THIRD EDITION

FIREFIGHTER'S HANDBOOK
Firefighter I and Firefighter II

Third Edition

FIREFIGHTER'S HANDBOOK
Firefighter I and Firefighter II

DELMAR
CENGAGE Learning

Australia • Brazil • Japan • Korea • Mexico • Singapore • Spain • United Kingdom • United States

Firefighter's Handbook: Firefighter I and Firefighter II
Delmar, Cengage Learning

Vice President, Career and Professional Editorial: Dave Garza

Director of Learning Solutions: Sandy Clark

Managing Editor: Larry Main

Product Development Manager: Janet Maker

Senior Product Manager: Jennifer A. Starr

Editorial Assistant: Amy Wetsel

Vice President, Career and Professional Marketing: Jennifer McAvey

Executive Marketing Manager: Deborah S. Yarnell

Senior Marketing Manager: Erin Coffin

Marketing Coordinator: Shanna Gibbs

Production Director: Wendy Troeger

Production Manager: Mark Bernard

Senior Content Project Manager: Jennifer Hanley

Art Director: Benj Gleeksman

Technology Project Manager: Christopher Catalina

Production Technology Analyst: Thomas Stover

Library of Congress Control Number: 2008940383

ISBN-13: 978-1-4283-3982-8
ISBN-10: 1-4283-3982-5

Delmar
Executive Woods
5 Maxwell Drive
Clifton Park, NY 12065-2919

Cengage Learning is a leading provider of customized learning solutions with office locations around the globe, including Singapore, the United Kingdom, Australia, Mexico, Brazil and Japan. Locate your local office at: **international.cengage.com/region**

Cengage Learning products are represented in Canada by Nelson Education, Ltd.

For your lifelong learning solutions, visit **delmar.cengage.com**
Visit our corporate Web site at **cengage.com**

Notice to the Reader

Publisher does not warrant or guarantee any of the products described herein or perform any independent analysis in connection with any of the product information contained herein. Publisher does not assume, and expressly disclaims, any obligation to obtain and include information other than that provided to it by the manufacturer. The reader is expressly warned to consider and adopt all safety precautions that might be indicated by the activities described herein and to avoid all potential hazards. By following the instructions contained herein, the reader willingly assumes all risks in connection with such instructions. The publisher makes no representations or warranties of any kind, including but not limited to, the warranties of fitness for particular purpose or merchantability, nor are any such representations implied with respect to the material set forth herein, and the publisher takes no responsibility with respect to such material. The publisher shall not be liable for any special, consequential, or exemplary damages resulting, in whole or part, from the readers' use of, or reliance upon, this material.

Printed in Canada
1 2 3 4 5 X X 10 09 08

*F*irefighter's Handbook is dedicated to the greathearted and courageous fire-fighters and emergency responders who have given of themselves the greatest sacrifice, their lives. Every year, the fire and emergency service community changes forever when the lives of firefighters are ended or damaged in the course of their duties. As we continue on we are left with the scars of these losses and the tears of many loved ones left behind. We share in the heartache of the loss of every fire-fighter and emergency responder. Let their lives shine on in the dedication and brav-ery of those left to respond when the tones drop, the bells ring, and the sirens blare.

This text is also dedicated to the driving force behind the continuation of fire-fighter heritage, the sharing of wisdom and experience, and the art of discovery and learning—trainers and educators. Every classroom session, practical scenario, and review session directly affects the quality of response the fire service provides. Never underestimate the power of positive change the training and education com-munity holds.

In honor and support of all Fire Service Educators, we are privileged to announce that Delmar, Cengage Learning will donate a portion of the proceeds to the National Fire Academy Alumni Association (NFAAA) for every copy of *Firefighter's Handbook* that we sell. The NFAAA was selected as the sole recipient of this contribution because of the similarities of our missions and our belief that NFAAA makes a positive difference in the education, safety, and welfare of firefighters.

And to every firefighter who has touched the life of someone in need and made a positive difference—you are truly the epitome of human compassion and selfless-ness. Don't ever stop caring

Contents

SECTION I
Firefighter I

1

Overview of the History and Development of the American Fire Service

2

Fire Department Organization, Command, and Control

3

Communications and Alarms

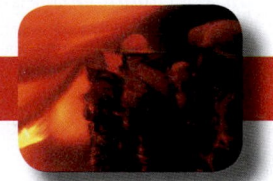

4

Fire Behavior

5

Firefighter Safety

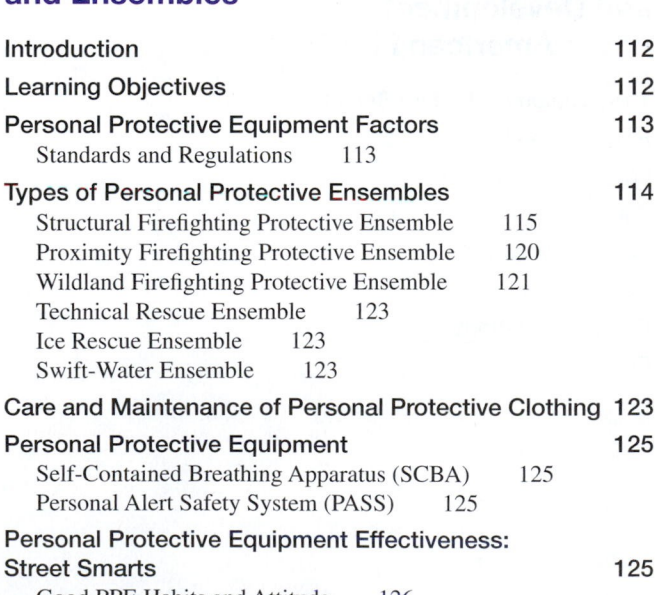

6

Personal Protective Clothing, Equipment, and Ensembles

7

Self-Contained Breathing Apparatus

8

Portable Fire Extinguishers

9

Water Supply

10

Fire Hose and Appliances

11

Nozzles and Fire Streams

12

Protective Systems

13

Building Construction

14

Ladders

xii   Contents

18

Ventilation

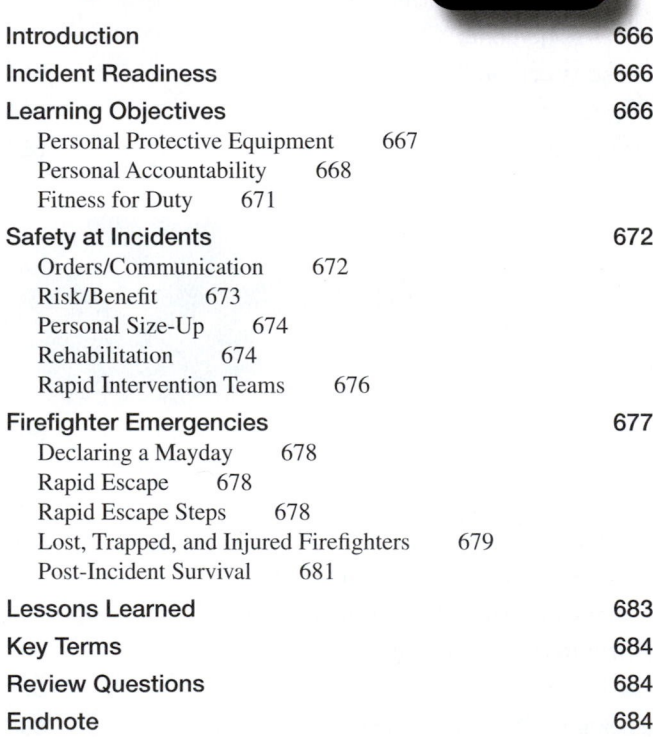

SECTION II
Firefighter II

2

Fire Department Organization, Command, and Control

3

Communications and Alarms

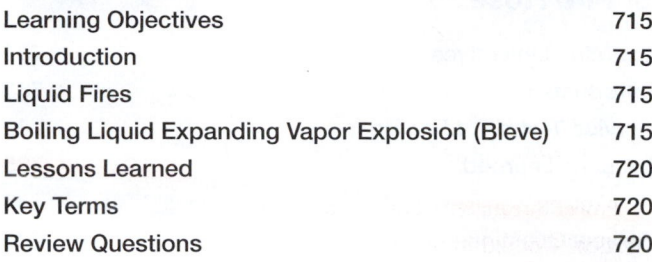

4

Boiling Liquid Expanding Vapor Explosion

5

Firefighter Safety

9

Water Supply

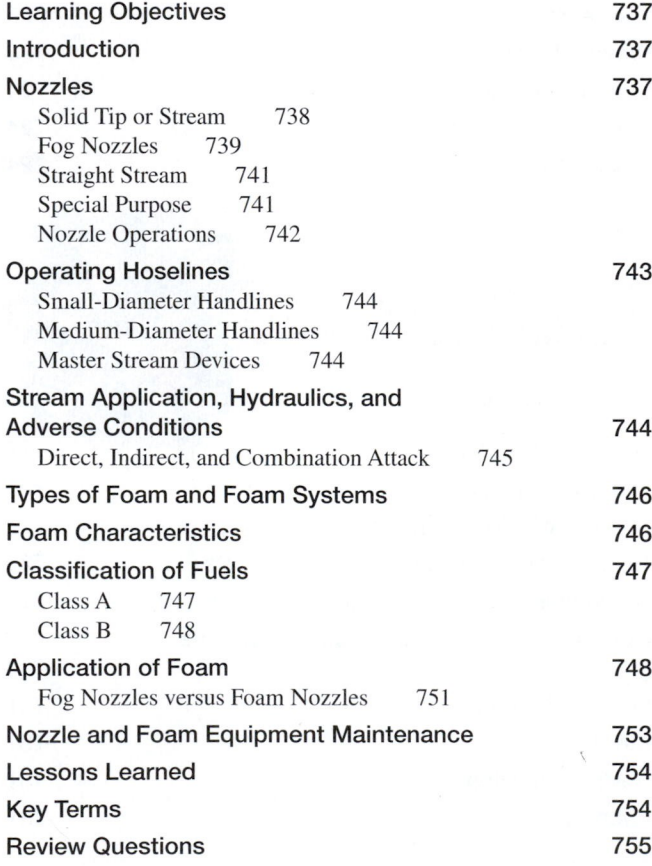

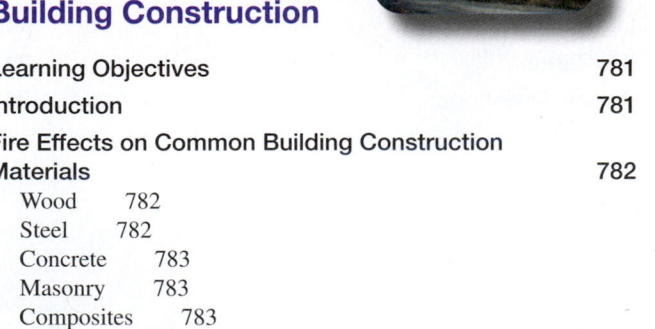
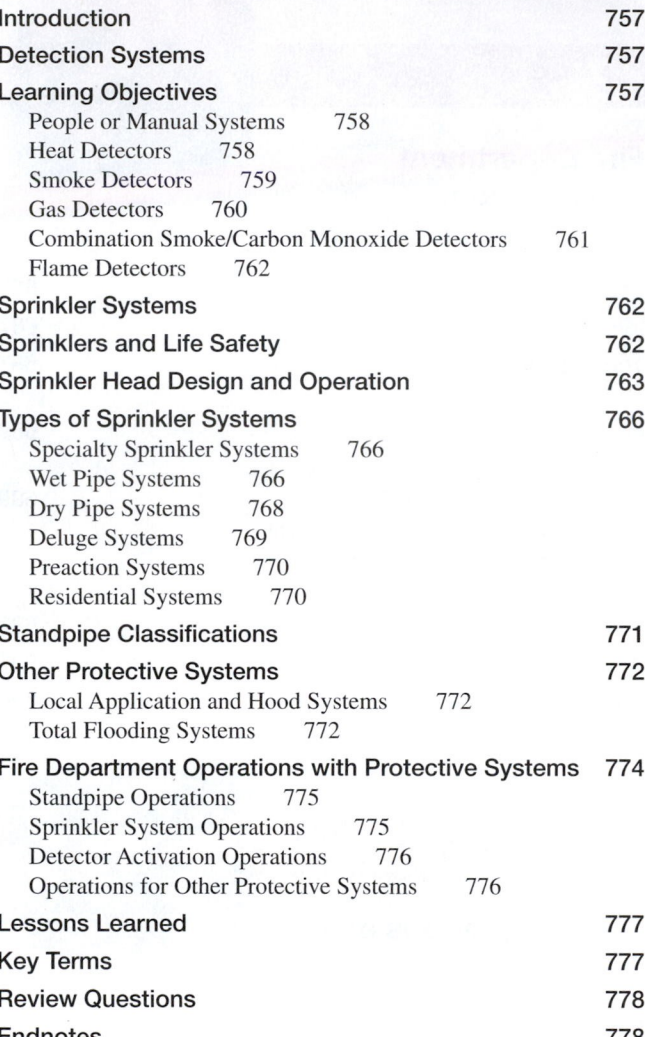

16

Vehicle Rescue and Extrication

17

Forcible Entry

18

Ventilation

19

Fire Suppression

20

Fire Cause Determination

21

Prevention, Public Education, and Pre-Incident Planning

23

Firefighter Survival

31

Disaster and Large Incident Response

SECTION III
Emergency Medical Response

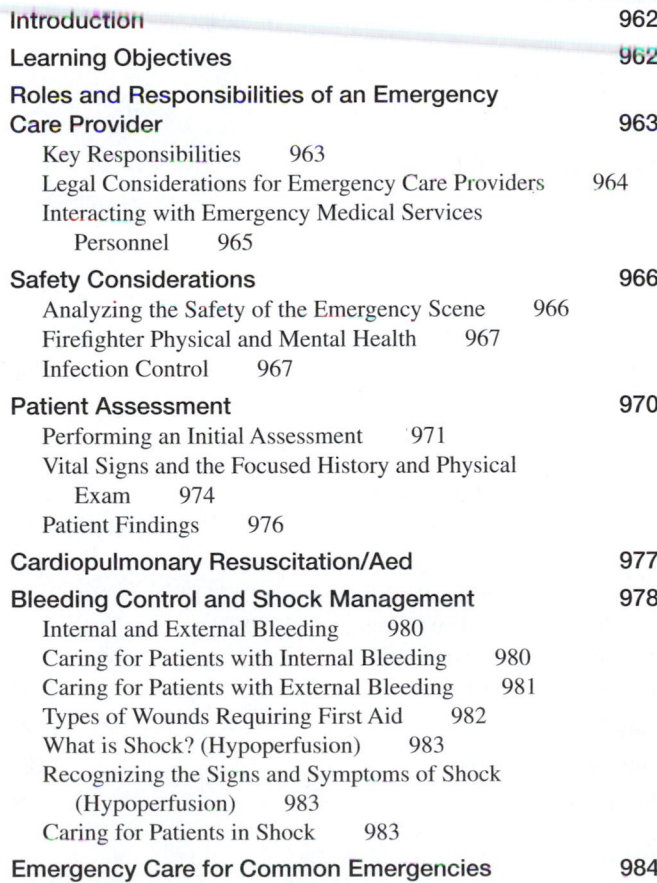

22

Emergency Medical Services

SECTION IV
Hazardous Materials and Terrorism Response

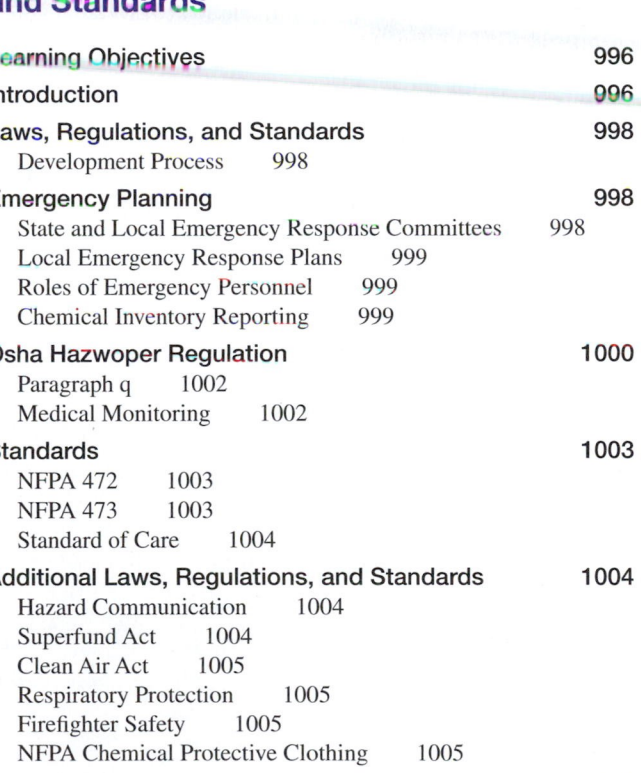

24

Hazardous Materials: Laws, Regulations, and Standards

25

Hazardous Materials: Recognition and Identification

26

Hazardous Materials: Information Resources

27

Hazardous Materials: Personal Protective Equipment

28

Hazardous Materials: Protective Actions

29

Product Control and Air Monitoring

30

Terrorism Awareness

Foreword

Once again the *Firefighter's Handbook* has been updated and released and is in the hands of firefighters, company officers, and chiefs across America. Few publications contain the volume and variety of information and material that you will find in this handbook. As always there are many chapters on tried and true firefighting standards such as engine operations, search and rescue, tools and equipment, and the history of our great service. But there is much more included that today's modern firefighters need to study and master to protect their community and themselves. Hazardous materials, terrorism issues, and the most recent strategic and tactical philosophies are all here for you to read, reread, and put to use on the streets of your city or town.

Americans across the country rely on their local fire department to handle just about every fire, emergency, disaster, and crisis that comes along. We still search for lost children and respond to water and gas leaks, floods and tornadoes, and of course both small and large fires involving buildings, vehicles, and the wildland interface. The chapters of *Firefighter's Handbook* are the rules of the road for firefighters. Just about every alarm you will respond to will present challenges and situations that you can find the solutions to in this handbook. Take your copy and read some every day. Examine the detailed graphics and photographs that are used to illustrate the vital and important aspects of the work that you do. Practice the skills and techniques that are outlined throughout the chapters and stay on top of your game. Your community is depending on you.

John J. Salka, Jr.
Battalion Chief
New York City Fire Department

Preface

OUR HISTORY

Delmar, a part of Cengage Learning, has been proudly serving the fire service for over ten years, and has been publishing trade-related training materials since the post-World War II era.

Delmar Publishers began in 1945 immediately following the end of World War II. Soldiers returning from the war were in need of jobs, and the industry needed people educated in the trades to fill available positions. It was an economic boom, and textbooks played a major role in educating these new workers. With this new educational need, Delmar was born as a publisher.

In the early 1980s, the Thomson Corporation bought Delmar Publishers, which quickly became part of the Thomson Learning branch of information providers. It has over the past few decades transitioned our company into what it is today – a provider of a variety of training solutions, including textbooks, CDs, videos and online courses.

In 2006, the growth and potential of Thomson Learning attracted the attention of both Apax Partners, a global company focused on investment, and Omers, a Canadian-based company offering pension services to retired municipal employees. Through a partnership, Apax and Omers bought Thomson Learning with the intention of inspiring further value and growth.

Today, as we continue to grow globally, we work to deliver even more training solutions, highly customized options and specialized content to colleges, universities, professors, students, reference centers, government agencies, corporations and professionals around the world.

As we proceed through the twenty-first century we hope that you will to turn to us to meet all your training needs as we continue to *Answer Your Call!*

OUR AUTHORS

We are proud to present a new version of *Firefighter's Handbook!* Complete with all the information that is found in the main version—*Firefighter's Handbook: Firefighting and Emergency Response, Third Edition*— this new version fully meets the requirements set forth by NFPA 1001 *Standard for Firefighter Professional Qualifications,* 2008 Edition, while dividing the content between Firefighter levels I and II as defined by the Standard.

As is the tradition in the fire service, this version has truly become a reality through team effort. These chapters are a result of many hours of collaboration to ensure that we offer a practical approach and deliver the most current information available, while meeting the needs of those programs that teach Firefighter I and II in separate courses.

We wish you the best as you enter the world of firefighting, and as you begin your training, we would like to share with you three guiding principles:

1. Build your skill sets.
2. Remain informed.
3. Above all—stay safe!

Firefighter's Handbook: Firefighter I and Firefighter II

The newest member of the *Firefighter's Handbook* team, author Jake Rhoades collaborated closely with our team, as well as members of the validation team, to meticulously separate the content within these pages into separate sections. At the same time, this author integrated review material within the FF II chapters to create a seamless transition between FF I and II. With this approach, cumbersome cross-referencing or supplemental handouts are not required when candidates move to the next level of their training.

Jake Rhoades

Jake Rhoades has been in the fire service for over 16 years and is currently the Assistant Chief of Jenks Fire Department in Jenks, Oklahoma. He is the former Training Officer of the Stillwater, Oklahoma Fire Department. He started his career as a firefighter with the department and has been promoted through the ranks to driver/operator in 1995, lieutenant in 1999, and training officer in 2005. His responsibilities as Training Officer included planning and coordinating all training activities; serving as Health and Safety Officer; overseeing all special operation programs that include hazardous materials and technical rescue; developing promotional testing and evaluation processes; and developing all recruit training. He currently serves as Planning Team Manager for the Oklahoma Task Force One where he manages the planning operations during emergency responses. He also serves Oklahoma Task Force 1 as an instructor for structural collapse and safety, and develops and performs administrative duties during emergency response of OK TF 1.

He is currently enrolled in the Executive Fire Officer program at the National Fire Academy and has certified to the following: FF I and II, FOI and II, Trench Rescue, Confined Space Rescue, Rope Rescue, Structural Collapse, Instructor I and II, Hazardous Materials Technicians well as numerous other certifications.

Jake was also recently named the advocate and representative for the state of Oklahoma for the Everyone Goes Home program. As part of the program, he delivers the Courage to be Safe program and encourages the adoption and commitment to the 16 Life Safety Initiatives.

As an author and developer for Delmar, Jake creates original materials, reviews materials as a subject matter expert, and serves on the committees to validate learning objectives for inclusion in fire related manuals and other materials.

Firefighter's Handbook: Firefighting and Emergency Response, Third Edition

Andrea A. Walter

Content Editor
Author: Chapter 31, Large Incident and Disaster Management
Andrea A. Walter is a firefighter with the Metropolitan Washington Airports Authority at the Washington Dulles International Airport and a life member and former officer of the Sterling Volunteer Rescue Squad, in Sterling, Virginia. Walter has been active in the fire and emergency services community for many years, serving as the Manager of the Commission on Fire Accreditation International for the International Association of Fire Chiefs, and assisting in a variety of projects with the National Volunteer Fire Council, Women in the Fire Service, and the United States Fire Administration. She has over eighteen years of experience in the fire and emergency services. In addition to being an author for this text, Walter serves as the project's content editor/lead author and has been with the text project since its inception. Walter is involved in a variety of other Delmar text projects and instructional supplements. She is also an author and the content editor for Delmar's *First Responder Handbook: Fire Service Edition* and *First Responder Handbook: Law Enforcement Edition,* as well as *Exam Preparation: Firefighter I and II,* and *Exam Preparation: Hazardous Materials Awareness and Operations.*

Richard Bonnett

Author: Chapter 22, Emergency Medical Services
Bonnett is an EMS Captain with the Metropolitan Washington Airport's Authority and a life member of the Millwood Volunteer Fire and Rescue Department. He has been actively involved with fire and EMS for over twenty years. He is a Nationally Registered Paramedic and is certified as an Advanced Life Support Coordinator. He has been an active instructor in the State of Virginia for over twelve years teaching at the state and local level. He is training center faculty in Advanced Cardiac Life Support at Winchester Medical Center and also teaches Pediatric Advanced life Support. Bonnett recently co-founded Comprehensive Safety and Health Training, a company offering basic and advanced life support classes in the tri-state area.

Willis T. Carter

Founding Author: Chapter 3, Communications and Alarms
Chief Willis Carter has been a member of the fire service for thirty-five years. He began his career in 1972 as a firefighter with the Shreveport (La.) Fire Department. In 1978, Carter performed a lateral transfer into what was then known as "Fire Alarm", where he was a call

taker and dispatcher until his promotion to Chief of Communications in 1986. Carter is responsible for the management and operations of the Fire Communications Division, which has a staff of 45 who operate the primary Public Safety Answering Point (PSAP) for the Caddo Parish (Louisiana) 9-1-1 system. In addition to PSAP operation, the Chief is also responsible for the Information Technology section of the Fire Department which supports the network, hardware, and software utilized by the 650 member Fire Department.

Carter is active in public safety communications outreach and education. He is a member of the International Fire Chief's Association and is an assessment-team leader for the Commission on Accreditation for Law Enforcement Agencies, Inc. (CALEA). Carter led the effort for the Shreveport Fire Department Communications Center to become the first public safety communications center in the state of Louisiana to achieve accredited status through CALEA and the only fire communications center in the nation to receive CALEA accreditation.

Carter currently serves as President Elect of the Association of Public Safety Communications Officials, International (APCO) Board of Officers. APCO is the world's oldest and largest not-for-profit public safety communications association with a membership of over 15,000 members world wide. He will be installed as President of this association at the close of the APCO annual training conference which will be held in Baltimore in August of 2007.

Dennis Childress

Author: Chapter 17, Forcible Entry; Chapter 18, Ventilation; Chapter 19, Fire Suppression

Childress, an author, lecturer and fire services instructor, recently retired from the Orange County Fire Authority in Southern California, and has been in the fire service for just over thirty-nine years. He is a Certified Chief Officer with the state of California, and he holds an Associate of Arts degree in Fire Science and a Bachelor of Science degree in Fire Protection Administration. He holds a seat on the Board of Directors for the Southern California Fire Training Officers Association, and sits on the Statewide Training and Education Advisory Committee for the State Fire Marshal's Office. He is a principal member of the NFPA 1500 Fire Service Occupational Safety and Health Committee, and the NFPA 1561 Standard for Emergency Services Incident Management System Committee. He has au-

thored a number of articles in fire service publications over the years, and he has also been an instructor in Fire Command and Management in the California State Fire Training System for over twenty years. He has also been a major part of the development of the Fire Officer Command curriculum for the State of CA over the last fifteen years.

Ronny J. Coleman

Author: Chapter 1, Overview of the History, Tradition and Development of the Amercian Fire Service

Chief Coleman is a nationally and internationally recognized member of the fire service who formerly served as the Chief Deputy Director, Department of Forestry and Fire Protection, and as California State Fire Marshal. He has served in the fire service for thirty-eight years. Previously he was Fire Chief for the Cities of Fullerton and San Clemente, California, and was the Operations Chief for the Costa Mesa Fire Department. Chief Coleman possesses a Master of Arts Degree in Vocational Education from Cal State Long Beach, a Bachelor of Science Degree in Political Science from Cal State Fullerton, and an Associate of Arts Degree in Fire Science from Rancho Santiago College. He has served in many elected positions in professional organizations, including President, International Association of Fire Chiefs; Vice President, International Committee for Prevention and Control of Fire; and President, California League of Cities, Fire Chiefs Department

David W. Dodson

Author: Chapter 4, Fire Behavior; Chapter 5, Firefighter Safety; Chapter 13, Building Construction; Chapter 23, Firefighter Survival

Dodson is the owner of and lead instructor for Response Solutions, LLC in Eastlake, Colorado. He is an independent contract instructor for fire departments and fire service trade conferences nationwide, and is a regular speaker at the annual Fire Department Instructor's Conference (FDIC) and Firehouse World Expo. He has many years in the fire service and specializes in firefighter safety issues.

He started his fire service career with the U.S. Air Force. He served at Elmendorf AFB in Alaska and spent two years teaching at the USAF Fire School. After the USAF, Dodson spent almost seven years as a Fire Officer and Training/Safety Officer for the Parker

Fire District in Parker, Colorado. He became the first Career Training Officer for Loveland Fire and Rescue in Colorado and rose through the ranks, including time as a HAZMAT Technician, Duty Safety Officer, and Emergency Manager for the city. He accepted a Shift Battalion Chief position for the Eagle River Fire District in Colorado before starting his current company, Response Solutions, which is dedicated to teaching firefighter safety and practical incident handling. Chief Dodson has served on numerous national boards including the NFPA Firefighter Occupational Safety Technician Committee and the International Society of Fire Service Instructors (ISFSI). He also served as president of the Fire Department Safety Officers' Association. In 1997, Dodson was awarded the ISFSI "George D. Post Fire Instructor of the Year."

Robert F. Hancock

Founding Author: Chapter 15, Ropes and Knots; Chapter 16, Rescue Procedures

Hancock is Assistant Chief/Administration with Hillsborough County Fire Rescue in Tampa, Florida, a department with 615 career personnel and 205 volunteers, He was hired in November 1974 as a firefighter and was promoted through the ranks to his present position in October 1993. He was awarded an Associate of Science Degree in Fire Science, with honors, from Hillsborough Community College. He graduated from the Executive Fire Officers Program at the National Fire Academy and has been certified as an instructor with the State of Florida since 1983. Hancock is chairman of the Florida Fire Chiefs' Disaster Response Communications Sub-Committee, charged with identifying short- and long-term solutions to the disaster response communication issue statewide.

Chris Hawley

Author: Chapter 24, Hazardous Materials: Laws, Regulations, and Standards; Chapter 25, Hazardous Materials: Recognition and Identification; Chapter 26, Hazardous Materials: Information Resources; Chapter 27, Hazardous Materials: Personal Protective Equipment; Chapter 28, Hazardous Materials: Protective Actions; Chapter 29, Product Control and Air Monitoring; Chapter 30, Terrorism Awareness

Chris Hawley is a Deputy Program Manager for Computer Sciences Corporation (CSC) and is responsible for several WMD courses within the DOD/FBI/DHS International Counterproliferation Program. This program provides threat assessment, HazMat and Anti-Terrorism training and full scale exercises worldwide and is focused on Eastern Europe and Central Asia. Previous to the International work, Chris retired as a Fire Specialist with the Baltimore County Fire Department. Prior to this assignment he was assigned as the Special Operations Coordinator. As the Special Operations Coordinator he reported to the Division Chief of Special Operations, and was responsible for the coordination of the Hazardous Materials Response Team and the Advanced Technical Rescue Team along with two team leaders. Prior to this assignment he was assigned to the Fire Rescue Academy as one of four shift instructors. As a shift instructor he was responsible for all County-wide training on his shift. He has been a HazMat responder for over 19 years. Chris has twenty-five years experience in the fire service and prior to Baltimore County he was a Hazardous Response Specialist with the City of Durham, NC Fire Department.

Chris has designed innovative programs in hazardous materials and anti-terrorism and has assisted in the development of many other training programs. He has assisted in the development of programs provided by the National Fire Academy, Federal Bureau of Investigation, U.S. Secret Service, and many others. Chris has presented at numerous, local, national and international conferences, and writes articles on a regular basis. He serves on a variety of pivotal committees and groups at the local state, and federal levels. He also works with local, state, and federal committees and task forces related to hazardous materials, safety, and terrorism.

Chris has published four texts on hazardous materials and terrorism response for Delmar. He co-authored *Special Operations for Terrorism and HazMat Crimes,* along with Mike Hildebrand and Greg Noll through Red Hat Publishing. He also assists a variety of publishers with the review and development of emergency services texts and publications.

T. R. (Ric) Koonce, III

Founding Author: Chapter 8, Portable Fire Extinguishers; Chapter 9, Water Supply; Chapter 10, Fire House and Appliances; Chapter 11, Nozzles, Fire Streams, and Foam; Chapter 12, Protective Systems

Koonce is an Assistant Professor and Program Head of Fire Science Technology at J. Sargeant Reynolds

Community College in Richmond, Virginia. He is a retired Battalion Chief with the Prince George's County (Maryland) Fire Department and has thirty five years of fire service experience. He is an adjunct instructor for the Virginia Department of Fire Programs. He holds two associate degrees, a Bachelor of Science degree in Fire Service Management from University College of the University of Maryland, and a Certificate of Public Management from Virginia Commonwealth University.

Frank J. Miale

Founding Author: Chapter 4, Fire Behavior; Chapter 14, Ladders; Chapter 18, Ventilation

Miale is a Battalion Chief (ret.) with over thirty years in the FDNY. A twenty-five-year active member in his local Volunteer Lake Carmel Fire Department, he maintains a busy role as treasurer and training instructor. A former high school teacher, he holds two Bachelor of Science degrees with several concentrations in Education, Biology, and Fire Administration. During his career in the FDNY, he taught at the NYC fire academy, participated in the introduction of a communication system using apparatus-mounted computers, and headed a special Emergency Command Unit while an active line officer. Formerly the Training Officer for the 27th Battalion in the FDNY, he taught many ladder company and ventilation courses throughout the country. His career was spent primarily in busy ladder companies in Brooklyn, Harlem, and the South Bronx sections of New York City prior to promotion to Chief Officer. He is the recipient of nine awards for courage and valor, including two department medals from the FDNY, and has been published many times in WNYF, Fire Command, and Fire Service Today.

Geoff Miller

Founding Author: Chapter 20, Salvage, Overhaul and Fire Cause Determination

Miller is a thirty-three year veteran of the fire service and is currently the Deputy Chief of Operations with the Sacramento Metropolitan Fire District in California. Previous assignments have included seven years as an Assistant Chief, six years as a Battalion Chief, four years as the district's Training Officer, ten years as a line Captain, and two years as an Inspector. He is assigned to a CAL FIRE Type 1 Incident Command Team and FEMA Incident Support Team as a Plans Chief. He is also a task Force Leader and Plans Manager for USAR CA TF 7 Sacramento. Through the FEMA assignment he responded to the Pentagon and World Trade Center disasters along with numerous hurricanes and National Security Special Events. He has been involved in several California Fire Fighter I and II curriculum development workshops as well as participating on the rewrite of Fire Command 1A and 1B.

Robert Morris

Founding Author: Chapter 17 Forcible Entry

Morris is a veteran of the New York City Fire Department and has been assigned to some of the busiest fire companies in New York City, including Ladder Company 42, Engine 60 in the Bronx, and Rescue Company 3 in Manhattan. After serving in the Bronx and Harlem, he served as Company Commander of Ladder Company 28. Captain Morris is currently Company Commander of Rescue Company 1 in Manhattan. Captain Morris is the recipient of seventeen meritorious awards, including three department medals.

Jeff Pindelski

Author: Chapter 2, Fire Department Organization, Command, and Control; Chapter 3, Communications and Alarms; Chapter 6, Personal Protective Clothing and Ensembles; Chapter 7, Self-Contained Breathing Apparatus; Chapter 16, Rescue Procedures

Jeffrey Pindelski is an eighteen-year-plus student of the fire service. Jeff is a Battalion Chief with the Downers Grove Fire Department in Illinois. He previously served for twelve years as a Firefighter and Lieutenant on the Truck and Heavy Rescue Company. In addition to his background in a career position, he has also served on departments in a volunteer and part time capacity.

Jeff is a staff instructor at the College of Du Page and also instructs courses at the Romeoville Fire Academy. He is a Certified Fire Officer III and Instructor III while also being certified as a Fire Suppression Incident Safety Officer. Chief Pindelski holds a Masters Degree in Public Safety Administration from Lewis University, a Graduate Certificate in Managerial Leadership and a Bachelors Degree from Western Illinois University.

He has been involved with the design of several training programs dedicated to firefighter safety and survival including R.I.C.O. (Rapid Intervention Company Operations) which is a forty-hour Rapid Intervention training program held on a national level. Jeff has served on review committees for Delmar and is coauthor of *Rapid Intervention Company Operations*. Chief Pindelski was a recipient of the State of Illinois Firefighting Medal of Valor in 1998 and has been published regularly in several trade journals on various fire service related topics.

Marty Rutledge

Author: Chapter 8, Portable Fire Extinguishers; Chapter 9, Water Supply; Chapter 10, Fire Hose and Appliances; Chapter 11, Nozzles, Fire Streams, and Foam; Chapter 12, Protective Systems; Chapter 14, Ladders; Chapter 15, Ropes and Knots; Chapter 20, Salvage, Overhaul, and Fire Cause Determination; Chapter 21, Prevention, Public Education, and Planning

Rutledge is a Firefighter/Engineer and ARFF Specialist, for Loveland Fire and Rescue in Loveland, Colorado. He is a member of the Fire Certification and Advisory Board to the Colorado Division of Fire Safety, as well as serving as the State First Responder program coordinator. He is also a member of the Colorado State Fire Fighter's Association and has over seventeen years of fire and emergency services experience in both volunteer and career ranks. He is a member of Wind and Fire Motorcycle Club, and has and ridden motorcycles with firefighters in different countries around the world. Rutledge has authored and served as technical expert for a supplementary firefighter training package for *Firefighter's Handbook* and has co-authored Delmar's *First Responder Handbook: Fire Service Edition, First Responder Handbook: Law Enforcement Edition,* as well as *Exam Preparation: Firefighter I and II,* and *Exam Preparation: Hazardous Materials Awareness Operations.*

Donald C. Tully

Founding Author: Chapter 21, Prevention, Public Education, and Pre-Incident Planning

Tully is a member of the Orange County, California, Fire Authority. With over thirty years in the fire service, he has also been a Division Chief/Fire Marshal in Buena Park and Westminster, California, for ten years, and a Fire Technology Instructor at Santa Ana College, California. He is Past President of the Orange County Fire Prevention Officers' Association.

He also served as a member of NFPA Committees 1221 (CAD Dispatch and Public Communications) and 72 (Fire Alarms), and as a member of the California State Fire Marshal Committees on Fire Sprinklers and Residential Care Facilities (ad hoc committees). He is a California State Certified Chief Officer, Fire Officer, Fire Investigator, Fire Prevention Officer, and Fire Service Instructor and Technical Rescue Specialist.

Thomas J. Wutz

Founding Author: Chapter 2, Fire Department Organization, Command, and Control; Chapter 7, Self-Contained Breathing Apparatus

As a Chief of the Fire Services Bureau, New York State Office of Fire Prevention and Control, Wutz manages the Fire Services Bureau's twenty full-time staff and 300 part time instructors as well as develops and delivers fire training outreach programs, training policies and procedures serving 1850 fire departments (career and volunteer), with 24,000 students annually. In addition, he provides program management for the New York State Fire Mobilization and Mutual Aid Plan. Chief Wutz is also certified as a Type II Incident Commander for the New York State Incident Management Assistance Team, most recently serving as deputy IC for the State IMAT assistance to Jackson County Mississippi, September 2005

Chief Wutz retired as base fire chief for the 109th Military Airlift Group, United States Air Force/ New York Air National Guard (twenty-eight years) with thirty-one years total military service and has served as a volunteer firefighter in various departments in New York state for thirty-five years.

OUR REVIEW AND VALIDATION COMMITTEES

Through the dedication of our authors, content and technical reviewers, as well as our Fire Advisory Board members and validation committee, the third edition of *Firefighter's Handbook* continues to remain up to date with the changing landscape of the fire service world. As part of the development process, each chapter is carefully reviewed by a select number of practicing individuals in the fire service who offer their expertise and insight as the authors revise the content. Additionally, technical reviewers thoroughly check the manuscript in detail for clarity and accuracy.

We are excited to announce that for the third edition of this book, we have created a validation committee to ensure that the content meets 100% of the NFPA Standard 1001. Learning objectives, which have been validated by this committee of subject-matter-experts, are included at the beginning of each essentials of firefighting chapter. The learning objectives tie all components of the learning solution (text, curriculum, test bank, supplementals) to the NFPA standard providing instructors and students a pathway to meet the intent of the Job Performance Requirements outlined in the Standard.

The Validation Process

The National Fire Protection Association (NFPA) professional qualifications standards identify the minimum Job Performance Requirements (JPRs) for fire service positions. JPR state the behaviors required to perform specific skill(s) on the job. The JPR statements must be converted into instructional objectives with behaviors, conditions, and standards that can be measured within the teaching/learning environment.

Our process includes the development of the learning objectives from the JPR by a subject matter expert. The learning objectives are then reviewed and validated by a committee. The committee reviews the learning objectives to ensure that the JPR was correctly interpreted and also makes recommendation for additional learning objectives within our materials. Our authors are provided with the validated learning objectives to develop the materials. This ensures that 100% of the standard is met within our materials.

The learning objectives are used throughout the entire development process. This includes the development of the book, curriculum, and the certification test question banks. This process ensures consistency among all materials and offers the best possible materials available on the market.

For a complete list of our review and validation committee members, as well as other contributors to this book, please refer to the Acknowledgments section.

ABOUT THIS BOOK

Welcome new recruits, volunteers and experienced firefighters to an inside look at the current practices, technology and initiatives in the fire service! In our tradition of offering quality firefighter training manuals, Delmar is proud to introduce this new addition to our *Firefighter's Handbook series*. Within the pages of this book you will find all the information you need to successfully complete the Firefighter I and II training courses which prepare student for certification testing as well as for excelling on promotional tests. It also serves well as a guide to refresh your knowledge and skills as you continue on as a firefighter.

Firefighter's Handbook: Firefighter I and Firefighter II is a comprehensive guide to the basic principles and fundamental concepts involved in firefighting, emergency medical services, and hazardous materials operations, as defined by the 2008 edition of the National Fire Protection Association 1001 *Standard for Firefighter Professional Qualifications*. With a practical, straight-forward approach and step by step sequences to explain knowledge and skill criteria, this book can be used by both new and experienced firefighters and hazardous materials responders to enable them to perform a wide variety of firefighting and emergency service activities.

While content in this book is conveniently divided into separate modules for the Firefighter I and Firefighter II levels, as defined by the Standard, it continues to maintain the quality of the other books in the series. It achieves this by focusing on the critical topics—keeping you and your team safe, explaining new initiatives, and highlighting new technology. This book is intended not only to train firefighters, but to ensure optimal proficiency on the job.

As a firefighter you will make a difference in the lives of many. Use your knowledge, practice your skills, and above all—be safe.

Organization of this Book

Firefighter's Handbook: Firefighter I and Firefighter II consists of the thirty-one chapters that make up the original *Firefighter's Handbook: Firefighting and Emergency Response, Third Edition*. The only difference is that the content within these thirty-one chapters is now reorganized and located in different portions of the book—depending on whether the content relates to Firefighter I (Section I), Firefighter II (Section II), EMS (Section III), or Hazardous Materials and Terrorism (Section IV) training requirements. All the essential information—from the history of the fire service to the governing laws and regulations, from the use of apparatus and equipment to the practice of procedures, from understanding fire behavior and building construction to effective planning and prevention measures, as well as special topics such as terrorism and large incident response—remains covered in this new book.

The chapters are set up to deliver a straightforward, systematic approach to training, and each includes an outline, learning objectives, introduction, lessons learned, key terms, review questions, and a list of additional resources.

Also included at the front of the book is an NFPA 1001 and 472 Correlation Guide that correlates the requirements outlined in the Standard to the content in this version of the *Firefighter's Handbook* by chapter and page references. These resources can be used as a quick reference and study guide.

Features of this Book

Firefighter's Handbook: Firefighter I and Firefighter II contains a number of features that make it unique. It offers a realistic approach to emergency response—it is comprehensive in coverage, including essential information, but presents the content in a clear and concise manner, so it is easy to read, follow, and understand. We also recognize that it is essential in this field to not only acquire knowledge, but to put it into practice as well.

The following recommends how you can best utilize the features in this book to gain competence and confidence in learning firefighting essentials.

NFPA Standard 1001 and 472 Correlation Guides

These grids provide a correlation between this book and the requirements for Firefighter I and II and Hazardous Materials Awareness and Operations for NFPA Standards 1001 and 472, 2008 Editions. Each performance requirement is linked to specific book chapters and page numbers for easy reference.

5.1.1 General Knowledge Requirements	Learning Objective(s)	Chapter(s)	Page(s)
Organization	2-1	2	22–25
Role	2-3	2	25–28
Mission	1-1 2-2	1, 2	2–3 24

	NFPA 472, Chapter 4 Awareness	Learning Objective(s)	Chapter(s)	Page(s)
4.1	General.			
4.1.1	Introduction.			
4.1.1.1	Awareness level personnel shall be persons who, in the course of their normal duties, could encounter an emergency involving	24-1, 24-2	24	846–849, 851–853

Street Stories

Introductory chapters open with a personal experience written by notable contributors from across the nation. These personal accounts engage you in the events and help to remind you of the importance of practicing the knowledge and skills presented in the chapter.

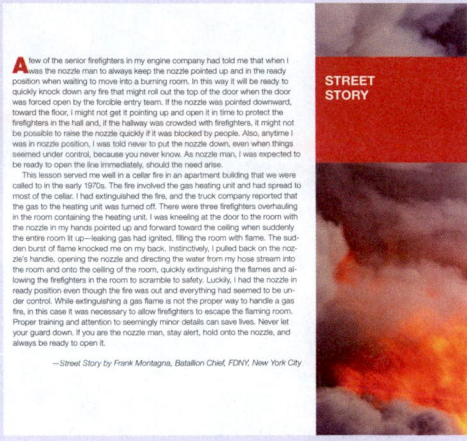

NEW! Validated Learning Objectives

The numbered learning objectives that appear in the beginning of each chapter were validated by a committee of experts and tie all the components of the learning solution (text, curriculum, test bank, and supplements) to the NFPA Standard. Each learning objective is colorized to indicate the level of training as dictated by the Standard—Firefighter I (black), Firefighter II (red). Information that goes above and beyond the Standard is also identified (blue).

LEARNING OBJECTIVES

After completing this chapter, the reader should be able to:

5-1 Describe potential accidents and injuries that may occur during structural firefighting operations.

5-2 List individual actions and attitudes that can help reduce firefighter injuries and death.

Job Performance Requirements

Step-by-step photo sequences illustrating important procedures are integrated throughout the chapters. These are intended to be used as a guide in mastering the job performance skills and to serve as important review.

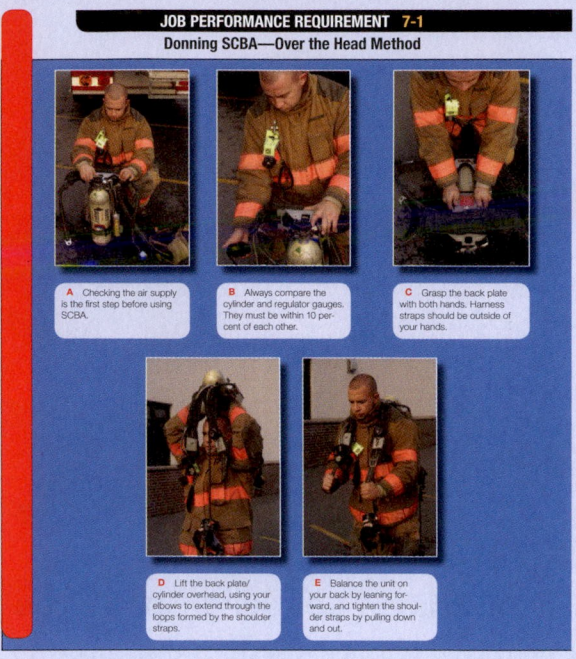

JOB PERFORMANCE REQUIREMENT 7-1
Donning SCBA—Over the Head Method

A   Checking the air supply is the first step before using SCBA.

B   Always compare the cylinder and regulator gauges. They must be within 10 percent of each other.

C   Grasp the back plate with both hands. Harness straps should be outside of your hands.

D   Lift the back plate/cylinder overhead, using your elbows to extend through the loops formed by the shoulder straps.

E   Balance the unit on your back by leaning forward, and tighten the shoulder straps by pulling down and out.

NEW! ViewPoint

These reports recognize variances in firefighter practices across the United States, providing a more informed view of the world of firefighting

VIEWPOINT

Often, residential sprinkler initiatives in communities are met with considerable opposition from homeowners and home builders for a variety of reasons.

StreetSmart Tip

As is true in any profession, sometimes experience can be the best teacher. These tips are power-packed with a wide variety of hints and strategies that will help you be a street-smart firefighter.

STREETSMART TIP

When connecting to a siamese, the firefighter should choose the far left outlet first. This will allow better access for the spanner wrench used to tighten the coupling. Also, some connections only have clappers installed on the right side and must be connected on the left side first.

Firefighter Fact

These boxes offer a detailed snapshot of facts based on firefighting history, experience, and recorded date to provide essential background information.

FIREFIGHTER FACT

New York City's tragic Triangle Shirtwaist Factory fire in 1911 killed 146 people. Most of the fatalities were women who jumped to their deaths trying to avoid being burned. This fire led to the concept of adding sprin-

Note

This feature highlights and outlines important points for you to learn and understand. Based on key concepts, this content is an excellent source for review.

NOTE

Firefighters must realize that protective systems are designed for specific purposes and will not completely defend a property unless they are designed with that goal in mind.

Safety and Caution

As a firefighting professional, you will face situations in which you will need to react immediately in order to ensure you safety and the safety of others. These tips provide this necessary advice.

SAFETY

All sprinkler system alarms should be treated as an actual fire until confirmed otherwise. Just because a building is fully sprinklered does not mean the fire department should prepare to fight fire differently.

CAUTION

Firefighters wishing to use a Class II or III standpipe system should not use the hose provided. It is not tested and may be single-jacketed unlined hose with a nozzle that has no shutoff. Firefighters using this system should replace the hoseline with fire service hose.

Lessons Learned

Lesson Learned summarize the main points presented in the chapter and is ideal for review purposes.

> ### LESSONS LEARNED
>
> Protective systems are designed to automatically detect or suppress a fire, or to assist in extinguishing the fire. They can apply water or other extinguishing agents. Protective systems have been credited with saving both lives and property and are essential fire protection tools. Firefighters need to understand the value and operation of these systems to protect their communities.

Review Questions

At the end of every chapter, the review questions assist you with the learning process and help you to evaluate your knowledge and ensure mastery of the content

> ### REVIEW QUESTIONS
>
> 1. How does an OS&Y valve work?
> 2. What is the FDC and how is it used?

New to this Edition of the Series

The third edition of the *Firefighter's Handbook* series, including this book, contains many new updates and additional information to address the needs of the fire service today:

New Initiatives in the Fire Service

- An all new chapter on large incident and disaster response prepares responders for current day events
- Thoroughly revised chapter on Fire Behavior provides a more practical approach to this topic and translates technical information into a easy to understand format
- Emphasis on the implementation of the National Incident Management System (NIMS) stresses the importance teamwork and a structured response to incidents
- New section on Fire Life Safety Initiatives seeks to help reduce firefighter injuries and line of duty deaths

- New JPR photo sequences covering special rescue and rapid intervention procedures to ensure that everyone goes home
- Reports on the latest terrorist and HazMat crimes, and how to respond to a criminal/terrorist activity keep responders informed of potential threats to the community
- Metric conversions accompany all U.S. Customary/English measurements for quick reference

Keeping You Safe

- Up to date statistics from the U.S. Fire Administration that highlight areas of concern related to dangers of the job
- New information on the dangers of cyanide poisoning to help you prevent this from occurring with firefighters in your department
- New section on predicting building collapse that highlight proactive actions to prevent catastrophes
- New practices to ensure firefighters remain safe on the fireground through proper use of personal protective equipment
- New mandates on managing air supply and performing emergency checks to avoid a life and death rescue situation
- New sections on securing building facilities, scene illumination, and establishing safe work zones
- New information on radiation to prepare hazardous materials responders for this potential threat

New Technology and Equipment

- New section on Voice Over Internet Protocol (VOIP) highlights the advantages of use of this technology in departments.
- Global Positioning Technology (GPS) technology and Automatic Vehicle Location (AVL) systems are explained in order to keep emergency responders aware of these new deployment and tracking systems should their jurisdiction use them.
- A new section on thermal imaging camera use in search and rescue explains when and how to use this technology, as well as its advantages and drawbacks, so that emergency responders are prepared should it be required at an incident.
- Current information on air monitoring helps hazardous materials responders keep informed of new technology in this field.
- New photos reflect current apparatus, equipment, tools and procedures.

OUR CUSTOM OPTIONS

As we look toward the future, Delmar is proud to be the first to offer custom solutions to our firefighters. We recognize that the fire service is a complex and changing world, and it is our duty to provide you with options for training. We are committed to providing you with a package to meet your specific needs. Listed here are the four versions of the third edition of *Firefighter's Handbook* that we offer to the fire service:

Firefighter's Handbook: Firefighting and Emergency Response

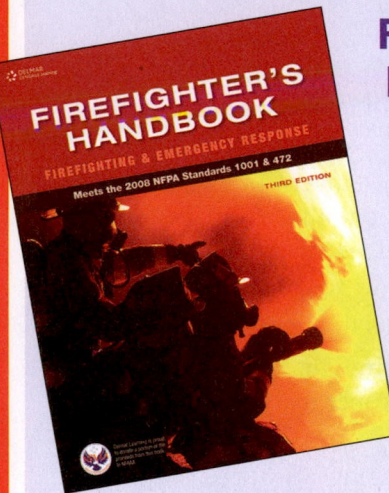

This comprehensive book includes all the need-to-know information for the firefighter candidate. It not only meets 100% of the 1001 Standard, but it also presents additional information that is essential for proficiency on the job.

- Meets 100% of 2008 NFPA Standard 1001 and 472
- Includes basic firefighter training, EMS and hazardous materials operations
- For those who combine training on all Firefighter I and II topics in preparing candidates for certification

Order#: 978-1-4180-7320-6

Firefighter's Handbook: Firefighter I and Firefighter II

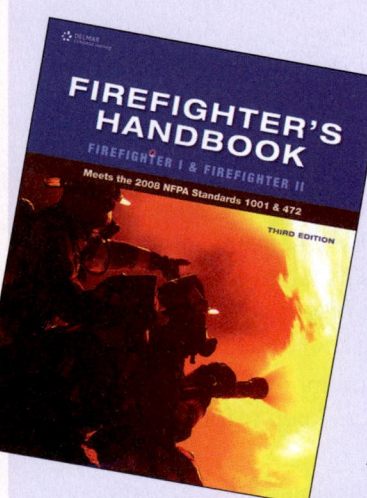

Firefighter I and II levels are divided into separate sections in this brand-new offering, providing convenience to those who teach these two levels separately.

- One book that actually divides out the content between Firefighter levels I and II
- Meets 100% of the 2008 NFPA Standard 1001 and 472
- Includes basic firefighter training, EMS and hazardous materials response

Order#: 978-1-4283-3982-8

Firefighter's Handbook: Essentials of Firefighting

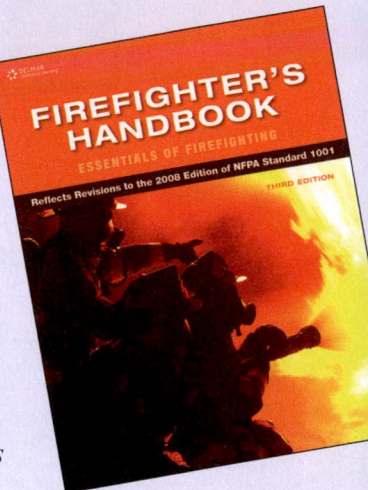

For those who teach hazardous materials in a separate course, this book provides all the need-to-know information for firefighting, exclusive of the hazardous materials content in the third edition of *Firefighter's Handbook*.

- Updated to the 2008 NFPA 1001 Standard
- Includes the same chapters and content as the original third edition, without the hazardous materials response chapters
- For those who conduct hazardous materials training separate from firefighter training

Order#: 978-1-4180-7324-4

Hazardous Materials Handbook: Awareness and Operations Levels

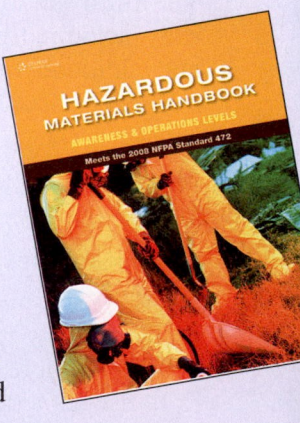

A handy resource for hazardous materials training, this book provides all the information included in the hazardous materials chapters of the third edition of *Firefighter's Handbook*.

- Meets the 2008 NFPA Standard 472
- Includes exclusively the hazardous materials response content
- For those who conduct hazardous materials training separately

Order#: 978-1-4283-1971-4

Note: We are offering these four versions of the Firefighter's Handbook for this edition. Further custom options are available for special orders. To learn more, please contact your Delmar representative.

OUR CURRICULUM PACKAGE

NEW! Skill Sheets

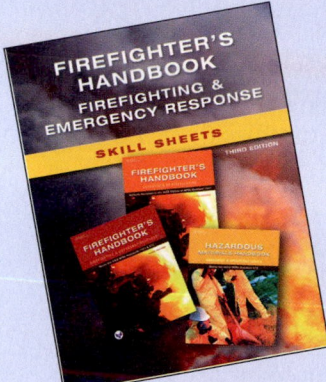

Drawn directly from the curriculum, the Job Performance Requirement and Supplement Skills sheets are also available separately in this soft cover book. Handy for tracking individual candidate progress, this book is three hole punched and perforated for ease of use on the training ground.

Order#: 978-1-4180-7326-8

Student Study Guide

Thoroughly revised for the third edition, this is helpful in the classroom setting as a guide for study and a tool for assessing progress. The study guide consists of questions in multiple formats, including new and revised questions to support the new edition of the book.

Order#: 978-1-4180-7322-0

New! Firefighter Skills DVD Series

Covering all the Skills outlined in the Instructor's Curriculum, this four part series combines live action video, graphics, and animations to illustrate important step-by-step skills essential to the job of a firefighter. Each 5 to 8 minute clip walks the viewer through the steps of the skill, while providing important review information and safety tips.

- Firefighter Skills DVD Series

Order#: 978-1-4283-1085-8

- Hazardous Materials Awareness and Operations

Order#: 978-1-4283-1089-6

ACKNOWLEDGMENTS

Firefighter's Handbook: Firefighting and Emergency Response, third edition, remains true to the Delmar tradition of remaining dedicated to the individuals we serve—among them, both aspiring and experienced firefighters. However, we would not be able to accomplish this without the contributions of many professionals whose passion, commitment, and hard work have helped shape a book of which we all can be proud.

We wish to thank those that contributed to the previous editions of this book. These founding reviewers and advisory board members offered valuable insight—they provided guidance to the content since inception and were our sounding boards in development. For their assistance, we extend our deepest gratitude.

In addition, we recognize the contributors to the third edition of this book:

Our Focus Group

A group of experts from across the United States gathered in a meeting focused on this *Firefighter's Handbook* series. Our sincere thanks go to those who traveled to meet with us and to provide us with thoughts, ideas, and recommendations for the new edition of *Firefighter's Handbook:*

James Dalton, Instructor, Chicago Fire Academy, Chicago, Illinois

Gary Fulton, Curriculum Specialist, Office of the State Fire Commissioner, Lewistown, Pennsylvania

Aaron Heller, Captain, Hamilton Township NJ Fire District #9, Township pf Hamilton, New Jersey

Jason Lloyd, Instructor/Industrial Fire, Texas Engineering Extension Service (TEEX), College Station, Texas

Gary McCarty, Battalion Chief, Fire Prevention/Community Relations, Salt Lake City Fire Department, Salt Lake City, Utah

Pat McAuliff, Director of Fire Science and EMS, Colin County Community College, McKinney, Texas

Frank Miratsky, Battalion Chief of Training, Omaha Fire & Rescue, Omaha, Nebraska

Howard Sykes, Assistant Chief, Lebanon Volunteer Fire Department/ Durham Technical Community College, Durham, North Carolina

Andrea Walter, lead author, firefighter, Note: List new job affiliation. Metropolitan Washington Airports Authority, Dulles Airport Fire Department, Washington D.C.

Our Reviewers

Our respect and appreciation to those who invested time in the review of the manuscript to provide us with insight and advice for the new edition of *Firefighter's Handbook:*

Charlie Brush, Standards Supervisor, State Fire Marshal, Bureau of Fire Standards and Training, Ocala, Florida

Andrew Byrnes, Hazardous Materials Program Coordinator, Institute of Emergency Services and Homeland Security, Utah Fire & Rescue Academy, Provo, Utah

Michael Dugan, Captain, FDNY Ladder Co.# 123, New York, New York

Jason Goodale, Engineer, Loveland Fire Rescue, Loveland, Colorado

Todd Haines, Fire Protection Engineer, Dallas Fort Worth International Airport, Texas

S.R. Hunter, Battalion Chief, Bethany Fire Department/Oklahoma State University/OKC, Bethany, Oklahoma

Richard Kelly, Major, Oklahoma City Fire Department, Oklahoma City, Oklahoma

Gary Kistner, Program Coordinator, Southern Illinois University, Carbondale, Illinois

William Klein, Program Coordinator, Fire Protection Technologies, Chemeketa Community College, Salem, Oregon

Philip Oakes, Assistant Chief of Training, Laramie County Fire District #6, Cheyenne, Wyoming

Jeremy Pope, Captain, Fire Operations , Metropolitan Washington Airports Authority, Dulles Airport Fire Department, Washington D.C.

Robert W. Royall, Jr., Chief of Emergency Operations, Harris County Fire Marshall's Office, Crosby, Texas

Tom Ruane, Fire and Life Safety Consultant, Peoria, Arizona

Glen Rudner, Hazardous Materials Officer, Virginia Department of Emergency Management, Dumfries, Virgina

Howard Sykes, Assistant Chief, Lebanon Volunteer Fire Department/ Durham Technical Community College, Durham, North Carolina

William Shouldis, Deputy Fire Chief, Philadelphia Fire Department, Philadelphia, Pennsylvania

Bruce Trego, Homeland Security Training Coordinator, Office of the State Fire Commissioner, Lewistown, Pennsylvania

Our Advisory Board Members

To those experts who work behind the scenes and provide us with continuing guidance on this book as well as other learning materials on our emergency services list, we extend our gratitude:

Steve Chikerotis, Battalion Chief, Chicago Fire Department, Chicago, Illinois

Tom Labelle, Executive Director, New York State Association of Fire Chiefs, East Schodack, New York

Jerry Laughlin, Deputy Director of Education Services, Alabama Fire College, Tuscaloosa, Alabama

Jeff Pindelski, Battalion Chief, Downers Grove Fire Department, Downers Grove, Illinois

Peter Sells, District Chief of Operations Training, Toronto Fire Services, Toronto, Canada

Billy Shelton, Executive Director, Virginia Department of Fire Programs, Glen Allen, Virginia

Doug Fry, Fire Chief , City of San Carlos Fire Department, San Carlos, California

Pat McCauliff, Director of Fire Science and EMS, Colin County Community College, McKinney, Texas

Michael Petroff, Western Director for Fire Department Safety Officers Association, St. Louis, Missouri

Adam Piskura, Director of Training, Connecticut Fire Academy, Windsor Locks, Connecticut

Theresa Staples, Program Manager, Firefighter and HazMat Certifications, Centennial, Colorado

Our Validation Committee Members

These committee members met and meticulously reviewed the 2008 Edition of the Standard 1001 along with the learning objectives of *Firefighter's Handbook*, created by a JPR subject matter expert for the third edition. The result of this meeting was a set of validated learning objectives that would tie the entire *Firefighter's Handbook* package together with the new edition of the Standard. We whole-heartedly thank those who attended this meeting and helped to initiate a new learning solution for this series:

Charlie Brush, Standards Supervisor, State Fire Marshal, Bureau of Fire Standards and Training, Ocala, Florida

Mollie C. Clakley, Training Approval and Testing Program Leader, Texas Commission on Fire Protection

Gary Fulton, Curriculum Specialist, Pennsylvania State Fire Academy, Lewistown, Pennsylvania

Ian Kraft, Fire Academy Trainer, State of Wyoming, Fire Marshal's Office, Riverton, Wyoming

William Klein, Program Chair, Fire Protection Technologies, Chemeketa Community College, Salem, Oregon

Ken Kroeker, Emergency Services Officer, Manitoba Emergency Services College, Brandon, Manitoba, Canada

Rachel Nix, Director, Arkansas Fire Academy, Camden, Arizona

Jake Rhoades, Training Officer, Stillwater Fire Department, Stillwater, Oklahoma

Derrick Warner, Training Officer, Round Lake Fire, Round Lake, Illinois

Photography

The majority of photographs shown in this book are the result of numerous days of photo shoots at various locations. A special acknowledgment is owed to the very patient firefighters who facilitated numerous days of shooting. Our appreciation is extended to the following individuals for their invaluable knowledge and willingness.

Kevin P. Terry, Chief, Fuller Road Fire Department (Colonie, New York), and Investigator, Town of Colonie Police Department

Brian Curran, Captain, Fuller Road Fire Department (Colonie, New York), and Patrolman, Town of Colonie Police Department

Kenneth Dott, Lieutenant, City of Albany Department of Fire and Emergency Services Albany, New York

Steven M. Leonardo, Past Chief, Shaker Road Loudonville Fire Department, and Investigator, Town of Colonie Police Department

Warren E. Carr, Jr., Past Chief, S.W. Pitts Hose Company, Inc. (Latham, New York)

Mike Kelleher, Captain, Troy Fire Department (Troy, New York)

Credit is also owed to the many departments, models, and photographers who shared their time, expertise, equipment, and photographs with us:

Fuller Road Fire Department, Colonie, New York

West Albany Fire Department, Colonie, New York

Fort Hunter Fire Department, Schenectady, New York

S.W. Pitts Hose Company, Inc., Latham, New York

Troy Fire Department, Troy, New York

Albany Fire Department, Albany, New York

Metropolitan Washington Airports Authority, Washington, DC

Loveland Fire and Rescue, Loveland, Colorado

Poudre Fire Authority, Fort Collins, Colorado

South Metro Fire and Rescue, Greenwood Village, Colorado

Hillsborough County Fire Rescue, Tampa, Florida

Baltimore County Fire Department, Baltimore, Maryland

Fairfax County Department of Fire and Rescue, Fairfax, Virginia

Sterling Park Rescue Squad and Sterling Volunteer Fire Department, Sterling, Virginia

Ashburn Volunteer Fire and Rescue Department, Ashburn, Virginia

Loudoun County Department of Fire Rescue Services, Leesburg, Virginia

Our photographers included:

Mike Galitelli, Metroland Photo, Inc., Loudonville, NY

Michael Dzaman, Dzaman Photography, Latham, New York

Rick Fulford, Cockeysville, Maryland

Rick Michalo and Brentt Sporn, California Fire Photos, Anaheim, California

Street Stories

Chapter 1: Andrea A. Walter, lead author, *Firefighter's Handbook*

Chapter 2: Mike Smith, Battalion Fire Chief, District of Columbia Fire and EMS Department

Chapter 3: James Angle, Chief, Palm Harbor Fire Department, Palm Harbor, Florida

Chapter 4: Mike Kelleher, Captain, Troy Fire Department, Troy, New York

Chapter 5: F. C. (Fred) Windisch, Fire Chief, Ponderosa VFD, Houston, Texas

Chapter 6: Randy Scheerer, Battalion Chief, Newport Beach Fire Department, Newport Beach, California

Chapter 7: William Tatsch, Engineer, Travis County Emergency Services District 6, Austin, Texas

Chapter 8: Mike Gala, Jr., Lieutenant, Ladder 148, FDNY, New York City

Chapter 9: Billy Goldfeder, E.F.O., Battalion Chief Loveland-Symmes Fire Department, Loveland, Ohio

Chapter 10: Michelle Steele, Lieutenant, Miami-Dade Fire and Rescue, Miami, Florida

Chapter 11: Frank Montagna, Batallion Chief, FDNY, New York City

Chapter 12: William Shouldis, Deputy Chief, Philadelphia Fire Department, Philadelphia, Pennsylvania

Chapter 13: Kevin P. Terry, Chief of Department, Fuller Road Fire Department, Town of Colonie, New York

Chapter 14: Peter F. Kertzie, Captain, Buffalo Fire Department, Buffalo, New York

Chapter 15: Joseph De Francesco, Coordinator, Madison County Fire and Rescue Services

Chapter 16: Jeffrey Pindelski, Battalion Chief, Downers Grove, Illinois Fire Department

Chapter 17: Randall F. Parr, EFO, Fire Chief, Tomball Fire Department, Tomball, Texas

Chapter 18: William Tatsch, Engineer, Travis County Emergency Services District 6, Austin, Texas

Chapter 19: James Angle, Chief, Palm Harbor Fire Department, Palm Harbor, Florida

Chapter 20: Julius Stanley, Lieutenant, Chicago Fire Department, Chicago, Illinois

Chapter 21: Mary K. Marchone, Program Manager, Montgomery County Fire and Rescue Services, Montgomery County, Maryland

Chapter 22: Gordon M. Sachs, Chief, Fairfield Fire and EMS, Fairfield, Pennsylvania

Chapter 23: Dave Dodson, Lead Instructor, Response Solutions, Colorado

Chapter 24: Rob Schnepp, Assistant Chief of Special Operations for the Alameda County (California) Fire Department

Chapter 25: Gregory G. Noll, Emergency Planning and Response Consultant, Hildebrand and Noll Associates, Lancaster, Pennsylvania

Chapter 26: Tom Creamer, Special Operations Coordinator, City of Worcester Fire Department, Worcester, Massachusetts

Chapter 27: Mike Callan, President, Callan and Company, Middlefield, Connecticut

Chapter 28: Gregory L. Socks, Captain, Montgomery County, Maryland, Hazmat Team

Chapter 29: Bill Hand, Houston Fire Department Hazardous Materials Response Team, Houston, Texas

Chapter 30: Christopher Hawley, Baltimore County Fire Department (Ret.)

Chapter 31: Captain Mark McAdams, Austin Fire Department Special Operations Battalion, Austin, Texas

Delmar Emergency Services Team

And to those we rarely take the time to recognize because this is their job, a special thanks. The Delmar team developed, produced, and marketed the third edition of *Firefighter's Handbook*, setting an example not only for getting the job done, but also having the creativity and fortitude to go above and beyond. Our appreciation goes to Alison Pase, Janet Maker, Rich Hall, Jennifer Starr, Jennifer Hanley, Amy Wetsel, Maria Conto, Erin Coffin, and Shanna Gibbs.

HOW TO CONTACT US

At Delmar, listening to what our customers have to say is the heart of our business. If you have any comments or feedback on *Firefighter's Handbook* or any of our products, you can e-mail us at delmar.fire@cengage.com, or fax us at 518-881-1262, Attention: Fire Rescue Editorial.

For additional information on other titles that may be of interest to you, or to request a catalog, visit www.delmarfire.cengage.com.

NFPA 1001, 2008 Edition Correlation Guide

5.1.1

General Knowledge Requirements	Learning Objective(s)	Chapter(s)	Page(s)
Organization	2–1	2	10–13
Role	2–3	2	13–16
Mission	1–1	1, 2	4–7
	2–2		12
SOPs	2–5	2	24–26
	2–6		24–26
Other agencies	2–7	2	27
Member Assistance Program	5–17	5	102
Physical fitness	5–14	5, 23	100–103
	5–15		100–103
	23–2		671–672
	23–3		671–672
NFPA 1500	5–3	5, 6	94–95
	5–4		94–95
	16–15		
Knot characteristics	15–5	15	428–435
	15–6		434–440, 450–461
	15–7		434–440, 450–461
Rope differences	15–2	15	428–434
	15–4		428–434
Rope service life	15–8	15	440–447
Knot types	15–14	15	447–449, 464–470
Hoisting	15–15	15	447–449, 464–470
Response usage	15–1	15	431–434

5.1.2

General Skills Requirements	Learning Objective(s)	Chapter(s)	Page(s)
PPC donning	6–14	6	112–129
	6–5		112–129
	6–1		112–129
	6–10		112–129
	6–11		123–129
PPC doffing	6–12	6	126–129
Hoisting	15–16	15	447–449, 464–470
Documentation	6–14	6	113–114, 125–129
	15–3	15	428–429, 433–434

5.2.1

Task Statement	Learning Objective(s)	Chapter(s)	Page(s)
Initiate the…to the dispatch center.			
(A) Requisite Knowledge			
Reporting emergencies	3–1	3	37–44
SOPs	3–2	3	49–55
Radio operations	3–2	3	49–55
Information	3–3	3	37–39
	3–4		37–49
(B) Requisite Skill			
Operate equipment	3–5	3	49–55
	3–6		37–49
Relay information	3–7	3	37–44, 49–54
Record information	3–8	3	37–39

5.2.2

Task Statement	Learning Objective(s)	Chapter(s)	Page(s)
Receive a…information is relayed.			
(A) Requisite Knowledge			
Non–emergency calls	3–9	3	37–39
(B) Requisite Skills			
Operate equipment	3–10	3	37–39
	3–11		37–39

5.2.3

Task Statement	Learning Objective(s)	Chapter(s)	Page(s)
Transmit and…by the AHJ.			
(A) Requisite Knowledge			
Radio procedures	3–12	3	49–55
Emergency traffic	3–12	3	49–55
Evacuation	3–12	3	49–55
(B) Requisite Skills			
Operate equipment	3–15	3	49–55

5.3.1 Task Statement	Learning Objective(s)	Chapter(s)	Page(s)
Use SCBA . . . to air depletion.	7–17	7, 23	159–164, 174–180
	23–16		676–681
	7–18		159–164, 174–180
	23–17		676–681
(A) Requisite Knowledge			
SCBA conditions and requirements	7–1	7	134–136
	7–3		136–139
	7–2		136–139
	7–4		139–140
Uses	7–8	7	146–153
	7–14		134–137
	7–5		140–146
Components	7–9	7	146–153
Donning	7–12	7	153–159, 170–173
Breathing	7–7	7	142–146
Emergency procedures	7–20	7	159–164, 174–180
Physical requirements	7–6	7	140–142
(B) Requisite Skills			
Control breathing	7–15	7	142–146
Cylinder replacement	7–26	7	165–167, 182–183
Restricted passage	7–16	7	159–164, 174–180
Emergency procedures	7–21	7	159–164, 174–180
Donning procedures	7–13	7	153–159, 170–173

5.3.2 Task Statement	Learning Objective(s)	Chapter(s)	Page(s)
Respond on . . . is correctly used.			
(A) Requisite Knowledge			
Riding apparatus	5–18	5	106–107
Hazards	5–19	5	106–107
Prohibited practices	5–19	5	106–107
PPE use	5–10	5, 6	106–107
	6–3		112–129
(B) Requisite Skills			
Safety equipment	5–10	5, 6	106–107
	6–3		112–129

5.3.3 Task Statement	Learning Objective(s)	Chapter(s)	Page(s)
Establish and . . . protected work areas.			
(A) Requisite Knowledge			
Emergency scene hazards	4–5	4, 13, 16, 19	76–84
	13–10		345–348, 356–365
	16–1		478–484
	19–1		607–636

5.3.3 Task Statement	Learning Objective(s)	Chapter(s)	Page(s)
Dismounting apparatus	16–43 19–77	16, 19	497–499 633–635
Operating at scene	5–21 16–6 19–78 23–1	5, 16, 19, 23	97–107 479, 497–499 633–635 666–677
Protective equipment	5–11 6–2 16–7 19–15	5, 6, 16, 19	97–100, 103–107 112–129 497–499 633–635
(B) Requisite Skills PPC	6–16 16–8 19–16	6, 16, 19	112–129 478–499 608–609, 633–635
Scene control	16–42 19–79	16, 19	497–499 633–635
Dismounting apparatus	16–44 19–80	16, 19	497–499 633–635
Scene operations	16–41 19–81	16, 19	497–499 633–635

5.3.4 Task Statement	Learning Objective(s)	Chapter(s)	Page(s)
Force entry . . . ready for entry.	17–1	17	514–523
(A) Requisite Knowledge Basic construction	17–19	17	526–553
Operation	17–21 17–26 17–29 17–23	17	526–553 542–545 545–547 532–536
Forcible entry dangers	17–22 17–27 17–30	17	523–525, 526–553 523–525, 542–545 523–525, 545–547
(B) Requisite Skills Tool usage	17–5	17	515–553
Force entry	17–24 17–25 17–31 17–28	17	526–542, 550–553 539–540, 552–553 545–547 542–545

The purpose of this Correlation, and the abbreviated text, is intended to be used in conjunction with the full version of the NFPA Standard to direct the user to the exact component of the JPR for which the listed learning objective(s) and page number(s) are referenced.

5.3.5 Task Statement	Learning Objective(s)	Chapter(s)	Page(s)
Exit a..integrity is maintained			
(A) Requisite Knowledge			
Accountability	23–7	23	668–669
	23–8		668–669
Communication	23–14	23	672, 676–678
Evacuation	23–15	23	676–678
Safety	16–3	16	479–480
Hazards	13–11		345–348, 356–365
	16–4		479–480
	19–2		608–636
Emergency procedures	7–19	7, 16, 23	159–164, 174–180
	16–23		480–484, 492–495, 500–504
	23–12		677–681
(B) Requisite Skills			
Poor visibility operations	16–14	16	478–492
Guideline	16–16	16	478–492
	16–17		478–492
Air management	7–27	7, 16	142–146
	16–24		
Hazards and safety	16–5	16	479–480

5.3.6 Task Statement	Learning Objective(s)	Chapter(s)	Page(s)
Set up…assignment is accomplished.			
(A) Requisite Knowledge			
Parts	14–1	14	371–375
Hazards	14–14	14	375–379
Placement	14–6	14	380–389
Angles	14–6	14	386–389
Safety	14–9	14	386–389, 393–398, 408–418
Top placement	14–11	14	386–389
(B) Requisite Skills			
Carries	14–17	14	391–399, 405–423
Raise	14–17	14	391–399, 405–423
Extend	14–7	14	391–399, 405–420
Support	14–12	14	380–389
Height requirements	14–8	14	380–389
Placement	14–13	14	380–389
	14–10		391–399, 405–420

5.3.7 Task Statement	Learning Objective(s)	Chapter(s)	Page(s)
Attack a..fire is extinguished.			
(A) Requisite Knowledge			
Fire streams	19–63	19	614–619, 630–635
Precautions	19–64 19–11	19	614–619, 630–635
Results	19–65	19	614–619, 630–635
Fuel hazards	19–76	19	630–635
Safety	19–10	19	614–619, 630–635
Accidents/Injuries	19–66	19	614–619, 630–635
Vehicle access	19–67	19	614–619, 630–635 630–635
Overhaul	19–68	19	630–635
(B) Requisite Skills			
Fuel types	19–69	19	630–635
Fuel leaks	19–70	19	630–635
Nozzle operations	11–10 19–71	11, 19	298–304 614–619, 630–635
Water application	19–72	19	614–619, 630–635
Hose advancement	11–17 19–74	11, 19	298–306 614–619, 630–635
Hidden fires	19–75	19	630–635

5.3.8 Task Statement	Learning Objective(s)	Chapter(s)	Page(s)
Extinguish fires...arson are preserved.			
(A) Requisite Knowledge			
Attack lines	19–95	19	615–619, 635–636
Dangers	19–96	19	635–636
Extinguishing agents	19–97	19	635–636
Overhaul	19–98	19	635–636
Extinguishment	19–99	19	635–636
Exposure protection	19–100	19	635–636
Hazardous materials	19–101	19	635–636
Origin/cause/preservation	20–31 20–32	20	655–657 655–657

5.3.8 Task Statement	Learning Objective(s)	Chapter(s)	Page(s)
(B) Requisite Skills			
Hazards	19–102	19	615–619, 635–636
Hose line operations	19–103	19	615–619, 635–636
Overhaul	19–104	19	615–619, 635–636
Extinguishment	19–105	19	615–619, 635–636
Water application	19–106	19	615–619, 635–636
Penetration	19–104	19	615–619, 635–636
Hidden fires	19–105	19	615–619, 635–636
Fire patterns	20–33	20	655–657
Extinguishment	19–105	19	615–619, 635–636

5.3.9 Task Statement	Learning Objective(s)	Chapter(s)	Page(s)
Conduct a…is not compromised.			
(A) Requisite Knowledge			
Tools	17–4	17	514–551
Ladders	14–15	14	383
Poor visibility operations	7–28	7	164–169, 184–185
Tenability	16–9	16	479–480
Search	16–12	16	478–485
Roles	16–11	16	478–485
Victim location	16–10 16–13	16	480–486 480–486
Victim removal	16–19	16	492–497, 500–510
Respiratory protection	16–18	16	478–479
(B) Requisite Skills			
SCBA use	7–16	7	159–164, 174–180
Ladders	14–16	14	380–383, 386–389, 391–399, 405–420
Firefighter rescue (scenario I)	7–22 16–21	7, 16	159–164, 174–180 486–495, 500–508,
Firefighter rescue (scenario II)	7–23 16–22	7, 16	159–164, 174–180 486–495, 500–508
Victim rescue	16–20	16	484–495, 500–508
Tenability	13–12 16–2	13, 16	345–348, 356–365 479–484

5.3.10 Task Statement	Learning Objective(s)	Chapter(s)	Page(s)
Attack an...brought under control.			
(A) Requisite Knowledge			
Fire streams	11–1	11, 19	298
	11–2		298–303
	11–3		298–303
	11–18		304–306
	19–17		615–619
	11–19		304–306
	19–18		615–619
	11–4		298–299, 307–311
Nozzles	11–7	11	298–304
	11–8		298–304
	11–9		298–306
	11–6		298–306
	11–5		298–306
Precautions	10–15	10	243–250, 274–286
Results	11–12	11	298–306
Hazards	4–6	4, 13, 19	76–86
	13–4		345–348, 356–365
			607–609
Exposure protection	19–54	19	619–627
Exposure consequences	4–7	4	76–80
Matter	4–1	4	68–71
	4–11		68–71
Accidents/Injuries	5–1	5	93–95
Attack lines	11–11	11	304–306
	19–21	19	615–619
Backup team	19–35	19	619–626
Attack and control	19–36	19	619–624
Expose hidden fires	20–18	20	652–655
(B) Requisite Skills			
Water hammer	11–13	11	298–304
Nozzles	11–14	11	298–304
Attack methods	11–20	11, 19	298–306
	19–20		619–636
Hose line advancement	10–16	10, 19	243–250, 274–286
	19–37		617–624
Extending hose	10–25	10	253, 289–291
Replace hose	10–26	10	254
Operate hose lines	10–17	10	248–250, 285–286
Hose coupling	10–11	10	234–235, 259–261
Carrying hose	10–14	10	236, 265–267
Fire attack	19–39	19, 20	617–624
	20–19		652–656

5.3.11 Task Statement	Learning Objective(s)	Chapter(s)	Page(s)
Perform horizontal..cleared of smoke.			
(A) Requisite Knowledge			
Types of ventilation	18–12	18	
	18–1		558–560
	18–8		560–563, 575–576
	18–13		571–574, 580–583
	18–15		574–575
	18–6		591–598
	18–2		591–598
Safety	18–9	18	590–591
Fire behavior	4–3	4, 8	63–65
	4–8		65–68
	4–15		65–71, 76–79
	4–18		80–84
	4–13		72–76
	4–14		76–79
	4–9		72
	8–1		190–192
	4–16		77
	4–4		63–65
Products of combustion	4–2	4	63
	4–17		79–81
Backdraft	4–21	4, 18	81, 84–86
	18–7		563–568
	4–20		81, 82–86
	18–11		563–569
	18–11		563–569
Fire growth	4–12	4	71
(B) Requisite Skills			
Use tools and equipment	14–18	14	399, 419–423
Safety procedures	18–16	18	569–580

5.3.12 Task Statement	Learning Objective(s)	Chapter(s)	Page(s)
Perform vertical…ventilation is accomplished.			
(A) Requisite Knowledge			
Heat transfer	4–13	4	72–76
Thermal layering	4–16	4	77
Ventilation techniques	18–18	18	560–563, 583–598
	18–10		560–563, 583–598
	18–19		560–563, 583–598
	18–3		560–563, 583–598
	18–14		560–563, 583–598
Collapse	13–14	13	345–348
	13–15		363
	13–16		363–365

5.3.12 Task Statement	Learning Objective(s)	Chapter(s)	Page(s)
Structural integrity	13–1	13	340–343
	13–2		340–345
	13–3		345–348
	13–6		349–356
	13–8		365–366
	13–7		363
Advantages/Disadvantages	18–21	18	583–598
	18–22		583–598
	18–23		583–598
(B) Requisite Skills			
Use tools and equipment	14–18	14	399, 419–423
Hoisting	15–16	15	447–449, 464–470
Cutting operations	18–24	18	583–598
Sounding	18–20	18	583–598
Opening	18–17	18	583–598
Ground ladder operations	14–19	14, 18	379–389, 400–401 655–664
	18–25		
Roof ladder operations	14–21	14	400–401
Carrying tools and equipment	14–19	14	399, 419

5.3.13 Task Statement	Learning Objective(s)	Chapter(s)	Page(s)
Overhaul a...fire is extinguished.			
(A) Requisite Knowledge			
Hose lines for overhaul	20–25	20	652–655
Limiting water damage	20–21	20	652–655
Hidden fires	20–16	20	652–655
	20–17		
Dangers	20–22	20	652–655
	20–23		
Area of origin	20–29	20	652–657
	20–31		
Scene protection	20–30	20	655–657
(B) Requisite Skills			
Attack line deployment	20–26	20	652–655
Void spaces	20–20	20	652–655
Water application	20–27	20	652–655
Hidden fires	20–24	20	652–657
Area of origin	20–24	20	652–657
Extinguishment	20–28	20	652–655

5.3.14 Task Statement	Learning Objective(s)	Chapter(s)	Page(s)
Conserve property...from further damage.			
(A) Requisite Knowledge			
Property conservation	20–1	20	641–655
Protection methods	20–2	20	641–652
Salvage covers	20–3	20	641–645
Sprinkler operations	12–5	12, 20	325–327, 333–335, 641–652
	20–15		
Flow stoppage	12–4	12	318–325
Main control valve	12–3	12	318–327
Forcible entry	17–20	17, 20	523–526
	20–10		646–647
(B) Requisite Skills			
Furniture clustering	20–11	20	645–652, 658–662
Coverings	20–11	20	645–652, 658–662
Salvage covers	20–9	20	645–652, 658–662
Water chutes	20–14	20	648–652, 662
Water removal	20–13	20	648–652, 662
Covering openings	20–4	20	641–652
Area of origin protection	20–12	20	641–652
Flow stoppage	12–6	12	325–327, 335
Main control valve	12–7	12	318–327

5.3.15 Task Statement	Learning Objective(s)	Chapter(s)	Page(s)
Connect a...flow is unobstructed.	9–1	9	210
	9–4		210–213
	9–11		212–213, 220–221
	9–8		217
(A) Requisite Knowledge			
Mobile water supply vehicle	9–6	9	217–220
	9–12		217–220
	9–13		217–220
Hydrants	9–9	9, 10	213–217
	10–19		250–252, 287
	10–20		250–252, 287
Static water supply	9–2	9	212–213
Water source connections	9–5	9, 21	213–220
	10–21		250–253, 287–288
(B) Requisite Skills			
Laying supply hose	10–28	10	250–252, 254–255, 287
Drafting operations	10–22	10	252–253, 288

5.3.15 Task Statement	Learning Objective(s)	Chapter(s)	Page(s)
Portable water tanks	9–7	9	217–220
Hose lay connections	10–27	10	254–255
Supply hose connections	10–23	10	250–252, 287
Hydrant operation	9–10 10–24	9, 10 250–252	213–217

5.3.16 Task Statement	Learning Objective(s)	Chapter(s)	Page(s)
Extinguish incipient…techniques are followed.	8–9	8	201–204
(A) Requisite Knowledge			
Classes of fire	4–10 8–2 8–4	4, 8	72 201–204 193–200
Ratings system	8–5 8–3	8	199–200 191–200
Extinguisher limitations	8–6 8–7	8	200 200–201, 205
(B) Requisite Skills			
Extinguisher operation	8–8	8	193–201, 205
Fire approach	8–8	8	193–201, 205
Selection	8–8	8	193–201, 205
Safety	8–8	8	193–201, 205

5.3.17 Task Statement	Learning Objective(s)	Chapter(s)	Page(s)
Illuminate the…listed safety precautions			
(A) Requisite Knowledge			
Safety	19–47	19	625–626
Power supply	19–49	19	625–626
Deployment	19–50	19	625–626
(B) Requisite Skills			
Lighting operations	19–51	19	625–626
Power cords and connections	19–52	19	625–626
GFI operations	19–53	19	625–626
Lighting placement	19–48	19	625–626

5.3.18

Task Statement	Learning Objective(s)	Chapter(s)	Page(s)
Turn off…is safely completed.			
(A) Requisite Knowledge			
Utilities and safety	19–41	19	625–626
Hazards	19–43	19	625–626
Safety equipment	19–44	19	625–626
(B) Requisite Skills			
Control devices	19–45	19	625–626
Operations	19–46	19	625–626
Hazards	19–42	19	625–626

5.3.19

Task Statement	Learning Objective(s)	Chapter(s)	Page(s)
Combat a…assignment is completed.			
(A) Requisite Knowledge			
Types	19–8	19	609–614, 627–630
Parts	19–55	19	609–614, 627–630
Methods	19–56	19	609–614, 627–630
Safety	19–7	19	609–614
	19–9		609–614
	19–57		609–614, 627–630
(B) Requisite Skills			
Exposure threats	19–58	19	609–614, 627–630
Exposure protection	19–60	19	609–614, 627–630
Hand tools	19–61	19	609–614, 627–630
Fire lines	19–62	19	609–614, 627–630
Suppression	19–59	19	609–614, 627–630

5.5.1

Task Statement	Learning Objective(s)	Chapter(s)	Page(s)
Clean and…or reported otherwise.			
(A) Requisite Knowledge			
Cleaning methods	6–6	5, 6, 7, 14,	123–129
	7–10	15, 17, 20	164–169, 181–182
	14–2		375–379
	15–10		440–447, 462–463
	20–5		641–645, 652–653
	5–13		98–100, 103–104
	10–6		
Solvents	6–7	6, 7, 10,	123–129
	7–24	14, 15, 17, 20	164–169, 181–182
	10–7		229–234
	14–3		375–379
	15–11		440–447, 462–463
	17–7		523–526
	20–6		

5.5.1 Task Statement	Learning Objective(s)	Chapter(s)	Page(s)
Solvents	17–7		525–526
	20–6		641–645, 652–653
Guidelines	6–8	6, 7, 8, 14,	123–129
	7–25	15, 20	164–169,181–182
	8–11		201–204
	14–4		375–379
	15–12		440–447, 462–463
	20–7		641–645, 652–653
(B) Requisite Skills			
Tool selection	6–9	6, 7, 10,	123–129
	7–11	14, 15, 17, 20	164–169, 181–182
	10–8		229–232
	14–5		375–379
	15–13		440–447, 462–463
	17–8		525–526
	20–8		641–645, 652–653
Guidelines	15–9	15, 17	440–447, 462–463
	17–6		525–526
Records/Reports	5–14	5	100, 104

5.5.2 Task Statement	Learning Objective(s)	Chapter(s)	Page(s)
Clean, Inspect…state for service.	10–9	10	227–232
(A) Requisite Knowledge			
Defective hose	10–1	10	226–232
	10–32		227–232
	10–33		227–232
Cleaning	10–3	10	229–230
Make ready for use	10–12	10	235–243, 261–273
(B) Requisite Skills			
Cleaning	10–4	10	229–230
Equipment operation	10–5	10	229–230
Hose marking	10–34	10	229–230
Gaskets	10–10	10	232
Make ready for use	10–13	10	235–243, 261–273

6.1.1

General Knowledge Requirements	Learning Objective(s)	Chapter(s)	Page(s)
Responsibilities of FF II	2–11	2, 19	689–692
	2–14		692–696
	2–8		689–692
	19–29		689
	2–13		689–697
	2–10		689–697
	19–30		689
Assigned duties	2–4	2	687–688
FF II role	2–4	2	687–688

6.1.2

General Skills Requirements	Learning Objective(s)	Chapter(s)	Page(s)
Command need	2–9	2, 19	689–697
	19–31		868–874
IMS operation	2–12	2, 19	689–697
	19–32		868–874
Function within IMS	2–15	2, 19	689–697
	19–33		868–874

6.2.1

Task Statement	Learning Objective(s)	Chapter(s)	Page(s)
Complete a . . . report is complete.			
(A) Requisite Knowledge			
Content	3–16	3	711–712
Accurate reports	3–17	3	706–712
Consequences	3–17	3	706–712
Information gathering	3–16	3	711–712
Coding	3–18	3	711–712
(B) Requisite Skills			
Codes	3–19	3	711–712
Proofing	3–20	3	711–712
Technology operation	3–19	3	711–712

6.2.2 Task Statement	Learning Objective(s)	Chapter(s)	Page(s)
Communicate the . . . is accomplished safely			
(A) Requisite Knowledge			
Standard Operating Procedures	3–13	3	701–712
(B) Requisite Skills			
Communications equipment	3–14	3	701–712

6.3.1 Task Statement	Learning Objective(s)	Chapter(s)	Page(s)
Extinguish an . . . haven is reached	11–26	11	746–753
(A) Requisite Knowledge			
Foam effects	11–25	11	746–748
Foam generation	11–25	11	746–748
Poor foam generation	11–33	11	746–748
	11–34		746–748
Hydrocarbon vs. Polar solvent	11–23	11	748
Foam characteristics	11–22	11	746–747
	11–24		746–747
Fog nozzles vs. foam nozzles	11–35	11	748–753
Foam application	11–28	11	748–753
Hazards	11–31	11	751
Hazard avoidance	11–32	11	748–753
(B) Requisite Skills			
Foam preparation	11–27	11	748–753
Component assembling	11–27	11	748–753
Application	11–29	11	748–753
Spill operations	11–30	11	748–753

6.3.2 Task Statement	Learning Objective(s)	Chapter(s)	Page(s)
Coordinate an . . . of changing conditions.	2–16	2, 5, 11, 19	687–697
	5–5	23	723–724
	19–25		869–874
	23–4		935–940
	11–21		744–745
	19–27		862–863, 866–874
(A) Requisite Knowledge			
Nozzle/hose selection	11–15	11, 19	737–743
	19–28		866–874
Adapter/appliance selection	11–16	11, 19	737–743, 748–753
	19–38		866–874
Dangerous conditions	13–19	13, 19	782–799
	19–4		862–863, 866–874
	13–4		781–799
	19–5		862–863, 866–874

The purpose of this Correlation, and the abbreviated text, is intended to be used in conjunction with the full version of the NFPA Standard to direct the user to the exact component of the JPR for which the listed learning objective(s) and page number(s) are referenced.

The purpose of this Correlation, and the abbreviated text, is intended to be used in conjunction with the full version of the NFPA Standard to direct the user to the exact component of the JPR for which the listed learning objective(s) and page number(s) are referenced.

6.3.4 Task Statement	Learning Objective(s)	Chapter(s)	Page(s)
Protect evidence . . . on the scene.			
(A) Requisite Knowledge			
Origin and cause	20–34	20	895–898
Evidence	20–35	20	895–898
Evidence protection	20–36	20	895–898
	20–37		895–898
FF II role	20–38	20	895–898
Property removal	20–39	20	895–898
(B) Requisite Skills			
Origin location	20–40	20	895–898
Causes	20–41	20	895–898
Evidence	20–42	20	895–898

6.4.1 Task Statement	Learning Objective(s)	Chapter(s)	Page(s)
Extricate a . . . hazards are managed.			
(A) Requisite Knowledge			
Fire department role	16–25	16	810–813
	16–45		808
	16–40		808–810
	16–37		808–810
	16–33		808–810
	16–48		810–814, 830–836
	16–51		812–813
	16–26		810–813
	16–35		810–813
Auto body construction	16–50	16	810–814, 830–836
	16–27		810–814, 830–836
Vehicle components	16–28	16	804–808
	16–36		810–813
	16–29		804–808
Extrication equipment	16–30	16	804–808
Safety	16–31	16	804–808
(B) Requisite Skills			
Tool operation	16–32	16	804–814, 830–836
Cribbing	16–47	16	810
Vehicle component removal	16–46	16	808–810
	16–38		808–810
	16–39		808–810
	16–34		808–810
	16–49		804–814, 830–836
	16–52		804–814, 830–836
	16–53		804–814, 830–836

6.4.2 Task Statement	Learning Objective(s)	Chapter(s)	Page(s)
Assist rescue . . . assignment is completed.			
(A) Requisite Knowledge			
Fire fighter's role	16–54	16	814–829
Hazards	16–55 16–56	16	814–829
Tools	16–58	16	814–829
Rescue	16–60	16	814–829
(B) Requisite Skills			
Rescue tools	16–59	16	814–829
Barriers	16–57	16	814–829
Provide assistance	16–61	16	814–829

6.5.1 Task Statement	Learning Objective(s)	Chapter(s)	Page(s)
Perform a . . . the proper authority.			
(A) Requisite Knowledge			
Policy	21–3	21	922–924
Fire causes and prevention	21–4	21	922–924
Safety survey	21–2 21–1	21	904, 922–924 922–929
Referrals	21–7	21	922–924
(B) Requisite Skills			
Forms	21–8	21	922–924
Hazards	21–5	21	922–924
Findings	21–6	21	922–924
Communication	21–6	21	922–924

6.5.2 Task Statement	Learning Objective(s)	Chapter(s)	Page(s)
Present fire . . . answered or referred.			
(A) Requisite Knowledge			
Informational materials	21–9	21	924–929
Presentation skills	21–10	21	924–929
Station tour	21–11	21	924–929
(B) Requisite Skills			
Documentation	21–12 21–13	21	924–929 924–929

6.5.3 Task Statement	Learning Objective(s)	Chapter(s)	Page(s)
Prepare a . . . diagrams are prepared.	9–14	9	727–730
(A) Requisite Knowledge			
Water supply	9–3 12–9	9, 21	727 762–777
System fundamentals	12–1 12–2	12	757–762 762–774
Diagramming	21–16	21	930–932
Preincident survey	21–14 21–15	21	929–932
Accuracy	21–17	21	929–932
(B) Requisite Skills			
System components	21–18	21	930–932
Sketching	21–19	21	929–932
Hazard detection	21–20	21	929–932
Forms	21–21	21	929–932

6.5.4 Task Statement	Learning Objective(s)	Chapter(s)	Page(s)
Maintain power . . . or reported otherwise.			
(A) Requisite Knowledge			
Cleaning methods	17–9	17	850–852
Solvents	17–10	17	850–852
Guidelines	17–11	17	850–852
Reporting	17–12	17	850–852
(B) Requisite Skills			
Tool selection	17–13	17	850–852
Guidelines	17–14	17	850–852
Recording/reporting	17–15	17	850–852
Equipment operation	17–16 17–17 17–18	17	850–852 850–852 850–852

6.5.5 Task Statement	Learning Objective(s)	Chapter(s)	Page(s)
Perform an . . . results are recorded.			
(A) Requisite Knowledge			
Hose testing	10–2 10–29	10	733–735 733–735
Hose removal	10–31	10	733–735
Test results	10–35	10	733–735
(B) Requisite Skills			
Equipment and records	10–30	10	733–735

NFPA 472, 2008 Edition Correlation Guide

	NFPA 472, Chapter 4 Awareness	Learning Objective(s)	Chapter(s)	Page(s)
4.1	**General.**			
4.1.1	**Introduction.**			
4.1.1.1	Awareness level personnel shall be persons who, in the course of their normal duties, could encounter an emergency involving hazardous materials/weapons of mass destruction (WMD) and who are expected to recognize the presence of the hazardous materials/WMD, protect themselves, call for trained personnel, and secure the area.	24-1, 24-2	24	996–1000, 996–1005
4.1.1.2	Awareness level personnel shall be trained to meet all competencies of this chapter.	24-2	24	996–1005
4.1.1.3	Awareness level personnel shall receive additional training to meet applicable governmental occupational health and safety regulations.	24-6	24	1002–1005
4.1.2	**Goal.**			
4.1.2.1	The goal of the competencies at the awareness level shall be to provide personnel already on the scene of a hazardous materials/WMD incident with the knowledge and skills to perform the tasks in 4.1.2.2 safely and effectively.	24-1, 24-2	24	996–999, 996–1005
4.1.2.2	When already on the scene of a hazardous materials/WMD incident, the awareness level personnel shall be able to perform the following tasks:	24-1, 24-2	24	996–999, 996–1005
(1)	Analyze the incident to determine both the hazardous material/WMD present and the basic hazard and response information for each hazardous material/WMD agent by completing the following tasks:	25-1, 25-2, 25-3, 25-4, 25-5, 25-6, 25-7, 25-8	25	1010–1055
(a)	Detect the presence of hazardous materials/WMD.	25-1, 25-3	25	1010–1055

	NFPA 472, Chapter 4 Awareness	Learning Objective(s)	Chapter(s)	Page(s)
(b)	Survey a hazardous materials/WMD incident from a safe location to identify the name, UN/NA identification number, type of placard, or other distinctive marking applied for the hazardous materials/WMD involved.	25-1, 25-2	25	1010–1028
(c)	Collect hazard information from the current edition of the DOT Emergency Response Guidebook.	26-5	26	1072–1089, 1099–1101
(2)	Implement actions consistent with the emergency response plan, the standard operating procedures, and the current edition of the DOT Emergency Response Guidebook by completing the following tasks:	26-5	26	1072–1089, 1099–1101
(a)	Initiate protective actions.	28-1	28	1130–1143, 1167
(b)	Initiate the notification process.	24-3	24	999
4.2	**Competencies—Analyzing the Incident.**			
4.2.1*	**Detecting the Presence of Hazardous Materials/WMD.** Given examples of various situations, awareness level personnel shall identify those situations where hazardous materials/WMD are present and shall meet the following requirements:	25-1, 25-2, 25-3	25	1010–1066
(1)*	Identify the definitions of both *hazardous material* (or *dangerous goods*, in Canada) and *WMD*.	25-1	25	1010–1066
(2)	Identify the UN/DOT hazard classes and divisions of hazardous materials/WMD and identify common examples of materials in each hazard class or division.	25-1, 25-2	25	1011–1028
(3)*	Identify the primary hazards associated with each UN/DOT hazard class and division.	25-2	25	1011–1028
(4)	Identify the difference between hazardous materials/WMD incidents and other emergencies.	24-1	24	996–999
(5)	Identify typical occupancies and locations in the community where hazardous materials/WMD are manufactured, transported, stored, used, or disposed of.	25-1, 25-3	25	1010–1055
(6)	Identify typical container shapes that can indicate the presence of hazardous materials/WMD.	25-4	25	1028–1055
(7)	Identify facility and transportation markings and colors that indicate hazardous materials/WMD, including the following:	25-5	25	1028–1055
(a)	Transportation markings, including UN/NA identification number marks, marine pollutant mark, elevated temperature (HOT) mark, commodity marking, and inhalation hazard mark	25-2, 25-5, 25-6, 25-7	25	1010–1055
(b)	NFPA 704, *Standard System for the Identification of the Hazards of Materials for Emergency Response*, markings	25-5	25	1023–1025
(c)*	Military hazardous materials/WMD markings	25-5	25	1025–1026
(d)	Special hazard communication markings for each hazard class	25-2, 25-5	25	1023–1028
(e)	Pipeline markings	25-5	25	1026–1027
(f)	Container markings	25-2, 25-6, 25-7	25	1011–1028
(8)	Given an NFPA 704 marking, describe the significance of the colors, numbers, and special symbols.	25-5	25	1023–1025
(9)	Identify U.S. and Canadian placards and labels that indicate hazardous materials/WMD.	25-2	25	1010–1028

	NFPA 472, Chapter 4 Awareness	Learning Objective(s)	Chapter(s)	Page(s)
(10)	Identify the following basic information on material safety data sheets (MSDS) and shipping papers for hazardous materials:	26-1, 26-2, 26-3	26	1089–1095, 1102
(a)	Identify where to find MSDS.	26-1, 26-2, 26-3	26	1089–1095, 1102
(b)	Identify major sections of an MSDS.	26-1, 26-2, 26-3	26	1089–1095, 1102
(c)	Identify the entries on shipping papers that indicate the presence of hazardous materials.	26-4	26	1093–1098
(d)	Match the name of the shipping papers found in transportation (air, highway, rail, and water) with the mode of transportation.	26-4	26	1093–1098
(e)	Identify the person responsible for having the shipping papers in each mode of transportation.	26-4	26	1093–1098
(f)	Identify where the shipping papers are found in each mode of transportation.	26-4	26	1093–1098
(g)	Identify where the papers can be found in an emergency in each mode of transportation.	26-4	26	1093–1098
(11)*	Identify examples of clues (other than occupancy/ location, container shape, markings/color, placards/ labels, MSDS, and shipping papers) the sight, sound, and odor of which indicate hazardous materials/WMD.	25-1, 28-6	25, 28	1010–1055, 1143–1157
(12)	Describe the limitations of using the senses in determining the presence or absence of hazardous materials/WMD.	25-1	25	1105
(13)*	Identify at least four types of locations that could be targets for criminal or terrorist activity using hazardous materials/WMD.	30-1	30	1207–1209
(14)*	Describe the difference between a chemical and a biological incident.	30-7	30	1216–1226
(15)*	Identify at least four indicators of possible criminal or terrorist activity involving chemical agents.	30-2	30	1200–1214
(16)*	Identify at least four indicators of possible criminal or terrorist activity involving biological agents.	30-2	30	1209–1223
(17)	Identify at least four indicators of possible criminal or terrorist activity involving radiological agents.	30-1, 30-2	30	1209–1223
(18)	Identify at least four indicators of possible criminal or terrorist activity involving illicit laboratories (clandestine laboratories, weapons lab, ricin lab).	30-1, 30-2, 30-4	30	1209–1223
(19)	Identify at least four indicators of possible criminal or terrorist activity involving explosives.	30-1, 30-2, 30-7	30	1209–1223
(20)*	Identify at least four indicators of secondary devices.	30-2	30	1209–1223
4.2.2	**Surveying Hazardous Materials/WMD Incidents.** Given examples of hazardous materials/WMD incidents, awareness level personnel shall, from a safe location, identify the hazardous material(s)/WMD involved in each situation by name, UN/NA identification number, or type placard applied and shall meet the following requirements:			
(1)	Identify difficulties encountered in determining the specific names of hazardous materials/WMD at facilities and in transportation.	25-2, 26-1	25	1010–1066, 1077
(2)	Identify sources for obtaining the names of, UN/NA identification numbers for, or types of placard associated with hazardous materials/WMD in transportation.	25-1, 25-2	25	1010–1055

NFPA 472, Chapter 4 Awareness		Learning Objective(s)	Chapter(s)	Page(s)
(3)	Identify sources for obtaining the names of hazardous materials/WMD at a facility.	24-4, 24-5, 26-8	24, 26	1089–1093, 1102, 1098
4.2.3*	**Collecting Hazard Information.** Given the identity of various hazardous materials/WMD (name, UN/NA identification number, or type placard), awareness level personnel shall identify the fire, explosion, and health hazard information for each material by using the current edition of the DOT *Emergency Response Guidebook* and shall meet the following requirements:			
(1)*	Identify the three methods for determining the guidebook page for a hazardous material/WMD.	26-5	26	1072–1089, 1099–1101
(2)	Identify the two general types of hazards found on each guidebook page.	26-5	26	1072–1089, 1099–1101
4.3*	**Competencies—Planning the Response. (Reserved)**			
4.4	**Competencies—Implementing the Planned Response.**			
4.4.1*	**Initiating Protective Actions.** Given examples of hazardous materials/WMD incidents, the emergency response plan, the standard operating procedures, and the current edition of the DOT *Emergency Response Guidebook*, awareness level personnel shall be able to identify the actions to be taken to protect themselves and others and to control access to the scene and shall meet the following requirements:			
(1)	Identify the location of both the emergency response plan and/or standard operating procedures.	24-4, 24-5	24	997–1000
(2)	Identify the role of the awareness level personnel during hazardous materials/WMD incidents.	24-2	24	1000–1005
(3)	Identify the following basic precautions to be taken to protect themselves and others in hazardous materials/WMD incidents:	27-1	27	1108–1118
(a)	Identify the precautions necessary when providing emergency medical care to victims of hazardous materials/WMD incidents.	28-1	28	1130–1143
(b)	Identify typical ignition sources found at the scene of hazardous materials/WMD incidents.	28-6	28	1147–1151
(c)*	Identify the ways hazardous materials/WMD are harmful to people, the environment, and property.	27-1, 27-2	27	1108–1118
(d)*	Identify the general routes of entry for human exposure to hazardous materials/WMD.	27-1, 27-2	27	1108–1118
(4)*	Given examples of hazardous materials/WMD and the identity of each hazardous material/WMD (name, UN/NA identification number, or type placard), identify the following response information:	25-1, 25-2	25	1010–1066
(a)	Emergency action (fire, spill, or leak and first aid)	26-5	26	1072–1089
(b)	Personal protective equipment necessary	27-4	27	1118–1126
(c)	Initial isolation and protective action distances	26-5	26	1072–1089
(5)	Given the name of a hazardous material, identify the recommended personal protective equipment from the following list:	27-4	27	1118–1126
(a)	Street clothing and work uniforms	27-4	27	1118–1126
(b)	Structural fire-fighting protective clothing	27-4	27	1118–1126
(c)	Positive pressure self-contained breathing apparatus	27-4	27	1118–1126

	NFPA 472, Chapter 5, Operations	Learning Objective(s)	Chapter(s)	Page(s)
5.1.1.3*	The operations level responder shall receive additional training to meet applicable governmental occupational health and safety regulations.	24-1, 24-2, 24-4	24	996–1005
5.1.2	**Goal**			
5.1.2.1	The goal of the competencies at this level shall be to provide operations level responders with the knowledge and skills to perform the core competencies in 5.1.2.2 safely.	24-2, 24-3	24	996–1005
5.1.2.2	When responding to hazardous materials/WMD incidents, operations level responders shall be able to perform the following tasks:	24-2	24	996–1005
(1)	Analyze a hazardous materials/WMD incident to determine the scope of the problem and potential outcomes by completing the following tasks:	24-2, 24-3, 24-4	24	996–1005
(a)	Survey a hazardous materials/WMD incident to identify the containers and materials involved, determine whether hazardous materials/WMD have been released, and evaluate the surrounding conditions.	25-1, 25-2	25	1010–1055
(b)	Collect hazard and response information from MSDS; CHEMTREC/CANUTEC/SETIQ; local, state, and federal authorities; and shipper/manufacturer contacts.	26-1, 26-2, 26-3, 26-4, 26-8	26	1072–1102
(c)	Predict the likely behavior of a hazardous material/WMD and its container.	28-6	28	1143–1157
(d)	Estimate the potential harm at a hazardous materials/WMD incident.	27-1	27	1108–1118
(2)	Plan an initial response to a hazardous materials/WMD incident within the capabilities and competencies of available personnel and personal protective equipment by completing the following tasks:	27-4	27	1108–1118
(a)	Describe the response objectives for the hazardous materials/WMD incident.	28-1	28	1130–1143
(b)	Describe the response options available for each objective.	28-1	28	1130–1143
(c)	Determine whether the personal protective equipment provided is appropriate for implementing each option.	27-4	27	1118–1126
(d)	Describe emergency decontamination procedures.	28-7	28	1157–1169
(e)	Develop a plan of action, including safety considerations.	28-1	28	1130–1143
(3)	Implement the planned response for a hazardous materials/WMD incident to favorably change the outcomes consistent with the emergency response plan and/or standard operating procedures by completing the following tasks:	28-1	28	1130–1143
(a)	Establish and enforce scene control procedures, including control zones, emergency decontamination, and communications.	28-1, 28-2, 28-3, 28-4	28	1130–1143
(b)	Where criminal or terrorist acts are suspected, establish means of evidence preservation.	30-3	30	1216
(c)	Initiate an incident command system (ICS) for hazardous materials/WMD incidents.	28-1	28	1130–1143
(d)	Perform tasks assigned as identified in the incident action plan.	28-1	28	1130–1143
(e)	Demonstrate emergency decontamination.	28-7	28	1157–1169

	NFPA 472, Chapter 5, Operations	Learning Objective(s)	Chapter(s)	Page(s)
(4)	Evaluate the progress of the actions taken at a hazardous materials/WMD incident to ensure that the response objectives are being met safely, effectively, and efficiently by completing the following tasks:	28-1	28	1130–1143
(a)	Evaluate the status of the actions taken in accomplishing the response objectives.	28-1	28	1130–1143
(b)	Communicate the status of the planned response.	28-1	28	1130–1143
5.2	**Core Competencies—Analyzing the Incident.**			
5.2.1*	**Surveying Hazardous Materials/WMD Incidents.** Given scenarios involving hazardous materials/WMD incidents, the operations level responder shall survey the incident to identify the containers and materials involved, determine whether hazardous materials/WMD have been released, and evaluate the surrounding conditions and shall meet the requirements of 5.2.1.1 through 5.2.1.6.			
5.2.1.1*	Given three examples each of liquid, gas, and solid hazardous material or WMD, including various hazard classes, operations level personnel shall identify the general shapes of containers in which the hazardous materials/WMD are typically found.	24-2, 25-1, 25-4, 25-5, 25-6, 26-6	24, 25, 26	995–1006, 1010–1055, 1072–1102
5.2.1.1.1	Given examples of the following tank cars, the operations level responder shall identify each tank car by type, as follows:	25-1, 25-6, 25-7	25	1048–1051
(1)	Cryogenic liquid tank cars	25-1, 25-6, 25-7	25	1048–1051
(2)	Nonpressure tank cars (general service or low pressure cars)	25-1, 25-6, 25-7	25	1048–1051
(3)	Pressure tank cars	25-1, 25-6, 25-7	25	1048–1051
5.2.1.1.2	Given examples of the following intermodal tanks, the operations level responder shall identify each intermodal tank by type, as follows:	25-1, 25-3, 25-5, 25-7	25	1046–1047
(1)	Nonpressure intermodal tanks	25-1, 25-3, 25-5, 25-7	25	1046–1047
(2)	Pressure intermodal tanks	25-1, 25-3, 25-5, 25-7	25	1046–1047
(3)	Specialized intermodal tanks, including the following:	25-1, 25-3, 25-5, 25-7	25	1046–1047
(a)	Cryogenic intermodal tanks	25-1, 25-3, 25-5, 25-7	25	1046–1047
(b)	Tube modules	25-1, 25-3, 25-5, 25-7	25	1046–1047
5.2.1.1.3	Given examples of the following cargo tanks, the operations level responder shall identify each cargo tank by type, as follows:	25-1, 25-3, 25-5, 25-7	25	1034–1036
(1)	Compressed gas tube trailers	25-1, 25-3, 25-5, 25-7	25	1044–1045
(2)	Corrosive liquid tanks	25-1, 25-3, 25-5, 25-7	25	1039–1040
(3)	Cryogenic liquid tanks	25-1, 25-3, 25-5, 25-7	25	1043–1044

	NFPA 472, Chapter 5, Operations	Learning Objective(s)	Chapter(s)	Page(s)
(1)	Highway transport vehicles, including cargo tanks	25-1, 25-3, 25-5, 25-7	25	1011–1025
(2)	Intermodal equipment, including tank containers	25-1, 25-3, 25-5, 25-7	25	1011–1025
(3)	Rail transport vehicles, including tank cars	25-1, 25-3, 25-5, 25-7	25	1011–1025
5.2.1.2.2	Given examples of facility containers, the operations level responder shall identify the markings indicating container size, product contained, and/or site identification numbers.	25-1, 25-3, 25-5, 25-7	25	1011–1025
5.2.1.3	Given examples of hazardous materials incidents, the operations level responder shall identify the name(s) of the hazardous material(s) in 5.2.1.3.1 through 5.2.1.3.3.	25-1, 25-3, 25-5, 25-7	25	1011–1028
5.2.1.3.1	The operations level responder shall identify the following information on a pipeline marker:	25-1, 25-3, 25-5, 25-7	25	1026–1027
(1)	Emergency telephone number	25-1, 25-3, 25-5, 25-7	25	1026–1027
(2)	Owner	25-1, 25-3, 25-5, 25-7	25	1026–1027
(3)	Product	25-1, 25-3, 25-5, 25-7	25	1026–1027
5.2.1.3.2	Given a pesticide label, the operations level responder shall identify each of the following pieces of information, then match the piece of information to its significance in surveying hazardous materials incidents:	25-1, 25-3, 25-5, 25-7	25	1027–1028
(1)	Active ingredient	25-1, 25-3, 25-5, 25-7	25	1027–1028
(2)	Hazard statement	25-1, 25-3, 25-5, 25-7	25	1027–1028
(3)	Name of pesticide	25-1, 25-3, 25-5, 25-7	25	1027–1028
(4)	Pest control product (PCP) number (in Canada)	25-1, 25-3, 25-5, 25-7	25	1027–1028
(5)	Precautionary statement	25-1, 25-3, 25-5, 25-7	25	1027–1028
(6)	Signal word	25-1, 25-3, 25-5, 25-7	25	1027–1028
5.2.1.3.3	Given a label for a radioactive material, the operations level responder shall identify the type or category of label, contents, activity, transport index, and criticality safety index as applicable.	25-1, 25-3, 25-5, 25-7	25	1019–1020, 1028
5.2.1.4*	The operations level responder shall identify and list the surrounding conditions that should be noted when a hazardous materials/WMD incident is surveyed.	26-5, 30-2, 30-3, 30-7	26, 30	1072–1102, 1200–1223
5.2.1.5	The operations level responder shall give examples of ways to verify information obtained from the survey of a hazardous materials/WMD incident.	26-1, 26-2, 26-3, 26-4, 26-5	26	1072–1102
5.2.1.6*	The operations level responder shall identify at least three additional hazards that could be associated with an incident involving terrorist or criminal activities.	30-2, 30-3, 30-4	30	1200–1223

	NFPA 472, Chapter 5, Operations	Learning Objective(s)	Chapter(s)	Page(s)
5.2.2	**Collecting Hazard and Response Information.** Given scenarios involving known hazardous materials/WMD, the operations level responder shall collect hazard and response information using MSDS, CHEMTREC/CANUTEC/SETIQ, governmental authorities, and shippers and manufacturers and shall meet the following requirements:			
(1)	Match the definitions associated with the UN/DOT hazard classes and divisions of hazardous materials/WMD, including refrigerated liquefied gases and cryogenic liquids, with the class or division.	25-2	25	1011–1028
(2)	Identify two ways to obtain an MSDS in an emergency.	26-1, 26-2, 26-3	26	1089–1093, 1102
(3)	Using an MSDS for a specified material, identify the following hazard and response information:	26-1, 26-2, 26-3	26	1089–1093, 1102
(a)	Physical and chemical characteristics	26-1, 26-2, 26-3	26	1089–1093, 1102
(b)	Physical hazards of the material	26-1, 26-2, 26-3	26	1089–1093, 1102
(c)	Health hazards of the material	26-1, 26-2, 26-3	26	1089–1093, 1102
(d)	Signs and symptoms of exposure	26-1, 26-2, 26-3	26	1089–1093, 1102
(e)	Routes of entry	26-1, 26-2, 26-3	26	1089–1093, 1102
(f)	Permissible exposure limits	26-1, 26-2, 26-3	26	1089–1093, 1102
(g)	Responsible party contact	26-1, 26-2, 26-3	26	1089–1093, 1102
(h)	Precautions for safe handling (including hygiene practices, protective measures, and procedures for cleanup of spills and leaks)	26-1, 26-2, 26-3	26	1089–1093, 1102
(i)	Applicable control measures, including personal protective equipment	26-1, 26-2, 26-3	26	1089–1093, 1102
(j)	Emergency and first-aid procedures	26-1, 26-2, 26-3	26	1089–1093, 1102
(4)	Identify the following:	26-8	26	1089–1093, 1102
(a)	Type of assistance provided by CHEMTREC/CANUTEC/SETIQ and governmental authorities	26-8	26	1089–1093, 1102
(b)	Procedure for contacting CHEMTREC/CANUTEC/SETIQ and governmental authorities	26-8	26	1089–1093, 1102
(c)	Information to be furnished to CHEMTREC/CANUTEC/SETIQ and governmental authorities	26-8	26	1089–1093, 1102
(5)	Identify two methods of contacting the manufacturer or shipper to obtain hazard and response information.	26-8	26	1089–1093, 1102
(6)	Identify the type of assistance provided by governmental authorities with respect to criminal or terrorist activities involving the release or potential release of hazardous materials/WMD.	26-8	26	1089–1093, 1102
(7)	Identify the procedure for contacting local, state, and federal authorities as specified in the emergency response plan and/or standard operating procedures.	26-8	26	1089–1093, 1102

	NFPA 472, Chapter 5, Operations	Learning Objective(s)	Chapter(s)	Page(s)
(8)*	Describe the properties and characteristics of the following:	25-8	25	1061–1065
(a)	Alpha radiation	25-8	25	1061–1065
(b)	Beta radiation	25-8	25	1061–1065
(c)	Gamma radiation	25-8	25	1061–1065
(d)	Neutron radiation	25-8	25	1061–1065
5.2.3*	**Predicting the Likely Behavior of a Material and Its Container.** Given scenarios involving hazardous materials/WMD incidents, each with a single hazardous material/WMD, the operations level responder shall predict the likely behavior of the material or agent and its container and shall meet the following requirements:			
(1)	Interpret the hazard and response information obtained from the current edition of the DOT *Emergency Response Guidebook*, MSDS, CHEMTREC/CANUTEC/SETIQ, governmental authorities, and shipper and manufacturer contacts, as follows:	26-5, 26-6, 26-7	26	1072–1089
(a)	Match the following chemical and physical properties with their significance and impact on the behavior of the container and its contents:	25-8, 28-6	25, 28	1061–1062 1143–1157
i.	Boiling point	25-8, 28-6	25, 28	1056, 1143–1157
ii.	Chemical reactivity	25-8, 28-6	25, 28	1060, 1143–1157
iii.	Corrosivity (pH)	25-8, 28-6	25, 28	1059, 1143–1157
iv.	Flammable (explosive) range [lower explosive limit (LEL) and upper explosive limit (UEL)]	25-8, 28-6	25, 28	1061, 1143–1157
v.	Flash point	25-8, 28-6	25, 28	1061, 1143–1157
vi.	Ignition (autoignition) temperature	25-8, 28-6	25, 28	1061, 1143–1157
vii.	Particle size	25-8, 28-6	25, 28	1061, 1143–1157
viii.	Persistence	25-8, 28-6	25, 28	1058, 1143–1157
ix.	Physical state (solid, liquid, gas)	25-8, 28-6	25, 28	1056, 1143–1157
x.	Radiation (ionizing and non-ionizing)	25-8, 28-6	25, 28	1062, 1143–1157
xi.	Specific gravity	25-8, 28-6	25, 28	1058, 1143–1157
xii.	Toxic products of combustion	25-8, 28-6	25, 28	1065, 1143–1157
xiii.	Vapor density	25-8, 28-6	25, 28	1058, 1143–1157
xiv.	Vapor pressure	25-8, 28-6	25, 28	1057, 1143–1157
xv.	Water solubility	25-8, 28-6	25, 28	1059, 1143–1157
(b)	Identify the differences between the following terms:	25-8, 28-6	25, 28	1055–1057

	NFPA 472, Chapter 5, Operations	Learning Objective(s)	Chapter(s)	Page(s)
i.	Contamination and secondary contamination	28-7	28	1157–1169
ii.	Exposure and contamination	28-7	28	1157–1169
iii.	Exposure and hazard	28-7	28	1157–1169
iv.	Infectious and contagious	28-7	28	1157–1169
v.	Acute effects and chronic effects	28-7	28	1157–1169
vi.	Acute exposures and chronic exposures	27-1	27	1108–1110
(2)*	Identify three types of stress that can cause a container system to release its contents.	28-2, 28-6	28	1143–1157
(3)*	Identify five ways in which containers can breach.	28-6	28	1143–1157
(4)*	Identify four ways in which containers can release their contents.	28-6	28	1143–1157
(5)*	Identify at least four dispersion patterns that can be created upon release of a hazardous material.	28-2	28	1143–1157
(6)*	Identify the time frames for estimating the duration that hazardous materials/WMD will present an exposure risk.	28-2, 28-3, 28-4, 28-5	28	1143–1157
(7)*	Identify the health and physical hazards that could cause harm.	27-1	27	1108–1116
(8)*	Identify the health hazards associated with the following terms:	27-1	27	1108–1116
(a)	Alpha, beta, gamma, and neutron radiation	27-1	27	1108–1116
(b)	Asphyxiant	27-1	27	1108–1116
(c)*	Carcinogen	27-1	27	1108–1116
(d)	Convulsant	27-1	27	1108–1116
(e)	Corrosive	27-1	27	1108–1116
(f)	Highly toxic	27-1	27	1108–1116
(g)	Irritant	27-1	27	1108–1116
(h)	Sensitizer, allergen	27-1	27	1108–1116
(i)	Target organ effects	27-1	27	1108–1116
(j)	Toxic	27-1	27	1108–1116
(9)*	Given the following, identify the corresponding UN/DOT hazard class and division:	30-7	30	1216–1223
(a)	Blood agents	30-7	30	1216–1223
(b)	Biological agents and biological toxins	30-7	30	1216–1223
(c)	Choking agents	30-7	30	1216–1223
(d)	Irritants (riot control agents)	30-7	30	1216–1223
(e)	Nerve agents	30-7	30	1216–1223
(f)	Radiological materials	30-7	30	1216–1223
(g)	Vesicants (blister agents)	30-7	30	1216–1223
5.2.4*	**Estimating Potential Harm.** Given scenarios involving hazardous materials/WMD incidents, the operations level responder shall estimate the potential harm within the endangered area at each incident and shall meet the following requirements:			
(1)*	Identify a resource for determining the size of an endangered area of a hazardous materials/WMD incident.	26-5	26	1072–1102
(2)	Given the dimensions of the endangered area and the surrounding conditions at a hazardous materials/WMD incident, estimate the number and type of exposures within that endangered area.	26-5	26	1072–1102

	NFPA 472, Chapter 5, Operations	Learning Objective(s)	Chapter(s)	Page(s)
(3)	Identify resources available for determining the concentrations of a released hazardous material/WMD within an endangered area.	26-5	26	1072–1102
(4)*	Given the concentrations of the released material, identify the factors for determining the extent of physical, health, and safety hazards within the endangered area of a hazardous materials/WMD incident.	26-5	26	1072–1102
(5)	Describe the impact that time, distance, and shielding have on exposure to radioactive materials specific to the expected dose rate.	27-1	27	1108–1116
5.3	**Core Competencies—Planning the Response.**			
5.3.1	**Describing Response Objectives.** Given at least two scenarios involving hazardous materials/WMD incidents, the operations level responder shall describe the response objectives for each example and shall meet the following requirements:			
(1)	Given an analysis of a hazardous materials/WMD incident and the exposures, determine the number of exposures that could be saved with the resources provided by the AHJ.	28-1, 29-2, 29-3, 30-3	28, 29, 30	1130–1143, 1174–1181, 1194, 1200–1266
(2)	Given an analysis of a hazardous materials/WMD incident, describe the steps for determining response objectives.	28-1, 29-2, 29-3, 30-3	28, 29, 30	1130–1143, 1174–1181, 1194, 1200–1266
(3)	Describe how to assess the risk to a responder for each hazard class in rescuing injured persons at a hazardous materials/WMD incident.	25-2, 28-1, 30-3	25, 28, 30	1130–1143, 1174–1181, 1194, 1200–1266
(4)*	Assess the potential for secondary attacks and devices at criminal or terrorist events.	30-3, 3-4, 30-7	30	1200–1266
5.3.2	**Identifying Action Options.** Given examples of hazardous materials/WMD incidents (facility and transportation), the operations level responder shall identify the options for each response objective and shall meet the following requirements:			
(1)	Identify the options to accomplish a given response objective.	28-1	28	1130–1143
(2)	Describe the prioritization of emergency medical care and removal of victims from the hazard area relative to exposure and contamination concerns.	28-7	28, 30	1130–1143, 1200–1226
5.3.3	**Determining Suitability of Personal Protective Equipment.** Given examples of hazardous materials/WMD incidents, including the name of the hazardous material/WMD involved and the anticipated type of exposure, the operations level responder shall determine whether available personal protective equipment is applicable to performing assigned tasks and shall meet the following requirements:			
(1)*	Identify the respiratory protection required for a given response option and the following:	27-4, 27-5, 27-6, 27-7	27	1118–1126
(a)	Describe the advantages, limitations, uses, and operational components of the following types of respiratory protection at hazardous materials/WMD incidents:	27-4, 27-5, 27-6, 27-7	27	1118–1126
i.	Positive pressure self-contained breathing apparatus (SCBA)	27-4, 27-5, 27-6, 27-7	27	1118–1126
ii.	Positive pressure air-line respirator with required escape unit	27-4, 27-5, 27-6, 27-7	27	1118–1126
iii.	Closed-circuit SCBA	27-4, 27-5, 27-6, 27-7	27	1118–1126

	NFPA 472, Chapter 5, Operations	Learning Objective(s)	Chapter(s)	Page(s)
iv.	Powered air-purifying respirator (PAPR)	27-4, 27-5, 27-6, 27-7	27	1118–1126
v.	Air-purifying respirator (APR)	27-4, 27-5, 27-6, 27-7	27	1118–1126
vi.	Particulate respirator	27-4, 27-5, 27-6, 27-7	27	1118–1126
(b)	Identify the required physical capabilities and limitations of personnel working in respiratory protection.	27-4, 27-5, 27-6, 27-7	27	1118–1126
(2)	Identify the personal protective clothing required for a given option and the following:	27-4, 27-5, 27-6, 27-7	27	1118–1126
(a)	Identify skin contact hazards encountered at hazardous materials/WMD incidents.	27-4, 27-5, 27-6, 27-7	27	1118–1126
(b)	Identify the purpose, advantages, and limitations of the following types of protective clothing at hazardous materials/WMD incidents:	27-4, 27-5, 27-6, 27-7	27	1118–1126
i.	Chemical-protective clothing: liquid splash–protective clothing and vapor-protective clothing	27-4, 27-5, 27-6, 27-7	27	1118–1126
ii.	High temperature–protective clothing: proximity suit and entry suits	27-4, 27-5, 27-6, 27-7	27	1118–1126
iii.	Structural fire-fighting protective clothing	27-4, 27-5, 27-6, 27-7	27	1118–1126
5.3.4*	**Identifying Decontamination Issues.** Given scenarios involving hazardous materials/WMD incidents, operations level responders shall identify when emergency decontamination is needed and shall meet the following requirements:			
(1)	Identify ways that people, personal protective equipment, apparatus, tools, and equipment become contaminated.	28-7	28	1157–1169
(2)	Describe how the potential for secondary contamination determines the need for decontamination.	28-7	28	1157–1169
(3)	Explain the importance and limitations of decontamination procedures at hazardous materials incidents.	28-7	28	1157–1169
(4)	Identify the purpose of emergency decontamination procedures at hazardous materials incidents.	28-7	28	1157–1169
(5)	Identify the factors that should be considered in emergency decontamination.	28-7	28	1157–1169
(6)	Identify the advantages and limitations of emergency decontamination procedures.	28-7	28	1157–1169
5.4	**Core Competencies—Implementing the Planned Response.**			
5.4.1	**Establishing and Enforcing Scene Control Procedures.** Given two scenarios involving hazardous materials/WMD incidents, the operations level responder shall identify how to establish and enforce scene control, including control zones and emergency decontamination, and communications between responders and to the public and shall meet the following requirements:			
(1)	Identify the procedures for establishing scene control through control zones.	28-4	28	1157–1169
(2)	Identify the criteria for determining the locations of the control zones at hazardous materials/WMD incidents.	28-4	28	1157–1169
(3)	Identify the basic techniques for the following protective actions at hazardous materials/WMD incidents:	28-3, 28-4, 28-5	28	1157–1169

	NFPA 472, Chapter 5, Operations	Learning Objective(s)	Chapter(s)	Page(s)
(5)	Identify the capabilities and limitations of personnel working in the personal protective equipment provided by the AHJ.	27-4, 27-7	27	1124–1126
(6)	Identify the procedures for cleaning, disinfecting, and inspecting personal protective equipment provided by the AHJ.	27-7	27	1120–1121
(7)	Describe the maintenance, testing, inspection, and storage procedures for personal protective equipment provided by the AHJ according to the manufacturer's specifications and recommendations.	27-7	27	1120–1121
5.5	**Core Competencies—Evaluating Progress.**			
5.5.1	**Evaluating the Status of Planned Response.** Given two scenarios involving hazardous materials/WMD incidents, including the incident action plan, the operations level responder shall evaluate the status of the actions taken in accomplishing the response objectives and shall meet the following requirements:			
(1)	Identify the considerations for evaluating whether actions taken were effective in accomplishing the objectives.	26-5, 28-1	26, 28	1072–1089, 1130–1143
(2)	Describe the circumstances under which it would be prudent to withdraw from a hazardous materials/WMD incident.	28-1, 28-6	28	1130–1157
5.5.2	**Communicating the Status of the Planned Response.** Given two scenarios involving hazardous materials/WMD incidents, including the incident action plan, the operations level responder shall communicate the status of the planned response through the normal chain of command and shall meet the following requirements:			
(1)	Identify the methods for communicating the status of the planned response through the normal chain of command.	28-1	28	1130–1143, 1167
(2)	Identify the methods for immediate notification of the incident commander and other response personnel about critical emergency conditions at the incident.	28-1, 28-5	28	1130–1143, 1167
5.6*	**Competencies—Terminating the Incident. (Reserved)**			

Operations Level, Chapter 6

	NFPA 472, Chapter 6, Product Control	Learning Objective(s)	Chapter(s)	Page(s)
6.6	**Mission-Specific Competencies: Product Control.**			
6.6.1	**General.**			
6.6.1.1	**Introduction.**			
6.6.1.1.1	The operations level responder assigned to perform product control shall be that person, competent at the operations level, who is assigned to implement product control measures at hazardous materials/WMD incidents.			
6.6.1.1.2	The operations level responder assigned to perform product control at hazardous materials/WMD incidents shall be trained to meet all competencies at the awareness level (Chapter 4), all core competencies at the operations level (Chapter 5), all mission-specific competencies for personal protective equipment (Section 6.2), and all competencies in this section.			

	NFPA 472, Chapter 6 Product Control	Learning Objective(s)	Chapter(s)	Page(s)
6.6.1.1.3	The operations level responder assigned to perform product control at hazardous materials/WMD incidents shall operate under the guidance of a hazardous materials technician, an allied professional, or standard operating procedures.			
6.6.1.1.4*	The operations level responder assigned to perform product control at hazardous materials/WMD incidents shall receive the additional training necessary to meet specific needs of the jurisdiction.			
6.6.1.2	**Goal.**			
6.6.1.2.1	The goal of the competencies in this section shall be to provide the operations level responder assigned to product control at hazardous materials/WMD incidents with the knowledge and skills to perform the tasks in 6.6.1.2.2 safely and effectively.			
6.6.1.2.2	When responding to hazardous materials/WMD incidents, the operations level responder assigned to perform product control shall be able to perform the following tasks:			
(1)	Plan an initial response within the capabilities and competencies of available personnel, personal protective equipment, and control equipment and in accordance with the emergency response plan or standard operating procedures by completing the following tasks:	28-1, 29-2	28, 29	1135–1140, 1143–1157, 1174–1181
(a)	Describe the control options available to the operations level responder.	28-1, 29-2	28, 29	1135-1140, 1143–1157
(b)	Describe the control options available for flammable liquid and flammable gas incidents.	28-1, 28-6, 29-2	28, 29	1135–1140, 1143–1157
(2)	Implement the planned response to a hazardous materials/WMD incident.	28-1, 29-2	28, 29	1135–1140, 1143–1157
6.6.2	**Competencies—Analyzing the Incident. (Reserved)**			
6.6.3	**Competencies—Planning the Response.**			
6.6.3.1	**Identifying Control Options.** Given examples of hazardous materials/WMD incidents, the operations level responder assigned to perform product control shall identify the options for each response objective and shall meet the following requirements as prescribed by the AHJ:			
(1)	Identify the options to accomplish a given response objective.	28-1, 29-2, 30-3	28, 29, 30	1135–1140, 1143–1157, 1174–1181, 1214–1216
(2)	Identify the purpose for and the procedures, equipment, and safety precautions associated with each of the following control techniques:	29-1	29	1174–1181
(a)	Absorption	29-1	29	1174–1181
(b)	Adsorption	29-1	29	1174–1181
(c)	Damming	29-1	29	1174–1181
(d)	Diking	29-1	29	1174–1181
(e)	Dilution	29-1	29	1174–1181
(f)	Diversion	29-1	29	1174–1181
(g)	Remote valve shutoff	29-1	29	1174–1181
(h)	Retention	29-1	29	1174–1181
(i)	Vapor dispersion	29-1	29	1174–1181
(j)	Vapor suppression	29-1	29	1174–1181

Firefighter I

Section I of this book is your introduction to the world of firefighting. These chapters cover the requirements outlined in Firefighter level I of NFPA 1001 *Standard for Firefighter Professional Qualifications,* 2008 Edition. It is the most comprehensive of your training—covering everything from the history and current state of the fire service, the role of a firefighter in the department, rules and policies of the department, and the importance of safety and fitness, to the practical knowledge and hands-on skills required of a firefighter on the fireground. Information is inclusive of the Standard requirements, but also goes above and beyond where necessary to ensure your proficiency on the job. Straight-forward and easy to follow, these chapters will serve as a foundation for your next level of training and the next section in this book—Firefighter II.

1

Overview of the History and Development of the American Fire Service

- Learning Objectives
- The Mission of the Fire Service
- Roots in the Past
- The Fire Service of Today

- Lessons Learned
- Key Terms
- Review Questions
- Endnote

When I began my work in the fire and emergency services, I was confused by the many traditions, symbols, and practices that have come straight out of history. In these high-tech times, when industry technology seems to change almost weekly, it seemed odd to find these traditions still alive and well in the fire service.

As I learned more about the history of the fire service, I realized how valuable these pieces of history are to what we do every day. It is from this history that the fire service derives its pride, honor, integrity, and courage. The symbolism ties our modern-day practices to those of the early firefighters and emergency responders. It provides firefighters and emergency responders with a unique sense of belonging, perspective, unity, and promise for the future.

The history of firefighting and emergency response also gives modern-day firefighters the safety and protection that we need to survive and thrive in the industry. This history, including knowledge of the incidents in which firefighters have fallen in the line of duty, provides us with the information and expertise we need to ensure that firefighters can perform their duties safely with safer personal protective equipment, technological advances, and improved operational practices.

The types of emergency calls may have changed with society, but our dedication to protecting our communities and our brother and sister firefighters has not. I hope it never does.

The history of the fire service helps modern firefighters learn from the successes and failures of the past and unites us in a common tradition of serving those in need.

—*Street Story by Andrea A. Walter, Lead Author,*
Firefighter's Handbook

THE MISSION OF THE FIRE SERVICE

In general, the tasks of firefighters are the same the world over: to save lives and property from fire and other emergencies. However, not all fire departments approach that goal in the same manner. Many variables are considered by the agencies that provide fire protection and emergency services. Some agencies only protect airports, whereas others do structural firefighting and provide emergency medical services (EMS). Fire departments that provide EMS may only perform basic life support (BLS), whereas others may provide advanced life support (ALS).

The term *mission*, used in the context of modern fire department management, is usually synonymous with an agency's purpose for existence, or even its legal authority to act in a certain manner. A **mission statement** is a written declaration by a fire agency describing the things that it intends to do to protect its citizenry or customers. When we look at the variety of situations under which the fire service functions, we begin to understand why one mission statement will not cover all agencies. The size and organizational structure of fire departments range from small volunteer companies that protect only a few hundred souls to metropolitan departments that protect millions.

Many fire agencies have a written mission statement. Firefighters should understand their departments' mission statements and how they contribute to their departments' successes. Mission statements vary from very general to very specific. Once written and posted, it is the responsibility of every individual in that department to attempt to achieve its goals. Each and every firefighter, from the beginning of civilization to the modern fire service, has contributed something to the accomplishment of an agency's mission. That is one of the legacies of the fire service.

ROOTS IN THE PAST

Somewhere, lost in the ancient past, is the name of the very first firefighter. That person was probably thrust into the job of fighting fire out of sheer desperation, to protect himself, his family, or his possessions. It was a courageous act then, and it has remained a courageous act as the fire problem has passed from those obscure ancient events to the streets of small towns and villages, suburban cities, metropolitan areas, airports, harbors, marinas, and wildlands, all of which are protected by contemporary firefighters.

> **NOTE**
>
> What is important to recognize is that the task of being a firefighter is an old, yet constantly evolving occupation.

The fire service has a great many traditions. These traditions are often in conflict with the constant changes required in the fire service to keep up with the task of combating fire in a modern society. Being a firefighter today is much different from being one in the past, but practically everything firefighters and fire service organizations do to save lives and protect property is rooted in events from the past.

> **FIREFIGHTER FACT**
>
> The fire service of today is a direct result of an evolution in the methodology, technology, and responsibility of a service that has been vital to communities since the beginning of civilization.

Present-day firefighters need to recognize that what they do today is going to be the basis for the processes future firefighters will use. The present will eventually become part of the past. Understanding the history of the fire service is a lot like climbing a ladder, as shown in **Figure 1-1.** The ladder has to be well grounded in order to be stable, and there must be strong beams to support the ladder. There are rungs to stand on as the ladder is climbed. The two beams of the fire service's history are courage and commitment. The bottom rungs represent the experience and knowledge gained from past years. The rung we are standing on is the present. The ones that are beyond our reach are the future. The ladder is being raised to achieve the goal of saving lives and protecting property. We anticipate that we will always be climbing that ladder in an unending quest to achieve that goal.

The history, traditions, and development of the fire service covers a long period of time. Significant events have shaped the development of fire service technology and methods. Individual people have actually shaped fire protection philosophy through their actions. Fire service history includes a colorful and exciting accumulation of personalities, tools, equipment, and apparatus that has been used to combat both the ordinary fire and the conflagration. The legacy and heritage of the fire service is a pageant that is rich with wisdom and experience.

Someone once made a joke that the fire service is "200 years of tradition, unhampered by progress." Nothing could be more inaccurate. For the last 350 years, the fire service in this country has advanced on all fronts. What has not changed a great deal is the nature of individuals who have chosen to follow a fire and emergency service career, either as a paid, career firefighter or as a volunteer. Granted, in the last few decades there have been more changes in the firefighters' working environment than at any other time in the history of firefight-

ing. We just have to search for the stepping stones and recognize that they are incremental steps of change.

> **NOTE**
>
> The fire service has evolved into what it is today because it is a service for people. What communities need and want from their fire service determines what firefighters should or could be providing them. That is what makes the role of the firefighter so exciting and fulfilling. It is a job that not everyone can do. It is not a job that remains static over time. It is a constantly challenging and stimulating environment where the past collides with the future in the actions of those who serve today.

Although many fire agencies have their own mission statement, that does not mean they are all different in every way. Some factors are common to all agencies. For example, fire agencies all have an organizational structure, and an inventory of facilities, apparatus, equipment, and methods for responding to emergencies. They almost all conduct a variety of programs to achieve their goal of protecting life and property. These programs, in general, fall into the categories of fire and emergency operations, fire prevention, arson investigation, training, emergency medical services, communications, and maintenance to name a few. These elements of the fire service are all founded on events and decisions from the past.

THE FIRE SERVICE OF TODAY

From the 1950s to the end of the twentieth century, the fire service saw more changes than in the previous 200 years. Most of the hardware and basic components of the service have evolved fairly slowly. The process of accelerating change has more to do with the actual duties of the firefighter and fire agency staffing. For example, the role of EMS has grown rapidly in the fire service; many departments are much more involved in **hazardous materials** response, **Figure 1-2,** than ever before; the role of **search and rescue** has grown in view of major earthquakes and structural collapses; and fire service agencies are becoming more and more involved as first responders in **terrorism** incidents. In this same era, residential sprinklers and smoke detectors were introduced as a means of mitigating risks.

Although fire protection remains essentially a function of local government, the second half of the twentieth century saw efforts to create a national focus on the fire problem in the United States and to bring together training and information sources at state and national levels to improve the business of fire

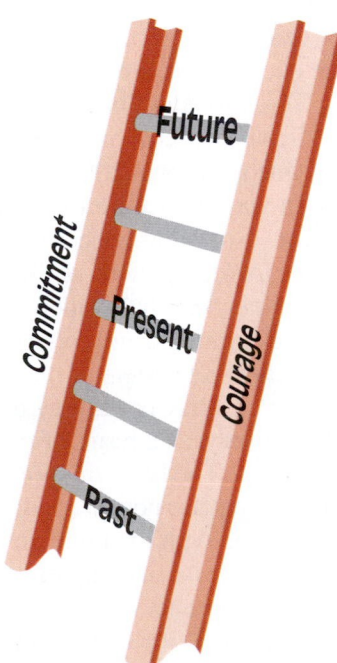

FIGURE 1-1 Understanding the history of the fire service is like climbing a ladder.

FIGURE 1-2 The fire service has expanded into many areas, including hazardous materials response.

protection. President Harry Truman created a National Conference on America's fire problem in 1947. The conference was overshadowed, however, by the occurrence of the Texas City Disaster. The explosion of the ammonium nitrate stored on the ship Grandcamp took the lives of 405 people, which included most of the Texas City Fire Department. It was truly the first catastrophic hazardous materials event in America.

In the 1960s, the Johnson Foundation sponsored an event called the "Wingspread Conference" that raised the level of awareness of the state of fire protection. During the 1970s, the federal government recognized that the country's fire problem was changing and commissioned a panel of experts to study it. This group, called the National Commission on Fire Protection and Control, published a document titled *America Burning*. As a result of that document, a lot of positive change occurred, including the creation of the United States Fire Administration and the development of the National Fire Academy. The 1970s also saw the creation of the incident command system (ICS) through the activity of the FireScope project in dealing with major wildland fires. Change was starting to become a natural part of the fire services evolution as it became increasingly more capable of coping with existing threats and hazards.

In the twenty-first century, the fire service continues to grow and develop. There are an estimated 30,635 fire departments in the United States with an estimated 1,140,900 firefighters. The actual numbers vary because new communities are created that require fire protection, and sometimes several fire organizations band together as a result of mergers or consolidations. Many fire departments are still totally volunteer, while others are growing larger due to population increases and increased density in some urban areas. In some areas, fire and rescue services may be provided by private, for-profit companies. Fire departments are increasingly complex in this day and age, with a diversification in the services they provide. In addition, fire departments are also more gender and ethnically diverse, offering new insights into the way they provide services to their communities.

According to the United States Fire Administration, a fire department in this country responds to a fire approximately every 20 seconds. Even with the incredible advancements in fire service tactics, training, and technology, the United States still has the fourth highest fire death rate of all industrialized countries, with someone dying in a fire an average of every two hours and someone injured by fire approximately every 23 minutes. Fire kills more Americans than all natural disasters combined. America is still burning, and the fire service continues to race to meet these ever-evolving challenges.[1]

The late twentieth century and the early twenty-first century also presented a new challenge to the fire service—the threat of terrorism in this country. The attacks on the World Trade Center and the Pentagon on September 11, 2001, shed new light on the fire service by demonstrating that firefighters are not immune to the horrible acts of terrorists. Although this event will remain fresh in every American's mind, it is very possible that the events of that day will shape the fire service in new and different ways well into the future. Then, too, there are equally challenging problems that are driven not by terrorists but by the nature of America's growing fire problems. For example, the return of the "urban conflagration" has occurred as a result of the urban wildland interface. The Oakland Fire in 1993 was only the first of many "firestorms" that have destroyed lives and property. Hurricane Katrina in 2005 tested the mettle of the fire service in ways that were totally different from events in the past, especially concerning the ability to sustain operations for long periods of time. These varied events have placed new emphasis upon the creation of a national system for command and control and a renewed focus on a comprehensive system of mutual aid.

The fire service of today has a very diverse, complicated system of delivery, as a result of the contributions of all past generations. The basic mission of saving lives and protecting property has not changed. Only the manner and means to cope with the problem have changed. Nonetheless, the fire service of today still contains a great deal of the past in its inventory of tools, equipment, and methods. If we go back to the beginning of this chapter and revisit the term mission, it may be possible to reconcile the past with the present. The fire service is today what it has become from past experiences.

What is most important is that each generation be more adequately prepared to do the job effectively and efficiently—and more safely. The National Fallen Firefighters Foundation has a goal of reducing the deaths of firefighters over the next few decades, and the individual firefighter, such as yourself, is an important part of making that happen. The culture of the fire service needs to heed a new motto: "Everyone goes home." This attitude must start at the entry level.

ETHICS

What is predictable about the future is that firefighters will likely be called on to deal with whatever dilemma or disaster strikes the community they serve. They must be prepared professionally to meet these challenges and to make accurate decisions for a positive outcome.

LESSONS LEARNED

What does the future hold for firefighters as they begin their journey? In one sense a firefighter's career is like climbing that ladder we described earlier in the chapter. There are a lot more rungs to be climbed, as firefighters pursue their careers. And, while we stand on only one rung at a time, we must be able to place our hands on the rungs that are higher.

No one should attempt to predict the future without giving due credit to the past. The future will contain a lot of challenges and opportunities. The future will contain some changes that will be difficult to deal with, just as they were for previous generations. The future will contain a requirement that firefighters continue to learn and develop skills that did not even exist the day they entered the fire service, not unlike those from the past.

KEY TERMS

Hazardous Material   A substance (either matter—solid, liquid, or gas—or energy) that, when released, is capable of harming people, the environment, and property. It includes weapons of mass destruction (WMD) as defined in 18 U.S. Code, Section 2332a, as well as any other criminal use of hazardous materials, such as illicit labs, environmental crimes, or industrial sabotage.

Mission Statement   A written declaration by a fire agency describing the things that it intends to do to protect its citizenry or customers.

Search and Rescue   Attempts by fire and emergency service personnel to coordinate and implement a search for a missing person and then effect a rescue.

Terrorism   Acts of violence that are arbitrarily committed against lives or property and intended to create fear and anxiety.

REVIEW QUESTIONS

1. What is a mission statement? Why do fire and emergency services departments have mission statements?

2. Describe three major fire service events that have occurred since you have been interested in a fire service career. What impact have they had on the fire service?

3. What do you think the future of the American fire service will look like in five years? In ten years? In 100 years?

4. Discuss what effects you think terrorism and large-scale national disasters have had on the fire service.

ENDNOTE

1. Statistics from "The United States Fire Service," Fire Analysis and Research Division, One-Stop Data Shop, Quincy, MA, www.nfpa.org.

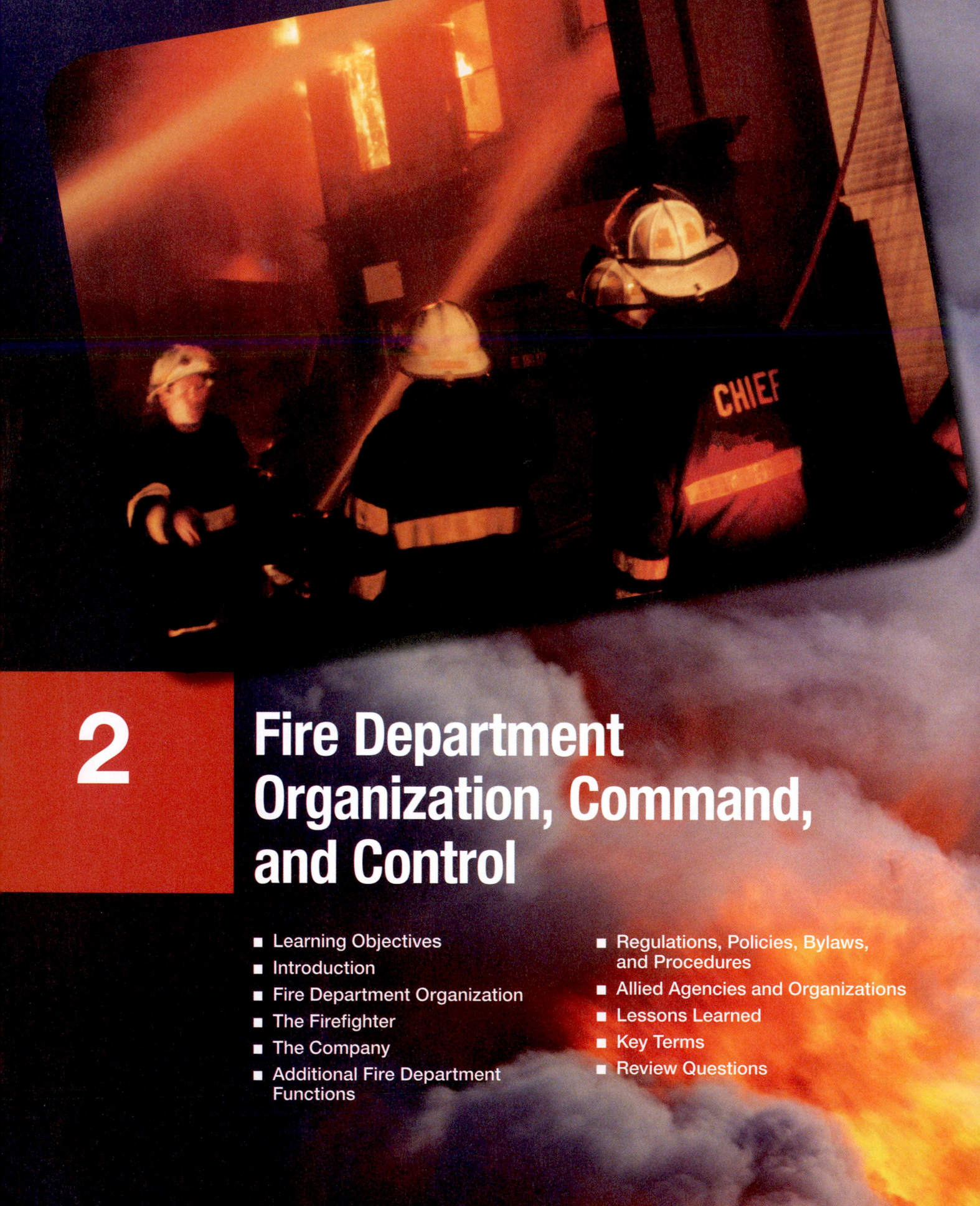

2

Fire Department Organization, Command, and Control

- Learning Objectives
- Introduction
- Fire Department Organization
- The Firefighter
- The Company
- Additional Fire Department Functions

- Regulations, Policies, Bylaws, and Procedures
- Allied Agencies and Organizations
- Lessons Learned
- Key Terms
- Review Questions

Winters in Washington, D.C., can be fickle. The temperature could be 70 degrees during January, or it could easily be 17 degrees with three feet of snow on the ground. On January 13, 1982, it was the latter. When people woke up that morning to blowing snow and sub-teen temperatures, they may have expected rough weather. Little did they realize, however, that by afternoon rush hour the city would experience a major subway crash in the tunnel with seven fatalities and that Air Florida flight 90 would collide, with seventy-nine souls on board, into the 14th Street Bridge that links Washington with Virginia across the Potomac River. Concepts such as command and control and tools such as the Incident Command System (ICS) were not readily employed in 1982. The incident was under the command of a senior fire department officer. The outcome would depend heavily on the abilities of that officer because he was responsible for all aspects of command. I was a young officer in command of a ladder company that day. Before the day was over, I would respond to both the subway and the airplane crashes. My clearest recollection of the incidents that day is that there was a great amount of confusion. Many times orders were issued only to be countermanded or withdrawn. Jurisdiction disputes existed among many of the responding entities. The number of casualties was compounded by the fact that many of the dead and injured had to be removed from the wreckage in the subway and at the 14th Street Bridge. The aircraft itself went into the water with no flames, but it took most of the victims with it. The weather was the final trump card played against us that day, making traveling and responding extremely difficult. This was a very taxing and difficult operational day for anyone to handle and should not reflect on the commanders who were in charge on that fateful day. Recently, however, a panel of chief officers who had experience in the implementation of ICS and the concepts of command and control revisited that day in! the form of an exercise. No facts were given except for the pertinent data of resources available, time of day, weather, and the scene as reported by the first arriving units. Needless to say, the outcome from a command point of view was much more successful this second time. The ICS can be considered one of the high points in tool development for the fire service of the twentieth century. The fire service continues to develop tools, and concepts continue to evolve as the need dictates. Some of these tools, such as the ICS, will be used well into the next millennium; that is, if we embrace and use them wisely. The leaders of the fire service must ensure that those who follow us maintain an open mind and continue to educate themselves in order to push the concept of command and control to levels where the safety of our people will be ensured and our operations will be successful in protecting civilians and property.

—Street Story by Mike Smith, Battalion Fire Chief,
District of Columbia Fire and EMS Department

INTRODUCTION

The American fire service has a history and tradition that is more than 250 years old, dating back to Ben Franklin who organized the Union Fire Company in 1736. That basic organizational structure is still in use today. Depending on the size and needs of a community, a fire department will consist of a number of companies in one or a number of stations throughout a community. The number of stations depends on the geography of the response area, travel time, and distance.

Companies are usually divided into functions such as engine, ladder, or truck companies and specialized units such as rescue or hazardous materials companies. In addition, many fire departments provide either basic or advanced life support emergency medical services. The organization is designed to establish a clear division of work assignments:

- To assign responsibility for a specific response area.
- To assign responsibility for a specific activity.
- To eliminate duplication of work and confusion.
- To establish an adequate level of equipment and personnel response to control the emergency scene or incident.

This chapter describes the many different roles a firefighter may have in the organization of a fire department. A well-trained firefighter is essential for the department to accomplish its mission.

FIRE DEPARTMENT ORGANIZATION

Fire departments, as shown in **Figure 2-1,** are like any other organization in that they have a reason for existing and a structure for operations. The mission statement should communicate the reason for being, and the organizational structure defines the chain of command and authority.

The Business of Fire Protection

It is up to a jurisdiction to determine the type and level of fire protection that it wishes to deliver. There are several types of fire department service delivery, each having its distinct advantages and disadvantages.

Career or paid departments consist of members who are employed in full-time positions with benefits. Volunteer departments are made up of members who may receive payment on a per call/hourly basis or no payment at all. Members of volunteer departments will often hold careers in different fields outside of emergency services. Volunteer fire departments hold a very important place in the history of the fire service and still make up the majority of its members.

Some jurisdictions utilize a system of both career and volunteer members to provide the services of the department. Although the majority of firefighters in the United States is volunteer, the majority of the population is protected by career or combination departments, **Figure 2-2**. Whichever type of department is utilized, the main objectives, goals, and purpose remain the same. The governing body for the fire department will be dependent on the type or classification of the department.

The public or municipal fire department is most commonly found in larger cities or densely populated areas. In the structure of this department, the Fire Chief is directly responsible for the operations of the department but reports to the City Manager or Mayor. Funding for this type of department is accomplished through taxes, for which the fire department must compete with other municipal departments such as public works or police.

Types of Fire Departments

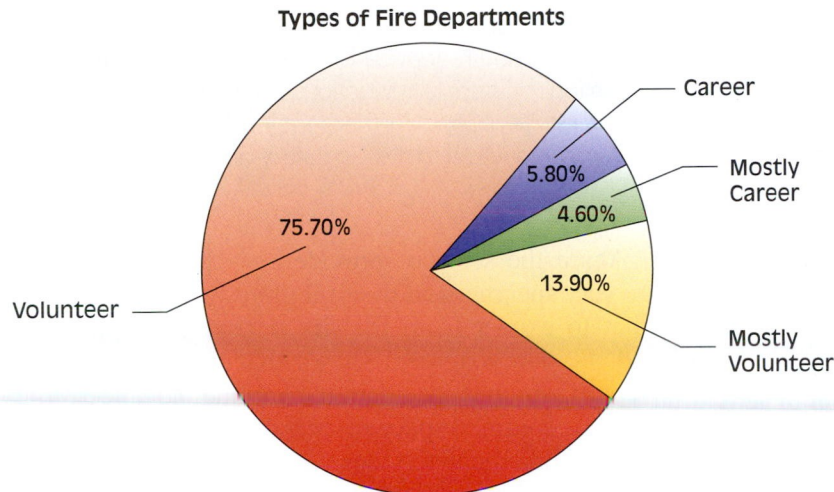

Career

5.80%

Mostly Career

4.60%

75.70%

13.90%

Volunteer

Mostly Volunteer

US Population Protected

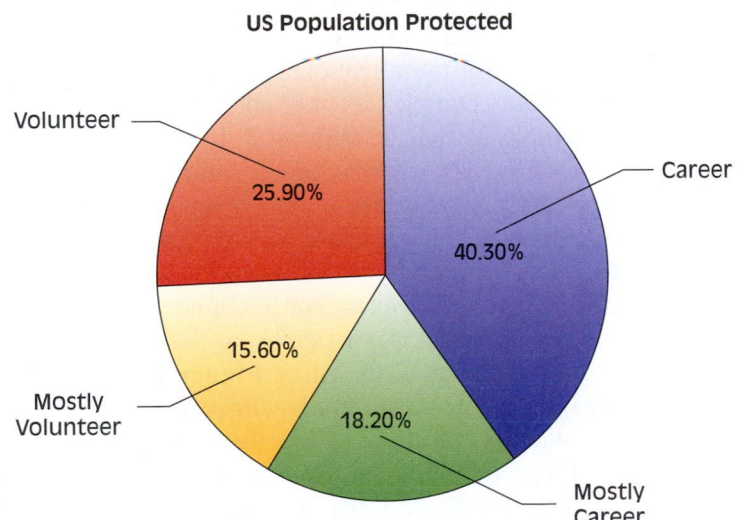

Volunteer

Career

25.90%

40.30%

15.60%

18.20%

Mostly Volunteer

Mostly Career

FIGURE 2-2 Although the majority of fire-fighters in the United States are volunteer, the majority of the population is protected by career or combination departments. *(Source: Adapted from* A Needs Assessment of the U.S. Fire Service, *published as a cooperative study by the U.S. Fire Administration and the National Fire Protection Agency.)*

The fire district or fire protection district is funded by a dedicated property tax that is used solely for fire department services. The Fire Chief is still responsible for department operations but reports to an elected board that administers the tax proceeds collected.

Volunteer fire departments can fit into either category or be supported by subscriptions for their services or fundraising activities. Federal fire departments are staffed by employees of the federal government and are most commonly found at government installations such as military bases. The U.S. Forest and Parks Service staffs one of the largest federal fire departments. Their funding and resources will be dependent on decisions made by the federal government.

There are also industrial fire departments, sometimes referred to as brigades, that provide emergency services at facilities under the same business management. These brigades are often specially trained to deal with potential hazards specific to the facility. They may be designed to function independently or in cooperation with the fire department that the facility property falls under. Their funding and resources will be dependent on the decisions made by the management of the facility.

Providing fire protection and life safety services is a business. Like businesses of all sizes and types, the fire service strives to provide the highest quality services as effectively and efficiently as possible while working to meet customer or community expectations. The fire service also works to meet these goals while protecting the safety of firefighters, emergency responders, and the public. Also similar to businesses, the fire and emergency services have adopted a quality mind-set and a focus on customer service in the provision of their emergency and nonemergency services.

Mission Statement

Each fire department should have a mission statement. The mission statement will provide both meaning and direction to the members of the organization. This will also provide the members of the organization with a clear and defined purpose of the type and level of service the department will provide. Also, the mission statement must be specific, so the public—the people depending on the entire department for help—know what to expect from the organization and understand what duties firefighters can and cannot perform.

STREETSMART TIP

The fire department must communicate its mission statement to the public. This can be accomplished during fire prevention inspections and fire safety information presentations to community groups.

The mission statement provides a clear and concise statement of what the organization intends to do or accomplish. The statement should contain the type of services or product to be delivered and who will receive this service or product.

SAMPLE MISSION STATEMENT 1

The Midway Fire Department is organized to deliver fire prevention, fire suppression (extinguishment), and rescue services to the citizens of its protection area. This will include response to conduct vehicle extrication, hazardous materials mitigation, and basic life support emergency medical services.

The sample mission statement describes what the Midway Fire Department does. The mission statement does not limit what an organization will do; it provides the focus for the service it will provide to the people in its community. This is especially true for an emergency response organization when a person calls for help.

If the Midway Fire Department decides to expand its services to include additional activities, its mission statement would read as follows:

SAMPLE MISSION STATEMENT 2

The Midway Fire Department is organized to deliver fire prevention, life safety, fire suppression (extinguishment), and rescue services to the citizens of its protection area. This will include response to conduct vehicle extrication, hazardous materials mitigation, confined space rescue, advanced life support emergency medical services, disaster response, and fire life safety code enforcement.

Again the services provided by this organization are very specific, and the firefighters and public know what to expect from the fire department. The second mission statement shows that the Midway Fire Department has decided to provide advanced life support, confined space rescue, disaster response, and code enforcement services.

Organizational Structure

To accomplish its tasks, a fire department must have some type of organizational structure. This structure may be as complex as that shown in **Figure 2-3** or as simple as that of **Figure 2-4.** Both figures show a structure for internal organization and how functions and responsibilities are delegated in the organization. Neither of these shows the receiver of the fire department's services, the citizens they protect, or how the fire department is connected to the community. Another way of defining the fire department organization could look like **Figure 2-5.** This structure shows circles of interdependence the organization has with its community. The fire department depends on the community and governing body for support. In turn, the community depends on the fire department for services and the governing body depends on the department to provide those services. In addition, the fire department must foster collaborative relationships and have knowledge of organizations that they must work closely with on the local, state, and federal levels. The firefighter is the key position in the organization and is essential for the department to fulfill its mission.

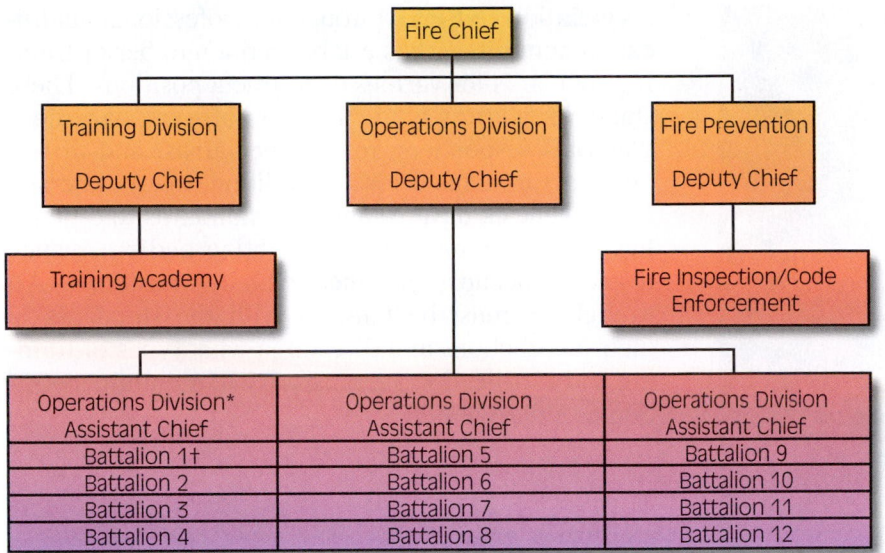

*Each division has one heavy rescue company.
† Battalion consists of four engine companies and one ladder company.

FIGURE 2-3 The organizational structure for a medium to large fire department shows the division of work assignments and chain of command. Most large departments will have a structure similar to this, with separate divisions for suppression, training, and fire prevention.

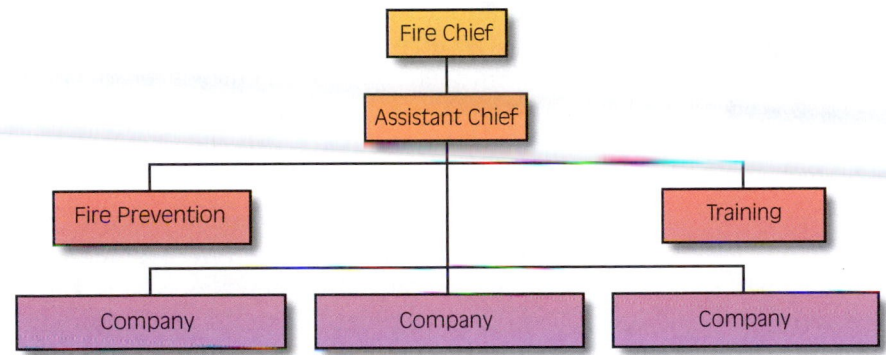

FIGURE 2-4 The organizational structure for a small fire department will have a more direct chain of command and structure. In this size organization, one individual may be responsible for a number of functions.

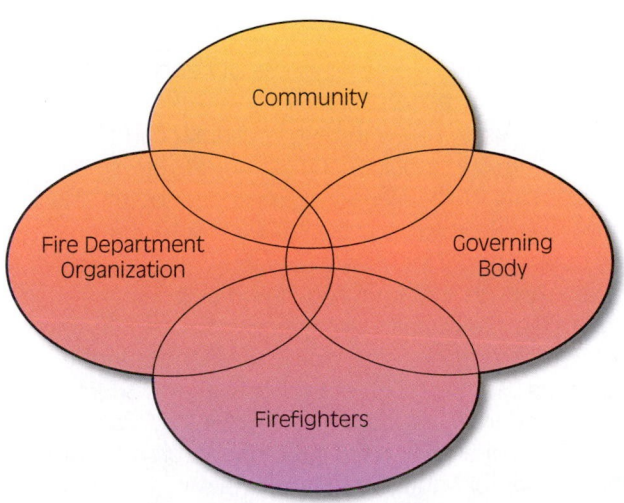

FIGURE 2-5 Different from most organizational charts, this organizational structure shows the interdependence of the community, fire department, governing body, and firefighters.

THE FIREFIGHTER

A firefighter is an individual trained to perform the function of fire prevention and suppression. Firefighters must possess both the knowledge and skills to be able to perform safely and effectively at an emergency incident, **Figure 2-6.**

Depending on the mission of the fire department or organization, there may be other areas that firefighters must be knowledgeable about in addition to firefighting. A firefighter may be required to be an **emergency medical technician (EMT)** or **paramedic (EMT-P);** a **hazardous materials technician;** a **rescue specialist, Figure 2-7;** a fire investigator; or a fire prevention officer. Many other positions are available in the firefighting community, and as firefighters begin their training they can start establishing personal and career goals. The only limitation to advancement and personal growth is a failure to take advantage of training and learning opportunities.

FIGURE 2-6 The firefighter is the individual who makes a department operate.

There are many organizations in the fire and emergency service community that assist in the determination of the necessary qualifications firefighters need to perform their jobs. The **National Fire Protection Association (NFPA),** through its professional qualification committee, has established a number of training standards for various department positions. These standards address firefighters, fire officers, hazardous materials responders, rescue specialists, and driver/operators to name a few. In addition, many state and local agencies have established standards and qualification requirements for firefighting and emergency response occupations and volunteer positions. All firefighters must be familiar with the standards in their jurisdictions and determine what types of training and certification are necessary to maintain a position as a firefighter.

> **FIREFIGHTER FACT**
>
> Two national organizations provide accreditation of state, provincial, territorial, and federal government entity certification testing processes. These are the International Fire Service Accreditation Congress (IFSAC) and the National Board on Fire Service Professional Qualifications (ProBoard). A firefighter who successfully completes a certification testing process through an IFSAC or ProBoard accredited entity is eligible for certification through that accredited certifying entity.

Through a consensus development process, the NFPA provides the Standard for Fire Fighter Professional Qualifications, also known as **NFPA 1001, Figure 2-8.** This standard outlines the minimum knowledge, skills, and abilities firefighters must have at two different levels, Firefighter I and Firefighter II.

ROLE OF FIREFIGHTER I

Firefighter training and education is based largely on NFPA 1001, Standard for Fire Fighter Professional Qualifications. This standard lists the critical skills and knowledge that firefighters must have to function safely and effectively at fires and emergency incidents.

NFPA 1001 splits basic firefighter education into two levels, Firefighter I and Firefighter II. Firefighter I outlines the most basic knowledge and skills that firefighters need, such as fire behavior, forcible entry, self-contained breathing apparatus (SCBA), personal protective equipment, ladders, ventilation, salvage, and overhaul. Firefighter II covers an expanded set of topics and skills to enhance the Firefighter I training and make firefighters more proficient at vehicle extrication, flammable liquid fires, flammable gas cylinder fires, evidence protection, fire safety surveys and public education presentations.

Some topics appear in both Firefighter I and Firefighter II, including communications, structure fire operations, and equipment maintenance. For these topics, the Firefighter I level provides the introduction to the concepts and how to apply them as a firefighter, and the Firefighter II level expands that knowledge and skill set.

Just like fire departments across the country differ in structure and operations, so do fire department training programs. Some firefighters will go through a Firefighter I course and then a separate Firefighter II course at a later time. Other firefighters will take one course that covers both Firefighter I and Firefighter II material.

The Firefighter I and Firefighter II levels allow firefighters to learn the basics first and then progress on to make themselves more versatile in firefighter and emergency operations.

(A)

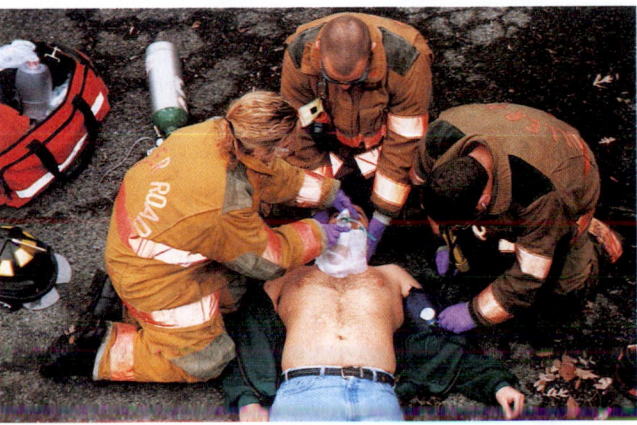

(B)

(C)

FIGURE 2-7 Some positions available to firefighters are (A) rescue specialist, (B) paramedic, and (C) hazardous materials technician.

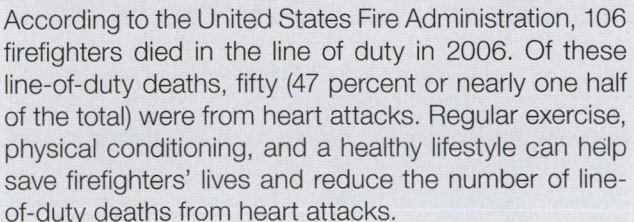

FIGURE 2-8 National Fire Protection Association 1001 Standard for Fire Fighter Professional Qualifications provides the basis for firefighter training.

In addition to knowledge and skills gained through training and experience, firefighters must also be in good physical condition to perform the demanding and strenuous tasks involved in firefighting and emergency response.

FIREFIGHTER FACT

According to the United States Fire Administration, 106 firefighters died in the line of duty in 2006. Of these line-of-duty deaths, fifty (47 percent or nearly one half of the total) were from heart attacks. Regular exercise, physical conditioning, and a healthy lifestyle can help save firefighters' lives and reduce the number of line-of-duty deaths from heart attacks.

Listed here are the typical requirements an individual trained in structural fire suppression would be expected to meet:

- Know the department's organizational structure and operating procedures.
- Perform all duties safely.

- Know the department's response area or district, including streets and hazards.
- Maintain firefighting equipment, especially personal protective equipment.
- Respond to alarms of fire as a member of a trained unit/company.
- Use self-contained breathing apparatus (SCBA) and personal protective equipment.
- Rescue people endangered by the fire.
- Use fire department tools to conduct forcible entry, ventilation, and fire extinguishment at structure, vehicle, and ground-cover fires.
- Conduct overhaul operations at a fire scene.
- Present fire safety information to the community.
- Provide basic life support activities such as CPR or control of bleeding.

Depending on the organization's mission statement and responsibilities, the firefighter may also be required to perform additional duties or functions. A sampling of these and the corresponding training standard are shown in **Table 2-1.**

THE COMPANY

Historically the basic unit of a fire department is the **company,** a team of firefighters with apparatus assigned to perform a specific function in a designated response area. A company is typically designated as an engine, ladder, truck, squad, or rescue depending on the type of apparatus, equipment, and firefighters on board. The company officer is responsible for all activities of the company.

TABLE 2-1 Additional Duty Training Standards	
Position Description	**Training Standard**
Fire apparatus driver/ operator	NFPA 1002
Airport firefighter	NFPA 1003
Hazardous Materials Technician	NFPA 1472
Public safety communications telecommunicator	NFPA 1061
Paramedic/emergency medical technician	State or local standards

STREETSMART TIP

Depending on history, traditions, and/or geography, the titles used to designate a position in the fire department may vary from those presented in this text. Not all organizations have the position of lieutenant and some use a military designation such as sergeant or private. Firefighters should be familiar with the rank structure and titles used in their own organization.

Company officers are supervisory-level positions and are responsible for both firefighters and administrative duties, **Figure 2-9A** and **B**. Requirements for the position of company officer are addressed in a separate training standard, NFPA 1021, Standard for Fire Officer Professional Qualifications.

Just as firefighters and company officers have standards that help in determining the knowledge, skills, and abilities to complete their jobs, so does the apparatus they ride to emergency incidents. NFPA 1901, Standard for Automotive Fire Apparatus, outlines minimum equipment, specifications, and capabilities for various types of firefighting apparatus. **Table 2-2** outlines some of the basic requirements for common types of fire apparatus.

The Engine Company

The **engine company,** as shown in **Figure 2-10,** is organized to provide firefighters who deliver water at the fire scene, deploy hoselines, and attack and extinguish fires in vehicles and structures.

An engine company often uses an apparatus referred to as an engine or pumper. NFPA 1901 states that a pumper should have a permanently mounted fire pump with a capacity of at least 750 gallons per minute (gpm). NFPA 1901 also states that a pumper should carry no less than 300 gallons of water. Other specific hose and ladder requirements are listed in Table 2-2.

In some organizations, the engine company may also perform rescue or squad functions, as detailed later in this section. This increases the amount of equipment necessary on the apparatus and may result in a different company designation, such as a rescue engine.

The Truck Company

The primary functions of the **truck company**, as shown in **Figure 2-11** (also known as a ladder company), are to carry firefighters for forcible entry, search and rescue, ventilation, provision of ladders, securing of utilities, and overhaul functions at a fire scene.

(A)

(B)

FIGURE 2-9   Company officers perform both (A) firefighting and (B) administrative duties.

There are three basic aerial devices that are utilized: the aerial ladder, the tower ladder, and the snorkel (Figure 2-11A, B, and C). Each has distinct features that give the device its advantages and disadvantages for fireground operations.

An aerial ladder consists of a power-driven ladder that is affixed to a truck. On the fireground, it can be used to rescue trapped civilians and firefighters, provide quick access to upper floors, and place master streams in advantageous positions. The disadvantage of an aerial ladder is that it does not provide abundant working room at the ladder tip and can thus be difficult to get victims onto for removal.

A tower ladder is the same as an aerial ladder, but a basket is mounted to the end of the ladder. This bas-

ket provides a working area for firefighters at the ladder tip. The use of a tower ladder is different from the operations of other aerial devices. The basket itself offers a solid working platform and railings to provide protection for the firefighters.

Snorkels are platforms or baskets attached to articulating booms mounted on a truck. They are very useful for situations requiring the maneuverability of a master stream device at an upper level. Snorkels are also useful in situations requiring movement around overhead obstacles. Disadvantages of the snorkel include the articulating action of the booms themselves. When the booms are raised, they project outward at the hinge, making it difficult to operate in tight areas. Because of their design, snorkels have a

TABLE 2-2   Overview of Basic Apparatus Requirements

Pumper

Fire pump with minimum rated capacity of 750 gpm

Water tank with a minimum of 300 gallons

Ground ladders, to include at least one attic ladder, one extension ladder, and one straight ladder with roof hooks

A minimum of 15 feet of soft suction hose or 20 feet of hard suction hose

A minimum of 800 feet of 2½-inch or larger diameter fire hose

A minimum of 400 feet of 1½-inch, 1¾-inch, or 2-inch fire hose

Aerial Fire Apparatus

A minimum of 115 feet of ground ladders, to include at least one attic ladder, two straight ladders with folding roof hooks, and two extension ladders

Apparatus may or may not have a fire pump and water tank

Mobile Water Supply Apparatus

A minimum of 1,000-gallon water tank

Apparatus may or may not have a fire pump. If the apparatus does have a pump, there are minimum requirements for the fire hose that must be carried

Quint

A permanent fire pump with a minimum rated capacity of 1,000 gpm

Water tank with a minimum of 300 gallons

A minimum of 85 feet of ground ladders, to include at least one attic ladder, one extension ladder, and one straight ladder with roof hooks

A minimum of 15 feet of soft suction hose or 20 feet of hard suction hose

A minimum of 800 feet of 2½-inch or larger diameter fire hose

A minimum of 400 feet of 1½-inch, 1¾-inch, or 2-inch fire hose

considerably diminished "scrub area" (area coverable by an aerial device) compared with the aerial ladder and tower ladder.

NFPA 1901 states that the apparatus used by truck companies, an aerial fire apparatus, should have a minimum of 115 feet of ground ladders, including at least one attic ladder, two straight ladders with folding roof hooks, and two extension ladders. Aerial fire apparatus may or may not be equipped with a fire pump. Aerial fire apparatus are also required to carry an extensive list of hand and power tools to accomplish the many tasks that are associated with truck company operations.

In addition to traditional truck company operations, in some jurisdictions, truck companies may also perform the functions of a rescue company, as detailed in the next subsection. Regardless of apparatus, all fire departments must be able to perform the functions associated with the truck company.

The Rescue Company

The **rescue company** (also known as a squad or squad company) is shown in **Figure 2-12A** and **B.** It is organized to provide specially trained firefighters and specialized rescue equipment at an incident scene. The apparatus designated as a rescue company carries tools to conduct forcible entry, and search and rescue at fire operations. Most rescue companies also carry specialized tools to conduct vehicle extrications, confined space rescue, rope rescue, and other forms of technical rescue operations. These specialized functions depend on local requirements and the mission of the organization.

FIGURE 2-10 The engine company provides the firefighters with the tools, equipment, and hose necessary for fire suppression operations. *(Courtesy of William Friedrich)*

(A)

(B)

(C)

FIGURE 2-11 The ladder or truck company is designed to carry firefighters and for forcible entry, search and rescue of building occupants, ventilation, ladder provision, and overhaul functions at a fire scene. Trucks can be classified as (A) aerials, (B) tower ladders, or (C) snorkels, and each truck has inidividual, distinct advantages. *(Courtesy of William Friedrich)*

(A)

(B)

FIGURE 2-12 The rescue company (A) is designed to carry specialized equipment for rescue operations and (B) may provide support equipment such as light towers.

Specialty/Combination Units

Because the fire service has a great amount of diversity in the types of services provided to communities, and an equally great diversity in the way these services are delivered, a wide variety of unique fire and emergency service apparatus are used in departments worldwide.

Combination units are typically a blend of two major company functions into one piece of apparatus. For example, many departments use a **quint** to deliver fire suppression services. A quint is a combination of an engine and a ladder company and can perform either function as necessary on a fire scene. According to NFPA 1901, a quint must have a permanently mounted fire pump and water tank, an area to store hose, an aerial device (ladder or platform) with a permanently mounted waterway, and at least 85 feet of ground ladders.

Another type of specialty unit, a mobile water supply apparatus, is used in areas where a water supply is not present, such as hydrants or natural water sources. Mobile water supply apparatus, also known as tenders, carry a minimum of 1,000 gallons of water per NFPA 1901, and they may or may not have a fire pump and fire hose.

Specialty units provide unique types of services at emergency scenes. Examples of specialty units include wildland fire or brush fire response units, **Figure 2-13A;** water tenders, Figure 2-13B; **aircraft rescue and firefighting (ARFF) apparatus,** Figure 2-13C; light and air units, Figure 2-13D; ventilation units, Figure 2-13E; hazardous materials units; mass casualty response units; and water rescue and firefighting apparatus, Figure 2-13F. These units are specially equipped to handle unique emergency situations, and the personnel who operate on these units are specially trained.

Emergency Medical Services

Many fire departments provide either basic life support (BLS) or advanced life support (ALS) **emergency medical services.** This may be an additional duty assigned to an existing engine, truck, or rescue company or a specific unit designated as a BLS/ALS response unit. Also, fire departments may operate ambulances to provide transport services.

The Chief Officers

The chief of the department, as shown in **Figure 2-14,** is the person ultimately responsible for the operations and administration of the fire department. To carry out these responsibilities, the chief may have a number of deputy, division, assistant, or battalion chiefs.

The rank structure and position designation depend on the size, needs, and in some cases the history of an individual fire department. Not all fire departments

(A)

(B)

(C)

(D)

(E)   (F)

FIGURE 2-13   Examples of specialty units include: (A) wildland or brush units, (B) water tenders, (C) aircraft rescue and firefighting (ARFF) units, (D) lighting units, (E) ventilation units, and (F) water rescue and firefighting apparatus. *(Photos B, C, E, and F courtesy of William Friedrich; Photo D courtesy Du Page Co. OEM)*

FIGURE 2-14 Chief officers are responsible for many functions, including command of emergency operations as well as planned events.

will have all of the positions listed next because some designations serve the same function:

- *Chief of department.* Responsible for all department functions.
- *Deputy chief.* A staff position designated by the chief of department for a specific function, such as personnel, training, or administration. Deputies usually have the authority to act in the absence of the chief.
- *Assistant chief.* Similar to a deputy chief; the assistant may be a staff position responsible for a functional area. In some departments, an assistant chief may be a higher rank than a deputy chief.
- *Division chief.* Usually an operational position responsible for a large geographic area or a number of battalions.
- *Battalion/district chief.* An operational position responsible for a specific number of companies.

The number of officers in any department depends on the size of the organization and the necessity to maintain unity of command, **span of control,** and division of task responsibilities. Chief officers are management-level positions and requirements for these positions are addressed in a separate training standard, NFPA 1021, Standard for Fire Officer Professional Qualifications.

ADDITIONAL FIRE DEPARTMENT FUNCTIONS

Because of the complex issues and tasks facing a fire department, many additional functions are assigned to its operations. Some of these, such as training or fire prevention, are necessary for day-to-day opera-

tions. Additional sections may be established to manage complex issues such as hazardous materials, urban/technical search and rescue, water rescue, or the delivery of emergency medical services.

Fire Prevention and Life Safety

One of the prime missions of all fire departments is the prevention of fires. Preventing fire not only reduces the risk to a community both in lives and property but also reduces the exposure of firefighters to a very dangerous and hazardous duty. The organization of a fire prevention office is usually divided into two functions: code enforcement/inspection services and fire/life safety education. A chief-level officer usually heads the fire prevention office.

Fire department personnel, as shown in **Figure 2-15,** should conduct fire prevention inspections. This provides the organization with increased knowledge of the hazards present and provides the opportunity to develop pre-incident surveys while performing a valuable service to the citizens.

FIGURE 2-15 The most important function of the fire prevention staff is to conduct inspections to prevent fires or ensure early control by fire suppression systems.

Public fire/life safety education activities are usually assigned to the fire prevention office and include fire prevention education as shown in **Figure 2-16** and fire survival programs as shown in **Figure 2-17.** In addition to code enforcement, fire prevention programs are designed to prevent the start of a fire. If an area were to experience a large number of fires starting because of careless disposal of smoking materials, an education program on proper disposal would be delivered to the public. Fire survival programs educate the public on what to do after a fire has started. An example of these activities includes education on installation and maintenance of smoke detectors or **exit drills in the home (EDITH).**

Staff assigned to fire prevention/fire inspection duties should meet the requirements as addressed in NFPA 1031, Standard for Professional Qualifications for Fire Inspector and Plan Examiner.

FIGURE 2-16 Some fire departments utilize unique characters such as E.D.I.T.H. the Clown to help children relate to the message of fire prevention.

FIGURE 2-17 Fire survival programs, such as fire extinguisher training, are routinely conducted by the fire department. *(Courtesy of Marsha Giesler)*

Training

Training begins with basic firefighter or probationary training, as shown in **Figure 2-18,** and continues with proficiency training as new tools, equipment, or techniques become available. Depending on the size, organizational structure, and other responsibilities, a chief-level officer usually heads the training division or group. Regardless of size, all fire departments must have a designated training officer or division.

The training division administers and documents all training activities for the department and will usually have a number of instructors assigned to present specific subjects. This division may also be responsible for conducting mandatory employee safety training required by the Occupational Safety and Health Administration (OSHA). This program requires firefighters to be trained to perform assigned tasks in a safe manner. Some of the required topics for firefighting are respiratory protection, response safety, and workplace safety.

Training officers and instructors should meet the requirements for these positions as addressed in NFPA 1041, Standard for Fire Service Instructor Professional Qualifications.

Emergency Medical Services

As mentioned previously in this chapter, and depending on the size of the organization or its jurisdiction, the emergency medical services (EMS) function may be a separate division within the fire department with a chief-level officer responsible for its activities.

FIGURE 2-18 Training must be a continuing function in all fire departments regardless of size or area served. *(Courtesy of Eastern Oklahoma Technology Center)*

FIGURE 2-19 Trench rescue is one of many specialized operations requiring additional equipment and training.

Apparatus Maintenance and Purchasing

Large departments may have a fire apparatus maintenance or repair shop that is usually responsible for all vehicle repairs, maintenance, and purchasing. Depending on local policy, this organization may be headed by a fire department officer or a nonuniform staff member knowledgeable in the areas of vehicles, heavy equipment, and fire apparatus.

Special Operations

Depending on the size of its community or potential hazards present, a fire department may establish a special operations section, which will deliver or support services such as hazardous materials mitigation; high-rise operations; air operations; confined space rescue, **Figure 2-19;** trench rescue; and swift water or ice rescue. A chief-level officer usually heads these specialized functions.

Individuals assigned to these areas must have specific knowledge and skills for their specialty, which usually requires many additional hours of training.

REGULATIONS, POLICIES, BYLAWS, AND PROCEDURES

All organizations must have regulations, policies, bylaws, and procedures to operate in an effective manner. This is especially true in a fire department to ensure an adequate and effective emergency response.

These regulations are implemented and used to establish the daily and emergency operations of the organization.

Regulations

Regulations are the rules that determine how an organization operates and are usually established by top-level management or a governing body. Regulations address topics such as time and attendance, uniform, and use of department equipment and vehicles. These are usually very specific in nature and may carry some type of disciplinary action for violation.

In addition, state or federal government agencies, such as the Occupational Safety and Health Administration (OSHA), may establish regulations governing fire department activities. These usually deal with such issues as health and safety, antidiscrimination laws, and work hours.

OSHA was created by the William Steiger Act signed into law by President Nixon in 1970. It is regulated by the U.S. Department of Labor and is charged with ensuring health and safety in the workplace. Violations of OSHA standards can carry fines up to $70,000.00 for each instance.

Public employees in state and local governments are not covered by OSHA, but most states have their own occupational safety and health plans. State plans covering private sector employees must also cover state and local government employees. Twenty-two states and territories have OSHA approved state occupational safety and health plans. An additional four, Connecticut, New Jersey, New York, and the Virgin Islands, have state plans that cover public employees only. These plans may be even more stringent than those set forth by the federal OSHA regulations

The federal program is the minimum set of standards that must be addressed in the individual state plans. Enforcement of the regulations in state plans is under state control. Unless a local government in a state without an OSHA approved plan for public safety employees adopts OSHA provisions, federal OSHA has no authority in enforcement. In these instances, enforcement will be carried out by the individual state's Department of Labor. OSHA considers NFPA standards as reference for the fire service when an issue is not addressed specifically in an OSHA regulation.

> **NOTE**
>
> Regulations established by the governing body of a fire department are internal in nature and design, whereas regulations established by state or federal agencies have the force of law, with some penalty or fine charged for violations.

Policies

Policies are formal statements or directives established by fire department managers to provide guidance for decision making. Policies will usually be general in nature and provide a framework for administering the day-to-day department activities. Topics addressed by policies include types of response levels, staffing of companies, and the release of information at an incident.

Bylaws

Volunteer fire departments provide a great deal of fire protection, both in the United States and Canada. These departments may be organized as previously described in this chapter, as independent corporations, or as some combination of both types of organizational structure. A fire corporation will usually be organized as a not-for-profit organization with a governing body of a president and vice president, or a board of directors. The board of directors or membership will establish bylaws in place of regulations or policies on how the department business structure is organized. Generally, fire and emergency operations are not covered as part of the bylaws.

Procedures

Standard operating procedures (SOPs) provide specific information and instructions on how a task or assignment is to be accomplished.

SOPs are established so that all members of a department will perform the same function with the same level of uniformity. Also, these procedures are generally tactical in nature because, in most instances, they address emergency operations. In addition, all SOPs must be distributed and understood for all department members. Some topics addressed by SOPs include cleaning and use of protective equipment, apparatus response policies, use of specific tactics or tools, and how to establish a water supply at a fire scene.

A note on terminology: This text uses the term *standard operating procedure* to describe documentation outlining actions or procedures to be followed. Many departments use the term standard operating guideline (SOG) to describe this documentation. SOGs are often thought of as providing guidance to company officers and firefighters when operating on emergency scenes as opposed to outlining strict parameters for actions. Each department is different in its use of SOP and SOG terminology and even different in its interpretation of the flexibility of the documents. It is up to the individual firefighter to understand what terms are used by his or her department.

There are a variety of ways in which SOPs are developed and documented. In general, the following should be true:

- SOPs address the who, what, when, where, and how of a topic.
- Firefighter safety is the first consideration in all procedures.
- SOPs are brief, clear, and concise.
- Lengthy SOPs are broken down into smaller sections.
- SOPs cover dynamic topics and must be reviewed often, at least once every three years.
- A method for changing SOPs exists if they should become obsolete, inaccurate, or ineffective.

Figure 2-20 shows a sample of an SOP that might be found in a typical fire service organization.

Sample Standard Operating Procedure

Subject: Emergency Vehicle Response   Section: 303
Date: 01/01/2007
Purpose: To provide requirements for emergency response of Delmar Fire Department apparatus and vehicles.
Scope: This procedure applies to all drivers/operators of Delmar Fire Department vehicles.

1. Apparatus will not exceed the posted speed limit at any time.
2. Seat belts must be used by all personnel riding on apparatus.
3. No riding is allowed on the rear step area nor standing in the crew compartment while the vehicle is in motion.
4. Apparatus must come to a complete stop before attempting to proceed in any of the following conditions:
   - Intersection against a red traffic light
   - All railroad grade crossings
   - Blind intersections
   - Stop signs
   - Any condition when the operator cannot account for all traffic lanes
5. A spotter shall be located at the rear of the apparatus at any time the vehicle moves in reverse.
6. Operators are responsible for any traffic citation issued to them.

FIGURE 2-20 A sample standard operating procedure.

ALLIED AGENCIES AND ORGANIZATIONS

During the course of any operation, the fire department and firefighter interact with many different people and organizations. Depending on the local jurisdiction, a few of these agencies could be police/law enforcement, public works or highway department, utility companies, environmental conservation/protection agencies, and private business. Depending on the nature of the incident, these organizations may be assisting with the operation or they may be receiving fire department services.

- *Police.* Provide assistance with traffic management and scene security at all incidents. May be the assisting or lead agency in fire investigations, bomb threats, or terrorism incidents.
- *Public works.* Provide assistance with traffic management and maintain public water system for fire protection. May maintain fire station facilities and provide assistance in securing materials (earth, sand) for the control of hazardous material incidents.
- *Utility companies.* Secure building and street utilities during or after an incident.
- *Environmental protection.* Depending on the incident they may be a lead or assisting agency at a hazardous materials spill.
- *Private business.* May provide supplies for emergency operations or may be the receiver of fire department services.

The agencies listed are by no means all-inclusive, as the nature or complexity of an incident may provide an opportunity for the fire department to work hand in hand with many additional agencies. Proper planning prior to an incident should be conducted to define these entities and establish the appropriate relationships and contacts as necessary.

> **NOTE**
>
> **Mutual aid or assistance agreements** are prearranged written agreements of the type and amount of assistance one jurisdiction will provide to another in the event of a large-scale fire or disaster. Mutual aid plans may include the fire department resources from a town, county, or entire state. The key to understanding mutual aid is that it is a reciprocal agreement. Many fire departments use automatic mutual aid to provide for immediate dispatch to an incident. This may be to supplement staffing or equipment required for high hazard facilities such as a hospital.

LESSONS LEARNED

To be proficient as a firefighter I, individuals must know their roles and responsibilities within the organization and how they apply to the overall mission of the fire service and their fire department. A firefighter must also know how the fire department command structure works, the application of standard operating procedures, and how to work with allied agencies during an incident.

KEY TERMS

Aircraft Rescue and Firefighting (ARFF)   Of or pertaining to firefighting operations involving fixed or rotary wing aircraft.

Company   A team of firefighters with apparatus assigned to perform a specific function in a designated response area.

Emergency Medical Services   The delivery of prehospital medical treatment.

Emergency Medical Technician (EMT)   An individual trained and certified to provide basic life support emergency medical care.

Engine Company   The unit designation of a group of firefighters assigned to a piece of apparatus designed to deliver water to the fire scene.

Exit Drills in the Home (EDITH)   A fire survival program to encourage people to practice fire drills from their home or residence.

Hazardous Materials Technician   An individual trained to meet the requirements of CFR OSHA 1910.120, Technician Level for Hazardous Materials Response.

Mutual Aid or **Assistance Agreements**   Prearranged written agreements of the type and amount of assistance one jurisdiction will provide to another in the event of a large-scale fire or disaster. The key to understanding mutual aid is that it is a reciprocal agreement.

National Fire Protection Association (NFPA)   A not-for-profit membership organization that uses a consensus process to develop model fire prevention codes and firefighting training standards.

NFPA 1001   Standard for Fire Fighter Professional Qualifications, a national consensus training standard establishing the job performance requirements of tasks to be performed by firefighters.

Paramedic (EMT-P)   An individual trained and certified to provide advanced life support emergency medical care, including drug therapy.

Quint   A combination fire service apparatus with components of both an engine company and a truck company.

Rescue Company   The unit designation of a group of firefighters assigned to perform specialized rescue work and/or tactics and functions such as forcible entry, search and rescue, ventilation, and so on.

Rescue Specialist   A firefighter with specialized training and experience in areas such as high angle rope rescue, confined space, trench, or structural collapse rescue.

Span of Control   The ability of one individual to supervise a number of other people or units. The normal range is three to seven units or individuals, with the ideal being five.

Standard Operating Procedure (SOP)   Specific information and instruction on how a task or assignment is to be accomplished.

Truck Company   The unit designation of a group of firefighters assigned to perform tactics and functions such as forcible entry, search and rescue, ventilation, and so on.

REVIEW QUESTIONS

1. Write a mission statement for a fire department including the level of services provided and to whom the services will be delivered.

2. Draw an organizational diagram for a fire department including the chain of command and who is responsible for training, fire prevention, apparatus maintenance, and fire scene operations.

3. List five functions or tasks performed by firefighters in a fire department.

4. Explain the reason for standard operating procedures.

5. List five areas addressed by standard operating procedures.

6. Explain where to find the rules and regulations for a particular firefighting organization and identify who is responsible for developing and enforcing them.

7. List five rules and regulations of a firefighting organization and describe how they apply to the firefighter.

8. Research a jurisdiction's mutual aid agreement and list the apparatus, specialized equipment, and personnel available for response.

9. List and describe the functions of five allied agencies that assist a fire department.

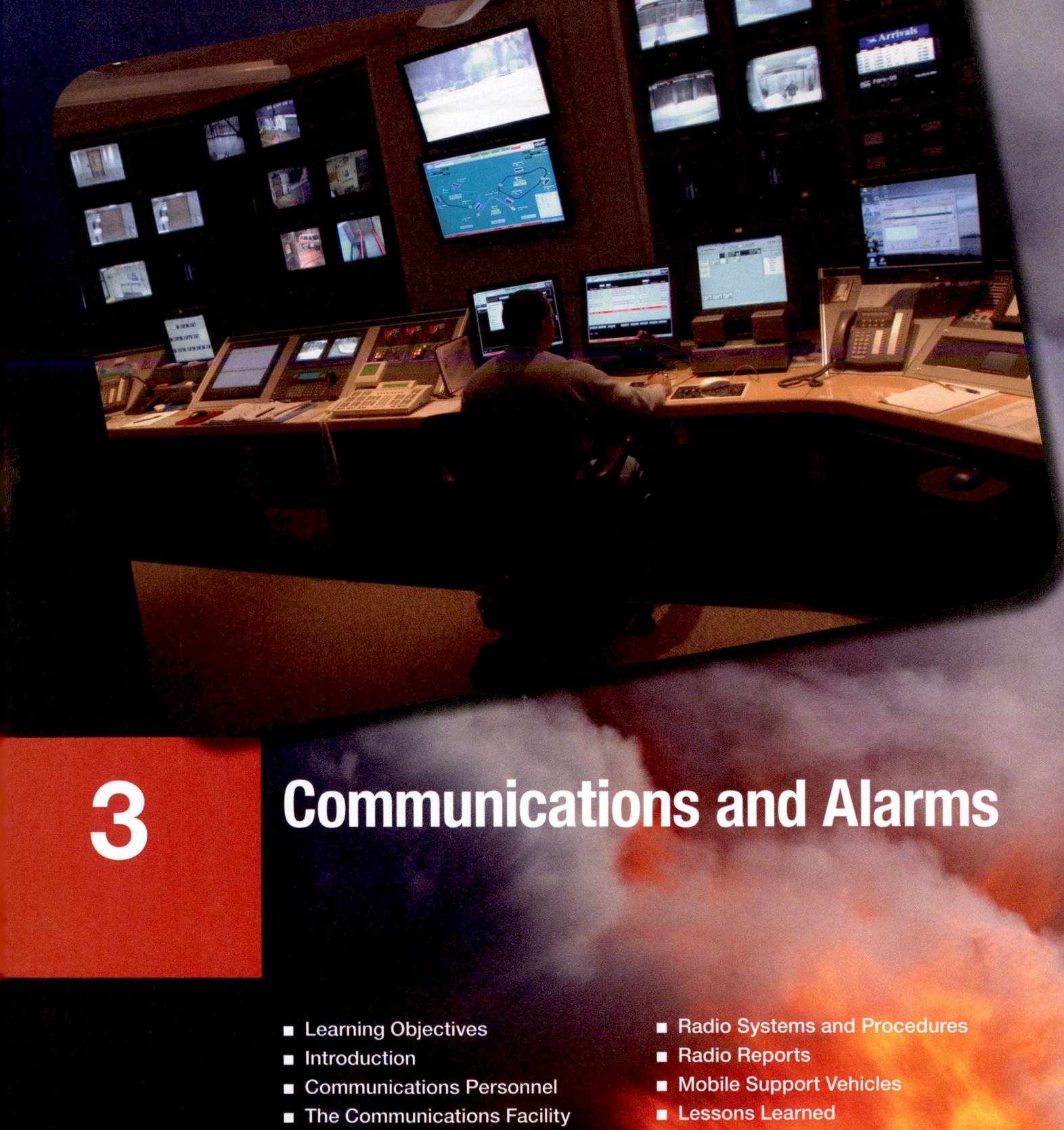

3 Communications and Alarms

everal years ago I had the unique opportunity to spend time working in the 9-1-1 communications center. It was a very eye-opening experience for me since I had always been a firefighter in the field. Spending time in the center allowed me to realize the role that communications personnel play in the overall fire and emergency services system. As a firefighter, I would get frustrated when calling dispatch on the phone for alarm numbers and times, and the dispatcher would say, "We are busy—can you call back?" I would think to myself, "How long can it take to give me times and a number? I have to get these reports done, and the dispatchers are holding me up." When responding to emergency calls, if the dispatch information was not completely accurate, the field personnel would often blame the dispatchers and communications personnel. We often did not consider that maybe the caller did not provide accurate information. After spending time in the communications center, I realized a new appreciation for what communications personnel do on a daily basis. While it may be frustrating to be told to call back for information, I realized that dispatchers were dealing with units calling in from twenty-four other stations. While they were dealing with these phone calls, they also had to answer and dispatch 9-1-1 emergency calls, answer inquiries from the media calling about incidents, and of course talk to people calling 9-1-1 for nonemergency information. I also came to understand that the communications personnel often have difficulty in obtaining accurate information from callers. Many times this is because the callers are upset, frantic, or just do not have good information to pass on to the communications center. The information given to the responding units is only as good as the information available to the dispatchers. My experience in the communications center taught me a great deal about the role of the communications personnel. They are truly the critical link between the public and the fire rescue system. They are a dedicated group who try to do the best job possible, even under difficult circumstances and situations. They take the 9-1-1 calls and dispatch the units and then try to anticipate the needs of the units throughout the incident. Do not make the same mistake that I did. Visit your communications center and experience what they do for yourself. And when they ask you to call back because they are busy, just do it with a smile.

—*Street Story by James Angle, Chief,*
Palm Harbor Fire Department, Palm Harbor, Florida

LEARNING OBJECTIVES

After completing this chapter, the reader should be able to:

3-1 Describe the process for an individual to report a fire or other emergency.

3-2 Explain the prescribed fire department radio procedures, including routine traffic, emergency traffic, and emergency evacuation signals.

3-3 List the critical pieces of information needed to properly process an emergency call.

3-4 Discuss alarm protocols for a dispatch center.

3-5 Demonstrate the prescribed fire department radio procedures, including routine traffic, emergency traffic, and emergency evacuation signals.

3-6 Demonstrate the purpose and function of all alarm-receiving instruments and personnel-alerting equipment.

3-7 Promptly relay information to the communications center.

3-8 Obtain complete information.

3-9 Explain the procedures for answering non-emergency telephone calls and operating an intercom system.

3-10 Demonstrate the procedures for answering non-emergency telephone calls and operating an intercom system.

3-11 Demonstrate the proper method for answering fire department telephones for nonemergency administrative calls and notifying the proper individual.

3-12 Identify proper protocol for operating fire department radios for routine communications, emergency traffic, and emergency evacuation procedures.

3-15 Demonstrate the proper method for operating both mobile and portable radios for both normal and emergency traffic.

*The FF I and II levels, as defined by the NFPA 1001 Standards, are identified in different colors:
FF I = black, FF II = red, additional information = blue.

INTRODUCTION

"Nine-one-one. What is your emergency?"

"Oh, please send help! There has been a terrible accident!"

"What is the location of the accident?"

"It's at the intersection of Main and...."

Terrifying, emotionally wrenching scenarios such as this unfold thousands of times each day in emergency communications centers all across America. To get the information emergency responders need to do their jobs effectively, **telecommunicators** must be prepared to communicate effectively with citizens and relay accurate information to first responders.

In this chapter, we discuss many aspects of fire service **communications.** It is safe to say that throughout history, effective communications have been essential to all successful endeavors. Public safety communications is certainly no exception to this. What is communications? Simply put, communications is the sending, giving, or exchanging of information. Although communications takes on many forms, conveyance of information through the medium of speech is considered one of the most universal functions. It is

estimated that interactions involving speech account for 74 percent of our communications time.[1] Yet effective communication is still a challenge for most of us in today's busy and complex workplace.

NOTE

From the standpoint of fire service **emergency communications center** operations, the communications process must include four basic elements: information from the caller must be received, understood, recorded accurately, and communicated to emergency responders.

The communications process is illustrated in **Figure 3-1.** Reports from citizens can usually be categorized as true emergencies, perceived emergencies, nonurgent reports, or requests for information.

Despite our knowledge of the importance of complete communications, in many cases complete communication does not occur between individuals. Ironically, it has not been until recently that many fire service leaders have come to realize the importance of good, reliable communications and the impact that

FIGURE 3-1 The communications process must be complete and clearly understood in order to be effective.

successful communication can have on the delivery of the services provided by their departments. As a result of a greater awareness of the need to maintain effective communications, many fire service leaders have started taking proactive measures to ensure that the quality of the communications process is maintained. These measures include:

1. Teaching communications skills to employees.
2. Upgrading communications systems.
3. Incorporating modern technology into the daily operational routines of their departments.

While it takes many special skills and various personnel to provide total service, the communications center is the heart of the operations. As a result, fire service leaders are putting more emphasis on having qualified staff managing and operating emergency dispatch systems. Due to this increased awareness, greater emphasis is being placed on the selection and training of the telecommunicators who staff communications centers in larger departments, and also on those smaller volunteer departments who rely on volunteers to perform these duties.

> **FIREFIGHTER FACT**
>
> A communications center may also be referred to in some instances as a PSAP, public safety answering point.

COMMUNICATIONS PERSONNEL

Although the names may have changed and the roles and responsibilities expanded, the basic role of the telecommunicator has remained remarkably the same. The primary role of the telecommunicator is to receive emergency requests from citizens, evaluate the need for a response, and ultimately sound the alarm that starts first responders on their way to the scene of an emergency.

Receiving calls is only the first challenge facing telecommunicators. This is the beginning of what can be a very complex process of gathering information and deploying emergency apparatus, personnel, and equipment. A myriad of tasks go on behind the scenes of communications centers. Telecommunicators provide a variety of ancillary services to support field operations. In some cases a telecommunicator serves as the public relations officer for the department. After all, a telecommunicator is the first person to speak to the caller in need of emergency services. A telecommunicator is often the first person contacted by the news media when updates are required. Some departments issue departmental news and updates from their communications center, whereas others require that this type of information be distributed by the chief or the person acting as the public information officer. In either case, the news media can be the department's best friend or worst enemy; they should be handled with care.

In some instances, local protocols may dictate that after initial contact by the citizen, the telecommunicator will maintain contact with the caller, enabling the telecommunicator to provide valuable **prearrival instructions.** As a result of the expanding role of the telecommunicator, fire service leaders are reconsidering the complexity of emerging technology and the importance of the individuals hired to operate these systems.

> **FIREFIGHTER FACT**
>
> A telecommunicator's job has been accurately defined by some as hours of boredom punctuated by moments of terror.

NFPA 1061, Standard for Professional Qualifications for Public Safety Telecommunicator, clearly outlines behavioral characteristics or traits that a person should possess to be a viable candidate for a position in the emergency communications center. Among these traits are the ability to perform multiple tasks, the ability to make decisions based on common sense and standard values, the ability to maintain composure in high stress conditions, and the ability to remember details and recall information easily. Candidates must also be able to exercise voice control, which includes the ability to maintain balanced tone, modulation, volume, and inflection while communicating. Only a tried, trusted, and complete training program will ensure that telecommunicators are armed with the knowledge and skills necessary to do the job.

A quality training program must be followed by a comprehensive work performance evaluation program to ensure fire service leaders that their communications centers are staffed by well-trained personnel who are able to apply the training they receive and develop the skills they need to perform the required tasks. One organization that has been a leader in the field of communications training for many years is the **Association of Public-Safety Communications Officials – International, Inc. (APCO).** APCO has established minimum training standards that are being adopted by many states throughout the country. Initial certification through a tested program coupled with a strong continuing education program is paramount to the efficient and effective operations of a modern communications center.

An adequate staffing level at communications centers is the next most significant concern of fire service leaders. Most communications center managers will readily attest that there are never enough personnel when a **mass casualty** or **multiple-alarm incident** occurs. Communications managers rely on historical data to produce staffing models that closely resemble actual staffing requirements. By using some widely accepted formulas, managers are able to closely replicate staffing needs.

NOTE

The NFPA states that "Ninety-five percent of alarms shall be answered within 30 seconds, and in no case shall the initial operator's response to an alarm exceed 60 seconds." It additionally recommends that "the dispatch of the appropriate fire services shall be made within 60 seconds after the completed receipt of an emergency alarm."

To meet this requirement, some departments with limited resources and low call volumes need only ensure that at least one person is on duty at all times to answer emergency phones. Regardless of size, fire departments must provide well-trained personnel to serve in the telecommunicator role because those individuals who answer calls from citizens have a direct impact on the overall response time of the agency.

THE COMMUNICATIONS FACILITY

The importance of staffing levels and having properly trained personnel in a communications center in order to process emergency calls within the time prescribed by the NFPA has been discussed. Now we turn our attention to the places where these telecommunicators work. If the facility is intended to house emergency communications operations, there are several important considerations.

FIREFIGHTER FACT

The first national emergency telephone number, 999, was introduced in Great Britain in 1930.

Communications centers have many different configurations. Some are served by **9-1-1** (pronounced nine, one, one) systems, and some rely on a regular seven-digit telephone number to receive calls. Many serve only a single agency, whereas others have joint authority over police, fire, and emergency medical services, and even utilities. Many communications centers are nothing more than a small alcove in a larger administrative office, with only one person responsible for communications duties for a single agency such as police. That same person might also book prisoners, answer phones, serve walk-up customers, and perform other combinations of job functions as assigned. Another example is a small volunteer fire department that has only one person on duty to answer calls and dispatch personnel. This is common in jurisdictions with low call volume activity. As shown in **Figure 3-2,** other facilities can be larger and constructed expressly for the purpose of serving as an emergency communications center. These are generally found in larger metropolitan areas with high volumes of emergency responses.

Regardless of the size or type of facility, all serve the one common goal of receiving and disseminating information that can be both emergency and nonemergency in nature.

The NFPA also has standards (NFPA 1221) for location and construction of emergency communications centers. Modern emergency communications facilities are typically constructed in areas where there is little risk of damage either by natural or manmade hazards. Communications centers should not be constructed in low-lying areas prone to flooding or along coastlines prone to hurricanes or high winds. In like terms, most agree that they should be located in areas of limited traffic and limited exposure to manmade hazards such as terrorism, arson, and so on. The NFPA suggests that communications centers that are

FIGURE 3-2 Exterior view of a 9-1-1 emergency communications center that houses fire/EMS, police, and sheriff communications operations. *(Courtesy of Caddo Parish 9-1-1)*

built closer than 150 feet to existing structures should receive special attention regarding protection from collapse. Therefore, parks and other open areas are well-suited locations for these types of facilities.

Special attention should also be given to protecting the facility from vandalism and the effects of civil disturbance. Only authorized personnel should be permitted access to the communications center. Communications centers should have a limited number of outside windows, and all outside entrances should be monitored by security systems. The facility should also be monitored by a fire alarm system that complies with **NFPA 72,** National Fire Alarm Code.

> **CAUTION**
>
> Every communications facility should be supported by a backup location in the event that the primary facility encounters problems that render it inoperable and result in evacuation.

Emergency communications centers should be equipped with backup power supplies that also meet the requirements set forth by the NFPA. An example of a backup power generator is shown in **Figure 3-3.** Backup power generators must be able to maintain the operations of a large communications center for prolonged periods of time. Some emergency generators are fueled by diesel fuel, which is stored on site as shown in **Figure 3-4.** It is also important to note that liquid petroleum or natural gas may fuel some power-generating equipment. Most backup power supplies are designed with electrical load distribution, which provides separation of power between emergency and nonemergency circuits. In this type of arrangement, where multiple generators are in use, they are typically installed with a switching device that automati-

cally transfers the emergency loading to the first generator to achieve operating speed.

Uninterruptible power supplies (UPS) are also used in some of the more modern communications centers. These units use battery power to provide uninterrupted power to the critical communications and computer equipment used in the communications center. Properly designed and regularly tested power backup systems that employ automatic switching devices that do not interrupt the operations are extremely important, especially when modern computerized systems are in use. A large UPS system is shown in **Figure 3-5.**

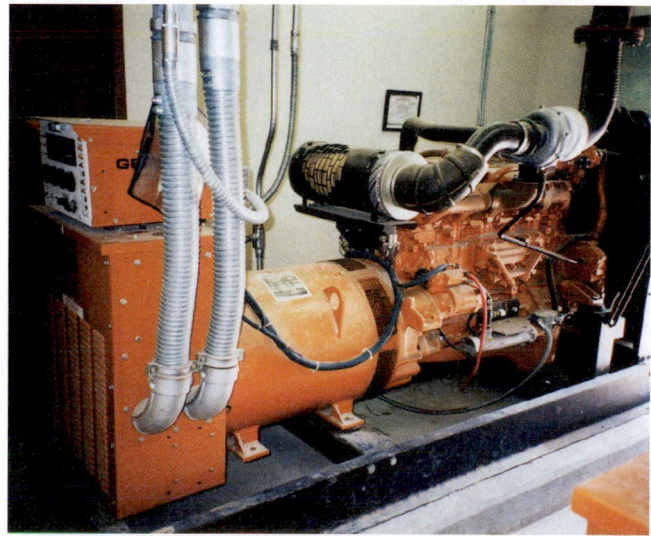

FIGURE 3-3 The emergency communications center is protected from power outages by use of backup power generators. Generators of this type can provide continuous power to operate the communications center during periods of power outages. *(Courtesy of Caddo Parish 9-1-1)*

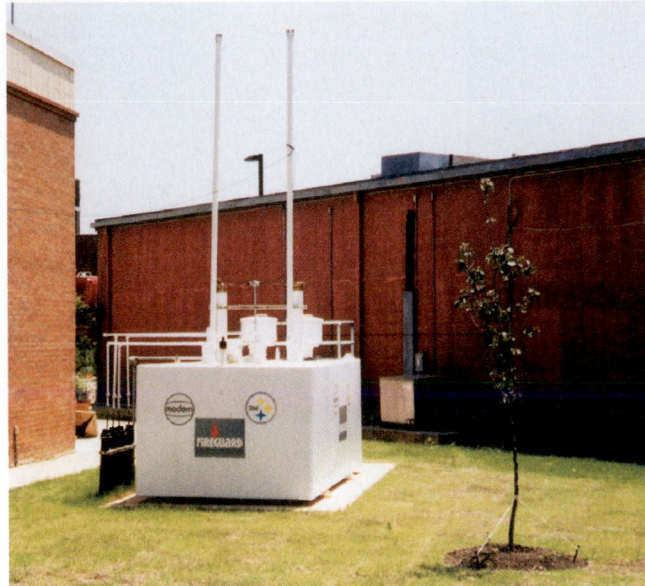

FIGURE 3-4 A 500-gallon supply of fuel is maintained to ensure that uninterrupted power can be provided for a minimum of twenty-four hours and for longer periods if necessary. Liquefied petroleum gas and natural gas (not shown) may also be used to power backup generators. *(Courtesy of Caddo Parish 9-1-1)*

FIGURE 3-5 UPS systems such as this help maintain uninterrupted service should a power outage occur.

COMPUTERS IN THE FIRE SERVICE

The proliferation of new and more technically advanced computer hardware, along with the development of highly specialized computer software designed especially for use by public safety agencies, has made it possible for many departments to incorporate the use of computers into their communications systems. Many fire departments have found that the use of **computer-aided dispatch (CAD)** systems,

FIGURE 3-6 Computer-aided dispatch workstations such as this one are found in many high-activity communications centers.

Figure 3-6, has made it easier to handle increased call volumes.

The computer can keep track of the location of active incidents and which units have been assigned to respond and therefore help telecommunicators to better manage their resources. At the same time, computers can create and store valuable records on each incident and other departmental activities. Computers have greatly improved the way in which statistical analyses of departmental activity are performed. Additionally, computers can provide remote locations with access to a variety of information that is stored in a main **database** that can be accessed when needed. This information may include maps, hazardous materials information, prefire plans, policies and procedures, mutual aid instructions, or any other information that may be useful to firefighters. This type of information is often stored on mobile data computers in all types of fire apparatus and **command vehicles.** Computers also allow fire personnel to access off-site databases that may be beneficial either for training or incident mitigation.

RECEIVING REPORTS OF EMERGENCIES

The call-taking process consists of receiving a report, interviewing, and referral or dispatch composition. **Figure 3-7** provides a visual representation of the various stages of call processing. Whether a highly trained telecommunicator or a firefighter is performing this duty, speed is very important during the interview portion of the call-taking process. Fire depart-

ment standard operating procedures (SOPs) regarding taking emergency calls should be well known by firefighters to ensure that the proper information is obtained from the caller as well as to prevent a delay in emergency response. Telecommunicators must be able to prioritize incoming calls in order to ensure that the most important call gets the fastest attention. Incoming telephone calls should be answered in the following priority: (1) 9-1-1 and other emergency lines, (2) direct lines, and (3) business or administrative lines. Telecommunicators should answer incoming lines promptly and determine if the caller has an emergency. If not, and another emergency line is ringing, the telecommunicator should put the caller on hold and answer the other incoming call. However, speed is of little consequence if accurate and complete information is not obtained from the caller and relayed to emergency response personnel.

When receiving reports of emergencies by telephone, telecommunicators should always speak clearly and slowly with good volume. If questions to the caller have to be repeated, valuable time is wasted. When speaking with a citizen who is reporting an emergency, the telecommunicator's voice should project authority and knowledge. The telecommunicator should use plain, everyday language at all times

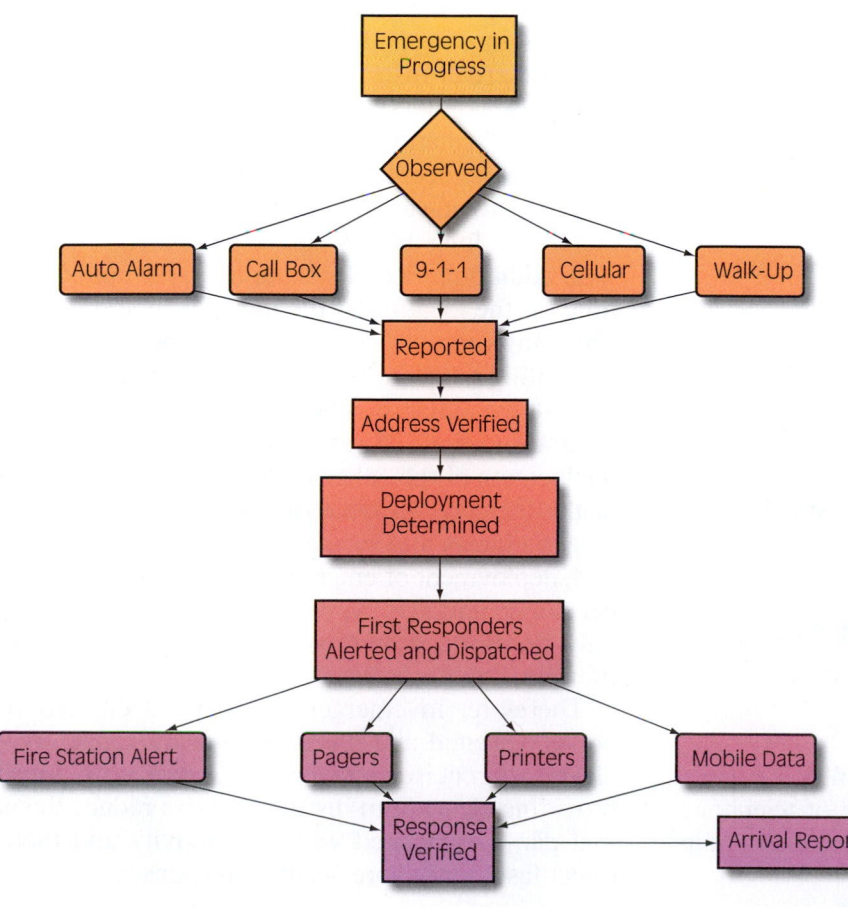

FIGURE 3-7 This figure illustrates the work flow of call processing by a public safety telecommunicator.

when speaking to the public and maintain a polite and friendly tone. It is never appropriate to argue with a caller. The citizen is the "customer."

Time is of the essence when answering and processing requests for service. The 9-1-1 and other emergency lines should always be the top priority. When answering incoming lines, the telecommunicator should always identify the department and ask, "Do you have an emergency?" or "What is your emergency?" It is critically important for the telecommunicator to control the conversation in order to obtain the necessary information in the least amount of time. In many cases, the caller knows what has happened but finds it difficult to relay key elements of the situation to emergency service providers. The key information must be extracted from the caller, even in cases where a caller is hysterical or in severe emotional distress. A calm and authoritative tone of voice is helpful. In most cases, asking short, specific questions (name, address, telephone number, location) is the most effective tactic.

In the case of nonemergency calls, the telecommunicator should make every effort to accommodate the requests of the callers, which includes transferring them to another agency that may be better suited to provide assistance. It is important from a public relations standpoint to always offer to transfer the caller if the telephone system is capable of this function. Prior to transferring the call, the telecommunicator should provide the number to the caller in case the transfer process is interrupted or fails.

> **STREETSMART TIP**
>
> A professional telecommunicator who speaks with confidence and authority will be able in most cases to gain the trust of the caller, who will then be more apt to follow the instructions issued by the call taker.

From time to time a caller may hang up, be disconnected, or simply drop the telephone before providing all of the needed information. This could be the result of a medical problem, an equipment failure, or simply the fact the caller realized he or she dialed a wrong number.

> **CAUTION**
>
> Telecommunicators should diligently follow up on all incomplete calls.

How this is done varies greatly from one department to another depending on the type of telephone system available, but the basic fundamentals remain the same.

When processing an emergency call, telecommunicators should always attempt to obtain the following minimum information:

- Location of the emergency
- Nature of emergency
- Callback number
- Caller's location and situation

Without a doubt, the most important piece of information is the location of the incident. What is happening is of little importance if the call taker doesn't know where to send help. Once the caller has provided the location, the telecommunicator should attempt to secure any additional information that the caller can provide with respect to landmarks such as cross-streets or easily recognizable buildings, if this can be done safely.

> **CAUTION**
>
> Getting and verifying callback numbers of the caller reporting the emergency is extremely important in the event that it is necessary to recontact the caller to obtain additional information concerning the incident.

Life safety is of primary importance. Therefore, the call taker should always try to determine if the caller is in danger. If so, the call taker should provide prearrival instructions in accordance with local protocols.

This information can be relayed to field units via radio as it is received and updated. Callers should be asked to provide all the information they can about an incident as long as they can do so safely.

In addition to a callback number, the caller's proximity to the location of the incident should be noted. This information is sometimes beneficial when attempting to locate incidents by following the directions of the caller. It is helpful if the first telecommunicator and the first responders are able to visualize the location of the caller in proximity to the locations of the incident. In most dispatch systems, once sufficient address and incident type information is verified, deployment of emergency apparatus and personnel can be initiated. As a rule, the average citizen will only find a need to report an emergency once in a lifetime.

Therefore, in emergency situations citizens may be very excited and unable to formulate and convey clear and concise information. A telecommunicator providing clear, calm instructions can reduce the normal panic associated with this activity and thereby effect faster and more accurate responses.

The information callers provide can be extremely valuable but only if the call taker is able to ask the right questions that will generate meaningful responses. After deployment of the appropriate responders, communications center personnel should retrieve any pre-incident plans, SOPs, information about hazardous conditions, or any other information that may be available on site at the communications center and relay all pertinent information to the emergency responders.

In the case of emergency medical calls, much more information may be requested from the caller in accordance with **emergency medical dispatch (EMD)** protocols. Emergency medical dispatch systems are designed to meet the needs of specific jurisdictions by providing the telecommunicators with a set of standard questions and predetermined actions used to evaluate the situation. A response to an emergency medical incident must be based on a telecommunicator's ability to recognize and react to the "symptoms" being displayed by the victim, rather than an attempt on the part of the telecommunicator to make a "diagnosis" of the injury or illness.

Methods of Receiving Reports of Emergencies

Reports of emergencies can be received in a variety of ways. We briefly discuss some of the most common means in this section:

- Conventional telephones
- Wireless or cellular telephones
- Emergency call boxes
- Automatic alarms
- TDD equipment for the hearing impaired
- Still alarms or walk-ups

Receiving Reports by Telephone

Throughout the country, conventional residential and business telephones are the most commonly used method of reporting emergencies. However, the use of cellular telephones is becoming much more popular as they become available to larger cross sections of citizens, as discussed in the next subsection.

According to the **National Emergency Number Association (NENA),** nearly 93 percent of the population of the United States is covered by some type of 9-1-1, and 95 percent of 9-1-1 coverage is enhanced

9-1-1. Approximately 96 percent of the geographic United States is covered by some type of 9-1-1.

Incidents have been reported in which a panicked citizen has cited problems locating the "eleven" on the telephone touch pad. Regardless of whether **basic 9-1-1** or **enhanced 9-1-1** is used, these systems provide access to emergency services via the use of a simple, easy-to-remember three-digit number.

In addition to voice communication, enhanced 9-1-1 service provides emergency communications centers with the telephone number and address from which the call is originating. This decreases the time necessary to determine the caller's location when callers are either unwilling or unable to provide this information. This feature eliminates the need to trace or research telephone company records to associate an address with a telephone number. Again, features like this provide valuable assistance quickly and effectively. They are extremely important to emergency call takers, especially in cases where the caller is very excited, the caller is not familiar with an area, or for other reasons the caller is unable to communicate clearly the location of an emergency.

Both basic and enhanced 9-1-1 service are available through conventional business and residential telephone service. In addition to conventional or wired telephones, wireless communications devices are growing more popular as a means of communicating emergencies.

Receiving Reports via Cellular Telephones

Wireless systems, such as cellular telephones, are proving very beneficial to the process of reporting emergencies. They are widely available to a large cross section of citizens and are very mobile. Any 9-1-1 calls initiated by a citizen using a cellular telephone are routed to a predetermined answering point for processing. In some areas, the state highway patrol or state police is designated to receive all 9-1-1 cellular calls for a specific area or jurisdiction. In other areas of the country, 9-1-1 calls that originate from cellular telephones are routed to a communications center that may serve either a single agency or multiple agencies and even multiple jurisdictions.

Although the use of cellular telephones is proving very beneficial, it also has some negative points. The use of cellular telephones has caused significant increases in communications center call volumes. For example, on today's busy highways, one accident usually generates multiple reports by motorists using cellular telephones to access 9-1-1. Cellular callers are also more likely not to know their exact location, because they may be traveling along busy highways in an unfamiliar area. In these types of situations, landmarks and direction of travel are very important to the telecommunicator. In the case of small departments, the increase in the number of calls generated by cellular phones can have a budgetary impact. Also, in some cases, the telecommunicator will be criticized unfairly for not getting "accurate" information when in fact some of the things discussed earlier in this chapter are the cause for inaccurate information. Government-mandated upgrades in technology require cellular manufacturers to provide a means by which the location of wireless callers can be determined and provided to the emergency call taker. This greatly reduces the risk of location errors during the processing of emergency reports received via cellular phones.

New technologies created for automobiles provide citizens with specialized communication systems built into vehicles that will contact emergency services with the press of a button or when certain safety systems, such as air bags, are activated within the vehicle. This new wireless emergency reporting technology does not provide much information about the nature of the emergency; however, using satellite technology it can provide the communications center with an exact position of the vehicle reporting the emergency. As this system is used and tested, its popularity in new automobiles may grow in the future.

> **NOTE**
>
> **Voice Over Internet Protocol (VOIP)** technology is becoming a very popular option for telephone service. This technology converts audio signals into digital data that is transmitted over the Internet. Just like accessing the Internet, the telephone uses an IP address. This Internet data allows a telephone to access any other telephone located throughout the world. Because the data is compressed, it allows for more calls to be passed through a system than traditional forms such as circuit switching. Because this system bypasses the telephone company, it offers considerable cost savings and is being used by many people for that reason.

Since the system is computer driven, it is susceptible to the same problems encountered with computers, such as viruses, bugs, and so on. Another downfall to this system is that there is presently no way to associate an IP address with a specific geographic location for emergency response as presented with enhanced 9-1-1. At this time, telephone service providers can only have 9-1-1 VOIP calls converted to numbers that ring in the public safety answering points (PSAP) serving the general area where the caller is located—specific address information is not attainable from the call itself as with enhanced 911. VOIP will definitely be an option in the future as the drawbacks are eliminated. Features such as preplans, critical incident dispatch data, and even access to a person's medical history may be provided to emergency responders in the future by use of VOIP.

Receiving Reports via Municipal Fire Alarm Systems

Municipal fire alarm systems allow a coded or voice message to be generated from an alarm box typically located in highly visible, easily accessible areas that are open to the general public. Systems of this type came into use during the late 1800s and many are still in use today with few upgrades and modifications.

> **FIREFIGHTER FACT**
>
> **Emergency call boxes** were first installed in the United States in 1852 and are still used in many parts of the country as a means of reporting emergencies.

According to the Boston Fire Department, the first emergency call box was placed in operation in Boston, Massachusetts, on April 28, 1852. This system is still in operation today and has seen only one upgrade since its installation. While Boston finds it beneficial to use a system of this type, some cities have discontinued them largely due to high rates of false alarms. As seen in **Figure 3-8,** call boxes can be operated via "hardwired" systems or they can be of the wireless solar-powered variety that can be installed in remote locations not serviced by electrical power or conventional telephone lines. An example of this type of technology is shown in **Figure 3-9.**

Some call boxes simply transmit a preset identification number to the communications center without providing a means for the reporting party to communicate verbally with the communications center. Others are equipped with signal switches that allow the caller to select the type of emergency being reported by pressing the appropriately labeled button, **Figure 3-10.**

FIGURE 3-8   "Fire alarm" boxes of this type came into being during the late 1800s, and some are still in use today. *(Courtesy of Shreveport Fire Department)*

FIGURE 3-9   More modern technology is being used as solar-powered cellular call boxes go up in areas with limited access to power and communications networks. This type of call box supports voice communications with emergency communications centers. *(Courtesy of Caddo Parish 9-1-1)*

FIGURE 3-10   Some call boxes are equipped with signal switches that allow the caller to select the type of emergency being reported.

Receiving Reports via Automatic Alarm Systems

There are two types of public alarm systems as defined by the NFPA. A **Type A reporting system**, **Figure 3-11,** is defined by NFPA as "a system in which an alarm from a fire alarm box is received and is retransmitted to fire stations either manually or automatically." A **Type B reporting system**, **Figure 3-12,** is defined by NFPA as "a system in which an alarm from a fire alarm box is automatically transmitted to fire stations and, if used, to outside alerting devices." Properly designed and installed automatic fire alarm systems can be the key element to any building's overall safety. Automatic alarm monitoring systems are typically comprised of five common types:

1. Local protective signaling system: An alarm system operating in the protected premises.
2. Auxiliary protective signaling system: An alarm system that utilizes a municipal fire alarm box to transmit a fire alarm from a protected property to a fire communications center.

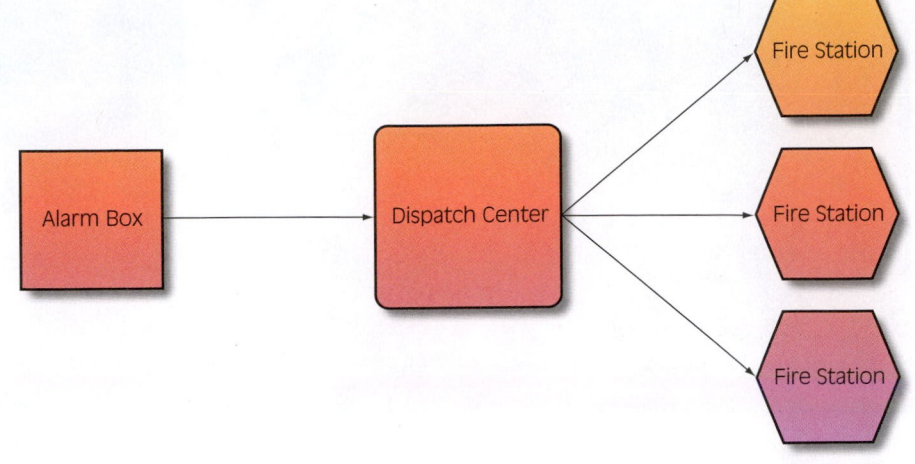

FIGURE 3-11 Type A municipal alarm systems typically transmit an alarm from a call box to a communications center where the alarm is retransmitted to emergency responders.

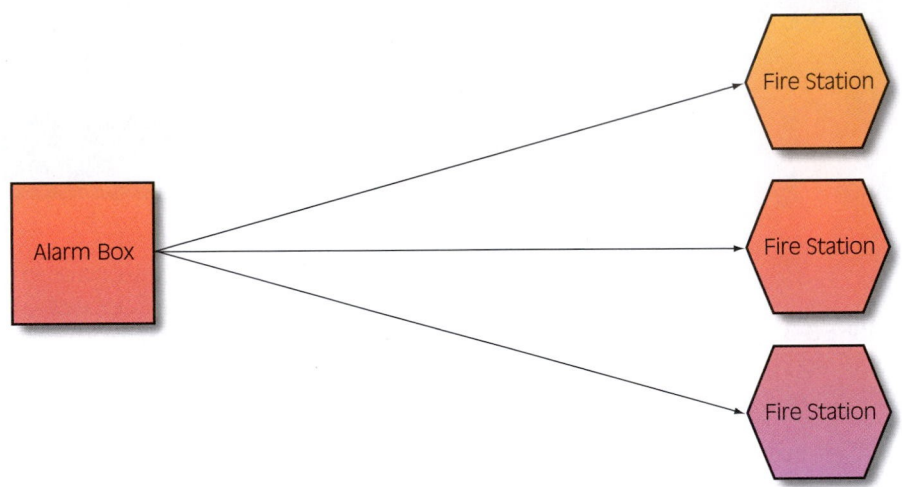

FIGURE 3-12 Type B municipal alarm systems typically transmit alarms directly from a call box to first responders.

3. Remote station protective signaling system: An alarm system that connects a protected premise over leased telephone lines to a remote station such as a fire communications center.
4. Central station protective signaling system: An alarm system that connects a protected premise to a privately owned monitoring site that monitors the lines constantly for any indication of fire or other trouble signals.
5. Proprietary protective signaling system: An alarm system that protects contiguous or noncontiguous properties with common ownership from a location on the protected property.

Alarm systems typically consist of a system of sensors designed to detect smoke, heat, or a combination of both, and also manual stations that can be activated by occupants in the event a fire is detected. Detecting devices protect specific and limited areas. Initiating the signal for any fire alarm system is accomplished by

many means: manual pull boxes, heat detectors, smoke detectors, flame detectors, and fire gas detectors.

Flame and fire gas detectors are used in very specific situations. Flame detectors (sometimes referred to as light detectors) are located in places where very rapid fire spread is possible or in very large open areas. Flame detectors work by detecting radiation, ultra violet, and infrared energy given off by a flame. Fire gas detectors use sensors to detect gases produced by combustion, such as carbon dioxide and carbon monoxide, and may be used in areas of chemical processes or mining, or in refineries.

Residential carbon monoxide detectors are often confused with smoke detectors by citizens calling in alarms. These detectors set off a local alarm based on the accumulation of carbon monoxide over a period of time. Carbon monoxide can be harmful in high levels over a short time period as well as lower levels over an extended period. Many jurisdictions are now making it mandatory that all residential buildings have CO

FIGURE 3-13 Manual pull box stations are used to activate local alarms.

(A)

(B)

FIGURE 3-14 (A) Heat detectors are designed to activate fire alarms and automatic extinguishing systems when ambient room temperatures reach certain levels. (B) Smoke detectors monitor air quality and can also activate automatic alarm systems. *(Courtesy of BRK Brands, INC., a Jarden Corporation (NYSE: JAH). All rights reserved.)*

FIREFIGHTER FACT

ADA regulation 35.162 states: "The legislative history of Title II specifically reflects congressional intent that public entities must ensure that telephone emergency services, including 9-1-1 services, be accessible to persons with impaired hearing and speech through telecommunication technology."

detectors installed. **Figure 3-13** shows a manual pull box, and **Figure 3-14** shows a heat and smoke detector. Fire alarms and detection devices are covered in greater detail in Chapter 12 (in Sections I and II).

Receiving Reports via TDD

Telecommunications Devices for the Deaf (TDDs) are becoming much more widely used by individuals with hearing disabilities. The **Americans with Disabilities Act (ADA)** entitles citizens to equal service from public agencies.

Therefore, communications centers are required to remain ready to receive calls via specialized equipment designed to allow citizens to communicate with telecommunicators through the use of a keyboard using text messages instead of voice communications, **Figure 3-15.** These devices essentially serve as a backup when sophisticated capabilities provided with enhanced 9-1-1 and Computer Aided Dispatch (CAD) are present.

Receiving Reports via Still Alarm or Walk-Ups

From time to time citizens may report an emergency directly to the personnel at the fire station. Receiving complete and accurate information is just as important

FIGURE 3-15 Devices such as this assist public safety agencies in communicating with citizens who have hearing impairments.

FIGURE 3-16 A firefighter relays information from the fire station to the communications center via direct telephone circuit.

in these cases as those that are received at the communications center by telecommunicators. The section titled Receiving Reports of Emergencies (earlier in this chapter) provides information that can be applied to deal with a citizen who is reporting an emergency in person. **Figure 3-16** shows a firefighter relaying information to the communications center that was received from a citizen who actually stopped at the fire station and reported an emergency. Although the protocols of different departments may vary with respect to how these types of reports are handled, it is important to always obtain as much information as possible from the person reporting the emergency. Once this is done, and following departmental protocols, the person taking the report should immediately activate the emergency notification system and relay all of the information to the communications center. How specific emergency notification systems are activated is covered in departmental protocols. The same is true in the case of notifying the communications center that a report of an emergency has been received. Some departments may use **ring-down circuits, base radio,** or **mobile radio** to communicate with their communications center. It is very important for firefighters always to notify the communications center of location and assignment when reports are received directly from citizens. Telecommunicators can then ensure that any necessary support and assistance are provided.

EMERGENCY SERVICES DEPLOYMENT

Once an emergency is recognized and subsequently reported or relayed to the communications center, the next step is to determine what action must be taken. A variety of methods is used to accomplish this throughout the country; however, some elements of the process are essential regardless of jurisdiction or geographic location. Identification of the situation, address verification, and unit selection must occur so that telecommunicators can deploy the appropriate types and numbers of emergency responders.

As stated earlier, the most important information to obtain from the caller is the address. However, to deploy the most effective emergency response, the telecommunicator must also know the nature of the emergency. Emergency response organizations typically identify the most common types of situations and preassign a standard response to each such situation. An example would be a fire department that predetermines that the routine response to all single-family dwelling fires will be three engines, one ladder, and one district chief. Based on this plan, the telecommunicators have baseline criteria from which to develop a **deployment plan.** In the modern fire service, deployment plans are based not only on apparatus types, but also take into consideration what equipment is carried on the apparatus, the number of personnel, and their skill levels.

The process of deploying apparatus and personnel varies greatly from one department to another. For instance, in smaller communities with a low volume of emergency response activity, a manual **run card system** may be sufficient to manage the emergency response deployment process. Manual run card systems, similar to the one shown in **Figure 3-17,** are typically comprised of a card file containing street and location information relating to a jurisdiction and predetermined unit assignments for each location.

STREET NAME: Delmar Ave.						BOX NUMBER: 1128	
BLOCK RANGE FROM: 1000 TO: 1500							

BLOCK	ENGINES	TRUCKS	RESCUE	D/CHIEF	A/CHIEF	MEDIC	OTHER
1000	1, 5, 4, 7, 10	1, 7, 10	1, 9	1, 2, 3	1	1, 4, 5	
1100	1, 5, 4, 7, 10	1, 7, 10	1, 9	1, 2, 3	1	1, 4, 5	
1200	1, 5, 4, 7, 10	4, 7, 10	1, 9	1, 2, 3	1	1, 4, 5	
1300	1, 4, 5, 7, 10	1, 7, 10	1, 9	1, 2, 3	1	1, 4, 5	
1400	1, 5, 4, 7, 10	1, 7, 10	1, 9	2, 1, 3	1	1, 5, 4	
1500	1, 5, 4, 7, 10	1, 7, 10	1, 9	2, 1, 3	1	1, 5, 4	

| STREET NAME: Delmar Ave | | | | | BLOCK RANGE FROM: 1000 TO: 1500 | | |

FIGURE 3-17 Run cards show the response assignment of a variety of apparatus to specific street addresses.

Enhanced 9-1-1 systems provide number identification and location information even in those cases where computer-aided dispatch is not being used.

As discussed earlier, jurisdictions with high volumes of activity may use automated systems to assist in the deployment process. CAD systems are widely used today and provide a very sophisticated method of assessing resources and making recommendations for deployment of equipment and personnel.

Some jurisdictions utilize **Global Positioning System (GPS)** technology to aid in the deployment of responders. Originally dedicated for military operations, GPS works off a system of twenty-four satellites placed into orbit by the U.S. Department of Defense. The satellites are used as reference points to calculate positions. As long as three of the twenty-four satellite signals are attainable, a two dimensional location such as latitude and longitude can be obtained. **Automatic Vehicle Location (AVL)** systems, as seen in **Figure 3-18,** utilize GPS technology to pinpoint the location of an emergency incident. AVL systems can also detect the closest response vehicle, which telecommunicators can view as one of several icons on a computerized map.

Other uses of GPS technology for the fire service include helping to coordinate street directions for units such as mutual aid who may not be familiar with a response area when assigned, as shown in **Figure 3-19;** providing the location of fire hydrants; and in some systems, even visualizing building footprints and access.

Regardless of which method of deployment is used—whether it is accomplished manually or through the use of a CAD system—a predetermined deployment of apparatus must exist. An example of a deployment table is shown in **Figure 3-20.**

The basic elements of the deployment process remain the same in either the manual or automated systems: Verify the location and nature of the emergency and determine what resources are available.

Once telecommunicators receive a report of an emergency situation, verify the location, and determine the appropriate deployment scheme, the next step is to notify the emergency responders. As is the case with deployment plans, this process also varies greatly from agency to agency. Volunteer departments may rely on personal pagers or use an automatic telephone system that "rings" multiple telephones on a common circuit and, in some cases, a system of sirens to alert them of an emergency. **Figure 3-21** shows a volunteer wearing a personal pager. Digital pagers may also be utilized to supplement VHF pagers and

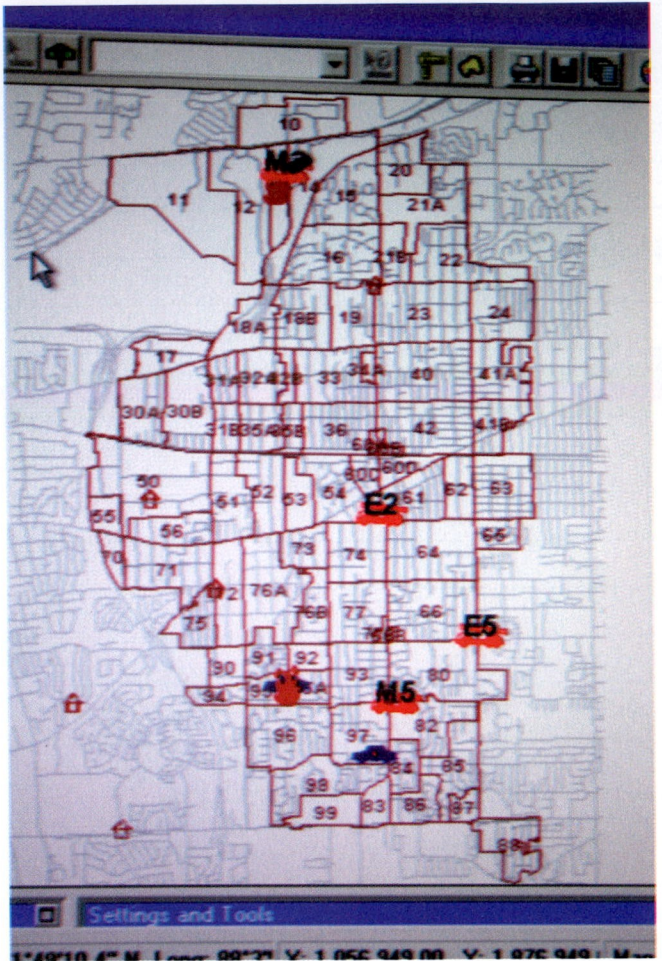

FIGURE 3-18 AVL systems help to locate the response unit closest to an incident location.

provide additional information on location, nature of the emergency, or progress reports concerning an incident.

A variety of **home alerting devices** are also available. These are used by some agencies that operate either an all-volunteer or combination volunteer and paid department. A common example of this type of alerting device is shown in **Figure 3-22.**

In departments where fire stations are staffed twenty-four hours a day, some type of **fire station alerting system** is usually employed. Fire station alerting can be accomplished in a variety of ways, but should

FIGURE 3-19 GPS mounted in apparatus can serve many functions such as routing information and hydrant locations.

TYPE	DISP. CODE	DESCRIPTION	RESPONSE
Fire	01	Single Company Response	1E
	11	Structure Fire/Alarm (Residential)	2E-1D-1R-2T
	21	Structure Fire/Alarm (Commercial)	3E-1D-1R-1T
	51	Hospitals	4E-2D-2R-2T
EMS	08	Medical (Noncritical)	1E-(COLD)
	18	MVA	1M-1E

Legend: E — Engine or Pumper
D — Chief Officer
R — Rescue Unit
T — Truck or Aerial
M — Medic Unit
(Cold) — Nonemergency

FIGURE 3-20 Deployment tables such as the example shown are used to identify the appropriate apparatus response for the type of dispatch.

FIGURE 3-21 Some departments use personal pagers to alert personnel of the need to respond to emergencies.

FIGURE 3-23 This is yet another device used to alert emergency responders. This unit is equipped with relays that activate station lights and open station doors in preparation for a response. *(Courtesy of Shreveport Fire Department)*

FIGURE 3-22 This device is typically found in fire stations and also in some private homes and is used to call out emergency responders.

FIGURE 3-24 Encoders such as this are used to control many paging systems.

always comply with NFPA standards. In some systems, a voice message is transmitted from the communications center to a fire station via a vocal alarm system. These systems typically operate either via some type of control unit connected to leased telephone circuits or through a radio transmitter. **Figure 3-23** shows

a receiver that would be located at the fire station. The telecommunicator, using the control device, which may be similar to the **encoder** shown in **Figure 3-24,** decides the appropriate fire stations to notify and activates the system. Normally some type of distinctive tone is transmitted via a public address system

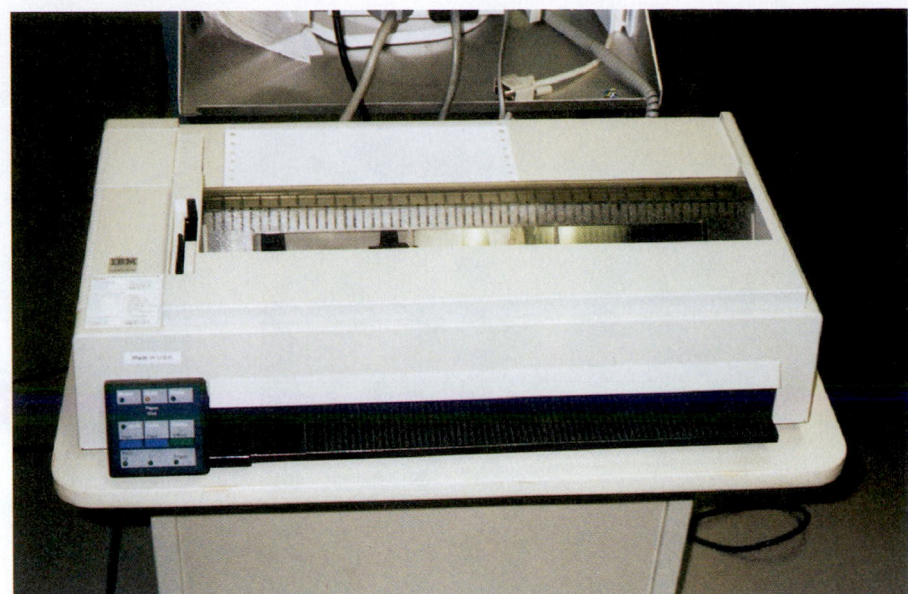

FIGURE 3-25 Printers such as this one are used to relay incident information to first responders at the time of dispatch. They are typically referred to as "tear and run" printers. *(Photo courtesy of Caddo Parish 9-1-1)*

within the fire station to alert personnel of an incoming message.

This alert tone is followed by voice instructions over the PA from the telecommunicator. Some fire station alerting systems perform additional functions such as turning on selected lights in the fire station, opening apparatus bays, turning off appliances, and controlling traffic signals. Some systems are capable of a zoned alert that can notify only specified areas of a station such as those areas that house both fire and EMS units, and personnel.

Departments that utilize CAD systems can enhance this method by installing "tear and run" printers in each fire station. These printers provide the responders with a hard copy printout showing details of the incident and location. A printer of this type is shown in **Figure 3-25**.

Some departments also employ **mobile data terminals** and **mobile data computers** in their dispatch and deployment process. These units allow dispatch information to be transmitted directly to the apparatus on a display screen or, in some cases, directly to mobile printers. Modern communications equipment such as this can provide a two-way flow of confidential information between the communications center and emergency responders. It also frees up a radio frequency from excessive radio traffic. **Figure 3-26** shows a mobile data terminal.

STREETSMART TIP

Remember that the response times of first responders are directly impacted by the amount of time required for telecommunicators to receive information, verify location, determine deployment, compose a dispatch, and transmit this information to first responders.

FIGURE 3-26 Mobile data terminals such as this one are used by some departments and provide information to emergency responders while en route to incidents. Unit status can also be managed with systems of this type.

Regardless of the type of fire station alerting system in use, it is critically important that accurate information pertaining to an emergency situation be transmitted from the communications center to first responders in a clear, concise, and expedient manner.

Again, speed and accuracy are of the utmost importance.

TRAFFIC CONTROL SYSTEMS

Careful planning and use of the technology described in this section can create a safer emergency response path without undue or prolonged disruption of normal traffic flows. However, there is no better way for emergency responders to reduce traffic accidents than by exercising prudent judgment and applying safe driving practices at all times.

To speed the response of emergency responders as well as reduce stress and increase the safety of both emergency responders and the general public, some jurisdictions utilize various types of emergency preemption systems to control traffic signals and provide a safe transition to a priority right-of-way for emergency vehicles. A system such as this is shown in **Figure 3-27A and B.** Systems of this type are designed to recognize an emergency response vehicle and actually allow it to change the traffic control signals on its route to allow clear passage for the emergency responders. A variety of these systems are in operation today and each uses slightly different technology. Some operate using radio-frequency while others communicate with remote signal detection devices through the use of microwave or laser technology mounted on the vehicle. The use of traffic control systems should not be taken for granted by responders and due caution should always be exercised when responding to calls. Emergency responders cannot be of help unless they first arrive on scene safely.

RADIO SYSTEMS AND PROCEDURES

Once apparatus and personnel are deployed to emergency situations, the function of fire communications personnel then becomes that of providing support for the field units deployed. The primary link between the communications center and field units is the radio system. The use of radios in the fire service serves to

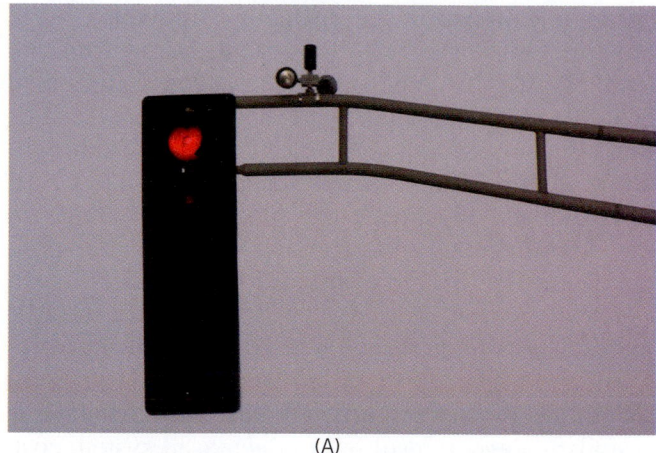

(A)

(B)

FIGURE 3-27 Systems that control traffic signals contain (A) a receiver on the traffic light and (B) a signal unit mounted to the vehicle.

carry both verbal and digital messages. Radio systems have various components; however, every system must have at minimum a base station and antenna capable of transmitting at a power or signal strength necessary to provide coverage to all parts of a jurisdiction.

The radio-frequencies that have commonly been used by the fire services are VHF low band frequencies, 33 to 46 MHz; the VHF high band frequencies 150 to 174 MHz; and the UHF frequencies, 450 to 460 MHz. These frequency ranges have provided reliable fire service communications for many years. However, as the result of growth, the fire service has experienced severe difficulties when attempting to add frequencies to their radio systems. The **Federal Communications Commission (FCC)** closely monitors frequency allocations, and in some cases additional frequencies in the ranges mentioned are simply not available. As the result of a need for additional frequency spectrum, other frequencies have been approved for use by the fire service. The 800-MHz frequency range is being used successfully by some departments for voice communications. Voice communications via the 800-MHz frequency range are limited by the FCC to systems using trunking technology. This type of radio system is discussed next. The FCC has mandated that by 2013 all radios must meet narrowband requirements, which will allow additional radio frequencies to be utilized. Narrowbanding will change the spacing of channels in the elctromagnetic spectrum and is estimated to double the available radio channels for responders.

One common radio system used by fire departments is a simplex system that uses only one frequency to transmit outgoing messages and to receive incoming messages, **Figure 3-28.** The advantage is simplistic design, resulting in decreased system cost. The primary disadvantage of systems of this type is the limited range and interference between multiple units in the same system attempting to access the base station simultaneously.

A more complex system, shown in **Figure 3-29,** is a duplex system that uses two frequencies per channel, transmitting outgoing messages on one and receiving incoming messages on the other. This system uses base repeater stations whereby a fixed control station transmits an outgoing message that is received at a repeater site and retransmitted to mobile and portable units in the field. The benefits of systems of this type are more range and the elimination of the self-interference found in simplex systems. The disadvantages are more complex system design, the need for multiple frequencies, system cost, and the ongoing maintenance costs associated with the system.

Multisite trunking systems as depicted in **Figure 3-30** use computer processors that make the most efficient use of radio spectrum. Multiple transmitters operating on different channels are controlled by microprocessors that sense available channels and reallocate their use as needed. In duplex systems of this type, user transmissions are transmitted on one frequency, received by the field unit on another, and retransmitted back to the communications center on the other. The user does not notice the fact that the system is changing frequencies with each transmission. However, this more efficient allocation of radio-frequency resources allows the use of fewer frequencies by individual agencies. Several agencies can operate simultaneously on the same trunked radio system and not interfere with each other. The benefits of systems of this type are expanded range, more efficient frequency use, and the ability of multiple agencies to operate on one system. Disadvantages are more complex system design and overall higher system cost.

STREETSMART TIP

The proper operation of the radio system, regardless of type, is of primary importance if information to and from the scene of the emergency is to be relayed in a timely and accurate manner.

CAUTION

On-scene personnel must listen before they talk. Routine or nonemergency traffic should not interfere with emergency communications. Transmissions should occur when airwaves are clear.

Proper radio discipline is very important during active incidents. When using any trunked two-way radio, it is important to depress and hold the "push to talk" button at least two seconds before talking to

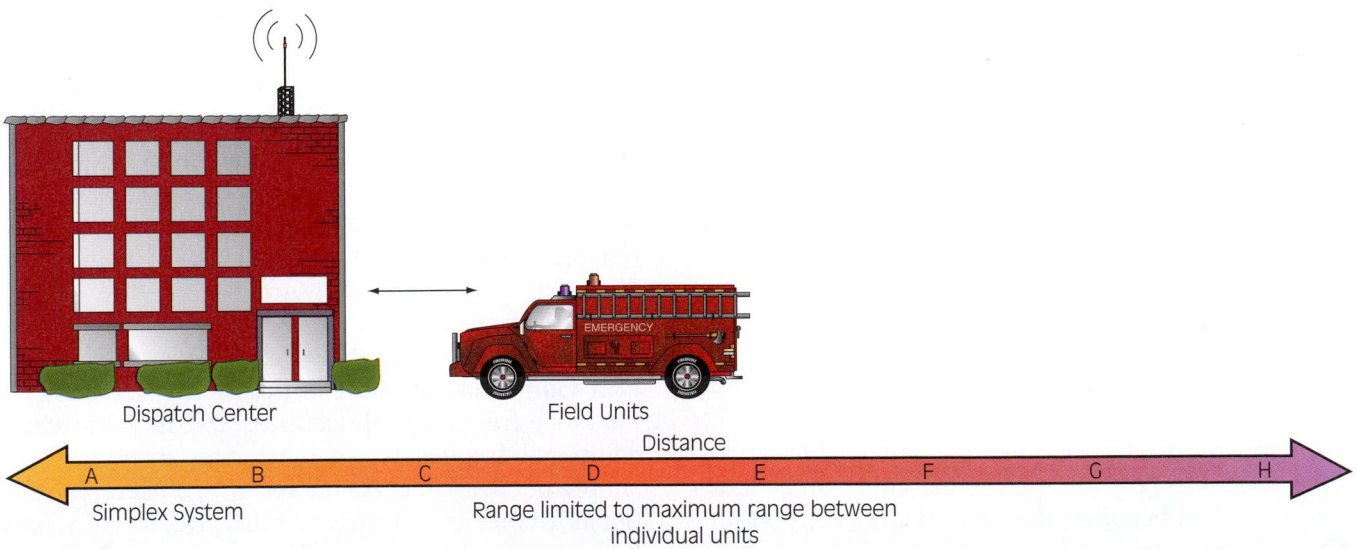

Dispatch Center   Field Units
Distance

A   B   C   D   E   F   G   H

Simplex System   Range limited to maximum range between individual units

FIGURE 3-28 Simplex radio system designs such as this one are reliable and relatively inexpensive to install. However, they are limited with respect to range of operations.

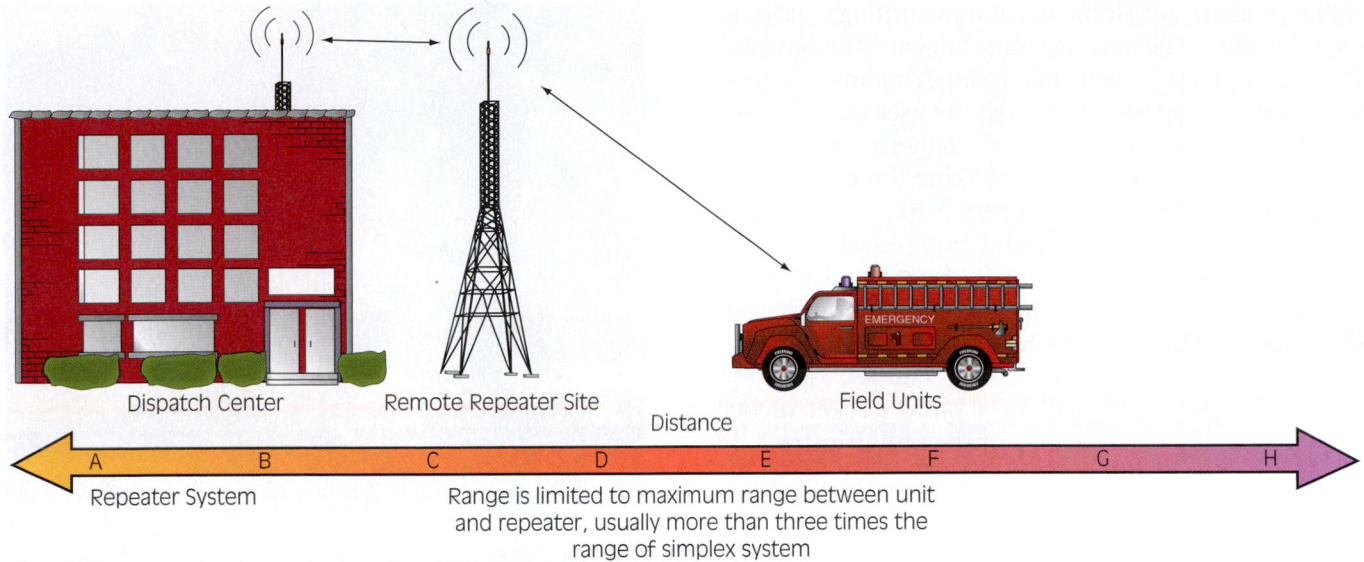

FIGURE 3-29 A slightly more advanced duplex design using multiple transmitters extends the operating range of the simplex system shown in Figure 3-28.

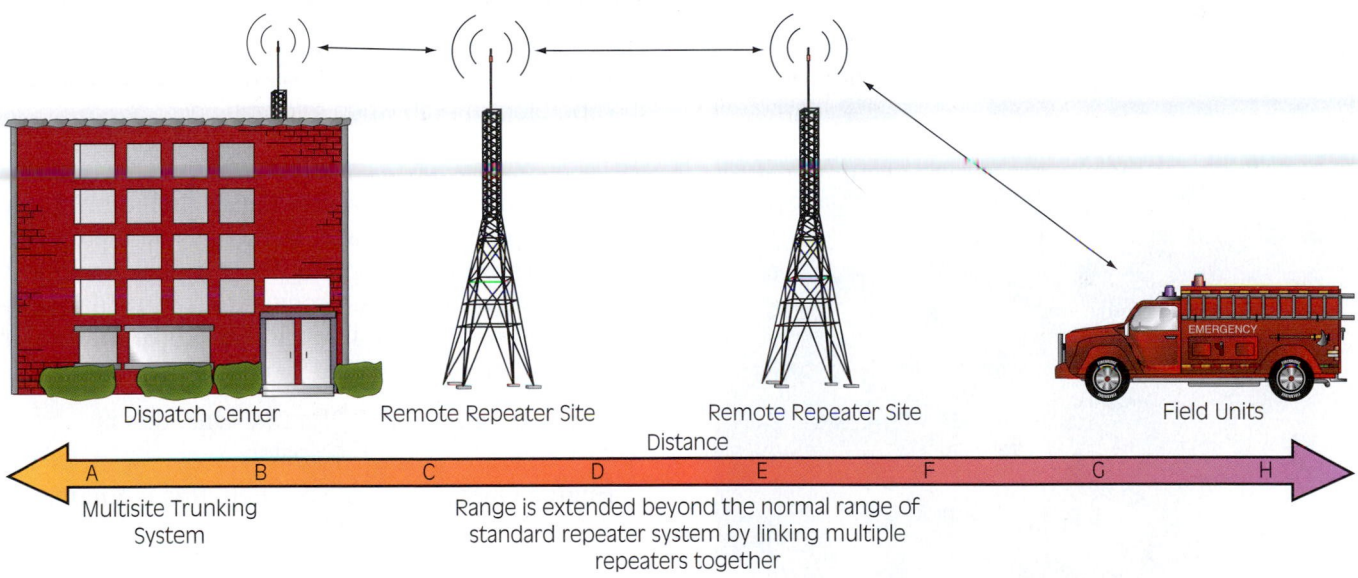

FIGURE 3-30 Multisite trunked radio systems provide perhaps the best coverage and also offer direct benefits associated with the most efficient use of radio resources.

avoid **clipping** the first part of the message. The same is true at the end of the message. The user should always pause briefly before releasing the "push to talk" button to avoid clipping the end of a message. Before keying any radio, users should know what they are going to say. The golden rule is to be brief but be concise. Firefighters should avoid touching any radio antenna during transmission to avoid burns that can result from radio-frequency energy.

FIREFIGHTER FACT

Some radios are equipped with a time-out feature that forces brevity by interrupting transmissions that are longer than the preset duration allowed. This feature also prevents long-term open microphones from accidental keying.

The user should never attempt to operate a radio while eating or chewing anything and should never use slang, profanity, or jargon when transmitting a message over the radio. The message must be clearly understood. During large-scale incidents, radio economy is important. When using mobile radios, the user should speak clearly (not shout) across the microphone, as shown in **Figure 3-31,** as opposed to speaking directly into the microphone as pictured in **Figure 3-32.**

When using portable radio equipment, the unit should be held perpendicular to the ground with the antenna facing skyward to allow for better radio wave distribution. The user should speak across the microphone as shown in **Figure 3-33.** The use of a portable radio that is located in a radio pocket or belt clip impairs the performance of the unit. **Figure 3-34** shows the improper position of the portable radio when transmitting. In transmitting on both mobile and portable radios, holding the microphone one to two inches from the mouth and at a 45° angle is important for a clear transmission.

Microphones should never be left on the seat or in other locations where they might accidentally be keyed

FIGURE 3-32 Improper use of a mobile microphone. The radio microphone should not be held directly in front of the operator's mouth when transmitting.

up. Microphones and portable radios should be placed in their appropriate storage locations or in a protected location. Local agency protocols and policy will dictate how radios are to be operated in a particular jurisdiction. However, the following is a simple illustration of the "call up" method that is used by some agencies:

Telecommunicator:	"Engine One, this is dispatch."
Engine One:	"Dispatch, this is Engine One."
Telecommunicator:	"Engine One, respond to 1234 Main Street for a trash fire."
Engine One:	"Dispatch, Engine One is responding for a trash fire at 1234 Main Street."
Engine One:	"Dispatch, Engine One is on the scene and establishing command. We have a small trash fire in the rear of the building, no exposures, no assistance needed."
Telecommunicator:	"Engine One, dispatch copies."

FIGURE 3-31 This figure shows the proper use of a mobile radio microphone.

FIGURE 3-33   In this figure, the user has positioned the portable radio properly and is speaking across the microphone.

FIGURE 3-34   This figure shows the improper positioning of a portable radio.

Ten codes are a set of radio signals preceded by the number "10" that make up a predetermined message. Considerable debate exists as to whether radio codes or clear speech should be used, but in most cases that decision is made based on the needs of each specific agency. Use of radio signal codes provides a somewhat more confidential and cryptic means of communicating. Confidentiality can be very important in situations where a responder may be unable to request assistance while in the presence of a combative or dangerous subject. However, such codes must be learned and remembered and are not usually standardized from department to department, making communications during multi-jurisdictional responses somewhat problematic. Under guidelines set by the National Incident Management System (NIMS), ten codes impede interoperability of communications and information transfer among different agencies. It is recommended that plain common language understood by all be used.

Clear speech, on the other hand, is exactly what the name implies. Clear speech is used to convey information and issue instructions. The use of clear speech eliminates much of the confusion associated with the use of radio codes and, although somewhat lengthier, in most cases is easily understood by all. However, even in the clear speech environment, the same phrase may carry a different meaning from agency to agency. For example, "in-service" for some agencies means that a crew is going to work at an incident. On the other hand, in some agencies "in-service" means that the crew is available for an assignment. Thus, it is important not to assume anything during radio communications.

Another important function associated with the radio systems is the issuance of emergency evacuation signals to on-scene personnel who may be subject to imminent danger. Some departments use the radio system to transmit electronic tones that are intended to attract the attention of firefighters and alert them of the need to evacuate to a safe area. This works relatively well in most cases. However, a typical emergency scene is very noisy and not everyone on the scene has a portable radio. Another system used by some departments employs the use of apparatus air horns. Three bursts on the air horn indicate the need to evacuate. Regardless of the type of system used, it is critically important that firefighters learn it and be familiar with how it is used. Firefighter safety is paramount in all operations.

The use of two-way radios has grown greatly over the years. Routine administrative traffic, which is necessary in the daily operations of a department, should never be allowed to interfere with emergency operations. In departments that have access to multiple radio channels, emergency operations should be assigned to a separate channel dedicated for use on that scene only. This greatly improves the ability for incident commanders to communicate with on-scene personnel and also minimizes the threat of interference from some other source.

Radios and radio systems are evolving as the advances in technology grow. Newer radio systems can identify radios assigned to a particular apparatus to help in tracking transmissions. As a result of the technology, some radios incorporate an emergency feature that alerts the emergency communications center when a crew is in trouble or encountering an emergency. Firefighters should spend time familiarizing themselves with the radio system in use in their department and the many features associated with the equipment and the system.

NOTE

In summary, firefighters must consider several important factors when transmitting messages across mobile or portable radios. First, the information provided to an emergency communications center or other emergency service units must be accurate, clear, and complete. The radio transmissions must also be within the time parameters established by the department or local jurisdiction if such policies and procedures exist. Firefighters should consult their department or jurisdiction's policies on time parameters for radio communications, specific terminology, and proper designations for units operating on a radio system.

NOTE

The initial incident commander remains in command of the incident until command is either assumed by a higher ranking officer or transferred to another officer.

RADIO REPORTS

Until field units arrive on the scene, for all practical purposes, the communications officer is the incident commander. The first unit arriving on the scene should establish command and give a brief initial report otherwise known as a size-up.

This initial report contains pertinent, but brief, information about the on-scene conditions. The initial report should be given using clear, precise, language. The report should contain, at minimum, the following information: (1) the correct address, (2) a situation evaluation, (3) where the emergency is located in the building, (4) some information about the building as well as its potential occupants, (5) a request for any other agency support such as law enforcement, (6) the location of the on-scene command post, and (7) the identity of the incident commander. Initial reports may also contain a very brief action plan for the incident.

In the case of most routine emergency medical incidents, this much detail is not necessary. However, in the event of a major accident or other incidents involving multiple patients, the same type of detailed information should be provided along with any additional information such as a quick assessment of the number of patients and general types of injuries involved.

Status or progress reports are very important to the overall success of any major scene operations. Most fire command officers agree that the first status report from the field should be made no longer than ten minutes into the incident, and every ten to fifteen minutes thereafter until the situation is brought under control. Communications centers may implement SOPs that call for "time marking" incidents at important thresholds for improved documentation and reporting purposes. Some CAD mobile data computer systems perform this function automatically.

Procedures for firefighters reporting possible life-threatening conditions or calling for assistance are a necessity for every fire department. Terminology such as **Emergency Traffic** or **Mayday** should automatically clear the radio channel and give that person the opportunity to relay his or her messege. Traditionally, the term emergency traffic has been utilized to signal that important information about a possibly life-threatening situation is to follow, and mayday has signified that a member is in trouble and in need of assistance. Firefighters must be trained to call a mayday the moment they even think they may be in trouble on the fireground because time is critical. A firefighter may communicate a mayday for himself or for any other member on the fireground who is in distress. When a mayday is given, it should receive top priority over the radio.

Procedures must be put into place for calling a mayday and firefighters must know their department's procedures and be well trained in them. Make certain that all members understand the meaning and use of the term. An example of a mayday procedure may be as follows:

1. Remain calm and call for the mayday: "Mayday, Mayday, Mayday." Activate the emergency button on the radio, if so equipped, at this time. (Some radios have a button that will distinguish the identifier on a screen in the command post or dispatch center.)

2. Transmit your unit number and name.

3. Give your last known location or assignment.

4. State what your emergency is and give the status of other members of your crew.

5. Secure acknowledgement by command.

6. Activate **Personal Accountability Safety System (PASS).** (An audible alarm signaling that a firefighter is in trouble. It may be activated by absence of movement or manually.)

MOBILE SUPPORT VEHICLES

Major events involving fire and EMS sometimes call for the use of **mobile support vehicles (MSVs)** or mobile communications units. These vehicles greatly enhance the overall effectiveness of the communications system in use at the scene of major incidents. Coordination of the communications process is absolutely necessary in order to manage large-scale operations effectively. MSVs provide an on-scene command post from which operations can be directed. The need to deploy vehicles of this type is usually determined by the size of the incident and the projected duration of activities. MSVs are normally highly specialized vehicles designed exclusively for use as on-scene command posts. MSV size depends greatly on the jurisdiction that it serves. The vehicle should have radios that allow communications on all of the frequencies that may be in use by the command agency and any additional jurisdictions that may be called in for mutual aid. Telephone service to vehicles such as these can be provided by the local telephone company via temporary connections or through the use of cellular technology. The MSV should be equipped to operate on both battery power as well as standard 120-V current. **Figure 3-35** shows an example of an MSV that is the result of modifications to a public transit bus. Conversions such as these, although often time consuming, are for the most part a much less expensive way to incorporate the use of this type of vehicle into the operations of a department. Custom-built units from the factory are also available, as shown in **Figure 3-36.** This particular unit was custom built to serve both fire and police operations through a cooperative agreement between the services.

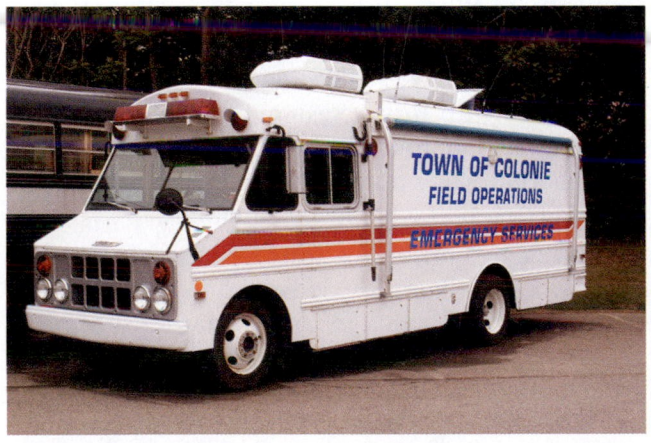

FIGURE 3-35 The vehicle pictured here is the result of a conversion project performed on a city transit system bus. The resulting mobile support vehicle costs much less than a custom-built factory unit.

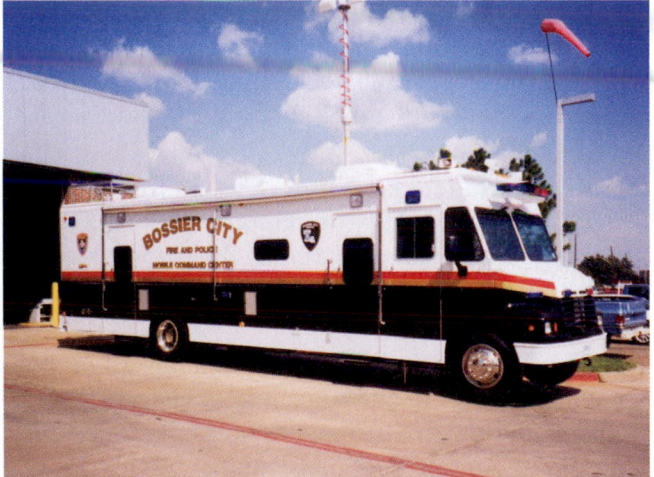

FIGURE 3-36 Unlike the vehicle shown in Figure 3-35, the vehicle shown here is a custom-built mobile support vehicle. *(Courtesy of Bossier City Fire Department)*

LESSONS LEARNED

Whether a telecommunicator is in a large emergency communications center or a firefighter is answering an emergency call at a fire station, the manner in which calls are answered and information processed has a direct impact on citizens' impressions of the department. In the case of emergency incidents, information that is accurately collected and rapidly transmitted to first responders is paramount to the successful resolution of the incident. The ability to answer incoming calls quickly, gain control of the call, and in some instances, calm the caller is a very important aspect of this process. Remember that the telecommunicator is the first person on the scene. Utilize the caller's eyes and ears. A telecommunicator's knowledge and skills, combined with the ability to make wise use of all available resources, will play a vital role in the successful outcome of an emergency incident.

KEY TERMS

9-1-1   Emergency telephone number that provides access to the public safety services in the community, region, and, ultimately, nation.

Americans with Disabilities Act   Public law that bars discrimination on the basis of disability in state and local services. Enacted in 1990.

Association of Public-Safety Communications Officials – International, Inc. (APCO)   International not-for-profit organization dedicated to the advancement of public safety communications. Membership is made up of public safety professionals from around the world.

Automatic Vehicle Location (AVL)   System that utilizes GPS technology to pinpoint the location of an emergency incident as well as the response vehicle that may be closest according to a computerized map.

Base Radio   Radio station that contains all of the antennas, receivers, and transmitters necessary to transmit and receive messages. Normally located in a station or communications center.

Basic 9-1-1   Telephone system that automatically connects a person dialing the digits "9-1-1" to a predetermined answering point through normal telephone service facilities. Number and location information is not normally provided in basic systems.

Clipping   Term associated with the use of two-way radios that is used to describe instances when either the first part of a message or the last part of a message is cut off as the result of either speaking before pressing the transmit key or releasing the transmit key prior to the end of a transmission.

Command Vehicle   Typically used by operations chief officers in the fire service.

Communications   Sending, giving, or exchanging of information.

Computer-Aided Dispatch (CAD)   Computer-based automated system that assists the telecommunicator in assessing dispatch information and then recommends responses.

Database   Organized collection of similar facts.

Deployment Plan   Predetermined response plan of apparatus and personnel for specific types of incidents and specific locations.

Emergency Call Box   System of telephones connected by private line telephone, radio-frequency, or cellular technology usually located in remote areas and used to report emergency situations.

Emergency Communications Center   Facility either wholly or partially dedicated to being able to receive emergency and, in some instances, nonemergency reports from citizens. Centers such as these are sometimes referred to as fire alarm, headquarters, dispatch, or a public safety answering point (PSAP).

Emergency Medical Dispatch (EMD)   System designed for use by telecommunicators to assist them in evaluating patient symptoms using predetermined criteria and responses.

Emergency Traffic   Term used to signify that a priority message is to follow on the radio.

Encoder   Device that converts an "entered" code into paging codes, which in turn activate a variety of paging devices.

Enhanced 9-1-1   Similar in nature to basic 9-1-1 but with the capability to provide the caller's telephone number and address.

Federal Communications Commission (FCC)   Government agency charged with administering the provisions of the Communications Act of 1934 and the revised Telecommunications Act of 1996 and is responsible for nonfederal radio-frequency users.

Fire Station Alerting System   System used to transmit emergency response information to fire station personnel via voice and/or digital transmissions.

Global Positioning System (GPS)   System of twenty-four satellites used as reference points to calculate positions. The satellites were placed into orbit by the U.S. Department of Defense and were originally dedicated for military operations.

Home Alerting Devices   Emergency alerting devices primarily used by volunteer department personnel to receive reports of emergency incidents.

Mass Casualty   EMS incidents that involve more than five patients.

Mayday   A term used *only* to signify that a person is in a life-threatening situation and needs immediate assistance. A mayday can be declared by anyone having knowledge of a person in distress.

Mobile Data Computer   Communications device that, unlike the mobile data terminal, does have information processing capabilities.

Mobile Data Terminal (MDT)   Communications device that in most cases has no information processing capabilities.

Mobile Radio   Complete receiver/transmitter unit that is designed for use in a vehicle.

Mobile Support Vehicle (MSV)   Vehicle designed exclusively for use as an on-scene communication center and command post.

Multiple-Alarm Incident   Involves the response of additional personnel.

National Emergency Number Association   Not-for-profit organization founded in 1982 and made up of more than 6,000 members. The association fosters technical advancement, availability, and implementation of a universal emergency telephone number system.

NFPA 72   National Fire Alarm Code.

Personal Accountability Safety System (PASS)   A small motion-sensitive device worn by firefighters. The motion device will go into an audible alarm signaling that the firefighter may be in trouble if there is no movement for a set period of time. It can also be manually activated by the firefighter if desired. Also known as a personal alert safety system.

Prearrival Instructions   Self-help instructions intended to enhance the overall safety of the citizen until first responders arrive on the scene.

Ringdown Circuits   Telephone connection between two points. Going "off-hook" on one end of the circuit causes the telephone on the other end of the circuit to "ring" without having to dial a number.

Run Card System   System of cards or other form of documentation that provides specific information on what apparatus and personnel respond to specific areas of a jurisdiction.

Telecommunications Devices for the Deaf (TDD)   Device that allows citizens to communicate with the telecommunicator through the use of a keyboard over telephone circuits instead of voice communications.

Telecommunicator   Individual whose primary responsibility is to receive emergency requests from citizens, evaluate the need for a response, and ultimately sound the alarm that sends first responders to the scene of an emergency.

Type A Reporting System   System in which an alarm from a fire alarm box is received and retransmitted to fire stations either manually or automatically.

Type B Reporting System   System in which an alarm from a fire alarm box is automatically transmitted to fire stations and, if used, to outside alerting devices.

Voice Over Internet Protocol (VOIP)   Technology used in telephones that converts audio signals into digital data that is transmitted over the Internet.

REVIEW QUESTIONS

1. List four basic elements of communications.
2. What is the primary role of the telecommunicator in relation to fire and EMS apparatus?
3. What is the most important part of call processing for a telecommunicator or firefighter?
4. List the order in which ringing lines should receive priority.

5. When speaking with citizens who are reporting emergencies, what should the call taker's voice project?

6. When processing a nonemergency call, if an emergency line rings, what steps should the telecommunicator take?

7. Telecommunicators or others who answer emergency telephone lines must extract key information from individuals who are, in some instances, hysterical and in severe emotional distress. How do telecommunicators extract information in such a case?

8. What is the most important information to learn from a caller reporting an emergency?

9. What other information is important to obtain in case it becomes necessary to recontact the caller?

10. What are emergency medical dispatch protocols designed to accomplish?

11. What is the primary benefit of basic or enhanced 9-1-1 service?

12. What sector of our society would have the most need for TDD equipment?

13. When operating radio equipment, emergency personnel should always do what before talking?

14. How should a user speak when using a mobile radio?

ENDNOTE

1. NFPA 1061. National Fire Protection Association, Quincy, MA.

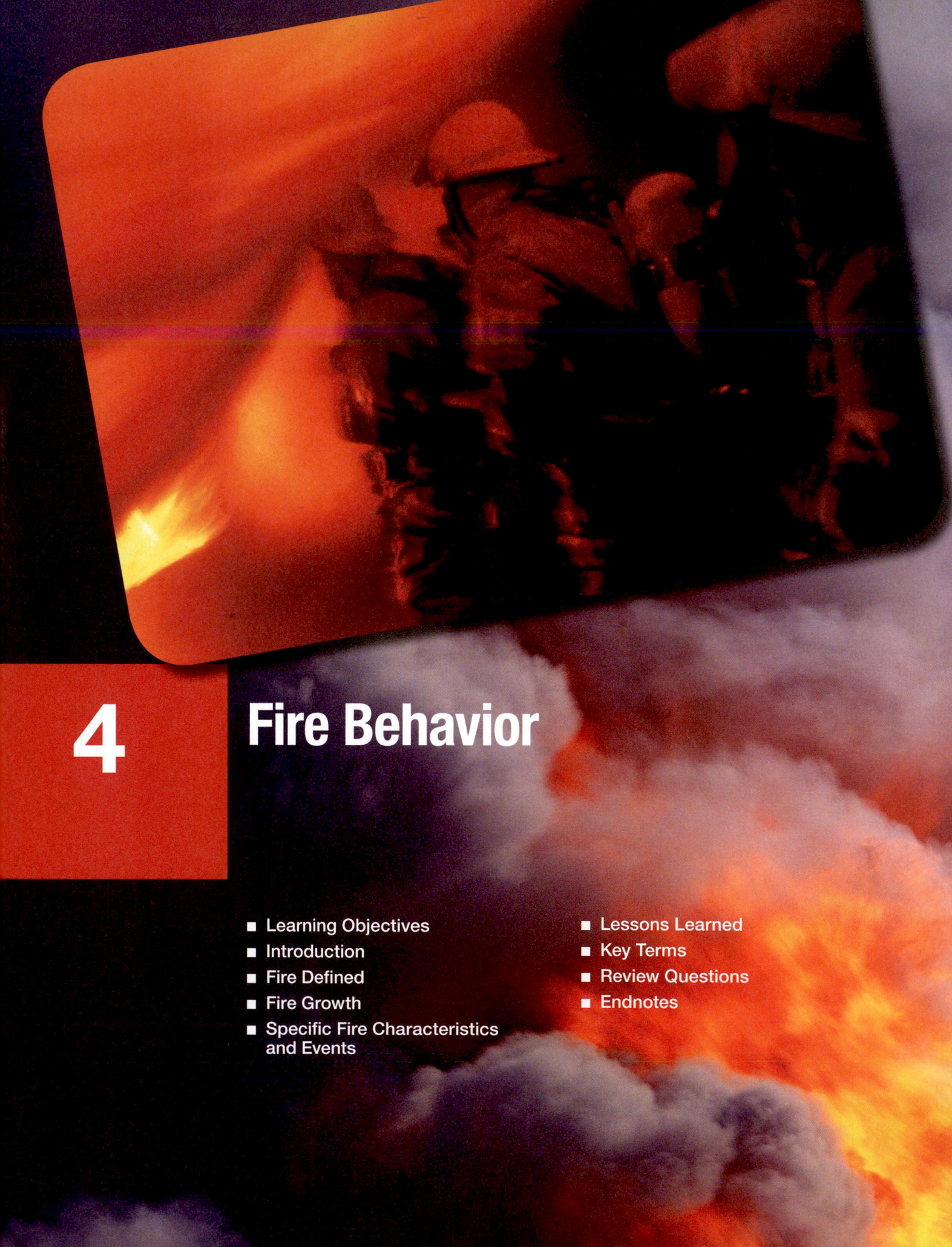

4

Fire Behavior

It had been a relatively quiet day. As soon as the box alarm came in we pulled out the door and the column of smoke was clearly visible in the late afternoon winter sky. We arrived with the first-due engine and truck companies. As we pulled in, I saw the fire building was a two-story wood frame structure with heavy brown and black smoke coming from both floors. A quick glance revealed that the windows on the first floor had been painted on the outside covering the glass. It appeared that this was a vacant structure probably being used by squatters or as a crack house.

After receiving an initial report, my crew and I headed toward the building. In my agency, one of the rescue squad's duties at structure fires is to perform the initial search unless otherwise directed by an incident commander. I saw that the engine company was laying their line while the truck was preparing to force entry into the building. During the size-up, a civilian ran over to us advising us that his father lived there and he could not find him.

At that point I split the crew up. I sent two men to assist the truck company in forcing entry through the front door, while the driver and I went to the rear of the building. As we arrived in the back we could hear some of the windows falling on both floors with flames venting from them. It was at this time we thought we heard moaning coming from the back porch area. We put on our SCBA (self-contained breathing apparatus) and forced entry through a door to the back porch. Inside we found the smoke had banked down to approximately ten feet from the kitchen door. We forced the kitchen door open and were only able to advance a few feet. Most of the kitchen was starting to erupt in flames. I thought to myself that it looked like a "Hollywood" fire. With everything burning, we started to back out because no one could have survived in there.

As we backed out of the kitchen my partner reached up to close the kitchen door. As he was doing this, the smoke became jet black and was now down to floor level. The temperature became unbelievably hot. I thought to myself, "I hope what could happen doesn't," but it did. The porch flashed over. It was incredible. It was like having a giant orange flashbulb go off in your face. The entire porch was in flames. As we turned and headed back toward the door, we kept the wall to the left because, on the way in, the wall had been on our right. We came to the back wall and turned right to where the door should have been. What we didn't realize was that the inside panel of the door was covered by plywood with no doorknobs, and it was secured with latches on the inside because the resident was concerned about break-ins. When we could not find the door, I actually became afraid that we would not get out alive. My partner made a Mayday call giving our location while I searched for the door. The pain from the heat was like being stung by a thousand bees. At one point I just wanted to stand up and try to run where I thought the door was but I told myself, "Stand up and you've really had it. Stay calm, stay together, and they'll get us out." I thought of my son who was only six weeks old at the time, my partner's kids, and our wives, and I said to myself, "I'm not dying in this place."

We stayed together in the area by the rear wall, when suddenly I felt someone grab me by the air-pack harness and throw me out the door. As I landed on the ground I turned and saw my partner come flying out the door. Thankfully, one of the captain/paramedics had seen us enter the building and heard our Mayday call.

Because we were wearing our full protective ensemble—boots, bunker pants, hoods, coats, gloves, helmets, and SCBA—we were fortunate enough to have sustained only minor burns. Our training also saved us by prompting us to create a search pattern that led us back to the door, make a Mayday call, and stay together.

—Street Story by Mike Kelleher, Captain, Troy Fire Department, Troy, New York

LEARNING OBJECTIVES

After completing this chapter, the reader should be able to:

4-1   Describe the properties of solid, liquid, and gas fuels associated with structural fires.

4-2   Define combustion.

4-3   Name the parts of the fire tetrahedron.

4-4   Describe the methods for extinguishing a fire.

4-5   Describe hazardous conditions found on an emergency scene.

4-6   Explain how the effects of fire can create a hazardous condition in a structure.

4-7   Explain the health and safety effects of prolonged exposure to the products of combustion.

4-8   Explain the four sources of heat.

4-9   Identify the classes of fire (A, B, C, D, and K).

4-10   Explain the five classes of fire and the risks associated with each class.

4-11   Identify the percentage of oxygen required to sustain human life.

4-12   Describe the effects of oxygen on fire.

4-13   List the three modes of heat transfer.

4-14   List the four phases of fires.

4-15   Discuss the chemistry and physics of fire through the stages of a fire in a structure.

4-16   Describe the principles of thermal layering in a structure fire.

4-17   Identify products of combustion.

4-18   Describe the specific characteristics of plastic, liquid, electrical, and metal fires.

4-20   Describe hostile fire events including flashover and backdraft, and the methods for preventing them from occurring.

4-21   Discuss the observations in reading smoke and the warning signs of hostile fire events.

*The FF I and II levels, as defined by the NFPA 1001 Standards, are identified in different colors: FF I = black, FF II = red, additional information = blue.

INTRODUCTION

Since ancient times, fire has been considered one of the most important life-sustaining components on earth—along with water, weather, and air. From their earliest memories, people have been warned against its dangers, entertained by its explosions and sound, comforted by its warmth, frightened by its power, mystified by its characteristics, and assisted by its light. In the history of mankind, fire has played a major role as a tool in the development of society. Sometimes an ally, sometimes an enemy, much has been learned about it, especially in the last thirty years, **Figure 4-1.**

FIGURE 4-1 Since its discovery, fire has been considered both an ally and an enemy.

- Once considered mysterious, fire has been studied endlessly—to this day. While scientists don't understand every nuance regarding fires, we have learned much. So what is **fire**? The simple answer would describe the bright flame one would see, and the accompanying heat one would feel when a material burns. While this answer may work for the lay person, a firefighter needs a much more specific understanding of fire. Actually, firefighter survival has never been more dependent on the understanding of fire behavior. Today's fires are hotter, more dynamic, and faster spreading than those of the past (thanks to a "plastic" society). To best understand fire behavior—and survive the fire extinguishment effort—a firefighter needs to combine three things:

- An understanding of fire dynamics.

- Some supervised experience and first-hand fire observations.

- An on-going commitment to learn more about fires.

This chapter will help provide an understanding of fire dynamics; your fire instructors and other fire officers need to provide the supervised fire observations; and the commitment to learn more about fires comes from, well, your commitment.

Before we get to the fire dynamics, please review the feature boxes, "Measurements" and "Chemistry 101"—they will help you put the science in perspective and help you understand concepts presented in other chapters of this book.

The study of fire dynamics starts with an understanding of the combustion and extinguishment process and then examines fire growth and the characteristics of various types of fires. Once we know what causes fire to begin, grow, and spread, the means employed to extinguish it will become more understandable.

FIRE DEFINED

In the context of this book, fire is burning, and burning is "combustion." **Combustion** is a chemical reaction that includes the self-sustaining rapid oxidation of a fuel accompanied by the release of heat and light.[1]

The term *combustion* can be further defined based on the speed (or rate) that a fuel oxidizes. A **deflagration** is combustion at a rate below the speed of sound (subsonic) and a **detonation** is combustion above the speed of sound (supersonic).

In order for any type of combustion to take place, certain ingredients need to come together. These ingredients include heat, fuel, oxygen, and a self-sustaining chemical reaction. The assembly of these ingredients is referred to as the **fire tetrahedron**, depicted in **Figure 4-2** as a pyramid-shaped diagram.

MEASUREMENTS

Units of measurements are referenced in this and other chapters of the book. There are two general systems of unit measurements that can be used. The English (British) system is used primarily in the United States. Canada and most other developed nations use the International System of Units (also called the metric system). Chemists and other scientists also tend to favor the International System of Units. This book will use the English system and provide International equivalents where appropriate.

Understanding what is being measured is also important. The table included here outlines the type of things that are measured and gives unit designations in English and International terms.

Measurement Type	English System	International System
Distance	Inches	Millimeters
	Feet	Centimeters
	Yards	Meters
	Miles	Kilometers
Volume	Ounces	Milliliters
	Quarts	
	Gallons	Liters
Weight	Ounces	Grams
	Pounds	Kilograms
	Tons	
Temperature	Degrees Fahrenheit	Degrees Celsius
Pressure	Pounds per Square Inch	Pascals or Kilopascals

CHEMISTRY 101

Chemistry can be described as a study of the composition of matter (anything that takes up space and can be measured) and the changes that take place with matter. Matter has physical properties (like color, density, and boiling point) and can be found in one or more of three physical states: solid, liquid, or gas. Matter also has chemical properties (composition and the potential to change). The burning process impacts matter by changing these physical and chemical properties.

The composition of matter can be divided into two groups: compounds and elements. A *compound* is a pure substance that can be broken down into more simple substances by various means. An *element* cannot be broken down into more simple substances. For example, each of the following compounds can be broken down into several elements:

- Water (hydrogen and oxygen)
- Sugar (a simple mix of carbon, hydrogen, and oxygen)
- Wood (a more complex mix of carbon, hydrogen, oxygen, and trace minerals like calcium)

Carrying this further, a compound is made up of **molecules**—the smallest particle of the compound that can still exist and retain the physical and chemical properties of the compound. In other words, a stick of wood is made up many molecules (linked carbon, hydrogen, oxygen, and minerals) that form the compound. As these compounds link together, cells are formed.

An element is made up of atoms. Atoms are the smallest particles of an element. An atom contains a nucleus (protons and neutrons) encircled by an orbit of electrons, as shown in the accompanying figure. These electrons are in motion and tend toward staying in balance with the protons in the nucleus. A group of atoms that have the same number of protons make up the element (and have a similar field of electrons trying to stay balanced).

Obviously, something has to bring together the atoms that create molecules. This is where the make-up of an atom comes in: atoms that have a balance between the protons and orbiting electrons are considered to be stable. Atoms lacking sufficient electrons will link up with other atoms and form molecules. Molecules can link up with even more atoms and other molecules in their constant attempt to balance electrons. In other words, molecules are always in a state of change. Sometimes, as in the case of a substance in a solid physical state, the change is quite slow. When atoms and molecules do link, a bond is created. In the case of the molecule illustrated here, four carbon elements have bonded to a carbon element forming a hydrocarbon molecule known as methane. When bonding occurs, a certain amount of heat is

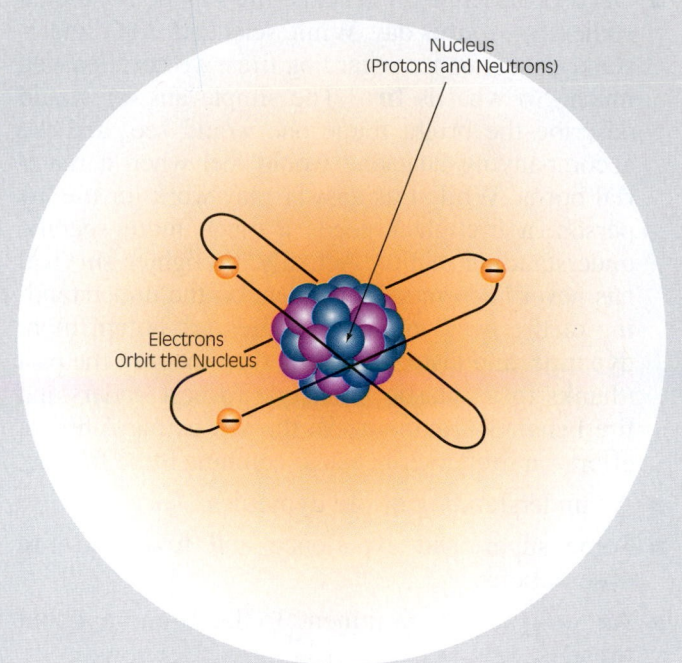

An atom is made of protons and neutrons orbited by electrons.

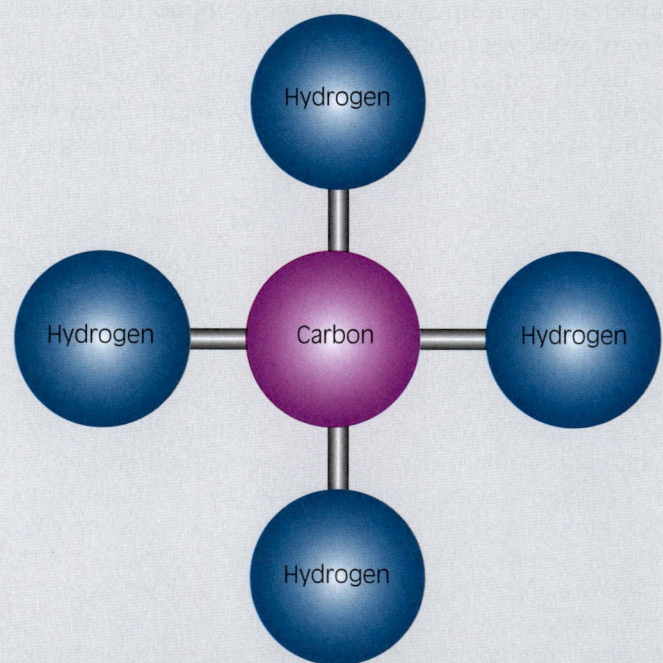

A methane molecule (CH^4—Hydrocarbon).

absorbed (called an endothermic reaction). If something breaks the bond, then it stands to reason that heat will be released (exothermic reaction).

Many events can cause atoms (and therefore molecules) to bond or separate: heat, pressure, or the exposure to another chemical (like oxygen). This is the foundation of understanding fires.

The Fire Triangle   The Fire Tetrahedron

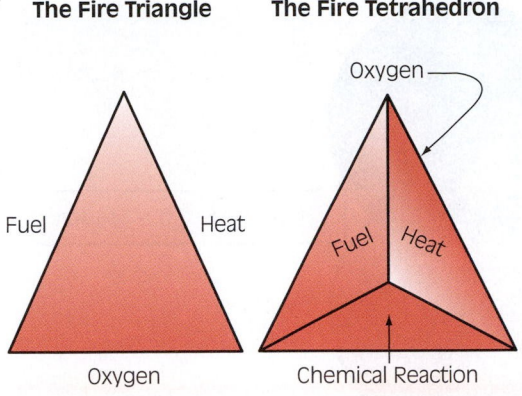

FIGURE 4-2 The old and new ways of visualizing the combustion process, the fire triangle and the fire tetrahedron.

NOTE

The combustion process was once depicted as a triangle with three sides. Each side represented an essential ingredient for fire. Heat, fuel, and oxygen were thought to be the essential elements. As the scientific study of fire progressed, it became evident that a fourth ingredient was necessary. That fourth element was the actual chemical chain reaction that permitted flame propagation. From this discovery, a new four-sided figure was used to represent the essential ingredients for fire—the fire tetrahedron.

Removal of one or more of the four ingredients is the key to fire extinguishment. Basic firefighting strategies and tactical practices are based on the principle of removing a tetrahedron ingredient. In its simplest form, fire extinguishment can be obtained by:

1. *Cooling the burning material.* This is the most common method used and is typically accomplished by using water, which has tremendous cooling properties.

2. *Removing the fuel (or allowing the fuel to be consumed).* Shutting off a pipeline valve that is allowing propane to fuel a fire is an example of removing the fuel.

3. *Excluding oxygen from the fire.* Placing a lid over a pot of burning food will eliminate oxygen. Some firefighting foams achieve the same result as they "coat" the fuel and exclude oxygen.

4. *Breaking the self-sustaining chemical reaction.* Several extinguishing agents have been developed that chemically inhibit the self-sustaining process.

The fire tetrahedron and fire extinguishment methods become our starting place for understanding fire behavior. Now we will build on this foundation and discuss each ingredient of the tetrahedron.

Heat

For a fire to begin, fuels need to be heated to the point that the fuel degrades and produces burnable gases. Heat stimulates the molecules that make up a given material. Eventually, the molecules will collide with one another and start to break apart (off-gas). In essence, heat is energy. The sources of heat energy are chemical, mechanical, electrical, and nuclear.

Chemical

Chemical heat is created when various chemicals react with each other. Oxygen (an **oxidizer**) is a powerful chemical in that it is always trying to react with other chemicals. **Oxidation** is defined as any process in which oxygen combines with an element or substance. When an oxidizer causes another material to break down, heat—and sometimes light—are created as molecular and atomic bonds are broken. Any chemical reaction that results in a heat release is known as an **exothermic reaction**, **Figure 4-3**. Rust is a slow form of oxidation that occurs when the oxygen in air breaks down steel and other metals. The heat generated is minimal and is dissipated in the environment. Exothermic reactions that happen quickly—or in an enclosed space—can create enough heat to cause the self-ignition of fuels. Once ignition occurs, combustion becomes a self-sustaining chemical heat source—the very definition of combustion.

A chemical reaction that absorbs heat or requires heat to bond atoms or molecules is called an **endothermic reaction**, **Figure 4-4**.

Mechanical

Mechanical heat is created from the friction of two materials rubbing against each other. Friction causes heat that can reach levels hot enough to ignite surrounding combustible material. Humankinds' discovery of how to make fire is tied to the discovery of creating friction (mechanical heat) with sticks or flint stones. The buildup of heat from friction is often the cause of fire in machinery. Because heat from friction can be produced whenever any rubbing or compression occurs, **Figure 4-5**, industrial processes that employ this type of action must have a means by which to carry the heat away from the source. These heat transfer mediums can be simply water or

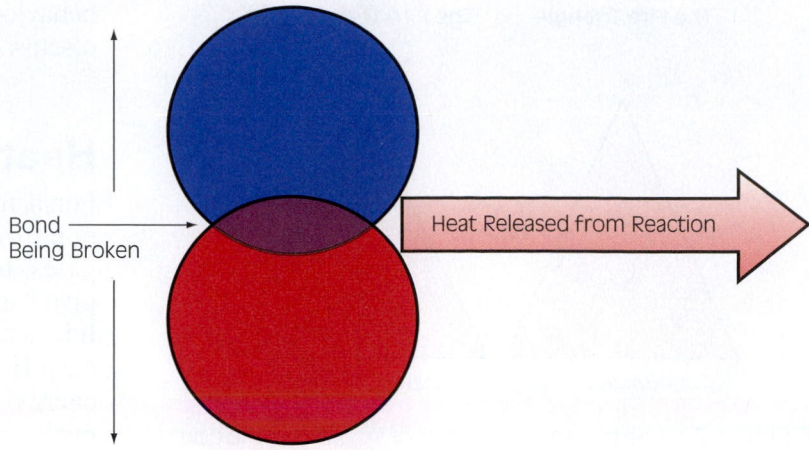

FIGURE 4-3   An exothermic reaction.

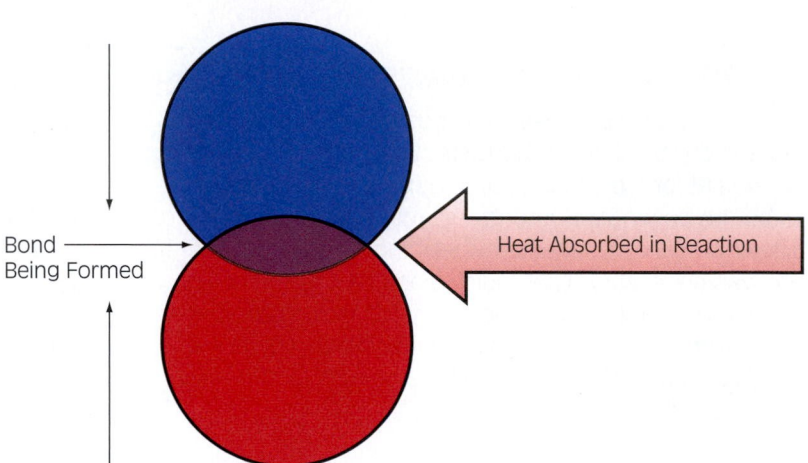

FIGURE 4-4   An endothermic reaction.

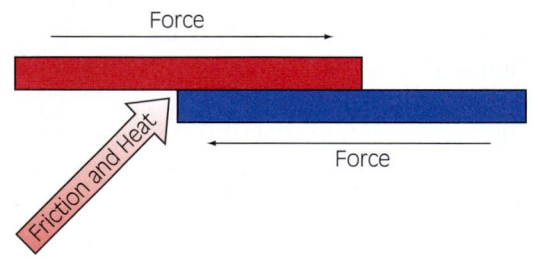

FIGURE 4-5   Heat from friction can be produced whenever any rubbing or compression occurs.

a chemical solution designed in various forms to act as a coolant.

The breakdown of a cooling mechanism will cause the two materials rubbing against each other to heat up to a point at which the surrounding materials can ignite. Understanding how this can result in a fire becomes important when inspections for fire hazards are taking place. When extinguishing fire, ensuring that the source of the heat has been stopped becomes significant. Although a fire can be extinguished even

if the source is still pumping in heat, the likelihood of keeping the fire extinguished is reduced.

Electrical

As a source of heat, electricity is probably the most recognized. As firefighters, it is important to recognize forms of electrical energy. Obviously, lightning, arcing, and electrical systems in buildings are sources of electrical heat but also included in this classification are static electricity and induction heating.

Electricity is simply a flow of electrons from a place where there are electrons to a place where they are lacking, **Figure 4-6.** Lightning is the sudden discharge of built-up electrons. In electrical systems, an electrical conductor permits the flow of electrons from one place to the other. As the electrons flow through a conductor, a certain amount of resistance to the electron flow is created based on the density and insulation of that conductor. Electrical heat is created in electrical equipment when the conductor attempts to carry more electrons than the material is capable of.

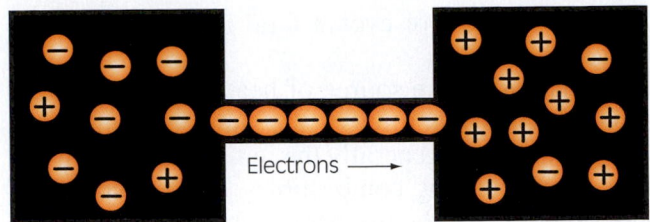

FIGURE 4-6   Electricity is simply a flow of electrons from a place where there are many to a place where they are lacking.

Arcing can be caused many ways but is mostly associated with the flash or spark created when two conductors actively carrying electrons across each other are suddenly separated.

Static electricity occurs when dissimilar materials are rubbed, scraped, or suddenly joined or separated. This action creates heat in the form of mechanical heat, but it also creates a different potential in electrical charges due to the tearing apart of the surface at the molecular level. Electrons from one substance are taken by another and when enough of them collect, they will attempt to equalize by jumping the gap in the form of a static discharge.

Induction heating takes place when electromagnetic waves stimulate the molecules of a given material. Another type of induction heat can be found in a microwave oven. In this case, food is subjected to waves of alternating electrical energy. Here the current consists of microwaves that alternate direction at high speeds resulting in molecular bombardment of the substances' internal electrons attempting to join the flow. The resultant collisions of the electrons release heat energy as the molecular bonds break apart and heat food.

Nuclear

The last heat source is nuclear, **Figure 4-7.** The means by which nuclear energy generates heat comes from radioactive materials that are very unstable and are

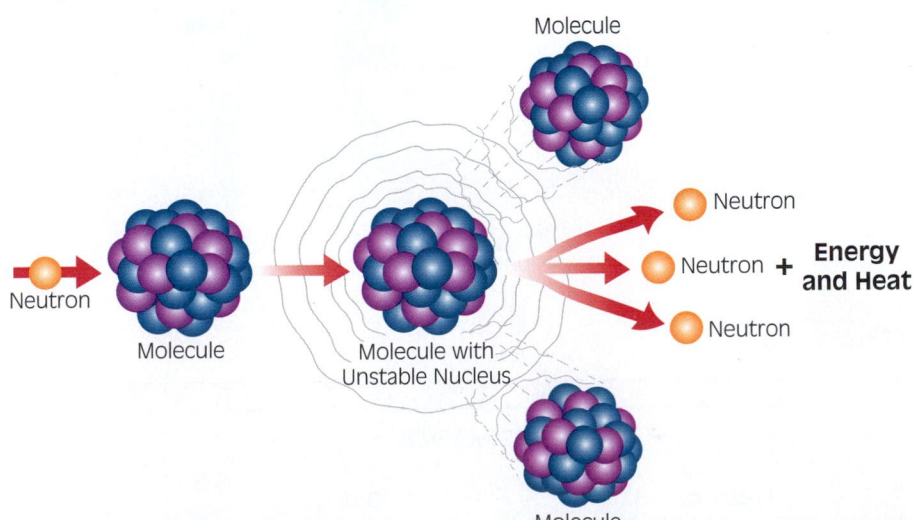

FIGURE 4-7   The process of creating nuclear energy and heat.

constantly breaking down as they seek to form a stable molecular composition (a balance of electrons). In this process, atomic particles are randomly flying in all directions. In the case of a nuclear power plant, the radioactive material is encased in a heavy shield called a core made of very dense materials such as lead. When heat is desired, the radioactive material, usually in the form of rods, is pulled out of the protective surrounding core, and a controlled amount of energy is permitted to be transferred into water surrounding the core. This heat release turns the water to steam. The steam then turns turbines that generate electricity. As the need for steam is increased or reduced, the rods are extracted from or reinserted into the protective core. When fully inserted, the nuclear bombardment is confined.

If a nuclear heat source causes a fire, there is little to do but protect uninvolved areas, evacuate, and let more skilled technicians handle the emergency. On-site personnel are trained to act in such an emergency. The real danger inherent in a nuclear fuel fire results from exposure to radiation. Without proper protection, a firefighter can sustain a serious long-term illness or even a fatal injury from radiation exposure.

In discussing the source of heat energy, one common thread winds through the subject. In all cases, the heat source is generally the initiator of the fire that ignites surrounding combustibles, be they solid, liquid, or gas. Once the fire begins, the combustion process creates additional chemical heat that becomes a self-sustaining process.

Fuel

Fuel refers to the material that is being consumed by the combustion process. Technically, fuel is **matter**—that is, some sort of material that can be touched, felt, or measured. Matter can be found in one of three physical states: solid, liquid, or gas, **Figure 4-8.** One law of physics says that matter is neither lost nor gained—it just changes form. The state of matter for some fuels is dependent on temperature, **Figure 4-9.** Water is a good example.

To take part in the combustion process, most fuels must be in their gaseous state. In other words, solid

FIGURE 4-8 States of matter. Solid materials have dense arrangements of molecules whereas gaseous molecules are more free-flowing.

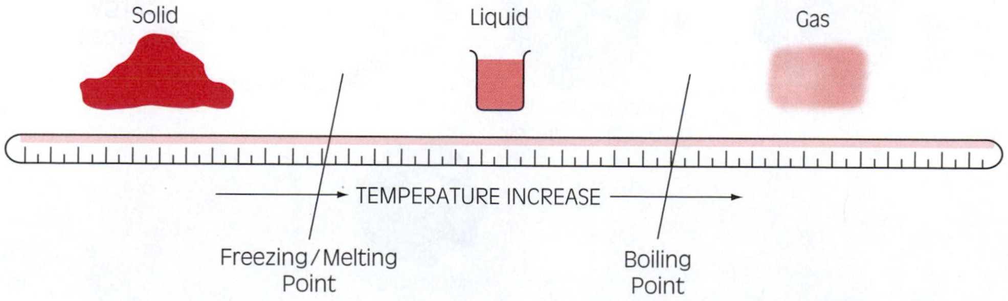

Solid   Liquid   Gas

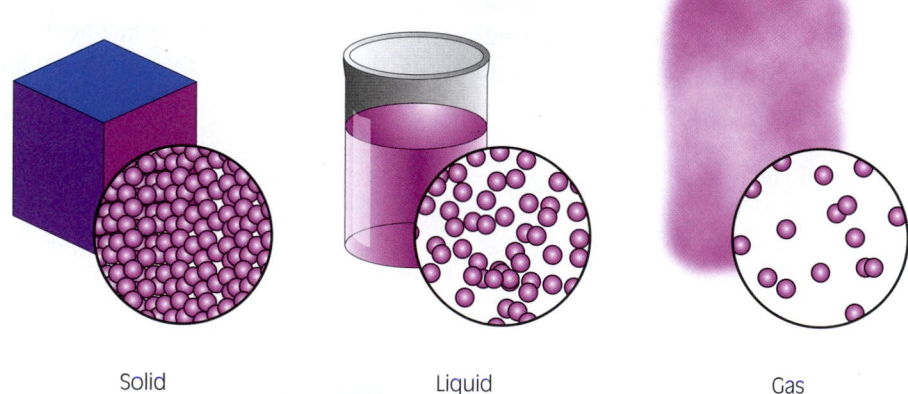

FIGURE 4-9 States of matter, in many cases, are temperature dependent.

Solid   Liquid   Gas

TEMPERATURE INCREASE

Freezing/Melting
Point

Boiling
Point

and liquid fuels must liberate gaseous molecules in order to react with oxygen. Heat energy accelerates the liberation of gaseous molecular matter from a given fuel. Let's now examine the specifics of solids, liquids, and gases, which will become the foundation for understanding fire growth.

Solids

The molecules in a solid material are packed closely together and bound in such a way as to give the solid shape. When heat is applied to a fuel, the molecules within become agitated and begin to collide with one another. Eventually, the applied heat energy will cause the molecules to break apart (called decomposition). **Pyrolysis** is the term used to describe the decomposition, or transformation of a compound caused by heat. Some of the molecules being broken apart are those that will readily combine with oxygen and become oxidized. As the bonds with oxygen are formed, a heat-producing (exothermic) reaction occurs, and the released heat causes the free-floating molecules to develop a greater affinity for the oxygen. These newly formed by-products of oxidation then release even more heat, and the chain reaction called combustion is under way.

The amount of heat a solid material can absorb before it begins to break down is directly proportional to its surface-to-mass ratio. The greater the surface-to-mass ratio, the easier it is to ignite. For example, a solid 8″ × 16″ heavy timber has incredible mass in relation to its surface area. The heavy timber can absorb lots of heat before it breaks down chemically —the internal mass of molecules serve as a heat sink. Conversely, the same heavy timber cut down into 2″ × 4″ lumber will have a high surface-to-mass ratio and much less capacity to absorb heat before it breaks down, Figure 4-10.

Solids that are in a fine particulate state (sawdust, grain flour, etc.) have extraordinarily high surface-to-mass ratios and little capacity to absorb heat. When heat energy is applied, the fine particulates will burn extremely quickly. In some cases, the combustion is so quick that a detonation can occur.

Liquids

In a liquid, the ability to burn is dependent on the substance's ability to place its molecules into suspension. A liquid cannot burn unless it is in suspension, also referred to as atomization. That liquid will then become engaged in the self-sustaining combustion

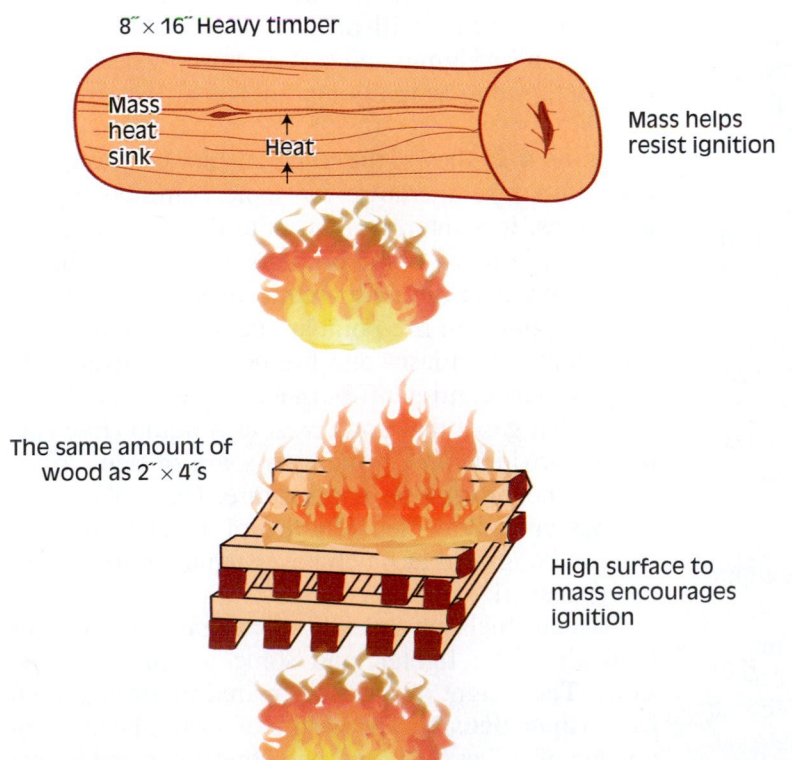

8″ × 16″ Heavy timber

Mass heat sink

Heat

Mass helps resist ignition

The same amount of wood as 2″ × 4″s

High surface to mass encourages ignition

FIGURE 4-10 Solid mass serves as a heat sink. High surface-to-mass solids encourage ignition.

process if it has combustible properties. In addition to atomization, several other concepts need to be discussed relative to liquid fires. **Vapor pressure** is the amount of force that is pushing vapors from a liquid. The higher the vapor pressure, the greater the amount of vapor (gas) being put into the air. When a burnable liquid has a high vapor pressure (like gasoline), the more intense the burning will be. Heating any liquid will increase its vapor pressure. **Boiling point** refers to the temperature at which a liquid will convert to a gas at a vapor pressure equal to or greater than atmospheric pressure.

Similar to a solid, a pooled liquid will act as a heat sink and dissipate the introduced heat into the cooler areas of the liquid. When the entire pool of liquid is heated or the ability to dissipate is overcome by the application of heat, a rise in the temperature of the liquid will occur. As heat increases, the liquid will liberate more molecules. If the molecules released are burnable, flashpoint (or self-ignition; defined in the next section) can occur.

Gases

Gaseous fuels are molecules already in a state of suspension and they are essentially primed for combustion. Fuels in a gaseous state can readily mix with the oxygen in air. The air that surrounds Earth is approximately 21 percent oxygen. The remainder of air is mostly nitrogen (78 percent) and a mixture of trace gases that make up the final 1 percent.

There are two triggers that must come together to allow a gaseous fuel to combust: the right temperature and the right mixture in air. The same is true for gases being released from solids and liquids. Individual gases have unique properties as it relates to temperature and mixture.

Through chemical analysis and scientific testing, it has been determined that gases can only ignite when certain concentrations of that gas are present in air. If enough combustible gas is not present, it is said that the mixture is too lean to burn. If there is too much gas, it is said to be too rich to burn. When a concentration of a gas falls into the range where it can ignite, it is said to be within its **flammable limits**. **Table 4-1** shows the flammable limits of some common fuels.

FIREFIGHTER FACT

Flammable limits of the materials in Table 4-1 demonstrate how different the flammable limits of different substances can be. Especially notable are the upper and lower limits of carbon monoxide, a substance found at every fire. It will ignite across a much greater range than natural gas.

TABLE 4–1   Flammable Limits of Some Materials

	Lower	Upper
Acetone	2.6	12.8
Butane	1.9	8.5
Kerosene	0.7	5
Natural gas	6.5	17
Gasoline (92 octane)	1.5	7.6
Carbon monoxide	12.4	74

The same scientific testing has determined that a given fuel must also be at a certain temperature before combustion can take place. These temperatures are known as **flashpoint**, **fire point**, and **ignition temperature**.

- *Flashpoint.* The lowest temperature at which a liquid will off-gas an ignitable mixture that will simply flash—but not sustain burning—given an outside spark or flame.
- *Fire Point.* The lowest temperature at which a liquid will off-gas an ignitable mixture that will ignite and continue to burn, given an outside spark or flame.
- *Ignition Temperature.* The lowest temperature at which any fuel will off-gas an ignitable mixture that can self-ignite and sustain burning. (May also be referred to as ignition point.)

Table 4-2 shows the relationship of flashpoint and ignition temperature for some common liquids and gases. It is interesting to note that the fire point isn't usually listed in the properties of a given fuel—usually because the temperature difference between the flashpoint and fire point are negligible. One other important note: Flash- and fire points are given only to the vapors coming off burnable liquids. That is not to say that gases that don't exist as a liquid (like carbon monoxide) can't ignite by a spark or flame prior to the known ignition temperature. Carbon monoxide has an ignition temperature of 1,128°F (609°C) but can be ignited at much lower temperatures by an external spark or flame.

Gaseous fuels have weight. Some gases are heavier than air, some lighter, and some weigh about the same. The weight of a gas compared to air is known as its **vapor density**. The weight of air has been given a value of "1" at 70°F and one atmosphere—14.7 psi (21°C and 101 kPa). Gases that weigh more than air

Product	Flashpoint	Ignition Temperature
Gasoline	−45°F (−43°C)	536°F (280°C)
Diesel Fuel	125°F (49°C)	494°F (256°C)
Propane	−155°F (−104°C)	842°F (480°C)
Natural Gas	−306°F (−188°C)	1076°F (580°C)

TABLE 4-2 Flashpoint and Ignition Temperatures of Common Liquids and Gases

All figures are approximate; sources vary in the values given to each product.

will have a value greater than one and those lighter will have a value less than one. In a fire environment, significant heat is present that can cause gases that are typically heavier than air to rise (become buoyant). Such gases will continue to rise until they lose their heat as they travel away from the heated environment—then fall back to Earth.

Understanding the properties and physical state of fuels is a strong foundation for selecting the best method and agent to suppress a fire. Obviously, you would not want to use water to extinguish a burning liquid or solid that reacts violently with water. In fact, the fire service has developed a fire classification system to better assist firefighters and non-firefighters in choosing the best extinguishing agent for the fuel involved (see the feature box, "Classes of Fire").

Oxygen

Oxygen is an important ingredient in the science of fire. It acts as a catalyst for the combustion process. The chemical reaction that occurs during combustion, called oxidation, is the process of oxygen bonding to other elements and compounds. The amount of oxygen present affects the process of oxidation, thus affecting the process of combustion.

FIGURE 4-11 The speed of combustion is affected by the amount of oxygen present.

the chemical reaction is accelerated, **Figure 4-11.** In addition, the presence of oxygen can affect a material's combustibility. Higher concentrations of oxygen can cause some materials to ignite spontaneously or permit materials to burn that would not burn under otherwise normal oxygen concentrations.

FIREFIGHTER FACT

The concentration of oxygen in the atmosphere is important to the combustion process. Oxygen occurs normally in the Earth's atmosphere in a concentration of approximately 21 percent. Combustion can occur in oxygen concentrations of 14 percent or greater. In concentrations less than 14 percent, there is not enough oxygen present to support combustion.

STREETSMART TIP

Oil in the presence of high levels of oxygen will ignite spontaneously. For this reason, any pipe, gauge, or fitting that carries oxygen posts the warning to "use no oil." Nomex, a material that is used as a fire protection component in fire protection equipment, will ignite and burn in high levels of oxygen.

With a diminished amount of oxygen, the combustion process is slowed. With an abundance of oxygen,

CLASSES OF FIRE

Five classifications of fire have been developed to help simplify the decision of choosing the best agent to extinguish a fire, based on the type of substance involved:

Class A: Fires made up of ordinary combustibles such as paper, wood, cloth, rubber, and other organic solids, including petrochemical solids (plastics).

Class B: Fires involving flammable and/or combustible liquids such as oil, gasoline, alcohol, and petroleum-based paints and stains.

Class C: Fires involving equipment and materials that are energized by electricity. Note that electricity is actually the heat source that propagates the fire and often transfers to other fuels of the Class A or B type to sustain the burning process.

Class D: Fires involving combustible metals such as magnesium, titanium, zirconium, lithium, sodium, potassium, and many others. Note that these materials are often difficult to extinguish.

Class K: Fires involving combustible cooking fuels such as vegetable and animal oils and fat. While similar to a class B fire, these oils and fats have characteristics that allow a range of extinguishing agents that differ from traditional Class B agents.

Each of the fire classes present unique considerations (and extinguishment options) which will be discussed later in this chapter as well as in Section I, Chapter 8.

Chemical Chain Reaction

When the ingredients of heat, fuel, and oxygen combine to start the combustion process, a chemical chain reaction is formed that makes the combustion self-sustaining as long as the three other ingredients remain present. As scientists studied this process, they discovered that there were many complex reactions taking place at the molecular level. They also discovered that certain chemicals introduced into the burning process could stop the flaming. These chemicals didn't actually remove heat, fuel, or oxygen—they simply broke the chemical reactions taking place. This discovery brought us the fire tetrahedron and led to the development of extinguishing agents (like dry chemicals) found in portable fire extinguishers.

The foundation for understanding how fires begin is grounded in the fire tetrahedron. Typically, some sort of event brought together the heat and fuel—and oxygen was present in air. Once the three ingredients came together, the self-sustaining chemical reaction took over. With that understanding, we can now look at fire growth.

FIRE GROWTH

Once a fire begins—usually a result of an event that brings together the ingredients of the fire tetrahedron—it will grow in a self-sustaining manner. The heat being released by the fire (exothermic energy) will be transferred to other fuels and begin heating them up. As more fuels become involved, the fire will transition through various phases. Along the way, the fire is releasing many by-products that can make the situation worse. The firefighter that understands fire growth can better predict fire behavior and make an appropriate intervention to stop that growth. This section will show how a fire grows through the discussion of heat transfer, phases of burning, and products of combustion.

Modes of Heat Transfer

Heat is a by-product of combustion that is of significant importance to the firefighter. It is heat that causes fire spread to other fuels. That exothermic energy can be transmitted in many ways. Knowing how heat is transmitted from one place to another is an important step in knowing how to control the extension of fire.

The three modes by which heat transfers its energy from one substance to another are conduction, convection, and radiation, **Figure 4-12.** The type of substance being heated, the distance covered by the material being heated, and the ability of the substance to retain the heat will be factors in the spread of fire.

Conduction

Conduction is the transfer of heat through a solid object. When an object heats up, the atoms become agitated and begin to collide with one another. A chain reaction of colliding molecules and atoms, like a wave of energy, occurs and causes the agitated molecules to pass the heat energy to areas of nonheat. As the heat increases, so do the waves of energy passing to the cooler areas, **Figure 4-13.**

When examining conduction as a heat transfer vehicle, it is important to understand that heat is conducted through materials at different rates. The rate of heat transfer will depend on several factors, the most significant being density. **Density** is the mass per unit volume of a substance under specified conditions of pressure and temperature. If a given volume of any substance is weighed, we can determine its relative density.

The more dense a material, the better conductor it will be. Because density is a function of weight, the heavy substances are generally better conductors. Metals, being among the most dense in the universe, are generally better conductors of heat. Steel can conduct heat very easily and actually cause substances with low ignition temperatures (like paper and wood) to ignite distally from the point where the steel is being heated. Conversely, wood is a poor conductor of heat. Because of its low ignition temperature, the wood is likely to ignite at the point of heat contact before it can conduct the heat elsewhere. Concrete is quite dense and has tremendous capacity to absorb and retain heat.

Convection

Convection is the transfer of heat through air and liquid currents. The air and gases that surround a fire and those gases coming off the flames are much hotter than those distal to a fire. These hot gases will become buoyant and rise. This upward movement creates a rising column that carries heat (known as a thermal column). The air and gas molecules in the thermal column are quite agitated and, because of their gaseous nature, will expand to accommodate the agitation. This forms a rising and expanding convection plume. This convection plume is also known as a **thermal plume**: a column of heat rising from a heat source.

As the plume rises, it will transfer heat to any medium it contacts. Additionally, the plume starts

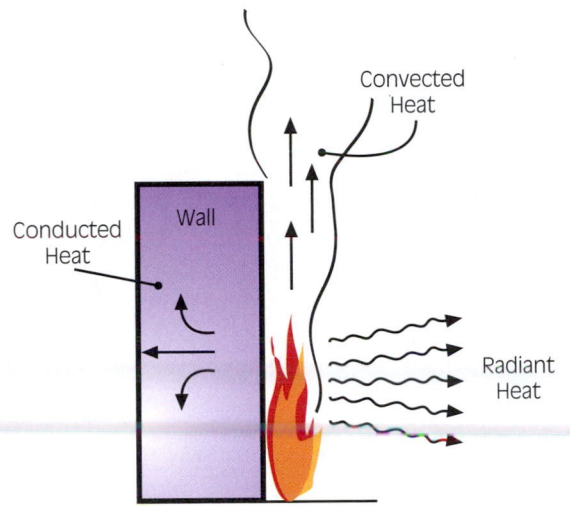

FIGURE 4-12 Examples of heat transfer in fire.

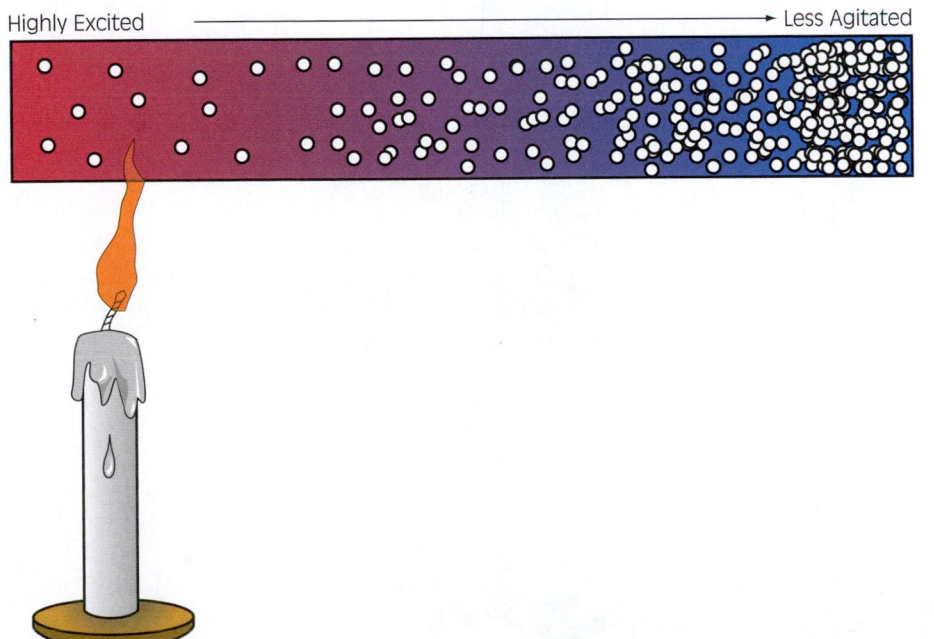

FIGURE 4-13 Conduction. A metal bar is heated at one end, causing the molecular activity that transfers the heat.

mixing with cooler air. Eventually, the heat will be lost and the plume will reach equilibrium with the atmosphere. As the molecules in the plume calm down, they start to return to their original density and become less buoyant. As the buoyancy decreases, that particular volume of air drops. If it is heated again by the rising heat energy, its agitation increases again. If it falls outside the thermal plume, it drops down to the bottom of the plume. A fully formed plume will resemble a mushroom as the upper level of the heat plume cools, stratifies, and begins to drop outside the rising column, **Figure 4-14.**

In theory, an unobstructed plume that is unaffected by any other influences such as horizontal air current would behave in the mushroom fashion. In a structure, the flow and rate of convection would be accelerated when the heated air is confined and must seek a "cooler" space after meeting a vertical obstruction such as a ceiling. The "mushrooming" can spread the heat much farther and faster horizontally than an unobstructed plume can, **Figure 4-15.** Hallways, stairways, and vertical shafts become serious convection pathways because they are restricting the expanding heat and gases.

Convection currents created by heat in the air occur in liquids in the same fashion. The process of evaporation is accelerated in a liquid heated in an open-topped container. **Evaporation** is defined as the process in which the molecules of a liquid are liberated into the atmosphere at a rate greater than the rate at which the molecules return to the liquid. Ultimately the liquid becomes fully airborne in a gaseous state.

Liquid molecules that are heated tend to bounce off the sides of the container and reintroduce the energy back into the liquid. The reintroduced energy then causes more molecules to collide until the energy forces them over the top of the container where they escape. This is also the same phenomenon that causes pressure in a closed container to rise—in some cases past the engineered design limits of the container.

Radiation

Radiation is the transfer of heat through invisible lightwaves. When combustion occurs, light is produced. These lightwaves range from ultraviolet to

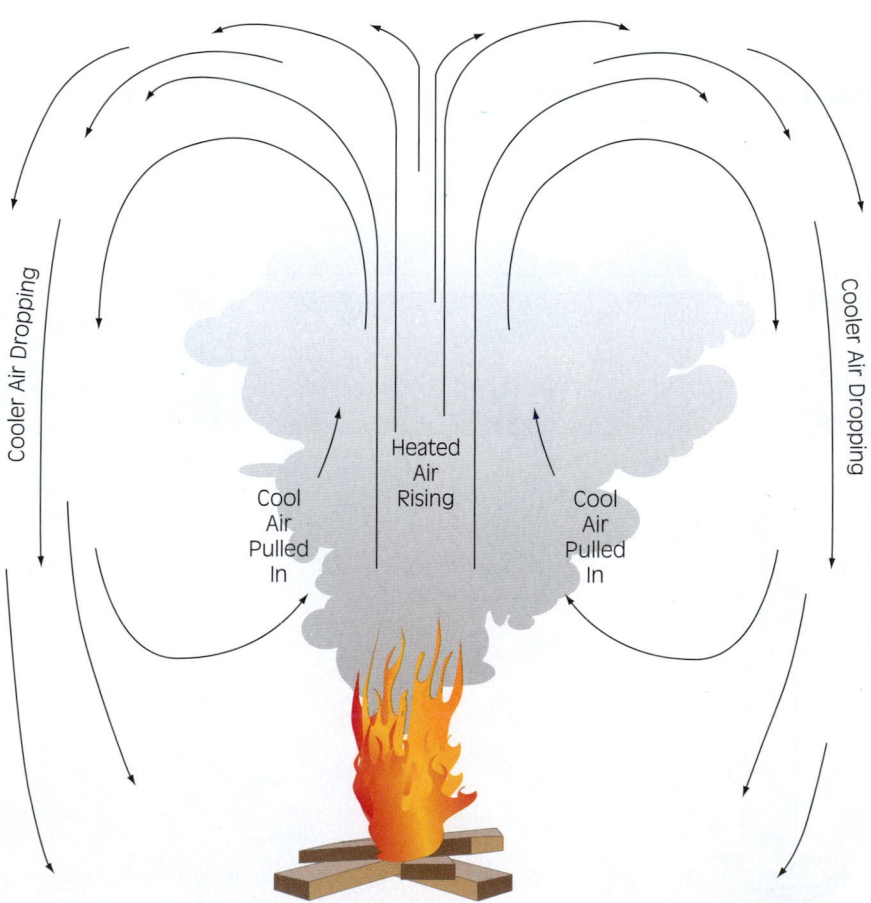

FIGURE 4-14 A natural fire plume in an unrestricted space.

infrared. Contained in the lightwaves are electromagnetic forces that can actually travel through air and deposit themselves on remote objects. An example of this is the sun and its ability to heat the Earth.

Fire produces the same infrared lightwaves and, with enough concentration, can permit fire to jump from the source to a distant object, heat it up, and if intense enough, cause it to ignite, **Figure 4-16A and B.** Several factors need to be in place before this can occur: the source must be strong enough to sustain the bombardment of the lightwaves; the recipient object must be able to absorb the heat energy; and that object must be able to build up the heat without it dissipating away through its own heat sink.

Firefighters must cope with this phenomenon because fire can extend through radiation as much as through conduction or convection. Water is used to absorb the heat and carry it away so it cannot heat surrounding materials. Water absorbs the accumulating heat from the exposed objects, thereby depriving them from reaching ignition temperature. By controlling the heat ingredient of the fire tetrahedron, the fire is prevented from extending. To a limited extent, water can also be employed to combat conduction extension. In that case, the water carries away the heat, keeping the unburned or unheated portion of the material from reaching the ignition temperature or permitting it to transfer the heat to another substance.

All three methods of heat transfer are usually present during fires. Some fire behaviorists have added a fourth method of heat transfer—direct flame contact. In practical terms this may seem obvious. Scientifically, direct flame contact is just a continuation of the self-sustaining combustion process. With

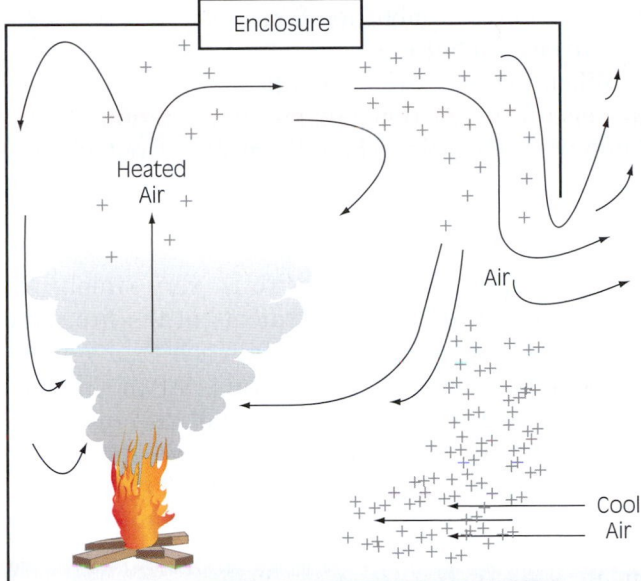

FIGURE 4-15 A convection fire plume in a restricted space. Mushrooming and heat transfer is accelerated in restricted spaces.

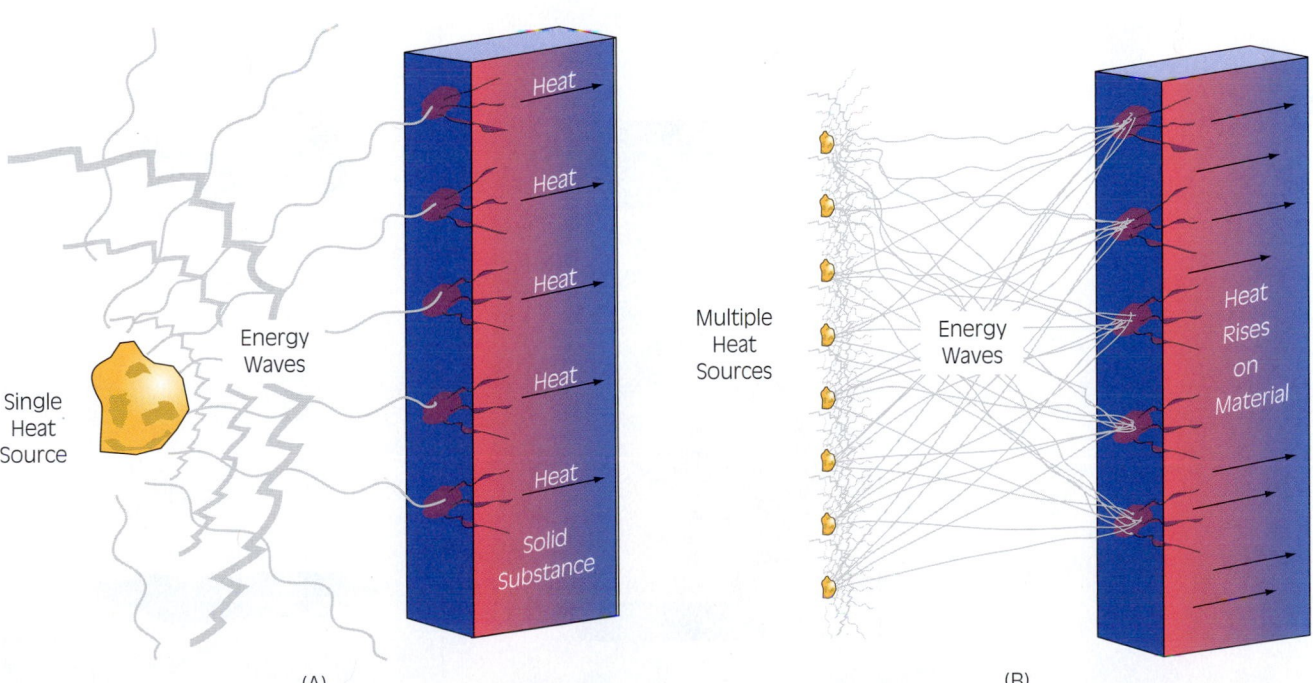

FIGURE 4-16 (A) Radiation, single heat source. (B) Radiation, multiple heat sources.

all modes, heat is being transferred and other fuels begin to burn. As more fuels burn, more heat is developed, and the fire will grow in defined phases. Let's look at those phases.

Fire Phases

The process of burning occurs in clearly defined phases. By recognizing the different phases, a firefighter can better understand the process of burning and fight the fire at different levels and with different tactics and tools. As is true in any combat situation, knowledge of the enemy's needs and practices leads to the use of different tactics and practices to attack that enemy successfully.

NOTE

Burning materials follow a specific sequence (stages or phases) from start to finish. Older texts refer to these phases as the incipient stage, the smoldering stage, and the free-burning stage. Today's firefighting science has redefined the fire phases. They are the ignition, growth, fully developed, and decay phases, **Figure 4-17**.

Ignition Phase

When a substance begins to heat up, it liberates gases that can burn. The preliminary heating usually comes from an outside source such as a match or spark from a fire already burning. Heating can also occur through conduction, convection, or radiation from another fire or heat source. At some point, the amount of heat being created exceeds the amount of heat that is dispersed and the combustion feeds itself the necessary heat to sustain burning.

When the necessary ingredients of a self-sustaining chemical reaction are present, ignition occurs. Ignition is that point where the need for outside heat application ceases and the ability of the material to sustain combustion comes from the heat generation of the material itself.

The ignition stage of the fire is, very simply, the point at which the four ingredients in the fire tetrahedron come together, materials reach their ignition temperatures, and a fire is started. At the ignition stage, the fire is typically very small and limited in area. Typically, plenty of oxygen is available for the fire in the ignition phase. Also, a definitive flame will be present and a thermal plume develops. Within a room, the thermal plume from a floor level fire will likely cool before it hits the ceiling.

Ignition

Growth

Fully Developed

Decay

FIGURE 4-17 The four phases of fire.

Growth Phase

From the point of ignition, fire begins to grow. Starting out as a small flame, other combustibles heat up, liberate flammable gases, and ignite, spreading the chain reaction to other flammables and resulting in an increase in fire size.

The speed of the growth and ultimately the size of the fire are dependent on several factors:

- *Oxygen supply:* The amount of oxygen will have a direct effect on the speed of growth and the size of the fire. Any limitation of the oxygen supply will curtail the growth and can even result in extinguishment.

- *Fuel:* The size of the fire will naturally depend on the amount of fuel available to burn and the heat release rate of those fuels.

- *Container size:* In a structure, the container would be the surrounding walls, ceiling, and any obstructions that make a room. A large container would permit dissipation of heat and slow the growth of the fire. Within a small room, heat can quickly build and radiate back from the ceiling and walls—causing other fuels to rapidly ignite.

- *Insulation:* Heat that is radiated back into unburned areas will accelerate growth. If the container wall and ceiling are insulated, the amount of heat kept trapped in the container and not permitted to escape serves to initially slow heat transfer. This insulation can hold only so much heat. Once that capacity is reached, the heat will be "reinvested" in the fire formula.

During the growth phase, an obvious thermal plume develops. Within a room, the thermal plume will reach the ceiling and begin moving laterally. This marks the beginning of thermal layering (see the feature box, "Thermal Layering"). A fire within a room will likely extend to other rooms and other parts of the building during the growth phase, **Figure 4-18.** This extension can happen within seconds depending on the preceding list of factors.

VIEWPOINT

The combination of better-insulated buildings and higher-temperature fires have caused the growth-phase fire to become rapid. Fires can double or even triple in size within seconds! Perhaps the growth-phase fire should be relabeled the *Rapid-Growth Phase.*

If plenty of air remains available, a growth-phase fire will rapidly spread and transition into the fully developed phase (covered next). If air is restricted or unavailable, the fire growth becomes vent-controlled; that is, it will grow only as air becomes available (also called an "underventilated" or "air-controlled" fire).

CAUTION

A vent-controlled fire in a building is extremely dangerous to firefighters because the burning process is incomplete: plenty of heat and fuel is available and the flames present are not as intense as they could be because they lack air. Introduction of air (like when a firefighter opens a door) can cause the fire to intensify explosively.

THERMAL LAYERING

Thermal layering is the stratification of air and fire gases into layers based on their temperatures. When confined to a structure, the hottest of the gases and air will accumulate near the ceiling of the compartment or room, and as the amount of by-products from fire increases, the gases, heated air, and smoke will bank down until they can find an escape route.

What kind of temperatures are we talking about here? To answer, it is important to note that fires are not absolute — there are so many factors that govern fire behavior and therefore temperatures. Given that, some general temperature ranges have been cited in research efforts that may give us some insight on the heat levels one can expect inside a burning room during the growth phase. Flame temperatures vary depending on the type of fuel burning but typically range from 1,700° to 2,200°F (927° to 1,204°C). In the growth phase, ceiling temperatures can reach 1,100° to 1,500°F (593° to 815°C). Near the floor, temperatures will hover between 100° and 200°F (38° to 93°C).[2]

Thermal layering is an important concept for firefighters because it affects how to enter and function in a room or area that is on fire or where the fire has just been extinguished. Firefighters must stay low to the floor when attacking structural fires because the hottest gases rise and stay at the highest layers in the compartment. One goal of interior firefighting is to keep the building in "thermal balance." In essence, thermal balance is the maintenance of natural thermal layering. Fire streams can mess up the thermal layer and, therefore, thermal balance. Efforts to keep the thermal layer up high (and cool air low) help make interior firefighting safer.

FIGURE 4-18 Room-to-room fire spread can be rapid during the growth phase. *(Photo by Keith Muratori from FIREGROUNDIMAGES .com)*

A vent-controlled fire—by nature of its incomplete burning—will be creating lots of smoke (see the section Products of Combustion). In many cases, the smoke being created fills the room and precludes air from entering the room. This will eventually choke the fire. If the room remains intact (doors and windows stay closed), the fire will skip the fully developed phase and transition into a decay phase.

One last note regarding growth-phase fires: Most hostile fire phenomena that can injure or kill firefighters are being set up during the growth phase (see the section Hostile Fire Events).

Fully Developed Phase

The fully developed phase is recognized as the point at which all contents within the perimeter of the fire's boundaries are burning. In a structure this would mean the entire contents of a room or all the rooms in a structure, **Figure 4-19.** In an outside fire, it would mean all combustible material within the fire's farthest reach. For a fire to become fully developed, plenty of air must be available. In a structure, the speed and extent of a fully developed phase is controlled or regulated by the amount of air that can be introduced or supplied to the fire area—it will only become fully developed if that air is available. A vent-controlled growth phase fire can rapidly become fully developed when some part of the structure fails (collapse, window or door failure, etc.) and unlimited air is allowed to reach the fire. In an outside fire, because the amount of air is unlimited, the amount of fuel will dictate the size of the fire, making it a fuel-dependent fire.

FIGURE 4-19 Full involvement of a structure is an example of a fully-developed-phase fire.

Decay Stage

When the point at which all fuel has been consumed is reached, the fire will begin to diminish in size. Ultimately, the fire will extinguish itself when the fuel or oxygen supply is exhausted. Obviously, this can take a considerable period of time. The bright array of flames will diminish, becoming a series of separate flame fronts, **Figure 4-20.** The flames will then disappear until only glowing embers are visible. Eventually, those too will disappear.

As a fire moves through its phases, many by-products of the burning process are being released. Let's look at these by-products and discuss how they impact firefighters.

FIGURE 4-20 The decay phase.

Products of Combustion

As stated earlier, matter is neither lost nor gained—it changes form. In the burning process, many chemical reactions are taking place and the fuels being burned are changing form rapidly. The by-products of this process are heat, light, and smoke.

- *Heat:* We've already discussed the role of heat at fires. As heat is transferred through the various modes, materials that are not yet burning start off-gassing. Some of these gases get drawn into the fire, but most of the gases join the thermal plume in the form of smoke.

- *Light:* The lightwaves produced by combustion are seen as the actual flames.

- *Smoke:* **Smoke** is defined as the product of incomplete combustion or decomposition heating (pyrolysis) and includes an aggregate of solids, liquids, and gases suspended in the thermal plume.

In all forms, the products of combustion, **Figure 4-21,** can be hazardous and deadly to fire victims *and* firefighters. An examination of each of the products can help us better understand their inherent dangers.

Heat

Heat, as a by-product of the combustion process, can be dangerous to firefighters, causing dehydration and heat exhaustion as well as burns. Each of these conditions can be mild to severe and ultimately life threatening. The amount of heat energy released over a cer-

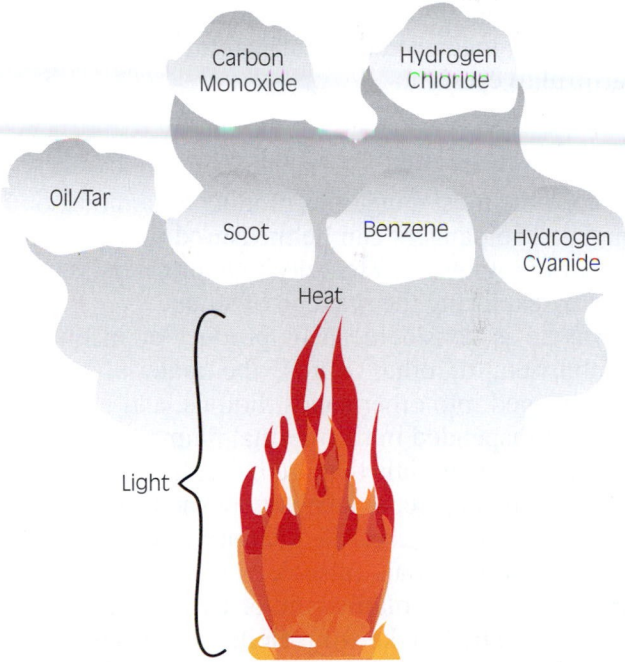

FIGURE 4-21 The products of combustion are deadly and include light, heat, and smoke.

tain period of time is called the heat release rate. Heat release rates are influenced by the quantity and type of burning materials. When a fire burns within a room, the walls and ceiling will absorb the released heat. When the room can no longer absorb heat, it begins reflecting that heat in a radiant form. This radiant heat feedback causes burning materials to release heat at a higher rate. The radiant heat feedback will also raise floor-level temperatures significantly. The greater the heat release rate for a particular fire, the greater the danger to firefighters for heat-related injuries.

> **NOTE**
>
> The unprotected human form can only withstand so much heat before death occurs. Most medical and fire life-safety references agree that resuscitation is not possible for unprotected humans that are exposed to 300°F (149°C) for one minute.[3] This temperature and time threshold can help firefighters understand the rescue profile for fire victims.

Light

The light associated with flaming may not seem dangerous to firefighters. In most cases, this is true: the eye's ability to constrict (pupil) and the reflex reaction to close one's eyes self-limit the danger of light. Even so, certain types of fuels burn with a very intense light that can cause temporary or permanent damage to eyesight. Examples include the burning of metals such as magnesium. Electrical arcing can also be harmful to eyesight.

Smoke

Perhaps the most dangerous product of combustion is smoke. The dangers can be measured in a multitude of ways ranging from toxicity to explosiveness. Let's start by exploring the makeup of smoke.

Smoke is considered a by-product of incomplete combustion. In other words, the material burning has changed into other solids, liquids, and gases that become suspended in the thermal plume. Smoke also comes from materials that are being heated—but not flaming. In essence, smoke is unburned fuel, **Figure 4-22.** With the increased quantity and complexity of synthetic materials in society, smoke has become amazingly toxic, volatile, and/or explosive. The feature box, "The Smoke from Hell," shows some common materials in smoke.

When no smoke is visible coming off flames, it is likely that the flames are well vented and the burning fuel is being reduced to water vapor, carbon dioxide, and trace amounts of carbon monoxide—all colorless. Some call this complete combustion. Complete combustion poses a lesser threat to firefighters, although carbon dioxide is an asphyxiant and carbon monoxide is a flammable toxin. Smoke management has never been more important than it is now. For most building fires, the smoke behavior will ultimately dictate the fire behavior.

FIGURE 4-22   Smoke is an aggregate of solids, aerosols, and gases. *(Photo by Keith Muratori from FIREGROUNDIMAGES .com)*

SPECIFIC FIRE CHARACTERISTICS AND EVENTS

In this final section of the chapter, we'll look at some characteristics of various types of fires (liquids, electrical, and metals). Additionally, we'll examine several unique fire events firefighters must be familiar with that can occur within a compartment or structure, including flashover and backdrafts. We'll finish the section with an introduction to the street-smart process of "reading smoke" to predict fire behavior within a building.

Liquid Fires

As presented earlier, liquid fuels must vaporize in order to burn. **Vaporization** is the process in which liquids are converted to a gas or vapor. The rate of

THE SMOKE FROM HELL

DAVE DODSON

Today's fire environment contains some of the most toxic, volatile, and explosive smoke that the fire service has ever faced. The chemical reactions taking place as synthetic materials burn or heat-degrade can challenge most chemistry professors. Although research continues, the scientific study of smoke production in today's fires has given us some useful information that can help firefighters understand why fires are hotter, more toxic, and more explosive than ever. The list that follows illustrates some of the more notable fuels in smoke, although the list of smoke products can literally fill volumes.

- *Solids.* Carbon (soot) is the most common solid in smoke. Other solids include ash (minerals like calcium that can't burn but can carry heat), natural and synthetic fibers, and dust.

- *Liquids.* The burning process can create atomized liquids, vapors, and microscopic droplets that become suspended. Basically, these can be called aerosols. Specifically, the aerosols in smoke include water vapor, oils, tar, polynuclear aromatic hydrocarbons, formaldehyde, and trace acids.

- *Gases.* Carbon monoxide leads the list of gases produced during incomplete burning. Also notable in quantity and flammability are hydrogen cyanide, benzene, and acrolein. Additionally, smoke can contain polyvinyl chloride, hydrogen chloride, hydrogen sulfide, phosgene, nitrogen dioxide, ammonia, and phenol.

Now we'll get personal. A review of firefighter news items, investigative reports, and incident data indicates that:

- Firefighter cancer rates are escalating while the rates for the general public are declining.

- Firefighters that get just a few breaths of zero-visibility smoke either die or face significant medical issues that can become chronic.

- Firefighters are being treated for cyanide poisoning from just a "little" smoke exposure outside of burning buildings.

- The rate of firefighter flashover (and rapid-fire-spread) deaths are going up in an environment where the number of fires is going down.

The bottom line: Smoke is fuel—and it has never been more deadly or explosive. A firefighter crawling through zero-visibility smoke is being saturated with an explosive poison. It's like swimming through a lake of warmed diesel fuel. Today's smoke is truly the smoke from hell.

---

vaporization is dependent on many factors but heat will always accelerate the rate. If the vapors are above their flashpoint—and within their flammable limits—they will ignite with the introduction of a spark or flame. If the vapors are above their ignition temperature and in the right mix with air, they will self-ignite. The growth of a liquid fire is dependent on the container holding the liquid. If burning liquid leaves the container, it will flow and spread the fire to other materials—called a running fuel fire.

Most class B liquid fires don't mix well with water. Using water can actually liberate more vapors and intensify the fire. Class B fires are typically extinguished by coating the liquid with an agent (foam) so that the liquid vapors cannot escape and mix with air. Specific foam types have been developed for various liquid fires, although the method of suppressing vapors is the same.

Electrical Fires

Recall that a Class C fire is one where electrical energy is creating heat. Electrical heat that causes arcing can create temperatures in excess of 2,000°F (1,093°C) and actually cause metals to melt or fuse. The same heat can obviously heat and ignite nearby Class A or B materials. The control of a Class C fire starts with a control of the electricity. Once the electrical power has been removed, then fire control efforts are aimed towards the involved Class A or B materials. Obviously, the use of water can present extreme shock hazards while the fire is still electrically energized. Occasionally—and only under the direct supervision of trained and experienced fire officers and/or power company representatives—water can be used to cool surrounding metals and other fuels until such time as the electricity is controlled.

Metal Fires

The Class D metal fire is a chemical reaction fire. Almost all metals will burn if conditions are correct. Some of these metals will burn with great violence. Among the combustible metals of significance are thorium, titanium, plutonium, hafnium, lithium, magnesium, zirconium, zinc, uranium, sodium, and potassium.

The combustible metals differ somewhat in their reactions under fire. In some cases, the mere presence of water will cause a violent reaction, releasing heat and brilliant light. In other cases, the mere presence of air will cause the reaction. Some metals will burn so hot that the water molecule will actually be broken down into its component hydrogen and oxygen atoms. Then the hydrogen burns away in the presence of oxygen creating a brilliant and hot fire. Each metal's characteristics must be evaluated. Fortunately, these metals are not found in great abundance in normal occupancies. They are usually found in industrial processes and their presence should be communicated in advance to responding firefighters.

Other metals are also subject to burning but usually only when the material is in fine shaving form. For example, steel will not ignite easily, but a piece of steel wool will ignite rapidly when flame is applied.

Control of Class D fires can be quite difficult—or amazingly simple. The shape, size, amount, and type of metal burning will dictate the best extinguishment method. For example, a small magnesium fire can be extinguished with an overwhelming volume of water. The same volume of water applied to a large magnesium fire can result in an explosion. For other metal fires, simply covering the metal with clean, dry sand will aid in extinguishment. Additional study (beyond the scope of this book) will acquaint the firefighter with Class D fire control methods.

Hostile Fire Events

Many believe that any uncontrolled fire in a building is a hostile fire event. This philosophy is right on. The firefighter, however, needs to add greater specificity to the phrase "hostile fire events." In firefighter terms, a **hostile fire event** can be described as unique fire phenomenon (such as flashover, backdraft, or rapid fire spread) that takes place within a building. Historically, hostile fire events have been defined and argued over by scientists, scholars, and fire experts. Most agree on the definition of flashovers and backdrafts. Less agreeable are the definitions of rapid fire spread and smoke explosions as phenomena.[4] Herein, we'll list and describe three hostile fire events: flashover, backdraft, and rapid fire spread.

Flashover

Flashover is a sudden event that occurs when all the contents of a container (room) reach their ignition temperature nearly simultaneously. As a fire continues to burn in a compartment, more and more heated gases are generated, and the hottest layer of gases becomes larger, expanding downward because of the ceiling. The exothermic energy is being absorbed by exposed contents in the room and the container itself (walls, ceiling). The walls and ceiling can absorb only so much heat—when they can't absorb any more heat, they reflect that heat back into the box interior (radiant feedback). This causes a surge in the flaming activity (higher rate of heat release). Additionally, the sudden radiant heat feedback raises the overall temperature of all the unburned contents of the compartment—even those at floor level. When the heat has brought the entire contents of a container to their ignition temperature, a flashover will occur if air is present. In a flashover, the entire contents of the compartment ignite almost simultaneously, generating intense heat and flames. This rapid change in the compartment occurs within seconds and is a dangerous and often deadly occurrence for firefighters.

Prior to flashover, smoke gases that have lower ignition temperatures may start to ignite intermittently within the upper thermal layer. The ignition of these lower ignition temperature gases is known as **rollover**. Rollover can appear as fingers of flames that come and go in the superheated upper layer. In essence, these are gases that either are low in volume and burn off or are flowing in and out of their flammable limits. Rollover is a proactive warning sign that flashover is imminent. Unfortunately, firefighters crawling in thick smoke may not see the rollover above their heads because smoke contains so many particulates (solids) that obscure vision. Rollover that becomes constant and begins to ignite all of the smoke is known as flame-over, which starts rapid fire spread (discussed later). Smoke leaving a room that is ready to flashover will be extremely agitated (turbulent)—another warning sign.

> **CAUTION**
>
> Rollover and turbulent smoke flow leaving a room are significant warning signs that flashover is imminent.

Backdraft

Backdraft is the sudden and explosive ignition of pressurized, superheated, and oxygen-deprived gases (within a closed space) caused by the reintroduction of oxygen. A backdraft can be considered a detonation of those oxygen-deprived, superheated, and pressurized fire gases. Reintroduced oxygen becomes the detonation trigger.

A fire burning in a closed container, such as a room in a structure, can use all the available oxygen in the compartment, slowing the combustion process and the fire. The introduction of a source of oxygen to the compartment, such as opening a door or window to a room, will bring together the four elements of the fire tetrahedron again, causing the room to violently burst into flames.

A room that is ready to backdraft will exhibit signs of heat that include:

- Smoke under pressure that will be seeping from cracks and seams.
- Yellowish-brown or yellowish-grey smoke coming from cracks and seams of the pressurized space (sulfurlike).
- Windows that are black stained and bowed outward (signs of heat and pressure).

Some texts list "puffing or sucking of air and smoke" as a warning sign of backdraft. If air is being allowed back into a pressurized compartment, it means the pressure is temporarily leaving elsewhere and air is being allowed into the closed environment. These are late warning signs. In other words, movement and airflow is taking place and the backdraft is in the process of being triggered.

The term *smoke explosion* is used by many synonymously with backdraft. Some fire service leaders argue against this and would like to see the term *smoke explosion* applied to situations where a spark or flame causes the ignition of trapped gases—as opposed to oxygen (see the feature box, "Backdraft versus Smoke Explosion").

Rapid Fire Spread

A rapid fire spread event is one where the accumulated smoke within a building ignites and suddenly spreads the fire. The term *flame-over* has been used to explain the same event but is more typically applied to the room where active fire is taking place. Let's explore the difference a bit more. With the advent of so many synthetic materials that off-gas lots of hydrocarbon aerosols, smoke has become increasingly contiguous. In other words, smoke creates continuity of fuel throughout a building. Once this smoke reaches a self-sustaining combustion temperature (temperatures above the aerosols' fire-point), it only needs the original flame to trigger its burning. When this happens, the fire can race violently to any place dense smoke has traveled.

Another way to look at this phenomenon is to compare the current fire spread rates to classic growth-phase teachings. In a simpler time, firefighters were taught that a growth-phase fire can double in size every minute. With the continuity of fuel that smoke now presents, a fire can grow ten-fold in mere seconds when the smoke ignites! When this happens,

BACKDRAFT VERSUS SMOKE EXPLOSION

DAVE DODSON

Fires inside of buildings can present a myriad of unusual events. Most experienced fire officers can recall a structural fire where some sort of unexplainable event took place: fire burning downward, fires that blew themselves out, easily burnable fuel that never ignited right next to something that was totally consumed. Occasionally, a fire will present itself in a way that defies all that is understood. The collection of these anecdotal experiences helps us confirm and dispute what we know. In the case of backdrafts, most in the fire service agree that it is an explosive event that occurs when oxygen is introduced into a confined, superheated, and pressurized environment that was previously oxygen starved. What then do you call an event where a smoke and air mixture, distal from the active fire, briefly ignites and expands with the introduction of spark or flame? Some have suggested that a spark-caused ignition of gases away from the active fire be called a smoke explosion. I agree with the latter.

In a typical interior fire, plenty of smoke is being created by the fire itself. Additional smoke is being produced as other materials (not burning) break down from heat. The collective smoke plume will migrate away from the fire and seek an upward path of least resistance. Typically, the smoke production is so great (thanks to synthetic materials) that smoke must flow in all directions because the "box" simply can't conduit the sheer volume of smoke in an upward path. As hot smoke flows through the building, it will heat up and cause additional materials to off-gas. These gases have never been exposed to a spark or flame! Some of this smoke will entrain air as it travels but may hit a dead end (like the top of a stairwell or within an HVAC duct system). An errant spark (like a glowing ember from the original fire, an electrical wire ground short, or even a bursting energized light bulb) introduces an ignition source for the gases. If the gases and aerosols in smoke are above their respective flashpoints (remember—smoke contains many aerosols) but below their ignition temperature, they will briefly flash—a flash accompanied by a shock wave. This shock wave can drop a suspended ceiling grid, blow out a wall, and displace firefighters.

Maybe the fire service needs to make some definition adjustments and embrace the term smoke explosion as an event separate from a backdraft.

firefighters crawling in the smoke are enveloped in the rapid fire spread. Those that are able to escape have reported that they never saw it coming (smoke contains many particulates that obscure visibility).

Reading Smoke at Structural Fires

When responding to structural fires, a firefighter can apply an understanding of basic fire behavior by "reading smoke." Smoke issuing from multiple openings of a structure is often the only clue as to what a fire is doing within the building. Reading smoke can help firefighters discover clues about the location of the fire within a building as well as the severity of the fire and the potential for a hostile fire event such as flashover or backdraft.

Typically, firefighters view smoke as "light" or "heavy." While this is fine for a rapid radio report, it is not descriptive enough when trying to understand what is actually happening with a fire. Smoke from a structural fire has four attributes that must be analyzed. These are volume, velocity, density, and color, **Figure 4-23.** These four attributes need to be compared from each opening that issues smoke. Taken collectively, the differences in attributes can paint a story about the behavior and location of fire within the building. Let's look at the attributes and what they are indicating.

Volume

Smoke volume is an indicator of the amount of fuels that are "off-gassing" within a given space. In itself, smoke volume tells very little about the fire. It can, however, establish relativity about the size of the fire event. For example, it doesn't take much of a fire event to fill a small fast-food restaurant with smoke. On the other hand, it would take a serious fire event to produce enough smoke to show even a small amount of smoke leaving a large warehouse or "big-box" store.

Velocity

The "speed" and flow characteristic of smoke that leaves a building is referred to as velocity. In actuality, smoke velocity is an indicator of pressure that has built up within the building. Only two things can create smoke pressure: heat and restricting the volume of smoke within a container (a room or building). If firefighters can define what is pushing the smoke (heat

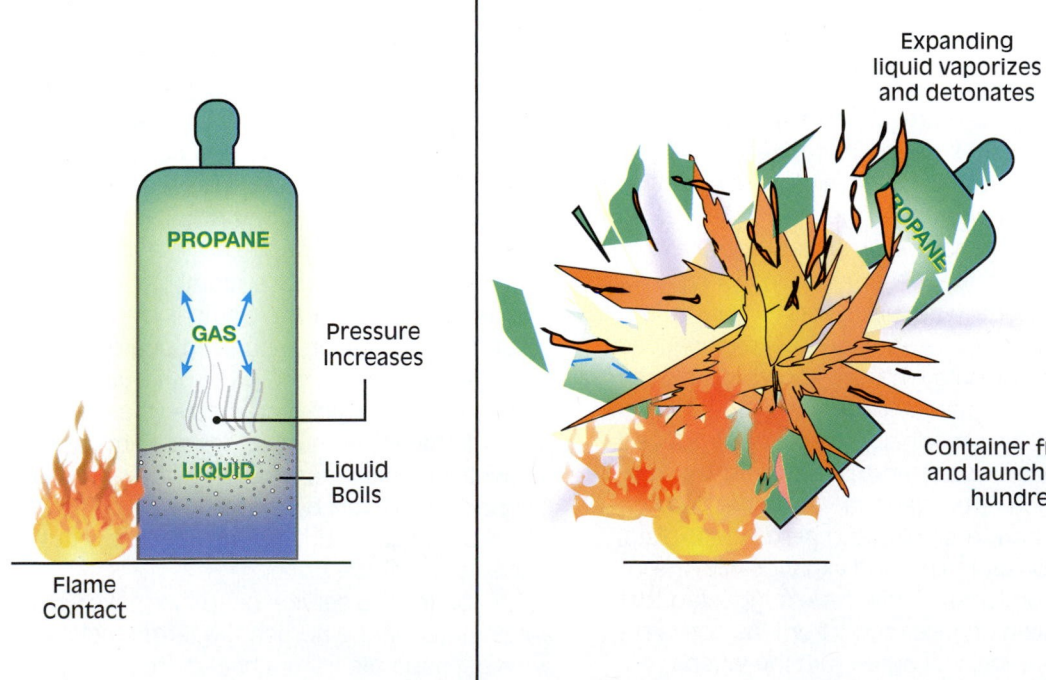

FIGURE 4-23   A boiling liquid vapor explosion (BLEVE).

or volume), they can achieve a better understanding of the conditions within a building. Heat-pushed smoke will leave the building and gradually slow as the smoke rises. Volume-pushed smoke will slow down immediately, and will likely sink upon contact with the outside atmosphere. The flow characteristic of smoke can also help firefighters understand interior conditions. If the flow characteristic of the smoke leaving an opening is agitated or turbulent (some call this boiling or rolling smoke), a flashover is likely to occur, **Figure 4-24.** This turbulent flow is caused by the rapid expansion of the smoke due to heat. In other words, the box cannot absorb any more heat and is unable to hold the rapid expansion. In these cases, the structure must be ventilated and cooled prior to entry. Victims in these conditions have little chance of surviving due to smoke toxicology and thermal exposure. A smooth or calm smoke flow characteristic (called "laminar flow") indicates that the box is still absorbing smoke heat. Comparing the velocity of smoke leaving various openings can help a firefighter locate the fire within the building.

Density

Incomplete burning causes smoke to thicken (become more dense with fuel). Smoke density is indicative of the amount of fuel that is laden within the smoke (solids, aerosols, and gases). The greater the smoke density, the more likely a hostile fire event, such as flashover or rapid fire spread, can occur. In essence, the thicker the smoke, the more spectacular the flashover or fire spread. Thick, black smoke within a compartment reduces the chance of life sustainability due to the toxicology of the smoke. A few breaths of thick,

black smoke will render a victim unconscious and cause death within minutes.

Color

For single-fuel fires, smoke color may indicate the type of material burning. In typical residential and commercial fires, it is rare that a single fuel source is emitting smoke. Smoke color can, however, tell the firefighter which stage of burning is taking place or where the fire is within a building. Virtually all solid materials will emit a white "smoke" when first heated. This white smoke is mostly moisture. As a material dries out and breaks down, the color of the smoke will change. Natural materials like unfinished wood will change to tan or brown, whereas plastics and painted surfaces will turn to gray. Gray is a result of moisture, carbon, and hydrocarbons (black) mixing. All materials will eventually off-gas a black smoke—flame contact will cause materials to give off black smoke right away. As smoke leaves a fuel that is ignited, it heats up other materials and the moisture from those objects can cause black smoke to turn gray. As smoke travels, carbon content from the smoke will deposit along surfaces and objects. This also lightens the smoke color.

By combining these smoke attributes, some basic observations about the fire can be made before firefighters even enter a structure. Smoke velocity and color differences from opening to opening help firefighers find the location of the fire. Faster/darker smoke is closer to the fire seat, whereas slower/lighter smoke is farther away. If smoke from multiple openings is a constant color and velocity, firefighters should start thinking that the fire is deep-seated within the

FIGURE 4-24 Smoke leaving a building has four attributes: volume, velocity, density, and color. *(Photo by Keith Muratori from FIREGROUNDIMAGES.com)*

TABLE 4-3   Reading Smoke Shortcuts

What You See	What It Can Mean
Black with push	Close to the seat of the fire
White smoke with velocity	Heat-pushed smoke that has traveled a distance or has had the carbon filtered (like smoke through a crack)
Same color (white/gray) and same velocity from multiple openings	Deep-seated fire, possibly located well within a building or in combustible voids
Brown smoke	Unfinished wood reaching late heating (can support flame); usually a sign that a contents fire is transitioning into a structural fire; when coming from structural spaces of lightweight wood structures, a warning sign of collapse
Turbulent smoke	Warning sign of impending flashover
Yellowish-gray smoke from cracks or seams	Warning sign of impending backdraft
Thin, black smoke moving fast	Flame-pushed smoke; nearby fire
Smoke moving faster than firefighters can crawl	Warning sign that rapid fire spread is imminent

building. In these cases, the smoke has traveled some distance or has been pressure-forced through closed doors or seams (walls/concealed spaces) prior to leaving the building (the carbon and hydrocarbons get scrubbed or filtered as the smoke travels). **Black fire** is a good term to describe smoke that is high volume, turbulent velocity, ultradense, and black.

CAUTION

Black fire is a sure sign of impending autoignition and flashover.

Other factors influence smoke and may cause attributes to change. Wind, poor thermal balance, fire streams, ventilation openings, and sprinkler systems change the appearance of smoke. With study and practice, the four attributes of smoke—volume, velocity, density, and color—can help firefighters refine their ability to read smoke and ultimately help protect their own safety and predict what fires will do next within a building. **Table 4-3** outlines some street-tested shortcuts to reading smoke. Remember, shortcuts are just that—they are far from absolute. Err on the conservative side when applying them.

LESSONS LEARNED

Firefighter survival and fire attack effectiveness are dependent on firefighters and fire officers understanding fire dynamics. Fire is combustion and combustion is best defined as a chemical reaction that includes the self-sustaining rapid oxidation of a fuel accompanied by the release of heat and light. Four ingredients need to come together to have combustion: heat, fuel, oxygen, and a chemical chain reaction. These ingredients make up the fire tetrahedron. Principles of fire extinguishment are grounded in the process of removing one or more of the ingredients of the tetrahedron.

Heat sources include chemical, mechanical, electrical, and nuclear. Fuels are found in solid, liquid, and gaseous states, although it is the gaseous molecules present in each of these states that burn. Oxygen is a powerful catalyst that must be present for com-

bustion—air typically contains 21 percent oxygen. Concentrations of oxygen below 21 percent will slow the burning process, whereas those concentrations above will accelerate burning. When some type of event brings the aforementioned ingredients together, a chemical chain reaction is started that becomes self-sustaining.

Once a fire begins, it will begin releasing lots of heat. This heat is transferred to other fuels through conduction, convection, and radiation—causing the fire to grow. The fire will grow in defined phases—ignition, growth, and fully developed. After a fire has fully developed, it will slowly slip into a decay phase. Along the way, the fire is releasing many by-products that can make the situation worse. These include heat, light, and smoke and all are dangerous to victims and firefighters.

Fires involving liquids, electricity, and metals require special considerations. Fires in structures can present hostile fire events like flashover, backdraft, and rapid fire spread. Learning how to read smoke (its volume, velocity, density, and color) can help a fire-fighter understand the size and location of fire within buildings as well as predict hostile fire events.

KEY TERMS

Backdraft   The sudden and explosive ignition of pressurized, superheated, and oxygen-deprived gases (within a closed space) caused by the reintroduction of oxygen.

Black Fire   A term used to describe smoke that is high volume, turbulent velocity (hot), ultradense, and black. Black fire is a sure sign of impending autoignition and flashover.

Boiling Point   The temperature where a liquid will convert to a gas at a vapor pressure equal to or greater than atmospheric pressure.

Combustion   A chemical reaction that includes the self-sustaining rapid oxidation of a fuel accompanied by the release of heat and light. Commonly referred to as *fire*.

Deflagration   Combustion rate below the speed of sound (subsonic).

Density   The mass per unit volume of a substance under specified conditions of pressure and temperature.

Detonation   Combustion rate above the speed of sound (super sonic).

Endothermic Reaction   Chemical reactions that absorb heat or require heat to bond atoms or molecules.

Evaporation   A process in which the molecules of a liquid are liberated into the atmosphere at a rate greater than the rate at which the molecules return to the liquid. Ultimately the liquid becomes fully airborne in a gaseous state.

Exothermic Reaction   A chemical reaction that results in heat release.

Fire   (see combustion)

Fire Point   The lowest temperature at which a fuel off-gases an ignitable mixture that, when introduced to a spark or flame, will ignite and sustain burning.

Fire Tetrahedron   Four-sided pyramid-like figure used to depict the four ingredients necessary for combustion: heat, fuel, oxygen, and chemical chain reaction.

Flammable Limits   The concentration level of a substance at which it will burn. Flammable limits are expressed as a percent range mixed in air.

Flashover   A sudden event that occurs when all the contents of a container reach their ignition temperature nearly simultaneously.

Flashpoint   The lowest temperature at which a fuel off-gases an ignitable mixture that, when introduced to a spark or flame, will briefly ignite, but not sustain burning.

Hostile Fire Event   A unique fire phenomenon (such as flashover, backdraft, and rapid fire spread) that takes place within a building.

Ignition Temperature   The lowest temperature that a fuel will off-gas an ignitable mixture that will self ignite and continue to burn (also known as autoignition). May also be referred to as ignition point.

Matter   Something that occupies space and can be perceived by one or more senses; a physical body, a physical substance, or the universe as a whole. Something that has mass and exists as a solid, liquid, or gas.

Molecule   The smallest particle into which an element or a compound can be divided without changing its chemical and physical properties; a group of like or different atoms held together by chemical forces.

Oxidation   Any process in which oxygen combines with an element or substance.

Oxidizer   A substance that readily releases oxygen, and by yielding oxygen, an oxidizer can easily cause or enhance the combustion of other materials. Oxidizers can dramatically increase the rate of burning when the combustible material is ignited.

Pyrolysis   Decomposition or transformation of a compound caused by heat.

Rollover   The intermittent ignition of lower-ignition-temperature gases within the upper thermal layer of a room. Rollover is a warning sign of impending flashover.

Smoke   The products of incomplete combustion or decomposition heating (pyrolysis) that include an aggregate of solids, liquids, and gases suspended in the thermal plume.

Thermal Layering   The stratification of gases produced by fire into layers based on their temperature.

Thermal Plume   A column of heat rising from a heat source. A fully formed plume will resemble a mushroom as the upper level of the heat plume cools, stratifies, and begins to drop outside the rising column.

Vapor Density   The weight of a gas compared to air. The weight of air has been given a value of "1" (at 70°F and one atmosphere or, equivalently, 14.7 psi [21°C and 101 kPa]). Gases that weigh more than air will have a value greater than one and those lighter will have a value less than one.

Vapor Pressure   The amount of force that is pushing vapors from a liquid.

Vaporization   The process in which liquids are converted to a gas or vapor.

REVIEW QUESTIONS

1. What is the definition of combustion?
2. What are the four ingredients of the fire tetrahedron?
3. List the four general methods to extinguish a fire.
4. What are the three states of physical matter and how does the physical state affect fire growth?
5. What are the three modes of heat transfer?
6. List and describe the four fire phases.
7. Which fire phase sets the stage for hostile fire events?
8. How are the terms *thermal layering* and *thermal balance* different?
9. Which of the products of combustion pose the greatest threat to firefighters? Why?
10. List three hostile fire events. What are the warning signs that accompany each?
11. What are four attributes that help firefighters "read smoke"?
12. What is meant by "black fire" and what does it mean to firefighters?

ENDNOTES

1. Research reveals that there are literally dozens of differing definitions for the word *combustion*. The one presented in this text is a simplified hybrid developed by the author to address the audience of recruit firefighters.
2. Temperature figures were averaged from several test fire data streams cited by the National Institute of Standards and Technology (NIST) and the Bureau of Alcohol, Tobacco, and Firearms (ATF).
3. The exact human life threshold temperatures differ among texts. The temperature given here is a conservative average among the texts.
4. Text and website research reveals that the debate on defining fire phenomena continues. For example, one text defines eleven types of flashover. The material presented in this chapter represents a simplified way to look at hostile events and predict them proactively.

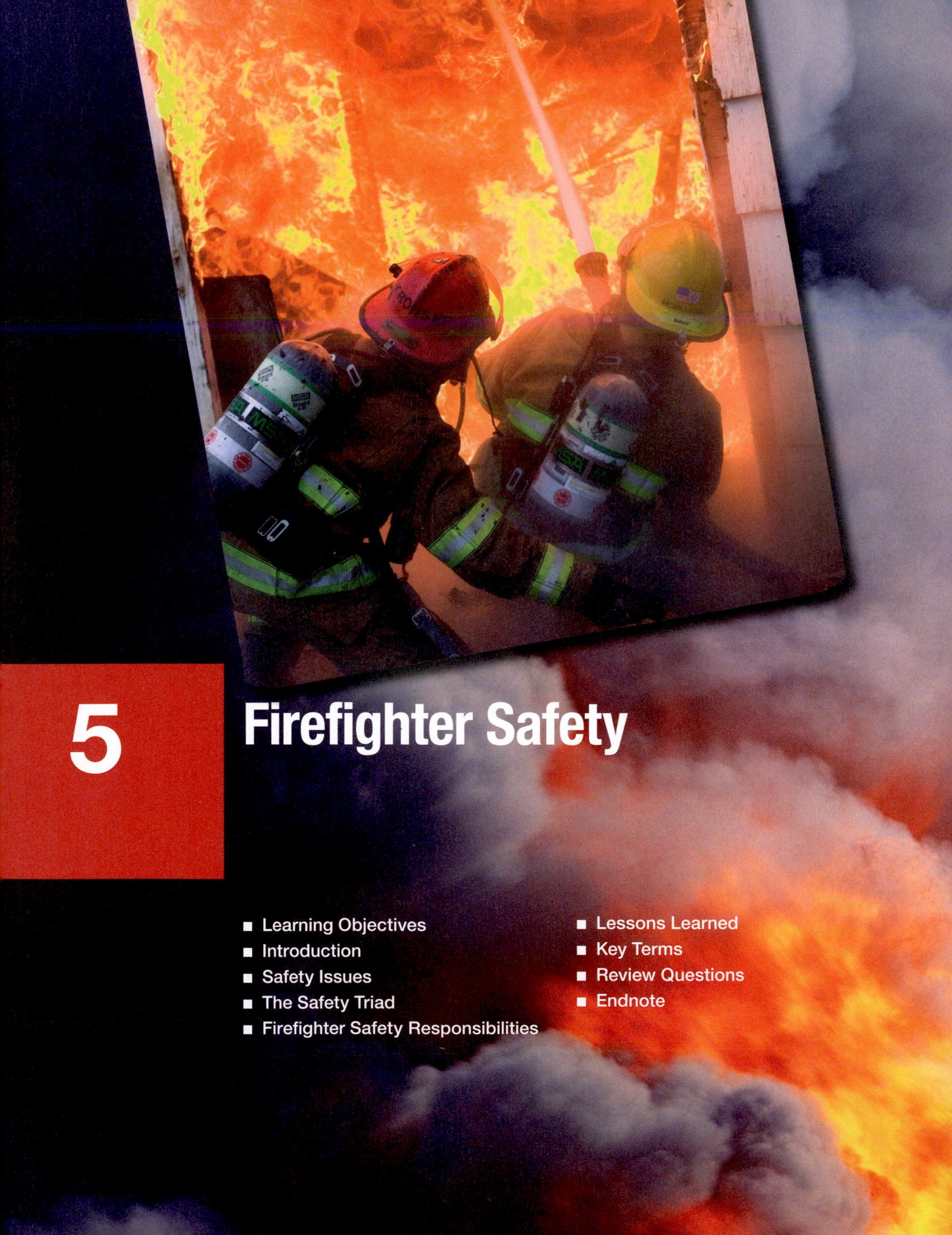

5

Firefighter Safety

I was at work one day when the phone rang. The caller stated that my son Scott, who happens to be both a career and volunteer firefighter, was injured at a fire on his career job. It is the phone call a parent always dreads. I immediately had a rush of adrenaline, an increase in respiration, and became flushed. I asked, "How bad?" The caller informed me that luckily the injuries appeared to be minor.

Of course I made a phone call or two to verify the information and to explain the situation to my wife. I was informed that Scott had some burns on his shoulders and hand, where there would be some scarring, and that he would miss a shift or two, but he would be O.K. "O.K.?" I guess that term takes on a new definition to a father who is 45 miles away and worried!

I left work early to coincide with Scott's arrival at home. He had large bandages around his shoulders and his hand, and he was beginning to feel some pain. We talked about the fire, and he told me it was a typical garden apartment-type fire and that his company had been assigned on automatic mutual aid. He and his lieutenant were in search mode because residents reported an eleven-year-old was trapped. Their backup line was right behind them, but it was not quite ready and they were; seconds became hours!

During their search, crawling on hands and knees to stay below the heat and smoke as they had learned, Scott felt something very hot on the back of his right hand, a sharp pain. He turned to his lieutenant and both of them shouted at the same time, "Let's get out of here now. It is too hot!" They exited quickly.

They were "greeted" at the door, and their coats were smoking and discolored around the shoulders. The firefighter at the door began patting the smoking gear, stating it was "on fire." The air barrier was now gone, and both the lieutenant and Scott felt the heat, big time! They both ended up with second-degree burns on both shoulders, and the hand burn resulted from the inside Nylon label on the gloves melting, even though they were compliant gloves!

The good news is that the properly designed and worn personal protective clothing prevented serious injuries, maybe even third-degree, debilitating burns. The coats, pants, and gloves could be replaced, but the scar tissue is still there. Scott received a very nice leather jacket from the personal protective equipment (PPE) outer shell material manufacturer, and the damaged PPE pictures can be seen at major fire service conferences across the country.

As Scott's parents, it was a difficult day for both of us, and emotions were running high. I think sometimes we forget how our choice to be a member of the fire service can have such an impact on our family members. His mother was very concerned about the inevitable scarring and strongly suggested that he really needed to quit doing that! Scott looked her in the eye and with deep conviction in his voice stated, "Mom, it's my job." Kind of made me shiver. But I know one thing. Scott and his lieutenant learned a lot that day—about safety, about PPE, and that on the fireground you never know what might come next. As he shares what he learned with other firefighters, foremost on his mind is what he might teach to his son, Cody, the next generation of Windisch firefighters.

—Street Story by F. C. (Fred) Windisch,
Fire Chief, Ponderosa VFD, Houston, Texas

INTRODUCTION

Collectively, the fire service knows what injures and kills firefighters. We know what the dangers are; we know what the injury and death data suggests; and we've created many standards, technological advances, and initiatives to help improve firefighter safety. Simply put, the firefighting profession is one of significant risk—that is, one filled with the potential to be seriously injured or killed. Typically, this potential for death has created the aura that firefighters are hero types. Unfortunately, the "hero badge" has led to several unnecessary injuries and deaths—which could be construed as recklessness or arbitrary aggressiveness. Today's firefighter understands that certain risks have no tangible benefit. In simple terms, this understanding is called risk/benefit-thinking, which is a form of risk management. Specifically, **risk** is defined as the chance of injury, damage, or loss. **Risk management** is the process of minimiz-ing the chance, degree, or probability of damage, loss, or injury. Firefighters practice risk management when they ask questions of risk/benefit: Why risk a team of firefighters for a building that is basically lost? Why cross over a collapse-zone barrier tape just to get a better angle for a fire stream?

Firefighter safety is grounded in understanding risks and risk/benefit thinking—as a starting point. Firefighter safety is so much more, however. "Being safe" is a state of mind, an attitude, and a way of life that needs to permeate each and every facet of our fire service involvement. Each chapter of this handbook includes specific information to help make you "more safe." In this chapter, we want to focus on some big-picture firefighter safety issues and show how individual firefighters fit into an entire firefighter safety system. With this knowledge, the firefighter can fill the "hero" role with intelligent aggressiveness and wisdom rather than recklessness.

SAFETY ISSUES

One way to keep firefighter injury and death events to a minimum is to understand what events and circumstances typically lead to injury and/or death.

A study of injury and death trends has inspired fire and safety professionals to create standards and procedures as well as many safety initiatives to help prevent injuries and deaths. These efforts directly affect some of the training and tactics the fire service employs today. The firefighter who is keenly aware of safety issues and practices good injury-prevention habits is actually helping the fire service address safety issues.

FIREFIGHTER FACT

Firefighter injury and death information is invaluable and has, in many notable cases, prevented additional injuries. For example, the Hackensack (NJ) Fire Department lost five firefighters after a roof collapsed on interior firefighters battling an attic fire in an automobile service shop. This shop had a bowstring-trussed roof. Following the incident and subsequent investigations, recommendations were made by various fire service organizations to help prevent a similar incident. In New Jersey, efforts succeeded in developing a building placard system to warn firefighters that a given building is built with a truss roof system.

One Oregon chief publicly thanked the fire service organizations for the New Jersey recommendations—he experienced a similar situation and withdrew his firefighters ten minutes before the roof collapsed.

Firefighter Injury and Death Trends

A study of firefighter injury and death statistics reveals that approximately 70 percent of all duty deaths and injuries occur during emergency activities (incident response and on-scene operations). The balance is split between training activities and other duties. This helps explain where firefighters get hurt and killed. The next question is what causes the injuries and deaths. **Figure 5-1** shows the leading causes of death, and **Figure 5-2** shows typical leading causes of injuries.

Firefighter death statistics also track the nature of the injury that caused the death. These statistics have shown that heart attacks (as a result of overexertion) are the leading type of death-producing injury. Internal trauma, crushing injuries, and asphyxiation follow heart attacks. Data suggest that firefighter duty deaths are not noticeably decreasing—hovering around 100 duty-related deaths each year in the United States. Although advances in equipment, training, and uniform procedures over the past decade have certainly prevented deaths and injuries, we still haven't made a notable reduction in the number of each.

According to the U.S. Fire Administration (USFA), firefighter fatalities as a result of fire-related causes (burns, asphyxiation, and structural collapse) have actually increased.[1] This trend suggests that even more emphasis needs to be placed on predicting fire behavior and structural collapse during actual fires. Additionally, there appears to be an increasing trend in the number of deaths associated with responding to and returning from

FIGURE 5-1 Firefighter fatalities by cause—2006 sample. *(Source: Firefighter Fatalities in the U.S.—2006, Rita F. Fahy, Paul R. LeBlanc, and Joseph L. Molis, June 2007. Reprinted with permission from NFPA 5000-2006, Building Construction and Safety Code, Copyright © 2006, National Fire Protection, Quincy, MA. This reprinted material is not the complete and official position of the NFPA on the referenced subject, which is represented only by the standard in its entirety.)*

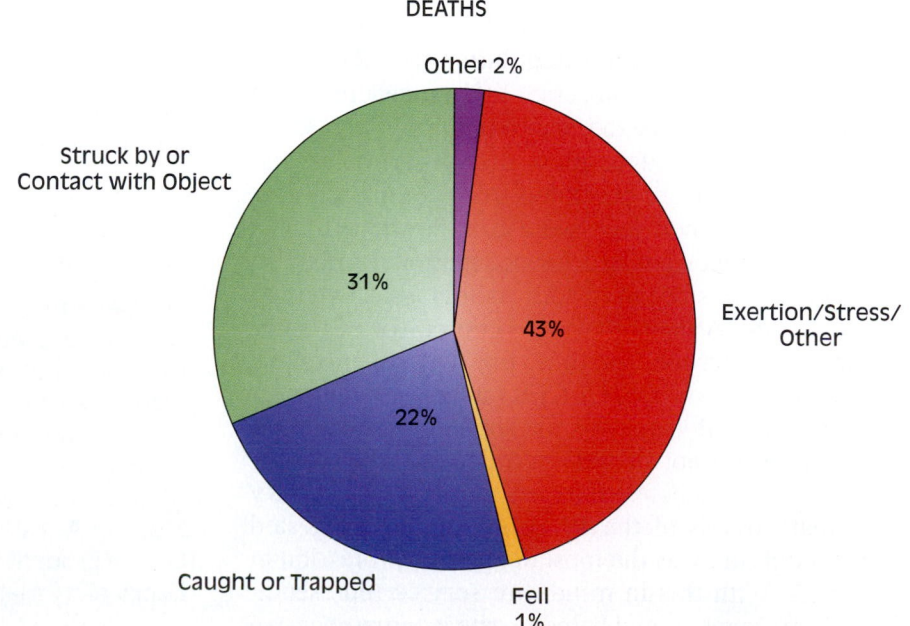

DEATHS

Other 2%

Struck by or Contact with Object   31%

43%   Exertion/Stress/Other

22%

Caught or Trapped

Fell 1%

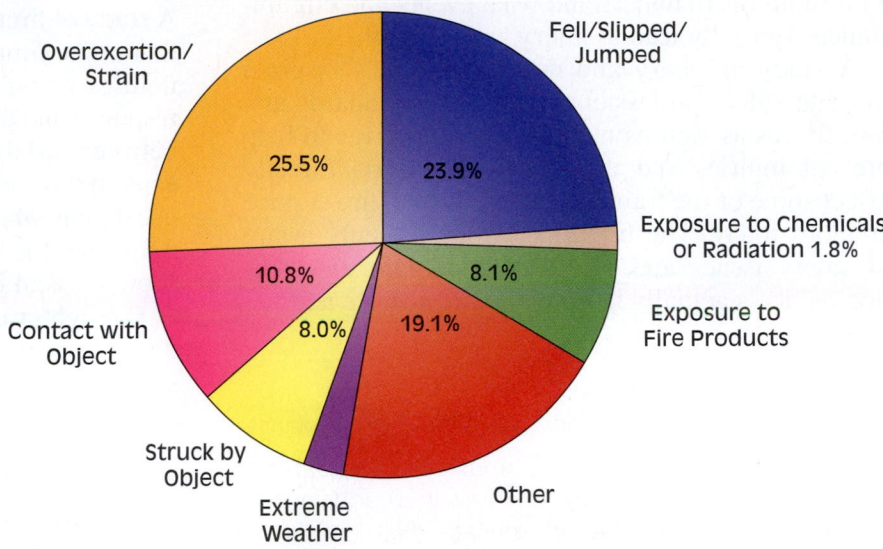

INJURIES

Fell/Slipped/Jumped 23.9%

Overexertion/Strain 25.5%

Exposure to Chemicals or Radiation 1.8%

Exposure to Fire Products 8.1%

Other 19.1%

Extreme Weather 2.8%

Struck by Object 8.0%

Contact with Object 10.8%

FIGURE 5-2 Firefighter injuries by cause—2006 sample. *(Source: U.S. Firefighter Injuries—2006, Michael J. Karter, Jr., November 2007. Reprinted with permission from NFPA 5000-2006, Building Construction and Safety Code, Copyright © 2006, National Fire Protection, Quincy, MA. This reprinted material is not the complete and official position of the NFPA on the referenced subject, which is represented only by the standard in its entirety.)*

incidents. Disturbingly, the failure to use seat belts has been cited as a major contributing factor in these deaths.

Understanding historical data and trends helps us understand why certain safety standards and regulations were developed to help reduce deaths and injuries, and reverse disturbing trends.

Safety Standards and Regulations

In 1970, the William Stieger Act was passed by Congress and signed into law by President Nixon. This act created the **Occupational Safety and Health Administration (OSHA),** which is part of the Department of Labor. OSHA is responsible for the enforcement of safety-related regulations in the workplace. These regulations are part of the **Code of Federal Regulations (CFR).** OSHA CFRs outline specific processes and procedures to be followed to help reduce the chance of injury.

Interestingly enough, most fire departments and public agencies were exempted from complying with OSHA CFRs when they were first developed. This created a double standard for occupational safety. In the 1980s, states began to address this issue by creating state OSHA plans that were compulsory for public agencies. For the first time, fire departments were obligated to follow OSHA CFRs adopted by an individual state. Unfortunately, the spontaneous nature of firefighting did not fit well into defined OSHA procedures and processes. Further, the firefighter injury and death trends of the 1970s and 1980s suggested that firefighting was the most dangerous profession in America. With this in mind, fire service representatives came together and helped write a comprehensive

occupational health and safety standard for the fire service that is now known as NFPA 1500.

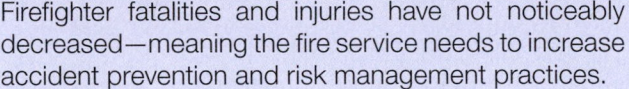

NOTE

Firefighter fatalities and injuries have not noticeably decreased—meaning the fire service needs to increase accident prevention and risk management practices.

NFPA 1500, Standard on Fire Department Occupational Safety and Health Program, outlines consensus requirements and procedures for the safety of fire service personnel—much like OSHA CFRs do for industry. The accompanying feature box, NFPA 1500 Topics, shows the topical areas addressed by NFPA 1500. As you can see, the standard was written to help fire departments address a whole host of safety issues. The NFPA 1500 standard (and most NFPA standards for that matter) doesn't have the enforcement power of an OSHA CFR unless a local jurisdiction adopts the standard as a law.

NFPA 1500 TOPICS

Fire Department Administration
Training, Education, and Professional Development
Fire Apparatus, Equipment, and Driver/Operators
Protective Clothing and Protective Equipment
Emergency Operations
Facility Safety
Medical and Physical Requirements
Member Assistance and Wellness Programs
Critical Incident Stress Program
Explanatory Material

NOTE

Currently, OSHA looks at NFPA standards as a guideline to address issues not directly covered by CFRs.

With OSHA looking at NFPA as a guideline, a quasi OSHA/NFPA alliance is formed that furthers the importance and accountability placed on firefighter safety. A classic example of this alliance is the requirement and clarification given to safe initial fire attack operations. The two-in/two-out rule is a concept underscored in OSHA 29 CFR 1910.134 and NFPA 1500. In essence, the rule states that before firefighters engage in interior fire attack operations, they must have two equipped and ready firefighters who form a team to make that attack. Additionally, two equipped and ready firefighters must be outside to monitor the progress of the interior crew and serve as an initial rapid intervention team should the interior firefighters experience an emergency. It is also important to note the outside team will likely need to perform some support activities (incident command or pump operations). The persons performing these tasks must be immediately available to rescue the interior team. These two are not considered a dedicated rapid intervention team—as more resources arrive, a dedicated rapid intervention team is required. This concept is covered more specifically in Chapters 19 and 23 (in Sections I and II).

NFPA has also created other standards that address safety. It can be argued that all NFPA standards address safety, but specifically, NFPA has created a 1500 series that focuses on safety issues (see the feature box, "NFPA Safety-Related Standards").

NFPA SAFETY-RELATED STANDARDS

1500	F.D. Occupational Safety and Health Program
1521	F.F. Safety Officer
1561	Emergency Services Incident Command System
1581	F.D. Infection Control Program
1582	Comprehensive Occupational Medical Program for F.D.s
1583	Health-Related Fitness Programs for Firefighters
1584	Rehabilitation Process for Members During Emergency Operations and Training Exercises*

*Currently a "recommended practice" but may become a standard in the future.

The **National Institute of Occupational Safety and Health (NIOSH)** has also entered the firefighter safety equation. NIOSH does not set regulations or standards—they write recommendations based on the investigation of firefighter fatalities. In the 1990s, President Clinton signed a directive that mandated that NIOSH investigate incident-related firefighter fatalities. Congress continues to fund the directive. NIOSH is tasked with preparing recommendations so that other fire departments can take preventive steps to reduce the chance of an injury or fatality. As it relates to standards and regulations, the results and recommendations of the NIOSH investigations are used to help update and write new standards and regulations to help reverse injury and death trends.

NOTE

NIOSH investigative reports should be mandatory reading. They can be found at http://www.cdc.gov/niosh/fire.

NIOSH *Alerts* are developed to address disturbing trends. The Alert is a publication that summarizes an injury or death trend and offers recommendations to help reduce the reoccurrence of similar events, as shown in **Figure 5-3.** Standards and regulations are a starting place to help reduce firefighter injuries and deaths. Additionally, fire service organizations have developed several initiatives to help reverse trends and prevent further injuries and deaths.

FIGURE 5-3 An example of a NIOSH *Alert* issued to address disturbing trends in firefighter injuries and deaths.

SIXTEEN FIREFIGHTER LIFE SAFETY INITIATIVES

1. Define and advocate the need for a cultural change within the fire service relating to safety; incorporating leadership, management, supervision, accountability, and personal responsibility.
2. Enhance the personal and organizational accountability for health and safety throughout the fire service.
3. Focus greater attention on the integration of risk management with incident management at all levels, including strategic, tactical, and planning responsibilities.
4. All firefighters must be empowered to stop unsafe practices.
5. Develop and implement national standards for training, qualifications, and certification (including regular recertification) that are equally applicable to all firefighters based on the duties they are expected to perform.
6. Develop and implement national medical and physical fitness standards that are equally applicable to all firefighters, based on the duties they are expected to perform.
7. Create a national research agenda and data collection system that relates to the initiatives.
8. Utilize available technology wherever it can produce higher levels of health and safety.
9. Thoroughly investigate all firefighter fatalities, injuries, and near misses.
10. Grant programs should support the implementation of safe practices and/or mandate safe practices as an eligibility requirement.
11. National standards for emergency response policies and procedures should be developed and championed.
12. National protocols for response to violent incidents should be developed and championed.
13. Firefighters and their families must have access to counseling and psychological support.
14. Public education must receive more resources and be championed as a critical fire and life safety program.
15. Advocacy must be strengthened for the enforcement of codes and the installation of home fire sprinklers.
16. Safety must be a primary consideration in the design of apparatus and equipment.

Firefighter Safety Initiatives

In 2004, the USFA and the National Fallen Firefighters Foundation (NFFF) hosted a summit to discuss firefighter death and injury trends. From that meeting, sixteen Firefighter Life Safety Initiatives were developed to help reach a stated goal of reducing firefighter fatalities by 50 percent in ten years (see the feature box, "Sixteen Firefighter Life Safety Initiatives").

Also in 2004, the International Association of Fire Chiefs received several grants to help develop a National Firefighter Near Miss Reporting System. The program was developed to help identify near-miss patterns that would be used to help develop strategies for reducing injuries and fatalities. Firefighters are encouraged to report their close call using an online system that takes about ten to fifteen minutes. Once submitted, fire service professionals analyze the report and then de-identify names before posting the report on a Web site for others to learn from. Annually, the data collected from the reports are analyzed, and reports and recommendations are offered. The system and reports can be found at http://www.firefighternearmiss.com.

In 2005, several national fire service organizations agreed that a collective "time-out" was needed following another year of disturbing firefighter injuries and deaths. The time-out took the form of a "stand-down" safety day. A stand-down refers to an event where routine activities are suspended so that focus can be drawn to a specific need—in this case, preventing firefighter deaths and injuries. The stand-down has become an annual event (usually several days in June) that incorporates activities that honor fallen firefighters, draws attention to a specific trend (for example, apparatus collisions), and offers specific training to help prevent further deaths.

In 2007, the Firefighter Life Safety Initiatives were revisited at another summit meeting. One theme that emerged from the second summit was the role of firefighter attitudes and discipline to "act safe." While fire service efforts will continue to address death and injury trends by improving equipment, communications, training, and procedures, in the end, it comes down to individual firefighters that embrace a commitment to reduce deaths and injuries through their preventive actions. Following the 2007 summit, the NFFF received sponsorship to launch a Whistle Stop tour of the United States to raise awareness of the Firefighter Life Safety Initiatives. Highlights of this tour can be found at http://www.everyonegoeshome.org.

It is significant to note that these efforts have not reduced the number of firefighter fatalities—indicating that time and more focused preventive energy are necessary to evaluate these initiatives.

Preventive Actions

The whole goal of exploring the preceding safety issues (trends, standards and regulations, and safety initiatives) is to reduce the potential for further firefighter injury and deaths. Once this understanding is in place, preventive actions need to be developed. Preventive actions are developed by addressing the "safety triad" (defined in the next section) and assigning action responsibilities at various organization levels. This combination creates the firefighter safety system.

Before we discuss the triad and responsibilities, it is important to note that *all firefighters should be empowered to stop unsafe actions—intervene—without fear of retribution or embarrassment*. Likewise, those who have their actions stopped should view the intervention as a positive—someone perceived that their actions were unsafe. A "thank you" is warranted.

(A)

(B)

THE SAFETY TRIAD

Most fire service operational environments are made up of three key components: procedures, equipment, and personnel. To mitigate injuries, each of these components needs to be addressed. This mitigation effort is called the safety triad, **Figure 5-4.**

Procedures

As stated, the three basic components or elements of the safety triad come together to form an operational environment like the one found at working structure fires or multicar extrications. If the operation is to be "safe," each component needs to be addressed.

The term *procedure* is used in a very generic form here to describe all sorts of formal (written) and informal processes that are in place in a fire department. Procedures or processes make up the structure for all the activity at an incident. The first-arriving engine at a fire alarm activation is the most likely to start a series of processes to investigate the alarm.

Formal processes are in writing and take on many forms: standard operating procedures (SOPs), standard operating guidelines (SOGs), departmental directives, temporary memorandums, and the like. In some departments, formal processes are derived from standard evolutions (such as a fire attack evolution) or lesson plans. These evolutions and lessons can be drilled periodically, on a rotating basis, to ensure that a crew's response to a given situation is appropriate. Some departments adopt training manuals as their operating standard. When a manual gives choices, for instance a hose load, the chosen load can just be circled in the manual. No matter what the source, the key component is that formal processes

(C)

FIGURE 5-4 The safety triad includes (A) procedures, (B) equipment *(Photo courtesy of Richard W. Davis)*, and (C) personnel.

and evolutions are in writing. In taking this approach, the department achieves consistency and preventive actions in its operations.

Some fire departments recognize both procedures and guidelines. In this context, procedures are strict processes with little or no flexibility, and guidelines

1. 1.1   Incident Command System
2. Emergency Ground Operations
   2.1   Rapid action group
   2.2   Gas/odor investigation
   2.3   Auto alarms
   2.4   Train fires
   2.5   Vehicle fire
   2.6   Fires at postal facilities
   2.7   Emergent driving procedure
   2.8   Kaneb pipeline response
   2.9   Volunteer and fire apparatus placement for motor vehicle accidents (MVAs)
   2.10   Operations involving Thompson Valley Ambulance
   2.11   Minimal staffing for interior firefighting
   2.12   Foreground formation and activation of companies
   2.13   Standard fire attack procedures/dwelling fires
3. Alarm Levels/Dispatching
   3.1   City alarm level assignments
   3.2   Rural alarm level assignments
   3.3   Fire resource officer
   3.4   Fire alarm panel operation and response policy
   3.5   Mutual/automatic aid agreement
   3.6   Staffing considerations during adverse weather conditions
   3.7   Cancellation procedures for emergency medical service (EMS) and MVA incidents
4. Hazardous Materials
   4.1   Hazardous materials operations
5. Emergency Medical Services
   5.1   Duties for non-EMS-certified personnel
6. Aircraft Rescue and Firefighting (ARFF)
   6.1   ARFF standby policy
7. Technical Rescue and Special Operations
   7.1   Vehicle extrication
   7.2   Rope
   7.3   Trench
   7.4   Collapse rescue
   7.5   Confined space
   7.6   Farm equipment and industrial rescue
   7.7   Loveland dive rescue standard operating procedures
   7.8   Use of Civil Air Patrol

FIGURE 5-5   Sample SOP index.

are adaptable templates that give wide application flexibility. **Figure 5-5** shows a typical SOP index and **Figure 5-6** shows a typical SOP format.

Informal procedures or processes are those that are obviously part of a department's routine but are not written. Informal processes are passed on through new member training as well as in day-to-day operations. An example of an informal process is the practice of having firefighters place a full-face hood across their bunker boots so that they are reminded to don the hood before placing their feet in the boots. Another example is the apparatus operator's use of a grease pencil to make status marks on pump panel gauges as a quick reference; at a glance the operator can see if changes have taken place.

SAFETY

Both formal and informal procedures can be responsible for the overall safety of a department.

Equipment

In the past few years, the fire service has seen the release of vast amounts of new equipment designed uniquely for improved "safety." Equipment helps make an operation more safe but is arguably the least important factor in the safety triad of procedures, equipment, and personnel.

Each of these is important as it relates to safety.

SAFETY

Making equipment safe is addressed in three ways: equipment selection, equipment inspection and maintenance, and equipment application.

Equipment Selection

Most critical equipment used in firefighting (PPE, ladders, hose, etc.) is designed and built to meet NFPA standards. NFPA standards are written with safety

Purpose:
To establish policy and direction to all department members regarding minimal staffing and resource allocation for safe and aggressive interior structural firefighting.

Responsibility:
It is the responsibility of all officers and firefighters engaged in firefighting operations to adhere to this policy. The Incident Commander is accountable for procedure included within this policy.

Procedure:
1. This policy is applicable to situations where the Incident Commander (IC) has made a tactical decision to initiate an *offensive fire attack*, by firefighters, inside the structure. Additionally, tactical firefighting assignments that expose firefighters to an atmosphere that is immediately dangerous to life and health (IDLH) dictate the application of this policy.[1]

2. Prior to initiating interior fire attack or exposure of firefighters to an IDLH atmosphere, a *minimum of four (4) firefighters shall assemble on scene*.[2] These four members shall utilize a "two-in, two-out" concept.

3. The *"two-in"* firefighters that enter the IDLH atmosphere shall remain as partners in close proximity to each other, generally fulfilling the operational role as the *FIRE ATTACK GROUP*. As a minimum, the *"two-in"* firefighters entering the IDLH atmosphere shall have full PPE, with SCBA and PASS devices engaged, and have between them a two-way portable radio, forcible entry tool, and flashlight or lantern.

[1] An *IDLH atmosphere* can be defined as an atmosphere that would cause immediate health risks to a person who did not have *Personal Protective Equipment (PPE)* and/or *Self-Contained Breathing Apparatus (SCBA)*. This includes smoke, fire gases, oxygen deficient atmospheres, or hazardous materials environments. For Loveland Fire and Rescue application, an IDLH atmosphere can be further defined as an environment that is *suspected* to be IDLH, has been *confirmed* to be IDLH, or *may rapidly become* IDLH. The use of full protective equipment including an activated SCBA and an armed PASS device is mandatory for anyone working in or near an IDLH atmosphere.

[2] The firefighters must be SCBA qualified and capable of operating inside fire buildings without immediate supervision.

FIGURE 5-6 Sample SOP format.

paramount to the topic. The fire department that purchases equipment that meets NFPA standards is buying equipment that has a certain safety element built into it.

Equipment Inspection and Maintenance

For equipment to be safe, it must be inspected and maintained. Firefighters spend many hours each week ensuring that their equipment is clean, properly functional, and ready for rough service during critical situations, **Figure 5-7.**

SAFETY

Firefighters must always prepare for the worst—including equipment failure.

Because many different firefighters may use and maintain a given piece of equipment, complete documentation of repairs and maintenance is essential. Further, a complete set of guidelines is often developed or adopted for essential equipment. These guidelines include:

- Selection
- Use
- Cleaning and decontamination
- Storage
- Inspection/documentation
- Repairs
- Criteria for retirement

The use, inspection, and maintenance of equipment must be done in accordance with instructions developed by the equipment manufacturer along with

FIGURE 5-7   Ensuring that equipment is clean, functional, and well maintained is essential to firefighter safety.

any policy and procedure developed by the local fire department. When caring for equipment, individual firefighters must initiate the appropriate documentation in accordance with department policy.

Equipment Application

Choosing the right tool for a given job is paramount for safety. It is imperative to use equipment in the manner for which it was designed. Using an ax as a prying tool can cause the handle to shatter. At times, the ingenuity and inventiveness of firefighters facing an unusual situation places equipment in a position for which it may not be intended. Extra care and insight are required here.

Personnel

Human factors are often cited as the cause of injuries and deaths. To mitigate these injuries and deaths, the safety triad must address personnel issues. A firefighter's training, fitness/health, and attitude all factor into the safety equation.

Training

Unfortunately, many firefighters become comfortable in their knowledge or position and let the basics slip away unknown. Don't fool yourself—knowledge and skill retention is lost if you don't use it on a frequent basis. Each and every skill, knowledge item, or behavior that is learned in basic academy should be drilled, tested, or confirmed every year. In a perfect example, the New York City Fire Department (FDNY) launched a huge "back-to-the basics" program after a series of firefighter duty deaths within a year. The FDNY line leadership felt that many of the deaths could have been avoided had firefighters and fire officers practiced the "basics." Many departments utilize company drills in order to ensure that the "basics" are practiced, **Figure 5-8.**

Because regular training is the single most important step in firefighter safety, the firefighter must strive to retain the information and skills that are presented in training sessions. The accompanying feature box, Strategies to Help Retain Skills and Knowledge, outlines some strategies for firefighters to help retain essential skills and information and thereby improve safety:

NOTE

Without a doubt, the single most important step that can be taken to reduce firefighter injuries is that of regular training. It is widely accepted that once a person enters the fire service, the training never ends.

Firefighter Fitness and Health

The safety and well-being of any firefighter increases with the health of the individual firefighter. Much has been written on the benefits of healthy firefighters—the most of which centers on physical health.

FIGURE 5-8   Basic skills must be practiced on a regular basis.

STRATEGIES TO HELP RETAIN SKILLS AND KNOWLEDGE

- *Train at every opportunity.* Test yourself to see what you recall and how well you perform—your life depends on it.
- *Envision the application of training.* Following a training session, visualize its application at an incident. This can be done internally or, preferably, with peers.
- *Create a personal reference library.* Always take notes and keep handouts—then file them in a system that you maintain throughout your fire service tenure. The process of taking notes, reading handouts, and filing them helps imprint information and can serve as a great future recall tool.
- *Acknowledge that skills will diminish over time if they are not used.* An old fire service adage says firefighters need to spend 95 percent of their training time practicing for what they do 5 percent of the time. If it has been some time since you've tied knots—go tie knots. In basic fire academies, numerous skills are learned in a short time period. You may be able to retain the skills to pass your academy finals, but try the same skills a year or two later and you'll be amazed at how many diminish without practice.

To handle the inherent stress of firefighting, each firefighter's body must be accustomed to and capable of handling stress. Additionally, firefighters need to protect themselves from, and prevent the spread of, communicable diseases and infections. Keys to improving physical health, and therefore firefighter safety, include these:

- *Annual health screening for all firefighters and line officers.* These physicals should also include vaccination and immunization offerings, blood screening, and stress testing.

FIGURE 5-9 Work hardening must become a habit—leading to increased safety and professional tenure.

CAUTION

"Overexertion/Stress" continues to lead the list of causes of firefighter duty deaths and contributes significantly to injuries.

- *Work hardening and mandatory ongoing fitness programs.* A personal devotion to physical fitness is essential to firefighter safety. Fitness includes improving flexibility, strength, and cardiovascular endurance. **Work hardening** is the effort and physical training put forth to better perform physical tasks without overstressing or injuring an individual. The firefighter who addresses these fitness areas will breeze through any "fit-for-duty" agility tests that a fire department may require as a condition of employment or association. Improper lifting techniques and slip and fall accidents are two of the most common causes of injury on duty. Work hardening helps prevent these injuries through strength, balance, and coordination improvement, **Figure 5-9.**
- *Firefighter fueling (nutrition) education.* Nutrition is important not only to lifestyle changes (like losing excessive fat) but also to incident readiness. This is especially true with the sporadic nature of the firefighter's energy demands. See Section I, Chapter 23, which more clearly defines appropriate fueling of the firefighter.

CAUTION

Some diet fads and meal-replacement aids may help an individual lose weight but often interfere with maintaining energy during working incidents.

Signs and Symptoms of Critical Incident Stress

Numbing and Withdrawal
Reexperiencing the Event
Flashbacks
Depression
Sleep Difficulties
Substance Abuse
Guilt
Family Problems
Low Job Efficiency
Fear of Next Incident
Anger and Hostility

FIGURE 5-10 Critical incident stress is inevitable and unpreventable. CISM can be addressed through recognition (know the signs and symptoms), peer support, and debriefings.

Attention to physical health is indeed important, but mental health is also important to firefighter safety. Keys to addressing mental health include the following:

- *Training and understanding of* **critical incident stress management (CISM).** Knowing the signs and symptoms of critical incident stress is important, **Figure 5-10.** Most departments have developed a process to have a **critical incident stress debriefing (CISD)** team activated or available for unusual events or at the request of one or more responders.

- *Utilizing available* **member assistance programs (MAP).** A MAP offers individuals a confidential approach to dealing with situations and problems that can affect job performance. Drug and alcohol dependence, depression, worker relationships, and job stress are a few of the issues that can be addressed through a MAP. These types of programs may also be referred to by local agencies as "employee assistance programs" or "EAP."

Attitude

Of all the "people" factors affecting safety, attitude is the hardest to address. Many factors affect the attitude of a given individual, not to mention the fact that attitudes are dynamic. Here are some factors that affect safety attitudes:

- *The fire department's safety "culture."* The culture of an organization is reflected in the ideas, skills, and customs that are passed through generations. If safety has not been given a high priority by administrators and members, a poor safety attitude will be reflected in daily activities. The inverse is also true.

- *The fire department's history.* A department that has experienced a firefighter duty-related death or serious injury is more likely than not to have increased safety awareness. Likewise, some departments have learned from others' experiences—this is known as having a **vicarious experience.** Realizing that a potential for injury or death exists is an important first step in developing a proactive safety attitude.

- *The example set by others.* Firefighters and line officers display their safety attitudes in what they do rather than what they say. These examples will shape the attitudes of others, **Figure 5-11.**

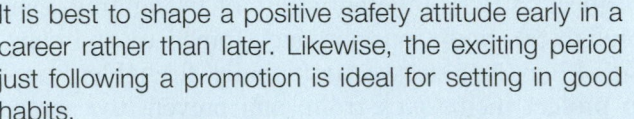

ETHICS

It is best to shape a positive safety attitude early in a career rather than later. Likewise, the exciting period just following a promotion is ideal for setting in good habits.

Previously we presented attitudes and actions to help reduce firefighter injuries and deaths. The fol-

FIGURE 5-11 Developing a positive safety attitude and practicing safe habits will demonstrate safe examples to others.

lowing are further steps a firefighter can take to create a positive safety attitude:

- *Practice good habits.* Each and every incident response and training activity results in opportunities to practice and reinforce good habits. PPE use, readiness, team formation, and assignment follow-through are all items that lead to a more positive safety attitude. These habits will serve the firefighter well when the incident becomes more complex or dangerous.

- *Learn from others.* Each month, trade magazines and investigative reports outline circumstances that have led to a firefighter injury or death. These serve as a constant reminder that an injury can happen. Firefighters can learn these lessons vicariously. Significant fatality and injury reports can be found at the NIOSH Web site, http://www.cdc.gov/niosh/fire.

- *Be vigilant.* Inactivity, complacency, and overconfidence have been cited as primary factors in many firefighter fatalities. Often, merely visualizing an injury will help firefighters regain focus and obtain vigilance in checking their safety attitude.

FIREFIGHTER SAFETY RESPONSIBILITIES

Firefighter safety is dependent on the efforts of everyone. Specifically, responsibility for firefighter safety rests in one of three areas: the department itself (administration), the working team, and the individual firefighter, **Figure 5-12.** Knowing these responsibilities—especially where individuals fit—will help achieve firefighter safety. Firefighter safety is achieved if, and only if, all "partners" hold up their end of the safety partnership.

The Department

The responsibility for firefighter safety ultimately rests with the department's leadership. This is an incredible weight to carry. Fire chiefs must create and enforce rules, procedures, and expectations in order to address the issue of firefighter safety. Additionally, a department's leadership must obtain budget appropriations to purchase equipment and apparatus that will help make firefighting safer. Often, a fire chief will appoint a health and safety officer to specifically address firefighter safety issues. In a nutshell, NFPA 1500 outlines what firefighter safety components are needed as part of an occupational safety and health program.

To hold up the department's end of the safety partnership, the fire chief, health and safety officer, or other administrative officer should create a health and safety committee, develop standard procedures and guidelines, and implement a risk management plan. Additionally, the department should research and purchase safety equipment, as well as develop and deliver hazard awareness training. The following paragraphs address each of these department responsibility areas.

Creation of a Health and Safety Committee

This group is usually made up of personnel from all levels. The focus here is to identify and create solutions for safety issues through a committee process.

(A)

(B)

(C)

FIGURE 5-12   Firefighter safety is dependent on all partners holding up their responsibilities: (A) administration, (B) teams, and (C) individual firefighters.

Development of SOPs

Defining a proper and expected level of procedure or behavior is an important step in addressing operations that can cause injury. It is common to have safety SOPs that address these areas:

- Personal protective equipment (PPE)
- Firefighter injuries
- Training safely
- Zoning
- Evacuation
- Emergency driving
- Slip and fall hazards
- Self-contained breathing apparatus (SCBA) use and care
- Accountability systems
- Incident command
- Riding apparatus
- Lifting heavy objects
- Power tool use
- Infection control
- Hazard reporting

Implementation of a Risk Management Plan

NFPA 1500 requires that the department develop, implement, and train all firefighters on a risk management plan. This plan is developed based on local needs and includes expectations from the community. The risk management plan can be quite comprehensive to address the many facets of the fire service (incident operations, facilities, equipment, health, etc.) Typically, though, the risk management plan acknowledges that certain risks are inherent in firefighting and that certain levels of risk-taking need to be defined. Defining risks based on the perceived benefit of the risk is known as risk/benefit thinking. An example of this risk/benefit thinking follows:

- Significant risk to the life of a firefighter shall be limited to those situations where the firefighter can potentially save endangered lives (risk a life to save a life).
- Situations endangering valued property shall cause firefighters to take a calculated and weighted risk (reduce risk for valued property).
- Where no life or valued property can be saved, no risk shall be taken by firefighters (take no risk for that lost).

- In situations where the risk to firefighters is excessive—or conditions are deteriorating quickly—firefighters should be limited to defensive operations (retreat).

Research and Purchase of Appropriate Apparatus and Equipment

Firefighting requires rugged, specially designed equipment that ensures a certain level of reliability and safety for the user. NFPA standards outline some of the requirements that make apparatus and equipment safer. This equipment is often very expensive and requires a significant effort to obtain funds. Once equipment has been purchased, the department needs to develop policies for the use, maintenance, and documentation of that equipment that is in accordance with manufacturers' recommendations.

Development and Delivery of Hazard Awareness Training

Virtually all training is designed to help firefighters operate in a safe manner. The training effort must also inform firefighters of the hazards they may face on any particular type of incident.

The Team

The department as a whole cannot be effective in its firefighter safety effort without the support of a team approach to handling incidents. Therefore, the team that handles an incident must hold up its part of the safety partnership. This team obviously includes the individual firefighter. To ensure safety, the team should follow these procedures:

1. Utilize an **incident command system (ICS).** The ICS should detail lines of authority and communication, be reasonable in its span of control, and include an action plan. Section chiefs, division/group supervisors, and team leaders need to communicate progress and status. Orders need to be clear and accomplished within an acceptable risk environment.

2. Work together and remain intact. The "buddy" concept is imperative for firefighter safety.

3. Look after each other. Team members who continuously watch each other will find and address conditions that will lead to injury. Aggressive rehabilitation will provide hydration, rest, and nourishment—thereby increasing the time the team can perform safely. At the team level, issues of PPE and SCBA use or misuse should be addressed. Like-

wise, the team must watch each other for signs of fatigue, dangerous overaggressiveness, unsafe acts, and freelancing.

> **CAUTION**
>
> The separation of members within a team is a contributing factor to firefighter fatalities.

The Individual Firefighter

> **SAFETY**
>
> A firefighter's awareness of his surroundings, proficiency in tasks, appropriate use of equipment, and most important, positive attitude about safety will ultimately prevent injuries.

The individual firefighter holds the final key to making the safety partnership work. The following paragraphs outline the individual contributions that a firefighter must make to the safety partnership.

Be Ready

Readiness is an attitude as well as a physical state. Reporting to duty or an incident with a physical limitation will increase everyone's danger. Firefighters working while sick or under the influence of certain over-the-counter and prescribed drugs can experience less than clear thinking, as well as injury. Likewise, mental barriers and "baggage" will likely impair a firefighter's ability to think and act.

> **STREETSMART TIP**
>
> A firefighter's readiness includes not only wearing PPE and SCBA as appropriate, but also being ready mentally and physically.

Understand and Act within the Chain of Command

For an incident command system to work, each individual must fill a role and not operate outside of it. Each person is responsible to someone else. Crossing these lines of responsibility causes confusion—a deadly proposal. In addition to understanding the chain of command, firefighters must clearly understand their orders (tasks). It is good practice to repeat the orders that have been given and to ask for clarification if those orders are not clear.

Perform as Trained

Incident task needs may place a firefighter in a position to perform a skill or task that the individual may not have been trained for. This should not be allowed to happen.

The incident scene is rarely the place for on-the-job training of critical skills. Firefighters should practice individual skills—knot tying, ladder raising, and tool use can deteriorate over time. Additionally, the department has outlined SOPs to be followed—firefighters should do so. Your ability to perform as trained is dependent on your discipline in maintaining skills and knowledge.

> **SAFETY**
>
> If a firefighter has not been trained for a task that is required, the firefighter should tell the direct on-scene supervisor.

Prevent Freelancing

Fire and rescue work is a team effort orchestrated to an action plan developed by the incident commander. Team leaders implement the plan and provide feedback to the IC. Firefighting teams follow the team leader's direction. Freelancing occurs when a team operates outside the action plan or when individuals work alone. Working alone or outside the action plan endangers individuals and the team, as **Figure 5-13** demonstrates. The individual firefighter can help prevent freelancing by not working alone and supporting the action plan when part of a team.

FIGURE 5-13 Freelancing endangers individuals and the team. This firefighter is working alone in a collapse zone—for what gain?

Incident Engagement Checklist

- Don personal protective equipment appropriate for the response.
- Mount apparatus using handholds.
- Buckle seat belt, don headset—report you are ready.
- Listen to radio reports and listen for details.
- Mentally run through tasks that might be required for the type of incident.
- Don SCBA or collect tools, gloves, etc.., if appropriate and if this can be done with seat belt on.
- Upon arrival, listen for orders.
- Unbuckle only when orders have been given.
- Prior to leaving the apparatus, look out the window and make sure the path is clear. Watch for traffic when dismounting.
- Close all doors and compartments after retrieving tools and/or equipment.
- Scan the environment for overhead wires, trip/fall hazards, and unusual circumstances.
- If the apparatus is in contact with electrical wires or other hazard, the firefighter may be required to leap clear of the apparatus when dismounting.
- Know who you are reporting to—make sure your accountability tag/passport is processed.
- Proceed with orders.
- Make sure you can be seen and heard by other firefighters at all times during an incident.

FIGURE 5-14 Firefighters should perform a mental incident engagement checklist for every response.

Use an Incident Engagement Checklist

Individual firefighters can also help achieve safety if they practice a standard approach to incidents—that is, doing things in a standard way all the time. Using an engagement checklist, **Figure 5-14,** helps achieve a standard approach to incidents.

Safety When Riding the Apparatus

Riding the fire apparatus has become one of the most common dangers the firefighter experiences. Alarmingly, firefighters are injured or killed while riding to and from incidents. A firefighter can minimize the chance of injury by following a few simple steps. The first is to always wear a seat belt! A firefighter should never attempt to stand or don PPE while the apparatus is in motion. Donning gear while the vehicle is in motion may require loosening of the seat belt, and the firefighter may be thrown around the enclosed cab in the event of a sudden stop or turn. Firefighters must always remain seated with seat belt fastened when the vehicle is moving.The firefighter must resist the temptation to ride the tailboard even during training drills and hose loading. Once the firefighter is in the cab of the apparatus, he should ensure that all doors are latched or the safety gate is in place. If a firefighter rides in fire apparatus that is not equipped with a fully enclosed crew compartment (open jump seat), it is important to use eye and ear protection as well as any installed safety bar or restraint device. In all cases, it is common practice for the firefighter to remain seated and belted until the officer or driver gives instructions to dismount.

Firefighters must pay careful attention when mounting and dismounting apparatus. Use handrails when mounting and dismounting the apparatus to prevent slipping and to help maintain balance especially when wearing PPE. The firefighter should always be alert to slip hazards such as water, foam, and ice. If the apparatus is in contact with electrical wires or other hazard, the firefighter may be required to leap clear of the apparatus when dismounting.

INDIVIDUAL ACTIONS AND ATTITUDES THAT HELP REDUCE INJURIES AND DEATHS

—DAVE DODSON

Actions:

- Strive to be always ready.
- Use your Protective Equipment as intended and NEVER short-cut its use.
- Always wear your seat belt—without fail.
- Always work as part of the team: be in the sight line (or physical touch in low visibility) of another firefighter at all incidents.

- Practice heart-healthy habits.
- Make a lifelong commitment to keep fit.
- Train and practice skills as if your life depended on it—it does!

Attitudes:

- Safety is a state of mind—apply it in all endeavors.
- Appreciate and reciprocate efforts to stop unsafe acts.
- Adopt an "everyone goes home" philosophy.
- Own up to your mistakes so that others may learn.

The incident engagement checklist will certainly help individual firefighters meet their responsibilities to the safety triad. Now we get personal. The firefighter safety system only works if everyone takes responsibility for their own safety. You can help achieve this goal by maintaining the desire and the discipline to act in safe ways with a safe-minded attitude.

LESSONS LEARNED

The issue of firefighter safety is dependent on many factors. The fact that roughly 70 percent of injuries and deaths occur during emergency activities demonstrates that firefighting is a risky profession. The majority of these deaths and injuries are a result of overexertion (heart attacks). Alarming trends include an increase in firefighter deaths and injuries due to burns, being caught by collapse, and responding to and returning from incidents. To help prevent injuries and deaths, fire service organizations have developed numerous initiatives to call attention to these trends and to help minimize further occurrences. Fire departments are required to follow federal regulations called OSHA CFRs. In addition, NFPA standards, like NFPA 1500, have been written and adopted to further the firefighter safety effort.

Accident prevention is actually the process of mitigating hazards or intervening with the accident chain. Within any operational environment, the safety triad is used to address issues. The triad includes efforts to address the operational environment of procedures, equipment, and personnel. To make sure the safety triad is effective, a partnership between the department administration, the working teams, and the individual firefighter is formed. Each of these partners carries an equal weight in creating a safe atmosphere. It is then the individual's challenge to develop safe habits and attitudes that help prevent injuries and deaths.

KEY TERMS

Code of Federal Regulations (CFR)   The documents that include federally promulgated regulations for all federal agencies.

Critical Incident Stress Debriefing (CISD)   A formal gathering of incident responders to help defuse and address stress from a given incident.

Critical Incident Stress Management (CISM)   A process for managing the short- and long-term effects of critical incident stress reactions.

Incident Command System (ICS)   An expandable management system used to deal with a myriad of incidents. Helps achieve the highest level of accountability and effectiveness for incident handling. Limits span of control and provides a framework to help make tasks manageable.

Member Assistance Program (MAP)   A defined program that offers professional mental health and other health services to employees.

National Institute for Occupational Safety and Health (NIOSH)   A federal institute tasked with investigating firefighter fatalities and making recommendations to prevent reoccurrence.

Occupational Safety and Health Administration (OSHA)   The federal agency, under the Department of Labor, that is responsible for employee occupational safety.

Risk   The chance of injury, damage, or loss; hazard.

Risk Management   The process of minimizing the chance, degree, or probability of damage, loss, or injury.

Vicarious Experience   A shared experience by imagined participation in another's experience.

Work Hardening   A phrase given to the effort and physical training designed to prepare an individual to better perform the physical tasks that are expected of the individual. Work hardening is key in preventing injuries from typical firefighting tasks.

REVIEW QUESTIONS

1. Define risk management and the concept of risk/benefit.

2. List the leading causes of death and injury in the fire service.

3. List the NFPA standards that affect and pertain to firefighter occupational safety.

4. What is the purpose of a NIOSH *Alert*?

5. List and briefly describe the three key components of the safety triad.

6. Explain the difference between formal and informal procedures.

7. Name the three factors that influence the equipment portion of the safety triad.

8. Name the three factors that influence the personnel portion of the safety triad.

9. Name the three partners that work together to achieve firefighter safety and give examples of how the individual firefighter can contribute to firefighter safety.

ENDNOTE

1. U.S. Fire Administration, Firefighter Fatality Retrospective Study, FA-220, USFA, Washington, DC, April 2002.

6

Personal Protective Clothing, Equipment, and Ensembles

As the department's training officer, I served as the safety officer on all major incidents. One afternoon while working in my office at our suppression headquarters station, I heard Engine 4 dispatched to a vegetation fire. Engine 4 arrived on scene to report approximately one acre involved, with the fire rapidly spreading. They requested two additional engines. The paramedics I was working with decided to drive down the street to observe, as we were on the opposite side of the involved area. I decided to ride with them. I did not plan on this being a major incident and did not take any protective clothing with me.

As we approached the area across from the fire, we noticed that the fire had spread across the ravine and was now threatening homes and utility services. An engine had laid lines, and one firefighter was attempting to put a 2½-inch line into service to protect the exposures. The incident commander was calling for two strike teams. The paramedics were given an assignment to help extend lines to the right flank. I jumped in to assist with the exposure line. My protective equipment consisted of slacks, a dress shirt, a tie, and shoes. Within minutes we were exposed to heat and smoke thick enough to cause difficulty breathing. Through the smoke an engineer brought me his turnout coat and helmet. He later gave me his turnout boots and stood at the pump panel in stockinged feet.

Not only was I completely ineffective as a safety officer, I was a safety hazard myself! By not having my protective gear, I created a safety hazard for myself and took away safety gear from another firefighter. The few minutes it would have taken for me to grab my gear would have been more than worthwhile. Our protective clothing is as essential to us as a scalpel is to a surgeon. The lesson I learned was never to respond to an incident without my safety gear and never to enter the hazard environment without it. Fortunately, no one was injured and the fire was extinguished with no loss of structures.

—Street Story by Randy Scheerer, Battalion Chief,
Newport Beach Fire Department, Newport Beach, California

LEARNING OBJECTIVES

After completing this chapter, the reader should be able to:

6-1   Describe the role of personal protective clothing (PPC) for firefighters.

6-2   Describe safety equipment used by emergency responders to safely function at an emergency scene.

6-3   Identify the various types of personal protective equipment (PPE) used in the fire service.

6-4   Select proper fire department documents, standards, or codes to find information.

6-5   List the components and unique elements of structural, proximity, and wildland PPC ensembles.

6-6   Given a specific piece of equipment, list and describe various tools and inspection and cleaning methods used to return the equipment back to service.

6-7   List and describe proper use of cleaning solvents.

6-8   Describe the proper manufacturers' or departments' procedures for tool and equipment cleaning.

6-9   Given manufacturer recommendations, departmental guidelines, tools, and cleaning materials, demonstrate the ability to properly clean, inspect, and service tools and equipment.

6-10   Describe a serviceability inspection of structural PPC.

6-11   Describe the conditions and damage that render structural PPC unserviceable.

6-12   Demonstrate proper doffing and reuse posture for personal protective clothing.

6-13   Demonstrate a team check following PPE donning.

6-14   Given a PPC ensemble, appropriately don the ensemble within one minute.

6-15   Demonstrate the proper use of PPE and safety equipment, including seat belts, used when riding fire apparatus.

6-16   Demonstrate emergency responder operations while wearing PPC for various emergency situations.

*The FF I and II levels, as defined by the NFPA 1001 Standards, are identified in different colors:
FF I = black, FF II = red, additional information = blue.

INTRODUCTION

Firefighters and emergency medical providers respond to incidents that are often immediately dangerous to life and health (IDLH). **Figure 6-1.** Often, the difference between injury and safety is determined by the responder's personal protective equipment (PPE). It is important to note, however, that PPE provides a minimum level of protection and should be considered the last resort of protection for firefighters and emergency responders operating at an incident. Proper risk/benefit analysis, fire streams, zoning (the establishment of specific hazard zones), and sound tactics and procedures should provide a greater measure of safety for teams, especially in IDLH atmospheres. Simply stated, PPE constitutes the primary layer of protection that a firefighter dons prior to an emergency response and is worn until the incident is successfully mitigated.

The NFPA states that personal protective clothing (PPC) for the firefighter includes turnout coat, protective trousers, firefighting boots, a helmet with eye protection, a protective hood, and firefighting gloves. Personal protective equipment (PPE) includes full protective clothing plus a self-contained breathing apparatus (SCBA) and a **personal alert safety systems (PASS)** device.

In addition to the basic ensemble, defined collections of personal protective clothing and equipment make up ensembles that are designed for different firefighter hazards. These ensembles include structural, proximity, and wildland firefighting ensembles. Other ensembles may include technical rescue, ice rescue, and swift-water rescue gear. Each of these ensembles is discussed later in this chapter. Ensembles for EMS infection control and hazardous materials PPE are used by many fire departments. Additional informa-

FIGURE 6-1   A hostile fire within a structure creates an IDLH environment. Personal protective equipment can help the firefighter work in an IDLH environment.

tion on these can be obtained locally or in Section IV, Chapter 27 of this book. Ensemble components are designed to meet a minimum level of safety. Each component has specific limitations that govern its performance and that of the entire ensemble.

Exceeding the limitations of an ensemble's components is not the only way injuries occur. Injuries and illnesses have also been suffered by the firefighter who fails to properly don and secure PPE—usually because the wearer skipped complete donning in the haste to perform a task.

CAUTION

Failure to operate within an ensemble component's designed limitations can lead to an injury, illness, or perhaps death of the user.

This chapter begins with a discussion of the factors that influence PPE design and use, including national standards and regulations. Next, the structural fire-fighting protective ensemble, including various components of personal protective clothing, is covered, followed by various other ensembles and PPC care and maintenance. The chapter then continues with a discussion of how firefighters can effectively use PPE and keep their first and last defense intact. SCBA—although an important component of PPE—is discussed in detail in Section I, Chapter 7. Other PPE items such as those for emergency medical operations and hazardous materials operations are discussed in other chapters.

PERSONAL PROTECTIVE EQUIPMENT FACTORS

Firefighter PPE has evolved significantly during the past two decades. The reasons for these recent improvements are varied. A study of the history of injuries, evolution of materials, and sound risk management practices have led to vast improvements in protective equipment. Modern PPE has been developed as a result of the direct efforts of labor groups (most notably, the International Association of Firefighters [IAFF]), other fire service membership associations, equipment manufacturers, and government entities such as NASA. These efforts usually come together as a result of a consensus process that establishes minimum standards. The National Fire Protection Association (NFPA) provides the forum for this consensus building. Although NFPA standards certainly influence PPE design and use, the federal government has also become involved in the PPE equation through the development of regulations and guidelines.

Standards and Regulations

The NFPA has developed standards for firefighter protective clothing, equipment, and ensembles, **Figure 6-2**. Typically, these standards cover component parts, manufacturers' quality assurance and labeling requirements, design requirements, user information, performance requirements, and test methods. The specific NFPA standards listed in Figure 6-2 primarily address the design, performance, and manufacturing of clothing and equipment. Additionally, NFPA has developed an important "use" standard. NFPA 1500, Standard on Fire Department Occupational Safety and Health Program, dedicates a whole chapter to the use, care, and maintenance of many forms of protective clothing and equipment.

**NFPA Standards That
Address PPE and Ensembles**

1500	Fire Department Occupational Safety and Health Program
1851	Standard on Selection, Care, and Maintenance of Protective Ensembles for Structural and Proximity Fire Fighting
1971	Protective Ensemble for Structural Firefighting
1975	Station/Work Uniforms for Firefighters
1977	Protective Clothing and Equipment for Wildland Firefighting
1981	Open-Circuit Self-Contained Breathing Apparatus for the Fire Service
1982	Personal Alert Safety Systems
1983	Life Safety Rope and System Components
1991	Vapor-Protective Hazardous Ensembles/Materials Emergencies
1999	Protective Clothing for Medical Emergency Operations

FIGURE 6-2 NFPA develops consensus standards that address PPE.

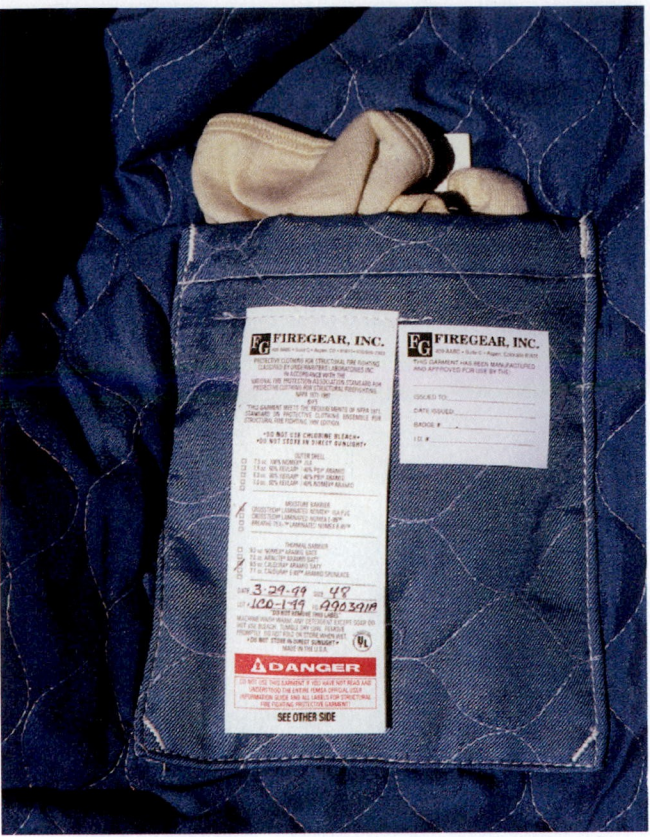

FIGURE 6-3 NFPA-compliant PPE components should have a permanently affixed label.

All protective clothing worn by a firefighter should meet current applicable standards. Firefighters should check protective clothing for a conspicuous and permanently attached label that verifies that a component meets an applicable NFPA standard and states what agency performed verification testing, **Figure 6-3.**

> **NOTE**
>
> Even if a piece of clothing or equipment meets all of the applicable standards, there is no guarantee that it will protect the wearer from the hazards that may be encountered while firefighting. It is the discretion of the wearer and the agency having jurisdiction to determine if it is appropriate for the situation being presented.

The federal government is also involved in protective clothing and equipment use. Of primary importance is the involvement of the Occupational Safety and Health Administration (OSHA). OSHA is responsible for the development and enforcement of regulations—namely, the Code of Federal Regulations (CFRs)—that govern safe work practices. At the firefighter level, this involvement essentially means that failure to use appropriate PPE in certain applications (e.g., IDLH atmospheres) means that a federal regulation has been violated, leading to a potentially expensive fine and disciplinary actions against the firefighter and the department. Additional information on OSHA can be found in Section I, Chapter 2. Other government and ancillary agencies that are involved in protective clothing and equipment issues include the Environmental Protection Agency (EPA), Centers for Disease Control and Prevention (CDC), American National Standards Institute (ANSI), American Society for Testing and Materials (ASTM), and the National Institute for Occupational Safety and Health (NIOSH).

TYPES OF PERSONAL PROTECTIVE ENSEMBLES

NFPA defines *ensemble* as multiple elements of compliant protective clothing and equipment that when worn together provide protection from some risks, but not all risks, of emergency incident operations. The NFPA has developed minimum ensemble standards for structural, proximity (commonly used for aircraft rescue and firefighting), and wildland firefighting. NFPA also addresses specialized ensembles that exist for technical rescue such as Urban Search and Rescue (USAR). The important point to underscore is the word *ensemble*. Frequently, firefighters short-circuit the concept of an ensemble in the use of their protective clothing and equipment. An action as simple as forgetting to secure the top fastener of a protective coat violates the concept of an ensemble and can lead to an injury.

> **SAFETY**
>
> To be effective, all components of a protective ensemble must be utilized as recommended by the manufacturer. Failure to complete an ensemble can cause an injury.

Structural Firefighting Protective Ensemble

The protective ensemble for structural firefighting includes all of the components listed and shown in **Figure 6-4.** For the sake of clarification, NFPA defines structural firefighting as "the activities of rescue, fire suppression, and property conservation in buildings, enclosed structures, aircraft interiors, vehi-

cles, vessels, or like properties that are involved in a fire or emergency situation."

Structural personal protective clothing is commonly referred to as **bunkers**. The term *bunkers* was originally associated with short boots and protective pants that duty firefighters would only wear at night—donned in the bunk room of a fire station. During the day, these same firefighters wore long coats and three-quarter pull-up boots—a practice no longer allowed by NFPA 1500 due to the inadequate protection offered to the wearer. (Many serious burns have been experienced by firefighters using this antiquated type of clothing, **Figure 6-5.**) Today, many terms are used to refer to the structural firefighting protective clothing. These terms include the previously mentioned *bunkers* as well as *turnouts* and *structural gear*. The components of structural PPC are discussed next.

Coats and Trousers

The heart of structural PPC is the coat and trouser combination. Both components rely on a layered protection system that includes a fire-resistive outer shell, vapor barrier, and thermal barrier.

FIGURE 6-4 A full structural firefighting ensemble includes more than the NFPA minimum required components.

FIGURE 6-5 Traditional ¾ boots and long coats are no longer acceptable PPE according to NFPA standards.

The three layers help the coat and trouser meet thermal protective criteria—insulation that minimizes the chance that the wearer will be burned. This criterion is called **thermal protective performance (TPP)**. The minimal TPP for structural coats and trousers is 35. The TPP number divided by two equals the number of seconds it will take for human skin to sustain a second degree burn when exposed to flame. Many variables such as dirt, moisture (including perspiration), age of the fabric, and fabric compression (from an SCBA or bending) can reduce the TPP of the clothing.

FIREFIGHTER FACT

Total Heat Loss (THL) provides a measurement of how well a garment allows heat and moisture to escape from an ensemble. This is important in regards to preventing heat stress for the firefighter. The higher the THL value, the better a garment is able to breathe. NFPA 1971 states that the minimum requirements for THL of coats and trousers should be 205. THL and TPP are in direct relation with each other, however. Generally, to raise the THL value of a garment, the TPP must be lowered because it is the insulation qualities provided in higher TPP numbers that provide heat protection. A balanced combination of the two must be found to offer the best protection.

The coat and trouser combination has reflective trim to increase the visibility of the wearer to others, **Figure 6-6.** Flaps, enclosures, wristlets, **Figure 6-7,** and fastening devices are all designed to seal the ensemble and to provide a protective interface with

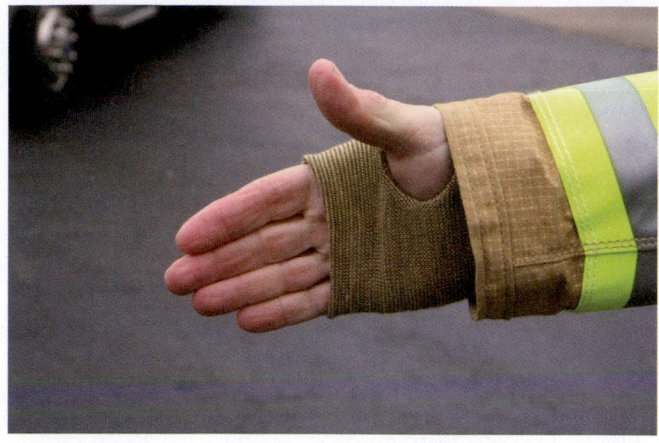

FIGURE 6-7 Wristlets help seal the ensemble and provide an interface between the gloves and coat.

FIGURE 6-8 The DRD is a system that forms a harness around a firefighter's upper body and enables another firefighter to drag them by the system if they become incapacitated.

FIGURE 6-6 Reflective trim increases firefighter visibility to others during low-light activities.

gloves, hoods, boots, and helmets. Failure to "seal" the ensemble invites injury in hazardous environments. Trousers should have heavy-duty suspenders to help keep them from sagging when they become wet with perspiration or water. Coats manufactured after March 1, 2007, are required to be outfitted with a drag rescue device (DRD), **Figure 6-8.** The DRD is a system that forms a harness around a firefighter's upper body and enables another firefighter to drag them by the system if they become incapacitated. DRDs are designed only for dragging the incapacitated firefighter and are not permitted to be used as a harness for vertical lifting or lowering.

Shoulder padding, knee reinforcements, and various types of pockets and accessory attachments can also be found on coats and trousers. Some firefighters have had webbing sewn into their structural trousers to create a "seat harness" for certain rescue situations. This customizing should only be performed by a certified manufacturer that understands and complies with NFPA standards for both structural PPC and life safety system components (NFPA 1983, Standard on Fire Service Life Safety Rope and System Components).

Helmets

The classic firefighter's helmet was designed to help shed water and to prevent hot embers from falling down on the firefighter's neck, back, and ears. This classic design remains in most styles of helmets, **Figure 6-9A–C.** Newer helmet safety features exceed those of previous designs in many ways and include impact resistance; thermal insulation; fire-resistant earflaps for a layered interface with hoods, coats, and SCBA face pieces; chin straps; and clear or tinted face shields and/or eye protection accessories. Earflaps in helmets must have a thermal protective performance (TPP) of 20 or more. The chin strap is an important part of the helmet—it must be utilized to take full advantage of the helmet's safety features. Failure to use the chin strap is an invitation to injury or loss of the helmet during critical operations.

Eye Protection

Firefighters use many different forms of eye protection. Goggles, safety glasses, and wrap-around shields are examples. An SCBA facepiece can provide primary eye protection, although a structural helmet face shield may not. The helmet face shield—because of its wide facial opening—may not prevent windblown particulate or liquid splashes from getting into the eyes, **Figure 6-10.**

Gloves

Hand protection is essential in the structural ensemble. Gloves meeting NFPA standards for structural firefighting must provide thermal protection as well as protection from cuts, punctures, and scrapes. Unfortunately, dexterity is almost always reduced when wearing structural gloves. This is a common complaint of firefighters.

(A)

(B)

(C)

FIGURE 6-9 Helmet styles range from traditional leather with long brims to sleek aerodynamic ones made from composite materials. Each style has unique advantages and disadvantages.

(A)   (B)

FIGURE 6-10   Different types of eye protection are available for firefighters. (A) Wearing safety glasses under a face shield improves eye protection. (B) A face shield alone may not protect the eyes.

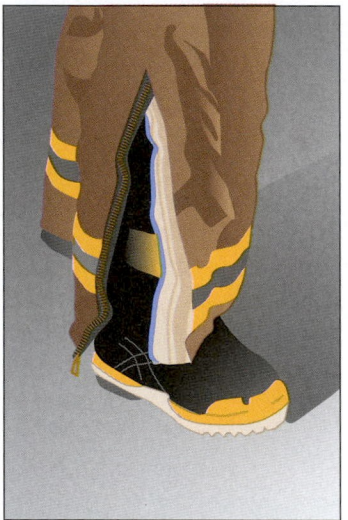

Common Rubber Boot

- Easy to Don
- Excellent Water Repellency
- Easy Decontamination
- Inexpensive
- Sloppy Fit

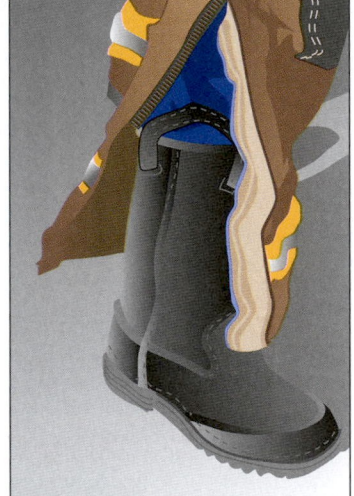

Leather Pull-Up Boot

- Lightweight
- Durable
- Comfortable
- Minimal Ankle Support

Leather Lace-Up Boot

- Tight Fit
- Ankle Support
- Durable
- Expensive

FIGURE 6-11   Firefighters have choices for structural firefighting footwear—each with its own advantages and disadvantages.

The simple act of practicing hands-on tool use while wearing gloves can increase muscle memory and reduce the frustrating loss of dexterity.

Footwear

The firefighter's choices for approved and effective footwear are growing. Now firefighters can choose from traditional rubber-like boots or leather-type boots.

Each has advantages and disadvantages, **Figure 6-11.** Both must meet NFPA standards for foot protection.

Protective Hoods

An important interface that creates an encapsulating link to the firefighter's helmet, coat, and SCBA face piece is the protective hood. Hoods are made of fire-resistive, form-fitting cloth (usually Nomex, PBI, or

Kevlar), that protects the face, ears, hair, and neck in areas not covered by the helmet, earflaps, and coat collar, **Figure 6-12.** It is important to note that structural protective hoods have a TPP less than that of a structural coat (20 for a hood versus 35 for a coat). Additionally, hoods are designed to be worn over the straps of an SCBA face piece. This practice helps protect the SCBA face piece straps and helps keep the hood from binding.

SAFETY

Wearing the hood over the facepiece allows for a proper seal of the facepiece to the wearer's head.

Hearing Protection

While not necessarily defined by the NFPA standards as part of the list of personal protective clothing, hearing protection is necessary for high noise areas. Many typical rescue and firefighting operations can expose firefighters to noise levels above 100 decibels (dB), which can cause hearing damage (including permanent damage) after a short-duration exposure. For this reason, firefighters should always have rapid access to hearing protection. Hearing protection takes on many forms—foam plugs, rigid earmuffs, and headsets are common examples. Many fire apparatus are equipped with a technology that combines hearing protection/intercom/radio microphone into a single headset. These combination systems make communication within the cab of a responding apparatus efficient and effective, **Figure 6-13.**

Station/Work Uniform

Another garment that is not listed as part of the personal protective clothing definition but which is of equal importance to firefighter safety is the station/work uniform. Firefighters assigned to emergency response duty are often required to wear a uniform that meets NFPA standards. NFPA 1975, Standard on Station/Work Uniforms for Fire Fighters, outlines manufacturers' guidelines and requirements for work uniforms. The underlying message is that a work uniform can add a protective measure to firefighters engaged in support activities or station duties. A work uniform that meets NFPA standards is not designed to protect the wearer from IDLH atmospheres, but can add another layer of reasonable protection under wildland, proximity, and structural ensembles.

FIGURE 6-12 A protective hood provides an interface layer that links the coat collar, helmet flaps, and opening for an SCBA mask.

FIGURE 6-13 Combination headsets protect the wearer's hearing and also increase communication effectiveness between crew members and other responders.

should be wearing their PPE properly each and every time they dismount an emergency apparatus on a call. Complacency on the fireground in regard to PPE has resulted in many firefighter line-of-duty fatalities.

Proximity Firefighting Protective Ensemble

The proximity firefighting protective ensemble is most often associated with aircraft rescue and firefighting (ARFF). The proximity turnout coat, protective trousers, hood, and gloves use an aluminized coating to help reflect radiant heat, **Figure 6-14.** Although similar in many ways to structural firefighting garments, proximity protective garments must meet more stringent heat reflection and wearer insulation standards.

The aluminized fabric allows the wearer to get closer to fires that emit extreme radiant heat such as that found with petroleum-based fires. Bulk fuel facilities, airports, and certain chemical plants are complexes where the assigned fire attack teams may be required to wear proximity ensembles.

(A)

(B)

FIGURE 6-14 Proximity firefighting ensembles can utilize (A) a special helmet or (B) a full hood. Either can interface with an SCBA. Note the gold-anodized visors.

In addition to the aluminized fabric, the proximity hood features a full face shield that is coated with an anodized gold material to help create a "mirrored" reflective surface. Without this special coating, the wearer could receive radiant facial burns, and the face shield could quite possibly melt.

Wildland Firefighting Protective Ensemble

Wildland firefighting conditions are unique in that firefighting operations are often outdoors, often require prolonged physical effort, and are typically accomplished when ambient temperatures are high. Fighting wildland fires with a structural ensemble can invite strained necks, heat stroke, and sprained ankles. Wildland ensembles (called **brush gear** by many) address the specific needs of the wildland firefighter: lightweight; provides breathability, firm ankle support, and hot ember protection, **Figure 6-15.**

> **CAUTION**
>
> Synthetic material should never be worn under wildland ensembles—most synthetic materials can melt and cause severe burns to the wearer.

FIGURE 6-16 Wildland PPE includes cotton or fire-resistive undergarments. Long-sleeve T-shirts are a *must!*

The wildland PPE ensemble is designed to be worn over undergarments. These undergarments (long-sleeve T-shirt, pants, and socks) should be 100 percent cotton or of a fire-resistive material, **Figure 6-16.**

Other highlights of the wildland PPE ensemble are discussed next.

Lightweight Jacket/Shirt and Trousers

These garments are typically made of a fire-resistive material or a treated cotton. Wool has also been used for wildland protective garments. Once again, these protective elements need to be worn over undergarments in order to increase thermal insulation.

Footwear

Lace-up leather boots that rise well above the ankle (8 to 10 inches) help protect the wearer from cuts, snakebites, and burns. Additionally, a good fitting, tightly laced boot can help prevent ankle sprains and reduce foot fatigue.

Fire Shelter

A **fire shelter** is another unique component in the wildland PPE ensemble, **Figure 6-17.** The fire shelter must be carried in a case that protects the aluminized

FIGURE 6-15 Wildland PPE is lightweight but still provides protection from hot, flying embers.

(A)

(B)

(C)

(D)

FIGURE 6-17 A wildland fire shelter is an essential PPE component. Training for rapid deployment can help save a firefighter's life. *(Photos courtesy of Nemo Nieman Photography)*

fabric from being crushed, yet still allows the shelter to be deployed quickly. A fire shelter is a last-resort protective device for firefighters caught or trapped in an environment where a firestorm or blow-up is imminent.

Web Gear

Although not listed in the NFPA standard for wildland PPE, **web gear** is essential for the wildland firefighter. Web gear consists of a belt (with or without shoulder support straps) that can carry a fire shel-

FIGURE 6-18 Wildland web gear is designed to keep personal items, water, flares, and the fire shelter within easy reach.

ter, water bottles, flares (fusees), a radio, and other assorted gear to help the wildland firefighter, **Figure 6-18.** Often, a web gear setup includes a detachable day sack that can carry meals and maps, or even overnight sleeping gear and personal items.

Technical Rescue Ensemble

Rescue operations such as confined space, structural collapse, rope, and trench do not necessarily require full structural PPE. In these cases, a lightweight yet durable ensemble is often used to protect rescuers. As **Figure 6-19A** shows, this ensemble typically consists of a durable pant and overshirt (or coverall) coupled with lace-up leather boots, tight-fitting durable gloves, a lightweight helmet, hearing protection, eye protection, and a harness.

Ice Rescue Ensemble

In areas where frozen lakes and recreation ponds exist, the fire department may offer an ice rescue service. The ice rescue ensemble includes a buoyant, insulated suit that protects the wearer from submersion and the freezing cold water should the rescuer break through the ice. The suit has a form-fitting face seal and hood to minimize the amount of water that can enter the suit. Typically one-piece, the suit has sealed gloves and boots built in. To complete the ensemble, the suit should be worn with a simple chest harness and lightweight helmet with face screen (similar to a football face mask), **Figure 6-19B.**

Swift-Water Ensemble

The use of structural PPE near swift water can introduce more danger to the firefighter if swept into the stream. While specific intense training can teach firefighters how to "float" in structural PPE, it is not to be used for swift-water environments. The swift-water ensemble includes a typical work uniform overlaid with a personal flotation device (PFD), harness with throw-rope bag, lightweight helmet with face cage, and no-slip gloves, **Figure 6-19C.**

CARE AND MAINTENANCE OF PERSONAL PROTECTIVE CLOTHING

The key to maintaining personal protective clothing in a high state of readiness is simple: Follow the specific instructions given by the manufacturer. NFPA requires manufacturers to clearly label care instructions for cleaning each piece of equipment. In addition, manufacturers should provide the user with specific instructions and information that address the following:

- Safety considerations
- Limitations of use
- Marking recommendations and restrictions
- Warranty information
- Sizing/adjustment procedures
- Recommended storage practices
- Inspection frequency and details
- Donning and doffing procedures
- Interface issues

Specific to cleaning and maintenance, the manufacturer will provide cleaning instructions and precautions, including a warning that the user should not wear PPC that is not thoroughly clean and dry.

Clothing exposed to biological and chemical contaminants should be decontaminated in accordance with manufacturers' instructions, **Figure 6-20.** NFPA 1581, Standard on Fire Department Infection Control Program, requires that clothing be cleaned at least once every six months . Traditionally, dirty and "battle-worn" PPC has been seen as a "badge of distinction" for firefighters, signifying that they are experienced or work in a busy company. In all reality, dirty, charred, or contaminated PPC is an invitation for trouble and can lead to numerous problems. Contaminants and dirt impair the heat reflective qualities of the PPC and have proved in studies to lead to certain types of cancer with repeated exposures over time.

(A)

(B)

(C)

FIGURE 6-19 Specialized ensembles include (A) the technical rescue ensemble, (B) the ice rescue ensemble, and (C) the swift-water ensemble. *(Courtesy of Ron Marley)*

Obviously, PPC that is dirty or contaminated should be cleaned immediately. Fire departments should have specific cleaning equipment or a cleaning service that can clean protective clothing in accordance with manufacturers' guidelines. Washing structural or wildland protective clothing with linens or other household items should be forbidden because cross-contamination can result.

PPC that exhibits damage—no matter how minor—should not be worn until it is properly repaired by a manufacturer-approved representative. Personal protective clothing must be routinely inspected and, when appropriate, retired and disposed of as suggested by the manufacturer. This last maintenance check is best accomplished through a team process that involves the individual firefighter, his or her officer, and the organization. This partnership process creates checks

FIGURE 6-20 Structural and wildland clothing should be washed in accordance with manufacturers' recommendations. Dedicated washers that are used for PPE only can help lower cross-contamination of other washables.

and balances that help ensure that PPC is serviceable for the user. Documentation should be kept on all PPC addressing age, cleaning, and any repairs made to it during its life cycle.

PERSONAL PROTECTIVE EQUIPMENT

As discussed at the beginning of this chapter, personal protective equipment includes personal protective clothing plus a self-contained breathing apparatus (SCBA), and a **personal alert safety system (PASS)** device.

Self-Contained Breathing Apparatus (SCBA)

The self-contained breathing apparatus is one of the most impotant items of personal protective equipment used by firefighters and rescue personnel. SCBA allows firefighters to enter hazardous atmospheres to perform critical interior operations including offensive fire attack; search, rescue, and removal of victims; ventilation; and overhaul. In addition, SCBA is used at non-fire incidents, such as hazardous material or confined space rescue situations involving toxic fumes or oxygen-deficient atmospheres. Because SCBA is such

an important piece of equipment, Section I, Chapter 7 is devoted entirely to that subject, including care and maintenance.

Personal Alert Safety System (PASS)

A PASS device is a small, motion-sensitive unit that is typically battery powered and includes a loud, audible warning signal. Some devices include a small flashing beacon or strobe. The PASS device, when worn on a firefighter, senses the firefighter's motion. As long as the device senses motion, the alarm will not sound. Inactivity for thirty seconds causes the device to send a pre-alert "chirp" or other reminder to the wearer. If the wearer fails to move, the device will go into alarm mode and emit a loud noise to signal or warn other firefighters that the wearer may be in trouble. Most new PASS devices are an integral part of the SCBA unit, although some fire departments issue PASS devices to individual firefighters, **Figure 6-21A–C.**

PASS devices have been shown to fail due to water immersion, high heat, and the position of a downed firefighter muffling its alarm. The 2007 edition of NFPA 1982 is addressing these issues with a much more stringent testing procedure for these devices prior to marketing. The most common problem with PASS devices results when wearers simply forget to turn their units on when the unit is not integrated into the SCBA. Oversights of this nature have contributed to numerous firefighter fatalities.The National Fire Protection Association now requires that PASS devices integrated with SCBA activate automatically when the SCBA air supply is turned on. Some PASS devices will also have the ability to track and locate firefighters on the fireground and, in some cases, even allow the Incident Commander the capability of monitoring air consumption by the user.

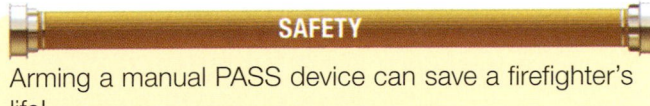

SAFETY

Arming a manual PASS device can save a firefighter's life!

PERSONAL PROTECTIVE EQUIPMENT EFFECTIVENESS: STREET SMARTS

To maximize the effectiveness of all PPE, firefighters must develop automatic behaviors.Simple steps can help achieve these behaviors. This final—and perhaps most important—section of the chapter contains suggestions to help firefighters develop good habits that will reduce the chance of injury.

(A)

(B)

(C)

FIGURE 6-21 PASS devices can help save a firefighter's life—but they must be activated. (A) A manual PASS device must be armed by the wearer. (B and C) An integrated PASS device on an SCBA, which is activated when the wearer opens the SCBA bottle.

Good PPE Habits and Attitude

Fire departments spend literally thousands of dollars equipping firefighters with PPE for the hazards they may face. Unfortunately, many firefighters are still injured—all too often because the individual firefighter has failed to properly utilize the PPE. This unfortunate result can be eliminated simply by good PPE habits and a positive attitude toward safety. The proper PPE attitude starts with a simple concept:

SAFETY

Firefighters should don all PPE necessary for the potential worst case scenario.

Granted, this approach may lead to "overdressing" for an incident. In these cases, the firefighters' company officer, incident commander, or incident safety officer may allow firefighters to "dress-down." The inverse, however, is unacceptable. Firefighters who take a "wait and see" attitude to decide what level of

(A)

(B)

FIGURE 6-22 Duty personnel should set up their gear for rapid—and complete—donning. Establish good habits to help eliminate shortcuts.

PPE they are going to need have set themselves up to shortcut their PPE ensemble if, on arrival at the scene, the situation requires immediate lifesaving actions. It is better to be prepared first—then act.

Another key to proper and effective use of PPE is the development of good habits that include fast, proper, and complete donning of the appropriate PPE ensemble, **Figure 6-22A** and **B.** Unfortunately, many firefighters begin shortcutting their use and completeness of PPE donning as days, weeks, months, and years of firefighting experience are accrued. This is one of the travesties that has kept injury rates high for firefighters.

Establishing good habits takes nothing more than self-discipline and practice. The benefits of self-discipline applied to PPE completeness pay a dividend in the form of **acclimation**. PPE acclimation simply refers to getting used to the wearing of PPE. The restrictive, encompassing, and stifling heat initially associated with most firefighter PPE simply becomes less of an issue and physically less taxing with acclimation. No firefighter particularly enjoys donning full structural PPE in 100°F (38°C) heat. The firefighter who is acclimated physically and mentally to the PPE will, however, accept the merits of the PPE and have less stress—and certainly less risk—than the firefighter who complains that the gear is too hot for the weather. The firefighter with good PPE habits and a positive PPE attitude knows that acclimation, rest, and on-scene **rehabilitation (rehab)** will beat the discomfort that some firefighters complain about.

When donning PPE, the firefighter should follow the manufacturer's instructions and ensure that shortcuts have been eliminated. Firefighters who have experienced a rash of false alarms (i.e., unintentional fire alarm activations in a building) seem to be particularly guilty when it comes to shortcutting their PPE. The firefighter with an appropriate PPE attitude will view each one of these incidents as an opportunity to fine-tune donning procedures and achieve **mastery** in PPE donning. Further, it makes sense that a firefighter who has appropriately donned the PPE is in a strong readiness position to be immediately assigned duties in IDLH atmospheres. Firefighters must resist the temptation to get caught in the "cry wolf" syndrome. They must always be ready!

Perfecting Practices in PPE Use

Daily, firefighters utilize PPE as their first and last defense against injury and illness. From these experiences come many "streetsmart" suggestions to help new firefighters:

- Keep PPE clean, inspected, and serviceable—an ownership attitude!

- Practice "team checks." Firefighters should check each others' PPE for readiness, **Figure 6-23.**

- When on duty in a station, position PPE for rapid, complete donning. A practical example is to place a structural hood across bunker pants and boots as a reminder to don the hood first so it will interface appropriately under the bunker coat.

- When doffing PPE, take the time to prepare it for the next response. Set the clothing up for rapid donning—preparation time spent now speeds the next donning.

STREETSMART TIP

Drinking water prior to wearing PPE ensembles can help prevent stress-related injuries. Firefighters should have a bottle of water available near their gear or at their seat on the apparatus, and take a few drinks as they respond to the emergency.

FIGURE 6-23 Firefighting teams should check each other's PPE for complete donning. This "team check" can help prevent a burn or other injury.

- Do not let the urgency of the situation override prudent judgment.
- The structural protective ensemble also includes a flashlight, personal tool, radio, earplugs, accountability tag (see Section I, Chapter 23), and a partner.
- Wear PPE properly—not wearing chinstraps, leaving coats open or skin exposed, and not fastening buckles and straps will only lead to problems on the fireground. PPE cannot adequately provide protection if it is not worn in the manner prescribed, **Figure 6-24.**
- Practice doesn't make perfect—perfect practice makes perfect! Firefighters need to use PPE the right way every time to cement good habits. Can firefighters don their structural PPC within one minute? Can they don and activate the SCBA within one minute (including PASS device if provided by AHJ)?
- Be the PPE success example—not an injury statistic.

FIGURE 6-24 Wearing equipment improperly is asking for trouble. The improperly fastened straps on this firefighter's SCBA create an entanglement hazard. In addition, it increases the firefighter's chances for back injuries because the weight of the unit is not properly dispersed without fastening the straps. Also note the entanglement hazard created by the improper use of the firefighter's suspenders.

- When wearing any protective ensemble, the firefighter must increase water consumption to help stay hydrated.

Mastery of PPE skills allows the firefighter to concentrate on the important task at hand.

LESSONS LEARNED

Firefighter personal protective equipment and ensembles are the first and last defense against injuries and illness in a profession that is extremely hazardous. The firefighter who develops good PPE habits and a positive PPE attitude is likely to be quicker, more prepared, and better acclimated to the real-world confines of the personal protective ensemble. Shortcutting the use of PPE is a dangerous practice that can contribute to injuries.

Through national consensus standards like the NFPA process, firefighter protective clothing and equipment has advanced significantly in the past two decades. NFPA standards address the performance requirements and manufacturing criteria for PPC and PPE as well as use standards for firefighters and fire departments. Federal government entities also influence PPC and PPE use and guidelines.

Different protective ensembles exist for firefighters, including structural, proximity, and wildland ensembles. The key with any ensemble is the simple concept that an entire ensemble must be donned completely and correctly in order for the ensemble to protect the wearer. A shortcut can open the door to an injury. SCBA, harnesses, hearing protection, eye protection, and other PPE components must also be considered to protect the firefighter from hazards.

Care and maintenance of PPC is essential to its readiness. The firefighter must understand manufacturers' instructions and information regarding the fit, use, cleaning, and serviceability of assigned PPC. A sense of ownership is important here.

Finally, PPE effectiveness is dependent on an individual firefighter's good PPE habits and positive PPE attitude. Donning PPE for the worst-case scenario prior to incident engagement is preferable to a wait and see approach. The firefighter must be ready!

KEY TERMS

Acclimation   The act of becoming accustomed or used to something. Typically achieved through repeated practice within a given set of conditions.

Brush Gear   Another term for a wildland personal protective ensemble.

Bunkers   A term that is used mostly to describe the components of a structural firefighting ensemble. The original use of the term *bunkers* referred only to the pant/boot combination that firefighters wore at night and placed next to their bunks for rapid donning.

Fire Shelter   A last-resort protective device for wildland firefighters caught or trapped in an environment where a firestorm or blowup is imminent.

Mastery   The concept that an individual can achieve 90 percent of an objective 90 percent of the time.

Personal Alert Safety System (PASS)   A device that emits a loud alert or warning that the wearer is motionless.

Rehabilitation (Rehab)   A shortened word meaning rehabilitation. Rehab typically consists of rest, medical evaluation, hydration, and nourishment.

Thermal Protective Performance (TPP)   A rating level, expressed in seconds, used to characterize the protective qualities of a PPE component before serious injury is experienced by the wearer.

Web Gear   The term given to a whole host of personal items carried on a belt/harness arrangement worn by wildland firefighters. Items include water bottles, a fire shelter, radio, and day sack.

Zoning   A term given to the establishment of specific hazard zones, that is, hot zone, warm zone, cold zone, and also collapse zones.

REVIEW QUESTIONS

1. Explain what is meant by the phrase "PPE is the first and last resort."

2. List four government and ancillary agencies or associations that influence PPE design or use.

3. Name the agency responsible for developing consensus firefighter PPE standards. Name the agency that writes PPE use regulations.

4. List the components of a structural PPE ensemble.

5. Describe the importance of TPP and "sealing" a structural ensemble.

6. Describe how a proximity ensemble differs from other ensembles.

7. List the components of a wildland PPE ensemble.

8. Describe the operational concept of a PASS device.

9. Explain the relationship of a work uniform and undergarments to PPE ensembles.

10. List three important guidelines relative to the washing of PPE garments.

11. Define the concept of acclimation in regards to PPE.

12. List five practices for improved effectiveness with PPE.

7

Self-Contained Breathing Apparatus

A call came in at around 2:00 A.M. for a vehicle fire. Four miles away from the scene we saw a large fireball. When we arrived on scene, we found a twenty-foot motor home fully involved. The fireball we saw was a propane tank rupturing—after the fire we found the tank more than twenty feet from the vehicle. The owners said they were on a hunting trip and just had the vehicle serviced.

This call was in a remote area and a second alarm was dropped for a second engine and tanker. We pulled a 13/4 line, donned our Self-Contained Breathing Apparatus (SCBA), and then started an attack. As we approached the vehicle, we could hear rounds of ammunition popping off. We were facing several safety concerns: one, we were a two-person engine; two, a lack of water while waiting for our second alarm; and three, the ammunition rounds that were popping off.

About ten minutes into the fire, my face mask sucked my face as though my tank were out of air. My low air alarm did not sound, and I had no warning of low air. Because of a lack of training and knowledge, my first instinct and action was to pull my mask off and breathe.

As I did this, the first thing I took in was the smoke from the vehicle. Luckily, it was a vehicle fire, and all I had to do was go upwind to get fresh air and move out of the hazard.

When the assistant chief arrived on scene I informed him of the faulty SCBA and said it needed to be taken out of service and inspected. Being a young and confident firefighter, I was convinced that it was a faulty pack and not operator error.

I was wrong. My lack of training and understanding of the equipment was a humbling experience. At the time of the incident, the type of SCBA used was a mine safety appliance (MSA), with a face piece and hose that connected to the regulator. With this type of SCBA, a missed step or an accidental occurrence can cause things to go wrong.

One common mistake is not making sure the connection is correctly sealed. For those who have never used this type of SCBA, it is important to note that when connecting the hose to the regulator, it is a blind connection, as the regulator is down on the waist. The other major concern is something pinching your hose and blocking the airflow. This type of face mask and hose is similar to having a garden hose hanging from your face—it can be pinched off very easily.

Given my inexperience, I did not realize these potential problems. Also, I panicked when I thought I was out of air. Nothing was wrong with my SCBA; I had simply pinched my hose. If I realized that, all I needed to do was take four to five steps away from the vehicle.

I was lucky that this happened during a vehicle fire and not a structure fire. Consequently, when I took off my mask, it was one lung full of smoke and not several. But it is highly possible that one lung full of smoke can kill you if you do not take the necessary steps to protect yourself. It is very important that you are extremely confident in the use of all of your protective equipment and that you know how to troubleshoot in an emergency situation.

—Street Story by William Tatsch, Engineer,
Travis County Emergency Services District 6, Austin, TX

LEARNING OBJECTIVES

After completing this chapter, the reader should be able to:

7-1   List conditions requiring respiratory protection.

7-2   Explain ways that smoke harms the body.

7-3   Describe the effects of oxygen deficiency and toxic gases on the human body.

7-4   List the legal requirement(s) for use of self-contained breathing apparatus (SCBA).

7-5   Describe the limitations of SCBA.

7-6   Identify the physical requirements of an individual wearing a SCBA.

7-7   Describe methods for conserving air at the emergency scene.

7-8   Identify the common types of SCBA used in the fire service.

7-9   Name the components of the SCBA used in the fire service.

7-10   Given a specific piece of equipment, list and describe various tools as well as inspection and cleaning methods used to return the equipment back to service.

7-11   Given manufacturer recommendations, departmental guidelines, tools, and cleaning materials, demonstrate the ability to properly clean, inspect, and service tools and equipment.

7-12   Identify the donning procedures for a SCBA.

7-13   Demonstrate the ability to don a SCBA and activate the system within 1 minute.

7-14   Identify the proper uses of SCBA.

7-15   Demonstrate proper breathing techniques while wearing a SCBA.

7-16   Demonstrate the ability to maneuver through restricted passages while wearing a SCBA

7-17   List the five steps that can lead to an organized rapid escape.

7-18   List the three steps that should be taken when entrapment occurs.

7-19   Explain emergency procedures that responders take when an air supply is depleted.

7-20   Given a situation and specified SCBA malfunction, list self-preservation and emergency actions.

7-21   Given a situation involving a specified SCBA malfunction, demonstrate self-preservation and emergency actions.

7-22   Demonstrate rescue techniques for locating and removing a downed firefighter with a functioning SCBA.

7-23   Demonstrate rescue techniques for locating and removing a downed firefighter with a malfunctioning SCBA.

7-24   List and describe proper use of cleaning solvents.

7-25   Describe the proper manufacturers' or departments' procedures for tool and equipment cleaning.

7-26   Demonstrate proper exchange of an SCBA air cylinder.

7-27   Demonstrate air conservation techniques.

7-28   Identify methods rescuers can use to minimize the psychological effects they may experience while operating in limited visibility.

*The FF I and II levels, as defined by the NFPA 1001 Standards, are identified in different colors:
FF I = black, FF II = red, additional information = blue.

INTRODUCTION

Failure to understand and know how to use **self-contained breathing apparatus (SCBA)** properly could result in injury or death to a firefighter, failed rescue attempts of victims, or deterioration of the emergency incident. In addition to the short-term effects experienced during the emergency incident, firefighters may suffer long-term health problems due to repeated exposure to toxic environments. The **respiratory system** of the human body is extremely vulnerable to injury, especially from toxic conditions and gases encountered during firefighting operations. To protect firefighters in this environment, fire departments must have a respiratory pro-

tection policy or "mask rule" as outlined in NFPA 1404, Standard for Fire Service Respiratory Protection Training.

This policy should require all personnel to not only wear, but use their SCBA during operations where an **immediate danger to life and health (IDLH)** atmosphere may be encountered. Because it is impossible to predetermine all IDLH atmospheres, this policy must include operations during interior or exterior fire attack such as structure, vehicle, and dumpster fires; below-grade or confined space rescue; and hazardous materials incidents. Respiratory protection provided by SCBA is necessary, even during exterior defensive operations as shown in **Figure 7-1.**

FIGURE 7-1 Large volumes of smoke require the use of SCBA, even for exterior operations as shown here at a tire storage facility.

FIGURE 7-2 These firefighters in full protective equipment, including SCBA, are ready to begin interior firefighting operations.

SAFETY

Any inhaled toxic gas can directly cause disease of the lung tissue. SCBA, keen senses, incident command, and a thorough safety program are vital to keep firefighters safe from these deadly toxins and other safety hazards.

On average, one in twelve firefighters is injured in the line of duty each year, with more than 4,000 occurrences of smoke inhalation being reported yearly. It accounts for roughly 18 percent of firefighter fatalities and 21 percent of fireground injuries. The tasks of firefighters can only be accomplished effectively through the proper use of personal protective equipment (PPE), **Figure 7-2.**

Respiratory protection has long been realized as important to the fire service in the rescue of fire victims. Crude attempts from using beards to filter smoke to having hoses connected to apparatus pumping air to a mask have existed in the history of the fire service. SCBA similar to what we see today were not introduced into the fire service until the early 1970s. Today's self-contained breathing apparatus come directly from technology utilized by the National Aeronautics and Space Administration (NASA) in the space exploration program.

Over the last thirty years, many firefirefighter line-of-duty deaths have been attributed to running out of air in hostile environments. During the investigation of these incidents, questions concerning the design, duration, and use of the SCBA were reviewed, along with the training that firefighters had received for its use. As a result of these tragic incidents, many changes have occurred in SCBA including design and use. The most notable of these changes follow:

- *SCBA weight.* The weight of a standard thirty-minute SCBA unit used in the fire service has been reduced by 15 to 17 pounds, approximately 40 percent since 1980. This reduces the physical stress on the firefighter.
- *Positive pressure.* A constant supply of air is delivered, pressurizing the face piece, keeping toxic gases from entering. This pressure, which is slightly above atmospheric pressure, also helps maintain face piece seals.

NOTE

All SCBA used or manufactured for fire service use must be positive pressure.

- *Improved design.* Breathing tube, regulator, and harness designs have been improved to limit catastrophic failure during firefighting operations.
- *Improved regulator design.* New regulator designs provide increased airflow and redundancy, providing a backup in the event of a regulator/pressure-reducing failure.
- *SCBA maintenance.* Field- and factory-level maintenance programs have been implemented to ensure proper operation of SCBA units.
- *PASS devices.* Personal Alert Safety Systems (PASS) are audible warning devices that incorporate a motion detector to sense movement. The PASS will automatically sound an alarm signal (and usually a small flashing strobe) if movement is undetected for thirty seconds. This alarm alerts other firefighters to an unconscious or incapacitated firefighter. The device can also be activated

manually to signal others that a firefighter is in trouble and needs help. Newer SCBA units have the PASS device incorporated in the SCBA design, while older removable PASS devices are attached to the SCBA unit, usually somewhere on the harness. These devices are in common use and are required by NFPA 1500. Newer technology being introduced with PASS devices is allowing the Incident Commander to monitor a firefighter's air supply and location on the fireground.

- *Training.* **Mask confidence training,** or **"smoke divers"** training programs develop the firefighter's knowledge of and confidence in using an SCBA unit.

- *Increased regulation.* Implementation of **respiratory protection programs** is required by OSHA 29 CFR 1910.134 or NFPA 1500 and NFPA 1404.

These improvements are only as effective as the training firefighters receive and the proficiency they develop using SCBA. For this reason, training must be continual and review of basic and advanced skills must be emphasized. Technology, regulations, and mandates are only as good as an individual firefighter's commitment to use SCBA.

CONDITIONS REQUIRING RESPIRATORY PROTECTION

Four conditions that present respiratory hazards are commonly found at fire or other emergency incidents:

- Oxygen deficiency
- High temperatures
- Smoke or other by-products of combustion
- Toxic environments

These conditions can be found separately, such as an oxygen-deficient atmosphere in a confined space situation, or combined in a fire incident. Regardless of the incident, any time the potential for these environments exists or develops, firefighters must use their SCBA.

Oxygen-Deficient Environments

The human body and fire are similar as both require oxygen to survive. Fire (combustion) consumes oxygen and produces toxic gases that may displace or dilute the available oxygen. Atmospheres with oxygen concentrations below 19.5 percent are classified as **oxygen-deficient atmospheres.** As shown in **Table 7-1,** decreased oxygen affects the human body

TABLE 7-1 Effects of Hypoxia (Reduced Oxygen)	
Oxygen Present/ Available (%)	Symptoms
21	Normal conditions, no effect
19.5	OSHA definition as oxygen deficient
17	Some muscular impairment, increased respiratory rate
12	Dizziness, headache, rapid fatigue
9	Unconsciousness
7 to 6	Death within a few minutes

by causing muscular impairment, mental confusion, and eventually death. This in combination with toxic gases is the cause of most fire deaths for individuals lacking the protection of SCBA.

STREETSMART TIP

Installed fire-extinguishing systems such as total flooding carbon dioxide or halon systems create an oxygen-deficient atmosphere. When entering a structure or area where this type of system has activated, even without a fire, the firefighter must have the protection of SCBA.

Elevated Temperatures

The respiratory system of the human body is extremely delicate and sensitive to elevated temperatures. Even air temperatures related to a recreational activity such as a sauna, commonly 120° to 130°F (49° to 54°C), taken into the lungs may cause a serious decrease in blood pressure and circulatory system problems. Air temperature as low as 165°F (74°C) has been shown to cause death within a time frame of one minute for a respiratory tract that is unprotected. Inhaling heated gases will cause an accumulation of fluid in the lungs, called **pulmonary edema,** which can cause death from **asphyxiation.** In addition, the damage caused by inhaling heated air or gases is long term and not reversible by treatment of fresh, cool air.

FIREFIGHTER FACT

Temperatures in a structure fire reach 1,000° to 2,400°F. One unprotected breath at this temperature level will cause death or severe damage to a firefighter's respiratory system.

Smoke

Smoke is the combination of unburned products of combustion—particles of carbon, tar, and associated gases such as carbon monoxide, carbon dioxide, sulfur dioxide, hydrogen cyanide, benzene, and acrolein. Large amounts of these gases are now contained in smoke due to the ever-increasing use of plastics in manufacturing products used in our daily lives. Inhalation of small amounts of any of these materials may be fatal.

Smoke can cause significant damage to the human body in four different ways:

- Simple asphyxiation through displacement of available oxygen in the air.
- Chemical irritation such as edema or inflammation caused by hindrance of the respiratory tract.
- Chemical asphyxiation whereby the body's capability to utilize oxygen at the cellular level is disrupted.
- Any combination of these.

EFFECTS OF TOXIC GASES AND TOXIC ENVIRONMENTS

The combustion process produces toxic gases and irritants that affect both the short- and long-term health of firefighters operating in hazardous environments. In addition, when products of combustion combine, they may form toxins that are lethal. A swimming pool at a residential occupancy, **Figure 7-3,** indicates storage of chemicals such as chlorine, which will pro-

duce a poisonous gas. Firefighters must understand that there is no routine fire. Even a light smoke condition at this type of structure could be deadly.

Some of the common gases produced in a fire affect not only the respiratory system, but the circulatory system as well, **Table 7-2.** In addition to the toxic products of combustion, chemicals present at commercial occupancies will produce many additional toxins, thus requiring a higher level of protection for firefighters, **Figure 7-4.**

Carbon Monoxide

Carbon monoxide (CO) is produced in great quantities during the combustion process and is one of the most lethal gases found in a fire. This colorless

FIGURE 7-3   The swimming pool is a warning to firefighters of the possible presence of chemicals in storage at this residence.

TABLE 7-2	Toxic Gases Formed as Products of Combustion		
Gas	**Toxicological Effect**	**Produced By**	**IDLH (ppm)***
Carbon dioxide	Displaces oxygen	Free burning	40,000
Carbon monoxide	Displaces oxygen	Incomplete combustion	1,200 to 1,500
Hydrogen cyanide	Chemical asphyxiant	Wool, silk, nylon, polyurethane	50
Hydrogen chloride	Respiratory irritant	Polyvinyl chloride (PVC), building materials, furnishings	50
Nitrogen dioxide	Pulmonary irritant	Small quantities from fabrics, large quantities from cellulose nitrate	20
Phosgene	Poison	Burning refrigerants; produces hydrochloric acid when inhaled	25

*ppm 5 parts per million, ratio of volume of gas compared to volume of air.

FIGURE 7-4 Agricultural and hardware store occupancies will usually have a large supply of chemicals that are extremely hazardous during firefighting operations.

TABLE 7-3 Symptoms of Carbon Monoxide Poisoning

Symptoms	Carbon Monoxide Concentration (ppm)*
Mild headache	1,000
Headache, nausea	1,300
Unconsciousness after 1 hour†	1,500
10-minute exposure, dizziness, nausea	3,200
Fatal, less than 1-hour exposure	4,000
Danger of death in 1 to 3 minutes	10,000

*ppm 5 parts per million, ratio of volume of gas compared to volume of air.

†IDLH level.

and odorless gas is always present, especially during incomplete combustion or in areas of poor ventilation. Carbon monoxide can also be found in homes with defective furnaces and clogged chimneys. Its presence is of great concern, especially during the overhaul stage when respiratory protection habits may be lax.

CAUTION

Firefighters should always use their SCBA, even when the fire is thought to be extinguished. Lethal amounts of carbon monoxide may be present.

Ordinarily, 98 percent of the oxygen carried by the blood is bound to the hemoglobin, which is contained in the red blood cells. CO attaches itself to the red blood cells and prevents oxygen from bonding with the hemoglobin. In fact, CO combines with blood over 200 times easier than oxygen. It does not act directly on the body, but prevents oxygen from being distributed to the body, causing **hypoxia,** followed by death. **Table 7-3** shows the stages and symptoms of this process.

Intense physical activities during firefighting operations, **Figure 7-5,** require the respiratory system to increase the respiratory rate to deliver increased oxygen to the muscles and organs of the body. If a firefighter is working unprotected (without the protection of SCBA) in an atmosphere of CO, the increased respiration augments the ability of CO to incapacitate the firefighter. CO also does not metabolize out of the blood very quickly. During repeated exposures to low levels of CO during a shift, or during a number of fires, the blood level of CO is compounded. CO symptoms range from a mild headache to death, Table 7-3.

Hydrogen Cyanide

Hydrogen cyanide is another toxin prevalent in large quantities in smoke. Hydrogen cyanide is colorless and is produced by the combustion both of natural products such as wool, silk, cotton, and paper as well as synthetics or plastics. Hydrogen cyanide can be present in an atmosphere long before the ignition temperature of a material is reached or after a fire is extinguished due to off-gassing, or **quantitative decomposition,** of materials. It is being argued in some studies that hydrogen cyanide poisoning may be the primary cause of death in smoke inhalation victims as opposed to carbon monoxide. It is often very hard to detect in the human body unless tested for immediately due to the toxin having a very short **half-life.** When combined with carbon monoxide, the effects of hydrogen cyanide are compounded significantly.

FIGURE 7-5 The light smoke condition present during overhaul will contain large amounts of carbon monoxide and hydrogen cyanide, requiring SCBA protection.

FIREFIGHTER FACT

Depending on the laws, rules, and regulations of a particular state or local government, OSHA standards may or may not be legal requirements. In general, the U.S. Department of Labor OSHA regulations do not apply to municipal employees, including firefighters. However, a number of states, **Figure 7-6,** have adopted their own plans, which must be as stringent as the federal requirements. In these states, the requirements do apply to municipal employees, including firefighters. Generally, volunteer firefighters are considered employees for the purposes of health and safety regulation. Simply said, firefighters must know the rules and regulations that apply to their organization. For more information on OSHA, see Section I, Chapter 2.

LEGAL REQUIREMENTS FOR SELF-CONTAINED BREATHING APPARATUS USE

Toxins are always present in the by-products of combustion. Common sense dictates that firefighters should use SCBA on every fire scene—from start to finish. For the firefighter's safety and health, a number of regulations have been developed for SCBA use. A number of organizations have also established regulations and standards concerning the design and use of SCBA, **Table 7-4.**

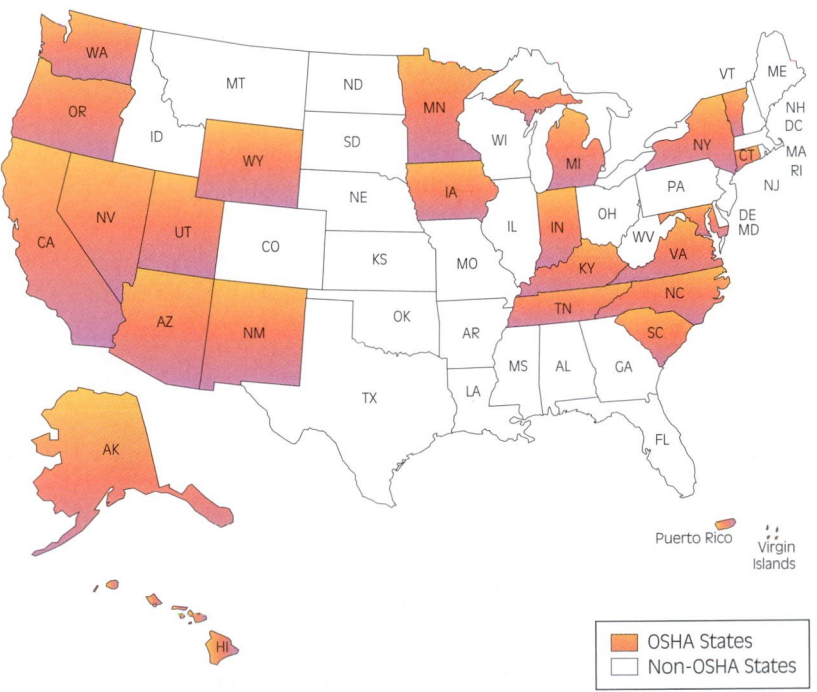

OSHA States
Non-OSHA States

FIGURE 7-6 States requiring respiratory protection plans in accordance with OSHA's 29 CFR 1910.134 regulations.

TABLE 7-4	Organizations Concerned with SCBA Design and Use	
Organization	**Standard**	**Application**
National Institute for Occupational Safety & Health (NIOSH)	42 CFR Part 84	Requirements for design, testing, and certifying SCBA
Occupational Safety & Health Administration	29 CFR 1910.134	Respiratory protection programs for SCBA use
Occupational Safety & Health Administration	29 CFR 1910.156	Fire Brigade Standard, references 1910.134
National Fire Protection Association	NFPA 1404	Standard for a Fire Department SCBA Program
National Fire Protection Association	NFPA 1500	Standard on Fire Department Occupational Safety and Health Program
National Fire Protection Association	NFPA 1981	Standard on Open-Circuit SCBA for the Fire Service

Title 29 Code of Federal Regulations, Section 1910.134

OSHA 29 CFR 1910.134 establishes the standard for all entries into IDLH atmospheres. The April 1998 revision contains requirements related to interior structural firefighting and defines interior structural firefighting as an IDLH atmosphere. In addition to requiring the use of SCBA, the standard also establishes requirements for a complete respiratory protection program and regular medical evaluation of employees designated to wear and use SCBA.

Although employers (fire departments) are responsible for providing a safe and healthy work environment, all firefighters have a duty to understand and follow these regulations.

NFPA 1500: Standard on Fire Department Occupational Safety and Health Program

The National Fire Protection Association has established **NFPA 1500,** Standard on Fire Department Occupational Safety and Health Program. This standard covers many safety and health-related issues for firefighters, including respiratory protection. The difference between this standard and the OSHA regulations is that a government authority (city, town, county, or state), called the **authority having jurisdiction (AHJ),** must adopt the standard as policy for the fire department.

The NFPA has three additional standards dealing with SCBA: **NFPA 1404,** Standard for a Fire Department Self-Contained Breathing Apparatus Program, **NFPA 1981,** Standard on Open-Circuit Self-Contained Breathing Apparatus for Fire Service, and **NFPA 1982,** Standard on Personal Alert Safety Systems (PASS). The NFPA 1404 standard establishes minimum requirements for fire department respiratory protection programs including the use of SCBA in the fire service, regarding training, safety, emergency procedures, maintenance, and breathing air. The NFPA 1981 standard establishes design and performance criteria, test methods, and certification for open-circuit SCBA intended for fire service use.

Both 29 CFR 1910.134 and NFPA 1500 contain similar requirements for respiratory protection. These are nationally recognized standards, and fire departments in jurisdictions without mandatory respiratory protection requirements should adopt or reference these for operations requiring the use of SCBA.

LIMITATIONS OF SELF-CONTAINED BREATHING APPARATUS

As with any other tool used by the fire service, SCBA has a number of limitations that firefighters must understand if they are to use the unit effectively and safely. These limitations are the SCBA unit itself (i.e., its size, weight, and limited air supply) and the physi-

FIGURE 7-7   Continuous training with SCBA is one of the keys to effective firefighting operations.

cal and physiological condition of the user. Firefighters must be well trained to successfully complete all tasks and assignments requiring the use of SCBA, **Figure 7-7.**

SCBA Design and Size

The design and size of SCBA units vary greatly by manufacturer and by the age of the unit. Older units placed in service as little as ten years ago may be heavier and bulkier than newer ones. Firefighters must be conditioned to the added weight and bulk of the SCBA and the personal protective equipment they must wear. Depending on the activity level required and the physical condition of the user, an SCBA cylinder will often be consumed more quickly than the length of time that it is rated for. These factors limit the distance a firefighter may advance into a building or fire and requires frequent crew rotations at large incidents. Other concerns include:

- Visibility is restricted, peripheral vision is reduced, and fogging of the face piece can occur.
- Depending on the manufacturer's style, the age of the unit, and the type of cylinder used, SCBA units will add 23 to 35 pounds of weight and 9 to 15 inches to the profile of the firefighter. This will limit the mobility of the firefighter.
- Unless the SCBA unit is equipped with a voice amplification accessory **Figure 7-8,** the firefighter's voice will be muffled and difficult for others to understand, especially if the firefighter is trying to communicate by radio.
- The quantity of air is limited. It is vital that firefighters know the status of the cylinder air supply prior to its use and be familiar with their own air consumption rate in a given situation.

FIGURE 7-8   A voice amplifier can help a user communicate while wearing an SCBA.

STREETSMART TIP

The Philadelphia Fire Department conducted extensive testing in a firefighting skills proficiency course with 1000 firefighters using SCBA. The average air consumption for SCBA rated for thirty minutes was less than fifteen minutes—from full cylinder to low air warning. Forcible entry and ventilation are two tasks requiring increased physical activity, **Figure 7-9A** and **B.**

Limitations of the SCBA User

Firefighters themselves can be limited in their use of SCBA. A firefighter's physical, mental, and emotional condition can cause usage problems. When these limitations are coupled with the confines of the SCBA unit there can be serious problems.

(A)

(B)

FIGURE 7-9 (A) Ventilation and (B) forcible entry are physically demanding. They produce increased respiration rates and air consumption.

Physical Limitations

- The total protective envelope including SCBA and PPE adds approximately 40 to 50 pounds of weight to the firefighter.

- The additional weight and bulk affect agility and mobility, requiring a high level of physical strength and endurance.

- Even though face pieces have been fitted and tested, weight loss or a twenty-four-hour growth of facial hair may affect the ability to obtain a good seal.

- In addition to strength and endurance, the stress of firefighting in elevated temperatures requires additional cardiovascular and respiratory conditioning. Lack of conditioning will increase stress on the body.

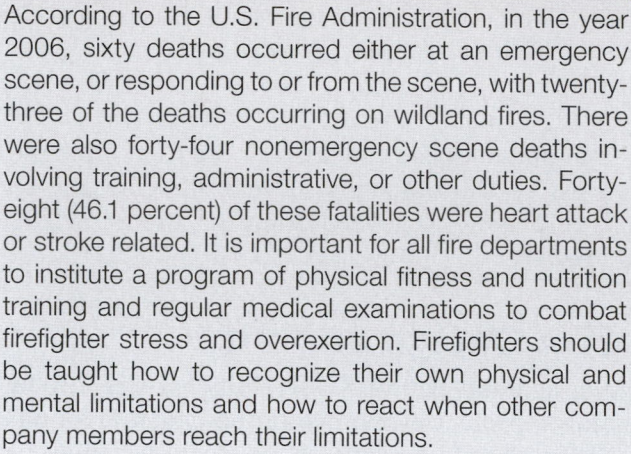

FIREFIGHTER FACT

According to the U.S. Fire Administration, in the year 2006, sixty deaths occurred either at an emergency scene, or responding to or from the scene, with twenty-three of the deaths occurring on wildland fires. There were also forty-four nonemergency scene deaths involving training, administrative, or other duties. Forty-eight (46.1 percent) of these fatalities were heart attack or stroke related. It is important for all fire departments to institute a program of physical fitness and nutrition training and regular medical examinations to combat firefighter stress and overexertion. Firefighters should be taught how to recognize their own physical and mental limitations and how to react when other company members reach their limitations.

Psychological Limitations

- Lack of confidence in the SCBA unit and its ability to protect the firefighter may cause anxiety, increasing the breathing rate.

- The degree of training or experience users have with SCBA affects their self-confidence and ability to function.

- Increased physical stress may cause anxiety.

- Emotional conditions, such as fear of being confined, excitement, or claustrophobia, may increase the user's breathing rate and air consumption.

These problems are addressed with proper training and education. Training in SCBA use is not a one-time deal; it must be continuous so firefighters maintain their proficiency and confidence with SCBA.

Air Supply Management

It is imperative that firefighters understand their air consumption rates to have an idea of how long they can work while on air. This information must be known and understood prior to a firefighter being put into a situation. A team making entry into a hazardous environment will only be as strong as the "weakest link" or lowest air cylinder pressure of the team members. All members of a team should always be aware of each other's air level.

A crew's effectiveness in the hazard zone on the fireground will be directly related to their ability to manage their air supply. Basic physiology tells us that different firefighters will consume the air in their SCBA at different rates. The working air supply will depend on the firefighter's training, physical condition, activity, and mental state experienced under the stressful conditions encountered during firefighting.

Firefighters are responsible for determining their individual **point of no return** whenever they enter a hazardous atmosphere. Per regulatory standards, the low pressure warning device on a firefighter's SCBA is to activate when the supply of air in the SCBA cylinder reaches below 25 percent of its rated capacity. Routinely, a firefighter will need more time to exit than would be allowed if relying solely on the low pressure warning device. Thinking in very simple terms: we know that the low pressure warning device activates when the air supply reaches a level of 25 percent of the cylinder's capacity. This means that it took 75 percent of the cylinder's capacity to get to the point that the firefighter is at, which can leave the firefighter with a deficit of as much as 50 percent of the cylinder's capacity to safely exit. It is very important that a firefighter **not** rely solely on the SCBA's low pressure warning device.

The 2002 edition of NFPA 1981 mandated the use of the **heads up display (HUD),** which is a visual signal signifying the available amount of an air cylinder's rated capacity. These parameters basically stated that SCBA were to have LED lights visible inside of the facepiece indicating when 50 percent of the air cylinder's capacity was remaining. Unfortunately, not all fire departments are operating with SCBA that are compliant with the latest edition of standards such as NFPA 1981.

Oftentimes, it is thought that an air bottle with a larger capacity is the answer to the problem of air management, and this can be very misleading. Utilizing a bottle with more air capacity will equate to a firefighter spending an increased amount of time operating inside of the hazard zone. Firefighting activity is very stressful on the human body and spending additional time working in the hazard zone will increase the amount of stress on the firefighter. This additional stress leads to the potential of both increasing injuries to firefighters as well as exposure to cardiac-related events such as heart attack. If cylinders with larger capacities are being used, extra attention must be given to air management in order to avoid these potential situations.

Determining a point of no return for a firefighter can be as simple as checking air level prior to entry into the hazard zone when donning the mask and checking it again when reaching the objective. This allows a firefighter to know exactly how much air it took to arrive at this point and indicates approximately how much will be needed, theoretically, to get back to the entry point. The disadvantage to this method, however, is that it leaves no safety cushion should the firefighter experience any problems while exiting—and if the firefighter does not check the level regularly, there may be insufficient air to exit safely. Also, more air may be needed for exit than entry because the firefighter will already have been working and may be stressed to the point of needing more.

Another air management method that has been practiced is the 10-10-10 rule, which simply means that crews are allowed ten minutes to enter and work, ten minutes to exit, and ten minutes as a safety cushion in case they experience problems. An advantage to this method is that it gives a clear definition of the working and withdrawal period. Disadvantages are that some firefighters may breathe down a bottle much faster than the allotted ten-minute time parameters in a given situation so that someone must be dedicated to watching and notifying the crew of their time periods as they pertain to the rule.

Simply stated, a good practice for firefighters is to exit the hazard zone before the low pressure warning device activates on their SCBA. The air remaining in an SCBA after the activation of the low pressure warning device is best left as a safety cushion to be used in an emergency situation.

NFPA 1404 specifically states that firefighters must be trained in the "identification of factors that affect the duration of air supply" and "determination of the point of no return." Calculating a consumption rate on a regular basis from a consumption test can help departments accomplish this. Consumption testing in this manner has advantages because a firefighter is working at a constant rate when determining it.

Consumption testing will involve a firefighter working on an obstacle course while experimenting with different breathing rates and techniques, **Figure 7-10A–C.** To get a true indication, the obstacle course must be set up exactly the same each time a consumption test is performed, **Figure 7-11.**

The test should be performed until the firefighter cannot draw any additional air from the SCBA. This will give the individual a true indication of exactly how much time is left to operate in the event that the firefighter becomes trapped or lost. It will also demonstrate to the firefighter how much air is still left in an SCBA even after the low pressure/25 percent alarm stops operating. Knowing this information can help keep firefighters calm and possibly enable them to save their own lives, **Table 7-5.**

(A)

(B)

(C)

FIGURE 7-10 Consumption testing will involve a firefighter going through an obstacle course of simulated fireground activities while experimenting with different breathing rates and techniques.

There are various methods of breathing that may help in reducing a firefighter's consumption rate. It will take experimentation on the firefighters' part to find which one works the best for them. When using any method, it is important to take normal breaths and exhale slowly to keep CO_2 in the lungs within proper balance. Firefighters should *never* hold their breath

in an attempt to save air. Due to the body's release of adrenaline when firefighting, oxygen is being consumed at a higher rate and holding the breath could cause unconsciousness.

A controlled breathing method can provide for the most efficient use of air. Breathing only through the mouth or nose is not the answer to conserving

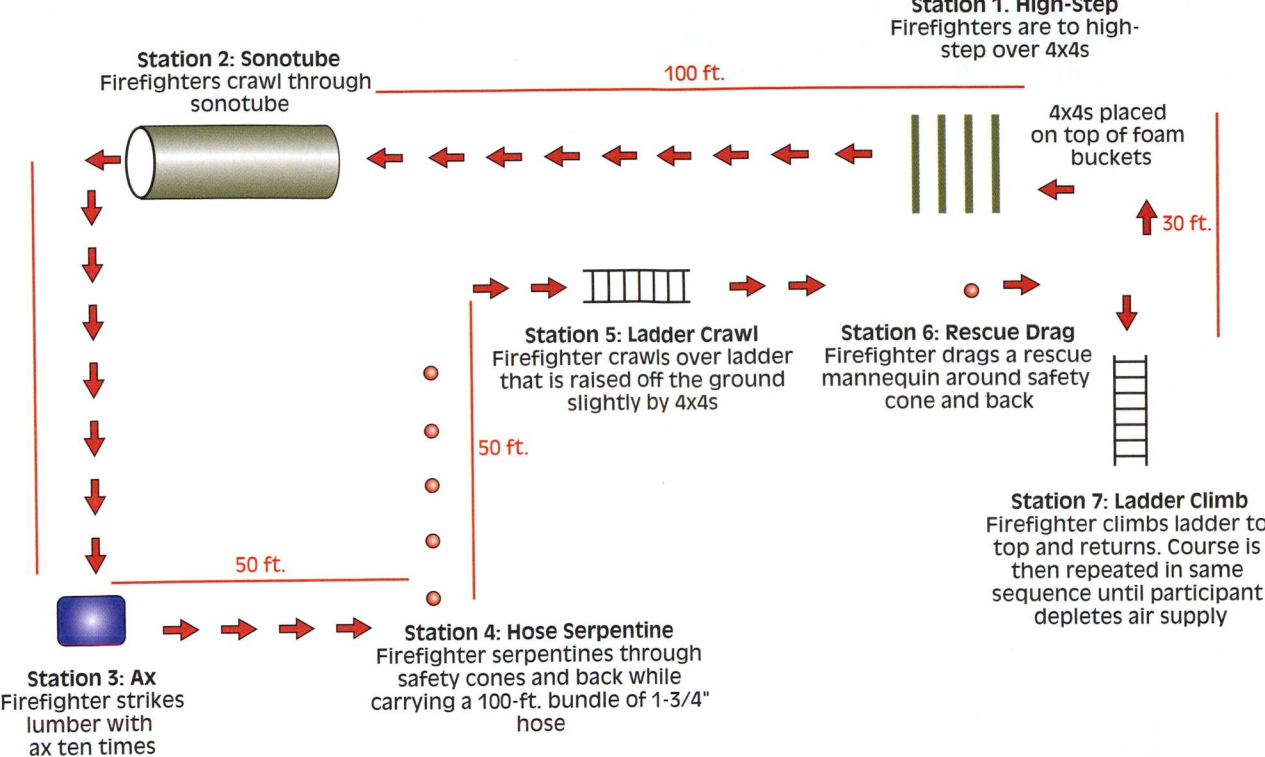

Sample Air Consumption Course
Distances between stations need to be the same each time the test is performed to obtain accurate results.
Firefighters should also have a partner to monitor their safety and record information at all times while on the course.

FIGURE 7-11 An example of an air consumption test.

TABLE 7-5	Air Consumption Test Record

Name _____

Date _____

Starting Cylinder Pressure _____ psi

Starting Time _____

Low Pressure Warning Device (25%) Activation Time _____

Minutes/Seconds Until Low Pressure Warning Device Sounded

_____ Minutes _____ Seconds

Air Cylinder Empty Time (No more air to mask) _____

Time Worked After Low Pressure Warning Device Sounded

_____ Minutes _____ Seconds

Total Time Worked From Start to Empty Cylinder

_____ Minutes _____ Seconds

air. Breathing only through the mouth results in an increased respiratory rate while preventing the body from utilizing all available oxygen before exhalation. Breathing exclusively through the nose results in short breaths that do not fill the lungs to their full capacity.

However, inhaling through the mouth and exhaling through the nose is a method that provides more than adequate air exchange and can be beneficial while engaged in a heavy work load. To reverse that pattern—inhale through the nose and exhale through the mouth—is another method that provides for good air exchange and is easy to remember. Slow, deliberate exhalation is the key to both approaches and any other breathing technique in helping to conserve air.

Whatever breathing techniques are used by a firefighter, it is paramount that they are performed calmly and efficiently. Training, such as regular consumption tests, will allow a firefighter to practice and perfect breathing techniques and will also help promote emotional stability of firefighters while wearing SCBA. A regular physical fitness program will also help firefighters to wear SCBA with less fatigue and minimum consumption rates.

TYPES OF SELF-CONTAINED BREATHING APPARATUS

Two different types of SCBA are in general use in today's fire service: open-circuit and closed-circuit systems. In an **open-circuit SCBA,** exhaled air is vented to the outside atmosphere, whereas in a **closed-circuit SCBA,** the exhaled air stays in the system for filtering, cleaning, and circulation. The open-circuit SCBA, **Figure 7-12A** and **B,** is the most commonly used for firefighting operations. The closed-circuit type is sometimes used for specialized rescue incidents.

Open-Circuit Self-Contained Breathing Apparatus

A number of manufacturers supply SCBA for fire service use. Regardless of the manufacturer, SCBAs for fire service use are designed and built in accordance with **NIOSH** and NFPA standards. Certain parts of an SCBA unit may be interchangeable with parts from another manufacturer; however, this practice will void the NIOSH and manufacturers' certifications.

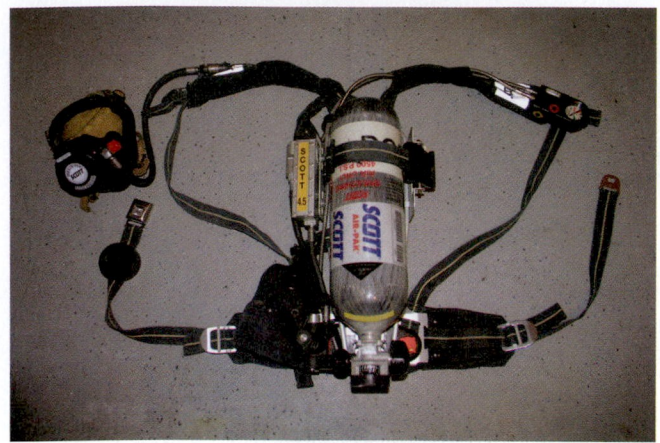

FIGURE 7-12   A number of manufacturers produce open-circuit SCBA for fire service use.

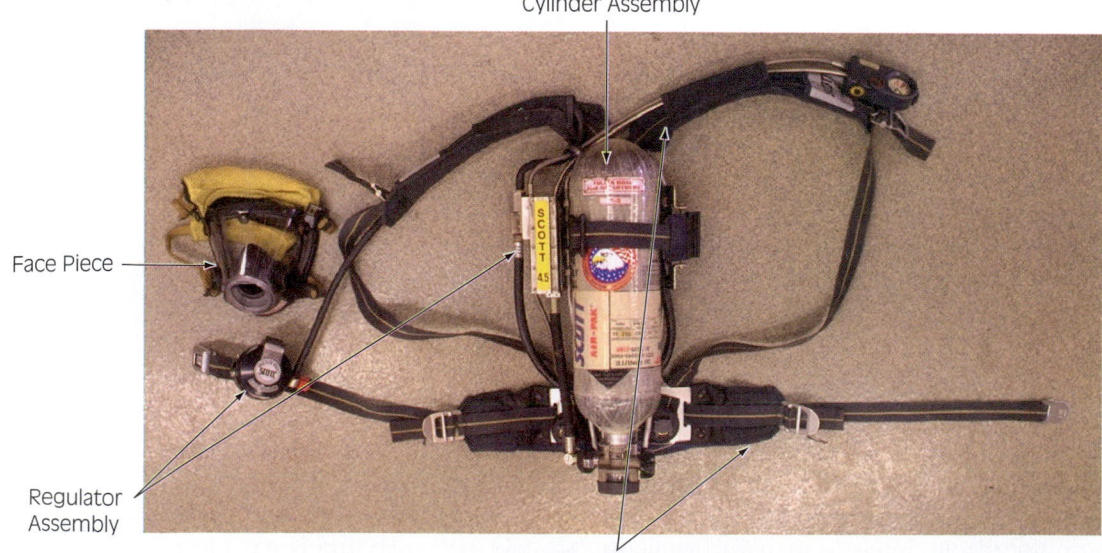

Cylinder Assembly

Face Piece

Regulator Assembly

Backpack/Harness Assembly

FIGURE 7-13   The four components of the open-circuit SCBA are the backpack/harness, cylinder, regulator, and face piece assemblies.

Generally, as shown in **Figure 7-13,** SCBA has four basic assembly components:

- *Backpack and harness:* The backpack holds the air cylinder, and the harness allows the unit to be worn.
- *Cylinder:* Includes cylinder, valve, and cylinder pressure gauge.
- *Regulator:* Includes the high-pressure hose from the cylinder, the regulator, and end of service time indicator (EOSTI).
- *Face piece assembly:* Includes face piece, exhalation valve, and head harness.

Backpack and Harness Assembly

The backpack assembly, **Figure 7-14,** is designed to hold the air cylinder and provide the straps for securing the SCBA unit to the firefighter. It will consist of a metal or a high-temperature-resistant plastic frame, a mechanism for securing the air cylinder, and shoulder and waist straps. Depending on the manufacturer's design and style, the regulator may be attached to the waist straps. The straps are adjustable to the size of the user and designed to distribute the weight of the unit. Most models in use today are designed to have the greatest portion of the unit's weight carried on the hips by the waist straps.

Cylinder Assembly

SCBA cylinders, **Figure 7-15,** contain the compressed air for breathing by the user. For this reason, the cylinder must be strong and durable; therefore, it constitutes most of the weight of the unit. Most cylinders in use today are aluminum, fiberglass/aluminum composite, and Kevlar or carbon composite materials. The change from steel to aluminum and then composite cylinders has reduced the weight of an SCBA cylinder by almost 50 percent. Typical cylinder capacities are listed in **Table 7-6.**

Types of Cylinders. The SCBA cylinders used in the fire service vary in material and type of manufacture. Listed here are general descriptions of the most common types of cylinders used in the fire service. Refer to manufacturers' information for exact specifications.

- *Steel.* Because of improvements in design and weight reduction using other materials, steel cylinders are generally not used for SCBA service. Most of the steel cylinders still in fire service use are used to provide air supply for various rescue tools. They have an unlimited service life as long as they pass a hydrostatic test every five years.
- *Aluminum.* Aluminum cylinders are used for thirty-minute (2,216-psi) rated SCBA units. This type of cylinder contains 44 cubic feet of compressed air and weighs about 22 pounds when charged. These cylinders have an unlimited service life as long as they pass a hydrostatic test every five years.
- *Fiberglass (hoop-wrapped).* Fiberglass cylinders are used for thirty-minute (2,216-psi) rated SCBA units and are constructed with an aluminum inner shell. The cylinder sides are wrapped with fiberglass for strength. This type of cylinder contains 44 cubic feet of compressed air and weighs about 16 pounds when charged. These cylinders

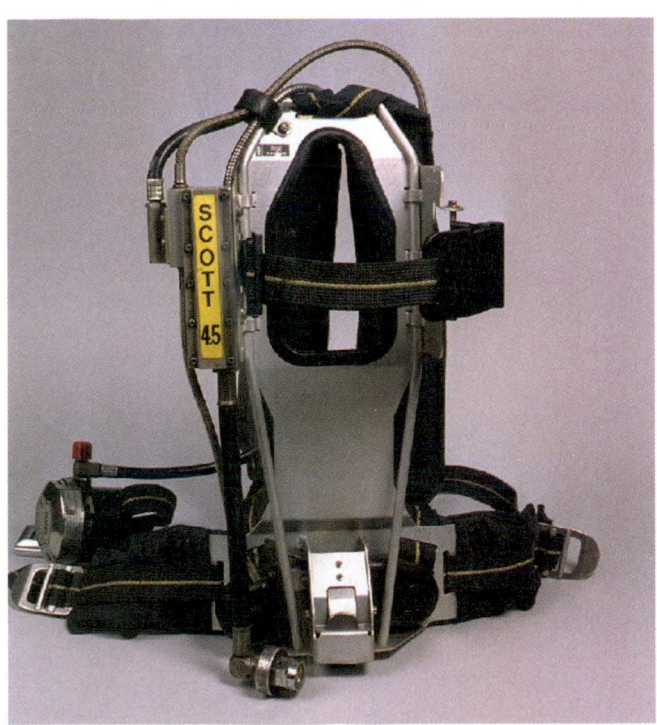

FIGURE 7-14 A typical SCBA backpack/harness assembly.

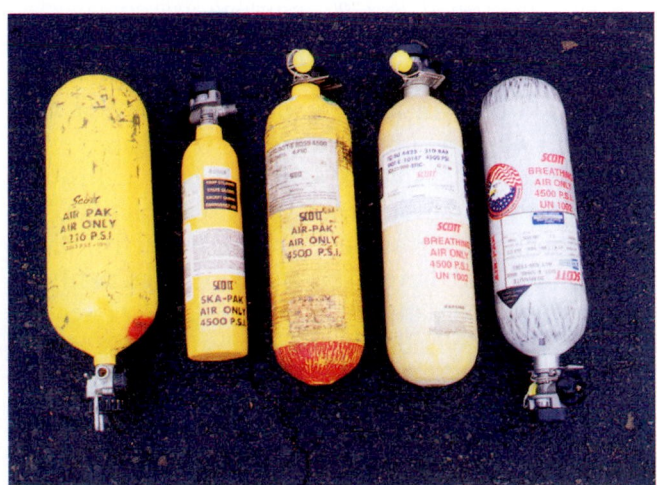

FIGURE 7-15 SCBA cylinders for fire service use, from left to right, are steel, aluminum, fiberglass, Kevlar composite, and carbon fiber composite.

TABLE 7-6   SCBA Air Cylinder Capacities

Rated Duration (minutes)	Material	Cylinder Pressure (psi)	Capacity-Compressed Air (ft³)
30	Aluminum or composite	2,216	44
30	Aluminum or composite	4,500	44
45	Carbon fiber	3,000	65
45	Aluminum or composite	2,216	65
45	Aluminum or composite	4,500	65
60	Aluminum or composite	4,500	88

have a service life limited to fifteen years and must have a hydrostatic test every three years. In accordance with current U.S. Department of Transportation (USDOT) regulations, these cylinders are destroyed at the end of their service life.

■ *Fiberglass (full-wrapped).* Full-wrapped fiberglass cylinders are used for thirty-, forty-five-, and sixty-minute (4,500-psi) rated SCBA units. They are constructed with an aluminum inner shell and fully wrapped with fiberglass for strength. These cylinders contain 44 to 88 cubic feet of compressed air and weigh 16 (44 ft³) to 26 (88 ft³) pounds when charged. They have a service life limited to fifteen years and must have a hydrostatic test every three years. In accordance with current USDOT regulations, these cylinders are destroyed at the end of their service life.

■ *Kevlar/carbon composites.* Composite cylinders are used for thirty-, forty-five-, and sixty-minute (4,500-psi) rated SCBA units and are constructed with an aluminum inner shell and fully wrapped with Kevlar or carbon fibers for strength. These cylinders contain 44 to 88 cubic feet of compressed air and weigh 12 (44 ft³) to 22 (88 ft³) pounds when charged. Kevlar cylinders must have a hydrostatic test performed every three years. Carbon fiber bottles must be hydrostatically tested every five years. Presently, both have a service life limited to fifteen years. In accordance with current USDOT regulations, these cylinders are destroyed at the end of their service life.

SCBA Cylinder Failures. The rated duration of all SCBA cylinders is based on laboratory testing; actual duration will vary significantly with each individual user. One of the most important factors affecting air supply duration is the physical and physiologic condition of the user. Firefighting is physically demanding work in a very harsh environment, and when combined with the stress of searching for victims, it usually leads to increased breathing and consumption of air. For this reason, even firefighters in top physical condition may only be able to operate for fifteen to twenty minutes with a unit rated for thirty minutes.

CAUTION

In the past few years, the failure of composite-type cylinders has occurred with increasing frequency. In three instances, cylinders failed while stored on fire apparatus. Investigations concluded that the failures were caused by stress-corrosion cracking of the fiberglass wraps resulting from exposure to a strong corrosive agent. Fiberglass composite cylinders are particularly at risk for stress-corrosion cracking because the fibers are under constant tension due to the internal pressure. When the structural integrity of the overwrap is weakened, a catastrophic failure of a cylinder can occur that may result in serious injury or death. Persons responsible for maintenance of composite cylinders should take measures to ensure that they do not come in contact with strong corrosive agents. In addition, cylinders should be washed only with a mild soap and water solution, and all recommendations of the cylinder manufacturer or distributor with regard to maintenance, requalification, and use must be carefully followed. In addition, during the 1980s a number of aluminum cylinders developed catastrophic failures at the neck area, near the connection of the cylinder valve. Even with regulations and testing, failures do occur and firefighters must be aware of the correct maintenance procedures for SCBA cylinders.

Cylinder Testing. The USDOT regulates compressed gas cylinders, including those used for SCBA, and requires hydrostatic testing. This test is done to ensure that the cylinder is capable of withstanding its rated pressure and capacity and the stress created when the cylinder is being filled. Cylinder testing is usually accomplished by an outside vendor; however, some large fire departments may have their own service unit that conducts these tests.

Once the test is complete, each cylinder is labeled or stamped as shown in **Figure 7-16.** This label must show the licenses of the testing organization and the date of the most recent test. One should never attempt to fill a cylinder with an out-of-date test label.

> **CAUTION**
>
> In 1993 a firefighter filled a composite-type SCBA cylinder that was beyond its useful service life of fifteen years. After filling the cylinder, he placed it back on the apparatus and then a catastrophic failure occurred. The cylinder valve exploded out of the cylinder and killed the firefighter instantly. One should never service an SCBA cylinder that is out of date for hydrostatic testing or a composite cylinder that has passed its fifteen-year service life.

Air Quality for SCBA Use

The quality of the compressed breathing gas used in open-circuit SCBA has a direct effect on the performance of this equipment. Many standards address the quality or purity of compressed air used for breathing purposes. The most generally accepted are those established by the Compressed Gas Association in its pamphlet G-7.1-1989, which are incorporated by reference in OSHA 29 CFR 1910.134 and NFPA 1500. Briefly, the quality of air used for SCBA must be Grade D or better, as established in the pamphlet. Through research, it has been found that the minimum qualities in this grade air are required to allow for the best operation and maintenance of SCBA components. This classification establishes maximum allowable quantities of impurities allowed in breathing air. In accordance with this specification the supply source (purifier/compressor) must be tested at least every three months and certified as shown on the certificate in **Figure 7-17.** NFPA 1404 has one exception to the Grade D breathing air requirement and that is a lower maximum moisture content of 24 ppm or drier.

This air is obtained from an air cylinder refill system owned by a fire department or from a commercial vendor. These systems, commonly called cascade systems, consist of a compressor, purifier, air quality

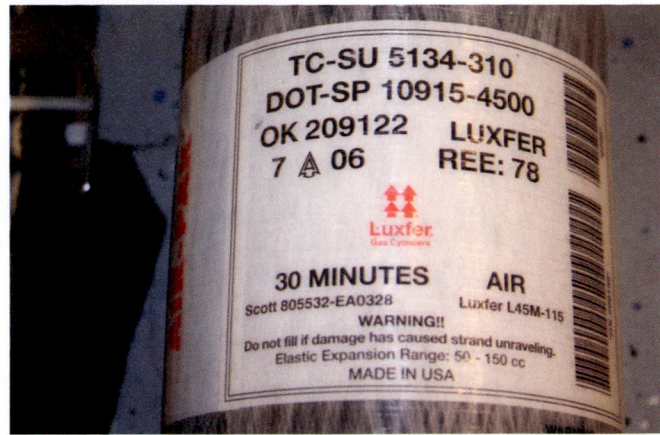

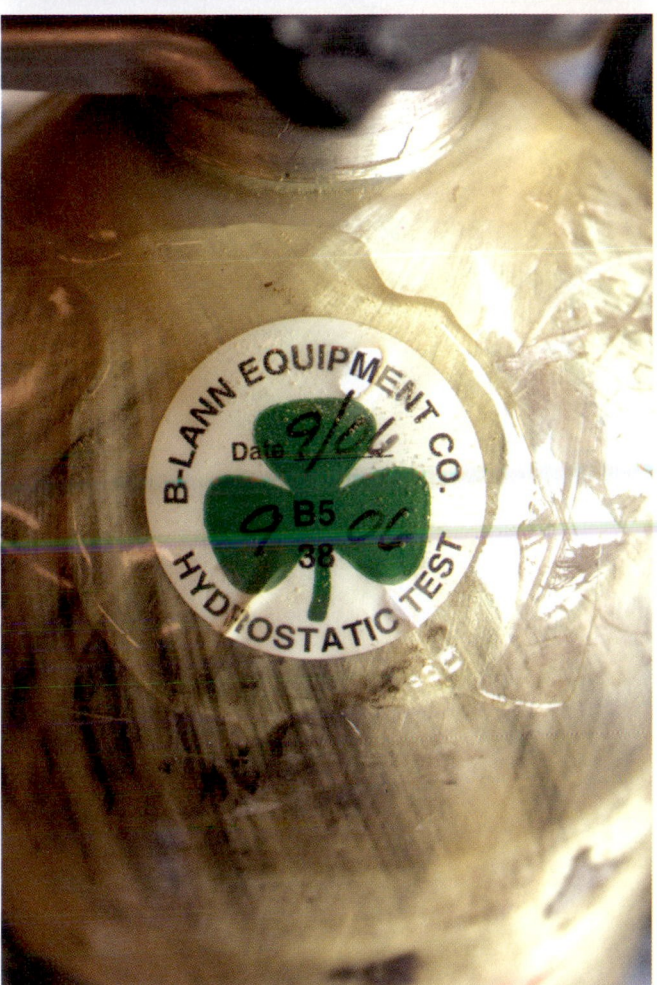

FIGURE 7-16 Depending on the type of material, SCBA cylinders are required to be hydrostatically tested every three to five years.

monitor, storage system, booster compressor, control panel, and cylinder fill containers **Figure 7-18.**

Regulator Assembly

Depending on the manufacture and style of the SCBA unit, the regulator assembly, **Figures 7-19A** and **B,** is attached to the face piece or waist straps. The regulator

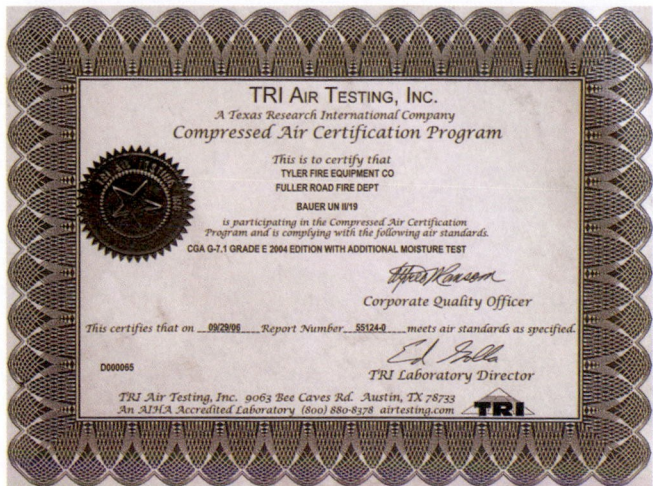

FIGURE 7-17 All air systems for SCBA use must be tested every three months and certified as shown by this air test quality certificate.

(A)

(B)

FIGURE 7-19 SCBA designed for the fire service will use either (A) a face piece–mounted regulator or (B) a waist-mounted regulator.

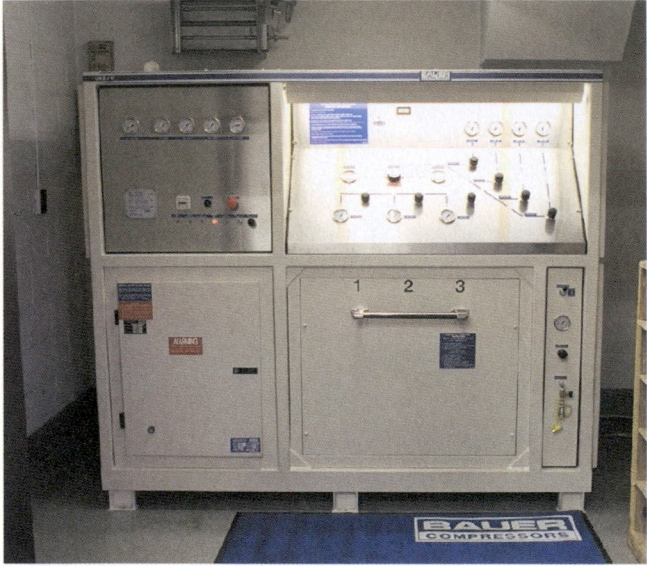

FIGURE 7-18 A compressor is the second type of system used to service SCBA cylinders. These are usually located at stations or other fixed facilities; however, larger departments may have mobile units.

reduces the high-pressure air from the cylinder to a low pressure that is slightly above atmospheric pressure, and controls the flow of air to the face piece. The regulator contains a diaphragm that is activated by the breathing action of the user. This action creates a pressure differential, which opens the diaphragm and allows air to flow to the face piece. In addition, all SCBA units designed for fire service use also maintain a constant positive pressure airflow to the face piece. The exhalation action moves the diaphragm to the closed position and opens exhalation valves, and the exhaled air is vented outside the face piece.

Again depending on the manufacturer and style of the SCBA unit, the regulator will have color-coded valves to control both normal and emergency operation. One is the main line valve, usually colored yellow for normal operation; the second is the bypass valve, usually colored red for emergency operation. On some units, this red knob may not be a true bypass and may function only as a purge valve to get additional airflow to the facepiece; the increased airflow aids in clearing mask fogging and any debris from the regulator if left unprotected. It is imperative that firefighters understand the particular unit that they are using and consult the manufacturer's instructions for that unit.

During normal operation in units with a bypass, the main line valve is fully open, allowing maximum air to flow, and the bypass valve is in the fully closed position. The valves should remain in these positions for proper operation of the unit. In the event of a malfunction of the regulator or high-pressure reducer, the bypass valve must be manually operated.

> ### STREETSMART TIP
>
> When using SCBA, firefighters must work in teams of a minimum of two, and all team members must exit the hazardous environment when a low air supply alarm sounds. A firefighter must never leave a partner or allow a partner to leave alone.

The regulator may have a remote pressure gauge, **Figure 7-20,** that is part of the regulator body or attached to the shoulder straps. This gauge should provide the same readings as those shown on the cyl-

FIGURE 7-20   The regulator may have a remote pressure gauge as part of the regulator body, or it may be attached to the shoulder straps. This gauge should provide the same readings as those shown on the cylinder valve.

inder valve. There may be some difference, but the two gauges must read within 10 percent (some organizations use 100 psi) of each other or the unit should be taken out of service. Depending on the manufacturer, the increments shown on the gauge may be in percentages of a full cylinder or will display the actual psi remaining in the cylinder.

SCBA units designed for fire service use have a low air supply warning alarm, which is usually part of the regulator assembly. This alarm sounds when the cylinder pressure decreases to approximately one-fourth of the rated capacity of the cylinder. Depending on the manufacturer, some alarms on face piece-mounted regulators will also vibrate the face piece assembly to warn the firefighter of the low air supply. NFPA 1981 requires SCBA units to be equipped with two different types of low air alarms. The two alarms must function independently—failure of one alarm must not affect the operation of the second—and each must alert different senses, for instance, an audible alarm accompanied by a visual or tactile (vibration) alarm.

Face Piece Assembly

The face piece assembly, **Figures 7-21A** and **B,** provides fresh air to the firefighter wearing an SCBA unit. In addition, protection from the hazardous environment is provided to the face and eyes. The assembly consists of a flexible rubber or silicon mask with a lens, exhalation valves, and a harness with adjustable straps. Most manufacturers provide a number of different sizes for proper fit without leakage. As mentioned later in this chapter, 29 CFR 1910.134 and NFPA 1500 require an annual fit test to ensure proper sizing of the face piece. Depending on manufacturer and style, the face piece will have a low-pressure air hose to connect to the regulator or the regulator will connect directly to the face piece.

The exhalation valve is the outlet for exhaled breath and prevents toxic gases from entering the face piece. These valves must be inspected regularly because dirt or moisture in a cold climate may keep it partially open, allowing toxic gases to enter the face piece. The following section on donning procedures explains how to check the operation of the exhalation valves.

The face piece assembly may have a number of options depending on manufacturer and style. These include a "nose cup" to prevent fogging and "voice amplifiers" to facilitate communication. The NIOSH regulation requires the nose-cup option in cold climates.

The last part of the face piece assembly is the harness and straps. This provides a tight fit to the head and prevents the face piece from loosening during firefighting operations. Again there are variations

(A)

(B)

FIGURE 7-21   Two types of SCBA face pieces. The regulator connects to (A) the face piece, and (B) the hose connects to the regulator.

among different manufacturers, but the types in use today are a web style or headnet style. When the face piece is stored, the harness straps must be in the full out position to reduce wear on the straps and facilitate quick donning. Also, regardless of manufacturer, when donning the face piece, the straps must be pulled straight back from the head to ensure proper fit.

Each SCBA manufacturer has specific instructions for cleaning individual face pieces. Failure to follow these instructions may result in damage to the face piece and potential failure during use. An SCBA face piece must be cleaned after each use or regularly to remove dust and particles and to prevent the spread of communicable diseases. To minimize this problem, many fire departments issue each individual an SCBA face piece and in some circumstances a regulator for individual use.

Closed-Circuit Self-Contained Breathing Apparatus

Generally, closed-circuit SCBA is not used for firefighting operations. For services that do utilize this system, the most common use is for hazardous materials incidents. Below-grade rescue use requires an extended air supply. These units are available with air

supplies that range from thirty minutes to approximately four hours.

Closed-circuit SCBAs contain a cylinder of oxygen, a filter system, a regulator, and valves. They work on the principle of cleaning and filtering exhaled breath and adding pure oxygen to continue operation. The duration of the air supply is based on the filtering/cleaning and oxygen capacity of the unit.

Open-Circuit Supplied Air Respirators

Open-circuit **supplied air respirators (SARs),** also called airline respirators, **Figure 7-22,** are similar to SCBA units, except that the air supply cylinder is remote from the user. Air is supplied in the same manner as for a regular SCBA, but the hose connecting the cylinder and the SCBA unit may be up to 300 feet long, **Figure 17-22A.** These types of units are not used for firefighting operations, but are commonly used for hazardous materials incidents or confined space rescues. They provide the user with a long-duration air supply with mobility and agility. This type of unit must be equipped with an SCBA escape unit with duration of approximately five to ten minutes, **Figure 7-22B.**

(A)

(B)

FIGURE 7-22 (A) A supplied air respirator with face piece, hose, and air supply. (B) Note the emergency escape cylinder.

DONNING AND DOFFING SELF-CONTAINED BREATHING APPARATUS

Depending on apparatus seat belt position and department procedures, SCBA may be placed in service in a number of ways. The most common donning procedures are from a seat-mounted position in the apparatus, from a side compartment on the apparatus, or from a storage case. Regardless of how the SCBA unit is stored or mounted, firefighters should always refer to the manufacturers' instructions for specific procedures.

General Considerations

Before using any SCBA unit, regardless of manufacturer, model, or donning method, operational safety checks must be performed. As shown in **Table 7-7** and **Figure 7-23**, these checks must be conducted on a daily or regular basis, or immediately prior to using the SCBA unit.

Once these checks are completed and the unit operates properly, the SCBA is ready for use. If any of the components listed in Table 7-7 does not operate properly, is broken, or is damaged, the unit is taken out of service immediately.

> **CAUTION**
>
> On some units, if the high-pressure line was not bled down after the last use, the low air alarm may not activate. If the low air alarm does not activate, the firefighter should close the cylinder valve, bleed all air from the system, close all valves, again open the cylinder valve two to three turns, listen for the low air alarm to activate, and follow the rest of step 2 as described.

Storage Case

Generally, two methods are used to don SCBA units that are stored in their cases. These are the "over the head" and the "coat" method. The method used is usually a matter of personal preference and training.

TABLE 7-7	Standard Daily Checks for SCBA
SCBA Component	**Check for Operation**
Cylinder gauge	Cylinder is 100% full; usually a green shaded area on gauge.
Cylinder valve (slowly open two to three full turns)	Listen for audible low alarm to activate. If it does not activate or continues to sound, place unit out of service.
Regulator or remote gauge and cylinder gauge	Compare gauges. They should read within 10% of each other.
PASS device	If unit has an integral PASS device, check for operation. If PASS device is a separate unit, check operation.
Valves	Check all valves for operation and proper position. Inhale and exhale through the face piece, making certain that air is flowing properly.
Valves and regulator	Close cylinder valve; bleed down system.
Face piece	Check that face piece is clean, free from cracks or deterioration; check condition of hose.
Backpack and harness	Check harness assembly for deterioration and that straps are fully extended. Check condition of straps and buckles. Make sure cylinder is secured.

FIGURE 7-23 Firefighters must perform regular checks of SCBA to ensure the unit's ability to operate.

For safety, SCBA units stored in cases should only be donned once the apparatus has arrived on scene.

STREETSMART TIP

It is good practice to form the habit of readying the SCBA unit for use as it is put into storage on the apparatus. Fully extending and dressing all straps on the unit can save time on an emergency scene. Once the firefighter is tangled in an SCBA, it might take precious time and possibly the assistance of another firefighter to untangle the mess.

CAUTION

Firefighters should always refer to the specific donning instructions for the particular SCBA unit used by a fire department. Some units have a "chest" or "cross" strap, which should be fastened before the shoulder straps are tightened. In addition, some manufacturers recommend loosening the shoulder straps slightly so the weight of the SCBA is carried on the hips. All straps must be worn in accordance with manufacturer's instructions to ensure that the weight of the SCBA is distributed as engineered and to eliminate the hazards of entanglement or loss of balance on the fireground.

JPR 7-1: Donning Self-Contained Breathing Apparatus, Over the Head Method

(For step-by-step photos of this skill sequence, see page 170)

These instructions and figures show the use of a face piece-mounted regulator. The procedure for a waist strap-mounted regulator is generally the same, with the exception of donning the face piece. Refer to the face piece instructions for specific details. In addition, the hand position referred to may vary for left-handed users.

1. Check the air supply of the SCBA unit. The cylinder gauge must indicate the cylinder is at least 90 percent full or greater.

2. Slowly open the cylinder valve two to three full turns and listen for the low air alarm to activate as the regulator pressurizes. After the alarm sounds,

open the cylinder valve fully. If it does not sound, or does not stop sounding, place the unit out of service (follow local SOP) and use another unit, again completing the first two steps.

3. Compare the cylinder gauge to the regulator or remote gauge. These gauges must be within 10 percent of each other.

4. Spread and fully extend the harness assembly straps (shoulder and waist straps).

5. With the cylinder valve pointing away from you, grasp the back plate and/or cylinder with both hands. All straps should be outside of your hands.

6. With the regulator hanging freely, lower your head and lift the back plate/cylinder overhead. Using your elbows to extend through the loops formed by the shoulder straps, pull the shoulder straps into the body and grasp them with the hands. Let the shoulder straps slide through the hands as the unit slides into place.

7. Balance the unit on your back by leaning forward, and tighten the shoulder straps by pulling out and down. Grasp the waist strap and buckle the belt, adjusting the belt for a firm fit on the hips.

Once the SCBA unit is in place, refer to the section on the SCBA face piece for detailed donning instructions.

JPR 7-2: Donning Self-Contained Breathing Apparatus, Coat Method

(For step-by-step photos of this skill sequence, see page 171)

These instructions and figures show the use of a face piece-mounted regulator. The procedure for a waist strap-mounted regulator is generally the same, with the exception of donning the face piece. Refer to the face piece instructions for specific details. In addition, the hand position referred to may vary for left-handed users.

1. Check the air supply of the SCBA unit. The cylinder gauge must indicate the cylinder is at least 90 percent full or greater.

2. Slowly open the cylinder valve two to three full turns and listen for the low air alarm to activate as the regulator pressurizes. After the alarm sounds, open the cylinder valve fully. If it does not sound, or does not stop sounding, place the unit out of service (follow local SOP) and use another unit, again completing the first two steps.

On some units, if the high-pressure line was not bled down after the last use, the low air alarm may not activate. If the low air alarm does not activate, close the cylinder valve, bleed all air from the system, close all valves, again open the cylinder valve two to three turns, listen for the low air alarm to activate, and follow the rest of step 2 as described.

3. Compare the cylinder gauge to the regulator or remote gauge. These gauges must be within 10 percent of each other.

4. Spread and fully extend the harness assembly straps (shoulder and waist straps).

5. Position the SCBA unit with the cylinder valve toward you. Using your left hand, grasp the left shoulder strap at the top of the back plate/harness assembly and grasp the lower portion of the same strap with your right hand. When the unit is positioned correctly, the left shoulder strap will be on your right as you face the SCBA unit.

6. Lift the SCBA unit, swing it around the left shoulder and onto your back. Both hands are still grasping the shoulder strap.

7. Holding the left shoulder strap with your left hand, release your right hand and put your right arm through the right shoulder strap.

8. Balance the unit on your back by leaning forward. Tighten the shoulder straps by pulling out and down on the straps. Grasp the waist strap and buckle the belt, adjusting the belt for a firm fit on the hips.

Once the SCBA unit is in place, refer to the section on the SCBA face piece for detailed donning instructions.

Remember, always refer to the specific donning instructions for the particular SCBA unit used by a fire department.

Seat-Mounted Apparatus

With the use of enclosed firefighter riding positions becoming more common, many fire departments are mounting SCBA units at the seat position, as shown in **Figure 7-24**. This method allows for quick donning and the unit is readily available for regular inspection. Various types of mounting brackets and hardware are used for this type of mount, and firefighters must be familiar with the style used in their department. The specific steps for donning the SCBA unit from this position are detailed in the skill given later.

Three important safety requirements must be observed if SCBA units are mounted at the seat position. These are storing of the face piece, donning the unit while the vehicle is moving, and checking the cylinder gauge. The SCBA face piece should never be left connected to the regulator for storage. Ideally, the face piece should be stored in a bag to maintain cleanliness and keep the unit free of dust particles, which could affect operation or injure the user's eyes.

FIGURE 7-24 Many fire departments are mounting SCBA units at the seat position, allowing for easy access for inspection and quick donning.

In addition, a firefighter should never attempt to stand or don the SCBA unit while the apparatus is in motion. Donning gear while the vehicle is in motion may require loosening of the seat belt, and the firefighter may be thrown around the enclosed cab in the event of a sudden stop or turn. Firefighters must always remain seated with seat belt fastened when the vehicle is moving.

> **CAUTION**
>
> Firefighters should always be properly seat-belted when the vehicle is in motion. It is a good rule of thumb for all firefighters to remain seated with seat belts fastened until the driver/operator of the apparatus has stopped the vehicle and set the parking brake.

From the seated position, it is extremely difficult, if not impossible, to check the cylinder gauge and compare it to the regulator gauge. If this is not accomplished prior to response, then after dismounting the apparatus, firefighters should use the buddy system to check each other's gauges, **Figure 7-25.** No excep-

FIGURE 7-25 After dismounting the apparatus, firefighters use the buddy system to check each other's cylinder gauges.

tions are permissible: firefighters should always be properly seated with seat belts fastened at all times when the vehicle is moving.

> **CAUTION**
>
> Fire apparatus should never be in motion unless all personnel are properly seated with their seat belts fastened—NO EXCEPTIONS.

JPR 7-3: Donning Self-Contained Breathing Apparatus, Seat-Mounted Apparatus

(For step-by-step photos of this skill sequence, see page 172)

These instructions and figures show the use of a face piece-mounted regulator. The procedure for a waist strap-mounted regulator is generally the same, with the exception of donning the face piece. Refer to the face piece instructions for specific details. In addition, the hand position referred to may vary for left-handed users.

1. Check the cylinder pressure prior to response, if possible. Do not attempt to check the unit when the

vehicle is in motion. If this is not possible, use the buddy system, turn the unit on, and check the cylinder gauge when you dismount the apparatus.

2. With straps fully extended, place the right arm between the right shoulder strap and back plate assembly. Repeat this action for the left arm.

3. Tighten the shoulder straps by pulling out and down on the straps. Grasp the waist strap and buckle the belt, adjusting the belt for a firm fit on the hips. Do not entangle the waist belt with the vehicle seat belt.

4. Fasten the seat belt properly—*the apparatus should not be in motion until this step is completed.*

5. Open the cylinder valve two to three full turns and listen for the low air alarm to activate as the regulator pressurizes. After the alarm sounds, open the cylinder valve fully. If the alarm does not sound or does not stop sounding, place the unit out of service (follow local SOP) and use another unit once the vehicle is stopped, again completing the first two steps. Compare the cylinder gauge to the regulator or remote gauge, making sure that they are within 10 percent of each other.

6. Dismount the apparatus and recheck shoulder and waist straps.

Always refer to the specific donning instructions for the particular SCBA unit used by a fire department. Some units have a "chest" or "cross" strap, which should be fastened before the shoulder straps are tightened. In addition, some manufacturers recommend loosening the shoulder straps slightly so the weight of the SCBA is carried on the hips.

7. If not able to check the cylinder pressure prior to response, with a partner, use the buddy system to compare the cylinder gauge to the regulator or remote gauge, making sure that they are within 10 percent of each other.

On some units, if the high-pressure line was not bled down after the last use, the low air alarm may not activate. If the low air alarm does not activate, close the cylinder valve, bleed all air from the system, close all valves, again open the cylinder valve two to three turns, listen for the low air alarm to activate, and follow the rest of the procedure as described.

Once the SCBA unit is in place, refer to the section on the SCBA face piece for detailed donning instructions.

Compartment or Side-Mounted Apparatus

Some fire departments carry their SCBA units mounted on the side of apparatus or in compartments. Depending on the height of the apparatus compartment and the bracket used to mount the SCBA, the donning method may be similar to that for the seat-mounted position except that the firefighter is standing rather than sitting. If the cabinet height or mounting bracket position of the SCBA does not allow for ease of donning while standing, the firefighter should remove the SCBA unit and use the "coat" or "over the head" method of donning.

To don the SCBA unit from a compartment or side-mounted storage position, it is recommended to follow the donning instructions for the method best suited for the particular mounting style.

Donning the SCBA Face Piece

Most SCBA face pieces are donned in a similar manner, with the difference being in the style of head straps and the location of the regulator. Proper donning of the SCBA face piece is essential to protect the firefighter from the effects of toxic gases and hazardous atmospheres. Each firefighter must be fitted for the face piece to be used with a particular manufacturer's SCBA as required by OSHA 29 CFR 1910.134:

29 CFR 1910.134 (f)(1)
The employer shall ensure that employees using a tight-fitting facepiece respirator pass an appropriate qualitative fit test (QLFT) or quantitative fit test (QNFT) as stated in this paragraph.
29 CFR 1910.134(f)(2)
The employer shall ensure that an employee using a tight-fitting facepiece respirator is fit tested prior to initial use of the respirator, whenever a different respirator facepiece (size, style, model, or make) is used, and at least annually thereafter.[1]

In addition, 29 CFR 1910.134, NFPA 1500, and SCBA manufacturers' recommendations prohibit *anything* that may interfere with proper fit and seal of the face piece. This specifically includes eyeglasses and facial hair such as beards, moustaches, and long sideburns, all of which are listed in these standards. To allow anything such as the items listed to come between the sealing surface of the face piece and skin is in direct violation of these regulations.

> **CAUTION**
>
> No glasses or facial hair are allowed between the skin and the sealing surface of the SCBA.

JPR 7-4: Donning Self-Contained Breathing Apparatus, Face Piece

(For step-by-step photos of this skill sequence, see page 173)

1. With the head straps/harness in the full out position, hold the head harness with one hand while

placing the face piece on the face, with the chin properly located in the chin pocket.

2. Fit the face piece to the face, pull the head harness over the head, and make sure straps are lying flat against the head with no twists.

3. Tighten the neck or lower straps by simultaneously pulling both straps to the rear.

4. Check the fit of the head harness by stroking the harness down the back of the head using one or both hands. Retighten the neck straps at this time.

CAUTION

It is important to tighten either neck or lower straps first, pulling straight back. Pulling straps outward may damage the straps and prevent proper seal. The temple straps are adjusted by pulling the strap ends toward the rear of the head. Regardless of the type of regulator used, one should always check for proper face piece seal. A firefighter who does not get a leak-free seal should not enter the hazardous environment.

---

5. Adjust the temple straps by simultaneously pulling both straps to the rear.

6. Check for proper seal by attaching the regulator and inhaling a breath, activating the regulator. Hold that breath for about five seconds and listen and feel for any air leaks. If leaks occur, repeat steps 2 through 5. Forcefully exhale at this time also to make certain that the exhalation valve is working properly. In addition, once the regulator or breathing tube is connected, listen for air leaks from the positive pressure mode. If leaks occur, repeat steps 2 through 5.

7. Pull the protective hood on over the entire head to cover exposed skin. Ensure all long hair is inside the protective hood. Also ensure that the protective hood is tucked completely down inside the bunker coat. Pull up and fasten the bunker coat collar. Place the helmet on, and adjust it for the increased size of the head, due to the face piece and protective hood. Fasten the helmet strap. Make sure the helmet shroud is pulled over the neck and coat collar. Place protective gloves on.

8. As you approach the hazardous atmosphere, install regulator on face piece or connect the hose to the regulator for a waist-strap-mounted regulator. Airflow to the face piece will begin with the breathing action. Again, listen for air leaks from the positive pressure mode. If leaks occur, repeat steps 2 to 5.

Removing/Doffing the SCBA Unit

Upon exiting the hazardous atmosphere, firefighters should remove the SCBA unit and rest. Generally, to remove the unit, the donning process is reversed following manufacturers' instructions for the type or model of unit used. If awaiting another assignment, the face piece should be removed to allow normal breathing and to conserve air. Doing so will not only help to conserve air but will avoid a number of other problems as well.

Wearing the mask without air flowing into it obstructs the wearer's peripheral vision, causes the mask to fog up, and increases the CO_2 intake of the firefighter. The latter occurs because the firefighter is trying to breathe through the small or restricted opening that acts as the attachment point for the regulator, and rebreathes expelled CO_2 causing hyperventilation. In turn, hyperventilation can lead to impaired decision making and increased air consumption once the firefighter hooks back into the air supply.

This advice on removing and donning the SCBA should also be heeded when donning the unit prior to making entry. The regulator or face piece must not be contaminated by laying it on the ground or other dirty environment. This may cause injury if firefighters need to use the unit again. Once removed, the cylinder should be refilled and the unit should be placed in a "ready" position, **Figure 7-26,** with the straps extended to facilitate a trouble-free donning if the firefighter must return to work quickly.

Depending on local SOP, once an assignment with SCBA is complete, firefighters should report to rehabilitation for rest, fluids, and monitoring of vital statistics. Many respiratory protection programs limit firefighters to consumption of two SCBA cylinders, after which mandatory rehabilitation is required.

FIGURE 7-26 The "ready" position; straps extended to facilitate a trouble-free donning if the firefighter must return to work quickly.

Again, this rest period for fluids and monitoring of vital statistics reduces the physical and mental stress encountered during firefighting operations.

SELF-CONTAINED BREATHING APPARATUS OPERATION AND EMERGENCY PROCEDURES

In accordance with 29 CFR 1910.134, NFPA 1404, and NFPA 1500, fire departments must establish respiratory protection programs for firefighters using SCBA. In addition, the firefighters must be proficient in the safe use of SCBA, donning and doffing procedures, individual limitations, and limitations of the SCBA unit.

Safe Use of SCBA

Safe use of SCBA is essential to firefighter survival during operations requiring its use. Firefighting tasks are both mentally and physically demanding and firefighters must be in excellent physical condition. The SCBA unit and protective equipment add weight and bulk to the firefighter, causing increased exertion with loss of body fluids through perspiration. These actions increase during firefighting operations and firefighters must be aware of them and of the symptoms of heat stress and their own limitations and abilities.

The following items are essential for maximum safety while using SCBA:

- All firefighters using SCBA must be certified physically fit for respirator use in accordance with local policy or 29 CFR 1910.134.
- Fire departments must establish an accountability system, **Figure 7-27,** to track personnel entering an IDLH atmosphere.

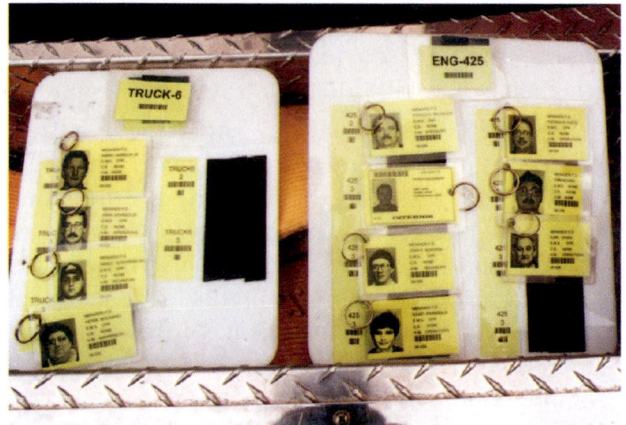

FIGURE 7-27 A typical accountability system to track personnel entering an IDLH atmosphere.

- Firefighters must work in teams of two as a minimum. If one member of the team must leave for any reason, all members of the team must exit the IDLH.

- In accordance with OSHA CFR 29 1910.134, the "two in/two out" rule states that a rescue team of two firefighters must be available and ready to rescue or assist the firefighters operating in the IDLH.
- PASS devices, **Figures 7-28A** and **B,** must be activated in order to function. If the PASS is an integral part of the SCBA unit, it must be activated before entering a hazardous environment.
- Fire departments should establish policies for firefighter rehabilitation during operations requiring SCBA use. This policy should include the maximum number of cylinders a firefighter can use before mandatory rehabilitation.
- During rehabilitation, EMS personnel should monitor vital signs, **Figure 7-29,** and firefighters must hydrate to replace body fluids.
- Air consumption will vary with each individual's physical condition, the level of training, the task performed, and the environment.
- The SCBA face piece should never be removed in a contaminated environment.
- Depending on an individual's air consumption and the amount of time required to exit a hostile environment, the low air alarm may not provide adequate time to exit. The individual should not panic, but move deliberately to a safe environment. All team members must exit together, even if only one low air alarm is sounding.

Operating in a Hostile Environment

Firefighters will use SCBA in hostile environments and toxic atmospheres that will have limited visibility. During firefighting operations requiring the use of SCBA, teams conducting search or fire attack must be able to function effectively. Firefighters should follow these general rules:

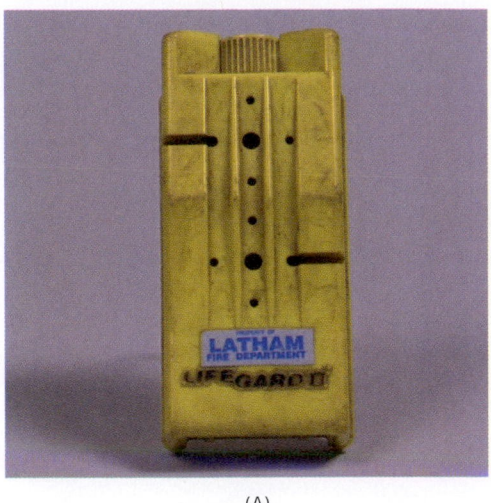

(A)

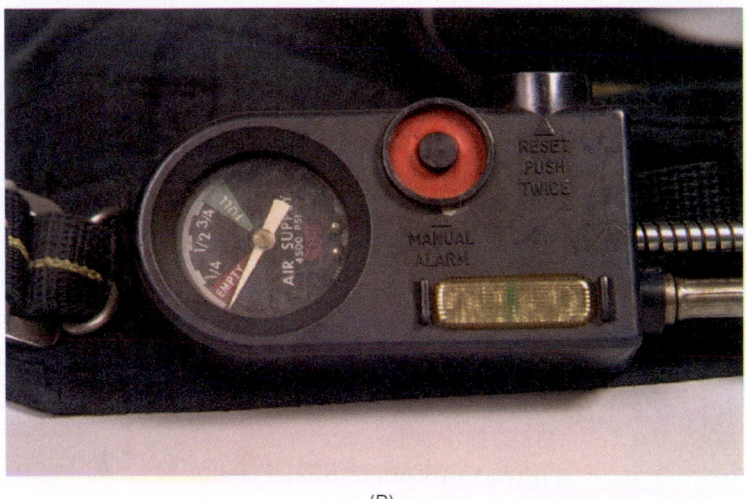

(B)

FIGURE 7-28 PASS devices used with SCBA include (A) the independent self-contained unit that attaches to the harness and (B) the integral type that is part of the SCBA unit.

FIGURE 7-29 EMS personnel should monitor vital signs, and firefighters must hydrate to replace body fluids during firefighting operations.

FIGURE 7-30 Firefighters should always remain low. Heat and smoke from the fire will rise; floor level provides the best visibility.

■ Always check in with the accountability/company officer when entering or exiting a building or when an assignment is complete.

■ Always remain low, **Figure 7-30.** Heat and smoke from the fire will rise, so floor level provides the best visibility. This will also reduce the risk of injury to a firefighter from walking into a hole in the floor or falling into a pit or shaft.

■ Check the environment and closely monitor conditions for changes through the use of thermal imaging technology when available.

■ Never remove the face piece.

■ Maintain an awareness of location, floor level, front or rear of building.

■ Ventilate as you advance, as long as it will not spread the fire. Ventilation allows the products of combustion to escape and provides a better environment.

■ Check for outside openings such as windows and doors. This will provide a means of escape in an emergency and provide the firefighters' location to outside personnel.

■ Always maintain direct contact with other team members. It is also acceptable for both firefighters to remain in contact by holding on to the same tool or a short piece of rope. There are also times where firefighters will be following a rope or hoseline into or out of a hostile environment. In this case, or in cases where direct contact is broken, firefighters should remain in constant voice communication until direct contact can be reestablished or until they are safely out of the IDLH atmosphere.

■ Never enter a hostile environment alone.

Restricted Openings

As a matter of survival, firefighters caught in a collapse or cut off from their means of egress may have to fit through a tight spot to get themselves to safety while wearing their SCBA. At no other time should a firefighter place themselves in a precarious position such as this. If a tight spot must be searched, probing it with a tool is the recommended procedure. Whenever going through obstacles, it is paramount that firefighters make certain that conditions on the other side are safe. It is also highly recommended that firefighters train on these techniques with their firefighting gloves in place and vision obstructed so that they become confident in working with the different components of their SCBA in less than ideal conditions without compromising their safety.

One conventional way for firefighters to get through an obstacle is to simply shift their pack to the left side. The SCBA will be shifted to the firefighter's left side to allow protection of the regulator or pressure reducer. It will also allow more freedom of movement for the firefighter because lines from these devices to the face piece are located on the left side of the SCBA on most units. (If lines from the pressure reducer are on the right, simply substitute going to the right in place of the left in the procedures described.)

JPR 7-5: Shifting the SCBA

(For step-by-step photos of this skill sequence, see page 174)

1. The left-side shoulder strap is loosened first followed by the extension of the waist strap
2. The SCBA is now shifted over to the left as far as possible allowing the firefighter's profile to be reduced. It may be necessary for a larger-framed firefighter to remove the right arm completely from the shoulder strap to accomplish this maneuver.
3. After the obstruction is cleared, the SCBA is shifted back into the center of the back and the straps are tightened once again.

Another method to get through a tight space such as wall studs with an SCBA is to "swim" through the obstacle backwards. Advantages of this technique are that a firefighter is not required to remove or loosen any part of their SCBA, which can slow an escape considerably.

JPR 7-6: Performing the Backwards "Swim"

(For step-by-step photos of this skill sequence, see page 176)

1. The firefighter sits down with the back against the wall and the SCBA centered between the two obstacles. The firefighter's feet should be positioned out in front of the body as far as possible with the but-

tocks raised off the ground. This is the key to make this technique work.
2. The firefighter then pulls shoulders and elbows inward towards the center of the body while pushing backwards through the space.
3. Once the neck of the SCBA cylinder is clear of the wall stud, the firefighter should rotate the body, "dipping" the right shoulder and employing a "swim" type maneuver to bring the rest of the body through the obstacle.

Another simple method for firefighters to reduce their profile to clear an obstacle is to perform a "forward dive" technique.

JPR 7-7: Executing the Forward Dive

(For step-by-step photos of this skill sequence, see page 178)

1. Centering the body between the wall studs or other obstacle, the firefighter places both arms in front through the obstacle attempting to get the elbows past the obstacle.
2. As the firefighter exhales, arms are pulled inward towards the center of the body allowing the firefighter to fall forward through the obstacle.
3. Once the SCBA is clear of the obstacle, the firefighter can use the arms to pull the rest of the way through. A slight turn of the hips may be required to clear the firefighter's waist.

Firefighters may have to get beneath an obstacle to facilitate their escape, which may also require them to lower their profile. The SCBA can be shifted to the left side by loosening the shoulder straps just as if the firefighter were advancing in an upright position, **Figure 7-31.**

FIGURE 7-31 Firefighters may have to get beneath an obstacle to facilitate their escape, which may also require them to lower their profile.

In a very extreme circumstance, a firefighter may have to resort to removing the SCBA from the body to facilitate clearing an obstacle. A firefighter must be very cautious when removing the SCBA as it further complicates the nature of the situation at hand. The SCBA should not be removed from the firefighter's back unless absolutely necessary.

JPR 7-8: Full Escape Procedure

(For step-by-step photos of this skill sequence, see page 179)

To remove the SCBA in a constricted area or if heat conditions dictate that a firefighter be as low as possible, the following steps can be followed:

1. While lying to the side, the firefighter will have to loosen the shoulder straps and remove the waist belt of the SCBA.

2. The firefighter should next "roll" out of the pack by rolling over to the left. Going to the left will allow the firefighter increased freedom of movement. At no time should the firefighter lose control of the left SCBA strap.

3. To further facilitate freedom of movement, the SCBA should be rotated so that the cylinder valve is facing away from the firefighter. All straps will need to be placed in a neat, organized manner on top of the SCBA to facilitate an easy redonning.

4. The firefighter should then move with the SCBA in front but keeping it close to the body to protect it and prevent the face piece from being pulled off. The firefighter should never lose contact with the SCBA. It is imperative that the firefighter make certain that no holes or elevation changes exist in the floor as the firefighter moves forward. When clear of the obstacle, the SCBA can be redonned by laying out the straps and rolling back into the pack.

Emergency Procedures

SCBA units can be damaged and malfunction, or the firefighter may become tangled in debris while operating. Situations such as these will require the calling of a Mayday to get immediate assistance. A number of emergency procedures exist that will assist a firefighter in the safe escape from the hazardous environment. Above all, firefighters must remain calm and rely on their training and knowledge.

It is imperative that a firefighter become thoroughly familiar with the breathing apparatus and possess a basic knowledge about preventive field maintenance for the particular unit they are using. Minor failures such as free flow of air or improper connections are very common on the fireground. These are often the result of operator error or improper preventive maintenance.

When a failure occurs in a hostile environment, those firefighters who are familiar with their SCBA unit will be able to remain calm and provide a remedy to the situation while exiting the area. The most important rule to remember in the case of a malfunction or depletion of air is to never remove the face piece of the SCBA. The face piece itself will afford the firefighter some protection for the face, eyes, and respiratory area while leaving the hazardous conditions.

To find and remedy a failure, a standardized emergency check procedure is stressed. This ensures that a firefighter will find the problem and execute the proper procedure that will enable exit from the hostile environment. The procedures listed can be modified to accommodate the different types of SCBA that may be in use, **Figure 7-32.**

How to Perform an Emergency Procedure Check

1. Determine need. Is there a problem?

2. Notify partner/Call for help (Mayday) and activate PASS.

3. Place left hand on face piece.

4. Slide hand down mask—check regulator.

5. Check air saver switch.

6. Check by-pass or purge valve. Is it open or closed? Can it provide air in the open position to allow for your escape?

7. Follow line from regulator to pressure reducer—check for problems.

8. Check if cylinder valve is in open position.

9. Check if cylinder is securely connected to high-pressure line.

10. Correct any problems found in check as you find them.

11. If not able to correct problem, immediately leave area with assistance to safe area.

Many SCBA units are equipped with an emergency escape breathing support system (EBSS) or a "buddy breathing" attachment, **Figure 7-33.** Current NIOSH and NFPA standards do not recognize the use of this option, because it utilizes one air source to support two users. Fire departments should review SOPs concerning "buddy breathing" and determine the best option for their needs. Practice these procedures during training, as an emergency is not the time to try something new.

Many cables, wires, and forms of ductwork run through ceiling spaces and are very vulnerable to collapse under fire conditions, causing entanglement haz-

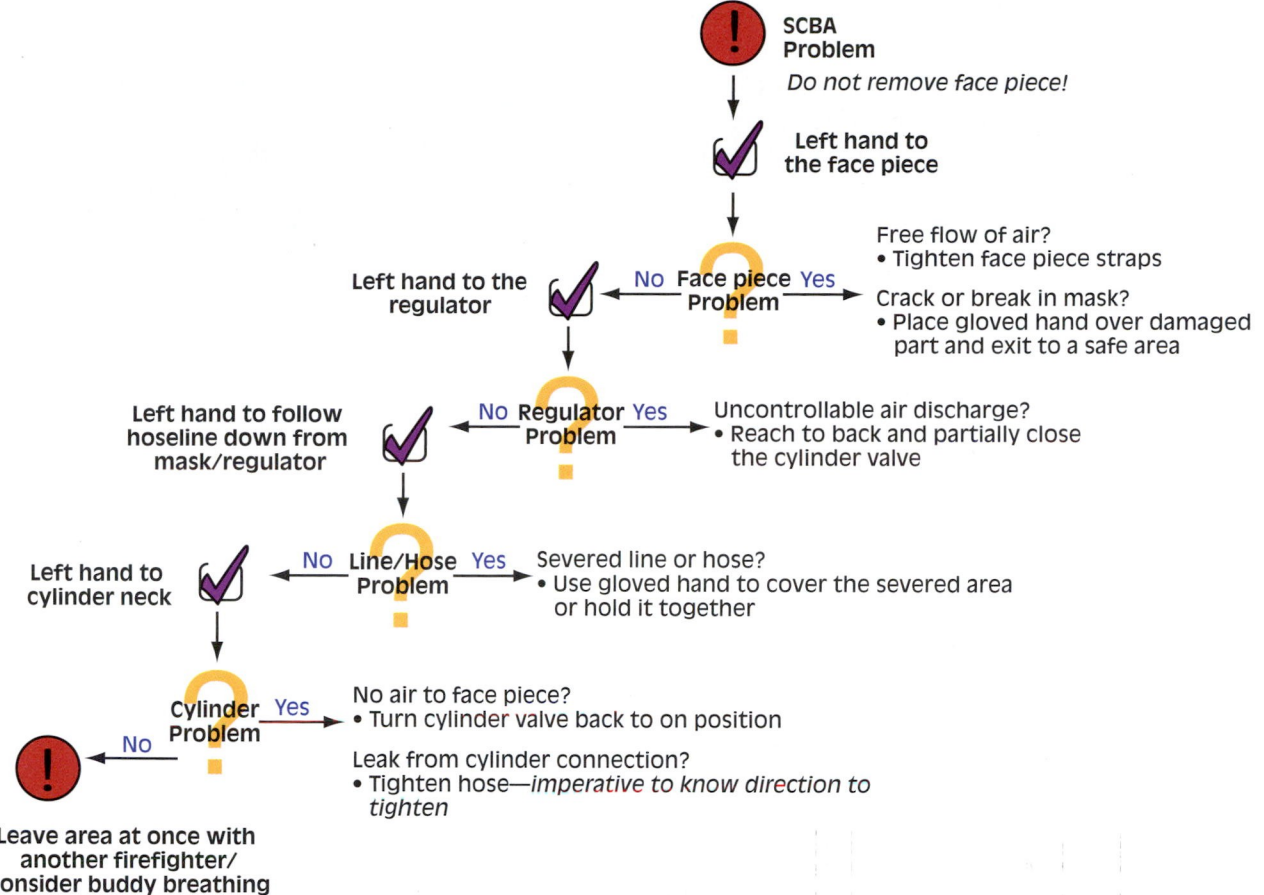

SCBA Problem
Do not remove face piece!

Left hand to the face piece

No ← **Face piece Problem** → Yes
Left hand to the regulator

Free flow of air?
• Tighten face piece straps

Crack or break in mask?
• Place gloved hand over damaged part and exit to a safe area

No ← **Regulator Problem** → Yes
Left hand to follow hoseline down from mask/regulator

Uncontrollable air discharge?
• Reach to back and partially close the cylinder valve

No ← **Line/Hose Problem** → Yes
Left hand to cylinder neck

Severed line or hose?
• Use gloved hand to cover the severed area or hold it together

Cylinder Problem → Yes
No

No air to face piece?
• Turn cylinder valve back to on position

Leak from cylinder connection?
• Tighten hose—*imperative to know direction to tighten*

Leave area at once with another firefighter/ Consider buddy breathing

FIGURE 7-32 Emergency procedures check.

FIGURE 7-33 EBSS, or a "buddy breathing" attachment that is common on newer SCBA.

ards for firefighters. If a firefighter becomes entangled, it is imperative to stop forward movement because trying to "muscle" forward will only pull an entanglement tighter on the SCBA and possibly wedge the entanglement into an unreachable position on the unit. It is strongly recommended that firefighters carry a knife or wire cutters to free themselves from an entanglement. When cutting wires in which a firefighter is entangled, it is important that only one wire at a time is cut; this will

decrease the possibility of injury in the event that utilities have not been fully controlled. For this same reason, firefighters should brush wires out of their path with an open hand or the back of their hand. A firefighter may be able to navigate through an entanglement by employing a "swimming" technique or a "football carry" technique.

JPR 7-9: Swim Method for Entanglement

(For step-by-step photos of this skill sequence, see page 180)

1. The firefighter should identify the entanglement and stop forward movement. An attempt should be made to get low and possibly back out of the hazard.

2. A firefighter should adopt a face-up position with the SCBA against a wall if at all possible.

3. While slowly moving forward, the firefighter should reach one arm across the body and slide an open hand across the floor surface, catching any entanglement hazards. Using the same arm, the hazard should then be lifted above the head while the firefighter moves under it.

4. The firefighter's opposite arm should now be slid across the floor repeating the same motion and

holding any additional entanglement hazard above the firefighter.

5. The first arm can now be brought down to the waist where the technique can be repeated. Firefighters should work together as a team to navigate the hazard if possible.

6. If the firefighter is entangled and cannot move, the firefighter should rotate the body about one quarter of the way to the side and attempt to reach the entanglement by sliding an arm down the leg and "swimming" it back up. It is important that the firefighter does not roll too far in one direction as this could cause the entanglement to wrap more tightly around the firefighter's body. This action should be repeated until contact is made with the entanglement and it can be removed.

FIREFIGHTER FACT

New SCBA designs incorporate a variety of innovative technologies to provide maximum safety for the user. Some SCBAs incorporate the PASS device or distress alert as an integral component of the unit. These systems offer redundant visual and audible warning alarms, with the audible alarm located on the harness assembly and the visual warning, a flashing LED, located on the regulator or remote gauge.

Other new designs include a technology that projects a display of the level of air in the cylinder to the face piece, straight ahead in the field of vision of the firefighter. This "Head's Up Display (HUD)" allows a firefighter to keep the head up, look straight, and continue to perform firefighting or other duties, while still being able to see the level of air remaining in the SCBA. The HUD uses an LED indication for air levels of full, three-quarter, one-half, and one-quarter. Further, the HUD will flash visual notification of one-half and one-quarter remaining air supply. The HUD may also include a visual indicator of temperature. It may be retrofitted into older SCBA units. According to the 2002 revision of NFPA 1981 and all other revisions thereafter, all brand new SCBA are to include the HUD in order to meet the specifications outlined in the standard.

Another new design feature is the **Rapid Intervention Crew/Universal Air Connection (RIC/UAC).** The RIC/UAC is a common style connector for use in emergency situations, such as rescuing a trapped firefighter. Rapid intervention teams (RIT) can carry an extra SCBA cylinder and recharge the trapped firefighter's air, as well as their own, during the rescue mission. The universal air connection is a requirement (according to revisions of NFPA 1981 made after 2002) and is mounted within 4 inches of the cylinder valve outlet with a relief valve to prevent overfilling of lower-pressure cylinders, such as a 2,216-psi cylinder if a 4,500-psi cylinder is used as a filling source, **Figure 7-34.**

FIGURE 7-34 Universal Air Coupling or Rapid Intervention Crew Coupling, UAC/RIC.

CAUTION

Removing the SCBA from the user's body should only be performed as a last resort in a life or death situation. Removing the SCBA from the users back can lead to further problems, such as loss of control, loss of valuable time to redon, and entanglement.

STREETSMART TIP

Many firefighters carry a small tool or wire cutters to free themselves from entanglements. Make certain that this tool is able to cut through the materials that will be encountered and is large enough to be manipulated with a gloved hand.

INSPECTION AND MAINTENANCE OF SELF-CONTAINED BREATHING APPARATUS

As with any piece of equipment used in the fire service, SCBA units must be ready to go on a moment's notice. Considering the importance of SCBA to the firefighter, inspection must be completed on a daily or regular basis.

There are a number of variations of maintenance procedures for inspection and servicing of SCBAs. Always follow the manufacturer's instructions and recommendations provided with the unit. The step-by-step instructions given in this chapter are intended for training purposes and may differ from the procedures recommended for the SCBA used by a particular organization.

Daily Maintenance

SCBA units should be checked daily to ensure they are secured and ready for operation. When an SCBA is used during an emergency scene or a training exercise, the unit should be serviced and checked in the same manner.

JPR 7-10: Daily Inspection of SCBA

(For step-by-step photos of this skill sequence, see page 181)

1. Check to ensure the cylinder is full.
2. Open the cylinder valve slowly, two to three full turns, until the low air alarm is activated. When the alarm stops sounding, open the cylinder valve completely. If the alarm does not activate or continues to sound, place the unit out of service following department procedures.

Depending on the manufacturer, some units—if the high-pressure line was not bled down after use—will not have activation of the low air alarm. In this case, close the cylinder valve, bleed all air from the system, close all valves and open the cylinder valve as described.

3. Compare the cylinder and regulator gauges. Gauge readings should be within 10 percent of each other.
4. Check to see that all hose connections are tight and free from leaks.
5. Check the bypass or purge valve and the main line valves for operations.
6. Close the cylinder valve and (depending on manufacturer's instructions) drain all air from the system. Note that the low air warning alarm activates when the regulator gauge reads about 20 to 25 percent. Also make certain that the lights on the heads-up display (HUD) are working properly.
7. Ensure the main line valve (if the unit has one) is open and that the bypass or purge valve is fully closed.
8. Check the condition of the face piece, if one is assigned to the unit. Ensure cleanliness and check for cracks and for missing or broken straps. Make sure that the diaphragm is in place. Clean face piece per local department policy, normally using a 1 percent bleach water solution.
9. Check the harness assembly, condition of belts, and cylinder fastening device.
10. Check the operation of the PASS device.

Monthly Maintenance

The monthly SCBA check contains all elements of the daily check but adds several checks of the mechanics of the system. Firefighters should be trained by their department to conduct the monthly maintenance. An example of a monthly check sheet is shown in **Figure 7-35.** Any irregularities should be noted and repaired,

or the SCBA should be pulled from service until a certified technician can repair the unit.

Annual and Biannual Maintenance

NIOSH and SCBA manufacturers require a number of different functional tests of SCBA units. Only a manufacturer's authorized or trained service personnel shall conduct these tests. Firefighters should refer to the instructions for the SCBA units used.

Changing SCBA Cylinders

SCBA cylinders must be changed after use, following local SOPs. Depending on the size and air consumption rate of a firefighter, allowing a cylinder at 90 percent full to remain in service could mean a loss of two to five minutes of air supply. This time may be the difference in being able to successfully exit or escape from a hazardous situation.

The firefighter should follow these procedures for cylinder replacement:

JPR 7-11: Cylinder Replacement Procedure

(For step-by-step photos of this skill sequence, see page 182)

1. Have full air cylinder ready for use.
2. Place the SCBA unit on the ground with the cylinder valve toward you.
3. Push in the knob on the cylinder valve and close the cylinder valve.
4. Bleed residual air in the high-pressure line by slowly opening the bypass or purge valve. When the flow of air from the face piece stops, close the valve. On some units, the pressure must be released by "breathing down" the regulator or opening the main line valve. Refer to the manufacturer's instructions for the unit in use.
5. Disconnect the high-pressure coupling from the cylinder. If the knob is difficult to turn, the high-pressure line may still be pressurized; repeat step 5 and attempt to disconnect the coupling again. Lay the coupling on the back plate to prevent dirt or grit from contaminating the threads or seat.
6. Release the cylinder clamp or locking mechanism used on the unit and slide the empty cylinder down out of the harness back plate assembly.
7. Inspect the O-ring on the seat of the high-pressure hose for any damage, scratches, or foreign matter. If the O-ring is damaged, replace it in accordance with the manufacturer's instructions.

SCBA FIELD MAINTENANCE SHEET

DATE: _____   REDUCER #: _____   LOCATION: _____

Air-Pak shows no use; visual and function checked only _____

2.2 / 4.5   30 DAY / REPAIR

(REGULATOR ASSY.)

REG. COVER:	OK	Y / N	RETAINING RING:	OK	Y / N
DIAPHRAGM:	OK	Y / N	REG. GASKET:	OK	Y / N
PURGE VALVE:	OK	Y / N	DONNING SWITCH:	OK	Y / N
THUMB LATCH:	OK	Y / N	REG. HOSE:	OK	Y / N

(BACKFRAME ASSY.)

SHOULDER HARNESS:	OK	Y / N	WAIST BELTS:	OK	Y / N
REMOTE GAUGE:	OK	Y / N	ALL HOSES:	OK	Y / N
EBSS EQUIPMENT:	OK	Y / N	PACK ALERT:	OK	Y / N

(FUNCTION TEST & LEAK TEST)

BREATHING:	OK	Y / N	PURGE VALVE:	OK	Y / N
LOW AIR ALARM:	OK	Y / N	REMOTE GAUGE:	OK	Y / N
DONNING SWITCH:	OK	Y / N	ANY LEAKAGE:	OK	Y / N

(BOTTLE CONDITION)

Note any excessive wear or damage to bottle and/or valve. Bottle # _____

*LIST ANY PROBLEMS NEEDING REPAIRS: _____

_____

_____

*LIST ANY PARTS REPLACED: _____

_____

_____

Rev: 07-04-02   EQUIP. # _____

FIGURE 7-35   SCBA Field Maintenance Sheet.

8. Replace with a fully charged cylinder. Reverse the process in step 6 by sliding the cylinder into the back plate assembly. Do not secure the cylinder clamp until the high-pressure line is aligned and connected.

9. Align the high-pressure hose with the cylinder valve assembly and connect the hose. This connection should be hand tightened.

10. Lock the cylinder into place on the back plate assembly.

11. Open the cylinder valve, check for air leaks, and compare the cylinder valve reading with the regulator reading. These should be within 10 percent of each other.

12. Close the cylinder valve, bleed all air from the system, ensure that the main line and bypass

valves are in the proper position, and place the unit in its mounting bracket or storage case.

Other units may have an integrated PASS device; once the pressure is off the system, the PASS device should be shut off.

JPR 7-12: Additional Steps for Two-Person SCBA Cylinder Replacement

(For step-by-step photos of this skill sequence, see page 183)

Many times on a fire scene, firefighters will require other firefighters to change out a cylinder while they are wearing the SCBA. This can be done safely using the same sequence of events, with the following exceptions:

1. While wearing the SCBA, stop and lean forward, or kneel on the ground. Ensure that your head is pointed down and out of the way of the cylinder as it is changed.

2. The firefighter changing out the cylinder should follow the steps for removing the cylinder as noted in steps 1 through 7 above, taking care not to hit the head of the firefighter wearing the pack while removing the cylinder.

3. Before replacing a full cylinder, the firefighter changing the cylinder should ensure that it is full. It is good practice to quickly show the bottle gauge to the firefighter wearing the SCBA unit; this way both firefighters have seen the gauge and know the cylinder is full.

4. The firefighter changing the cylinder should replace the cylinder as noted in steps 8 through 11 of JPR 7-11.

Servicing SCBA Cylinders

When the cylinder capacity is below full, the cylinder must be serviced. As noted earlier in this chapter, filling SCBA cylinders is completed using a cascade system, **Figure 7-36A**, or compressor/purifier system, **Figure 7-36B**. Regardless of how this is accomplished, the following safety precautions must be followed:

- The air source being used must be tested and certified meeting the requirements of Compressed Gas Association Pamphlet G-7.1-1989, OSHA 29 CFR 1910.134, and NFPA 1500.

- All cylinders must have a current hydrostatic test date: no more than five years for aluminum/ steel or carbon cylinders, or three years for Kevlar or fiberglass composite cylinders. In addition, Kevlar and fiberglass composite cylinders cannot be more than fifteen years old. Presently, carbon cylinders also fall under the fifteen-year lifespan.

(A)

(B)

FIGURE 7-36   A cascade system is one of the systems available to service SCBA cylinders. These may be fixed or mobile units.

- All fill stations must have fragmentation containment devices in case of cylinder failure.

- All manufacturers' recommendations should be followed, especially for recalls or safety notices concerning cylinder capacity.

- Fill rate may vary; 300 to 600 psi per minute is considered an acceptable range.

CAUTION

A firefighter should never attempt to fill an SCBA cylinder without fragmentation protection. It is important never to fill a composite SCBA cylinder in a water-filled fragmentation protection device. Water may infiltrate between the composite layers, causing a weakness and failure.

JPR 7-13: Servicing an SCBA Cylinder Using a Cascade System

(For step-by-step photos of this skill sequence, see page 184)

Many organizations require certification before using a cascade system or air compressor/purifier system. Do not attempt this skill without training on the specific system used by a particular fire department and follow the procedures recommended by the system manufacturer. Never attempt to fill a cylinder above its rated capacity and never attempt to fill bottles with different rated capacities at the same time.

1. Check the hydrostatic test date of the cylinder. Remove from service if test is out of date or usable service life for composite cylinders has passed.

2. Inspect the cylinder for any signs of physical damage, discoloration from heat, gouges, or nicks in the cylinder surface. Never fill an SCBA cylinder that has visible signs of damage or an out-of-date hydrostatic test.

3. Place the cylinder in the fragmentation containment device.

4. Connect the fill hose to the cylinder, and if it is equipped with a bleed valve, close it.

5. Open the SCBA cylinder valve.

6. Open the valve at the cascade system manifold or the fill hose valve, depending on how the system is designed. Open both valves if the system has both.

7. Open the valve of the cascade cylinder with the lowest pressure. This pressure must be higher than the pressure in the cylinder being filled. Control the fill rate of air to avoid excessive heating or chatter in the cylinder. If the cylinder heats or chatters, reduce the fill rate.

8. Observe the cylinder gauge and watch that the cylinder fills at approximately 300 to 600 psi per minute.

9. Close the cascade cylinder valve when the pressure equalizes with the pressure in the SCBA cylinder. If the SCBA cylinder is not full, open the valve on the cascade system cylinder with next highest pressure. Repeat this step until the SCBA cylinder is full.

10. Close all valves on the cascade system, then close the SCBA cylinder valve.

11. Open the cascade system fill hose bleeder valve and bleed off excess line pressure. Failure to follow this step could result in injury from the fill line as it discharges high-pressure air or blows off the O-ring seal on the end of the fill hose.

12. Disconnect the cascade system fill hose from the cylinder valve, remove the cylinder from the fragmentation container, and place in storage.

JPR 7-14: Servicing an SCBA Cylinder Using a Compressor/Purifier System

(For step-by-step photos of this skill sequence, see page 185)

Many organizations require certification before using either a cascade system or air compressor/purifier system. Do not attempt this skill without training on the specific system used by a particular fire department, and follow the procedures recommended by the system manufacturer. Never attempt to fill a cylinder above its rated capacity.

1. Check the hydrostatic test date of the cylinder. Remove from service if the cylinder test date is not current or, if a composite cylinder, it is beyond the fifteen-year service life.

2. Inspect the cylinder for any signs of physical damage, discoloration from heat, gouges, or nicks in the cylinder surface. Never fill an SCBA cylinder that has visible signs of damage or an out-of-date hydrostatic test.

3. Place the cylinder in the fragmentation containment device.

4. Connect the fill hose to the cylinder, and if equipped with a bleed valve, close valve.

5. Open the SCBA cylinder valve.

6. Operate the compressor/purifier system in accordance with the manufacturer's instructions and open the outlet valve.

7. Set the cylinder pressure adjustment on the compressor/purifier system to the correct pressure to fill the cylinder. Never attempt to fill a cylinder above its rated capacity.

8. Open the manifold valve and again check the fill pressure.

9. Open the fill station valve to begin filling the cylinder.

10. Observe the cylinder gauge and watch that the cylinder fills at approximately 300 to 600 psi per minute. Control the fill rate of air to avoid excessive heating or chatter in the cylinder. If the cylinder heats or chatters, reduce the fill rate.

11. Close the fill station valve when the cylinder is full.

12. Close the SCBA cylinder valve.

13. Open the fill hose bleeder valve and bleed off excess line pressure. Failure to follow this step could result in injury from the fill line as it discharges excess air pressure or damages the O-ring seal.

14. Disconnect the fill hose from the cylinder valve, remove the cylinder from the fragmentation container, and place in storage.

JOB PERFORMANCE REQUIREMENT 7-1
Donning SCBA—Over the Head Method

A Checking the air supply is the first step before using SCBA.

B Always compare the cylinder and regulator gauges. They must be within 10 percent of each other.

C Grasp the back plate with both hands. Harness straps should be outside of your hands.

D Lift the back plate/cylinder overhead, using your elbows to extend through the loops formed by the shoulder straps.

E Balance the unit on your back by leaning forward, and tighten the shoulder straps by pulling down and out.

Hands on Left Shoulder Strap of SCBA

A With your left hand, grasp the left shoulder strap at the top of the back plate/harness assembly and grasp the lower portion of the same strap with your right hand. When the unit is positioned correctly, the left shoulder strap will be on your right as you face the SCBA unit.

B Lift the SCBA unit, then swing it around the left shoulder and onto your back. Both hands are still grasping the shoulder strap.

C Balance the unit on your back by leaning forward. Tighten the shoulder straps by pulling out and down. Grasp the waist strap and buckle the belt, adjusting the belt for a firm fit on the hips.

JOB PERFORMANCE REQUIREMENT 7-3
Donning SCBA—Seat-Mounted Apparatus

A Check the cylinder pressure prior to response, if possible. Do not attempt to check the unit when the vehicle is in motion. If this is not possible, use the buddy system, turn the unit on and check the cylinder gauge when you dismount the apparatus.

B With straps fully extended, place the right arm between the right shoulder strap and back plate assembly. Repeat this action for the left arm. Tighten the shoulder straps by pulling out and down on the straps.

C Grasp the waist strap and buckle the belt, adjusting the belt for a firm fit on the hips. Do not entangle the waist belt with the vehicle seat belt.

D Fasten the seat belt properly; the apparatus should not be in motion until this step is completed.

E Dismount the apparatus and recheck shoulder and waist straps.

Donning SCBA—Face Piece

A With the head straps/harness in the full out position, hold the head harness with one hand while placing the face piece on the face, with the chin properly located in the chin pocket.

B Tighten the neck or lower straps by simultaneously pulling both straps to the rear.

C Check the fit of the head harness by stroking the harness down the back of the head. Retighten the neck straps as necessary.

D Check for proper seal by attaching the regulator and inhaling a breath, activating the regulator. Hold that breath for about five seconds and listen and feel for any air leaks.

E Pull on protective hood to cover exposed skin and head. Place helmet on head with earflaps down. Pull up coat collar and buckle. Snug or buckle helmet straps. Place gloves on hands.

JOB PERFORMANCE REQUIREMENT   7-5
Shifting the SCBA

A   The left side shoulder strap is loosened first followed by the waist strap being extended.

B   The SCBA is now shifted over to the left as far as possible allowing the firefighter's profile to be reduced. It may be necessary for a larger-framed firefighter to remove the right arm completely from the shoulder strap to accomplish this maneuver.

C After the obstruction is cleared, the SCBA is shifted back into the center of the back and the straps are tightened once again.

JOB PERFORMANCE REQUIREMENT 7-6
Backwards Swim Technique

A The firefighter sits down with the back against the wall with the SCBA centered between the two obstacles. The firefighter's feet should be positioned out in front of the body as far as possible with the buttocks raised off the ground. This is the key to this technique working.

B The firefighter then pulls shoulders and elbows inward toward the center of the body while pushing backwards through the space.

C Once the neck of the SCBA cylinder is clear of the wall stud, the firefighter should rotate the body and employ a "swim" type maneuver to bring the rest of the body through the obstacle.

A Centering the body between the wall studs or obstacle, the firefighter places both arms in front through the obstacle, attempting to get the elbows past the obstacle.

B As the firefighter exhales, the arms are pulled inward toward the center of the body, allowing the firefighter to fall forward through the obstacle.

C Once the SCBA is clear of the obstacle, the firefighter can use the arms to pull the rest of the way through. A slight turn of the hips may be required to clear the firefighter's waist.

A While lying to the side, the firefighter will have to loosen the shoulder straps and remove the waist belt of the SCBA. The firefighter should next "roll" out of the pack by rolling over to the left. Going to the left will allow the firefighter increased freedom of movement.

B To further facilitate freedom of movement, the SCBA should be rotated so that the cylinder valve is facing away from the firefighter. All straps will need to be placed in a neat organized manner on top of the SCBA to facilitate an easy redonning.

C The firefighter should then move with the SCBA in front but keeping it close to the body to protect it and prevent the face piece from being pulled off. The firefighter should never lose contact with the SCBA. It is imperative that the firefighter makes certain that no holes or elevation changes exist in the floor as the firefighter moves forward. When clear of the obstacle, the firefighter can redonn the SCBA by laying out the straps and rolling back into the pack.

Swim Method for Entanglement

A Identify entanglement. Stop forward movement. Try to get low and possibly back out of the hazard.

B Firefighters should position themselves face up with their SCBA against a wall if at all possible. While slowly moving forward, the firefighter should reach one arm across the body and slide an open hand across the floor surface, catching any entanglement hazards. This arm should then be lifted to hold the hazard above the head while the firefighter moves under it.

C The firefighter's opposite arm should now be slid across the floor repeating the same motion, holding any additional entanglement hazard above the firefighter. The first arm can now be brought down to the waist where the technique can be repeated. Firefighters should work together as a team to navigate the hazard if possible.

D If the firefighter is entangled and cannot move, the firefighter should rotate the body about one quarter of the way to its side and attempt to reach the entanglement by sliding the arm down the leg and swimming it back up. It is important that the firefighter does not roll too far in one direction as this could cause the entanglement to wrap around the firefighter tighter. This should be repeated until contact is made with the entanglement and it can be removed.

Daily Inspection of SCBA

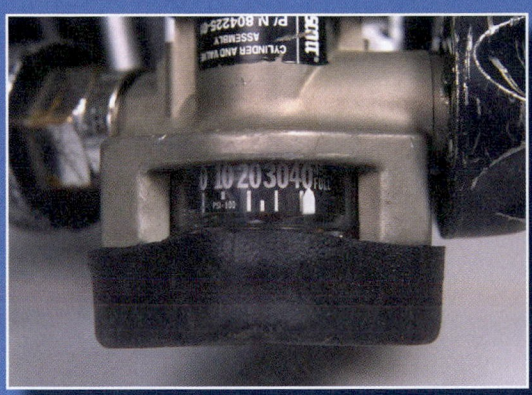

A Check to ensure that the cylinder is full.

B Compare cylinder gauge and regulator gauge. Gauge readings should be within 10 percent of each other.

C and **D** After checking for proper operation, ensure air is purged from the system, the main line valve (if the unit has one) is open, and the bypass valve is fully closed.

JOB PERFORMANCE REQUIREMENT 7-11
Cylinder Replacement Procedure

A   Place the SCBA unit on the ground with the cylinder valve toward you.

B   Disconnect the high-pressure coupling from the cylinder. If the knob is difficult to turn, the high-pressure line may still be pressurized.

C   Release the cylinder clamp or locking mechanism used on the unit and slide the empty cylinder down out of the harness back plate assembly.

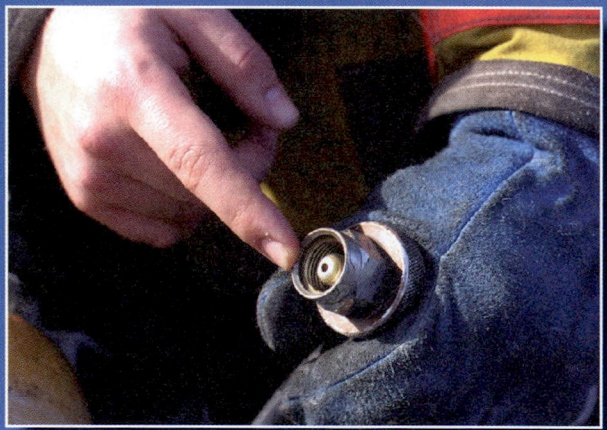

D   Inspect the O-ring on the seat of the high-pressure hose for any damage, scratches, or foreign matter.

JOB PERFORMANCE REQUIREMENT 7-12

Additional Steps for Two-person SCBA Cylinder Replacement

A   Stop and lean forward in a stable position.

B   In some cases, it will be easier to kneel on the ground.

C   Always know the amount of air in your cylinder.

JOB PERFORMANCE REQUIREMENT   7-13
Servicing a SCBA Cylinder Using a Cascade System

A   Check the hydrostatic test date of the cylinder. Remove from service if test is out of date or usable service life for composite cylinders has passed.

C   Open the valve at the cascade system manifold or the fill hose valve. Open the valve of the cascade cylinder with the lowest pressure. This pressure must be higher than the pressure in the cylinder being filled. Observe the cylinder gauge and watch that the cylinder fills at approximately 300 to 600 psi per minute. Close the cascade cylinder valve when the pressure equalizes with the pressure in the SCBA cylinder. If the SCBA cylinder is not full, open the valve on the cascade system cylinder with next highest pressure. Repeat this step until the SCBA cylinder is full.

B   Place the cylinder in the fragmentation containment device, connect the fill hose to the cylinder, and if it is equipped with a bleed valve, close it. Open the SBCA cylinder valve.

JOB PERFORMANCE REQUIREMENT   7-14
Servicing a SCBA Cylinder Using a Compressor/Purifier System

A   Check the hydrostatic test date of the cylinder. Remove from service if the cylinder test date is not current or, if a composite cylinder, it is beyond the fifteen-year service life.

B   Place the cylinder in the fragmentation containment device.

C   Connect the fill hose to the cylinder and close the bleed valve.

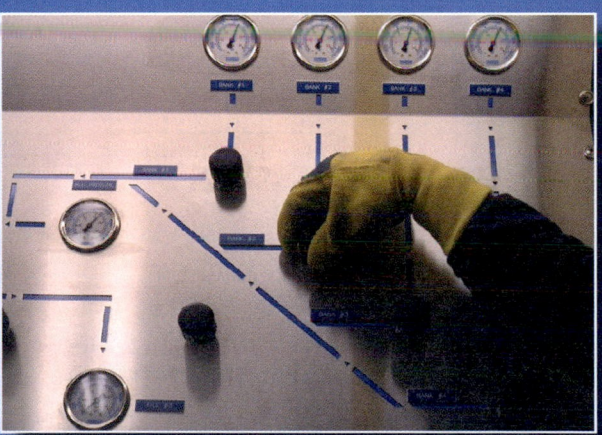

D   Operate the compressor/purifier system in accordance with the manufacturer's instructions.

LESSONS LEARNED

The SCBA unit is to a firefighter what a weapon is to a soldier. There are many events in military history where soldiers have lost their lives and battles because of lack of knowledge of their weapons. The same holds true for firefighting.

As with any firefighter skill, there is no substitute for proper training with SCBA. In addition to initial training and certification in SCBA use, continued practice and advanced training are necessary to ensure that firefighters maintain proficiency. The same is true for maintenance because some firefighters have a tendency to treat SCBA as a tool, like a ladder or ax, rather than as the vital piece of protective equipment that it is. To prevent failures from occurring and endangering lives, it is imperative that firefighters thoroughly inspect and test the function of SCBA as often as possible.

Each year a number of line-of-duty deaths are documented where firefighters for various reasons have died because they did not rely on their training with self-contained breathing apparatus. Regardless of the nature of the alarm, firefighters must always be prepared to go in harm's way; to do so safely, they must be knowledgeable and proficient in the use of SCBA.

STREETSMART TIP

Firefighters should be encouraged to perform a "buddy check" with their partner or team members before entering a hazardous environment. Firefighting is teamwork and this teamwork can be a lifesaver.

KEY TERMS

Asphyxiation   Condition that causes death due to lack of oxygen and an excessive amount of carbon monoxide or other gases in the blood.

Authority Having Jurisdiction (AHJ)   The responsible governing organization or body having legal jurisdiction.

Carbon Monoxide (CO)   Colorless, odorless, poisonous gas that when inhaled combines with the red blood cells excluding oxygen.

Closed-Circuit SCBA   A type of SCBA unit in which the exhaled air remains in the system to be filtered and mixed with oxygen for reuse.

Half-life   The time required for a drug or other substance deposited in a living organism to be metabolized or eliminated by normal biological processes.

Heads Up Display (HUD)   A device that displays—in the user's field of vision—the amount of air remaining in the SCBA cylinder .

Hydrogen Cyanide (HCN)   A prevalent toxin that is readily found in large quantities in smoke. Hydrogen cyanide is colorless and is produced by the combustion of natural products such as wool, silk, cotton, and paper as well as synthetics or plastics.

Hypoxia   A deficiency of oxygen.

Immediately Dangerous to Life and Health (IDLH)   The maximum level of a dangerous atmo-sphere one could be exposed to and still escape without experiencing any effects that may impair escape or cause irreversible health effects.

Mask Confidence, or "Smoke Divers," Training   Training courses designed to develop a firefighter's skills and confidence for using SCBA.

NFPA 1404   National Fire Protection Association standard created by the Fire Service Training Committee detailing the requirements for fire service SCBA programs, including training and maintenance procedures.

NFPA 1500   National Fire Protection Association standard created by the Technical Committee on Fire Service Occupational Safety and Health that addresses a number of issues concerning safety and protective equipment.

NFPA 1981   National Fire Protection Association standard specific to open-circuit SCBA for fire service use that contains additional requirements above the NIOSH certification.

NFPA 1982   National Fire Protection Association standard specific to Personal Alert Safety Systems (PASS).

NIOSH   National Institute for Occupational Safety and Health, 42 CFR Part 84, sole responsibility for testing and certification of respiratory protection including fire service SCBA.

Open-Circuit SCBA   A type of SCBA unit in which the exhaled air is vented to the outside atmosphere.

OSHA 29 CFR 1910.134   Standard establishing minimum medical, training, and equipment levels for respiratory protection programs.

Oxygen-Deficient Atmosphere   An atmosphere with an oxygen content below 19.5 percent by volume.

Point of No Return   Refers to the firefighter's individual management of their air supply. It is determined by the amount of air consumed going into an IDLH environment versus the amount of residual air needed to safely exit the environment.

Positive Pressure   A feature of SCBA providing a continuous supply of air, delivered by the regulator to the face piece, keeping toxic gases from entering. This pressure (1½ to 2 psi, depending on the manufacturer) is slightly above atmospheric pressure.

Pulmonary Edema   Fluid filling the lungs causing death by drowning.

Quantitative Decomposition   The temperature at which a material begins to decompose and release quantities of gases prior to reaching its ignition temperature.

Rapid Intervention Crew/Universal Air Connection (RIC/UAC)   A special coupling that will allow an SCBA cylinder that is low on air to be transfilled from another SCBA cylinder regardless of the manufacturer's make or model.

Respiratory Protection Program   Management program designed to ensure employee respiratory protection as required by OSHA 29 CFR 1910.134 and NFPA 1500.

Respiratory System   The system of the human body that exchanges oxygen and waste gases to and from the circulatory system.

Self-Contained Breathing Apparatus (SCBA)   A type of respiratory protection in which a self-contained air supply and related equipment are worn or attached to the user. Fire service SCBA is required to be of the positive pressure type.

Supplied Air Respirator (SAR)   A type of SCBA in which the self-contained air supply is remote from the user, and the air is supplied by means of air hoses.

REVIEW QUESTIONS

1. List two conditions requiring respiratory protection and explain the effects of these conditions on the human body.

2. List and explain the effects of oxygen deficiency on the human body.

3. List and explain one legal requirement for use of SCBA.

4. List two different types of operation of self-contained breathing apparatus and describe each.

5. List the four basic assemblies of a SCBA unit.

6. Explain and demonstrate two different SCBA donning procedures at 100 percent accuracy in the time limit established by the authority having jurisdiction.

7. Explain and demonstrate routine inspection procedures for SCBA used by the authority having jurisdiction in accordance with the manufacturer's instructions.

8. Explain and demonstrate maintenance and servicing procedures for the SCBA used by the authority having jurisdiction in accordance with the manufacturer's instructions.

9. Explain the proper procedure to troubleshoot and remedy a problem encountered with an SCBA while operating in a hostile environment.

10. Explain and demonstrate the proper procedures for refilling an empty SCBA air cylinder.

ENDNOTE

1. U.S. Department of Labor, Washington, DC, 29 CFR 1910.134(f).

8

Portable Fire Extinguishers

I remember going to a fire several years ago as a firefighter assigned to a ladder company. I was the can man for the tour, which meant my tool assignment was a 6-foot wooden hook and a 2½-gallon water extinguisher. We were out on building inspection duty, and an alarm came in that we were not assigned to; however, because we were only six blocks away the dispatcher told us to respond. As we were responding, we were informed of children possibly trapped in a rear bedroom.

We were the first company to arrive, and heavy smoke was pushing from the second-floor front window of a four-story apartment house. We proceeded to the fire apartment, and as we forced entry, heavy smoke was pushing out of the apartment. Without the luxury of a hoseline, we proceeded to crawl down a narrow smoke- and heat-filled hallway. The heat was tremendous, and the lieutenant kept pushing me down the hall, telling me we were getting closer to the fire.

At this time, we were able to see fire lapping out of a bedroom door. I was instructed to use my water extinguisher to hit the fire both around the top of the door and the ceiling, driving the fire back into the bedroom. The lieutenant then took my 6-foot hook, grabbed the bottom of the bedroom door with it, and pulled it closed. Fire was now lapping from around the door, and I was able to give quick short blasts of the extinguisher and knock it down. The forcible entry man climbed over our backs and with the lieutenant was able to search the rear bedrooms, finding no one. My job now was to stay at the door to the fire and protect lives until a hoseline was stretched and operating. It only took about two minutes to get the hoseline operating and for me to join my company in the search effort, but it felt like a lifetime. I have been to many fires since that day, both as a firefighter and an officer, and I have seen much accomplished with a little 2½-gallon water extinguisher. It is an underrated piece of equipment that when properly used is worth its weight in gold.

—*Street Story by Mike Gala, Jr., Lieutenant, Ladder 148, FDNY*

LEARNING OBJECTIVES

After completing this chapter, the reader should be able to:

8-1   Identify the five classes of fire (A, B, C, D, and K).

8-2   Explain the five classes of fire and the risks associated with each class.

8-3   Discuss the risks associated with various types of fires and the potential damage associated with various types of fire extinguishers.

8-4   List and describe the kinds of fire extinguishers used for each class.

8-5   Explain how the rating systems of portable extinguishers are utilized for proper extinguisher selection.

8-6   Discuss the limitations of portable extinguishers.

8-7   Given a specific type of extinguisher, describe its operation.

8-8   Given a Class A, B, or C fire, select the appropriate extinguisher, make a safe approach and completely extinguish the fire.

8-9   Discuss the inspection requirements of portable fire extinguishers.

*The FF I and II levels, as defined by the NFPA 1001 Standards, are identified in different colors: FF I = black, FF II = red, additional information = blue.

INTRODUCTION

Portable fire extinguishers are designed to fight small fires, unusual fires that are not easily extinguished with water, or fires that cannot be reached quickly with hoselines. The definitions of these types of fires should be in the local Fire Department Standard Operating Procedures. A portable fire extinguisher can knock down a small fire or control a larger one until firefighters can stretch, or advance, a hoseline to the fire area. Citizens have also used fire extinguishers to control a small fire until the fire department arrives. When used by untrained persons, fire extinguishers have created deadly delays in alerting the fire department. Therefore, it is very important that firefighters are proficient in using fire extinguishers and in teaching proper extinguisher use to any interested citizens.

VIEWPOINT

Defining a "small" fire or a fire that would necessitate the use of a portable fire extinguisher is sometimes difficult. Normally, firefighters should consider the size of the extinguisher to be the determining factor in choosing the proper extinguisher for the job. However, there are no hard and fast rules and, in most cases, firefighter experience will afford the ability to make this decision.

Historically, firefighters have used hose streams to fight fires, and have used extinguishers only occasionally. This trend is changing. Truck company firefighters now regularly carry a pressurized water extinguisher into structures as part of truck company operations. Because fire extinguishers are a valuable tool, firefighters should be proficient in their use. A simple acronym, PASS, will help users with basic extinguisher operation. The four basic PASS steps are: Pull the pin, Aim the nozzle, Squeeze the handle, and Sweep the *base of the fire* with the extinguishing agent.

STREETSMART TIP

When using a portable fire extinguisher, it is common for members of the general public—and in some cases trained fire professionals with tunnel vision—to waste agent by discharging the extinguisher before reaching the base of the fire. Often the agent is discharged on the smoke before the base of the fire is reached. Besides wasting agent, this method of fighting fires directly affects life safety and property conservation. Firefighters must be proficient in both fire extinguisher use and teaching fire extinguisher skills to members of the general public.

While these four steps are used in all types of extinguishers, they can be efficiently practiced using a pressurized water extinguisher, which is easily refilled and cost effective to use. Firefighters must be knowledgeable about fire extinguisher types, extinguishing agents, and the five classes of fire. Throughout the fire service, the phrase "Use the right tool for the right job" becomes particularly meaningful when choosing the correct fire extinguisher.

Fire extinguishers come in a variety of types and sizes. Firefighters should know the extinguishers carried on each apparatus, as well as those inside the various buildings of the response district. During building inspections, firefighters should create a pre-incident plan of special hazards and locations where an extinguisher would be a valuable tool to control a fire. Fire prevention will dictate the correct extinguishers and extinguisher location for the occupancy. Firefighting skill will dictate the use of the right extinguisher for the conditions present during a fire. The general public will often ask firefighters questions about firefighting and fire extinguisher use. Firefighters must be prepared to answer questions and train citizens to use portable fire extinguishers, when requested.

FIRE CLASSIFICATION AND RISK

The type or nature of the material burning, or fuel, defines the class of fire. Fuel is the key ingredient because, as it varies, so does the fire. The different classes of fires are used to identify the type of extinguishers and extinguishing agents used to put them out. There are four traditional classes of fire with an additional fire class having been added in recent years. It is important for firefighters to pre-incident plan for the various fuel types and locations within their response district, as this will give fire crews a head start in combating a fire or potential fire. The Fire Prevention Bureau will set requirements for the size, type, and locations of fire extinguishers or fire protective systems for use by the building occupants. (For more information on fire classes, see Section I, Chapter 4.)

NOTE

Standards dictating the requirements for fire extinguisher location have changed and no longer require fire extinguishers in some classes of buildings, such as educational and assembly facilities. Research has shown that fire sprinkler systems activate much more quickly than an individual retrieving and putting into use a fire extinguisher. The emphasis is now on early evacuation and notification along with quick response by fire suppression units. There are still applications in these facilities for extinguishers, such as kitchens and other locations, and the Fire Prevention Bureau will set these standards according to the fire code. Sprinkler systems have proved to be a faster and more effective means of fire suppression and life safety. Fire Department engine companies will need to work closely with the Fire Prevention Bureau to ensure that standards are adhered to.

Class A

Class A fires involve ordinary combustibles such as wood, paper, cloth, plastics, and rubber. These fuels can be extinguished with water, water-based agents or foam, and multipurpose **dry chemicals.** Because of its availability and cost effectiveness, water is usually the agent chosen by the fire department.

Class B

Class B fires involve flammable and combustible liquids, gases, and greases. Common products are gasoline, oils, alcohol, propane, and cooking oils. Flammable liquids that are flowing and dripping horizontally or overflowing their container and spilling vertically (creating a three-dimensional flow), such as an overflowing tank, are considered special hazards. Special hazards refers to situations where fire extinguishers have not been tested and may be inadequate.[1] It is important to carefully evaluate the situation prior to attacking these types of fires. Some solids under fire conditions may melt and act like flammable liquids. Common extinguishing agents for Class B fires are carbon dioxide (CO_2), regular and multipurpose dry chemical, and foam.

SAFETY

Pressurized flammable liquids and gases should be extinguished by stopping the flow of the fuel. In some cases, firefighters need to extinguish the fire to shut off control valves. These situations are extremely dangerous because firefighters have to work in a flammable atmosphere and any source of ignition may reignite that atmosphere with catastrophic results. This entry into a flammable atmosphere should be the last available option.

Class C

Class C fires involve *energized* electrical equipment. As water is a conductor of electricity, water-based agents cannot be used as extinguishing agents. The recommended method of fighting these fires is to turn off or disconnect the electrical power and then use an appropriate extinguisher, depending on the remaining fuel source. Class C extinguishers have extinguishing agents and hoses with nozzles that will not conduct electricity. Exclusively Class C extinguishers are not made, but are categorized with another class of extinguisher: a BC extinguisher or an ABC extinguisher. Class C agents include carbon dioxide (CO_2) and regular and multipurpose dry chemicals.

Class D

Class D fires involve combustible metals and alloys such as magnesium, sodium, lithium, and potassium. These metals, once found in factories or other industrial areas, can now be found in some lightweight motor vehicle engine components or lawn mower bodies. Great care must be used when attempting to extinguish a fire in these types of fuels, and firefighters should always be on the lookout for erratic fire behavior when applying water to fires involving motor vehicles or storage facilities. Water and other extinguishing agents can react violently when applied to burning combustible metals and can endanger nearby firefighters. Fires involving Class D materials will almost appear to explode when water is applied to them. Also, because of the differences in the metal and alloy fuels, there is no universal Class D extinguishing agent that works on all Class D materials; what works well on one metal may be relatively ineffective on a different alloy.

Personnel must use the correct and uncontaminated (clean, dry, and without any other foreign materials in it—basically as it comes from the factory) extinguishing agent for each different Class D material. Facilities that use or store these materials are required to maintain adequate amounts of the proper extinguishing agents to combat any potential fire situation. Class D agents are called **dry powders** and should not be confused with dry chemicals, which, although dry and powdery, are not the same. (Some of these agents are dry sand, phosphate salts, or silica.) Other special agents such as Lith-X and Met-Ex are not commonly used and are only mentioned here. Firefighters using these special agents locally should seek additional information and training in their use.

Class K

Class K involves fires in cooking appliances using combustible cooking fuels such as vegetable or animal oils and fats. Its fuels are similar to Class B fuels but involve high-temperature cooking oils. Typically, firefighters have used Class B extinguishers on these types of fires, but they have been less effective on deep layers of cooking oils. Class K agents are usually **wet chemicals,** water-based solutions of potassium carbonate-based chemical, potassium acetate-based chemical, potassium citrate-based chemical, or a combination.[2] These agents are usually used in fixed systems, **Figure 8-1,** but some portable extinguishers are available.

FIGURE 8-1   Class K equipment.

TYPES OF FIRE EXTINGUISHERS

Many types of fire extinguishers are available for purchase and use, **Figure 8-2.** The best type of extinguisher depends on many factors and each should be considered prior to placing an extinguisher in use. The wrong extinguisher can be worse than no extinguisher.

Factors for selecting an extinguisher are the type of fuel, the person using the extinguisher, and the building or environment where it will be used. The first factor to consider when choosing an extinguisher is the type and amount of fuel present. This will provide clues as to the type and size of fire to anticipate. The amount of fuel determines fire intensity, and the wrong sized extinguisher will not completely knock down the fire. The user of the extinguisher and the occupancy in which it will be used represent another factor. Are potential users trained to combat a fire effectively or will they use improper technique and exhaust the extinguisher before the fire is knocked down? What will the people in the building do with or to the extinguisher while it is stored? Will the extinguisher be tampered with, stolen, or otherwise ineffective when needed?

Other factors are the type of building construction and occupancy. What hazards or other conditions exist? What building areas need to be protected to avoid further problems? These potential hazards set the fire code requirements for selecting and placing the extinguishers. Environmental conditions must be examined for effects on the fire and the extinguishing agent. Temperature may eliminate using water-based agents that may freeze; corrosive atmospheres may require special protection for the user; the wind may blow away the extinguishing agent before it reaches the fire; and a confined space may create an unsafe or nonsurvivable atmosphere for the user.

A final factor would be the type of equipment protected. Often this last factor is given too much consideration. Delicate equipment and high-value items may require special considerations as they may be irreparably damaged in an extinguishing operation. It should be stressed, however, that the main objective is to extinguish the fire completely and effectively. Most extinguishing agents can be cleaned from equipment, whereas damage from a fire can render equipment permanently destroyed. It is important to remember that certain extinguishing agents are corrosive to certain substances. This should be a factor when choosing an extinguisher for the occupancy.

Types of Extinguishing Agents

Because of its ability to absorb heat, water is the basic fire-extinguishing agent for Class A materials. However, water is subject to freezing and can sometimes be ineffective—and sometimes dangerous—on other classes of fuels. To combat the freezing problems, an alkali salt can be added as an antifreeze agent; this type of extinguisher is called a **loaded stream extinguisher.** Water-based foam extinguishers for use on Class B fires have either **Aqueous Film-Forming Foam (AFFF)** or **Film-Forming Fluoroprotein Foam (FFFP)** of both the regular and **polar solvent type.** These agents are effective in cooling and smothering the fire and creating a vapor barrier. (See Section II, Chapter 11 for more information on foam.) **Alcohol-resistant foam (AR foam)** contains a polymer that will form protective layering between the burning surface and the foam. This layering prevents foam breakdown due to alcohol being present in the burning fuel. Alcohol-resistant foams are used for fighting fuel fires that contain alcohol additives, such as the E85 gasoline blend.

Carbon dioxide (CO_2) is an inert gas stored under pressure as a liquid capable of being self-expelled. The colorless and odorless gas is effective in smothering a Class B or C fire. CO_2 works best in enclosed or semi-enclosed areas as the agent can be easily blown away by the wind. Personnel should take caution to avoid oxygen deprivation when CO_2 is used in small enclosed areas. This clean agent is still popular with those concerned about property damage, although in well-ventilated areas it can be ineffective. Some sensitive electrical equipment can be thermoshocked by CO_2.

Dry chemical extinguishing agents are particles propelled by a gaseous medium for distribution. There are three general categories: sodium bicarbonate-based, potassium-based, and multipurpose dry chemicals. The first two categories work on Class B and C fires, and are called regular dry chemical agents. Sodium

FIGURE 8-2   Various types of fire extinguishers.

bicarbonate, similar to baking soda, is the agent in the first category and is highly effective for grease fires while cooking. The second category includes potassium bicarbonate, potassium chloride, and urea-based potassium bicarbonate—all of which are usually more effective than sodium bicarbonate. The third category is known as a multipurpose dry chemical agent—effective on Class A, B, and C fires. These multipurpose dry chemicals are monoammonium phosphate.[3] Dry chemicals are very effective due to their coating action, which reduces the chances of reignition. The coating action is a drawback, however, when protecting sensitive items such as computer mainframes.

Wet chemical agents are water-based solutions of potassium carbonate-based chemical, potassium acetate-based chemical, potassium citrate-based chemical, or a combination of those chemicals.[4] They are used for special applications, particularly Class K fires.

Clean agents are the agents used to replace halon or halogenated hydrocarbon extinguishers. These agents are thought to cause damage to the Earth's ozone layer and all halon production has been banned by an international treaty: the Montreal Protocol on Substances that Deplete the Ozone Layer. These agents may still

be in limited use, but must be phased out and replaced with clean agent systems. Clean agents are gases that do not conduct electricity or leave a residue, and are nonvolatile. They are divided into two classes—halocarbon agents and inerting gases—and are currently rated as somewhat ineffective for local application. Research on their application continues.[5]

Kinds of Extinguishers

Many types of portable fire extinguishers are in use today. Some are small and handheld, while others are so large that they require a wheeled cart to move them. Most extinguishers operate on the same basic principle of storing and expelling an extinguishing agent. Fire extinguishers are labeled to make their firefighting rating quick and easy to identify. The older versions of fire extinguishers are labeled with colored geometrical shapes with letter designations, **Figure 8-3A.** Newer fire extinguishers are labeled with a picture label system, **Figure 8-3B.** Many fire extinguishers are multiuse, that is, a single extinguisher may be used to fight more than one class of fire, **Figure 8-3C.** Class A and B fire extinguishers have a numerical rating discussed

(A)

(B)

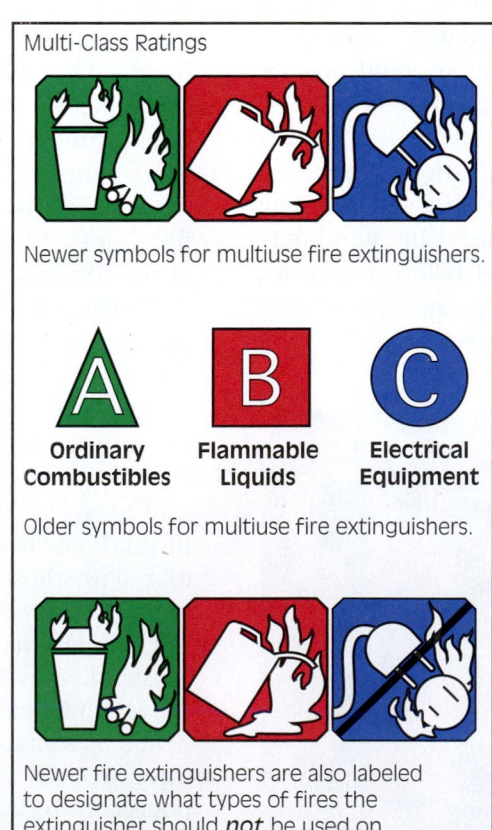

(C)

FIGURE 8-3 (A) Older versions of fire extinguishers are labeled with colored geometrical shapes with letter designations. (B) Newer fire extinguishers are labeled with a picture label system. (C) Many fire extinguishers can be used to fight more than one type of fire.

later in this chapter. Listed here are current kinds of fire extinguishers for each type of agent:[6]

1. Water type for Class A fires
   a. Pump type
   b. Pressurized water
   c. Pressurized loaded stream
2. Foam extinguishers for Class A and B fires—stored pressure
3. Carbon dioxide for Class B and C fires—self-expelling gas, stored pressure
4. Halon and clean agents for Class B and C, or Class A, B, and C
5. Dry chemical for Class B and C or Class A, B, and C fires (also some dry powder, Class D extinguishers)
   a. Stored pressure
   b. Cartridge-operated type
6. Wet chemical for Class K fires—stored pressure

Table 8-1 highlights the appropriate type of extinguisher to use for each type of fire.

Pump-type extinguishers are hand-pumped devices of two designs, depending on whether the pump is internal or external to the tank. The pump tank extinguisher has the pump housed inside the water tank area and each stroke pumps water out of the tank, **Figure 8-4**. This type of extinguisher is simple to operate and maintain. The second type is designed for wildfires and is a backpack pump that has the tank carried on the back with the hose and pump in front of the operator, **Figure 8-5A** and **Figure 8-5B**. Both usually come in a 2½-gallon size.

Pressurized water, pressurized loaded stream, and stored pressure extinguishers operate by means of an expelling gas that propels the agent out of the container. The differences are seen in the type of container and hose and nozzle assemblies. The container differences result from the various pressures required to store the agent. The hose and nozzle differences

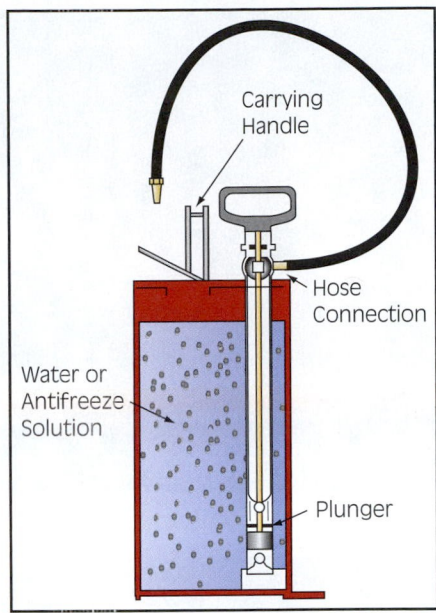

FIGURE 8-4   A pump tank fire extinguisher.

relate to the agent and its special need for application. Some agents may be dispensed over a wide area, others over a smaller area; some foam needs an air-aspirating nozzle or may require the nozzle to be nonconductive.

The basic principle is that the agent is inside the container with a pressurizing gas above it under constant pressure. Some use air or nitrogen as an expelling gas that is added to the container with the agent. Other agents, like carbon dioxide, are their own expelling gases. Most extinguishers have a gauge to measure the pressure of the gas. Some extinguishers, CO_2 extinguishers for example, do not have gauges. These extinguishers are weighed regularly to ensure their readiness. **Figures 8-6** through **8-15** show the various extinguishers and how they store their product.

Cartridge-operated extinguishers are used for some regular and multipurpose dry chemical and most

Extinguisher Type	Class A Fire	Class B Fire	Class C Fire	Class D Fire	Class K Fire
Type A Extinguisher	×				
Type BC Extinguisher		×	×		×
Type ABC Extinguisher	×	×	×		×
Metal/Sand (Class D Extinguisher)				×	
Type K Extinguisher		×			×
CO_2		×	×		

TABLE 8-1   It is important to use the proper extinguisher on the particular type of fire.

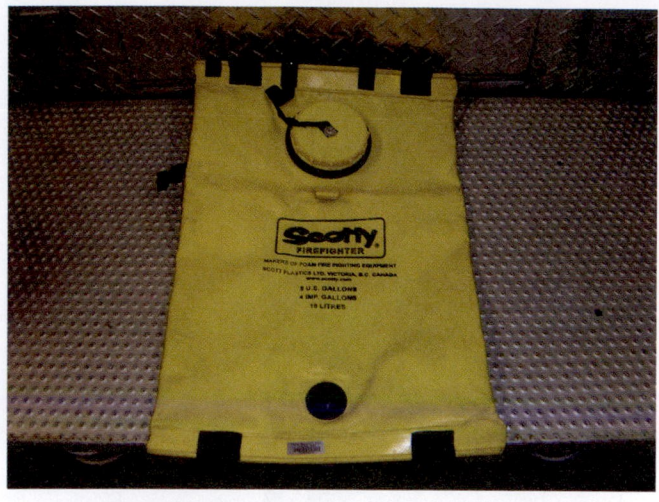

(A)

(B)

FIGURE 8-5 (A) A backpack tank made of flexible material. (B) A traditional backpack pump tank fire extinguisher made of metal. *(Courtesy of Fred Schall)*

FIGURE 8-7 Stored pressure water extinguisher. *(Courtesy of Fred Schall)*

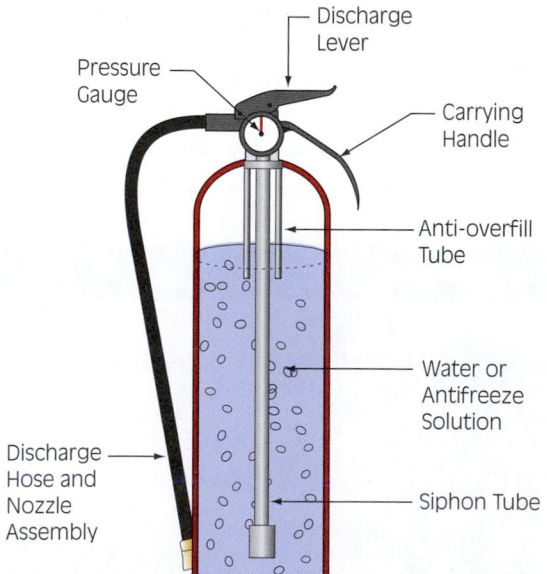

FIGURE 8-6 Inner workings of a stored pressure water extinguisher.

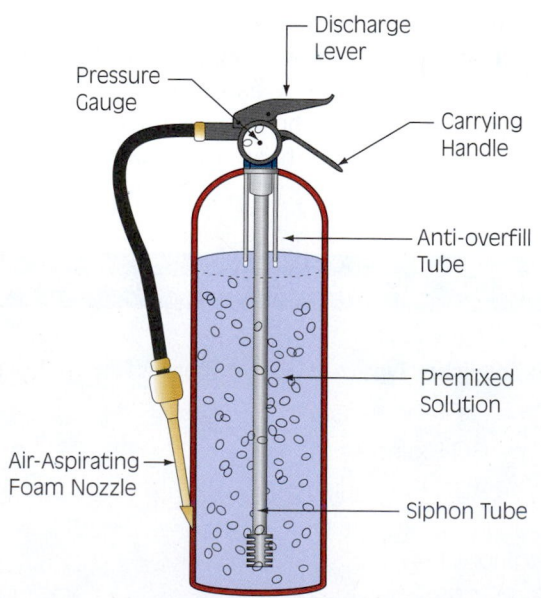

FIGURE 8-8 Inner workings of a stored pressure AFFF or FFFP extinguisher with air-aspirating nozzle.

FIGURE 8-9 Stored pressure foam extinguisher.

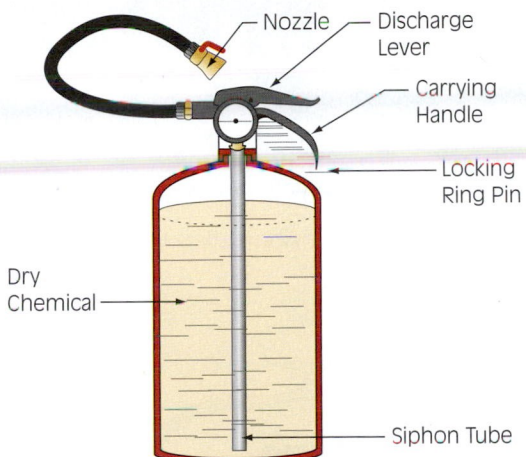

FIGURE 8-10 Inner workings of a stored pressure dry chemical extinguisher.

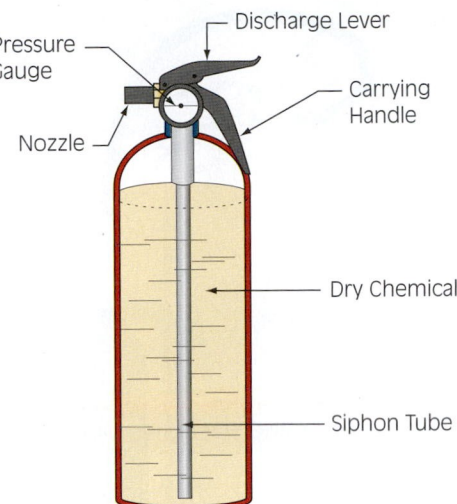

FIGURE 8-11 Inner workings of a stored pressure dry chemical extinguisher with fixed nozzle.

FIGURE 8-12 Stored pressure dry chemical extinguisher.

dry powder Class D extinguishers. They are similar to stored pressure extinguishers except that instead of being under constant pressure, the expelling gas is stored in a cartridge on the side of the container. When the puncturing lever is depressed, it ruptures a disk and the gas is expelled into the tank, which pressurizes it. The operator controls the flow of agent by a nozzle mounted at the end of the hose. Cartridge-type extinguishers are used when the agent may cake excessively and needs to be stirred. The agent is accessible without charging the pressure or dumping the agent. This type is also easy to refill by adding the agent into the tank and a new cylinder to the charging mechanism, **Figure 8-16** and **Figure 8-17.**

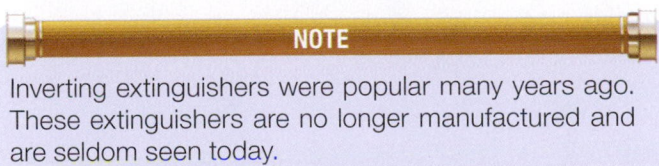

NOTE

Inverting extinguishers were popular many years ago. These extinguishers are no longer manufactured and are seldom seen today.

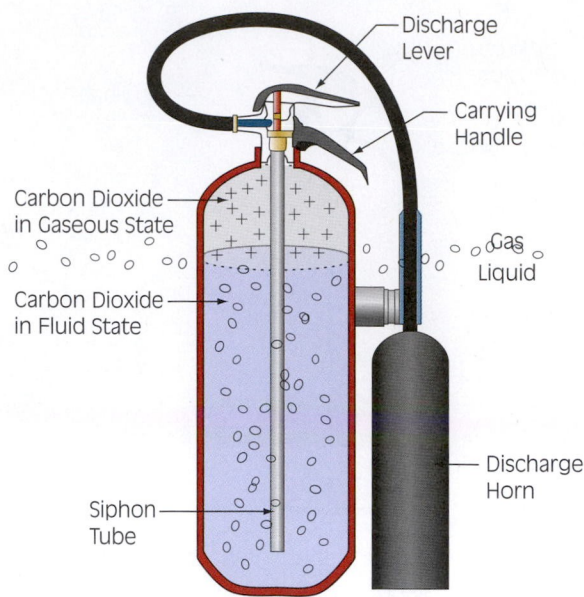

FIGURE 8-13 Inner workings of a carbon dioxide extinguisher.

FIGURE 8-15 Carbon dioxide extinguishers.

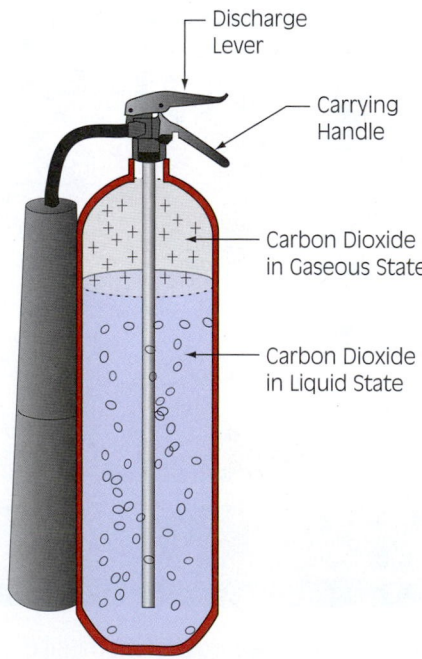

FIGURE 8-14 Inner workings of a carbon dioxide extinguisher with a fixed nozzle.

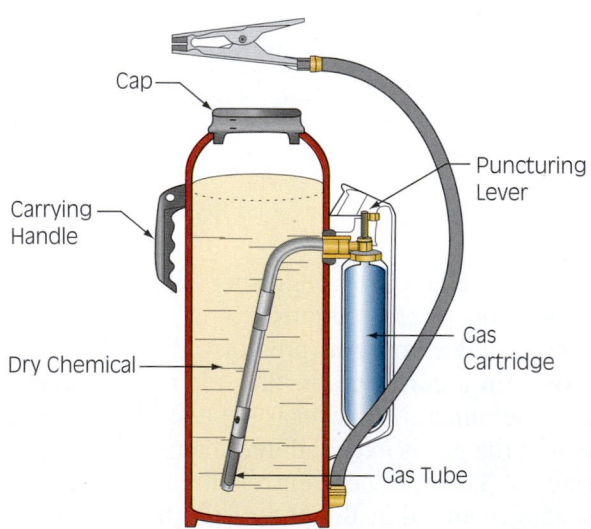

FIGURE 8-16 Inner workings of a cartridge-operated dry chemical extinguisher.

FIGURE 8-17 Cartridge-operated dry chemical extinguisher.

Class A

Testing of Class A extinguishers and agents utilizes wood cribbing. The cribbing is ignited, allowed a preburn period, and then attacked with the extinguisher. For a 1-A rating, the extinguisher should extinguish about 1 cubic foot (0.03 cubic meters) of wood cribbing, **Figure 8-18.** The ratings increase as the amount of fire suppressed increases; for instance, a 2-A extinguisher will put out twice the fire of a 1-A, and so on. In a wood paneling and excelsior test, the excelsior is ignited and allowed to spread to wood paneling before it is extinguished. Larger scale tests use only the wood cribbing and are conducted outside, while smaller tests are done indoors.

Class B

The test for Class B extinguishers involves igniting a pan of a flammable liquid (heptane), allowing a preburn period, and attacking the fire. The size of the pan determines the rating, **Figure 8-19.** For example, a 4-square-foot (0.37-square-meter) pan that is extinguished should yield a rating of 4-B. The rating numbers follow the size of the pan fire and are based on an inexperienced extinguisher operator. For example, an inexperienced operator using a 20-B rated extinguisher on a 20-square-foot fire should find the results adequate and should be able to extinguish the fire. The number rating compares approximately to the square footage to be extinguished, although larger units will not offer a direct relationship, as larger fires require more agent per area than the smaller ones.

RATING SYSTEMS FOR PORTABLE EXTINGUISHERS

Each class of fuel is subjected to a separate type of extinguisher test for its class. The testing of extinguishers is usually conducted by an independent testing agency such as Underwriters Laboratories, Factory Mutual Research Corporation, or a government agency such as the Coast Guard. This section covers the tests for Classes A, B, and C. Classes D and K have special tests that are more specific to their fuels or application. The appropriate ratings and symbols are noted on the label of the extinguisher. (See Figure 8-3.)

FIGURE 8-18 Wood cribbing for Class A extinguisher test.

Class C

The testing of Class C extinguishers and agents tests only the conductivity of the agent and the nozzle or hose and nozzle combination, **Figure 8-20.** There is no actual fire test for Class C agents or extinguishers. No numbers are assigned with the Class C rating: the extinguisher is simply referred to as a Class C extinguisher.

FIGURE 8-19 Flammable pan for Class B extinguisher test.

LIMITATIONS OF PORTABLE EXTINGUISHERS

Fire extinguishers have limited capabilities, and attempting to exceed those capabilities can increase damage and cause injuries. They are designed for specific purposes and are usually a first-aid method for fire extinguishment. Fire extinguishers are designed and rated with certain types and sizes of fires in mind; using the wrong class or the wrong size extinguisher may cause problems. When thinking of size, it is usually best to pick the larger size. However, picking the largest fire extinguisher available to put out a small fire can result in the waste of unused agent and additional expense. The wrong class extinguisher may not do the job and may cause a reaction or electrical shock.

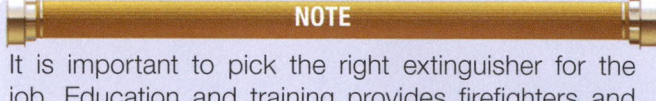

NOTE

It is important to pick the right extinguisher for the job. Education and training provides firefighters and the general public with the ability to make the correct choice.

PORTABLE EXTINGUISHER OPERATION

Suppose a wastebasket is on fire and the firefighter has chosen the correct extinguisher for extinguishment. Now, what is the right *way* to use it? For firefighters and the general public alike, the operation of most fire extinguishers is covered by the acronym PASS. PASS outlines the four simple steps for extinguisher

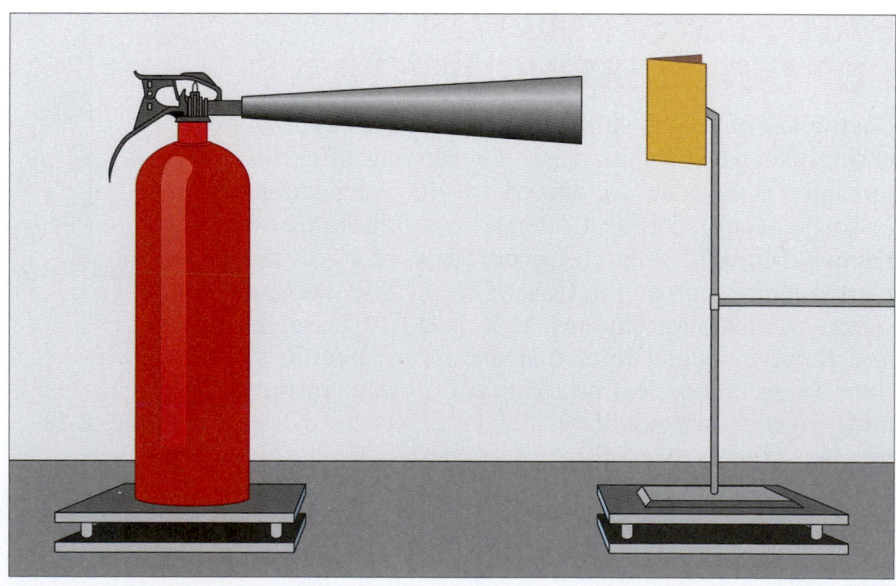

FIGURE 8-20 Class C test for electrical conductivity of agent.

use: P—pull the pin; A—aim the nozzle; S—squeeze the operating handle; and the second S—sweep the nozzle across the *base* of the fire.

> **NOTE**
>
> PASS, the four steps for using a fire extinguisher:
>
> **P:** *Pull* the pin.
> **A:** *Aim* the nozzle.
> **S:** *Squeeze* the handle.
> **S:** *Sweep* the base of the fire.

If the extinguisher does not extinguish the fire, the firefighter should not deploy a second extinguisher. Multiple extinguishers used at the same time, however, are more effective than using one at a time. Unfortunately, many buildings have been lost due to improper "one at a time" use of multiple extinguishers, and the untrained operators are quickly forced to exit the building due to the fire's increasing intensity. In these situations, the fire department usually is not notified until the fire has increased greatly in size. Extinguishers are meant to combat small fires in the growth stage. Large fires are beyond the capabilities of portable extinguishers.

> **SAFETY**
>
> When teaching the general public fire extinguisher use, it is paramount that they be taught to think of themselves as a critical link in the firefighting system. Extinguisher use should not preclude dispatch of fire departments or brigades. If a layperson is able to extinguish a fire, dispatched fire crews can always stand down. In most cases, at least one engine company will proceed to the scene to verify the fire is extinguished with no possibility of reignition.

Firefighters must be adept at using all types of firefighting equipment. This includes portable fire extinguishers. Even though extinguisher use is increasing in firefighting operations, in many services this tool is not used as often as a hoseline. It is, therefore, extremely important that firefighters train using portable fire extinguishers to stay proficient with the use of this tool and to effectively teach firefighting techniques to the general public.

> **NOTE**
>
> Stored-pressure water extinguishers are designed to be carried in an upright position when approaching a fire. The firefighter is cautioned to keep the hose in hand at all times.

> **STREETSMART TIP**
>
> Civilians are not immune from tunnel vision.[7] Often their plan of attack is to attempt fire extinguishment using one extinguisher, and then they plan to call the fire department. They almost extinguish the fire, but run out of agent and go for another extinguisher, which again fails to put out the fire. One group of maintenance personnel used more than twenty extinguishers on an apartment fire: when the first arriving officer, on a 1,000-gpm pumper, found fire coming through the roof, he called for a third-alarm assignment. The civilians were caught up in fighting the fire. This tunnel vision facilitated great loss of property. Education and training could have changed this scenario.

> **STREETSMART TIP**
>
> In an emergency, a firefighter may use a CO_2 extinguisher on scenes where dangerous dogs or other animals are present. Normally the noise of the CO_2 extinguisher discharging will scare the animal. The cold cloud of agent will usually further frighten the animal into retreat. Note: Discharging the CO_2 extinguisher directly onto the animal can render permanent harm to the animal. The extinguisher should be thought of as a reasonable force tool. Only the amount of force (agent) to get the animal to retreat should be used and no more. This choice of protection should be a last resort for firefighters or other persons in danger: if possible, notify proper animal control resources and wait for them to secure the scene before proceeding.

CARE AND MAINTENANCE OF PORTABLE EXTINGUISHERS

Care and maintenance of fire extinguishers is fairly straightforward. Simple inspections (see next section) and careful storage prevent most problems. The extinguisher should be inspected carefully when first placed in service and properly stored in a cabinet or on a bracket. Keeping apparatus-mounted extinguishers in their brackets will prevent damage from falling or from striking other equipment. New apparatus standards require this and it is a good practice on older vehicles. Routine apparatus equipment checks should require the extinguisher to be visually examined for proper pressure and visible damage and to ensure the pin seal is in the proper place. If any problems are noted, the unit should be removed and sent for repairs.

Vehicle operators should periodically remove the unit from its bracket to hand test its weight and to do a better visual check. Dry chemical or dry powder extinguishers carried on apparatus should occasionally be rotated upside down and shaken to keep powders from packing at the bottom as a result of vehicle vibration. Cleaning of any dirt or grit will help keep the unit in good working order. Finally, any extinguisher that has been discharged, has low pressure on the gauge, or has a broken seal should be removed from the vehicle until thoroughly examined and/or recharged.

Recharging a water extingisher is a simple process that is usually performed in the fire station by any firefighter. As always, the manufacturer's recommendation should be consulted for the capacity of water for a particular extinguisher and for the proper refill procedures. Normally, portable water extinguishers contain 2.5 gallons of water. Some agencies prefer to add Class A foam or other additives to their portable water extinguishers. This mixture can assist in making an extinguisher more effective. Usually, reservicing an extinguisher is as simple as removing the top, **Figure 8-21A**; adding water, **Figure 8-21B**; adding foam, **Figure 8-21C**; replacing the top, **Figure 8-21D**; and pressurizing the extinguisher using compressed air from the station or apparatus to the manufacturer's recommended pressure, **Figure 8-21E**. Once an extinguisher is serviced, it should be checked for leaks and returned to service.

INSPECTION REQUIREMENTS

Some very popular fire extinguishers of the past have now been declared obsolete and should be removed from service. These obsolete extinguishers include soda acid, chemical foam (except film-forming foam), vaporizing liquids such as carbon tetrachloride, cartridge-operated water or loaded stream, and any copper or brass extinguisher with solder or rivets (except pump tanks). Some of these have been removed because of the hazards of the extinguishing agents. Most of these extinguishers have the potential to explode when pressure is applied during use. Firefighters should never attempt to operate an obsolete extinguisher and should instead have them decommissioned by trained technicians.

The inspection of fire extinguishers by firefighters is usually a visual inspection, **Figure 8-22**. If something does not look right (gauges inoperative or not reading correct pressure, unit or hose damaged, evidence of extinguisher leaking, etc.) the unit should be removed and replaced. Extinguishers on apparatus should be inspected each time the vehicle is inspected. Extinguishers in buildings should be inspected every 30 days, or must be electronically monitored. **Electronic monitoring** is defined as a method of electronic communication (data transmission) between an in-place fire extinguisher and an electronic monitoring device/system. The monitoring system is similar to a Tamper Alarm on a sprinkler system or a shut-off valve. For extinguishers, the system uses data transmission to notify the monitoring company that the extinguisher has been tampered with in some way or has been removed. Even with the monitoring system, the NFPA mandates that fire extinguishers be annually maintained and inspected.

Extinguishers that are new, or old ones being returned to service, should be examined prior to their placement back on apparatus. The visual inspection should verify a tag indicating its last service and/or annual inspection. Annual inspection and maintenance by a service technician is recommended. The pin seal should be firmly in place. If the unit has a gauge, it should register the proper pressure. If a gauge is not present, lifting the unit or weighing it will determine if it is properly filled. Some extinguishers such as carbon dioxide extinguishers can only be truly checked by weighing; the filled weight is stamped on the cylinder. A careful look at the unit should detect any damage, dents, or corrosion to the shell, hose, nozzle, and other parts. If any defects are present, the unit should be removed and replaced. An important check should be to see if the unit is within the time frame for its hydrostatic testing, which is a pressure test to ensure that the unit will not explode when operated. The test period ranges from five to twelve years; firefighters should consult their service center or the NFPA standard to verify that extinguishers are compliant.[8]

> **NOTE**
>
> Electronic monitoring of fire extinguishers does not take the place of a physical inspection; an annual physical inspection is still required. The monitoring company is notified if the extinguisher is tampered with or removed. The monitoring company will notify the building or business owner and a physical inspection would still be required for the affected extinguisher. The electronic monitoring will save the business or building owner the trouble of checking the extinguisher monthly, but the added cost of monitoring may not be cost effective.

(A)

(B)

(C)

(D)

(E)

FIGURE 8-21 (A) Unscrew and remove the top. (B) Add the manufacturer's recommendation of water. (C) Add foam, if required. (D) Replace the top. (E) Charge the extinguisher with the manufacturer's recommendation of air.

FIGURE 8-22 Conducting visual inspections.

and not allow the fire to get between them and the exit. The object of aiming at the base of the fire is to sweep the fire away from the firefighter, confining it and driving it back to its origin. Aiming at other points may cause the fire to spread beyond the capability of the extinguisher. The extinguisher must be aimed from the proper distance, which often is set by the room size and fire size. Various extinguishers can reach a variety of distances up to 50 feet (15 meters), but reaching and being effective are two different subjects. The practical effective range of a pressurized water or foam extinguisher is about 20 feet (6 meters), dry chemical unit range is 15 to 20 feet (4.5 to 6 meters), and a CO_2 type extinguisher range is 10 to 15 feet (3 to 4.5 meters). When using a CO_2 extinguisher, firefighters must ensure that the aiming hand is on the handle, not on the horn where the cold can freeze the skin.

> **SAFETY**
>
> Firefighters must remember to have a path of escape behind them; they cannot allow the fire to get between them and the exit.

> **SAFETY**
>
> Improper operation of a CO_2 extinguisher can result in injury to the operator.

JPR 8-1: Portable Extinguisher Operation

(For step-by-step photos of this skill sequence, see page 205)

1. Pull the safety pin that prevents accidental discharge. Cartridge-operated extinguishers have a pin at the activation lever for the cartridge, not on the operating handle at the nozzle. The pin is held in place by a wire or plastic tie; a simple tug removes this wire and the pin.

2. Aim the extinguisher nozzle toward the base of the fire closest to the firefighter. Firefighters must remember to have a path of escape behind them

3. Squeeze the two parts of the handle together to apply the agent. The firefighter should continue to squeeze the handle long enough to cover the whole fire area with agent. Short blasts are sometimes appropriate, but too many can empty the extinguisher without extinguishing the fire.

4. Lastly, the firefighter sweeps the fire by carefully keeping the nozzle aimed at the base of the fire, continuing to push the fire back. The sweeping motions help extinguish the entire fire instead of creating a pocket that may allow the fire to circle around the firefighter. Foam extinguishers can also use the methods of regular foam application that are covered in Section II, Chapter 11.

JOB PERFORMANCE REQUIREMENT 8-1
Portable Extinguisher Operation

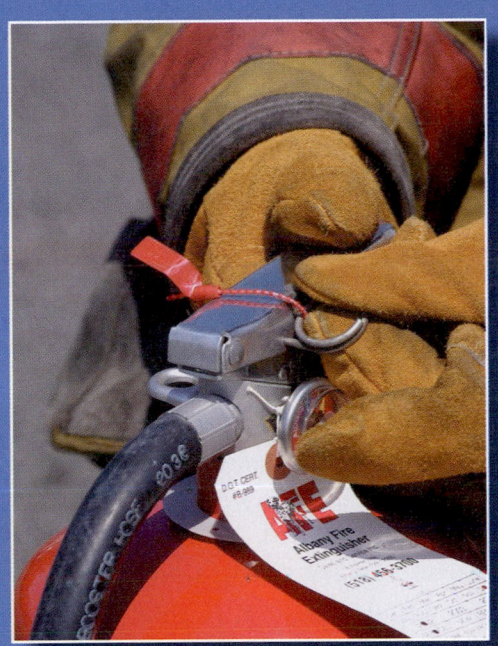

A Pull the pin.

B Aim the nozzle.

C Squeeze the handle.

D Sweep the base of the fire.

LESSONS LEARNED

Fire extinguishers can be used as initial response tools or to fight fires in special situations. Firefighters must know how to use fire extinguishers and to teach the public how to use them. Part of that knowledge includes classifying fires by their fuels, especially ordinary combustibles, flammable liquids and gases, and energized electrical equipment. Knowing the classes of fire is the first step toward choosing the right extinguisher and the correct size. The four-step process for using an extinguisher is PASS: pull, aim, squeeze, and sweep.

KEY TERMS

Alcohol-Resistant Foam (AR Foam)   Foam containing a polymer that will form protective layering between the burning surface and the foam. This layering prevents foam breakdown due to alcohol present in the burning fuel. Alcohol-resistant foams are used for fighting fuel fires that contain alcohol additives, such as the E85 gasoline blend.

Aqueous Film-Forming Foam (AFFF)   A synthetic foam that as it breaks down forms an aqueous layer or film over a flammable liquid.

Carbon Dioxide (CO₂)   An inert colorless and odorless gas that is stored under pressure as a liquid that is capable of being self-expelled and is effective in smothering Class B and C fires.

Class A   Classification of fire involving ordinary combustibles such as wood, paper, cloth, plastics, and rubber.

Class B   Classification of fire involving flammable and combustible liquids, gases, and greases. Common products are gasoline, oils, alcohol, propane, and cooking oils.

Class C   Classification of fire involving energized electrical equipment, which eliminates using water-based agents.

Class D   Classification of fire involving combustible metals and alloys such as magnesium, sodium, lithium, and potassium.

Class K   A classification of fire as of 1998 that involves fires in combustible cooking fuels such as vegetable or animal oils and fats.

Dry Chemicals   Dry extinguishing agents divided into two categories. Regular dry chemicals work on Class B and C fires; multipurpose dry chemicals work on Class A, B, and C fires.

Dry Powders   Extinguishing agents for Class D fires.

Electronic Monitoring   A method of electronic communication (data transmission) between an in-place fire extinguisher and an electronic monitoring device/system.

Film-Forming Fluoroprotein Foam (FFFP)   Foam that incorporates the features of AFFF and fluoroprotein foams with good resistance and the film-forming barrier.

Loaded Stream Extinguisher   Combats the water freezing problem by adding an alkali salt as an anti-freezing agent.

Polar Solvent Type of Foam   Foam that is compatible with alcohol and/or polar solvents by creating a polymeric barrier between the water in the foam and the polar solvent.

Wet Chemicals   Extinguishing agents that are water-based solutions of potassium carbonate-based chemicals, potassium acetate-based chemicals, potassium citrate-based chemicals, or a combination of these.

REVIEW QUESTIONS

1. Name the four traditional classes of fire and, for each class, give an example and name an extinguishing agent.
2. Name the new class of fire, describe its materials, and name an extinguishing agent.
3. What are the three categories of agents of dry chemicals?
4. What is a wet chemical?
5. How does carbon dioxide extinguish a fire?
6. Why are halons no longer produced for fire extinguishers?
7. How does a stored pressure extinguisher work?
8. What are the two kinds of pump-type extinguishers?

9. How are Class A extinguishers rated?

10. What does a 10-B rating mean?

11. How is a Class C extinguisher tested?

12. Describe the four steps of operating an extinguisher.

13. Describe a visual inspection process on a fire extinguisher.

14. What is the main focus concerning fire extinguishers and assembly occupancies?

ENDNOTES

1. "Special hazard" is a term that has several meanings. In this case, it denotes special fire situations that are so unique that it would be impossible to test and rate the effectiveness of a fire extinguisher for each special case.

2. NFPA 17A: *Standard for Wet Chemical Extinguishing Systems*, 1998 Edition (Quincy, MA: National Fire Protection Association), p. 17A-5.

3. Gagnon, Robert, *Design of Special Hazard and Fire Alarm Systems* (Clifton Park, NY: Delmar Learning, a part of the Thomson Corporation, 1998), pp. 149–150.

4. NFPA 17A: *Standard for Wet Chemical Extinguishing Systems*, 1998 Edition (Quincy, MA: National Fire Protection Association), p. 17A-5.

5. Gagnon, Robert, *Design of Special Hazard and Fire Alarm Systems* (Clifton Park, NY: Delmar Learning, a part of the Thomson Corporation, 1998), pp. 101–102.

6. Gagnon, Robert, *Design of Special Hazard and Fire Alarm Systems* (Clifton Park, NY: Delmar Learning, a part of the Thomson Corporation, 1998), p. 30.

7. OSHA regulations limit untrained or limited trained persons to fight only incipient fires.

8. NFPA 10: *Standard for Portable Fire Extinguishers*, 1998 Edition (Quincy, MA: National Fire Protection Association), Chs. 4 and 5.

9

Water Supply

S everal years ago, I was a brand new assistant fire chief in a rural area of Florida. Having come from the New York metro area, "rural water supply" was new to me. In New York, if you needed water, you just used a hydrant! But this was not the case in the new department.

We were dispatched to a reported building fire. As I approached the scene, I found a very large single-family dwelling with heavy fire involvement in approximately 30 percent of the structure. I knew that there were no hydrants in that part of the district, so I started to radio the Emergency Communications Center for multiple mobile water supply apparatus. I wanted every mobile water supply apparatus for miles around. And I got them. Eventually.

As the incident progressed, I was interrupted two or three times by some of our firefighters who were mumbling "something" about water supply. I was busy and made it clear that I had it covered. As one might expect, the fire progressed faster than the arrival of the mobile water supply apparatus, and for all intents and purposes, the building burned to the ground. Not a good night.

In the mop-up stages of the fire, I went up to some of the firefighters who were speaking to me earlier and I asked them what they were trying to tell me. What they'd been trying to tell me was that a major river was less than 100 feet away from the building—and they were going to suggest we draft out of that river for our water supply. Hmmm. Great idea.

I learned a few lessons from that fire. The first was that if you're a Yankee, and you burn down a 150-year-old "pride of the South" building, you will hear about it for 201 years! But more importantly, I learned there's no excuse not to be prepared when you know your district has areas without municipal water supply. You know what areas you protect right now—and certainly know where you have and don't have municipal water supply. Plan ahead for rural water supply before the fire comes in. Know what mobile water supply apparatus resources will be due to respond and have them respond on the first alarm. And drill with mutual and automatic aid companies before a fire to make sure that your radios and hose connections are all compatible. At the fire scene is too late to find out. Also, a few seconds of listening to the firefighters may have made a big difference in our tactical operation. This fire incident now prompts me to do a walk-around on almost any incident I'm involved with. The walk-around gives me a view of the big picture and allows me to determine our strategy—had I done it that night, it would have been clear to me that water, and lots of it, was flowing behind the building.

—*Street Story by Battalion Chief Billy Goldfeder, E.F.O.,*
Loveland-Symmes Fire Department, Loveland, Ohio

LEARNING OBJECTIVES

After completing this chapter, the reader should be able to:

9-1   Explain the value of proper water supply to other goals of firefighting.

9-2   Explain the features of a water distribution system and identify sources of water supply for firefighting.

9-4   List and describe the sources of water supply for fire protection.

9-5   Explain the various methods of connecting fire apparatus to various water sources.

9-6   Explain the purpose of a mobile water supply operation in the rural setting.

9-7   Demonstrate deployment and operation of portable water tanks and the required equipment

to transfer water from the water tank to a fire apparatus.

9-8   Identify the valves associated with water distribution systems.

9-9   Identify types of fire hydrants and their uses.

9-10   Demonstrate the correct operation of a fire hydrant.

9-11   Identify the proper pressures associated with water distribution systems.

9-12   Explain a portable water tank operation.

9-13   Describe the loading and offloading of tanks for mobile supply operations.

9-15   Identify the cause of obstructions and damage to fire hydrants and mains.

*The FF I and II levels, as defined by the NFPA 1001 Standards, are identified in different colors:
FF I = black, FF II = red, additional information = blue.

INTRODUCTION

Water supply is one of the most critical elements of firefighting because water is the most common extinguishing agent. In considering the expression "the first five minutes of fighting a fire sets the stage for the entire battle," remember: without water it is not even a fight. Water supply is important in areas with a water distribution system, and in areas without a distribution system one must be created. If there is no water supply, then hoses, appliances, and engines are useless.

Where the water comes from, how much is available, and how it is delivered are key questions. Water supply dictates the **fire flow capacity** or possible fire flow. **Pressure** is the force, or weight (lbs.) of water, measured over an area (sq. in.). The **fire flow requirement** is the amount of water required for putting out the fire.[1] Knowing the amount of water available and how much is needed lets firefighters select the right strategy, tactics, hose appliances, and fire streams. All are interrelated; a large fire without sufficient water leads to failure, and property and possibly lives will be lost as a result.

Water is normally economical, readily available, and abundant. It has the ability to absorb large quantities of heat. In most rural areas, and some urban areas, it is readily available to draft from a river or pond. The additional needs for fire protection make municipal water more expensive but not as costly as other extinguishing agents. Its availability and abun-

dance often depend on the fire's location. Desert or rural areas have limited water supplies or, at times, no water at all. A city can have areas where the availability of water is limited. In such situations, firefighters must transport the necessary water by water tender. Finally, water's ability to absorb large quantities of heat makes it extremely effective in fighting fire. Water absorbs such large amounts of heat from the fire that it cools the fuel below its ignition temperature. (More information on this can be found in Section I, Chapter 4.)

SOURCES OF WATER SUPPLY

An understanding of the water supply situation for firefighting requires knowledge of the source of the water and how it gets from that point to the fire scene. When using a lake as the water source, it is pretty obvious, but other cases are not as clear. How does the water get to a fire hydrant and what factors are involved in making sure that enough water is available for the needs? A wide range of natural and man-made factors affects water sources. The weather is probably the greatest of these factors. The Earth's water is in a constant cycle of change. The sun evaporates water into the atmosphere. It condenses into clouds that carry it to other places where it eventually falls as rain. Some areas get so much rain they are called rain forests, whereas other places get none and are deserts.

Some areas alternate between rainy and dry seasons; others have plenty of water but it is frozen some or all of the time. These are just some problems encountered with water supply.

Groundwater

Most of the Earth's freshwater supply is groundwater. Groundwater is water that seeps into the ground from rain and other surface sources. It collects in pockets of the Earth or permeates into layers of water-bearing soil. These pockets are called **aquifers** and some are large underground rivers. The level of the water under the Earth's surface is the **water table,** which can rise and fall. Springs are groundwater sources that naturally flow to the surface; others are under great pressure and when drilled may force their way to the surface. This pressure may decrease over the years such that pumps will be needed to draw the water to the surface, as is the case with most wells, **Figure 9-1.** Shallow wells are closer to the Earth's surface and are cheaper to drill but are more prone to changes in the water table and are subject to contamination and saltwater infiltration. Deep wells may penetrate through several layers of water before finding an aquifer. They are a more predictable water source with less chance of contamination. Wells used for municipal water supply and fire protection must be of sufficient volume; often multiple wells are employed. Farm irrigation systems can provide adequate water for firefighting operations. Some departments carry special adaptors and hardware to tap into these irrigation water supply systems.

Surface Water

Surface water is the world's most common source of water—almost 75 percent of the Earth is covered by water. Unfortunately, most is found in the oceans and seas with limited availability. Other natural sources of surface water are rivers, lakes, and ponds, **Figure 9-2.** Man-made surface sources include lakes, ponds, reservoirs, swimming pools, and water tanks. Surface water may be fresh, salt, or brackish water, and many surface water sources are influenced by tidal changes. **Tidal changes** are the rising and falling of the surface water levels due to the gravitational effects between the Earth and the moon. In some areas, these changes are insignificant but in others a more than 40-foot difference exists between high and low tide. When using a tidal water source, it is important to understand the effects of the tide and know when the changes will occur or the water source may not be available when it is needed.

Mobile Water Supply Apparatus

For many small fires and in those areas without a water distribution system, the water tank on fire apparatus supplies the water, having attained this water from another source. The tank sizes of fire apparatus vary, but most engines today have 500-gallon or larger tanks. If the vehicle has a tank of over 1,000 gallons and is primarily used to transport water, it is considered a mobile water supply apparatus, **Figure 9-3.** The term **water tender** is used to describe mobile water supply apparatus, although some jurisdictions use the term **tanker.** The fire service is still divided on this language to some degree. The National Fire Service Incident Management System Consortium uses the term water tender to describe land-based water supply apparatus in its Model Procedures Guide for Structural Firefighting. This guide has been endorsed by the International Association of Firefighters (IAFF), International Association of Fire Chiefs (IAFC), National Wildfire Coordinating Group (NWCG), and Fire Resources of

FIGURE 9-1 Well pump with storage tanks.

FIGURE 9-2 A surface water source. *(Photo courtesy of Donald Fischer)*

FIGURE 9-3   Mobile water supply apparatus.

Southern California Organized for Potential Emergencies (FIRESCOPE) groups. The guide refers to a tanker as an aircraft capable of carrying and dropping water for firefighting operations. Although regional, some fire departments use *tanker* to describe the land-based water supply apparatus and **tender** to describe hose-carrying apparatus. This chapter will use the term tender, because it is becoming the preferred standard term for ground vehicle units, while tankers are understood to be air units, such as helicopters and airplanes. Tenders might or might not have pumps and hose. Mobile water supply apparatus range from 1,000 to over 8,000 gallons with some units being tractor-trailer units. Tenders combined with **portable water tanks** can efficiently provide large volumes of water to a fireground operation. Tenders may have a small booster or attack pump of at least 250 gpm, a fire pump of at least 750 gpm, or a transfer pump of at least 250 gpm. Transfer pumps use a vacuum-operated pump to draw and expel their load.

SAFETY

Mobile water supply apparatus have been involved in a large percentage of fire department vehicle accidents and firefighter injuries and fatalities. The rural environment that requires these large, heavy vehicles has older, narrower, crowned roads that are often driven by persons with less experience, training, and supervision. Some fire departments cannot afford a new vehicle so they modify an old delivery tank truck into a water supply unit without the necessary safety modifications—such as extra baffles and heavier brakes. Another problem is that a gasoline tank truck with 7,600 gallons of water has a 20 percent heavier load than if carrying lighter weight gasoline. Fire departments using this type of apparatus need to examine all aspects of this problem and work to provide safe and effective fire protection.[2]

Tanks, Ponds, and Cisterns

An additional source of water can come from developed water tanks, ponds, and cisterns. Water tanks may be underground, ground level, or elevated and may be connected to a piping system, have a **dry hydrant** or other connection, or have just a drafting point. Ponds may be developed for fire protection reasons and may be lined or unlined with or without dry hydrants. A **cistern** is an underground water tank made from natural rock or concrete. Cisterns can store large quantities of water—30,000 gallons or more—in areas without other water supplies or as backup. Connections are dry hydrants, drafting pits, or other types. Firefighters should also be aware of other sources of water available to them such as swimming pools and any other body of water that can be accessed for firefighting.

WATER DISTRIBUTION SYSTEMS

How does the water get to the fire? This is a good question in areas with fire hydrants because the main water source may be many miles away. Water distribution systems have several components, including a method of getting the water, filtration or treatment processes, and a method of storage, supply, and distribution. The important water system components that concern firefighters are the supply and distribution system, including storage facilities. The water can be obtained from a surface or groundwater source. Water supply and distribution systems should be designed to meet the community's domestic, industrial, and fire protection needs at peak periods.

Small groundwater systems require a well with a pumping station that can also treat and store the water. This system will connect to a local supply system. In larger well or surface systems, multiple supply, processing, and storage units with massive feeder and distribution lines are often used.

Water is supplied in three ways. The first is gravity fed, **Figure 9-4A,** in which the water source is at a higher elevation than where the water is used, such as in a mountain reservoir feeding a valley city. The next is a pumped system, **Figure 9-4B,** in which pumps are used to draw water from a well, river, or reservoir and pump it through the system. The third type is a combination gravity-pumped system, **Figure 9-4C,** where part of the system relies on pumps and the remainder on gravity. This is the most common type because even good gravity systems find areas where distance or elevation requires additional pumping and storage for maintaining pressure. It is considered to

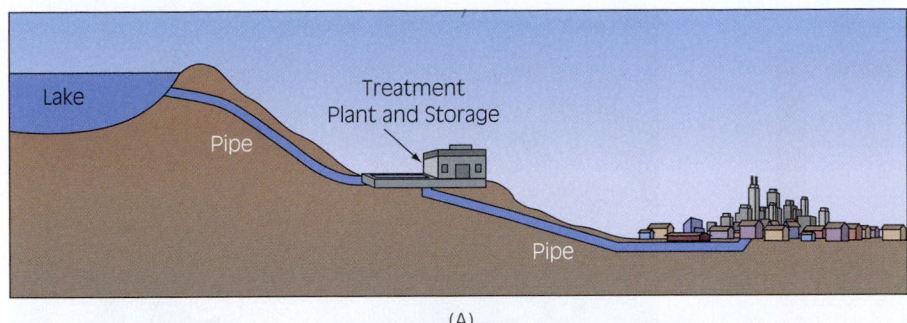

(A)

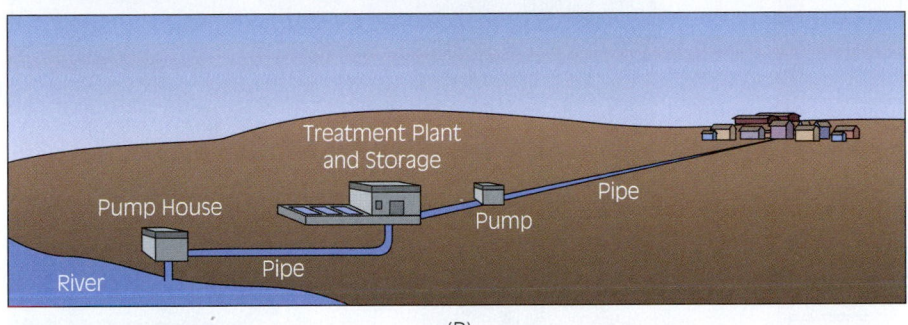

(B)

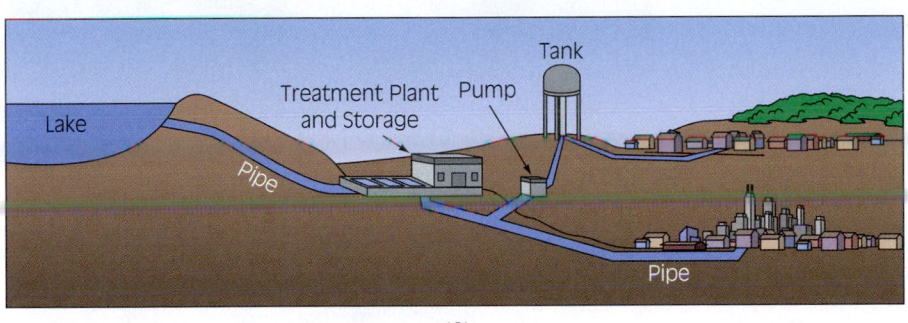

(C)

FIGURE 9-4 (A) Gravity-fed water distribution system. (B) A direct pump water distribution system. (C) A combination gravity-pumped water distribution system.

be a combined system because elevated water tanks that are filled by pumps in turn supply the piping system by gravity.

After treatment for drinking, water goes into the distribution system or water mains. The mains are divided into feeders and distribution lines. Primary feeders supply secondary feeders and then distribution lines. Primary and secondary feeder lines in large cities can be measured in feet; some of New York City's aqueducts are 20 feet in diameter. Some older and larger cities have separate water systems just for fire protection. Distributor lines are the water mains which supply building connections and fire hydrants and are used for fire protection; they range in size from 6 to 16 inches. Good water distribution systems are interconnected into loops and grids that allow multidirectional supply and reduce dead-end mains that cannot provide adequate pressure or volume, **Figure 9-5.**

STREETSMART TIP

Firefighters should know the location of dead-end water mains and, where possible, avoid using them. Connecting to a dead-end water main may not provide adequate water, or if another unit connects to another hydrant on the same dead-end main, it may rob the original unit of what was a good water supply.

FIRE HYDRANTS

Fire hydrants allow the fire department to access water supply systems. A fire hydrant is also known as the "plug." The term *plug* is a throwback to days when firefighters drilled a hole in wooden water mains, then capped the hole with a plug when they were finished. The two major hydrant types are wet and dry barrel hydrants. Another hydrant type is a dry hydrant that is neither dry nor a true hydrant but a pipe system for

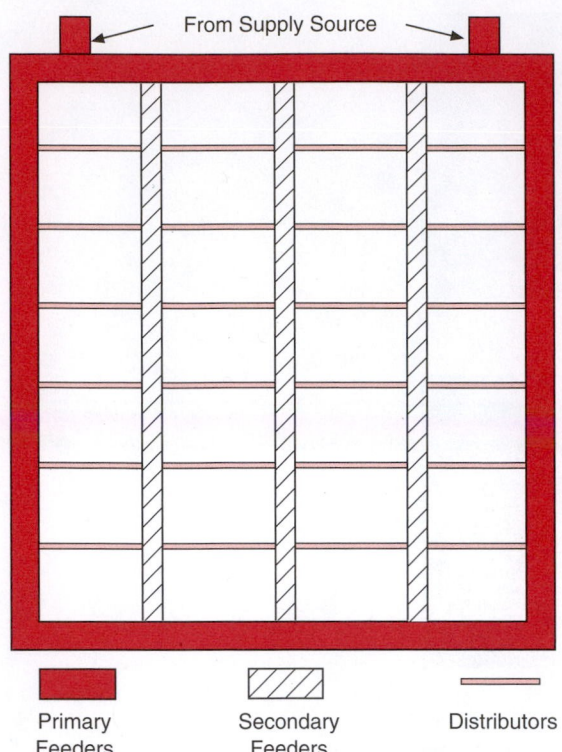

From Supply Source

Primary
Feeders

Secondary
Feeders

Distributors

FIGURE 9-5   Grid or looped system.

drafting from a static water source. Some specialty hydrants are also used.

Wet Barrel

Wet barrel hydrants, **Figure 9-6** and **Figure 9-7,** have water in the barrel up to the valves of each outlet. They are used in areas that are not subject to freezing temperatures, primarily California and Florida. The wet barrel hydrant allows each outlet to be controlled by a separate valve and can have additional lines taken off or supplied if an outlet is available. This additional connection does not require the flow through other outlets to be stopped. Some also have a main control valve that can control the flow to all outlets.

Dry Barrel

Dry barrel hydrants, **Figure 9-8** and **Figure 9-9,** are used in areas where freezing temperatures could damage the hydrant. It uses a valve at the base of the hydrant to control water flow to all outlets. The base and this valve are below ground at the level of the water main. Connecting additional lines to a flowing dry barrel hydrant requires the entire hydrant to be shut down until the new connect is made and then reopening the valve or adding a gate on one of the outlets prior to charging the hydrant. When the dry barrel hydrant valve is fully closed or partially

FIGURE 9-6   Multivalved wet barrel hydrant.

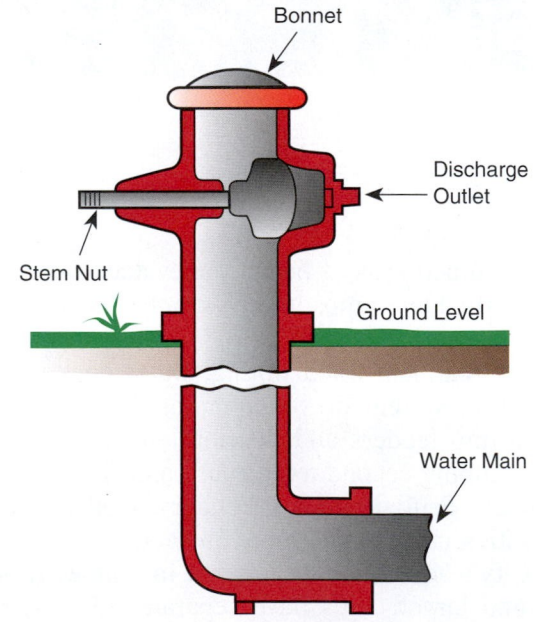

Bonnet

Discharge
Outlet

Stem Nut

Ground Level

Water Main

FIGURE 9-7   Typical schematic of wet barrel hydrant.

opened, a drain allows the water to drain from the hydrant preventing damage from freezing. Firefighters should ensure that the hydrant is operated and shut down with the valve in either the fully opened or closed position. Operation of a dry barrel hydrant in a

FIGURE 9-8   Dry barrel hydrant.

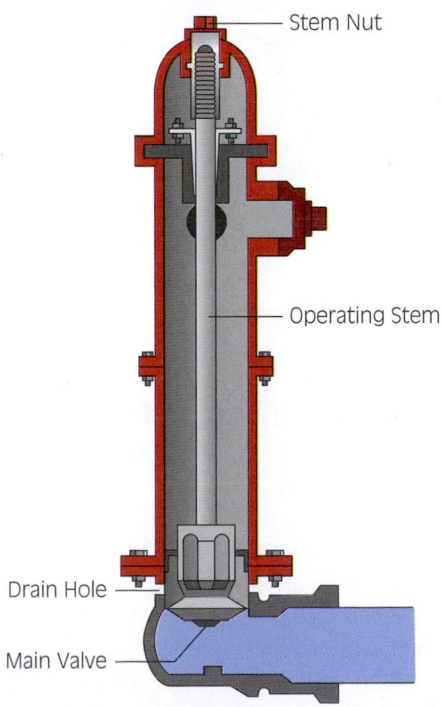

FIGURE 9-9   Typical schematic of dry barrel hydrant.

Stem Nut

Operating Stem

Drain Hole

Main Valve

partially opened position causes undermining of the hydrant and ground, with imminent or eventual damage to nearby roads, buildings, and fire apparatus.

> **CAUTION**
>
> Firefighters should ensure that dry barrel fire hydrants are operated and shut down with the valve in either the fully opened or closed position. When the valve is not fully opened or closed, undermining of the hydrant and ground with imminent or eventual damage to nearby roads, buildings, and fire apparatus can occur, plus it may restrict volume. Also, anytime a valve is operated, it should be done slowly to prevent a sudden surge of pressure referred to as a **water hammer.**

Dry Hydrant

A dry hydrant is not really a fire hydrant but a connection point for **drafting** from a static water source such as a pond or stream. A dry hydrant is a pipe system with a pumper suction connection at one end and a strainer at the other, **Figure 9-10** and **Figure 9-11.** They are used primarily in rural areas with no water distribution system. Dry hydrants may be found in urban and suburban areas as a backup water supply or where certain buildings rely on a nearby pond or stream for a water source. Fire departments benefit by having prepared and well-planned access to the water source; some even have graveled areas for parking and a turnaround point for the apparatus. Landowners benefit by having a ready source of water in case of fire and may even receive a discounted cost for property insurance.

FIGURE 9-10   Dry hydrant.

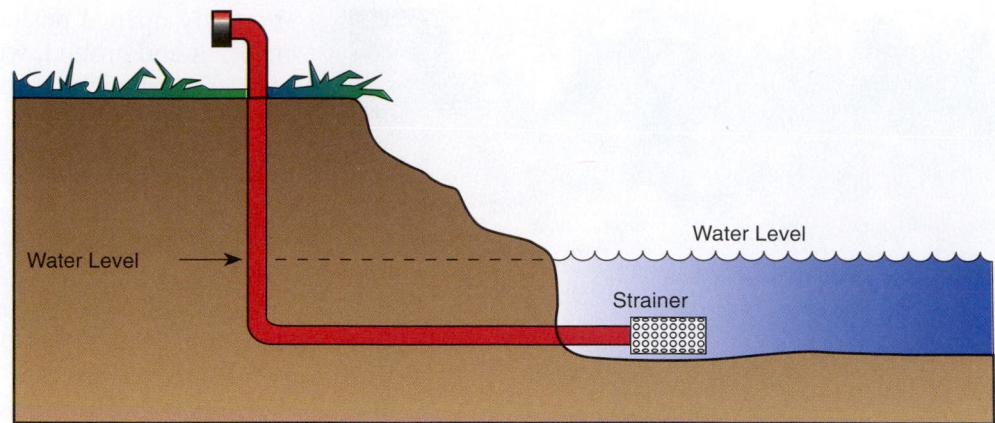

FIGURE 9-11 Schematic of dry hydrant installation.

Specialty Hydrants

Wall hydrants are mounted on the wall of a building, **Figure 9-12.** Firefighters should make sure that they are connecting to a wall hydrant and not the fire pump test connection outlets. A wall hydrant has a direct connection to the water supply system, whereas a fire pump test connection is an inspector's way to test the flow of the fire pump. The test connection will not supply any additional water; it will just rob the water from the fire pump.

A flush-type hydrant is mounted below grade level and is usually found in a pit, vault, or valve box. The purpose of mounting the hydrant below grade allows access to a water source in areas where an above-grade hydrant would interfere with operation of the facility. Airports, shipping terminals, and many European cities utilize flush-mounted hydrants.

High-pressure hydrants are connected to a separate high-pressure water system used only for fire protection purposes. Some larger cities, San Francisco is one, have these types of systems installed in the downtown commercial and industrial areas to provide high water volume and pressure for large fires. Some of these systems draw their water directly from the city harbor, and the pumps are often maintained by the fire department. These hydrants are either of the wet or dry barrel design.

Hydrant Protective Devices

Hydrant protective devices or valves are designed to protect the water system associated with a fire hydrant. Most fire hydrants are connected directly with the domestic water supply for a city, town, or municipality. With the increased security resulting from the events of 9/11, many water suppliers have decided to add measures to protect the water supply. Interior valves and exterior protective devices, **Figure 9-13,**

FIGURE 9-12 Wall fire hydrant.

have been installed on some hydrants in some water systems. In most cases, the exterior systems require a lock and key. In some jurisdictions, firefighters may encounter these locking hydrant protection devices that prohibit unauthorized use of the hydrants. Their main purpose is to prohibit work crews from using water from these hydrants. They are most commonly found on privately owned property such as industrial plants and airports rather than in municipal jurisdictions. Firefighters who may be operating in these areas should be familiar with these devices and know

FIGURE 9-13 External hydrant protective device.

the location of the key or tool that makes the hydrant operational. Firefighters subject to encountering these devices should train on their rapid removal to increase tactical effectiveness.

VIEWPOINT

By their design, water systems are vulnerable to terrorist attack. Water is attained either by surface runoff or by wells. The storage for either is by tank or by surface lakes or ponds. Delivery is by underground pipes all of which surface at some point. Fire hydrants are part of these system surfaces. As stated, many water companies—mostly private entities—have added protective measures for the hydrants in their systems, for protection against either terrorism or water theft. Firefighters have been charged with the responsibility of removing these exterior devices in the event of a fire. Often, firefighters feel these devices are unnecessary as they add time to the firefighting operation. A key or custom tool is necessary in most cases to unlock the protective device and must be kept on the apparatus and accounted for by firefighters. While most engine companies have the ability to cut off the locks or devices, it takes precious time away from the initial operation. The interior hydrant protective devices seem to be effective and simple to use. However, retrofit or initial purchase costs can be prohibitive in some fire systems, taking dollars away from apparatus, equipment, and manpower needs. It is often argued that the weakest link in the water supply system is not the hydrants but the water storage system or treatment facility. It would seem that every portion should be equally protected. Whether or not these devices serve an adequate, cost-effective purpose can be debated. The use of protective devices and fire service concerns should be discussed thoroughly with the water supply system's administration.

VALVES ASSOCIATED WITH WATER DISTRIBUTION SYSTEMS

The valves in public water systems are usually nonindicating type gate valves and check valves. The **gate valves** or butterfly valves are opened and closed to control water flow, **Figure 9-14.** Nonindicating gate valves are installed at interconnections of water mains, at intermediate points on long sections of water mains, and before each hydrant and major building connection. **Check valves** are installed to control water flow in one direction, typically when different systems are interconnected. **Backflow preventers** are check valves or a pair of check valves that prevent a backflow of water from one system into another. They are mostly required where a building water or fire protection system connects with the public water system. Backflow preventers are required for environmental and health reasons. The backflow preventer is installed on the building's private water piping—after the building's main water shutoff. Private water valves are of the indicating type, for example, a post indicator valve (PIV), a wall indicator valve (WIV), or an outside stem and yoke valve (OS&Y). These valves are commonly associated with standpipe and sprinkler systems.

RURAL WATER SUPPLY

Rural water supply or mobile apparatus water supply operations can occur anywhere, not just in rural areas. Urban and suburban areas often have undeveloped areas or other places where the hydrants are just too far away to use effectively. Limited access highways and large bridges have water supply problems. Even redundant water supply systems can suffer a catastrophic event that knocks out or limits the water supply. Firefighters must understand the basics of rural water supply. Rural firefighters find it easier to adjust to water supply operations in urban areas. Conversely, urban firefighters may not fully understand rural systems.

Rural water supply operations require careful coordination and control. A water supply group supervisor should be part of the incident command system with full authority over mobile water apparatus operations. Firefighters assigned to water supply operations may find the work less glamorous than actual firefighting, but the nozzle person cannot put out the fire without this vital support.

Portable Water Tanks

Since mobile water supply apparatuses are designed to transport water, they need to be able to quickly offload water and return to the **fill site** as soon as possible.

FIGURE 9-14 Hydrant with plumbing. Note the location of a gate valve between the water main and hydrant.

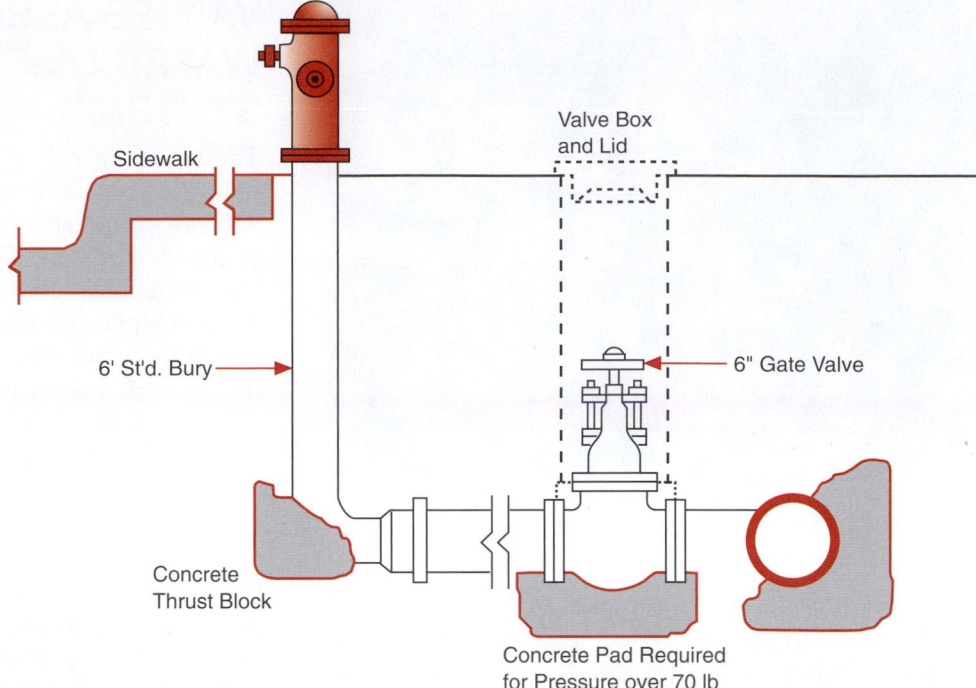

Valve Box and Lid

Sidewalk

6' St'd. Bury

6" Gate Valve

Concrete Thrust Block

Concrete Pad Required for Pressure over 70 lb

FIGURE 9-15   Portable water tanks are an essential piece of equipment for shuttle operations.

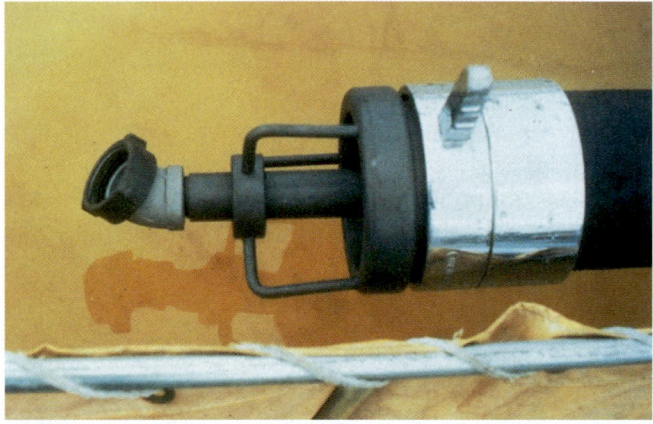

FIGURE 9-16   Jet siphon valve. *(Photo courtesy of Donald Fischer)*

To speed the operation, each mobile apparatus should have a portable water tank with a capacity equal to or greater than its tank size, **Figure 9-15.** The collapsible or inflatable tanks are either set up next to the attack engine or, if space is not available, set up at a relay engine that drafts from the tank and supplies the attack engines. Depending on the size of the fire and availability of tanks and mobile apparatus, several tanks may be used to increase the fire flow. Many large modern tenders use a **jet dump** or **jet siphon** to speed the unloading of tanks, **Figure 9-16.** The jet dump or siphon draws the water out as opposed to just using gravity in a nonassisted system.

The portable tank is erected or inflated at the **dump site** in a location where the loaded mobile apparatus can offload water. Using suction hose, the attack engine will draft water for suppression operations. Placing a

heavy tarp on the ground prior to opening the portable tank can help protect the tank liner from damage. In larger operations, multiple tanks may be set up together. Departments that use portable tanks should regularly practice mobile water supply shuttle operations.

Mobile Water Apparatus Operation

A mobile water apparatus operation is a **shuttle operation** that involves mobile water supply apparatus moving large quantities of water between a dump site and a fill site. The dump site is where the water is delivered for quick unloading. As the mobile apparatuses arrive already full, the dump site is set up first. Using portable tanks and a supply engine to draft from the tanks speeds the process and allows the mobile apparatus to return to the fill site for another load of water. Personnel at the dump site connect hoselines or operate dump valves to fill the portable tanks and help set up the drafting operation for supplying the attack pumpers, **Figure 9-17.** Dump sites should be selected for availability to unload multiple mobile apparatus, turnaround area for the mobile apparatus, operational area, continued access to the fireground, and safety of personnel. The fill site is the location of the water source where the mobile apparatus are filled, **Figure 9-18.** The fill site should be staffed with enough personnel to quickly connect hoselines to fill the mobile apparatus and may require a pumper to draft from a static source, **Figure 9-19.** The fill site should have an adequate water source—enough to cover the contingency of supplying multiple units—and it should have sufficient space for vehicles to turn around. The optimum fill site area would allow vehicles to safely and expeditiously drive *through* the fill site without the need to turn around to exit.

Shuttle operations control the fire flow capacity of the incident, **Figure 9-20.** Fire flow rates can be cal-

FIGURE 9-18 Tender at fill site with pressurized water source.

FIGURE 9-17 Tender at dump site dropping water directly into portable tank.

FIGURE 9-19 A pumper at fill site drafting water. *(Photo courtesy of Donald Fischer)*

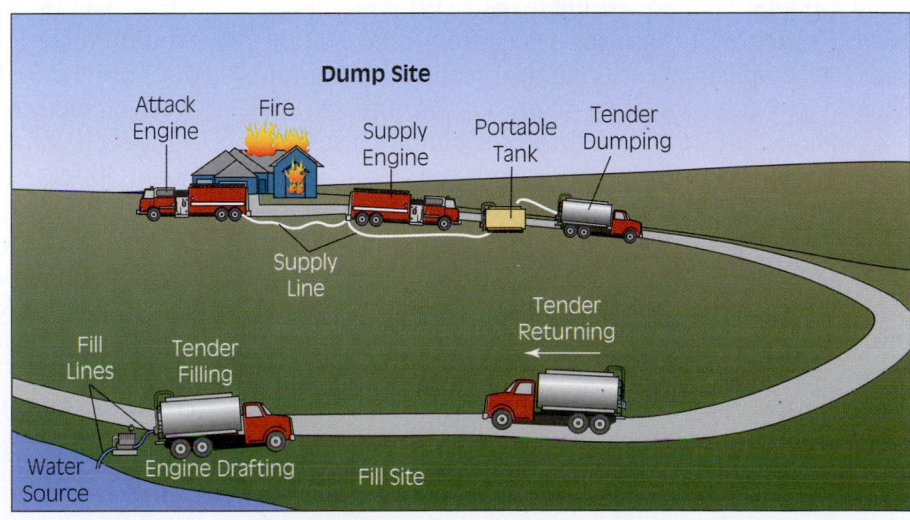

FIGURE 9-20 Tender shuttle operation.

culated by adding the time to fill a mobile apparatus, turn the unit around and return to the dump site, drop the water, and turn around and return to the fill site divided by the quantity of water carried. In this formula, a 10 percent residual is figured. Normally, due to tank construction and for time purposes, 10 percent of the tank water remains in the tank after the dump operation. If a 1,000-gallon (3,785-L) mobile water supply apparatus has a 10-minute shuttle time, the flow rate cannot exceed 90 gpm (340 L/min) [1,000 gallons (3,785 L) – 100 gallons or 10 percent (378 L) = 900 gallons (3406 L) / 10 minutes].[3] Reducing the shuttle time or adding more mobile apparatus or larger tanks would increase the flow rate. With five 2,000-gallon (7,570-L) units operating on 10-minute cycles, the flow rate would be 900 gpm (3,406 L/min) [10,000 gallons (37,850 L) – 1000 gallons (3785 L) /10 minutes]. Mobile water shuttle operations time cannot be decreased by increasing vehicle speed on the highway. Time should only be gained at dump and fill sites by increased efficiency of personnel and equipment.

SAFETY

It is important to note that attempts to lower shuttle time cannot and should not be done by increasing vehicle speed. Time should only be gained at dump and fill sites by increased efficiency of personnel and equipment. This efficiency is gained through strategy and tactics covered in more advanced engineer or officer training programs.

PRESSURE ASSOCIATED WITH WATER DISTRIBUTION SYSTEMS

All of the Earth's water is under pressure. See Chapter 11 (in Section I) for more discussion on pressure. Even seawater is under **atmospheric pressure,** which is 14.7 pounds per square inch (psi) (101 kPa) at sea level, **Figure 9-21.** Drafting water from a lake or the sea is accomplished by taking advantage of this atmospheric pressure. Creating a partial vacuum or low atmospheric pressure area inside a pump causes the atmospheric pressure on the water's surface to force the water up the suction hose and into the pump, which adds pressure and pumps it out. Pressure is the force, or weight, of water measured over an area, **Figure 9-22.**

At sea level, a 1-foot (0.305-meter) column of water 1 inch square weighs 0.433 pounds (0.19 kg), creating a pressure at the bottom of 0.433 psi (3 kPa). Pressure can also be expressed as feet of head with the 1-foot column having a head of 1 foot (0.305 meters) and 2.31 feet (0.7 meter) of head equaling 1 psi (6.895

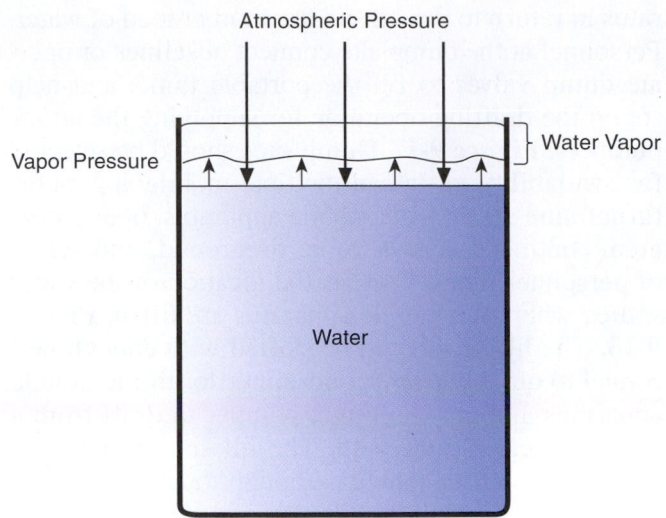

FIGURE 9-21   Atmospheric pressure being exerted on container of water.

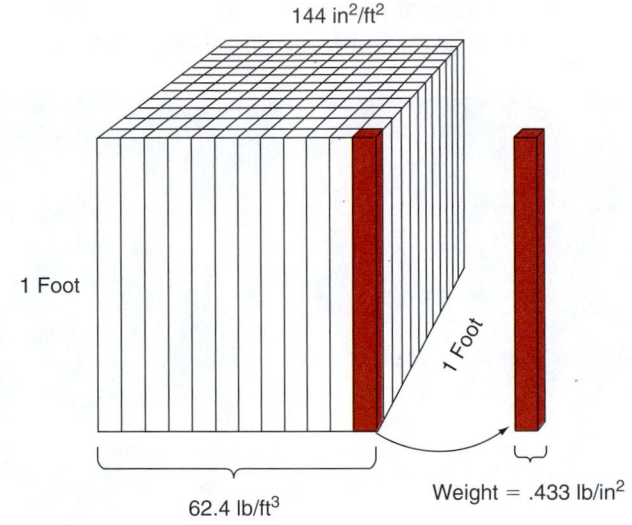

144 in²/ft²

1 Foot

1 Foot

62.4 lb/ft³

Weight = .433 lb/in²

FIGURE 9-22   Pressure is expressed in pounds per unit area (psi).

kPa). An atmospheric pressure of 14.7 psi (101 kPa) would create a head of almost 34 feet (10.4 meters) [14.7 psi 3 2.31 ft/lbs 5 33.9 feet], if a perfect vacuum were created, or the ability to lift a column of water 33 feet (10.1 meters). Practicality limits this ability to about 30 feet (9 meters).

Atmospheric pressure reduces 0.5 psi per 1,000 feet of elevation. Therefore, a drafting operation in the mountains of Colorado would not produce as much lift as a drafting operation near the ocean in California. A tank in a distribution system that stores water 100 feet (30 meters) high would create 100 feet (30 meters) of head or 43 psi (296 kPa) [100 feet 4 2.31 ft/lbs 5 43.3 psi or 100 feet 3 0.433 lbs/ft 5 43.3 psi]. Pressure in a nonflowing closed system is equal at all points but, while flowing, it is reduced by friction created by its movement and the loss of pressure at the opening. Pressure in an open system is subject to atmospheric pressure.[4]

Water distribution systems are supplied under pressure to meet domestic and fire protection needs. This dual use can cause some firefighting concerns: the system pressure is never as high as firefighters would prefer as this type of pressure could cause leaks in the plumbing of private homes. The average pressures in the United States are between 65 and 80 psi (448 to 551 kPa) with a typical minimum of 20 psi (137 kPa).[5]

Pressures reaching 150 psi (1,034 kPa) or higher are usually damaging to plumbing. The 20-psi (137-kPa) figure is the recommended low **residual pressure** when pumping from a hydrant; pressures below this may create a vacuum in part of the supply system that could cause damage to the system. Fire departments should know the normal and flowing pressures and the capacity of the water distribution system—including areas that are high or low in either category. High-flow and/or high-pressure areas are an advantage, whereas low-pressure and/or low-flow areas should be avoided if possible.

LESSONS LEARNED

Water is the most common fire extinguishing agent used by firefighters and it must be supplied in sufficient quantity to accomplish extinguishment. Firefighters must understand the relationship between the amount of water that can be supplied and the amount needed. If they do not have enough water, their efforts will not be effective.

Supplying the water needed requires an understanding of the water source and the valves and hydrants in the distribution system. Firefighters should also know about the different pressures available in the system, along with the likely problems.

Firefighters in areas without a distribution system must create one using mobile supply apparatus, portable water tanks, and shuttle operations. These firefighters quickly learn the value of water as it can be difficult to sustain adequate fire flow with this type of operation.

KEY TERMS

Aquifer   A formation of permeable rock, gravel, or sand holding water or allowing water to flow through it.

Atmospheric Pressure   The pressure exerted by the atmosphere, which for Earth is 14.7 pounds per square inch at sea level.

Backflow Preventers   A check valve or set of valves used to prevent a backflow of water from one system into another.

Check Valve   Valve installed to control water flow in one direction, typically when different systems are interconnected.

Cistern   An underground water tank made from natural rock or concrete.

Drafting   The pumping of water from a static source by taking advantage of atmospheric pressure to force water from the source into the pump.

Dry Hydrant   A piping system for drafting from a static water source with a fire department connection at one end and a strainer at the water end.

Dump Site   The area where tenders are unloaded or their load dumped.

Fill Site   The area where tenders are filled or attain their water.

Fire Flow Capacity   The amount of water available or amount that the water distribution system is capable of flowing.

Fire Flow Requirement   A measure comparing the amount of heat the fire is capable of generating versus the amount of water required for cooling the fuels below their ignition temperature.

Gate Valve   Indicating or nonindicating valve that is opened and closed to control water flow.

Jet Dump   A device that speeds the process of dumping a load of water from a tanker/tender.

Jet Siphon   A device that speeds the process of transferring water from one tank to another.

Portable Water Tanks   Collapsible or inflatable temporary tanks for the storage of water that is dumped from tankers or tenders. Usually carried by the tender to set up a dump site.

Pressure   The force, or weight, of a substance, usually water, measured over an area.

Residual Pressure   The pressure in a system after water has begun flowing.

Shuttle Operation   The cycle in which mobile water supply apparatus is dumped, moves to a fill site for refilling, and is returned to the dump site.

Tanker   The term given to aircraft capable of carrying and dropping water or fire retardant. Some departments still use the term to describe land-based water apparatus.

Tender   The abbreviated term for water tender. A water tender is defined as a land-based mobile water supply apparatus. Some departments still use

the term *tender* to describe a hose-carrying support apparatus.

Tidal Changes   The rising and falling of the surface water levels due to the gravitational effects between the Earth and the moon.

Water Hammer   A sudden surge of pressure created by the quick opening or closing of valves in a water system. The surge is capable of damaging piping and valves.

Water Table   The level of groundwater under the surface.

Water Tender   The term given to land-based water supply apparatus.

REVIEW QUESTIONS

1. How does the fire flow capacity relate to the fire flow requirements?
2. What are the four reasons for using water for firefighting?
3. Where is most freshwater found?
4. Name three surface water sources.
5. What are the two names for mobile water supply apparatus?
6. Water is distributed or supplied through a supply system in what three ways?
7. What is a dead-end main and why should it be avoided?
8. Explain the two main types of fire hydrants and where they are found.

9. Why is a dry hydrant different from other hydrants?
10. What purpose does a backflow preventer serve?
11. What does a gate valve do?
12. Describe a portable water tank.
13. Explain a tender shuttle.
14. What is pressure?
15. What is the difference between static and residual pressure?
16. If a hydrant's water pressure drops when flowing a line by 9 percent, how many more lines can be taken from that hydrant?

ENDNOTES

1. For more information, see the following: Thomas Sturtevant, *Introduction to Fire Pump Operations* (Clifton Park, NY: Thomson Delmar Learning, 1997), Chapter 11; NFPA 1231, *Standard on Suburban and Rural Water Supply*; or *Fire Protection Handbook*, 18th ed. (Quincy, MA: National Fire Protection Association, 1997), Chapter 6-5.

2. See many of the fire service publications about this safety issue. To examine closely the accident, injury, and fatality statistics, see the *NFPA Fire Journal* issues containing annual firefighter injuries and fatalities surveys.

3. Tank size is the total capacity of the tank, but often the tender cannot actually offload its capacity or the time to do so would

not be productive. Departments operating tenders should conduct tests in which the tender is loaded and offloaded as if under fireground conditions and weigh the vehicle and divide the offloaded amount by 8.33 lb/gallon (or 1 kg/L) to determine its operating capacity.

4. For more information, see Thomas Sturtevant, *Introduction to Fire Pump Operations* (Clifton Park, NY: Thomson Delmar Learning, 1997), Chapter 11.

5. *Fire Protection Handbook*, 18th ed. (Quincy, MA: National Fire Protection Association, 1997), pp. 6–72.

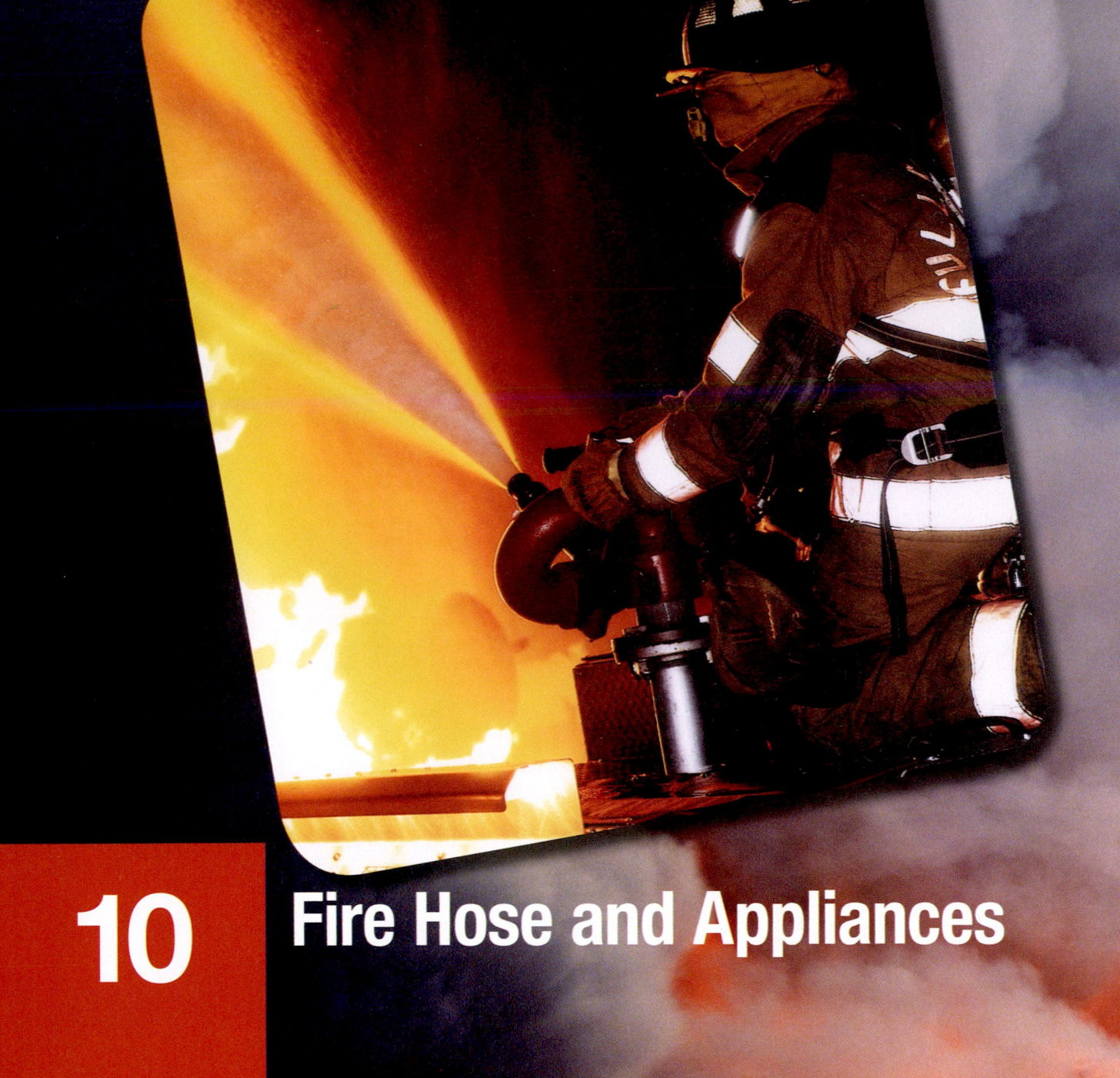

10

Fire Hose and Appliances

Our station covers an area that has lots of old abandoned homes with wood construction. One evening a call came in as a house fire, and as we approached from several blocks away, we could see the smoke. When we arrived, the house was fully involved. We were told that a homeless person had taken residency in this shack—it had no power. As we attempted to go in, we saw that he was a major pack rat. There were several appliances, including washing machines and refrigerators, and parts throughout the front of the house, and that made it nearly impossible to gain access. We had to squeeze through the door sideways and walk over things. Inside there were more piles: candles, cans, papers, etc.

It was an incredible obstacle just for the rescue crew to get to the victim. (Eventually they found him behind a locked door. He wasn't breathing but he did have a pulse. He also had a gun in one hand and a knife in the other!) In terms of advancing the hose, we decided to do it as a team. We needed an action plan. Normally it's a three-person job, but in this case it took two entire crews. Instead of one person at the front, we had somebody every 4 or 5 feet, because otherwise as soon as we pulled, we'd get stuck on something.

That day was really long and by the end of the incident, more than half of the house was destroyed. But we really learned to adapt to a difficult situation. Most important is remembering and applying the basic skills you know to each specific incident. I remember learning hoseline rolling in training, but it's not something we use much in application. In this case, though, it was the best way to get the job accomplished. I also learned the importance of teamwork. A firefighter may come in on a fire assigned to one job, but when there's an issue like this, we all have to work together as a team.

—*Street Story by Michelle Steele, Lieutenant,*
Miami-Dade Fire and Rescue, Miami, Florida

LEARNING OBJECTIVES

After completing this chapter, the reader should be able to:

10-1   Describe the construction, care, and maintenance of fire hose and hose couplings.

10-3   Explain fire hose cleaning techniques.

10-4   Demonstrate the proper method of fire hose cleaning.

10-5   Demonstrate the proper operation of hose washers and drying equipment.

10-6   Given a specific piece of equipment, list and describe various tools as well as inspection and cleaning methods used to return the equipment back to service.

10-7   List and describe proper use of cleaning solvents.

10-8   Given manufacturer recommendations, departmental guidelines, tools, and cleaning materials. Demonstrate the ability to properly clean, inspect, and service tools and equipment.

10-9   Identify the types of hose couplings and threads.

10-10   Demonstrate replacing fire hose couplings gasket.

10-11   Demonstrate connecting and disconnecting various sizes and types of couplings on hose.

10-12   List and describe fire hose rolls and loads and their use.

10-13   Demonstrate the proper methods of rolling fire hose and loading fire hose back on apparatus.

10-14   Demonstrate hose carries.

10-15   Explain the precautions to be followed when advancing a hoseline to a fire in terms of communications, equipment, personnel, and safety.

10-16   Demonstrate the ability to advance charged and uncharged hoselines in various firefighting situations to include ladders and stairwells.

10-17   Demonstrate discharging an attack line while secured to a ground ladder.

10-18   Demonstrate how to connect supply and attack lines to standpipe connections.

10-19   Identify hydrant connections and their proper operation.

10-20   List and describe tools and equipment required to establish a hydrant water supply.

10-21   Explain the various methods of connecting fire apparatus to various water sources.

10-22   Demonstrate the proper method of connecting and placing hard-sleeve hose for a drafting operation.

10-23   Demonstrate connecting large-diameter hose to a fire hydrant.

10-24   Demonstrate the correct operation of a fire hydrant.

10-25   Demonstrate extending a hoseline.

10-26   Demonstrate replacing a burst section of hoseline.

10-27   Demonstrate forward and reverse hose lays to a hydrant.

10-28   Demonstrate the proper way to hand lay supply hose to a water source from a fire department apparatus over a given distance.

10-32   Describe a procedure for removing a defective hose from service.

10-33   Describe a procedure for reporting a defective hose being taken out of service.

10-34   Demonstrate the proper way to identify and mark defective fire hose.

*The FF I and II levels, as defined by the NFPA 1001 Standards, are identified in different colors: FF I = black, FF II = red, additional information = blue.

INTRODUCTION

This chapter introduces the firefighter to one of the most important tools for fighting fires: hose. Hose is the tool used to move water to the fire. Firefighters should know about this tool, how it is made, and how to care for it. More importantly, firefighters must know how to properly store hose on the apparatus and how to quickly move it from the apparatus to the location where firefighting will take place.

The hose techniques learned in this chapter are basic to the techniques of fire control. These techniques, along with many others in this book, are the foundation of firefighting in a safe, efficient, and effective manner. Firefighters should learn them well and practice them often. In situations where things do not go well on the fireground, many experienced fire officers often comment about the need to "get back to the basics."

Fire hose is a flexible conduit used to move water or another agent from a source to the fire. Early firefighters used bucket brigades to supply the water to the engines while top-mounted nozzles sprayed the water onto the fire. Leather hose was invented to allow firefighters to advance nozzles from the engine to the fire and to supply the engine with water. Today, many different materials have replaced leather hose; however the basic tasks performed for supply and attack remain the same. Firefighters, fire brigade members, wildland firefighters, and some building occupants continue to use hose today.

Couplings, adapters, and appliances are used in modern firefighting to connect hose and adapt it to different sizes required for specific fire ground operations. Hose couplings were found to be a problem at major fires early in the twentieth century. The different fire departments fighting these fires discovered that their hose couplings were not compatable. Today, most departments use National Standard Hose Threads. Departments that do not use National Standard carry adaptors to make connections with mutual aid departments. Newer types of metals and composites have made it possible to construct lighter-weight adapters. This helps make the job of firefighting somewhat easier.

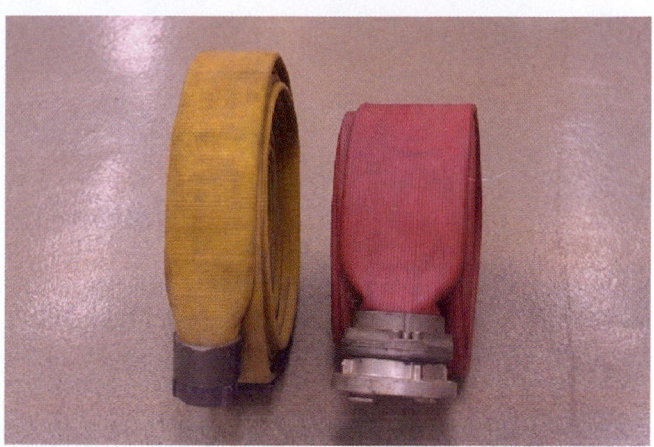

FIGURE 10-1 Woven and rubber-coated fire hose.

wrapped construction, a fabric or mesh material is impregnated with rubber or plastic, wrapped around a rubber **liner** and covered with rubber or plastic. Braided construction uses a yarn braided over the rubber liner, with a rubber or plastic layer separating it from the next layer. A rubber or plastic cover encases the entire hose. Woven hose has a rubber liner and one or more outer layers called a **jacket,** which is usually woven of cotton, synthetic, or blended materials. Some hoses have jackets of rubber or plastic-impregnated woven construction. The jacket is designed to protect the lining from heat and mechanical damage. **Hard suction hose** has a plastic or wire **helix** to provide addition support and prevent collapse under a vacuum when drafting. Some newer synthetic types of hose combine the jacket and liner. Fire hose is divided into categories: attack, supply, occupant use, forestry, and hard suction hose. Each type of hose has certain requirements for construction, size, and pressure. Most hose comes in 50- or 100-foot (15- or 30-m) lengths except as noted.

> **FIREFIGHTER FACT**
>
> Large conflagrations in Boston in 1872 and Baltimore in 1904 resulted in many large cities sending fire units to help fight the fire only to discover that their hose couplings were not compatible and could not be connected.

> **FIREFIGHTER FACT**
>
> Cotton and other natural fibers replaced leather hose. Rubber liners were added in the late 1800s. Prior to the rubber liner, firefighters had to wait for the cotton or natural fibers to absorb enough water to cause them to swell and slow the leakage of water. Some fire departments and building occupant standpipe hoses use unlined single-jacket hose. Current standards now recommend lined hose.

CONSTRUCTION OF FIRE HOSE

Fire hose has two components: the hose itself and the couplings that connect the hose sections or appliances. Fire hose is made in three types of construction: wrapped, braided, and woven, **Figure 10-1.** In

Booster hose (sometimes called a "trash line") is smaller diameter, rubber-coated hose of ¾- or 1-inch (19- to 25-mm) diameter size. Booster hose is usually

FIGURE 10-2 Booster hose on reel.

FIGURE 10-3 Various hoselines in a hose bed.

mounted on a reel and can be used for outside fires or overhaul operations after the fire is extinguished, **Figure 10-2.** It is not designed to be used for structural firefighting. Booster hose has a limited flow rate of up to 30 gpm (114 L/min) and, although easier to maneuver and control, is unsafe for structural firefighting due to the maximum gallonage possible and length shortcomings. Once the entire hose is deployed off the reel, it is not feasible to add additional sections to it in a timely manner. Many departments have stopped purchasing booster hose.

Attack hose is used by trained personnel to fight fires. This hose is connected to nozzles and distributors, master stream appliances, standpipes and sprinkler systems, hydrants, and fire pumps. Attack hose diameter is a minimum of 1½ inches (38 mm) and can include large-diameter hose when supplying master stream devices. The NFPA requires pumpers to carry 400 feet of attack hose. Attack hose is usually service tested annually at 250 to 400 psi (1720 kPa). Any use above this pressure may result in hose and equipment damage or personnel injury. Attack hose can be either single or, more commonly, double-jacketed hose, **Figure 10-3.** Standpipe packs often use single-jacketed hose to reduce the weight of the pack and allow easier manuevering. **Medium-diameter hose (MDH)** is either 2½- or 3-inch (63- or 75-mm) hose. Hoses that are smaller than 2½ inches are called **small lines** or **small-diameter hose.**

Supply hose or **large-diameter hose (LDH)** is larger hose (3½ inches [90 mm] or bigger) used to move water from one engine to another, from an engine to a fire department connection (FDC), or from the water source to a portable hydrant for distribution. Common sizes are 4 and 5 inches (100 to 125 mm) with some 6 inch (150 mm). The diameter of supply hose allows water to move at lower pressure.

NFPA requires pumpers to carry at least 800 feet of 2 ½-inch or larger supply line.

Hard suction hose comes in sizes from 2½ to 6 inches (65 to 150 mm) in diameter and is usually matched to the size of the main intake of the pump for which it will be used. Because of its more rigid construction, it is less flexible or movable than other hose. Hard suction hose is tested on its ability to withstand vacuum. Hard suction hose is standard in 10-foot (3-m) lengths.

Soft suction hose (also known as soft sleeve) is large-diameter woven hose used to connect a pumper to a hydrant, **Figure 10-4.** It is also sized according to the size of the intake of the pumper and usually comes in sizes from 4½ to 6 inches (112 to 150 mm). The lengths of a section vary: 20, 25, and 30 feet (6, 8, and 9 m). NFPA requires pumpers to carry at least 15 feet of soft sleeve intake hose or 20 feet of hard suction hose.

Occupant use hose is hose used in standpipe systems for building occupants to fight incipient fires. It is usually 1½-inch (38-mm) diameter single-jacket hose similar to attack hose. Older types of this hose may be unlined. Most fire departments do not use the occupant use hose, and if they use the standpipe system, they replace it with their own hoselines.

Forestry hose is specially designed for use in forestry and wildland firefighting, **Figure 10-5.** It comes

FIGURE 10-4 A soft suction hose.

in 1- and 1½-inch (25- and 38-mm) sizes and should meet U.S. Forestry Service specifications. Forestry hose is woven hose and comes in 50, 75, and 100 foot (15, 23, and 30 m) lengths.

CARE AND MAINTENANCE OF FIRE HOSE

The care of hose begins with the careful folding and placement of the dry hose on the apparatus. Properly placed hose should not have sharp turns or bends that can damage the liner (see Hose Loads section). Hose that is not new should always be folded at different places from the previous folds to prevent over-stressing one area of the hose.

NOTE

Hose standards still recommend that fire hose be changed out every thirty days with a visual inspection at the time of change. With the new synthetic hoses, many departments no longer use this standard.

The **hose bed** should be designed to facilitate the circulation of air to allow the hose to dry while it is stored on the apparatus. This prevents mildew of natural fiber hose and vehicle rusting. The use of a hose bed cover will prevent rain and other road contaminants from getting into hose, as well as blocking sunlight, which can affect rubber and plastic-coated hose. Synthetic hose can be loaded while wet.

NOTE

The most common damage to hose is wear and tear due to dragging hose—which is especially damaging to couplings. Excessive pressure and water hammers should be avoided, as this can cause sudden rupture of the hose or separation of the hose from the coupling.

When used on the fireground or for training, several steps can be taken to reduce damage to hose. Avoid laying hose over sharp or rough corners; use a hose roller or a salvage cover to prevent damage. Remove any kinks in hose prior to charging it or soon thereafter. Do not allow vehicle traffic to drive over

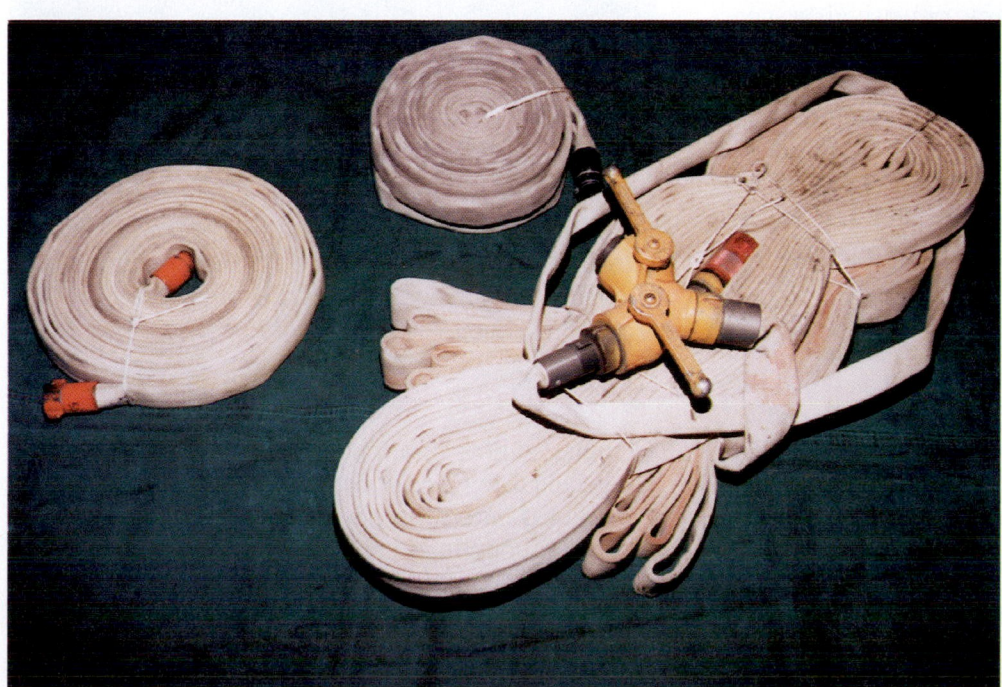

FIGURE 10-5 Hose packs and rolls for wildland firefighting.

the hose, which could damage the couplings. Vehicles cause more damage to uncharged hose than charged hose because the inner liner is not protected when uncharged. Clean dirt and grit from hose with a brush or plain water to prevent these materials from creating abrasions on the hose. Firefighting exposes hose to heat and burning embers as well as broken glass, nails, and other sharp objects that can quickly damage it. Avoiding excessive heat, burning embers, and chemicals—especially gasoline and oil—can keep the hose in good and usable condition, and can save the fire district unnecessary costs. Mild chemicals can be removed using detergent, but large amounts of strong chemicals should be removed using hazardous materials decontamination procedures, if possible. Care should be taken to prevent hose from freezing. Frozen hose should be carefully moved to a safe place to thaw prior to any folding or bending. Any hose that has been subjected to damage of any kind should be service-tested prior to returning to service.

Water hammer commonly damages fire hose. Water hammer is the pressure surge that occurs when water valves are rapidly opened or closed. Water that is not flowing has its pressure equalized throughout the whole system. Slamming a valve shut or opening it quickly allows this pressure to rapidly act on the parts of the system and damage hose, valves, pumps, and the water supply system including water mains, hydrants, and valves. Water hammers result from mistakes made by people operating valves and nozzles too quickly. The correct action is to operate all valves slowly and let the pressure either bleed off or build up slowly without the surge.

Dirty or contaminated hose should be rolled and returned to the station for cleaning or washed on scene using a portable washer. Cleaning may also be done by placing the hose on a flat surface, rinsing with water, and scrubbing with mild detergent. When detergents are used, the hose should be thoroughly rinsed and hung in a hose tower or storage rack until dry. Commercial hose washers are also available but are not capable of adequately cleaning the threads of a coupling. Some work by connecting the hose washer to a hydrant and the hose is rinsed as it passes through, **Figure 10-6.** This type of washer still requires hand scrubbing by personnel. Another type has the ability to scrub, wash, rinse, and reroll the hose, **Figure 10-7.** Heated hose drying cabinets are also available. Care must be taken to prevent overheating and damaging the hose. Stored hose also requires maintenance: occasionally water should be flowed through stored sections to prevent the liner from drying and cracking, and then it should be redried and stored.

The final part of the care of hose involves a regular visual inspection by firefighters after each use. Fire-

FIGURE 10-6   Hose washer on hydrant.

FIGURE 10-7   An automatic hose washer.

fighters should carefully inspect the outer cover and couplings for any damage. Coupling care is discussed later but the outer cover should be checked as hose is reloaded for discoloration and abrasions. Further care is conducted during the annual service test.

TYPES OF HOSE COUPLING

Couplings allow hose and appliances to be joined or connected using a set of connection devices. Couplings are divided into two types: threaded and nonthreaded. Threaded couplings use a screw thread that secures the two sections of hose together, while nonthreaded couplings use locks or cams. Couplings are made of brass, aluminum, or an alloy called pyrolite, which is lighter than brass and more resistant to bending.

Threaded couplings are further separated into two different types: one with external threads, the male, and the other with internal threads, the female, with a matching thread type. A threaded hose coupling set is typically of three-piece construction with a hose bowl on each hose and a swivel attached to the female coupling. The bowls are attached to the hose using an expansion ring with a tail gasket to prevent leaks. The internal swivel also has a thread or swivel gasket to stop leakage where the threads connect, **Figure 10-8A.**

Threaded couplings are of common design but with many nonstandard types of threads. Today, although a National Standard Thread (NST)—also called National Hose Thread, NHT—exists, there are other "standard" hose threads. Some include National Pipe Thread, Eastern Hose Thread, Pacific Coast Thread, Quebec Standard Thread, and Canadian Standards Association. New York and Chicago each have both a fire department thread and a city hose thread. Personnel checking new hose and connection equipment should ensure the correct threads are used on all hose, appliances, connections, and so on. Hose threads allow the couplings to screw or fasten together. National Standard hose threads and other types have blunt end threads called a **Higbee cut, Figure 10-8B.** The blunt end threads keep hose from cross-threading since they can be started at only one point. A groove is notched into coupling lugs to help firefighters align the Higbee cuts. This notch is called the **Higbee indicator.**

> **CAUTION**
>
> Personnel checking new hose connection equipment should ensure that the correct threads are used on all hose, appliances, connections, and so on. An emergency fire scene is not the correct time to discover that threads on equipment are not compatible.

Nonthreaded couplings use locks or cams to secure the connection. Both couplings are identical and either end can be used to connect to another coupling. The couplings are aligned and twisted one-quarter turn, locking them together. **Storz couplings, Figures 10-8 C** and **D,** are the most popular of these types but

(A)

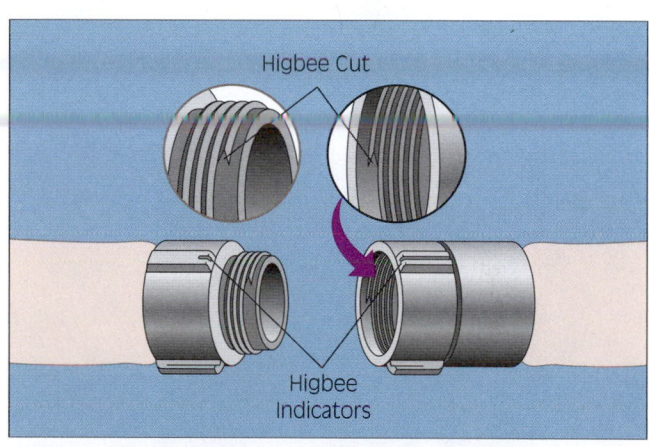

(B)

(C)

(D)

FIGURE 10-8 (A) Male and female threaded couplings. (B) A Higbee or blunt end cut on the thread of a coupling. (C) Storz coupling. (D) Quarter-turn thread. *(Photos A, C, and D by Fred Schall)*

there are other kinds. Some nonthreaded couplings use expansion rings and others are attached by wire or straps on the exterior of the hose that bind the hose to the couplings.

To assist firefighters in making, tightening, or breaking the connection of a coupling set, lugs or handles are placed on the couplings. The most common type of lug is the rocker lug, but pins and recessed pins are also used. Handles are used mostly on hard and soft sleeves. Spanner wrenches are used with the lugs to assist in tightening and loosening the couplings.

CARE AND MAINTENANCE OF COUPLINGS

Care and maintenance of hose couplings requires keeping them clean and preventing mechanical damage. Proper storage and rolling, and simply paying attention to where the coupling is placed can solve most problems. Prevention of mechanical damage begins with caution being taken against dropping the couplings, which damages the threads or the coupling itself. Couplings should not be dragged. Whenever possible, personnel should carry hose near the couplings to avoid damaging them. Firefighters should avoid the placement of hose couplings in vehicle traffic areas where the couplings may be struck or rolled over. Good storage and rolling methods also prevents damage to exposed threads, **Figure 10-9.**

The care of couplings should include a visual inspection each time the hose is reloaded. The couplings should be checked for any damage to the coupling and threads, the proper operation of the swivel, and the presence and condition of the gasket. The gasket should be pulled out and checked for cracks and pliability. One last check would be for any movement or slippage of the hose from the coupling. If dirt or grit is detected in the coupling, rinse it and take a brush to the coupling and threads to clean it. Remember, dirt and grit may prevent tightening or loosening of a connection.

HOSE TOOLS AND APPLIANCES

> **NOTE**
>
> Hose tools are accessories that help firefighters move or operate hoselines. Appliances are devices that water flows through, including adapters and connectors. Many of these devices must be the correct size to work with the size(s) of hose being used. Appliances allow firefighters to connect various hoses and nozzles, **Figure 10-10.** Nozzles and foam equipment are covered in Chapter 11 (in Sections I and II), and master stream devices are covered later in this chapter.

FIGURE 10-9 Hose storage rack.

FIGURE 10-10 Various hose appliances.

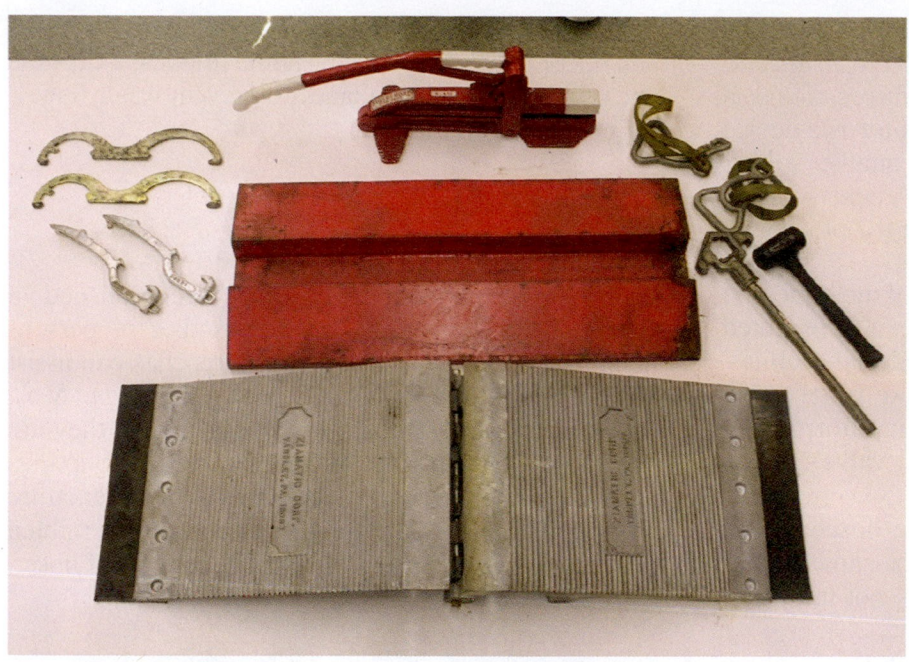

FIGURE 10-11   Various hose tools.

Tools include rope hose tools, wrenches, rollers, clamps, and other items, **Figure 10-11.** A **rope hose tool** is about 6 feet (2 m) of ½-inch (13-mm) rope spliced into a loop with a large metal hook at one end and a 2-inch (50-mm) ring at the other. It is used to tie in hose and ladders, for carrying hose, and for many other tasks requiring a short piece of rope. Some departments have an engine or spanner belt to accomplish the same tasks.

A **hose strap** is a variation of the rope hose tool, although shorter in length. The hose strap is perhaps the most useful tool in handling a charged hoseline due to its effective design of a strap, cinch clip, and forged handle.

Spanner wrenches come in several sizes and are used to tighten or loosen couplings. They may also be useful as a pry bar, door chock, gas valve control, and so on.

A **hydrant wrench** is the tool used to operate the valves on a hydrant and may also be used as a spanner wrench. Some are plain wrenches and others have a ratchet feature to speed the operation of the valve.

A **hose roller** or **hoist** has a metal frame, with a securing rope, shaped to fit over a windowsill or edge of a roof. It has two rollers that allow the hose to roll over the edge, preventing chafe, **Figure 10-12.**

A **hose clamp** is a device used to control the flow of water by squeezing or clamping the hose shut. Some work by pushing a lever, which closes the jaws of the device, and others have a screw mechanism or hydraulic pump that closes the jaws. Hose clamps create a pressure buildup in the hose, so it is important to place the clamp at least 5 feet from any coupling (toward the supply side) and at least 20 feet from an apparatus.

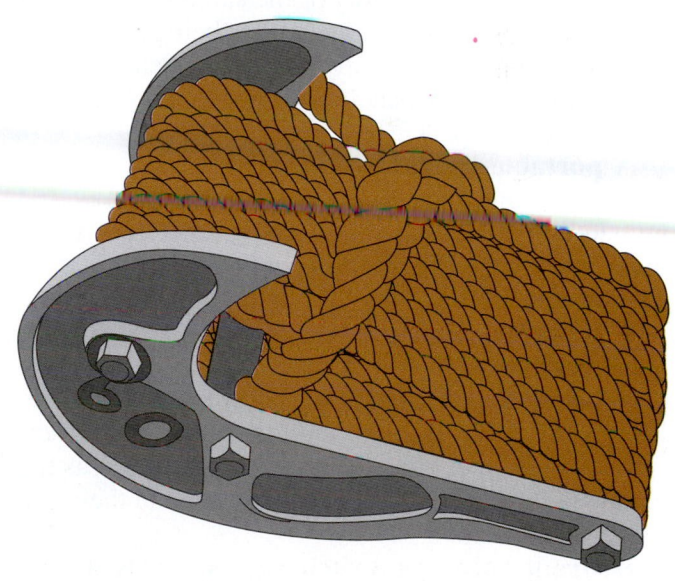

FIGURE 10-12   Hose roller.

Another device used for stopping leaks without shutting down the line is a **hose jacket.** Hose jackets may be a metal or leather device that is fitted over a leaking area in the hose and is either clamped or strapped to control the leak.

A **hose bridge** is a device that allows vehicles to pass over a section of hose without damaging it.

A **hose cart** is a handcart or flat cart modified to be able to carry hose and other equipment around large buildings. Some departments use them for high-rise situations.

A **double female** is an appliance that allows two male ends of hose to be connected, and a **double male** does the same for two female ends. A double male is used to connect the two female thread couplings when doing a split lay. An **increaser** is used to connect a smaller hose to a larger one, and a **reducer** connects a larger hose to a smaller one. An **adapter** is a device that changes one type of hose thread to another, allowing connection of two different lines. Adapters have a male end on one side and a female on the other with each side being a different thread type.

Gate valves have an inlet and outlet of the same size with a gate to control water flow; they can be mounted on apparatus or appliances connected to hoselines. Large-diameter hose requires an **intake relief valve** at the receiving engine. This valve may function as a combined overpressurization relief valve, a gate valve, and an air bleed-off.

A **wye** is a device that divides one hoseline into two or more. The wye lines may be the same size or smaller size, and the wye may or may not have gate control valves to control the water flow.

A variation of the wye is the **water thief,** which has one inlet and one outlet of the same size plus two smaller outlets with all of the outlets being gated. The standard water thief usually has a 2½-inch (65-mm) inlet with one 2½-inch (65-mm) and two 1½-inch (38-mm) outlets.

A **portable hydrant** or **manifold** is like a large water thief and may have one or more intakes and numerous outlets to allow multiple hoselines to be utilized with or without a pumper being directly at the fire location.

A **siamese** is a device that connects two or more hoselines into one line. Siamese valves either have clapper valves or gate valves to prevent loss of water if only one line is connected. Clapper valves are spring-loaded and operate automatically, whereas the gate valve requires that a firefighter operate the valve to the open position.

Hydrant valves or **switch valves** are used on a hydrant to allow an engine to connect and charge its supply line immediately but also allow an additional engine to connect to the same hydrant without shutting down the hydrant.

A **strainer** is placed over the end of a suction hose to prevent debris from being sucked into the pump. Some strainers have a float attached to keep them at or near the water's surface. A strainer designed to draw water from a shallow, flat surface is typically called a "duck's foot." The duck's foot strainer is most often used for drafting from portable water tanks. A different style of strainer or screen is located on each intake of a pump.

A **distributor pipe** or **extension pipe** allows a nozzle or other device to be directed into holes to reach basements, attics, and inaccessible areas. The distributor pipe has self-supporting brackets that help hold it in place when in use.

A **hose cap** is not an appliance, but it is a tool that can be used with hose. As the name implies, it caps the end of a hoseline or appliance to prevent water flow.

COUPLING AND UNCOUPLING HOSE

Connecting hose couplings can be accomplished in several ways, depending on the number of personnel available. The use of spanner wrenches can assist in tightening and loosening hose couplings. Most threaded couplings should be tightened until they are "hand tight." Arbitrarily tightening threaded couplings with a spanner wrench causes stress and premature wear on gaskets. Spanners should be used to tighten leaking couplings or to assist the uncoupling of hose.

> **NOTE**
>
> Remember, when making a connection: "righty tighty-lefty loosey." Turning a wrench to the right tightens the connection, while turning the wrench to the left loosens the connection.

The first coupling technique is a foot-tilt method for one person.

JPR 10-1:   One-Person Foot-Tilt Method

(For step-by-step photos of this skill sequence, see page 259)

1. With the male coupling lying flat on the ground, place the left foot directly behind the coupling, stabilizing it and raising it upward.
2. With the female coupling in the right hand, bend down and bring the two couplings together. When aligned, turn the swivels to the right.

Another one-person method is the over-the-hip method.

JPR 10-2:   One-Person Over-the-Hip Method

(For step-by-step photos of this skill sequence, see page 259)

1. Hold the hose with the female coupling slightly below the waist.
2. Pick up the male coupling, align the two couplings, and swivel the female coupling to make the connection.

The over-the-hip method can be accomplished with two firefighters.

JPR 10-3: Two-Person Over-the-Hip Method

(For step-by-step photos of this skill sequence, see page 260)

1. The two firefighters face each other—each holding a coupling. The couplings are brought together, aligned, and swiveled. The firefighter with the female coupling watches and adjusts the alignment of the couplings.

Uncoupling hose is usually done in the opposite manner in which it was connected. Occasionally, a connection is too tight to easily break. After ensuring the hose is not charged and the pressure has been bled off, the first recommendation is to use spanner wrenches to loosen the connection.

JPR 10-4: Uncoupling Hose with Spanners

(For step-by-step photos of this skill sequence, see page 260)

1. With the hose lying flat on the ground, place the spanners on the lugs going in opposite directions.
2. Push the spanners downward to break the connection.

If spanners are unavailable, one of the following methods can be used.
The one-person uncoupling method is known as the knee-press method.

JPR 10-5 Knee-Press One-Person Uncoupling Method

(For step-by-step photos of this skill sequence, see page 261)

1. Fold the hose coupling over on itself on the ground and press your knee down on the coupling.
2. Try to twist the coupling to the left, breaking the connection.

The two-person method is called the stiff-arm method and is similar to the two-person coupling, but the firefighters brace themselves and twist the coupling while holding their arms stiff with elbows slightly bent, **Figure 10-13.**

HOSE ROLLS

The type of hose roll is normally dictated by department policy and the application for which it will be used. Firefighters should practice to be proficient in all types of hose rolls.

FIGURE 10-13   The two-person method is called the stiff-arm method and is similar to the two-person coupling, but the firefighters brace themselves and twist the coupling while holding their arms stiff, with elbows slightly bent.

Straight/Storage

The straight or storage hose roll is the easiest to work with. This is called a storage roll because of its common use, and it is often used when picking up after a fire. An exception used by some departments is to store damaged hose with the male end out and/or a knot tied at the end to identify it as unuseable hose.

JPR 10-6: Straight or Storage Hose Roll

(For step-by-step photos of this skill sequence, see page 261)

1. Start with the hose flat on the ground. From the male end, to protect the threads, roll it straight to the opposite end.
2. Once the roll is finished, it is ready to be moved to storage.

Single Donut

The single-donut hose roll is used when access to either or both couplings may be needed. There are several ways to do a donut roll.

JPR 10-7: Single-Donut Hose Roll

(For step-by-step photos of this skill sequence, see page 262)

1. Lay the hose lying flat. Fold the hose on top of itself with the male coupling on top about 3 feet (1 m) short of the female coupling.

2. Starting at the fold, roll the hose toward the couplings with the extra hose on the female end protecting the male coupling. A second firefighter can assist in guiding the hose and adjusting the slack as it is rolled.

3. Leave a small space at the center of the roll to provide a handhold.

JPR 10-8: Single-Donut Hose Roll (alternate method)

(For step-by-step photos of this skill sequence, see page 263)

1. Lay the hose out flat. Starting at a point off-center about 6 feet (2m) toward the male coupling, roll the hose toward the female coupling. The extra hose will protect the male coupling and the female coupling will be exposed.

2. The handhold space and the extra firefighter are also useful with this method.

Twin or Double Donut

The twin or double-donut roll is used for special applications and works best with 1½- to 2-inch (38- to 50-mm) hose.

JPR 10-9: Twin-Donut Hose Roll

(For step-by-step photos of this skill sequence, see page 264)

1. First the hose is laid flat with both couplings at one end and each half lying parallel to the other.

2. At the center, the loop is folded over the top of both halves.

3. The roll is started toward the couplings at the same time.

4. At the end, the roll may be tied together for carrying.

5. The twin donut can be secured by using the hose itself. This is called a self-locking roll. To accomplish this, extend the amount of hose that is used for the starting fold and loop. Allow this excessive hose to "flop" as the twin donuts are rolled. When finished, adjust the loop size to one large and one small loop. Place the large loop through the small loop to secure. The large loop can also be used as a carrying strap.

HOSE CARRIES

The type of hose carry is dictated by user preference and on-scene conditions. Firefighters should practice to be proficient in all types of hose carries.

Drain and Carry

The drain and carry method is used to combine the two steps of draining and carrying a section of hose into one operation. This is usually done with one section of hose.

JPR 10-10: Drain and Carry

(For step-by-step photos of this skill sequence, see page 265)

1. The firefighter starts at one end of the hose and with the coupling held waist height feeds the hose over the shoulder and back down to the waist.

2. A fold is created and the hose is laid on itself back to the front.

3. The firefighter continues to walk forward folding and refolding the hose at the waist until finished. The hose can then be carried to the new location.

Shoulder Loop Carry

This carry is similar to rolling an electrical cord around one's arm but with bigger loops.

JPR 10-11: Shoulder Loop Carry

(For step-by-step photos of this skill sequence, see page 266)

1. Place the nozzle or end of hose over the shoulder resting against the back.

2. Walk forward about 3 feet (1 m), pick up the hose, and form a bight to bring the hose back up and over the shoulder, creating a loop.

3. Continue as each section is picked up and carried forward.

4. If you need to move in the opposite direction, the loops are collected and raised with your hands and then rotated to the opposite direction.

5. Return the hose to the opposite shoulder moving forward in the new direction.

Single-Section Street Drag

The simple hose drag technique can move one or two hoselines.

JPR 10-12: Single-Section Street Drag

(For step-by-step photos of this skill sequence, see page 267)

1. Put the end of a section of hose over your shoulder with the coupling in front at waist height and walk away dragging the line.

2. Place a line over each shoulder and pull two lines.

3. If additional sections are needed, additional firefighters can do the same with the following sections until the desired amount of hose is stretched.

> **NOTE**
>
> The simple hose drag technique can cause damage to the hose couplings. Firefighters must always use common sense and be protective of all firefighting equipment.

HOSE LOADS

Apparatus hose load designs are dependant on the types of firefighting operations a company will employ. The components of fire attack, including water supply, small- and medium-sized attack lines, protective systems, and master stream devices must be efficiently incorporated onto an apparatus for rapid deployment. Attack lines may be prepared in advance and may be pre-connected to the discharges, or the attack line may be "pulled" according to the requirements of the fire scene.

A well-trained company should be able to perform any of the required fire scene tasks quickly and efficiently. Efficiency will be largely based upon how (and how often) the company trains with the hose load on their apparatus. With some thought, the hose loads can be designed to deploy with the hose couplings, adaptors, and appliances at the proper place when the lines are deployed. Training is the key; no matter how good the hose loads and attack theory are, they are useless if they cannot be proficiently and professionally utilized.

> **NOTE**
>
> Time and care taken during the loading of hose will be repaid many times over when pulling the hose loads on an incident.

A **dutchman** is a short fold of hose or a reverse fold that is used allowing for coupling placement on the load. When loading a hose onto an apparatus, there are times when a coupling will fall where a fold should be. A dutchman moves the coupling to another point in the load, **Figure 10-14.**

Accordion Load

Accordion hose loads can be used for preconnected hoselines or for providing additional supply line. The accordion load is ideal for making up **shoulder loads.**

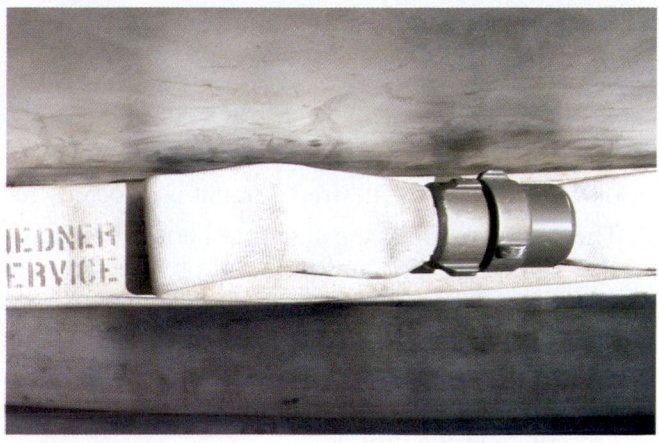

FIGURE 10-14   A dutchman is a short fold of hose or a reverse fold that is used when loading hose and a coupling comes at a point where a fold should take place or when two couplings end up on top of each other. The dutchman moves the coupling to another point in the load.

Care should be used with the accordion load because overpacking the hose bed can stress and crack the liner of the hose.

JPR 10-13: Accordion Load

(For step-by-step photos of this skill sequence, see page 268)

1. For use with preconnected lines, the accordion load should be started with the female coupling at the pump outlet or the front of the bed.

2. The hose is placed in the bed on its side rather than lying flat. The hose is laid out toward the rear of the bed, where it is folded and returned to the front where the process is repeated.

3. When the layer of the hose fills the bed, the hose is brought up at the rear and folded back on itself starting the next layer, again on the left side.

4. Coupling placement is important as couplings require a little bit of room to spin around in the bed. It is wise to use a dutchman to stagger the couplings in the bed to keep a coupling from lying close to another coupling. This load should be continued until the hose bed is full or until the desired amount of hose is loaded. If the line is pre-connected, a nozzle can also be pre-connected. When loading supply line, the hose should have adaptors available to allow for flexibility: normally it is not possible to pre-plan a forward or reverse lay. Adaptors will allow the hose couplings to be converted as required. A blind cap, can be added to protect threads.

Building a hoseline from an accordion hose load is an easy task.

JPR 10-14: Advancing an Accordion Load

(For step-by-step photos of this skill sequence, see page 269)

1. A nozzle or other appliance is attached to the end of the hose and the desired amount is selected.

2. The firefighter pulls the amount of hose about half-way out of the bed and places it on one shoulder with the nozzle at the bottom. Use care not to have too much of the hose in front, which could cause tripping. It is better to drag the excess behind.

3. Step away from the bed. Pull additional folds and drop them to the ground.

4. When the point is reached where the shoulder load is to be deployed, the firefighter allows the hose to flake off one fold at a time while moving forward.

If additional hose is needed, another firefighter may repeat this process with the attached hose instead of the nozzle.

Flat Load

The flat load is used for supply lines and some attack lines. It involves simply laying the hose flat. The intended use of the line will dictate whether the female or male end remains exposed when the line is loaded.

JPR 10-15: Flat Load

(For step-by-step photos of this skill sequence, see page 270)

1. A water supply flat load starts with the male end of the hose on the right side of the bed at the rear. This will allow this hose load to be connected to another hose load for long hose lays.

2. The hose is laid flat until the front is reached where the hose is folded over and brought to the rear. When back at the rear, turn the hose off the first layer and lay it flat next to the first row.

3. When the edge of the hose bed is reached, a second layer is started with a double layer, moving back to the right side of the bed.

4. An appliance, such as a ball valve, can be placed on the female end on top of the load.

JPR 10-16: Advancing a Flat Load from a Supply Bed

(For step-by-step photos of this skill sequence, see page 271)

1. Attach the nozzle or appliance and determine the amount of hose needed to be placed on the shoulder.

2. If possible, separate these folds from the bed by twisting them on their sides. Place the folds over your shoulder with the nozzle in front of you.

3. If twisting the folds is not possible (possibly due to the amount of hose in the bed), pull a section of hose off and place it over your shoulder, with the nozzle in front of you. Pull additional sections off and place them on top of the previous section on your shoulder.

4. When the desired amount of hose is out of the bed and on your shoulder, step away from the bed and again pull some extra folds to place on the ground. The hose should now deploy as an accordion.

Horseshoe Load

The horseshoe load is normally used for supply line. It is relatively simple to load and usually deploys well. It is especially useful for operations that require the entire hoseload to be deployed at once.

JPR 10-17: Horseshoe Load

(For step-by-step photos of this skill sequence, see page 273)

1. Start the load at a rear corner of the hose bed with the hose on its side.

2. Lay the hose toward the front wall of the hose bed and then along the wall to the other side and back toward the rear.

3. At the rear, fold the hose alongside itself and head back to where it started. Fold it again and another "U" is formed.

4. When the bed is filled, start a new layer by laying the hose along the side of the bed on top of the hose load and continue to the front of the bed. Continue the hose across the front of the bed, and down the opposite side of the bed toward the rear.

5. Lay a few flakes of hose on top longways for quick access. Attach a hose strap, ball valve, hose bag, or other set of appliances or devices deemed necessary for quick deployment.

Making a shoulder load from the horseshoe bed is somewhat similar to the accordion load. Care should be taken when reaching the outer edges of the hose bed because the increasing length of hose between folds may cause problems advancing the line.

JPR 10-18: Advancing a Horseshoe Load

(For step-by-step photos of this skill sequence, see page 274)

1. Place the nozzle on the hose and select the desired amount of hose to deploy.

2. Pull the hose and place it on your shoulder.

3. Step away to pull the hose out of the bed.

Finish Loads and Preconnected Loads

Finish loads and preconnected loads can utilize the three methods of loading hose previously discussed. Some use combinations of these loads in different layers to assist firefighters in advancing hose quickly from an engine. Finish loads can be used for assisting in laying supply line or attack lines. Preconnected loads are usually used for attack lines; therefore, the hose must be loaded in such a way that the *entire* hose load can be easily deployed. Failure to clear the hose from the bed will cause kinking and jamming of the hose in the bed when the pump operator charges the line. Apparatus and hose damage may occur if hose loaded in a bed is charged prematurely.

A straight finish load, **Figure 10-15,** is usually used with a straight hose lay and simply involves taking the final length or two of a load and laying it flat across the top of the load. A rope with adapters, a spanner wrench, and a hydrant wrench attached allows the layout person to quickly have all the necessary tools and enough hose to make the hydrant connection. For laying out, a horseshoe load is made at the top of the final hose load layer. At the center point of the "U," the last few wraps of the hose are brought to the top of the load in a flat load and the rope with adapters and wrenches is attached. Placing a twist in the line at the point where it changes from a horseshoe load to a flat load, as shown in the reverse horseshoe load in **Figure 10-16** will help the hose deploy. A reverse horseshoe load is used for an attack line. It is loaded

backwards—from the front of the bed to the rear—with a few sections of hose loaded flat on top. The nozzle is attached at the rear for quick deployment.

Similarly, an attack line can be attached to the end of a hose load. This works when a **backstretch** or **flying stretch** will be utilized. A wye or gate valve can be installed at the end of the hose load before the horseshoe. (One will have to be used if changing hose size.)

FIGURE 10-15 A straight finish load simply involves taking the final length or two of a load and laying it flat across the top of the load. A rope with adapters, a spanner wrench, and a hydrant wrench attached allows the layout person quick access to all the necessary tools and enough hose to make the hydrant connection.

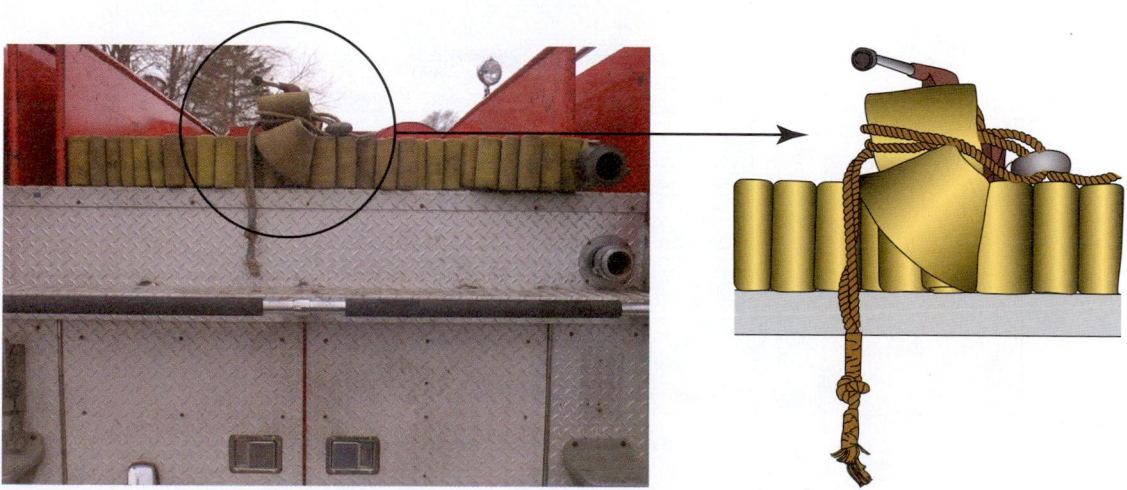

FIGURE 10-16 A reverse horseshoe load for laying out is made on top of the hose load but in the reverse direction (front to back), and at the center point of the "U" of the horseshoe the rope with adapters and wrenches is attached. The first portion of the hose may need a twist in it to get it to change direction.

JPR 10-19: Reverse Horseshoe Load for an Attack Line

(For step-by-step photos of this skill sequence, see page 276)

1. For an attack line, create the reverse horseshoe again on top of the hose bed. When the amount of hose is loaded, attach a nozzle and lay it in the center of the horseshoe with the nozzle at the rear of the hose bed.

2. Grab the nozzle and place a few of the hose folds over your shoulder. Grab the horseshoe load at the bottom of the "U," pulling it out of the hose bed and allowing it to flake out while moving to the fire.

Preconnected lines can be made up using any number of loads or combinations. Since they are preconnected to a discharge they will always start with the female end where the discharge is located. It is possible to use preconnected loads with all accordion or horseshoe layers. An example of a combination load would be a horseshoe bottom layer, with the top layer accordioned, or a flat load bottom layer and a gasner top, **Figure 10-17A** and **B**. The combinations are often designed based on a specific need at a certain pre-planned property within the response district. When using multiple layers or various hose loads, it is common to place **ears** on the hose to assist in pulling the layer, **Figure 10-18**.

Flat Load, Minuteman Load, and Triple-Layer Load

Preconnected loads must allow rapid removal of the hose from the slot or bed. Many preconnected loads have been invented based on need and ingenuity of fire crews. The following loads are among the more popular.

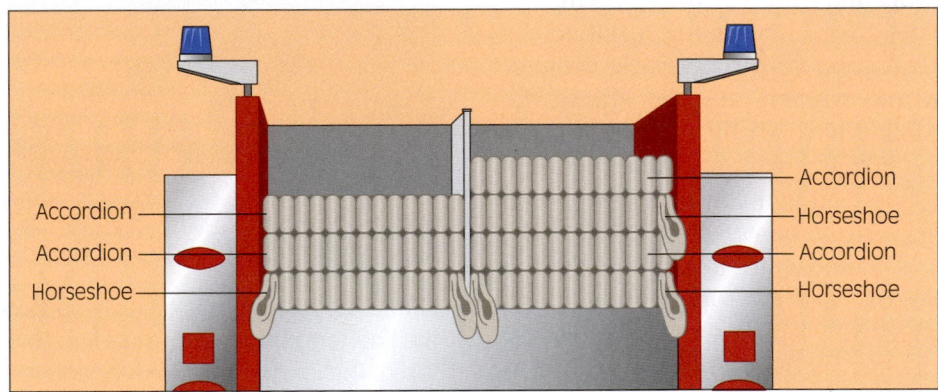

Accordion
Accordion
Accordion
Horseshoe

Accordion
Horseshoe
Accordion
Horseshoe

View from End of Hose Bed

(A)

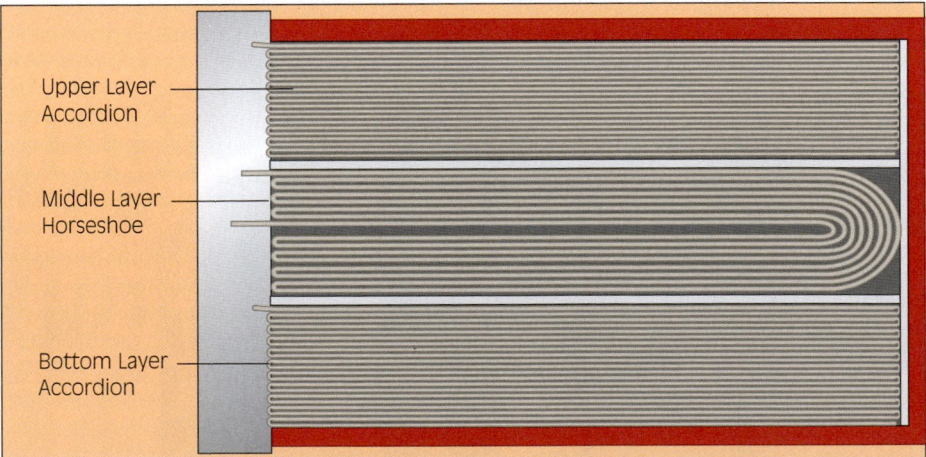

Upper Layer Accordion

Middle Layer Horseshoe

Bottom Layer Accordion

View from Top of Hose Bed

(B)

FIGURE 10-17 Preconnected combination loads include horseshoe, accordion, accordion layers or alternating horseshoe and accordion layers. (A) Horseshoe, accordion, accordion layers. (B) Alternating horseshoe and accordion layers.

FIGURE 10-18   Ears on hose load.

The flat load, as a preconnect, is also based on the flat load described earlier and starts at the discharge.

JPR 10-20: Advancing the Flat Load from a Preconnect Bed

(For step-by-step photos of this skill sequence, see page 277)

1. Start the flat load at the discharge with the hose laid out. At a point from one-third to one-half the length of the line, an ear or row of ears should be added to assist in pulling the line.

2. To advance the line, grab the nozzle with one hand and place it over the shoulder with the other hand reaching around and pulling the ear.

3. Walk away, pulling the line behind.

Some departments have placed ears in the load to mark the lengths of hose—such as an ear every 100 feet. The ears mark the hose according to their size. The smaller-sized ears mark the hose from the nozzle, while the larger ears mark the length closer to the bed, **Figure 10-19.** This arrangement allows additional flexibility: if 100 feet of hose is deemed necessary, a firefighter can grab the nozzle and the smallest ear. The hose is broken at the next connection pulled from the bed and connected to an available discharge. This allows efficient use of the hose and doesn't leave a "spaghetti mess" of unnecessary hose to straighten.

The minuteman load is a preconnected load using a narrower section of the hose bed. This narrower section is called a **slot load** and can allow more hoselines to be placed in the same space. Some longer length versions may use more than one slot, and all require attention when being loaded. These loads have become more popular because they are quick and efficient, **Figure 10-20.**

FIGURE 10-19   Smaller-sized ears mark the hose from the nozzle, while larger ears mark the length closer to the bed. *(Courtesy Loveland Fire and Rescue)*

FIGURE 10-20   The minuteman load or slot load is a preconnected load using a narrower section of the hose bed, which allows more hoselines to be placed in the same space. Some longer length versions may use more than one slot.

JPR 10-21: Minuteman or Slot Load

(For step-by-step photos of this skill sequence, see page 278)

1. The minuteman load has the female end connected to the discharge in the slot, and several sections of hose are flat loaded into the slot, with the last few feet (about a meter) placed aside out of the bed at the front of the slot. An ear may be placed one or two folds from the bottom to assist in pulling.
2. Next start a section of hose with the nozzle and load the hose flat above that. The nozzle will therefore be in the middle of the slot at the point where it will be pulled.
3. When the designated number of sections is finished, connect the set-aside piece to the pieces with the nozzle attached. The connected sections are on the top layer.

JPR 10-22: Advancing the Minuteman Load

(For step-by-step photos of this skill sequence, see page 280)

1. Lift up the nozzle and layers above it while pulling them out and placing them midway on the shoulder.
2. Step away to remove the remainder of the top layers.
3. Turn around and pull the ear to remove the remaining hose.
4. When the bottom sections are fully stretched out, allow the shoulder load to flake out toward the fire.

JPR 10-23: Triple-Layer Load

(For step-by-step photos of this skill sequence, see page 281)

1. Connect all the hose to the discharge and attach the nozzle at the other end with the hose stretched out to its full length.
2. At a point two-thirds of the way from the hose bed, make a fold and double the hose on itself back toward the engine. Make another fold, the last part with the nozzle, to create the third layer. These three layers are treated as one for loading on the engine.
3. Load the hose in the slot in a layer and make a bend, creating another layer, until all hose is loaded. The nozzle's layer should be at the edge of the bed for pulling.

JPR 10-24: Advancing the Triple-Layer Load

(For step-by-step photos of this skill sequence, see page 282)

1. Grab the layer with the nozzle and place it on the shoulder.
2. Pull the layers out of the slot, or another firefighter can grab the next layer.
3. Stretch the hose to the fire.

Many departments use the loads presented here. Sometimes these loads are combined with each other, and sometimes they are combined with new loads. Any hose load used should serve the needs of the department and should be quick and efficient to load and deploy. Firefighters should consult their department SOPs and be well versed in the hose loads most commonly used by their agency.

Stored Hose Loads/Packs

Apparatuses typically carry stored hose rolls and special application hose packs. Hose rolls are just that—extra sections of rolled hose for replacing damaged hose or short sections to help assemble an evolution. These hose rolls can be stored as a straight roll, donut roll, or double donut. Hose packs can be numerous in design and makeup. High-rise (or standpipe) packs are hose loads that are preassembled and bundled to be easily carried into a building, **Figure 10-21.** High-rise packs may also include a tool and appliance bag containing adapters, spanners, and hose straps.

Wildland Firefighting Hose Loads

Wildland firefighting often requires firefighters to stretch hoselines a great distance from the engine while fighting the fire. To accomplish this task, the hose is rolled and bundled together to allow it to be carried on the firefighters' backs while carrying tools or stretching and advancing the line. Placing two bundles together allows each firefighter to carry 200 feet (60 m) of 1-inch (25-mm) or 1½-inch (38-mm) hose.

JPR 10-25 Modified Gasner Bar Pack

(For step-by-step photos of this skill sequence, see page 283)

1. Start at the male end of the hose (100 feet [30m]) and form a roll 30 inches (0.8m) in diameter on flat ground.

FIGURE 10-21 High-rise (standpipe) hose packs are pre-assembled in a bundle for deployment.

2. Continue to roll the hose in the direction of the rows and make the rolls as tight as possible.

3. The completed field roll, with the female coupling on the outside, is ready for packing. Grab the outside of the hose roll near the female coupling and pull toward the center of the roll.

4. The fold should now protect the female end. The top layer of hose can now be tied to the bottom layer using a piece of rope 32 inches long (0.9 m). Two rolls of hose may be placed together and tied off with rope on the outside of the rolls with a shoulder strap to allow carrying.

A modification would be to add a gated wye or tee at the end of each roll of hose. To do this merely requires that when the roll is closed together, the female end is left to one side, on the outside, and the gated wye or tee is attached after tying the roll off. When placing the two rolls together, they should be placed so that the gated wyes are on opposite sides of the pack, **Figure 10-22.**

NOTE

The gasner load is also a useful load for attack lines in structural firefighting.

ADVANCING HOSELINES— CHARGED/UNCHARGED

The engine company's purpose is to advance hoselines to the seat of the fire and to apply water for extinguishment. These tasks should be accomplished in the most efficient manner, whether around, onto, or

Top Layer

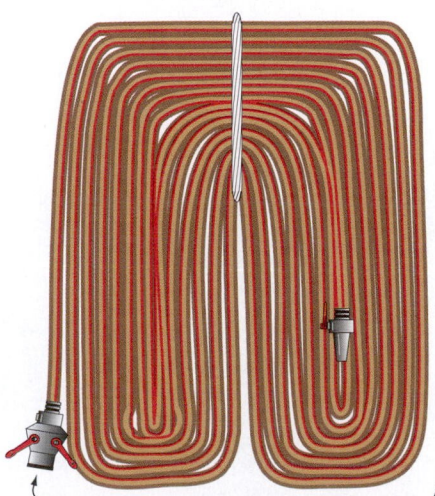

Add Gated Wye (Closed) and
Remove Hose Clamp

Bottom Layer

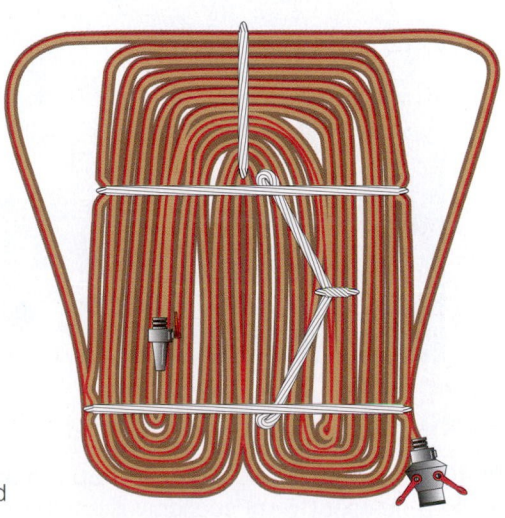

FIGURE 10-22 Modified gasner bar pack with added gated wyes.

into buildings. The following evolutions are based on a four-person crew with one person being at the layout position and connecting to the water source. As crew situations change, the crew positions may also change. Remember that two in/two out firefighter safety rules apply, and conducting any fireground operation without sufficient personnel for a firefighter rescue team can be dangerous and unlawful. See Chapters 2 and 5 (in Sections I and II) for more information.

When advancing hoselines in any of the evolutions described, the nozzleperson will advance the first shoulder load with the nozzle. The officer takes the second position on the line if it is a two- or three-person line. The engine driver will take the third position on a three-person line. If additional personnel are available, they will be substituted for the officer and driver. The driver is responsible for clearing the hose bed prior to charging any of the lines.

The officer determines the length of the hoseline by figuring the distance from the engine to the location of the fire, and ascertains the conditions to make a determination of whether a dry line or wet line will be advanced. Upon reaching the fire area, there should be about 50 feet (15 m) of hose left for movement around the fire area. When hoisting a hoseline up the outside of a building, a 50-foot (15-m) section will reach four floors. Advancing the same section inside the building via a stairwell will reach only one individual floor, **Figure 10-23.**

When operations require moving a charged hoseline to another locale—such as another floor—it is best in terms of time and efficiency to shut down, drain, and advance the line as a dry line if fire conditions allow. When the new location is reached, the line is then recharged and the fire attack operations will continue.

Into Structures

Advancing a hoseline into a structure requires careful placement of the pumper and hoseline, proper selection of the correct size and length hoseline, and skillful execution by the hose crew. The engine should be positioned in a safe location that allows access for hoselines without hindering any other apparatus placement, especially a ladder company, and allows for any changes in fire conditions and equipment use, **Figure 10-24.** The hoseline must be advanced to attack the fire while also being at a position to prevent any fire spread. The hoseline should be large enough to suppress the fire, a minimum of 1½ inches (38 mm). Finally, it is the skillful crew that carefully and safely advances the line, ensuring that they are ready to combat the fire.

To advance the line into a structure, the crew selects a hoseline and properly removes it from the engine, deploying it toward the entrance. A careful evaluation of the fire behavior and building conditions is required prior to entering any structure or room. Structural firefighting involves the possibility of flashover, backdrafts, and building collapse. Entering a burning building without having a safe plan of attack and emergency escape can be fatal.

Firefighters should ensure that there is adequate hose available at the entry point. If the fire is not near this door, the line may be extended into the building on a shoulder load. However, when reaching the entrance to the area of the fire, the excess hose is removed from the shoulder and carefully laid in a

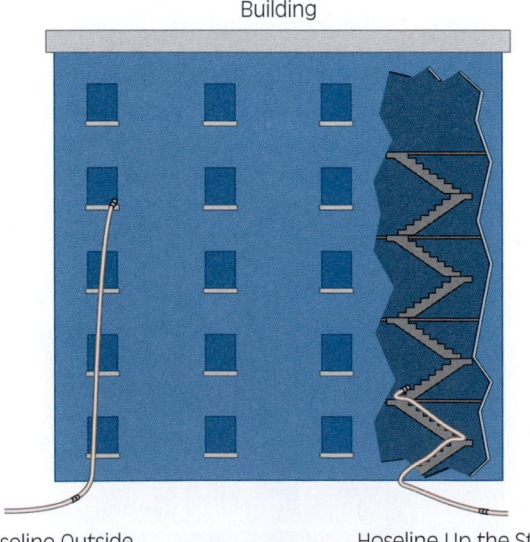

Building

Hoseline Outside   Hoseline Up the Stairs

FIGURE 10-23   When taking a hoseline up the outside of a building a 50-foot (15-m) section will reach four floors, while taking the same section inside the building via a stairwell will reach only one individual floor.

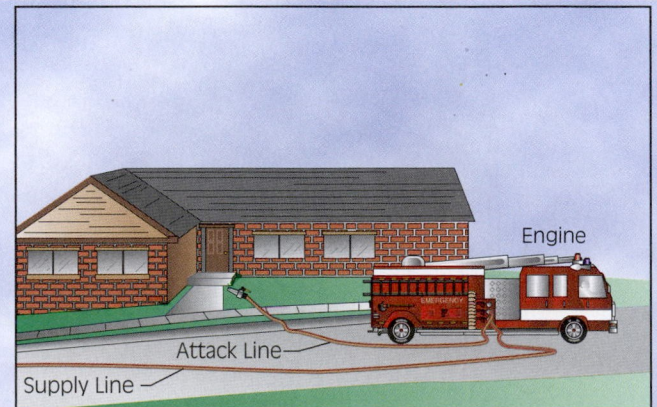

Engine

Attack Line

Supply Line

FIGURE 10-24   Park the engine to allow good access for hoselines without hindering any other apparatus placement.

manner to allow it to be easily pulled into the building or room. This would include straightening the hose and, if necessary, looping it back and forth so that it will pull without kinking. The nozzleperson and other firefighters should all be on the same side of the hoseline and properly spaced on the line for easier movement of the hose. The nozzleperson is at the nozzle, the second firefighter a few feet (a meter or so) back, and the next firefighter, if extra firefighters are available, between the midpoint of the first section and the coupling, **Figure 10-25.**

Prior to entering the structure, the door is quickly checked for heat using the back of the hand with the glove peeled back and away to expose the skin. The firefighter will then ensure the glove is replaced to cover the exposed skin. With the line charged, the firefighter briefly opens the nozzle to bleed off air from the hose and selects the proper pattern. See nozzle operation in Chapter 11 (in Sections I and II). Crouching in a low or kneeling position while positioned to the side of the door, the firefighter opens the door and moves the line forward. When advancing the line past doorways, around corners, and furniture, remember that the extra time spent ensuring that the line can move smoothly past obstacles will be time saved. Allowing the line to run short wastes time if the crew has to send someone back to free it.

Up and Down Stairs

When the fire is on a level other than the ground floor, firefighters will need to advance the line up or down stairs. If the fire does not involve the stairs, the best method is to advance an uncharged line to the fire floor. Make sure that the proper length and size of hoseline is chosen, considering the length and size of the structure. The hoseline is advanced from the engine as described and the crew carries the shoulder loads into the building and up the stairs, allowing it to play out as needed, **Figure 10-26.** Use caution so that the hose does not slip back down the stairs, and do not allow it to get caught on stair handrails where it can become wedged. To help avoid this and the resulting kinking of the hose, lay the hose against the outside edge or wall of the stairs. If it is necessary to run the hoseline up between the handrails, it should be done in a vertical manner and tied off with a rope hose tool or strap at the upper level. The hose tool is passed around the hose and pulled snugly, near a coupling if possible.

The rope or strap is wrapped around the railing several times and secured back on itself. When the desired floor is reached, the shoulder load is carefully laid out and deployed to avoid kinks (pile of spaghetti) once it is charged. One ideal solution for the extra hose, if possible, is to run the line up the stairs toward the next level and back down to the fire floor.

FIGURE 10-25   The nozzleperson is at the nozzle, the officer or second firefighter a few feet (a meter or so) back, and the next firefighter, if there is one, between the midpoint of the first section and the coupling.

FIGURE 10-26   A crew advancing an uncharged hoseline up stairs.

This creates a large loop that will be easy to pull as gravity will help it move down, **Figure 10-27.** With the hoseline charged and bled it is ready to advance.

If the stairwell or the landing on the desired floor is involved with fire, the crew will need to advance a charged line. The line is brought to the entrance or landing as described earlier, charged, and advanced at the point where the attack is to be made. The nozzle is operated to darken down the fire at the next landing, then shut down and quickly advanced to that level, where it is opened again to secure the landing and moved forward to maintain the floor.

JPR 10-26: Advancing a Charged Hoseline Up a Stairwell

(For step-by-step photos of this skill sequence, see page 284)

1. Apply water from bottom of stairs to darken down fire.
2. Shut down nozzle and advance quickly up stairs.
3. At top of stairs, open nozzle and continue fire attack.

Care is taken to ensure that sufficient hose and personnel are available to move to the next floor without stopping on the stairs.

FIGURE 10-27 Extra hose on stairs creates a large loop that will be easy to pull.

While advancing a hoseline on a stairwell is always a challenge, it is more so when going down the stairs. Firefighters going down the stairs will encounter the chimney effect of rising heat, hot gases, and possibly fire. "Making the floor" when the fire is on a lower level is a true test of a fire attack crew.

Using a Standpipe System

Advancing hoselines using a standpipe system involves two different hoseline evolutions. The first is the engine driver connecting to the fire department connection on the structure, and the second is the hose crew connecting to the standpipe outlet and advancing the hoseline to attack the fire.

The pumper first establishes a water supply from the nearest hydrant whenever possible. Pre-emergency planning should have determined the water supply for each standpipe system, including any necessary hose evolutions if the fire department connection is not near the building entrance. The engine driver advances a supply line of medium or large diameter hose to the fire department connection. Local standard operating procedures (SOPs) should outline what size hose should be connected. Depending on the length of the lay and the style of hose load, the driver can drag the hose or make up a shoulder load.

The hose load and type of coupling will determine the need for any adapters as the male end is needed at the siamese connection on the building. Reaching the siamese, the driver removes all caps, usually two, and checks the operation of the clapper valve, **Figure 10-28.** If the clapper is inoperative but open, then advance a second line to the connection prior to charging the system. The driver connects the hose first to the left outlet, as the rotation of the coupling will be easier to do when a second line is connected, **Figure 10-29.** Returning to the engine, straightening any kinks, the driver disconnects the hose from the bed and connects to the pump's discharge outlet. The line(s) are charged to the proper pressure.

Occasionally, the fire department connection might be damaged or inoperable. Several steps can be taken to use this system. Pick the one that gets the system into service the quickest. If the siamese swivel does not turn, place a double male into the siamese and add a double female to connect the hose, **Figure 10-30.** If the siamese cannot be used at all, the supply line must be extended into the building to a first floor outlet. To connect the hose, a double female will be required, **Figure 10-31.** After connecting the hose but before opening the outlet, the driver must return to the engine and break the supply line and connect to the discharge outlet and charge the line to the proper pressure. The driver then returns to the outlet

FIGURE 10-28 Driver checking standpipe connection and clapper valve.

FIGURE 10-30 If siamese swivel is inoperable, place a double male into the siamese and add a double female to connect the hose.

FIGURE 10-29 Driver attaching second supply line to FD connection. Note the first line is on left inlet.

FIGURE 10-31 If the siamese cannot be used at all, the supply line must be extended into the building to a first floor outlet. To connect the hose a double female will be required.

and opens the valve, **Figure 10-32.** Opening the outlet valve prior to breaking, connecting, and charging the line will result in the standpipe draining into the hoseline and charging the hose bed. A second line should then be run to the next most convenient outlet following the same steps.

The advancing of the interior hoseline to attack the fire should use a standpipe pack that includes the attack hose and nozzle, a gated wye, a short section of supply hose, and necessary wrenches and adapters. The crew proceeds first to the floor below the fire or, depending on fire conditions and location of the standpipe outlets in relation to the fire, an outlet on another lower floor can be chosen. The crew disassembles the hose pack and connects to the outlet, **Figure 10-33** and **Figure 10-34.** The crew stretches the hose either in the stairwell or hallway depending on fire location and conditions. The crew then advances the line.

Working Hose Off Ladders

Advancing a hoseline up a ladder can be done with the line charged or uncharged. The best and safest manner is to advance an uncharged hoseline up the ladder and into the building or onto a fire escape. The other method

FIGURE 10-33 Engine crew with standpipe equipment.

FIGURE 10-32 After connecting the hose at the outlet and the discharge outlet of the pump, the line is charged to the proper pressure. The driver then returns to the outlet and opens the valve.

FIGURE 10-34 Engine crew deployed with hose connected to standpipe.

advances a charged hoseline up a ladder for operation inside the building or from the ladder. This is the least preferred option and should be done using great caution.

SAFETY

The best and safest manner is to advance an uncharged hoseline up the ladder and into the building or onto a fire escape before it is charged.

When advancing an uncharged hoseline over a ladder, start with the nozzleperson and/or firefighters removing and advancing the shoulder loads, as described earlier, to the base of the ladder.

JPR 10-27: Advancing a Charged Hoseline Over a Ladder

(For step-by-step photos of this skill sequence, see page 285)

1. The shoulder load is laid on the left side of the ladder.
2. The nozzleperson takes the nozzle and pulls it under the left armpit, across the chest, and over the right shoulder, allowing the nozzle to rest in the small of the back.
3. The nozzleperson, with both hands free for climbing, begins climbing the ladder to a point about 20 feet (6 m) up and stops. The next firefighter at the bottom of the ladder places the hoseline over the left shoulder at the next coupling and begins to climb the ladder. The two firefighters coordinate and maintain their distance on the ladder and hose.

Additional lengths and firefighters are added as needed and space allows.

JPR 10-28: Advancing an Uncharged Hoseline Over a Ladder

(For step-by-step photos of this skill sequence, see page 286)

1. Upon reaching the entry point of the building—a window, balcony, or the roof—the firefighters stop in their respective positions on the ladder. The nozzleperson removes the nozzle and places it over the tip of the ladder into the building.
2. The nozzleperson climbs into the building, using the top rung of the ladder as a hose roller, and pulls the hoseline up as the next firefighter climbs.
3. The nozzleperson must complete pulling up the section as the next firefighter arrives at the tip. The second firefighter dismounts into the building and pulls the remainder of the hose into the building while the nozzleperson advances the line. To remove the

line from the building after use, the line is drained and the process of advancing the hose is reversed.

Another method to advance the line over a ladder involves using rope hose tools or straps wrapped around the line just below the nozzle and at the next coupling, and the hose tool looped over the shoulder of the firefighter carrying the line up the ladder, **Figure 10-35**. Additional firefighters would drape the hose over the left shoulder and assist the advance of the line as described previously.

Advancing a charged hoseline over a ladder requires multiple firefighters as the line is heavier and climbing the ladder with a charged line is dangerous. It is usually done when the line is to be operated from the ladder. The hoseline is actually passed by the firefighters rather than advanced by one or more firefighters. The hoseline is brought to the base of the ladder and it is charged and then air in the line is bled. Firefighters climb the ladder spacing themselves at the opening and then at the feet of

FIGURE 10-35 Another method to advance the line over a ladder involves using rope hose tools or straps wrapped around the line just below the nozzle and at the next coupling. The hose tool or strap is looped over the shoulder of the firefighter.

the next firefighter. When they are in place, the hoseline is passed up the ladder from one firefighter to the next until it reaches the opening, **Figure 10-36.**

> ### CAUTION
>
> Advancing a charged hoseline over a ladder requires multiple firefighters to advance the hoseline, because the line is heavier and climbing a ladder with the line is dangerous.

If the line is to be used from the ladder, the nozzle-person puts the nozzle through the top two rungs of the ladder allowing 1 foot (0.3 m) to be extended beyond the rung for movement and the line is secured with a rope hose tool. The rope hose tool is wrapped around the hoseline snugly. Two or three wraps are taken around the second rung and the hose tool secured on itself. The hoseline is also secured by rope hose tools every 20 feet (6 m) and at the base of the ladder, **Figure 10-37.** Operating a hoseline from a ground ladder requires the ladder to be securely tied in and heeled. (See Section I, Chapter 14.)

FIGURE 10-36 Firefighters passing a charged hoseline up a ladder from one firefighter to the next until it reaches the opening.

FIGURE 10-37 A charged hoseline on a ladder. The hoseline is also secured by rope hose tools every 20 feet (6 m) and at the base of the ladder.

> ### SAFETY
>
> Operating a hoseline from a ground ladder requires the ladder to be securely tied in and heeled.

An uncharged hoseline can also be advanced to the tip of a ladder, secured in the same manner as the charged line, and then operated. The same safety precautions apply plus the line should be charged in a very careful and slow manner.

ESTABLISHING A WATER SUPPLY CONNECTION

Several different methods exist for establishing a water supply depending on the type of water source (static or hydrant), style of hydrant, the hose lays used (discussed later), and whether a pumper will be used at the water source. Static water sources obviously require a pumper at the source, but a hydrant source may not, depending on the volume and pressure of the hydrant and its distance from the fire. A supported fire hydrant is one that has a pumper at the fire and another pumper at the hydrant.

Firefighters need to be able to connect directly to a fire hydrant or assist the engine driver in making the connections to hydrant and static sources.

From Hydrants

Using an unsupported hydrant requires a hoseline to be connected to a fire hydrant without an engine at the hydrant. This can be done using a single hoseline, multiple lines, or with adapters or valves to allow additional lines or a pumper to be connected later. Most departments use some type of finished hose load for a layout load. This load has a roll or several folds of hose to make any connections, a hydrant wrench and necessary adapters, and a rope to tie it all together and to allow the hydrant to be wrapped. "Wrapping the hydrant" allows the hose to be pulled far enough to reach the hydrant and helps prevent the hose from being dragged by the engine as it pulls away. The engine stops at the hydrant or layout point and the officer gives instruction for the hose lay. The layout person dismounts the vehicle and removes the layout load from the hose bed. Any additional adapters should also be removed from the engine. The layout person pulls sufficient hose and advances it to the hydrant and wraps either the hose or rope around the hydrant to secure it. The firefighter should be positioned to prevent being pinned between the hose and the hydrant. At this point the engine can move forward completing the hose lay, **Figure 10-38.**

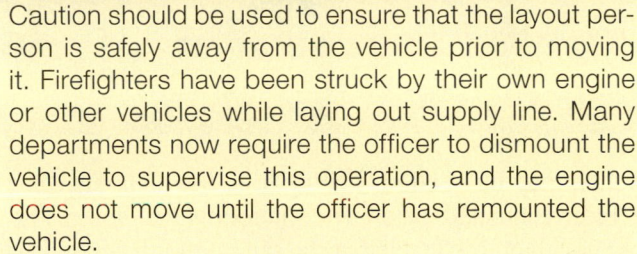

After unwrapping the hydrant and removing the rope and tools, the hose is stretched out to remove any kinks. The layout person picks the proper hydrant outlet and removes the cap. This outlet depends on hydrant style and size and SOPs. Some departments utilize a gate valve or a switch valve for additional lines or a pumper to be connected without shutting down the hydrant, **Figure 10-39.** To remove the cap, the hydrant

FIGURE 10-38 The layout person pulls the layout section and enough hose to reach and wrap the hydrant.

FIGURE 10-39 A supply line with a switch valve connected to hydrant.

FIGURE 10-40 The layout person making the hose connection to a hydrant.

wrench may be needed. The hydrant itself should then be briefly flushed prior to connecting the hose. The hose or adapter is leveled at the outlet and the coupling is screwed on, carefully ensuring that the threads are not crossed. The coupling should be hand tightened; although, to stop leakage of water from the connection, wrench tightening may be needed after turning the hydrant on, **Figure 10-40.** The hydrant valve is opened when water is called for and leaks and kinks are removed as the layout person returns to the engine.

An engine can be connected directly to a fire hydrant or to a switch valve already in service to supply attack lines or to support another engine. A soft sleeve hose is typically connected to the large hydrant outlet or the switch valve. The hard sleeve may also be used as well as other supply hose. Remember that the larger the hose, the more water that can move through it. The engine should be set up with gate valves at the inlet to speed up the process; in fact, some fire departments carry the soft sleeve preconnected. Care should be taken to make sure that the hose is connected to an inlet rather than a discharge valve of the pumper.

JPR 10-29: Soft-Sleeve Hydrant Connection

(For step-by-step photos of this skill sequence, see page 287)

1. Connecting a soft sleeve to a hydrant starts with proper parking of the engine. The charged line should be without kinks and undue stress on the hose or couplings. If the hose rests on the ground, there should be a chafing block between it and the ground. The hose is removed from its compartment or the hose bed, and it is stretched from the engine to the hydrant.
2. The outlet cap is removed, the hydrant flushed, and the coupling connected.

3. If the line is not preconnected to the engine, the hose must be attached to the engine's inlet. If a gate valve is on the inlet, the hose is connected directly to it. If not preconnected, the inlet cap must be removed. This will cause water in the pump to flow out, so the engine driver/operator should ensure the tank-to-pump valve is closed. The sleeve is then connected to the inlet. The hydrant valve can now be opened slowly to avoid a water hammer. The gate valve on the engine or switch valve on the hydrant must also be opened, if used.

STREETSMART TIP

Care should be taken while removing kinks in large-diameter hose (LDH). For safety reasons, the firefighter should always try to "kick" the kink out of LDH. If it is necessary to lift the LDH to remove a kink, it should be done on the charged and pressurized side of the hose (toward the hydrant). It is important to use a strong lifting position (using the legs and a tight "core" posture). Firefighters must be cautious to avoid pinch-type injuries to the hands and fingers. If possible, a hose strap should be wrapped around the hose so that it can be dragged to remove the kink.

From Static Water Supplies

The supply of water from a static source requires use of an engine and its hard sleeves to draft the water. The engine must be positioned close enough to place the hard sleeve and strainer into the water or to reach the dry hydrant connection. Connecting a hard sleeve to a dry hydrant is done using the same procedure as connecting to a regular hydrant without the normal concerns surrounding creation of a vacuum. A vacuum has to be created, and it is important to make a tight connection to prevent leakage and loss of the vacuum.

JPR 10-30: Setting Up for Drafting Operation

(For step-by-step photos of this skill sequence, see page 288)

1. The placement of a hard sleeve and strainer directly into a water source requires engine placement near the water and slightly at an angle.
2. Connect the hard sleeve to the engine inlet.
3. Attach a strainer and guide rope to the other end of the sleeve.
4. The hard sleeve and strainer are attached to the pumper's inlet and the engine moves forward as the crew positions the sleeve into the water.
5. If a nonfloating strainer is used, the rope also keeps the strainer from resting on the bottom. Nonfloating strainers should be well under the water's surface to prevent drawing in of air. When the strainer and hose are in place, the engine is stopped, the line is secured, and drafting operations begin.

EXTENDING HOSELINES

Despite the best efforts of judging the needed length of a hoseline, there will be occasions when the line comes up short and will need to be extended. Wildland firefighters encounter this often, not because of misjudgment, but because they have successfully driven the fire back beyond the length of their hoselines. Whatever the case, all firefighters should be familiar with techniques used to extend their hoselines. While the pump operator can shut down the line when the additional hose is needed, it is easier to extend the line using either a break-apart nozzle or a hose clamp.

JPR 10-31: Extending a Hoseline with a Break-Apart Nozzle

(For step-by-step photos of this skill sequence, see page 289)

1. The preferred method of extending a hoseline is to use a break-apart nozzle. The additional hose is brought to the nozzle end of the hoseline, and the line is shut down with the control handle.
2. The nozzle tip is removed and the additional hose added if it has 1½-inch (38-mm) couplings.
3. If extending a line with hose couplings larger than 1½ inch (38 mm), then a 1½-inch (38-mm) to 2½-inch (65-mm) increaser will be required at the control handle.
4. The additional line with the nozzle tip attached is advanced and charged from the control valve.

JPR 10-32: Extending a Hoseline Using a Hose Clamp

(For step-by-step photos of this skill sequence, see page 290)

1. The clamp and additional hose are brought up to the nozzle end of the hose. Clamping a hose can be dangerous, so firefighters need to ensure that the clamp is safely operated and is fully locked before releasing the clamp. The firefighter is positioned to the side of the clamp to prevent injury if it slips. The hoseline is clamped just before the nozzle.
2. Remove the nozzle. The additional line is attached as is the nozzle and the line is then advanced. When water is called for, a firefighter removes the hose clamp.

SAFETY

If an operation requires the extension of a hoseline and a clamp is used, the firefighter operating the clamp should stay with it to ensure it is locked. This firefighter should have the responsibility of holding the clamped position in case of equipment malfunction or other problems. The clamp firefighter should not leave the clamp until the line is extended and the nozzleperson is ready for water in the new section of hose.

Wildland firefighters often extend their hoselines as they advance on a fire. Some of their hose loads plan for this with the addition of a gated wye or small hose clamp at each layer of hose.

JPR 10-33: Wildland Hose Advancing and Extension

(For step-by-step photos of this skill sequence, see page 291)

1. A preconnected line is pulled to begin the hose lay. As the line is moved forward, it is operated for suppression and crew protection.
2. When the line is almost fully extended, a second line is rolled back in preparation for the next length.
3. With the line clamped, the nozzle is removed and placed on the new length and the hoses are connected.
4. The clamp is released and the line advanced. When the line is again fully extended, the team repeats the clamping and extending process. As wildland hose is typically of smaller diameter than structural firefighting hose, it is easier to clamp.

REPLACING SECTIONS OF BURST HOSE

The bursting of a section of hose is a dangerous situation that stops the flow of water to the fire and can injure firefighters and bystanders while causing property or water damage. It is a situation that requires immediate attention, especially when firefighters are operating under fire conditions. The hoseline must immediately be shut down.

The other firefighter should then remove the burst section and replace the section or if enough hose is available, reconnect the two adjacent sections. If enough hose is not available, a new section must quickly be brought and connected. The line may then be recharged.

NOTE

It is important for firefighters to remember the dynamics of hoselines when replacing a burst section. Fire hose, due to the pressure and water inside, will extend its length when wet. If replacing a section, the dry section will most likely be found to be shorter than the section to be replaced. For this reason, a good rule to follow is to replace a burst section of hose with two sections of hose.

Typically, the pump operator will shut down the line. If this is not possible, a hose clamp can be used to stop the water flow. If no clamp is available, a firefighter can fold the hose twice over itself and kneel down to hold pressure buildup in the kinks. This will stop the flow sufficiently to have another firefighter replace the burst section.

HOSE LAY PROCEDURES

Hose lay procedures bring the water to the fire location by placing supply hose between the water source and the attack engine. SOPs should cover preferred hose lays and water supply operations, and new apparatus is then designed for these preferences. The direction of the engine's travel to or from the water source gives each procedure its name. Supply lines and the hose beds on apparatus are designed to use at least one of the following three hose lay techniques; many are designed to allow all three. Any needed adapters are either attached to the supply line or readily accessible to the layout person. Each of these may work with either a hydrant or static source of water with the static source needing a supply engine at the source.

Forward Lay

Forward lay refers to the engine stopping first at the water source to drop off a supply line(s) and then advancing to the location of the fire, **Figure 10-41.** The forward lay is a preferred lay because a water supply can be established by connecting and charging the hydrant, and it may need no additional engines for the water supply. The attack engine is closer to the fire with access to additional hoselines and equipment. The forward lay works best when a water source is located in the approach path of the engine. An additional engine may still be required at the water source if needed to support the pressure or volume and will be required if the source is a static water source. The forward lay can also be used with a portable water tank operation where the fire is off the main road and the tank dump site is on the main road.

Reverse Lay

A reverse lay is the opposite of the forward lay with the supply line being dropped off at the fire location and the engine laying the hose toward the water source, **Figure 10-42.** Typically, the first arriving engine is used as the attack engine with the second engine doing the reverse lay and being the support or supply engine. Reverse lays can be used where a manifold, attack lines, and equipment can be dropped off at the

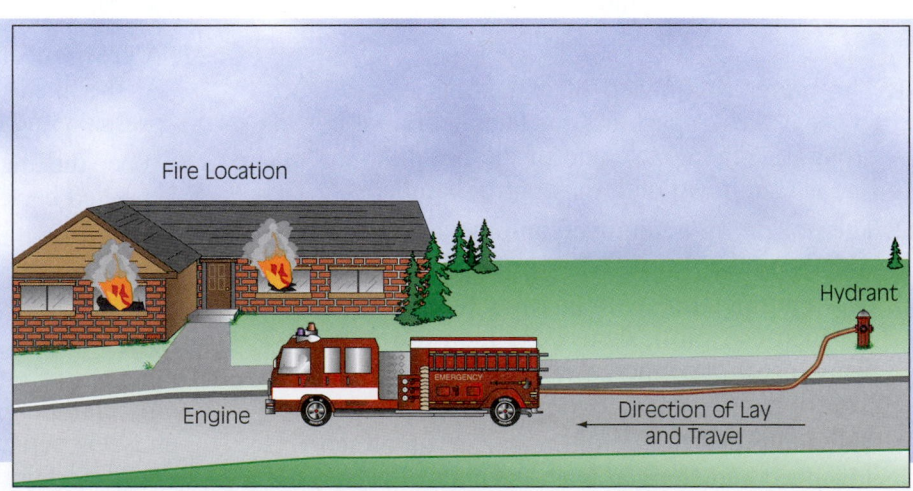

FIGURE 10-41 The forward or straight hose lay.

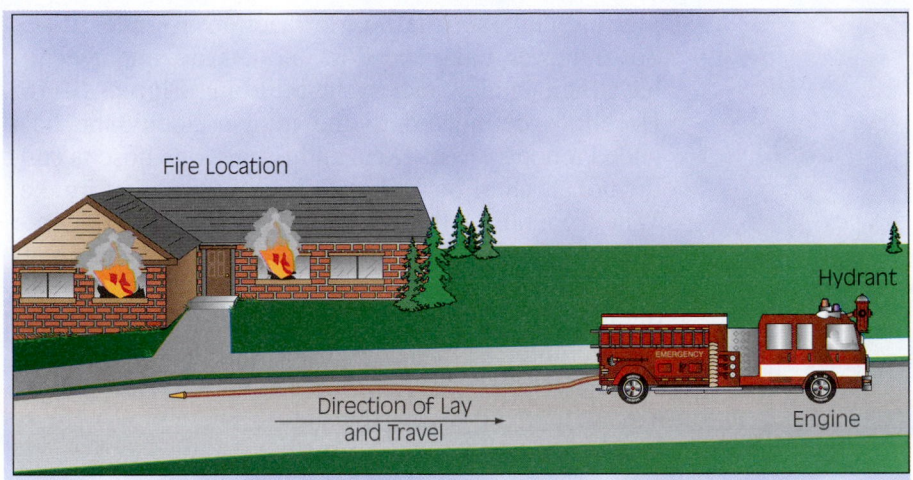

FIGURE 10-42   The reverse hose lay.

FIGURE 10-43   The split hose lay.

fire with the engine going to the water source. This is less preferred but may be necessary in areas with few responding units and poor or static water sources.

Split Lay

The split lay is used where the fire and the water source are in two different directions, such as on two different streets, thus needing to split the lay between two engines. The split lay has the first engine laying its line from a point or intersection to the fire location and the second engine laying its line from the point to the water source, **Figure 10-43.**

DEPLOYING MASTER STREAM DEVICES

Master streams or **heavy appliances** are non-hand-held water applicators capable of flowing over 350 gallons of water per minute (1325 L/min). Four basic types of master stream devices are used, although the names are sometimes interchanged.

The wagon pipe is a permanently mounted master stream device on an engine that either has a prepiped water connection or needs a short section of hose to connect it to the pump, **Figure 10-44A.** Some departments refer to wagon pipes as a "deck gun." If it is without permanent piping, then either short or regular sections of supply hose can be run from the discharge to the inlets of the device. It is about 9 feet (2.8 m) in the air and can reach about four stories with a solid stream nozzle. A wagon pipe can also be directed at lower angles than a ground-mounted deluge set, **Figure 10-44B.**

A similar looking device called a deluge set is not permanently mounted so it can also be removed and operated from the ground. Different models of deluge sets have one base that can be used both mounted on the apparatus or on the ground while others have separate bases for each mounting. Deluge sets, when operated from the top of an engine, may have prepiped water connections to a pump and need no supply lines attached. Some newer ones come with an extension device or are able to be adjusted in several

(A)

(B)

FIGURE 10-44   (A) A wagon pipe. (B) A deluge set and a wagon pipe operating.

positions that extend their reach by several feet. Supply lines can be run up from the discharge to the inlet if no pre-pipe water connection is available. When securely attached to the engine, they can also be operated at low angles or can reach four stories. Portable deluge sets operated from the ground may not be operated at an angle lower than 25 degrees. Most deluge sets have a pin or other device that prevents them from going below this angle, and this pin should only be removed when the set is bolted down to the apparatus. Going below this angle can cause the set to become unstable.

When operating a portable deluge set on the ground, the intakes normally should be facing the fire build-

ing to counteract the nozzle reaction, **Figure 10-45.** Some newer balanced-flow models do not need to have their intakes facing the building, **Figure 10-46.** Hoselines connected to the intakes should be balanced among the intakes, and if only one hose is run, the nozzle shall be operated directly over that intake. When operating the set at the minimum 25-degree angle, the nozzle should not pass beyond the two outboard intakes. The advancing of medium- or large-diameter hoselines to the deluge set is also similar to the standpipe siamese described earlier.

> ### SAFETY
>
> Portable deluge sets operated from the ground may not be operated at an angle lower than 25 degrees. Most deluge sets have a pin that prevents them from going below this angle, and this pin should only be removed when the set is bolted down to the apparatus. Going below this angle can cause the set to become very unstable.

A monitor pipe is a permanently attached master stream device with a prepiped waterway on an aerial device such as an aerial ladder or platform. The monitor pipe has either a direct discharge valve if the unit also has a pump or needs a supply hoseline(s) if not equipped with a pump. If a hoseline from an engine is needed, the monitor pipe has a siamese or manifold that can be supplied similarly to a standpipe system, as described earlier.

A ladder pipe is a non-permanently mounted device that needs a hoseline for its waterway on an aerial ladder or platform. Some ladder trucks have both a monitor pipe and a ladder pipe.

The ladder pipe is not permanently mounted and therefore needs a hoseline for rigging it up the ladder and another for supply. The ladder pipe is often assembled, completely or partially, with its siamese, hose for the ladder, guide ropes, and the pipe itself. The ladder pipe is removed from its bracket and mounted on the aerial ladder in the center. The nozzle and **stream straightener** are adjusted to the operating position, and guide ropes are attached, **Figure 10-47A.** If the hose is also attached, it is deployed down the aerial to the turntable and the ground. If the hose is not attached, it is often run up the ladder from the turntable and attached to the pipe. In either case, the hose is run in the center of the ladder and secured at the top with a rope hose tool. The bottom of the hose and the siamese are placed on the ground and the aerial is elevated to its operating position. Another rope hose tool is attached at the base of the aerial, **Figure 10-47B.** A supply line(s) is run from an engine to the siamese and attached. When water is

Proper Operating Procedure for Portable Deluge Set with Two Supply Lines

Proper Operating Procedure for Portable Deluge Set with One Supply Line

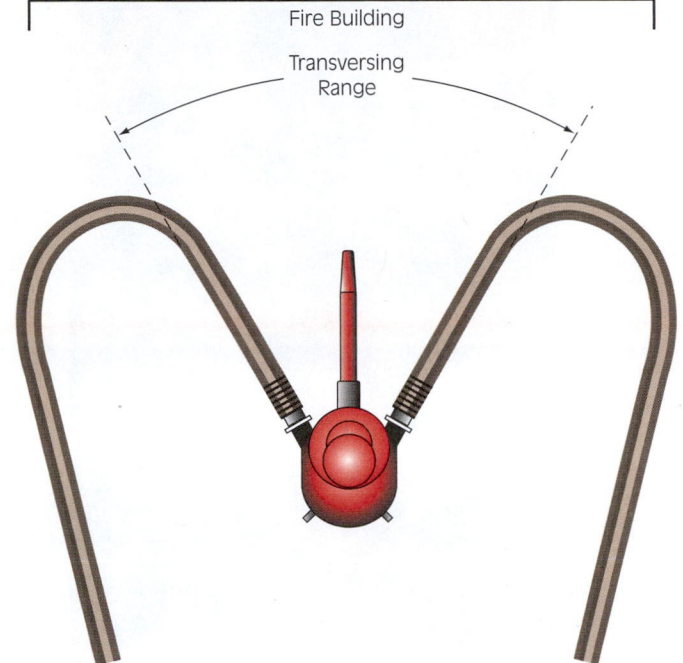

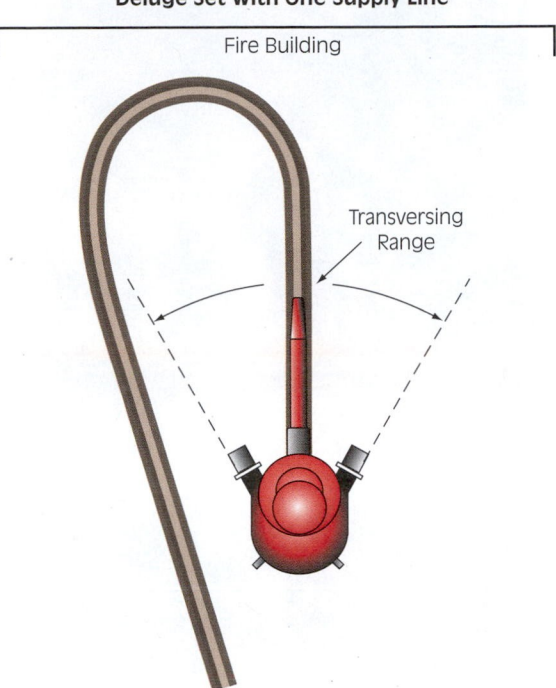

FIGURE 10-45 Proper operations of portable deluge sets.

FIGURE 10-46 A portable deluge set in operation.

called for, the lines are charged slowly and carefully to ensure everything is properly connected.

When operating master streams with solid stream tips, the following rules apply for figuring reach, vertically and horizontally. For every foot (0.3 m) of vertical reach needed, move the device 1 foot (0.3 m) away. For horizontal reach, each pound of pressure (6.89 kPa) equals 1 foot (0.3 m) of reach. An angle of 45 degrees is about the maximum angle that can produce an efficient stream. The reach is generally limited to three floors above the nozzle's height.

(A)

(B)

FIGURE 10-47   Ladder pipe operations. (A) Ladder pipe on ladder. (B) Ladder pipe hose with rope hose tools secured to the hose. Notice the hose is run in the center of the ladder.

FIRE HOSE INSPECTION

Fire hose needs to be tested prior to being placed in use and then retested annually during its lifetime. Hose also should be tested after being damaged and after repairs have been made. To ensure accuracy of the testing program and for personnel safety, a record-keeping system must be utilized. The record system should have identification numbers, dates of testing, repairs, and other important notations about each section.[1]

The actual testing and servicing of fire hose will be covered as a firefighter II but it is important for all personnel to know how to remove a defective hose from service and what procedures are to be used for reporting a defective hose once it is taken out of service. Any damaged or suspect sections should be removed from service until tested or repaired.

After the test is completed, the hose has the current date written on it, and it is cleaned, dried, and returned to service or rolled and put in storage.

Hose that is worn, damaged, or does not pass service testing should be removed from service. Consult department policies on removing hose from service. Many departments have firefighters mark the hose as "out of service" with a tag, marker, or other identifying method. Out-of-service hose should be kept separate from sections of hose that are "in service" and safe for operational use.

When conducting service testing of hose, firefighters should ensure that the results of the hose testing are recorded. Consult department policies and procedures for the appropriate method of recording the hose testing results.

SAFETY

Hose testing is a destructive process that identifies weak hose by causing it to burst. Firefighters should secure the nozzle end of the hose and operate hose valves from a distance to prevent personal injury if the hoseline breaks and begins to whip about.

JOB PERFORMANCE REQUIREMENT   10-1
One-Person Foot-Tilt Method

A With the male coupling lying flat on the ground, place the left foot directly behind the coupling, stabilizing it and raising it upward.

B With the female coupling in the right hand, bend down and bring the two couplings together. When aligned, turn the swivels to the right.

JOB PERFORMANCE REQUIREMENT   10-2
One-Person Over-the-Hip Method

A Hold the hose with the female coupling slightly below the waist.

B Pick up the male coupling, align the two couplings, and swivel the female coupling to make the connection.

JOB PERFORMANCE REQUIREMENT   10-3
Two-Person Over-the-Hip Method

The two firefighters face each other—each holding a coupling. The couplings are brought together, aligned, and swiveled. The firefighter with the female coupling watches and adjusts the alignment of the couplings.

JOB PERFORMANCE REQUIREMENT   10-4
Uncoupling Hose with Spanners

A   With the hose lying flat on the ground, place the spanners on the lugs going in opposite directions.

B   Push the spanners downward to break the connection.

JOB PERFORMANCE REQUIREMENT 10-5
One-Person Knee-Press Uncoupling Method

A Fold the hose coupling over on itself on the ground and press your knee down on the coupling.

B Try to twist the coupling to the left, breaking the connection.

JOB PERFORMANCE REQUIREMENT 10-6
Straight or Storage Hose Roll

A Start with the hose flat on the ground. From the male end, to protect the threads, roll it straight to the opposite end.

B Once the roll is finished, it is ready to be moved to storage.

JOB PERFORMANCE REQUIREMENT   10-7
Single-Donut Hose Roll

A   Lay the hose lying flat. Fold the hose on top of itself with the male coupling on top about 3 feet (1 m) short of the female coupling.

B   Starting at the fold, roll the hose toward the couplings with the extra hose on the female end protecting the male coupling. A second firefighter can assist in guiding the hose and adjusting the slack as it is rolled.

C   Leave a small space at the center of the roll to provide a handhold.

JOB PERFORMANCE REQUIREMENT   10-8
Single-Donut Hose Roll (Alternate Method)

A   Lay the hose out flat. Starting at a point off-center about 6 feet (2m) toward the male coupling, roll the hose toward the female coupling. The extra hose will protect the male coupling and the female coupling will be exposed.

B   The handhold space and the extra firefighter are also useful with this method.

JOB PERFORMANCE REQUIREMENT 10-9
Twin-Donut Hose Roll

A First the hose is laid flat with both couplings at one end and each half lying parallel to the other.

B At the center, the loop is folded over the top of both halves.

C The roll is started toward the couplings at the same time.

D At the end, the roll may be tied together for carrying.

E The twin donut can be secured by using the hose itself. This is called a self-locking roll. To accomplish this, extend the amount of hose that is used for the starting fold and loop. Allow this excessive hose to "flop" as the twin donuts are rolled. When finished, use the extra hose at the center to form a bight around the two end couplings.
(Photo courtesy Loveland Fire and Rescue)

JOB PERFORMANCE REQUIREMENT   10-10
Drain and Carry

A   The firefighter starts at one end of the hose and with the coupling held waist height feeds the hose over the shoulder and back down to the waist.

B   A fold is created and the hose is laid on itself back to the front.

C   The firefighter continues to walk forward folding and refolding the hose at the waist until finished. The hose can then be carried to the new location.

JOB PERFORMANCE REQUIREMENT 10-11
Shoulder Loop Carry

A   Place the nozzle or end of hose over the shoulder resting against the back.

B   Walk forward about 3 feet (1 m), pick up the hose, and form a bight to bring the hose back up and over the shoulder, creating a loop.

C   Continue as each section is picked up and carried forward.

D   If you need to move in the opposite direction, the loops are collected and raised with your hands and then rotated to the opposite direction.

E   Return the hose to the opposite shoulder moving forward in the new direction.

JOB PERFORMANCE REQUIREMENT 10-12
Single-Section Street Drag

A Put the end of a section of hose over your shoulder with the coupling in front at waist height and walk away dragging the line.

B Place a line over each shoulder and pull two lines.

C If additional sections are needed, additional firefighters can do the same with the following sections until the desired amount of hose is stretched.

JOB PERFORMANCE REQUIREMENT 10-13
Accordion Load

A   For use with preconnected lines, the accordion load should be started with the female coupling at the pump outlet or the front of the bed.

B   The hose is placed in the bed on its side rather than lying flat. The hose is laid out toward the opposite end of the bed, where it is folded and returned to the front where the process is repeated.

C   When the layer of the hose fills the bed, the hose is brought up at the rear and folded back on itself starting the next layer, again on the left side.

D   Coupling placement is important as couplings require a little bit of room to spin around in the bed. It is wise to use a dutchman to stagger the couplings in the bed to keep a coupling from lying close to another coupling. This load should be continued until the hose bed is full or until the desired amount of hose is loaded. If the line is pre-connected a nozzle can be pre-connected, also. When loading supply line, the hose should have adapters available to allow for flexibility: normally it is not possible to preplan a forward or reverse lay. Adapters will allow the hose couplings to be converted as required. A blind cap can be added to protect threads.

JOB PERFORMANCE REQUIREMENT 10-14
Advancing an Accordion Load

A A nozzle or other appliance is attached to the end of the hose and the desired amount is selected.

B The firefighter pulls the amount of hose about halfway out of the bed and places it on one shoulder with the nozzle at the bottom. Use care not to have too much of the hose in front, which could cause tripping—it is better to drag the excess behind.

C Step away from the bed. Pull additional folds and drop them to the ground.

D When the point is reached where the shoulder load is to be deployed, the firefighter allows the hose to flake off one fold at a time while moving forward.

JOB PERFORMANCE REQUIREMENT   10-15
Flat Load

A   A water supply flat load starts with the male end of the hose on the right side of the bed at the rear. This will allow this hose load to be connected to another hose load for long hose lays.

B   The hose is laid flat until the front is reached where the hose is folded over and brought to the rear. When back at the rear, turn the hose off the first layer and lay it flat next to the first row.

C   When the edge of the hose bed is reached, a second layer is started with a double layer, moving back to the right side of the bed.

D   An appliance, such as a ball valve, can be placed on the female end on top of the load.

JOB PERFORMANCE REQUIREMENT 10-16
Advancing a Flat Load from a Supply Bed

A Attach the nozzle or appliance and determine the amount of hose needed to be placed on the shoulder.

B If possible, separate these folds from the bed by twisting them on their sides. Place these folds over your shoulder with the nozzle in front of you.

(Continues)

JOB PERFORMANCE REQUIREMENT 10-16

Advancing a Flat Load from a Supply Bed (*Continued*)

C   If twisting the folds is not possible (possibly due to the amount of hose in the bed), pull a section of hose off and place it over your shoulder, with the nozzle in front of you. Pull additional sections off and place them on top of the previous section on your shoulder.

D   When the desired amount of hose is out of the bed and on your shoulder, step away from the bed and again pull some extra folds to place on the ground. The hose should now deploy as an accordion load.

JOB PERFORMANCE REQUIREMENT 10-17
Horseshoe Load

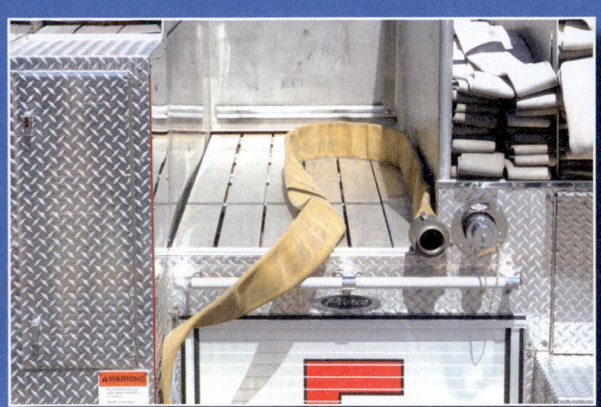

A Start the load at a rear corner of the hose bed with the hose on its side.

B Lay the hose toward the front wall of the hose bed and then along the wall to the other side and back toward the rear.

C At the rear, fold the hose alongside itself and head back to where it started. Fold it again and another "U" is formed.

D When the bed is filled, start a new layer by laying the hose flat on its side and continue to the front of the bed. Continue the hose down the side of the bed toward the rear, or

E Lay a few flakes of hose on top longways for quick access. Attach a hose strap, ball valve, hose bag, or other set of appliances or devices deemed necessary for quick deployment.

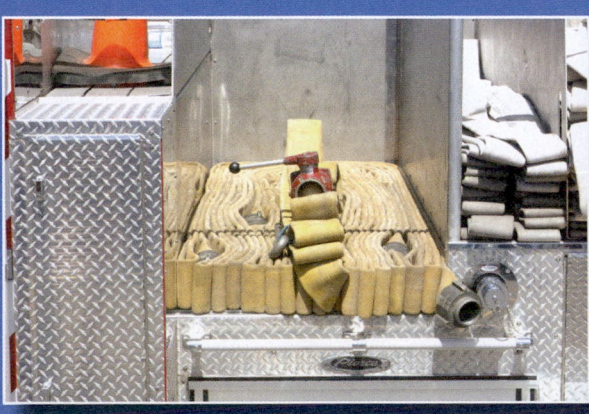

JOB PERFORMANCE REQUIREMENT 10-18
Advancing a Horseshoe Load

A Place the nozzle on the hose and select the desired amount of hose to deploy.

B Pull the hose and place it on your shoulder.

C Step away to pull the hose out of the bed.

JOB PERFORMANCE REQUIREMENT 10-19
Reverse Horseshoe Load for an Attack Line

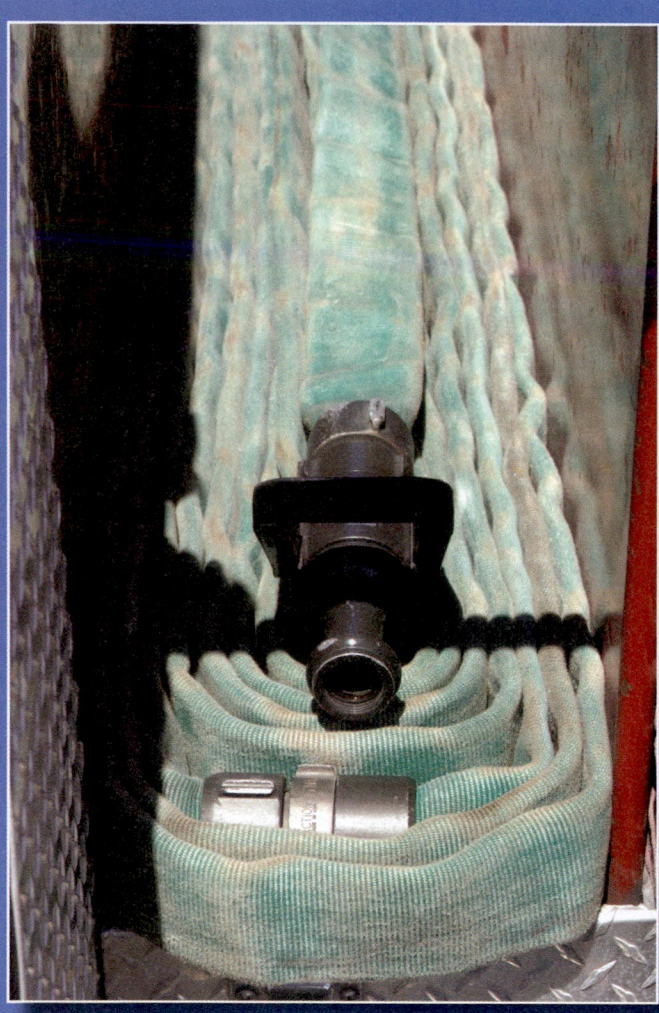

A For an attack line, create the reverse horseshoe again on top of the hose bed. When the amount of hose is loaded, attach a nozzle and lay it in the center of the horseshoe with the nozzle at the rear of the hose bed.

B Grab the nozzle and place a few of the hose folds over your shoulder. Grab the horseshoe load at the bottom of the "U," pulling it out of the hose bed and allowing it to flake out while moving to the fire.

JOB PERFORMANCE REQUIREMENT   10-20
Advancing the Flat Load from a Preconnect Bed

A Start the flat load at the discharge with the hose laid. At a point from one-third to one-half the length of the line, an ear or row of ears should be added to assist in pulling the line.

B To advance the line, grab the nozzle and place it over the shoulder with the other hand reaching around and pulling the ear.

C Walk away, pulling the line behind.

JOB PERFORMANCE REQUIREMENT   10-21
Minuteman or Slot Load

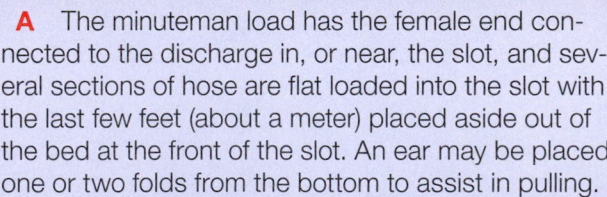

A   The minuteman load has the female end connected to the discharge in, or near, the slot, and several sections of hose are flat loaded into the slot with the last few feet (about a meter) placed aside out of the bed at the front of the slot. An ear may be placed one or two folds from the bottom to assist in pulling.

B   Next start a section of hose with the nozzle and load the hose flat above that. The nozzle will therefore be in the middle of the slot at the point where it will be pulled.

C When the designated number of sections is finished, connect the set-aside piece to the pieces with the nozzle attached. The connected sections are on the top layer.

JOB PERFORMANCE REQUIREMENT   10-22
Advancing the Minuteman Load

A   Lift up the nozzle and layers above it while pulling them out and placing them midway on the shoulder.

B   Step away to remove the remainder of the top layers.

C   Turn around and pull the ear to remove the remaining hose.

D   When the bottom sections are fully stretched out, allow the shoulder load to flake out toward the fire.

JOB PERFORMANCE REQUIREMENT 10-23
Triple-Layer Load

A Connect all the hose to the discharge and attach the nozzle at the other end with the hose stretched out its full length.

B At a point two-thirds of the way from the hose bed, make a fold and double the hose on itself back toward the engine. Make another fold, the last part with the nozzle, to create the third layer. These three layers are now treated as one for loading on the engine.

C Load the hose in the slot in a layer and make a bend, creating another layer, until all hose is loaded. The nozzle's layer should be at the edge of the bed for pulling.

JOB PERFORMANCE REQUIREMENT 10-24
Advancing the Triple-Layer Load

A Grab the layer with the nozzle and place it on the shoulder.

B Pull the layers out of the slot, or another firefighter can grab the next layer.

C Stretch the hose to the fire.

JOB PERFORMANCE REQUIREMENT 10-25
Modified Gasner Bar Pack

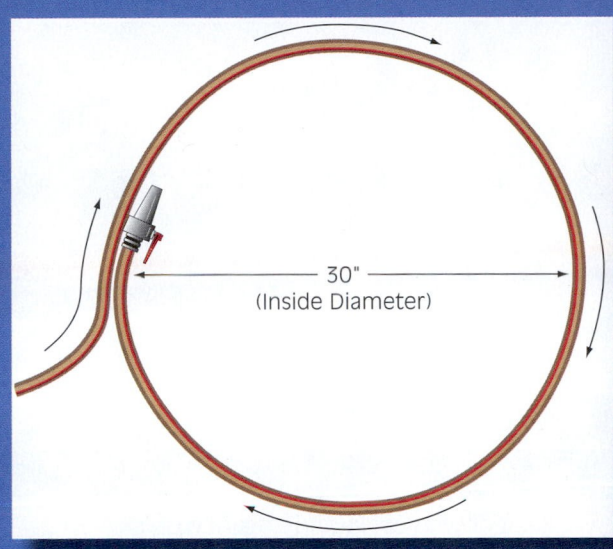

A Start at the male end of the hose (100 feet [30m]) and form a roll 30 inches (.8m) in diameter on flat ground.

B Continue to roll the hose in the direction of the rows and make the rolls as tight as possible.

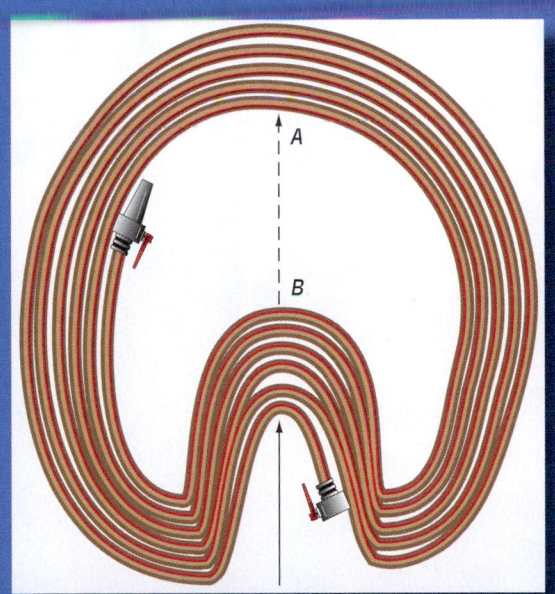

C The completed field roll, with the female coupling on the outside, is ready for packing. Grab the outside of the hose roll near the female coupling and pull toward the center of the roll.

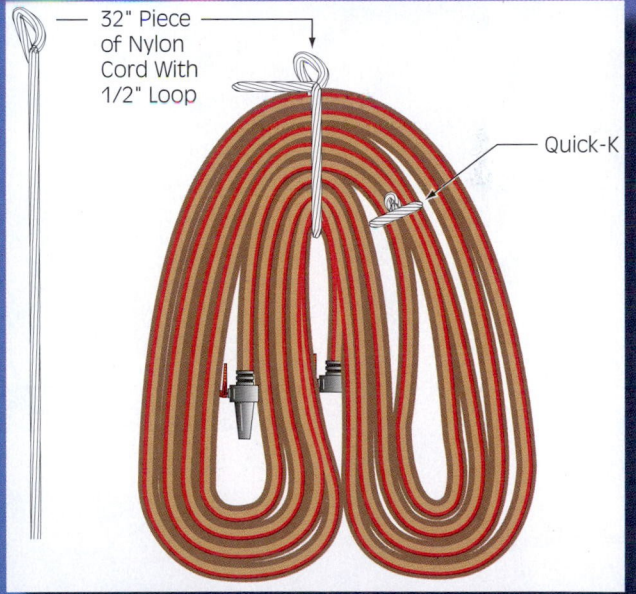

D The fold should now protect the female end. The top layer of hose can now be tied to the bottom layer using a piece of rope 32 inches long (0.9 m). Two rolls of hose may be placed together and tied off with rope on the outside of the rolls with a shoulder strap to allow carrying.

JOB PERFORMANCE REQUIREMENT   10-26
Advancing a Charged Hoseline Up a Stairwell

A   Apply water from bottom of stairs to darken down fire.

B   Shut down nozzle and advance quickly up stairs.

C   At top of stairs, open nozzle and continue fire attack.

JOB PERFORMANCE REQUIREMENT 10-27
Advancing a Charged Hoseline Over a Ladder

A The shoulder load is laid on the left side of the ladder.

B The nozzleperson takes the nozzle and pulls it under the left armpit, across the chest, and over the right shoulder, allowing the nozzle to rest in the small of the back.

C The nozzleperson, with both hands free for climbing, begins climbing the ladder to a point about 20 feet (6 m) up and stops. The next firefighter places the hoseline over the left shoulder at the next coupling and begins to climb the ladder. The two firefighters coordinate and maintain their distance on the ladder and hose.

JOB PERFORMANCE REQUIREMENT 10-28
Advancing an Uncharged Hoseline Over a Ladder

A   Upon reaching the entry point of the building—a window, balcony, or the roof—the firefighters stop in their respective positions on the ladder. The nozzleperson removes the nozzle and places it over the tip of the ladder into the building.

B   The nozzleperson climbs into the building, using the top rung of the ladder as a hose roller, and pulls the hoseline up as the next firefighter climbs.

C   The nozzleperson must complete pulling up the section as the next firefighter arrives at the tip. The second firefighter dismounts into the building and pulls the remainder of the hose into the building while the nozzleperson advances the line. To remove the line from the building after use, the line is drained and the process of advancing the hose is reversed.

JOB PERFORMANCE REQUIREMENT   10-29
Soft-Sleeve Hydrant Connection

A   Connecting a soft sleeve to a hydrant starts with proper parking of the engine. The charged line should be without kinks and undue stress on the hose or couplings. If the hose rests on the ground, there should be a chafing block between it and the ground.

B   The hose is removed from its compartment or the hose bed and it is stretched from the engine to the hydrant. The outlet cap is removed, the hydrant flushed, and the coupling connected.

C   If the line is not preconnected, the hose must be attached to the engine's inlet. If a gate valve is on the inlet, the hose is connected directly to it. If not preconnected, the inlet cap must be removed. This will cause water in the pump to flow out so the engine driver/operator should ensure the tank-to-pump valve is closed. The sleeve is then connected to the inlet. The hydrant valve can now be opened slowly to avoid a water hammer. The gate valve on the engine or switch valve on the hydrant must also be opened, if used.

JOB PERFORMANCE REQUIREMENT 10-30
Setting Up for a Drafting Operation

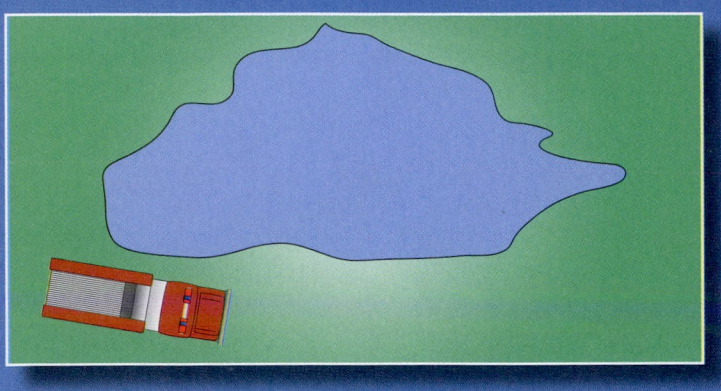

A The placement of a hard sleeve and strainer directly into a water source requires engine placement near the water source—slightly at an angle.

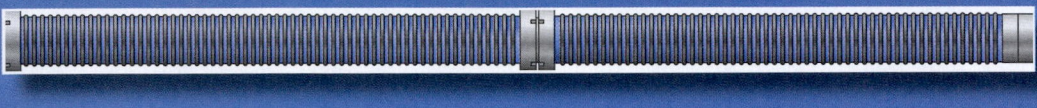

B Connect the hard sleeve to the engine inlet.

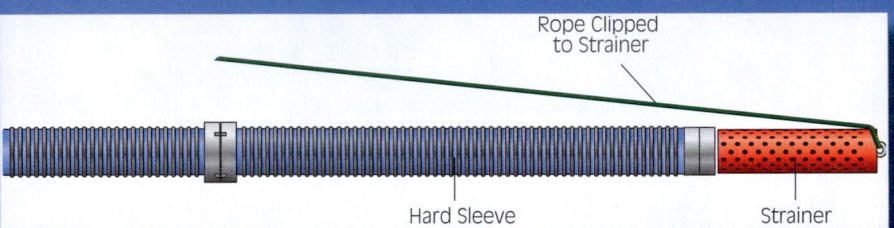

Rope Clipped to Strainer

Hard Sleeve

Strainer

C Attach a strainer and guide rope to the other end of the sleeve.

D The hard sleeve and strainer are attached to the pumper's inlet and the engine moves forward as the crew positions the sleeve into the water.

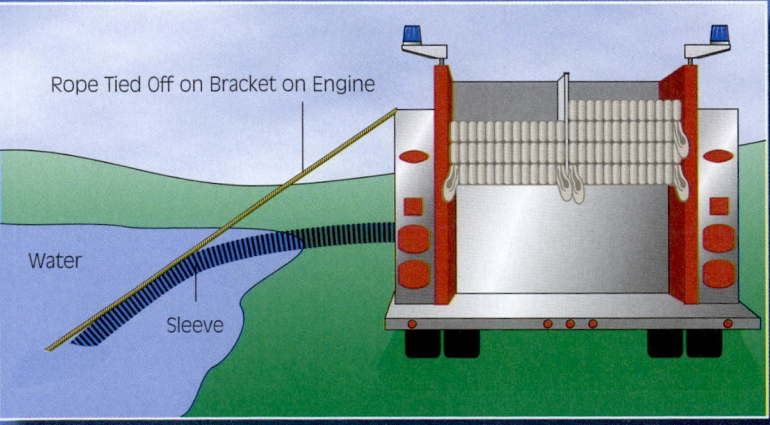

Rope Tied Off on Bracket on Engine

Water

Sleeve

E If a nonfloating strainer is used, the rope also keeps the strainer from resting on the bottom. Nonfloating strainers should be well under the water's surface to prevent drawing in of air. When in place, the engine is stopped, the line is secured, and drafting operations begin.

Extending a Hoseline with a Break-Apart Nozzle

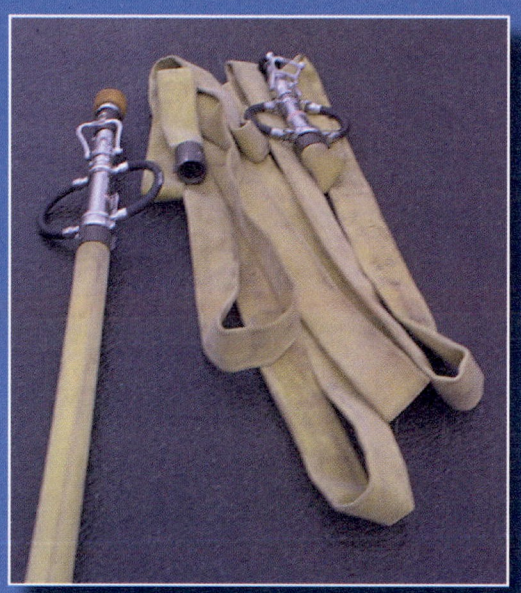

A The preferred method of extending a hoseline is to use a break-apart nozzle. The additional hose is brought to the nozzle end of the hoseline and the line shut down with the control handle. Bring additional hose to the end of the hoseline and shut down the line.

B The nozzle tip is removed and the additional hose added if it has 1½-inch (38-mm) couplings.

C If extending a line with hose couplings larger than 1½ inches (38 mm), then a 1½-inch (38-mm) to 2½-inch (65-mm) increaser will be required at the control handle.

D The additional line with the nozzle tip attached is advanced and charged from the control valve.

JOB PERFORMANCE REQUIREMENT   10-32
Extending a Hoseline Using a Hose Clamp

A   The clamp and additional hose are brought up to the nozzle end of the hose. Clamping a hose can be dangerous, so firefighters need to ensure that the clamp is safely operated and is fully locked before releasing the clamp. The firefighter is positioned to the side of the clamp to prevent injury if it slips. The hoseline is clamped just before the nozzle.

B   Remove the nozzle. The additional line is attached as is the nozzle and the line is then advanced. When water is called for, a firefighter removes the hose clamp.

JOB PERFORMANCE REQUIREMENT 10-33
Wildland Hose Advancing and Extension

A A preconnected line is pulled to begin the hose lay. As the line is moved forward, it is operated for suppression and crew protection.

B When the line is almost fully extended, a second line is rolled back in preparation for the next length.

C With the line clamped, the nozzle is removed and placed on the new length, and the hoses are connected.

D The clamp is released and the line advanced. When the line is again fully extended, the team repeats the clamping and extending process. As wildland hose is typically of smaller diameter than structural firefighting hose, it is easier to clamp.

LESSONS LEARNED

Fire hose, adapters, and appliances allow firefighters to move water from the source to the pumper and then onto the fire incident scene. Without these valuable tools, firefighters would be extremely limited in the ability to move water and to conduct most fire suppression operations—especially interior fire attack. Firefighters must understand the proper use and care of these tools and how to connect, advance, and operate these tools. These are the basics of firefighting, and the best method of learning these basics is practical application.

KEY TERMS

Adapter   A device that adapts or changes one type of hose thread to another, allowing connection of two different lines. Adapters have a male end on one side and a female on the other with each side being a different thread type, for example, an iron pipe to national standard adapter.

Attack Hose   Small- to large-diameter hose used to supply nozzles and other applicators or protective system for fire attack. Attack hose commonly means handheld hoselines from 1½ to 2½ inches (38 or 63 mm) in diameter.

Backstretch or **Flying Stretch**   An attack line lay where the engine is at the hydrant and the line is stretched back from the engine to the fire. The flying stretch is a version of the backstretch where the engine stops in front of the fire, the attack portion is removed, and the engine proceeds to the hydrant.

Booster Hose   Smaller diameter, flexible hard-rubber-coated hose of ¾- or 1-inch (19- to 25-mm) size usually mounted on reel that can be used for small trash and grass fires or overhaul operations after the fire is out.

Distributor Pipe or **Extension Pipe**   Devices that allow a nozzle or other device to be directed into holes to reach basements, attic, and floors that cannot be accessed by personnel. The distributor pipe has self-supporting brackets that help hold it into place when in use.

Double Female   Allows the two male ends of hose to be connected.

Double Male   Used to connect two female thread couplings.

Dutchman   A short fold of hose or a reverse fold that is used when loading hose and a coupling comes at a point where a fold should take place or when two sets of couplings end up on top of or next to each other. The dutchman moves the coupling to another point in the load.

Ears   Elongated folds or flaps at the ends of a layer of hose to assist in pulling that layer.

Fire Hose   A flexible conduit used to convey water or other agent from a water source to the fire.

Forestry Hose   Specially designed hose for use in forestry and wildland firefighting. It comes in 1- and 1½-inch (25- and 38-mm) sizes and should meet U.S. Forestry Service specifications.

Hard Suction Hose   A special type of hose that does not collapse when used for drafting.

Helix   The metal or plastic bands or rings used in hard suction hose to prevent its collapse under drafting conditions.

Higbee Cut   The blunt ending of the threads of fire hose couplings that allows the threads to be properly matched, avoiding cross-threading.

Higbee Indicator   A groove notched into coupling lugs to help firefighters align the Higbee cuts.

Hose Bed   The portion or compartment of fire apparatus that carries the hose.

Hose Bridge   Device that allows vehicles to pass over a section of hose without damaging it.

Hose Cap   A device used to cap the end of a hoseline or appliance to prevent water flow.

Hose Cart   A handcart or flat cart modified to be able to carry hose and other equipment around large buildings. Some departments use them for high-rise situations.

Hose Clamp   A device to control the flow of water by squeezing or clamping the hose shut. Some work by pushing a lever that closes the jaws of the device, and others have a screw mechanism or hydraulic pump that closes the jaws.

Hose Jacket   Metal or leather device used for stopping leaks without shutting down the line that is fitted over the leaking area and either clamped or strapped together to control the leak.

Hose Roller or **Hoist**   A metal frame, with a securing rope, shaped to fit over a windowsill or edge of a roof with two rollers to allow the hose to roll over the edge, preventing chafe.

Hose Strap   A short strap with a forged handle and cinch clip attached. Used to help maneuver hose and attach hose to ladders and stair rails.

Hydrant Valve or **Switch Valve**   Valve used on a hydrant that allows an engine to connect and charge its supply line immediately but also allows an additional engine to connect to the same hydrant without shutting down the hydrant, and increases the flow of the hydrant.

Hydrant Wrench   Tool used to operate the valves on a hydrant. May also be used as a spanner wrench. Some are plain wrenches and others have a ratchet feature to speed the operation of the valve.

Increaser   Used to connect a smaller hose to a larger one.

Intake Relief Valve   Required on large-diameter hose at the receiving engine that functions as a combined overpressurization relief valve, a gate valve, and an air bleed-off.

Jacket   The outer part of the hose, often a woven cloth or rubberized material, which protects the hose from mechanical and other damage.

Liner   The inner layer of fire hose, usually made of rubber or a plastic material, that keeps the water in the tubing of the hose.

Master Stream or **Heavy Appliances**   Non-hand-held water applicator capable of flowing over 350 gallons of water per minute (1325 L/min).

Medium-Diameter Hose (MDH)   Either 2½- or 3-inch (63- or 75-mm) hose.

Occupant Use Hose   Hose that is used in standpipe systems for building occupants to fight incipient fires. It is usually 1½-inch (38-mm) single-jacket hose similar to attack hose.

Portable Hydrant or **Manifold**   Like a large water thief and may have one or more intakes and numerous outlets to allow multiple hoselines to be utilized with or without a pumper at the fire location.

Reducers   Used to connect a larger hose to a smaller one.

Rope Hose Tool   About 6 feet (2 m) of 1½-inch (13-mm) rope spliced into a loop with a large metal hook at one end and a 2-inch (50-mm) ring at the other. Used to tie in hose and ladders, carry hose, and perform many other tasks requiring a short piece of rope.

Shoulder Load   Hose load designed to be carried on the shoulders of firefighters.

Siamese   A device that connects two or more hose-lines into one line with either a clapper valve or gate valve to prevent loss of water if only one line is connected.

Slot Loads   Narrow section of a hose bed where hose is flat loaded in the slot.

Small Lines or **Small-Diameter Hose**   Hose less than 2½ inches (63 mm) in diameter.

Soft Suction Hose   Large-diameter woven hose used to connect a pumper to a hydrant. Also known as a soft sleeve.

Spanner Wrench   Used to tighten or loosen couplings. It may also be useful as a pry bar, door chock, gas valve control, and so on.

Storz Couplings   The most popular of the non-threaded hose couplings.

Strainers   Placed over the end of a suction hose to prevent debris from being sucked into the pump. Some strainers have a float attached to keep them at or near the water's surface. A different style of strainer or screen is located on each intake of a pump.

Stream Straightener   A metal tube, commonly with metal vanes inside it, between a master stream appliance and its solid nozzle tip. The purpose is to reduce any turbulence in the stream, allowing it to flow straighter.

Supply Hose or **Large-Diameter Hose (LDH)**   Larger hose (3½ inches [90 mm] or bigger) used to move water from the water source to attack units. Common sizes are 4 and 5 inches (100 to 125 mm).

Water Hammer   The pressure surge that occurs when water valves are rapidly opened or closed.

Water Thief   A variation of the wye that has one inlet and one outlet of the same size plus two smaller outlets with all of the outlets being gated. The standard water thief usually has a 2½-inch (65-mm) inlet with one 2½-inch (65-mm) and two 1½-inch (38-mm) outlets.

Wye   A device that divides one hoseline into two or more. The wye lines may be the same size or smaller size, and the wye may or may not have gate control valves to control the water flow.

REVIEW QUESTIONS

1. Describe the various types of hose construction and how each type is used by firefighters.
2. What is the difference between single- and double-jacketed hose?
3. What are the different types of hose couplings important to firefighters?
4. What is a hose clamp?
5. What types of wrenches are used with hose?
6. What are double males and double females used for?
7. Describe the foot-tilt method of coupling hose.
8. What is a dutchman?
9. Describe flat, accordion, and horseshoe loads.
10. How is a hoseline advanced up a stairwell?
11. Describe connecting a supply line to a hydrant.
12. How can a hoseline be extended without shutting down the line at the pump?
13. What is a forward lay?
14. What is a deluge set?
15. What is a triple-layer load?

ENDNOTE

1. NFPA 1962, *Standard for Care, Use and Testing of Fire Hose Including Couplings and Nozzle*s (Quincy, MA: National Fire Protection Association, 2003), pp. 1962–7.

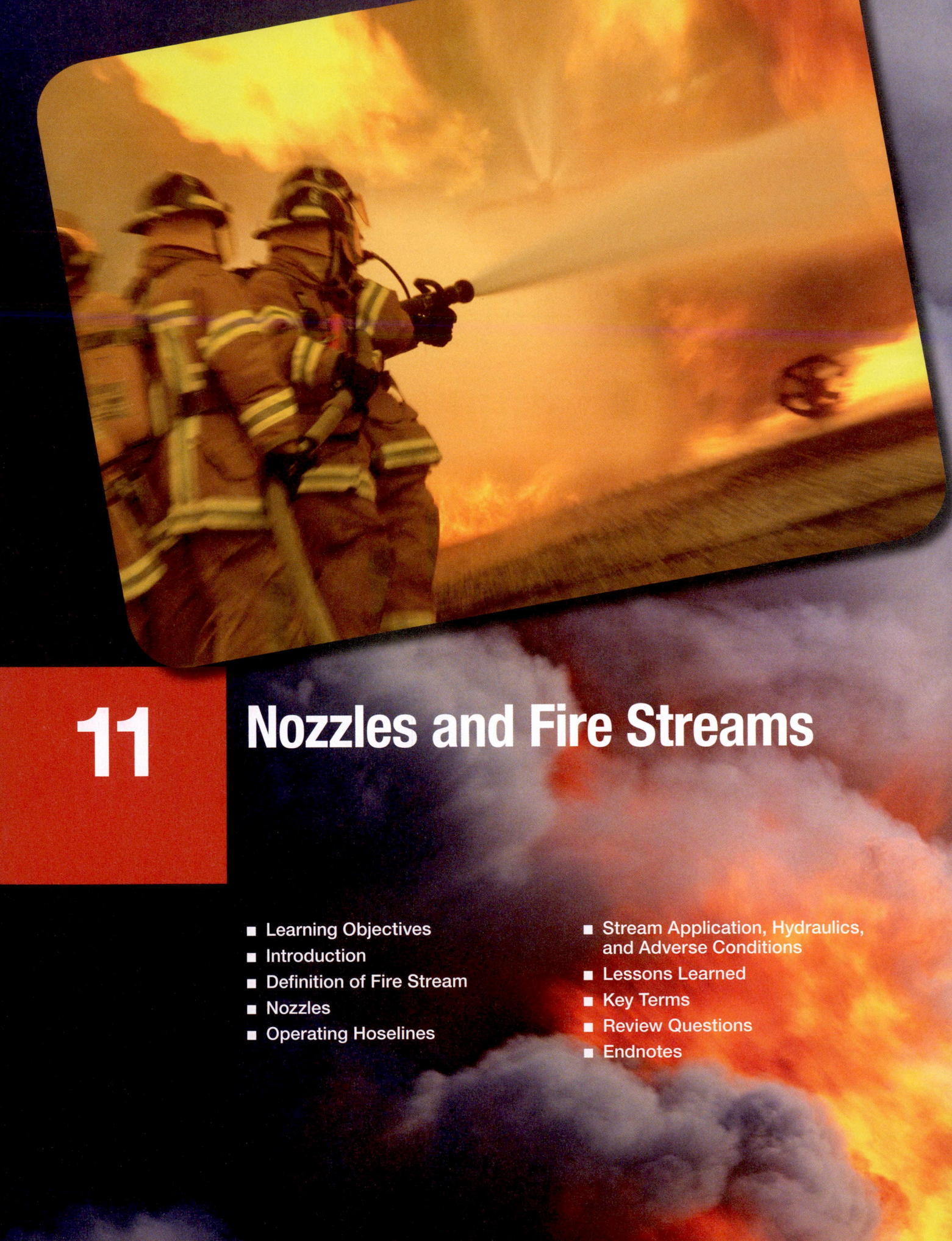

11 Nozzles and Fire Streams

A few of the senior firefighters in my engine company had told me that when I was the nozzle man to always keep the nozzle pointed up and in the ready position when waiting to move into a burning room. In this way it will be ready to quickly knock down any fire that might roll out the top of the door when the door was forced open by the forcible entry team. If the nozzle was pointed downward, toward the floor, I might not get it pointing up and open it in time to protect the firefighters in the hall and, if the hallway was crowded with firefighters, it might not be possible to raise the nozzle quickly if it was blocked by people. Also, anytime I was in nozzle position, I was told never to put the nozzle down, even when things seemed under control, because you never know. As nozzle man, I was expected to be ready to open the line immediately, should the need arise.

This lesson served me well in a cellar fire in an apartment building that we were called to in the early 1970s. The fire involved the gas heating unit and had spread to most of the cellar. I had extinguished the fire, and the truck company reported that the gas to the heating unit was turned off. There were three firefighters overhauling in the room containing the heating unit. I was kneeling at the door to the room with the nozzle in my hands pointed up and forward toward the ceiling when suddenly the entire room lit up—leaking gas had ignited, filling the room with flame. The sudden burst of flame knocked me on my back. Instinctively, I pulled back on the nozzle's handle, opening the nozzle and directing the water from my hose stream into the room and onto the ceiling of the room, quickly extinguishing the flames and allowing the firefighters in the room to scramble to safety. Luckily, I had the nozzle in ready position even though the fire was out and everything had seemed to be under control. While extinguishing a gas flame is not the proper way to handle a gas fire, in this case it was necessary to allow firefighters to escape the flaming room. Proper training and attention to seemingly minor details can save lives. Never let your guard down. If you are the nozzle man, stay alert, hold onto the nozzle, and always be ready to open it.

—Street Story by Frank Montagna, Batallion Chief, FDNY, New York City

LEARNING OBJECTIVES

After completing this chapter, the reader should be able to:

11-1   Define a fire stream.

11-2   List and explain the properties of a fire stream.

11-3   Describe the various types of fire streams and their purposes.

11-4   Describe the principles of fire streams as they relate to friction loss, nozzle pressure, nozzle reaction, and elevation.

11-5   Describe the capabilities in terms of flow rate, pattern, and reach for various nozzles.

11-6   Describe the effect that low or excessive nozzle pressure has on fire stream application.

11-7   Explain the differences between a manually adjusted and automatic nozzle design.

11-8   Given a selection of nozzle types, identify a solid, fog, and broken stream nozzle.

11-9   Explain how to safely operate and control various nozzles.

11-10   Demonstrate the operation of various pattern options while flowing water through a nozzle.

11-11   Describe the characteristics of the various sizes of hoselines.

11-12   List the observable interior and exterior changes that indicate proper application and effect of a fire stream on a fire.

11-13   Demonstrate the opening and closing of nozzles to prevent a water hammer from occurring.

11-14   Demonstrate the operation of adjustable nozzles.

11-17   Demonstrate the ability to advance a 1½-inch (38-mm) or larger hoseline while applying the proper fire suppression techniques.

11-18   Describe the methods of fire stream application: direct, indirect, combination.

11-19   List the advantages and disadvantages for each type of application: direct, indirect, combination.

11-20   Demonstrate the methods for accomplishing each type of fire attack.

*The FF I and II levels, as defined by the NFPA 1001 Standards, are identified in different colors:
FF I = black, FF II = red, additional information = blue.

INTRODUCTION

Fires are usually extinguished by using water to cool the heat that is produced. Foam is added to improve water's extinguishment ability or for fires involving fuels where plain water is ineffective. Water, in proper quantities, is delivered using nozzles and fire streams to reach the seat of the fire. This chapter examines fire streams of water, various nozzles and appliances, attack applications, and basic hydraulic principles. Proper nozzle selection provides the attack crew with the proper tool required to fight the fire successfully. Nozzle selection is important, because each fire situation may require a different appliance.

DEFINITION OF FIRE STREAM

A **fire stream** is the extinguishing agent—usually water—that leaves the nozzle and flows toward its target, usually the fire. The four elements affecting a fire stream are the pump, water, hose, and nozzle. Properly executed fire streams are successful in extinguishing fires, while poorly developed or improperly aimed streams allow fires to burn and grow in size. Proper fire streams are the result of the knowledge, skills, and abilities of the nozzle firefighter, the company officer, and the pump operator. A proper fire stream has sufficient volume, pressure, and direction to reach the target in the desired shape or form and pattern. It is important that firefighters understand fire streams and how they are applied in various firefighting situations. (See Sections I and II, Chapter 19 for situations.)

NOZZLES

Nozzles are the appliances that allow application of the extinguishing agent. The two basic types of nozzles are solid stream (also called a smooth bore, straight bore, or solid tip) and fog nozzles. Each type, particularly fog nozzles, are available in different styles. For example, **combination nozzles** are capable of providing a straight stream or adjustable spray patterns. The type of nozzle used at a particular fire department is based entirely on user preference and each has advantages and disadvantages.

Important factors in nozzle selection are nozzle pressure, flow, reach, stream shape, and reaction.

Nozzle pressure is the pressure required for effective nozzle operation and relates to flow and reach. Nozzles are designed to operate at a specific pressure, usually 50, 75, 80, or 100 psi (345, 517, 552, or 690 kPa). Pressure is measured in pounds per square inch (psi) or kilopascals (kPa).

Nozzle flow is the amount or volume of water that a nozzle will provide at a given pressure. Flow is critical because the amount of water provided determines the amount of heat absorbed or cooled. Some nozzles only flow a set volume at a set pressure, while others can be adjusted manually or automatically adjust their flow. Flow is measured in gallons per minute (gpm) or liters per minute (L/min).

Nozzle reach is the distance the water will travel after leaving the nozzle. Greater reach is important in large rooms or during exterior fire operations. Reach is a function of the pressure, which is converted to velocity or speed, of the water leaving the nozzle. Reach is measured in feet or meters. Reach is affected by the factors of stream shape, water pressure, wind direction, gravity, and friction of the air. The angle of the nozzle can affect the reach: Maximum horizontal reach occurs at 32 degrees, while maximum vertical reach is obtained at 65 to 70 degrees. The objective is

to reach the fire with maximum effect, not to be at the maximum distance from the fire.

Stream shape, also called stream pattern, is the arrangement or configuration of the droplets of water or foam as they leave the nozzle, **Figure 11-1.** The shape of the pattern helps determine the reach of the fire stream.

Nozzle reaction is the force of nature that makes the nozzle move in the opposite direction of the water flow, **Figure 11-2.** The nozzle operator must counteract or fight the backward thrust exerted by the nozzle to maintain control of the nozzle and to direct it to the correct location. The nozzle pressure and stream shape affect nozzle reaction.

Solid Tip or Stream

Solid tip, solid stream, or smooth bore nozzles deliver an unbroken or solid stream of water at the tip and toward the fire, **Figure 11-3** and **Figure 11-4.** The **solid stream nozzle** can deliver its water as a solid mass or cone of water or, when bounced off a ceiling, wall, or other object, as large water droplets. This solid mass breaks or shears apart the farther the water travels. The solid stream nozzle flow is a factor of the tip size at a certain nozzle pressure. Excessive or reduced nozzle pressures have adverse effects on stream performance. Handlines use tips from ¾ to 1¼ inches (19 to 32 mm) at 50 psi (345 kPa), and master streams use tip sizes of 1¼ inches (32 mm) and larger at 80 psi (552 kPa).

Solid stream handlines can reach over 70 feet (21 m) and master streams about 100 feet (33 m). They have the ability to penetrate through the fire's heat without absorbing that heat before reaching the target. A solid stream has less effect on a room's thermal balance, that is, the layers of heated gases in a burning room (see Section I, Chapter

FIGURE 11-1 Nozzles showing the stream shape for straight, solid, and wide pattern streams.

FIGURE 11-2 Nozzle reaction is the force of the water as it escapes and pushes back toward the nozzle and nozzleperson.

FIGURE 11-3 Various solid tips (stacked).

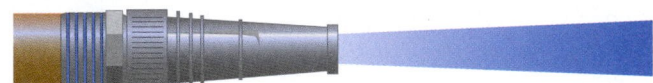

FIGURE 11-4 The flow pattern of a solid tip nozzle.

4), and produces minimal amounts of steam compared with narrow and wide fog patterns. Disruption of the thermal balance of a room can cause steam burns to firefighters as heavier water vapor descends onto firefighters after discharging the water. Many fire departments have tried to minimize the possibility of thermal burns by utilizing solid stream nozzles for aggressive, interior fire attack. A solid stream provides good penetration into burning or smoldering piles of materials. Smooth bore nozzles are more durable and easier to maintain than fog nozzles as they contain fewer parts. The disadvantages of a solid stream are the lack of volume control (other than changing tip sizes or adjusting the shutoff), the lack of fog protection when working close to the fire, and a higher nozzle reaction at the same pressure than a fog nozzle.

Fog Nozzles

Fog nozzles deliver either a fixed spray pattern or a variable combination pattern with both **straight stream** and spray patterns that can be adjusted by

the nozzleperson. Fixed spray pattern nozzles are of the impinging design in which the nozzle has a series of holes at the end that creates a water spray, **Figure 11-5.** The variable fog patterns vary from the straight stream pattern, which is similar to a solid stream pattern, to a wide-angle fog pattern of at least 100 degrees, **Figure 11-6** and **Figure 11-7.** The different types of combination fog nozzles depend on the variations allowed and include constant volume or set gallonage nozzles; variable, adjustable, or selectable gallonage nozzles; and automatic or constant pressure nozzles, **Figure 11-8.** The **constant** or **set volume nozzle** has one set volume at a set pressure, for example, 60 gpm at 100 psi (227 L/min at 690 kPa), and the only adjustment is the pattern. **Variable, adjustable,** or **selectable gallonage nozzles** allow the nozzleperson to select from two or three flow choices and to adjust the flow pattern; the flow choice is made at the end of the hose. To maintain the correct pressure, the nozzleperson or company officer should ensure the pump operator is aware of the choice.

The **automatic** or **constant pressure nozzle** has a flow that is adjusted by the **pump operator,** who increases the pressure, which increases the gallonage. The pattern can be adjusted by the nozzleperson. Some newer automatic nozzles have a dual-pressure option for normal and low-pressure situations that is controlled by the nozzleperson. The automatic feature has a spring mechanism built around the baffle that reacts to increased pressure and adjusts the flow and reach of the nozzle, **Figure 11-9.** The disadvantage of

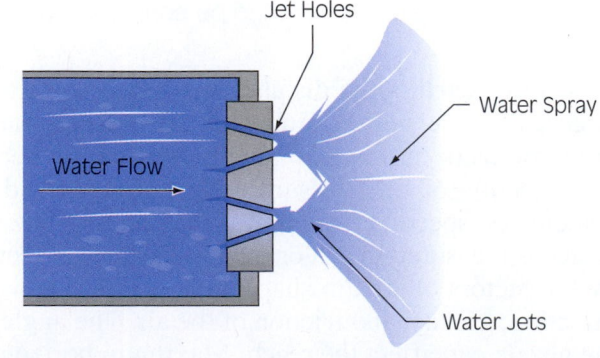

FIGURE 11-5 A fixed fog nozzle pattern of impinging design.

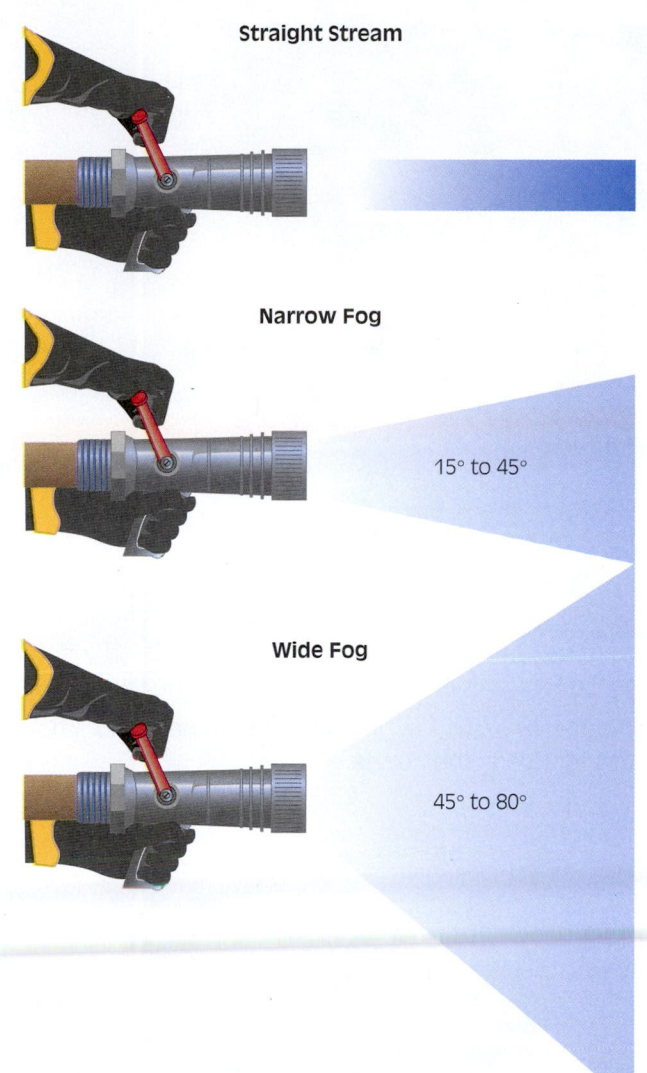

Straight Stream

Narrow Fog

15° to 45°

Wide Fog

45° to 80°

FIGURE 11-6 Variable combination fog nozzle patterns. From top to bottom: straight stream, narrow fog, and wide fog.

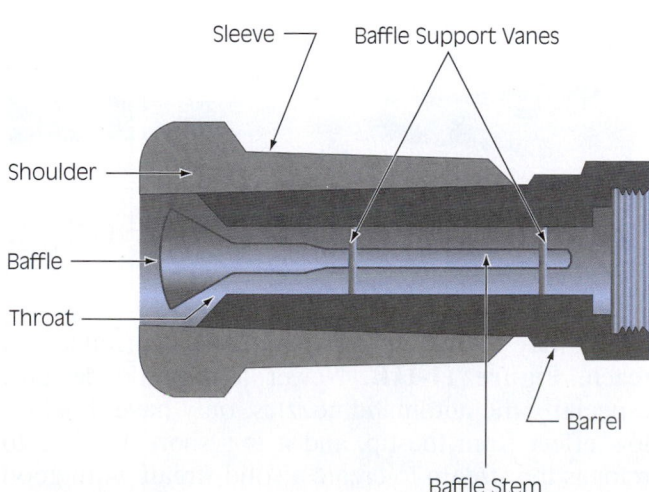

Sleeve

Baffle Support Vanes

Shoulder

Baffle

Throat

Barrel

Baffle Stem

FIGURE 11-7 Parts of a fog nozzle.

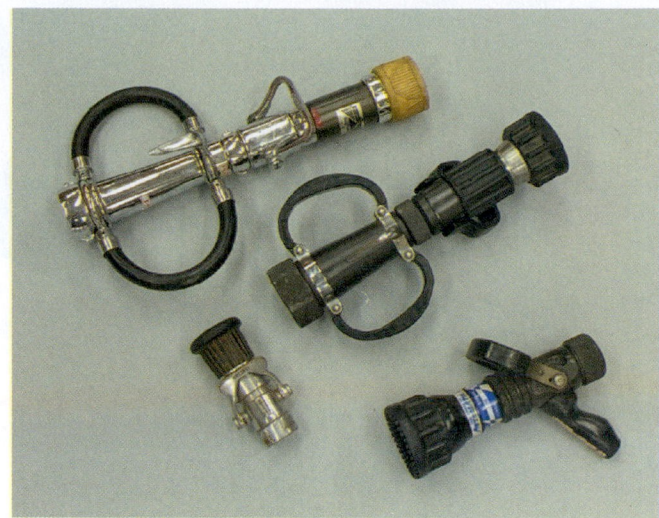

FIGURE 11-8 Various styles of fog nozzles.

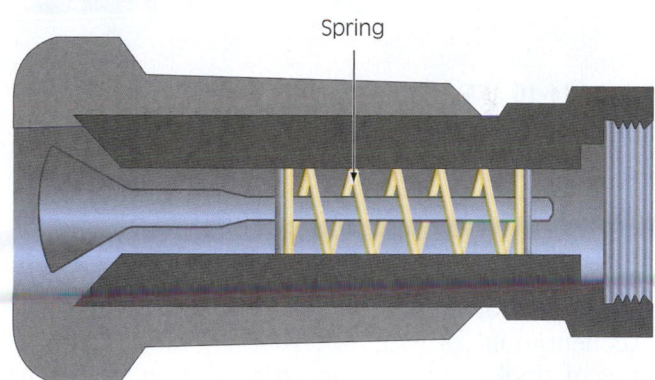

Spring

FIGURE 11-9 An automatic or constant pressure fog nozzle.

these nozzles is that the pump operator, who is farther away from the fire, controls the maximum flow. This makes it important for the company officer or nozzleperson to be in contact with the pump operator to control this flow. Fog nozzles have traditionally operated at 100 psi (690 kPa) but new low-pressure nozzles operating at 50 or 75 psi (345 or 517 kPa) have been approved. The lower nozzle pressure gives the same volume, but nozzle reaction is reduced and longer lengths of hose can be used when at the maximum pump pressure. This type of nozzle can also be beneficial in areas with poor water pressure. On the low-pressure nozzle settings, additional flow is available at higher pump pressures.

Fog nozzles provide good reach that varies with the pattern from 25 to over 100 feet (7.5 to 133 m). They also provide good penetration. Adjustable fog nozzles provide personal protection for the nozzleperson by utilizing a screening effect from the convected and radiated heat. Fog provides better heat absorption, but can change to steam before reaching the seat of the fire.

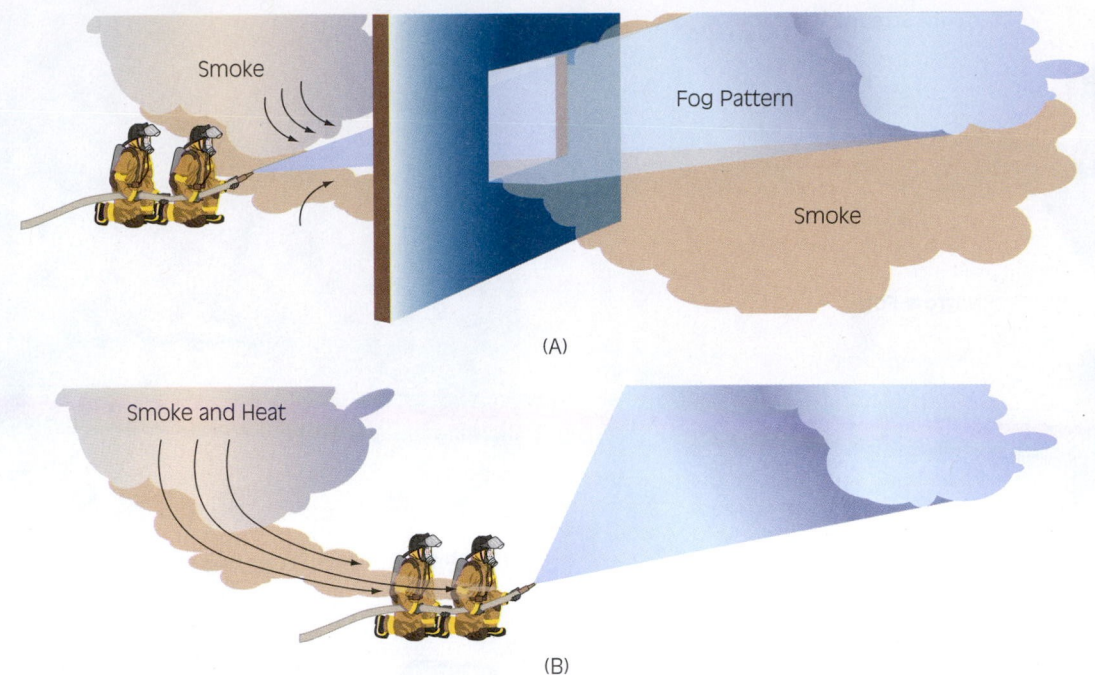

Smoke

Fog Pattern

Smoke

(A)

Smoke and Heat

(B)

FIGURE 11-10 (A) One advantage of a fog nozzle is that, with ventilation at a window, it draws out the heat and smoke. (B) The disadvantage is that heat and gases are drawn onto the firefighters at the nozzle.

However, this steam can extinguish hidden fire and works well for indirect attack. Fog streams are also excellent tools for hydraulic ventilation. Large quantities of smoke can be removed by aiming a fog cone out an open window or other opening in the structure. Firefighters must be aware that this method of removing smoke can also draw heat from the fire, **Figures 11-10A** and **B.** A 1½-inch (38-mm) fog nozzle moves more air than a 14-inch (0.36-m) smoke ejector, and a 2½-inch (63-mm) fog nozzle moves more air than a 24-inch (0.6-m) smoke ejector.

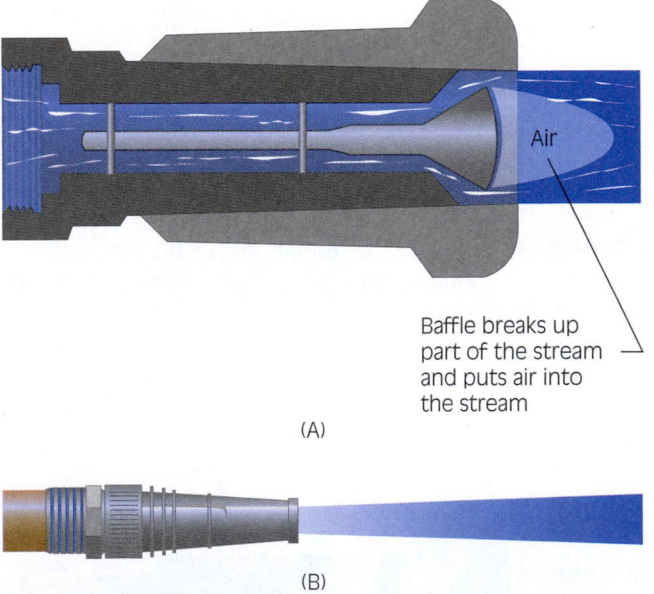

Air

Baffle breaks up part of the stream and puts air into the stream

(A)

(B)

FIGURE 11-11 Comparison of (A) straight and (B) solid streams at tip.

> **SAFETY**
>
> A fog nozzle, by nature of its design, will also cause air swirling and movement, which can disrupt the thermal balance of a room. To avoid hindering firefighting operations, as well as to protect fire crews from thermal burns, firefighters should use straight streams when attacking fires in high-heat environments or when trying to cool interior compartments while advancing on a fire.

Straight Stream

The straight stream nozzle pattern on a fog nozzle, **Figure 11-11A** creates a hollow type stream that is similar in shape to the solid stream pattern, but the straight stream pattern must pass around the baffle of the nozzle. This creates an opening in the pattern,

which may allow air into the stream and reduce its reach, **Figure 11-11B.** Newer fog nozzle designs, especially the automatic nozzles, only have this hollow effect from the tip, and it is a short distance to refocus the stream to create a solid stream with good reach and penetration abilities. In fact, some are better than solid stream nozzles.

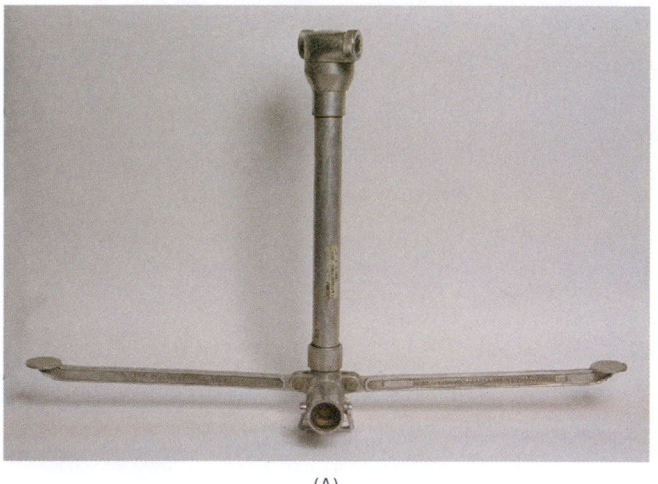

(A)

(B)

FIGURE 11-12 (A) Cellar nozzle and (B) Bresnan distributor.

Special Purpose

Special-purpose nozzles were developed for use in limited types of situations by some fire departments. Special-purpose nozzles are not often used, but firefighters should know when and how to use them. **Cellar nozzles** and **Bresnan distributors** can be used to fight localized fires in basements or cellars when firefighters cannot make a direct attack on the fire, **Figures 11-12A** and **B.** The cellar nozzle has four spray nozzles, and the Bresnan distributor has six or nine solid tips or broken stream openings that are designed to rotate in a circular spray pattern.

Piercing nozzles were originally designed to penetrate the skin of aircraft and now have been modified to pierce through building walls and floors, **Figure 11-13.** Some have striking points that allow them to be driven through the material. Other nozzles include high-pressure fog nozzles that can operate at 1,000 psi (6,895 kPa) and industrial and forestry nozzles that are combination fog nozzles with a shutoff built into the design.

Another special nozzle is a **water curtain nozzle,** which is designed to spray water to protect against exposure to heat, **Figure 11-14.** When using a water curtain, its spray should be directed on the object being protected to absorb the heat, not just up into the air as radiant heat passes through clear materials, such as water, without being absorbed.

Nozzle Operations

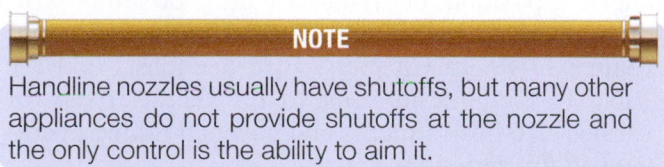

NOTE

Handline nozzles usually have shutoffs, but many other appliances do not provide shutoffs at the nozzle and the only control is the ability to aim it.

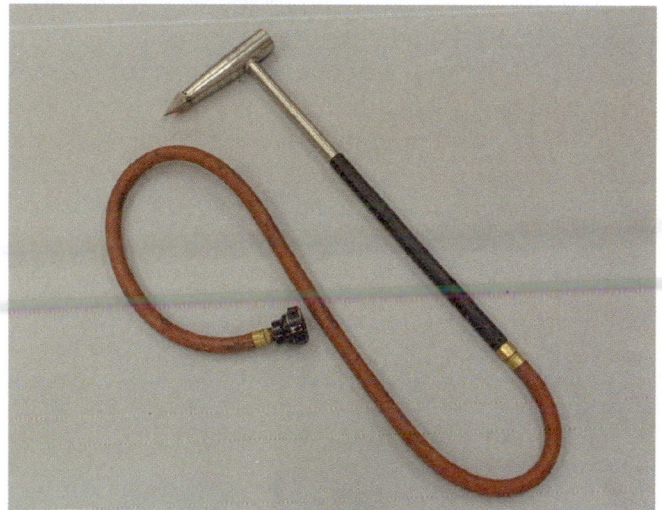

FIGURE 11-13 Piercing nozzle.

FIGURE 11-14 Water curtain nozzle.

Solid stream nozzles are simple to operate with the nozzleperson controlling the shutoff, tip size, and aim. The typical shutoff for a solid tip nozzle is a lever, or bale, that controls a ball valve inside the nozzle. This type of lever allows the nozzle to be fully or partially opened or closed. For proper gallonage and flow, a nozzle size and nozzle tip are selected to match the desired flow. This is coordinated with the pump operator for proper pump pressure and safety. The tips are usually carried stacked: the smaller sizes are screwed onto the larger sizes. When choosing a tip size, firefighters should carry the smaller-sized tips in their pocket to facilitate firefighting operation.

Fog nozzles have either the lever-type open/close shutoff, which is the most common, or the rotating type of valve. Nozzles with rotating valves are commonly found on wildland firefighting hose. For both types of nozzles, the fog pattern can be adjusted by rotating the nozzle barrel counterclockwise to move from straight stream to narrow fog to wide-angle fog, **Figure 11-15.** Turning it counterclockwise also closes the nozzle for the rotating valves. Clockwise rotation adjusts the pattern the opposite way and opens the valve fully for the rotating valves. Fog nozzles with variable gallonage have an additional ring on the col-

FIGURE 11-16   Close-up of combination fog nozzle showing the adjustments for selecting patterns and for selecting gallonage.

lar that rotates from one gallonage setting to the next, **Figure 11-16.** Both gallonage and pattern adjustments can be detected in the dark because the nozzle clicks at each position. Fog nozzles have more applications than smooth bore nozzles and many firefighters consider them more effective.

OPERATING HOSELINES

Advancing hoselines was covered in Section I, Chapter 10 including some of the initial operation of the nozzle. Included was straightening the hose, having firefighters properly spaced on the same side of the line, and with the line charged, bleeding off air from the hose and nozzle, and selecting the proper pattern. Most hoselines are operated from a crouching or kneeling position, but lying, standing, or sitting positions are also used, **Figure 11-17, Figure 11-18,** and **Figure 11-19.** In Chapter 19 (in Sections I and II), Table 19-1 highlights some of the characteristics of various size hose streams and covers hoseline operations in fire situations.

FIGURE 11-15   Close-up of rotating type of nozzle shutoff.

FIGURE 11-17 Firefighter operating nozzle in the crouching or kneeling position.

FIGURE 11-18 Firefighter operating nozzle in the lying position.

FIGURE 11-19 Firefighter operating nozzle in the sitting position. Notice that the hoseline is placed in a loop to counter nozzle reaction.

Small-Diameter Handlines

Small-diameter handlines are typically 1½, 1¾, or 2 inches (38, 44, or 50 mm) in diameter and flow from 100 to over 250 gpm (378 to over 946 L/min).[1] When flowing at the lower volumes, these lines can be operated by one person; larger volumes require two persons.[2] Both fog and solid tip nozzles can be used for small lines. Small lines are popular because of their ease of mobility, the number of personnel required to operate them, and their ability to extinguish one to three typical rooms of fire with their flow.

Medium-Diameter Handlines

Medium-diameter hose for handlines is 2½-inch or 3 inch (63 or 75-mm) hose and can be used with solid tip and fog nozzles, flowing from 165 to 325 gpm (625 to 1,230 L/min). The 2½-inch hose has been the standard size hoseline for the fire service. Many fire departments abandoned the 2½-inch hose as a fire attack tool, advocating 1¾-inch and 2-inch lines because of increased maneuverability. However, increased Btu development of fires due to plastics and low-mass fuels has caused many to reintroduce the 2½-inch hoseline to increase gpm flow to quench the higher heat. This is especially important in large, commercial structures or buildings with high fire loading (like an auto body repair shop or high rack storage warehouse-type stores).

Medium-sized hoselines usually require two or more personnel to operate them because they are less mobile. Some departments use 3-inch (75-mm) hoses with 2½-inch (65-mm) couplings for hoselines or masterstream devices.

Master Stream Devices

Master stream devices (see Section I, Chapter 10) are capable of flowing over 350 gpm (1,325 L/min), and some have capabilities of many thousands of gallons per minute. This is the main artillery of the fire service and used when large volumes of water are required. These devices must be apparatus-mounted or secured properly. Safety should be a major concern during their operation. Specific operating instructions are necessary for each type of device, so training and firefighter familiarity is a major concern. Master stream devices require only one person to operate them, but a major drawback lies in their poor or complete lack of mobility.

NOTE

The method of stream application or fire attack depends on the fire's fuel, its location, and the suppression equipment of the fire department. The size and type of hoseline and nozzle figure prominently in fire attack.

STREAM APPLICATION, HYDRAULICS, AND ADVERSE CONDITIONS

Application of fire streams vary according the method of fire attack and conditions encountered, including environmental factors and water supply. The fire stream must have the proper pressure and flow. Firefighters—particularly the pump operator—must have a thorough understanding of hydraulics. Improper hydraulic calculations are the leading causes of poor fire streams.

Direct, Indirect, and Combination Attack

Three methods of fire attack exist: direct, indirect, and combination. **Direct fire attack** is used to attack the fire by aiming the flow of water directly at the seat of the fire, **Figure 11-20.** Direct fire attack is used on most deep-seated fires that require penetration by the hose stream.

Indirect fire attack is used to attack interior fires by applying a fog stream into a closed room or compartment, converting the water into steam to extinguish the fire, **Figure 11-21.** Firefighters apply the water at the doorway and then close the door, allowing the steam to put the fire out. The closed compartment is needed to contain the steam, thereby smothering the fire. The estimated quantity of water applied is the amount needed for total conversion to steam to fill the room. As water is converted to steam at 212°F (100°C), it expands 1,600 to 1,700 times its volume; 1 ft^3 of water (0.028 m^3 or 7.48 gallons or 28 L) would fill a 1,700-ft^3 (48.1-m^3) room, At 1,000°F (500°C), this expansion is over 4,000 times. Because the entire space is filled with steam, the indirect attack should not be used when people are in, or suspected to be in, the space. It is also important to remember that this water-to-steam expansion can damage windows and walls containing the fire. Too much water applied into the room can cause the weaker portions of the structure to fail, allowing the "contained" fire to escape.

A **combination fire attack** uses a blend of the direct and indirect fire attacks, with firefighters applying water to both the fuel and the atmosphere of the room, **Figure 11-22,** using a circular, "Z," or "T" motion. The straight stream is directed to the seat of the fire and then the nozzle is quickly changed to a fog and the upper areas of the room are covered with a quick shot into the thermal cover. This puts water on the seat of the fire and the atmosphere, while creating steam to extinguish the fire in any hidden areas. Ventilation with this combination attack controls the flow of fire gases and steam, improving the survival chances of victims, and makes this the typical attack in structural firefighting.

FIGURE 11-20   Firefighter directly attacking a fire.

Applying an effective fire stream will yield noticeable results when applied to a fire:

■ Heat reduction
■ Flame reduction
■ Smoke reduction
■ Fire location becomes tenable

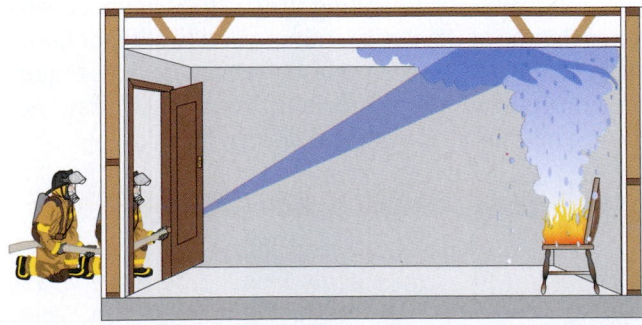

FIGURE 11-21   Firefighter using indirect attack by applying water into room and then closing the door.

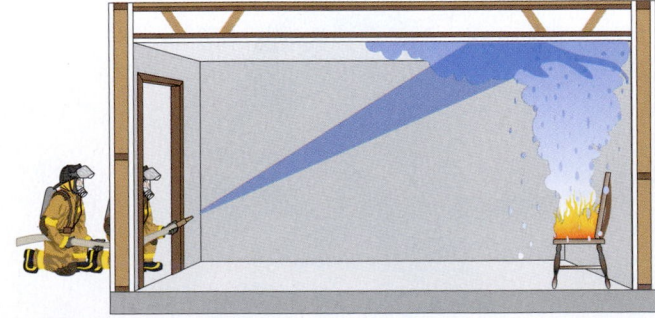

FIGURE 11-22   Firefighter using combination fire attack directing the stream from the ceiling to the fire with a circular, "Z" or "T" motion.

Basic Hydraulics, Friction Loss, and Pressure Losses in Hoselines

Hydraulics is the study of fluids at rest and in motion. Water moving through a hoseline and nozzle is an example of water in motion, while use of the nozzle shutoff leads to water at rest. An effective fire stream must have sufficient volume and pressure and be delivered in the correct form. This relationship begins with the water supply into the fire pump, through the lines to the nozzle tip and onto the fire. See Chapter 9 in Section I for more information.

NOTE

It is important for firefighters to have a basic understanding of hydraulics. Most of the formulas used in this section are necessary information for those firefighters serving in a driver/operator or engineer capacity.

Pressure is the force divided over an area, usually expressed in pounds per square inch (psi) or kilopascals (kPa) where 1 psi = 6.895 kPa. Pressure is required to lift, push, or move water. Force is a measurement of weight and is measured in pounds or kilograms. The pull of gravity on a mass of water creates a force. Other external pushing and pulling, like the actions of a fire pump, applied to a substance (water) can also create motion. Water weighs 62.4 pounds per cubic foot (1,000 kg/m³), which creates a force of 62.4 pounds (28.3 kg). This would also create a pressure of 62.4 pounds per square foot (or 1,000 kg/m²) or 0.434 pounds per square inch or psi of 3 kPa, **Figure 11-23.** Pressure can also be measured in feet or meters of head or the height of a column of water. Thus, a column of water 100-feet (30-m) high would create a head of 100 feet (30 m) or a pressure of 43.4 psi (300 kPa).

Firefighters should be familiar with the types of pressure: Atmospheric pressure is pressure exerted by the atmosphere or, simply, the weight of the air at the Earth's surface. At sea level, this pressure is 14.7 psi (101 kPa), and gauges reading this pressure show **absolute pressure,** which is indicated as pounds per square inch absolute (psia). **Gauge pressure** normally measures pressure minus atmospheric pressure and is measured in psi or psig (pounds per square inch gauge). Most fire department readings are from gauge pressure and begin at zero, **Figure 11-24. Vacuum (negative) pressure** is the measurement of pressure less than atmospheric pressure, which is usually read in inches of mercury (in. Hg or mm Hg). Fire apparatus capable of drafting have at least one gauge that measures vacuum pressure, called a compound gauge. Today, it is common for all gauges on apparatus to be

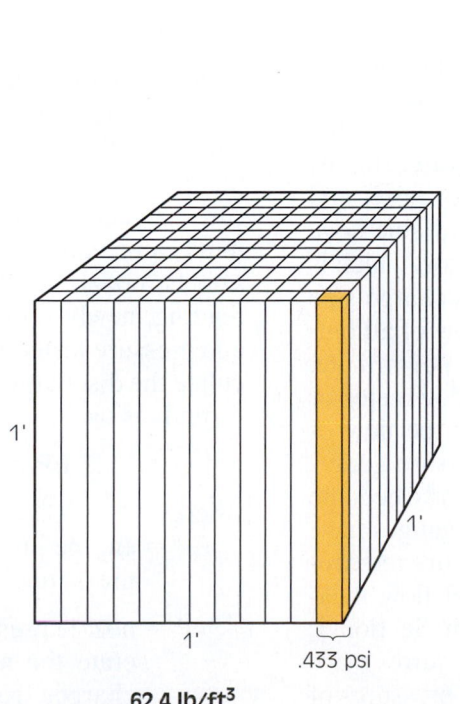

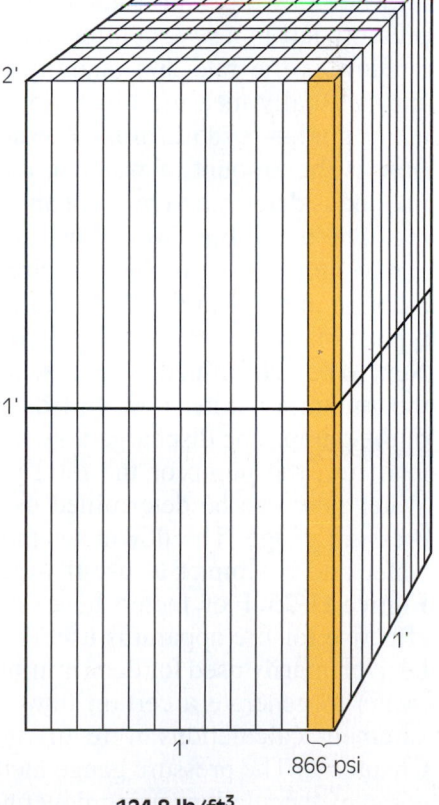

FIGURE 11-23 Water in a container (like a hose) exerts pressure at the lowest point.

1'

1'

1'

.433 psi

62.4 lb/ft³

2'

1'

1'

1'

.866 psi

124.8 lb/ft³

FIGURE 11-24 Typical pump pressure gauges.

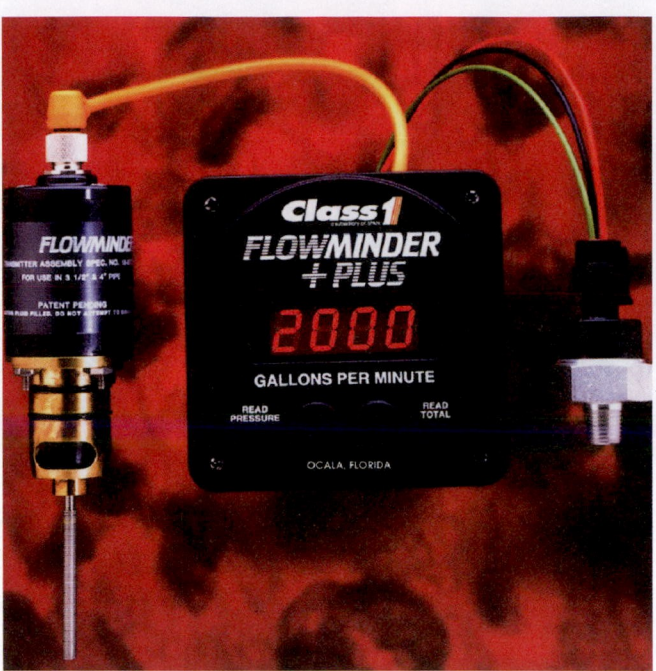

FIGURE 11-25 Flowmeters eliminate the need for friction loss calculations during pump operations. *(Photo courtesy of Class One)*

of the compound type. **Head pressure** measures the pressure at the bottom of a column of water in feet (meters) (see example given earlier) and head pressure can be gained or lost when water is being pumped above or below the level of the pump. A head of 2.31 feet (0.7 m) would equal 1 psi (6.895 kPa). This is also called elevation pressure and is usually rounded to 5 psi (35 kPa) per floor (or 10 feet) when pumping water up or down in a building. **Velocity pressure** is the pressure in a hose being converted to velocity or speed as it leaves an opening. When leaving a nozzle, this becomes the nozzle pressure. Section II, Chapter 9 discussed static pressure, which is the pressure with no water flowing, and residual pressure, which is the pressure left in a system after water is flowing.

Flow is the rate and quantity of water delivered and is usually measured in gallons or liters per minute (1 gpm = 3.785 L/min). The **needed** or **required flow** is the amount of water required to put out the fire and is determined by what and how much is burning. The **available flow** is the amount of water that can be moved to extinguish the fire. The water supply, pump(s) and capability, size and length of hose, and type(s) of nozzles used will determine available flow. Next to be determined is the **discharge flow,** which is the amount of water flowing from the discharge side of the pump. The discharge flow is the flow of all outlets from the pump or the total gallons per minute. Water flow can be determined using a flowmeter or pressure gauge. The flowmeter measures the flow of water and is simpler to use than the pressure gauge, **Figure 11-25.** Flowmeters have only recently become effective for fire apparatus use. Pressure gauges have been primarily used for determining pressure requirements to generate a certain flow. (Actual flow measurement calculations were discussed in Section I, Chapter 9.) The pressure gauge method requires more complex calculations using flow rate and pressures of

the nozzle and any **friction** produced in the system. This friction causes some of the pressure to be lost and is known as **friction loss.** Note that friction loss in hose and appliances, including nozzle pressures, and elevation losses or gains only occur when water is in motion or flowing. When all water flow stops, the system equalizes at the highest pressure in the system or hoseline. A sudden stop of water flow creates a water hammer or pressure surge that could damage equipment, piping, and hose. (See Section I, Chapter 10 for more on water hammer.)

After determining the needed flow, the pump operator adds the pressures of the various parts of the pump to the nozzle system and increases engine pressure to generate that flow. Errors made in this formula can create too much or too little pressure, which affects nozzle performance. To ensure consistent reading, newly purchased equipment should be flow and pressure tested to validate these formulas. To calculate the discharge pressure of a pump, the following formula is used:

$$EP = NP + FL \pm E + SA$$

where:
  EP = engine pressure, also called pump discharge pressure.

  NP = nozzle pressure, the pressure required to operate the nozzle and the pressure as it discharges from the nozzle.

FL = friction loss in the hose(s). (There can be multiple cases of friction losses, such as the hose between the pump and a standpipe connection and also the attack line hose.)

E = elevation gain or loss from moving the weight of the water up or down in relation to the pump.

SA = friction loss in any special appliance, not hose or elevation. It is also called appliance friction loss.

Each of the formula's components is either a given number or must be calculated. Engine pressure is the addition of the other components and is typically the highest value for any of the hoselines. If more than one hoseline is pumped, the engine pressure is figured for each line and set to the highest psi requirement. The lower pressure line(s) are gated down; that is, the valve is slowly closed. This reduces the size of the opening and lowers the pressure in that line. Nozzle pressure is usually a given value for each type of nozzle. **Table 11-1** gives the typical values.

Nozzle pressure pushes the water out of the nozzle, but it also pushes the nozzle in the opposite direction against the firefighter or whatever is holding it. A higher nozzle reaction means more effort is required by the nozzleperson and hose crew to control the nozzle and hoseline to keep them from getting away or when advancing an opened line. Now it is easier to understand why operating a nozzle in the sitting or lying position is less stressful than the standing position—the ground helps absorb this reaction. Nozzle reaction can make an unsecured hoseline unsafe and dance like a wild snake. It is important to keep a firm grip. Nozzle reaction is a relationship between the nozzle pressure and the type of nozzle. Nozzle reac-

tion can be figured for a smooth bore nozzle by using the following formula:

$$NR = 1.57 \times d^2 \times NP$$

where

NR = nozzle reaction

1.57 = a constant (some round to 1.5)

d = diameter of the nozzle tip in inches

NP = nozzle pressure

For example, a 1-inch nozzle with 50 psi of nozzle pressure flowing at 210 gpm would have a nozzle reaction of 78.5 lb.

$$\begin{aligned} NR &= 1.57 \times d^2 \times NP \\ &= 1.57 \times (1 \text{ in.})^2 \times 50 \text{ psi} \\ &= 1.57 \times 1 \text{ in.}^2 \times 50 \text{ psi or } 1.57 \times 50 \text{ psi} \\ &= 78.5 \text{ lb} \end{aligned}$$

The nozzle reaction for a fog nozzle is figured using the following formula:

$$NR = gpm = \sqrt{NP} \times 0.0505$$

where

NR = nozzle reaction

gpm = gallons per minute

$\sqrt{NP}$ = square root of the nozzle pressure

0.0505 = constant

A fog nozzle flowing 200 gpm at 100 psi of nozzle pressure would have a nozzle reaction as follows:

$$\begin{aligned} NR &= gpm \times \sqrt{NP} \times 0.0505 \\ &= 200 \text{ gpm} \times \sqrt{100} \text{ psi} \times 0.0505 \\ &= 200 \text{ gpm} \times 10 \times 0.0505 \\ &= 101 \text{ lbs.} \end{aligned}$$

A quick fireground formula for fog nozzles operating *with a nozzle pressure of 100 psi* poses the nozzle reaction is half the gpm.

$$NR = gpm \times 0.5$$

With this formula, a fog nozzle flowing 200 gpm at 100 psi of nozzle pressure would have a nozzle reaction as follows:

$$\begin{aligned} NR &= 200 \text{ gpm} \times 0.5 \\ &= 100 \text{ lbs.} \end{aligned}$$

This formula has a small margin of error (in this case 1 lb) but can allow for quicker calculations when needed.

When using these formulas, firefighters should remember that the lower the NR the easier a hoseline is for firefighters to handle. High nozzle reactions can

TABLE 11-1 Nozzle Types and Pressure

Type of Nozzle	Nozzle Pressure, psi (kPa)
Smooth bore handline	50 (345)
Fog handline, normal	100 (690)
Fog handline, mid pressure	75 (517)
Fog handline, low pressure	50 (345)
Smooth bore master stream	80 (552)
Fog master stream	100 (690)

require more firefighters on the hoseline to control it and can even drive firefighters backwards.

NOTE

At the same pressure and relatively same gpm, the smooth bore nozzle has a higher nozzle reaction than the fog nozzle. However, most fire departments operate fog nozzles at 100 psi, and the nozzle pressure for fog nozzles becomes greater than the smooth bore nozzle at this pressure.

Friction loss is the loss of energy from the turbulence, or rubbing, of the moving water through the hose. The pump operator compensates for friction loss by increasing the pump pressure for the correct pressure to the nozzle. Friction loss is based on four principles.

The first is that the friction loss varies directly with the length of the hose if all other variables are held constant. This means that if the length of the hose doubles, so does the friction loss.

Second, with all other variables held constant, friction loss varies approximately with the square of the flow. Therefore, if the flow rate doubles, then the friction loss will increase four times or be squared ($2^2 = 2 \times 2 = 4$).

Third, when the flow remains constant, friction loss varies inversely with the hose diameter. This means that at the same flow, the friction loss will decrease as the size of the hose diameter gets larger.

The fourth and final principle states that for any given velocity, the friction loss will be about the same regardless of the water pressure. This says that the speed of the water flow rather than the pressure is what determines friction loss.[3] Friction loss can be calculated as follows, where these principles are factored into the formula:

$$FL = Q^2 \times c \times L$$

where
  FL = friction loss of a hoseline

  Q = quantity of water in hundreds of gallons per minute, that is, the flow:

  Q = gpm/100

  c = friction loss coefficient for the diameter of the hose with 2½-inch (65-mm) hose being used as the standard (see **Table 11-2**)

  L = length of hose in hundreds of feet:

  L = hose length/100

TABLE 11-2 Friction Loss Coefficients

Hose Diameter, inches (mm)	Coefficient Value (*c*)
1½ (38)	24
1¾ (44)	15.5
2 (50)	8
2½ (65)	2
3 (76) with 2½-inch (65-mm) couplings	0.8
4 (101)	0.2
5 (127)	0.08
6 (152)	0.05

Note: Use caution because the hose diameter required may not be the same as the labeled hose diameter.

An example calculating the friction loss formula is given next, using a 2-inch hoseline that is 150 feet long with a flow of 200 gpm, **Figure 11-26.**

$$FL = Q^2 \times c \times L$$
$$= (200/100)^2 \times 8 \times (150/100)$$
$$= (2)^2 \times 8 \times 1.5$$
$$= (4) \times 8 \times 1.5$$
$$= 48 \text{ psi}$$

STREETSMART TIP

When discussing friction loss formulas and the values of the coefficient *c*, many firefighters will use different formulas. The formula FL = $Q^2 \times c \times L$ was once given as FL = $2Q^2 \times c \times L$ but with different values for *c*. The different values depended on whether the coefficient had already been multiplied by 2. Many older firefighters remember the formula as FL = ($2Q^2 = Q$) $\times c \times L$, which has changed over time as hose construction has reduced the friction created in the line.

Elevation can be a positive number if the nozzle is above the level of the pump or a negative number if below the pump. If even, elevation is not a factor. Elevation is equal to head pressure of 0.434 psi per foot (9.81 kPa/m) of height.

NOTE

Elevation is usually rounded to 0.5 psi per foot or 5 psi per floor or 10 feet (10 kPa/m) for practical fireground purposes.

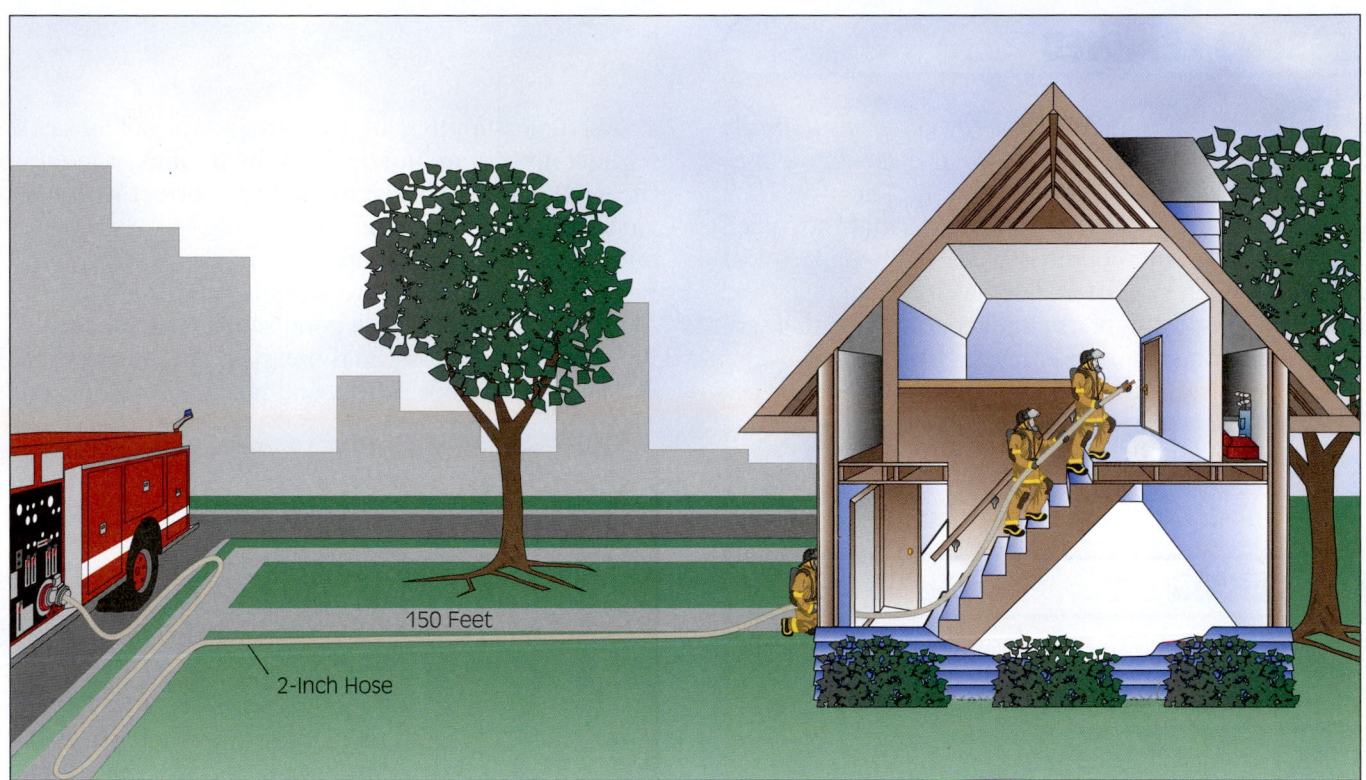

FIGURE 11-26   Example for friction loss and engine pressure calculations.

Appliance friction loss is the friction loss created by movement of water through the valves and turns of the appliances. Appliance friction loss varies with each device. Examples include wagon pipe or deluge set, 20 psi; ladder pipe, 50 psi; and 2½ to 2½ siamese or wye, 5 psi. Manufacturer's specifications or department SOPs should be consulted for guidance in determining these friction loss factors.

The next example uses the preceding example's 2-inch hoseline that is 150 feet long flowing at 200 gpm, with a fog nozzle and operating on the second floor. Using that friction loss equation, engine pressure (EP) would be:

$$EP = NP = FL \pm E + SA$$
$$= 100 + 48 + 5 + 0$$
$$= 153 \text{ psi}$$

STREETSMART TIP

Improper pump or nozzle operation is usually the cause of ineffective fire streams.

Adverse Conditions

Adverse conditions that affect fire streams are of two types: natural and man-made. The major natural factor affecting a fire stream is the wind and wind direction. Anyone who has tried to wash a car on a windy day knows the wind will break up a stream and deflect it from its target unless the flow is being aimed downwind. This also occurs with fire streams. Rain, snow, hail, and objects such as tree branches and wires are solid or semi-solid objects that deflect and break up hose streams. Gravity and air friction are also natural factors, especially when attempting to "lob" water a great distance. These natural causes cannot be removed. Moving the stream closer to its target or in a better position allow these effects to be reduced.

LESSONS LEARNED

Fire streams are made of water that leaves a nozzle and heads toward the target. The nozzle defines the characteristics of the fire stream. The two main types of nozzles are solid tip and fog, and they need to match the fire conditions and fire department resources. Each type of nozzle has its advantages and disadvantages; firefighters should use the one that will do the best job of suppressing the fire. Proper use of nozzles requires shutoffs and even some special types of nozzles. Fire streams and fire conditions determine how firefighters will attack a fire.

An understanding of fire streams is not possible without understanding the basic hydraulics of moving the water from a source to the fire. Correct hydraulic calculations require an understanding of the pressures and the amount of friction loss in the system. They result in an adequate fire stream being delivered in the proper amount with a good shape. Calculations of the flow, friction loss, and pressure are combined as required for the hoseline(s) and nozzle(s) being used.

KEY TERMS

Absolute Pressure   The measurement of pressure, including atmospheric pressure. Measured in pounds per square inch absolute.

Automatic or **Constant Pressure Nozzle**   Nozzle with a spring mechanism built in that reacts to pressure changes and adjusts the flow and resultant reach of the nozzle.

Available Flow   Amount of water that can be moved to extinguish the fire. Depends on the water supply, pump(s) and their capabilities, and the size and length of hose.

Bresnan Distributor.   A device used to fight fire in basements or cellars when firefighters cannot make a direct attack on the fire; has six or nine solid tips or broken stream openings designed to rotate in a circular spray pattern.

Cellar Nozzle   A device with four spray nozzles designed to rotate in a circular spray pattern for fighting fires in basements or cellars when firefighters cannot make a direct attack on the fire.

Combination Fire Attack   A blend of the direct and indirect fire attack methods, with firefighters applying water to both the fuel and the atmosphere of the room.

Combination Nozzle   A spray nozzle that is capable of providing straight stream and spray patterns, which are adjustable or variable by the operator. Most fog nozzles used today are combination nozzles.

Constant or **Set Volume Nozzle**   Nozzle with one set volume at a set pressure. For example, 60 gpm at 100 psi (227 L/min 690 kPa). The only adjustment is the pattern.

Direct Fire Attack   An attack on the fire made by aiming the flow of water directly at the material on fire.

Discharge Flow   Total amount of water flowing from the discharge side of the pump.

Fire Stream   The water or other agent as it leaves the hose and nozzle toward its objective, usually the fire.

Flow   The rate or quantity of water delivered, usually measured in gallons per minute or liters per minute (1 gpm = 3.785 L/min).

Fog Nozzle   Delivers either a fixed spray pattern or variable combination of straight stream and spray patterns.

Friction   Caused by the rubbing of materials against each other while in movement and converts or robs some of the movement energy into heat energy.

Friction Loss   Measurement of friction in a system such as a hoseline.

Gauge Pressure   Measures pressure without atmospheric pressure. Normally fire department gauges do not measure atmospheric pressure. Gauge pressure is measured in psi or psig.

Head Pressure   Measures the pressure of a column of water in feet (meters). Head pressure gain or loss results when water is being pumped above or below the level of the pump. A head of 2.31 feet (0.7 m) would equal 1 psi (6.895 kPa).

Hydraulics   The study of fluids at rest and in motion.

Indirect Fire Attack   An attack made on interior fires by applying a fog stream into a closed room or compartment, thus converting the water into steam to extinguish the fire.

Needed or **Required Flow**   Estimate of the amount of water required to extinguish a fire in a certain time period. Based on the type and amount of fuel burning.

Nozzle   A tapered or constricted tube used to increase the speed or change the direction of water or other fluids.

Nozzle Flow   The amount or volume of water that a nozzle will provide. Flow is measured in gallons per minute or liters per minute.

Nozzle Pressure   The pressure required to effectively operate a nozzle. Pressure is measured in pounds per square inch or kilopascals.

Nozzle Reach   The distance the water will travel after leaving the nozzle. Reach is a function of the pressure, which is converted to velocity or speed of the water leaving the nozzle.

Nozzle Reaction   The force of that makes the nozzle move in the opposite direction of the water flow. The nozzle operator must counteract the thrust exerted by the nozzle to maintain control.

Piercing Nozzle   Nozzle originally designed to penetrate the skin of aircraft and now has been modified to pierce through building walls and floors.

Pump Operator   A generic term to describe the person responsible for operating a fire apparatus pump. Other commonly used titles include motor pump operator, engineer, technician, chauffeur, and driver/operator.

Solid Stream Nozzles   Type of nozzle that delivers an unbroken or solid stream of water to the fire. Also called solid tip, straight bore, or smooth bore.

Straight Stream   A nozzle pattern that creates a hollow stream, similar in shape to the solid stream pattern, but the straight stream pattern must pass around the baffle of the nozzle. Newer fog nozzle designs, especially the automatic nozzles, only have this hollow effect from the tip on and, hence, create a solid stream with good reach and penetration abilities, some better than solid stream nozzles.

Stream Shape   The arrangement or configuration of the water or other agent droplets as they leave the nozzle.

Vacuum (Negative) Pressure   The measurement of the pressure less than atmospheric pressure, which is usually read in inches of mercury (in. Hg or mm Hg) on a compound gauge.

Variable, Adjustable, or Selectable Gallonage Nozzle   Nozzle that allows the nozzleperson to select the flow, with usually two or three choices, and the pattern.

Velocity Pressure   The forward pressure of water as it leaves an opening.

Water Curtain Nozzle   Designed to spray water to protect exposures against heat by wetting the exposure's surface.

REVIEW QUESTIONS

1. Define a fire stream.
2. Explain what a nozzle is, how it works, and how nozzle reaction affects its operation.
3. What is the amount or volume of water leaving the nozzle called?
4. Explain the difference between a solid tip nozzle pattern and a straight stream nozzle pattern.
5. A selectable or variable nozzle allows the operator to adjust which feature?
6. What is the purpose of a Bresnan distributor?
7. Why should the pattern of a water curtain be applied onto a solid object?
8. What is the most common type of shutoff device for a nozzle and how does it work?
9. Explain the difference between a direct and indirect attack.
10. What is the difference between atmospheric and gauge pressure?
11. Identify common nozzle pressures for solid tip and fog nozzles.
12. To reduce the friction loss in hose, what two things can be done?

ENDNOTES

1. A 1½-inch (38-mm) hoseline is the minimum size hoseline for structural firefighting.
2. Although the hoseline can be operated by one person, safety regulations require operation with a second team member.
3. Thomas Sturtevant, *Introduction to Fire Pump Operations,* 2nd ed. (Clifton Park, NY: Thomson Delmar Learning, 2005).

12 Protective Systems

One winter night in the heart of downtown Philadelphia an entire fire protection system failed; a trio of firefighters was lost, a thirty-eight-story skyscraper was destroyed, and the confidence of those dependent on fixed fire protection systems was shaken. It was the devastating No. 1 Meridian Plaza fire that changed my view on active and passive protective devices.

Upon arrival of the first-alarm companies, fire was venting from the twenty-second floor. At the onset it was a controllable fire. Then the fire protection systems began to quickly fail. Initially, it was the electrical systems. The primary and secondary wires shorted and placed the entire high-rise office building in total darkness. With the elevator stalled, access to the upper floors was limited to climbing the three fire towers. After firefighters hauled their equipment to the twentieth floor they discovered the pressure-reducing valves on the standpipe system were improperly set. Generating an effective fire streams system was impossible. Later, without an operable HVAC the core temperature in the building began to escalate. Pipe chases for the bathroom plumbing were never sealed. These "poke-holes" permitted the toxic smoke to spread in all directions. Smoke began to bank down, and the ventilation group became disoriented and lost and perished on a smoke-charged upper floor. With conditions rapidly deteriorating, a partially sprinklered structure, crews working with only hand lights, and practically no way to ventilate the building, the odds did not favor the men and women of the Philadelphia Fire Department.

After twelve alarms were struck, bringing more than 300 firefighters to the scene, and nineteen long hours, the blaze was contained to eight floors. In the end, ten sprinkler heads stopped the upward progress of the flames on the thirtieth floor. The lessons from No. 1 Meridian Plaza are plentiful, but one undisputed fact is that fire protection systems must be properly designed, installed, and maintained.

Recently, at the Canarsia Senior Citizen Center on Vandalla Avenue in New York City, a self-closing steel fire door remained ajar and a sprinkler system did not operate in a tenth-floor common hallway. When heavy winds pushed flames like a blowtorch from an upper floor apartment into the hallway, a search and rescue group was engulfed in the fireball. Tragically, again, three firefighters were killed because essential components of the fire protection system did not properly function.

Daily across this nation, emergency responders are witness to incidents that reinforce the need for knowledge of these basic systems. Somewhere in the nation this week a rate-of-rise alarm in a commercial building warned of an incipient fire, a sprinkler head activated in a factory and extinguished a minor fire before it spread to a manufacturing process, a closed fire-rated door in a hospital contained the smoke to a particular room before a massive evacuation was needed, a damper over a kitchen grill closed to prevent the involvement of a metal duct that crisscrossed vital structural supports in a restaurant, and the annunciator panel in an enclosed shopping mall pinpointed the exact location of a potential problem. Unfortunately, when these fire protection systems malfunction, the lives and safety of the occupants and emergency responders are at a much higher risk.

What is the price of success? In the fire rescue service, personnel safety and operational efficiency are surely two of the keys; often both hinge on fire protection systems. Firefighters often see firsthand the economic value of a properly maintained and tested fire protection system. I know both the economic and emotional value of these auxiliary appliances. I believe that an in-depth knowledge of sprinkler and standpipe systems will help to "save our own."

—Street Story by William Shouldis, Deputy Chief,
Philadelphia Fire Department, Philadelphia, Pennsylvania

INTRODUCTION

Protective systems help to guard lives and property by detecting and/or suppressing fires. Detection systems detect the presence of a fire and alert building occupants and/or the fire department, and may also activate a suppression system. Suppression systems are devices that help firefighters in controlling fires. Some systems are designed to automatically detect and suppress a fire, while others assist firefighters and occupants in extinguishing the fire. They are also called **auxiliary appliances.** Regardless of the term used, firefighters must know about these systems and how to use them. Due to the implementation and maintenance costs, great efforts have been expended to require these systems in building and fire prevention codes. Therefore, to the builders, owners, and occupants, the expectation is that the fire department will use them correctly.

NOTE

Remember that auxiliary appliance devices can make the job of firefighting safer and easier.

Detection systems are varied. Some require people to detect the fire and manually activate an alarm. Others are highly complex systems that can detect a fire almost at its ignition and sound an alarm. Some may also activate suppression systems.

The two main suppression systems are sprinklers and standpipes. These are devices that allow water to be used as an extinguishing agent. When the situation dictates that another agent may be more effective, other types of protective systems are required. Some use the agents found in portable fire extinguishers to cover room or larger areas. These systems include dry and wet chemicals, inerting gases, and halogenated agents.

Detection systems, suppression systems, and alarm systems (covered briefly in Section I, Chapter 3) can all be tied together to form a protective ensemble for building occupants. This chapter describes individual detection and suppression components as well as the interface of each. It also covers fire department operations using protective systems.

DETECTION SYSTEMS

Detection systems are designed to notify people of a fire or other problem. Simple detection systems can actually be viewed as a warning system requiring a person to recognize a danger. More complex systems use a series of devices to automatically detect a fire or event and initiate a warning alarm. Fire detection and warning can also come from the initiation of a suppression system. For example, a sprinkler activates, tripping a water flow alarm that notifies occupants. Regardless of how complex or simple, detection systems are designed to notify people that a potentially life-endangering event is happening. By understanding the basic principles of operation of various detection systems, firefighters more effectively utilize these systems to perform their job. However, until firefighters are on scene to verify the type and extent of emergency, they should remain cautious and take all actions to prepare and safeguard themselves. For firefighting, all detection system alarms should be considered to house a hostile atmosphere until proved otherwise by the fire companies.

SPRINKLER SYSTEMS

Sprinkler systems are designed to automatically distribute water through sprinklers placed at set intervals on a system of piping, usually in the ceiling area, to extinguish or control the spread of fires. Most sprinkler heads detect the heat of a fire and begin to apply water directly over the source of the heat. Sprinkler heads, unless deluge-type heads, are heat-sensitive devices that react to a fixed temperature and disperse water in a specific pattern and quantity over a set area. Sprinkler systems are highly effective. In fact, some people describe the benefit of sprinklers as similar to having a firefighter constantly on duty in the protected building. The NFPA publishes standards for installation of sprinkler systems and their inspection and maintenance.[1]

SPRINKLERS AND LIFE SAFETY

Sprinkler systems were originally designed in the late 1800s. Their intent was to protect property, especially businesses and factories, from total loss from fires. They are almost 100 percent effective. Improper action by humans is usually the catalyst when sprinkler systems malfunction. Improper maintenance or inadequate water supply due to human error are two common mistakes leading to sprinkler failure. In the early 1900s, the idea that sprinklers might be able to save lives was beginning to take shape.

America's fire loss statistics show that most structure fires, fire damage, and, of greatest importance, fire injuries and fatalities occur in residential properties.

Yet the American home is one of the least protected properties when it comes to fire safety. Homes in the United States are usually built of wood and other combustible materials. Until the mid-1970s, most American homes did not have smoke detectors and only a few non-high-rise residential units had sprinklers. Recognizing the potential benefit of these types of systems, the U.S. Fire Administration, National Fire Protection Association, International Association of Fire Chiefs, and Factory Mutual, among many others, led the research efforts to develop faster responding and less expensive residential sprinkler systems. The first standard on residential sprinkler systems was developed in 1975.[2] Some cities and counties now have mandatory sprinkler requirements in all newly constructed residential structures, and many states require them in multifamily dwellings. There is still, however, a long way to go to place them in all new residential buildings. The American heritage of protecting personal property rights often allows citizens the right to die in their own homes, the least protected and inspected of any occupancy classes.

Residential sprinklers are designed for **life safety** and not necessarily to protect property. A residential sprinkler system may completely extinguish a fire in a protected area as unprotected areas are allowed in the design of such systems. Fires that may occur in these unprotected areas can gain great headway or move to other unprotected areas, and the system may not be able to prevent a total loss. Fires have occurred in apartment buildings, motels, and hotels protected by a life safety sprinkler system, yet have resulted in a total loss of the property. When analyzing these fires and rating the effectiveness of the protection system, the presence of occupant or firefighter fatalities or injuries will determine the effectiveness of the system. Absence of these fatalities or injuries validates that the system did what it was designed to do.

Sprinkler systems are also not totally effective in life safety in some types of fires. These include slow burning and smothering fires that produce fatal quantities of smoke without generating sufficient heat to activate the system, flash fires that rapidly engulf a person, or a fire in the immediate vicinity of the victim where the fire inflicts the injuries prior to system activation. Victims who are disabled or unable to remove themselves from the fire area also factor into fire system inadequacy. Protecting against all situations is impossible, but residential sprinkler systems combined with effective smoke detection systems and coupled with aggressive firefighting tactics and strategies allow maximum life safety in the home.

TYPES OF SPRINKLER SYSTEMS

Many types of occupancies and businesses exist, each with specific attributes when it comes to fire suppression. Some sprinkler systems protect homes or apartment buildings. Others protect businesses with highly sensitive electronic equipment. Still others must protect homes and businesses in areas subject to harsh winter conditions. As a result, many types of specialty sprinkler systems have been engineered to perform fire suppression in different circumstances.

Specialty Sprinkler Systems

Specialty sprinkler systems include some combination-type sprinkler systems and systems that cannot meet the standards for some reasons. They may have an inadequate water source or supply or may be a partial or outside system. Even if a system does not completely meet a standard, it provides a higher measure of protection than if no protection were available. Fire departments with in-district properties using specialty sprinkler systems should be familiar with their intended protection strategy and any of their limitations.

Wet Pipe Systems

A **wet pipe sprinkler system** has automatic sprinklers attached to pipes with water under constant pressure, **Figure 12-1.** This allows quick response when the head is opened. The wet pipe system is the simplest sprinkler system in design and operation. The main or alarm valve is a one-way check or clapper valve that prevents water from reentering the water supply and, when closed, shuts off the water flow to the alarm line. Both sides of the alarm valve have pressure gauges that register the water pressure of the supply and the system. The system side gauge should read a slightly higher pressure as the reclosing clapper valve will trap any pressure surges. The alarm line piping usually has a **retard chamber** that acts to prevent false alarms from a sudden pressure surge in the water supply, **Figure 12-2.** The chamber collects a small volume of water before allowing a continued flow to the alarm device. The water from a surge is drained from a small hole in the bottom of the collection chamber. A waterflow indicator, a vane or paddle in the waterway, detects the water flow and activates an alarm signaling system.

A wet pipe system has three more valves. The first is the control valve, which is used to shut off the supply of water to the system and is usually an **outside stem and yoke (OS&Y) valve, Figure 12-3.** The second valve is the main drain, which allows the system to be drained for maintenance or to be restored from a fire, and the last valve is an inspector test valve. The test valve is at the farthest end of the system and is used to simulate the flow of a single head and to measure the response time of the system.

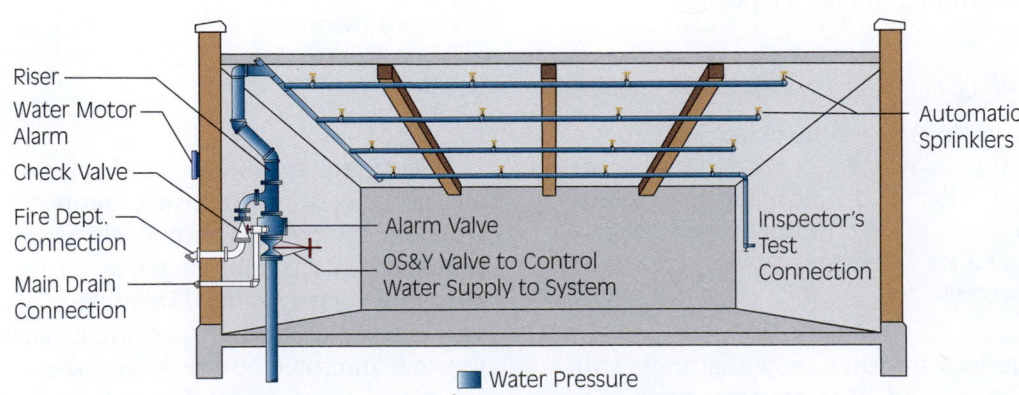

Riser
Water Motor Alarm
Check Valve
Fire Dept. Connection
Main Drain Connection

Alarm Valve
OS&Y Valve to Control Water Supply to System

Automatic Sprinklers

Inspector's Test Connection

Water Pressure

FIGURE 12-1 Wet pipe sprinkler system.

The operation of a wet pipe system starts with the fusing or bursting of a sprinkler head, which causes it to begin applying water to the fire. As the water pressure in the system begins to fall, the main check valve opens and water flows into the system and the alarm line, filling the retard chamber, and then activating the automatic alarm and water motor gong. The alarm signal may be used to notify the fire department or an alarm company. *After ensuring the fire is out or completely under control,* fire personnel will close the control valve. Sprinkler maintenance personnel replace the head and restore the system. When the system is shut down, a firefighter with a radio should be posted at the control valve and be ready to reopen the valve.

FIGURE 12-2   Retard chamber.

> **NOTE**
>
> Some wet pipe systems are located in or have parts of the system located in areas subject to freezing temperatures. To protect these systems, an antifreeze solution can be added to the water in the system. These systems require special attention to restore and maintain.

> **STREETSMART TIP**
>
> Building personnel are often overly afraid of water damage to property, which is one of the reasons that sprinklers have not been 100 percent effective at extinguishing fires. These personnel, or sometimes firefighting personnel, shut down the control valve prior to complete fire extinguishment. The fire is allowed to rekindle and causes additional loss. Most water-damaged materials can be salvaged but fire-damaged materials are usually destroyed. Firefighters should not close down any sprinkler system until ensuring the fire is extinguished or completely under control. They should also keep hoselines in place until a fire is declared extinguished.

Dry Pipe Systems

In **dry pipe systems,** air under pressure replaces water in the system to protect against freezing temperatures. The system uses a dry pipe valve to maintain pressurized air above and supply water under pressure below the valve, **Figure 12-4** and **Figure 12-5.** A small amount of water at the seat of the valve, called priming water, maintains the seal at the valve and is filled to the priming level. The clapper valve has a locking mechanism that keeps the clapper open until it is manually reset to prevent **water columning.**

Dry pipe valves use an air differential system with a small amount of air pressure over the larger head surface of the clapper valve, which keeps back the

FIGURE 12-3   OS&Y valves chained in the opened position.

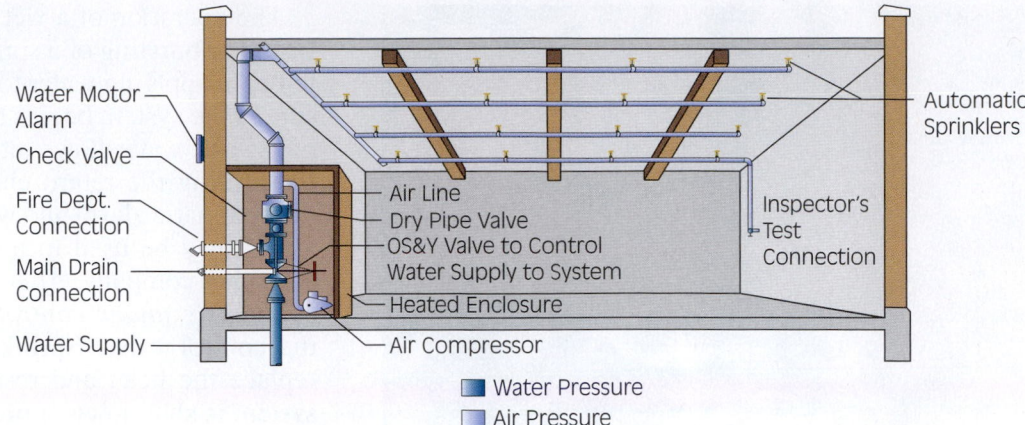

FIGURE 12-4 Dry pipe system schematic.

Water Motor Alarm
Check Valve
Fire Dept. Connection
Main Drain Connection
Water Supply

Air Line
Dry Pipe Valve
OS&Y Valve to Control Water Supply to System
Heated Enclosure
Air Compressor

Automatic Sprinklers
Inspector's Test Connection

■ Water Pressure
□ Air Pressure

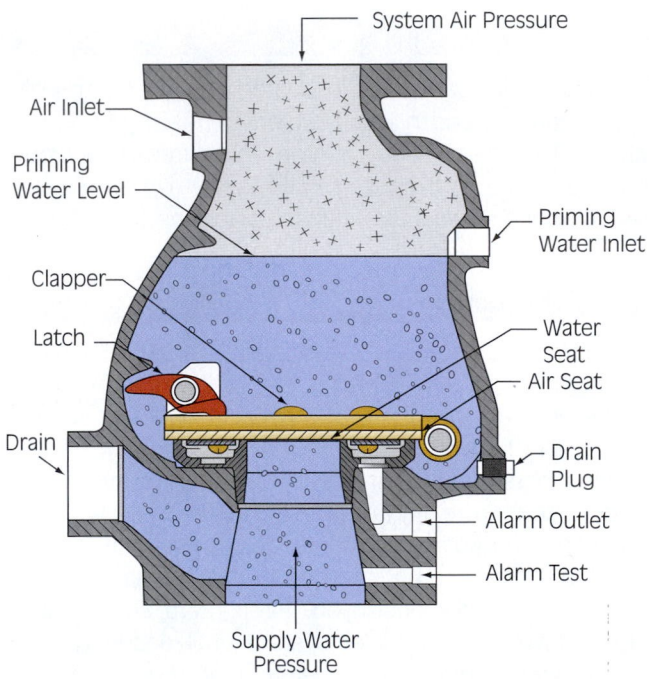

FIGURE 12-5 Dry pipe valve holds back the water with air pressure.

System Air Pressure
Air Inlet
Priming Water Level
Clapper
Latch
Drain
Supply Water Pressure
Priming Water Inlet
Water Seat
Air Seat
Drain Plug
Alarm Outlet
Alarm Test

higher water pressure exerted on the smaller water side of the clapper valve. If water were allowed to fill the riser above the clapper, the water column's weight would never allow the clapper to be forced open and would make the system inoperative. When a sprinkler head is fused by heat, the air is discharged. As the air pressure drops below the pressure of the supply water, the clapper valve locks open. Since air, as opposed to water, is in the system, dry pipe systems are slightly slower to flow water than wet pipe systems. Most systems have either an **exhauster** or an **accelerator** to speed the operation of the dry pipe valve. The exhauster detects the decrease in air pressure and helps bleed off air. The accelerator detects

the decrease in air pressure and pipes air pressure below the clapper valve, speeding its opening. Drain and alarm valves in a dry pipe system are similar to wet pipe systems. Dry pipe systems are normally used in unheated buildings, in buildings that refrigerate or freeze materials, or in outdoor applications where freezing temperatures occur—these systems require a heated valve room.

The dry pipe system is more complex in design than a wet pipe system and also harder to return to service after activation. This system requires the dry pipe valve cover to be opened after draining the system, and the lock on the valve must be reset. The valve must be primed, air pressure must charge the lines, and the control valve must be opened carefully to prevent retripping of the system or creation of a water column.

Deluge Systems

Deluge systems are designed to protect areas that may have a fast-spreading fire that could engulf an entire area. Petroleum-handling facilities, aircraft hangars, some manufacturing facilities, and hazardous materials storage areas are all examples of occupancies that may have a deluge system. The essential difference between a standard wet system and a deluge system is the individual sprinkler head. Deluge systems utilize open heads, without any fusible elements. When a fire is detected, the deluge system delivers water (or foam solution) to all heads in the area—allowing total coverage of the area.

Operationally, the deluge system must interface with a detection system. Once the detection system activates, it sends a signal to a "deluge valve," which opens, sending water or foam solution to all the open heads, **Figure 12-6.** Most municipal water systems lack the pressure to effectively supply the numerous open heads in the deluge system. For this reason, del-

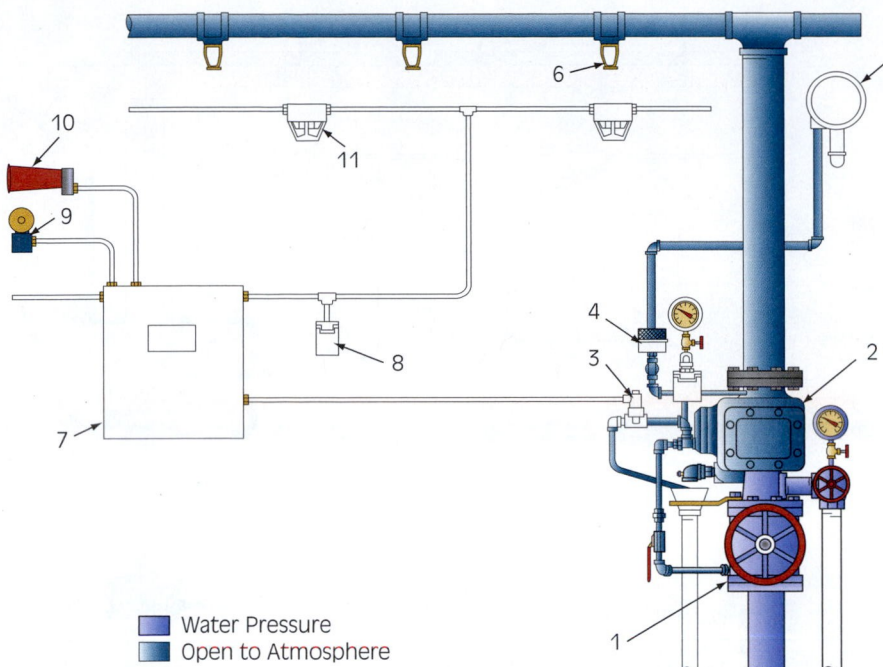

FIGURE 12-6 Deluge system schematic: 1, OS&Y valve; 2, deluge valve with basic trim; 3, solenoid valve and electric actuation trim; 4, pressure alarm switch; 5, water motor alarm; 6, spray nozzles or open sprinklers; 7, deluge releasing panel; 8, electric manual control stations; 9, fire alarm bell; 10, trouble horn; 11, heat detectors.

Water Pressure
Open to Atmosphere

uge systems also incorporate a fire pump to boost pressure in the system. Likewise, deluge systems that deliver foam solution require a foam supply tank, foam pump, and foam mixing device.

As can be imagined, deluge system activation will cause tremendous quantities of water to flow. While this is essential to fire control of high hazards, activation of the system in the absence of a fire will likely cause water damage and a significant cleanup effort. For this reason, deluge system activation usually requires activation of several detectors versus a single detector. For example, several flame detectors must activate before a signal is sent to the deluge valve. Some systems require detection from different kinds of devices (i.e., activation of a flame detector **and** a heat detector). To prevent unneeded damage from activation, the deluge system may utilize a manual override alarm and "deadman" switch. In these systems, an alarm is sent telling occupants that the deluge system is ready to activate. If someone does not push a button in a prescribed time, the system will deluge the area. A person who does push the hold button cannot release the button until the activated detectors are reset—which requires additional persons in the system. If the person does release the button prior to system reset, the deluge valve will open and the system will be fully activated—hence the deadman label.

Preaction Systems

Preaction systems are similar to the dry pipe and deluge system. The system has closed piping and heads with air under no or little pressure, but the water does not flow until signaled from a separate fire detection system, **Figure 12-7.** The preaction valve then opens and allows water to flow through the system to the closed heads. When an individual head is heat activated, it opens and water attacks the fire. Preaction systems are used in areas where the materials protected are of high value and water damage would be expensive. Computer rooms, museums, or buildings storing historical items are examples of where a preaction system would be installed.

Residential Systems

Residential sprinkler systems are smaller and more affordable versions of wet or dry pipe sprinkler systems, **Figure 12-8.** They are designed to control the level of fire involvement while residents escape. The sprinkler system water supply is combined with the domestic water supply, and flow rates are designed for one or a few heads in operation. Residential sprinkler heads have a faster response time and use a lighter and smaller piping than commercial wet/dry pipe systems. Residential systems were the first suppression

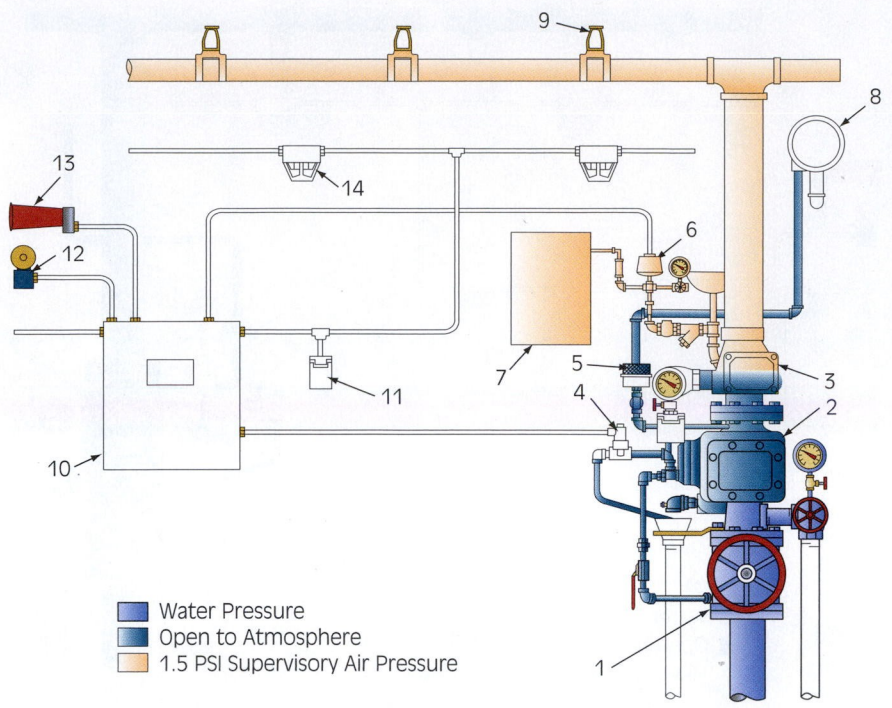

FIGURE 12-7 Preaction system schematic: 1, OS&Y valve; 2, deluge valve with basic trim; 3, check valve; 4, solenoid valve and electric actuation trim; 5, water pressure alarm switch; 6, 1.5-psi low air pressure alarm switch; 7, 1.5-psi supervisory air pressure control; 8, water motor alarm; 9, automatic sprinklers; 10, deluge releasing panel; 11, electric manual control stations; 12, fire alarm bell; 13, trouble horn; 14, heat detectors.

■ Water Pressure
■ Open to Atmosphere
■ 1.5 PSI Supervisory Air Pressure

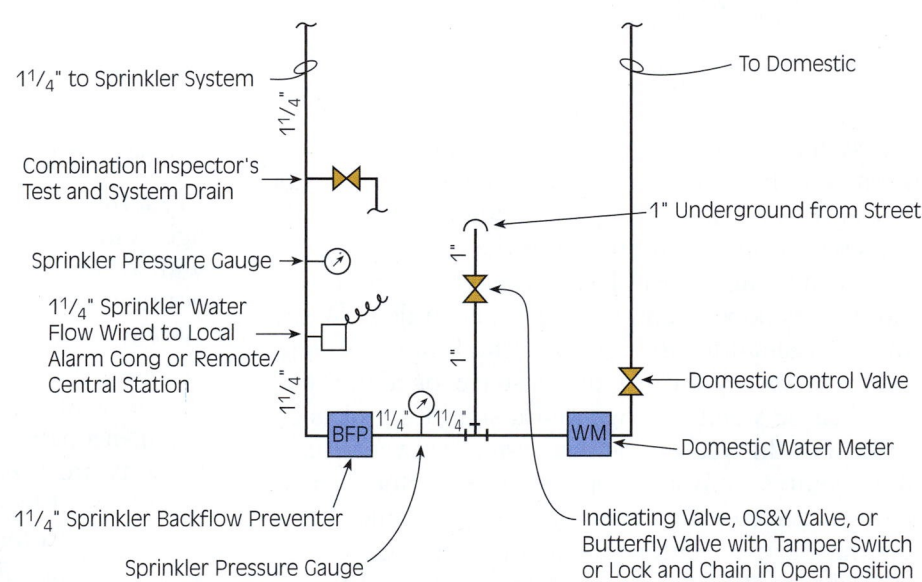

FIGURE 12-8 Schematic of residential sprinkler system.

SPRINKLER SYSTEM CONNECTIONS AND PIPING

The connections and piping comprise most of the components of the sprinkler system and provide the water from its source to the heads The main water supply can come from public or private water. Depending on the requirements of the system and the water source, a fire pump may be included. A secondary water source is supplied via a fire department connection, which allows pumpers to supplement the water supply, **Figure 12-9.**

systems to use plastic piping. Residential systems use a check valve, waterflow alarms, and drains similar to the bigger systems and are not required to have a fire department connection, although some do. Some residential systems also use antifreeze to protect all or part of the system.

> **NOTE**
>
> When pre-planning a building, serious consideration should be given to the source of the fire department supply water. If a water supply robs water flow from the sprinkler system, it is not a good choice and alternates should be considered.

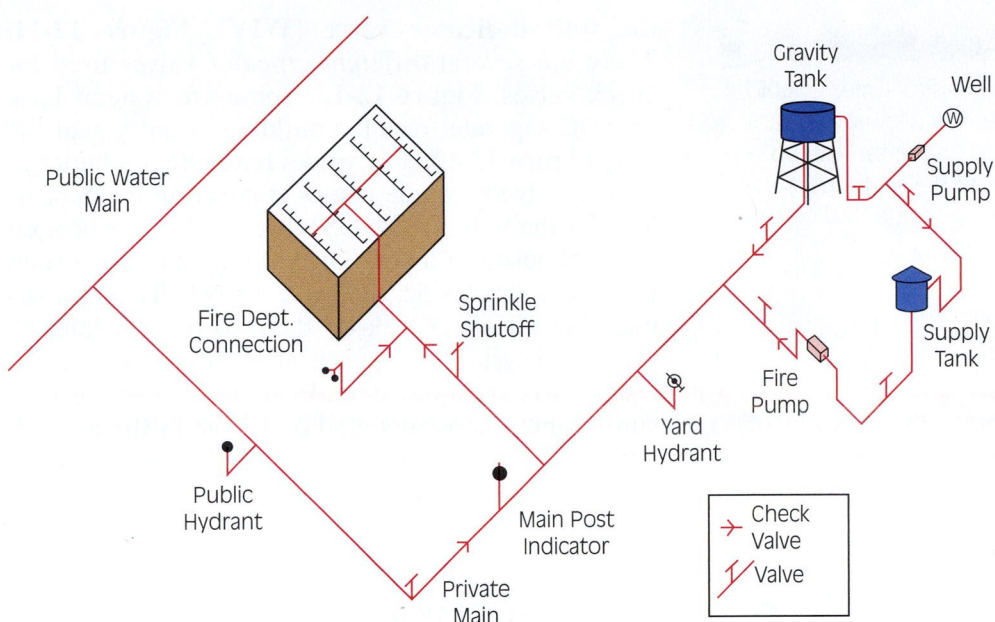

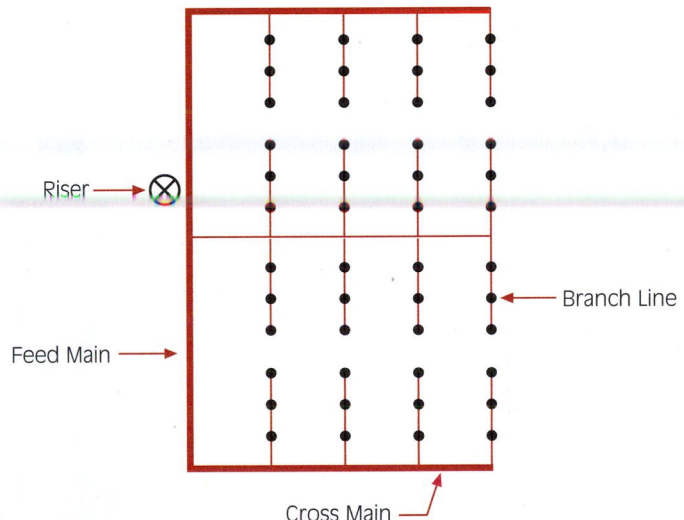

FIGURE 12-9 Various water supply systems to a sprinkler system.

It is best to use a different water main or supply source than the one being used by the sprinkler system. The fire department connection is usually piped in past the main control valve and can supply the system even if the main control valve is closed.

Another required valve in most systems is a backflow preventer, a set of one-way check valves designed to keep water in a sprinkler system from reentering the public water supply. These valves are an environmental requirement, and most codes require and mandate them even on old fire protection systems, as well as landscaping sprinklers and the like. The system may have other control valves and check valves and will have a water control valve if connected to a public water supply.

Next in line is the main control valve, which is of the wet, dry, deluge, or preaction type as described earlier. Above that is the riser. The riser pipe feeds the mains that supply the cross mains to the branch lines, **Figure 12-10.** The branch lines have the sprinkler heads attached to them. Some larger systems may have sectional valves that divide the system into floors or other sub-areas of the system. The location of these valves can be important when shutting down the system for maintenance. They should be kept locked and supervised to prevent tampering. Tamper alarms alert the alarm monitoring company whenever someone operates the valve, and unauthorized valve operations are quickly checked to prevent an arsonist from disabling the system prior to setting a fire.

Firefighters should be able to properly connect a supply line to a fire department connection for either a sprinkler or standpipe. This requires the correct amount of hose to reach from the pumper's discharge

FIGURE 12-10 Sprinkler piping diagram.

gate to the connection, a spanner wrench to tighten the couplings, and any adapters needed to complete the connection. The firefighter should check the connection for damage, remove the cap cover, and check inside the siamese or stortz connection for debris or damage, and should check the operation of the clapper valve (if present) and an O-ring gasket. If no clapper valve is provided, all necessary connections must be made prior to charging the system. When connecting to a siamese, the outlet on the far left should be chosen first, because this will allow better access for using the spanner wrench to tighten the coupling. After completing the connection, the pump operator should be advised that the line is ready for charging.

CONTROL DEVICES FOR SPRINKLER SYSTEMS

The three main control devices for sprinkler and standpipe systems are the outside stem and yoke valve (OS&Y), the **post indicator valve (PIV),** and the **wall indicator valve (WIV), Figure 12-11.** There are several different types of valves used for check valves, **Figure 12-12.** Some fire systems have an FDC separate from the building, usually near the PIV, **Figure 12-13.** The valves have either a butterfly or gate valve type: the names come from the appearance of the valves. The OS&Y valve has a wheel on a stem housed in a yoke or housing. When the stem is exposed or outside, the valve is open. These valves must have a chain lock on them to prevent tampering. If the system is supervised, a tamper switch will also be found on the stem. PIVs and WIVs are very similar, but one is mounted on a post in the ground, and the other is mounted on a wall. Both valves are housed in a metal case with a small window, reading either "OPEN" or "SHUT." A wrench or a wheel

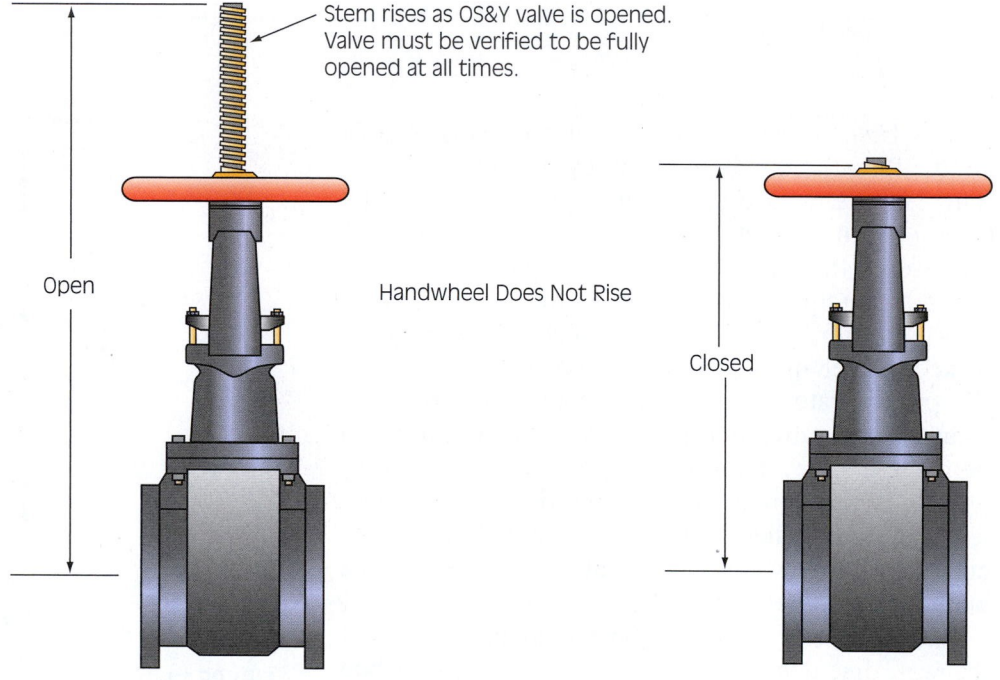

Stem rises as OS&Y valve is opened. Valve must be verified to be fully opened at all times.

Open

Handwheel Does Not Rise

Closed

OS&Y Valve Visual Indication

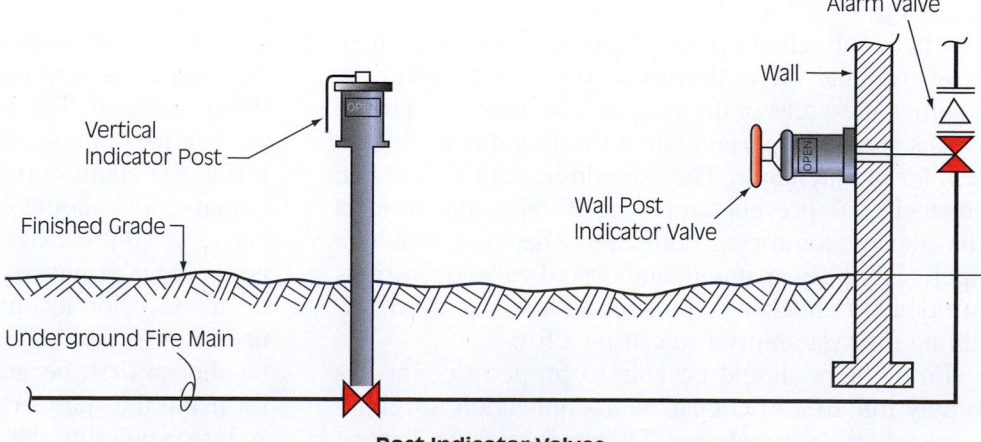

Alarm Valve

Wall

Vertical Indicator Post

Wall Post Indicator Valve

Finished Grade

Underground Fire Main

Post Indicator Valves Visual Indication

FIGURE 12-11 OS&Y valves and post indicator valves.

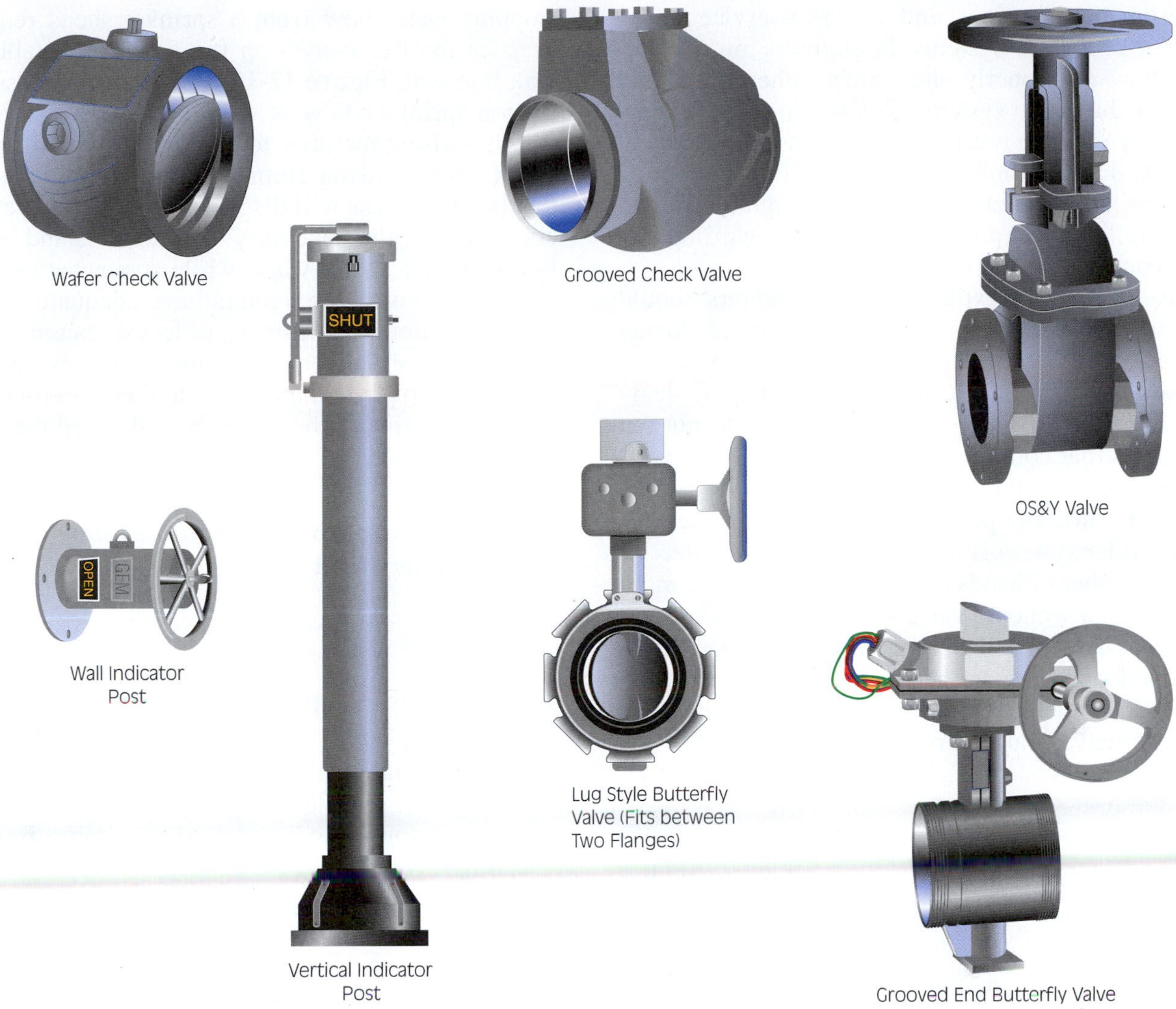

Wafer Check Valve

Grooved Check Valve

SHUT

OS&Y Valve

OPEN GEM

Wall Indicator
Post

Lug Style Butterfly
Valve (Fits between
Two Flanges)

Vertical Indicator
Post

Grooved End Butterfly Valve

FIGURE 12-12   Fire protection valves.

FIGURE 12-13   Fire department connection and post indicator valve. The box on the side of the PIV is a tamper alarm switch box.

controls these valves; a padlock and chain are used to lock them in the open position. Some water control valves—such as public water system valves, for instance—are typically of the nonindicating type.

RETURNING SPRINKLER SYSTEMS TO SERVICE

Fire departments used to regularly test and/or restore sprinkler and other protective systems to service for building owners. Many engine and ladder companies carried a sprinkler kit with wrenches, extra heads, stops, and other equipment to do this. Today most departments no longer provide this service. For liability purposes, departments either stop the flowing head or shut down the system, letting the owner contact a fire protection service company for system restoration. Fire departments should have local SOPs to address the response, operations, and return-to-service guidelines for protective systems. Firefighters must understand how to properly shut down either individual heads or the entire system. This will mitigate excessive water damage until the heads can be completely restored. This portion of the text will cover returning a sprinkler system to service: firefighters should always follow local protocol and SOPs regarding any fire scene activity and ensure compliance.

Once the fire is extinguished, firefighters should leave at least one charged hoseline in place during overhaul. With possible reignition in mind, hoselines may be pulled from the scene and broken down—leaving the hoselines that would be necessary to put out small fires that could rekindle. When the fire is verified completely out, the first step to restore the system is to shut down any pumper supplying the system. If the sprinkler system is to be restored immediately or a large number of heads have opened, either the main sprinkler or a sectional valve should be shut down and drained. The sprinkler head is replaced with an identical one from the spare heads that are required to be kept in a cabinet in the sprinkler control room. The sprinkler valves must be reset or reopened.

ETHICS

Due to increased liability, most fire departments do not test or restore sprinkler systems. These tasks are now left to commercial sprinkler system companies. Most fire departments will stop the flow of water or shut down the system. Obviously, this would also involve the closure of business operations or establishing a fire watch until the system can be restored.

The simplest—and often quickest—way to stop water flow from an individual sprinkler head is to insert a stop. Sprinkler stops may include manufactured sprinkler tongs or improvised wedges, dowels, or clamps, **Figure 12-14.** While it may seem easy,

FIGURE 12-14 Sprinkler tongs and wedges.

stopping water flow from a sprinkler head requires practice to effectively stop the flow and establish a leak-free seal, **Figure 12-15.** The firefighter assigned to stop sprinkler flow at the head will get seriously wet. It is also imperative to be familiar with the physics of water flowing from the system. It is possible that the water flow will displace oxygen, and firefighters must take the necessary steps of PPE and SCBA to remain safe. SCBA use, while somewhat cumbersome, will ensure firefighters have adequate air and face protection. It is important to use caution when climbing a wet ladder and to make sure eye protection is used to prevent injury. Following a step-by-step plan will minimize the exposure and speed the water stoppage.

JPR 12-1: Using Stops to Stem the Flow of a Sprinkler Head

(For step-by-step photos of this skill sequence, see page 335)

1. Empty your PPE pockets of radios or other items that can be damaged from getting wet.
2. Select the right stop for the head discharging water. Remember, simple wedges may not fit into the recessed sprinkler heads found in some finished ceilings.
3. Choose a stable ladder, desk, or other platform to reach the head.
4. Put your face shield down or don eye protection. Climb the platform and secure the wedges in place.
5. Wedges should be placed as shown.

The other way to shut off the water flow from opened sprinklers is to shut down either the main sprinkler valve or a floor or sectional valve. This task will require a key or set of bolt cutters because the main valves are locked, as discussed earlier. If the main valve is an OS&Y valve, it is turned until the stem is no longer exposed. If it is a PIV or WIV valve, it is turned until it reads "SHUT." Personnel with a portable radio should stand by the valves until the system is recharged or until the fire is determined and verified as extinguished. If the entire system is shut down or if there is a sectional drain, that valve can be opened. After draining the system, this valve should be closed. Once the system is shut down and drained, the heads may be replaced. Replacement of the heads is recommended as soon as possible because in the event of an additional fire, even with the system shut down, the fire department can pump through the fire department connection or turn the system back on and have it operational for firefighting. Refilling a system should be done slowly to prevent damaging the system.

FIGURE 12-15   Sprinkler tongs and wood wedges stopping sprinkler flow.

To avoid liability of restoring sprinkler systems, some fire departments *do not* replace heads and restore the system. In these cases, the fire department should require that a **fire watch** be established until the system has been restored by a qualified technician. Some occupancies such as hotels, public assemblies, and high-rises *require* a sprinkler system for life safety. In these cases, the fire department should not allow the building to be occupied without an approved fire watch system or sprinkler restoration.

STANDPIPE CLASSIFICATIONS

Standpipe systems are designed to allow firefighters to fight fires in larger buildings by prepiping water supply lines through the building. Some systems allow both occupants and firefighters to use the system. The systems for building occupant use have hoselines and nozzles attached. Standpipe systems are not automatic; they must be operated by people. They are used in high-rise buildings; large commercial, retail, and industrial buildings; places of public assembly; and other areas where advancing hoselines from hydrants or fire apparatus would be difficult due to the size of the building. Tunnel or subway systems and shopping malls have horizontal standpipe systems

Standpipe systems are classified according to the intended user and each class sets requirements for volume and size of outlets. *Class I systems* are designed for use by the fire department or trained personnel such as a fire brigade, **Figure 12-16.** The systems use 2½-inch (65-mm) outlets. A single standpipe should flow a minimum of 500 gpm (1,890 L/min) plus an additional 250 gpm (945 L/min) for each additional riser. No hose is provided for this class. Outlets are found in stairwells on each floor, near the exits, or in the hallways on each floor near the stairs.

A *Class II system* is designed for use by untrained building occupants and has a 1½-inch (38-mm) outlet with a minimum flow of 100 gpm (378 L/min), **Figure 12-17.** Hose is provided and is either 1 or 1½ inch (25 or 38 mm). Firefighters wishing to use a Class II system should not use the hose provided, because it is not tested and may be single-jacketed, unlined hose

FIGURE 12-16   Class I standpipe system.

FIGURE 12-17   Class II standpipe system.

with a nozzle that has no shutoff. Firefighters using this type system should replace the hoseline with fire service hose.

A *Class III system* is used by either of the users of the other two classes but meets the Class I requirements for flow. It has both 1½-inch (38-mm) and 2½-inch (65-mm) outlets or may have a 2½-inch (65-mm) outlet with a 2½- to 1½-inch (65- to 38-mm) reducer. All of the flows given are at the top outlet with a residual pressure of 65 psi (448 kPa).[3] Any of these systems may have a **pressure-regulating device,** which is designed to maintain or reduce the pressure to a set amount—regardless of the height of the building. Standpipes have height limits on each section (about 280 feet [88 m]), but the pressure difference in an unregulated system could vary by almost 140 psi (966 kPa) from top to bottom.

The types of standpipe systems are differentiated based on water supply. An automatic wet pipe system has a water supply that is ready to conduct firefighting, whereas automatic dry and semi-automatic dry pipe systems use dry pipe until a hose station (an outlet) is opened to allow water to flow into the system. The semi-automatic system requires an activation valve to be opened manually. The manual dry pipe is dry and relies on the fire department connection for water supply, whereas a manual wet pipe system has the piping filled with water but needs the fire department connection for firefighting supply.[4]

STANDPIPE SYSTEM CONNECTIONS AND PIPING

Standpipe systems, which allow for the manual application of water in large buildings, can range from very simple to highly complex with multiple connections, risers, and sources of water supply, and many outlets on each floor or area. A manual dry system can be a single riser having a fire department connection and a drain at the bottom, and an outlet on each floor that is capped at the top. The wet manual type may just add a small water line to maintain water in the piping. The other systems become more complex as a water supply is added and are found in larger buildings with more fire flow requirements. Some standpipe systems are combined with the building sprinkler system and are designed with outlets coming off a branch line.

The components of a standpipe system are the piping, outlets with hose and other attachments, the valves, the fire department connection, and any monitoring devices. The piping includes the riser or risers and any attachments needed to plumb between the water sources and the uppermost or furthest outlet. The piping must be capable of discharging the required volume and pressure with minimal friction loss. The outlets are usually placed in the building stairwell or may be in a wall-mounted cabinet. Any required pressure-regulating devices would be attached to the outlet. A Class II or III system would have a hoseline, and a nozzle would be attached. The hose connection usually has a wheel-type handle to operate the valve and a cap covering the threads. The nozzles may be a combination fog–straight stream or a solid tip nozzle, and quite often have no shutoff, which could confuse the untrained user.

Standpipe valves are similar to the valves used on sprinkler systems, including various types of gate and check valves, with the addition of hose valves at the outlets. If a water supply is provided, an indicating gate valve is required as well as check valves or backflow preventers. A check valve is also added with a drain on the fire department connection line to allow drainage and prevent freezing. Gate valves should also be installed on each section or separate riser to allow

maintenance work to be performed without shutting down the entire system. Drain valves are installed on each section. The fire department connection(s) is an inlet or siamese device with protective caps and a raised letter sign marked "Standpipe." If there are multiple connections that are not interconnected, the sign should state the zone covered. Monitoring systems are covered in the next section.

ALARMS FOR STANDPIPES AND SPRINKLERS

Alarm and monitoring systems are found in most sprinkler and many standpipe systems, especially Class II and III systems. Sprinkler systems are both fire detection and suppression devices and are designed to notify at least the building occupants. Occupant-used standpipe systems should be monitored, as occupants might not notify 9-1-1 prior to fighting the fire. Most protective systems require monitoring to prevent tampering or deliberate attempts to disable the system. These may be electronically or mechanically activated by the movement of a gate valve. Waterflow alarms are also electrical or mechanical. The most familiar mechanical one is the water motor gong, which operates like a waterwheel with the flow of the water causing it to turn and strike a gong, **Figure 12-18.**

FIGURE 12-18   Water motor gong.

The water motor gong, unless connected to another alarm, is a local alarm only. Other waterflow devices are hooked to an electronic sensor that is connected to a paddle or vane inside the riser. A monitoring alarm company notifies the fire department about fires, but its own personnel respond to tamper alarms.

OTHER PROTECTIVE SYSTEMS

Many other types of protective systems are used today. Some are rather simple, as in those designed to protect the grill, fryer, and ductwork of a local restaurant. Others are extremely complex and are designed to prevent or suppress an explosion in highly hazardous locations. Firefighters need to understand the most common types of protective systems: fire departments responding to the more complex systems should ensure that all firefighters are familiar with the operation of the system, the hazards of the fires they control, and extinguishing agents.

Local Application and Hood Systems

One of the most common types of protective systems is a **local application system,** especially the systems protecting the cooking areas of restaurants, **Figure 12-19.** Local application means that the system is designed to protect only a certain portion of the building, usually directly where the hazard will occur or spread. Local applications are also used in laboratory hoods, paint booths, and other small hazardous locations, **Figure 12-20.** Most of the local application systems use dry or wet chemical agents with the most popular being an ABC dry chemical. Higher temperature cooking units and oils have created the need for a new Class K classification, and these systems also use dry or wet chemicals. Hood systems use a heat-sensitive device or a manual switch for activation.

Total Flooding Systems

Total flooding systems are used to protect an entire area, room, or building. The total flooding system discharges an extinguishing agent that completely fills or

FIGURE 12-19   Local application system (notice nozzles) protecting cooking area and ductwork.

floods the area with the extinguishing agent to smother or cool the fire or break the chain reaction. Total flooding systems can use carbon dioxide or other inert gases, halogenated or clean agents, dry chemicals, or foam as extinguishing agents (see Section I, Chapter 8 for a discussion of these agents). They are effective as long as the proper amount discharges, the area is contained to prevent loss of agent, or the fire goes out prior to ventilation of the agent. There are hazards associated with each type of application with which firefighters and building personnel should be familiar.

tem activation. Firefighters, using SCBA, may enter the area to ensure the fire is extinguished and all persons have been safely removed. This agent is very clean and is often used in computer and electronics areas. If the fire is out, ventilation of the area and restoration of the system are the only cleanup activities required.

CAUTION

Halon and the newer clean agents are also used for total flooding applications. Depending on the toxicity and concentration of agent needed, some of these systems may even be used when people are still occupying the area. Firefighters should still use SCBA when entering areas where even safe agents have been discharged to ensure that the fire is out. As with CO_2, electronics and other high-cost or sensitive valuables are protected by these systems.

NOTE

The use of halon is being phased out. To protect the ozone layer, the Montreal Protocol treaty, enacted in 1989, bans production of substances believed to be harmful to the Earth's ozone layer. Halon falls into this category of substances, and no further production of halon is permitted. However, there is halon in storage and its use continues. There are some applications where halon is considered the best and most effective extinguishing agent, such as the cargo bins of airliners. For this use, there is a "critical use exemption" to the Montreal Protocol. Other halon fire suppression systems may still use the product, but its use must be discontinued as the systems are upgraded or replaced. Although it is no longer produced, halon may still be found in many places and firefighters should be familiar with its use in their response areas.

FIGURE 12-20 Dry chemical nozzles are located above the island and at ground level.

FIGURE 12-21 Low-pressure carbon dioxide extinguishing system.

FIRE DEPARTMENT OPERATIONS WITH PROTECTIVE SYSTEMS

Standpipe, sprinkler, and other protective systems should be part of the fire department overall strategic plan to provide fire protection for its community. This strategic plan recognizes the community's hazards and tries to keep them within certain limits. When properties or processes create hazards beyond those limits, protective systems are required. The fire department must survey its hazards and also its resources, both public and private. A program of maintenance, inspections, and pre-emergency planning keeps the department in compliance with the strategic plan. The property owner does maintenance, the fire prevention bureau does the inspections, and fire companies do the preplanning. Preplanning identifies hazards and resources, and some key resources are protective systems. Fire companies should know where every protective system is and how it operates. For standpipe and sprinkler systems, fire company personnel should know the type of system, the location of all fire department connections, two closest water supply points, and the location of key valves. This information should be noted on area maps, pre-emergency plans, fire department laptop software, and dispatch information. Knowing where the systems are and how to use them is the first step in coordinating a proper action plan. The next step is having SOPs that give a recommended course of action in buildings with these systems.

From an operational point of view, protective systems can be separate components requiring the fire department to look in different areas to understand what has been activated. In newer systems, the protective systems are integrated into a single "smart" system that includes an annunciator panel, fire alarm control panel, and system override controls. Typically, annunciator panels are near the primary entrance to the building and simply "announce" what has been activated and where. The fire alarm control panel and other protective system controls are in a locked room and allow the fire department to reset, silence, or otherwise control the system. Large occupancies such as a high-rise may have a complete fire command center where the fire department can monitor and/or control all protective systems as well as HVAC systems and building intercom/communications systems.

Standpipe Operations

Standpipe operations start with establishing a water supply to the fire department connection with a minimum of one hoseline. Additional hoselines are added as needed. The pump operator should immediately charge the standpipe system for the pressure required at the reported fire level. The first arriving unit officer should have personnel check the annunciator panel and go to the reported location of the fire. Methods of proceeding to this location will vary depending on the situation, but may involve use of stairs or elevators and should be covered in SOPs. Firefighters using elevators should do so only if they are equipped with firefighter service control systems. They should not take the elevator directly to the reported fire floor because the doors could open directly into the fire. Fire crews should stop at least two floors below the reported fire level.

Personnel should have full personal protective equipment, standpipe pack, and forcible entry equipment. The suggested standpipe pack equipment is a minimum of 150 feet (46 m) of 1½-inch (38-mm) hose or larger (many departments are now using 1¾ or 2 inch [45 and 50 mm]), a 2½- to 1½-inch (65- to 38-mm) gated wye with 1½- to 2½-inch (38- to 65-mm) increaser, 5 to 10 feet (1½ to 3 m) of 2½-inch (65-mm) hose, spanner wrench, adjustable wrench,

FIGURE 12-22   Hose pack for standpipe use.

and valve wheel. Additional sections of hose are recommended if personnel allow, **Figure 12-22.** Personnel should connect their standpipe pack to the nearest and safest outlet in relation to the fire location. This may require using the floor below and advancing the hoseline toward the fire location. Depending on the fire's location, the line may be charged prior to entering the hallway or delayed until reaching the door of the room or unit on fire. Additional lines may be run off the gated wye, or off another standpipe outlet on that floor or another floor.

Sprinkler System Operations

Fire incident operations in buildings with sprinkler systems begin with an investigation of the building to determine if water is flowing or if a fire is present. Typically, the first arriving officer checks the annunciator panel and then directs the crew to pull fire attack lines (or begin standpipe operations). Typically, SOPs require the first arriving crew to start fire attack while the pump operator sets up to support the sprinkler system. Establish a water supply to the fire department connection with a minimum of one medium-diameter hoseline. Additional hoselines are added as needed. The pump operator should charge the sprinkler system when smoke or fire is showing, when the water motor gong is operating, or when ordered by the officer-in-charge. The system should be charged immediately if it is combined with a standpipe system. Sprinkler systems are normally charged and maintained at 150 psi (1035 kPa). Personnel should advance hose-

lines into the fire area and conduct extinguishment and overhaul operations. If no fire is found, the officer may direct firefighters to stop sprinkler flow as discussed earlier in this chapter. It is important to note that all sprinkler system alarms should be treated as an actual fire until confirmed otherwise. Just because a building is fully sprinkled does not mean the fire department should prepare to fight fire differently. Being fully prepared with PPE, tools, radios, and so forth, should take precedence.

Detector Activation Operations

Operationally, the generic "fire alarm activation" response can range from investigating an alarm malfunction to a full-blown fire operation. Histories of false alarms have caused some firefighters to take a less than ready approach to their operations. Not only is this dangerous, it can be fatal. Most experienced firefighters can tell a story of arriving at a fire alarm activation, only to be surprised that an actual fire existed. These firefighters now respond to all alarms just as if an actual fire were awaiting them. It cannot be overemphasized: Firefighters must treat all fire alarm activations as an actual fire and prepare for the worst.

First arriving firefighters at an alarm activation should dismount the apparatus wearing full PPE with SCBA, and carry necessary tools and a radio. Typically, the first step at an alarm activation is to investigate. At residential structures, firefighters should meet with the building occupant if no fire signs or conditions exist. In cases where nobody is awaiting fire department arrival, the officer or designee should conduct a "360," which is a systematic walk around the structure to check for smoke or other indicators of an internal fire. Local SOPs should address further steps if no visible signs exist. These SOPs can involve notifying neighbors, watching the structure for evidence of fire or smoke, or entering the structure and conducting an interior search. If access can be made to the structure, caution should be employed. Smoke detector and CO detector activation may sound similar, but the CO can incapacitate unprepared firefighters with little warning.

At commercial structures, the first arriving crew should first check the annunciator panel to ascertain which device has activated. If no annunciator exists, firefighters may have to access the fire alarm control panel if no obvious signs are found. After business hours, firefighters should perform a "360" to look for signs. If the building is so equipped, the officer will then access the lock or knox box and enter the structure. Anytime a lockbox is utilized, the officer should radio dispatch that locked entry is being made. This will allow dispatch to notate the event with a time stamp of when it occurred. After checking the annunciator panel, the firefighter should proceed to the indicated zone and investigate. If any signs of smoke or fire are visible, the operation transitions from investigation to fire attack mode.

Buildings equipped with strobes and loud audible warning devices can be annoying to occupants if they see no signs of fire. Firefighters must not get trapped in a "false alarm mentality" and arbitrarily silence these alarms to investigate the activated zone. Premature silencing may send a message to occupants that they can return to their business. Using simple foam earplugs (which should be in the pocket of firefighters' PPE) can reduce the potential of temporary hearing loss while investigating the alarm. Once the activated detector is found, the immediate area can be checked for potential causes. If no fire signs are found, the firefighters can then take steps to clear the detector and attempt to reset the alarm. Correction of systems that will not reset are the property owner or manager's responsibility. Often, the fire department needs to advise the building representatives to contact their alarm company or a repair technician. All fire department actions must be documented on the incident report.

Operations for Other Protective Systems

Local SOPs should address the operations for total flooding, foam, dry chemical, and other unique systems. Total flooding systems that have discharged present a hazard to occupants, and firefighters must never enter this environment without first engaging their SCBA. The introduction of air into total flooding environments can reignite the fire. Firefighters should also check ceiling spaces and the rooftop ventilator to make sure the fire has not extended out of the hood and ducts. Hood systems that have discharged will require an overhaul effort to ensure that the fire is totally extinguished. Dry chemical systems are quite effective for fire control but require a large cleanup effort. This cleanup is not the responsibility of firefighters: it is important to notify occupants that breathing dry chemical can cause lung irritation. Activation of suppression systems may cause secondary damage to electrical circuits and other building infrastructure. It is important to be aware of this potential and protect firefighters and occupants accordingly.

JOB PERFORMANCE REQUIREMENT 12-1
Using Stops to Stem the Flow of a Sprinkler Head

A Select the right stop for the head discharging water. Remember, simple wedges may not fit into recessed sprinkler heads found in some finished ceilings.

B Ensure proper PPE and SCBA protection. Choose a stable ladder, desk, or other platform to reach the head.

C Put your face shield down or don eye protection.

D Climb the platform and secure the wedges in place.

E Wedges should be placed as shown.

LESSONS LEARNED

Protective systems are designed to automatically detect or suppress a fire, or to assist in extinguishing the fire. They can apply water or other extinguishing agents. Protective systems have been credited with saving both lives and property and are essential fire protection tools. Firefighters need to understand the value and operation of these systems to protect their communities.

Sprinkler systems are used for detection and suppression and can apply water or foam to extinguish the fire. Each type of system uses a different alarm valve but many of the other features are similar. Firefighters are still needed in sprinkler-protected property to finish extinguishing any fire that the system could not.

Standpipe systems facilitate manual fire suppression in which people do the firefighting. Through the use of a system of pipes with water supply and discharge connections, standpipes supply water in large buildings to facilitate fire suppression.

Other protective systems detect fires and apply extinguishing agents to fires in proximity to the hazard. Restaurants and kitchens usually have a hood system protecting the cooking areas, while other systems are used to protect high hazards or high value items. These protective systems are highly specialized and firefighters must understand their operation and any hazards of the agents.

Fire department operations at buildings with protective systems are typically outlined in local SOPs. Firefighters should treat all fire alarm activations as the worst-case scenario, meaning investigation with full PPE, tools, and radios. Operations involving standpipe and sprinkler systems require the fire department to support the system with a water supply and fire attack hoselines.

KEY TERMS

Accelerator A device to speed the operation of the dry pipe valve by detecting the decrease in air pressure. It pipes air pressure below the clapper valve, speeding its opening.

Auxiliary Appliances Another term for protective devices, particularly sprinkler and standpipe systems.

Deluge System Protective system designed to protect areas that may have a fast-spreading fire engulfing the entire area. All of its sprinkler heads are already open, and the piping contains atmospheric air. When the system operates, water flows to all heads, allowing total coverage. The system uses a deluge valve that opens when a separate fire detection system senses the fire and signals to trip the valve open.

Dry Pipe System Air under pressure replaces the water in the system to protect against freezing temperatures. The sprinkler control valve uses a dry pipe valve to keep pressurized air maintained above with the supply water under pressure below the valve.

Exhauster A device to speed the operation of the dry pipe valve by detecting the decrease in air pressure. It helps bleed off air.

Fire Watch An organized patrol of a protected property when the sprinkler or other protection system is down for maintenance. Personnel from the property regularly check to make sure a fire has not started and assist in evacuation and prompt notification of the fire department.

Life Safety Term applied to the fire protection concept in which buildings are designed to allow for the escape of building occupants without injuries. Life safety usually makes the building more fire resistant, but this is not the main goal.

Local Application System Designed to protect only a certain or local portion of the building, usually directly where the hazard will occur or spread.

Outside Stem and Yoke (OS&Y) Valve Has a wheel on a stem housed in a yoke or housing. When the stem is exposed or outside, the valve is open. Also called an outside screw and yoke valve.

Post Indicator Valve (PIV) A control valve that is mounted on a post case with a small window, reading either "OPEN" or "SHUT."

Preaction System Similar to the dry pipe and deluge systems. The system has closed piping and heads with air under no or little pressure, but the water does not flow until signaled open from a separate fire detection system. The preaction valve then opens and allows water to flow through the system to the closed heads. When an individual head is heat activated, it opens and water attacks the fire. Usually used when water can cause a large dollar loss.

Pressure-Regulating Device   Designed to control the head pressure at the outlet of a standpipe system to prevent excessive nozzle pressures in hoselines. Some are field adjustable, whereas others are preset at the factory.

Residential Sprinkler System   Smaller and more affordable version of a wet or dry pipe sprinkler system designed to control the level of fire involvement such that residents can escape.

Retard Chamber   Acts to prevent false alarms from a sudden pressure surge in the water supply by collecting a small volume of water before allowing a continued flow to the alarm device. The water from a surge is drained from a small hole in the bottom of the collection chamber.

Sprinkler System   Protective system designed to automatically distribute water through sprinklers placed at set intervals on a system of piping, usu-ally in the ceiling area, to extinguish or control the spread of fires.

Standpipe System   Piping system that allows for the manual application of water in large buildings.

Total Flooding System   Used to protect an entire area, room, or building by discharging an extin-guishing agent that completely fills or floods the area with the extinguishing agent to smother or cool the fire or break the chain reaction.

Wall Indicator Valve (WIV)   A control valve that is mounted on a wall in a metal case with a small window, reading either "OPEN" or "SHUT."

Water Columning   A condition in a dry pipe sprinkler system in which the weight of the water column in the riser prevents the operation of the dry pipe valve.

Wet Pipe Sprinkler System   Has automatic sprin-klers attached to pipes with water under pressure all the time.

REVIEW QUESTIONS

1. A person has fallen asleep on a sofa while smok-ing and a small fire has erupted. Explain the events that will transpire regarding the sprinkler system.

2. How does an OS&Y valve work?

3. What is the FDC and how is it used?

4. What are the steps to restoring a wet pipe sprin-kler system?

5. Explain the procedure to stop the flow from a sprinkler head.

6. Name the three classes of standpipes. Can these standpipes be used by the general public? Why or why not?

7. What equipment should be carried in a high-rise fire pack?

8. For firefighters, what is the hazard with a carbon dioxide, total flooding extinguishing system?

9. What is the difference between an annunciator panel and a fire alarm control panel?

10. Can firefighters use an elevator to access upper floors in a building that is on fire?

11. When should firefighters silence the audible fire alarm devices in a building?

ENDNOTES

1. See NFPA Standards 13, 13D, 13R, and 25 for more information.

2. NFPA 13D: *Standard for the Installation of Sprinkler Systems in One- and Two-Family Dwellings and Manufactured Homes* (Quincy, MA: National Fire Protection Association, 2007).

3. NFPA, *Fire Protection Handbook*, 20th ed. (Quincy, MA: National Fire Protection Association, 2008).

4. NFPA, *Fire Protection Handbook*, 20th ed. (Quincy, MA: National Fire Protection Association, 2008).

13 Building Construction

On February 14, 2007, the Northeast was hit with a massive winter storm that left more than twenty inches of snow in the Capital District. At about 11:00 P.M., my fire department was dispatched to a water flow alarm in the industrial/commercial area of our district. Because of poor road conditions, a single engine company and an assistant chief were sent to investigate. I knew the building to be one of a row of similar one-story warehouses approximately 150 feet by 300 feet with concrete block exterior walls and a wood roof supported by large timber trusses. The building was divided into two occupancies of equal size.

I was monitoring the radio transmissions at the alarm and heard that the personnel had found a broken sprinkler pipe with a significant water flow but was having difficulty securing the water supply to the system. Because I had previously been inside the building and knew the general layout, I responded to the scene to assist.

As I pulled up to the scene, I noticed that the 30 mph winds had caused large snow drifts on the roofs of many of the buildings in the area. I conferred with the Assistant Chief, then we went into the building to try to shut down the sprinkler system. We went in through the rear door of the warehouse and proceeded to the front of the building where the outside stem and yoke (OS & Y) valves were located. We closed the valves, and as the sound of the water spraying began to subside, we all heard rather loud creaking noises coming from the building.

Upon walking back out into the general warehouse area, I shined my hand light up to the ceiling area and saw that the original leak in the sprinkler system was caused by the bottom chord of one of the roof trusses breaking and falling on the sprinkler pipe. Apparently, the snow load on the roof overloaded this truss and caused it to fail. Once this was detected, I ordered an immediate evacuation of the building. As I exited the building, I requested the Town building department to send a representative to the scene as well as a responder for the building owner or occupants.

The engine company returned to service and I moved my vehicle across the street to wait for the building department and the responder to arrive. Within three minutes of evacuation, half of the roof of the warehouse collapsed into the building where we had been operating. This collapse caused the rear block wall to fall outward into the parking lot and sever the natural gas lines that ran to the ceiling heaters in the building. I immediately transmitted the box for a full structural assignment to be dispatched to the scene. We established two water supplies and set up a tower ladder to recon the damage to the building from above. After two hours, the utility company was able to secure natural gas service to the building. There had been no ignition or any secondary collapse and all companies returned to service.

It is important that all firefighters and fire officers be familiar with building construction. As a fellow firefighter always said, "The building is our enemy." Routine calls, if not observed properly, can result in the death or serious injury of firefighters.

—*Street Story by Kevin P. Terry, Chief of Department, Fuller Road Fire Department, Town of Colonie, New York*

LEARNING OBJECTIVES

After completing this chapter, the reader should be able to:

13-1   Describe the relationship between loads, imposition of loads, and forces.

13-2   List and describe structural elements.

13-3   Describe the effects of fire on five common building materials.

13-6   List and describe the types of building construction.

13-7   Identify the effects of time on various construction types when exposed to fire.

13-8   List and describe hazards associated with alternative (hybrid) building construction types.

13-9   Explain how the effects of a fire can create a hazardous condition in a structure.

13-10   Describe hazardous conditions found on an emergency scene.

13-11   Describe the critical elements alerting responders to a hazardous situation.

13-12   Perform an assessment to determine whether an area of a structure is safe for rescue operations.

13-13   Identify the effects that fire suppression operations have on a structure.

13-14   Describe building collapse hazards associated with fire suppression operations.

13-15   Explain indicators of collapse or structural failure.

13-16   Explain the process for predicting structural collapse or roof failure.

*The FF I and II levels, as defined by the NFPA 1001 Standards, are identified in different colors:
FF I = black, FF II = red, additional information = blue.

INTRODUCTION

Many fire departments pride themselves in their ability to launch aggressive interior structural fire attacks. Unfortunately, many firefighters are injured and killed when that same structure collapses, **Figure 13-1.** Often, buildings collapse without a "visual" warning such as sagging floors and roofs, leaning walls, and cracks. To keep from getting trapped in a collapse, firefighters must understand the types of structures they enter from the perspective of how the buildings are assembled, what materials are used, and how buildings react to fire. Additionally, firefighters must understand how fire travels through a building and choose appropriate tactics to stop the fire before key structural elements are attacked by the fire. Many firefighter fatality investigations conclude that fire departments need more training and education on building construction and the effects of fire on buildings. This chapter introduces several key topics regarding building construction and how fire affects buildings. It is important to note, however, that this chapter is merely an introduction. Firefighters must bridge the information in this chapter with a long-term commitment to study and research building construction and, more importantly, to explore the buildings within their jurisdiction.

This chapter begins by exploring some basic terms and mechanics of building construction, and then examines structural hierarchy and fire effects on materials. That information is then applied to classic and new construction types. The chapter concludes with a look at collapse hazards associated with structural fires.

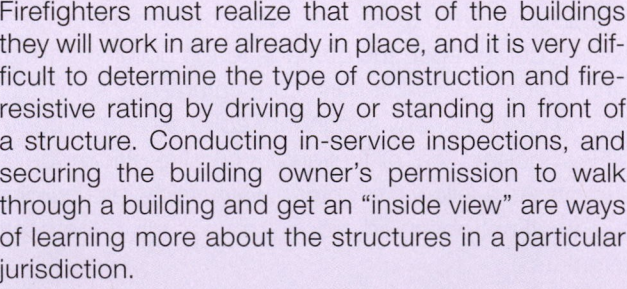

STREETSMART TIP

Firefighters must realize that most of the buildings they will work in are already in place, and it is very difficult to determine the type of construction and fire-resistive rating by driving by or standing in front of a structure. Conducting in-service inspections, and securing the building owner's permission to walk through a building and get an "inside view" are ways of learning more about the structures in a particular jurisdiction.

If new construction is being conducted in a firefighter's response area or jurisdiction, contact the building department and secure a set of the plans, and arrange to visit the construction site. Also, take photographs of how the building is built and what materials are used for future training references. Many new materials and construction methods are introduced regularly.

Types of Loads

Loads can be divided into two broad categories as it relates to building construction: dead loads and live loads. **Dead loads** include the weight of all materials and equipment that are permanently attached to the building. **Live loads** include equipment, people, movement, and materials not attached to the structure. Dead loads and live loads can be more specifically described using the following terms:

- **Concentrated load.** A concentrated load is a load that is applied to a small area, **Figure 13-2.** An example of a concentrated load is a heating, ventilation, and air-conditioning (**HVAC**) unit on a roof.

- **Distributed load.** A distributed load is a load applied equally over a broad area, **Figure 13-3.** Examples of this include snow on a roof or a hoist attached to numerous roof supports.

FIGURE 13-1 This collapse happened seconds after firefighters were repositioned.

FIGURE 13-2 The steel stairs and air-conditioning unit apply a concentrated load on this roof structure. Also note the potential instability of the air-conditioning unit placed on cement blocks.

BUILDING CONSTRUCTION TERMS AND MECHANICS

Firefighters need a basic understanding of certain terms and concepts associated with building construction. Obviously, buildings are constructed to provide a protected space to shield occupants from elements. The building must be built to resist wind, snow, and rain, and still resist the force of gravity. Additionally, the intended use of the building can add a tremendous amount of weight, placing more stress on the building's ability to resist gravity. In building terms, these elements create building **loading.** Loads are then imposed on building materials. This imposition causes stress on the materials, called force. Forces must be delivered to the earth in order for the building to be structurally sound. With this basic understanding, we can start to define terms and mechanics.

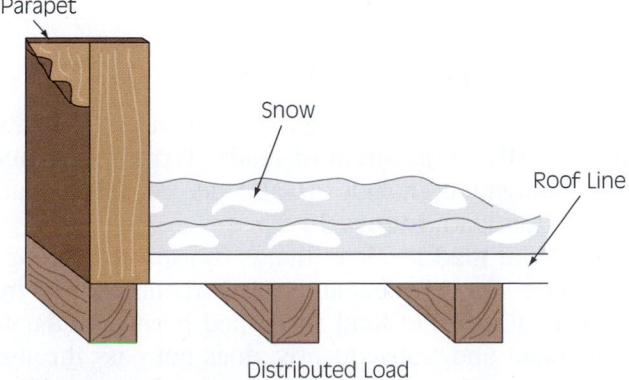

FIGURE 13-3 The weight of snow is a distributed load on this roof structure.

FIGURE 13-4 This ladder pipe operation is applying an impact load to the wall of this structure. As the wall weakens, it will eventually collapse. *(Courtesy of William H. Schmitt, Jr.)*

- **Impact load.** Impact load is a load that is in motion when applied, **Figure 13-4.** Crowds of people, fire streams, and wind gusts are examples of impact loads.
- **Design load.** Design loads are loads that an engineer has planned for or anticipated in the structural design.
- **Undesigned load.** Undesigned load is a load that was not planned for or anticipated. Buildings that are altered or are being used for occupancy other than original intent create an undesigned load. One common example is a residential structure that is converted to a print shop or legal office. These buildings were not designed to hold the additional live loads caused by the change in occupancy.
- **Fire load.** A fire load is the amount of heat generated when the building and its contents burn. It is important to note that the construction industry personnel (including architects and engineers) do not recognize fire load in their vocabulary—it is a fire engineering term.

Imposition of Loads

Loads must be transmitted to structural elements. This is called imposition of loads. Terms associated with imposition include axial load, eccentric load, and torsion load, **Figure 13-5.**

An **axial load** is a load that is transmitted through the center of an element and runs perpendicular to the element. **Eccentric load** is applied perpendicular to an element and, subsequently, does not pass through the center of the element. **Torsion load** is a load that is applied offset to an element, causing a twisting stress to the material.

Forces

Loads imposed on materials create stress and strain on the materials used to make the element. Stress and strain are defined as forces applied to materials. These forces are further defined as compression, tension, and shear, **Figure 13-6.** In **compression,** forces tend to push materials together. **Tension** occurs when forces tend to pull a material apart. **Shear** occurs when a force tends to "tear" a material apart—the molecules of the material are sliding past each other.

All loads—and the forces they create—must eventually pass through the structure and be delivered to the earth through the foundation of the building. Under normal conditions, structures will resist failure. Under fire conditions, the materials used to resist forces start breaking down. Eventually, gravity takes over and pushes the building to the earth.

> **STREETSMART TIP**
>
> As a building burns, the structural elements decompose and lose their strength. This causes a change in the forces and the way the design loads are applied, leading to structural failure and collapse.

The time it takes for gravity to overcome the structure during a fire is not predictable. A number of variables determine the amount of time a material can resist gravity and fire degradation. These include:

- Material type and mass
- Surface-to-mass ratio
- Overall load being imposed
- BTU development (fire load)
- Type of construction (assembly method)
- Alterations (undesigned loading)
- Age deterioration/care and maintenance of the structure
- Firefighting impact loads
- Condition of fire-resistive barriers

> **STREETSMART TIP**
>
> Surface-to-Mass Ratio: **Surface-to-mass ratio** is defined as the exposed exterior surface area of a material divided by its weight. In simple terms, smaller, lighter structural members will have a large surface with small mass when compared with larger structural members capable of carrying an equal load. A 3 × 14-inch solid wood beam may carry the same design load as a parallel chord truss made of 2 × 4s. The trusses have much more wood surface exposed and are more likely to ignite and burn rapidly.

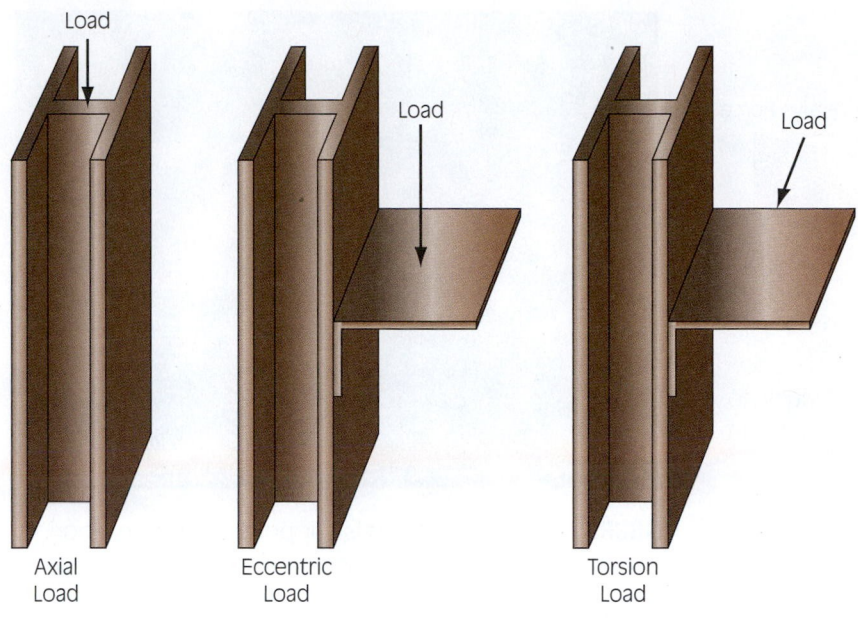

Axial
Load

Eccentric
Load

Torsion
Load

Application of Loads

FIGURE 13-5 There are three types of loads that can be transmitted through a structural member: axial, eccentric, and torsion.

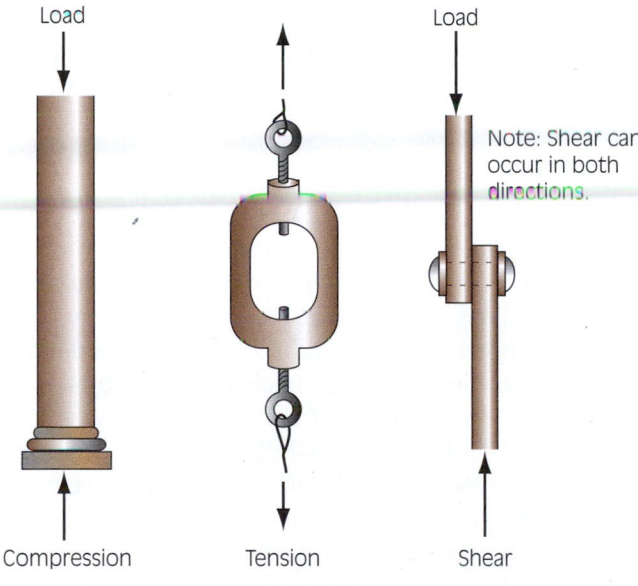

Compression

Tension

Note: Shear can occur in both directions.

Shear

Types of Loads

FIGURE 13-6 Loads are applied to a structural member as compression, tension, and shear forces.

The larger the surface area, the smaller the mass, the quicker it will burn or fail. Also, in combustible construction, the large surface area provides more fuel for the fire. Surface-to-mass ratio may also be applied to lightweight steel or pre-engineered buildings. These lightweight structural steel members will absorb heat quickly, and the steel elements will lose their strength and fail, causing collapse.

STRUCTURAL ELEMENTS

Buildings are an assembly of structural elements designed to transfer loads to the earth. Structural elements can be defined simply as beams, columns, walls, and connections. Each of these elements work together as an assembly to effectively make the load transfers to the building foundation, which delivers the building's live and dead loads to earth.

Beams

A **beam** is a structural element that transfers loads perpendicular to its length. Obviously, something must support the beam—usually a wall or column. It stands to reason that beams are used to create a covered space. In doing so, the beam is subjected to loads that causes the beam to deflect. The top of the beam becomes subjected to a compressive force whereas the bottom of the beam is subjected to tension, **Figure 13-7.** The distance between the top of the beam and the bottom of the beam dictates the amount of load the beam can carry. "I beams" are very typical and usually refer to the use of steel or engineered wood products to form the beam. The top of the I is known as the top **chord;** the bottom of the I is called the bottom chord. Chords are sometimes referred to as "flanges." The material in between the chords is known as the **web.** There are numerous types of beams, although the principal method of load transfer remains the same. A few types of beams include:

■ **Simple beam.** A beam supported at the two points near its ends.

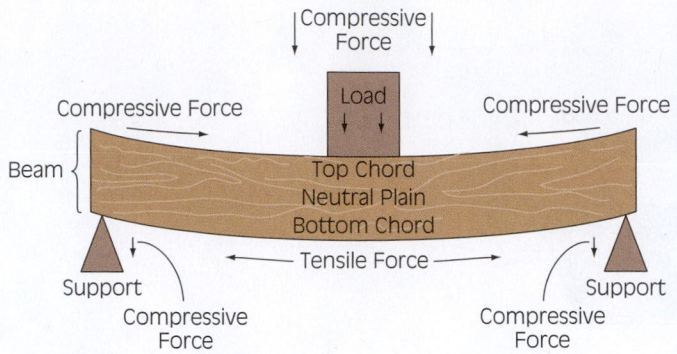

FIGURE 13-7 A beam transfers a load perpendicular to the load—creating compressive and tensile forces within the beam.

- **Continuous beam.** A beam that is supported in three or more places.
- **Cantilever beam.** A beam that is supported at only one end—or a beam that extends over a support in such a way that the unsupported overhang places the top of the beam in tension and the bottom in compression.
- **Lintel.** A beam that spans an opening in a load-bearing masonry wall—such as over a garage door opening. Lintels are often called "headers"—a slang term.
- **Girder.** A beam that supports other beams.
- **Joist.** A wood framing member used to support floors or roof decking. A **rafter** is a joist that is attached to a ridge board to help form a peak.
- **Truss.** A series of triangles used to form a structural element that in many ways is really a "fake" beam. That is, a truss uses geometric shapes, lightweight materials, and connections to transfer loads just like a beam. Trusses will be covered in detail later in this chapter.
- **Purlin.** A series of wood beams placed perpendicular to trusses or other beams to help support roof decking.

Columns

A **column** is any structural component that transmits a compressive force parallel through its center. Columns typically support beams and other columns, **Figure 13-8.** Columns are typically viewed as the vertical supports of a building; however, columns can be diagonal or even horizontal. The guiding principle is that a column is totally in compression.

Walls

A wall is also a component that transmits compressive force through its center. Simply put, a wall is a really long, but slender, column. Walls are subdivided

FIGURE 13-8 This column is supporting a beam, flooring, and another column. Columns are subjected to compressive forces.

into two categories: load-bearing walls and non-load-bearing walls. A **load-bearing wall** carries the weight of beams, other walls, floors, roofs, or other structural elements as well as the weight of the wall itself. A non-load-bearing wall need only support its own weight. A partition wall between two stores in a strip mall is an example of a non-load-bearing wall.

Connections

As mentioned previously, beams, columns, and walls must be connected in some fashion in order to effectively transfer loads. Often, the connection is the weak link as it relates to structural failure during fires. The connection point is often a small, low-mass material that lacks the capacity to absorb much heat, thereby failing quicker than an element that has more mass such as a column or wall. Connections fall into three categories: pinned, rigid, and gravity. Pinned connections use bolts, screws, nails, rivets, and similar devices to transfer load. Rigid connections refer to a system where the elements are bonded together such that all the columns (or load-bearing walls) are bonded to all the beams (usually by welds, poured concrete, or glue). Typically, failure of one element will cause the loads to be transferred to other elements. Gravity connections are just that—the load from an element is held in place by gravity alone.

Together, structural elements defy gravity and make a building sound. A series of columns and beams used to hold up a building are often referred to as the skeletal frame. Post and beam describes the same concept. Beams resting on walls are simply called wall-bearing buildings. One factor that has not been discussed is the suitability of the materials used to form structural elements. The next section covers these materials and how they act during fires.

TRIBUTE TO THE OL' PROFESSOR
DAVE DODSON

Francis L. Brannigan is a true friend to the fire service. For over thirty-five years Mr. Brannigan has shared his knowledge of fire and, more specifically, effects of fire on buildings. His book, *Building Construction for the Fire Service,* is a must-read for any firefighter and critical reading for anyone wanting to promote into fireground decision-making positions.

The fire service has affectionately called Mr. Brannigan the "Ol' Professor." His teachings have saved untold numbers of firefighters. I will never forget my first exposure to Mr. Brannigan. It was at a national conference in Cincinnati. The Ol' Professor was teaching a daylong class on steel buildings. As a young and inexperienced firefighter, I was all ears. The class taught me two things. First, I had much to learn about fire effects on building construction. I was way behind in my knowledge, and the Ol' Professor motivated me to make a never-ending knowledge quest to understand buildings. Second, I realized that reading buildings and reading smoke were the keys to rescuing people and putting the "wet stuff on the red stuff" before the building fails.

Over the years, Mr. Brannigan has coined many powerful—and lifesaving—phrases. Among my favorites:

Trusses

"BEWARE the TRUSS!"

"A truss is a truss, is a truss . . ."

(in reply to the notion that a bowstring truss is more dangerous than other trusses)

"The bottom chord of a truss is under tension—It's like you hanging on a rope. If the rope gets cut, you will fall. So it is with a truss."

"Failure of one element of a truss may cause the entire truss to fail—failure of one truss can cause other trusses to fail."

Columns

"The failure of a column is likely to be more sudden than failure of a beam."

"The slightest indication of column failure should cause the building to be cleared immediately."

On Collapses

"There is a tendency among those concerned about building stability to make light of partial collapse... a partial collapse is very important to at least two groups—those under it and those on top of it!"

"From an engineering point of view, (lightweight, trussed) buildings are made to be disposable... we don't make disposable firefighters!"

In response to those who claim a building collapsed without warning during a fire:

"The warning is the brain—in your ability to understand buildings and anticipate how they will react to a fire."

The fire service is indebted to the Ol' Professor, who died in 2006. In memory, we must give thanks to all he has done.

FIRE EFFECTS ON COMMON BUILDING CONSTRUCTION MATERIALS

Many factors determine which material is used to form structural elements. Quality, cost, application, engineering capabilities, and adaptability all play in the suitability of a material. In some cases, the material chosen for a structural application needs to meet fire resilience criteria. Regardless, the firefighter needs to understand how these materials react to fire. In the past, the fire service looked at the characteristics of four basic material types: wood, steel, concrete, and masonry. Each of these materials can be found together or separately. Each material reacts to fire in a different way, **Table 13-1.** Now, advanced material technology has found its way into structural elements. Buildings are being assembled using plastics, graphites, wood derivatives, and other composites. This section covers the four basic building materials as well as some of the new composites.

Wood

Wood is perhaps the most common building material. It is used in millions of residential and commercial buildings. Wood is relatively inexpensive, easy to manipulate, and a replenishable natural resource. Wood has marginal resistance to forces compared to

TABLE 13-1 Performance of Common Building Materials under Stress and Fire

Material	Compression	Tension	Shear	Fire Exposure
Brick	Good	Poor	Poor	Fractures, spalls, crumbles
Masonry block	Good	Poor	Poor	Fractures, spalls
Concrete	Good	Poor	Poor	Spalls
Reinforced concrete	Good	Fair	Fair	Spalls
Stone	Good	Poor	Fair	Fractures, spalls
Wood	Good w/grain; poor across grain	Marginal	Poor	Burns, loss of material
Structural steel	Good	Good	Good	Softens, bends, loses strength
Cast iron*	Good	Poor	Poor	Fractures

*Some cast iron may be ornamental in nature and not part of the structure or load bearing.

its weight, but it does the job for most residential and small commercial buildings. Wood also burns—and in doing so gives away its mass. The more mass a section of wood has, the more material it must burn away before strength is lost. This is true of native wood—that is, wood that has been cut from a tree. Engineered wood can react differently when exposed to heat from a fire. Engineered wood includes a host of products that take many pieces of native wood and glue them together to make a sheet, longer beam (trees only grow so tall!), or stronger column. Plywood delaminates when exposed to fire. Some newer wood products such as composites, which are discussed later in this chapter, can fail just through exposure to heat (they don't necessarily have to burn).

Steel

Steel is a mixture of carbon and iron ore heated and rolled into structural shapes to form elements for a building. Steel has excellent tensile, shear, and compressive strength. For this reason, steel is a popular choice for girders, lintels, cantilevered beams, and columns. Additionally, steel has high factory control. It is easy to change its shape, increase its strength, and otherwise manipulate it during production.

As it relates to fires, steel loses strength as temperatures increase. The specific range of temperatures depends on how the steel was manufactured. Cold drawn steel, like cables, bolts, rebar, and lightweight fasteners, loses 55 percent of its strength at 800°F (427°C). Extruded structural steel used for beams and columns loses 50 percent of its strength at 1,100°F (538°C). Structural steel will also elongate or expand as temperatures rise. At 1,000°F (583°C), a 100-foot-

long beam (30 meters) will elongate 10 inches (23 centimeters). Imagine what that could do to a building. If a beam is fixed at two ends, it will try to expand—and likely deform, buckle, and collapse. If the beam sits in a pocket of a masonry wall, it will stretch outward and place a shear force on the wall—which was designed only for a compressive force. This could knock down the whole wall! Steel structural elements must stay in their original shape. Any deflection, sagging, or stretching takes the steel out of its engineered shape and failure is likely to be quick.

> **CAUTION**
>
> Steel softens, elongates, and sags when heated, leading to collapse. Cooling structural steel with fire streams is just as important as attacking the fire.

Because steel is an excellent conductor of heat, it will carry heat of a fire to other combustibles. This can cause additional fire spread, sometimes a considerable distance from the original fire.

Concrete

Concrete is a mixture of portland cement, sand, gravel, and water. It has excellent compressive strength when cured. The curing process creates a chemical reaction that bonds the mixture to achieve strength. The final strength of concrete depends on the ratio of these materials, especially the ratio of water to portland cement. Because concrete has poor tensile and shear strength, steel is added as reinforcement. Steel can be added to concrete in many ways. Concrete can be poured over steel rebar and become part of the concrete mass when cured. Cables can be placed through the plane of concrete and be tensioned, compressing the concrete to give it required strength. Cables can be pretensioned (at a factory) or posttensioned (at the job site). Precast concrete refers to slabs of concrete that are poured at a factory and then shipped to a job site. Precast slabs are "tilted up" to form load-bearing walls—thus the term tilt-up construction.

All concrete contains some moisture and continues to absorb moisture (humidity) as it ages. When heated, this moisture content will expand, causing the concrete to crack or spall. **Spalling** refers to a large pocket of concrete that has basically crumbled into fine particles, taking away the mass of the concrete. Reinforcing steel that becomes exposed to a fire can transmit heat within the concrete, causing catastrophic spalling and failure of the structure. Unlike steel, concrete is a heat sink and tends to absorb and retain heat rather than conduct it. This heat is not easily reduced. Concrete can stay hot long after the fire is out, causing additional thermal stress to firefighters performing overhaul.

Masonry

Masonry is a common term that refers to brick, concrete block, and stone. Masonry is used to form load-bearing walls because of its compressive strength. Masonry can also be used to build a veneer wall. A **veneer wall** supports only its own weight and is most commonly used as a decorative finish. Masonry units (blocks, bricks, and stone) are held together using **mortar.** Mortar mixes are varied but usually contain a mixture of lime, portland cement, water, and sand. These mixes have little to no tensile or shear strength. They rely on compressive forces to give a masonry wall strength. A lateral force that exceeds the compressive forces within a masonry wall will cause quick collapse of the wall.

STREETSMART TIP

Masonry Walls Collapse: Masonry has very little lateral stability, and in many cases the roof or floor structure of a building holds the walls in place. Steel beams or joists will expand during a fire, creating lateral loads that the walls were not designed to withstand. In addition, wood roof structures will burn away or collapse during a fire, leaving little lateral support. The effects of the fire, pulling forces of the collapsing wood structure, or the force of a hose stream may cause a free-standing masonry wall to collapse.

Brick, concrete block, and stone have excellent fire-resistive qualities when taken individually. Many masonry walls are typically still standing after a fire has ravaged the interior of the building. Unfortunately, the mortar used to bond the masonry is subject to spalling, age deterioration, and washout. Whether from age, water, or fire, the loss of bond will cause a masonry wall to be very unstable, **Figure 13-9.**

Composites

New material technologies have introduced some interesting challenges for the firefighting community. Composites are a combination of the four basic materials listed previously as well as various plastics, glues, and assembly techniques. Of particular interest are the many wood products that are widely used for structural elements.

Engineered wooden I beams (slang term is "I-joists") are nothing more than wood chips and veneers that are press-glued together into the shape of an I beam, **Figure 13-10.** The chords are typically made from laminated veneer lumber (LVL) while the web is a piece of oriented strand board (OSB) sheeting. While structurally strong (stronger than a comparable solid wood joist), the wooden I beam fails quickly when heated. Actually, no fire contact is required. Ambient heating can cause the binding glue to fail, leading to a quick collapse. The bottom of a beam is under tensile forces. If the bottom of the beam falls off, due to glue failure, the beam will immediately snap and collapse.

New products, known as FiRP (fiber-reinforced products) are becoming common in the construction industry. FiRP can be plastic fibers mixed with wood to give the wood increased tensile strength. As with most plastics, fire exposure can cause quick failure as the plastic melts. The mixture of steel and wood as a structural element can cause rapid collapse because steel expands faster than wood. This causes stress at the intersection of the two materials, **Figure 13-11.**

Structural insulated panels (SIP) are another interesting composite. This technique is characterized by

FIGURE 13-9 Prior to a fire, the effects of age will take their toll on masonry walls. How stable is this wall with joint deterioration and lack of full mortar bond? Actually, this is an unsupported masonry wall which is inherently unstable—made more unstable with the deteriorated masonry.

FIGURE 13-10 To save on materials and cost, the use of composites or engineered wood structural members is becoming popular. Shown here is a typical wood I beam with finger-glued chords (also called flanges). The web is OSB. The glues used in this assembly will degrade when heated—and fail quickly.

FIGURE 13-11 A composite truss. Rapid heating will cause the stamped-steel to separate from the wood chords.

large wall and roof panels made of expanded polystyrene sandwiched between two sheets of OSB, which is a sheet of wood chips bonded by glue, **Figure 13-12A** and **B.** The OSB is covered with a typical wall finish. It is anticipated that a fire will cause rapid deterioration of the load-bearing panel.

(A)

(B)

FIGURE 13-12 These are SIP panels where expanded polystyrene is sandwiched between OSB sheets (A). The panels will make up the load-bearing walls and roof (building in background). Fire and heat can easily enter the wall space. (B) Failure of the wall panel will cause instability and rapid collapse.

To review, this chapter has so far explored the basic terminology, mechanics, elements, and materials used in the construction of buildings. It has also discussed a bit about how fire attacks materials and causes failure. The next section explores the various methods used to assemble buildings.

TYPES OF BUILDING CONSTRUCTION

Over time, five broad categories of building construction types have been developed to help classify structures. These categories give firefighters a basic understanding of the arrangement of structural elements and the materials used to construct the building. Unfortunately, these broad classifications are dangerously incomplete for firefighters and may lead to deadly assumptions about the makeup of a building. As stated before, firefighters need to explore the buildings within their jurisdiction to determine how buildings are assembled.

It is important to note that buildings are built to meet certain codes. These codes are designed to give occupants time to escape during a fire. Concrete is **fire resistive**—meaning it has some capacity to withstand the effects of fire. Other materials, like steel and wood, need fire-resistive assistance to give occupants a chance to escape. Building codes outline **fire-resistive ratings,** occupancy classifications, and means of egress based on five general types of buildings. The features of each type of construction will be discussed shortly, but first it is important to understand fire resistance for structural elements. **Table 13-2** outlines the number of hours that a structural element needs to be protected for the five

TABLE 13-2   Types of Construction from NFPA 220

	Type I		Type II		Type III		Type IV		Type V	
	442	332	222	111	000	211	200	2HH	111	000
Exterior Bearing Walls—										
Supporting more than one floor, columns, or other bearing walls	4	3	2	1	0	2	2	2	1	0
Supporting one floor only	4	3	2	1	0	2	2	2	1	0
Supporting a roof only	4	3	1	1	0	2	2	2	1	0
Interior Bearing Walls—										
Supporting more than one floor, columns, or other bearing walls	4	3	2	1	0	1	0	2	1	0
Supporting one floor only	3	2	2	1	0	1	0	1	1	0
Supporting roofs only	3	2	1	1	0	1	0	1	1	0
Columns—										
Supporting more than one floor, columns, or other bearing walls	4	3	2	1	0	1	0	H[1]	1	0
Supporting one floor only	3	2	2	1	0	1	0	H[1]	1	0
Supporting roofs only	3	2	1	1	0	1	0	H[1]	1	0
Beams, Girders, Trusses & Arches—										
Supporting more than one floor, columns, or other bearing walls	4	3	2	1	0	1	0	H[1]	1	0
Supporting one floor only	2	2	2	1	0	1	0	H[1]	1	0
Supporting roofs only	2	2	1	1	0	1	0	H[1]	1	0
Floor-Ceiling Assemblies	2	2	2	1	0	1	0	H[1]	1	0
Roof-Ceiling Assemblies	2	1½	1	1	0	1	0	H[1]	1	0
Exterior Nonbearing Walls	0	0	0	0	0	0	0	0	0	0

▓ Those members that shall be permitted to be of approved combustible material.

[1]"H" indicates heavy timber members; see text for requirements.

Source: Copyright © 2008, National Fire Protection Association, Quincy, MA.

FIGURE 13-13 This parking garage is of Type II construction. The protective coating applied to the structural steel may increase the fire-resistive rating, but note that the unprotected corrugated metal flooring and interior steel structure are not protected.

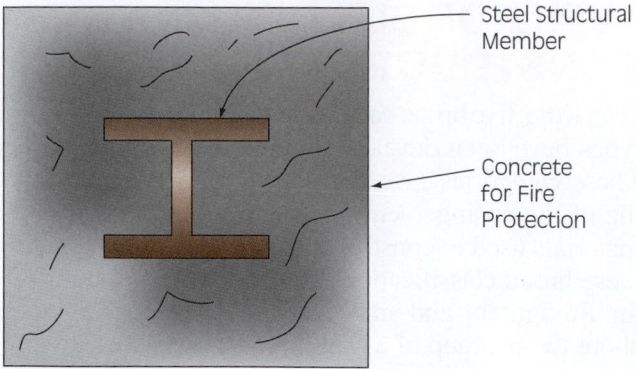

FIGURE 13-14 To achieve a Type I fire-resistive rating, structural steel members are encased with concrete to prevent failure from the effects of a fire.

types of construction. Simply put, firefighting time is not part of the fire-resistive and building construction equation. Fire-resistive ratings are established in a laboratory. In the real world, fire resistance ratings could "underperform" due to many factors. For example, a structural element with a two-hour fire rating may fail in thirty minutes if it was not assembled correctly or if it has been damaged by some event (fire, explosion, or other collapses). Fire resistance for structural members can be achieved by various methods including drywall (gypsum wallboard), spray-on coatings, and concrete, **Figure 13-13.** Aging, alterations, and wear can damage fire-resistive methods to the point that structural elements have no fire resistance protection.

The following paragraphs outline the basic definition of each building type, its general configuration, and some historical fire spread problems associated with each. Also included are some construction methods that do not fit into the five common types.

Type I: Fire-Resistive

Type I fire-resistive construction is a type in which structural elements are of an approved noncombustible or limited combustible material with sufficient fire-resistive rating to withstand the effects of fire and prevent its spread from story to story. Concrete-encased steel, **Figure 13-14,** monolithic-poured cement, and steel with spray-on fire protection coatings are typical of Type I, **Figure 13-15.** Generally, the fire-resistive rating must be three to four hours depending on the specific structural element. Fire-resistive construction is used for high-rises, large sporting arenas, and other buildings where a high volume of people are expected to occupy the building.

FIGURE 13-15 A typical Type I building, with structural members designed to resist the effects of fire for three to four hours. This building is of reinforced concrete construction.

Most Type I buildings are typically large, multistoried structures with multiple exit points. Fires are difficult to fight due to the large size of the building and the subsequent high fire load. Type I buildings rely on protective systems to rapidly detect and extinguish

FIGURE 13-16 Buildings of Type II construction will have structural elements with little or no protection from the effects of fire. Remember, in the event of a fire, these unprotected steel structural members may fail and collapse quickly.

fires. If these systems do not contain the fire, a difficult firefight will be required. Fire can spread from floor to floor on high-rises as windows break and the next floor windows fail, allowing the fire to jump. Fire can also make vertical runs through utility and elevator shafts. Regardless, firefighters are relying on the fire-resistive methods to protect the structure from collapsing. The collapse of fire-resistive structures can be massive, as we are reminded from the World Trade Center collapses in New York City.

Type II: Noncombustible

Type II noncombustible construction is a type in which structural elements do not qualify for Type I construction and are of an approved noncombustible or limited combustible material with sufficient fire-resistive rating to withstand the effects of fire and prevent its spread from story to story. More often than not, Type II buildings are steel, **Figure 13-16.** Modern warehouses, small arenas, and newer churches and schools are built as noncombustible. Because the steel is not required to have significant fire-resistive coatings, Type II buildings are susceptible to steel deformation and resulting collapse. Fire spread in Type II buildings is influenced by the contents. While the structure itself will not burn, rapid collapse is possible from the content Btu release stressing the steel.

Suburban strip malls with concrete block load-bearing walls and steel roof structures can be classified as Type II. Fires can spread from store to store through wall openings and shared ceiling and roof support spaces. The roof structure is often of lightweight steel that fails rapidly. More often than not, the fire-resistive device used to protect the roof structure is a dropped-in ceiling. Missing ceiling tiles, damaged drywall, alterations, and utility penetrations can ren-

der the steel unprotected. These buildings may have combustible attachments such as facades and signs as well as significant content fire loading.

Type III: Ordinary

The term **Type III ordinary construction** is often misapplied to wood frame buildings. By definition, ordinary construction includes buildings where the load-bearing walls are noncombustible, and the roof and floor assemblies are wood. Most commonly, this is load-bearing brick or concrete block with wood roofs and floors. Ordinary construction is prevalent in most downtown or "main street" areas of older towns and villages, **Figure 13-17.** Firefighters have long called

FIGURE 13-17 Buildings of Type III, ordinary construction, are common throughout North America. These typical "Downtown USA" buildings provide many challenges to firefighters, such as void spaces and common walls allowing rapid fire extension and little structural protection.

ordinary construction "taxpayers." This slang is derived from landlords who built buildings with shops and/or restaurants on the first floor and apartments above in order to maximize income to help pay property taxes. Newer Type III buildings include strip malls with block walls and wood truss roofs, **Figure 13-18.**

FIREFIGHTER FACT

Sagging or bowing load-bearing walls are often pulled back in alignment by tightening a steel rod that runs through the building from wall to wall. A small interior fire can elongate this steel and cause catastrophic wall failure. These buildings can be spotted by decorative stars or ornaments (called spreaders) on the outside brick wall.

Ordinary construction presents many challenges to firefighters. In older buildings, numerous remodels, restorations, and repairs have created suspect wall stability and hidden dangers.

Ordinary construction has many void spaces where fire can spread undetected. Common hallways, utilities, and attic spaces can communicate fire rapidly. Masonry walls hold heat inside, making for difficult firefighting. Wood floors and roof beams are often gravity fit within the masonry walls. These can release quickly and cause a general collapse, leaving an unsupported masonry wall. Older Type III buildings have structural mass; therefore, they burn for a long time.

Type IV: Heavy Timber

Type IV heavy timber construction can be defined as those buildings that have block or brick exterior load-bearing walls and interior structural members, roofs, floors, and arches of solid or laminated wood

without concealed spaces. The minimum dimensions for structural wood must meet the criteria in **Table 13-3.** Heavy timber buildings, as the name suggests, are quite stout and are used for warehouses, manufacturing buildings, and some older churches, **Figure 13-19.** In many ways, a Type IV building is like a Type III—just larger dimension lumber instead of common wood beams and trusses. Some firefighters mistakenly call Type IV buildings "mill construction." Mill construction is a much more stout, collapse-resistive building that may or may not have block walls. A new Type IV building is hard to find. The cost of large-dimension lumber and laminated wood beams makes this type of construction rare in modern construction.

Fire spread in a heavy timber building can be fast due to wide-open areas and content exposure. The exposed timbers contribute Btus to the fire. Because of the mass and large quantity of exposed structural wood, fires burn a long time. If the building housed machinery at one time, oil-soaked floors will add more heat to the fire and accelerate collapse. Once floors and roofs start to sag, heavy timber beams may release from the walls. This is accomplished by making a fire-cut on the beam; then the beam is gravity fit into a pocket within the exterior load-bearing masonry wall, **Figure 13-20.** As the floor sags, it loses its contact point with the wall and simply slides out of its pocket without damage to the wall. It is important to recall that a free-standing masonry wall has little lateral support and requires compressive weight from floors and roofs to make it sound.

Type V: Wood Frame

Type V wood frame construction is perhaps the most common construction type. Homes, newer small businesses, and even chain hotels are built pri-

FIGURE 13-18 One of the most common uses of Type III, ordinary construction, is the "strip mall" with masonry walls and lightweight steel or wood trusses. Common problems associated with this type of construction are void spaces allowing for rapid fire extension and collapse of lightweight structural elements.

TABLE 13-3   Heavy Timber Dimensions

Type of Element	Use	Size
Column	Supporting floor load	8- × 8-in. minimum any dimension
Column	Supporting roof load	6-in. smallest dimension, 8-in. depth minimum
Beams and girders	Supporting floor load	6-in. width and 10-in. depth minimum
Beams, girders, and roof framing	Supporting roof loads only	4-in. width minimum, 6-in. depth minimum
Framed or laminated arches	As designed	8-in. minimum dimension
Tongued and grooved planks	Floor systems	3-in. minimum thickness with additional 1-in. boards at right angles
Tongued and grooved planks	Roof decking	2-in. minimum thickness

marily with wood, **Figure 13-21.** Older wood frame buildings were built as **balloon frame**—that is, wood studs ran from the foundation to the roof and floors were "hung" on the studs. As can be envisioned, fire could enter the wall space and run straight to the attic. In the early 1950s, builders started using a

FIGURE 13-19   Type IV buildings, heavy timber construction, have large wood structural elements with great mass. The mass of these structural members requires a long burn time for failure. The connections, usually steel, are the weak points in this type of construction.

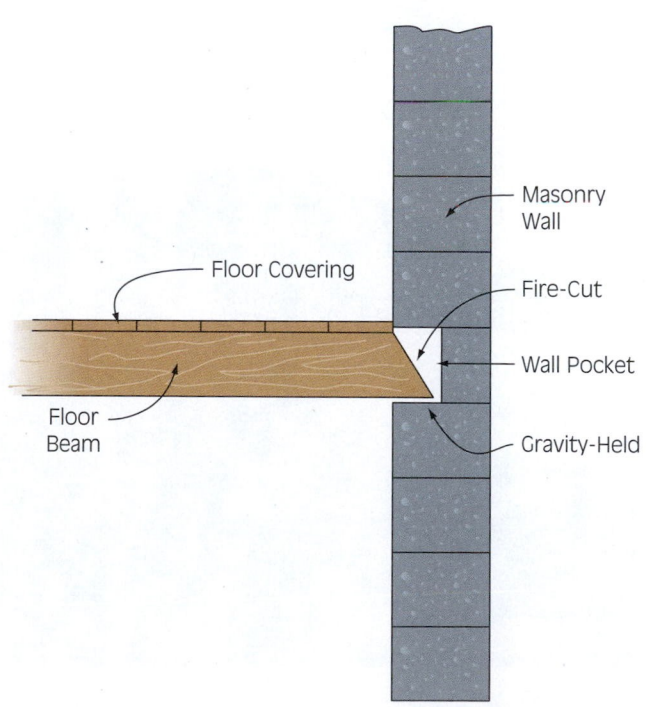

FIGURE 13-20   Wood and heavy timber beams were often "fire-cut" so that a fire-damaged, sagging floor would simply slide out of the wall pocket in order to preserve the wall.

FIGURE 13-21 The wood frame structure, Type V construction, is the most common type of construction in North America.

platform framing arrangement where one floor was built as a platform for the next floor. This created **fire stopping** to help minimize fire spread. Newer wood frame buildings utilize lightweight wood trusses for roofs and floors. This is akin to a "horizontal balloon frame" that can allow quicker lateral fire spread. Coupled with high surface-to-mass wood exposure, collapse becomes a real possibility, **Figures 13-22 A–C.** Some codes require truss spaces to have fire stopping every 500 square feet (46 sq. meters). Even with this fire stopping, it remains dangerous to step onto the 500 square feet where the fire is. Wood

(A)

(B)

(C)

FIGURE 13-22 (A, B) Note the void spaces at the first and second floor levels and in the attic area that are created by the use of truss systems. (C) The building after it was completed. Note the sloping grade—firefighters need to communicate this observation to avoid confusion on which is the first floor.

frame structures may appear more like a Type III ordinary building because of a brick-wall appearance. Remember, brickwork may be a simple veneer to add aesthetics.

To protect structural members from a fire, wood frame construction typically uses gypsum board (drywall or the brand name Sheetrock). Gypsum protects wood frame members by absorbing heat. When heated, the gypsum goes through a chemical process called "calcification." Drywall that has been calcified is very fragile and crumbles—allowing fire spread.

Once finished, wood frame buildings typically have many rooms that can help compartmentalize content fires. Fire that penetrates wall, floor, or attic spaces becomes a significant collapse threat, especially in newer buildings. Often, the only warning that fire has penetrated these spaces is the issuance of smoke from crawl space vents, gable end vents, and eaves.

Other Construction Types (Hybrids)

As mentioned earlier, the five broad building types can actually lead to dangerous assumptions. Newer construction and alternative building methods may not fit cleanly into one of the above types. Although not an official term, we can call these "hybrid buildings." Some hybrid buildings are actually two types of construction. For example, a particular restaurant may be built as a Type II noncombustible, yet it is topped with a large wood frame structure to hide rooftop HVACs and cooking vent hoods, **Figure 13-23A** and **B.** The combustible square feet space of the false dormers and wood frame structure can exceed that of most homes.

Lightweight Steel Frame

New lightweight steel homes resemble and are built like "stick" wood-frame homes. These buildings are actually a "post and beam" steel building with lightweight steel studs to help partition the home. OSB is added to the studs to help make the house more "stiff" and increase wind-load strength, **Figure 13-24.**

Insulated Concrete Formed (ICF)

Another interesting construction type uses polystyrene foam blocks or panels to make a form for a lightweight concrete mud mixture. The concrete is not contiguous—there are many voids, utility runs, and foam block spacers (made of plastic or galvanized

(A)

(B)

FIGURE 13-23 (A) The decorative roof assembly is a Type V wood frame structure while the occupancy space is Type II noncombustible. (B) This building uses two types of construction.

steel), **Figures 13-25 A** and **B.** These structures are called "ICF" or insulated concrete formed. However, it is important not to be fooled by any claims that these buildings are concrete or are less combustible. In reality, these composite buildings are assem-

FIGURE 13-24 This lightweight steel home is built similar to a Type V. OSB sheeting gives the steel rigidity to torsional loads such as wind.

bled with plastics, polystyrene, lightweight steel, and lightweight concrete. When finished, these buildings may resemble wood frame or even ordinary construction. Extended window and door jambs are clues that indicate the wall is thicker than that of typical wood or masonry built buildings.

Structural Insulated Panel (SIP) Wall

As mentioned in the preceding composite material section, SIP wall uses panels that are expanded polystyrene sandwiched between sheets of OSB. The entire load-bearing wall and roof structure are made from these panels (with no structural framing). The building is assembled like the house of playing cards you built as a child. These panels rely on drywall (interior) and exterior finishes to protect the assembly from fire. Heat alone can cause the panels to lose integrity.

Other hybrid buildings include, but are not limited to, hay bales, aerated concrete, steel/styrene, and gypsum/Portland lime blocks. The phrase "green building" is being used more often. Although no official definition exists, "green" can be applied to buildings that are assembled using energy and environmentally friendly methods or to buildings that have earthen roof or wall assemblies. Some of the earthen buildings have soil, grass, shrubs, and other living plants used as insulation and to help the carbon "footprint" of the building. The fire service has very little research information on the stability of these new types of buildings during fires. One thing is certain: firefighters should expect rapid collapse due to the low-mass, high surface-to-mass exposure of structural elements.

Manufactured buildings can be defined as those structures that are built at a factory and then trucked to a job site. These building are quite light with little mass. Where a stick-built home uses 2×4 or 2×6 lumber, the manufactured home uses 1×2 and 2×2 lumber. These buildings use galvanized strapping to give required strength. In any case, these buildings burn quickly and collapse equally fast.

Relationship of Construction Type to Occupancy Use

Before considering the basic types of construction, many officials and builders first look at the anticipated use of the building—its occupancy type. **Occupancy classifications** are called many different names around the country, but they are usually broken down into five basic arenas: residential, commercial, business, industrial, and educational. Each of these general occupancies has a number of hazards that firefighters must understand, **Table 13-4.** Remember, a building may have been built for one type of occupancy only to be sold and converted to another occupancy type for which it may not have been designed. Firefighters should go out and explore the buildings in their community.

COLLAPSE HAZARDS AT STRUCTURE FIRES

It cannot be overstated that firefighters have to understand the buildings in their jurisdiction. Constant study and site visits will help them "read" buildings. Reading buildings is essential to anticipating collapse proactively. This section addresses some specific collapse threats that the fire service has experienced throughout history (see the feature box, "Historically Significant Building Collapses") and the importance of understanding buildings and how they react at structure fires.

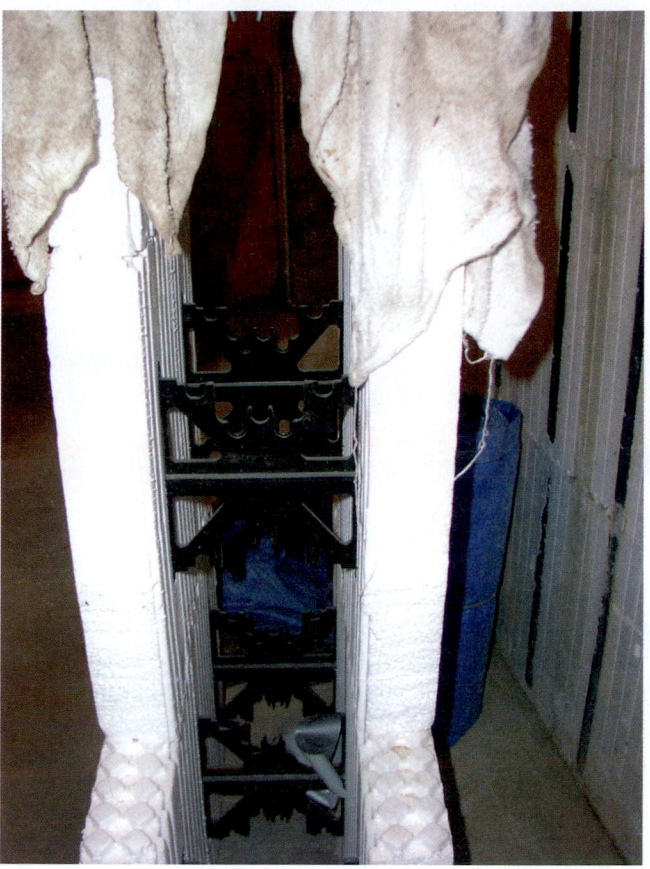

(A)   (B)

FIGURE 13-25   (A) This wall is a load-bearing foam block unit filled with a lightweight concrete mud mix. (B) Note the black plastic spacers that will melt early in a fire.

TABLE 13-4   Typical Hazards Associated with Occupancies

Occupancy	Type of Construction	Hazards
Residential	Type V, most common	Fire loading, truss construction, owner alterations, rapid fire extension in void spaces
Commercial	Type III, most common	Fire loading, truss construction, rapid fire extension in void spaces, unknown occupancy change
Educational	Type II, most common	Unprotected structural steel, collapse, high fire load in some areas
Business	Types II and III, most common	Unknown change in occupancy, high fire load, difficult to ventilate
Industrial	Types I and II, most common	Hazardous materials, difficult to ventilate

Trusses

Truss roof collapses have killed many firefighters. As stated previously, a truss is actually a fake beam. A truss uses geometric shapes (the triangle) to create a structural element similar to a beam. A wood truss can actually be stronger than a like-sized solid wood beam, and does so with less material, **Figure 13-26.** It is this loss of material and subsequent increase in exposed surface area that make them so vulnerable during fires. Trusses rely on each and every part of the truss to carry a portion of the imposed load. Like

HISTORICALLY SIGNIFICANT BUILDING COLLAPSES

DAVE DODSON

Many firefighters have been killed as a result of building collapse during firefighting operations. With each of these tragic losses, lessons can be learned. The following is a brief look at some of the more significant collapses. Each of these events should be researched to find all the contributing factors that led to the event. The italicized lessons are perceptions that are shared in the spirit of preventing firefighter injuries and death.

Wonder Drug Store Fire, New York City, 1966

Firefighters responded to a commercial structure fire only to discover that the building shared a basement with another building. The concealed fire advanced rapidly and undetected. The ensuing firefight trapped and killed twelve firefighters. *It is vitally important to pre-plan buildings prior to a fire event. Older buildings may have access ways sealed from adjoining buildings. Shared utilities and other hidden voids can facilitate fire spread.*

The Vendome Hotel Fire, Boston, 1972

Nine firefighters died during overhaul of a fire in an old, remodeled hotel. The investigation revealed that a masonry wall had been breached to make way for an air duct. Just above the breach, a column carried the load from floors above. A corner of the five-story building collapsed, trapping the firefighters. There were no obvious signs of impending collapse. The Vendome building was brought back from disrepair in 1971, and many alterations were made that were unknown to the fire department. *Firefighters should take an interest in the construction activities in and around buildings. Remodeling and restoration can compromise structural elements.*

Abandoned Building Fire, Detroit, 1980

A fire was reported in a large, abandoned building that was scheduled for demolition. Responding firefighters found "light smoke" showing. The fire escalated rapidly due to the poor interior conditions and wide-open spaces. One firefighter died while trying to escape. Two other firefighters died when a firewall collapsed. *Abandoned buildings are much like a building under construction. Firefighters cannot take anything for granted. Being prepared to encounter rapid fire spread and suspect integrity should be the order of the day. Defensive operations must respect collapse zones.*

Hackensack Ford Dealership Fire, Hackensack, New Jersey, 1988

A fire was discovered in the attic space above an automobile repair garage. Responding firefighters launched an aggressive attack. The bowstring truss roof space was being used as a parts storage area, placing additional load on the structure. During firefighting operations, the roof collapsed, trapping and killing five firefighters. *Fighting fires in truss spaces is like playing Russian roulette. Trusses create a wide, open space beneath them. Clear spans are a warning sign of quick collapse should the truss space be involved in fire. Where there are no occupants to rescue, firefighters should reduce their risk and fight fire from safe attack points.*

Sun Rise Gift Shop Fire, Orange County, Florida, 1989

Firefighters responded to a fire in a single-story commercial structure. Interior conditions were described as light smoke and no heat. The fire had gained headway in a truss space above the ceiling. The tile-covered roof collapsed twelve minutes after firefighters arrived and killed two firefighters. *A fire can be roaring over firefighters' heads without them being aware. Firefighters should routinely inspect the ceiling space above their heads for fire. Once fire or heavy, dark smoke conditions are found in truss spaces, tactics should change. Tile roof coverings are quite attractive but show very little signs that the roof supporting them is about to collapse.*

Western Interior Services Building Fire, Brackenridge, Pennsylvania, 1991

Firefighters were attempting to attack a fire in the basement of a large commercial building with concrete floors supported by steel columns. During the attack, the floor collapsed, trapping and killing four firefighters. *Basement fires present many difficult challenges to firefighters. Limited access, trapped heat and smoke, and the storage nature of basements must be factored into fire attack. Unfinished basements allow the fire to attack the floor above and the floor supports rather quickly. Unprotected steel exposed to fire will soften quickly, leading to rapid collapse. The so-called "20-minute rule" for interior firefighting is a dangerous assumption.*

One Meridian Plaza High Rise Fire, Philadelphia, 1991

Three firefighters died when they became disoriented and ran out of air while fighting fire in the high-rise building. The fire started on the twentieth floor and ran up to the thirtieth floor, where a sprinkler system extinguished the fire. Although the firefighters did not die from a collapse, this event is significant in that the Incident Commander feared a catastrophic collapse due to stress cracks found in the concrete stair towers and withdrew the firefighters. *There are many lessons learned from this event. The fire investigation details and the many building construction issues, as well as the potential for fatalities, associated with high-rise firefighting are available from the U.S. Fire Administration (www.fema.gov/usfa).*

Mary Pang Fire, Seattle, Washington, 1995

An arson fire in a multi-use commercial building caused the deaths of four firefighters when a portion of the floor collapsed. The building had been altered several times, and a lightweight "pony wall" had been used to replace a portion of a load-bearing wall. The building had a confusing layout, including entry points at two different elevations (like a walkout basement). The advanced fire was not apparent from the "front" side of the building. An aggressive interior attack was underway for approximately thirty minutes when the floor collapsed. *Buildings that have gone through several owners and occupancy changes should always be suspect. Pre-fire planning helps uncover hazards that could change firefighting tactics. Firefighters should make a habit of reporting fire conditions, unique building features, and visual observations to their direct fireground supervisor.*

West Mendocino Avenue House Fire, Stockton, California, 1997

Two firefighters died when a home addition collapsed during interior firefighting operations. The homeowners had built a large, two-story, clear-span dance studio ad-

dition on the back of their home. Firefighters entered the front of the building and found a heavy fire and dense smoke conditions. The clear-span portion collapsed, killing the firefighters. *Homeowners may not follow established building practices and codes when making additions. A "360" check prior to fire attack can uncover significant hazards when "reading" the building. Unusual or significant additions should be communicated to all working firefighters. Firefighters cannot pre-plan every home, so they must rely on their ability to read buildings and read smoke conditions.*

World Trade Center, New York City, 2001

Terrorists hijacked two large airliners and hit the twin towers of the World Trade Center. The Fire Department of New York (FDNY) responded and began the biggest rescue effort in fire service history. The high-rise towers collapsed, killing 343 of FDNY's bravest. Thousands of civilians also died in the collapse. Steel high-rise construction relies on fire-resistive coatings to protect the steel. The combination of burning jet fuel and the trauma of the aircraft strikes rendered the steel unprotected. Failure came much more quickly than the four-hour time limit prescribed for the Type I fire-resistive construction found in high-rises. *Firefighters should never rely on fire protection time ratings when making decisions for fire attack. History will always remember that the FDNY firefighters died trying to rescue trapped civilians. They have the undying respect of people throughout the world.*

FIGURE 13-26 Wood trusses provide a large surface-to-mass ratio, fuel load, and void spaces—three of the worst structural collapse contributors a firefighter will encounter during structural firefighting operations.

a beam, the top of the truss (called the top chord) is typically under a compressive force. The bottom chord is under tension. In between the two chords, connecting members (the web) transfer the two forces creat-

ing stress and strain. Failure of one part of the truss will likely cause the whole truss to fail. This distributes the weight of the failed truss to other trusses—which may not have the capacity to take that weight—thereby starting a domino effect collapse. Trusses come in many styles and shapes. Bowstring truss, parallel chord truss, and open web joists are some of the more common names. Trusses are also classified by the type of material used in assembly.

Wood Trusses

Wood trusses are commonly used for roof assemblies but are increasingly being used for floor support as well. Wood trusses are an assembly of many pieces of wood. Some may even be press-glued particles. These pieces are connected using **gusset plates.** A gusset plate is a simple galvanized steel plate (very thin) with perforations punched into the plate. The perforations are used to pierce into wood fibers to hold pieces together. These perforations only penetrate the wood a fraction of an inch (⅜ inch is typical), **Figure 13-27.** During fires, the steel gusset heats up and

FIGURE 13-27 A typical parallel chord truss. The gusset plates on this truss are pressed into the wood. In addition to the decomposing of the wood element, the light-gauge steel plates will deform and pull away from the wood under fire conditions.

FIGURE 13-28 Unprotected open web steel joists present a large surface area to absorb the heat of a fire, expand, and collapse. Structural steel will lose 50 percent of its strength at temperatures of 1,000°F (538°C).

transfers heat into the very wood fibers that are being held. If the heating is slow (like a smoke-filled attic), the wood decomposes, allowing the gusset plate to fall out. If the heating is fast, like sudden exposure to flame, the steel expands too quickly for the wood and the gusset simply pops out. Either way, truss failure is imminent. Sometimes the truss gingerly stays together because of its connection to roofing or flooring materials, yet a sudden force, like a walking firefighter, will cause the truss to disassemble and suddenly collapse.

Some wood trusses are being assembled using glue instead of gusset plates. Some manufacturers claim that a glued truss is actually stronger than a gusseted one. This may be true in the absence of heat or fire. Remember, heat can cause glue failure—it doesn't have to burn.

Wood trusses are mass-produced at a factory where quality control may not be adequate. Further, the truss gusset plates or glue may vibrate or be damaged while being delivered to a job site. Once on the job site, contractors may use shortcuts to lift the truss into position, furthering the damage to gusset plates or glued joints.

Steel Trusses

Steel trusses are no less susceptible to collapse than wood trusses. Like wood, steel trusses are an assembly of pieces—typically angle iron for the chords and cold-drawn round stock for the web. The pieces are tack-welded together to form the truss unit. While not a true joist by definition, many call the common steel truss an open web steel joist, **Figure 13-28.** The term bar joist is also used to describe an open web steel joist. These trusses expose a large surface area (but low mass) to heat during fires. Given the lack of mass, the truss heats quickly and will soften and expand. The expansion can cause wall movement. (Remember, masonry walls must be loaded axial with compressive force.) Lateral movement can cause wall collapse. If the wall does not move, the steel truss will twist and buckle to allow expansion. It is very important to keep steel trusses cool.

Although the focus here is buildings, many other structures (like antennae, bridges, water towers, and large signs) use steel trusses as part of their assembly. A tall cell phone relay tower is actually a vertical, cantilevered, steel truss (beam).

In 2005, NIOSH issued an *Alert* regarding firefighting operations in buildings with truss construction. The alert cites an alarming trend in the number of firefighters killed due to truss collapse and recommends that firefighting operations immediately resort to a defensive (outside) posture once fire has entered a truss space.

ETHICS

Many firefighters have died because of truss collapse. The best way to honor those that have died is to identify those buildings in your jurisdiction that have truss construction and develop pre-plans so that a similar occurrence can be prevented.

Void Spaces

Trusses create large void spaces. The area between the chords of trusses will allow fires to spread horizontally, **Figure 13-29.** Some codes require fire stopping in floor truss spaces but may allow wide-open attic spaces. Fires can start in void spaces due to electrical and other utility problems. In Type III ordinary construction, voids are numerous. Some voids may pass through masonry walls, causing fire spread from one store to the next in a row of buildings. The obvious collapse danger with void spaces is that the fire may be undetected with simultaneous destruction of structural elements.

FIGURE 13-29 Lightweight floor truss systems have many void spaces. This could be called "horizontal balloon frame."

Roof Structures

The roof of a building can be flat, pitched, or inverted. Many factors help determine how and why the roof is built the way it is. Sometimes the roof is designed to hide rooftop HVACs. Other times the roof shape is designed to shed snow, accommodate a vaulted ceiling, or merely give the building character, **Figure 13-30.** As it relates to structural collapse, the roof style may allow a large volume of fire to develop. Other roofs, like the mansard, have many concealed spaces. Dormers are protrusions from a roof structure. Dormers can be used to introduce daylight into a roof space that is converted into a living space, **Figure 13-31.** Other dormers are actually aesthetic (false) and can fool ventilation crews attempting to relieve heat from a roof space.

Stairs

First arriving firefighting crews rely on internal stairways to help gain access for rescue and fire attack. For years, firefighters have found stairways to be durable and a bit stronger than other interior components. This is a dangerous assumption in newer wood frame buildings. Stairs are now being built offsite and simply hung in place using light metal strapping, **Figure 13-32.** Additionally, stairs are being made using lightweight engineered wood products that fail quickly when heated. Remember, press-glued wood chip products can fail from the heat of smoke—no flame is required.

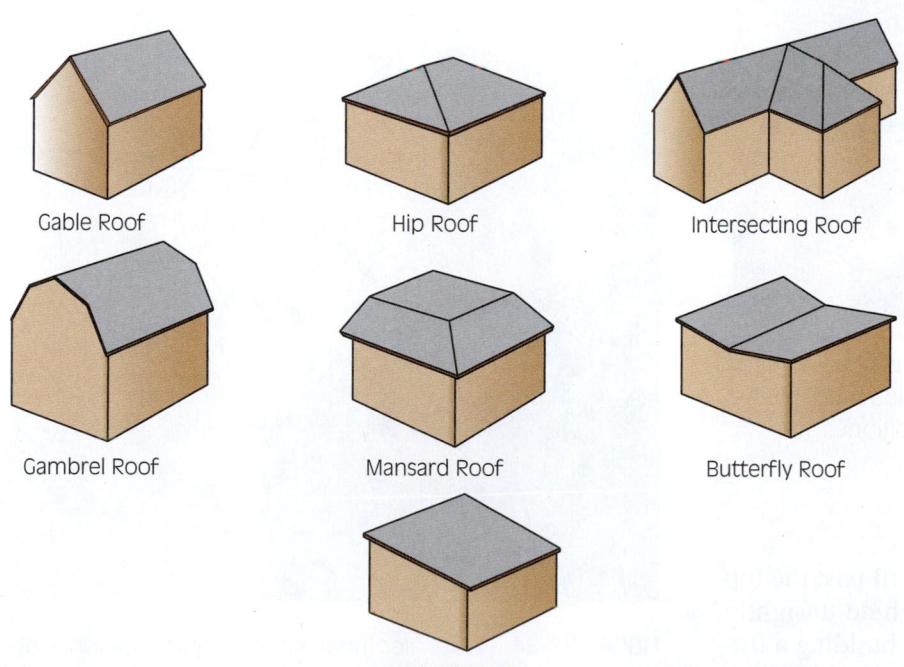

Gable Roof

Hip Roof

Intersecting Roof

Gambrel Roof

Mansard Roof

Butterfly Roof

Shed Roof

FIGURE 13-30 Some common roof framing styles used in wood frame or ordinary construction.

FIGURE 13-31 An internal view of dormers in a mansard roof. Fire can run through the many voids between the exterior and the roof.

FIGURE 13-32 This prefab stair assembly is hung in place by thin metal strapping. Note the staples and plastic shims that can quickly fail under fire conditions.

Parapet Walls

A **parapet** wall is the extension of a wall past the top of the roof. Parapets are used to help hide unsightly roof equipment and HVACs and give a building a finished look. Typically masonry, these walls are free-standing with little stability because they have no

compressive loading to give the masonry strength. Collapse of parapets typically happen when the roof structure starts to sag—this lifts up the parapet in an eccentric way, causing it to fall, **Figure 13-33**. Business owners hang signs, utility connections, and other loads on the parapet. During a fire, the steel cables and bolts holding these will weaken or change shape and subsequently pull down the parapet, **Figure 13-34**. Historically, dozens of firefighters have been killed by collapsing parapets.

FIGURE 13-33 This is the scene of a typical "parapet" wall failure common in Type III ordinary construction.

FIGURE 13-34 This electrical service entrance and attached sign may be the eccentric load causing an early failure of this parapet wall.

Collapse Warning Signs

Firefighters must rely on building material knowledge, building construction principles, and an understanding of fire effects on buildings in order to predict or anticipate collapse. Waiting for a visual sign that a building will collapse is dangerous, especially in newer buildings. There are, however, some factors and observations that can be used to help anticipate collapse. These include:

- Deterioration of mortar joints and masonry
- Overall age and condition of the building
- Cracks, in anything
- Signs of building repair including reinforcing cables and tie-rods
- Large open spans
- Bulges and bowing of walls
- Sagging floors
- Abandoned buildings
- Large volume of fire
- Long firefighting operations—remember gravity?
- Smoke coming from cracks in walls
- Dark smoke coming from truss roof or floor spaces (Brown smoke indicates that wood is being heated significantly; black smoke means combustibles have ignited or are near ignition.)
- Multiple fires in the same building or damage from previous fires

Buildings under Construction

Buildings are especially unsafe during construction, remodeling, and restoration. The word *unsafe* applies not only to fire operations but also to rescues, odor investigations, and on-site inspections. Buildings need only meet fire and life safety codes when they are completed. During construction, many of the protective features and fire-resistive components are incomplete. Additionally, stacked construction material may overload other structural components. This is not to say that contractors are using unsafe practices, but rather underscore the fact that exposed structural elements, incomplete assemblies, and material stacks will contribute to a rapid collapse if a fire were to develop.

STREETSMART TIP

Beware of Buildings under Construction: A building as a complete unit has a number of interdependent parts. During construction, these parts may not be fully connected (steel), may not be at their full design strength (concrete), and may lack any type of fire protection (gypsum board or concrete). Also, many structural elements may be held in place by scaffoldings, false work, or forms. Because of this, a building under construction is exposed to early collapse potential due to the effects of fire or other elements such as high winds.

Historical building restoration and general remodel projects in buildings are similar to buildings under construction. Firefighters may find temporary shoring of walls, floors, and roofs while other structural components are being updated, replaced, or strengthened. Contractors may use simple 2 × 4s to temporarily shore up heavy timber, leading to disastrous results during fire conditions. The best approach for firefighters to take when responding to fires in buildings under construction is to be defensive. They should make sure everyone is out and accounted for and then attack the fire from a safe location. A building under construction is likely insured and can be replaced—a firefighter's life cannot.

Time

There are no time limits for firefighting operations within a building. Tests have shown that the age-old "twenty-minute rule" used by some fire officers is no longer accurate. Roofs and walls can collapse within minutes of fire involvement given certain conditions. An overloaded truss (due to improper storage or other factors) can collapse immediately when heated.

The passage of time during a building fire works against the firefighter teams. In fact, the window of opportunity to perform an aggressive interior fire attack has diminished significantly due to lighter-weight construction and higher BTU heat release of plastics. As it relates to predicting firefighting time within a structure, a few truisms have emerged:

- The lighter the structural element, the faster it comes down.
- The heavier the imposed load, the faster it comes down.
- Wet (cooled) steel buys time.
- Gravity and time are constant, resistance is not.
- There is no *window* of time for interior operations when a building is under construction, being renovated, or being disassembled.
- Brown or dark smoke coming from lightweight structural components means time is up.

Preparing for Collapse

At building fires, the incident commander (and/or assigned safety officer) needs to predict collapse *PROACTIVELY.* Communicated information and

PREDICTING COLLAPSE

There is no perfect formula for predicting collapse. A simple approach is presented here to help you understand the mental process that incident commanders and safety officers use to help predict collapse. All firefighters need to be aware of the process and should communicate conditions and observations to ensure the process is proactive. The approach explained here can be applied to buildings and other structures such as bridges, cranes, towers, signs, mine shafts, wells, or any "engineered" construction attacked by fire.

Step 1: Classify the construction type.

Step 2: Determine structural involvement or heating (read the smoke and flames).

Step 3: Visualize and trace loads (determine the weak link).

Step 4: Evaluate time.

Step 5: Predict and communicate the collapse potential (establish a collapse zone).

observations from working firefighter teams helps with predicting collapse. Therefore, all firefighters need to understand the mental process that is used to proactively predict collapse (see the feature box, "Predicting Collapse").

STREETSMART TIP

Collapse: Every firefighter must understand two rules about structural collapse during fire operations. The first is that the potential for structural failure during a fire always exists. Do not set artificial time limits based on experience. The second rule is to establish a **collapse zone,** as shown in **Figure 13-35,** which is an area around and away from a building where debris will land if the building fails. As an absolute minimum, this distance must be at least 1½ times the height of the building. The walls may crumble into a pile or they may tip out the full height of the building. Also you need to provide extra room for cascading debris.

Once a building has been searched for occupants, the risks firefighters take to control the fire should be reduced—after all, it is now a property issue. Many firefighters have been killed fighting interior fires, only to have the building torn down after the investigation. Outside (defensive) firefighting operations can be equally dangerous if firefighters wander into the collapse zone, **Figure 13-36.**

In all cases, firefighters who witness collapse indicators or an actual collapse should be empowered to report the condition. If an imminent threat to firefighters exists, the firefighting team should declare a firefighter emergency ("Mayday. Mayday. Mayday.") and initiate an immediate withdrawal.

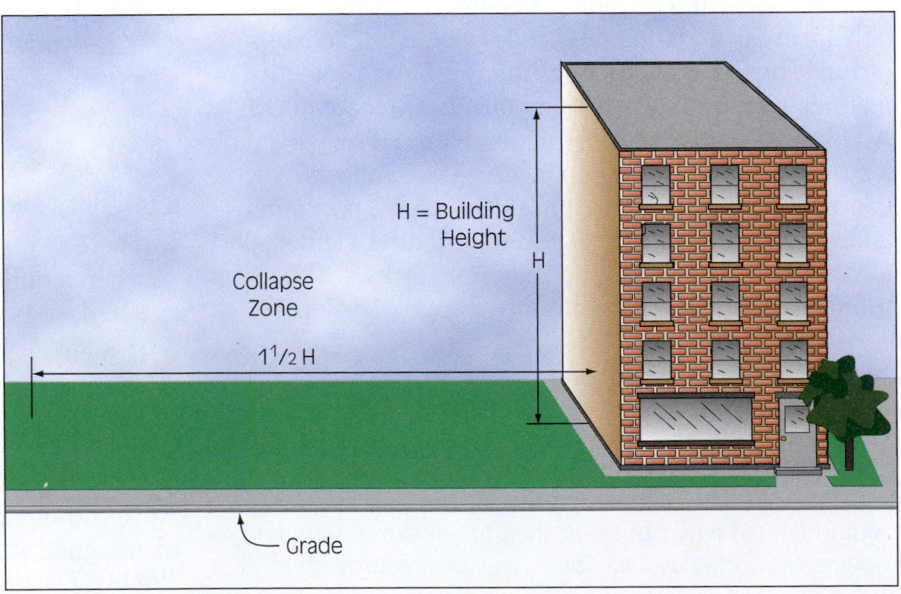

FIGURE 13-35 A minimum collapse zone should be 1½ times the height of the building.

FIGURE 13-36 These photos show the effects of fire on a masonry wall. Note the debris and distance the bricks fell away from the building. Firefighters should always establish a collapse zone, as shown in Figure 13-35.

LESSONS LEARNED

Many firefighters have been killed as a result of building collapse from structural fires. To prevent future deaths, firefighters must understand the buildings in which they fight fires. This understanding comes from a long-term commitment to read and study building construction information. Additionally, firefighters must get into buildings within their jurisdictions to survey and explore the way buildings are assembled, remodeled, and used in the real world. Knowledge of building construction starts with an understanding of loads, forces, and materials found in the structural makeup of buildings. Firefighters also study the effects of fires on materials and construction types. The five classic types of construction are being challenged by new construction methods. Trusses are used in virtually all new buildings. Trusses have high surface-to-mass characteristics that rapidly absorb heat and subsequently fail quickly. Failure of one truss can cause failure of other trusses. There are no rules for how long a building will last while on fire. Many factors determine when materials and construction design fail and gravity pushes down the building. Buildings under construction are losers from a firefighting point of view—they collapse quickly. Understanding the process to proactively predict collapse helps firefighters communicate conditions and observations that can save lives.

KEY TERMS

Axial Load   A load passing through the center of the mass of the supporting element, perpendicular to its cross section.

Balloon Frame   A style of wood frame construction in which studs are continuous the full height of a building.

Beam   A structural member subjected to loads perpendicular to its length.

Cantilever Beam   A beam that is supported at only one end.

Chord   The top and bottom components of a beam or truss (sometimes referred to as *flanges*). The top chord is subjected to compressive force; the bottom chord is subjected to tensile force.

Collapse Zone   The area around a building where debris will land when it falls. As an absolute minimum this distance must be at least 1½ times the height of the building.

Column   A structural element that is subjected to compressive forces—typically a vertical member.

Compression   A force that tends to push materials together.

Concentrated Load   A load applied to a small area.

Continuous Beam   A beam that is supported in three or more places.

Dead Load   The weight of the building materials and any part of the building permanently attached or built in.

Design Load   A load the engineer planned for or anticipated in the structural design.

Distributed Load   A load applied equally over a broad area.

Eccentric Load   A load perpendicular to the cross section of the supporting element that does not pass through the center of mass.

Fire load   The amount of heat generated when the building and its contents burn.

Fire Resistive   The capacity of a material to withstand the effects of fire.

Fire-Resistive Rating   The time in hours that a material or assembly can withstand fire exposure. Fire-resistive ratings are usually provided for testing organizations. The ratings are expressed in a time frame, usually hours or portions thereof.

Fire Stopping   Pieces of material, usually wood or masonry, placed in stud or joist channels to slow the extension of fire.

Girder   A large structural member used to support beams or joists—that is, a beam that supports beams.

Gusset Plate   A connecting plate used in truss construction. In steel trusses, these plates are flat steel stock. In wood trusses, the plates are either light-gauge metal or plywood.

HVAC   Acronym for heating, ventilation, and air-conditioning unit. HVACs are typically a rooftop unit on commercial buildings. Buildings may have one or dozens of these units.

Impact Load   A load that is in motion when it is applied.

Joist   A wood framing member that supports floor or roof decking.

Lintel   A beam that spans an opening in a load-bearing masonry wall.

Live Load   The weight of all materials and people associated with but not part of a structure.

Load-Bearing Wall   Any wall that supports other walls, floors, or roofs.

Loading   The weight of building materials or objects in a building.

Mortar   Mixture of sand, lime, and portland cement used as a bonding material in masonry construction.

Occupancy Classifications   The use for which a building or structure is designed.

Parapet   The projection of a wall above the roofline of a building.

Platform Framing   A style of wood frame construction in which each story is built on a platform, providing fire stopping at each level.

Purlins   A series of wood beams placed perpendicular to trusses to help support roof decking.

Rafter   A wood joist that is attached to a ridge board to help form a peak.

Shear   A force that tends to tear a material by causing its molecules to slide past each other.

Simple Beam   A beam supported at the two points near its end.

Spalling   Deterioration of concrete by the loss of surface material due to the expansion of moisture when exposed to heat.

Surface-to-Mass Ratio   Exposed exterior surface area of a material divided by its weight.

Tension   A force that pulls materials apart.

Torsion Load   A load parallel to the cross section of the supporting member that does not pass through the long axis. A torsion load tries to "twist" a structural element.

Truss   A rigid framework using the triangle as its basic shape to emulate a beam.

Type I, Fire-Resistive Construction   Type in which the structural members, including walls, columns, beams, girders, trusses, arches, floors, and roofs, are of approved noncombustible or limited combustible materials with sufficient fire-resistive rating to withstand the effects of fire and prevent its spread from story to story.

Type II, Noncombustible Construction   Type not qualifying as Type I construction, in which the structural members, including walls, columns, beams, girders, trusses, arches, floors, and roofs, are of approved noncombustible or limited combustible materials with sufficient fire-resistive rating to withstand the effects of fire and prevent its spread from story to story.

Type III, Ordinary Construction   Type in which the exterior walls and structural members that are portions of exterior walls are of approved noncombustible or limited combustible materials, and interior structural members, including wall, columns, beams, girders, trusses, arches, floors, and roofs, are entirely or partially of wood of smaller dimension than required for Type IV construction or of approved noncombustible or limited combustible materials.

Type IV, Heavy Timber Construction Type in which exterior and interior walls and structural members that are portions of such walls are of approved noncombustible or limited combustible materials. Other interior structural members, including columns, beams, girders, trusses, arches, floors, and roofs, shall be of solid or laminated wood without concealed spaces.

Type V, Wood Frame Construction Type in which the exterior walls, bearing walls, columns, beams, girders, trusses, arches, floors, and roofs are entirely or partially of wood or other approved combustible material smaller than the material required for Type IV construction.

Undesigned Load A load not planned for or anticipated.

Veneer Wall A covering or facing, not a load-bearing wall, usually with brick or stone.

Web The portion of a truss or I beam that connects the top chord with the bottom chord.

REVIEW QUESTIONS

1. What are three ways loads are imposed on materials?

2. List the three types of forces created when loads are imposed on materials.

3. Name three kinds of beams.

4. Explain the effects of fire on steel structural elements.

5. How does a masonry wall achieve strength?

6. List and define the five common types of building construction. Give an example of each type that is located in your district or response area.

7. List the three parts of a truss and explain what forces are being applied to each.

8. List three buildings in your district or response area that have truss construction.

9. List how fire affects the four more common building materials in use today.

10. Diagram and label four different roof shapes.

11. List and describe eight conditions or observations that might indicate potential structural collapse.

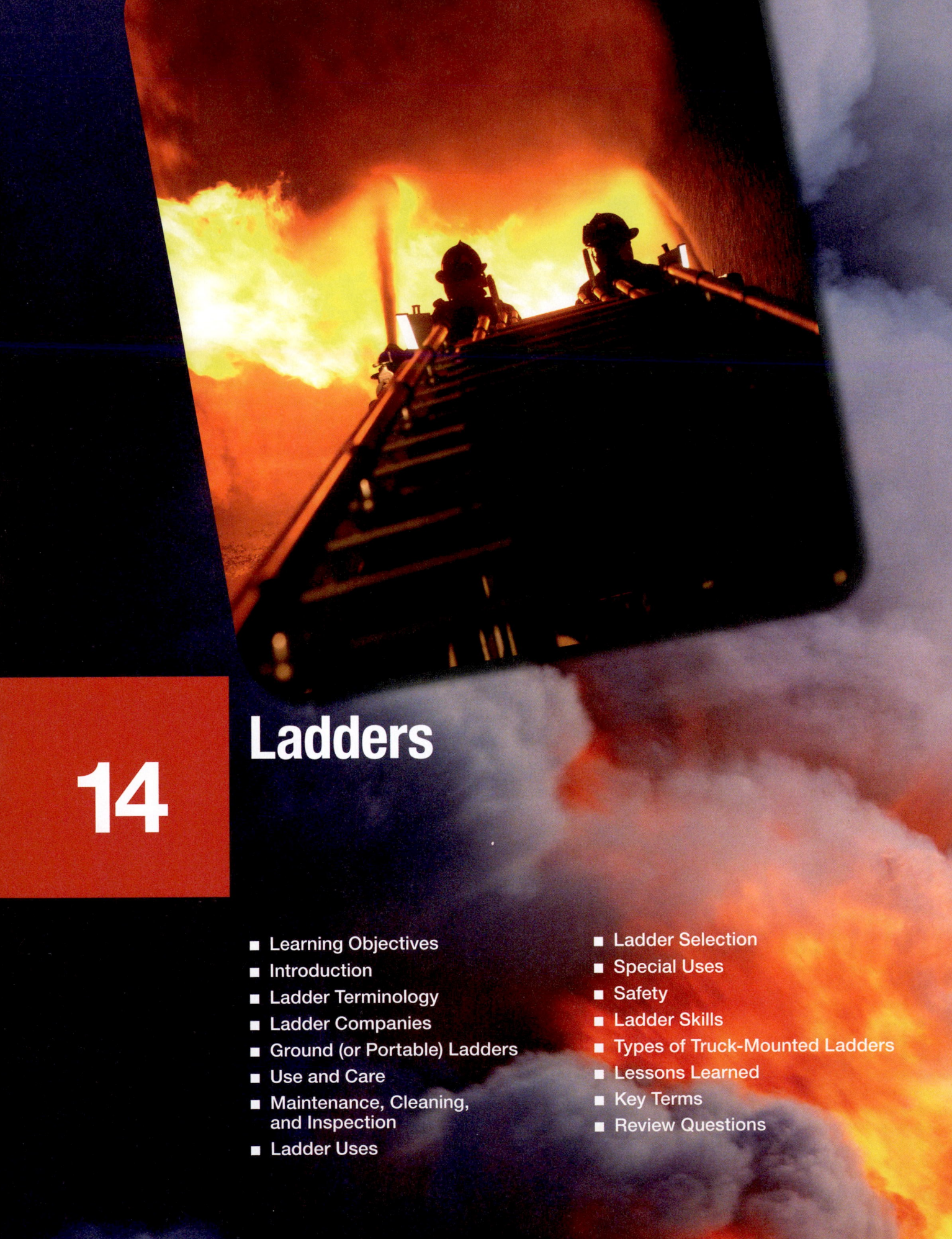

14

Ladders

With newer and more technical equipment being introduced to the fire service on an almost daily basis, it is hard to stay focused on basic firefighting. Ground ladders are as basic as you can get; however, they are still as important today as they ever were. When you place a ladder to a window, you have an instant exit for both occupants and firefighters. Place a roof ladder on a roof and you have a safe place to start ventilation. Of course, do not forget about the occasional cat stuck in a tree. As long as rescuing cats does not interfere with more serious firefighting duties, this type of rescue provides training and good public relations.

New firefighters often wonder why they must practice raising ground ladders over and over, almost to the point where they no longer have to consciously think about what they are doing. That is exactly the point. Any time, day or night, snow or rain, heat or cold, firefighters must use ground ladders with practiced skill and an understanding of their features and limitations.

This concept became apparent to me several years ago at the scene of a 2½-story wood-frame, occupied structure fire. As my crew approached the fireground, we noticed a man visible through the lower sash of a double-hung window on the second floor. His outline was obscured by thick black smoke tinged with orange, which meant that his seconds were numbered. From a different vantage point, another crew observed the same scene. In an instant, both groups bolted for their respective ladder trucks to retrieve the appropriate ground ladders needed to effect the rescue. Both crews returned with exactly the same sizes and types of ground ladders for the job. This was no coincidence. The members of both crews had received their training at different times and by different instructors, but the proper use of ground ladders had been embedded in the minds of both crews, so they responded to the task in the same way.

Keep in mind that the rescue was successful because of ground ladders, not a thermal imaging camera, not an articulated telescoping tower ladder, not any other specialized, expensive equipment—just ground ladders and firefighters who knew what they were doing. Of course, there are situations that call for special equipment, but the proper knowledge and use of ground ladders will save lives and reduce property damage at almost every fire.

—Street Story by Peter F. Kertzie, Captain,
Buffalo Fire Department, Buffalo, New York

LEARNING OBJECTIVES

After completing this chapter, the reader should be able to:

14-1   Name the parts of different types of ground ladders and describe their functions.

14-2   Given a specific piece of equipment, list and describe various tools, as well as inspection and cleaning methods, used to return the equipment back to service.

14-3   List and describe proper use of cleaning solvents.

14-4   Describe the proper manufacturers' or departments' procedures for tool and equipment cleaning.

14-5   Given manufacturer recommendations, departmental guidelines, tools, and cleaning materials, demonstrate the ability to properly clean, inspect, and service tools and equipment.

14-6   Describe the ladder selection process and the functions for which different types of ground ladders can be used.

14-7   Demonstrate the ability to place a ladder at the proper height and angle.

14-8   Choose the proper ladder length needed for various job functions or situations in an emergency operation.

14-9   Identify proper ladder placement and raising techniques.

14-10 Place the ladder while demonstrating safe placement practices and avoiding hazards.

14-11 Describe the qualities that are suitable for placement and stabilization of a ground ladder.

14-12 Evaluate the structural capability of walls and roofs in supporting ladder operations.

14-13 Recognize hazards prior to placement of the ladder.

14-14 Identify hazards associated with carrying, placing, raising, and lowering ladders.

14-15 Identify proper ladder techniques used for rescue operations.

14-16 Demonstrate the proper selection, placement, and raising of ground ladders for rescue operations.

14-17 Demonstrate skills associated with ladders, to include carrying and raising ladders and extending and locking flies.

14-18 Demonstrate the proper carry and operation of ventilation tools, equipment, and various ladders.

14-19 Demonstrate the safe carry of vertical ventilation equipment while ascending and descending ladders.

14-20 Demonstrate the proper tool selection, tool use, and ladder positioning for vertical ventilation.

14-21 Demonstrate safe and proper use of a roof ladder on a pitched roof.

*The FF I and II levels, as defined by the NFPA 1001 Standards, are identified in different colors:
FF I = black, FF II = red, additional information = blue.

INTRODUCTION

Ladders are primarily used in the fire service for providing access to elevated or below-grade locations. Originally, ladders were constructed of wood; however, as the need for greater height became more evident, the supporting solid wood beams were replaced with newer technology, truss-type beams. This truss construction is a design that removes the nonsupport portions of a solid structural member without reducing strength but significantly reducing weight. While this design permitted longer ladders to be constructed, there were still size limits. Introduction of lightweight high-strength aluminum replaced the wood, first in solid beam construction, then in truss construction.

Later, higher strength aluminum alloys were developed and later still, fiberglass ladders were developed. Along with the change in the ladder's material makeup, new design technology continued to meet ladder needs. For instance, several smaller ladders were arranged to be adjustable by sliding them against one another through connecting channels. These extension ladders were designed to reach a range of heights. As ladders got longer and longer, additional features such as tormentor poles were designed into the ladder to assist in raising.

Ladders can be used for many purposes besides climbing. This chapter describes many innovations that have evolved from hands-on practical use.

LADDER TERMINOLOGY

A ladder is defined as "a structure consisting of two long sides crossed by parallel rungs, used to climb up and down" and as "a means of ascent and descent." Today, fire departments boast many different ladder types and lengths.

There are many parts to a ladder. It is important for every firefighter to know each part by its name. See **Figures 14-1 A–C.**

VIEWPOINT

Different parts of the ladder might have multiple names. For example, one department might use the term *cleats,* while another might call the same part of the ladder a *shoe.* Different ladder skill actions might also have different names. One department might use the phrase "raise ladder," while another might say "ladder up." Normally, these terms evolve from history and tradition, or sometimes just from geographic location. The specific command or ladder part name is not as important as the need for all department members to use common terminology.

SAFETY

Common terminology usage will reduce miscommunication when passing along information or requesting assistance.

Parts of a Ladder

- *Beam.* The side of the ladder. It is the rail that runs the full length of the ladder from top to bottom from which rungs span, **Figure 14-1A.**

- *Bed section.* The part of the ladder that is the foundation, usually the part of the ladder that is in touch with the ground or attached to the body of an aerial ladder truck. It is the section from which all other sections (see Fly section entry) are raised in extension ladders. This is also called the **bed ladder.**

- *Heel (also called foot, base, or butt).* The bottommost part of a ladder. It is usually a reinforced section with points, spurs, or rubber pads to reduce slipping on various surfaces.

- *Spurs, spikes, cleats, shoes, and butt plates.* The pointed shoes that are attached to the base of a ladder to dig into the surface and prevent slippage during use, **Figure 14-1B.**

- *Dogs, pawls, rung locks, or ladder locks.* The mechanisms in an **extension ladder** that ride up along with a fly section and engage a rung of the section from which they extend to prevent retraction through the use of a spring-loaded lock.

- *Fly section.* The section (or sections) of an extension ladder that is (are) raised. Also called the **fly ladder.**

- *Pads.* Nonslip pieces of rubber or plastic that attach to the bottom of a ladder, usually in a swivel-type foot designed to lie flat against the ground or floor to prevent slippage on smooth surfaces, **Figure 14-1C.**

- *Guides/channels.* Channels of the **bed ladder** that permit the fly sections to ride up and maintain stability.

- *Halyard.* The cable or rope, made of nylon, hemp, or steel, that is used to raise or lower the fly section out of or into the bed section through the use of pulleys.

- *Sensor label.* A heat-sensitive label affixed to the ladder to alert firefighters that the ladder has been exposed to a potentially damaging heat level and that testing should be performed before it is used again.

- *Hooks.* Retractable hooks that permit certain types of ladders to be placed on a slanted roof surface and used for footing stability.

- *Rung locks.* see Dogs.

- *Pawls.* see Dogs.

- *Protection plates.* Reinforced metal that is built up at chafing points to avoid weakening created by rubbing and friction wear.

- *Pulley.* A wheel with a groove through which the **halyard** passes. It is used for raising or lowering the fly sections of extension ladders.

- *Rails.* Running lengthwise, the upper and lower surfaces of the beams.

- *Rungs.* The "steps" of a ladder that connect one beam to the other. Generally round, they can also be flattened on the upper side for more secure footing.

- *Stops.* Limiters built into the bed section to prevent the extension fly sections from being overextended. They are found in the shape of solid blocks or angled metal.

- *Tie-rods.* Found on wooden ladders, these metal rods secure the two beams and prevent them from spreading apart when the rungs are doweled into the beams. These rods prevent the rungs from pulling out, which would result in a ladder collapse.

- *Tip.* The top of the ladder. In an extension ladder, it would be the top of the fly section that attains the greatest height.

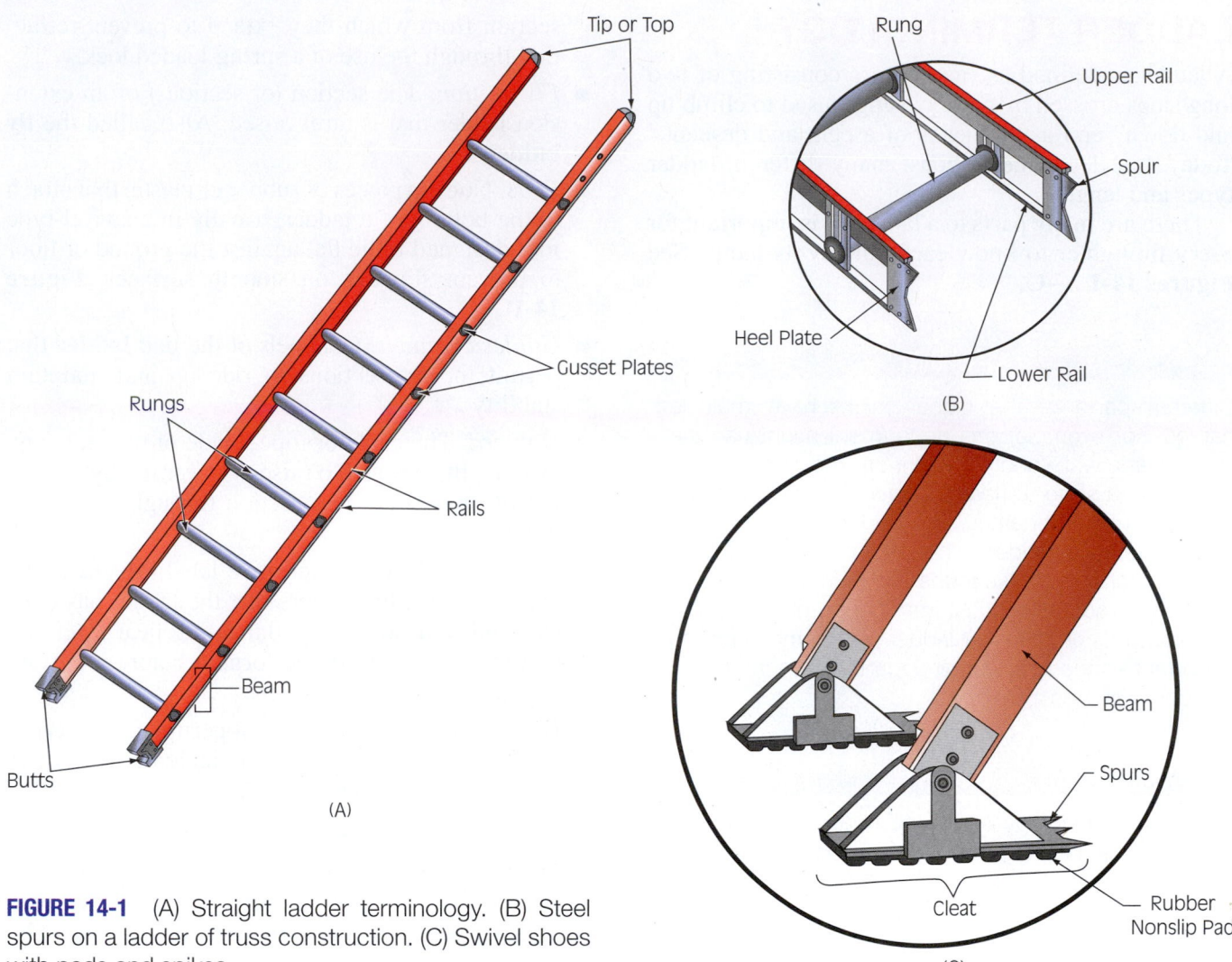

FIGURE 14-1 (A) Straight ladder terminology. (B) Steel spurs on a ladder of truss construction. (C) Swivel shoes with pads and spikes.

LADDER COMPANIES

In the past, most fire departments employed a single apparatus and *it* was the "fire department." Eventually as the number of fire companies grew, they became specialized to perform specific duties. The required tasks were logically associated with the apparatus. Ladder companies are the apparatus that carry ladders and other devices, tools, and personnel to upper levels. Tower ladders and articulating boom ladders are included in this category, although technically, they are not ladders. These units, however, carry ladders that are more traditional in construction and use.

As more and more fire departments created special units to serve particular functions, the personnel assigned to those apparatus were trained to operate the equipment on that vehicle. With specialized tools designed to complete a specific function, a certain "division of labor" developed in which firefighters on certain types of apparatus performed designated functions. Engine companies carried hose, nozzles, and appliances that served the water delivery function. Ladder companies, originally developed to gain access to upper floors, became responsible for tasks associated with entry. That included forcible entry of the fire area, access to the roof for ventilation operations, and access to upper floors for entry, search, and rescue. Eventually, the need for access to higher levels of the building became evident. Higher reaching ladders, better designed tools, and a wider measure of knowledge became the trademark of the ladder company and its personnel.

GROUND (OR PORTABLE) LADDERS

Carried on all ladder truck (and engine) apparatus is a complement of ground ladders, sometimes referred to as portable ladders. The most common types are straight ladders, extension ladders, and various types of specialized ladders. Although standard ladder lengths are often found on an apparatus, each might contain different lengths depending on the department and its needs. Ladders are also carried on engine company mobile water supply apparatus for limited applications.

Straight Ladder

Also referred to as a wall ladder, the straight ladder is a fixed length ladder. Usually found in lengths between 12 and 20 feet, they are generally long enough to gain access into first-floor and second-floor windows, **Figure 14-2.** The surrounding topography of the ground will affect this capability. Beyond the height of the fixed ladders, extension ladders may be employed. The straight ladder is generally light, can be carried by one person, and can often be raised by one person using special techniques designed to perform that function. Straight ladders can be used for access, ventilation of upper floor windows, and escape.

In departments without ladder companies, several ladders chosen to serve the needs of the response area are carried on the pumping engines or mobile water supply apparatus. Several lengths of ladders can be carried **nested** into one another and secured to the apparatus by restraining devices. This arrangement maximizes carry capability with minimal storage space use, **Figure 14-3.**

Extension Ladder

An **extension ladder** consists of two or more ladders that operate as a unit. The bed ladder acts as the nest for the movable fly ladder(s). In the two-piece extension ladder, the **fly ladder** slides in channels built into the bed ladder, **Figure 14-4.** Some ladders that reach beyond 25 feet can have two or more fly ladders with each ladder running through channels of the ladder beneath it. Extended by use of a rope **halyard,** these ladders are designed to maximize mechanical advantage through the use of pulleys. These pulleys are arranged so that as the fly ladders extend, the rungs of the various sections come to rest in alignment at each stop. Each level that the ladder can reach is locked into place on the bed ladder by the use of rung locks, also called dogs or pawls. These locks secure the unit into place. The value of an extension ladder is that one ladder can be used to reach various heights by moving the fly ladders up and out from the bed ladder. One adjustable ladder can serve the function of several fixed length ladders. In addition, if the need for different heights is required at a remote location from the apparatus, the same ladder can be used. Access to a second-story window and then to a third or even a roof can be accomplished without the need for a second ladder. Primarily carried on the ladder truck,

FIGURE 14-2 Straight wall ladders.

FIGURE 14-3 Ground ladder nested on the side of a ladder truck.

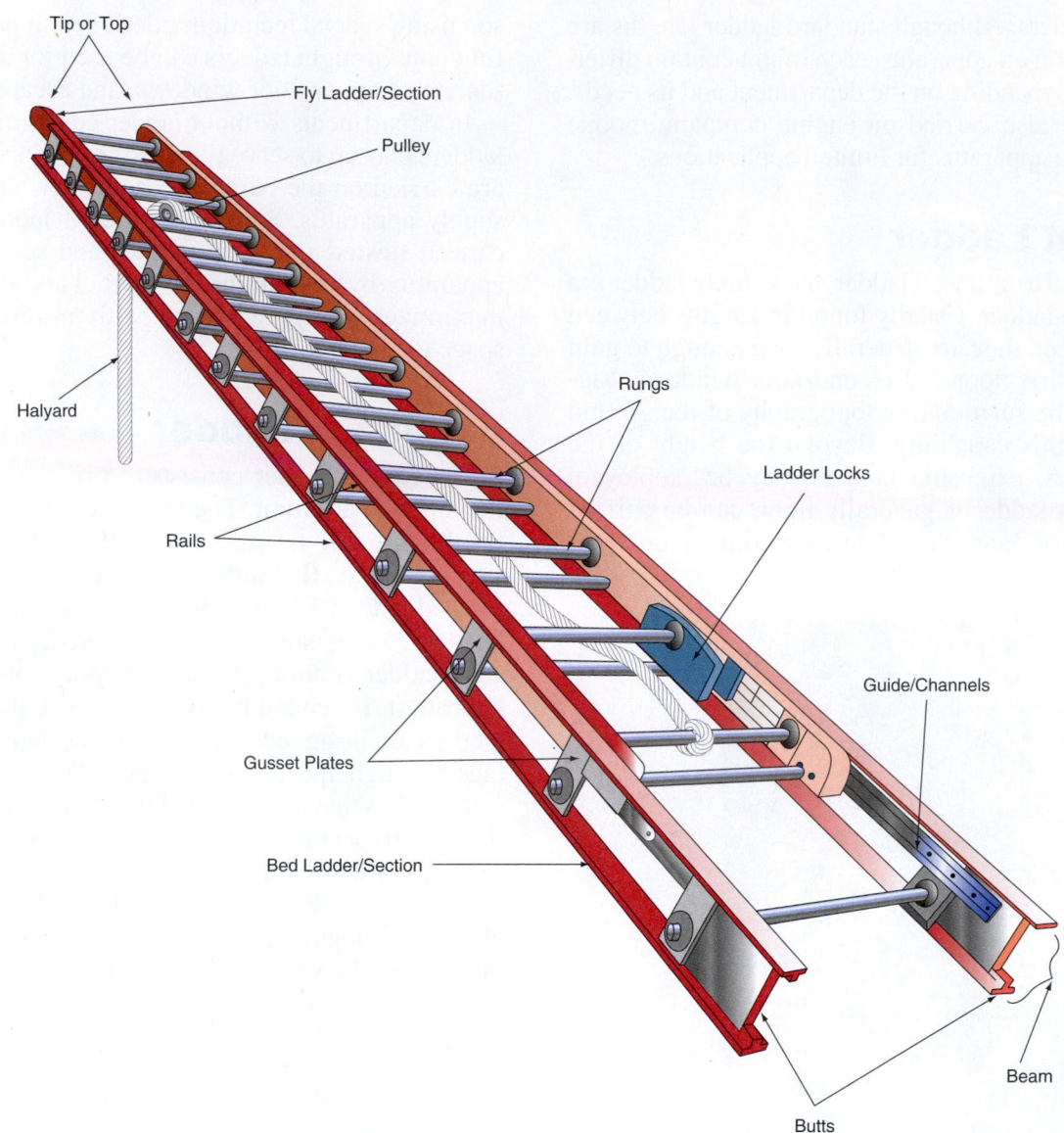

FIGURE 14-4 Extension ladder terminology.

they are also found on pumpers for use when a ladder company is not on the scene.

When an extension ladder exceeds 40 feet, it is required to be equipped with **staypoles,** also called tormentor poles. Called Bangor (pronounced "Bang-gore") ladders or pole ladders, these ground ladders generally do not exceed 50 feet. The staypoles are used only for raising, and, once the ladder is in a raised position, they are not used to support any weight on the ladder. Because of the weight and size of these ladders, the staypoles are needed to push up the tip of the ladder in the initial stages of raising. Used primarily for elevated access in remote locations where aerial apparatus cannot be placed, they require up to six firefighters and are generally used when no other means of access is available, **Figure 14-5.**

Roof or Hook Ladder

The roof or hook ladder is basically a straight wall ladder that possesses a set of retractable hooks at the tip end, **Figure 14-6.** Used when operating on a sloped roof, it enables a firefighter to work with more secure footing. When extended, the hooks are placed over the ridge of a peaked roof while the ladder rests on the sloped roof. The rungs are then used by the firefighter to gain a foothold where it would otherwise be impossible. In the retracted position, the hooks remain out of the way, enabling this ladder to be used as a standard straight wall ladder. A roof or hook ladder is usually found in lengths of from 12 to 24 feet. A hook ladder is not designed to be used as a hanging ladder.

FIGURE 14-5 Bangor (*Bang-gore*) ladder raise. This ladder is also called a Pole Ladder. The staypoles are used only for raising.

Folding Ladder

The folding ladder is known by many names. Also called a suitcase ladder, an attic ladder, or a closet ladder, its function is to enable transport into narrow and confined places where a ladder is required. Although usually carried in a 10-foot model, it is available in lengths from 8 to 16 feet. It is a straight ladder that can provide access to attic spaces through hatches from the top floor of a structure. Its collapsible feature is very useful when attic access hatches are found in closets. When bedded (or folded), it is very portable through tight corners and into narrow spaces, **Figure 14-7A.** It can be used to remove trapped occupants through an elevator car's roof hatch. It is an ideal ladder to use to reach a sprinkler shutoff located high off the floor in a large open area of a warehouse. Foot-pads are provided as a safety feature to prevent slipping because this ladder is very often used indoors on structure floors, **Figure 14-7B.**

A-Frame Combination Ladder

An A-frame ladder is a combination ladder that can be used in various configurations. When nested, it is easily stored and acts as a mini-extension ladder, **Figure 14-8A.** Without the use of halyards, the fly ladder

FIGURE 14-6 A hook ladder, with retracted hooks, being used as a straight ladder.

can be manually raised to the desired level and locked into place, **Figure 14-8B.** When fully articulated and extended, it can be locked into place as a full fixed straight ladder. When articulated into an A shape, it becomes a stepladder, **Figure 14-8C.** Each position is provided with locking mechanisms to secure the ladder into that configuration. Like the folding ladder, its greatest asset is its mobility in tight places because it can reach locations that would otherwise be unsafe or difficult with conventional length ladders.

USE AND CARE

Because of the harsh environment in which ladders will be used on a regular basis, as much care as possible must be employed to prevent the avoidable causes of ladder damage. See **Table 14-1** for a list of some tips that should be second nature to all firefighters. **NFPA Standard 1932,** Use, Maintenance and Service Testing of Fire Department Ground Ladders,

(A)

(B)

FIGURE 14-7   Folding ladder (also called a suitcase or attic ladder). (A) Folded. (B) Opened.

covers the use, maintenance, and service testing of ground ladders.

MAINTENANCE, CLEANING, AND INSPECTION

Because of the nature of firefighting, ladders will be subjected to a great deal of misuse, for instance, in emergency situations where on-site adaptations are employed when time is short. As a result, extra vigilance must be exercised when inspecting ladders, **Table 14-2.** With high levels of heat, overloading, and general rough handling, fire service ladders experience abuse that warrants special inspection attention. Ladders should be inspected at regular intervals and after each use—before they are stowed on the apparatus. **NFPA Standard 1931** sets out the standards to which fire service ladders must conform. Also required is a certification label affixed to the ladder that provides

evidence that the ladder is in conformance with the NFPA standard. Any ladder that needs repair should be immediately taken out of service.

Defects are uncovered through regular inspection and maintenance, **Table 14-3.** When a ladder is put through regular maintenance routines, it is generally verified that it is in a state of readiness. Any work beyond general maintenance should be performed by trained ladder repair technicians. In addition, detailed maintenance records need to be kept to confirm that required inspections are being performed regularly.

Cleaning Ladders

The act of cleaning a ladder is not just a function of improving its appearance. Dirt can cake on moving parts and prevent proper operation. Dirt and caustic substances can also act as an abrasive, causing accelerated wear of moving parts. Generally, warm, soapy

(A)

(B)

(C)

FIGURE 14-8 Combination A-frame ladder. (A) Used as a short extension ladder. (B) Being converted from an extension ladder to an A-frame stepladder. (C) In the A-frame stepladder mode.

TABLE 14-1   Ground Ladder Tips

- Store ladders in clean, dry places.
- Store ladders with multiple support points to prevent bowing or sagging. Support points should be inspected regularly to prevent damage due to ladder wear.
- Avoid moisture and keep ladders as dry as possible.
- Do not place ladders where they are subjected to heat or exhaust.
- Do not lean ladders on movable objects such as apparatus.
- Do not place ladders in out-of-sight areas where they can become tripping hazards.
- When left unattended, secure ladders at their tips to prevent them from toppling.
- Ladder position is important. If at all possible, a ladder should not be placed over a doorway or a window.
- Paint will tend to hide defects. Paint only 14 inches of the tips or heels for identification.

TABLE 14-2   General Inspection Guidelines

- Check halyard ropes for undue wear. Replace if rope is kinked or frayed.
- Check rivets for tightness.
- Check for length indicators.
- Check that ladder has additional warnings required by NFPA (electrical warning, positioning warning, etc.).
- Make sure that rub plates are secure, without burrs, and not worn out.
- Check support plates for tightness.
- Check the heat sensor label for presence and condition.
- Check for splitting or cracks, which are cause for repair or replacement.
- Make sure that bolts are secure and not overtightened.
- Make sure that rungs have no play or movement.
- Examine any discoloration closely for possibility of heat damage.
- Check for cracks in welds, which are cause for concern.
- Check for any wavy or deformed areas on the surface of the ladder that might indicate damage.
- Examine contact points where a ladder might develop wear from rubbing during transport.
- Ensure that moving parts are well lubricated and clean, as per department policy.

TABLE 14-3   Extension Ladder Inspection Guidelines

- Ensure that dogs are freely operating and springs are operational and in place.
- Make sure rope halyard is not knotted, kinked, or worn.
- Make sure there is no excessive play in the halyard.
- Ensure that pulley is not out of round and operates freely.
- Ensure that fly ladders operate and slide freely in channels.
- Ensure that any operating latches on Bangor ladders do not bind.
- Ensure that the halyard cable of multiple-fly extension ladders is snug when in bedded position. Inspect and replace cable if it is kinked or frayed.
- In the bedded and extended positions, ensure that all rungs line up when pawls engage.
- With any of the special folding type or articulating ladders, check that the hinges are secure.

mild detergent and a scrub brush will remove most dirt. During the cleaning process, closer inspection of the ladder will occur as opposed to a general overview inspection. The nooks and crannies of the ladder will be observed, and defects are more likely to be uncovered. If a buildup of tar or grease is discovered, proper solvents and thorough cleaning are indicated with renewed lubrication of moving parts if warranted. Manufacturer's recommendations should be consulted for the proper choice of a cleaning agent for use with ladder material.

LADDER USES

VIEWPOINT

Ladders are designed and tested for one function—climbing. However, there are departments who choose to use ladders for many different applications. Firefighters should always operate within the scope of the local department protocol and policy. Knowledge of allowable ladder uses will become even more important when responding on mutual aid calls. One department's ladder practices might not necessarily be acceptable to another department. Communication and training are key to solving this dilemma.

By design, ladders are used primarily for climbing. However, ladders are used for many additional purposes in the fire service. Many departments use them as a shoring tool, a fence, a means to hold back loose debris, or a chute to channel water with a tarp—and these are just some of the other available applications. Some of the more exotic uses are covered later in the Special Uses section.

NOTE

It is important to remember that ladders are tools that have an intended purpose. All ladders should be inspected after use. However, if ladders are used for any purpose other than access and rescue, they should be thoroughly and carefully inspected before returning them to service.

Access

The most obvious use of a ladder is for access. A ladder can provide a path to an otherwise inaccessible opening or height. The first image often pictured is an elevated climb, but a ladder can be used to descend into an opening or as a bridge between two points at the same level.

Rescue

This is probably the most recognizable use at a fire scene. The drama in extracting a victim from the "jaws of death" is played out in most daily newspapers and television news programs. Not to diminish the importance of this role, but it is probably the use of ladders that is employed least often!

Salvage Operations

There are times when a ladder can be used as a tool to support salvage covers in salvage operations or to protect hoselines from falling glass.

Stability

The use of a roof or hook ladder will provide footing stability to a firefighter working on a sloped roof.

Ventilation

Using a ladder for ventilation can take place in one of two ways. A firefighter can use the ladder to remove glass with a tool from an elevated position. Also, the ladder itself can be used as the tool that takes out the glass. As always, safety must be paramount in either use.

Bridging

A ladder can be an effective bridge between two points. It can be used to support weight over a weakened floor or afford stable footing over a collapsed stairway. It has been successfully used to reach victims through windows from across a shaft, such as an alley. When bridging, a bedded extension ladder is the safest to use. By using it in this fashion, the firefighter is afforded the benefit of two ladders that support one another. When bridging, the portion of the ladder that rests on the support points should be at least 5 feet on each end of the ladder for each 10 feet spanned. In other words, the portion of the ladder that rests should be one-half of the distance spanned.

Elevated Streams

The use of a ground ladder for an elevated stream has lost its place in the fire service over the years with the introduction of the **tower ladder** and **ladder pipe.** It is still, however, an option if needed. The application of water from an exterior location off a ground ladder may be employed when no other approach to the fire is available. This certainly does not advocate the practice of exterior firefighting over the interior approach, but in some cases there might be an advantage to such a practice. For example, in a structure that has had its interior stairs burned away with no other access, firefighters might find the elevated handline stream an asset. Apparatus-mounted appliances in the form of tower ladders, ladder pipes, or deck guns might be unable to be positioned for remote position water application.

Elevated Work Position

A ladder can serve as an exterior work platform. Just as a painter uses a ladder for reach, so can a firefighter. There may be a need to remove something or check for heat during overhauling.

LADDER SELECTION

Once a general tactical attack plan is formulated, ladder selection can take place. Among the many ladder choices will be straight, extension, aerial, and tower. Once the target is identified, a secondary set of considerations should be entertained. What length ladder is necessary, and what will be done with the ladder after the initial use? When used as an access path to a fire or search area it must be left in place. Or will the firefighter be using it at several locations, as would be the case for ventilating a series of upper level locations? Whereas a straight ladder would be ideal in the first application, an extension ladder might be more appropriate if different heights will be necessary after the first goal is attained. The ease of portability and use of the straight ladder might have to be sacrificed in order to meet the needs of the goal.

NOTE

One of the first considerations that will have to be made is where the ladder will be placed. The second will be the length of the ladder needed to reach the objective. A general rule of thumb for ladder selection involves the normal heights of the floors of residential and commercial structures. Residences measure approximately 8 to 10 feet from floor to floor. The average commercial story will average approximately 10 to 12 feet from floor to floor. This information is essential for choosing the correct length ladder to use. When in doubt, firefighters should use a longer ladder.

Some additional considerations that generally need to be entertained once potential locations have been identified are:

- *Ground condition.* Is the ground area soft, hard, muddy, gravel, concrete, or slippery?

- *Height needed.* Is the distance single height or several heights?

- *Purpose.* Access will require stationary placement. Rescue might entail extracting a number of people or moving the ladder for multiple rescue operations. In conjunction with rescue, other tasks (e.g., ventilation) may be required of the ladder.

- *Slope of ground.* At some locations, the ground slope may prevent the safe and level deployment of a ladder.

- *Accessibility of location.* Just getting the ladder to the objective may present a challenge. Ladder portability through passageways to rear or sides, over the roof and down the rear of a structure, or over fences could challenge the objective.

- *Available personnel.* The number of personnel available for assistance will greatly affect the choice of ladders.

- *Overhead considerations.* Electrical wires, porch overhangs, tree limbs, clotheslines, signs, and other obstacles can hinder ladder operations.

- *Raising space considerations.* **Tip arc** clearance will be necessary for raising, or the ladder might need to be raised first and then carried to the location in vertical mode.

- *Stability.* Is the structure capable of supporting the weight of the ladder?

Because ladder work usually requires two people and not always the same two people, specific guidelines must be established in the raising of a department's ladders. There are many ways to raise a ladder, and the individual past history of each firefighter who may have worked for a utility company, construction, or in general household duties will present an array of different raising techniques. One standard technique should be established so that each firefighter will be able to perform the same function, same foot location, same hand placement, and so on. At any time, the firefighter can be interchangeable but the routine will remain the same. This eliminates wasted effort and unnecessary discussions on how to perform the skill at the scene.

Ladder-raising skills should be practiced and become second nature to all firefighters. This in no way implies that discussion and communication should be curtailed, but, on operations such as this, discussion should not be necessary beyond the identification of the target and a general "thinking out loud" so each firefighter knows what the other is going to do.

Butt Section

If the butt or heel of the ladder will be placed on flat ground, there generally will not be a problem deploying a ladder. However, if the ground slopes, raising a ladder might be impossible. The practice of chocking one beam with fillers such as wood or chocks is dangerous and should not be employed under any circumstances.

> **CAUTION**
>
> Ladder chocking is dangerous! One shift of the ladder as weight is being distributed can cause the entire ladder to topple. The operation is too dangerous and should not be permitted. Firefighters should attempt to ladder in an alternate location if sloping ground is encountered. In cases where sloping ground must be dealt with, the use of an aerial-mounted ladder would be called for.

The point where the butt is positioned should be directly under the target with an appropriate distance from the wall to create the proper climbing angle of about 75 degrees—usually one fourth of the working distance of the ladder. It is easier to adjust the ladder's angle to the building than to move the raised ladder laterally. When transporting a ladder, the heel or butt should be carried in the direction of the target. Once the target is reached, the butt can be planted and the ladder raised almost without hesitation. If the butt were carried in reverse, the lead firefighter (carrying the tip end) would have to pass the target before the butt could be planted. This practice wastes valuable time and requires additional effort.

Fly Section

Just as the butt of the ladder dictates the ladder placement, the tip of the fly dictates how the ladder will be used. There are several specific locations where the placement of the tip will be important and contribute to the success of the operation and achievement of the goals.

Windows

When the tip is placed at a window, two general guidelines should be followed for placement:

1. If the ladder will be used for access or escape, the tip should not extend into the window frame, **Figure 14-9A.** For every inch the ladder penetrates the opening, the size of the opening is reduced.

Furthermore, the higher the extended ladder is into the opening, the more the victims or rescuers will be exposed to the heat escaping the window at the upper level as they attempt to mount the ladder. The ideal location for access is for the ladder tip to be level or slightly below (no more than a few inches) the windowsill. This permits the firefighter to climb in face first over the windowsill below any heat. In addition, this ladder placement will provide some measure of safety to a victim or firefighter if immediate escape is required. The ladder location will at least provide a place on which to land if immediate escape is absolutely required.

2. Placement of the ladder with the tip at the top of the window frame to either side of the window, **Figure 14-9B,** has several positive and negative aspects. On the positive side, it affords the firefighter a relatively safe location from which to perform ventilation. The firefighter can remove windows with a tool while remaining below and to the side of the venting heat. However, to make access once the window is opened, the firefighter must negotiate the distance from the window to the ladder that is created by the angle of the window to the wall. As the distance from the tip increases, the distance of the beam from the building increases. At the sill level, depending on the height of the ladder, the distance could be from 18 to 24 inches. This gap can create difficulty in victim removal or access to a window that is spewing heat and smoke.

> **NOTE**
>
> Some ladder companies have installed mini-strobe lights on the tips of aerial ladders that only illuminate in the raised position for visibility at night or in smoke. Other department members carry portable magnetic strobes or strobes with a hook to attach to the tips of ladders to make them more visible in an emergency egress.

Roof Level

When placed for access to a roof, the tip should extend above the roof level approximately five rungs. That will provide about 5 feet of visible ladder over the roof. There are good reasons for this over-extension.

1. The firefighter should climb onto the roof without overreaching and causing imbalance. The extended tip affords a handhold that can be used while dismounting the ladder.

(A)   (B)

FIGURE 14-9   (A) Ladder placed with the tip below the windowsill. (B) Ladder placed with the tip at the top of the windowsill to either side.

2. The tip will be visible to any roof occupants that need to find it, especially if it is an escape route.

3. The upper rungs of the ladder might be needed as a tie-off point for a rope where one is needed and the roof structures offer no alternatives. This practice should be employed only when absolutely necessary, such as when a lifesaving effort is involved. It should not be used merely as a tie-off point for rope in any other application because it would immobilize the ladder, which might be needed for a rescue effort elsewhere.

Fire Escapes

When raised to a fire escape, the position of the ladder will be guided by the intended use for the given situation:

1. If the ladder tip is on the fire escape itself, the tip should be level with or below the upper rail so it can be mounted by straddling the rail and stepping onto the ladder.

2. If the ladder tip can be placed against a wall adjoining the fire escape, it should be several rungs above the fire escape rail so a person can swing a leg over the rail and step onto a rung while holding onto an upper rung. This is the preferred method. It will provide a victim with a stronger sense of stability.

When placing a ladder on a fire escape for victim removal, the priority and sequence should be that the first ladder is placed opposite the drop ladder. The second ladder is placed on the same side as the drop ladder on the floor above, **Figure 14-10.** In this manner, the evacuees can mount ladders at three points without having to wait for the person in front to clear the area. Additionally, if all the escape ladders can only be served by a single file access, removal will only occur as fast as one person can climb on a ladder. That speed is usually slow.

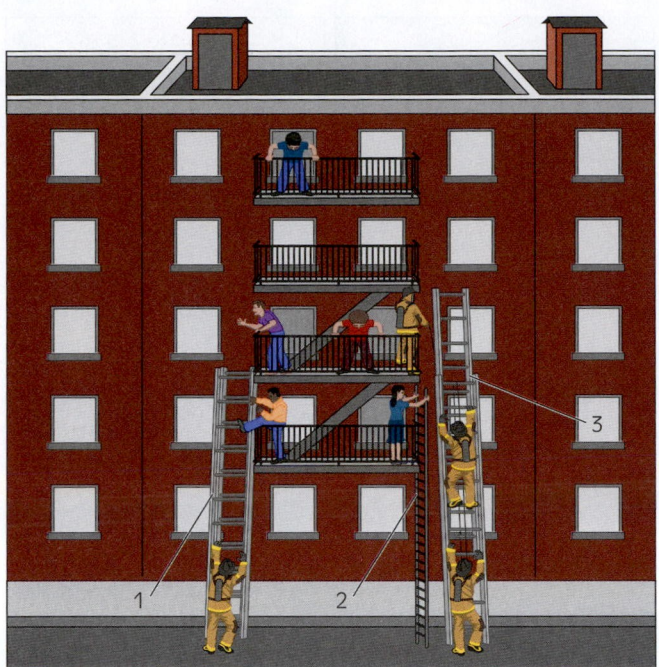

FIGURE 14-10 Ladder placement at fire escapes: (1) Drop ladder or stairway. (2) Opposite drop ladder or stairway. (3) Balcony above drop ladder on the same side of balcony to provide for maximum escape flow.

SPECIAL USES

Ladders can be used as tools or as portable stairs as long as ladder integrity is not compromised. Whether for venting upper floor windows with the tip or as a fence, firefighters should be alert to visualizing adaptive uses for a ladder to achieve the firefighting goal.

SAFETY

Ladders were designed for a specific purpose. Using a ladder for any purpose other than access to above- or below-grade areas must result in the ladder being removed from service and thoroughly inspected.

Removal of Numerous Victims

The usual method of removing a victim is to raise the ladder, ascend and secure the victim onto the ladder, and then descend escorting the victim. When multiple victims need removal, taking one person at a time can be a very time-consuming process, even if several ladders and firefighters are available. A method to "keep the flow going" is to place two or more ladders at the escape point. This will usually be a large area where victims might be collecting such as a roof of a lower building, an area of escape refuge, or perhaps a school classroom where children have been brought. One ladder is used by the firefighters strictly to ascend, and the others are used to descend escorting a victim. The one "supply" ladder can service many escape ladders, **Figure 14-11.**

Chute with a Tarp

If water from a leak or part of an upper floor firefighting operation is threatening lower floors with damage, a ladder with a tarp draped over it can be used as a makeshift water chute to direct water coming through the ceiling out a window, **Figure 14-12.**

Over a Fence

Two short ladders tied together at the tip in an A-frame format can be used to climb over a fence. There are times when a fence cannot be cut and operations are impeded by the obstacle, **Figure 14-13.**

Elevated Hose Streams

In a location that could not be approached by conventional apparatus with elevated stream capability, a handline from a portable ground ladder was a standard firefighting technique in the days before mechanized master streams. It can still be effective when

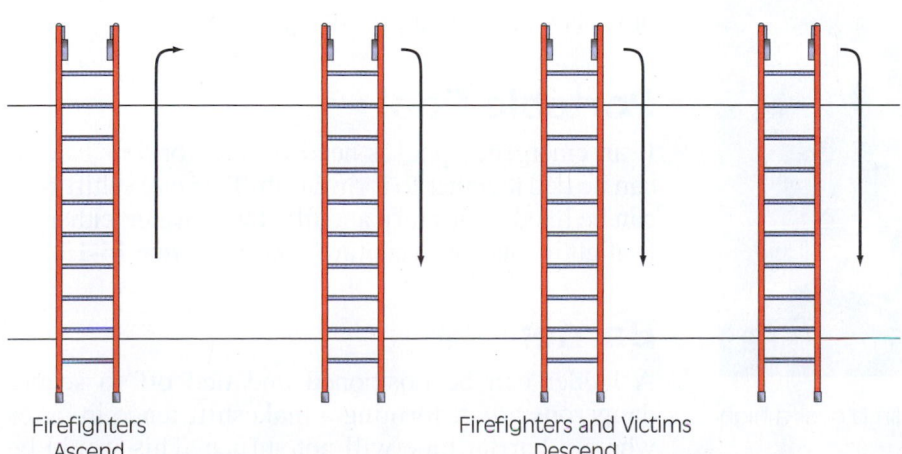

Firefighters
Ascend

Firefighters and Victims
Descend

FIGURE 14-11 Multiple ladder rescue technique.

FIGURE 14-12 Ladder with salvage cover, plastic sheet, or tarpaulin used as a chute to divert and discharge water.

FIGURE 14-14 A handline can be used off a ground ladder for difficult-to-reach areas. Note the use of a ladder belt.

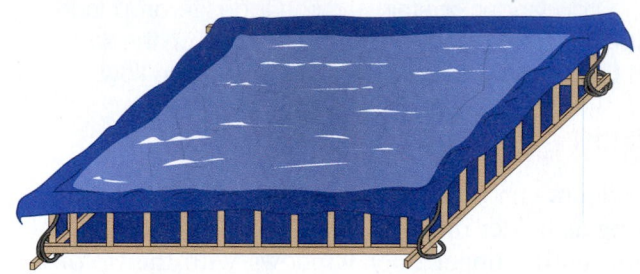

FIGURE 14-15 Ladders can be turned into an emergency water pool or collection area.

FIGURE 14-13 Ladders can be used to climb over a high fence.

no other approach is available, **Figure 14-14.** When using this technique, basic safety practices must be employed. The firefighter and hose must be secured to the ladder, and the ladder must be stabilized at the base or at the tip. Each department must develop this operation in a form that uses its individual inventory of tools without compromising safety.

Portable Pool

If an emergency pool is needed, three or four ladders can be tied together to form a crib. This makeshift pool can be lined with a tarp and filled with water, either for firefighting use or to capture runoff, **Figure 14-15.**

Barrier

A ladder can be positioned and tied off to secure dangerous areas, forming a makeshift fence in cases where a barrier tape will not suffice. This would be especially useful where, for example, a large hole needs to be barricaded. Distracted individuals might accidentally move through the presence of flimsy barrier tape. The ladder barrier will offer a positive visual deterrent as well as a physical mechanism to prevent a pedestrian from passing, **Figure 14-16.**

Support

A dangling sign or cornice that needs to be supported across a span and that is beyond the capability of a mere rope can be supported by a ladder that spans the unsafe structural component. Using ropes, each end can be tied off to secure objects only as an emergency structural stabilizer. This temporary emergency measure should be replaced as soon as possible with tools designed for the situation. The ladder must be thoroughly inspected before being placed back into service, **Figure 14-17.**

Hoist Point

A set of ladders tied off at the tip and at the base into an A-frame configuration can be used as an emergency hoist point if a pulley and rope are attached to the apex of the ladder triangle. Ropes are lashed at the beam intersections and where the rungs run parallel. In addition, the support for the pulley is lashed to the rungs, creating a stable base, **Figure 14-18.** Special care should be used to make sure the ladder weight limits are not exceeded.

Ventilation Fan Supports

A short ladder spanning an opening—such as a cellar opening or a hole made in a floor—can support a ventilation fan or blower. It can also be used to hang a fan at an upper level opening if such fan placement is desired, **Figure 14-19.** The fan can be placed on the ladder, in front of the opening, and operated from an exterior location.

These are just some of the examples of optional ladder usage. As is always the case, innovation must be tempered with safety, and the weight limits of a ladder should never be exceeded. In addition, whenever a ladder is used in a fashion other than that for

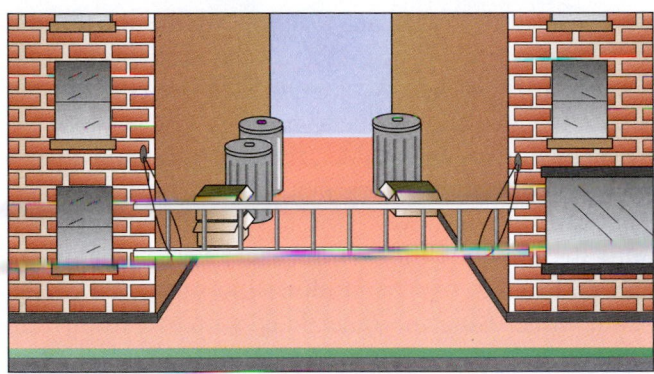

FIGURE 14-16   A ladder can be used as a barrier.

FIGURE 14-17   Ladder used as a shoring tool. A ladder secured to substantial objects by ropes can assist in stabilizing a structural defect as an emergency measure.

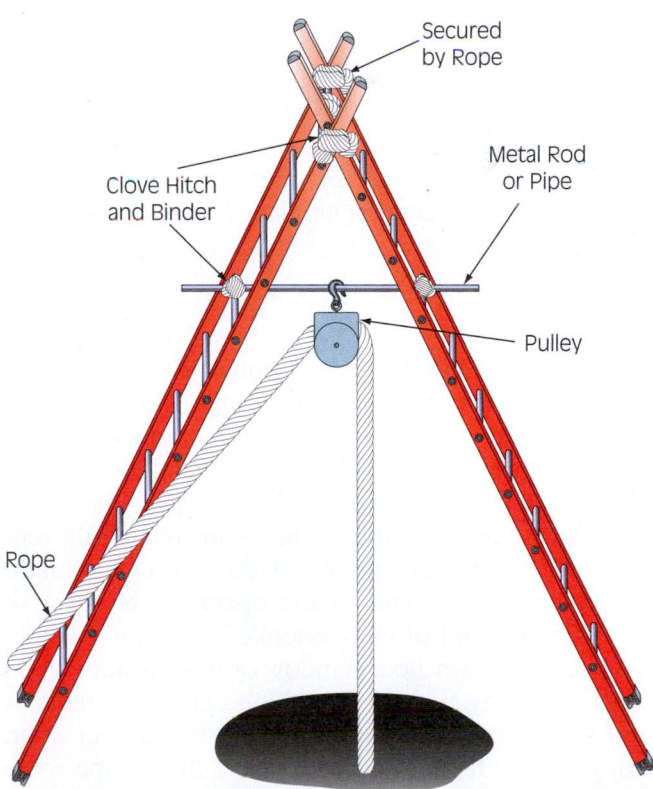

FIGURE 14-18   A-frame hoist.

FIGURE 14-19 A ladder can be used to support a fan in a doorway.

which it was designed, a thorough inspection is essential before the ladder can be placed back into service.

SAFETY

In most cases, ladder safety is equated with common sense. Overreaching, overbalancing, or overloading ladder limits are the most common causes of ladder-related injuries in general use. At fires, a host of new dangers is added to the list. Ladder placement is a critical element. Obvious dangers are generally recognized, but placement of a ladder where it might become a hazard later in the operation is one risk that is easily and often overlooked. Placing a ladder in front of a lower floor window or door is not a good practice at any time. It becomes a critical danger if and when fire erupts out that lower window or door, exposing the ladder and cutting off the escape route of firefighters using the ladder. Also, persons emerging from the door might topple the ladder and close

an effective escape route for firefighters or cause a firefighter to fall from the ladder.

> **SAFETY**
>
> An act or operation that would not be done on the ground should not be done on a ladder.

The use of gloves reduces the possibility of pinching fingers or skin, which can occur with articulating, folding, or extension ladders. When climbing, placing, or positioning a ladder with debris falling, the firefighter must be looking toward the target. Eye protection is essential.

The use of the correct ladder is important. With an extension ladder, the length is adjustable. With a fixed straight ladder, a target that is beyond the ladder length should not be reached by decreasing the angle to the building or by standing on a set of rungs closer to the tip.

All ladders are electrical conductors despite their construction material. Moisture on the surface negates any composite insulation. Overhead wires must always be considered "live," and improperly insulated, and should be avoided. The quality of the insulation should never be trusted. Age, weather, or heat from the fire can reduce the insulation value to nothing.

When moving on a ladder, firefighters should always keep at least three limbs in contact with the ladder. They should not move to the next rung (hand or foot) until the previous limb is in place. Some firefighters choose to slide one hand along the underside of the beam while using the other hand to move from rung to rung. This movement makes the three-limb contact rule easier for them. Others choose to move their hands alternately from rung to rung. No matter which method a firefighter uses, movement up or down a ladder should be rhythmic, safe, and smooth. Firefighters carrying a tool up or down a ladder should slide their free hand along the underside of the beam while ascending or descending, as just described. A ladder should always be butted (also referred to as footed, heeled, or tended) at the bottom by another firefighter. In the event a ladder must be left unattended, it should be secured by a short rope or cord at the tip to prevent toppling. Finally, whenever the job requires reaching and working off a ladder, the firefighter should be secured to the ladder with a ladder belt or the firefighter should utilize a leg lock. (The leg lock is described later in this chapter in the Ladder Skills section.)

> **SAFETY**
>
> When ascending ladders, firefighters should keep their eyes straight ahead, occasionally looking up the climbing path for hazards and to view the objective.

Even though ladder use is common in the fire services, firefighters must not become complacent to safety concerns and potentially dangerous situations. Dangerous situations change from one fire scene to another. Firefighters must learn to quickly scan the area in which a ladder will be deployed. Anything considered a hazard should be removed. If removal is not possible, the ladder should be deployed elsewhere.

Overhead Obstructions

Tree limbs, structural overhangs, television and telephone wires, and of course overhead electrical lines carry injury potential. The path in which the ladder will be raised must be visualized before setting the ladder base and before the ladder is lowered.

Overhead electrical wires are always a hazard. Although the wires are insulated when installed, weathering can create cracks, which in turn can provide exposure to a bare current-carrying cable underneath. There are many other dangers to be alert for when raising a ladder. If a fire has caused the insulation of a wire to be burned away and the wire is in contact with electrically conductive material (aluminum siding, a fire escape railing, television cable wires, etc.), this material could be electrically charged, **Figure 14-20.** If a ladder is placed against an energized material, the ladder could become an immediate path to ground. If the firefighter is in contact with the ladder, the ground path may course electricity through the firefighter's body.

Climbing Path

The climbing path is the imaginary passageway a firefighter climbs through while ascending a ladder. The path is not only the straight line created by the ladder butt to the ladder tip, but also the pass-through "tunnel" area that is necessary for a person to climb, **Figure 14-21.** If a firefighter is required to alter the normal climbing angle or squeeze through a tight space created by wires, tree limbs, or structural components, the climbing path is obstructed. The space a breathing apparatus cylinder will occupy must be accounted for when estimating the space needed for the climbing path.

Ground Considerations

An assurance that the ladder will be stable must be confirmed before raising. The ground under the ladder must be level and not create a dangerous lateral lean. The longer the ladder, the more a lateral lean will be exaggerated, **Figure 14-22.** What might be an acceptable lateral lean at a very low level might be dangerously beyond a safe position at an extended height.

Ladder Load

The number of people permitted on a ladder at one time will vary. The load capacity is based on weight, not the number of people. The load limits of a ladder should always be understood, known, and not

Aluminum Siding

FIGURE 14-20 A ladder can make an electrical connection to ground.

FIGURE 14-21 "Climbing path" pass-through area.

exceeded, **Figure 14-23.** Because each type of ladder can have different construction and weight-carrying capacity or be built differently by various manufacturers, firefighters should be aware of their department's equipment or at least know where to look for the capacity statement. The recommended maximum load will be found on the manufacturer's label affixed to the ladder, **Figure 14-24.**

Working Off a Ladder

When working off a ladder, it is imperative that the firefighter be secured to the ladder. Ladder belts or safety harnesses have the ability to hook onto a rung to secure the firefighter should balance be lost. Another method of locking into a ladder is by use of a leg lock, **Figure 14-25.** (The leg lock is described later in this chapter in the Ladder Skills section.)

Ladder Storage

When ladders are stored for extended periods it is important that they be supported by more than just two support points. Ideally, ladders should be stored on a flat surface with the ladder on its beam or flat on both beams. By avoiding specific contact points, the tendency for a ladder to sag is reduced. For ladders that are often removed and replaced, a storage technique not using the same contact points will reduce any wear and point weakening. A ladder that is continually hung on its upper beam and constantly slid on and off along a hanger could develop wear spots and weak points at those locations on the beam.

Apparatus Ladder Storage

Under ideal conditions, ladders should be stored under cover in compartments, **Figures 14-26 A–D.** With protection, ladders will not be subjected to weather that could accelerate deterioration of halyards or permit caking of grime or debris on moving parts. It will also prevent accumulation of ice, snow, water,

FIGURE 14-22 Uneven ground effect is magnified as the ladder increases in height.

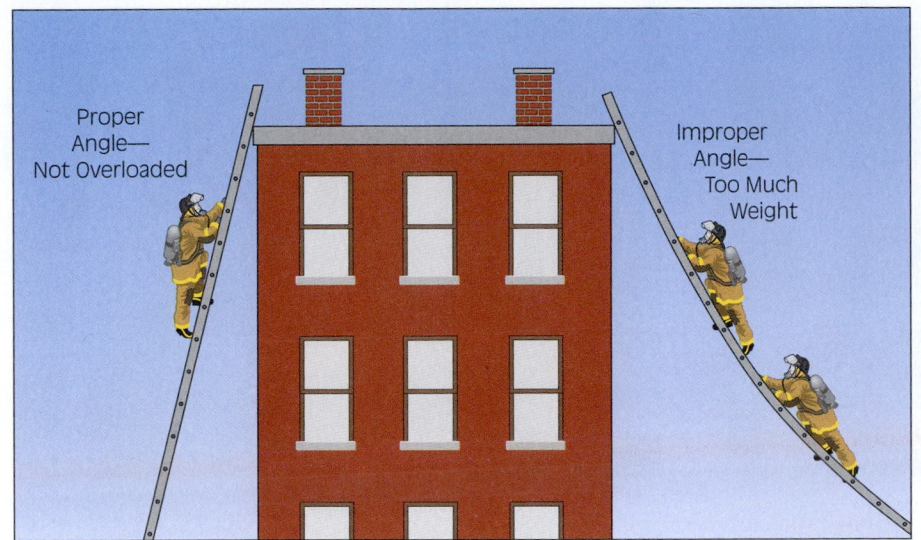

FIGURE 14-23 Ladders must be positioned properly and not overloaded.

FIGURE 14-24 Ladder load limit is found on the label attached to the ladder.

or winter de-icing products on the ladders during inclement weather. This is especially relevant during below-freezing conditions when ice can accumulate on ladder rungs or extension ladder operating parts.

Ladder Apparatus Parking

Most new ladder apparatus have ladder storage compartments that unload at the rear of the apparatus. Very often, units responding to the emergency place their apparatus directly behind a parked ladder apparatus, impeding ground ladder access, and, in some cases, preventing use of ladders all together, **Figure 14-27A.** The best way to

FIGURE 14-25 Leg lock.

prevent the actions of another from affecting the operation is to be proactive. When parking the ladder apparatus, the ladder apparatus driver should place the apparatus at an angle to the fire building. The ground ladders will still be able to be removed, and the turntable ladder position will remain unaffected, **Figure 14-27B.** In addition, all

(A)

(B)

(C)

(D)

FIGURE 14-26　Various ground ladder storage practices.

department apparatus drivers should be made aware of this so that, in their haste to place their own company into operation, they do not block the activities of another.

Ladder Painting

Ladders should never be painted as a means of maintenance, especially wooden ladders. With good intent, ladders are painted in their entirety with a specific color for unit identification or for general appearance, but paint obscures imperfections and hides defects that would otherwise be discovered during inspections. However, small areas of ladders should be painted for identification, visibility, and quick reference.

- *Identification.* Some departments use a color code for individual companies to quickly identify equipment during the postoperation fireground take-up (breakdown) period. It is acceptable to paint the base of a ladder up to the first rung for this purpose, but no part of any structural connecting plates, operating parts, or rivets should be painted.

- *Visibility.* The ladder may be painted from the tip down to the top rung for visibility. Yellow, white, and/or reflective paint helps make the tip visible in dark conditions and can be a genuine aid in raising. It is also a valuable feature for the firefighters using a raised ladder as an escape path from a roof or upper level location. Aerial ladder tips can also be painted to serve this safety function.

- *Quick reference.* The ladder size should be affixed on places that would be instantly visible. In addition to enabling the size to be ascertained immediately when obtaining a ladder, it would also be visible from the ground when the ladder is in a raised position. The use of ladder length identification numbers on the butt (or base) end of the ladder will assist in choosing the correct ladder from the ladder storage bed.

- *Hoist points.* One type of rope hoist method employs using a bowline loop inserted through a specific set of rungs. Small painted identifier strips may encircle the rungs involved for quick recognition (see Ladder Skills section).

(A)

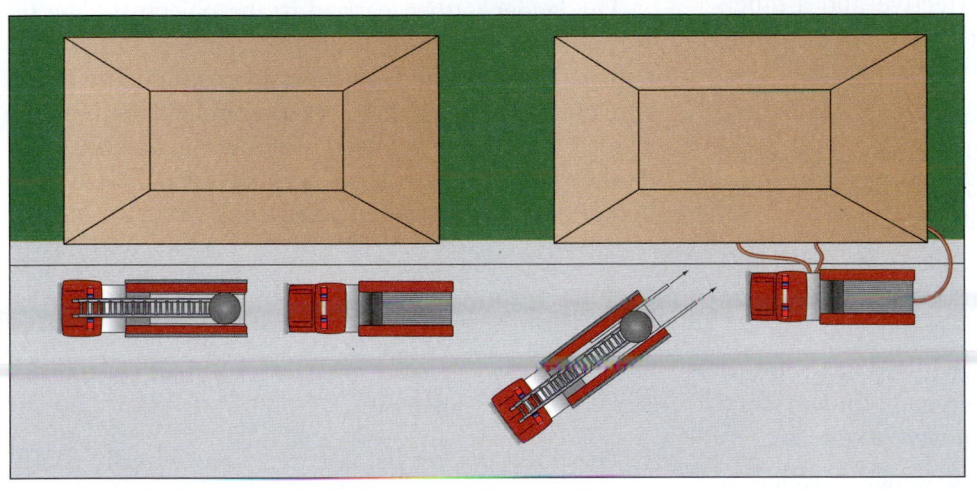

(B)

FIGURE 14-27 (A) It is important to leave room to remove portable ground ladders. (B) Parking apparatus can impede ladder access: parking on an angle can be a simple solution.

Certification and Testing Procedures

NFPA Standard 1931, Design and Verification Tests for Fire Department Ground Ladders, outlines specific procedures for ladder testing and certification. This standard requires that a ladder manufacturer attach a certification label to each ladder verifying that the ladder was manufactured according to the NFPA standard and to OSHA guidelines.

When a ladder model is designed, it is subjected to rigid testing to ensure it will withstand the rigorous tasks required of it on the fireground. Ladder weight loads are tested while the ladder is in a horizontal position and resting on one beam, supported only at the tip and the butt, and again while the ladder is in the climbing position of 75 degrees. The ladder is tested to ensure that both butt spurs remain on the ground under simulated climbing conditions and that the rungs do not fail under simulated fireground operations. The ladder is tested to ensure that swaying and twisting will not result and that the butt spurs will not move across

a surface. Special-use ladders (roof, pole, combination ladders, etc.) are tested to ensure they are capable of accomplishing their specialized functions. It is important to note that a design model ladder is tested and certified, not each individual ladder. After the testing process, the ladder is discarded. The ladder label will attest that the model of ladder is in compliance with the NFPA standard and OSHA guidelines. For more detailed information on ladder certification and testing, please consult NFPA Standard 1931.

LADDER SKILLS

Moving a ladder from point A to point B and setting it in position may seem to be a simple task. However, without common terminology and technique, the moving and positioning of a ladder can be chaotic. It is important that firefighters understand commands given to carry, position, and raise a ladder. These commands should be established by local department policy. Simple and basic commands relying on common sense should be the norm.

Carrying Ladders

Several techniques are used for carrying ladders. Each technique has its own advantages and disadvantages. For short distances, the suitcase carry is sufficient. If longer distances are involved, the shoulder carry is more suitable to reduce fatigue. The butt (or base) firefighter should be the lead and the tip firefighter should be in the rear. Once the raising point is reached, the firefighters plant the base and raise the ladder in almost one motion.

JPR 14-1: Suitcase Carry

(For step-by-step photos of this skill sequence, see page 405)

The suitcase carry is used primarily as a short carry maneuver. Although it is very effective and a quick method of moving a ladder, carrying a ladder in this manner for long distances can be fatiguing.

1. With the butt of the ladder facing the direction of travel, the firefighters position themselves on the same side of the ladder facing in the direction of travel; one firefighter is at the tip and one is at the butt.

2. The tip firefighter controls the operation and gives the commands. On command, both firefighters raise the ladder to stand on one beam.

3. Grasping the upper ladder beam with the hand closest to the ladder, in unison, the firefighters lift and carry the ladder to the objective as if it were a suitcase.

4. If the ladder is heavy or the distance traveled is lengthy, a shoulder carry should be used. However, if the suitcase carry is selected, a third firefighter can assist, with the additional person taking position in the middle of the ladder.

The firefighter at the butt is leading and can use the other hand to push away minor obstructions, open gates, and so on. The firefighter at the tip is the guide. It is up to this guide firefighter to ensure that turns are not too sharp or that angles are not created that cannot be negotiated by the tip firefighter. Communication is key, and the guide should dictate an alternate route if the travel path is unacceptable or if another carry method is necessary.

JPR 14-2: Shoulder Carry

(For step-by-step photos of this skill sequence, see page 406)

1. The shoulder carry is useful for operations where firefighters will have to carry a ladder a good distance. The tip firefighter controls the operation and gives the commands. With the ladder flat on the ground, the two firefighters position themselves, facing opposite the direction of travel (facing the tip) on one side of the ladder beam. One is at the tip and one is at the butt.

2. On command, both firefighters squat down and grasp a rung and raise the ladder to stand on one beam.

3. In unison, the firefighters lift the ladder using their legs for the power maneuver.

4. Both firefighters pivot in unison toward the direction of travel (facing the butt), place their free arm through the space between two rungs, and place the underside of the upper beam to rest on their shoulders.

5. The hand of the arm that was inserted through the rungs reaches to the next rung forward and grasps it.

The ladder is then carried to the objective. The tip firefighter is the guide. The free hand of either firefighter may be used to push away obstructions. In the case of large ladders, three or more firefighters might be needed to perform the carry. The firefighters must use on-the-spot adjustments to account for different firefighter body heights when positioning along the ladder.

JPR 14-3: Flat Carry

(For step-by-step photos of this skill sequence, see page 407)

1. The three-person flat carry is a particularly useful technique when great disparity exists in the height of the firefighters. It also helps to evenly distribute the weight of the ladder. With the ladder flat on the ground, the two firefighters who are about the same height will position themselves, facing the direction of travel, on one side of the ladder beam. One is at the tip and one is at the butt. The firefighter who is taller or shorter locates midway on the other beam. The tip firefighter controls the operation and gives the commands. On command, all three firefighters squat down and grasp a beam with the hand nearest the ladder. Using their legs for the power maneuver, the firefighters stand with the ladder. The ladder will slant up or down toward the odd-sized person, but the weight will be evenly distributed. As with the other carries, the tip firefighter is the guide. The butt firefighter leads the way to the objective.

2. The sequence is basically the same for the four-person flat carry, except that two firefighters are at the tip and two firefighters are at the butt. Either the beam or the rung can be grasped during the carry, depending on personal choice.

Single Carry

A single firefighter can carry a small ladder individually by hoisting it up onto the shoulder and inserting an arm through the ladder rungs and grasping the next forward rung set, or over the upper beam for weight balance, **Figure 14-28.** When initially placed on the shoulder, the tip should be slightly elevated with the greater weight on the butt (or base) of the ladder. The ladder should be balanced so that it can be adjusted by moving the hand on the rung or the beam to make a final weight equalization adjustment. When being carried, the butt (or base) of the ladder should be slightly dipped for maximum visibility and control.

Carrying Victims

Many types of carry techniques are used when dealing with a victim on a ladder. The level of consciousness, fear on the part of the victim, and trust in the firefighter are factors that will aid in removing a victim down a ladder. This is covered in greater detail in Section I, Chapter 16.

Raising Ladders

Several considerations must be weighed when raising a ladder. The heel or foot of the ladder must be a calculated distance from the building for stability. The distance will depend on the target height. With the correct distance from the building for the given height, the ladder will be at a safe climbing angle, ideally about 75 degrees. For most people, this angle will be naturally comfortable and not require excessive bending or reaching to accomplish tasks performed while on the ladder. In addition, the ladder is designed to carry its rated weight at this angle. At lower angles, the ladder will actually bend due to the weight it is carrying. At higher angles, climbing and attempting any tasks while on the ladder may be unsafe. The distance between the ground and the point of contact with the structure is called the **working length, Figure 14-29.** If the ladder tip goes beyond the building when it is placed against the roof, the **working length** is from the ground to the point where the ladder contacts the roof level.

The number of firefighters needed to raise a ladder will vary. Depending on its size, as many as six firefighters may be needed to raise one ladder. Usually one, two, or three firefighters will be able to raise almost any of the ladders carried by a specific department. When conducting ladder operations, common terminology and practice is the key. There is no specific set of commands or one correct way to raise a ladder. Ladder operations depend on factors such as the specific goal to be attained, the size of the ladder, the manpower available, the topography of the area, and so on. However, it is very important that every firefighter from the same department use like methods and common ladder terminology. All firefighters must attempt to accomplish the same objective in the same way.

FIGURE 14-28 Single-person ladder carry.

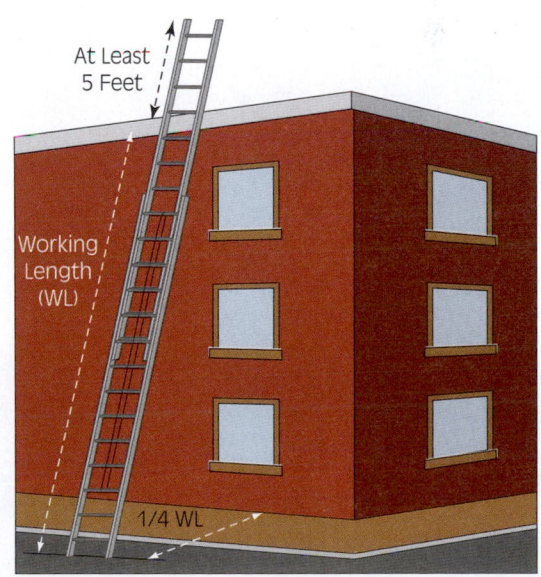

FIGURE 14-29 The working length is the distance from the ground to the point where the ladder contacts the building. The ladder is placed at a point approximately one-quarter of the working length from the building.

Most firefighting situations call for two firefighters to raise a ladder. The average ladder used at a fire is in the 18- to 35-foot range. The two people required are referred to by the functions they perform. The pivot firefighter is referred to as the heel, butt, or footer, while the firefighter raising the ladder is called the raiser, tip, or fly.

NOTE

There are many different ladder commands used. The actual verbiage is not important. What is paramount is that common terminology be used. Ladder commands must be consistent for each agency, and interagency operations should be considered when deciding what actual commands the department will choose. When a ladder command is given, usually by the tip firefighter, every member of the operation must be thinking of the same potential actions required to accomplish the goal.

Rung and Beam Raises

There are two methods of raising a ladder to the vertical position, **Figures 14-30 A** and **B.** In one method, called the beam raise, the raiser brings the ladder up using the beam in a hand-over-hand motion. In the rung (or flat) raise, the raiser uses the same approach but turns the ladder 90 degrees and utilizes the rungs for raising in a hand over hand motion. The beam raise is a good technique when the approach to the objective is parallel to the wall to be scaled. The beam raise is especially helpful when buildings are in close proximity with limited access. This would also

be true where the presence of overhead obstructions would impede the operation. The rung raise is a good technique to use when the approach to the objective is perpendicular to the wall to be scaled.

STREETSMART TIP

The standard formula for determining the proper distance the foot or butt should stand out from the building is one-quarter of the working distance of the ladder from the base of the wall—resulting in an approximate 75-degree working angle. A good way to estimate this is to stand with the toes against the foot of the beams and then, with outstretched arms, reach for a rung at about shoulder height. If the firefighter can grasp the rung, the angle is within a range that is acceptable. Remember that this is a rule of thumb for the average firefighter. An exceptionally tall or short firefighter might have to make some adjustments. The time for this adjustment calculation is during drills, not during the fireground operations.

JPR 14-4: Two-Person Rung Raise

(For step-by-step photos of this skill sequence, see page 408)

1. With the rung raise, the ladder is brought to the raising point and positioned perpendicular and out from the wall approximately one-quarter of the working length from the wall or objective. The footer positions the ladder on the ground and places a foot at each beam to plant the pivot point of the raise. The toes of the boots are snubbed up against the heel of the ladder for stability. The raiser is positioned at the tip. If necessary, the raiser can position down the ladder toward the footer. The raiser position on the ladder depends on the length of the ladder and the comfort level of the firefighter.

2. When ready, the raiser grasps the beams or the rungs and raises the ladder to hip level using the strength of the legs as lifting mechanisms.

3. Using the upper body with the back straight and legs slightly bent, the raiser lifts the ladder overhead while stepping under the ladder and pivoting to face the footer in one fluid motion. Using a hand-over-hand motion, the raiser with outstretched arms, moves from rung to rung while walking toward the footer until the ladder is in the vertical position. The footer constantly surveys the area around the ladder for hazards that could impede the operation.

4. Once the ladder is vertical, the footer accepts some of the weight and the ladder is then lowered onto the objective.

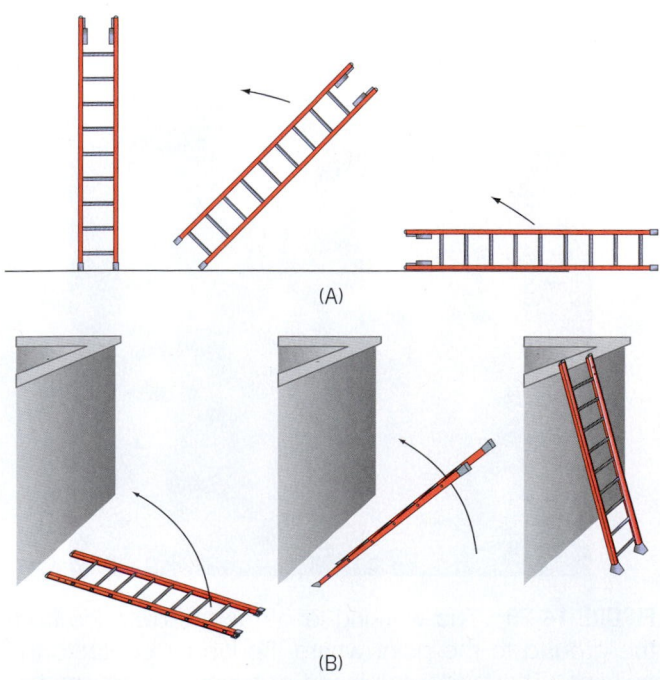

(A)

(B)

FIGURE 14-30   (A) Beam raise. (B) Rung raise.

JPR 14-5: Two-Person Beam Raise

(For step-by-step photos of this skill sequence, see page 409)

1. The ladder is placed down on the beam parallel to the building approximately one-quarter the working distance of the ladder. The footer is positioned in such a way that the firefighter's back is to the structure. The footer places a foot on the beam of the ladder to plant the pivot point of the raise. The toe of the boot is snubbed up against the heel of the ladder for stability and to secure the pivot point. The raiser is positioned approximately two-thirds of the length of the ladder away from the base and toward the tip, adjusting this position as necessary.

2. When ready, the raiser grasps the beams or the rungs and raises the ladder to hip level using the strength of the legs as lifting mechanisms.

3. Using the upper body with the back straight and legs slightly bent, the raiser then lifts the ladder overhead while stepping under the ladder and pivoting to face the footer in one fluid motion.

4. Using a hand-over-hand motion, the raiser moves along the beam with outstretched arms while walking toward the footer, until the ladder is in the vertical position.

5. Simultaneously, the footer uses both hands to steady the ladder by moving up the beam in a hand-over-hand method as the ladder ascends to the vertical position.

The ladder is then lowered into the building.

If the ladder is an extension ladder, there are additional factors and steps to consider before lowering the ladder into the building. Any extension ladder over 25 feet will usually require three persons to carry it to the objective because of the sheer weight of the ladder and its contribution to instability. The third firefighter can assume a raiser position if the ladder is too heavy for one firefighter to raise. When positioning an extension ladder, the fly section is positioned away from the building, **Figure 14-31.**

JPR 14-6: Fly Extension Raise

(For step-by-step photos of this skill sequence, see page 411)

1. Once the ladder is positioned and vertical, the footer unties the halyard while the raiser accepts the weight and stabilizes the ladder.

2. The footer and raiser use a foot to toe or support the same ladder beam to keep it from moving during the raising operation. The footer extends the ladder to the desired height by pulling down on the rope.

3. The footer must watch to ensure the fly section is locked and secured in place once the desired height is reached.

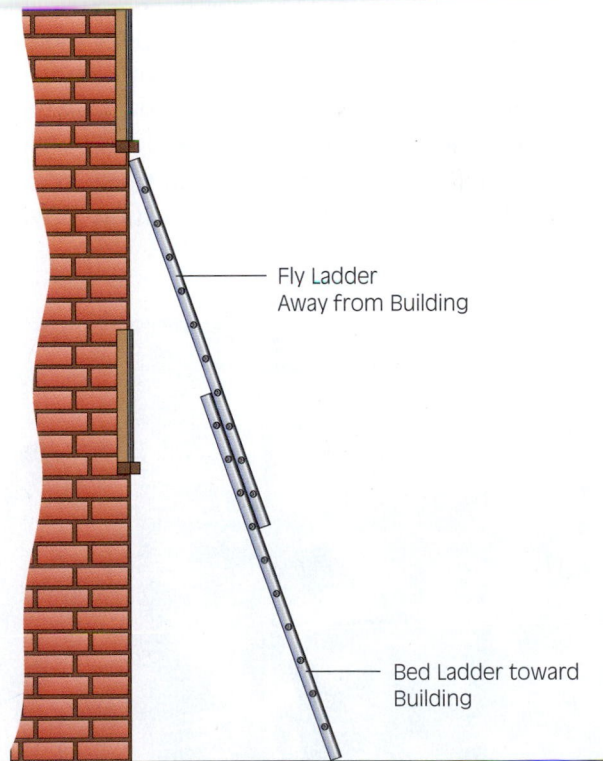

Fly Ladder Away from Building

Bed Ladder toward Building

FIGURE 14-31 Extension ladder. The fly ladder is positioned away from the building, the bed ladder toward the building.

4. The ladder is lowered to the building and the footer then secures the rope around the rungs and ties a knot in the halyard as a safety precaution, leaving the rope centered on the rung.

5. For personal safety, it is important that the raiser hold the weight of the ladder by grasping the bed section of the ladder on the beams where the sliding fly section can operate free of any hand contact.

SAFETY

When raising the fly ladder with the halyard, the firefighter should grasp the rope so that the heel of the hand is the uppermost part of the hand on the rope. The hand is twisted so that it creates a 90-degree angle with the rope with the balance of the rope passing through the fist and out between the firefighter and the thumb, **Figure 14-32.**

JPR 14-7: Three-Person Raise

(For step-by-step photos of this skill sequence, see page 412)

1. Firefighters will find it necessary to use additional firefighters to raise the larger ladders, due to ladder weight and size. With a three-person ladder raise, a rung raise should be used. The ladder may be carried using any one of the carries, but once the ladder butt is placed on the ground, the ladder should be pivoted to the rung raise position with the fly section out prior to the raise. (The ladder can be pivoted before raising the fly section, if necessary.)

2. The heeler remains at the butt end to heel the ladder. The firefighter in the mid position moves to

FIGURE 14-32 Firefighter grasps the rope with the heel of the hand at 90 degrees to the rope for a better grip.

the tip end and takes one of the beams from the raiser. Both of these firefighters are now raisers.

3. The raisers lift the ladder and each places a beam upon their shoulder. The heeler crouches down and grabs a rung with both hands and gives the command and the raisers lift the ladder up and position themselves under their respective beam.

4. With their arms extended straight, each raiser proceeds toward the building, using the beam to lift the ladder. (If the fly section remains in, the ladder can be pivoted at this point to move it to the outside.) The raisers now accept the weight and stability responsibility of the ladder. They each move to a position on the sides of the ladder and use their hands and feet to steady it.

5. It is important for these firefighters to keep their hands away from the moving fly section and to keep them on the outside of the beam. The heeler grasps the halyard and extends the ladder to the proper position. All three firefighters watch the operation and watch to ensure the ladder is at the proper position and locked.

6. The heeler then grasps the ladder with one hand on each beam. The raising firefighters then move back to the outside of the ladder. Each places a foot on the bottom rung, while the inside firefighter uses both feet to steady the ladder on the ground. All three firefighters lower the ladder toward the objective until the tip is resting on the objective.

If the ladder requires adjusting for climbing angle, all three firefighters will adjust. Lowering the ladder will result in these steps performed in reverse. As always, lowering a ladder is not normally a critical step, and additional firefighters should assist, if possible.

JPR 14-8: Four-Person Ladder Raise

(For step-by-step photos of this skill sequence, see page 414)

1. Whenever possible, a four-person ladder raise should be used, as an additional firefighter can assist with the weight and leverage of the ladder. This operation is usually safer for the scene and will help prevent injuries that could result. The four-person raise is basically the same as the three-person raise, but with an additional heeler. As with the three-person raise, a rung raise should be used. The ladder may be carried using any one of the carries, but once the ladder butt is placed on the ground, the ladder should be pivoted to the rung raise position with the fly section out prior to the raise. (The ladder can be pivoted before raising the fly section, if necessary).

2. The heelers remain at the butt end to heel the ladder. The firefighters at the tip are the raisers, and they lift the ladder and place a beam upon their shoulders. The heelers crouch down and grab a rung and beam with each hand. The command to raise is given and the raisers lift the ladder up and position themselves under their respective beam.

3. With their arms extended straight, each raiser proceeds toward the building, using the beam to lift the ladder.

4. If the fly section remains in, the ladder can be pivoted at this point to move it to the outside. Once the ladder is vertical, the firefighters shift position so that there is a firefighter on the outside of each beam, one firefighter on the outside of the rungs, and a firefighter on the inside of the rungs. The three outside firefighters steady the ladder, while the inside firefighter grasps the halyard and raises the ladder. All firefighters ensure that hands are free of the moveable fly section. All four firefighters watch the operation and watch to ensure that the ladder is at the proper position and locked.

5. The outsider firefighters assume the job of heeling the ladder using their feet to steady the ladder on the ground and their hands to control the descent. All four firefighters lower the ladder toward the objective. If the ladder needs adjusting for climbing angle, all four firefighters help adjust.

Taking down the ladder will result in these steps performed in reverse.

JPR 14-9: Lower Ladder into Building

(For step-by-step photos of this skill sequence, see page 416)

1. From the raising procedure, the footer is left standing between the ladder and the structure, while the raiser is standing on the outside of the ladder facing the building. The raiser grasps a set of rungs at shoulder height and places one foot on the lowest rung. The footer grasps both beams at shoulder height.

2. As the raiser lowers the ladder into the structure, the footer accepts the weight of the ladder in a controlled maneuver until the ladder is resting against the building.

If any angle adjustment is necessary, the firefighters can lift the ladder off the ground by each grasping a rung while positioned on either side of the ladder beams. The ladder can then be safely adjusted by lifting and pulling it away from the building until

the desired angle is attained. Fine adjustments can be made by moving the ladder out and letting the ladder tip ride down the wall. In like manner, the tip can be pushed up the wall if the angle needs to be decreased.

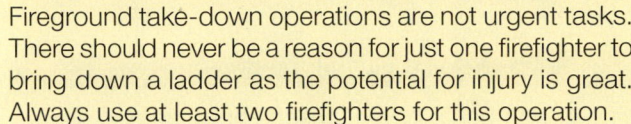

SAFETY

Fireground take-down operations are not urgent tasks. There should never be a reason for just one firefighter to bring down a ladder as the potential for injury is great. Always use at least two firefighters for this operation.

In some cases, the situation might call for only one person to raise the ladder. The condition in which a firefighter would be working alone is rare and generally *not recommended* by NFPA standards and prohibited by some fire departments. However, it is possible for a ladder to be raised by one person provided the ladder is of a length and weight that is not beyond the capacity of that person—and provided that the local Authority Having Jurisdiction (AHJ) allows this practice. Under normal two-person raising conditions, one firefighter will secure the pivot of the ladder foot or butt while the other raises the tip section. With only one person, the base pivot firefighter is absent. A substitute can be employed through the use of a wall, fence, bush, or anything that is substantial enough to act as a foot for the ladder.

JPR 14-10: One-Person Raise

(For step-by-step photos of this skill sequence, see page 417)

1. The ladder is placed on the ground in a rung-raise position with the ladder heel against the substitute foot.

2. The firefighter stands at the tip of the ladder facing the building and raises the tip with a beam in each hand to the hip level.

3. With an upward swing, the firefighter raises the ladder to an overhead position and walks the ladder into the building using a sliding hand motion along the beams to support the weight of the ladder as the ladder ascends into a vertical position.

4. When the ladder is in a vertical position, it will be against the building from base to tip.

5. The firefighter must continue to support the weight of the ladder. The firefighter then sets the ladder angle by lifting the ladder off the ground and backing away, while the tip rests against the building until the desired angle is attained. The ladder butt is then placed on the ground. The ladder will be at

approximately the correct angle when the firefighter's toes are at the base and the outstretched arms are able to reach the beams at shoulder height.

As normal fireground take-down operations are not urgent, there should never be a reason for a firefighter to bring down a ladder alone. Take-down should be accomplished as described in the previous ladder raise sections.

> ### VIEWPOINT
> Some departments have devised shortcuts and alternate ways of performing the various fireground functions. Tactics, such as the one-firefighter ladder deployment, can be accomplished in many different ways, **Figure 14-33.** Any methods used should be approved by the fire department and should be practiced by all firefighters to ensure proficiency.

Leg Lock

The leg lock is used to secure the firefighter to the ladder when both hands must be used to perform a task and a ladder belt is not available.

JPR 14-11: The Leg Lock

(For step-by-step photos of this skill sequence, see page 418)

1. The firefighter first chooses the desired height and climbs up one additional rung.
2. Then the locking leg is inserted into the space between the rungs and over the rung that was at knee height.
3. Using the supporting leg as an assist, the firefighter lowers the entire body until the leg inserted between the rungs can be inserted through the next lower set

of rungs such that the foot projects beyond the ladder beam. Twisting the ankle slightly, the toes are projected beyond the ladder beam. Then the instep of the foot becomes the locking mechanism to the ladder or, alternatively, just hooks a lower rung of the ladder. Then, the desired operation may be performed with the stability of the firefighter in place. To remove the leg lock, perform the leg lock steps in reverse.

> ### VIEWPOINT
> Many fire service agencies spend an immense amount of time training, which produces additional ways of accomplishing tasks. Some agencies find the currently accepted method of performing a leg lock cumbersome and difficult, others utilize it regularly but find that it is not easy for a few of their firefighters to accomplish due to their build or other circumstances. Therefore, these agencies have come up with a different way of performing this task. Firefighters should always ensure they are following the accepted methods of performing firefighter tasks, as outlined in the department policies, procedures, and SOPs. An alternate leg lock is known as the hook-in leg lock. A firefighter places one foot on the chosen rung. The firefighter then places the other leg over the top of the next rung and places the heel of the other foot on the tip of the boot of the bottom foot, **Figure 14-34.** This leg lock is said to be advantageous for firefighters who have somewhat thicker or shorter legs.

The second alternate leg lock is known as the hyper-extended leg lock and works well for taller firefighters performing quick manuevers. A firefighter hyperextends one leg up and into the next

FIGURE 14-33 An alternate one-firefighter ladder raise tactic.

FIGURE 14-34 Firefighter using an alternative leg lock maneuver.

FIGURE 14-35 Firefighter using an alternative arm lock maneuver.

rung. Firefighters may also use alternate arm lock manuevers while performing routine ladder operations, **Figure 14-35.** As always, firefighters should follow procedures, train extensively, and find out what works well for them, personally and in different circumstances.

Carrying Tools

When carrying tools on a ladder, a certain amount of security is sacrificed. The positive grip that would be afforded by the hand holding the tool is negated. When possible, tools should be passed up to another firefighter first. Another technique is to hang the tool on an upper rung, and climb to it.

JPR 14-12: Carrying Tools

(For step-by-step photos of this skill sequence, see page 419)

1. Hang the tool on an upper rung.
2. Climb up to it.
3. Reposition the tool to another upper rung and again climb up to it.

One last—but not the best—technique is to ride the tool up the rails while climbing. This last technique is suited for climbing an aerial ladder where high side rails are available. A ground ladder is not a good candidate for this method, and special care must be exercised.

Mounting and Dismounting

Getting on and off a ladder is the most difficult action for the uninitiated. It requires releasing a grip on the very thing that is supporting the climber off

the ground. As height increases, so does the fear of falling. A person who has a fear of heights dreads the very act of moving from one rung to another.

Before a firefighter ascends a ladder it should be secured. If a ladder is not secured with a rope, it should be heeled by another firefighter.

JPR 14-13: Heeling a Ladder

(For step-by-step photos of this skill sequence, see page 420)

1. A firefighter should stand on the outside of the ladder, facing the building or objective. After another firefighter begins to ascend the ladder, the heeler moves into position, grasps a beam with each hand at about eye level, and chocks the ladder with a foot by snubbing the toe of the boot against the butt,
2. or by placing a foot on the lower rung. The heeler pushes the ladder toward the objective. This firefighter must be prepared to move out of the way as a descending firefighter nears the bottom of the ladder.

> **SAFETY**
>
> Some departments use a heeler underneath or on the inside of the ladder. This practice is considered dangerous by many. To accomplish this task in this manner, the heeler will be facing away from the objective. This, in effect, blinds the heeler to the emergency situation and puts the heeler at risk of being struck by falling objects.

Again, to be as safe as possible while climbing, at least three limbs should be in contact with either the ladder or the objective at all times if practical. Only one limb should move at a time, and the next limb to move should not do so until the previous one is secure. Whether mounting a ladder from an elevated location or dismounting the ladder to a structural item such as a fire escape or a window, a few simple common-sense actions must be followed. First, the firefighter should make sure to step onto a secure and stable objective. Using the hand-move first will permit the climber to get an immediate "feel" for whether the target is unstable, shaky, or questionable. This action will occur while the full weight of the firefighter is still on the ladder. The next foot movement permits the firefighter to gradually apply some weight onto the target while one hand and foot are still on the ladder. If at any time the objective becomes unstable as more weight is shifted, the firefighter can still pull back and retreat to the safety of the stable ladder. Once the second foot is on the objective and secure, the second hand can release its grip from the ladder.

If the dismount will be made onto a balcony, a parapet, or a roof where there is no structural element to grasp, the firefighter should first check the structural stability with a tool. If stability seems to be intact, the firefighter can move one foot onto the objective and, while still grasping the ladder with both hands, shift weight to that foot. If any unstableness develops, immediate retreat is possible without a loss of balance. If stable, the weight shift can continue until the majority of the firefighter's weight is on the objective. When ensured that stability is present, the firefighter can move the other foot to the objective and let go of the ladder.

Window Dismount

When climbing into a window from a ladder, two methods are used. The method chosen should be used based on the conditions. When the ladder is placed on the side of the window, a step-in method is satisfactory provided the ladder is close enough to the building. It is always better to step down onto the windowsill so that if there is structural instability, weight can be withdrawn. If the step-up method is used, the firefighter's weight is shifting onto the structure, and, if failure occurs, the firefighter might already be committed and fall from the ladder. Mounting the ladder is essentially the same process in reverse. It is best to step up to and on the ladder rung rather than step down to the rung. This permits a more positive grip and less chance of loss of balance.

If entry into the window must be made under heat and smoke conditions, the ladder's tip is placed under the windowsill and the firefighter actually climbs right over the tip of the ladder and into the window. This method keeps the firefighter low. Exit from the window would require a somewhat different approach. This exiting method should be consistent with the accepted department policies and practices.

Ornamental Works

When mounting or dismounting a ladder from an ornamental works such as a fire escape or a raised parapet wall at the front of the building, special care should always be observed to account for structural weakening from fire, weather, or just plain deterioration from lack of maintenance. The weight shift from the ladder should be slow and calculated to permit quick withdrawal should structural stability become questionable.

Roof Ladder Deployment

A ladder of any size or type is first raised to the eave of the roof directly under the desired access point. The hook ladder (with hooks engaged) is raised alongside the prepositioned ladder using the beam-raise method. The ladder is pivoted to rest on its rungs on the edge of the roof. One firefighter climbs the non-hook ladder. The footer maintains the stability of the climbing ladder and assists with the hoisting of the hook ladder if a third firefighter is not available or if the climbing ladder is secured. The hook ladder is pulled by the firefighter on the ladder and pushed by the firefighter on the ground, on its beam, until it reaches its balance point of the edge of the roof. The firefighter on the ladder pushes the ladder up the slope of the roof until the hooks clear the peak. Once over the peak, a fire tug is employed to secure the hook ladder into the roof.

JPR 14-14: Engaging the Hook on a Hook Ladder

(For step-by-step photos of this skill sequence, see page 421)

1. The hook (or roof) ladder is held hip high.

2. Using the palm of the hand, the hook is depressed against its spring-loaded resistance.

3. Once at the limit of the hook's travel, it is rotated 90 degrees to the perpendicular position from the ladder. The hook is able to rotate in either direction.

4. When fully rotated, the pressure on the hook is gradually released, and it is permitted to return to its limit in the up/open position. The same procedure is used for the other hook.

VIEWPOINT

Some fire services choose to deploy a roof ladder across the roof on the beam—pivoting it at the top of the sliding move to catch the hooks on the roof peak. Others choose to engage the hooks once the ladder is raised and before it is slid across the roof. All of these methods are acceptable methods. As in any deployment, the procedures must be accepted by the AHJ and performed in a safe manner.

Hoisting Ladders by Rope

On occasion, the need to use a ladder from an elevated location might arise. For example, there might be a window that can be reached from the flat roof of a one-story building, or the only way to get to the roof of a four-story building is to climb up a ladder from the roof of a three-story building. In any case, the need to get a ladder to an elevated location might be present. All ladders are treated the same, and the same technique can be used for any ladder that has two beams and rungs. When lifting an extension ladder, the fly ladder should be **nested** or bedded before raising. A utility rope of sufficient strength is used for this operation.

JPR 14-15: Hoisting Ladders by Rope

(For step-by-step photos of this skill sequence, see page 422)

1. A bowline or a large figure eight on a byte knot is tied at one end at a point that will create a loop that is 9 feet in circumference or at least sufficient for the knot to rest above the next set of rungs.

2. The middle of the ladder is ascertained by counting rungs. Once the middle is established, the firefighter counts up two rungs and passes the rope through the hole that the rungs create. Strips of electrical tape or painted bands can be used to semi-

permanently identify these rungs. With the ladder on the ground, the butt (or base) of the ladder is gently lifted and the rope slid up to the marked rungs while the slack rope is taken up.

3. The knot should be approximately in the center of the ladder between the beams. The ladder is then rolled over so that the rope runs under the ladder rungs and is arranged so that it is between the ladder and the building.

4. The command to hoist may be given. The tip will tend to lean away from the building and prevent the ladder tip from snagging on any projections caused by windowsills, protruding brickwork, or any other structural element that might cause difficulty. Once the rope knot is at the point of raising, the ladder is pivoted on the edge of the structure and gently pulled down on the roof. Then the remaining portion of the ladder is pulled onto the roof.

When lowering the ladder, the procedure is reversed. However, during lowering, the ladder is placed between the rope and the building to avoid any snagging on building protrusions. The tip will drag along the building, and the base or heel of the ladder will tend to stay out from the wall a few inches and ride over any obstructions.

TYPES OF TRUCK-MOUNTED LADDERS

Many types of ladder trucks are used by the fire service today. Each was invented and designed to serve a particular function and has been named by its function.

Aerial Ladder

An aerial ladder is an apparatus-mounted ladder capable of reaching heights of generally up to 100 feet, with some that even go beyond 110 feet, **Figure 14-36**. An aerial ladder with a reach capability of 144 feet was once tested in field service. It was taken out of service after a short time because of limited use, safety concerns, and maintenance.

Because of the heavy-duty use and reach, most of these ladders are very heavy and are constructed of some combination of steel, aluminum, and other metallic alloys designed for strength using a truss-style construction for maximum strength.

Aerial ladders are designed so that various sections slide out from one another to produce greater reach, **Figure 14-37**. Each fly ladder section is designed to overlap the section below sufficiently to maintain stability. It is usually made of a steel/aluminum alloy for lighter weight. The lowest section of the ladder is called the bed ladder. It is usually made of steel and is attached to the main body of the apparatus by a combination of pins and pistons that allow it to be raised from the horizontal to the vertical positions. The bed ladder is attached to a **turntable,** a 360-degree rotatable platform that is attached to the framework of the apparatus. The fly sections of an aerial ladder are the sections that extend out from the bed ladder and one another to reach whatever height is desired.

The bed ladder, including the nested fly sections, is raised out of the bed of the apparatus through the use of **hydraulic pistons, Figure 14-38.** The turntable rotates the ladder to align with the target, and the fly sections are extended to reach the target.

Proper terminology for use of these ladders is very important, **Figure 14-39.** The same word must mean the same thing to all people. **Table 14-4** describes the terms that should be used to identify the operation desired.

FIGURE 14-36   Rear-mount aerial ladder.

FIGURE 14-37   Extended aerial ladder. Note the sliding sections.

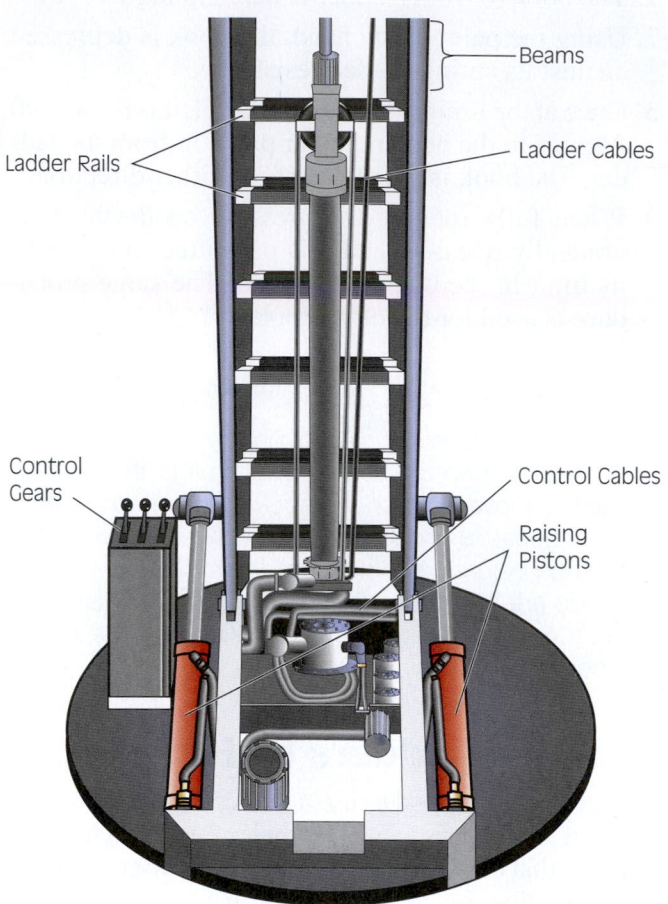

FIGURE 14-38   Aerial ladder raising mechanisms as seen from under a raised bed ladder.

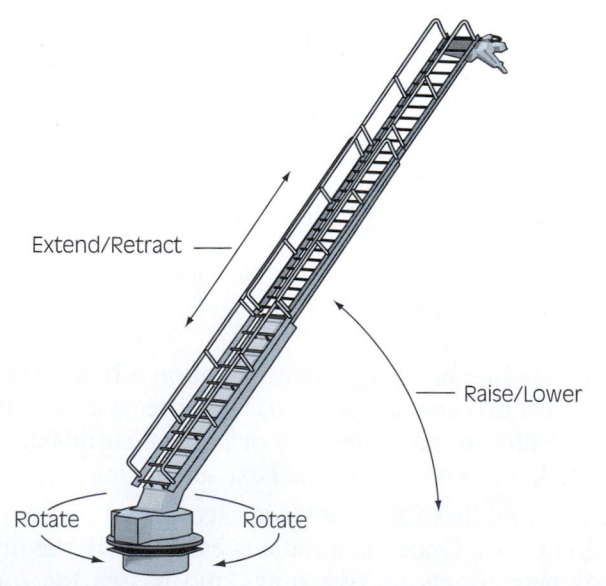

FIGURE 14-39   Ladder positioning terminology.

TABLE 14-4	Terminology for Apparatus-Mounted Ladder Raising
Raise:	Lifts the ladder from horizontal to vertical angles.
Lower:	Lowers the ladder from vertical angles to horizontal.
Rotate:	Turns ladder right and left.
Extend:	Increases the length and reach by extending fly sections.
Retract:	Decreases length and reach by nesting the fly sections.

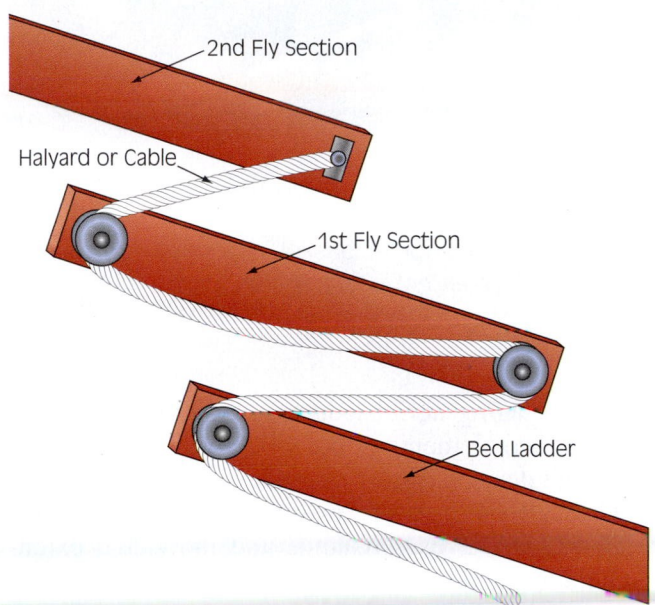

FIGURE 14-40 Typical multifly halyard hoisting pulley arrangement. As the halyard is pulled, the fly sections are hoisted in unison with all the rungs aligning at each locking position.

The word raise should refer only to the raising of the bed ladder out of the bed and not to making the ladder longer. The order to make the ladder longer would be to extend the ladder.

SAFETY

To avoid confusion, firefighters should use standard terminology when working with ladders.

NOTE

In many cases, the operator of the ladder might not be in visual contact with the target and is being directed by a forward observer. In cases such as this, common terminology becomes essential.

The ladder's reach is achieved through the use of cables and pulleys. Using the physics of mechanical advantage associated with block and tackle, a continuous cable is spun off a drum and woven through the various fly sections of the ladder, **Figure 14-40.** As the first fly section moves out from the bed ladder, the cable and pulley configuration pulls out the second and third sections at the same time. When in operation, each fly section moves faster than its parent section. This permits quick placement of the ladder at the target. Depending on the pulley and cable arrangement, the upper fly section can be moving three times faster than the lowest fly section.

Once the mainstay of a fire department's high elevation needs, the aerial ladder has become only one of the fire service's high elevation tools. Today's fire service has other apparatus such as tower ladders and **articulating boom ladders** that are equally capable and, in some cases, better designed for increased versatility. Newer designs incorporate features of both the tower and the articulating boom.

Tower Ladder

The **tower ladder** is often found as a standard piece of equipment in any moderate to large fire department. It is still referred to as a ladder company because it carries a complement of ground and portable ladders. Performing many of the same functions as the aerial ladder, in many departments the tower ladder has replaced older aerial ladders as they are retired from service.

The tower ladder, **Figure 14-41,** has a telescopic boom with a mounted basket capable of holding from 750 to more than 1,000 pounds. Like the aerial ladder, it is capable of rotation, extension, and retraction. The fly sections of the boom telescope out of the bed boom section. It has an advantage over the aerial ladder in that it can hold many people at the same time and is capable of many uses, including use as a work platform, an observation vantage point, and an elevated water stream appliance. The tower ladder can remove many trapped victims at the same time with limited commitment of resources. One firefighter at the turntable as a safety person and one in the basket can safely remove many victims, including children and small animals from an elevated location such as a

FIGURE 14-41 Tower ladder.

FIGURE 14-42 A snorkel basket can reach places not possible with other types of apparatus.

window or rooftop. A ground ladder would generally require the use of a separate firefighter for each victim removed to assist in climbing down the ladder.

In addition, the tower ladder affords people who suffer from a fear of heights, psychologically paralyzed victims, and persons with physical and mental disabilities a safe escape.

The tower ladder is a very versatile piece of equipment that achieves many fireground elevation needs. It has a tendency, however, to take slightly longer to place into operation and position than the aerial ladder.

Articulating Boom Ladder

The articulating boom (sometimes known as a snorkel ladder) truck was among the first designs for elevated platform use in the fire service. First appearing in the early 1960s, it broke new ground that eventually led to changes in fire service tactics and equipment design. Some of the original designs had a ladder attached to the booms for escape, but its use proved very impractical because of the angles and accessibility shortcomings. Although not a true ladder, it has been used

to replace aging aerial ladders. Newer designs have incorporated elements of the telescopic and articulating boom designs, **Figure 14-42.**

Through the use of several articulating booms, a snorkel ladder uses balance and individual extension and retraction capability to place the bucket into places that are not reachable by tower ladders due to obstructions. By positioning the booms at different angles, the basket can be lowered behind an obstruction. Tower and aerial ladders would be unable to reach the same objective.

An **articulating boom ladder** can also be used as an elevated water application platform or observation point and can simultaneously remove several victims. The disadvantage of a snorkel-type ladder is that it requires observation of many possible points of contact. Each articulation joint is capable of striking an object, building, electric wires, or other obstacle.

JOB PERFORMANCE REQUIREMENTS   14-1
Suitcase Carry

A   With the butt of the ladder facing the direction of travel the firefighters position themselves on the same side of the ladder facing in the direction of travel; one firefighter is at the tip and one is at the butt.

B   The tip firefighter controls the operation and gives the commands. On command, both firefighters raise the ladder to stand on one beam.

C   Grasping the upper ladder beam with the hand closest to the ladder, in unison, the firefighters lift and carry the ladder to the objective as if it were a suitcase.

D   The concept is the same for a three-person carry, with the additional person taking position in the middle of the ladder.

A The shoulder carry is useful for operations when firefighters will carry a ladder a good distance. The tip firefighter controls the operation and gives the commands. With the ladder flat on the ground, the two firefighters position themselves, facing opposite the direction of travel (facing the tip) on one side of the ladder beam. One is at the tip and one is at the butt.

B On command, both firefighters squat down and grasp a rung and raise the ladder to stand on one beam.

C In unison, the firefighters lift the ladder using their legs for the power maneuver.

D Both firefighters pivot in unison toward the direction of travel (facing the butt), place their free arm through the space between two rungs, and place the underside of the upper beam to rest on their shoulders.

E The hand of the arm that was inserted through the rungs reaches to the next rung forward and grasps it.

Flat Carry

A   With the ladder flat on the ground, the two firefighters who are about the same height will position themselves, facing the direction of travel, on one side of the ladder beam. One is at the tip and one is at the butt. The firefighter who is taller or shorter locates midway on the other beam. (The tip firefighter controls the operation and gives the commands.) On command, all three firefighters squat down and grasp a beam with the hand nearest the ladder. Using their legs for the power maneuver, the firefighters stand with the ladder. The ladder will slant up or down toward the odd-sized person, but the weight will be evenly distributed. As with the other carries, the tip firefighter is the guide. The butt firefighter leads the way to the objective.

B   The sequence is basically the same for the four-person flat carry, except that two firefighters are at the tip and two firefighters are at the butt. Either the beam or the rung can be grasped during the carry, depending on personal choice.

JOB PERFORMANCE REQUIREMENTS 14-4
Two-Person Rung Raise

A With the rung raise, the ladder is brought to the raising point and positioned perpendicular and out from the wall approximately one-quarter of the working length from the wall or objective. The footer positions the ladder on the ground and places a foot at each beam to plant the pivot point of the raise. The toes of the boots are snubbed up against the heel of the ladder for stability. The raiser is positioned at the tip. The raiser can position down the ladder toward the footer, if necessary. The raiser position on the ladder depends on the length of the ladder and the comfort level of the firefighter.

B When ready, the raiser grasps the beams or the rungs and raises the ladder to hip level using the strength of the legs as lifting mechanisms.

C Using the upper body, with the back straight and legs slightly bent, the raiser lifts the ladder overhead while stepping under the ladder and pivoting to face the footer in one fluid motion. Using a hand-over-hand motion, the raiser, with outstretched arms, moves from rung to rung while walking toward the footer until the ladder is in the vertical position. The footer constantly surveys the area around the ladder for hazards that could impede the operation.

D Once the ladder is vertical, the footer accepts some of the weight and the ladder is then lowered onto the objective.

Two-Person Beam Raise

A The ladder is placed down on the beam parallel to the building approximately one-quarter the working distance of the ladder. The footer is positioned in such a way that the firefighter's back is to the structure. The footer places a foot on the beam of the ladder to plant the pivot point of the raise. The toe of the boot is snubbed up against the heel of the ladder for stability and to secure the pivot point. The raiser is positioned approximately two-thirds of the length of the ladder away from the base and toward the tip, adjusting this position as necessary.

B When ready, the raiser grasps the beams or the rungs and raises the ladder to hip level using the strength of the legs as lifting mechanisms.

C Using the upper body with the back straight and legs slightly bent, the raiser then lifts the ladder overhead while stepping under the ladder and pivoting to face the footer in one fluid motion.

(*Continues*)

D Using a hand-over-hand motion, the raiser moves along the beam with outstretched arms while walking toward the footer, until the ladder is in the vertical position.

E Simultaneously, the footer uses both hands to steady the ladder by moving up the beam in a hand-over-hand method as the ladder ascends to the vertical position. The ladder is then lowered into the objective.

Fly Extension Raise

A Once the ladder is positioned and vertical, the footer unties the halyard while the raiser accepts the weight and stabilizes the ladder.

B The footer and raiser use a foot to toe or support the same ladder beam to keep it from moving during the raising operation. The footer extends the ladder to the desired height by pulling down on the rope.

C The footer must watch to ensure the fly section is locked and secured in place once the desired height is reached.

D The ladder is lowered to the building and the footer then secures the rope around the rungs and ties a knot in the halyard as a safety precaution, leaving the rope centered on the rung.

E For personal safety, it is important that the raiser hold the weight of the ladder by grasping the bed section of the ladder on the beams where the sliding fly section can operate free of any hand contact.

JOB PERFORMANCE REQUIREMENTS   14-7
Three-Person Raise

A Firefighters will find it necessary to use additional firefighters to raise the larger ladders, due to ladder weight and size. With a three-person ladder raise, a rung raise should be used. The ladder may be carried using any one of the carries, but, once the ladder butt is placed on the ground, the ladder should be pivoted to the rung raise position with the fly section out prior to the raise. (The ladder can be pivoted before raising the fly section, if necessary.)

B The heeler remains at the butt end to heel the ladder. The firefighter in the mid position moves to the tip end and takes one of the beams from the raiser. Both of these firefighters are now raisers.

C The raisers lift the ladder and each places a beam upon their shoulder. The heeler crouches down and grabs a rung with both hands and gives the command and the raisers lift the ladder up and position under their respective beam.

D   With their arms extended straight, each raiser proceeds toward the building, using the beam to lift the ladder. (If the fly section remains in, the ladder can be pivoted at this point to move it to the outside.) The raisers now accept the weight and stability responsibility of the ladder. They each move to a position on the sides of the ladder and use their hands and feet to steady the ladder.

E   It is important for these firefighters to keep their hands away from the moving fly section and keep them on the outside of the beam. The heeler grasps the halyard and extends the ladder to the proper position. All three firefighters watch the operation and watch to ensure the ladder is at the proper position and locked.

F   The heeler then grasps the ladder with one hand on each beam. The raising firefighters then move back to the outside of the ladder. Each places a foot on the bottom rung, while the inside firefighter uses both feet to steady the ladder on the ground. All three firefighters lower the ladder into the objective until the tip is resting on the objective.

JOB PERFORMANCE REQUIREMENTS   14-8
Four-Person Ladder Raise

A   Whenever possible, a four-person ladder raise should be used, as one more firefighter will assist with the weight and leverage of the ladder. This operation is usually safer for the scene and will help prevent injuries. The four-person raise is basically the same as the three-person raise, but with an additional heeler. As with the three-person raise, a rung raise should be used. The ladder may be carried using any one of the carries, but, once the ladder butt is placed on the ground, the ladder should be pivoted to the rung raise position with the fly section out prior to the raise. (The ladder can be pivoted before raising the fly section, if necessary.)

C   With their arms extended straight, each raiser proceeds toward the building, using the beam to lift the ladder.

B   The heelers remain at the butt end to heel the ladder. The firefighters at the tip are the raisers, and they lift the ladder and place a beam upon their shoulders. The heelers crouch down and grab a rung and beam with each hand. The command to raise is given and the raisers lift the ladder up and position under their respective beam.

D   If the fly section remains in, the ladder can be pivoted at this point to move it to the outside. Once the ladder is vertical, the firefighters shift position so that there is a firefighter on the outside of each beam, one firefighter on the outside of the rungs and a firefighter on the inside of the rungs. The three outside firefighters steady the ladder, while the inside firefighter grasps the halyard and raises the ladder. All firefighters ensure that hands are free of the moveable fly section. All four firefighters watch the operation and watch to ensure the ladder is at the proper position and locked.

E   The outsider firefighters assume the job of heeling the ladder using their feet to steady the ladder on the ground and their hands to control the descent. All four firefighters lower the ladder into the objective. If the ladder needs adjusting for climbing angle, all four firefighters help adjust. Taking down the ladder will result in these steps performed in reverse.

JOB PERFORMANCE REQUIREMENTS   14-9
Lower Ladder into Building

A   From the raising procedure, the footer is left standing between the ladder and the structure, while the raiser is standing on the outside of the ladder facing the building. The raiser grasps a set of rungs at shoulder height and places one foot on the lowest rung. The footer grasps both beams at shoulder height.

B   As the raiser lowers the ladder into the structure, the footer accepts the weight of the ladder in a controlled maneuver until the ladder is resting against the building.

JOB PERFORMANCE REQUIREMENTS 14-10
One-Person Raise

A The ladder is placed on the ground in a rung raise position with the ladder heel against the substitute foot.

B The firefighter stands at the tip of the ladder facing the building and raises the tip with a beam in each hand to the hip level.

C With an upward swing the firefighter raises the ladder to an overhead position and walks the ladder into the building using a sliding hand motion along the beams to support the weight of the ladder as the ladder ascends into a vertical position. When the ladder is in a vertical position it will be against the building from base to tip.

D The firefighter must continue to support the weight of the ladder. The firefighter then sets the ladder angle by lifting the ladder off the ground and backing away while the tip rests against the building until the desired angle is attained. The ladder butt is then placed on the ground. The ladder will be at approximately the correct angle when the firefighter's toes are at the base and the outstretched arms are able to reach the beams at shoulder length.

JOB PERFORMANCE REQUIREMENTS 14-11
The Leg Lock

A The firefighter first chooses the desired height and climbs up one additional rung.

B Then the locking leg is inserted into the space between the rungs and over the rung that was at knee height.

C Using the supporting leg as an assist, the firefighter lowers the entire body until the leg inserted between the rungs can be inserted through the next lower set of rungs such that the foot projects beyond the ladder beam. Twisting the ankle slightly, the toes are projected beyond the ladder beam. Then the instep of the foot becomes the locking mechanism to the ladder or, alternatively, just hooking a lower rung of the ladder. Then, the desired operation may be performed with the stability of the firefighter in place. To remove the leg lock, perform the leg lock steps in reverse.

JOB PERFORMANCE REQUIREMENTS 14-12
Carrying Tools

A Hang the tool on an upper rung.

B Climb up to it.

C Reposition the tool to another upper rung and again climb up to it.

D One last—but not the best—technique is to ride the tool up the rails while climbing.

JOB PERFORMANCE REQUIREMENTS   14-13
Heeling a Ladder

A   A firefighter should stand on the outside of the ladder, facing the building or objective. After another firefighter begins to ascend the ladder, the heeler moves into position, grasps a beam with each hand at about eye level, and chocks the ladder with a foot by snubbing the toe of the boot against the butt,

B   or by placing a foot on the lower rung. The heeler pushes the ladder toward the objective. This firefighter must be prepared to move out of the way as a descending firefighter nears the bottom of the ladder.

JOB PERFORMANCE REQUIREMENTS 14-14

Engaging the Hook on a Hook Ladder

A The hook (or roof) ladder is held hip high.

B Using the palm of the hand,

C The hook is depressed against its spring-loaded resistance. Once at the limit of the hook's travel, it is rotated 90 degrees to the perpendicular position from the ladder. The hook is able to rotate in either direction.

D When fully rotated, the pressure on the hook is gradually released, and it is permitted to return to its limit in the up/open position. The same procedure is used for the other hook.

JOB PERFORMANCE REQUIREMENTS 14-15
Hoisting Ladders by Rope

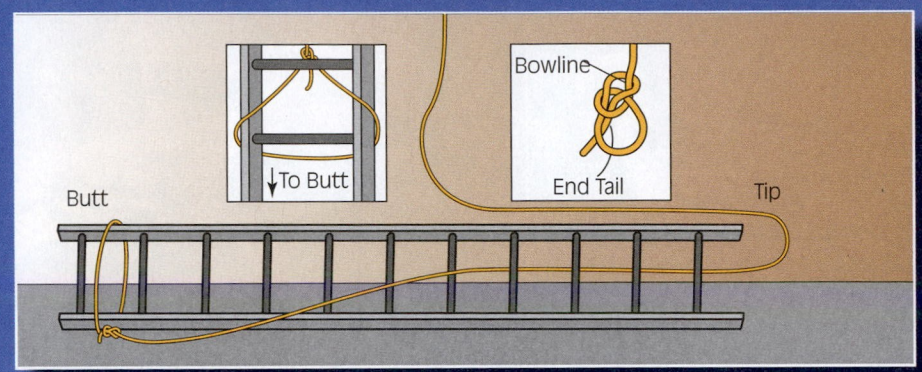

A A bowline or a large figure of eight on a byte knot is tied at one end at a point that will create a loop that is 9 feet in circumference or at least sufficient for the bowline knot to rest above the next set of rungs.

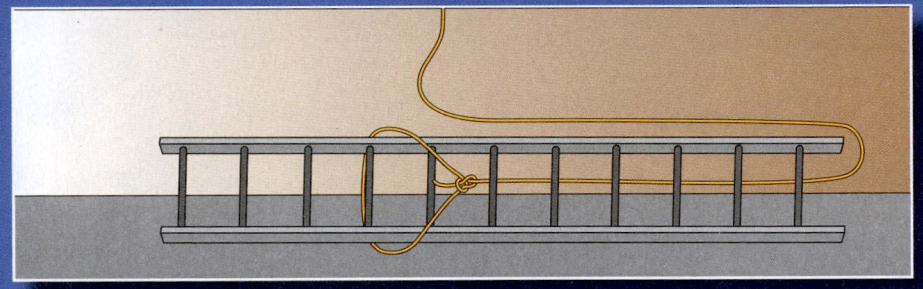

B The middle of the ladder is ascertained by counting rungs. Once the middle is established, the firefighter counts up two rungs and passes the rope through the hole that the rungs create. Strips of electrical tape or painted bands can be used to semipermanently identify these rungs. With the ladder on the ground, the butt (or base) of the ladder is gently lifted and the rope slid up to the marked rungs while the slack rope is taken up.

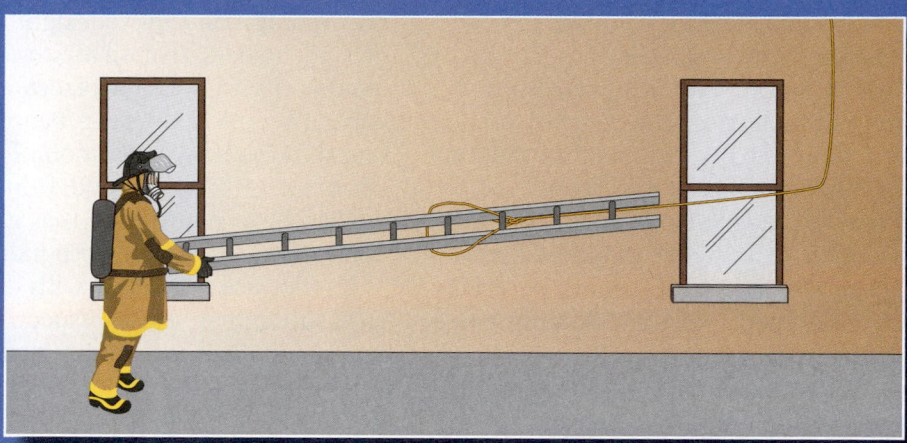

C   The knot should be approximately in the center of the ladder between the beams. The ladder is then rolled over so that the rope runs under the ladder rungs and is arranged to be between the ladder and the building.

D   The command to hoist may be given. The tip will tend to lean away from the building and prevent the ladder tip from snagging on any projections caused by windowsills, protruding brickwork, or any other structural element that might cause difficulty. Once the rope knot is at the point of raising, the ladder is pivoted on the edge of the structure and gently pulled down on the roof. Then the remaining portion of the ladder is pulled onto the roof.

LESSONS LEARNED

Ladders have many applications for use in the fire service beyond just climbing. With safety always in the forefront, innovative techniques can provide greater uses for ladders. However, it is important to remember what use a ladder was designed for and where the strengths and weaknesses lie. Any use other than such requires a thorough inspection of the ladder before returning it to service on the apparatus.

The different types of ladders, from truck-mounted aerial and tower ladders to portable ground ladders, are designed for a specific use. Roof ladders are used to stabilize footing on sloped roofs, and folding ladders are used for tight space application. New designs are always being developed. Some new designs in truck-mounted ladders include combination aerial ladder/bucket combinations and telescopic boom/articulating boom combinations. Ground ladders are also undergoing constant redesign. Collapsible telescopic ground ladders that save storage space and can be hand-carried into tight places and then extended are currently on the drawing board and might be introduced in the new generation of ground ladders. From climbing to bridging and from hoisting to shoring, ladders have many uses that can provide a superior application when used with judicious common sense.

Ladder use is packed with additional dangers that each firefighter must respect. Maintenance duties, inspection, and recording results all have a place where safety is the underlying motive and the overlying concern.

KEY TERMS

Articulating Boom Ladder   An apparatus with a series of booms and a platform on the end. It is maneuvered into position by adjusting the various boom sections into place to position the platform at the desired location.

Bed Ladder   The nonextending part of an extension ladder.

Extension Ladder   A ladder consisting of two or more sections that has the ability to be extended to a desired height through the use of a halyard.

Fly Ladder   That portion of a ladder that extends out from the bed ladder. Also called fly section.

Halyard   A rope or cable that is used to raise the fly ladders of an extension ladder.

Hydraulic Pistons   Mechanical rams that operate by pressure exerted through the use of a liquid, usually some form of oil.

Ladder Pipe   An appliance that is attached to the underside of an aerial ladder for an elevated water application.

Nested   The state when all the ladders of an extension ladder are unextended.

NFPA Standard 1931   Standard for manufacturer's design of fire department ground ladders.

NFPA Standard 1932   Standard on use, maintenance, and service testing of in-service fire department ground ladders.

Staypoles   The stabilizer poles attached to the sides of Bangor ladders that are used to assist in the raising of this type of ladder. Once raised, they are not used to support the extended ladder.

Tip Arc   The path that a ladder's tip will take while being raised.

Tower Ladder   An apparatus with a telescopic boom that has a platform on the end of the boom or ladder. It can be extended or retracted and rotated like an aerial ladder.

Turntable   The rotating platform of a ladder that affords an elevating ladder device the ability to turn to any target from a fixed position.

Working Length   The length of the ladder that spans the distance from the ground to the point of contact with the structure. This does not include any distance the ladder might go beyond the point of contact as would be the case when the tip extends beyond the roof.

REVIEW QUESTIONS

1. What is the main difference between a wall ladder and an extension ladder?

2. What is the function of the hooks on a roof ladder?

3. What is the main function of cleaning and maintenance duties?

4. What is the main use of ladders at fire operations?

5. Under what conditions is a ground ladder used as an elevated water stream application device?

6. Describe the rules for mounting and dismounting a ladder.

7. When using a ladder for access, why should it be left in position?

8. What is the "climbing path"?

9. How far away from the building should the ladder be placed?

10. Describe the two techniques a firefighter can use to stabilize a ladder.

11. When stabilizing an extension ladder, the raiser who is stabilizing the ladder must take care to ensure proper hand placement. Why?

12. What are the hazards a firefighter should look for when raising a ladder? What additional resource is advisable when a ladder must be raised in the area of charged power lines or other overhead hazards?

15

Ropes and Knots

For many rookie firefighters, the prospect of memorizing the array of knots taught in recruit school seems daunting and relatively unimportant. Advancing hoselines into burning buildings and performing other fireground operations are much more exciting and seemingly more important skills to master.

However, a number of experiences in my twenty-five-year career as a firefighter have convinced me of the necessity of not just learning the knots but practicing them over and over until their tying becomes almost instinctual.

As a young chief I once had the experience of ordering a vent saw to the roof of a three-story building. I watched as a young firefighter started and briefly warmed up a K12 saw according to department protocols and then shut off the saw and tied it into the center of the rope hanging down from the roof so that the tail of the rope could be used as a tag line. Confident that the saw would soon arrive at the roof to cut the vent hole that would allow the hose team to advance to the top floor, I turned to answer a question. The startled expression etched on this captain's face was matched only by the obvious anguish of the firefighter whose knot had slipped loose, sending the saw crashing to the ground from the eaves of the roof. Luckily no one was in the saw's path, and that rookie survived his lesson to become a department chief and a master of fire service knots!

My first high-angle rescue also taught a similar lesson. The call was for a teenager who had fallen to the bottom of a 110-foot gorge. After reaching the unconscious but still breathing patient and stabilizing his multiple fractures and bleeding, a decision was made to secure him to a backboard and haul him back up the cliff face in a Stokes basket.

The necessary ropes and webbing slings were lowered to our position, and I proceeded to diamond lash the young man into the litter and attach haul lines to the litter bridle in the way that I had been trained. As the litter was being hauled up the cliff face, my partner and I used tag lines affixed to each end of the litter to pull the basket away from the ledges and snags. Everything was going according to plan until the litter was approximately halfway up the cliff. The patient suddenly regained consciousness. The combination of his being disoriented by his surroundings coupled with multiple facial injuries that further compromised his airway caused him to begin thrashing around quite violently against his restraints. Visions of this panicked young man clawing his way out of the litter and plunging to his death on rocks at the bottom of the falls made the few minutes that it took to complete the haul to the top of the cliff an eternity. I replayed the tying of the knots that were used to secure the patient over and over again in my mind as the litter inched its way up the cliff face.

Fortunately, the knots had all been tied correctly and safetied in just the way I had been taught, and the patient survived both the fall and the rescue, eventually making a full recovery. Since that day, a rope short has hung from the top drawer of the desk in my office, and I will often throw knots when some phone caller has me on hold.

Mastery of the unique basic skills associated with an occupation is one of the defining characteristics of any profession. Like suturing for physicians or writing legal briefs for lawyers, knot tying is one of the most basic and perhaps most critical of the core skills associated with our profession. Take pride in your ability to tie knots correctly, safely, and efficiently, for you might be asked to on your next call.

—*Street Story by Joseph De Francesco,*
Coordinator, Madison County Fire and Rescue Services

LEARNING OBJECTIVES

After completing this chapter, the reader should be able to:

15-1   List various uses for rope in support of response activities.

15-2   Identify materials and styles of construction for fire service rope.

15-3   Select proper fire department documents, standards, or codes to find information.

15-4   Describe the characteristics of life safety and utility ropes.

15-5   Define the basic terminology used when discussing ropes, knots, and hitches.

15-6   Identify the basic knots and hitches used in the fire service.

15-7   Describe the uses of selected knots and hitches.

15-8   List conditions that would cause a rope to be placed out of service.

15-9   Demonstrate the techniques of inspecting, cleaning, maintaining, and storing rope in accordance with the department's types of rope and procedures or guidelines.

15-10 Given a specific piece of equipment, list and describe various tools, as well as inspection and cleaning methods, used to return the equipment back to service.

15-11 List and describe proper cleaning solvents.

15-12 Describe the proper manufacturers' or departments' procedures for tool and equipment cleaning.

15-13 Given manufacturers' recommendations, departmental guidelines, tools, and cleaning materials, demonstrate the ability to properly clean, inspect, and service tools and equipment.

15-14 Identify and describe the types of knots and hitches that may be used for raising and lowering various tools and equipment.

15-15 Describe the method of securing firefighting tools or equipment to be hoisted.

15-16 Demonstrate the proper raising and lowering of various tools and equipment using fire department ropes, knots, and hitches.

*The FF I and II levels, as defined by the NFPA 1001 Standards, are identified in different colors:
FF I = black, FF II = red, additional information = blue.

INTRODUCTION

Rope is one of the most important and routinely used tools in the fire service. In this chapter, the firefighter will learn how to select the proper rope for a given application based on its construction material and method. Also the firefighter will learn the primary uses of rope in the fire service, how to tie the knots utilized in the fire service, and how to properly inspect, maintain, and store rope.

STREETSMART TIP

Most firefighters will advise that tying knots is a skill that requires constant practice to remain proficient. It is also a needed skill that can be easily overlooked in the day-to-day operations of firefighting. Well-rounded firefighters will use some down time to practice tying knots. Keeping small sections of rope in common areas of the fire station is a handy reminder of the importance of knot-tying ability.

ROPE MATERIALS AND THEIR CHARACTERISTICS

Rope is constructed of a wide variety of natural and synthetic materials. Each material has a different set of characteristics that impact its appropriateness (or inappropriateness) for utilization in the fire/rescue service. Some materials present characteristics that make them perfect for specific applications while rendering them useless for other applications.

The earliest ropes made by man and used in the fire service were made of natural materials. The use of natural material ropes by the vast majority of departments continued until the 1980s. Unfortunately, as it happened, a series of incidents occurred that resulted in firefighters being killed or seriously injured while using natural fiber ropes. These incidents forced the fire service to reexamine the type of rope materials being utilized as **life safety lines.** This review ultimately resulted in the drafting of NFPA 1983, **Figure 15-1,** which deals with life safety ropes, harnesses, and hardware. NFPA 1983 established, for the first

FIGURE 15-1   2006 Edition of NFPA 1983 Standard.

FIGURE 15-2   Type 1 manila rope.

are twisted tightly together during the manufacturing process. This twisting creates the bond, through friction, that gives the rope its strength. Although the rope appears to be one continuous length, it is actually innumerable short fibers twisted together.

Manila rope is available in different types, with Type 1 being the higher quality and the one that was used most often by fire departments. Type 1 manila rope is manufactured from the higher quality inner fiber of the abaca plant leafstalk and can be identified by the colored string or ribbon that is twisted into the strands of the rope.

For years it was a common practice to soak new manila rope in water to make it limber. In actuality, this soaking reduced its strength by approximately 50 percent (which it never regained). This same reduction in strength occurs when a manila rope is stored in a compartment or area that has a high degree of humidity.

time (effective date June 6, 1985), minimum standards for this type of equipment if it is to be used by firefighters during the performance of their duties.

The following section presents the types of materials and characteristics of the different ropes used in the fire service.

Natural Materials

The materials that fall into the natural materials category include manila, sisal, and cotton. Because they are all natural materials, they share some of the same poor characteristics with regard to rot, mildew, abrasion resistance, natural deterioration/degradation due to age, a very low strength-to-weight ratio (when compared with synthetic materials), and a low **shock load** absorption capability.

SAFETY

Any ropes manufactured from natural materials should be used only as utility lines (ladder halyards, hoisting lines, etc.) and never in a life safety situation.

Manila

Manila rope, **Figure 15-2,** is made from the fibers that grow in the leafstalk of the abaca plant, which grows predominantly in the Philippines (Manila), hence the name. These are relatively short fibers that

Sisal

Sisal is another fiber obtained from plant leaves, in this case the agave plant, native to southern Mexico. Ropes made of sisal fiber have approximately 25 percent less tensile strength than manila fiber ropes of similar diameter.

Cotton

Cotton fiber rope is made from the seed hairs obtained from the cotton boll. As can be expected, this rope tends to be very soft and pliable. It is also much lower in tensile strength (approximately 50 percent) when compared with manila, due in part to the shortness of the fibers relative to manila or sisal. It is very susceptible to damage from abrasion.

Synthetic Materials

The primary synthetic materials utilized in the manufacture of ropes are nylon, polypropylene, polyethylene, and polyester. In the fire service today, ropes

FIGURE 15-3 The manila fibers can be seen sticking out, while the nylon kernmantle rope is smooth and unbroken.

FIGURE 15-4 Fishing line is the simplest example of nylon line.

TABLE 15–1	Rope Strengths	
	Temperature at Which There Is a Loss of Strength (°F)	Char or Melting Temperature (°F)
Manila	180°	375°
Nylon	300°	400°–500°
Polypropylene	200°	275°–300°
Polyethylene	230°	285°
Polyester	300°	450°–650°

Nylon

The material nylon was introduced in 1938 by E. I. Du Pont de Nemours and Company. It began to be heavily used to make ropes during World War II due to the shortage of natural fibers. The simplest example of nylon line is monofilament (single filament) fishing line, **Figure 15-4.** Nylon ropes are constructed of this same type of filament configured in multifilament bundles.

Negative properties of nylon include susceptibility to damage by acids (particularly mineral acids), loss of up to 25 percent strength when wet and/or frozen, stretching (elongation) when under load (use caution when utilizing to stabilize objects), and inability to float on water.

> **NOTE**
>
> Nylon has the following positive properties: high melting point (400–500°F), excellent abrasion resistance, can be bent sharply, high **tensile strength** (3 to 3½ times that of equal size manila), and a resistance to most chemicals (not acids).

Polyester

The most significant differences between nylon and polyester materials are that polyester has a good resistance to both acids and alkalis (however, an 80%+ acid and 10%+ base will affect it), it has low elongation under load, and its strength is not negatively affected by being wet or frozen. However, polyester does not handle shock loading very well. A correlation can be drawn between its low elongation characteristic and its ability to handle shock loading.

made of these materials are the rule rather than the exception. These materials have excellent properties resisting rot, mildew, and natural degradation due to age. They are also much more resistant to physical damage and damage from abrasion. One of the major advantages/differences between these materials and the natural materials is that the fibers making up the rope are continuous from end to end, as opposed to the short fibers inherent to natural material ropes. This difference is readily identifiable in **Figure 15-3.**

Some of the other significant advantages that ropes manufactured from synthetic materials have over those manufactured from natural materials are a high strength-to-weight ratio, much greater shock load absorption, higher resistance to acids (polypropylene and polyethylene—not nylon), and no permanent loss of significant strength when they become wet. The temperature at which the various rope materials begin to lose strength and either melt or char is shown in **Table 15-1.**

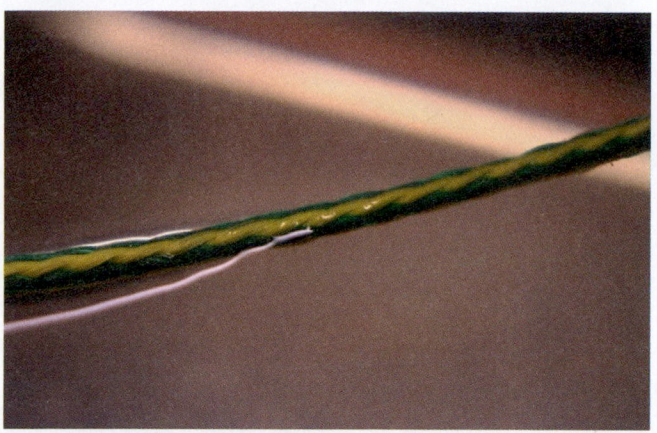

FIGURE 15-5   Polypropylene rope floating on water.

FIGURE 15-6   Brightly colored polyethylene rope floating on water.

Polypropylene

Ropes constructed from polypropylene, **Figure 15-5,** are primarily used for water rescue operations by the fire service. This is due to the fact that water has no effect on their strength, and even more beneficial is the fact that they will float. Ropes of this material may be utilized in industrial operations where there is a high likelihood of exposure to chemicals and/or acids due to their high resistance to these materials. Polypropylene also has good resistance to rot and mildew.

Polypropylene ropes have a low melting point, have low resistance to abrasion, are susceptible to damage from sunlight, and have a relatively low breaking strength.

Polyethylene

Polyethylene ropes, **Figure 15-6,** have properties that are very similar to those of polypropylene. However, they will float indefinitely and can be purchased in bright, highly visible colors if desired.

CONSTRUCTION METHODS AND THEIR CHARACTERISTICS

When natural fiber ropes dominated the fire service, they were basically constructed utilizing the laid (twisted) method, **Figure 15-7.** When synthetic materials were introduced, a variety of other construction methods were also introduced.

Modern ropes utilize a number of different construction techniques (braided, braid-on-braid, and kernmantle), each with its own particular set of characteristics, which in conjunction with the type of fiber used combines to determine a given rope's properties. The two broad categories that rope falls into are **static** or **dynamic.** Static ropes have very little (less than 2 percent) elongation at normal safe working loads. This characteristic is very beneficial in the rescue field and the primary reason why most departments use static rope for rescue operations. Dynamic rope on the other hand has a much higher degree of elongation (10 to 15 percent) at normal safe working loads. This is the primary reason why this is the type of rope most preferred by mountaineers and others where fall protection is the primary purpose of a rope.

Laid (Twisted)

As previously mentioned, the laid method is the most common type of construction for natural fiber ropes. It is also utilized with synthetic fibers, but is by no means the construction method of choice for rescue rope. Laid ropes are formed by twisting individual fibers together to form strands or bundles. These strands (three or more for larger ropes) are then laid

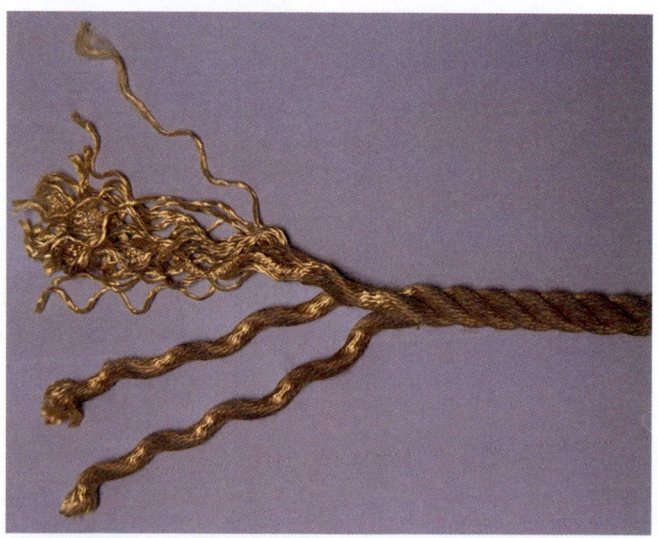

FIGURE 15-7   Example of laid construction method.

(twisted) together to form the finished rope. One of the major drawbacks to this type of construction method is that with all of the twisting to form the strands every fiber is exposed on the outside of the rope. This means that every fiber has potential to be damaged by abrasion, sunlight, chemicals, and so on.

Another type of laid rope is used primarily for mountaineering and called hard laid. It is generally much stiffer than other types and tends to get stiffer the more it is used. It is also difficult to form knots in, and one has to be very careful to use safety knots, especially when using a bowline knot. This stiffness also makes it difficult to handle and store.

A characteristic that is common to both types of laid ropes is that it tends to impart and/or accentuate spinning and subsequent twisting and possibly knotting at the bottom end of the rope when used for a rappel rope.

The one real advantage of laid rope is that, because all fibers are exposed, it is easy to inspect. Again, it is important to remember that this also makes all of the exposed fibers susceptible to damage.

Braided

The braided method of rope construction, **Figure 15-8,** is utilized predominantly with synthetic fibers, although there are some natural fiber braided ropes. Braided ropes are formed by weaving small bundles (not twisted) of fibers together, much the same as braiding hair, uniformly and systematically. These ropes are generally smooth to the touch and have a good degree of flexibility.

Although braided rope does not induce or accentuate spinning during rappelling, the braiding process exposes every fiber to abrasion, sunlight, and other forms of physical damage.

Braid-on-Braid

As can be assumed from the name, this type of rope is formed by braiding a sheath over a smaller braided core, **Figure 15-9.** This type of construction often results in an approximately 50/50 split between the core and the sheath in the total strength of the rope. There is, however, at least one manufacturer that has designed braid-on-braid rope so that the core maintains 80 percent of the strength. Ropes of this con-

Braided

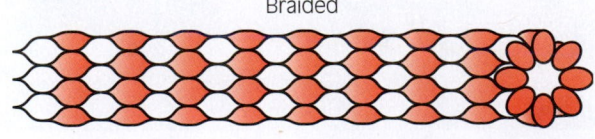

FIGURE 15-8   Example of braided construction method.

Braid-on-Braid

FIGURE 15-9   Example of braid-on-braid construction method.

FIGURE 15-10   Example of kernmantle construction method with the outer kern cut and pulled back at the end, showing the inner mantle section.

struction method tend to be quite dynamic (stretchy) and most have a soft sheath that is more susceptible to damage from abrasion.

Kernmantle

The term kernmantle, when separated, very accurately describes this construction method, **Figure 15-10.** The first part, **kern,** is a derivative of the term kernel, which is defined as "the central, most important part of something; core; essence." The second part of the word, **mantle,** is defined as "anything that cloaks, envelops, covers, or conceals." In a kernmantle rope, the kern generally carries the vast majority of the load, accounting for approximately 75 percent of the strength of the rope, with the mantle making up the remaining 25 percent. The mantle's main purpose is to protect the kern from physical damage.

Kernmantle ropes can be either dynamic or static depending on the configuration of the fibers of the kern. When the kern fibers are twisted together and the mantle woven over them, the result is a dynamic kernmantle rope. When the kern fibers run parallel continuously and the mantle is woven over them, the result is a static kernmantle rope.

Kernmantle ropes tend to be quite resistant to abrasion and other forms of physical damage since the main strength of the rope, the kern, is protected by the mantle. However, since the vast majority of the kernmantle ropes used in the fire service are manufactured from nylon, they are subject to the possible sources of damage listed previously in the synthetic materials section.

CAUTION

Care must be taken that lifelines are not stored in an area that has the potential of exposing them to acids, such as near the batteries or battery box vent of an apparatus.

PRIMARY USES

Although ropes have many uses, the fire service tends to utilize them on a regular basis for a few tasks and operations that can be divided into two classifications: utility and life safety. These classifications are discussed in detail in the following section. If a department/area has a particular evolution not shown here as a part of their standard operating procedures, this is by no means intended to intimate that they are wrong or improper.

A firefighter should be familiar with department standard rope use, as well as a broad range of possible uses. Firefighters are faced with new and different challenges and unusual situations every day. Experience and a broad knowledge base are vital assets of firefighting.

Utility

Rope used for utility purposes has no governing standards; therefore, it is the responsibility of the AHJ (authority having jurisdiction) to determine its type and construction. Because of the nongoverned status, firefighters must take extra precautions when using **utility rope.** A strong basic understanding of the capabilities of utility rope is required to avoid property damage and injury.

Some of the tasks for which utility ropes are used are shown in **Figures 15-11 A–C.** There are many uses too numerous to list, but firefighters should become intimately familiar with the common uses within their own department.

Firefighting and Rescue Uses

Ropes used for structural search and rescue guide ropes and other non-life-supporting operations do not fall into the category of life safety ropes according to NFPA 1983. However, it would be prudent to utilize either a life safety rope or a rope used specifically for

(A)   (B)   (C)

FIGURE 15-11   (A) Utility rope hoisting a pike pole. (B) Hoisting a ladder. (C) Hoisting a charged hoseline.

these purposes, as opposed to utilizing everyday utility rope.

Ropes, harnesses, and hardware utilized anywhere that there will be a life supported must comply with NFPA 1983, which sets minimum standards for equipment used in these types of situations. NFPA 1983 categorizes life safety ropes as **light-use (or one-person) or general-use (or two-person) ropes** and sets minimum tensile strength requirements for each. While this chapter does not go into depth on life safety rope operations, firefighters need to know the basics in order not to endanger themselves, other firefighters, or the persons they are attempting to rescue.

FIRE SERVICE KNOTS

All fire services have experience with some type of rescue or operation that requires ropes and knots. Each service will have preferred methods and knots to use. Members of the various fire services will sometimes argue the merits of one knot over another: this argument usually boils down to user preference. Given the number of knots possible and the various types of rope available, there is sometimes no correct black-and-white answer to the question "Which knot is the best?" Therefore, the NFPA has left this knowledge to the AHJ. Firefighters should consult the local protocol, policies, and SOPs to find which knots are expected for the various situations in their agency. A good working knowledge of the more common knots is important, and these knots are covered in this chapter.

While some of the knots covered here have been around for as long as people have been tying vines together, some were introduced along with the switch to synthetic fibers and modern construction techniques; some of the knots that have been used for years in the fire service do not hold well in these new ropes. The bowline is an excellent example. It is possible, once tension is released, for a bowline knot to release.

Cavers and mountaineers used synthetic ropes for many years. Because of synthetic rope characteristics and life safety applications, these mountaineers began to utilize and perfect their knot-tying technique. With the fire service switch to synthetic ropes, these updated and revised knot-tying techniques were incorporated into department applications.

Terms used for Rope and Knots

Parts of a Rope

Figure 15-12 shows the three separate and distinct parts of a rope: (1) the working end, (2) the standing end, and (3) the running end.

Elements of a Knot

Just as firefighters need to know the basic terms used to describe the parts of a rope, they also need to know the terms used to describe the elements that are combined to form a knot, **Figure 15-13.** A **round turn** is formed by continuing the loop on around until the

Working End

Standing Part

Running End

FIGURE 15-12 The three parts of a rope: working end, standing part, and running end.

FIGURE 15-13 Left to right: a round turn, a bight, and a loop. Take the loose end of the working end after tying the primary knot, and secure it by making a round turn around the standing knot and bringing the loose end through. Make a round turn in the standing portion of the rope, and slide the round turn down over the object being hoisted.

sections of the standing part on either side of the round turn are parallel to one another. A **bight** is a doubled section of rope, usually made along the standing part. A bight forms a U-turn in the rope that does not cross itself. A **loop** is a turn in the standing part that crosses itself and results in the standing part continuing on in the original direction of travel.

Knots

The following pages provide step-by-step instructions for tying the knots that are utilized most often by the fire service. They are tried and true and straightforward. Fellow firefighters should not have any prob-

lem recognizing the knot when it needs to be untied. Later, this chapter shows how each of these knots can be used for various tasks that may be required at any given incident.

It is important to point out that the step-by-step instructions are intended to teach firefighters the steps necessary to tie the various knots. Seldom are firefighters called upon to tie any of these knots while standing in a brightly lighted, air-conditioned classroom or apparatus bay.

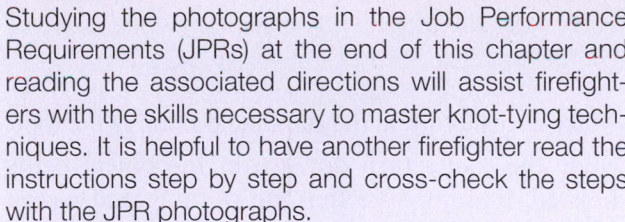

STREETSMART TIP

Studying the photographs in the Job Performance Requirements (JPRs) at the end of this chapter and reading the associated directions will assist firefighters with the skills necessary to master knot-tying techniques. It is helpful to have another firefighter read the instructions step by step and cross-check the steps with the JPR photographs.

All knots should be neat and tight. The terms used to describe this practice are to dress and set a knot. **Dressing** a knot is the practice of making sure that all parts of the knot are lying in the proper orientation to the other parts and look exactly as the pictures indicated. **Setting** a knot is the finishing step of making sure the knot is snug in all directions of pull. This is an important step because a perfect knot, if not set, will slip when a load is applied. This is due to the lack of appropriate friction and cinching effects that cause knots to hold. **Figures 15-14 A** and B show a knot both incorrectly set (A) and correctly dressed and set (B).

(A)

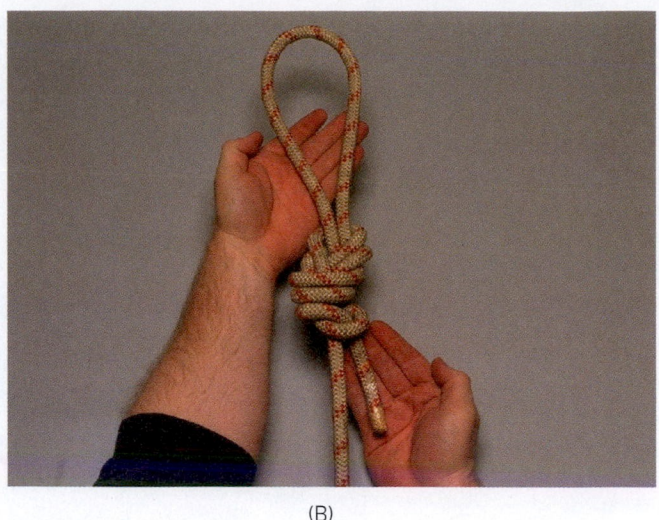

(B)

FIGURE 15-14   (A) A loose and sloppy knot. (B) A knot properly dressed and set.

Half Hitch

A half hitch is almost always utilized in conjunction with some other knot and is used to maintain the proper orientation of the object being hoisted.

JPR 15-1: Half Hitch

(For step-by-step photos of this skill sequence, see page 450)

1. Make a round turn in the standing portion of the rope.
2. Slide the round turn down over the object being hoisted, making sure that the running end passes under the working end. If necessary, on particularly long objects such as pike poles more than one half hitch can be used.

Overhand (Safety) Knot

An overhand knot is generally used to secure the loose end of the working end after tying a knot. All of the knots described in this chapter should be secure in and of themselves, but the addition of this simple knot can provide an extra measure of security.

JPR 15-2: Overhand (Safety) Knot

(For step-by-step photos of this skill sequence, see page 450)

1. Take the loose end of the working end after tying your primary knot and secure it by making a round turn around the standing part and bringing the loose end through this round turn and the primary knot.

Clove Hitch

The clove hitch is used to attach a rope to an object, such as a pole, tree, or fence post, or to a tool, such as a pike pole or hoseline. A clove hitch can be tied anywhere along the rope if needed. It will hold equally well when tension is applied from either direction—if tied correctly.

There are basically two ways of tying a clove hitch. Step-by-step instructions and figures are provided for each method. The first way, which is fastest, is to tie a clove hitch in hand (sometimes referred to as in the open) and then slip it over the object. In many cases, however, this is not possible, so it is important that firefighters also know how to tie the clove hitch around an object.

JPR 15-3: Clove Hitch Tied in the Open

(For step-by-step photos of this skill sequence, see page 451)

1. Hold the rope in your hands with the working end to your right.

2. Form a loop with the working end passing in front of the standing part (between you and the rope).

3. Move the rope down to the right and form another loop exactly like the previous loop.

4. Place the loop formed in step 3 behind the loop formed in step 2. You now have a clove hitch.

5. Keep the clove hitch loops in the proper orientation to one another and slide them over the object to be secured.

6. Finish with a safety knot.

JPR 15-4: Clove Hitch Tied around an Object

(For step-by-step photos of this skill sequence, see page 452)

1. With the working end, make a loop around the object.

2. Cross the working end over the standing part and make another loop around the object.

3. Bring the working end under at the point you crossed with the loop. At this point your loops should be side by side with the standing part and the loose end of the working end opposing one another.

4. Finish with a safety knot.

Becket Bend and Double Becket Bend

Also known as the sheet bend and double sheet bend, the becket bend and the double becket bend are very useful knots for tying ropes together. The becket bend is utilized to tie ropes of equal diameter together, while the double becket bend is used most often when tying ropes of unequal diameter.

JPR 15-5: Tying a Becket Bend

(For step-by-step photos of this skill sequence, see page 453)

1. Form a bight in the running end of the line that needs to be extended.

2. Utilizing the working end of the rope being added, bring it up through the bight.

3. Take this same end around both sides of the bight, forming a loop around the bight.

4. Still utilizing this same end, pass it between itself and the rope in which you formed the bight.

5. When dressed and set, the end you were working with should be at approximately 90 degrees to the rope.

6. Finish with a safety knot.

JPR 15-6: Tying a Double Becket Bend

(For step-by-step photos of this skill sequence, see page 454)

1. Form a bight in the running end of the larger line.

2. Utilizing the working end of the smaller rope, bring it up through the bight.

3. Take this same end all the way around both sides of the bight forming a round turn around the bight.

4. Continue around the bight a second time.

5. Still utilizing this same end, pass it between itself and the rope in which you formed the bight. At this point the rope you have been working with should have formed what looks like the number eight (8).

6. When dressed and set, the end you were working with should be at approximately 90 degrees to the rope. Finish with a safety knot.

Bowline Knot

Although it was the mainstay of fire service knots for years, the advent of synthetic fiber ropes has greatly reduced the utilization of the bowline knot. This is due in large part to the fact that the bowline is an inherently loose knot, which is beneficial when it needs to be untied but has a dangerous tendency to slip when used on synthetic fiber ropes.

CAUTION

If a bowline is not properly dressed, set, and secured with a safety knot there is a possibility of it inverting and becoming a slip knot.

JPR 15-7: Tying a Bowline

(For step-by-step photos of this skill sequence, see page 455)

1. Holding the standing part in your hand with the working end coming toward you, form a loop in the rope with the working end on your side when it passes the standing part.

2. Still utilizing the working end, bring it up through the loop formed in step 1.

3. Pass this same end around behind the standing part and then bring it back down through the loop you brought it up through.

4. Be sure that the working end is on the inside of the loop that you have just formed with the completed

knot. If left on the outside there is a much greater chance of it getting caught and the knot inverting to a slip knot.

5. Finish with a safety knot.

Figure Eight Knots

The figure eight knots came into use in the fire service during the same time line as synthetic fiber rope. This was a natural transition because cavers and mountaineers, from whom the fire service adopted synthetic fiber ropes, had been utilizing figure eight knots for years. Figure eight knots have several good features. They are very simple to tie; they form a knot that will not slip or pull from the rope (as a bowline knot will); and they place less stress on the rope, as the turns in the figure eight are not as sharp as in the other knots. Figure eight knots are the preferred knots when working with synthetic rope.

The name *figure eight knot* is used to describe many different knots; a modifier is added to delineate one knot from the other. This section covers the *basic* figure eight, the *follow-through* figure eight, and the figure eight *on a bight*.

JPR 15-8: Tying a Basic Figure Eight

(For step-by-step photos of this skill sequence, see page 456)

The basic figure eight knot is useful when an "end of the line" knot is needed, such as when a rappel rope is not long enough to reach the ground or solid landing. This knot is also useful when joining two ropes of equal diameter together.

1. Form a bight in the working end of the rope, holding the bight out in front of you by the apex so that both sides of the bight are hanging parallel.
2. Take the short end (working end) of the rope and wrap it around the standing part from front to rear.
3. After the working end comes around the standing part, you will have formed an eye. Now take the working end through the eye from front to rear.
4. If you are joining two ropes, at this point you will take the working end of the second rope and, coming from the opposite direction, trace the path of the original rope as it flows through the knot. You should end up with both working ends lying parallel to the opposite rope's standing part.
5. Finish with a safety knot with each working end around the opposite rope's standing part.

JPR 15-9: Tying a Follow-Through Figure Eight

(For step-by-step photos of this skill sequence, see page 457)

The follow-through figure eight knot, sometimes called a *traced eight knot,* is very useful when attaching a utility or life safety line rope to an object that does not have a free end available.

1. Follow the directions for tying a basic figure eight; however, you need to make sure that you have an adequate amount of rope left on the working end side of the basic knot to go around the object you are securing the rope to.
2. If you are securing the rope to an object, wrap the working end around the object at this point. (From this point on the procedure is very similar to the procedure for joining two ropes together.)
3. As the name implies, you now follow the other rope through the knot, coming from the opposite direction. In effect, you have formed a figure eight on a bight (the next knot described) around an object.
4. You should conclude with the end of the working end parallel to the standing part.
5. Finish with a safety knot.

JPR 15-10: Tying a Figure Eight on a Bight

(For step-by-step photos of this skill sequence, see page 458)

The figure eight on a bight is utilized when the object being attached to has a free end available to place the rope over and is very useful in forming a loop in the line that will not slip or cinch up. This knot can also be used when a loop is needed in the middle of a rope.

1. Start by forming a bight. The size of the bight needs to be relative to the object the finished loop is going to be placed over or around.
2. Grab both ropes so that the bight is hanging down from one side of your hand while the standing part and the end of the working part are hanging out of the opposite side of your hand. Basically you are forming a second bight of doubled ropes.
3. Hold this bight out in front of you by the apex so that both sides (four parts of the same rope) of the bight are hanging parallel.
4. Take the first bight and wrap it around the standing part from front to rear.
5. After the bight comes around the standing part, you will have formed an eye. Now take the bight through the eye from rear to front.

6. You have now formed an easy, reliable loop that will not slip. Do not forget to dress and set your knot and apply a safety with the end of the working end.

Rescue Knot

Almost every department and/or recruit school has a knot or combination of knots that the agency recognizes as the best knot to be used as a rescue knot. The rescue knot described in this section is by no means intended to imply that an organization's choice is incorrect. The knot presented here is one of the easiest, fastest, and surest among the many that have been tried over the years. It should come as no surprise that the primary knot utilized is a member of the figure eight family, a figure eight on a bight.

This rescue knot can be used on oneself, a conscious patient, or an unconscious patient. The basic method of applying this rescue knot does not change with the patient; however, the difficulty factor does go up if the patient is unconscious. The following instructions are based on applying the rescue knot to oneself. However, it is also important to practice the rescue knot on other people to simulate victims needing rescue. An emergency scene is not the proper venue for learning skills.

JPR 15-11: Tying a Rescue Knot

(For step-by-step photos of this skill sequence, see page 459)

1. Tie a figure eight on a bight. Your initial bight should be approximately 6 feet long. This will allow you to tie the figure eight on a bight, including a safety at the working end, and still have a finished loop approximately 4 to 5 feet long. This loop is going to need to be long enough to go around the victim's waist and back up between the legs. The loop does not need a precise measurement, but it is important to remember that the length cannot be adjusted once the knot is tied. Time should be spent sizing and applying this knot on different size victims in practice scenarios. Firefighters should rehearse this skill until they feel comfortable and can be consistently successful with this maneuver on the first try.

2. Place the loop on the floor or step into it with the actual figure eight knot behind you. Raise the loop to waist level, being careful not to get above the top of the hip bones.

3. Reach down between your legs, pulling the standing part of the rope up in front of you. The standing part should not be between your body and the loop but on the outside of the loop.

4. Continuing to hold the rope in front of you, raise your hand holding the rope above your head. Reach out with the other hand and grasp the ropes, both the one coming from between your legs and the one hanging down from the hand over your head.

5. With the hand over your head make three full twists, forming a loop. Place the loop over your head and shoulders so that it comes to rest in your underarms.

6. Keeping the standing part pointed generally down, adjust what slack you can out of the rope between the figure eight knot coming from between your legs and the twists forming the loop that is now around your chest. Raise the standing part up over your head, allowing the twists to wrap around themselves and cinching down on the rope forming the loop around your chest. This is important to keep the loop from cinching down on your chest. Arranging the ropes that come up between your legs as your weight is lifted by the rope will help ease the discomfort somewhat.

Remember this tip when lifting a conscious or unconscious patient.

Water Knot

Webbing has become very popular and is carried by many firefighters in their personal protective clothing for use as an emergency harness or hose holding strap.

> **SAFETY**
>
> The water knot is the only knot that is recommended for use to form a length of webbing into a load bearing loop.

This is one knot that absolutely must be kept neat and be tightly set.

JPR 15-12: Tying a Water Knot

(For step-by-step photos of this skill sequence, see page 461)

1. Tie a simple overhand knot in one end of the webbing, making sure the webbing lies flat as it crosses back over itself. Leave enough tail beyond the overhand knot to be able to place a safety knot on that side of the water knot when finished.

2. Take the other end of the webbing and thread it through in the opposite direction, again making sure that the webbing lies flat as it crosses itself. Make sure you feed enough webbing through so that you are able to place a safety knot on this side of the water knot.

INSPECTION

As with any emergency service tool, all ropes must be inspected and properly maintained to ensure they are in good shape for use during an emergency incident. These inspections should be a matter of department policy and done on a regular basis; many departments conduct monthly inspections. It is also recommended that all inspections of life safety ropes be logged on a form specifically designed for this purpose. These forms can be purchased from most rescue supply houses or the department can develop its own. An example is shown in **Figure 15-15.** In order for an inspection log to be useful, the ropes themselves need to be individually identified.

If a life safety rope is found to be damaged or suspect it should be immediately removed from service. NFPA 1983 lists re-use requirements for a life safety rope used during an actual emergency operation. This is mainly due to the unavoidable stresses that can be placed on life safety rope during emergency operations. Emergency use can subject the rope to the forces of shock loading and overloading and can result in chafing, abrasion, and exposure to chemicals/acids. These ropes should be inspected and either returned to or removed from service.

Ropes should be inspected along their entire length. This can be done when they are being placed back in storage, coiled, or bagged, **Figure 15-16.** It is best to inspect ropes after they have been cleaned. Generally, all ropes should be visually inspected for abrasion,

DELMAR ROPE USAGE & HISTORY

SERIAL NUMBER	I.D. MARKING	155' LENGTH	1/2" DIAMETER

Kernmantle FIBER	Red COLOR	E-2-Bend CONSTRUCTION	06F94421KB MFG'S LOT NUMBER

6-6-94 DATE OF MFG. — 3-11-95 ISSUE DATE — 3-11-95 DATE IN SERVICE

INSPECT ROPE FOR DAMAGE OR EXCESSIVE WEAR EACH TIME IT IS DEPLOYED AND AGAIN AFTER EACH USE. IMMEDIATELY RETIRE ALL SUSPECT ROPES.

DATE USED	INCIDENT LOCATION	TYPE OF USE	ROPE EXPOSURE	DATE INSPECTED	INSPECTOR'S INITIALS	ROPE CONDITION & COMMENTS
9/16/95	Shaker High	Training	Ground, Grass + Dirt Some Sunlight	9/16/95	V.G. F. Carr, Jr.	Good No Damage
11/18/95	Vecky	Training	Ground, Dirt, Grass Rocks, MLN	11/18/95	F. Carr, Jr.	Good No Damage
7/15/96	Boght	Training	Dirt, Grass, Rocks	7/15/96	F. Carr, Jr.	Good No Damage

FIGURE 15-15 A standard life safety rope inspection form.

FIGURE 15-16 It is very important that rope is inspected as it is being put back into a rope bag.

laceration, chemical exposure, melting, and excessive fuzzing, many of which can be seen in **Figure 15-17.** They should also be inspected **tactilely** by running the rope through the hands. For this reason, firefighters should not wear gloves when inspecting and storing

FIGURE 15-17   Damaged rope.

FIGURE 15-18   It is important to twist apart a laid rope to inspect between strands.

a rope, because they are feeling for foreign material embedded in the rope, slippery spots (possible chemical damage), voids in the center fibers (kernmantle and braid-on-braid), stiff or hard spots, and soft spots. For safety purposes, care should be taken to visually inspect for this foreign material. Firefighters do not want to injure themselves unnecessarily with cuts and abrasions.

Laid (Twisted)

When inspecting laid ropes, firefighters should look and feel for all of the items mentioned above. In addition, laid ropes should be untwisted at random intervals to inspect the inside between the strands, **Figure 15-18.** When untwisting the strands of a natural fiber rope, it is important to watch for small particles such as sand or gravel that may damage the rope fibers. If a rope is stored wet or stored in a very humid location there may be signs of mold, mildew, or rot. The presence of mildew is not necessarily a terminal problem if found early enough. If these signs are present, the rope should be removed from service and the cause of the mold should be addressed as soon as possible.

Braided

Braided rope should be inspected visually and tactilely as described previously. It cannot be twisted apart, so inspecting the inside is not possible. With braided rope, all strands will appear on the surface somewhere along the rope.

Braid-on-Braid

It is very important to remember, when inspecting braid-on-braid rope, that there is no way to see the inside braided rope. The person doing the inspection needs to pay particular attention to the tactile part of the inspection, watching and feeling for possible types of damage previously listed. Another problem with this type of rope is that the outside braid will sometimes slip over the inner braid, causing the rope to invert on itself. A rope with this problem should be immediately removed from service. The outside braid typically represents 50 percent of the strength, so any damage to the outside braid will have a significant impact on the remaining strength.

Kernmantle

As with braid-on-braid rope, there is no way to see the kern portion, which typically represents 75 percent of the total strength of a kernmantle rope. The tactile inspection is the best and only way to discover damage to the kern. It is possible (but unusual) for the kern to be damaged without damaging the mantle. One should be alert for any change in the regular uniform feel of the rope as it is run through ungloved hands. Irregularities that can be discovered in a thorough tactile inspection include flat spots, voids, bunches, stiffness, and limpness. The key is to search for a different feel. If a minor imperfection seems to have been located, it may not necessitate the removal of the rope, but would certainly indicate the need for another thorough inspection.

STREETSMART TIP

Another way to inspect the rope, especially if a minor imperfection seems to have been located, is to tie it off to an object and do a thorough tactile inspection while placing slight tension on it. This may cause the minor imperfections to be more pronounced and therefore easier to feel.

MAINTENANCE

As with all firefighting and life safety tools and equipment, proper maintenance of utility and life safety ropes is required in order to ensure that they are available, safe, and ready for immediate use at an emergency scene. The maintenance of ropes is not difficult or complicated; it is generally limited to cleaning, inspecting, and storing.

Occasionally the firefighter may be called on to assist with placing new rope in service. If the rope was purchased in bulk, it is important to carefully adhere to the manufacturer's instructions with regard to cutting, sealing, whipping, and so on. The same goes for cleaning and storage of ropes. Most new ropes (especially life safety ropes) purchased from a rescue equipment dealer come with instructions and warnings, either attached to or packaged with the rope. An example is shown in **Figure 15-19.**

Cleaning

As mentioned in the previous section, the best policy for cleaning is to follow the manufacturer's instructions; however, since not all rope comes with instructions, the following are some general guidelines for the cleaning of rope.

Natural Materials

The cleaning of natural fiber rope is very difficult because water cannot be utilized. Remember from the earlier discussion that natural fiber ropes lose approximately 50 percent of their strength when wet and do not regain the lost strength when they dry. Firefighters are limited to brushing off the loose dirt and foreign materials with a stiff broom or brush, **Figure 15-20.** This is another reason why the fire service is moving away from the use of natural fiber rope, even as utility rope.

Synthetic Materials

Ropes manufactured from synthetic materials can be cleaned in a number of ways. Once again it is always the best policy to follow the manufacturer's instructions. In the absence of these, however, the following procedures can be utilized:

1. Use tap/cold water (high-temperature hot water may damage rope).
2. If detergent use is necessary, use only mild detergent that has been well diluted. (Use absolutely no bleach or any cleaning product containing bleach.)

Hand Washing

The most basic method of cleaning rope is to wash it by hand, using a large utility sink, bucket, or hose and scrub brush. The first two methods (sink or bucket) require immersing the rope, letting it soak for a few minutes, and then agitating and scrubbing it with the hands or small brushes, **Figure 15-21.** The third method (hose and scrub brush) described involves laying the rope out on the apparatus bay floor or apron, wetting it with a garden hose, and scrubbing it with a brush or broom. If detergent is needed, it can be mixed in a bucket and the brush or broom can be dipped into it.

If detergent is used with any of these methods, all residue must be removed after the washing is complete. If the sink or bucket method is used, this can be accomplished by changing the water or using other sinks or buckets full of fresh water to thoroughly

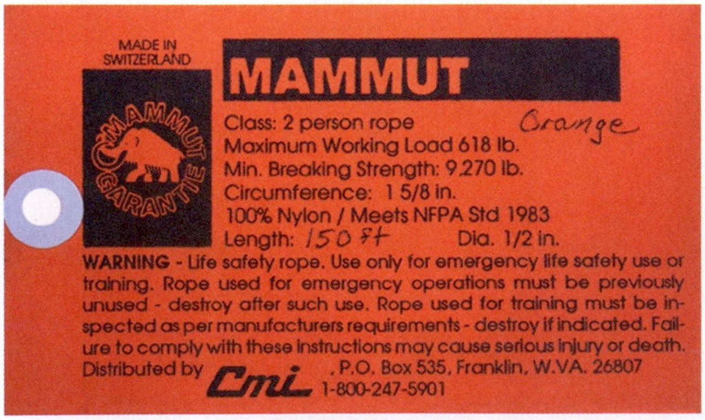

FIGURE 15-19 A standard manufacturer's warning/instruction label.

FIGURE 15-20 Manila rope is best cleaned by laying it out and brushing it off.

rinse the rope. With the hose method, the rope must be rinsed thoroughly with clean water; it helps to occasionally move the rope once or twice so that the soap trapped underneath it can be rinsed away.

Rope Washer

A number of different rope washers are available for purchase from rescue equipment suppliers. These are generally small PVC devices that fasten directly to a hose bib or to the male end of a garden hose, **Figure 15-22.** They are designed so that the water is directed at the rope from all different directions at the same time. The rope is inserted in one end (the direction of travel is usually marked) of the washer and slowly pulled out the other end. Without some method of injecting detergent into the hose stream, detergent cannot be used with rope washers.

Washing Machine

If using a clothes washing machine to wash rope, it is very important to make sure that the machine itself cannot damage the rope. Only a front-loading machine (with a glass window) should be used. These machines do not have an agitator for the rope to get caught on or wrapped around. Placing the rope in a large mesh bag before placing it in a machine, **Figure 15-23,** is the best way to keep it from getting fouled in the machinery and terribly knotted. If a large mesh bag is not available, the rope can be "chained," **Figure 15-24,** to try to keep it from getting fouled and/or knotted. Once again the previous directions regarding use of detergents should be followed and all residue rinsed out of the rope.

FIGURE 15-21 A kernmantle (nylon) rope being washed in a bucket.

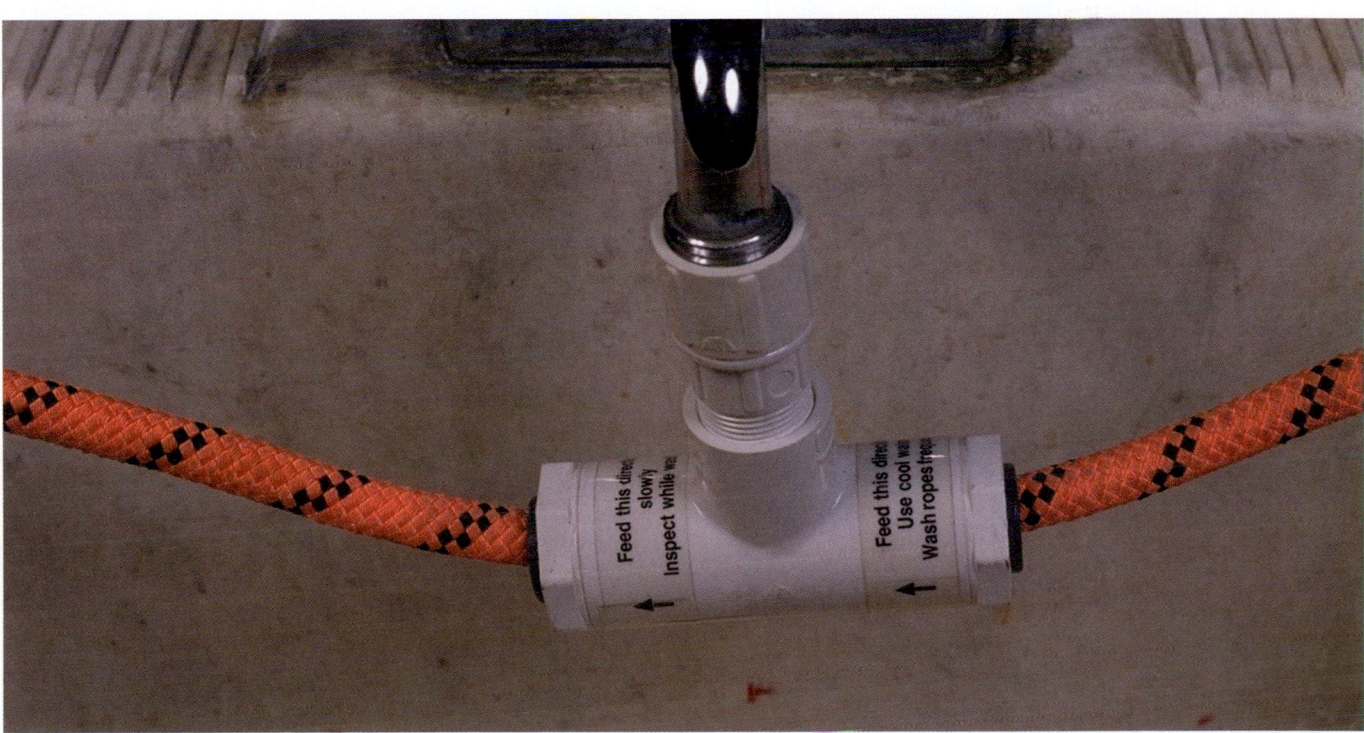

FIGURE 15-22 A commercial (PVC) rope washer being used to wash a kernmantle rope.

FIGURE 15-23   A kernmantle rope can be placed in a mesh bag and washed in a front-loading washing machine.

FIGURE 15-24   A kernmantle rope can also be "chained" and washed in a front-loading washing machine.

Drying

No matter what cleaning method is utilized, the rope must be completely dry prior to being stored. A few different methods of drying a rope are discussed next.

Laying Flat to Dry

The rope can be laid out on the apparatus bay floor or any other area that is clean, dry, and out of direct sunlight. Rope should be turned as it dries so it dries uniformly throughout.

Hanging to Dry

This is one of the easier methods of drying ropes. They can be hung in a hose tower, from the bar joists/trusses of the apparatus bay, or from any other place that is clean, dry, and out of direct sunlight. The rope must not come in contact with roofing tar, paint, bug spray, or other contaminants that may have been sprayed on bar joists or trusses. It is important to use care when removing the rope so that it does not get snagged or otherwise damaged by the truss plates on wood trusses or slag from welds on steel bar joists.

CAUTION

Extreme caution is necessary if a clothes dryer is to be used. Some departments utilize and some texts list clothes dryers as a viable alternative for drying rope. Not only is there the same issue as with the washer (i.e., the possibility of the rope becoming entangled in the machinery), but with a dryer (gas or electric) it is very difficult to tell what the interior temperature is, even if there is a "low-temperature" setting. The temperature could surpass the manufacturer's recommendations.

NOTE

It is very important that rope be stored in a manner that allows for quick identification, access, and deployment at emergency operations, **Figure 15-25.**

Storage

Quick identification is very important so that a firefighter can rapidly select the proper type (utility or life safety) and length required for a given application.

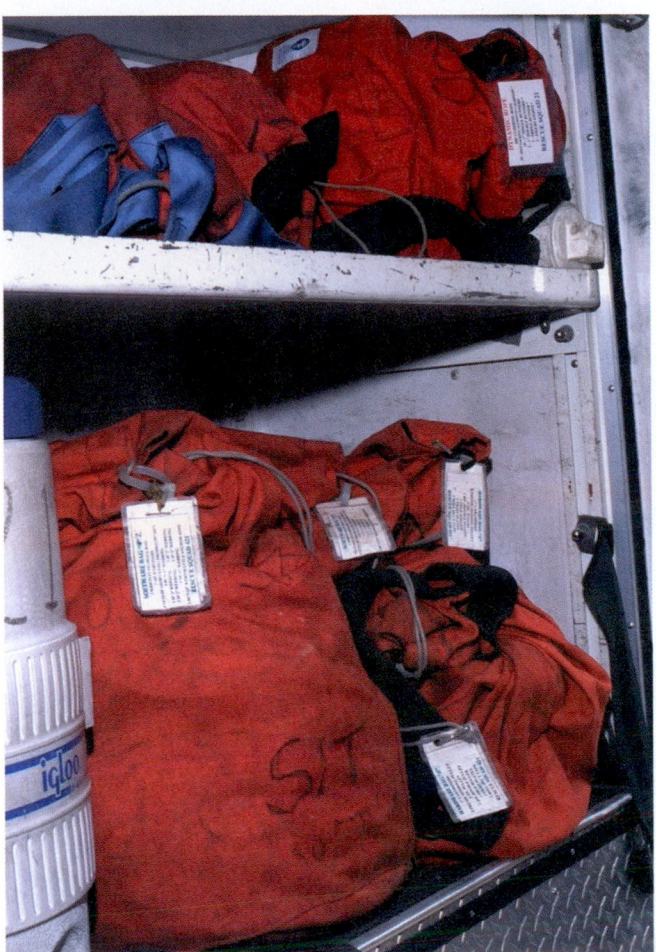

FIGURE 15-25 Properly identified, bagged, and stored life safety ropes.

FIGURE 15-26 Different types, sizes, and colors of bags are useful for quickly identifying different sizes and types of ropes.

There are many ways to accomplish quick identification, **Figure 15-26.** Many departments use different material types, sizes, or colors of bags or tags to differentiate among utility ropes, life safety ropes, and ropes of different lengths. Each department needs to establish a policy or defined procedure to quickly identify the different types of ropes.

SAFETY

It is imperative that fire departments establish a system to quickly differentiate life safety rope from utility rope.

Quick access is usually not a problem on fire or rescue apparatus, especially if there is a routine inspection policy in place that keeps the ropes from being buried behind other equipment. Remember that ropes must be stored away from areas where they might be exposed to battery acid, fuel (including fumes from either), or sunlight.

The issue of quick deployment has been addressed for the most part by the storage of ropes in specially designed bags, although some departments still utilize the coiling method to store their ropes. While either method, **Figure 15-27,** will work to keep the rope neatly stored, the bags have some significant advantages. Depending on its construction, a bag offers protection from dirt, liquid contaminants, sunlight, abrasion, and so on. If a rope is properly bagged, it is easier to deploy than a rope that is coiled.

Coiling

The standard coil used by fire departments for years will work with either natural or synthetic ropes equally well. If the coiled method of storage is selected by a department, it is imperative that all members of the department be familiar with the process of both coiling and deploying from the coil. Coiling is not the storage method of choice for life safety rope.

Some departments have built special apparatus to assist with the coiling of their ropes, but this is not

FIGURE 15-27 The two most common methods of storing life safety ropes (bagged or coiled).

really necessary. It is possible to use improvised standards and common items to coil a rope. The heel or tip of a ground ladder (do not use the tip of a roof ladder), a four-legged kitchen chair, or any item that has parallel posts (legs) approximately 18 to 24 inches apart will work for rope coiling. The distance between the standards and the number of wraps per layer are dependent on the size and length of the rope being coiled.

JPR 15-13: Coiling a Rope

(For step-by-step photos of this skill sequence, see page 452)

1. Measure off an amount of rope equal to approximately three times the distance between the standards to be used in tying the finished coil.
2. Starting with this point measured at one of the standards, wrap the rope around the standards, forming the loops and moving toward you until you have nine to twelve loops on the standards. Then form a second layer on top of the first, moving away from you. Continue in this manner until approximately twice the amount measured off in step 1 remains.
3. Begin wrapping this remaining rope around the coils that you formed in step 2. Finish these wraps off with a clove hitch.
4. Utilizing the rope measured off in step 1, form a bight leaving the loose end of the rope coming out toward you. Feed the bight from the back side to the front side on the wraps. You should now have the loose end and the bight sticking out toward you.
5. Place the loose end through the bight and adjust the bight until it is snug where it passes through the bight.

FIGURE 15-28 A coiled rope in the process of being deployed.

To deploy a rope that has been coiled in this manner, the firefighter pulls the loose end out of the bight and pulls the bight out from between the loops. Next, the firefighter pulls a couple of loops out of the center and drops the coil if in an elevated position or feeds the rope out, pulling it from the center, as needed. A firefighter must be aware that if the coil is dropped, it may hang up before it reaches the ground, either on itself or on some structural member. If extending the rope horizontally, **Figure 15-28,** it is usually easier if one person holds the coil on the ground while another person pulls the rope out of the coil.

Bagging

The utilization of special bags for the storage of rope is another of the things that came about with the switch to synthetic fiber ropes. Ideally, rope bags should have some type of attached watertight storage for the rope log. If they do not, the rope and bag need to be clearly identified and associated with a rope log on file in the apparatus or at the station.

The process for placing a rope in a rope bag is very simple (especially when compared to coiling a rope) and can be easily accomplished by one or two persons. Many bags have holes in the bottom for rope deployment and drainage purposes.

VIEWPOINT

Deciding whether to deploy a rope from the top or bottom opening of a bag remains the decision of the AHJ. Deploying a rope from the rope bag's bottom hole can be problematic: if the rope inside the bag becomes tangled, the problem is not as easily solved if the rope is deployed in this manner. Firefighters should ensure compliance with AHJ guidelines when reloading rope bags.

JPR 15-14: Bagging a Rope

(For step-by-step photos of this skill sequence, see page 463)

1. If policy is for the rope to be fed out the bottom hole, start by feeding enough rope out the bottom to tie a basic figure eight knot. If the rope is deployed from the top opening in the bag, tie a basic figure eight knot in the end of the rope and place it inside at the bottom of the bag.

2. Begin placing the rope in the bag by sliding your hand approximately 12 to 18 inches up the rope at a time and "stuffing" it in the bag. Do not coil the rope in the bag, because it increases the possibility that the rope will hang up as it is deployed.

3. Continue stuffing until the end of the rope is reached. Tie a basic figure eight knot in the end and place it on top of the rope in the bag. Securely close the bag.

The process for deploying a rope stored in a bag is very straightforward. If department policy is to deploy through the bottom hole, the firefighter grasps the figure eight tied just outside the hole and feeds the rope out of the bag as needed. Feeding the rope out allows the firefighter to keep the bag in case a problem develops, **Figure 15-29,** and also protects the integrity of the rope. If deploying the rope horizontally, the firefighter can either have someone take the end of the rope or the bag and walk in the desired direction of deployment. The rope will feed out as the person walks.

SAFETY

A life safety rope should never be dropped from an elevated position. This action can result in mechanical damage to the rope.

NOTE

If a firefighter has a good working knowledge of ropes and is skilled in tying the knots presented in this chapter, there should be no problem with hoisting almost any type of tool or equipment.

FIGURE 15-29 A bagged lifeline in the process of being deployed.

RIGGING FOR HOISTING

One of the primary uses of rope on an emergency scene is for hoisting or lowering of tools and equipment to the needed location. This is not the type of job that requires a life safety rope. Instead, a much smaller diameter rope can be utilized, which translates to a lighter, easier-to-carry rope. Ropes used for hoisting can be stored either coiled or bagged. As previously mentioned, it is usually much easier and more practical if the rope is fed out by the person who will be hoisting the item, **Figure 15-30,** as opposed to dropping the whole bag or trying to pass the end of the rope up to the person at the higher elevation.

This section presents examples of how to hoist a few specific tools and equipment and the proper knots to utilize when doing so.

STREETSMART TIP

In general, anything that has a closed handle can be hoisted with a figure eight or bowline, while longer cylindrical tools (i.e., ax and pike pole) can be hoisted using a clove hitch and half hitches.

Some departments have policies requiring the use of **tag/guide lines,** which are guide ropes held and controlled by firefighters on the ground, **Figure 15-31.** There are

FIGURE 15-30 A firefighter deploying a rope from an elevated position in order to hoist a tool.

FIGURE 15-31 Use of a tag/guide line. A rotary saw (closed handle) being hoisted with a tag line.

some circumstances in which tag lines should be used whether required or not. Examples of these circumstances are when an overhang(s) exists that the item is likely to get caught on, when the item rubbing against the side of the structure may be damaged, when there is a strong wind that may cause the item to blow out of control, or any time firefighters think a tag/guide line is necessary. All knots must be dressed, be set, and have safeties.

Specific Tools and Equipment

Ax

A small figure eight on a bight with a half hitch up the handle is the easiest and quickest way to hoist an ax.

JPR 15-15: Hoisting an Ax

(For step-by-step photos of this skill sequence, see page 464)

1. Tie a figure eight on a bight forming a small loop. Drop the loop over the ax handle.
2. Take the loop around the head of the ax, bringing it back up, paralleling the handle.
3. Place a half hitch approximately 6 to 8 inches below the handle.

Pike Pole

Pike poles should be hoisted point up.

JPR 15-16: Hoisting a Pike Pole

(For step-by-step photos of this skill sequence, see page 465)

1. To hoist a pike pole point up, place a clove hitch near the end of the handle.
2. Place two half hitches around the handle between the clove and the point, with the last one being located immediately below the head.
3. Raise the pole with the head at the highest point.

Hoselines

Hoselines can be hoisted either charged or uncharged. A charged hoseline is going to be dramatically heavier than an uncharged one.

JPR 15-17: Hoisting a Charged Hoseline

(For step-by-step photos of this skill sequence, see page 466)

1. With the nozzle bale in the closed (forward) position, tie a clove hitch around the hoseline 18 to 24 inches behind the nozzle.

2. Form a bight in the rope and feed it through the bale from the coupling side.

3. Form a half hitch and slip it over the tip of the nozzle.

4. The rope will ensure the bale remains closed.

5. The hose is ready to be hoisted.

JPR 15-18: Hoisting an Uncharged Hoseline

(For step-by-step photos of this skill sequence, see page 468)

1. With the nozzle bale in the closed (forward) position, tie a clove hitch around the hoseline 18 to 24 inches behind the nozzle.

2. Form a bight in the rope and feed it through the bale from the coupling side.

3. Form a half hitch and slip it over the tip of the nozzle.

4. The rope will ensure the bale remains closed.

5. The hose is ready to be hoisted.

Smoke Ejector, Chain Saw, Rotary Saw

All of these items and a host of others used on emergency scenes have closed handles or support pieces that can have rope tied around them for hoisting.

JPR 15-19: Hoisting Small Equipment

(For step-by-step photos of this skill sequence, see page 469)

1. Tie a follow-through figure eight through the closed handle.

2. The use of a tag line is highly recommended for items of this type, which tend to be heavy and hard to control with only the hoisting line from above.

Ladders

Both ground ladders and roof ladders are hoisted on a regular basis at emergency scenes. The hoisting procedure is the same for both types. Once again the use of a tag line is recommended, especially in those cases where the firefighter on the ground will not be able to guide the bottom of the ladder as it is hoisted.

JPR 15-20: Hoisting a Ladder

(For step-by-step photos of this skill sequence, see page 470)

1. Tie a large figure eight on a bight forming a loop approximately 3 to 4 feet long.

2. Go approximately one-third the length down the ladder and put the loop through the rungs.

3. Pull the loop up and slip it around the top of the ladder allowing it to slide back down, securing the ladder.

Securing a Rope between Two Objects

While the use of rope(s) to cordon off an area is not a common practice in many departments today, the need to secure a rope between two objects may arise at any emergency scene—even if only temporarily. If this need arises, a rope may be used as a barrier using one of two methods. A figure eight on a bight may be used to secure an anchor point.

JPR 15-21: Tying a Rope between Two Objects

(For step-by-step photos of this skill sequence, see page 471)

1. Starting at one end, secure the rope to a solid object (anchor) by forming a figure eight on a bight and sliding the rope over the anchor.

2. If the rope cannot be placed over the anchor, utilize a follow-through figure eight to secure the anchor point.

3. Lay the rope out, keeping it as straight as possible to minimize slack, until reaching the other objective to be tied.

4. Measure approximately one-third of the standing part of the rope toward the anchor and tie a figure eight on a bight resulting in a loop 6 to 12 inches long.

5. Wrap the running end of the rope around the objective and bring it to the figure eight on a bight tied in step 2.

6. Thread the running end through the loop and pull it back toward the objective, tightening the rope as necessary.

7. Using the running end, tie three or four consecutive half hitches around both sections of rope.

JOB PERFORMANCE REQUIREMENT   15-1
Half Hitch

Make a round turn in the standing portion of the rope. Slide the round turn down over the object being hoisted, making sure that the running end passes under the working end.

JOB PERFORMANCE REQUIREMENT   15-2
Overhand (Safety) Knot

Take the loose end of the working end after tying your primary knot and secure it by making a round turn around the standing part and bringing the loose end through between this round turn and the primary knot.

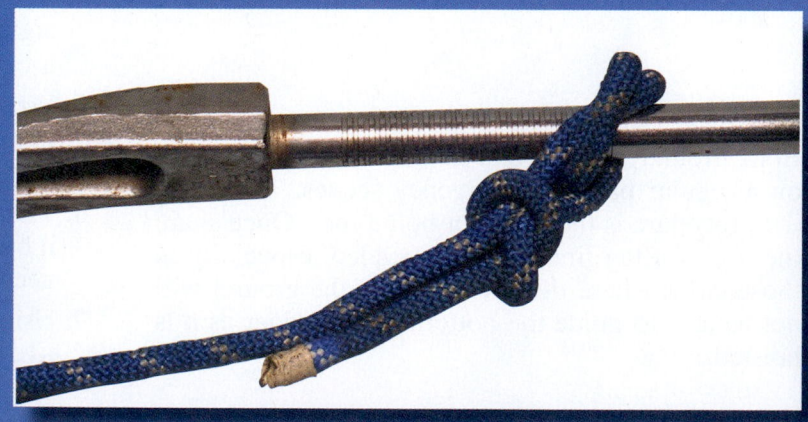

JOB PERFORMANCE REQUIREMENT 15-3
Clove Hitch Tied in the Open

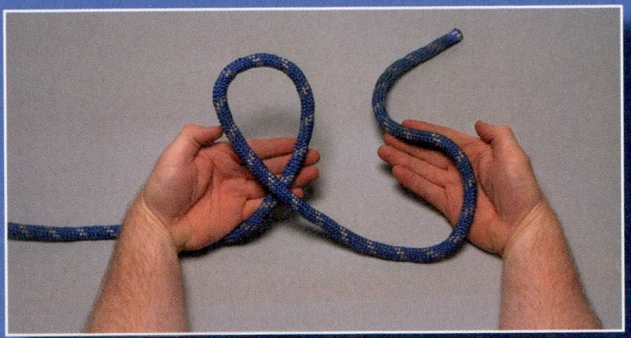

A Hold the rope in your hands with the working end to your right. Form a loop with the working end passing in front of the standing part (between you and the rope).

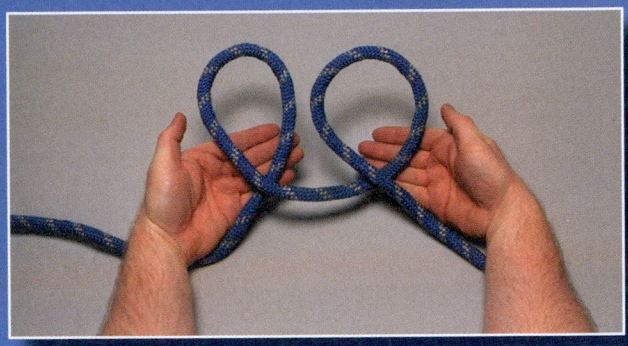

B Move the rope down to the right and form another loop exactly like the previous loop.

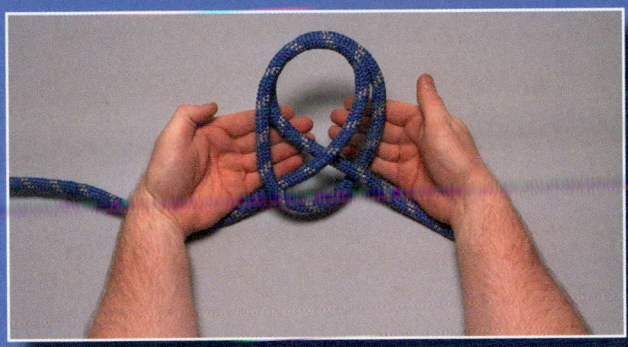

C Place the loop formed in step 3 behind the loop formed in step 2. You now have a clove hitch.

D Keep the clove hitch loops in the proper orientation to one another and slide them over the object to be secured.

E Finish with a safety knot.

JOB PERFORMANCE REQUIREMENT 15-4
Clove Hitch Tied around an Object

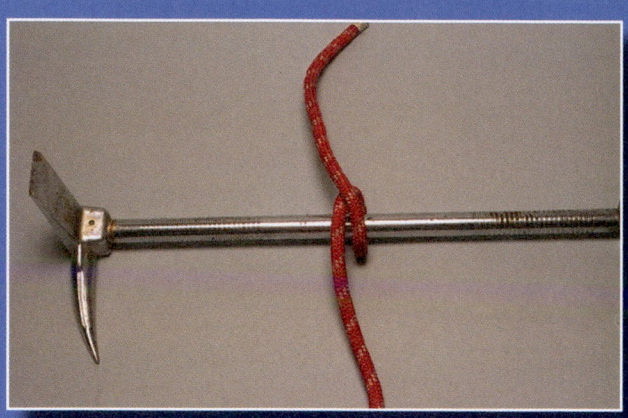

A With the working end, make a loop around the object. Cross the working end over the standing part and make another loop around the object.

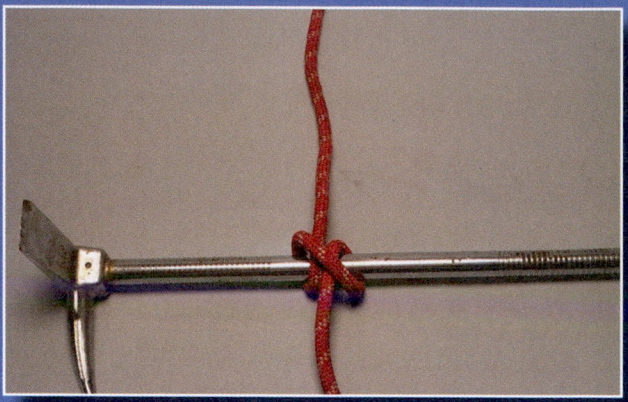

B Bring the working end under at the point you crossed with the loop. At this point your loops should be side by side with the standing part and the loose end of the working end opposing one another.

C Finish with a safety knot.

JOB PERFORMANCE REQUIREMENT 15-5
Tying a Becket Bend

A Form a bight in the running end of the line that needs to be extended.

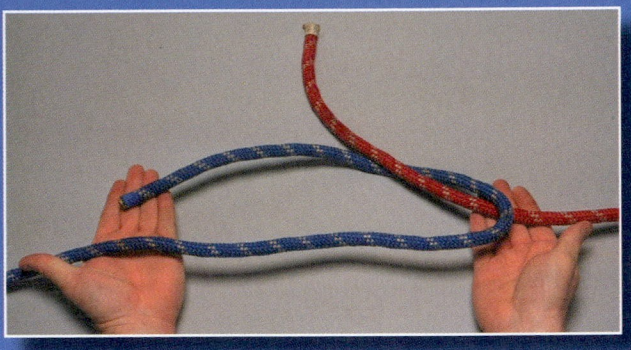

B Utilizing the working end of the rope being added, bring it up through the bight.

C Take this same end around both sides of the bight, forming a loop around the bight.

D Still utilizing this same end, pass it between itself and the rope in which you formed the bight. When dressed and set, the end you were working with should be at approximately 90 degrees to the rope.

E Finish with a safety knot.

JOB PERFORMANCE REQUIREMENT   15-6
Tying a Double Becket Bend

A   Form a bight in the running end of the larger line.

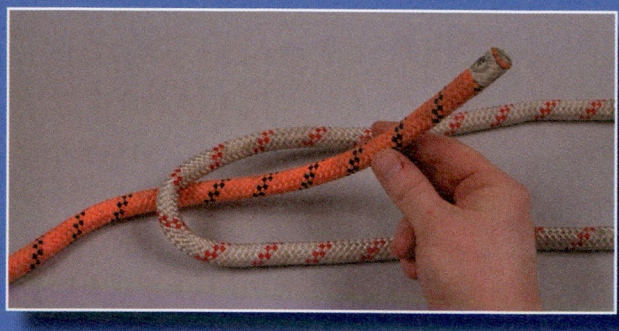

B   Utilizing the working end of the smaller rope, bring it up through the bight.

C   Take this same end all the way around both sides of the bight forming a round turn around the bight.

D   Continue around the bight a second time.

E   Still utilizing this same end, pass it between itself and the rope in which you formed the bight. At this point the rope you have been working with should have formed what looks like the number eight (8).

F   When dressed and set, the end you were working with should be at approximately 90 degrees to the rope. Finish with a safety knot.

JOB PERFORMANCE REQUIREMENT 15-7
Tying a Bowline

A Holding the standing part in your hand with the working end coming toward you, form a loop in the rope with the working end on your side when it passes the standing part.

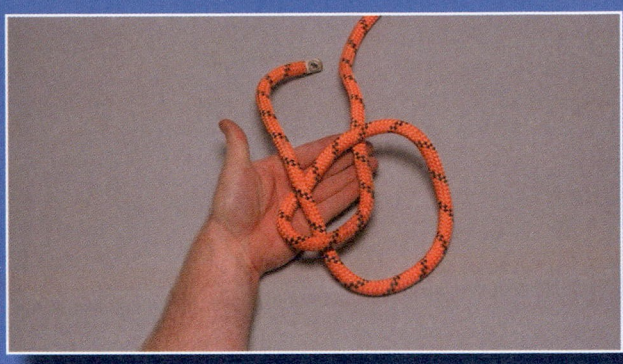

B Still utilizing the working end, bring it up through the loop formed in step 1.

C Pass this same end around behind the standing part and then bring it back down through the loop you brought it up through. Be sure that the working end is on the inside of the loop that you have just formed with the completed knot. If left on the outside there is a much greater chance of it getting caught and the knot inverting to a slip knot.

D Finish with a safety knot.

JOB PERFORMANCE REQUIREMENT   15-8
Tying a Basic Figure Eight

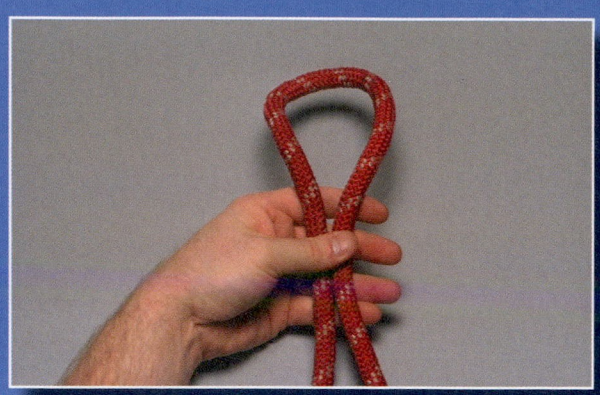

A   Form a bight in the working end of the rope, holding the bight out in front of you by the apex so that both sides of the bight are hanging parallel.

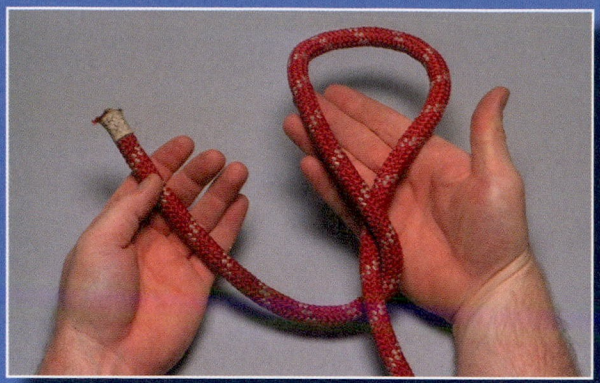

B   Take the short end (working end) of the rope and wrap it around the standing part from front to rear, JPR 15-8B.

C   After the working end comes around the standing part, you will have formed an eye. Now take the working end through the eye from front to rear.

D   If you are joining two ropes, at this point you will take the working end of the second rope and, coming from the opposite direction, trace the path of the original rope as it flows through the knot. You should end up with both working ends lying parallel to the opposite rope's standing part.

E   Finish with a safety knot with each working end around the opposite rope's standing part.

JOB PERFORMANCE REQUIREMENT   15-9
Tying a Follow-Through Figure Eight

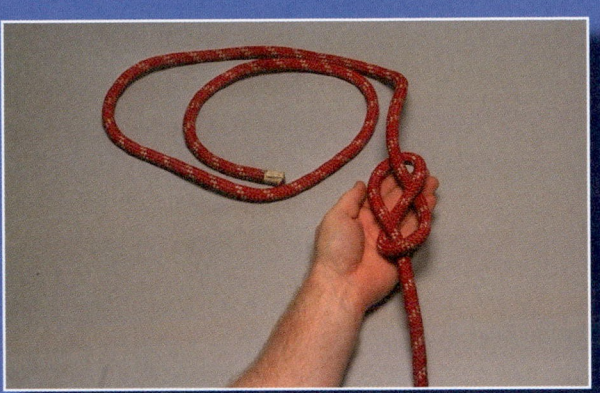

A Follow the directions for tying a basic figure eight; however, you need to make sure that you have an adequate amount of rope left on the working end side of the basic knot to go around the object you are securing the rope to.

B If you are securing the rope to an object, wrap the working end around the object at this point. (From this point on the procedure is very similar to the procedure for joining two ropes together.)

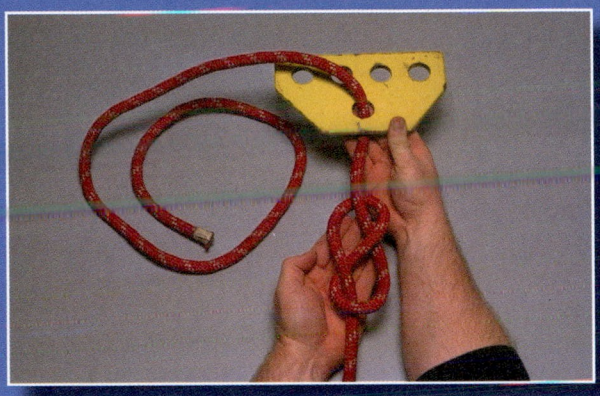

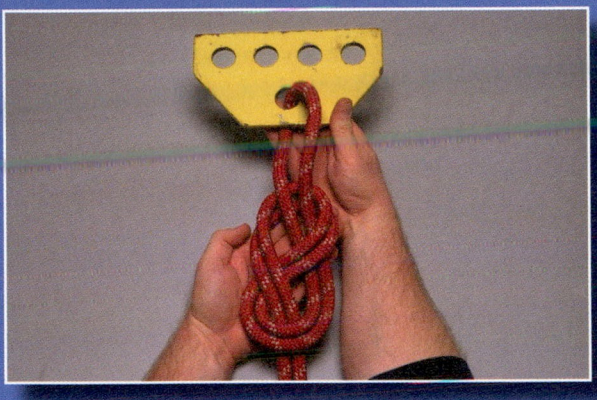

C As the name implies, you now follow the other rope through the knot, coming from the opposite direction. In effect, you have formed a figure eight on a bight (the next knot described) around an object.

D You should conclude with the end of the working end parallel to the standing part.

E Finish with a safety knot.

JOB PERFORMANCE REQUIREMENT   15-10
Tying a Figure Eight on a Bight

A   Start by forming a bight. The size of the bight needs to be relative to the object the finished loop is going to be placed over or around.

B   Grab both ropes so that the bight is hanging down from one side of your hand while the standing part and the end of the working part are hanging out of the opposite side of your hand. Basically you are forming a second bight of doubled ropes.

C   Hold this bight out in front of you by the apex so that both sides (four parts of the same rope) of the bight are hanging parallel.

D   Take the first bight and wrap it around the standing part from front to rear.

E   After the bight comes around the standing part, you will have formed an eye. Now take the bight through the eye from rear to front.

F   You have now formed an easy, reliable loop that will not slip. Do not forget to dress and set your knot and apply a safety with the end of the working end.

JOB PERFORMANCE REQUIREMENT 15-11
Tying a Rescue Knot

A Tie a figure eight on a bight. Your initial bight should be approximately 6 feet long. This will allow you to tie the figure eight on a bight, including a safety at the working end, and still have a finished loop approximately 4 to 5 feet long. This loop is going to need to be long enough to go around the victim's waist and back up between the legs. The loop does not need a precise measurement, but it is important to remember that the length cannot be adjusted once the knot is tied. Time should be spent sizing and applying this knot on different size victims in practice scenarios. Firefighters should rehearse this skill until they feel comfortable and can be consistently successful with this maneuver on the first try.

B Place the loop on the floor or step into it with the actual figure eight knot behind you. Raise the loop to waist level, being careful not to get above the top of the hip bones.

C Reach down between your legs, pulling the standing part of the rope up in front of you. The standing part should not be between your body and the loop but on the outside of the loop.

(*Continues*)

JOB PERFORMANCE REQUIREMENT 15-11
Tying a Rescue Knot (*Continued*)

D   Continuing to hold the rope in front of you, raise your hand holding the rope above your head. Reach out with the other hand and grasp the ropes, both the one coming from between your legs and the one hanging down from the hand over your head.

E   With the hand over your head make three full twists, forming a loop. Place the loop over your head and shoulders so that it comes to rest in your underarms.

F   Keeping the standing part pointed generally down, adjust what slack you can out of the rope between the figure eight knot coming from between your legs and the twists forming the loop that is now around your chest. Raise the standing part up over your head, allowing the twists to wrap around themselves and cinching down on the rope forming the loop around your chest. This is important to keep the loop from cinching down on your chest.

JOB PERFORMANCE REQUIREMENT   15-12
Tying a Water Knot

A Tie a simple overhand knot in one end of the webbing, making sure the webbing lies flat as it crosses back over itself. Leave enough tail beyond the overhand knot to be able to place a safety knot on that side of the water knot when finished.

B Take the other end of the webbing and thread it through in the opposite direction, again making sure that the webbing lies flat as it crosses itself. Make sure you feed enough webbing through so that you are able to place a safety knot on this side of the water knot.

JOB PERFORMANCE REQUIREMENT 15-13
Coiling a Rope

A   Measure off an amount of rope equal to approximately three times the distance between the standards to be used in tying the finished coil.

B   Starting with this point measured at one of the standards, wrap the rope around the standards, forming the loops and moving toward you until you have nine to twelve loops on the standards. Then form a second layer on top of the first, moving away from you. Continue in this manner until approximately twice the amount measured off in step 1 remains.

C   Begin wrapping this remaining rope around the coils that you formed in step 2. Finish these wraps off with a clove hitch.

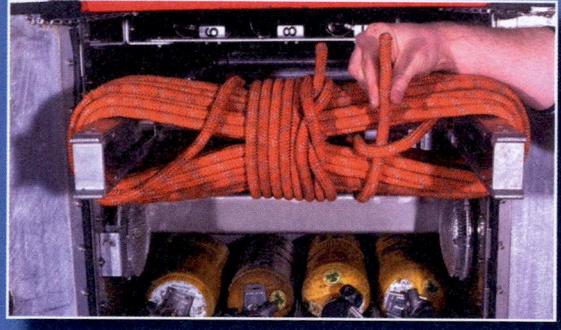

D   Utilizing the rope measured off in step 1, form a bight leaving the loose end of the rope coming out toward you. Feed the bight from the back side to the front side on the wraps. You should now have the loose end and the bight sticking out toward you.

E   Place the loose end through the bight and adjust the bight until it is snug where it passes through the bight.

JOB PERFORMANCE REQUIREMENT   15-14
Bagging a Rope

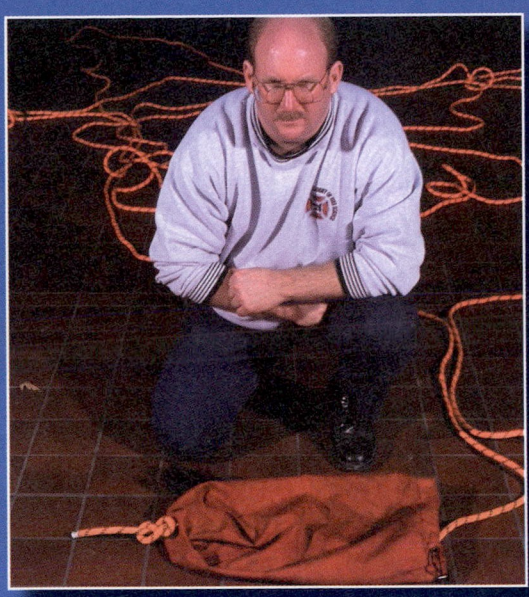

A   If policy is for the rope to be fed out the bottom hole, start by feeding enough rope out the bottom to tie a basic figure eight knot. If the rope is deployed from the top opening in the bag, tie a basic figure eight knot in the end of the rope and place it inside at the bottom of the bag.

B   Begin placing the rope in the bag by sliding your hand approximately 12 to 18 inches up the rope at a time and "stuffing" it in the bag. Do not coil the rope in the bag as it increases the possibility that the rope will hang up as it is deployed.

C   Continue stuffing until the end of the rope is reached. Tie a basic figure eight knot in the end and place it on top of the rope in the bag. Securely close the bag.

JOB PERFORMANCE REQUIREMENT   15-15
Hoisting an Ax

A Tie a figure eight and slide it around the ax handle all the way to the head of the ax.

B Wrap a loop around the head of the ax and back up the handle. Tie a half hitch on the handle a few inches above the clove hitch toward the butt end of the ax handle.

C Tie an additional half hitch on the handle at the butt end of the ax handle.

JOB PERFORMANCE REQUIREMENT   15-16
Hoisting a Pike Pole

A   To hoist a pike pole point up, place a clove hitch near the end of the handle.

B   Place two half hitches around the handle between the clove and the point, with the last one being located immediately below the head.

C   Raise the pole with the head at the highest point.

JOB PERFORMANCE REQUIREMENT 15-17
Hoisting a Charged Hoseline

A With the nozzle bale in the closed (forward) position, tie a clove hitch around the charged hoseline 18 to 24 inches behind the nozzle.

B Form a bight in the rope and feed it through the bale from the coupling end.

C Form a half hitch.

D Slip the half hitch over the nozzle.

E The charged hoseline is rigged for hoisting.

JOB PERFORMANCE REQUIREMENT   15-18
Hoisting an Uncharged Hoseline

A Fold the nozzle back on the hose, approximately three to four feet. Tie a clove hitch around the hose and nozzle together.

B Place a half hitch around the end of the hose, approximately six inches below the bend.

C The uncharged hoseline is rigged for hoisting.

JOB PERFORMANCE REQUIREMENT 15-19
Hoisting Small Equipment

A Tie a follow-through figure eight through the closed handle.

B The use of a tag line is highly recommended for items of this type, which tend to be heavy and hard to control with only the hoisting line from above.

JOB PERFORMANCE REQUIREMENT 15-20
Hoisting a Ladder

A Tie a large figure eight on a bight forming a loop approximately 3 to 4 feet long.

B Go approximately one-third the length down the ladder and put the loop through the rungs.

C Pull the loop up and slip it around the top of the ladder allowing it to slide back down, securing the ladder.

D Attach a tag line to the ladder, and it is ready to be hoisted.

JOB PERFORMANCE REQUIREMENT   15-21
Tying a Rope Between Two Objects

A Starting at one end, secure the rope to a solid object (anchor) by forming a figure eight on a bight and sliding the rope over the anchor.

B If the rope cannot be placed over the anchor, utilize a follow-through figure eight to secure the anchor point.

C Lay the rope out, keeping it as straight as possible to minimize slack, until reaching the other objective to be tied.

D Measure approximately one-third of the standing part of the rope toward the anchor and tie a figure eight on a bight resulting in a loop 6 to 12 inches long.

(Continues)

JOB PERFORMANCE REQUIREMENT 15-21

Tying a Rope Between Two Objects *(Continued)*

E   Wrap the running end of the rope around the objective and bring it to the figure eight on a bight tied in step 2.

F   Thread the running end through the loop and pull it back toward the objective, tightening the rope as necessary.

G   Using the running end, tie three or four consecutive half hitches around both sections of rope.

LESSONS LEARNED

This chapter presented information and introduced some skills to accomplish basic tasks required at emergency scenes. Although ropes have many uses, the fire service tends to utilize them on a regular basis for a few tasks and operations that can be divided into two classifications: utility and life safety. It is not possible to present all of the information about skills that firefighters may need as they progress in their careers and become involved in more technical and complex emergency operations. However, the lessons in this chapter can be an important link for day-to-day training exercises. Experience often is the most effective teacher, and in many cases, firefighters must utilize practical knowledge to retain the information presented here. However, firefighters are encouraged to regularly practice the basics learned in this chapter and pursue further information and training to expand their knowledge, skills, and abilities in this very interesting and exciting subject. The maintenance of ropes is not difficult or complicated; it is generally limited to cleaning, inspecting, and storing. Given the number of knots possible and the various types of rope available, there is sometimes no correct black-and-white answer to the question "Which knot is the best?" Therefore, the NFPA has left this knowledge to the authority having jurisdiction, AHJ.

KEY TERMS

Bight   A doubled section of rope, usually made along the standing part, that forms a U-turn in the rope that does not cross itself.

Dressing   The practice of making sure that all parts of a knot are lying in the proper orientation to the other parts and look exactly as the pictures herein indicate.

Dynamic   A rope having a high degree of elongation (10 to 15 percent) at normal safe working loads.

Kern   A derivative of the term kernel, which is defined as "the central, most important part of something; core; essence."

Life Safety Line   According to NFPA 1983, rope dedicated solely to the purpose of constructing lines for supporting people during rescue, firefighting, or other emergency operations, or during training evolutions.

Loop   A turn in the standing part that crosses itself and results in the standing part continuing on in the original direction of travel.

Mantle   Anything that cloaks, envelops, covers, or conceals.

Light-use (one-person) or **general-use (two-person) rope**   According to NFPA 1983, a one-person rope requires a minimum tensile strength of 4,500 pounds, and a two-person rope requires a minimum tensile strength of 9,000 pounds.

Round Turn   Formed by continuing the loop on around until the sections of the standing part on either side of the round turn are parallel to one another.

Running End   End of the rope that is not rigged or tied off.

Setting   The finishing step, making sure that the knot is snug in all directions of pull.

Shock Load   A load or impact being transferred to a rope suddenly and all at one time.

Standing Part   The part of a rope that is not used to tie off.

Static   A rope having very little (less than 2 percent) elongation at normal safe working loads.

Tactilely   Using the sense of touch to feel for any differences or abnormality.

Tag/Guide Lines   Tag lines are ropes held and controlled by firefighters on the ground or lower elevations in order to keep items being hoisted from banging against or getting caught on the structure as they are being hoisted.

Tensile Strength   Breaking strength of a rope when a load is applied along the direction of the length, generally measured in pounds per square inch.

Utility Rope   Rope used for utility purposes only. Some of the tasks utility ropes are used for in most every fire department are hoisting tools and equipment, cordoning off areas, and stabilizing objects. Also used as ladder halyards.

Webbing   Nylon strapping, available in tubular and flat construction methods.

Working End   The end of the rope that is utilized to secure/tie off the rope.

REVIEW QUESTIONS

1. What NFPA standard addresses life safety ropes and system components?

2. Polypropylene and polyethylene are two of the four synthetic materials discussed in this chapter. What is the major distinguishing feature between these two materials and nylon and polyester, the other two discussed?

3. Modern ropes manufactured from synthetic materials fall into two broad categories, static and dynamic. What is the major difference between the categories?

4. In a kernmantle rope, the kern carries the vast majority of the load. What is the role of the mantle?

5. According to NFPA 1983, there are two classifications of life safety ropes. What are they and what are the minimum requirements for each?

6. Three terms are commonly utilized to describe parts of a rope. What are they and what does each mean?

7. To "dress" and to "set" a knot means to do what to it?

8. What is the best feature of figure eight knots?

9. What is the only knot recognized for tying webbing?

10. According to NFPA 1983, how many times can a life safety rope be utilized for emergency service after once having been used during an emergency?

11. When storing ropes, it is important that the rope be stored in a manner that allows for what three things at emergency scenes?

12. If a rope, coiled or bagged, is dropped from an elevated position, what are some of the negative things that could occur?

16 Rescue Procedures

In August of 2001, the Chicago area experienced a heat wave that resulted in above average heat indexes for twelve consecutive days. The temperature on August 9 registered 95°F with a heat index of 105°.

At 6:56 that morning, the Fire Department responded with a Tower Ladder, Heavy Rescue, and Medic to the train station for a report of a pedestrian struck by a train.

Upon arrival at 6:58 a.m., the Tower Ladder officer was met by the conductor of the train who stated that there was a woman pinned beneath the train and that she was still alive. According to witnesses it is believed that the woman had fainted from the heat while waiting for the train and had fallen onto the tracks as the train was approaching. The engineer of the train was unable to stop before hitting her.

In the Chicago area, locomotives on the METRA line pull or lead the train when it is leaving the city. When traveling back into the city, the locomotive pushes the train from the rear with a coach car leading the way. The engineer is seated in the cab on the upper level of the coach car. It is beneath this lead coach car that the woman was pinned between the rails and braking components approximately 30 inches in front of the wheels. Now conscious, the patient was in extreme pain and understandably very frightened. Train crew members were prepared to back the train off the patient but were promptly stopped by rescue personnel.

After assuring scene safety, a firefighter from the Tower Ladder was able to crawl beneath the train to assess the patient's condition and reassure her that the fire department was there and was going to help her. The officer of the Tower Ladder assumed command and upgraded the response to include a Battalion Chief and engine company.

The set of tracks that this incident took place on serves as one of the major lines into the city of Chicago and services 120 passenger trains daily in addition to 40 freight trains. During peak times, passenger trains are scheduled to run approximately every six minutes.

For access and rescuer safety and because of movement caused by vibrations, the railroad was instructed to shut down all rail traffic that would pass through the affected area. As they arrived on scene, the Medic unit was assigned to patient care and promptly undertook prehospital trauma life support measures. A major concern at this point was that it was not known if the patient was suffering from crush syndrome—a condition that could cause an abundance of toxic substances to be released into the bloodstream once the weight constricting bloodflow from the entrapment is removed. If so, it was feared that the release of the weight of the train car could be fatal to the patient. For this reason and in an attempt to calm the patient, paramedics utilized a cellular phone to allow the patient to speak with her husband during the extrication process.

The Battalion Chief arrived on the scene and command was transferred. The officer from the Tower Ladder was reassigned as the Rescue Officer. Realizing the possibility of an extended operation, an additional heavy rescue company was requested from a neighboring department. Four options were considered to extricate the patient.

Option one was to cut the components beneath the rail car that were pinning the patient. After further assessment, this option was considered not viable. The case-hardened steel beneath the train would need to be cut with a torch. Working room did not allow for this and there would be no way to support the weight of the braking assembly once it was cut.

Option two was to disassemble the braking mechanism under the train piece by piece. The problem with this option was that it would require an extended amount of time and rescuers were battling to stay within the "golden hour."

Option three was to have a crane respond to the scene that would be capable of lifting the train car. Time was again considered to be the downfall of this option. In addition, the movement of the train car using this method may have caused the car to move in a manner that was beyond the control of the rescuers, causing further damage.

Option four was to attempt lifting the train car utilizing high-pressure air bags. This option presented itself as the most time-efficient and was immediately assigned to the Heavy Rescue and Engine company. As a precaution, the Engine company also began to disassemble the braking components beneath the train using pneumatic tools. At the same time, a crane from a local contractor was being summoned as a last resort in the event the operations put into place were unsuccessful.

The largest air bag on the Rescue was 36" x 36" and was the airbag of choice. Rescuers attempted to slowly tilt the train by utilizing a single-point lift technique. As the train was slowly lifted, a box crib was built to support the weight. Six inches of movement was all that the paramedics required to remove the patient. Extrication was complete at 7:32 a.m. (34 minutes elapsed time) and the patient was on her way to the trauma center, where it was determined that she had suffered a pelvic fracture in six locations accompanied by soft tissue trauma.

When the bell sounds, the fire department can be called for any type of emergency situation. As with any out of the ordinary rescue call, teamwork and training must prevail to give the victim the best chances for survival. Rescue situations often require that several different plans or alternatives be ready to go at a moment's notice if the original plan is not working. These factors all came into play on this morning, along with extensive knowledge of the equipment capabilities and resources available.

—*Street Story by Battalion Chief Jeffrey Pindelski, Downers Grove Fire Department, Illinois*

LEARNING OBJECTIVES

After completing this chapter, the reader should be able to:

16-1   Describe hazardous conditions found on an emergency scene.

16-2   Perform an assessment to determine if an area of a structure may or may not be safe for rescue operations.

16-4   Describe the critical elements alerting responders to a hazardous situation.

16-5   Demonstrate the ability to locate hazards and identify safe havens during emergency situations.

16-6   Describe procedures for safely functioning on an emergency scene.

16-7   Describe safety equipment used by emergency responders to safely function at an emergency scene.

16-8   Demonstrate emergency responder operations while wearing personal protective clothing for various emergency situations.

16-9   Identify the methods used to determine if a search area in a structure is safe for operations.

16-10   List indicators of the presence of victims in a rescue situation.

16-11   Describe the firefighter's role in meeting the search team's objectives.

16-12   Describe the purpose and operational characteristics of the primary and secondary search.

16-13   Identify signs of occupant location and methods for conducting interior search and rescue operations.

16-14   Demonstrate operating at the FF I level in limited visibility conditions.

16-15   Explain the immediate-assistance crew requirements (OSHA two-in/two-out) described in NFPA 1500.

16-16   React to an emergency situation by applying an identified emergency guideline.

16-17   Demonstrate the ability to locate and follow various guidelines.

16-18   Identify respiratory protection hazards in rescue operations.

16-19   Describe the proper procedures for victim drags and carries.

16-20   Demonstrate rescue techniques for locating and removing a victim with no self-contained breathing apparatus.

16-21   Demonstrate rescue techniques for locating and removing a downed firefighter with a functioning self-contained breathing apparatus.

16-22   Demonstrate rescue techniques for locating and removing a downed firefighter with a malfunctioning self-contained breathing apparatus.

16-23   Explain emergency procedures responders take when an air supply is depleted.

16-24   Demonstrate air conservation techniques.

16-41   Demonstrate the ability to perform work assignments in protected areas while following appropriate safety procedures.

16-42   Establish control zones using scene and traffic control devices.

16-43   Describe procedures for safely dismounting fire apparatus in traffic.

16-44   Demonstrate mounting and dismounting procedures for fire apparatus.

*The FF I and II levels, as defined by the NFPA 1001 Standards, are identified in different colors: FF I = black, FF II = red, additional information = blue.

INTRODUCTION

The term *rescue* in the emergency services has many meanings. In this chapter, rescue describes the actions that trained firefighters perform at emergency scenes to remove someone from imminent danger or to extricate them if they are already entrapped. Because of this danger, firefighters are often put at high risk during rescue operations. It is imperative that firefighters recognize these existent dangers and be cognizant of the methods available to minimize or control the associated risk. Rescue is a very broad subject and this chapter is not all-inclusive. Training must take place at higher levels on a consistent basis to keep the firefighter up to date with the most current obstacles that prevent them from performing their job. This chapter is only going to touch the surface of various rescue situations firefighters may find themselves confronted with. It is not the intent to make the reader an "expert" in all rescue situations discussed, but rather to bring the reader to an awareness level in order to recognize a situation for what it is, be aware of the dangers associated with it, and apply safe procedures to any potential rescue situation that may be encountered. The chapter does, however, cover building search and victim removal in more detail since these are the areas that are most common to daily firefighting operations.

HAZARDS ASSOCIATED WITH RESCUE OPERATIONS

Hazards are associated with every type of rescue operation. When firefighters are involved in a rescue operation, one of the biggest dangers that they must be aware of is the focusing of attention on a particular problem without proper regard for possible consequences or alternative approaches. This is generally referred to as **tunnel vision,** Figure 16-1. It is very easy to develop tunnel vision when a rescuer is involved in an unusually complex and/or lengthy rescue. Tunnel vision can keep, and in many cases has kept, the rescuer from seeing an obvious solution or more often an impending danger.

A thorough **risk/benefit analysis** needs to be performed each time personnel are committed to rescue operations and must be continual throughout the incident to prevent and overcome tunnel vision. The probability of success of an operation must be considered in relation to the degree of risk presented. Incident size-up is the first step in establishing this assessment. This initial assessment will serve as the foundation for all decisions made.

> **SAFETY**
>
> No building or life already lost is ever worth the life of a firefighter!

FIGURE 16-1 When a member of the rescue crew is injured, it is difficult to maintain focus on the overall task of maintaining responder safety.

Safe Haven

Firefighters operating within hazardous conditions must identify safe areas to retreat to while waiting to be rescued or until another method of escape is identified. The safe area is referred to as a safe haven and is defined as an area of refuge that can be utilized while waiting to be rescued or until you are able to escape the hazardous conditions. Safe havens are utilized when you can no longer exit the area in the same manner in which you entered because of a collapse, flashover, or other dangerous conditions. Prior to entry, the firefighter should identify at least two safe havens. Each and every firefighter should develop basic survival techniques that will allow them to escape dangerous or life threatening situations.

Characteristics of a safe haven are fundamentally the same for all emergency incidents: temporary safe area, away from the hazard, in a tenable environment, identifiable by rescuers, and self rescue may be initiated. Following are examples of safe havens for specific emergency incidents:

Structure Fire

- Rooms adjacent to or away from the fire
- Next to wall away from the hazard, ideally exterior wall
- In a doorway
- Close to a window
- In a void created by collapse

Wildland Fire

- Downhill, upwind
- In the burned area

Hazardous Materials Incidents

- Away from hazardous exposure
- Away from catastrophic failure areas of a tank or cylinder
- Shielded from heat sources

Upon entering the safe haven, the following safety and survival techniques should be followed:

- If lost or trapped, immediately use the "mayday" procedure.
- Maintain constant communication with Incident Commander or Safety Officer to let them know your location.

- Position yourself next to a wall, in a doorway, or close to a window to allow rescuers to find you.
- Lay horizontally with audible pass device positioned for maximum effectiveness.
- Stay calm and conserve air by controlling breathing.
- If lost or trapped, create audible signals such as tapping flashlights or other tools, and visible signals such as shining your flashlight to aid rescuers in locating you.
- Maintain constant contact with team members by maintaining team integrity.

Safe havens are identified to provide temporary safety during a dangerous situation. Firefighting is a team environment but all firefighters should be prepared to use the skills necessary for self-survival and rescue.

Psychological Limitations

Rescuers must minimize the psychological effects that they may experience while operating in limited visibility. The degree of training or experience users have with SCBA affects their self-confidence and ability to function. The following are examples of the psychological effects that firefighters may encounter.

- Lack of confidence in the SCBA unit and its ability to protect the firefighter may cause anxiety, increasing the breathing rate.
- The degree of training or experience users have with SCBA affects their self-confidence and ability to function.
- Increased physical stress may cause anxiety.
- Emotional conditions, such as fear of being confined, excitement, or claustrophobia, may increase the user's breathing rate and air consumption.

These problems are addressed with proper training and education. Training in SCBA use is not a one-time deal; it must be continuous so firefighters maintain their proficiency and confidence with SCBA.

SEARCH OF BURNING STRUCTURES

Searching burning structures is one of the most dangerous rescue situations regularly faced by firefighters. The best way to reduce the danger while searching an involved structure is through training, practice, and planning. It is paramount that search activities be closely coordinated with suppression and ventilation efforts on the fireground to avoid further hostile conditions being forced upon the search team.

Firefighters must always work in teams of two or more when entering an involved structure for any reason (i.e., interior firefighting, search and rescue, ventilation). In addition to this search team, a minimum of two firefighters must be standing by in full protective PPE with a charged hoseline ready to come in and assist the search team should a problem develop. This is commonly referred to as the two in/two out rule (OSHA 1910.134). Even if the search team also has a hoseline with them, this recommendation remains in effect. This team is referred to as the **Initial Raid Intervention Crew (IRIC)** and should be replaced with a team of at least four firefighters when resources arrive. The latter team is referred to as the Rapid Intervention Team (RIT) and is required under the parameters outlined in OSHA 190.134, NFPA 1500, NFPA 1710, and NFPA 1720.

As firefighters approach a structure that is going to be searched, they should perform a size-up to determine a "rescue profile." The rescue profile helps the firefighters prioritize the probability, location, and status of potential victims as well as the avenues of access and egress. While life, safety, and rescue always remain a priority, there are structural fire environments that are a "recovery" environment. Simply stated, a recovery environment means there is little to no chance of saving a victim due to fire and smoke conditions and/or building collapse potential. Factors that should be evaluated to determine the rescue profile include:

- *Occupancy type/time of day.* Residential structures can be occupied at any time, although typical sleeping hours present the greatest indication that victims may need rescue. Commercial occupancies seldom have sleeping occupants and, therefore, a greater chance exists that the occupants have self-evacuated. In some high-cost resort areas, employees may sleep on the premises due to the high cost of domicile, therefore increasing the rescue profile. Most adults will attempt to escape by means they normally use to exit. Children are likely to hide from the smoke and fire—and may be in closets, in cabinets, toy boxes, under beds, and even in their parents' bed.
- *Fire/smoke conditions.* Turbulent, black smoke (pressurized with high heat) is an unsurvivable environment for occupants due to pain threshold and toxicity—a recovery environment versus a rescue environment. Areas with less dense and lighter colored smoke indicate a higher rescue profile. Post-flashover compartments (rooms) are a recovery environment. Rooms adjacent and above an involved room have a higher rescue priority, assuming the smoke in these areas is not tur-

bulent. Remember, smoke is comprised mainly of unburned fuel—flashover of one room is likely to cause rapid ignition of dense smoke. Fire and smoke conditions need to be continually reevaluated by the search team if they make entry. These conditions will also need to be monitored and communicated by exterior fire companies.

- *Activity clues.* Cars in the driveway or garage, toys strewn about, shoveled snow and footprints, and open windows provide clues that a home is occupied, **Figure 16-2.** If home occupants are not out waiting for the fire department, the rescue profile is high. Conversely, no cars, boarded-up windows, signs of neglect, and vacancy signs should indicate a lower rescue profile.

In addition to the protective clothing and equipment listed earlier, firefighters should carry with them a forcible entry/egress tool (ax, maul, Halligan tool, etc.), two flashlights, portable radio, and thermal imaging camera, **Figure 16-3.** The forcible entry tool can be useful in gaining access to locked or blocked rooms within the structure and is also useful in extending the searching firefighters' reach under and behind objects such as beds, dressers, and tables and can assist the firefighter in creating an emergency egress if necessary. The flashlights are useful in searching if the smoke is not too thick; serving as possible orientation points, they can be useful to signal a rescue/backup crew should a firefighter get into trouble. The portable radio is helpful for keeping the incident commander informed of progress, fire/smoke conditions at a location, and the results of a search. It is very useful in communicating with a rescue/backup crew if a firefighter becomes disoriented or lost. Thermal imaging cameras and devices also

allow the search team to navigate through the smoke, thereby increasing the speed in which searches can be accomplished.

In many single-family residential structures, it may be possible to conduct a search thoroughly and safely without a **guideline/lifeline** or hoseline by using the wall as a reference, **Figure 16-4.** Utilizing a wall in this manner is referred to as conducting a "right-hand" or a "left-hand" search, meaning that the firefighter maintains constant contact between the wall and that side of the body. By doing this, the firefighter will return to the point of entry while searching all the way around the room.

In addition to specific fire conditions, firefighters should always maintain awareness of their position within a building, along with the location of objects and furniture encountered inside. Know where the fire has been, where it is at, and where it is going. If a search pattern starts to the left, continue to the left. Do not change to a right-handed pattern once started.

FIGURE 16-2 A streetside view of a typical residential occupancy. Note the clues that can be spotted in this photo: cars in driveway, toys, bicycles, and so on.

FIGURE 16-3 Well-equipped interior structural firefighting/search and rescue crews need a minimum of full PPE, SCBA, PASS, forcible entry tool, flashlight, portable radio, and thermal imaging camera.

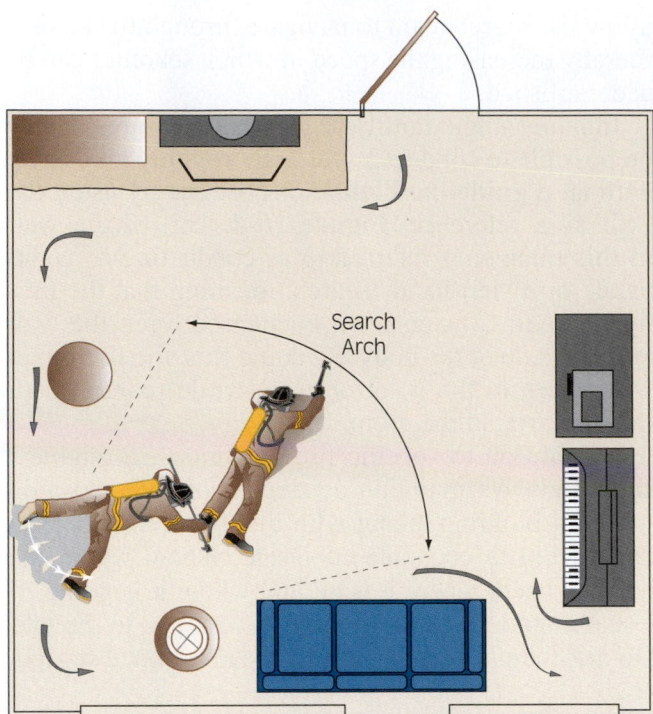

FIGURE 16-4 During an interior search, firefighters should stay in contact with a wall. If visibility is hampered, firefighters can reach into the center of the room using a tool or a "human chain" technique.

Actions such as this will lead to disorientation. Doors and windows are survival landmarks that are important to every firefighter operating inside structures. For this reason, it is important that firefighters sweep high enough on the wall to locate the door handle or window sill and make a mental note of its location, **Figure 16-5.** Conditions can deteriorate rapidly, sometimes without warning, so knowing the location of the closest door or window is a self-survival basic.

Firefighters should draw a picture in their mind of their surroundings as they move through a building with limited visibility. Some ways of accomplishing this are through features that are unique to areas within a building, such as furnishings, floor coverings, and door swings. Certain furnishings will be unique to specific areas—refrigerators and tables to kitchens, beds to bedrooms, and so forth. Floor coverings such as tile are generally in bathrooms or kitchens, while carpeting can be pinpointed to areas such as the living room or bedrooms. In a commercial structure, tile may be located in main aisles that could lead to exits, and carpeting could be located in work or display areas. Door swings are also possible indicators for orientation points. Generally speaking, doors to bedrooms and bathrooms will swing inward, as will entrance and exit doors in residential structures, whereas closet doors will swing outward to allow more space for storage. In a commercial structure,

FIGURE 16-5 Firefighters need to make certain that they "sweep" high enough on walls when searching to feel windows and door handles for possible exit.

doors to entrances and exits should swing outward. Again, these are generalizations and may not always be true depending on what has been changed or renovated in a building and whether it was inspected under a specific set of building codes.

STREETSMART TIP

Another way for firefighters to find their way back to where they entered a structure is to follow the hoseline back out. To be able to do this in firefighting conditions, it must be practiced during training until firefighters can tell which way the hose is going by feeling the couplings. The female coupling will have a smooth surface connected to a swivel with small rocker lugs. The male coupling will have larger rocker lugs. These larger rocker lugs are located on the pumper side of the coupling connection. The firefighter must locate these larger rocker lugs and follow the hose towards the pumper. This skill is imperative for firefighter survival and the firefighter should be able to perform this task in limited visibility, **Figure 16-6.**

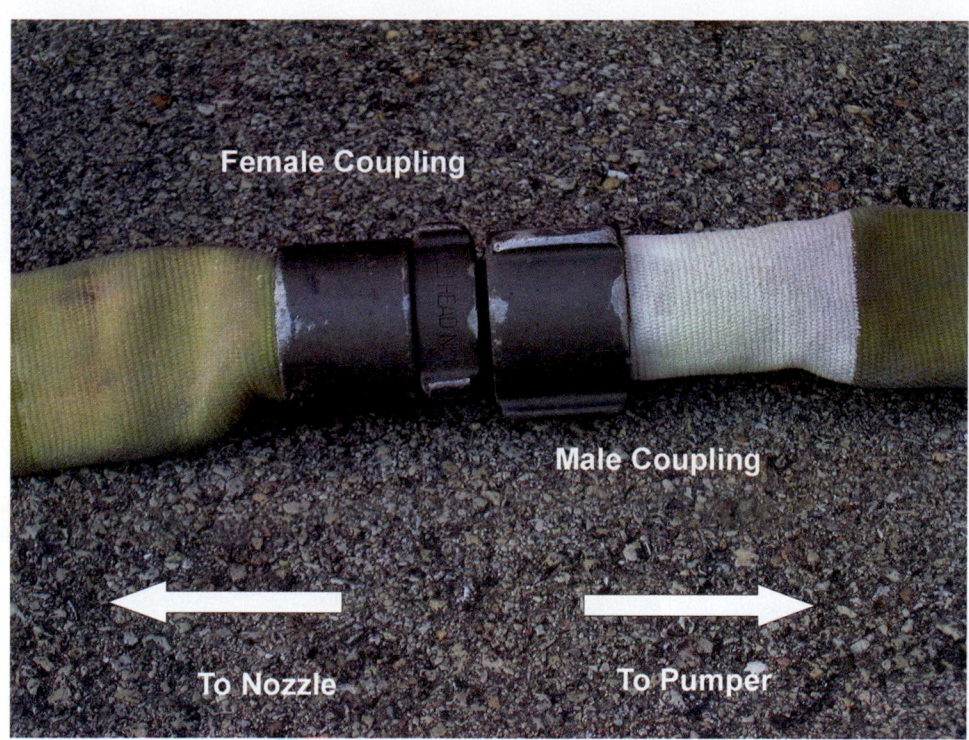

FIGURE 16-6 Lugs on the female coupling are much smaller than the lugs on the male ones.

The biggest asset to conducting a safe and successful search is to have practiced and trained together prior to the actual emergency and to have a plan prior to entering the structure. "Having a plan" means that the search team members know who is in charge, in what direction they are going, on what side (left or right) to keep the wall, and any other pertinent information.

Team members must stay together. The important point is to remain within visual and voice contact with one another. If searching a small room or area with decent visibility, one team member may remain at the entrance but must be able to stay in constant voice communication with the other member so that a reference point can be maintained.

A second effective search method is to leave a light at an entry point (a good reason to carry two flashlights) and have one firefighter search right upon entry into a room and the other left. Both firefighters should cross at a midway point in the room and meet back at the entry point. Advantages of this type of search are that it is quick and it basically provides for areas in that room to be searched twice by different firefighters.

Another way of covering more area is for one member to remain in contact with the wall while the other member holds onto the first member's arm, leg, or forcible entry tool, effectively doubling the distance that can be reached without leaving the wall, **Figures 16-7 A** and **B.** A disadvantage to this technique, however, is that firefighters holding onto each other may have a tendency to slow down a search. It is important that firefighters do not simply follow each other in a straight line when conducting a search using this method as it will leave many areas unsearched that could be easily accessed.

If a room is too hot to be entered, a firefighter can use a forcible entry tool to probe through the doorway or window. Tools can also be utilized to aid the firefighter in maintaining orientation and egress points while searching, **Figure 16-8.**

Searching a building is completed in two different operations: the primary search and the secondary search. These operations are also two of the tactical benchmarks utilized by the incident commander.

(A)

(B)

FIGURE 16-7 Crawling, holding on to one another in a straight line (A) is not very productive when searching. Extending off one another toward the center of an area being searched (B) will allow more area to be covered in a quick manner.

FIGURE 16-8 A tool such as a pike pole or hook can be used to help a firefighter keep his or her orientation while extending the search pattern.

STREETSMART TIP

The most effective life-saving effort is to advance a hoseline and attack the fire. This may include protecting egress paths for trapped victims as well as eliminating the danger of the fire itself.

Primary Search

The primary search is a search for both life and fire and is usually conducted prior to a fire being controlled. It is considered to be one of the most dangerous activities to perform on the fireground. During the primary search, the team is often ahead of the attack lines and may be above the fire (the most dangerous place). They search the areas that are most likely to have victims in a rapid, but thorough, manner. In residential occupancies, these would be bedrooms and routes of escape. In commercial occupancies victims would be expected near the exits or in offices during operating hours. Speed is critical when conducting a primary search—fire and IDLH (immediately dan-

gerous to life and health) conditions are increasing exponentially. As a general rule, search efforts should start closest to the fire working back towards the point of entry, but conditions or the given situation may dictate otherwise. Victims located nearest the fire are in the most danger and must be removed quickly to have the best chances of survival.

When conducting a search of the fire floor, firefighters should go to the fire area and search back towards their entry point. Following in order, victims trapped above the area of fire are in the next greatest degree of danger. Firefighters should begin their search immediately upon entering the floor above and work their search towards the fire since conditions can change very rapidly and necessitate their egress. Some important points to consider when operating on the floor above the fire are:

■ Make certain that the Incident Commander and interior crews know you are going above the fire.

- Secure an area of refuge that can be utilized if access to the stairs is lost.
- The team searching above the fire must be continuously updated on fire conditions.
- The stairwell should be protected by interior crews at all costs.

Often, it is advantageous for firefighters to ladder and break out a window and enter for a primary search versus a blind crawl through thick smoke and heat to find a stairway. This technique, also referred to as **Vent Enter Search (VES),** can also be carried out when access to a particular room is not attainable through ordinary routes of entrance due to fire conditions, **Figure 16-9.** VES is an extremely high-risk search technique that allows firefighters to search beyond fire conditions, oftentimes prior to a hoseline being put in place. It is considered an advanced fire-fighting skill and should only be carried out by well-trained and experienced firefighters.

It is important to remember that "opening" the structure can increase fire intensity and draw conditions towards the opening. Because of this, once the window is broken and cleared, the firefighter should give the smoke a chance to lift and the fire to light up prior to entry. Once inside, the firefighter should stay low and attempt to close the door to the room in order to isolate smoke and fire conditions from the room as quickly as possible. The room should be searched quickly with the firefighter returning to the ladder. At no time should firefighters move beyond the room entered by the ladder to search; doing so may lead to disorientation and fire conditions cutting them off from their means of egress. They should descend the ladder and move to the next room to be searched while repeating the same procedure. Remember, VES is not a substitute for a conventional interior search, because it will not cover all areas.

When conducting the primary search, visibility is often obscured by smoke conditions and darkness. The best visibility, portable light penetration and temperature conditions will be closer to the floor. Additionally, the noise created by SCBA, radios, crawling, running into furniture and obstacles, as well as general fireground activities can mask the simple whimper of a child or the tapping of a trapped occupant who has little or no strength left. Firefighters should occasionally pause and listen for cries or signals for help.

Once the primary search is completed, crews should notify the incident commander that the assigned search is complete and provide them with the results. Once all search crews report a completed search, the incident commander will broadcast an all clear. This is an incident tactical benchmark. Following an all clear, firefighters should reduce the degree of risk they take to help keep risk/benefit in balance.

Secondary Search

The secondary search is usually conducted once the fire is out or at least well under control. This search can be much more thorough since there is no immediate fire

FIGURE 16-9   Vent Enter Search techniques allow firefighters to search into areas that may not be readily accessible due to fire conditions.

danger, and the visibility is much better once the smoke is markedly reduced. Areas still needing extinguishment may be discovered during the secondary search phase. During the secondary search, it is imperative that firefighters search through debris that has fallen or been knocked over. It is also important that areas on the exterior of the fire building be searched for victims who have jumped or escaped and are injured.

It is highly recommended that crews different than the ones who performed the primary search of an area be assigned to the secondary search of that area to prevent complacency and to provide a fresh look at the area, which may produce a missed victim. Secondary searches must be very thorough and leave no area unsearched that could hide a victim. Many victims have been missed on search efforts and their bodies have been discovered when firefighters leave the scene. Beyond the tragedy itself, this is a professional embarrassment and can hold possible legal ramifications for fire departments. In an effort to help alleviate this, some departments require a third and final search to be completed prior to releasing control of the scene.

STREETSMART TIP

Many people act in unpredictable ways when a fire occurs. Because they are scared, they will try to hide from the fire, which is why firefighters must thoroughly check under beds and in closets and restrooms.
    Important points to remember include:

- Search under the protection of a hoseline when possible.
- Stay low.
- Evaluate conditions.
- Check behind doors.
- Call out and momentarily cease breathing.
- Listen.
- Isolate fire and conditions when possible.
- Vent as you go if it does not draw fire (see the Venting for Life section later in the chapter).
- Paint a picture in your mind of your surroundings.
- If you find one victim, search the immediate area for another.
- Check bathtubs and showers and closets.
- Do not throw furniture and objects.
- Provide progress reports, including conditions, actions being taken, and needs of the search crew.

Thermal Imaging Cameras and Search

Thermal imaging technology is one of the most valuable and advanced technological tools within the fire service to date. Thermal imaging technology has been available and used by the military since the 1950s.

FIGURE 16-10   A Thermal Imaging Camera (TIC) being utilized by a search team.

As the technology became declassified, it was passed down to the fire service. The first fire departments to use thermal imaging cameras (TICs) did so in the late 1980s. The use of TICs is now an important function for firefighters, not only in aiding search efforts and identifying potential hazards on the fire ground but for other uses as well, **Figure 16-10.** Some of these include but are not limited to:

- Initial size-up
- Overhaul
- Rapid Intervention Team (RIT) operations
- Hazardous material situations
- Pre-planning
- Fire inspections
- Urban Search and Rescue (USAR)
- Assisting law enforcement

How Does It Work?

TICs are of no use unless the user understands the images being represented on the screen. Thermal imaging technology is based on infrared energy,

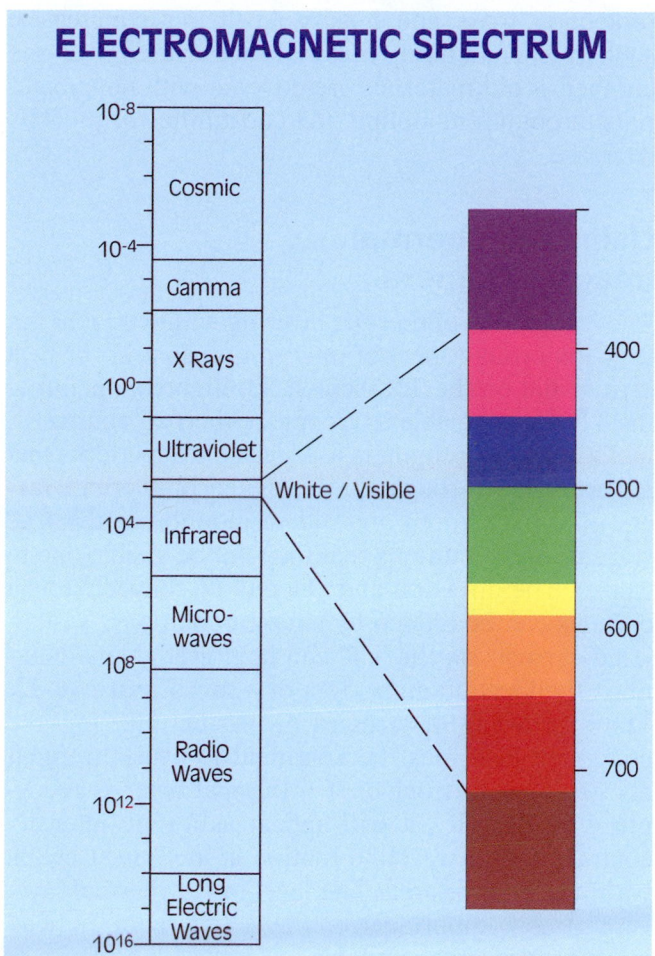

FIGURE 16-11 Infrared energy is not visible but is part of the electromagnetic spectrum.

FIGURE 16-12 Visual representation on a TIC screen.

Figure 16-11. All objects located in an area above absolute zero temperature (0° Kelvin) will emit infrared energy to some degree. Heat is classified as a form of infrared energy. Our unaided eye is unable to see this energy. Visible light is also considered infrared energy, but it is on a different wavelength than heat. The amount of heat energy given off by objects varies. The picture on the viewing screen of a TIC is a visual representation of temperature differences within an area. Even though visual impairment has taken place for the firefighter, the heat of combustion continues to exist. It is the heat that the technological function of thermal imaging capitalizes on. This is why a firefighter can get a visual representation of an area on the viewing screen when products of combustion have brought visibility levels near zero, **Figure 16-12.**

Even though visible light (which is blocked by smoke) is not required for use, the TIC cannot see through objects. A user may be able to obtain a skeletal view of a stud wall behind drywall, but this will only occur if heat or differentiation in temperatures

exists behind the drywall in order to illuminate the objects or wood framing behind them. The mass and density of an object will have a direct effect on the image visualized on the screen of the TIC. The technology of thermal imaging cannot see through the density of solid objects provided in any given space within a structure.

When speaking of infrared energy, there are three types of emitters, which helps explain how mass and density affect the image visualized on the screen.

Passive emitters are inanimate objects whose temperatures will vary depending on the environment and time limit that they are exposed. Basic physics tells us that heat moves from hot to cold. Heat will move to a cold object until the object is the same temperature as the surrounding environment. A passive emitter will absorb or reflect heat in the same manner.

Active emitters are objects that generate their own thermal energy, such as human beings and animals. These objects can be hidden or "masked" very easily when searching with the TIC. The density of clothing, blankets, or debris (passive emitters) covering a victim may prevent their body (active emitter) from being picked up by the TIC. It is important that the user understands this and is able to recognize shapes or objects that may be a part of a victim protruding from underneath a debris pile or the like.

Direct source emitters will give off the most thermal energy and will be easily detectable with the TIC. The fire itself can be classified as a direct source emitter.

Drawbacks and Limitations

The biggest drawback with TICs presently is that they may be too expensive for a large number of departments that are regulated by budgetary constraints. Many departments that could not have otherwise afforded a TIC have been able to do so through community fundraising and grants. If a department does not presently have a TIC, it is highly recommended that these options be aggressively pursued. Not having this technology readily available can mean the difference between life and death on the fireground.

It should also be mentioned that TICs, like any other piece of firefighting equipment, have their limitations for use. Most important, the use of thermal imaging cannot replace a secure and basic foundation in fireground search techniques.

The use of TICs can result in firefighters developing an overconfidence and a dependability on the use of the camera. The camera allows firefighters to see in otherwise near zero visibility environments, which can cause them to stand or walk into a structure without recognizing prominent reference points or keeping aware of obstacles such as holes in floors. This tunnel vision of concentrating on the camera and not the surroundings can also lead to disorientation if the camera fails and the firefighter must now exit.

The components of the camera are intricate. The camera is being used in a hostile environment. These intricate components are subjected to impact, heat, and water exposure once inside the fire building. Condensation and fogging will alter images seen. In addition, the TIC is powered by a battery. These are all factors that can lead to its failure.

> **SAFETY**
>
> The TIC should be used to supplement the firefighter's search techniques—not replace them!

TICs will not see through glass or water. Shiny objects such as mirrors, glass, aluminum, steel, and water will reflect infrared energy showing multiple images on the camera viewing screen. Glass and water are transparent to the unaided eye but are opaque to infrared wavelengths. Reflective surfaces within a structure such as doors or windows may not be discernible as exits or openings due to the high heat and contrasting information provided to the camera.

Each TIC will vary in the size viewing screen available according to manufacturer specifications. It should be understood that the field of view and depth perception is very limited with a TIC. Again, using the TIC to aid search techniques and not replace them will help to avoid this problem.

As mentioned earlier, density will have a direct effect on the capability of the TIC. Certain materials and construction features provide different levels of protection, mass, and density. As the mass increases, such as when building materials are layered over one another or old materials are covered with new materials through remodeling, the capabilities of the TIC decrease.

Using the Thermal Imaging Camera

The recognition of thermal layering within a room is a very important aspect of interpreting contrast or heat differential on the TIC screen. Firefighters operating the TIC must be able to recognize thermal contrast.

TICs allow firefighters to identify heat patterns that are available through the existence of thermal layers in a given room and aid in monitoring the true stage of fire conditions that may not be visible due to smoke. The direction and velocity of convected heat currents can be hidden by smoke conditions. Visualizing currents on the TIC can help firefighters determine the location and extent of a fire, **Figure 16-13.** The monitoring of changes in thermal layering and convected heat must be continual to avoid firefighters getting into trouble. If a thermal layer is recognized by the TIC, it will appear as a high intensity contrast with a wavelike motion at any given height within a room or area. The level of the thermal layer is of utmost importance to the search team, indicating the given time available to search for a victim or to not search at all.

Essentially, contrast creates the images that are provided for us through the camera's lens onto the visual screen that we are looking at. The images appear in different contrast because of the differences in temperatures from one object to another within an area. It is difficult for the TIC to provide a clear image if there is no heat contrast within the area. In

Convected Heat Currents

FIGURE 16-13 Visualizing convected heat currents on the TIC can help firefighters determine the location and extent of a fire.

the absence of contrast, the images are barely visible or appear as either all black or intensely white. Dark or black contrast indicates that the object is producing or holding very little heat in comparison to the surroundings, whereas white contrasts are objects and figures that are giving off more heat than the surroundings. The less heat emitted by an object, the darker the image will appear on the screen. The amount of heat within a space and the effect of heat on objects located in that space will determine the ease of distinguishing one object from another on the viewing screen.

Firefighters should also be aware that a victim may not be easily discernable on the viewing screen inside of a fire environment. Because of the high ambient temperature, a victim may appear the same color as the surroundings or even darker. This is known as image inversion. This is very unlike what is seen on the camera screen during training on the TIC, in an environment using nontoxic theatrical smoke or normal temperature environments. Thus, to locate a victim will require the TIC operator to be able to interpret shapes of objects as opposed to relying on only variations in shade on the screen.

The firefighter operating a camera should begin a systematic approach at the entryway of any area to be searched. To view an area properly, apply a specific pattern of scanning the room starting at the initial point of entry. This will help to recognize any hazards as well as victims. Scanning the room will give a firefighter the information needed to establish a plan of direction and pattern of search.

When scanning an area to search, perform a quick scan of the floor area first—the victim may be located near the door or you may determine that the floor is not structurally sound to advance any farther. When scanning with the TIC at an entryway to an area, point the camera at the top right or left at the ceiling area and scan slowly across it to the opposite corner from which you started. The reason for scanning the ceiling area after a quick scan of the floor is to determine any possible threatening conditions that may be over the intended rescue teams' search area before entry into the space. By also scanning the ceiling, a firefighter will be able to tell if heated gases, fire, or potential collapses are present. After the ceiling area has been viewed, bring the camera back to the original corner of the ceiling area that was the starting point. Then bring the camera downward from that corner to about midlevel on the wall and begin scanning at that level in the same way when viewing the ceiling. Repeat the process at floor level, pointing the camera from one side of the area or space to the other, **Figure 16-14.**

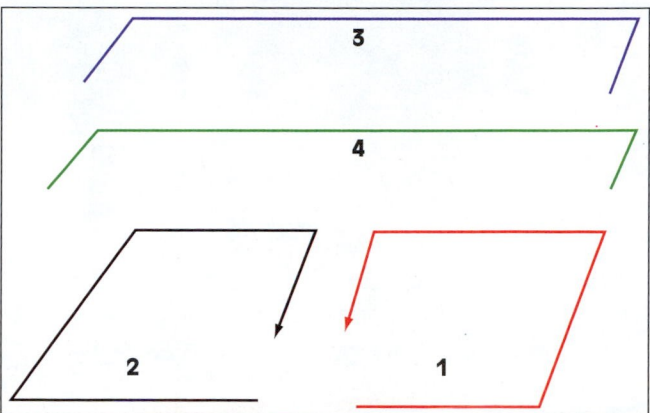

TIC Scanning Technique

FIGURE 16-14   Scanning with the TIC should follow a systematic order.

Large Area or Rope-Assisted Team Search

Conventional search techniques will not suffice when dealing with large structures. In these types of occupancies, it is mandatory that a guideline/lifeline be utilized. It is impossible to conduct a thorough search with an acceptable level of safety without one. This type of search is referred to as large area or "team" search. It is very easy to become disoriented and completely turned around when in a large open area such as a department store, warehouse, or industrial plant. As shown in **Figure 16-15,** these types of occupancies generally not only have large open spaces but often have obstructions (shelf units, machines, displays, etc.) at various and random locations within the open spaces. Even if no fire is present, **cold smoke** (smoke that remains after a fire is extinguished by a sprinkler system) can present a very serious hazard to firefighters in these structures.

FIGURE 16-15 Large area occupancies generally have not only large open spaces but may also incorporate obstructions (shelf units, machines, displays, etc.) at various and random locations within the open spaces.

FIGURE 16-16 The search rope system should always be anchored to an immovable object outside the hazard area.

If the proper procedures are followed, team search can be very safe and effective. The concept is simple: a rope is anchored to a stationary point outside the hazardous environment and firefighters lead out the search line as the team crawls into the structure following the line. There are many different procedures involving the use of main search lines. Some departments utilize short ropes or tethers to attach searchers to one another. Some use ropes or webbing from 15 to 20 feet in length attached to the main line while searching in patterns. Firefighters should be trained in the specific use and deployment of these lines and make certain they are familiar with the procedure their department chooses.

Five to seven firefighters will be required to conduct a large area search using a rope system. Several positions with specific responsibilities will have to be integrated:

- *Team Leader.* The team leader is responsible for the overall operations of the search team, which includes control and deployment of the main line. This individual will also be responsible for directing the search on the interior and maintaining communications with the control/entry supervisor. Tools essential for the team leader will include the main line search rope, two-way radio, hand light, and TIC.

- *Control/Entry Supervisor.* This supervisor is responsible for tracking the team's progress and, most important, monitoring and logging the time in and out of the structure as it pertains to the team's air supply. The control supervisor should have a worksheet and stopwatch to monitor and record the team's air supply and is responsible for accountability and monitoring conditions from outside the team's entry point. The team leader will communicate progress reports back to the control/entry supervisor regarding their progress and air supply status. The control/entry supervisor's main responsibility is to act as a safety officer for the interior rescuers.

- *Remaining Firefighters.* Remaining firefighters will need to bring in forcible entry/exit tools, hand lights, two-way radios, personal rope or tethers, and an additional air supply.

The search rope system should always be anchored to an outside area whether conditions are tenuous or not, **Figure 16-16.** At any time, visibility can go from clear to heavy smoke conditions, which will require the rope to provide the firefighters an immediate way out. It should also be stressed that the control/entry supervisor at the entry point will ensure the security of the main guideline to its attachment.

A team and the control man should have a set amount of time allowed for the team to be inside. Once this time is reached, the control/entry supervisor should alert the interior rescuers to make their way back out. The search team will only be as effective as the member with the lowest pressured SCBA cylinder.

The team should work from a search rope bag that can be shouldered while controlling the rope as it is deployed from the bag, **Figure 16-17.** The rope bag should never be attached to the SCBA of a rescuer. This will avoid the rope bag becoming entangled or caught on something and trapping the rescuer. Another important reason not to attach the bag to a rescuer comes into play should the need arise to leave the rope and bag, in which case the rescuer will be able to anchor it quickly and follow the line back out. It's a good idea to provide a quick-release strap (such as a buckle-and-clasp seatbelt-style connection) for the rope bag so it can be disengaged quickly when necessary. This also limits the possibility of entanglement problems if they occur with the team member holding the rope bag, in which case a quick-release strap provides a safe method for freeing oneself if needed. Another important reason for providing this feature is that all rope-assisted searches deployed as a main search line from a rope bag should be left in place, and preferably anchored, when the team exits the structure. Deployed search ropes should never be collected while traveling out of the structure.

The team members following the team leader should remain several feet behind on the line to allow the team leader to stop occasionally and draw the line tight, thus ensuring accountability and facilitating any need for a change of direction. The line should ideally be kept several inches off the floor to make it easier for a firefighter to find it with a gloved hand.

The team should occasionally stop and listen for any noises that might be made by the victim.

When changing direction (as going around a corner), the main line should be tied off and secured to prevent the rope from moving as the team travels through the building. This is important during exit so that the team is not going over areas not covered originally. Hazards such as balconies or holes in the floor could cause problems if the line is not secured. If the rope is not anchored properly, the line can be pulled back or through the hazard area, putting rescuers in harm's way as well as changing the route to safety. Tying the line off and leaving the rope bag when the team reaches their point of no return will allow the next team to resume the search from where the previous team left off. A hand light can be left next to the bag to allow the second team to find their way to it quickly.

A main line could have a series of knots tied into it in order to identify how deep a team has traveled within a structure. One common application of these knots is to tie them every 20 to 25 feet. Any tethers being utilized by individual members working off the main line may not exceed the length of the knot system that is supplied into the rope. For example, if a knot is tied into the rope every 15 feet, then all tethers shall not exceed 15 feet. The reason for this is simple. When a rescuer is searching off the main line using a tether, he or she will be searching in a semi-circular pattern off to one side of the main line while another rescuer is searching on the other side of the main line, with the team leader stationed at the knot in the main line, **Figure 16-18.** This causes the rescuers on tethers to always end up at the next knot in the main line if they have attached themselves to the previous knot.

The team leader may have to act as an anchor point on the main line in order for other members to search off it with their tethers. If the knot system is being used, the rescuers—after having extended themselves from the main line using tethers—shall return to the main line after completing a specific area and advance to the next knot. The advantage of using this type of system is that it can help determine how deep the team is into a structure and reference that information to their remaining air in their SCBA to determine a point of no return.

FIGURE 16-17 Firefighters should work from a search rope bag that can be shouldered while controlling the rope as it is deployed from the bag.

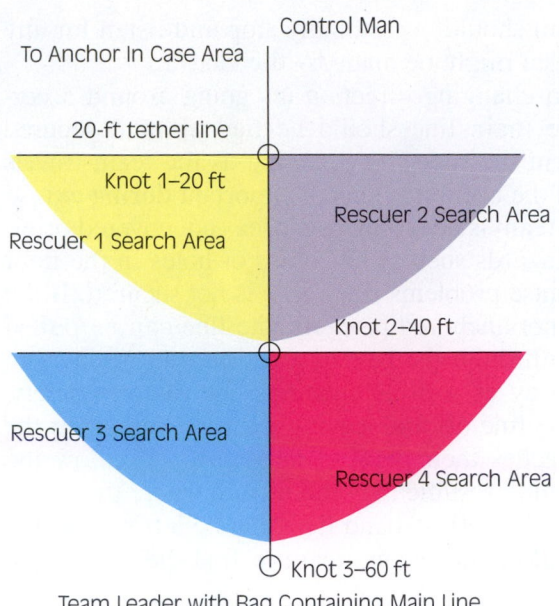

FIGURE 16-18 Semicircular main line search pattern.

As many as 20 percent of firefighters participating as rescuers in the mentioned studies would become victims themselves prior to completion of the exercise. Because of many factors, RIT teams should operate with five goals in mind when they are put into action on the fireground:

1. Locate the firefighter in trouble.
2. Assess the firefighter's condition and environment.
3. Provide an emergency air supply to the trapped or downed firefighter.
4. Call for additional teams and resources.
5. Attempt to remove the firefighter to safety.

Once these goals are accomplished, additional teams can focus on the rescue and actual removal.

RIT teams can also have a very positive impact on performing proactive safety functions such as placing additional ladders, controlling utilities, and removing security devices on the fireground prior to being called into action, dependent on the operating guidelines and procedures utilized by their department.

VICTIM REMOVAL, DRAGS, AND CARRIES

Victims must be removed as carefully and expeditiously as possible. The goal should be not to cause any further injury or aggravation of existing injuries during the rescue process. Many times, it is not possible to utilize all of the patient handling and immobilization skills that firefighters have learned due to the imminent danger presented by the heat, smoke, and gases in a structural fire. Other types of rescue situations also sometimes prevent the rescuer from using all the care that the person would like due to a continuing hazard, or being in a confining area or other hostile environment.

All carries and drags place additional stress on the rescuer's musculoskeletal system. Training and work hardening prepare the firefighter to perform these victim removal techniques. Of particular importance with all drags and carries is the need to keep a tight core. Simply put, firefighters performing drags and carries need to do so by tightening the core muscles around the hips, back, and torso. This creates a "power center" that aids in balance, strength, and injury prevention. Ideally, the firefighter should keep the spine (back) in a neutral position (straight and tight) and use the legs and buttocks for leverage and lifting power, **Figure 16-19.**

Many variations of rope-assisted search procedures are available and each has its distinct advantages and disadvantages. It is imperative that firefighters understand and practice the procedures adopted by their department. Without proper planning and practice, the probability of success and efficiency is greatly decreased while the risk and dangers increase dramatically.

Rapid Intervention Teams

A Rapid Intervention Team (RIT) is a team of specially trained firefighters who are solely designated to provide for the safety, search, and rescue of trapped or lost firefighters at an emergency incident. The formation of this team is mandated by NFPA 1500, NFPA 1710 and 1720, as well as OSHA 1910.134. Firefighters selected to attend RIT training should first be proficient in all basic firefighting skills. It is one of the most important positions on the fireground and should be filled with firefighters capable of performing advanced skills under the most adverse conditions. Rescuing a fellow firefighter will be very different than performing a civilian rescue. RIT operations are very labor and resource intensive. It has been shown through studies that it takes an average of twelve firefighters to remove one downed firefighter completely from a building.

3. The rescuer at the patient's head squats behind the patient, sliding arms under the patient's armpits and grasping the wrist of the patient's opposite arm (if possible). At the same time the rescuer at the patient's feet squats between the patient's feet and grasps the patient's legs under the knees (if possible) or as close as possible below the knees.

4. When ready to lift, the rescuers need to communicate clearly with each other so that they can stand as one and carry the patient to safety.

JPR 16-2: Seat Carry

(For step-by-step photos of this skill sequence, see page 501)

The seat carry can be utilized on conscious patients only and requires two rescuers.

1. The rescuers face each other. Each rescuer grasps his own right forearm just above the wrist. The rescuers then grasp each other's left forearm just above the wrist, forming a square "seat."

2. The rescuers lower the "seat" that they have just formed, allowing the patient to sit on the seat with arms across the shoulders of each rescuer.

3. When ready to lift, the rescuers need to communicate clearly with each other so that they can stand as one and carry the patient to safety.

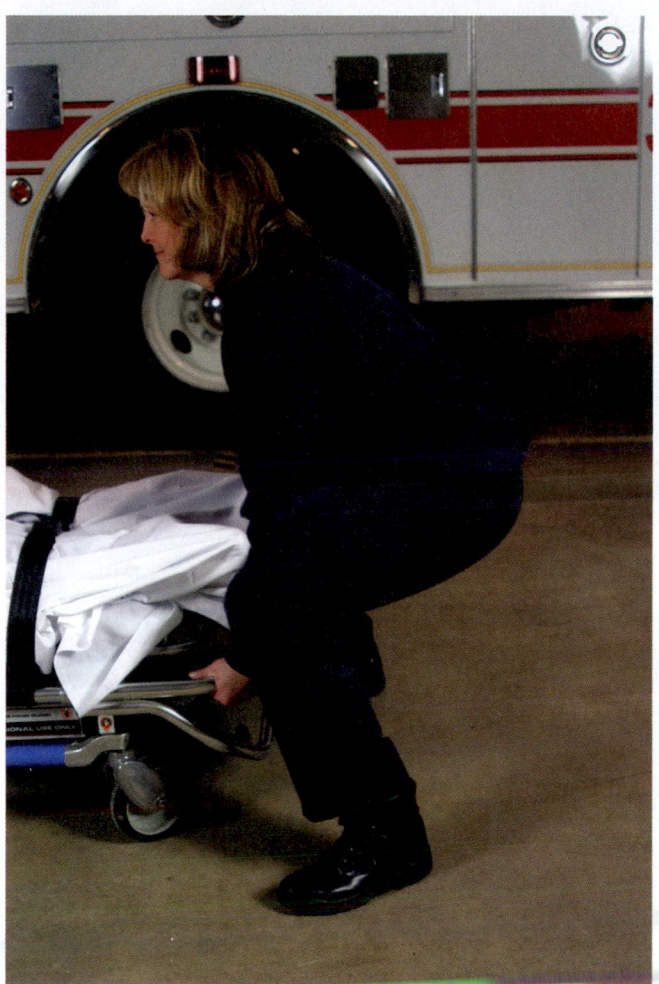

FIGURE 16-19   To avoid injury, firefighters should lift heavy objects using a "tight core" and leverage from the legs and buttocks.

Carries

JPR 16-1: Extremity Carry

(For step-by-step photos of this skill sequence, see page 500)

The extremity carry (sometimes referred to as the cross-arm carry) can be utilized on both conscious and unconscious patients and requires two rescuers.

1. Lay the patient on the back, with arms laid across the torso, knees bent, and feet pushed back approximately halfway to the buttocks.

2. For an unconscious patient, it may be easiest to have the rescuer stationed at the patient's feet reach down, grasp the patient's hands, and pull the patient into a sitting position using the rescuer's weight. Have a conscious patient assume a sitting posture with feet pulled back, lifting the knees above the floor.

Drags

Dragging a victim may be necessary when conditions dictate that rescuers remain low or manpower is limited. Certain types of drags will require the rescuers to stand, while others will allow the victim to be moved from the crawling position. Dragging a victim from the upright position will be easier than dragging the victim from a crawl. Using the leg muscles and principles of physics will make the task of dragging a victim more manageable.

Rescuers can move a victim by placing him or her on a blanket or salvage cover, using the victim's clothing as a handhold, or using a webbing sling, **Figure 16-20.** Each of these various types of drags is described here. All of the drags shown here can be carried out by a single rescuer.

> **STREETSMART TIP**
>
> A rescuer should always drag a victim head first in an effort to protect and avoid further injury to the victim.

FIGURE 16-20 Webbing and premanufactured slings can be valuable for dragging a victim.

JPR 16-3: Blanket Drag

(For step-by-step photos of this skill sequence, see page 502)

As mentioned, the blanket drag can also be accomplished utilizing a salvage cover or any other type of material of adequate size. The steps and procedures are the same regardless of the material being used.

1. With the patient lying face up, lay the material the patient is to be placed on along one side of the patient's body with just over one-half of the material gathered close to the patient's side.

2. Kneel by the patient on the opposite side from the blanket, extend the patient's arm (on your side) above the head if injuries permit, reach across the patient with one hand just above the waist and one just below the hips, and roll the patient toward you.

3. While supporting the patient, now on the side, tuck the gathered blanket material close to the body.

4. Roll the patient back onto the blanket material, return the arm to the side, and wrap the blanket snugly around the patient, supporting the head and neck as much as possible.

5. The patient is now ready to be dragged head first with the head and shoulders raised slightly off the floor.

JPR 16-4: Lift and Drag

(For step-by-step photos of this skill sequence, see page 503)

The lift and drag is a drag that utilizes one firefighter to drag the victim. To use this technique, conditions must allow standing up.

1. The rescuer will locate and assess the victim, placing the individual on his or her back.

2. The rescuer will then take up position at the head of the victim.

3. The rescuer will then wrap both arms around the victim's torso and lock hands together.

4. Using the leg muscles will allow the rescuer to stand up while lifting the victim.

5. Once the victim's torso is lifted, the individual can be dragged to safety.

JPR 16-5: Rescue of a Firefighter Wearing an SCBA—Push and Pull Drag

(For step-by-step photos of this skill sequence, see page 504)

This method works very well for removing an unconscious or incapacitated firefighter who is wearing an SCBA. The push and pull drag requires two rescuers. As the name implies, one rescuer will be pulling using upper body strength, while the other will utilize both legs to push the downed firefighter. This allows the rescuers and downed firefighter to remain low and work in narrow spaces.

1. The rescuer will locate and assess the downed firefighter, placing the firefighter on his or her back.

2. The waist strap of the victim's SCBA should be unfastened, and one strap should be passed through the crotch area and refastened. This will prevent the SCBA from riding up on the downed firefighter.

3. Rescuer 1 will take up position at the head of the downed firefighter, taking hold of the victim's SCBA shoulder strap.

4. Rescuer 2 will take up position between the legs of the downed firefighter, lifting one of the victim's legs over the rescuer's shoulder. The rescuer's head should be positioned high into the groin area with the shoulder driving into the buttocks of the downed firefighter. Correct positioning of Rescuer 2 is essential for this drag to be effective.

5. On command, Rescuer 1 should pull the downed firefighter while Rescuer 2 uses both legs to push simultaneously.

STREETSMART TIP

The rescuer must remember that none of these drags provides spinal immobilization and are intended to be utilized only in situations where greater harm will come to the patient if not immediately moved. If the patient is wearing a functioning SCBA, the rescuer needs to be careful not to break the face piece seal.

Ladder Removals

Using ladders for rescue is a twofold process involving rescuers on the interior as well as the exterior. To bring a victim down a ground ladder will require a minimum of at least four to six members, including both rescuers on the inside of the structure with the victim and team members on the outside. It will be the responsibility of the rescuers inside to get the victim to and over the windowsill, while the exterior team will need to get the victim down safely. Whichever presentation of a victim is shown at the window will determine the technique utilized by the rescuers on the ladder for removal. Communication between the teams will play a very important role. In this particular situation, the interior team can present the victim out the window head or feet first, facing towards or away from the rescuer.

JPR 16-6: Cradle Carry When Victim Is Presented Head First

(For step-by-step photos of this skill sequence, see page 505)

1. The exterior rescuer on the ladder will take up position at or below the windowsill ready to receive the victim head first, with the victims head off to one side or the other. This allows the remaining portion of the victim's torso and legs to come across the ladder horizontally with his or her back facing the rescuer on the ladder.

2. The rescuer on the ladder will position the victim by placing one forearm and hand underneath the victim's armpit in order to hold the weight of the upper torso and grasp the beam of the ladder while the other arm is placed under the top leg into the groin area of the victim. This will help to hold the weight of the victim's pelvic area and lower extremities.

3. The rescuer will descend the ladder grasping the rails while leaning inward to control the victim until reaching the base of the ladder, where additional rescuers can assist in moving the downed firefighter to waiting EMS personnel.

JPR 16-7: Cradle Carry When Victim Is Presented Feet First

(For step-by-step photos of this skill sequence, see page 507)

1. The exterior rescuer on the ladder will take up position at or below the windowsill ready to receive the victim.

2. The rescuer on the ladder will position the victim by placing the victim's legs to the outside of the ladder beams and hooking one arm into the groin area of the victim. Both hands should grasp the beam of the ladder firmly.

3. The upper body of the victim should then be placed onto the rescuer's opposite arm.

4. The rescuer will descend the ladder grasping the rails while leaning inward to control the victim until reaching the base of the ladder, where additional rescuers can assist in moving the victim to waiting EMS personnel.

STREETSMART TIP

If the rescuer at any time feels a loss of control, leaning into the ladder will stop the victim from moving.

Backboard, Stretcher, and Litter Uses

While the carries and drags discussed are emergency methods of moving patients, it is much preferred to utilize a backboard, stretcher, or litter to transport patients who must be moved. These pieces of equipment are designed to provide the protection, immobilization, and safety needed by injured patients. Each of these items has specific characteristics that make a particular piece of equipment more appropriate for a given situation than the other.

Backboards

Backboards (also known as long spine boards) are designed to provide the maximum in spinal immobilization. They are manufactured (commercially and homemade) from many different types of materials, the most common of which is plywood. However, the more modern backboards are manufactured using materials that allow the patient to remain on the backboard while being x-rayed. This minimizes the amount of movement the patient is subjected to and therefore reduces the possibility of aggravation of injuries. These newer materials also simplify the cleanup and decontamination process.

Placing a patient who is suspected of having a spinal injury onto a backboard takes a coordinated team effort. It is best to have four personnel available for this task. The rescuer at the patient's head is responsible for maintaining traction (in line with spine) on the patient's cervical spine and is also the person in charge of directing the process of placing the patient onto the backboard. The other three rescuers kneel along the side of the patient, one at the upper torso area, one at the hip area, and one at the knee/shin area.

JPR 16-8: Placing a Patient on a Backboard

(For step-by-step photos of this skill sequence, see page 509)

1. While the rescuer at the patient's head is maintaining traction, the rescuer at the upper torso places a cervical collar (or some other immobilization device) on the patient. Even with an immobilization device in place, traction must be maintained at all times during this process.

2. The backboard is laid alongside the patient on the opposite side from the three kneeling rescuers. When ready, the rescuer at the patient's head directs the others to "prepare to roll." The three rescuers reach across and grasp the patient at the appropriate locations: The rescuer at the torso grasps the patient at the shoulder and upper arm area; the rescuer at the hips grasps the patient just above and below the hip area; the rescuer at the knee/shin area grasps the patient just above and below the knee.

3. When ready, the rescuer at the head directs the others to "roll patient." At this time the three others roll the patient toward them onto the side. During all of these maneuvers, it is essential for the rescuer at the head to rotate the patient's head along with the body, maintaining traction.

4. With the patient on the side, one of the rescuers (usually the hip area rescuer) reaches over the patient and slides the backboard up tight against the patient's body. When this is completed, the rescuer at the head directs the others to "prepare to lower" and then "lower patient." At this command the three rescuers allow the patient to roll back down on top of the backboard.

5. At this point the patient should be on the backboard; however, it is usually necessary to move the patient in order to center the patient on the board. This needs to be done very carefully so as not to aggravate any injuries. The patient must be moved as a unit, meaning the patient's body is moved all at the same time. The safest way to accomplish moving the patient to the center of the backboard is to move the patient along the long axis of the body rather than try to move sideways.

   The rescuer at the head remains in charge. When ready to move the patient, the rescuer at the head directs the others to "prepare to slide." At this point, the rescuers reach across or straddle the patient, grasping the upper arm/chest, waist, and lower legs. When ready to slide, the rescuer at the head directs the others to "slide." All of the rescuers then slide the patient down toward the foot of the board while moving slightly toward the center. When the patient has been moved enough, the rescuer at the head directs them to "stop." The patient now needs to be moved back up on the board to the proper location both top-to-bottom and side-to-side. This is accomplished by reversing the just completed movement. The patient's distal pulses, movement, and sensation are then reassessed.

6. With the patient now in the proper location, the head is supported on both sides by sand-bags, towels, or commercial head blocks, or in some other acceptable manner. The actual method of strapping the patient to the board varies widely from department to department, but whatever the particular method, the patient must be securely fastened to the board when complete. The strapping method is one that has been found to work very well and can be accomplished in an expeditious manner when practiced.

Stretchers

Stretcher is a universal term that can be applied to the patient-carrying device for ambulances as well as the common army litter. Other terms and slang words are used for the ambulance-type stretcher. Perhaps most used is the term cot, although pram and wheels are still prevalent. The venerable army litter is still used in some situations.

A variety of different types of stretchers are manufactured by various companies and are in use today. It would be very difficult and unnecessary to describe in detail the operation of all of the different units available. However, it is imperative that firefighters know the correct method of operating the type of stretcher(s) that is used in their community.

The most common methods of placing patients onto these types of stretchers are by the extremity carry, by utilizing a backboard, or by having the patient lie directly onto the stretcher. Regardless of which of these methods is utilized, the rescuers need to assist and support the patient onto the stretcher. Once on the stretcher, the patient must be secured as soon as possible.

JPR 16-9: Safety Rules forAmbulance Stretchers

(For step-by-step photos of this skill sequence, see page 510)

1. Always make sure the patient is strapped securely prior to lifting, lowering, or moving.
2. Make sure that your partner is ready and understands what movement is desired.
3. Any time that a stretcher is being moved, a minimum of two rescuers should work together.
4. Do not attempt to roll a stretcher across uneven or rough terrain. It should be carried by an adequate number of personnel to make it safe. If at all possible, no rescuers should be backing up as they transport it.
5. The stretcher should be placed in the transport unit by the personnel responsible for operating the unit. Many times when firefighters or others attempt to assist, the stretcher gets unbalanced or in worst cases turns over.

The army stretcher is most commonly utilized in mass casualty incidents or at events where transport units are either not available or their use is not practical (large crowds, stadiums, evacuations, etc.). A patient may be placed on an army stretcher using the extremity carry, by having the patient lie directly on the stretcher, or using the method just given for placing a patient on a backboard. Whichever method is utilized, the patient must be securely strapped to the stretcher as soon as possible. Rescuers should not back up when carrying a patient on an army stretcher. The handles allow room for both rescuers to face and walk in the same direction.

Scene Assessment (Size-Up)

Scene assessment should be a predetermined sequence of steps or actions that is used in evaluating a crash scene. This evaluation is usually carried out by the officer; however, a good firefighter who conducts an assessment will be able to foresee many of the needs and requests of the officer.

NOTE

While it is important that this assessment be done quickly, it is more important that it be done accurately and completely.

Scene Safety

Good scene assessment considers many facts and probabilities. Specifically, the following items need to be taken into consideration:

- Traffic (see safety box)
- Number and type of vehicles involved
- Potential number and apparent extent of patient injuries
- Hazardous conditions (fire, HazMat, electrical, water, structural integrity, weather, crowds)
- Degree of entrapment

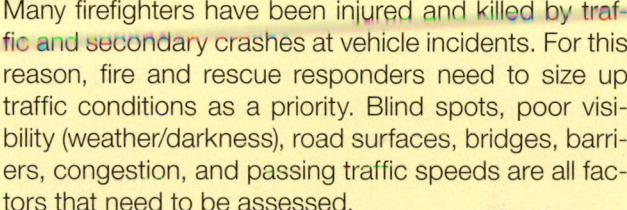

SAFETY

Many firefighters have been injured and killed by traffic and secondary crashes at vehicle incidents. For this reason, fire and rescue responders need to size up traffic conditions as a priority. Blind spots, poor visibility (weather/darkness), road surfaces, bridges, barriers, congestion, and passing traffic speeds are all factors that need to be assessed.

The results of this assessment will help determine the need for additional resources or assistance that might include more ambulances, law enforcement, specialized equipment, power company response, or other ancillary needs. It is important to mention that the scene assessment is an ongoing process.

ESTABLISHMENT OF WORK AREAS

Ideally, the fire department would like to shut down all traffic in and around the area of the vehicle incident. Unfortunately, the congestion this may cause can create secondary hazards and reduce the ability of additional responders to reach the scene. Many state and local law-enforcement agencies will work at all costs to keep traffic flowing. Because of this, it is essential that the fire department create protected work areas

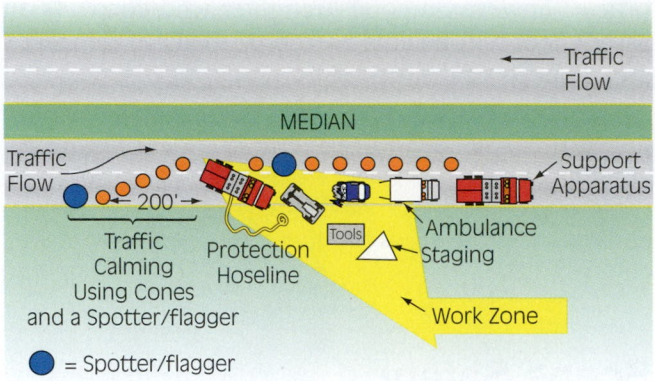

FIGURE 16-21 The first-arriving large fire apparatus should be positioned to create a traffic barrier and work zone. Cones and a spotter/flagger can help re-route traffic.

for rescuers and responders. Work areas are established using a combination of traffic barriers, traffic-calming strategies, and hazard zoning. Based on the scene assessment, the first arriving apparatus should be positioned to create a traffic barrier to help shield the greatest number of rescuers. Additional apparatus can increase the size of the barrier or be positioned past the scene so that they are also screened from traffic by the first arriving apparatus. While various positioning strategies exist, firefighters need to understand the logic and process of vehicle positioning so that they can work within the protection and limitations afforded by the barrier, **Figure 16-21.**

Traffic Barrier

A large fire apparatus such as an engine or heavy rescue can form the initial barrier by stopping in a position that shields the area where firefighters need to work. The apparatus should park with a slight diagonal angle to help increase the work area and make the apparatus appear larger to approaching traffic. The diagonal park also helps approaching traffic recognize that the fire apparatus is not moving. Even with a barrier in place, firefighters should mentally preplan an escape route should the barrier apparatus get hit by traffic.

NOTE

Traffic cones should never be used to create a barrier. Cones should be placed well behind the traffic barrier (apparatus). Some departments still use flares (fusees) to help with traffic calming. These devices are incendiary. It is important to make sure they cannot roll and will not ignite flammable gases, liquids, or vegetation.

Traffic Calming

Efforts expended to warn approaching traffic of an upcoming hazard are known as traffic-calming strategies. Typically, law-enforcement agencies and the Department of Transportation will handle this task. However, in the early stages of the incident, the fire department can begin some simple traffic-calming strategies to slow traffic, **Figure 16-22.** Placing traffic cones on the street to help guide traffic is a popular strategy.

Other traffic-calming strategies include:

- Deploying portable Accident Ahead signs.
- Switching off white lights and strobe warning lights that cause night blindness to approaching traffic.
- Utilizing arrow sticks (flashing amber lights that signal in which direction traffic needs to move).
- Positioning a firefighter to wave down traffic speeds. This is done using a wand light or other highly visible device. The firefighter waves the wand up and

FIGURE 16-22 Signs and lights are some devices that can be utilized for traffic calming.

down in a full arm length waving motion. This firefighter is exposed to traffic, but can also serve as a spotter for other rescuers. It is important to make sure an escape route is planned.

Hazards Zoning

Hazards that are identified in the scene assessment must be addressed. Initially, the fastest way to address the hazard is to create an exclusion zone (verbally or using flagging and/or traffic cones). Electrical wires will arc and move as automated systems try to reenergize the line. Pad-mounted transformers and electrical boxes can create a ground gradient of energy when they are damaged. Firefighters should treat them all as if they are energized and keep a 10-foot ring clear around the box. Damaged electrical poles can suddenly fall, dropping energized lines. Placing a lookout and preplanning an escape route can help zone these hazards. Hazardous materials should be treated as worst-case and zoned accordingly.

Most fire departments require that a fire attack line be pulled and charged as a standby measure whenever a vehicle extrication operation is under way. The firefighter assigned to this protective task can also serve as a lookout and remind other firefighters when they wander into an exclusion zone.

The vehicle(s) involved in the incident will likely present additional hazards. Fuel, electrical hazards, cargo, fire risk, and vehicle systems can be compromised in a motor vehicle incident. Understanding all the potential hazards a vehicle incident can present is well past the scope of this book; however, some hazards deserve a mention. Fuel systems on most vehicles are pressurized. If part of the fuel system is compromised, the fuel will escape under pressure initially.

Newer hybrid and alternative fuel systems present unique hazards ranging from **asphyxiation** to electrocution to toxicity. Alternative fuels can include compressed natural gas and propane. Hybrid systems include hydrogen and high-voltage battery systems combined with traditional fuels to increase engine efficiency. Each of these presents a unique set of hazards to responders. Firefighters should be encouraged to pursue understanding of new technologies through constant study of trade journals and extrication-specific training.

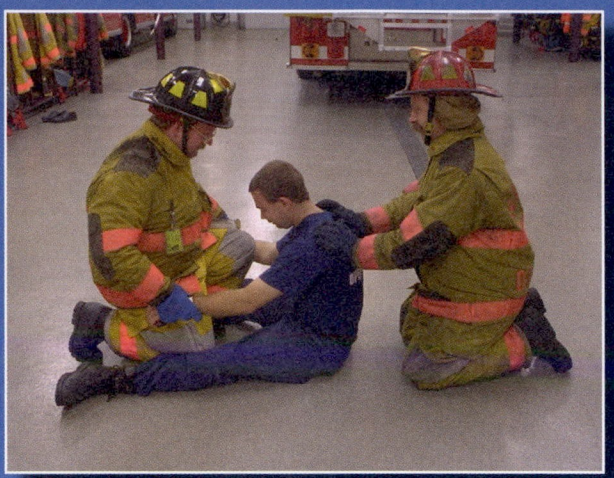

A Lay the patient on his back, with arms laid across torso, knees bent, and feet pushed back approximately halfway to buttocks. For an unconscious patient, it may be easiest to have the rescuer at the patient's feet reach down, grasp the patient's hands, and pull him into a sitting position using the rescuer's weight. Have a conscious patient assume a sitting posture with feet pulled back, lifting the knees above the floor.

B The rescuer at the patient's head squats behind the patient, sliding arms under the patient's armpits and grasping the wrist of the patient's opposite arm (if possible). At the same time, the rescuer at the patient's feet squats between the patient's feet and grasps the patient's legs under the knees (if possible) or as close as possible below the knees.

C When ready to lift, the rescuers need to communicate clearly with each other so that they can stand as one and carry the patient to safety.

Seat Carry

A   The rescuers face each other. Each rescuer grasps his own right forearm just above the wrist. The rescuers then grasp each other's left forearm just above the wrist, forming a square "seat."

B   The rescuers lower the "seat" that they have just formed, allowing the patient to sit on the seat with arms across the shoulders of each rescuer. When ready to lift, the rescuers need to communicate clearly with each other so that they can stand as one and carry the patient to safety.

JOB PERFORMANCE REQUIREMENT   16-3
Blanket Drag

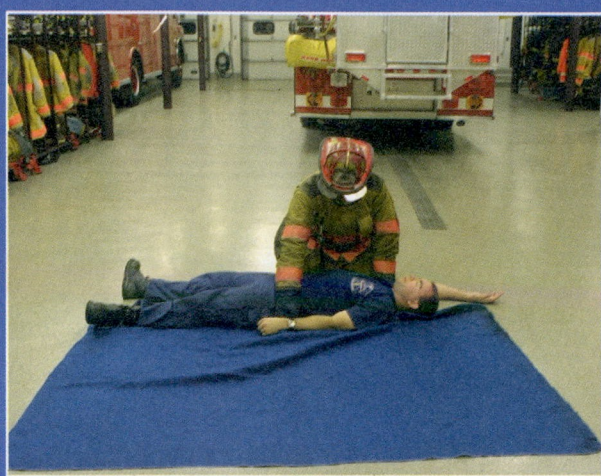

A   With the patient lying face up, lay the material the patient is to be placed on along one side of the patient's body with just over one-half of the material gathered close to the patient's side.

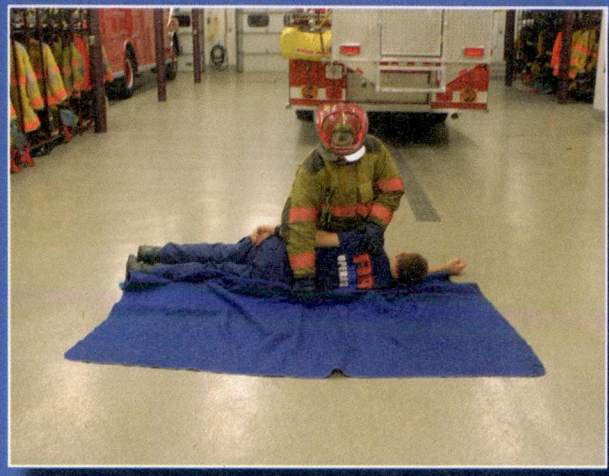

B   Kneel on the opposite side (from the blanket) of the patient, extend the patient's arm (on your side) above the head if injuries permit, reach across the patient with one hand just above the waist and one just below the hips, and roll the patient toward you. While supporting the patient, now on the side, tuck the gathered blanket material close to the body.

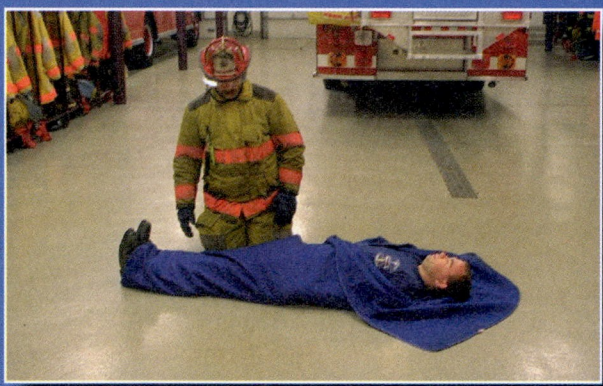

C   Roll the patient back onto the blanket material, return the arm to the side, and wrap the blanket snugly around the patient, supporting the head and neck as much as possible.

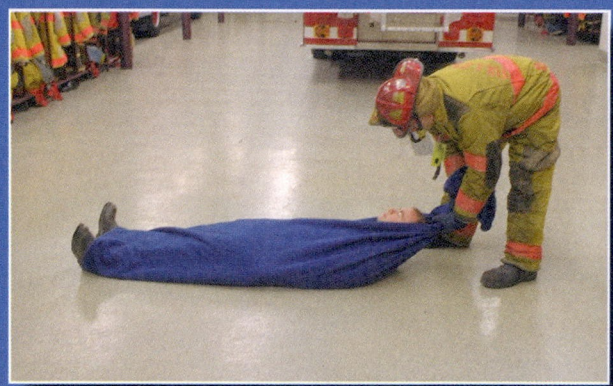

D   The patient is now ready to be dragged headfirst with the head and shoulders raised slightly off the floor.

JOB PERFORMANCE REQUIREMENT 16-4
Lift and Drag

A Locate and assess the victim and place the individual on his or her back. Take a position at the head of the victim. The rescuer will wrap his or her arms around the victim's torso.

B The rescuer will utilize his or her legs to stand up while lifting the victim. Once the victim's torso is lifted, he or she can be dragged to safety.

JOB PERFORMANCE REQUIREMENT   16-5

Rescue of a Firefighter Wearing an SCBA—Push and Pull Drag

A   Locate and assess the downed firefighter and place the firefighter on his or her back. The waist strap of the victim's SCBA should be unfastened, and one strap should be passed through the crotch area and refastened. This will prevent the SCBA from riding up on the victim firefighter.

B   One rescuer (Rescuer 1) will locate him- or herself at the head of the downed firefighter, taking hold of the downed firefighter's SCBA shoulder strap. Rescuer 2 will take a position inside the legs of the downed firefighter, lifting one of the downed firefighter's legs over his or her shoulder. Rescuer 2's head should be positioned high into the groin area with the shoulder driving into the buttocks of the downed firefighter. Correct positioning of Rescuer 2 is essential for this drag to be effective.

C   On command, Rescuer 1 should pull the downed firefighter while Rescuer 2 uses his or her legs to push simultaneously.

JOB PERFORMANCE REQUIREMENT 16-6
Cradle Carry When Victim Is Presented Head First

A The exterior rescuer on the ladder will be positioned at or below the windowsill, ready to receive the victim headfirst with the victim's head off to one side or the other. This allows the remaining portion of the victim's torso and legs to come across the ladder horizontally with the back facing the rescuer on the ladder.

B The rescuer on the ladder will position the victim by placing one forearm and hand underneath the victim's armpit in order to hold the weight of the upper torso and grasp the beam of the ladder while the other arm is placed under the top leg into the groin area of the victim. This will help to hold the weight of the victim's pelvic area and lower extremities.

(Continues)

JOB PERFORMANCE REQUIREMENT   16-6
Cradle Carry When Victim Is Presented Head First (*Continued*)

C   The rescuer will descend the ladder grasping the rails while leaning inward to control the victim until they reach the base of the ladder, and the rescuer is assisted by additional firefighters in moving the downed victim to waiting EMS personnel.

JOB PERFORMANCE REQUIREMENT 16-7
Cradle Carry When Victim Is Presented Feet First

A The exterior rescuer on the ladder will position him- or herself at or below the windowsill ready to receive the victim. The rescuer on the ladder will position the victim by placing the victim's legs to the outside of the ladder beams and hooking one arm into the groin of the victim. Both hands should grasp the beam of the ladder firmly.

B The upper body of the victim should then be placed onto the rescuer's opposite arm.

(Continues)

JOB PERFORMANCE REQUIREMENT 16-7

Cradle Carry When Victim Is Presented Feet First (*Continued*)

C The rescuer will descend the ladder grasping the rails while leaning inward to control the victim until they reach the base of the ladder, and the rescuer is assisted by additional firefighters in moving the victim to waiting EMS personnel.

JOB PERFORMANCE REQUIREMENT 16-8

Placing a Patient on a Backboard

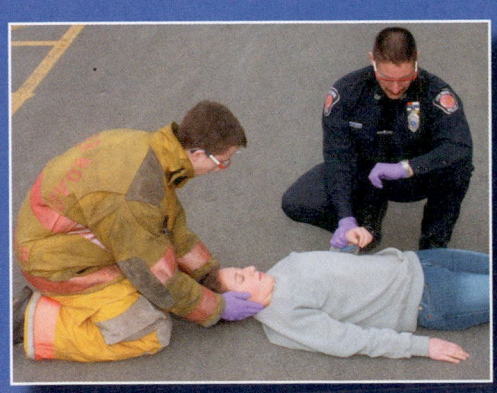

A A firefighter maintains manual stabilization while another first responder checks pulse, movement, and sensation.

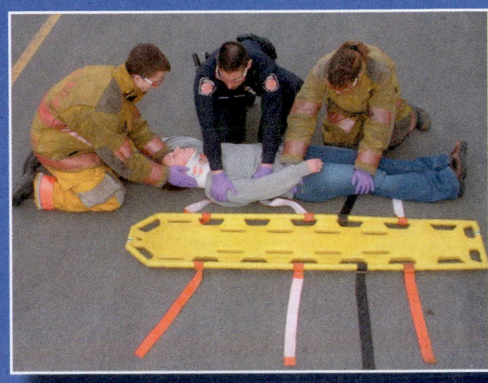

B A cervical collar is applied. While a firefighter maintains manual stabilization, two other firefighters take positions at the patient's shoulders and pelvis, reaching across the patient and grasping the patient's shoulders and pelvis, respectively.

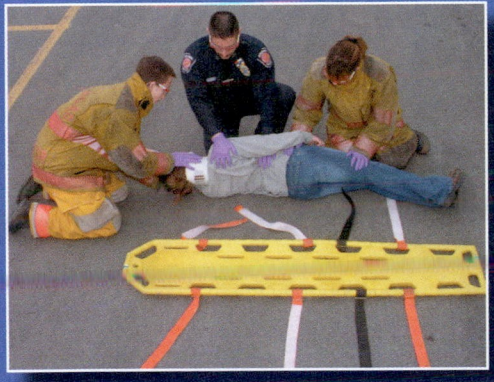

C On the command of the firefighter, while maintaining manual stabilization, the team rolls the patient onto the patient's side.

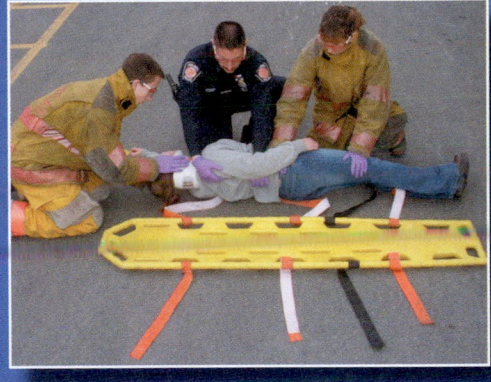

D One firefighter places the backboard under the patient with the bottom of the backboard at the patient's knees.

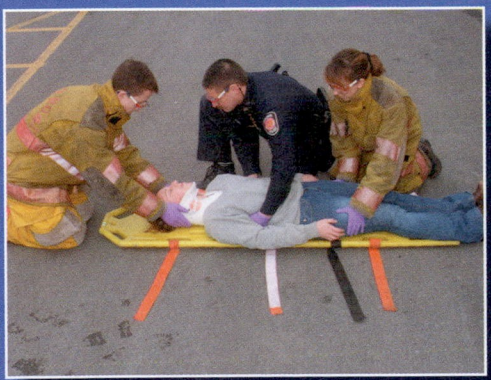

E On command, the team rolls the patient back onto the backboard, and the patient is pulled up to the center of the board using a long axis drag.

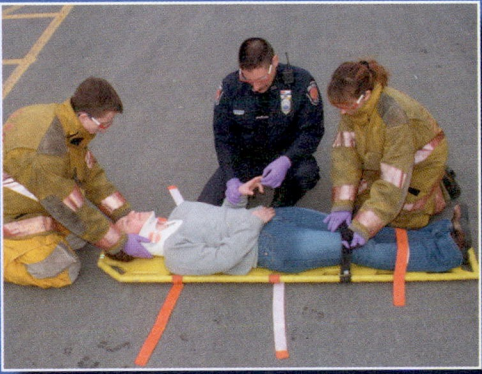

F Once the patient is centered on the backboard, the firefighter secures the patient to the backboard and reassesses distal pulses, movement, and sensation.

JOB PERFORMANCE REQUIREMENT   16-9
Safety Rules forAmbulance Stretchers

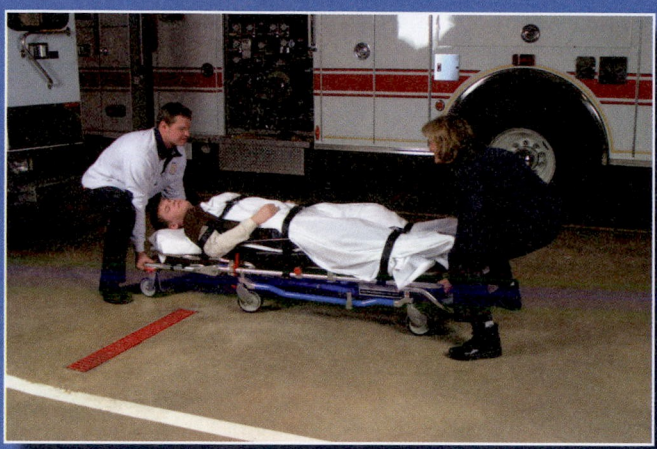

A   Always make sure the patient is strapped securely prior to lifting, lowering, or moving.

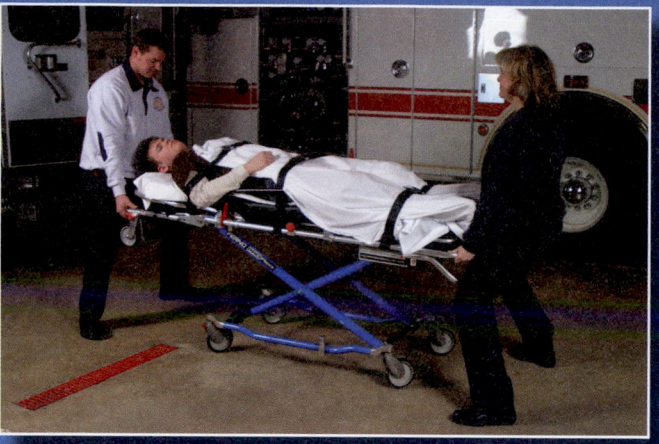

B   Any time that a stretcher is being moved a minimum of two rescuers should work together.

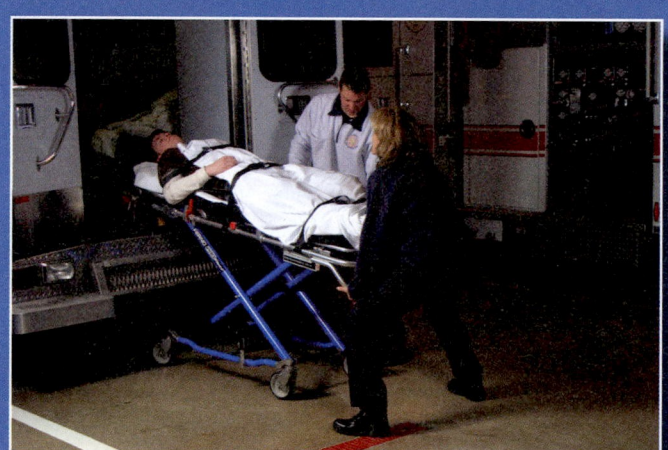

C   The stretcher should be placed in the transport unit by the personnel responsible for operating the unit. Many times when firefighters or others attempt to assist, the stretcher gets unbalanced or in worst cases turns over.

LESSONS LEARNED

This chapter covered a variety of the more common rescue situations that firefighters can expect to be confronted with from the day that they first step onto a unit. However, this does not even scratch the surface of the great variety of real-life rescue situations that firefighters confront on a daily basis. Rescue situations are classified as low-frequency (as far as occurrence) events that hold a very high probability of risk and danger to the firefighters involved. Risk/benefit analysis must be an ongoing process at each incident regardless of its nature. At no time should a firefighter be put in a hazardous position to save something that is already lost. It is imperative that firefighters stay aware of the "big picture" presented at every incident to stay safe and avoid tunnel vision. The risks faced by today's firefighters are ever changing. Firefighters need to keep the skills, knowledge, and abilities learned in this chapter in their mental toolbox, building on them throughout their careers and constantly updating them as new technology and equipment is developed and put into use.

KEY TERMS

Active Emitter Objects   Objects that generate their own thermal energy, for example, animals and humans. Active emitters can vary in relation to mass and density. They can also be suppressed by barriers such as clothing and construction features inside structures.

Asphyxiation   Loss of consciousness or death caused by too little oxygen and too much carbon dioxide.

Cold Smoke   Smoke that remains after a fire is extinguished by a sprinkler system.

Direct Source Emitter   An object that provides the most thermal energy and can be easily visualized through thermal imaging. Examples are flame, moving fire gases, or pre-flashover thermal layering within a room or structure.

Guideline/Lifeline   Rope used as a crew is searching a structure to assist them in finding their way back out.

Initial Rapid Intervention Crew (IRIC)   As outlined in NFPA 1710, two members of the initial attack crew must be assigned as a rapid deployment rescue team for the purposes of rescuing lost or trapped firefighters.

Passive Emitter   A inanimate object whose temperature will vary depending upon the environment and time frame that it is exposed to a heat source.

Tunnel Vision   The focus of attention on a particular problem without proper regard for possible consequences or alternative approaches.

Two In/Two Out   The procedure of having a crew standing by completely prepared to immediately enter a structure to rescue the interior crew should a problem develop.

Vent Enter Search (VES)   Extremely high-risk search technique of breaking out a window and entering for a primary search versus a blind crawl through thick smoke and heat to find a stairway. Oftentimes performed prior to hoselines being put into place.

REVIEW QUESTIONS

1. What is the biggest danger that firefighters must be aware of when involved in rescue operations?
2. Name at least three of the things firefighters should be looking for when approaching a building they will be searching.
3. Describe the difference between the primary search and the secondary search.
4. Describe the approach that firefighters should take when utilizing the thermal imaging camera for search operations.
5. Describe the basic procedures for conducting a large area or "team" search.
6. List the five goals that a Rapid Intervention Team should strive for when put into action on the fireground.

17 Forcible Entry

It was a late night call for a fire in a commercial strip center. One of our engines and a ladder company were on the automatic aid initial box assignment and, based on the dispatch information, we knew we were going to a working fire in a familiar area. The L-shaped strip center contained multiple occupancies including a restaurant, professional offices, stores, and a bank. The fire originated in the antique store in the middle of the approximately 30,000-square-foot structure and had good control of that occupancy. The store fronts were glass set in heavy aluminum frames. The entry doors were glass with heavy aluminum door jams representing a very easy, but very expensive, opportunity to break windows and doors to gain entry to each of the occupancies.

As we were arriving, the Incident Command Officer called for a second alarm. Our two units were given multiple assignments. The ladder company was assigned to roof ventilation and the engine company was ordered to open up all occupancies in the building and, with the assistance of another engine company, search for extension.

Only days before, our companies had completed forcible entry training on the lock method of our door assembly training system. The engine company officer broke his company into two teams and had each retrieve a K-Tool Kit from our apparatus. Within eight minutes, the crews had opened the remaining fifteen occupancies without breaking one window.

Because the crew reported to the Incident Commander so quickly, the Incident Commander assumed they had broken out doors and windows. He was thinking of all the unwarranted damage caused by our team. When he discovered, in actuality, that there was no collateral damage, other than the locks, he was amazed the team had completed the task so quickly. He began quizzing the firefighters on their tactics.

A very proud group of firefighters were happy to demonstrate their prowess by pulling and picking the lock on the door of the store where the fire originated. A lock can be replaced for $80, but a door assembly can cost in excess of $750; collateral damage was minimized through the use of the training.

The lesson learned at the event was twofold. First, firefighters were amazed their forcible entry training transferred to real-life situations. Second, because our team quickly reported completed tasks, including the bank, back to the Incident Commander, the other departments were able to realize the value in continuous training to improve response.

—Street Story by Randall F. Parr, EFO, Fire Chief,
Tomball Fire Department, Tomball, TX

LEARNING OBJECTIVES

After completing this chapter, the reader should be able to:

17-1   Identify forcible entry tools by common name and use.

17-4   Identify how and when to use hand and power tools during forcible entry in rescue operations.

17-5   Demonstrate proper techniques for carrying and operating selected hand and power tools.

17-6   Given a selection of hand and power forcible entry tools, protective equipment, and an assignment, select the proper tool and demonstrate the proper care, inspection, and maintenance after use.

17-7   List and describe proper cleaning solvents

17-8   Given manufacturer recommendations, department guidelines, tools, and cleaning materials, demonstrate the ability to properly clean, inspect, and service tools and equipment.

17-19   Describe building features and methods of forcible entry for doors, windows, walls, and other construction barriers.

17-20   Describe issues associated with forcible entry during salvage operations.

17-21   List the basic door construction types.

17-22   List hazards associated with forcing entry through doors.

17-23   Identify the various types of locks.

17-24   Demonstrate forcible entry techniques on a variety of doors.

17-25   Demonstrate the through-the-lock forcible entry method.

17-26   List the basic window construction types.

17-27   List hazards associated with forcing entry through windows.

17-28   Demonstrate the procedures for gaining entry through windows.

17-29   List the types of wall construction.

17-30   List hazards associated with forcing entry through walls.

17-31   Demonstrate procedures for breaching both interior and exterior walls and floors.

*The FF I and II levels, as defined by the NFPA 1001 Standards, are identified in different colors:
FF I = black, FF II = red, additional information = blue.

INTRODUCTION

According to statistics from the Federal Bureau of Investigation, over 2½ million homes are broken into every year, approximately 8,600 each day, or one break-in every 13 seconds. Because of these incidences, residents and owners install a variety of **locking devices** to protect both their homes and businesses. These devices present a number of challenges to firefighters who must gain access to a structure to conduct firefighting operations.

Forcible entry is often one of the first operations conducted at a scene to gain entry into secured areas and buildings whether in a fire, medical aid, or other type of emergency. At structural fires, a delay in gaining entry will reduce the ability to mount an aggressive fire attack, allowing the fire to extend out of control. Also, rescue operations require forcible entry to gain access to areas where victims are located. In addition, the renewed emphasis on rapid intervention teams, for the rescue of trapped firefighters, requires all firefighters to be knowledgeable and proficient with the forcible entry tools, **Figure 17-1.**

Forcible entry is a combination of knowledge and skills used to gain entry to a structure when the windows and doors are locked, blocked, or in some cases do not exist. To perform this essential task, firefighters must be knowledgeable about building construction and the operation of locking devices to break, remove, or bypass these elements. In addition, the firefighter must have the skills, gained by training and experience, to apply this knowledge using a variety of tools.

Knowledge

A working knowledge of the many types of locks, hardware, doors, and other assemblies is essential to successful forcible entry operations. Firefighters must be able to "size up" the quickest and easiest way to gain access to buildings, such as the doors shown in **Figure 17-2.** In addition, the firefighter must know which type of tool to use and the best method to gain access through the door to the building.

STREETSMART TIP

As with all firefighting skills, forcible entry must be studied and practiced on a continuing basis. As security concerns increase, the types and number of locks used will also increase. Firefighters should study the locks used in their area during the development of pre-incident plans. They can visit a locksmith to learn of new types of locks or check out the Web sites of lock manufacturers. Forcible entry is more brain power than it is muscle power.

FIGURE 17-1 A typical assortment of forcible entry tools used by fire departments.

FIGURE 17-2 Firefighters should size-up the way to gain entry, "try before you pry," and think about how they would force these doors.

Skill

The element of skill involves a firefighter's ability to apply knowledge of building construction, lock assemblies, tools, and techniques to accomplish the necessary tasks of forcible entry. This means choosing the proper tools and applying the best techniques when using the tools. Again, skills are developed by repeated practice and experimenting with new tools, locks, and techniques.

Experience

Experience can be acquired by three means. One is through drills and practice at training sessions and another is at the scene of actual fires and emergencies. Both are the means by which skill is developed and knowledge is gained as well as reinforced. A third and very good method of aquiring experience is

learning about other's experiences through case studies and reports. Still, the most important experience is gained from field operations, where firefighters' skills and knowledge are put to the true test.

FORCIBLE ENTRY TOOLS

Firefighters must have an understanding and knowledge of the tools available to conduct forcible entry. The selection and use of the "right" tool are essential if the task of forcible entry is to be completed as quickly as possible. Although any number of tools can be used to accomplish a task, the right tool for the right job is the quickest and easiest way to complete the operation. Because of geography or local tradition, many of the tools shown or listed may have another name or nomenclature. Firefighters must know the tools and their names as used by their particular department. Tools used for forcible entry operations are divided into several families or groups based on their intended uses as shown in **Table 17-1.**

This chapter does not attempt to show or describe all of the tools used for forcible entry because some tools are obsolete or used infrequently. The tools described in this chapter are the most common forcible entry tools used today.

Striking Tools

The group or family of **striking tools, Figure 17-3,** is used to deliver impact to other tools, such as a Halligan tool, in order to drive it into place. Striking tools are used for impact delivery to the lock or the door itself. They may force the door or even break it down.

Flathead Ax

The most common of the striking tools is the flathead ax. Although this tool may be used for cutting, it is generally used to drive the Halligan tool. Together, the

FIGURE 17-3 The group or family of striking tools includes the maul, small hammer, flathead ax, and Denver tools.

ax and Halligan tool are married together forming the **irons, Figure 17-4,** and are one of the most important and useful of all forcible entry tools. The flathead ax is available in 6- and 8-pound sizes, the weight being at the head of the ax. The 8-pound ax is best for striking purposes and is commonly known as the forcible entry ax. The handle of the ax can be comprised of wood or fiberglass. Fiberglass is best for all purposes, due to its high strength and ease of maintenance.

TABLE 17-1	Forcible Entry Tools	
Family or Group	**Tools**	**Operation**
Striking	Flathead ax, maul, sledgehammer, battering ram, hammer, punch & chisel, lock breaker	Deliver impact force to break lock or drive another tool
Prying	Crowbar, Halligan tool, hux, claw tool, pry bar, hydraulic spreader	Provide mechanical advantage, leverage
Cutting	Axes, saws, torches, bolt cutters	Cut material away; cut around locking devices
Pulling	Hooks, pike poles	Limited for forcible entry; breaks glass, gypsum board, Sheetrock
Through-the-Lock	K-tool, A-tool, picks & key tool, bam bam, vise grip or channel lock pliers, REX tool	Remove lock cylinder

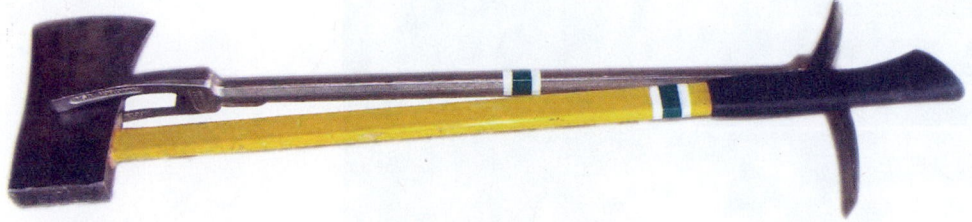

FIGURE 17-4 The flathead ax and Halligan tool form the tool set known as the *irons*. (*Courtesy of Ralph Restadius*)

STREETSMART TIP

A firefighter should never free-swing any of the striking tools. The firefighter using the striking tool, usually to apply force to a prying tool, should hold the tool firmly and use a strong and controlled stroke for power and accuracy.

Maul/Sledge

The maul or sledge is generally found in several sizes with the 10-pound maul being the most common and versatile. Other sizes include the 8-, 12-, and 16-pound versions. The 10-pound maul is sometimes carried with the Halligan tool to again form a set of "irons." The 16-pound maul is considered too heavy to drive other tools such as the Halligan and is sometimes used to break in doors or locks. Once again, fiberglass handles are recommended for strength and safety. Some new tool styles combine mauls with other tools, **Figure 17-5.**

The power a striking tool applies is explained by this equation: Force = Weight × Speed. A tool that is too heavy cannot be moved fast enough to develop proper force. Also, problems with accuracy and control can develop. On the other hand, a tool that is too light will not deliver enough force, which in turn will require delivering more strikes. Drilling and practice will help determine which tool is best for you.

Ram

Battering rams, **Figure 17-6,** are used by two or more firefighters to break through a door or wall. The desired result is to break or push the object in. There are several types, but the type most commonly used is the old-style fire department ram. This tool is made of steel and has four handles and a round end for battering and a forked end for breaking and penetrating. Another type of battering ram is smaller and is intended for use by one or two firefighters. This tool weighs approximately 20 to 25 pounds and is used to break doors and breach walls.

Prying and Spreading Tools

The group or family of **prying tools, Figure 17-7,** is used to spread apart a door from its **jamb,** move objects, or expose a locking device.

Halligan Tool

The original **Halligan tool, Figure 17-8,** designed by Hugh Halligan of the Fire Department of the City of New York, has proven to be the most important single forcible entry tool used in the fire service. Although the original style tool is not widely used today, the same basic concept with some improvements is capable of many forcible entry tasks.

Halligan-type tools are made in several weights and lengths, and of different materials. The best type for forcible entry and structural firefighting is a 30-inch forged steel tool. The basic Halligan design is a tool that has three primary parts as shown in Figure 17-8. These are the adz end, the pike end, and the fork end. The size and shape of the parts are important because they determine the mechanical advantage the tool will apply to the spreading force. As noted earlier,

FIGURE 17-5 The Denver tool is a maul with a cutting edge, prying end, and a pulling hook.

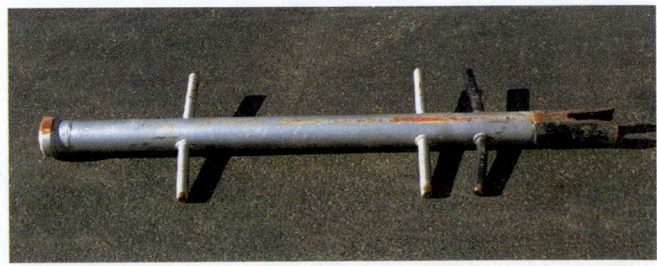

FIGURE 17-6 The battering ram requires brute strength to break a door, lock, or wall.

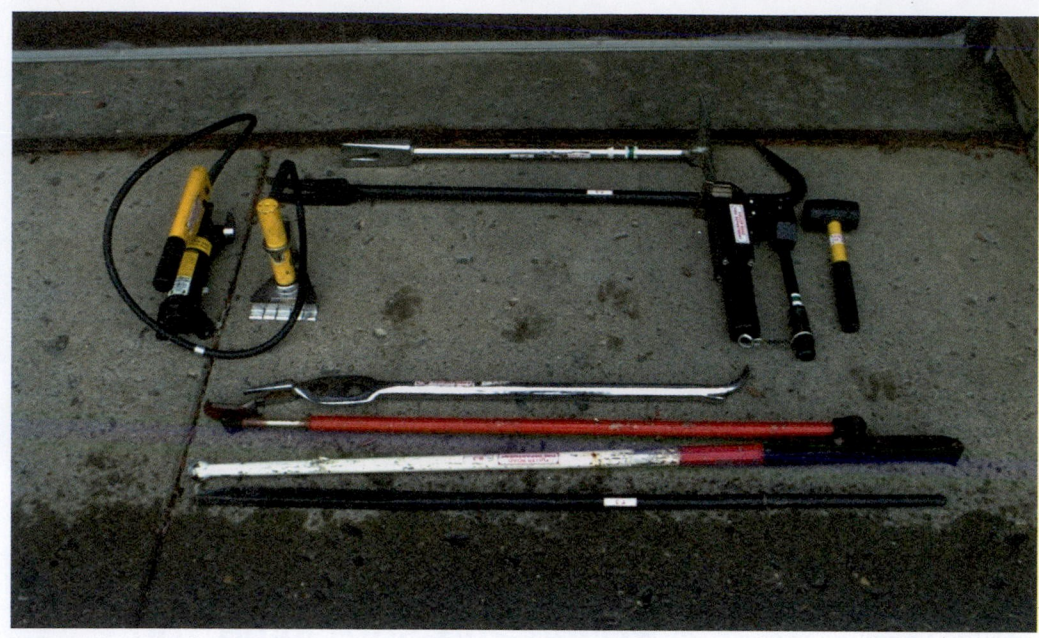

FIGURE 17-7 The group or family of prying tools includes the Halligan, claw tool, hux bar, Detroit door opener, pry bar, and hydraulic spreaders.

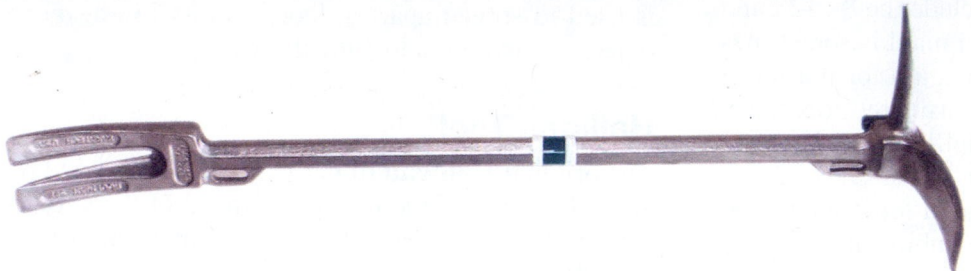

FIGURE 17-8 The Halligan tool is one of the most useful forcible entry tools. *(Courtesy of Ralph Restadius)*

the Halligan tool is often nested with the flathead ax, forming the irons.

Claw Tool

The claw tool, a forerunner to the Halligan tool, was the mainstay of forcible entry operations for the previous generation of firefighters. The claw tool had two major parts: the fork end, which had a hollow, curved profile, and the hook end, which was curved and ended with a sharp point. Generally, the tool was made in two sizes. The heavy claw is 42 inches long and the standard claw is 32 inches long. The standard claw was usually carried with a striking tool such as the ax or the kelly tool. The claw tool is obsolete now, having been replaced by the Halligan tool.

Kelly Tool

The Kelly tool is a steel bar, 28 inches (71.1 centimeters) long that has two main features. One end has a large adz, approximately 3 inches (7.6 centimeters) wide, and the other end has a large chisel or fork. This is an older tool and has since been replaced by the Halligan tool.

Hydraulic Spreaders

Handheld hydraulic spreaders, **Figure 17-9,** are available in two different styles. One is a hydraulic pump with a spreading device attached by a length of hose, and the second is a similar pump with the spreading devices attached directly to it. These tools are often called rabbit or rabbet tools. The spreading end is forced between the door and jamb at the location of the locking device and the pump is operated to spread the door and jamb apart. A tool is used to force the spreaders into the space between the door and jamb.

Power hydraulic spreaders can also be used to spread objects. Their application is generally for rescue work; however, they may be used to force entry into a building.

Miscellaneous Prying Tools

There are many other types of prying tools such as the crowbar, pry bar, and others, but most are of limited use. As with any tool, its use depends on the type of task to be accomplished and the method of forcible entry.

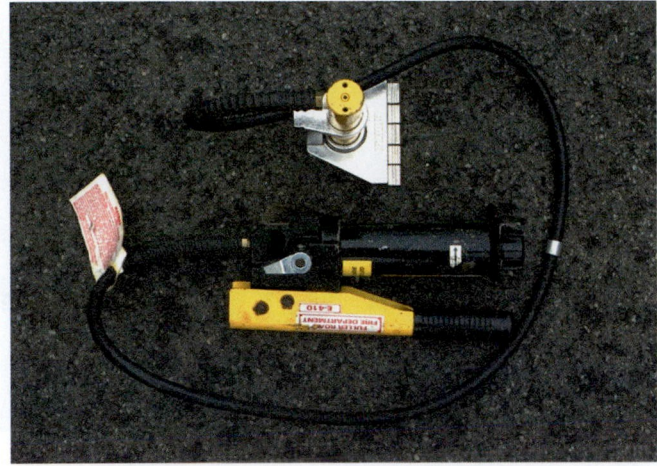

(A)

(B)

FIGURE 17-9 Handheld hydraulic spreaders are available in two different styles but operate on the same principle. (A) Handheld pump with remote spreader. (B) Handheld unit with integral built-in spreader.

Cutting Tools

The group or family of **cutting tools,** **Figure 17-10,** is used to cut away materials and expose the locking device or cut through a door or wall to accomplish forcible entry.

Ax

The fire department ax is seen in two basic configurations in **Figure 17-11:** the flathead ax and pick head ax. The flathead ax is most commonly used as a striking tool for forcible entry. In addition to cutting, this tool may be used as a prying tool, especially when equipped with a fiberglass handle. The pick head ax has a point that can be used for puncturing, pulling, and, to a limited extent, prying. It is recommended that all axes be equipped with fiberglass handles for strength and safety.

Handsaws

The most commonly used handsaw is the metal cutting hacksaw. A strong, high-quality frame with a supply of good-quality blades is used for many firefighting operations, especially vehicle rescue. Carpenter saws, both rip and crosscut, may also be carried on fire apparatus for use in areas where power saws are too large, are too noisy, or present a safety hazard.

Bolt Cutters

Bolt cutters are used to cut metal bars, cables, wires, and other hardware. These tools come in a variety of sizes; the most common for forcible entry is 36 inches

FIGURE 17-10 The group or family of cutting tools includes axes, saws (both power and manual), and bolt and wire cutters.

(A)

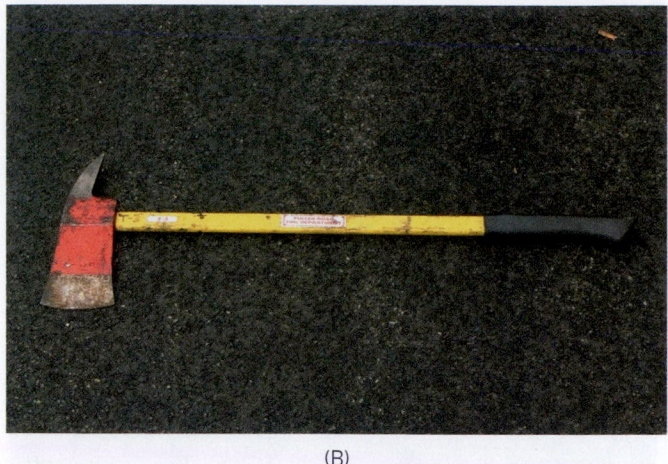

(B)

FIGURE 17-11   The fire department ax is seen in two basic configurations: (A) flathead and (B) pick-head.

in length. This size will cut steel up to ¾ inches thick. Firefighters should avoid using bolt cutters to cut heavy-duty padlocks with case-hardened shackles. The hardness and the thickness of the metal make them difficult to cut and will damage the cutting edge of the tool.

Power Cutting Tools—Saws

Power saws, **Figure 17-12,** are the most common power cutting tools used for forcible entry work. They are available in two basic types: the rotary saw with a circular blade and the chain saw. The rotary saw in Figure 17-12A has a carbide-tipped blade. Figure 17-12B shows a rotary saw with a metal-cutting disc and Figure 17-12C shows a chain saw. Rotary and chain saws are usually powered by a two-cycle gasoline engine and require two firefighters for safe operation.

Carbide-Tipped Blades

The rotary power saw is widely used and can be adapted to cut a variety of materials. Wood and composition materials such as roofs and flooring are cut with a carbide-tipped blade, which has 12 to 24 teeth. Light-gauge metals and polycarbonate plastic may also be cut with this blade. When cutting with a carbide-tipped blade, it is important to maintain full rpms to avoid having the saw bind in the material being cut.

Metal Cutting Blades

The most common metal cutting blade is an abrasive disc made of aluminum oxide. This blade is used to cut locks, hardware, steel doors, and roll-down gates, and for other forcible entry applications.

NOTE

With all blades, care must be taken not to bind or bend the blade while cutting. This is particularly important with abrasive discs because they can shatter or disintegrate.

Masonry Cutting Blades

Masonry materials such as concrete, brick, block, and stone can be cut with an abrasive disc made of silicon carbide or a steel with a diamond matrix blade. When cutting with these blades, a spray of water on the blade will cut down on the production of dust and keep the blade cool, thus prolonging the life of the blade.

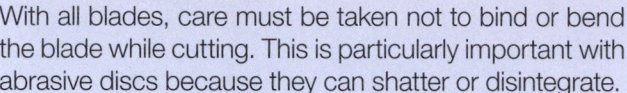

STREETSMART TIP

Gasoline or other hydrocarbon fuels will break down the bonding material used in the manufacture of abrasive disc blades. To avoid this, blades must be stored separately from the fuel can for the saw. One should always refer to the manufacturer's recommendations on use and storage. Checking the fuel supply and starting the engine of gasoline-powered equipment before taking it to the location of the cutting operation will save time and frustration. Some departments start and run the saw each tour to avoid problems.

Chain Saws

Chain saws are used primarily for ventilation purposes; however, they have application as a forcible entry tool depending on the type of building construction and opening required. Several types of saws and chains are available, and many departments use a carbide-tipped chain. The chain saw can also be used to cut through wood siding, wood frame walls, certain doors, and light-gauge metal.

(A)

(B)

Reciprocating Saws

The reciprocating saw, **Figure 17-13,** is an extremely versatile tool. These saws are powered by electricity, either stationary (house current) or portable (generator). Cordless, battery-powered saws are now available and can be operated remotely from a power source, thus increasing versatility. A variety of blades are available for this type of saw depending on the material to be cut.

Cutting Torch

The cutting torch, **Figure 17-14,** uses a fuel such as acetylene mixed with oxygen to produce a flame to heat metal. A jet of pure oxygen is then applied to intensify the heat, creating a flame temperature in excess of 5,000°F (2,760°C), melting the metal. Many departments use these tools for heavy forcible entry in addition to other cutting tools. Use of cutting torches requires specialized training in addition to following manufacturer's recommendations and department procedures.

CAUTION

Acetylene is an unstable and extremely flammable gas. It is essential to always follow manufacturer's recommendations for storage and use of this material. In addition, cutting torches should never be used in a flammable environment.

(C)

FIGURE 17-12 Gasoline-powered saws are available in two types: the rotary saws shown in (A) with the wood cutting blade and (B) with the abrasive disc, for either masonry or metal cutting, or the chain saw shown in (C). The chain saw is usually used for roof ventilation.

FIGURE 17-13 The reciprocating saw is a versatile tool for forcible entry and rescue work.

FIGURE 17-14 Cutting torches use a mixture of acetylene and oxygen to generate a high-temperature flame. They cut metal by melting it.

Pulling Tools

The most common type of **pulling tool** is the **hook** or **pike pole, Figure 17-15.** These tools are grouped by the type of head and handle length and are used to open up walls and ceilings, to vent windows, and to pull up roof boards or other building materials.

Special Tools

A number of specialized tools are available to assist with or conduct forcible operations. Many of these have been developed by firefighters after years of experimentation and trial and error.

Bam Bam or Dent Puller

Originally designed as an automotive body work tool, the bam bam is used to pull lock cylinders. It consists of a shaft, case-hardened screw, and slide hammer. The screw is turned into the lock cylinder and the slide hammer is operated to pull the cylinder out.

Duck Bill Lock Breaker

The duck bill lock breaker is a wedge designed to break open heavy-duty padlocks, **Figure 17-16.** The long, tapered head is placed into the shackle of the lock and driven down with a flathead ax (8-pound minimum) or a maul. This wedging action will pull the shackle out of the body of the lock.

K-Tool and Lock Pullers

The K-tool, **Figure 17-17,** is used to perform through-the-lock forcible entry. The K-tool is designed to pull out lock cylinders and expose the mechanism in order to open the lock with the various key tools. The back of the tool is shaped like the letter *K* and slides over the lock cylinder. The front of the tool has a loop for the adz of the Halligan tool.

The K-tool is placed on the lock cylinder and tapped into place to gain a good purchase or bite on the lock. The Halligan tool is then used to pry/pull the cylinder

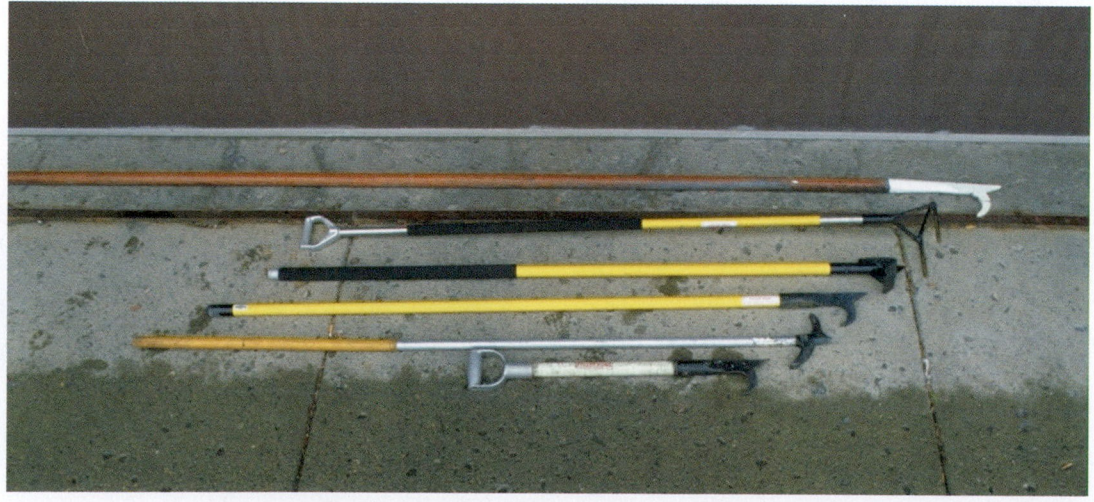

FIGURE 17-15 The most common type of pulling tool is the hook or pike pole, available in various styles and lengths.

out of the lock mechanism. This exposes the lock mechanism, which is operated with the proper key tool. Key tools include the bent, square, and screwdriver types, as well as a pick to slide open the shutter on some rim locks. A locking-type pliers is a good addition to the K-tool kit.

A number of other tools similar to the K-tool are available that can be used to perform the same operation of through-the-lock forcible entry. These are the A-tool, officer's tool, or REX tool, **Figure 17-18,** to name a few.

Combinations of Tools

To perform the task of forcible entry most of the tools or groups of tools discussed are used in combination with other tools. Striking tools are used with prying

FIGURE 17-16 Duck bill lock breaker forcing a heavy-duty padlock.

tools, such as the flathead ax and Halligan tool, and these tools should be carried or stored together on apparatus. In addition, the firefighter must be able to size up the forcible entry task and choose the right combination of tools to provide adequate leverage or force. Experience will provide the firefighter with the knowledge to choose the right combination and to apply this knowledge to complete the task.

SAFETY WITH FORCIBLE ENTRY TOOLS

As with all tools and equipment, if misused or used for the wrong task, forcible entry tools will create safety hazards. Firefighters will become familiar with the tools, their operation, and their maintenance during training, which should result in safe operation. A number of general rules apply to all operations as follows:

- Always wear proper personal protective equipment including hand and eye protection.
- Follow manufacturer's guidelines for proper operations.
- Do not attempt to cut material other than that for which a blade or tool was designed.
- Operate with regard to the safety of others in the immediate work area.
- Make sure tools are in proper operating condition before use.

FIGURE 17-17   The K-tool (shown with key tools) is designed to perform through-the-lock forcible entry. *(Courtesy of Ralph Restadius)*

FIGURE 17-18 The REX tool is used to pull lock cylinders in the through-the-lock method of forcible entry.

■ Most forcible entry operations require teamwork. Never attempt to use tools alone that require two firefighters.

■ When the task is complete and if the tool is no longer needed, secure it to prevent tripping or other hazards.

■ Tools should be stored and easily accessible, **Figure 17-19.**

> ### NOTE
>
> The safety guidelines provided are general in nature and firefighters are reminded to read manufacturer's operating instructions for the specific tools used in their department. In addition, they should not use any power tools or other spark-producing equipment in an explosive or flammable atmosphere.

> ### SAFETY
>
> Whenever forcing entry into an unfamiliar area, firefighters must be very cautious as to what they may face when creating the opening. Immediately dangerous to life and health (IDLH) conditions, extreme fire exposure, drug lab chemicals, booby traps, or other hazards may be present.

Rotary and Chain Saws

As the use of security gates and overhead doors increases, the power saw has become the tool of choice to remove the door or gate. The rotary saw with a metal cutting blade is an effective tool for these operations. These saws present a number of hazards,

FIGURE 17-19 These slide-out trays provide easy access to forcible entry tools. Note the mounting brackets that hold the tools in place.

and firefighters should follow these guidelines to complete the operation safely:

■ Always follow the manufacturer's instructions.

■ Conduct daily checks for operation and blade condition.

■ Check the saw for fuel and proper operation before proceeding to the entry location.

■ Equip the saw with a carry strap (standard equipment with some manufacturers).

■ Use the right blade for the material being cut, **Figures 17-20 A** and **B.**

■ Never carry a running saw up a ladder or through a crowd of firefighters.

■ Power saws require two firefighters: the saw operator and a guide firefighter.

■ Eye protection is required when running any power equipment, especially power saws.

Carrying Tools

Many forcible entry tools have sharp or pointed ends and must be carried safely from fire apparatus to the fire scene. Firefighters should always be aware of their safety in addition to the safety of other firefighters.

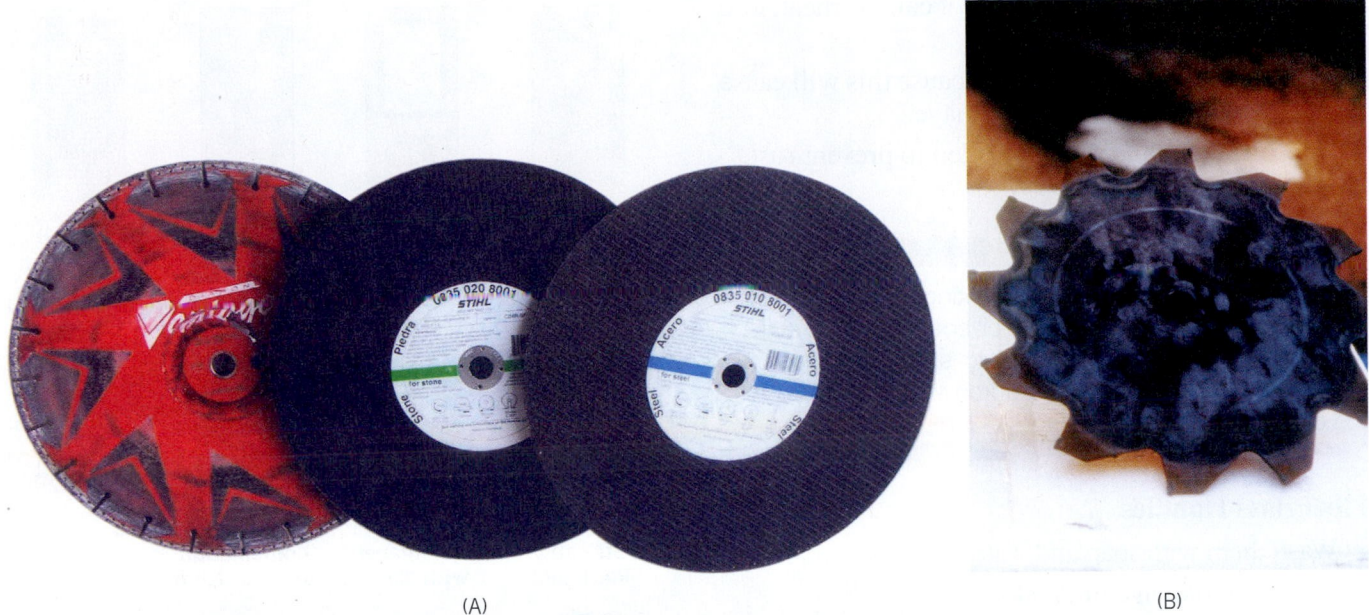

FIGURE 17-20   (A) Rotary saws have three types of blades: wood, masonry, and metal. Several examples are shown here. *(Courtesy of Ralph Restadius)* (B) This is a new style of wood and roofing blade for the rotary saw.

- *Axes.* Carry the ax with the blade away from the body and the pick head covered or in its scabbard. Never carry an ax over the shoulder.
- *Prying tools.* Similar to the ax, pointed and sharp edges should be carried away from the body and covered if possible.
- *Striking tools.* These tools tend to be heavy so the head should be carried close to the ground and not allowed to swing freely. When using these tools, do not use a free-swing motion. Firmly grasp the tool with two hands and use a controlled and accurate stroke to move the tool.

Hand Tools

Hand tools used for forcible entry are constructed of metal, wood, fiberglass, or some combination of these materials. All tools must be inspected regularly for condition, cracks in the handles, burrs in the metal, and loose heads.

MAINTENANCE OF FORCIBLE ENTRY TOOLS

Proper tool maintenance is the first step to tool safety. Tools must be inspected and cleaned on a regular basis and checked for wear and damaged parts.

General Guidelines

- Inspect tools regularly and keep them clean and operational. Follow any manufacturer or department guidelines for inspection and maintenance of all forcible entry tools.
- Documentation of tool maintenance is important. Follow department procedures for proper documentation of tool maintenance and service, and for placing tools that are unsafe or not working out of service. Be sure to report unsafe or nonfunctional forcible entry tools to the proper authority within the department so the situation can be corrected.
- Follow manufacturer guidelines for battery-operated tools and ensure that batteries are functional for operational equipment. Follow manufacturer and department procedures for rotating or cycling batteries in the tools for maximum operational use.

Metal Heads and Parts

- Remove any dirt or rust with steel wool or emery cloth.
- Use a metal file to maintain the proper profile and cutting edge.
- Sharpen edges and remove burrs with a file.
- Do not keep the blade edge too sharp; this may cause it to chip when in use.

- Do not grind the blade because it can overheat, lose temper, and become soft.
- Do not paint the metal parts because this will cause sticking and also hide any damage.
- Keep all metal parts lightly oiled to prevent rust.

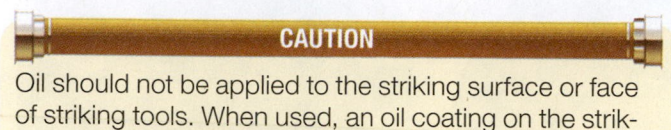

CAUTION

Oil should not be applied to the striking surface or face of striking tools. When used, an oil coating on the striking surface may cause the tool to slip or glance off, causing injury, and little or no force will be applied to the object being hit.

Fiberglass Handles

- Wash them with soap and water and dry completely.
- Check for damage or cracks.
- Make sure metal parts are secure.

Wood Handles

- Clean with soap and water, rinse, then dry completely.
- Check for damage and sand off any splinters.
- Do not paint or varnish the handles. A coat of boiled linseed oil may be applied if necessary.
- Ensure that the head is securely fastened to the handle.

CONSTRUCTION AND FORCIBLE ENTRY

The type and construction of the many different features of buildings, such as doors, windows, gates, walls, floors, and roofs, must be recognized and understood by firefighters to force entry. A thorough knowledge of the construction of these building features will lead to successful forcible entry operations.

Door Construction

Passage doors are manufactured in many styles as shown in **Figure 17-21** and are manufactured from plastic, wood, metal, and/or glass. The door assembly consists of the door itself, the frame or jamb, **mounting hardware,** and locking device mounted in a jamb, which is **rabbeted** or has a stop attached to it, **Figure 17-22.** A rabbeted jamb is formed or milled into the casing that the door closes against to form a seal. The stopped jamb has a piece of molding nailed or attached to the casing for the door to close against. The stopped jamb can be removed, allowing access to the lock assembly and hence easier forcible entry.

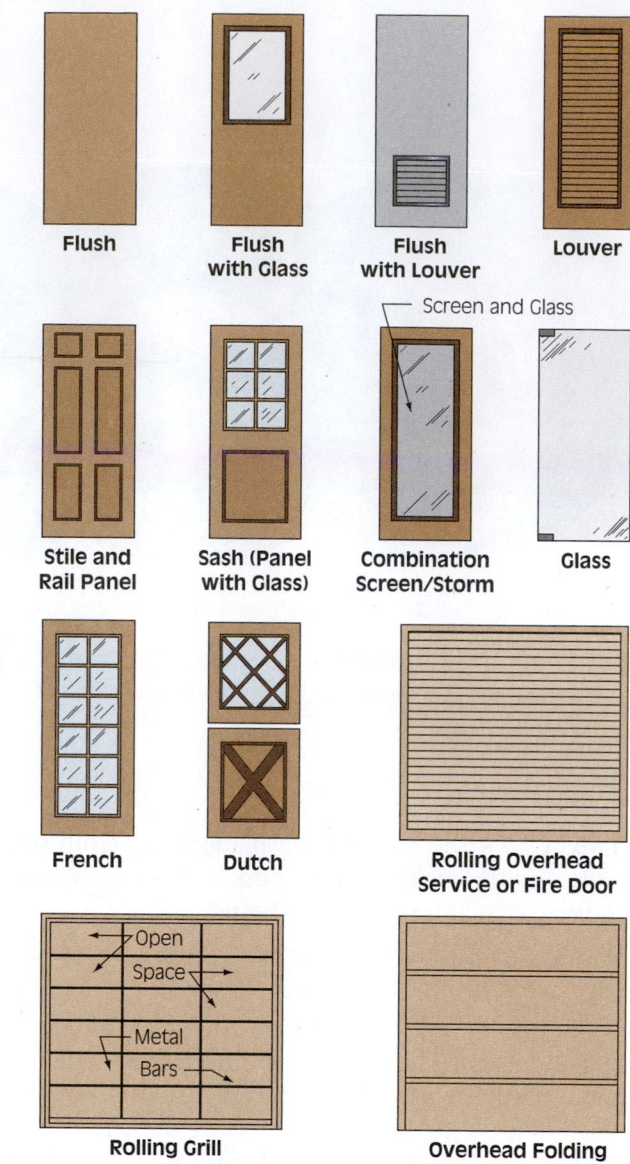

FIGURE 17-21 Doors are manufactured in a variety of styles and materials.

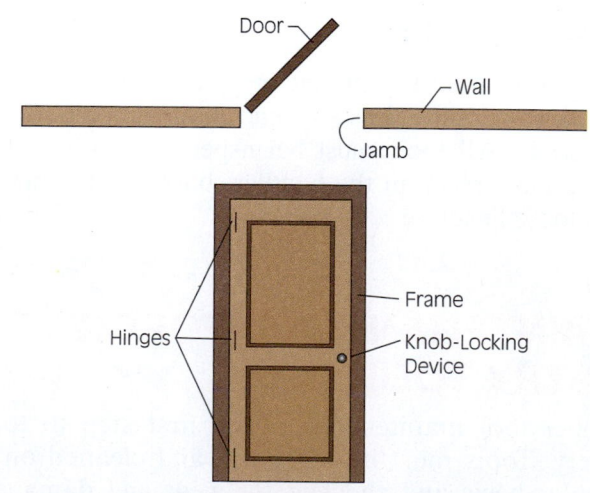

FIGURE 17-22 The parts of a door assembly.

Types of Doors

Wood Doors

There are three types of wood doors. **Panel doors** have a solid stile and rails with panels made of wood or glass or other materials. **Flush or slab doors** are flat or have a smooth surface and may be of either **hollow-core** or solid-core construction. **Solid-core doors** are more resistant to fire and more secure, usually making them more difficult to force. **Ledge doors** are built with solid material, usually individual boards, and are common in barns and warehouses. Wood doors are used primarily in residential construction and are installed on a wood frame or jamb secured to wall framing with nails or screws.

Metal Doors

Metal doors manufactured either as hollow-core or metal clad, **Figure 17-23,** are common in new construction. These are usually installed in metal frames and can be very secure. The metal clad door, which has a steel surface with a wood core, is manufactured in a number of designs and architectural finishes and some may have a fire-resistive rating. Generally metal doors are used in commercial construction and as exterior doors in residential construction. Metal doors are more substantial and are secured to the wall construction, **Figure 17-24.** Forcible entry may

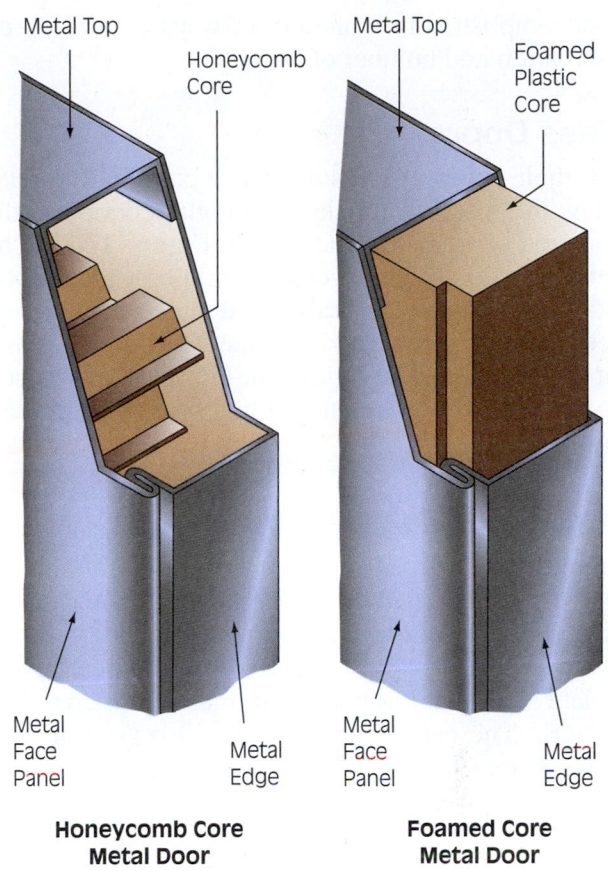

Honeycomb Core Metal Door

Foamed Core Metal Door

FIGURE 17-23 Metal doors are of both solid- and hollow-core design.

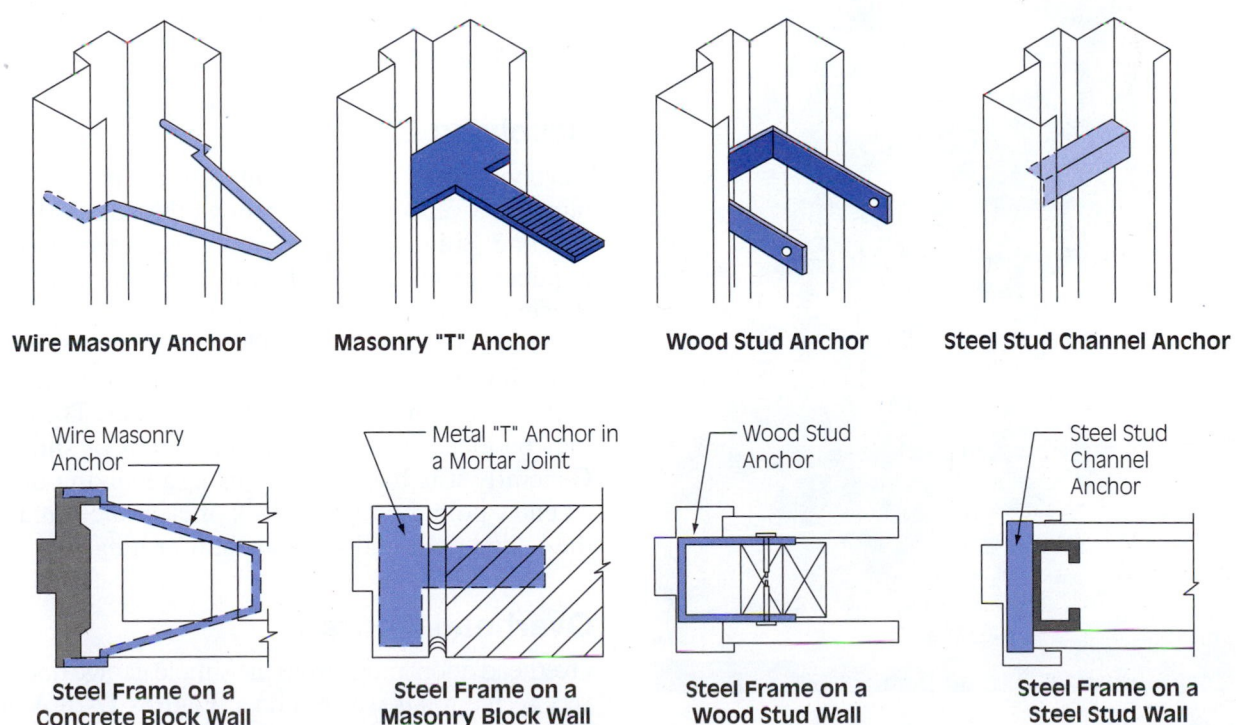

Wire Masonry Anchor

Masonry "T" Anchor

Wood Stud Anchor

Steel Stud Channel Anchor

Steel Frame on a Concrete Block Wall

Steel Frame on a Masonry Block Wall

Steel Frame on a Wood Stud Wall

Steel Frame on a Steel Stud Wall

FIGURE 17-24 Metal doors are secured to walls in a number of ways.

be accomplished in a number of ways depending on installation and number of locks.

Glass Doors

Two main types of glass doors are commonly in use today: the metal or tubular frame glass door and the tempered or frameless glass door, **Figure 17-25.** The metal frame is the most common type of glass door, and both styles are generally used in commercial and mercantile occupancies. Most have **tempered glass,** but a laminated safety glass and polycarbonate glazing (Lexan) are common and can be found in these doors.

Tempered Glass Doors

These doors are made of thick tempered glass, hung on a tubular metal frame, with hinges located at the top and bottom of the door and a locking device set in a metal rail at the bottom of the door. Tempered glass is plate glass that has been heat treated to increase its strength. The best method to force this type of door is the through-the-lock technique. If it is necessary

FIGURE 17-25 Typical commercial glass door. *(Courtesy of Ralph Restadius)*

to break tempered glass, firefighters should strike a sharp blow with a pointed tool such as the pike of a Halligan tool. Firefighters must be aware that the glass will shatter so they should stand to one side and strike the glass in a lower corner.

Door Swing Direction

Doors are hung in jambs with hinges, and forcible entry is accomplished by working with the direction of swing. The firefighter must determine which way the door swings, and the common way of describing this is in relationship to the forcible entry team. Doors with exposed or visible hinges will swing toward the forcible entry team, **Figure 17-26A.** If no hinges are visible and there is a stop on the door frame, the door will swing away from the team, **Figure 17-26B.**

Sliding Doors

Sliding doors are usually found in residential occupancies and consist of sliding and fixed sections of tempered glass in a wood or metal frame. The locking device is usually very light with a strike screwed into the jamb, **Figure 17-27.** These doors are forced by prying the door away from the jamb with a Halligan tool, breaking the lock striker out of the stationary frame. To prevent entry by intruders, many occupants will place a pipe, broomstick, or other solid object in the track to prevent the door from sliding. With these objects used as additional security devices, breaking the glass may be the only option for entry, but it should be used as a last resort.

Revolving Doors

Revolving doors are made up of four sections (doors) hung on a vertical shaft or hinge that allows the door to rotate. Building and life safety codes require revolving doors to collapse and allow occupants a rapid exit if needed. This is accomplished by releasing the arms that hold the door apart. The "panic-proof" type of revolving door will automatically collapse when two sections are moved in opposite directions. These doors may be locked with key cylinder locks or slide bolts. Generally it is best to avoid using revolving doors as an entry point because of lack of clearance created by the folded door and vertical shaft or hinge.

Overhead Doors

Overhead doors range from the simple garage door to the very secure roll-down steel door, **Figures 17-28 A** and **B.** These doors are built from wood, steel, aluminum, and fiberglass and may have a solid or insulated core.

(A)

(B)

FIGURE 17-26 Hinges indicate the direction of door swing and forcible entry. An outward-swinging door is shown in (A) and an inward-swinging door in (B).

Depending on the occupancy and security requirements, they may have windows of glass or some type of synthetic material.

Residential Garage Doors

Overhead garage doors used in residential construction are typically three- to five-section folding doors of wood or metal construction, Figure 17-28A. Older style doors may be one-piece slabs that tilt up into the garage. Overhead doors are installed on tracks with rollers and balance springs to assist in opening. A folding overhead garage door may be forced by any of several methods:

- Break a panel or window, reach in, and unlock the securing device.
- Pull the lock cylinder and utilize through-the-lock forcible entry.
- Automatic openers hold the door in the closed position. To disconnect the opener, break out a panel near the attachment mechanism, reach in with a tool to grab the release cord, and pull as shown in **Figure 17-29.**

> **CAUTION**
>
> At a recent fire, the initial attack lines advanced through the garage area of a house. The overhead door, which was equipped with an automatic operator, closed either because the balance springs were weakened by fire or the door operator shorted out. This trapped the attack team inside the garage with no means of escape. Once an overhead door is opened, a tool should be placed (a six-foot hook works well) in the track or a pair of locking pliers should be used on the track to prevent the door from closing.

Commercial Garage Doors

Commercial overhead doors are similar in operation and construction to residential doors, as previously shown in Figure 17-28B. The exception to this is the type of locking and security devices. These doors may be forced using the same methods mentioned for residential doors or by cutting the door with a rotary saw, **Figure 17-30.** After the door is opened, the same precautions used for residential overhead doors must be taken to ensure that it does not close, thus trapping firefighters.

FIGURE 17-27 The strike of a residential sliding glass door is secured with two small screws and can be easily forced.

(A)

(B)

FIGURE 17-28 Typical overhead doors: (A) residential and (B) commercial.

Roll-Down Steel Doors and Gates

Roll-down steel doors, **Figures 17-31A** and **B,** are of heavy steel construction to provide a higher level of security or as a rated door in a firewall or separation. Roll-down doors can be divided into two types—ectional rolling steel doors and continuous steel doors. Sectional rolling steel doors are made up of individual sections that are attached one to the next to make a door. Continuous steel doors are made of large sections of metal that are connected together to make one continuous piece. Roll-down doors generally use three different methods of opening and closing:

- *Manually operated:* The steel gate is lifted by hand with the assistance of springs.
- *Chain operated:* A chain hoist mechanism is used to lift the door.
- *Electrically powered:* An electric motor connected to a switch is used to raise and lower the door.

The method used to force these doors will depend on the type and number of locks and security devices. Generally the best way to accomplish forcible entry is to:

- Cut or force the locks.
- Attack the hardware.
- Cut through the gate in a manner similar to that shown in Figure 17-30.

There are different methods of cutting through these gftes. Some are very complex and do not work on different types of doors. An inverted "V" cut is simple, quick and works well on all roll-down doors.

Security gates used in commercial occupancies provide security while allowing for the display of merchandise. The most common means of securing these gates is to use padlocks. These padlocks can be cut or forced by the use of several types of tools; the

FIGURE 17-29 To disable a motorized residential overhead door opener, the firefighter can break out a pane of glass or door panel and pull the release rope.

FIGURE 17-30 On older commercial overhead doors, a large inverted V-shaped cut is made in the gate with a power saw. A slat on each side is pulled toward the center and removed. This allows all slats to be removed and a large opening created.

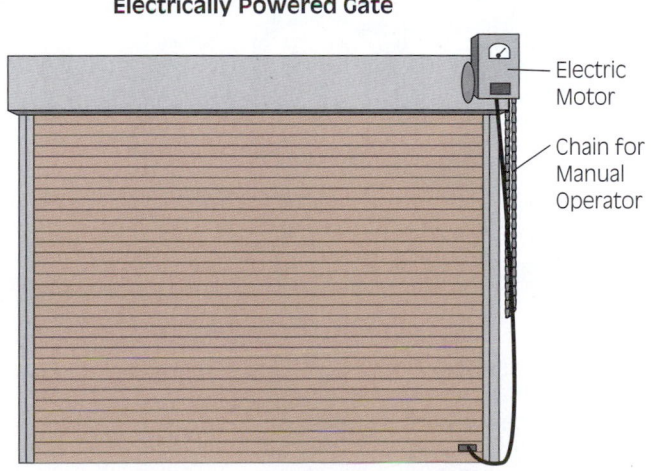

Electrically Powered Gate

Electric Motor

Chain for Manual Operator

(A)

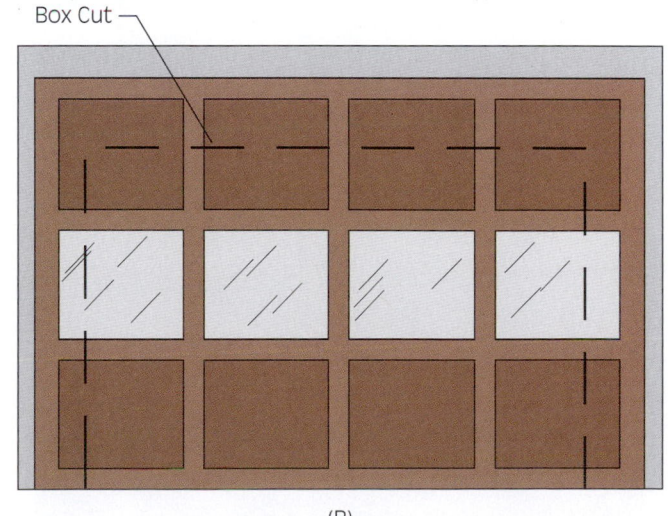

Box Cut

(B)

FIGURE 17-31 (A) Roll-down steel doors are of heavy steel construction to provide a higher level of security or as a rated door in a firewall. (B) Box cut used for this door.

most reliable method is the power saw with a metal cutting disc. Cutting or otherwise forcing the hardware that secures the gate is a method that can be used with some success. An example of this is cutting the hasp that secures the padlock to the gate.

Locks

Locks are designed and intended to keep unwanted or unwelcome visitors out of a building or occupancy. Firefighters—although not unwelcome visitors—will face the problem of locked door and entry access on a regular basis. As security requirements increase or break-ins occur, additional locks will be placed on a door as shown in **Figure 17-32.**

The most important part of forcible entry is to know the type of lock, how the lock operates, and how to disable it. Knowing this will save valuable time and energy on the fireground and provide quicker access, rescue, and fire extinguishment. The basic nomenclature of locks as shown in **Figure 17-33** will assist the firefighter in understanding the operation of most locks. Lock mechanisms used to provide security for doors and other openings fall into the following general categories.

FIGURE 17-32 To gain access through this door, the firefighter must know how to force these three different styles of locks.

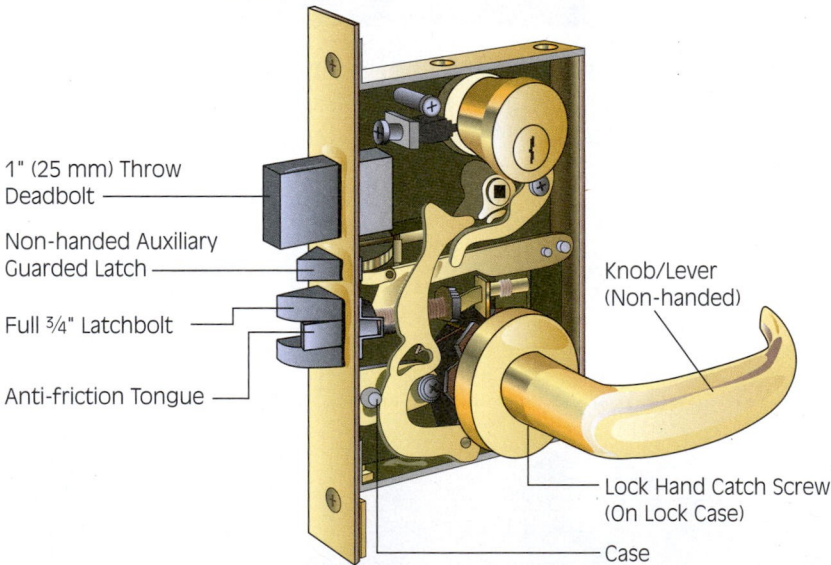

1" (25 mm) Throw Deadbolt

Non-handed Auxiliary Guarded Latch

Full ¾" Latchbolt

Anti-friction Tongue

Knob/Lever (Non-handed)

Lock Hand Catch Screw (On Lock Case)

Case

P

Cylinder Suffix
P = Standard Mortise Cylinder
C = Concealed Mortise Cylinder
R = Interchangeable Core Cylinder
L = Loss Cylinder
J = Interchangeable Core Cylinder Less Core
F = Interchangeable Core Cylinder Less Schlage Logo
W = Less Concealed Cylinder
T = Temporary Interchangeable Core Construction Core Cylinder

FIGURE 17-33 The parts of a lock device.

Key in the Knob Lock

This lock, **Figure 17-34,** is most common in residential occupancies and on interior doors in commercial occupancies. The outside of the lock will have a keyway for operation and the inside will usually have a keyway or button. The bolt on this type of lock is either a latch or dead latch bolt type. The latch bolt has a throw of approximately ½ inch, making this an easy door to force.

Mortise Lock

The mortise lock, **Figure 17-35A,** is designed to fit into a cavity in the edge of a door and is usually found in commercial occupancies. The mortise lock is designed with three types of operating latches: the dead bolt, the dead bolt and latch, and the pivoting dead bolt. The pivoting dead bolt may also be known as an Adams Rite lock, which is used in narrow frame/stile aluminum doors. Most mortise locks operate with a key that, when inserted in the keyway and turned, turns a cam, which moves the dead bolt or latch and opens the lock.

- *Dead bolt.* This lock has one sliding bolt and is locked or unlocked by one complete turn of the key (**Figure 17-35B**) in the cylinder. This will only operate the dead bolt; additional latches or locks are operated separately.
- *Dead bolt and latch.* (Figure 17-35B) Similar to the dead bolt, but with an additional latch operated by a doorknob.
- *Pivoting dead bolt.* This lock, **Figure 17-35C,** is used on metal and glass doors and has a bolt that is housed vertically when retracted and pivots up to the horizontal when placed to lock. The **bolt throw** projects approximately 1½ inches into the **strike plate** when locked.

FIGURE 17-34 A key in the knob-type lock usually found in residential occupancies. *(Courtesy of Master Lock Company, LLC)*

(A) (B) (C)

FIGURE 17-35 A typical mortise style lock: (A) the exterior, or normal view, (B) the locking mechanism, and (C) the pivoting dead bolt (Adams Rite lock).

Rim Locks

Rim locks, also known as surface locks, **Figure 17-36,** attach on the inside of a door with the cylinder extending through the door and with a keyway visible on the outside. There are many variations of this lock but they are all mounted on the surface of the door.

Tubular Locks

Tubular locks are mounted in a hole that has been bored into the door and are best described as a combination of the key in the knob lock and the mortise lock. However, instead of a knob, the tubular lock uses a cylinder to operate the bolt and is recognized by the cylinder's cover which protrudes about 1½–2 inches on each side of the door. A typical tubular lock is shown in **Figure 17-37.**

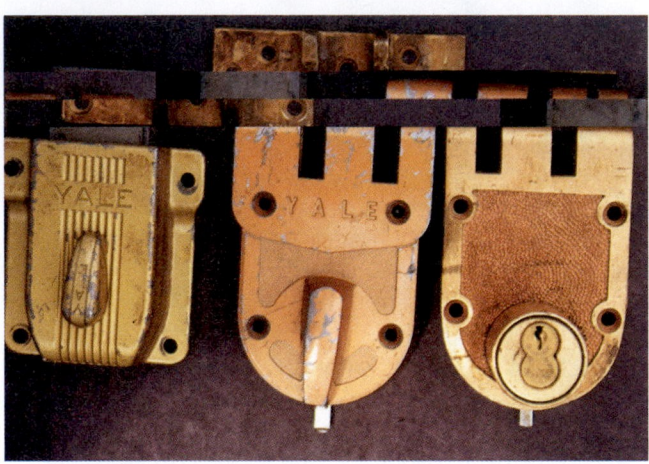

FIGURE 17-36 Rim locks (from left to right): a dead bolt, a vertical bolt and striker plate, and a vertical bolt with key cylinder.

FIGURE 17-37 Tubular dead bolts. *(Courtesy of Master Lock Company, LLC.)*

Padlocks

Padlocks are portable or detachable locking devices that are manufactured for regular and heavy-duty service. This type of locking device has a movable shackle that locks into the body of the lock and is used to secure a door or gate using a hasp or chain. Padlocks come in a wide variety of shapes and sizes, **Figure 17-38.**

Regular padlocks have a shackle of less than 1¼ inch in diameter and are not usually made of hardened steel. These may be cut with a bolt cutter or broken with a lock breaker. Heavy-duty padlocks have shackles larger than ¼ inch in diameter and are made of hardened steel, **Figure 17-39.** In addition, the shackles on both the heel and toe are locked when the shackle is depressed into the lock case, and both sides of the shackle must be cut to open the lock.

Special Locks

Included here are all devices that do not fit into other categories. Examples of these devices are overhead door locks, magnetic locks, or card key entry systems, **Figures 17-40 A** and **B.** Overhead door locks are similar to rim locks and may be forced by gaining access through the lock. Card key entry systems are regular locking devices operated by an electric actuator or solenoid. Again, conventional or through-the-lock forcible entry is the key to access.

> **STREETSMART TIP**
>
> Many office occupancies and hotels/motels are using electronic locks for security. The fire department should contact the facility management people to arrange a procedure for obtaining a master card key if available. This reduces the amount of work for the forcible entry crew and could reduce property damage.

Magnetic locks use a magnetic force to hold a door secure and the system is usually disabled from the interior of the building. In most instances it is quicker and easier to find another means of entry.

Additional Security Devices

As security requirements increase, home and business owners have begun to install many varieties of locks and security devices. These may be as simple as a broom handle in the track of a sliding door or

(A)

(B)

(C)

(D)

FIGURE 17-38 Padlocks are available in many shapes and sizes. *(Photos courtesy of Master Lock Company, LLC)*

as complicated as a number of additional locks on a door. These additional devices may not provide visible signs or indication of their use to the forcible entry team, **Figures 17-41 A–D.** If these types of devices are in use, the forcible entry team may need to find an alternate means of entry or use a rotary saw to gain access.

When buildings are being remodeled or demolished, the fire department should check with the owner or contractor to gain permission to salvage unwanted door locks or security devices. These make excellent training aids, especially if similar locks are in use in other buildings in the jurisdiction. In addition, new construction should be surveyed to determine the types of locking devices being installed.

FIGURE 17-39 Cutting a padlock with a power saw. Note that both shackles are being cut at once.

(A)

(B)

FIGURE 17-40 (A) Residential overhead door locks are a type of rim lock and are forced easily using the through-the-lock method. (B) Card key entry systems are standard locks with electric actuators.

METHODS OF FORCIBLE ENTRY

For fire department operation there are three standard methods of forcible entry:

1. Conventional
2. Through-the-lock
3. Power tools

These methods involve the use of certain tools and the application of many techniques.

> **CAUTION**
>
> All forcible entry operations must be coordinated with fire attack and ventilation. Lack of coordination may result in rapid fire spread or a backdraft.

Conventional Forcible Entry

This is an old and reliable method involving the use of leverage, force, and impact. The primary tools used in this method are the irons, consisting of the Halligan tool and the flathead ax. This technique requires procedures that will accomplish one or more of the following:

- Force the door away from the jamb, pulling the bolt away and free from the strike plate.
- Break the lock or striker.
- Break the door and/or the frame.
- Force or remove the hinges.
- Breach the wall or door.

> **STREETSMART TIP**
>
> Always size up the forcible entry task. Regardless of the type of lock, at a quick glance, the fastest way to force entry to the door shown in **Figure 17-42** is to break the glass and unlock the door from the inside. But be aware that there may be additional locking devices out of view that may not be easily unlocked. At a fire situation this could magnify the problems associated with heat, smoke, and fire possibly rolling out through the broken glass. On the contrary, once the door in **Figure 17-43** is forced, the firefighter may encounter the household goods stored in front of the door.

In general, conventional forcible entry is reliable and is the primary method at structural fires. Conventional forcible entry may also involve the use of the hydraulic forcible entry tools.

(A)

(B)

(C)

(D)

FIGURE 17-41 These auxiliary locking devices are usually not detected by the forcible entry team: (A) floor-mounted stop plate, (B) door blocker, (C) steel bar and brackets, and (D) sliding bolt. Note that the bar and sliding bolt are often homemade devices.

JPR 17-1: Door Swinging Away from the Forcible Entry Team

(For step-by-step photos of this skill sequence, see page 548)

Remember to always try the door to determine if it is locked or secured and forcible entry is actually necessary.

1. Size up the door to determine swing; the number, type, and location of locking devices; and the type of door and frame. With this type of door swing, the hinge pins are not visible. Also check to make sure the door is locked/secured.

2. Gap the door by slipping the adz between the door and the door stop 6 inches (15.2 centimeters) above

FIGURE 17-42 Firefighters cannot assume that the quickest way to force entry to this type of door is to break the glass and unlock the door from the inside. Additional locking devices may be out of view.

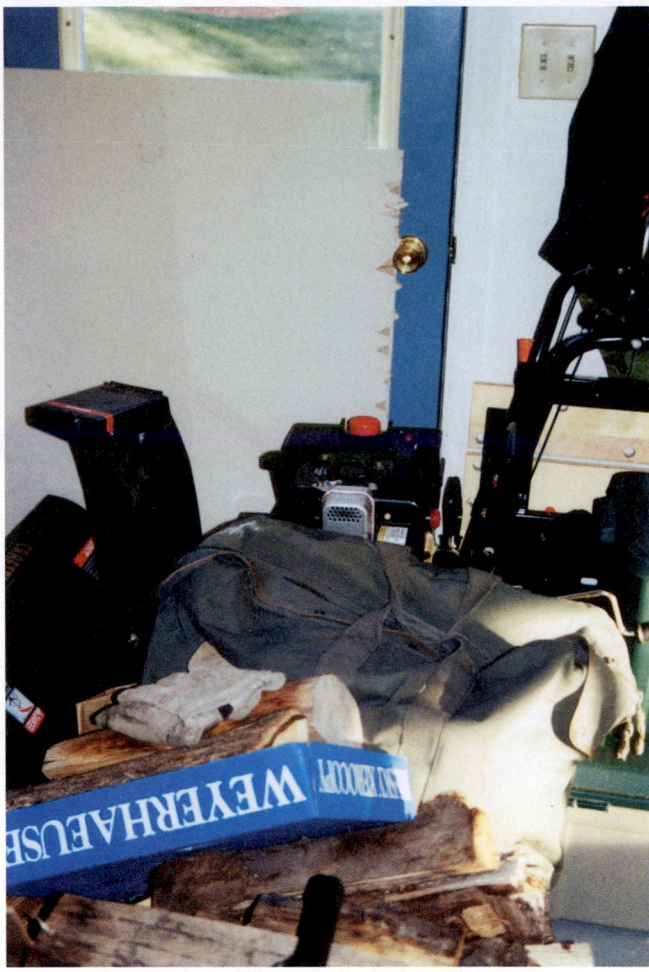

FIGURE 17-43 Once a door is forced, the firefighter may encounter many obstacles.

or below center of the lock and push up or down to spread the door from the frame. This will allow for placement of the fork end of the Halligan tool. Note that on some doors this may cause the door to open. On door frames with a nailed doorstop, this method will break the stop and allow easy entry of the fork.

3. Set the fork of the Halligan tool into the gapped area with the bevel of the fork toward the door no more than 6 inches above or below the center lock. As the Halligan tool is struck with the flathead ax or maul, gradually bring the tool away from and at an approximately 90-degree angle to the door.

4. Set the tool using the flathead ax or maul, driving the Halligan tool in so that the tips are around the door and locked in. If difficulty is experienced, the fork may be inserted with the bevel toward the frame.

5. Force the door. With the tool set, the door is spread from the frame. The Halligan tool is pushed sharply in the direction of the door swing and opens as shown in. This procedure is repeated as necessary for each lock.

On doors that swing away from you, try to control the door as it is forced. Generally, the door is

controlled by the firefighter with the Halligan tool. As the door swings in, the firefighter must reach in with the tool and bring the door back under control.

CAUTION

Doors leading into the fire area should be blocked or chocked open when firefighters are conducting a search or attacking the fire in that area in addition to other activities. This will prevent the door from closing behind crews and will allow rapid egress if needed.

JPR 17-2: Door Swinging Toward the Forcible Entry Team

(For step-by-step photos of this skill sequence, see page 550)

1. Size up the door to determine swing; the number, type, and location of locking devices; and type of door frame and hinges. Also check to make sure the door is locked/secured.

2. Gap the door by driving the adz or fork end of the Halligan tool between the door and frame 6 inches above or below the lock. Work the tool in until it contacts the doorstop. The adz end is preferred, but on some doors, especially steel, it will have a tight fit between the door and jamb and the gap may be started by using the fork end. If door is tight, work tool up and down to open up gap.

3. Set the tool by driving the adz or fork end past the door around the doorstop. Drive in the tool until it is "locked in" around the inside of the door. Due to the configuration of the door and wall, using the adz end will allow the tool to clear the wall.

4. Force the door by pulling the tool away from the door. This will cause the bolt to pull out of the strike, the lock to break, or the door frame assembly to fail and the door will open. When dealing with several locks, this procedure may have to be repeated for each lock.

Forcing Doors with Hydraulic Forcible Entry Tools

The hydraulic forcible entry tool (HFT) is primarily used on doors that swing away and have strong metal frames.

1. Place the spreader jaw of the tool between the door and jamb as close as possible to the lock.

2. Set the jaw into the gap to the doorstop using a tool if necessary. On tight-fitting doors, a driving tool may be needed to set the jaw into the doorstop.

3. Pump the tool until the bolt is pulled out of the strike and push the door open.

> **STREETSMART TIP**
>
> If there is more than one lock present and they are relatively close together, place the jaw of the tool between the locks.

Through-the-Lock Forcible Entry

The **through-the-lock method** of forcible entry involves attacking the locking mechanism by removing the key cylinder and then operating the lock with alternative means. This method of forcible entry is best applied to mortise, rim, or tubular-type lock cylinders.

In general, the choice of through-the-lock forcible entry is made when entry needs to be gained with damage kept to a minimum. Sometimes it is also the quick-est means of entry and is the best method when forcing entry of metal framed glass and all glass doors.

JPR 17-3: Unscrewing or Wrenching the Locking Cylinder

(For step-by-step photos of this skill sequence, see page 552)

This method is not as quick as pulling the cylinder with a lock puller and should be used only if time allows. In addition, trim work or walls may not allow the free rotation of the pliers.

1. Size up the lock to determine the type of lock and the feasibility of utilizing this method. Lock cylinders with protective collars may not be able to be unscrewed.

2. Using locking-type pliers, lock the pliers onto the cylinder.

3. Turn the lock cylinder counterclockwise to unscrew the cylinder.

4. Remove the cylinder and insert the proper end of the key tool to operate the locking mechanism as shown later in this chapter.

> **STREETSMART TIP**
>
> Through-the-Lock Technique Modified with Rim Locks: If the cylinder crumbles and will not pull out or the lock will not unlock, the firefighter can drive the lock off the back of the door with the pike of the Halligan tool or the handle of the lock puller.

JPR 17-4: Through-the-Lock Entry Using the K-Tool

(For step-by-step photos of this skill sequence, see page 553)

1. Size up the lock to determine the type of lock and the feasibility of utilizing this method.

2. Force the blades of the K-tool over the cylinder and decorative ring. Tap the K-tool into place until firmly set.

3. Place the adz of the Halligan tool into the loop of the K-tool. If necessary, the Halligan can be tapped to firmly set the K-tool.

4. Pull up on the Halligan tool and pull the cylinder.

> **STREETSMART TIP**
>
> The K-tool is designed to be able to pull cylinders that are very close to the bottom or edge of the door. This is done by placing the narrow, straight blade of the K-tool toward the bottom of the door to allow room for the tool.

A number of different lock-pulling tools may be used in place of the K-tool. The A-tool, officer's, and REX tool, **Figure 17-44,** are designed to be driven behind the lock cylinder to get a "bite" or "purchase" on a substantial part of the cylinder. These tools are most effective on cylinders that are recessed into a door and on odd-shaped cylinders such as tubular locks.

Operating Lock Mechanisms

The final step using through-the-lock forcible entry is to manipulate the lock by using the proper key tool. Many types of key tools are used, but the most common is the two-sided, flat steel key or square key tool, **Figure 17-45.** This tool has a bent end used for mortise locks and a straight end used for rim and tubular locks.

The correct key tool to operate the locking mechanism is determined by examining the cylinder after it is removed. Lock cylinders fall into two different categories, the mortise having a cam device on the back of the cylinder and the rim lock with a flat or square blade on the back of the cylinder, **Figure 17-46.** The tubular lock has a tailpiece similar to the rim lock.

To open a mortise lock, visualize that the key hole where the cylinder was prior to pulling it out is that of the face of a clock. The keyway of the cylinder prior to removal was at the six o'clock position and the unlocking action will occur within the five to seven or seven to five o'clock area, **Figure 17-47.** Place the bent end of the key tool into the keyway and move it to the five or seven o'clock position and engage the locking mechanism. Move it to the opposite side to retract the dead bolt. If the lock mechanism has a spring lock, push down with the key tool to depress it and then move it. To unlock the spring latch, move the key tool up to the three o'clock or nine o'clock position and engage the latch mechanism, push, and hold it in place. This will pull back the spring latch and allow the door to open.

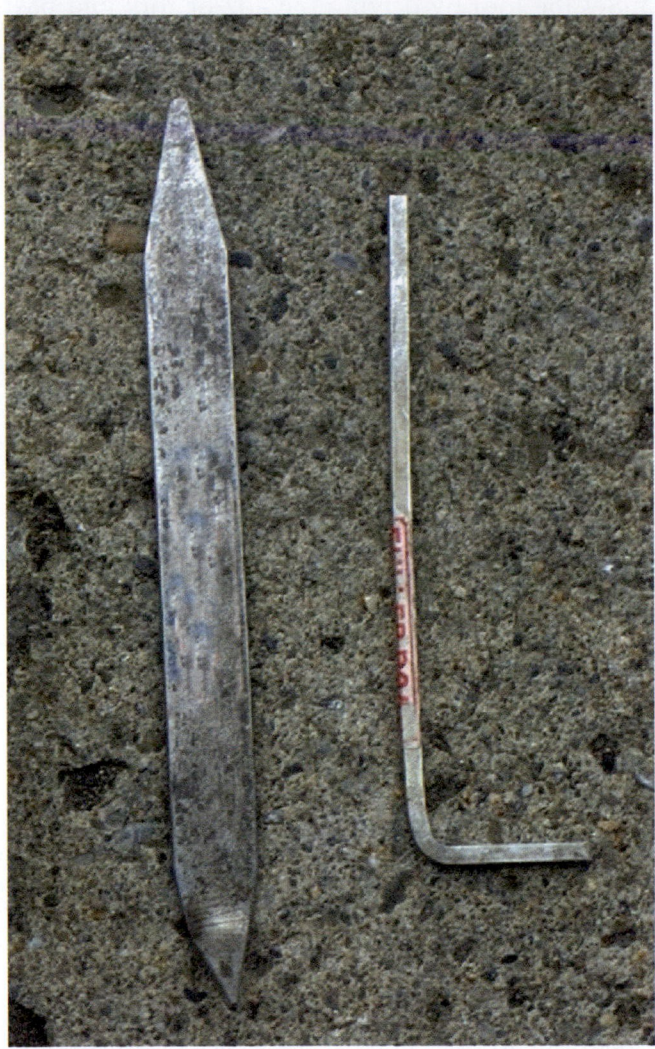

FIGURE 17-45 The two types of key tools used for through-the-lock forcible entry.

FIGURE 17-44 The REX tool is used to pull lock cylinders in the through-the-lock method of forcible entry.

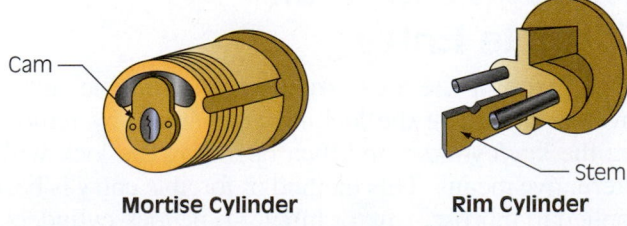

Mortise Cylinder   Rim Cylinder

FIGURE 17-46 To operate the lock, look at the back of the cylinder to determine the right key tool to use.

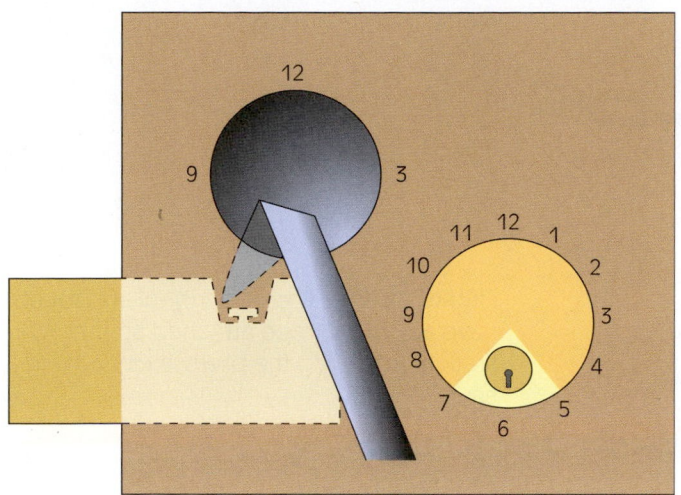

FIGURE 17-47 The unlocking action will occur within the five to seven or seven to five o'clock area.

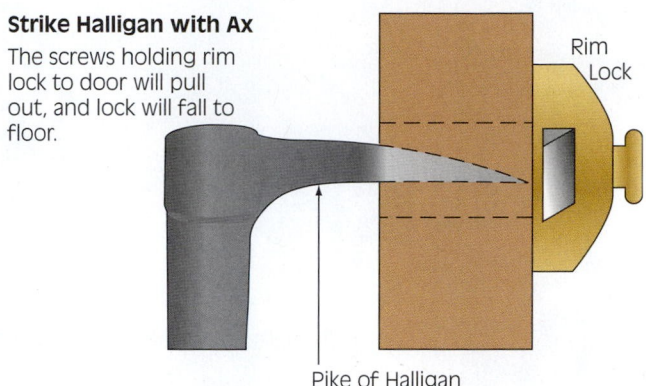

Strike Halligan with Ax
The screws holding rim lock to door will pull out, and lock will fall to floor.

Pike of Halligan

FIGURE 17-48 Strike the Halligan tool with the ax and drive the lock off the door. This method is used for rim locks only.

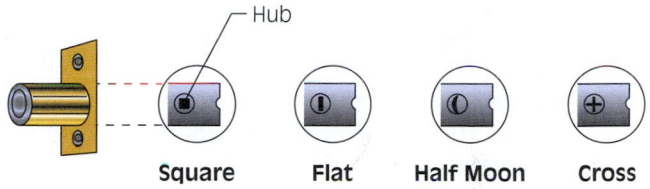

FIGURE 17-49 The bolt of a tubular lock is operated with a key tool matching one of the shapes shown.

If, after removing the cylinder, a tailpiece is visible, then the lock is a rim lock. To open, insert the straight blade of the key tool into the slot on the rim lock and rotate the tool to unlock it.

Lock Variations

Shutter Guard

Many locks have a spring-loaded shutter guard that will close when the cylinder is pulled. This will not allow a key tool to open the lock. A shaped tool such as a pick is used to slide the shutter open. Slide the shutter horizontally toward the edge of the door and hold it open until the key tool is in place. Then turn the key tool to open.

Night Latch

A rim lock with the night latch engaged can only be unlocked from the inside.

Through-the-Lock Technique Modified

When the mechanism of a rim lock cannot be operated because of the presence of a night latch or shutter guard, or in any situation where the key tool cannot open the lock, firefighters should use the following procedure. Place the pike of the Halligan tool into the hole and against the back of the rim lock. Strike the Halligan tool with an ax and drive the rim lock off the door, **Figure 17-48.**

Tubular Locks

Locks such as the key in the knob and tubular dead bolts are opened in a manner similar to the rim lock. The knob is removed with a lock puller or knocked off with a heavy tool such as the ax or Halligan tool.

This will expose the latch mechanism, which can then be operated by the key tool or screwdriver.

Removing the cylinder with a lock puller exposes the tubular dead bolt lock. The adz of the Halligan tool can be used to shear the screws, and the cylinder can sometimes be removed in this manner. The bolt mechanism is then operated with a key tool or screwdriver, depending on the shape of the hub, **Figure 17-49.**

Forcing a Padlock with a Lock Breaker

1. Place the wedge end of the lock breaker through the shackle. Depending on the size of the lock a Halligan tool may also be used for this task.
2. Strike the back or head of the lock breaker with a maul or heavy striking tool.
3. Continue step 2 until both shackles break.

Forcing a Padlock by Cutting with a Rotary Saw

1. Attach locking-type pliers to the lock case and lock the jaws. The locking pliers must have a chain or rope attached so the firefighter can hold the pliers clear of the saw.
2. A rotary saw with a metal cutting blade is used to cut the lock shackles, **Figure 17-50.**

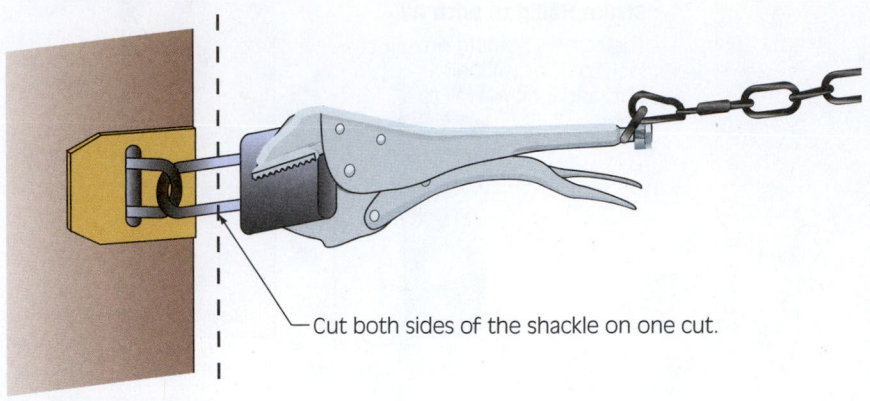

Cut both sides of the shackle on one cut.

FIGURE 17-50 The locking pliers must have a chain or rope attached so that the firefighter holding the pliers is clear of the saw.

WINDOWS

Windows are found in many types, architectural styles, sizes, and construction as shown in **Figure 17-51.** These are installed in buildings to provide light and natural ventilation. Usually firefighters will force entry into a structure using door openings; however, windows may be used as an alternate means of entry especially for rapid intervention team operations. To successfully conduct forcible entry operations through windows, firefighters must know about the four construction features of windows: glazing, sash, frame, and security devices.

Forcible Entry of Windows

There are many types of windows and each individual one requires a specialized technique to force entry through it. There are two general reasons to force a window: to gain entry and for horizontal ventilation.

STREETSMART TIP

When forcing windows, an opening must be created to allow for safe entry and exit with full protective equipment in place. It is important to remove all obstructions such as air-conditioning units, security bars, or child protective gates. The opening created may be needed to make a rapid exit if conditions deteriorate.

To make a large enough opening for entry, it is often necessary to break the glazing and the sash to create the largest possible opening. A common adage heard from experienced firefighters is "make the window a door." At structure fires, the speed of operations is more important than damage done to windows. With this goal in mind, the quickest method is to break the glass and the sash to provide a quick opening for entry, rescue, and ventilation. At the same time, firefighters must remember not to break glass unnecessarily and create a safety hazard.

Glazing

The glass or other clear material portion of the window that allows light to enter is the **glazing.** The most common glazing material is glass. There are different types of glass, of which these are just a few:

- Regular or plate glass
- Tempered glass
- Laminated (safety) glass
- Wire glass

Regular/Plate Glass

Regular glass is relatively weak and easy to break, and when struck with a tool, it breaks into very sharp, knife-like shards. Plate glass is used in larger windows and is generally thicker than regular glass, with thicknesses of ¼ to 3¾ inch being the most common. This type of glass will break into large, heavy, and sharp pieces, which can be very dangerous. A long-handled tool such as a hook or pike pole works best, **Figure 17-52.** With both types of this glass, the firefighter should stand to the side and strike the window. All shards of glass that remain must be cleaned out of the window opening.

CAUTION

Firefighters should always use a tool such as a hook (pike pole) or other long-handled tool to clean all glass out of a window. Care must be taken that glass shards do not slide down the tool's handle.

Double Hung Window has two movable vertical sashes.

Single Hung Window has single movable vertical sash.

Jalousie Windows have horizontal glass slats that pivot in unison with a single control.

Casement Windows are hinged on the side, operate individually, and swing outward.

Vertical Pivoting Windows swing on center pivots.

Projected Windows may project in or out.

A Top-Hinged in-swinging sash

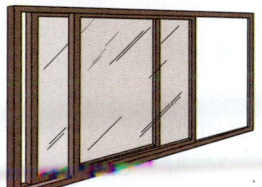

Horizontal Sliding Windows can have one or more moving sashes.

Specialty Windows

Circle Top

Awning Windows are top hinged and open together with a single control.

Quarter Circle

Bull's-eye

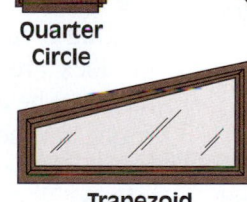

Trapezoid

FIGURE 17-51 Windows are found in many types, architectural styles, sizes, and construction.

Tempered Glass

This material is glass that has been heat treated to give it additional strength. The technique to use when breaking tempered glass is to strike the glass with a pointed tool, such as the pike of the Halligan tool or a pick ax. Best results are obtained when the glass is struck near a corner where the glass is more rigid than at the center of the pane.

STREETSMART TIP

A unique property of tempered glass is that when it is broken, the whole pane will fracture and fragment into small pieces. Certain panes, even those that have been fractured, may remain in place. These will require the use of a tool to clear out the remaining glass.

Laminated Glass

Laminated glass is also known as safety glass. It is commonly composed of two or more sheets of glass with a plastic sheet between them. The purpose of the plastic sheet is to hold the glass together if broken, thus reducing the hazard of flying glass. Laminated glass is most commonly found in automobile windshields; however, it is also used for glazing in windows and doors.

Wire Glass

Wire glass has a wire mesh embedded between two or more layers of glass, **Figure 17-53.** The primary purpose is to give the glass increased fire resistance. When exposed to high heat, the glass will break. The

FIGURE 17-52   Use a long-handled tool such as a hook or pike pole to break glass out of a window.

FIGURE 17-53   Wire glass has a wire mesh embedded between two or more layers of glass.

wire will then hold the glass together and allow the glazing to remain intact to stop the horizontal spread of the products of combustion.

To break the wire glass, the firefighter strikes with a tool, an ax or Halligan, and then cleans the glass out of the frame. This is best accomplished by carefully chopping through the wire with the blade of an ax. The firefighter must use caution because the entire piece of wire glass will fall as one complete unit when cut free.

Polycarbonate Glazing

These are plastic products produced to provide additional strength as compared with regular glass. One common type is Lexan. Lexan is approximately 250 times stronger than regular glass and is available in thicknesses up to 4 inches. Polycarbonate glazing can be removed by cutting it with a carbide-tipped circular saw. It can also be removed by striking it near a corner and driving it out. A heavy striking tool is best, and progressively driving it in may work it free from the sash.

Types of Windows

Double Hung/Check Rail Windows

These windows have upper and lower sashes that both slide vertically. The sash may be composed of wood or metal and will have a locking device in place to lock the sashes to one another or to the frame. To force open this type of window, the firefighter should use a tool to force the bottom sash up and break the locking mechanism. Wood sash windows with a single lock are easier to force than those composed of metal sashes/frames or other heavily secured window frames. When this type of window is encountered, breaking the glass and sash may be the most prudent option.

Energy Efficient Windows

Most new windows and replacement windows are equipped with double and triple layers of glass. An inert gas is placed between the panes of glass and the windows are very well sealed. These energy-efficient windows tend not to fail in a fire as single-pane windows can. This in turn will keep products of combustion trapped inside the structure longer. From the exterior point of view, they will give little indication of the severe conditions inside. Extra care must be taken when using these windows for entry. Because of their construction, energy-efficient windows are usually more difficult to break than a single-pane window. With the individual layers of glass and inert gas sandwiched together, they form a single, stronger unit.

Casement Windows

Casement windows are hinged on the side of the sash and generally open outward. They commonly have a metal or wood sash and can be secured by one or more latches and a crank for opening and closing. The way to open this type of window is to break the glass with a tool, reach inside, and operate the crank. The screen may have to be cut out of the way to reach inside.

STREETSMART TIP

The firefighter needs to size up the window and decide if it is wide enough to enter safely and exit. Often the window sash and center mullion must be removed to provide sufficient room to enter.

Awning Windows

Awning windows are hinged so that they may swing out and may have a crank-operated mechanism to move the sash. They may have a wood or metal sash. The procedure used to open this type of window is similar to that of a casement window. The window must be broken to operate the mechanism. Many awning windows are too small to enter and the sash will have to be removed or broken out to allow for entry.

Jalousie Windows

Similar to the awning window, the jalousie window has small sections of glass that are operated with a crank mechanism and overlap when closed.

Projected Windows

This type of window will pivot at the top, bottom, or center and is most commonly found in commercial buildings. Pivoting windows rotate on the top or at the sides.

Fixed Windows

Fixed windows are nonoperable windows used primarily for aesthetics and the introduction of light into a structure. They can be found in inaccessible areas, such as the top of a wall near the roof of a commercial building.

CAUTION

Firefighters must be in full protective equipment before breaking any windows. In addition to the hazard of sharp glass and other materials, the firefighter may have to contend with the smoke, heat, and fire that can vent from a broken window.

Bars and Gates

Windows and other openings that require security measures are often fitted with gates or bars, presenting a unique forcible entry situation for firefighters. These bars and gates must be removed or forced out of the way to allow for entry and exit into the structure. The following procedures may be necessary to accomplish this task:

- *Force the locking devices.* Gates that have exposed locks can be opened by forcing or cutting the lock.
- *Attack the fastenings.* Gates and bars can be removed by breaking or cutting through the bolts or attacking the point where the bars or gates are set.
- *Cutting the gates/bars.* The firefighter can use the rotary saw with metal cutting disc (aluminum oxide) or the sawsall to cut the gate or bars.

BREACHING WALLS AND FLOORS

Emergency situations often dictate that the walls of a structure must be opened to allow entry or to remove trapped firefighters or victims. This is especially true in the event of a collapse or blocked exit. There are two main considerations when breaching walls:

- *The type of construction of the building.* Wood-framed buildings with lath and plaster or drywall are usually easy to breach. Solid brick and reinforced concrete buildings are more formidable and will be more difficult to get through and hence require more effort.
- *Tools available.* During the initial stages of an operation, regular hand tools or possibly a power saw may be the only tools available. These tools may be adequate for numerous operations, but solid masonry or reinforced concrete construction will require specialized tools and equipment.

Techniques for Breaching Walls
Breaching Wood-Framed Walls

1. Size up the wall, trying to avoid the area around doors and corners due to narrow stud spaces. It may be beneficial to create an inspection hole using an ax handle or Halligan bar, **Figure 17-54,** to verify the size-up, and check for obstructions, barriers, and

FIGURE 17-54 The firefighter uses an ax handle or Halligan tool to poke through a wall to determine if there are any obstructions on the other side.

fire conditions. The outcome of the inspection will determine whether to move to another location.

2. Remove the wall covering from your side first to prevent the extension of heat and smoke into the area.

3. If plumbing pipes are encountered, try to bend or break the pipes (plastic and cast iron will shatter, copper will bend).

4. When a large enough hole is made, push in the wall cover on the other side to complete the hole. If necessary, a stud can be removed by attacking the connection at the sill or plate. If the wall is a bearing wall, do not remove more than one stud. If a larger opening is required, shoring or some other supporting techniques will be necessary for long-term operations.

Breaching Masonry Walls— Block or Brick

To breach these walls without utilizing power tools, the only option may be the 12-pound maul, the Halligan tool, the flathead ax, or a battering ram.

1. Start by removing a single unit of block or brick. Work at the mortar joints because this is usually the weak point.

2. Once the joint is weakened, use the largest striking tool available and break the masonry unit.

3. Proceed by knocking out the surrounding units or release them at the mortar joint.

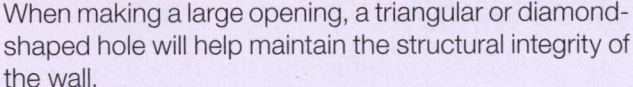

STREETSMART TIP

When making a large opening, a triangular or diamond-shaped hole will help maintain the structural integrity of the wall.

When opening masonry walls with power tools such as a power saw with a masonry disc, cut a triangle by making two angled cuts followed by a cut parallel to the mortar line. It may be quicker to use the maul on the mortar line to break the bricks free.

Breaching Reinforced Concrete Walls and Floors

Solid concrete reinforced walls with steel reinforcing rods are extremely difficult to open up even with the proper tools. The most common tools utilized are the jackhammer, the rotary power saw with masonry blade, and the diamond-tipped chain saw. An oxyacetylene torch may be needed to cut through the reinforcing rods.

CAUTION

Cutting pre- or posttensioned concrete may cause the structural member to lose its strength and collapse. Wall or floor breaching operations in buildings with this type of structural members should be managed by specialized rescue teams with structural engineering assistance.

Breaching Metal Walls

Before any cutting operations are started, the location of heavy structural members, such as columns, should be determined. The tool of choice for this operation will be the rotary saw with a metal cutting disc. The quickest cut is a triangle cut large enough to allow for safe entry, similar to the procedure for overhead doors shown earlier in Figure 17-30. Also, depending on the type of metal siding, it may be cut by making two vertical cuts to loosen the wall material. The metal siding can then be pushed in or pulled out depending on the situation. If necessary, the first two cuts can be joined to remove the entire piece. When light-gauge metal is encountered, the wall can be opened up with hand tools (i.e., the blade of the ax).

Techniques for Breaching Floors

Cutting Wood Floors with a Power Saw

The rotary saw with a 12-inch-diameter (30.4 centimeters) carbide-tipped blade will cut a maximum depth of 4 inches (10.1 centimeters), **Figure 17-55**. This should be sufficient to cut through most floors in one cut. The hole must be of sufficient size and proper shape. A rectangle, square, or triangle can be cut followed by the removal of the finished flooring and the subflooring, **Figure 17-56**. Carpeting and ceramic tiling should be removed before using the power saw.

Cutting Wood Floors with an Ax

When breaching a wood floor using an ax, locate the floor joists and make cuts close to them. Cut along the (parallel) joist and cut the finish floor on a bias (angle). Pull up the finish flooring to expose the sub-

FIGURE 17-55 The rotary saw with a 12-inch-diameter carbide-tipped blade will cut a maximum depth of 4 inches.

FIGURE 17-56 A rectangle, square, or triangle can be cut, followed by the removal of the finished flooring and the subflooring.

flooring. Cut the subfloor and pull up to expose the area below. Make all cuts on the subfloor first, then pull it up to confine the heat and smoke. Push down any ceiling or other obstructions.

Note these important things to consider when breaching floors:

- Always take precautions to avoid cutting wires, pipes, and conduit when performing breaching operations.
- Maintain the structural stability of the building when opening walls and floors.
- Beware of sparks produced by metal cutting tools.
- Always operate with proper protective equipment including eye, ear, and respiratory protection.

TOOL ASSIGNMENTS

The necessary tools to accomplish the tactics of structural firefighting must be carried in with the first on-scene and then later arriving units. The timely arrival of primary tools is as important as the placement and operation of the first hoseline and cannot be left to chance.

Tool assignments are based on the occupancy and construction of the building (i.e., multiple dwelling, private dwelling, commercial, etc.), position or task assigned, and department standard operating procedures or policies.

JOB PERFORMANCE REQUIREMENT 17-1

Conventional Forcible Entry—Door Swinging Away from the Forcible Entry Team

A Size up the door to determine swing; the number, type, and location of locking devices; and the type of door and frame.

B Place the adz between the door and the door stop 6 inches above or below center of the lock and push up or down to spread the door from the frame.

C Place the fork of the Halligan tool into the gapped area with the bevel of the fork toward the door no more than 6 inches above or below the center lock.

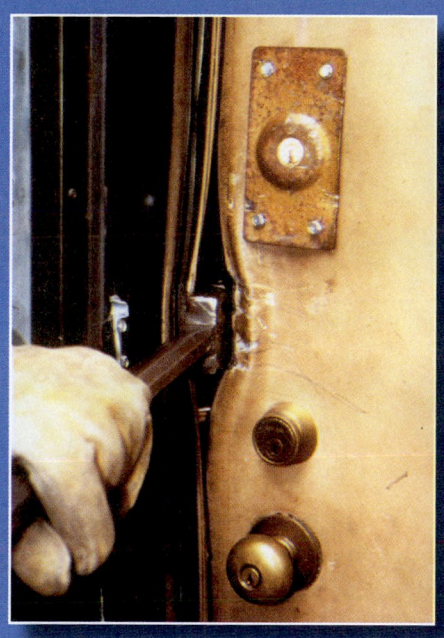

D Set the tool using the flathead ax or maul. Drive the Halligan tool in so that the tips are around the door and locked in.

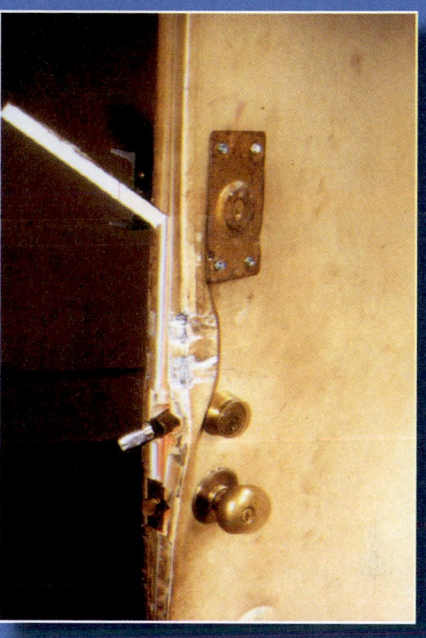

E If difficulty is experienced in setting the tool, drive fork in with bevel toward frame to allow easier entry.

F This procedure is repeated for each lock.

JOB PERFORMANCE REQUIREMENT   17-2

Conventional Forcible Entry—Door Swinging Toward the Forcible Entry Team

A   Check the door to determine swing; the number, type, and location of locking devices; and the type of door frame and hinges.

B   Create a gap between the door and jamb by driving the adz or fork end of the Halligan tool between the door and frame 6 inches above or below the lock.

C   On tight doors, the gap may be started with the fork end.

D If door is tight, work tool up and down to open up gap.

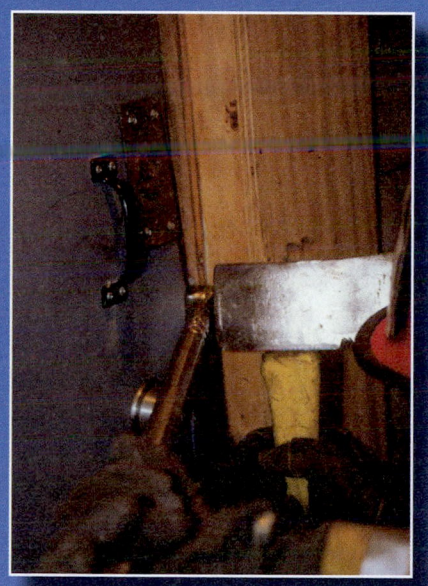

E Set the tool by driving the adz or fork end past the door around the doorstop.

F This inside view shows how the adz is set.

G Force the door by pulling the tool away from the door.

JOB PERFORMANCE REQUIREMENT   17-3

Through-the-Lock Forcible Entry—Unscrewing or Wrenching the Locking Cylinder

A   Using locking-type pliers, lock the pliers onto the cylinder. Turn the lock cylinder counterclockwise to unscrew the cylinder.

B   Use proper key tool to open lock.

JOB PERFORMANCE REQUIREMENT   17-4

Through-the-Lock Forcible Entry—Using the K-Tool

A   Force the blades of the K-tool over the cylinder.

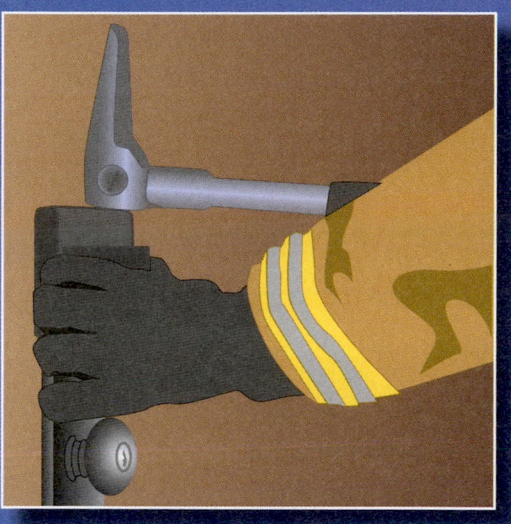

B   Place the adz of the Halligan tool into the loop of the K-tool.

C   Pull up on the Halligan tool, and pull the cylinder out.

LESSONS LEARNED

Forcible entry is a key tactic in structural firefighting and emergency operations, and firefighters must understand the tools, equipment, and methods used for forcible entry. As with all other fireground tactics, teamwork is an essential element. Failure to quickly conduct effective forcible entry may result in delayed search and rescue operations and unnecessary fire spread. In addition, with the renewed emphasis on rapid intervention crews, the art and skill of forcible entry techniques is a basic element for these operations.

Firefighters must continually size up buildings for firefighting and rescue operations, including how to force entry into a building.

KEY TERMS

Bolt Throw   The distance the bolt of a lock travels into the jamb or strike plate. Usually ½ to 1½ inches.

Cutting Tools   The group of tools used to cut through or around materials.

Flush or **Slab Doors**   Doors that are flat or have a smooth surface and may be of either hollow-core or solid-core construction.

Forcible Entry   The fire scene task of gaining entry to a building or secured area by disabling, breaking, or going around locking and security devices.

Glazing   The glass or other clear material portion of the window that allows light to enter.

Halligan Tool   From the prying group, a 30-inch forged steel tool with three primary parts: the adz end, the pike end, and the fork end.

Hollow-Core Door   Any door that is not solid, usually with some type of filler material between face panels.

Hook   A tool with a 32-inch to 12-foot handle with a pike and hook on one end. Used for pulling ceilings or separating other materials. Also known as a pike pole.

Irons   The combination of a Halligan tool and flathead ax or maul.

Jamb   The mounting frame for a door.

Laminated Glass   Glass composed of two or more sheets of glass with a plastic sheet between them. The purpose of the plastic sheet is to hold the glass together if broken, thus reducing the hazard of flying glass.

Ledge Door   Door built with solid material, usually individual boards, common in barns and warehouses.

Locking Devices   A mechanical device or mechanism used to secure a door or window.

Mounting Hardware   Hinges, tracks, or other means of attaching a door to the frame or jamb.

Panel Doors   Doors with a solid stile and rails with panels made of wood or glass or other materials.

Pike Pole   See hook.

Prying Tools   The group of tools used to separate objects by means of a mechanical advantage.

Pulling Tools   The group of tools used to pull away materials.

Rabbeted   A door stop that is cut (rabbeted) into the door frame. On metal door frames the stop is an integral part of the frame.

Slab Door   See flush or slab door.

Solid-Core Doors   Doors made of solid material such as wood or having a core of solid material between face panels.

Strike Plate   The metal piece attached to a door jamb into which the lock bolt slides. Also called a strike or striker.

Striking Tools   The group of tools designed to deliver impact forces to break locks or drive another tool.

Tempered Glass   Plate glass that has been heat treated to increase its strength.

Through-the-Lock Method   A method of forcible entry in which the lock cylinder is removed by unscrewing or pulling and the internal lock mechanism is operated to open a door. Also, the family of tools used to perform this operation.

Wire Glass   Glass with a wire mesh embedded between two or more layers to give increased fire resistance.

REVIEW QUESTIONS

1. Choose an engine or truck company in a fire department and identify five forcible entry tools and describe their use.

2. List the inspection and maintenance procedures for five forcible entry tools.

3. List three different types of doors and describe a method of forcible entry for each.

4. List five types of locks and describe their operation.

5. List the steps for the three types of conventional forcible entry.

6. Demonstrate conventional forcible entry on a variety of doors.

7. Describe or demonstrate the through-the-lock forcible entry method.

8. List four construction features of windows, and methods of gaining entry.

9. Describe three considerations when breaching walls.

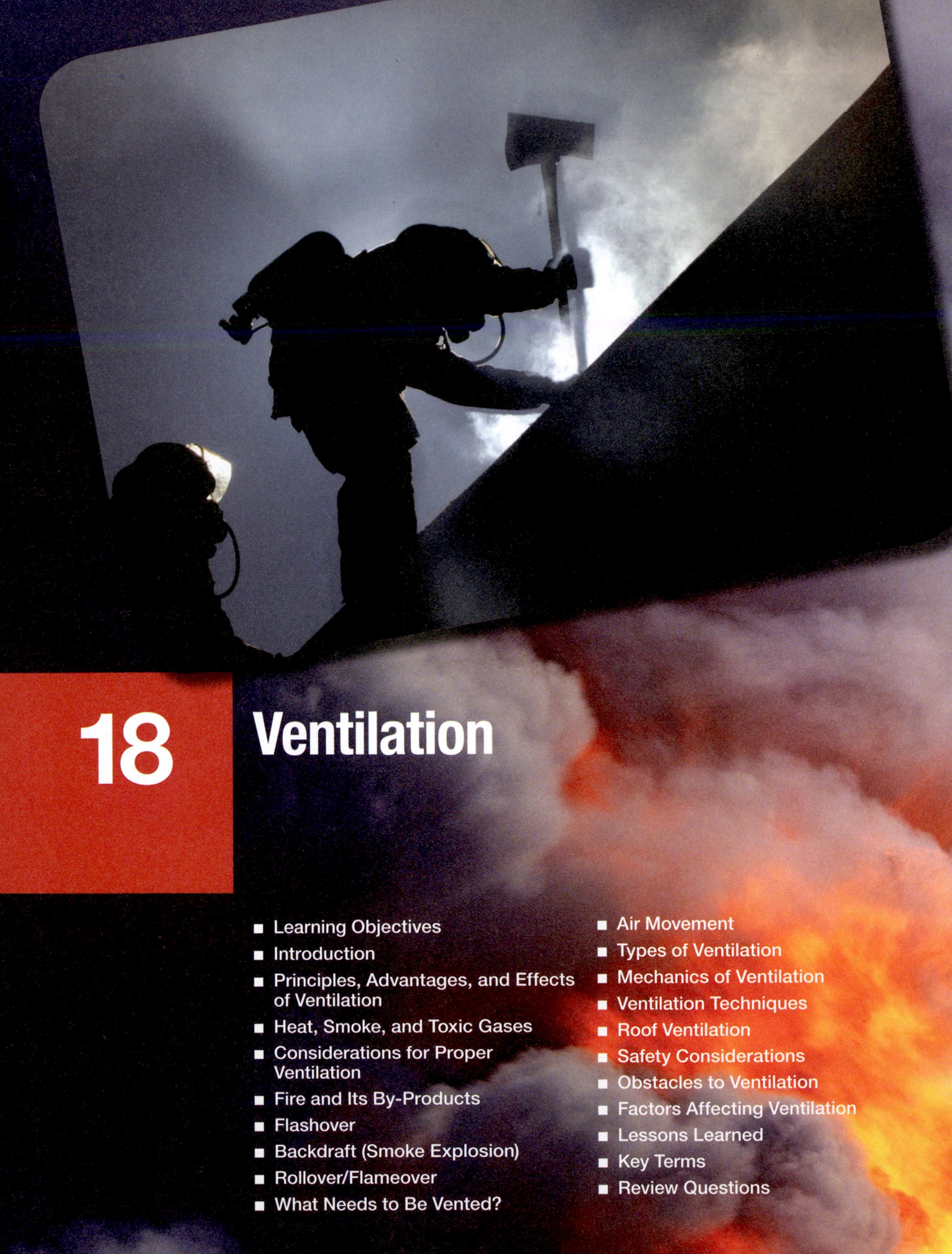

18 Ventilation

The call came out mid-afternoon in August for a reported structure fire. This was our third structure fire in the past four days. The first two looked suspicious, and we had contacted the state fire marshal office to investigate the fires. Both structures had burned to the ground, so finding evidence was going to be difficult.

The fire department I was with at the time was a combination department; the fire chief and I were the only two paid firefighters and the rest of the department was volunteer. The Chief responded in the command vehicle, and I responded in one of the engines with a volunteer. A second engine responded about five minutes after us with a crew of four.

The Chief arrived first on scene and reported that the structure had just flashed over and that if we hurried, we could stop the fire and save the structure. When the volunteer and I arrived on scene, we put on our bunker gear and self contained breathing apparatus, grabbed a hose, and went to the front door. When we tried to make entry into the structure, something was blocking the door. We also noticed that the door was unlocked. The firefighter and I both pushed on the door, it moved, and we were able to make entry.

Upon entering the structure, we encountered several problems. The first was that we could not see more than six feet in front of us, as the smoke was very thick and black. The second problem was the heat. It was so hot that we had to enter on our stomachs to advance. The third problem was making entry into an unoccupied structure without waiting for our second engine to arrive on the scene.

We advanced about five feet into the structure, but because it was so hot and dark, we had to stop and listen for the fire. We could hear it crackling all around but were unable to see it. The heat was so intense that I told the firefighter that we should back out. It was at that moment we faced our next problem: The carpet had melted to our bunker gear and we were stuck to the floor. We rocked back and fourth, and with an adrenaline rush, we were able to pull ourselves free from the floor.

As we exited the structure, I looked over my shoulder and saw the flames pouring out of the doorway we had been in. After the second engine company arrived, they said they had seen the house, and it had a lot of black smoke and flames coming from the doorway.

I ordered the second engine company to set up a ventilation fan. We set up a rapid intervention team, grabbed a fresh firefighter (the Chief) and re-entered the structure. It was still very hot but the ventilation fan had started to clear the smoke. We were making entry to the seat of the fire when out of nowhere we found ourselves engulfed in flames again. We fell to the floor, used a straight stream setting, and broke out several windows for additional ventilation.

We made it to the kitchen, where the fire had started, knocked down the fire, and then pulled ourselves out to let the next teams come in to start the overhaul and look for any possible fire extinction.

When the fire was out, we started our investigation. First, the home was empty. Second, the item blocking the door was the stove from the kitchen; it had been moved to the living room. Third, the fire was started by an accelerant—the fuel load was too small to cause the amount of heat that we had experienced. So with this information we contacted the state fire marshal's office for an investigator. We hoped that we had preserved enough evidence for them.

We had been careless, however. Because we had experienced a couple of structure fires in a short time, our mindset was not on safety but on trying to save evidence. Instead of making a rapid entry into an empty structure, we should have set up the ventilation fan and started the positive pressure ventilation first. Had we done this, we would have prevented firefighter injury. All three of us had steam burns to our chests and backs and second degree burns to our ears.

This fire showed the importance of ventilation and how well it works when used correctly. Once we set up the ventilation van, it took out some of the heat and smoke and we were able to make entry to the fire. When we did the horizontal ventilation it removed the rest of the heat and smoke from the fire.

—Street Story by William Tatsch, Engineer,
Travis County Emergency Services District 6, Austin, TX

LEARNING OBJECTIVES

After completing this chapter, the reader should be able to:

18-1   Identify the principles, advantages, and effects of ventilation.

18-2   Explain the factors affecting ventilation.

18-3   Discuss the factors affecting vertical ventilation.

18-6   Describe limitations when performing ventilation.

18-7   Discuss the observations in reading smoke and the warning signs of hostile fire events.

18-8   Describe the effects of air movement in horizontal ventilation.

18-9   Discuss safety considerations when venting operations are in progress.

18-10   List and describe safety considerations required for ventilation operations.

18-11   Describe hostile fire events, including flashover and backdraft, and the methods for preventing them from occurring.

18-12   Identify the various types of ventilation.

18-13   List the equipment used in mechanical ventilation.

18-14   Discuss specific tactics for below-grade ventilation to include the use of positive pressure.

18-15   Describe hydraulic ventilation techniques.

18-16   Demonstrate the proper procedures for breaking windows and door glass while accomplishing horizontal ventilation.

18-17   Demonstrate how to clear vertical and other ventilation openings using hand tools.

18-18   Describe the need for roof ventilation and explain ventilation techniques given different roof and basement types.

18-19   Describe the obstacles to vertical ventilation.

18-20   Demonstrate proper sounding of a roof during roof operations.

18-21   Describe the use of and how to accomplish trench or strip-cut ventilation.

18-22   Describe differences between trench or strip ventilation and normal vertical or horizontal ventilation.

18-23   Describe advantages and disadvantages of vertical and trench or strip ventilation.

18-24   Demonstrate proper vertical ventilation for flat and pitched roofs and floors utilizing hand and power tools.

18-25   Demonstrate the proper tool selection, tool use, and ladder positioning for vertical ventilation.

*The FF I and II levels, as defined by the NFPA 1001 Standards, are identified in different colors: FF I = black, FF II = red, additional information = blue.

INTRODUCTION

Ventilation can be defined as the planned, methodical, and systematic removal of pressure, heat, smoke, gases, and in some cases, even flame from an enclosed area through predetermined paths. Ventilation is an essential part of the tactical and strategic objectives of modern fire extinguishment. Nevertheless, ventilation sometimes isn't employed until late in the initial fire attack phase or only when initial efforts of fire attack have been unsuccessful. The late application of proper ventilation has subjected firefighters to extreme and unforgivable circumstances while attacking fires. Ventilation is a very complex subject area with many facets. Additional information can be found in books that have been written solely on the subject of ventilation.

PRINCIPLES, ADVANTAGES, AND EFFECTS OF VENTILATION

Ventilation is the relief of the products of combustion from an enclosed area. Combustion of organic material produces heat, smoke, pressure, and fire gases that quickly fill up an enclosed structure. That structure could be a one-room shed, a ten-room private dwelling, or a twenty-story building. In every case, the need to relieve the structure of these products of combustion is a very essential part of the fire suppression effort. The practice provides benefits that go beyond merely suppression to extinguish the fire.

First, by relieving the structure of heat through channeling it into the atmosphere, the fire is deprived of the ability to heat up other parts of the structure. Fire burns when the gases of a combustible substance are liberated. This liberation occurs when heat from an existing fire is applied to the unburned material. The unburned material heats up, liberates gases, and then ignites, permitting the fire to spread. Using ventilation, the heat is exhausted and dissipated into the atmosphere where its ability to spread fire through the structure is reduced.

Second, ventilation channels smoke out of the structure. Smoke is a combination of material, mostly unburned hydrocarbons that have a tarry consistency. Because most of these substances are microscopic in size, they are very light and can stay suspended in air. The heat emitted by the fire carries the smoke to all parts of the structure. Because smoke is made up of only partially burned solid bits of microscopic material, it obscures vision, and the more dense the smoke is, the more it will obscure. Light smoke can be like a fog in which shapes can be seen from several feet away. A heavy smoke condition can obscure light so completely that even a powerful light is rendered totally ineffective. Not only does lack of vision seriously hamper firefighting operations, but it also prevents victims from escaping.

Additionally, the unburned tarry hydrocarbons contain substances that irritate the eyes. In the natural function of the eye's protective action, anytime a foreign material is introduced, the tear ducts attempt to flush the eye with body fluids (tears) to rid it of the irritation. These tears blur vision so badly that it is nearly impossible to see.

Smoke contains many other products of combustion that are deadly substances. Many harmful compounds are mixed in with smoke, and a person that is exposed to this material will suffer ill and potentially lethal effects, **Table 18-1.** When people fall victim to smoke or its components, they usually fall into unconsciousness, with death coming a short time later. The removal of smoke, heat, and toxic gases will add survival time to a potential victim who is unconscious, increasing the chance of successful rescue.

HEAT, SMOKE, AND TOXIC GASES

When fire burns, air heats, expands, becomes lighter, and rises. It also begins to exert pressure on anything that surrounds it. The rising heated air becomes a means by which fire is spread or communicated to surrounding materials by convection, and, to some extent, radiation from the upper levels (see Section I, Chapter 4). Fire can also spread into other places within a structure by the pressure that is created. Pressure will take the path of least resistance within a confined area as it tries to become equalized. If that path leads to more combustible material, the heated air will spread its heat to that substance and permit it to ignite, thereby spreading the fire. Fire gases consist of many products of combustion that can contribute to the spread of fire, render a living being unconscious leading to death, or contribute the necessary ingredients for an explosion. Many substances found as the by-products of combustion are deadly.

Because of the desire to conserve energy, structures being built today and those being renovated are outfitted with heavy insulation and tight, weatherproof seams. These features make ventilation even more important because the high heat generated by the inability of the heat to escape is radiated into the compartment. These tight construction practices lead to hotter fires, early failure of structural components, and greater incidences of flashover and backdraft.

TABLE 18-1	Gases Produced by Fire
Carbon monoxide	Takes the place of oxygen in the blood
Carbon dioxide	Overstimulates the rate of breathing
Hydrogen sulfide	Causes respiratory paralysis
Sulfur dioxide	Extremely irritating to eyes and respiratory tract
Ammonia	Extremely irritating to eyes, nose, throat, and lungs
Hydrogen cyanide	Highly toxic; used commercially as a vermin fumigant
Hydrogen chloride	Can become hydrochloric acid in mucous membranes
Nitrogen dioxide	Causes respiratory distress as a delayed reaction
Acrolein, phosgene	Gases found in certain kinds of fires; lethal in small doses

These phenomena are examined in greater detail later in this chapter.

Last, in today's climate-controlled buildings that are equipped with windows that are unopenable or that are even windowless, reliance on ventilation is greater in order to move the products of combustion out of the structure. An overview of mechanical ventilation is given later.

CONSIDERATIONS FOR PROPER VENTILATION

In order for ventilation to assist in the fire extinguishment operation, the firefighter must first understand the behavior of fire gases, which include smoke, in a building. Because heat is lighter than air, it tends to rise. Smoke, when mixed with heat, also rises. As the smoke rises, it collects under any vertical obstruction and mushrooms in all directions when it meets that obstruction. Gradually, the smoke and heat fills the structure, starting from the highest point in the structure, and "banks down" until the entire structure is filled, **Figure 18-1.** To fully understand the concept, imagine a building. Then, pull it out of the ground and turn it upside down. While it is upside down, begin filling the cellar with water. Any channel that the water finds as it drains out of the inverted

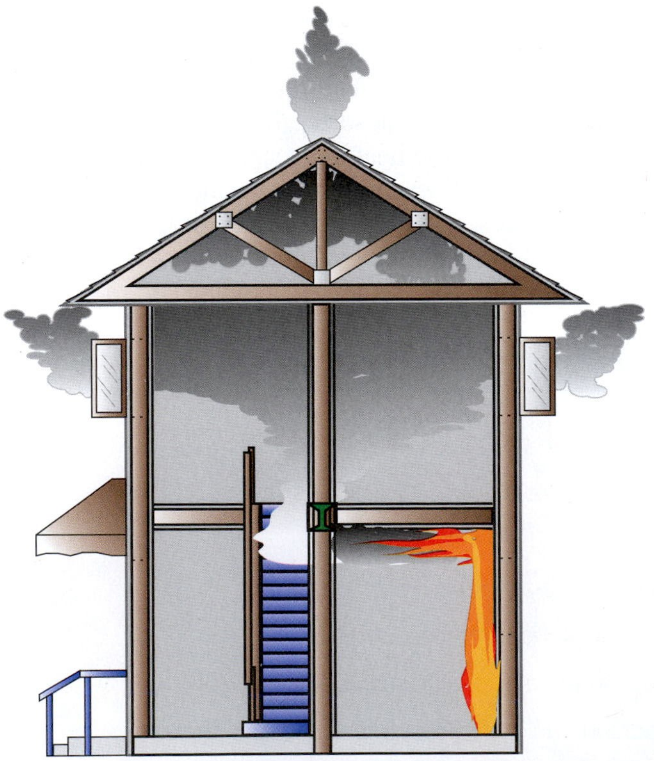

FIGURE 18-1 Heat, smoke, and fire will follow the path of least resistance and find their way through any available opening.

cellar into the inverted attic will be the same channels that smoke and heat will use to "drain up" out of that cellar, and, just as water will leak out of small holes in the windows or siding in the inverted structure, so too will smoke leak out in the structure that is right side up. So even though smoke is coming out of the lower floors of a structure, it is only a portion of the smoke and heat that is collecting in the upper part of the structure.

There are advantages and disadvantages to horizontal and vertical ventilation.

Advantages

Allows smoke and heated gases to escape into the atmosphere
Increased visibility allows firefighters to enter the structure
Improves tenability for both firefighters and victims
Reduces the potential for hostile fire conditions
Reduces the potential of fire spread
Reduces property and structural damage

Disadvantages

If not performed properly, it may cause adverse fire heat, and smoke spread
Roof ventilation openings cause further structural damage
Firefighter safety during roof ventilation operations

Vertical ventilation is defined as the removal of gases and smoke through vertical channels. This will prevent fire extension by convection from occurring at a remote part of the building. The opening of the structure at the top will permit the fire gases and smoke to exit the building. Another type of ventilation that is often associated with fire suppression is **horizontal ventilation,** which is the channeling of smoke and heat out of the structure through horizontal openings such as windows and doors. There is a distinct difference between the needs and results of vertical and horizontal ventilation. Horizontal ventilation permits the fire's by-products to be pushed out of the structure by the advancement of the fire suppression crews. In areas remote from the fire, it will permit a reduction of smoke, and, to a lesser degree, built-up heat that will aid in a search for victims and any fire extension. Without horizontal ventilation in front of an advancing hose team, the heat, smoke, and now a new element, steam, have nowhere to go. With no outlet for the water to push the smoke, steam, and heat out of, it will be pushed over the hose team at the ceiling level. When water is applied to a fire, the fire is extinguished by the removal of heat from the fire pyramid. When raised to 212°F (100°C), water turns to steam. It takes a tremendous amount of heat to turn water into steam, and, when it does, it expands 1,700 times. That means that when 1 gallon of water is heated to the proper level, it creates 1,700 gallons of steam. If the hose team is about to apply water to a fire in an enclosed room without a ventilation hole opposite their position, the pressure created will push back out through the same opening from which the hose team entered. The fire, heat, smoke, and fire gases will push right over the hose team and super-heat the area behind the nozzle team, surrounding them with the heat of the fire.

By opening a channel for the products of combustion to exit as the nozzle team moves in, the heated smoke and steam will take the path of least resistance and vent from the fire room. It is very important for the hose team to move in with a wall of water in front of them and push the fire out the window or another opening, **Figure 18-2.**

Horizontal ventilation requires some forethought before performance. If performed at the wrong time or at the wrong place, it can accelerate fire spread. For example, if the venting is about to be performed in a part of the structure that is two rooms away from the fire room, the air currents that are created could pull the fire into a room that might otherwise not have been involved, **Figure 18-3,** but this will not occur all the time. There are many reasons why this might or might not happen. Some of the factors that will influence the air currents in a structure fire are vertical vent openings, horizontal openings, outside wind direction, the direction the hose attack team is using in relation to the fire, and the room being vented.

When properly performed, the ventilation operation can be as critical as the nozzle team applying water to the fire. It is important to note that in the confusion of a fireground operation, it may not be possible to follow a manual's every line and sentence. Tempered

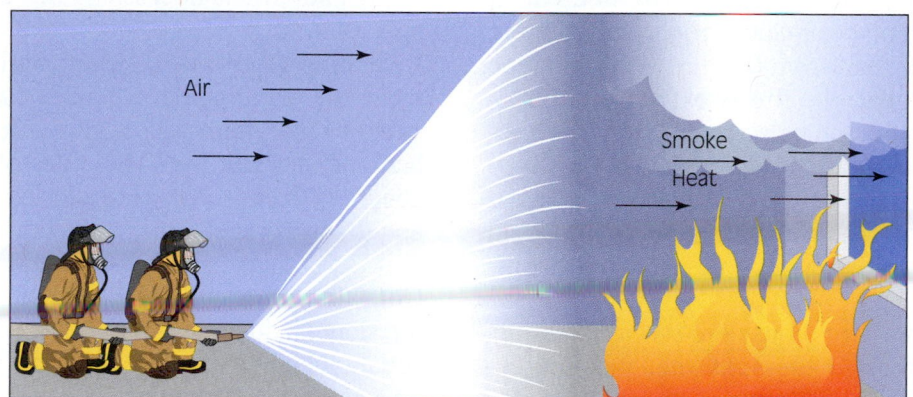

FIGURE 18-2   Heat, smoke, and fire will always take the path of least resistance. When pressurized by the air created by an advancing hose team, the products of combustion will seek a path that will equalize that pressure. An opening on the opposite side of the advancing nozzle team is essential.

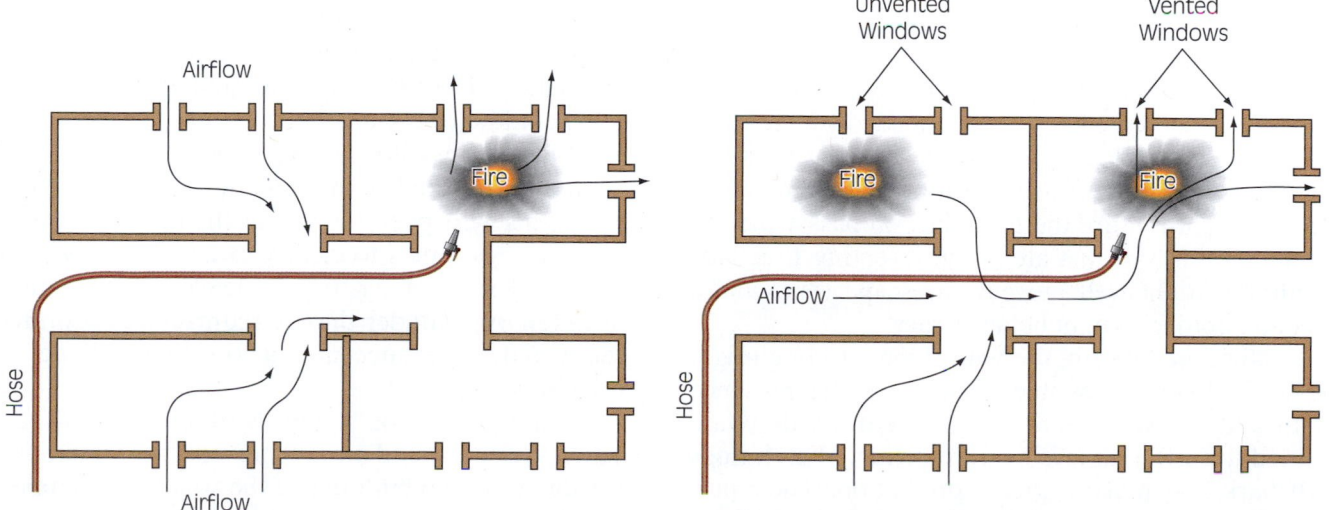

FIGURE 18-3   Air movement is created by water application. Openings in back of the nozzle team will create airflow from behind in the direction of the hose team. It can be a source of fresh cool air, or it can pull fire to the nozzle from behind. Indiscriminate ventilation can be a liability. Careful assessment and proper timing are important.

by situational awareness, the firefighter must utilize experience, knowledge, and training to determine the appropriateness of the ventilation operations.

In rare situations, no ventilation at all is performed at the fire extinguishment operation. Some occupancies are protected by systems that discharge a fire-inhibiting gas such as halon into the room. In this type of occupancy, an identifying placard usually warns that this gas is present. A system such as this might be used where delicate materials are found, such as a museum, computer rooms, biological labs, or archival storage. This type of occupancy might suffer more damage from water than from the fire. With this type of extinguishing agent, the fire is extinguished by a chemical reaction with the flame production (see Section I, Chapter 4). In a structure protected by this type of system, ventilation must be avoided until ordered by the incident commander. Venting this room would permit the fire-inhibiting gases to escape and reignition to occur.

Many factors must be considered when venting. These include access to the vent site, wind direction, weather conditions, exposures, the given material burning, the height of the building, the potential for fire spread, and the escape route. Not every factor will have the same impact at every fire. Each incident is unique, and the specifics of occupancy group and type, building features, and systems associated with the structure must be considered.

Access to the vent site can be tricky. Access to a window to be vented might be impeded by the presence of electrical wires, sloping ground, or sheer height. A window off a porch roof that needs to be vented might require the firefighter to pass windows on that porch roof that are already venting fire. This condition might well prevent an escape route and a second approach might be necessary.

Ventilation duties in the rear of the building might be difficult or impossible because of the presence of locked gates or guard animals. **Guard dogs** are trained for different levels of protection. **Watch dogs** will bark and make a great commotion. Their purpose is to scare an intruder. Guard dogs will bark and assault an intruder. Their function is to warn and injure. Attack dogs are trained not to bark and will attack an intruder with ferocious intent.

A hose stream from a hoseline or even a water extinguisher will send a guard dog running. In all cases, the best approach is to guide the dog into some kind of confined area and leave it there. Occasionally, police might have to be called to deal with the dogs by using tranquilizers or by destroying them.

The presence of electric supply lines to the structure can impede access to a vent site. Unless specifically designed, all fire service ladders, including those made of wood, are electrically conductive, and that presents a great risk when raising or positioning them in the vicinity of electricity. When the window needing ventilation is close to electric power lines, extra care must be employed. Live wires might have had their insulation burned off and energized surrounding metallic substances such as aluminum siding. A ladder resting against that energized siding can result in electrocution. When sizing up the approach, consideration must include the presence of overhead wires and, when necessary, the use of an alternate approach.

Sloping ground will significantly affect the placement of ground ladders. This is especially true if the street or road **frontage** is at a different level than the back of the building. The sides of the building will generally have some kind of slope so that the front and the back meet. Unless the landscaping is terraced with level steps, the ladders will have to be raised from uneven ground. Depending on the severity of the slope, ladders might be unavailable for use in this situation and ventilation of certain windows might be delayed.

The height of the building will also have an impact on access to a window. If the window is out of the reach of ground ladders and not reachable by aerial or tower ladders, venting might have to be performed by using adjoining windows or from windows above or below.

Wind is a factor that can dramatically alter the ventilation. The removal of a window that faces the wind could cause a complete reversal of the air current flow. If the wind is strong enough, the attack hose team can be faced with a blowtorch or wind tunnel effect that will push the fire at them. There are just too many variables to clearly establish a "go/no go" set of rules about whether to leave the glass intact or to remove it under these circumstances. Common sense and experience are factors that will help to solve this judgment call.

Humid, rainy, or foggy weather tends to cool smoke-laden, heated air and prevents the smoke from **lifting** out of the building. If the weather is humid or rainy, the smoke from the chimney does not rise in a column but tends to spread out horizontally. On a clear day, the smoke rises in a crisp column into the upper atmosphere, where it dissipates. Other atmospheric conditions such as pressure or temperature inversions alter natural air movement, affecting ventilation and hindering operations.

Section I, Chapter 4 explains how combustion is a process that creates its own life. If the heat produced is not high enough or is dissipated quickly, the fire self-extinguishes. Through the use of ventilation, firefighters can remove some of the fire's ability to extend. If the products of combustion, especially heat and smoke, are confined in a structure, two of the three essential ingredients (heat and fuel) are present and ready to be consumed. By venting, those ingredients are removed harmlessly into the atmosphere and permitted to burn where they will do no harm to advancing firefighters.

FIRE AND ITS BY-PRODUCTS

During the pure combustion process, energy is released from an exothermic reaction as heat and light (flame), **Figure 18-4A.** Heat without smoke can be observed as wavy lines emanating from the fire. Other gases that are produced by the combustion process are not visible. The reuniting of molecules to form new substances is part of the endothermic reaction, **Figure 18-4B.** This process actually takes heat from its surroundings and uses it to bond the loose ends of other molecules.

CAUTION

Substances created by fire can be very harmful to a human. The body reacts to the invasion of foreign substances by attempting to rid itself of them. Unburned particulate matter that settles into the mucous membranes of the eye and nasal passages results in a flow of liquid from the tear ducts and nasal passages to naturally flush out the foreign material. When inhaled into the lungs, the body attempts to isolate the foreign material by coating it with phlegm. Then, the body attempts to take the collected phlegm and expel the foreign material by coughing.

When properly performed, ventilation will remove some of the harmful agents associated with the by-products of combustion. Additionally, the presence of certain chemicals and chemical compounds serves to accelerate the human respiration rate. The human body uses two mechanisms to regulate the rate of breathing. One is the level of oxygen. If the oxygen level diminishes, the rate of breathing increases. The oxygen-reading capability of the brain is actually the alternate monitor that regulates the breathing rate. The primary monitor that regulates breathing is the level of carbon dioxide in the blood. Under normal

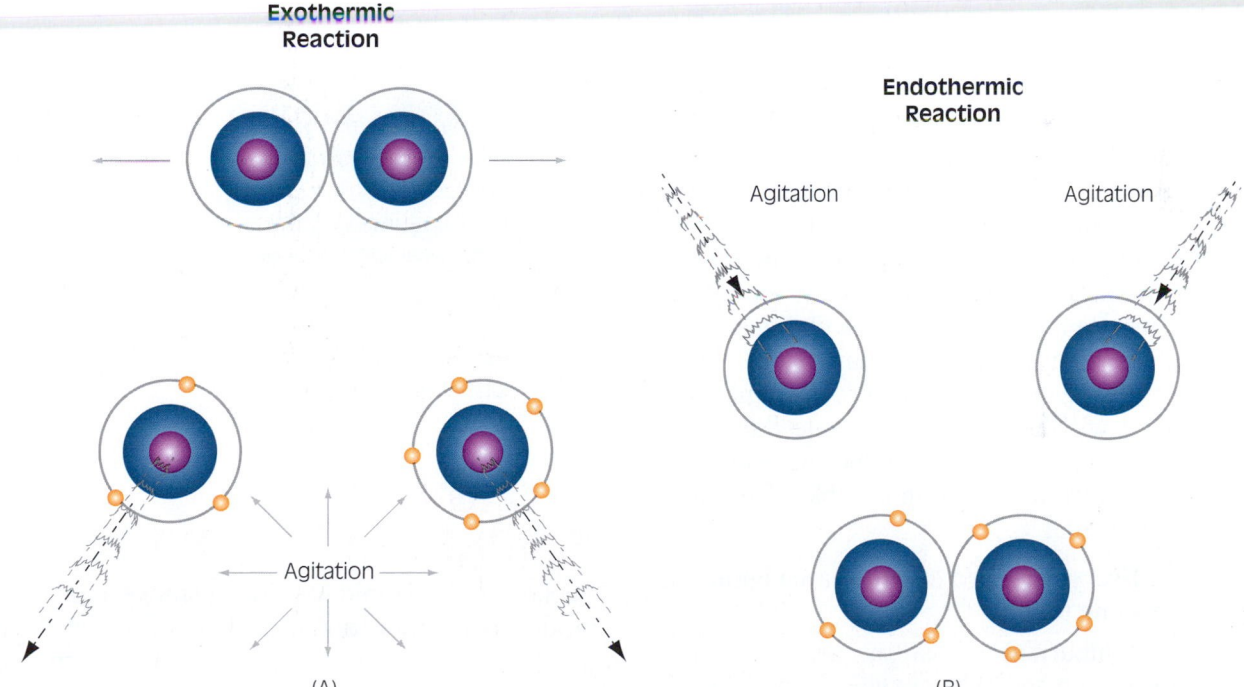

FIGURE 18-4 (A) In an exothermic reaction, the bond broken between two atoms or molecules creates vibration or agitation, and heat is created by that action. There are times when bonds are broken, heat is created, and the atoms or molecules find stability in another combination minus that bond. In that case, a chemical reaction releases heat, but it stabilizes as another substance. (B) Heat is present when atoms or molecules are agitated or vibrating. Very simply stated, if the loose ends of the atoms or molecules attach to another and become balanced, the agitation ceases and the heat that was being generated is absorbed by the newly created bond.

conditions, the body takes in oxygen, uses it, and produces carbon dioxide as a by-product of energy usage by the various muscles of the body. When the brain sensors read an increasing level of carbon dioxide, the body speeds up the breathing rate to increase the oxygen/carbon dioxide exchange in the lungs.

Because the by-product of combustion is carbon dioxide, the body inhales it along with the other poisonous gases produced. As the brain senses the increasing levels of carbon dioxide, it increases the breathing rate to expel the building level of CO_2. In the case of carbon monoxide, another product of combustion, which the blood cells absorb over 200 times faster than oxygen, the result is a faster and faster rate of poison absorption.

> **FIREFIGHTER FACT**
>
> The faster the structure is opened and ventilated, the more quickly harmful gases are replaced with fresh air and the safer the environment as a consequence of pressure reduction, increased visibility, and increased tenability.

Since all firefighters should be wearing self-contained breathing apparatus, one might ask if ventilation is important since a fresh air supply is available to the wearer. The answer is an emphatic yes! Ventilation increases the survival time of a trapped victim who might be overcome or a firefighter who has become disoriented and run out of air.

Ventilation has many other benefits beyond limiting the inhalation of contaminated air. Removal of smoke improves visibility, aiding in the search and rescue of victims. Reducing the danger from a hostile event is a key element in today's rapidly developing fires. Dangers such as holes in floors are easier to spot, and avenues of fire extension become more obvious. With better visibility, the use of tools becomes safer.

If heat is not permitted to linger on a material long enough, it will be unable to liberate the material's combustible gases. By venting the enclosure, the heat level is kept from becoming capable of producing these phenomena:

- *Flashover.* Everything in a confined area ignites at almost the same time.
- *Backdraft.* Unburned smoke is heated in the absence of oxygen and, when oxygen is introduced, produces an explosive force.
- *Smoke explosion.* Unburned products of smoke and gases accumulate in areas of the structure, and when oxygen is introduced the gases rapidly ignite with explosive force. However, sustained combustion is generally not indicated.

- *Rollover/Flameover.* Fire begins to ignite smoke overhead in "fingers of fire" that reach out and begin to consume fuel in the gaseous state.

The mechanics associated with each of these phenomena and prevention techniques are discussed as each relates to ventilation.

FLASHOVER

Light, smoke, and heat are liberated as part of the combustion process. As the trapped heat collects at the ceiling level, its cumulative temperature increases. Across the upper area of the room, heat begins to radiate downward, heating the contents of the room. When the overall temperature reaches the ignition point of another substance in the room, a new chain reaction combustion site occurs and additional heat is added beyond the initial source of fire. As each item in the room follows suit, more and more heat is created and more and more items ignite. In a very short time, the entire room and all of its contents are on fire. This happens very rapidly. Slow-motion photography shows that it is not an instantaneous occurrence but a very rapid fire spread. It is important to know the mechanics of a flashover in order to recognize its development. It is even more important to know what course of action to take before, during, and after a flashover.

> **FIREFIGHTER FACT**
>
> In a hostile environment, at best, the survival time of a firefighter in bunker gear and breathing apparatus, fully encapsulated with gloves, hood, and helmet flaps down, is estimated to be between ten and fifteen seconds.

The best survival skill here is Hostile Event Recognition and avoidance.

BACKDRAFT (SMOKE EXPLOSION)

Backdraft is the rapid ignition of smoke and unburned products of combustion as a result of the introduction of oxygen into the environment. This introduction can come in many forms, such as the misuse of PPV or via an air pocket as a result of fog nozzle water application during fire attack. It can also occur with the sudden introduction of oxygen from opening windows and doors at horizontal levels of the structure. Sometimes confused with a flashover, backdraft is a very different event.

When fire burns organic materials in the presence of oxygen, carbonaceous materials are transformed into carbon, carbon dioxide, and water. When there is a shortage of oxygen, incomplete combustion occurs, and, instead of forming carbon dioxide (CO_2), a less stable compound called carbon monoxide (CO) is formed, **Figure 18-5.** During this process, two of the three classic substances for a complete burn are present: fuel and heat.

Incomplete combustion occurs as the oxygen level decreases. Visually, this can be observed by the amount of yellow flame present. With adequate oxygen, the flame is bright yellow. As the oxygen level decreases, the yellow becomes bright orange, then dull orange, then red. As the flame approaches red, the level of oxygen is getting so low that if continued the fire would self-extinguish for lack of oxygen. As discussed in Section I, Chapter 4, carbon monoxide has a very wide range of flammability. When oxygen is introduced into the mix, combustion will again occur to complete the process by converting the CO into CO_2. Under the proper conditions, this can occur with a violent reaction.

Some form of backdraft occurs at almost every fire that remains unvented prior to the arrival of firefighters. Its violence and damage will be directly commensurate with the distance from where the oxygen is introduced and the location of the flames, **Figures 18-6 A** and **B.**

As the fire consumes greater amounts of the oxygen in the room, the production of carbon monoxide increases. With the heat, pressure builds in the confined space. Smoke can be seen puffing from openings, and black smoke condenses on the surface of any cooler surface, such as glass or a wall. Initially the entire room is filled with heat, carbon monoxide, and other fire gases. When an opening occurs, a billow of smoke escapes as the pressure from inside attempts to equalize. Then, what is described as "sucking" occurs where, once a large majority of the pressure in the space is released, cool outside air is introduced into the room. The cooler air causes the volume of air to contract and pull air in to attempt to equalize again. Being cool, it stays low to the floor and at some **thermal level** a mix of highly concentrated CO is mixed with the fresh, 21 percent oxygen-laden air. As the cool air snakes its way to the source of the fire to replace air that is being heated and driven upward, what could be described as a tunnel of fresh air finds its way to the fire, **Figure 18-7.** An observer might even see the smoke lift off the floor and just before ignition, actually see the fire burning across the expanse of the floor below the smoke level. In the meantime, in terms of the oxygen level on the floor, with each few inches the smoke rises, the CO and O_2 mix more. This mixture of gases at some point will permit the concentration of CO to be in the range of 14.5 to 74 percent. Once that concentration reaches the flame, the three components for combustion are present and primed for ignition. The burn begins at the seat of the fire and, like the fuse on a firecracker, burns back to the opening in the space along the perimeter of the tunnel of oxygen-laden air. This occurs so rapidly that it seems like an explosion. The burn causes the heat, which in turn causes rapid expansion of the surrounding air with such rapidity and force that it can blow out windows, knock down walls, and hurl firefighters out openings and across the street. The degree of force is dependent on several factors.

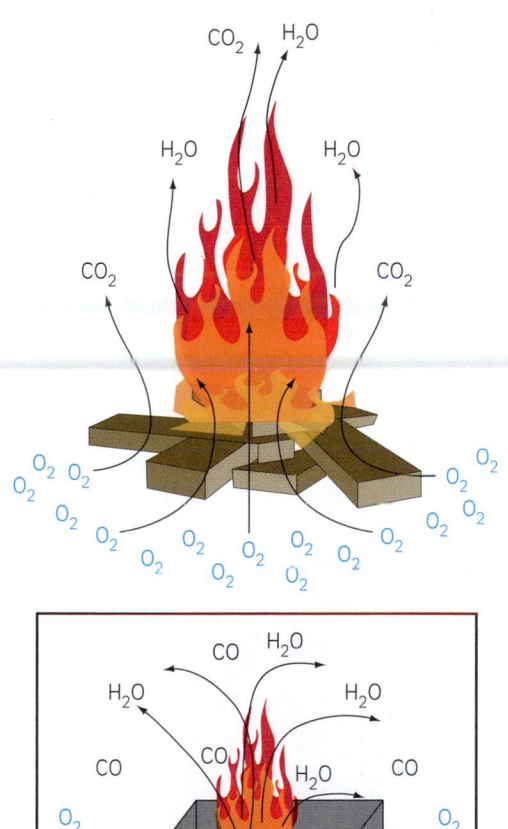

FIGURE 18-5 When combustion occurs, oxygen is combined with an organic material, and, through a series of bond breaking and recombination processes, carbon dioxide and water are driven off. When oxygen is lacking, instead of two oxygen atoms combining with a carbon atom to form carbon dioxide (*dioxide* meaning "two oxygens"), the carbon atom combines with only one oxygen atom, forming carbon monoxide (*monoxide* meaning "one oxygen"). In the absence of adequate oxygen, carbon monoxide is generated instead of carbon dioxide.

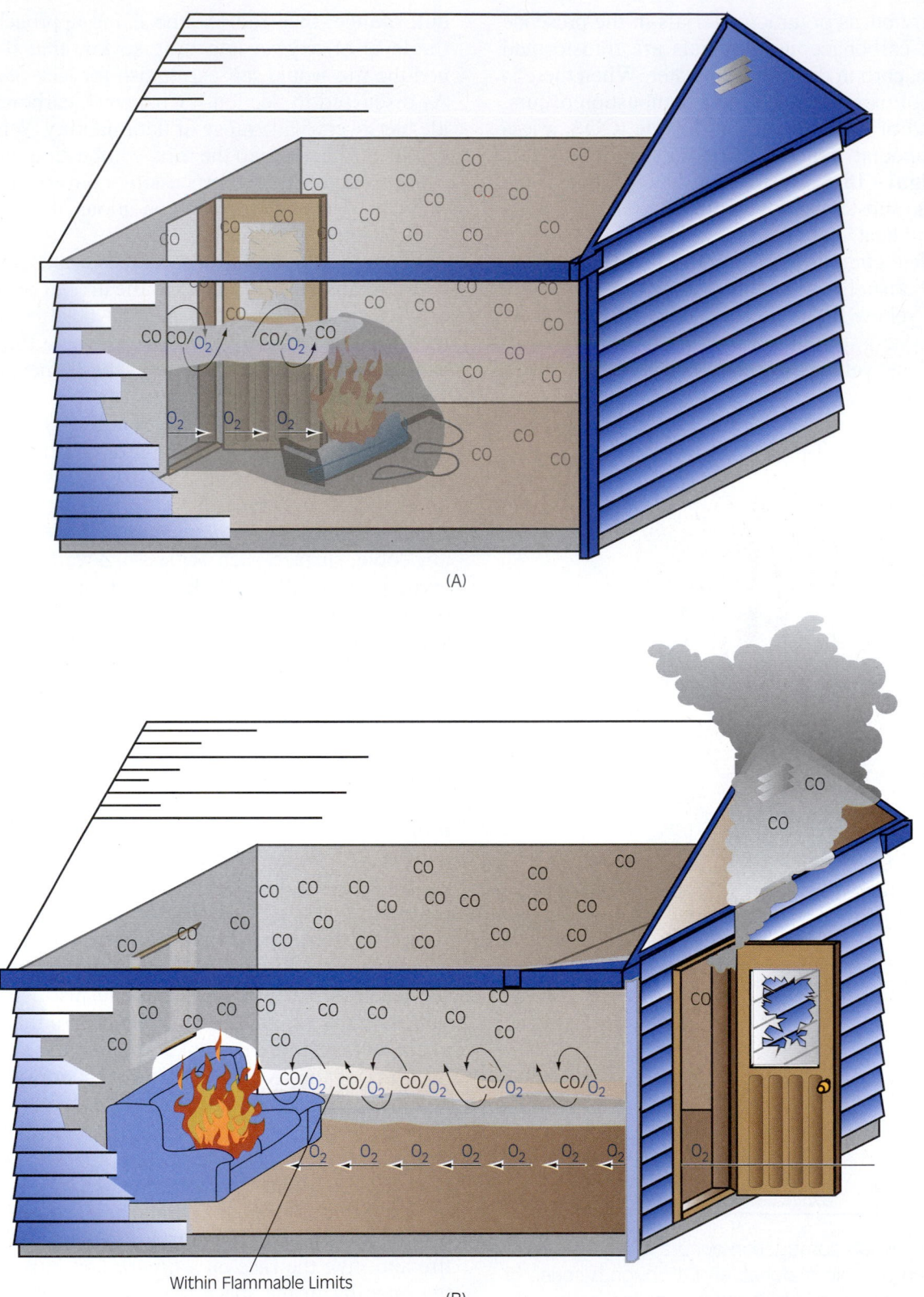

(A)

Within Flammable Limits

(B)

FIGURE 18-6 (A) When the distance between the opening and the source of ignition is short, there is little opportunity for the gases to circulate and form a mixture that is within carbon monoxide's flammable limits. When the mixture does ignite, it appears more like a flame flare-up than an explosion-like ignition. (B) A large distance between the opening and the location of the fire will permit a greater mixing of carbon monoxide and cool oxygen-laden air. The greater distance will permit the carbon monoxide concentration to be in its flammable limits over a greater area before it reaches an ignition source. A larger instantaneous ignition will result with great force. The greater the distance, the greater will be the resultant ignition force.

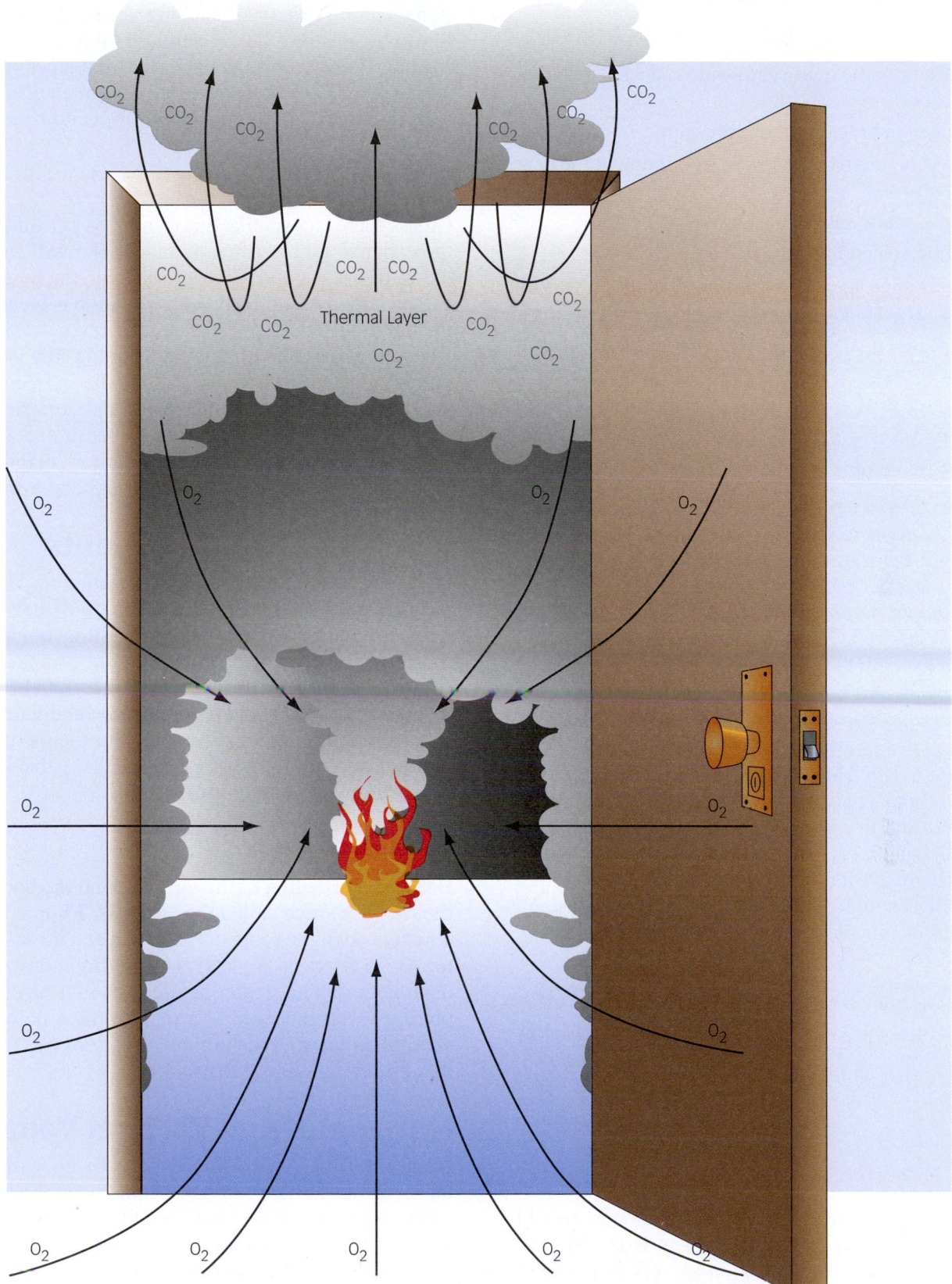

FIGURE 18-7 Through the open door, you can see the pressurized smoke containing carbon monoxide venting outside the upper level of the door as the oxygen-laden cooler air enters at the lower level of the thermal layer. Looking down a "tunnel" you can see the cool oxygen-laden air mixing with the oxygen-starved carbon monoxide around the perimeter of the tunnel. When it reaches the fire, the mixture will ignite and, in chain reaction fashion, follow the ignitable mixture in the tunnel right out to the opening.

Signs of a Potential Backdraft

- Smoke-stained windows
- Puffing of smoke at seams and cracks of windows and doors
- Smoke pushing out under pressure
- No visible flame or very dull red flame in the depth of the smoke
- Heavy, black smoke
- Tightly sealed building
- Large, open area structure (supermarket, bowling alley, department store)
- Large, open void (cockloft, between-space of hanging ceiling)
- Extreme heat

Consider these two backdraft conditions and the difference in the results:

- *Short distance to opening.* The atmosphere is laden with carbon monoxide. As the fresh air works in toward the thermal plume, it brings fresh oxygen. The outer perimeter mixes with the CO and achieves the proper flammability range. When the seat of the fire is reached, the tunnel perimeter ignites and follows the path to the opening. Since it is limited and relatively short, the results are unspectacular and appear merely as the fire flaring up.

- *Long distance to opening.* The same initial condition exists. However, in this case the amount of the proper concentration from the opening to the seat of the fire is much greater. When the ignition occurs, it travels a greater distance and amasses much greater force. Although technically it is moving in a chain reaction fashion, the almost instantaneous chemical reaction is like an explosion.

NOTE

The greater the distance between the smoldering fire and the opening, the greater the explosive mixture involved, and consequently, the greater the force created.

ROLLOVER/FLAMEOVER

Rollover/flameover was described in greater detail in Section I, Chapter 4, but the subject is briefly revisited here because ventilation is so very important to prevent this phenomenon. When products of combustion are produced, the heat brings them to higher levels, usually accumulating near the ceiling. The heated gases accumulating at the upper levels of the compartments reach their ignition point and begin to spread across the room at the ceiling level. Fingers of flame can be observed reaching across the room followed by a wall of flame behind it. When an advancing hoseline disrupts the upper thermal layer, the heat at that layer is forced down into an unbalanced state. This action can disrupt the rollover phenomenon.

WHAT NEEDS TO BE VENTED?

Unless there is a ventilation opening to permit escape, the expanding heated steam and smoke will roll over the wall of water and drop down behind the hose team, **Figures 18-8 A** and **B.** When ventilation is mentioned, the first image that appears is a fully involved building fire emitting smoke from every opening. It appears to be well vented. However, long before the entire building requires venting, the smaller voids and compartments need to exhaust the increasing pressure and intensifying heat condition. If done in a timely manner, the involvement of the entire building might be avoided.

Voids and Compartments

When examining the subject of ventilation, all compartments must be treated with the same understanding. For example, a residential building is merely a very large compartment with many subcompartments (i.e., apartments). Each subcompartment can be subdivided further (i.e., rooms), then again subdivided (i.e., closets), **Figure 18-9A.** Rounding out the picture, the building might also have eaves, peaks, gables, cocklofts, and other voids, **Figure 18-9B.**

Cocklofts

Between the ceiling of the top floor and the bottom of the roof is a space called the **cockloft, Figure 18-9C,** which is often the route of major fire extension. Chapter 13 (in Sections I and II) on building construction dealt with this in greater detail. It cannot be overemphasized that this void is a major attack point for a ventilation crew, especially in a top-floor fire or a fire that has extended into that space.

Horizontal and Vertical Voids

Because all heat will follow the path of least resistance, an unobstructed channel could be in the form of a horizontal or vertical void. By following **pipe chases** or electrical wire pathways, heat and fire may extend without being seen. Ventilating these voids by opening them at the highest points can expose any extending fire for extinguishment. In addition, by opening the voids, the trapped heat is diffused into unheated spaces, thus minimizing the chance of heating up a new, uninvolved portion of the building, **Figure 18-10.**

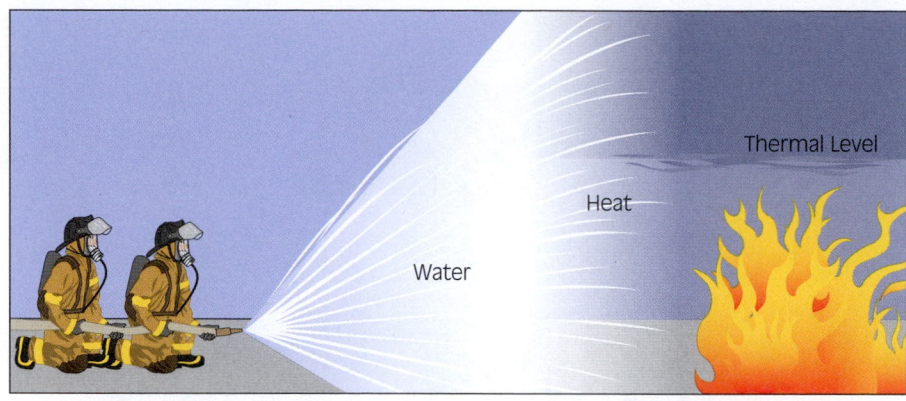

(A)

(B)

FIGURE 18-8   (A) Applying water to the upper levels of a thermal layer will cool and disrupt the rollover effect that is apt to occur with the proper conditions. Ventilation is critical when this is done. (B) As a hose team advances into the fire and sprays water in droplet form, it creates a wall of water and disrupts the high-heat thermal layer and cools the upper levels of the compartment. Water absorbs heat as it turns to steam, expanding 1,700 times as it does. If there is no path for the expanding water/steam conversion, it will take the path of least resistance, in this case over the wall of water and the nozzle team. The water movement will then pull any heat from the back of the nozzle team and roll over on top of it.

AIR MOVEMENT

Defined and discussed in greater depth in Section I, Chapter 4, the attributes of the modes of heat transmission that relate directly to ventilation are revisited here:

- *Convection*. The normal travel path of heat as it follows the convection cycle can be interrupted and challenged by opening a hole in the upper portions of the compartment. Then, instead of traveling up, across, and back down, the heated air will travel up, across, and out of the structure. By providing a path of least resistance in the normal path of heated air, the heat is carried into the atmosphere where it harmlessly dilutes with cooler air.

- *Conduction*. Heat tends to travel along dense material. It can cause remote combustibles hidden from view to ignite. By exposing the heat conductor and ventilating the void where it exists, the trapped heat is provided with another less resistant path, thereby preventing the heat from traveling to uninvolved areas of the structure.

- *Radiation*. In most cases, radiation is an outside function. Under normal conditions, there would be no real need to ventilate. However, in smaller applications, such as an overheated wire, the heat from the wire can radiate to nearby combustible building components. While a combination of the three modes of heat transfer would be at work here, radiation is contributing. Ventilation would have its desired effect as heat is carried away from the recipient of the radiant heat, thereby carrying away and preventing, or at least delaying, the exposed material from reaching its ignition temperature.

TYPES OF VENTILATION

Ventilation can be performed using one of several methods either singularly or in combination with one another. Natural ventilation merely requires opening doors and windows and letting physics take care of

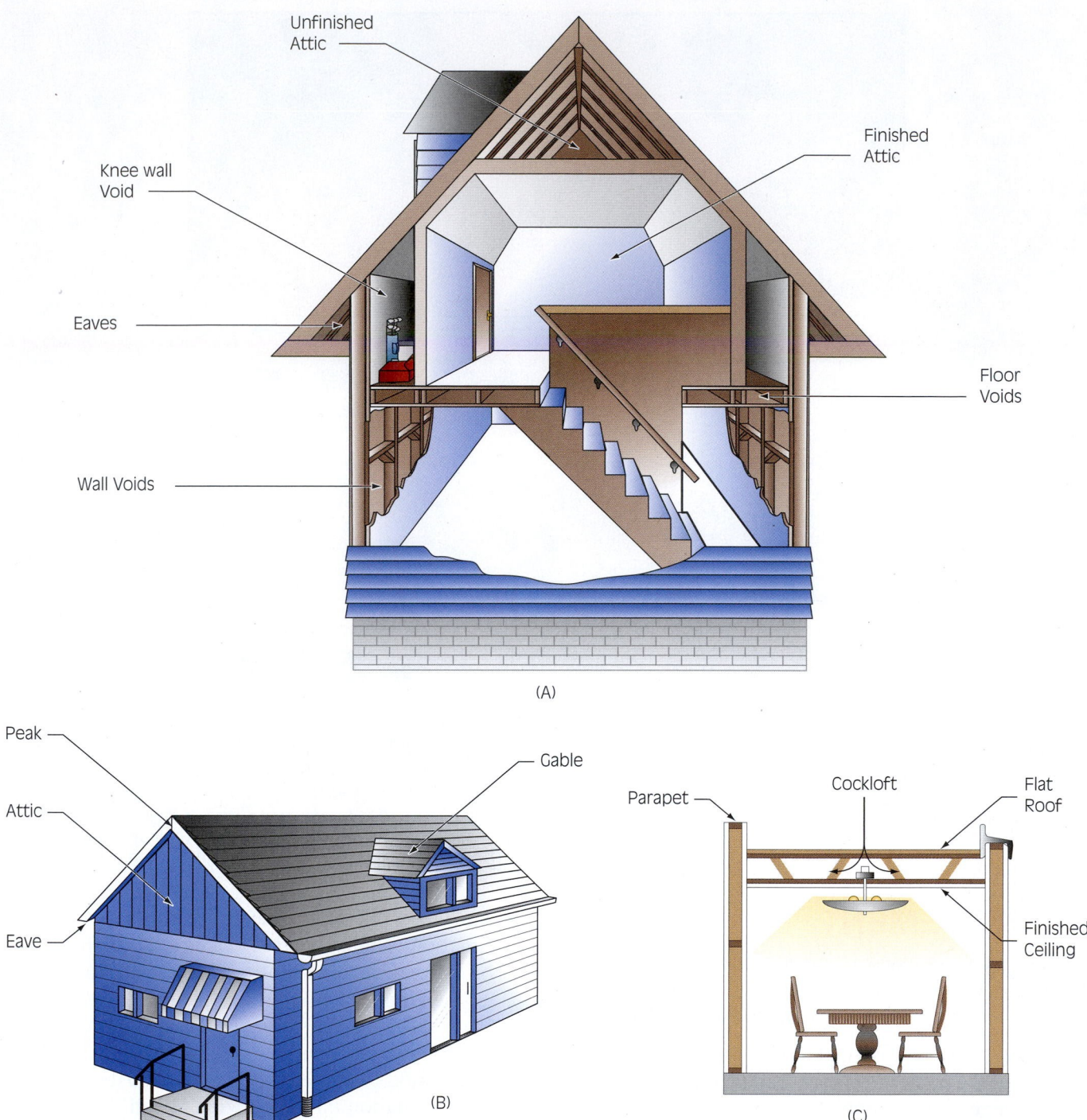

(A)

(B)

(C)

FIGURE 18-9   Voids in a typical structure that can trap heat and permit fire extension.

the rest, **Figure 18-11.** Among the mechanical means are the use of smoke fans consisting of ejectors and blowers, and the use of water to create air movement.

Natural

Firefighters still use natural methods for evacuating smoke from a building. Merely opening windows and doors will provide natural ventilation. This choice is

the most appropriate for incidents where time is not essential and there is no urgency to the emergency, permitting a slower venting operation. Diffusion of smoke and heat is a slow process, but it will eventually be complete when equilibrium is achieved. The natural movement of air currents is the action in this type of ventilation. This method would be appropriate for a light smoke condition where the incident is under control. It is the method of choice for removal

FIGURE 18-10   Ventilation openings are necessary to permit fire gases and steam to vent into the open air. On the left, with no openings provided, the fire is pushed into uninvolved areas of the building, and water is having no effect on the fire in the wall. On the right, an opening provides a path out through the roof to vent heat and fire gases, thus limiting extension.

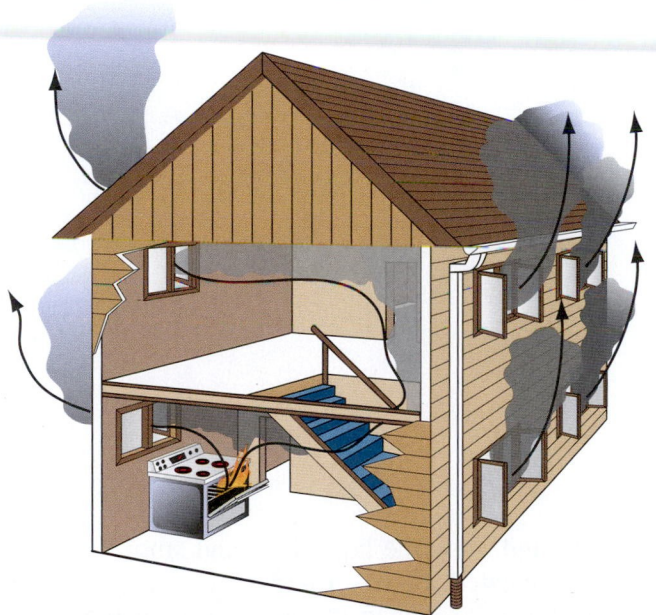

FIGURE 18-11   Smoke will be carried throughout the building to upper floors by normal air currents mixed in with the heat.

of residual smoke from an incident such as food on the stove, overheated motor, leftover smoke from a fire some distance away in the building, or similar situations. A combination of air movement and diffusion is employed in the unassisted natural ventilation process.

Cutting a hole in the roof or breaking out windows is another type of natural ventilation when greater speed is needed or increased volumes of smoke and heat need to be removed from a structure, **Figure 18-12**. The rising heat will take smoke and fire gases to upper levels following the path of least resistance. A large opening in the roof with an unobstructed path from the lower floor affords an express channel for ventilation using the natural method.

Mechanical

The use of mechanical devices can speed the ventilation process, especially in compartments deep within a structure or under conditions where natural air movement is not satisfactory. Mechanical aids can accelerate the air movement and even reverse the airflow against natural air current movement. Use of fans or blowers in a positive or negative mode and water from a nozzle can provide a large air movement volume. Some buildings may have built-in smoke and heat vents that can also be utilized when needed.

Heating, Ventilation, and Air Conditioning

Heating, ventilation, and air conditioning (HVAC) systems can be used effectively for ventilation in sealed, climate-controlled buildings. Closing intake dampers that would draw in smoke from the outside

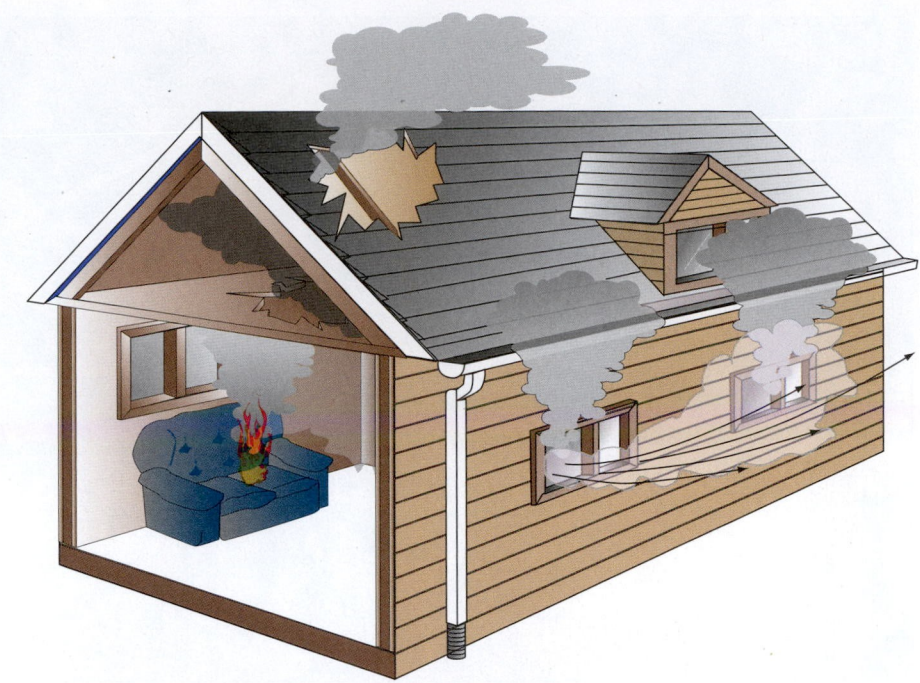

FIGURE 18-12 The hotter the air, the faster it will circulate throughout the building. Without the roof opening, the smoke will "bank down" and fill the structure. The openings provide a path to the outside for the heat and smoke.

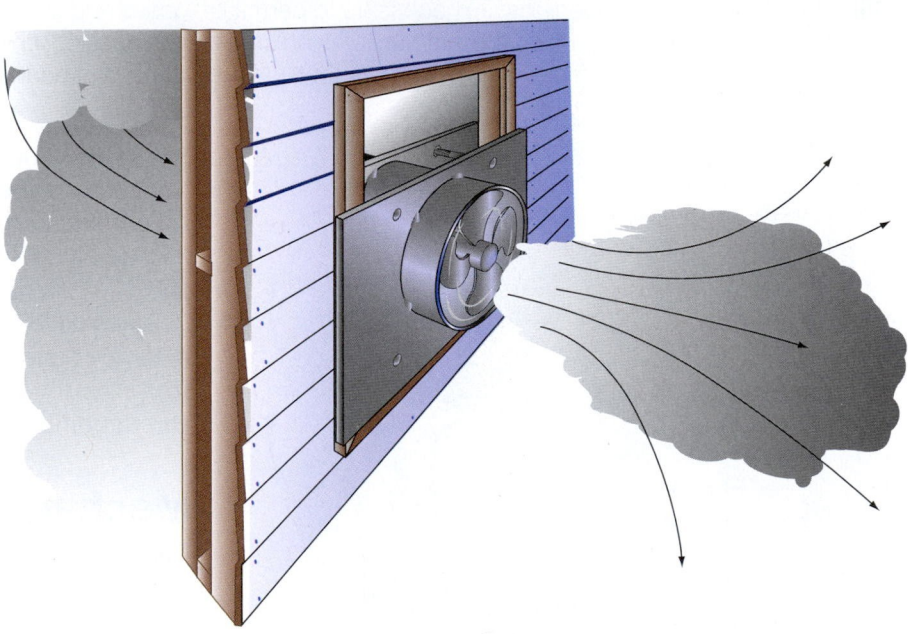

FIGURE 18-13 A smoke ejector exhaust fan placed in an opening will pull air through the fan as it ejects air out of the structure.

robs the fire of fresh air. With the opening and activation of exhaust fans, smoke and heat are removed from the compartment. Uninvolved floors can be vented mechanically, and air movement in the building in general can be controlled.

Smoke Fans

Two types of fans are used for smoke and heat removal. Ejectors were among the first fans used by the fire service and were employed primarily to suck out the smoke and heat. Placed in an opening, the fan developed a negative pressure on the structure side of

the unit and drew the heated air and smoke into the fan to be ejected, **Figure 18-13.**

The fan is placed in the window with the arrow for airflow leading to the outside and safely secured. The openings that could cause the ejected smoke to be pulled back in and recirculated are sealed with tarps or plastic. During this time, establishment of an unobstructed path of airflow from the outside opposite the fan to facilitate fresh air introduction would be beneficial.

The disadvantage of this type of smoke removal is that the firefighting crew positioning the fan has to work exposed to the smoke and heat, **Figure 18-14.**

In addition, the smoke must be pulled through the unit, subjecting it to caustic substances and deposits that could eventually cause mechanical difficulties to the motor. Deposits on fan blades can also create inefficiencies. Positioning a fan on upper level floors from the exterior is even more hazardous. In these situations, the fan is usually placed on the inside.

Positive Pressure

Positive pressure ventilation (PPV) has been around for over thirty years but only recently has it become an active tool in the arsenal. Air is introduced into the smoke-filled area through the use of PPV fans or blowers. The compartment becomes positively charged with air, and the heat and smoke are displaced through chosen avenues to the exterior, **Figure 18-15.** In addition to being used in the later stages of a fire to clear the smoke to ease overhauling, PPV has been employed with some success during the attack stage of fire suppression. In today's rapidly developing fires, which are the result of increased fuel loading and the dramatic rise in polycarbonates (plastics) in contemporary furnishings, the use of PPV in the inital stages of fire attack have narrowed in applicability. Although PPV can be used in a Positive Pressure Attack (PPA) mode, the incident commander must adhere to the following contraindications of PPV use:

1. Imminent or confirmed rescue of a firefighter or civilian
2. Working fire attack
3. Unknown location of fire
4. Inability to provide a proper exhaust point
5. Structure that presents an overpressurization or hostile event indicator

FIGURE 18-14 Placement of a smoke ejector will subject the firefighters to products of combustion because in order to operate it, the firefighter must stand in the path of the airflow. In addition, it forces the firefighters to work in a hostile environment in which it is difficult to see and breathe.

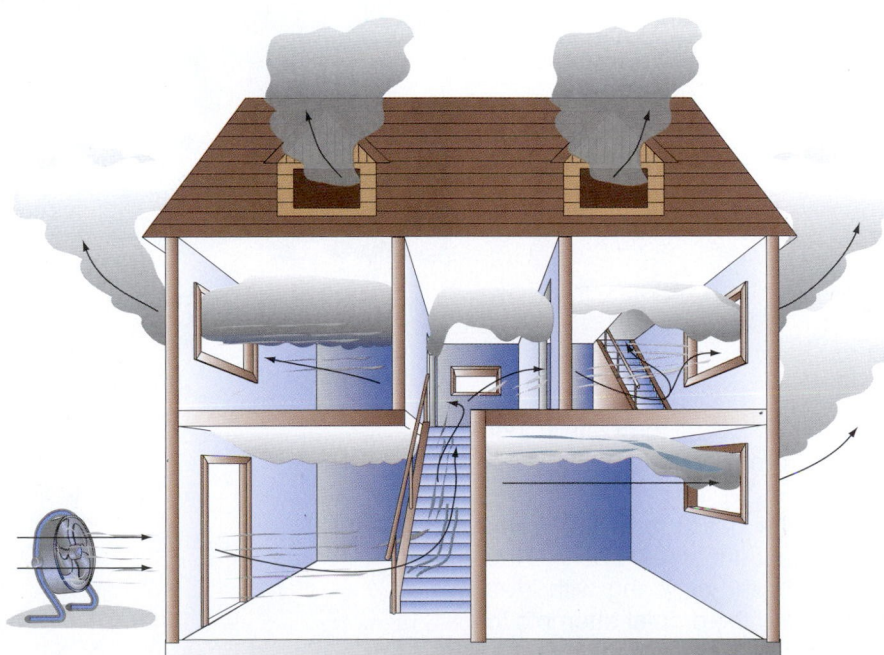

FIGURE 18-15 Positive pressure literally pressurizes the structure and forces smoke out any path of least resistance. Almost the same effect would occur if a light breeze were blowing directly into the structure from one side and venting out the other side.

However, when used properly, the smoke and heat are totally removed from the access paths of the nozzle team and search team. Careful consideration must be taken when selecting positive pressure ventilation in certain types of fires. For example, a basement or below-grade fire may have no window or ventilation openings that would allow smoke to escape after pressure is applied. Firefighters must overcome this by creating paths or holes for smoke to escape if positive pressure ventilation is used.

A distinct advantage of the PPV blower is ease of setup. The fan merely needs to be positioned at the entrance to the structure with no exposure of the setup crew to dangers from within the structure. Furthermore, some of the blowers are designed for a single-person setup. This is a particularly distinct advantage where personnel are limited. If properly applied, the blower can do the job of several firefighters attempting to open multiple holes with ladders at different locations. While not a cure-all for ventilation problems, PPV can be a material asset to an operation. The placement of the blower will depend on the fan's size and capacity and the size of the opening.

Hydraulic

The last of the mechanical modes of ventilation to use is water. Known as hydraulic ventilation, water is employed to create air movement, **Figure 18-16.** It is a quick method for expulsion of smoke and heat without the need for additional tools. Use of water can cause water damage, but it has some assets, too. If the need to evacuate a smoke-filled room is essential to search for a victim, the time needed to set up an alter-

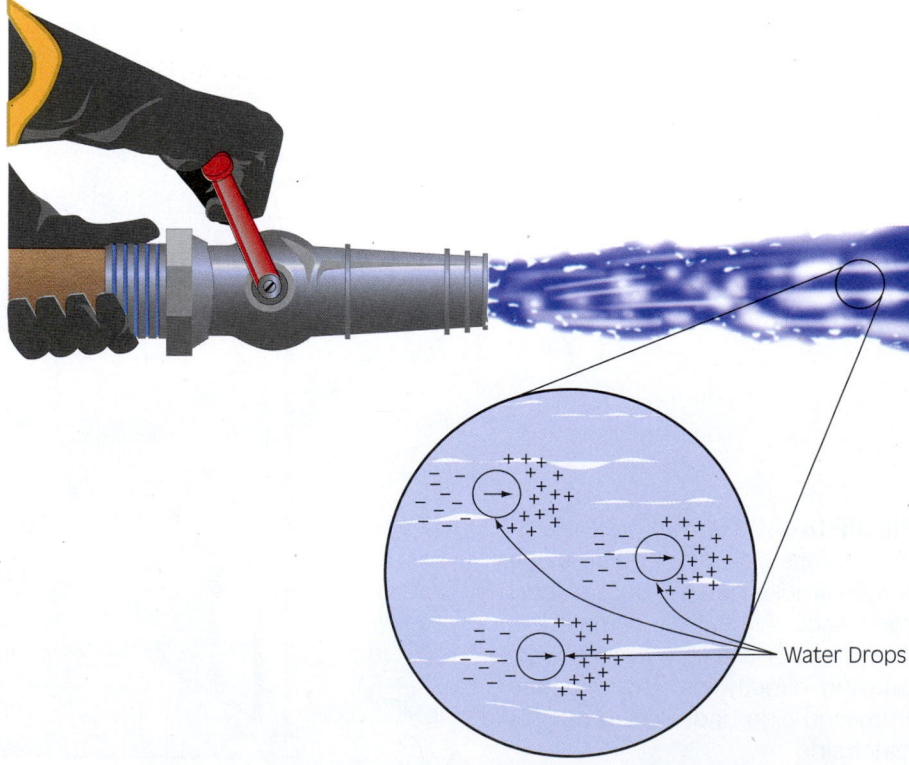

FIGURE 18-16 Water droplets under pressure from a hose (or even gravity) compress air in front, thus pushing air in the direction of the water flow. The vacuum created behind the droplet pulls the air along with it as the air behind the droplet attempts to fill the low-pressure void.

Water Drops

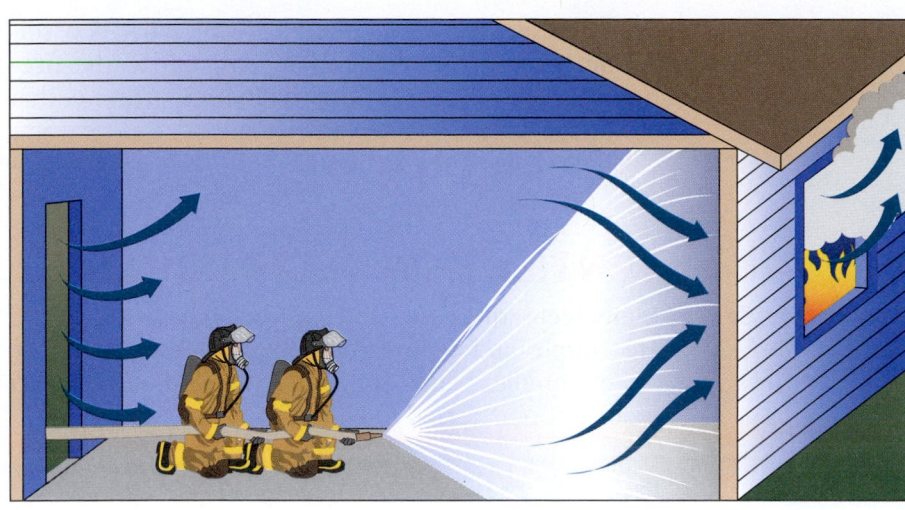

FIGURE 18-17   A coarse stream from a solid-bore nozzle or a spray or fog stream will create the airflow necessary to move smoke out an opening.

native mechanical ventilation tool might be unacceptable. Water can also be used to quickly cool down a heated atmosphere in a structure by sucking and pushing the heated smoke out of the structure.

The nozzle team will quickly feel the benefits as the cooler air from the uninvolved portion of the structure is pulled to them from behind. At the same time, the air movement evacuates the heated upper portion of the compartment, **Figure 18-17.**

This method of ventilation is the most rudimentary. Tests have shown that when properly positioned, the mechanical blowers move a greater amount of cubic feet per minute (cfm) than smoke ejectors, and the use of a nozzle to move air is the most inefficient.

Using a Water Stream to Ventilate. Ventilation by hydraulic means is as simple as directing the hose stream out of the window to remove smoke and heat. Using a fog or spray stream is most effective because the water movement actually pushes and pulls the air surrounding the droplets created by the nozzle. A solid stream nozzle can also be effective by opening the nozzle partially to create a coarse, broken stream.

Whatever type of stream is used, it is placed so that the water is being directed out of the opening. A spray stream setting should be between a 30- to 60-degree angle. Fog and broken solid streams are placed so that the water discharge is entirely outside and not quite touching the sides of the opening. Ideally, the stream should fill 90 percent of the opening. The nozzle stream distance from the opening will depend on the width of the stream. The nozzle team can then move closer or back off until the maximum beneficial effect is achieved. Because of the many variables that are always present, a certain amount of trial and error is involved with finding the best position for maximum effectiveness.

MECHANICS OF VENTILATION

The entire ventilation process, regardless of how it is achieved, is simply the movement of air from a high-pressure location to a place where lower pressure exists. Whether created artificially or naturally, air will move from a high-pressure area to a low-pressure one. Knowing this natural tendency of air to move will assist the firefighter assigned to the task of ventilation.

Vertical Ventilation

Based on the "heat rises" rule of physics, the collection of heat at the upper levels of a structure will spread fire to those upper levels if the heat level rises to the ignition temperature of the structure. By opening vertical arteries to release any pent-up gases, the odds of fire spreading to involve other parts of the structure or contents are reduced. In addition, the heated air being released will be replaced with cooler air at the lower levels. This improvement of the conditions will assist fire suppression crews attempting to locate the fire, search for possible victims, and position for suppression activities, **Figure 18-18** and **Figure 18-19.**

Horizontal Ventilation

Horizontal ventilation conforms to the same rules of vertical ventilation, which are governed by physics. Both types are actually a form of what is called **diffusion,** a naturally occurring event in which molecules keep spilling excess levels of high concentrations to areas of low concentrations. In addition to the vertical gravitation, the molecules also move laterally. Given a wall, the molecules bounce back into the mix. Given an opening, the molecules will pass into the less con-

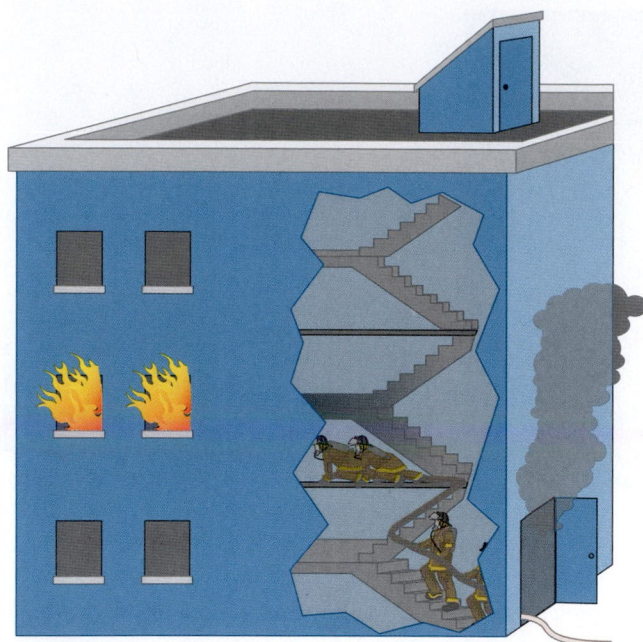

FIGURE 18-18 Smoke in an unvented building impedes progress. Searches are slowed. Discovery of the fire location and deployment of hoselines are delayed.

FIGURE 18-19 While a vented building will still have smoke in it, the difference is profound. The processes of searching, hose deployment, and fire discovery are markedly improved. In addition, the survival time of overcome victims is increased. Last but not least, potential victims might be able to self-evacuate, freeing up personnel for firefighting efforts.

centrated environment, **Figure 18-20.** That is a scientific description of what happens, but to the firefighter on the fireground, all that is important is that the openings are made and the smoke and heat are channeled out of the structure.

VENTILATION TECHNIQUES

Many techniques are used to effect ventilation. Some are simple and require no special tools; others are more complex, are dangerous to implement, and require sophisticated tools.

Break Glass

The quickest way to open a building for ventilation is to break glass. While not always the technique of choice, it is the best investment of time for results gained if done properly. Many windows can be broken in the time it takes to force open one locked door.

SAFETY
When breaking glass, a fully equipped firefighter wearing proper protective equipment is mandatory. Slivers of broken glass can become lodged in eyes and shards of glass can act as airborne daggers.

Glass can and has penetrated skin deep enough to sever arteries and veins. The wearing of protective equipment will reduce the severity or perhaps even completely eliminate an injury.

Open Doors

Merely opening a door will usually exhaust huge volumes of smoke and heat built up on the other side by pressure from the combustion. Other doors will also be helpful in opening the compartment. Keeping the door on its hinges is a good practice because the need might arise to close that door to limit fire extension or to employ a PPV blower with prechosen ventilation channels.

Effects of Glass Panes

Many windows have several panes of glass separated by wood or aluminum dividers, **Figures 18-21 A–C.** When the need to remove glass is evident, the entire window should be removed. An adage exists in the fire service vernacular: "Turn every window into a door." This refers to the possibility that the window may be needed for escape. Removal of the sashes, cross members, and any other obstacle that would impede access or egress is necessary.

Heated

Normal

FIGURE 18-20 Heated air has more agitation in its molecules, causing internal pressure in a compartment. This will, in turn, create greater velocity when air exits an opening. Normal diffusion takes much longer to occur when only natural air movement and currents are employed.

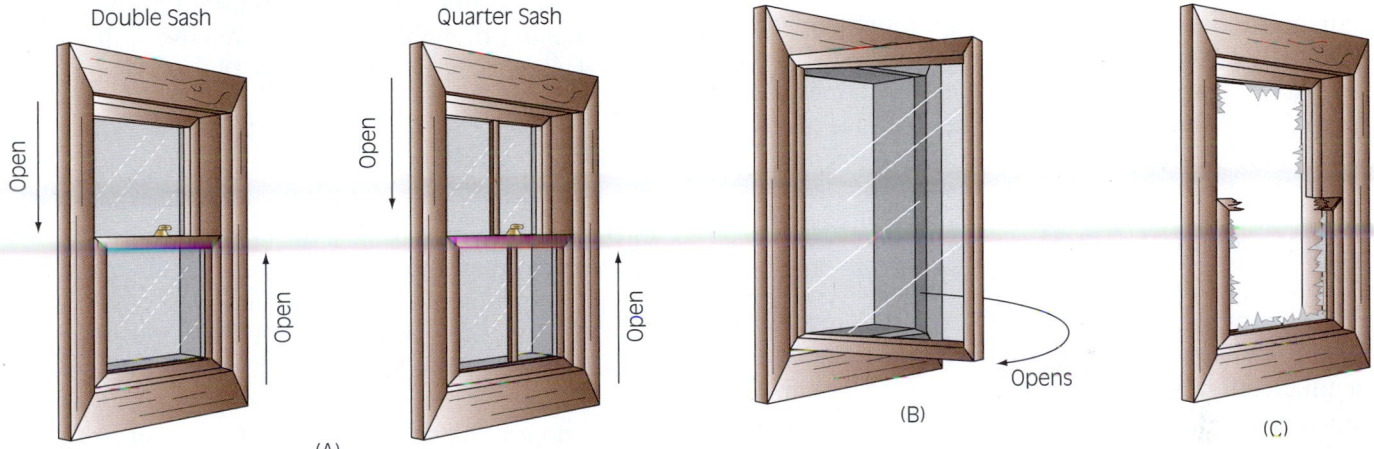

Double Sash

Open

Open

Quarter Sash

Open

Open

Opens

(A)

(B)

(C)

FIGURE 18-21 (A) Sash windows open by pulling the lower section up or the upper section down. The number of panes of glass will not change the number of sashes. (B) A casement window opens like a door, from right to left or from left to right, usually through the use of a crank or handle. (C) Regardless of the type of window, when it is vented, all the glass, including cross members, should be removed to maximize airflow with a minimum of obstructions.

When removing glass for ventilation, it is important to remove all glass. Shards that remain in the sash will impede airflow and reduce ventilation. A sweep around the perimeter of the sash with a tool will often produce quick, satisfactory results.

Using a Tool to Take Out a Window

Regardless of the tool used, basic principles should be followed. Except for special types of glass (i.e., tempered, bulletproof), glass will break regardless of the method employed. What is important is the amount of open area created once the glass is removed.

SAFETY

Once glass is broken, whatever was on the other side of that glass will be coming out. Therefore, when removing glass, it is imperative that the firefighter not be in the path created by its removal. Firefighters should position themselves off to one side and crouch underneath the window level; this is far preferable and safer than standing directly in front of the window while removing the glass. It is imperative that the firefighter be fully clothed in PPE (personal protective equipment). See Section I, Chapter 6.

Using the tool to its maximum length to protect the firefighter, the tool is brought down from the upper sash through the window's midsupport and into the lower sash. Usually, this will remove a large portion of the glass. Then, the tool is used to sweep around the perimeter of the window to remove any shards that remain in the sash. Finally, the tool is used to reach into the window and unhitch any shade, blinds, or curtains that will impede airflow. If the firefighter is not able to capture the fastening, the tool can be used to hook the curtains or blinds and pull the entire unit out of the window and down until it releases. These units are held in place by small nails or screws and forcible removal is not difficult. It is essential to use care when dropping window treatments inside the structure. A victim might be lying under the window and could be inadvertently covered up. This area should be thoroughly searched as soon as possible.

The glass removal should be a two-step movement: (1) breaking through the glass from upper to lower sashes and then (2) sweeping the perimeter to remove remaining glass. Of course, not every single piece of glass will disengage. The remaining small shards will not materially affect the airflow. Good judgment must be employed when deciding whether to move on to the next window or continue to work on that opening.

Rope and a Tool

A technique that has produced satisfactory results when operating off a flat roof beyond the reach of a ladder is to use a tool and a rope. Many windows can be taken out in a short period with this operation. The rope must be secured to the tool so there is no chance it will detach.

STREETSMART TIP

Some firefighters have had small rings welded to the tool and use a positive latch rope clip for quick attachment.

JPR 18-1: Using a Rope and Tool to Ventilate a Window

(For step-by-step photos of this skill sequence, see page 599)

1. The rope is securely attached to the tool using an approved slip-free knot. The tool is then lowered to the window the firefighter desires to ventilate.
2. A turn of rope is taken around the firefighter's hand. The tool is then tossed out as far as possible from the building in a horizontal direction. As gravity takes the tool down and the end of the fully

extended rope is reached, the tool will swing into the glass at the predetermined length of rope and break out the window.

This technique is not the best method because it leaves shards of glass in place and does not remove screens or window shades or curtains, but it is effective when no other approach can be easily made.

Hook or Pike Pole

The use of a hook, also called a pike pole, is another effective way to take out glass. The length of the pole keeps the firefighter a safe distance from the falling glass, and it affords the firefighter the opportunity to stand off to the side of the opening. Pike poles or hooks come in assorted lengths. The use of a longer hook enables access to windows that would otherwise be out of reach without a ladder. A hook can also be used to extend the reach of a firefighter attempting to open or close a door, as a pry tool to remove roof material while standing out of the way of the opening, and for other useful purposes that will extend a firefighter's reach or remove a firefighter from dangerous positions, **Figures 18-22 A–F.**

Iron or Halligan

The iron or Halligan tool is brought down diagonally through the glass starting at the upper corner. The tool should be brought through the cross members to create one large opening. Then the tool is used to sweep around the perimeter to clean out any large shards. A disadvantage of this tool is its short length. It places the firefighter in proximity to flying glass. If prying open a door or hatch, it places the firefighter almost in the path that the sudden gush of heat and smoke will take. Careful planning on the part of the firefighter must be employed to minimize exposure.

Ax

Similar to the iron or Halligan tool, the ax affords limited reach and places the firefighter in a potentially hazardous location. When using an ax for venting glass, the side of the ax is used to break the glass. Use of the blade portion or the striking head might cause the firefighter to lose a grip on the handle and lose the tool due to the manner in which a hand grasps the handle. This technique will not break tempered glass, for which a sharp pointed tool is required.

Portable Ladder

A very effective method of taking out glass in upper level windows, especially in private dwellings, is by the use of a portable ground ladder. The first thing

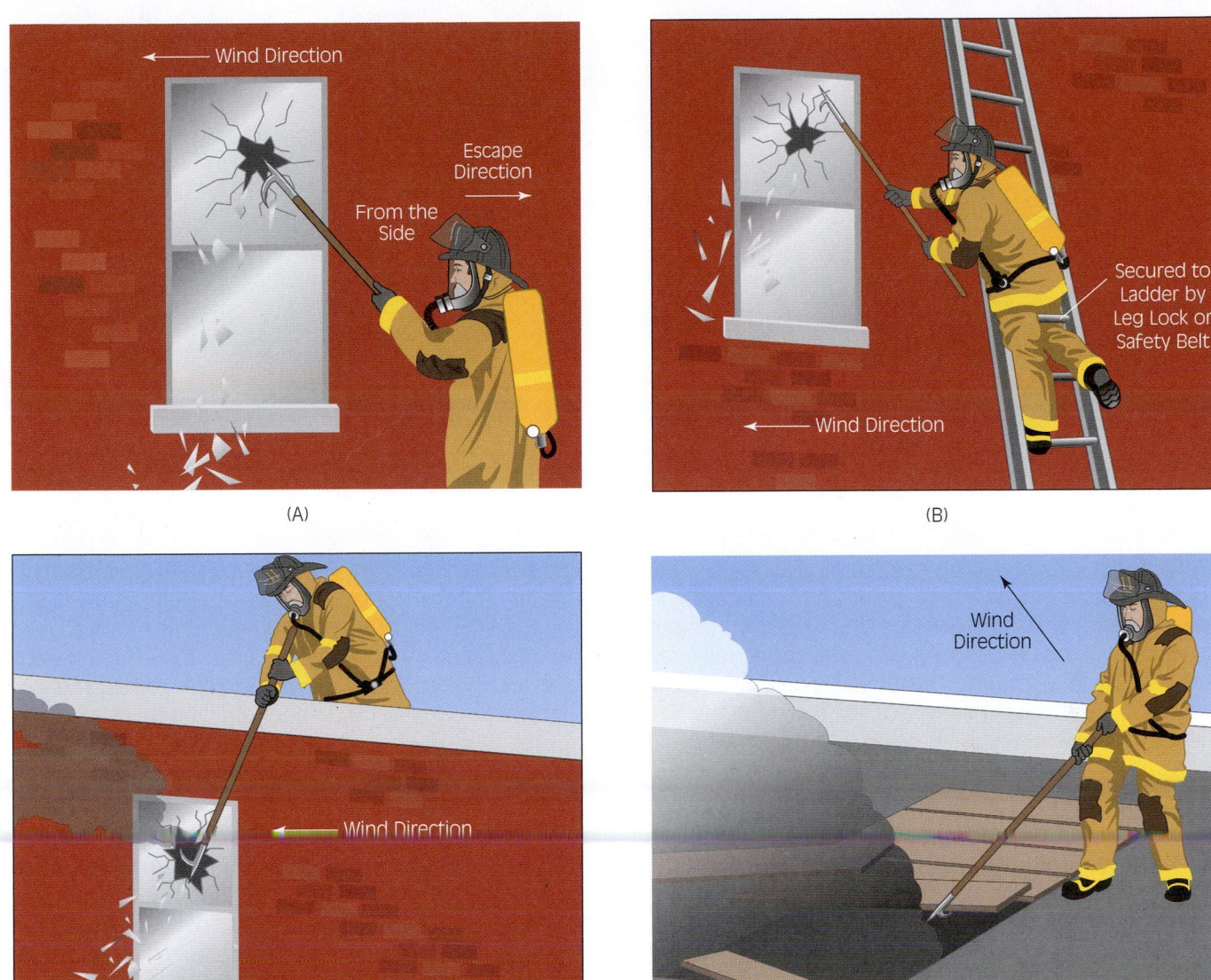

FIGURE 18-22 Making a ventilation hole requires some preplanning. (A) Firefighters should make the hole so that heat, smoke, and possibly flame do not envelop them. They must not cut off escape routes, and they should work so that the smoke is blowing away from firefighters' positions. (B) When working off a ladder, the same general precautions are necessary. Firefighters must be secured to the ladder before performing any action. This could be through use of a safety belt or a leg lock. (C) When venting from above, firefighters use the wind to their advantage and stand off to one side so that they are not standing in the path of any initial billow of heat. (D) When pulling off roof boards, firefighters should work in the clear air with the wind blowing smoke away, and be careful with roof debris. It will most likely be hidden in the smoke.

to do is check for overhead obstructions. The ladder is placed to the side of the window against the siding of the house. Then the base (or heel) is arranged and measured so that the tip will fall into the glass at about two-thirds the height of the window. The ladder is then repositioned in front of the window perpendicular to the ground and shoved into the building. The tip of the ladder crashes through the upper pane and often comes down through the sash into the lower pane. The disadvantage of this technique is that sometimes only the top sash window breaks and window hangings remain in place that impede airflow. It is important to know how the interior attack is progressing and the location of the interior team before performing this operation.

Although far from totally effective, it does afford a quick method of getting some ventilation into operation in several upper level windows in a short period

FIGURE 18-22 (Continued) Making a ventilation hole requires some preplanning. (E) When removing a skylight, firefighters work with the wind at their backs. It is sometimes less work to lift off the entire housing than to break out each individual pane of glass. In addition, the glass for these units is usually reinforced. Depending on the amount of tar used to waterproof the unit, it might be easier to just remove the panes of glass by bending back the frame and letting each individual pane slide out. (F) When using an ax to remove window glass, the flat side of the ax head should be used, not the point or the striking surface. The position of the hands on the handle will offer better control of the ax as it breaks through the glass.

of time. When completed with the initial openings, ladders can be repositioned and firefighters can climb them to clean out the rest of the glass, window dressings, blinds, and so on, and to enter the building for a search if necessary.

> **SAFETY**
>
> When using a tool or a ladder to break glass, large glass shards can ride down the tool or the ladder into the firefighter. This shard can shatter in the face of the firefighter or slice through protective equipment into flesh.

> **SAFETY**
>
> When operating on a roof, there should always be a second escape route provided opposite the entry point.

Negative Pressure Ventilation

Areas of high pressure will flow to areas of low pressure. A fan moves air through the use of this physics principle. In ventilation, positive pressure is created by a fan that compresses air through the use of fan blades. That will tend to flow to an area of low pressure in the direction in which the fan is pointed. On the back side

of the fan blade, negative pressure is being created. When placed in a window facing the outward flow, the positive pressure is forced out into the area outside the compartment. Along with that flow comes the heat and smoke trapped in the compartment. That is the principle under which a smoke ejector works.

Knowing this principle, it is better understood that the closing of all avenues of pressure equalization from the discharge side of the fan to the input side is essential. If not properly sealed, the air will churn or recirculate and merely circle around to the intake side of the fan, **Figures 18-23 A–D.** For effective air movement, a barrier must be created that separates the positively charged air from the negatively charged air sectors. In a structure where many windows are broken and there are holes in roofs and walls, this is not easily done. Therein lies the major weakness with this technique. Drapes of plastic and tarps can be employed to limit churning.

This type of setup can be very effective in limited access compartments such as cellars or basements where few openings exist. Fans can be placed over the opening in the cellar stairs or in a hole made in the first floor near a window and used to draw out the smoke through suction. This tool is primarily an overhauling aid and should not be used during the attack mode. The fan could bring fire to the location, placing the operators in a dangerous position.

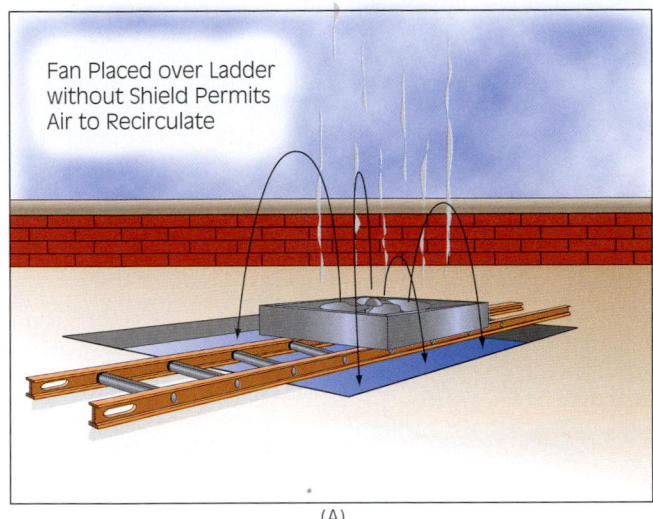

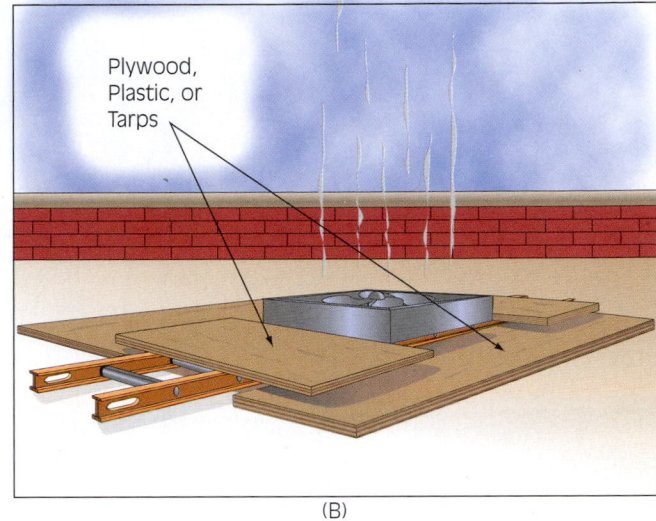

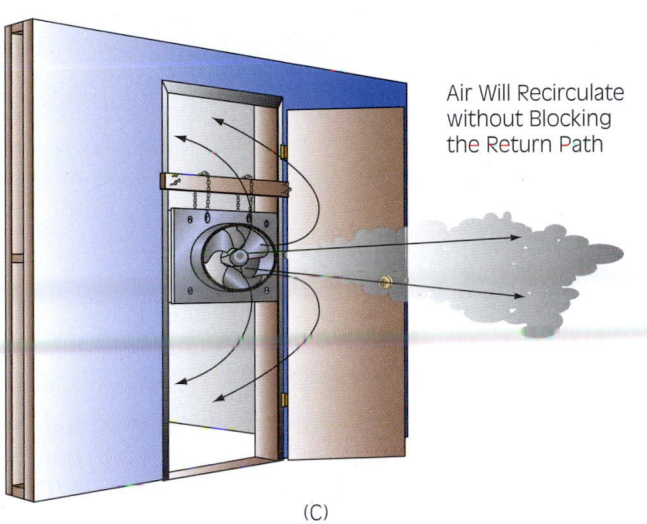

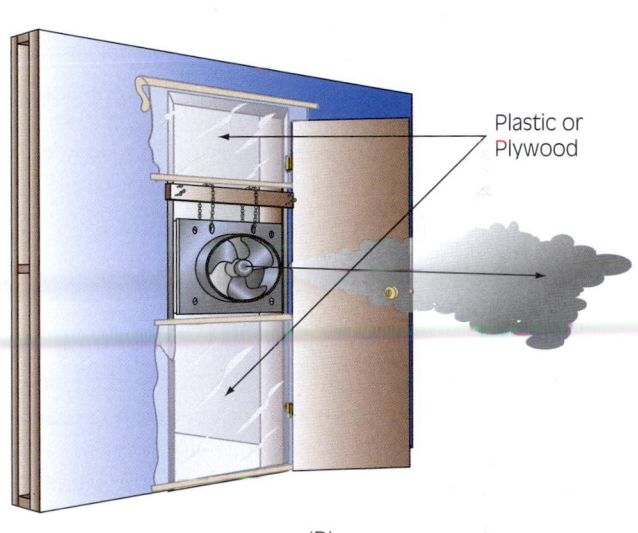

FIGURE 18-23 (A) When using an exhaust fan, it is important to cover the openings around the unit. Without a means to block the air from returning, it will merely circulate, reducing efficiency. (B) When covering any opening around the exhaust fan, the vacuum necessary to operate efficiently will be created and the exhausted air will not be sucked back around the fan. (C) When using an exhaust fan in a door, the air will circulate from the exhaust side into the intake side if no provision is made to block that flow. (D) Through the use of plastic, tarps, or even a piece of plywood, the air is prevented from being pulled back into the intake side of the fan.

Positive Pressure Ventilation

PPV uses an entirely different approach than negative pressure ventilation, but with the same principle. The positive pressure technique actually injects air into the compartment and pressurizes it. In an attempt to equalize, the smoke and heat are carried out into the areas of lower pressure outside the structure. There are many variations of this practice. The basic principle is to create a cone of air and force it from the outside to the inside, **Figure 18-24.**

Depending on the circumstances, fans can be set up to augment one another. Two fans can be placed side by side to cover a wide opening. They may be stacked to create an elongated cone of air. If positioned one behind the other, the rear fan can supercharge the front fan for greater efficiency. The fans can be placed in remote locations to augment the airflow from one part of the structure to another, and three or more fans can be used in conjunction with one another: one or two at the opening and others

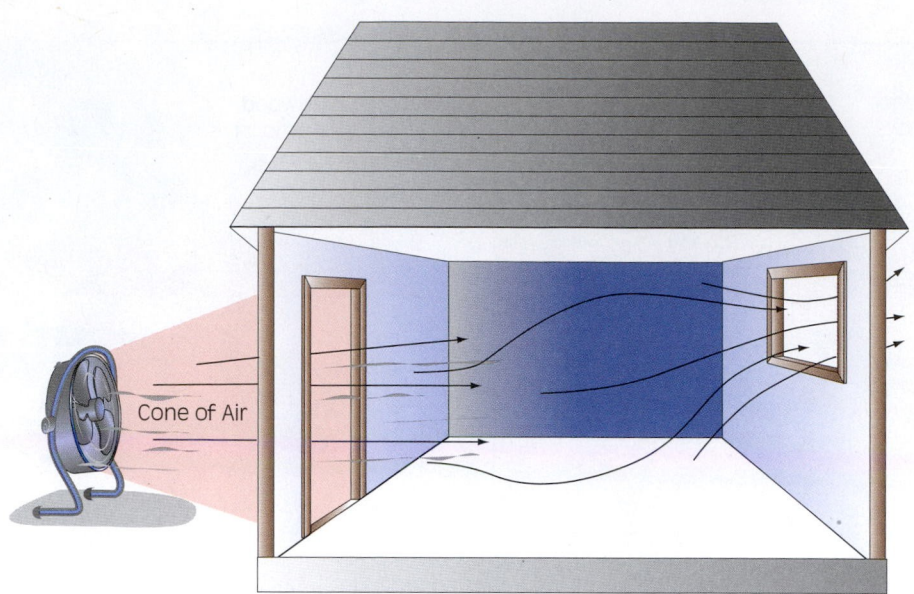

Cone of Air

FIGURE 18-24   With positive pressure ventilation, the theory is to actually pressurize the compartment and then the smoke and heat will actually be pushed out another opening. To be effective, certain actions must be taken. (1) The blower or fan must be placed a short distance from the opening so that a "cone of air" is created that just barely exceeds the opening being used. (2) The exhaust opening should be smaller than the introduction opening for maximum efficiency. That opening size depends on the number of blowers and their capacity. There are many variations where this practice can be effective. Stacking fans, using them in tandem, or placing one behind the other can have varying degrees of success depending on the structure and its configuration.

spaced throughout the structure in upper floors and along long corridors or split in halls that fork. Understanding the principle is all that is needed to create a custom application to fit an unusual situation.

Tests have shown that effective pressures can be generated up to twenty-five stories from the street opening location of a PPV blower. Augmentation beyond that level can still provide effective pressure generation. Intermediate fans will also assist in maintaining the flow.

Generally speaking, for every cubic foot of air injected into a compartment, a cubic foot of air must be ejected. The pressure loss experienced is due to the many openings in any run that will permit some air pressure to escape.

STREETSMART TIP

Practical experimentation of positive pressure ventilation can be employed in any structure (even the fire station) through the use of tissue paper. With a fan or blower set up, tissue paper strips are hung in openings throughout the structure. When the fan is activated, the airflow can be demonstrated, redirected to different openings, and observed. The velocity of the air movement can be observed by the degree to which the tissues in the openings are affected.

Positive Pressure Fan/Blower Setup and Operating Principles

To effectuate positive pressure, the blower should be set up outside the structure. The cone of air that is produced must completely cover the opening. This can be achieved by moving the fan closer to or farther away from the building opening. Generally, the size of the exhaust opening must be larger than the air introduction opening. Positive pressure is most efficient when the exhaust opening is ¾ to 1½ times the size of the entrance opening. Communication is essential between the intake side personnel and the exhaust side personnel. All members engaged in a positive pressure operation must be properly trained and understand the principles of the evolution.

Many different combinations of blowers and fans can be utilized. They can be placed in tandem, stacked, side by side, in-line, or in any configuration that accomplishes the goal of pressurizing the compartment. Depending on the size, shape, and number of cubic feet in the structure, many different layouts will be successful.

- Positive pressure evacuation of smoke is successful for distances of up to 1,000 feet.
- For each cubic foot in a building, a cubic foot needs to be exhausted. Multiple floors and multiple rooms

can be systematically vented by opening up a single room and window until it is vented, then closing the room off and moving to the next room. This technique is particularly effective in a school classroom-type building, motel, hotel, or similar room layout.

Depending on the size of the blower and the department's practice in using it in overhauling and/or attack modes, the novice firefighter should become familiar with the many different aspects of positive pressure techniques. The basic rules are as follows:

1. The introduction opening should be larger than the exhaust opening.

2. The air can be channeled into different parts of a building by opening and closing windows and doors.

3. A cone larger than the introduction opening is necessary to prevent recirculation of air back out of the structure.

4. Communication at the point of entry and the point of exhaust is important.

5. Practice will provide the understanding of the basic principles that can then be applied to actual fire scene operations.

ROOF VENTILATION

Vertical ventilation can be attained using several methods. One such method is to use the existing building features that allow for removal of heat, smoke, and the by-products of combustion. Those features include penthouse doors, skylights, and ventilation shafts. Care should be taken to ensure that use of these features would not spread the fire, **Figure 18-25.**

In terms of the tactical function, there are two types of vertical ventilation openings: offensive and defensive. The primary offensive vertical ventilation openings are generally referred to as offensive heat holes. These openings are typically placed over the fire or as close to the main body of the fire as is reasonable or necessary. The primary defensive vertical ventilation openings are generally considered strip or trench ventilation operations and are broken down into subtypes from there. The goal is to place offensive openings into the structure and evaluate the need for a defensive ventilation opening. Communication with interior forces will assist in the evaluation process for vertical ventilation operations.

The primary offensive heat hole should be cut directly over the fire if possible or in the area of greatest pressurization. The more direct the heat travel to the outer air, the less potential for extension of the fire. Based upon the amount of fire or pressurization, expanding the initial offensive heat hole may be necessary.

One example of defensive operations is the strip cut, also called a trench cut. Based upon the operation's objective and the overall operational strategy, these can be employed as a primary ventilation hole. The primary purpose of the defensive operation is to create a line of defense and, as a result, it is not generally ordered until the primary hole has been completed.

SAFETY

Firefighters who operate on the roof of any structure should sound the roof prior to stepping onto it and then sound it constantly as they move around the roof area. To do this, firefighters simply tap a tool onto the roof, listen for a sound, and make sure the tool bounces back off the roof. If the tool sinks into the roof and does not bounce back, that area is potentially unsafe for a firefighter to walk on.

Expandable Cut

An expandable cut is the most efficient type of cut for the time expended, **Figure 18-26.** It can produce a hole as large as is needed. Because ventilation is such a necessary operation, the roof must be opened quickly. After planning the cut so that the wind will tend to blow the smoke away from the work area, an initial cut line about 4 feet long is made. The second cut is a small triangular purchase point. Then a third, fourth, and fifth cut are made. Using a hook (pike pole), the purchase hole is used to hook the roof material and, like a giant lid, the top is lifted off. This type of cut can produce a hole of about 16 square feet (1.4 square meters).

Care must be taken to prevent damage to support rafters or cross members when cutting a roof. A support rafter can be felt when the saw suddenly begins to slow down as it tries to cut a thickness greater than the roof sheathing. The speed of the saw will increase and the operator will actually be able to feel the difference when the rafter or support beam is passed.

If a larger opening is necessary, making three additional cuts using one of the sides of the original hole as the first cut expands the hole. With only three more cuts, the size of the hole is doubled to 32 square feet 2.9 square meters. Again, if an even larger opening is required, another three cuts will increase the hole size by another 16 square feet (1.4 square meters) to a total of 48 square feet (4.4 square meters). If a larger opening is still required, the large hole can be finished off with only two cuts, **Figure 18-27.**

Occasionally, a close approach to a roof opening is impossible because of the heat condition. In this

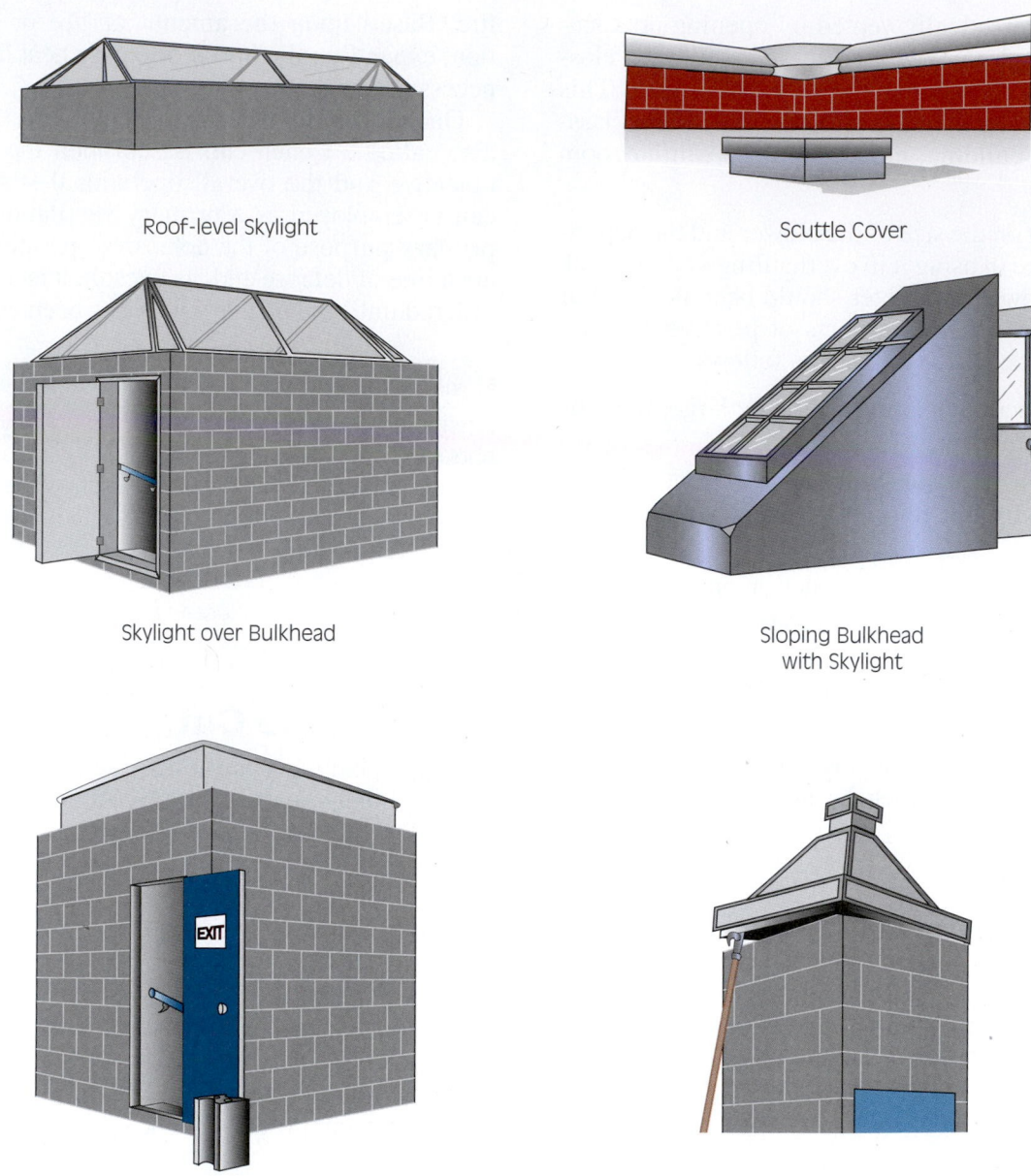

Roof-level Skylight

Scuttle Cover

Skylight over Bulkhead

Sloping Bulkhead
with Skylight

Bulkhead

Dumbwaiter Bulkhead

FIGURE 18-25   Many types of roof structures afford quick and effective initial ventilation. As is always the case, the time spent in creating a hole should be balanced by the amount of efficiency that will be gained. For example, opening a bulkhead door will be quick and effective. It should be performed before a roof cut is initiated.

case, the expanded cut might have to be performed by making as large a hole as possible using whatever part of the hole is approachable, **Figure 18-27** and **Figure 18-28.**

By comparing the numbers, the use of this type of cut will produce a hole that is 64 square feet. In terms of manual labor, if each hole were cut individually, it would take five cuts to produce each hole. A total of four holes would require twenty-five cuts to produce 64 square feet of opening. Using the expandable cut, the same opening is produced with only thirteen cuts. In addition, one large opening produces more airflow than several smaller holes of comparable area.

Among the disadvantages of this type of cut is the amount of debris produced. With each section that is opened, roofing material, roof boards, and wood with protruding nails are left lying about. Additionally, the hole produced is large and a firefighter can fall into the opening, even with the joists that are spanning the opening after the roof material is removed.

Two-Panel Louver Operation

The two-panel louver operation cut is primarily used as an offensive heat hole. It is often used on building construction that utilizes plywood sheeting under

Expandable Roof Cut Sequence
(Each Section Is Pulled as Cuts are Completed)

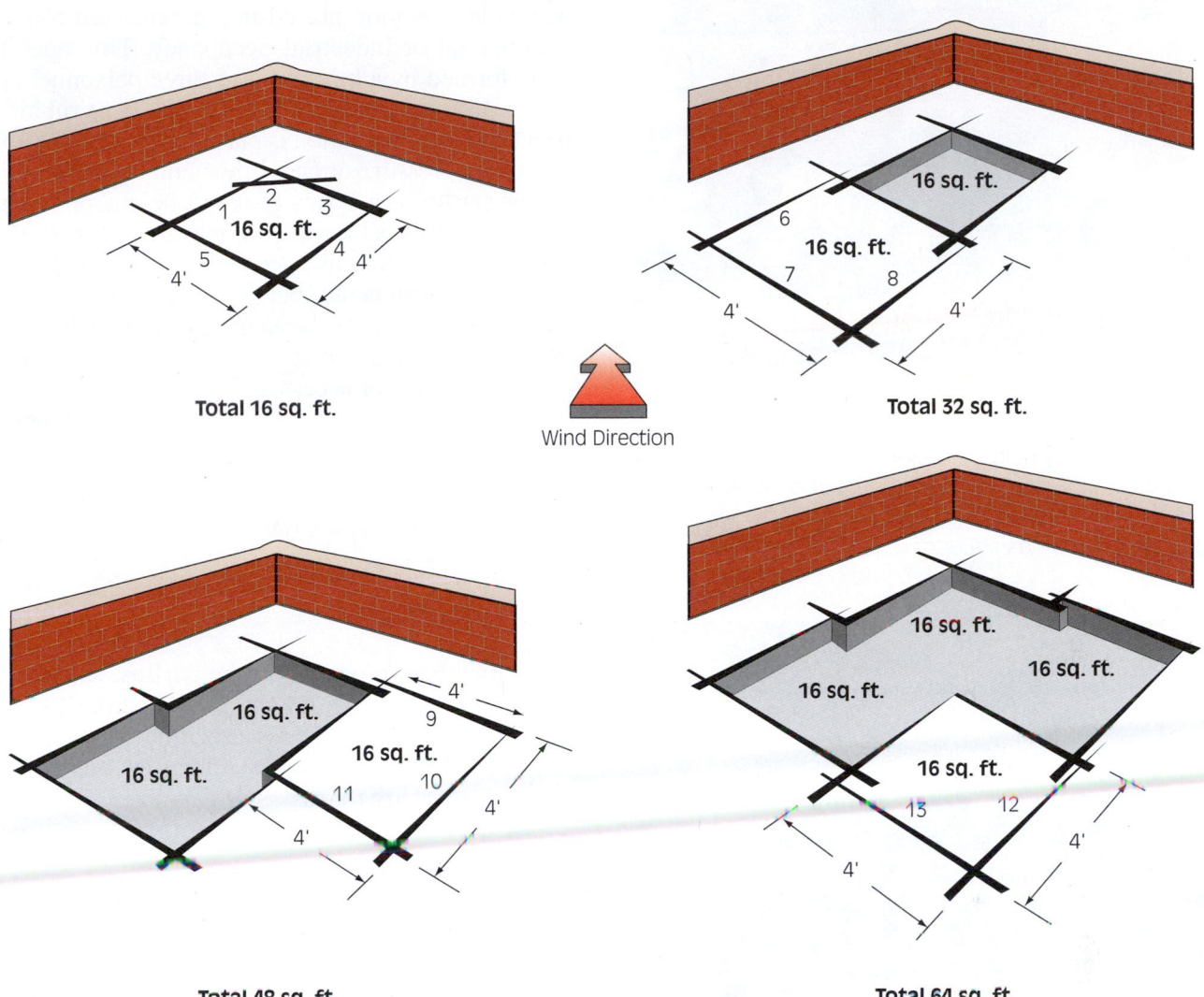

Wind Direction

Total 16 sq. ft.

Total 32 sq. ft.

Total 48 sq. ft.

Total 64 sq. ft.

FIGURE 18-26   When cutting open a roof is necessary, the expandable cut is the most effective, quickest, and most efficient way to get a large opening for vertical ventilation. Each new cut is merely an extension of the previous opening. The overall number of cuts required is dramatically fewer than if each hole were individually cut.

paper as a roof cover. It can also be used on roofs with 1 × 6 sheeting.

The two-panel louver operation is based on a series of cuts beginning with an outside cut, **Figure 18-29A.** The saw is placed into the roof decking and a cut is made toward the fire to indicate the outside #1 rafter with the nose of the chain saw bar. The saw is turned opposite the fire and the head cut is made away from the inital #1 rafter. The head cut continues to the next rafter #2, which is then rolled until the following rafter #3 is indicated. This head cut is stopped when the #3 rafter is contacted. Once the #3 rafter is contacted, the firefighter will make a downward cut toward his or her feet along the inside of the #3 rafter approximately 4 feet.

Next, the saw is moved over 4 to 6 inches and the same function is performed on the opposite side of the #3 rafter. The firefighter will then continue the head cut, moving away from the fire, rolling the next #4 rafter, and stopping at the last and final #5 rafter. The firefighter will then move back to the #1 rafter and make a downward cut approximately four feet. The bottom skim cut is established by tying into the fire side cut and skimming the bottom cut toward the means of access and egress. The final cut ties the 8th and final cut on the inside of the 5 rafter to the bottom skim cut. Once completed, the opening should be two 4 × 4-foot panels for a total of 32 square feet.

The opening should be evaluated for its potential effectiveness in pressure reduction and appropriate

FIGURE 18-27 Notice here that the opening had to be made in a linear fashion. In this fire, the heat was so intense that a rectangular cut was impossible. However, using the existing hole to make the next series of cuts could expand the opening. Also notice the amount of debris that accumulates. Visible in clear conditions, a severe tripping factor exists under smoke conditions.

FIGURE 18-28 The initial hole was cut and the "lid" lifted off in one piece. In the foreground, the cuts of the second hole can be seen along the right side of the hole. This type of roof, called an inverted roof, consists of a flimsy under-roof support system while the actual weight-carrying members exist at the occupancy ceiling level. This photo illustrates how little support is available.

ventilation. If the need to expand this opening is evident, the same steps can be repeated, working toward an area of safety and egress. The last step is to push in the roofing pieces creating louvers along the cut and allowing the by-products of combustion to escape.

STREETSMART TIP

Do not forget to clear all ceilings within the structure after performing vertical ventilation operations.

Louver in Lieu

A louver-in-lieu operation is typically an offensive heat hole operation placed into a panelized roof of a commercial or industrial occupancy. This operation is performed by a minimum of three personnel operating with a minimum of two saws and one rubbish or roof hook. The operation is placed in the area of greatest smoke pressurization while members are standing on the purlins—a series of wood beams placed perpendicular to steel trusses to help support roof decking—which in this instance run approximately 8 feet (2.44 meters) on center between beams.

Members access the area of operation and split into two teams, with the Company Officer on the team closest to the means of access and egress. This allows the officer the opportunity to observe the safety zones and escape routes as well as the ventilation operations.

JPR 18-2: Louver in Lieu

(For step-by-step photos of this skill sequence, see page 600)

1. As members progress up the purlins, smoke indicaters are placed into the roof decking to determine the amount of pressurization or involvment. It should be noted that if members encounter direct flame exposure to trusses or panelized roof assembly, they are to immediately retreat to an area of noninvolvement.

2. Once they have determined the area to perform their operations, the outside team member will place the saw into the roof decking and cut toward the fire to indicate the top side rafter or member. The next cut will be on the backside of the top side member and will establish the head cut.

3. The opposite side member will mirror the operations of the outside team member. Both members will cut back toward the purlin they are standing on and will stop once they have contacted the purlin.

4. Then, establishing a parallel cut 4 inches inside the purlin, the firefighters will cut back toward their means of egress until they encounter the first rafter.

5. Firefighters will roll this first rafter and will stop cutting at the next or second rafter.

6. They will then reach out and cut on the upside of the second rafter to establish a bottom cut.

7. Once the first louver in lieu is in place, the team will reestablish the head cut on the second louver by cutting a new head cut on the backside of the second rafter and following the same sequence of cuts for the second and all subsequent louvers in the determined operation.

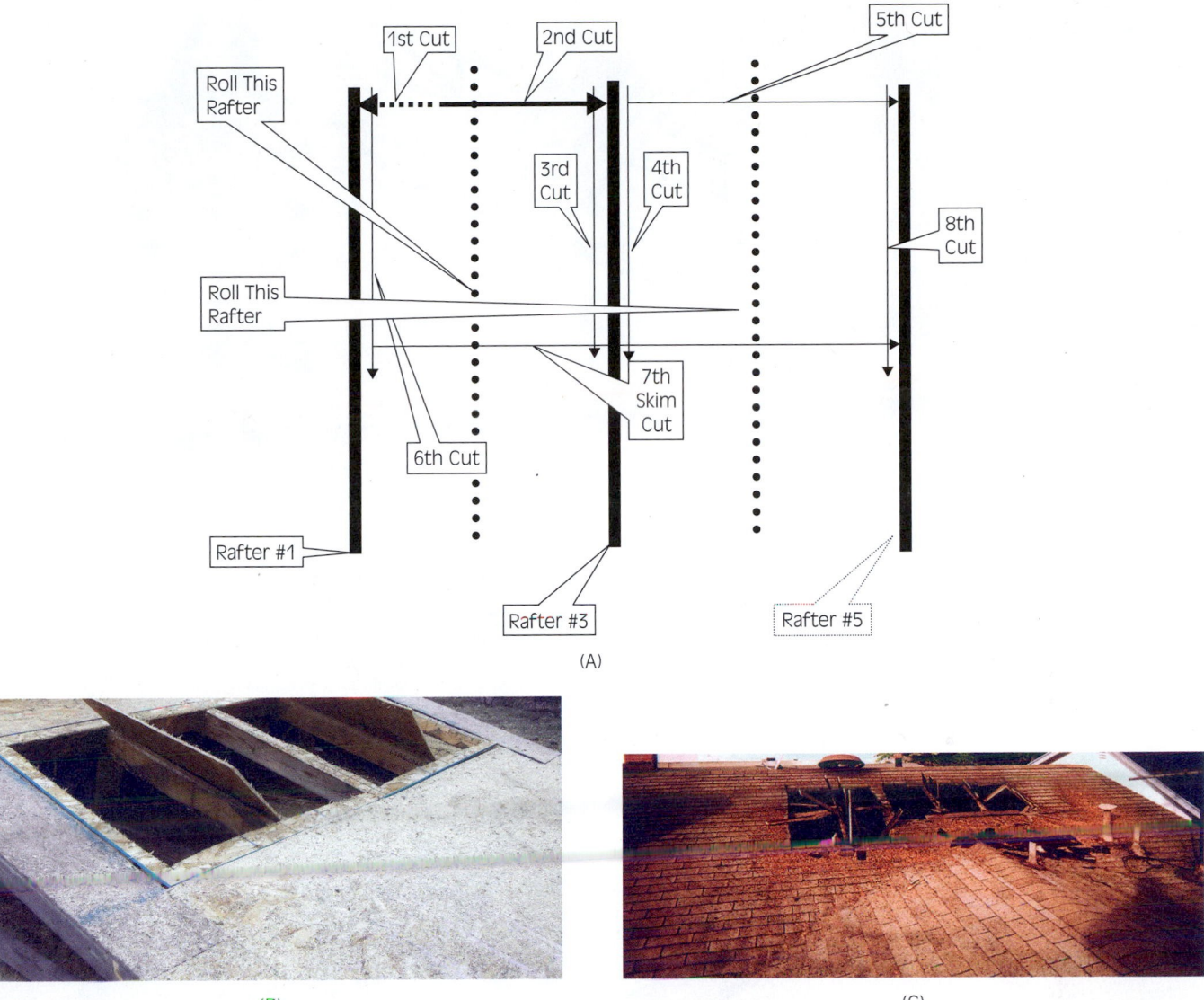

FIGURE 18-29 (A) Cutting sequence for a two-panel residential louver. (B) A completed two-panel residential louver will provide an approximate 4x8-foot opening in the deck of the occupancy. This maneuver should be the starting point of all residential operations. Once complete, personnel should evaluate the need to increase the opening while working toward their means of egress, or create another opening in a safe location toward the means of egress. (C) This photo shows a two-panel residential louver that has been extended to accommodate the need for further ventilation of the structure. The need for further ventilation is based on the interior conditions reported by interior crews operating within the structure and the amount of turbulent, pressurized smoke or fire issuing from the ventilation opening.

Triangular Cut

On some types of roofs, the use of the triangular cut technique is favored. A roof supported by an open web bar joist with Q-decking is the best candidate for this type of cut. Because the span of the web bar joist often exceeds 24 inches (60.9 centemeters), the opening can become a funnel if opened using conventional square cuts, **Figure 18-30.**

The use of a triangular cut will help support the underlying Q-decking because it is interlocked. By cutting across the panel on a diagonal, the tendency to sag will be reduced.

Of necessity, these holes will be relatively small and might require several cuts to adequately perform the ventilation mission.

Strip Cut or Trench Cut

Purely defensive in design and execution, the trench cut is a roof opening that ventilates the cockloft area or open attic space of a building where the fire is spreading under the roof, **Figure 18-31.** After the offensive heat hole opening is made, the strategy might call for making a stop at a particular location. It is usually predicated on structural elements inherent in the

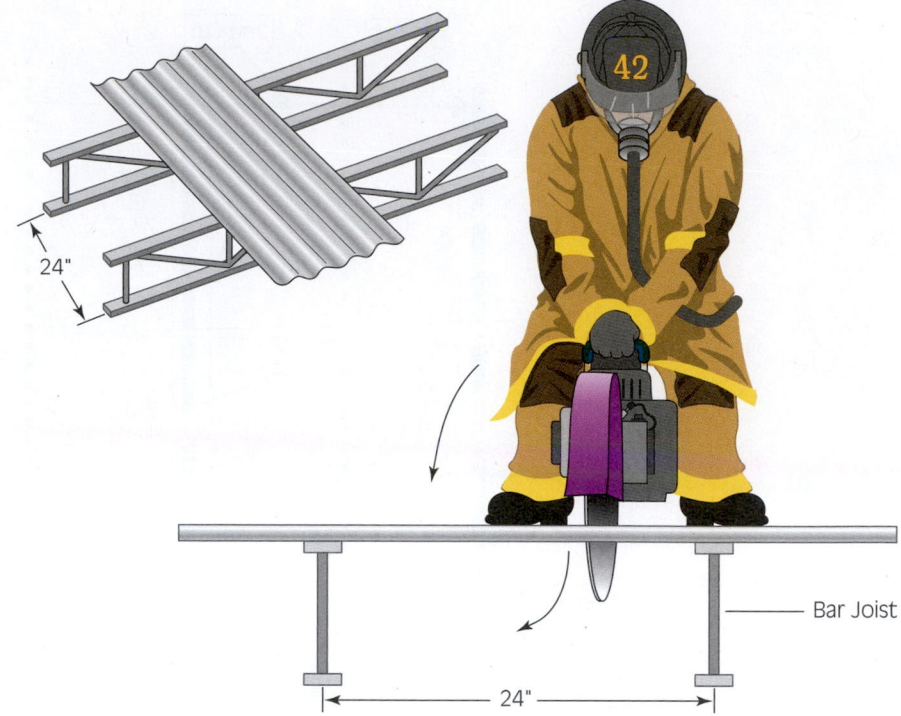

FIGURE 18-30 The open web bar joist-type roof structure requires extra vigilance. It is possible to cut away the supporting panel on which the firefighter is standing. Because of the large span between the supporting joists, the member can topple into the newly created hole.

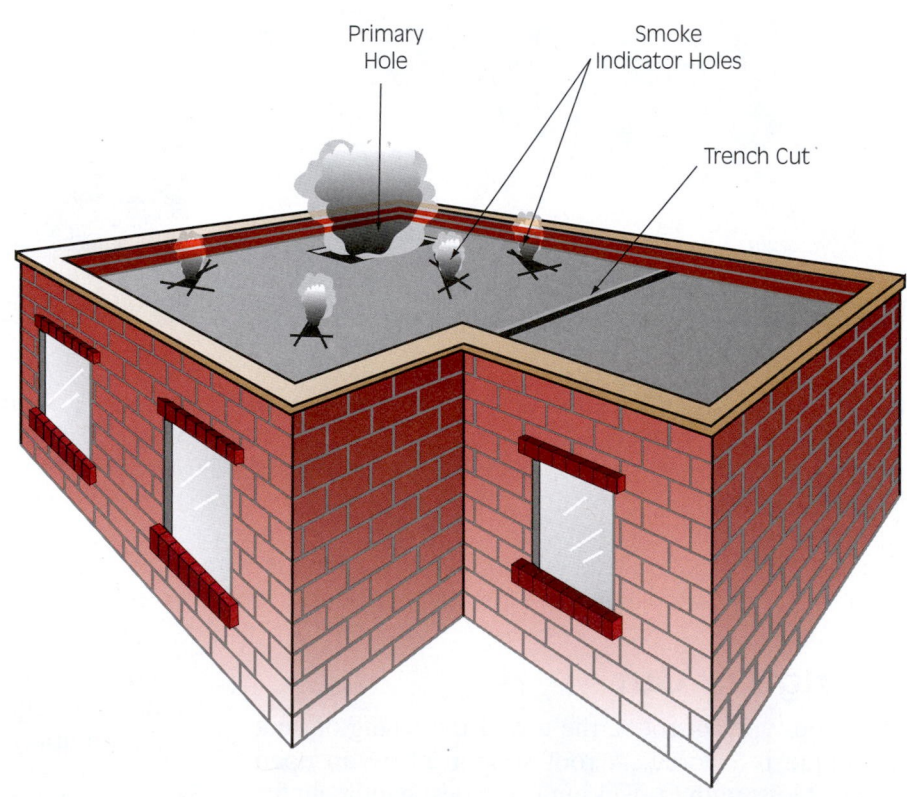

FIGURE 18-31 A strip cut is used to cut off fire extension. A totally defensive action, its use means that the fire side of the cut is being given up. Once opened, no firefighters should be operating on the fire side of the cut. A strip cut is (1) 2 to 3 feet wide, (2) not opened until the entire cut is complete, (3) not opened until all personnel are on the "safe side" of the cut, and (4) made using any available building features.

building itself, such as the connecting point of several wings or a place where the roof area narrows due to the presence of structural elements.

Along with the advantages and disadvantages of horizontal and vertical ventilation, the main advantage of the strip cut is that with the strip cut, the gases that are expanding horizontally under the roof are vented to the atmosphere and not permitted to pass a chosen defensive line, **Figure 18-32.** A disadvantage is that when the decision is made to use a trench cut, it is understood that the building on the fire side of the cut has been determined to be unsavable.

Firefighters must understand and determine the rate of fire spread in order to effectively determine

Primary Hole

Strip Cut

FIGURE 18-32 A trench cut is a defensive move. Ceilings should be pulled below the cut to promote vertical airflow through the trench. Additionally, a handline should be in place below the opening to cut off any horizontal extension.

the location of the strip cut. Improper location of the trench cut could result in unnecessary structural damage or increased firefighter risk during trenching operations.When a strip cut is being properly utilized, a hoseline should be positioned below the cut with ceilings pulled down to expose the opening. Any fire that actually reaches the opening and threatens to jump the gap should be extinguished with the hose from below. Under no circumstance should a hoseline be operated into a hole from the roof side. It will only serve to drive the heat and smoke down into the top floor environment. The purpose of this hole is to permit the laterally extending fire and heat to escape into the atmosphere where it poses little threat. A hose may be operated on the roof to extinguish a roof fire, but it should never be operated into the ventilation hole.

Below Grade or Basement Ventilation

Basement ventilation creates many challenges for firefighters. These include limited access, limited ventilation openings, collapse hazard, and superheated gases upon entry. These may be overcome by applying the ventilation principles discussed throughout this chapter.

Firefighters should attempt to locate any exterior basement openings, such as windows or cellar doors, prior to entry. Opening these will allow the firefighter to combine both horizontal and vertical ventilation techniques in order to release the smoke and heated gases prior to entry.

If no exterior openings exist, the firefighters must treat the ground floor much like a roof and apply roof ventilation techniques. The same safety considerations that are used when firefighters are operating on a roof also apply when firefighters are operating on the ground floor above a basement fire. As with roof operations, firefighters should always use sounding techniques, ensure adequate escape routes, and avoid directing fire streams through ventilation openings with attack crews present.

Ventilation openings must be made in the ground floor as close as possible to a window or other exterior openings. A fan can be placed at the top of a stairway with the cone of air covering the basement doorway creating positive pressure ventilation. To assist in firefighter safety, this method is ideal as it pushes heat and smoke away from the path of the firefighters as they descend the stairway. It is critical to begin ventilation, whether vertical or positive pressure, as early as possible because firefighter crews will descend down the very same opening where the heat and fire gases are traveling.

Inspection Cut

The first operation that needs to be accomplished on a flat roof is the inspection cut. The inspection cut is placed into the roof to determine the following:

- Roof covering and depth of covering
- Roof sheeting material

- Rafter direction
- Conditions directly below firefighters
- Type of operations the ventilation crews are going to accomplish

The first cut in establishing the inspection cut is made at 45 degrees to a bearing wall, followed by another cut opposite the first. Then the triangular inspection cut is completed with another cut, **Figure 18-33.** The goal is to identify the rafter and roll it when coming in contact with it. The inspection cut is larger than the smoke indicator hole and could pose a danger if placed in a path of access and egress to the ladder.

Smoke Indicator Hole

The smoke indicator hole is the only opening that will adequately determine the conditions directly below firefighters operating on the roof. The smoke indicator hole is a small triangular opening approximately large enough to push the D-handle of a roof hook through but small enough that the firefighters operating on the roof will not put a foot through the roof itself. A smoke indicator hole should be placed into the path of access and egress approximately every 15 to 20 feet of travel.

FIGURE 18-33 In this photo, personnel have accessed the roof and established the primary inspection hole opposite their primary means of access to the operational area and their primary means of egress to their ground ladder. The inspection hole is used to determine five key objectives: (1) roof covering, (2) roof decking, (3) rafter direction or roof assembly type, (4) conditions and pressurization immediately beneath the inspection hole, (5) operational objectives required to accomplish vertical ventilation.

SAFETY CONSIDERATIONS

When considering the placement of a hole for ventilation, whether it is horizontal or vertical, it is of paramount importance to consider the benefit gained against the possible liability created. In some cases the best choice will be not to vent at that particular location, for example, if venting would expose a victim and rescuer on a ladder to danger.

Will Ventilation Permit the Fire to Extend?

There is no justification for permitting a fire to extend in order to complete a task that is assigned. On occasion, an order may be given without full understanding of the consequences of that order because early reconnaissance is incomplete and fact gathering is still taking place. Some action may still be required but the firefighter should consider the possible consequences of that action and report to the incident commander for reevaluation if conditions are changing out of sight of that incident commander. Most incident commanders in the field would gladly countermand an order given when new information shows the original order to be unsound or unsafe.

Will the Escape Route Be Cut Off?

Of paramount importance to firefighters is their individual safety. When performing a ventilation activity, an eye toward escape must always be in the forefront. When venting a series of windows, the firefighter must work toward the escape point, **Figure 18-34.** There should always be two easily recognizable ways off a roof.

Escape routes should be lighted at night if possible and the presence of such escape routes made known to everyone on the roof. Rooftop LCES (Lookouts, Communications, Escape Routes, and Safety Zones) are the primary responsibility of the person assigned to the Roof Division or Ventilation Group.

Will Ventilation Endanger Others?

An overview of the operation in progress must always be in place when operating. Careful consideration must be given to what is called **mission vision,** a "blinder" view of the work being performed because the mission of the individual is overriding all other activities. The activity of one firefighter performing a venting task must never endanger the position of another person.

FIGURE 18-34 Ventilate in the direction of the escape route so escape is not cut off.

When opening a roof, care must be taken to advise everyone else on the roof where the holes are. In dark, smoky conditions, it is easy to fall into the fire.

In addition, the torn-up roofing material should be cleared away as soon as possible to remove the tripping hazards. As the roof team pulls off the roof, they should pile it up in one place if possible. Thinking ahead of possible safety problems and working to remove them before they present themselves is safety in action.

Work in Teams

A firefighter should never work alone. The presence of a team member will afford a second set of eyes and another mind at work. It is very difficult for any one person to be aware of everything that is taking place at the same time. At times the presence of another firefighter will afford the opportunity to quickly discuss the manner in which to attack the problem at hand. A team member might make the difference between being located when a floor collapses or making a misjudgment resulting in a fall down a shaft. The presence of a team member could make the difference between being able to pry open a bulkhead door or lift a heavy skylight and not being able to do it because it is too difficult for one person.

Proper Supervision

Unfortunately, in some places, the presence of supervision is all too often considered to be a waste of personnel. There should be a person on the scene to make decisions when conflicting options are available. The presence of recognized supervision will ensure that the effort is unified. Supervision also typically brings a certain amount of experience to the activity, because supervisors are usually the more experienced personnel.

A supervisor's presence also helps to prevent team members from becoming too focused on the mission itself, because the nature of a supervisor's job is to oversee the whole operation of the sector and coordinate subordinates' activities to meet the objectives of the teams under the supervisor's command.

OBSTACLES TO VENTILATION

The importance of ventilation cannot be overstated. Because of the unpredictability of fire operations, firefighters will be confronted with many unforeseen circumstances that will delay ventilation activities. Listed next are some of the obstacles that might be encountered.

Access

Access should be one of the initial size-up considerations when arriving at the scene. The firefighter should first assess the needs of the ventilation objective, then determine the route to be employed to reach the location of the job performance task. It might be via the adjoining building, if present, by use of a ladder, or perhaps from the ground. Roof access might be delayed because of surrounding impediments such as overhead wires, elevated trains or roadways, light pole placement, or any other obstruction that might delay placement of a ladder or apparatus, **Figure 18-35.**

Access to a rear yard might be impeded by the presence of high, locked fences or by the presence of watch dogs. Alternative measures must be employed and thought out in advance before the obstacles are encountered. In routine nonemergency responses or activities, it is a good practice to map out an access strategy for a particular building and then assume that such a route would not be available. The constant practice of formulating a second and third plan on a routine basis will permit the ability to do so at a real emergency with greater ease. It is simply a matter of practicing the thinking process routine until it becomes a reflex action.

Security Devices

The presence of security devices can impede ventilation both in access and in timing. Building owners may have barricaded themselves and their possessions behind gates, screens, steel doors, and closed-up windows. In some cases, window openings have been blocked off and sealed, **Figure 18-36.**

FIGURE 18-35 Roof access by ladder can be slowed down when obstacles are encountered. Aerial ladder placement must take into account overhead wires, structures, and presence of utility poles before being set into place.

FIGURE 18-36 In the interest of security, occupants will place steel security screens or completely close off an opening that might serve as an entry point for a would-be burglar. Notice the structural material behind the steel screen on the window to the left. Any obstructions to a full opening will reduce ventilation.

In high-crime areas, some shop owners have even removed skylights and replaced the hole in the roof with cheap plywood attached to inferior structural support, all held in place by a few nails. Once covered with roofing, the former hole is no longer discernible and has been the cause of firefighter deaths as members walked on the covered-over holes and fell through.

> **STREETSMART TIP**
>
> These unknown alterations mandate a very cautious approach to roof operations and often result in delay of roof ventilation procedures.

Height

From multistory skyscrapers to one-story garages, the firefighter must be alert to the structure's ventilation needs. Sometimes the need to cut a hole in an area that is out of reach of the tool at hand or from a particular position will be encountered. The need for a longer tool for reach or a rope tied to a tool or a ladder for access should be part of the initial size-up routine for the member assigned to ventilation duties. It must be assumed that reach will be a problem. What appears to be a single-story building from the street side can possess a topographical drop-off in the rear, placing the same floor that is at street level in the front several stories off the ground in the rear. The firefighter must practice thinking in a proactive fashion rather than reactive.

Poor Planning

One great obstacle to ventilation that is addressable is planning. Time is not a luxury. Proper on-site planning must occur without delay. A quick size-up and implementation of the plan are essential for timely ventilation. If ventilation is delayed, the interior team will suffer. Backdrafts or flashovers, decreased survival time of trapped victims, and arduous working conditions will result from delayed ventilation.

Personnel Assignment

A task that is assigned to a shorthanded or inexperienced crew will delay ventilation. If a ladder is needed to reach a setback in order to vent windows, and only one person is assigned to the task, not only will OSHA rules be violated, but objective attainment will be delayed.

Of course, there are times when adequate personnel are not yet on the scene. It is at this time that the ingenuity of the firefighter comes into play. With safety as an overriding concern the heads-up firefighter might be able to effect some ventilation while waiting for reinforcements. In a roof operation, a two-person team can open many openings relatively quickly while awaiting assistance. Structural components such as skylights, doors, roof access hatches, and ventilators can be easily opened or removed by individuals working together. A roof-cutting operation might need additional assistance but at least some openings are being produced to vent. Or two firefighters might be able to tie their tools together to get to a window that is out of the reach of just one tool. However, doing this while five other reachable windows are still not vented would be a poor use of resources.

Unfamiliar Building Layout

Especially in large buildings, the floor layout can be very confusing to the firefighter outside the building attempting to figure out which set of windows serves the fire area. This is particularly true of occupancy floor layouts that form L-shaped or U-shaped configurations in a building that appears to be square on the outside.

The building layout can also be an obstacle to a firefighter attempting to reach the rear of the building. Building wings, fences, lower floor extensions, or unusual configurations of the building's dimensions can confuse the firefighter and delay access. Some occupancies such as contemporary cluster-type construction townhouses can have door entrances on the same floor but serve three different levels. For example, there could be three entrances on the same floor. One door opens to a stairway that leads to the apartment on the floor below, one opens on the same floor as the entrance doors, and the last opens to a stairway that serves the apartment on the floor above. For the firefighter attempting to get to adjoining occupancy

on the floor above the fire to vent, the need to force open three different doors might be necessary and will delay ventilation.

Observing structures while under construction, requiring building owners to identify the occupancy served, conducting familiarization drills, entering pertinent information on response ticket critical information (if such a feature is available), or simply entering an uninvolved lower floor occupancy to look at the floor layouts are just some of the ways to address this problem. Often the same floor plan exists throughout a multistory building to reduce construction costs. Utility supply voids will often dictate the "rubber stamp" floor layout of multistory occupancies. The counter to this is those buildings where the tenants' preferences are addressed during construction and each floor is different. Again, here is another very strong reason for "walk throughs" and inspections.

Ventilation Timing

Venting too early can lead to fire extension; venting too late causes unnecessary punishment to the interior forces and can even prevent forward progress. The possibilities of backdraft and flashover have already been discussed. The timing is dependent on the type of ventilation being performed.

Vertical ventilation of stairways, hallways, and any paths of egress and ingress are paramount and must be effected without delay. These openings are commonly referred to as Firefighter Access Holes and are paramount in occupancies with a central hallway. If fire is known to be in the area under the roof boards or on a top floor extending into the cockloft, opening of the roof must commence without delay.

NOTE

The importance of early removal of the heat and smoke through vertical channels cannot be overstated.

As with all facets of firefighting, however, nothing can be written in stone. The opening of an entrance door to a fire occupancy on a lower floor that will vent up the stairs and out the roof opening must be carefully planned. There may be situations where escaping building occupants are coming down the stairs or firefighters are moving up the stairs. In such situations, vertical ventilation through this channel must be timed properly and may even have to be delayed. Delay might be necessary when opening a roof cut because the opening might expose an adjoining building to extension. A hoseline might have to be put into place to protect the exposure before the cut can be opened.

Cut a Roof—Open a Roof

Cutting a roof and opening a roof are different processes and this distinction should be clear to firefighters. Cutting a roof is the process of making the necessary cuts to perform the ventilation. Once the cuts are made, the opening of the roof can begin. The opening refers to the actual removal of the roof material. Therefore, a roof can be cut and not yet opened. There might be situations where a delay is necessary in opening the roof because of exposures or line placement or some other reason. This does not necessarily stop the cut. The cuts can be made but the opening delayed until the time is right.

FACTORS AFFECTING VENTILATION

Several factors can affect the attainment of proper ventilation. Partial openings, screens, type of roof material, wind direction, weather, building size, and construction features are some of the more prominent elements.

Partial Openings

Research has demonstrated that one single opening of a given size has greater ventilation capability than several holes that equal the same opening area. Much of this has to do with airflow characteristics. Like water, airflow is not like a chain of molecules that stay in line and move along in single file. Air is made up of many individual molecules that will flow in generally the same direction, but will tend to drift randomly in all directions. Air will move in the direction of less resistance. The eddy currents that can be observed when smoke is released into an unconfined and unchannelled area demonstrate this. If under enough pressure, smoke appears to billow.

Looking at a cross section of a chimney, it is clear that smoke rubs along the edges but tends to flow in a more uniform manner in the center. The greater the circumference of the vessel (i.e., smokestack) the less friction on the sides and the greater the flow. Once released into the atmosphere, the column of smoke continues to rise until the central core of the column is finally exposed to the reduced outside temperatures and begins to stratify. However, along the outside perimeter there is a constant peeling off of the smoke in eddy currents. The eddy currents will tend to reduce the flow, **Figure 18-37.**

Transforming the large picture of the smokestack into a roof opening, the same principles can be seen at work. With each opening, the outside perimeter of the

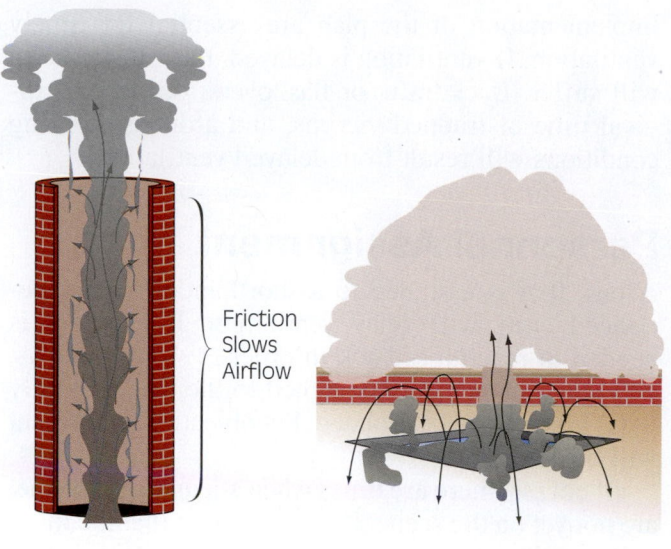

Friction Slows Airflow

Smoke Stack   **Roof Opening**

FIGURE 18-37   Airflow is reduced by friction.

opening will afford the airflow opportunities to eddy and slow down the overall flow. This is the underlying factor in the principle that many small holes will not be as efficient as one large hole. The large hole will have one perimeter of a given size. The many small holes will have a greater overall length of perimeters. The greater the length of perimeter, the more opportunity to reduce the speed of the airflow.

Partially Broken Windows

The same principle just described comes into play again. If a windowpane that is square is broken and many crescent-shaped shards of glass are left in place, not only is the area of the opening reduced but the presence of the shards creates more perimeter opening distance, **Figure 18-38.**

Screens

The presence of insect screens in a window that has been broken out is magnified when one considers that a screen is like a solid panel with hundreds of holes.

NOTE

Conservative estimates say that the presence of a screen in a window reduces the airflow by approximately 50 percent.

Put into real-time numbers, failure to remove a screen after taking out the window is like opening 100 windows, then permitting a shutter to close on

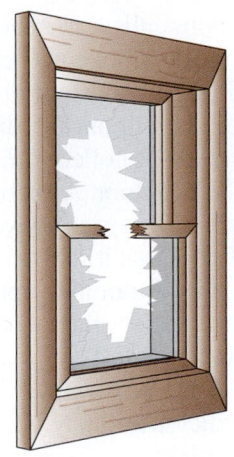 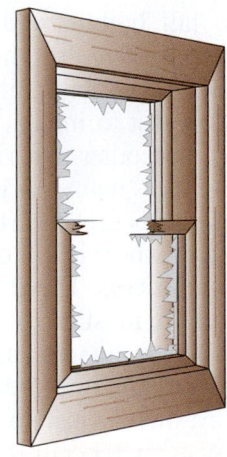

FIGURE 18-38 Airflow is greatest through a window where glass is fully removed. Screens, shades, curtains, and window cross members should also be removed.

44 of them. The effort is wasted and the ventilation is greatly reduced. Removal of the window treatments, curtains, shades, or blinds is also important. Any obstruction in the opening will reduce the airflow.

Roof Material

In some buildings, particularly those of older vintage, the roof material could be several layers thick and several layers might need to be removed before the cut is opened. When planning the cut, the firefighter must be aware that the kerf cut of the saw will usually permit a thin line of smoke to vent. If several layers of roofing must be removed, the original cut should not penetrate the under roof area. The cut should only go as deep as the roofing material that will be removed on the first pass, **Figure 18-39**. The smoke from the kerf cut can be enough to obscure vision and impede the initial roof material removal.

In some types of construction, the roof material might actually prevent the opening of a roof. The purpose of a roof is to shed water. Some materials are well designed for that purpose but are deadly in a fire. In the case of a corrugated metal roof over an open bar joist support structure, the heat from the fire melts the underside of the waterproofing material, usually some form of petrochemical product. Flammable gases are created and trapped below the cooler upper roof and tend to collect in the troughs of the corrugated roof. When the heat generates enough pressure, the gases find a way into the underside of the roof through the joints of the metal. The presence of this flammable gas in the cockloft area can be explosive when reached by fire. Additionally, the unsuspecting firefighter might be providing the missing oxygen as the roof is cut. A kerf cut is all that is necessary to

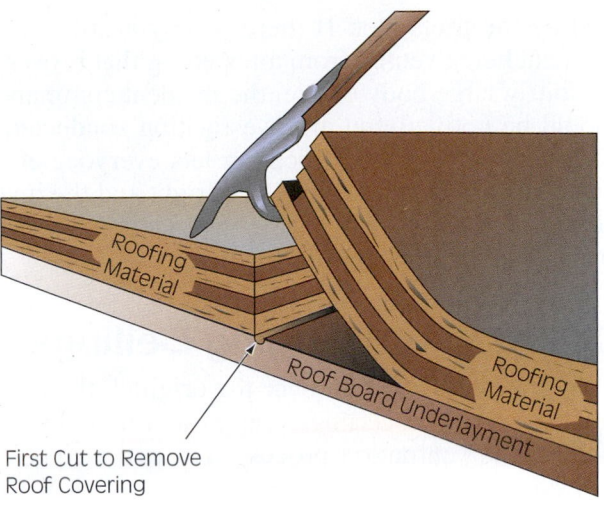

First Cut to Remove Roof Covering

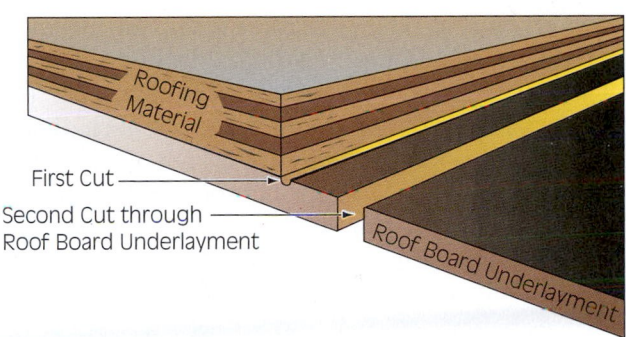

First Cut — Second Cut through Roof Board Underlayment

FIGURE 18-39 If a roof is thick with roof covering, two cuts might have to be made. The first cut is made shallow to remove the covering. It should not pierce the sheathing to prevent smoke from interfering with the completion of the first cut. With the second cut, the roof sheathing is cut and then removed.

open an oxygen supply route. Usually without warning the roof will erupt into fire at the place of least resistance—most often the place where the roof has just been weakened by the saw cuts.

Special care must be employed and constant study of the latest innovations and techniques is essential for the firefighter to maintain and improve a margin of safety in a known hostile environment.

Other types of roofs also have their particular strong and weak points. Any type of truss roof is always suspect because of the nature of its collapse potential. Placement of any personnel on this type of roof must be very carefully considered. The full extent of the fire must be known before commitment of personnel to this type of roof for ventilation is ordered.

In many cases, one roof has been built over another, sometimes with several feet between them. Obviously, the opening of the upper roof will have no effect on the ventilation of any spaces below the under roof. The effects produced by the opening made must be evaluated by the firefighter. If there is only a little smoke and heat being vented from an opening that is over an obviously large body of fire, the incident commander should be notified and an investigation conducted to determine why. The notification lets everyone at the scene know that something is not right and the investigation determines why it is so in order to address the problem and complete the objective.

Dropped or Hanging Ceilings

Like a second roof built over the original, the installation of dropped ceilings creates a similar impediment to the ventilation process. Trapped air pockets will conceal fire and smoke, and the raging fire on the floor will not be able to vent through the opening in the roof. In older buildings, especially of the commercial type that have a regular occupant turnover, several hanging ceilings may be in place.

The space between ceilings is a haven for fire gases to collect. Venting them can be a real challenge because the dropped ceilings are (1) not always obvious, (2) often out of reach of the roof firefighter, and (3) difficult to open from either above or below. The space is usually large and is ripe for a backdraft that will cause the entire ceiling to drop into the firefighting forces below.

Advancing Under a Dropped Ceiling

The airflow in front of a nozzle can work its way into the space above a hanging ceiling, especially if it is of the grid and insert type. The fire is pushed into the space when the heat lifts the tiles like trap doors. Heat and fire gases fill the space and, when conditions are right, can ignite and blow down the whole grid onto the advancing firefighters, trapping them in a web of metal grid.

STREETSMART TIP

When entering an occupancy with a suspended ceiling, the first thing that should be done is to have a tile pushed up from the safety of the doorway into the area to determine the conditions above the ceiling.

If fire is already in the space, the tiles can usually be unseated with the hoseline merely by directing the stream upward and extinguishing fire in the ceiling while advancing.

Building Size

The building size will affect attempts to ventilate. A building's height and width both have profound effects on the ability to move air.

In tall buildings, a phenomenon called a neutral plane can occur. It is the location where smoke will collect instead of continuing to rise. Fireground experience has found it to inhabit just a couple of floors in some cases, and other experiences have had it collect on up to ten floors many stories above the fire floor. Although certain texts give a particular number of stories above a fire floor where it will occur, the fact is that it will occur where it occurs. There is no good rule of thumb to apply because in most cases, the numbers are laboratory generated or results of carefully structured tests in existing buildings. In a fire, because so many variables are at play, there is just no way to predict accurately where this plane will occur if it occurs at all.

Factors that affect the development of a neutral plane include the presence or absence of HVAC systems and their associated ducts; wind direction around the building; presence of other buildings in the general area and height in relation to the structure on fire; outside temperature and its relative difference to the structure temperature inside; and presence or absence of smoke shafts in the building. The only way to determine what the smoke is doing in the upper floors of a high-rise building is to have reconnaissance teams report on their findings.

In large buildings, air movement can defy normal expectations. Wind direction on the outside of the building can create a negative pressure on one side of the building and cause the smoke to gravitate toward that section. This can become very confusing when smoke is being reported on a side of the building that is remote from the fire. Ventilating for such a fire can be a monumental task, and careful analysis of air movement as reported by reconnaissance teams must be conducted in order to maximize a coordinated ventilation effort.

The smaller the structure, the greater the similarity the air movement will be between buildings of the same type. Light breezes will not usually be much of a factor in a residential 2½-story wood frame building surrounded by trees and other dwellings. However, the same breeze in the same structure that sits high on an unprotected hill might exhibit very different characteristics. The same goes for a building that is facing a wide river or ravine as opposed to a structure that is surrounded by buildings of similar height, even if they are 70, 80, (21.3, 24.3 meters) or more feet high. There are no absolutes in this area. So many variables factor into the formula that only experience, observation, and common sense can prepare the firefighter to observe and predict what will most likely be the result of a given situation.

Weather

Some general observation can be made about how weather will affect air movement. On cool dry days, smoke-filled structures will vent quickly. On rainy humid days, the smoke will lift slowly. Snow affects the air movement and a light breeze will tend to "pull" the smoke out of the upper vertical openings. There is no magic to this knowledge. It goes back to fire behavior and what makes smoke rise.

The heated, smoke-filled air rises because the heated air weighs less than the surrounding air sample. In humid environments, the air is actually cooled and that slows the vertical velocity of the plume. While the fire is still pumping smoke out, the column flattens quickly as it cools and might not even leave the structure at the upper levels. That will tend to make it stay low and not lift.

Rain will have the same effect with an added element. The droplets of water will, in addition to cooling the superheated gases, actually impede the vertical flow physically. There are two forces at work: (1) The heat is lifting the smoke particles, and (2) the water droplet is colliding with the airflow as gravity pulls it down resulting in a diminished vertical airflow. The ultimate result is a slowing of the venting process.

Snow affects the rising column of smoke even more profoundly. If the rising air is hot enough, the snow melts, robbing the column of heat and cooling off the air volume. The result is water that further cools down the thermal plume. Snow adds an additional factor of air resistance. Since it has weight, the snowflake will actually force the thermal plume to work harder to rise. The combination of the additional weight to lift, the cooling of the snowmelt, and the water that does further cooling works against the venting process. One can expect difficulty when trying to ventilate vertically in snow.

Horizontal venting, however, is not necessarily affected the same way because gravity or thermal plume generation does not generally affect it. Heated, pressurized air will look for a path of least resistance and will find an opening in a wall (i.e., window or door) regardless of the weather conditions. However, once outside the structure, the horizontal venting turns to a vertical direction and up the side of the structure. At that point the smoke will be affected as described earlier. Once the fire is extinguished and the source of the heat is removed, the ambient atmospheric conditions will tend to move indoors. When the humidity reaches the inside of the structure, the smoke will then tend to stay low and not lift.

Great success can be achieved with positive pressure ventilation or with the use of a hoseline stream to create negative pressure and pull the smoke out. A fog or spray stream can be used to move air by opening the nozzle and directing the stream out the window or door. The droplets of water actually push the air in front of it (positive pressure) and pull the air behind it (negative pressure). The overall effect is the creation of an artificial airflow that can be very successful.

Opening Windows

Opening windows is the simplest way to open a compartment. Not every fire requires the removal of glass. Sometimes areas unaffected by fire can become filled with smoke and merely need to be vented to remove smoke. For many years firefighters were taught to open windows two-thirds from the top and one-third from the bottom to effect ventilation. That concept has changed. Based on what was previously discussed it is now known that one large hole is more efficient than several small holes, so one full sash opening is better than two equal in area. The same holds for windows with some exceptions.

Generally, it is better to open the top sash fully and let the airflow out. The replacement air will come from the open door and the lower levels of the room. There are exceptions to this rule of thumb, **Figures 18-40 A–E.**

If the smoke condition from the door opening will make the room conditions worse, then the choice would be to close the door and open the window. If there is one window, the two-thirds/one-third rule is appropriate. The heated upper air will vent from the upper level and the cooler fresh air will enter through the lower opening.

If there are two windows, one should be fully opened at the upper sash and the other opened fully from the bottom sash. Obviously, the greater the opening at the upper level, the larger the heated air volume that can be allowed to escape.

If there are three windows, two can be opened at the top and one at the bottom. If there is a breeze or wind blowing into the outside wall on which there is a window, the window facing the breeze should be opened from the bottom. The forced air will then actually pressurize the room and the vent/exit opening will be at the upper levels where the heat has collected.

The preceding scenarios represent ideal conditions. The firefighter must evaluate the effect that is being created and be prepared to adjust the plan if the desired effect is not forthcoming. Again, the many variables that tend to be present can change the rules. Nothing can be assumed and everything must be monitored until the incident is over.

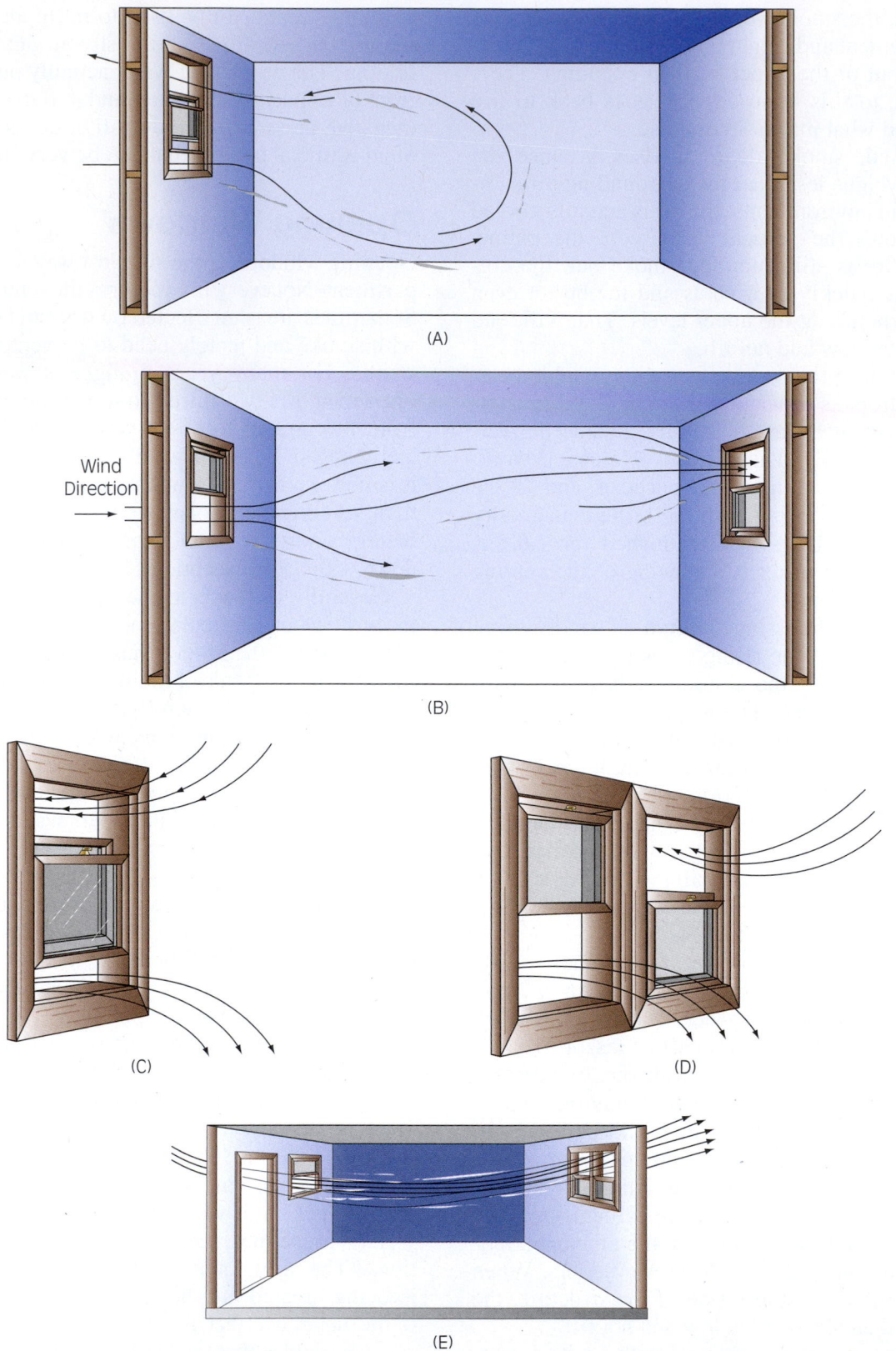

FIGURE 18-40 Sash windows should be opened to maximize natural airflow characteristics while maintaining the principle that large openings are more efficient than many small openings. Heat rises and cool air sinks. Depending on the number of windows and their placement, various different combinations will be effective. (A) Single window in a room. Open it two-thirds from the top, one-third from the bottom. (B) Two windows in a room. Open bottom sash of window facing wind fully. Open opposite window fully from the top sash. (C) Single window; another view. (D) Double window. Open one sash fully from the top and the other sash fully from the bottom. (E) Cross ventilation. Open window sash facing wind fully from the bottom and the others across the room fully open from the top.

JOB PERFORMANCE REQUIREMENT 18-1
Using a Rope and Tool to Ventilate a Window

A Make sure the rope is securely attached to the tool using an approved slipfree knot. The tool is then lowered to the window.

B Wrap a turn of rope around your hand and toss the tool out as far as possible in a horizontal direction. The tool is then swung into the glass and will break out the window.

Note: Before performing this task, make sure there are no firefighters or victims in the building or below the building.

JOB PERFORMANCE REQUIREMENT   18-2
Louver in Lieu

A   Detailed drawing identifying cutting terms used in louver-in-lieu operation.

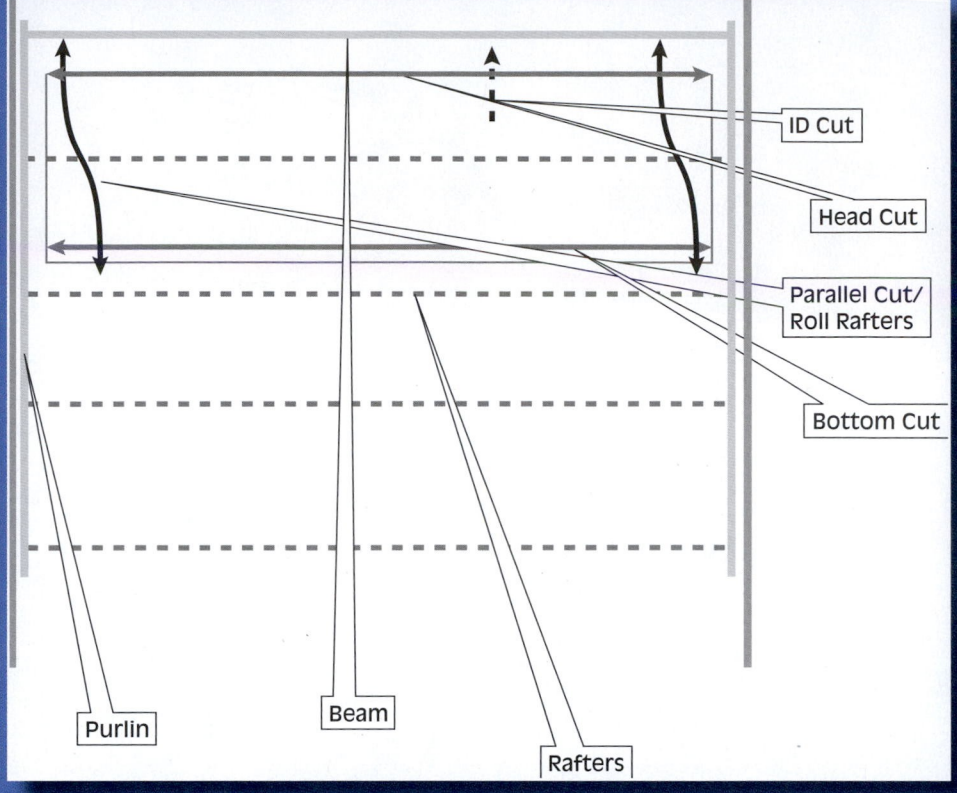

ID Cut

Head Cut

Parallel Cut/ Roll Rafters

Bottom Cut

Purlin

Beam

Rafters

B   Once the lead saw (firefighter on the right) has established the ID cut and the primary head cut, he will begin the parallel cuts 4 inches inside the purlin. The secondary saw (firefighter on the left) operates in unison to establish the head cut, followed by the parallel cuts.

C   Here both saws work in unison back toward their means of egress, extending the parallel cuts and rolling the first rafter with which they come in contact.

D Another view of the ventilation team making the parallel cuts approximately four inches inside the purlin, working back toward their means of egress.

E Once the ventilation team has made contact with the second rafter or support member, they stop cutting any further and establish the bottom cut. In this photo, the ventilation team has come in contact with the second support member and are stopping the parallel cuts.

F This photo shows the fourth and final cuts in the sequence. The bottom cut is made approximately four inches on the top side of the second support member.

G This photo shows a successful louver-in-lieu operation, in which a tactically offensive heat hole was established in a commercial occupancy heavily involved with fire. The louver in lieu is approximately 32 x 8 feet and quickly increased the visibility and tenability for interior companies.

LESSONS LEARNED

Ventilation is a tool that is used in firefighting just as any other tool is used. It must be understood, manipulated to the greatest advantage, and used carefully. Its proper use can make the difference between extinguishing a fire and creating a conflagration. Its judicious use can enable a firefighter to enter a structure to make a rescue. Used inappropriately it can permit a fire to extend into uninvolved portions of a structure.

There are no hard and fast rules beyond some very simple truisms. Heat rises and cold air drops. Heated air expands and cooled air contracts. Natural air movement will follow the path of least resistance. Airflow can be artificially generated by mechanical means. Beyond these, the rest depends on the conditions.

KEY TERMS

Cockloft The area between the roof and the ceiling.

Diffusion A naturally occurring event in which molecules travel from levels of high concentration to areas of low concentration.

Frontage The portion of a property that faces and actually touches the street.

Guard Dogs Trained animals that will bark and attack an intruder.

Horizontal Ventilation Channeled pathway for fire ventilation via horizontal openings.

Lifting A term used to describe the removal of upper level smoke and heat when cool air replaces the upper level hot air that is escaping.

Mission Vision A term used to describe a condition in which a person becomes so focused on an objective that peripheral conditions are not noticed, as if the person is wearing blinders.

Pipe Chase A construction term used to describe a void designed to house building water supply and waste pipes. The term electrical chase is used for wiring.

Thermal Level A layer of air that is of the same approximate temperature.

Vertical Ventilation Channeled pathway for fire ventilation via vertical openings.

Watch Dogs Trained dogs that will bark and create a commotion, but will not attack.

REVIEW QUESTIONS

1. Define ventilation.
2. What are the advantages of venting a structure?
3. Name the two types of ventilation.
4. Define the terms backdraft, flashover, and rollover.
5. Describe an expandable roof cut.
6. Describe the concept of the two types of forced ventilation, negative and positive.
7. What is the difference between opening a roof and cutting a roof?
8. A defensive strip or trench cut must be done on which side of the fire wall and why?
9. The louver-in-lieu operation is an important part of the ventilation process. Is it defensive or offensive, and how many people must be assigned to make it happen?
10. When breaking glass for horizontal ventilation there are two distinct steps. Describe them.

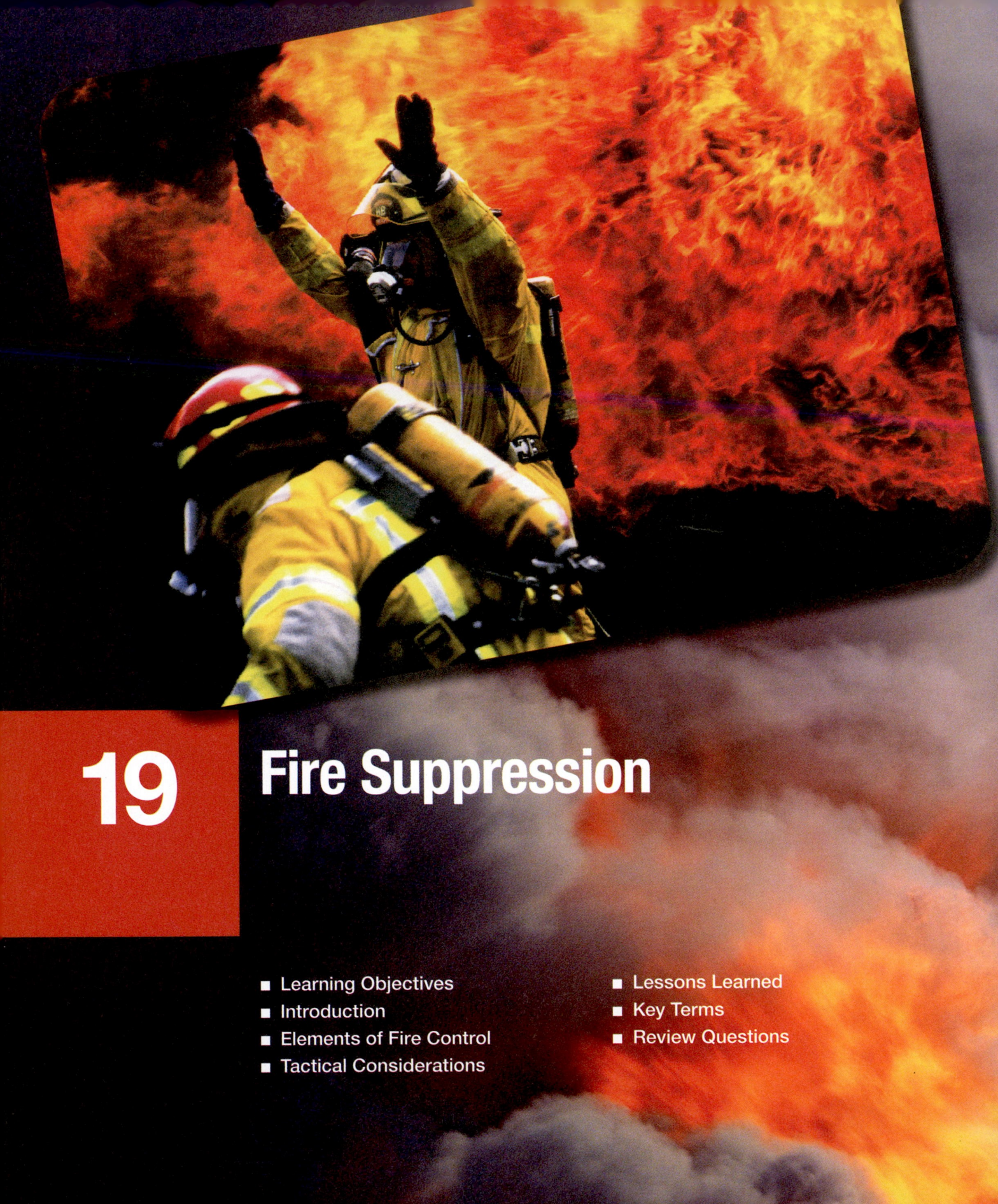

19 Fire Suppression

- Learning Objectives
- Introduction
- Elements of Fire Control
- Tactical Considerations
- Lessons Learned
- Key Terms
- Review Questions

The Monroeville Fire Department's Rescue 4 arrived at the scene of an apartment building in the early morning hours to find heavy smoke and fire venting out of a basement apartment. The building was a **garden apartment** structure with two stories in the front and three in the rear because of the topography, since it was built into a hillside. Each floor had four apartments between stairwells. Each **stairwell** was protected by firewalls or separation walls. The basement apartment was fully involved in the rear. The crew stretched a preconnected 1½-inch hoseline through the front door and down the steps to attack the fire and protect the interior stairwell.

The other first-alarm units arrived and began normal fireground operations, search and rescue, water supply, ventilation, and secondary fire attack lines. The arriving chief quickly ordered the second alarm.

Upon my arrival, on a second-alarm engine, the fire had **autoextended** up the rear of the building, entering the cockloft, and was venting through the roof in several areas. The third alarm was sounded. We were assigned the task of stretching a 2½-inch handline up the southern stairwell, pulling the ceiling, and supporting the firewall. By supporting the firewall, I mean we were to ensure that the fire did not get past us into the next set of apartments. We were assisted by a second engine company with a 1½-inch handline. Although an uncommon procedure today, we operated for the duration of our SCBA bottles, exited for fresh bottles twice, and returned.

After our third SCBA bottle, we exited the building and were told to take a break. By this time the fire was under control, forward extension stopped, and the fire would just require overhaul. It was daybreak. The scene in the daylight was impressive. At the height of the fire, three elevated master streams had operated over the main body of fire while six handlines supported interior firewalls, preventing further horizontal fire spread.

What I saw and experienced reinforced everything I had learned from fire academy instructors and through training. For a safe and effective outcome, the suppression of fires requires a coordinated attack, with crews working under a controlled command system and all personnel carrying out the assignments given them. Firefighters must understand the importance of fire suppression techniques and understand the role they play on the fire scene. At this fire, spread was prevented by the incident commander knowing the proper application of fireground tactics and the firefighters understanding suppression methods, application of water, and working as a team.

—*Street Story by James Angle, Chief, Palm Harbor Fire Department, Palm Harbor, Florida*

LEARNING OBJECTIVES

After completing this chapter, the reader should be able to:

19-1 Describe hazardous conditions found on an emergency scene.

19-2 Describe the critical elements alerting responders to a hazardous situation.

19-3 Explain how the effects of fire can create a hazardous condition in a structure.

19-7 Explain the wildland fire triangle in relation to wildland or ground cover firefighting.

19-8 List and describe types of ground cover fires.

19-9 Discuss some of the features of topography to be considered in wildland firefighting.

19-10 Identify hazards that may develop while fighting a vehicle fire.

19-11 Identify the safe approach to a vehicle fire to avoid hazardous situations.

19-15 Describe safety equipment used by emergency responders to safely function at an emergency scene.

19-16 Demonstrate emergency responder operations while wearing personal protective clothing for various emergency situations.

19-17 Describe the methods of fire stream application: direct, indirect, combination.

19-18 List the advantages and disadvantages for each type of application: direct, indirect, combination.

19-20 Demonstrate the methods for accomplishing each type of fire attack.

19-21 Identify the proper type of attack lines and fire streams to use when extinguishing interior structural fires.

19-34 Explain the difference between offensive and defensive modes of fire attack.

19-35 Define the role of a secondary entry team as it relates to interior structural firefighting operations.

19-36 List tactical objectives when fighting structural fires at grade level and above and below grade levels of a structure.

19-37 Demonstrate the ability to advance charged and uncharged hoselines in various firefighting situations to include ladders and stairwells.

19-39 Demonstrate the methods for fighting interior structural fires at grade level and above and below grade levels in a structure.

19-41 Discuss the properties, principles, and safety concerns associated with utilities.

19-42 Recognize hazards associated with various utilities.

19-43 Describe proper methods for disconnecting utilities and the hazards associated with each.

19-44 Specify the proper safety equipment required for disconnecting utilities and its use.

19-45 Demonstrate recognition of utility control devices.

19-46 Demonstrate the operation of utility control valves and switches.

19-47 Describe safety principles and practices associated with illuminating an incident site.

19-48 Demonstrate the ability to illuminate an incident scene with available equipment.

19-49 Describe power supply capacity and limitations when illuminating an incident site.

19-50 Explain various illumination deployment methods.

19-51 Demonstrate the operation of fire department generators and lighting systems.

19-52 Demonstrate the ability to connect power supply cords.

19-53 Demonstrate the ability to reset ground fault interrupters.

19-54 List considerations and techniques for protecting exposures during structure fires.

19-55 List and describe parts of a ground cover fire.

19-56 Explain the methods used to contain or suppress ground cover fires.

19-57 Summarize safety principles and practices while fighting ground cover fires.

19-58 Demonstrate the ability to recognize and identify potential exposures during ground cover fires.

19-59 Demonstrate the ability to suppress a ground cover fire using water.

19-60 Demonstrate the ability to protect exposures during ground fires.

19-61 Demonstrate procedures for constructing a fire line and extinguishing a ground fire with appropriate hand tools.

19-62 Demonstrate procedures to maintain the integrity of an established fire line during ground fire operations.

19-63 Identify proper fire stream techniques used in fighting vehicle fires.

19-64 Identify hazards affiliated with advancing hoselines at a vehicle fire.

19-65 Describe signs of an effective firefighting stream when extinguishing a vehicle fire.

19-66 Identify fire suppression techniques to avoid accidents and injuries at vehicle fires.

19-67 Identify forcible entry techniques used for vehicles.

19-68 Identify the methods for overhauling an automobile fire.

19-69 Recognize various types of fuel used in automobiles.

19-70 Demonstrate how to control fuel leaks in automobiles.

19-71 Demonstrate the operation of various pattern options while flowing water through a nozzle.

19-72 Demonstrate effective vehicle fire extinguishment.

19-73 Apply proper suppression techniques to minimize flash fires at vehicle incidents.

19-74 Demonstrate the ability to advance a 1-inch (38-mm) or larger hoseline while applying the proper fire suppression techniques.

19-75 Demonstrate the ability to overhaul a vehicle fire in order to locate and expose concealed fires.

19-76 Identify hazards associated with alternative vehicle fuels.

19-77 Describe procedures for safely dismounting fire apparatus in traffic.

19-78 Describe procedures for safely functioning on an emergency scene.

19-79 Establish control zones using scene and traffic control devices.

19-80 Demonstrate mounting and dismounting procedures for fire apparatus.

19-81 Demonstrate the ability to perform work assignments in protected areas while following appropriate safety procedures.

19-95 Identify the proper type of attack lines and fire streams to use when extinguishing stacked, piled, or other outdoor Class A materials.

19-96 Identify hazards presented with stacked or piled burning Class A materials.

19-97 Explain how and when Class A extinguishing agents should be used on various burning material configurations.

19-98 Identify the tools and techniques used for breaking up Class A materials.

19-99 Explain why stacked and piled materials may not be easily extinguished.

19-100 Identify fire suppression techniques used to protect exposures and extinguish Class A fires.

19-101 Identify safety hazards associated with storage buildings and container fires.

19-102 Demonstrate safety procedures during inherent hazardous situations.

19-103 Demonstrate extinguishing Class A fires using the proper hoseline or master stream.

19-104 Demonstrate penetrating and breaking up Class A material with hand tools and water streams, evaluate its effectiveness, and modify if necessary.

19-105 Demonstrate extinguishment and overhaul procedures for exterior Class A fires in order to expose hidden fires.

19-106 Demonstrate the use of water application devices while fighting exterior Class A materials fires.

*The FF I and II levels, as defined by the NFPA 1001 Standards, are identified in different colors: FF I = black, FF II = red, additional information = blue.

INTRODUCTION

This chapter gives the reader information that pulls together all of the other information presented to this point. Like a recipe, which lists ingredients and then explains how to blend and cook them, this chapter explains how to put into action the techniques and methods that have already been discussed.

This chapter looks at some of the common types of fire as well as some common techniques of firefighting in different disciplines. It opens with the more common elements of fire control, detailing most of the areas with which average firefighters will be involved during their careers. In this section, seven subjects provide information needed when working in various situations common to the average fire department. A second section details tactical considerations. At the conclusion of this chapter, the reader will have a good foundation of knowledge in fighting fire and tactical situations involving a number of elements common to today's fire service.

ELEMENTS OF FIRE CONTROL

Before firefighters can respond appropriately to an emergency involving fire suppression, they must know the basic principles involved in the processes that create and sustain fire. This section deals with those areas common to a fire department. It discusses structural firefighting elements, ground cover or wildland firefighting elements, vehicle fires, flammable liquid and gas fires, the process of fire extinguishment, and proper stream selection.

In a previous chapter the fire tetrahedron and its importance in the suppression of fire was discussed. The elements of that tetrahedron are mentioned again here in an effort to instruct firefighters in safe, efficient, expedient fire control. A simple rule is that the earlier the fire department arrives on scene, and the more knowledge they bring with them in the fire suppression process, the lower the losses and the less risk taken.

Structural Fire Components and Considerations

When discussing structural fire components and considerations, it makes sense to look again at the fire tetrahedron, **Figure 19-1.** Most structural firefighting involves the suppression of Class A materials within the structure or as part of the structure itself. These elements are commonly suppressed by removing one side of the fire tetrahedron. The typical fire department will respond with water as an extinguishing agent and suppress the fire by removing the heat, oxygen, or both as quickly as possible. This may not seem difficult in its purest form but can be extremely complicated or hazardous if not done properly and safely. A number of structural firefighting considerations must be made prior to the extinguishment, or a great number of things can go wrong.

Listed are a number of factors that must be taken into consideration.

1. Length of time the fire has been burning

2. Building construction materials

3. Occupancy type and contents

4. Resources available (amount of water, staffing, equipment, etc.)

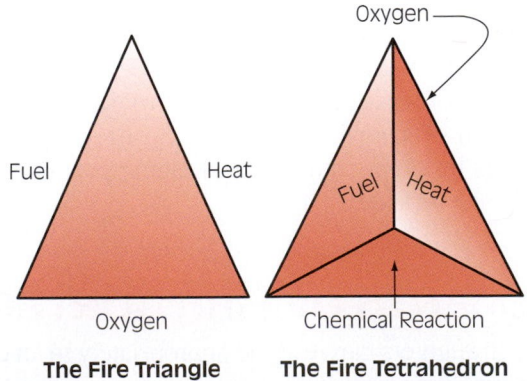

FIGURE 19-1 The old and new ways of visualizing the combustion process, the fire triangle and the fire tetrahedron.

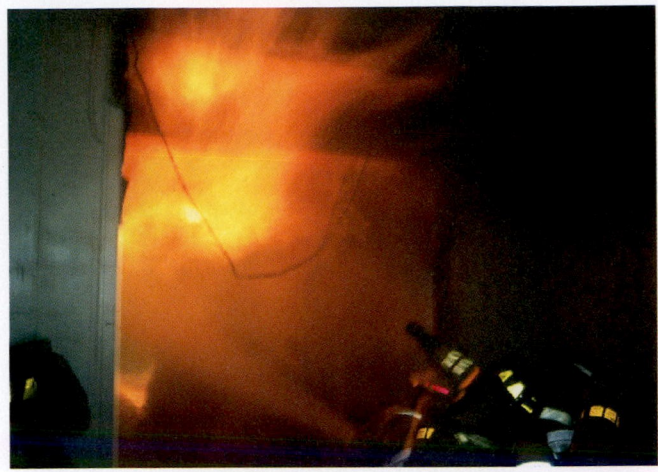

FIGURE 19-2   A flashover may occur in the early stages of a structure fire. *(Courtesy of Phill Queen)*

Taking these factors in order of listing and not in order of importance, recommendations can be made as to how a structure fire should be fought.

The length of time the fire has been burning is an important factor because it can determine the stability of the structure and the amount of fire the responding units will face. It will also determine which stage of involvement the fire is presenting.

As discussed in previous chapters, flashover and backdraft are extremely dangerous fire phenomena that are predicated on the stages of fire in structures. In **Figure 19-2** a flashover is shown as it occurs over the heads of a couple of firefighters. Based on length of burn, the integrity of the structure itself can be in jeopardy.

> **CAUTION**
>
> Most structure fires begin in contents and then spread to the structure itself. If not contained, this structural involvement can create concerns about collapse, both inside as well as outside.

This brings up the second factor: the building materials themselves. Is the structure made of highly combustible or noncombustible materials? How long will it take for the fire to become involved in the structure itself? **Figure 19-3** shows an apartment fire that began in its contents and then burned into the structure. Chapter 13 (in Sections I and II) details the types of construction materials and demonstrates the dangers involved in structural firefighting.

The third factor listed is the building occupancy and contents. Building occupancy means the proposed use of the structure. Is it designed to be an automobile repair shop or a preschool? Is it a retail store or a board and care facility? Each of these occupancies presents different fire attack protocols. The life hazards, the

Structural fire components and considerations are a large part of the responding agency's responsibility. Fire service personnel must be well schooled in all of these factors and situations so as to ensure efficiency, safety, and responsibility.

Ground Cover Fire Components and Considerations

In some areas of the country, ground cover fires are a very large part of the fire service's work. Not all departments have jurisdictions with open areas that burn, but those that do find it to be a responsibility that is not taken lightly. In fact, large campaign fires are usually wildland (ground cover) fires. These fires, being so large and involving so many resources, usually cause the first-in departments to utilize other departments from miles around, and in some cases out-of-state departments, to assist in the attack. **Figure 19-4** shows a large wildland fire and its impact on the surrounding area. So, even if a department does not have any open land to burn in its response area, any firefighter in any department could be summoned to help in another agency's wildland area. This being the case, it is a good idea for all firefighters to know the components and considerations involved in ground cover firefighting.

In Section I, Chapter 4 the fire tetrahedron was discussed. In the previous section of this chapter the same tetrahedron was mentioned, but in the field of ground cover firefighting a fire triangle is used, **Figure 19-5.** In wildland firefighting this triangle reflects a different set of principles related to a different set of conditions when compared with structural firefighting. Each side of the triangle can be broken down into smaller components in order to understand the components and considerations of ground cover or wildland firefighting.

FIGURE 19-3 A contents fire has burned into the structure itself. *(Courtesy of Rocco Di Francesco)*

entrance accesses, and the ventilation components are among many details associated with the occupancy types that have to be taken into consideration by the responding units. Along with the occupancy, the building contents must also be considered. In most cases, the contents will be predicated by the occupancy type. The preschool will have dramatically different contents than the repair garage. These contents will have a bearing on the suppression efforts and control problems encountered by responding units.

The fourth factor listed is resources available. Fire suppression involves all of the factors listed, as well as the resources available to the firefighting team. Some areas of the country have limited water supplies while others have practically unlimited water. Some departments have large numbers of responding apparatus while others have only one engine with the second responding unit a very long distance away. Some areas of the country have staffing problems while others do not. Some areas have unlimited training opportunities while others do not. These and many other situations are considered when dealing with response resources. These resources—or the lack thereof—have a great bearing on the tactics and strategy used in fighting structure fires.

FIGURE 19-4 The Baker fire in Southern California lays out a large footprint on the area. *(Courtesy of Craig Covey)*

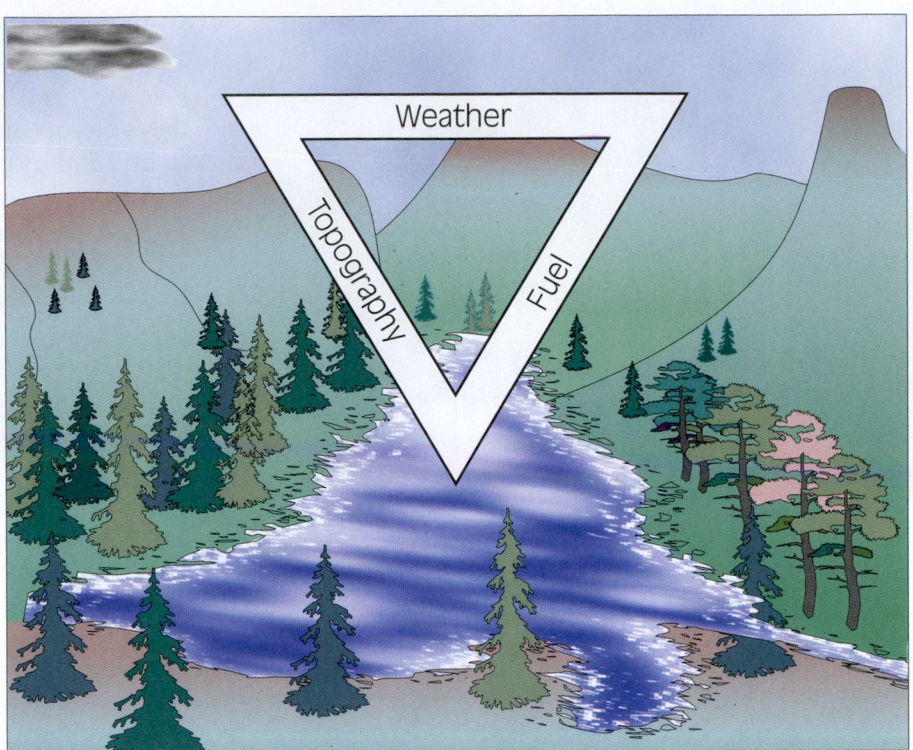

FIGURE 19-5 The wildland fire triangle differs from the structural fire triangle.

Weather

The weather component of the wildland fire triangle is the most dynamic of the group and hence is discussed first, **Figure 19-6.** This is because of its dramatic effect on wildland fire conditions. It changes all the time, including any number of times in a given day. Given similar fuels and topography, wildland fires will burn in a radically different way in different weather conditions.

> **CAUTION**
>
> Weather is one of the main causes of firefighter fatalities on wildland fires. In fact, whole teams of firefighters have been caught and killed on wildland fires a number of times over the years because of weather changes that were not recognized in time to change tactics or locations.

> **CAUTION**
>
> Frontal changes (cold and hot), high and low pressures, winds, and storms all have direct and dangerous effects on wildland fires.

Each of these elements can change in very little time on any given day. It is also a fact that some ground cover fires grow so large that they can create their own weather.

A number of safety rules or laws apply to wildland firefighting, based on years of study and practice. A great number of them pertain to the weather and its dangers to the fire environment because this area of wildland firefighting has caused so many injuries and deaths among the firefighter ranks over the years.

Fuels

In the wildland environment, there are a great number of fuel types, **Figure 19-7.** From fine grasses to large timber, the fuels burn differently. There are two main variables in fuels and their effects on fire. These are **rate of spread** and **fire intensity.**

The rate of spread is the speed with which the fire travels across the ground or through the bushes and trees. This rate of spread is affected by a number of variables, all of which are in the wildland fire triangle. In the case of fuels, it can be said that the lighter or finer the fuel, the faster the rate of spread. Given similar weather and topography, fire will run much faster in grasses and light vegetation than in heavy brush or timber. **Figure 19-8** shows a grassy fuel fire moving quickly across the terrain. In fact, this deceptiveness has caused a great number of firefighter deaths and injuries over the years. Numbers of firefighters have been overrun in grassy fuels thinking they could "outrun" the fire when things went bad. It is a sad fact that many firefighters have been caught and burned when running from "light fuels" wildland fires.

FIGURE 19-6 The components that make up the weather side of the wildland fire triangle.

FIGURE 19-7 The components that make up the fuels side of the wildland fire triangle.

FIGURE 19-8   Grasses and light fuels burn fast and must be respected. *(Courtesy of Rocco Di Francesco)*

The second term used here is the fire intensity. This is the amount of heat and flame production given off by a given fuel type. Rate of spread has an inverse relationship to fire intensity. Whereas lighter fuels have a higher rate of spread, they have a lower fire intensity. Heavier fuels have a lower rate of spread and a greater fire intensity. Light grassy fuels burn fast, but not with great flame lengths, while heavy timber burns slower, but with much more intensity, releasing higher flame lengths and Btus. **Figure 19-9** shows a timber fire with long flame heights and high intensity. Flames can run up to 100 feet or more into the air over a timber or heavy brush fire.

The last fuel concept discussed here is fuel spacing or continuity. The fuel can be spaced tightly or loosely. Depending on its compactness, the fuel may or may not burn well. If air can circulate around the

fuels and they are close enough to each other that they practically touch, then the fire will burn better than if the fuels are so densely packed that no air can circulate in and around them. At the other extreme is fuel that is so sparse that the fire cannot travel from plant to plant. In this case the fire may not run as well. It could even put itself out.

Topography

The last side of the wildland fire triangle is the side called topography, **Figure 19-10.** This term deals with the layout of the land. The **steepness of slope,** the direction the slope faces, the **drainages** and **river bottoms,** and the **ridges** and **saddles** are all critical to the behavior of fire in the wildland.

It is a fact of nature that fire runs uphill faster than downhill. This is because the flame will run vertically and because the slope places the uphill vegetation close to the flame, **Figure 19-11.** The plants are preheated and burn quicker as the heat travels upward. Conversely, it is said that when a fire is burning down a hill, it is "backing down" because the flame is leaning away from its direction of travel. When the fire backs down a hill, it naturally travels slower because the fuel ahead of it is not being preheated and has to catch fire by conduction rather than convection. It then naturally follows that the steeper the hill, the faster the fire will travel upward and the slower it will travel downward.

> **STREETSMART TIP**
>
> A concept the wildland firefighters follow here is that from flat to a 30 percent slope the fire will double its rate of spread and at a 55 percent slope it will double again.

The direction the slope faces is also very important because vegetation grows differently on different **aspects.** In the western United States, for example, the sun warms southern-facing slopes more than it does others. Because of this, the vegetation on the southern-facing slopes is dryer and generally lighter in consistency. This causes the southern-facing slopes to burn with higher rates of spread than the other aspects. Also, with the sun preheating the fuel, the chance of fire activity on these southern-facing slopes is much higher. In other parts of the country, a different aspect will be dryer, but the principle is the same. Firefighters should know what aspect gets more sun in an area before planning their strategy in fighting a ground cover fire. **Figure 19-12** shows a team of firefighters entering an area that has mul-

FIGURE 19-9   Heavy timber fires burn with high intensity and flame length. *(Courtesy of Phill Queen)*

FIGURE 19-10 The components that make up the topography side of the wildland fire triangle.

FIGURE 19-11 The angle of the terrain will have an effect on a running fire. Note how the flame "leans" into the hill, preheating the fuel above it. If this fire was burning or "backing" down the hill, it would move much slower because the fuel is not preheated.

FIGURE 19-12 A fire crew enters an area with complex topographic features and stops to study the terrain. *(Courtesy of Rocco Di Francesco)*

tiple topographic features. All must be understood and monitored.

Some more common wildland firefighting terms related to topography are saddle, ridge, drainage or **chimney, box canyon,** and **midslope.** These are described in the Key Terms section at the end of

this chapter. They all have great relevance in wildland firefighting. Each one has associated dangers when involved in fire. All but the ridge tend to draw fire to themselves. If a ground cover fire is anywhere near these topographic features, the danger level is very high because each will draw the fire to itself in a rate-of-spread increase that will astound the firefighter. These features are responsible for many firefighter deaths in wildland fires.

When discussing ground cover fire components and considerations, it becomes clear that it is a very complex science. Firefighters must understand these components to better ensure the safety and survival of those who fight fires in this environment. The numbers of firefighter deaths and injuries in wildland firefighting are very high, second only to residential firefighting. It is for this reason that ground cover firefighting must be understood and respected by firefighters the world over.

Vehicular Fire Components and Considerations

Studying vehicular fire components and considerations has become a lot more complex over the years. A number of variables can be involved in this practice. This section deals with the more common variables associated with today's automotive industry.

The most logical way to discuss the automobile and its dangers when burning will be to start at the front and work toward the rear. In past years automobile models had a metal bumper held on to the car with brackets. When hit, the bumper nearly always damaged the car. The industry then developed hydraulic supports for the bumpers in order to lessen the chance of car damage when lightly struck. Today's cars have hydraulic supports in both the front and rear to protect the car. The point here is to avoid standing in front of any bumper after it is been heated.

SAFETY

Firefighters should understand that when heated, as in an automobile fire, these hydraulic fluid-filled bumper systems undergo great stress as the fluid expands. This stress can cause the bumper to be propelled off the car, traveling up to 40 feet or more. Persons standing in front of the bumper (front or rear) when this happens can be severely injured.

The engine compartment has a mass of very hot metal filled with flammable liquids (fuel, oil, and hydraulic fluids). Any fuel leak can ignite quickly because of the heat from the engine. In many cases the fuel fire is extinguished and then keeps reigniting as the heat from the engine contacts new fuel leaking from burned lines or melted parts. These are Class B fires as opposed to Class A fires. In some cases, the engine itself is made of flammable metals such as magnesium. This is a Class D fire. Another danger

in this area is the battery and its associated wiring. A battery is a plastic box filled with a hazardous material (sulfuric acid) that creates a flammable gas (hydrogen) as well as electricity. There is the potential for a Class C fire.

The recent addition of electric vehicles to the nation's highways adds another component to the firefighter's information needs. The firefighter must be able to recognize an electric vehicle and be knowledgeable about the differences it presents on the fireground. The first thing to look for will be an insignia of some type, designating electric power. The insignia will be located on the sides or rear of the body. The next step in identification comes when the hood is opened. The firefighter will not find items normally found on a gasoline engine automobile, such as an engine, a transmission, cylinder heads, an alternator, fuel injectors or spark plugs, an air filter, or accessory drive belts. Instead, the firefighter will find high-voltage warning labels on components. Although the chances of finding an electrical (Class C) fire are remote due to numerous manufacturer safeguards, firefighters may have to alter their tactics when fighting fire in this type of vehicle.

The primary dangers in the passenger compartment are the materials used in construction. Most materials found in the passenger compartment are plastic, a form of polyvinyl chloride (PVC). PVC is extremely toxic when burned. A great number of studies have been performed on this material during the past few years and all have come to the same conclusion.

SAFETY

Plastic PVC is many times more toxic than wood and cloth when burned. Firefighters must wear SCBA when working around this material in any stage of combustion. A second danger in the passenger compartment that few consider is that of the hydraulic pistons used to open the hatchback and hoods in some models. These pistons will react to heat in the same way as bumpers do.

Another item that must be mentioned here is the air bag mounted in the dash on either or both sides, and also on the side of the front and rear seats in some models. Great care must be taken when working around these items because of the chance of an accidental discharge. Each model has its own idiosyncrasies that must be studied individually to truly understand the mechanics involved in its design. In most cases, they can be disarmed by disconnecting the battery to the car and waiting some period of time before working around the bag. The side air bag can be disconnected by cutting the power cord running up the side of the seat from the floor to the bag. It is suggested that even after disconnecting the power to air bags, caution should be used when working around them.

The next area to consider is the trunk. At first, a car's trunk might not seem to pose much of a problem but that is not the case. The trunk is used for storage. Firefighters have no idea what is stored in the trunk (without being told by the owner) until it is opened. It could be any type of flammable liquid or gas container, ammunition, explosive commodity, or other dangerous cargo. It has been known to even contain humans. It may be prudent to also have a hoseline charged and ready when accessing this area of the automobile just as is recommended for all other compartments.

SAFETY

When opening the trunk, great care must be taken to avoid exposure to potential harm. The firefighter should stand to the side, or crouch very low, and wear full protective clothing with the eyes protected by a helmet shield pulled down or by goggles.

The last area of the automobile to consider as dangerous is the underside or fuel storage area. Action movies almost never stage a car crash without the fuel tank rupturing and exploding. In real life this is extremely rare; however, firefighters must approach with caution. What can and usually does happen in car accidents is that the fuel tank will be damaged and be leaking raw fuel on the ground around the vehicle. In this case all precautions must be taken to secure possible ignition sources until the fuel leak is stopped. Dirt or any nonflammable absorbent material will do the job of recovery until something more efficient can be obtained.

When discussing the fuel tank it is important to remember that gasoline may not be the only fuel present. Cars are run on a number of fuels today. Gasoline, diesel, propane, methanol, electricity, compressed natural gas (CNG), hydrogen, and steam are but a few of the fuel systems in use today. Each one carries its own risk and will dictate the tactics used in securing the fuel source.

The last consideration to be mentioned here is that of another type of vehicle, a freight truck or tractor and trailer combination. For the purposes of fire components and considerations, trucks (all but the trailer) are simplified as short vehicles. The compartments are the same, only compressed. The materials used are relatively the same and in the same locations. The engine compartment is larger but contains just about the same materials and hazards associated with automobiles. The same precautions are necessary.

The trailer on the truck is another story and can also be dealt with by simplifying it. For the purposes of this chapter, the trailer body is treated as a structure off the ground. It can contain just about anything a structure on the ground contains and can be just as dangerous. All of the principles of caution with structure fires will hold true here. In many cases these containers will be placarded giving the responder clues to the contents and how fire may be fought using them.

Studying vehicular fire components and considerations is not an easy task. A vehicle contains nearly all hazards associated with other areas of firefighting, but they are compressed into an area approximately 8 by 20 feet (2.4 by 6.0 meters) with possible occupants. The risks are there and the firefighter must be aware of every possibility.

Process of Fire Extinguishment

This section begins to develop thought processes based on the previous sections within this and other chapters. The concepts and considerations of the differing situations and classes of fire previously discussed begin to be put into the context of fighting fire.

NOTE

Before firefighters take on the beast of flame, they must first understand that fire is a multisided three-dimensional presentation of heat and chemistry. It must be considered from six sides. It has a top, a bottom, and four sides. It will follow laws of physics and applied science that only the most seasoned firefighter will understand, and even then things will go wrong on occasion. The goal in defeating this beast can only be met by extending an aggressive, fast, well-placed, adequate fire attack.

To accomplish this goal, a number of steps must be taken. The first step is to create a plan of attack. This is done by first locating the fire and determining its extent. Generally speaking, the longer it takes to find the fire, the longer it will take to put it out. Good communication becomes very important here. As Section II, Chapter 2 describes, this report is built on careful and complete observation of the fire and its development to this point. Each department will usually have SOPs that determine who is first on scene and where each unit should be located. These policies are not set in concrete because each fire is different. However, they have been set up in order to maintain order and efficiency. The attack plan will be built on the observations, training, and opinions of the first-in officer after considering the facts as perceived and predicting a course of action.

The second step is then to apply the plan of action as quickly, efficiently, and safely as possible. Firefighters will take their skills, knowledge, and ability into the face of danger in an effort to mitigate the situation as quickly as possible. Three methods of fire attack are possible: the direct attack, the indirect attack, and the combination attack of fighting fire. The direct attack is quite simply the act of putting water directly onto the seat of the fire, **Figure 19-13**. This method is the most efficient use of water on free-burning fires. A solid or straight stream of water is applied in short bursts directly onto the burning fuels (Class A) until the fire darkens or is extinguished. **Figure 19-14** shows a firefighting team about to enter a burning structure and apply a direct attack on the seat of this fire. If used indoors and done properly, the thermal balance of the room may still hold, giving the firefighting team a clear view of the fire and contents of the structure.

Another method is that of the indirect attack, **Figure 19-15**. This attack is usually utilized when the firefighting team does not see the seat of the fire. A fog stream is applied to the upper areas of the room or above the fire seat in an effort to "steam" the fire out. **Figure 19-16** shows a firefighting team applying a fog stream into the upper area of a structure in an indirect attack method. Directing the stream into the superheated upper atmosphere above the fire causes the water to "boil" or "vaporize," steaming the area and causing the fire to die. This method, if used indoors, will usually greatly disturb the thermal balance, causing a loss of visibility in the structure as heat and products of combustion are circulated downward. When using this method, the firefighting team should attempt to shield themselves from the steam vapor by backing out of the area after applying the water. They can reenter as soon as the vapor begins to dissipate.

> **CAUTION**
>
> Note also that the indirect method is not appropriate for occupancies with victims in need of rescue, because the steam created by the attack will burn the victims quickly.

The last method is that of a combination attack. This attack is a blend of both the direct and indirect methods. The straight stream is directed to the seat of the fire and then the nozzle is quickly changed to a fog and the upper areas of the room are covered with a quick shot into the thermal cover. This method will darken the fire and stop the growth of flashover.

In each of these methods of applying water, firefighters must remember that the goal of the fire department is to save property. If water is not applied in proportion to the amount of fire, two things can

FIGURE 19-14   A firefighting team prepares for a direct attack on a contents fire. *(Courtesy of Phill Queen)*

FIGURE 19-13   A direct attack with firefighters applying a straight stream onto the seat of a fire.

FIGURE 19-15   The indirect attack has the firefighters applying a 30-degree fog into the upper heat layer of the fire in order to create a steam that will extinguish the fire.

FIGURE 19-16 A team of firefighters applies a fog stream in an indirect attack on a fire. *(Courtesy of Central Net Fire)*

FIGURE 19-17 Firefighters use a fog pattern to shield themselves from radiant heat. *(Courtesy of Phill Queen)*

happen. If not enough water is used, extinguishment will not take place. If too much is used, then the damage by water may exceed that done by fire. The lesson here is to use only what is needed, saving the rest for overhaul if needed.

Proper Stream Selection

Fire streams were discussed at length in Section I, Chapter 11 with information regarding a number of things that are mentioned again here. The purpose of this section is to transition from knowledge of the burning characteristics of various occupancies and hazards to actual fireground use of the information.

In the case of fighting fire with water, it is important to understand that in order to be successful, sufficient water must be applied directly to the fire in order to control it. That statement may sound simple until one begins to take into account all of the mechanics of performing that task—not only the physical movement but the mental as well. Success will be based on a number of factors in the selection of the stream alone. Some of those factors, in no particular order, are proper stream type, stream size, stream placement, timing, water supply or quantity of water, stream reach, mobility needs, tactics required, speed of deployment, and personnel available.

Consideration of some of these factors provides a better understanding of the firefighter's role in proper stream selection. The first one was the proper stream type. Is a fog stream the correct choice? If so, which width? **Figure 19-17** shows firefighters using a fog stream to protect themselves from radiant heat. Or is the straight stream a better choice? Each situation calls for differing stream requirements. The most basic answer to stream type may be in the type of attack required: direct, indirect, or combination. Each has its place in given circumstances.

FIGURE 19-18 A firefighter applies a single straight stream onto a structure in an attempt to reduce some of the fire's heat. *(Courtesy of Phill Queen)*

Next to be considered is stream size. The typical response to this question may very well be "Big water, big fire and little water, little fire." It was once said that firefighting is like going to battle. The weapon must match the target. **Figure 19-18** and **Figure 19-19** show examples of the need to have the stream size match the fire size to be effective.

CAUTION

If a large line is needed, then it should be pulled right away so as not to play "catch up" with the fire. Some firefighters think the stream to start with should be the stream that will eventually be used, going smaller later, rather than ever having to go larger later.

The placement of the fire stream can make or break an attack plan. The fireground must be coordinated at all times without exception. This is a battle with

FIGURE 19-19   If the line selected does not match the size needed, the fire will burn longer and hotter and can jeopardize the operation.

FIGURE 19-20   The first line pulled should be positioned between the potential victims and the fire.

FIGURE 19-21   Firefighters use a deck gun to hold the fire until more lines can be put into place. *(Courtesy of Rocco Di Francesco)*

FIGURE 19-22   Large stream appliances at work on a large fire. *(Courtesy of Rocco DiFrancesco)*

an enemy that can injure or kill. When an order is given to place a fire stream in a particular location, it is part of a plan. Without that stream, the plan may fail, so this factor can be very important. An example may be a fire in a dwelling where the first line goes between the trapped occupants and the fire, with the second stream going to the seat of the fire, **Figure 19-20.** This operation will not work without both streams being coordinated and controlled. Care must also be taken to avoid opposing streams, because these streams may drive heat and products of combustion toward the opposing firefighting teams. Still another related factor is that of timing, which will go with speed of deployment. This is self-explanatory. Timing is everything on the fireground. Stopping the progress of the fire as quickly as possible will allow the firefighter to work toward the goals of rescue and property conservation.

Many differing types of nozzles are used in the fire service. One of the commonly used types has the ability to control the amount of water it puts out. In cases of water conservation because of poor sources, a nozzle that conserves water would be advantageous. In such a case, water supply or quantity of water available plays a significant role. A different set of rules applies when the water is limited.

Another factor to consider is the mobility of the stream that will be needed. In some cases, lines are advanced quickly into the fire, whereas in other cases the line is placed in a fixed position holding the fire until another more mobile line can be set up to take over. **Figure 19-21** shows a deck gun attack on a large fire in an effort to bring the fire into handline size by reducing some of its heat. In this same area are the factors of stream reach and tactics. A line may be placed on an exposed structure that is large and unmoving, while a smaller, more mobile line is extended into the burning occupancy to fight the fire.

Everything taken to a fire is a resource operated by the fire department. The personnel, water, equipment, and knowledge are all resources. The selection of the proper fire stream is a significant decision. **Figure 19-22** shows the proper use of large streams on a

TABLE 19-1 Hose Stream Characteristics

Type or Size of Line	Reach (ft)	Mobility	GPM	Common Use
1-inch or greater booster or reel line	25–50	Excellent	10–40	Very small nonstructural fires or overhaul
1½ to 1¾ inches	25–50	Good	40–175	Quick attack, one to three rooms, vehicle fires
2½ inches	50–100	Fair to poor	125–350	One floor or more, personnel permitting
Master stream	100–200	Poor to none	350–2,000	Large, fully involved structures or exposure protection

fully involved mercantile fire. **Table 19-1** shows the differing sizes of nozzles and the effort required in order to work properly.

Firefighters carry out their part of the attack plan using water and a good basic knowledge of the requirements being predicted by the fire they are fighting. Knowledge of the factors listed, as well as many others, will arm the firefighter with an arsenal when going into battle against a fire.

TACTICAL CONSIDERATIONS

Earlier in this chapter the elements of fire control were discussed in some detail regarding the most common fire expectations of the average firefighter. Those sections detailed the components and considerations a firefighter must be knowledgeable in to be safe and successful on the fireground. This section deals with those same types of fires in a tactical firefighting capacity. It builds on the information presented in the previous section in an effort to actually use the information in fighting fire.

The actual fighting of fire on the fireground is broken down into tactical objectives. Each of these objectives must be accomplished so that the safety of the firefighters, occupants, and others is a primary consideration. If there are no occupants, the overall plan still remains in order, keeping the firefighters safe and efficient. These goals are universal in the fire service, although different departments may use different terms. The priority was first set in place a great many years ago by Lloyd Layman, with the following acronym: **RECEO**—Rescue, Exposures, Confinement, Extinguishment, Overhaul.

VIEWPOINT

Over the years a number of variations have been used for fire attack acronyms. Some currently popular are:

RECEO—Rescue, Exposures, Confinement, Extinguishment, and Overhaul
REVAS—Rescue, Exposures, Ventilation, Attack, and Salvage
SPARS—Size-up, Position Apparatus, Attack Fire, Rescue, Support Functions
LOUVERS (truck companies)—Ladders, Overhaul, Utilities, Ventilation, Entry, Rescue, and Salvage

There are many more fire attack acronyms that firefighters may encounter, but the point is well made that the fire service has set itself a series of goals that must be accomplished in order that the fireground remains as safe and orderly as possible. All firefighters should refer to local or departmental policies and procedures for the tasks and order of fire tactics.

NOTE

A second point that must be made is that the command structure must be solid and well defined in order to accomplish the set objectives in each of the goal areas.

Just as an army employs a rank structure to fight a battle, the fire service uses a rank structure to fight fire. One person will be in charge and appoint people who will command different parts of the operation under that person. **Figure 19-23** shows a battalion chief conferring with his captains in planning his objectives. This system is called the incident com-

FIGURE 19-23 A battalion chief meets with his captains to set operational plans into action. *(Courtesy of Rocco Di Francesco)*

mand system **(ICS)** and it is mandated by NFPA, OSHA, and NIMS. The ICS is basically designed to maintain order in any emergency operation. It can be fire, flood, rescue, first aid, mass casualty, or anything demanding a number of emergency responders. A single person will be in charge of the incident and this person is called the incident commander (IC). This person works from the goals listed earlier utilizing strategy and tactics just as a general would in the military. The IC will delegate assignments to arriving units in an effort to accomplish the goals in a safe and efficient order.

To explain the system and complexities of fighting a fire, this discussion sets the scene with an average, generic fire, because the methods and operating principles will apply to any fireground operation.

As soon as the fire is reported, and before units ever pull from the station, mental preparation takes place. These are the recall factors of past similar incidents, the recall factors of knowledge about the area where the alarm is reported to be, the recall factors regarding the fire prevention inspections of that area, the recall factors of time of day and route to be taken in response, and the factors regarding the condition of the responding units as to their capabilities. Are they fully staffed today? Are the water tanks full and the pumps working properly? Are the crews fresh and ready to go to work in possibly very dangerous conditions? All of these factors, and many more, will race through the minds of the responders. Each of these factors will eventually dictate the strategies to be considered.

Upon arrival another set of factors must be considered. All of them have to do with the environment of the response. **Table 19-2** demonstrates that the number of factors having to do with the operation can be almost overwhelming. The seasoned firefighter will go through all of these and possibly more in calculating the environment presented at the emergency. And each of them can greatly affect the outcome of the incident depending on how they are handled.

Based on the situation presenting itself to the arriving firefighters, three methods or modes of attack are

TABLE 19-2   Fireground Factors

Building

Size	Construction type	Condition
Age	Openings	Utilities
Concealed spaces	Access	Effect of fire
Extent of fire	Interior fuel load	Exterior fuel load

Fire

Size	Location	Direction of travel
Time since ignition	Extent	Materials involved
Material left to burn	Fire load	Stage of involvement

Occupancy

Type	Value	Fire load
Status (used/vacant)	Hazards of occupancy	Life hazard
Arrangement	Obstructions	

possible: the **offensive attack,** the **defensive attack,** and the **combination attack.** These modes are predicated on the resources available at the emergency. If the arriving units have adequate resources to handle the situation, and the structure is safe to enter, then they will fight the fire aggressively and offensively. They will attack the problem head-on and, following department standards, will accomplish their objectives efficiently, effectively, and safely. If they do not have adequate resources to aggressively handle the situation, then they will have to fight the fire in a defensive mode of attack. This mode will be continued until enough resources can be massed to then change to an aggressive, offensive attack. The combination mode is most often utilized when a rescue must be accomplished but not enough resources are on scene to handle the operation entirely. An example would be when a line is pulled to defend potential victims until they are removed, knowing that the line will do nothing to fight fire because it is not large enough or practical for the size of the fire.

As the operation begins to unfold, the IC calls for an attack on the fire. The IC's call will be based on the conditions presented earlier. Hoselines will be pulled, utilizing methods prescribed by the responding department's SOPs. As mentioned earlier, rescue will be the first priority, and all actions will be directed toward accomplishing that task. A line may be pulled and placed between the fire and the people needing rescue, or the situation may call for rescue first and no line pulled until it is completed, **Figure 19-24.** Each situation will dictate its own need.

A recent rule by the Occupational Safety and Health Administration sets safety standards on the fireground for rescuing occupants from burning buildings.

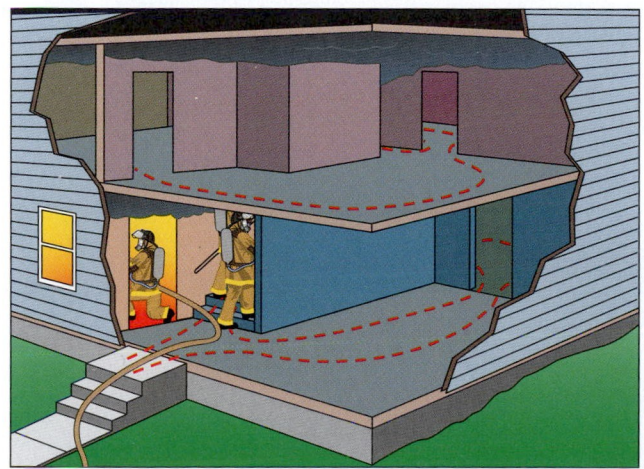

FIGURE 19-24 If it is not known whether victims exist, then the first line into the structure should assist the search and rescue functions.

> **NOTE**
>
> The two in/two out rule tells the fire department that the teamwork concept must be followed on the fireground at all times.

This law is designed to protect the firefighter from being injured in rescue attempts where fire conditions are critical.

> **FIREFIGHTER FACT**
>
> In 1998 the Federal Occupational Safety and Health Administration issued a Final Rule titled 29 CFR Parts 1910 and 1926 Respiratory Protection. This regulation requires both private industry and the fire service to practice the safe use of self-contained breathing apparatus (SCBA) by mandating a number of rules regarding face piece fit testing, physical requirements for wearers, apparatus requirements, and more. A part of this Final Rule also mandates the number of firefighters required on the fire scene when interior operations are to take place in hazardous situations. The two in/two out rule dictates that during interior structural firefighting in atmospheres that are immediately dangerous to life and health (IDLH), and that do not have lives requiring immediate rescue, at least two firefighters must enter together and remain in visual and voice contact with each other at all times. Two more firefighters must be on standby for the potential rescue of the interior team. If there is a confirmed rescue to be made, the rule does not apply but the team **must** have good evidence of that need for rescue.
>
> The 1997 NFPA 1500 Standard for Occupational Safety and Health also requires this same two in/two out rule as part of its requirements. This was created in an effort to further follow OSHA laws and protect the safety of firefighters the world over.

If the situation does not call for rescue because there are no dangers to people or animals, then the second priority is placed into action. That priority is exposure. Almost anything in the path of the fire can be called an exposure. The most common exposure is a structure next to the burning building where the fire is heating it to a point where it is close to catching on fire, **Figure 19-25.** Another example would be a car fire that is about to ignite nearby brush or tall grass. If an unburned object (i.e., exposure) is about to catch on fire, it is the responsibility of the responding firefighter to stop the spread of the fire to that object. The logic here is that the item burning has already sustained damage, but the exposed item has not. So firefighters save the exposed item before concentrating on the burning item. Exposure protection is accomplished by either removing the exposed item from the area or wetting it to cool its surface below the point of ignition, **Figure 19-26.** Radiant heat is the main cause of exposure fire, so wetting the exposed item removes that heat at the point of contact.

As soon as all exposures are deemed safe, it is time to consider the next step in the firefighting pro-

FIGURE 19-25   The most common exposure problem is the closest building to the building on fire.

cess, which is fire confinement. The fire's parameters are not always clearly defined. Visibility may be difficult at best. The members of the firefighting team must use all of their senses in order to locate the fire. Once located, water or another extinguishing agent is applied in a method that will confine the fire to its area of burn. Stopping the fire's forward progress must be done first, before the fire can be extinguished. It was mentioned earlier that after rescue came fire control. The confinement of the fire is controlling the fire. Some departments will even broadcast a "fire is under control" call to the IC when this step is accomplished. In the wildland, this point becomes even more dominant. An "under control" term may be issued days before actual fire extinguishment is accomplished. When the fire is surrounded and can burn no further outside of this confinement, then it is considered to be under control.

In many cases, to bring the fire under control in structure firefighting, another task may have to be added. Ventilation (discussed in detail in Section I, Chapter 18) coordinated with water application may be the only way to achieve the confinement results safely. This will be described more later in the discussion of particular structure fires.

It is common practice to attack fires from the unburned side, **Figure 19-27.** That is to say that the line will be brought into action between the fire and

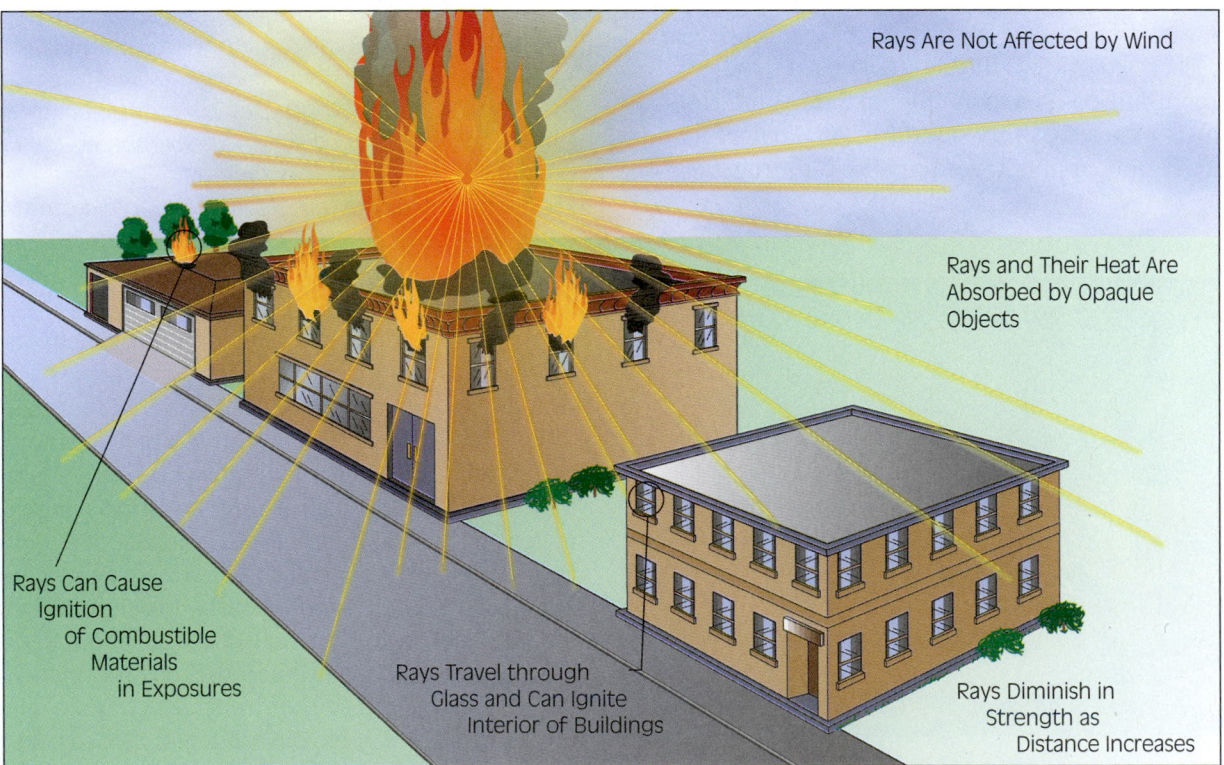

FIGURE 19-26   Radiant heat is the main cause of exposure fires within short distances. Radiant heat will travel in straight lines from the heat source to nearby objects.

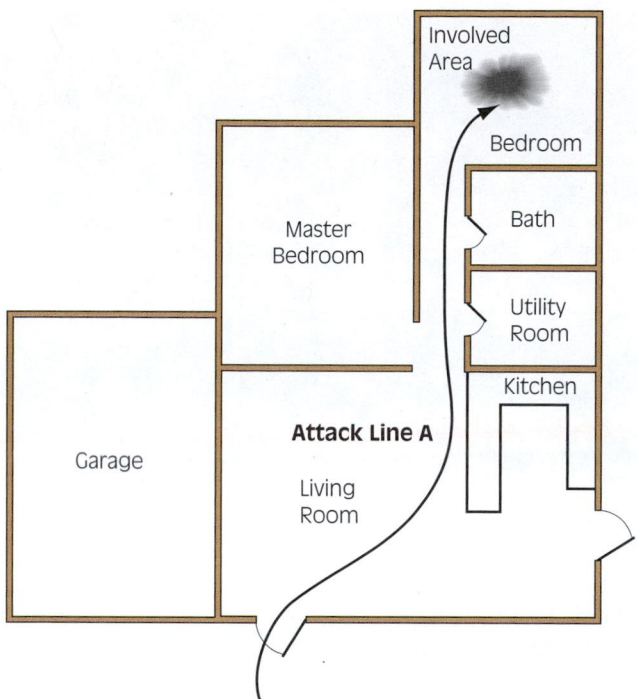

FIGURE 19-27   Fires most commonly should be attacked from their unburned side, pushing the products of combustion away from unburned materials and areas.

its intended direction of travel. The opposite of this tactic would be to attack the fire from the rear, pushing the heat and other products of combustion forward into unburned areas, causing far more damage. This is a tactical consideration in almost all applications of firefighting from structure to wildland or ground cover to flammable liquids and gases.

When the fire is controlled or confined, then the actual extinguishment begins. This is accomplished with direct or indirect attack strategies and resource use. This can also be thought of as property conservation. Saving the property not damaged is a very high priority here. The fire has been confined and now is extinguished utilizing the latest methods and practices in order to save the undamaged property. The use of water is restricted to only as much as is necessary, because water can cause as much and, in some cases, more damage than the fire.

As soon as the fire is extinguished, the overhaul begins. This is a methodical system of ensuring that all embers and chances of reignition are removed. The area is carefully worked by moving every item and wetting all combustibles that could carry new life to the extinguished fire. During this phase of the operation, the cause of the fire is usually determined. By carefully sifting through the fire debris the investigator looks for clues that will lead to the cause of ignition. A closely coordinated effort between these people will result in the call being closed out so the

firefighters can return to their respective stations to prepare for the next call.

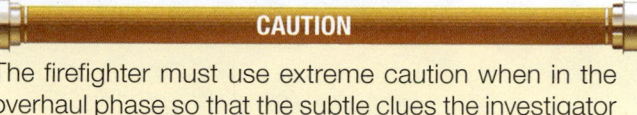

CAUTION

The firefighter must use extreme caution when in the overhaul phase so that the subtle clues the investigator is working with are not disturbed.

One more point must be made about firefighter safety. That point is about acts of terrorism. The fire service is called upon in nearly every conceivable emergency, representing local authority in the form of a fire department, fire district, or emergency services agency. Unfortunately the firefighter may represent the authority being targeted by an act of terrorism. For this reason it is important that the firefighter be made aware of the possibility of danger from cowardly acts of aggressiveness. Some recognized threats to firefighters are:

- Undetonated explosives: secondary devices, low-order explosions, malfunctions
- Structural instability: collapsed buildings, falling/thrown debris, live electrical lines, leaking gas mains
- Fire
- Hazardous materials (poisons, etc.)
- Biohazards

Further training in this area is highly recommended for all firefighters. Local jurisdictions, police, FBI, the State Fire Marshal, and the National Fire Academy offer this type of training. More information on terrorism is given in Chapter 30.

All of the information given in this section must be applied to tactical considerations in fighting fire in all emergency situations, bringing the information in this chapter into a fine focus for practical use.

Below-Ground Structures or Basements

Fires below a structure, such as in basements, can be some of the hardest fires to fight due to the harsh conditions the firefighter may have to face. Fire and its combustion products travel upward with heat by convection and radiation. To fight one of these fires, the firefighters must travel down through this superheated air and smoke in order to reach the fire itself, **Figure 19-28**. In just about any other firefighting attack the firefighter can access the fire from below or the side, keeping low and out of the products of combustion. That is not true here.

FIGURE 19-28 Firefighters must travel down through superheated gases and toxic products of combustion in order to fight a basement fire.

> **NOTE**
>
> The key to fighting a basement fire is to ventilate as soon as possible, releasing the heat and smoke, making access and attack easier for the firefighting team.

The basement can also be very difficult to ventilate. Section I, Chapter 18 tells of a number of ventilation techniques, and that information will prove invaluable in a below-ground fire such as a basement.

The role of the first-in engine company in these fires is to get water onto the fire as quickly as possible. This is primarily because the fire itself will be traveling upward and outward, creating a very poor situation for anything or anybody above the fire in the structure. The second-in unit must quickly assist with the ventilation as well as assist with the line already pulled. A backup line must be pulled as quickly as is practical for the safety of the team that has entered the basement. The attack team must be backed up as soon as possible, and this is a situation where a single firefighter will never enter the area alone. The team concept must be strictly followed.

In cases where access to the basement cannot be made, it may be possible to punch through the floor above the room on fire and flood the room with water or high expansion foam, **Figure 19-29.** Whatever the attack method, it must be done quickly because everything above the fire is exposed.

Structures Equipped with Sprinklers or Standpipes

Fighting fire in occupancies equipped with sprinkler systems or standpipes has been a boon to the fire service over the last decade or two. Before the mandates of these systems, firefighters were not nearly as successful as they are today fighting fires in these types of occupancies.

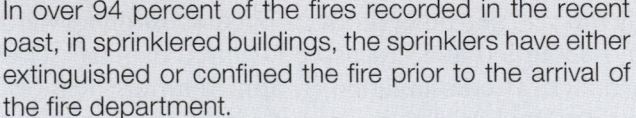

FIGURE 19-29 If access to the below-ground fire is too dangerous or not practical, then other means must be used to extinguish the fire through the ceiling of the burning room.

The sprinklered building creates a unique situation for the firefighter.

> **FIREFIGHTER FACT**
>
> In over 94 percent of the fires recorded in the recent past, in sprinklered buildings, the sprinklers have either extinguished or confined the fire prior to the arrival of the fire department.

The team entering the structure has but to find the seat of the fire and complete extinguishment. This will still not be easy because the smoke and heat will be held down over the combustibles by the cooling action of the water spray from the system. The firefighting team will have to enter the structure very carefully without shutting down the protection afforded by the operating sprinklers, and they will have to find the seat of the fire through very limited visibility to complete the extinguishment of the fire. After this is done, the system can be shut down for a short time during overhaul. Chapter 12 (in Sections I and II) details the operation of sprinkler systems.

The standpipe system is somewhat different. This pipe system will carry water to the firefighting team, easing their dependence on hoselines over great distances. This system is invaluable in fighting fires in high-rise structures. It saves a great amount of time by having the water ready for the team as it arrives at the floor on fire. Team members simply hook the hose carried aloft to the system and then charge it by turning on a valve at the standpipe. Section I, Chapter 10 describes this operation in some detail.

WORKING SMART AND SAFE ON THE JOB

A common thread throughout discussions on structure firefighting is the task of securing utilities. By securing the utilities on scene, you can ensure the safety of all the firefighters working the incident.

Securing Utilities

One of the first steps in going to work at a structure fire is to control the various utilities that are active within the building. For most structures, this will include electrical service, gas service, and water.

Controlling the utilities is important in fire operations because it helps to protect the firefighters working both inside and outside the structure. Fire can damage routes and controls of utilities and cause gas or electricity to escape, further compounding the fire problem and creating a more dangerous environment. Firefighters performing operations inside the structure, such as fire attack or overhaul, could come into contact with electrical lines or gas lines. It is essential that these utilities be shut down to avoid unnecessary firefighter injury. Uncontrolled utilities can also contribute to the spread of fire to uninvolved areas of the structure, causing more fire damage to the building.

Each jurisdiction will have different SOPs regarding the control of building utilities, so firefighters should be familiar with their department's guidelines for this activity. Additionally, the control mechanisms that allow firefighters to shut off the various utilities are different depending on the service provider. These providers typically offer utility-control training to fire departments at little or no cost.

In general, securing building utilities includes the following steps:

1. Wearing full PPE (personal protective equipment), gather any necessary tools for the types of utilities at the structure.
2. Control the electric service at the meter if possible.
3. Locate the main breaker box; note which circuits have already tripped; then shut off power to individual breakers first. After the individual breakers are off, shut off the main breaker. (Noting tripped circuits before shut down procedures assists the fire investigators in their work after the fire.)
4. Check for auxiliary power supply sources. Some buildings have multiple power feeds or circuits wired to an uninterrupted power supply (batteries and/or generator) that are separate from the main breaker. If found, communicate to the Incident Commander (IC) their presence and shut off power only if directed.
5. Locate the natural gas meter and shut off the gas service. (Most are quarter turn valves.) In the case of propane-fed buildings, locate the propane storage tank and shut off the gas by opening the valve cover and turning the valve to the fully closed position. Carefully close the cover once the valve is shut off.
6. Locate the main domestic water valve and, using the appropriate tool, rotate valve until closed.

In the case of each utility that is shut off or controlled, a lock-out/tag-out system should be used. This means the utility should be locked into the "off" position and tagged by the person who controlled the utility. This is done to prevent anyone from turning the service back on without authorization from the IC or his or her designee. Firefighters should consult with their department to learn proper lock-out/tag-out procedures.

Emergency Scene Lighting

As was mentioned many times, firefighter safety is of paramount concern. Securing electrical power on scene does two things. It creates a safer firefighting environment by removing the chance of accidental electrocution, but at the same time darkens that environment by shutting off all power to lights. As the fire is darkened, some form of alternative lighting must be brought forth so the fire team can see their work site while they complete their mission safely, even all the way through overhaul.

There are a number of forms of emergency lighting used by the fire service, including:

- Truck- or engine-mounted generators with electric cord reels and portable lights.
- Portable generators with electric cord reels and portable lights.
- Apparatus-mounted lighting on telescoping poles or high on the sides of the units. Specialty apparatus used for scene lighting. These units are usually dual purpose, for example, an apparatus used for both an air-refilling station and scene lighting, **Figures 19-30A** and **B.**

While these and other forms of emergency scene lighting are used in structural firefighting, they are used in all other forms of emergency work, as well. These include:

- Automobile accidents
- Trench rescue
- Confined space rescue
- Hazardous materials scenes
- Mass casualty scenes
- EMS (emergency medical services) scenes
- Command Post operations, as well as other on-scene needs
- Any night work requiring detail and firefighter safety

It is the responsibility of every firefighter to know the equipment he or she uses. Lighting equipment is no exception. A firefighter must understand the safety aspects of the equipment, the limitations and exceptions of the equipment, and how to deploy the best lighting possible, with the equipment available.

(Continues)

FIGURE 19-30 (A) Specialty apparatus for scene lighting and SCBA filling. (B) Scene lighting and SCBA refilling taking place on the fire scene.

The following steps should be followed when operating portable lighting equipment:

1. Properly position lighting by extending and properly adjusting height.
2. Position lighting above ground and away from water.
3. Position light to adequately light work zone.
4. Properly position light in area away from traffic.
5. Position power source (generator) within reach of lighting equipment.
6. Ensure fuel level prior to starting portable unit.
7. Properly position junction box and power cords out of traffic areas.
8. Ensure connections from generator to lighting equipment.
9. Start the generator and supply power.
10. Ensure operation of lighting equipment.

As with all fire department electrical equipment, there are safeguards and precautions that firefighters should be aware of when operating electrical equipment, including lighting, on an emergency scene. **GFI (ground fault interrupters)** are one such device built into the electrical cords and connectors used by the fire service. These devices are simply "automatic shutoffs" built into the cords and connectors that will shut down all power to the appliances used in the event of any short or grounding. The firefighter then must assume that there is a problem with the devices and discontinue their use. When safe to use again, each device usually will be equipped with a "switch" for resetting the GFI and operations can continue.

All electrical equipment must be maintained in order to avoid electrical shock. Avoid extending any light poles or equipment into overhead hazards or obstructions as well as avoiding tripping hazards from electrical supply cords. Finally, when setting up any electrical equipment, avoid setting up equipment in any water. Ideal placement of any electrical equipment is on higher ground and in a way that scene safety is maintained.

Exposure Fires

An **exposure fire** can be described as any combustible item being threatened by something burning in another area. The fire may be carried to the exposure by three ways. Section I, Chapter 4 described these as conduction, convection, and radiation.

As stated earlier in this chapter, the second priority after rescue, in most fire scenarios, is exposure protection. This form of property protection is one of the primary focus points in most fires. A number of factors are involved in most exposure fires. In no order of preference they are:

- *Wind.* Convection carrying embers can be a problem.
- *Distance.* The nearer the exposure, the greater chance of fire.
- *Material.* The makeup of the exposed surface determines its combustibility.
- *Intensity of fire.* The intensity of the primary fire factors into its spread.

A couple of basic methods are used to defend an exposure from fire, and they are based on the methods of heat transfer. In conditions of conduction, a cooling medium must be applied to the material being heated by the conduction. An example would be a water stream applied to a tank exposed to a fire burning under or on it. **Figure 19-31** shows foam being applied to a tanker that is exposed to a fire next to it. In cases of convection, a downwind patrol must be provided so as to ensure that embers do not ignite other materials. An example would be a downwind house with a wood shake roof, near a fire with flying brands being put into the air. Next is radiation as a medium for ignition of an exposure. The best method would be to apply water to the exposed area to cool it, **Figure 19-32.** An example would be, as a structure is free burning, the first hoseline is placed between the burning structure and the next door unit. The water is applied to the exposed structure until another line can be trained onto the burning structure itself.

Exposure fires are common and dangerous, especially when the firefighting teams do not see the second fire and are put in danger by it.

FIGURE 19-31 Firefighters apply foam to a tank truck exposed to a nearby fire. *(Courtesy of Central Net Fire)*

Nonstructural Fires

This subsection discusses fighting fire in nonstructural situations. It opens with wildland or ground cover firefighting, then transitions to vehicle fires. The section finishes with a short discussion on trash and dumpster fires.

Ground Cover or Wildland Fire

The ground cover or wildland fire can be either a short pickup-type situation or a campaign fire situation lasting days to weeks or more. The types of ground fire include woodlands, forests, grasslands, brush, prairies, and other such vegetation. It can be a short, easy, nonthreatening call or a hard-fought, killer-of-firefighters call. In each of these cases the objectives remain basically the same:

- Confine the spread of the fire by surrounding it with hoselines, by removing the fuel, or by another means of stopping its growth.

- Guide the fire by control measures in order to keep it from burning into areas of higher intensity or faster growth.

- Operate from a position of strength and safety, never giving the fire the advantage of jeopardizing the safety of the fire crews or the citizens being protected.

The initial size-up is very important in that this type of firefighting can be very time consuming and

FIGURE 19-32 Radiation will travel through water or any opaque material. In order for water to be effective, it must be applied to the exposed surface in order to cool it.

exhausting. A number of basic facts must be considered on arrival that will dictate the attack plan for this type of fire. These facts are based on the wildland fire triangle discussed earlier and are used to determine which attack method will be most successful in fighting this fire. Based on the facts, the officer will ask two questions: What is the fire doing now and what will it be doing in the future? From this, the attack plan will be formed and the fight will begin.

The direct attack is a form of attacking the fire itself, in some fashion. It will usually involve hoselines, but not always. Hand tools or aircraft can be used in many cases as successfully as handlines. A point of attack is made on the fire called an **anchor point.** From there, the team will work up one of the **flanks of the fire** working toward the **head of the fire** (the progressive end), **Figure 19-33.** Based on the factors listed in the subsection on ground cover fire components and considerations, the firefighting team will progress up the most active side of the fire until eventually the fire is surrounded and extinguished. **Figure 19-34** shows a firefighting team in direct attack of a grass fire by working up the flank while staying in the burn.

SAFETY

By staying in or near the burn (walking with one foot in the black, or burn, and one foot in the green, or unburned area) a firefighter's safety is greatly enhanced.

If a team is progressing up the opposite flank, that is called a parallel method of attack. Both teams will meet at the head or near there and join lines, controlling

the fire. Note that there is a difference between controlling and extinguishing the wildland fire. Controlling is just that. The fire is controlled and no longer burns into threatened areas. Extinguishment is the full extinguishment of all embers and fires contained within the perimeter of the control line. Extinguishment may take days to accomplish after the fire has been controlled.

Once a fire has reached such proportions that a direct attack is impractical, an indirect attack is utilized. This type of attack is common on large campaign fires that sometimes burn for days or even weeks. The indirect attack involves getting ahead of the fire in order to remove fuel from its path, **Figure 19-35.** This can be done by any of several methods. A few

FIGURE 19-34   A fire crew in the direct attack mode works on a grass fire. *(Courtesy of Phill Queen)*

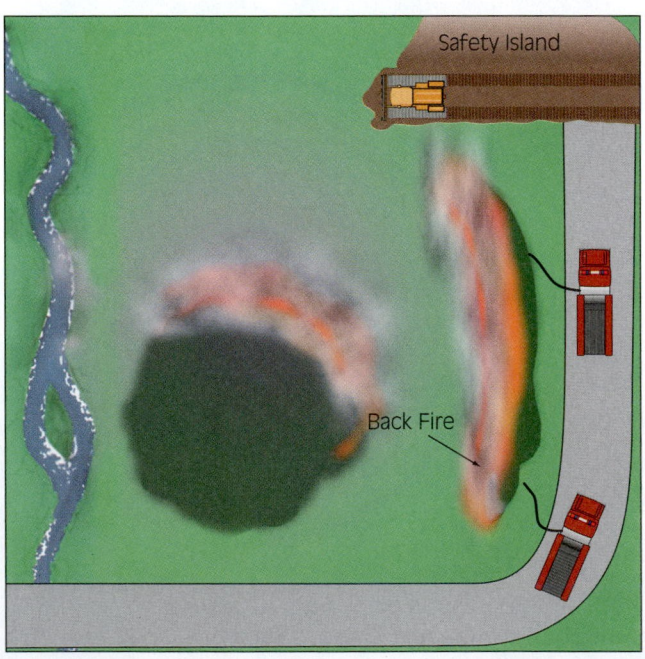

FIGURE 19-35   An indirect attack on a ground cover fire with two engines and a dozer.

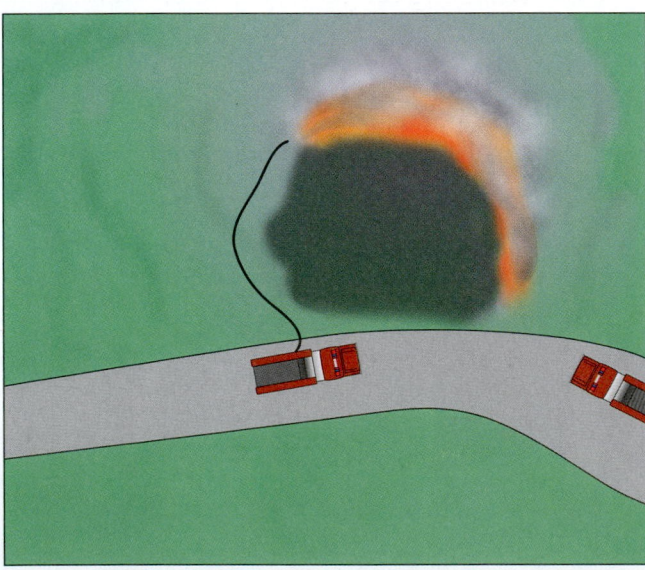

FIGURE 19-33   A direct attack on a ground cover fire.

of the more common methods are removing fuel by hand or with bulldozers, removing fuel by utilizing natural barriers such as roads and rock outcroppings, or burning off the fuel in controlled conditions. **Figure 19-36** shows the last method as a firefighter burns

FIGURE 19-36 An indirect attack is accomplished by burning off fuel just before the main fire arrives. *(Courtesy of Phill Queen)*

off fuel before the main body of the fire can reach it. This method of firefighting is very scientific and practiced by only the most experienced fire teams. It requires the ability to read the weather, the fuels, and topography very carefully in order to remain in control of the attack. Consideration of safety and all of the safety lessons and rules must be followed very carefully on this type of firefighting.

Another type of wildland firefighting is now being utilized quite commonly in the fire service today and that is **interface firefighting.** This type of firefighting involves crews that are often oriented more toward structural firefighting. They are asked to go into the rural areas in an effort to protect structures from wildland fires, **Figure 19-37.** The skills involved with this type of firefighting are very different from urban firefighting. A number of classes and curricula are being promoted in these areas for fire service personnel that are designed to create a much safer and smarter firefighter as society continues to move out of the cities.

Ground cover or wildland firefighting is much different than structural firefighting. It requires different tools, clothing, methods of attack, and knowledge.

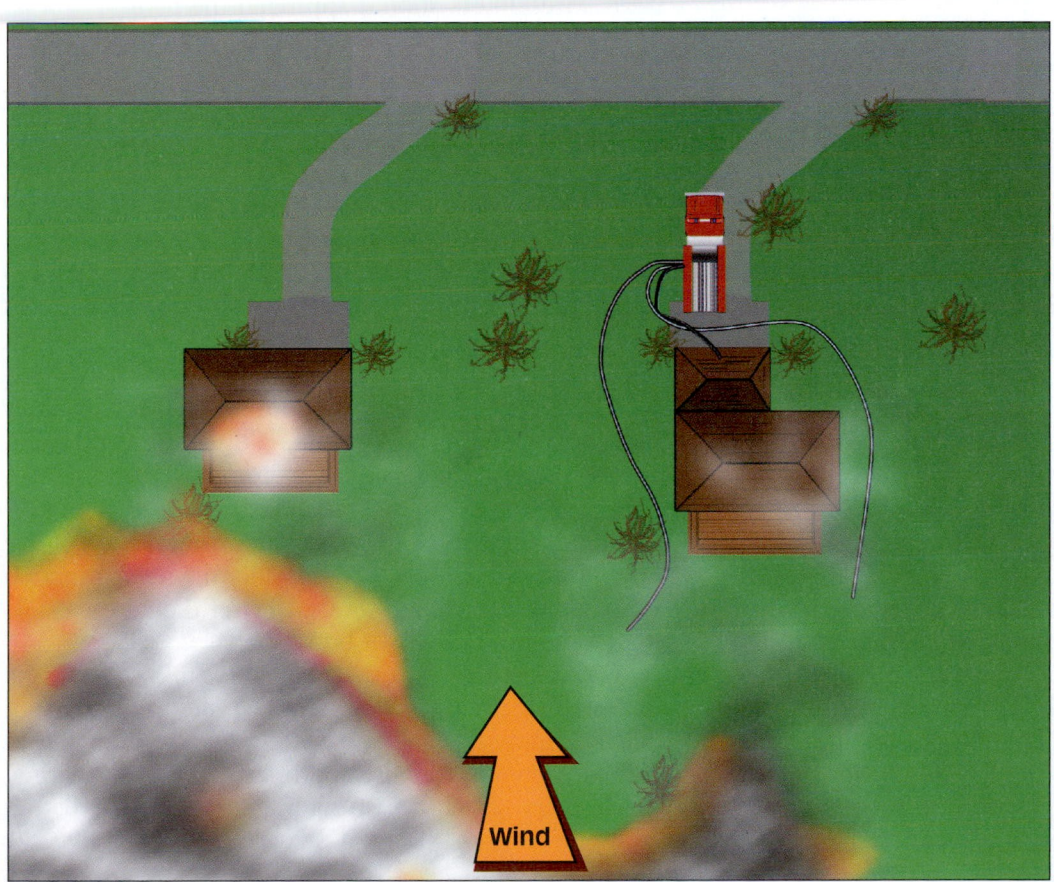

FIGURE 19-37 Protecting homes in the interface is a growing practice across the country for engine companies used to fighting fire in the cities.

FIGURE 19-38   A helicopter is readied for wildland fire-fighting. *(Courtesy of Brentt Sporn)*

FIGURE 19-39   The firefighters' first attack point is the passenger compartment in order to protect life. *(Courtesy of Brentt Sporn)*

Figure 19-38 shows a common tool used in wildland firefighting: a helicopter with a water dropping tank. To be safe, efficient, and effective, the firefighter of today must be educated in this area as well as that of the structural fire world.

Vehicular Firefighting

Vehicle fires are very common in today's fire service. Some of the more common reasons for vehicle fires are:

- Collision resulting in leaking fuel and sparks
- Electrical problems causing short circuits
- Overheated components such as tires, engines, or brakes
- Muffler and catalytic converter problems
- Discarded smoking materials
- Arson

For vehicular firefighting, the NFPA recommends at least a 1½-inch hoseline and preferably a combination nozzle with a partial fog of approximately a 30-degree pattern initially. The initial attack should protect the occupants if they are still in the vehicle, **Figure 19-39.** Once the occupants have been rescued, the attack on the fire itself may begin. The suggested method of attack is direct. The attack should be from the upwind and uphill side of the vehicle in order to reduce the firefighter's exposure to heat and the products of combustion. This is also important in situations where leaking fuel is under the vehicle and can run toward the firefighters. Consideration must also

be applied to chocking the tires on any vehicle burning on unlevel ground. It could be disastrous if the vehicle's brake lines burned through and the vehicle began to roll away or toward the fire engine. Full PPE must be worn when fighting vehicle fires because the combustibles are toxic and sometimes explosive.

Fire attack should begin with a 30 degree fog pattern. Firefighters should approach the vehicle at a 45 degree angle to avoid dangers presented by hydraulic fluid-filled bumper systems and exploding tires. During approach, the firefighters should sweep the bottom of the vehicle to extinguish any visible flames and remove any fuel runoff. This will also cool the tires, fuel tank, and any hydraulic fluids affected by the fire. Effective fire streams will reduce the amount of heat, smoke, and fire. This will allow firefighters to continue firefighting operations. Once the firefighters are close enough, they can locate and attack the seat of the fire.

The vehicle fire can also pose situations in areas not encountered in other fires—closed hood and trunk areas. Some of the reasons for gaining access include; the complete extinguishment of the fire, the investigation of the fire's origin, the securing of the vehicle and all of its contents, and possibly the determination of whether the vehicle was carrying illegal materials that factored into why it burned.

Gaining entry into the hood or trunk area can be extremely difficult. In these cases, the firefighter will have to apply forcible entry techniques to gain access to these compartments. As in structural firefighting, firefighters should always remember to "try before you pry." In most of today's automobiles a steel cable runs from the interior dash area to the latch at the front of the engine compartment. Unfortunately, this cable burns through quickly and access to the engine compartment must be by force, as the cable mechanism will be out of service. The trunk area is very similar as there is a cable from the interior driver area to the trunk in most new vehicles, which may be of no use due to the fire.

A Halligan bar may be used to knock out windows as well as to open hoods and trunks. Before attempting to gain full access to the trunk or engine compartment, it may be necessary to create an opening large enough to direct a fire stream in order to extinguish the fire.

Power saws may be used to cut locks in order to gain entry. Hydraulic tools may be used if the fire conditions allow as detailed in Section II, Chapter 16. A variety of forcible entry tools may be used when forcing entry into a vehicle; however, the firefighter should also apply common sense and ingenuity as well as consider the vehicle's design and construction in forcible entry techniques.

CAUTION

As mentioned earlier in the book, backdraft and flashover are two structure fire phenomena that firefighters must be aware of. This is also true with automobile fires. A closed passenger compartment or trunk may be the perfect place for a backdraft to be set up. Or, that same area can see a flashover if the conditions are right. Care must be taken when working in these environments.

Overhaul of a vehicle must be accomplished to locate and extinguish any hidden or smoldering fires. The battery to the vehicle must be disconnected by removing the cables before any overhaul begins to ensure firefighter safety. As discussed in the "Vehicular Fire Components and Considerations" subsection, care must be exercised around the bumpers, hatchback struts, air bag assemblies, tires, and fuel areas especially if any of these areas have been exposed to fire or heat.

Firefighters should be very thorough and take their time during overhaul operations to ensure complete fire extinguishment. This may require them to remove vehicle components such as the dash and seat covering in order to effectively accomplish this task. Overhaul operations should be conducted only to the extent necessary to ensure complete extinguishment. Depending upon circumstances, law enforcement or fire investigator may need to be called to the scene.

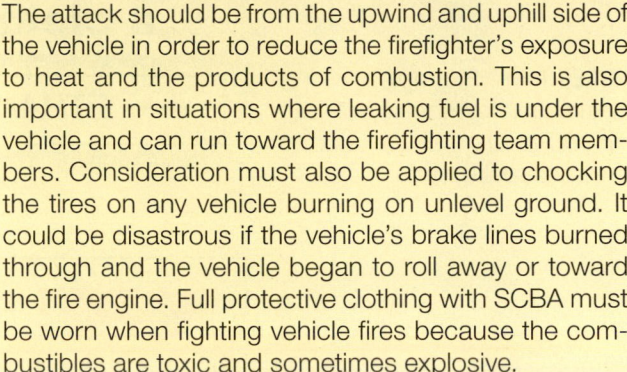

SAFETY

The attack should be from the upwind and uphill side of the vehicle in order to reduce the firefighter's exposure to heat and the products of combustion. This is also important in situations where leaking fuel is under the vehicle and can run toward the firefighting team members. Consideration must also be applied to chocking the tires on any vehicle burning on unlevel ground. It could be disastrous if the vehicle's brake lines burned through and the vehicle began to roll away or toward the fire engine. Full protective clothing with SCBA must be worn when fighting vehicle fires because the combustibles are toxic and sometimes explosive.

Alternative-Fueled Vehicles

Today's firefighters are faced with a much more complex task when fighting vehicle fires than they were a few years ago. In past years, they were faced with a gasoline-driven automobile and the tactics associated with it provided only a few options. On today's highways, we are faced with a great number of alternative fuel sources for our vehicles. The fuels used are categorized into a number of vehicular types:

- Flex-fuel vehicles
- Bi-fuel vehicles
- Compressed natural gas (CNG) and liquefied petroleum gas (LPG) vehicles
- Electric vehicles

Fighting fire in these newer types of fueled vehicles will be tactically different when their fuels are involved. If the fuels are not involved, the tactics won't change as nothing else in the manufacture of vehicles has changed as it relates to firefighting. So, let's take a look at each and compare them to the standard gasoline-driven automobile of the past.

Flex-Fuel Vehicles

These vehicles have a standard fuel tank, fuel system, and engine and are designed to run on regular unleaded gasoline and an alcohol fuel (either ethanol or methanol) in any mixture. For example, they could be 85 percent ethanol and 15 percent gasoline (E85) or any other mixture. In California, ethanol has already been blended into standard gasoline as a component because of clean air demands on the industry. The blend is a 6 percent blend of ethanol to gasoline. A quick way to identify vehicles with a flex-fuel system is by the identification symbols used on stickers on or in the vehicle, **Figure 19-40.**

Fighting a fuel fire in a flex-fuel vehicle is much like the standard gasoline-powered car fire. You are

Powered By ETHANOL
...a clean burning fuel

ETHANOL POWERED

FIGURE 19-40 Examples of identification symbols that can help firefighters identify vehicles with a flex-fuel system.

Responding to Incidents Involving Ethanol and Gasoline Fuel Mixtures

The Pipeline and Hazardous Materials Safety Administration (PHMSA) is alerting emergency responders to appropriate emergency response guidance for responding to incidents involving fuel mixtures composed of ethanol (or "ethyl alcohol") and gasoline in various concentrations. The most common of these fuels, designated E85 (85% ethanol and 15% gasoline), recently has begun to be used in volume in the Midwest, primarily in the states of Illinois and Minnesota.

Fires involving E85 and other ethanol/gasoline mixtures containing more than 10% ethanol should be treated differently than traditional gasoline fires because these mixtures are polar/water-miscible flammable liquids (i.e., they mix readily with water) and will degrade the effectiveness of fire-fighting foam which is not alcohol-resistant. For this reason, PHMSA recommends use of alcohol-resistant foam to fight fires involving these fuel mixtures. Properties of ethanol/gasoline fuels that may be of interest to emergency responders are provided in the chart below.

Properties of Fuel Ethanol

Property	Comment
Vapor density	Ethanol vapor, like gasoline vapor, is denser than air and tends to settle in low areas. However, ethanol vapor disperses rapidly.
Solubility in water	Fuel ethanol will mix with water, but at high enough concentrations of water, the ethanol will separate from the gasoline.
Flame visibility	A fuel ethanol flame is less bright than a gasoline flame but is easily visible in daylight.
Specific gravity	Pure ethanol and ethanol blends are heavier than gasoline.
Conductivity	Ethanol and ethanol blends conduct electricity. Gasoline, by contrast, is an electrical insulator.
Toxicity	Ethanol is less toxic than gasoline or methanol. Carcinogenic compounds are not present in pure ethanol; however, because gasoline is used in the blend, E85 is considered to be potentially carcinogenic.
Flammability	At low temperature (32°), E85 vapor is more flammable than gasoline vapor. However at normal temperatures, E85 vapor is less flammable than gasoline, because of the higher autoignition temperature of E85.

April 26, 2006

SAFETY ALERT

U.S. Department
of Transportation

Pipeline and
Hazardous Materials
Safety Administration

FIGURE 19-41 DOT guidance on incidents involving flex-fuel vehicles.

FIGURE 19-42 Examples of identification symbols that can help firefighters identify vehicles with an alternative fuel system.

PROPANE/FLEET LPG

battling a flammable liquid fire. The difference comes into play when you have fuel burning in an open area (under or around the car), in which case you will have to use foam, **Figure 19-41.**

Bi-Fuel Vehicles

The bi-fuel vehicle has two separate fuel systems, with the capability to switch from one to the other; the vehicle can be fueled by either system. One system will usually run on gasoline or diesel, while the other may be CNG or LPG. We will discuss fighting fire with these fuels in the next section of this chapter since one is a flammable liquid and the other a flammable gas.

CNG and LPG Vehicles

CNG- and LPG-powered vehicles will have compressed gas tanks much like a barbeque tank, only larger. All of the fuel lines will be high-pressure lines. Unlike the alternative-fueled vehicles discussed so far, these vehicles will not give you a choice as to how to fight a fire when the fuel system has been compromised with an engine fire or fully involved vehicle fire. You will truly have a compressed flammable gas fire. Fighting a fire of compressed gas is discussed in the next section of this chapter. Identifying these vehicles will depend on decals or stickers, such as those shown in **Figure 19-42,** or by noticing the fuel tank itself. In most cases these compressed fuel tanks will be in the trunk area; under a side panel of a van, truck, or school bus; or in the bed of a pick-up truck.

CAUTION

In two recent cases in California when vehicles fueled by CNG burned, the fuel tanks located in the trunk areas BLEVE'd sending tank parts over 50 feet away from the vehicles. Approaching these vehicles will be very hazardous when fully involved.

CAUTION

Vehicles powered by CNG and/or LPG involved in accidents may not be on fire upon the arrival of the fire department. It must be recognized that any of the high-pressure lines containing this gas (fuel) may be ruptured. The scene must be contained due to the possibility of escaping gas. An approach from upwind and uphill should be attempted and a line pulled, charged, and manned as a precaution.

Electric-Fueled Vehicles

The procedures around electric vehicles may differ due to the nature of the power source and its effect on the automobile itself. The automobile industry has gone to great lengths to educate firefighters in safe methods of fighting electric vehicle fires. The most recommended method to date consists of seven steps:

1. Approach vehicle at an angle to front or rear.
2. Identify vehicle type.
3. Secure and stabilize vehicle.
4. Power off electric vehicles.
5. Make the initial attack on vehicle fires with water and/or foam.
6. Use standard tools and cut-in areas for victim extrication.
7. Do not cut into high-voltage components. These are usually color coded by law to stand out from the rest of the components.

The action taken by an attack team on a vehicle fire in step 5 is to extinguish electric vehicle fires with water and/or foam. If it appears that the fire has extended into the battery pack, the firefighter should attempt to extinguish it following the instructions for the specific battery type. This information changes as technology changes, so it must be obtained from in-service training on an ongoing basis. If the vehicle is in a safe area, under some circumstances, firefighters may allow the fire to self-extinguish. Vehicle manufacturers are now beginning to use a number of alternative fuels in an effort to become more fuel efficient. A firefighter may find CNG, LPG, and others. For fighting fire involving these fuels, refer to "Flammable Liquids and Gases Fires" content in Section II, Chapter 19.

Establishing Safe Work Areas

Most vehicle fires will be found in places of travel, roadways, freeways, turnpikes, and so on. This will most probably create another safety hazard for the firefighting team. We've already considered the hazards associated with fighting vehicle fires, but now we are faced with traffic hazards at the scene. It has been an unfortunate statistic for many years that firefighters are injured and/or killed at vehicle fires and other roadway emergencies when other vehicles penetrate the working areas where these firefighters are abating the emergency. Great care must be taken to block these areas from outside penetration. Whenever possible the first arriving apparatus should position itself between approaching traffic and the scene,

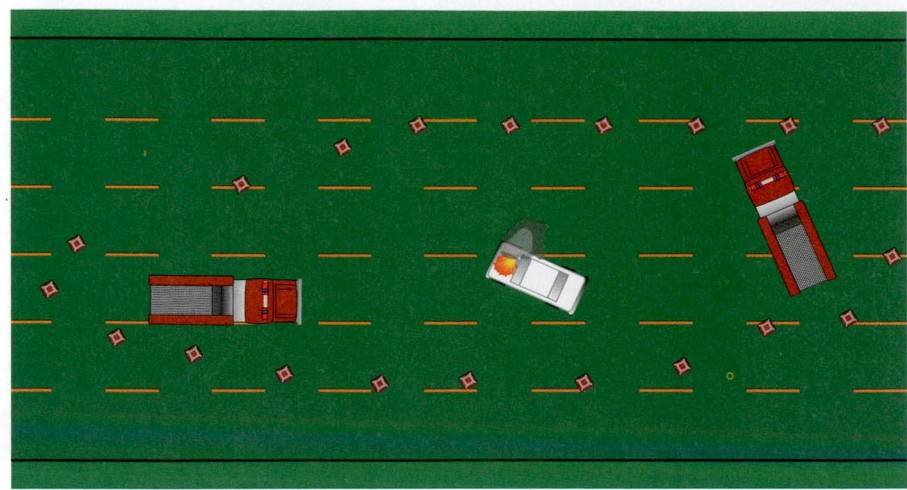

FIGURE 19-43 To protect firefighters on scene in the street, the engine is typically used to block traffic. Cones are used quite often to assist in this endeavor.

Figures 19-43 and **19-44.** It should be placed in a manner that gives approaching and working firefighters full protection from outside traffic. Some sources recommend this position be up to 50 feet from the accident itself. The second-arriving apparatus should be placed in a manner that would assist in the blockage of traffic if possible. If that is not practical or possible, then it should be positioned just beyond the accident or fire, utilizing the blocking effect of the first apparatus. It is recommended that traffic cones and/or road flares be used to enhance the safety zone being established.

FIGURE 19-44 The fire apparatus is used to block traffic from the firefighters' working area. *(Courtesy of Phill Queen)*

CAUTION

Some roadway emergency scenes may not be conducive to the use of road flares because the flare would bring an outside hazard to the scene. Hybrid automobiles with leaking flammable gases, flammable liquid fuels spilled but not on fire, dry roadside brush, as well as other items could ignite if preventa, not itive measures are not taken.

ommended that the use of vehicle lights be reduced to parking lights augmented with the flashing red lights that are already being used.

STREETSMART TIP

Highway safety research has shown that the placement of roadway flares is most visible when placed as follows: The length of the flare line should be twice that of the posted vehicle speed on that roadway. So, 70 feet of flares would be posted on a roadway posted at 35 mph, and 130 feet of flares would be placed on a roadway of 65 mph. The flares should be placed approximately every 15 feet or every three strides taken by the firefighter placing the flares.

Another method that must be used to enhance scene safety is emergency lighting. This is accomplished with the emergency apparatus itself, other emergency vehicles, and road flares. Some departments are even experimenting with portable lighted signs that are taken from the apparatus and placed in the roadway between approaching traffic and the scene. Apparatus lighting is standard in the industry based on NFPA Apparatus Standards. The use of that lighting is where some problems can surface. One problem concerns the use of headlights, and sometimes high-beam headlights, facing approaching traffic at the emergency scene. It is quite possible that these lights may temporarily blind approaching drivers, causing them to drive into firefighters. It is rec-

As the firefighters work on the vehicle fire, caution must be considered regarding other traffic in the area. Unfortunately, a great number of firefighter injuries occur each year at vehicle fires when passing motorists strike firefighters working on the accident scene.

Great care must be taken knowing that passing motorists will be looking at the accident more than the people working in the area. It is also important to realize that flashing lights and excitement will attract some drivers to the scene. Figure 19-44 shows an engine being used to block traffic from the vehicle fire scene. Traffic cones, road flares, or fire engines are to be used to block traffic from the work area in order to better protect firefighters working an accident.

While fighting vehicle fires may at first seem simple, this is far from the truth. Vehicle fires are complex, dangerous situations, requiring skill in fighting each of the four classes of fires. Safety must always be considered in these situations.

Trash and Dumpster Fires

A trash fire, whether contained in a bin (dumpster) or a truck, or spread on the ground, can surprise the most seasoned firefighter with a burning rate or intensity not expected. Trash can be anything discarded by anyone. Trash can produce toxic gases and, in many cases, items discarded can be hazardous and sometimes even explosive. The unprepared firefighter can be injured or killed fighting these seemingly innocent fires.

As the fire engine arrives on scene, it is important for the team members to be fully dressed in their full protective clothing including SCBA. Water is most usually the agent used in the extinguishment of these typically Class A combustibles.

SAFETY

When first applying water, the firefighter must be protected from the possible explosion of heated containers such as aerosol cans, bottles, and glass items. Using the protection of the side of the bin or the side of the engine should be considered when first attacking a trash fire.

Figure 19-45 shows a firefighter applying water while being protected from exposure.

Many times these fires will deeply seat themselves, not allowing the water to penetrate the materials. In these situations a number of things can be done, including the following:

- Use a tool such as a pike pole or rubbish hook to stir the material as water is applied.
- Use a tractor or dozer to move larger masses of trash in order to penetrate piles.
- If the fire is in a truck, consider having it dumped while water is applied.

FIGURE 19-45   A firefighter approaches a dumpster fire, keeping below the top edge of the container. *(Courtesy of Brentt Sporn)*

- The use of water agents (light water) can be an effective method of achieving deeper penetration into the materials.
- In some cases the bin or dumpster may have to be filled with water to achieve better results.

The trash or rubbish fire can pose extinguishment problems as well as hazards not considered in such nonemergency-type calls. The firefighter must be knowledgeable about the possibilities and proceed with caution when dealing with the unknown contents of these fires.

Outside Stacked or Piled Materials Fires

Many times the firefighter is asked to extinguish a fire in outside storage. These may be lumber piles, used tire piles, stored pallets, bundled paper or cardboard, or just about anything else that is flammable and stored outside. **Figure 19-46** shows a typical yard storage of high piled pallets. The hose size, stream used, amount of water, appliances for water delivery, and tools for overhaul will either make the job easy or, if not selected properly, make the job difficult.

SAFETY

Working around any high piled storage is dangerous. This danger is magnified if the materials are on fire, due to the high fire load, or even wet after extinguished, due to the pile stability. Piled storage is held up by the integrity of the items stored, and as that changes with either fire or water damage, those piles may topple. Collapse zones must be maintained whenever working around high piled storage.

FIGURE 19-46 Typical high piled storage that can jeopardize firefighter safety.

Logically, large piles with large fires will require large amounts of water. The delivery of that water will be the key to success. Large-diameter lines with the proper nozzles, usually smooth bore or straight stream capable, will most often be utilized for quick knockdown. "Big Fire equals Big Water" is a good rule of thumb as firefighters may employ monitors or even ladderpipes to apply adequate amounts of water during suppression. As this knockdown is taking place, it will be most important to keep an eye on any exposures that may be threatened. Those may consist of nearby piles of the same material, small

buildings, or even distant exposures threatened by flying fire brands. Then, after initial knockdown, the fog and/or foam nozzles will come into effect. Many times the streams will require additives, such as those described in Section II, Chapter 11, in order to add to the penetrating power of the water. Then the overhaul begins. In most cases, firefighters can expect to have to pull all the materials out of every pile while water is being applied. In the case of used tire storage, it is not uncommon to have to bring in large tractors, cranes with buckets, or other large hauling or moving equipment to break down the piles. Hand tools, such as pike poles and shovels, will also be used as the piles are brought down to manageable sizes. Water will continue to be applied as firefighters dismantle the piled materials staying alert for any pockets of hidden fires or other materials or hazards that may also be hidden within the materials. Foam may also be a valuable tool during overhaul due to its penetrating and smothering ability. In many cases the overhaul will last far longer than fighting the fire itself.

CAUTION

The products of combustion will continue to develop during the overhaul phase of the work, and it is important to continue to wear SCBA or some type of respiratory protection during this time.

LESSONS LEARNED

The basic principles of firefighting are based on sound scientific laws as well as years of firefighting experience. Every combustible item and occupancy must be carefully studied and understood in order for firefighters to be effective, efficient, and safe in their work.

Just as both synthetic and natural items differ, so will the tactics and strategies necessary to extin-

guish them differ when they burn. With newer automobile fuels and technology advances, the job of the firefighter is a constantly changing and challenging one. Knowing the basic elements of fire control, and applying them safely, will result in a better environment for all.

KEY TERMS

Anchor Point   A safe location from which to begin line construction on a wildland fire.

Aspect   The direction a slope faces given in compass directions.

Autoextended   When a fire goes out the window on one floor, up the side of the building, which is often noncombustible, and extends through the window or cockloft directly above.

Box Canyon   A canyon open on one end and closed on the other. They become very dangerous when wildfire enters them.

Chimney   A topographic feature on the side of a hill or mountain that naturally collects water runoff, channeling it to the bottom of the rise. Fire is attracted to this feature. Also referred to as drainage.

Combination Attack   A combined attack based on partial use of both offensive and defensive attack modes.

Defensive Attack   A calculated attack on part of a problem or situation in an effort to hold ground until sufficient resources are available to convert to an offensive form of attack.

Drainage   A topographic feature on the side of a hill or mountain that naturally collects water run-off, channeling it to the bottom of the rise. Fire is attracted to this feature. Also referred to as chimney or fire chimney.

Exposure Fire   Any combustible item threatened by something burning nearby that has caught on fire.

Fire Intensity   A measurement of Btus produced by a fire. Sometimes measured in flame length in the wildland environment.

Flanks of the Fire   The sides of a wildland fire running from the start point up each side to the end of the fire running into unburned areas.

Garden Apartment   A two- or three-story apartment building with common entryways and layouts on each floor, surrounded by greenery and landscaping, sometimes having porches and patios.

Ground Fault Interrupter (GFI)   An electrical switch that will trip when electrical current is interrupted or shorted. It may be reset when the cause of the interruption is corrected.

Head of the Fire   The running top or aggressive end of the fire away from the start point.

Interface Firefighting   Fighting wildland fire and protecting exposed structures in rural settings.

Midslope   An area partway up a slope. Any location not on the bottom or top of a slope, as in a mid-slope road crossing the slope horizontally.

Offensive Attack   An aggressive attack on a situation where resources are adequate and capable of handling the situation.

Rate of Spread   A ground cover fire's forward movement or spread speed. Usually expressed in chains or acres per hour.

RECEO   Acronym coined by Lloyd Layman standing for Rescue, Exposures, Confinement, Extinguishment, and Overhaul.

Ridge   The land running between mountain peaks or along a wide peak. A high area separating two drainages running parallel with them.

River Bottom   Topographic feature where water runs from higher elevations to lower. Can be dry or wet depending on season or recent rains.

Saddle   A pass between two peaks that has a lower elevation than the peaks. Wind will pass through this area faster than over the peaks, so fire is drawn into this feature.

Stairwell   An enclosed stairway attached to the side of a high-rise building or in the center core of same.

Steepness of Slope   The degree of incline or vertical rise to a given piece of land.

REVIEW QUESTIONS

1. Explain how water suppresses fire in relation to the fire tetrahedron.

2. Give at least two factors that must be taken into consideration prior to an attack plan being formed and extinguishment of a structure fire.

3. Ground cover firefighting involves a number of principles very different from those used for structural firefighting. Explain how the wildland fire triangle differs from the fire tetrahedron.

4. Vehicle firefighting involves a number of built-in dangers. Name an exterior feature and an interior feature that can injure a firefighter who is not very careful.

5. Why does a basement fire have the potential to be so much more hazardous to a firefighter than a ground-level fire in the same occupancy?

6. OSHA has enacted the two in/two out ruling. Explain this rule and its impact on the fireground during structural operations.

7. Advances in automobile technology have caused some major changes in fuel systems. Explain how a firefighter can tell the differences in these types of vehicles.

8. When working an incident in heavy traffic, a number of safety precautions must be considered. Give two examples of possible precautions we can deploy.

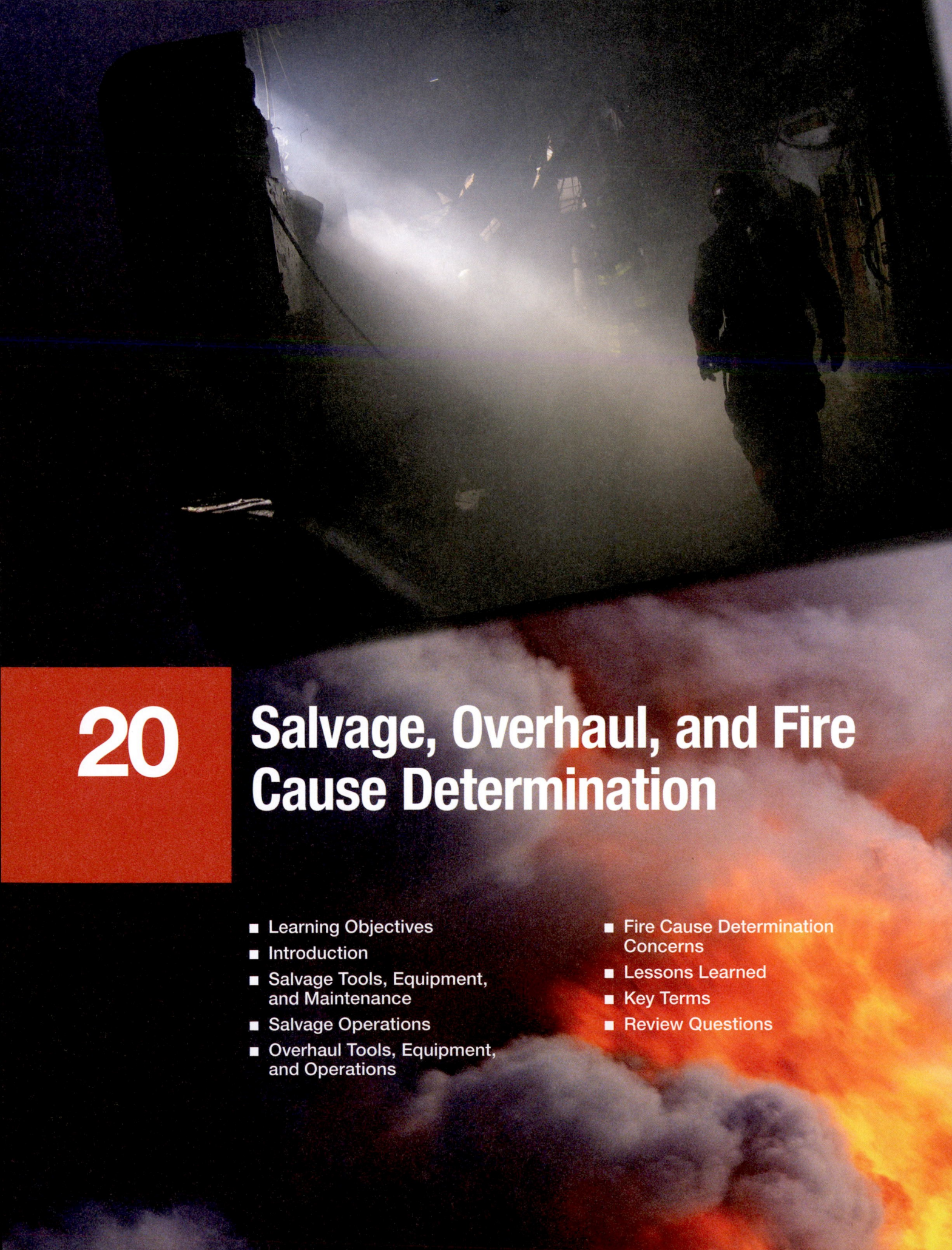

20

Salvage, Overhaul, and Fire Cause Determination

Many firefighters take more of an interest in going in and attacking the fire. But salvage is important too. Most of the damage to a person's property is caused by smoke and water, not by the actual fire. And those contents can have so much meaning for people—all they want is their private possessions. Beyond that, if there's a possibility of arson, it's your job to preserve the evidence.

I remember receiving a call for a fire at The Museum of Science and Industry—it came in through enhanced 9-1-1. By the time we responded, the museum had white smoke curling from the roof. The initial companies were locating the seat of the fire (not easy since the museum occupies four city blocks). They figured out it was in a void in the roof, and we started doing salvage. I acted as coordinator. In a house fire, we'd be likely to rope the belongings in the center of the room and put salvage covers over everything. But in this case, there were fixed displays, so we couldn't move them, but we still covered everything. We built chutes for the stairwells so that the water could be channeled without damaging the carpets, and we diked the elevators. If we can prevent water damage and channel out some of the smoke, we'll have a better goodwill image with citizens. In this case, you may not gain the admiration of fellow firefighters, but what we did was save a lot of artifacts that couldn't have been replaced.

—Street Story by Julius Stanley, Lieutenant,
Chicago Fire Department, Chicago, Illinois

LEARNING OBJECTIVES

After completing this chapter, the reader should be able to:

20-1   Explain the purpose of salvage and overhaul operations and how "customer service" factors into the procedures.

20-2   Describe methods of property conservations during salvage and overhaul operations.

20-3   List the various types and usage of salvage covers during salvage and overhaul operations.

20-4   Demonstrate the proper method for covering doors, windows, and floor or roof openings during salvage operations.

20-5   Given a specific piece of equipment, list and describe various tools, inspection, and cleaning methods used to return the equipment back to service.

20-6   List and describe the use of cleaning solvents

20-7   Describe the proper manufacturer's or department's procedures for tool and equipment cleaning.

20-8   Given manufacturer recommendations, departmental guidelines, tools, and cleaning materials demonstrate the ability to properly clean, inspect, and service tools and equipment.

20-9   Demonstrate proper salvage cover rolls and folds so the cover is ready for immediate deployment.

20-10   Describe issues associated with forcible entry during salvage operations.

20-11   Arrange a room's contents into a salvageable position and perform salvage cover throws, folds, and rolls.

20-12   Demonstrate how to separate damaged items to a safe location during salvage operations while protecting fire investigation paths.

20-13   Demonstrate methods of water removal during salvage operations.

20-14   Demonstrate how to construct water chutes and catch-alls to reduce damage during salvage operations.

20-15   Describe how salvage operations are accomplished when automatic sprinklers are in use.

20-16   Identify tools and equipment and describe how they are used in overhaul operations.

20-17   Describe the various types of tools, and the methods to use them in order to locate and expose hidden fires.

20-18   Discuss extinguishment and overhaul procedures for interior structure fires.

20-19   Demonstrate methods of locating and suppressing hidden fires in walls and subfloors.

20-20   Demonstrate the safe opening of void spaces to expose hidden fires while maintaining structural integrity.

20-21   Explain how fire attack and ventilation affects the overhaul effort.

20-22   Explain how to look for structural stability and where to search buildings for hidden fires.

20-23   Describe safety considerations when performing overhaul operations.

20-24   Demonstrate proper methods for exposing hidden fires while preserving any obvious signs of the area of origin.

20-25   List the water application for various attack lines in order to enhance overhaul operations.

20-26   Demonstrate proper attack line selection and its use during overhaul operations.

20-27   Demonstrate the proper methods for applying water to get maximum effectiveness.

20-28   Demonstrate the procedures used to verify that complete extinguishment of a fire has occurred.

20-29   List and describe obvious signs of arson.

20-30   Explain the importance of fire scene preservation.

20-31   Describe visual indicators used in determining area of origin.

20-32   Identify methods to protect the area of origin, potential fire cause, and arson evidence.

20-33   Demonstrate locating and assessing burn patterns to assist in determining the point of origin of an exterior Class A fire.

*The FF I and II levels, as defined by the NFPA 1001 Standards, are identified in different colors:
FF I = black, FF II = red, additional information = blue.

INTRODUCTION

Salvage and overhaul operations are not often viewed as critical tasks within the fire service. They are not associated with the excitement of fire attack, rescue, or ventilation. This perception leads to a downplaying of importance and an "afterthought" mentality. Fire cause determination is something new firefighters will not be actively involved in; however, all firefighters need to understand this task. During fire attack, with proper training, it is possible for firefighters to combat the fire while watching for and protecting potential signs and conditions that will assist the fire investigator. This chapter explains the reasons for and importance of salvage, overhaul, and fire cause determination and describes the necessary firefighter skills and abilities needed to effectively perform these operations.

SALVAGE TOOLS, EQUIPMENT, AND MAINTENANCE

All fire incidents involve a potential loss of material goods. Whether they are easily replaceable articles of clothing or furniture or irreplaceable photographs, heirlooms, antiques, or memorabilia, they all have meaning to their owner. It is for this reason that firefighters can make a tremendous impact on an incident by aggressively salvaging the occupancy, **Figure 20-1A** and making sure not to cause more damage than the emergency itself has, **Figure 20-1B.**

When considering salvage, people usually think of the interior of a structure. Items on and around the exterior of a house can be just as valuable. While first-in company officers are making their size-up to determine the extent of the involvement, they may be able to quickly move some patio furniture or arrange to have vehicle(s) removed or covered with a salvage cover. It is important that the company officer "triage" the entire scene. If quick action can save property without endangering the operation, then the officer should make an effort to do so. It is possible for firefighters to have tunnel vision and to waste time trying to save one portion of the scene while endangering the integrity of the whole scene. Quick thinking and experience are vital traits of good company officers.

(A)

(B)

FIGURE 20-1 (A) Good and (B) bad salvage operations.

It is rewarding to report to owners who fear the worst possible loss that salvage crews were able to save a particular special item for them. Special items make a home unique, and firefighters should always consider this while working any fire scene. A quick scan followed by consideration of what the homeowner might want to save can be a great service to a person who suffers the tragedy of a fire.

STREETSMART TIP

Company officers who are in command of a scene can often perform a small necessary or safety-oriented task if there are no other firefighters in staging and if the task will not interfere with the safety of the on-scene crews or the command of the scene. Commanding the scene is the primary role of that officer.

The basic premise of salvage operations is to remove the harmful atmosphere from the material or to protect the material from the harmful atmosphere. Several techniques are used to accomplish this goal and will be discussed later. Any technique is rendered useless unless the proper tools are used, and there are many tools and different pieces of equipment for salvage work. Some operations might require complex equipment while others demand a hammer and a few nails.

Salvage Covers

The mainstay of salvage operations is the salvage cover or a variation. Salvage covers are made out of several different materials, most commonly plastic, canvas, and treated canvas. A plastic cover has advantages in weight and water resistance. A canvas cover is durable and, if treated, can compare to the plastic cover for water resistance capability.

The size of the salvage cover can vary from 10×12 feet to 12×16 feet. The perimeter of the cover is ringed with grommets spaced at intervals, **Figure 20-2.** The ideal interval is 16 inches to match up with the studs in a building wall.

Salvage covers are the main materials used for building water chutes and catch-alls, which are discussed later.

A simple tool that can replace or supplement the salvage cover is **Visqueen** or black plastic, which can be carried easily by a firefighter, **Figure 20-3.** The Visqueen comes in rolls 120 feet long, with an unfolded width of 20 feet. An entire roll of plastic can be carried by putting a dowel in the center tube (with a cord attached) and hanging the cord over the firefighter's neck so the roll is in front of the firefighter. The only additional tool needed for plastic use is a razor knife or scissors to cut the desired lengths. In some cases the plastic is precut into various sizes for quick deployment. By precutting some lengths, the weight

FIGURE 20-3 Firefighter carrying Visqueen roll.

of the roll is reduced, thus minimizing the stress on the firefighter carrying it. Visqueen is also relatively inexpensive and can be left with the property owner.

Floor Runner

Another item used for salvage operations is the floor runner, which is usually approximately 3 feet wide and 20 feet long. It is normally made of a lightweight canvas-type material for easy deployment. It is used just like its name implies—to cover the floor down a hallway or along a traffic area, **Figure 20-4.** If there is a fire in one area of the building, firefighters should try not to damage the flooring in another. Remember, customer service really makes the difference and is a sign of professionalism.

Water Vacuum

A very useful tool in salvage operations is the water vacuum, **Figure 20-5,** which is available in two basic types. One is worn like a backpack and is used to

FIGURE 20-2 Salvage cover grommets.

FIGURE 20-4 Floor runner in place.

FIGURE 20-5 Firefighter using water vacuum.

remove water from areas where access prohibits squeegee use; the other is usually larger and is moved around on wheels. If SCBAs are being worn, the "backpack" water vacuum cannot also be worn. In this case, other methods will have to be considered, or firefighters will have to wait until the air quality is safe and the area is safe to work in without SCBA.

The water vacuum can also be used to quickly drain catch-alls so that they do not overflow. The water vacuum usually has a limited capacity of around 5 gallons. If a large amount of water must be removed by a water vacuum, it is advisable to have the owner contract with a professional restoration company.

Another option carried by some fire agencies is either a submersible pump or "float-a-pump." The submersible pump can be dropped directly into water and used to pump water out through an attached line. This is a good tool for flooded basements. The float-a-pump simply floats on top of the water, drafting water and pumping it through a hoseline. The submersible needs electrical power, while the float-a-pump is gasoline-driven.

Miscellaneous Salvage Tools

Basic hammers, nails, and staple guns also have a place in salvage work. Having these tools available allows firefighters the ability to weatherproof doors, windows, ventilation holes, or other openings to further protect the occupant's property from weather. Visqueen is usually the salvage cover of choice as it can be left with the structure at little cost to the fire department.

Doors and windows need to be secured with plywood or some other sturdy material to offer any resistance to vandals. This task is handled by board-up crews or restoration companies and is discussed later.

Sometimes salvage covers need to be used to prevent a leak from damaging an item to be salvaged or an item that is too large to move to a safer location. A hammer and a few nails can be used to nail the salvage cover to the wall and cover the exposed item. Firefighters also use a small fastening tool called an S-hook that can be hammered into the wall to secure a sturdy place to hang the cover, **Figure 20-6.**

A thorough knowledge of available tools to be used in salvage work will make the task much easier to accomplish. As in all projects, with the right tools the job will be done correctly while minimizing damage.

Maintenance of Salvage Tools and Equipment

After salvage and overhaul tools and equipment are used, they must be cleaned and inspected like any other piece of firefighting equipment to make sure they are ready for the next emergency. A written log should be kept on the maintenance and use of the tools so proper maintenance is done.

Salvage tools and equipment are often exposed to hazardous materials and should be placed out of service if damaged. If, through the course of a salvage operation, the equipment suffers damage, the fire department should seek reimbursement from the property owners through their insurance company. Just as the insurance company is liable for the owners' property, it may also be liable for fire department equipment if it is damaged. The department should document the damage thoroughly and then contact the insurance company through the **chain of command** to determine the means to secure reimbursement. This reimbursement may not apply in all states, so it is important to contact a local insurance commissioner to determine if this is an option.

When salvage covers are used at an emergency scene, they will end up with material or debris on them. This material should not be spilled on clean areas after it has been collected. The best way to keep this material in the cover is to do a "loose fold and roll," **Figure 20-7.** Essentially, with this technique the cover is folded into itself so the debris caught is trapped inside. This is done by bringing both sides to the middle once and then once again, and then rolling the cover. This will keep all of the material inside until it can be disposed of. Once the material is disposed of and the cover is brushed off, it must be washed using a mild soap solution, hung to dry, inspected for tears or holes, repaired, folded, and placed back on

FIGURE 20-6   Various fastening tools.

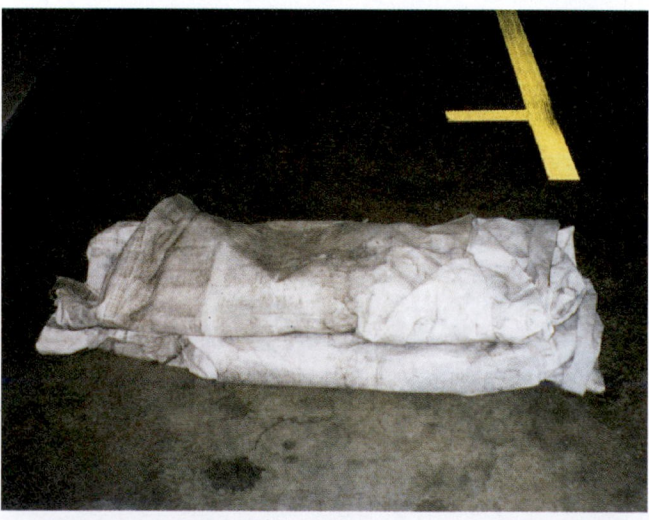

FIGURE 20-7   Loose fold and roll.

FIGURE 20-8 Salvage cover drying rack.

the apparatus. (Note that this reveals another benefit of Visqueen—no maintenance.)

Drying the cover can prove difficult due to its size. Sometimes hooks are placed along high exterior roof-lines or interior ceilings to hang the cover to dry. Another method is to rig a system to haul the cover up alongside of a building to dry, **Figure 20-8.** If neither of these is available, it can be laid out on the apparatus floor, but it must be turned over so both sides dry.

Once dry, the cover should be inspected for holes or tears. A simple way to inspect it is for firefighters to drape the cover over themsleves and look for light shining through any holes or rips. Holes or tears should be marked with chalk and repaired.

Salvage Cover Folds and Rolls

After they are cleaned and inspected for damage, salvage covers are either rolled or folded for storage. The available compartment space often dictates which method to use. The roll is accomplished by placing the cover on a large flat surface; if it has a treated side, it is placed up. Sprinkling some baby powder on the cover and spreading it around with a broom will keep the cover from sticking to itself if it is not used for long periods.

JPR 20-1: Salvage Cover Roll

(For step-by-step photos of this skill sequence, see page 658)

Two firefighters are needed to roll a cover.

1. Place one firefighter at either end of the cover. Both firefighters place one hand on the edge of the cover one-quarter of the way in and the other hand on the outside edge. Each firefighter should be on the same outside edge facing each other.

2. While grasping the cover, with a quick flipping motion, bring the outside edge to the middle.

3. The firefighter's inside hand should be at the fold. Repeat this fold again and again until the fold is at the middle of the cover. Fold the opposite end of the cover in the same manner.

4. The cover should now be about 3 feet wide and ready to be rolled. The treated side is rolled inside and the cover is ready for deployment.

Many salvage covers are folded for deployment by one firefighter, while others are folded for deployment by two or more firefighters. Folding a cover is similar to folding a bed sheet. It should be neat and orderly and take up minimal space on the apparatus. Some departments, to keep the folds neat and tidy, use a plywood template to adjust the size, **Figures 20-9A** and **B.** JPR 20-2 shows the proper skills to prepare a salvage cover for a one-person deployment. JPR 20-3 shows the proper skills to prepare a salvage cover for a two-person deployment. When folding a salvage cover, it is vital that the deployment method be considered. These covers often must be deployed quickly.

JPR 20-2: Salvage Cover—One-Person Deployment

(For step-by-step photos of this skill sequence, see page 659)

1. With the cover flat on the ground, fold the outside edge to the center point of the cover. Smooth out wrinkles and align edges.

2. Fold the same edge to the center point again. One fold is now beneath the other fold. Smooth out wrinkles and align edges.

3. Fold the opposite side in a like manner. Smooth out wrinkles and align edges.

4. Fold one end to a point just short of the center point. Smooth out wrinkles and align edges.

5. Fold the same end again. One fold is now beneath the other fold. Smooth out wrinkles and align edges.

6. Repeat steps to fold the opposite end. Smooth out wrinkles and align edges. The cover is now ready for a one-person deployment.

JPR 20-3: Salvage Cover—Two-Person Deployment

(For step-by-step photos of this skill sequence, see page 660)

Two firefighters are needed to roll a cover.

1. Fold the cover in half as in the one-person deployment fold. Smooth out wrinkles and align edges.

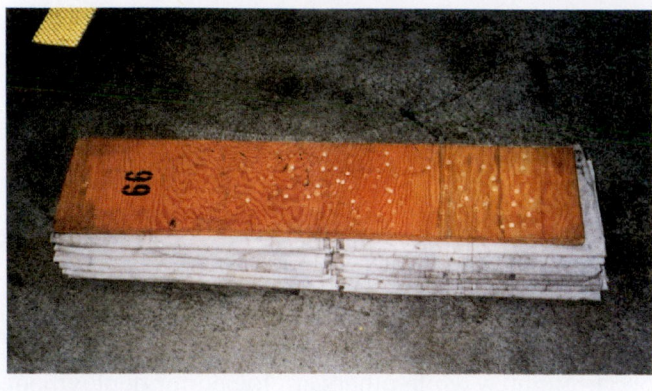

(A)

(B)

FIGURE 20-9 Salvage cover fold. (A) Salvage cover fold template. (B) Paper towel between salvage covers in compartment.

2. In similar fashion, fold the cover in half (lengthwise) again. Smooth out wrinkles and align edges.
3. Fold the cover in half (widthwise). Smooth out wrinkles and align edges.
4. Fold the cover end over again (widthwise). Smooth out wrinkles and align edges.
5. Fold the cover in half two more times (widthwise). The cover is ready for a two-person deployment.

JPR 20-4: Shoulder Toss

(For step-by-step photos of this skill sequence, see page 661)

Another way to deploy a salvage cover is to use the shoulder toss. This method is done by a single firefighter and is reserved for covering large unbreakable items. Rack storage in a warehouse is one example where this could be used.

1. Place a folded salvage cover over one arm.
2. Using a straight arm movement, toss the cover over the items being salvaged.

All tools must be thoroughly cleaned, dried, and checked for damage. The water vacuum must be rinsed several times to remove all contaminants from the tank. Chain saws will require more than the normal amount of maintenance and be taken apart and cleaned thoroughly. In most cases, chainsaws are run in atmospheres they are not designed to run in and used to cut things they are not really designed to cut. They must be inspected and cleaned from top to bottom. When chainsaws are put back into service, they should start within a few pulls and should run smoothly. Chainsaws are crucial to fire attack, rescue, ventilation, salvage, and overhaul. All firefighting tools need to be in good working order every time they are used.

SALVAGE OPERATIONS

The basic goal behind salvage operations is property conservation. Firefighters want to reduce the possibility of damage from a harmful substance, which in this case is fire. This may entail moving the item out of harm's way, covering the item, or removing the harmful substance from the area.

Safety Considerations

As with all emergency scene operations, the salvage group members must be aware of their surroundings and work as safely as possible. If salvage operations are going on while active suppression operations are taking place—as is often the case—the salvage crew must be appropriately dressed for the conditions; this means dressed the same as the attack crew—full protective clothing with self-contained breathing apparatus. (The individual components of full protective clothing are discussed in Section I, Chapter 6.)

The same is true for forcible entry tools. The salvage group is also sent to areas where fire attack is not taking place, such as the apartment below the involved apartment. In this case, it is necessary for group members to carry forcible entry tools for entry into the exposure apartment, or to gain entry through any interior doors within a business or home that are locked for security reasons.

CAUTION

The crew members must be aware of their surroundings at all times to ensure they do not find themselves in the wrong place at the wrong time. Although the content of a structure is important to its owner, it is not worth injury or a firefighter's life!

One of the most common hazards of salvage work is ceiling collapse. The salvage group may be tasked with covering homeowner possessions or other materials to protect them from water that is accumulating in the ceiling area. The roof could have leaks or be under repair, sprinkler heads in the attic could have fused, the attack crew might not have done a good job of monitoring water flow, or the fire might have warranted a tremendous amount of water to suppress it. For these reasons, the plasterboard (**Sheetrock**) will become weak and sometimes fail. In most cases, there will be warning signs such as the plasterboard seams beginning to show as water seeps into them, **Figure 20-10.** With this sign comes the knowledge that water is building up in the insulation. At a weight of 8.33 lbs per gallon, the weight will make the ceiling fail. Another sign of failure is plasterboard that sags as if it were not nailed properly.

If these signs appear, the water must be drained before the ceiling collapses. The easiest mitigation is to stop the flow and remove the water. This can be accomplished by quickly salvaging the area below the ceiling and creating a drain hole in the plasterboard to relieve the water buildup. If it is not possible to salvage the area, a quick decision must be made: Is water damage more serious than ceiling collapse? In most cases, water damage will be the lesser of the two evils.

A safe scene involves knowing what other operations are ongoing and where they are taking place. Ventilation crews working above the salvage operation can create a dangerous situation if the cut roof or ceiling drops. Firefighters should be listening for noise and radio traffic and should be aware of crews working above and below them. The Incident Commander (IC) or Operations Section Chief should be aware of all crews and their location in the building and should be contacted if the members of the salvage crew have any questions.

Stopping Water Flowing from Sprinkler Heads

Stopping the flow of water from sprinkler heads is often the main step in a salvage operation. Several types of sprinkler stops are available—the most common being the sprinkler wedge. This is usually a wooden wedge, like a doorstop, that can be jammed into place in the opening of a flowing sprinkler to stop the flow of water. The firefighter must be careful not to break the head off the sprinkler while hammering the wedge in place. This job is not easily accomplished as it is usually done from a ladder while water is flowing under a considerable amount of pressure.

> **CAUTION**
>
> A flowing sprinkler head can displace oxygen within a 3 to 5 foot area surrounding the head. It is important for personal safety that firefighters wear SCBA and are "on air" when attempting to stop flow from a sprinkler head or attempting to perform any duties near a flowing sprinkler head.

The firefighter assigned this task must wear eye protection and full protective clothing. The firefighter will need to lock into the ladder in order to get some force behind the wedge and to eliminate the possibility of falling. Then it is merely a matter of patience. As long as the main drain has been opened and the system is shut down, the firefighter is only working against residual pressure within the system. Getting this assignment is a task most firefighters rarely forget!

Other types of sprinkler stops work on more of a mechanical theory: The stop is placed in the opening and either screwed or locked into place by the operator. The same protective clothing and safety procedures apply to all types of sprinkler stops.

Methods of Protecting Material Goods

The "removing items from harm's way" approach can be determined very quickly by asking one simple question: Can an item be moved more quickly than it can be covered? If the answer is yes, firefighters should go for it! This will usually apply to quickly

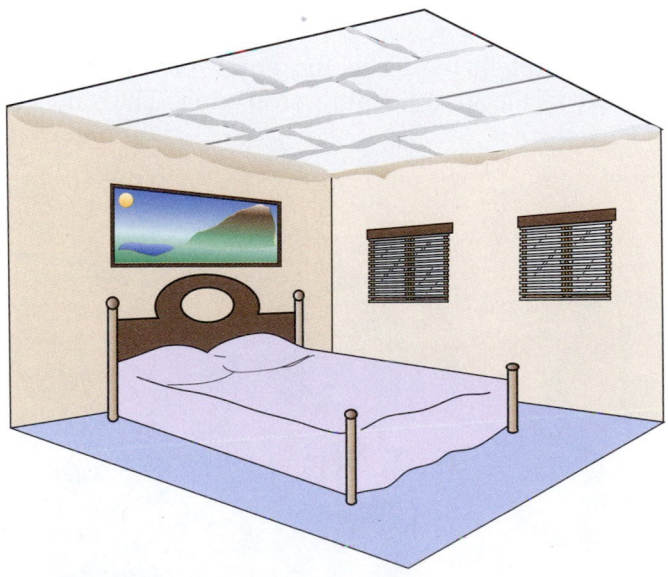

FIGURE 20-10   Plasterboard (Sheetrock) with water seeping through the seams.

accessed items like patio furniture and garage items. These are easily moved from harm's way as they usually do not have to be moved very far to get them out of the smoke and heat. Items that are in the way of the fire attack crews should be moved to assist the operation and also accomplish salvage.

Arranging of Furnishings and Salvage Cover Deployment

Covering the item is usually the best choice when the item is either too large to be moved or the movement would be too time consuming. When the decision is made to cover items with salvage covers or Visqueen, the room should be surveyed for the largest item or central piece. All smaller items should be moved to this area, **Figure 20-11.** Consider a bedroom for example. Items should be removed from the closets and placed on the bed, dressers should be pushed up close to the bed, and pictures should be removed from the walls and placed on the bed. If dressers are too large to be moved, then the drawers can be placed on the bed to protect their contents.

Once everything is ready to be covered, the salvage cover or Visqueen should be deployed. This is done by one or two firefighters. With one firefighter holding the cover or Visqueen firmly by the bottom fold, the other pulls the top fold or leading edge, if it is rolled, away from the firefighter holding the cover, **Figure 20-12.** It is easier and will save time if the cover is rolled or unfolded directly onto the items needing protection. When the cover is extended, it is then placed on the furnishing and the sides are unfolded. At this point, the reason for having the treated side out is obvious—so that it is facing the ceiling to deflect any water dripping down.

If there are breakables under the cover, extra care must be taken. Depending on the room being salvaged, a glass piece should be placed on top of the cover after the room is salvaged to designate that valuables

are under the cover. This signals other crews that the cover should be removed gently. The salvage cover edges should be "rolled" under so that they won't collect any water or debris. If this is a one-firefighter operation, the cover is just laid on the items and the firefighter plays it out individually.

ETHICS

Salvage operations can provide temptations to firefighters who would like a souvenir of an incident. Sometimes items are slightly damaged and crews will determine the value to be zero and take the item back to the firehouse. This is stealing and should never be allowed to take place. The consequences to this type of behavior are obvious. Even the allegation of theft will bring a very unprofessional image upon the department.

Covering items does not just apply to furnishings; it can also apply to flooring. When attacking an attic fire, the attack crew can bring with them a salvage cover and quickly spread it out prior to pulling ceilings or making attic access cuts. This is also where the floor runner comes into play. This again is just one of those little customer service efforts that create good publicity for the fire department.

Balloon Toss

If two firefighters are available, the balloon toss is another method of deploying a salvage cover. This method is used if the object to be covered is slightly taller than the firefighters.

JPR 20-5: Balloon Toss

(For step-by-step photos of this skill sequence, see page 662)

1. Two firefighters place the cover on the ground next to the item needing protection. The salvage

FIGURE 20-11 Furniture arranged for salvaging.

FIGURE 20-12 Two firefighters deploying a salvage cover.

cover is unfolded out of the accordion fold so that the cover is across the base of the item at its full length.

2. The firefighters grab the top inside edge of the cover and pick it up to about waist height. Both firefighters simultaneously move the cover's edge in an upward motion so that it catches air, and they then move to the other side of the item while the cover is still in the air.

3. The firefighters guide the cover to the floor on the opposite side. Once the items are covered, the firefighters dress the cover by folding the edges under to keep water from pooling and to lessen the chance for the edges of the cover to be disturbed by accident.

Working with Visqueen

Working with Visqueen or a rolled salvage cover is essentially the same as working with a folded cover. The limiting factor is the size of the rolled cover. In both cases, the roll is held by one firefighter and the other pays out the remaining portion. Once the material is in place, the sides are folded out and the edges are rolled under. A good example of where Visqueen is much better than a salvage cover is in a supermarket where the long rows of food would need to be covered. By starting at one end of the row, the proper length of Visqueen can be determined and cut. The Visqueen is then placed over the row and everything is covered.

Another use is covering the ventilation holes in a roof. The Visqueen can be nailed to the roof using wood strips. This is normally a task for the restoration company or board-up crew, and it is discussed later.

Limiting Exposure to Smoke

As discussed in Section I, Chapter 18, firefighters should always consider the by-products of combustion—heat, smoke, and toxic gases—as a danger. These by-products should be contained from spreading into undamaged portions of a structure and should be removed from the entire structure using forced air ventilation, **Figure 20-13.** Containment of the by-products can be an action as simple as closing doors after a room has been searched or closing any nearby doors during the fire attack advance. This action can save the contents of rooms from additional smoke, fire, and water damage. Adding positive pressure ventilation not only allows firefighters better visibility and safety from the products of combustion but also affords an additional measure of protection against damaging the contents of the unburned portion of the structure.

FIGURE 20-13 Positive pressure ventilation in place.

Water Removal

Water from fire attack operations can leak through the floor and damage rooms or entire floors below the fire. It is critical to capture, divert, or remove the water to successfully salvage these areas. If there is a floor drain, the operation becomes a little easier. If no drain exists, firefighters should examine their options. If a toilet can be removed, a drain hole can be created. Firefighters can squeegee water down this created drain as they would a floor drain. Firefighters can be creative and use salvage covers to create dikes to channel water into the drain or out of the building. Lining a stairwell with a salvage cover can create a chute to channel the water out of the area.

Catch-All

An example of the need to use a catch-all would be a multiple-story office building. A fire may occur in an office and be extinguished by the sprinkler system. This could cause a major salvage problem on a lower floor.

Catch-alls are used to contain the water dripping through the ceiling until a system is set up to remove the water from the building (chutes, water vacuums, squeegees, etc.). Catch-alls have sometimes been constructed using ladders and salvage covers to form a containment area. However, firefighters should consider the weight contained inside this catch-all. At 8.33 pounds per gallon, this weight can soon cause a structural integrity problem in an already damaged building, as the weight is all focused on a small area of the floor. It is much safer to consider a catch-all as a mitigation effort and construct it using a single salvage cover. A system to remove the water from the area should be constructed or devised as quickly as possible.

Catch-alls are easy to make. The firefighter places the cover with the treated side down and rolls the sides in to the desired size of the catch-all. The next step is to fold the ends as shown in **Figure 20-14,** roll the ends in, and tuck them tightly so they do not unroll. The cover is then flipped over, **Figure 20-15,** so as the catch-all fills up it will not unroll the sides or ends. As the water depth will only be 2 to 3 inches, a water vacuum may be necessary to adjust the water level until the water can be directed away with a chute or other method.

FIGURE 20-14   Catch-all end folds.

FIGURE 20-15   Catch-all.

Water Chute

If the water volume is too much for the catch-all, then a water chute is used so the water can be diverted out of the building through a window or doorway or down a stairwell. The water chutes are a combination of ladders or pike poles and salvage covers. The salvage covers are draped on the ladders so they work like a trough and the water flows down them. Office furnishings or other equipment is used to support the ladders. The salvage cover can also be stretched out and a pike pole placed on either side. The cover is wrapped around the poles and the ends rolled in until the desired width of the chute is attained.

If the length is longer than a single cover, a leakproof seal must join the two, **Figure 20-16.** Once the direction of the flow is determined, the upper cover is put in place with the last 6 inches folded back on itself. The next cover is laid on the next portion of the ladder with its top 6 inches laid on top of the bottom 6 inches of the cover above. The covers are then rolled together toward the bottom cover and flattened out when there is no more to roll.

Post Indicator Valve and Outside Screw and Yoke Valve

Many buildings are now equipped with sprinkler systems. These systems are beneficial at stopping or containing a fire but can cause damage to the unburned portions of the buildings if allowed to flow freely. It is important that firefighters know how to shut these systems down to prevent unnecessary damage. The IC will determine when the sprinkler is no longer needed for fire suppression.

Essentially two types of valves are used to shut down a sprinkler system: the post indicator valve

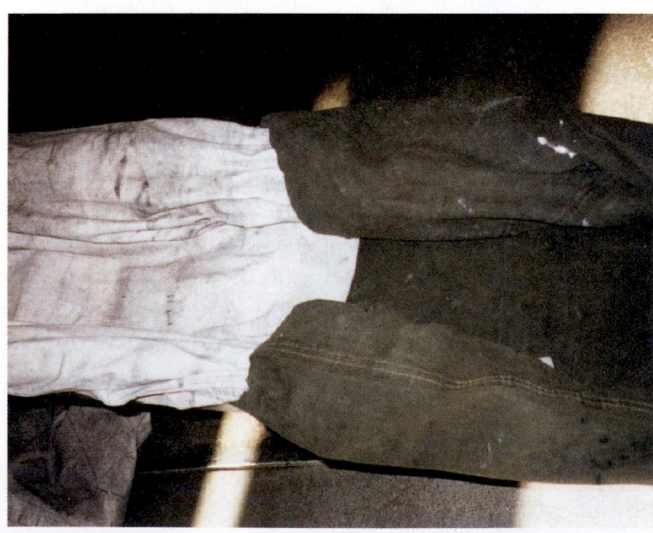

FIGURE 20-16   Leakproof fold of water chute.

(PIV), **Figure 20-17,** and the outside screw and yoke (OS&Y) valve, **Figure 20-18.** The PIV is generally located near the sprinkler connection for the building, but it may be just about anywhere around the building. The PIV is about 3 feet tall with a window near the top that states either "Open" or "Shut." This indicates the status of the valve. The PIV should be locked in the open position. Most PIVs are monitored so that if someone attempts to close them, an alarm sounds. If a PIV is found in the closed position it should be noted to the IC and turned to the open position if active firefighting is still taking place. To close the valve, the firefighter should break the lock, which is usually a knockoff-type lock, and remove the handle, which is usually locked on the valve.

STREETSMART TIP

A firefighter should always take bolt cutters when sent to close a sprinkler valve in case it is locked with a regular lock rather than a knockoff lock.

FIGURE 20-17 Post indicator valve. Note that the PIV is shut and unlocked. If found in this position at an incident, it should be opened and the condition reported to the incident commander and the fire investigator.

FIGURE 20-18 Outside screw and yoke valve.

The handle is used to close the valve. It helps to remember, "Righty, tighty—lefty, loosey," or clockwise to tighten and counterclockwise to loosen. Once the PIV is closed, the firefighter should also locate the system drain and open it. This will assist in bleeding the system and reducing the amount of water that will drain through the sprinkler heads.

The OS&Y valve will be located either on an exterior wall of the building or just inside the building. It looks like a steering wheel with a large screw sticking out of the middle of it. If the screw is out then the valve is open; if it is in then it is closed. The wheel may be locked so it cannot be turned to close the valve, or it may be a supervised system that sends an alarm if attempts are made to close the wheel. Bolt cutters are used to cut the lock to shut this valve. Again, if the valve is found in the closed position, the firefighter should notify the IC and turn it on if active firefighting is still taking place. If there is any possibility of having the valve reopened due to fire spread then it should not be turned off. The system should only be shut down when all active firefighting is completed.

Sprinkler Stops

If no valves can be located, sprinkler stops will have to be used. In some cases, sprinkler stops will need to be put in place to minimize the amount of water that will flow out of the head. An A-frame ladder will reach most ceilings in sprinklered buildings. When placing a sprinkler stop from an A-frame ladder, the ladder is steadied by another firefighter as the firefighter climbs it to provide a good platform.

The Importance of Salvage Operations

Salvage operations can be as important as any actions a fire service agency can perform on an emergency scene. In most cases, a majority of the damage has already been done to the structure and the main job is to make sure the emergency does not get any worse after arrival. This is where salvage operations can be very important to the success of the incident. Firefighters will learn throughout their career that the ability to make something good out of a very bad situation becomes paramount in order to find some value in what they do. With fires, that outcome is in doubt until the end of the incident. Firefighters apply all of their knowledge, skills, and abilities to aggressively attack the problem, but if all that is left is a smoldering parking lot, what have they accomplished? Salvage can immediately give firefighters that sense of purpose. It is no longer acceptable to believe that merely saving the house next door means a successful operation. Firefighters need to preserve as much of the salvageable property as possible. The operation must ensure life safety, exposure protection, property conservation, and scene stabilization. With this thought process directing firefighters' efforts, the "customer" will agree that they accomplished what they set out to do.

OVERHAUL TOOLS, EQUIPMENT, AND OPERATIONS

The process of overhaul is as important as the initial extinguishment of the fire. The process of making sure the fire is completely extinguished can be worthwhile. The ability to check walls, floors, ceilings, and attics—especially with blown-in insulation and dead air spaces—is a skill that needs to be learned and practiced whenever firefighters attend a working incident. Due to many factors, including successful fire prevention programs, the fire service is responding to fewer working fire incidents. The need to use good and thorough firefighting practices and make them habits is paramount to successful firefighting.

Tearing into buildings, cutting through floors, pulling ceilings, and searching for hidden fires are tasks that most firefighters enjoy. The tools used for overhaul are built for these tasks, so these tools are among firefighters' favorites. In most cases, firefighting requires precision and attention to details. In overhaul, the mission is simple—leave no area unsearched if there is potential exposure. A firefighter's knowledge of building construction and fire behavior is an asset in property conservation efforts.

Common Tools

Common overhaul tools are pike poles, pitchforks, rubbish hooks, shovels, axes, chain saws, carry-alls, and wheelbarrows, **Figure 20-19.** All of these tools meet the goals of getting into areas to determine if they were exposed to fire and removing necessary items so they will not cause any more fire. Most of these tools are common and their purpose does not need to be explained. The few exceptions are the rubbish hook and the carry-all. The rubbish hook is a heavy tool with a D-handle that is between 4 and 6 feet in length. It has two tongs spaced about 6 inches apart that run perpendicular to the handle: It looks like a rake with only two teeth. The tool is heavy enough to easily break through ceilings or strip shingles from a roof. It is a mainstay for ventilation crews on roof operations. Because of its weight and the D-handle, it is easily used to breach and pull plasterboard.

Carry-All

The other tool needing a little explanation is the carry-all. Its name makes it pretty clear as to its purpose. It is an approximately 6-foot-square piece of heavy canvas with a rope strung through the grommets for handles that is used to carry debris out of the building once it is loaded, **Figure 20-20.** It is used in areas where wheelbarrows cannot be taken, due to access or other restrictions.

FIGURE 20-19   Various overhaul tools.

FIGURE 20-20 Firefighters carrying debris in a carry-all.

Operations

Once the determination has been made that fire investigation concerns have been met, overhaul can commence. Fire investigation concerns are discussed later in this chapter. Overhaul operations consist of breaching walls, floors, ceilings, and dead air spaces in order to confirm that fire is not present in these areas. All areas directly exposed to the fire should be stripped of coverings such as plasterboard or paneling.

Firefighters can look for obvious signs of hidden fire and utilize their senses by feeling the wall for heat, smelling around outlets and other openings for a scent of anything actively burning, and listening for sounds of items burning. There are cases where buildings have been remodeled, leaving void spaces in the walls or subfloors. For asthetic purposes, ceilings are sometimes lowered, floors raised, or walls extended or doubled. These situations leave many places for fire to travel and can cause serious problems for the overhaul operations. Firefighters should be aware of construction methods and notice inconsistencies that could spell trouble for firefighting operations. For example, a 100-year-old home with interior drywall board should cause firefighters to question whether or not a fire is fully extinguished. Homes built in this era were constructed with lath and plaster interiors, so it should be obvious that a remodel of some kind has occurred. The lath might have been stripped and replaced with no void spaces or new walls might have been constructed over the old walls. Firefighters should also be aware of the typical and atypical signs of structural stability problems: sagging joists, floor, or ceilings; leaning walls; a spongy roof; structural stress creaking, cracking, or other abnormal noises; or any other situation that *any* firefighter finds suspect or abnormal. Safety is the job of all firefighters.

The use of a thermal imaging camera (TIC) can find fires and hotspots in areas that firefighters are not able to see or smell. The bottom line is, if there is a chance something could be burning behind a wall covering, the wall covering must be removed to make sure it is all clear. It is much better to be safe than sorry in this instance. In the overhaul process, there should be at least one charged hoseline with a dedicated firefighter on the nozzle for overhaul operations. Once uncovered, hidden fires may be easy to extinguish, but they may also blow up and cause heavy damage and injury or death to firefighters working the overhaul process. Firefighters may be called upon to use various types of nozzles to combat possible or confirmed hidden fires discovered during the overhaul process. In addition to straight stream and fog nozzles, cellar nozzles, Bresnan Distributers, or piercing nozzles may be utilized. Firefighters must keep a watchful eye out for changing conditions, and experience will help firefighters ascertain the situations where hazards might erupt.

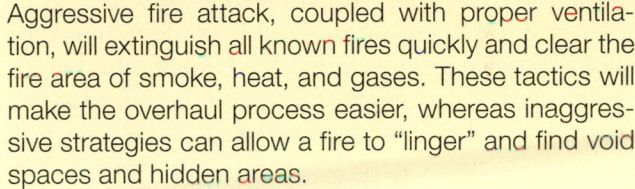

SAFETY

Aggressive fire attack, coupled with proper ventilation, will extinguish all known fires quickly and clear the fire area of smoke, heat, and gases. These tactics will make the overhaul process easier, whereas inaggressive strategies can allow a fire to "linger" and find void spaces and hidden areas.

Firefighters must visually observe the wooden structural members to make sure there is no smoldering fire behind the coverings. The building structural members should also be visualized in order to confirm that the building's structural integrity has not been compromised by the overhaul activities. Fire can get into these areas through electrical outlets; via conducted heat from metallic pipes or wiring; through breaches in the walls or floors where conduit runs; via heating, ventilation, and air conditioning (HVAC) ductwork; or because of a lack of fire stops in improperly constructed buildings.

Once these areas are breached, all of the insulation, if any is in place, needs to be removed so the entire space can be visualized. This is extremely important in attic spaces. Several different types of blown-in insulation can smolder for many hours prior to igniting, **Figure 20-21.** Some of the more common are cellulose or redwood insulation, which is often found in older homes. If this type of blown-in insulation is found, then the task of removing all of the exposed material becomes paramount. The smallest of embers can travel to remote areas within the attic and smolder for many hours prior to growing to a free burning state. The entire attic must be checked, and revisits are recommended, as discussed later in the chapter. Newer homes have rolled-in fiberglass bats that are easily removed and for the most part noncombustible, although the paper backing that the fiberglass is attached to will burn

FIGURE 20-21   Blown-in insulation in an attic.

Overhauling Roofs

Overhauling roofs can be a very long and tedious process. Sometimes there can be multiple roofs under the most visible one and each one needs to be overhauled for possible fire extension. Sometimes flat roofs are left in place while pitched roofs are built over them, thus creating two attic spaces and a real overhaul nightmare. As with all overhaul operations, the material exposed to fire that could smolder for some time needs to be removed. This is why roofs are particularly difficult. During a hot summer day there is nothing worse than working on a roof to overhaul it because of the heat, access problems, the sheer magnitude of the operation, and the inherent safety concerns of working above ground.

CAUTION

During overhaul operations, although there may not be any visible smoke in the structure, there is still a possibility of the presence of carbon monoxide and other toxic gases that pose a respiratory hazard to firefighters. All firefighters conducting overhaul operations must wear SCBA until it is determined that the atmosphere is free of hazards.

Electronic Heat Sensors

A relatively new way to check buildings for hot spots is to use electronic heat sensors or thermal imaging cameras. These sensors determine where heat is higher as compared with the surrounding areas. The sensors' alerting mechanisms differ depending on the sophistication of the equipment. Some simply sound different when they are pointing at an area that is hotter than the surrounding areas, whereas others give the operator the ability to see and pinpoint the exact area that is showing a heat signature. Some of the more sophisticated devices are also used during searches to find people through the smoke, underwater, or lost in the wilderness, and for other applications, **Figures 20-22A** and **B.**

Revisits of the Involved Structure

Even after all areas have been checked, it is important to revisit the scene of the incident. Crews should never leave the scene if there is any active evidence that any part of the building is not completely extinguished. As with all good operations and focusing on customer service, it is always wise to return to make additional checks. The crews are checking for any smoldering areas and to see if the occupants have any questions or need assistance. A good rule is to revisit 2 hours after the last company or personnel leaves the scene and then again within 10 hours. This way the building is visited twice within 12 hours of incident termination.

In most cases, crews from the next shift can go by the incident for the revisit and also learn from it by looking at the areas of involvement and problems encountered. Volunteer companies will have to work around their normal work schedules. Another option is to have the business owner hire a security company to maintain a fire watch, or local law enforcement may be able to assist if the fire service agency is unable.

Debris Removal

The need to remove the debris while searching for hidden fires is important, but as discussed later in this chapter this needs to be done in cooperation with the fire investigator. All of the loose material damaged in the fire must be removed and further extinguished outside the structure. As walls or ceilings are pulled, the material needs to be removed from the building. This is where carry-alls or wheelbarrows come in handy, **Figure 20-23.** The material needs to be placed in an area outside the structure where cleanup crews can easily access it after the firefighters leave.

The material is placed in a pile. It is lightly hosed down to keep from creating a bigger problem down the street with a substantial amount of runoff. Firefighters must be certain the material is extinguished. Debris placed on driveways is the easiest to clean up after the event. Although placing debris in front yards outside a main window may seem convenient for firefighters, customer service should always be considered.

STREETSMART TIP

Sometimes the debris needs to be placed outside a window or other opening, instead of taking it to the driveway or other hard surface. Visqueen should be placed outside the opening to protect any plants or lawn under the plastic, and to facilitate the clean-up process. Remember: customer service! This effort will be appreciated by the building owner.

(A)

(B)

FIGURE 20-22 (A) Helmet with thermal imaging camera attached. (B) Firefighter using a handheld thermal imaging camera.

After the overhaul crews have completed the removal of material and are sure it is completely extinguished, they can cover the debris with Visqueen to reduce the eyesore.

The Importance of Overhaul Operations

As with salvage operations, the customer service aspects of overhaul are many. The detailed search for hidden fires, the ability to limit the amount of uncov-

FIGURE 20-23 Firefighter using a wheelbarrow to remove debris.

ered debris left behind, and the overall ability to leave the building in a state where it is safe for the owners to get in and start looking for lost articles are very important processes. Overhaul work is often the only thing the owners get to see the result of. Making sure overhaul works in concert with the fire investigator is another way to ensure the operation is successful.

FIRE CAUSE DETERMINATION CONCERNS

As the incident progresses through the aspects of initial dispatch, response, arrival, rescue, containment, extinguishment, ventilation, salvage, and overhaul, firefighters need to be aware of possible clues as to how the incident started. The information Dispatch receives can be vital to the rescue/attack operation. Although the Fire Officer will be gathering the information and formulating a plan of attack, it should be the responsibility of all firefighters to listen to the reported information, if possible, and to build a mental picture of the situation. During dispatch, all available information should be documented. Details such as weather, time of day, address, type of dispatch (i.e., garage fire versus bedroom fire), and the number of phone calls received for the fire can give firefighters—and later, investigators—a clue as to what the fire was doing prior to arrival, **Figure 20-24**.

During response, additional information may be gathered by answering these types of questions: Is the structure evacuated or still occupied? Are callers

FIGURE 20-24 Scene size-up begins with the dispatch of the call, and continues with information from dispatch en route to the call.

reporting explosions or odd colors of smoke? Were citizens seen running or driving from the incident at a high rate of speed? Upon arrival on scene, firefighters may encounter witnesses who have pertinent information concerning the fire scene. Witnesses can be great sources of information, but they can also be part of a crime scene. Firefighters should note if any bystanders seem overly interested in the fire or the firefighter activities. It is possible for firefighters to notice observers that are present at more than one fire scene. Any pertinent information should always be relayed to the IC and ultimately to the Fire Investigator.

For exterior fires and fires involving high-pile class A combustibles—stacked lumber, pallets, rolls of paper, and so on—firefighters must ensure that any suspicious evidence is protected. It is basic instinct to scatter the piled contents and completely douse them with foam and water to ensure the fire is extinguished—and this quite possibly might be the extinguishment scenario. It is a firefighter's main job to ensure that the fire is completely extinguished, and this may cause extensive overhaul practices to be employed. However, whenever possible, any evidence of the point of origin (V-patterns, deep charring, etc.) evidence of suspicious liquids or accelerants, matches, wadded paper, or any suspicious situations should be protected or noted by the firefighters. Observant firefighters on scene can provide fire investigators with reliable information as to where the fire was, what spectators said about the fire, and how the fire reacted during fire attack. Inside the structure, firefighters should observe exactly where the fire was located in the building, how the fire reacted to suppression efforts, and the conditions under which people were found in the building. The TIC can be used to find hidden fires and areas of intense heat that could reignite. Firefighters should also pay attention to how the contents of the building were arranged, whether

there were signs of break-in, and the position of interior doors. It is also important to note the building's electrical system, including the condition and position of circuit breakers and appliances. The inspectors should be briefed about this information, **Figure 20-25,** and it should be noted in the company officer's report. The smallest of items that seem out of place need to be reported because these may be the missing pieces of the puzzle for the investigator.

During overhaul, firefighters must ensure that the investigator preapproves the removal of any part of the structure or any piece of furnishing. Firefighters must not throw things out of the building for expediency. Each step of overhaul has to be taken with knowledge and care so as not to make the investigation more difficult. It may be appropriate to consider conducting overhaul from the unburned side of the wall. This action will preserve burn patterns critical to a fire investigation. As investigators work through a structure, they will be able to release the different rooms for more thorough overhaul. Smoke markings on the plasterboard may be important clues for the investigator. Experienced investigators are only as good as the clues they observe. If they arrive on scene and all they have is a bare wood floor and wooden studs for walls, they are at a disadvantage.

FIGURE 20-25 Fire investigators should be thoroughly briefed about fire and building conditions existing during fire attack. *(Courtesy of Fuller Road Fire Department, Colonie, New York)*

Preservation of Evidence

In the process of their work, firefighters may come upon something out of the ordinary and will need to make every effort to preserve it until the fire investigator can observe it—even if this means stopping the entire fire suppression operation. Firefighters need to leave items where they found them and protect and preserve the area as found until investigators can examine it. In this case, the area should be cordoned off with fire line tape to secure the affected area. Fire ground activities that may damage items in the protected area should be avoided. All possible steps should be taken to ensure that firefighters do not become involved in a legal matter because they made simple, preventable mistakes.

If a room is overhauled and cleaned of all contents prior to the arrival of investigators and it is later found that an incendiary device was located in that room, it could result in a legal matter for the firefighter and the department.

Basics of Point of Origin Determination

The determination of the point or area of origin can be a very scientific pursuit in which exotic testing is done and the exact point is determined—or it can be as simple as looking at a room and knowing where the fire started based on the indicators present. Fire investigators employ the scientific approach to every fire they investigate. The basic clues a firefighter should watch for include notation of where the fire was first noticed or reported; where the heaviest damage was; or any smoke, heat, or other types of markings on the walls, ceilings, or floor indicating where the fire started. The basic premise behind fire behavior is that it travels the path of least resistance. As fires grow, smoke and heat are forced up until they reach an obstruction, usually the ceiling. The smoke and heat will branch out along the ceiling. As the smoke and heat go in opposite directions the wall will be marked with a "V pattern" marking, **Figure 20-26.** This is a common term in the fire investigation field. The V will start at the bottom nearest the area of origin and proceed up the wall.

Another simple manner to determine the fire's starting point and where it spread is the **depth of char.** Wherever the fire burned the longest is where the fire's depth of char will be the greatest, **Figure 20-27.** This can be illustrated by looking at a log in the fireplace. Usually the center of the log has the most charring and it diminishes toward each end. Wherever the fire started would have had the longest period of time to burn the wood members in that area, thus the

FIGURE 20-26   "V" pattern in a structure fire.

FIGURE 20-27   Depth of char.

charring would be greater in that area. The only misleading part of this indicator is when an area of the fire took longer to extinguish than others. This type of attack crew information can assist the investigator in determining the point of origin. These signs could be indicative of possible incendiary, or arson, fires. A TIC can be useful for tracing heat patterns. However, if the investigators are called to court, the evidence will not be admissable unless the TIC images are recorded.

Extent of fire damage is just one of many indicators of fire behavior and this damage can be influenced by many other factors. It is a straightforward way to determine the area of origin for simple or large-scale incidents alike. The key to assisting the fire investigators is not to rush the overhaul process. It is imperative that firefighters pay attention to what is going on around them during all aspects of an incident and report items that fall outside the norm.

JOB PERFORMANCE REQUIREMENT 20-1
Salvage Cover Roll

A Place one firefighter at either end of the cover. Both firefighters place one hand on the edge of the cover one-quarter of the way in and the other hand on the outside edge. Each firefighter should be on the same outside edge facing each other.

B While grasping the cover, with a quick flipping motion, bring the outside edge to the middle.

C The firefighter's inside hand should be at the fold. Repeat this fold again and again until the fold is at the middle of the cover. Fold the opposite end of the cover in the same manner.

D The cover should now be about three feet wide and ready to be rolled. The treated side is rolled inside and the cover is ready for deployment.

Salvage Cover—One-Person Deployment

A With the cover flat on the ground, fold the outside edge to the center point of the cover. Smooth out wrinkles and align edges.

B Fold the same edge to the center point again. One fold is now beneath the other fold. Smooth out wrinkles and align edges.

C Fold the opposite side in a like manner. Smooth out wrinkles and align edges.

D Fold one end to a point just short of the center point. Smooth out wrinkles and align edges.

E Fold the same end again. One fold is now beneath the other fold. Smooth out wrinkles and align edges.

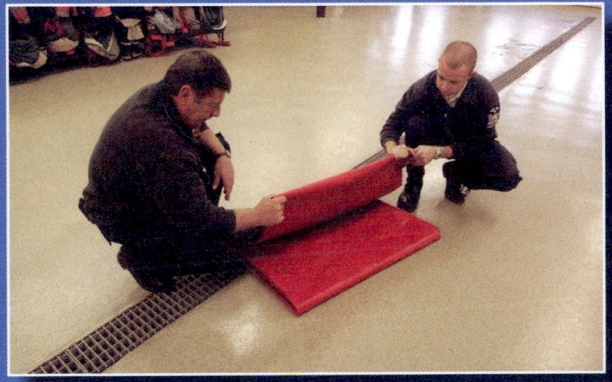

F Repeat steps to fold the opposite end. Smooth out wrinkles and align edges. The cover is now ready for a one-person deployment.

JOB PERFORMANCE REQUIREMENT 20-3
Salvage Cover—Two-Person Deployment

A Fold the cover in half as in the one-person deployment fold. Smooth out wrinkles and align edges.

B In similar fashion, fold the cover in half (lengthwise) again. Smooth out wrinkles and align edges.

C Fold the cover in half (widthwise). Smooth out wrinkles and align edges.

D Fold the cover end over again (widthwise). Smooth out wrinkles and align edges.

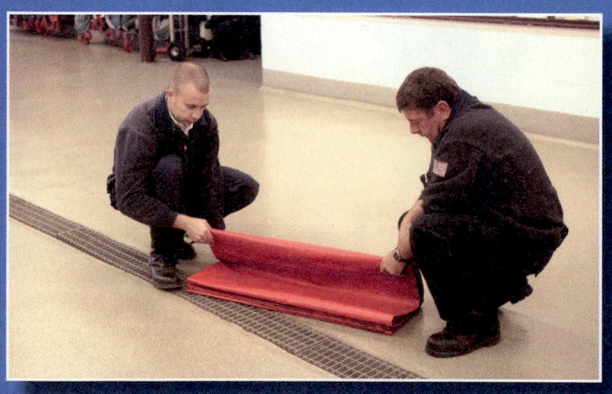

E Fold the cover in half two more times (widthwise). The cover is ready for a two-person deployment.

JOB PERFORMANCE REQUIREMENT 20-4
Shoulder Toss

A Place a folded salvage cover over one arm.

B Using a straight arm movement, toss the cover over the items being salvaged.

JOB PERFORMANCE REQUIREMENT 20-5
Balloon Toss

A   Two firefighters place the cover on the ground next to the item needing protection. The salvage cover is unfolded out of the accordion fold so that the cover is across the base of the item at its full length.

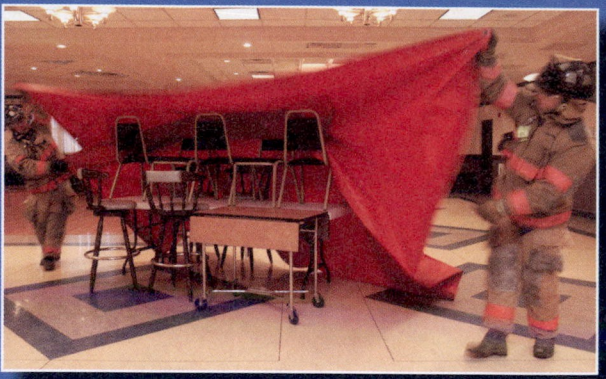

B   The firefighters grab the top inside edge of the cover and pick it up to about waist height. Both firefighters simultaneously move the cover's edge in an upward motion so that it catches air, and they then move to the other side of the item while the cover is still in the air.

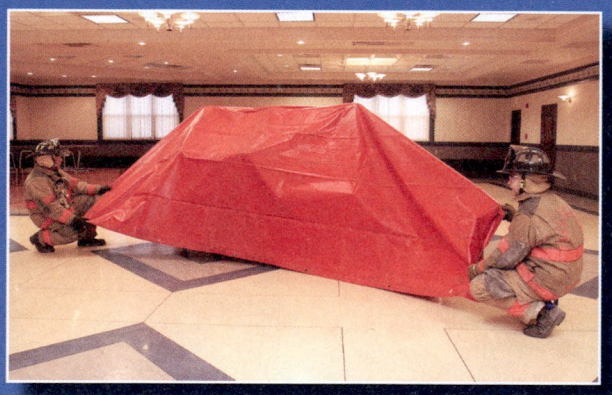

C   The firefighters guide the cover to the floor on the opposite side. Once the items are covered, the firefighters dress the cover by folding the edges under to keep water from pooling and to lessen the chance for the edges of the cover to be disturbed by accident.

LESSONS LEARNED

This chapter dealt with areas not commonly reinforced through training. Most fire training consists of the vital roles of fire attack, building construction, search and rescue, and so on. Salvage and overhaul duties, while not glamorous, are vital to the functions of fire investigation and property conservation. These duties can often go unnoticed as firefighters are focused on the more exciting aspects of an incident. Most firefighters would rather talk about the great "stop" they made instead of the minimal damage the lower floor of the structure received. However, salvage and overhaul duties do have a direct impact on the success of an incident. They are also a great customer service for the homeowner, even if they are never recognized.

KEY TERMS

Chain of Command Common fire service term that means to always work through one's direct supervisor. The fire service is viewed as a paramilitary organization and because of this all requests for information outside the assigned workplace should go through the supervisor.

Depth of Char A term commonly used by fire investigators to describe the amount of time wooden material had burned. The deeper the char, the longer the material was burning or exposed to direct flame.

Sheetrock A trademark and another name for plasterboard.

Visqueen A trade name for black plastic. It can be used very effectively in salvage and overhaul operations. Many examples are discussed in this chapter.

REVIEW QUESTIONS

1. Why is the exterior of a building important to salvage operations?
2. What are the three types of material that salvage covers are made of?
3. What advantages does Visqueen have over customary salvage covers?
4. How can a fire department be reimbursed for equipment damaged in the course of working the incident?
5. How do firefighters check for holes in the salvage covers?
6. What are the basics of furniture arranging for salvage operations?
7. What are some indicators that a Sheetrock ceiling has a potential to fall?
8. What does PIV stand for?
9. How are overhaul and fire investigation linked?
10. What does a revisit of a structure fire accomplish?
11. What is a "V" pattern?
12. Why is knowing the "depth of char" important?

23 Firefighter Survival

I t is one thing to talk about the life and death nature of firefighting—yet another when you are actually experiencing a situation that is about to kill you. A million thoughts go through your head and time seems to slow down. You rely on training, instinct, awareness, and even luck to get you through. My own experience happened one cold winter night—the kind of night where you didn't want to go outside. I was the second-arriving company officer to a rural, single-story farmhouse with a well-involved attic fire. The older home had a sturdy, high-mass roof structure so we elected an aggressive interior attack—that is, pulling ceilings and hitting it from underneath. Conditions inside the home were clear, although we knew there was significant fire over our heads in the attic space. In short time, we had two lines operating inside and were making good progress. I noticed that the ceiling was sagging in the large "great-room" area of the house—and advised all crews to stay near walls and doorways and away from any open-span areas in case the roof collapsed. Seeing this potential increase, I thought we could simply force the collapse by pulling down the sagging area. I told my crew to take cover behind me in a doorway while I yanked down the ceiling. All was "safe" in my mind because we were eliminating a hazard and making access to the burning attic easier. Finding a good leverage point became troublesome so I moved into the room just a few feet. The roof then let go behind me.

As I turned to escape out the doorway, the roof hit me and flipped my 230 pounds over a couch. The first thought to cross my mind was dying. The second thought was a question: Where did my crew go? Training then kicked in. I took inventory of my environment: I was on my back (with SCBA—ouch), still breathing, legs working, no flames, light smoke, and a ceiling/roof structure pinning me to the floor and couch. I called for my crew—no reply. I called out for anyone—no reply. I couldn't get to my radio or PASS device manual activator. I thought about lying motionless to help activate the device but blew off that idea and went into self-help mode. I reasoned that I should try to self-extricate and get to my crew somehow. My instinct told me to move. I knew the fire was right above me with only drywall and framing separating me from it. With an energy burst, I was able to spin myself over to my side and began to crawl through a slight void created by the furniture in the room. Earlier, I was aware of three exits to the area. The way I came in was now blocked but I recalled a side door and a rear sliding glass door. Luckily, the void led to the sliding glass door. I exited and went to the entry point of my crew. They were shielded from the collapse by the doorway and they simply exited to begin looking for another way in to find me. What seemed like ten minutes under that roof was more like seconds according to my crew. They didn't even have time to report the collapse before they were out and I met them.

Training, instinct, awareness, and luck figured into this survival. Use this chapter to help build the training, instinct, and awareness to help you survive. Don't rely on the luck.

—Street Story by Dave Dodson, Lead Instructor, Response Solutions, Colorado

LEARNING OBJECTIVES

After completing this chapter, the reader should be able to:

23-1   Describe procedures for safely functioning on an emergency scene.

23-2   Define a healthy lifestyle.

23-3   Explain how a healthy lifestyle can promote well-being and assist the firefighters in their day-to-day responsibilities.

23-5   List the three main components that lead to incident readiness.

23-6   Describe safety equipment used by emergency responders to safely function at an emergency scene.

23-7   Identify characteristics of the various types of personnel accountability systems.

23-8   Identify a firefighter's personal responsibility to the overall personnel accountability system.

23-9   Describe the three components that lead to "fitness for duty."

23-11   Define personal size-up.

23-12   Explain emergency procedures responders take when an air supply is depleted.

23-13   List and briefly describe the four components of rehabilitation.

23-14   Describe the implementation of procedures for emergency communication protocols.

23-15   Explain the application of emergency evacuation procedures to an incident.

23-16   List the five steps that can lead to an organized rapid escape.

23-17   List the three steps that should be taken when entrapment occurs.

*The FF I and II levels, as defined by the NFPA 1001 Standards, are identified in different colors: FF I = black, FF II = red, additional information = blue.

INTRODUCTION

The excited radio transmission of "Mayday, mayday, mayday, firefighters injured and trapped" will present perhaps the most stressful circumstances that a firefighter will ever face. While no firefighter or fire officer expects this to happen, it would be negligence if training efforts failed to address firefighter survival and the handling of situations in which firefighters are injured, trapped, missing, disoriented, low on air, or imperiled—all firefighter emergencies.

Firefighter survival is codependent on two focus areas: Prevention through readiness and training on firefighter emergency procedures. Obviously, it makes sense to attempt to prevent a firefighter emergency from happening. Daily, firefighters can help prevent firefighter emergencies through incident readiness—that is, the efforts to ensure that all firefighters are mentally and physically ready to respond. Personal protective equipment (PPE), task accountability, and fitness for duty are all issues that help prevent firefighter emergencies. At an incident, firefighter actions, inaction (knowing when to hold back), and attention to hazards will help prevent a firefighter emergency. Likewise, attention to team continuity, orders and communication, rapid intervention planning, and rehabilitation all help prevent firefighter emergencies.

If a firefighter emergency were to occur, a planned, systematic process of self-rescue or rapid intervention would have to be established to avoid compounding the seriousness of the emergency. Having procedures in place for rapid escape and self-extrication, accountability recall, rapid intervention, and post-incident emotional processing is essential. This chapter discusses all of these areas and gives firefighters survival tools to help prevent firefighter emergencies and to assist them when such an unfortunate event does occur.

INCIDENT READINESS

Firefighters might well represent the only profession that requires peak mental and physical performance with no notice or warm-up. Constant readiness is a challenge, but when the alarm sounds, it is imperative that firefighters engage quickly. Preparing for incident response involves more than firefighters merely making sure they know where they are going. Incident readiness is a mental process that answers a few questions:

- Am I in a position to respond?
- Is my personal protective equipment available?

- What is my relationship to the response?
- Physically, can I respond?
- Mentally, can I check out of my current thoughts and focus on response?

A negative answer to any of these questions may set a firefighter up for an injury. To avoid this risk, firefighters must take steps to ensure that their "system" is assembled and ready for response. Some readiness system items are addressed before an alarm is received, while others are addressed on receipt of the alarm. This "system" is the key to survival and includes some important components.

> **NOTE**
>
> Namely, the readiness system includes PPE, personal accountability, and fit-for-duty status.

FIGURE 23-1 PPE is part of the survival system. Is it ready?

Personal Protective Equipment

The personal protective ensemble can be considered the first step in one's ability to focus on incident handling. Specifically, PPE is the first thing firefighters put on for protection and the last thing they want to to utilize to protect them, **Figure 23-1.** If firefighters rely on PPE to protect them from some hazard, they have in effect used up all other protective measures. As an example, if firefighters find that their PPE prevented a serious burn, that means the PPE became the last means of prevention. Firefighting streams, ventilation, zoning, and access/egress efforts have likely failed if the firefighter got to the point that PPE made the difference between injury and no injury.

> **SAFETY**
>
> One fire service adage claims that PPE is the first and last means of defense from risks.

> **CAUTION**
>
> At the beginning of a duty tour or shift, firefighters must ensure that their protective equipment is dry, serviceable, and ready for quick donning.

In many departments, firefighters must check all their various ensembles. This could mean that a firefighter needs to prepare their structural ensemble, technical rescue personal gear, aluminized proximity suits, wildland gear, and (EMS) PPE each and every shift. The readiness check ensures that the firefighters are ready for any type of incident that is likely to occur within a response district, which could include an airport, commercial and residential structures, rural areas and fields, and EMS responsibilities.

When preparing PPE for readiness, firefighters should ensure the following:

1. All clothing materials are dry. Wet clothing reduces the protective insulation of the clothing and may lead to steam burns in heated environments.

2. All PPE is present and positioned so that it may be rapidly donned. This is important for station-assigned firefighters and those who respond to a station to staff an apparatus. Firefighters who respond from home directly to an incident must pack their dry gear in such a way that it can be donned quickly and appropriately.

3. Essential "pocket tools" are available and in working order. It is not uncommon for firefighters to carry an assortment of personal tools that they feel are important. Small flashlights, trauma shears, doorstops, nylon webbing, carabiners, grease pencils, chalk, sprinkler wedges, multiple-tool pliers, knives, foam earplugs, self-escape rope, and two-way radios are all common pocket tools found in firefighters' PPE.

4. Alternative PPE items are appropriately packed and ready for use. As stated earlier, many firefighters must use (EMS) PPE, wildland PPE, proximity PPE, or lightweight technical rescue PPE. These should also be checked and made ready.

5. In addition to the preceding PPE items, firefighters must also check those ancillary protective equipment items that are assigned to their individual riding position:

- *Radio.* Check to make sure the battery is charged (or is charging) and is set on the appropriate response or typically assigned tactical frequency.

- *SCBA.* Check to see that the air cylinder is full, mask is clean and ready, and the straps are let out to facilitate rapid donning. Most fire departments require that firefighters do a quick operational check to assure the regulator, PASS device, and low-air alarm is functional.

- *Thermal imaging camera.* Ensure that the device is in a readiness state as recommended by the manufacturer and the battery is charged.

- *EMS PPE.* Take a few seconds to make sure there are multiple pairs of the correct size of disposable gloves available for quick donning. Also check to make sure that protective eye wear is clean and ready.

When an alarm is received, the time spent donning gear *before* arrival at the incident is time well invested. People waiting until incident arrival to don PPE are setting themselves up to "shortcut" their protective ensemble. These shortcuts are caused by firefighters who are concerned and distracted by the situation and the actions that need to be started rather than by completion of their ensemble. Donning structural PPE prior to incident response is preferable; however, driving in full PPE can introduce some of its own safety issues. First, wearing PPE while driving or riding in an apparatus during a response does NOT eliminate the need for seat belts. Vehicles equipped with seat belts that do not fit around donned PPE can set the stage to eliminate their use. This should not be allowed. In these cases, it is better to don PPE upon arrival versus the choice to not wear the seat belt. Large fire apparatus typically allow more seat belt webbing to fit around PPE. When the seat belts do fit around PPE, some drivers argue that donned PPE may interfere with the driver's ability to safely operate in-cab controls. Training can overcome this argument.

SAFETY

If a driver makes the choice to don PPE after driving to an incident, strict discipline must be exercised to fully don PPE before engaging in the incident.

There are times when a firefighter may be dispatched to an incident while riding in the apparatus. In these cases, the urge to unbuckle the seat belt and start donning PPE may present itself. Resist this urge and don only what you can with the seat belt on! Finish donning on arrival.

CAUTION

Firefighters should not find themselves at an incident without their PPE fully donned; they must practice strict self-discipline to complete the ensemble. They must not let the urgency of the situation override prudent judgment!

Following an incident, the same attention to PPE readiness needs to be reinvested. In many cases, the gear may have to be decontaminated. Gear that is wet from perspiration or fire streams should be dried.

VIEWPOINT

The evolving trend of issuing firefighters two sets of gear is encouraged and shows an overdue respect for the need to have "ready gear" after an incident.

Remember, firefighters who have been operating in smoke at typical building and vehicle fires are wearing gear contaminated with toxins—many of which are carcinogens that have long half-life values that can be inhaled or skin-absorbed several days after the fire (benzene is a notable example).

Personal Accountability

Personal accountability is an essential and required part of firefighter readiness. Personal accountability refers not only to the established accountability system used by a particular fire department, but also to the firefighter's relationship to the response and ability to perform as trained.

Accountability System

The accountability system used by a department can take on many forms. As technology advances, firefighters may soon be tracked electronically using sophisticated locator and vital sign monitors. Typically, however, accountability systems will fall into one of three general types:

1. *Passport.* This is a crew-card system that is tracked on a status board by a monitor. Members of a crew give a name chip to the company officer or team leader, who places all of the team names on a card or **passport.** When given an assignment, the passport is given to the monitor, who tracks all crews. Team leaders must report any changes in their teams' location or assignment.

2. *Tag.* In this system, individual firefighters report to staging and give an identification tag to the staging manager. The staging manager groups the tags into teams and tracks the progress of each. First arriv-

ing crews and personnel will not have a staging/ accountability manager to collect their tags. In these cases, individual tags are usually hooked on a ring of the first-due apparatus, and the staging/account-ability manager, once assigned, will collect these.

3. *Company officer.* This system is perhaps the old-est and most used system. Here, the company offi-cer, team leader, or other supervisor is responsi-ble for keeping track of the crew. This system is the least preferred system in that many firefighters have been lost and many crews broken up due to the lack of formality of the system.

Regardless of the system used, each firefighter must be aware of how the system works and follow its specific guidelines. Failure to be accounted for on the incident is akin to **freelancing**—that is, performing a task that has not been assigned or performing a task alone—both are fire service taboos, **Figure 23-2.**

STREETSMART TIP

Not only can firefighters endanger themselves through freelancing, but may also indirectly injure others when they attempt to find the freelancer.

Worse, if freelancers become lost or trapped, nobody will come to assist if they do not know free-lancers are on scene or where they are working.

Relationship to the Response

Another part of the accountability equation is atten-tion to each individual's relationship to the response. Specifically, firefighters' relationship to the response includes their assignments and their own personal size-ups.

FIGURE 23-2 Accountability systems take on many forms—firefighters must know how to check in.

■ *Assignment.* In most cases, firefighters are given readiness assignments—that is, being assigned as given crew members for given positions on the appa-ratus. With these assignments come responsibilities to ensure that they perform preassigned or officer-directed duties. In cases where firefighters have no preassigned duty, firefighters must be prepared to perform a host of duties. The readiness elements entail a mental preparation to perform the tasks of that position. If firefighters are assigned positions that they have not performed recently, they may not be ready to act—leaving an invitation to injury. In some departments, a list of crew duties for a given seated position on the apparatus is listed on a plas-tic card—this is a great reminder. Firefighters should take the time to reacquaint themselves with the expected tasks, tools, and procedures required. This improves readiness and reduces injury potential. In those cases where the firefighter is working with an unfamiliar crew and/or officer, the firefighter should take the time to review expectations and talk through crew assignments with the officer to make sure that he or she is on the same page as the new crew.

■ *Personal Size-Up.* **Personal size-up** can be defined as the continuous situational awareness and mental evaluation of firefighters' immediate environments, facts, and probabilities. This mental evaluation should have them evaluating the weather, time of day, current chain of command, and likely assign-ment. Once again, they are mentally preparing themselves—thereby reducing injury potential.

Work as Trained

STREETSMART TIP

Personal accountability is achieved when all firefighters and fire officers are able to perform assigned essential tasks—tasks they have been trained for—and keep the chain of command advised of their progress.

Of significant importance here is everyone's ability to perform as trained. A common thread discovered in post-incident investigations is the fact that many of the injured were performing tasks they had never experienced before. In some cases, the training had been given by the individual's department, but the individual either forgot how to accomplish the task or chose a method that was inappropriate for the situa-tion. How can this be prevented? Listed next are two key points to assist firefighters.

1. *Firefighters should perform as trained.* This sounds easier than it is. Most departments invest consider-able cost to train firefighters, so firefighters should do their part and constantly practice and perfect their skills. For career firefighters, this should be

TEST YOURSELF

DAVE DODSON

The first company officer I worked for had little interest in making sure his crew practiced their academy skills on a frequent basis. I guess he figured we knew our stuff. I certainly believed that I knew my stuff—and spent most of my station time and new-firefighter energy to learn things that the academy did not teach us. Six months later, I was reassigned to a different company officer who was quite the opposite. He drilled our crew constantly. I realized immediately that some of my skills had deteriorated (embarrassingly so). The drills were tough—and, at times, I felt that nervous stress of being tested. As if the drills weren't enough, the boss would ask spontaneous questions and test our knowledge of fire service facts, figures, and processes. Talk about stress! I began to wonder why he was so relentless in testing us. Was he trying to prove something? Was he that unsure of our abilities? Was he showing off how much he knew? Why was he trying to embarrass me in front my peers?

After a few months, I privately approached the new skipper to share my developing fear of being tested and to ask him why he was always trying to expose our weaknesses. After a brief period of silence, he provided a perspective that became my new fire service focus.

In a friendly tone, the company officer explained that it is human nature to have test-anxiety. Each incident we respond to is a test—and we needed to meet that challenge head on. He went on to explain that there are two keys to overcoming any test anxiety:

1. Accept tests as a personal tool to discover what you know and what you don't.
2. Never pass up a test—any test. It doesn't matter if the test is in the form of an evolution drill, promotion-practice exam, public speaking challenge, actual incident, or even the geography quiz in the travel section of the Sunday paper—don't pass it up.

Being humbled in discovering what you don't know is, well, humbling, but provides a ready status check of where you stand.

Further, he explained that the fire service is a team environment, and everyone brings strengths and weaknesses to that environment. The point of his testing was not about exposing weaknesses but to create an environment where the crew would become comfortable with each other. The more you know about each others' strengths and weaknesses, the better.

To this day, I'm always testing myself—and remain humbled by what I don't know or can't do. It motivates me to discover and learn more. Taking this approach has served me well in the fire service: I am less fearful of tests, which has helped my promotional opportunities, drill proficiency, public speaking opportunities, incident-handling abilities, and even those coffee-talk trivia sessions. Additionally, I find that I am more apt and comfortable in admitting what I don't know—but willing to discover. Never pass up a test. Go ahead—test yourself!

no challenge—evenings around the station are perfect times for doing this—it does, however, take motivation. Volunteers have a bigger challenge. They should take the initiative to go to the station and practice, and get others to join in the effort.

2. *Firefighters need to know their strengths and weaknesses.* The achievement of a firefighter certificate does not mean a firefighter has all of the necessary skills to be an effective emergency responder. It is merely a minimum. Training never ends for a firefighter—ever! Firefighters must constantly test themselves and try to improve those areas where they discover weaknesses (see the box Test Yourself). As a matter of fact, firefighters should strive for mastery of their assigned tasks, **Figure 23-3.** Mastery can be defined as the ability to achieve 90 percent of an objective 90 percent of the time. Further, firefighters should expand their abilities.

CAUTION

Firefighters should not let the incident scene become an avenue for discovery and on-the-job training.

FIGURE 23-3 Firefighters achieve mastery of tasks through repeated training. Mastery reduces the chance of injury.

Fitness for Duty

Fitness for duty is more than just being "in good shape."

> **NOTE**
>
> Fitness for duty includes mental fitness, physical fitness, wellness, energy, and rest, **Figure 23-4.**

If just one of these is less than adequate, the window to being injured opens.

Mental Fitness

Being mentally ready to respond to an incident at all times rarely happens—firefighters' lives are complex, busy, and full of mental and, in some cases, emotional challenges. As has been stated often, a firefighter's size-up begins on receipt of the incident. If a firefighter is involved in an emotional event during this phase, key size-up information may be missed. The key is for the firefighter to "check out" of the environment and "check in" to size-up. The firefighter's non-firefighting acquaintances may not understand the rapid mental departure—it is common, however, for them to forgive the firefighter.

Once again, it is easier to say a firefighter should "check in" to the incident than to actually do it. Many firefighters have a mental ritual to help them achieve the check-in state every time. They create a ritual—a physical or mental act that reminds them to check out of their emotional or current mental state. For some, it comes when they buckle their seat belt and respond. Others take a deep breath and force the exhalation. Still others perform a rapid mental checklist: What is the call? Where is the call? Am I prepared to respond?

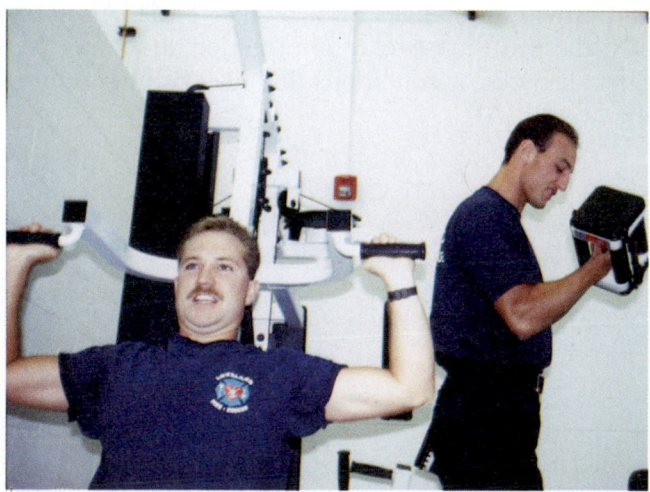

FIGURE 23-4 Fitness for duty includes more than just being "in shape." It also includes wellness, rest, energy, and mental fitness.

What is my relationship to the response—first due, second due?

Physical Fitness and Wellness

There is no doubt that fit firefighters are less injury prone than those who are unfit. Fitness is a lifestyle commitment. On receipt of the alarm, physical fitness becomes more than an issue of how much firefighters "work out." Are there lingering injuries? Are they recovering from a cold or the flu? Are prescriptions or over-the-counter drugs on-board? Did they just finish a work out? What is the status of their hydration and nourishment? Any of these conditions can affect performance and fitness. The simple act of going from a resting heart rate at alarm time to a maximal heart rate on arrival at a working fire can stress the body to the point of collapse if it is not capable of accommodating the spike. A firefighter who is in "great shape" can collapse quickly at a fire when fighting off the infection of a flu. Likewise, persons taking prescription drugs may experience rapid loss of strength or fatigue easier. The lasting effects of alcohol can also impair the firefighter (and endanger the team) well after consumption.

Physical fitness and wellness are highly individual—that is, firefighters are all different and have varied fitness needs based on genetics, gender, metabolic rate, and personal history (injuries, tolerance levels, etc.). However, firefighting contains some universal demands that require a certain level of fitness and wellness. Specifically, firefighters need to create a fitness and wellness lifestyle that includes the following:

- *Cardiovascular conditioning.* Without a doubt, this is critical. Firefighters seldom operate at incidents with a resting heartbeat. Repeated conditioning at elevated heart/breathing rates strengthens the heart/lung relationship and reduces the chance of overexertion at incidents.

- *Core strengthening.* The human core can be defined as the muscle structure that supports the hips, back, and thoracic cavity. This "power center" is used in virtually all physical activity at an incident. Tight-core lunges, crunches, and balance-oriented exercises (like using a fitness ball) help strengthen this core.

- *Flexibility improvement.* The stress on joints, tendons, and ligaments can be brutal in the firefighting world—strains and sprains are the leading firefighter injury. Following a brief cardiovascular warm-up, firefighters should apply stretching and flexing routines to help condition joints, tendons, and ligaments.

- *Resistance training.* Firefighting is labor intensive. Resistance training (weight lifting) helps build muscle that makes the firefighting labor demand easier on the body. While many strategies exist, resistance training that systematically works all major muscle groups is best and, combined with core strengthening, will help keep postural balance.

- *Nutritional balance.* Perhaps the most neglected area of fitness and wellness of firefighters is the daily intake of food. Although this highly personal topic is fodder for much debate, most fitness experts agree that the key is to eat less but eat more often. Eating a balance of carbohydrates, protein, and fats along with a programmed intake of supplements helps the firefighter find nutritional balance.

- *Hydration maintenance.* Firefighters working a shift should stay well hydrated by drinking plenty of water throughout the shift. Firefighters who are on call or in volunteer systems should keep drinking water close and hydrate en route to the fire station or incident. If it is anticipated that work will be done in high temperatures and/or high humidity, begin hydrating in advance.

Energy and Rest

Safety is directly affected by the energy potential and rest a firefighter has stored. Obviously, if the firefighter is short on sleep, the physical and mental capacities are reduced. Likewise, if it has been more than three hours since a meal or the food (energy) in the stomach has been used up, the firefighter will reach a fatigued state quicker.

CAUTION

For most firefighters, the worst time to respond to an emergency is just after midnight, when they have not received much sleep after a long day, it has been five or six hours since the last meal, and the mind and body are in repair mode.

What is the solution? The answer lies in the firefighter's awareness. The firefighter should be attentive to energy and rest levels and plan to communicate an early rehabilitation need to an officer during operations.

STREETSMART TIP

One simple suggestion: Firefighters should keep a balanced energy bar (protein, carbohydrate, and fat) in the apparatus or car and eat part of it during their response. A half a bar that is balanced with protein, carbohydrate, and fat will serve the body well if it has been more than three hours since the last meal. Hydrating prior to incident engagement, however, is essential. Drinking some water while en route helps maintain heart stroke volume when sudden work demands are required.

STREETSMART TIP

Quality sleep plays a major role in preparing the body and brain for readiness. If it has been more than 16 hours since you've had quality sleep, you are at risk for injury. It is medically indisputable that a quick 30 to 40 minute power nap during waking hours helps improve performance! This daytime "safety nap" is beneficial for those who work 24-hour shifts where the likelihood of sleep interruption is looming.

SAFETY AT INCIDENTS

With roughly half of all duty-related injuries and deaths happening at the incident scene, it makes sense for each firefighter to develop a system to follow to minimize risk of injury or death. It is important to note here that nobody tries to get injured at an incident—the circumstances of the incident have created the potential. In most cases, the individual receiving the injury or the other firefighters and fire officers have failed to proactively "see" the events that led to the injury. Additionally, injuries and death occur when firefighters allow the urgency of the situation to override prudent judgment. What is being said here is valid—injuries can be prevented. At the individual level, a firefighter can prevent injuries and even death through a series of mental and physical actions. This section shows these mental and physical actions.

Orders/Communication

The incident commander is responsible for assembling an **incident action plan (IAP)** that is implemented by teams performing tasks. These tasks are assigned to organized teams in the form of orders. Each team is responsible for carrying out the order and providing updates on a regular basis. Additionally, the team must relay information about any pertinent hazards or conditions that may be important to the overall IAP.

SAFETY

As individuals, firefighters must keep their team leader advised of conditions and hazards they find as work is performed.

The discovery of a weak stairway, a significant weight load in an attic or on the roof, worsening smoke conditions, or fire in a void space are all important facts that should be relayed to the team leader and subsequently up the chain of command.

Occasionally, a team performing an order is given a different order by someone else in the incident command system. In this case, the team leader needs to inform the person giving the new order that they are already under orders. The person giving the new order then must either countermand the original order or find someone else to perform the new order. In these cases, it is imperative that both the team and the person giving the new order communicate any changes to the person giving the original order.

Typically, the first arriving group of teams and apparatus performs a prescribed set of tasks or orders based on a pre-incident survey or standard set of procedures. In these cases, the teams should know not only the tasks, but also the tools required and any safety considerations. An official safety briefing is usually not held and teams carry on their task. This does not mean that team members should not vocalize safety thoughts among themselves—in fact, this should be encouraged. Building construction considerations, fire behavior and smoke observations, and access/egress routes should be discussed as the team approaches its task.

Later-arriving teams are given assignments based on the IAP. These teams should ascertain essential information and hold a quick, formal safety briefing based on the conditions present and the hazards that are faced, **Figure 23-5**. This practice is not new—wildland fire crews are required to perform a crew briefing prior to task accomplishment. Most proactive fire departments are initiating these safety briefings as a matter of habit. **Figure 23-6** lists good habits regarding orders and communications.

FIGURE 23-5 Later-arriving teams can benefit from a quick, formal safety briefing prior to performing tasks.

Good Habits Regarding Orders and Communications

- Clearly understand any order; ask questions and seek clarity.

- Assemble tools and openly discuss procedure as you approach tasks.

- Relay all pertinent hazards found during the course of task completion.

- Update crew status on a regular basis.

- Advise when tasks are complete.

- Communicate rotation to rehab and/or staging.

FIGURE 23-6 Practicing good reporting habits enhances communications.

Risk/Benefit

Another key to firefighter survival is the concept of risk/benefit. Simply stated, **risk/benefit** is an evaluation of the potential benefit that a task will accomplish in relationship to the hazards that will be faced while completing the task. As an example, the task of vertically ventilating a residential structure fire will benefit the fire suppression and lifesaving effort by increasing visibility, reducing heat buildup, and reducing overall damage. That benefit, however, is at the risk of roof collapse, operating on a potentially steep incline, or operating without quick egress. The team operating on the roof is often in the best place to judge this relationship. Signs of collapse, difficult footing, and obscured visibility may present a significant risk to the team. The risk/benefit evaluation in this example may cause the team leader to withdraw the team.

Risk/benefit evaluations take place at different levels within an ICS. At the command level, an overall evaluation takes place, as reflected in the IAP. At the team level, crews evaluate their immediate environment and report this back to the person giving them their orders, **Figure 23-7**. Regardless of the level of evaluation, some basic guidelines can be used to help make risk/benefit decisions:

- Firefighters will take a significant risk to save a known life.

- Firefighters will take a calculated risk, and provide for additional safety, to save valuable property or reduce the potential for civilian and firefighter injuries.

FIGURE 23-7 Solid risk/benefit analysis means taking no risk for that which is already lost. *(Courtesy of Richard W. Davis)*

- Firefighters will take no risk to their safety to save what is already lost.
- In situations where conditions are deteriorating quickly, firefighters should retreat to a defensive position.

These guidelines are common within the fire service and are more clearly spelled out in NFPA Standard 1500, Fire Department Occupational Safety and Health Program.

Personal Size-Up

As mentioned several times in this book, firefighters must perform a personal size-up. A size-up is the continuous situational awareness and mental evaluation process. Specifically, firefighters should continually evaluate the safety of their environment by staying aware of the following:

- *Established work areas.* Firefighters have to be aware of collapse zones, barriers to traffic, and hazard isolation zones.
- *Hazardous energy.* Electrical equipment, pressure vessels, chemicals, and even springs and cables can suddenly release—causing injury and death to firefighters. It is important to scan the environment and mentally log all forms of hazardous energy.
- *Smoke conditions.* Smoke is fuel waiting to ignite. Firefighters crawling in low-visibility, hot smoke are flirting with an explosive environment. Cooling and ventilation are key.
- *Escape routes.* The firefighter who continually plans two or three escape routes while performing tasks is investing in survivability if the established work area breaks down. The firefighter should count doors and windows, scan for traffic barriers, and plot pathways to "safe havens."

- *Air management.* When utilizing SCBA, be aware of the amount of air that is being used. As a general guideline: the first half of the bottle air volume is used for entry and work, and the second half is for egress and to provide a safety margin in case of disorientation or other firefighter emergency.

Rehabilitation

A study of firefighter duty-related injury and death statistics shows that stress and overexertion consistently rank as leading causes.

Firefighting is hot, arduous work performed in PPE that does not allow body heat to evaporate through sweat. A key health and safety concern is controlling the heat stress that occurs when a firefighter's internal core temperature rises above its normal level during incident activities. Core temperatures that have risen above normal can lead to heat- and heart-related injuries and death. Proper rehabilitation efforts (rehab for short) will help keep firefighters safe by reducing heat stress. Additionally, good rehab practices will provide the rest, hydration, and nourishment needed for sustained operations.

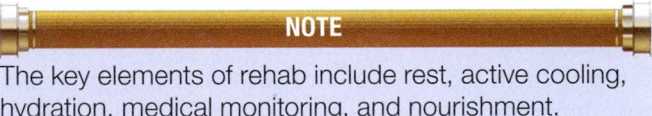

The key elements of rehab include rest, active cooling, hydration, medical monitoring, and nourishment.

Rest

At an incident, rest is achieved during crew rotation. When a firefighter is rotated to rehabilitation, the firefighter should maximize the opportunity to rest by sitting down, by allowing medical personnel (which should be assigned) to do vital sign check, **Figure 23-8,** and by mentally disengaging from the event (which rests the mind). The firefighter who achieves these three important steps can actually work longer at any given incident—it allows a brief recovery from physical and mental stress and allows other rehabilitation elements (hydration and nourishment) to repair and prepare the firefighter for further work. The medical evaluation by a trained medical provider is an important part of rest. Many signs and symptoms indicate the need for further rest—or removal from the incident scene.

One additional point can be made here. Some firefighters do not feel the need for rehabilitation and may be reluctant to spend time in rehabilitation while others are aggressively engaged in the incident—these firefighters are potential candidates for overexertion.

FIGURE 23-8 Effective rehabilitation includes a medical evaluation by BLS or ALS personnel.

FIGURE 23-9 Rehabilitation should start well before a firefighter is thirsty or tired. Failure to rehabilitate "early and often" opens the door to injury.

Remember, the time to rehabilitate is before a firefighter gets tired. "Rehab early and often" should be the fire service creed, especially if the incident duration will span hours, **Figure 23-9.**

Active Cooling

As mentioned at the start of this section, increased core temperatures can cause heat- and heart-related injuries and death. During rehab, firefighters should engage in active cooling to reduce their core body temperature. Typically, firefighters have used passive cooling as a strategy to reduce core temperatures. Passive cooling includes the use of shade, air movement, and rest to bring down core temperatures. Medical personnel have typically monitored heart rate and a firefighter's perceived comfort to determine when a person has recovered sufficiently to don PPE and resume firefighting operations. New studies have shown that this strategy is not adequate and that heart rate and perceived comfort are not good indicators of sufficient core temperature cooling.[1] Effective core temperature cooling is achieved using an "active cooling" strategy. Active cooling is best achieved using a forearm immersion technique. Simply put, firefighters should doff their coats and submerge their hands and forearms in a basin of cool water. This technique is more effective than misting fans (also considered active cooling) and has a reduced tendency to cause the chills that can happen when firefighters go immediately from hot environments to cold environments. Some firefighters "dunk" their heads in cool water to cool off. This strategy is inherently flawed in that the heat-regulating part of the brain (hypothalamus) is instantly cooled—and sends a signal that all is good and the body's natural cooling mechanisms slow down. Unfortunately, the body core and blood flow is still hot. Forearm submersion acts as a heat exchanger that will bring down core temperatures in a systematic way.

Hydration

Hydration cannot be overemphasized in any incident environment, whether it is one of heat stress, cold stress, heavy workload, light workload, mentally taxing events, or long-duration events. Water is vital to the peak operation of virtually every body system from transport of nutrients, to blood flow, to waste removal, to temperature regulation. When the body becomes dehydrated, these systems start to shut down in order to protect themselves. With this shutdown comes fatigue, reduced mental ability, and, in extreme cases, medical emergencies such as renal (kidney) failure, shock, and death. The working firefighter must accommodate for the wearing of heavy clothing that does not allow evaporation of sweat. Additionally, the firefighter must account for strenuous physical activity under stressful situations. Hydration of firefighters should become paramount—even to the point of excess. Firefighters must not wait until they are thirsty to drink water! Some anecdotal history and studies have suggested that it is possible to over-hydrate. In most of these cases, some other condition existed (drug use or other medical condition).

A working, sweating firefighter is usually behind the hydration curve.

Substituting carbonated and/or sugared beverages or other liquids for water can slow the absorption of water into the system. For this reason, just water should be given for the first hour. For activities lasting longer than an hour, some consideration can be given to adding essential electrolytes and nutrients along with the water. Many sports drinks are available that can achieve this.

Nourishment

While dehydration and thermal stress can lead to energy depletion, most firefighters associate energy depletion with the need for food. Nutrition or "fueling" of the firefighter can have the effect of rejuvenating the firefighter or putting the firefighter to sleep. Too often, rehabilitation-feeding efforts accomplish the latter. A firefighter properly nourished will work smarter and safer. The firefighter who is improperly fed will not only want to "crash," but will likely make sluggish mental calculations leading to injury.

So what is the proper way to nourish firefighters? A brief study of essential nourishment theory can help answer this question. In basic terms, maximizing energy from the human machine takes a balance of four essential elements: oxygen, water, blood sugar (from food), and insulin. When this balance is present, other essential hormones and enzymes are created that make for a well-running human machine. The element that is often misprescribed is food— the foundation for building balance. True, all food is fuel for the firefighter. That is, all food is converted to glucose (blood sugar) for the muscles to use as fuel. Insulin is released into the bloodstream to help convert the blood sugar into energy for muscle use. The key to providing quick energy to the firefighter is to find a balance of protein, fat, and low-glycemic-index carbohydrates. Ideally, this balance should be a 30/30/40 mix—that is, 30 percent protein, 30 percent fat, and 40 percent carbohydrate.[3] This balance provides essential elements from three food groups. A balanced approach achieves a few benefits. One, the balance will stabilize insulin release into the bloodstream, helping to reduce the rollercoaster of blood sugar levels that often lead to sporadic activity, chemical imbalance, and fatigue. Second, the balance approach stimulates the release of hormones and enzymes that optimize human performance—both physical and mental.

Choosing the best protein, carbohydrate, and fat also promotes steady, sustained performance. Protein is best derived from low-fat meats such as turkey, chicken, and fish. Eggs and cheese also offer protein. Fats should be of the monounsaturated type like olive oil, nuts, and peanut butter. Often, carbohydrates are dangerous in that so many of the foods typically found on the fire scene are rich in unfavorable carbohydrates. Candy, bread, potatoes, and bananas are all carbohydrates that are quite rich and have a tendency to slow down the worker—and spike blood sugar levels. Low-glycemic carbohydrates include green vegetables, apples, tomatoes, and whole grain breads.

A good example of a quick, balanced rehabilitation meal would include water, sliced turkey sandwiches on thin-sliced whole grain bread with mustard or ketchup, apples, and a handful of peanuts. Low-sugar energy bars can be stocked on apparatus and used as a meal replacement. It is best to use energy bars that are balanced 30/30/40.

Rapid Intervention Teams

Section I, Chapter 16 introduced the use of a rapid intervention team (RIT) to rescue firefighters experiencing an emergency. While the ultimate incident goal should be to *avoid* the need for rapid intervention, it is prudent (and in many cases, mandatory) to establish a RIT when firefighters are engaged in risky environments. The concept of RITs has become commonplace in many fire departments, although they may be titled differently. Other examples include: Rapid intervention company (RIC), firefighter assist and support team (FAST), and rescue assist team (RAT).

The concept of rapid intervention is started early at an incident. The two-in/two-out rule (known as initial rapid intervention or IRIT) is designed to provide for immediate intervention if needed. This requirement is specific to interior fire attack but should be conceptually used for all incidents where firefighters are taking risks. As more resources arrive, the RIT becomes a designated team for a defined task. This RIT should have no other assignment than to prepare for the rapid deployment in support of a search and/or rescue. The dedicated RIT should be assembled using well-trained and experienced firefighters and officers. Specific RIT techniques and duties are considered "advanced skills" past the scope of this book, although some general RIT information and responsibilities are included here.

TABLE 23-1 RIT Responsibilities Checklist	
Tool Assignment	**Rescue**
RIT Leader/Officer	■ Immediate location of companies
Radio	■ Identify hazards and safety issues
Rope bag	
Thermal imaging camera	
Hand light	
Firefighter 1	**Escape**
Radio	■ Open up and remove blocked or secured doors and windows
Forcible entry (irons)	■ Raise ladders to windows and roofs
Hand light	
Firefighter 2	**Circle Survey**
Radio	■ Building size and special features
RIT pack	■ Dimensions
Hand light	■ Construction and stability
Firefighter 3	■ Exposures within and without
Radio	
Specialty tools	
Hand light	
Additional Tools	**Hazards**
Webbing	■ Information placards
Attic ladder	■ Electrical power and downed wires
Rescue litter	■ Hazardous materials and hazard conditions
Power saws	■ Gases: natural/propane/other
Tarps for tool staging	
Extra SCBA cylinders	
Handline	
EMS jump kit	

STREETSMART TIP

Firefighters assigned to the RIT should be prepared to act immediately on orders to rescue firefighters in need. This means the crew assigned the RIT task should be equipped and prepared for the task.

The RIT assignment is not a passive one—the members assigned should proactively prepare for engagement and assist in providing additional egress options for committed crews. **Table 23-1** outlines responsibilities for the RIT.

When activated for an actual firefighter emergency, the RIT will engage the emergency as trained. It is important that other firefighters resist the urge to self-assign themselves to help out the RIT. While it may be likely that others will be reassigned to help, it is important to remain focused and not abandon previous assignments.

CAUTION

Upon hearing a Mayday or RIT activation, *DO NOT* abandon your non-RIT assignment until reassigned by the Incident Commander or Operations Section Chief.

FIREFIGHTER EMERGENCIES

So far, this chapter has focused on preventing injuries and deaths through proactive actions. Occasionally, preventive measures do not address every circumstance or the evolution of the incident presents the working firefighters with an emergency that threatens their lives. In these cases, each firefighter must rely on instincts and training to help survive the situation. To help understand the actions to be taken during an actual or potential firefighter emergency, the firefighter must study procedures for rapid escape and declaring a **Mayday** for lost and trapped situations. The word *mayday* is a universal call for help. Further, the firefighter must understand that survival includes processes for the rescue of trapped and lost firefighters as well as long-term mental survival. The steps presented here should be second nature to firefighters—requiring lots of training and continual practice.

Declaring a Mayday

The specific procedures for declaring any mayday should be developed at the local fire department level. For the scope of this book, we are presenting some general procedures that fire departments may use for declaring a mayday.

Once a firefighter or firefighting team discovers an emergency need for assistance, regardless of the cause, a mayday should be declared. To declare a mayday, the firefighter should:

1. Transmit the declaration over the radio by repeating the word mayday three times followed by the firefighter's ICS assignment. Example: "Mayday, mayday, mayday. Fire Attack Team 2 is declaring a mayday."

2. Wait for an acknowledgment. If no acknowledgment is made within a few seconds, repeat the declaration. If no acknowledgment is gained, follow the next step.

3. Once an acknowledgment is made (or if no acknowledgement is made), report the nature of the mayday and current location. If lost or disoriented, report the best guess or last known location. Example: "Fire Attack Team 2 is disoriented on the second floor. Zero visibility and worsening heat conditions."

4. Manually activate PASS device and proceed with the steps outlined next for the specific type of emergency being experienced.

Rapid Escape

During the course of task completion, firefighting teams may discover a situation that requires immediate escape or may be ordered to do so via radio and fireground warnings. The firefighting teams must then take immediate steps to exit the building or environment.

A building being attacked by fire is always subject to collapse. Once this attack degrades the load-bearing portions of the structure, a collapse will occur. Teams that witness these events must take steps to initiate an evacuation. Teams that hear an established evacuation signal must also immediately egress. Some typical evacuation signals include a repeated series of air horn blasts from apparatus or a special radio tone followed by the evacuation order. The next section discusses some important steps that will assist in achieving immediate escape.

Rapid Escape Steps

1. *Preplan the escape.* All firefighters operating in a hazardous environment should continually look for multiple escape routes. Even in routine or non-threatening environments, the firefighter should be constantly evaluating escape routes. This process develops good habits that will serve the firefighter well when a threatening situation does develop.

2. *Immediately report the need for rapid evacuation.* Actual collapse or signs of impending collapse can be communicated via radio while exiting the area. A delay in reporting these conditions could

FIGURE 23-10 Rapid fire spread and partial collapse are likely to trigger the need for rapid escape.

endanger other crews. Likewise, the witnessing of a flashover (or signs of an impending flashover or other hostile fire event) should be cause for immediate reporting.

3. *Acknowledge rapid evacuation or escape signals.* In cases where the team has not initiated the rapid escape, the team must acknowledge that it has heard the signal. Signals vary from department to department but typically consist of an audible fireground signal (a defined air horn signal or a barrage of sirens and air horns) as well as a radio broadcast (an alert tone followed by the evacuation order).

4. *Rapidly escape.* Rapid escape means just that—firefighters should leave the area immediately. They should leave hoselines and heavy tools and escape. Trying to bring these types of items with them has resulted in injuries to many firefighters. Often, it is faster for the team to use an egress that is different from the way it entered. This is where the team's escape preplan observations pay off. At times, firefighters will be faced with a situation that requires a temporary escape—in these cases, the priority is to protect the crew and plan for further escape. Here are some examples:

   - *Room escape.* It may be possible to enter an uninvolved room, shut the door, and use a window for further escape. If the window is above the first floor, the firefighter may have to clear the window and wait for a ladder. These actions should be communicated if it is not obvious to crews outside. If the room has no window, the firefighter should think through the options. Often, a simple wall breach (breaking out the drywall over wood studs) can get the team to rooms with alternative escape routes.

   - *Roof/balcony escape.* When fire or collapse has cut off stairwells and hallways, it may be easier to get to the roof or balcony and await ladders. In these cases, it is best to close doors or create barriers that separate the team from the fire. It is essential to communicate the team's needs if the firefighters assisting the team are not in visual contact.

   - *Self-rescue.* In some extreme cases, it might be possible to escape through a system of self-rescue. Many firefighters carry a personal rescue rope or strap so they can lower themselves to another floor. Firefighters equipped and trained to do this need to establish a barrier to fire spread, secure a reliable anchor point, have a device or method to descend the rope, and have a method to break windows on lower floors.

   - *Ladder escape.* When a ladder is in place and available for rapid escape, firefighters should use safe

and common methods to mount and descend the ladder. In hurried situations (like collapse and flashover conditions), firefighters may opt for dangerous headfirst mounting and ladder slides. It is important to note that these techniques can lead to serious injuries or death. Locally, fire departments may choose to instruct firefighters on these techniques. These skills are highly technical and require close supervision and instruction for practice. A firefighter should not attempt to practice these without an experienced instructor who uses protective systems.

5. *Report successful escape.* Once the team has escaped the building or area, members should report that the team is safe and accounted for. This process is usually accomplished with a fireground **personnel accountability report (PAR).** A PAR is an organized approach to accounting for everyone on scene, **Figure 23-11.** Typically, the accountability manager runs through all the assignments given thus far and asks (usually via radio) for each crew to report its status. From this, unaccounted teams are identified and rapid intervention is initiated. Many departments perform a PAR at given intervals, such as when dispatch announces time intervals (every 10 to 30 minutes). Most departments perform a PAR after an emergency or when the incident operation changes modes or strategies (for instance, switches from offensive to defensive operations).

Lost, Trapped, and Injured Firefighters

Firefighting operations place crews in environments in which they have potentially never been—large buildings with unusual floorplans, small confined

FIGURE 23-11 A PAR is a personnel accountability report organized to check the status of all crews working an incident. PARs should take place every half-hour or after an evacuation or any firefighter emergency.

areas with minimal escape routes, and a whole host of stock, furniture, and arrangement obstacles. With this comes the potential to be trapped or lost during fire suppression, search, or rescue assignments. While most firefighters will agree that an entrapment is a true firefighter emergency, fewer believe that being lost or disoriented is an emergency. They believe that, eventually, they will find a way out. This thinking needs to be altered. Likewise, firefighters that are experiencing an SCBA low-air warning within an IDLH environment are experiencing a firefighter emergency. Again, this is an attitude shift.

CAUTION

A firefighter or crew that has lost spatial bearings in an IDLH atmosphere is experiencing a firefighter emergency!

With low air situations, entrapments, and lost situations, it is important to have a mental process to follow to avoid complicating the situation.

Low-Air Situations

Good SCBA air management techniques should provide a margin of safety for firefighters to enter, work, and egress an IDLH environment *before* the low-air warning device on the SCBA activates. When a firefighter experiences a low-air warning while in an IDLH environment, immediate communication and action is warranted:

1. Confirm the low-air warning with the assigned team.
2. Evaluate the air-status of other team members.
3. Size up the path and distance to a visible egress point.
4. If the egress point can't be seen, communicate via radio that a low-air warning is being experienced and report the team location within the building (some FD procedures require that a Mayday be declared).
5. Stop, try to control breathing rate, and proceed calmly to the closest egress. Keep the rapid intervention team advised of location and progress.

Entrapments

The first step a firefighter should take in an entrapment is to get assistance. Activation of a PASS device is warranted and the declaration of a "Mayday" should be made over the radio. Some radios are equipped with an emergency assistance button. It should be activated if the radio is so equipped. The Mayday will be followed up with radio procedures and communications to get assistance to the trapped firefighter—often via the rapid intervention team. The firefighter should follow this up with other noise-making efforts. Banging on a pipe or throwing debris may be helpful, but the firefighter must be careful not to use up energy or excessive air in doing this. Using visual signals such as a flashlight may also prove helpful. The second step in dealing with the entrapment is to size up the situation and develop a plan. Some key questions to answer include:

- What exactly is causing the entrapment?
- What is the exposure to fire/smoke/further collapse?
- How much breathing air is left in the SCBA bottle?
- What are the extent of injuries?
- Is there anything that can be done to self-extricate?
- Is there any self-first-aid that can be performed?
- How can remaining air be conserved?

The trapped firefighter should attempt self-extrication. This must be planned and systematic as opposed to reckless and panic driven. All of these steps accomplish two important points. First, a process is created to help with rescue. Second, the firefighter's mind is kept active to help ward off panic.

Lost/Disoriented Firefighters

It cannot be overemphasized that a firefighter or team lost in an IDLH atmosphere is in fact experiencing a firefighter emergency. The steps to overcome this emergency are simple and can result in a quick resolution prior to an injury. First, the firefighter or team must report the fact that they are lost. This is also a Mayday situation and should be transmitted as such over the radio. The Mayday will lead to radio traffic trying to ascertain last known position and any clues that might help a RIT locate the firefighters. Second, the lost firefighters need to manually activate their PASS devices. Finally, the firefighters need to stop, take a few deep breaths, and calm down to help ward off panic. Once calmed, the firefighters should take inventory of their surroundings. From this, they can establish direction, door and window locations, and potential paths. As with firefighter entrapments, some questions should be answered:

- What is the exposure to fire/smoke/further collapse?
- How much breathing air is left in the SCBA bottle?
- How can remaining air be conserved?
- Are there other options that have not been explored?

It is important to maintain radio contact with the RIT members and the incident commander. A team that is lost needs to help each other remain calm and conserve air. Any member that appears to panic can have a negative effect on the others—the team should exercise steps to keep minds open.

Injured Firefighters

Perhaps the most stressful situation that will ever face a firefighter is the realization that a fellow firefighter has become seriously injured. Regardless of what caused the injury, firefighters will typically drop whatever task they have been assigned and rush to aid their fellow firefighter, **Figure 23-12**. In some cases, this rapid focus on assisting a coworker becomes heroic due to extreme dangers faced by the now-rescuers and the firefighter victim. It should also be said that the potential to injure or kill many firefighters could be risked during firefighter rescues. One example of this unnecessary risk is the all-too-often scenario where firefighters rush to a debris pile to start rescue of firefighters caught under a collapse. A secondary collapse occurs and compounds the event. To minimize this risk, firefighters must trust RITs. The RIT is designed to attempt the rescue and remove the victim firefighter from the hazard so appropriate medical care can begin.

Once a firefighter is found to be trapped or injured, the RIT should be activated. It is natural for another team that is in proximity of the trapped firefighter or crew to engage in the rescue—this must be communicated! Collapse, exposure to the fire, and smoky conditions may exist and make the rescue difficult. Furthermore, the trapped firefighter or crew will be in an area that is hard to access by rescuers. Practicing the removal of firefighters from tight enclosed areas or through small windows is an investment that can save firefighters' lives.

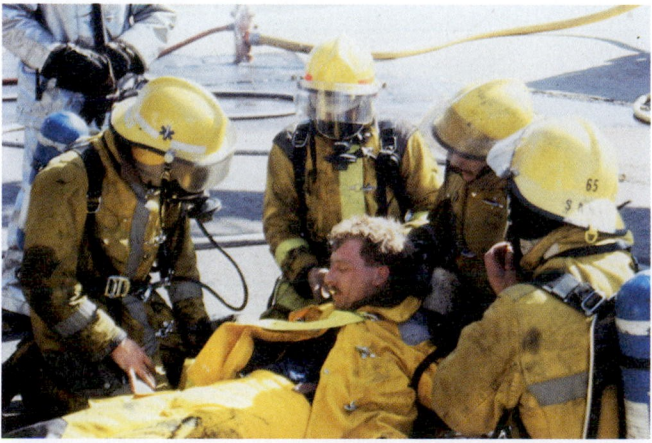

FIGURE 23-12   A serious firefighter injury or fatality will cause significant incident stress. Focus and use of RITs will minimize unnecessary risk during firefighter rescue and help maintain incident control.

Firefighters who are not on the RIT or in the immediate vicinity of the rescue should resist the urge to rush in and help. History shows that most firefighter rescues require the assignment of many teams. Firefighters should let the RIT take the lead and wait for their assignment. In this way, they are part of the plan—not the problem.

Post-Incident Survival

Once an incident has reached the point where crews are starting to clean up and be released, a firefighter's mind may begin to relax as the "buzz" of adrenaline fades. This phenomenon has been called **post-incident thought patterns** and is responsible for many injuries and deaths. Common post-incident injuries include strains, sprains, and being struck by objects.

> **NOTE**
>
> The causes of these post-incident injuries seem almost ironic in a profession where aggressive and calculated risk taking is a hallmark.

Incidents that are especially gruesome or involve significant human tragedy can easily impact firefighters in an emotional sense, leading to long-term mental and health issues. This is called critical incident stress (CIS). It is important that firefighters address post-incident thought patterns as well as CIS.

Post-Incident Thought Patterns

One cause of post-incident injuries has to do with the little-studied notion of post-incident thought patterns. In essence, this is inattentiveness. In cases of especially difficult, unusually spectacular, or particularly challenging incidents, firefighters will tend to reflect on their actions. The replay of the incident starts almost instantly when the order is given to "pick up." This introspection is normal. The switch from activities requiring brainpower and physical energy to an activity that is so routine as to be dull is a hard one to adapt to. Herein lies the problem. Some signs that post-incident thought patterns are affecting crews include faraway stares, firefighters wanting to be alone, and firefighters who stop and look about as if they have forgotten their task.

> **NOTE**
>
> Simple safety reminders or appropriate jocularity can help firefighters regain focus and reduce injury potential.

FIGURE 23-13   A quick huddle during post-incident activities can serve to remind firefighters to continue to work safe and smart during pickup activities.

One method to reduce the impact of these thought patterns is to take a time-out and have everyone gather for a quick incident summary and safety reminder, **Figure 23-13.** These huddles can be effective for everyone or just small groups.

Another factor to consider that leads to inattentiveness is chemical imbalance. Even the most successful rehabilitation program cannot prevent firefighters from experiencing fatigue and mental drain. With the end of an incident, especially one requiring major physical effort, comes the relaxation of the firefighter's mind, which, in turn, starts shutting down protective chemicals that stimulate performance. The adrenaline "rush" is over and the firefighter's metabolism will return to a "repair" state. This causes a mental slowdown that can lead to unclear thinking and result in injuries.

Another way to see this chemical and mind imbalance is to look at a firefighter's tools from a layperson's point of view. Most laypersons would literally have to concentrate on carrying an ax, pike pole, chain saw, roof ladder, or hose length in order to not hurt themselves or anyone around. Consider also doing these tasks in bulky, restrictive clothing and heavy boots. The firefighter does these things after incredible energy bursts under frightening conditions. Familiarity plays a certain role here, but concentration is still required. If the mind has been taxed, the body has been fatigued, and the signal to "relax" has been given—yet the concentration required to do the task has remained the same—the potential for injury rises.

Whether the issue is chemical imbalance or post-incident thought patterns, the firefighter needs to stay alert and try to pick up signs of potential injury and take steps to "survive" without injury.

Critical Incident Stress

Firefighters are expected to tolerate a certain level of incident stress given the nature of the environment that firefighting brings. Dealing with community hazards, injuries, and death can even be classified as normal for the profession. Some events, however, can trigger a significant emotional response from responders. These emotional responses may not always be external. In fact, many firefighters harbor the reaction internally. Events that typically induce an emotional response include:

- Death or serious injury to a coworker or a person known to the responder
- Mass casualty incidents
- Death or serious injury to a child, especially from an act of violence or crime
- A prolonged rescue attempt that results in a death

Firefighters will exhibit signs of CIS in many ways. **Figure 23-14** outlines a few of these signs and symptoms. It is important to note that none of these signs is abnormal—emergency responders will continue to feel the stress of critical incidents. What is important is to "survive" the CIS through a process of critical incident stress management (CISM).

CISM can take on many forms, both formal and informal. Formal CISM is usually accomplished following an incident of significant magnitude—one affecting numerous responders. Typically, formal CISM involves a critical incident stress debriefing

Selected Signs and Symptoms of Critical Incident Stress	
Behavioral:	Increased irritation, aggressiveness, withdrawal, flashbacks, inattentiveness, alcohol/drug abuse, and memory loss
Physical:	Loss of appetite, insomnia, fatigue, headaches, muscle stiffness, and hypertension
Psychological:	Guilt, sadness, depression, career introspection, claimed burnout, fear, and unsociability

FIGURE 23-14   Signs of critical incident stress are natural following "trigger" incidents.

(A)

(B)

FIGURE 23-15   Critical incident stress management sessions can be (A) informal such as a "defusing" or (B) formal such as a process that includes peer support and mental health professionals.

(CISD) process that is led by peers and/or health professionals. An informal process usually includes a "defusing" that allows persons to deal with an ongoing rescue or to voice concerns and thoughts of the incident in a peer environment. Defusing should take place just prior to releasing firefighters from the scene, or back at the station just after the incident, **Figures 23-15 A** and **B.**

To help manage critical incident stress as well as daily incident stress, the firefighter should use CIS processes and tested stress-reduction principles such as exercise, adequate and appropriate nutrition, deep breathing, deep relaxation (massage, meditation, etc.), hobby development, and positive social interaction. With these steps, the firefighter can survive post-incident stress in the long term.

LESSONS LEARNED

Firefighter survival is dependent on many proactive and preventive actions. These actions include incident readiness, safe operations at incidents, and appropriate preparation and response to firefighter emergencies. Incident readiness includes those efforts to prepare the firefighter's personal protective equipment, personal accountability, and individual fitness for duty. Of particular note, fitness for duty relies on the firefighter's mental and physical wellness as well as the firefighter's attention to energy and rest.

Safe operations and, therefore, survival are dependent on team continuity and the elimination of freelancing. Additionally, safe incident operations are achieved when teams perform orders with attention to communications and timely updates. While operations are under way, individual teams must practice solid risk/benefit analysis for their specific job. During the incident, crews must take advantage of the rest, active cooling, hydration, and nourishment offered through established rehabilitation processes.

Rapid intervention teams are formed for immediate deployment should a firefighter emergency take place. To meet this challenge, RITs must assemble tools and be briefed on essential information that will assist in a more rapid reaction should an emergency occur. Firefighter emergencies require the individual firefighter to practice a clear and concise approach to dealing with the emergency. Using the term *Mayday* can signal everyone of the need for firefighter rescue. Whether trapped or lost, the firefighter must remain calm and follow a defined series of steps to mitigate the situation.

Finally, firefighters must survive long term through an understanding of post-incident thought patterns and the effects of critical incident stress. Understanding the factors that lead to each is a useful tool in dealing with the reality of emergency responder stress and can help reduce the potential for firefighter injury or death.

KEY TERMS

Freelancing   The act of working alone or performing a task for which the firefighter has not been assigned.

Incident Action Plan (IAP)   A strategic and tactical plan developed by the incident commander.

Mayday   A universal call for help. A Mayday indicates that an individual or team is in extreme danger. When declaring a mayday, it is best to repeat the word three times (Mayday, mayday, mayday).

Passport   A term given to a specific accountability system where crews are tracked using a card (passport) with all members listed. An accountability manager tracks the passports on an accountability board.

Personal Size-Up   A continuous situational awareness and mental evaluation of an individual's immediate environment, facts, and probabilities.

Personnel Accountability Report (PAR)   This is an organized roll call of all units assigned to an incident.

Post-Incident Thought Patterns   A phenomenon that describes an individual's inattentiveness following a significant incident. Post-incident thought patterns can lead to injuries or even death.

Risk/Benefit   An evaluation of the potential benefit that a task will accomplish in relationship to the hazards that will be faced while completing the task.

REVIEW QUESTIONS

1. List the three main components that lead to incident readiness.
2. Define the four key checks to ensure that individual personal protective equipment is ready for response.
3. Describe the advantages and disadvantages of three kinds of personnel accountability systems.
4. Define personal size-up.
5. Briefly describe the three components that lead to fitness for duty.
6. Name three practices that lead to team continuity. Describe how each can increase firefighter survival.
7. Define risk/benefit.
8. List and briefly describe the four components of on-scene rehabilitation.
9. Describe the procedures that should be taken to establish and prepare for the assignment of a rapid intervention team.
10. List the five steps that can lead to an organized rapid escape.
11. Describe the reasons why a lost firefighter is considered an emergency.
12. List the three steps that should be taken when entrapment occurs.
13. Compare and contrast post-incident thought patterns and critical incident stress.

ENDNOTES

1. Thomas McLellan, PhD, "Safe Work Limits While Wearing Firefighting Protective Clothing," Toronto Fire Department Grant Study Report (Toronto, Ontario, Canada: Defense Research and Development Council, 2002).

2. U.S. Fire Administration, Emergency Incident Rehabilitation, FA-114 (Washington, DC: USFA Publications, July 1992).

3. Barry Sears, *Enter the Zone* (New York, NY: HarperCollins Publishers, 1995).

Firefighter II

Once you have successfully completed the Firefighter level I training that is included in the first section of this book, you are ready to embark on your next step toward qualification as Firefighter level II. To become a level II firefighter, it is necessary that you do fully achieve qualification at level I prior to initiating training at level II. This requires meeting the knowledge and skill requirements defined in NFPA 1001 *Standard for Firefighter Professional Qualifications,* 2008 Edition.

Section II will now take you to this higher level of training. As defined by the NFPA 1001 Standard, 2008 Edition, the Firefighter II is empowered to assume and transfer command within an incident command system by having the necessary knowledge and skills to act as a leader among your fellow firefighters and maintain direct communications with a supervisor. This responsibility is not to be taken lightly and having knowledge of safety regulations and procedures within your Authority Having Jurisdiction, as well as a complete understanding of your role within the department, will be required.

Each chapter within this section continues where the corresponding chapter in Section I concluded, and provides a higher level of technical training. However, some chapters are identified as containing exclusively Firefighter I content. For this reason, those chapters will *not* appear again in this section and numbering of the chapters in this section will *not* be sequential in all cases. Where applicable, chapters in this section are numbered according to their corresponding chapter in Section I. For example, Chapter 2 is identified as having *both* Firefighter I and II content. Therefore, the content within this chapter is split between Section I, Chapter 2 (Firefighter I) and Section II, Chapter 2 (Firefighter II). Cross-references are provided for you along the way to help you with the transition, and review material has been included within the chapters of Section II to help bridge learning between the two levels.

2

Fire Department Organization, Command, and Control

- Learning Objectives
- Introduction
- The Firefighter
- Incident Management
- Incident Command System
- Lessons Learned
- Key Terms
- Review Questions

LEARNING OBJECTIVES

After completing this chapter, the reader should be able to:

2-4 Describe the role of the firefighter II in the fire service while performing duties within required standards and regulations.

2-8 Explain the need for command at incident locations.

2-9 Recognize the need for command at incident locations.

2-10 Discuss the incident command system as it relates to an emergency operation.

2-11 Describe duties and responsibilities in establishing and transferring command within the incident command system.

2-12 Initiate incident command and transfer command at an incident scene.

2-13 List and describe the components of an incident command system.

2-14 List the function of each assigned role in an incident command system.

2-15 Perform an assigned role within an incident command system.

2-16 Discuss teamwork and its part in firefighting.

*The FF I and II levels, as defined by the NFPA 1001 Standards, are identified in different colors: FF I = black, FF II = red, additional information = blue.

INTRODUCTION

This section describes the many different roles a firefighter II may have in the organization of a fire department. A firefighter II is defined by NFPA 1001, Standard for Fire Fighter Professional Qualifications, as a person at the second level of progression "who has demonstrated the skills and depth of knowledge to function under general supervision." A well-trained firefighter is essential for the department to accomplish its mission.

THE FIREFIGHTER

A firefighter is an individual trained to perform the function of fire prevention and suppression. Firefighters must possess both the knowledge and skills to be able to perform safely and effectively at an emergency incident, **Figure 2-1.**

The Team

The department as a whole cannot be effective in its firefighter safety effort without the support of a team approach to handling incidents. Therefore, the team that handles an incident must hold up its part of the safety partnership. This team obviously includes the individual firefighter. To ensure safety, the team should follow these procedures:

FIGURE 2-1 The firefighter is the individual who makes a department operate.

ROLE OF THE FIREFIGHTER II

Firefighter training and education is based largely on NFPA 1001, Standard for Fire Fighter Professional Qualifications. This standard lists the critical skills and knowledge that firefighters must have to function safely and effectively at fires and emergency incidents.

NFPA 1001 splits basic firefighter education into two levels, Firefighter I and Firefighter II. Firefighter I outlines the most basic of knowledge and skills that firefighters need, such as fire behavior, forcible entry, self-contained breathing apparatus (SCBA), personal protective equipment, ladders, ventilation, salvage, and overhaul. Firefighter II covers an expanded set of topics and skills to enhance the Firefighter I training and make firefighters more proficient at vehicle extrication, flammable liquid fires, flammable gas cylinder fires, evidence protection, fire safety and pre-incident surveys, and public education presentations.

Some topics appear in both Firefighter I and Firefighter II, including communications, structure fire operations, and equipment maintenance. For these topics, the Firefighter I level provides the introduction to the concepts and how to apply them as a firefighter, and the Firefighter II level expands that knowledge and skill set.

Just like fire departments across the country differ in structure and operations, so do fire department training programs. Some firefighters will go through a Firefighter I course and then a separate Firefighter II course at a later time. Other firefighters will take one course that covers both Firefighter I and Firefighter II material.

The Firefighter I and Firefighter II levels allow firefighters to learn the basics first and then progress on to make themselves more versatile in firefighter and emergency operations.

1. *Utilize an* **incident command system (ICS).** The ICS should detail lines of authority and communication, be reasonable in its span of control, and include an action plan. Section chiefs, division/group supervisors, and team leaders need to communicate progress and status. Orders need to be clear and accomplished within an acceptable risk environment.

2. *Work together and remain intact.* The "buddy" concept is imperative for firefighter safety.

3. *Look after each other.* Team members who continuously watch each other will find and address conditions that will lead to injury. Aggressive rehabilitation will provide hydration, rest, and nourishment—thereby increasing the time the team can perform safely. At the team level, issues of PPE and SCBA use or misuse should be addressed. Likewise, the team must watch each other for signs of fatigue, overaggressiveness that is dangerous, unsafe acts, and freelancing.

> **CAUTION**
>
> The separation of members within a team is a contributing factor to firefighter fatalities.

The individual firefighter holds the final key to making the safety partnership work. The following list includes areas that the firefighter must remember.

- Be ready.
- Understand and act within the chain of command.
- Perform as trained.
- Use an incident engagement checklist (See Section I, Chapter 2).

Prevent Freelancing

Fire and rescue work is a team effort orchestrated to an action plan developed by the incident commander (IC). Team leaders implement the plan and provide feedback to the IC. Firefighting teams follow the team leader's direction. Freelancing occurs when a team operates outside the action plan or when individuals work alone. Working alone or outside the action plan endangers individuals and the team, as **Figure 2-2** demonstrates. The individual firefighter can help prevent freelancing by *not* working alone and supporting the action plan when part of a team.

FIGURE 2-2 Freelancing endangers individuals and the team. This firefighter is working alone in a collapse zone—for what gain?

INCIDENT MANAGEMENT

Fire departments respond to millions of emergency incidents every year. These incidents are the operations for which a firefighter trains. Some incidents require a single company; some require multiple companies. Extremely large incidents such as wildfires or natural disasters may require mutual aid assistance from outside the authority having jurisdiction and involve fire agencies on a national basis.

Command and Control

To manage incidents, firefighters must understand the concept of command and control and how it is applied at the emergency incident. The basic tactical unit at any incident is the company or single resource. As the size or complexity of the incident grows, additional companies and resources respond. To maintain command and control, the command officer must have the ability to manage effectively the increasing number of companies operating at an incident. This is accomplished by using unity of command and span of control.

Unity of command means having one designated leader or officer in charge of an operation, company, or single resource. Every operating resource has a single designated supervisor. On the initial response this could be a company officer or a firefighter. All units responding to the incident then report to command and receive instructions on operations. As the incident becomes larger, command may transfer to a high ranking officer. This is accomplished by briefing the new command on the incident, actions taken, and units operating.

Span of control refers to the number of resources (people, companies, etc.) that any one person supervises. For emergency operations the span of control is usually three to seven, with the optimum number being five. This allows for effective management and safety of all operations.

STREETSMART TIP

Local and regional policies may dictate the format of the radio report, and firefighters should be familiar with this format and structure. In addition, the locality or region may have special terminology for radio reports, such as location designations. Firefighters should be familiar with these specific requirements and local terminology.

Establishing Command

The first unit arriving at an incident should establish command of the incident and give a brief initial report known as a size up. This is accomplished with a radio announcement, for example: "Engine 438 arriving, establishing Maple Avenue Command. We have a two-story wood frame residential structure with smoke showing from the second floor. Two occupants are reported in the structure." This advises all other responding units of a fire condition in an occupied residential building. Generally, standard operating procedures (SOPs) will dictate the tasks assigned to other responding units.

Transfer of Command

Transfer of command is the process of briefing an individual of equal or higher experience or authority so that person can assume command of an incident. As noted earlier, the first arriving unit must establish command of an incident and as the incident becomes more complex, the command structure will expand and the incident commander (IC) will usually change. The information necessary for a complete transfer of command is known as a status or situation status report and must include the following information:

- What has happened or the type of incident
- Incident action plan (strategy and tactics)
- Current units/resources operating, status, and location
- Civilians injured or trapped
- Firefighters injured or trapped
- Actions that have been accomplished
- Ability to control the incident with available resources

Transfer of command should occur during a face-to-face meeting, but under extreme conditions transfer may be accomplished by radio or telephone. Transfer can only be accomplished when the person assuming command is at the incident scene. Also, the individual assuming command must acknowledge receipt of the information, confirm its accuracy, and announce the transfer of command to avoid any confusion with the previous IC. The more complex an incident is, the more complex this briefing will be, possibly involving additional staff or technical experts. The use of tactical worksheets can enhance the ability of transferring command.

INCIDENT COMMAND SYSTEM

The **incident command system (ICS)** is exactly what the name implies: a systematic approach for the command, control, and management of an emergency incident. It provides a command structure and designated responsibilities for the functions that must be

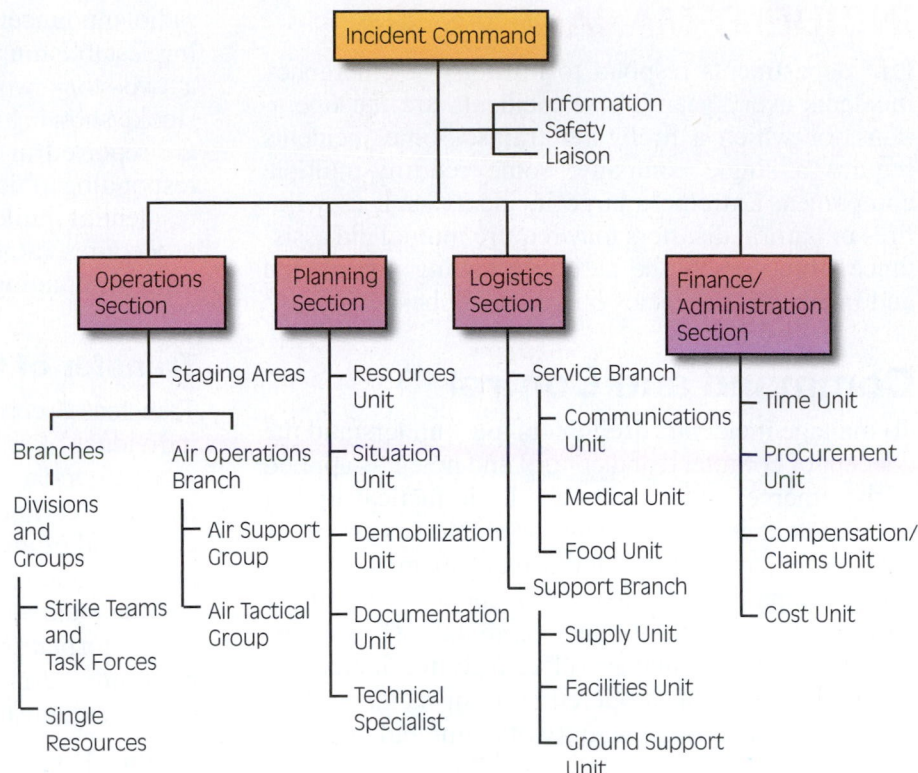

FIGURE 2-3 This chart of a typical incident management system shows the modular organization necessary to manage an incident.

addressed to stabilize any incident, from a room and contents fire to a large-scale natural disaster.

The concept of incident command, as shown in **Figure 2-3,** has evolved over the years and is based on the FireScope (Firefighting Resources of California Organized for Potential Emergencies) Project from California. Firescope was developed as a result of numerous large wildland fires that occurred during 1970. The California Department of Forestry, working with the Los Angeles City Fire Department, the Los Angeles County Fire Department, the Santa Barbara County Fire Department, the Governors Office of Emergency Services, and the U.S. Forest Service, developed Firescope to provide a unified system for use by all agencies.

Although intended for wildland firefighting, its broader value for handling a variety of incidents was realized. Other variations of this system, such as Fire Ground Command (FGC) developed by the Phoenix Fire Department, were established. The main difference that existed in the systems was in the area of terminology. In 1990, the National Fire Service Incident Command Sytem (ICS) Consortium was formed in an effort to finalize a common system of incident command no matter where an incident took place. The ICS consortium released a document in 1993 titled *Model Procedures Guide for Structural Firefighting* that reflected aspects of each system. This material was adopted by the National Fire Academy and presented in its courses.

To function properly, any ICS must contain the following components:

- *Common terminology.* The designation of a term that is the same throughout an ICS. Common terminology is instrumental in making certain that communications are easily understood.

- *Modular organization.* The ability to start small and expand if an incident becomes more complex.

- *Integrated communications.* The ability of all units or agencies to communicate at an incident. This may be as simple as assigning a tactical radio channel to an incident or developing a communications plan for use by many agencies at a large-scale incident.

- *Consolidated incident action plan.* The strategic goals to eliminate the hazard or control the incident. All companies or agencies working an incident must have the same objective.

- *Span of control.* The ability of one individual to supervise a number of other people or units. At an emergency incident the range for safe supervision is usually three to seven people or units. The ideal number is five.

- *Designated incident facilities.* These may be as simple as a command post established at a chief's vehicle **Figure 2-4A,** or complex enough to include a staging area, rehabilitation area for firefighters, feed-

(A)   (B)

FIGURE 2-4   Depending on an incident's complexity, a command post may be as simple as (A) the command officer's vehicle or (B) a specially designed mobile command post. *(Photo B courtesy Du Page Co. OEM)*

ing facilities, and office space for other agencies. A specially designed mobile command post is shown in **Figure 2-4B.** In some instances, it is possible to designate these facilities prior to an incident.

The events that took place on September 11, 2001, further illustrated the need for a standardized system of operating at and controlling resources during an incident where multiple agencies may interact. In an effort to provide a common model for this standardization, President Bush issued Homeland Security Presidential Directive 5 in February of 2003. This directive called for the development of a **National Incident Management System (NIMS).** NIMS was designed to lead agencies through the tasks of preventing, preparing, responding, and recovering from incidents by concentrating on several main components applicable to all four areas. These components are:

- *Command and management.* Definition of the operating mechanisms and structure needed to run an incident, coordination of different agencies and processes for communicating information to the public.
- *Preparedness.* Preplanning and training exercises.
- *Resource management.* Systems in place to document, deploy, track, and recover the associated costs of resources throughout an incident.
- *Communications and information management.* Interoperability of communications systems and information transfer through a common method that is understood.
- *Supporting technologies.* Technology-based items that help facilitate the ongoing operations during an incident that may be beyond the ordinary.

- *Continual maintenance and management.* A system for review and modification to ensure the system better meets the needs of responders.

The goal of this system is to provide the necessary tools for governmental and private agencies to work jointly at all levels. The need for this system was further demonstrated by the events that took place during Hurricane Katrina in the Southern Gulf Area in 2005.

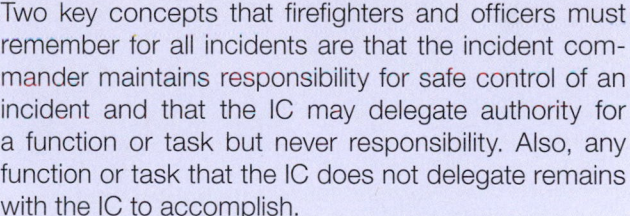

NOTE

Two key concepts that firefighters and officers must remember for all incidents are that the incident commander maintains responsibility for safe control of an incident and that the IC may delegate authority for a function or task but never responsibility. Also, any function or task that the IC does not delegate remains with the IC to accomplish.

Common designators are used for all resources assigned to an incident. Resources are assigned and managed in one of the following ways:

- *Single resource.* Personnel or vehicle and required equipment.
- *Task force.* A task force is any combination of single resources within the span of control guidelines, assembled for a common task or assignment. A structural assignment task force may consist of three engines, two trucks, and one rescue company.
- *Strike team.* A combination of a specific number of units of the same kind and type. A water

supply strike team would be five Type 1 engine companies.

The command organization at an incident must be developed at a pace that enables management of the tactical deployment of firefighters as that incident becomes more complex, as **Figure 2-5** demonstrates. Command is handled on three levels:

1. *Strategic,* which includes establishing overall objectives, setting priorities, predicting outcomes, and allocating resources.
2. *Tactical,* which is assigned to division or group supervisors who direct activities toward specific objectives such as ventilation, overhaul, salvage, and so forth.
3. *Task,* which is assigned to individual company or group leaders.

The strategic and tactical levels form the foundation for achieving the incident commander's overall objectives. The task level is the actual work required to accomplish the tactical objectives set by the division or group supervisors.

> **NOTE**
>
> A **Rapid Intervention Team (RIT)** is a team of specially trained firefighters who are solely designated to provide for the safety, search, and rescue of trapped or lost firefighters at an emergency incident, as shown in **Figure 2-6**. These teams are referred to by many names and descriptions, such as **Rapid Intervention Crew (or Company, RIC), Firefighter Assist and Support Team (FAST)**, GO Team, Rescue Assist Team (RAT), and many other denominations that have been given to them by various departments across the globe. The names and related jargon are not important—it is the task and purpose of these teams on the fireground that is.

Five Major Functions of an Incident Command System

An ICS has five major functional areas, four of which may or may not be established depending on incident complexity. All incidents will always have an incident commander, who may establish the operations,

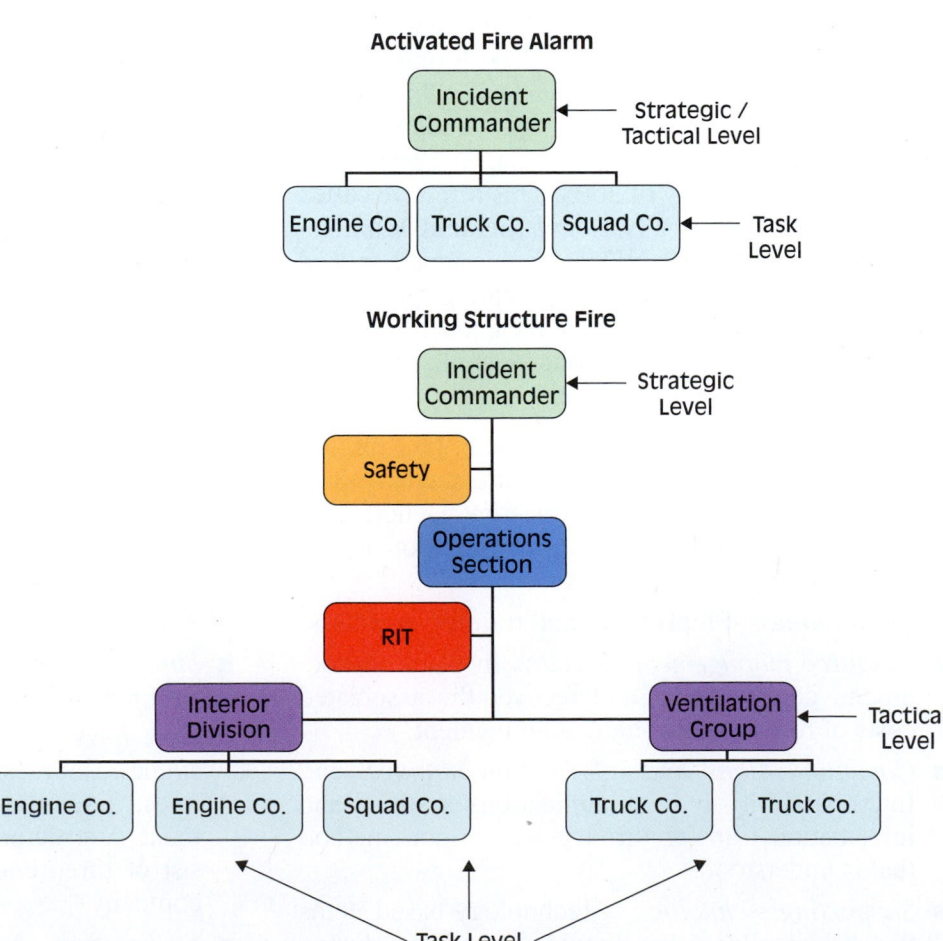

FIGURE 2-5 Command organization must be expanded as an incident becomes more complex in nature.

planning, logistics, and finance/administration sections. When these positions are established, an individual is designated as the chief and reports directly to the IC. In addition, the IC may establish three command staff positions—safety officer, liaison officer, and public information officer (PIO)—to assist with the incident management. These positions are not considered to be under the span of control guidelines. During incidents involving long operational periods or other agencies, these functions should be established. The managerial level of the organization structure used should follow a format similar to that shown in **Figure 2-7.**

Command

The incident commander is responsible for developing the **strategic goals** for control of an incident. This may be a chief-level officer, captain, or firefighter, depending on the staffing of the fire department that responds and who arrives first. The IC has the authority to request and assign resources to an incident, the power to establish functional areas to control the incident, and the responsibility for the safety of all responders. The IC manages the incident, not the **tactics.** Tactics are the actions taken to achieve the strategic goals of the incidents.

Operations

The operations section chief is responsible for implementing the tactical assignments to meet the strategic goals established by the IC. The operations chief

FIGURE 2-6   The Rapid Intervention Team's purpose is to provide for the safety or search and rescue of trapped or lost firefighters at an emergency incident.

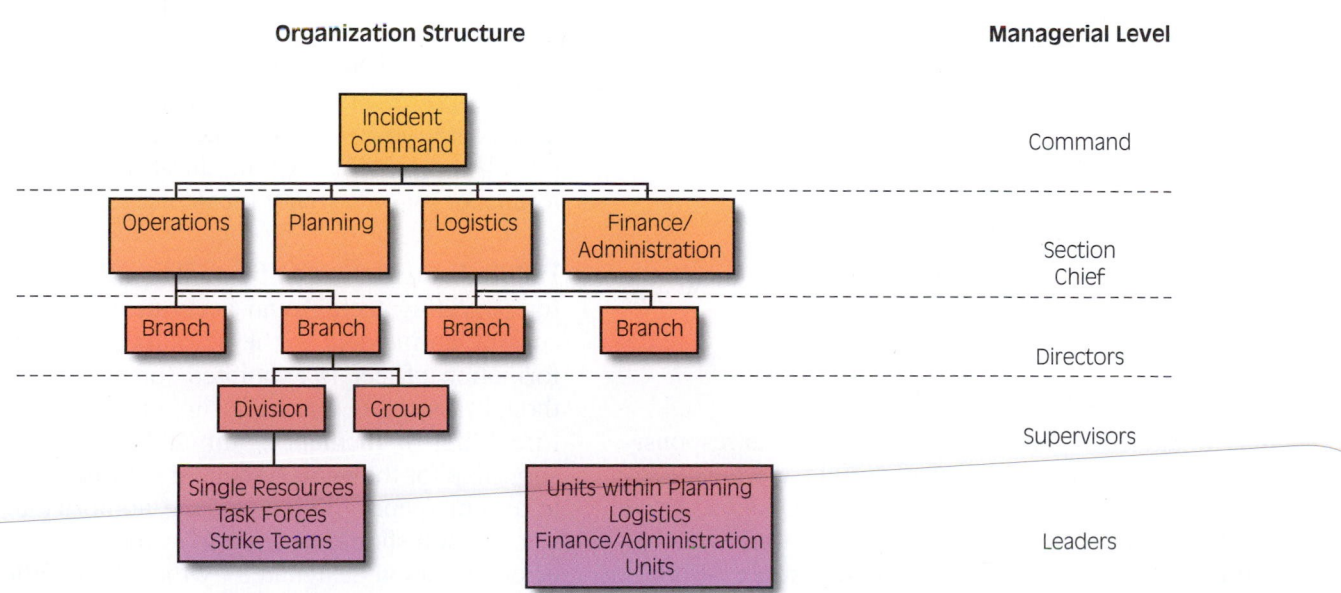

FIGURE 2-7   Organizational structure showing the managerial level for an incident.

reports directly to the IC. Depending on the size and complexity of the incident, the operations section may be a single unit or a number of **branches** or **divisions.**

Staging is part of the operations section under the direct control of the operations chief with the assistance of the staging area manager. All apparatus and personnel assigned to staging are committed to the incident and must be available for deployment within three minutes.

Planning

The planning section chief is responsible for the development of the incident action plan. This plan is based on evaluation of the information and particulars of an incident. The planning section is also responsible for tracking the status of all resources used at an incident.

The planning section provides the IC with situation status report updates. These reports detail what has happened at the incident, injuries, how the incident is being controlled, resources in use, and suggested actions for stabilization of the incident. For large-scale incidents or incidents of long duration, the planning section will develop an Incident Action Plan (IAP) for approval and use by the IC. IAPs provide a consistent manner to inform all entities of what overall objectives need to be accomplished. IAPs are formulated for time blocks or operational periods. The results of the tasks performed during an operational period are documented and utilized in the planning process for the next operational period.

Logistics

The logistics section chief is responsible for securing the facilities, services, equipment, and materials for an incident. Usually this is accomplished by using a support and service section. The support section is responsible for medical support for incident responders, communication, and food services. The services section is responsible for supplies (food, medical, and specialized extinguishing agent), facilities, and the resources to deliver these items.

Finance/Administration

The finance/administration section chief is responsible for documenting cost of materials and personnel for the incident. This is an important function, especially when the potential for reimbursement exists from areas such as state or federal disaster declarations and insurance companies. Generally this function will be established for large-scale or long-duration incidents such as natural disasters.

Command Staff Positions

An incident commander has the ability to create staff positions to assist with an incident. These individuals report directly to the IC.

> **STREETSMART TIP**
>
> The PIO position is very important for incidents involving evacuation of people or an unusual event. The news media is a valuable resource for conducting evacuation operations. Another benefit from this is the development of a relationship with the media for publicizing other events such as fire safety education or recruitment campaigns.

> **FIREFIGHTER FACT**
>
> To implement an incident command system, many fire departments use tactical worksheets like that shown in **Figure 2-8.**

Safety Officer. The safety officer is responsible for the safety of all responders at an incident. The incident safety officer reports directly to the IC and has the authority to stop any activities that pose an immediate danger to incident responders. All other safety issues are channeled through the IC. For large-scale incidents the safety officer will develop an incident safety plan to be incorporated as part of the IAP, which will cover the overall incident objectives and strategies.

Liaison Officer. The liaison officer is responsible for communication and contact between other agencies that respond. These may include environmental protection, police, highway or transportation, and state and federal government agencies. This position is especially important for long-duration, hazardous materials, or mass-casualty incidents.

Public Information Officer. The PIO is responsible for providing factual and accurate information concerning an incident to the news media. Only one information officer is appointed for each incident, although assistants can be assigned as necessary. The Fire Chief or Incident Commander is ultimately responsible for the management of information released at an emergency. Therefore, the authority to release information should be limited to the Fire Chief, PIO, or other person designated by the Chief. When information is released, it is imperative that it preserves the integrity and dignity of any party that may be involved.

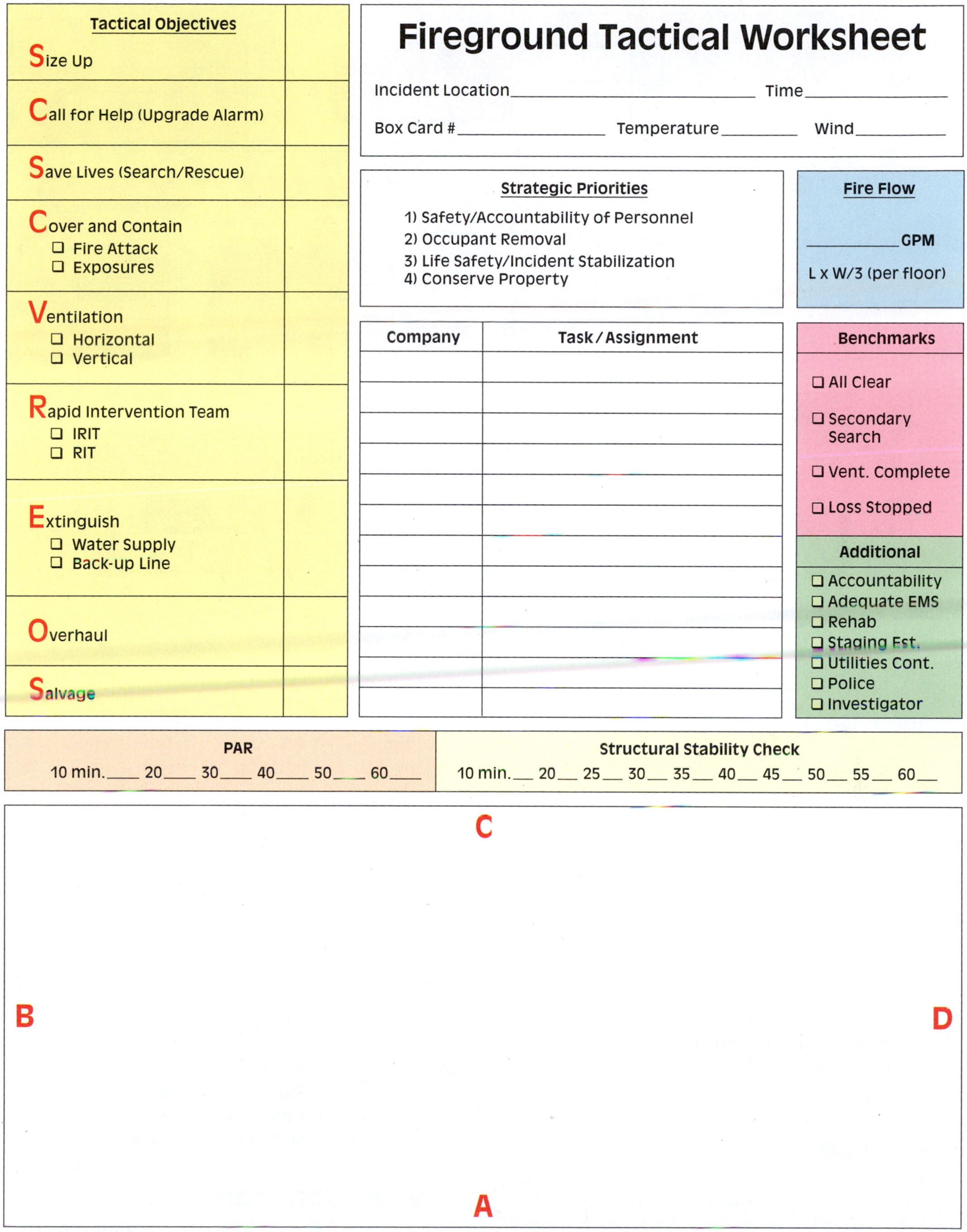

Tactical Objectives	
Size Up	
Call for Help (Upgrade Alarm)	
Save Lives (Search/Rescue)	
Cover and Contain ☐ Fire Attack ☐ Exposures	
Ventilation ☐ Horizontal ☐ Vertical	
Rapid Intervention Team ☐ IRIT ☐ RIT	
Extinguish ☐ Water Supply ☐ Back-up Line	
Overhaul	
Salvage	

Fireground Tactical Worksheet

Incident Location_____ Time_____

Box Card #_____ Temperature_____ Wind_____

Strategic Priorities

1) Safety/Accountability of Personnel
2) Occupant Removal
3) Life Safety/Incident Stabilization
4) Conserve Property

Fire Flow

_____GPM

L x W/3 (per floor)

Company	Task / Assignment

Benchmarks

☐ All Clear

☐ Secondary Search

☐ Vent. Complete

☐ Loss Stopped

Additional

☐ Accountability
☐ Adequate EMS
☐ Rehab
☐ Staging Est.
☐ Utilities Cont.
☐ Police
☐ Investigator

PAR	Structural Stability Check
10 min.____ 20____ 30____ 40____ 50____ 60____	10 min.___ 20___ 25___ 30___ 35___ 40___ 45___ 50___ 55___ 60___

C

B

D

A

FIGURE 2-8 Tactical worksheets provide the incident commander with a guide for managing an incident.

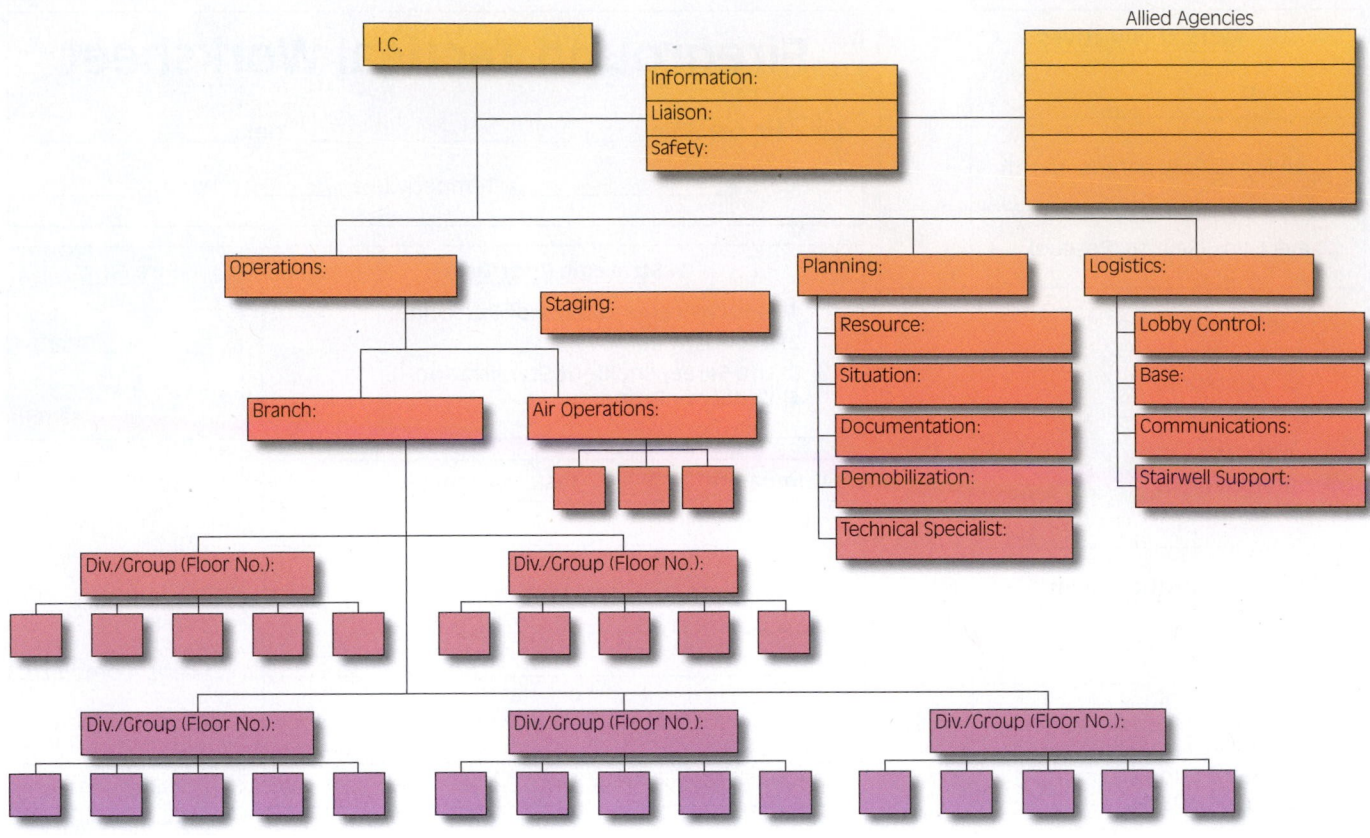

FIGURE 2-8   (Continued)

Incident Command Designations

■ *Division.* The division is responsible for all operations within an assigned geographic area. At structural incidents a division may be designated for each floor or level of a building.

■ *Group.* A functional designation to conduct a specific task such as search and rescue or ventilation. A group can operate in coordination with any division or sector.

■ *Section.* The organizational level with responsibility for a major functional area of the incident.

■ *Branch.* As an incident expands, branches are established to maintain span of control over a number of divisions or groups. A branch must have at least two divisions or groups.

> **STREETSMART TIP**
>
> Incident command terms and designations may vary by jurisdiction. For example, *sector* and *division* are interchangeable terms used for tactical assignments. Firefighters must know the terms used in their organization!

■ *Type.* A resource that has specific capabilities or equipment. As an example, a Type 1 engine company has a minimum 1,000-gpm pump, a 400-gallon water tank, 1,200 feet of $2\frac{1}{2}$ inch hose, and a 500-gpm master stream appliance with four firefighters.

■ *Task force.* Any combination of single resources assembled for an assignment. As an example, a task force for a structure fire assignment may be three engines, two ladder trucks, and one rescue company.

■ *Strike teams.* Designation for a set number of resources of the same type and kind. A water supply strike team would be five Type 1 engine companies.

■ *Crew.* A specific number of firefighters with an assigned task, but usually without apparatus. A ventilation crew may consist of six firefighters with tools but not assigned to a truck company.

Unified Command

One of the benefits of any ICS is the ability to establish what is known as a unified command structure. The **unified command** structure is used to manage

INCIDENT MANAGEMENT TEAMS

Incident Management Teams are made up of predetermined individuals to fill the command and general staff positions when an incident takes place. This allows for properly trained and qualified individuals to be available for any type of incident that may occur. Training and experience will have a bearing on the type of designation given to the team. Teams will be classified as one of five types, and the designation will determine the level or extent of an incident that the team is prepared to handle.

The five type designations are:

Type 5: Local level—this team type classification will consist of fire personnel who are trained in command and general staff functions from an immediate area. Their training and resources will allow them to have the capability to carry out the operations of an initial **operational period** of six to twelve hours.

Type 4: City or County level—a team that fits the criteria for this type designation may have members included from fire, police, and EMS from a densely populated area. Their training and capabilities will enable them to carry out operations for the initial operational period as well as during the transition to a higher-level team designation.

Type 3: State or Metropolitan Area level—teams meeting this type designation will possibly have members from response organizations such as fire, police, and EMS, as well as groups within an area determined by the Department of Homeland Security. A Type 3 team may have the capability to handle larger incidents on a state level.

Type 2: National and State level—Type 2 teams can be federally or state certified. Typically their experience, training, and resources will allow them the ability to conduct operations on state or less significant national incidents. These teams are regulated by the U.S. Forest Service.

Type 1: National and State level—these teams are able to sustain operations with minimal or no outside support mechanisms. The individuals on these teams have the highest level of training and experience. These teams are also regulated by the U.S. Forest Service.

an incident involving multiple response agencies or when multiple jurisdictions have responsibility for control of an incident. Unified command has only one IC and IAP, but it allows for agencies with jurisdiction to be part of the command structure or team.

For example, a natural disaster may involve a large number of agencies from multiple levels of government. The initial incident commander may be the chief of the fire department; however, due to the complex nature of the incident, law enforcement, building code, and social service officials may be needed for legal reasons or for their expertise. Individuals from these agencies may be designated deputy ICs or may assume command as incident problems and priorities involve their specific discipline. Regardless of how complex or large an incident is, the incident will have one and only one IC. The use of unified command structure requires knowledge and training before the incident!

LESSONS LEARNED

To survive on the fire scene, firefighters must know the roles and responsibilities of the personnel, how their fire department command structure works, and how to function as part of that command structure.

According to NIOSH case studies conducted after firefighter fatalities, some of the most common threads listed as contributing factors to firefighter deaths in the line of duty are command and organization related.

KEY TERMS

Branch   The command designation established to maintain span of control over a number of divisions, sectors, or groups.

Common Terminology   The designation of a term that is the same throughout an IMS.

Division   Command designation responsible for operations within an assigned geographic area.

Firefighter Assist and Support Team (FAST)   A company designated to search for and rescue trapped or lost firefighters. May also be called a rapid intervention team (RIT).

Incident Command System (ICS)   An organized, systematic method for the command, control, and management of an emergency incident.

Incident Management Team   A team of specially trained individuals that fill the command and general staff positions of the Incident Command System when an incident takes place.

Integrated Communications   The ability of all units or agencies to communicate at an incident.

Modular Organization   The ability to start small and expand if an incident becomes more complex.

National Incident Management System (NIMS)   A standardized system of operating at and controlling resources during an incident where multiple agencies may interact.

Operational Period   The time frames for operations at an incident.

Rapid Intervention Crew (RIC)   A company designated to search for and rescue trapped or lost firefighters. May also be called a **Firefighter Assist and Support Team (FAST).**

Rapid Intervention Team (RIT)   A company designated to search for and rescue trapped or lost firefighters. May also be called a **Firefighter Assist and Support Team (FAST).**

Span of Control   The ability of one individual to supervise a number of other people or units. The normal range is three to seven units or individuals, with the ideal being five.

Staging   Part of the operations section where apparatus and personnel assigned to the incident are available for deployment within three minutes.

Strategic Goals   The overall plan developed and used to control an incident.

Tactics   The specific operations performed to accomplish the strategic goals for an incident.

Unified Command   The structure used to manage an incident involving multiple response agencies or when multiple jurisdictions have responsibility for control of an incident.

Unity of Command   Principle that there is one designated leader or officer to command an incident.

REVIEW QUESTIONS

1. List and define the five major components of the incident command system.
2. Describe duties and responsibilities of assuming and transferring command for the ICS used by a particular organization.
3. Describe the importance of the incident command system and the origin of the system.
4. What is the difference between unified command and unity of command?
5. What are the five different types of incident command teams?
6. What personnel make up the types of incident command teams?

3

Communications and Alarms

- Learning Objectives
- Introduction
- Computers in the Fire Service
- Emergency Services Deployment
- Radio Systems and Procedures

- Records
- Lessons Learned
- Key Terms
- Review Questions

INTRODUCTION

Once the initial emergency call is received, the firefighter II must be able to complete alarm assignments using communications equipment available at the fire department. Once completed the information must be recorded accurately in an incident report. This chapter will outline the methods for receiving and carrying out an alarm as well as describing the documentation process necessary for an incident report.

COMPUTERS IN THE FIRE SERVICE

The proliferation of new and more technically advanced computer hardware, along with the development of highly specialized computer software designed especially for use by public safety agencies, has made it possible for many departments to incorporate the use of computers into their communications systems. Many fire departments have found that the use of **computer-aided dispatch (CAD)** systems, **Figure 3-1,** has made it easier to handle increased call volumes.

The computer can keep track of the location of active incidents and which units have been assigned to respond and therefore help telecommunicators to better manage their resources. At the same time, computers can create and store valuable records on each incident and other departmental activities. Computers have greatly improved the way in which statistical analyses of departmental activity are performed. Additionally, computers can provide remote locations with access to a variety of information that is stored in a main database that can be accessed when needed. This information may include maps, hazardous materials information, prefire plans, policies and procedures, mutual aid instructions, or any other information that may be useful to firefighters. This type of information is often stored

FIGURE 3-1 Computer-aided dispatch workstations such as this one are found in many high-activity communications centers.

on **mobile data computers** in all types of fire apparatus and command vehicles. Computers also allow fire personnel to access off-site databases that may be beneficial either for training or incident mitigation.

EMERGENCY SERVICES DEPLOYMENT

Once an emergency is recognized and subsequently reported or relayed to the communications center, the next step is to determine what action must be taken. A variety of methods is used to accomplish this throughout the country; however, some elements of the process are essential regardless of jurisdiction or geographic location. Identification of the situation, address verification, and unit selection must occur in order for telecommunicators to deploy the appropriate types and numbers of emergency responders.

As stated earlier, the most important information to obtain from the caller is the address. However, to deploy the most effective emergency response, the telecommunicator must also know the nature of the emergency. Emergency response organizations typically identify the most common types of situations and pre-assign a standard response to each such situation. An example would be a fire department that predetermines the routine response to all single-family dwelling fires will be three engines, one ladder, and one district chief. Based on this plan, the telecommunicators have baseline criteria from which to develop a deployment plan. In the modern fire service, deployment plans are based not only on apparatus types, but also take into consideration what equipment is carried on the apparatus, the number of personnel, and their skill levels.

The process of deploying apparatus and personnel varies greatly from one department to another. For instance, in smaller communities with a low volume of emergency response activity, a manual **run card system** may be sufficient to manage the emergency response deployment process. Manual run card systems, similar to the one shown in **Figure 3-2,** are typically comprised of a card file containing street and location information relating to a jurisdiction and predetermined unit assignments for each location.

FIREFIGHTER FACT

Enhanced 9-1-1 systems provide number identification and location information even in those cases where computer-aided dispatch is not being used.

As discussed earlier, jurisdictions with high volumes of activity may use automated systems to assist in the deployment process. CAD systems are widely used today and provide a very sophisticated method of assessing resources and making recommendations for deployment of equipment and personnel.

Some jurisdictions utilize **Global Positioning System (GPS)** technology to aid in the deployment of responders. Originally dedicated for military operations, GPS works off a system of twenty-four satellites placed into orbit by the U.S. Department of Defense. The satellites are used as reference points to calculate

STREET NAME: _Delmar Ave._   BOX NUMBER: _1128_

BLOCK RANGE FROM: _1000_ TO: _1500_

BLOCK	ENGINES	TRUCKS	RESCUE	D/CHIEF	A/CHIEF	MEDIC	OTHER
1000	1, 5, 4, 7, 10	1, 7, 10	1, 9	1, 2, 3	1	1, 4, 5	
1100	1, 5, 4, 7, 10	1, 7, 10	1, 9	1, 2, 3	1	1, 4, 5	
1200	1, 5, 4, 7, 10	1, 7, 10	1, 9	1, 2, 3	1	1, 4, 5	
1300	1, 4, 5, 7, 10	1, 7, 10	1, 9	1, 2, 3	1	1, 4, 5	
1400	1, 5, 4, 7, 10	1, 7, 10	1, 9	2, 1, 3	1	1, 5, 4	
1500	1, 5, 4, 7, 10	1, 7, 10	1, 9	2, 1, 3	1	1, 5, 4	

STREET NAME: _Delmar Ave_   BLOCK RANGE FROM: _1000_ TO: _1500_

FIGURE 3-2 Run cards show the response assignment of a variety of apparatus to specific street addresses.

positions. As long as three of the twenty-four satellite signals are attainable, a two dimensional location such as latitude and longitude can be obtained.

Automatic Vehicle Location (AVL) systems, as seen in **Figure 3-3,** utilize GPS technology to pinpoint the location of an emergency incident. AVL systems can also detect the closest response vehicle, which telecommunicators can view as one of several icons on a computerized map.

Other uses of GPS technology for the fire service include helping to coordinate street directions for units such as mutual aid who may not be familiar with a response area when assigned, as shown in **Figure 3-4;** providing the location of fire hydrants; and in some systems, even visualizing building footprints and access.

Regardless of which method of deployment is used, whether it is accomplished manually or through the use of a CAD system, a predetermined deployment of apparatus must exist. An example of a deployment table is shown in **Figure 3-5.**

FIGURE 3-4 GPS mounted in apparatus can serve many functions such as routing information and hydrant locations.

The basic elements of the deployment process remain the same in either the manual or automated systems: Verify the location and nature of the emergency and determine what resources are available.

Once telecommunicators receive a report of an emergency situation, verify the location, and determine the appropriate deployment scheme, the next step is to notify the emergency responders. As is the case with deployment plans, this process also varies greatly from agency to agency. Volunteer departments may rely on personal pagers or use an automatic telephone system that "rings" multiple telephones on a common circuit and, in some cases, a system of sirens to alert them of an emergency. **Figure 3-6** shows a volunteer wearing a personal pager. Digital pagers may also be utilized to supplement VHF pagers and provide additional information on location, nature of the emergency, or progress reports concerning an incident.

A variety of **home alerting devices** are also available. These are used by some agencies that operate either an all-volunteer or combination volunteer and paid department. A common example of this type of alerting device is shown in **Figure 3-7.**

In departments where fire stations are staffed twenty-four hours a day, some type of **fire station alerting system** is usually employed. Fire station alerting can be accomplished in a variety of ways, but should always comply with NFPA standards. In some systems, a voice message is transmitted from the communications center to a fire station via the use of a vocal alarm system. These systems typically operate via either some type of control unit connected to leased telephone circuits or a radio transmitter. **Figure 3-8** shows a receiver that would be located at the fire station. The telecommunicator, using the control device, which may be similar to the encoder shown in **Figure 3-9,** decides the appropriate fire stations to

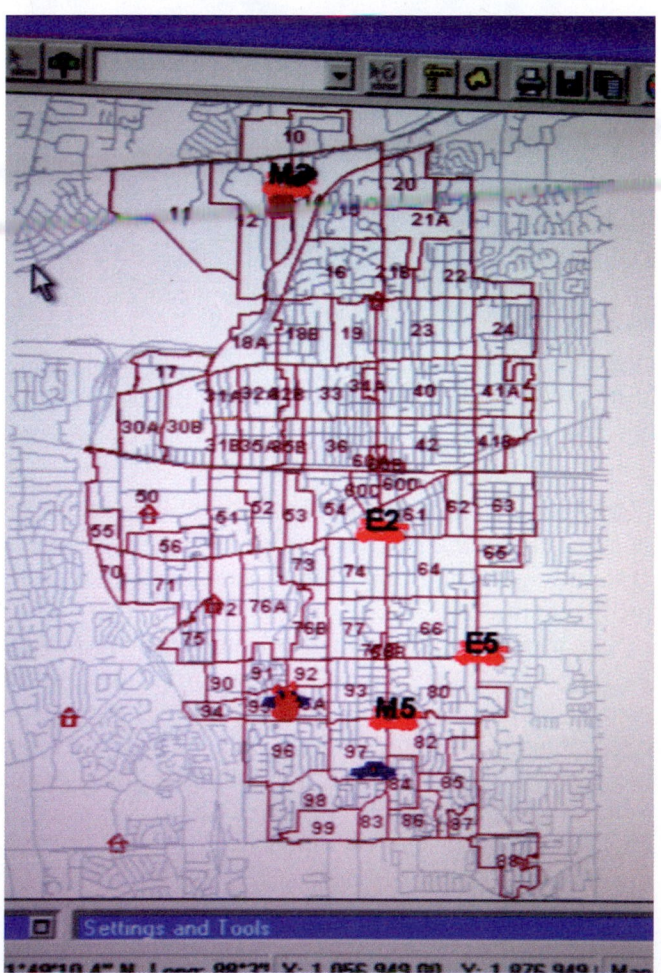

FIGURE 3-3 AVL systems help to locate the response unit closest to an incident location.

TYPE	DISP. CODE	DESCRIPTION	RESPONSE
Fire	01	Single Company Response	1E
	11	Structure Fire/Alarm (Residential)	2E-1D-1R-2T
	21	Structure Fire/Alarm (Commercial)	3E-1D-1R-1T
	51	Hospitals	4E-2D-2R-2T
EMS	08	Medical (Noncritical)	1E-(COLD)
	18	MVA	1M-1E

Legend: E — Engine or Pumper
D — Chief Officer
R — Rescue Unit
T — Truck or Aerial
M — Medic Unit
(Cold) — Nonemergency

FIGURE 3-5 Deployment tables such as the example shown are used to identify the appropriate apparatus response for the type of dispatch.

FIGURE 3-6 Some departments use personal pagers to alert personnel of the need to respond to emergencies.

FIGURE 3-7 This device is typically found in fire stations and also in some private homes and is used to call out emergency responders.

notify and activates the system. Normally some type of distinctive tone is transmitted via a public address system within the fire station to alert personnel of an incoming message.

This alert tone is followed by voice instructions over the PA from the telecommunicator. Some fire station alerting systems perform additional functions such as turning on selected lights in the fire station, opening apparatus bays, turning off appliances, and controlling traffic signals. Some systems are capable of a zoned alert that can notify only specified areas of a station such as those stations that house both fire and EMS units and personnel.

Departments that utilize CAD systems can enhance this method by installing "tear and run" printers in each fire station. These printers provide the responders with a hardcopy printout showing details of the incident and location. A printer of this type is shown in **Figure 3-10.**

FIGURE 3-8   This is yet another device used to alert emergency responders. This unit is equipped with relays that activate station lights and open station doors in preparation for a response. *(Courtesy of Shreveport Fire Department)*

FIGURE 3-9   Encoders such as this are used to control many paging systems.

Some departments also employ **mobile data terminals (MDT)** and mobile data computers in their dispatch and deployment process. These units allow dispatch information to be transmitted directly to the apparatus on a display screen or, in some cases, directly to mobile printers. Modern communications equipment such as this can provide two-way confidential information flows between the communications center and emergency responders. It also frees up a radio frequency from excessive radio traffic. **Figure 3-11** shows a mobile data terminal.

> ### STREETSMART TIP
>
> Remember that the response times of first responders are directly impacted by the amount of time required for telecommunicators to receive information, verify location, determine deployment, compose a dispatch, and transmit this information to first responders.

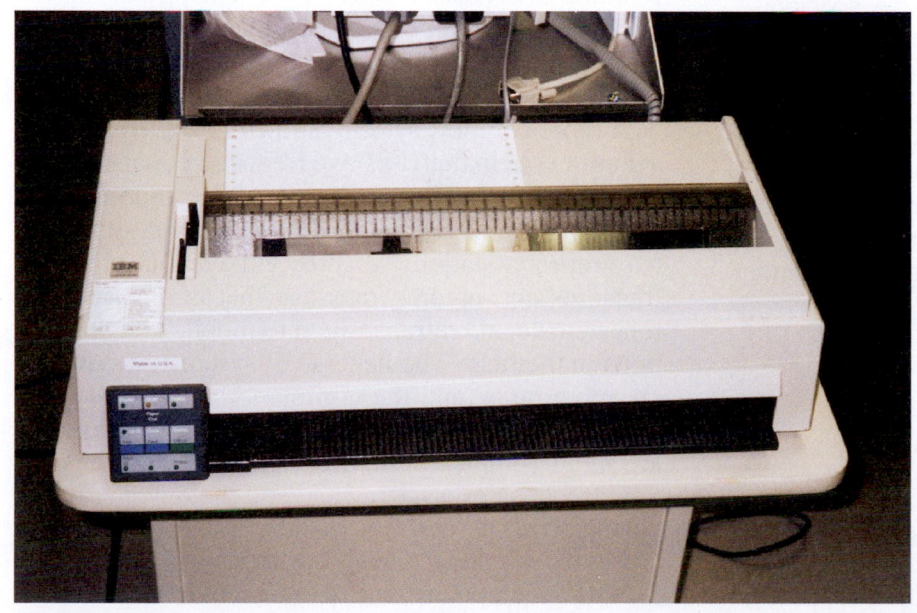

FIGURE 3-10   Printers such as this one are used to relay incident information to first responders at the time of dispatch. They are typically referred to as "tear and run" printers. *(Photo courtesy of Caddo Parish 9-1-1)*

FIGURE 3-11 Mobile data terminals such as this one are used by some departments and provide information to emergency responders while en route to incidents. Unit status can also be managed with systems of this type.

> **CAUTION**
>
> Thousands of collisions resulting in tragic accidents involving emergency response vehicles are evidence of the dangerous nature of emergency responses.

Regardless of the type of fire station alerting system in use, it is critically important that accurate information pertaining to an emergency situation be transmitted from the communications center to first responders in a clear, concise, and expedient manner.

Again, speed and accuracy are of the utmost importance.

RADIO SYSTEMS AND PROCEDURES

Once apparatus and personnel are deployed to emergency situations, the function of fire communications personnel then becomes that of providing support for the field units deployed. The primary link between the communications center and field units is the radio system. In the fire service, radios carry both verbal and digital messages. Radio systems have various components; however, every system must have at minimum a base station and antenna capable of transmitting at a power or signal strength necessary to provide coverage to all parts of a jurisdiction.

The radio frequencies that have commonly been used by the fire services are VHF low band frequencies, 33 to 46 MHz; the VHF high band frequencies 150 to 174 MHz; and the UHF frequencies, 450 to 460 MHz. These frequency ranges have provided reliable fire service communications for many years. However, as the result of growth, the fire service has experienced severe difficulties when attempting to add frequencies to their radio systems. The Federal Communications Commission (FCC) closely monitors frequency allocations, and in some cases additional frequencies in the ranges mentioned are simply not available. As the result of a need for additional frequency spectrum, other frequencies have been approved for use by the fire service. The 800-MHz frequency range is being used successfully by some departments for voice communications. Voice communications via the 800-MHz frequency range are limited by the FCC to systems using trunking technology. This type of radio system is discussed next. The FCC has mandated that by 2013 all radios must meet narrowband requirements, which will allow additional radio frequencies to be utilized. Narrowbanding will change the spacing of channels in the electromagnetic spectrum and is estimated to double the available radio channels for responders.

One common radio system used by fire departments is a simplex system that uses only one frequency to transmit outgoing messages and to receive incoming messages, **Figure 3-12.** The advantage is simplistic design, resulting in decreased system cost. The primary disadvantage of systems of this type is the limited range and interference between multiple units in the same system attempting to access the base station simultaneously.

A more complex system, shown in **Figure 3-13,** is a duplex system that uses two frequencies per channel, transmitting outgoing messages on one and receiving incoming messages on the other. This system uses base repeater stations whereby a fixed control station transmits an outgoing message that is received at a repeater site and retransmitted to mobile and portable units in the field. The benefits of systems of this type are more range and the elimination of the self-interference found in simplex systems. The disadvantages are more complex system design, the need for multiple frequencies, system cost, and the ongoing maintenance costs associated with the system.

Multisite trunking systems as depicted in **Figure 3-14** use computer processors that make the most efficient use of radio spectrum. Multiple transmitters

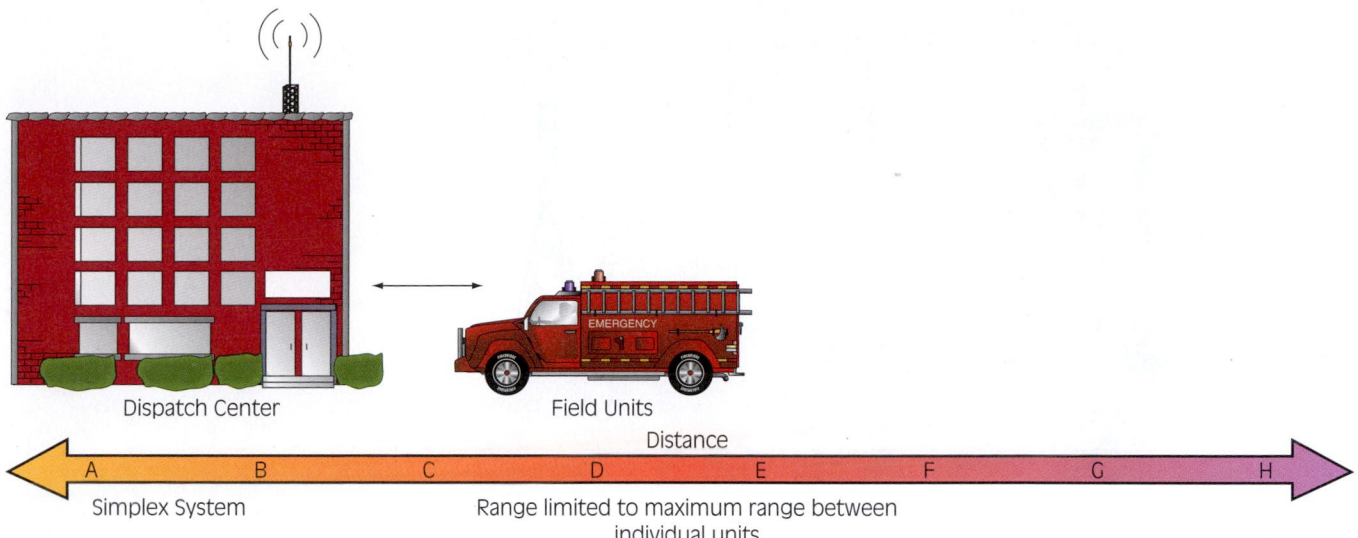

FIGURE 3-12 Simplex radio system designs such as this one are reliable and relatively inexpensive to install. However, they are limited with respect to range of operations.

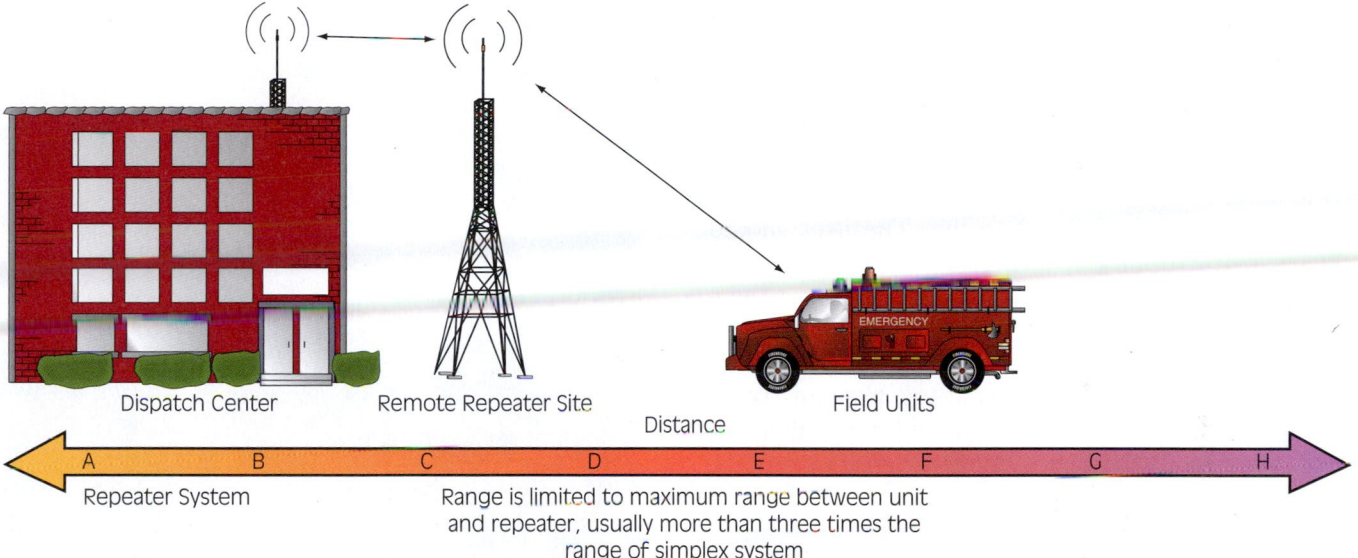

FIGURE 3-13 A slightly more advanced duplex design using multiple transmitters extends the operating range of the simplex system shown in Figure 3-12.

operating on different channels are controlled by microprocessors that sense available channels and reallocate their use as needed. In duplex systems of this type, user transmissions are transmitted on one frequency, received by the field unit on another, and retransmitted back to the communications center on the other. The user does not notice that the system is "changing" frequencies with each transmission. However, this more efficient allocation of radio-frequency resources allows the use of fewer frequencies by individual agencies. Several agencies can operate simultaneously on the same trunked radio system and not interfere with each other. The benefits of systems of this type are expanded range, more efficient frequency use, and the ability of multiple agencies to

operate on one system. Disadvantages are more complex system design and overall higher system cost.

STREETSMART TIP

The proper operation of the radio system, regardless of type, is of primary importance if information to and from the scene of the emergency is to be relayed in a timely and accurate manner.

CAUTION

On-scene personnel must listen before they talk. Routine or nonemergency traffic should not interfere with emergency communications. Transmissions should occur when airwaves are clear.

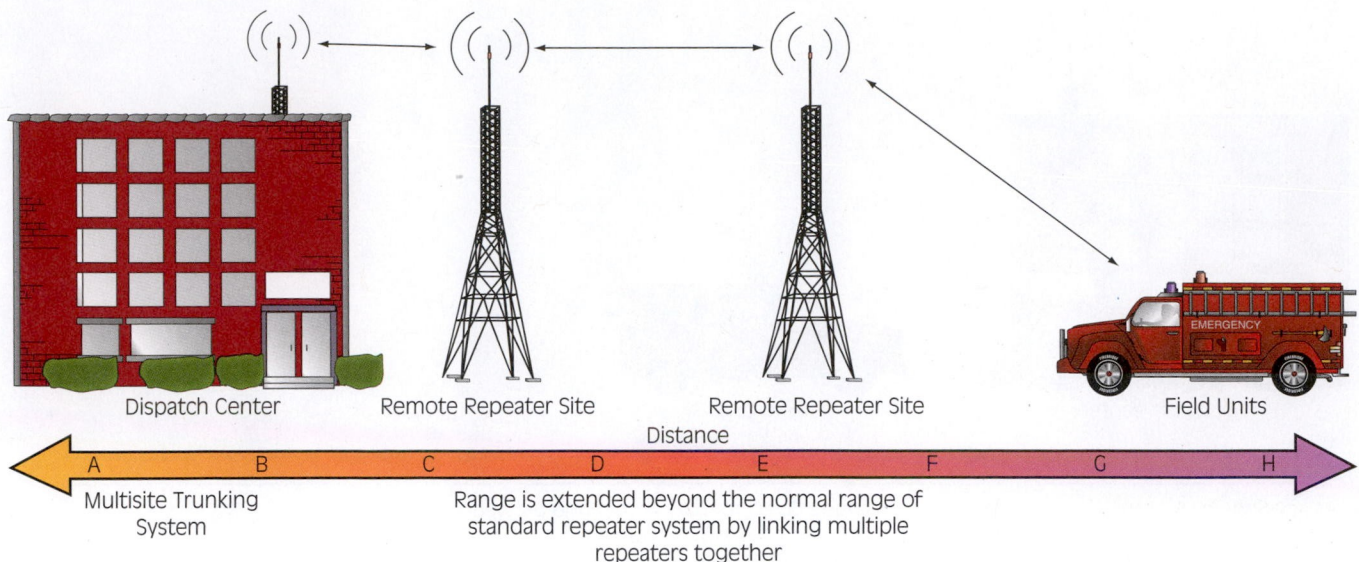

Dispatch Center   Remote Repeater Site   Remote Repeater Site   Field Units

Distance

A   B   C   D   E   F   G   H

Multisite Trunking
System

Range is extended beyond the normal range of
standard repeater system by linking multiple
repeaters together

FIGURE 3-14 Multisite trunked radio systems provide perhaps the best coverage and also offer direct benefits associated with the most efficient use of radio resources.

Proper radio discipline is very important during active incidents. When using any trunked two-way radio, it is important to depress and hold the "push to talk" button at least two seconds before talking to avoid clipping the first part of the message. The same is true at the end of the message. The user should always pause briefly before releasing the "push to talk" button to avoid clipping the end of a message. Before keying any radio, users should know what they are going to say. The golden rule is to be brief but be concise. Firefighters should avoid touching any radio antenna during transmission to avoid burns that can result from radio-frequency energy.

FIREFIGHTER FACT

Some radios are equipped with a time-out feature that forces brevity by interrupting transmissions that are longer than the preset duration allowed. This feature also prevents long-term open microphones from accidental keying.

The user should never attempt to operate a radio while eating or chewing anything and should never use slang, profanity, or jargon when transmitting a message over the radio. The message must be clearly understood. During large-scale incidents, radio economy is important. When using **mobile radios,** the user should speak clearly (not shout) across the microphone, as shown in **Figure 3-15,** as opposed to speaking directly into the microphone as pictured in **Figure 3-16.**

When using portable radio equipment, the unit should be held perpendicular to the ground with the

FIGURE 3-15 This figure shows the proper use of a mobile radio microphone.

FIGURE 3-16 Improper use of a mobile microphone. The radio microphone should not be held directly in front of the operator's mouth when transmitting.

FIGURE 3-17 In this figure, the user has positioned the portable radio properly and is speaking across the microphone.

antenna facing skyward to allow for better radio wave distribution. The user should speak across the microphone as shown in **Figure 3-17.** The use of a portable radio that is located in a radio pocket or belt clip impairs the performance of the unit. **Figure 3-18** shows the improper position of the portable radio when transmitting. In transmitting on both mobile and portable radios holding the microphone one to two inches from the mouth and at a 45° angle is important for a clear transmission.

Microphones should never be left on the seat or in other locations where they might accidentally be keyed up. Microphones and portable radios should be placed in their appropriate storage locations or in a protected location. Local agency protocols and policy will dictate how radios are to be operated in a particular jurisdiction. However, the following is a simple illustration of the "call up" method that is used by some agencies:

Telecommunicator:	"Engine One, this is dispatch."
Engine One:	"Dispatch, this is Engine One."

FIGURE 3-18 This figure shows the improper positioning of a portable radio.

Telecommunicator:	"Engine One, respond to 1234 Main Street for a trash fire."
Engine One:	"Dispatch, Engine One is responding for a trash fire at 1234 Main Street."
Engine One:	"Dispatch, Engine One is on the scene and establishing command. We have a small trash fire in the rear of the building, no exposures, no assistance needed."
Telecommunicator:	"Engine One, dispatch copies."

Ten codes are a set of radio signals preceded by the number "10" that make up a predetermined message. Considerable debate exists as to whether radio codes or clear speech should be used, but in most cases that decision is made based on the needs of each specific agency. Use of radio signal codes provides a somewhat more confidential and cryptic means of communicating. Confidentiality can be very important in situations where a responder may be unable to request assistance while in the presence of a combative or dangerous subject. However, such codes must be learned and remembered and are not usually standardized from department to department, making communications during multi-jurisdictional responses somewhat problematic. Under guidelines set by the National Incident Management System (NIMS), ten codes impede interoperability of communications and information transfer among different agencies. It is recommended that plain, common language that is understood by all be used.

Clear speech, on the other hand, is exactly what the name implies. Clear speech is used to convey information and issue instructions. The use of clear speech eliminates much of the confusion associated with the use of radio codes and, although somewhat lengthier, in most cases is easily understood by all. However, even in the "clear speech" environment, the same phrase may carry a different meaning from agency to agency. For example, "in-service" for some agencies means that a crew is going to work at an incident. On the other hand, in some agencies "in-service" means that the crew is available for an assignment. Thus, it is important not to assume anything during radio communications.

A firefighter at the scene of an emergency may not always perform tasks or assignments under direct supervision. While performing these duties, the firefighter may recognize the need for additional support or resources. The firefighter must be capable of identifying the needed resources and effectively communicating through the chain of command in accordance with local operating guidelines and the incident management system. The firefighter is expected to be able to communicate clearly and accurately when using any communications device during an emergency incident.

Another important function associated with the radio systems is the issuance of emergency evacuation signals to on-scene personnel who may be subject to imminent danger. Some departments use the radio system to transmit electronic tones that are intended to attract the attention of firefighters and alert them of the need to evacuate to a safe area. This works relatively well in most cases. However, a typical emergency scene is very noisy and not everyone on the scene has a portable radio. Another system used by some departments employs the use of apparatus air horns. Three bursts on the air horn indicate the need to evacuate. Regardless of the type of system used, it is critically important that firefighters learn it and be familiar with how it is used. Firefighter safety is paramount in all operations.

The use of two-way radios has grown greatly over the years. Routine administrative traffic, which is necessary in the daily operations of a department, should never be allowed to interfere with emergency operations. In departments that have access to multiple radio channels, emergency operations should be assigned to a separate channel dedicated for use on that scene only. This greatly improves the ability for incident commanders to communicate with on-scene personnel and also minimizes the threat of interference from some other source.

Radios and radio systems are evolving as the advances in technology grow. Newer radio systems can identify radios assigned to a particular apparatus to help in tracking transmissions. As a result of the technology, some radios incorporate an emergency feature that alerts the emergency communications center when a crew is in trouble or encountering an emergency. Firefighters should spend time familiarizing themselves with the radio system in use in their department and the many features associated with the equipment and the system.

NOTE

In summary, firefighters must consider several important factors when transmitting messages across mobile or portable radios. First, the information provided to an emergency communications center or to other emergency service units must be accurate, clear, and complete. The radio transmissions must also be within the time parameters established by the department or local jurisdiction if such policies and procedures exist. Firefighters should consult their department or jurisdiction's policies on time parameters for radio communications, specific terminology, and proper designations for units operating on a radio system.

RECORDS

Complete and accurate communications center records should be maintained on all responses. It is considered routine practice in most communications centers to record all emergency telephone and radio traffic either in some type of manual log book or on magnetic or digital recording devices for future reference. Additional information may be obtained from bystanders as well as occupants and should be recorded for later use.

In most states, fire reports are considered public record and as such are available to newspaper reporters, insurance adjusters, lawyers, and others. Records of this type speak for the department. A well-written report will be very valuable when years later a firefighter is called to testify at a deposition regarding legal action resulting from an incident.

The consequences of inaccurate reporting include:

- *Information inaccuracies in local reporting contribute to inaccurate data used in national reports.*
- *Conflicts between the report itself and the recollection of the actions taken by on-scene responders.*
- *Inaccurate reporting has a negative effect on the credibility of the reporting agency.*

The following is the minimum information that should be recorded and maintained as the legal record of an incident.

- *Time call received.* The time the telecommunicator received the call from the caller.
- *Units dispatched.* The unit numbers of the units dispatched to the incident.
- *Dispatch times.* The time of the initial dispatch and the times of any additional units subsequently dispatched.
- *Arrival times.* The arrival times of all units at the scene and the unit initiating the incident management system.

- *Command post information.* Commander, location, and so on.
- *Requests.* For example, for additional fire department units or for other agencies or services.
- *All clear time.*
- *Under control time.*
- *Back in service times for all units.*

This information can be used to complete a more detailed National Fire Incident Report or other appropriate reports approved for use by a particular state or jurisdiction.

National Fire Incident Reporting System (NFIRS)

A National Fire Incident Reporting System was put into place by the U.S. Fire Administration in 1976 to be used as a tool for assessing the fire problem in the United States. It serves as a database of fire incident information as it pertains to the United States. The uses of this data vary and are widespread depending on the end user. Along with the U.S. Fire Administration, a volunteer group known as the National Fire Information Council (NFIC) helps to administer the program. Local fire departments submit NFIRS data to their state office (usually the state Fire Marshal) who in turn consolidates the data. All fifty states, as well as Washington, DC, report NFIRS data. A common and consistent format of recording data is integral to the accurateness of the reports. For this reason, a common coding procedure has been developed for recording information on fire reports. This coding procedure is not only consistent with NFIRS, but it is also the format presented in NFPA 901, Uniform Coding for Fire Protection. Firefighters should be familiar with the coding procedures used by their department or agency, such as the NFIRS, as these codes are important in fire incident reports and documentation.

Figure 3-19 shows an example of a "manual" dispatch card that is used to record incident information. This type of logging information is typically found in smaller departments and is also used as backup systems for departments that rely on CAD and other automatic systems.

Inc. Number: 9807670				Under Control Time: 10:20		
Address: 1234 Delmar Ave						
Location: CITY HALL				Dispatch Code:		
Complaint: FIRE ALARM (21)						
Reported By: Joe Smith				Call Back Number: 555-5555		
UNIT	I/ROUTE	ARRIVE	TO HOSP	AT HOSP	COMP	I/SERVE
E01	1005	1007			1025	—
E04	1005	1008			1020	—
E05	1005	1006			1020	—
D01	1005	1015			1020	—
T01	1005	1007			1020	—
R01	1005	1007			1020	—

FIGURE 3-19 A manual incident card that is used to maintain a record of event history.

LESSONS LEARNED

The manner in which calls are answered and information processed has a direct impact on citizens' impressions of the department. In the case of emergency incidents, information that is accurately collected and rapidly transmitted to first responders is paramount to the successful resolution of the incident.

The information transmitted to the firefighter will play a vital role in the successful response to an emergency incident. Ultimately, an accurate incident report can be documented from notification through response and mitigation.

KEY TERMS

Automatic Vehicle Location (AVL) System that utilizes GPS technology to pinpoint the location of an emergency incident as well as the response vehicle that may be closest according to a computerized map.

Computer-Aided Dispatch (CAD) Computer-based automated system that assists the telecommunicator in assessing dispatch information and then recommends responses.

Fire Station Alerting System System used to transmit emergency response information to fire station personnel via voice and/or digital transmissions.

Global Positioning System (GPS) System of twenty-four satellites used as reference points to calculate positions. The satellites were placed into orbit by the U.S. Department of Defense and were originally dedicated for military operations.

Home Alerting Devices Emergency alerting devices primarily used by volunteer department personnel to receive reports of emergency incidents.

Mobile Data Computer Communications device that, unlike the mobile data terminal, does have information processing capabilities.

Mobile Data Terminal (MDT) Communications device that in most cases has no information processing capabilities.

Mobile Radio Complete receiver/transmitter unit that is designed for use in a vehicle.

Run Card System System of cards or other form of documentation that provides specific information on what apparatus and personnel respond to specific areas of a jurisdiction

REVIEW QUESTIONS

1. What baseline criteria must telecommunicators utilize when developing a deployment plan?

2. List and describe the minimum information that should be recorded and maintained as the legal record of an incident.

3. When operating radio equipment, emergency personnel should always do what before talking?

4. How should a user speak when using a mobile radio?

5. Define NFIRS and describe its importance in today's fire service

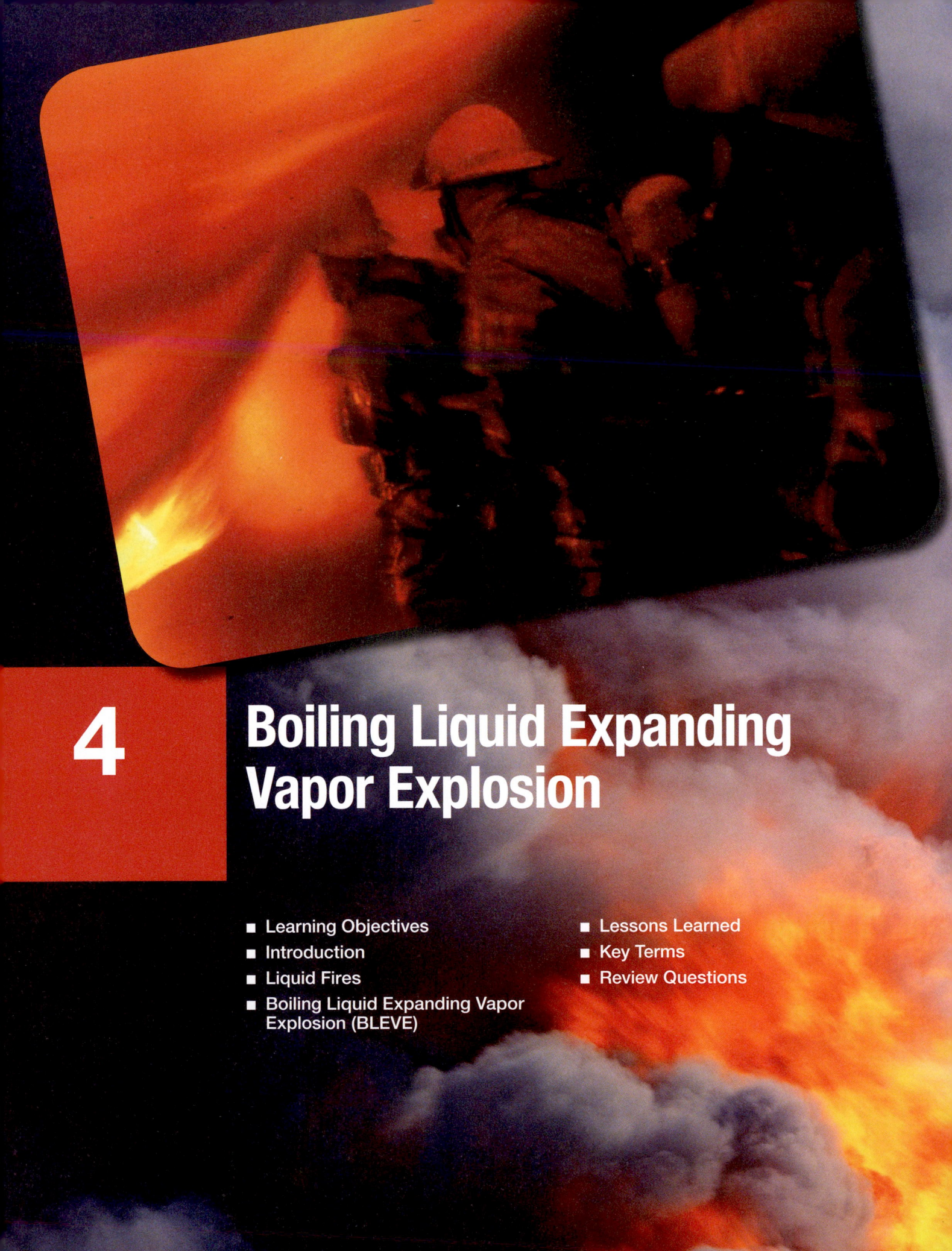

4

Boiling Liquid Expanding Vapor Explosion

LEARNING OBJECTIVES

After completing this chapter, the reader should be able to:

4-19   Describe the conditions leading to a boiling liquid expanding vapor explosion.

*The FF I and II levels, as defined by the NFPA 1001 Standards, are identified in different colors:
FF I = black, FF II = red, additional information = blue.*

INTRODUCTION

Prior to reading this chapter, review the following section/sub-sections in Section I, Chapter 4:

- Fire Defined
  - Fuel
    - Liquids
    - Gases

Also, review the following key terms in Section I, Chapter 4: *boiling point* and *evaporation*.

During an emergency response, a boiling liquid expanding vapor explosion (BLEVE) can prove to be a catastrophic incident. This chapter provides an understanding of the fire dynamics of liquid fires and BLEVEs.

LIQUID FIRES

Liquid fuels must vaporize in order to burn. **Vaporization** is the process in which liquids are converted to a gas or vapor. The rate of vaporization is dependent on many factors, but heat will always accelerate the rate. If the vapors are above their flashpoint—and within their flammable limits—they will ignite with the introduction of a spark or flame. If the vapors are above their ignition temperature and in the right mix with air, they will self-ignite. The growth of a liquid fire is dependent on the container holding the liquid. If burning liquid leaves the container, it will flow and spread the fire to other materials—called a running fuel fire.

Most class B liquid fires don't mix well with water. Using water can actually liberate more vapors and intensify the fire. Class B fires are typically extinguished by coating the liquid with an agent (foam) so that the liquid vapors cannot escape and mix with air. Specific foam types have been developed for various liquid fires, although the method of suppressing vapors is the same.

When liquids are held in a closed vessel that is being heated, the liquid inside will eventually boil. The boiling creates **vapor pressure** in the closed container. If

the pressure is increased until it reaches a point where the vessel can no longer withstand the pressure exerted against it, the vessel will fail. When the vessel fails, the ensuing release of vapor and liquid can be very violent, causing the explosive release of tremendous forces. This explosive release is called a **BLEVE** (boiling liquid expanding vapor explosion), **Figure 4-1.**

> **FIREFIGHTER FACT**
>
> The most common example of a BLEVE is cooking popcorn. The liquid inside the hard kernel shell heats, boils, and exerts pressure against its container until the shell fails. The result is a cooked kernel that escapes its container as the pressure inside equalizes with outside pressure.

If the liquid in a vessel is not flammable, the container rupture can still be violent, resulting in a force that can send container fragments great distances accompanied by a shock wave. If the liquid is flammable, the container failure will exhibit a fireball that adds a fire extension element to the problem. The BLEVE of a railroad tank car containing propane can exhibit a fireball and shockwave that can destroy everything for several square blocks. If the liquid is a hazardous material, still another set of variables must be addressed. BLEVE prevention includes the use of adequate cooling streams to keep the vessel and liquid cool.

BOILING LIQUID EXPANDING VAPOR EXPLOSION (BLEVE)

When tanks, trucks, tank cars, or other containers are involved in a fire situation, there are a number of hazards. One very deadly hazard is a BLEVE, **Figure 4-2.** A large number of firefighters have been killed by propane tank BLEVEs, and when a BLEVE occurs it usually results in more than one firefighter being killed at a single incident. The type of container and the product within the container will dictate how

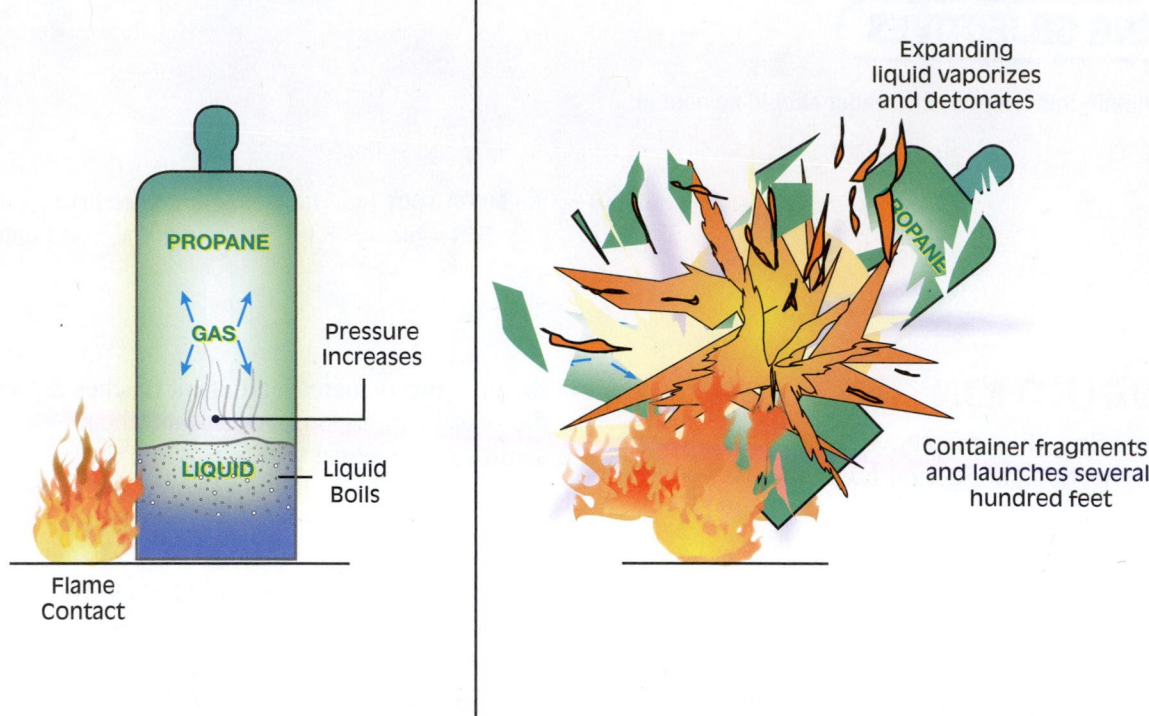

FIGURE 4-1 A boiling liquid vapor explosion (BLEVE).

severe a BLEVE may be. The basis of a BLEVE is the fact that the pressure inside the container increases and exceeds the maximum pressure the container was designed to handle. The contents are violently released, and if the material is flammable, an explosion or large fireball occurs. In the recent past there have been several incidents involving BLEVEs that resulted in emergency responder deaths and injuries, thus emphasizing the need to recognize and prevent this event before it occurs.

Another phenomenon that transpires with containers is known as violent tank rupture (VTR), which occurs with nonflammable materials. The concept is the same as a BLEVE, but there is not a characteristic fireball and explosion. With both a BLEVE and a VTR there is some form of heat increase inside the container, typically from a nearby fire. The fire heats the container, which heats the contents. The contents will boil, which creates expanding vapors, which in turn increase the pressure inside the container. In some containers, the relief valve will activate, relieving the pressure. In some cases the pressure inside the tank is greater than the relief valve can handle and the pressure continues to increase.

One of two possibilities can occur with a BLEVE or a VTR. One is that the relief valve will not be able to handle the increase in pressure and the tank will fail. The other is that the fire or heat source that is creating the problem will weaken the container shell,

and the resulting increasing pressure will vent at this weakened portion of the container. If the heat source is a fire and the container product is flammable, when the relief valve activates the raw product coming out of the container typically ignites. This may increase the temperature of the tank, increasing the pressure. It is never advisable to extinguish the fire coming from a relief valve, as that is a safety mechanism. It is possible to cool the top of the tank near the relief valve with an unstaffed hose stream.

The difference between a BLEVE and a VTR occurs when the container fails. A BLEVE occurs with flammable liquids, such as propane. When the container fails, the vapors of the flammable liquid ignite, creating the explosion. How the container ruptures and the amount of product released will determine how severe the explosion will be. With a VTR, the container will fail. Since the product is nonflammable, the product will not ignite. The only event is a rupture of the container, spilling its contents. The container can still rocket, and the resulting release of pressure can be violent. A VTR can occur with a container of water or other "nonhazardous material." As an example, a 55-gallon (208-liter) steel drum of water, when heated, can travel several hundred feet depending on where the release point is on the container. In most cases, the bungs (screw-top caps) will release, flying a considerable distance, and the container will remain mostly intact.

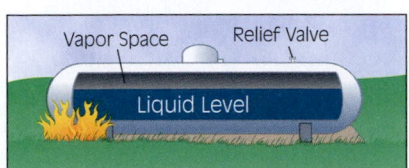

Fire impinging on a propane tank.

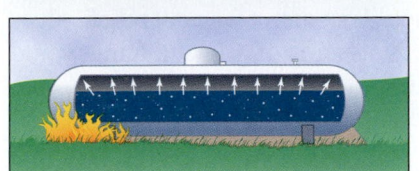

As heat increases inside of tank, the pressure also increases. The liquid will begin to boil.

As the pressure increases, the relief valve will open, releasing propane. The propane, being heavier than air, will sink.

The vapor will reach the fire and ignite.

The relief valve will ignite, also causing heat to be on the tank by that flame. The pressure will increase in the tank.

As the pressure in the tank is increasing, the tank may discolor and the pitch of the relief valve will get higher. Eventually the tank will rupture. This is known as a BLEVE.

FIGURE 4-2   Diagram of a BLEVE.

The failure of the container when impinged from a fire usually occurs as the fire is impacting the tank in the vapor space. When heat is applied to a section of the tank that has no internal mechanism to provide cooling, failure of the steel can occur. When fire impinges on the liquid portion of the tank, the liquid will distribute the heat spread internally and the steel tank typically will not weaken. The problem is that responding firefighters do not usually know the liquid level, and one cannot easily predict when a tank may fail. Any fire impingement on a tank is a serious problem, and withdrawing from the scene may be the best course of action.

This brings up another interesting note: A casual observation of BLEVEs indicates that most BLEVEs and firefighter deaths occur within the first few minutes of arrival. The clock starts ticking on the BLEVE time bomb from the first minute heat is applied to the tank. The clock does not start with the 9-1-1 call or the arrival of firefighters. The critical time for safety may have already passed before firefighters arrive on the scene. If firefighters arrive at an incident involving a propane tank on fire or being impinged by fire, they are in extreme danger. If the relief valve is not operating, the danger is even more pronounced. Operating in close proximity to a tank in this situation can be a fatal mistake. Firefighters should follow this risk/benefit analysis: *Risk a lot to save a lot, and risk a little to save a little.*

The dangers associated with a BLEVE are:

- The fireball can engulf responders and exposures.
- Metal parts of the tank can fly considerable distances.
- Liquid propane can be released into the surrounding area and be ignited.
- The shock wave, air blast, or flying metal parts created by a BLEVE can collapse buildings or move responders and equipment.

JPR 4-1: Fighting a Fire for Tanks and Other Containers Under Pressure

A tank or container can fail at any time, and it is impossible to determine the exact moment a tank is going to fail.

(For step-by-step photos of this skill sequence, see page 719)

1. Firefighters should withdraw immediately in the case of rising sound from venting relief valves or discoloration of the tank.

2. Fire must be fought from a distance with unstaffed or unmanned hose holders or monitor nozzles.

   ■ The tank should be cooled with flooding quantities long after the fire is out. A minimum of 500 gpm (1,893 liters per minute) at the point of flame impingement is recommended by the NFPA.

   ■ If the water is vaporizing on contact, firefighters are not putting enough water on the tank. Water should be running off the tank if it is being cooled.

   ■ Firefighters should not direct water at relief valves or safety devices, as icing may occur. Icing would block the venting material, which could cause an increase in pressure inside the tank.

   ■ The tank may fail from any direction, and any tank that is being exposed to a fire can fail at any moment.

   ■ For massive fire, it is recommended to use unstaffed or unmanned hose holders or monitor nozzles. If this is impossible, firefighters should withdraw from the area and let the fire burn.

JOB PERFORMANCE REQUIREMENT 4-1

Fighting a Fire for Tanks and Other Containers Under Pressure

B   Fire must be fought from a distance with unstaffed hose holders or monitor nozzles. In this photo, a liquid propane tank has overturned and the propane is being flared (burned) off before the truck can be righted. *(Courtesy of Maryland Department of the Environment Emergency Response Division)*

A   A tank or container can fail at any time, and it is impossible to determine the exact moment that failure will occur. Firefighters should withdraw immediately when increasing sound from venting relief valves or a discoloration of the tank is noticed. In this photo, a high-pressure tube trailer carrying compressed hydrogen is on fire. *(Courtesy of Maryland Department of the Environment Emergency Response Division)*

LESSONS LEARNED

Fires involving liquids require special considerations. Firefighter survival and fire attack effectiveness are dependent on firefighters understanding the fire dynamics of a BLEVE. Firefighter safety is paramount to the correct application of their understanding and knowledge of liquid fires that may evolve into a boiling liquid expanding vapor explosion.

KEY TERMS

BLEVE   Boiling liquid expanding vapor explosion. Describes the rupture of a container when a confined liquid boils and creates a vapor pressure that exceeds the container's ability to hold it.

Vapor Pressure   The amount of force that is pushing vapors from a liquid.

Vaporization   The process in which liquids are converted to a gas or vapor.

REVIEW QUESTIONS

1. What does BLEVE stand for?
2. What effects do the boiling point and vapor pressure have on the possibility of a BLEVE?
3. What are the dangers associated with a BLEVE?
4. Define ignition temperature.
5. What is the difference between a BLEVE and a VTR?

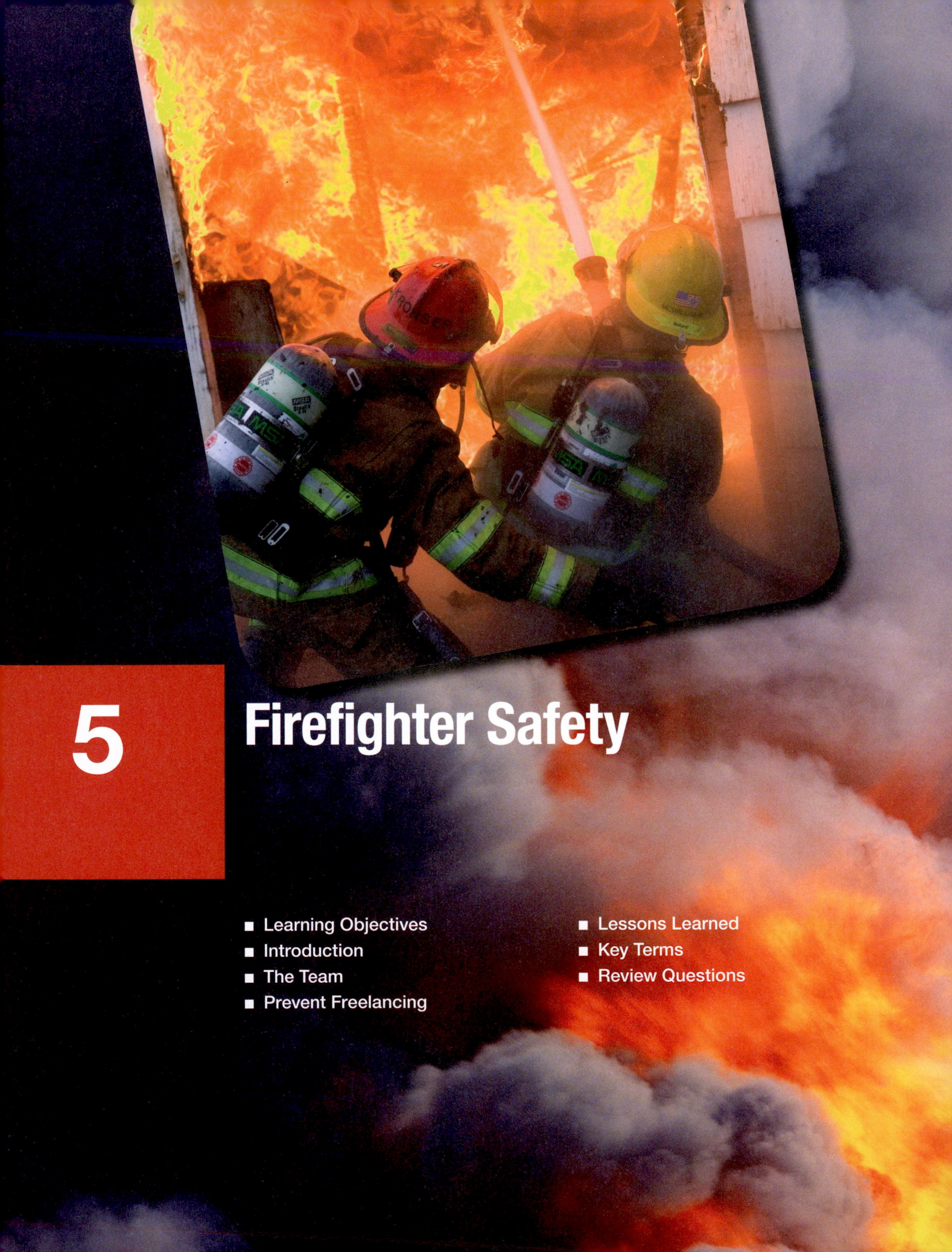

5

Firefighter Safety

- Learning Objectives
- Introduction
- The Team
- Prevent Freelancing
- Lessons Learned
- Key Terms
- Review Questions

INTRODUCTION

A firefighter should never work alone or freelance on the fireground. The presence of a team member will afford a second set of eyes and another mind at work. It is very difficult for any one person to be aware of everything that is taking place at the same time. At times the presence of another firefighter will afford the opportunity to quickly discuss the manner in which to attack the problem at hand. A team member might make the difference between being located when a floor collapses or making a misjudgment resulting in a fall down a shaft. The presence of a team member could make the difference between being able to pry open a bulkhead door or lift a heavy skylight and not being able to do it because it is too difficult for one person.

THE TEAM

The department as a whole cannot be effective in its firefighter safety effort without the support of a team approach to handling incidents. Therefore, the team that handles an incident must hold up its part of the safety partnership. This team obviously includes the individual firefighter. To ensure safety, the team should follow these procedures:

1. *Utilize an* **incident command system (ICS).** The ICS should detail lines of authority and communication, be reasonable in its span of control, and include an action plan. Section chiefs, division/group supervisors, and team leaders need to communicate progress and status. Orders need to be clear and accomplished within an acceptable risk environment.

2. *Work together and remain intact.* The "buddy" concept is imperative for firefighter safety.

3. *Look after each other.* Team members who continuously watch each other will find and address conditions that will lead to injury. Aggressive rehabilitation will provide hydration, rest, and nourishment—thereby increasing the time the team can perform safely. At the team level, issues of PPE and SCBA use or misuse should be addressed. Likewise, the team must watch each other for signs of fatigue, overaggressiveness that is dangerous, unsafe acts, and freelancing.

CAUTION

The separation of members within a team is a contributing factor to firefighter fatalities.

The individual firefighter holds the final key to making the safety partnership work. The following list includes areas that the firefighter must remember.

■ Be ready.

■ Understand and act within the chain of command.

■ Perform as trained.

■ Use an incident engagement checklist. (See Section I, Chapter 5.)

PREVENT FREELANCING

Fire and rescue work is a team effort orchestrated to an action plan developed by the incident commander (IC). Team leaders implement the plan and provide feedback to the IC. Firefighting teams follow the team leader's direction. Freelancing occurs when a team operates outside the action plan or when individuals work alone. Working alone or outside the action plan endangers individuals and the team, as **Figure 5-1** demonstrates. The individual firefighter can help prevent freelancing by *not* working alone and by supporting the action plan when part of a team.

FIGURE 5-1 Freelancing endangers individuals and the team. This firefighter is working alone in a collapse zone— for what gain?

LESSONS LEARNED

To ensure the safety of all personnel during an emergency operation, a partnership must be formed between the department administration, the working teams, and the individual firefighter. Each of these partners carries an equal weight in creating a safe atmosphere. It is then the individual's challenge to develop safe habits and attitudes that help prevent injuries and deaths.

KEY TERMS

Incident Command System (ICS)   An expandable command system used to deal with a myriad of incidents. Helps achieve the highest level of accountability and effectiveness for incident handling. Limits span of control and provides a framework to help make tasks manageable.

REVIEW QUESTIONS

1. What key concepts must an individual remember and apply to make the safety partnership work?
2. Explain how the incident management system relies on tcamwork.
3. Describe the "buddy" concept in relation to firefighter safety.
4. Name the three partners that work together to achieve firefighter safety and give examples of how the individual firefighter can contribute to firefighter safety.
5. How can freelancing be prevented?

9

Water Supply

INTRODUCTION

In Section I, Chapter 9, you learned that water supply is one of the most critical elements of firefighting because water is the most common extinguishing agent. Water supply is important in areas with a water distribution system, and in areas without a distribution system one must be created. If there is no water supply, then hoses, appliances, and engines are useless.

Where the water comes from, how much is available, and how it is delivered are key questions. Water supply dictates the **fire flow capacity** or possible fire flow. **Pressure** is the force, or weight (lbs.) of water, measured over an area (sq. in.). The **fire flow requirement** is the amount of water required for putting out the fire. Knowing the amount of water available (through hydrant testing) and how much is needed lets firefighters select the right strategy, tactics, hose appliances, and fire streams. All are interrelated; a large fire without sufficient water leads to failure, and property and possibly lives will be lost as a result.

SOURCES OF WATER SUPPLY

Review the following sections in Section I, Chapter 9 prior to progressing through the material in this chapter.

- Sources of Water Supply
  - ☐ Ground Water
  - ☐ Surface Water
  - ☐ Mobile Water Supply Apparatus
  - ☐ Tanks, Ponds, and Cisterns
- Water Distribution Systems
- Rural Water Supply
  - ☐ Portable Water Tanks
  - ☐ Mobile Water Apparatus Operation

TESTING OPERABILITY AND FLOW OF HYDRANTS

Testing should be conducted on fire hydrants periodically to ensure they are operable and to determine the flow rate of the hydrants, **Figure 9-1.** Regular testing

FIGURE 9-1 Firefighters inspecting and servicing a hydrant.

can save firefighters time and frustration by identifying inadequate hydrants in a nonemergency situation. Testing of the hydrant may be the responsibility of the water department, but the fire department should at least have access to the results of the testing. Some fire departments do the testing and maintenance of the hydrants, whereas others have an agency policy involving strong joint programs with the water department. All testing should be coordinated between the two agencies.

The first set of tests involves the operability of the hydrants. On wet and dry barrel hydrants, the

FIGURE 9-2 Typical pump panel gauges are Bourdon gauges. *(Courtesy Loveland Fire and Rescue)*

test starts with a visual inspection for damage. Then after ensuring the hydrant valve is closed, inspectors remove all of the caps and check the threads and gaskets for damage. The inspector should also inspect for objects—cans, toys, trash, and the like are sometimes placed inside a hydrant as a "joke." The next task would be to check that all valves on the hydrant allow water to flow. A dry barrel hydrant should be checked to ensure the drain valve is working properly by determining that the water level is dropping inside the barrel after the hydrant has been shut off. For a dry hydrant, a visual inspection of the piping, caps, and gaskets should be conducted. A flow test would be the next step to ensure water is available, followed by backflushing of the hydrant to clear any debris from the strainer. Backflushing involves pumping into the outlet of the hydrant so that water flows out of its intake. This should not be done on other types of hydrants as it will contaminate the water supply. Once the hydrant is inspected, replace the caps and oil any moving parts as needed. Painting the hydrant, if necessary, will complete the inspection process.

SAFETY

It is important to inspect hydrants for objects that are placed inside the hydrant as a "joke." If not removed, these objects can cause serious damage to a fire pump, adversely affecting fire department operations and firefighter safety.

The next tests are flow tests that should be conducted during the normal operations of the supply system. There are two ways to do a flow test in a pressurized system; one is a fireground method and the other is more extensive and is done for fire department planning purposes and insurance ratings. The fireground method uses the gauges on the pumper that is using the hydrant, **Figure 9-2,** while the other requires a 2½-inch (63-mm) cap with a **Bourdon gauge, Figure 9-3,** hydrant wrenches, **Pitot gauges,** a ruler, and paper to record the data.[1] The tests are discussed in the next section.

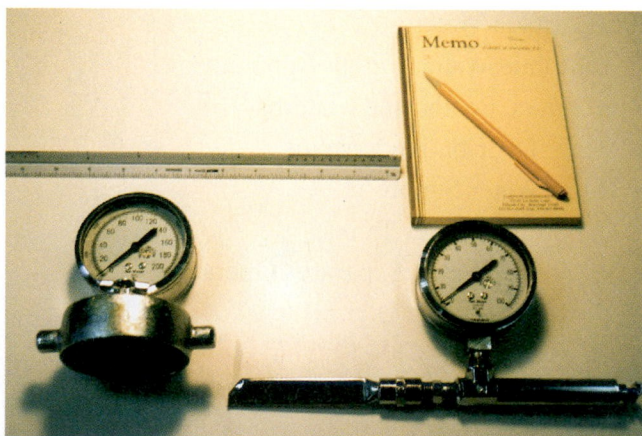

FIGURE 9-3 Pitot gauge and cap gauge with Bourdon gauge.

DETERMINING STATIC, RESIDUAL, AND FLOW PRESSURES

The fireground method of flow testing involves connecting a pumper to a hydrant and turning it on. Prior to charging any lines, the pump operator checks the compound gauge for the static pressure. **Static pressure** is the pressure in the system with no hydrants or water flowing. The pump operator then charges the first line with the desired volume, checking the compound gauge for the residual pressure. Residual pressure is the pressure left in the system after the flow and friction loss from the flow. The pump operator then compares the percentage of pressure drop from static to residual. With this information, the pump operator can determine how much water is available for additional hoselines. Two comparisons are used for percent drop as shown in **Table 9-1.**

Consider this example using the first column of drop: a static pressure of 50 psi (345 kPa) with a flow of 500 gpm (1,893 L/min) with a residual pressure of 45 psi (310 kPa). Percentage drop is 50 (static) − 45 (residual) = 5 psi drop. 50/5 = 10 percent. With a 10 percent drop, the first column would allow an additional three times or 1,500 gpm (56,775 L/min) to come from the hydrant.

Some departments prefer to use a more conservative chart due to anomolies in their water system. Some water systems may have an inconsistent sustained water flow or other problems. Therefore, to be safe, the department is more careful with the amount of water pulled from a single hydrant. The second column is more conservative and would allow only an additional 1,000 gpm (3,785 L/min) to be taken. Department SOPs should define percentage drop measurements. It is important that the residual pressure should never drop below 20 psi (137 kPa).

The second test involves testing multiple hydrants and is not conducted during fire operations. The first hydrant selected is tested with the cap gauge to take a static pressure with no hydrants flowing, **Figure 9-4.** The residual pressure is also measured on this hydrant. The second hydrant is opened and the flow measured, preferably at a 2½-inch (63-mm) outlet, with a Pitot gauge while the residual reading at the first hydrant is taken. The Pitot gauge is inserted into the stream's midpoint and out about half the distance of the diameter, **Figure 9-5.** This process would continue if additional hydrants are to be tested, gathering each

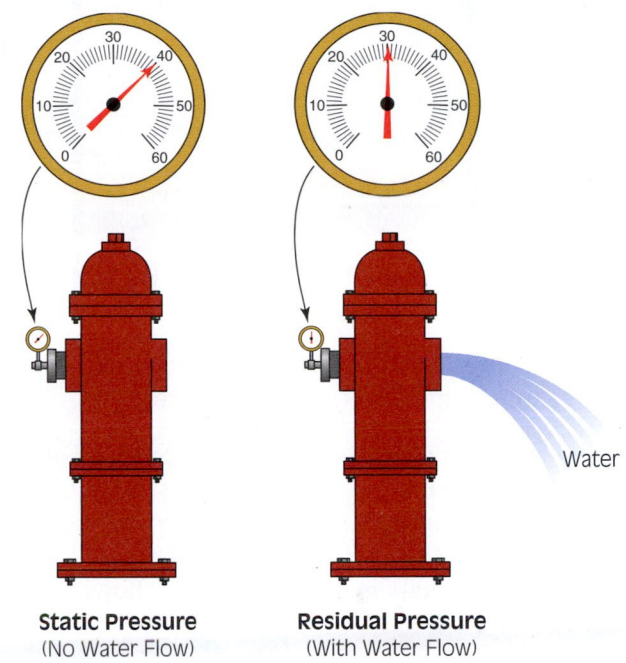

Static Pressure (No Water Flow)   **Residual Pressure** (With Water Flow)

FIGURE 9-4   Static and residual pressures.

FIGURE 9-5   Hydrant testing with a Pitot gauge.

TABLE 9-1	Percentage Drop Measurements	
Percentage Drop 1	**Percentage Drop 2**	**Amount of Additional Water**
0–10% drop	0–5% drop	Three times the amount
11–15% drop	6–10% drop	Two times the amount
16–25% drop	11–20% drop	One time the amount

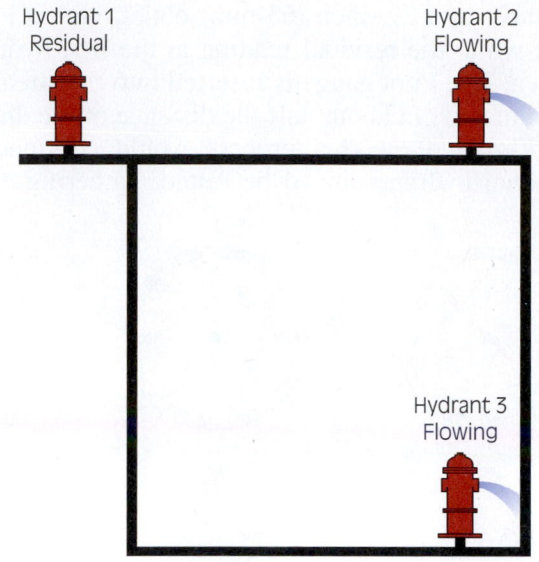

Hydrant 1
Residual

Hydrant 2
Flowing

Hydrant 3
Flowing

FIGURE 9-6 Flow testing diagram showing flowing of two hydrants with another used for residual pressures.

flowing hydrant pressure on the Pitot gauge and at the same time gathering the residual on the first hydrant, **Figure 9-6.** The next step involves the calculations of discharge for the second hydrant and beyond. Using a precalculated chart is easiest, or the following formula can be used:

$$Q = 29.83 \times c \times d^2 \times \sqrt{p}$$

where Q is quantity in gallons, c is the coefficient of the outlet, d^2 is the diameter of the outlet in inches squared, and $\sqrt{p}$ is the square root of the pressure. Each type of outlet or nozzle has a coefficient with most hydrants having a number of .90, .80, or .70.[2] The way to determine the correct coefficient for hydrants is simply to feel the inside edge of the discharge orifice. A smooth or rounded edge uses the .9 coefficient. A square edge requires a .8, and a rough lip or edge uses .7. For example, a hydrant with a 2½-inch outlet, a coefficient of .90, and a Pitot reading of 49 psi:

$$
\begin{aligned}
Q &= 29.83 \times c \times d^2 \times \sqrt{p} \\
&= 29.83 \times (0.9) \times (2.52)^2 \times (\sqrt{49}) \\
&= 29.83 \times (0.9) \times (6.25) \times (7) \\
&= 1{,}175 \text{ gpm}
\end{aligned}
$$

OBSTRUCTIONS AND DAMAGE TO FIRE HYDRANTS AND MAINS

Obstructions and damage can occur to fire hydrants and water mains as they can to any other type of system. Nature, vandals, accidents, and improper actions by members of the fire department can cause problems. Sudden events such as an earthquake or flood can undermine or break mains or hydrants. As a water system matures, small leaks can develop. Tree roots will often grow toward a water source and can eventually interfere with the piping system. Minerals or organisms can survive the water treatment process and build up inside the water pipes. Eventually, this buildup will impede water flow. Vandals can do damage by opening or closing valves that can shut off the flow of water or undermine a hydrant. Opening a hydrant in freezing weather without allowing water to flow from the hydrant can cause the hydrant to freeze and crack. Vandals can also put debris into the system damaging the hydrant or the pump.

SAFETY

Firefighters should always check the hydrant outlet, flushing it if necessary, prior to connecting a supply hose.

Accidents can crack pipes or break off a hydrant. It is possible for entire buildings to become severely damaged due to a broken hydrant and main undermining the foundation. Street damage can also result. Damage from improper actions by firefighters or others includes opening and closing valves or hydrants too quickly—creating a water hammer. Firefighters should also ensure a hydrant is opened or closed fully and should always ensure that hoses are not cross-threaded when connecting a hydrant. If any damage is suspected, the water company should be notified immediately.

LESSONS LEARNED

Firefighters must understand the relationship between the amount of water that can be supplied and the amount needed. If they do not have enough water, their efforts will not be effective. Through proper testing of hydrants, firefighters should be able to determine the different pressures available in the system, along with any likely problems. Without the proper testing of hydrants, the fire department will not know how much water may be available and at what presssure upon arrival at the incident scene.

KEY TERMS

Bourdon Gauge The type of gauge found on most fire apparatus that operates by pressure in a curved tube moving an indicating needle.

Fire Flow Capacity The amount of water available or amount that the water distribution system is capable of flowing.

Fire Flow Requirement A measure comparing the amount of heat the fire is capable of generating versus the amount of water required for cooling the fuels below their ignition temperature.

Pitot Gauge A device with an opening in its blade-shaped section that allows water to flow to a Bourdon gauge and register the flowing discharge pressure of an orifice.

Pressure The force, or weight, of a substance, usually water, measured over an area.

Residual Pressure The pressure in a system after water has begun flowing.

Static Pressure The pressure in the system with no hydrants or water flowing.

REVIEW QUESTIONS

1. Where is most freshwater found?
2. Name three surface water sources.
3. List some of the common causes of obstructions and damage to fire hydrants and mains.
4. What is the difference between static and residual pressure?
5. What is the purpose of periodic testing and operation of fire hydrants?

ENDNOTES

1. NFPA 291, *Recommended Practices for Fire Flow Testing and Marking of Fire Hydrants,* 2002 Edition.

2. NFPA 291, *Recommended Practices for Fire Flow Testing and Marking of Fire Hydrants,* 2002 Edition.

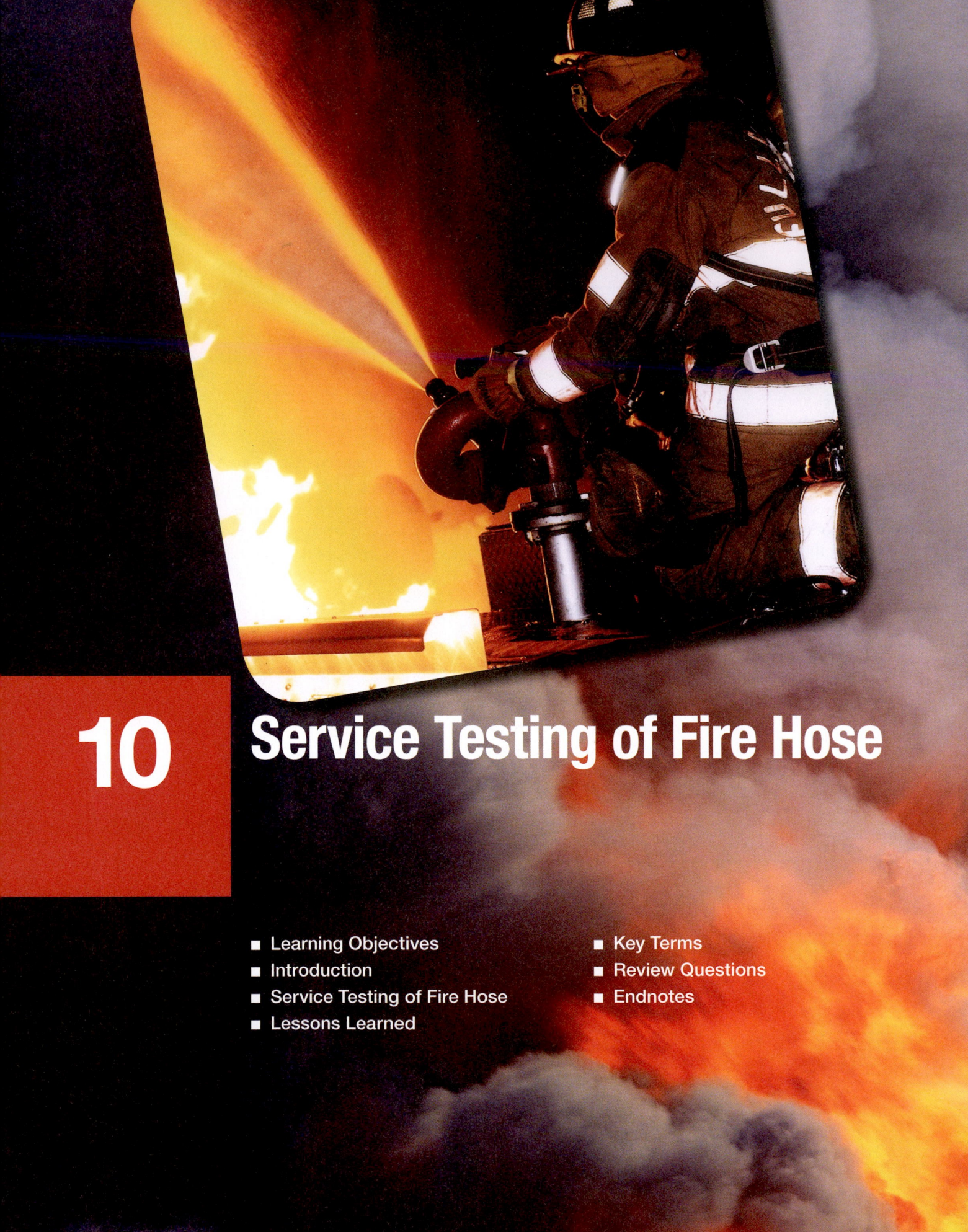

10

Service Testing of Fire Hose

INTRODUCTION

Fire hose needs to be tested prior to being placed in use and then retested annually during its lifetime. Hose also should be tested after being damaged and after repairs have been made. To ensure accuracy of the testing program and for personnel safety, a record-keeping system must be utilized. The record system should have identification numbers, dates of testing, repairs, and other important notations about each section.[1]

SERVICE TESTING OF FIRE HOSE

The testing of hose begins with a visual inspection of the hose coupling. This inspection should be done with the annual test, during routine reloading and reconnection of hose sections, and any time the hose is washed. The inspection should check thread damage. The coupling should not be misshapen and it should swivel freely. Missing, worn, or damaged gaskets should be replaced and the inspector should note any missing lugs, slippage of the hose, or loose collars.[2] As the hose is loaded, it should also be visually inspected for any type of damage. Any damaged or suspect sections should be removed from service until tested or repaired.

SAFETY

Hose testing is a destructive process that identifies weak hose by causing it to burst. Firefighters should secure the nozzle end of the hose and operate hose valves from a distance to prevent personal injury if the hoseline breaks and begins to whip about.

The service test consists of testing the hose under pressure or, in the case of a hard sleeve, under a vacuum. These tests can be done with a hose testing machine, a fire apparatus pump, or a stationary pump. Pressure testing is designed to check for hose failure, and proper safety precautions should always be taken. Firefighters should follow manufacturers' instructions or NFPA standards. The hose is laid out in a straight line and a visual inspection conducted. The hose is marked, using a new color each year, at the point where the couplings are connected to check for any slippage of the coupling under pressure, **Figure 10-1.** If the couplings do slip during service testing, the section of hose should be removed from service and marked in accordance with department policies. A hose test valve, a gate valve with a ¼-inch (6.4-mm) opening drilled into it, is placed on the discharge of the pump. This hose test valve is designed to limit the flow rate of the water into the hose to reduce chances of injury if the hose fails during the test.

The hoselines are attached to the pumping device with no more than 300 feet (91 m) of length per line and a nozzle or test cap with a bleeder valve at the discharge end. The hoseline is charged to about 45 psi (310 kPa) with the nozzle open to bleed off any air and then closed. The hoseline and couplings are checked for leaks with couplings tightened with a spanner if necessary. The hose is now ready for the pressure testing.

Each new length of hose made after 1987 has its service test pressure rating stamped on it, **Figure 10-2.** In most cases, this is about 110 percent above its maximum operating pressure and maintained for at least three minutes.[3] SOPs should define maximum operating pressures for hoselines and pumps for each department. After the test is completed the hose has the current date written on it, and it is cleaned, dried, and returned to service or rolled and put in storage.

Hose that is worn, damaged, or does not pass service testing should be removed from service. Consult department policies on removing hose from service. Many departments have firefighters mark the hose as "out of service" with a tag, marker, or other identifying method. Out-of-service hose should be kept separate from sections of hose that are "in service" and safe for operational use.

When conducting service testing of hose, firefighters should ensure that the results of the hose testing are recorded. Consult department policies and procedures for the appropriate method of recording the hose testing results.

Hard sleeves are tested by being connected to a suction source and capped at the opposite end with a transparent cap. A vacuum of 22 inches of mercury (74.5 kPa) must be reached and maintained for 10 minutes while the inner lining is viewed for any collapse. If the hard sleeve is used under positive pressure, it must pass a service test of 165 psi (1138 kPa).[4]

FIGURE 10-1 The hose is marked at the point where the couplings are connected to check for any slippage of the coupling under pressure.

FIGURE 10-2 Each new length of hose made after 1987 has its pressure testing rating stamped on it.

LESSONS LEARNED

Firefighters must understand proper procedures for safely conducting fire hose testing and the equipment that is needed to complete the tests. Documentation is one of the most important aspects during the testing process and all firefighters must know the documentation process, including proper recording of results as well as what must be recorded when removing hose from service.

KEY TERMS

Fire Hose   A flexible conduit used to convey water or other agent from a water source to the fire.

REVIEW QUESTIONS

1. Describe the testing of fire hose.
2. Explain the importance of documenting fire hose that is being taken out of service.
3. Describe each of the steps associated with fire hose testing.

ENDNOTES

1. NFPA 1962, *Standard for Care, Use and Testing of Fire Hose Including Couplings and Nozzles,* pp. 1962–7 (Quincy, MA: National Fire Protection Association, 2003).

2. NFPA 1962, *Standard for Care, Use and Testing of Fire Hose Including Couplings and Nozzles,* pp. 1962–7 (Quincy, MA: National Fire Protection Association, 2003).

3. NFPA 1962, *Standard for Care, Use and Testing of Fire Hose Including Couplings and Nozzles,* pp. 1962–9 (Quincy, MA: National Fire Protection Association, 2003).

4. NFPA 1962, *Standard for Care, Use and Testing of Fire Hose Including Couplings and Nozzles,* pp. 1962–9 (Quincy, MA: National Fire Protection Association, 2003).

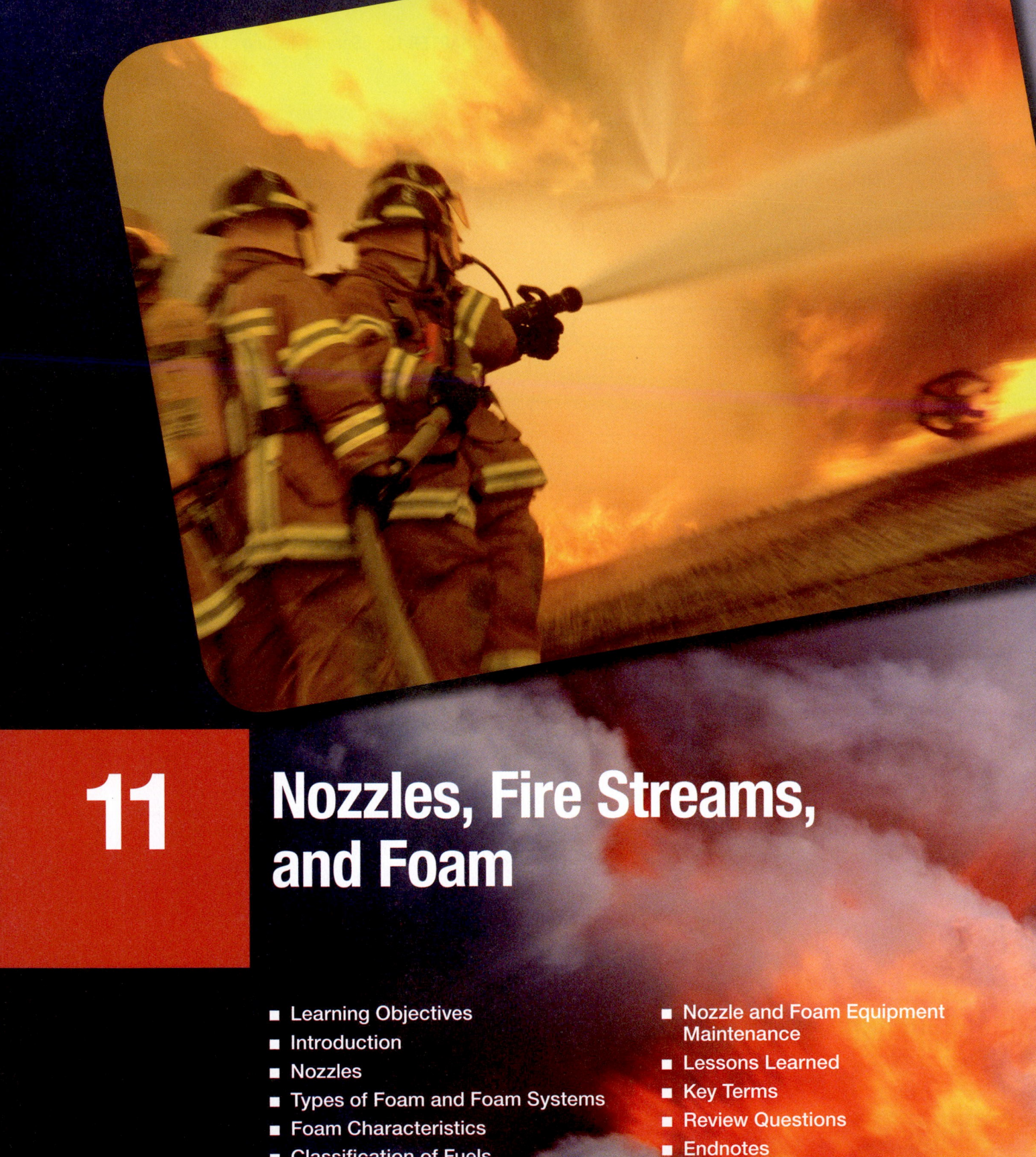

11

Nozzles, Fire Streams, and Foam

- Learning Objectives
- Introduction
- Nozzles
- Types of Foam and Foam Systems
- Foam Characteristics
- Classification of Fuels
- Application of Foam

- Nozzle and Foam Equipment Maintenance
- Lessons Learned
- Key Terms
- Review Questions
- Endnotes

LEARNING OBJECTIVES

After completing this chapter, the reader should be able to:

11-15   Identify the correct nozzle and hose for various structural fire operations.

11-16   Identify the different types of adapters and appliances used for various fireground applications.

11-21   Define a plan of action for fighting fire in terms of attack modes and styles.

11-22   Define *foam*.

11-23   Describe the types of fuel and the foam concentrates that work on each type of fuel.

11-24   Describe the various types of foam concentrates and their limitations.

11-25   Explain the principles of foam and foam making for fire suppression and incident mitigation.

11-26   Demonstrate the ability to disassemble, inspect, and clean foam-generating equipment.

11-27   Demonstrate the process for preparing the foam concentrate and assembling foam-generating equipment required for foam operations.

11-28   Describe the foam application methods used for various fuel incidents.

11-29   Demonstrate foam application techniques for fighting various fuel incidents.

11-30   Approach and retreat from hazardous spills while maintaining team integrity.

11-31   Describe the hazards associated with firefighting foam use.

11-32   Explain the methods used to mitigate hazards when applying foam.

11-33   Identify the causes of ineffective foam generation.

11-34   Describe measures that should be taken to reduce the possibility of ineffective foam generation.

11-35   Differentiate between the use of a fog nozzle and a foam nozzle for foam application.

*The FF I and II levels, as defined by the NFPA 1001 Standards, are identified in different colors:
FF I = black, FF II = red, additional information = blue.

INTRODUCTION

Fires are usually extinguished by using water to cool the heat that is produced. **Foam** is added to improve water's extinguishment ability or for fires involving fuels where plain water is ineffective. Water and foam, in proper quantities, are delivered using hose and nozzles to reach the seat of the fire. This chapter examines foam, various nozzles and appliances, and attack applications. Proper nozzle selection provides the attack crew with the proper tool required to fight the fire successfully. Nozzle selection is important,because each fire situation may require a different appliance.

NOZZLES

Nozzles are the appliances that allow application of the extinguishing agent. The two basic types of nozzles are solid stream (also called a smooth bore, straight bore, or solid tip) and fog nozzles. Each type, particularly fog nozzles, are available in different styles. For example, **combination nozzles** are capable of providing a straight stream or adjustable spray patterns. The type of nozzle used at a particular fire department is based entirely on user preference and each has advantages and disadvantages.

VIEWPOINT

The use of straight stream versus fog nozzles has been an ongoing controversy in the fire service for many years. Solid stream, fog, or automatic fog nozzles of various designs are all available to fire departments. Some departments choose to use one type, others use two, and still others all three. Nozzle selection results from careful consideration of the types of local fires and firefighting conditions, along with testing and experimentation. Equipment evolves and improves regularly, so fire departments must reevaluate equipment choices and fire tactics to ensure the best equipment is chosen. Nozzles are seldom used at their maximum performance capabilities, but are often used to maximize the performance of their operators. Firefighters should train to maximize the use of the nozzle.

Important factors in nozzle selection are nozzle pressure, flow, reach, stream shape, and reaction.

Nozzle pressure is the pressure required for effective nozzle operation and relates to flow and reach. Nozzles are designed to operate at a specific pressure, usually 50, 75, 80, or 100 psi (345, 517, 552, or 690 kPa). Pressure is measured in pounds per square inch (psi) or kilopascals (kPa).

Nozzle flow is the amount or volume of water that a nozzle will provide at a given pressure. Flow is critical because the amount of water provided determines the amount of heat absorbed or cooled. Some nozzles only flow a set volume at a set pressure, while others can be adjusted manually or automatically adjust their flow. Flow is measured in gallons per minute (gpm) or liters per minute (L/min).

SAFETY

Selecting the proper nozzle for the fire situation improves firefighting operations and personnel safety.

Nozzle reach is the distance the water will travel after leaving the nozzle. Greater reach is important in large rooms or during exterior fire operations. Reach is a function of the pressure, which is converted to velocity or speed, of the water leaving the nozzle. Reach is measured in feet or meters. Reach is affected by the factors of stream shape, water pressure, wind direction, gravity, and friction of the air. The angle of the nozzle can affect the reach: Maximum horizontal

reach is achieved at 32 degrees, while maximum vertical reach is obtained at 65 to 70 degrees. The objective is to reach the fire with maximum effect, not to be at the maximum distance from the fire.

Stream shape, also called stream pattern, is the arrangement or configuration of the droplets of water or foam as they leave the nozzle, **Figure 11-1.** The shape of the pattern helps determine the reach of the fire stream.

Nozzle reaction is the force of nature that makes the nozzle move in the opposite direction of the water flow, **Figure 11-2.** The nozzle operator must counteract or fight the backward thrust exerted by the nozzle to maintain control of the nozzle and to direct it to the correct location. The nozzle pressure and stream shape affect nozzle reaction.

Solid Tip or Stream

Solid tip, solid stream, or smooth bore nozzles deliver an unbroken or solid stream of water at the tip and toward the fire, **Figure 11-3** and **Figure 11-4.** The solid stream nozzle can deliver its water as a solid mass or cone of water or, when bounced off a ceiling, wall, or other object, as large water droplets. This solid mass breaks or shears apart the farther the water travels. The solid stream nozzle flow is a factor of the tip size at a certain nozzle pressure. Excessive or reduced nozzle pressures have adverse effects on stream performance. Handlines use tips from ¾ to 1¼ inches (19 to 32 mm) at 50 psi (345 kPa), and master streams use tip sizes of 1¼ inches (32 mm) and larger at 80 psi (552 kPa).

Solid stream handlines can reach over 70 feet (21 m) and master streams about 100 feet (33 m). They have the ability to penetrate through the fire's heat without absorbing that heat before reaching the target. A solid stream has less effect on a room's thermal balance, that is, the layers of heated gases in a burning room (see Section I, Chapter 4), and produces minimal amounts of steam compared with narrow and wide fog patterns. Disruption of the thermal balance of a room can cause steam burns to firefighters as heavier water vapor descends onto firefighters after discharging the water. Many fire departments have tried to minimize the possibility of thermal burns by

FIGURE 11-1 Nozzles showing the stream shape for straight, solid, and wide pattern streams.

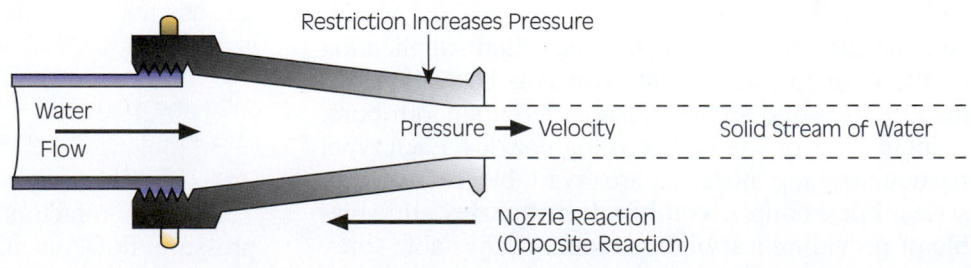

FIGURE 11-2 Nozzle reaction is the force of the water as it escapes and pushes back toward the nozzle and nozzleperson.

Restriction Increases Pressure

Water Flow

Pressure → Velocity   Solid Stream of Water

Nozzle Reaction (Opposite Reaction)

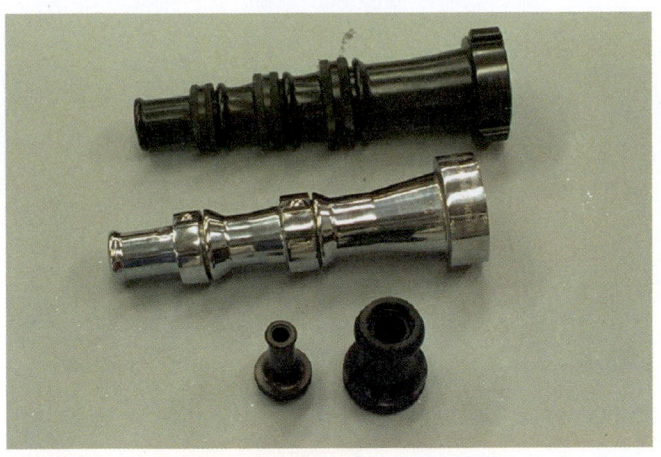

FIGURE 11-3 Various solid tips (stacked).

Solid Stream

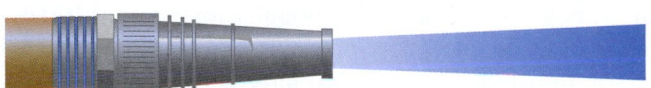

FIGURE 11-4 The flow pattern of a solid tip nozzle.

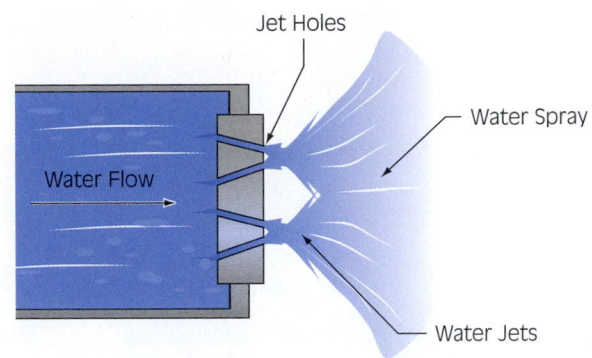

FIGURE 11-5 A fixed fog nozzle pattern of impinging design.

utilizing solid stream nozzles for aggressive, interior fire attack. A solid stream provides good penetration into burning or smoldering piles of materials. Smooth bore nozzles are more durable and easier to maintain than fog nozzles as they contain fewer parts. The disadvantages of a solid stream are the lack of volume control (other than changing tip sizes or adjusting the shutoff), the lack of fog protection when working close to the fire, and a higher nozzle reaction at the same pressure than a fog nozzle.

Fog Nozzles

Fog nozzles deliver either a fixed spray pattern or a variable combination pattern with both **straight stream** and spray patterns that can be adjusted by the nozzleperson. Fixed spray pattern nozzles are of the impinging design in which the nozzle has a series of holes at the end that creates a water spray, **Figure 11-5**. The variable fog patterns vary from the straight stream pattern, which is similar to a solid stream pattern, to a wide-angle fog pattern of at least 100 degrees, **Figure 11-6** and **Figure 11-7.** The different types of combination fog nozzles depend on the variations allowed and include constant volume or set gallonage nozzles; variable, adjustable, or selectable gallonage nozzles; and automatic or constant pressure nozzles, **Figure 11-8.** The **constant** or **set volume nozzle** has one set volume at a set pressure, for example, 60 gpm at 100 psi (227 L/min at 690 kPa), and the

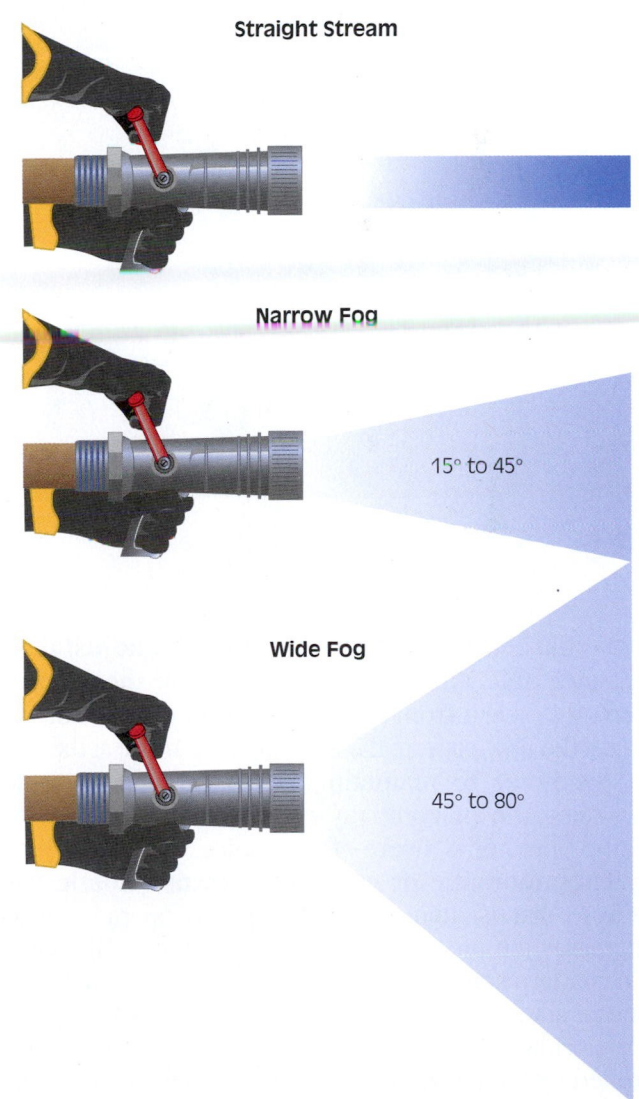

FIGURE 11-6 Variable combination fog nozzle patterns. From top to bottom: straight stream, narrow fog, and wide fog.

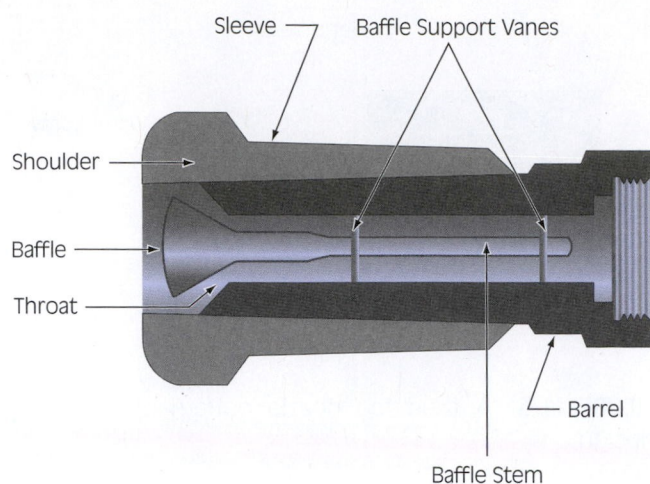

FIGURE 11-7   Parts of a fog nozzle.

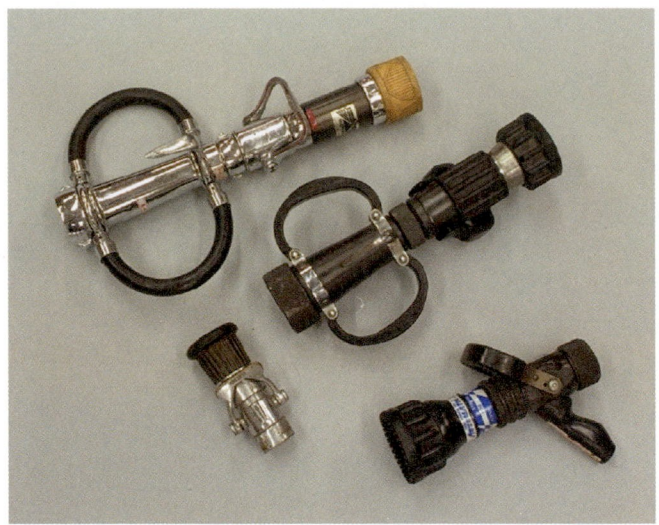

FIGURE 11-8   Various styles of fog nozzles.

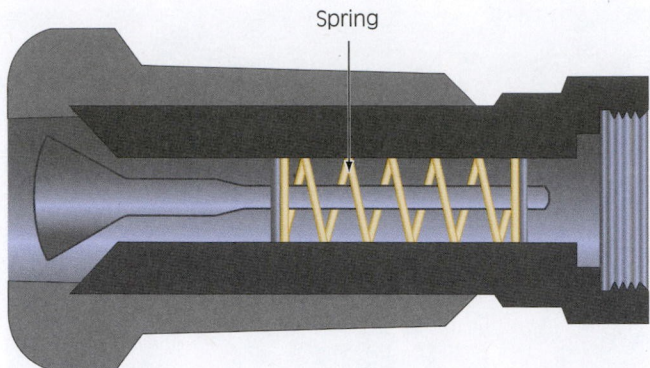

FIGURE 11-9   An automatic or constant pressure fog nozzle.

only adjustment is the pattern. **Variable, adjustable,** or **selectable gallonage nozzles** allow the nozzleperson to select from two or three flow choices and to adjust the flow pattern; the flow choice is made at the end of the hose. To maintain the correct pressure, the nozzleperson or company officer should ensure the pump operator is aware of the choice.

The **automatic** or **constant pressure nozzle** has a flow that is adjusted by the pump operator, who increases the pressure, which increases the gallonage. The pattern can be adjusted by the nozzleperson. Some newer automatic nozzles have a dual-pressure option for normal and low-pressure situations that is controlled by the nozzleperson. The automatic feature has a spring mechanism built around the baffle that reacts to increased pressure and adjusts the flow and reach of the nozzle, **Figure 11-9.** The disadvantage of these nozzles is that the pump operator, who is farther away

from the fire, controls the maximum flow. This makes it important for the company officer or nozzleperson to be in contact with the pump operator to control this flow. Fog nozzles have traditionally operated at 100 psi (690 kPa) but new low-pressure nozzles operating at 50 or 75 psi (345 or 517 kPa) have been approved. The lower nozzle pressure gives the same volume, but nozzle reaction is reduced and longer lengths of hose can be used when at the maximum pump pressure. This type of nozzle can also be beneficial in areas with poor water pressure. On the low-pressure nozzle settings, additional flow is available at higher pump pressures.

Fog nozzles provide good reach that varies with the pattern from 25 to over 100 feet (7.5 to 133 m). They also provide good penetration. Adjustable fog nozzles provide personal protection for the nozzleperson by utilizing a screening effect from the convected and radiated heat. Fog provides better heat absorption, but can change to steam before reaching the seat of the fire. However, this steam can extinguish hidden fire and works well for indirect attack. Fog streams are also excellent tools for hydraulic ventilation. Large quantities of smoke can be removed by aiming a fog cone out an open window or other opening in the structure. Firefighters must be aware that this method of removing smoke can also draw heat from the fire, **Figures 11-10A** and **B.** A 1½-inch (38-mm) fog nozzle moves more air than a 14-inch (0.36-m) smoke ejector, and a 2½-inch (63-mm) fog nozzle moves more air than a 24-inch (0.6-m) smoke ejector.

SAFETY

A fog nozzle, by nature of its design, will also cause air swirling and movement, which can disrupt the thermal balance of a room. To avoid hindering firefighting operations, as well as to protect fire crews from thermal burns, firefighters should use straight streams when attacking fires in high-heat environments or when trying to cool interior compartments while advancing on a fire.

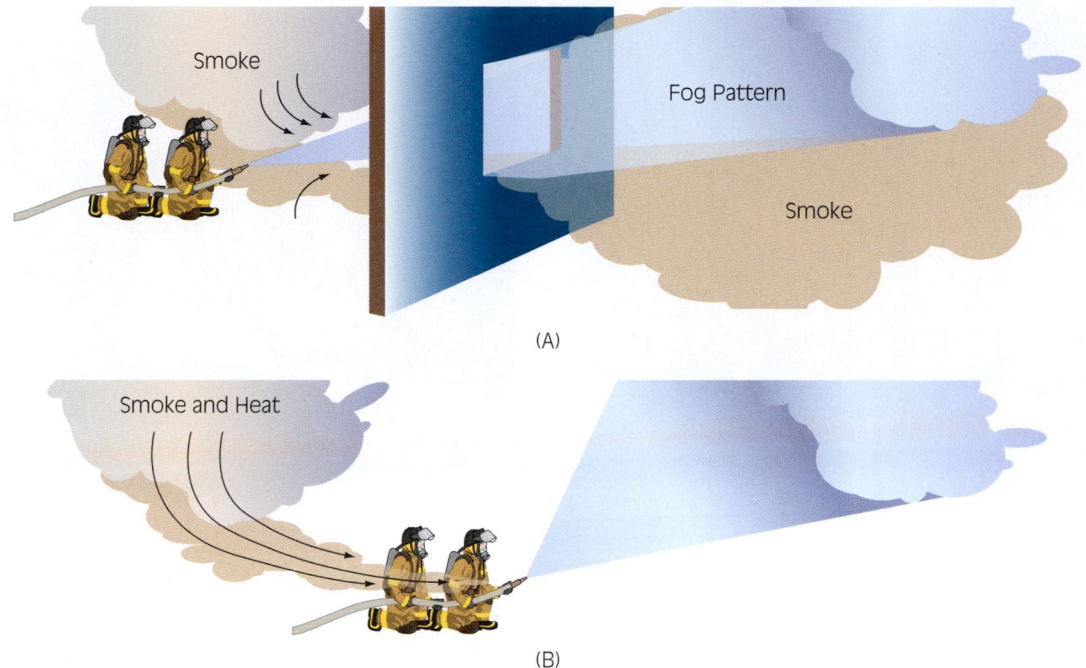

FIGURE 11-10   (A) One advantage of a fog nozzle is that, with ventilation at a window, it draws out the heat and smoke. (B) The disadvantage is that heat and gases are drawn onto the firefighters at the nozzle.

Straight Stream

The straight stream nozzle pattern on a fog nozzle, **Figure 11-11A,** creates a hollow type stream that is similar in shape to the solid stream pattern, but the straight stream pattern must pass around the baffle of the nozzle. This creates an opening in the pattern, which may allow air into the stream and reduce its reach, **Figure 11-11B.** Newer fog nozzle designs, especially the automatic nozzles, only have this hollow effect from the tip, and it is a short distance to refocus the stream to create a solid stream with good reach and penetration abilities. In fact, some are better than solid stream nozzles.

Special Purpose

Special-purpose nozzles were developed for use in limited types of situations by some fire departments. Special-purpose nozzles are not often used but firefighters should know when and how to use them. **Cellar nozzles** and **Bresnan distributors** can be used to fight localized fires in basements or cellars when firefighters cannot make a direct attack on the fire, **Figures 11-12A** and **B.** The cellar nozzle has four spray nozzles, and the Bresnan distributor has six or nine solid tips or broken stream openings that are designed to rotate in a circular spray pattern.

Piercing nozzles were originally designed to penetrate the skin of aircraft and now have been modified

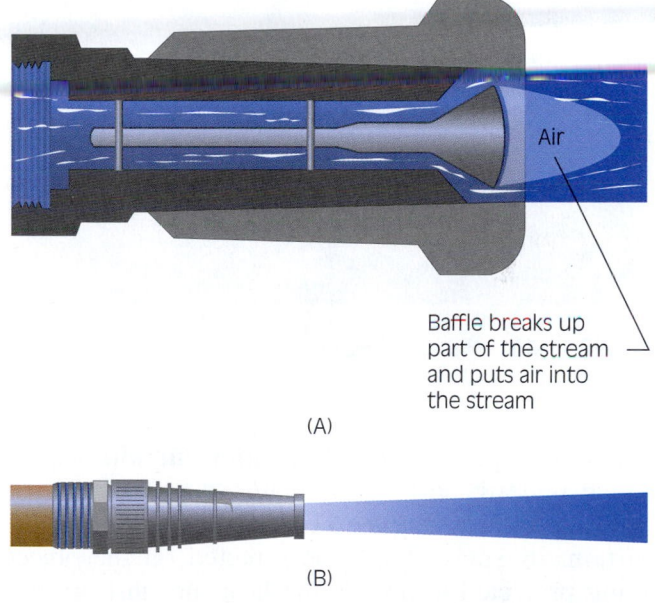

Baffle breaks up part of the stream and puts air into the stream

(A)

(B)

FIGURE 11-11   Comparison of (A) straight and (B) solid streams at tip.

to pierce through building walls and floors, **Figure 11-13.** Some have striking points that allow them to be driven through the material. Other nozzles include high-pressure fog nozzles that can operate at 1,000 psi (6,895 kPa) and industrial and forestry nozzles that are combination fog nozzles with a shutoff built into the design.

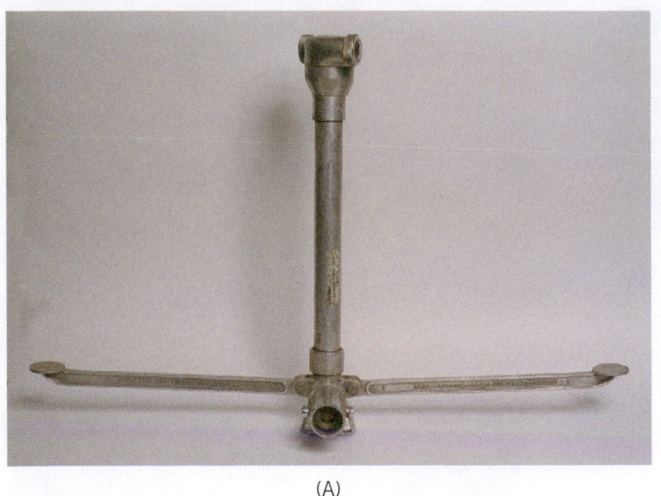

(A)

(B)

FIGURE 11-12 (A) Cellar nozzle and (B) Bresnan distributor.

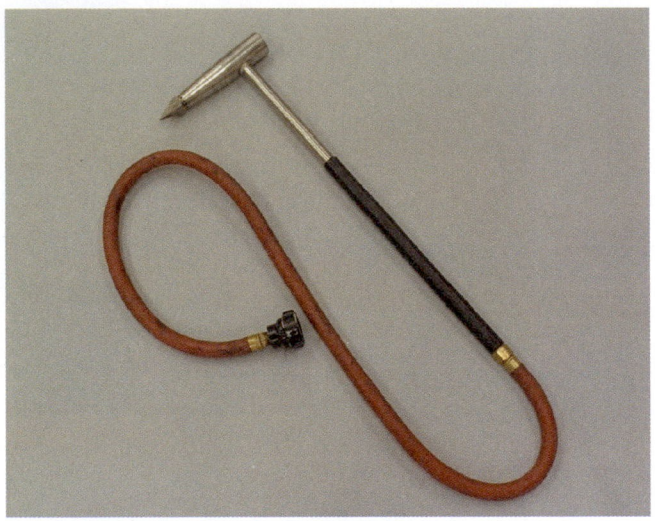

FIGURE 11-13 Piercing nozzle.

FIGURE 11-14 Water curtain nozzle.

Another special nozzle is a **water curtain nozzle,** which is designed to spray water to protect against exposure to heat, **Figure 11-14.** When using a water curtain, its spray should be directed on the object being protected to absorb the heat, not just up into the air as radiant heat passes through clear materials, such as water, without being absorbed.

Nozzle Operations

NOTE

Handline nozzles usually have shutoffs, but many other appliances do not provide shutoffs at the nozzle and the only control is the ability to aim it.

Solid stream nozzles are simple to operate with the nozzleperson controlling the shutoff, tip size, and aim. The typical shutoff for a solid tip nozzle is a lever, or bale, that controls a ball valve inside the nozzle. This type of lever allows the nozzle to be fully or partially opened or closed. For proper gallonage and flow, a nozzle size and nozzle tip are selected to match the desired flow. This is coordinated with the pump operator for proper pump pressure and safety. The tips are usually carried stacked: the smaller sizes are screwed onto the larger sizes. When choosing a tip size, firefighters should carry the smaller-sized tips in their pocket to facilitate firefighting operation.

Fog nozzles have either the lever-type open/close shutoff, which is the most common, or the rotating type of valve. Nozzles with rotating valves are commonly found

FIGURE 11-15 Close-up of rotating type of nozzle shutoff.

FIGURE 11-16 Close-up of combination fog nozzle showing the adjustments for selecting patterns and for selecting gallonage.

on wildland firefighting hose. For both types of nozzles, the fog pattern can be adjusted by rotating the nozzle barrel counterclockwise to move from straight stream to narrow fog to wide-angle fog, **Figure 11-15.** Turning it counterclockwise also closes the nozzle for the rotating valves. Clockwise rotation adjusts the pattern the opposite way and opens the valve fully for the rotating valves. Fog nozzles with variable gallonage have an additional ring on the collar that rotates from one gallonage setting to the next, **Figure 11-16.** Both gallonage and pattern adjustments can be detected in the dark because the nozzle clicks at each position. Fog nozzles have more applications than smooth bore nozzles and many firefighters consider them more effective.

OPERATING HOSELINES

Advancing hoselines was covered in Section I, Chapter 10 including some of the initial operation of the nozzle. Included was straightening the hose, having firefighters properly spaced on the same side of the line, and with the line charged, bleeding off air from the hose and nozzle, and selecting the proper pattern. Most hoselines are operated from a crouching or kneeling position, but lying, standing, or sitting positions are also used, **Figure 11-17, Figure 11-18,** and

FIGURE 11-17 Firefighter operating nozzle in the crouching or kneeling position.

FIGURE 11-18 Firefighter operating nozzle in the lying position.

FIGURE 11-19 Firefighter operating nozzle in the sitting position. Notice that the hoseline is placed in a loop to counter nozzle reaction.

Figure 11-19. In Chapter 19 (in Sections I and II), Table 19-1 highlights some of the characteristics of various size hose streams, and covers hoseline operations in fire situations.

Small-Diameter Handlines

Small-diameter handlines are typically 1½, 1¾, or 2 inches (38, 44, or 50 mm) in diameter and flow from 100 to over 250 gpm (378 to over 946 L/min).[1] When flowing at the lower volumes, these lines can be operated by one person; larger volumes require two persons.[2] Both fog and solid tip nozzles can be used for small lines. Small lines are popular because of their ease of mobility, the number of personnel required to operate them, and their ability to extinguish one to three typical rooms of fire with their flow.

Medium-Diameter Handlines

Medium-diameter hose for handlines is 2½-inch or 3-inch (63- or 75-mm) hose and can be used with solid tip and fog nozzles, flowing from 165 to 325 gpm (625 to 1,230 L/min). The 2½-inch hose has been the standard size hoseline for the fire service. Many fire departments abandoned the 2½-inch hose as a fire attack tool, advocating 1¾-inch and 2-inch lines because of increased maneuverability. However, increased Btu development of fires due to plastics and low-mass fuels has caused many to reintroduce the 2½-inch hoseline to increase gpm flow to quench the higher heat. This is especially important in large, commercial structures or buildings with high fire loading (like an auto body repair shop or high rack storage warehouse-type stores).

Medium-sized hoselines usually require two or more personnel to operate them because they are less mobile. Some departments use 3-inch (75-mm) hoses with 2½-inch (65-mm) couplings for hoselines or master stream devices.

Master Stream Devices

Master stream devices (see Section I, Chapter 10) are capable of flowing over 350 gpm (1,325 L/min), and some have capabilities of many thousands of gallons per minute. This is the main artillery of the fire service and used when large volumes of water are required. These devices must be apparatus-mounted or secured properly. Safety should be a major concern during their operation. Specific operating instructions are necessary for each type of device, so training and firefighter familiarity is a major concern. Master stream devices require only one person to operate them, but a major drawback lies in their poor or complete lack of mobility.

NOTE

The method of stream application or fire attack depends on the fire's fuel, its location, and the suppression equipment of the fire department. The size and type of hoseline and nozzle figure prominently in fire attack.

STREAM APPLICATION, HYDRAULICS, AND ADVERSE CONDITIONS

Application of fire streams vary according to the method of fire attack and conditions encountered, including environmental factors and water supply. The fire stream must have the proper pressure and flow. Firefighters—particularly the pump operator—must have a thorough understanding of hydraulics. Improper hydraulic calculations are the leading causes of poor fire streams.

Direct, Indirect, and Combination Attack

Three methods of fire attack exist: direct, indirect, and combination. **Direct fire attack** is used to attack the fire by aiming the flow of water directly at the seat of the fire, **Figure 11-20.** Direct fire attack is used on most deep-seated fires that require penetration by the hose stream.

Indirect fire attack is used to attack interior fires by applying a fog stream into a closed room or compartment, converting the water into steam to extinguish the fire, **Figure 11-21.** Firefighters apply the water at the doorway and then close the door, allowing the steam to put the fire out. The closed compartment is needed to contain the steam, thereby smothering the fire. The estimated quantity of water applied is the amount needed for total conversion to steam to fill the room. As water is converted to steam at 212°F (100°C), it expands 1,600 to 1,700 times its volume; 1 ft³ of water (0.028 m³ or 7.48 gallons or 28 L) would fill a 1,700-ft³ (48.1-m³) room, **Figure 11-22.**

At 1,000°F (500°C), this expansion is over 4,000 times. Because the entire space is filled with steam, the indirect attack should not be used when people are in, or suspected to be in, the space. It is also important to remember that this water-to-steam expansion can damage windows and walls containing the fire. Too much water applied into the room can cause the weaker portions of the structure to fail, allowing the "contained" fire to escape.

A **combination fire attack** uses a blend of the direct and indirect fire attacks, with firefighters applying water to both the fuel and the atmosphere of the room, **Figure 11-23.** The nozzleperson opens the nozzle and directs the stream toward the ceiling and then the fire with a circular, "Z," or "T" motion. This puts water on the seat of the fire and the atmosphere, while creating steam to extinguish the fire in any hidden areas. Ventilation with this combination attack controls the flow of fire gases and steam, improving the survival chances of victims, and makes this the typical attack in structural firefighting.

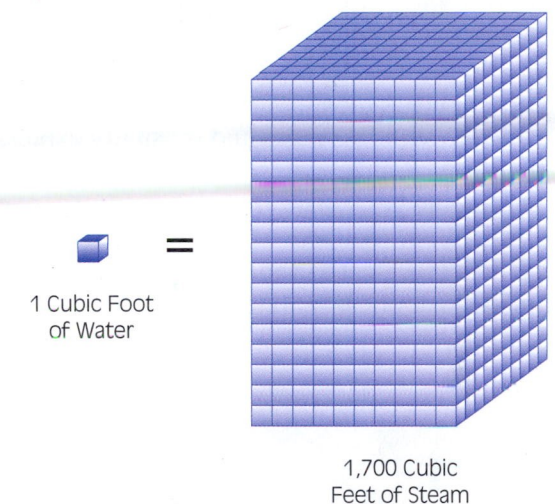

1 Cubic Foot
of Water

=

1,700 Cubic
Feet of Steam

FIGURE 11-22   One cubic foot of water in liquid form expands 1,700 times when converted to steam at 212°F.

FIGURE 11-20   Firefighter directly attacking a fire.

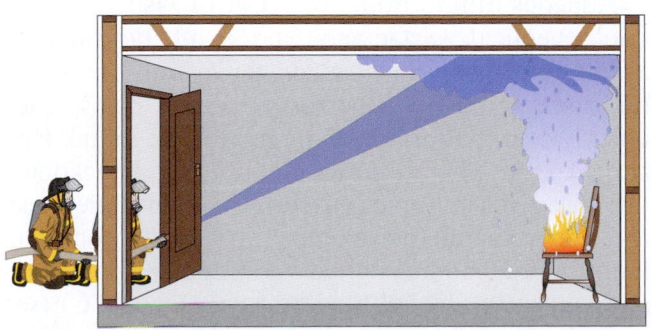

FIGURE 11-21   Firefighter using indirect attack by applying water into room and then closing the door.

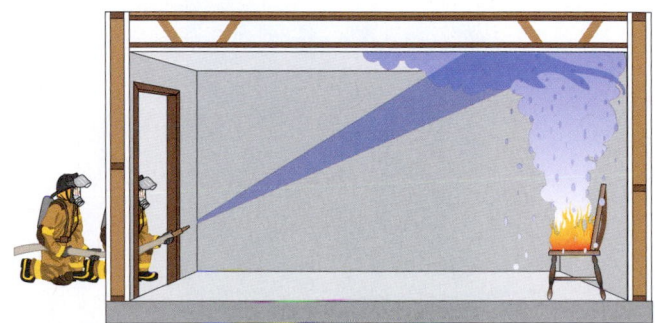

FIGURE 11-23   Firefighter using combination fire attack directing the stream from the ceiling to the fire with a circular, "Z" or "T" motion.

TYPES OF FOAM AND FOAM SYSTEMS

Class B foam is an aggregate of gas-filled bubbles formed from aqueous solutions of specially formulated concentrated liquid foaming agents.[3] The bubbles are filled with a gas, usually air, creating a blanket over the surface of the fuel. This blanket cools and smothers the fire while sealing in the vapors. The mechanical action of mixing the foam concentrate in the water makes a foam solution to which air is added. Combining these three components makes the foam lighter than the flammable liquids and gives it the ability to float over their surface, **Figure 11-24.**

Class A foam is an aggregate of gas-filled bubbles formed from aqueous solutions of detergent or soap-based surfactants used to penetrate ordinary combustible materials. This penetrating action keeps the fuel wet and reduces its ability to burn.

FOAM CHARACTERISTICS

Foam's ability to extinguish fires is based on several characteristics. The various foams available differ in their abilities to supply these characteristics. The **application rate** is the amount of foam or foam solution needed to suppress a fire and is usually expressed in gallons per minute per square foot (gpm/ft^2) or liters per minute per square meter (L/min/m^2). Application rates vary depending on the fuel type and severity and fuel depth. See NFPA Standard 11 for recommended application rates.

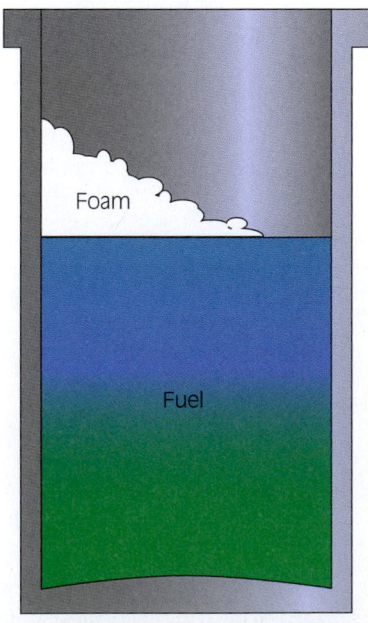

FIGURE 11-24 Foam applied on Class B fuel.

Application rates must be maintained for a minimum amount of time, and additional foam will need to be reapplied to any remaining fuel to prevent reignition. The use of application rates allows firefighters to preplan for probable incidents or assess needs for foam prior to beginning a foam attack. For example, a shallow spill fire of 1,000 ft^2 (90 m^2) of regular gasoline would require an application rate of 0.1 gpm/ft^2 using a 3 percent foam concentrate with a minimum run time of 15 minutes. This would equal a need for 100 gpm of foam solution or 3 gallons of foam concentrate for 15 minutes for a total of 45 gallons of foam concentrate required.

Heat resistance is the ability of the foam to stand up to the heat of the fire or to hot surfaces near the fire. **Knockdown speed** is how fast the foam spreads across the surface of a fuel. **Fuel resistance** is the ability to tolerate the fuel and to avoid becoming saturated with the fuel. **Vapor suppression** is the ability to contain or control the production of fuel vapors.

Protein foam is made from chemically broken down natural protein materials, such as animal blood, that have metallic salts added for foaming. The foam created is stiff, elastic, and has a high water retention rate and heat resistance. It is highly effective in vapor suppression but can be blown away by the wind, does not spread easily, and has a limited shelf life. It cannot be saturated or dipped into the fuel (poor fuel resistance), it has slow knockdown time and breaks down when used with dry chemical extinguishing agents. It is applied in 3 and 6 percent concentrations.

Fluoroprotein foam was designed as improved protein foam and has a fluorinated surfactant added. This surfactant allows the foam to be dipped into the fuel. In fact, fluoroprotein can be sub-surface injected into a fuel tank and will rise through the fuel to the surface. It also has the ability to work with dry chemical extinguishing agents. It has high heat resistance and vapor suppression capabilities and a better knockdown time than protein foam. Fluoroprotein foam is popular at fuel processing and storage facilities and is applied in 1, 3, and 6 percent concentrations.

Aqueous film forming foam (AFFF) is made from fluorochemical surfactants and synthetic foaming agents that have a quick drain-down time. This feature creates a liquid that forms a film that spreads quickly over the surface of the flammable liquid, **Figure 11-25A.** This film provides a faster knockdown time but gives up some heat resistance, fuel tolerance, and vapor suppression compared with fluoroprotein foam. AFFF can be applied with regular fog nozzles and comes in alcohol-type concentrates (ATC) (see later discussion). Because of its effectiveness, ease of use, and reduced need for special applicators, it is carried by most fire departments. AFFF can also

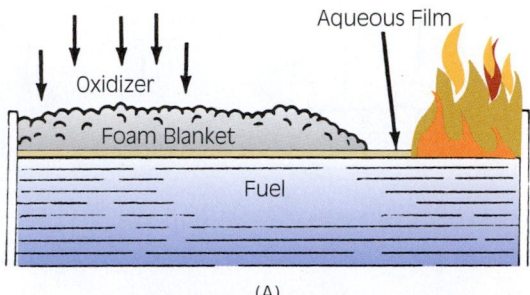

FIGURE 11-25 (A) AFFF applied on Class B fuel. Note the film barrier on the surface. (B) Alcohol-resistant AFFF foam.

be used with dry chemical extinguishing agents. The quick drain-down does require continued application with a poorer resistance to burnback than fluoroprotein foam. It is also applied in 1, 3, and 6 percent concentrations.

Alcohol-resistant foams contain a polymer that forms a protective layer between the burning surface and the foam, preventing the foam breakdown by alcohols present in the burning fuel. Alternative fuels are becoming more popular and fire tactics are changing accordingly. One alternative fuel, E85, consists of 85 percent ethanol and 15 percent gasoline and is a polar/water-miscible flammable liquid. Highly flammable, E85 will be easily ignited by heat, sparks, or flames. Alcohol-resistant foams will have the letters *AR* preceding the type of foam, such as AR AFFF and AR FFFP, **Figure 11-25B.**

Fluoroprotein film-forming foam (FFFP) combines protein with the film-forming fluorosurfactants of AFFF to improve on the qualities of both types of foam. It has the fast film-forming capabilities of AFFF and the burnback and heat resistance of protein foams. It has some alcohol resistance.

Detergent-type foams use synthetic surfactants to break down the surface tension of water and create a foaming blanket. These foams originally were good at penetrating into Class A materials but lacked heat resistance and stability. Newer types have overcome these drawbacks, and these types are used both as an active firefighting agent and as a protective barrier against fire spread. A special type of detergent foam is used for high expansion foam, which is used to fill up entire areas such as mine shafts or buildings.

CLASSIFICATION OF FUELS

Foams are used for Class A and B fires, and some specific considerations affect their use.

Class A

Sometimes piles of Class A materials are extinguished using a wetting agent, often a detergent-like substance, to help soak through the piles. Today, a new class of foams is available for Class A materials that uses a detergent base that extinguishes by reducing the surface tension of the water, allowing greater penetration into the materials. The foamy water solution has the ability to cling to the sides of objects, thus enhancing protection. This clinging ability is now being used to protect homes in urban interface areas during wildland fires, **Figure 11-26.** Urban fire departments are using it for interior firefighting and have reported faster fire extinguishment with less water. Class A foams use application ranges of 0.03 to 1.0 percent, which require a separate and more accurate proportioning system than Class B foams. Some Class B foams like AFFF have a detergent action that can be used on Class A materials.

Disadvantages of Class A foams include cost of equipment and agent, additional possibilities of equipment failure, possible effects on the environment and fire investigation laboratory tests.[4] Other

FIGURE 11-26   Firefighters applying Class A foam in wildland–urban interface. *(Photo courtesy of Ansul, Inc.)*

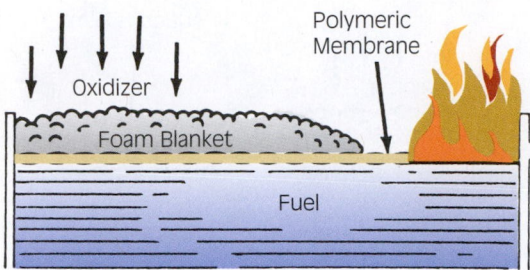

FIGURE 11-27   Polar solvent or alcohol-type foam applied on Class B polar solvent fuel. Note the polymeric film barrier on the surface.

potential problems are more difficult salvage operations and a customer service problem of explaining the milky residue on furniture and possessions to the homeowner.

Class B

Class B fuels include flammable liquids in two categories: hydrocarbons and polar solvents. Because gases should be extinguished by shutting off the flow of the fuel, firefighters do not use foam on them.

Hydrocarbons

Hydrocarbons cover a wide range of substances in forms ranging from gaseous to liquid to semisolid to solid. Common examples are heating oil, diesel fuel, gasoline, kerosene, petroleum jelly, paraffin, and asphalt. These fuels do not mix with water; they are not **miscible** or water soluble and the best method to extinguish large quantities is foam. These foams work by cooling and smothering to extinguish the fire and provide a vapor barrier.

Polar Solvents

Polar solvents mix with water and this ability causes either a breakdown of the foam or mixing of flammable vapors into the bubbles of ordinary foams. These liquids include alcohols—both methyl and ethyl—

lacquer thinners, acetone, ketones, acrylonitrile, and acetates. Perhaps the most commonly used is ethyl alcohol. When added to water with some flavoring, ethyl alcohol becomes an alcoholic beverage such as whiskey. Many gasoline blends today have polar solvents in a solution, such as Methyl Tert-Butyl Ether (MTBE), and require alcohol-resistant foams for effective extinguishment. As mentioned, E85 presents unique firefighting challenges by forming a mixture of 85 percent polar solvent and 15 percent hydrocarbon. This fuel seems to be a probable alternative to normal gasoline and its presence is becoming notable—with more areas of the country becoming receptive to it. Normal foams will break down when used on fires involving these types of mixtures. To prevent the breakdown of ordinary foams, special foams called polar solvent-type, alcohol-resistant concentrate, or alcohol-type foams have been developed. These polar solvent-type foams create a **polymeric barrier** that separates the polar solvent from the liquid of the foam, **Figure 11-27.**

APPLICATION OF FOAM

Foam is a mixture that requires a device to proportion, meter, or mix the foam concentrate into the water. Air must then be added to the foam solution to complete the process. There are several ways of adding air to the foam concentrate. Concentrations are usually expressed as a percentage of foam concentration to water in the solution. For example, a 3 percent concentration would have three parts of foam concentrate to every ninety-seven parts of water. In premixed systems, the foam concentrate is added directly to the tank, and the solution is pumped through the hoselines. The problems are the resulting limited supply, the size of the tank, and the possible damage caused to the pump and valves by the foam solution. This process also uses a tremendous amount of water to flush the tank, pump, and hoses.

One common proportioner is an **eductor, Figure 11-28,** which works on the **Venturi principle.** In a venturi eductor, water enters its inlet, which is then reduced down in a piping system that is narrower, causing the water to move faster through the smaller opening. As the speed increases, the pressure drops and creates suction. At the point of suction, the foam concentrate pick-up tube is attached and the concentrate is drawn up the tube to mix with the water, **Figure 11-29. Figure 11-30** shows how to connect and operate an in-line eductor.

Several types of eductors can be permanently piped into or added to a hoseline. One that is always piped through the venturi is an **in-line eductor,** and one that has a separate waterway and valve to allow plain water to pass by the venturi is called a **bypass eductor, Figure 11-31.**

Eductors must have the proper gpm flow, correct pressure, be clean, and not have any back-pressure situations such as hose kinks, excess hose, a partially opened nozzle, or too much elevation to work properly. Firefighters must ensure the proper setup is used to match the eductor, nozzle, and hoseline to give the correct flow. The eductor and nozzle must be clean of dried foam concentrate and other debris. Things as simple as a few pebbles in the nozzle baffle or kinks in the hose can reduce the flow and prevent foam pickup. The eductor must be set at the correct percentage for the concentrate.

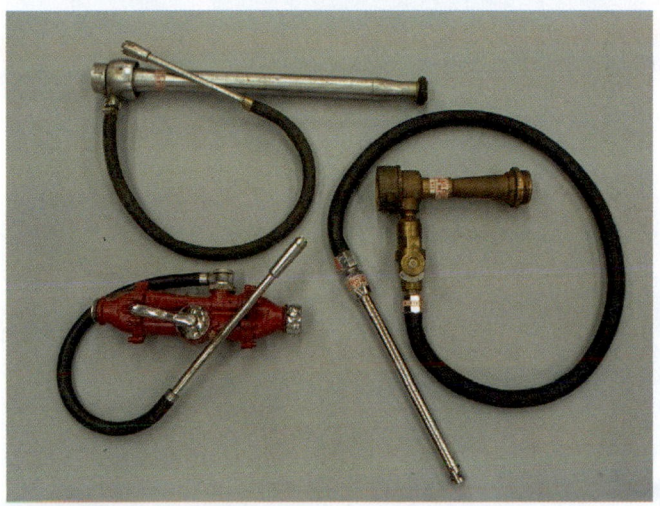

FIGURE 11-28 From right to top left clockwise: an in-line eductor, bypass eductor, and a foam eductor nozzle.

> **STREETSMART TIP**
>
> Practice and maintenance are often keys to good foam operations.

A non-recirculating, direct inject foam proportioner has a foam pump that injects foam concentrate, on demand, into the water stream on the discharge side of the water tank. The on-demand design ensures no foam is recirculated back into the water tank.

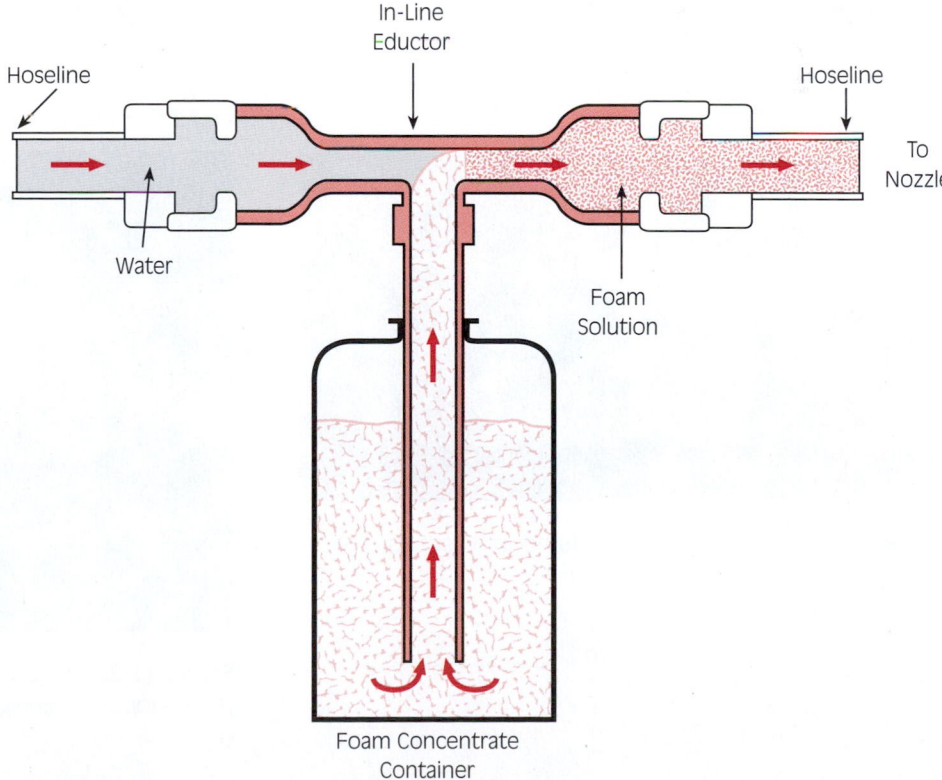

FIGURE 11-29 Foam eductor using the venturi principle.

FIGURE 11-30 The in-line eductor is connected either directly to the pump discharge or to a supply hoseline. Eductors require a specific amount of hose between the eductor and the nozzle, plus the gallonage of the nozzle must match the eductor.

FIGURE 11-31   Close-up of bypass eductor bypass valve.

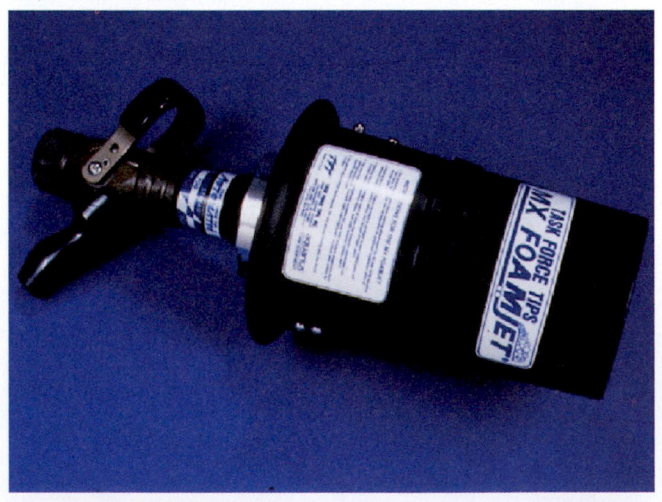

FIGURE 11-32   Handline-operated medium expansion foam generator. *(Photo courtesy of Task Force Tips)*

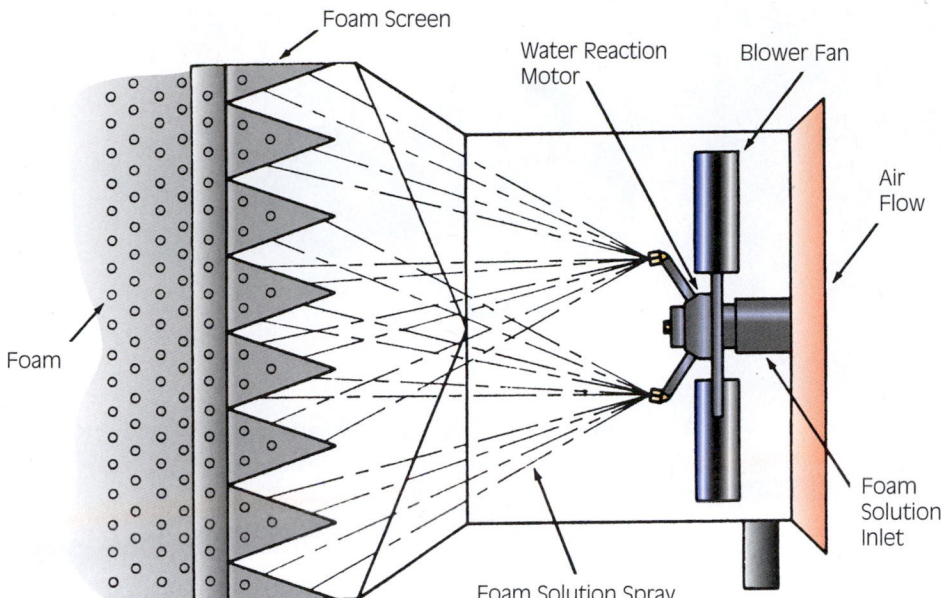

Foam Screen

Water Reaction Motor

Blower Fan

Air Flow

Foam

Foam Solution Inlet

Foam Solution Spray

FIGURE 11-33 High expansion foam generator.

An around-the-pump proportioner uses an eductor between the intake and discharge side of the pump. The discharged water pressure is pumped through the eductor to siphon the concentrate into the intake side of the pump and through the pump. This system allows the pump to operate at full flow, but may cause damage to the pump and valves without proper back-flushing of the entire pump.

Balanced pressure systems use one of two foam pumping methods, in which the foam is pumped under pressure into a metering chamber that balances the pressure of the concentrate and water controlling the flow of the foam solution.

In **compressed air foam systems (CAFS)** or dual-injection systems, the concentrate is in a separate foam tank and a foam pump pumps the concentrate directly into the hoseline, which is metered by a flow-metered microprocessor. After the foam solution is created, an air compressor line injects air into the hoseline to create a light and fluffy foam. These systems are used for Class A foams.

To finish making the foam, air must be added to the foam solution, normally at the nozzle. The various types of foam nozzles, regular fog nozzles, and foam generators have the ability to aspirate various quantities of air into the foam solution. The goal is to get the correct expansion ratio, that is, amount of air added to the solution. The expansion ratio is expressed as the volume of air added to the volume of solution. For example, 10:1 refers to 10 volumes of air added to 1 volume of solution. Foam's expansion ratios are in three ranges: low, medium, and high. Low expansion foam has an air to solution ratio of up to 20 to 1, or as much as 20 ft³ (0.56 m³) of finished foam for each cubic foot (0.028 m³) of foam solution. Medium expansion foam has ratios ranging from 20:1 to 200:1, and high expansion foam is rated 200:1 and higher (the

upper limit is about 1,000:1). The creation of medium and high expansion foams requires a special foam generator, **Figure 11-32** and **Figure 11-33.** This special unit has a nozzle that sprays the foam solution onto a screen or mesh and a fan that pushes large volumes of air to create millions of bubbles. These foam generators may be gas, electric, hydraulic, or water powered.

HAZARDS ASSOCIATED WITH FOAM APPLICATION

■ Mismatching of foam to fuel type may cause breakdown of foam blanket.
■ Disturbing the foam blanket will allow fuel vapors to escape and possibly reignite.
■ Improper application may cause fuel and fire to spread to uninvolved areas.
■ Insufficient foam application may compromise team safety.
■ Foam may hide trip and fall hazards.

In order to alleviate the hazards that are associated with applying foam, each of the aforementioned hazards should be assessed and appropriate actions applied as explained throughout this chapter.

Fog Nozzles versus Foam Nozzles

Originally, foam operations required a special foam nozzle to properly aspirate air into the foam solution. Some of these foam nozzles incorporated the eductor into the nozzle, combining all of the foam-making steps. Some of these combined-eductor nozzles are still in use. As these nozzles require having the

FIGURE 11-34   Typical foam nozzles.

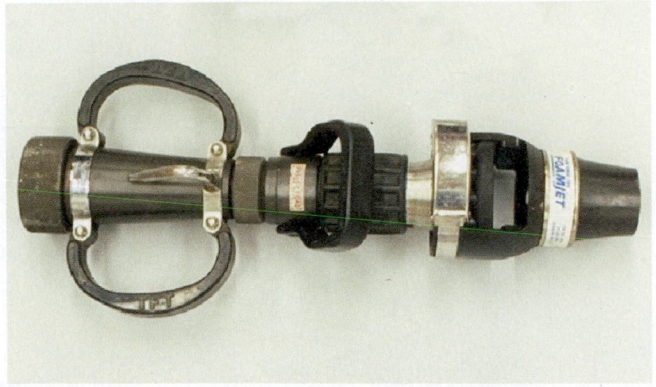

FIGURE 11-35   Clip-on foam attachment for fog nozzle.

FIGURE 11-36   The bank-in technique of foam application.

foam concentrate containers, and personnel to move the containers at the nozzle, they are not very popular. Foam nozzles are designed today to aspirate the proper amounts of air and apply the foam to the fuel. They do this by having air vents or ports built into the nozzle. The movement of foam solution past these ports draws air into the solution, both before and after the tip. Foam nozzles, **Figure 11-34,** are designed for low and medium expansion foams with expansion ratios of 8:1 to 20:1 in the low range, and up to 50:1 in the medium range. Foam nozzles are recommended with protein and fluoroprotein foams.

When AFFF was introduced, one of its advantages was that it did not require a special foam nozzle for application. Fog nozzles typically used by the fire department could be used. This helped reduce the costs of purchasing an effective foam system, and many departments began carrying foam concentrate and foam eductors on all pumpers. While the cost is lower, so is the expansion ratio, with 8:1 being at the high end. To increase this expansion efficiency while keeping costs lower, some manufacturers now have clip-on or snap-on foam nozzle adapters that attach to the fog nozzle and make it a foam nozzle, **Figure 11-35.** These are effective for increasing the expansion ratio and are almost as good as a specially designed foam nozzle. CAFSs, by preinduction of air, are effective with a fog nozzle and even with solid tip nozzles.

Foam from nozzles is applied using one of three techniques. The first is the bank-in technique, in which the foam strikes the ground before the fire and rolls into the fire, **Figure 11-36.** The second is the bank-back or bounce-off technique, in which the foam is banked off a wall or other object and rolls back into the fire, **Figure 11-37.** The third technique is the raindown or snowflake technique, in which the foam is sprayed high into the air over the fire and it floats down onto it, **Figure 11-38.**

FIGURE 11-37 The bank-back technique of foam application.

FIGURE 11-38 The rain-down technique of foam application.

NOZZLE AND FOAM EQUIPMENT MAINTENANCE

Just like other equipment used in the fire service, nozzles and foam appliances must be cleaned and maintained regularly to ensure proper operation. The regular maintenance of this equipment usually coincides with the maintenance schedule of other equipment on the apparatus or in the station. When maintaining and cleaning this equipment, all manufacturer's guidelines and departmental policies should be followed. Together, these guidelines and policies should include information such as the cleaning and maintenance schedule, the method and materials used for cleaning, the necessary skill level of firefighters tasked with cleaning and maintenance, documentation, tagging or identification procedures, the process to place appliances or other equipment out of service for repair and who should be responsible for the repair work, and any pertinent special instructions. Effective hose streams depend on equipment that is properly serviced and in good working order. Maintenance and cleaning of all fire service equipment is vital to effective firefighting operations.

LESSONS LEARNED

Fire streams are made of water that leaves a nozzle and heads toward the target. The nozzle defines the characteristics of the fire stream. The two main types of nozzles are solid tip and fog, and they need to match the fire conditions and fire department resources. Each type of nozzle has its advantages and disadvantages; firefighters should use the one that will do the best job of suppressing the fire. Proper use of nozzles requires shutoffs and even some special types of nozzles. Fire streams and fire conditions determine how firefighters will attack a fire.

When the fuels involved in a fire are not compatible with water, other agents must be used to fight the fire and, commonly, this other agent is foam. Firefighters should know the different kinds of fuels and types of foams available. Foam requires special equipment to create and apply it, and some special application techniques are used.

KEY TERMS

Alcohol-Resistant Foams   These foams contain a polymer that forms a protective layer between the burning surface and the foam, preventing the foam breakdown by alcohols present in the burning fuel.

Application Rate   Amount of foam or foam solution needed to extinguish a fire. Usually expressed in gallons per minute per square foot or liters per minute per square meter.

Automatic or Constant Pressure Nozzle   Nozzle with a spring mechanism built in that reacts to pressure changes and adjusts the flow and resultant reach of the nozzle.

Bresnan Distributor   A device used to fight fire in basements or cellars when firefighters cannot make a direct attack on the fire; has six or nine solid tips or broken stream openings designed to rotate in a circular spray pattern.

Bypass Eductor   Eductor with two waterways and a valve that allows plain water to pass by the venturi or through the venturi to create foam solution.

Cellar Nozzle   A device with four spray nozzles designed to rotate in a circular spray pattern for fighting fires in basements or cellars when firefighters cannot make a direct attack on the fire.

Combination Fire Attack   A blend of the direct and indirect fire attack methods, with firefighters applying water to both the fuel and the atmosphere of the room.

Combination Nozzle   A spray nozzle that is capable of providing straight stream and spray patterns, which are adjustable or variable by the operator. Most fog nozzles used today are combination nozzles.

Compressed Air Foam Systems (CAFS)   A foam system where compressed air is injected into the foam solution prior to entering any hoselines. The fluffy foam created needs no further aspiration of air by the nozzle.

Constant or Set Volume Nozzle   Nozzle with one set volume at a set pressure. For example, 60 gpm at 100 psi (227 L/min 690 kPa). The only adjustment is the pattern.

Detergent-Type Foams   Use synthetic surfactants to break down the surface tension of water to create a foaming blanket.

Direct Fire Attack   An attack on the fire made by aiming the flow of water directly at the material on fire.

Eductor   Device that siphons a liquid from a container into a moving stream.

Fluoroprotein Film-Forming Foam (FFFP)   Combines protein with the film-forming fluorinated surfactants of AFFF to improve on the qualities of both types of foam.

Fluoroprotein Foam   Designed as an improved protein foam with a fluorinated surfactant added.

Foam   An aggregate of gas-filled bubbles formed from aqueous solutions of specially formulated concentrated liquid foaming agents.

Fog Nozzle   Delivers either a fixed spray pattern or variable combination of straight stream and spray patterns.

Fuel Resistance   Ability to tolerate the fuel and to avoid being saturated by or picking up the fuel.

Heat Resistance   The ability of foam to stand up to the heat of the fire or to hot surfaces near the fire.

Indirect Fire Attack   An attack made on interior fires by applying a fog stream into a closed room or compartment, thus converting the water into steam to extinguish the fire.

In-Line Eductor   Eductor in which the waterway is always piped through a venturi.

Knockdown Speed   Speed with which foam spreads across the surface of a fuel.

Miscible   Having the ability to mix with water.

Nozzle   A tapered or constricted tube used to increase the speed or change the direction of water or other fluids.

Piercing Nozzle   Nozzle originally designed to penetrate the skin of aircraft and now has been modified to pierce through building walls and floors.

Polymeric Barrier   A separation barrier made up of polymer or a chain of molecules linked in a series of long strands. This separates a polar solvent from an ATC foam blanket.

Protein Foam   Made from chemically broken down natural protein materials, such as animal blood, that have metallic salts added for foaming.

Solid Stream Nozzles   Type of nozzle that delivers an unbroken or solid stream of water to the fire. Also called solid tip, straight bore, or smooth bore.

Straight Stream   A nozzle pattern that creates a hollow stream, similar in shape to the solid stream pattern, but the straight stream pattern must pass around the baffle of the nozzle. Newer fog nozzle designs, especially the automatic nozzles, only have this hollow effect from the tip on and, hence, create a solid stream with good reach and penetration abilities, some better than solid stream nozzles.

Vapor Suppression   Ability to contain or control the production of fuel vapors.

Variable, Adjustable, or Selectable Gallonage Nozzle   Nozzle that allows the nozzleperson to select the flow, with usually two or three choices, and the pattern.

Venturi Principle   A process that creates a low-pressure area in the induction chamber of the eductor and allows the foam concentrate to be drawn into and mixed with the water stream.

Water Curtain Nozzle   Designed to spray water to protect exposures against heat by wetting the exposure's surface.

REVIEW QUESTIONS

1. Explain what a nozzle is, how it works, and how nozzle reaction affects its operation.

2. Explain the difference between a solid tip nozzle pattern and a straight stream nozzle pattern.

3. List and describe the types of foams, their characteristics, and their purpose.

4. Define foam and name its three major components.

5. How does AFFF work on a flammable liquid fire?

6. If a foam concentration were 6 percent, how many gallons of concentrate would be added to how many gallons of water to equal 100 gallons?

ENDNOTES

1. A 1½-inch (38-mm) hoseline is the minimum size hoseline for structural firefighting.

2. Although the hoseline can be operated by one person, safety regulations require operation with a second team member.

3. NFPA, *Fire Protection Handbook,* 18th ed. (Quincy, MA: National Fire Protection Association, 1997), pp. 6–349.

4. Jeff Stern and J. Gordon Routley, *Class A Foam for Structural Firefighting* (Emmittsburg, MD: U.S. Fire Administration, December 1996).

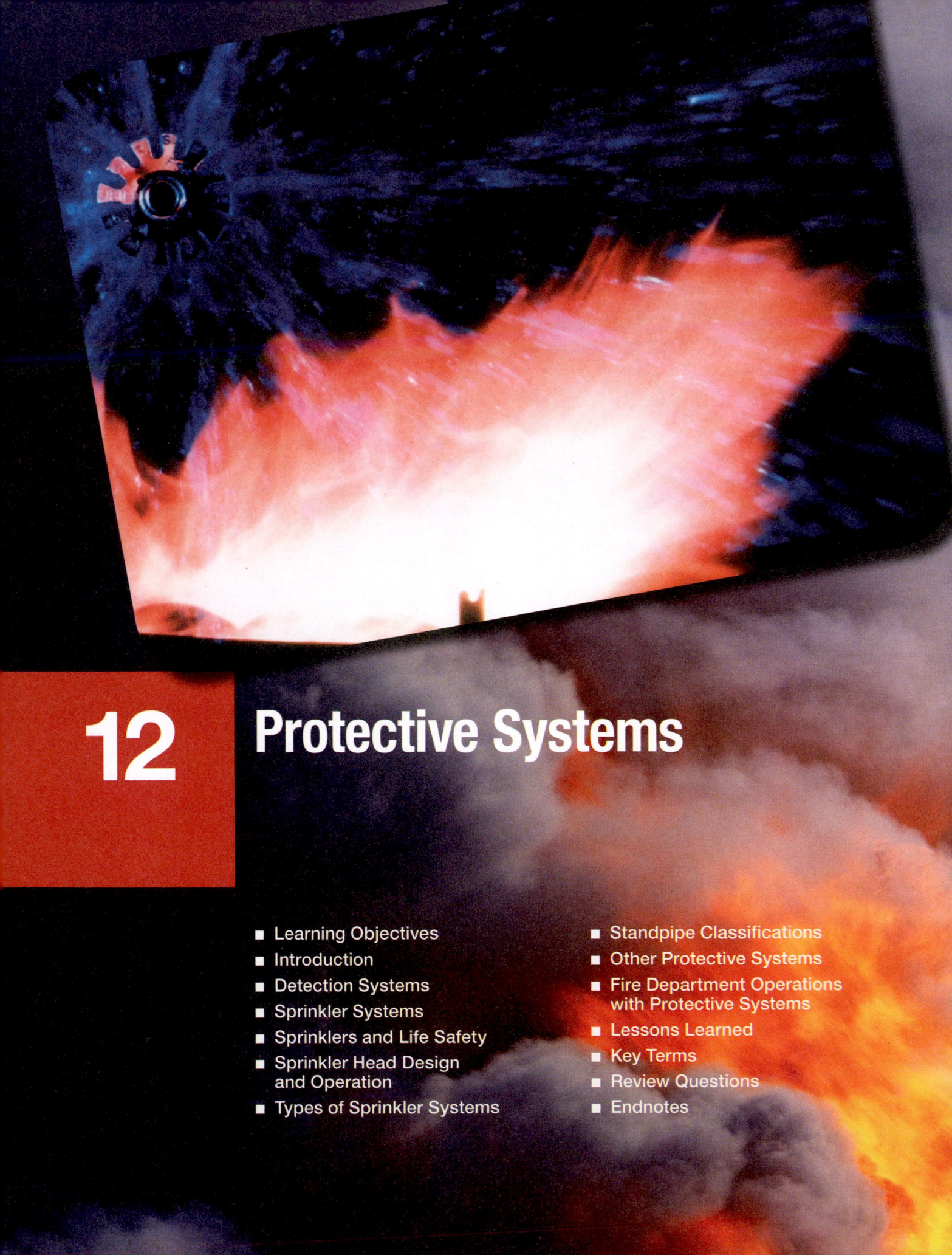

12 Protective Systems

INTRODUCTION

Protective systems help to guard lives and property by detecting and/or suppressing fires. Detection systems detect the presence of a fire and alert building occupants and/or the fire department, and may also activate a suppression system. Suppression systems are devices that help firefighters in controlling fires. Some systems are designed to automatically detect and suppress a fire, while others assist firefighters and occupants in extinguishing the fire. They are also called **auxiliary appliances.** Regardless of the term used, firefighters must know about these systems and how to use them. Due to the implementation and maintenance costs, great efforts have been expended to require these systems in building and fire prevention codes. Therefore, to the builders, owners, and occupants, the expectation is that the fire department will use them correctly.

NOTE

Remember that auxiliary appliance devices can make the job of firefighting safer and easier.

Detection systems are varied. Some require people to detect the fire and manually activate an alarm. Others are highly complex systems that can detect a fire almost at its ignition and sound an alarm. Some may also activate suppression systems.

The two main suppression systems are sprinklers and standpipes. These are devices that allow water to be used as an extinguishing agent. When the situation dictates that another agent may be more effective, other types of protective systems are required. Some use the agents found in portable fire extinguishers to cover a room or larger areas. These systems include dry and wet chemicals, inerting gases, and halogenated agents.

Detection systems, suppression systems, and alarm systems (covered briefly in Section I, Chapter 3) can all be tied together to form a protective ensemble for building occupants. This chapter describes individual detection and suppression components as well as the interface of each.

DETECTION SYSTEMS

Detection systems are designed to notify people of a fire or other problem. Simple detection systems can actually be viewed as a warning system requiring a person to recognize a danger. More complex systems use a series of devices to automatically detect a fire or event and initiate a warning alarm. Fire detection and warning can also come from the initiation of a suppression system. For example, a sprinkler activates, tripping a water flow alarm that notifies occupants. Regardless of how complex or simple, detection systems are designed to notify people that a potentially life-endangering event is happening. By understanding the basic principles of operation of various detection systems, firefighters more effectively utilize these systems to perform their job. However, until firefighters are on scene to verify the type and extent of emergency, they should remain cautious and take all actions to prepare and safeguard themselves. For firefighting, all detection system alarms should be considered to house a hostile atmosphere until proved otherwise by the fire companies.

CAUTION

Fire companies should take somewhat of a "distrustful" attitude when it comes to detection system activations and suspect the worst-case scenario until fire department units are on scene and prove otherwise. Detection systems are sophisticated and the people operating them are normally proficient; however, personal safety should always dictate the required personal protective equipment (PPE) and actions on a fire call.

People or Manual Systems

People can alert other building occupants and can call the fire department after discovering a fire. Some buildings are equipped with a manual fire alarm system that requires a person to discover the fire and then pull a lever or push a button. A restaurant cooking area or a laboratory hood with a local application system that sounds an alarm and activates the suppression system can have a manual switch for people to operate. Street box systems are also manual fire alarm systems.

Manual systems suffer from two typical problems. A person must be present, awake, and alert to discover the fire and then activate the system. Unfortunately, sleeping people and pets make poor fire detectors, because all of their detection senses are also asleep. The second problem is that many manual fire detection systems are local only, meaning that they will alert any building occupants but will not summon any additional help. A public education program and proper signs at the pull station can help teach people about the need to call the fire department.

Heat Detectors

Heat detectors operate by detecting the heat of a fire at a fixed temperature or as the rising temperature builds at a rapid rate. These can be used as part of a suppression system such as a sprinkler system or can operate a fire protection device such as closing a door. Some are merely detection devices that operate an alarm. Heat detectors are slow to detect fire and hence should not be used for life safety. They are, however, inexpensive and have a low rate of false alarms. Heat detectors can be of the spot type, which covers just one area, or of the line type, which covers a larger area.

The *rate-of-rise heat detector* measures temperature increases above a predetermined rate. The pneumatic or air-sensitive device uses a diaphragm with a small hole or orifice that acts as a relief valve for slow temperature increases. When a fast temperature increase occurs, the diaphragm is pushed toward a contact point and the alarm is activated, **Figure 12-1.**

The *fixed temperature heat detector* is usually electrically operated with a bimetallic strip of two metals that expand at different rates, eventually bending to touch a contact point and complete the alarm circuit, **Figure 12-2.** Another fixed temperature type uses a metal element similar to solder that melts at a fixed temperature and breaks a circuit or opens a cap. Line-type detectors can be either pneumatic or electric, **Figure 12-3.**

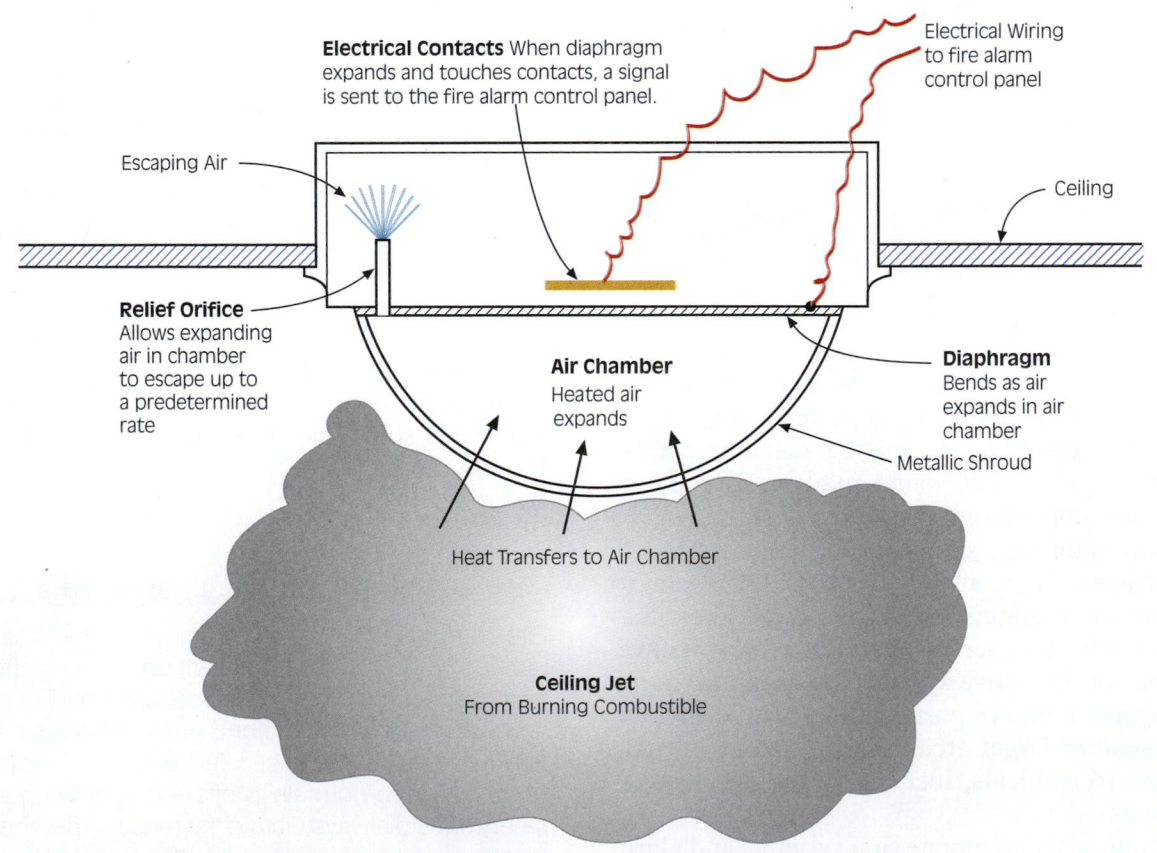

FIGURE 12-1 Rate-of-rise heat detector.

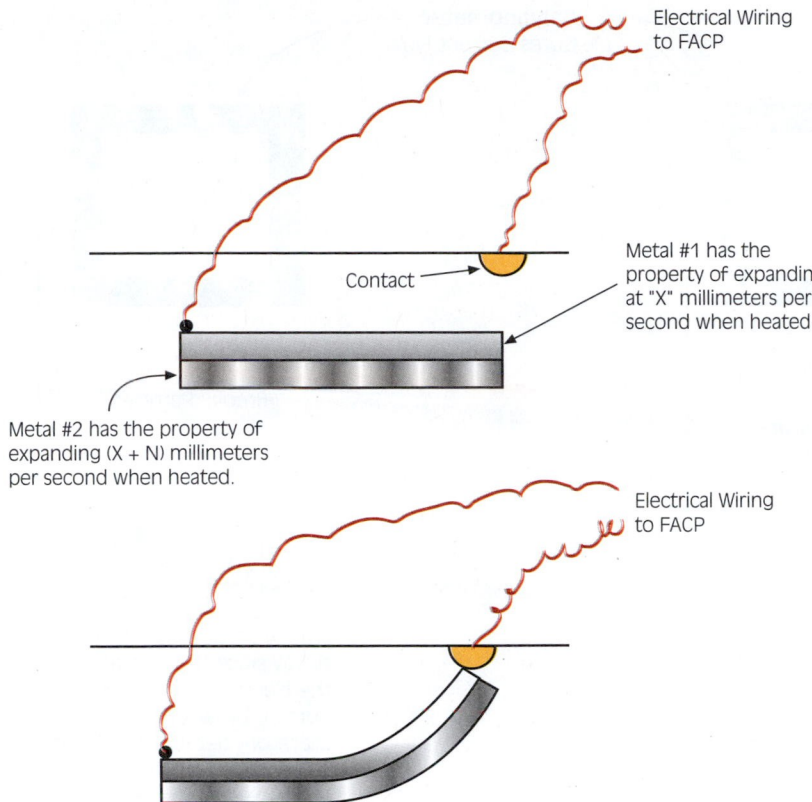

Electrical Wiring to FACP

Contact

Metal #1 has the property of expanding at "X" millimeters per second when heated

Metal #2 has the property of expanding (X + N) millimeters per second when heated.

Electrical Wiring to FACP

Since metal #2 expands at a rate faster than metal #1, the bimetallic strip will deform toward the contact when heated until the circuit is closed, and a signal is sent to the FACP.

FIGURE 12-2 Fixed temperature heat detector.

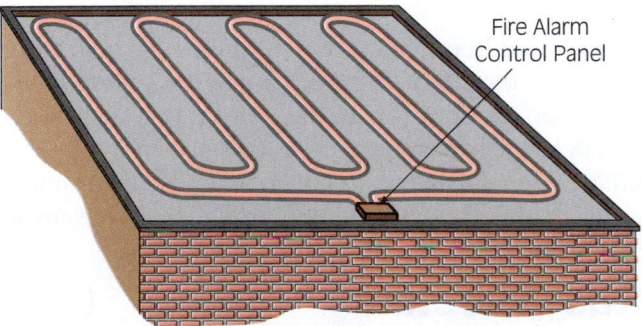

Fire Alarm Control Panel

FIGURE 12-3 Line-type heat detector layout.

Smoke Detectors

Smoke and toxic gases are the leading killers of people in residential structure emergencies. Smoke detectors have greatly improved residential life safety by detecting and alerting building occupants to the harmful effects of fire. Smoke detectors are the most prevalent automatic detection system. Even with the abundance of smoke detectors installed in homes across North America, many occupants do not understand how they work—or how to fix problems. Therefore, it is common for citizens to request assistance from the fire department when the detectors malfunction. It is important that firefighters understand the opera-

tion of smoke detectors and have the ability to explain the problems with the detector and suggest solutions to the building owner or occupant.

Smoke detectors are hard (permanently) wired, battery operated, or a combination of hard wiring with a battery backup. Smoke detectors work primarily on two principles: ionization and photoelectric.

Ionization detectors are the most common type of smoke detector and have been installed in most homes and apartments since the early 1980s. These detectors use a radioactive element that emits ions into a chamber. The positive and negative ions are measured on an electrically charged electrode. When smoke enters the chamber, the flow of ions to the electrode changes and the alarm is activated, **Figure 12-4.** Ionization detectors are spot-type devices (usually functioning independent of other detectors) that are inexpensive and are quick to react to small smoke particles from flaming fires. Ionization smoke detectors can also detect invisible smoke; certain fire gases that are colorless can disrupt ion flow within the chamber. Ionization smoke detectors are reliable but can be activated by cooking, steam (even from a hot shower), electrostatic discharge (from a thunderstorm), and dust buildup. Some people may classify these activations as false alarms, but in actuality, these activations indicate that the detector is working as designed. Most ioniza-

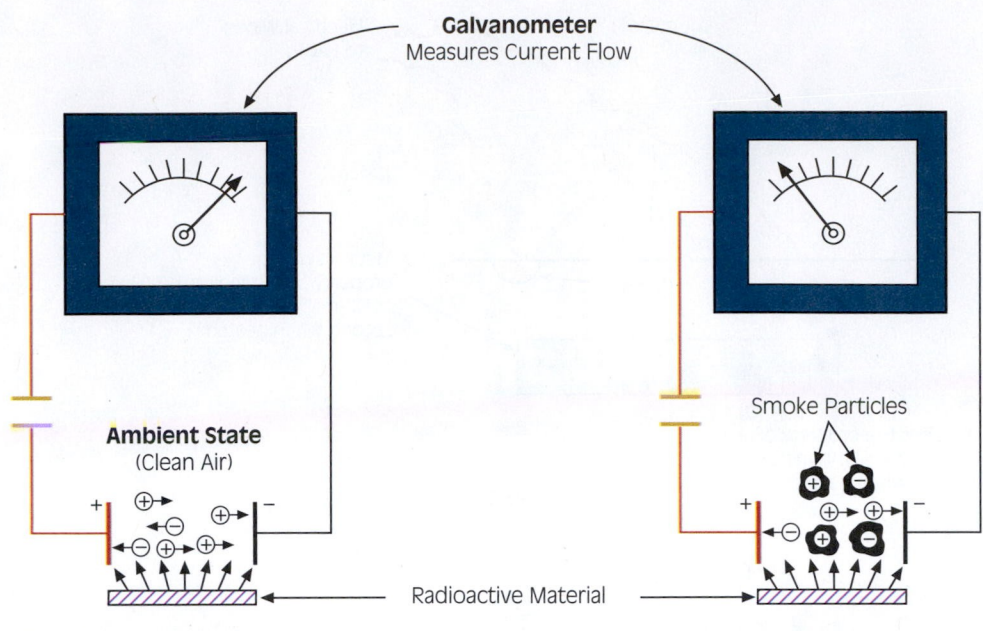

Galvanometer
Measures Current Flow

Ambient State
(Clean Air)

Radioactive Material

Smoke Particles

A small amount of radioactive material ionizes the air, giving a charge to particles within the small space shown. Positively charged particles are attracted to the electrode at right, and negatively charged particles are drawn to the electrode at left.

Smoke particles move more slowly since their attraction to the electrodes is diminished. Current between the electrodes therefore decreases. A signal is sent to the FACP when current drops below a predetermined level.

FIGURE 12-4 Ionization-type smoke detector.

tion detectors will reset once the smoke has cleared. Sometimes it is necessary to "blow out" the detector to clear the internal chamber of smoke products and dust or spider webs. Ionization detectors that are activated by an electrostatic discharge may not reset and the detector will likely need to be replaced. Smoke detectors that "chirp" typically indicate that the battery is low.

Photoelectric detectors use two different methods to detect the smoke of a fire and can be spot or line detectors—which function in conjunction with other detectors: when one detector activates, all detectors in the line will alarm. The first method used by a photoelectric detector is the *light obscuration detector,* which has a light beam aimed at a light sensor. When the smoke particles obscure or block the light beam, the light sensor notes the loss of light and activates the alarm, **Figure 12-5.** *The light-scattering detector* operates by having a light beam aimed at the end of the chamber with a light sensor in an angled-off chamber, **Figure 12-6.** When the smoke enters the chamber, the smoke particles scatter the light, which strikes the sensor, activating the alarm. Photoelectric detectors are activated by visible smoke and are appropriate for areas where smoke may develop but dissipate visually, such as in cooking areas and where

fuel-powered equipment is working. Like ionization detectors, dust and steam can cause activation of the photoelectric detector. If the smoke that causes the photoelectric detector to activate is oily (as from a grease fire), the detector may not reset once the smoke clears as the film will likely obscure the light sensor.

Gas Detectors

Gas-sensing detectors are designed to find the presence of a certain gas or gases prior to their reaching a concentration that can cause danger. A flammable gas detector would identify the presence of the gas, such as a petroleum product, before it reached its ignitable concentration. Other gases, such as carbon monoxide, would be detected prior to reaching a concentration that could cause death or injury. Gas detectors can be permanently mounted devices or portable units.

Carbon monoxide (CO) detectors for the home are popular and are even required in some communities and, like smoke detectors, have saved many lives. However, this popularity has placed an additional service load on fire departments. CO detectors function using several different methods. Consulting manufacturers' literature is perhaps the best approach to understand the method of operation and requirements

Light Obscuration Type

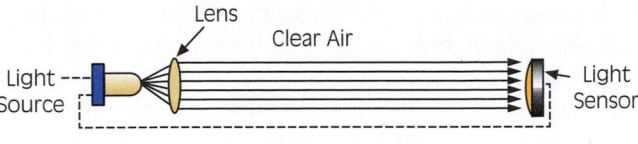

Principle of Operation

Light Obscuration Type

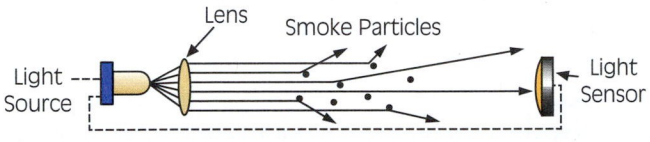

Principle of Operation

FIGURE 12-5   Light obscuration photoelectric smoke detector.

Light-Scattering Type

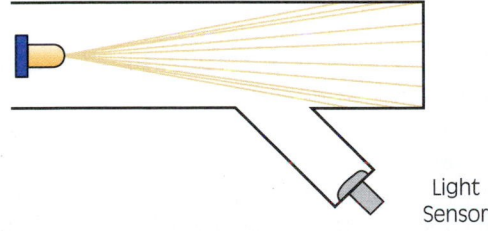

Typical Photoelectric Detector Chamber with Clean Air

Light-Scattering Type

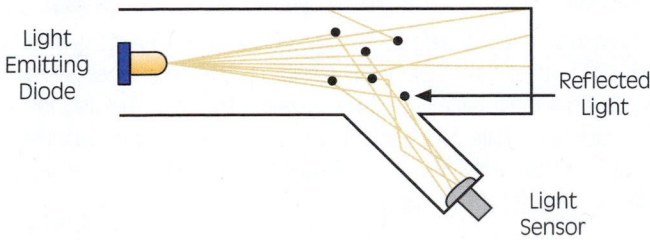

Typical Photoelectric Detector Chamber with Smoke

FIGURE 12-6   Light-scattering photoelectric smoke detector.

to reset the device. Most CO detectors, however, provide an early warning alarm (usually an intermittent beep,) when low-level CO is detected. Some CO detectors have an LED display that indicates the level (in parts per million) of CO being detected. Persons reporting smoke detector activation may actually be hearing a CO detector activation—and vice versa. CO is odorless and colorless and even though smoke is not visible, dangerous levels of CO may be present. Therefore, firefighters must be ready for a fire or CO situation when responding to any residential alarm. Many fire departments carry gas detectors to check for flammable gases, carbon monoxide, other hazardous materials, and the oxygen content of the air.

Combination Smoke/Carbon Monoxide Detectors

New and improved detection systems are always being manufactured. Some companies are now producing combination smoke/carbon monoxide detectors that function as the name describes. These detectors, while simplifying things for the builder and homeowner, can make discerning the situation on a particular fire incident a little more difficult for firefighters. As citizens are normally somewhat excited when calling in a home-safety-device alarm, they may not correctly deduce which alarm has activated. Therefore, when responding to residential detection alarms, firefighters must be prepared for a smoke detector activation to actually be a carbon monoxide activation and vice versa—or possibly both.

To help alleviate this type of situation, detector manufacturers are adding different alarms and different colored lights for each type of situation. This alarm could be three beeps, a pause, and repeat for smoke alarms, and four beeps, a pause, and repeat for carbon monoxide alarms. Again, while this appears helpful, an excited homeowner may not correctly distinguish between the two. It might be possible for 9-1-1 Dispatch to ask scripted questions to gain more information to ascertain the alarm type; however, a homeowner should never be prompted to remain in a structure that has potentially unsafe CO levels or the possibility of fire.

Other manufacturers have added voice prompts. When these smoke detectors sense smoke, they will alarm with an audible voice stating, "Fire, fire, fire." Some detectors can be programmed with information to describe the location, for example, "Warning, smoke detected in kitchen. Evacuate." When they sense carbon monoxide, the alarm will respond with the CO information in the same manner. Specific information is much more helpful. However, seasoned firefighters can undoubtedly tell stories about fire calls where citizens have confused this type of information or have been frightened and unable to repeat any information correctly. Preparing for all situations is key to firefighter readiness.

Flame Detectors

Flame detection devices detecting flames or lightwaves are mainly of three types: ultraviolet (UV), infrared (IR), or a combined ultraviolet-infrared (UV-IR) detector. These detectors are very sensitive and quick to alarm. They are used to protect petroleum and chemical facilities where fast and high temperature fires can occur.

Ultraviolet flame detectors sense the radiation of high-intensity flaming fires by detecting the lightwaves emitted in the UV spectrum (below 4,000 angstroms), which is *below* the visual light wavelengths. The UV detectors must have proper shielding from light sources, especially sunlight and welding. Dust or moisture can also fog the lens of the UV detector and cause detection problems.

Infrared flame detectors sense the infrared radiation (8,500 to 12,000 angstroms) of a fire using a photocell, which is *above* the visual light wavelengths. IR detectors are effective in rapidly spreading fires, but must be screened from sunlight and have a limited distance range.

Combined *ultraviolet-infrared flame detectors* are used to rapidly detect the flames of a fire but without the false alarm problems of either a UV or IR detector. This detector requires both the UV and IR components to be activated before an alarm is sent. Because the weakness of one does not affect the other, the double positive usually and accurately indicates a fire.

STREETSMART TIP

It is common for firefighting companies to respond to a detector activation with a full alarm assignment, only to have it canceled when the detection company receives "proper code" or verification that there is not a fire or other dangerous situation. Fire companies should cancel according to local SOPs; however, the best choice in terms of best practice and customer service would be for the "first due" apparatus or a fire department support vehicle to continue to respond on a nonemergent basis. This company or support person can verify that the fire response is not necessary and can offer support to business persons on scene or answer questions about the fire-detection system. If nothing else, this is a great way for fire crews to see and pre-plan some of the structures in their first-due area.

CAUTION

When responding to ANY type of residential detector activation, firefighters should always prepare for the possibility of fire, carbon monoxide, and other dangerous atmosphere situations. Firefighters should wear complete and proper PPE with SCBA until the atmosphere is proved safe.

SPRINKLER SYSTEMS

Sprinkler systems are designed to automatically distribute water through sprinklers placed at set intervals on a system of piping, usually in the ceiling area, to extinguish or control the spread of fires. Most sprinkler heads detect the heat of a fire and begin to apply water directly over the source of the heat. Sprinkler heads, unless deluge-type heads, are heat-sensitive devices that react to a fixed temperature and disperse water in a specific pattern and quantity over a set area. Sprinkler systems are highly effective. In fact, some people describe the benefit of sprinklers as similar to having a firefighter constantly on duty in the protected building. The NFPA publishes standards for installation of sprinkler systems and their inspection and maintenance.[1]

SPRINKLERS AND LIFE SAFETY

Sprinkler systems were originally designed in the late 1800s. Their intent was to protect property, especially businesses and factories, from total loss from fires. They are almost 100 percent effective. Improper action by humans is usually the catalyst when sprinkler systems malfunction. Improper maintenance or inadequate water supply due to human error are two common mistakes leading to sprinkler failure. In the early 1900s, the idea that sprinklers might be able to save lives was beginning to take shape.

FIREFIGHTER FACT

New York City's tragic Triangle Shirtwaist Factory fire in 1911 killed 146 people. Most of the fatalities were women who jumped to their deaths trying to avoid being burned. This fire led to the concept of adding sprinklers to buildings for life safety and providing occupational safety for workers.

America's fire loss statistics show that most structure fires, fire damage, and, of greatest importance, fire injuries and fatalities occur in residential properties.

Yet the American home is one of the least protected properties when it comes to fire safety. Homes in the United States are usually built of wood and other combustible materials. Until the mid-1970s, most American homes did not have smoke detectors and only a few non-high-rise residential units had sprinklers. Recognizing the potential benefit of these types of systems, the U.S. Fire Administration, National Fire Protection Association, International Association of Fire Chiefs, and Factory Mutual, among many others, led the research efforts to develop faster responding and less expensive residential sprinkler systems. The first standard on residential sprinkler systems was developed in 1975.[2] Some cities and counties now have mandatory sprinkler requirements in all newly constructed residential structures, and many states require them in multifamily dwellings. There is still, however, a long way to go to place them in all new residential buildings. The American heritage of protecting personal property rights often allows citizens the right to die in their own homes, the least protected and inspected of any occupancy classes.

NOTE

Firefighters must realize that protective systems are designed for specific purposes and will not completely defend a property unless they are designed with that goal in mind. These types of systems are of great benefit to firefighting strategy and tactics but are only one of many firefighting tools.

VIEWPOINT

Often, residential sprinkler initiatives in communities are met with considerable opposition from homeowners and home builders for a variety of reasons. While sprinklers can save lives, many people are not willing to pay the additional costs of installing sprinkler systems in the home. Fire departments are often active in supporting residential sprinkler initiatives and educating the public about the benefits of protective systems. For more information about residential sprinklers, consult the Residential Fire Safety Institute at www.firesafehome.org.

Residential sprinklers are designed for life safety and not necessarily to protect property. A residential sprinkler system may completely extinguish a fire in a protected area as unprotected areas are allowed in the design of such systems. Fires that may occur in these unprotected areas can gain great headway or move to other unprotected areas and the system may not be able to prevent a total loss. Fires have occurred in apartment buildings, motels, and hotels protected by

a life safety sprinkler system that have resulted in a total loss of the property. When analyzing these fires and rating the effectiveness of the protection system, the presence of occupant or firefighter fatalities or injuries will determine the effectiveness of the system. Absence of these fatalities or injuries validates that the system did what it was designed to do.

Sprinkler systems are also not totally effective in life safety in some types of fires. These include slow burning and smothering fires that produce fatal quantities of smoke without generating sufficient heat to activate the system, flash fires that rapidly engulf a person, or a fire in the immediate vicinity of the victim where the fire inflicts the injuries prior to system activation. Victims who are disabled or unable to remove themselves from the fire area also factor into fire system inadequacy. Protecting against all situations is impossible, but residential sprinkler systems combined with effective smoke detection systems and coupled with aggressive firefighting tactics and strategies allow maximum life safety in the home.

SPRINKLER HEAD DESIGN AND OPERATION

Sprinklers or sprinkler heads are the key components of a sprinkler system, **Figure 12-7**. Most important are the heat-sensitive parts that usually detect heat

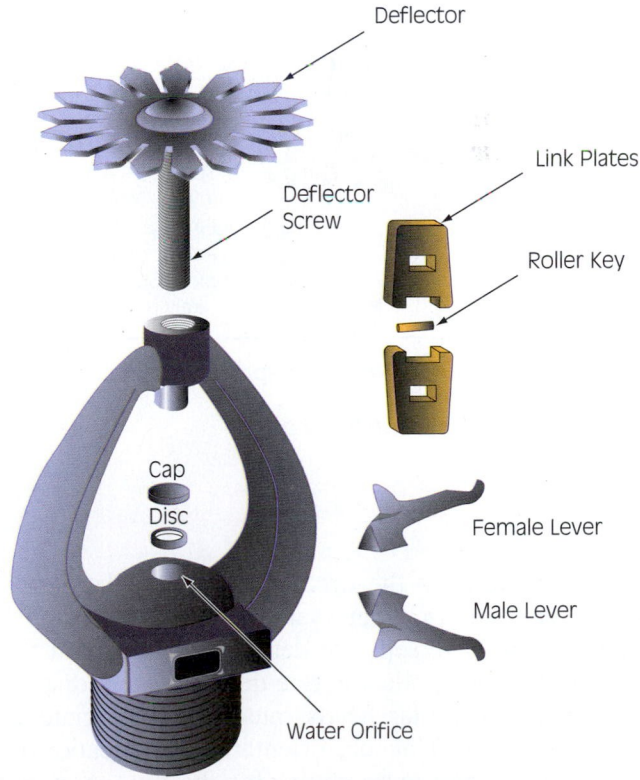

FIGURE 12-7 Sprinkler head parts.

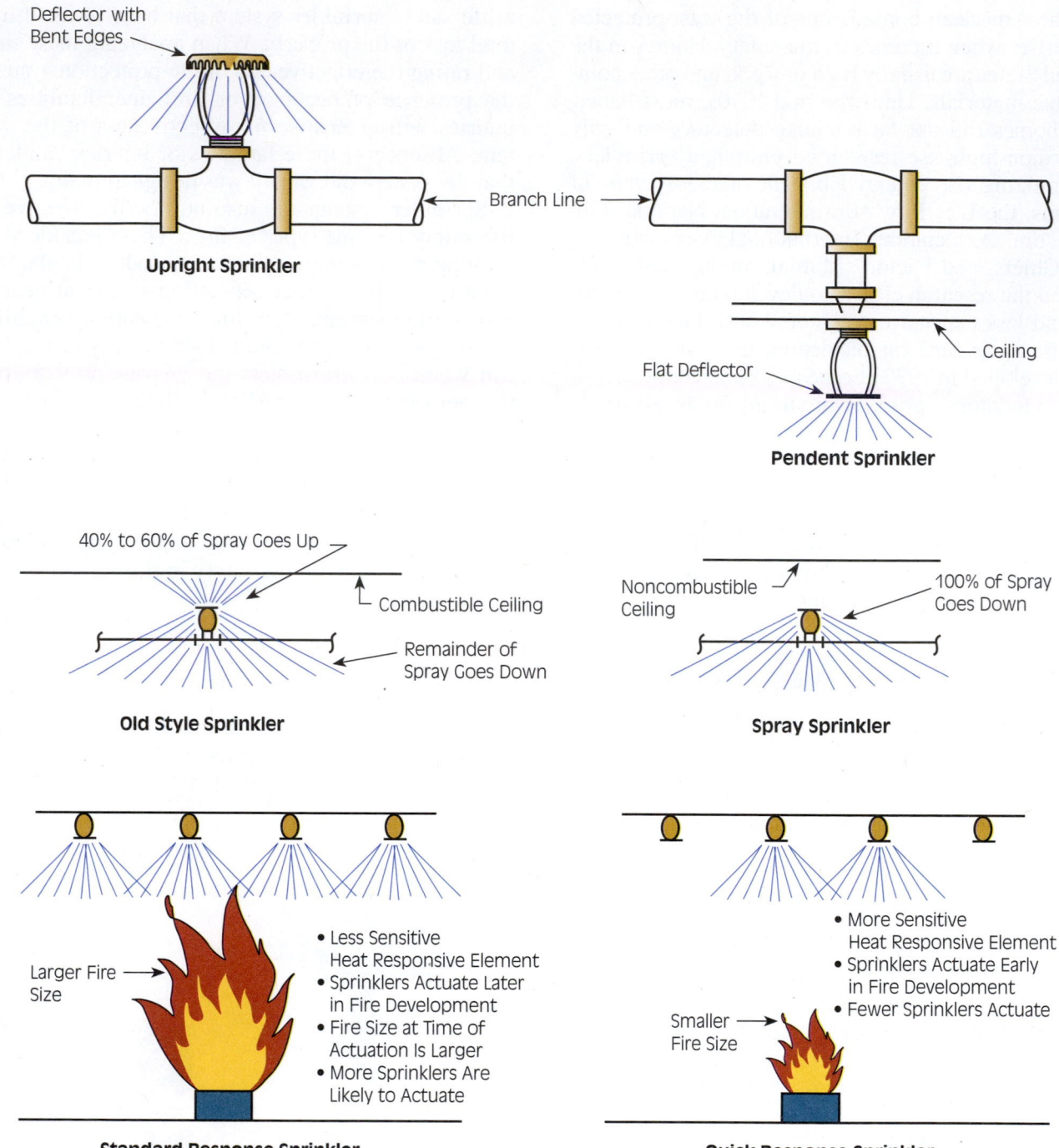

FIGURE 12-8 Sprinkler differences.

and apply water to the fire. (Note the word *usually.* Some types of sprinkler systems, such as a deluge-type system, have separate detection devices but the head is still the applicator. See deluge system discussion later). Sprinkler heads must be appropriate in design and performance, orientation, application or environment, and temperature for the type of property to be protected.

Sprinkler heads come in many designs. These differences affect their performance. One main difference in sprinkler heads is the new and old-style sprinkler head, **Figure 12-8.** Sprinklers designed and manufactured up to the early 1950s are called *old-style sprinklers.* They deflect the majority of the water toward the ceiling to further break up its pattern. The effect was that the water was concentrated in a smaller area.

The *new style* or *standard sprinklers* have a much more even flow across the coverage area and do not bounce the water off the ceiling, Figure 12-8. Standard sprinklers are marked with SSU (Standard Sprinkler Upright) or SSP (Standard Sprinkler Pendent) on the deflector. Orientation means up, down, or sideways. Upright sprinklers have the head vertically above the piping with the deflector at the top, while pendents have the head suspended below the piping with the deflector at the lowest point, **Figure 12-9A–C.** The design of sidewall sprinklers allows them to be placed near the wall and, although the deflectors look bent, they provide the correct pattern. Sidewall heads may be pendent, upright, or horizontal. Each must be properly positioned.

Application and environment may require corrosion-resistant heads, special dry head with extension piping, rack storage head, or decorative head. The orifice or the size of the water opening varies from ¼ to ¾ inch (6.4 to 19 mm) depending on the occupancy protected. Sizes other than ½ inch and 17/32 inch (12.7 and 13.5 mm) are noted on the sprinkler frame.[3] Temperature ratings range from ordinary at 135°F (57°C) to ultra high at 625°F (343°C) with the most common temperature being 165°F (74°C).[4] See **Table 12-1.**

The operation of a sprinkler head begins with the heat of the fire raising the temperature of the sprinkler head and bringing its fusible element to the fusing point. The fusible elements can be of three types. The first is a fusible link that has a metal or solder that melts at a fixed temperature. The second is a bulb filled with a liquid, leaving only a small air bubble, which expands and bursts the bulb at a fixed temperature. The third

(A)

(B)

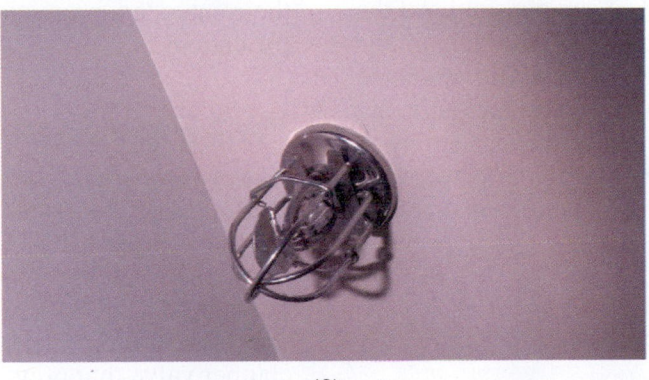

(C)

FIGURE 12-9 Types of sprinklers: (A) upright, (B) pendent (without guard), and (C) pendent sidewall (with guard). *(Photos by Fred Schall)*

TABLE 12-1   Temperature Ratings, Classifications, and Color Coding

| Temperature Rating | | Temperature Classification | Color Coding | Glass Bulb Colors |
°F	°C			
135–170	57–77	Ordinary	Uncolored or black	Orange or red
175–225	79–107	Intermediate	White	Yellow or green
250–300	121–149	High	Blue	Blue
325–375	163–191	Extra high	Red	Purple
400–475	204–246	Very extra high	Green	Black
500–575	260–302	Ultra high	Orange	Black
625	343	Ultra high	Orange	Black

Note: A portion of this table is reprinted with permission from the Fire Protection Handbook, 19th edition. Copyright © 2003, National Fire Protection Association, Quincy, MA 02269.

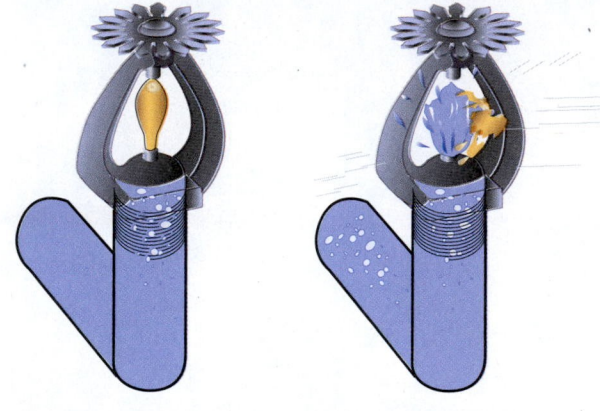

FIGURE 12-10 Sprinkler head closed and opened or operating.

type uses a chemical pellet that melts at a fixed temperature. The fusible element melts or bursts and levers holding the cap then fall out. The cap pops out and the water flows, striking the deflector and spraying into the designed pattern, **Figure 12-10.** A newer innovation is an on/off head that operates at a fixed temperature, but when the fire's temperature drops, it allows a spring to close the waterway. This type of sprinkler can reopen and close as necessary, depending on the temperature.

TYPES OF SPRINKLER SYSTEMS

Many types of occupancies and businesses exist, each with specific attributes when it comes to fire suppression. Some sprinkler systems protect homes or apartment buildings. Others protect businesses with highly sensitive electronic equipment. Still others must protect homes and businesses in areas subject to harsh winter conditions. As a result, many types of specialty sprinkler systems have been engineered to perform fire suppression in different circumstances.

Specialty Sprinkler Systems

Specialty sprinkler systems include some combination-type sprinkler systems and systems that cannot meet the standards for some reasons. They may have an inadequate water source or supply or may be a partial or outside system. Even if a system does not completely meet a standard, it provides a higher measure of protection than if no protection were available. Fire departments with in-district properties using specialty sprinkler systems should be familiar with their intended protection strategy and any of their limitations.

Wet Pipe Systems

A **wet pipe sprinkler system** has automatic sprinklers attached to pipes with water under constant pressure, **Figure 12-11.** This allows quick response when the head is opened. The wet pipe system is the simplest sprinkler system in design and operation. The main or alarm valve is a one-way check or clapper valve that prevents water from reentering the water supply and, when closed, shuts off the water flow to the alarm line. Both sides of the alarm valve have pressure gauges that register the water pressure

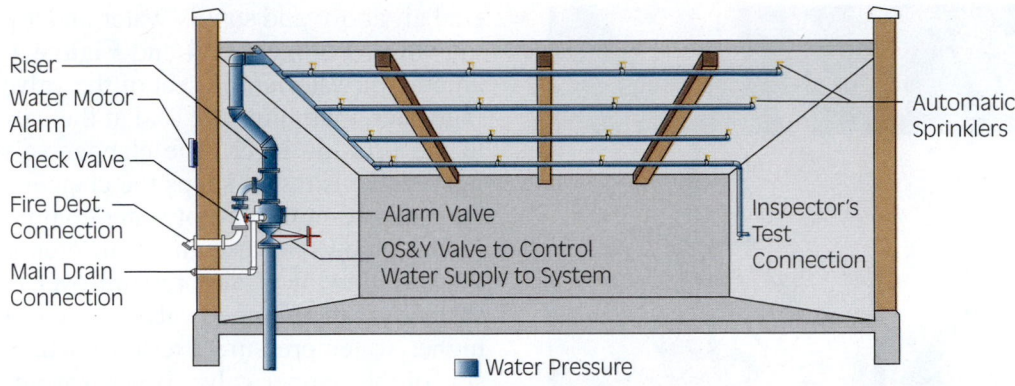

Riser
Water Motor Alarm
Check Valve
Fire Dept. Connection
Main Drain Connection

Automatic Sprinklers

Alarm Valve
OS&Y Valve to Control Water Supply to System

Inspector's Test Connection

■ Water Pressure

FIGURE 12-11 Wet pipe sprinkler system.

of the supply and the system. The system side gauge should read a slightly higher pressure as the reclosing clapper valve will trap any pressure surges. The alarm line piping usually has a retard chamber that acts to prevent false alarms from a sudden pressure surge in the water supply, **Figure 12-12.** The chamber collects a small volume of water before allowing a continued flow to the alarm device. The water from a surge is drained from a small hole in the bottom of the collection chamber. A waterflow indicator, a vane or paddle in the waterway, detects the water flow and activates an alarm signaling system.

A wet pipe system has three more valves. The first is the control valve, which is used to shut off the supply of water to the system and is usually an outside stem and yoke (OS&Y) valve, **Figure 12-13.** The second valve is the main drain, which allows the system to be drained for maintenance or to be restored from a fire, and the last valve is an inspector test valve. The test valve is at the farthest end of the system and is used to simulate the flow of a single head and to measure the response time of the system.

The operation of a wet pipe system starts with the fusing or bursting of a sprinkler head, which causes it to begin applying water to the fire. As the water pressure in the system begins to fall, the main check valve opens and water flows into the system and the alarm line, filling the retard chamber, and then activating the automatic alarm and water motor gong. The alarm signal may be used to notify the fire department or an alarm company. *After ensuring the fire is out or completely under control,* fire personnel will close the control valve. Sprinkler maintenance personnel replace the head and restore the system. When the system is shut down, a firefighter with a radio should be posted at the control valve and be ready to reopen the valve.

FIGURE 12-12 Retard chamber.

NOTE

Some wet pipe systems are located in or have parts of the system located in areas subject to freezing temperatures. To protect these systems, an antifreeze solution can be added to the water in the system. These systems require special attention to restore and maintain.

FIGURE 12-13 OS&Y valves chained in the opened position.

Dry Pipe Systems

In **dry pipe systems,** air under pressure replaces water in the system to protect against freezing temperatures. The system uses a dry pipe valve to maintain pressur- ized air above and supply water under pressure below the valve, **Figure 12-14** and **Figure 12-15.** A small amount of water at the seat of the valve, called prim- ing water, maintains the seal at the valve and is filled to the priming level. The clapper valve has a lock- ing mechanism that keeps the clapper open until it is manually reset to prevent water columning.

Dry pipe valves use an air differential system with a small amount of air pressure over the larger head surface of the clapper valve, which keeps back the higher water pressure exerted on the smaller water side of the clapper valve. If water were allowed to fill the riser above the clapper, the water column's weight would never allow the clapper to be forced open and would make the system inoperative. When a sprin- kler head is fused by heat, the air is discharged. As the air pressure drops below the pressure of the sup- ply water, the clapper valve locks open. Since air, as opposed to water, is in the system, dry pipe systems are slightly slower to flow water than wet pipe sys- tems. Most systems have either an exhauster or an accelerator to speed the operation of the dry pipe valve. The exhauster detects the decrease in air pres- sure and helps bleed off air. The accelerator detects the decrease in air pressure and pipes air pressure below the clapper valve, speeding its opening. Drain and alarm valves in a dry pipe system are similar to wet pipe systems. Dry pipe systems are normally used in unheated buildings, in buildings that refrigerate or freeze materials, or in outdoor applications where freezing temperatures occur—these systems require a heated valve room.

The dry pipe system is more complex in design than a wet pipe system and also harder to return to service after activation. This system requires the dry pipe valve cover to be opened after draining the sys- tem, and the lock on the valve must be reset. The valve must be primed, air pressure must charge the lines, and the control valve must be opened carefully to prevent retripping of the system or creation of a water column.

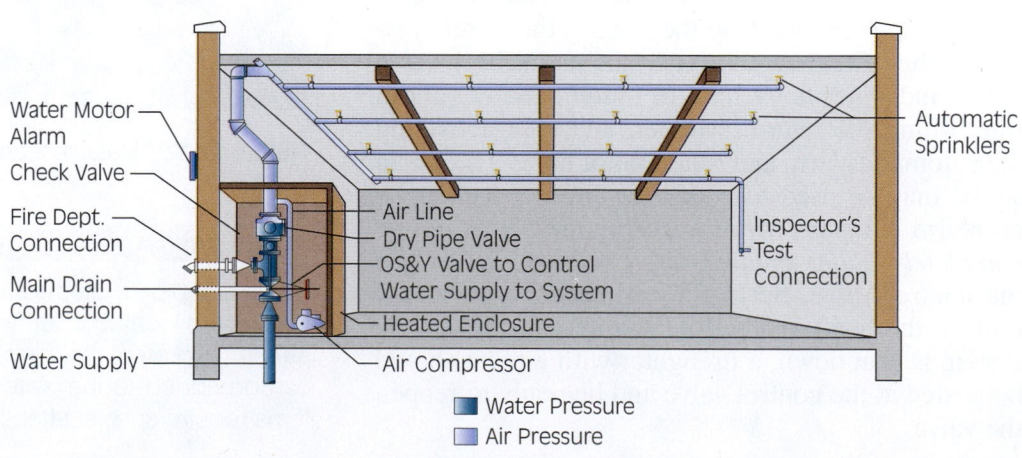

Water Motor Alarm
Check Valve
Fire Dept. Connection
Main Drain Connection
Water Supply
Air Line
Dry Pipe Valve
OS&Y Valve to Control Water Supply to System
Heated Enclosure
Air Compressor
Automatic Sprinklers
Inspector's Test Connection

■ Water Pressure
■ Air Pressure

FIGURE 12-14 Dry pipe system schematic.

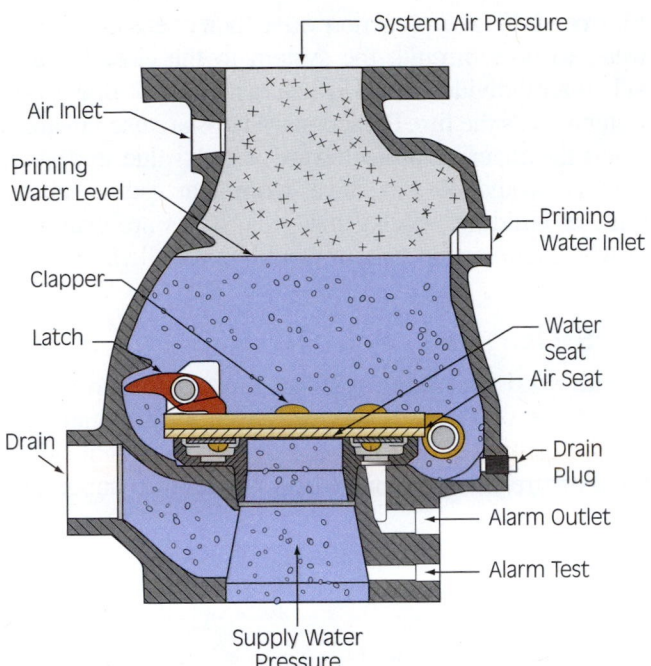

FIGURE 12-15 Dry pipe valve holds back the water with air pressure.

Deluge Systems

Deluge systems are designed to protect areas that may have a fast-spreading fire that could engulf an entire area. Petroleum-handling facilities, aircraft hangars, some manufacturing facilities, and hazardous materials storage areas are all examples of occupancies that may have a deluge system. The essential difference between a standard wet system and a deluge system is the individual sprinkler head. Deluge systems utilize open heads, without any fusible elements. When a fire is detected, the deluge system delivers water (or foam solution) to all heads in the area—allowing total coverage of the area.

Operationally, the deluge system must interface with a detection system. Once the detection system activates, it sends a signal to a "deluge valve," which opens, sending water or foam solution to all the open heads, **Figure 12-16.** Most municipal water systems lack the pressure to effectively supply the numerous open heads in the deluge system. For this reason, deluge systems also incorporate a fire pump to boost pressure in the system. Likewise, deluge systems that deliver foam solution require a foam supply tank, foam pump, and foam mixing device.

As can be imagined, deluge system activation will cause tremendous quantities of water to flow. While this is essential to fire control of high hazards, activation of the system in absence of a fire will likely cause water damage and a significant cleanup effort. For this reason, deluge system activation usually requires activation of several detectors versus a single detector. For example, several flame detectors must activate before a signal is sent to the deluge valve. Some systems require detection from different kinds of devices (i.e., activation of a flame detector and a heat detector). To prevent unneeded damage from activation, the deluge system may utilize a manual override alarm and "deadman" switch. In these systems, an alarm is sent telling occupants that the deluge system

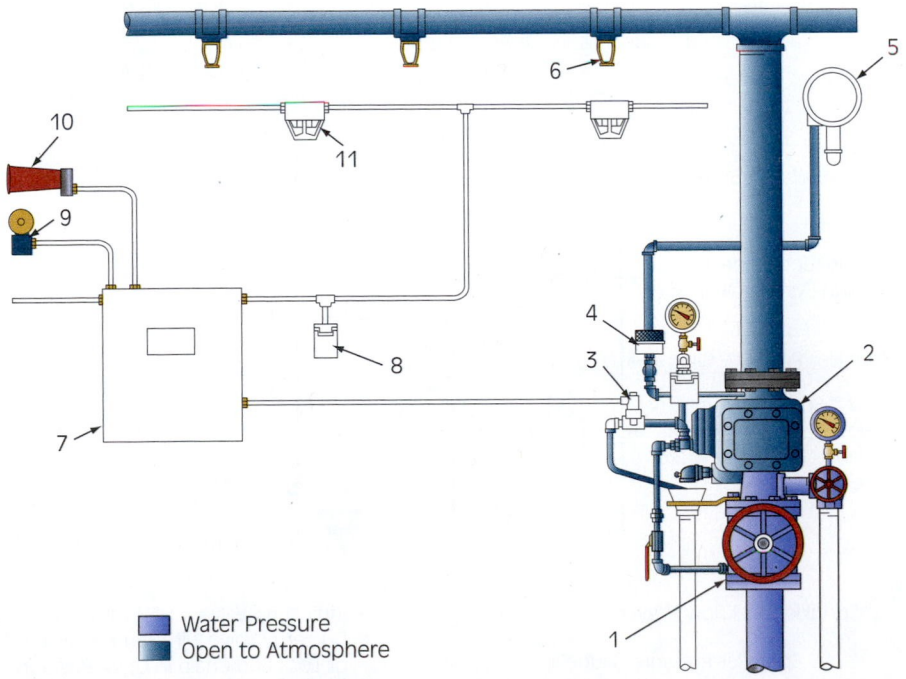

Water Pressure
Open to Atmosphere

FIGURE 12-16 Deluge system schematic: 1, OS&Y valve; 2, deluge valve with basic trim; 3, solenoid valve and electric actuation trim; 4, pressure alarm switch; 5, water motor alarm; 6, spray nozzles or open sprinklers; 7, deluge releasing panel; 8, electric manual control stations; 9, fire alarm bell; 10, trouble horn; 11, heat detectors.

is ready to activate. If someone does not push a button in a prescribed time, the system will deluge the area. A person who does push the hold button cannot release the button until the activated detectors are reset—which requires additional persons in the system. If the person does release the button prior to system reset, the deluge valve will open and the system will be fully activated—hence the deadman label.

Preaction Systems

Preaction systems are similar to the dry pipe and deluge system. The system has closed piping and heads with air under no or little pressure, but the water does not flow until signaled from a separate fire detection system,

Figure 12-17. The preaction valve then opens and allows water to flow through the system to the closed heads. When an individual head is heat activated, it opens and water attacks the fire. Preaction systems are used in areas where the materials protected are of high value, and water damage would be expensive. Computer rooms, museums, or buildings storing historical items are examples of where a preaction system would be installed.

Residential Systems

Residential sprinkler systems are smaller and more affordable versions of wet or dry pipe sprinkler systems, **Figure 12-18.** They are designed to control the level of fire involvement while residents escape. The

FIGURE 12-17 Preaction system schematic: 1, OS&Y valve; 2, deluge valve with basic trim; 3, check valve; 4, solenoid valve and electric actuation trim; 5, water pressure alarm switch; 6, 1.5-psi low air pressure alarm switch; 7, 1.5-psi supervisory air pressure control; 8, water motor alarm; 9, automatic sprinklers; 10, deluge releasing panel; 11, electric manual control stations; 12, fire alarm bell; 13, trouble horn; 14, heat detectors. *(© Copyright Simplex Grinnell. All rights reserved.)*

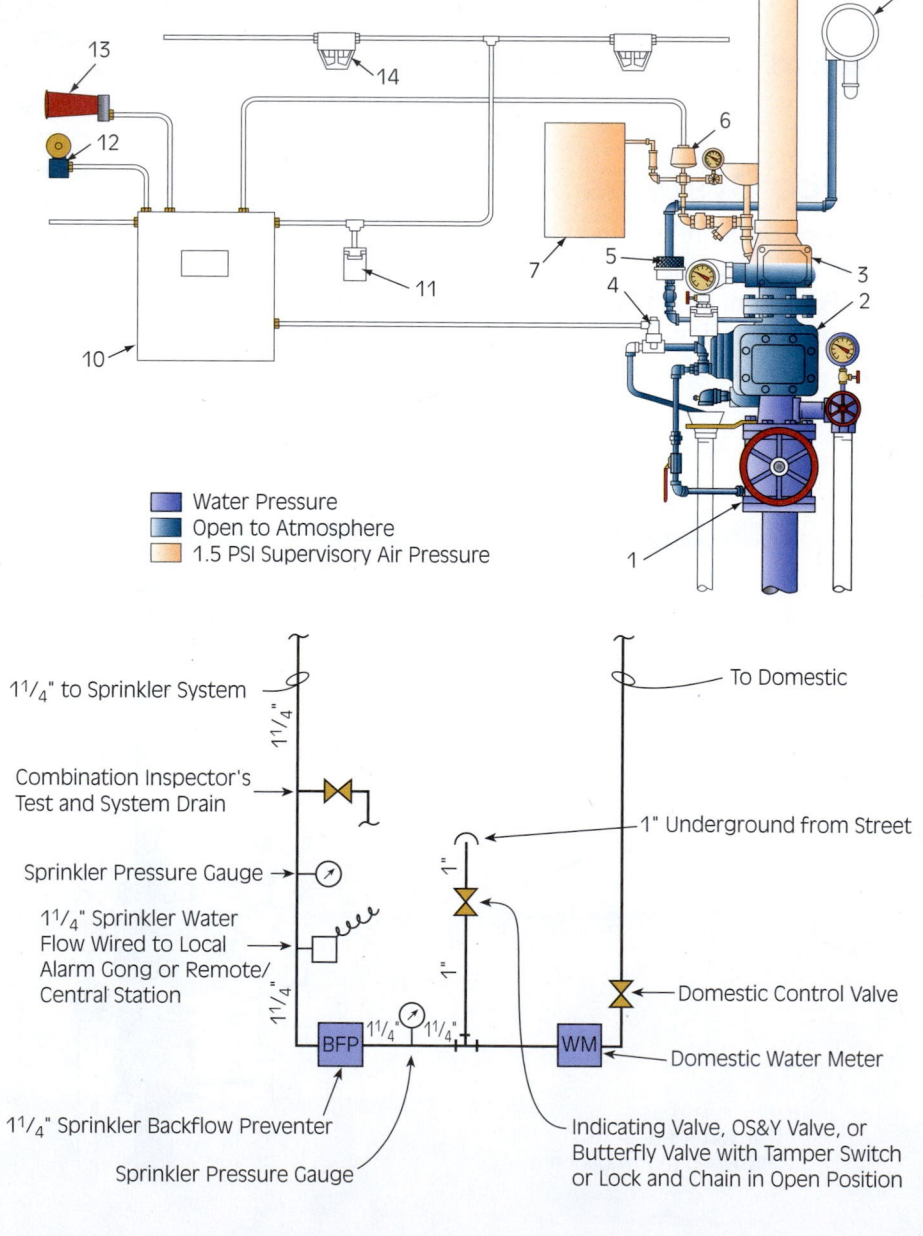

- Water Pressure
- Open to Atmosphere
- 1.5 PSI Supervisory Air Pressure

FIGURE 12-18 Schematic of residential sprinkler system.

sprinkler system water supply is combined with the domestic water supply, and flow rates are designed for one or a few heads in operation. Residential sprinkler heads have a faster response time and use a lighter and smaller piping than commercial wet/dry pipe systems. Residential systems were the first suppression systems to use plastic piping. Residential systems use a check valve, waterflow alarms, and drains similar to the bigger systems and are not required to have a fire department connection, although some do. Some residential systems also use antifreeze to protect all or part of the system.

STANDPIPE CLASSIFICATIONS

Standpipe systems are designed to allow firefighters to fight fires in larger buildings by prepiping water supply lines through the building. Some systems allow both occupants and firefighters to use the system. The systems for building occupant use have hoselines and nozzles attached. Standpipe systems are not automatic; they must be operated by people. They are used in high-rise buildings; large commercial, retail, and industrial buildings; places of public assembly; and other areas where advancing hoselines from hydrants or fire apparatus would be difficult due to the size of the building. Tunnel or subway systems and shopping malls have horizontal standpipe systems.

Standpipe systems are classified according to the intended user and each class sets requirements for volume and size of outlets. *Class I systems* are designed for use by the fire department or trained personnel such as a fire brigade, **Figure 12-19.** The systems use 2½-inch (65-mm) outlets. A single standpipe should flow a minimum of 500 gpm (1,890 L/min) plus an additional 250 gpm (945 L/min) for each additional riser. No hose is provided for this class. Outlets are found in stairwells on each floor, near the exits, or in the hallways on each floor near the stairs.

A *Class II system* is designed for use by untrained building occupants and has a 1½-inch (38-mm) outlet with a minimum flow of 100 gpm (378 L/min), **Figure 12-20.** Hose is provided and is either 1 or 1½ inch (25 or 38 mm). Firefighters wishing to use a Class II system should not use the hose provided, because it is not tested and may be single-jacketed, unlined hose with a nozzle that has no shutoff. Firefighters using this type system should replace the hoseline with fire service hose.

CAUTION

Firefighters wishing to use a Class II or III standpipe system should not use the hose provided. It is not tested and may be single-jacketed unlined hose with a nozzle that has no shutoff. Firefighters using this system should replace the hoseline with fire service hose.

FIGURE 12-19 Class I standpipe system.

FIGURE 12-20   Class II standpipe system.

A *Class III system* is used by either of the users of the other two classes but meets the Class I requirements for flow. It has both 1½-inch (38-mm) and 2½-inch (65-mm) outlets or may have a 2½-inch (65-mm) outlet with a 2½- to 1½-inch (65- to 38-mm) reducer. All of the flows given are at the top outlet with a residual pressure of 65 psi (448 kPa).[5] Any of these systems may have a pressure-regulating device, which is designed to maintain or reduce the pressure to a set amount—regardless of the height of the building. Standpipes have height limits on each section (about 280 feet [88 m]), but the pressure difference in an unregulated system could vary by almost 140 psi (966 kPa) from top to bottom.

> **CAUTION**
>
> Pressure-regulating valves on standpipe systems may cause firefighter problems if improperly installed or reset. Some are preset at the factory and must be installed on the correct floor to properly regulate the pressure. Others are adjustable and can be reset after installation. Improperly installed pressure-regulating devices have contributed to some firefighter fatalities in high-rise firefighting. Regular and thorough inspection and testing of these devices by a qualified technician is highly recommended due to maintenance problems.

The types of standpipe systems are differentiated based on water supply. An automatic wet pipe system has a water supply that is ready to conduct firefighting, whereas automatic dry and semi-automatic dry pipe systems use dry pipe until a hose station (an outlet) is opened to allow water to flow into the system. The semi-automatic system requires an activation valve to be opened manually. The manual dry pipe is dry and relies on the fire department connection for water supply, whereas a manual wet pipe system has the piping filled with water but needs the fire department connection for firefighting supply.[6]

OTHER PROTECTIVE SYSTEMS

Many other types of protective systems are used today. Some are rather simple, as in those designed to protect the grill, fryer, and ductwork of a local restaurant. Others are extremely complex and are designed to prevent or suppress an explosion in highly hazardous locations. Firefighters need to understand the most common types of protective systems: fire departments responding to the more complex systems should ensure that all firefighters are familiar with the operation of the system, the hazards of the fires they control, and extinguishing agents.

Local Application and Hood Systems

One of the most common types of protective systems is a local application system, especially the systems protecting the cooking areas of restaurants, **Figure 12-21.** Local application means that the system is designed to protect only a certain portion of the building, usually directly where the hazard will occur or spread. Local applications are also used in laboratory hoods, paint booths, and other small hazardous locations, **Figure 12-22.** Most of the local application systems use dry or wet chemical agents with the most popular being an ABC dry chemical. Higher temperature cooking units and oils have created the need for a new Class K classification, and these systems also use dry or wet chemicals. Hood systems use a heat-sensitive device or a manual switch for activation.

Total Flooding Systems

Total flooding systems are used to protect an entire area, room, or building. The total flooding system discharges an extinguishing agent that completely fills or floods the area with the extinguishing agent to smother or cool the fire or break the chain reaction. Total flooding systems can use carbon dioxide or other inert gases, halogenated or clean agents, dry chemicals, or foam as extinguishing agents (see Section I, Chapter 8 for a discussion of these agents). They are effective as long as the proper amount discharges, the area is contained to prevent loss of agent, or the fire goes out prior to ventilation of the agent. There are hazards associated with each type of application, with which firefighters and building personnel should be familiar.

FIGURE 12-21 Local application system (notice nozzles) protecting cooking area and ductwork.

FIGURE 12-22 Dry chemical nozzles are located above the island and at ground level.

<table>
<tr><td>

CAUTION

Carbon dioxide (CO_2) and other inert gases such as nitrogen can be used to smother a fire. CO_2 is most often used for firefighting, while nitrogen is used to inert an atmosphere to allow welding or flammable activities to take place near combustible materials. CO_2 systems can be either high- or low-pressure storage systems and are designed to fill the area with a concentration of the gas that extinguishes the fire, **Figure 12-23.** Unfortunately, this smothering effect is also fatal to any living beings in the area, and an evacuation period is built into the system activation. Firefighters, using SCBA, may enter the area to ensure the fire is extinguished and all persons have been safely removed. This agent is very clean and is often used in computer and electronics areas. If the fire is out, ventilation of the area and restoration of the system are the only cleanup activities required.

</td><td>

CAUTION

Halon and the newer clean agents are also used for total flooding applications. Depending on the toxicity and concentration of agent needed, some of these systems may even be used when people are still occupying the area. Firefighters should still use SCBA when entering areas where even safe agents have been discharged to ensure that the fire is out. As with CO_2, electronics and other high-cost or sensitive valuables are protected by these systems.

</td></tr>
</table>

FIRE DEPARTMENT OPERATIONS WITH PROTECTIVE SYSTEMS

Standpipe, sprinkler, and other protective systems should be part of the fire department overall strategic plan to provide fire protection for its community. This strategic plan recognizes the community's hazards and tries to keep them within certain limits. When properties or processes create hazards beyond those limits, protective systems are required. The fire department must survey its hazards and also its resources, both public and private. A program of maintenance, inspections, and pre-emergency planning keeps the department in compliance with the strategic plan. The property owner does maintenance, the fire prevention bureau does the inspections, and fire companies do the preplanning. Preplanning identifies hazards and resources, and some key resources are protective systems. Fire companies should know where every protective system is and how it operates. For standpipe and sprinkler systems, fire company personnel should know the type of system, the location of all fire department connections, the two closest water supply points, and the location of key valves. This information should be noted on area maps, pre-emergency plans, fire department laptop software, and dispatch information. Knowing where the systems are and how to use them is the first step in coordinating a proper action plan. The next step is having SOPs that give a recommended course of action in buildings with these systems.

From an operational point of view, protective systems can be separate components requiring the fire department to look in different areas to understand what has been activated. In newer systems, the protective systems are integrated into a single "smart"

FIGURE 12-23 Low-pressure carbon dioxide extinguishing system.

system that includes an annunciator panel, fire alarm control panel, and system override controls. Typically, annunciator panels are near the primary entrance to the building and simply "announce" what has been activated and where. The fire alarm control panel and other protective system controls are in a locked room and allow the fire department to reset, silence, or otherwise control the system. Large occupancies such as a high-rise may have a complete fire command center where the fire department can monitor and/or control all protective systems as well as HVAC systems and building intercom/communications systems.

Standpipe Operations

Standpipe operations start with establishing a water supply to the fire department connection with a minimum of one hoseline. Additional hoselines are added as needed. The pump operator should immediately charge the standpipe system for the pressure required at the reported fire level. The first arriving unit officer should have personnel check the annunicator panel and go to the reported location of the fire. Methods of proceeding to this location will vary depending on the situation, but may involve the use of stairs or elevators and should be covered in SOPs. Firefighters using elevators should do so only if they are equipped with firefighter service control systems. They should not take the elevator directly to the reported fire floor because the doors could open directly into the fire. Fire crews should stop at least two floors below the reported fire level.

FIGURE 12-24   Hose pack for standpipe use.

nearest and safest outlet in relation to the fire location. This may require using the floor below and advancing the hoseline toward the fire location. Depending on the fire's location, the line may be charged prior to entering the hallway or delayed until reaching the door of the room or unit on fire. Additional lines may be run off the gated wye, another standpipe outlet on that floor, or another floor.

> **STREETSMART TIP**
>
> When advancing a hoseline in a stairwell, extra hose should be looped up the stairs and then down so that when pulled, gravity will assist in pulling the hose down instead of pulling it up the stairs.

> **CAUTION**
>
> Firefighters using elevators should do so only if they are equipped with firefighter service control systems. They should not take the elevator directly to the reported fire floor because the doors could open directly into the fire. Fire crews should stop at least two floors below the reported fire level. If a fire is reported on or below the sixth floor, it would be wise to use the stairs for entry, if possible.

Personnel should have full personal protective equipment, standpipe pack, and forcible entry equipment. The suggested standpipe pack equipment is a minimum of 150 feet (46 m) of 1½-inch (38-mm) hose or larger (many departments are now using 1¾ or 2 inch [45 and 50 mm]), a 2½- to 1½-inch (65- to 38-mm) gated wye with 1½- to 2½-inch (38- to 65-mm) increaser, 5 to 10 feet (1½ to 3 m) of 2½-inch (65-mm) hose, spanner wrench, adjustable wrench, and valve wheel. Additional sections of hose are recommended if personnel allow, **Figure 12-24**. Personnel should connect their standpipe pack to the

Sprinkler System Operations

Fire incident operations in buildings with sprinkler systems begin with an investigation of the building to determine if water is flowing or if a fire is present. Typically, the first arriving officer checks the annunciator panel and then directs the crew to pull fire attack lines (or begin standpipe operations). Typically, SOPs require the first arriving crew to start fire attack while the pump operator sets up to support the sprinkler system. Establish a water supply to the fire department connection with a minimum of one medium-diameter hoseline. Additional hoselines are added as needed. The pump operator should charge the sprinkler system when smoke or fire is showing, the water motor gong is operating, or when ordered by the officer-in-charge. The system should be charged immediately if it is combined with a standpipe system. Sprinkler systems are normally charged and maintained at 150 psi (1035 kPa). Personnel should advance hoselines into the fire area and conduct extinguishment and overhaul operations. If no fire is found, the officer may direct firefighters to stop sprinkler flow as discussed earlier in this chapter. It is important to note that all sprinkler system alarms should be treated as an actual fire until confirmed otherwise. Just because a building is fully sprinkled does not mean the fire department should prepare to fight fire differently. Being fully prepared with PPE, tools, radios, and so forth, should take precedence.

Detector Activation Operations

Operationally, the generic "fire alarm activation" response can range from investigating an alarm malfunction to a full-blown fire operation. Histories of false alarms have caused some firefighters to take a less than ready approach to their operations. Not only is this dangerous, it can be fatal. Most experienced firefighters can tell a story of arriving at a fire alarm activation, only to be surprised that an actual fire existed. These firefighters now respond to all alarms just as if an actual fire were awaiting them. It cannot be overemphasized: Firefighters must treat all fire alarm activations as an actual fire and prepare for the worst.

First arriving firefighters at an alarm activation should dismount the apparatus wearing full PPE with SCBA, and carry necessary tools and a radio. Typically, the first step at an alarm activation is to investigate. At residential structures, firefighters should meet with the building occupant if no fire signs or conditions exist. In cases where nobody is awaiting fire department arrival, the officer or designee should conduct a "360," which is a systematic walk around the structure to check for smoke or other indicators of an internal fire. Local SOPs should address further steps if no visible signs exist. These SOPs can involve notifying neighbors, watching the structure for evidence of fire or smoke, or entering the structure and conducting an interior search. If access can be made to the structure, caution should be employed. Smoke

detector and CO detector activation may sound similar, but the CO can incapacitate unprepared firefighters with little warning.

At commercial structures, the first arriving crew should first check the annunciator panel to ascertain which device has activated. If no annunciator exists, firefighters may have to access the fire alarm control panel if no obvious signs are found. After business hours, firefighters should perform a "360" to look for signs. If the building is so equipped, the officer will then access the lock or knox box and enter the structure. Anytime a lockbox is utilized, the officer should radio dispatch that locked entry is being made. This will allow dispatch to notate the event with a time stamp of when it occurred. After checking the annunciator panel, the firefighter should proceed to the indicated zone and investigate. If any signs of smoke or fire are visible, the operation transitions from investigation to fire attack mode.

Buildings equipped with strobes and loud audible warning devices can be annoying to occupants if they see no signs of fire. Firefighters must not get trapped in a "false alarm mentality" and arbitrarily silence these alarms to investigate the activated zone. Premature silencing may send a message to occupants that they can return to their business. Using simple foam earplugs (which should be in the pocket of firefighters' PPE) can reduce the potential of temporary hearing loss while investigating the alarm. Once the activated detector is found, the immediate area can be checked for potential causes. If no fire signs are found, the firefighters can then take steps to clear the detector and attempt to reset the alarm. Correction of systems that will not reset are the property owner or manager's responsibility. Often, the fire department needs to advise the building representatives to contact their alarm company or a repair technician. All fire department actions must be documented on the incident report.

Operations for Other Protective Systems

Local SOPs should address the operations for total flooding, foam, dry chemical, and other unique systems. Total flooding systems that have discharged present a hazard to occupants, and firefighters must never enter this environment without first engaging their SCBA. The introduction of air into total flooding environments can reignite the fire. Firefighters should also check ceiling spaces and the rooftop ventilator to make sure the fire has not extended out of the hood and ducts. Hood systems that have discharged will require an overhaul effort to ensure that the fire is totally extinguished. Dry chemical systems are quite effective for fire control but require a large cleanup effort. This cleanup is not the responsibility of fire-fighters: it is important to notify occupants that breathing dry chemical can cause lung irritation. Activation of suppression systems may cause secondary damage to electrical circuits and other building infrastructure. It is important to be aware of this potential and protect firefighters and occupants accordingly.

LESSONS LEARNED

Protective systems are designed to automatically detect or suppress a fire, or to assist in extinguishing the fire. They can apply water or other extinguishing agents. Protective systems have been credited with saving both lives and property and are essential fire protection tools. Firefighters need to understand the value and operation of these systems to protect their communities.

Sprinkler systems are used for detection and suppression and can apply water or foam to extinguish the fire. Each type of system uses a different alarm valve but many of the other features are similar. Firefighters are still needed in sprinkler-protected property to finish extinguishing any fire that the system could not.

Standpipe systems facilitate manual fire suppression in which people do the firefighting. Through the use of a system of pipes with water supply and discharge connections, standpipes supply water in large buildings to facilitate fire suppression.

Other protective systems detect fires and apply extinguishing agents to fires in proximity to the hazard.. Restaurants and kitchens usually have a hood system protecting the cooking areas, while other systems are used to protect high hazards or high value items. These protective systems are highly specialized and firefighters must understand their operation and any hazards of the agents.

KEY TERMS

Auxiliary Appliances   Another term for protective devices, particularly sprinkler and standpipe systems.

Deluge System   Protective system designed to protect areas that may have a fast-spreading fire engulfing the entire area. All of its sprinkler heads are already open, and the piping contains atmospheric air. When the system operates, water flows to all heads, allowing total coverage. The system uses a deluge valve that opens when a separate fire detection system senses the fire and signals to trip the valve open.

Dry Pipe System   Air under pressure replaces the water in the system to protect against freezing temperatures. The sprinkler control valve uses a dry pipe valve to keep pressurized air maintained above with the supply water under pressure below the valve.

Local Application System   Designed to protect only a certain or local portion of the building, usually directly where the hazard will occur or spread.

Preaction System   Similar to the dry pipe and deluge systems. The system has closed piping and heads with air under no or little pressure, but the water does not flow until signaled open from a separate fire detection system. The preaction valve then opens and allows water to flow through the system to the closed heads. When an individual head is heat activated, it opens and water attacks the fire. Usually used when water can cause a large dollar loss.

Residential Sprinkler System   Smaller and more affordable version of a wet or dry pipe sprinkler system designed to control the level of fire involvement such that residents can escape.

Sprinkler System   Protective system designed to automatically distribute water through sprinklers placed at set intervals on a system of piping, usually in the ceiling area, to extinguish or control the spread of fires.

Standpipe System   Piping system that allows for the manual application of water in large buildings.

Total Flooding System   Used to protect an entire area, room, or building by discharging an extinguishing agent that completely fills or floods the area with the extinguishing agent to smother or cool the fire or break the chain reaction.

Wet Pipe Sprinkler System   Has automatic sprinklers attached to pipes with water under pressure all the time.

REVIEW QUESTIONS

1. What are the two most common styles of smoke detectors?

2. How does an ionization detector operate?

3. Describe an upright and pendent sprinkler?

4. A person has fallen asleep on a sofa while smoking and a small fire has erupted. Explain the events that will transpire regarding the sprinkler system.

5. Name the three classes of standpipes. Can these standpipes be used by the general public? Why or why not?

6. What is the difference between an annunciator panel and a fire alarm control panel?

ENDNOTES

1. See NFPA Standards 13, 13D, 13R, and 25 for more information.

2. NFPA 13D, *Standard for the Installation of Sprinkler Sytems in One- and Two-Family Dwellings and Manufactured Homes* (Quincy, MA: National Fire Protection Association, 2007).

3. NFPA, *Fire Protection Handbook,* 20th ed. (Quincy, MA: National Fire Protection Association, 2008).

4. NFPA 13, *Standard for the Installation of Sprinkler Systems,* 2007 Edition (Quincy, MA: National Fire Protection Association).

5. NFPA, *Fire Protection Handbook,* 20th ed. (Quincy, MA: National Fire Protection Association, 2008).

6. NFPA, *Fire Protection Handbook,* 20th ed. (Quincy, MA: National Fire Protection Association, 2008).

13

Building Construction

LEARNING OBJECTIVES

After completing this chapter, the reader should be able to:

13-4 Identify the hazards imposed on a structure by fire.

13-5 List the effects of fire on various building structure materials and assemblies.

13-12 Perform an assessment to determine whether an area of a structure is safe for rescue operations.

13-17 Describe structural instability and signs of possible structural failure.

13-18 Demonstrate the ability to analyze changing fire conditions and their effects on structural integrity.

13-19 Identify structural fire considerations to be made prior to extinguishment

*The FF I and II levels, as defined by the NFPA 1001 Standards, are identified in different colors: FF I = black, FF II = red, additional information = blue.

INTRODUCTION

Many firefighters are injured and killed when fire or suppression activities weaken a structure, **Figure 13-1.** Often, buildings collapse without a "visual" warning such as sagging floors and roofs, leaning walls, and cracks. Firefighters must understand the types of structures they enter from the perspective of how the buildings are assembled, what materials are used, and how buildings react to fire. Additionally, firefighters must understand how fire travels through a building and choose appropriate tactics to stop the fire before key structural elements are affected by fire and suppression activities. Many firefighter fatality investigations conclude that fire departments need more training and education on building construction and the effects of fire on buildings. This chapter reinforces several key topics introduced in Section I, Chapter 13 regarding building construction and how fire affects buildings.

FIGURE 13-1 This collapse happened seconds after firefighters were repositioned.

STREETSMART TIP

Firefighters must realize that most of the buildings they will work in are already in place, and it is very difficult to determine the type of construction and fire-resistive rating by driving by or standing in front of a structure. Conducting in-service inspections, and securing the building owner's permission to walk through a building and get an "inside view" are ways of learning more about the structures in a particular jurisdiction.

If new construction is being conducted in a firefighter's response area or jurisdiction, contact the building department and secure a set of the plans, and arrange to visit the construction site. Also, take photographs of how the building is built and what materials are used for future training references. Many new materials and construction methods are introduced regularly.

FIRE EFFECTS ON COMMON BUILDING CONSTRUCTION MATERIALS

Many factors determine which material is used to form structural elements. Quality, cost, application, engineering capabilities, and adaptability all play in the suitability of a material. In some cases, the material chosen for a structural application needs to meet fire resilience criteria. Regardless, the firefighter needs to understand how these materials react to fire. In the past, the fire service looked at the characteristics of four basic material types: wood, steel, concrete, and masonry. Each of these materials can be found together or separately. Each material reacts to fire in a different way, **Table 13-1.** Now, advanced material technology has found its way into structural elements. Buildings are being assembled using plastics, graphites, wood derivatives, and other composites. This section covers the four basic building materials as well as some of the new composites.

Wood

Wood is perhaps the most common building material. It is used in millions of residential and commercial buildings. Wood is relatively inexpensive, easy to manipulate, and a replenishable natural resource. Wood has marginal resistance to forces compared to its weight, but it does the job for most residential and small commercial buildings. Wood also burns—and in doing so gives away its mass. The more mass a section of wood has, the more material it must burn away

before strength is lost. This is true of native wood—that is, wood that has been cut from a tree. Engineered wood can react differently when exposed to heat from a fire. Engineered wood includes a host of products that take many pieces of native wood and glue them together to make a sheet, longer **beam** (trees only grow so tall!), or stronger **column.** Plywood delaminates when exposed to fire. Some newer wood products such as composites, which are discussed later in this chapter, can fail just through exposure to heat (they don't necessarily have to burn).

Steel

Steel is a mixture of carbon and iron ore heated and rolled into structural shapes to form elements for a building. Steel has excellent tensile, shear, and compressive strength. For this reason, steel is a popular choice for **girders,** lintels, cantilevered beams, and columns. Additionally, steel has high factory control. It is easy to change its shape, increase its strength, and otherwise manipulate it during production.

As it relates to fires, steel loses strength as temperatures increase. The specific range of temperatures depends on how the steel was manufactured. Cold drawn steel, like cables, bolts, rebar, and lightweight fasteners, loses 55 percent of its strength at 800°F (427°C). Extruded structural steel used for beams and columns loses 50 percent of its strength at 1,100°F (538°C). Structural steel will also elongate or expand as temperatures rise. At 1,000°F (583°C), a 100-foot-long beam (30 meters) will elongate 10 inches (23 centimeters). Imagine what that could do to a building. If

TABLE 13-1	Performance of Common Building Materials under Stress and Fire			
Material	**Compression**	**Tension**	**Shear**	**Fire Exposure**
Brick	Good	Poor	Poor	Fractures, spalls, crumbles
Masonry block	Good	Poor	Poor	Fractures, spalls
Concrete	Good	Poor	Poor	Spalls
Reinforced concrete	Good	Fair	Fair	Spalls
Stone	Good	Poor	Fair	Fractures, spalls
Wood	Good w/grain; poor across grain	Marginal	Poor	Burns, loss of material
Structural steel	Good	Good	Good	Softens, bends, loses strength
Cast iron*	Good	Poor	Poor	Fractures
*Some cast iron may be ornamental in nature and not part of the structure or load bearing.				

a beam is fixed at two ends, it will try to expand—and likely deform, buckle, and collapse. If the beam sits in a pocket of a masonry wall, it will stretch outward and place a shear force on the wall—which was designed only for a compressive force. This could knock down the whole wall! Steel structural elements must stay in their original shape. Any deflection, sagging, or stretching takes the steel out of its engineered shape, and failure is likely to be quick.

> **CAUTION**
>
> Steel softens, elongates, and sags when heated, leading to collapse. Cooling structural steel with fire streams is just as important as attacking the fire.

Because steel is an excellent conductor of heat, it will carry heat of a fire to other combustibles. This can cause additional fire spread, sometimes a considerable distance from the original fire.

Concrete

Concrete is a mixture of portland cement, sand, gravel, and water. It has excellent compressive strength when cured. The curing process creates a chemical reaction that bonds the mixture to achieve strength. The final strength of concrete depends on the ratio of these materials, especially the ratio of water to portland cement. Because concrete has poor tensile and shear strength, steel is added as reinforcement. Steel can be added to concrete in many ways. Concrete can be poured over steel rebar and become part of the concrete mass when cured. Cables can be placed through the plane of concrete and be tensioned, compressing the concrete to give it required strength. Cables can be pretensioned (at a factory) or posttensioned (at the job site). Precast concrete refers to slabs of concrete that are poured at a factory and then shipped to a job site. Precast slabs are "tilted up" to form **load-bearing walls**—thus the term tilt-up construction.

All concrete contains some moisture and continues to absorb moisture (humidity) as it ages. When heated, this moisture content will expand, causing the concrete to crack or spall. **Spalling** refers to a large pocket of concrete that has basically crumbled into fine particles, taking away the mass of the concrete. Reinforcing steel that becomes exposed to a fire can transmit heat within the concrete, causing catastrophic spalling and failure of the structure. Unlike steel, concrete is a heat sink and tends to absorb and retain heat rather than conduct it. This heat is not easily reduced. Concrete can stay hot long after the fire is out, causing additional thermal stress to firefighters performing overhaul.

Masonry

Masonry is a common term that refers to brick, concrete block, and stone. Masonry is used to form load-bearing walls because of its compressive strength. Masonry can also be used to build a veneer wall. A veneer wall supports only its own weight and is most commonly used as a decorative finish. Masonry units (blocks, bricks, and stone) are held together using mortar. Mortar mixes are varied but usually contain a mixture of lime, portland cement, water, and sand. These mixes have little to no tensile or shear strength. They rely on compressive forces to give a masonry wall strength. A lateral force that exceeds the compressive forces within a masonry wall will cause quick collapse of the wall.

> **STREETSMART TIP**
>
> Masonry Walls Collapse: Masonry has very little lateral stability, and in many cases the roof or floor structure of a building holds the walls in place. Steel beams or **joists** will expand during a fire, creating lateral loads that the walls were not designed to withstand. In addition, wood roof structures will burn away or collapse during a fire, leaving little lateral support. The effects of the fire, pulling forces of the collapsing wood structure, or the force of a hose stream may cause a free-standing masonry wall to collapse.

Brick, concrete block, and stone have excellent fire-resistive qualities when taken individually. Many masonry walls are typically still standing after a fire has ravaged the interior of the building. Unfortunately, the mortar used to bond the masonry is subject to spalling, age deterioration, and washout. Whether from age, water, or fire, the loss of bond will cause a masonry wall to be very unstable, **Figure 13-2.**

Composites

New material technologies have introduced some interesting challenges for the firefighting community. Composites are a combination of the four basic materials listed previously as well as various plastics, glues, and assembly techniques. Of particular interest are the many wood products that are widely used for structural elements.

Engineered wooden I beams (slang term is "I-joists") are nothing more than wood chips and veneers that are press-glued together into the shape of an I beam, **Figure 13-3.** The chords are typically made from laminated veneer lumber (LVL) while the **web** is a piece of oriented strand board (OSB) sheeting. While

FIGURE 13-2 Prior to a fire, the effects of age will take their toll on masonry walls. How stable is this wall with joint deterioration and lack of full mortar bond? Actually, this is an unsupported masonry wall which is inherently unstable—made more unstable with the deteriorated masonry.

FIGURE 13-4 A composite truss. Rapid heating will cause the stamped-steel to separate from the wood chords.

FIGURE 13-3 To save on materials and cost, the use of composites or engineered wood structural members is becoming popular. Shown here is a typical wood I beam with finger-glued chords (also called flanges). The web is OSB. The glues used in this assembly will degrade when heated—and fail quickly.

structurally strong (stronger than a comparable solid wood joist), the wooden I beam fails quickly when heated. Actually, no fire contact is required. Ambient heating can cause the binding glue to fail, leading to a quick collapse. The bottom of a beam is under tensile forces. If the bottom of the beam falls off, due to glue failure, the beam will immediately snap and collapse.

New products, known as FiRP (fiber-reinforced products) are becoming common in the construction industry. FiRP can be plastic fibers mixed with wood to give the wood increased tensile strength. As with most plastics, fire exposure can cause quick failure as the plastic melts.

The mixture of steel and wood as a structural element can cause rapid collapse because steel expands faster than wood. This causes stress at the intersection of the two materials, **Figure 13-4.**

Structural insulated panels (SIP) are another interesting composite. This technique is characterized by large wall and roof panels made of expanded polystyrene sandwiched between two sheets of OSB, which is a sheet of wood chips bonded by glue, Figure 13-12A and B. See Section I, Chapter 13. The OSB is covered with a typical wall finish. It is anticipated that a fire will cause rapid deterioration of the load-bearing panel.

TYPES OF BUILDING CONSTRUCTION

Over time, five broad categories of building construction types have been developed to help classify structures. These categories give firefighters a basic understanding of the arrangement of structural elements and the materials used to construct the building. Unfortunately, these broad classifications are dangerously incomplete for firefighters and may lead to deadly assumptions about the makeup of a building. As stated before, firefighters need to explore the buildings within their jurisdiction to determine how buildings are assembled.

It is important to note that buildings are built to meet certain codes. These codes are designed to give occupants time to escape during a fire. Concrete is **fire resistive**—meaning it has some capacity to withstand the effects of fire. Other materials, like steel and wood, need fire-resistive assistance to give occupants a chance to escape. Building codes outline **fire-resistive ratings,** occupancy classifications, and means of egress based on five general types of buildings. The features of each type of construction will be discussed shortly, but first it is important to understand fire resistance for structural elements. **Table 13-2** outlines the number of hours that a structural element needs to be protected for the five types of construction. Simply put, firefighting time is not part of the fire-resistive and building construction equation. Fire-resistive ratings are established in a laboratory. In the real world, fire resistance ratings could "under-

TABLE 13-2   Types of Construction from NFPA 220

	Type I		Type II		Type III		Type IV		Type V	
	442	332	222	111	000	211	200	2HH	111	000
Exterior Bearing Walls—										
Supporting more than one floor, columns, or other bearing walls	4	3	2	1	0	2	2	2	1	0
Supporting one floor only	4	3	2	1	0	2	2	2	1	0
Supporting a roof only	4	3	1	1	0	2	2	2	1	0
Interior Bearing Walls—										
Supporting more than one floor, columns, or other bearing walls	4	3	2	1	0	1	0	2	1	0
Supporting one floor only	3	2	2	1	0	1	0	1	1	0
Supporting roofs only	3	2	1	1	0	1	0	1	1	0
Columns—										
Supporting more than one floor, columns, or other bearing walls	4	3	2	1	0	1	0	H[1]	1	0
Supporting one floor only	3	2	2	1	0	1	0	H[1]	1	0
Supporting roofs only	3	2	1	1	0	1	0	H[1]	1	0
Beams, Girders, Trusses & Arches—										
Supporting more than one floor, columns, or other bearing walls	4	3	2	1	0	1	0	H[1]	1	0
Supporting one floor only	2	2	2	1	0	1	0	H[1]	1	0
Supporting roofs only	2	2	1	1	0	1	0	H[1]	1	0
Floor-Ceiling Assemblies	2	2	2	1	0	1	0	H[1]	1	0
Roof-Ceiling Assemblies	2	1½	1	1	0	1	0	H[1]	1	0
Exterior Nonbearing Walls	0	0	0	0	0	0	0	0	0	0

[shaded bar]   Those members that shall be permitted to be of approved combustible material.

[1]"H" indicates heavy timber members; see text for requirements.

Source: Copyright © 2008, National Fire Protection Association, Quincy, MA.

FIGURE 13-5 This parking garage is of Type II construction. The protective coating applied to the structural steel may increase the fire-resistive rating, but note that the unprotected corrugated metal flooring and interior steel structure are not protected.

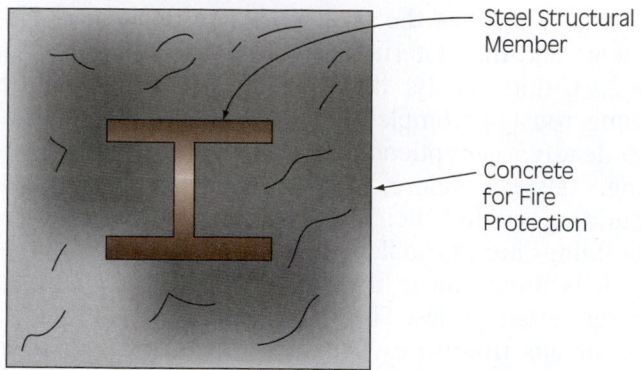

Steel Structural Member

Concrete for Fire Protection

FIGURE 13-6 To achieve a Type I fire-resistive rating, structural steel members are encased with concrete to prevent failure from the effects of a fire.

perform" due to many factors. For example, a structural element with a two-hour fire rating may fail in thirty minutes if it was not assembled correctly or if it has been damaged by some event (fire, explosion, or other collapses). Fire resistance for structural members can be achieved by various methods including drywall (gypsum wallboard), spray-on coatings, and concrete, **Figure 13-5.** Aging, alterations, and wear can damage fire-resistive methods to the point that structural elements have no fire resistance protection.

The following paragraphs outline the basic definition of each building type, its general configuration, and some historical fire spread problems associated with each. Also included are some construction methods that do not fit into the five common types.

Type I: Fire-Resistive

Type I fire-resistive construction is a type in which structural elements are of an approved noncombustible or limited combustible material with sufficient fire-resistive rating to withstand the effects of fire and prevent its spread from story to story. Concrete-encased steel, **Figure 13-6,** monolithic-poured cement, and steel with spray-on fire protection coatings are typical of Type I, **Figure 13-7.** Generally, the fire-resistive rating must be three to four hours depending on the specific structural element. Fire-resistive construction is used for high-rises, large sporting arenas, and other buildings where a high volume of people are expected to occupy the building.

Most Type I buildings are typically large, multistoried structures with multiple exit points. Fires are difficult to fight due to the large size of the building and the subsequent high **fire load.** Type I buildings rely

FIGURE 13-7 A typical Type I building, with structural members designed to resist the effects of fire for three to four hours. This building is of reinforced concrete construction.

on protective systems to rapidly detect and extinguish fires. If these systems do not contain the fire, a difficult firefight will be required. Fire can spread from floor to floor on high-rises as windows break and the next floor windows fail, allowing the fire to jump. Fire

FIGURE 13-8 Buildings of Type II construction will have structural elements with little or no protection from the effects of fire. Remember, in the event of a fire, these unprotected steel structural members may fail and collapse quickly.

can also make vertical runs through utility and elevator shafts. Regardless, firefighters are relying on the fire-resistive methods to protect the structure from collapsing. The collapse of fire-resistive structures can be massive, as we are reminded from the World Trade Center collapses in New York City.

Type II: Noncombustible

Type II noncombustible construction is a type in which structural elements do not qualify for Type I construction and are of an approved noncombustible or limited combustible material with sufficient fire-resistive rating to withstand the effects of fire and prevent its spread from story to story. More often than not, Type II buildings are steel, **Figure 13-8.** Modern warehouses, small arenas, and newer churches and schools are built as noncombustible. Because the steel is not required to have significant fire-resistive coatings, Type II buildings are susceptible to steel deformation and resulting collapse. Fire spread in Type II buildings is influenced by the contents. While the structure itself will not burn, rapid collapse is possible from the content Btu release stressing the steel.

Suburban strip malls with concrete block load-bearing walls and steel roof structures can be classified as Type II. Fires can spread from store to store through wall openings and shared ceiling and roof support spaces. The roof structure is often of lightweight steel that fails rapidly. More often than not, the fire-resistive device used to protect the roof structure is a dropped-in ceiling. Missing ceiling tiles, damaged drywall, alterations, and utility penetrations can render the steel unprotected. These buildings may have combustible attachments such as facades and signs as well as significant content fire loading.

FIGURE 13-9 Buildings of Type III, ordinary construction, are common throughout North America. These typical "Downtown USA" buildings provide many challenges to firefighters, such as void spaces and common walls allowing rapid fire extension and little structural protection.

Type III: Ordinary

The term **Type III ordinary construction** is often misapplied to wood frame buildings. By definition, ordinary construction includes buildings where the load-bearing walls are noncombustible, and the roof and floor assemblies are wood. Most commonly, this is load-bearing brick or concrete block with wood roofs and floors. Ordinary construction is prevalent in most downtown or "main street" areas of older towns and villages, **Figure 13-9.** Firefighters have long called ordinary construction "taxpayers." This slang is derived from landlords who built buildings with shops and/or restaurants on the first floor with apartments above in order to maximize income to help pay property taxes.

Newer Type III buildings include strip malls with block walls and wood truss roofs, **Figure 13-10.**

Ordinary construction presents many challenges to firefighters. In older buildings, numerous remodels, restorations, and repairs have created suspect wall stability and hidden dangers.

Ordinary construction has many void spaces where fire can spread undetected. Common hallways, utilities, and attic spaces can communicate fire rapidly. Masonry walls hold heat inside, making for difficult firefighting. Wood floors and roof beams are often gravity fit within the masonry walls. These can release quickly and cause a general collapse, leaving an unsupported masonry wall. Older Type III buildings have structural mass; therefore, they burn for a long time.

Type IV: Heavy Timber

Type IV heavy timber construction can be defined as those buildings that have block or brick exterior load-bearing walls and interior structural members, roofs, floors, and arches of solid or laminated wood without concealed spaces. The minimum dimensions for structural wood must meet the criteria in **Table 13-3.** Heavy timber buildings, as the name suggests, are quite stout and are used for warehouses, manufacturing buildings, and some older

FIGURE 13-10 One of the most common uses of Type III, ordinary construction, is the "strip mall" with masonry walls and lightweight steel or wood trusses. Common problems associated with this type of construction are void spaces allowing for rapid fire extension and collapse of lightweight structural elements.

TABLE 13-3   Heavy Timber Dimensions

Type of Element	Use	Size
Column	Supporting floor load	8- × 8-in. minimum any dimension
Column	Supporting roof load	6-in. smallest dimension, 8-in. depth minimum
Beams and girders	Supporting floor load	6-in. width and 10-in. depth minimum
Beams, girders, and roof framing	Supporting roof loads only	4-in. width minimum, 6-in. depth minimum
Framed or laminated arches	As designed	8-in. minimum dimension
Tongued and grooved planks	Floor systems	3-in. minimum thickness with additional 1-in. boards at right angles
Tongued and grooved planks	Roof decking	2-in. minimum thickness

churches, **Figure 13-11.** In many ways, a Type IV building is like a Type III—just larger dimension lumber instead of common wood beams and trusses. Some firefighters mistakenly call Type IV buildings "mill construction." Mill construction is a much more stout, collapse-resistive building that may or may not have block walls. A new Type IV building is hard to find. The cost of large-dimension lumber and laminated wood beams makes this type of construction rare in modern construction.

Fire spread in a heavy timber building can be fast due to wide-open areas and content exposure. The exposed timbers contribute Btus to the fire. Because of the mass and large quantity of exposed structural wood, fires burn a long time. If the building housed machinery at one time, oil-soaked floors will add more heat to the fire and accelerate collapse. Once floors and roofs start to sag, heavy timber beams may release from the walls. This is accomplished by mak-

ing a fire-cut on the beam; then the beam is gravity fit into a pocket within the exterior load-bearing masonry wall, **Figure 13-12.** As the floor sags, it loses its contact point with the wall and simply slides out of its pocket without damage to the wall. It is important to recall that a free-standing masonry wall has little lateral support and requires compressive weight from floors and roofs to make it sound.

Type V: Wood Frame

Type V wood frame construction is perhaps the most common construction type. Homes, newer small businesses, and even chain hotels are built primarily with wood, **Figure 13-13.** Older wood frame buildings were built as **balloon frame**—that is, wood studs ran from the foundation to the roof and floors

FIGURE 13-11 Type IV buildings, heavy timber construction, have large wood structural elements with great mass. The mass of these structural members requires a long burn time for failure. The connections, usually steel, are the weak points in this type of construction.

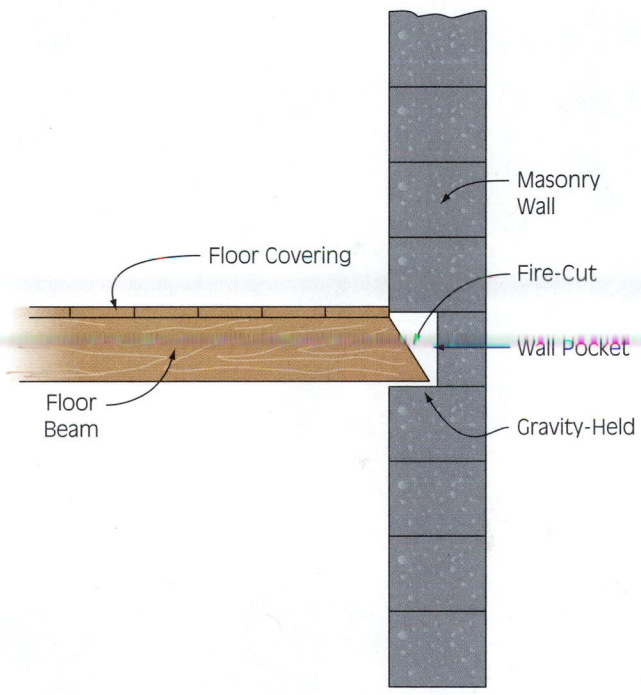

FIGURE 13-12 Wood and heavy timber beams were often "fire-cut" so that a fire-damaged, sagging floor would simply slide out of the wall pocket in order to preserve the wall.

FIGURE 13-13 The wood frame structure, Type V construction, is the most common type of construction in North America.

were "hung" on the studs. As can be envisioned, fire could enter the wall space and run straight to the attic. In the early 1950s, builders started using a **platform framing** arrangement where one floor was built as a platform for the next floor. This created **fire stopping** to help minimize fire spread. Newer wood frame buildings utilize lightweight wood trusses for roofs and floors. This is akin to a "horizontal balloon frame" that can allow quicker lateral fire spread. Coupled with high surface-to-mass wood exposure, collapse becomes a real possibility, **Figures 13-14 A–C.** Some codes require truss spaces to have fire stopping every 500 square feet (46 sq. meters). Even with this fire stopping, it remains dangerous to step onto the 500 square feet where the fire is. Wood frame structures may appear more like a Type III ordinary building because of a brick-wall appearance. Remember, brickwork may be a simple veneer to add aesthetics.

To protect structural members from a fire, wood frame construction typically uses gypsum board (drywall or the brand name Sheetrock). Gypsum protects wood frame members by absorbing heat. When heated, the gypsum goes through a chemical process called "calcification." Drywall that has been calcified is very fragile and crumbles—allowing fire spread.

Once finished, wood frame buildings typically have many rooms that can help compartmentalize content fires. Fire that penetrates wall, floor, or attic spaces becomes a significant collapse threat, especially in newer buildings. Often, the only warning that fire has penetrated these spaces is the issuance of smoke from crawl space vents, gable end vents, and eaves.

(A)

(B)

(C)

FIGURE 13-14   (A, B) Note the void spaces at the first and second floor levels and in the attic area that are created by the use of truss systems. (C) The building after it was completed. Note the sloping grade—firefighters need to communicate this observation to avoid confusion on which is the first floor.

Other Construction Types (Hybrids)

As mentioned earlier, the five broad building types can actually lead to dangerous assumptions. Newer construction and alternative building methods may not fit cleanly into one of the above types. Although not an official term, we can call these "hybrid buildings." Some hybrid buildings are actually two types of construction. For example, a particular restaurant may be built as a Type II noncombustible, yet it is topped with a large wood frame structure to hide rooftop **HVACs** and cooking vent hoods, **Figures 13-15A and B.** The combustible square feet space of the false dormers and wood frame structure can exceed that of most homes.

Lightweight Steel Frame

New lightweight steel homes resemble and are built like "stick" wood-frame homes. These buildings are actually a "post and beam" steel building with lightweight steel studs to help partition the home. OSB is added to the studs to help make the house more "stiff" and increase wind-load strength, **Figure 13-16.**

(A)

(B)

FIGURE 13-15   (A) The decorative roof assembly is a Type V wood frame structure while the occupancy space is Type II noncombustible. (B) This building uses two types of construction.

FIGURE 13-16   This lightweight steel home is built similar to a Type V. OSB sheeting gives the steel rigidity to torsional loads such as wind.

Insulated Concrete Formed (ICF)

Another interesting construction type uses polystyrene foam blocks or panels to make a form for a lightweight concrete mud mixture. The concrete is not contiguous—there are many voids, utility runs, and foam block spacers (made of plastic or galvanized steel), **Figures 13-17 A** and **B.** These structures are called "ICF" or insulated concrete formed. However, it is important not to be fooled by any claims that these buildings are concrete or are less combustible. In reality, these composite buildings are assembled with plastics, polystyrene, lightweight steel, and lightweight concrete. When finished, these buildings may resemble wood frame or even ordinary construction. Extended window and door jambs are clues that indicate the wall is thicker than that of typical wood or masonry built buildings.

Structural Insulated Panel (SIP) Wall

As mentioned in the preceding composite material section, SIP wall uses panels that are expanded polystyrene sandwiched between sheets of OSB. The entire load-bearing wall and roof structure are made from these panels (with no structural framing). The building is assembled like the house of playing cards you built as a child. These panels rely on drywall (interior) and exterior finishes to protect the asembly from fire. Heat alone can cause the panels to lose integrity.

Other hybrid buildings include, but are not limited to, hay bales, aerated concrete, steel/styrene, and gypsum/Portland lime blocks. The phrase "green building" is being used more often. Although no official definition exists, "green" can be applied to buildings that are assembled using energy and environmentally friendly methods or to buildings that have earthen roof or wall assemblies. Some of the earthen buildings have soil, grass, shrubs, and other living plants used as insulation and to help the carbon "footprint" of the building. The fire service has very little research information on the stability of these new types of buildings during fires. One thing is certain: firefighters should expect rapid collapse due to the low-mass, high surface-to-mass exposure of structural elements.

Manufactured buildings can be defined as those structures that are built at a factory and then trucked to a job site. These building are quite light with little mass. Where a stick-built home uses 2×4 or 2×6 lumber, the manufactured home uses 1×2 and 2×2 lumber. These buildings use galvanized strapping to give required strength. In any case, these buildings burn quickly and collapse equally fast.

Relationship of Construction Type to Occupancy Use

Before considering the basic types of construction, many officials and builders first look at the anticipated use of the building—its occupancy type. **Occupancy classifications** are called many different names around the country, but they are usually broken down into five basic arenas: residential, commercial, business, industrial, and educational. Each of these general occupancies has a number of hazards that firefighters must understand, **Table 13-4.** Remember, a building may have been built for one type of occupancy only to be sold and converted to another occupancy type for which it may not have been designed. Firefighters should go out and explore the buildings in their community.

COLLAPSE HAZARDS AT STRUCTURE FIRES

It cannot be overstated that firefighters have to understand the buildings in their jurisdiction. Constant study and site visits will help them "read" buildings. Reading buildings is essential to anticipating collapse proactively. This section addresses some specific collapse threats that the fire service has experienced throughout history (refer to the Section I, Chapter 13 feature box, "Historically Significant Building Collapses") and the importance of understanding buildings and how they react at structure fires.

Trusses

Truss roof collapses have killed many firefighters. As stated previously, a truss is actually a fake beam. A truss uses geometric shapes (the triangle) to create a structural element similar to a beam. A wood truss can actually be stronger than a like-sized solid wood beam, and does so with less material, **Figure 13-18.** It is this loss of material and subsequent increase in exposed surface area that make them so vulnerable during fires. Trusses rely on each and every part of the truss to carry a portion of the imposed load. Like a beam, the top of the truss (called the top **chord**) is typically under a compressive force. The bottom chord is under tension. In between the two chords, connecting members (the web) transfer the two forces creating stress and strain. Failure of one part of the truss will likely cause the whole truss to fail. This distributes the weight of the failed truss to other trusses—which may not have the capacity to take that weight—thereby starting a domino effect

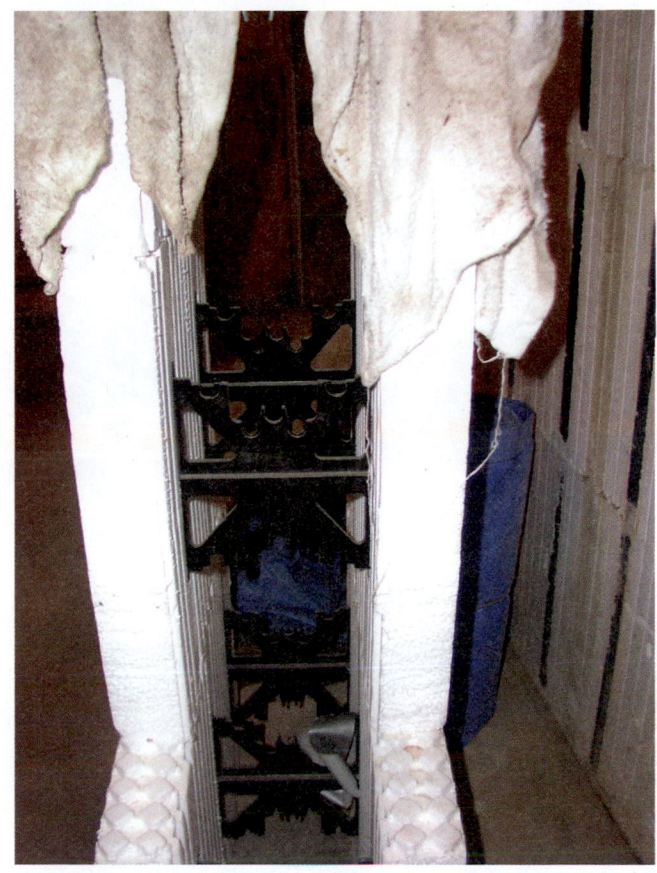

(A)   (B)

FIGURE 13-17   (A) This wall is a load-bearing foam block unit filled with a lightweight concrete mud mix. (B) Note the black plastic spacers that will melt early in a fire.

TABLE 13-4	Typical Hazards Associated with Occupancies	
Occupancy	**Type of Construction**	**Hazards**
Residential	Type V, most common	Fire loading, truss construction, owner alterations, rapid fire extension in void spaces
Commercial	Type III, most common	Fire loading, truss construction, rapid fire extension in void spaces, unknown occupancy change
Educational	Type II, most common	Unprotected structural steel, collapse, high fire load in some areas
Business	Types II and III, most common	Unknown change in occupancy, high fire load, difficult to ventilate
Industrial	Types I and II, most common	Hazardous materials, difficult to ventilate

collapse. Trusses come in many styles and shapes. Bowstring truss, parallel chord truss, and open web joists are some of the more common names. Trusses are also classified by the type of material used in assembly.

Wood Trusses

Wood trusses are commonly used for roof assemblies but are increasingly being used for floor support as well. Wood trusses are an assembly of many pieces of wood. Some may even be press-glued particles. These

FIGURE 13-18 Wood trusses provide a large surface-to-mass ratio, fuel load, and void spaces—three of the worst structural collapse contributors a firefighter will encounter during structural firefighting operations.

FIGURE 13-19 A typical parallel chord truss. The gusset plates on this truss are pressed into the wood. In addition to the decomposing of the wood element, the light-gauge steel plates will deform and pull away from the wood under fire conditions.

pieces are connected using gusset plates. A **gusset plate** is a simple galvanized steel plate (very thin) with perforations punched into the plate. The perforations are used to pierce into wood fibers to hold pieces together. These perforations only penetrate the wood a fraction of an inch (3/8 inch is typical), **Figure 13-19.** During fires, the steel gusset heats up and transfers heat into the very wood fibers that are being held. If the heating is slow (like a smoke-filled attic), the wood decomposes, allowing the gusset plate to fall out. If the heating is fast, like sudden exposure to flame, the steel expands too quickly for the wood and the gusset simply pops out. Either way, truss failure is imminent. Sometimes the truss gingerly stays together because of its connection to roofing or floor-

ing materials, yet a sudden force, like a walking fire-fighter, will cause the truss to disassemble and suddenly collapse.

Some wood trusses are being assembled using glue instead of gusset plates. Some manufacturers claim that a glued truss is actually stronger than a gusseted one. This may be true in the absence of heat or fire. Remember, heat can cause glue failure—it doesn't have to burn.

Wood trusses are mass-produced at a factory where quality control may not be adequate. Further, the truss gusset plates or glue may vibrate or be damaged while being delivered to a job site. Once on the job site, contractors may use shortcuts to lift the truss into position, furthering the damage to gusset plates or glued joints.

Steel Trusses

Steel trusses are no less susceptible to collapse than wood trusses. Like wood, steel trusses are an assembly of pieces—typically angle iron for the chords and cold-drawn round stock for the web. The pieces are tack-welded together to form the truss unit. While not a true joist by definition, many call the common steel truss an open web steel joist, **Figure 13-20.** The term *bar joist* is also used to describe an open web steel joist. These trusses expose a large surface area (but low mass) to heat during fires. Given the lack of mass, the truss heats quickly and will soften and expand. The expansion can cause wall movement. (Remember, masonry walls must be loaded axial with compressive force.) Lateral movement can cause wall collapse. If the wall does not move, the steel truss will twist and buckle to allow expansion. It is very important to keep steel trusses cool.

FIGURE 13-20 Unprotected open web steel joists present a large surface area to absorb the heat of a fire, expand, and collapse. Structural steel will lose 50 percent of its strength at temperatures of 1,000°F (538°C).

Although the focus here is buildings, many other structures (like antennae, bridges, water towers, and large signs) use steel trusses as part of their assembly. A tall cell phone relay tower is actually a vertical, cantilevered, steel truss (beam).

In 2005, NIOSH issued an *Alert* regarding firefighting operations in buildings with truss construction. The alert cites an alarming trend in the number of firefighters killed due to truss collapse and recommends that firefighting operations immediately resort to a defensive (outside) posture once fire has entered a truss space.

ETHICS

Many firefighters have died because of truss collapse. The best way to honor those that have died is to identify those buildings in your jurisdiction that have truss construction and develop pre-plans so that a similar occurrence can be prevented.

Void Spaces

Trusses create large void spaces. The area between the chords of trusses will allow fires to spread horizontally, **Figure 13-21.** Some codes require fire stopping in floor truss spaces but may allow wide-open attic spaces. Fires can start in void spaces due to electrical and other utility problems. In Type III ordinary construction, voids are numerous. Some voids may pass through masonry walls, causing fire spread from one store to the next in a row of buildings. The obvious collapse danger with void spaces is that the fire may be undetected with simultaneous destruction of structural elements.

Roof Structures

The roof of a building can be flat, pitched, or inverted. Many factors help determine how and why the roof is built the way it is. Sometimes the roof is designed to hide rooftop HVACs. Other times the roof shape is designed to shed snow, accommodate a vaulted ceiling, or merely give the building character, **Figure 13-22.** As it relates to structural collapse, the roof style may allow a large volume of fire to develop. Other roofs, like the mansard, have many concealed spaces. Dormers are protrusions from a roof structure. Dormers can be used to introduce daylight into a roof space that is converted into a living space, **Figure 13-23.** Other dormers are

FIGURE 13-21 Lightweight floor truss systems have many void spaces. This could be called "horizontal balloon frame."

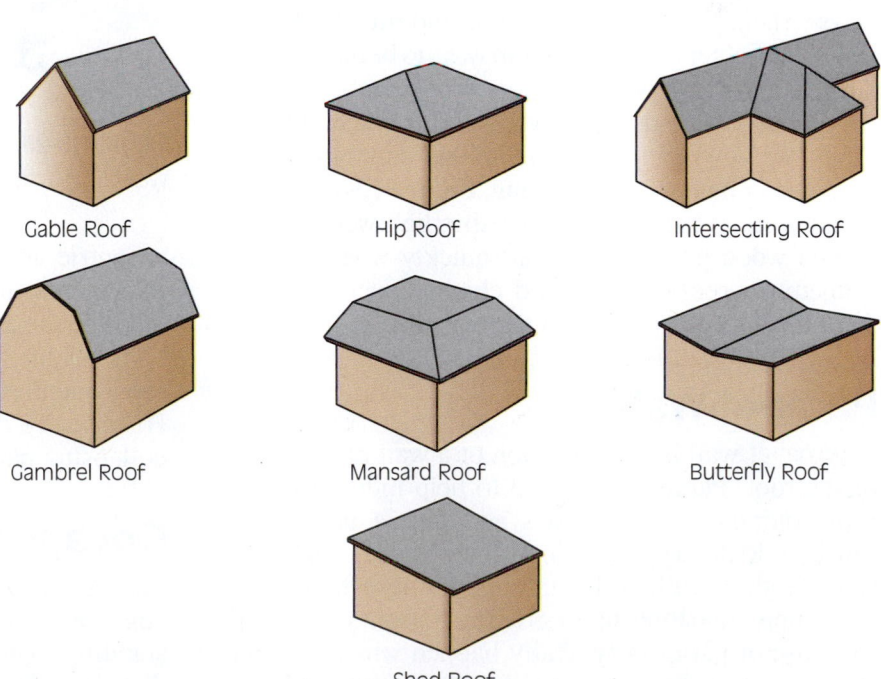

Gable Roof

Hip Roof

Intersecting Roof

Gambrel Roof

Mansard Roof

Butterfly Roof

Shed Roof

FIGURE 13-22 Some common roof framing styles used in wood frame or ordinary construction.

FIGURE 13-23   An internal view of dormers in a mansard roof. Fire can run through the many voids between the exterior and the roof.

FIGURE 13-24   This prefab stair assembly is hung in place by thin metal strapping. Note the staples and plastic shims that can quickly fail under fire conditions.

FIGURE 13-25   This is the scene of a typical "parapet" wall failure common in Type III ordinary construction.

actually aesthetic (false) and can fool ventilation crews attempting to relieve heat from a roof space.

Stairs

First arriving firefighting crews rely on internal stairways to help gain access for rescue and fire attack. For years, firefighters have found stairways to be durable and a bit stronger than other interior components. This is a dangerous assumption in newer wood frame buildings. Stairs are now being built offsite and simply hung in place using light metal strapping, **Figure 13-24.** Additionally, stairs are being made using lightweight engineered wood products that fail quickly when heated. Remember, press-glued wood chip products can fail from the heat of smoke—no flame is required.

Parapet Walls

A **parapet** wall is the extension of a wall past the top of the roof. Parapets are used to help hide unsightly roof equipment and HVACs and give a building a finished look. Typically masonry, these walls are free-standing with little stability because they have no compressive **loading** to give the masonry strength. Collapse of parapets typically happen when the roof structure starts to sag—this lifts up the parapet in an

eccentric way, causing it to fall, **Figure 13-25.** Business owners hang signs, utility connections, and other loads on the parapet. During a fire, the steel cables and bolts holding these will weaken or change shape and subsequently pull down the parapet, **Figure 13-26.** Historically, dozens of firefighters have been killed by collapsing parapets.

Collapse Warning Signs

Firefighters must rely on building material knowledge, building construction principles, and an understanding of fire effects on buildings in order to predict or anticipate collapse. Waiting for a visual sign

FIGURE 13-26 This electrical service entrance and attached sign may be the eccentric load causing an early failure of this parapet wall.

that a building will collapse is dangerous, especially in newer buildings. There are, however, some factors and observations that can be used to help anticipate collapse. These include:

- Deterioration of mortar joints and masonry
- Overall age and condition of the building
- Cracks, in anything
- Signs of building repair including reinforcing cables and tie-rods
- Large open spans
- Bulges and bowing of walls
- Sagging floors
- Abandoned buildings
- Large volume of fire
- Long firefighting operations—remember gravity?
- Smoke coming from cracks in walls
- Dark smoke coming from truss roof or floor spaces (Brown smoke indicates that wood is being heated significantly; black smoke means combustibles have ignited or are near ignition.)
- Multiple fires in the same building or damage from previous fires

Buildings under Construction

Buildings are especially unsafe during construction, remodeling, and restoration. The word unsafe applies not only to fire operations but also to rescues, odor investigations, and on-site inspections. Buildings need only meet fire and life safety codes when they are

completed. During construction, many of the protective features and fire-resistive components are incomplete. Additionally, stacked construction material may overload other structural components. This is not to say that contractors are using unsafe practices, but rather to underscore the fact that exposed structural elements, incomplete assemblies, and material stacks will contribute to a rapid collapse if a fire were to develop.

> **STREETSMART TIP**
>
> Beware of Buildings under Construction: A building as a complete unit has a number of interdependent parts. During construction, these parts may not be fully connected (steel), may not be at their full design strength (concrete), and may lack any type of fire protection (gypsum board or concrete). Also, many structural elements may be held in place by scaffoldings, false work, or forms. Because of this, a building under construction is exposed to early collapse potential due to the effects of fire or other elements such as high winds.

Historical building restoration and general remodel projects in buildings are similar to buildings under construction. Firefighters may find temporary shoring of walls, floors, and roofs while other structural components are being updated, replaced, or strengthened. Contractors may use simple 2 × 4s to temporarily shore up heavy timber, leading to disastrous results during fire conditions. The best approach for firefighters to take when responding to fires in buildings under construction is to be defensive. They should make sure everyone is out and accounted for and then attack the fire from a safe location. A building under construction is likely insured and can be replaced—a firefighter's life cannot.

Time

There are no time limits for firefighting operations within a building. Tests have shown that the age-old "twenty-minute rule" used by some fire officers is no longer accurate. Roofs and walls can collapse within minutes of fire involvement given certain conditions. An overloaded truss (due to improper storage or other factors) can collapse immediately when heated.

The passage of time during a building fire works against the firefighter teams. In fact, the window of opportunity to perform an aggressive interior fire attack has diminished significantly due to lighter-weight construction and higher BTU heat release of plastics. As it relates to predicting firefighting time within a structure, a few truisms have emerged:

- The lighter the structural element, the faster it comes down.

PREDICTING COLLAPSE

There is no perfect formula for predicting collapse. A simple approach is presented here to help you understand the mental process that incident commanders and safety officers use to help predict collapse. All firefighters need to be aware of the process and should communicate conditions and observations to ensure the process is proactive. The approach explained here can be applied to buildings and other structures such as bridges, cranes, towers, signs, mine shafts, wells, or any "engineered" construction attacked by fire.

Step 1: Classify the construction type.

Step 2: Determine structural involvement or heating (read the smoke and flames).

Step 3: Visualize and trace loads (determine the weak link).

Step 4: Evaluate time.

Step 5: Predict and communicate the collapse potential (establish a collapse zone).

- The heavier the imposed load, the faster it comes down.
- Wet (cooled) steel buys time.
- Gravity and time are constant, resistance is not.
- There is no *window* of time for interior operations when a building is under construction, being renovated, or being disassembled.
- Brown or dark smoke coming from lightweight structural components means time is up.

Preparing for Collapse

At building fires, the incident commander (and/or assigned safety officer) needs to predict collapse *PROACTIVELY.* Communicated information and observations from working firefighter teams helps with predicting collapse. Therefore, all firefighters need to understand the mental process that is used to proactively predict collapse (see the feature box, "Predicting Collapse").

STREETSMART TIP

Collapse: Every firefighter must understand two rules about structural collapse during fire operations. The first is that the potential for structural failure during a fire always exists. Do not set artificial time limits based on experience. The second rule is to establish a **collapse zone,** as shown in **Figure 13-27,** which is an area around and away from a building where debris will land if the building fails. As an absolute minimum, this distance must be at least 1½ times the height of the building. The walls may crumble into a pile or they may tip out the full height of the building. Also you need to provide extra room for cascading debris.

Once a building has been searched for occupants, the risks firefighters take to control the fire should be reduced—after all, it is now a property issue. Many firefighters have been killed fighting interior fires, only to have the building torn down after the investigation. Outside (defensive) firefighting operations can

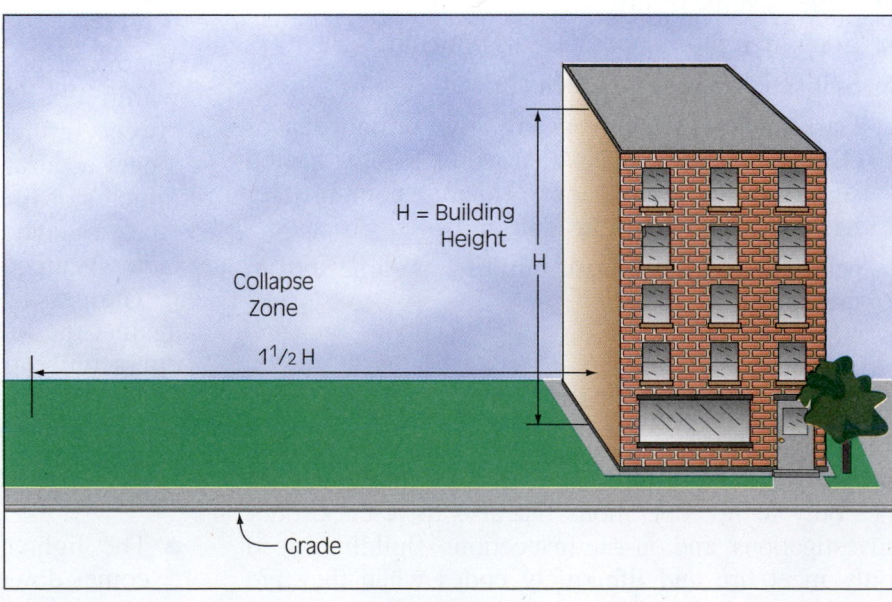

FIGURE 13-27 A minimum collapse zone should be 1½ times the height of the building.

FIGURE 13-28 These photos show the effects of fire on a masonry wall. Note the debris and distance the bricks fell away from the building. Firefighters should always establish a collapse zone, as shown in Figure 13-27.

be equally dangerous if firefighters wander into the collapse zone, **Figure 13-28.**

In all cases, firefighters who witness collapse indicators or an actual collapse should be empowered to report the condition. If an imminent threat to firefighters exists, the firefighting team should declare a firefighter emergency ("Mayday. Mayday. Mayday.") and initiate an immediate withdrawal.

LESSONS LEARNED

To prevent collapse-related firefighter fatalities, firefighters must understand the buildings in which they fight fires. This understanding comes from a long-term commitment to read and study building construction information. Additionally, firefighters must get into buildings within their jurisdictions to survey and explore the way buildings are assembled, remodeled, and used in the real world. Firefighters also study the effects of fires on materials and construction types. The five classic types of construction are being challenged by new construction methods. Trusses are used in virtually all new buildings. Trusses have high surface-to-mass characteristics that rapidly absorb heat and subsequently fail quickly. Failure of one truss can cause failure of other trusses. There are no rules for how long a building will last while on fire. Many factors determine when materials and construction design fail and gravity pushes down the building. Buildings under construction have a much higher collapse risk factor from a firefighting point of view. Understanding the process to proactively predict collapse helps firefighters communicate conditions and observations that can save lives.

KEY TERMS

Balloon Frame   A style of wood frame construction in which studs are continuous the full height of a building.

Beam   A structural member subjected to loads perpendicular to its length.

Chord   The top and bottom components of a beam or truss (sometimes referred to as *flanges*). The top chord is subjected to compressive force; the bottom chord is subjected to tensile force.

Collapse Zone   The area around a building where debris will land when it falls. As an absolute minimum this distance must be at least 1 ½ times the height of the building.

Column   A structural element that is subjected to compressive forces—typically a vertical member.

Fire Load   The amount of heat generated when the building and its contents burn.

Fire Resistive   The capacity of a material to withstand the effects of fire.

Fire-Resistive Rating   The time in hours that a material or assembly can withstand fire exposure. Fire-resistive ratings are usually provided for testing organizations. The ratings are expressed in a time frame, usually hours or portions thereof.

Fire Stopping   Pieces of material, usually wood or masonry, placed in stud or joist channels to slow the extension of fire.

Girder   A large structural member used to support beams or joists—that is, a beam that supports beams.

Gusset Plate   A connecting plate used in truss construction. In steel trusses, these plates are flat steel stock. In wood trusses, the plates are either light-gauge metal or plywood.

HVAC   Acronym for heating, ventilation, and air-conditioning unit. HVACs are typically a rooftop unit on commercial buildings. Buildings may have one or dozens of these units.

Joist   A wood framing member that supports floor or roof decking.

Load-Bearing Wall   Any wall that supports other walls, floors, or roofs.

Loading   The weight of building materials or objects in a building.

Occupancy Classifications   The use for which a building or structure is designed.

Parapet   The projection of a wall above the roofline of a building.

Platform Framing   A style of wood frame construction in which each story is built on a platform, providing fire stopping at each level.

Spalling   Deterioration of concrete by the loss of surface material due to the expansion of moisture when exposed to heat.

Truss   A rigid framework using the triangle as its basic shape to emulate a beam.

Type I, Fire-Resistive Construction   Type in which the structural members, including walls, columns, beams, girders, trusses, arches, floors, and roofs, are of approved noncombustible or limited combustible materials with sufficient fire-resistive rating to withstand the effects of fire and prevent its spread from story to story.

Type II, Noncombustible Construction   Type not qualifying as Type I construction, in which the structural members, including walls, columns, beams, girders, trusses, arches, floors, and roofs, are of approved noncombustible or limited combustible materials with sufficient fire-resistive rating to withstand the effects of fire and prevent its spread from story to story.

Type III, Ordinary Construction   Type in which the exterior walls and structural members that are portions of exterior walls are of approved noncombustible or limited combustible materials, and interior structural members, including wall, columns, beams, girders, trusses, arches, floors, and roofs, are entirely or partially of wood of smaller dimension than required for Type IV construction or of approved noncombustible or limited combustible materials.

Type IV, Heavy Timber Construction   Type in which exterior and interior walls and structural members that are portions of such walls are of approved noncombustible or limited combustible materials. Other interior structural members, including columns, beams, girders, trusses, arches, floors, and roofs, shall be of solid or laminated wood without concealed spaces.

Type V, Wood Frame Construction   Type in which the exterior walls, bearing walls, columns, beams, girders, trusses, arches, floors, and roofs are entirely or partially of wood or other approved combustible material smaller than the material required for Type IV construction.

Web   The portion of a truss or I beam that connects the top chord with the bottom chord.

REVIEW QUESTIONS

1. What are three ways loads are imposed on materials?

2. List the three types of forces created when loads are imposed on materials.

3. Name three kinds of beams.

4. Explain the effects of fire on steel structural elements.

5. How does a masonry wall achieve strength?

6. List and define the five common types of building construction. Give an example of each type that is located in your district or response area.

7. List the three parts of a truss and explain what forces are being applied to each.

8. List three buildings in your district or response area that have truss construction.

9. List how fire affects the four more common building materials in use today.

10. Diagram and label four different roof shapes.

11. List and describe eight conditions or observations that might indicate potential structural collapse.

16

Vehicle Rescue and Extrication

LEARNING OBJECTIVES

After completing this chapter, the reader should be able to:

16-3 Identify the characteristics commonly associated with a safe haven.

16-25 Describe the fire department's responsibility at a vehicle accident.

16-26 Identify the medical needs of a vehicle accident victim.

16-27 Define the terminology associated with motor vehicle extrication operations.

16-28 Describe the safe operation of vehicle extrication, tools and equipment.

16-29 List dangers associated with vehicle components and systems encountered during a vehicle extrication.

16-30 Describe the use and limitations of hand and power equipment in relation to vehicle extrication.

16-31 List the safety procedures required when using vehicle extrication equipment.

16-32 Demonstrate the use and operation of hand and power tools designed for forcible entry and vehicle extrication.

16-33 Identify the hazards associated with an automobile incident.

16-34 Recognize and mitigate the hazards found at an automobile incident.

16-35 Identify the considerations for mitigating hazards at a vehicle incident.

16-36 Identify informational resources related to vehicle extrication to identify hazards by type, year, and manufacturer.

16-37 Describe highway traffic safety considerations at an automobile incident.

16-38 Demonstrate vehicle positioning at an automobile incident.

16-39 Demonstrate highway traffic safety considerations at an automobile incident.

16-40 Describe vehicle positioning at an automobile incident.

16-45 Describe scene size-up for an automobile incident.

16-46 Demonstrate scene size-up for an automobile incident.

16-47 Demonstrate proper vehicle stabilization using cribbing and shoring materials.

16-48 Describe the procedures for gaining access into a vehicle.

16-49 Gain access into a vehicle.

16-50 Identify common access points based on given auto body construction and manufacturer.

16-51 Describe the procedures for disentanglement.

16-52 Demonstrate disentanglement of an accident victim.

16-53 Remove an accident victim from a vehicle without further injury.

16-54 Describe a firefighter II's role during a technical rescue situation.

16-55 Explain the various types of specialized rescue situations presented and the specific hazards associated with each of them.

16-56 Describe the safety control zones used in a technical rescue operation.

16-57 Demonstrate the ability to keep the public away from a technical rescue incident by maintaining safety control zones.

16-58 Identify various rescue tools and their intended use.

16-59 Demonstrate the ability to differentiate between various types of rescue tools.

16-60 Describe rescue practices and the goals associated with them.

16-61 Demonstrate the ability to assist a rescue team as requested or directed during a technical rescue situation.

*The FF I and II levels, as defined by the NFPA 1001 Standards, are identified in different colors: FF I = black, FF II = red, additional information = blue.

INTRODUCTION

In this chapter, **rescue** describes the actions that trained firefighters perform at emergency scenes to remove someone from imminent danger or to **extricate** them if they are already entrapped. Because of this danger, firefighters are often put at high risk during rescue operations. It is imperative that firefighters recognize the existing dangers with vehicle extrication and be cognizant of the methods availailable to minimize or control the associated risk. Training must take place at higher levels on a consistent basis to keep the firefighter up to date with the most current obstacles that prevent them from performing their job. This chapter is only going to touch the surface of vehicle rescue and extrication situations that may confront the firefighter. This chapter discusses victim removal and vehicle extrication in detail since these are the areas that firefighters may encounter in daily operations.

HAZARDS ASSOCIATED WITH RESCUE OPERATIONS

A thorough **risk/benefit analysis** needs to be performed each time personnel are committed to rescue operations and must be continual throughout the incident to prevent and overcome tunnel vision. The probability of the success of an operation must be considered in relation to the degree of risk presented. Incident size-up is the first step in establishing this assessment. This initial assessment will serve as the foundation for all decisions made. Prior to initiating any rescue operations, firefighters must also establish emergency exit points and areas of refuge known as safe havens in case conditions change and no longer allow them to exit an area in the manner in which they entered.

Safe Haven

Firefighters operating within hazardous conditions must identify safe areas to retreat to while waiting to be rescued or until another method of escape is identified. The safe area is referred to as a safe haven and is defined as an area of refuge that can be utilized while waiting to be rescued or until you are able to escape the hazardous conditions. Safe havens are utilized when you can no longer exit the area in the same manner in which you entered because of a collapse, flashover, or other dangerous conditions. Prior to entry, the firefighter should identify at least two safe havens. Each and every firefighter should develop basic survival techniques that will allow them to escape dangerous or life threatening situations.

Characteristics of a safe haven are fundamentally the same for all emergency incidents: temporary safe area, away from the hazard, in a tenable environment, identifiable by rescuers, and self rescue may be ini-

tiated. Safe havens for specific emergency incidents may include the following:

Structure Fire

- Rooms adjacent to or away from the fire
- Next to wall away from the hazard, ideally exterior wall
- In a doorway
- Close to a window
- In a void created by collapse

Wildland Fire

- Downhill, upwind
- In the burned area

Hazardous Materials Incidents

- Away from hazardous exposure
- Away from catastrophic failure areas of a tank or cylinder
- Shielded from heat sources

Upon entering the safe haven, the following safety and survival techniques should be followed:

- If lost or trapped, immediately use the "mayday" procedure.
- Maintain constant communication with Incident Commander or Safety Officer to let them know your location.
- Position yourself next to a wall, in a doorway, or close to a window to allow rescuers to find you.
- Lay horizontally with audible pass device positioned for maximum effectiveness.
- Stay calm and conserve air by controlling breathing.
- If lost or trapped, create audible signals such as tapping flashlights or other tools, and visible signals such as shining your flashlight to aid rescuers in locating you.
- Maintain constant contact with team members by maintaining team integrity.

Safe havens are identified to provide temporary safety during a dangerous situation. Firefighting is a team environment but all firefighters should be prepared to use the skills necessary for self survival and rescue.

MOTOR VEHICLE ACCIDENTS

As was discussed in the introduction to this chapter, motor vehicle crashes are probably the most common rescue situation that today's firefighters respond to. The firefighters' most valuable tool at vehicle crash incidents is their knowledge, experience, and skill—

not, as is often believed, more powerful tools. For the purpose of this section, a good working definition of extrication is "to set free, release, or disentangle a patient from an entrapment."

Operations at motor vehicle accidents should follow a predetermined sequence of events. This list outlines the fire department's role at a motor vehicle accident and ensures that none of the essential procedures or operations are overlooked:

1. Scene assessment (size-up)
2. Establishment of work areas
3. Vehicle stabilization
4. Patient access and stabilization
5. Disentanglement
6. Patient removal
7. Scene stabilization

These procedures/operations are discussed in further detail later in this section.

Tools and Equipment

Tools and equipment utilized at vehicle crash incidents range from the most basic of firefighting tools, such as axes and pry bars, to the much more complex and specialized power hydraulic tools, air bags (low and high pressure), and battery-powered saws. A statement made earlier bears repeating here: The firefighters' most valuable tool at vehicle crash incidents is their knowledge, experience, and skill—not, as is often believed, more powerful tools. While power hydraulic tools, air bags, and the other advanced tools and equipment available today have certainly increased a firefighter's capabilities, they cannot replace the knowledge, skill, and ability developed through training and experience.

Power Hydraulic Tools

Many different companies manufacture power hydraulic tools today, but they all operate on basically the same principle. A hydraulic pump is powered by a gasoline engine, an electric motor, an air-driven motor, or the apparatus engine itself through a power take-off. Some manufacturers offer a manual hydraulic

FIGURE 16-1   Gasoline engine-powered hydraulic pump for extrication equipment.

pump as a backup should the primary power hydraulic pump fail. The hydraulic pump, **Figure 16-1,** provides the required fluid and pressure to operate the variety of spreaders, cutters, and rams available. Because different companies' tools operate at different pressures and use different types of fluid, most of the manufacturers use hoses that will not connect to anything other than a compatible tool.

If a department has more than one type of power tool, personnel adding hydraulic fluid must be sure they have the right hydraulic fluid for the specific tool. The wrong type of fluid can cause serious damage.

Spreaders, **Figure 16-2,** were the first power hydraulic tool that became available to firefighters, and they are still widely used today. Spreaders can be used to both push and pull (if they are equipped with a chain attachment); they can be used to spread things apart or squeeze them together. Generally the pulling forces are less than the pushing forces for the same tool.

Cutters (also known as shears) were first introduced as a separate attachment to go on the spreader arms, either at the ends sticking out or coming back between the arms. All cutters of this type should be removed from service since there is a possibility of the blades becoming crossed and shattering. Modern cutters, **Figure 16-3,** are completely separate tools and come in a variety of sizes also. The smaller cutters are designed for working in close places to cut pedal supports or steel rods.

Rams were the next hydraulic tool to appear with limited available sizes, but that has changed such that rams that measure as small as 10 inches but extend up to more than 60 inches are available, **Figure 16-4.** Some rams are designed to both push and pull; however, the pulling strength is generally one-half that of the pushing strength.

Combination tools combine the functions of the spreader and the cutter. This type of tool incorporates the cutting blades on the inside of the spreader arms,

FIGURE 16-2   Power hydraulic spreaders.

FIGURE 16-3   Power hydraulic cutters.

Figure 16-5. These tools are generally less powerful than the individual stand-alone spreader or cutter.

FIGURE 16-4   Power hydraulic rams.

Air Bags

Air bags used in rescue operations come in high-pressure and low/medium-pressure styles, **Figures 16-6A** and **B.** Each style has operational properties that are particular to that style, and therefore, each style serves a specific purpose in the rescue field.

High-pressure bags operate at a maximum inflation pressure of approximately 130 psi. They come in sizes ranging from 6 by 6 inches up to the largest ones measuring upwards of 36 by 36 inches. The larger bags are capable of lifting up to approxi-

FIGURE 16-5 Power hydraulic combination tool.

FIGURE 16-7 A high-pressure bag can be placed in a very small space at a vehicle extrication scene.

(A)

(B)

FIGURE 16-6 (A) A typical high-pressure air bag set. (B) A typical low-pressure air bag set. *(Courtesy of Rick Michalo)*

mately 80 tons. One of the major advantages of these bags is that when deflated they are only 1 inch thick, allowing them to be inserted into very small spaces when necessary, **Figure 16-7.** The biggest drawback is that even the largest bag will lift a maximum of approximately only 20 inches.

> ### CAUTION
>
> Although it is possible to stack high-pressure bags, extreme caution must be exercised to keep them from shifting under a load and being expelled at high velocity.

Low/medium-pressure bags operate at 7–10 psi/12–15 psi inflation pressure, respectively, depending on specific type and brand. These bags are bulky when deflated, not getting much less than 3 to 6 inches thick, but can lift heavy loads a considerable distance.

> ### SAFETY
>
> Firefighters should remember these important safety precautions when operating air bags:
>
> - Bags must be on or against a solid base.
> - Never inflate bags against sharp or hot objects.
> - Make sure that air supply is adequate. (Do not use oxygen.)
> - You must crib as you lift.
> - Do not stack more than two bags, and if they are different sizes, the smaller bag goes on top.
> - Operate controls carefully, filling bags in a controlled manner.

Air Chisels and Reciprocating Electric Saws

Standard air chisels have been used in the rescue field for quite some time. These chisels are designed to be operated at between 100 and 150 psi. However, newer versions of the air chisel, **Figure 16-8,** have been designed specifically for rescue operations and operate at up to 300 psi. The performance of these new air chisels is markedly improved.

Reciprocating electric saws are one of the newer tools making their way into the extrication field. One of the drawbacks with the earlier versions was that it was necessary to have 120 volts to operate them. However, with the evolution of the high-capacity, battery-powered saws on the market today, they are finding widespread acceptance and usage in the rescue field, **Figure 16-9.**

SAFETY

Firefighters should remember these important safety precautions when operating air chisels and reciprocating electric saws:

- Do not use in hazardous or flammable atmospheres because they can cause sparks.
- Make sure air supply is adequate. (Do not use oxygen.)
- Make sure that no victims or rescuers are in the way when cutting.
- Use caution cutting hardened steel because blades may chip and/or break.

Scene Assessment (Size-Up)

Scene assessment should be a predetermined sequence of steps or actions that is used in evaluating a crash scene. This evaluation is usually carried out by the

FIGURE 16-8 A high-pressure air chisel kit. *(Courtesy of Rick Michalo)*

FIGURE 16-9 A battery-powered reciprocating saw.

officer; however, a good firefighter who conducts an assessment will be able to foresee many of the needs and requests of the officer.

NOTE

While it is important that this assessment be done quickly, it is more important that it be done accurately and completely.

Scene Safety

Good scene assessment considers many facts and probabilities. Specifically, the following items need to be taken into consideration:

- Traffic (see boxed safety note following this list)
- Number and type of vehicles involved
- Potential number and apparent extent of patient injuries
- Hazardous conditions (fire, HazMat, electrical, water, structural integrity, weather, crowds)
- Degree of entrapment

SAFETY

Many firefighters have been injured and killed by traffic and secondary crashes at vehicle incidents. For this reason, fire and rescue responders need to size up traffic conditions as a priority. Blind spots, poor visibility (weather/darkness), road surfaces, bridges, barriers, congestion, and passing traffic speeds are all factors that need to be assessed.

The results of this assessment will help determine the need for additional resources or assistance that might include more ambulances, law enforcement, specialized equipment, power company response, or other ancillary needs. It is important to mention that the scene assessment is an ongoing process.

Establishment of Work Areas

Ideally, the fire department would like to shut down all traffic in and around the area of the vehicle incident. Unfortunately, the congestion this may cause can create secondary hazards and reduce the ability of additional responders to reach the scene. Many state and local law-enforcement agencies will work at all costs to keep traffic flowing. Because of this, it is essential that the fire department create protected work areas for rescuers and responders. Work areas are established using a combination of traffic barriers, traffic-calming strategies, and hazard zoning. Based on the scene assessment, the first arriving apparatus should be positioned to create a traffic barrier to help shield the greatest number of rescuers. Additional apparatus can increase the size of the barrier or be positioned past the scene so that they are also screened from traffic by the first arriving apparatus. While various positioning strategies exist, firefighters need to understand the logic and process of vehicle positioning so that they can work within the protection and limitations afforded by the barrier, **Figure 16-10.**

Traffic Barrier

A large fire apparatus such as an engine or heavy rescue can form the initial barrier by stopping in a position that shields the area where firefighters need to work. The apparatus should park with a slight diagonal angle to help increase the work area and make the apparatus appear larger to approaching traffic. The diagonal park also helps approaching traffic recognize that the fire apparatus is not moving. Even with a barrier in place, firefighters should mentally preplan an escape route should the barrier apparatus get hit by traffic.

> **NOTE**
>
> Traffic cones should never be used to create a barrier. Cones should be placed well behind the traffic barrier (apparatus). Some departments still use flares (fusees) to help with traffic calming. These devices are incendiary. It is important to make sure they cannot roll and will not ignite flammable gases, liquids, or vegetation.

Traffic Calming

Efforts expended to warn approaching traffic of an upcoming hazard are known as traffic-calming strategies. Typically, law-enforcement agencies and the Department of Transportation will handle this task. However, in the early stages of the incident, the fire department can begin some simple traffic-calming strategies to slow traffic, **Figure 16-11.** Placing traffic cones on the street to help guide traffic is a popular strategy.

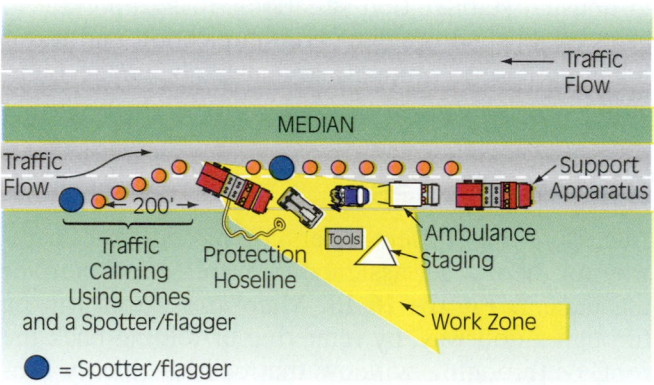

FIGURE 16-10 The first-arriving large fire apparatus should be positioned to create a traffic barrier and work zone. Cones and a spotter/flagger can help re-route traffic.

FIGURE 16-11 Signs and lights are some devices that can be utilized for traffic calming.

Other traffic-calming strategies include:

- Deploying portable Accident Ahead signs.
- Switching off white lights and strobe warning lights that cause night blindness to approaching traffic.
- Utilizing arrow sticks (flashing amber lights that signal in which direction traffic needs to move).
- Positioning a firefighter to wave down traffic speeds. This is done using a wand light or other highly visible device. The firefighter waves the wand up and down in a full arm length waving motion. This firefighter is exposed to traffic, but can also serve as a spotter for other rescuers. It is important to make sure an escape route is planned.

Vehicle Stabilization

Vehicle stabilization is required on all accidents. It can be as simple as putting the transmission in gear and completely chocking a wheel, **Figure 16-12,** or extremely complex, requiring cribbing, air bags, ropes, or other tools and equipment. The rescuer needs to use the information gathered during the assessment process to determine the amount of stabilization required. In any case, if a vehicle has an injured person inside, it must be stabilized by taking the weight off the vehicle's **suspension system.** (Note: Deflating the tires alone will not accomplish this.) This can be accomplished by using **cribbing, Figure 16-13.** Once cribbing is in place with a vehicle on flat ground, deflating the tires will transfer the weight of the vehicle onto the cribbing and off the vehicle's suspension system. In more complex situations, stabilization may require the use of a variety of tools and equipment. Stabilization struts, cables, "step chocks," **Figure 16-14,** winches, and booms are just a few examples of the many tools that can be

FIGURE 16-13 Box cribbing can be used to take the weight off the suspension of the vehicle.

FIGURE 16-14 Step chocks are also a quick method of stabilizing a vehicle.

used to help stabilize vehicles. Tow-vehicle and vehicle-recovery businesses may have equipment that can help with vehicle stabilization. Utilizing the expertise of vehicle-recovery technicians can be helpful in unusual stabilization situations. It is important to have established previous communication and training with these technicians so that safety and accountability issues are clear.

Patient Access

Accessing the patient refers to providing a pathway for the rescuer assigned to evaluate and care for the patient, **Figure 16-15.** Many times this can be through a doorway, by removing or folding back the roof, or through a window that can be quickly broken or removed. The medical needs of the patient will dictate the extrication techniques that will be applied for that particular vehicle accident.

Proper patient assessment and triage must be performed in order for rescuers to provide the proper

FIGURE 16-12 To be properly stabilized, a wheel should be chocked both in front and in back.

FIGURE 16-15 Many times access to the patient can be made by removing the rear window. Note that the vehicle is properly cribbed, and the glass edges the patient attendant has to crawl over are covered. *(Courtesy of Rick Michalo)*

emergency care. The treatment of a patient involved in a motor vehicle accident should begin as soon as safe access is available. Treatment should take into account the type of collision, the patient position within the vehicle, and the signs and symptoms presented by the victim. The needs of each patient will be different dependent upon the type of collision that has occurred; however, no injury, regardless of severity, should be discounted due to the amount of force involved in a motor vehicle collision.

Firefighters should focus their attention to life threatening injuries involving the airway, breathing, and circulation, while simultaneously stabilizing the spine. Trauma may be evident from the force of the collision and cause broken bones and lacerations that should be treated as soon as possible. Firefighters should deliver emergency care up to their level of ability until EMS arrives and patient care is transferred. Refer to Section III, Chapter 22 for the proper assessment and care of the patient.

Once access is gained:

- The patient can be evaluated.
- Life support activities can be initiated.
- The patient's position as it relates to the extrication activities can be evaluated.
- The patient can be protected from further injuries.
- Patient **packaging** (bandaging and preparation for removal by stretcher) can be initiated.

Removing Glass

Glass may need to be removed to allow a rescuer access to a patient. The glass used in vehicle windshields is laminated safety glass and is actually several layers of

glass and plastic sandwiched together. It is designed this way to crack or "spider web" under impact but remain together in one piece. Glass utilized in side and rear windows is tempered safety glass. The process of forming tempered glass leaves tension across the surface of the finished piece. When struck hard enough in a concentrated area, tempered glass will break into many small pieces or "nuggets."

Tempered glass can be easily removed by striking the glass with the point of a tool such as a halligan or pick-head axe but can penetrate into the passenger compartment if the rescuer loses control. A spring-loaded center punch applied to a corner of the glass will allow it to be broken in a much more controlled manner. Sometimes it is important that the nuggets be contained from scattering about the passenger compartment. The following can be utilized in these situations, but it is important to note that this technique will not work if the glass is wet since the duct tape will not adhere to the glass.

JPR 16-1: Removing Tempered Glass

(For step-by-step photos of this skill sequence, see page 830)

1. Make certain that the vehicle is stabilized.
2. Place several strips of duct tape across the window to be broken, leaving an excess amount in the middle of each strip. The excess will be pressed together and will form handles to remove the glass.
3. Apply several other strips of duct tape to the window in a criss-cross or "H" pattern. The tape should only be placed on the glass itself.
4. A spring-loaded center punch is pressed into the corner of the glass. The firefighter should make certain that the hand not operating the punch is positioned in a manner that will help prevent the hand operating the tool from penetrating into the passenger compartment.
5. Once the glass is broken, it can be removed in one piece by pulling out on the handles formed in the tape in step 1.

Windshield removal is most often performed simultaneously with removal or folding of the roof of the vehicle. The major advantage to this technique is the amount of time that can be saved in both gaining access for treatment of the patient as well as removal.

SAFETY

Cutting laminate glass will produce very fine shards of glass as well as dust.

JPR 16-2: Removing the Windshield and Roof

(For step-by-step photos of this skill sequence, see page 831)

1. Make certain that the vehicle is stabilized.

2. The windshield should be "chopped" along the bottom edge with a pick-head axe. Placing a hand on the pick of the axe will help control it and make certain that only one corner of the blade penetrates the glass.

3. The front roof posts on both the passenger side and driver side should be cut through next, using cutters or hand tools such as a reciprocating saw or hacksaw.

4. Relief cuts will next be placed in the roof on both sides of the vehicle, allowing ample space to gain access when the roof is folded.

5. A tool such as a pike pole will next be placed across the rear section of roof, with downward pressure, to form a fulcrum for the windshield and roof portion being folded over.

6. The windshield and attached assembly will be folded over the rear section.

7. Once folded over, easy access is attained and patient stabilization can begin as removal is carried out. The folded-over portion should also be secured so that a wind gust cannot blow it back over onto the patient or firefighters.

Disentanglement

The process of disentangling a patient begins with the rescuer who is tending the patient advising the incident commander as to the extent of injuries and mechanism of entrapment. The best pathway for patient removal is determined, keeping in mind that the pathway must provide ample working space and adequate space for removal of the packaged patient.

Once the pathway is selected, the method or combination of methods for disentanglement must be selected. The options are:

- **Disassembly.** The actual taking apart of the vehicle components.

- **Distortion.** The bending of sheet metal or components.

- **Displacement.** The relocating of major parts (i.e., doors, roof, dash, steering column).

- **Severance.** The cutting off of components (i.e., brake pedal, steering wheel).

The method of disentanglement selected from those given here will provide information to the rescuer in the selection of the proper tools to accomplish the desired goal. The safety of the rescuer(s) and the patient must always be at the forefront.

Extrications can be as simple as removing a window, unlocking a door, or sliding a seat back, or as complicated as forcing and/or removing vehicle doors, roofs, dashboards, seats, and so on. In general, as vehicles become larger and heavier (i.e., buses, tractor trailers, trains) extrication becomes more difficult due to the heavier structural components and severity of the crashes.

It may be necessary to remove a door to remove a patient or gain access. Door removal can be difficult if the door is severely damaged. Do not overlook the obvious when it comes to removing doors: Is the door locked? Can an alternative door be used? The following method is used to remove a door using spreaders.

JPR 16-3: Door Removal Using Hydraulic Spreaders

(For step-by-step photos of this skill sequence, see page 833)

1. Make certain that the vehicle is stabilized.

2. Insert a Halligan between the door handle and the body of the vehicle, moving the bar up and down to create a purchase point where the tips of the spreaders can be inserted.

3. Position the spreader tips above or just below the safety bolt or "nader bolt" of the door. All vehicles manufactured after 1973 will have one.

4. Open the spreaders trying to roll or "pop" the door latching mechanism off the safety bolt. If metal begins to tear, stop spreading, reposition the tool, and start over.

5. Once the door is free from the safety bolt, enough room may be available to remove the patient. If not, or if the door is to be completely taken off, a safety strap will be applied to the door to allow a second rescuer to control its movement as it is separated from the vehicle. This will prevent any injuries to feet and legs as well as prevent the door from damaging equipment such as hoses or power cords.

VIEWPOINT

Another variation of this windshield and roof removal technique is to remove the rear window glass and make additional cuts through the rear posts allowing total roof removal. Local protocols and operating procedures will dictate which technique should be utilized.

6. The spreader tips should be placed into the top hinge of the door, widening it enough so that it breaks.

7. The bottom hinge should be spread until it breaks next. As the door is released, the second firefighter should firmly hold the strap and pull the door away from the firefighter operating the tool.

Many times during frontal impact collisions, the dashboard of the vehicle gets forced upon the occupants leaving them entrapped beneath it.

JPR 16-4: Displacing a Dashboard

(For step-by-step photos of this skill sequence, see page 835)

1. Make certain that the vehicle is stabilized. Cribbing must be placed beneath the rear base of the front door jamb on both sides of the vehicle.

2. Relief cuts are placed 5 to 6 inches below the intersection of the front support at the rocker panel, as well as 8 to 10 inches back on the horizontal rocker panel, on both sides of the vehicle.

3. Hydraulic rams are inserted between the rear of the door jam and dashboard on both sides of the vehicle.

4. Both rams will be extended at a steady and equal rate until enough room is available for patient removal.

5. Firefighters will need to place cribbing in the lower relief cuts as the dashboard is folded over to maintain its position.

SAFETY

Do not stand in front of doors when hinges are being broken. If released, the springs on the hinges can shoot out at a high rate of speed and cause injury, **Figure 16-16.**

FIGURE 16-16 Springs on door hinges can shoot out at a high rate of speed and cause injury to rescuers.

Patient Removal

Once the pathway has been created and made as safe as possible by covering sharp edges left by cut or torn metal, covering any broken glass that the patient could come in contact with, and removing all tools that may be in the way, the properly packaged patient should be carefully removed, **Figure 16-17,** again remembering that the goal is to minimize any aggravation of existing injuries.

Patient removal should be made only with the direct supervision of a certified EMS responder—typically a paramedic or experienced EMT. Appropriately, the supervising EMS provider should be focused on patient care. Often, the removal of the patient and subsequent movement toward an ambulance takes the rescuers and patient out of the work area protected by traffic barriers. Spotters should watch for traffic during patient movement, and escape routes must be planned.

Scene Stabilization

After patient removal, firefighters must refocus their energy toward securing the incident scene. Although this is not as exciting as the actual rescue, firefight-

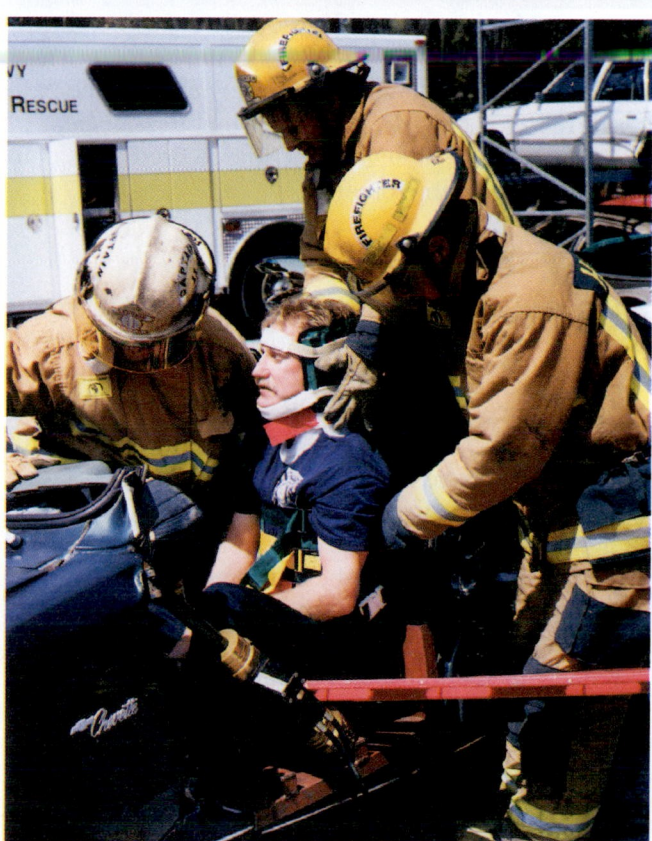

FIGURE 16-17 A properly packaged patient being carefully removed from a two-vehicle accident and placed onto a long spine board. *(Courtesy of Rick Michalo)*

ers must continue to exercise safe practices. Complacency must not take place once the patient is removed. Hazard assessment must remain continual; many firefighters have been injured during scene "pick-up" operations. Extreme attention is required to release tools that are holding the vehicle—many components may "spring back" when the force of the rescue tool is relieved. Cribbing used to shore and support the vehicle might need to be retrieved, creating a potentially unstable vehicle again. At times, it may be best to allow tow vehicles to connect to the vehicle prior to cribbing removal. As tools are being removed from the wreckage, firefighters should inspect the cutting/contact surfaces, hydraulic lines, and safety features for damage and report any problems or observations to their officer. Additionally, the scene stabilization effort may include the following:

Vehicle Recovery

Fire resources are often asked to stand by for the tow vehicle to overturn or "hook" the damaged vehicle. The tow-vehicle operator must be advised of vehicle hazards such as fuel leaks and supplement restraint systems (SRS air bags) that have not deployed. Local SOPs may require that the battery of a damaged vehicle be disconnected. Only trained personnel should perform this disconnect. Tensioned tow cables introduce another hazard. It is a good practice to "envision" the tow cable breaking and stay clear of the path it will recoil. Likewise, it is important to stay clear of the vehicle as it is being moved and to predict its path should the tow cable break.

Fluid/Parts Cleanup

The use of a dry absorbent is typical for oils, engine coolant, and diesel fuel. Gasoline is best absorbed with specially designated "fuel pads" and disposed of in accordance with local SOPs. When handling fuel pads, it is best to use gloves that can be decontaminated. Battery acids are extremely caustic and should be handled with diligence. Vehicle parts that have been strewn about can be sharp. Law-enforcement officials may want strewn parts to be left in place to assist with accident reconstruction investigations. If not, carefully place all loose parts back into the wrecked vehicle.

Once the scene has been stabilized, the responding crews must secure all deployed equipment and pick up traffic-calming devices. This exposes rescuers to traffic hazards again. As a rule, they should not put their back to traffic—they should pick up traffic cones while facing traffic. Often, spotters are needed to help apparatus back out of position and into traffic. These spotters must also have a traffic lookout watching their back.

SPECIALIZED RESCUE SITUATIONS AND TOOLS

The fire department is considered an "all service" organization and is often called upon to mitigate emergency situations that may be outside the normal realm of training for most of its members. Specialized rescue calls are all the more dangerous because their respective training programs are often underfunded since such calls are few in number, yet they hold a high degree of risk for rescuers by their very nature. Because of this, many times rescuer fatalities outnumber the initial victims of the incident. It is important that firefighters recognize these specialized situations for what they are and call for the necessary assistance, at the same time ensuring that the situation does not worsen for the victim and they do not become victims themselves. With any specialized rescue incident, certain elements are imperative. These are:

- *Continual hazard assessment and risk benefit analysis.* As with all incidents, is the amount of risk exposure to the firefighters justified? Are the members performing a true rescue or is it a clear recovery operation? Specialized rescue incident conditions change rapidly and must be evaluated continuously to avoid missing important information and to overcome tunnel vision.

- *Operating guidelines and procedures.* NFPA 1670, Operations and Training for Technical Rescue Incidents, requires that fire departments identify possible hazards in their jurisdiction that may require specialized rescue. Basic pre-planning can establish needs prior to receiving a call for help and establish the minimums for personal protective equipment. Firefighters must be aware of their role, evacuation procedures, and the resources available to them.

- *Scene control.* Specialized rescue incidents will draw a lot of attention; therefore, onlookers as well as media will need to be contained. Zones will need to be established to prevent non-essential personnel, bystanders, and patients' family members or coworkers from getting in harm's way. An unauthorized, untrained or nonessential person in the hazard area is an invitation for trouble. Setting up control zones will help limit access and maintain accountability. Zones that need to be established include a hot zone, warm zone, cold zone, and possibly a no entry zone. The hot zone contains the greatest degree of risk or hazard to the rescue workers and is the area where rescue operations will be taking place. The warm zone is a limited access area where operations that support actions being taken in the hot zone should be performed. The cold zone is set up for staging, nonessential personnel and the public. Incident Command for an incident will also be set up in the cold zone. A no entry zone is an area that is not allowed to be entered by anyone due to an extreme or unpredictable hazard condition.

- Information released should be provided through the use of a Public Information Officer, and the use of outside resources will need to be closely regulated. *Incident management and accountability.* Command and control must be established for all incidents and a personnel accountability system must be in place. It is not necessary for the Incident Commander to be trained to the level of the incident, but it is necessary to have trained resources readily available to help them make decisions. At a minimum, a Safety Officer trained to the level of the incident and a Rescue Group Leader responsible for rescue plan development should be present along with the Incident Commander on specialized rescue situations.

Rope Rescue

Rope rescue deals with a victim who is either above or below normal ground level and beyond practical means of rescue, thereby requiring the use of rope and related equipment. Vertical or high-angle rescue, **Figure 16-18,** entails the weight of the victim and rescuer relying solely on rope for support. Rope rescue can also include low-angle or slope evacuation where rescuers must bring the victim up a steep incline. Some form of rope rescue is relevant in all disciplines of specialized rescue to a degree.

If a patient is to be lowered or raised using a life safety rope, it is imperative that the personnel on scene have the training, skill, and equipment to do so safely. Many fire/rescue departments have developed special teams to respond to these types of incidents. The teams have the proper equipment and specialized training and are required to maintain proficiency. However, firefighters should be familiar with the basic equipment and techniques utilized by their department at this type of incident.

Fall hazards are the obvious danger at rope rescue incidents but others that may exist include:

- Bad or slippery footing
- Falling objects
- Hazardous atmosphere
- Equipment misuse
- Utilities
- Trip hazards

Responsibilities of first-arriving firefighters to rope rescue incidents may include scene size-up, obtaining necessary resources, securing an area of at least 300 feet, limiting traffic and sources of vibration within the hazard area, and making contact with the victim if it is safe.

FIGURE 16-18 Vertical or high angle rescue entails the weight of the victim and rescuer relying solely on rope for support. *(Courtesy of James Pelliterri)*

Water Rescue

Water rescue operations can be very dangerous for the rescuers. In many cases, the victim(s) can be seen, which results in the emotions of the bystanders being very high. They may expect the firefighters to jump in the water immediately to rescue the person in trouble. That is the last thing a rescuer should do, and it should never be done without a personal flotation device (PFD).

SAFETY

Every firefighter at the scene of a water rescue incident should have a PFD on if at all possible. If not, those without PFDs should be kept well back from the water's edge. If they are allowed to be near the water's edge and something happens in front of them, they may go into the water without the proper equipment.

Following in priority order are the methods and procedures that firefighters should utilize to rescue a victim from static (lake) water and slow-moving rivers and streams:

1. *Reach.* If the victim can be reached without the rescuer entering the water, a pike pole, pool rescue pole, or anything that can be used to extend the rescuer's reach can and should be utilized, **Figure 16-19.**

2. *Throw.* If the victim cannot be reached directly but is within throwing range of a rescue throw bag or flotation device attached to a rope, then one should be thrown and the rope utilized to pull the victim to safety, **Figure 16-20.**

3. *Row.* If the victim cannot be reached by either of the preceding methods and there is a suitable boat available, the rescuer(s) should utilize it. Extreme caution should be exercised on approaching the victim. In a panic, the victim could capsize the boat, resulting in the rescuers now becoming potential victims. The best course of action is not to try lifting the victim into the boat at all (unless it is designed for rescue operations), but to hold onto the person and return to shore, **Figure 16-21.**

4. *Go.* If none of the preceding options is possible, the absolute last method is for the rescuer to enter the water, **Figure 16-22.** Extreme caution must be exercised when approaching the victim. In a panic, the victim could injure the rescuer or make it diffi-

cult for the rescuer to assist him or her. If possible, the rescuer should carry an additional PFD to the victim. In still or extremely slow-moving water, it is a good idea for the rescuer to swim out holding a **tether line.** This will allow the rescuers on shore to pull the rescuer and victim back to shore.

SAFETY

A rescuer should not use an attached tether line in moderate to fast moving water. It can pull the rescuer under.

FIGURE 16-20 A firefighter using the throw method to rescue a victim in the water. Note the PFD and the underhand throwing technique.

FIGURE 16-21 A firefighter using the row method in a small boat to rescue a victim in the water. Note the extra PFD in the boat for the victim. Do not attempt to lift the victim into a small boat; instead have the victim hold on to the side.

FIGURE 16-19 A firefighter using the reach method to rescue a victim in the water. Note the PFD and the use of a pole to extend the firefighter's reach.

FIGURE 16-22   A firefighter using the go method to rescue a victim in the water. Note that the firefighter is carrying an extra PFD for the victim.

FIGURE 16-23   Obstacles in the water such as low head dams also produce currents that can pull rescuers under and cavitate boat motors. *(Courtesy of James Pelliterri)*

Swift-water rescue is a specialized field requiring additional training, practice, and protective equipment. The force of a fast moving stream is considerable, even in shallow currents. Obstacles in the water, such as low head dams, also produce currents that can pull rescuers under and cavitate boat motors, **Figure 16-23.**

As mentioned before, tether lines increase the danger to rescuers. Special rigging and procedures are required for a successful rescue. Local SOPs and training are required for swift-water rescues; however, some basic rescuer safety procedures can be discussed here. First, all rescuers must wear PFDs. They should try to encourage the victim to keep afloat until a swift-water team can arrive and set up. If an immediate rescue attempt must take place, rescuers can

place themselves at the shore and attempt the throw method, making sure a float is attached to the end of the rope. A PFD is a great float and will help calm the victim. Rescuers should throw the rope well upstream of the victim and allow the current to drift the rope toward the victim. Once the victim has the rope, rescuers can encourage the victim to don the PFD. The rope is not used as a tether; it acts merely as a guide rope. The victim holds the rope, and the rescuers allow plenty of slack in the line. As the victim drifts into the stream, the rescuers release just enough slack to let the victim swing toward the shore. Once again, this is a last-ditch method. Properly trained swift-water technicians are the key to a successful rescue.

Ice rescue is considered another aspect of water rescue. Many of the safety precautions and rescue procedures just presented with regard to water rescue apply to ice rescue as well. The procedures and hazards are very similar with the added element of extreme cold. In place of or in addition to PFDs, rescue personnel working at an ice rescue scene should have thermal rescue suits, which provide the required personal flotation as well as protection from the extreme cold. The rescuers in **Figure 16-24** are wearing thermal suits.

If rescuers must go out onto the ice themselves, they need to use whatever is available to distribute their weight over as large an area as possible. Some items that work well for this are ladders, sheets of plywood, planks of wood, and backboards.

The rescuer should use the reach (with extension) or throw method to keep from having to be right up next to the hole. Ice rescue victims may not have the strength or determination to grab a rope or pole. The effects of the cold water quickly rob the human body of energy. In these cases, firefighters with appropriate thermal flotation PPE should try to approach the victim in such a way as not to disturb the ice the victim is clinging to. The best approach is to come in behind the victim and use a rescue collar or other rope device to secure the victim. It is essential to coach the victim to hold tight to the rope but not to the rescuer. Again, these rescue skills are beyond the scope of this book. All firefighters should train and practice under the supervision of a competent instructor.

Responsibilities of first-arriving firefighters to water rescue incidents where the victim may not be apparent may include scene size-up, obtaining necessary resources such as divers, securing the area, and establishing the point where the victim was last seen through witness interviews.

CAUTION

A rescuer should not approach too close to the hole itself. The ice is obviously not stable in that area.

(A)   (B)

FIGURE 16-24   (A) A ladder, a sheet of plywood, or a specialized ice rescue boat can be used to spread a rescuer's weight over a larger area. (B) Using a throw bag or pike pole to reach out to a victim who has fallen through the ice prevents the rescuer from having to approach the weakened area too closely. *(Courtesy of Halfmoon-Waterford Fire District 1)*

Structural Collapse Rescue

Structural collapse may occur for any number of reasons: weakening due to age or fire, environmental causes (earthquake, tornado, hurricane, flooding, rain, or snow buildup on roofs), or an explosion (accidental or intentional). Whatever the cause, the result can be any number of victims trapped, injured, or killed. One of the firefighter's first priorities at this type of incident is to determine the number of possible victims. This information along with the construction type, size, and occupancy of the collapsed structure will have a direct impact on the assistance required. Operations at collapse incidents will take place in five stages:

1. Reconnaissance and rescue of surface victims
2. **Void** search
3. Selected debris removal
4. General debris removal
5. Debriefing

Rescue of victims who are not trapped or are lightly trapped should be done as soon as safely possible. Initial responding firefighters may participate in the reconnaissance and rescue of surface victims. It is very common

FIGURE 16-25   Firefighters need to remember that victims may be buried under the debris that they are climbing on and over at a collapse scene.

for this process to already be under way by civilians from the immediate area when a rescue unit arrives. It is important to remember that as civilians and rescuers are climbing on and over debris there is a possibility that other victims may be trapped under that debris, **Figure 16-25.** As operations become more involved, adequately trained resources will be a necessity to carry on throughout the remaining incident stages.

SAFETY

Safety of the rescuers needs to be the number one consideration as in all rescue operations. Some of the particular hazards to be aware of at collapse scenes are secondary collapse, live electrical wires, and gas leaks.

SAFETY

Every action taken near a collapsed building will have a reaction.

Structural collapses are generally classified into three different types, which are described next. However, it is not uncommon to encounter more than one type at any given collapse incident.

Pancake Collapse

A pancake collapse occurs when both sides of the supporting walls or the floor anchoring system fails. It is characterized by the roof or upper floor(s) falling parallel onto the one below, often causing a domino effect, which may result in several or all of the floors collapsing one upon the other until the lowest level is reached. This type of collapse often results in many small voids being created by debris between floors supporting the upper floor, **Figure 16-26.**

Lean-To Collapse

A lean-to collapse occurs when only one side of the supporting walls or floor anchoring system fails. It is characterized by one side of the collapsed roof or floor remaining attached or supported by the wall or anchoring system that did not fail. This type of collapse usually results in a significant void being created near the remaining wall, **Figure 16-27.**

V-Type Collapse

A V-type collapse occurs when the center of the floor or roof support system is overloaded (improper placement of stock, buildup of snow or rain, etc.) or becomes weakened for some reason (fire, rot, termites, improper removal of support beams, etc.). It is characterized by both sides remaining attached or supported by the walls or anchoring system but collapsing in the center. This type of collapse will usu-

ally result in voids being left on each side near the supporting walls, **Figure 16-28.**

It is important for the firefighter to know where these voids can be anticipated in each of the different types of collapses since this is where the greatest likelihood of finding survivors exists.

Reconnaissance and Rescue of Surface Victims

The initial size-up by first-arriving firefighters will set the tone for the entire operation. The initial size-up should include a six-sided approach to surveying the structure. The six-sided approach will consist of the front, rear, both

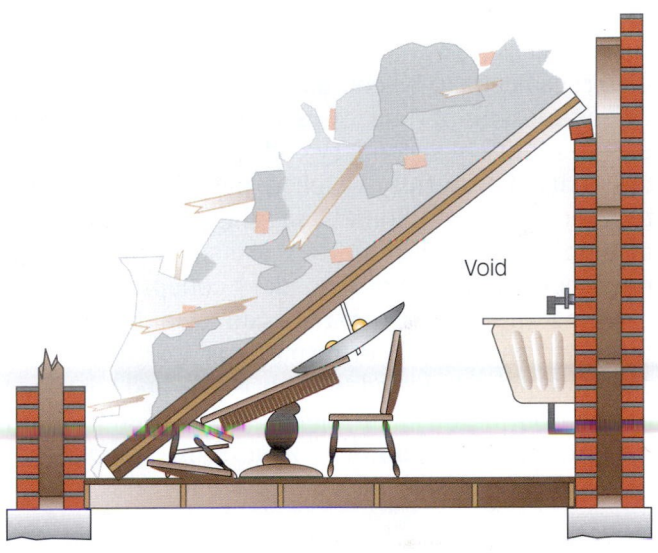

FIGURE 16-27   A lean-to collapse. Note voids, where survivors may be located, that have been created by debris during structural collapse.

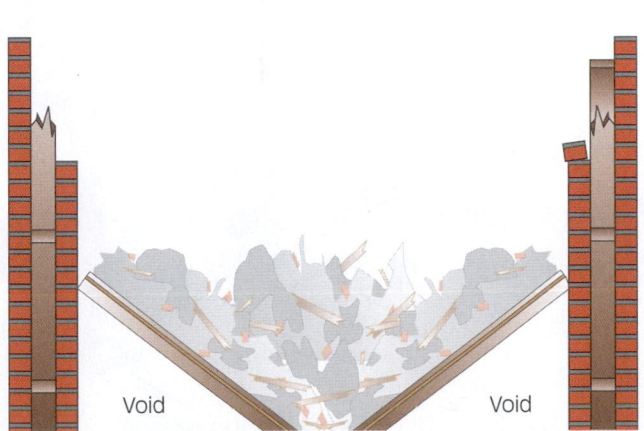

FIGURE 16-28   A V-type collapse. Note voids, where survivors may be located, that have been created by debris during structural collapse.

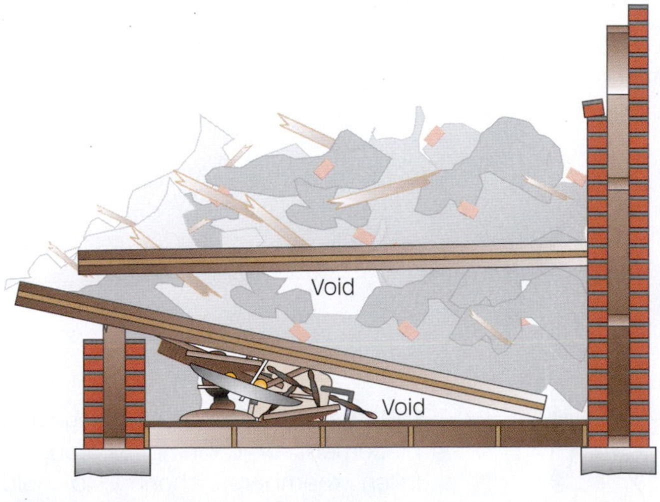

FIGURE 16-26   A pancake collapse. Note voids, where survivors may be located, that have been created by debris during structural collapse.

sides, top, and bottom. An aerial device set up outside of the collapse zone can provide a valuable location in which to size up the collapse area. Once acknowledged, hazards should be handled in one of four ways:

1. Avoid the hazard area
2. Remove the hazard
3. Shore or support the hazard
4. Monitor the hazard for deterioration over time

An important observation to be made is the status of utilities. With collapse, live wires and broken pipes can be distributed among the debris pile. It may not be possible to control utilities in the immediate vicinity of the building due to the collapse. This may require the control of utilities at other points remote from the incident site. Monitoring the collapse area for hazards such as natural gas, flammable vapors, and hazardous materials needs to be continuous.

The most important assessment is recognition of the possibility of secondary collapse and making certain that it is carefully weighed and considered prior to rescue attempts. Sounds, sagging floors, walls out of plumb, or material shifting are some of the most prominent indicators of secondary collapse.

Using the **hailing system** can also help locate those victims capable of response at this stage. Rescuers are placed around the debris pile and silence is ordered on the incident scene. One rescuer will call for some sort of acknowledgment from a trapped victim if they can hear the call. After the first rescuer takes a turn, silence should be exercised by the incident personnel for a time period to give the trapped victim a chance to reply. After that time period, the next rescuer should repeat the process. If a reply is heard, the rescuers should attempt another call from a different location or angle to help pinpoint the location of the trapped victim. It is important to note that a reply may not be voice communication but sounds such as tapping objects or other noise.

By pinpointing information, the initial responders can save technical rescue teams critical time in initiating advanced rescue efforts. Knowing where the survivors may be is just part of the problem. The firefighter must also know how to make the area as safe as possible prior to entering. Having dealt with the issues previously discussed (live wires, gas leaks, etc.), the firefighter needs to have a basic knowledge of cribbing, **shoring,** and tunneling. **Figure 16-29** shows examples of each technique.

- *Cribbing.* As previously defined, cribbing is the use of various dimensions (2×4, 2×6, 4×4, 6×6, etc.) of lumber arranged in systematic stacks (pyramid,

---

SAFETY

It is important to remember in all of these operations that the firefighter is not lifting the fallen members, but merely supporting them where they are. To lift them may cause further collapse.

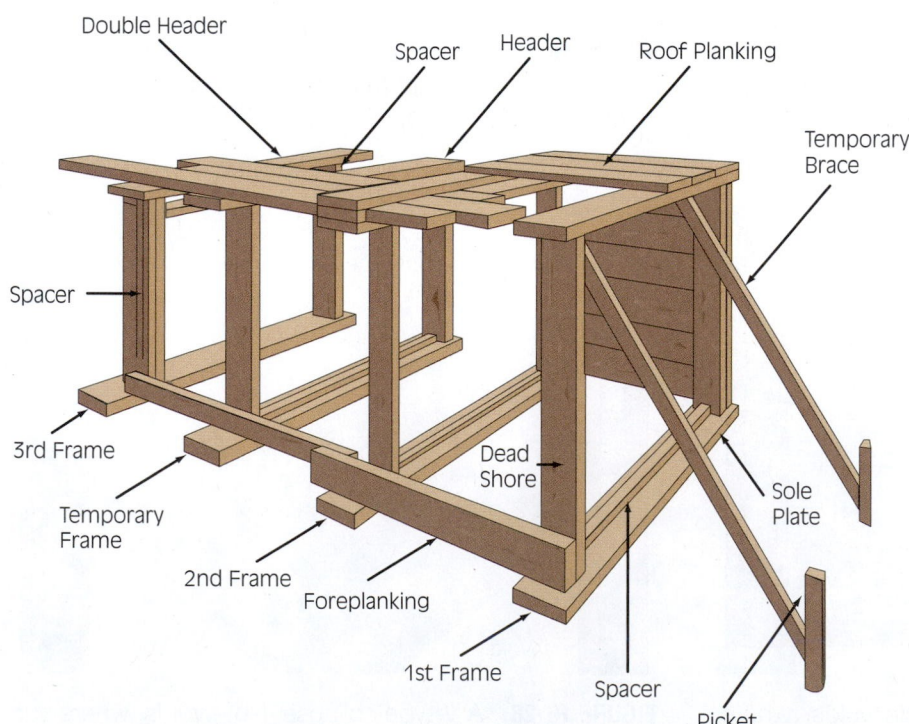

Perspective of Timbering

FIGURE 16-29 The proper usage and placement of cribbing to support fallen members, shoring to hold back the sides and roof debris, and tunneling supports are required prior to entering trench or collapse areas.

box, step, etc.) to support an unstable load. The same principle applies to the use of cribbing in collapse operations; however, it is common to use timbers of larger dimensions.

- *Shoring.* Shoring is the use of timbers to support and/or strengthen weakened structural members (roofs, floors, walls, etc.) in order to avoid a secondary collapse during the rescue operation. These can be very complex operations and should not be undertaken without the input, advice, and guidance of experts in the engineering field, preferably someone with experience with collapsed structures.

- *Tunneling.* Tunneling may be required to reach survivors located in the voids created during the collapse. This is a very dangerous and time-consuming process and should not be undertaken unless all other avenues have been tried. Tunneling should not be utilized as a general search process; a specific destination is necessary along with a good indication that survivor(s) will be located when the destination is reached.

Trench and Below-Grade Rescue

OSHA 29CFR 1926 regulates operations specific to trenches. A **trench** is defined as an excavation that is deeper than it is wide, with 15 feet being the maximum width. Most trench or below-grade incidents occur at construction sites, utilities (gas, water, sewer, etc.), maintenance sites, or well digging sites.

In a trench collapse, the most immediate danger to the victim is usually a compromising of the ability to breathe. It follows then that the first priority on reaching a victim is to uncover the victim's head and as much of the chest as possible. Supplementing respiration with oxygen should also occur as soon as safely possible.

Even with the need for haste as described in the previous paragraph, rescuers should assume that, because the fire department has been called to the scene for a rescue, proper safety procedures were not being utilized to begin with, such as the trench box being above ground and not in the hole, shown in **Figure 16-30.** With this in mind, the rescue crew must approach the scene very carefully keeping their safety in the forefront. The rescuers should use **ground pads** or something else to distribute their weight over a larger area (the same as with ice rescue discussed earlier) when approaching the caved-in area, to prevent causing more material to fall in on the victim or others. Any fissures in the soil or sides of the trench should be noted as they are an indication of additional collapse, **Figure 16-31.** If the

FIGURE 16-30 Proper safety procedures such as the use of a trench box are often not being utilized to begin with when rescuers are called to a scene. *(Courtesy of James Pelliterri)*

FIGURE 16-31 Any fissures in the soil or sides of the trench should be noted as they are an indication of additional collapse. *(Courtesy of James Pelliterri)*

victim's location can be determined, it should be marked on the ground outside of the trench, and a rope should be thrown to the victim if his arms are free so that in the case of a secondary collapse he can be located quickly.

For a trench or below-grade incident to be resolved as safely as possible, the previously mentioned ground pads

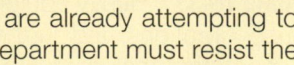

SAFETY

Even if coworkers or others are already attempting to assist the victim(s), the fire department must resist the urge to jump in and help until the area is made safe.

should be in place, all nonessential personnel and equipment should be kept away from the site, the cave-in area must be sho red up, the air quality needs to be monitored, fresh air needs to be introduced into the area, and adequate egress needs to be provided for the rescuers.

With the exception of shoring, initial responders should be able to perform the other necessary actions.

Entering the caved-in area will require extensive equipment and training, **Figure 16-32.** More than likely, this equipment may not be available on the initial responding unit. In the case of a trench collapse, the safest (for victim and rescuers) and most effective method of freeing a victim is to dig with small hand tools. Equipment such as a backhoe, **Figure 16-33,**

FIGURE 16-32 Rescuers entering a trench must have the necessary training and equipment put into place prior to entry. *(Courtesy of James Pelliterri)*

FIGURE 16-33 Heavy equipment such as a backhoe can cause vibrations and further collapse near a trench. It is imperative that rescuers get control of these items. *(Courtesy of James Pelliterri)*

should never be utilized no matter how skilled the operator may be. Sources of vibration can cause secondary collapse and should be stopped within an area of at least 300 feet by responders when establishing control zones.

While one rescue team is in the process of excavating the victim, there should be another team topside addressing the need to remove the victim once freed. It can be anticipated that the victim is going to have injuries such as trauma to the head, spine, torso, and extremities. The removal team needs to be factoring these possible injuries into the removal plan and be prepared to provide adequate immobilization and hauling systems.

Confined Space Rescue

OSHA 29 CFR 1910.146 regulates operations regarding confined spaces. A **confined space** is defined as a space that is large enough to be entered for working, has limited entry and exit points, and is not designed for continuous occupancy. Typical confined spaces are utility vaults, storm water or sewer culverts, septic tanks, industrial boilers, storage tanks, grain silos, cement hoppers, sawdust collectors, fertilizer hoppers, and areas aboard ships, **Figure 16-34.** As can be seen from this list, confined spaces can and do exist in every department's jurisdiction.

> **NOTE**
>
> It is estimated that 60 to 80 percent of the people who die each year in confined space incidents do so while attempting to "rescue" a victim. Although the majority of these "rescuers" are untrained civilians or coworkers, surprisingly some are firefighters.

FIGURE 16-34 Confined spaces are present in every community. *(Courtesy of James Pelliterri)*

Many tasks must be accomplished before an entry into the confined space is permissible for rescue. Size-up and information gathering is once again critical by initial responders. A qualified rescue team will need to be contacted. Securing the scene is an important task that initial responders may be assigned. Isolating or eliminating hazards through lock out/tag out procedures will need to be accomplished. Site workers may be able to provide valuable information regarding lock out/tag out and alternative entry points into the space, as well as providing a permit for entry that could disclose valuable information to the responders.

The most common hazard encountered at a confined space incident is an oxygen-deficient atmosphere, followed closely by toxic and/or hazardous vapors. Entry into a confined space by qualified rescuers should never be attempted without having first sampled the atmosphere and confirmed it as tenable or identified the hazards. Initial responders can provide atmospheric monitoring of the confined space and ventilation by the use of an air blower and flexible duct, **Figure 16-35.**

Whenever members of a rescue team enter a confined space, they must utilize appropriate personal protective clothing and equipment (for the known and possible hazards), including lifelines for each member and a radio or other communication system between the rescue team and the operations chief. The space must be constantly monitored, preferably by individual monitors on each entry team member, but at a minimum by one member of the entry team assigned this task. A backup crew must be fully equipped and dressed out to act as a rescue team for the entry team if a problem develops or the entry team needs assistance for any reason.

FIGURE 16-35   Ventilation is a task that can be completed by initial responding firefighters. *(Courtesy of James Pelliterri)*

Rescue from Electrical Situations

Rescuers responding or arriving at an incident scene involving electrical wires, transformers, transfer stations, and so on, must assume these items are energized until advised otherwise by the utility company representative at the scene. Very often wires will appear to be de-energized because they are lying still on the ground and no arcing is visible. This is not necessarily the case; the wire may be well grounded at that time or there may be a time delay reset breaker on the line that will reset itself after a set amount of time and may do so a number of times before finally staying in the off position. Any victim who is in contact with an electrical wire must also be considered energized and cannot be touched. Firefighters need to remember that none of their protective clothing and/or equipment is designed to protect them from electrical current.

Vehicle accidents involving electrical equipment like pad transformers and downed wires are especially dangerous. It is best to leave victims in the vehicle and not approach. Often, the vehicle may be energized—yet the passengers are safe as long as they do not exit. A victim in an energized car is like a bird on an overhead power wire—that is, as long as the ground is not touched, no electrocution will take place. Anything or anyone who touches the vehicle and ground at the same time will cause an easier path for electricity to seek ground. Firefighters attempting to approach involved electrical equipment may feel a tingling sensation in their boots. This is a dangerous sign known as ground gradient. When the tingling is felt, firefighters must back away using a shuffle foot motion to keep their feet in contact with the ground. In all cases, the fire department shall verify that power company technicians have been dispatched.

If the electrical contact is within a building, it should be possible to shut the power off at the main disconnect. It is recommended that the main disconnect be utilized as opposed to individual breakers because sometimes circuits or equipment may be receiving electricity from more than one breaker. If this (shutting off main disconnect) method is used, a firefighter should be stationed at the disconnect, **Figure 16-36,** to ensure that no one turns the power back on either accidentally or because the person did not know a rescue was in progress.

Firefighters should never enter an enclosed electrical transfer station, substation, or vault without a company representative on scene to ensure that all electrical power is shut off. Every wire, bracket, frame, transformer, and so on, within the enclosed area should be assumed to be energized prior to the on-scene company representative confirming otherwise.

FIGURE 16-36 Rescuers must take control of the electrical disconnect and remain in contact with the rest of the team via portable radio.

Some departments carry specialized equipment such as hot sticks, insulated cutters, and insulated lineman's gloves on their apparatus. If these items are to be utilized they must be stored, tested, and certified in accordance with the very strict standards that apply to such equipment. The personnel assigned to utilize this equipment must also receive specialized training in the use and care of this equipment.

Most fire departments have recognized that this is a highly specialized and dangerous operation and best left to the professionals with the utility company.

Any time large electrical wires are going to be cut (energized or not), one of the dangers to be aware of is what is known as **reel coil.** This is memory that the wire develops from being placed on the wooden spool as it is being manufactured. This reel coil will cause the wire to recoil (spring away) from the point at which it was cut, attempting to resume the coiled form again. This recoiling can be very quick and powerful in heavier wire and can cause severe injury from the impact even if it is de-energized.

Industrial Entrapment Rescue

Extricating a victim from an industrial entrapment is usually a very complex process. In most cases if the entrapment is relatively minor, the victims' coworkers will have already extricated the victim by the time the fire department arrives. Due to the number and complexity of machines used in industrial plants today, it is impossible to have a specific plan for each possible entrapment.

The operations at an industrial entrapment incident should follow a predetermined sequence of events. The process presented previously in the vehicle crash extrication section is applicable at an industrial entrapment with very minor changes.

The following order of procedures works very well and ensures that none of the essential procedures or operations is overlooked:

1. Assess the incident (size-up).
2. Identify on-scene or available experts.
3. Stabilize entrapping machinery.
4. Access and stabilize of the patient.
5. Disentangle the patient.
6. Remove the patient.

Incident assessment should be a predetermined sequence of steps or actions that is used in evaluating an entrapment. This evaluation is usually carried out by the officer. However, at an industrial entrapment incident, there may be a firefighter on the crew with knowledge and/or experience that could be invaluable. A good officer should know this ahead of time and never hesitate to utilize it. As previously stated, while it is important that this assessment be done quickly, it is more important that it be done accurately and completely.

Although identifying on-scene or available experts is shown as the second step it is vitally important that this expert advice be available during the assessment phase if at all possible. It is quite possible that the expert may be able to provide a simple and quick extrication method that is not obvious. It is also possible that the expert can prevent further injury to the victim or injury to the rescuers by preventing the rescuers from taking some action that, although appearing to be logical and effective, would cause the machine to react in an unexpected manner. A good example of this is that while turning off the power to a machine would appear to be the logical thing to do, on some punch press-type machines this would cause them to finish out their cycle.

Personnel conducting an assessment need to take the following into consideration as part of a good incident assessment:

- The apparent extent of injuries
- Current and anticipated hazards to the victim and/or rescuers
- Disentanglement requirements
- Support needs

The results of the assessment will reveal whether any additional assistance is required, such as:

- Additional medical units
- Additional or specialized extrication equipment
- Advanced medical assistance, such as an on-scene doctor
- Other specialized equipment, such as a light unit
- Hazardous materials team

Stabilization of the entrapping machinery should begin with shutting down the power (electrical, hydraulic, and pneumatic) and releasing stored energy (air or hydraulic pressure) to the machine, as soon as it is determined that this will not cause any further movement. Once the power is shut off to the machine, a firefighter should be assigned to stand by the shut-off switch to ensure that it is not turned back on inadvertently.

Generally, the next step in stabilizing the machine is to crib or block the entrapping part so that there will be no further unwanted or unplanned movement. If the machine is in midpoint of a movement, it may be necessary to crib both above and below the moving part. Although in many cases it is quite appropriate and convenient to use the same cribbing materials utilized for stabilizing vehicles, heavier and stronger materials are sometimes required since the machine may be capable of crushing regular cribbing. It may also be possible to use chains or cables to assist in stabilizing the moving part.

Accessing the patient is usually not that difficult in these types of incidents; however, accessing the entrapped appendage may be very difficult or impossible in some cases. It is very beneficial, from the medical point of view, if access can be gained to the entrapped appendage.

The three overriding factors when deciding on the method to be employed to disentangle the patient are (1) the medical condition of the patient, (2) the amount of time required to complete the extrication, and (3) the effect the extrication activities will have on the patient. While the damage to the entrapping machinery should be considered, it has a very low priority when compared with the patient's well-being.

> **NOTE**
>
> Amputation is a viable option in industrial entrapment incidents more than in any other particular incident. This is due to the severe damage that is often done to the entrapped appendage and concern for the medical condition of a patient who may not be able to tolerate a complex and lengthy extrication operation.

When it is possible to do so without causing further injury to the patient, the simplest and fastest method of disentanglement is to operate the machinery manually through the rest of its normal cycle or to manually operate the machinery in reverse, backing the entrapped appendage out of the machine.

When these methods are not possible or practical, the entrapping machinery must be disassembled, forced, or cut. Obviously the use of on-scene expert advice and assistance is highly recommended during this phase of the operation. The tools and equipment discussed in the vehicle extrication section may also be very useful in industrial entrapment operations. However, in many instances it might be impossible to force the machinery without causing greater harm to the patient or impossible to cut the machinery components due to their size or strength. In these cases, the machinery must be disassembled piece by piece.

Elevator and Escalator Rescue

When a call is received for an elevator or escalator incident, the responding fire department unit should immediately request that the dispatcher confirm with the calling party that the service company has already been called and a technician is en route; if not, one should be requested immediately. The elevators being installed today and even those that were installed many years ago are very complex. Unless there is a compelling medical emergency requiring that the rescuers access or remove the passengers from the elevator immediately, they should await the arrival of the service technician. In many cases the service technician will be able to repair the problem immediately or at least be able to bring the car to a landing so that the passengers can be removed through normal methods. At incidents where there is no immediate need to access or remove the passengers, the fire department's responsibility is to establish communications with the passengers, ensuring them that they are perfectly safe and help is on the way.

The two basic types of elevator operating systems are hydraulic and electric/cable. Hydraulic elevators are raised and lowered by use of a hydraulic pump connected via a valve system to a multisection ram, on top of which the elevator car sits. A very basic example of this type can be found in many garages where they are utilized to lift vehicles to be worked on. The power unit for a hydraulic elevator is usually found near the lowest floor served by the elevator, **Figure 16-37.** This type of elevator is not generally used in buildings of more than five stories.

Electric/cable elevators are raised and lowered by the use of an electric motor connected to a drum or cable sheave. The cable(s) that is taken up (to raise the car) or let out (to lower the car) by the drum/sheave unit is connected to the frame of the elevator car. The motor drum/sheave unit for this type of elevator is generally located directly above the elevator shaft, **Figure 16-38,** though it can be located at the bottom or in rare cases in an adjacent room. Electric/cable elevators are in use at all high-rise buildings. Extremely tall buildings may have two or more sets of elevators, with one set serving the lower floors and another set serving the upper floors.

Elevators have a variety of doors, openings, and safety interlocks that need to be understood by rescu-

FIGURE 16-37 A hydraulic elevator pump and valve assembly located in the mechanical room, usually on the first floor.

FIGURE 16-38 An electric/cable elevator motor and drum/cable sheave assembly in the equipment room usually located above the hoistway.

ers. Starting with the **hoistway** door, these doors are locked in the closed position unless the car is at the correct level (the landing) in the hoistway to release the locking devices. These doors can be opened by the use of special hoistway door keys, which come in a wide variety of key shapes and styles. Unless the responding unit has an assortment of these keys, the rescuers will have to rely on the building management to have the proper key available when needed. When a hoistway door is opened, an electrical safety interlock will not allow the car to move.

The elevator car doors themselves do not have a locking mechanism and can be opened from either the inside or the outside of the car by pushing them apart. These doors also have an electrical safety interlock that will not allow the car to move if the slightest opening is detected.

All electric/cable elevator cars have emergency access panels either in the roof or through the side (in multiple car hoistways), **Figure 16-39.** Hydraulic elevators that have an emergency valve that can be used (by an elevator technician) to lower the car to a landing are not required to have these emergency access panels. These emergency panels are equipped with the same type of electrical safety interlock that will not allow the car to move when they are open.

Modern in-car control panels, **Figure 16-40,** generally have the floor selection buttons, an emergency stop button, and a "Fire Service" key slot. Some will have an "Independent Service" switch located behind a locked panel or a separate key slot. Many will also have a telephone or intercom-type system for emergency usage.

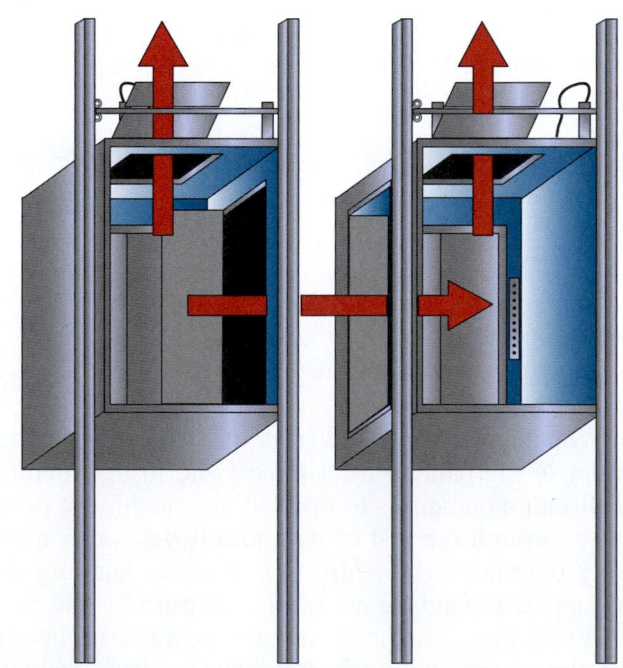

FIGURE 16-39 Elevator cars may have top emergency access panels, side emergency access panels, or both.

FIGURE 16-40 Firefighters must know and understand the operation of the various types of elevator control panels that exist within a particular response area.

The Fire Service key slot is for use by firefighters who may want to utilize the elevator car during the investigation of a possible structure fire. When this key switch is operated, the car will respond only to commands from the in-car control panel and not to any calls from other floors. The car door will not automatically open on reaching the selected floor; the operator in the car must push and hold the Door Open button.

When the Independent Service switch is operated, the car will respond only to commands from the in-car control panel and will not respond to calls from other floors. However, when the selected floor is reached the door will automatically open.

If it becomes necessary for rescuers to access or remove elevator passengers, this should only be

attempted by rescue personnel with special training and experience. This can be a very dangerous process for both rescuers and passengers.

The first step in accessing the passengers is to determine the location of the car in the hoistway, which can often be a time-consuming process. If rescuers are able to communicate with the passengers, they will be able to assist by telling rescuers at what floor they got on the elevator or possibly the last floor they saw indicated by the in-car floor indicator. If rescuers cannot communicate with passengers or they cannot assist, the quickest method is to go to the elevator pit below the elevator and look up, or to the equipment room above the elevator looking down, and then estimate the car's location.

While personnel are determining the car's location, a firefighter should locate the electrical control panel, which should be in the equipment room, and shut off the power to the elevator. (As with industrial entrapment procedures, this firefighter must maintain control of the electrical power.) Once an estimated location is determined, rescuers open the closest hoistway door to confirm the exact location. An open hoistway door is extremely dangerous to rescuers, passengers, and bystanders, so access to the open hoistway must be controlled. Use of a ladder or other barrier is strongly recommended.

If the car is at or very close to a landing, the hoistway doors can be opened as previously discussed, the car doors can be pushed open, and the passengers assisted out of the car. Once the hoistway doors are opened, if enough of the top part of the car doors is accessible from the landing for passengers to be removed through, a ladder needs to placed through the opening down into the car, and a rescuer needs to enter the car to assist the passengers in climbing out, **Figure 16-41**. If the open hoistway doors reveal that enough of the bottom part of the car doors is accessible from the landing, a ladder needs to placed from the landing to the car, and a rescuer needs to climb into the car to assist the passengers, **Figure 16-42**. This is very dangerous; the open hoistway must be blocked.

If the car doors are not accessible from a landing, the rescuers need to go to the landing above and open the hoistway doors so that they can use a ladder to gain access to the top of the car.

Once on top of the car, the rescuer can open the top emergency panel and gain access to the passengers

FIGURE 16-41   Passengers can be rescued from an elevator with only the top half of the car door opening visible from the landing by extending a ladder down into the car. Firefighters then assist the passengers out.

FIGURE 16-42   Passengers can be rescued from an elevator with only the bottom half of the car door opening visible from the landing by extending a ladder up into the car. Firefighters then assist the passengers out. Note that the hoistway opening needs to be protected.

FIGURE 16-43   Passengers can be rescued from an elevator with the top of the car 6 to 10 feet below the landing by extending a ladder down to the top of the car and a firefighter assisting a passenger out the top emergency access panel. A Class 2 harness and safety lines need to be connected to both firefighter and passenger.

through it. If the car is close enough for a ladder to be placed from the landing above to the top of the car, the passengers can be removed up the ladder, **Figure 16-43.** If this method is used, passengers need to be placed in a life safety harness and attached to a lifeline as they begin exiting the car.

If the car is not close enough for a ladder to reach it and it is in a common hoistway with other elevator cars, it may be possible to bring one of the other cars alongside and transfer the passengers through the side emergency panel. If this method is chosen, the rescuers will need to build a makeshift bridge between the cars. Short ladders, backboards, wooden planks, and similar items work well for this. Once firefighters are ready to transfer the passengers, one rescuer needs to join the passengers in the stuck car in order to assist in the transfer, **Figure 16-44.**

Though not common, incidents involving escalators do occur. They usually involve a passenger getting feet caught in the area where the steps disappear into the **landing plate** or getting fingers caught under the moving hand rail. As with an elevator incident, the responding unit should request that a service technician be dispatched immediately. The other type of incident that can be encountered involving escalators

FIGURE 16-44 Passengers can be rescued by using a working elevator stopped alongside of the stuck elevator. Passengers are transferred out through the side emergency access panel. Note the bridging in place and that a firefighter assists the passengers from both cars.

FIGURE 16-45 The top landing of an escalator with the landing plate removed. The motor/gear assembly can be seen.

is when a service technician becomes caught in the gear mechanism while working on the unit.

As with all of the other incidents that involve machinery, one of the first things to be done is to shut off and control the power supply. Most modern escalators have emergency shutoff switches located at the bottom landing on one of the end plates adjacent to the hand rail. The motor and drive gear assembly for escalators is usually located under the top landing plate, **Figure 16-45,** or near the top under the escalator itself, behind a removable panel if this area is accessible.

In cases of foot entrapment at the landing plate, the plate itself can be taken up by removing the screws that hold it down. Figure 16-45 shows one of these plates removed. It is very difficult to reverse the direction of travel of an escalator due to a number of safety devices designed to prevent the escalator from reversing if power is lost. In the case of finger/hand entrapment, the firefighters can loosen the wheel that drives the hand rail. It is located in the same area as the motor and drive gear. If more extensive disassembly is required to extricate an entrapped passenger or

repair person, it will be necessary to wait for a service technician.

Farm Equipment Rescue

Rescue of victims who are trapped, pinned, impaled, or otherwise injured by farm equipment can be very challenging. While on the surface it may appear that these types of extrications are little different from vehicle extrications, this is not the case. Everything about the farm equipment itself inherently makes extrication a much more difficult process. Farm equipment is built to withstand the rigors of very hard use on a daily basis and not bend, break, or otherwise fail. Therefore, it is routine in extrications involving this type of equipment to have to disassemble the equipment rather than use the other methods discussed in the vehicle extrication section (distortion, displacement, and severing). Couple this with the fact that these events are often located far from paved roads that allow apparatus with all their rescue equipment to be parked close, and rescue becomes all the more difficult.

If a fire department's district has the potential to be required to respond to this type of incident, specialized training should be provided to firefighters. An excellent resource for information and assistance should an incident occur is the local equipment dealer; the department should work on developing a working relationship with dealers prior to an actual incident.

JOB PERFORMANCE REQUIREMENT   16-1
Removing Tempered Glass

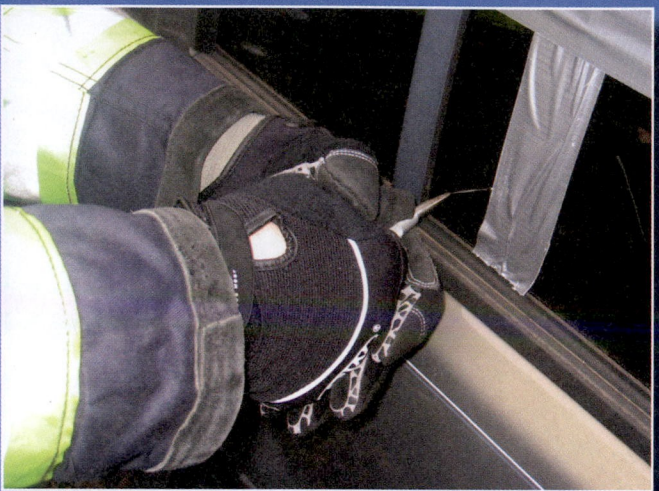

A   Make certain the vehicle is stabilized. Place duct tape across the window to be broken leaving an excess amount in the middle. The excess will be pressed together and will form handles to remove the glass. Apply several other strips of duct tape to the window in a crisscross or "H" pattern. The tape should be placed on the glass only.

B   A spring-loaded center punch is pressed into the corner of the glass. The firefighter should make certain that the hand not operating the punch is positioned in a manner that will help prevent the operating hand from penetrating into the passenger compartment.

C   Once the glass is broken, it can be removed in one piece by pulling out on the handles formed by the duct tape.

JOB PERFORMANCE REQUIREMENT 16-2
Removing the Windshield and Roof

A Make certain that the vehicle is stabilized. The windshield should be "chopped" along the bottom edge with a pick-head axe. Placing a hand on the pick of the axe will help control it, making certain that only one corner of the blade penetrates the glass.

B The front roof posts on both the passenger side and driver side should be cut through next, using cutters or hand tools such as a reciprocating saw or hacksaw.

C Relief cuts are placed in the roof on both sides of the vehicle, allowing ample space to gain access when the roof is folded.

(Continues)

JOB PERFORMANCE REQUIREMENT 16-2
Removing the Windshield and Roof (Continued)

D   A tool such as a pike pole is placed across the rear section of roof. Apply downward pressure to form a fulcrum for the windshield and roof portion being folded over.

E   The windshield and attached assembly will be folded over the rear section.

F   Once folded over, easy access is attained and patient stabilization can begin as removal is carried out. The folded-over portion should also be secured so that a wind gust cannot blow it back over onto the patient or firefighters.

JOB PERFORMANCE REQUIREMENT 16-3
Door Removal Using Hydraulic Spreaders

A   Make certain that the vehicle is stabilized. Insert a halligan between the door handle and body, moving the bar up and down to create a purchase point where the tips of the spreaders can be inserted.

B   Open the spreaders, trying to roll or "pop" the door latching mechanism off the safety bolt. If metal begins to tear, stop spreading, reposition the tool and start over again.

(Continues)

JOB PERFORMANCE REQUIREMENT   16-3
Door Removal Using Hydraulic Spreaders *(Continued)*

C   Once the door is free from the safety bolt, it's possible that enough room will be available to remove the patient. If not, or if the door is to be completely taken off, a safety strap should be applied to the door to allow a second rescuer to control its movement as it is separated from the vehicle. This will prevent injuries to feet and legs as well as prevent the door from damaging equipment such as hoses or power cords. The spreader tips should be placed into the top hinge of the door, widening it enough so that it breaks.

JOB PERFORMANCE REQUIREMENT 16-4
Displacing a Dashboard

A Make certain that the vehicle is stabilized. Cribbing must be placed beneath the rear base of the front door jam on both sides of the vehicle.

B Relief cuts are placed 5 to 6 inches below the intersection of the front support at the rocker panel.

(Continues)

JOB PERFORMANCE REQUIREMENT 16-4
Displacing a Dashboard *(Continued)*

C   Relief cuts are also placed 8 to 10 inches back on the horizontal rocker panel on both sides of the vehicle.

D   Hydraulic rams are inserted between the rear of the door jam and dashboard on both sides of the vehicle. The rams will both be extended at a steady and equal rate until enough room is available for patient removal.

E   Firefighters will need to place cribbing in the lower relief cuts as the dashboard is folded over to maintain its position.

LESSONS LEARNED

The risks faced by today's firefighters involved in vehicle rescue and extrication are ever changing. Firefighters need to keep the skills, knowledge, and abilities learned in this chapter in their mental toolbox, building on them throughout their careers and constantly updating them as new technology and equipment is developed and put into use.

KEY TERMS

Confined Space A space that is large enough to be entered but is not designed for continuous occupancy.

Cribbing The use of various dimensions of lumber arranged in systematic stacks (pyramid, box, step, etc.) to support an unstable load.

Disassembly The actual taking apart of vehicle components.

Displacement The relocating of major parts (i.e., doors, roof, dash, steering column) of a vehicle.

Distortion The bending of sheet metal or components.

Extricate To set free, release, or disentangle a patient from an entrapment situation.

Golden Hour The first hour after the onset of trauma to a person. If surgical intervention can be reached within that hour, chances for survival are greatly increased.

Ground Pads Sheets of plywood, planks, aluminum sheets, and so on, used to distribute weight over a larger area.

Hailing System A system used to help locate a victim at a collapse situation. Rescuers are placed at strategic areas of a debris pile while ordering complete silence of people and machinery. One rescuer will then call out for a response from the victim, hoping to receive an acknowledgement. If none is given, the next positioned rescuer will call out. If a reply is heard, another rescuer at a different position should call out in order to pinpoint the position of the response.

Harnesses Webbing sewn together to form a belt, seat harness, or seat and chest harness combination.

Hoistway The shaft in which an elevator or a number of elevators travel.

Landing Plate The plate at the top or bottom of an escalator where the steps disappear into the floor.

Packaging The bandaging and preparing of a patient to be moved from the place of injury to a stretcher.

Reel Coil Memory that wire develops from having been placed on a wooden spool as it is being manufactured.

Rescue Those actions that firefighters perform at emergency scenes to remove victims from imminent danger or to extricate them if they are already entrapped.

Risk/Benefit Analysis Weighing the probable outcomes in comparison to the level of risk that rescuers will be exposed to.

Severance The cutting off of components (i.e., brake pedal, steering wheel) in a vehicle.

Shoring The use of timbers to support and/or strengthen weakened structural members (roofs, floors, walls, etc.) in order to avoid a secondary collapse during the rescue operation.

Suspension System The springs, shock absorbers, tires, and so on, of a vehicle.

Tether Line A rope that is held by a team on shore during a water rescue to be used to haul the rescuer and victim back to shore.

Trench An excavation that is deeper than it is wide, with 15 feet being the maximum width.

Voids Spaces within a collapsed area that are open and may be an area where someone could survive a building collapse.

REVIEW QUESTIONS

1. What is the biggest danger that firefighters must be aware of when involved in rescue operations?

2. List the five goals that a Rapid Intervention Team should strive for when put into action on the fireground.

3. What elements must exist at all specialized rescue incidents to ensure safety and efficiency?

4. List in order and describe the four methods presented for carrying out a water rescue.

5. List and describe the three types of structural collapses covered.

6. List the four ways in which hazards can be controlled at collapse incidents.

7. Describe the actions and procedures that must be taken to make a trench collapse safe prior to rescuers entering.

8. What is the most common hazard encountered at confined space incidents?

9. Describe the difference in how an elevator operates when in the "Fire Service" and "Independent Service" modes.

10. Describe the basic procedures for conducting a large area or "team" search.

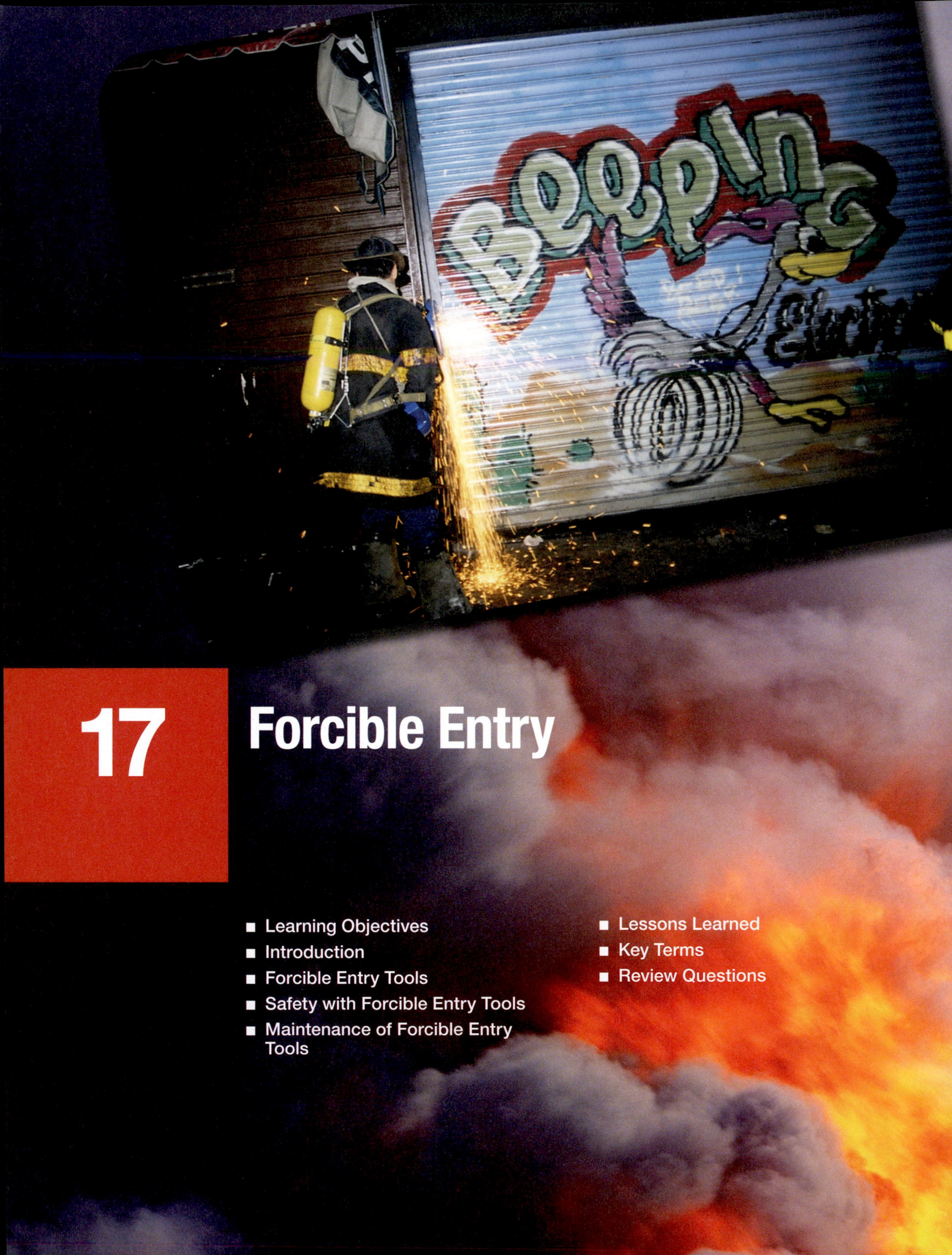

17

Forcible Entry

LEARNING OBJECTIVES

After completing this chapter, the reader should be able to:

17-2   Name the appropriate hand and power tool(s) to be used for specific forcible entry operations.

17-3   Demonstrate the ability to select tools for forcible entry.

17-9   Identify cleaning methods given manufacturer or department guidelines.

17-10   List the correct cleaning solvents for various equipment following manufacturer or department guidelines.

17-11   List the procedures for documenting equipment maintenance in accordance with manufacturer or department guidelines.

17-12   Identify procedures for reporting problems with tools and equipment.

17-13   Select the correct service and maintenance tool for a given situation.

17-14   Demonstrate the use of manufacturer guidelines in maintaining power plants, power tools, and lighting equipment.

17-15   Demonstrate the proper completion of recording and reporting forms as described in department guidelines.

17-16   Demonstrate the operation of power plants, power tools, and lighting equipment.

17-17   Demonstrate the maintenance of power plants, power tools, and lighting equipment.

17-18   Demonstrate the proper cycling of batteries from battery-operated power tools.

*The FF I and II levels, as defined by the NFPA 1001 Standards, are identified in different colors: FF I = black, FF II = red, additional information = blue.

INTRODUCTION

Forcible entry is often one of the first operations conducted at a scene to gain entry into secured areas and buildings whether in a fire, medical aid, or other type of emergency. It is a combination of knowledge and skills used to gain entry to a structure when the windows and doors are locked, blocked, or in some cases do not exist. To perform this essential task, firefighters must be knowledgeable in the tools used in forcible entry. In addition, the firefighter must have the skills, gained by training and experience, to apply this knowledge using a variety of tools.

Knowledge

A working knowledge of the many types of locks, hardware, doors, and other assemblies is essential to successful forcible entry operations. Firefighters must be able to "size-up" the quickest and easiest way to gain access to buildings, such as through the doors shown in **Figure 17-1.** In addition, the firefighter must know which type of tool to use and the best method to gain access through the door to the building.

FIGURE 17-1 Firefighters should size-up the way to gain entry, "try before you pry," and think about how they would force these doors.

Skill

The element of skill involves a firefighter's ability to apply knowledge of building construction, lock assemblies, tools, and techniques to accomplish the necessary tasks of forcible entry. This means choosing the proper tools and applying the best techniques when using the tools. Again, skills are developed by repeated practice and experimenting with new tools, locks, and techniques.

Experience

Experience can be acquired by three means. One is through drills and practice at training sessions and another is at the scene of actual fires and emergencies. Both are the means by which skill is developed and knowledge is gained as well as reinforced. A third and very good method of aquiring experience is learning about other's experiences through case studies and reports. Still, the most important experience is gained from field operations, where firefighters' skills and knowledge are put to the true test.

FORCIBLE ENTRY TOOLS

Firefighters must have an understanding and knowledge of the tools available to conduct forcible entry. The selection and use of the "right" tool is essential if the task of forcible entry is to be completed as quickly as possible. Although any number of tools can be used to accomplish a task, the right tool for the right job is the quickest and easiest way to complete the operation. Because of geography or local tradition, many of the tools shown or listed may have another name or nomenclature. Firefighters must know the tools and their names as used by their particular department. Tools used for forcible entry operations are divided into several families or groups based on their intended uses as shown in **Table 17-1.**

This chapter does not attempt to show or describe all of the tools used for forcible entry because some tools are obsolete or used infrequently. The tools described in this chapter are the most common forcible entry tools used today.

TABLE 17-1   Forcible Entry Tools

Family or Group	Tools	Operation
Striking	Flathead ax, maul, sledgehammer, battering ram, hammer, punch & chisel, lock breaker	Deliver impact force to break lock or drive another tool
Prying	Crowbar, Halligan tool, hux, claw tool, pry bar, hydraulic	Provide mechanical advantage, leverage
Cutting	Axes, saws, torches, bolt cutters	Cut material away; cut around locking devices
Pulling	Hooks, pike poles	Limited for forcible entry; breaks glass, gypsum board, Sheetrock
Through-the-lock	K-tool, A-tool, picks & key tool, bam bam, vise grip or channel lock pliers, REX tool	Remove lock cylinder

Striking Tools

The group or family of **striking tools, Figure 17-2,** is used to deliver impact to other tools, such as a Halligan tool, in order to drive it into place. Striking tools are used for impact delivery to the lock or the door itself. They may force the door or even break it down.

Flathead Ax

The most common of the striking tools is the flathead ax. Although this tool may be used for cutting, it is generally used to drive the Halligan tool. Together, the ax and Halligan tool are married together forming the **irons, Figure 17-3,** and are one of the most important and useful of all forcible entry tools. The flathead ax is available in 6- and 8-pound sizes, the weight being at the head of the ax. The 8-pound ax is best for striking purposes and is commonly known as the forcible entry ax. The handle of the ax can be comprised of wood or fiberglass. Fiberglass is best for all purposes, due to its high strength and ease of maintenance.

FIGURE 17-2 The group or family of striking tools includes the maul, small hammer, flathead ax, and Denver tools.

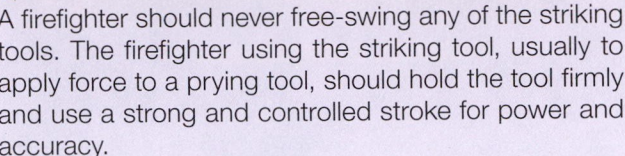

Maul/Sledge

The maul or sledge is generally found in several sizes with the 10-pound maul being the most common and versatile. Other sizes include the 8-, 12-, and 16-pound versions. The 10-pound maul is sometimes carried with the Halligan tool to again form a set of "irons." The 16-pound maul is considered too heavy to drive other tools such as the Halligan and is sometimes used to break in doors or locks. Once again, fiberglass handles are recommended for strength and safety. Some new tool styles combine mauls with other tools. **Figure 17-4.**

The power a striking tool applies is explained by this equation:

$$F = m \times a$$

A tool that is too heavy cannot be moved fast enough to develop proper force. Also, problems with accuracy and control can develop. On the other hand, a tool that is too light will not deliver enough force, which in turn will require delivering more strikes. Drilling and practice will help determine which tool is best for you.

Ram

Battering rams, **Figure 17-5,** are used by two or more firefighters to break through a door or wall. The desired result is to break or push the object in. There are several types, but the type most commonly used is the old-style fire department ram. This tool is made of steel and has four handles and a round end for battering and a forked end for breaking and penetrating. Another type of battering ram is smaller and is intended for use by one or two firefighters. This tool weighs approximately 20 to 25 pounds and is used to break doors and breach walls.

FIGURE 17-3 The flathead ax and Halligan tool form the tool set known as the *irons. (Courtesy of Ralph Restadius)*

FIGURE 17-4 The Denver tool is a maul with a cutting edge, prying end, and a pulling hook.

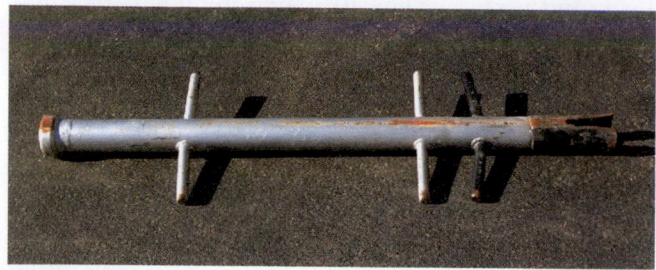

FIGURE 17-5 The battering ram requires brute strength to break a door, lock, or wall.

Prying and Spreading Tools

The group or family of **prying tools, Figure 17-6,** is used to spread apart a door from its jamb, move objects, or expose a locking device.

Halligan Tool

The original **Halligan tool, Figure 17-7,** designed by Hugh Halligan of the Fire Department of the City of New York, has proven to be the most important single forcible entry tool used in the fire service. Although the original style tool is not widely used today, the same basic concept with some improvements is capable of many forcible entry tasks.

Halligan-type tools are made in several weights and lengths, and of different materials. The best type for forcible entry and structural firefighting is a 30-inch forged steel tool. The basic Halligan design is a tool that has three primary parts as shown in Figure 17-7. These are the adz end, the pike end, and the fork end. The size and shape of the parts are important because they determine the mechanical advantage the tool will apply to the spreading force. As noted earlier,

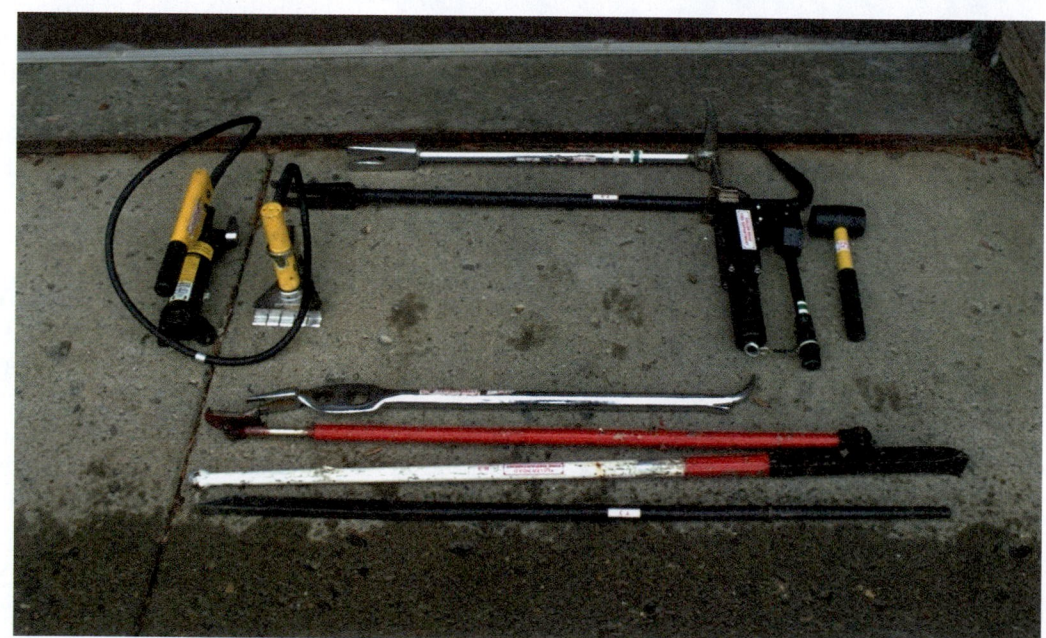

FIGURE 17-6 The group or family of prying tools includes the Halligan, claw tool, hux bar, Detroit door opener, pry bar, and hydraulic spreaders.

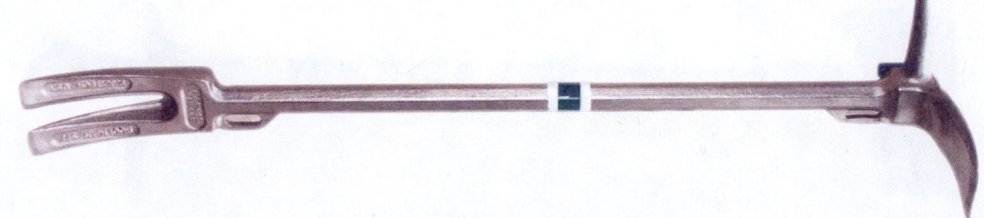

FIGURE 17-7 The Halligan tool is one of the most useful forcible entry tools. *(Courtesy of Ralph Restadius)*

the Halligan tool is often nested with the flathead ax, forming the irons.

Claw Tool

The claw tool, a forerunner to the Halligan tool, was the mainstay of forcible entry operations for the previous generation of firefighters. The claw tool had two major parts: the fork end, which had a hollow, curved profile, and the hook end, which was curved and ended with a sharp point. Generally, the tool was made in two sizes. The heavy claw is 42 inches long and the standard claw is 32 inches long. The standard claw was usually carried with a striking tool such as the ax or the kelly tool. The claw tool is obsolete now, having been replaced by the Halligan tool.

Kelly Tool

The Kelly tool is a steel bar, 28 inches (71.1 centimeters) long, that has two main features. One end has a large adz, approximately 3 inches (7.6 centimeters)

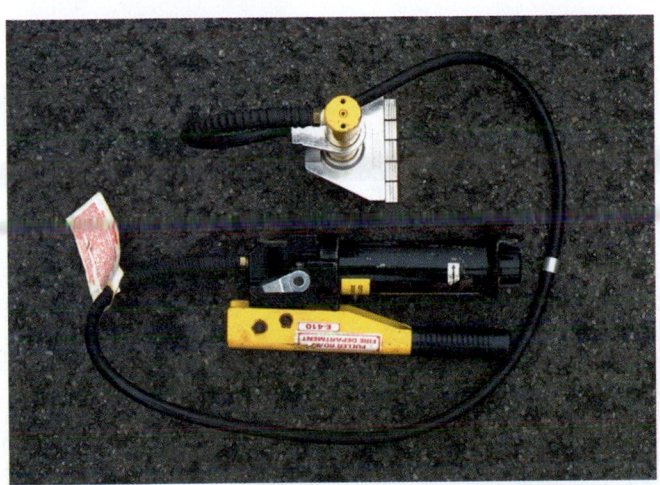

(A)

(B)

FIGURE 17-8 Handheld hydraulic spreaders are available in two different styles but operate on the same principle. (A) Handheld pump with remote spreader. (B) Handheld unit with integral built-in spreader.

wide, and the other end has a large chisel or fork. This is an older tool and has since been replaced by the Halligan tool.

Hydraulic Spreaders

Handheld hydraulic spreaders, **Figure 17-8,** are available in two different styles. One is a hydraulic pump with a spreading device attached by a length of hose, and the second is a similar pump with the spreading devices attached directly to it. These tools are often called rabbit or rabbet tools. The spreading end is forced between the door and jamb at the location of the locking device and the pump is operated to spread the door and jamb apart. A tool is used to force the spreaders into the space between the door and jamb.

Power hydraulic spreaders can also be used to spread objects. Their application is generally for rescue work; however, they may be used to force entry into a building.

Miscellaneous Prying Tools

There are many other types of prying tools such as the crowbar, pry bar, and others, but most are of limited use. As with any tool, its use depends on the type of task to be accomplished and the method of forcible entry.

Cutting Tools

The group or family of **cutting tools, Figure 17-9,** is used to cut away materials and expose the locking device or cut through a door or wall to accomplish forcible entry.

Ax

The fire department ax is seen in two basic configurations in **Figure 17-10:** the flathead ax and pick-head ax. The flathead ax is most commonly used as a striking tool for forcible entry. In addition to cutting, this tool may be used as a prying tool, especially when equipped with a fiberglass handle. The pick-head ax has a point that can be used for puncturing, pulling, and, to a limited extent, prying. It is recommended that all axes be equipped with fiberglass handles for strength and safety.

Handsaws

The most commonly used handsaw is the metal cutting hacksaw. A strong, high-quality frame with a supply of good-quality blades is used for many firefighting operations, especially vehicle rescue. Carpenter saws, both rip and crosscut, may also be carried on fire apparatus for use in areas where power saws are too large, too noisy, or present a safety hazard.

FIGURE 17-9   The group or family of cutting tools includes axes, saws (both power and manual), and bolt and wire cutters.

(A)

(B)

FIGURE 17-10   The fire department ax is seen in two basic configurations: (A) flathead and (B) pick-head.

Bolt Cutters

Bolt cutters are used to cut metal bars, cables, wires, and other hardware. These tools come in a variety of sizes; the most common for forcible entry is 36 inches in length. This size will cut steel up to $^3/_8$ inches thick. Firefighters should avoid using bolt cutters to cut heavy-duty padlocks with case-hardened shackles. The hardness and the thickness of the metal make them difficult to cut and will damage the cutting edge of the tool.

Power Cutting Tools—Saws

Power saws, **Figure 17-11,** are the most common power cutting tools used for forcible entry work. They are available in two basic types: the rotary saw with a circular blade and the chain saw. The rotary saw in Figure 17-11A has a carbide-tipped blade. Figure 17-11B shows a rotary saw with a metal-cutting disc and Figure 17-11C shows a chain saw. Rotary and chain saws are usually powered by a two-cycle gasoline engine and require two firefighters for safe operation.

(A)

(B)

(C)

FIGURE 17-11 Gasoline-powered saws are available in two types: the rotary saws shown in (A) with the wood cutting blade and (B) with the abrasive disc, for either masonry or metal cutting, or the chain saw shown in (C). The chain saw is usually used for roof ventilation.

Carbide-Tipped Blades

The rotary power saw is widely used and can be adapted to cut a variety of materials. Wood and composition materials such as roofs and flooring are cut with a carbide-tipped blade, which has 12 to 24 teeth. Light-gauge metals and polycarbonate plastic may also be cut with this blade. When cutting with a carbide-tipped blade, it is important to maintain full rpms to avoid having the saw bind in the material being cut.

Metal Cutting Blades

The most common metal cutting blade is an abrasive disc made of aluminum oxide. This blade is used to cut locks, hardware, steel doors, and roll-down gates, and in other forcible entry applications.

> **NOTE**
>
> With all blades, care must be taken not to bind or bend the blade while cutting. This is particularly important with abrasive discs because they can shatter or disintegrate.

Masonry Cutting Blades

Masonry materials such as concrete, brick, block, and stone can be cut with an abrasive disc made of silicon carbide or a steel with a diamond matrix blade. When cutting with these blades, a spray of water on the blade will cut down on the production of dust and keep the blade cool, thus prolonging the life of the blade.

> **STREETSMART TIP**
>
> Gasoline or other hydrocarbon fuels will break down the bonding material used in the manufacture of abrasive disc blades. To avoid this, blades must be stored separately from the fuel can for the saw. One should always refer to the manufacturer's recommendations on use and storage. Checking the fuel supply and starting the engine of gasoline-powered equipment before taking it to the location of the cutting operation will save time and frustration. Some departments start and run the saw each tour to avoid problems.

Chain Saws

Chain saws are used primarily for ventilation purposes; however, they have application as a forcible entry tool depending on the type of building construction and opening required. Several types of saws and chains are available, and many departments use a carbide-tipped chain. The chain saw can also be used to cut through wood siding, wood frame walls, certain doors, and light-gauge metal.

Reciprocating Saws

The reciprocating saw, **Figure 17-12,** is an extremely versatile tool. These saws are powered by electricity, either stationary (house current) or portable (generator). Cordless, battery-powered saws are now available and can be operated remotely from a power source, thus increasing versatility. A variety of blades are available for this type of saw depending on the material to be cut.

Cutting Torch

The cutting torch, **Figure 17-13,** uses a fuel such as acetylene mixed with oxygen to produce a flame to heat metal. A jet of pure oxygen is then applied to intensify the heat, creating a flame temperature in excess of 5,000°F (2,760°C), melting the metal. Many departments use these tools for heavy forcible entry in addition to other cutting tools. Use of cutting torches requires specialized training in addition to following manufacturer's recommendations and department procedures.

CAUTION

Acetylene is an unstable and extremely flammable gas. It is essential to always follow manufacturer's recommendations for storage and use of this material. In addition, cutting torches should never be used in a flammable environment.

Pulling Tools

The most common type of **pulling tool** is the **hook** or **pike pole, Figure 17-14.** These tools are grouped by the type of head and handle length and are used to open up walls and ceilings, to vent windows, and to pull up roof boards or other building materials.

Special Tools

A number of specialized tools are available to assist with or conduct forcible operations. Many of these have been developed by firefighters after years of experimentation and trial and error.

Bam Bam or Dent Puller

Originally designed as an automotive body work tool, the bam bam is used to pull lock cylinders. It consists of a shaft, case-hardened screw, and slide hammer. The screw is turned into the lock cylinder and the slide hammer is operated to pull the cylinder out.

Duck Bill Lock Breaker

The duck bill lock breaker is a wedge designed to break open heavy-duty padlocks, **Figure 17-15.** The long, tapered head is placed into the shackle of the

FIGURE 17-12 The reciprocating saw is a versatile tool for forcible entry and rescue work.

FIGURE 17-13 Cutting torches use a mixture of acetylene and oxygen to generate a high-temperature flame. They cut metal by melting it.

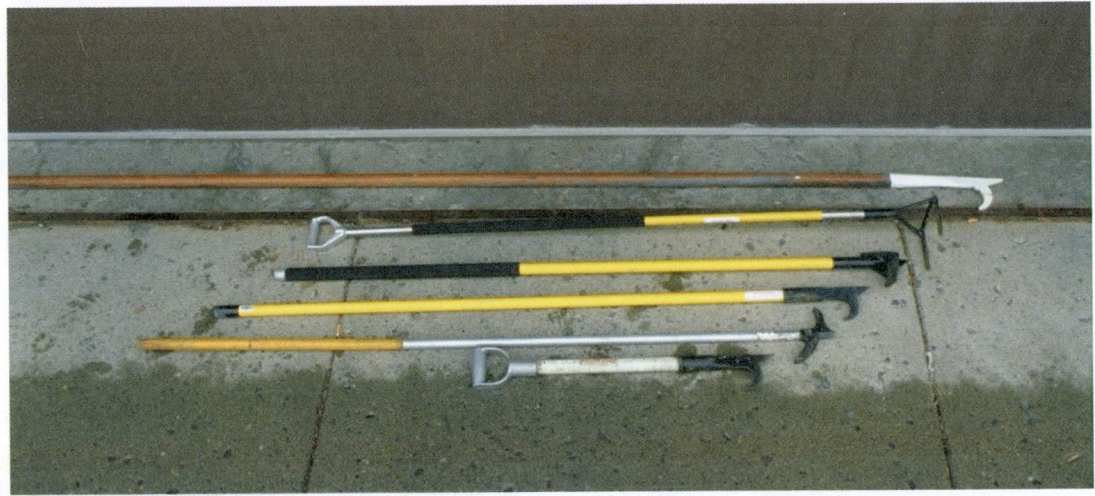

FIGURE 17-14 The most common type of pulling tool is the hook or pike pole, available in various styles and lengths.

FIGURE 17-15 Duck bill lock breaker forcing a heavy-duty padlock.

lock and driven down with a flathead ax (8-pound minimum) or a maul. This wedging action will pull the shackle out of the body of the lock.

K-Tool and Lock Pullers

The K-tool, **Figure 17-16,** is used to perform through-the-lock forcible entry. The K-tool is designed to pull out lock cylinders and expose the mechanism in order to open the lock with the various key tools. The back of the tool is shaped like the letter *K* and slides over the lock cylinder. The front of the tool has a loop for the adz of the Halligan tool.

The K-tool is placed on the lock cylinder and tapped into place to gain a good purchase or bite on the lock. The Halligan tool is then used to pry/pull the cylinder out of the lock mechanism. This exposes the

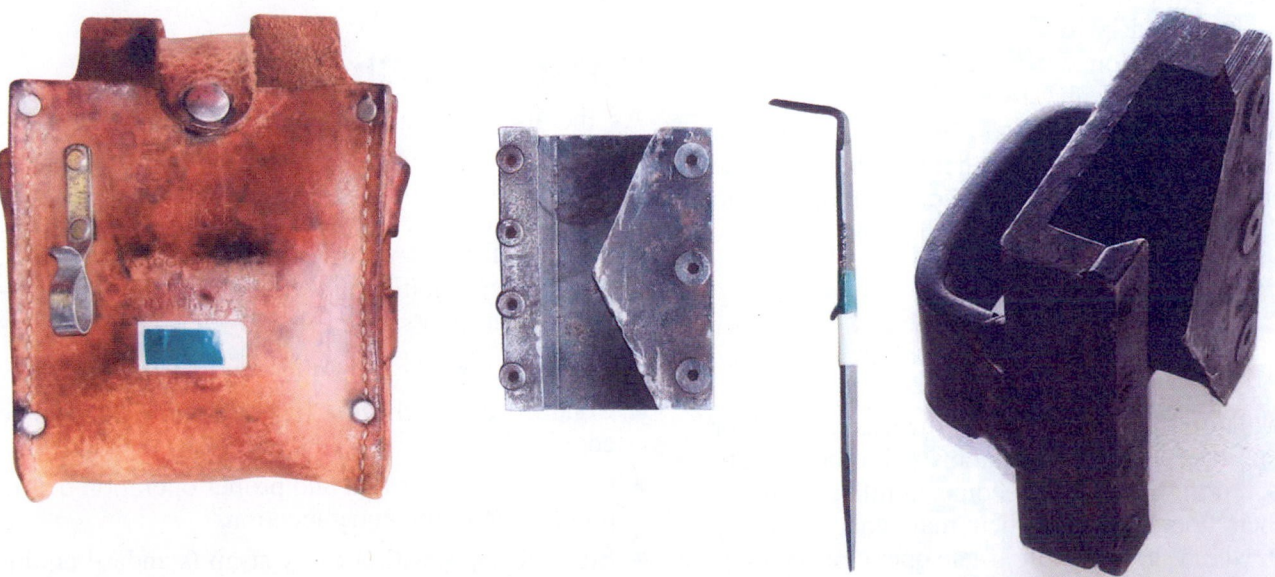

FIGURE 17-16 The K-tool (shown with key tools) is designed to perform through-the-lock forcible entry. *(Courtesy of Ralph Restadius)*

FIGURE 17-17   The REX tool is used to pull lock cylinders in the through-the-lock method of forcible entry.

lock mechanism, which is operated with the proper key tool. Key tools include the bent, square, and screwdriver types, as well as a pick to slide open the shutter on some rim locks. A locking-type pliers is a good addition to the K-tool kit.

A number of other tools similar to the K-tool are available that can be used to perform the same operation of through-the-lock forcible entry. These are the A-tool, officer's tool, or REX tool, **Figure 17-17,** to name a few.

Combinations of Tools

To perform the task of forcible entry, most of the tools or groups of tools discussed are used in combination with other tools. Striking tools are used with prying tools, such as the flathead ax and Halligan tool, and these tools should be carried or stored together on apparatus. In addition, the firefighter must be able to size up the forcible entry task and choose the right combination of tools to provide adequate leverage or force. Experience will provide the firefighter with the knowledge to choose the right combination and to apply this knowledge to complete the task.

SAFETY WITH FORCIBLE ENTRY TOOLS

As with all tools and equipment, if misused or used for the wrong task, forcible entry tools will create safety hazards. Firefighters will become familiar with the tools, their operation, and their maintenance during training, which should result in safe operation. A number of general rules apply to all operations as follows:

- Always wear proper personal protective equipment including hand and eye protection.
- Follow manufacturer's guidelines for proper operations.
- Do not attempt to cut material other than that for which a blade or tool was designed.
- Operate with regard to the safety of others in the immediate work area.
- Make sure tools are in proper operating condition before use.
- Most forcible entry operations require teamwork. Never attempt to use tools alone that require two firefighters.
- When the task is complete and if the tool is no longer needed, secure it to prevent tripping or other hazards.
- Tools should be stored and easily accessible, **Figure 17-18.**

NOTE

The safety guidelines provided are general in nature and firefighters are reminded to read manufacturer's operating instructions for the specific tools used in their department. In addition, they should not use any power tools or other spark-producing equipment in an explosive or flammable atmosphere.

SAFETY

Whenever forcing entry into an unfamiliar area, firefighters must be very cautious as to what they may face when creating the opening. Immediately dangerous to life and health (IDLH) conditions, extreme fire exposure, drug lab chemicals, booby traps, or other hazards may be present.

Rotary and Chain Saws

As the use of security gates and overhead doors increases, the power saw has become the tool of choice to remove the door or gate. The rotary saw with a metal cutting blade is an effective tool for these operations. These saws present a number of hazards, and firefighters should follow these guidelines to complete the operation safely:

- Always follow the manufacturer's instructions.
- Conduct daily checks for operation and blade condition.
- Check the saw for fuel and proper operation before proceeding to the entry location.
- Equip the saw with a carry strap (standard equipment with some manufacturers).

- Power saws require two firefighters: the saw operator and a guide firefighter.
- Eye protection is required when running any power equipment, especially power saws.

Hand Tools

Hand tools used for forcible entry are constructed of metal, wood, fiberglass, or some combination of these materials. All tools must be inspected regularly for condition, cracks in the handles, burrs in the metal, and loose heads.

MAINTENANCE OF FORCIBLE ENTRY TOOLS

Proper tool maintenance is the first step to tool safety. Tools must be inspected and cleaned on a regular basis and checked for wear and damaged parts.

General Guidelines

- Inspect tools regularly and keep them clean and operational. Follow any manufacturer or department guidelines for inspection and maintenance of all forcible entry tools.
- Documentation of tool maintenance is important. Each tool repair should be documented for future reference. Also document any tools that have been taken out of service. All repairs and unserviceable tools should be communicated to co-workers and through the chain of command. Out of service

FIGURE 17-18 These slide-out trays provide easy access to forcible entry tools. Note the mounting brackets that hold the tools in place.

- Use the right blade for the material being cut, **Figures 17-19A** and **B.**
- Never carry a running saw up a ladder or through a crowd of firefighters.

(A)

(B)

FIGURE 17-19 (A) Rotary saws have three types of blades: wood, masonry, and metal. Several examples are shown here. *(Courtesy of Ralph Restadius)* (B) This is a new style of wood and roofing blade for the rotary saw.

equipment should be removed from the apparatus and clearly marked according to departmental guidelines.

- Follow department procedures for proper documentation of tool maintenance, service, and placing tools that are unsafe or not working out of service. Be sure to report unsafe or nonfunctional forcible entry tools to the proper authority within the department so the situation can be corrected.

- Follow manufacturer guidelines for battery-operated tools and ensure that batteries are functional for operational equipment. Follow manufacturer and department procedures for rotating or cycling batteries in the tools for maximum operational use.

Metal Heads and Parts

- Remove any dirt or rust with steel wool or emery cloth.
- Use a metal file to maintain the proper profile and cutting edge.
- Sharpen edges and remove burrs with a file.
- Do not keep the blade edge too sharp; this may cause it to chip when in use.
- Do not grind the blade because it can overheat, lose temper, and become soft.

- Do not paint the metal parts because this will cause sticking and also hide any damage.
- Keep all metal parts lightly oiled to prevent rust.

> **CAUTION**
>
> Oil should not be applied to the striking surface or face of striking tools. When used, an oil coating on the striking surface may cause the tool to slip or glance off, causing injury, and causing little or no force to be applied to the object being hit.

Fiberglass Handles

- Wash them with soap and water and dry completely.
- Check for damage or cracks.
- Make sure metal parts are secure.

Wood Handles

- Clean with soap and water, rinse, then dry completely.
- Check for damage and sand off any splinters.
- Do not paint or varnish the handles. A coat of boiled linseed oil may be applied if necessary.
- Ensure that the head is securely fastened to the handle.

LESSONS LEARNED

Forcible entry is a key tactic in structural firefighting and emergency operations, and firefighters must understand the tools, equipment, and methods used for forcible entry. As with all other fireground tactics, teamwork is an essential element. Failure to quickly conduct effective forcible entry may result in delayed search and rescue operations and unnecessary fire spread.

KEY TERMS

Cutting Tools   The group of tools used to cut through or around materials.

Forcible Entry   The fire scene task of gaining entry to a building or secured area by disabling, breaking, or going around locking and security devices.

Halligan Tool   From the prying group, a 30-inch forged steel tool with three primary parts: the adz end, the pike end, and the fork end.

Hook   A tool with a 32-inch to 12-foot handle with a pike and hook on one end. Used for pulling ceilings or separating other materials. Also known as a pike pole.

Irons   The combination of a Halligan tool and flat-head ax or maul.

Pike Pole   See hook.

Prying Tools   The group of tools used to separate objects by means of a mechanical advantage.

Pulling Tools   The group of tools used to pull away materials.

Striking Tools   The group of tools designed to deliver impact forces to break locks or drive another tool.

REVIEW QUESTIONS

1. Choose an engine or truck company in a fire department and identify five forcible entry tools and describe their use.

2. List the inspection and maintenance procedures for five forcible entry tools.

3. Describe the following types of tools: prying tools, pulling tools, striking tools, and cutting tools.

4. Describe the documentation process for a broken or defective tool.

5. List the safety precautions that must be taken with forcible entry tools.

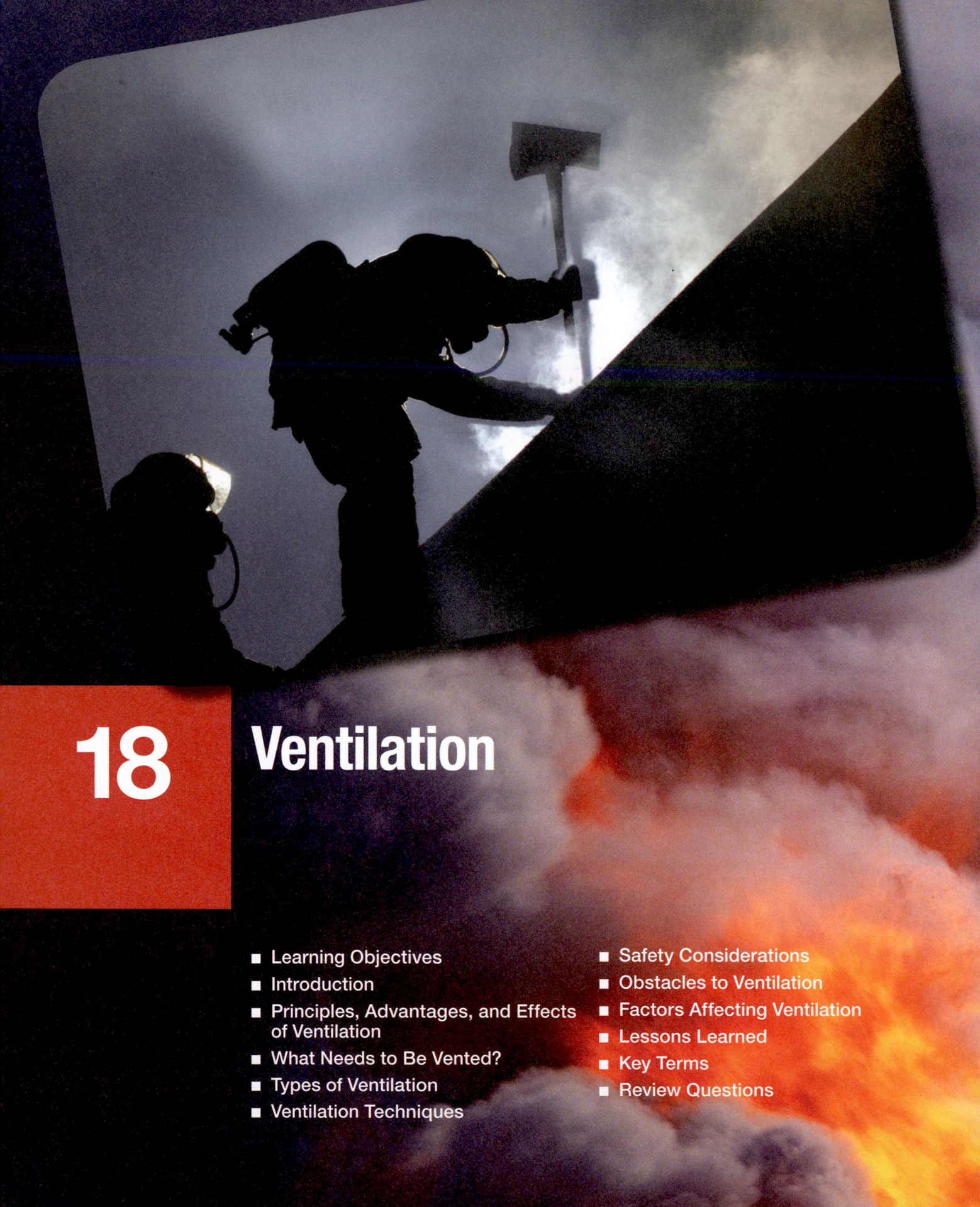

18 Ventilation

INTRODUCTION

In Section I, Chapter 18, ventilation was explained in detail. Ventilation is an essential part of the tactical and strategic objectives of modern fire extinguishment. Nevertheless, ventilation sometimes isn't employed until late in the initial fire attack phase or only when initial efforts of fire attack have been unsuccessful. The application of proper ventilation has allowed firefighters to successfully perform search and rescue operations during emergency incidents. Ventilation is a very complex subject area with many facets. Additional information can be found in Section I, Chapter 18, as well as texts written solely on the subject of ventilation.

PRINCIPLES, ADVANTAGES, AND EFFECTS OF VENTILATION

Ventilation is the relief of the products of combustion from an enclosed area. Combustion of organic material produces heat, smoke, pressure, and fire gases that quickly fill up an enclosed structure. That structure could be a one-room shed, a ten-room private dwelling, or a twenty-story building. In every case, the need to relieve the structure of these products of combustion is a very essential part of the fire suppression effort. The practice provides benefits that go beyond merely suppression to extinguish the fire.

First, by relieving the structure of heat through channeling it into the atmosphere, the fire is deprived of the ability to heat up other parts of the structure. Second, ventilation channels smoke out of the structure. Additionally, the unburned tarry hydrocarbons contain substances that irritate the eyes. Smoke contains many other products of combustion that are deadly substances. Many harmful compounds are mixed in with smoke, and a person that is exposed to this material will suffer ill and potentially lethal effects, **Table 18-1.** When people fall victim to smoke or its components, they usually fall into unconsciousness, with death coming a short time later. The removal of smoke, heat, and toxic gases will add survival time to a potential victim who is unconscious, increasing the chance of successful rescue.

When properly performed, the ventilation operation can be as critical as the nozzle team applying water to the fire. It is important to note that in the confusion of a fireground operation, it may not be possible to follow a manual's every line and sentence. Tempered by situational awareness, the firefighter must utilize experience, knowledge, and training to determine the appropriateness of the ventilation operations.

TABLE 18-1	Gases Produced by Fire
Carbon monoxide	Takes the place of oxygen in the blood
Carbon dioxide	Overstimulates the rate of breathing
Hydrogen sulfide	Causes respiratory paralysis
Sulfur dioxide	Extremely irritating to eyes and respiratory tract
Ammonia	Extremely irritating to eyes, nose, throat, and lungs
Hydrogen cyanide	Highly toxic; used commercially as a vermin fumigant
Hydrogen chloride	Can become hydrochloric acid in mucous membranes
Nitrogen dioxide	Causes respiratory distress as a delayed reaction
Acrolein, phosgene	Gases found in certain kinds of fires; lethal in small doses

A coordinated effort between fire attack teams, ventilation teams, and rescue teams is required. Miscommunication can lead to improper tactics leading to an increase in fire spread, decreased visibility, disruption of the thermal balance, and ultimately an unsafe or untenable situation for search and rescue teams as well as victims.

Ventilation increases the survival time of a trapped victim who might be overcome or a firefighter who has become disoriented and run out of air. Removal of smoke improves visibility, aiding in the search and rescue of victims. Reducing the danger from a hostile event is a key element in today's rapidly developing fires. Dangers such as holes in floors are easier to spot, and avenues of fire extension become more obvious. With better visibility, the use of tools becomes safer.

If heat is not permitted to linger on a material long enough, it will be unable to liberate the material's combustible gases. By venting the enclosure, the heat level is kept from becoming capable of producing these phenomena:

- *Flashover.* Everything in a confined area ignites at almost the same time.
- *Backdraft.* Unburned smoke is heated in the absence of oxygen and, when oxygen is introduced, produces an explosive force.
- *Smoke explosion.* Unburned products of smoke and gases accumulate in areas of the structure, and when oxygen is introduced the gases rapidly ignite with explosive force. However, sustained combustion is generally not indicated.
- *Rollover/Flameover.* Fire begins to ignite smoke overhead in "fingers of fire" that reach out and begin to consume fuel in the gaseous state.

The mechanics associated with each of these phenomena and prevention techniques are discussed as each relates to ventilation.

WHAT NEEDS TO BE VENTED?

Unless there is a ventilation opening to permit escape, the expanding heated steam and smoke will roll over the wall of water and drop down behind the hose team, **Figures 18-1A** and **B.** When ventilation is mentioned, the first image that appears is a fully involved building fire emitting smoke from every opening. It appears to be well vented. However, long before the entire building requires venting, the smaller voids and compartments need to exhaust the increasing pressure and intensifying heat condition. If done in a timely manner, the involvement of the entire building might be avoided.

TYPES OF VENTILATION

Ventilation can be performed using one of several methods either singularly or in combination with one another. Natural ventilation merely requires opening doors and windows and letting physics take care of the rest, **Figure 18-2.** Among the mechanical means are the use of smoke fans consisting of ejectors and blowers, and the use of water to create air movement.

Types of ventilation include the following:

- Natural
- Mechanical
- **Vertical**
- **Horizontal**

VENTILATION TECHNIQUES

Many techniques are used to effect ventilation. Some are simple and require no special tools; others are more complex, are dangerous to implement, and require sophisticated tools. Refer to Section I, Chapter 18 for a detailed discussion of each of the following ventilation techniques:

- Break glass
- Open doors
- Negative pressure ventilation
- Positive pressure ventilation
- Roof ventilation

SAFETY CONSIDERATIONS

When considering the placement of a hole for ventilation, whether it is horizontal or vertical, it is of paramount importance to consider the benefit gained against the possible liability created. For example, if venting would expose a victim and rescuer on a ladder to danger, the best choice would be *not* to vent. Consider the following questions:

- Will ventilation permit the fire to extend?
- Will the escape route be cut off?
- Will ventilation endanger others?

OBSTACLES TO VENTILATION

The importance of ventilation cannot be overstated. Because of the unpredictability of fire operations, firefighters will be confronted with many unforeseen circumstances that will delay ventilation activities.

(A)

(B)

FIGURE 18-1   (A) Applying water to the upper levels of a thermal layer will cool and disrupt the rollover effect that is apt to occur with the proper conditions. Ventilation is critical when this is done. (B) As a hose team advances into the fire and sprays water in droplet form, it creates a wall of water and disrupts the high-heat thermal layer and cools the upper levels of the compartment. Water absorbs heat as it turns to steam, expanding 1,700 times as it does. If there is no path for the expanding water/steam conversion, it will take the path of least resistance, in this case over the wall of water and the nozzle team. The water movement will then pull any heat from the back of the nozzle team and roll over on top of it.

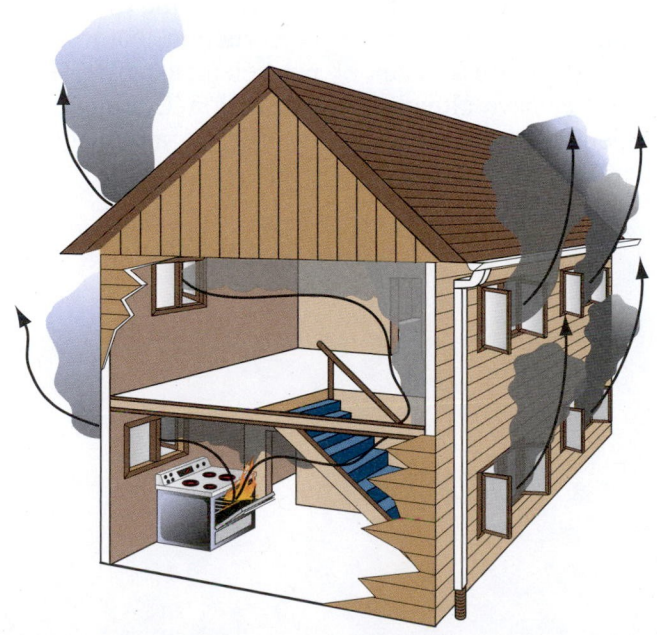

FIGURE 18-2   Smoke will be carried throughout the building to upper floors by normal air currents mixed in with the heat.

Listed next are some of the obstacles that might be encountered.

- *Access* should be one of the initial size-up considerations when arriving at the scene. The firefighter should first assess the needs of the ventilation objective, then determine the route to be employed to reach the location of the job performance task.

- The presence of *security devices* can impede ventilation both in access and in timing. Building owners may have barricaded themselves and their possessions behind gates, screens, steel doors, and closed-up windows.

- From multistory skyscrapers to one-story garages, the firefighter must be alert to the structure's ventilation needs. The *height* of a building presents many challenges that range from reach to tool availability.

- Another obstacle to ventilation that is addressable is *planning*. A quick size-up and implementation of the plan are essential for timely ventilation.

- A task that is assigned to a shorthanded or inexperienced crew will delay ventilation. Of course, there are times when adequate *personnel* are not yet on the scene.

- Especially in large buildings, the floor *layout* can be very confusing to the firefighter outside the building attempting to figure out which set of windows serves the fire area. The building layout can also be an obstacle to a firefighter attempting to reach the rear of the building.

Ventilation Timing

Venting too early can lead to fire extension; venting too late causes unnecessary punishment to the interior forces and can even prevent forward progress. The possibilities of backdraft and flashover have already been discussed. The timing is dependent on the type of ventilation being performed.

Vertical ventilation of stairways, hallways, and any paths of egress and ingress are paramount and must be effected without delay. These openings are commonly referred to as firefighter access holes and are paramount in occupancies with a central hallway. If fire is known to be in the area under the roof boards or on a top floor extending into the **cockloft,** opening of the roof must commence without delay.

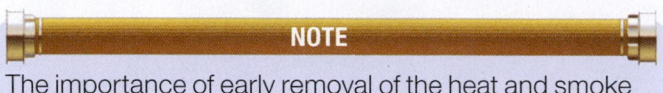

NOTE

The importance of early removal of the heat and smoke through vertical channels cannot be overstated.

As with all facets of firefighting, however, nothing can be written in stone. The opening of an entrance door to a fire occupancy on a lower floor that will vent up the stairs and out the roof opening must be carefully planned. There may be situations where escaping building occupants are coming down the stairs or firefighters are moving up the stairs. In such situations, vertical ventilation through this channel must be timed properly and may even have to be delayed. Delay might be necessary when opening a roof cut because the opening might expose an adjoining building to extension. A hoseline might have to be put into place to protect the exposure before the cut can be opened.

FACTORS AFFECTING VENTILATION

Several factors can affect the attainment of proper ventilation. Partial openings, screens, type of roof material, wind direction, weather, building size, and construction features are some of the more prominent elements. All of these factors were described in Section I, Chapter 18 and should be reviewed in order to apply the effects to search and rescue operations.

LESSONS LEARNED

Ventilation is a tool that is used in firefighting just as any other tool is used. Its judicious use can enable a firefighter to enter a structure to make a rescue. Used inappropriately it can permit a fire to extend into uninvolved portions of a structure and endanger the lives of victims as well as firefighters.

A complete understanding of the concepts that affect ventilation is necessary during operations but there are no hard and fast rules. Each incident will be different based upon the factors presented during the emergency, and the firefighter must consider each and every factor in order to use ventilation successfully to perform search and rescue.

KEY TERMS

Cockloft   The area between the roof and the ceiling.

Horizontal Ventilation   Channeled pathway for fire ventilation via horizontal openings.

Vertical Ventilation   Channeled pathway for fire ventilation via vertical openings.

REVIEW QUESTIONS

1. What are the types of ventilation that may be utilized during emergency operations?
2. What are the advantages to ventilating a structure?
3. Define the terms backdraft, flashover, and rollover.
4. Define and describe vertical and horizontal ventilation.
5. List the factors that affect ventilation and the methods that may be used to overcome these obstacles.

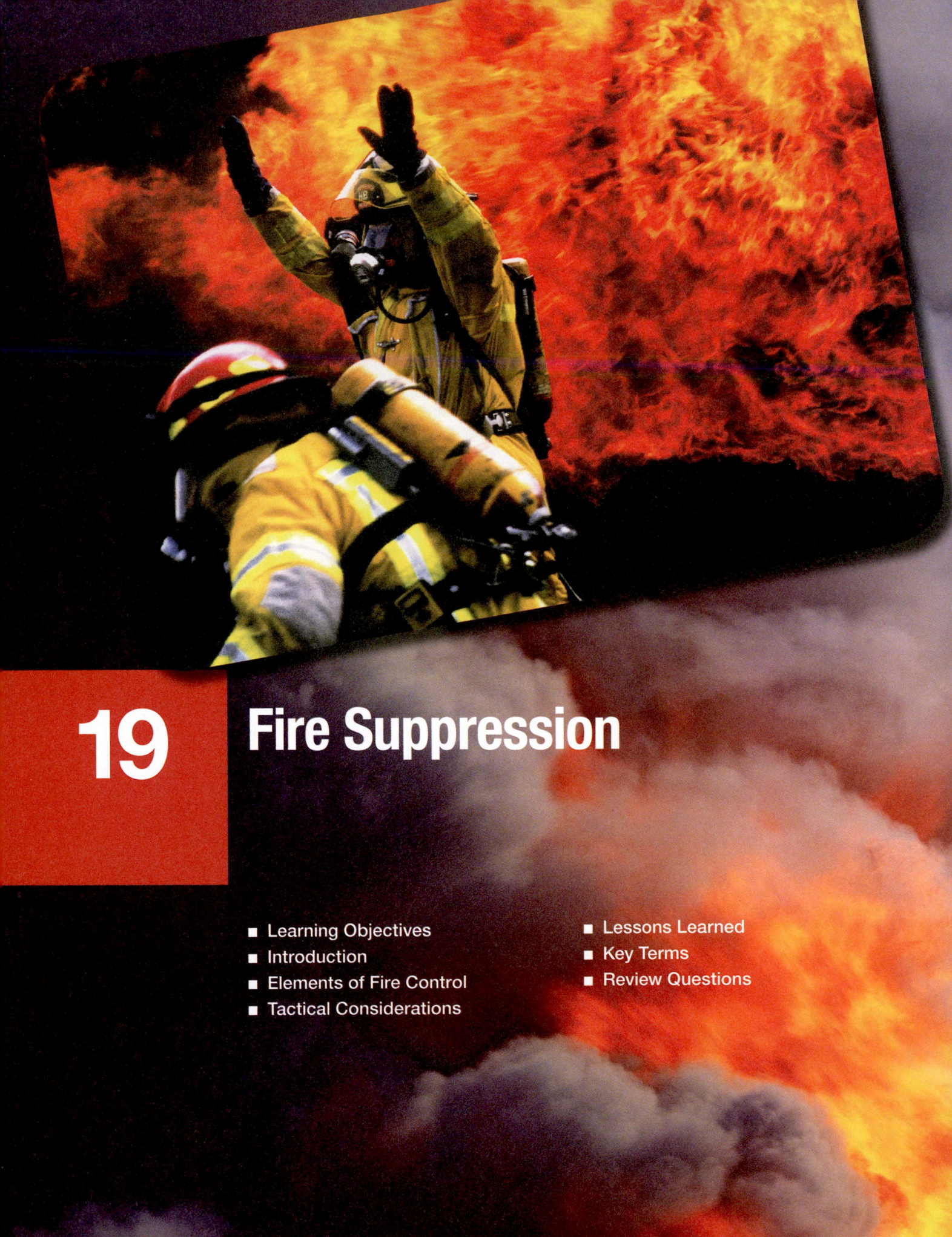

19 Fire Suppression

- Learning Objectives
- Introduction
- Elements of Fire Control
- Tactical Considerations

- Lessons Learned
- Key Terms
- Review Questions

LEARNING OBJECTIVES

After completing this chapter, the reader should be able to:

19-4 Identify structural fire considerations to be made prior to extinguishment.

19-5 Identify the hazards imposed on a structure by fire.

19-6 Identify the effects that fire suppression operations have on a structure.

19-12 Specify the appropriate actions to take when hazardous situations present themselves.

19-13 Describe the properties of pressurized flammable gases and their effects on firefighting.

19-14 List the effects of fire on closed pressurized cylinders.

19-19 Explain the difference between the modes of fire attack.

19-22 Demonstrate the ability to evaluate fire progression.

19-23 Demonstrate the ability to analyze changing fire conditions and their effects on structural integrity.

19-24 Describe fire suppression techniques and their application for various types of structural fires.

19-25 Discuss teamwork and its part in firefighting.

19-26 Demonstrate assembling a team for an interior structural firefighting operation.

19-27 Define a plan of action for fighting fire regarding attack modes and styles.

19-28 Identify the correct nozzle and hose for various structural fire operations.

19-29 Explain the need for command at incident locations.

19-30 Discuss the incident command system as it relates to an emergency operation.

19-31 Recognize the need for command at incident locations.

19-32 Demonstrate the ability to maintain effective communications as needed during an emergency operation.

19-33 Perform an assigned role within an incident command system.

19-38 Identify the different types of adapters and appliances used for various fireground applications.

19-40 Demonstrate the fire attack for grade level, above grade, and below grade.

19-82 Identify the contents of a cylinder by its characteristics.

19-83 Name the methods for identifying the contents of a container.

19-84 List components of a gas cylinder.

19-85 Describe water application types and water flow rates for pressurized cylinder fires.

19-86 State the procedures for premature fire extinguishment of a pressurized cylinder.

19-87 List pressurized cylinder valve types and their operation.

19-88 Demonstrate proper advance and retreat techniques for fighting a flammable gas cylinder fire.

19-89 Demonstrate proper water application techniques for fighting a flammable gas cylinder fire.

19-90 Demonstrate the ability to assess a flammable gas cylinder's integrity.

19-91 Recognize the effect a fire is having on a flammable gas cylinder.

19-92 Detect and communicate changing conditions of a flammable gas cylinder during fire fighting operations.

19-93 Demonstrate the ability to operate flammable gas cylinder control valves.

19-94 Recognize and adapt to changing conditions in a flammable gas cylinder fire fighting operation.

*The FF I and II levels, as defined by the NFPA 1001 Standards, are identified in different colors:
FF I = black, FF II = red, additional information = blue.*

INTRODUCTION

This chapter looks at some of the common types of fire as well as some common techniques of firefighting in different disciplines and expands into specialized incidents like those involving cylinders. It opens with the more common elements of fire control, detailing most of the areas with which average firefighters will be involved during their careers. At the conclusion of this chapter, the reader will have a good foundation of knowledge in fighting fire and tactical situations involving the elements common to today's fire service.

ELEMENTS OF FIRE CONTROL

Before firefighters can respond appropriately to an emergency involving fire suppression, they must know the basic principles involved in the processes that create and sustain fire. This section deals with those areas common to a fire department. It discusses structural firefighting elements, ground cover or wildland firefighting elements, vehicle fires, flammable liquid and gas fires, the process of fire extinguishment, and proper stream selection.

In a previous chapter the fire tetrahedron and its importance in the suppression of fire was discussed. The elements of that tetrahedron are mentioned again here in an effort to instruct firefighters in safe, efficient, expedient fire control. A simple rule is that the earlier the fire department arrives on scene, and the more knowledge they bring with them in the fire suppression process, the lower the losses and the less risk taken.

Structural Fire Components and Considerations

When discussing structural fire components and considerations, it makes sense to look again at the fire tetrahedron, **Figure 19-1.** Most structural firefighting involves the suppression of Class A materials within the structure or as part of the structure itself. These elements are commonly suppressed by removing one side of the fire tetrahedron. The typical fire department will respond with water as an extinguishing agent and suppress the fire by removing the heat, oxygen, or both as quickly as possible. This may not seem difficult in its purest form but can be extremely complicated or hazardous if not done properly and safely. A number of structural firefighting considerations must be made prior to the extinguishment, or a great number of things can go wrong.

Listed are a number of factors that must be taken into consideration.

1. Length of time the fire has been burning
2. Building construction materials
3. Occupancy type and contents
4. Resources available (amount of water, staffing, equipment, etc.)

Taking these factors in order of listing, and not in order of importance, recommendations can be made as to how a structure fire should be fought.

The *length of time* the fire has been burning is an important factor because it can determine the stability of the structure and the amount of fire the responding units will face. It will also determine which stage of involvement the fire is presenting.

As discussed in previous chapters, flashover and backdraft are extremely dangerous fire phenomena that are predicated on the stages of fire in structures. In **Figure 19-2,** a flashover is shown as it occurs over the heads of a couple of firefighters. Based on length of burn, the integrity of the structure itself can be in jeopardy.

> **CAUTION**
>
> Most structure fires begin in contents and then spread to the structure itself. If not contained, this structural involvement can create concerns about collapse, both inside as well as outside.

This brings up the second factor: the *building materials* themselves. Is the structure made of highly combustible or noncombustible materials? How long will it take for the fire to become involved in the structure itself? **Figure 19-3** shows an apartment fire that began in its contents and then burned into the

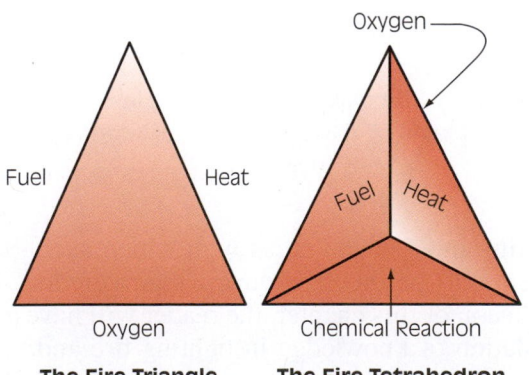

FIGURE 19-1 The old and new ways of visualizing the combustion process, the fire triangle and the fire tetrahedron.

FIGURE 19-2 A flashover may occur in the early stages of a structure fire. *(Courtesy of Phill Queen)*

FIGURE 19-3   A contents fire has burned into the structure itself. *(Courtesy of Rocco Di Francesco)*

departments have large numbers of responding apparatus while others have only one engine with the second responding unit a very long distance away. Some areas of the country have staffing problems while others do not. Some areas have unlimited training opportunities while others do not. These and many other situations are considered when dealing with response resources. These resources—or the lack thereof—have a great bearing on the tactics and strategy used in fighting structure fires.

Structural fire components and considerations are a large part of the responding agency's responsibility. Fire service personnel must be well schooled in all of these factors and situations so as to ensure efficiency, safety, and responsibility.

Flammable Liquids Fire Components and Considerations

Unlike most Class A fires that either go out when water is applied or continue to burn at a given pace if water is not applied, flammable liquids fires can actually be complicated by improper actions of firefighters. Whereas Class A fires are somewhat predictable and follow given laws of nature when burning or when water is applied to them, flammable liquids fires are not. Each flammable liquid has its own specifications that will affect its burning characteristics and extinguishment.

Section I, Chapter 4 introduced a number of terms related to flammable liquids. Knowledge of these terms is critical. Flash point was mentioned, and that characteristic of every flammable liquid will make a very big difference in a material's storage, handling, and extinguishment.

> **CAUTION**
>
> The vapors given off by a flammable liquid can be that material's most hazardous profile, sometimes traveling great distances before finding an ignition source.

For instance, an automobile accident with leaking fuel and no fire can turn into a very dangerous situation if the vapors traveling outward find a distant ignition source bringing fire back to the leaking area.

Another term used in the field of flammable liquids is **solubility.** This is the degree with which a liquid will mix with water. In chemistry, the materials that are soluble in water are called **polar solvents.** Materials that are polar solvent, such as alcohol, will dilute or mix with water quite easily. Hydrocarbon-based liquids (gasoline, oil, etc.) are nonpolar solvents and do not mix with water. They either float on water or

structure. Chapter 13 (in Sections I and II) details the types of construction materials and demonstrates the dangers involved in structural firefighting.

The third factor listed is the *building occupancy and contents.* Building occupancy means the proposed use of the structure. Is it designed to be an automobile repair shop or a preschool? Is it a retail store or a board and care facility? Each of these occupancies presents different fire attack protocols. The life hazards, the entrance accesses, and the ventilation components are among many details associated with the occupancy types that have to be taken into consideration by the responding units. Along with the occupancy, the building contents must also be considered. In most cases, the contents will be predicated by the occupancy type. The preschool will have dramatically different contents than the repair garage. These contents will have a bearing on the suppression efforts and control problems encountered by responding units.

The fourth factor listed is *resources available.* Fire suppression involves all of the factors listed, as well as the resources available to the firefighting team. Some areas of the country have limited water supplies while others have practically unlimited water. Some

FIGURE 19-4   Firefighters gently apply foam to a flammable liquid fire. *(Courtesy of Phill Queen)*

FIGURE 19-5   Firefighters apply water fog to cool a venting propane tank. *(Courtesy of Phill Queen)*

sink to the bottom of it. Most flammable liquids are not water soluble, but knowing which ones are can go a long way toward the prevention or extinguishment of a fire involving these materials. Liquids that are not water soluble do not mix with water and either float on water or sink below it. This is termed the liquid's specific gravity. Knowing the specific gravity of the liquid in question will tell the firefighter which method of extinguishment should be utilized. If the liquid is heavier than water and sinks, then we know water can be used to extinguish it, if applied properly. If the liquid is lighter than water, and floats on the water, then an alternate means of extinguishment must be used. **Figure 19-4** shows firefighters applying Class B foam to a flammable liquid fire by gently covering its burning surface.

Knowing the material that has spilled, or is burning, and some of its characteristics will greatly benefit the fire company, providing them with a margin of safety and the ability to effectively accomplish their goal of mitigating the hazard.

Flammable Gas Fire Components and Considerations

Flammable gases, in many cases, can be grouped with the flammable liquids category. In fact, a great number of flammable gases are products of flammable liquids. Often when shipping a flammable gas, the manufacturer will compress the gas in order to ship a larger quantity, turning the gas into liquid under pressure.

As with flammable liquids, it is very important for the firefighter to understand the hazards associated with gases. Gases are typically stored under pressure. Therefore, when a container is ruptured or leaking,

the gas escapes immediately, creating a hazard as large as the gas can spread. An example is propane. **Figure 19-5** shows a propane tank fire with firefighters cooling and containing the flames. Propane is stored as a liquid and the gas coming off the liquid is used for a multitude of purposes. Should a propane container rupture, the liquid turns into a gas, expanding over 250 times. That can be a really large problem for the responding firefighter.

The firefighter will want to know the properties of the flammable liquids so a plan of action can be determined and the hazard can be abated. One of those properties will be the gas's **flammable range.** This range will determine whether the fuel is too rich or too lean to burn in the given space it occupies. If the spill or leak is inside a structure with no windows or doors open, then there is a chance that the gas is too thick or rich to burn, as the ratio between oxygen and the gas will not support combustion. Conversely, if the spill is in a structure with open windows and a stiff breeze is blowing through the building, the gas-to-oxygen ratio may be too thin or too low to support the burning of the gas.

Another term common to flammable gases is **vapor density.** This term is very similar to the specific gravity term used with liquids. Basically, it is the weight of the gas as compared with air. Is the gas lighter or heavier than air? Knowing this fact about the material will predicate the plan of action taken because the gas may be hanging low and heavy, looking for an ignition source, or it may be light and dissipate quickly, causing far less concern.

The last term discussed here is possibly the most important: **toxicity.** Simply put, will the product kill a person who encounters it? This contact can be by a means other than breathing. It can be through skin absorption or ingestion also. Knowing this fact will determine what the responders should wear. Will turnout clothing be enough or will a hazardous materials suit have to be worn? And, in some cases, will that hazardous materials suit be enough?

The flammable gas fire components and considerations are many. Just as there are literally thousands of places where these gases can be found, there are a number of ways to deal with them when they are found. The firefighter interested in longevity will pay close attention to the details involved with this firefighting hazard.

Process of Fire Extinguishment

This section begins to develop thought processes based on the previous sections within this and other chapters. The concepts and considerations of the differing situations and classes of fire previously discussed begin to be put into the context of fighting fire.

> **NOTE**
>
> Before firefighters take on the beast of flame, they must first understand that fire is a multisided three-dimensional presentation of heat and chemistry. It must be considered from six sides. It has a top, a bottom, and four sides. It will follow laws of physics and applied science that only the most seasoned firefighter will understand, and even then things will go wrong on occasion. The goal in defeating this beast can only be met by extending an aggressive, fast, well-placed, adequate fire attack.

To accomplish this goal, a number of steps must be taken. The first step is to create a plan of attack. This is done by first locating the fire and determining its extent. Generally speaking, the longer it takes to find the fire, the longer it will take to put it out. Good communication becomes very important here. As Section II, Chapter 2 described, this report is built on careful

and complete observation of the fire and its development to this point. Each department will usually have SOPs that determine who is first on scene and where each unit should be located. These policies are not set in concrete because each fire is different. However, they have been set up in order to maintain order and efficiency. The attack plan will be built on the observations, training, and opinions of the first-in officer after considering the facts as perceived and predicting a course of action.

The second step is then to apply the plan of action as quickly, efficiently, and safely as possible. Firefighters will take their skills, knowledge, and ability into the face of danger in an effort to mitigate the situation as quickly as possible. Three methods of fire attack are possible: the direct attack, the indirect attack, and the combination attack of fighting fire. The direct attack is quite simply the act of putting water directly onto the seat of the fire, **Figure 19-6.** This method is the most efficient use of water on free-burning fires. A solid or straight stream of water is applied in short bursts directly onto the burning fuels (Class A) until the fire darkens or is extinguished. **Figure 19-7** shows a firefighting team about to enter a burning structure and apply a direct attack on the seat of this fire. If used indoors and done properly, the thermal balance of the room may still hold, giving the firefighting team a clear view of the fire and contents of the structure.

Another method is that of the indirect attack, **Figure 19-8.** This attack is usually utilized when the firefighting team does not see the seat of the fire. A fog stream is applied to the upper areas of the room or above the fire seat in an effort to "steam" the fire out. **Figure 19-9** shows a firefighting team applying a fog stream into the upper area of a structure in an indirect attack method. Directing the stream into the superheated upper atmosphere above the fire causes the water to "boil" or "vaporize," steaming the area and causing the fire to die. This method, if used indoors, will usually greatly disturb the thermal balance, causing a loss of visibility in the structure as heat and products of combustion are circulated downward.

FIGURE 19-6   A direct attack with firefighters applying a straight stream onto the seat of a fire.

FIGURE 19-7   A firefighting team prepares for a direct attack on a contents fire. *(Courtesy of Phill Queen)*

FIGURE 19-8   The indirect attack has the firefighters applying a 30-degree fog into the upper heat layer of the fire in order to create a steam that will extinguish the fire.

When using this method, the firefighting team should attempt to shield themselves from the steam vapor by backing out of the area after applying the water. They can reenter as soon as the vapor begins to dissipate.

CAUTION

Note also that the indirect method is not appropriate for occupancies with victims in need of rescue, because the steam created by the attack will burn the victims quickly.

FIGURE 19-9   A team of firefighters applies a fog stream in an indirect attack on a fire. *(Courtesy of Central Net Fire)*

The last method is that of a combination attack. This attack is a blend of both the direct and indirect methods. The straight stream is directed to the seat of the fire and then the nozzle is quickly changed to a fog and the upper areas of the room are covered with a quick shot into the thermal cover. This method will darken the fire and stop the growth of flashover.

In each of these methods of applying water, firefighters must remember that the goal of the fire department is to save property. If water is not applied in proportion to the amount of fire, two things can happen. If not enough water is used, extinguishment will not take place. If too much is used, then the damage by water may exceed that done by fire. The lesson here is to use only what is needed, saving the rest for overhaul if needed.

Firefighters should never limit their nozzle selection during any emergency. Different nozzles have specific purposes that are intended to work better during different incidents. Please review Chapter 11 (in Section I) for more information and the advantages and disadvantages of specific nozzles. This knowledge is important when applying to the suppression activities that are presented in this chapter.

Proper Stream Selection

Fire streams were discussed at length in Section I, Chapter 11 with information regarding a number of things that are mentioned again here. The purpose of this section is to transition from knowledge of the burning characteristics of various occupancies and hazards to actual fireground use of the information.

In the case of fighting fire with water, it is important to understand that in order to be successful, sufficient water must be applied directly to the fire in order to control it. That statement may sound simple until one begins to take into account all of the mechanics of performing that task—not only the physical

FIGURE 19-10 Firefighters use a fog pattern to shield themselves from radiant heat. *(Courtesy of Phill Queen)*

FIGURE 19-11 A firefighter applies a single straight stream onto a structure in an attempt to reduce some of the fire's heat. *(Courtesy of Phill Queen)*

movement but the mental as well. Success will be based on a number of factors in the selection of the stream alone. Some of those factors, in no particular order, are proper stream type, stream size, stream placement, timing, water supply or quantity of water, stream reach, mobility needs, tactics required, speed of deployment, and personnel available.

Consideration of some of these factors provides a better understanding of the firefighter's role in proper stream selection. The first one was the proper stream type. Is a fog stream the correct choice? If so, which width? **Figure 19-10** shows firefighters using a fog stream to protect themselves from radiant heat. Or is the straight stream a better choice? Each situation calls for differing stream requirements. The most basic answer to stream type may be in the type of attack required: direct, indirect, or combination. Each has its place in given circumstances.

Next to be considered is stream size. The typical response to this question may very well be "Big water, big fire and little water, little fire." It was once said that firefighting is like going to battle. The weapon must match the target. **Figure 19-11** and **Figure 19-12** show examples of the need to have the stream size match the fire size to be effective.

FIGURE 19-12 If the line selected does not match the size needed, the fire will burn longer and hotter and can jeopardize the operation.

FIGURE 19-13 The first line pulled should be positioned between the potential victims and the fire.

CAUTION

If a large line is needed, then it should be pulled right away so as not to play "catch up" with the fire. Some firefighters think the stream to start with should be the stream that will eventually be used, going smaller later, rather than ever having to go larger later.

The placement of the fire stream can make or break an attack plan. The fireground must be coordinated at all times without exception. This is a battle with an enemy that can injure or kill. When an order is given

to place a fire stream in a particular location, it is part of a plan. Without that stream, the plan may fail, so this factor can be very important. An example may be a fire in a dwelling where the first line goes between the trapped occupants and the fire, with the second stream going to the seat of the fire, **Figure 19-13**. This operation will not work without both streams being coordinated and controlled. Care must also be taken to avoid opposing streams, because these streams may drive heat and products of combustion toward the opposing firefighting teams. Still another related factor is that of timing, which will go with speed of deployment. This is self-explanatory. Timing is everything on the fireground. Stopping the progress of the fire as quickly as possible will allow the firefighter to work toward the goals of rescue and property conservation.

Many differing types of nozzles are used in the fire service. One of the commonly used types has the ability to control the amount of water it puts out. In cases of water conservation because of poor sources, a nozzle that conserves water would be advantageous. In such a case, water supply or quantity of water available plays a significant role. A different set of rules applies when the water is limited.

Another factor to consider is the mobility of the stream that will be needed. In some cases, lines are advanced quickly into the fire, whereas in other cases the line is placed in a fixed position holding the fire until another more mobile line can be set up to take over. **Figure 19-14** shows a deck gun attack on a large fire in an effort to bring the fire into handline size by reducing some of its heat. In this same area are the factors of stream reach and tactics. A line may be placed on an exposed structure that is large and unmoving, while a smaller, more mobile line is extended into the burning occupancy to fight the fire.

Everything taken to a fire is a resource operated by the fire department. The personnel, water, equipment, and knowledge are all resources. The selection of the proper fire stream is a significant decision.

FIGURE 19-15 Large stream appliances at work on a large fire. *(Courtesy of Rocco DiFrancesco)*

Figure 19-15 shows the proper use of large streams on a fully involved mercantile fire. **Table 19-1** shows the differing sizes of nozzles and the effort required in order to work properly.

Firefighters carry out their part of the attack plan using water and a good basic knowledge of the requirements being predicted by the fire they are fighting. Knowledge of the factors listed, as well as many others, will arm the firefighter with an arsenal when going into battle against a fire.

TACTICAL CONSIDERATIONS

Earlier in this chapter the elements of fire control were discussed in some detail regarding the most common fire expectations of the average firefighter. Those sections detailed the components and considerations a firefighter must be knowledgeable in to be safe and successful on the fireground. This section deals with those same types of fires in a tactical firefighting capacity. It builds on the information presented in the previous section in an effort to actually use the information in fighting fire.

The actual fighting of fire on the fireground is broken down into tactical objectives. Each of these objectives must be accomplished in order that the safety of the firefighters, occupants, and others is a primary consideration. If there are no occupants, the overall plan still remains in order, keeping the firefighters safe and efficient. These goals are universal in the fire service, although different departments may use different terms. The priority was first set in place a great many years ago by Lloyd Layman, with the following acronym: **RECEO**—Rescue, Exposures, Confinement, Extinguishment, Overhaul.

FIGURE 19-14 Firefighters use a deck gun to hold the fire until more lines can be put into place. *(Courtesy of Rocco Di Francesco)*

TABLE 19-1   Hose Stream Characteristics

Type or Size of Line	Reach (ft)	Mobility	GPM	Common Use
1-inch or greater booster or reel line	25–50	Excellent	10–40	Very small nonstructural fires or overhaul
1½ to 1¾ inches	25–50	Good	40–175	Quick attack, one to three rooms, vehicle fires
2½ inches	50–100	Fair to poor	125–350	One floor or more, personnel permitting
Master stream	100–200	Poor to none	350–2,000	Large, fully involved structures or exposure protection

VIEWPOINT

Over the years a number of variations have been used for fire attack acronyms. Some currently popular are:

RECEO—Rescue, Exposures, Confinement, Extinguishment, and Overhaul

REVAS—Rescue, Exposures, Ventilation, Attack, and Salvage

SPARS—Size-up, Position Apparatus, Attack Fire, Rescue, Support Functions

LOUVERS (truck companies)—Ladders, Overhaul, Utilities, Ventilation, Entry, Rescue, and Salvage

There are many more fire attack acronyms that firefighters may encounter, but the point is well made that the fire service has set itself a series of goals that must be accomplished in order that the fireground remains as safe and orderly as possible. All firefighters should refer to local or departmental policies and procedures for the tasks and order of fire tactics.

FIGURE 19-16 A battalion chief meets with his captains to set operational plans into action. *(Courtesy of Rocco Di Francesco)*

NOTE

A second point that must be made is that the command structure must be solid and well defined in order to accomplish the set objectives in each of the goal areas.

Just as an army employs a rank structure to fight a battle, the fire service uses a rank structure to fight fire. One person will be in charge and appoint people who will command different parts of the operation under that person. **Figure 19-16** shows a battalion chief conferring with his captains in planning his objectives. This system is called the **incident command system (ICS)** and it is mandated by NFPA, OSHA, and NIMS. The ICS is basically designed to maintain order in any emergency operation. It can be fire, flood, rescue, first aid, mass casualty, or anything demanding a number of emergency responders. A single person will be in charge of the incident and this person is called the incident commander (IC). This person works from the goals listed earlier utilizing strategy and tactics just as a general would in the military. The IC will delegate assignments to arriving units in an effort to accomplish the goals in a safe and efficient order. Each firefighter must have a thorough knowledge of the ICS system and its components in order for an emergency to run smoothly and effectively. The ICS system is discussed in detail in Section II, Chapter 2 and should be reviewed on a frequent basis to remain familiar with the system and where each person fits into the system.

To explain the system and complexities of fighting a fire, this discussion sets the scene with an average, generic fire, because the methods and operating principles will apply to any fireground operation.

As soon as the fire is reported, and before units ever pull from the station, mental preparation takes place. These are the recall factors of past similar incidents,

TABLE 19-2	Fireground Factors	
Building		
Size	Construction type	Condition
Age	Openings	Utilities
Concealed spaces	Access	Effect of fire
Extent of fire	Interior fuel load	Exterior fuel load
Fire		
Size	Location	Direction of travel
Time since ignition	Extent	Materials involved
Material left to burn	Fire load	Stage of involvement
Occupancy		
Type	Value	Fire load
Status (used/vacant)	Hazards of occupancy	Life hazard
Arrangement	Obstructions	

the recall factors of knowledge about the area where the alarm is reported to be, the recall factors regarding the fire prevention inspections of that area, the recall factors of time of day and route to be taken in response, and the factors regarding the condition of the responding units as to their capabilities. Are they fully staffed today? Are the water tanks full and the pumps working properly? Are the crews fresh and ready to go to work in possibly very dangerous conditions? All of these factors, and many more, will race through the minds of the responders. Each of these factors will eventually dictate the strategies to be considered.

Upon arrival, another set of factors must be considered. All of them have to do with the environment of the response. **Table 19-2** demonstrates that the number of factors having to do with the operation can be almost overwhelming. The seasoned firefighter will go through all of these and possibly more in calculating the environment presented at the emergency. And each of them can greatly affect the outcome of the incident depending on how they are handled.

Based on the situation presenting itself to the arriving firefighters, three methods or modes of attack are possible: the **offensive attack,** the **defensive attack,** and the **combination attack.** These modes are predicated on the resources available at the emergency. If the arriving units have adequate resources to handle the situation, and the structure is safe to enter, then they will fight the fire aggressively and offensively. They will attack the problem head-on and, following

department standards, will accomplish their objectives efficiently, effectively, and safely. If they do not have adequate resources to aggressively handle the situation, then they will have to fight the fire in a defensive mode of attack. This mode will be continued until enough resources can be massed to then change to an aggressive, offensive attack. The combination mode is most often utilized when a rescue must be accomplished but not enough resources are on scene to handle the operation entirely. An example would be when a line is pulled to defend potential victims until they are removed, knowing that the line will do nothing to fight fire because it is not large enough or practical for the size of the fire.

As the operation begins to unfold, the IC calls for an attack on the fire. The IC's call will be based on the conditions presented earlier. Hoselines will be pulled, utilizing methods prescribed by the responding department's SOPs. Some common hose evolutions were covered in Section I, Chapter 10. The concept of **teamwork** will dominate the scene as each person will be responsible for his or her own assigned details of the operation. This teamwork will continue through the entire operation just as it would in a military operation. Each team member will be given an assignment based on the priority of the incident. As mentioned earlier, rescue will be the first priority, and all actions will be directed toward accomplishing that task. A line may be pulled and placed between the fire and the people needing rescue, or the situation

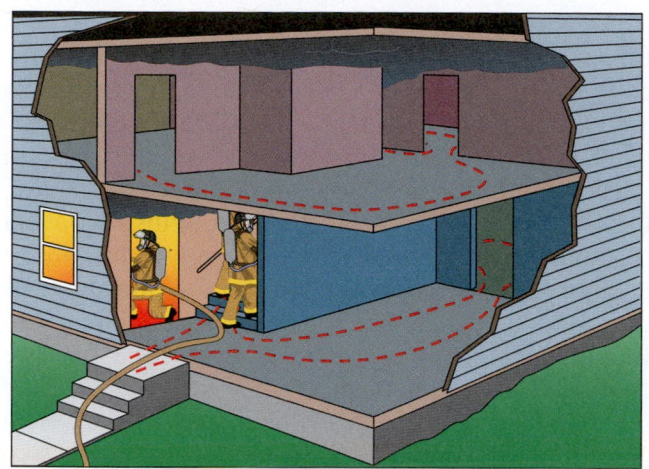

FIGURE 19-17 If it is not known whether victims exist, then the first line into the structure should assist the search and rescue functions.

may call for rescue first and no line pulled until it is completed, **Figure 19-17.** Each situation will dictate its own need.

The department as a whole cannot be effective in its firefighter safety effort without the support of a *team approach* to handling incidents. Therefore, the team that handles an incident must hold up its part of the safety partnership. This team obviously includes the individual firefighter. To ensure safety, the team should follow these procedures:

1. Utilize an incident command system (ICS).
2. Work together and remain intact.
3. Look after each other.

CAUTION

The separation of members within a team is a contributing factor to firefighter fatalities.

The *individual firefighter* holds the final key to making the safety partnership work. The following list includes areas that the firefighter must remember.

- Be ready.
- Understand and act within the chain of command.
- Perform as trained.
- Use an incident engagement checklist. (See Section I, Chapter 5.)

Fire and rescue work is a team effort orchestrated to an action plan developed by the incident commander. Team leaders implement the plan and provide feedback to the IC. Firefighting teams follow the team leader's direction. *Freelancing* occurs when a team operates outside the action plan or when individuals work alone. Working alone or outside the action plan

endangers individuals and the team, as **Figure 19-18** demonstrates. The individual firefighter can help prevent freelancing by not working alone, and supporting the action plan when part of a team.

The team process is the basis for successful operations and firefighter safety when responding to the various emergencies described throughout this chapter. The team must function together at all times and on all levels for complete success at any incident.

A recent rule by the Occupational Safety and Health Administration sets safety standards on the fireground for rescuing occupants from burning buildings.

NOTE

The two in/two out rule tells the fire department that the teamwork concept must be followed on the fireground at all times.

This law is designed to protect the firefighter from being injured in rescue attempts where fire conditions are critical.

FIREFIGHTER FACT

In 1998 the Federal Occupational Safety and Health Administration issued a Final Rule titled 29 CFR Parts 1910 and 1926 Respiratory Protection. This regulation requires both private industry and the fire service to practice the safe use of self-contained breathing apparatus (SCBA) by mandating a number of rules regarding face piece fit testing, physical requirements for wearers, apparatus requirements, and more. A part of this Final Rule also mandates the number of firefighters required on the fire scene when interior operations are to take place in hazardous situations. The two in/two out rule dictates that during interior structural firefighting in atmospheres that are immediately dangerous to life and health (IDLH), and that do not have lives requiring immediate rescue, at least two firefighters must enter together and remain in visual and voice contact with each other at all times. Two more firefighters must be on standby for the potential rescue of the interior team. If there is a confirmed rescue to be made, the rule does not apply but the team **must** have good evidence of that need for rescue.

The 1997 NFPA 1500 Standard for Occupational Safety and Health also requires this same two in/two out rule as part of its requirements. This was created in an effort to further follow OSHA laws and protect the safety of firefighters the world over.

If the situation does not call for rescue because there are no dangers to people or animals, then the second priority is placed into action. That priority is *exposure*. Almost anything in the path of the fire can

FIGURE 19-18 Freelancing endangers individuals and the team. This firefighter is working alone in a collapse zone—for what gain?

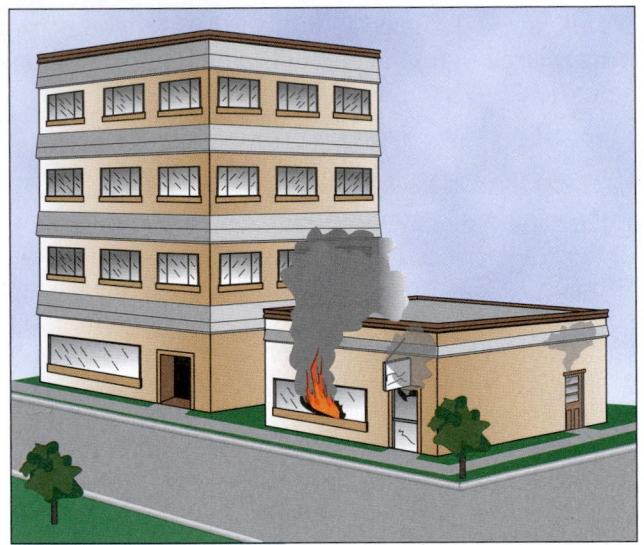

FIGURE 19-19 The most common exposure problem is the closest building to the building on fire.

be called an exposure. The most common exposure is a structure next to the burning building where the fire is heating it to a point where it is close to catching on fire, **Figure 19-19.** Another example would be a car fire that is about to ignite nearby brush or tall grass. If an unburned object (i.e., exposure) is about to catch on fire, it is the responsibility of the responding firefighter to stop the spread of the fire to that object. The logic here is that the item burning has already sustained damage, but the exposed item has not. So firefighters save the exposed item before concentrating on the burning item. Exposure protection is accomplished by either removing the exposed item from the area or wetting it to cool its surface below the point of ignition, **Figure 19-20.** Radiant heat is the main cause of exposure fire, so wetting the exposed item removes that heat at the point of contact.

As soon as all exposures are deemed safe, it is time to consider the next step in the firefighting process, which is fire *confinement.* The fire's parameters are not always clearly defined. Visibility may be difficult at best. The members of the firefighting team must use all of their senses in order to locate the fire. Once located, water or another extinguishing agent is applied in a method that will confine the fire to its area of burn. Stopping the fire's forward progress must be done first, before the fire can be extinguished. It was mentioned earlier that after rescue came fire control. The confinement of the fire is controlling the fire. Some departments will even broadcast a "fire is under control" call to the IC when this step is accomplished. In the wildland, this point becomes even more dominant. An "under control" term may be issued days before actual fire extinguishment is accomplished. When the fire is surrounded and can burn no further outside of this confinement, then it is considered to be under control.

In many cases, to bring the fire under control in structure firefighting another task may have to be added. Ventilation (see Chapter 18, in Sections I and II) coordinated with water application may be the only way to achieve the confinement results safely. This will be described more later in the discussion of particular structure fires.

It is common practice to attack fires from the unburned side, **Figure 19-21.** That is to say that the line will be brought into action between the fire and its intended direction of travel. The opposite of this tactic would be to attack the fire from the rear, pushing the heat and other products of combustion forward into unburned areas, causing far more damage. This is a tactical consideration in almost all applications of firefighting from structure to wildland or ground cover to flammable liquids and gases.

When the fire is controlled or confined, then the actual *extinguishment* begins. This is accomplished with direct or indirect attack strategies and resource use. This can also be thought of as property conservation. Saving the property not damaged is a very high priority here. The fire has been confined and now is extinguished utilizing the latest methods and practices in order to save the undamaged property. The use of water is restricted to only as much as is necessary, because water can cause as much and, in some cases, more damage than the fire.

As soon as the fire is extinguished, the *overhaul* begins. This is a methodical system of ensuring that all embers and chances of reignition are removed. The area is carefully worked by moving every item and wetting all combustibles that could carry new life to the extinguished fire. During this phase of the operation, the cause of the fire is usually determined. By carefully

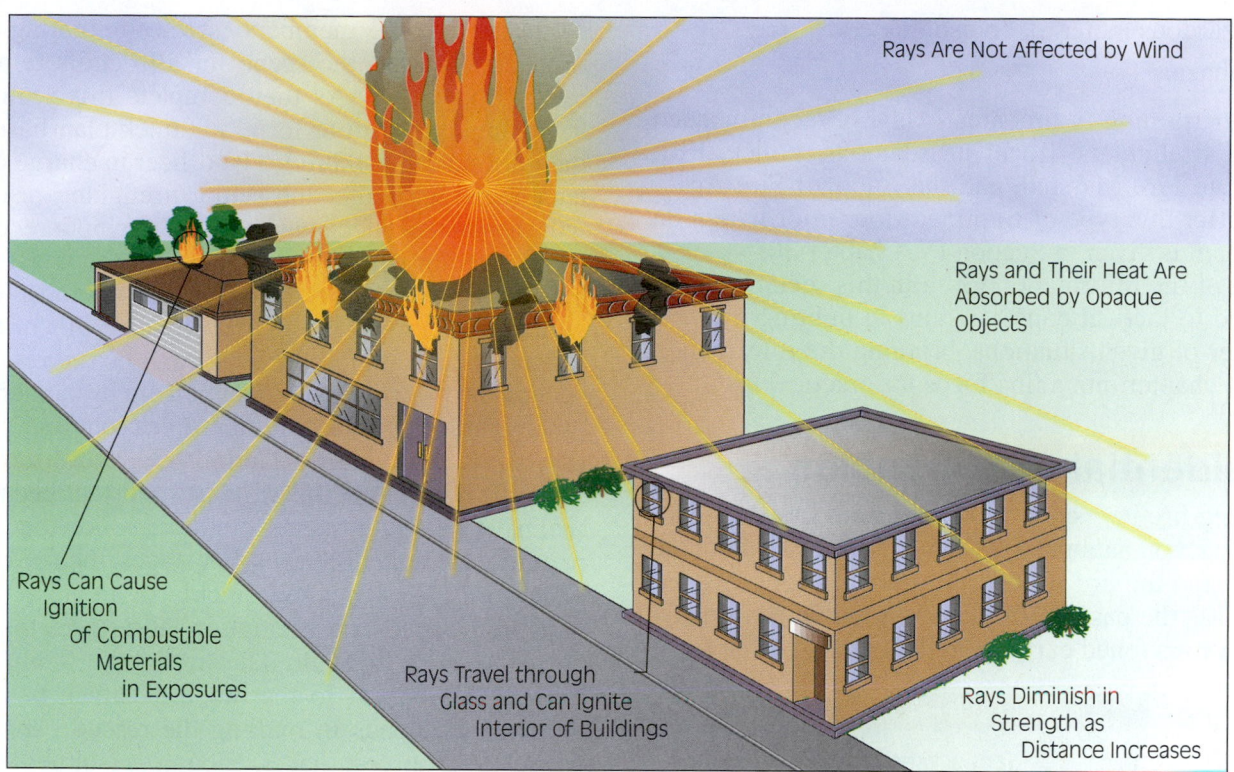

Rays Are Not Affected by Wind

Rays and Their Heat Are Absorbed by Opaque Objects

Rays Can Cause Ignition of Combustible Materials in Exposures

Rays Travel through Glass and Can Ignite Interior of Buildings

Rays Diminish in Strength as Distance Increases

FIGURE 19-20 Radiant heat is the main cause of exposure fires within short distances. Radiant heat will travel in straight lines from the heat source to nearby objects.

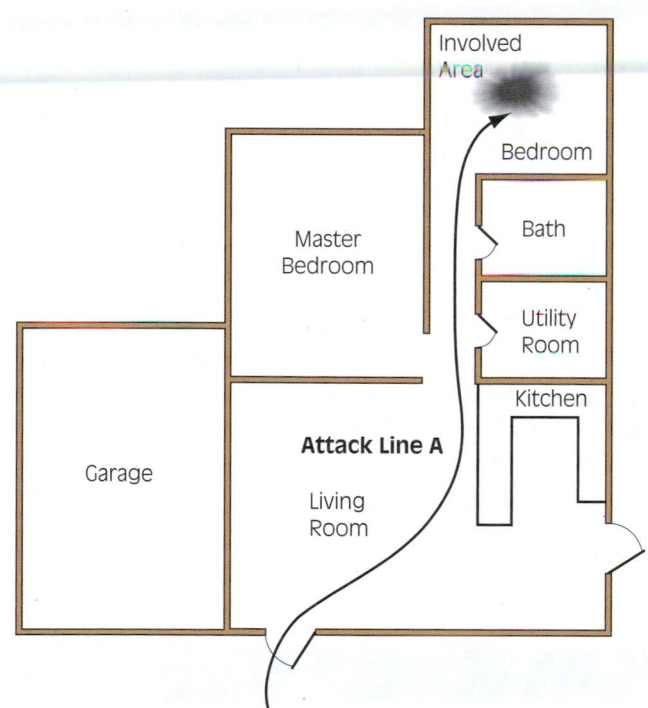

Involved Area

Bedroom

Master Bedroom

Bath

Utility Room

Kitchen

Attack Line A

Garage

Living Room

FIGURE 19-21 Fires most commonly should be attacked from their unburned side, pushing the products of combustion away from unburned materials and areas.

sifting through the fire debris the investigator looks for clues that will lead to the cause of ignition. A closely coordinated effort among these people will result in the call being closed out so the firefighters can return to their respective stations to prepare for the next call.

> **CAUTION**
>
> The firefighter must use extreme caution when in the overhaul phase so that the subtle clues the investigator is working with are not disturbed.

One more point must be made about firefighter safety. That point is about acts of terrorism. The fire service is called upon in nearly every conceivable emergency, representing local authority in the form of a fire department, fire district, or emergency services agency. Unfortunately the firefighter may represent the authority being targeted by an act of terrorism. For this reason it is important that the firefighter be made aware of the possibility of danger from cowardly acts of aggressiveness. Some recognized threats to firefighters are:

- Undetonated explosives: secondary devices, low-order explosions, malfunctions
- Structural instability: collapsed buildings, falling/thrown debris, live electrical lines, leaking gas mains
- Fire

- Hazardous materials (poisons, etc.)
- Biohazards

Further training in this area is highly recommended for all firefighters. Local jurisdictions, police, FBI, the State Fire Marshal, and the National Fire Academy offer this type of training. More information on terrorism is given in Section IV, Chapter 30.

All of the information given in this section is next applied to tactical considerations in fighting fire in a number of given situations, bringing the information in this chapter into a fine focus for practical use.

Residential Occupancies

Fighting fire in residential occupancies has the greatest impact on national fire fatality figures, for both the public and the fire service. A direct result of this statistic was the passage of the two in/two out ruling by OSHA mentioned earlier and in Section I, Chapter 5.

> **FIREFIGHTER FACT**
>
> More people die in residential fires than in any other type of fire in the United States. This is true for both the public and the firefighters.

For this reason, the fire service must place a much greater emphasis on safety here than almost anywhere else. This subsection deals with firefighting in residential occupancies in both single-family and multifamily dwellings.

Before discussing particular residential occupancy types, it is important to discuss a number of generic considerations that hold for all residential fire emergencies. The best way to do this is to go on scene and discuss each step of consideration for the firefighter. Then, subsequent sections will detail single-family and multifamily dwelling operations.

En route to the alarm, the firefighting teams begin the size-up process discussed earlier. The team considers all factors, both silently and in discussions with each other. This practice prepares the team members for the multiple details that will be thrust on them as soon as they arrive on scene.

On arrival, the members of the firefighting team immediately scan the scene for all details remotely related to the fire. As they complete this scan they have already begun to form an attack plan based on the information gathered. The officer in charge of the team (captain, lieutenant, senior firefighter, or other) will then make a decision based on the input received, and the action will begin.

As discussed earlier, the order of attack must follow guidelines set forth in the department SOPs or training mandates. The first priority will be the rescue of occupants. Information is sought from bystanders and other sources as to the number, location, and condition of potential victims, **Table 19-3.** Determining the stage the fire is in at this time is important because it will have a direct influence on the direction the rescue or attack will take. In the earlier stages, the fire can be attacked aggressively and quickly in an effort to cut it off from the rescue attempt. In later stages that will be much more difficult. The firefighting team must make a decision as to the mode of attack based on the resource needs in making the rescue. Are there adequate resources, that is, enough personnel, equipment, and water to accomplish the order of attack selected? Or are the resources on scene limited, causing the team to hold as it waits for adequate help to arrive?

The two in/two out rule is important here. If a rescue is being considered, but there are no facts to substantiate that one is needed, then the rule applies that adequate manpower must be on scene before entering the structure. If the facts presented show that a definite rescue must be executed, then the rule is relaxed and that rescue can take place. As described in Section I, Chapter 16, a primary search for victims is conducted very early, as soon as the amount of personnel on scene will allow.

The sequence of priority of multiple rescues is another factor to consider. The sequence shown in **Figure 19-22,** or order of priority, is most generally as follows:

1. Those closest to the fire
2. Largest grouping of threatened people

TABLE 19-3	Rescue Factors	
Number of occupants	Location of occupants	Mobility of occupants
Verification by bystanders	Condition of occupants	Firefighting required
EMS/ALS required	Manpower for search	Fire's burn time
Hazards to firefighters	Access to victims	Escape after rescue

3. Anyone else in the fire area

4. Those in the areas that will eventually be exposed

After any necessary rescues have been performed, the next priority will be exposure control. This was discussed earlier when tactical considerations were covered. Following exposure control comes confinement of the fire, then the actual firefighting itself. As these events begin, a number of simultaneous events must take place to ensure the safety of the firefighting team members. The utilities to the occupancy must be secured and ventilation must be considered. The utilities will consist of the electrical power, the water, and gas supplying the occupancy. A firefighter is typically assigned the task of utility shutoff. This person must find the electrical panel and shut off the breaker switches or unscrew the fuses in older occupancies. Then the gas must be secured by going to the meter or tank and turning off the flow of gas supplying the structure. After that, the company commander will decide if the domestic water to the occupancy should be shut off also. This is usually done if there has been any chance of the water lines breaking during the operation.

CAUTION

The firefighting force does not want to be electrocuted, blown up, or have water running that they do not control. Water can cause as much—or more—damage than the fire.

During this part of the operation, ventilation must be considered. Ventilation is discussed in Chapter 18

FIGURE 19-22 When numerous rescues face the first team they must prioritize their actions based on victim exposure.

(in Sections I and II). A well-coordinated ventilation effort will greatly assist the fire attack team in almost all circumstances. The extinguishment and overhaul parts of the fire attack then will take place bringing the situation to a close.

At this point, many fire departments will offer services to the occupants of the burned structure. These services consist of housing for the night, clothing, food, and, in some cases, counseling. These services are offered by the fire department through a number of volunteer and charity organizations all over the country. Two primary organizations are the American Red Cross and the Salvation Army, but there are a great many others.

This section dealt generically with residential occupancy firefighting. The next topic is the finer details of fighting fire in single-family and multifamily occupancies.

Single-Family Occupancies

The most common element of single-family dwellings is the layout. Granted, floor plans will differ more than porcupines have quills, but there is still commonality. In the one-story home, the bedrooms will generally be on one side of the dwelling, and the living, cooking, and laundry facilities on the other. In the two-story home, the bedrooms will generally be on the upper floor (not always) with possibly one of them downstairs. The other functional areas of the home will be downstairs. In the two-story home it is usually true that there are fewer rooms downstairs than upstairs also, because the functional rooms tend to be larger than the bedrooms. Why is this important? Rescue! In the dark, smoke-filled, hot confines of the fire environment, it is nice to have a little bearing on where the search may be most productive.

Another common consideration of two-story homes is that of heat travel. The heated, toxic products of combustion will tend to travel upward because they are usually lighter than air. They will travel toward the sleeping areas of the common two-story dwelling, many times moving up the open stairways.

It is typically the responsibility of the first-in engine to either take command of the incident and control the operation or to immediately attack the fire. As the engine arrives, the driver pulls to the far side of the structure passing by the front, giving firefighters a view of both sides and the front, and making room for other responding units, **Figure 19-23.** This gives the team three of four sides and quite a lot of information from which to draw an attack plan. Then more often than not, lines are pulled to either attack the fire or protect rescue team members. **Figure 19-24** shows an engine parked with lines pulled back to the structure

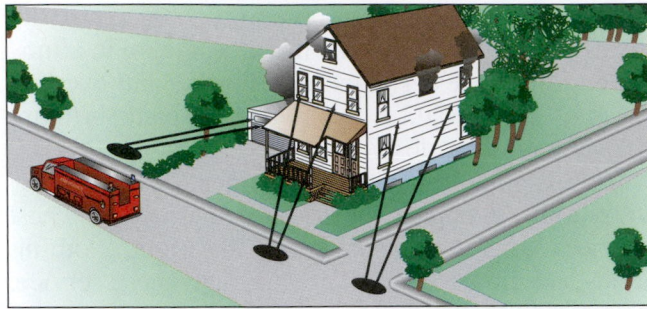

FIGURE 19-23 As the first-in unit arrives, it should pull past the building, giving firefighters a good look at up to three different sides or views of the fire.

FIGURE 19-24 An engine has pulled past the involved structure to pull working lines. *(Courtesy of Phill Queen)*

it passed on arrival. In rural areas the engine may have laid a line from a water source to the fire, or they may have brought water with them in the form of a very large tank on the apparatus or a water tank truck. In metropolitan areas it is common for the first-in unit to attack the fire with the tank water in the unit and the second-in unit to bring water from the hydrant or

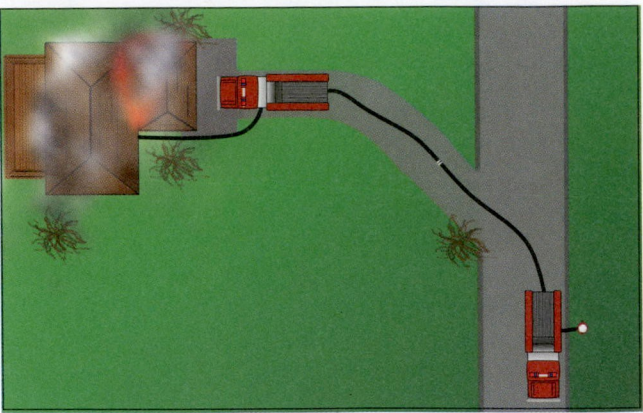

FIGURE 19-25 The second-due engine may bring water to the first-due unit that went into a "quick attack" using tank water.

FIGURE 19-26 Here the second-in engine has laid hose from the hydrant and is assisting the first-in unit. *(Courtesy of Phill Queen)*

water source, **Figure 19-25.** Departments' SOPs will differ in this area based on experience and resource availability.

As the second-in team arrives at the fire, they are commonly required to offer backup to the first-in unit. **Figure 19-26** shows two engines at a fire scene where the second-in unit has laid hose and is assisting the first-in unit. Some of the common tasks they are responsible for include these:

- Back up the initial lines laid by the first-in unit
- Protect the access and egress of the first-in company
- Assist with attack while the first-in works on exposures
- Assist with exposure protection if the situation warrants more lines
- Back up the initial team in any requests they make

The truck company in most areas will arrive and secure the occupancy's utilities and begin ventilation.

FIGURE 19-27 A ventilation operation on a single-family dwelling. *(Courtesy of Ed Hatfield)*

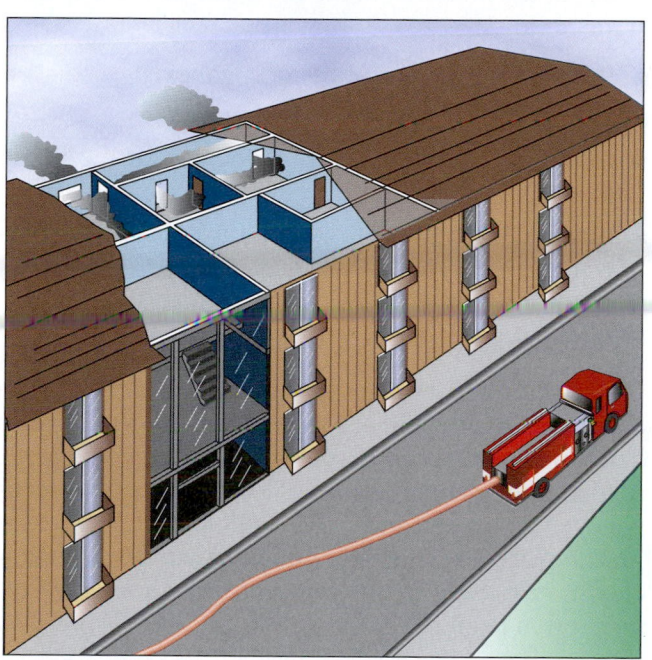

FIGURE 19-28 In many cases, in larger occupancies or complexes, the smoke may be the best indicator of the fire's location.

In occupancies that require entry tools not common to engine companies, the truck assists with building entry, and the ladders will be used for a great many purposes as mentioned in Section I, Chapter 14. **Figure 19-27** shows a ventilation operation on a single-family residence.

The role of the firefighting teams in single-family dwellings is basic and, as mentioned before, potentially hazardous because of the need to rescue occupants in dangerous situations.

Multifamily Occupancies

Multifamily occupancy firefighting is also potentially dangerous and many times frustrating.

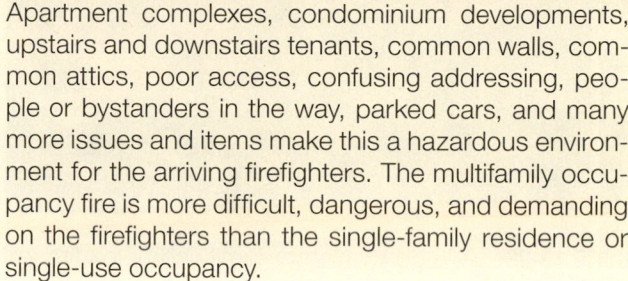

CAUTION

Apartment complexes, condominium developments, upstairs and downstairs tenants, common walls, common attics, poor access, confusing addressing, people or bystanders in the way, parked cars, and many more issues and items make this a hazardous environment for the arriving firefighters. The multifamily occupancy fire is more difficult, dangerous, and demanding on the firefighters than the single-family residence or single-use occupancy.

In large developments, very often the smoke may be the best indicator of the fire's location, giving the firefighters a better mark on which to focus, **Figure 19-28.**

The role of the first-in unit is basically the same as for a single-family occupancy. Size-up and the action must be quick and decisive. It will be based on the factors as perceived by the arriving team and whatever they have gathered from other sources. If the structure has a common means of egress for the occupants, they must be protected by the team, allowing escape. This can be considered a form of rescue, which is the first priority of firefighting (RECEO).

The resources needed at a multifamily occupancy fire are going to have to be upgraded because of the logistics of moving personnel and equipment into areas of need. The distance from the street alone can be great and complex. The greater number of occupants will pose a rescue situation demanding more effort. The structural design lends itself to exposure problems, and confinement will be more difficult due to access problems and larger running areas for the fire to travel. Extinguishment may be difficult because of hidden areas in which the fire will run between floors and units. **Figure 19-29** shows a large multifamily complex on fire during its construction and the size of the pending firefight.

Business and Mercantile Occupancies

The business or mercantile fire is fought somewhat differently than the residential fire. Here, depending on the time of day, the life hazard potential for fire victims is not as great, and the fire and life hazards to the firefighter can be greater.

In most cases the business having the fire is either closed with no occupants, or open with the occupants

FIGURE 19-29 A large multistory family complex fire will entail a lot more resources than a single-family dwelling fire. *(Courtesy of AP Images/The Tennessean, John Partipilo)*

either evacuated or fighting the fire. In either case, the chance of having to save lives is far less than for residential firefighting. On the other hand, however, the hazards associated with the occupancy itself can be greater to the firefighter.

CAUTION

Many businesses utilize hazardous materials or dangerous processes, and sometimes storage can be piled high. These can pose a serious risk to the firefighters entering these occupancies under dark, smoky, hot conditions.

Typically, the first-in engine will go to the front of the building (allowing appropriate space for the first-due truck company to position in front of the building as well) in an effort to locate occupants who can assist with information about the fire. They will ask

FIGURE 19-30 The first-in unit at a fire should take a position in front of the building, making certain that access is still allowed to put a truck company into operation. *(Courtesy of Orange County Fire Authority)*

about its location, spread potential (what is burning), lives endangered, time of ignition and source, hazards to watch out for (see Chapter 24 for hazardous materials placarding and recognition), most important items or processes to protect from fire or water, and where access can be made safely. The officer will then decide on an attack plan based on these and other factors known about the area or business, **Figure 19-30.** If the officer has decided to attack the fire, then the firefighter is told to pull lines from the apparatus following department SOPs in preparation for attack. If the fire can be attacked without endangering crew members, then an attack is made on the fire. If entering the building will put the crew in danger (per the two in/two out rule), the crew will prepare the attack by laying lines and placing equipment, but wait for reinforcements to assist in the interior attack.

The second-in engine has a number of options based on the conditions of the fire, the department's SOPs, or the direction of the IC. Some of these options are as follows:

- Respond to the rear of the occupancy.
- Supply the first-in engine with water.
- Join the first-in unit with manpower for the initial attack.
- Lay into the sprinkler connection and supply added water to the system.
- Stage at the street or hydrant waiting for instructions.

If the first-in engine has not made entry due to either the business being closed or not being able to determine the extent or possibility of fire, then the second-in engine will respond to the rear of the occupancy. The second-in engine will often find the fire first, **Figure 19-31,** because many business occupancies have storage in the rear and this is where fires most often begin.

FIGURE 19-31 Many times the second-in unit will find the greater hazards when they respond to the rear of the structure.

The first-due truck also has a number of options based on SOPs or need. Some of the uses for the first-in truck can be any of the following:

- Forcible entry
- Ventilation of structure
- Search and rescue (primarily multistory ladder work)
- Ground ladder operations
- Securing of utilities
- Salvage
- Aerial streams

This is only a partial list and if one truck cannot accomplish all of these tasks in a very short time by itself, then a second-in truck company may be called for assistance.

In these types of fires, firefighter injuries and fatalities can occur from falls and building collapse. The falls can occur from roofs either into the building or off the roof on the side of the building. Chapters 13 (in Sections I and II) and Section I, Chapter 18 describe safety on roofs and collapse potential.

The business or mercantile fire can be a potentially dangerous operation based on the hazards associated with the products manufactured and stored, and the structural integrity during a fire. This type of firefighting must be taken very seriously and studied well in order for the firefighter to be successful and safe.

Multistory Occupancies

The preceding discussions about fighting fires in structures apply in this section as well. Every tactic and factor considered can be utilized in attacking fires in multistory or high-rise structures. In the case of a multistory fire, it must also be expected that the resources needed for this task will be much greater than for any single-story structure. **Figure 19-32** shows a staging area for a multistory fire beginning to fill with responding units. The support structure of the firefighting system will require multiple layers because the fatigue factors alone can overwhelm smaller systems. The National Fire Academy suggests that, depending on the size of the fire, up to one hundred firefighters may be needed for full extinguishment. In downtown Los Angeles, high-rise responses have been known to contain over seventy-five engines, not counting truck companies and other units.

The number of units responding, as well as the workload, will be predicated on the height of the structure, the location of the fire, and local protocol. If the structure is four stories and the fire is on the third floor, not as many units will be required as for a fire on the fifty-sixth floor of a seventy-five-floor building.

This section is intended to give a basic overview of the high-rise fire operation, because an in-depth study is beyond the scope of this book. Common courses in multistory or high-rise firefighting often run up to forty hours in length. Multiple courses are available and a number of books have been written on the subject.

In the typical high-rise response, a greater number of units will respond due to the nature of the work involved in these fires. Common practice is for the first-in unit to go to the **control room.** This room contains items that will assist in the determination of the

FIGURE 19-32 A staging area for a high-rise fire begins to fill with resources. *(Courtesy of Getty Images/Jamie Rector)*

location and severity of the fire. Some of the items found in a typical control room are:

- Alarm panels giving alarm details as to location, and so on
- Building utility controls
- Keys to all areas of the building including the elevators
- Communications to all parts of the building
- Sprinkler system activation details and controls
- Controls to auxiliary water pumping systems to supplement the building systems
- Maps and diagrams of the building layout and systems

Communications will be extremely important in this operation because the next unit to arrive typically will go to the fire's reported location. As this unit climbs the stairs or accesses the elevators they must be in communication with the first-in unit getting feedback as to that unit's findings regarding location and fire severity.

STREETSMART TIP

In many jurisdictions, a general rule in high-rise fire-fighting is not to use the elevator. There is too great a chance that the crew would be taken to the fire floor, and as the door opened, they would be exposed to fire and conditions that could be dangerous to their survival, **Figure 19-33.** Because of the extreme height of many of today's high-rise structures, some jurisdictions may take elevators under certain circumstances. Two of those are if the elevator has firefighter lock-out controls built for fire department access and control or if they are part of a split bank of elevators that does not access the reported fire floor.

VIEWPOINT

There are a number of views on the use of stairways or elevators in high-rise firefighting. One view is that elevators should not be used unless they are a split bank and the bank used does not go near the reported fire floor. Another view holds that stairs must be used no matter how high the fire. After the fire is located, controlled elevators may then be used to a level at least two floors below the fire. Still another view is that elevators should always be used to save the fire attack team energy to fight the fire. The bottom line is that every firefighter must know and follow their department SOPs on the use of elevators in a firefighting incident. In every case, the elevator will not be taken to the fire floor. All fire attack will be from the stairwell, keeping the unburned safety of the stairwell at the team's back.

FIGURE 19-33 Firefighters must only use elevators that are secured so that accidental fire exposure will never occur.

If an elevator is utilized, the firefighters will attempt to stop and get off the elevator two or more floors below the reported fire floor and walk the stairs the rest of the way. This way they can remain in the **stairwell** safely before deploying onto the fire floor. This crew will typically take a number of items with them for setting up an attack on the fire. The typical list that an engine or truck company may use to select what to take with them is shown in **Table 19-4.**

Upon reaching the fire floor the attack team will set up an operation using the standpipe for water supply to the hose they brought with them. They will go to the floor below the fire and connect the hose to the standpipe, **Figure 19-34,** bringing it up to the fire floor in a layout that will play out as they enter the fire floor to attack the fire. Depending on the size of the fire, they may either attack or wait for a second unit for support. As in residential fire attack, if occupants are on the fire floor, then the first line must go between them and the fire in an effort to create an escape path or at least to contain the fire until sufficient personnel are on scene to focus on the rescue of the people.

The first-in truck will usually go into a ventilation mode because ventilation will be extremely important in this situation. The products of combustion will be concentrated on the fire floor and possibly others, making rescue and firefighting very difficult at best. Their first task will be to capture the stairwells before the fire does, giving the fire teams access to the fire as well as exits for the evacuees, **19-35.** The ventilation of a high-rise building can be extremely complex. For more on this issue refer to Section I, Chapter 18.

TABLE 19-4   Suggested High-Rise Equipment to Be Carried Aloft

Engine Company

One spare SCBA cylinder per member

One large hand lantern or light per company

One standpipe hose bundle to include gated wye and nozzle (see Chapter 12)

Two spanner wrenches and a small pipe wrench or adjustable pliers

One ax

One Halligan tool

One pike pole

Portable radio for each team member

Thermal Imaging Camera (TIC)

Truck Company (four-person unit)

One spare SCBA cylinder per member

Two large hand lanterns or lights

Two axes (one flathead, or one flathead ax and one sledgehammer/maul)

One Halligan tool

One pike pole

Circular saw with fuel can (or for interior operations, a hydraulic forcible entry tool)

One 100-foot drop line (3⅛-inch rope if possible)

Portable radio for each team member

Thermal Imaging Camera (TIC)

Large-area search kit

FIGURE 19-34   The high-rise fire is usually attacked from the stairwells. Laying hose up from the floor below the fire floor will give the firefighting team a safer and more efficient operation.

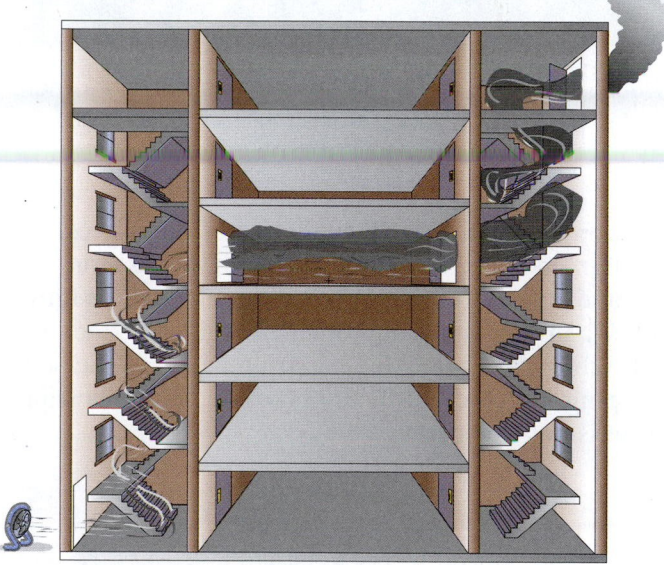

FIGURE 19-35   Truck company begins ventilation process.

As other units arrive, they will begin filling in a great number of important operational positions. Some of these assignments could include search, attack team assistance, elevator control, equipment staging two floors below the fire floor (SCBAs, hose, lighting, etc.), ventilation team assistance, water supply to sprinkler or standpipe connections, salvage below the fire floor, and many more.

The multistory or high-rise fire can be one of the most taxing to personnel in the fire service. It will require more personnel, planning, practice, and other resources than almost any other type of operation. The only fire that will exceed it in resource needs will be the wildland conflagration,. The firefighter assigned to a unit with the possibility of responding to a high-rise fire must be prepared for this assignment, because it is dangerous and extremely taxing

physically. **Figure 19-36** demonstrates that a multi-story fire requires more resources than a smaller single-story structure.

Below-Ground Structures or Basements

Fires below a structure, such as in basements, can be some of the hardest fires to fight due to the harsh conditions the firefighter may have to face. Fire and its combustion products travel upward with heat by con-

FIGURE 19-37 Firefighters must travel down through superheated gases and toxic products of combustion in order to fight a basement fire.

vection and radiation. To fight one of these fires, the firefighters must travel down through this superheated air and smoke in order to reach the fire itself, **Figure 19-37.** In just about any other firefighting attack the firefighter can access the fire from below or the side, keeping low and out of the products of combustion. That is not true here.

NOTE

The key to fighting a basement fire is to ventilate as soon as possible, releasing the heat and smoke, making access and attack easier for the firefighting team.

The basement can also be very difficult to ventilate. Section I, Chapter 18 tells of a number of ventilation techniques. That information will prove invaluable in a below-ground fire such as a basement.

The role of the first-in engine company in these fires is to get water onto the fire as quickly as possible. This is primarily because the fire itself will be traveling upward and outward creating a very poor situation for anything or anybody above the fire in the structure. The second-in unit must quickly assist with

FIGURE 19-38 If access to the below-ground fire is too dangerous or not practical, then other means must be used to extinguish the fire through the ceiling of the burning room.

the ventilation as well as assist with the line already pulled. A backup line must be pulled as quickly as is practical for the safety of the team that has entered the basement. The attack team must be backed up as soon as possible and this is a situation where a single firefighter will never enter the area alone. The team concept must be strictly followed.

In cases where access to the basement cannot be made, it may be possible to punch through the floor above the room on fire and flood the room with water or high expansion foam, **Figure 19-38.** Whatever the attack method, it must be done quickly because everything above the fire is exposed.

Structures Equipped with Sprinklers or Standpipes

Fighting fire in occupancies equipped with sprinkler systems or standpipes has been a boon to the fire service over the last decade or two. Before the mandates of these systems, firefighters were not nearly as successful as they are today fighting fires in these types of occupancies.

The sprinklered building creates a unique situation for the firefighter.

FIREFIGHTER FACT

In over 94 percent of the fires recorded in the recent past, in sprinklered buildings, the sprinklers have either extinguished or confined the fire prior to the arrival of the fire department.

The team entering the structure has but to find the seat of the fire and complete extinguishment. This will still not be easy because the smoke and heat will be held down over the combustibles by the cooling action of the water spray from the system. The firefighting team will have to enter the structure very carefully without shutting down the protection afforded by the operating sprinklers, and they will have to find the seat of the fire through very limited visibility to complete the extinguishment of the fire. After this is done, the system can be shut down for a short time during overhaul. Chapter 12 (in Sections I and II) details the operation of sprinkler systems.

The standpipe system is somewhat different. This pipe system will carry water to the firefighting team, easing their dependence on hoselines over great distances. This system is invaluable in fighting fires in high-rise structures. It saves a great amount of time by having the water ready for the team as it arrives at the floor on fire. Team members simply hook the hose carried aloft to the system and then charge it by turning on a valve at the standpipe. Section I, Chapter 10 describes this operation in some detail.

WORKING SMART AND SAFE ON THE JOB

A common thread throughout discussions on structure firefighting is the task of securing utilities. By securing the utilities on scene, you can ensure the safety of all the firefighters working the incident.

Securing Utilities

One of the first steps in going to work at a structure fire is to control the various utilities that are active within the building. For most structures, this will include electrical service, gas service, and water.

Controlling the utilities is important in fire operations because it helps to protect the firefighters working both inside and outside the structure. Fire can damage routes and controls of utilities and cause gas or electricity to escape, further compounding the fire problem and creating a more dangerous environment. Firefighters performing operations inside the structure, such as fire attack or overhaul, could come into contact with electrical lines or gas lines. It is essential that these utilities be shut down to avoid unnecessary firefighter injury. Uncontrolled utilities can also contribute to the spread of fire to uninvolved areas of the structure, causing more fire damage to the building.

Each jurisdiction will have different SOPs regarding the control of building utilities, so firefighters should be familiar with their department's guidelines for this activity. Additionally, the control mechanisms that allow firefighters to shut off the various utilities are different depending on the service provider. These providers typically offer utility-control training to fire departments at little or no cost.

In general, securing building utilities includes the following steps:

1. Wearing full PPE (personal protective equipment), gather any necessary tools for the types of utilities at the structure.
2. Control the electric service at the meter if possible.
3. Locate the main breaker box; note which circuits have already tripped; then shut off power to individual breakers first. After the individual breakers are off, shut off the main breaker. (Noting tripped circuits before shut down procedures assists the fire investigators in their work after the fire.)
4. Check for auxiliary power supply sources. Some buildings have multiple power feeds or circuits wired to an uninterrupted power supply (batteries and/or generator) that are separate from the main breaker. If found, communicate to the Incident Commander (IC) their presence and shut off power only if directed.
5. Locate the natural gas meter and shut off the service. (Most are quarter turn valves.) In the case of propane-fed buildings, locate the propane storage tank and shut off the gas by opening the valve cover and turning the valve to the fully closed position. Carefully close the cover once the valve is shut off.
6. Locate the main domestic water valve and, using the appropriate tool, rotate valve until closed.

In the case of each utility that is shut off or controlled, a lock-out/tag-out system should be used. This means

(Continues)

the utility should be locked into the "off" position and tagged by the person that had controlled the utility. This is done to prevent anyone from turning the service back on without authorization from the IC or his or her designee. Firefighters should consult with their department to learn proper lock-out/tag-out procedures.

Emergency Scene Lighting

As was mentioned many times, firefighter safety is of paramount concern. Securing electrical power on scene does two things. It creates a safer firefighting environment by removing the chance of accidental electrocution, but at the same time darkens that environment by shutting off all power to lights. As the fire is darkened, some form of alternative lighting must be brought forth so the fire team can see their work site while they complete their mission safely, even all the way through overhaul.

There are a number of forms of emergency lighting used by the fire service, including:

- Truck- or engine-mounted generators with electric cord reels and portable lights.
- Portable generators with electric cord reels and portable lights.
- Apparatus-mounted lighting on telescoping poles or high on the sides of the units. Specialty apparatus used for scene lighting. These units are usually dual purpose, for example, an apparatus used for both an air-refilling station and scene lighting, **Figures 19-38A** and **B.**

While these and other forms of emergency scene lighting are used in structural firefighting, they are used in all other forms of emergency work, as well. These include:

- Automobile accidents
- Trench rescue
- Confined space rescue
- Hazardous materials scenes
- Mass casualty scenes
- EMS (emergency medical services) scenes

- Command Post operations, as well as other on-scene needs
- Any night work requiring detail and firefighter safety

It is the responsibility of every firefighter to know the equipment he or she uses. Lighting equipment is no exception. A firefighter must understand the safety aspects of the equipment, the limitations and exceptions of the equipment, and how to deploy the best lighting possible, with the equipment available.

The following steps should be followed when operating portable lighting equipment:

1. Properly position lighting by extending and properly adjusting height.
2. Position lighting above ground and away from water.
3. Position light to adequately light work zone.
4. Properly position light in area away from traffic.
5. Position power source (generator) within reach of lighting equipment.
6. Ensure fuel level prior to starting portable unit.
7. Properly position junction box and power cords out of traffic areas.
8. Ensure connections from generator to lighting equipment.
9. Start the generator and supply power.
10. Ensure operation of lighting equipment.

As with all fire department electrical equipment, there are safeguards in place for fireground operations built into the devices used. GFI (ground fault interrupters) are one such device built into the electrical cords and connectors used by the fire service. These devices are simply "automatic shutoffs" built into the cords and connectors that will shut down all power to the applicances used in the event of any short or grounding. The firefighter then must assume that there is a problem with the devices and discontinue their use. When safe to use again, each device usually will be equipped with a "switch" for resetting the GFI and operations can continue.

(A)

(B)

FIGURE 19-38 (A) Specialty apparatus for scene lighting and SCBA filling. (B) Scene lighting and SCBA refilling taking place on the fire scene.

Exposure Fires

An **exposure fire** can be described as any combustible item being threatened by something burning in another area. The fire may be carried to the exposure by three ways. Section I, Chapter 4 described these as *conduction, convection,* and *radiation.*

As stated earlier in this chapter, the second priority after rescue, in most fire scenarios, is exposure protection. This form of property protection is one of the primary focus points in most fires. A number of factors are involved in most exposure fires. In no order of preference they are:

- *Wind.* Convection carrying embers can be a problem.
- *Distance.* The nearer the exposure, the greater chance of fire.
- *Material.* The makeup of the exposed surface determines its combustibility.
- *Intensity of fire.* The **fire intensity** of the primary fire factors into its spread.

FIGURE 19-39 Firefighters apply foam to a tank truck exposed to a nearby fire. *(Courtesy of Central Net Fire)*

A couple of basic methods are used to defend an exposure from fire, and they are based on the methods of heat transfer. In conditions of conduction, a cooling medium must be applied to the material being heated by the conduction. An example would be a water stream applied to a tank exposed to a fire burning under or on it. **Figure 19-39** shows foam being applied to a tanker that is exposed to a fire next to it. In cases of convection a downwind patrol must be provided so as to ensure that embers do not ignite other materials. An example would be a downwind house with a wood shake roof, near a fire with flying brands being put into the air. Next is radiation as a medium for ignition of an exposure. The best method would be to apply water to the exposed area to cool it, **Figure 19-40.** An example would be, as a structure is free burning, the first hoseline is placed between the burning structure and the next-door unit. The water is applied to the exposed structure until another line can be trained onto the burning structure itself.

Exposure fires are common and dangerous, especially when the firefighting teams do not see the second fire and are put in danger by it.

SAFETY

It is important to put lookouts in service in order to maintain safety whenever there is any possibility of an exposure fire.

Flammable Liquids and Gases Fires

Flammable liquids and gases have a number of things in common that the firefighter can use to control a fire in these commodities. The mechanisms of extinguishment are well known to the fire service. These are some of them:

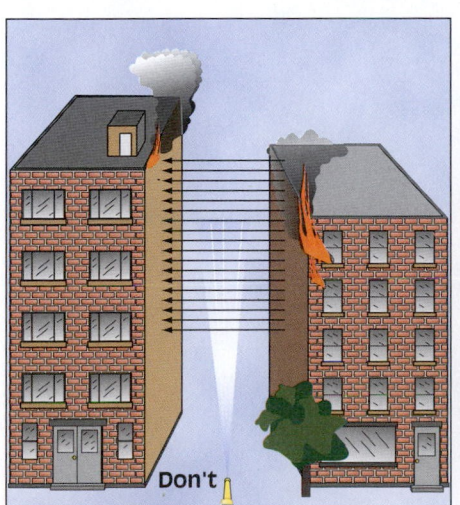
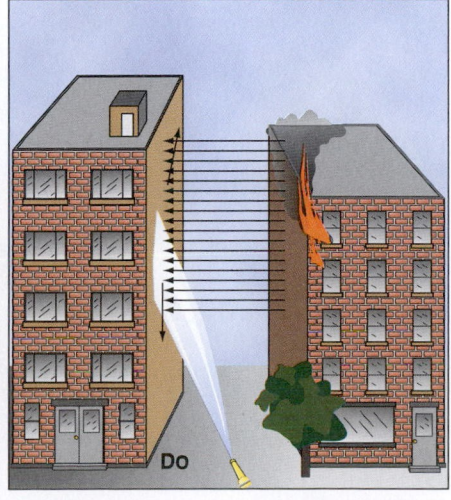

FIGURE 19-40 Radiation will travel through water or any opaque material. In order for water to be effective, it must be applied to the exposed surface in order to cool it.

- Smother the fuel, cutting off the supply of oxygen.
- Starve the fuel by removing unburned material from the area.
- Interrupt the chain reaction process with chemical agents.
- Cool the fuel, reducing vapor pressure.

A great range of possible situations can arise involving combustible and flammable liquids and gases and their burning. As the first-in unit arrives, a number of facts must be obtained in order to handle the situation effectively. Finding answers to the questions asked may be difficult or sometimes impossible. In many situations, hazardous materials teams may be needed in order to identify the materials and the recommended method of handling them. Hazardous materials are discussed in more detail in Chapter 24.

As opposed to most of the other fire scenarios discussed thus far, flammable liquids and gases do not follow the laws of extinguishment that are basic to the fire service. Water may not be the choice best suited for firefighting here. Knowing this, it is possible to set up some very generic principles for fighting most flammable liquid and gas fires as follows:

- Identify the material involved and its hazards.
- Utilize written assistance with the DOT Emergency Response Guidebook and other sources.
- Evaluate the threat to nearby similar commodities such as gas cylinders and tanks. For instance, get water on exposed tanks to reduce internal pressure buildup and prevent the possibility of a BLEVE (see Section II, Chapter 4), **Figure 19-41.**
- Determine which extinguishment agent and principle is best.

- For small fires consider CO_2 or dry chemicals.
- For larger fires consider foam suited for that particular commodity.
- For some fires consider letting the fire burn. This is particularly true in cases where the burning product from the leak is contained. In the event of premature fire extinguishment, the remaining product release may cause a greater hazard than that caused by the fire. While burning, the fire consumes the product. If the fire is extinguished, the remaining product may create a vapor cloud or it may spill onto the ground. The unburned product or its vapors may travel long distances, find an ignition source, and reignite or create an explosion hazard. Two methods firefighters can use to overcome this situation are to apply hose streams to disperse a vapor cloud or apply a foam blanket to suppress the vapors from the product on the ground. The method used to control vapors or the spill depends on the characteristics of the product being released. Diking and damming must also be considered for product run-off.
- Look for shutoff valves or switches to shut down the flow of material to the fire.

Incidents involving combustible and flammable liquids or gases are very dangerous and have caused the fire service to bring in the concept of hazardous materials teams to deal with these special situations. The number of materials considered hazardous in today's society pose a very great threat to the fire service and its ability to fight fire. Knowledge and safety are the keys in working with these materials.

FIGURE 19-41 Water is applied to the heated metal surface in order to keep it cool, slowing the pressure buildup inside the exposed tank car. This will reduce the possibility of a BLEVE.

FIGURE 19-42 Cylinders present additional risks to responders because not only can the contents be hazardous, but if the cylinder is involved in a fire it may explode.

Containers

Much like the containers discussed in Section I, Chapter 19, cylinders present additional risks to responders because not only can the contents be hazardous, but if the cylinder is involved in a fire it may explode. It is important for firefighters to understand the characteristics of cylinders and the dangers that they present during an emergency situation.

Cylinders

Cylinders, like those shown in **Figure 19-42,** come in 1-pound (0.45 kg) sizes up to several thousand pounds and carry a variety of chemical products. The product and its chemical and physical properties will determine in what type of container the product is stored.

CAUTION

Other than the hazard of the chemical itself, the big hazard of all cylinders is that they are pressurized.

The pressures range from a low of 200 psi to a high of 5,000 psi (14–345 bar). One of the most common cylinders firefighters run across is the propane tank, which can range in size from 1 pound to millions of gallons. In residential homes, firefighters will find everything from the 20-gallon (75.7 L) cylinder for barbecues to the 100- to 250-pound (45–113 kg) cylinders used as a fuel source for the home. In some areas it is not uncommon to find 1,000-pound (454 kg) cylinders, often buried underground.

Identifying the contents of a cylinder is critical in emergency incidents. Firefighters must recognize the potential hazards of transporting, storing, and using compressed gases and cylinders involved in fire. Some clues to assist the firefighter in identifying cylinders and their contents include location and occupancy; placards, labels and markings; and cylinder shapes:

- *Location and occupancy.* Compressed gases are used in many situations including homes, garages, industry, retail businesses, and illegal processes. It is not unusual to find cryogenic cylinders in convenience stores or restaurants. They typically store carbon dioxide (CO_2) for the soda-dispensing machines. In rural communities, farms present unique risks due to the storage of pesticides and fertilizers. In general, the more industrialized the community, the more compressed gases and other hazardous materials the community will contain. A challenge to firefighters is identifying a particular cylinder in an occupancy that uses multiple chemicals. Another significant hazard is the use of cylinders for purposes other than their intended use. In methamphetamine labs it is common to find propane tanks being used to store anhydrous ammonia. The corrosive nature of ammonia reacts with the cylinder valves making the cylinder extremely hazardous.

- *Placards, labels, and markings.* Placards, which are required by the Department of Transportation (DOT) for vehicles transporting hazardous materials, are used to warn responders of the hazards of the materials being carried on the vehicle. Responders must be cautious as vehicles may be missing placards or improperly placarded. Individual cylinders should be marked with labels that provide positive identification of the contents. Other than their size, labels

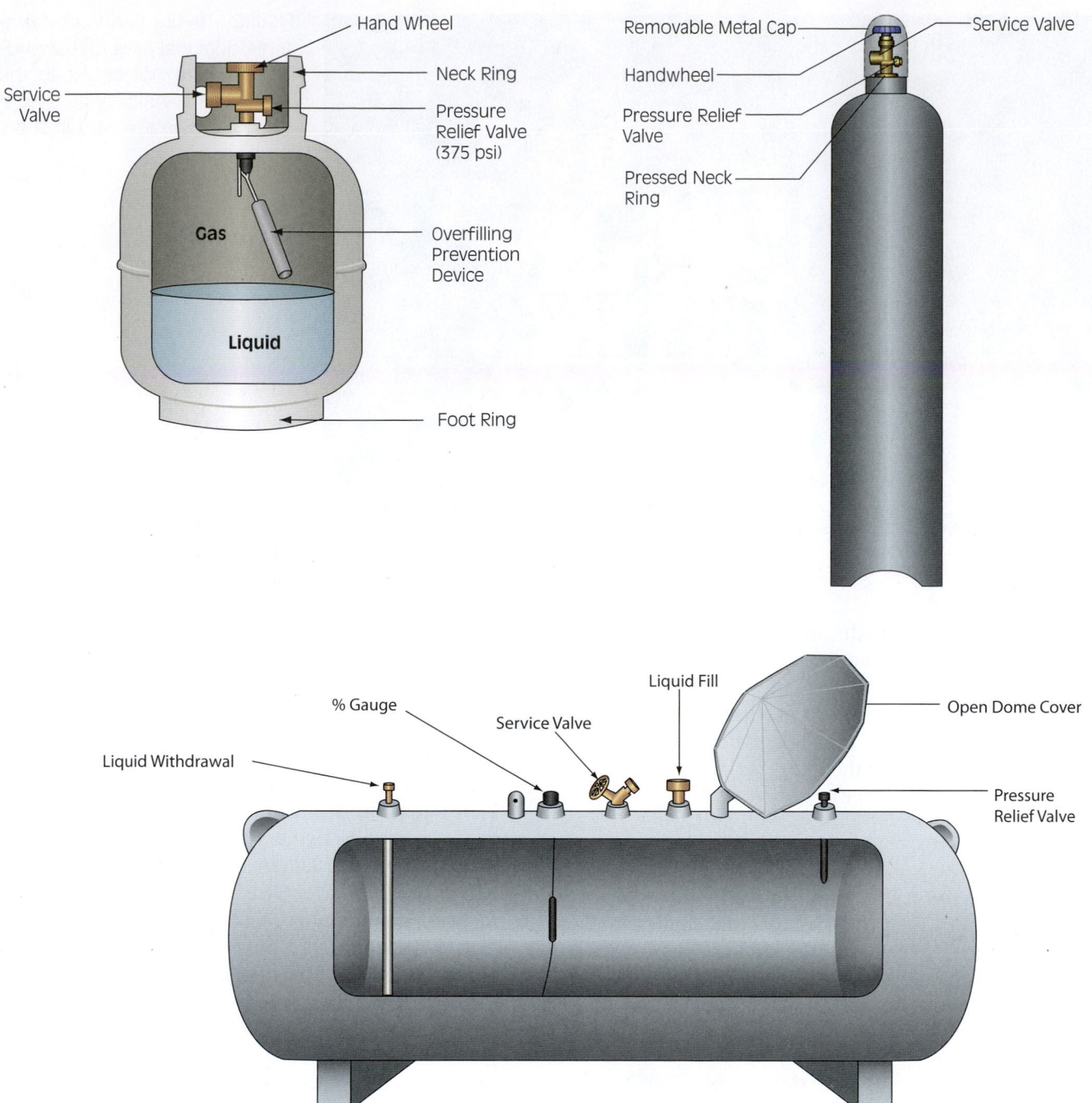

FIGURE 19-43 Compressed gas cylinders come in many shapes and sizes. Firefighters must be aware of the components of each cylinder type prior to the emergency incident. These three cylinders are commonplace in society yet have many of the same components.

are identical to placards. The color coding of a cylinder; such as green for oxygen, may be unreliable as color coding varies from manufacturer to manufacturer. Color coding of caps also has no real value in identification because caps may be interchangeable.

■ *Cylinder shapes*. Low-pressure cylinders are usually short and broad. High pressure cylinders are usually

tall and thin. Specialized cylinders that hold cryogenic gases (extremely cold) appear to be high pressure, but in reality are low pressure. The bulkiness of the cylinder is a result of the large amount of insulation required to keep the material cold. When in transport, compressed gases are usually shipped in cylinder shaped containers with rounded ends.

Cylinder Components and Valves

Cylinders usually have relief valves or frangible disks in the event they are over pressurized or are involved in a fire, **Figure 19-43.** These safety devices are designed to relieve excess pressure. A powerful spring holds the relief valve in the closed position. The valve will remain in the closed position as long as the pressure inside the tank is less than the pressure on the spring. A significant rise in tank pressure caused by heating will result in the activation of the valve. A loud hissing or whistling sound accompanies the activation of the relief valve. A loud pop followed by a release of product will occur when the relief valve is fully opened. As pressure falls within the cylinder, the valve will close. If heating continues, pressure will rise and the relief valve will then reactivate.

Upon over-pressurization, the disk in a cylinder equipped with frangible disks is designed to fail at a predetermined pressure. It will result in the release of all product as it has no reset capabilities like those of a relief valve.

The United States is one of the few countries that mandates relief valves. Most cylinders used in other countries do not have this feature. Incidents aboard ships may involve these types of cylinders as the cylinders are being trans-shipped and are destined for delivery in another country. Most communities, regardless of their size, have cylinders of chlorine and sulfur dioxide used in water treatment. These cylinders come in 100- to 150-pound (45–68 kg) and 1-ton (907 kg) cylinders, which could create a major incident if they were ruptured or suffered a release.

Boiling Liquid Expanding Vapor Explosion (BLEVE)

When tanks, trucks, tank cars, or other containers are involved in a fire situation there are a number of hazards. One very deadly hazard is known as a boiling liquid expanding vapor explosion (BLEVE), **Figure 19-44.** A large number of firefighters have been killed by propane tank BLEVEs. When a BLEVE occurs, it usually results in more than one firefighter being killed at a single incident. The type of container and the product within the container will dictate how severe a BLEVE may be. The basis of a BLEVE is the fact that the pressure inside the container increases and exceeds the maximum pressure the container was designed to handle. The contents are violently released, and if the material is flammable, an explosion or large fireball occurs. In the recent past there have been several incidents involving BLEVEs that resulted in emergency responder deaths and injuries, thus emphasizing the need to recognize and prevent this event before it occurs.

Another phenomenon that transpires with containers is known as violent tank rupture (VTR), which occurs with nonflammable materials. The concept is the same as a BLEVE, but there is not a characteristic fireball and explosion. With both a BLEVE and a VTR there is some form of heat increase inside the container, typically from a nearby fire. The fire heats the container, which heats the contents. The contents will boil, which creates expanding vapors, which in turn increase the pressure inside the container. In some containers the relief valve will activate, relieving the pressure. In some cases the pressure inside the tank is greater than the relief valve can handle and the pressure continues to increase.

One of two possibilities can occur with a BLEVE or a VTR. One possibility is that the relief valve will not be able to handle the increase in pressure and the tank will fail. The other possibility is that the fire or heat source that is creating the problem will weaken the container shell, and the resulting increasing pressure will vent at this weakened portion of the container. If the heat source is a fire and the container product is flammable, when the relief valve activates the raw product coming out of the container typically ignites. This may increase the temperature of the tank, increasing the pressure. It is never advisable to extinguish the fire coming from a relief valve, as that is a safety mechanism. It is possible to cool the top of the tank near the relief valve with an unstaffed hose stream.

The difference between a BLEVE and a VTR occurs when the container fails. A BLEVE occurs with flammable liquids, such as propane. When the container fails, the vapors of the flammable liquid ignite, creating the explosion. How the container ruptures and the amount of product released will determine how severe the explosion will be. With a VTR the container will fail. Since the product is nonflammable, the product will not ignite. The only event is a rupture of the container, spilling its contents. The container can still rocket, and the resulting release of pressure can be violent. A VTR can occur with a container of water or other "nonhazardous" material. As an example, a 55-gallon (208 liter) steel drum of water, when heated, can travel several hundred feet depending on where the release point is on the container. In most cases the bungs (screw-top caps) will release, flying a considerable distance, and the container will remain mostly intact.

The failure of the container when impinged from a fire usually occurs as the fire is impacting the tank in the vapor space. When heat is applied to a section of the tank that has no internal mechanism to provide cooling, failure of the steel can occur. When fire impinges on the liquid portion of the tank, the liquid

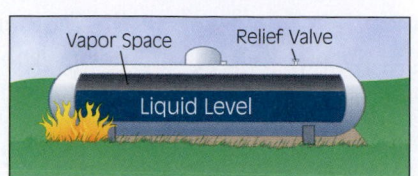
Vapor Space   Relief Valve
Liquid Level

Fire impinging on a propane tank.

As heat increases inside of tank, the pressure also increases. The liquid will begin to boil.

Lower Level

As the pressure increases, the relief valve will open, releasing propane. The propane, being heavier than air, will sink.

The vapor will reach the fire and ignite.

The relief valve will ignite, also causing heat to be on the tank by that flame. The pressure will increase in the tank.

As the pressure in the tank is increasing, the tank may discolor and the pitch of the relief valve will get higher. Eventually the tank will rupture. This is known as a BLEVE.

FIGURE 19-44   Diagram of a BLEVE.

will distribute the heat spread internally and the steel tank typically will not weaken. The problem is that responding firefighters do not usually know the liquid level, and one cannot easily predict when a tank may fail. Any fire impingement on a tank is a serious problem, and withdrawing from the scene may be the best course of action.

This brings up another interesting note: A casual observation of BLEVEs indicates that most BLEVEs and firefighter deaths occur within the first few minutes of arrival. The clock starts ticking on the BLEVE time bomb from the first minute heat is applied to the tank. The clock does not start with the 9-1-1 call or the arrival of firefighters. The critical time for safety may have already passed before firefighters arrive on the scene. If firefighters arrive at an incident involving a propane tank on fire or being impinged by fire, they are in extreme danger. If the relief valve is not operating, the danger is even more pronounced. Oper-

ating in close proximity to a tank in this situation can be a fatal mistake.

An example: one recent event involved the death of two firefighters and injuries to seven other emergency responders. The location was a farm at which a propane tank was on fire. The relief valve on the tank was operating and the vapors from the relief valve were on fire. About eight minutes after the firefighters had arrived, the tank exploded into four separate parts. The four parts went in four different directions. The two firefighters who died were 105 feet away and were struck by one piece of the tank, dying instantly.

BLEVEs and VTRs can result in injuries and fatalities. Any time containers are under stress, such as during a fire, they can fail. Many times they fail violently and with severe consequences. Some containers have relief valves, while others do not. Materials that are highly poisonous such as chlorine will not have a relief valve; they will have a fusible metal plug that vents the pressure of the tank. This fusible metal plug is not like a relief valve that shuts off when the pressure inside the tank is decreased. Once a fusible metal plug melts, the contents of the tank come out, no matter the pressure. When a tank is on fire or is being impinged by fire, the tank is being weakened and the contents are being heated, creating increased pressure.

Fire Suppression Involving Cylinders

Fire suppression of cylinders is dangerous and sometimes unpredictable with results that can be catastrophic. Firefighters should follow this risk/benefit analysis: *Risk a lot to save a lot, and risk a little to save a little.* Incident command should conduct a thorough risk benefit analysis prior to determining operational methods when fighting a fire for tanks and other containers under pressure.

- How long has the tank been exposed to heat and/or flames?
- Where is the tank leaking—relief valve or piping leak?
- Is the relief valve operating?
- Is the sound from the relief valve increasing?
- Is there an accessible shut off valve?
- What are the advantages of approaching the tank?
- What are the advantages of going defensive with unmanned monitors?
- What is the advantage of evacuating the area?

A tank or container can fail at any time; it is impossible to determine the exact moment a tank is going to fail. Once operational methods have been determined, safety should dictate practices due to the volatility of the cylinder. All personnel should use extreme caution during suppression activities. The dangers associated with a BLEVE are:

- The fireball can engulf responders and exposures.
- Metal parts of the tank can fly considerable distances.
- Liquid propane can be released into the surrounding area and be ignited.
- The shock wave, air blast, or flying metal parts created by a BLEVE can collapse buildings or move responders and equipment.

Defensive operations may be employed due to the unpredictability of the situation. If so, firefighters may employ the following techniques. See Section IV, Chapter 25.

1. Firefighters should withdraw immediately in the case of rising sound from venting relief valves or discoloration of the tank.

2. Fire must be fought from a distance with unstaffed or unmanned hose holders or monitor nozzles.

- The tank should be cooled with flooding quantities long after the fire is out. A minimum of 500 gpm (1,893 liters per minute) at the point of flame impingement is recommended by the NFPA.
- If the water is vaporizing on contact, firefighters are not putting enough water on the tank. Water should be running off the tank if it is being cooled.
- Firefighters should not direct water at relief valves or safety devices, as icing may occur. Icing would block the venting material, which could cause an increase in pressure inside the tank.
- The tank may fail from any direction, and any tank that is being exposed to a fire can fail at any moment.
- For massive fire, it is recommended to use unstaffed or unmanned hose holders or monitor nozzles. If this is impossible, firefighters should withdraw from the area and let the fire burn.

In order to approach a fire of this nature, a team effort must take place to ensure the safety of all firefighters. All personnel should be aware of the hazards and changing conditions when approaching a cylinder.

If the situation dictates that firefighters approach the cylinder, constant communication and awareness should be used to watch for changing conditions or signs of a BLEVE. A minimum of two 1½-inch hoselines flowing a minimum of 125 gallons per minute is required before approaching the relief valve in order to cut off the fuel to the fire. Firefighters should approach the cylinder in the following manner:

- Cool the tank using master streams until the relief valve resets.
- Use two teams of firefighters, one per hoseline. Use wide fog pattern to deflect heat and protect firefighters. A team leader will be positioned in the middle to direct both hoseline teams.
- Position each team approximately five feet apart so that when the fog pattern is open the two overlap and remain in constant contact with the ground, creating a protective wall.
- While flowing water, both teams simultaneously approach the tank from a 45 degree angle or from the side. NEVER approach a cylinder from the ends.
- Approach the shut off valve that has been identified slowly and using the fog pattern for protection.

- The fog pattern should overlap the valve, which will allow a member from one team to turn off the valve while remaining protected.
- Once the fuel is cut off, the fire should extinguish and the firefighters should retreat in the same manner they approached, continuing to apply water to the tank. Care must be taken to avoid reignition or other hazards during retreat.

Incidents involving compressed gas cylinders are very dangerous situations and firefighters must be aware of the dangers and warning signs associated with each of these situations

LESSONS LEARNED

The basic principles of firefighting are based on sound scientific laws as well as years of firefighting experience. Every combustible item and occupancy must be carefully studied and understood in order for firefighters to be effective, efficient, and safe in their work.

Flammable liquids and compressed gases, such as those contained in cylinders, must be treated with extreme caution during an emergency incident due to the extreme life safety hazard that they pose to firefighters. Applying the knowledge of the basic elements of fire suppression combined with safety techniques necessary for each individual emergency incident, will result in successful fireground operations.

KEY TERMS

Combination Attack   A combined attack based on partial use of both offensive and defensive attack modes.

Control Room   A room on the ground floor of a high-rise building where all building systems controls are located.

Defensive Attack   A calculated attack on part of a problem or situation in an effort to hold ground until sufficient resources are available to convert to an offensive form of attack.

Exposure Fire   Any combustible item threatened by something burning nearby that has caught on fire.

Fire Intensity   A measurement of Btus produced by a fire. Sometimes measured in flame length in the wildland environment.

Flammable Range   Ratio of gas to air that will sustain fire if exposed to flame or spark.

Incident Command System (ICS)   A command system utilized on the emergency scene that is designed to keep order and follow a sequence of set guidelines.

Offensive Attack   An aggressive attack on a situation where resources are adequate and capable of handling the situation.

Polar Solvent   A material that will mix with water, diluting itself.

RECEO   Acronym coined by Lloyd Layman standing for Rescue, Exposures, Confinement, Extinguishment, and Overhaul.

Solubility   A liquid's ability to mix with another liquid.

Stairwell   An enclosed stairway attached to the side of a high-rise building or in the center core of same.

Teamwork   A number of persons working together in an effort to reach a common goal.

Toxicity   Poisonous level of a substance.

Vapor Density   Weight of a gas in relation to air. Air is rated 1.

REVIEW QUESTIONS

1. Explain how water suppresses fire in relation to the fire tetrahedron.

2. Give at least two factors that must be taken into consideration prior to an attack plan being formed and extinguishment of a structure fire.

3. Explain why firefighting tactics will differ based on a flammable liquid's specific gravity.

4. Why does a basement fire have the potential to be so much more hazardous to a firefighter than a ground-level fire in the same occupancy?

5. Explain the procedures recommended if a firefighter is considering taking an elevator to the fire floor in a high-rise or above-ground fire. What does your jurisdiction recommend?

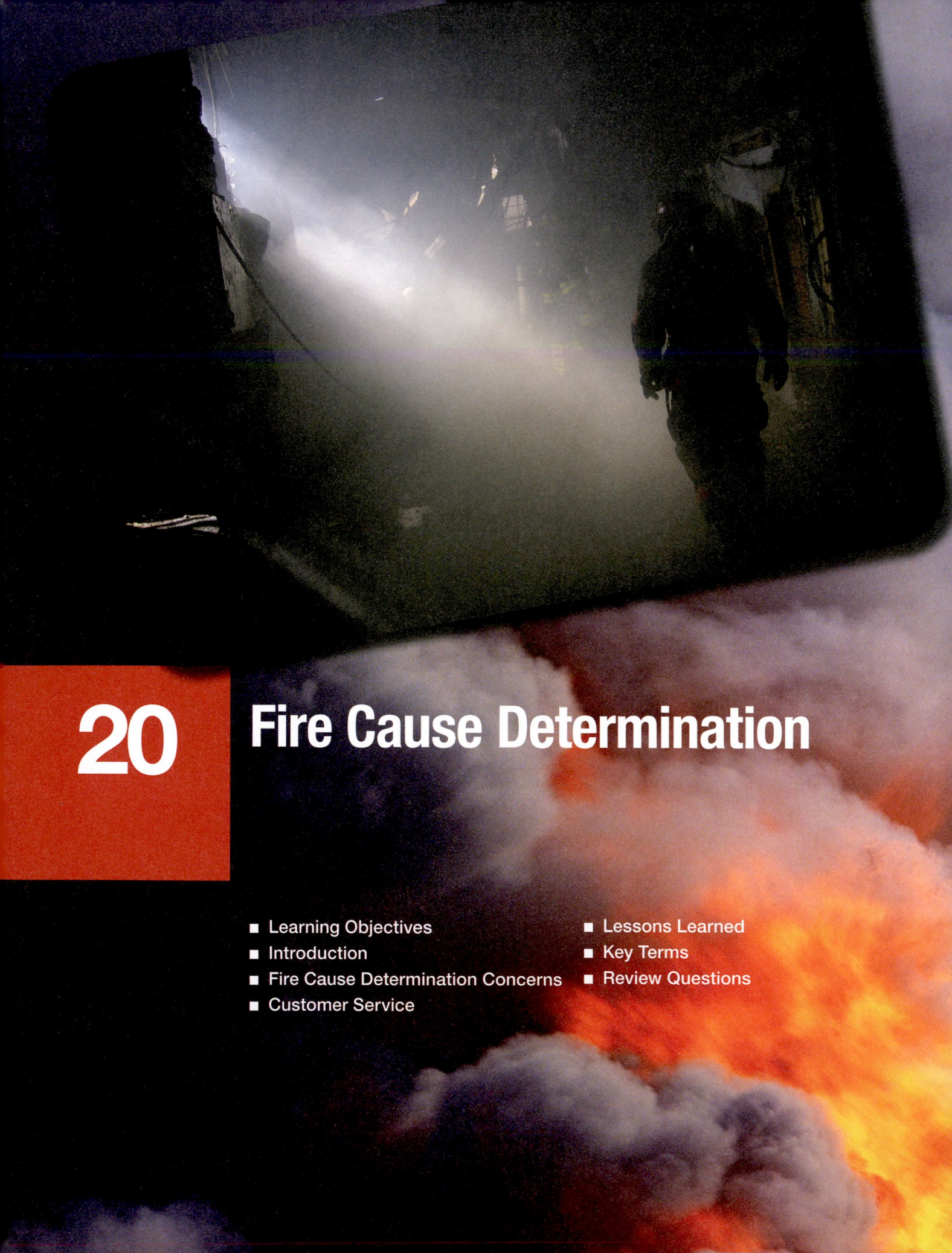

20 Fire Cause Determination

- Learning Objectives
- Introduction
- Fire Cause Determination Concerns
- Customer Service
- Lessons Learned
- Key Terms
- Review Questions

INTRODUCTION

Fire cause determination is something new firefighters will not be actively involved in; however, all firefighters need to understand this task. During fire attack, with proper training, it is possible for firefighters to combat the fire while watching for and protecting potential signs and conditions that will assist the fire investigator. This chapter explains the reasons for and importance of fire cause determination and describes the necessary firefighter skills and abilities needed to effectively perform this operation.

FIRE CAUSE DETERMINATION CONCERNS

As the incident progresses through the aspects of initial dispatch, response, arrival, rescue, containment, extinguishment, ventilation, salvage, and overhaul, firefighters need to be aware of possible clues as to how the incident started. The information that dispatch receives can be vital to the rescue/attack operation. Although the Fire Officer will be gathering the information and formulating a plan of attack, it should be the responsibility of all firefighters to listen to the reported information, if possible, and to build a mental picture of the situation. During dispatch, all available information should be documented. Details such as weather, time of day, address, type of dispatch (i.e., garage fire versus bedroom fire), and the number of phone calls received for the fire can give firefighters—and later, investigators—a clue as to what the fire was doing prior to arrival, **Figure 20-1.**

FIGURE 20-1   Scene size-up begins with the dispatch of the call, and continues with information from dispatch en route to the call.

During response, additional information may be gathered by answering these types of questions: Is the structure evacuated or still occupied? Are callers reporting explosions or odd colors of smoke? Were citizens seen running or driving from the incident at a high rate of speed? Upon arrival on scene, firefighters may encounter witnesses who have pertinent information concerning the fire scene. Witnesses can be great sources of information, but they can also be part of a crime scene. Firefighters should note if any bystanders seem overly interested in the fire or the firefighter activities. It is possible for firefighters to notice observers that are present at more than one fire scene. Any pertinent information should always be relayed to the IC and ultimately to the Fire Investigator.

For exterior fires and fires involving high-pile class A combustibles—stacked lumber, pallets, rolls of paper, and so on—firefighters must ensure that any suspicious evidence is protected. It is basic instinct to scatter the piled contents and completely douse them with foam and water to ensure the fire is extinguished—and this quite possibly might be the extinguishment scenario. It is a firefighter's main job to ensure the fire is completely extinguished, and this may cause extensive overhaul practices to be employed. However, whenever possible, any evidence of the point of origin (V patterns, deep charring, etc.), evidence of suspicious liquids or accelerants, matches, wadded paper, or any suspicious situations should be protected or noted by the firefighters. Observant firefighters on scene can provide fire investigators with reliable information as to where the fire was, what spectators said about the fire, and how the fire reacted during fire attack. Inside the structure, firefighters should observe exactly where the fire was located in the building, how the fire reacted to suppression efforts, and the conditions under which people were found in the building. The TIC can be used to find hidden fires and areas of intense heat that could reignite. Firefighters should also pay attention to how the contents of the building were arranged, whether there were signs of break-in, and the position of interior doors. It is also important to note the building's electrical system, including the condition and position of circuit breakers and appliances. The inspectors should be briefed about this information, **Figure 20-2,** and it should be noted in the company officer's report. The smallest of items that seem out of place need to be reported because these may be the missing pieces of the puzzle for the investigator.

During overhaul, firefighters must ensure that the investigator preapproves the removal of any part of the structure or any piece of furnishing. Firefighters must not throw things out of the building for expediency. Each step of overhaul has to be taken with knowledge and care so as not to make the investigation more difficult. It may be appropriate to consider conducting overhaul from the unburned side of the wall. This action will preserve burn patterns critical to a fire investigation. As investigators work through a structure, they will be able to release the different rooms for more thorough overhaul. Smoke markings on the plasterboard may be important clues for the investigator. Experienced investigators are only as good as the clues they observe. If they arrive on scene and all they have is a bare wood floor and wooden studs for walls, they are at a disadvantage.

FIGURE 20-2 Fire investigators should be thoroughly briefed about fire and building conditions existing during fire attack. *(Courtesy of Fuller Road Fire Department, Colonie, New York)*

Preservation of Evidence

Firefighters must be aware of the consequences of disturbing or removing physical evidence and also that any potential evidence must be tracked through a chain of custody. The chain of custody means that evidence is secured from the time that it is found through the investigation process. Although firefighters will not be responsible for the final disposition of the evidence, they must know what they need to do when potential evidence is discovered. The types of evidence a firefighter may discover may include burn patterns, faulty wiring, incendiary devices, fuel containers, and accelerants.

In the process of their work, firefighters may come upon something out of the ordinary and will need to make every effort to preserve it until the fire investigator can observe it—even if this means using hose stream control to avoid disturbing any evidence from its original location or stopping the entire fire suppression operation. When potential evidence has been discovered, firefighters need to leave items where they found them and protect and preserve the area as found until investigators can examine it. In this case, the area should be cordoned off with fire line tape to secure the affected area. Fire ground activities that may damage items in the protected area should be avoided. All possible steps should be taken to ensure that firefighters do not become involved in a legal matter because they made simple, preventable mistakes. If a room is overhauled and cleaned of all contents prior to the arrival of investigators and it is later found that an incendiary device was located in that room, it could result in a legal matter for the firefighter and the department.

A term used in the fire investigation community is *spoliation*. NFPA defines spoliation as the loss, destruction, or material alteration of an object or document that is evidence or potential evidence in a legal proceeding by one who has the responsibility for its preservation. The most important term in this definition is **potential evidence.** Almost any item within and around the structure could potentially become evidence in the determination of the area of origin, fire cause, or identification of who or what may be responsible for the event.

Every effort must be made to minimize the alteration of the scene before the investigator has an opportunity to examine the scene. Depending on the size and scope of the investigation, the investigator can be suppression personnel collecting data to complete their fire report of a more complex scene where a full time investigator has been called to the scene by the Officer in Charge.

Spoliation can have a dramatic impact on everyone associated with the fire or explosion scene. Destruction of evidence can affect the ability of both the public and private investigator to come up with an accurate determination of the area or origin and cause. The home or business owner is impacted by not truly knowing what happened and preventing them from seeking remedy should this fire have been caused by a third party like an appliance manufacturer. Like the property owners, the insurance companies representing everyone involved are impacted financially from being able to recover funds they paid out for the claim, when in fact they could have taken civil actions against the responsible party for the incident. The spoliation of evidence could also lead to key evidence not being able to be submitted allowing the arsonist to go free to commit their crime again to another innocent victim.

When firefighters arrive on the scene and interact with individuals such as occupants, owners, or witnesses, it is important to remember what they see and hear. Some statements may be crucial to the investigation. If it is unusual enough, the firefighter should attempt to document any comment heard so it can more easily be recalled at a later time. At all times the firefighter must maintain a professional demeanor and treat all individuals on the emergency scene with respect. Any statement regarding the incident should be avoided by the firefighter and directed through the incident commander.

Basics of Point of Origin Determination

The determination of the point or area of origin can be a very scientific pursuit in which exotic testing is done and the exact point is determined—or it can be as simple as looking at a room and knowing where the fire started based on the indicators present. Fire investigators employ the scientific approach to every fire they investigate. The basic clues a firefighter should watch for include notation of where the fire was first noticed or reported; where the heaviest damage was; and any smoke, heat, or other types of markings on the walls, ceilings, or floor indicating where the fire started. The basic premise behind fire behavior is that it travels the path of least resistance. As fires grow, smoke and heat are forced up until they reach an obstruction, usually the ceiling. The smoke and heat will branch out along the ceiling. As the smoke and heat go in opposite directions the wall will be marked with a "V pattern" marking, **Figure 20-3.** This is a common term in the fire investigation field. The V will start at the bottom nearest the area of origin and proceed up the wall.

Another simple manner to determine the fire's starting point and where it spread is the **depth of char.** Wherever the fire burned the longest is where the fire's depth of char will be the greatest, **Figure 20-4.** This can be illustrated by looking at a log in the fireplace. Usually the center of the log has the most

FIGURE 20-3   "V" pattern in a structure fire.

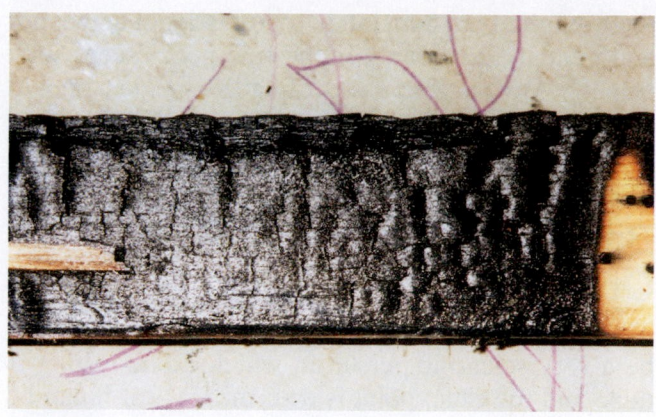

FIGURE 20-4   Depth of char.

charring and it diminishes toward each end. Wherever the fire started would have had the longest period of time to burn the wood members in that area, thus the charring would be greater in that area. The only misleading part of this indicator is when an area of the fire took longer to extinguish than others. This type of attack crew information can assist the investigator in determining the point of origin. These signs could be indicative of possible incendiary, or arson, fires. A TIC can be useful for tracing heat patterns. However, if the investigators are called to court, the evidence may not be admissable unless the TIC images are recorded.

Extent of fire damage is just one of many indicators of fire behavior and this damage can be influenced by many other factors. It is a straightforward way to determine the area of origin for simple or large-scale incidents alike. The key to assisting the fire investigators is not to rush the overhaul process. It is imperative that firefighters pay attention to what is going on around them during all aspects of an incident and report items that fall outside the norm.

The Importance of Fire Cause Determination

Assisting the fire investigator is a skill that should be second nature to all firefighters. The gathering of information from the time of dispatch throughout the incident should become automatic. Every detail should be noted and passed along, at some point, to the IC and the fire investigator. Fire investigators are just another part of the team who are attempting to make this tragedy as tolerable as possible for the owner. Determining the cause and origin of the incident is the responsibility of the fire department, and every firefighter will play a roll in the process. Firefighter actions can either help or hinder the investigation.

CUSTOMER SERVICE

Firefighters have an obligation to make sure the building is secured after they have completed all of their operations. This would include the revisits discussed in Section I, Chapter 20, boarding up all openings—including roof openings—so the building is secure, and securing the utilities. Some jurisdictions rely on board-up crews or restoration companies to handle this function, and the insurance carrier will pay those crews. If a building is not insured, then the local public works or a local lumberyard might donate some plywood to help the owner defray costs. Again, customer service will leave a lasting positive experience for the homeowner.

These are simple items to consider prior to leaving an incident scene. But what about the displaced owners? Where are they going to stay or do business? What are they going to wear for clothes if everything was destroyed? Do they have the means to contact their insurance company? Do they have insurance? Does the IC need to contact the Red Cross or other charitable organization in order to get them a place to stay or some clothing? All of these things should be considered and, if at all possible, the fire service should play an active role in assisting with these issues for the owners.

LESSONS LEARNED

Assisting the fire investigator is a skill that should be second nature to firefighters. The gathering of information from the time of dispatch throughout the incident should become automatic. Fire investigators are just another part of the team who are attempting to make this tragedy as tolerable as possible for the property owner.

Determining the cause and origin of the incident is every firefighter's duty and, like all other parts of the job, a very important factor. Firefighters always want to assist an investigation or criminal prosecution.

KEY TERMS

Depth of Char A term commonly used by fire investigators to describe the amount of time wooden material had burned. The deeper the char, the longer the material was burning or exposed to direct flame.

Potential Evidence Any item within and around the structure could potentially become evidence in the determination of the area of origin, fire cause, or identification of who or what may be responsible for the event.

REVIEW QUESTIONS

1. What is a "V" pattern?
2. Why is knowing the "depth of char" important?
3. What are the steps that must be accomplished when securing a building?
4. Name the individuals in your organization who are responsible for determining the cause of a fire.
5. What steps should be taken to protect evidence during an investigation?

As a fire service educator for many years, the one thing I have noticed that sets apart people in prevention from people in suppression is the time it takes to see results. In suppression, the effects of the interventions are seen in minutes, hours, or at the most several days. Prevention is different. The results may not be seen for months, or years, for that matter.

I have been fortunate in my career to see fire prevention pay big dividends. Fire deaths and severe fires have been drastically reduced in Montgomery County, Maryland, while the population and rate of building construction continue to grow. This is true not only here but throughout the country. Safety campaigns, smoke detectors, sprinklers, child-resistant lighters, school programs, and the fact that people are eating out more and smoking less are all contributing factors.

A particular case stands out in which a school fire safety program made the difference between life and death. Alena Tune of Rockville, Maryland, paid attention to the advice she got in school on fire safety and used it to save herself and her four-year-old sister when a fire broke out in their home. It was 3 A.M. on November 17, 1995. Alena woke up to the screams of her sister. She put to practice the survival techniques she learned in school. She got down on her hands and knees, got to her sister, and put her on her back. She crawled 30 feet to the front door and safety. This happened thirteen years after firefighters began teaching fire safety to fourth graders.

So, remember that your prevention message may not pay off today or tomorrow, but armed with lifesaving information, people can use it during their lifetime to save lives and property.

—*Street Story by Mary K. Marchone, Program Manager, Montgomery County Fire and Rescue Services, Montgomery County, Maryland*

STREET STORY

LEARNING OBJECTIVES

After completing this chapter, the reader should be able to:

21-1 Identify reasons public education programs are vital to public relations and the community.

21-2 Describe the purpose, value, and goals of a quality fire safety survey program.

21-3 List organization policies and procedures for conducting fire and life safety surveys in private dwellings.

21-4 Identify common causes of fire and fire prevention practices to follow in private dwellings.

21-5 Demonstrate the ability to recognize hazards during a private dwelling survey.

21-6 Communicate findings and provide residents with information concerning fire and life safety hazards according to departmental guidelines.

21-7 List referral procedures for discrepancy items found during a private dwelling survey.

21-8 Demonstrate completion of a private dwelling survey.

21-9 Describe the parts of and the purpose of fire safety information.

21-10 Identify presentation skills required to deliver a fire safety brief.

21-11 Describe guidelines for conducting a fire station tour.

21-12 Conduct a fire safety presentation using prepared materials.

21-13 Demonstrate completion of fire safety presentation reports.

21-14 List the basic components of a pre-incident survey and form completion.

21-15 Describe the typical safety problems that are encountered during a fire safety survey.

21-16 Identify the common symbols used in diagramming construction features, utilities, hazards, and fire protection systems.

21-17 Explain the importance of accurate diagrams in pre-incident surveys.

21-18 Given pictures or graphics of various fire suppression and detection system components, identify the system and the component.

21-19 Given a structure, procedures, equipment, and an assignment, perform a pre-incident survey to include a sketch of the site, buildings, and special features.

21-20 Given a structure or visuals of a structure with hazards and special considerations, identify the hazards and/or special considerations that should be added to a pre-incident sketch.

21-21 Given pre-incident forms and directions, complete all forms in accordance with the directions.

*The FF I and II levels, as defined by the NFPA 1001 Standards, are identified in different colors: **FF I = black, FF II = red, additional information = blue.**

INTRODUCTION

What is the cardinal mission of the fire service in America? Unquestionably, it is the preservation of life and property. The scope of this lofty task is seemingly ever expanding.

FIREFIGHTER FACT

As recently as thirty years ago firefighters were generally expected to respond only to fires—structure, vehicle, wildland. Today's firefighters are called on to be experts in all aspects of emergency activities: medical emergencies, hazardous materials incidents, urban search and rescue, swift water rescue, ice rescue, high-angle and confined space rescue, possible terrorist situations—and the list goes on.

With all the demands placed on firefighters by the increasing expectations of the public to perform in relatively new areas of technical expertise, it is easy to lose sight of the simplest and most effective method of achieving the goal of the preservation of life and property: prevention. Any physician will attest to the benefits of physical conditioning. Keeping in shape now helps to prevent devastating injury and illness in the future. Why do fire departments do preventive maintenance on their apparatus and equipment? One reason is that changing the oil regularly prevents costly engine rebuilds in the future. So it is with fire prevention and education. In an enlightened world, the firefighter's job must be to save lives and property from the effects of fire and other emergencies through the effective application of time-honored and innovative fire prevention and education techniques.

This chapter explains, through the application of the three Es, how the level of fire safety in the community can be enhanced. This information is presented to assist the firefighter/inspector in conducting a quality fire safety survey of both residential and nonresidential occupancies. Furthermore, the chapter highlights programs available for instructing communities in some proper fire prevention activities and appropriate responses to emergency situations. Finally, a discussion is included on the need and methodology for **pre-incident surveys** or **pre-incident plans,** by firefighters for emergencies at target hazards.

FIGURE 21-1   Fire prevention officer reviewing plans.

FIGURE 21-2   Fire prevention inspector walking through new construction.

ADMINISTRATION OF THE FIRE PREVENTION DIVISION

The duties and responsibilities of fire prevention officers are often not well understood by suppression firefighters. These men and women are seen around the station or at headquarters, but little is known about their daily routines. Fire prevention officers play a key role in ensuring that the fire department meets its goals of life safety and property conservation in the community it serves.

The Fire Prevention Division is responsible for all aspects of the fire and life safety of buildings and occupants prior to an emergency incident. This includes the "engineering" aspect of fire prevention such as interaction with architects and builders, new construction plans review and approval, **Figure 21-1,**

installation of fire detection and suppression systems, inspection of newly constructed or remodeled occupancies, **Figure 21-2,** administration of hazardous materials disclosure programs, inspection of high-risk occupancies requiring specialized expertise, and public education and information programs. After a fire or hazardous materials incident has occurred, the Fire Prevention Division is often called on to conduct the "origin and cause" investigation, **Figure 21-3,** and coordinate the follow-up activities of other agencies, when appropriate.

Fire prevention duties may be the responsibility of one fire prevention officer, or specialized assignments may be given to a number of officers, depending on local conditions. Whichever approach is taken, it is certain that the effectiveness of the Fire Prevention Division's efforts is greatly minimized without the cooperation and assistance of suppression firefighters.

FIGURE 21-3 Fire prevention officers inspecting a fire scene. *(Courtesy of Pam Rodriguez, Loveland Fire and Rescue)*

FIRE COMPANY FIRE SAFETY SURVEY PROGRAM

The events of 9/11 put firefighters in the status of "hero" for most American citizens. However, that kind of support can wane over time. To keep citizen support high, firefighters must remain in the public eye and must continue to show that they are indeed the heroes who will risk a life for a saveable life and will save homes and property in the event of a fire or emergency. Citizens should be confident that, once they have activated the emergency response system, firefighters will respond and take every possible action to mitigate the emergency they are experiencing.

An excellent way to maintain a higher profile with the public is to enact a proactive Fire Company Fire Safety program. This type of program is positive on multiple levels. It allows firefighters the opportunity to enter into buildings and businesses to see and plan the business layout and mitigate many hazards that could lead to a fire or life safety incident, while keeping the fire crews in the public eye, showing that firefighters are focused on customer service and are taking extra steps to keep the public safe. An effective fire safety survey program will reap immediate and future benefits for the community and fire departments. The "corrective action to problems" part of the program should be approached with the following objectives in mind: reduction of fire hazards, opportunity to increase public awareness of the dangers of fire, positive public contact, and building familiarization and pre-incident survey.

Fire Safety Survey responsibilities are generally assigned to fire suppression companies by the Fire Prevention Division. For obvious reasons, every effort must be made to ensure that such assignments do not require companies to leave their first-due response areas. The number of surveys assigned to particular fire companies will vary depending on the size and complexity of the occupancies and the level of emergency activity of the fire company. There are so many positive aspects of a good and well-planned survey program that they far outweigh the few negative aspects. Some crews will find it difficult to keep up with assigned surveys due to heavy call volume in their particular district or for other valid reasons. The fire administration should plan the survey program with such factors in mind, monitor the suppression company survey load, and adjust the program accordingly. At the very least, the identified target hazards should be visited by fire crews—even if an official fire survey cannot take place due to time constraints. Fire crews must know their in-district businesses and must remain positive in the public eye.

Equipment

Any job, done right, requires the proper tools and equipment. A minimum complement of tools and equipment for the firefighter/inspector is suggested here:

- Complete standard-issue uniform, clean and pressed
- Occupancy fire safety survey files
- Clipboard with notepad and pencil
- Flashlight
- Fire safety survey forms
- Standardized information bulletins
- Violation notices
- Fire code reference manual

Many company officers find it convenient to keep their forms and equipment organized in a catalog case (large briefcase), **Figure 21-4.**

Preparation for Fire Safety Surveys

With ever-increasing demands on firefighters' time, it is prudent for fire company officers to routinely designate specific days and times for conducting fire safety surveys. Planning and organizing an afternoon's surveys is essential. Preparation should include these steps:

1. Plan the area to be surveyed. When practical, all occupancies within a given shopping center, office building, or block should be visited on the same day.

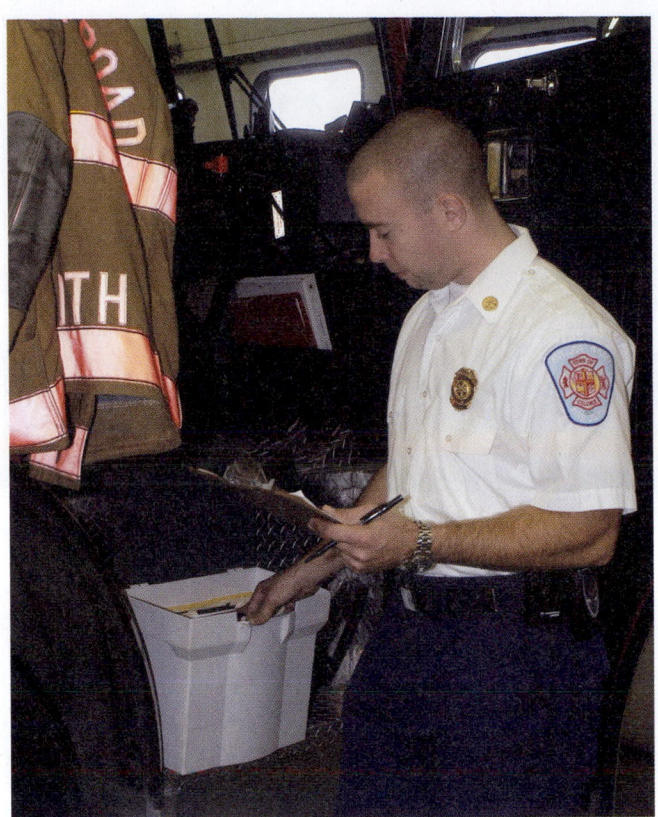

FIGURE 21-4   Company officer preparing for inspections.

FIGURE 21-5   An officer reviews occupancy files before leaving the station.

2. Review occupancy files prior to leaving the station. Note past problems, corrections, and recommendations to check for compliance or reoccurrence, **Figure 21-5.**

3. Give consideration to the type of activities conducted at the business relative to the time of day chosen for the survey. For example, it is usually not good practice to visit a restaurant during lunch hour or a small manufacturing plant after 3:30 P.M. In the case of the restaurant, the staff may be too busy to assist engine companies, and it is not unusual for a small manufacturing plant to shut down after 3:30 P.M.

4. Should the fire company make an appointment with the business owner or arrive unannounced? There is no specific rule in this regard. Generally, fire companies do not make surprise visits. Many business owners appreciate a telephone call prior to the fire safety survey; others feel it is not necessary. Writing a note in the survey file indicating the owner's preference is good practice. Note, too, that given the unpredictable nature of the firefighting business, keeping an appointment is not always possible. When making an appointment, remember to explain this to the business owner.

5. When possible, complete as much of the paperwork as possible prior to leaving the station. A sur-vey can demand important time from the business owner/manager: having paperwork in order prior to the safety visit saves time and presents a professional image.

Conducting the Fire Safety Survey

How many firefighters are necessary to conduct a fire safety survey? This is a judgment call for the company officer. As building familiarization is an important aspect of the fire safety survey program, it would seem appropriate to include the entire company. However, consideration must be given to the business owner's perspective. Too many firefighters may be seen as a "show of force" or an intimidation. Too few inspectors may result in a less than thorough or inefficient survey. Care must also be taken to minimize the disruption to business activities. It certainly does not take three or four inspectors to inspect a small barbershop, for example, but this may be the right number for a large warehouse or manufacturing occupancy.

After arriving at the occupancy to be surveyed, the crew should proceed directly to the front office, **Figure 21-6.** They should not linger in front or immediately begin inspecting the exterior of the building. Although the fire department has a statutory right to survey all businesses, permission must always be obtained from the owners or their representatives prior to beginning any safety visit.

Firefighters should introduce themselves to the first employee encountered, state their business, and request to see the person responsible for general building and employee safety. This may be the business

FIGURE 21-6   Upon arrival at the site to be inspected, inspectors should proceed directly to the office.

FIGURE 21-7   Introduction of a company officer to a business owner.

owner, plant safety engineer, operations manager, or a maintenance engineer, **Figure 21-7.** Whoever is designated should have the authority to speak for management and be able to correct any safety problems noted. It should be stressed that the survey or "safety visit" is for the benefit of the business.

NOTE

The company officer should insist that the crew be escorted during the fire safety survey visit. The escort will be able to grant access to all areas of the building and answer any questions that may arise. Additionally, the possibility of unfounded questions or impropriety on the part of the fire company will be eliminated.

Occasionally, but fortunately infrequently, a business owner will deny permission to perform a fire safety survey. If not immediately offered by the business representative, the company officer should ask for an explanation for the denial. Perhaps it is as simple as the time is not convenient for the business owner, the owner is too busy, or the safety manager is not present. Rarely is it a matter of the owner just refusing to be inspected with no rational explanation. This does occur, though. Under no circumstances should company officers engage in an argument with owners or attempt to persuade them to change their minds through the use of threats of authority or intimidation. A calm and reassuring approach is always the best tactic. "Would another day be more convenient?" or "We are not here to disrupt your business, only to help you maintain a safe working environment for you, your employees, and your customers." If all else

fails, the matter should be referred to the Fire Prevention Division. Ultimately, an **administrative warrant** may have to be obtained from the local magistrate.

NOTE

There are several schools of thought regarding the proper methodology for routing through a building when conducting the fire safety survey. Outside to inside, top to bottom, left to right are all acceptable. It is generally of little consequence in what order the occupancy is inspected as long as the method is efficient, systematic, and thorough.

Typical Safety Problems

A primary objective of the fire company fire safety survey program is, of course, the elimination of fire and safety hazards. A **fire hazard** can be defined as any condition, situation, or operation that could lead to the unwanted ignition of combustibles or result in proper combustion becoming uncontrolled. Fire safety problems in all occupancies can be categorized in general terms: exiting, fire protection equipment, use and storage of hazardous materials, and electrical and general fire safety.

Exiting

Most fire prevention officials would agree that a building's **means of egress, Figure 21-8,** provides the most basic level of life safety to its occupants.

NOTE

Exits are the most often abused and neglected system in a building.

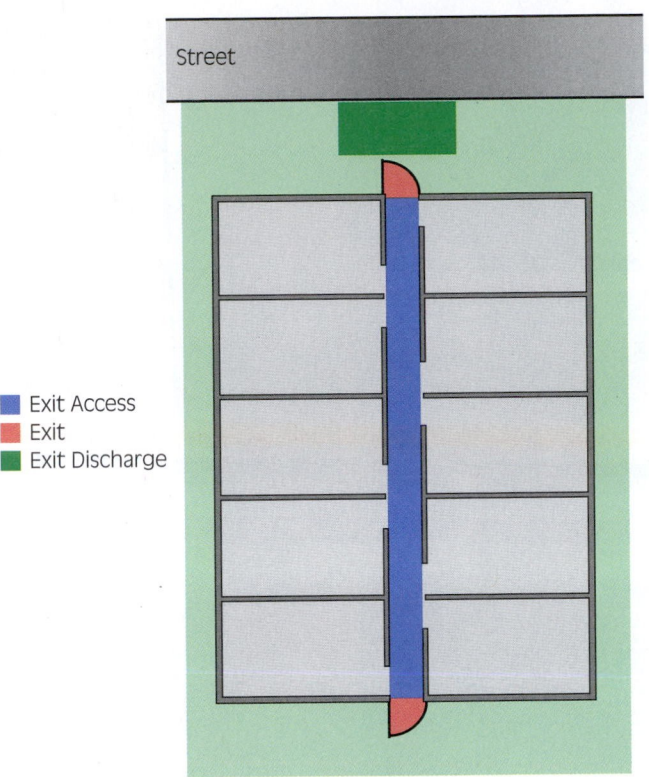

- Exit Access
- Exit
- Exit Discharge

FIGURE 21-8   Means of egress.

FIGURE 21-9   Exit corridors must be kept clear of all storage.

The inspector must verify that no exits are compromised by ensuring the following:

- Clear and unobstructed access is provided to all exits.
- Exits are identified and well lit.
- The proper type of door is used and it opens in the direction of travel.
- Exits are equipped with the proper opening and well-maintained locking hardware.
- Exit discharges to a public way are clear.

Exit accesses must be kept clear and unobstructed by storage or other materials that may hinder their use. Service corridors in malls and hotels are easy targets for storage of recently delivered merchandise or unused tables, chairs, or other items, **Figure 21-9.**

All exit accesses and exits are required to be identified with an exit sign, **Figure 21-10.** An exception is made for the main entrance to an occupancy because it is assumed that it is obvious even to visitors. Generally, exit signs are required to be illuminated only in assembly occupancies or high-rise buildings. When the exit access is not straightforward or is confusing, as in a warehouse or office with many corridors, intermittent directional signs should be used. If the fire inspector becomes lost or misdirected while walking through the building, then it can be assumed that visitors may become lost as well. This situation should be addressed by the use of intermittent directional exit signs.

FIGURE 21-10   All exits are required to be clearly identified.

STREETSMART TIP

The use of a "This Is Not An Exit" sign is not considered good practice because people in a hurry to leave the building under emergency circumstances are likely to focus on the word "Exit" and may become disoriented as a result.

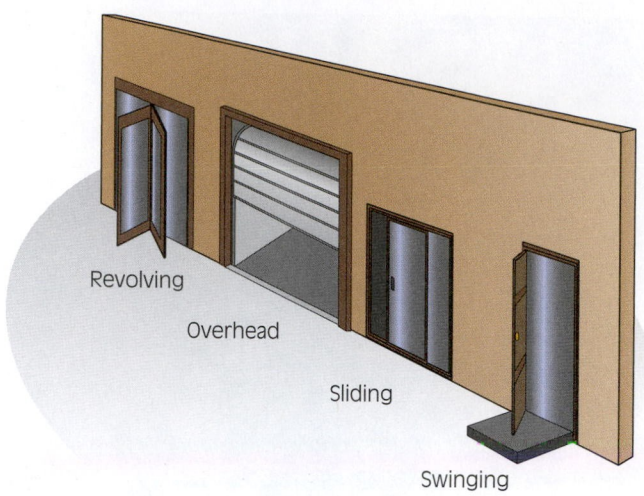

FIGURE 21-11 Of the four types of doors, only the swinging door can be used for a required exit.

The inspector must ensure that the proper type of door is used as an exit. Of the four classifications of doors—revolving, overhead, sliding, and swinging—only the swinging door is permitted to be used for a required exit. **Figure 21-11** shows the four classifications of doors.

Exits are required to be equipped with specific types of opening and latching devices depending on the occupancy classification. Approved exit hardware can be grouped into three categories: no-knowledge, panic, and special egress devices.

No-knowledge hardware, Figure 21-12, is a broad category of locking devices that require no key or special knowledge to operate. Lock sets that combine the latching mechanism with a dead bolt are an example, **Figure 21-13.** The dead bolt is retracted simultaneously with the normal latching device when the knob or lever is turned. This device provides the security of a dead bolt while maintaining simplicity of operation. No-knowledge hardware is required on all exits, except the main entrance, in all occupancies when panic hardware is not required.

NOTE

A double-cylinder lock set is key operated on both sides. This type of hardware is permitted only on the main entrance to most occupancies provided a sign reading "This Door to Remain Unlocked During Business Hours" is displayed above the door. Double-cylinder dead bolts are also permitted in private residential occupancies.

FIGURE 21-12 No-knowledge hardware is required on all exits not requiring panic hardware, except on the main entrance. *(Courtesy of Loveland Fire and Rescue)*

STREETSMART TIP

Although double-cylinder dead bolts are permitted in residential occupancies by most building and fire codes, firefighters should strongly discourage their use. During an emergency evacuation of a home, there is no time to fumble for a key.

Panic hardware is required on exits in all assembly, educational, and institutional occupancies. It usually consists of a locking mechanism activated by a bar across the door, **Figure 21-14.** The mechanism must operate and the door must open with no more than 15 pounds of force. This is to allow even small children and frail adults to open the door.

Occasionally, for security purposes, a business owner will request the use of a **special egress control device.** The operators of occupancies such as jewelry or electronic equipment stores, secured areas of an airport, weapons dealers, and the like often feel the need for the additional security provided by locked exits. This, of course, is an obvious violation of the fire code. To address the concerns of these owners

FIGURE 21-13   Interconnected lock set and dead bolt.

FIGURE 21-14   Exit doors should be opened fully to ensure their function.

while ensuring the safety of their employees and customers, fire and building officials have approved the limited use of special egress control devices. This type of device is allowed, provided that an approved automatic fire sprinkler system and an approved automatic smoke detection system protect the building. The door-release will unlock the door in a maximum of fifteen seconds. Pushing on the bar, **Figure 21-15,** activates the device. With the specific approval of the fire marshal, these devices are permitted in the following occupancies: general businesses, factories, institutions, mercantile businesses, storage facilities, and group care facilities and residential care facilities housing clients with various forms of dementia.

Special egress control devices must automatically deactivate whenever the sprinkler or smoke detection systems activate or whenever there is a loss of electrical power to the building or the device. There must also be posted on the door a sign that reads "Keep Pushing. Door Will Open in 15 Seconds. Alarm Will Sound" or similar wording.

Exit discharges must be kept clear to allow for occupants to continue their travel to a safe position away from the structure, **Figure 21-16.** Fire codes usually define a "safe distance" as the nearest public way, that is, a street, alley, sidewalk, or major parking lot.

NOTE

Fire protection equipment, whether portable or fixed, is the building occupants' first line of defense against unwanted fire. Firefighters must ensure that this equipment is not only present and maintained in proper working condition, but that the building occupants know when and how to use it.

Protection Equipment

During a fire safety survey, it is most appropriate for the inspector to question the business representative: "Do your employees know where your fire extinguishers are located and how to use them?" "Do you

FIGURE 21-15 Special egress control devices often look like normal panic hardware, but must be identified with a sign.

FIGURE 21-16 Firefighters should inspect the exit discharge to ensure that occupants have clear access to the public street or parking area.

provide any training in their proper use?" "Do you know how to manually activate your kitchen cooking equipment fire suppression system?" "Do you know where the fire sprinkler control valves are located?" "Do you know how a fire sprinkler system operates?" Many people are still under the misunderstanding that all fire sprinkler heads operate simultaneously or that they are activated by smoke!

Portable Fire Extinguishers. Firefighters should inspect for and identify the types of fire extinguishers in an occupancy, ascertain that they conform to the fire prevention requirements for that occupancy, **Figure 21-17,** and ensure the following:

- Fire extinguishers are located per local fire code.
- The extinguisher has the proper classification and rating for its location.
- The fire extinguisher is mounted on the wall and is easily visible and accessible.
- The pressure gauge indicates that the extinguisher is fully charged.

- The tag on the extinguisher indicates that it has been serviced in a time frame conforming to local requirements. (Some jurisdictions require annual servicing, whereas others permit a time span as long as five years between services.)

Automatic Fire Sprinkler Systems. Firefighters should check fire sprinkler systems for the following:

- All water supply valves are open and secured.
- The fire department connection should be free of obstructions and capped, **Figure 21-18.**
- Access should be provided to system risers, valves, and gauges, **Figure 21-19.**
- Pressure gauges on the system riser should show appropriate pressure.
- All areas of the building must be properly protected.
- Sprinkler heads should not be damaged, painted, or corroded.

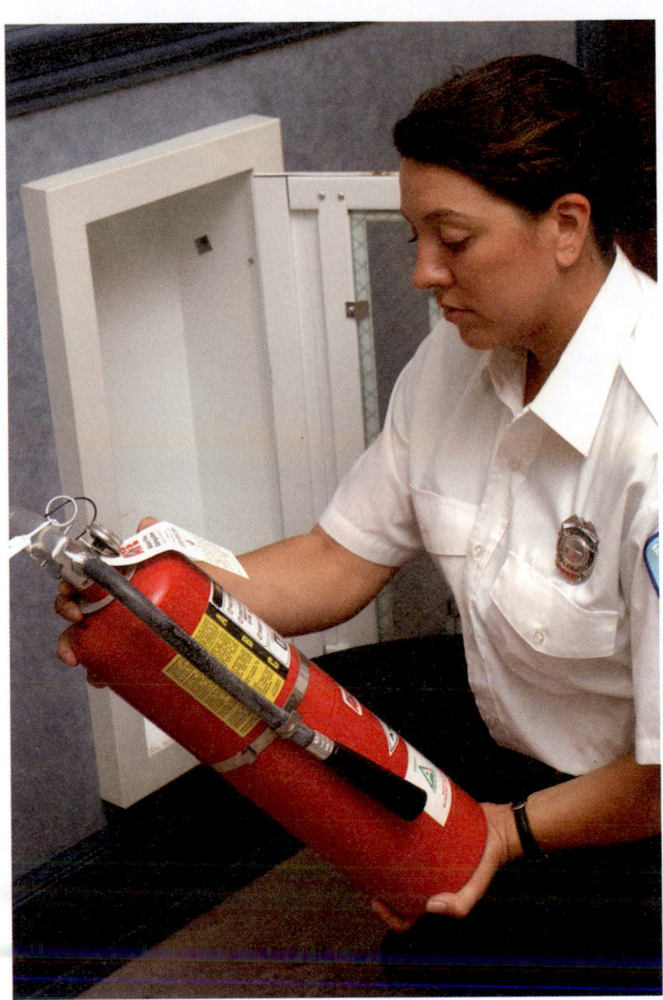

FIGURE 21-17   Firefighter inspecting a fire extinguisher.

FIGURE 21-18   Outside control valves must be open and secured. The fire department connection must be clear of obstructions and capped.

Standpipes should be inspected to ensure:

- Proper water level is maintained in supply tanks—if present. Freezing precautions should be present in areas susceptible to frost.
- Fire pumps start and operate.
- Water control valves are open.
- Fire cabinets are accessible and no other objects or materials are stored in or blocking them.
- All gaskets are in good condition.
- Individual discharge valves operate properly.
- Discharge outlets will accommodate fire department hose through proper hose threads or adaptors.
- Any hose present is in good condition.
- Nozzles are operable, adequate for the system, and stored properly.
- Any doors or hose reels or racks are operational.
- Drains are free of all dirt and are operational.

FIGURE 21-19   Access should be provided to fire sprinkler control valves and gauges.

- Fire Department Connections (FDCs) have been inspected, paying special attention to the threads and valves. Firefighters should ensure there is no refuse or any other object placed inside the FDC that would hinder operation. Last, but of great importance, firefighters should check to make sure fences or other objects are not blocking FDC access and that vegetation is not allowed to grow wild, hindering access. FDCs should always be marked and visible.

FIGURE 21-20 The sprinkler system should be flow tested regularly. Some inspectors flow the system themselves, while others ask the business owner to do so.

FIGURE 21-21 Many departments allow plastic sandwich-type bags to be placed over sprinkler heads in spray paint booths to prevent the accumulation of overspray on heads.

The system should be flow tested regularly, preferably quarterly, and records kept, **Figure 21-20.**

The system should be serviced in a timely manner, usually at five-year intervals.

Sprinkler systems in spray paint booths deserve special mention. By the very nature of the operation, sprinkler heads in spray booths quickly become covered with overspray. This condition will not only delay activation of the head but will most likely prevent its operation altogether. Therefore, keeping the heads clean is essential. To protect sprinkler heads that are likely to be ruined by overspray residue, NFPA 25, Inspection, Testing and Maintenance of Water-Based Fire Protection Systems, requires that they be covered with thin plastic or small paper bags, as shown in **Figure 21-21.** These bags should be changed daily.

Restaurant Cooking Surface and Exhaust Hood Suppression Systems. Fixed extinguishing systems protecting the cooking surfaces and exhaust hood and ducts in restaurants are required whenever the cooking process produces grease-laden vapors. Pizza and bread ovens and stoves that are used for heating water or soup do not require a system; however, nearly all other commercial cooking do. Such systems are essentially large fire extinguishers mounted on the wall, **Figure 21-22.** They contain either a wet or dry extinguishing agent.

When inspecting cooking surfaces and hood and duct systems, firefighters should ensure the following:

■ The extinguishing agent cylinders are charged and armed.

■ Nozzles are free of grease and are capped.

■ Fusible links are free of grease.

FIGURE 21-22 Fixed fire extinguishing system for cooking surfaces.

■ Hood filters are clean and in place, **Figure 21-23.** (Filters not only trap grease, but also restrict a surface fire from spreading to the duct.)

■ The manual activation control is accessible, **Figure 21-24.**

■ The tag on the system indicates that it has been serviced annually.

FIGURE 21-23   Hood filters should be clean and in place.

FIGURE 21-24   Restaurant workers should know the location of the manual release and how to activate the extinguishing system.

Similar systems are found in automotive, furniture finishing, and spray paint booths. They should be inspected in the same manner as restaurant systems.

Heat and Smoke Detection Systems. Beginning at the system control panel, **Figure 21-25,** the firefighter should ensure the following:

- All indicators on the panel are in a "normal" condition and not in "trouble" mode.
- The panel is receiving AC power.

- All detection devices are present and properly mounted—not hanging from the ceiling.
- Records are provided indicating that the system has been tested periodically.

It is often a good idea for firefighters to request that the system be tested in their presence. While not always practical given the nature of the occupancy, witnessing the activation of the system will not only ensure that it functions but will stress to the owner the importance of the system.

Fixed Gaseous Extinguishing Systems.   These types of extinguishing systems protect building areas where water from a sprinkler system would cause massive amounts of damage to sensitive equipment, such as computers or electrical systems. The extinguishing agent is generally CO_2, halon, or halon replacement. In essence, these protection systems serve the purposes of detection and suppression.

FIREFIGHTER FACT

Because it is thought to be a danger to the environment, halon has virtually been replaced by other agents that are more environmentally friendly. Existing halon systems are "grandfathered" and can continue operation. Halon is no longer produced, and only remaining stockpiles of halon are available. Recharging systems with halon is generally cost-prohibitive, and as a result, most systems are modified to use more environmentally safe extinguishing agents.

CAUTION

The detection system should be inspected as any other heat and smoke detection system would be, but the function test should be omitted. Function tests should never be done by firefighters because an accidental discharge may result and the department could be liable for damages.

The survey of the suppression portion of the system involves checking the cylinder gauges for proper pressure, ensuring that all discharge nozzles are unobstructed either by foreign objects or storage, and making sure access to the manual discharge control is provided.

Use and Storage of Hazardous Materials

Ensuring the proper use and storage of hazardous materials is a subject that often requires a great deal of knowledge, expertise, training, and experience. Many fire prevention officers dedicate large portions of their careers to such endeavors. However, from the firefighters' perspective, some general principles apply.

FIGURE 21-25 Firefighter checking fire alarm system control panel.

Inside Storage and Use

- Storage and use areas must be well ventilated.
- Sources of ignition must be eliminated in any area using or storing flammable or combustible liquids or gases. These include open flames and welding or grinding of metal.
- The maximum quantity of flammable and combustible liquids permitted varies according to the occupancy.
- Quantities of less than 5 gallons must be stored in approved **safety containers, Figure 21-26.**
- Small quantities should be stored in approved storage cabinets, **Figure 21-27.**
- Larger quantities must be stored in rooms specially designed for such purposes. These rooms are constructed of fire-resistant material, have raised door sills to prevent leakage, have liquid-tight floors and walls, have explosion-proof electrical fixtures, and are well ventilated.
- Containers with a capacity greater than 30 gallons may not be stacked.
- Dispensing must be done through an approved pump or self-closing faucet, **Figure 21-28.**
- Dispensing containers must be properly bonded and grounded to protect against the discharge of static electricity, **Figure 21-29.**
- Compressed gas cylinders, whether full or empty, should be stored with protective caps in place and secured with chains or straps to prevent falling, **Figure 21-30.**

FIGURE 21-26 Approved flammable liquid safety can.

Outside Storage

- The selected storage area should be properly distanced from buildings and property lines to minimize the potential for fire spread or exposure.
- Storage areas must have **secondary containment.** This typically means a 6-inch curb around the storage area to contain spilled liquid and is typically built to withstand 1.5 times the volume of the largest container, **Figure 21-31A.**

FIGURE 21-27   Small quantities of hazardous material should be stored in approved storage cabinets.

FIGURE 21-29   Properly bonded and grounded barrels.

FIGURE 21-28   Approved self-closing faucet.

FIGURE 21-30   Compressed gas cylinders secured with a chain and with caps in place.

(A)

Red:	Flammability	0 through 4
Blue:	Health Hazards	0 through 4
Yellow:	Reactivity of Chemical	0 through 4
White:	Special Hazards (See examples below)	

Ratings are 0 through 4, with 4 being the highest risk.

No Special Hazards   Reacts violently to water   Reacts violently to water and oxygen

(B)

FIGURE 21-31 (A) Storage of flammable liquids with curb around storage area. (B) The NFPA 704 placard system identifies hazards in fixed storage facilities. The colors and numbers quickly give first responders vital information.

■ Sources of ignition must be eliminated, keeping in mind the potential for the migration of vapors from spilled liquids.

Hazardous materials that are located in fixed storage facilities are not governed by the same regulations as the transportation industry. Local jurisdictions may develop a system of labeling for hazardous materials. Most local jurisdictions choose to adopt the NFPA 704 (Standard System for the Identification of the Fire Hazards of Materials) placard system, **Figure 21-31B.** This system provides valuable information to first responders and notifies employees and other persons using the facilities of the contents. The placard system is simple to read and provides information about health hazards, flammability, and reactivity. Firefighters inspecting facilities with these types of placards should speak to the appropriate persons at the facility to review the types of materials contained on the property and the associated hazards. These hazardous materials should be documented in the fire safety survey report and first-due responders should be advised.

Electrical Hazards

Building electrical systems require specialized knowledge to ensure proper design and installation. The firefighter should call on the expertise of a local electrical inspector whenever questionable installation practices are noted. However, a firefighter can be effective in eliminating electrical hazards by observing the following:

■ Check all fuse and breaker panels to verify that overcurrent protection devices have not been defeated and that all spaces on the panel are equipped with a breaker switch or blank to cover the opening, **Figure 21-32.**

■ Access to the panels should be maintained free of storage and debris.

■ All outlets and junction boxes should have an approved cover in place.

■ Ensure that outlets, switches, lights, and appliances are approved for the location installed. Remember that ordinary electrical devices are not permitted in rooms designed for the use and storage of flammable or combustible liquids or explosives.

■ Extension cords are not permitted to be used in place of permanent wiring. This problem is widespread and becomes more acute when light-duty "zip cords" are utilized. If the installation of additional permanent outlets is not possible, multiplug adapters with built-in circuit breakers are generally an acceptable alternative.

■ No exposed wiring, approved or otherwise, should be permitted to run through doorways or under floor coverings or be stapled to a wall.

FIGURE 21-32 Circuit breakers must never be prevented from tripping by the use of tape or other objects. All spaces on the panel must be covered with either a breaker switch or a blank.

General Fire Safety

> **NOTE**
>
> Poor housekeeping practices, while seemingly a simple issue, are often the root cause of many fire hazards and eventually fires and injuries. Accumulated trash and debris are often found blocking exits or access to fire protection equipment. Haphazard storage of combustibles near open flame or hot surface devices such as water heaters or space heaters is another cause of accidental fires.

Dust accumulation is a common problem in woodworking and textile manufacturing occupancies. Care must be taken to prevent the buildup of dust on all horizontal surfaces, especially those overhead. Dust explosions are caused when combustible dust particles suspended in the air contact an ignition source. Fortunately these devastating explosions are an infrequent occurrence. **Figures 21-33A** and **B** shows the ductwork for a dust collection system.

Proper disposal of rags used with flammable or combustible fuels, solvents, and oils is easily accomplished but so often neglected. Autoignition of such rags due to their casual treatment has resulted in more than a few fires. Oily rags must be stored or discarded in an approved safety can, **Figure 21-34**.

Fire codes restrict smoking in certain occupancies. General smoking in buildings should be discouraged by firefighters. Smoking and nonsmoking areas should be clearly identified and proper containers provided for the disposal of smoking materials.

Building Exterior

No fire prevention fire safety survey is complete until the inspector has walked around the outside of the building. The following observations should be noted:

- Address numerals of sufficient size should be posted where responding emergency personnel can easily see them, **Figure 21-35**.
- All access roads should be considered fire lanes and kept clear of vehicles and storage, **Figures 21-36A** and **B**.
- Secured key boxes (also called Knox Boxes), when provided, should be checked for current keys, **Figure 21-37**. Firefighters should encourage, if not require, the use of such boxes for after-hours entry into buildings by emergency personnel.
- Trash disposal areas should be away from buildings and free of debris, **Figure 21-38**.
- Outside storage of flammable and combustible liquids should be as described earlier.
- Accumulated dried vegetation near buildings should be eliminated, **Figure 21-39**.
- Recognize any deficiencies in the chimneys or flues and be aware of situations that could possibly be a hazard, **Figure 21-40**.

Concluding the Fire Safety Survey

At the conclusion of the fire safety survey, the firefighter should convey the findings to the building representative. This "wrap-up meeting," as shown in **Figure 21-41,** need not be a formal event. In fact, the efficient inspector will have pointed out any deficiencies or suggestions noted while walking through the business at the time observed. In this way, at the wrap-up, the firefighter can reiterate the findings in summary form prior to leaving the premises. The inspector should remember to thank occupants for their time and praise them for their fire safety efforts. If criticism is necessary, the inspecting officer should exercise tact.

(A)

(B)

FIGURE 21-33   (A) Woodworking occupancy with a dust collection system. (B) The exterior of a dust collection system.

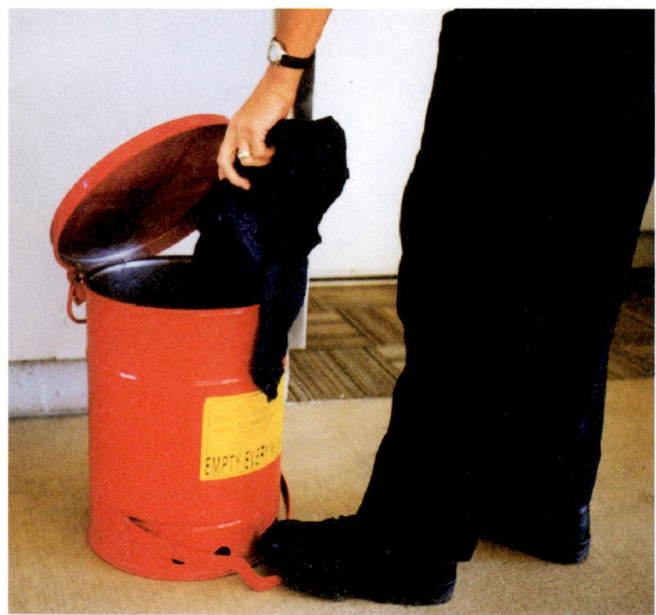

FIGURE 21-34   Oily rags must be stored in approved safety cans with self-closing lids.

FIGURE 21-35   All occupancies should have plainly visible address numerals.

(A)

(B)

FIGURE 21-36   (A) Unblocked fire lane. Concrete blocks under the turf are often used to allow fire apparatus to drive in landscaped areas. (B) Blocked fire lane. Fire lanes must be kept open at all times.

FIGURE 21-37   Key boxes on the outside of a building allow firefighters access to a locked building. They should be checked for current keys.

FIGURE 21-38   Trash storage areas should be free of debris and away from buildings.

All deficiencies, safety problems, and suggestions must be documented, even those corrected in the presence of the firefighter. It is important for all firefighters, regardless of rank, to have a working knowledge of the building fire safety survey reports. Firefighters should be able to complete a fire safety survey report and explain all parts of it to the building occupant.

> **NOTE**
>
> Safety problems that pose an imminent life hazard, in the opinion of the firefighter, must be corrected before the firefighter leaves the premises. Examples include locked or blocked exits, welding operations in the vicinity of open flammable liquids, and electrical wiring that presents an obvious shock hazard.

FIGURE 21-39 Dried grass at exterior of building should be eliminated.

FIGURE 21-40 Exterior conditions can reveal clues as to fire cause, rate of spread, etc.

FIGURE 21-41 Inspector discussing findings with business owner.

Many different styles of notices are available. An example of a Fire and Life Safety Inspection Notice is shown in **Figure 21-42.** At a minimum, the notice should include the name and address of the business, the date of the fire safety survey, the building representative, safety problems noted with their corresponding code references, the fire department inspector's name, and a compliance date. Most often, notices are completed prior to leaving the building. If this is not possible because further research is required, it is perfectly acceptable to return later with the completed notice. The original should be given to the building occupant and a copy should be filed in the occupancy file. Remember, this is a legal document and must be treated as such.

Re-Visits

Few things can negate the validity of a fire safety survey faster than the lack of follow-up. All documented safety problems (and they should all be documented) must be reinspected in some manner to emphasize their significance and ensure compliance. If it is important enough to document, then it is important

FIRE SAFETY CHECKLIST

BUSINESS NAME

PHONE #

STREET ADDRESS   SUITE   ZIP CODE

BUSINESS OWNER   HOME PHONE

OWNER ADDRESS

EMERGENCY CONTACT

ADDRESS   EMERGENCY CONTACT PHONE #

INSPECTION CONTACT	INSPECTION DATE		INSPECTION COMPLETED	
	#1	#2	( ) YES	( ) NO

#1	#2	

ARTICLE 1: ADMINISTRATION
0136: No space heaters allowed in occupancy
0137: Notices & orders shall be complied with

ARTICLE 9: ACCESS
0901: Install approved address identification
0902: Maintain & post fire department access lanes
0903: Lock Box required for access along with current keys

ARTICLE 10: FIRE PROTECTION SYSTEMS
1001: Maintain fire protection systems in an operable condition
1002: Inspection & test reports to be made available
1004: FDC shall remain clear of obstructions
1005: 3ft. clear space to be maintained around hydrants
1009: All control valves & flow switches shall be monitored
1011: All commercial cooking equipment shall have automatic fire protection
1012: Provide a 40 BC fire extinguisher within 30 feet of cooking equipment
1013: Hood & duct system shall be free of grease
1014: Hood extinguishing system to be serviced and links replaced every 6 months
1015: Fire alarm equipment shall not be obstructed
1017: "Local Alarm Only" signage shall be provided
1018: Maximum travel distance for fire extinguishers exceeded
1019: Properly mount fire extinguisher
1020: Service fire extinguisher annually

ARTICLE 11: GENERAL
1103: Dumpsters shall be 5 feet from building
1104: Maintain good housekeeping
1105: Maintain storage 18 in. below sprinkler heads
1106: Maintain storage 24 in. below ceiling
1107: No storage in boiler, mechanical, or electrical room
1108: Drapes and other decorative material shall be fire resistive
1109: Aboveground gas meters & piping shall be protected
1110: Maintain clearance from combustibles and heat producing devices
1114: Maintain fire resistive construction
1115: Maintain fire doors and dampers
1117: Remove accumulation of exterior waste

#1	#2	

ARTICLE 12: EGRESS
1201: Remove all obstructions from path of exit
1202: Exit door shall swing out when occupancy exceeds 50
1203: Remove all unapproved locking devices from exit doors
1205: Emergency power equipment for egress illumination shall be maintained
1206: Exit paths shall be identified by exit signs
1207: Exit signs shall be illuminated
1208: No storage under unprotected stairways
1213: Dead end corridors over 20 feet in length

ARTICLE 25: PLACES OF ASSEMBLY
2501: Maximum occupant load signs to be posted in occupancies over 50

ARTICLE 74: COMPRESSED GAS
7401: Compressed gas cylinders shall be secured

ARTICLE 80: HAZARDOUS MATERIALS
8002: Tanks, piping & valves shall be protected from vehicular damage

ARTICLE 85: ELECTRICAL EQUIPMENT AND WIRING
8501: Install cover plate on all electrical outlets, switches and junction boxes
8503: Discontinue the use of extension cords
8504: Cords shall not be subject to damage
8505: Cords shall be maintained in good condition
8506: Discontinue the use of multi-plug adapters
8507: Maintain 30 inch clearance around electrical panels
8508: Electrical panels shall be labeled as to area served
8509: Doors into electrical rooms shall be labeled

MISCELLANEOUS:
SEE ATTACHED SUPPLEMENTAL

COMMENTS:

REQUIRED PERMIT	ASSOCIATED FEE	NOT CURRENT	CURRENT

A routine fire inspection has been made of these premises to determine if any life or fire hazard exists. The inspection also was made to determine if any violations of the Uniform Fire Code exist. If violations have been found, this shall serve as your official notice. All violations must be corrected within the time allowed. A re-inspection will be made at a later date to verify the required corrections. Should you need assistance regarding this inspection, contact the Fire Prevention Bureau at 962-2537.

I, the undersigned, am in receipt of a copy of this inspection report and am aware of the hazards noted. I am also aware that this is a routine inspection for correction of violations and **may not encompass every possible violation.**

SIGNATURE OF BUSINESS OWNER, MANAGER, OR RESPONSIBLE PARTY   TITLE   DATE

SIGNATURE OF INSPECTOR

FIRE DEPARTMENT: WHITE   1ST INSPECTION: CANARY/BUSINESS OWNER   2ND INSPECTION: PINK/BUSINESS OWNER

FIGURE 21-42 Inspection notice should be given to business representatives at the conclusion of the inspection.
(Courtesy of Loveland Fire and Rescue, Loveland, Colorado)

enough to follow up. However, not all safety problems require the time and effort of another trip to the business site. Many departments use a self-clearing card, an example of which is shown in **Figure 21-43**. It may have different names, but the concept is the same. For a limited number of very specific infractions, the business owner affirms that the violation has been corrected by signing a postcard and mailing it to the fire department. Examples of qualifying infractions are address numerals that have been enlarged so they are clearly visible, removal of an extension cord, or recharging a fire extinguisher.

More serious problems require a re-visit. The time allotted to make the corrections will vary with the complexity and seriousness of the violation. However, once a re-visit date has been established and agreed to by the business owner, every effort should be made by the inspector to return on the specified day. Additional compliance time may be granted at the discretion of the inspector. Failure on the part of the business owner to make a reasonable effort to comply with violation notices should be referred to the Fire Prevention Division.

HOME FIRE SAFETY SURVEYS

FIREFIGHTER FACT

National fire statistics prove that year after year roughly 80 percent of all injuries and deaths in America occur in the home. To the credit of the fire service and the public, actual numbers of injuries and deaths have been decreasing in recent years. But the fact remains that eight of ten people killed or injured by fire are in the "safety" of their homes.

The overwhelming majority of fire prevention efforts are concentrated on the business community. The reasons are deep seated and fundamental but are also centered around privacy issues. Americans simply will not allow authorities legislated access to their homes, no matter how noble the reason.

This does not preclude the fire service from making an effort to conduct voluntary home fire safety survey programs. Such programs, whether available to entire jurisdictions or only targeted areas, can be beneficial to the community as well as the fire depart-

---

	Dept. Use
	Batt. _____
	Unit _____ Shift_____
	Insp. _____
	(Print name)

DELMAR COUNTY FIRE AUTHORITY
INSPECTION NOTICE

INSP. # _____   Received by _____

Violation

Business Name: _____Address:_____
As a result of an inspection by the Delmar County Fire Authority on (date _____
the violation(s) listed below were noted.

☐   Vio. Code <u>25</u>   1. State law requires that all fire extinguishers be serviced annually CCR TITLE 19
      A. Service accomplished by (Co.) _____
      B. State license number _____

☐   Vio. Code <u>25</u>   2. Provide a type 2A 10BC fire extinguisher. Extinguisher shall be installed on the hanger,
      conspicuously located, readily accessible in the event of fire. UFC 10.301

☐   Vio. Code <u>01</u>   3. Address numbers shall be placed on all buildings, plainly visible from the street. Numbers
      shall contrast with their background. UFC 10.208

☐   Vio. Code <u>09</u>   4. Discontinue use of extension cords and/or multiplug adapters. UFC 85.106

☐   Vio Code ___   5. Other: _____
_____

_____ I certify under penalty of perjury that the Violations noted above have been corrected. _____

Signature_____ Date corrected_____

Sign and return this Notice. Failure to correct the identified Violation(s)
within 14 days after inspection date will result in reinspection or other
appropriate legal action to ensure compliance.

➡ F121-16.2 (R3/90)   **FILE COPY**

FIGURE 21-43 A business owner may affirm that minor code violations have been corrected by signing and returning a self-clearing card.

ment. Not only will the threat of fire in family homes be reduced, but much goodwill for the fire department will be fostered. No homeowners should ever be refused the opportunity for a fire safety survey of their home. Homeowners who take the initiative to ask fire crews for a home visit will often be vocal about their support for the fire service, especially if the fire department is facing budget cuts or other restructuring that could have a negative impact on the citizens of the community.

Voluntary residential fire safety surveys can be used to point out fire and life safety hazards, check and install smoke detectors and carbon monoxide detectors, and instruct families in proper emergency preparedness techniques. When granted permission to enter a home, firefighters are cautioned to limit their movement in the home to the "less private" areas. In other words, stay out of the bedrooms.

Typical hazards found in the home are not unlike those discussed previously. Exiting problems are usually centered around double-cylinder dead bolts on doors, security bars on doors and windows, **Figure 21-44,** or rooms constructed or remodeled without proper exiting requirements. Electrical problems involve overloaded circuits and extension cords. Storage around furnaces and water heaters is often noted. Combustibles left on top of floor furnaces continue to cause many fires at the start of each heating season. Improperly stored flammables and combustibles in basements and garages are regularly observed. Garages are normally huge "catchalls" for small quantities of hazardous materials of all kinds.

Attention should be given to issues of safety to small children. Accessible electrical outlets without protective caps, unsecured household chemicals, swimming pools without security fences, and electrical appliances in bathrooms are areas that deserve the serious attention of parents and child care providers. Firefighters should consider that making any repairs within a private residence could lead to legal liability. Firefighters may perform repairs within the bounds of those authorized within the fire department such as hanging smoke detectors or replacing batteries. A referral list could be maintained by the fire department so that when deficiencies are found, the occupant may be advised of who to contact to remedy the problem. The referral procedures may include advising the occupant to contact the proper repair company, utility company, landlord, etc., and document all deficiencies as well as the referral to the occupant.

Smoke and carbon monoxide detectors should be installed, **Figure 21-45.** The importance of these devices cannot be overstated and must be stressed to homeowners. Firefighters should remind residents that their chance for survival in case of a fire increases by 50 percent if a smoke detector is installed and functioning. However, correct installation is only a good first step. Detectors must be maintained by changing their batteries on a regular basis, **Figure 21-46.**

It is becoming more and more beneficial for fire departments when administration officials think "outside the box." Trying new and different ways of doing business with the citizen in mind can result in a positive public perception of the fire service and the local fire department in particular.

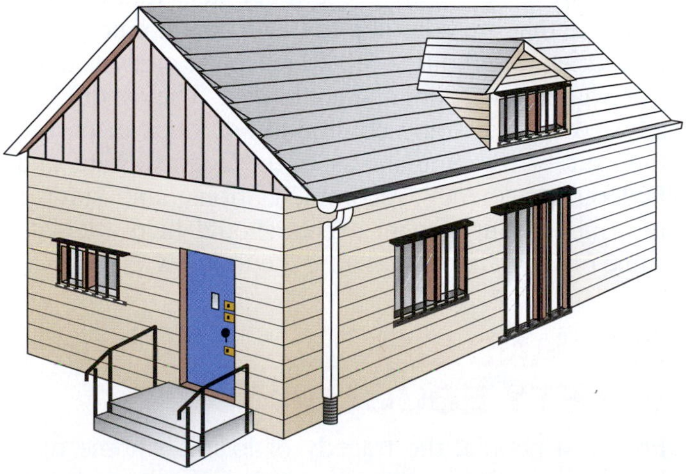

FIGURE 21-44 House with security bars.

FIGURE 21-45 Smoke and carbon monoxide detectors are available separately or as one unit. (*Copyright 2008 BRK Brands, Inc., a Jarden Corporation (NYSE: JAH). All rights reserved.*)

FIGURE 21-46 Batteries in smoke detectors should be checked monthly and replaced every six months. Some detectors have lithium batteries that do not require replacement.

FIGURE 21-47 Firefighter showing others how to place a lid on a burning pan.

STREETSMART TIP

In Loveland, Colorado, the Fire Prevention Bureau of Loveland Fire and Rescue is randomly accompanying local pizza delivery persons on a delivery. If the patrons ordering the pizza have working smoke detectors in their home, the Pizza delivery company provides the pizza free of charge. If not, the accompanying firefighter will install free detectors or replace the batteries.

Programs like these can cost little or nothing to the department. Detectors and batteries can be purchased with grant monies or donated by local businesses or fund raising entities. With creative scheduling, the bureau members' salary costs will not increase. When the press is contacted and the story hits the paper, the publicity is enormous for the involved agencies and especially the fire department.

Other departments have instituted programs of Life Safety visits. After a fire with a fatality, a fire company may visit a target audience, such as a two-block area surrounding the fire. The presentation of fire safety information may be a formal or a casual meeting. The focus of these types of presentations is answering the public's questions concerning the general reasons for the fatality and discussing the preparations necessary to avoid a repeat of the scenario. The public can be quite an ally for the fire department when it comes to needed apparatus or equipment upgrades or replacement, and public service programs can go a long way in gaining citizen support for the fire department. With just a little innova-

tive thought, there can be an endless supply of fresh ideas for these types of programs.

Finally, firefighters should not fail to seize the opportunity to discuss emergency preparedness with the home's occupants. This should include first-aid fire-fighting techniques such as fire extinguisher placement and use, stove top fire extinguishment, **Figure 21-47,** smoke and carbon monoxide detector placement, and barbecue fires. Instruction should be given in proper home evacuation in case of fire. But this area could be expanded to include preparedness for natural disasters such as earthquakes, tornadoes, and flooding.

STREETSMART TIP

In this era of terrorism awareness, many citizens may be in a panic mode when asking for safety advice and emergency preparedness. Firefighters should attempt to calm such individuals when answering their questions about safety and survival.

There are many scenarios that could become reality and preparing for each individual incident is impossible. However, there are preparatory steps that citizens can take that are common to the different emergency categories. Firefighters should teach the citizens to be motivated by wisdom and not by fear. They should be advised to take some commonsense steps to ensure their safety. Keeping abreast of the County, State, and Federal emergency recommendations and staying informed about current events should help alleviate some of the panic concerning the unknown.

FIRE AND LIFE SAFETY EDUCATION

For most people, the tragedy of losing a home or a loved one to fire is something that happens to someone else in some other neighborhood or town. How many post-incident reports have appeared on television where the neighbor remarks, "I can't believe this happened on our block. This is such a quiet street"? Teaching citizens to recognize life safety hazards and to react appropriately is clearly a fire department function and responsibility. The problem for fire safety educators is how to make fire and safety personal issues in the community.

> **NOTE**
>
> National and state statistics do little to impress people about the immediacy of a problem or its potential impact on their lives. All attempts should be made to enlighten citizens in regard to fire safety issues in their community. Citizens should be taught what they can personally do to prevent becoming a statistic.

Some departments have personalized the tragedy of fire by conducting tours of burned-out homes where occupants have been injured or killed. Specially built fire safety houses, as shown in **Figure 21-48,** allow more intensive training for the general public. These fire safety houses are designed to simulate fire conditions in a safe and controlled environment. They allow citizens to experience firsthand some of the conditions present during smoke or fire conditions. The citizens learn that a basic evaluation can become a confusing scenario under adverse conditions. Educational efforts must be tailored to meet the specific needs of the community and not be limited strictly to fire-related issues. For example, water safety classes are frequently taught by fire service personnel in communities with a high concentration of backyard swimming pools.

FIGURE 21-48 This school bus was redesigned and specially equipped to become a fire safety house.

Fire and Life Safety Program Presentations

Firefighters are frequently asked to speak to groups of all ages regarding fire and life safety issues. For a presentation to be worthwhile and effective, it should contain three main components: preparation, presentation, and practice.

The preparation phase involves the firefighter and the audience. The firefighter must be well versed with current knowledge on the subject matter to be presented and come prepared with appropriate audiovisual equipment, training aids, and literature. Once in front of the audience, the firefighter should prepare the group for the information they are about to receive. An effective method for gaining the audience's attention is to make the subject matter personal to them. For example, when speaking on the topic of drownproofing to a group of parents, it might be beneficial to recount a recent incident in the community, minus the graphic details, of course. The firefighter should let them know that they are about to receive information they can use to ensure the incident is not repeated at their homes. Getting the audience interested and personally involved is conducive to learning and the goal of the preparation phase.

Now that the audience is primed to learn, it is time to present the information. How the information is transferred from the firefighter to the individuals in the group is left to the creativity of the speaker. Showing a video and answering questions may be easy for the firefighter, but generally not much learning takes place. Professional educators know that the level of learning increases proportionally with the number of senses affected. In other words, the more the audience can see, hear, touch, or smell the subject matter, the more the information will be retained. When speaking to a group of hotel workers about evacuation of their guests during an emergency, it is important to let them hear the fire alarm sounding and see the strobe lights flashing. When speaking to children, demonstrating specialized equipment used in fire prevention and education helps to keep the children's attention, **Figures 21-49A** and **B.**

In the final phase of the program, the participants are asked to apply, practice, or demonstrate what they have learned. A group of small children being taught how to report an emergency could practice by using fire station telephones with a firefighter on an extension playing the role of a dispatcher receiving the call. An industrial fire brigade, having been instructed in the proper selection and use of fire extinguishers, could be allowed to extinguish small pan fires. (Appropriate safety equipment is mandatory. Firefighters must always verify local department policy regarding live

(A)

(B)

FIGURE 21-49 These (A) fire safety robots and (B) fire safety characters keep the children focused and help to get the message across.

fire training.) Firefighters should observe this phase closely, ensuring that the information presented is being applied properly and making correction where necessary.

STREETSMART TIP

When conducting tours for children, a nice touch—that parents and caregivers usually appreciate—is to stress the importance of "putting things back where you found them." Fire apparatus have equipment stored in specific locations and every firefighter knows where to go for that particular piece of equipment on a particular apparatus. Stressing this point to children helps the adults teach the importance of returning items to their proper location in a practical way.

Forms of Fire and Life Safety Programs

Public education programs usually take three general forms: public service announcements, school programs, and adult/homeowner programs. Fire station tours also provide an excellent opportunity to heighten the public's awareness of fire safety issues.

Public Service Announcements (PSAs)

A simple, inexpensive, and timely method of spreading fire safety messages is to use the media. Radio, television, and newspapers are always willing to assist public agencies with their efforts. Short five- to ten-second television or radio spots or small news articles can be used for delivering safety messages on such seasonal topics as drowning prevention, fireworks safety, smoke detector maintenance, Halloween costumes and candy, and Christmas tree safety.

School Programs

Fire safety and education programs for elementary school children can be valuable and productive, **Figure 21-50.** They are designed to teach students the dangers of fire, how to prevent fires from happening, and what to do if they are caught in a fire. In many cases, the children will take the information home to their parents or caregivers and become instructors to the adults. These programs can take many forms: two of the more popular are NFPA's Learn Not to Burn and the Junior Firefighter Program.

The Learn Not to Burn program is designed for children in preschool through the third grade. It teaches children safety lessons using educational material and technical support provided by NFPA. Teachers attend workshops where they are instructed in course material by fire educators. The teachers then teach the program over a three-month span at their schools. Incorporated into the Learn Not to Burn curriculum are lessons on home evacuation using material from Operation EDITH (Exit Drills In The Home) and the ever-popular Stop, Drop, and Roll.

The Junior Firefighter Program can take many forms. It is generally presented to fifth-grade students in three stages. Phase 1 is usually an oral presentation and tabletop demonstration by a fire education specialist. Using an instructor's manual and student workbooks provided by the fire department, teachers complete phase 2 in about four weeks. Phase 3 consists of a visit to the school by an engine company. Firefighters describe their jobs and demonstrate their equipment. Additionally, they present Junior Firefighter patches or badges to students who have completed the program.

School Evacuation Drills

An effective fire prevention program in schools should be a matter of high priority not only to the fire department and school staff but to the community as a whole. A proper program will include regular and thorough fire safety surveys of the school premises by school staff and the fire department as described throughout this chapter. Additionally, the program must include emergency evacuation drills. Although such drills are usually referred to as "fire" drills, it must be emphasized that emergency evacuations can be necessary for other occurrences as well—bomb threats and natural disasters, to name two. Many states and local communities have adopted statutes requiring evacuation drills to be conducted at designated intervals, generally twice per school year. Where no such mandatory requirement exists, the fire department should work with school administrators to ensure that drills are conducted voluntarily.

Emergency exit drills are necessary to ensure that students and staff can safely, efficiently, and effectively evacuate their buildings in times of stress. Order and control are the key elements that will help to prevent injuries during the evacuation. A good emergency evacuation plan and drill will include the following points:

- Evacuation drills should be conducted as often as necessary to ensure that all occupants of the building are thoroughly familiar with the process. The fire department should be present whenever possible but at least once per school year.

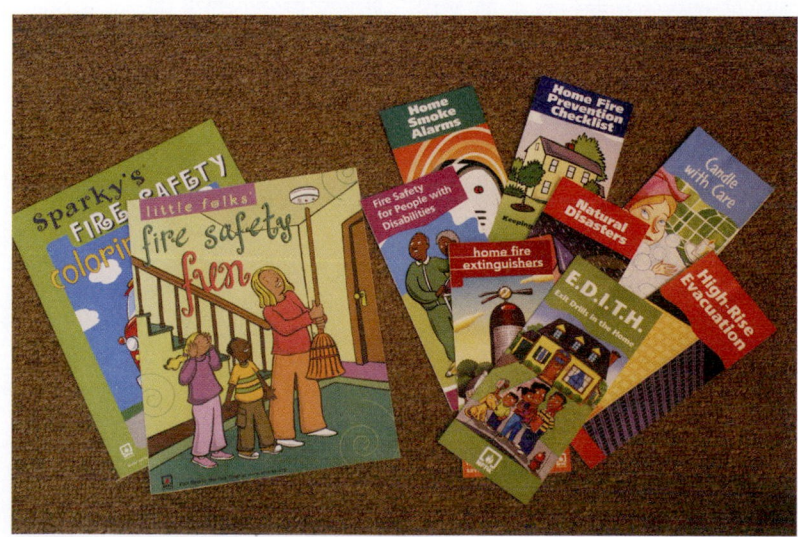

FIGURE 21-50 Popular public education programs are available through the NFPA.

- The focus of the drill should be placed on disciplined control and order. Speed, while important, should not be overemphasized. Speed is a by-product of properly planned and supervised evacuation.

- Specific exits should be assigned to groups of classrooms; for example, Classrooms 1, 2, and 3 should use Exit A and Classrooms 4, 5, and 6 should use Exit B. All building exits should be utilized to provide for the even distribution of exiting occupants. Alternative exits should also be designated.

- Students and teachers should proceed to predesignated assembly points outside and away from the building. Playgrounds are appropriate; parking lots are not. Teachers should be required to account for all students in their classes. Taking roll call once assembled outside is the preferred method. They should report the status of their classes to the school administrator in charge. Missing and unaccounted for children must be reported to responding firefighters.

- Emergency evacuation plans should be drawn in graphic form and posted in each classroom and at various locations throughout the school. **Figure 21-51** shows such a plan.

- To familiarize students and staff with the sound, the fire alarm system should be used whenever conducting an evacuation drill.

Adult Programs

The types of programs and lessons taught to adults are limited only by the imagination of fire service educators. Common themes include:

- Home fire safety
- Smoke detector placement and maintenance
- Drownproofing
- Fire extinguisher demonstrations and instruction for use
- Operation EDITH (Exit Drills In The Home)
- CPR instruction
- General first aid
- Hazardous materials awareness
- Earthquake, tornado, and flooding preparedness

Fire Station Tours

Visiting a local fire station has always been, and continues to be, one of the most popular field trips for all types of clubs, organizations, and schools. Station tours not only give the visitors a chance to see firefighting equipment and apparatus up close, but also give them a glimpse into life in a fire station. From the firefighters' perspective, tours provide an excel-

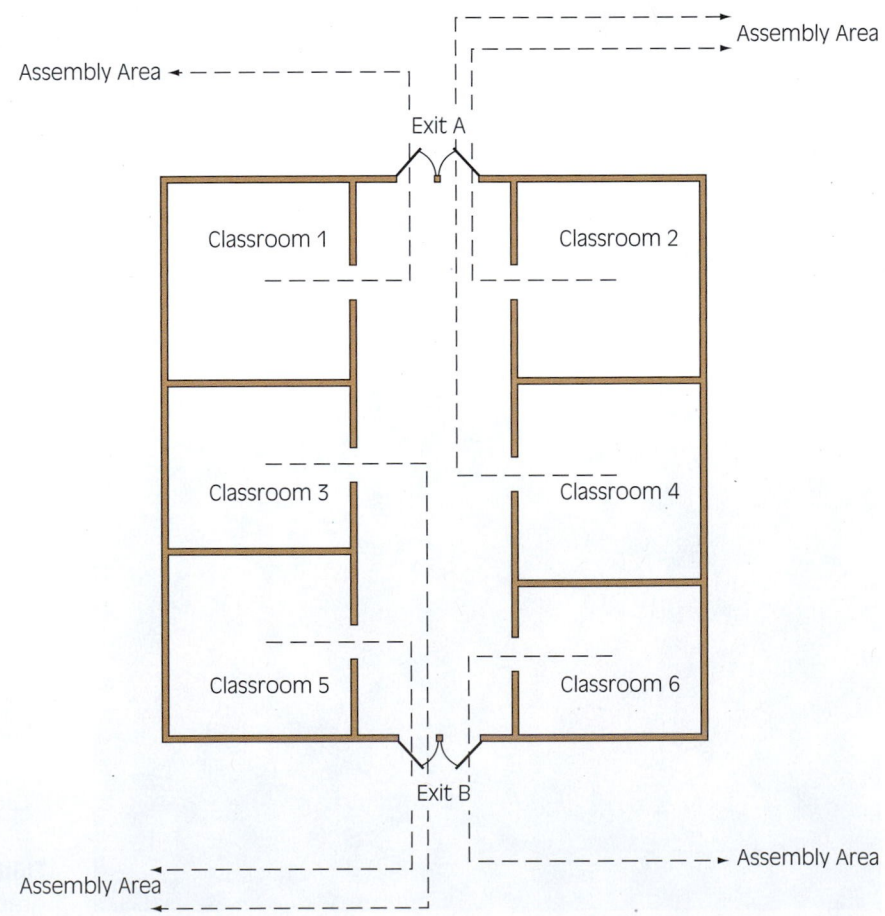

FIGURE 21-51 A graphic view of the school building showing exits, evacuation routes, and assembly areas should be posted in all classrooms and throughout the building.

lent opportunity to "spread the word" about fire and life safety.

Fire station tours should be tailored to fit the needs and interests of the visitors. Young children may be interested in trying on a firefighter's safety equipment.

CAUTION

Helmets should not be placed on very small children; neck injuries could result.

Children might also enjoy sitting in the engine's cab or watching a firefighter "slide the pole." Older children and adults may be interested in the operation of a pumper or dispatch and response procedures.

Fire safety messages should be interwoven into the station tour and directed at the appropriate level. Small children should be asked or taught how to call for emergency assistance, to "stop, drop, and roll," or how to properly react to an operating smoke detector and exit their homes safely. Older visitors should be given information regarding such issues as smoke detector and carbon monoxide placement and maintenance, home evacuation planning and drills, and general fire safety. This information should be presented orally during the tour and reinforced with written literature distributed at the end of the visit, **Figure 21-52.** It is generally not good practice to give visitors brochures and flyers during the tour because this will serve as a distraction. Firefighters should also take the opportunity to inform visitors about the specialized educational programs offered by their department. Giving visitors a printed schedule of upcoming events and a telephone number for registration or further information is an excellent method for encouraging participation.

PRE-INCIDENT SURVEY PROCESS

Fire prevention activities and pre-incident surveys, often referred to as preplanning, are related in some ways and different in others. From a fire prevention perspective, one assumes an incident can occur and seeks to take measures to ensure that it does not. Pre-incident surveys assume an incident has occurred and use tactics, strategies, and coordination of resources to minimize the impact on lives and property. The link between the two is education. Fire service educators strive to teach the public how to prevent fires and other emergencies, while at the same time teaching the proper action to take should an emergency occur.

FIGURE 21-52 Fire safety brochures available through the NFPA.

Pre-incident surveys can be as simple as company officers deciding from which hydrant they will lay hose for a particular building, or as complex as a coordinated effort among many agencies and many jurisdictions. Regardless of its scope, pre-incident surveys must be a collaborative effort between all divisions of the fire department and involve other agencies when appropriate.

Deciding to Preplan

In a perfect world, all structures in a given jurisdiction would be preplanned with the information stored in a massive database. That information would be available to emergency responders via mobile data terminals located on all apparatus. GPS (global positioning system) units would guide apparatus to the scene where flashing strobe lights would identify the affected structure. While such systems currently exist, they are often out of the financial reach of many fire agencies—at least for the immediate future. Therefore, most data gathered during the pre-incident survey process is stored in "hardcopy" form. For practical reasons—storage, staffing, and time constraints—fire agencies must prioritize occupancies for pre-incident surveys. In so doing, consideration should be given to the following:

- Type of occupancy
- Type of incident expected
- Life hazards—civilian and firefighter
- Nature of activities conducted at the occupancy
- Exposure to surrounding areas
- Complexity of firefighting operations
- Resources required

Structures such as high-rise buildings, hotels, malls, large industrial buildings, and multistory and multibuilding apartment complexes should be given high priority. Such occupancies are often referred to as **target hazards,** indicating that a greater than average life hazard or complexity of firefighting operations can be expected.

Site Visit

Generally, after deciding to preplan a structure, the first step is the on-site visit or fire safety survey. The site visit for pre-incident survey purposes differs from a fire prevention fire safety survey in that the firefighters' purpose is fact gathering from an operational strategic perspective as opposed to ensuring compliance with fire code regulations. Certainly if code safety problems are noted, they should not be ignored but rather dealt with in the usual manner.

During the pre-incident survey site visit, firefighters should be gathering information that will enable emergency responders to deal effectively with all levels of situations at the site. At a minimum, the following information should be obtained and documented:

- Occupancy classification
- Construction type and method
- Structure size, height, and number of stories
- Exiting systems
- Built-in fire protection
- Access points to the site and interior of the structure
- Exposure problems
- Hazardous materials usage and storage areas
- Personnel—civilian and firefighter—safety issues and features
- General firefighting concerns

Diagrams

An essential element of the pre-incident survey process is the diagram of the site, structure, or occupancy. Here is where the information gathered during the site visit is displayed in graphic form. Diagrams of the site and floor plans should be included and should be presented in both a **plan view** and a **sectional view** using a standardized set of symbols and drawn to scale, **Figure 21-53.** Most fire departments are using CAD drawings from the builder that can be imported into a fire department's computerized records. Other departments use purchased software to draw and store the building plans. Those departments that require fire personnel to draw the plans by hand should have a set of standard map symbols and all drawings should be to scale. The prevention bureau or officer from these departments should be consulted for the symbols used and the methods to draw the plans.

Diagrams can be hand drawn but the preferable method is to produce them using a computer. Sophisticated computer-aided drawing software need not be used. A simple paint or draw application will do the job. Updating the diagram will be a simple process if the original is produced in this manner.

The site plan should include the perimeter of the building, surrounding roadways, access points to the site and structures, fire hydrants, water main sectional valves, fire sprinkler control valves and connections, cross fencing, gates or other obstacles that may impede the movement of emergency vehicles, staging areas, and likely apparatus placement locations.

Floor plans should include a general layout of the interior of the building, floor by floor. Areas of high life hazard, exiting systems, fire protection features, hazardous materials use and storage areas, type of construction, roof openings, stairs and elevators, and other pertinent information can be shown on this drawing.

Seek Input from Others

The pre-incident survey process is a collaborative effort involving all levels and divisions of the fire department and often other agencies. All relevant information obtained should be assembled in narrative and graphic form and routed as appropriate to obtain the perspective, expertise, and input of others. For example, in urban areas, the water department may suggest that, due to anticipated volume of fire streams during a full-scale firefighting operation, the water pressure in city mains be boosted. Law enforcement may have input regarding anticipated

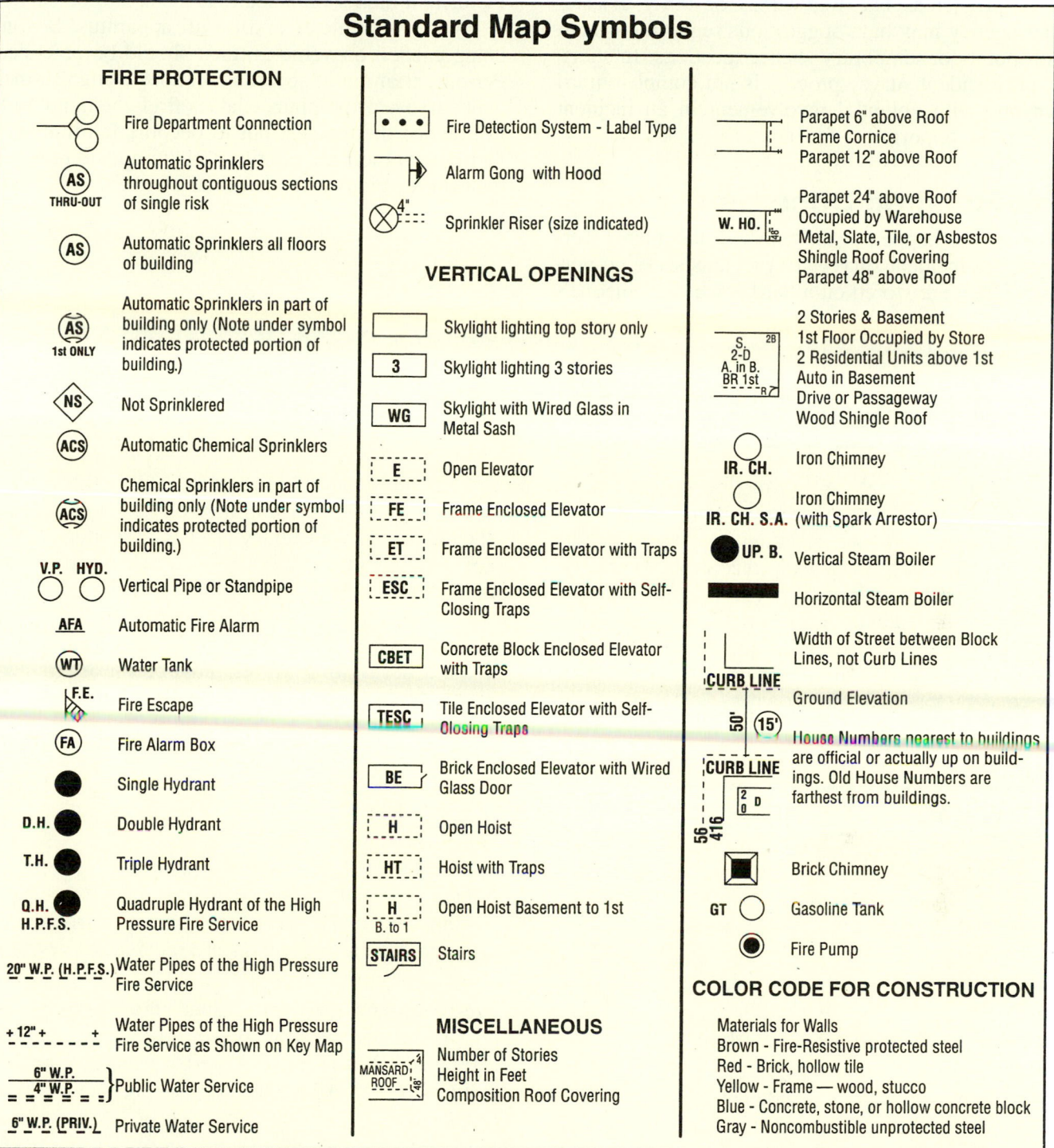

FIGURE 21-53 Standardized symbols for pre-planning.

traffic problems. The Red Cross or other disaster relief agency may have suggestions regarding evacuation points or temporary shelter locations. In short, the pre-incident survey process is not complete until everyone with potential involvement in an incident has had the opportunity for input.

The Finished Document

What should happen to the finished document? Certainly, pre-incident surveys or preplans serve no purpose if they are stored on a hard drive on someone's PC or in a binder on a shelf. They should be carried in command vehicles and on all apparatus. Beyond being carried on vehicles, they should be practiced. Periodic training at specific locations using pre-incident surveys or preplans enables firefighters not only to practice their skills, but to become familiar with facilities in their jurisdictions and to update pre-incident survey plans when necessary.

LESSONS LEARNED

It is hard for firefighters to imagine anything more exhilarating and satisfying than aggressively entering a burning structure, rescuing its occupants, and defeating the sworn enemy—uncontrolled fire. Or perhaps pulling a drowning child from a swimming pool and watching him respond as another firefighter applies lifesaving techniques. Unless it would be responding to a "fire out on arrival" call because smoke detectors that homeowners installed at the fire department's recommendation had alerted their sleeping families to the fire downstairs in the kitchen. Or reading in the morning newspaper about children who escaped serious injury when a fireworks sparkler ignited their clothes and they knew to "stop, drop, and roll" because the firefighters who came to their school had taught them how. There will always be emergencies and disasters. Preventing or lessening the impact of these devastating events must be the firefighter's constant goal.

KEY TERMS

Administrative Warrant   An order issued by a magistrate that grants authority for fire personnel to enter private property for the purpose of conducting a fire safety survey.

Fire Hazard   Any condition, situation, or operation that could lead to the ignition of unwanted combustion or result in proper combustion becoming uncontrolled.

Means of Egress   A safe and continuous path of travel from any point in a structure leading to a public way. Composed of three parts: the exit access, the exit, and the exit discharge.

No-Knowledge Hardware   Locking devices that require no key or special knowledge to operate.

Panic Hardware   Hardware mounted on doors that enable them to be opened by pushing from the inside.

Plan View   A drawing or diagram of a building or area as seen from directly overhead. May include a site plan or a floor plan.

Pre-Incident Plan   Advance planning of firefighting tactics and strategies or other emergency activities that can be anticipated to occur at a particular location. Also referred to as pre-incident survey or pre-plan.

Pre-Incident Survey   Advance planning of firefighting tactics and strategies or other emergency activities that can be anticipated to occur at a particular location. Also referred to as pre-incident plan or pre-plan.

Safety Container   A storage can that eliminates vapor release by using a self-closing lid. Also contains a flame arrestor in the dispenser opening.

Secondary Containment   Any approved method that will prevent the runoff of spilled hazardous materials and confine it to the storage area.

Sectional View   A vertical view of a structure as if it were cut in two pieces. Each piece is a cross section of the structure showing roof, wall, horizontal floor construction, and the location of stairs, balconies, and mezzanines.

Special Egress Control Device   Door hardware that will release and unlock the door a maximum of 15 seconds after it has been activated by pushing on the bar.

Target Hazard   An occupancy that has been determined to have a greater than average life hazard or complexity of firefighting operations. Such occupancies receive a high priority in the pre-incident survey process and often a higher level of first-alarm response assignment.

REVIEW QUESTIONS

1. List the four Es of fire prevention.
2. What are the objectives of a fire company fire safety survey program?
3. Identify the first step in conducting a fire safety survey after arrival at the occupancy to be surveyed.
4. Identify the steps to be taken in the event that permission to inspect is denied by the business owner or representative.
5. Define a fire hazard.
6. Identify the three components of a means of egress.
7. Where is panic hardware required?
8. Discuss the key areas of concern to the fire inspector when inspecting a fire sprinkler system.
9. When are commercial cooking surfaces and equipment required to be protected by a fixed fire extinguishing system?
10. Is it necessary to document a deficiency noted during a fire safety survey if it was corrected in the presence of the inspector?
11. When conducting a home fire safety survey, the firefighter should be aware of several typical fire hazards found in the home. Identify three.
12. Should fire department public education efforts be tailored to meet the needs of the community and not be limited to fire-related issues?

23

Firefighter Survival

LEARNING OBJECTIVES

After completing this chapter, the reader should be able to:

23-4 Discuss teamwork and its part in firefighting.

23-10 Demonstrate assembling a team for an interior structural firefighting operation.

*The FF I and II levels, as defined by the NFPA 1001 Standards, are identified in different colors:
FF I = black, FF II = red, additional information = blue.

INTRODUCTION

Teamwork is a critical factor in the successful operations at any emergency incident. This is especially critical when things go wrong during the incident.

If a firefighter emergency were to occur, a planned, systematic process of self-rescue or rapid intervention would have to be established to avoid compounding the seriousness of the emergency. Having procedures in place for rapid escape and self-extrication, accountability recall, rapid intervention, and post-incident emotional processing is essential.

This chapter discusses the individual's preparation as part of the team and the team approach to emergency response.

TEAMWORK

Firefighter survival is dependent on two focus areas: Prevention through readiness and training on firefighter emergency procedures. Obviously, it makes sense to attempt to prevent a firefighter emergency from happening.

Firefighters can help prevent responder emergencies through incident readiness—that is, the efforts to ensure that all firefighters are mentally and physically ready to respond. Attention to the following issues will assist in preventing responder emergencies:

- Fitness for duty
- Personal protective equipment (PPE)
- Team continuity
- Task accountability
- Hazards at incident
- Orders and communication
- Rapid intervention planning
- Firefighter actions and inaction (knowing when to hold back)
- Rehabilitation

> **NOTE**
>
> The success of all incident handling is dependent on trained, assembled teams accomplishing organized tasks, **Figure 23-1**.

The department as a whole cannot be effective in its firefighter safety effort without the support of a team approach to handling incidents. This team includes

FIGURE 23-1 Freelancing is eliminated and incident success is gained when specific tasks are assigned to teams of two or more firefighters operating from a single incident action plan.

the individual firefighter. To ensure safety, the team should follow these procedures:

1. Utilize an incident command system (ICS). The ICS should detail lines of authority and communication, be reasonable in its span of control, and include an action plan. Section chiefs, division/group supervisors, and team leaders need to communicate progress and status. Orders need to be clear and accomplished within an acceptable risk environment.

2. Work together and remain intact. The "buddy" concept is imperative for firefighter safety.

3. Look after each other. Team members who continuously watch each other will find and address conditions that will lead to injury. Aggressive rehabilitation will provide hydration, rest, and nourishment—thereby increasing the time the team can perform safely. At the team level, issues of PPE misuse should be addressed. Likewise, the team must watch each other for signs of fatigue, overaggressiveness that is dangerous, unsafe acts, and freelancing.

CAUTION

The separation of members within a team is a contributing factor to firefighter fatalities.

Continuity

Departments that allow well-meaning firefighters to perform as each sees fit, **freelancing,** may ultimately experience a significant injury or duty death. Freelancing is dangerous. Investigative reports cite freelancing as a significant contributing event to death or injury. To eliminate freelancing, every firefighter must be assigned to a team and be given specific assignments. On completion of a task, the team must report that the task is done and also report any update information that could be useful to the incident commander. The incident commander may then reassign the team or have the team report to staging or rehabilitation. Fire and rescue work is a team effort in support of an incident action plan developed by the incident commander. Team leaders implement the plan and provide feedback to the incident commander. Firefighting teams follow the team leader's direction. Freelancing also occurs when a team operates outside the action plan or when individuals work alone. Working alone or outside the action plan endangers individuals and the team, as **Figure 23-2** demonstrates. The individual firefighter can help prevent freelancing by *not* working alone and supporting the action plan when part of a team.

FIGURE 23-2 Freelancing endangers individuals and the team. This firefighter is working alone in a collapse zone—for what gain?

The team must exercise good judgment when completing a task. For example, a team performing overhaul may decide that an additional tool is needed. The simple decision is to send a lone firefighter to retrieve the tool, but this sets up a situation that breaks team continuity. Rather than sending a lone firefighter, it may be prudent to send two, leaving two to continue task completion. In the case of a three-person crew, the team leader should weigh the potential of breaking the team. In most cases, it is best to keep the team together and ask for a shuttle team to bring the needed tool or to take a momentary break and have the team exit the area together, get the tool, and then return. This process may seem inefficient, but it underscores a safe approach to team continuity. **Figure 23-3** outlines practices that contribute to team continuity.

Relationship to the Response

Another integral part of the team process is attention to each individual's relationship to the response. Specifically, firefighters' relationship to the response includes their assignments and their own personal size-ups.

■ *Assignment.* In most cases, firefighters are given readiness assignments—that is, being assigned as given crew members for given positions on the apparatus. With these assignments come responsibilities to ensure that they perform preassigned or officer-directed duties. In cases where firefighters have no preassigned duty, firefighters must be prepared to perform a host of duties. The readiness elements entail a mental preparation to perform the tasks of that position. If firefighters are assigned positions that they have not performed recently, they may not be ready to act—leaving

Keys to Team Continuity

- Know who is on your team and who is the team leader.

- Keep your team in *visual* contact at all times. If visibility is obscured, be within touch or voice distance of your partner.

- Communicate your needs and observations to the team leader.

- Rotate to rehab and/or staging as a team.

- Watch your team members—practice a strong "buddy-care" approach.

FIGURE 23-3 Team continuity reduces the chance for injuries.

FIGURE 23-4 Firefighters achieve mastery of tasks through repeated training. Mastery reduces the chance of injury.

an invitation to injury. In some departments, a list of crew duties for a given seated position on the apparatus is listed on a plastic card—this is a great reminder. Firefighters should take the time to reacquaint themselves with the expected tasks, tools, and procedures required. This improves readiness and reduces injury potential. In those cases where the firefighter is working with an unfamiliar crew and/or officer, the firefighter should take the time to review expectations and talk through crew assignments with the officer to make sure that he or she is on the same page as the new crew.

- *Personal Size-Up.* Personal size-up can be defined as the continuous situational awareness and mental evaluation of firefighters' immediate environments, facts, and probabilities. This mental evaluation should have them evaluating the weather, time of day, current chain of command, and likely assignment. Once again, they are mentally preparing themselves—thereby reducing injury potential.

Work as Trained

> **STREETSMART TIP**
>
> Personal accountability is achieved when all firefighters and fire officers are able to perform assigned essential tasks—tasks they have been trained for—and keep the chain of command advised of their progress.

Of significant importance here is everyone's ability to perform as trained. A common thread discovered in post-incident investigations is the fact that many of the injured were performing tasks they had never

been trained for. In some cases, the training had been given by the individual's department, but the individual either forgot how to accomplish the task or chose a method that was inappropriate for the situation. How can this be prevented? Listed next are two key points to assist firefighters.

1. *Firefighters should perform as trained.* Most departments invest considerable cost to train firefighters. Firefighters must do their part and constantly practice and perfect their skills.

2. *Firefighters need to know their strengths and weaknesses.* Training never ends for a firefighter! They must constantly test themselves and try to improve those areas where they discover weaknesses Firefighters should strive for mastery of their skills and continually expand their abilities, **Figure 23-4.**

> **CAUTION**
>
> Firefighters should not let the incident scene become an avenue for discovery and on-the-job training.

The individual firefighter holds the final key to incident safety. Remember the following:

- Be ready.
- Understand and act within the chain of command.
- Perform as trained.
- Use an incident engagement checklist (See Section I, Chapter 5).

ORDERS AND COMMUNICATION

The incident commander is responsible for assembling an **incident action plan (IAP)** that is implemented by teams performing tasks. These tasks are assigned to organized teams in the form of orders. Each team is responsible for carrying out the order and providing updates on a regular basis. Additionally, the team must relay information about any pertinent hazards or conditions that may be important to the overall IAP.

SAFETY

As individuals, firefighters must keep their team leader advised of conditions and hazards they find as work is performed.

The discovery of a weak stairway, significant weight load in an attic or on the roof, worsening smoke conditions, or fire in a void space are all important facts that should be relayed to the team leader and subsequently up the chain of command.

Occasionally, a team performing an order is given a different order by someone else in the incident command system. In this case, the team leader needs to inform the person giving the new order that they are already under orders. The person giving the new order then must either countermand the original order or find someone else to perform the new order. In these cases, it is imperative that both the team and the person giving the new order communicate any changes to the person giving the original order.

Typically, the first arriving company performs a prescribed set of tasks or orders based on a pre-incident survey or standard operating procedures/guidelines. In these cases, the teams should not only know the tasks, but also the tools required and any safety considerations. Team members should vocalize safety thoughts among themselves—in fact, this should be encouraged. Building construction considerations, fire behavior and smoke observations, and access/egress routes should be discussed as the team approaches its task.

Later-arriving companies are given assignments based on the incident action plan. These companies should gather essential information and discuss safety issues based on the hazards and conditions presented. **Figure 23-5.** This practice is not new—wildland fire crews are required to perform a crew briefing prior to task accomplishment. **Figure 23-6** lists good habits regarding orders and communications.

FIGURE 23-5 Later-arriving teams can benefit from a quick, formal safety briefing prior to performing tasks.

Good Habits Regarding Orders and Communications

- Clearly understand any order; ask questions and seek clarity.
- Assemble tools and openly discuss procedure as you approach tasks.
- Relay all pertinent hazards found during the course of task completion.
- Update crew status on a regular basis.
- Advise when tasks are complete.
- Communicate rotation to rehab and/or staging.

FIGURE 23-6 Practicing good reporting habits enhances communications.

RISK/BENEFIT

Another key to firefighter survival is the concept of risk/benefit. **Risk/benefit** is an evaluation of the potential benefit that a task will accomplish in relationship to the hazards that will be faced while completing the task. As an example, the task of vertically ventilating a residential structure fire will benefit the fire suppression and lifesaving effort by increasing visibility, reducing heat buildup, and reducing overall damage. That benefit, however, is at the risk of roof collapse, operating on a potentially steep incline, or operating without quick egress. The team operating on the roof is often in the best place to judge this relationship. Signs of collapse, difficult footing, and obscured visibility may present a significant risk to

FIGURE 23-7 Solid risk/benefit analysis means taking no risk for that which is already lost. *(Courtesy of Richard W. Davis)*

the team. The risk/benefit evaluation in this example may cause the team leader to withdraw the team.

Risk/benefit evaluations take place at different levels within an ICS. At the command level, an overall evaluation takes place, as reflected in the IAP. At the team level, crews evaluate their immediate environment and report this back to the person giving them their orders, **Figure 23-7.** Regardless of the level of evaluation, some basic guidelines can be used to help make risk/benefit decisions:

- Firefighters will take a significant risk to save a known life.

- Firefighters will take a calculated risk, and provide for additional safety, to save valuable property or reduce the potential for civilian and firefighter injuries.

- Firefighters will take no risk to their safety to save what is already lost.

- In situations where conditions are deteriorating quickly, firefighters should retreat to a defensive position.

These guidelines are common within the fire service and are more clearly spelled out in NFPA Standard 1500, Fire Department Occupational Safety and Health Program.

RAPID INTERVENTION TEAMS

Section I, Chapter 16 introduced the use of a rapid intervention team (RIT) to rescue firefighters experiencing an emergency. While the ultimate incident goal should be to *avoid* the need for rapid intervention, it is prudent (and in many cases, mandatory) to establish a RIT when firefighters are engaged in risky environments. The concept of RITs has become commonplace in many fire departments, although they may be titled differently. Other examples include: rapid intervention company (RIC), firefighter assist and support team (FAST), and rescue assist team (RAT).

The concept of rapid intervention is started early at an incident. The two-in/two-out rule (known as initial rapid intervention or IRIT) is designed to provide for immediate intervention if needed. This requirement is specific to interior fire attack but should be conceptually used for all incidents where firefighters are taking risks. As more resources arrive, the RIT becomes a designated team for a defined task. This RIT should have no other assignment than to prepare for the rapid deployment in support of a search and/or rescue. The dedicated RIT should be assembled using well-trained and experienced firefighters and officers. Specific RIT techniques and duties are considered "advanced skills" past the scope of this book, although some general RIT information and responsibilities are included here.

> **STREETSMART TIP**
>
> Firefighters assigned to the RIT should be prepared to act immediately on orders to rescue firefighters in need. This means the crew assigned the RIT task should be equipped and prepared for the task.

The RIT assignment is not a passive one—the members assigned should proactively prepare for engagement and assist in providing additional egress options for committed crews. **Table 23-1** outlines responsibilities for the RIT.

When activated for an actual firefighter emergency, the RIT will engage the emergency as trained. It is important that other firefighters resist the urge to self-assign themselves to help out the RIT. While it may be likely that others will be reassigned to help, it is important to remain focused and not abandon previous assignments.

> **CAUTION**
>
> Upon hearing a Mayday or RIT activation, *DO NOT* abandon your non-RIT assignment until reassigned by the Incident Commander or Operations Section Chief.

TABLE 23-1   RIT Responsibilities Checklist

Tool Assignment

RIT Leader/Officer

Radio

Rope bag

Thermal imaging camera

Hand light

Firefighter 1

Radio

Forcible entry (irons)

Hand light

Firefighter 2

Radio

RIT pack

Hand light

Firefighter 3

Radio

Specialty tools

Hand light

Additional Tools

Webbing

Attic ladder

Rescue litter

Power saws

Tarps for tool staging

Extra SCBA cylinders

Handline

EMS jump kit

Rescue

- Immediate location of companies
- Identify hazards and safety issues

Escape

- Open up and remove blocked or secured doors and windows
- Raise ladders to windows and roofs

Circle Survey

- Building size and special features
- Dimensions
- Construction and stability
- Exposures within and without

Hazards

- Information placards
- Electrical power and downed wires
- Hazardous materials and hazard conditions
- Gases: natural/propane/other

LESSONS LEARNED

Firefighter survival is dependent on many proactive and preventive actions. Safe operations and, therefore, survival are dependent on team continuity and the elimination of freelancing. Additionally, safe incident operations are achieved when teams perform orders with attention to communications and timely updates. While operations are under way, individual teams must practice solid risk/benefit analysis for their specific job.

Rapid intervention teams are formed for immediate deployment should a firefighter emergency take place. Firefighter emergencies require the individual firefighter to practice a clear and concise approach to dealing with the emergency.

KEY TERMS

Freelancing   The act of working alone or performing a task for which the firefighter has not been assigned.

Incident Action Plan (IAP)   A strategic and tactical plan developed by the incident commander.

Risk/Benefit   An evaluation of the potential benefit that a task will accomplish in relation to the hazards that will be faced while completing the task.

REVIEW QUESTIONS

1. List the three main components that lead to incident readiness.
2. Briefly describe the three components that lead to fitness for duty.
3. Name three practices that lead to team continuity. Describe how each can increase firefighter survival.
4. Define risk/benefit.
5. Describe the procedures that should be taken to establish and prepare for the assignment of a rapid intervention team.

31 Disaster and Large Incident Response

The 9-1-1 dispatchers in Austin, Texas, began receiving 9-1-1 calls in the early morning hours from commuters traveling along a major U.S. highway running through the city. The callers described a tanker with "green smoke" coming from the top. The fire department dispatched a full hazardous materials alarm assignment to the scene along with police and EMS. Upon arrival, the first-in companies saw they had a large chlorine tanker leaking from the top.

The fire department immediately set up a hot zone and began evacuation of downwind structures, including densely populated residential areas as well as commercial businesses near the highway.

The hazardous material team immediately began area monitoring operations as well as preparing a team to perform a reconnaissance entry into the hot zone. Once a hot zone was set up using standard reference guides, the fire department requested a unified command post with fire, police, EMS, state department of transportation, and the state department of public safety.

After downwind monitoring had been accomplished, law enforcement agencies were briefed, and they began evacuation procedures beyond the initial hot zone. Seeing that this was sizing up to possibly be an event that would cover at least one 12-hour operational period, the fire department activated the city/county Emergency Operations Center (EOC) and began requesting resources that were, or possibly may be, needed.

The command structure was set up using the national incident command system protocol and a mobile command post was dispatched to the scene to support command and research capabilities.

After using a software plume modeling program to further develop the evacuation perimeters, command developed a public information statement and requested that EOC initiate a public service broadcast giving instructions to the public through radio and television media. Command also communicated the location of the public information briefing location to all public media organizations.

Command initiated call back of off duty hazardous materials technicians after reconnaissance showed that the chlorine relief valve was functioning as it should and that the continuing loss of the product could be predicted.

A logistical support plan was developed for the different agencies involved. This was crucial for support of the hazardous materials team in regard to technical monitor support, research, and personal protective equipment.

Formal contacts were made with the owner of the tanker and data related to the cargo was obtained. At this time a private contractor with chlorine tanker expertise was contacted and requested to respond.

The fire department hazardous materials branch officer developed a plan to move the tanker approximately one mile to the location of a city-owned property. This was coordinated through the command post where law enforcement expanded the traffic control perimeter and fire/hazardous materials team recalculated the plume model changes created by moving the tanker.

A hazardous material technician with experience driving semi-trucks was briefed, and the vehicle was moved to the designated area—after coordinating with law enforcement and insuring that the site where the vehicle was being moved was evacuated. The media release of this move was coordinated through the EOC.

In the following hours, unified command developed a plan to continue long range monitoring with call back personnel using ATV's. A fully functional rehab area was set up to support all personnel and a fire department shift rotation was planned and executed through the fire department representative in the EOC.

The state environmental regulatory agency was notified and worked within the unified command structure through the EOC.

The chlorine tanker continued to vent and the fire/hazardous materials team continued monitoring. The decisions regarding the re-opening of roads, highways, and neighborhoods were constantly reviewed. Using the NIMS model and taking full advantage of the unified command structure, several options for mitigation were discussed and an action plan was developed.

The action plan developed required the off loading of the leaking tanker into another tanker provided by the owner. As this operation takes considerable time, the department developed a relief and rotation plan for companies.

The plan for the re-occupation of neighborhoods and the re-opening of roads, highways, and businesses was developed using constant input of monitoring data from hazardous material team members. The incident carried into a third operational period using resources from local, county, and state agencies. This incident was carried out with no injuries to citizens and responders by using a step-by-step approach. This was made possible by area pre-planning, multi-agency training, and adherence to the unified command system.

What began as a seemingly standard hazardous material alarm became an incident where thousands of citizens were affected, around one hundred responders were used, and multiple local and state agencies were involved. Without a well-planned response set up by these agencies, a larger disaster could have erupted and chaos would have overtaken the scene. It is important that you—as a firefighter—understand and support your department's, and your local, state, and even national agencies' plan for large disaster response.

—*Street Story by Captain Mark McAdams, Austin Fire Department*
Special Operations Battalion, Austin, Texas

INTRODUCTION

Every community is vulnerable to a large incident or disaster. A large incident or disaster cannot be easily defined to suit every occurrence. A disaster is commonly thought of as an incident that overwhelms a community's resources. In a small town, a bus accident could easily be classified as a disaster, because it could very well overwhelm the emergency resources available to handle the many patients, rescue efforts, and hazards entailed. In a larger community, this same scenario might have a minimal impact on the emergency resources. A large incident is another description of a disaster, yet it is a broader definition, encompassing many types of incidents that might not commonly be thought of as "disasters" in a traditional sense. An example would be a large warehouse fire or a hazardous materials incident at a school. This chapter discusses disaster and large-incident responses together, since the concepts and strategies used to manage them are the same.

TYPES OF DISASTERS AND LARGE INCIDENTS

Disasters may occur as a result of natural events, such as weather, or they may be a result of human actions. There are many different types of disasters and large-scale incidents that can occur. Disasters and large-scale incidents can have short-term problems for a community, such as a massive power outage, or long-term consequences for an area or region, as in the case of a nuclear power plant release that affects people over decades. Disasters can also lead to other disasters, such as a weather event that causes flooding, a dam break, power outages, and public health problems.

Firefighters should be familiar with what types of disasters may affect their communities, and most fire departments have specific plans for dealing with the disasters that are most likely to occur in their jurisdictions. These plans are often connected with a local government's disaster or emergency management plans. The emergency services are typically an integral part of these emergency plans and responders should be aware of their responsibilities during a disaster.

The following section outlines a number of disasters and large scale incidents that firefighters might face.

Water-Related Emergencies

Water-related emergencies may result from flooding and/or weather conditions or failure of critical infrastructure in a community, such as a dam failure, **Figure 31-1.** Flooding is often a result of prolonged or severe rain conditions, tsunamis, hurricanes, and the melting of large amounts of snow and ice. Flooding can strand residents in homes and businesses, cause

FIGURE 31-1   This photo shows a portion of New Orleans, Louisiana, after Hurricane Katrina hit in September 2005. The widespread flooding resulted from both the weather conditions and the failure of dams. *(Courtesy of FEMA/Michael Rieger)*

(A)

(B)

FIGURE 31-2   Far inland areas of Sumatra experienced minor flooding during the 2004 Indian Ocean tsunami (A); however, the coastlines experienced massive devastation (B). *(Courtesy of U.S. Geological Survey/Guy Gelfenbaum)*

massive damage to buildings and community infrastructure, create fires and hazardous materials emergencies, and prohibit emergency responders from reaching calls for help.

As already mentioned, one water-related emergency is a **tsunami.** A tsunami is a series of enormous waves that are caused by an event that actually occurs underwater, such as an earthquake or volcanic eruption. Tsunami waves can travel at incredibly fast speeds and generate waves as high as 100 feet. Tsunamis can result in structural damage, flooding, and water contamination.

One of the deadliest natural disasters that has ever occurred was a tsunami that resulted from a powerful earthquake underneath the Indian Ocean in December of 2004, **Figure 31-2.** The tsunami killed hundreds of thousands of people in many countries bordering the Indian Ocean.

Earthquakes and Landslides

Earthquakes result from shifts in plates of the Earth's crust along geographic fault lines, **Figure 31-3,** causing large areas of the Earth's surface to shake violently

FIGURE 31-3 The San Andreas fault in California. *(Courtesy of National Geophysical Data Center)*

FIGURE 31-4 Major damage to community infrastructure as a result of an earthquake. *(Courtesy of U.S. Geological Survey/E. V. Leyendecker)*

in waves of vibrations. Earthquakes typically result in injuries, fires, damage to buildings and community infrastructure, and hazardous materials emergencies, **Figure 31-4.** Earthquakes typically require substantial resources as there are multiple events taking place at the same time. Rescue teams are required to search damaged and collapsed buildings, which is not a quick process. Earthquakes may also result in an inability of first responders to reach locations in their communities.

Landslides are large areas of rock, earth, and/or debris that shift violently and unexpectedly down a slope. Landslides can result from earthquakes or prolonged rain conditions. Landslides can also result from human alterations to land areas. Landslides that occur near populated areas can cause injuries, damage to structures and infrastructure, fires, and hazardous materials emergencies.

Severe Weather Incidents

Severe weather incidents can include hurricanes, tornados, severe storms, and extremes in temperature. The aftereffects of such weather incidents can include multiple rescue scenarios, fires, collapsed buildings, damage to infrastructure, and hazardous materials releases. **Hurricanes** are low-pressure weather systems that form in the tropics, resulting in harsh, spinning winds accompanied by thunderstorms that move across the oceans and onto land. Hurricanes have winds that move from 74 miles per hour to greater than 155 miles per hour. Hurricanes can cause tornadoes to form on land. They can also create a great deal of rain, resulting in flooding and possible landslides. Major structural damage can result.

In the late summer of 2005, Hurricanes Katrina and Rita devastated portions of the southern United States, **Figure 31-5.** Hurricane Katrina was a Category 5 storm that did tremendous structural damage when it made landfall. Hurricanes Katrina and Rita also caused severe flooding in many areas. As a result of flooding and structural damage, large fires occurred and hazardous materials emergencies became frequent. This is an example of how one disaster can actually lead to or become many disasters. The hurricanes by themselves caused enough damage to overwhelm local responders, but the additional problems resulting from the hurricanes generated such a catastrophic condition that many state and federal resources were called in to help as well as teams of responders from all over the country. Years later, residents and businesses are still trying to recover, demonstrating that a disaster can last much longer than the initial severe weather event.

FIREFIGHTER FACT

Hurricanes are classified by the Saffir-Simpson Hurricane Scale, which is outlined in **Table 31-1.**

FIGURE 31-5   Examples of the wind and flood damage caused by Hurricane Katrina in 2005. In the top left photo the president surveys the flood damage in New Orleans. *(Photos courtesy of District Chief Chris E. Mickal/N.O.F.D. Photo Unit)*

TABLE 31-1   Saffir-Simpson Hurricane Scale

Scale Number (Category)	Sustained Winds (in mph)	Damage Likely
1	74–95	Minimal: Unanchored mobile homes, vegetation, and signs
2	96–110	Moderate: All mobile homes, roofs, small crafts, flooding
3	111–130	Extensive: Small buildings, low-lying roads cut off.
4	131–155	Extreme: Roofs destroyed, trees down, roads blocked, mobile homes destroyed, beach homes flooded
5	Greater than 155	Catastrophic: Most buildings destroyed, vegetation destroyed, major roads blocked, homes flooded

Source: FEMA, www.fema.gov.

TABLE 31-2   **Tornado Classifications**

Classification Number	Classification	Types of Damage Found
F0	Light	Chimneys damaged, tree branches broken, shallow-rooted trees toppled.
F1	Moderate	Roof surfaces peeled off, windows broken, some tree trunks snapped, unanchored. Manufactured homes overturned, attached garages may be destroyed.
F2	Considerable	Roof structures damaged, manufactured homes destroyed, debris becomes airborne (missiles generated), large trees snapped or uprooted.
F3	Severe	Roofs and some walls torn from structures, some small buildings destroyed, unreinforced masonry buildings destroyed, most trees in forest uprooted.
F4	Devastating	Well-constructed houses destroyed, other houses lifted from foundations and blown some distance, cars blown some distance, large debris becomes airborne.
F5	Incredible	Strong frame houses lifted from foundations, reinforced concrete structures damaged, automobile-sized debris becomes airborne, trees completely debarked.

Source: FEMA's *Design and Construction Guidance for Emergency Shelters. July 2000*

Tornados result from thunderstorms as violent winds, rotating in a funnel-shaped cloud that can eventually make land contact at speeds as high as 300 miles per hour, causing significant damage to structures and the environment, **Figure 31-6.** Scientists have been studying tornados to find ways of predicting their occurrence and warning residents before it is too late, yet tornados can still occur with little or no warning at all. Tornados typically result in major structural damage and can prohibit responders from accessing community areas that need help.

FIREFIGHTER FACT

Like hurricanes, tornados are classified using a scale based on their strength and ability to cause damage as opposed to their wind speeds. The scale used to measure tornados is the Fujita Scale, as shown in **Table 31-2.**

Severe storms may also be considered a large-scale incident or disaster, even though they may not be classified as a hurricane or tornado event. Each of these types of incidents can cause injuries, damage to buildings and infrastructure, and flooding (in the case of hurricanes and severe storms) and can prohibit emergency responders from reaching calls for help.

Large-scale incidents can also result from severe temperature extremes, such as prolonged heat or cold conditions and winter storms. Prolonged heat conditions can be taxing to emergency medical services because of the potential for a large number of

FIGURE 31-6   A tornado funnel cloud. *(Courtesy of National Oceanic and Atmospheric Administration/Department of Commerce)*

FIGURE 31-7 Winter storms can quickly evolve into large incidents or disasters. *(Courtesy of FEMA/Michael Rieger)*

victims experiencing heat-related emergencies. Winter storms and severe cold conditions can generate injuries but can also result in damage to buildings and infrastructure and, once temperatures rise, can eventually cause flooding with the melting of ice and snow, **Figure 31-7.**

Fires and Explosions

Although a fire in a single-family dwelling may not be considered a large-scale incident, a major fire involving several structures or a structure containing many people can result in a significant response by emergency services and government resources. Wildfires are another type of fire that can rapidly become a disaster, traversing many acres of land and threatening homes, businesses, and infrastructure, **Figures 31-8A and B.** Explosions may be a result of a weather emergency (such as an earthquake or tornado) or may be caused by equipment failures. Explosions can also be intentional as a result of criminal acts or terrorism.

Large-scale incident fires, wildfires, and explosions can cause structural and environmental damage and threaten many lives.

Power Outages

Humans rely a great deal on power that is supplied to their homes and businesses, and a mass power outage or a prolonged outage can lead rapidly to a large incident or disaster for emergency responders and local governments. Imagine the result of a prolonged power outage affecting a hospital or medical care facility and how taxing that could be to the local emergency response community. Power outages can adversely affect fire stations and fire department resources as well, making it more difficult to respond to calls for help.

Hazardous Materials Emergencies

Hazardous materials emergencies can quickly become disasters depending on the type of product that is leaking or spilled. These types of incidents can result in mass evacuations of people from homes and businesses and can result in infrastructure damage, explosions, fires, and large numbers of sick or injured victims, **Figure 31-9.** There are chemical scenarios that can impact citizens 15 to 20 miles (24 to 32 kilometers) away from a chemical release. Some chemical releases have the potential to kill hundreds of citizens and injure thousands. Understanding the chemical risk in your community is key to preparing for this type of disaster.

Volcanic Emergencies

Volcanos are mountains that open beneath the Earth's crust to the molten rock that lies underneath. The pressure from gases in this molten rock can build, resulting in its upward expulsion through the center of the mountain, **Figure 31-10.** This expulsion of molten rock, known as lava, can occur either gently, where lava oozes out of the top and slowly travels down the mountain, or it can burst violently into the sky.

Lava flows are destructive, taking down everything in their path, and generate fires along the periphery of the flow. Volcanic eruptions also release large quantities of ash into the air, which can result in medical emergencies and damage to structures and equipment. For this reason, a key consideration of response is just how far-reaching the incident scene is, because the area of the lava flow is not the only place where damage and injuries can occur. Volcanic eruptions can also spawn earthquakes, landslides, fires, and tsunamis.

(A)

(B)

FIGURE 31-8 (A) Wildfires can rapidly become a disaster, traversing many acres of land and threatening homes, businesses, and infrastructure. (B) Wildfire evacuations can be complicated by smoke and damaged or blocked highways, and they can limit the response paths for additional resources to get to the fire locations. *(Courtesy of FEMA/Bryan Dahlberg)*

FIGURE 31-9 Hazardous materials emergencies can quickly become large incidents or disasters. This incident may appear isolated to the truck, but the fact that it is located on a major highway can quickly complicate the emergency and the local community's ability to respond. *(Courtesy of Mark E. Brady, PGFD)*

FIGURE 31-10 Eruption in 1980 of Mount St. Helens in Washington. *(Courtesy of FEMA/NOAA News Photo)*

Nuclear Power Plant Emergencies

Although very rare, an emergency involving a nuclear power plant can be a very devastating situation, **Figure 31-11.** Nuclear power plants are closely monitored by government agencies and employ several "safety" measures in terms of equipment and practices, yet an emergency is still possible. The greatest risk in a nuclear power plant emergency is human exposure to radiation for those in the path of radioactive materials and gases known as a *plume*. Exposure to high levels of radiation can cause serious illnesses and possibly death.

FIGURE 31-11 Although rare, emergencies involving nuclear power plants can be devastating to the surrounding areas and the environment. *(Courtesy of NNS/Landov)*

Terrorism

Acts of terrorism are often large-scale emergencies or disasters because of the number of victims, the significant structural and infrastructural damage, and the complexity of responding agencies. One key factor to remember is that terrorists' main objective is to terrorize the community, and the best way to do so is to cause a disaster—an event that overwhelms the community's ability to respond effectively. Section IV, Chapter 30 details potential types of terrorist acts.

RESPONDING TO DISASTERS AND LARGE INCIDENTS

First responders represent the front lines of disaster and large-incident response. Local governments have emergency plans for dealing with disasters in general and hazard-specific annexes to the plan that go into more detail for specific types of incidents.

> **NOTE**
>
> The fire department has a role in every emergency plan because they are the first to arrive when disasters strike the community. The time to understand their role is before the disaster, not after it has occurred.

Mutual Aid

In disasters and large incidents, local emergency resources will quickly be overwhelmed, and mutual aid units will be required to assist in the response.

Mutual aid can occur in a number of ways, through mutual aid agreements or inter-jurisdictional plans. In disaster situations, mutual aid may be requested from beyond the normal scope of agreements, sometimes coming from neighboring states or even from across the country, such as what occured wtih the reponse to Hurricane Katrina.

In some areas of the country, there are specific regional resources designed for certain aspects of an incident, such as communications, command and management, or emergency medical care. These teams may respond to disaster-affected areas to provide expertise or equipment for one aspect of the emergency.

Expanding Response

A disaster or large incident will likely involve many more agencies and individuals than firefighters are commonly used to working with on an incident scene. In disasters expanding beyond regional capabilities, state resources may be called upon to respond, including the National Guard, state agencies, and specialized state emergency response teams.

The name of the agency responsible for coordinating large-scale incidents may vary from state to state. However, the agency's purpose remains essentially the same. Each state has a state Emergency Management Agency or Emergency Response Commission. This agency operates a command/coordinating center that is the state focal point for all disaster coordination. It is from this center that executive officials such as the governor are kept informed of the situation, and from this center the activities of different response elements are coordinated. For instance, a barge runs aground in a major river and three states down river are expected to see over a million gallons of oil impinge on their jurisdictions; that will trigger an activation of the State **Emergency Operations Center (EOC).** In turn, the EOCs will notify the counties expected to be effected. In another instance, a storm surge causes flooding in five counties of a given state. This flooding has severely impacted local resources, and adjoining counties are now requested to assist. Such a situation would also trigger the State EOC to activate.

The states have their EOCs that organize their response to a disaster situation, **Figure 31-12.** At the local level, typically the county or city, there are also EOCs whose emergency services play a major role in disaster response. In many communities, the fire service is the agency that runs the EOC and the emergency management function. The local EOC is staffed by representatives of the major government agencies who have a role in disaster response and recovery. Although emergency plans typically call for the EOC to command the disaster, in many cases they act as a coordinating body. The local EOC is the government body in command of the overall disaster, but they wouldn't, for instance, be in command of a local house fire that might occur at the same time as the large-scale emergency.

The type of event that opens an EOC varies, but there are a number of emergencies that would dedicate resources to the EOC. Many EOCs are in central locations and the required agencies report to the EOC to represent their agency. For some emergencies, only a few agencies may work the EOC, such as the fire department, police department, and public works. Some EOCs are staffed for snow emergencies, wildfires, hurricanes, flooding, hazardous materials

releases, or any other emergency requiring more than the usual resources to handle an emergency.

One way to think about the EOC is as a "giant phone directory" or resource guide. It is a misconception that the EOC, or even the Federal Emergency Management Agency (FEMA), has the mass resources to handle any disaster. They themselves do not have the actual resources but coordinate the response of the resources. If you need a crane, the EOC should have a contact to arrange a crane, through the local public works, state public works, or a private contractor. They can arrange for National Guard resources through the state and federal resources as well. One of the major tasks of the EOC is to ensure that a formal declaration of disaster is made.

There is a proper chain of events that needs to be followed in order to receive assistance, from the state or the federal government. In disaster situations, FEMA generally reimburses the local government for the expenses of responding to a declared disaster. The key to reimbursement is the declaration of a disaster by the local and the state government. The reimbursement is only for actions taken to respond to the disaster and falls outside of normal costs handled by the local government. It may seen odd to think that in the middle of a disaster someone has to be responsible for a formal (written) declaration of disaster at the local and state levels, but it is required to put the resources of the federal government into action. This is the primary role of the EOC—to make sure that the formal procedures are followed while emergency responders are saving lives. For some situations, like a hurricane where it is known in advance that a storm is coming, some communities will declare a state of disaster prior to the storm hitting, which allows them to recover the costs from the very beginning of the storm.

The following is a listing of agencies and individuals that may be involved in a state EOC:

- Health Department
- Environmental Agencies
- Treasury Department
- Highways and transportation officials
- Police/law enforcement
- The Red Cross and disaster relief groups
- Department of education
- Fire and EMS departments and agencies
- Emergency managers (local, city, county, regional, state)
- Public works agencies
- The National Guard
- Utilities representatives

FIGURE 31-12 A state EOC in operation.

When the need for resources and funds rises above the normal state response, a request will likely be formulated to the governor to request federal assistance. The governor of the state involved in the disaster can request a major disaster declaration to begin the process of marshalling federal resources and funds to the incident.

Federal Disaster Resources

Once a state governor requests a major disaster declaration and commits state funds and resources to a long-term recovery effort, the **Federal Emergency Management Agency (FEMA)** evaluates the request and recommends action to the White House based on the disaster, assessments of damage, committed resources, and the ability of the state to recover from the incident. The president can then approve or deny the declaration.

If a major disaster declaration is approved, the **Department of Homeland Security (DHS),** through FEMA, becomes intimately involved with the incident. FEMA can assist with search and rescue, electrical power, food, water, shelter, and other needs of affected citizens. FEMA also coordinates disaster aid programs for affected families and businesses in the disaster area.

The federal response to a disaster is structured by the National Response Plan (NRP). The following excerpt from FEMA describes this plan:

> The National Response Plan establishes a comprehensive all-hazards approach to enhance the ability of the United States to manage domestic incidents. The plan incorporates best practices and procedures from incident management disciplines—homeland security, emergency management, law enforcement, firefighting, public works, public health, responder and recovery worker health and safety, emergency medical services, and the private sector—and integrates them into a unified structure. It forms the basis of how the federal government coordinates with state, local, and tribal governments and the private sector during incidents.

There are a variety of federal agencies that may be involved in a disaster response, although the most involved will be the Department of Homeland Security, through FEMA. Incidents with a significant law enforcement component, such as disasters involving terrorism, will see a major involvement of the Federal Bureau of Investigation (FBI). The FBI has the responsibility to investigate all acts of potential terrorism and acts as the lead in the investigation. The FBI can also assist when the crime meets the criteria for the FBI's involvment, such as an interstate crime ring. Terrorist incidents may also get a response from National Guard Civil Support Teams (CST); these are 24-person teams that are equipped to respond to a weapons of mass destruction (WMD) incident much like a hazardous materials team.

There are also specialized teams at the federal level that offer expertise and equipment for key functions at a disaster scene. One of the successful initiatives supported by the fire service is seen in FEMA's **Urban Search and Rescue (USAR) teams, Figures 31-13A** and **B.** Groups of technical rescue specialists have been formed in strategic areas around the coun-

(A)

(B)

FIGURE 31-13 A USAR team in action. *(Courtesy of Fairfax County, Virginia, Fire and Rescue Department's Virginia Task Force 1 USAR Team)*

try to capitalize on the experience of many departments and personnel. This regional philosophy greatly benefits local departments by not depleting an area completely when the specialists are deployed. Also, regional deployment allows for the deployment of a team located hundreds of miles away while leaving other resources in place. The USAR teams are composed of technicians with such specialties as structural collapse, confined space rescue, trench/excavation rescue, swift water rescue, hazardous materials, and WMD. These teams go to great lengths to maintain enhanced skill levels and state-of-the-art equipment for their skill set. The regional concept continues to be of value in the acquisition of equipment. Where one department could not begin to afford the high costs of this technology, regionally the team can pool its resources for purchase. The deployment of USAR teams at a disaster site can greatly enhance the rescue capabilities of the local responders.

FIREFIGHTER FACT

There are currently 28 FEMA USAR teams across the country. For more information about the FEMA USAR program and teams, go to http://www.fema.gov/emergency/usr/index.shtm.

Another specialized federal resource is the **Disaster Medical Assistance Team (DMAT), Figure 31-14.** Currently, this responsibility lies with the Department of Health and Human Services, Office of Preparedness and Response. DMAT mobilizes medical resources to an affected area in the event of a disaster or mass casualty incident. DMAT teams are comprised of medical resources including physicians, nurses, therapists, pharmacists, emergency medical technicians (EMTs), and administrative staff. Understanding that a community compromised by disaster may be completely incapacitated, these teams are self-sufficient for 72 hours. The function of the activation depends on the disaster at hand. DMATs may serve as a primary means of medical care, backup to existing resources, or even a remote medical unit where victims may be received for treatment. This is a reactive force designed for rapid deployment and medical intervention as a stopgap measure until the situation is resolved or other resources can be used. Assisting the DMAT mission are several other response groups. They include:

- Disaster Mortuary Operational Response Team (DMORT)
- Veterinary Medical Assistance Response Team (VMAT)
- National Nurse Response Team (NNRT)

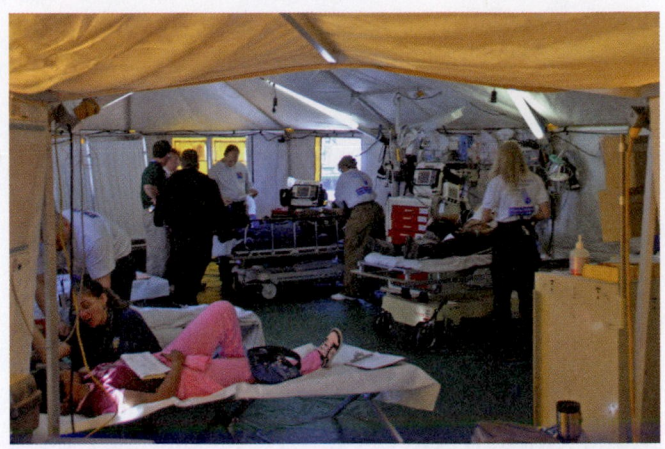

FIGURE 31-14 DMAT members care for patients in an acute care tent. *(Courtesy of FEMA/Marvin Nauman)*

- National Pharmacy Response Team (NPRT)
- Disaster Portable Morgue Units Team (DPMU)

Other Resources

Large incidents and disasters can exhaust many resources, and often help is needed faster than governments can mobilize the resources to act. Private sector organizations often play a role in disaster response at all levels—local, state, and federal. One of the most commonly seen resources is the Red Cross, a nonprofit organization that provides assistance with food, water, shelter, and basic necessities for victims of a disaster. These types of organizations are critical to protecting the livelihood of the victims involved in a disaster. Many local communities and state agencies include private sector and nonprofit organizations in their emergency plans to supplement municipal resources. For example, a local business may offer to shelter victims left homeless in a tornado, or a group of area restaurants may provide food for emergency workers kept on the incident scene for extended periods.

NATIONAL INCIDENT MANAGEMENT SYSTEM

Firefighters are very familiar with the Incident Command System (ICS) in managing everyday incidents. ICS is a very effective and efficient way of managing people and resources on an incident scene. As the United States began dealing with a number of large-scale incidents, it became apparent that the ICS concept could be expanded to include all agencies and individuals that respond to a disaster. In 2003, the president of the United States charged the Department of Homeland Security with developing and implementing a system that could be used to manage large incident response. This resulted in the creation of the National Incident Management System (NIMS).

NIMS has two key concepts at its core, flexibility and standardization. NIMS is designed to be flexible enough to adapt to the wide variety of disasters and responders. In addition, NIMS aims to provide standardized organizational structures and processes to improve interoperability between agencies and governments involved in the disaster. This standardization not only applies to processes but also establishes a set of terminology so that everyone working at an incident can speak the same language.

THE FIREFIGHTER IN DISASTER RESPONSE

All this discussion on the many layers of disaster response can be overwhelming to a firefighter. It may be hard to imagine how a firefighter fits into the complex response at a disaster, but in many respects, the firefighter is the critical link in the chain of emergency response to disaster.

There are several things that firefighters should consider concerning disaster and large incident response.

As a starting point, training is important for every firefighter in disaster response, which means learning about department policies, local emergency plans, local hazards, and the ICS and NIMS. Communications is also a critical factor in large incident response, and mastering department communication equipment and policies also provides a valuable skill in disaster response.

Also, firefighters should recall their knowledge of the department accountability and safety procedures, as these are just as important in large incident response as in everyday fires and emergencies. Firefighters should recognize that large incidents and disasters can be very physically and emotionally difficult events. Accordingly, firefighters should always look out for each other and watch for signs of critical incident stress in themselves and those around them, **Figure 31-15.**

The last important consideration is that firefighters must understand that freelancing can be dangerous in large incident response and may damage the effectiveness of an operation. There have been several instances of widespread freelancing in recent disasters in the United States that have caused additional problems for local responders beyond the damage brought by the disaster itself. It is in the nature of firefighters to want to help where help is needed. Often individuals and companies will arrive at an incident scene, sometimes hundreds of miles from their jurisdictions, because they want to help. One of the most difficult aspects of managing a large incident is controlling resources and using them where they are needed. Firefighters that arrive without being summoned can complicate this process, especially when they arrive in large numbers.

FIGURE 31-15   New York City firefighters search through the rubble for victims of the World Trade Center collapse in September 2001. *(Courtesy of FEMA/Michael Rieger)*

LESSONS LEARNED

Disasters and large incidents are a great burden on the fire and emergency response system, and firefighters are often called upon to act for extended operational periods and to perform tasks outside their usual daily routine. A disaster can bring resources from a wide variety of local, state, and federal agencies and the private sector. Disasters can affect communities for hours, weeks, and even years. The role of the firefighter is always an important one: a vital connection between the disaster victim and a potentially overwhelming army of responders.

KEY TERMS

Department of Homeland Security (DHS)   Federal agency created as a result of The National Strategy for Homeland Security and the Homeland Security Act of 2002 that leads the unified national effort to secure America by preventing and deterring terrorist attacks and protecting against and responding to threats and hazards to the nation. DHS ensures safe and secure borders, welcomes lawful immigrants and visitors, and promotes the free flow of commerce.

Disaster Medical Assistance Teams (DMAT) Teams that mobilize medical resources to an affected area in the event of a disaster or mass casualty incident; DMAT teams are composed of medical resources including physicians, nurses, therapists, pharmacists, EMTs, and administrative staff.

Earthquakes   The result of shifts in plates of the Earth's crust along geographic fault lines causing large areas of the Earth's surface to shake violently in waves of vibrations.

Emergency Operations Center (EOC)   A communication and coordination center placed in service during major emergencies or disasters that is composed of representatives from many different agencies.

Federal Emergency Management Agency (FEMA)   On March 1, 2003, the Federal Emergency Management Agency (FEMA) became part of the U.S. Department of Homeland Security (DHS). The primary mission of the Federal Emergency Management Agency is to reduce the loss of life and property and protect the nation from all hazards, including natural disasters, acts of terrorism, and other man-made disasters, by leading and supporting the nation in a risk-based, comprehensive emergency management system of preparedness, protection, response, recovery, and mitigation.

Hurricanes   Low-pressure weather systems that form in the tropics, resulting in harsh, spinning winds accompanied by thunderstorms that move across the oceans and onto land.

Landslides   Large areas of rock, earth, and/or debris that shift violently and unexpectedly down a slope.

Tornados   Weather events that result from thunderstorms as violent winds, rotating in a funnel-shaped cloud that can eventually make land contact at speeds as high as 300 miles per hour.

Tsunami   A series of enormous waves that are actually caused by an event that occurs underwater, such as an earthquake or volcanic eruption.

Urban Search and Rescue (USAR) Teams Groups of technical rescue specialists formed in strategic areas around the country to respond to disasters and provide technical rescue expertise and equipment; the USAR program is managed by FEMA.

Volcanos   Mountains that open beneath the Earth's crust to the molten rock, known as lava, that lies underneath; the pressure from gases in the lava can build, resulting in an upward expulsion of lava through the center of the mountain.

REVIEW QUESTIONS

1. What is a disaster? What are two major categories of disasters?

2. Compare and contrast the effects of a disaster involving a hurricane and one involving a tornado.

3. Compare and contrast the effects of a disaster involving an earthquake and one involving a volcanic eruption.

4. What is an Emergency Operations Center? Where is your local EOC? Where is your state EOC?

5. What agencies might be involved in a state EOC?

6. Give an example of a federal agency that would be involved in a disaster response after a hurricane.

7. Give an example of a federal agency that would be involved following a terrorist incident.

8. Give an example of a specialized federal response team that might be called upon to respond to a disaster involving an earthquake.

9. What kinds of disaster and large incident threats does your community or region face?

10. Where can you access your local, regional, or state emergency plans to review your agency's role in a disaster or large incident?

Emergency Medical Response

SECTION

III

Training requirements for emergency medical response are also mandatory as part of your Firefighter I and II training. Many departments and/or regions may recommend or even require a high level of EMS training. The chapter within this section provides basic first responder training as outlined in NFPA 1001 *Standard for Firefighter Professional Qualifications,* 2008 Edition. It includes coverage of these following topics:

- Roles and responsibilities of the emergency care provider
- Safety considerations, such as infection control
- Patient assessment
- The basics of CPR
- Bleeding control and shock management
- Emergency care for common emergencies

Make sure you check with your Authority Having Jurisdiction on local EMS requirements, so that you are properly prepared for emergency response within your area.

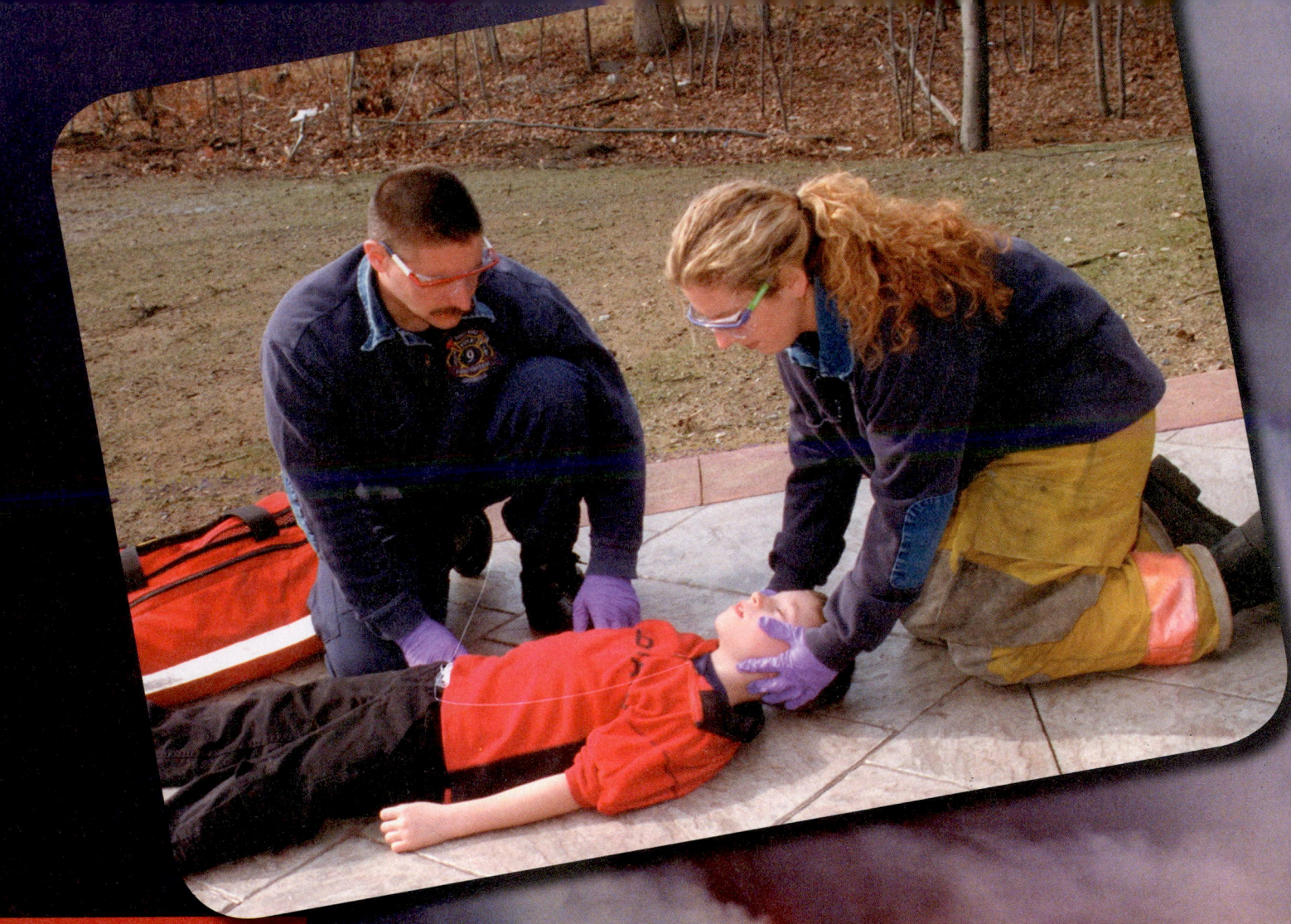

22 Emergency Medical Services

During the past twenty-two years, I have seen many occasions where EMS training of firefighters has been essential to the success of incidents—especially to the patients involved. There have been many times where an engine I was on arrived at the scene of a collision before an ambulance. There were times when we pulled victims from buildings and no EMS units were available. There have been other rescue situations—in a storm drain, in a steam tunnel, in overturned vehicles—where EMS was on the scene but did not have the proper personal protective equipment to enter the hazardous area to provide care. Still, one incident stands out in my mind as a clear example of why firefighters should have EMS training.

It was a clear weekday afternoon in Fairfax County, Virginia. It had been a typical day with typical calls. It was rush hour—traffic was at a standstill; no one seemed to be moving. A vehicle fire was dispatched, a short distance from Fire Station 34. I knew that Engine 34 and Ambulance 34 were out of service for HAZMAT training and that the next due engine (Fairfax City Engine 33) was about 4 miles away and would have to battle traffic. I was about 3 miles away, just leaving headquarters, but could see the smoke—a billowing, black loom-up. I decided to roll in, just to watch.

Squad 18, a heavy squad with a crew of three, responded from Fire Station 34 and arrived within a couple of minutes. As I recall, their size-up was "a stake body truck well involved." No big deal, I thought. Then, after a minute, they requested an ambulance for a burned subject. Then they requested another, along with two medic units. Then they requested a helicopter. What seemed at the start to be "no big deal" ended up to be a serious situation with five civilians critically burned.

The crew from Squad 18 began treating the patients. I assisted when I arrived, as did the crew from Engine 33. As ambulances and medic units began to arrive, they took over patient care, but the lifesaving measures had already been taken. Yes, the fire was eventually extinguished, but our job as firefighters is life safety first, then property conservation. Thanks to our efforts, all five patients survived.

If we had not been trained to provide EMS care, there could easily have been five fatalities at that scene. Further, with rush hour traffic, there were hundreds (if not thousands) of people who could have perceived that firefighters are only trained to put "wet stuff on red stuff" and that injured victims are of no concern to us. That is not the image the fire service should portray.

The truck was a total loss even before the fire department was called. The five victims, employed by a landscaping company, were riding in the back of the truck when one tried to light a cigarette. Unable to do so because of the wind, he knelt down or bent down and lit it—right over a can of gasoline. The victims were burned from the flash fire that occurred.

Ever since this incident, I have made sure that I have kept my EMS training current, and as a chief officer I have made sure that all personnel under my command have had the opportunity and encouragement to become EMS-trained. Patient care is a key part of what we do as firefighters, and we cannot do it effectively if we are not trained.

—Street Story by Gordon M. Sachs, Chief,
Fairfield Fire and EMS, Fairfield, Pennsylvania

INTRODUCTION

"When I became a firefighter, I expected to be fighting fires on every shift. I never anticipated that my engine company would respond to more emergency medical calls than fires last year. I am certainly glad that I received training in emergency care in firefighter school, because I have really helped save lives. There is more to being a firefighter than putting water on fire!"

It is true. There is more to being a firefighter than "putting the wet stuff on the red stuff." Firefighters are community defenders and trusted public saviors. When people do not know who else to turn to, they call the fire department. The community relies on firefighters to be skilled at fighting fires and to be creative problem solvers. That means firefighters need to know more than just basic firefighting techniques. That is where emergency medical care comes into the picture, **Figure 22-1.**

Emergency medical responses constitute more than 50 percent of total emergency responses for many fire departments all across the country. In some jurisdictions, emergency medical calls make up 75 to 80 percent of the fire department's total emergency responses per year.

This chapter covers very basic material that firefighters will need to know based on the NFPA 1001 standard and also covers standard first-aid practices. A wide variety of training courses in emergency medical services (EMS) are available that allow firefighters to increase their EMS knowledge and skills, and all firefighters are encouraged to continue their EMS training.

FIGURE 22-1 Emergency medical services star of life.

> **FIREFIGHTER FACT**
>
> Emergency medical care has evolved into an essential and intregal part of the fire service.

With the technology available today, firefighters can deliver lifesaving techniques to stabilize patients until emergency medical technicians and paramedics arrive, **Figure 22-2.**

> **FIREFIGHTER FACT**
>
> Firefighters are a critical, lifesaving link in the emergency response community and in many cases work side by side with emergency medical services personnel.

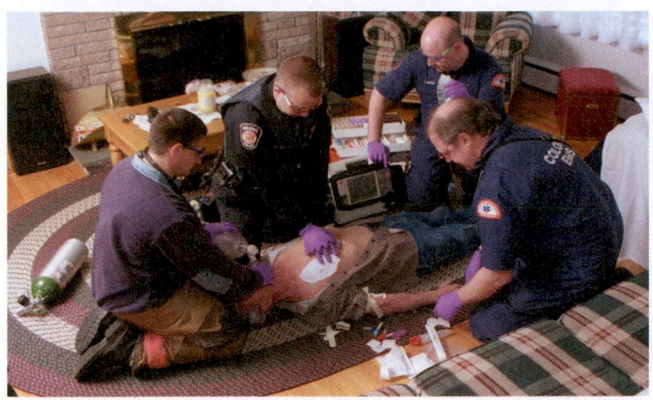

FIGURE 22-2 Firefighters are often called on to assist EMS crews with patient care.

(A)

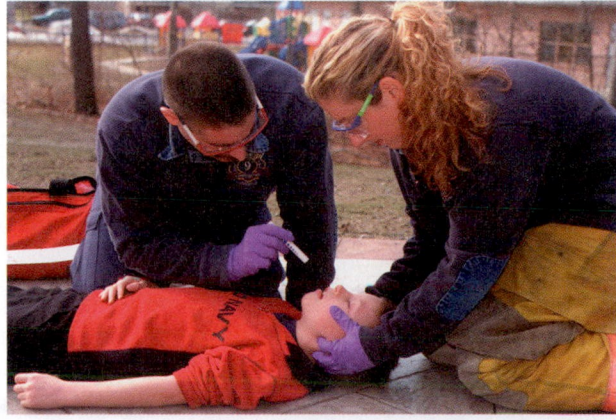

(B)

ROLES AND RESPONSIBILITIES OF AN EMERGENCY CARE PROVIDER

As a part of the emergency response system, firefighters often act as providers of emergency care and first aid. Every emergency service delivery system is different, yet many call on firefighters to assist as the first arriving emergency unit on the scene or to assist emergency medical technicians and paramedics in providing patient care. In any case, a working knowledge of the basics of emergency medical care is a very important part of a firefighter's training.

Key Responsibilities

Firefighters have several key responsibilities when responding to emergency medical incidents:

- To ensure their own safety, the safety of their team, and the safety of the patient.

SAFETY

The most important factor to consider when firefighters respond to emergency medical calls is safety, for themselves, their team members, and their patient.

- To act safely from the minute they step on the fire engine until every piece of equipment is clean or sterilized and placed back on the fire engine, **Figures 22-3 A–C.**
- To act in a professional manner at all times. People who have called for assistance are counting on firefighters to help them or their friends or family

(C)

FIGURE 22-3 (A) EMS calls begin when firefighters leave the station, (B) continue throughout the time on the scene, and (C) do not end until firefighters have cleaned the equipment and restocked the supplies.

FIGURE 22-4 Firefighters should always treat patients with respect. They should treat patients as they would want to be treated in the same situation.

members in a time of need. Firefighters must always be respectful and considerate, **Figure 22-4.**

■ To never cause a situation to become worse or act beyond the bounds of their training. Sometimes the greatest care that can be provided is compassion.

ETHICS

Emergency medical training is not only about what firefighters can do to assist a patient in crisis, it is also about recognizing what is beyond an individual firefighter's training and ability. If firefighters recognize that a situation is beyond their capabilities, and they call for the appropriate assistance and act within the limits of their training, they have done the best job they can for their patient.

■ To practice and update emergency care and first-aid skills with continued training. The emergency medical field is constantly changing and evolving. That means firefighters will have to learn new skills, as well as practice the old skills, **Figure 22-5.** Training for emergency medical responses and maintaining current knowledge and skills will allow firefighters to be ready for any emergency medical incident that occurs.

■ To know and practice the use of EMS equipment and maintain it properly. While emergency medical care may require the use of a firefighter's brain and hands, it also requires specialized equipment and tools. Firefighters should be familiar with each piece of EMS equipment and all supplies they carry. Also, equipment needs to be maintained properly. It should be cleaned after each use (or dis-

carded properly if it is disposable) and checked at the beginning of each shift, **Figure 22-6.**

■ To gather and document important patient information. Information about the patient on an emergency medical call can be critical for treatment. The firefighter must collect all important information about a patient before deciding on a course of action. Then, when EMS personnel arrive, the firefighter should transfer that information so that the EMS personnel can continue with appropriate care in the ambulance and in the hospital. Collecting patient information is discussed later in this chapter.

Legal Considerations for Emergency Care Providers

Emergency care providers should understand several important legal issues. First is the principle called **standard of care,** which is a legal term that means for every emergency medical incident, an emergency responder should treat the patient in the same manner as would another emergency responder with the same training. In short, this legal principle is what is used in court to demonstrate that the actions of an emergency care provider were adequate or inadequate. This legal issue emphasizes the importance of firefighters acting within the bounds of their emergency care and first-aid training and doing what they have been trained to do, and *only* what they have been trained to do.

Another important term to understand is **consent.** Emergency responders should always attempt to gain consent from the patient before beginning treatment. In the case of children or minors, consent should be obtained from a parent or legal guardian. Emergency

FIGURE 22-5 Firefighters should train on new EMS skills and continue to practice skills they have already learned. Practice makes skilled responders.

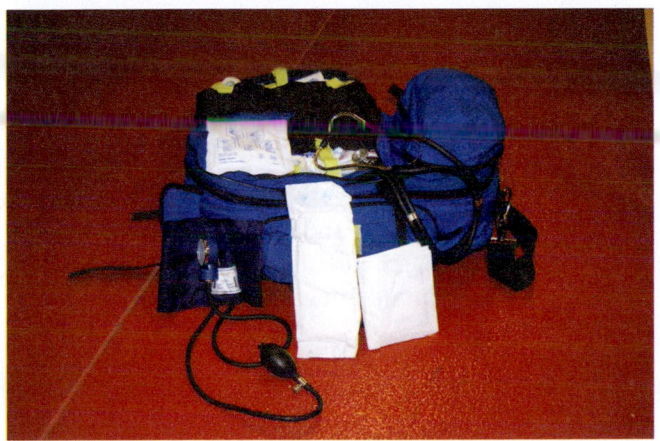

FIGURE 22-6 Equipment must be maintained properly.

responders should not withhold treatment from minors when a parent or guardian is not present. Emergency responders can begin treatment while making every effort to contact the parents or guardians to obtain permission.

Patients who are unconscious when emergency responders arrive are considered to be giving **implied consent** for emergency medical treatment. Patients who are not unconscious, and are mentally capable, have the right to refuse treatment.

Abandonment occurs when an emergency care provider begins treatment of a patient, and then leaves the patient or discontinues treatment prior to the arrival of an equally or higher trained responder.

CAUTION

If firefighters begin patient care, they must remain with the patient until another emergency responder arrives whose training is equal to or higher than theirs.

Interacting with Emergency Medical Services Personnel

As emergency care providers, firefighters interact with emergency medical service personnel while on the scene of an emergency. Emergency medical transport is done in a variety of ways. An ambulance staffed with EMTs and paramedics may come from the firefighters' own fire department, it may come from a separate rescue squad or EMS organization, or it may even come from a private transport company or a hospital.

In some cases, firefighters may be interacting with emergency medical personnel who have arrived by helicopter, commonly referred to as a medi-vac. **Medi-vac** units are staffed by nurses, paramedics, and EMTs who are usually skilled in treating patients with traumatic or serious injuries. A medi-vac unit can be seen in **Figure 22-7.**

It is also important to understand the different levels of care that are provided by EMS crews who may arrive on the scene. An ambulance may provide basic life support (BLS) care, which is the primary level of EMS care, or it may provide advanced life support

FIGURE 22-7 Medi-vac helicopters are used to transport patients to specialty care hospitals.

FIGURE 22-8 Firefighters should record important information for arriving EMS personnel and for their agency's patient information reports. Effective documentation is critical. *(Courtesy of Fred Schall)*

(ALS) care, which involves much more aggressive treatments for patients and specialized monitoring.

In any case, the arriving EMS personnel will rely on the firefighters on scene to give them critical patient information. What is important patient information? It varies from situation to situation and requires firefighters to use their best judgment, but a few basic pieces of information are important for every call: the patient's age, the patient's medical history, known drug allergies of the patient, medications the patient is taking, the circumstances surrounding the illness or injury the patient is reporting, the patient's vital signs and chief complaint, **Figure 22-8.**

Many other pieces of information can be helpful, and firefighters should consider gathering such information during an emergency medical call, if feasible: the patient's name and address, what condition the patient was found in, the patient's surroundings (example: condition of an automobile in a car accident), and the treatment that was given to the patient prior to the EMS personnel arriving on the scene.

All firefighters need to check with their department about their responsibility for documenting an emergency medical response. Many fire departments require that a patient care report or an emergency call record be completed for every emergency medical response. Thorough and accurate documentation of emergency medical responses will help protect the department and the firefighter in the event of legal action.

SAFETY CONSIDERATIONS

One of the most important considerations when performing emergency care and applying first-aid skills is the safety of the caregiver or individual firefighter

performing patient care and the safety of the emergency response team, such as an engine company crew responding on an emergency medical call.

> ### SAFETY
> Safety, including **infection control,** should be the primary consideration of every emergency responder.

Safety and infection control should also remain a primary consideration for the duration of the emergency run and even afterwards in the cleanup phase of the incident. *Remember:* Emergency responders who get sick or injured are no longer a part of the solution; they are a part of the problem. Safety is the primary consideration!

Analyzing the Safety of the Emergency Scene

Firefighting training teaches firefighters to observe each and every scene for situations that may be unsafe to the public or to the firefighters present. For example, firefighters must check for overhead obstructions whenever raising a ladder to a structure and check the floor of a structure for stability before stepping on to it. Engine company officers must survey the condition of a burning building to make sure it is not on the verge of collapse before sending firefighters inside for an interior fire attack.

While functioning on emergency medical calls, firefighters must also check for scene safety prior to entering. There are a few important considerations for EMS scene safety. Firefighters should listen carefully to all information given prior to arrival on the scene. It is essential to plan actions ahead of time! Also, firefighters should listen to the direction of the officer or the EMS provider in charge of the incident so the team is functioning as one unit, **Figure 22-9.** It may be necessary to stage away from the scene until it can be stabilized by law-enforcement personnel.

Firefighters should carefully observe the emergency scene and people present before entering. Indications of violent behavior, the use or presence of weapons, or signs of controlled substance or alcohol use may all be initial indicators of a safety concern. If after entering the emergency scene a safety concern is discovered, it is acceptable to take the team out of the situation until law-enforcement officers have secured the scene or the dangerous individual or condition has left the scene.

The firefighter should be sure to ask for additional assistance when safety concerns are present, whether in the form of additional fire and emergency medical units or a law-enforcement officer. It is also important to keep the communications center informed of the situation and the team's status to ensure continued safety throughout the incident.

Firefighter Physical and Mental Health

Firefighting is a strenuous occupation, and firefighters are encouraged to maintain a healthy lifestyle to meet the demands of firefighting work.

Emergency medical calls for assistance can also be strenuous work, both physically and, more often, mentally. A "healthy lifestyle" is different for each firefighter but generally includes regular exercise, a proper diet, and getting the right amount of sleep. Firefighters should contact their personal physician or department health care provider for assistance with developing a healthy lifestyle that meets their needs.

> ### NOTE
>
> A healthy lifestyle will help firefighters in their ability to respond to emergency medical calls for assistance as well as in typical firefighting activities.

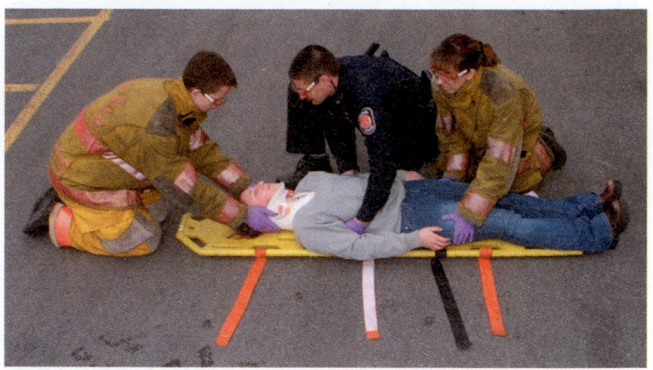

FIGURE 22-9   Team working together in patient care.

One of the most common problems for firefighters on emergency medical responses is back injuries. Firefighters should learn and practice proper lifting techniques (lifting with the legs, keeping the weight close to the body, etc.) when moving patients or emergency medical equipment.

Emergency medical calls can also take their toll on the mental health of a firefighter or emergency medical responder. Mental stress is a common reaction to some emergency medical calls. Firefighters should consult their department's policies on critical incident stress debriefing or other intervention techniques available for firefighters who feel adversely affected by emergency medical responses.

Infection Control

Safety for the firefighter responding to emergency medical incidents involves not only scene safety and the prevention of injury but also protection from contracting **communicable** or **infectious diseases.**

Prevention of Exposure to Infectious Diseases

The Centers for Disease Control and Prevention (CDC) is a federal organization that monitors outbreaks of infections and advises agencies on how to handle the situation and control the disease. The CDC, the Occupational Safety and Health Administration (OSHA), and the National Fire Protection Association (NFPA) have all written standards of practice or operational guidelines that firefighters should use when providing medical care.

Many laws and regulations have been enacted on the local, state, and national levels that assist in protecting emergency responders and health care providers from infectious diseases. In addition, first responders should follow any local or organizational policies for infection control.

Any time a firefighter treats a sick or injured patient, the risk of **exposure** to disease is present, both for the firefighter and for the patient. Firefighters should protect themselves and the patient from potential exposure to a disease. The most effective means of reducing the risk of spreading disease is hand washing, **Figure 22-10.**

Proper hand washing is a necessity in the medical profession, and many types of antimicrobial soaps are available for this purpose. Although soap is a great tool, the technique of washing is more important than the type of soap.

Frequent hand washing, although critical, may have a drying effect on the skin, and a moisturizing lotion may be necessary to combat the discomfort.

Firefighters must also be aware of the possibility of transmission of disease and sickness to their family and friends. Firefighters who do not practice proper infection control procedures can carry disease-causing microorganisms home with them to their families. As a general practice, firefighters should be extra careful never to wear soiled duty uniforms home. Washing soiled items in a separate station machine can help lower cross-contamination to other washables. It is advisable to shower and change at the station before leaving for home. Likewise, firefighters who respond from a station should keep soiled uniforms and personal protective equipment out of living areas of the station.

Body Substance Isolation

Simple precautions, called **body substance isolation (BSI) precautions,** can reduce the odds of disease transmission significantly. BSI involves wearing proper protective equipment on every call. A person

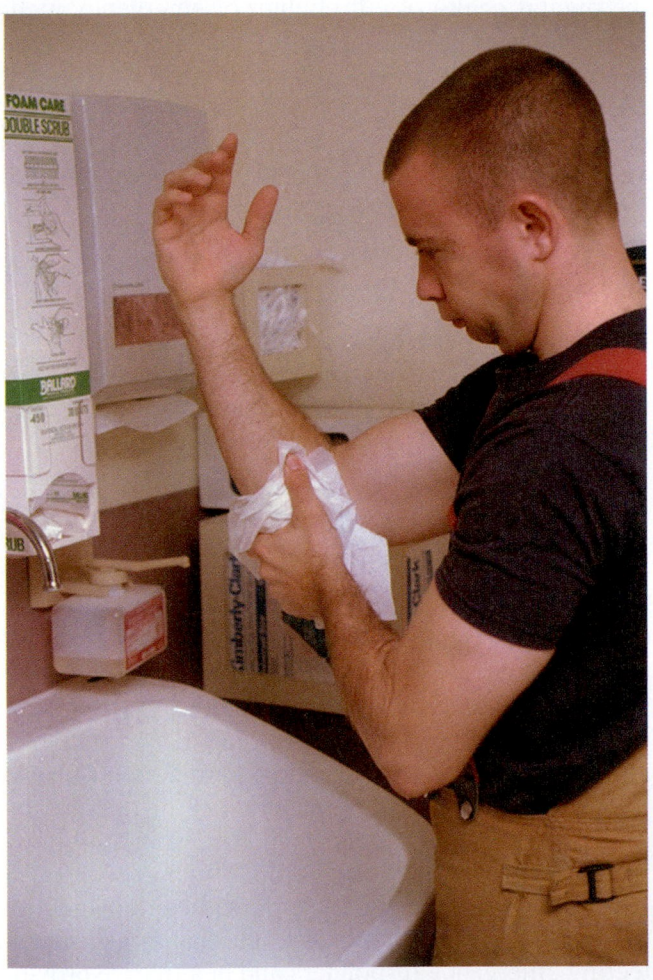

FIGURE 22-10 The most effective means of reducing the risk of spreading disease is hand washing.

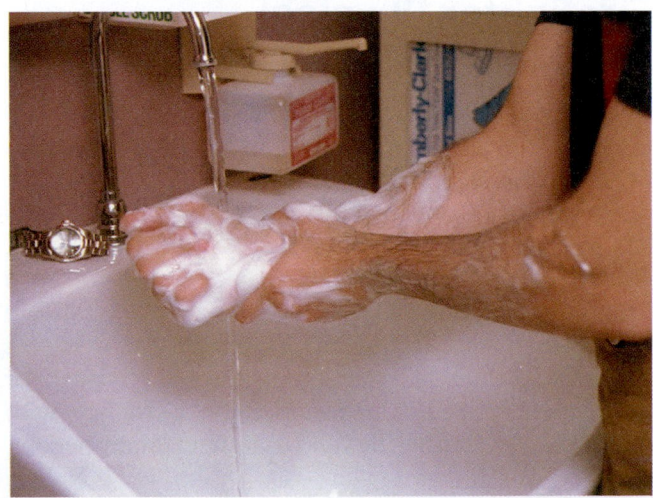

FIGURE 22-11 Proper hand washing involves vigorous lathering with soap and water for fifteen seconds (or longer) followed by thorough rinsing.

who is ill with a communicable disease will not necessarily appear ill and may not want to make the disease known to others. Therefore, firefighters should have an agency-approved face mask available and should consider wearing the minimum of goggles or safety glasses and protective gloves any time a medical scene is entered.

SAFETY

Appropriate BSI precautions must be taken for every patient.

The mainstay rule of emergency medical care is to treat all body fluids as if they were infectious, **Table 22-1.** If a substance is wet, it has the potential to be an infectious substance. Again, a firefighter should consider wearing a minimum of gloves and eye protection on every call. A face mask can be added, if necessary, for splash potential. A nonabsorbent or plastic gown may be added for medical scenes with gross contamination potential, **Figure 22-12.**

Protective gloves are an important barrier device for obvious reasons but are effective only if they fit and are used properly. Firefighters must don new gloves for each patient contact. They must never use the same pair of gloves on more than one patient. Double gloving can be used for incidents with gross

TABLE 22-1   Potentially Infectious Body Fluids
Blood
Amniotic fluid
Vaginal discharge
Semen
Cerebrospinal fluid
Pleural fluid
Synovial fluid
Peritoneal fluid
Pericardial fluid
Fluids with little potential to transmit bloodborne diseases:
■ Tears
■ Nasal discharge
■ Vomitus
■ Sputum
■ Saliva
■ Feces
■ Urine

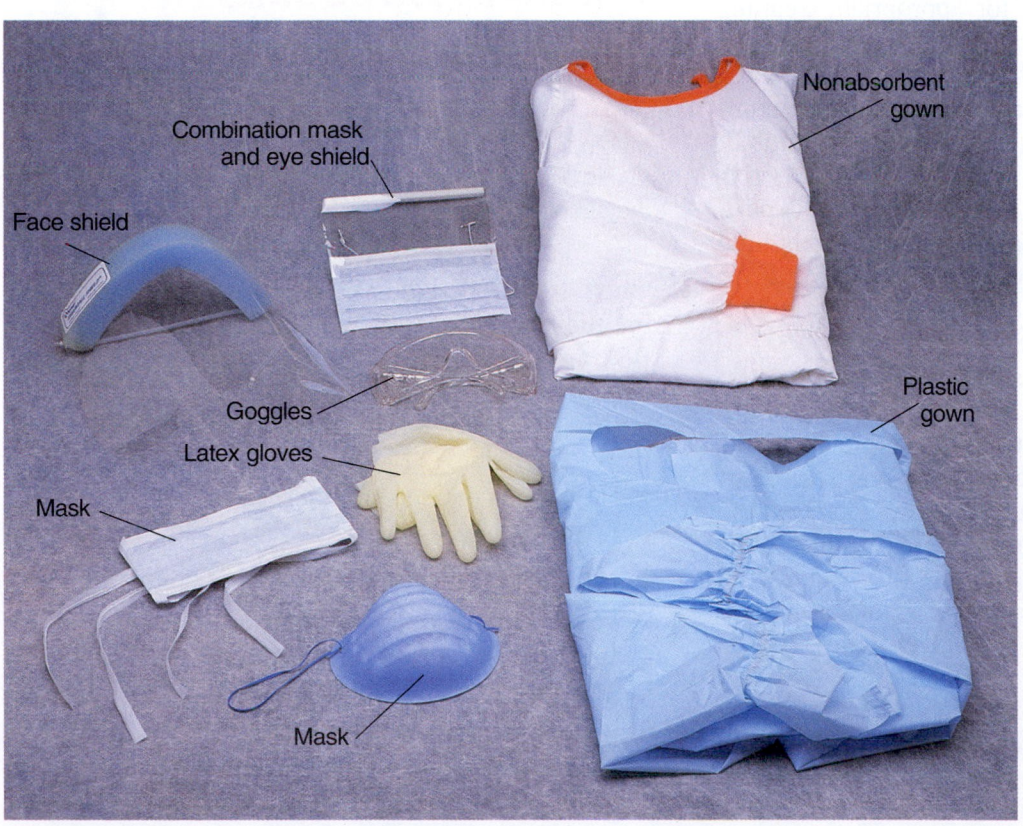

FIGURE 22-12 Types of personal protective equipment.

contamination potential. If gloves fail during use, fire-fighters should stop immediately, wash their hands or use alcohol foam or the like, and don another pair of gloves before returning to patient care.

JPR 22-1: Removing Gloves

For step-by-step photos of this skill sequence, see page 989)

1. Gloves should only be removed after all patient contact possibilities are finished and should be removed without touching the skin with the outer layer of the glove.
2. Firefighters should dispose of gloves in a container clearly marked with the biohazard label if there is gross blood present and wash their hands after every use of gloves.

Splashes of body fluid or chemicals can occur on emergency medical scenes, which is the purpose for eye protection in the BSI precautions. The eyes can provide a route for a disease microorganism to enter the firefighter's body. As stated earlier, goggles or safety glasses should also be used on every emergency call.

Any supplies or equipment used for treating a patient should either be disposed of in an appropriate container or cleansed and disinfected properly, **Figure 22-13.** Firefighters should wear gloves and safety glasses when disinfecting emergency medical equipment and should consult departmental policy for appropriate cleaning and disinfecting methods and procedures.

Firefighters who regularly participate in emergency medical incidents should consider receiving immunization against both common diseases and hepatitis B, an infectious disease that is the most dangerous risk to medical care providers. Firefighters who experience a significant exposure to blood or any body fluid should contact their supervisor immediately. Each agency will have policies concerning infection control and procedures, including exposures.

In short, BSI precautions are important because they protect the safety of the firefighters and their families and friends. BSI precautions also protect patients with compromised immune systems from disease-causing microorganisms that can be spread from the firefighter to the patient. Something as simple as the common cold can be life threatening to a patient with a compromised immune system. A review of precautions for infection control is covered in **Table 22-2.**

For more information on infection control, BSI, and exposure to infectious diseases, firefighters can contact the CDC or OSHA. Both agencies are a part of the federal government and will provide information as requested. More information is also available on their Web sites at http://www.cdc.gov and http://www.osha.gov.

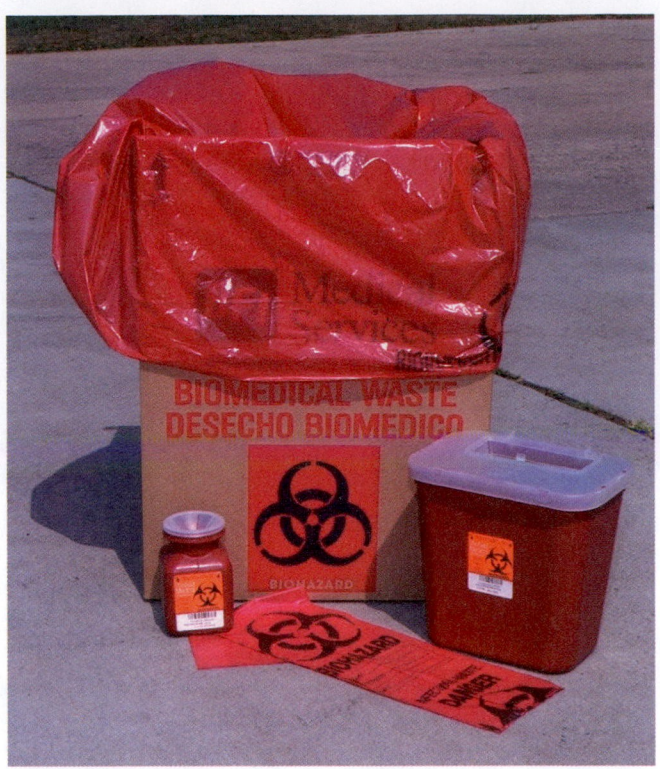

FIGURE 22-13 Examples of biohazard containers for disposal of infectious waste.

PATIENT ASSESSMENT

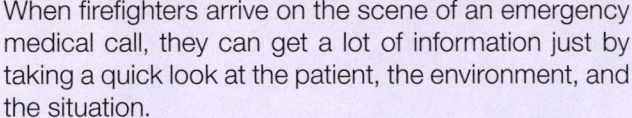

STREETSMART TIP

When firefighters arrive on the scene of an emergency medical call, they can get a lot of information just by taking a quick look at the patient, the environment, and the situation.

Once a firefighter has made sure the scene is safe for the team to enter and has donned appropriate BSI, an initial visual size-up of the patient(s) should be conducted. A quick visual assessment should reveal:

- How many patients are there?
- Is the patient awake?
- Is the patient in a harmful situation or environment?
- What position is the patient in?
- What people or objects are in the emergency scene that may have contributed to the patient's illness or injury?
- What is the skin color of the patient (for example, pale, red, bluish, etc.)?
- Is the patient having any trouble breathing?

TABLE 22-2	**Standard Precautions for Infection Control**

Wash Hands (Plain Soap or Alcohol-Based Foam or Substitute)

Wash after touching **blood, body fluids, secretions, excretions,** and **contaminated items.**

Wash immediately **after gloves are removed** and **between patient contacts.**

Avoid transfer of microorganisms to other patients or environments.

Wear Gloves

Wear when touching **blood, body fluids, secretions, excretions,** and **contaminated items.**

Put on **clean** gloves just **before touching mucous membranes** and **nonintact skin.**

Change gloves between tasks and procedures on the same patient after contact with material that may contain high concentrations of microorganisms. Remove gloves promptly after use, before touching noncontaminated items and environmental surfaces, and before going to another patient; then wash hands immediately (or use an alcohol-based foam or substitute in the absence of soap and water) to avoid transfer of microorganisms to other patients or environments.

Wear Mask and Eye Protection or Face Shield

Protect mucous membranes of the eyes, nose, and mouth during procedures and patient care activities that are likely to generate **splashes** or **sprays of blood, body fluids, secretions,** or **excretions.**

Wear Gown

Protect skin and prevent soiling of clothing during procedures that are likely to generate **splashes** or **sprays** of **blood, body fluids, secretions,** or **excretions.** Remove a soiled gown as promptly as possible and wash hands to avoid transfer of microorganisms to other patients or environments.

Patient Care Equipment

Handle used patient care equipment soiled with **blood, body fluids, secretions,** or **excretions** in a manner that prevents skin and mucous membrane exposures, contamination of clothing, and transfer of microorganisms to other patients or environments. Ensure that reusable equipment is not used for the care of another patient until it has been appropriately cleaned and reprocessed and that single-use items are properly discarded.

Linen

Handle, transport, and process used linen soiled with **blood, body fluids, secretions,** or **excretions** in a manner that prevents exposures and contamination of clothing and avoids transfer of microorganisms to other patients or environments.

Use resuscitation devices as an alternative to mouth-to-mouth resuscitation.

Firefighters with many years of experience can often use these quick visual surveys to determine how serious a patient's condition may be. These quick visual surveys can provide important patient information in the first few seconds after an emergency responder's arrival, as well as recognize any safety concerns for the patient and the care providers.

Performing an Initial Assessment

After a quick visual survey of the patient and the scene is done, the next step is an **initial assessment.** An initial assessment is the initial investigative action taken by care providers to determine if the patient has the basic signs of life as well as any serious, life-threatening injuries. The initial assessment covers the following:

1. Level of consciousness of the patient (Is the patient awake?)
2. Airway (Does the patient have an open airway?)
3. Breathing (Is the patient breathing adequately?)
4. Circulation (Is the patient's heart pumping blood to the body adequately?)
5. Major bleeding (Is there any major bleeding?)

The initial assessment can be done very quickly on a conscious patient who is alert and talking. Patients who are unconscious will require closer examination to conduct the initial assessment.

When performing an initial assessment on a patient, firefighters should consider the following:

Level of Consciousness

Is the patient awake? If not, is the patient responsive to loud verbal commands or painful stimulation? After the firefighter determines the level of consciousness of the patient, another member of the team should maintain stabilization of the head and neck if the patient is either unconscious or has experienced a traumatic injury that may endanger the spine (automobile accident, fall, industrial accident, etc.).

To stabilize the head and neck of the patient, the firefighter places one hand on either side of the patient's head and holds firmly so that the head and neck are in a straight line with the body, **Figures 22-14 A** and **B.** The purpose of stabilizing the head and neck is to keep them from moving so that if there is a neck injury, no further damage will be done to the neck or spinal cord.

Airway

The firefighter must ensure that the patient has an open airway. To do this for a patient without traumatic injury, the patient's airway is opened with the head tilt/chin lift by placing the palm of one hand on the patient's forehead and the fingers of the other hand underneath the chin. The firefighter lifts the chin with the fingers and presses lightly on the forehead to roll the head into the open airway position, **Figures 22-15 A** and **B.**

If the patient has a traumatic injury and there could be an injury to the patient's neck, the firefighter will need to use a different method for opening the airway, the jaw thrust. To do this, one hand is placed on either side of the patient's head, placing the fingers along the curve in the jaw bone near the ear. Using the fingers, the jaw bone is pressed out to open the airway, **Figures 22-16 A** and **B.**

The act of breathing brings air into the lungs where oxygen is passed into the bloodstream. Oxygen is essential to sustaining life. Many fire departments carry oxygen on their apparatus. Almost all patients experiencing difficulty breathing may benefit from the use of supplemental oxygen from cylinders that firefighters and EMS providers carry on their equipment. Note, however, that in very rare circumstances, the use of supplemental oxygen may worsen the patient's condition. Firefighters should consult their department's policies on the use of supplemental oxygen. Oxygen administration and the devices used to deliver oxygen to a patient are part of the curriculum of most higher level emergency medical training courses.

Breathing

When it is established that the patient has an open airway, or a firefighter has taken measures to open the airway, the next step is to check to see if the patient is

(A)

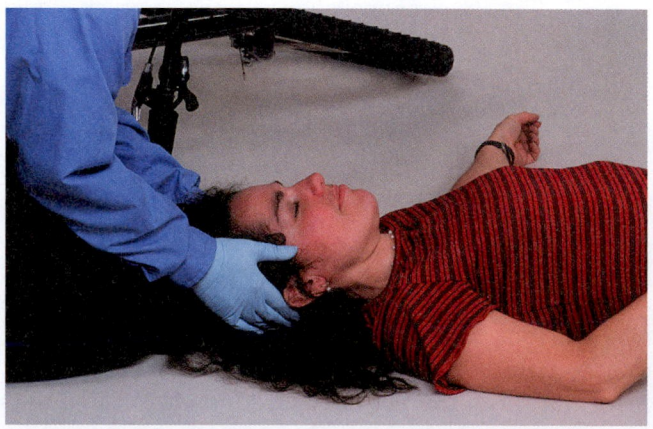

(B)

FIGURE 22-14   Firefighters stabilizing the head and neck of a (A) seated victim and (B) a victim lying on the ground.

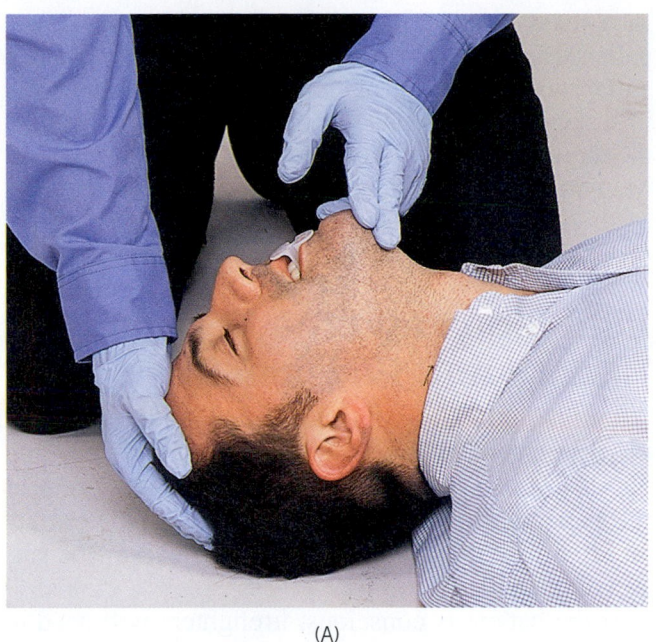

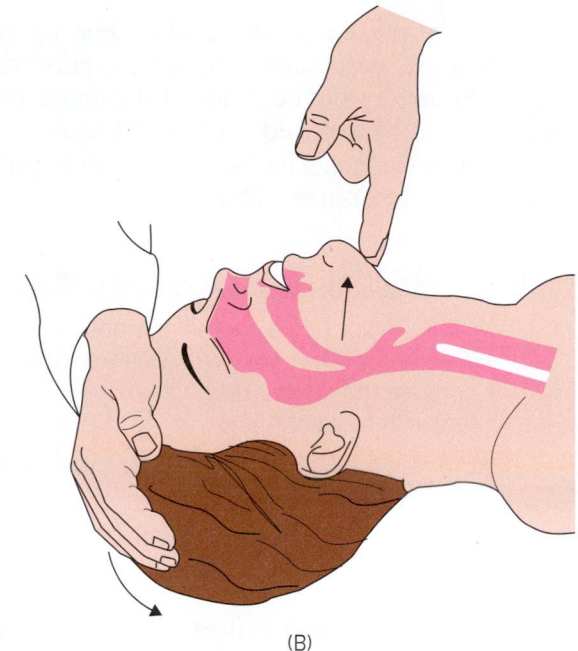

(A)   (B)

FIGURE 22-15 Opening the airway of a patient without traumatic injury using the head-tilt, chin-lift method.

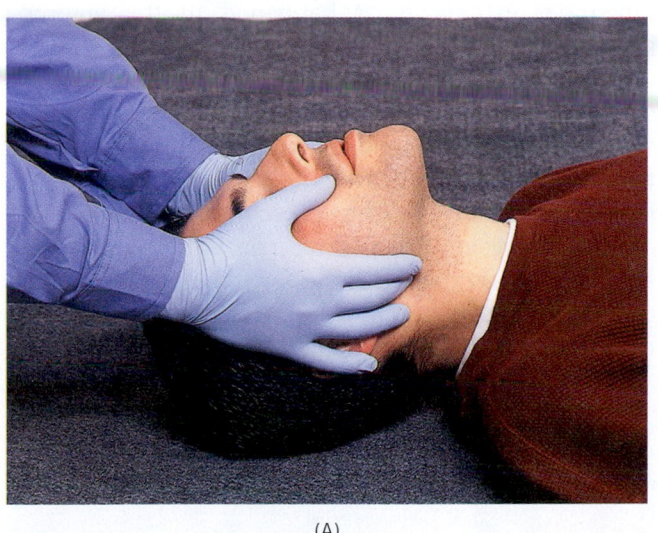

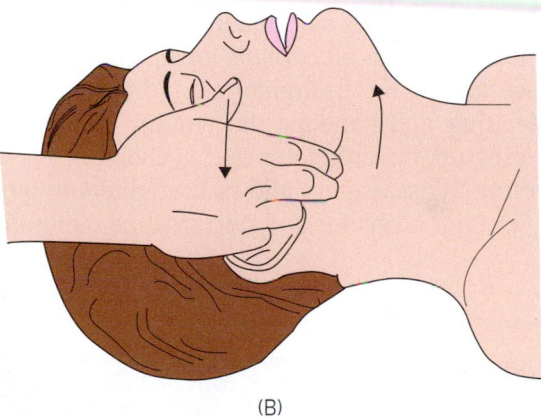

(A)   (B)

FIGURE 22-16 Opening the airway of a patient with traumatic injury using a jaw thrust.

breathing. To check for breathing, you need to look, listen, and feel for air movement. This is done by placing your ear over the patient's mouth while watching for the chest to rise and fall. If there is no breathing, the firefighter should provide rescue breathing for the patient while using a barrier device such as a pocket mask.

Circulation

After the patient has an open airway and is breathing (or someone is breathing for the patient) the firefighter should check for a pulse in the patient. For an unconscious victim, the firefighter should locate the **carotid pulse** in the neck. On a conscious patient, the firefighter can use the **radial pulse** in the wrist to determine the pulse rate and quality. For a small child or infant, the best place to locate the pulse rate is the **brachial artery** on the inside of the upper arm. The pulse should be checked for 5 to 10 seconds to determine if it is present. The locations of these pulse points are shown in **Figures 22-17 A–C.**

Major Bleeding

After determining that the patient has a pulse, the firefighter checks the patient from head to toe, front and back, for major bleeding. A sweeping motion is used with both hands along the entire length of the patient's body to check for major bleeding. If you see blood-soaked clothing, you need to remove the clothing to determine the source and severity of the bleeding. If the patient is lying down, it is important to check underneath the patient's body. This can be done without moving the patient by simply reaching under the patient's body when doing the sweeping motion, and checking the gloves frequently for signs of blood. If a source of major bleeding is located during the initial assessment, the firefighter should attempt to stop the bleeding and bandage the wound. Bleeding and bandaging are covered later in this chapter.

The initial assessment allows firefighters to quickly identify major problems and start treatment of patients.

In a mass casualty situation, where there are many patients, a triage system may be used. **Triage** is a quick and systematic method of identifying which patients are in serious condition and which patients are not, so that the more seriously injured patients can be treated first. Many triage systems are based on the same principles as the initial assessment. Firefighters can learn more about triage systems in higher levels of EMS training.

Vital Signs and the Focused History and Physical Exam

After the initial assessment of the patient is complete, the firefighter continues on to a focused history and physical exam, which is a thorough examination of the patient. This consists of three parts: (1) patient fact-finding, (2) vital signs, and (3) a head-to-toe survey of the patient.

If the patient is conscious, firefighters will need to go on a fact-finding mission for information about the patient. This includes trying to determine the age of the patient, important medical history, and allergies to medications. Firefighters should also try to establish what the patient's current situation is. Has the patient had chest pain for the past two hours? Does the patient remember exactly what happened in the automobile accident? Also, firefighters should always check patients for a bracelet, necklace, or watch that may contain vital medical information, **Figure 22-18.**

If the patient is unconscious, the firefighter can try to do fact-finding by asking questions of family members, friends, or bystanders present on the scene. If no one else is present or no one can provide medical facts about the patient, the firefighter may be able to find clues on the emergency scene to help establish some information about the patient.

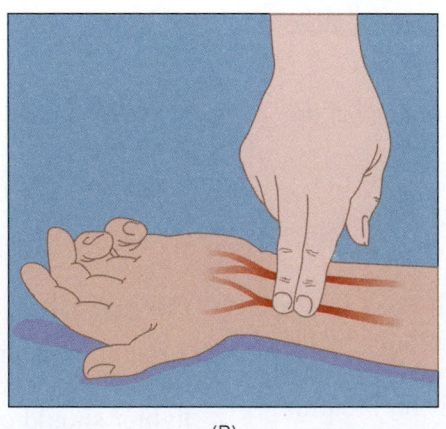
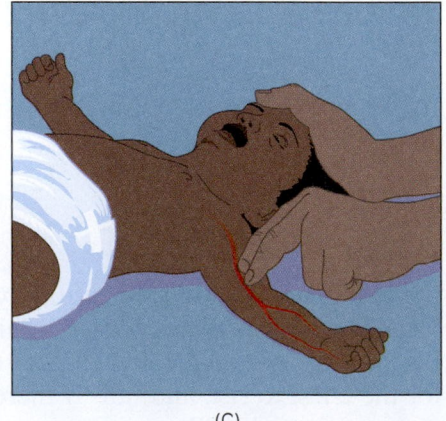

(A)   (B)   (C)

FIGURE 22-17   Locating pulses in (A) the carotid artery in the neck, (B) the radial artery in the wrist, and (C) the brachial artery in an infant's upper arm.

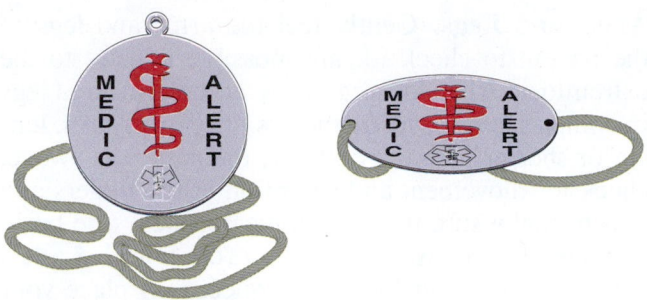

FIGURE 22-18 Examples of medical alert tags.

patient's chest and abdomen for the characteristic rise and fall that occur with each breath for one minute. The firefighter should also try to observe the quality of the respirations (labored, shallow, normal). Labored breathing is indicated by a patient's increased effort to breathe.

TABLE 22-3	Average Vital Sign Ranges, by Age	
Age	**Pulse**	**Respirations**
Newborn	120–160	40–60
1 year	80–140	30–40
3 years	80–120	25–30
5 years	70–115	20–25
7 years	70–115	20–25
10 years	70–115	15–20
15 years	70–90	5–20
Adult	60–80	2–20

After the fact-finding is done, the next step is to get a set of vital signs on the patient. For firefighters providing emergency first aid, the set of vital signs should include a pulse rate, a respiratory rate, and the color and temperature of the skin. **Table 22-3** shows average vital sign ranges by age of the patient.

Pulse Rate

The pulse rate is the number of times the heart is beating per minute. To establish the pulse rate, the firefighter should first locate the carotid pulse in the neck or the radial pulse in the wrist. Using the second hand on a watch or clock, the firefighter should count the number of beats for a minute to get the patient's pulse rate. With experience and practice, firefighters may choose to count the beats over a thirty-second period and multiply the number by two for the pulse rate per minute.

Respiratory Rate

The respiratory rate, like the pulse rate, is the number of times the patient is breathing in one minute. To get the respiratory rate, the firefighter observes the

Color and Temperature of the Skin

The firefighter should carefully observe the skin color and temperature of the patient, because this can be an important indicator of the patient's condition. The skin color may be normal, ashen (gray in color), blue or cyanotic, red, yellow, or pale (white). The skin temperature may be normal, cool and clammy, hot and dry, or hot and moist. These are important signs, and should be noted and recorded on all patients.

After the patient's vital signs are recorded, it is time to do a head-to-toe examination of the patient. It is important to check the patient's vital signs periodically. A good practice to follow is obtaining vital signs every three to five minutes. To conduct a head-to-toe survey of the patient, the firefighter should follow the procedure outlined here:

Head and Neck. Gently palpate the head and neck to check for any structural damage to the bones in the head, face, and neck.

Face and Eyes. Gently palpate the bones in the face to check for swelling or deformities. Examine the mouth, nose, and ears for the presence of foreign objects, bleeding, or discharge of fluid.

Carefully examine the patient's eyes and note the size of the pupil: normal, **constricted,** or **dilated.** Also, check to make sure that the pupils are equal in size for both the right and left eye. Using a small flashlight, quickly shine the light in each eye briefly and note whether or not the pupil is "reactive" to light. If the pupil is reactive, it will shrink with the presence of light and grow larger with the absence of the light source. Refer to **Figure 22-19** for examples of different pupil sizes.

Upper Torso and Chest. Check the bones of the rib cage and breastbone (sternum) for structural integrity by placing your hands on the chest and feeling for equal chest rise and fall or areas of instability. Also, check to see if the chest rise and fall from breathing is equal on both the right and left sides. An unequal chest rise and fall may indicate a serious breathing problem, such as a collapsed lung.

Lower Torso, Abdomen, and Pelvis. Check the abdomen and pelvic area for any obvious signs of traumatic injury, like bruising, swelling, or pulsating masses. A pulsating mass is a sign of a rupture, or impending rupture, of the body's largest artery, the aorta, and is a very serious condition. Using one hand on top of the other, press down in each of the four quadrants of the abdomen to check for pain, tenderness, or abdominal rigidity. Next, place one hand on each hip on the side of the body, squeeze together gently to check the stability of the pelvis, then press down on the top of the hips gently to continue to check the structural integrity of the pelvis.

Arms and Legs. Gently feel the arms and legs of the patient to check for any possible injuries to the extremities. Also, look carefully at the arms and legs in comparison to one another. Is one leg, or arm, longer or shorter than the other? If the patient is awake, check for movement and sensation of the fingers and toes to make sure the neurological system (the brain and spinal cord) is functioning. You can also ask a patient to grip your hand and squeeze, or place your hands at the bottom of the patient's feet and ask the patient to push against your hands. Check for the pulse in both arms and both legs to be sure that the heart is properly circulating blood to the extremities. The pulse in the foot can be felt on the top of the foot and the inside of the ankle underneath the protruding bone structure in the ankle, **Figure 22-20.**

Another very effective way of checking for circulation in the arms and legs of children under six years of age is by checking capillary refill. To do this, pinch the end of the fingers, or toes, until the skin underneath the nailbed becomes white. When you release the pressure, watch to see how quickly the skin becomes pink or normal again. Normally, the skin will return to its original color right away. If it takes longer than a second or two for the skin under the nail to return to its original color, there may be a circulation problem in the patient. **Figure 22-21** demonstrates the capillary refill technique.

Patient Findings

If the fire department requires that firefighters complete a call record for emergency medical patients, they should be sure to document the findings of the assessment and the patient's vital signs.

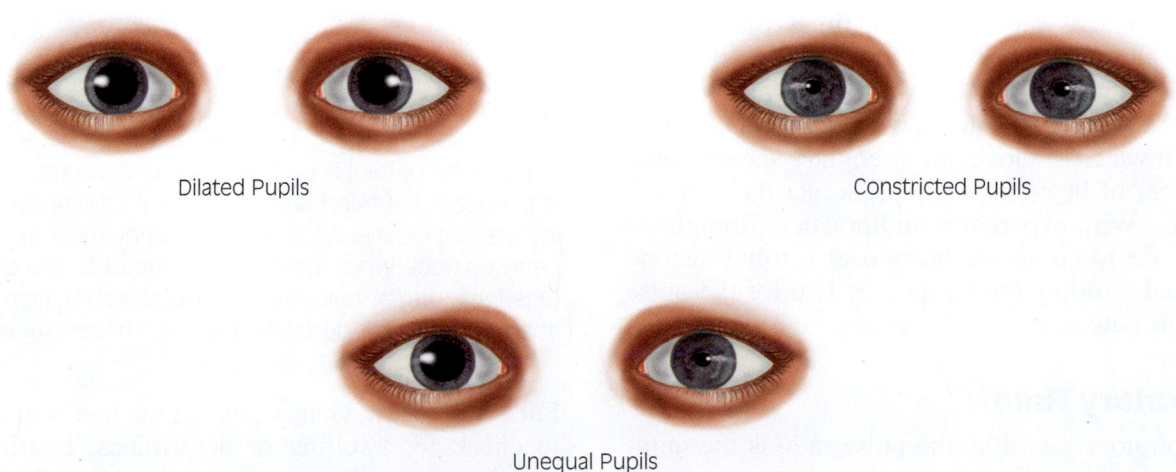

Dilated Pupils

Constricted Pupils

Unequal Pupils

FIGURE 22-19 Examples of pupil size.

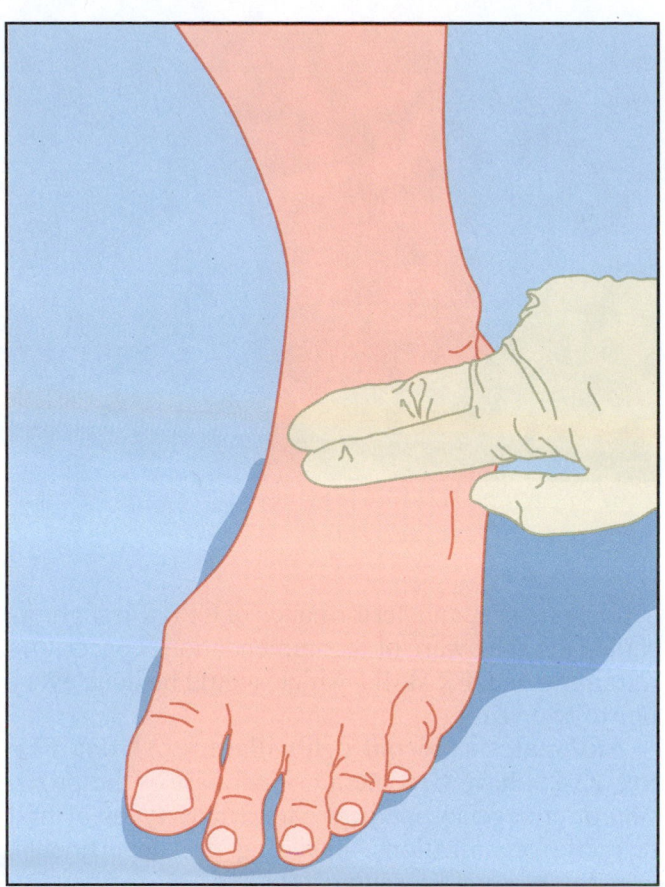
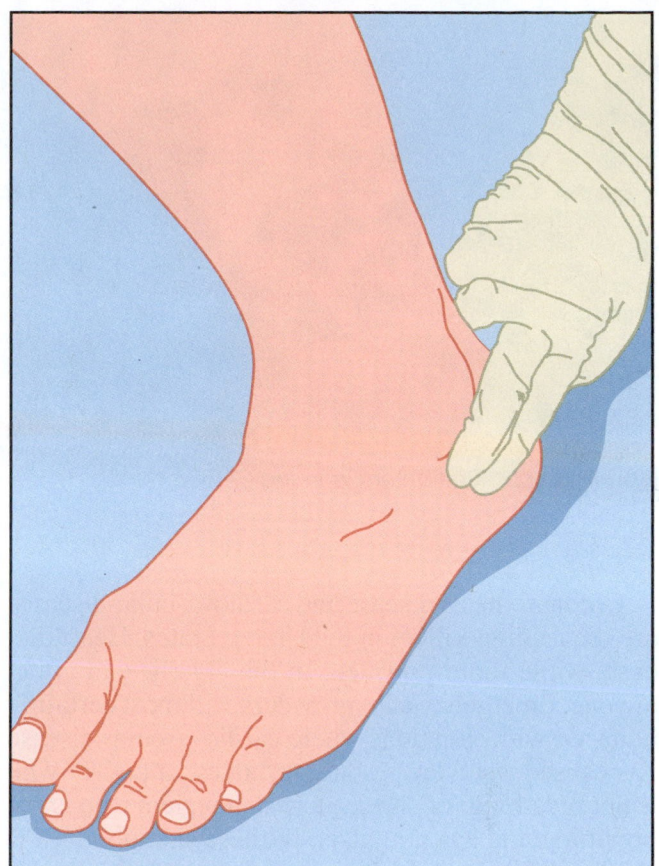

FIGURE 22-20 Locating pulses in the foot.

> **NOTE**
>
> Documenting patient information is very important, and firefighters should be familiar with their department's policies concerning the recording of patient and emergency response information.

Firefighters may have found something in the survey that is important to the treatment of the patient, both by an ambulance crew and in the hospital. Everything that is found must be passed on to the emergency medical provider taking over care of the patient. That includes the initial assessment, the focused history and physical exam, the patient's vital signs, and all patient fact-finding that was done while on the scene.

CARDIOPULMONARY RESUSCITATION/AED

One of the most basic and widely learned emergency response skills is cardiopulmonary resuscitation, or CPR. This skill is learned by all types of people, from lifeguards to physicians. CPR is a critical, lifesaving

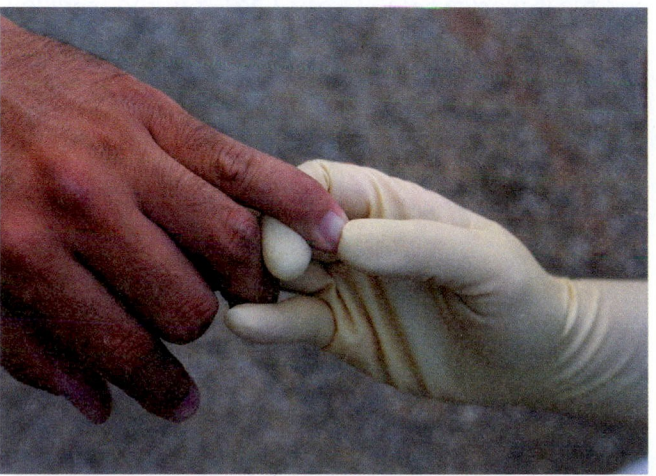

FIGURE 22-21 Capillary refill.

skill for firefighters and emergency responders to learn and practice, **Figure 22-22.**

> **NOTE**
>
> CPR and techniques for helping choking victims are extremely important knowledge to have not only while on duty with the fire department but also anywhere a firefighter may go.

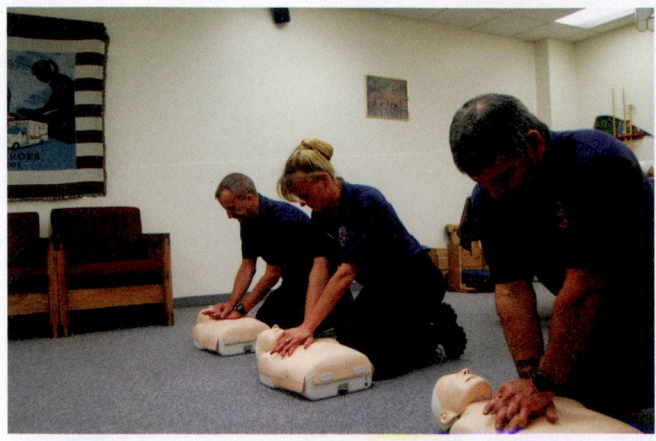

FIGURE 22-22 Firefighters learning CPR.

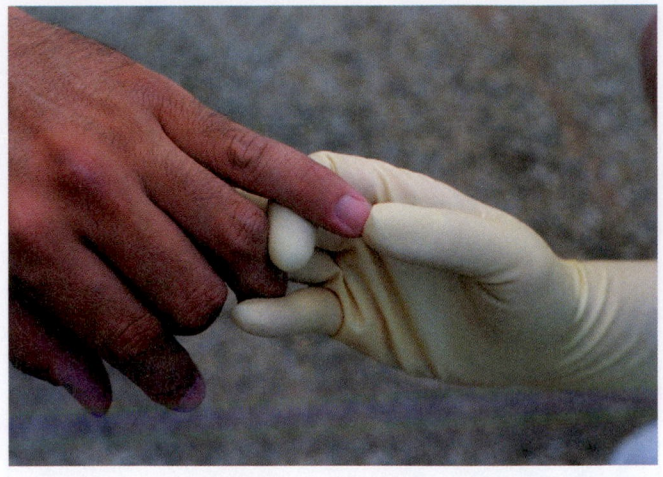

Coronary heart disease and cardiovascular diseases are the leading killers in the United States. Heart disease, while more prevalent in the elderly, can affect anyone. Emergency care providers will most certainly be faced with situations where cardiovascular disease has caused a sudden respiratory arrest (a patient who is not breathing) or cardiac arrest (a patient who is not breathing and has no pulse). In these situations, CPR is a basic lifesaving skill.

Several different organizations support CPR training, education, and research. The two largest organizations are the American Heart Association (AHA) and the American Red Cross. CPR training is widely available in the United States. If a particular department or agency does not have CPR classes available, firefighters can contact a local chapter of the AHA or the American Red Cross and ask them about CPR training.

SAFETY

CPR is a critical skill for firefighters to know and practice. They need this skill not only to provide care to patients on emergency medical calls but also to be prepared to help fellow firefighters on the scene of a fire or emergency incident. Heart disease is a leading killer of firefighters operating on emergency scenes every year. Firefighters should have the skills necessary to help the members of their team in need, including CPR and first-aid training.

As research on cardiovascular disease and cardiac arrest procedures continues, the skills and information regarding CPR may be changed or revised. Firefighters should follow the most current standards set forth by the AHA or the American Red Cross for edu-

cation, practice, and performance of CPR. Firefighters should also be aware of recommendations concerning retraining on CPR skills, which should be done every one to two years.

Automated external defibrillators (AEDs), Figure 22-23, have become a critical part of the provision of emergency medical care. Many types of first response organizations, including fire departments, use AEDs as a lifesaving tool. AEDs are becoming more common in nonemergency settings as well, including shopping malls, casinos, airplanes, airports, and even private homes. AEDs are becoming more common because they dramatically improve the outcome of cardiac arrest.

An AED is a small machine with an internal computer that analyzes a patient's heart rhythm and delivers electrical shocks, or defibrillations, as necessary to correct a heart with a malfunctioning rhythm. It requires some basic training to place the AED on a patient and use it appropriately. Many first response agencies are training firefighters to use AEDs at the same time they are training them on CPR techniques. Firefighters should consult their agency's policies and explore training courses for AED use by firefighters.

BLEEDING CONTROL AND SHOCK MANAGEMENT

The blood in the body brings oxygen and nutrients to the cells in the body in order for the cells to survive, grow, and reproduce. The heart and the system of tubes through which blood travels (*arteries, veins,* and *capillaries*) are very complex and vital systems, **Figure 22-24.** The **cardiovascular system** is a closed system with a pump (the heart) and specialized tubing (the blood vessels) that provide the body

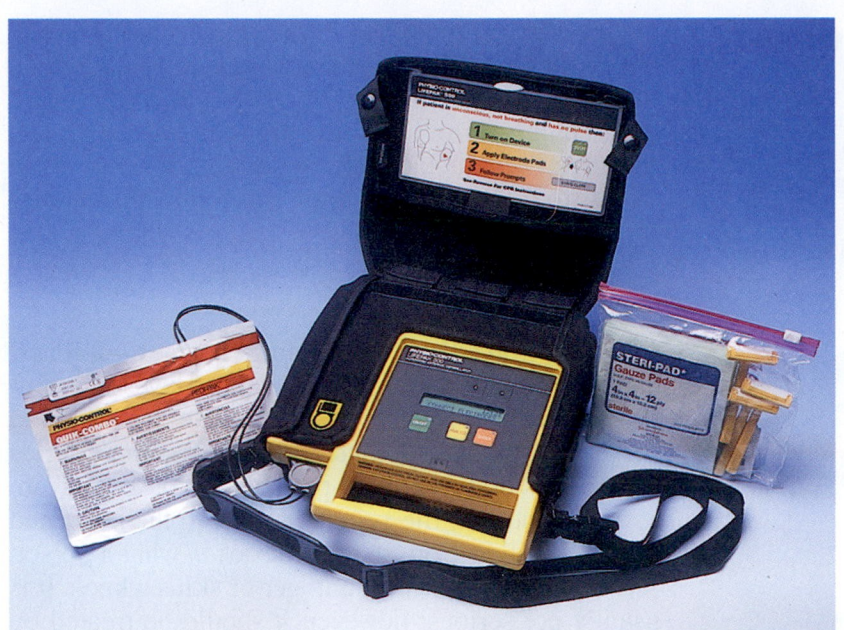

FIGURE 22-23   An AED is a small machine with a computer that analyzes a patient's heart rhythm and delivers electrical shocks, or defibrillations, as necessary to correct a heart with a malfunctioning rhythm.

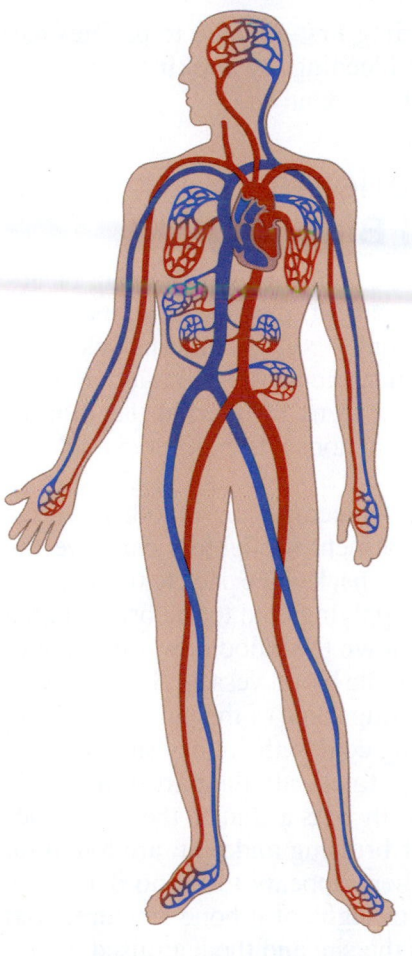

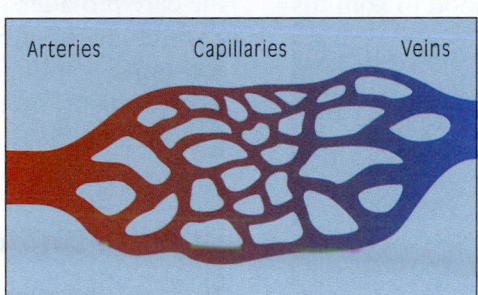

Arteries   Capillaries   Veins

FIGURE 22-24   The blood vessel system in the body, which is composed of arteries, veins, and capillaries.

with blood containing the elements humans need to live. Damage to the heart or the blood vessels may result in a lower volume of blood or an ineffective pumping system, which are serious, if not life-threatening, conditions that can cause a patient to go into shock, or hypoperfusion (a lack of oxygen and nutrients to the tissues).

CAUTION

Bleeding control and shock (hypoperfusion) management are important skills to learn in first-aid training. The firefighter should use BSI, as with all patients, but especially so in this situation because exposure to blood and body fluids is very likely.

Internal and External Bleeding

Internal bleeding occurs when there is bleeding within the body and no visible open wound is present. Internal bleeding can occur because of trauma to the body or because of illness. Internal bleeding can be a very serious condition for a patient.

Firefighters should look for the following signs, which may indicate internal bleeding:

- Bruising of the skin
- Pale skin
- Cold and clammy skin
- Dilated pupils
- Obvious deformities to major bones (like the pelvis, the upper leg, etc.)
- A rigid and tender abdomen
- Blood in the urine or from the rectum
- Blood from the mouth or nose, or blood in vomitus

Patients who the firefighter suspects are suffering from internal bleeding should be treated for shock (hypoperfusion), as covered later in this chapter.

External bleeding is bleeding that is coming from an open wound on the body. The three types of external bleeding are discussed next.

Arterial Bleeding

Arterial bleeding is bleeding from **arteries,** which are the blood vessels that carry blood that has been oxygenated away from the heart to the cells in the body. Arterial blood, because it has just been supercharged with oxygen, is bright red in color. Bleeding from the arteries can be very serious. In order for the oxygenated blood to get to the cells, the heart pushes it through the arteries under high pressure. When an artery is severed or damaged, this pressure can force a large quantity of blood out of the blood vessel system, depleting the system and possibly causing shock. If a larger artery is involved, the bleeding may be life threatening.

Firefighters providing first-aid care to patients can recognize arterial bleeding by looking for bright red blood from the wound site or blood spurting or pulsating from the wound site.

Venous Bleeding

Venous bleeding is bleeding from a vein. **Veins** are the blood vessels that carry blood from the cells in the body, after the oxygen and nutrients are used, back to the heart to be reoxygenated in the lungs. Venous blood, because it has been stripped of the oxygen it was carrying, is bluish red in color. Bleeding from the veins can also be very serious, especially if the bleeding is from a large vein. Unlike the arteries, however, the pressure on the blood in the veins is not as great, but if a larger vein is involved, venous bleeding can still be life threatening.

Firefighters providing first-aid care to patients can recognize venous bleeding by looking for bluish red blood from the wound site or blood flowing steadily from the wound site.

Capillary Bleeding

Capillary bleeding is bleeding from a capillary. **Capillaries** are the very small blood vessels connecting the arteries and the veins. These small blood vessels filter the oxygenated blood from the arteries to the cells and then take the used blood back from the cells and into the veins to return to the heart and lungs. Bleeding from the capillaries is what a person sees when he or she cuts a finger or skins a knee. It is usually not serious; however, it should be treated by the care provider.

Firefighters providing first-aid care to patients can recognize capillary bleeding by looking for blood slowly oozing from the wound site.

Caring for Patients with Internal Bleeding

Once firefighters have established what kind of bleeding is occurring in a patient, they can take measures to get the bleeding under control. Internal bleeding cannot be brought under control by first-aid actions. For patients exhibiting signs of internal bleeding or shock, there are other actions a firefighter can take to provide assistance.

For minor internal bleeding or, in other words, bruising, the best treatment firefighters can give the patient is to apply a cold pack or ice pack to the affected site. The purpose of applying cold to the bruised area is to attempt to slow down the blood flow seeping into the skin by narrowing the blood vessels. The cold will also help with alleviating some of the pain of bruising as well. After applying cold to the injury site, the firefighter should attempt to elevate the part of the body that is injured, unless there is a chance there is a fracture. For example, if bruising and pain are found on the lower leg, and there appears to be no deformity of the leg structure or signs of a bone fracture, cold should be applied to the site and the leg raised.

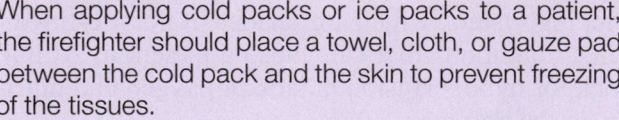

STREETSMART TIP

When applying cold packs or ice packs to a patient, the firefighter should place a towel, cloth, or gauze pad between the cold pack and the skin to prevent freezing of the tissues.

Major internal bleeding is a serious problem and requires rapid medical attention. The loss of a large volume of blood can cause a patient to go into shock. The treatment for patients in shock is covered in a later section.

Caring for Patients with External Bleeding

Three methods are used when dealing with a patient experiencing external bleeding, as explained next. Note that, before acting, the firefighter should remember to wear protective gloves, since exposure to blood and body fluids is likely.

Pressure on the Site of the Bleeding, or Direct Pressure

Capillary and some venous bleeding can usually be controlled by direct pressure. Direct pressure can also help in the case of arterial bleeding. Firefighters controlling bleeding with direct pressure should first locate the wound and source of the bleeding.

(They must look carefully, because a large volume of blood may be coming from a small cut or puncture.) The firefighter should place a sterile dressing or bandage over the site of the wound and apply pressure by pressing a hand against the wound site. Using a roll of gauze (preferably sterile) or a roll of bandage material, the firefighter wraps the site with the material, pulling it tight to create moderate pressure to the wound site. It is important to use a piece of medical or first-aid tape to secure the bandage and keep pressure on the wound site. If the bleeding soaks through the bandaging, it should not be removed to redo the bandage. Once a bandage is placed on a patient, it should not be removed. Instead, the firefighter should place another sterile dressing on the bandaged wound and wrap it again with a roll or gauze, this time applying a little more pressure with the bandaging, **Figure 22-25.**

Elevate the Site of the Bleeding

After applying direct pressure and a bandage to the wound that is bleeding, the firefighter should elevate the extremity or area that is bleeding. If direct pres-

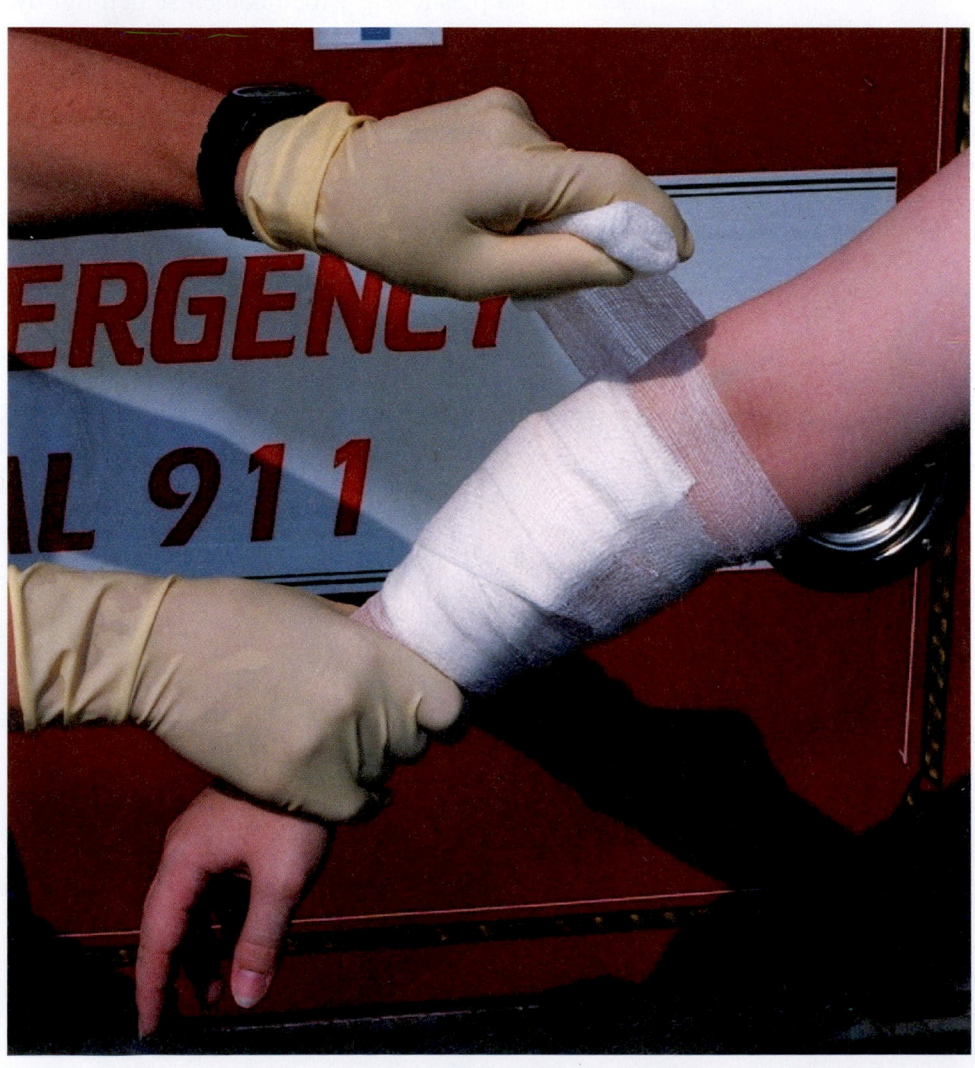

FIGURE 22-25 Applying a pressure dressing to a bleeding wound.

sure stops the bleeding, elevation may not be necessary. Elevation will make the blood flow to the site slower because it must travel "uphill" and, therefore, may slow down the bleeding. If the bleeding is from an arm or leg, it should be simply propped or raised above the level of the heart. If the bleeding is from the head or face, the firefighter should make sure the patient sits upright, but only if there is no back or neck injury.

Use Pressure Points

The blood vessel system in the body has some main vessels, or tubes, that supply certain areas of the body. If direct pressure and elevation do not help stop or slow the bleeding, applying pressure to these "pressure points" may slow the flow of blood to the body area, thus slowing the flow of blood from the wound, **Figure 22-26.** The best pressure points are in the upper arm (brachial artery) and in the hip/pelvic area (**femoral artery.**) In both of these areas, the major artery is near a bone where it can be compressed to slow blood flow.

Types of Wounds Requiring First Aid

There are several types of wounds that firefighters providing first aid should be able to recognize and treat.

Abrasion

An **abrasion** is a scrape or brush of the skin usually making it reddish in color and resulting in minor capillary bleeding. An abrasion is what results when

someone "skins a knee," a common event hood. Abrasions should be treated by applying bandages and, if the bleeding has not stopped, pressure.

Avulsion

An **avulsion** is an injury where a part of the skin torn away, but still attached, leaving a flap or loose area hanging. Avulsions should be treated by applying sterile bandages and, if the bleeding has not stopped, direct pressure.

Amputation

An **amputation** occurs when a part of the body is severed completely as a result of an injury. This can happen to fingers and toes and even arms and legs. Firefighters should first apply sterile bandages to the site where the amputation occurred with direct pressure, using elevation or pressure points as needed. Firefighters should then locate the severed body part and wrap it completely with a sterile dressing. Once it is wrapped in the dressing, the body part should be placed into a plastic bag which is then sealed tightly. The severed body part must be kept cold and transported to the hospital with the patient. Firefighters can place the severed body part that is sealed in a plastic bag into a second plastic bag filled with ice to keep the body part cold for transport.

Laceration or Incision

A **laceration** is a cut to the skin and underlying tissues that leaves an irregular, uneven pattern. An **incision** is a cut to the skin that leaves a straight, even pattern. In

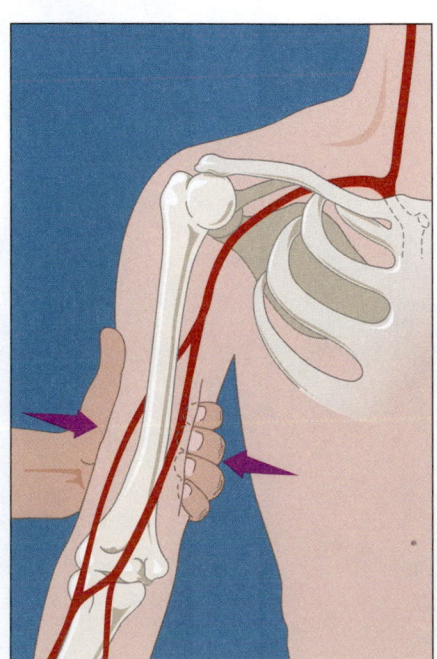
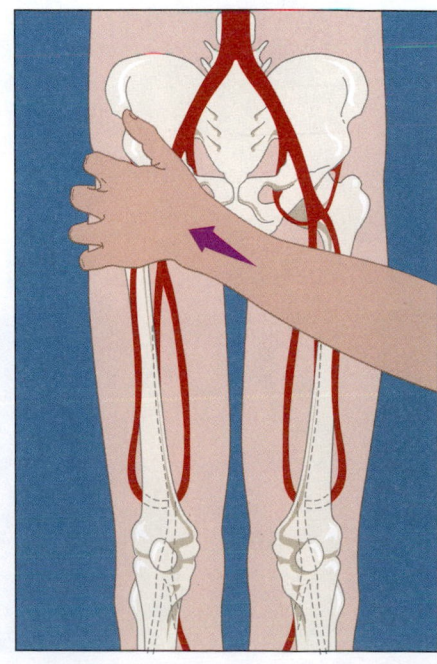

FIGURE 22-26 Using pressure points in the upper arm and the leg to decrease blood flow to a wound.

both cases, the firefighter should apply sterile bandages to the wound and, if the bleeding has not stopped, direct pressure. Elevation and pressure points may be necessary for a large or deep laceration or incision.

Puncture

A **puncture** is caused by an object that has stabbed the body. If the object is no longer in the wound, firefighters can treat the puncture wound by applying a sterile bandage and, if necessary, direct pressure, elevation, or pressure points. If the object causing the puncture remains in the body, firefighters should pack the sterile dressings around the wound site, leaving the object in the body. They should not attempt to remove an object imbedded in a wound, because that could cause more harm to the patient. The object must be secured so that it will not move around and cause more damage to the skin and tissues.

What Is Shock? (Hypoperfusion)

Shock, or **hypoperfusion,** is a condition caused by a problem with or failure of the circulatory system that results in a decreased supply of oxygen and vital nutrients to the body's tissues. In other words, it is a lack of blood and oxygen to the organs and tissues of the body. The body requires a regular and constant supply of oxygen and nutrients to function properly.

> **CAUTION**
>
> Shock (hypoperfusion) is a very serious condition and can cause death.

Shock (hypoperfusion) can result from a problem with the heart, which pumps the blood to the tissues. This can happen when a patient is experiencing a heart attack, and the heart has lost its ability to pump the blood adequately. Shock can also result from a failure in the blood vessels that carry the blood to and from the heart. This can be caused by a severe allergic reaction, which causes the blood vessels to expand rapidly, or by a trauma to the head or spinal cord, also causing a rapid expansion, or dilation, of the blood vessels.

Lastly, shock can result from a problem within the blood vessels, namely, a drop in the quantity of blood contained in the vessels. This can occur when the patient has an injury that has caused major internal or external bleeding. As the blood is lost from the body, the vessels and the heart have less oxygen carrying material to work with, resulting in the shock state. This type of shock can also occur in a patient who

has experienced an illness and is experiencing **dehydration.** A severely dehydrated state can cause shock. Dehydration can occur due to a variety of reasons. A patient experiencing vomiting and diarrhea for even a short period of time can become dehydrated quickly.

Recognizing the Signs and Symptoms of Shock (Hypoperfusion)

There are many signs and symptoms of shock, or hypoperfusion, that firefighters should look for in a patient, **Table 22-4.**

- Pale color to the skin, or a bluish tint to the skin, especially around the lips and in the nailbeds
- A cool temperature of the skin
- Sweating, or moist skin
- Pupils larger than normal, or dilated
- Rapid, shallow breathing
- A rapid, weak pulse
- Complaints of nausea and or vomiting
- Thirst
- Restlessness or anxiety
- Unconsciousness, or an altered level of consciousness

Firefighters should look for signs of shock (hypoperfusion) in all patients. Many different traumatic injuries or severe illnesses can result in shock, so firefighters should always be prepared for a shock state in a patient.

Caring for Patients in Shock

Firefighters who find a patient with the signs and symptoms of shock (hypoperfusion) should follow the basic treatment outlined here. Firefighters should also

TABLE 22-4   Signs of Shock
- Pale or bluish tint to skin color
- Cool skin temperature
- Moist or sweating skin
- Dilated pupils
- Rapid, shallow breathing
- Rapid, weak pulse
- Nausea, thirst, or vomiting
- Unconsciousness or an altered level of consciousness

treat patients in this manner if they suspect that they could go into shock.

Treating a Patient for Shock (Hypoperfusion)

1. Ensure scene safety and BSI precautions.

2. Assess the patient's level of consciousness and, if needed, maintain head and neck stabilization.

3. Make sure the airway of the victim is open. This is especially important in unconscious patients.

4. Ensure that the patient is breathing adequately and has a pulse. If either is absent, follow the procedures for cardiopulmonary resuscitation and/or rescue breathing.

5. Treat the injuries present and control any major bleeding.

6. Keep the patient warm. Place some blankets or a covering over the patient to prevent any body heat loss or exposure to cold in the environment.

7. Position the patient properly. The best possible position for a patient in shock is lying on the back. If head and neck stabilization is initiated during the initial assessment it should be continued throughout patient treatment. Placing a patient on the back may not be possible if the patient is having difficulty breathing, so the patient can be placed in a reclined position. A patient who is conscious and vomiting is best placed on the side so that the vomitus does not obstruct the airway.

8. Raise the legs of the patient to allow more blood to reach the heart. The legs should be raised together and no more than a foot above the level of the heart. Do not raise the legs of a patient who has a traumatic injury to the legs or pelvis.

9. Provide reassurance to the patient, and monitor vital signs every few minutes.

EMERGENCY CARE FOR COMMON EMERGENCIES

No two emergency medical calls are the same. Firefighters must learn the basics of first-aid treatment and then adapt them to each emergency. This section lists some common emergencies, things to look out for, and first-aid treatment.

Trouble Breathing

Trouble breathing is a very common emergency that firefighters may be called on for first-aid care. There are many reasons why a patient may be experiencing trouble breathing, anything from anxiety to a major respiratory disease. Common signs and symptoms of trouble breathing include wheezing, gasping for air, shallow breathing, pale or bluish skin color, anxiety, or even unconsciousness if the patient is having serious difficulty breathing.

Care Guidelines

1. After assessing scene safety and donning BSI, perform an initial assessment and attend to any major problems found with the airway, breathing, or circulation.

2. Reassure the patient, keep him or her calm, and monitor vital signs frequently because the patient's condition may deteriorate rapidly.

3. Recognize that this is a serious emergency and emergency medical care is needed right away.

Chest Pain

Chest pain is another common emergency that firefighters may encounter, as shown in **Figure 22-27.** Chest pain related to a heart problem or heart attack may be described as a tightness or squeezing feeling in the chest, and the pain may be felt not only in the chest but also in the jaw, abdomen, or arm (often the left arm). Chest pain can also result from some respiratory disorders, traumatic injuries, or even stomach problems such as indigestion.

Care Guidelines

1. After assessing scene safety and donning BSI, perform an initial assessment and attend to any major problems found with the airway, breathing, or circulation.

2. Reassure the patient, keep him or her calm, and monitor vital signs frequently because the patient's condition may deteriorate rapidly.

3. Recognize that this is a serious emergency and emergency medical care is needed right away.

4. Because chest pain may be a sign of a problem with the heart, there is a possibility the patient could go into shock or, in the worst case, cardiac arrest. Look for signs and symptoms of shock and treat the patient accordingly.

Medical Illnesses

A firefighter may be called on to assist a patient with a medical illness, which can range from stomach problems like nausea and vomiting to general illnesses like fevers, the flu, the common cold, or a diabetic emergency.

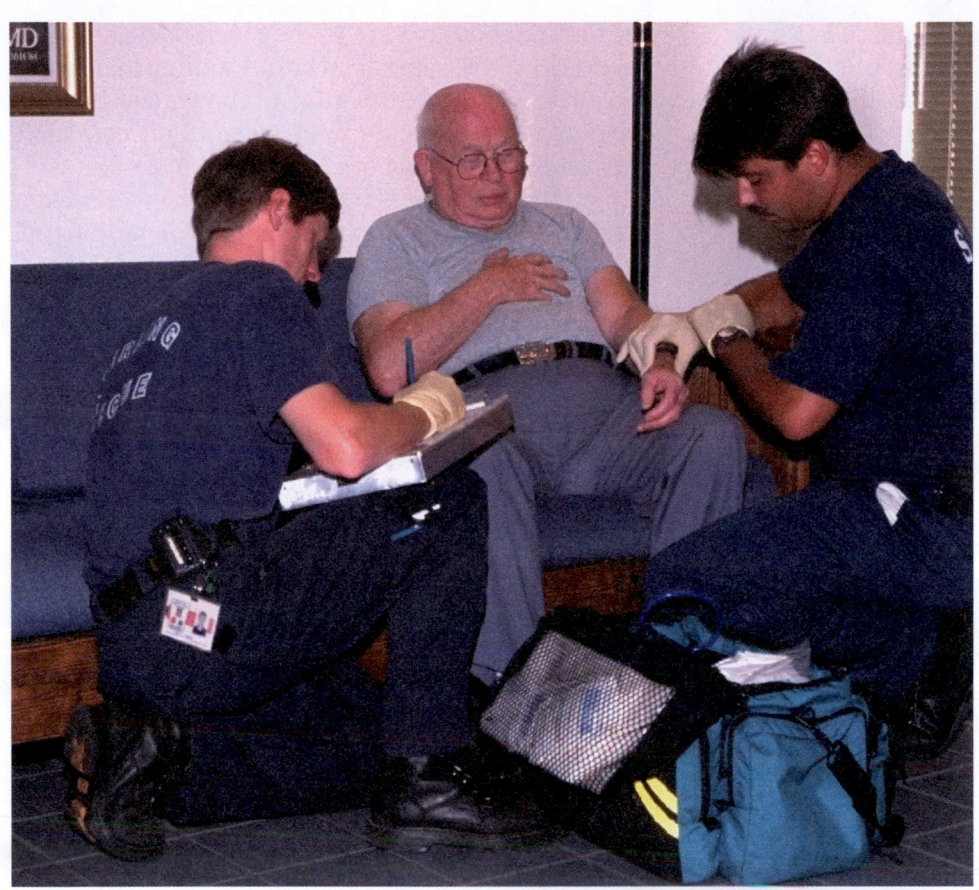

FIGURE 22-27 Firefighters treating a patient complaining of chest pain. *(Photo courtesy of Fred Schall)*

Care Guidelines

1. After assessing scene safety and donning BSI, perform an initial assessment and attend to any major problems found with the airway, breathing, or circulation.

2. Reassure the patient, keep the patient calm, and monitor vital signs frequently.

3. Recognize that this could be a serious emergency and emergency medical care is needed right away.

4. Because a medical illness may cause dehydration (lack of fluids) in a patient, there is a possibility the patient could go into shock. Look for signs and symptoms of shock and treat the patient accordingly.

Allergic Reactions

Allergic reactions result from the body's reaction to a substance to which there is an allergy. Many people experience allergy problems associated with hay fever or bee stings. In some cases, allergic reactions can be very severe, even life threatening. Allergic reactions can result from a variety of substances, such as types of food, things in the environment, bites and stings, and medications. Allergic reactions may be localized

TABLE 22-5	Signs of a Serious Allergic Reaction

- Trouble breathing
- Rapid pulse
- Shallow breathing
- Hives
- Anxiety

at the part of the body where the substance was introduced, and present as localized swelling, redness, and itching. Allergic reactions involving the entire body system may include trouble breathing, rapid pulse, shallow breathing, hives (red splotches or bumps on the body, typically concentrated on the torso), anxiety, and even shock, **Table 22-5**.

Care Guidelines

1. After assessing scene safety and donning BSI, perform an initial assessment and attend to any major problems found with the airway, breathing, or circulation. Severe allergic reactions can compromise

the airway and breathing of a patient because the reaction tightens the breathing passages in the lungs and airway. Be prepared to breathe for the patient if necessary.

2. Reassure the patient, keep the patient calm, and monitor vital signs frequently. Also, if the patient is still in contact with or near the substance causing the allergic reaction, remove the patient from the area or isolate and dispose of the substance. If a patient has been stung by a bee, make sure that the stinger has been removed by using a credit card or other straight edge to scrape the stinger away. Be extra careful not to squeeze the venom sac when removing a stinger.

3. Recognize that in an allergic reaction involving the body system, this is a serious emergency and emergency medical care is needed right away.

4. Because severe allergic reactions often cause the blood vessels to widen, there is a possibility the patient could go into shock. Look for signs and symptoms of shock and treat the patient accordingly.

Thermal Burns

Thermal burns, or burns caused by heat, can be as simple as a sunburn or as life threatening as severe burns across much of the body. Thermal burns are very painful and can cause the patient to lose body heat and fluids, causing dehydration.

Burns are divided into three categories based on their severity. **Superficial burns,** sometimes referred to as first-degree burns, occur when the outer layer or layers of skin are burned, **Figure 22-28A.** The skin will turn red and feel hot to the touch. Depending on the severity of the superficial burn, swelling may result. These burns are often painful to the patient but will generally heal in about a week, even if left untreated. Sunburn is an example of a superficial burn.

In **partial thickness burns,** sometimes called second-degree burns, additional layers of skin are burned, but the deeper layers of the skin remain undamaged, **Figure 22-28B.** The skin will be very red, and blisters will develop. These burns can cause the patient intense pain because the nerves under the skin are sometimes affected. The major sign of a partial thickness burn is the blistering of the skin; however, this blistering does not always occur immediately. It may take several hours before blisters develop. These types of burns require medical attention, and with treatment they generally heal within a few weeks.

The last category of burn is a **full thickness burn,** often referred to as a third-degree burn, **Figure 22-28C.** In full thickness burns, all layers of skin are burned. These burns will have the characteristics of partial thickness burns, but will also have areas of charred or blackened skin, which is the definitive sign of a full thickness burn. These burns are the most serious of the three types because the entire area of skin has been damaged, and the body is no longer able to hold fluids inside or keep bacteria and dirt outside. Many times nerves are also damaged or destroyed, and the patient may not complain of pain in the area of the full thickness burn. (However, the accompanying partial thickness burn areas will still be very painful.) Full thickness burns require medical attention, and these patients are extremely susceptible to infections. The healing process of a full

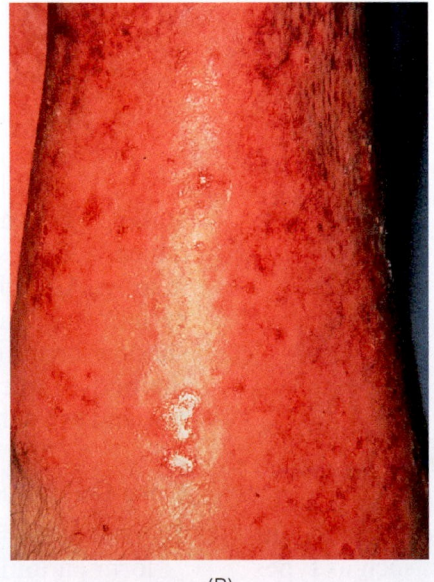
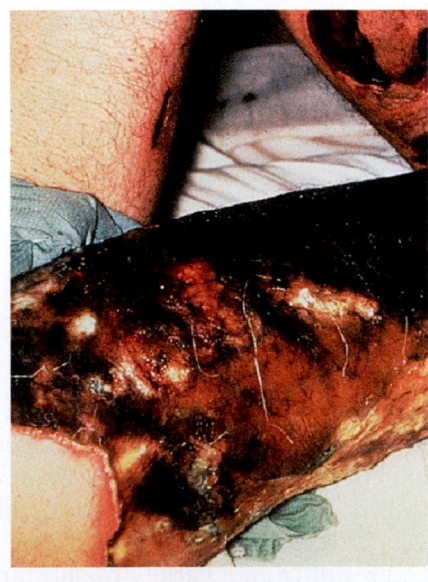

(A)   (B)   (C)

FIGURE 22-28   (A) Superficial burns, (B) partial thickness burns, and (C) full thickness burns.

thickness burn is usually long and painful and can be debilitating.

Care Guidelines

1. After assessing scene safety and donning BSI, perform an initial assessment and attend to any major problems found with the airway, breathing, or circulation. Be sure to remove the patient from the environment to stop the burning process.

2. Reassure the patient, keep the patient calm, and monitor vital signs frequently.

3. Cover the burned area with dry, sterile dressings to prevent infection. (Firefighters should consult their agency for specific treatment for burns. Some fire departments may have a different treatment for burns.)

4. Monitor the patient's airway and breathing constantly. Burn victims may also have inhaled smoke or hot gases, which can cause damage in the lungs and create airway or breathing problems. Constant monitoring of the patient is critical due to the possibility of airway swelling or other injury.

5. Also, monitor the patient for signs and symptoms of shock and treat accordingly.

Chemical Burns

Chemical burns are caused by chemical substances that come into contact with the skin or tissues of the body, creating a caustic reaction. A variety of chemical substances can cause a chemical burn. They can be found in many different locations, from large, industrial plants to family residences.

Care Guidelines

1. After assessing scene safety and donning BSI, perform an initial assessment and attend to any major problems found with the airway, breathing, or circulation.

2. Reassure the patient, keep the patient calm, and monitor vital signs frequently.

3. If the chemical substance is dry or in powder form, brush off as much of the substance as you can from the patient, and remove the patient's clothing, watches, and jewelry. (Firefighters should make sure they protect themselves when removing the chemical from the patient. Do not use bare hands or brush the substance into the wind where it may become airborne and harm others. Consider additional protective gear, such as masks and gowns.)

4. If the chemical substance is not in dry or powder form (liquid chemicals, acids, etc.), flush the sub-

stance off the patient with large volumes of gently flowing water. Use a garden hose without a pressure nozzle, bottles of water, or water from a faucet to flush the substance off the patient. Remove all clothing, watches, and jewelry from the patient.

5. Monitor the patient's airway and breathing constantly. Chemical burn victims may also have inhaled a substance that can cause damage to the lungs and create airway or breathing problems. Constantly monitor the patient because they can deteriorate rapidly.

6. If the chemical has gotten into the patient's eyes, flush the eyes with large amounts of gently flowing water for at least fifteen to twenty minutes.

Poisoning

The ingestion or inhalation of a caustic substance is considered poisoning. Poisoning can also result from the ingestion of a large quantity of a normally harmless substance, like over-the-counter medications or household substances as shown in **Figure 22-29.**

Care Guidelines

1. After assessing scene safety and donning BSI, perform an initial assessment and attend to any major problems found with the airway, breathing, or circulation.

2. Reassure the patient, keep the patient calm, and monitor vital signs frequently.

3. Do some investigative work to find out *exactly* what substance was ingested or inhaled. Keep that information handy.

4. *Do not* give the patient any liquid or food to ingest.

5. Contact the local poison control center. Check with your department or agency and find out what your local poison control center is and the telephone number for the center. Tell the poison control center all essential information about the patient, such as age, sex, medical history, and symptoms. Also, tell the poison control center *exactly* what substance was ingested or inhaled, and how much of it was ingested or inhaled. The poison control center will be able to suggest additional treatment for the patient before the EMS unit arrives.

STREETSMART TIP

Local poison control centers are very valuable and helpful resources. Even if the exact name or dosage of the substance is unknown, they may be able to help the emergency responder figure out what it is.

FIGURE 22-29 Many types of substances and materials can be harmful to people and cause poisoning. When caring for a patient who has possible poisoning, the firefighter should make every attempt to find out exactly what substance was ingested or inhaled and how much.

Poison control centers have special references to assist them in figuring out "mystery" substances. For example, if a pill is found, a firefighter can describe it to the poison control center, and they may be able to find out what the name of the drug is and even the dosage.

Fractures and Sprains

A **fracture** is a medical term for a broken bone. A **sprain** is an injury to the ligaments that hold joints in the body together and allow them to move. Firefighters may notice swelling and tenderness to the area of a fracture or sprain. In the case of a fracture, there may be deformity to the bone and body structure.

Care Guidelines

1. After assessing scene safety and donning BSI, perform an initial assessment and attend to any major problems found with the airway, breathing, or circulation.

2. Reassure the patient, keep the patient calm, and monitor vital signs frequently.

3. Do not move the patient, and protect the cervical spine area in cases where the patient has fallen or been involved in a moving vehicle accident.

4. Do not move the injured part of the body. Do not attempt to straighten an arm or leg that is deformed.

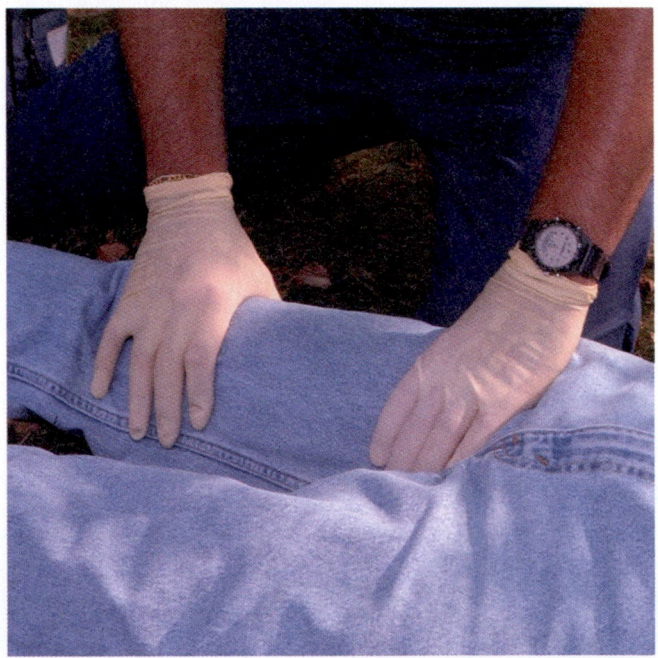

FIGURE 22-30 Firefighter stabilizing a leg injury until the EMS unit arrives.

Carefully protect the injured area and be sure it does not move or shift, **Figure 22-30**.

5. As needed, apply cold packs or ice packs to the injured area to help relieve some of the pain and swelling.

JOB PERFORMANCE REQUIREMENT 22-1
Removing Gloves

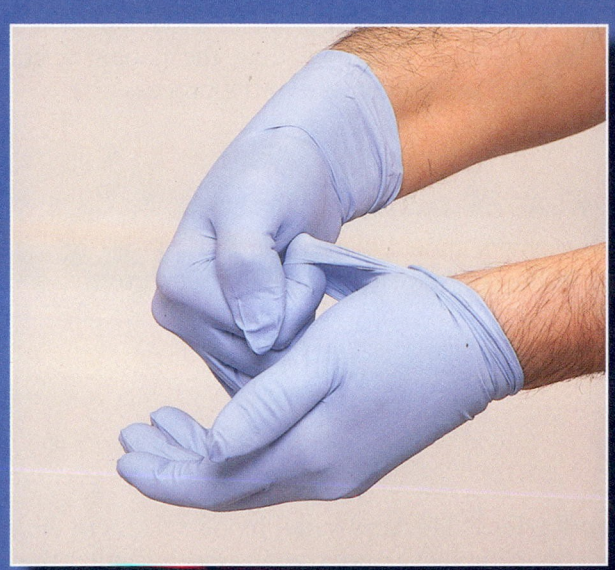

A Grasp the palm or outside cuff of the left glove with the gloved right hand.

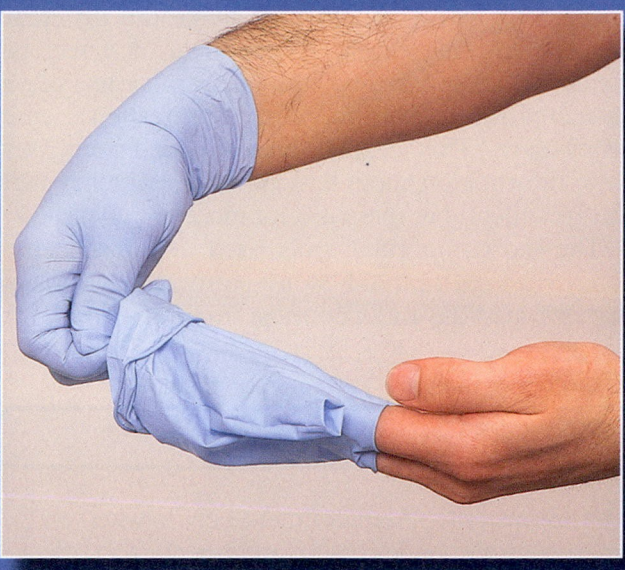

B Pull the left glove toward the fingertips. The glove should turn inside out as it is removed.

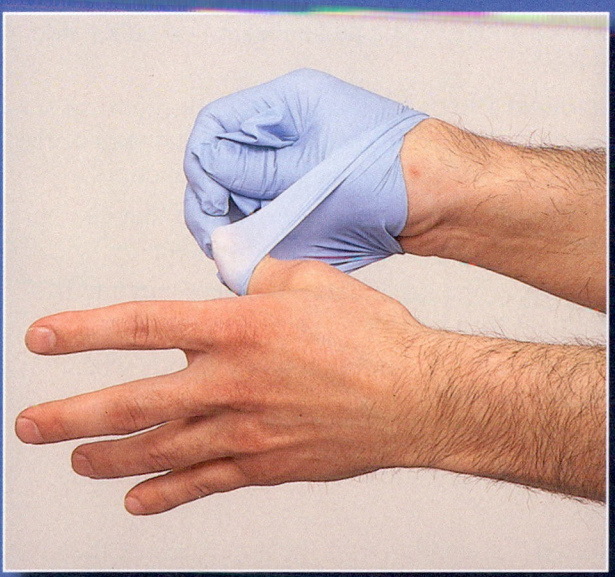

C Hold the removed glove in the still-gloved right hand. Insert the thumb or two fingers of the ungloved left hand under the cuff of the right glove, carefully avoiding any contaminated areas. Pull the right glove toward the fingertips, turning the glove inside out as it is removed. The soiled left glove should remain in the palm of the right glove as it is removed.

Bio-Hazard

D Dispose of the gloves in a container clearly marked with the biohazard label and wash hands thoroughly.

LESSONS LEARNED

Emergency medical care is a major part of most firefighters' responsibilities, and the treatment a firefighter gives a patient may make the difference between life and death. As the field of emergency medicine has progressed, the fire service has played an increasing role in providing first response emergency care. This chapter has provided some very basic information about first aid and emergency medical treatment, but there is a lot more to be learned.

The most important tools used to provide emergency medical first aid do not come from a trauma bag or a first-aid kit. The most important tool is the firefighter's brain. The job of a first-aid provider is a lot like that of a detective. Firefighters should use their senses to discover all that they can about a patient and the environment, and then treat the patient accordingly. A life could be saved in the process!

STREETSMART TIP

All firefighters should consider furthering their education to become emergency medical technicians and even paramedics.

KEY TERMS

Abandonment   Abandonment occurs when an emergency responder begins treatment of a patient, and then leaves the patient or discontinues treatment prior to the arrival of an equally or higher trained responder.

Abrasion   A scrape or brush of the skin usually making it reddish in color and resulting in minor capillary bleeding.

Allergic Reaction   The body's reaction to a substance to which there is an allergy.

Amputation   Occurs when part of the body is severed completely as a result of an injury.

Arterial Bleeding   Bleeding from an artery.

Arteries   The blood vessels, or tubes, within the body that carry blood rich with oxygen and nutrients away from the heart.

Automated External Defibrillator (AED) A portable computer-driven device that analyzes a patient's heart rhythm and delivers defibrillation shocks when necessary.

Avulsion   An injury where a part of the skin is torn away, but still attached, leaving a flap or loose area hanging.

Body Substance Isolation (BSI) Precautions A set of precautions for emergency responders designed to prevent exposure to any body fluid or substance.

Brachial Artery   A major artery in the inside of the upper arm that supplies blood to the arm. Can be used as a pressure point for controlling bleeding and for locating a pulse on an infant.

Capillaries   The very small blood vessels in the body that connect arteries and veins, and filter the oxygen and nutrients from the blood into the tissues of the body.

Capillary Bleeding   Bleeding from a capillary.

Cardiovascular System   The heart, blood vessels, and blood within the body.

Carotid Pulse   The pulse located on either side of the neck.

Chemical Burn   A burn caused by chemical substances that come into contact with the skin or tissues of the body, creating a caustic reaction.

Communicable Disease   A disease that can be transmitted from one person to another.

Consent   The acceptance of emergency medical treatment by a patient or victim.

Constricted   A condition of the pupils where they are much smaller than normal and may appear almost like a "pinpoint."

Dehydration   A loss of water and vital fluids in the body.

Dilated   A condition of the pupils where they are much larger than normal and can take up almost the whole colored portion of the eye.

Exposure   A contact with a potentially disease-producing organism; the contact does not necessarily produce the disease in the exposed individual.

External Bleeding   Bleeding that is coming from an open wound on the body.

Femoral Artery   A major artery in the lower body near the groin that supplies the leg with blood. Can be used as a pressure point for controlling bleeding in the lower extremities.

Fracture   A medical term for a broken or cracked bone in the body.

Full Thickness Burns   Burns affecting not only the skin structure but the tissues and muscles underneath. Full thickness burns may be red, white, or charred in color and will appear dry because the blood vessels in the skin are damaged extensively and are not supplying fluids to the area.

Hypoperfusion   A serious condition caused by a problem or failure of the circulatory system that results in a decrease of oxygen and vital nutrients to the body's tissues. Also known as shock.

Implied Consent   The assumption of acceptance of emergency medical treatment by an unconscious patient or a child with no parents or legal guardians present.

Incision   A cut to the skin that leaves a straight, even pattern.

Infection Control   Procedures and practices for firefighters and emergency medical care providers to follow to prevent the transmission of diseases and germs from a patient to themselves or other patients.

Infectious Disease   See *communicable disease*.

Initial Assessment   The initial investigative action taken by care providers to determine if the patient has the basic signs of life as well as any serious, life-threatening injuries.

Internal Bleeding   Bleeding within the body when no visible open wound is present.

Laceration   A cut to the skin and underlying tissues that leaves an irregular, even pattern.

Medi-Vac   An ambulance that transports patients by air. Typically, medi-vac units are helicopters with highly trained EMS personnel and nurses.

Partial Thickness Burns   Burns affecting the entire skin structure that lies over the top of the fatty tissues and muscles causing skin to turn red and blister.

Puncture   An injury caused by an object that has stabbed the body.

Radial Pulse   The pulse located in either wrist.

Shock   A serious condition caused by a problem or failure of the circulatory system that results in a decrease of oxygen and vital nutrients to the body's tissues. Also known as hypoperfusion.

Sprain   Injury to the ligaments that hold the body's joints together and allow them to move.

Standard of Care   A legal term that means for every emergency medical incident, an emergency responder should treat the patient in the same manner as another emergency responder with the same training.

Superficial Burns   Burns affecting the outermost layer of skin, which typically cause redness of the skin, swelling, and pain.

Thermal Burns   Burns caused by heat or hot objects.

Triage   A quick and systematic method of identifying which patients are in serious condition and which patients are not, so that the more seriously injured patients can be treated first.

Veins   The blood vessels, or tubes, within the body that carry blood lacking oxygen and nutrients back to the heart.

Venous Bleeding   Bleeding from a vein.

REVIEW QUESTIONS

1. What are the basic elements of an emergency medical system?

2. What is the firefighter's role in the emergency medical system?

3. Describe the principles of BSI. Discuss how you would use BSI as a firefighter providing emergency medical care to a patient. When are BSI precautions needed?

4. What are the five major parts of an initial assessment?

5. What are the three types of external bleeding and the characteristics of each?

6. What is shock (hypoperfusion)?

7. What are the three types of thermal burns, and how would a firefighter treat a patient with thermal burns?

8. What is the difference in treatment between a patient with thermal burns and a patient with chemical burns?

9. How do you contact your local poison control center?

10. What information do you give to the poison control center when contacting them for assistance?

Hazardous Materials and Terrorism Response

Also critical to firefighter training is knowledge of the principles and skills involved in responding to a hazardous materials or terrorist incident. The chapters within this section comply with both NFPA 1001 *Standard for Firefighter Professional Qualifications,* and NFPA 472 *Professional Competence of Responders to Hazardous Materials Incidents,* 2008 Editions.

The chapters within this section bring to light even more the concerns of the firefighter of today. It is absolutely essential that firefighters be trained to respond to these events, so that they can effectively prepare for any threat to the communities in which they serve. These chapters report on the latest in terrorist and hazardous materials crimes, offer new information on the dangers of radiation and other potential CBRNE response, provide current information on air monitoring and new technologies in this field, and more—all to ensure that you, as a firefighter, are aware of the dangers of our society, and can be ready to respond to hazardous materials and terrorist incidents.

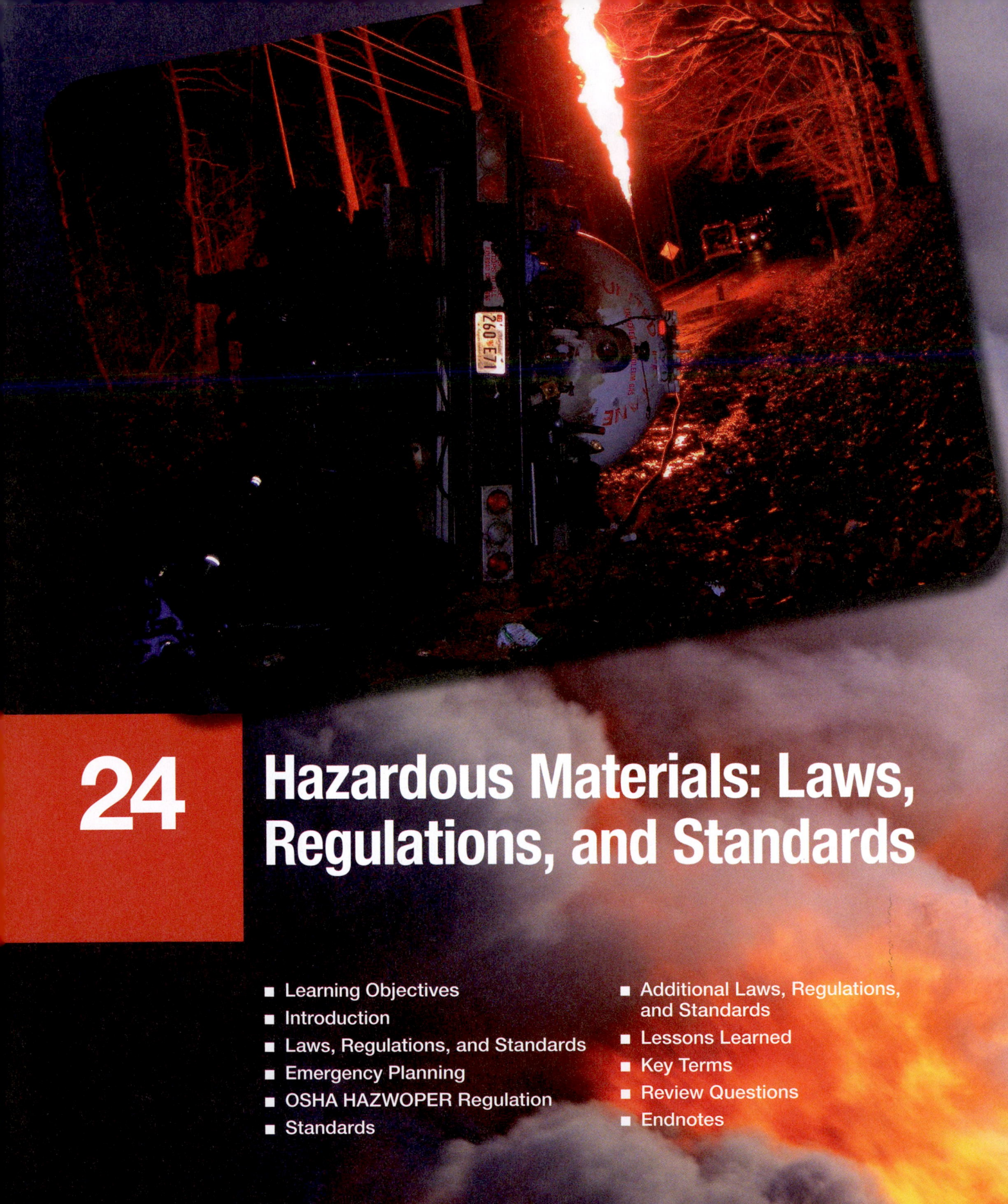

24 Hazardous Materials: Laws, Regulations, and Standards

We arrived on scene to find the patient lying beside the hot dusty road, baking in the summer sun. He was a male in his mid-thirties, wearing torn blue jeans, old tennis shoes, and no shirt. A local farmer was on the scene. He was driving home when he noticed the man beside the road. It was obvious the man was in distress, so the farmer called 9-1-1 from a cellular phone.

When we got off the fire engine, it was apparent that the man was struggling to breathe. He was sunburned, sweaty, and unable to respond to any of our questions. He was shaking slightly but it just did not look like the typical epileptic seizure. We put the man on high-flow oxygen and began to assess his vital signs. It was the size of his pupils that tipped us off. In addition to the pinpoint pupils, the patient was incontinent and producing volumes of frothy sputum. We stripped off his clothes, decontaminated him with water, and alerted the incoming ambulance by radio that we might have an organophosphate poisoning. Fortunately, we had recently had some hazardous materials training on pesticide exposures. Had we not been armed with that information, we may not have picked up on the clues so quickly.

The transport ambulance arrived what seemed like hours later, and by that time we were assisting his respirations with a bag valve mask and suctioning his airway. An IV was started, and atropine was administered in large doses to counteract the effects of a suspected pesticide. Unfortunately, we were never able to identify the substance he was exposed to or even know the duration of the exposure.

Sometimes, despite the greatest efforts and the best technology available, people do not survive. He fought hard to breathe and even harder to live, but in the end, neither we nor the hospital staff were able to save his life. He died an hour or so after we first saw him, probably never knowing why he became so sick.

I think about him from time to time, wondering if the call could have gone smoother or we could have done something different, but I always arrive at the same conclusion: You do your best, learn from any mistakes or situations you find yourself in, and just hope you have a better outcome the next time around.

We could have suffered a secondary exposure because we were not properly protected, but we did not. We could have had an inhalation exposure from his contaminated clothing, but we did not. We could all have ended up as patients, but we did not. It was only luck that brought us through without harm. Would I do it differently the next time? Absolutely. Our first responsibility as firefighters is to not make the problem worse when we show up! You are no good to anyone if you become part of the problem or end up being a victim yourself. Use common sense, wear the right PPE for the right situation, and, above all, think before you act. Do not count on luck to bring you back to the station.

—Street Story by Rob Schnepp, Assistant Chief,
Alameda County, CA, Fire Department

LEARNING OBJECTIVES

After completing this chapter, the reader should be able to:

24-1 Explain the local emergency response plan and standard operating guidelines.

24-2 Explain the student's role within these documents at the awareness level.

24-3 Explain the notification process to request assistance.

24-4 Explain the role of the local emergency planning committee.

24-5 Explain the role of the SARA Title III regulation and emergency response.

24-6 Explain other regulations that have an effect on fire department activities.

*The FF I and II levels, as defined by the NFPA 1001 Standards, are identified in different colors: FF I = black, FF II = red, additional information = blue.

INTRODUCTION

Hazardous materials response, as shown in **Figure 24-1,** is typically one of the specialty fields within the fire service. When the fire department is called to a hazardous materials release, the actual handling of chemical releases is usually done by a well-trained team whose function it is to handle such emergencies, but in almost all cases firefighters are required to assist in the effort.

NOTE

Many times at hazardous materials incidents the actions that the first-in firefighters take in the first five minutes determine how the next five hours/days/months will go.

Firefighters and EMS providers are bombarded with exposures to hazardous materials each and every day in their regular fire suppression and EMS duties. A large number of firefighters and EMS providers are exposed to bloodborne pathogen materials each day, sometimes suffering fatal effects from this exposure. By the nature of their jobs they are exposed to numerous toxic and cancer-causing agents, many times in non-emergency situations. When confronted with emergency situations, the exposure and risks increase as well. This chapter covers some hazardous materials risks and some fundamental response profiles that will help keep firefighters safe. As the fire service furthers its efforts to improve firefighter health and safety, the response to chemical accidents is one area in which extra precautions are needed.

SAFETY

In a fire situation, the injuries can be acute, for instance, if a building collapses during firefighting operations. A chemical exposure can also kill a firefighter immediately with a deadly dose, or it can cause serious illness or injury that prolongs death for ten to twenty years.

FIGURE 24-1   A hazardous material team member surveys a chemical agent lab using air monitors.

NOTE

There are two options for hazardous materials emergency response training. One can be trained to an Occupational Safety & Health Administration (OSHA) level or to meet the National Fire Protection Association levels. The requirements vary from state to state, but the minimum level is the OSHA training. The term *certification* is commonly used, and typically only the employer can certify the training. Neither OSHA nor the NFPA certifiy responders; their fire department certifies that they have met the training objectives outlined by OSHA or the NFPA. The NFPA requirements are much more stringent but better reflect the current emergency responders' actual duties. If trained to meet the NFPA training objectives, a responder would meet the OSHA requirements.

When firefighters receive their basic training, it is typically based on the NFPA training objectives for firefighters. The NFPA Standard 1001, also known as the Standard for Firefighter Professional Qualifications, has been revised and the 2008 edition was published in July 2007. As part of the revision to be certified as a Firefighter I, the student will also have to be certified at the Operations level according to NFPA 472, which is the Competence of Responders to Hazardous Materials/Weapons of Mass Destruction Incidents Standard. All of this information is important for firefighters' survival since no matter which state the training is providedin, firefighters are required to have some form of hazardous materials training. The training can be based on the OSHA regulations or the NFPA standards. The information in this text exceeds the requirements for OSHA training and covers both awareness and operations levels, and in some cases exceeds the operations level of the NFPA requirements. See the discussion on NFPA standards and OSHA's HAZWOPER for more information on awareness and operations level training.

As society has become more environmentally aware, so has the fire service. Hazardous materials response is still a very young field, and most communities have had fire department–based hazardous materials response teams for only nineteen years or less. The Jacksonville, Florida, team is the oldest hazardous materials response team in the country, and it is only twenty-six years old. This field is constantly changing and dynamic, and the response to hazardous materials emergencies has undergone three distinct changes in response methodology. Tactics used to handle incidents thirteen years ago certainly were appropriate then but might bring criticism if used today.

Technology is rapidly changing as well. **Air monitoring devices** that were used in the past strictly by hazardous materials response teams are now common on engine, truck, medic, and squad companies. On the hazardous materials response team side, technology has also increased to a level that requires a high degree of training and expertise. Response to chemical accidents is a rapidly changing field and brings new experiences every day. From building fires, tank truck fires, **clandestine drug labs,** shipboard incidents, pressurized cylinder leaks, and overturned tank trucks to the emerging incidents of terrorism, hazardous materials response teams face new challenges on every response. This chapter provides firefighters with a practical framework that can help keep them and their crews alive in situations that involve hazardous materials.

The next few sections describe in detail the many aspects of hazardous materials and appropriate actions when working those incidents. To establish the groundwork for these sections, it is necessary to provide a definition for hazardous materials. The text that follows provides these standard definitions.

Out of these definitions, the following may be considered the best definition: "A hazardous material is any substance that jumps out of its container when something goes wrong, and hurts and harms the things it touches." Ludwig Benner Jr. who worked for the National Transportation Safety Board, provided this early definition. It is a perfect definition for this field.

> **CAUTION**
>
> Almost everything is hazardous to human health if it is used improperly or escapes its container. This is demonstrated in **Figure 24-2.**

Even the most toxic material is not hazardous as long as it is used correctly and stays in its intended container. The other principle that is important here is that even if the material escapes its container, it must have the ability to cause harm to humans, property, or the environment. Every day hazardous materials are packaged to be transported and are intended to stay in that package. When firefighters become involved in an incident that involves a hazardous material, an increased level of concern results, but as long as the

FIGURE 24-2 The material shown here is an example of one that ignites when it escapes its container and comes in contact with the air. A material that is air reactive is known as *pyrophoric.*

material remains packaged, and the firefighters do not open the package, the risk to them is very minimal. The material may be toxic, flammable, corrosive, or reactive, and these properties remain no matter what happens—but to reach out and harm people it must escape or be forced from the container. The following are some standard definitions:

- *Hazardous materials.* A substance (either matter—solid, liquid, or gas—or energy) that when released is capable of creating harm to people, the environment, and property, including weapons of mass destruction (WMD) as defined in 18 U.S. Code, Section 2332a, as well as other criminal use of hazardous materials, such as illicit labs, environmental crimes, or industrial sabotage. Some specific agency definitions follow.

- *DOT hazardous material.* Any substance or material in any form or quantity that poses an unreasonable risk to the safety, health, and property when transported in commerce. The DOT also provides more exacting definitions with each hazard class.

- *EPA hazardous substances* (40 CFR300.5). A chemical released into the environment that could be potentially harmful to the public's health or welfare.

- *OSHA hazardous chemicals* (29 CFR1910.1200). Those chemicals that would be a risk to employees if exposed in the workplace.

LAWS, REGULATIONS, AND STANDARDS

It is important for the first responder to have a basic understanding of the legislative history of hazardous materials. It was in this area that fire departments first became regulated as to how they would respond and how they would be trained. Many environmental and safety regulations affect how firefighters respond to emergencies. This section provides an overview of the major federal laws, regulations, and standards. Readers should consult their local environmental and OSHA offices to learn about state and local laws and regulations.

Development Process

It is important to understand the differences among laws, regulations, and standards. **Laws** result from legislation passed by Congress and signed by the president. **Regulations,** on the other hand, are developed by government agencies, like OSHA or the EPA, and have the weight of law but are not passed by

Congress nor signed by the president. In some cases, laws require the development of regulations and provide a framework for the government agency to follow. **Standards** are developed and reviewed by a nongovernmental consensus committee, such as the NFPA. These do not have the weight of law but could be applied by a regulating agency or in court. As citizens and as members of an industry, it is important for firefighters to participate in the development and review of laws, regulations, and standards.

NOTE

As time goes on, standards are being applied with the weight of law, typically by OSHA.

EMERGENCY PLANNING

The first law that regulated how fire departments respond to emergencies was the **Superfund Amendments and Reauthorization Act,** commonly referred to as **SARA.** This law was passed in 1986 for the protection of emergency responders and the community. Its intent was to inform emergency responders of chemical hazards within their community. The main component of the law is called **Emergency Planning and Community Right to Know Act (EPCRA).** It is divided into two sections: planning for emergencies and providing a mechanism to get chemical storage information to emergency responders.

Standard operating procedures (SOPs) provide specific information and instructions on how a task or assignment is to be accomplished. SOPs are established so that all members of a department will perform the same function with the same level of uniformity. Also, these procedures are generally tactical in nature because, in most instances, they address emergency operations. In addition, all SOPs must be distributed and understood for all department members.

State and Local Emergency Response Committees

The planning portion established a requirement to provide a **State Emergency Response Committee (SERC)** in each state. It is the responsibility of the SERC to ensure the state has the resources necessary to respond safely and effectively to chemical releases. The SERC also had to provide the framework to implement **Local Emergency Planning Committees** known as **LEPCs.** The LEPC is a group composed of the representatives of the community, emergency responders, industry, hospitals, media, and other government agencies. Most LEPCs are set up on a county basis, although larger communities may have their own.[1] The LEPC is responsible for the development of a hazardous materials emergency plan and its annual revision.

Local Emergency Response Plans

The emergency plan is an important component of a successful response. The emergency plan should outline emergency contacts and procedures. Target facilities such as those that store extremely hazardous substances are to be included, along with specific information and response tactics. A decision tree for evacuation and in-place sheltering is usually included. If the decision is made to evacuate citizens, the plan provides the mechanism to shelter and take care of these evacuees. It is important for personnel to have an understanding of this plan and know how to access its resources during an emergency.

The LEPC is also responsible for ensuring that local resources are adequate to handle a chemical release in the community. Ensuring that local responders receive the proper amount of training and that the emergency plan is evaluated annually by conducting an exercise (drill), is also part of their responsibility. It is important for the emergency services to be an integral player in the LEPC. Decisions regarding a fire department's response to chemical incidents may be planned at the LEPC level. Many of the facilities using hazardous materials will have representatives at these meetings, and before an emergency incident is a better time to meet these representatives. Funding for training and planning is also available through some LEPCs, since most federal HAZMAT grants are provided through the LEPC.

Roles of Emergency Personnel

Awareness level personnel shall be persons who, in the course of their normal duties, could encounter an emergency involving hazardous materials/weapons of mass destruction (WMD) and who are expected to recognize the presence of the hazardous materials/WMD, protect themselves, call for trained personnel, and secure the area.

Operations level responders are persons who respond to hazardous materials/WMD incidents for the purpose of implementing or supporting actions to protect nearby persons, the environment, or property from the effects of the release.

The awareness level and operation level personnel must know their role in a hazardous materials inci-

dent based on the local emergency response plan and standard operating procedures. These will be determined by the authority having jurisdiction.

Chemical Inventory Reporting

The other section of EPCRA, usually referred to as SARA Title III, is the chemical reporting portion of the act. It requires some facilities to report chemical information to the state, LEPC, and local fire department. Failure to report this information on an annual basis can result in fines of $25,000 a day.

To qualify as a reporting facility, the facility has two methods of reaching a reporting threshold. Most facilities meet the reporting threshold of storing more than 10,000 pounds of a chemical. The 10,000-pound number is for a single chemical and is not the total of all chemicals on site. A facility that has more than 1,300 gallons of acetone is required to report because the amount represents more than 10,000 pounds of acetone. A facility that stores drums of ammonia and water is only required to report the ammonia, not the total weight of the drum. They do have to account for all of the ammonia at the facility, regardless of the location or type of container.

Another method of meeting the threshold of reporting is to store one of 366 chemicals that the EPA considers an **extremely hazardous substances (EHS).** The EHSs have separate reporting requirements and lower reporting thresholds. Some must be reported at 100 pounds. If these chemicals are released, then the facility has a whole host of responsibilities to comply with, including immediate notification of the LEPC, usually by dialing 9-1-1 or some other listed emergency contact number as determined by the LEPC. The common EHSs are chlorine, sulfur dioxide, anhydrous ammonia, and sulfuric acid. Because EHS materials can present an extreme threat to the community, it is important for emergency responders to become familiar with the facilities that use them and to use caution when responding to potential releases involving these materials.

Facilities that are required to report as SARA facilities are required to submit either a Tier 1 or Tier 2 Chemical Inventory Report. The Tier 2 report, **Figure 24-3,** is the most common and lists the chemical name, storage amount in a range, storage location and information, and emergency contact information, including twenty-four hour phone numbers. The facility is also required to submit a list of chemicals or **Material Safety Data Sheets (MSDS).** A site plan may also be required that outlines the storage areas. EHS facilities are also required to submit a copy of their emergency plan. Depending on the state or locality, reporting requirements may be more strin-

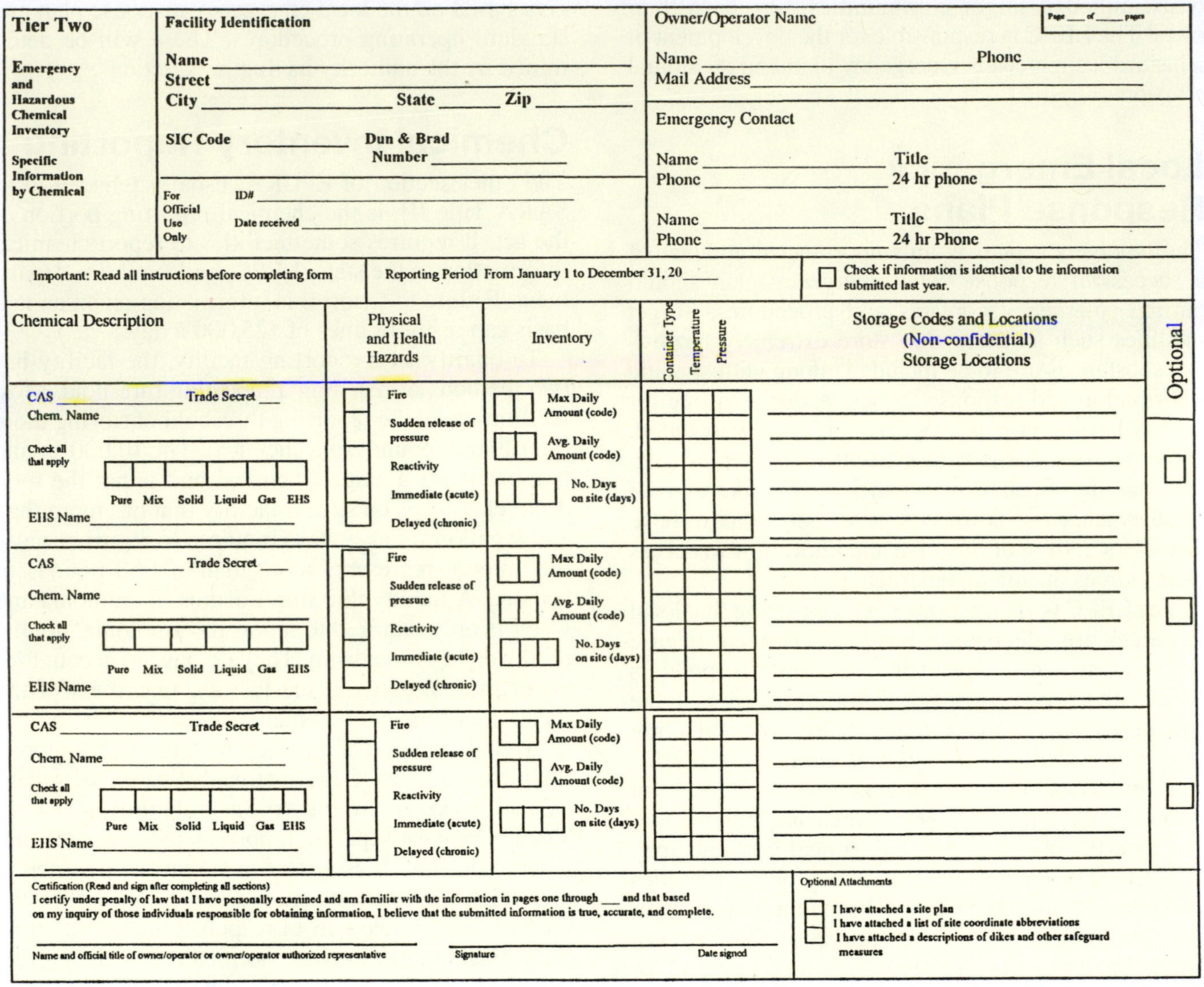

FIGURE 24-3 This form is an example of what facilities are required to submit to the fire department and the Local Emergency Planning Committee on an annual basis.

gent, although all have to meet the minimum listed. Note, however, that the information contained on a Tier 2 form regarding the chemicals that are located on site is for the prior year. Facilities report from January 1 to March 1 for the prior year.

The basis for this reporting is to inform the emergency responders of the hazard of responding to a SARA facility. It also allows the emergency responder to make informed decisions as to site entry, tactical decisions, and potential evacuation. MSDS are a valuable tool in obtaining chemical information, and SARA requires their availability to emergency responders.[2]

OSHA HAZWOPER REGULATION

Another part of SARA was the requirement for OSHA to develop a regulation covering activities that involve hazardous materials. The regulation, known as **Hazardous Waste Operations and Emergency Response,** became final on March 6, 1989. It is referred to as **HAZWOPER** or by its identification number of 29 CFR 1910.120.[3] There was considerable discussion within the fire service when this regulation was established as to whether or not it applied to all of the fire service. The primary unknown was whether

Tier Two	Facility Identification	Owner/Operator Name	Page____ of ____ pages

Tier Two

Emergency and Hazardous Chemical Inventory

Specific Information by Chemical

Facility Identification

Name _____
Street _____
City _____ State _____ Zip _____

SIC Code _____, ___ Dun & Brad Number _____

For Official Use Only — ID# _____
Date received _____

Owner/Operator Name

Name _____ Phone _____
Mail Address _____

Emergency Contact

Name _____ Title _____
Phone _____ 24 hr phone _____

Name _____ Title _____
Phone _____ 24 hr Phone _____

Important: Read all instructions before completing form	Reporting Period From January 1 to December 31, 20____	☐ Check if information is identical to the information submitted last year.

Confidential Location Information Sheet

Container Type | Temperature | Pressure

Storage Codes and Locations
(Confidential)
Storage Locations

Optional

CAS # _____

Chemical Name _____

☐

CAS # _____

Chemical Name _____

☐

CAS # _____

Chemical Name _____

☐

Certification (Read and sign after completing all sections)
I certify under penalty of law that I have personally examined and am familiar with the information in pages one through ___ and that based on my inquiry of those individuals responsible for obtaining information, I believe that the submitted information is true, accurate, and complete.

Name and official title of owner/operator or owner/operator authorized representative   Signature   Date signed

Optional Attachments
☐ I have attached a site plan
☐ I have attached a list of site coordinate abbreviations
☐ I have attached a descriptions of dikes and other safeguard measures

(B)

FIGURE 24-3 (Continued)

this regulation applied to volunteer firefighters. In some states Hazwoper only affected the private sector and not career firefighters, while in other states only the public sector was subject to this regulation. In many cases, volunteer firefighters were determined to be employees of the local government since they received workers' compensation benefits or received safety equipment from the employer, and so this regulation applied. To end this argument and confusion, the EPA adopted the same regulation and issued it on the EPA's behalf. The EPA regulation is also called HAZWOPER and is referenced by 40 CFR 311. It is interesting to note that OSHA now puts a provision in some of its regulations that specifically mandates the coverage of volunteer employees.

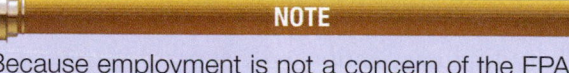

NOTE

Because employment is not a concern of the EPA, as it is with OSHA, all persons are covered by this regulation regardless of their employment status.

This regulation has had far-reaching effects for the fire service. It has required that certain training be provided, requires the development of standard operating procedures, and mandates certain requirements when handling chemical releases. It only allows persons trained in the handling of chemical releases to respond to chemical incidents. Only certain levels allow operation in and around chemical releases. Even just to be present at a chemical accident requires training, and as firefighters' activity levels increase, so does their need for further training.

Paragraph q

The majority of this OSHA regulation covers employers' responsibilities at hazardous waste sites. The last section, **paragraph q,** covers emergency response and applies to the fire service. It established five levels of training, as well as a requirement for annual refresher training. It requires that an incident command system be used and that there be an incident commander at chemical releases. At larger incidents, a safety officer must be appointed. As firefighters progress up the training levels the requirements become more detailed. The two in/two out rule, as discussed in other chapters, originated from this OSHA regulation.

The five training levels established by OSHA mirror the four levels established by the NFPA. Their basic responsibilities are as follows:

1. **Awareness Level.** Persons trained at the awareness level respond to possible hazardous materials/ WMD incidents, identify the potential for a hazardous materials release, call for trained assistance, and stand by isolating the area and denying entry to other persons. Persons trained at the awareness level cannot take any action beyond this. This level is intended for police officers, public works employees, and other government employees.

2. **Operations Level.** The next level of training above awareness level provides the foundation that allows for the responder to perform defensive activities at a hazardous materials/WMD incident. Acting defensively means that the responder does not enter a hazardous area, but can set up dikes, dams, and other confinement measures. Training at the operations level allows the firefighter to assist technicians in setting up the various activities that are required at an incident. Operations training can be expanded to include specialized activities such as decontamination, so that persons trained at this level can assist with this activity. This level is intended for the fire and EMS service.

3. **Incident Commander Level.** This level is designed for a person who has received operations level training and has received training in incident command procedures. This person will be in charge of the incident. To be the incident commander does not mean that this person has the highest level of chemical response training but is the senior response official. The IC must rely on the expertise of the other responders, such as the hazardous materials team, facility officials, or other technical specialists, to make strategic and tactical decisions.

4. **Technician Level.** This is the level at which offensive activities can be completed in the hazard area. Other than some specific restrictions outlined in HAZWOPER, there are no general restrictions as to the activities technicians can perform, as long as the activities fit within the scope of their training. This is the level at which leaks can be stopped and mitigation of the incident can be completed. Hazardous materials technicians are expected to mitigate or stop the incident from progressing.

5. **Specialist Level.** This level is identified only in HAZWOPER and refers to an individual who has received a higher level of training than a technician, or may be someone who specializes in a specific chemical or area of expertise. The training concentrates on chemistry and the identification of unknown materials. In some instances, the specialist supervises the technicians. The NFPA removed this level in 1992.

Medical Monitoring

Other sections of this regulation include a requirement for medical monitoring of certain employees. An individual's responsibilities mandate whether the employer must provide that individual with an annual physical. A physical is required if an individual meets any of these requirements:

- Was exposed to a chemical above the permissible exposure limit.
- Wears a respirator or is covered by the OSHA respiratory regulation (29 CFR 1910.134).
- Was injured due to a chemical exposure.
- Is a member of a hazardous materials team.

The physician determines the extent of the exam and can establish that the exams be given every two years if that time period is determined to be acceptable. For the fire service, the only persons required to be given a physical are the members of a hazardous materials team, although OSHA 1910.134 (Respiratory Protection) has a requirement that a medical survey questionnaire be answered by every firefighter. Depending on the answers to the questions, a physical may or may not be required. If a firefighter is exposed to a hazardous material above the OSHA-specified level, then the department is required to provide a physical exam to the firefighter. The physician determines the extent of the exam and any other future visit or tests.

OSHA requires that the medical records be kept by the employer for a period of thirty years past the last date of employment. This applies to the HAZWOPER regulation as well as other OSHA regulations. In simple terms any medical record generated by the employer, or for the employer, must be kept and must be available for employees if they request it.

STANDARDS

The group that establishes most of the standards for the fire service is the National Fire Protection Association. The NFPA establishes a variety of committees that develop standards, which are then made available for review and comment by the public. After the review process, each standard is voted on by the NFPA committee and, if passed, then goes to the next public meeting where it is voted on by the group in attendance. Once passed it then is sent to the Standards Council, where it is voted on again. If it passes there, it becomes a standard. As with emergency medical services in which an acceptable standard of care applies, typically based on a national or regional standard, the NFPA establishes a standard of care. A person cannot be held criminally liable for violating an NFPA standard, but may be held civilly liable.

> **NOTE**
>
> OSHA has used what is known as the general duty clause, which means that all employers have a general duty or an obligation to provide a workplace free from hazards.

One of the areas in which the NFPA standards have been applied as having the weight of a regulation is in the hazardous materials arena. OSHA has used this clause to cite employers for violating an NFPA standard.

NFPA 472

Prior to the 2008 edition, the NFPA had three hazardous materials response standards: NFPA 471, 472, and 473. In the 2008 edition, the NFPA eliminated the 471 Standard that covered the response to hazardous materials emergencies and moved much of its information into 472. The NFPA Hazardous Materials committee felt that training standards adequately covered the response objectives, and that only two standard documents were required. NFPA 472, Professional Competence of Responders to Hazardous Materials/Weapons of Mass Destruction Incidents is the listing of objectives required to meet the training levels established by the NFPA. OSHA established only basic criteria to meet in order for the employer to certify their employees as to their ability to respond to hazardous materials incidents. This NFPA document expands the requirements in order for the employer to certify their employees. The levels mirror those of the HAZWOPER regulation, but in 1992 the NFPA removed the specialist category. In the 1997 edition they added the following competencies: private sector specialist employee, hazardous materials branch officer, hazardous materials branch safety officer, tank car specialist, cargo tank specialist, and inter-modal specialist.

In the 2002 edition, the NFPA added objectives related to terrorism response. These changes were part of the 1997 edition but were listed as tentative interim amendments (TIAs).[4] The 2002 version adopts them as part of the standard, and they are covered by this text. The 2008 edition adds several law enforcement competencies for the collection of evidence in hazardous situations. There are several criminal scenarios that involve the potential use of hazardous materials or weapons of mass destruction and present significant risk to investigators. The revision to this standard provides training objectives for response personnel who may be collecting evidence when hazardous situations exist. Also in the 2008 revision is the addition of **mission-specific competencies,** which used to be reserved for hazardous materials technicians. These competencies reflect the realities of real-world incidents. Many firefighters have access to and the ability to use air monitoring devices, which used to be a technican-level skill. The revision to the standard provides that, if properly trained and equipped, an operations-level responder can use some air monitoring devices.

NFPA 473

The third standard is NFPA 473, Competencies for EMS Personnel Responding to Hazardous Materials/Weapons of Mass Destruction Incidents. This standard adds some additional competencies above NFPA 472 with regard to EMS issues. It provides EMS Level I and Level II training levels. The competencies in this edition have been updated and use common terminology known to the EMS response community. Instead of Level I and II, the standard now relies on BLS and ALS providers. To meet the competencies of the standard, all EMS personnel will meet the core competencies of Hazardous Materials Operations and some of the Mission Specific Competencies based on their on-scene responsibilities. *Emergency Medical Response to Hazardous Materials Incidents,* authored by Richard Stilp and Armando Bevelaqua, offers more information on this topic.

Standard of Care

As mentioned before, emergency responders have to abide by a standard of care. In years past it was deemed acceptable to wash spilled gasoline down a storm drain. To do that now is illegal, and personnel could face state and federal charges for violating the Clean Water Act. This standard of care is composed of the laws, regulations, standards, local protocols, and experience detailed earlier. Violations of this standard of care are based on three theories: **liability, negligence,** and **gross negligence.**

When operating in and around hazardous materials, liability becomes a very real issue. Other than EMS responses like those shown in **Figure 24-4,** firefighters at a chemical release are likely to find themselves personally liable for any wrongdoing. Liability is being responsible for personal actions. In the example of the individual who made the decision to wash gasoline down a storm drain, that individual could end up in jail or have to pay a substantial fine.

NOTE

Environmental and safety regulations are designed to place blame on the person who made the decision, not on the organization itself, unless the organization has policies that violate these regulations.

FIGURE 24-4 Just as EMS responders have to follow a standard of care so that the patient is provided an appropriate level of care, HAZMAT response has a similar standard of care. *(Courtesy of Cambria County, Pennsylvania, Emergency Services)*

Negligence is not following the standard of care or an accepted practice. Gross negligence is the willful disregard for the standard of care. In other words, a conscious decision was made not to follow the standard of care. To avoid becoming liable, firefighters should receive the training that is required for their positions and act only to that level of training. If firefighters are not sure what the next step should be, they should consult with their local HAZMAT team or other local or state environmental protection agency.

ADDITIONAL LAWS, REGULATIONS, AND STANDARDS

The items discussed next are ones that firefighters should be aware of, because they are commonly encountered or applied in chemical releases.

Hazard Communication

The hazard communication regulation issued by OSHA (29 CFR 1910.1200) requires that employers provide an MSDS for all chemicals located at a facility at quantities above "household quantities." The term household quantities allows for small amounts, such as those that would be purchased for home use, for which MSDS are not required. An example would be a gallon of household ammonia for use in an office. This would not require an MSDS, because this is a normal household quantity. If, however, a case of household ammonia were brought in, an MSDS would be required because this is not a normal household quantity.

The employer must provide a list of these MSDS materials and when they arrived on site. The employer must provide training on these MSDS materials and the hazard communication program. One of the important components of this program is the requirement for the facility to make these available for emergency responders. Fire departments themselves are responsible for following this regulation.[5]

Superfund Act

The Comprehensive Environmental Response, Compensation and Liability Act (CERCLA) is commonly referred to as the Superfund law. This law was established for the cleanup of toxic waste sites around the country. It was the first law to set the groundwork for the regulating of the fire service with regard to response to chemical emergencies. When responding to a Superfund site, some additional concerns and requirements must be followed.

The site has an existing emergency response plan, of which the local fire department should have a copy. The site should have its access limited, hazard zones established, and, depending on the nature of the work involved, may have a decontamination corridor created. Prior to work beginning on a Superfund site, the local fire department should meet with the site supervisor to learn the hazards of the site and any protection measures that may be in place. Because Superfund sites can vary from old landfills to mercury dumping sites, it is not possible to provide here a breakdown of the potential site hazards and precautions that the site may hold. Prior to becoming a Superfund site, the area may be listed as a National Priority List (NPL) site. Firefighters should contact their state EPA or federal regional EPA office to learn of any Superfund or NPL sites in their area.

Clean Air Act

The Clean Air Act Amendments (CAAA), passed in 1990, have some language that affects the fire service. It requires that certain facilities file additional planning documents and that the LEPC and the local fire service be involved in training and exercises. As of June 1999, facilities were required to submit emergency plans, many of which must be coordinated with the local fire department. Other requirements, such as exercises and worst-case scenarios, may also involve the fire department.

Respiratory Protection

OSHA's respiratory protection regulation (29 CFR 1910.134) also impacts the fire service, because it adds additional requirements that were previously exempted. The biggest changes are the inclusion of the two in/two out rule, which was in place for chemical incidents but now also includes structural fire situations. The fire service is now required to fit test all firefighters and provide them with a medical survey or a physical exam.

Record keeping has also changed. Specific records must be kept by the fire department regarding the employees and the respiratory protection program. Records include the listing of personnel training, daily equipment checks, periodic maintenance, routine service, review of the respiratory protection program, and other items.

Firefighter Safety

NFPA Standard 1500, Fire Department Occupational Safety and Health Program, is sometimes referred to when discussing hazardous materials issues. This standard is a "broad-based" program for providing a safe workplace for firefighters. It has been applied by OSHA using the general duty clause, specifically regarding the two in/two out requirement.

NFPA Chemical Protective Clothing

NFPA Standards 1991 and 1992 are standards for chemical protective clothing ensembles, including encapsulated suits, splash-protective suits, and support garments. They establish design and use requirements.

NFPA 1994, Standard on Protective Ensembles for Chemical/Biological Terrorism Incidents, has three levels of protective equipment that can be used in the event of a chemical or biological attack.

LESSONS LEARNED

The maze of laws, regulations, and standards can be confusing and sometimes overwhelming. Most laws and regulations are not easy to read and it can take years to comprehend their true intent. They are subject to interpretation and change frequently. Emergency responders must keep abreast of the ones that affect their everyday jobs. In larger departments, staff may be assigned to monitor health, safety, and environmental laws and regulations. In smaller departments there may not be assigned personnel. As a result, laws and regulations may be overlooked. The fire service is an industry the same as the chemical plant or gas station in any community. The fire service has to follow the same health, safety, and environmental laws and regulations. For everyone's health and safety, firefighters should keep abreast of these items, so they can help make communities safer places to live and work.

KEY TERMS

Air Monitoring Devices   Used to determine oxygen, explosive, or toxic levels of gases in air.

Awareness Level   The basic level of training for emergency response to an incident involving hazardous materials/weapons of mass destruction (WMDs), the basis of which is the firefighters' ability to recognize a hazardous situation, protect themselves and call for trained assistance, and secure the scene.

Clandestine Drug Labs   Illegal labs set up to manufacture street drugs.

Emergency Planning and Community Right to Know Act (EPCRA)   The portion of SARA that specifically outlines how industries report their chemical inventory to the community.

Extremely Hazardous Substances (EHS)   A list of 366 substances that the EPA has determined present an extreme risk to the community if released.

Gross Negligence   Occurs when an individual disregards training and continues to act in a manner without regard for others.

Hazardous Waste Operations and Emergency Response (HAZWOPER)   The OSHA regulation that covers safety and health issues at hazardous waste sites, as well as response to hazardous materials incidents.

Incident Commander Level   A training level that encompasses the operations level with the addition of incident command training. Intended to be the person who may command a chemical incident.

Laws   Legislation that is passed by the House and Senate, and signed by the president.

Liability   The possibility of being held responsible for individual actions.

Local Emergency Planning Committee (LEPC)   A group composed of members of the community, industry, and emergency responders to plan for a chemical incident, and to ensure that local resources are adequate to handle an incident.

Material Safety Data Sheet (MSDS)   Information sheet for employees that provides specific information about a chemical, with attention to health effects, handling, and emergency procedures.

Mission-specific Competencies   The knowledge, skills, and abilities to perform the mission of the organization. Examples are provided in NFPA 472.

Negligence   Acting in an irresponsible manner, or different from the way in which someone was trained; that is, differing from the standard of care.

Operations Level   The next level of training above awareness that provides the foundation which allows for the responder to perform defensive activities at a hazardous materials incident.

Paragraph q   The paragraph within HAZWOPER that outlines the regulations that govern emergency response to hazardous materials incidents.

Regulations   Developed and issued by a governmental agency and have the weight of law.

Specialist Level   A level of training that provides for a specific type of training, such as railcar specialist; someone who has a higher level of training than a technician.

Standards   Usually developed by consensus groups establishing a recommended practice or standard to follow.

State Emergency Response Committee (SERC)   A group that ensures that the state has adequate training and resources to respond to a chemical incident.

Superfund Amendments and Reauthorization Act (SARA)   A law that regulates a number of environmental issues, but is primarily for chemical inventory reporting by industry to the local community.

Technician Level   A high level of training that allows specific offensive activities to take place, to stop or handle a hazardous materials incident.

REVIEW QUESTIONS

1. How does the response to hazardous materials differ from a structural fire response?
2. How can hazardous materials be easily defined?
3. Who is to receive the chemical reporting required under SARA Title III?
4. Who is responsible for ensuring the proper response to a hazardous materials emergency?
5. Which regulation covers all emergency responders who respond to chemical accidents?
6. What is the difference between negligence and gross negligence?
7. Which regulation requires employers to maintain Material Safety Data Sheets?
8. What was the largest change to the 2008 Edition of NFPA 472?
9. Which training regulations are most firefighters required to follow?
10. To perform offensive activities at a hazardous materials release, what level of training is required?

ENDNOTES

1. Some LEPCs are established on a city basis; some states may only have one. The format varies from state to state.
2. The OSHA hazard communication regulation (29 CFR 1910.1200) also makes this requirement.
3. Federal regulations are published in the Federal Register, which comes out daily. Annually, they are published in a text called the Code of Federal Regulations (CFR). Each federal agency is identified by a two-digit number (29 for OSHA, 40 for EPA, 49 for DOT), and each regulation is given a number to identify it.
4. A TIA is tentative because it has been established outside of the normal NFPA process, in between revisions. When the standard comes up for revision it will be subject to the normal review process.
5. The applicability varies from state to state, depending on OSHA's jurisdiction. Some states may have more stringent regulations.

25

Hazardous Materials: Recognition and Identification

- Learning Objectives
- Introduction
- Location and Occupancy
- Placards, Labels, and Markings
- Other Identification Systems
- Containers
- Senses
- Chemical and Physical Properties
- Containers and Properties
- Lessons Learned
- Key Terms
- Review Questions
- Endnotes

While the news media tends to focus on the large and spectacular, the reality is that every day the fire service responds to literally hundreds of hazardous materials–related incidents. These incidents usually do not get much attention because they are small in scope and are handled with a minimal amount of fire department resources. Common examples include natural gas leaks and flammable and combustible liquid spills.

We have an old saying in the hazardous materials community that the initial ten minutes of an incident will dictate the tone for the first hour of an incident. Clearly, the actions—or inactions—of the first responders will set the tone for an incident. I think back to two incidents in my career that reinforce this point.

The first incident involved an engine company being sent to investigate a call regarding an unknown liquid spilled along a highway. Upon arrival, the engine company officer found a very viscous red liquid spilled along a long distance of a road, apparently from the rear of a tractor-trailer. Unsure of the identity and the potential hazard of the liquid, the officer requested a hazardous materials unit to respond to the scene and provide assistance. Using their monitoring and detection equipment, responders were eventually able to determine that the liquid posed no hazard to the community. In fact, the unknown liquid was eventually identified as strawberry syrup! Although the officer took some ribbing from his peers, he clearly made the proper decisions to ensure the safety of both his personnel and the community.

In the second incident, a police officer was called to provide assistance to a public works road crew that had discovered what appeared to be a 5-pound portable fire extinguisher wrapped in duct tape with a fuse on the top. Although he believed it to be a hoax, the officer still requested the response of the fire department bomb squad. After conducting a thorough risk assessment and requesting the additional assistance of a hazardous materials response team, the bomb squad was able to determine that the perceived hoax was, in fact, an actual explosive device. When the device was disrupted (i.e., blown up) by the bomb squad, it made a lasting impression on the police officer!

In both of these instances, recognition and identification by the first responders set the tone for the incident. If you do not know what to do, isolate the area, deny entry, and call for help.

—*Street Story by Gregory G. Noll, Emergency Planning and Response Consultant, Hildebrand and Noll Associates, Lancaster, Pennsylvania*

LEARNING OBJECTIVES

After completing this chapter, the reader should be able to:

25-1 Explain the various tools for recognition and identification of hazardous materials.

25-2 Identify the hazards associated with the nine hazard classes as defined by DOT.

25-3 Identify the standard occupancies where hazardous materials may be used or stored.

25-4 Identify the standard container shapes and sizes and common products.

25-5 Identify both facility- and transportation-related markings and warning signs.

25-6 Identify the standard transportation types for highway and rail.

25-7 Explain the use of transportation containers in identifying possible contents.

25-8 Explain the importance of understanding chemical and physical properties of hazardous materials.

*The FF I and II levels, as defined by the NFPA 1001 Standards, are identified in different colors: FF I = black, FF II = red, additional information = blue.

INTRODUCTION

This chapter is of primary importance to the emergency responder.

SAFETY

It is through recognition and identification (R&I) that firefighters can impact their ability to stay alive.

The inability to recognize the potential for chemicals to be present and the inability to identify the chemical hazard can place firefighters in severe danger. Firefighting is inherently dangerous, but the response to a hazardous materials release creates an additional risk. Not only can there be immediate effects from some materials, but multiple exposures can have far-reaching effects. As depicted in the opening photo, firefighters are using proper PPE but taking a considerable risk to rescue a live victim. More information on this rescue and risk is provided in the case study in Section IV, Chapter 28. Although fires have killed hundreds before, these types of fires are rare. Hazardous materials incidents have killed thousands and injured countless more. In 1984, a release of methyl isocyanate in Bhopal, India, killed several thousand and injured tens of thousands more. This incident was the basis for the Emergency Planning and Community Right to Know Act (EPCRA) because several facilities in the United States use this material.

NOTE

In the hazardous materials section, NFPA 1001 will require that the student receive hazardous materials training at the Operations level. The information in this text covers both Awareness level and Operations level. In some cases, it exceeds the Operations level. All of this information is important for firefighters' survival. See the discussion on NFPA standards and OSHA's HAZWOPER for more information on Awareness and Operations level training.

FIGURE 25-1 The four basic clues to recognition and identification are location and occupancy; placards, labels, and markings; container types; and senses.

Four basic clues to recognition and identification are (1) location and occupancy; (2) placards, labels, and markings; (3) container types; and (4) the senses. A mere suspicion in any of these areas should be enough to place a first responder on guard for the possibilities of a chemical release and its associated hazards, **Figure 25-1.**

LOCATION AND OCCUPANCY

CAUTION

The size of the community does not impact the potential for hazardous materials; every community has hazardous materials.

FIGURE 25-2 Agricultural supply stores have a large quantity of hazardous materials, including pesticides, herbicides, and fertilizers. In many cases they also have fuels, including propane.

FIGURE 25-3   If a community has a road, the potential for a hazardous materials incident exists. One of the most common chemical releases is a gasoline spill.

Most communities have a gasoline station or a hardware store. The average home has a large amount of hazardous materials that can cause enormous problems during a response. In rural communities, farms present unique risks due to the storage of pesticides and fertilizers, **Figure 25-2.** All of these locations and occupancies provide the potential for the storage of hazardous materials. In general, the more industrialized a community is, the more hazardous materials the community will contain. Communities adjacent to industrialized areas or along major transportation corridors (interstate highways, rail, water) may also have the same hazards because these materials can travel through the community, **Figure 25-3.** Buildings that typically store hazardous materials include hardware stores, hospitals, auto part supply stores, dry cleaners, manufacturing facilities, print

shops, doctors' offices, photo labs, agricultural supply stores, semiconductor manufacturing facilities, electronics manufacturing facilities, light to heavy industrial facilities, marine terminals, rail yards, airport terminals and fueling areas, pool chemical stores, paint stores, hotels, swimming pools, and food manufacturing facilities.

PLACARDS, LABELS, AND MARKINGS

This section examines the first concrete evidence of the presence of hazardous materials. A number of systems are used to mark hazardous materials containers, buildings, and transportation vehicles. The systems result from laws, regulations, and standards, and in some cases from a combination of the three. As an example, the **Building Officials Conference Association (BOCA)** code, which has been adopted as a regulation in local communities, requires the use of the NFPA 704M marking system, **Figure 25-4,** for certain occupancies.

Placards

The most commonly seen item for identifying the location of hazardous materials is the placard. The Department of Transportation (DOT) regulates the movement of hazardous materials ("dangerous goods" in Canada) by air, rail, water, roadway, and pipeline by means of 49 CFR 170-180. After meeting certain guidelines, a shipper must placard a vehicle to warn of the storage of chemicals on the vehicle, an example of which is shown in **Figure 25-5. Table 25-1** and **Table 25-2** provide further explanations of the placarding system.

NOTE

The quantity of hazardous materials that must be carried in order to require placarding is 1,001 pounds (454 kg), unless it is one of six classes of materials that require placarding at any amount.

The DOT has established a system of nine hazard classes that uses more than twenty-seven placards to identify a shipment. The idea behind these hazard classes is to provide a general grouping to a shipment and to provide some basic information regarding the potential hazards. The placards, like the one shown in **Figure 25-6,** are 10¾ inches by 10¾ inches and are to be placed on four sides of the vehicle. Labels are 3⁹⁄₁₀ inches by 3⁹⁄₁₀ inches and are affixed near the shipping name on the container. Labels, for the most part, are smaller versions of placards and are designed to provide warnings about the package contents. There are labels for some materials that do not have or require placards.

The system is designed so that materials designated by the DOT as potentially harmful to the environment, humans, and animals are easily identified. Such a material has to present an unreasonable risk to the safety, health, or property upon contact. The DOT has two placarding tables, which are called Table 1 and Table 2. The materials that are on Table 1 (as shown in Table 25-1) are those that are most

FIGURE 25-4 This NFPA 704M symbol is used to warn of potential chemical dangers in the building. It warns of fire, health, reactivity, and special hazards.

FIGURE 25-5 The DOT requires some shippers of hazardous materials to provide placards to warn responders of chemicals that may be on the truck.

TABLE 25-1   Materials That Require Placarding at Any Amount (DOT Table 1)

Hazard Class or Division	Placard Type
1.1	Explosives 1.1
1.2	Explosives 1.2
1.3	Explosives 1.3
2.3	Poison gas
4.3	Dangerous when wet
5.2 (Organic peroxide, Type B, liquid or solid, temperature controlled)	Organic peroxide
6.1 (Inhalation hazard, Zone A or B)	Poison inhalation hazard
7 (Radioactive label III only)	Radioactive

FIGURE 25-6   "Corrosive" placard. Not to scale.

hazardous and require the use of a placard no matter what the quantity being shipped. The materials in DOT Table 2 (as shown in Table 25-2) are those that require placarding at 1,001 pounds. The criteria establishing this 1,001-pound (454 kg) rule are not clearly defined, but responders should be aware that a spill of 999 pounds (453 kg) of a material can be as hazardous as 1,001 pounds (454 kg). The shipper uses the hazardous materials tables (49 CFR 172.504) to determine which labels and placards are required. The tables may also list a packing group for the material, which indicates the danger associated with the material being transported.

- Packaging Group I: Greatest danger
- Packaging Group II: Medium danger
- Packaging Group III: Minor danger

Packaging groups are only assigned to classes 1, 3–6, 8, and 9 as shown in **Table 25-3** through **Table 25-6.** These are determined based on flash points, boiling points, and toxicity.

TABLE 25-2   Materials That Require Placarding at 1,001 Pounds (454 Kg) (DOT Table 2)

Class or Division	Placard Type
1.4	Explosives 1.4
1.5	Explosives 1.5
1.6	Explosives 1.6
2.1	Flammable gas
2.2	Nonflammable gas
3	Flammable
Combustible Liquid	**Combustible**
4.1	Flammable solid
4.2	Spontaneously combustible
5.1	Oxidizer
5.2 (Other than organic peroxide, Type B, liquid or solid, temperature controlled)	Organic peroxide
6.1 (Other than Inhalation hazard, Zone A or B)	Poison
6.1 (PG III)	PG III
6.2	None
8	Corrosive
9	Class 9
ORM-D	None

> **NOTE**
>
> Placards are useful because they take advantage of four ways to communicate the hazard class they represent.

TABLE 25-3   Packing Groups: Flammability

Flammability— Class 3 Packing Group (PG)	Flash Point	Boiling Point
I		≤ 95°F (35°C)
II	≤ 73°F (23°C)	> 95°F (35°C)
III	≥ 73°F (23°C) and ≤ 141°F (60.5°C)	> 95°F (35°C)

NOTE

The DOT uses several terms to provide a measure of toxicity. The DOT establishes ranges for the shipment of toxic materials. The determination of how toxic a material is can be based upon exposure limits established for many chemicals. LC50 and LD50 are terms used to describe a concentration or amount of a material that causes lethal effects to 50 percent of an exposed population. LC is lethal concentration, used to measure toxicity of gases, and LD is a lethal dose of solids and liquids. To determine an LC50 or LD50, researchers expose a population of test animals to a given amount of a substance. At the level that 50 percent of the test population dies, this establishes the LC50 or LD50.

When discussing toxicity, two common units of measure for the amount of material required to cause toxic effects are parts per million (ppm) and milligrams per cubic meter cubed (mg/M3). The use of ppm usually is in reference to a gas and represents a part of a million units, as one marble is one part per million of a million marbles. The use of mg/M3 usually refers to an amount of a solid or liquid material. It is a comparison of the amount of chemical in milligrams (mg) as compared with a cubic meter of air, meaning the amount of the material that could cause toxic effects within that cubic meter. Similarly, the use of mg/L is a reference to the amount (mg) of a given toxic chemical within the space of a liter. More information related to toxicity is provided in Section IV, Chapter 27.

TABLE 25-4   Packing Groups: Poisonous Gas, Division 2.3

Hazard Zone	Inhalation Toxicity—Lethal Concentration (LC_{50})*
Hazard Zone A	LC_{50} less than or equal to 200 ppm
Hazard Zone B	LC_{50} greater than 200 ppm and less than or equal to 1,000 ppm
Hazard Zone C	LC_{50} greater than 1,000 ppm and less than or equal to 3,000 ppm
Hazard Zone D	LC_{50} greater than 3,000 ppm and less than or equal to 5,000 ppm

*LC_{50} is the lethal concentration (gases) to 50 percent of an exposed population.

of the placard, and they display the hazard class and division number in the bottom triangle.

As an example, the placard shown in Figure 25-6 is black and white, which represents the corrosive class. The top of the placard has a picture of a hand and a steel bar being eaten away by the corrosive material being poured on it, the middle of the placard states corrosive, and at the bottom the class number 8 shows.

The DOT also requires the addition of a four-digit number, known as the United Nations/North America (UN/NA) identification number, either on

Placards have distinct colors, they have a picture at the top of the triangle depicting a representation of the hazard, they state the hazard class in the middle

TABLE 25-5   Packing Group Division 6.1

For packing materials that are toxic via a route other than inhalation, the following are used:

Packing Group	Oral Toxicity—Lethal Dose (LD_{50})* (mg/kg)	Dermal Toxicity—Lethal Dose (LD_{50}) (mg/kg)	Inhalation via Dusts and Mists (LC_{50}) (mg/L)
I	≤ 5	≤ 40	≤ 0.5
II	> 5 and ≤ 200	> 40 and ≤ 200	> 0.5 and ≤ 2
III	Solids: > 50 and ≤ 200; Liquids: > 50 and ≤ 500	> 200 and ≤ 1000	> 2 and ≤ 10

*(LD_{50}) Lethal dose (solids and liquids) to 50 percent of the exposed population.

TABLE 25-6   Division 6.1, Toxic Materials Poisonous by Inhalation

Packing Group	Vapor Concentration and Toxicity
I (Hazard Zone A)	$V \geq 500$ LC_{50} and LC_{50} 200 mL/M^3

FIGURE 25-7 There are three ways to signify the four-digit ID number for bulk shipments. The most common is for the four-digit DOT identification number to be placed in the middle of the placard.

a placard or on an adjacent orange rectangle. This identifies a bulk shipment of over 119 gallons (450 liters) and provides an identity to the material. A tank truck carrying gasoline, which would be considered a bulk shipment, would display a Flammable Liquid placard with the number 1203 either in the middle of the placard or on an orange rectangle adjacent to the placard, as shown in **Figure 25-7.**[1] This provides an additional bit of information to the

responder, because without the UN/NA number, the only information provided would be that a flammable liquid was on board.

The nine hazard classes and subdivisions are discussed next.[2]

Class 1, Explosives (Figure 25-8)

- *Division 1.1.* Mass explosion hazard, such as black powder, dynamite, ammonium perchlorate, detonators for blasting, and RDX explosives.
- *Division 1.2.* Projectile hazard, such as aerial flares, detonating cord, detonators for ammunition, and power device cartridges.
- *Division 1.3.* Fire hazard or minor blast hazard. Examples include liquid-fuel rocket motors and propellant explosives.
- *Division 1.4.* Minor explosion hazard, which includes line throwing rockets, practice ammunition, detonation cord, and signal cartridges.
- *Division 1.5.* Very insensitive explosives, which do have mass explosion potential but during normal shipping would not present a risk. Ammonium nitrate and fuel oil (ANFO) mixtures are an example of this division.
- *Division 1.6.* Also very insensitive explosives that do not have mass explosion potential. Materials that present an unlikely chance of ignition are part of this grouping.

The placards for divisions 1.1, 1.2, or 1.3 will have the division number and compatibility group located just above the class number for any quantity. For divisions 1.4, 1.5, and 1.6, the placard will have a compatibility group number when a vehicle is transporting 1,001 pounds (454 kg) or more. The DOT

FIGURE 25-8 "Explosive" placards and labels.

provides compatibility charts that dictate that certain high-hazard shipments cannot be mixed with other hazardous materials. The compatibility charts are designed to prevent the mixing of certain materials in the event of an accident. Some materials when mixed can create substantial issues such as explosions, fires, or release of toxic gases.

SAFETY

Incidents involving explosives can be very dangerous, especially when involved in fire.

Making a tactical decision to attack a fire involving explosives can endanger the responders, especially if the fire has reached the cargo area of the vehicle. The recommendations in the DOT **Emergency Response Guidebook** (ERG—Guide page 112) should be followed, paying particular attention to the isolation and evacuation distances. When explosives are involved in traffic accidents not involving fire, the actual threat is minimized depending on the circumstances. As long as the explosives were transported legally and as they were intended to be transported, the responders should face little danger. Some cities require an escort and that the explosives be transported at nonpeak hours.

Spilled explosive materials may present a health hazard if inhaled or absorbed through the skin. When explosives are involved in a fire situation, the best course of action would be the immediate withdrawal from the area as well as isolating and denying entry. The area should be evacuated following recommendations from the current version of the DOT ERG. Many chemicals used in explosives are powerful oxidizers that if ignited can be very aggressive and will burn with considerable energy, and in some cases are explosive in nature.

ANFO EXPLOSION

On November 29, 1988, an engine company from the Kansas City, Missouri, Fire Department was dispatched to a reported pickup truck fire. While en route to the incident the engine company was told to use caution because explosives were reportedly involved. When they arrived they found that they had two separate fires, one in the pickup truck and the other in a trailer. The first engine began to extinguish the fire in the pickup truck and requested a second engine for assistance as well as the district battalion chief. They attempted to contact the second engine to warn them of the explosives on fire on top of the hill. The second engine arrived and began to attack the fire on top of the hill. They requested the assistance of the first engine and also requested a squad for water. From the radio communications with the battalion chief the crews thought that the explosives had already detonated.

The battalion chief was a quarter of a mile away where he had stopped to talk with the security guards when the explosives detonated. The explosion moved the chief's car 50 feet (15 m) and blew in the windshield. The blast was heard for 60 miles (87 Km) and damaged homes within 15 miles (24 Km). The chief requested additional assistance and staged the responding companies. There was a report that there were more explosives on the hill that had not yet detonated. Luckily the chief did not let any other responders into the scene because shortly after pulling back, a second explosion went off, reportedly larger than the first one.

The next morning a team of investigators went to the site to begin the investigation. They discovered that six firefighters were killed in the blast and the explosion was very devastating. Only one engine was recognizable; the other was reduced to the frame rail. The first blast was from 17,000 pounds (7,711 kg) of an ammonium nitrate, fuel oil, and aluminum mixture. It also had 3,500 pounds (1,588 kg) of ANFO. The second explosion involved 30,000 pounds (13,608 kg) of the mixture. The use of ANFO is common throughout the United States for blasting purposes. Ammonium nitrate is commonly used as a fertilizer in the agricultural business and is used on residential lawns. The Oklahoma City Alfred P. Murrah Federal Building explosive was devised of ammonium nitrate and nitromethane, very similar to ANFO. The World Trade Center bombing in 1993 used an explosive made up primarily of urea nitrate and three cylinders of hydrogen. Both explosives are comparable to each other, and when they are used improperly or for criminal purposes the results can be devastating.

FIGURE 25-9   "Gas" placards and labels.

Class 2, Gases (Figure 25-9)

■ *Division 2.1.* Flammable gases that are ignitable at 14.7 psi (1 bar) in a mixture of 13 percent or less in air, or have a flammable range with air of at least 12 percent regardless of the lower explosive limit (LEL). Propane and isobutylene are examples of this division.

■ *Division 2.2.* Nonflammable, nonpoisonous, compressed gas, including liquefied gas, pressurized cryogenic gas, and compressed gas in solution. Carbon dioxide, liquid argon, and nitrogen are examples.

■ *Division 2.3.* Poisonous gases that are known to be toxic to humans and would pose a threat during transportation. Chlorine and liquid cyanogen are common examples of this division. Gases assigned to this division are also assigned a letter code identifying the material's toxicity levels. These levels are discussed further in the section on toxicology. The hazard zones associated with this division are:

  ■ *Hazard Zone A.* LC_{50} less than or equal to 200 ppm
  ■ *Hazard Zone B.* LC_{50} greater than 200 ppm and less than or equal to 1,000 ppm
  ■ *Hazard Zone C.* LC_{50} greater than 1,000 ppm and less than or equal to 3,000 ppm
  ■ *Hazard Zone D.* LC_{50} greater than 3,000 ppm and less than or equal to 5,000 ppm

The hazard zones are a quick way to determine how toxic a material is. Hazard Zone A materials are more toxic than Hazard Zone B materials, and so forth. The shipping papers will identify these by the addition of Poison Inhalation Hazard (PIH) Zone A, B, C, or D.

> **NOTE**
>
> Remember that many of these gases have multiple hazards. Consider that based on information that can be obtained from a **Material Safety Data Sheet (MSDS)** and the DOT Emergency Response Guide (ERG), some of these products such as 2.2 can also be flammable and cause unforeseen problems.

Class 3, Flammable Liquids (Figure 25-10)

■ Flammable liquids have a flash point of less than 141°F (60.5°C). Gasoline, acetone, and methyl alcohol are examples.

■ Combustible liquids are those with flash points above 100°F (37.7°C) and below 200°F (93°C). The DOT allows liquids with a flash point of 100°F (37.7°C) to be shipped as a combustible liquid. Examples include diesel fuel, kerosene, and various oils.

FIGURE 25-10   "Flammable" and "Combustible" placards and label.

CAUTION

The difference between a flammable liquid and a combustible liquid is based on the material's flash point. The flash point is the temperature of the liquid at which there could be a flash fire if an ignition source is present. The fire service usually refers to a flammable liquid as one that has a flash point of 100°F (37.7°C) or lower. The DOT classifies any liquid with a flash point of 141°F (60.5°C) or lower as being a flammable liquid. Any liquid with a flash point greater than 141°F (60.5°C) is considered combustible by the DOT. It can be confusing when using the terms flammable and combustible to describe a liquid material. Most references, with the exception of the DOT, use 100°F (37.7°C) as the criteria for flammable and combustible.

Class 4, Flammable Solids (Figure 25-11)

- *Division 4.1.* Includes wetted explosives, self-reactive materials, and readily combustible solids. Examples include magnesium ribbons, picric acid, explosives wetted with water or alcohol, or plasticized explosives.

- *Division 4.2.* Composed of spontaneously combustible materials including pyrophoric materials or self-heating materials. An example is zirconium powder.

- *Division 4.3.* Dangerous-when-wet materials are those that when in contact with water can ignite or give off flammable or toxic gas. Calcium carbide when mixed with water makes acetylene gas, which is very flammable as well as unstable in this form.

Sodium is another example of a material that when wet can ignite explosively. Lithium and magnesium are not as explosive as sodium but will react with water. Magnesium if on fire will react violently if water is used in an attempt to extinguish the fire.

Class 5, Oxidizers and Organic Peroxides (Figure 25-12)

- *Division 5.1.* The class assigned to materials that readily release oxygen; by yielding oxygen, an **oxidizer** can easily cause or enhance the combustion of other materials. Oxidizers can dramatically increase the rate of burning when combustible material is ignited. Ammonium nitrate and calcium hypochlorite are examples.

- *Division 5.2.* The organic peroxides, which have the ability to explode or polymerize; if contained, it is an explosive reaction. These are further subdivided into seven types:
  - *Type A* can explode upon packaging. These are DOT forbidden, which means they cannot be transported and must instead be produced on site.
  - *Type B* can thermally explode, considered a very slow explosion.
  - *Type C* neither detonates nor **deflagrates** rapidly, and will not thermally explode.
  - *Type D* only detonates partially or deflagrates slowly, and has medium or no effect when heated and confined.
  - *Type E* shows low or no effect when heated and confined.
  - *Type F* shows low or no effect when heated and confined, and has low or no explosive power.
  - *Type G* is thermally stable and is desensitized.

FIGURE 25-11 "Flammable solids" placards and labels (Class 4).

FIGURE 25-12 "Oxidizers and organic peroxides" placards and labels (Class 5).

A chemical that polymerizes is one that presents an extreme danger to firefighters. Polymerization is a runaway chain reaction that once started cannot be stopped. If the material is contained, it will rupture the container in a violent manner, sending pieces of the container for a considerable distance. Polymerization can be best described as a chain. It starts off as one link, and once the reaction is started, many other links develop and attach themselves to the other links. Each link develops a great deal of heat and for each 41°F (5°C) of heat that develops, the reaction rate doubles. Thus, they quickly duplicate themselves, eventually rupturing the container. The reaction can result in a fire or explosion, since many of the chemicals that can polymerize are also flammable liquids. A container of spray foam insulation that is used for sealing spaces in walls around doors and windows is a good example. When the foam hits the air, it reacts and rapidly expands to a much greater volume. This reaction is uncontrollable and will continue to expand for quite some time until the polymerization is completed. A small container that holds a few ounces does not present much risk, but a railcar full of a chemical that may polymerize presents an extreme risk to the community.

FIGURE 25-13 "Poisonous materials" placards and labels (Class 6).

Class 6, Poisonous Materials (Figure 25-13)

- *Division 6.1.* Materials that are so toxic to humans that they would present a risk during transportation. Examples include arsenic and aniline.
- *Division 6.2.* Composed of microorganisms or their toxins, which can cause disease to humans or animals. Anthrax, rabies, tetanus, and botulism are examples.

Hazard zones are associated with Class 6 materials:

- *Hazard Zone A.* LC_{50} less than or equal to 200 ppm
- *Hazard Zone B.* LC_{50} greater than 200 ppm and less than or equal to 1,000 ppm.

Class 7, Radioactive Materials (Figure 25-14)

- Those materials determined to have radioactive activity at certain levels.
- Although there is only one placard, the labels shown in Figure 25-14 are further subdivided into Radioactive I, II, and III, with level III being the highest hazard. The designation I, II, or III is dependent on two criteria: the transport index and the radiation level coming from the package.

1. The transport index, which is comparable to the United Nations and IAEA criticality safety index is the degree of control the shipper is to use and is based on a calculation of the radiation threat the package presents.
2. Responders should understand that a package labeled radioactive may be emitting radiation. These emissions of radiation are legal within certain guidelines:
   - *Radioactive I label.* Less than 0.005 mSv/hr (0.5 mR/hr)
   - *Radioactive II label.* More than Radioactive I and less than 0.5 mSv/hr (50 mR/hr)
   - *Radioactive III label.* More than Radioactive II and less than 2 mSv/hr (200 mR/hr)
   - *Fissile material label.* For radioactive material that can be used as a fuel—common examples include uranium-233, uranium-235, uranium-239, and uranium-241

   Packages emitting more than 2 mSv/hr require special handling and transportation and are subject to additional regulations.

FIGURE 25-14 "Radioactive materials" placards and labels (Class 7).

FIGURE 25-15 "Corrosives" placards and labels (Class 8).

Class 8, Corrosives (Figure 25-15)

■ Includes both acids and bases, and is described by the DOT as a material capable of causing visible destruction in skin or corroding steel or aluminum. Examples include sulfuric acid and sodium hydroxide.

Class 9, Miscellaneous Hazardous Materials (Figure 25-16)

■ A general grouping that is composed of mostly hazardous waste. Dry ice, molten sulfur, liquid asphalt, and polymeric beads are examples that would use a Class 9 placard.

■ This is known as a catch-all category. If a substance does not fit into any other category and presents a risk during transportation then it becomes Class 9.

Other Placards and Labels

■ The Dangerous placard is used when the shipper is shipping a mixed load of hazardous materials. If the shipper ships 2,000 pounds (907 kg) of corrosives and 2,000 pounds (907 kg) of a flammable liquid, then instead of displaying two placards the

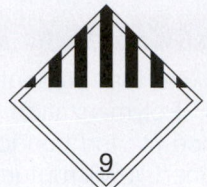 

FIGURE 25-16 "Miscellaneous hazardous materials" placards and labels (Class 9).

FIGURE 25-17 "Dangerous" placard.

FIGURE 25-18 "Inhalation" placard.

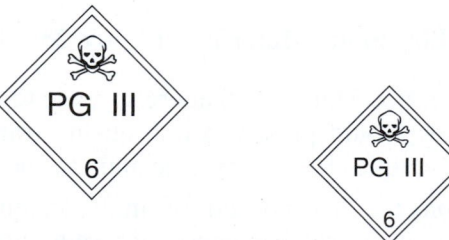

FIGURE 25-19 Another form of a poison placard and label.

shipper can display a Dangerous placard, as shown in **Figure 25-17.** If any of the items exceeds 2,205 pounds (1,000 kg) and is picked up at one location, then in addition to the Dangerous placard the shipper is to display the placard for the material that exceeds the 2,205 pounds (1,000 kg).

■ Specific-name placards or placards with "Inhalation Hazard," as shown in **Figure 25-18,** are sometimes used in place of the Poison Gas placard.

■ The PG III placard and label has replaced the Stow Away from Foodstuffs placard, which indicates that a poisonous material is being transported, but it is not poisonous enough to meet the rules to be placarded as a poison. Chloroform is an example that would use a placard like the one shown in **Figure 25-19.**

TABLE 25-7   Examples of Materials of Trade (MOT)

Class or Division	Example
Flammable gases (Division 2.1)	Acetylene, propane
Non-flammable gases (Division 2.2)	Oxygen, nitrogen
Flammable or combustible liquids (Class 3)	Paint, paint thinners, gasoline
Flammable solids (Division 4.1)	Charcoal
Dangerous-when-wet materials (Division 4.2)	Some fumigants
Oxidizers (Division 5.1)	Bleaching compounds
Organic peroxides (Division 5.2)	Benzoyl peroxides
Poisons (Division 6.1)	Pesticides
Some infectious substances (Division 6.2)	Diagnostic specimens
Corrosive materials (Class 8)	Muratic acid, drain cleaners, battery acid
Miscellaneous hazardous materials (Class 9)	Asbestos, self-inflating life boats
Consumer commodities (ORM-D)	Hair spray, spray paints

- "Other Regulated Material—Class D" (ORM-D) is a classification that is left over from a previous DOT regulation. It is a subdivision that includes ammunition and consumer commodities, such as cases of hair spray. The package will have the printing "ORM-D" on the outside of the package. The previous regulation used to have ORM-A through ORM-E, but these are now grouped together in Class 9, Miscellaneous Hazardous Materials.

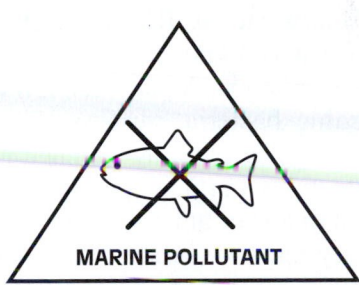

FIGURE 25-20   "Marine Pollutant" marking.

> **NOTE**
>
> The DOT has changed some terminology for Other Regulated Materials, commonly referred to as ORMs. Typically, these were shipments of consumer goods, which are now known as Materials of Trade (MOT). Some common examples are shown in **Table 25-7.** The DOT has changed some of the rules when transporting smaller quantities of hazardous materials when used in a business. With the exception of tanks containing diluted mixtures of class 9 materials, a person cannot transport more than 440 pounds (200 kg) of a material. If the material is a high-hazard material (packing group I), the maximum amount of material allowed in one package is one pound (0.5 kg) for solids and one pint (0.5 liters) for liquids.

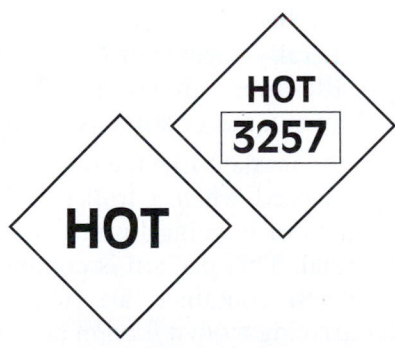

FIGURE 25-21   "Hot" placard.

- "Marine Pollutant" is displayed on shipments that, if the material were released into a waterway, would damage the marine life, **Figure 25-20.**

- Elevated temperature material will have a "HOT" label either to the side of or on the placard, as shown in **Figure 25-21,** if it meets one of the following criteria:

  - Is a liquid above 212°F (100°C).
  - Is a liquid that is intentionally heated and has a flash point above 100°F (37.7°C).
  - Is a solid at 464°F (240°C) or above.

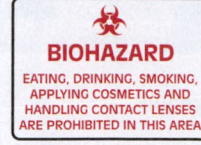

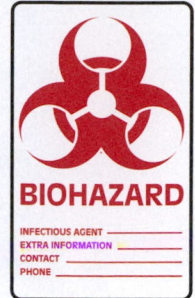

FIGURE 25-22 "BioHazard" labels.

- "Infectious Substances" or "BioHazard" are labels like those shown in **Figure 25-22** that are sometimes used on the outside of trucks. It is not required by the DOT but may be required by other agencies such as Health and Human Services or a state agency. The DOT now calls medical waste "regulated medical waste" or RMW. An infectious substance is a material known or reasonably expected to contain a pathogen. A pathogen is a microorganism (including bacteria, viruses, ricksettia, parasites, and fungi) or other agent, such as a proteinaceous infectious particle (prion), that can cause disease in humans or animals. They have also split this category into two subdivisions. Category A is an infectious substance in a form capable of causing permanent disability or life-threatening or fatal disease in otherwise healthy humans or animals when an exposure occurs. A category B material presents a lesser hazard, but the infectious substance is not in a form generally capable of causing permanent disability or life-threatening or fatal disease in otherwise healthy humans or animals when exposed.

- The Fumigated marker, like the one shown in **Figure 25-23,** is used when a trailer or railcar has been fumigated or is being fumigated with a poisonous material. This placard is commonly found near ports where containers are frequently fumigated after arriving from a foreign port.

- A white square background as shown in **Figure 25-24** is used in the following situations:

  - On the highway for controlled Radioactive III shipments

  - On rail for:

    - Explosives 1.1 or 1.2

    - Division 2.3 Hazard Zone A (poison gas) materials

FIGURE 25-23 "Fumigation" marking.

FIGURE 25-24 White-squared "Flammable Gas" placard.

- Division 6.1 Packing Group I Hazard Zone A (poison)

- Division 2.1 (flammable gas) in a DOT 113 tank car

- Division 1.1 or 1.2, which is chemical ammunition that also meets the definition of a material that is poisonous by inhalation

Problems with the Placarding System

The placarding system relies on a human to determine the extent of the load, determine the appropriate hazard classes, and interpret difficult regulations to determine if a placard is required. The placard then must be affixed to all four sides of the vehicle before shipment. Placards are only required for shipments that exceed 1,001 pounds (454 kg), except for materials listed in Table 25-1, which require placarding at any amount. It is suspected that 10 to 20 percent of the trucks traveling the highway are not placarded at all or not placarded correctly.

CAUTION

Given the restrictions of many cities, bridges, and tunnels where hazardous materials are not allowed, many trucks are probably not carrying the proper designations.

A placard can come off during transport and legally may not have to be immediately replaced. The fact that an incident involves a truck or train should alert the first responder to the potential for hazardous materials, and when a placard is involved extra precautions should be taken.

Labels

STREETSMART TIP

A tractor trailer was involved in an accident and had jackknifed in Baltimore County, Maryland. The hazardous materials team was called to assist with the fuel spill. Upon arrival, the hazardous materials team inquired as to the contents of the truck and was told by the first responders that the truck was carrying rocking chairs. While working to control the fuel leak the hazardous materials team noticed a greenish blue liquid coming from the front of the trailer, where it had been damaged in the accident. They located the driver and asked him about the contents of the trailer, which did not display a placard. The driver stated he was carrying rocking chairs, which is what the shipping papers showed the load to be.

The crews opened the back of the trailer to inspect the cargo and found rocking chairs. Upon closer inspection of the front of the truck, several drums were noticed in the very front of the trailer. When the driver was confronted by the hazardous materials team and a state trooper, the driver confessed he had picked up a load of unknown waste from his first stop. He did not know the contents of the drums and it took several days to determine the exact contents of the drums. Beware; it is unlikely that someone trying to evade the law will mark the shipment appropriately.

Labeling and Marking Specifics

Package markings must include the shipping name of the material, the UN/NA identification number, and the shipping and receiving companies' names and addresses. Packages that contain more than a **Reportable Quantity (RQ)** of a material must also be marked with an RQ near the shipping name. Packages that are listed as ORM-D materials should be marked as such. Some packages with liquids in them must use orientation arrows. Materials that pose inhalation hazards must affix an Inhalation Hazard label next to the shipping name, as shown in **Figure 25-25.** Hazardous wastes will be marked "Waste" or will use the EPA labeling system to identify these packages.

Labels are identical to placards, other than their size.

NOTE

Materials that have more than one hazard may be required to display a primary hazard label and a subsidiary label.

The primary label will have the class and division number in the bottom triangle, while the subsidiary label will not have the number at all, as shown in **Figure 25-26.** As an example, the material acrylonitrile, inhibited, is required to be labeled "Flammable" with a subsidiary label of "Poison."

OTHER IDENTIFICATION SYSTEMS

There are several other identification systems that are used in private industry to mark facilities and containers. Military shipments and pipelines are also marked to provide a warning as to the potential for hazardous materials. Much like the transportation system, the warnings are a clue to the potential presence of hazardous materials that could cause harm to the responders.

NFPA 704 System

One of the other more common systems used to identify the presence of hazardous materials is the NFPA 704 system. This system is designed for buildings, not transportation, and alerts the first responders to the potential hazards in and around a facility. The system is much like the placarding system and relies on a triangular sign that is divided into four areas, as shown in **Figure 25-27.** The four areas are divided by color as well and use a ranking system to identify severity. The four areas and colors are:

1. Health hazard—blue
2. Fire hazard—red

FIGURE 25-25 The DOT adds the "Poison—Inhalation Hazard" label to those materials that present severe toxic hazards.

3. Reactivity hazard—yellow

4. Special hazards—white

The system uses a ranking of 0–4 with 0 presenting no risk and a ranking of 4 indicating severe risk. The specific listings are discussed next.

Health

This listing is based on a limited exposure to the materials using standard firefighting protective clothing as the protective clothing for the exposure.

    4—Severe health hazard

    3—Serious health hazard

    2—Moderate hazard

    1—Slight hazard

    0—No hazard

Flammability

This listing pertains to the ability of the material to burn or be ignited.

    4—Flammable gases, volatile liquids, pyrophoric materials

    3—Ignites at room temperature

    2—Ignites when slightly heated

    1—Needs to be preheated to burn

    0—Will not burn

Reactivity

This listing is based on the material's ability to react, especially when shocked or placed under pressure.

FIGURE 25-26 Primary and subsidiary placards.

FIGURE 25-27 NFPA 704 system marking.

    4—Can detonate or explode at normal conditions

    3—Can detonate or explode if strong initiating source is used

    2—Violent chemical change if temperature and pressure are elevated

    1—Unstable if heated

    0—Normally stable

Special Hazards

This listing is used to indicate water reactivity and oxidizers, which are included in the NFPA 704 system. In some cases, other symbols may be used such as the tri-foil for radiation hazards, "ALK" for alkalis, and "CORR" for corrosives. In the presence of the slashed W there is also an accompanying ranking structure for water reactivity in addition to the hazards listed in the other triangles:

4—Not used with the slashed W and a reactivity ranking of 4

3—Can react explosively with water

2—May react with water or form explosive mixtures with water

1—May react vigorously with water

0—Slashed W is not used with a reactivity ranking of 0

Some potential problems exist with the NFPA 704 system, because it groups all of the chemical hazards listed in a building into one sign. If the sign is placed on a tank that contains one material, the system does a good job of warning about the contents of the tank, but does not provide the name of the product. For a facility that has hundreds—if not thousands—of materials, the system will only warn of the worst-case scenario. As an example, dramatically different tactics are used to handle a flammable gas incident versus a flammable liquids incident, but both can be classified as fire hazard 4. The system is best used to alert the first responder to the presence of hazardous materials and to warn of the worst-case scenario.

Hazardous Materials Information System

Commonly referred to as HMIS, the Hazardous Materials Information System was designed to provide a mechanism to comply with the federal hazard communication regulation, which requires that all containers be marked with the appropriate hazard warnings and the ingredients be provided on the label, **Figure 25-28.** Many products that come into the workplace are missing adequate warning labels. The HMIS is not a uniform system. It can be developed by the facility or by the manufacturer of the labels, so one system may vary from another. Most systems are similar to the NFPA 704 system and use blue, red, and yellow colors with a numbering system that provides an indication of hazard. The colors may be used in a triangle format or, in most cases, as stacked bars. The numbers are usually 0–4, the same as the NFPA system, but in rare cases may differ from the NFPA system. The facility manager or other representative should have the key to the symbols, or a chart should be provided somewhere in the facility indicating what the symbols and the warning levels are. The chart is usually stored with the MSDS. In most cases, a central

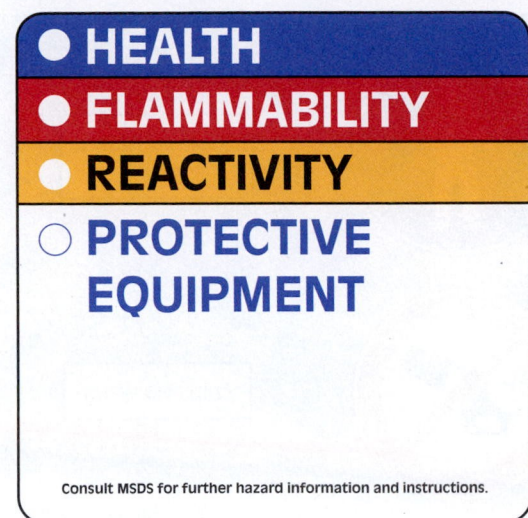

FIGURE 25-28   HMIS label.

location should be chosen for the MSDS and other hazard communication information.

In some systems, a picture is provided of the level of PPE required for the substance. Each HMIS is different, and responders should not assume any particular hazard level until the warning levels can be determined.

Military Warning System

The military uses the DOT placarding system when possible, but in some cases may use its own system. Within the DOT's Emergency Response Guidebook, an emergency contact number is given when responding to an incident involving a military shipment.

In most cases, for extremely hazardous materials, arms, explosives, or secret shipments, firefighters can assume prior to their arrival that the military is already aware of the incident and probably already responding. The higher the hazard the more likely there will be an escort for the shipment. There may be shipments in which the driver of the truck is not allowed to leave the cab of the truck and may provide warnings to stay away from the truck.

SAFETY

For high-security shipments the driver is armed, as are the personnel in the escort vehicle. If an incident occurs involving one of these vehicles, firefighters must obey the commands of the escorting personnel and determine if they have made the appropriate notifications.

If the driver and escort crew are killed or seriously injured in the accident, it would be advisable to notify the military about the incident, although with satellite tracking help is probably already on the way. The

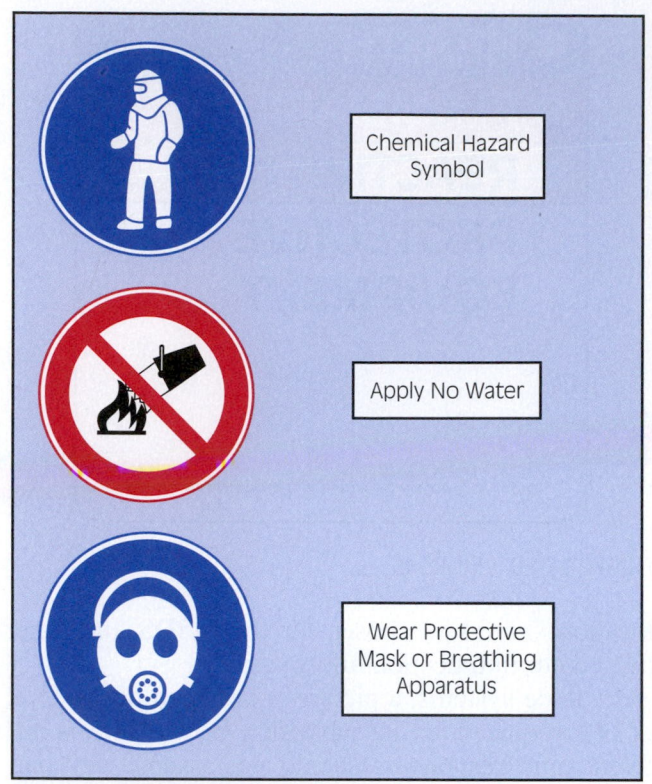

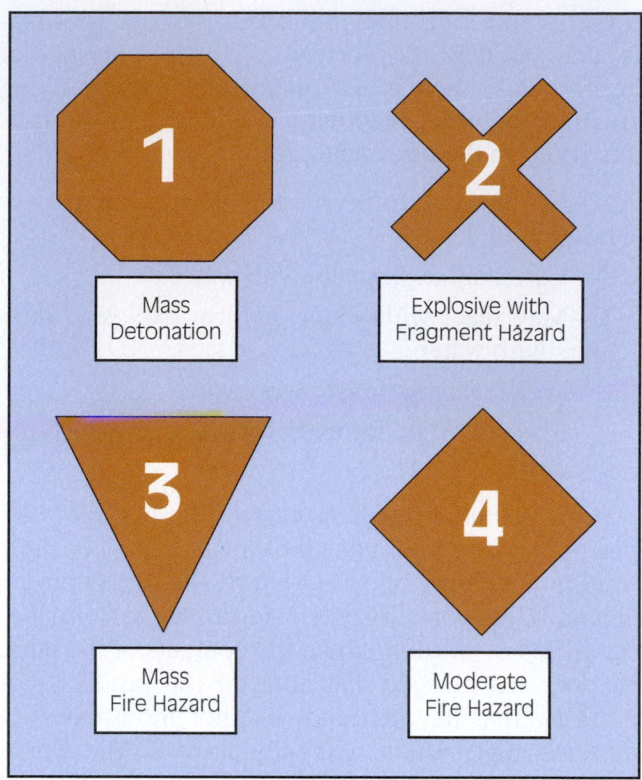

FIGURE 25-29   Military placards.

phone number to contact the military is in the DOT ERG, along with **Chemtrec's** and other emergency contact numbers.

Other incidents involving fuels, food, or military equipment may require notification of the military.

The military typically uses its own marking system at its facilities to mark the buildings. The military uses a series of symbols and a numerical ranking system as shown in **Figure 25-29**.

Pipeline Markings

Any place an underground pipeline crosses a mode of transportation, the pipeline owner is required to place a sign like the one shown in **Figure 25-30** that displays the wording "Warning, Caution, or Danger," the words "Petroleum Pipeline," or the hazardous contents of the pipe, as well as the owners name and telephone number. The pipeline contents may be general, as in "petroleum products," because the same pipeline may be used to ship fuel oil, gasoline, motor oil, or other products. Dedicated pipelines that carry only one product will be marked with the specific product that it carries. The pipeline should be buried a minimum of 3 feet and should be adequately marked.

Many of the larger pipelines, such as the Colonial pipeline that originates in Texas and ends in New York, are 26 inches in diameter along the main pipeline. In the event of a release involving the pipeline,

FIGURE 25-30   The owner of a pipeline is required to provide a sign with the words "Warning," "Caution," or "Danger"; the words "Petroleum Pipeline" or the hazardous contents of the pipe; and the owner's name and telephone number.

the line will be shut down immediately. Even with this immediate shutdown, the potential exists to lose several hundred thousand gallons of hazardous materials because the distance between the shutoffs is substantial.

An incident involving a pipeline can be a serious event—firefighters should not underestimate the need for considerable local, state, and federal resources. Within the fire department alone, considerable resources may be required such as command staff, logistic support, communications, and tactical units. A fuel oil pipeline rupture in Reston, Virginia, resulted in the loss of more than 400,000 gallons (1,514,164 liters) of fuel, requiring considerable resources from several states to control the spill. The resources included emergency response organizations; local, state, and federal assistance; and a considerable number of private cleanup companies.

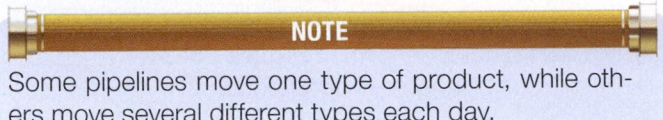

NOTE

Some pipelines move one type of product, while others move several different types each day.

The product in the pipelines varies from liquefied gases and petroleum products to slurried material. Pipeline companies are required to conduct in-service training and tours for the emergency responders in the communities their pipelines transverse. When firefighters have an incident on or near any pipeline, it is advisable to notify the pipeline owner of the incident, even if they are pretty sure the pipeline was not damaged. A train derailment in California caused a pipeline to shift, and the pipeline did not release any product until several days after the original derailment occurred. Most pipeline operators would like to have the opportunity to check the line as opposed to having a catastrophic release several days later because the line was not checked.

Container Markings

Most containers such as drums are marked with the contents of the drum, **Figure 25-31,** while cylinders have the name of the product stenciled on the side of the cylinder. In bulk shipments the bulk container will have the name of the product stenciled on the side. Trucks that are dedicated haulers will also stencil or mark the product name on two sides of the vehicle.

Pesticide Container Markings

Due to their toxicity, pesticides are regulated by the EPA as to how they are to be marked. The label on a pesticide container, such as the one shown in **Figure 25-32,** will have the manufacturer's name for the pesticide, which is not usually the chemical name for the product. The product name will usually not offer any clue as to the chemical makeup of the product, and in some cases may be difficult to research. In Section IV, Chapter 26, there is a discussion of chlordane and the multiple names it's referred to by. The label will also contain a signal word such as "Danger," "Warning,"

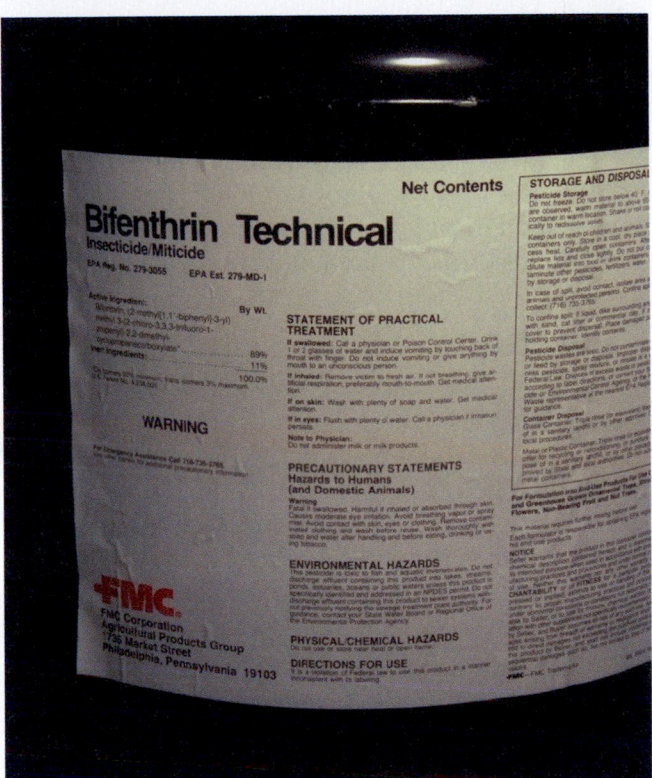

FIGURE 25-31 The label describes the contents of the drum.

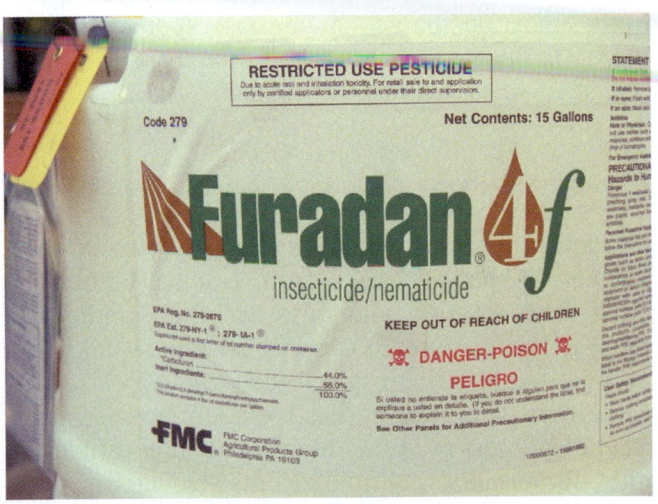

FIGURE 25-32 "Pesticide" labels.

or "Caution." The assignment of the signal word is the clue to how dangerous the material actually is. If the label indicates "danger" then extreme care should be used as the material is highly toxic. The designations of "warning" and "caution" present lesser hazards, but still should be considered very hazardous.

In the United States, the EPA issues an EPA registration number and in Canada the label will have a pest control number. The label will also include a precautionary statement and a hazard statement, examples of which include "Keep from Waterways," and

"Keep Away from Children." The active ingredients will be listed by name and percentage; in most cases the active ingredients are usually a small percentage of the product. Inert ingredients are also listed but not specifically named. It is the active ingredient that presents the major hazard. For liquid pesticides, the "inert" ingredient is usually water, or petroleum solvents such as xylene, plus emulsifying agents.

Radiation Source Labeling

In 2007, the International Atomic Energy Association (IAEA), a group that sets nuclear and radiation standards worldwide, issued a new radiation warning label. The old label using the trefoil commonly referred to as a "propeller" isn't really associated with radiation or nuclear material. The label shows energy waves coming from a trefoil with skull and cross bones, with a human running away. The label shown in **Figure 25-33** more accurately represents the hazards associated with radiation. It is not mandated for use by DOT, and the DOT will still require the DOT placards and labels, but the new label will be more common in the future. Devices or containers that store radiation sources will be using this type of label. On devices that use a radiation source and are relatively large, the label will be placed near the access panel to the radiation source. Obviously, responders should not try to access the source and should request radiation specialists to assist them.

CONTAINERS

Hazardous materials come in a variety of containers of many shapes and sizes, from 1-ounce bottles and larger bags to tanks and ships carrying hundreds of thousands of gallons. A survey of the materials in the average home will reveal a wide variety of storage containers. Compressed gas cylinders hold propane; steel containers hold flammable and combustible liquids; bottles, jars, and small drums hold various products. Plastic-lined cardboard boxes, and bags of various types are also used for chemical storage.

The type of material and the end use for the product determine the packaging. Packaging used to store household or consumer commodities is usually different than the industrial version. In some cases the industrial version may be full-strength undiluted product, whereas the household version is only a small percentage of that strength mixed with a less hazardous substance such as water. The type of container usually provides a good clue as to the contents of the package.

STREETSMART TIP

The more substantial, durable, and fortified a container is, like the container shown in **Figure 25-34,** the more likely the material inside is dangerous.

On the other hand, materials transported in fiberboard drums usually have no significance with regard to human health, although they may pose a risk to the environment.

When looking at the recognition and identification process, **first responders** should be alert for anything

FIGURE 25-34 The type of container can provide some clues as to the contents of the container. Because this drum is reinforced, it has a high likelihood of containing an extremely hazardous material.

FIGURE 25-33 "IAEA Radiation" label.

unusual when arriving at an incident. When on an EMS call to a residential home, it would be unusual to find a 55-gallon (208 liter) drum in a bedroom along with glassware associated with a lab environment. These types of recognition and identification clues should alert the first responders that additional assistance may be required. Arriving at an auto repair garage and finding 55-gallon (208 liter) drums and compressed gas cylinders should not be unexpected, however.

General

Containers come in a variety of sizes and shapes and the general category of containers is not exceptional. Most of the general containers are designed for household use but will be carried in large quantities when moved in transportation. When moving to bags and into drums and cylinders the move is made from household to industrial use. All of these types can be used in the home, but a super sack that can hold thousands of pounds of materials is not usually considered household.

Cardboard Boxes

With the popularity of shopping clubs and discount warehouses, more and more homeowners are buying materials by the case, when in the past they bought in much smaller quantities. Cardboard boxes are used to ship and contain hazardous materials. They can hold glass, metal, or plastic bottles. In some cases they may have a plastic lining, such as the box shown in **Figure 25-35,** which holds sulfuric acid. Many household pesticides, insecticides, and fertilizers are contained in cardboard boxes. With the exception of these products and sulfuric acid, most products contained in boxes are usually not extremely toxic to humans, but may present an environmental threat. Materials in transport to suppliers may be transported in larger cardboard boxes and then broken down at the retail level. Responders should note any labels on these packages, but the absence of any labels does not indicate that hazardous materials are not present.

Bottles

From 1-ounce bottles to 1-gallon (3.8 liter) bottles, the variety of containers is endless and the types of products contained in them too numerous to mention. In recent times manufacturers have begun to take precautions when packaging their materials for transport and use, especially when glass bottles are used. Nowadays, when chemicals are shipped in glass bottles, the bottles are usually packed in cardboard boxes and insulated from potential damage. One-gallon glass containers are usually shipped in what is known as carboys, like the one shown in **Figure 25-36.** Carboys provide

FIGURE 25-35 Typically, chemicals that can cause harm are not packaged in cardboard. This sulfuric acid is one example of a material that can cause harm, but unfortunately is packaged in cardboard.

FIGURE 25-36 To protect the glass bottle, which has a corrosive in it, a carboy is used in case the bottle is dropped.

a protective cover to protect against potential damage during transportation. If the container is dropped, the bottle should survive the fall. Carboys are usually seen in laboratories and in smaller chemical production facilities.

Ensuring the material's compatibility with the container it will be stored in is important, but the one area that usually results in a release is the use of an improper cap. The chemical must not only be compatible with the glass, it must also be compatible with the material the lid is composed of. A variety of materials are used in the manufacturing of lids. Many new glass containers, like the one shown in **Figure 25-37,**

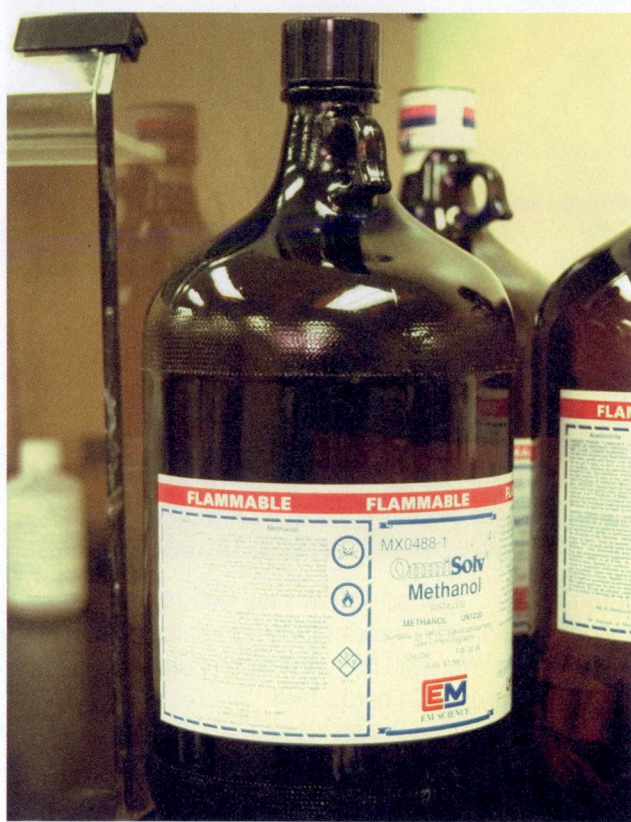

FIGURE 25-37   This glass jar is coated with a plastic coating that will not allow the liquid to spill out if the glass is broken or dropped.

are coated with plastic to avoid the bottle being broken when dropped. Even if the bottle is cracked, the contents are supposed to remain sealed within the plastic coating.

Bags

Bags are also commonly used as containers for chemicals. Bags can be as simple as paper bags or plastic-lined paper bags to fiber bags, plastic bags, and the reinforced super sacks or tote bags. It might be a surprise to open the back of a trailer and find four super sacks like those shown in **Figure 25-38** carrying a material that is classified as a poison. Bags carry anything from food items to poisonous pesticides, and the method of transportation varies widely.

Drums

When discussing hazardous materials, drums are the containers with which most responders are familiar. They vary from 1-gallon (3.8 L) sizes up to a 95-gallon (359.6 liter) overpack drum. The construction varies from fiberboard to stainless steel. The typical drum holds 55 gallons (208 liters) and weighs 400 to 1,000 pounds (181–454 kg). It is possible to get an idea of what a drum may contain by the construction of the drum. **Table 25-8** provides an indication of potential drum contents, but this is not an absolute listing; contents can and do vary from drum to drum.

Dewars Flask

Dewars, **Figure 25-39,** are vacuum flasks that are basically an insulated shipping container. Dewar flasks are nonpressurized, heavily insulated contain-

FIGURE 25-38   It can be quite surprising to open the back of a tractor trailer and find these super sacks. They can hold solid materials, some of which can be toxic.

TABLE 25-8   Drum Contents

Type of Drum	Possible Contents (in Order of Likelihood)
Fiberboard (cardboard), unlined	Dry, granular material such as floor sweep, sawdust, fertilizers, plastic pellets, grain, etc.
Fiberboard, plastic lined	Wetted material, slurries, foodstuffs, material that may affect cardboard or could permeate the cardboard
Plastic (poly)	Corrosives such as hydrochloric acid and sodium hydroxide, some combustibles, foodstuffs such as pig intestines
Steel	Flammable materials such as methyl alcohol, combustible materials such as fuel oil, motor oil, mild corrosives, and liquid materials used in food production
Stainless steel	More hazardous corrosives such as oleum (concentrated sulfuric acid)
Aluminum	Pesticides or materials that react with steel and cannot be shipped in a poly drum (e.g., concentrated hydrogen peroxide or beer)

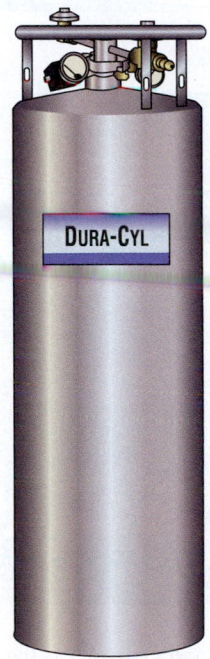

FIGURE 25-39 An example of a Dewars vacuum flask that is non-pressurized and heavily insulated.

ers used for carrying cryogenic materials such as liquefied nitrogen and oxygen. These containers usually have an insulated vacuum space between the inner and outer shell that is used to provide thermal insulation between the contents and the environment.

Cylinders

Cylinders, like those shown in **Figure 25-40,** come in 1-pound (0.45 kg) sizes up to several thousand pounds and carry a variety of chemical products. The product and its chemical and physical properties will determine in what type of container the product is stored.

CAUTION

Other than the hazard of the chemical itself, the big hazard of all cylinders is that they are pressurized.

The pressures range from a low of 200 psi to a high of 5,000 psi (14–345 bar). One of the most common cylinders firefighters run across is the propane tank, which ranges from 1 pound up to millions of gallons. In residential homes, firefighters will find everything from the 20-gallon (75.7 L) cylinder for barbecues to the 100- to 250-pound (45–113 kg) cylinders used as a fuel source for the home. In some areas it is not uncommon to find 1,000-pound (454 kg) cylinders, often buried underground.

Specialized cylinders that hold cryogenic gases (extremely cold) appear to be high pressure, but in reality are low pressure. The bulkiness of the cylinders is a result of the large amount of insulation required to keep the material cold. It is not uncommon to find cryogenic cylinders in convenience stores or fast food restaurants, which presents an unusual hazard. They typically are storing carbon dioxide (CO_2) for the soda-dispensing machines. If the system were to leak, a large amount of CO_2 would be released presenting an asphyxiation risk. Cylinders usually have **relief valves** or **frangible disks** in the event they are overpressurized or are involved in a fire. The United States is one of the few countries that mandates relief valves; most cylinders used in other countries do not have this feature. Incidents aboard ships may involve these types of cylinders as the cylinders are being trans-shipped and are destined for delivery in another country. Most communities, regardless of their size, have cylinders of chlorine and sulfur dioxide used in water treatment. These cylinders come in 100- to 150-

FIGURE 25-40 Cylinders present additional risks to responders because not only can the contents be hazardous, but if the cylinder is involved in a fire it may explode.

FIGURE 25-41 The type of vapor cloud commonly referred to as a plume varies with the terrain and buildings in the vicinity. The plume here represents one of the most common types.

pound (45–68 kg) and 1-ton (907 kg) cylinders, which could create a major incident if they were ruptured or suffered a release.

> **NOTE**
>
> The area affected by a 100-pound (45 kg) chlorine cylinder release can be several miles, **Figure 25-41,** causing serious injuries if not fatalities.

There was an attempt to allow the use of composite propane cylinders, both inside and outside of buildings. The draft version of NFPA 58, which is the Liquefied Petroleum Gas code, had language that would have allowed composite propane cylinders to be used in small portable heaters. The final version only allows the use of composite cylinders outside of buildings.

When this standard comes up for revision in 2010 this issue will come up again. These cylinders, which are constructed much like composite breathing apparatus cylinders, can be used in transportation through a special DOT permit. Composite cylinders can carry flammable gases, oxygen, and many other gaseous materials. Composite cylinders are in common use in Europe and are starting to become more popular in the United States. Two major tests have been conducted on these cylinders to determine the outcome if the cylinders were involved in a fire situation. From the initial testing, composite cylinders appear to react to fire situations better than steel cylinders. As is discussed later in this section, BLEVEs are a major life threat to responders and are common with steel cylinders. None of the composite cylinders BLEVE'd or had a violent rupture during the testing process.

FIGURE 25-42 A tote that holds approximately 300 gallons. This particular one is plastic with a steel frame for reinforcement.

FIGURE 25-43 The most common type of spill occurs when a valve is knocked off, releasing the contents.

Totes and Bulk Tanks

Both **totes** and **bulk tanks** are becoming more common, sized as they are between drums and tank trucks, and are used for a variety of purposes. These totes and tanks are also called intermediate bulk containers and have capacities between 119 gallons and 793 gallons (450–3,001 liters). Used in industrial and food applications, they hold flammable, combustible, toxic, and corrosive materials. They are constructed of steel, aluminum, stainless steel, lined materials, poly tanks, and other products, **Figure 25-42.** The usual capacity is approximately 300 gallons (1,136 liters). They are transported on flatbed tractor trailers or in box-type tractor trailers.

A common incident with totes can occur during offloading. Tanks are offloaded from the bottom through a swinging valve, such as the one shown in **Figure 25-43.** It is a common occurrence for this valve to swing out during transport and get knocked off during movement.

One unusual tote is made to transport calcium carbide, a material that when it gets wet forms acetylene gas, which is reactive and very flammable.

Pipelines

Pipelines vary in size and pressure, but can be sized between ½ inch (1.27 cm) and more than 6 feet (1.9 meters). They are commonly buried underground. The most common products they carry are natural gas, propane, and assorted liquid petroleum products. The larger petroleum pipelines originate in Texas and Louisiana and then proceed up throughout the East Coast. The West Coast also has its share of large pipelines, with Alaska having a majority. Pipelines can originate from any bulk storage facility and can cross many states; Some type of pipeline system is found in every state. Because the amount in the pipelines varies, it is important that first responders know the location of the pipelines and emergency contact names and phone numbers so if there is a suspected problem they can notify the pipeline owner immediately.

Radioactive Material Containers

The transportation of radioactive or nuclear materials is highly regulated by the DOT and the **Nuclear Regulatory Agency (NRC).** There are several types of containers that are used to transport these types of materials, all strengthened to keep the material inside the container. Radioactive material that is classified as low-level radioactive material, typically waste material, is required to be transported in what is called a strong, tight container. This container can be made of metal or wood but should be reinforced to prevent damage to an inner container holding the radioactive material. These materials would present little risk if they escaped the container and only emit radiation just above background levels.

Another type of package is the **excepted packaging** shown in **Figure 25-44,** which is designed to survive the normal conditions of transport. This packaging is designed to hold materials that have **low specific activity** (emitting low levels of radiation), also called **surface contaminated objects (SCO).** These are normally small-quantity shipments that contain natural or depleted uranium, or natural thorium. In some cases, the package does not have to be labeled since the material does not meet the DOT requirement for labeling and placarding. Industrial packaging is a container that is designed to be more substantial and has to pass some basic testing procedures, such as the drop test.

FIGURE 25-44 Excepted packaging is used to transport low-risk radioactive materials; the packaging is exempted from DOT regulations. *(Courtesy of IAEA)*

FIGURE 25-45 Industrial packaging is used to transport low-risk radioactive materials, typically waste. *(Courtesy of Kirstie Hansen, IAEA)*

FIGURE 25-46 Type A packaging is designed to hold low-risk radioactive material and offers moderate protection against accidents. *(Courtesy of John Mairs, IAEA)*

Industrial packaging, shown in **Figure 25-45,** usually consists of metal boxes or drums. Industrial packaging is also used to transport low specific activity (LSA) or SCO) materials. Commonly shipped materials are slightly contaminated tools, medical and research isotopes, and soil waste. For higher levels of radiation, a container known as a **Type A container,** shown in **Figure 25-46,** is used. The Type A container can be made of cardboard, wood, or most commonly, metal drums. The Type A containers are used to transport radioactive material (RAM) and are tested to withstand minor accidents. The level of radiation is still low because only materials that would not present a significant health risk can be placed in a Type A container. Materials transported in Type A packages are usually medical isotopes or industrial, agricultural, or research materials. The highest level of protection is offered by a **Type B container,** shown in **Figure 25-47,** and is used for the transportation of larger or more danger-

ous types of radioactive materials. The Type B containers must meet all types of testing procedures and must be able to withstand a severe accident. The Type B container is either a metal drum or a large, heavily shielded transport container. The Type B container may have 10 inches of lead shielding. **Spent nuclear fuel** is transported in Type B containers.

Type C packages are rarely used and are constructed to contain and transport the most hazardous amounts of radiation. Life threatening conditions exist if the contents are released or package shielding fails. Type C containers must pass very stringent testing criteria due to the life threatening products that they carry and are designed to withstand major impact without container failure.

Highway Transportation Containers

The type of vehicle provides some important clues as to the possible contents of the vehicle. The most common truck is a tractor trailer or a box truck. There are four

FIGURE 25-47 Type B containers, such as this container being loaded on a ship, are designed to hold high-risk radioactive materials. They are hardened transport cases capable of withstanding severe accidents. *(Courtesy of Pat Pavlicek, IAEA)*

FIGURE 25-48 Although in most cases refrigerated trailers are carrying food, there exists the possibility that they may have chemicals that require refrigeration to remain stable.

basic tank truck types that carry hazardous materials, with some additional specialized containers. Tractor trailers carry the whole variety of hazardous materials and portable containers. They can carry loose material that is not contained in any fashion other than by the truck itself. They can carry portable tanks that hold 500 gallons (1,893 liters) or bulk bags that weigh several tons.

CAUTION

When dealing with tractor trailers, the rule is to expect the unexpected.

Nothing is routine. Until the driver has been interviewed, the shipping papers looked at, and the cargo actually examined, a firefighter cannot confirm or deny the presence of hazardous materials. Sometimes the signage on the trailer is an indication of the possible

contents, and a trailer that has several placard holders is a likely candidate for hazardous materials transport. If a tractor trailer is refrigerated like the one shown in **Figure 25-48** and is carrying hazardous materials, extra precautions must be taken because the materials may require the cold temperature to remain stable.

Leakage is often found in containers known as **intermodal containers** or, more commonly, **sea containers.** These types of containers are typically used on ships, then offloaded onto a tractor trailer or loaded directly onto a flatbed railcar. These containers come from all over the world and can contain any imaginable commodity.

NOTE

The types of containers that are shipped in these trailers vary from bags, boxes, and drums to bulk tanks and cylinders.

Although the driver is supposed to have the shipping papers for the contents, on occasion the paperwork is missing or is sealed in the back of the trailer.

FIGURE 25-49 Many of the differences between a 306 and a 406 tank truck are internal. The biggest difference is that the dome covers on a 406 are less likely to open during a rollover, although the skin of the tank is thinner. *(Courtesy of Maryland Department of the Environment)*

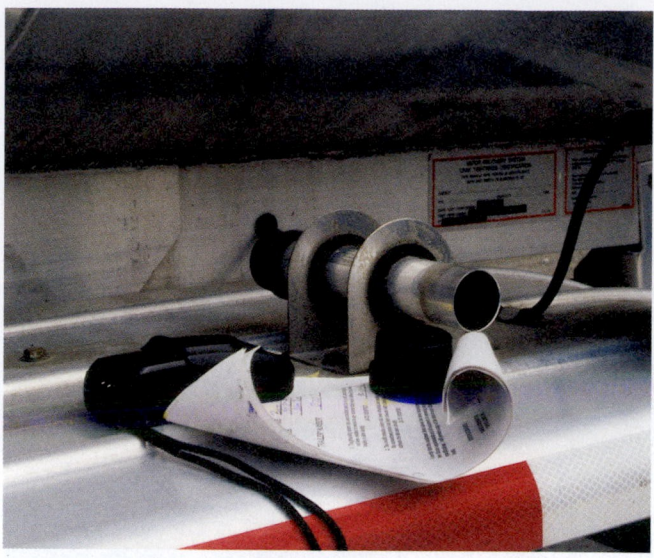

FIGURE 25-50 In most cases the shipping papers will be with the driver in the cab of the truck, but they may be in a special tube located on the trailer.

Determining the contents of a trailer can be very difficult and frustrating.

Tank trucks carry several hundred gallons up to a maximum of 10,000 gallons (37,854 liters). The DOT allows maximum loads by weight not by gallons, so the actual capacity varies state to state. The most common tank truck is the gasoline tank truck, which usually carries 5,000 to 10,000 gallons (18,927 to 37,854 liters). In September 1995 the DOT changed the regulations covering tank trucks, so two systems are used for identifying tank trucks, **Figure 25-49.** In the past the DOT wrote specifications as to how the manufacturer should build a tank truck. Today, they have established performance-based standards for the construction. The DOT allows trucks that were manufactured before the new regulation to remain on the road, as long as they meet the applicable inspection requirements. The four basic types of tank trucks are:

1. DOT-406/MC-306 gasoline tank truck
2. DOT-407/MC-307 chemical hauler
3. DOT-412/MC-312 corrosive tanker
4. MC-331 pressurized tanker

The DOT numbers are the more recently manufactured tanks, and the MC (motor carrier) numbers identify those tanks manufactured prior to September 1995. There are some differences in construction of the two types—some of which favor the emergency responders, others favor the shipping company—but overall the newer tanks hold up better during accidents and rollovers. If unsure of the type of tank, all tank trucks have **specification (spec) plates,** which list all the pertinent information regarding that tank. The spec plate in many cases is located on the drivers side of the tank near the front of the tank. In some

cases, shipping papers or MSDS are located in a paper holder as shown in **Figure 25-50** (tube).

DOT-406/MC-306

This is the most common tank truck on the road today, **Figure 25-51** and for that reason, along with the large number of shipments, suffers the most accidents. Although the most common products carried on these trucks are gasoline and diesel fuel, almost any flammable or combustible liquid can be found on these types of trucks. This truck is known for its elliptical shape and is usually made of aluminum. Older style tanks used to be made of steel, which presented an explosive situation known as a BLEVE. The tanks generally have three or four separate compartments, but two to five compartments are not uncommon. Newer style tanks have considerable vapor recovery systems as well as rollover protection, and in some cases these features are combined. The valving and piping are contained on the bottom of the tank, as shown in **Figure 25-52,** and the number of pots is indicated by the number of outlets as well as the number of manhole assemblies located on top of the tank. The maximum pressure that this truck can hold is 4.2 psi. The compartments are separated by bulkheads, which usually fail during a rollover situation. In the event that the shipper wishes to carry mixed loads, they may add a double bulkhead to protect against the mixing of chemical products. During a violent rollover, even this double bulkhead is likely to fail. Although in most states the shipper is not allowed to ship a mixed load of flammable and combustible

FIGURE 25-51 The DOT-406/MC-306 is the most common truck on the highway today and is referred to as a gasoline tanker. It can and frequently does carry other types of flammable and combustible liquids. *(Courtesy of Maryland Department of the Environment)*

FIGURE 25-52 Most 406/306 trucks have more than one tank (pots) and the number of tanks is indicated by the number of valves or the number of dome covers on top of the truck. *(Courtesy of Maryland Department of the Environment)*

materials, widely differing loads can be encountered from compartment to compartment.

> **CAUTION**
>
> During a rollover, the bulkheads may shift, allowing all products to mix.

On initial examination it may appear that only one tank has ruptured and is leaking, but it is possible to lose the entire contents of the tank through a leak in one compartment, like that shown in **Figure 25-53**. In the majority of rollovers experience has shown that at least one bulkhead has separated in almost every accident, resulting in product mixing. In most cases this is not a big problem because it may only be the mixing of different grades of gasoline. Within the individual tanks

themselves, baffles limit the movement of the product within the tank. The emergency shutoff valves, **Figure 25-54,** are located on the driver's side near the front of the tank and near the piping on the passenger side.

DOT-407/MC-307

The DOT-407/MC-307 tankers are the workhorses of the chemical industry, **Figure 25-55.** They carry a variety of materials including flammable, combustible, corrosive, poisonous, and food products. The two basic types of chemical tanks are insulated and uninsulated. The insulated tank can have a number of additional concerns that do not apply to an uninsulated tank. These tanks usually hold 2,000 to 7,000 gallons (7,571 to 26,498 liters), lower amounts than the 406/306 because most of the products they carry are heavier than petroleum products. The average amount found in these tanks is 5,000 gallons (18, 927 liters).

The uninsulated tank is round and has stiffening rings around the tank. The offloading piping is located on the bottom or off the rear of the tank. These tanks are composed of only one compartment, and its loading piping and manhole are usually on the top in the middle. These tanks generally do not hold up as well during rollovers and accidents as the insulated version. The shell is made of stainless steel and can hold pressures up to 40 psi (2.8 bar).

The insulated tank, **Figure 25-56,** is a covered version of the uninsulated tank, although in some cases it has a slightly smaller inner tank. The inner tank is made of stainless steel with about 6 inches (15.24 cm) of insulation, and the outer shell is made of aluminum. The inner tank may also be lined with a fiberglass or other liner depending on the chemical that is carried. Due to the aluminum and insulation these tanks hold up remarkably well during rollovers.

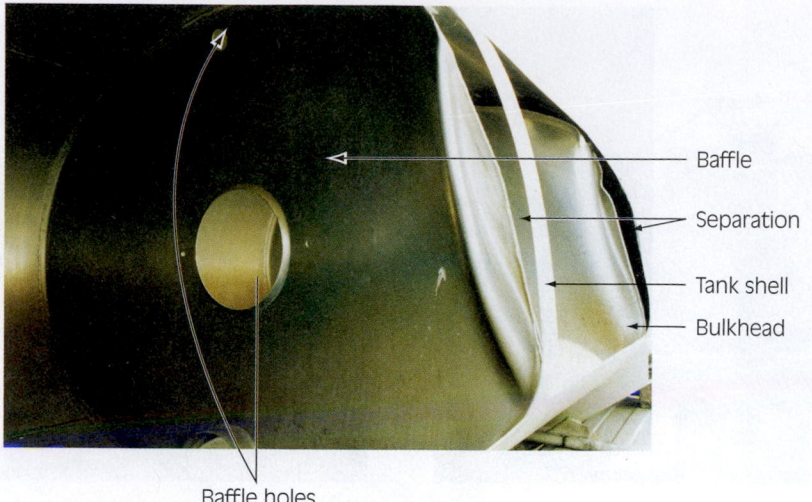

Baffle

Separation

Tank shell

Bulkhead

Baffle holes

FIGURE 25-53 When a truck rolls over on its side, the internal baffles and bulkheads may shift. The internal baffles reduce the amount of sloshing the liquid will do when the truck starts and stops. The baffles will allow product to move through the holes in the baffle wall. A bulkhead separates the compartments and does not allow the products to mix. Shown in the photograph is an MC-306 that rolled on its side. The side wall of the tank has been cut away, revealing a baffle and a bulkhead. Also shown is the separation that took place between the tank shell and the baffles/bulkhead. When the tank wall was cut away, strips of the tank shell were left where they should have connected to a baffle or bulkhead.

FIGURE 25-54 On most trucks there is a minimum of one emergency shutoff, and with most there are two. The most common location is near the driver's door; the other is usually located near the valve area.

FIGURE 25-55 On the right is an insulated MC-307/DOT-407, and on the left is an uninsulated MC-307/DOT-407. Both trucks hold comparable products, but the insulated one holds products that are heated or may require heating to offload.

SAFETY

One of the major problems with this insulated tank is that in the event of a leak, the location where the material leaks out of the outer shell is usually nowhere near the leak on the inner shell.

Within the insulation there can be heating and cooling lines depending on the product being carried, **Figure 25-57.** Products such as paint are shipped at

170°F (77°C) and need to be heated to that temperature for offloading. Some products need to remain at certain temperatures to remain stable, and first responders need to be aware of any special requirements.

In general, both types of tanks have rollover protection, similar piping, and relief valves that serve two purposes: overpressurization and vacuum protection. The emergency shutoffs are located near the front of the tank on the driver's side and near the offloading piping.

FIGURE 25-56 This insulated version is identical to the uninsulated tank but has an aluminum cover and insulation. Note the differences in these two trucks. The one on the right has safety railings around the manhole. Although not an absolute rule, the truck on the right would carry more dangerous products and would have other added safety features. Most of these items are not required but were added by the trucking company.

FIGURE 25-57 Products carried in an insulated 407/307 need to remain either heated or cooled. Some products may need to be heated for offloading. The heater coils that run around the tank heat the product up, allowing it to be offloaded.

FIGURE 25-58 The DOT-412/MC-312 is designed to carry corrosives and is similar in design to the uninsulated 307, although smaller. The inner tank may be lined with a variety of materials to prevent the corrosive from attacking the tank.

DOT-412/MC-312

These tankers, **Figure 25-58,** carry a wide variety of corrosives, both acids and bases. These tankers are round and are smaller in diameter than the 306s and 307s due to the weight of the corrosives they carry. Most petroleum products weigh about 8 pounds per gallon (1 kg per liter), while some corrosives weigh up to 15 pounds per gallon (1.8 kg per liter). Because of the weight, the stiffening rings used are generally bulkier than the ones used on DOT-407 tanks. These tankers are constructed of a single tank that carries up to 7,000 gallons (26,498 liters), with most tanks holding 5,000 or fewer gallons (18,927 liters). The tanks are made of stainless steel and are usually lined with a rubber or ceramic coating to protect against corrosion. The piping can be on top of the tank located in

the middle, but is usually located on the end of the tanker. The piping is usually contained within a housing that includes the manhole and offload piping. This housing protects the piping in the event of a rollover, **Figure 25-59.** The area around the manhole is usually coated with a material that resists the chemical being carried and is usually a black, tar-like coating.

MC-331

MC-331 tanks look like bullets and are noted for their rounded ends and smooth exterior, as shown in **Figure 25-60.** They carry liquefied gases that are liquefied by pressure. One of the most common products carried in this type of tank is propane. They also carry ammonia, butane, and other flammable and corrosive gases. These tanks carry up to 11,500 gallons

(45,532 liters) and have a general pressure of 200 psi (14 bar), although it can be as high as 500 psi (34.4 bar). The relief valves are located on top of the tank at the rear of the trailer, and they sometimes malfunction during a rollover. The tanks are made of steel, are uninsulated, and are heavily fortified with heavy bolts used in the piping and manholes. The tanks are usually painted white or silver to reduce the potential heating by sunlight.

The tanks contain a liquid along with a certain amount of vapor. The most liquid the tank is supposed to have is 80 percent to allow for expansion when heated. The liquid in the tanks is at atmospheric temperature but on release can go below 0°F (–18°C) and could cause frostbite upon contact. The pressure

in these tanks is of concern when responding to incidents involving these tanks.

FIGURE 25-59   The black coating around the manhole indicates that a corrosive is being carried. It is used to protect the tank from spillage. It is not required nor will it be found on all corrosive tanks.

FIREFIGHTER FACT

When a propane tank is emptied and, hence, the pressure reduced, the temperature of the propane drops below 0°F (–18°C). Temperature, pressure, and volume are interrelated. Think of an SCBA bottle. When it gets filled it becomes hot because the pressure and volume are increasing. When the SCBA bottle is used, it becomes cold because it is losing pressure and volume. Any time one of the parameters is changed, there is a corresponding change in the other properties. When a pressurized gas is pressurized to a point that it becomes a liquid, as is the case with propane, it allows for a lot of propane to be stored in a small container. When released, however, the temperature will drop because the pressure is decreasing in the container. This is known as autorefrigeration.

SAFETY

If the pressure increases at a rate higher than the relief valve can handle, the tank will explode. These explosions have been known to send pieces of the tank up to a mile, with the ends of the tank typically traveling the farthest, although any part is subject to becoming a projectile.

Boiling Liquid Expanding Vapor Explosion (BLEVE)

When tanks, trucks, tank cars, or other containers are involved in a fire situation there are a number of hazards. One very deadly hazard is known as a

FIGURE 25-60   MC-331 tanks carry liquefied compressed gases such as propane and ammonia. They are made of steel and are designed to carry a variety of products.

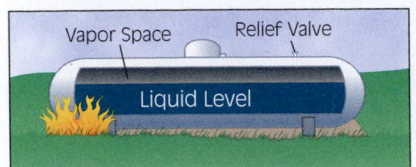
Fire impinging on a propane tank.

As heat increases inside of tank, the pressure also increases. The liquid will begin to boil.

As the pressure increases, the relief valve will open, releasing propane. The propane, being heavier than air, will sink.

The vapor will reach the fire and ignite.

The relief valve will ignite, also causing heat to be on the tank by that flame. The pressure will increase in the tank.

As the pressure in the tank is increasing, the tank may discolor and the pitch of the relief valve will get higher. Eventually the tank will rupture. This is known as a BLEVE.

FIGURE 25-61   Diagram of a BLEVE.

boiling liquid expanding vapor explosion (BLEVE), **Figure 25-61.** A large number of firefighters have been killed by propane tank BLEVEs, and when a BLEVE occurs it usually results in more than one firefighter being killed at a single incident. The type of container and the product within the container will dictate how severe a BLEVE may be. The basis of a BLEVE is the fact that the pressure inside the container increases and exceeds the maximum pressure the container was designed to handle. The contents are violently released, and if the material is flammable, an explosion or large fireball occurs. In the recent past there have been several incidents involving BLEVEs that resulted in emergency responder deaths and injuries, thus emphasizing the need to recognize and prevent this event before it occurs.

Another phenomenon that transpires with containers is known as violent tank rupture (VTR), which occurs with nonflammable materials. The concept is the same as a BLEVE, but there is not a characteristic fireball and explosion. With both a BLEVE and a VTR there is some form of heat increase inside the container, typically from a nearby fire. The fire heats the container, which heats the contents. The contents will boil, which creates expanding vapors, which in turn increases the pressure inside the container. In some containers the relief valve will activate, relieving the pressure. In some cases the pressure inside the tank is greater than the relief valve can handle and the pressure continues to increase.

One of two possibilities can occur with a BLEVE or a VTR. One possibility is that the relief valve will not be able to handle the increase in pressure and the tank will fail. The other possibility is that the fire or heat source that is creating the problem will weaken the container shell, and the resulting increasing pressure will vent at this weakened portion of the container. If the heat source is a fire and the container product is flammable, when the relief valve activates, the raw product coming out of the container typically ignites. This may increase the temperature of the tank, increasing the pressure. It is never advisable to extinguish the fire coming from a relief valve, as that is a safety mechanism. It is possible to cool the top of the tank near the relief valve with an unstaffed hose stream.

The difference between a BLEVE and a VTR occurs when the container fails. A BLEVE occurs with flammable liquids, such as propane. When the container fails, the vapors of the flammable liquid ignite, creating the explosion. How the container ruptures and the amount of product released will determine how severe the explosion will be. With a VTR the container will fail. Since the product is nonflammable, the product will not ignite. The only event is a rupture of the container, spilling its contents. The container can still rocket, and the resulting release of pressure can be violent. A VTR can occur with a container of water or other "nonhazardous" material. As an example a 55-gallon (208 liter) steel drum of water, when heated, can travel several hundred feet depending on where the release point is on the container. In most cases the bungs (screw-top caps) will release, flying a considerable distance, and the container will remain mostly intact.

The failure of the container when impinged from a fire usually occurs as the fire is impacting the tank in the vapor space. When heat is applied to a section of the tank that has no internal mechanism to provide cooling, failure of the steel can occur. When fire impinges on the liquid portion of the tank, the liquid will distribute the heat spread internally and the steel tank typically will not weaken. The problem is that responding firefighters do not usually know the liquid level, and one cannot easily predict when a tank may fail. Any fire impingement on a tank is a serious problem, and withdrawing from the scene may be the best course of action.

This brings up another interesting note: A casual observation of BLEVEs indicates that most BLEVEs and firefighter deaths occur within the first few minutes of arrival. The clock starts ticking on the BLEVE time bomb from the first minute heat is applied to the tank. The clock does not start with the 9-1-1 call or the arrival of firefighters. The critical time for safety may have already passed before firefighters arrive on the scene. If firefighters arrive at an incident involving a propane tank on fire or being impinged by fire, they are in extreme danger. If the relief valve is not operating, the danger is even more pronounced. Operating in close proximity to a tank in this situation can be a fatal mistake. Firefighters should follow this risk/benefit analysis: *Risk a lot to save a lot, and risk a little to save a little.*

The dangers associated with a BLEVE are:

- The fireball can engulf responders and exposures.
- Metal parts of the tank can fly considerable distances.
- Liquid propane can be released into the surrounding area and be ignited.
- The shock wave, air blast, or flying metal parts created by a BLEVE can collapse buildings or move responders and equipment.

The following stories illustrate the consequences of BLEVEs and VTRs:

- One recent event involved the death of two firefighters and injuries to seven other emergency responders. The location was a farm, at which a propane tank was on fire. The relief valve on the tank was operating and the vapors from the relief valve were on fire. About eight minutes after the firefighters had arrived, the tank exploded into four separate parts. The four parts went in four different directions. The two firefighters who died were 105 feet away and were struck by one piece of the tank, dying instantly.

- In two other farm incidents, firefighters lost their lives. In 1993 a fire and BLEVE in Ste. Elisabeth de Warwick, Quebec, Canada, killed three firefighters. In 1997 in Burnside, Illinois, two firefighters lost their lives in a fire and BLEVE. In all three of these fatal BLEVE events, the relief valves were operating upon arrival of the fire department.

- Another event worth noting is the 1984 PEMEX LPG Terminal fire in Mexico City, Mexico. An 8-inch (20 cm) pipeline broke while filling a tank at the terminal. The resulting vapor cloud, which was 650 feet by 572 feet by 8 feet (198 × 174 × 2.4 meters), ignited. This resulted in an explosion, which included a ground shock, and a major fire. The terminal had two large spheres, four small spheres, and forty-eight horizontal tanks. About fifteen minutes into the fire the first BLEVE occurred, and for the next ninety minutes tanks continued to BLEVE. Tanks and liquid propane rained down on the adjacent community. The death toll exceeded 500 people, and thousands were injured.

- In 1983 five Buffalo, New York, firefighters lost their lives when a three-story building collapsed. Nine other firefighters were injured in the massive explosion. The force of the explosion caused the first arriving apparatus, including a ladder truck, to be blown across the street. A propane tank inside the building had been struck and was leaking. An unknown ignition source sparked the explosion minutes after fire crews arrived, which caused the building to collapse.

■ In January 2003 a propane truck (MC-331) went over a guardrail and fell to the ground 35 feet (10.6 meters) below. The tank ruptured and the leaking propane ignited, resulting in a 600-foot-high (183 meter) fireball. The resulting explosion moved the truck several hundred feet away from the first impact area. The driver of the truck was killed in the accident.

As can be noted in these stories, BLEVEs and VTRs can result in injuries and fatalities. Any time containers are under stress, such as during a fire, they can fail. Many times they fail violently and with severe consequences. Some containers have relief valves, while others do not. Materials that are highly poisonous such as chlorine will not have a relief valve; they will have a fusible metal plug that vents the pressure of the tank. This fusible metal plug is not like a relief valve that shuts off when the pressure inside the tank is decreased. Once a fusible metal plug melts, the contents of the tank come out, no matter the pressure. When a tank is on fire or is being impinged by fire, the tank is being weakened and the contents are being heated, creating increased pressure. Some of the general rules of firefighting and propane tanks (or other containers under pressure) are outlined in JPR 25-1.

JPR 25-1: Fighting a Fire for Tanks and Other Containers Under Pressure

(For step-by-step photos of this skill sequence, see page 1066)

A tank or container can fail at any time, and it is impossible to determine the exact moment it is going to happen.

1. Firefighters should withdraw immediately in the case of rising sound from venting relief valves or discoloration of the tank.

2. Fire must be fought from a distance with unstaffed or unmanned hose holders or monitor nozzles.

■ The tank should be cooled with flooding quantities long after the fire is out. A minimum of 500 gpm (1,893 liters per minute) at the point of flame impingement is recommended by the NFPA.

■ If the water is vaporizing on contact, firefighters are not putting enough water on the tank. Water should be running off the tank if it is being cooled.

■ Firefighters should not direct water at relief valves or safety devices, as icing may occur. Icing would block the venting material, which could cause an increase in pressure inside the tank.

■ The tank may fail from any direction, and any tank that is being exposed to a fire can fail at any moment.

■ For massive fire, it is recommended to use unstaffed or unmanned hose holders or monitor nozzles. If this is impossible, firefighters should withdraw from the area and let the fire burn.

Specialized Tank Trucks

These types of tank trucks are used to carry unique chemicals or chemicals that have to be transported in a certain fashion. When gases are transported, they are transported as liquefied gases, as was described with the MC-331 tank trucks. They can also be transported as refrigerated gases or as compressed gases as will be described in this section. Other trucks are dry bulk which can carry a variety of products from grain to explosives. Materials that required high temperatures are transported in special vehicles to keep them hot. The intermodal series of tanks are cousins to their full size highway tanks but carry the chemicals in a comparable fashion.

MC-338 Cryogenic Tank Trucks

MC-338 tankers are uniquely constructed like the one shown in **Figure 25-62.** They have a tank with an outer shell. The inner container is steel or nickel, with a substantial layer of insulation; the exterior is aluminum or mild steel. The space between the shells is placed under a vacuum to assist in the cooling process. The ends of the tank are flat, and the piping is contained usually at the end of the tanker in a double door box. Relief valves are located on top of the tank, to the rear of the tank. The best way to describe this tanker is to compare it to a Thermos bottle on wheels. Cryogenic materials are gases that have been refrigerated to a temperature that converts them to liquids. Unlike liquefied gases, which use pressure to reduce them to liquids, these are cooled to the point of becoming liquid. To remain liquids, the material must be kept cool. To be a cryogenic material the liquid has to be at least $-150°F$ ($-101°C$) and can be as cold as $-456°F$ ($-271°C$).

The most common products are nitrogen, carbon dioxide, oxygen (liquid oxygen or LOX), argon, and hydrogen. The material inside is kept liquid by vacuum, and when on the road the tank will have a maximum of 25 psi (1.7 bar). As the truck travels, the sun will heat the material, and the pressure will

FIGURE 25-62   The MC-338 carries cryogenic liquefied gases. The tank resembles a rolling Thermos bottle.

increase. As the pressure increases the relief valve will open up, relieving any excess pressure into the atmosphere.

> **NOTE**
>
> It is not uncommon to see a white vapor cloud like that shown in **Figure 25-63** coming from the relief valves while the truck is traveling on the highway or sitting alongside the road. This is a normal occurrence and is not cause for alarm.

When the truck makes a delivery, the pressure needs to be increased to push the liquid out of the tank. The driver opens the piping and allows the material to flow into a heat exchanger, which is located just in front of the rear wheels of the tank. Once in the exchanger, the material will heat up and expand, increasing the pressure in the tank and forcing the liquid into the receiving tank. During transportation and offloading, a large amount of ice will build up on the piping and valves.

Tube Trailers

Tube trailers, **Figure 25-64,** contain several pressurized vessels, constructed much like the MC-331 tank. The DOT now refers to these as Multiple-Element Gas Containers (MEGC). They are constructed of steel and have pressures ranging from 3,000–5,000 psi (138–414 bar). They hold pressurized gases such as air, helium, and oxygen. The piping and controls are usually located on the rear of the trailer, but could be in the front. The typical delivery mode is that the driver will drop a full trailer off at a facility and pick up an empty one for refilling. Although

FIGURE 25-63   When transporting liquids it is not uncommon to see vapors coming from the relief valves. The truck can only travel with the tank at 25 psi or less. As it heats, the pressure increases, triggering the relief valve. This is a normal situation, and the tank will vent until below the 25 psi. When offloading the pressure can be increased to assist with the offloading procedure.

they are not subject to BLEVEs because they only contain a gas, if involved in a fire, tube trailers can experience a VTR and rocket in the same fashion as a BLEVE.

Dry Bulk Tanks

Dry bulk tanks resemble large, uninsulated MC-307s in shape, with bottom hoppers to unload the product, as shown in **Figure 25-65.** The tanks hold dry products The most common products are fertilizers, lime, flour, grain, and other common dry products.

FIGURE 25-64 The tubes on this trailor contain pressurized gases. The pressure can vary from 3,000 to 5,000 psi.

The potential hazard when dealing with these tankers is predominantly environmental but at times these tankers contain toxic materials. They are usually offloaded using air pressure either from the truck itself or at the facility.

Hot Materials Tanker

Hot materials tankers vary in that they can be modified MC-306s, 307s, 406s, 407s, or dry bulk containers, **Figure 25-66.** They may have a mechanism to keep the material hot or it may be loaded hot. It may require heating prior to offloading. Common products are tar, asphalt, molten sulfur, and fuel oil #7 and #8.

FIGURE 25-65 Dry bulk tanks carry a variety of projects. Some examples included fertillizers, explosives, and Portland Cement.

FIGURE 25-66 These trucks carry molten products and can be heating the product while driving. This practice is illegal but is found on occasion. The fuels used to heat the product are either diesel/kerosene or propane.

If the material is allowed to cool, it can cause problems for the responders or the shipping company. Tar trucks may be transported with a propane or fuel oil flame ignited to heat the product en route to the job site, although this is illegal in most states.

Intermodal Tanks

Intermodal tanks are increasing in use and carry the same types of products as their highway and rail companions, **Table 25-9** and **Figures 25-67, 25-68,** and **25-69.** They are called intermodal (IM) because they can be used on ships, railways, or highways.

FIGURE 25-67 This is an IMO-101 tank. Like the totes, these are bulk tanks capable of carrying a large quantity of product. These are normally placed on ships, then delivered locally by a truck, although trains can also be used.

NOTE

Smaller intermodal tanks may be found inside of box-type tractor trailers.

Intermodals follow three basic types: nonpressurized, pressurized, and highly pressurized. They are built in two ways. In one, they sit inside a steel frame, called a box-type framework; in the other, the tank is part of the framework, called a beam-type intermodal. Like tank trucks, they are assigned specification numbers, IM-101, IM-102, Spec 51 (specification 51). When used internationally they are called

IMOs. They are made to be dropped off at a facility and when empty picked up for refilling.

Intermodal tube modules are similar in design to the tube trailer previously discussed. These modules carry several pressurized vessels within a steel frame that is built to rigid specifications. Each compressed gas cylinder is independent with its own piping and valves and is capable of holding different specialized gases within a pressure range of 3,000–5,000 psi. These modules operate similar to a cascade system and transport nonliquefied gases under pressure, such as helium, oxygen, and carbon dioxide.

TABLE 25-9	IMO Containers		
Container	**Maximum Capacity (Gal)**	**Pressures (Psig)**	**Products**
IM-101 or IMO Type 1	6,340 (24,000 liters)	25.4–100 (1.8–7 bar)	Non-flammable liquids, mild corrosives, foods, and other products
IM-102 or IMO Type 2	6,340 (24,000 liters)	14.5–25.4 (1–1.7 bar)	Flammable materials, corrosives, and other industrial materials
Spec 51 or IMO Type 5	6,418 (24,295 liters)	100–500 (7–34.4 bar)	Liquefied gases, much like an MC-331
Specialized tanks (Cryogenic tanks are also known as Type 7 tanks)	Varies	Varies	Includes cryogenic tanks and tube banks (small tube trailers) that carry the same products as their highway counterparts

FIGURE 25-68 This is an intermodal tank, commonly referred to as an IM or IMO. This is a bulk tank that carries an average of 3,000 to 5,000 gallons. This tank is an IMO-101, which is an atmospheric tank. The orange panel indicates that it has a United Nations (UN) hazard code of 60 and a UN number of 2572. The code of 60 indicates that the product is a toxic material. The UN number indicates that the product is phenylhydrazine. The name is also stenciled on the sides of the container, and there are toxic placards.

FIGURE 25-69 This is another IMO-101, and the photo shows that the orange panel indicates a hazard code of 66 and a UN number of 2810. The hazard code indicates that the product is a highly toxic material, and there are fifty-two different materials listed for UN 2810. One would have to look at the shipping papers to figure out what is being carried in this tank. The tank is marked "foodstuff only," which means that it is holding a product used in food. It is interesting to note that many of the chemicals listed for UN 2810 are chemical warfare agents such as sarin nerve agent and comparable materials. The most likely candidate, considering the "foodstuff only" label, would be a medical-type product.

A tank truck is designed to function with its wheels down and when it is wheels up, the strength of the tank is in question. The stresses on a tank during a rollover cannot be seen and when rolled over there are more stresses placed on the tank. Gasoline tankers (MC306/DOT 406) are especially susceptible to failure during a righting operation. The principle responsible party is required by law to provide response experts, either from within their company or an outside contractor, to mitigate this type of incident. In some cases, the local public sector hazardous materials team would remain in command and monitor the scene until the scene is no longer an emergency. In some parts of the country, the local hazardous material team is aggressive in this mitigation and performs the offload. No matter who handles the incident, the product needs to be offloaded from the truck.

There are several methods of removing the product from the tank, some easier than others. One method is to remove the product through existing valves and piping, which can sometimes be awkward when the truck is upside down. Another method is to attach a valve assembly to the existing valves and then offload the product. Comparable to this method are flexible bladders that attach to the valves and are connected to piping. Another method is to drill the tank to allow for the offload of the product. Each of the methods has benefits and pitfalls and may not apply in every case. First responders can speed up this process by calling for another compatible tank truck to transfer product into and calling for a vacuum truck or pump to conduct the transfer. These requests should be made in consultation with the responsible party and the hazardous materials team. In some states, box trailers that have any quantity of chemicals, including household commodities, are required to offload prior to righting, as shown in **Figure 25-70.**

FIGURE 25-70   Righting an overturned truck can be dangerous because of fuel and other motor fluids, and there is potential for the cargo to be further damaged. Any hazardous materials present could be released, endangering those in the vicinity and causing environmental damage. *(Courtesy of Maryland Department of the Environment Emergency Response Division)*

Rail Transportation

As with highway transportation there are only a few types of railcars, and they are similar to their highway counterparts. The piping and shape may be the same but that is the extent of the similarities.

> **NOTE**
>
> In rail transportation, the quantities are greatly increased—up to 30,000 gallons (113,562 liters) for hazardous materials and up to 45,000 gallons (170,344 liters) for nonhazardous materials.

Rail incidents usually involve multiple railcars, whereas highway incidents usually involve one or two trucks. The incidents may occur in rural areas, away from water supplies and easy access. Rail incidents will involve multiple agencies, and the local community can expect a large contingent of assistance coming from the state and federal government, which in itself will be difficult to manage.

Railcars come in three basic types: nonpressurized, pressurized, and specialized cars. Although they are categorized in this fashion, the commodities they carry will determine the ultimate use of the car. As with highway transportation, there are dedicated railcars that will be marked with the products they carry.

> **CAUTION**
>
> The term nonpressurized car is actually misapplied because it can have up to 100 psi (7 bar) in the tank.

A nonpressurized car carries chemicals, combustible and flammable liquids, corrosives, and slurries. The easiest way to determine if the car is nonpressurized, such as the one shown in **Figure 25-71,** is to observe whether the valves, piping, and other appliances are located outside of a protective housing. In some cases, the car will be bottom unloaded, with the ability to load the car through the top, or the car may be top loaded and unloaded. The cars will usually have a small dome cover, but the relief valves and other piping are located outside of this dome. Most of the piping is located on the catwalk on top of the car.

Some cars have unique paint schemes such as those shown in **Figure 25-72.** Nonpressurized cars can be insulated and may have heating and cooling lines around the tank. Prior to offloading, the tank may need to be heated to ease the offloading process.

FIGURE 25-71   The indication that this is a nonpressurized railcar is given by the fact that all of the valves are on the outside of the tank and not contained within any protective housing.

FIGURE 25-72   This railcar is white with a red stripe, which is used to indicate hydrogen cyanide but on occasion is used to indicate dangerous materials. This car holds hydrogen fluoride anhydrous, which is a severe inhalation hazard and is corrosive. The term *anhydrous* indicates "no water" and means the hydrogen fluoride is pure.

FIGURE 25-73   The pressurized car is indicated by the valves being contained within the protective housing of the railcar.

Some nonpressurized railcars have an expansion dome. The piping valves and fittings sit on top of this dome, which was constructed to hold any potential expansion. These cars are not in regular service but may be seen in larger industrial facilities that have their own railroad service on site.

Pressurized tank cars, **Figure 25-73,** also carry a wide variety of products, including flammable gases like propane, and poisonous gases like chlorine and sulfur dioxide. The pressurized cars will have pressures in excess of 100 psi up to 600 psi (7–41.3 bar). Most pressure cars that carry flammable materials will be insulated with a spray-on insulation or may be thermally insulated. This insulation is 1 to 2 inches (2.5–5 cm) thick and helps reduce the chance of a

BLEVE during a fire situation. To determine if the railcar is pressurized, the firefighter can look at the spec plate located under the protective housing on top of the railcar, which will list all of the valves, piping, and fittings. A catwalk around the protective housing provides relatively easy access.

Specialized Railcars

Specialized railcars have the same characteristics as highway vehicles, and, in fact, in some cases highway box trailers are loaded onto flatbed railcars. When transported in this fashion they are referred to as trailers on flat cars (TOFC). Regular box trailers as well as refrigerated trailers can be found on flat cars. Much like highway trailers, there are freight boxcars that haul the same products as their highway trailers, with the exception of carrying much larger quantities. Examples of these cars are shown in **Figure 25-74.** The boxcars can carry boxes, cylinders, bulk tanks, totes, and super sacks. Refrigerated railcars are similar to highway and intermodal boxes and are referred to as reefers. They contain their own fuel source. Dry bulk closed railcars are common, and open hopper cars can also be found. Tube trailers for rail are also found, but they are rarely used; the most common is a highway tube trailer set on a flat car. Cryogenics are also carried in railcars and have the same low pressure (25 psi/1.7 bar) as in highway transportation, with the characteristic venting when the pressure increases.

Markings on Railcars

Railroads use the same placarding system as highways, with the exceptions noted in the placard section. The placards are the same size, but additional information on the railcars is more extensive than their counterparts on the highway. The information on

FIGURE 25-74   This specialized railcar holds liquefied carbon dioxide, a cryogenic material.

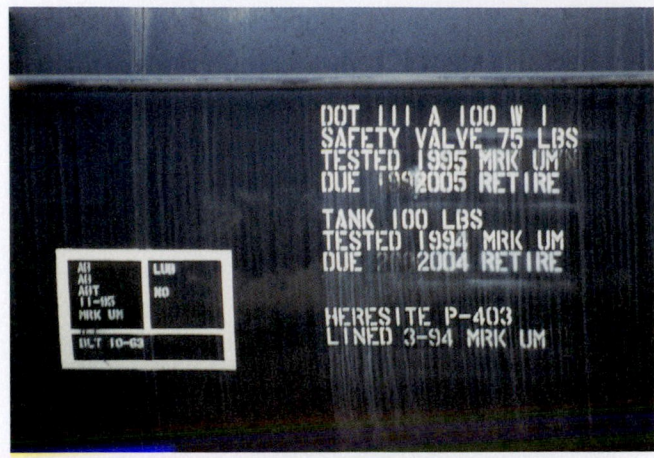

FIGURE 25-75 A railcar has a lot of information stenciled on the side. The FMLX identifies who owns the car. The X at the end indicates that the car is privately owned, that is, the railroad does not own the car. The 15020 is the car number and can be used to cross-reference the shipping papers. The photo on the left is the certification and testing data. The top line, DOT 111 A 100 W 1, provides most of the information. The tank is a DOT 111 specification, which means that it is a nonpressurized car. The A indicates the type of couplers, and the 100 is the maximum pressure for the car. The W and the 1 indicate the type of welds and other tank construction information. The last line is also important, and it shows that the tank was lined on 3-94, which means that it has an internal lining. In a derailment the lining could separate from the tank and the chemical could react with the tank metal.

the car itself is printed larger as compared to highway transport. By looking at the sides of a railcar, you can tell the specification type, maximum quantities, test pressures, relief valve settings, and other pertinent information. These types of markings are shown in **Figure 25-75.** In addition to a placard, the name of the hazardous materials will be stenciled on two sides of a dedicated railcar, as shown in **Figure 25-76.**

> **NOTE**
>
> Certain railcars may be painted in a configuration to identify their hazardous loads.

FIGURE 25-76 Dedicated railcar stencil.

It was once common to paint a car carrying hydrogen cyanide white with a red stripe running around the middle; it was called a "candy-striper." The specification information listed on the car is coded to reveal a lot of information about the car. The tank's owner can be determined by the code on the top line, which consists of four letters followed by a series of numbers. If the fourth letter is an *X*, then the car is owned by someone other than the railroad. The numbers indicate the car number and can be cross-referenced to determine the tank's contents when contacting the railroad. An example of the coding on a railcar follows.

DOT 111 A 60 AL W 1

- *DOT* is the authorizing agency.
- *111* is the class or specification number.
- The following markings in conjunction with the preceding markings will be found only on pressure rail cars. The letters will be found between the authorizing agency and the pressure markings:
  - A—has top and bottom shelf couplers.
  - S—has features of an A car and head puncture resistance.
  - J—has features of A and S and jacketed thermal protection.
  - T—has features of A, S, and J plus spray-on insulation.
- *60* is the pressure rating of the tank.

- *AL* is the material that the tank is made up of:
  - ☐ No letter—carbon steel
  - ☐ AL—aluminum alloy
  - ☐ A—carbon steel
  - ☐ A-AL—aluminum alloy
  - ☐ AN—nickel
  - ☐ B—carbon steel, elastomer lined
  - ☐ C, D, or E—alloy steel
- *W* is the construction type of the tank.
  - ☐ W—fusion welded
  - ☐ F—forged welded
  - ☐ X—longitudinally welded
- *1* is a number that describes fittings, materials, and linings, which can be retained in rail-specific text.

Bulk Storage Tanks

Bulk storage tanks range in size from 250 gallons (946 liters) up to millions of gallons and store a variety of products, the most common being petroleum products. These tanks are seen in residential homes and rural areas, and are common at an industrial facility, such as the one shown in **Figure 25-77.** In residential homes, the most common tank is a 250-gallon (946 liters) home heating oil tank, and some homes or small businesses may have gasoline or diesel fuel tanks that vary from 250 to 500 gallons (946 to 1,893 liters).

The two basic groupings of tanks are in-ground and aboveground. Since the passage of the EPA underground storage regulations, there has been considerable movement to remove **underground storage tanks (USTs)** and replace them with **aboveground storage tanks (ASTs).** If the facility owner elects to keep its tanks underground then additional requirements are placed on them for testing, containment, and leak prevention.

CAUTION

Leaks from these types of tanks can be overwhelming because the leak may not be detected for a period of time, giving it a chance to spread throughout the area, contaminating a large area.

A substantial release from a million-gallon tank of gasoline that escapes a facility would require an enormous response from the emergency responders and environmental contractors.

Underground tanks are usually constructed of fiberglass or are steel coated with an anticorrosion material. New tanks are usually double walled, that is, a tank within a tank, to prevent any spillage into the environment. Ethanol gasoline has created some problems because the ethanol may not be compatible with the fiberglass tanks and may cause a release. As new formulations for gasoline are developed, the true effect on the storage and piping systems may not be known for many years. The piping from the underground system comes up through the ground and its use will determine its route. The offload piping for an UST at a gasoline station comes up through the pump, which **Figure 25-78** demonstrates. The piping is manufactured so that in the event that a car knocks off the pump it will shut off the flow of gasoline and will snap the piping off at or near the ground level. The only spillage of gasoline would be from the amount in the piping aboveground and in the hose, if everything works properly.

FIGURE 25-77   The sizes of tanks vary from a few hundred gallons up to several million gallons. A catastrophic failure of the tank may overwhelm the responders' ability to handle the incident.

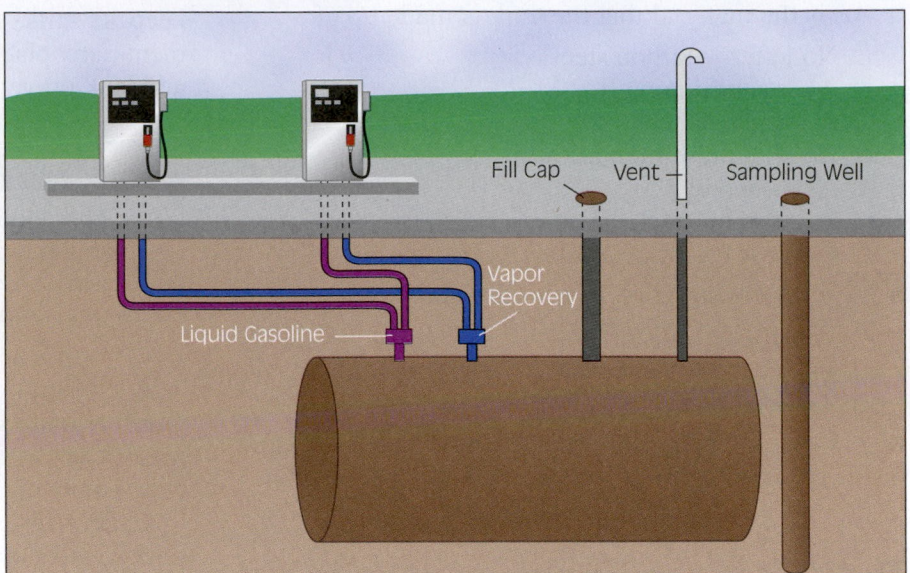

FIGURE 25-78 Piping system of a gas station.

The loading piping is located separately and all of the fill pipes are located in the same area for easy transfer from a tank truck. The fill pipes are usually color coded and marked so that the driver can differentiate between unleaded, unleaded super, and diesel fuel, although the color coding is not standard across the country and varies from company to company. At some location on the property there will be vent pipes for the tanks, generally away from the pumps and near the property line. There will also be other manhole covers approximately 6 to 8 inches (15.2–20 cm) in diameter that have a triangle on the top of the cover. These are inspection wells and typically surround the tank. The holes are drilled at various depths so that leaks can be easily detected by air monitoring of the well, or if water is in the well, by taking a sample. If a facility has had a leak, there may be a large number of these wells on and around the property. Most gas station tanks are 10,000 to 25,000 gallons (37,854–94,635 liters) in size and, if not properly monitored, can slowly release a substantial amount of product over a short period of time.

> **NOTE**
>
> It is not uncommon to find gasoline bubbling up in a basement miles away from the gas station and to find that the release occurred several years ago.

Another common problem arises when farms are redeveloped into housing developments. If unknown USTs are located on the property, they may or may not be discovered during the construction and can eventually leak, causing problems. The tank and environmental industry refers to a **leaking underground storage tank** as a **LUST.**

> **SAFETY**
>
> A recent incident in Baltimore County, Maryland, demonstrates that regardless of how many protection systems may be in place, a release off the property is still possible. One afternoon a gas station was getting a tank filled with fuel, as per normal procedure, and the tank filling went as expected with no problems. The gas station had installed a state-of-the-art alarm system that would indicate a fault in the system and alert the owners to any potential releases. The system goes through a system check after each sale, if no pump is operated for a period of thirty seconds. If a pump is on, or the period between sales is less than thirty seconds, then the system check does not begin to work. During a busy afternoon, it could be estimated that the system check would not function for a long period of time, because at least one of the pumps is operating all the time.
>
> In addition to the system check of each of the storage tank locations, some piping and the pumps all have electronic monitors that detect the presence of a liquid. When a liquid is detected, an alarm activates—the same type of alarm as if the system check fails the system. One of the potential problems with many self-service gas stations is that there is only one attendant, who for many reasons cannot leave the work area to either check on a problem outside or, depending on the alarm panel location, cannot check any alarms. In some gas stations the alarms are in other rooms and cannot be seen or heard by the only attendant. Another problem is that the liquid alarms will activate when it rains, tripping for rainwater.
>
> The pumps that sit on top of the storage tanks have the ability to supply upward of 80 psi (5.5 bar) of pressure to all pump station nozzles, although this is

governed down to an actual working pressure of 10 psi (0.7 bar) at all pumps. In this particular incident, a flange that comes out of the tank pump failed and separated from the tank pump. From the pump discharge there is a 2-inch- (5-cm) diameter opening, which at 10 to 80 psi (0.7–5.5 bar) could pump a considerable amount of fuel in a short time. One theory is that a customer may have set a nozzle down after not being able to pump fuel, or the amount of fuel being pumped was inadequate. This action would have resulted in the tank pump continuing to pump. A liquid alarm sounded for a number of areas at the gas station. The attendant notified the pump repair company as per protocol, but a response was delayed due to a number of extenuating circumstances.

Once the repair company arrived they found the tank empty; it was at this time the clerk informed them of the recent delivery. It was later found that an estimated 4,500 gallons (17,034 liters) were lost and were causing flammable vapor readings in a several-block area, although all the gasoline was later recovered. Within the gas delivery system, this station had the best protection in place, and had just retrofitted the station with a new piping system, but it shows that no matter how many protection systems are in place, it is still possible to have a release.

FIGURE 25-79 Due to increased environmental concerns, many tanks are being placed aboveground and are called aboveground storage tanks or ASTs.

FIGURE 25-80 This is a cone roof tank. It has a weak roof-to-shell seam so that in the event of an explosion the roof will come off, but the tank should remain intact.

Aboveground tanks are becoming more and more common, although they have been used for many years, **Figure 25-79.** They are of two basic construction types—upright and horizontal—and some are nonpressurized or atmospheric tanks and some are pressurized tanks. ASTs hold a wider variety of chemicals as opposed to their UST counterparts, but petroleum products are still a leading commodity stored in these type of tanks. They vary in size from the 250-gallon (946 liters) home heating oil tank to the several million gallon oil tank. In industrial and commercial applications, a containment area is required around the tank. The containment area must be able to hold the contents of the largest tank within the containment. Regular inspection of these areas is required to ensure that rainwater, snow, or ice does not cause a buildup of liquid i n the containment area, which would reduce the containment's ability to hold its intended amount. Depending on the weather conditions, the containment area's gate valve may be left in an open position, which leaves the facility at risk for product to escape the facility through the open drain or gate valve.

Upright storage tanks come in three basic construction types: **ordinary tank, external floating roof tank,** and **internal floating roof tank.** The ordinary tank, **Figure 25-80,** is constructed of steel and typically has a sloped or cone roof to shed rainwater, snow, and ice. The roof and tank shell seam is purposely made weak so that in the event of an explosion, only the top of the tank is relocated. One of the major problems with this type of tank is that when the tank is not full, there is room for vapors to accumulate.

CAUTION

Any time vapors are allowed to accumulate, the potential exists for a fire or explosion.

Some ordinary tanks are purposely constructed without a roof in place. These are typically used in safety vent situations and water treatment areas where a roof may cause additional problems. It is important to identify these tanks during pre-incident surveys so as not to misjudge the severity of an incident. In many chemical processes it is not uncommon to have a storage tank with piping into the tank just to catch overflow or the contents of a system in the event of overpressurization or failure. The materials can be hot and produce large amount of vapors, which, if a roof were present, would allow a buildup of pressure, causing a catastrophic failure of the tank and roof.

External floating roof tanks are used to eliminate the buildup of vapors. They ride on top of the liquid, **Figure 25-81,** and since there is no space between the roof and the liquid, vapors cannot accumulate and create problems. External floaters can be seen from the top of the tanks, and a ladder is affixed to the roof to allow access. A common incident with these types of tanks results from a lightning strike, which can cause a fire in the roof/shell interface area. This type of fire is difficult to extinguish and can result in roof/tank failure if not controlled quickly. It is also possible that water used for firefighting could also sink the roof, causing further problems.

Internal floaters are constructed in the same manner as external floaters but have an additional roof over the top of the tank. An example is shown in **Figure 25-82.** The type of roof varies, but is usually a slightly coned roof with vent holes along the outer edge of the tank shell, or a geodesic type of roof that is usually made of fiberglass. The internal floater suffers from the same type of fire problem as the external, although it is a reduced risk. If a fire were to start in the roof/shell interface, the roof makes it more difficult to extinguish.

FIGURE 25-81   This is an open floating roof tank, in which the roof floats on top of the product. This reduces the release of vapors, as there is no vapor space, and reduces the fire potential.

FIGURE 25-82   This is a covered floating roof tank, which is the same as an open floating roof tank but has a cover to keep out snow, rain, and debris. Another term for this type of tank is *geodesic domed tank.*

Specialized Tanks

Specialized tanks are a combination of the tank types discussed earlier in this section and include pressurized tanks and cryogenic tanks. The larger pressurized vessels are divided into two categories: low-pressure and high-pressure tanks. The low-pressure tanks hold flammable liquids, corrosive liquids, and some gases, up to 15 psig (1 bar). Common high-pressure commodities are liquefied propane, liquefied natural gas, or other gaseous or liquefied petroleum gases. An acid like hydrochloric acid, which has a high vapor pressure, would not be uncommon in a tank like this.

These types of tanks may have an external cover, which appears to be a tank within a tank. The pressurized tanks are larger versions of the propane tanks discussed earlier and can have a capacity of up to 9 million gallons (34,068,706 liters) although less than 40,000 gallons (151,416 liters) is typical. Liquefied petroleum gases are not only stored in pressurized tanks such as the one shown in **Figure 25-83;** these types of gases are also stored in other locations such as caves carved out of mountains, although this is rare. Because these facilities are not required to report under the SARA Title III regulations, firefighters will have to contact their local gas suppliers to find out how they store their gas products.

Upright cryogenic storage tanks, **Figure 25-84,** are located in almost every community, especially if the community has a hospital or medical center. Most hospitals are supplied with liquid oxygen (LOX) through the cryogenic tank. Facilities that sell or distribute compressed gases are likely to have cryogenic tanks. Fast-food restaurants and convenience stores are now using cryogenic tanks to supply carbon dioxide to the soda dispensing machines.

FIGURE 25-83   The specialized tank such as the propane tank shown here has some of the same properties as its transportation equivalents.

FIGURE 25-84   This tank holds cryogenic liquid oxygen and is typical of a cryogenic upright tank.

SENSES

Although the use of senses is listed as one of the four major recognition and identification categories, it is one that presents the most risk. Responders should never use smell, taste, or touch to assist in the hazardous materials identification process. If victims were unfortunately exposed to a material, responders should use the information gleaned from these persons, but only after they have been decontaminated. Responders can use hearing and vision, as well as other senses, to assist in the recognition and identification process. The ability to recognize potentially hazardous situations is key to survival, and there are typically visual clues available to responders. Each of the other recognition and identification categories relies on the use of the visual sense. The venting of a relief valve may only present an audible indication that there is a problem. As the pitch of the relief valve increases, this is an indication that the pressure is increasing inside of the container. Responders should never place themselves in harm's way to smell, taste, or touch hazardous materials; doing so may be the last act they perform.

Many chemicals are in a category known as desensitizers, which overwhelm the olfactory (sense of smell) system. Hydrogen sulfide, a toxic material, has a unique and offensive, rotten egg–like odor. After a few minutes of exposure, a person breathing in hydrogen sulfide will no longer have the ability to smell the toxic gas. The gas is still present, and could even be present in higher levels, but the exposed person no longer has the ability to smell, as their brain has told the olfactory system to ignore the smell. The odorant in natural gas, known as mercaptan, has the same effect.

> **CAUTION**
>
> The lack of an odor cannot be equated with a lack of toxicity. Many severely toxic materials are colorless and odorless.

CHEMICAL AND PHYSICAL PROPERTIES

Although not intended to be a full chemistry lesson, the material in this section is a key component of safety. The chemical and physical properties outlined here are appropriate for a firefighter's level of response. The identification and use of these key terms can determine the outcome of an incident and firefighter well-being. As a firefighter progresses up through the response levels the need for additional chemistry also increases. When in doubt, the firefighter should consult with a hazardous materials team or other resources such as Chemtrec or a local specialist. A lot of the terms to be discussed next can be applied throughout the entire firefighter text. The basis of a fire is a chemical reaction. The better that firefighters understand this chemical reaction, the better off they will be, which will also benefit the citizens of their communities.

States of Matter

The basic chemical and physical properties that are important to understand are the **states of matter: solid, liquid,** and **gas** as shown in **Figure 25-85.** The severity of an incident can be determined by knowing if the material is a solid chunk, a pool of liquid, or an invisible gas.

NOTE
The level of concern rises with each change of state; relatively speaking, a release of a solid material is much easier to handle than a liquid release, and it is nearly impossible to control a gas.

The control methodology for each state increases in difficulty from simple controls with a solid, to difficult with a gas. Evacuations have to be larger for releases involving gases, whereas a minimal evacuation would be required for most solid materials.

How the chemicals can hurt someone also varies with the state of material. Solids usually can only enter the body through contact or ingestion, although inhalation of dusts is possible. Liquids can be ingested, absorbed through the skin, and, if evaporating, inhaled. Gases on the other hand can be absorbed through the skin, and inhaled, and to some extent ingested.

Adjoining the states of matter are melting point, freezing point, **Figure 25-86,** boiling point, **Figure 25-87,** and condensation point. All of these are related because they are the points at which a material changes its state. The **melting point** is the temperature at which solid must be heated to transform the solid to the liquid state. Ice, for instance, has a melting point of 32°F (0°C). The **freezing point** is the temperature of a liquid when it is transformed into a solid. For water, the freezing point is 32°F (0°C). The actual temperatures vary by the tenths of a degree, but are very close. The **boiling point** is reached when the liquid is heated to the point at which evaporization takes place, that is, the liquid is being changed into a gas. Another way of defining boiling point is the temperature of a liquid when the vapor pressure exceeds the atmospheric pressure and a gas is produced. Water boils at 212°F (100°C) and changes into a gaseous state. The important thing to remember about boiling points is the fact that when the liquid approaches this temperature, vapors are being produced that can cause larger problems.

Vapor Pressure

Out of all of the chemical and physical properties, vapor pressure is one of the most important to a hazardous materials responder. If a product has a high **vapor pressure,** it can be very dangerous. A material with a low vapor pressure is typically not a major concern. Vapor pressure has to do with the amount

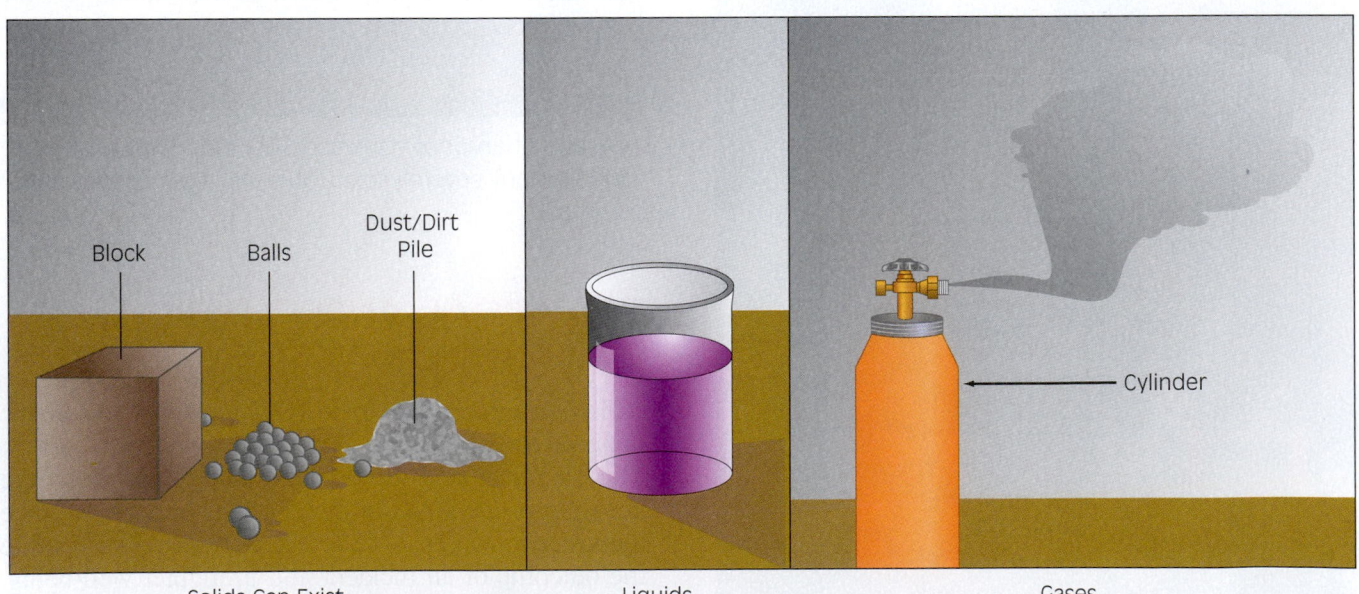

Block   Balls   Dust/Dirt Pile

Cylinder

Solids Can Exist in a Variety of Forms   Liquids   Gases

FIGURE 25-85 States of matter.

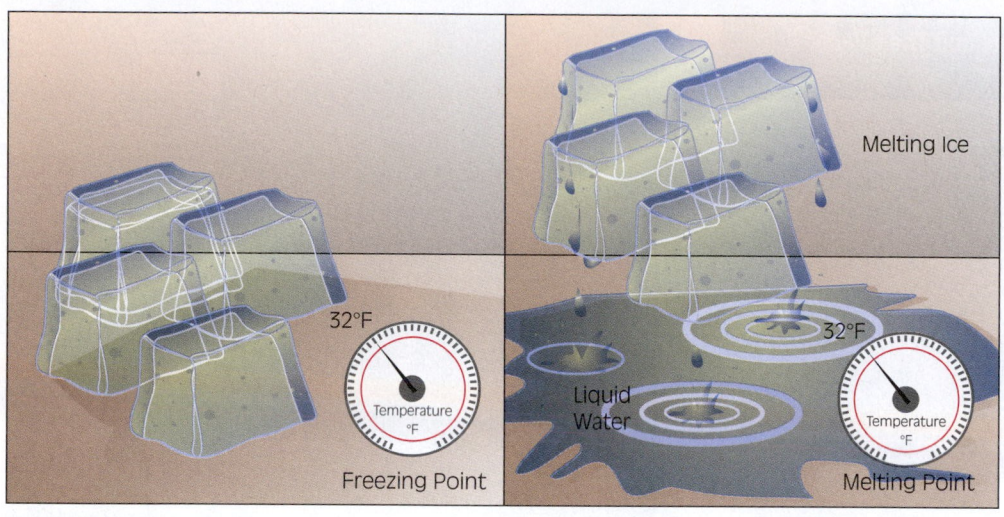

FIGURE 25-86 Melting and freezing points.

of vapors released from a liquid or a solid, **Figure 25-88.** The true definition is the pressure that is exerted on a closed container by the vapors coming from the liquid or solid. Vapor pressure can be related with the ability of a material to evaporate, in that the material is not really disappearing, it is just moving to another state of matter. Chemicals like gasoline, acetone, and alcohol all have high vapor pressures, whereas diesel fuel, motor oil, and water all have low vapor pressures.

Vapor pressures are measured in three ways: millimeters of mercury (mm Hg), pounds per square inch (psi), and atmospheres (atm). Normal or average vapor pressures are 760 mm Hg, 14.7 psi, and 1 atm. Although these figures are used to describe normal vapor pressure, they best describe atmospheric pressure. The temperature used is 68°F (20°C), which is the standard temperature. If a temperature is not provided, then it can be assumed it is 68°F. Chemicals that have a true vapor hazard are those in excess of 40 mm Hg, and they are considered volatile. Chemicals with a vapor pressure of less than 40 mm Hg do not travel throughout the area, but if close enough can harm through inhalation; they still present extreme toxicity through skin absorption. Chemicals with a vapor pressure above 40 mm Hg can be considered

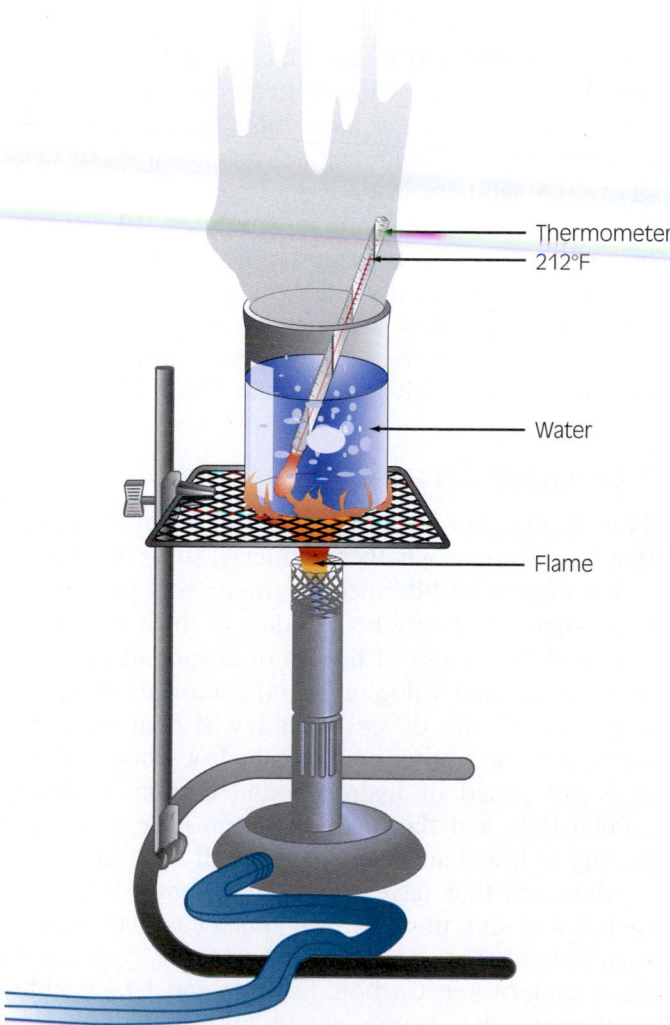

FIGURE 25-87 Boiling point.

A material with a vapor pressure is pushing against the sides of the container. If the vapor pressure is high and the material is in the wrong type of container, the container could fail.

FIGURE 25-88 Vapor pressure.

TABLE 25-10   Common Products and Their Vapor Pressures

Name	Vapor Pressure @ 68°F/20°C (mm Hg)	Boiling Point (°F)
Water	25*	212 (100°C)
Acetone	180	134 (57°C)
Gasoline	300–400	399 (204°C)
Diesel fuel	2–5	304–574 (151–301°C)
Methyl alcohol	100	149 (65°C)
Ethion (pesticide)	0.0000015	304 (151°C) (decomposes)
Sarin nerve agent	2.1	316 (158°C)

*The vapor pressure of water has been reported in various texts as between 17 and 25 mmHg. This text uses 25 mmHg, as it is the highest reported vapor pressure for water.

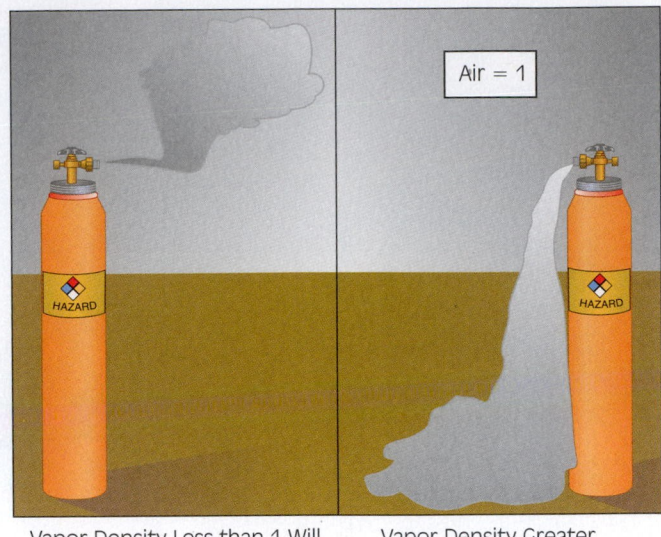

Vapor Density Less than 1 Will Rise in Air   Vapor Density Greater than 1 Will Sink

FIGURE 25-89   Vapor density.

inhalation hazards in addition to any other route of exposure they may possess. **Table 25-10** lists vapor pressures for some common products. The military uses the term **persistence,** which is an indication of the time that a material will remain as a liquid, and is directly related to the vapor pressure of the material. A material that is known to be persistant is one that has a low vapor pressure and will remain a liquid.

A unique chemical phenomenon called **sublimation** occurs when a chemical goes from a solid state of matter to a gas. The material never enters the liquid phase. The sublimation ability means that some solids have a vapor pressure and can move directly to the gaseous stage. Some solids such as dry ice (carbon dioxide) move quickly to the gaseous stage, while others, such as mothballs (naphthalene or paradichlorobenzene), move slowly.

Vapor Density

It is easy to confuse vapor pressure with vapor density, but they have two entirely different meanings. Vapor density determines whether the vapors will rise or fall, **Figure 25-89.** When deciding on potential evacuations and other tactical objectives (e.g., sampling and monitoring), this is an important consideration. One of the primary reasons that natural gas leaks do not ignite or flash back more often is the vapor density of natural gas. Air is given a value of 1, and all other

gases are compared to air. Gases that have a vapor density of less than 1 will rise in air and will dissipate, whereas gases with a vapor density greater than 1 will stay low to the ground. Natural gas has a vapor density of 0.5, while propane has a vapor density of 1.56, which causes it to seek out any low spots, like gullies or sewers. Propane is more likely to ignite or flash back because it has a greater potential of finding an ignition source. Most gases and vapors will stay low to the ground. There are only eleven common gases that rise in air, which are shown in **Table 25-11.**

Specific Gravity

This chemical property is similar to vapor density in that it determines whether a material sinks or floats in water, **Figure 25-90.** Specific gravity is of prime concern when efforts are being taken to limit the spread of a spill by the use of booms or absorbent material. Water is given a value of 1, and chemicals that have a specific gravity of less than 1 will float on water. Fuels, oils, and other common hydrocarbons (chemicals composed of hydrogen and carbon, typically combustible and flammable liquids) have a specific gravity of less than 1 and, hence, will float on water.

Materials that have a specific gravity of greater than 1 will sink in water. It is much easier to recover materials floating on top of the water as opposed to those underwater. Carbon disulfide and 1,1,1-trichloroethane are two common materials that sink in water. Any material that sinks in water is especially troublesome if it has the potential to reach any groundwater, because remediation (cleanup) efforts are difficult and expensive.

| **TABLE 25-11** | **Common Gases That Rise** | | | |

Gas Name	Vapor Density	Gas Name	Vapor Density
Diborane (B_2H_6)	0.97	Ammonia (NH_3)	0.60
Methane (CH_4)	0.55	Neon (Ne)	0.7
Hydrogen (H)	0.1	Helium (He)	0.138
Hydrogen cyanide (HCN)	0.9	Carbon monoxide (CO)	0.97
Acetylene (C_2H_2)	0.91	Ethylene (C_2H_4)	0.98
		Nitrogen (N)	0.967

Also of concern are materials that are water soluble, that is, have the ability to mix with water, not sink or float. Corrosives and many poisons are water soluble and are difficult to remove from a water source. Examples are provided in the following Firefighter Fact. Alcohol is also water soluble. Materials that are water soluble are difficult to extinguish.

FIREFIGHTER FACT

When dealing with water-soluble materials it is important to protect bodies of water, because the cleanup can be difficult or impossible if the material enters the water. In one incident more than 10,000 gallons (37,854 liters) of sodium hydroxide, a very corrosive material, entered a small stream. The creek had to be dammed, and a neutralizing agent was put into the stream. Any life existing in this mile of stream was killed by the sodium hydroxide, but the environmental damage could have spread for many more miles if it had not been neutralized. Once neutral, the water was released, which eventually led to a larger body of water, where no damage occurred. This took considerable resources and time to accomplish, and luckily the necessary neutralizing agent was located at the site of the spill.

In another incident a very small amount of a pesticide was sprayed near a pond. After a rainfall the pesticide ran into the pond, killing all of the fish. The only method of removing this pesticide is the addition of other chemicals that are also hazardous. It will be years before the pond is able to sustain life.

In another incident in Baltimore City a railcar was involved in an accident and released 18,000 gallons (68,137 liters) of hydrochloric acid, which eventually ran into a local stream. The addition of the acid actually raised the pH level of the stream to near acceptable levels, because this stream was already heavily contaminated from a number of other sources.

Note that nothing should be flushed into any body of water without the express permission of the local or state environmental agencies.

Corrosivity

This text has already discussed tanks and containers that hold corrosives, as well as the concerns of dealing with corrosives. Corrosive is a term that is applied to both acids and bases, and is used to describe a material that has the potential to corrode or eat away skin or metal. Some examples are shown in **Table 25-12.** People deal with corrosives every day in that the human body is naturally acidic and they use many corrosive materials in their everyday lives. Acids are sometimes referred to as corrosives, whereas bases are also known as alkalis or caustics. The accurate way to describe a corrosive is to identify the material's pH, which provides some measure of corrosiveness. The designation pH is an abbreviation for a number of terms, including potential of hydrogen, power of hydrogen, and percentage of hydrogen ions. It is used to designate the corrosive nature of a material. Acids have hydronium ions, and bases have hydroxide ions. It is the percent (ratio) of these ions that makes up the pH number. Materials having a pH of 0 to 6.9 are considered acids, and materials with a pH of 7.1 to 14 are considered bases. A material having a pH of 7 is considered neutral. The pH scale is a logarithmic scale, meaning the movement from 0 to 1 is an increase of 10. The movement from 0 to 2 is an increase of 100.

When dealing with a corrosive response, one of the common methods to mitigate the release is to neutralize the corrosive. One thought is that water can be used to dilute and thereby neutralize the spill. This presents two major issues: one, the mixing of water (chemical reaction), and two, the runoff from the reaction. Corrosives and water can be a dangerous combination. One should never add a corrosive to water, as it presents great risk. There may be some spattering and heat generation.

If 1 gallon (3.8 liters) of an acid had a pH of 0, it would take 10 gallons (38 liters) of water to move the material to a pH of 1. To move it to a pH of 2 would

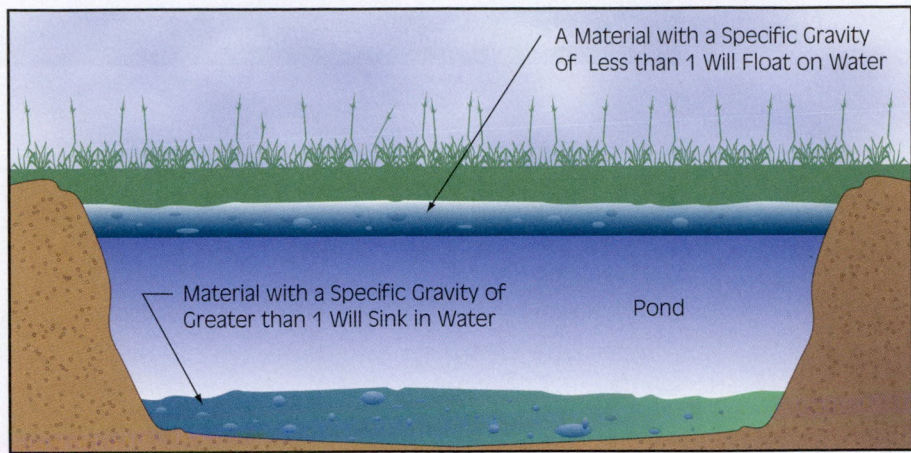

A Material with a Specific Gravity of Less than 1 Will Float on Water

Material with a Specific Gravity of Greater than 1 Will Sink in Water

Pond

FIGURE 25-90   Specific gravity.

TABLE 25-12   pH of Common Materials

Material	pH	Material	pH
Water*	7	Sulfuric acid	1
Stomach acid	2	Gasoline	7
Orange juice†	2	Hydrochloric acid	0
Drain cleaner**	14	Pepsi‡	2
Potassium hydroxide	14	Household ammonia***	14

*Tap water is usually a 7, while rainwater in the Northeast can be 3–6 (acid rain).

†Citric acid is the main ingredient.

**The main ingredient is sodium hydroxide (lye).

‡Phosphoric acid is the main ingredient.

***This is a 5 percent solution of ammonium hydroxide and water.

take 100 gallons (380 liters) of water, and to change the pH to 6 would require 1 million gallons (3,785,412 liters) of water—all for a 1-gallon (3.8 liters) spill. Chemically neutralizing a corrosive spill is the better choice, but even that can present some issues. When neutralizing a strong acid, a weak base should be used to perform the neutralization. A street method of calculation for neutralization is that it takes more than 8,800 pounds of potash to bring 1,000 gallons (3,785 liters) of 50 percent sulfuric acid to neutral, which is more realistic than controlling 1 million gallons (3,785,412 liters) of runoff. These examples are provided for information only; the neutralization of corrosives is a technician-level skill and can be very dangerous.

If the skin or eyes are exposed to a corrosive material, they should be immediately flushed with large quantities of water. This flushing should continue for at least twenty minutes uninterrupted. Some corrosive materials will cause immediate blindness and skin burns, and water should be used to prevent further injury.

Chemical Reactivity

When chemicals mix, they will have one of three types of reactions: exothermic, endothermic, or no reaction. The most common is the **exothermic reaction,** meaning the release of heat.

As discussed in Section I, Chapter 4, fire is a rapid oxidation reaction (exothermic) accompanied by heat and light. When most chemicals mix and provide an exothermic reaction, there usually is not a lot of light but there can be substantial heat. By mixing one tablespoon each of vinegar (acetic acid) and ammonia (base) at the same temperature, an immediate rise in temperature of 10°F (12°C) will occur. When handling oleum (concentrated sulfuric acid) and applying water to a spill, the resulting mixture will bubble, fume, boil, and heat to over 300°F (149°C) the instant the water hits the acid. An **endothermic reaction** is one in which the energy created by the reaction is absorbed and cooling occurs.

Flash Point

A flash point is described as the lowest temperature at which a fuel off-gases an ignitable mixture and that when introduced to a spark or flame will briefly ignite but not sustain burning, **Table 25-13.** This resulting flash fire will ignite just the vapors and self-extinguish once those vapors are burned up. The liquid itself does not burn; it is the mixture of vapors and air that ignites. Following closely behind the flash point is the fire point of a liquid. The fire point is the lowest temperature at which a fuel off-gases an ignitable mixture and, when introduced to a spark or flame, will ignite and sustain burning. In the laboratory, scientists can replicate flash points and fire points; on the street, however, they are usually one and the same.

Autoignition Temperature

The autoignition temperature, sometimes referred to as ignition point, is the lowest temperature at which a fuel will off-gas an ignitable mixture and at which the fuel will self-ignite and continue to burn. Ignition points are much higher than flash points and represent a potential hazard level depending on the temperature. Depending on the context, the term **self-accelerating decomposition temperature (SADT)** may be used, which is essentially the same thing as the autoignition temperature. Regardless of what it is called, it is a level to avoid.

Flammable Range

The two main areas within a flammable range are the lower explosive limit and upper explosive limit. The flammable range is the difference between the two extremes. A fire or explosion needs an ignition source, a fuel (vapors), and air. The proper vapor-to-air ratio is also required or ignition cannot occur. The **lower explosive** (flammable) **limit (LEL)** is the lowest amount of vapor mixed with air that can provide the proper mixture for a fire or explosion. The air monitor

that is used to detect combustible gases is designed to read this level. The **upper explosive** (flammable) **limit (UEL)** is the highest amount of vapor mixed with air that will sustain a fire or explosion.

Each year many natural gas explosions occur throughout the United States. The LEL of natural gas is 5 percent, so if it is mixed with 95 percent air with an ignition source present, an explosion or fire is possible, as shown in **Figure 25-91.** The UEL for natural gas is 15 percent, so 85 percent air is required to result in a fire or explosion. The flammable range for natural gas is 5 to 15 percent, and a fire or explosion can occur at any point in that range, as long as there is enough air to complete the mixture. With 4 percent methane and 96 percent air, there would not be any fire or explosion, nor would there be at 16 percent natural gas and 84 percent air. Acetylene, which is a common gas, has an LEL of 2.5 percent and UEL of 100 percent, which means that when 2.5 percent LEL is reached, the potential for a fire exists. Having less than the LEL is referred to as too lean, and amounts in excess of the UEL are called too rich. Levels less than the LEL are much safer than levels above the UEL.

> **SAFETY**
>
> When confronted with a situation that has a level higher than the UEL, the situation is very dangerous.

Ventilation should be carried out using non-spark-inducing devices, and great care should be taken to minimize any potential electrical arcs such as not using light switches or doorbells.

CONTAINERS AND PROPERTIES

The interrelationship with chemical and physical properties and the containers that hold hazardous materials is important. When there are chemical releases or incidents involving containers, knowing how the materials may react is important for responder safety. The lower the boiling point that a chemical has means

TABLE 25-13	Flash Points of Some Common Materials		
Material	**Flash Point**	**Material**	**Flash Point (°F)**
Gasoline	−45°F (−43°C)	Diesel fuel	> 100°F (38°C)
Isopropyl alcohol	53°F (11.6°C)	Motor oil	300–450°F (149–232°C)
Acetone	−4°F (−20°C)	Xylene	90°F (32°C)

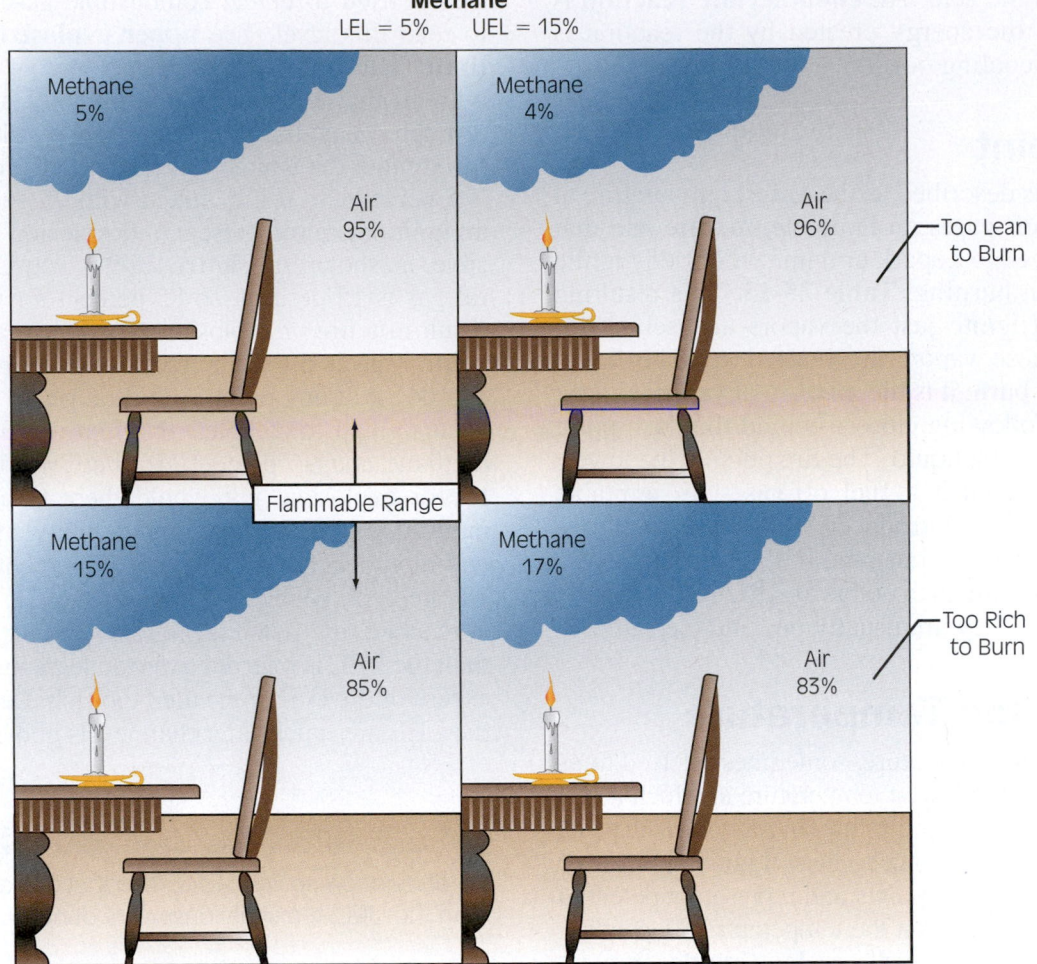

Methane
LEL = 5%   UEL = 15%

Methane 5%

Air 95%

Flammable Range

Methane 15%

Air 85%

Methane 4%

Air 96%

Too Lean to Burn

Methane 17%

Air 83%

Too Rich to Burn

To have a fire or explosion the lower explosive limit must be reached. Each gas has a flammable range in which there can be a fire or explosion. Below the LEL or above the UEL means there cannot be a fire.

FIGURE 25-91   Flammable range.

that the container is going to be under more pressure and, if involved in a fire situation, the heat will cause an increase in pressure within the container. Materials that react or are corrosive can also create pressure within a container or, if the corrosive is placed in the wrong container, can cause the container to fail. When containers are under pressure, there is a good chance that the venting or rupture of the container will be violent. Understanding flash point and flammable range are important as this response to potentially flammable materials is common. The lower the flash point, the greater the fire risk, and the more likely the container will be under some pressure, since the vapor pressure is generally more forceful as the flash point decreases.

Radiation

We are subjected to radiation exposure in various forms every day. Our bodies have radioactive substances in our makeup. We eat various forms of radiation sources,

and we breathe in radiation without any harm every day. Our exposure to these everyday radiation sources far exceeds those that would be found if we worked at a nuclear power plant. Through television, medical tests, and elevation we are subjected to levels of radiation that under normal circumstances cause us no harm. Some common radiation sources are listed in **Table 25-14.** To understand how radiation can hurt us, we must have an understanding of radioactivity. The basis of radioactivity is the makeup of the basic atom, specifically the nucleus. Having knowledge of radiation is important to understand not only how radiation affects the human body, but also how radiation monitors work.

An atom is comprised of electrons, neutrons, and protons. Protons and neutrons reside in the nucleus in the center of the atom, while electrons orbit the nucleus. Protons have a positive charge and determine the element or type of atom. Neutrons are the same size as protons but are neutral. Each element, with a given number of protons, can have several forms, termed **iso-**

TABLE 25-14   Common Radioactive Sources

Name	Energy Emitted	Half-Life	Use	Decays to Form
Polonium-214	α	164 microseconds	Dust-removing brushes	Lead-210
Lead-214	β and γ	27 min	Industrial	Bismuth-214
Cobalt-60	β and γ	5.3 years	X-ray machines, industrial applications	Nickel-60
Cesium-137	β	30 years	Medical field, industrial imaging, check source	Barium-137
Radon-222	α	3.8 days	Naturally occurring gas	Polonium-218
Uranium-238	α	4.5 billion years	Ore	Thorium-234
Plutonium-239	α	24,100 years	Nuclear weapons	Uranium-235
Americium-241	α	433 years	Smoke detectors	Neptunium-237

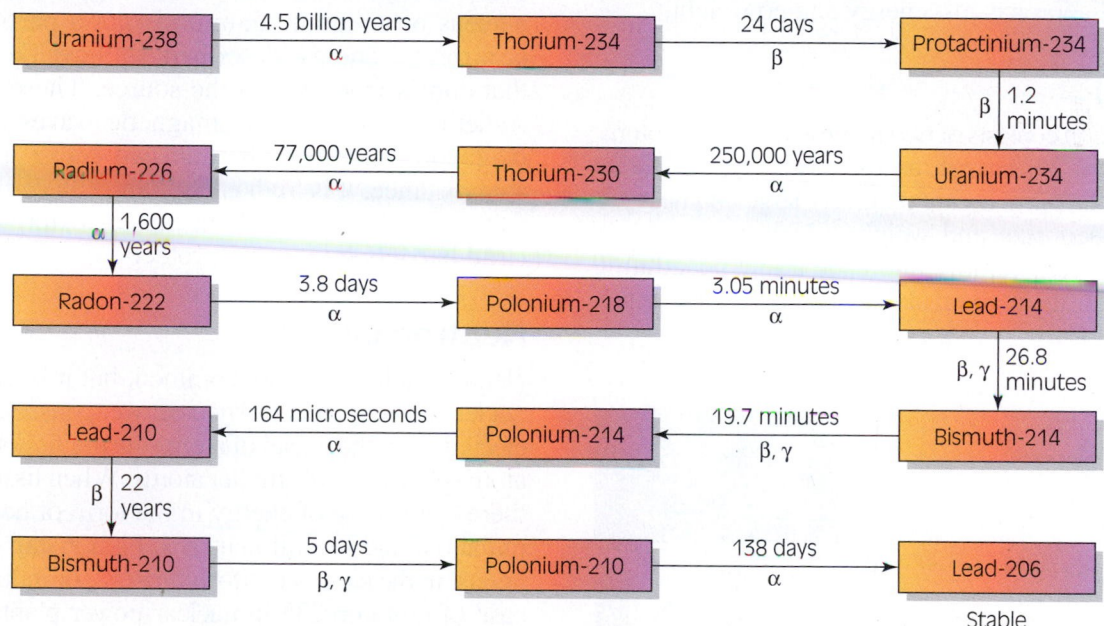

FIGURE 25-92   Half-life chart for uranium 238.

topes, which are determined by the number of neutrons in the nucleus. The chemical properties of each isotope of an element are the same—you cannot chemically or physically distinguish one from the other. If there are too few or too many neutrons, the nucleus becomes unstable. **Radioisotopes,** isotopes whose nuclei are unstable, are radioactive, and emit radiation to become more stable. This emission of radiation is known as **radioactive decay** and usually is in the form of gamma radiation but may also be alpha, beta, or neutron radiation, as shown in Table 25-14.

Some unstable material may become stable after one or two decays, but others may take many decay cycles. If a radioisotope decays by emission of an alpha or beta particle, the number of protons in the nucleus is changed, and the radioisotope becomes a different element. Uranium is the base for the development of **radon,** a common radioactive gas found in homes. Eventually, the radon will decay into lead, as shown in **Figure 25-92.** Each radioisotope has a constant **half life.** Half-life is defined as the amount of time for half of a given radioactive source to decay. The **activity** of a source of radioactive material is a measure of the number of decays per second that occur within the source. Source activity is measured in **becquerel (Bq)** or **curies (Ci).** One becquerel is equal to one radioactive decay per second. One curie is equal to 37 gigabecquerels, or 37 billion disintegrations per second, in the same way that one kilometer is equal to 1,000 meters. The physical size of a radioactive source is not an indicator of radioactive strength or activity. One curie is also the activity of exactly one gram of radium-222.

Types of Radiation

Radiation is comprised of two basic categories: ionizing radiation and nonionizing radiation. Some examples of nonionizing radiation include radio waves, microwaves, infrared, visible light, and ultraviolet light. Alpha, beta, gamma, and X-rays are forms of ionizing radiation, and some characteristics of each are provided in **Figure 25-93**. There are two subcategories of ionizing radiation, one with energy and weight, and the other comprising just energy. Alpha and beta are known as particulates and have weight and energy. Gamma radiation has just energy and no weight.

Alpha (α)

Alpha radiation consists of two neutrons and two protons, which carry a 2+ charge, and is identical to a helium nucleus. Alpha particles are relatively large and heavy.

Due to their size and weight, alpha particles lose their energy very rapidly and have a low penetrating ability resulting in their ability to only move a few

feet in air. It's primary hazard is through inhalation or ingestion. Street clothing or other protective clothing provides ample protection against alpha radiation.

Beta (ß)

Beta particles are electrons (−) or positrons (+) and weigh 1/1,836 the weight of a proton. Beta radiation is one of two forms, low or high energy, but both are particulates. The low energy is comparable to alpha but their smaller size and weight allows them to move a little farther in air, penetrate more than alpha particles, and cause more damage. High-energy beta moves even farther and can cause greater harm. Beta can move several feet and higher levels of protective clothing are required, but structural firefighters' clothing offers some protection

Gamma (γ)

Gamma rays come from the energy changes in the nucleus of an atom. Gamma is not a particulate but is airborne energy described as wavelike radiation that comes from within the source. These waves are sometimes called electromagnetic waves or electromagnetic radiation. Gamma can move a considerable distance. It is high energy and can cause internal damage to the body without causing external damage. Lead barriers offer some protection.

Neutron (n)

Neutron radiation is not common, but it is the basis for nuclear power plants and nuclear weapons. Neutrons are ejected from the nuclei of atoms during fission, when an atom splits into two smaller atoms. When fission occurs, there is a release of energy in the form of heat, gamma radiation, and several neutrons. Fission can be spontaneous, in the case of californium-252, or induced, in the case of uranium-235 in nuclear power plants. Isotopes that can be induced to fission are called **fissile.**

Neutrons can travel great distances but are stopped by several feet of water, concrete, or hydrogen-rich materials. Neutron radiation is dangerous because it readily transfers its energy to water, and on average the human body is about 68 to 75 percent water. One unusual aspect of neutron radiation is its ability to activate nonradioactive isotopes, in other words, to make them radioactive. The nucleus of the targeted atom can capture a neutron, which makes it a different isotope. If the nucleus of the new isotope is unstable, the isotope will be radioactive. For example, when natural gold, gold-197, is bombarded with neutrons, it can be activated and become gold-198, which emits beta and gamma radiation. Activation is the reason that the materials used in a nuclear reactor become radioac-

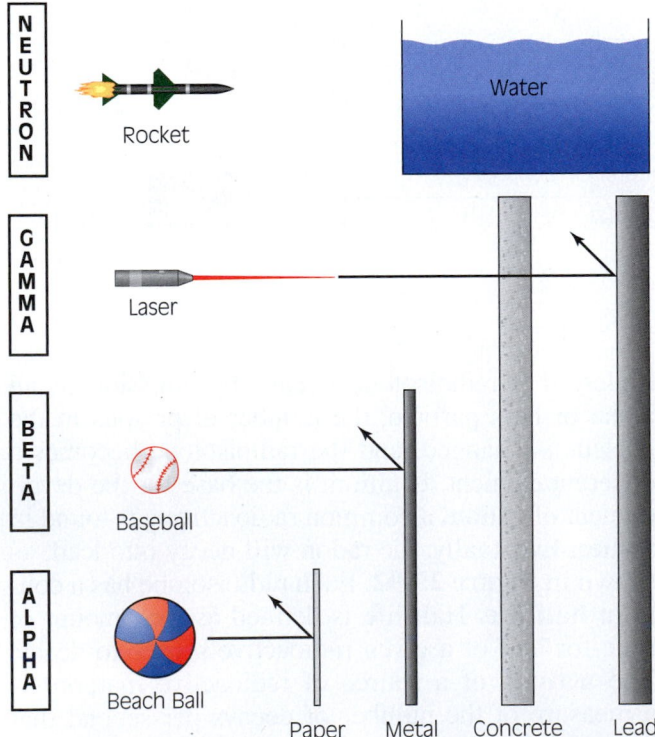

FIGURE 25-93 Examples of risks for ionizing radiation.

tive and must be disposed of as nuclear waste. Activation is accomplished only by neutron radiation.

X-rays

X-rays are comparable to gamma radiation waves, as they are wavelike, but X-rays are only dangerous when the X-ray machine in question is energized.

Toxic Products of Combustion

This is the one area in which firefighters suffer considerable chemical exposures. Any time a person is in smoke or breathes smoke, the body is being bombarded with toxic chemicals. Many toxic chemicals, such as carbon monoxide, carbon dioxide, hydrogen cyanide, hydrochloric acid, and phosgene, are produced in a fire. The worst type of chemical accident a firefighter can respond to is a house, car, or dumpster fire. Even brush fires are not exempt from the toxic products of combustion, because the field may have been sprayed with pesticides, herbicides, or other chemicals. Even burning wool or hay produces extremely toxic gases. Due to this constant exposure to these and other materials, it is important for firefighters to wear all of their protective clothing, especially the SCBA.

JOB PERFORMANCE REQUIREMENT   25-1

Fighting a Fire for Tanks and Other Containers Under Pressure

A   A tank or container can fail at any time, and it is impossible to determine the exact moment that failure will occur. Firefighters should withdraw immediately when increasing sound from venting relief valves or a discoloration of the tank is noticed. In this photo, a high-pressure tube trailer carrying compressed hydrogen is on fire. *(Courtesy of Maryland Department of the Environment Emergency Response Division)*

B   Fire must be fought from a distance with unstaffed hose holders or monitor nozzles. In this photo, a liquid propane tank has overturned and the propane is being flared (burned) off before the truck can be righted. *(Courtesy of Maryland Department of the Environment Emergency Response Division)*

LESSONS LEARNED

The ability to recognize and identify the potential for hazardous materials to be present at an incident is important for the first responder. The numbers of tank trucks, tank cars, and containers can easily overwhelm the beginning student. At any incident, there is always a factor that relates to the recognition and identification process, whether it is the location, placards, container type, or physical senses. It could even be that sixth sense that alerts people to a potential problem. When that occurs, it is important to proceed with caution.

One of the important lessons for responders to remember is that they do not have to commit everything to memory, but they should know where to access hazardous materials information. It is not expected that each community have a railcar expert in their fire department. What is expected is that each fire department have a contact person available around the clock to obtain that resource. It is possible that every material on this earth has the ability to cause harm in some fashion, but the chemical and physical properties play a factor in the type of harm that can be caused.

Materials that have low vapor pressures present little risk to the responders unless touched or eaten. Materials that have high vapor pressures do present a great risk to responders and to the community and should be treated with caution. Vapor pressure is only one of the terms with which responders should become familiar. Many responders have a fear of radiation, but this fear comes from a lack of education and experience. Understanding the harms from radiation is an important safety consideration. Local hazardous materials responders are a good resource and should be contacted early in an incident and whenever assistance dealing with hazardous materials is needed.

KEY TERMS

Aboveground Storage Tank (AST)   Tank that is stored above the ground in a horizontal or vertical position. Smaller quantities of fuels are often stored in this fashion.

Activity   A measure of the number of decays per second that occur with the source.

Autoignition Temperature   The temperature of a liquid at which it will ignite on its own without an ignition source. Can be compared to SADT (Self-Accelerating Decomposition Temperature).

Becquerel (Bq)   A measure of radioactivity; the metric version of Curies.

Boiling Point   The temperature to which a liquid must be heated in order to turn into a gas.

Building Officials Conference Association (BOCA)   A group that establishes minimum building and fire safety standards.

Bulk Tank   A large transportable tank, comparable to a tote, but considered to be the larger of the two.

Chemtrec   The Chemical Transportation Emergency Center, which provides technical assistance and guidance in the event of a chemical emergency; a network of chemical manufacturers that provide emergency information and response teams if necessary.

Cryogenic Gas   Any gas that exists as a liquid at a very cold temperature, always below $-150°F$ ($-101°C$)

Curies (Ci)   The measure of activity level for radiation sources.

Deflagrates   Rapid burning, which in reality with regard to explosions can be considered a slow explosion, but is traveling at a lesser speed than a detonation.

Emergency Response Guidebook (ERG)   Book provided by the DOT that assists the first responder in making decisions primarily at transportation-related hazardous materials incidents.

Endothermic Reaction   A chemical reaction in which heat is absorbed, and the resulting mixture is cold.

Excepted Packaging   Type of packaging used to transport low-risk radioactive materials; the packaging is exempted from DOT regulations.

Exothermic Reaction   A chemical reaction that releases heat, such as when two chemicals are mixed and the resulting mixture is hot.

External Floating Roof Tank   Tank with the roof exposed on the outside that covers the liquid within the tank. The roof floats on the top of the liquid, which does not allow for vapors to build up.

First Responders   A group designated by the community as those who may be the first to arrive at a chemical incident. This group is usually composed of police officers, EMS providers, and firefighters.

Fissile   Isotopes that can be induced into fission, which creates a release of energy.

Frangible Disk   A type of pressure-relieving device that actually ruptures in order to vent the excess pressure. Once opened the disk remains open; it does not close after the pressure is released.

Freezing Point   The temperature at which liquids become solids.

Gas   A state of matter that describes the material in a form that moves freely about and is difficult to control. Steam is an example.

Half Life   The amount of time for a given radiation source to decay, which means it is emitting radioactivity.

Intermodal Containers   These are constructed in a fashion so that they can be transported by highway, rail, or ship. Intermodal containers exist for solids, liquids, and gases.

Internal Floating Roof Tank   Tank with a roof that floats on the surface of the stored liquid, but also has a cover on top of the tank, so as to protect the top of the floating roof.

Isotope   A material that has a different form due to the number of neutrons that are in the nucleus.

Leaking Underground Storage Tank (LUST)   Describes a leaking tank that is underground.

Liquid   A state of matter that implies fluidity, which means a material has the ability to move as water would. There are varying states of being a liquid from moving very quickly to very slowly. Water is an example.

Low Specific Activity (LSA)   Designation that indicates that a material is emitting low levels of radiation.

Lower Explosive Limit (LEL)   The lower part of the flammable range; the minimum required to have a fire or explosion.

Material Safety Data Sheet (MSDS)   An information sheet that provides chemical-specific safety information.

Melting Point   The temperature at which solids become liquids.

Nuclear Regulatory Agency (NRC)   The government regulatory agency responsible for oversight of nuclear materials.

Ordinary Tank   A horizontal or vertical tank that usually contains combustible or other less hazardous chemicals. Flammable materials and other hazardous chemicals may be stored in smaller quantities in these types of tanks.

Oxidizer   Materials that readily release oxygen; by yielding oxygen, an oxidizer can easily cause or enhance the combustion of other materials. Oxidizers can dramatically increase the rate of burning when combustible material is ignited.

Persistence   An indication of the time that a material will remain as a liquid, and is related to vapor pressure.

Radioactive Decay   Process whereby, as a material emits radiation, it decays, changing its form; for example, uranium eventually becomes lead after it decays.

Radioisotope   A radioactive form of an element that is unstable due to the number of neutrons.

Radon   A radioactive gas that is emitted from the earth, sometimes collecting in basement areas.

Relief Valve   A device designed to vent pressure in a tank, so that the tank itself does not rupture due to an increase in pressure. In most cases these devices are spring loaded so that when the pressure decreases the valve shuts, keeping the chemical inside the tank.

Reportable Quantity (RQ)   Both the EPA and DOT use the term. It is a quantity of chemicals that may require some type of action, such as reporting an inventory or reporting an accident involving a certain amount of the chemical.

Sea Containers   Shipping boxes that were designed to be stacked on a ship, then placed onto a truck or railcar.

Self-Accelerating Decomposition Temperature (SADT)   Temperature at which a material will ignite itself without an ignition source present. Can be compared to ignition temperature.

Solid   A state of matter that describes materials that may exist in chunks, blocks, chips, crystals, powders, dusts, and other forms. Ice is an example.

Specification (Spec) Plates   All trucks and tanks have a specification plate that outlines the type of tank, capacity, construction, and testing information.

Spent Nuclear Fuel   Radioactive fuel that was used in a nuclear reactor.

States of Matter   Describes in what form matter exists, such as solids, liquids, or gases.

Sublimation   The ability of a solid to go to the gas phase without being liquid.

Surface Contaminated Object (SCO)   Materials that may be contaminated with radioactive waste and are usually low hazard.

Tote   A large tank usually 250 to 500 gallons (946–1893 liters), constructed to be transported to a facility and dropped for use.

Type A Container   A container designed to hold low-risk radioactive material and offers moderate protection against accidents.

Type B Container   A container designed to hold high-risk radioactive materials. It is a hardened transport case capable of withstanding severe accidents.

Underground Storage Tank (UST)   Tank that is buried under the ground. The most common are gasoline and other fuel tanks.

Upper Explosive Limit (UEL)   The upper part of the flammable range. Above the UEL, fire or an explosion cannot occur because there is too much fuel and not enough oxygen.

Vapor Pressure   The amount of force that is pushing vapors from a liquid. The higher the force the more vapors (gas) being put into the air.

REVIEW QUESTIONS

1. What are the nine hazard classes as defined by DOT?

2. An explosive placard is what color?

3. What three things on a placard indicate the potential hazards?

4. A DOT-406/MC-306 tank truck commonly carries what product?

5. A DOT-406 tank truck has a characteristic shape from the rear. What is it?

6. An MC-331 carries what type of gases?

7. What does the blue section of the NFPA 704 system refer to?

8. What are the locations of emergency shutoffs on an MC-331?

9. What are the four basic clues to recognition and identification for hazardous materials?

10. The placards shown in Table 25-1 are required for which quantities?

11. A tractor trailer carrying Division 1.1 materials is well involved in fire. What should be the firefighter's initial tactics?

12. When propane tanks are involved in a fire, what is a potential consequence?

13. What types of radiation are particles?

14. Which one type of radiation has the ability to activate other materials and make them radioactive?

15. What impact does half-life have with regard to a radiation source?

ENDNOTES

1. A tank truck placarded for gasoline could also haul diesel fuel or fuel oil without having to change placards.

2. The requirements for the subdivisions are presented in general. Other requirements may need to be met for a material to be assigned to a subdivision. Refer to 49 CFR 170–180 for more information.

26

Hazardous Materials: Information Resources

had been on the job for several years, and there was nothing unusual about the call when it first came in—nothing remarkable that would lead to any heightened sense of awareness. "Engine 29, check out the smell of smoke in the building at 211 Mason Street." It would be a simple food on the stove call or something to that effect.

When we arrived on the scene the local police department had already made initial entry to the facility (which turned out to be a restaurant) and reported a slight haze with a strange pungent odor but no visible fire. We made entry into the vacant facility and immediately noticed the haze and pungent odor, but because it was a restaurant we thought it was simply something to do with the food preparation process. As we made our way through the facility, we began to notice a strange taste in our mouths and several of those present began to experience irritation to the eyes and throat, yet still nothing clicked. It was, after all, simply a restaurant, perhaps some burnt wiring or bad cooking, no big deal. As we continued our investigation, several of those present began to complain of tightness in the chest, headaches, and nausea. I too was feeling strange but did not want to throw in the towel so another responder and I continued on.

Then suddenly as the haze began to clear via the doors and windows that had been opened for ventilation, we began to notice an abundance of dead insects and rodents throughout the facility. And now on the radio we heard the call for an ambulance as one of the responders who had been on scene for some time began to experience more than simply mild irritations. We began to realize that the haze was in fact some type of aerosolized insecticide/pesticide and that we had been exposed and contaminated. Now and only now did we begin to comprehend the error of our ways, so to speak, and implement the proper response and protocols.

As the weeks followed, I replayed the incident over and over and came to realize all of the mistakes that we had made, and how imperative it was to seek the proper training to ensure that I never put myself or those with me in this position again. Two of the responders present were unable to return to work for a period of time (in one case several months). They had to be treated for the effects of exposure to organophosphates. Both are fine today, but much has been written about the possible long-term effects of exposure to organophosphates, and one can only wonder what course these individuals' health could have taken had they been further exposed. And that I do, all too often.

Speaking only to my errors, there was no attempt to obtain the proper information from the appropriate resources or the owners. (After we realized we had no fire, we should have consulted the owners before further entry, because there was no life hazard.) Nor was subsequent contact made with the other tenants of this row structure who may have been able to provide information, or to even take basic actions to ensure the health and safety of those involved. The failure to properly research and gather information on this call could have resulted in much more severe health problems for all involved had the concentrations been higher or ventilation unsuccessful. But the fact that two responders were unnecessarily exposed still remains with me today.

—Street Story by Tom Creamer, Special Operations Coordinator (ret.),
City of Worcester Fire Department, Worcester, Massachusetts

INTRODUCTION

CAUTION

Specific and current chemical information is essential if the first responder is going to protect lives and property.

Chemical information is available through a variety of sources, including those carried on emergency apparatus and information sources that can be reached via telephone, fax, or computer. The shipper and the facility are required to maintain certain documents that will assist first responders in determining the chemical hazards they may face. Knowing what information is available and how to interpret this information is a valuable tool. This chapter discusses the most common sources of information.

EMERGENCY RESPONSE GUIDEBOOK

The DOT's ERG is a well-known book for emergency responders, **Figure 26-1.** The DOT makes one copy available for every emergency response apparatus in the country.

The ERG is commonly referred to as the DOT book or the orange book. This book, which is published about every three years, contains information regarding the most commonly transported chemicals as regulated by the DOT. It is intended as a guide-book for first responders during the initial phase of a hazardous materials incident. It provides information regarding the potential hazards of responding to these materials and is one of the only books that provides specific evacuation recommendations. Although an

excellent book for first responders, it does have limitations for the more advanced responder, who requires more specific chemical information.

The book consists of these major sections:

- Placard information
- **ADR/RID** marking system information
- Listing by DOT identification number
- Alphabetical listing by shipping name

FIGURE 26-1   The DOT *Emergency Response Guidebook* should be found in every emergency vehicle in the United States. It provides chemical emergency response information that is valuable to the first responder.

RESIST RUSHING IN !
APPROACH INCIDENT FROM UPWIND
STAY CLEAR OF ALL SPILLS, VAPORS, FUMES AND SMOKE

HOW TO USE THIS GUIDEBOOK DURING AN INCIDENT INVOLVING DANGEROUS GOODS

ONE **IDENTIFY THE MATERIAL** BY FINDING ANY **ONE** OF THE FOLLOWING:

> THE 4-DIGIT ID NUMBER ON A PLACARD OR ORANGE PANEL
>
> THE 4-DIGIT ID NUMBER (after UN/NA) ON A SHIPPING DOCUMENT OR PACKAGE
>
> THE NAME OF THE MATERIAL ON A SHIPPING DOCUMENT, PLACARD OR PACKAGE

IF AN **ID NUMBER** OR THE **NAME OF THE MATERIAL** CANNOT BE FOUND, SKIP TO THE NOTES BELOW.

TWO **LOOK UP THE MATERIAL'S 3-DIGIT GUIDE NUMBER** IN EITHER:

> THE ID NUMBER INDEX..(the yellow-bordered pages of the guidebook)
>
> THE NAME OF MATERIAL INDEX..(the blue-bordered pages of the guidebook)

If the guide number is supplemented with the letter "**P**", it indicates that the material may undergo violent polymerization if subjected to heat or contamination.

If the index entry is highlighted (in either yellow or blue), it is a TIH (Toxic Inhalation Hazard) material, a chemical warfare agent or a Dangerous Water Reactive Material (produces toxic gas upon contact with water). **LOOK FOR THE ID NUMBER AND NAME OF THE MATERIAL** IN THE TABLE OF INITIAL ISOLATION AND PROTECTIVE ACTION DISTANCES (the green-bordered pages). Then, if necessary, **BEGIN PROTECTIVE ACTIONS IMMEDIATELY** (see Protective Actions on page 298). If protective action is not required, use the information jointly with the 3-digit guide.

USE GUIDE 112 FOR ALL EXPLOSIVES EXCEPT FOR EXPLOSIVES 1.4 (EXPLOSIVES C) WHERE GUIDE 114 IS TO BE CONSULTED.

THREE **TURN TO THE NUMBERED GUIDE** (the orange-bordered pages) **AND READ CAREFULLY.**

NOTES IF A NUMBERED GUIDE CANNOT BE OBTAINED BY FOLLOWING THE ABOVE STEPS, AND A PLACARD CAN BE SEEN, LOCATE THE PLACARD IN THE TABLE OF PLACARDS (pages 16-17), THEN GO TO THE 3-DIGIT GUIDE SHOWN NEXT TO THE SAMPLE PLACARD.

IF A REFERENCE TO A GUIDE CANNOT BE FOUND AND THIS INCIDENT IS BELIEVED TO INVOLVE DANGEROUS GOODS, TURN TO GUIDE 111 NOW, AND USE IT UNTIL ADDITIONAL INFORMATION BECOMES AVAILABLE. If the shipping document lists an emergency response telephone number, call that number. If the shipping document is not available, or no emergency response telephone number is listed, IMMEDIATELY CALL the appropriate **emergency response agency listed on the inside back cover of this guidebook.** Provide as much information as possible, such as the name of the carrier (trucking company or railroad) and vehicle number. AS A LAST RESORT, CONSULT THE TABLE OF RAIL CAR AND ROAD TRAILER IDENTIFICATION CHART (pages 18-19). IF THE CONTAINER CAN BE IDENTIFIED, REMEMBER THAT THE INFORMATION ASSOCIATED WITH THESE CONTAINERS IS FOR THE WORST CASE POSSIBLE.

Page 1

FIGURE 26-2 The first page of the ERG is a step-by-step listing of how to use the book. When not sure how to proceed, the responder should turn to this page.

- Response guides
- Table of initial isolation and protective action distances
- List of dangerous water-reactive materials

On the inside cover, the book provides an example of a shipping document used in truck transportation and provides information on how the shipping document should be written. It also provides an example of a placard with an identification number panel. The first page, which is shown in **Figure 26-2,** is an essential page for responder safety because it outlines all of the actions the first responder should take. It provides a decision tree process by taking the reader through the incident step by step. It also has an important listing of the guides for explosives.

The next few pages of the ERG provide information on safety precautions and whom to call for assis-

HAZARD CLASSIFICATION SYSTEM

The hazard class of dangerous goods is indicated either by its class (or division) number or name. For a placard corresponding to the primary hazard class of a material, the hazard class or division number must be displayed in the lower corner of the placard. However, no hazard class or division number may be displayed on a placard representing the subsidiary hazard of a material. For other than Class 7 or the OXYGEN placard, text indicating a hazard (for example, "CORROSIVE") is not required. Text is shown only in the U.S. The hazard class or division number must appear on the shipping document after each shipping name.

Class 1 - Explosives

Division 1.1	Explosives with a mass explosion hazard
Division 1.2	Explosives with a projection hazard
Division 1.3	Explosives with predominantly a fire hazard
Division 1.4	Explosives with no significant blast hazard
Division 1.5	Very insensitive explosives with a mass explosion hazard
Division 1.6	Extremely insensitive articles

Class 2 - Gases

Division 2.1	Flammable gases
Division 2.2	Non-flammable, non-toxic* gases
Division 2.3	Toxic* gases

Class 3 - Flammable liquids (and Combustible liquids [U.S.])

Class 4 - Flammable solids; Spontaneously combustible materials; and Dangerous when wet materials/Water-reactive substances

Division 4.1	Flammable solids
Division 4.2	Spontaneously combustible materials
Division 4.3	Water-reactive substances/Dangerous when wet materials

Class 5 - Oxidizing substances and Organic peroxides

Division 5.1	Oxidizing substances
Division 5.2	Organic peroxides

Class 6 - Toxic* substances and Infectious substances

Division 6.1	Toxic* substances
Division 6.2	Infectious substances

Class 7 - Radioactive materials

Class 8 - Corrosive substances

Class 9 - Miscellaneous hazardous materials/Products, Substances or Organisms

* The words "poison" or "poisonous" are synonymous with the word "toxic".

Page 14

FIGURE 26-3 The DOT uses nine classes of hazardous materials, as described in Chapter 25, that are listed in the ERG. Occasionally the only information that is available is the hazard class.

tance. The first responder should already know where the closest local assistance will be coming from and where to contact state assistance if required. The DOT book provides a contact number for federal assistance, although responders should proceed by requesting local, state, and then federal assistance. The contact numbers and agencies are divided among Canada, the United States, and Mexico. It is important for the first responder to read the DOT book prior

to an incident because it provides a large amount of background material that could not be effectively read during an emergency.

The book also provides a listing of the hazard class system, **Figure 26-3,** and provides a reference point for the hazard classes that may be listed on placards or shipping papers. This listing precedes the placard section, which is valuable if the only information available is the placard. The placard section, **Figure 26-4,**

TABLE OF PLACARDS AND INITIAL
USE THIS TABLE ONLY IF MATERIALS CANNOT BE SPECIFICALLY IDENTIFIED BY

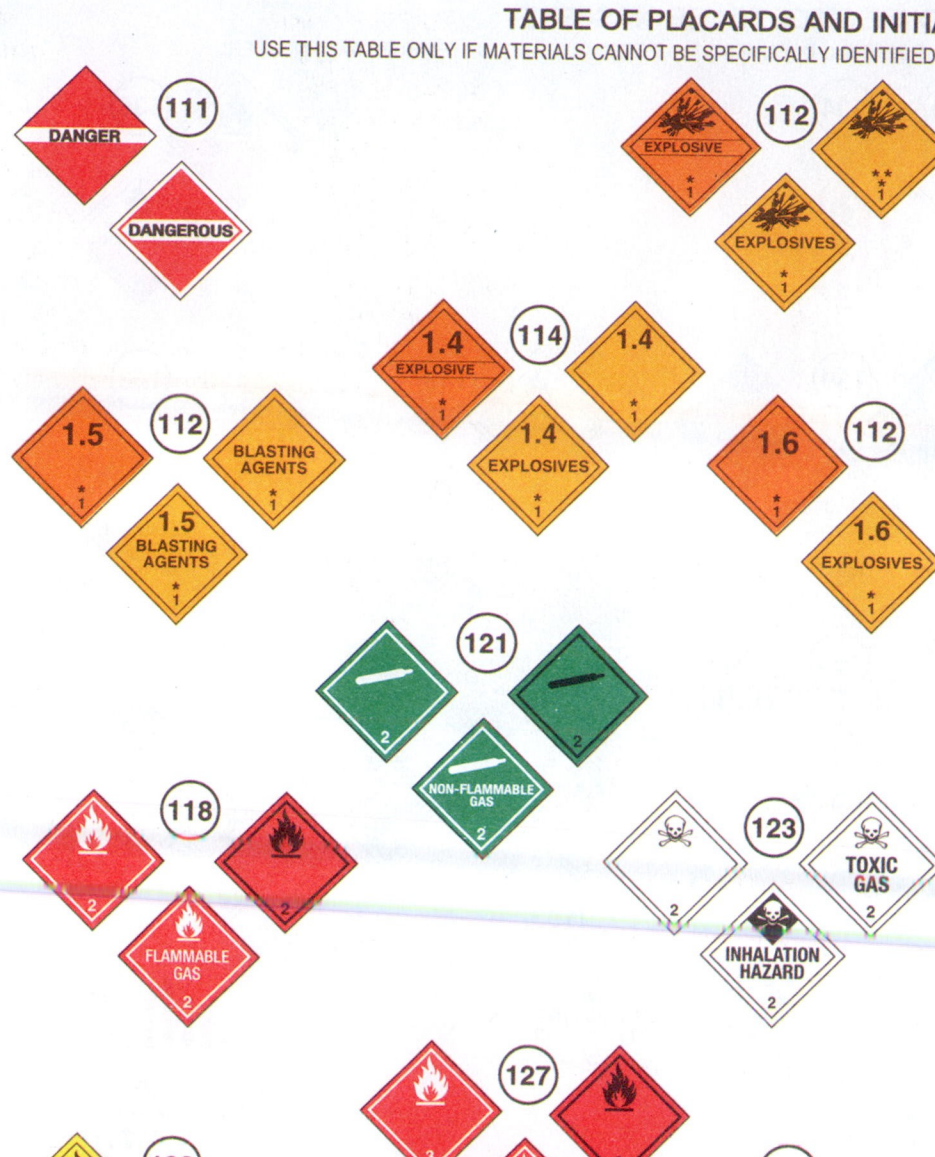

Page 16

FIGURE 26-4 If the only information that is available is a placard, the ERG can provide assistance. Listed below each placard is a guide number that provides response information based on the placard information.

provides information about how to proceed at an incident where the only information is a placard. All of the possible current placards are listed with their pictures, and the reader is referred to the accompanying guide page. Much like the placard section, included in the ERG are silhouettes of tank trucks, shown in **Figure 26-5,** and of rail cars, shown in **Figure 26-6.** Some international shipments use additional hazard markings as shown in **Figure 26-7.**

The yellow section, **Figure 26-8,** is a numerical listing by the identification, or ID, number. This is a

number assigned by the DOT that identifies a material being shipped. This number takes two forms, a North America (NA) number or a United Nations (UN) number, but it is always a four-digit number. Although one would think that this number would specifically identify a material, in some cases it may not. Depending on the material, it may be lumped into a general category such as Hazardous Waste, Liquid, or NOS (Not Otherwise Specified), but in most cases it will identify a specific substance. The yellow section lists the ID number, shipping name, and guide

RESPONSE GUIDE TO USE ON-SCENE
USING THE SHIPPING DOCUMENT, NUMBERED PLACARD, OR ORANGE PANEL NUMBER

FIGURE 26-4   (Continued).

Page 17

reference number. If the material is highlighted as shown in **Figure 26-9,** then the reader must refer to the guide page listed as well as the Table of Initial Isolation and Protective Action Distances, located at the back of the book in the green section.

The following abbreviations are used in the DOT ERG:

- NOS—not otherwise specified
- PIH—poison by inhalation
- P—polymerization hazard
- SCA—surface contaminated articles (radiation)
- LSA—low specific activity (radiation)
- PG I, II, or III—Packing Group I, II, or III
- ORM—other regulated material (listed as ORM A-E)
- LC_{50}—lethal concentration to 50 percent of the population
- LD_{50}—lethal dose to 50 percent of the population

ROAD TRAILER IDENTIFICATION CHART*

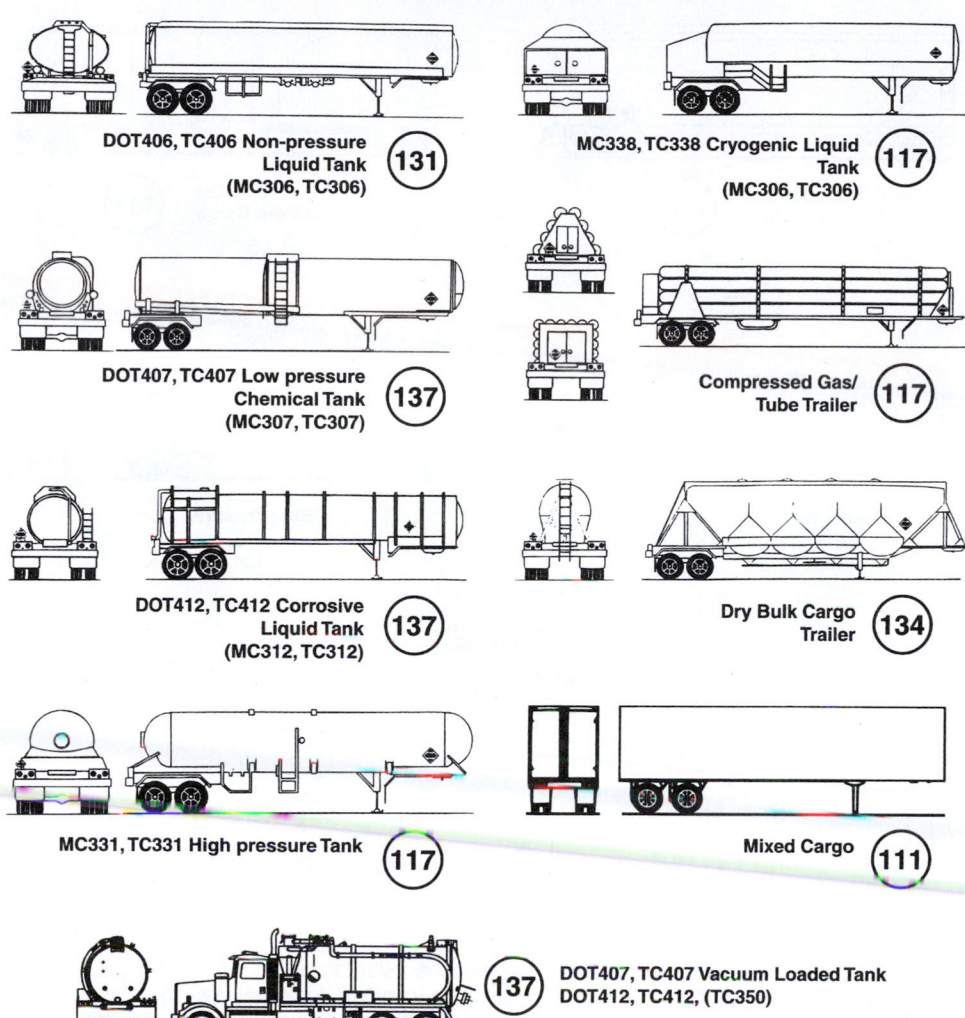

DOT406, TC406 Non-pressure Liquid Tank (MC306, TC306) (131)

MC338, TC338 Cryogenic Liquid Tank (MC306, TC306) (117)

DOT407, TC407 Low pressure Chemical Tank (MC307, TC307) (137)

Compressed Gas/ Tube Trailer (117)

DOT412, TC412 Corrosive Liquid Tank (MC312, TC312) (137)

Dry Bulk Cargo Trailer (134)

MC331, TC331 High pressure Tank (117)

Mixed Cargo (111)

DOT407, TC407 Vacuum Loaded Tank DOT412, TC412, (TC350) (137)

CAUTION: This chart depicts only the most general shapes of road trailers. Emergency response personnel must be aware that there are many variations of road trailers, not illustrated above, that are used for shipping chemical products. The suggested guides are for the most hazardous products that may be transported in these trailer types.

*   **The recommended guides should be considered as last resort if product cannot be identified by any other means.**

Page 19

FIGURE 26-5 Truck silhouettes are provided with the appropriate guide page for each of the truck styles.

The blue section of the book mirrors the yellow section except that it is alphabetical by shipping name, **Figure 26-10.** These shipping names are assigned by the DOT and for the most part are identical to the chemical names, but in some cases they can vary. A lot of names are assigned to chemicals, including the actual chemical name, synonyms, trade names, and shipping names.

SAFETY

Some chemicals can have more than sixty synonyms, which can create a confusing situation. **Table 26-1** provides one example.

The chemical known by the DOT as chlordane is also called by the names given in Table 26-1. This

RAIL CAR IDENTIFICATION CHART*

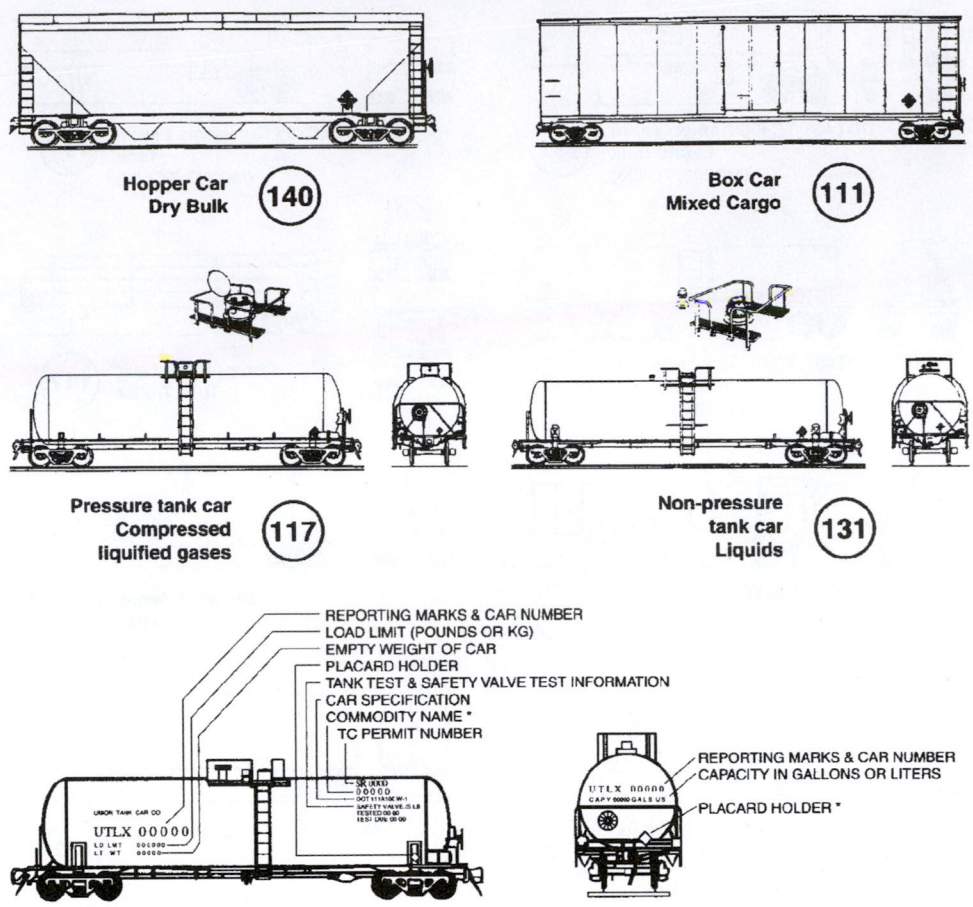

CAUTION: Emergency response personnel must be aware that rail tank cars vary widely in construction, fittings and purpose. Tank cars could transport products that may be solids, liquids or gases. The products may be under pressure. It is essential that products be identified by consulting shipping documents or train consist or contacting dispatch centers before emergency response is initiated.

The information stenciled on the sides or ends of tank cars, as illustrated above, may be used to identify the product utilizing:

a.   the commodity name shown; or

b.   the other information shown, especially reporting marks and car number which, when supplied to a dispatch center, will facilitate the identification of the product.

*   **The recommended guides should be considered as last resort if product cannot be identified by any other means.**

Page 18

FIGURE 26-6 Rail car silhouettes are provided with the appropriate guide page for each of the rail car styles.

section lists the ID number, shipping name, and guide reference number. If the material is highlighted in blue, the reader should refer to the guide page listed as well as the Table of Initial Isolation and Protective Action Distances, located at the back of the book in the green section.

The middle of the book, the orange section, makes up the actual guide pages, **Figure 26-11.** A total of sixty-one guides is given for the more than 4,000

chemicals listed by the DOT. It is this generalization that makes this book valuable to the initial responder but less so to an advanced responder, such as a hazardous materials technician. A hazardous materials technician needs more specific information, which the ERG does not provide.

Each guide takes up two pages and is divided into three major sections: potential hazards, public safety, and emergency response. The guide may also provide

HAZARD IDENTIFICATION CODES
DISPLAYED ON SOME INTERMODAL CONTAINERS

Hazard identification codes, referred to as "hazard identification numbers" under European and some South American regulations, may be found in the top half of an orange panel on some intermodal bulk containers. The 4-digit identification number is in the bottom half of the orange panel.

The hazard identification code in the top half of the orange panel consists of two or three digits. In general, the digits indicate the following hazards:

2 - EMISSION OF GAS DUE TO PRESSURE OR CHEMICAL REACTION

3 - FLAMMABILITY OF LIQUIDS (VAPORS) AND GASES OR SELF-HEATING LIQUID

4 - FLAMMABILITY OF SOLIDS OR SELF-HEATING SOLID

5 - OXIDIZING (FIRE-INTENSIFYING) EFFECT

6 - TOXICITY OR RISK OF INFECTION

7 - RADIOACTIVITY

8 - CORROSIVITY

9 - MISCELLANEOUS DANGEROUS SUBSTANCE

- Doubling of a digit indicates an intensification of that particular hazard (i.e. 33, 66, 88).
- Where the hazard associated with a material can be adequately indicated by a single digit, the digit is followed by a zero (i.e. 30, 40, 50).
- A hazard identification code prefixed by the letter "X" indicates that the material will react dangerously with water (i.e. X88).
- When 9 appears as a 2nd or 3rd digit, this may present a risk of spontaneous violent reaction.

FIGURE 26-7 Intermodal containers will have an orange panel near the placard like that shown in the figure. The four-digit number is the DOT ID number, and the one- or two-digit number above that is a classification code, providing information on the basic hazards.

some additional information about a specific chemical. It is important for the responder to read the entire guide, and, if referred to the Table of Initial Isolation and Protective Action Distances, to read that section also.

The potential hazards section lists the predominant hazard on the top line. If fire is the major concern, then it will be listed on top; if health is the major concern, then it will be listed on top. The information regarding the health effects provides the route of exposure and any major symptoms of exposure. It also details other potential health and environmental concerns with fire products and runoff. The fire section is not as detailed as the health section but does provide some assistance in identifying potential hazards. It provides information related to the flammability of the product and will state whether the vapors will stay low to the ground or will rise in air.

ID No.	Guide No.	Name of Material	ID No.	Guide No.	Name of Material
1027	115	Cyclopropane	1043	125	Fertilizer, ammoniating solution, with free Ammonia
1027	115	Cyclopropane, liquefied	1044	126	Fire extinguishers with compressed gas
1028	126	Dichlorodifluoromethane	1044	126	Fire extinguishers with liquefied gas
1028	126	Refrigerant gas R-12	1045	124	Fluorine
1029	126	Dichlorofluoromethane	1045	124	Fluorine, compressed
1029	126	Refrigerant gas R-21	1046	121	Helium
1030	115	1,1-Difluoroethane	1046	121	Helium, compressed
1030	115	Difluoroethane	1048	125	Hydrogen bromide, anhydrous
1030	115	Refrigerant gas R-152a	1049	115	Hydrogen
1032	118	Dimethylamine, anhydrous	1049	115	Hydrogen, compressed
1033	115	Dimethyl ether	1050	125	Hydrogen chloride, anhydrous
1035	115	Ethane	1051	117	AC
1035	115	Ethane, compressed	1051	117	Hydrocyanic acid, aqueous solutions, with more than 20% Hydrogen cyanide
1036	118	Ethylamine	1051	117	Hydrocyanic acid, liquefied
1037	115	Ethyl chloride	1051	117	Hydrogen cyanide, anhydrous, stabilized
1038	115	Ethylene, refrigerated liquid (cryogenic liquid)	1051	117	Hydrogen cyanide, stabilized
1039	115	Ethyl methyl ether	1052	125	Hydrogen fluoride, anhydrous
1039	115	Methyl ethyl ether	1053	117	Hydrogen sulfide
1040	119P	Ethylene oxide	1053	117	Hydrogen sulfide, liquefied
1040	119P	Ethylene oxide with Nitrogen	1053	117	Hydrogen sulphide
1041	115	Carbon dioxide and Ethylene oxide mixture, with more than 9% but not more than 87% Ethylene oxide	1053	117	Hydrogen sulphide, liquefied
1041	115	Carbon dioxide and Ethylene oxide mixtures, with more than 6% Ethylene oxide	1055	115	Isobutylene
1041	115	Ethylene oxide and Carbon dioxide mixture, with more than 9% but not more than 87% Ethylene oxide	1056	121	Krypton
1041	115	Ethylene oxide and Carbon dioxide mixtures, with more than 6 % Ethylene oxide	1056	121	Krypton, compressed
			1057	115	Lighter refills (cigarettes) (flammable gas)
			1057	115	Lighters (cigarettes) (flammable gas)
			1058	120	Liquefied gas (nonflammable)

Page 26

FIGURE 26-8 The yellow pages are a numeric listing of the chemicals regulated by the DOT. This listing is by four-digit United Nations/North America identification number and also provides the shipping name and the emergency response guide page information.

Depending on the product, it may also describe the material's ability to float on water. Products shipped at an elevated temperature as well as those materials that have the ability to **polymerize** will have a notation.

One of the problems with the ERG is the fact that gasoline, which is highly flammable, is in the same guide that is provided for diesel fuel, which to the fire service is combustible. Although the DOT uses 141°F (60.5°C) as the difference between flammable and combustible, the fire service still uses 100°F (38°C) as the difference. A major difference is seen when handling a gasoline spill as opposed to a diesel fuel spill under normal conditions. By only using

ID No.	Guide No.	Name of Material
2474	157	Thiophosgene
2475	157	Vanadium trichloride
2477	131	Methyl isothiocyanate
2478	155	Isocyanate solution, flammable, poisonous, n.o.s.
2478	155	Isocyanate solution, flammable, toxic, n.o.s.
2478	155	Isocyanate solutions, n.o.s.
2478	155	Isocyanates, flammable, poisonous, n.o.s.
2478	155	Isocyanates, flammable, toxic, n.o.s.
2478	155	Isocyanates, n.o.s.
2480	155	Methyl isocyanate
2481	155	Ethyl isocyanate
2482	155	n-Propyl isocyanate
2483	155	Isopropyl isocyanate
2484	155	tert-Butyl isocyanate
2485	155	n-Butyl isocyanate
2486	155	Isobutyl isocyanate
2487	155	Phenyl isocyanate
2488	155	Cyclohexyl isocyanate
2490	153	Dichloroisopropyl ether
2491	153	Ethanolamine
2491	153	Ethanolamine, solution
2491	153	Monoethanolamine
2493	132	Hexamethyleneimine
2495	144	Iodine pentafluoride
2496	156	Propionic anhydride
2498	129	1,2,3,6-Tetrahydrobenzaldehyde
2501	152	1-Aziridinyl phosphine oxide (Tris)
2501	152	Tri-(1-aziridinyl)phosphine oxide, solution

FIGURE 26-9   If the listing in the yellow pages or the blue pages is highlighted, as is the listing of ID number 2481, Ethyl isocyanate, then the reader must turn to the orange guide and to the green section. The green section is the Table of Initial Isolation and Protective Action Distances.

the guide as intended, the reader has no way of differentiating between the two products. Although taking this route might be safer, because it assumes

a worst-case scenario, there are times when this action is inappropriate and can lead to other problems, such as an unnecessary evacuation for a diesel fuel spill.

> **NOTE**
>
> Only the public safety section (orange pages) of the ERG provides recommendations about how far to isolate the scene and provides some initial strategic goals for the first arriving company. If the chemical is highlighted, then responders should use both the orange response guides and the information provided in the green section.

The orange section lists safety precautions and initial isolation distances for emergency responders and the public. The green section contains detailed evacuation information that includes small and large spills and also includes information that differentiates between daytime and nighttime operations. This detailed table may be used by responders to further protect the community during evacuation.

It is this section as well as the Table of Initial Isolation and Protective Action Distances (green section), **Figure 26-12,** that make this book one of the necessary tools in the mitigation of a chemical release. The green section provides detailed isolation and evacuation information. This information relates to extremely hazardous substances, ones that can present a significant risk to the community. The orange section lists isolation distances for the emergency responders, and the green section is for the community. It also recommends that responders contact the emergency contact provided on the shipping papers or, if these are not available, then to use one of the contacts listed on the inside back cover. It provides some tactical objectives for handling radioactive substances and some additional considerations regarding these types of incidents.

The public safety section also lists personal protective equipment (PPE) recommendations and provides four basic PPE scenarios. It always recommends the use of positive-pressure SCBA, which is always a good recommendation when responding to chemical releases. The next level in severity is one in which chemical protective equipment is required because turnout gear will provide limited protection. In some cases it will state that turnout gear will provide limited protection, without any recommendation for other PPE. PPE is discussed in Section IV, Chapter 27 of this text, and many of these same issues will be covered in that chapter. The last recommendation

Name of Material	Guide No.	ID No.	Name of Material	Guide No.	ID No.
Allyl alcohol	131	1098	Aluminum phosphide pesticide	157	3048
Allylamine	131	2334	Aluminum powder, coated	170	1309
Allyl bromide	131	1099	Aluminum powder, pyrophoric	135	1383
Allyl chloride	131	1100	Aluminum powder, uncoated	138	1396
Allyl chlorocarbonate	155	1722	Aluminum processing by-products	138	3170
Allyl chloroformate	155	1722			
Allyl ethyl ether	131	2335	Aluminum remelting by-products	138	3170
Allyl formate	131	2336	Aluminum resinate	133	2715
Allyl glycidyl ether	129	2219	Aluminum silicon powder, uncoated	138	1398
Allyl iodide	132	1723			
Allyl isothiocyanate, inhibited	155	1545	Aluminum smelting by-products	138	3170
Allyl isothiocyanate, stabilized	155	1545	Amines, flammable, corrosive, n.o.s.	132	2733
Allyltrichlorosilane, stabilized	155	1724	Amines, liquid, corrosive, flammable, n.o.s.	132	2734
Aluminum, molten	169	9260			
Aluminum alkyl halides	135	3052	Amines, liquid, corrosive, n.o.s.	153	2735
Aluminum alkyl halides, liquid	135	3052	Amines, solid, corrosive, n.o.s.	154	3259
Aluminum alkyl halides, solid	135	3052	2-Amino-4-chlorophenol	151	2673
Aluminum alkyl halides, solid	135	3461	2-Amino-5-diethylaminopentane	153	2946
Aluminum alkyl hydrides	138	3076	2-Amino-4,6-dinitrophenol, wetted with not less than 20% water	113	3317
Aluminum alkyls	135	3051			
Aluminum borohydride	135	2870			
Aluminum borohydride in devices	135	2870	2-(2-Aminoethoxy)ethanol	154	3055
			N-Aminoethylpiperazine	153	2815
Aluminum bromide, anhydrous	137	1725	Aminophenols	152	2512
Aluminum bromide, solution	154	2580	Aminopyridines	153	2671
Aluminum carbide	138	1394	Ammonia, anhydrous	125	1005
Aluminum chloride, anhydrous	137	1726	Ammonia, anhydrous, liquefied	125	1005
Aluminum chloride, solution	154	2581	Ammonia, solution, with more than 10% but not more than 35% Ammonia	154	2672
Aluminum dross	138	3170			
Aluminum ferrosilicon powder	139	1395	Ammonia, solution, with more than 35% but not more than 50% Ammonia	125	2073
Aluminum hydride	138	2463			
Aluminum nitrate	140	1438			
Aluminum phosphide	139	1397	Ammonia solution, with more than 50% Ammonia	125	1005

Page 99

FIGURE 26-10 The blue pages are alphabetical by shipping name. These listings also provide the UN/NA number and the emergency response guide page.

is the suggestion that in some cases chemical protective equipment is a good idea, but that turnout gear is acceptable for fire situations.

This section also provides information regarding initial evacuation distances for large spills and for fires. If the material is listed in the Table of Initial Isolation and Protective Action Distances, the reader will be referred to this section, located at the back of the book in the green section.

SAFETY

The initial isolation distances listed in the orange section of the DOT book vary from a minimum of 30 feet (9 meters) to a maximum of 800 feet (243 meters). The minimums are derived for materials that are solid or present little risk to the responder. The higher the toxicity or the risk, the greater the isolation distance. As more information becomes available, this isolation distance can increase or decrease as needed.

TABLE 26-1	Chemical Names
Aspon-Chlordane	Octachlorodihydrodicyclopentadiene
Belt	1,2,4,5,6,7,8,8-Octachloro-2,3,3a,4,7,7a-hexahydro-4,7-methanoindene
CD 68	1,2,4,5,6,7,8,8-Octachloro-2,3,3a,4,7,7a-hexahydro-4,7-ethano-1H-indene
Chloordaan	1,2,4,5,6,7,8,8-Octachloro-3a,4,7,7a-hexahydro-4,7-methylene indane
Chlordan	Octachloro-4,7-methanohydroindane
g-Chlordan	Octachloro-4,7-methanotetrahydroindane
Chlorindan	1,2,4,5,6,7,8,8-Octachloro-4,7-methano-3a,4,7,7a-tetrahydroindane
Chlor kil	1,2,4,5,6,7,8,8-Octachloro-3a,4,7,7a-tetrahydro-4,7-methanoindane
Chlorodane	1,2,4,5,6,7,8,8-Octachloro-4,7,7a-Tetrahydro-4,7-methanoindane
Chlortox	1,2,4,5,6,7,8,8-Octachlor-3a,4,7,7a-Tetrahydro-4,7-endo-methano-indan (German)
Clordan	Octa-Klor
Clorodane	Oktaterr
Cortilan-Neu	Ortho-Klor
Dichlorochlordene	1,2,4,5,6,7,8,8-Ottochloro-3A,4,7,7A-Tetraidro-4,7-endo-methano-indano (Italian)
Dowchlor	RCRA Waste Number U036
ENT 9,932	SD 5532
ENT 25,552-X	Shell SD-5532
HCS 3260	Synklor
Kypchlor	Tat Chlor 4
M 140	Topichlor 20
NCI-C00099	Topiclor
Niran	Topiclor 20
1,2,4,5,6,7,8,8-Octachloor-3a,4,7,7a-tetrahydro-4,7-endo-methano-indaan (Dutch)	Toxichlor
1,2,4,5,6,7,10,10-Octachloro-4,7,8,9-tetrahydro-4,7-methyleneindane	Velsicol 1068

Excerpted from Richard J. Lewis, Sr., *Dangerous Properties of Industrial Materials,* 8th ed. Van Nostrand Reinhold Co., New York, 1994.

The emergency response section provides information regarding fires, spills, and first aid. The fire section is divided between small fires and large fires, and potential tactics that can be used for both. For both types of fires, there is a listing of what type of extinguishing agent may be needed such as water, foam, dry chemical, halon, or CO_2. Specific materials may also list special agents that may be needed. Listed under the large fire section are some tactical considerations as well as some recommendations for specific substances. If the material is carried by tank truck or railcar, the book provides some additional information on fighting those types of fires. In this section some helpful hints are given that apply in many

GUIDE 111   MIXED LOAD/UNIDENTIFIED CARGO   ERG2004

POTENTIAL HAZARDS

FIRE OR EXPLOSION

- May explode from heat, shock, friction or contamination.
- May react violently or explosively on contact with air, water or foam.
- May be ignited by heat, sparks or flames.
- Vapors may travel to source of ignition and flash back.
- Containers may explode when heated.
- Ruptured cylinders may rocket.

HEALTH

- Inhalation, ingestion or contact with substance may cause severe injury, infection, disease or death.
- High concentration of gas may cause asphyxiation without warning.
- Contact may cause burns to skin and eyes.
- Fire or contact with water may produce irritating, toxic and/or corrosive gases.
- Runoff from fire control may cause pollution.

PUBLIC SAFETY

- CALL Emergency Response Telephone Number on Shipping Paper first. If Shipping Paper not available or no answer, refer to appropriate telephone number listed on the inside back cover.
- As an immediate precautionary measure, isolate spill or leak area for at least 100 meters (330 feet) in all directions.
- Keep unauthorized personnel away.
- Stay upwind.
- Keep out of low areas.

PROTECTIVE CLOTHING

- Wear positive pressure self-contained breathing apparatus (SCBA).
- Structural firefighters' protective clothing provides limited protection in fire situations ONLY; it may not be effective in spill situations.

EVACUATION

Fire

- If tank, rail car or tank truck is involved in a fire, ISOLATE for 800 meters (1/2 mile) in all directions; also, consider initial evacuation for 800 meters (1/2 mile) in all directions.

Page 170

FIGURE 26-11   The response guide in the orange section provides the responder with basic information regarding the health, fire, and public safety issues. Each part needs to be read completely before taking any action.

cases involving fires and these types of containers. The hints include the following:

- Fight fire from a distance using unstaffed monitors.
- Withdraw immediately if the sound level from the venting safety device rises or the tank begins to discolor.

- Cool containers with flooding quantities of water until well after the fire is out.

The spill or leak section lists some general tactical objectives and provides some specific information on some substances. It is important to be aware that this section could lead responders into an action that may be beyond their level of training if not used

ERG2004   MIXED LOAD/UNIDENTIFIED CARGO   **GUIDE 111**

EMERGENCY RESPONSE

FIRE
CAUTION: Material may react with extinguishing agent.

Small Fires
- Dry chemical, CO_2, water spray or regular foam.

Large Fires
- Water spray, fog or regular foam.
- Move containers from fire area if you can do it without risk.

Fire involving Tanks
- Cool containers with flooding quantities of water until well after fire is out.
- Do not get water inside containers.
- Withdraw immediately in case of rising sound from venting safety devices or discoloration of tank.
- ALWAYS stay away from tanks engulfed in fire.

SPILL OR LEAK
- Do not touch or walk through spilled material.
- ELIMINATE all ignition sources (no smoking, flares, sparks or flames in immediate area).
- All equipment used when handling the product must be grounded.
- Keep combustibles (wood, paper, oil, etc.) away from spilled material.
- Use water spray to reduce vapors or divert vapor cloud drift. Avoid allowing water runoff to contact spilled material.
- Prevent entry into waterways, sewers, basements or confined areas.

Small Spills • Take up with sand or other non-combustible absorbent material and place into containers for later disposal.

Large Spills • Dike far ahead of liquid spill for later disposal.

FIRST AID
- Move victim to fresh air. • Call 911 or emergency medical service.
- Give artificial respiration if victim is not breathing.
- **Do not use mouth-to-mouth method if victim ingested or inhaled the substance; give artificial respiration with the aid of a pocket mask equipped with a one-way valve or other proper respiratory medical device.**
- Administer oxygen if breathing is difficult.
- Remove and isolate contaminated clothing and shoes.
- In case of contact with substance, immediately flush skin or eyes with running water for at least 20 minutes.
- Shower and wash with soap and water.
- Keep victim warm and quiet.
- Effects of exposure (inhalation, ingestion or skin contact) to substance may be delayed.
- Ensure that medical personnel are aware of the material(s) involved and take precautions to protect themselves.

Page 171

FIGURE 26-11 (Continued).

correctly. As an example, most of the guides recommend that the reader can stop a leak if it can be done without risk. The intention is for a responder trained to the operations level to be able to close a remote shutoff valve if available. Most readers may incorrectly follow the recommendation and try to stop the leak regardless of the method, a tactic that requires a hazardous materials technician to accomplish safely. Another area of concern is the suggestion that a water spray be used to knock down vapors. If necessary to save lives, using a water spray is an acceptable tactic.

Page 320

TABLE OF INITIAL ISOLATION AND PROTECTIVE ACTION DISTANCES

ID No.	NAME OF MATERIAL	SMALL SPILLS (From a small package or small leak from a large package)				LARGE SPILLS (From a large package or from many small packages)			
		First ISOLATE in all Directions Meters (Feet)		Then PROTECT persons Downwind during- DAY Kilometers (Miles)	NIGHT Kilometers (Miles)	First ISOLATE in all Directions Meters (Feet)		Then PROTECT persons Downwind during- DAY Kilometers (Miles)	NIGHT Kilometers (Miles)
2204 2204	Carbonyl sulfide Carbonyl sulphide	30 m	(100 ft)	0.1 km (0.1 mi)	0.6 km (0.4 mi)	300 m	(1000 ft)	3.0 km (1.9 mi)	8.1 km (5.0 mi)
2232 2232	Chloroacetaldehyde 2-Chloroethanal	30 m	(100 ft)	0.2 km (0.1 mi)	0.3 km (0.2 mi)	90 m	(300 ft)	0.8 km (0.5 mi)	1.6 km (1.0 mi)
2334	Allylamine	30 m	(100 ft)	0.1 km (0.1 mi)	0.5 km (0.3 mi)	120 m	(400 ft)	1.1 km (0.7 mi)	2.5 km (1.5 mi)
2337	Phenyl mercaptan	30 m	(100 ft)	0.1 km (0.1 mi)	0.1 km (0.1 mi)	60 m	(200 ft)	0.4 km (0.2 mi)	0.6 km (0.4 mi)
2382 2382	1,2-Dimethylhydrazine Dimethylhydrazine, symmetrical	30 m	(100 ft)	0.1 km (0.1 mi)	0.2 km (0.1 mi)	60 m	(200 ft)	0.6 km (0.4 mi)	1.2 km (0.8 mi)
2407	Isopropyl chloroformate	30 m	(100 ft)	0.1 km (0.1 mi)	0.3 km (0.2 mi)	90 m	(300 ft)	0.7 km (0.5 mi)	1.5 km (0.9 mi)
2417 2417	Carbonyl fluoride Carbonyl fluoride, compressed	30 m	(100 ft)	0.2 km (0.1 mi)	1.1 km (0.7 mi)	90 m	(300 ft)	1.0 km (0.6 mi)	3.6 km (2.3 mi)
2418 2418	Sulfur tetrafluoride Sulphur tetrafluoride	60 m	(200 ft)	0.7 km (0.4 mi)	3.2 km (2.0 mi)	500 m	(1600 ft)	4.7 km (2.9 mi)	10.6 km (6.6 mi)
2420	Hexafluoroacetone	30 m	(100 ft)	0.3 km (0.2 mi)	1.3 km (0.8 mi)	800 m	(2500 ft)	7.2 km (4.5 mi)	11.0+ km (7.0+ mi)
2421	Nitrogen trioxide	30 m	(100 ft)	0.1 km (0.1 mi)	0.5 km (0.3 mi)	60 m	(200 ft)	0.4 km (0.3 mi)	1.9 km (1.2 mi)
2437	Methylphenyldichlorosilane (when spilled in water)	30 m	(100 ft)	0.1 km (0.1 mi)	0.1 km (0.1 mi)	30 m	(100 ft)	0.3 km (0.2 mi)	1.1 km (0.7 mi)
2438	Trimethylacetyl chloride	30 m	(100 ft)	0.1 km (0.1 mi)	0.2 km (0.1 mi)	60 m	(200 ft)	0.5 km (0.3 mi)	0.8 km (0.5 mi)
2442	Trichloroacetyl chloride	30 m	(100 ft)	0.2 km (0.2 mi)	0.8 km (0.5 mi)	120 m	(400 ft)	1.2 km (0.8 mi)	2.2 km (1.4 mi)
2474	Thiophosgene	90 m	(300 ft)	0.8 km (0.5 mi)	2.4 km (1.5 mi)	360 m	(1200 ft)	3.6 km (2.3 mi)	6.8 km (4.2 mi)
2477	Methyl isothiocyanate	30 m	(100 ft)	0.1 km (0.1 mi)	0.2 km (0.1 mi)	60 m	(200 ft)	0.5 km (0.3 mi)	1.0 km (0.7 mi)
2480	Methyl isocyanate	60 m	(200 ft)	0.5 km (0.3 mi)	1.9 km (1.2 mi)	600 m	(1800 ft)	5.4 km (3.3 mi)	11.0+ km (7.0+ mi)

FIGURE 26-12 The Initial Isolation and Protective Action Distances table provides recommended isolation distances for the materials that were highlighted in the yellow or blue sections. The table is divided into small spills and large spills.

SAFETY

The use of water indiscriminately to knock down "vapors" can create many other problems. These problems include increasing the spill size, creating runoff problems, possible chemical reactions, and adding an additional hazard to other responders who will try to mitigate the leak. If necessary, as in the case of knocking down anhydrous ammonia vapors that may affect a neighborhood, it is an acceptable tactic. To flow water into the back of a trailer that has a leaking drum, in which the contents have not been identified, is inappropriate and could cause further problems.

The first-aid section provides basic medical treatment and some basic decontamination recommendations for chemical burns. The information is basic and if confronted with a patient, the responder should contact the local hospital or Chemtrec for further assistance. The Table of Initial Isolation and Protective Action Distances is the green section in the back of the ERG, Figure 26-12. It provides specific isolation and evacuation distances for the materials that were high-

lighted in the yellow or blue section. An example of how to use the isolation distances is shown in **Figure 26-13.** In order to use the initial isolation protection, firefighters should identify the product by name and identification number. Firefighters must also identify the wind direction in relation to the incident location. The next step is to look in Table 1 (the green-bordered pages) for the ID number and name of the material involved in the incident. When more than one material is identified, use the entry with the largest protective action distance. Next, determine if the incident involves a small or large spill and the time of occurrence, day or night. Generally, a small spill is one that involves a single, small package (e.g., a drum containing up to approximately 200 liters), a small cylinder, or a small leak from a large package. A large spill is one that involves a spill from a large package, or multiple spills from many small packages. Day is any time after sunrise and before sunset. Night is any time between sunset and sunrise.

The initial isolation area is then located and all persons should be directed to move, in a crosswind direc-

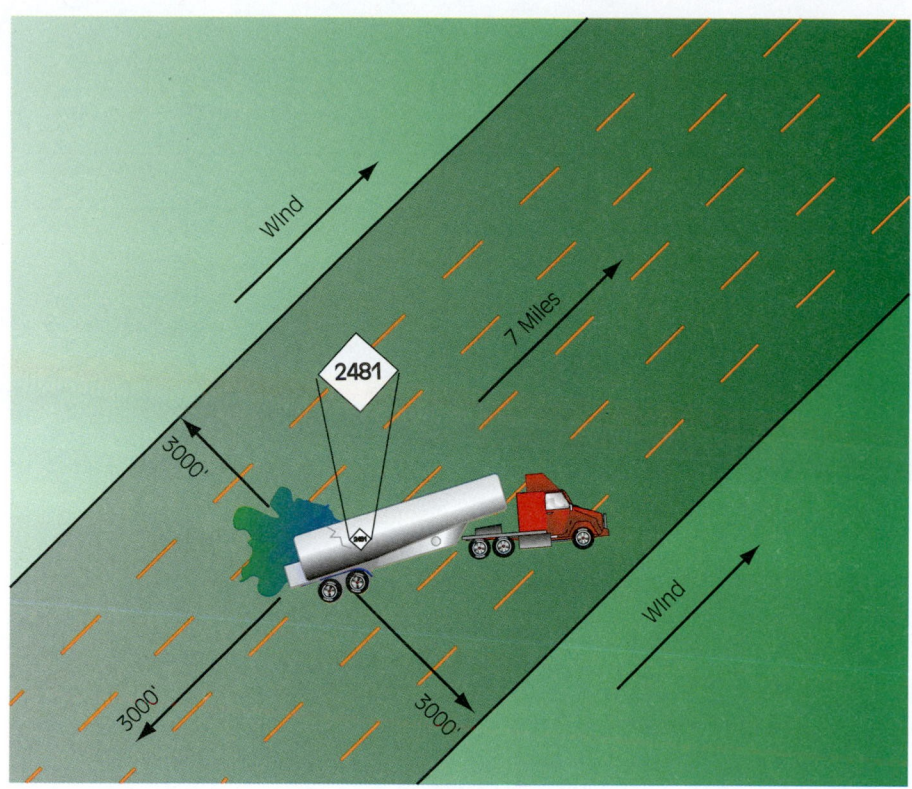

FIGURE 26-13 A large spill of ethyl isocyanate (ID 2481) is listed as having an isolation distance of 1,700 feet in all directions. In such a case, responders need to protect those persons downwind for 7+ miles.

tion, away from the spill to the distance specified. The protective action distance considers the material, spill size, and whether day or night, and gives the downwind distance for which protective actions should be considered. For practical purposes, the Protective Action Zone is a square, whose length and width are the same as the downwind. When initiating an evacuation of the protective action zone, firefighters should begin with those closest to the spill site and work away from the site in the downwind direction. The shape of the area in which protective actions should be taken into consideration are as shown in Figure 26-13. In the figure the spill is located at the center of the small circle. The larger circle represents the INITIAL ISOLATION zone around the spill.

> ### NOTE
>
> The distances are a guide for the first thirty minutes of an incident and use a typical day as defined by the DOT. The ERG is a guide to get an evacuation started. As soon as possible, air monitoring needs to be established to definitively define the evacuation distances. When establishing the protective action distances for a hazardous materials incident, exposures within the protective action zone must be considered by the firefighters. These exposures include people, property, and the environment (e.g., schools in session, office building, public waterways). Using proper risk management analysis, firefighters should use this information when formulating evacuation plans to protect the greatest number of exposures from potential harm.

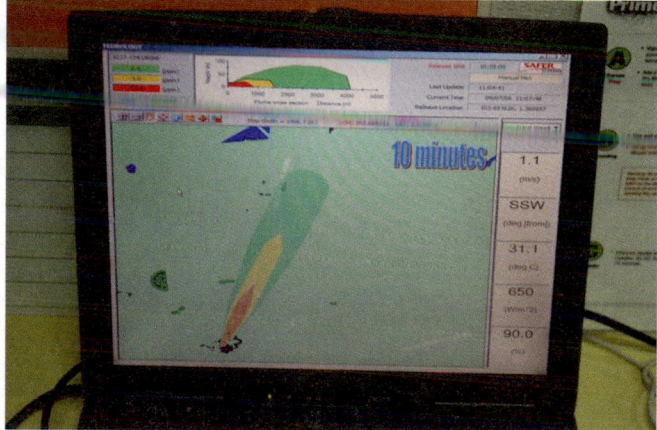

FIGURE 26-14 Software is available that can plot toxic gas cloud plumes, which can help determine isolation areas and guide evacuation decisions.

In addition to the protective action zone, responders may also utilize an emergency response software package known as CAMEO that uses a vapor cloud modeling program known as ALOHA. This modeling program is generally referred to as a plume dispersion model, an example of which is shown in **Figure 26-14,** and can assist with evaluating potential downwind evacuations, using real-time weather and data.

The DOT defines a small spill as a leaking container smaller than a 55-gallon (208 liters) drum or a leak from a small cylinder, whereas a large spill is

TABLE OF WATER-REACTIVE MATERIALS WHICH PRODUCE TOXIC GASES

Materials Which Produce Large Amounts of Toxic-by-Inhalation (TIH) Gas(es) When Spilled in Water

ID No.	Guide No.	Name of Material	TIH Gas(es) Produced	
1716	156	Acetyl bromide	HBr	
1717	155	Acetyl chloride	HCl	
1724	155	Allyltrichlorosilane, stabilized	HCl	
1725	137	Aluminum bromide, anhydrous	HBr	
1726	137	Aluminum chloride, anhydrous	HCl	
1728	155	Amyltrichlorosilane	HCl	
1732	157	Antimony pentafluoride	HF	
1745	144	Bromine pentafluoride	HF	Br$_2$
1746	144	Bromine trifluoride	HF	Br$_2$
1747	155	Butyltrichlorosilane	HCl	
1752	156	Chloroacetyl chloride	HCl	
1754	137	Chlorosulfonic acid	HCl	
1754	137	Chlorosulfonic acid and Sulfur trioxide mixture	HCl	
1754	137	Chlorosulphonic acid	HCl	
1754	137	Chlorosulphonic acid and Sulphur trioxide mixture	HCl	
1754	137	Sulfur trioxide and Chlorosulfonic acid	HCl	
1754	137	Sulphur trioxide and Chlorosulphonic acid	HCl	
1758	137	Chromium oxychloride	HCl	
1763	156	Cyclohexyltrichlorosilane	HCl	
1766	156	Dichlorophenyltrichlorosilane	HCl	
1767	155	Diethyldichlorosilane	HCl	
1769	156	Diphenyldichlorosilane	HCl	
1771	156	Dodecyltrichlorosilane	HCl	
1777	137	Fluorosulfonic acid	HF	

Chemical Symbols for TIH Gases:

Br$_2$	Bromine	HF	Hydrogen fluoride	PH$_3$	Phosphine
Cl$_2$	Chlorine	HI	Hydrogen iodide	SO$_2$	Sulfur dioxide
HBr	Hydrogen bromide	H$_2$S	Hydrogen sulfide	SO$_2$	Sulphur dioxide
HCl	Hydrogen chloride	H$_2$S	Hydrogen sulphide	SO$_3$	Sulfur trioxide
HCN	Hydrogen cyanide	NH$_3$	Ammonia	SO$_3$	Sulphur trioxide

Use this list only when material is spilled in water. *Page 345*

FIGURE 26-15 List of Dangerous Water-Reactive Materials from the green section of the ERG (1 of 4 pages).

defined as coming from a container larger than 55 gallons (208 liters) or a large leak from a cylinder. These are further explained in that a leak from several small containers may be considered a large leak. A small leak from a large container may also be considered a small leak.

The last section included in the green section is the List of Dangerous Water-Reactive Materials, **Figure 26-15.** This section provides the evacuation distances for these materials if they contact water. The distances vary from 0.3 (0.5 km) to 6 miles (9.7 km). Another helpful item that this section provides is the chemical that this material makes when it con-

tacts water, an unusual piece of information for this type of text.

The last pages are filled with definitions, glossary, and explanations. The inside back cover provides a listing of additional emergency contacts for the United States, Canada, and Mexico. Further discussion regarding these agencies is found later in this chapter.

The firefighter should consider these general reminders when using the DOT ERG:

- It is an excellent source for evacuation and isolation distances.

- It is an excellent source for water-reactive hazards.

- It is an excellent first responder document, one that will keep the responders safe.

- It is a great starting point for the first thirty minutes of the incident.

- The specific chemical information is limited, and in some cases too general for a hazardous materials team, although it is well suited for the first responder level.

- Specific identification of the product is essential to responder and citizen safety.

JPR 26-1: Using the DOT Emergency Response Guide (ERG)

(For step-by-step photos of this skill sequence, see page 1099)

JPR 26-1 outlines specific steps in using the ERG when responding to an incident involving dangerous goods. Please note that this JPR sequence is a basic guideline, and users should reference the first page of the ERG (as shown in Figure 26-2) for further information. Users should also practice using the ERG prior to an incident so that they are familiar with its contents and fully prepared to effectively mitigate the threat of hazardous materials.

1. Resist rushing in! Approach incidents from an upwind direction and stay clear of all spills, vapors, fumes, and smoke.

2. Identify the material by finding any one of the following:

   - 4-digit ID number on placard/ID panel.
   - 4-digit ID number on shipping document or package.
   - Name of material on shipping document, placard, or package.

3. Look up the material's 3-digit guide number in either:

   - ID number index
   - Name of material index

4. Turn to the numbered guide.

MATERIAL SAFETY DATA SHEETS

Material Safety Data Sheets, or MSDS, as they are commonly called, are a result of the hazard communication standard—OSHA regulation 29 CFR 1910.1200—which became effective in 1994. As stated in Section IV Chapter 24, this regulation requires employers who use chemicals above the household quantity to create MSDS. They are also required to develop a hazard communication plan, label all chemical containers, and provide training to employees on an annual basis. This training is based on the chemical hazards the employees will face in the workplace. An MSDS is required to have a variety of information. The amount, quality, and order of information is determined by the manufacturer. The original intent of the MSDS was to protect employees working at the facility, not emergency responders. Although many of the sheets do have applicability and serve a useful purpose, in some cases they may not have the necessary emergency response information. The information that may be found on an MSDS is listed in **Table 26-2.**

The quality of information varies from MSDS to MSDS and from manufacturer to manufacturer. The early MSDS had a lot of good information; however, as litigation has increased, the typical MSDS provides a worst-case scenario, which results in an extremely conservative approach to the handling of the chemical.

Given the choice of using the DOT ERG or an MSDS, the firefighter should rely more on the technical information on the MSDS. The DOT ERG places many chemicals into a few general categories. Although it may be conservative, the MSDS is specific to the chemical. It is important through the preplanning process that responders learn about the chemicals from a facility representative. Responders should work through the MSDS and the ERG and determine prior to the incident which information is most beneficial.

Although many attempts have been made to modify the format of the MSDS and to improve the quality of information, the MSDS has remained the same since inception. There are a couple of recommended formats, and a new format exists for the European chemical industry. It can be anticipated that the United States will have to conform to this format. The information provided in Table 26-2 is from this format. One of the improvements with this new MSDS format is a section specifically on emergency response procedures and considerations for firefighters. A sample MSDS is provided in **Figure 26-16.**

Using the MSDS Wisely

It is always recommended that responders use more than one source of information. The EPA issued a safety alert in June 1999 on the use of the MSDS, and the majority of that bulletin is reproduced here.

TABLE 26-2   Information Included on an MSDS

Information	Note
Chemical product and company identification	The name the chemical company uses to identify the product is used near the top of the MSDS. Information related to the manufacturer, such as the address and other contact information, is provided near the top of the first sheet. The MSDS is required to have a twenty-four-hour emergency contact number, although in most cases this number is Chemtrec's.
Chemical composition	The ingredients of the chemical are listed here. If the mixture is a trade secret, there is a provision to exclude this information. If this information is needed for medical reasons, the manufacturer is to provide this information to the physician. In some cases the true identity of the material may never be known, and only the hazards may be provided.
Hazards identification	In this section, an emergency overview is provided for both an employer and emergency responders. This section is lengthy and includes potential health effects, first-aid, and firefighting measures. The exposure levels are typically included with the health effects but may be listed separately. New to MSDS will be a section on accidental releases. This section will be beneficial to emergency responders. Handling and storage considerations, as well as engineering controls including proper PPE, are also provided in this section.
Physical and chemical properties	Specific chemical information such as boiling points, vapor pressures, flash points, and many other specific items are included in this section.
Stability and reactivity	Some chemicals become unstable after a period of time or as a result of poor storage conditions, or they may react with other chemicals. Any information regarding this type of detail is provided here.
Toxicology information	The long-term health effects and other concerns regarding acute and chronic exposures are provided in this section.
Ecological information	Information regarding any potential environmental effects is listed here.
Disposal considerations	Although in many cases this section states "follow local regulations," this is the section that outlines any regulatory requirements for disposal.
Transport information	Information regarding the DOT regulations is listed here—typically the hazard class and UN identification number.
Regulatory information	Any other regulations that apply to the use, storage, or disposal of the chemical are listed here. If the chemical is covered by any of the other regulations, this information is listed in this section.

CAUTION

A critical consideration when choosing a response strategy is the safety of emergency responders. Adequate information about on-site chemicals can make a big difference when choosing a safe response strategy. This information must include name, toxicity, physical and chemical characteristics, fire and reactivity hazards, emergency response procedures, spill control, and protective equipment. Generally, responders rely primarily on Material Safety Data Sheets (MSDS) maintained at the facility. However, MSDS may not provide sufficient information to effectively and safely respond to accidental releases. This Alert is designed to increase awareness of MSDS limitations, so that first responders can take proper precautions, and identify additional sources of chemical information, which could help prevent death or injury.

JPR 26-2: Determining an Action Plan Using the MSDS

(For step-by-step photos of this skill sequence, see page 1102)

A Material Safety Data Sheet can be a very effective tool when responding to an incident at a facility that houses chemicals. It can cut down on the response time and avoid injury or even death of those involved in the incident. When responding to such a facility, you would be well advised to follow these procedures:

1. Review the MSDS to determine the chemical threat and whether there are other products in the chemical that are of concern.

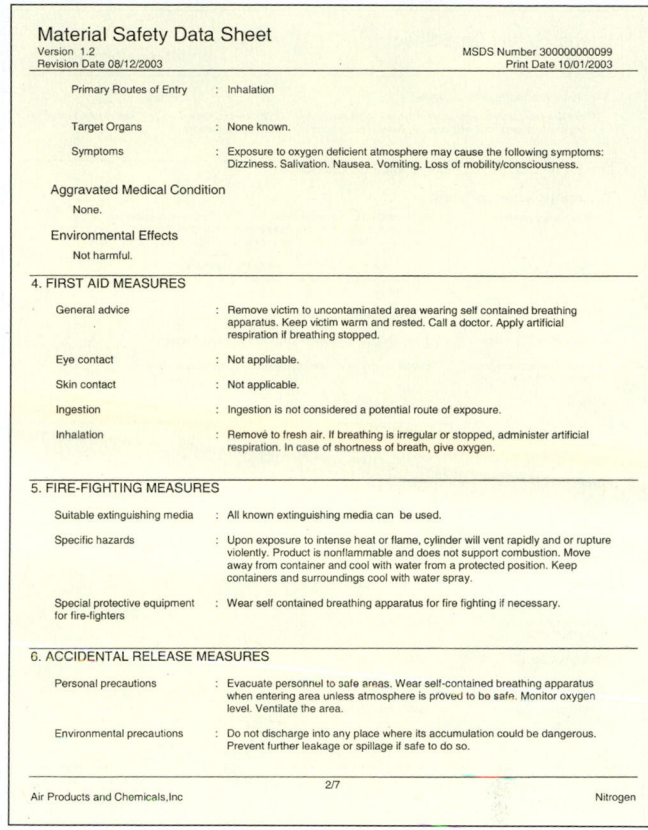

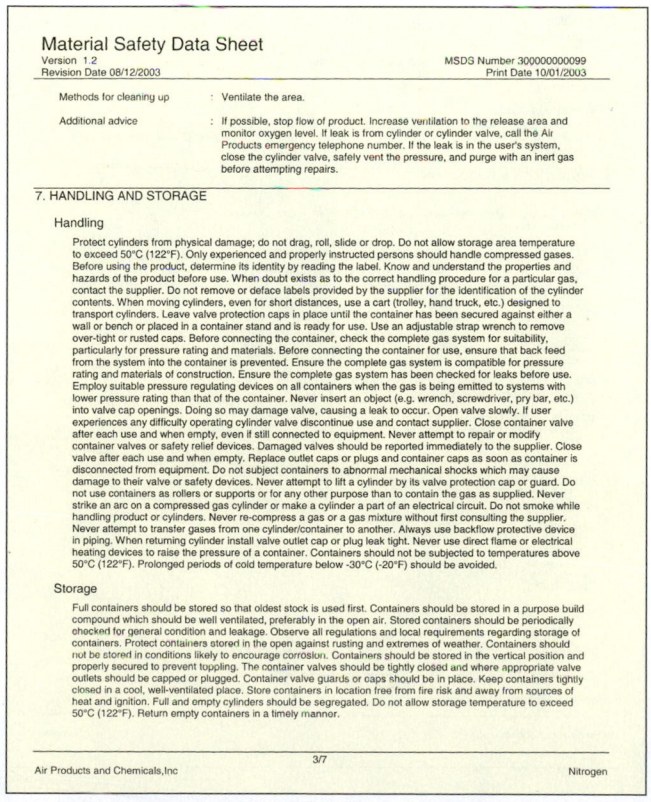

FIGURE 26-16 An example of an MSDS. *(Courtesy of Air Products and Chemicals, Inc. Presented for illustrative purposes only)*

Material Safety Data Sheet

Version 1.2
Revision Date 08/12/2003

MSDS Number 300000000099
Print Date 10/01/2003

Technical measures/Precautions

Containers should be segregated in the storage area according to the various categories (e.g. flammable, toxic, etc.) and in accordance with local regulations. Keep away from combustible material.

8. EXPOSURE CONTROLS / PERSONAL PROTECTION

Personal protective equipment

Respiratory protection	: Self contained breathing apparatus (SCBA) or positive pressure airline with mask are to be used in oxygen-deficient atmosphere. Air purifying respirators will not provide protection. Users of breathing apparatus must be trained.
Hand protection	: Sturdy work gloves are recommended for handling cylinders. The breakthrough time of the selected glove(s) must be greater than the intended use period.
Eye protection	: Safety glasses recommended when handling cylinders.
Skin and body protection	: Safety shoes are recommended when handling cylinders.
Special instructions for protection and hygiene	: Ensure adequate ventilation, especially in confined areas.
Remarks	: Simple asphyxiant.

9. PHYSICAL AND CHEMICAL PROPERTIES

Form	: Compressed gas.
Color	: Colorless gas
Odor	: No odor warning properties.
Molecular Weight	: 28 g/mol
Relative vapor density	: 0.97 (air = 1)
Density at 70 °F (21 °C)	: 0.075 lb/ft3 (0.0012 g/cm3) Note: (as vapor)
Specific Volume at 70 °F (21 °C)	: 13.80 ft3/lb (0.8615 m3/kg)
Boiling point/range	: -320.8 °F (-196 °C)
Critical temperature	: -232.6 °F (-147 °C)
Melting point/range	: -346.0 °F (-210 °C)
Water solubility	: 0.02 g/l

4/7

Air Products and Chemicals, Inc   Nitrogen

Material Safety Data Sheet

Version 1.2
Revision Date 08/12/2003

MSDS Number 300000000099
Print Date 10/01/2003

10. STABILITY AND REACTIVITY

Stability	: Stable under normal conditions.
Hazardous decomposition products	: None.

11. TOXICOLOGICAL INFORMATION

No known toxicological effects from this product.

12. ECOLOGICAL INFORMATION

Ecotoxicity effects

Aquatic toxicity	: No data available.
Toxicity to other organisms	: No data available.
Mobility	: No data available.
Bioaccumulation	: No data available.

Further information

No ecological damage caused by this product.

13. DISPOSAL CONSIDERATIONS

Waste from residues / unused products	: Contact supplier if guidance is required. Return unused product in orginal cylinder to supplier.
Contaminated packaging	: Return cylinder to supplier.

14. TRANSPORT INFORMATION

CFR

Proper shipping name	: Nitrogen, compressed
Class	: 2.2
UN/ID No.	: UN1066

IATA

Proper shipping name	: Nitrogen, compressed
Class	: 2.2
UN/ID No.	: UN1066

IMDG

Proper shipping name	: NITROGEN, COMPRESSED
Class	: 2.2
UN/ID No.	: UN1066

5/7

Air Products and Chemicals, Inc   Nitrogen

Material Safety Data Sheet

Version 1.2
Revision Date 08/12/2003

MSDS Number 300000000099
Print Date 10/01/2003

CTC

Proper shipping name	: NITROGEN, COMPRESSED
Class	: 2.2
UN/ID No.	: UN1066

Further Information

Avoid transport on vehicles where the load space is not separated from the driver's compartment. Ensure vehicle driver is aware of the potential hazards of the load and knows what to do in the event of an accident or an emergency.

15. REGULATORY INFORMATION

OSHA Hazard Communication Standard (29 CFR 1910.1200) Hazard Class(es)
Compressed Gas

Country	Regulatory list	Notification
USA	TSCA	Included on Inventory.
EU	EINECS	Included on Inventory.
Canada	DSL	Included on Inventory.
Australia	AICS	Included on Inventory.
South Korea	ECL	Included on Inventory.
China	SEPA	Included on Inventory.
Philippines	PICCS	Included on Inventory.
Japan	ENCS	Included on Inventory.

EPA SARA Title III Section 312 (40 CFR 370) Hazard Classification:
Sudden Release of Pressure Hazard

US. California Safe Drinking Water & Toxic Enforcement Act (Proposition 65)
This product does not contain any chemicals known to State of California to cause cancer, birth defects or any other harm.

16. OTHER INFORMATION

NFPA Rating

Health	: 0
Fire	: 0
Instability	: 0
Special	: SA

HMIS Rating

Health	: 0
Flammability	: 0
Physical hazard	: 3

Prepared by   : Air Products and Chemicals, Inc. Global EH&S Product Safety Department

For additional information, please visit our Product Stewardship web site at

http://www.airproducts.com/productstewardship/

6/7

Air Products and Chemicals, Inc   Nitrogen

FIGURE 26-16 (Continued) *(Courtesy of Air Products and Chemicals, Inc. Presented for illustrative purposes only)*

2. Don the appropriate level PPE for the chemical threat, particularly in cases where the material has escaped its container.

3. If the material has released into the building, follow procedures for evacuating and securing the building.

4. Determine which extinguishing agents are required for the incident and identify the container in which the material will be found.

Accidents and How the MSDS Relates

In May 1997 a massive explosion and fire occurred at an agricultural chemical packaging facility in eastern Arkansas. Prior to the explosion, employees observed smoke in a back warehouse and evacuated. The facility called local responders and asked for help to control smoldering inside a pesticide container. The local fire department rapidly responded and reviewed the smoldering product's MSDS. The MSDS lacked information on decomposition temperatures or explosion hazards. The firefighters decided to investigate the building. While they were approaching, a violent explosion occurred. Fragments from a collapsing cinder block wall killed three firefighters and seriously injured a fourth.

In April 1995, an explosion and fire at a manufacturing facility in Lodi, New Jersey, caused the death of five responders. The explosion occurred while the company was blending aluminum powder, sodium hydrosulfite, and other ingredients. Even though the material was water reactive, the MSDS for the product advised the use of a "water spray..to extinguish fire." The recommendation in the MSDS for "small fires" was to flood with water; however, "small fire" was not defined, the amount of water necessary was not specified, and no information dealt with how to respond to large fires (which can occur during blending processes).

The MSDS only described the hazards associated with the product. In this case, responders needed information on the hazards associated with the reactivity during the blending process (which was significantly different from the product).

Emergency responders should note that the chemical information provided on an MSDS usually presents the hazards associated with that particular product. Once the product is placed in a process some factors may change, resulting in the increase, decrease, or elimination of hazards. These factors may include reactions with other chemicals and changes in temperature, pressure, and physical/chemical characteristics.

MSDS in the Workplace

In 1988 the Occupational Safety and Health Administration (OSHA) required facilities storing or using hazardous chemicals to comply with the Hazard Communication Standard. This standard requires employers to provide employees with an MSDS for every hazardous chemical present on site and to train those employees to properly recognize the hazards of the chemicals and to handle them safely. An MSDS normally provides information on the physical/chemical characteristics and first-aid procedures. This information is valuable for employees to safely work with the chemical. However, the content for the MSDS on emergency response procedures, fire, and reactive hazards may be insufficient for local responder use in an emergency situation. Vagueness, technical jargon, understandability, product versus process concerns, and missing information on an MSDS may increase the risk to emergency responders. MSDS are provided by manufacturers, importers, and/or distributors. MSDS chemical hazard information can vary substantially depending on the provider. Sometimes this discrepancy is due to different testing procedures. However, whoever prepared the MSDS is responsible for ensuring the accuracy of the hazard information. **Table 26-3** summarizes information from various MSDS for the chemical azinphos methyl and illustrates how different sources can provide varied and conflicting information.

SHIPPING PAPERS

Besides the use of placards when chemicals are transported, the carrier is required to provide shipping papers for the cargo. The shipping papers generally provide the following information:

- Shipping company
- Destination of the packages
- Emergency contact information
- The number and weight of the packages
- The proper shipping name of the materials and the hazard class of the materials
- Special notations for hazardous materials

If the vehicle is not carrying hazardous materials, there is no requirement to have any specific information, which at times can be confusing for first responders. When carrying hazardous materials the shipping papers may include a packing group (PG) number listed as a I, II, or III. The lower the number, the worse the chemical is, and it may have special shipping requirements.

TABLE 26-3   Comparison of MSDS Data for Azinphos Methyl—AZM

MSDS Note	MSDS–A	MSDS–B	MSDS–C	MSDS–D	CAMEO
Hazard Rating	Health—2 Flammable—0 Reactivity—0	None listed	Health—3 Flammable—2 Reactivity—2	Health—4 Flammable—0 Reactivity—0	Health—3 (extremely hazardous) Flammable—2 (ignites when moderately heated) Reactivity—2 (violent chemical change possible)
Reactivity Hazards	Stable under normal conditions. Hazardous polymerization will not occur.	Depends on the characteristics of dust; decomposes under influence of acids and bases.	Stable material. Unstable above 100°F sustained temperature. Hazardous polymerization will not occur.	Releases toxic, corrosive, flammable, or explosive gases. Polymerization will not occur.	Will decompose.
Incompatibility	High Temperatures, oxidizers, alkaline substances	Acids and bases	Heat, moisture	Heat, flames, sparks, and other ignition sources	Heat, UV light
Fire Hazards	Vapors from fire are hazardous.	Combustible, gives off irritating or toxic fumes (or gases) in a fire.	Decomposes above 130°F with gas evolution and dense smoke. Dust explosion hazard for large dust cloud.	Containers may rupture or explode if exposed to heat.	Decomposes giving off ammonia, hydrogen, and CO.

Other information may include a reportable quantity (RQ) for a hazardous material. Some chemicals have a threshold of reporting when spilled much like in storage situations and SARA section 304. The RQ will be listed on the shipping papers. If the material is spilled and exceeds the quantity listed, the driver must report the spill to the **National Response Center (NRC).**

The driver/operator of the vehicle is supposed to keep the shipping papers with the vehicle at all times and should be able to provide them upon request.

CAUTION

Problems occur in accidents where there are multiple shipping papers from a variety of pickups. If the driver has kept all of the previous shipping papers from other days, it can be confusing sorting out what the actual cargo is.

If the driver provides information regarding the cargo, and the shipping papers agree with that information, the results should be confirmed with the shipping company. If all three are in agreement as to what is being shipped, it is probably accurate, but this is not guaranteed. A driver who picked up some illegal waste is not going to put it on a shipping paper, nor placard the vehicle, and probably will not admit to having it. It is obvious that the shipping company will not know about it, so until visually confirmed with appropriate levels of PPE, one must never assume what a cargo consists of.

Mode of Transportation

In highway transportation the shipping papers are called a bill of lading, or most commonly just shipping papers. The papers are supposed to be within arm's reach of the driver, most commonly in a pouch on the driver's door. On tank trucks a duplicate set of papers may be located in a tube attached near the landing gear on the driver's side. In most cases the driver will leave the papers in the truck, and they usually need to be retrieved. If the driver had many pickup stops there will be multiple papers, and some companies put each individual item on a ticket. The hazardous materials are sometimes color coded, as in the case of United Parcel Service (UPS), which uses red tabs to identify hazardous materials packages.

For rail, the shipping papers are called the **consist** or **waybill** and are in the control of the engineer, or the conductor if the train has one. There may be two sets of papers, one listing the contents of each car and one that is by car number, with the first car after the engine being number 1 (in some cases car number 1 is the last car). In a derailment this numbering system usually goes out the window but in some cases may be useful. Most railcars are identified well with car numbers, and as long as they are upright can usually be identified.

Rail uses a number known as a **Standard Transportation Commodity Code (STCC),** usually referred to as a "stick" number. It is a seven-digit number, and if it starts with a "49," the material is a hazardous material. Out of all of the modes of transportation, in rail incidents the engineer is not likely to surrender the papers and will want to accompany the papers anywhere they go. Rail rules require that the engineer be in possession of the papers at all times. If they are lost, engineers face serious personal penalties. There is a computer system in place called Operation Respond that tracks rail shipments. If the car number is known, the computer can identify what the cargo is in a particular car. The system is being slowly implemented in several cities and only involves a handful of carriers at this time.

On a ship, the papers are called the **dangerous cargo manifest,** or **DCM** for short. These papers are in the control of the ship's captain or the master who is second in command of the ship. Regardless of the location of the ship and the crew complement, there is always someone in charge of the ship. There may be a considerable number of crew aboard a ship, but there may be only one person who speaks English. The most difficult time to obtain information regarding cargo is when the ship is loading and unloading, especially for container ships. Although the papers may be on board for all of the containers, it can be some time before the crew actually knows where each container is and what the actual cargo is. One of the major problems is determining what is in each individual container, and this information may be limited or not available at all. Luckily each container is well marked with identification information and after some time the source can be traced to get information.

It is interesting to note that one of the biggest hazards from ships comes from the fuel on board. Even a cruise ship can have up to a million gallons of fuel on board. When dealing with ships, special training is required, because they are extremely hazardous to operate on, and responders can get easily lost and/or fall into deep holds in the ship.

In air, the shipping papers are called **Air Bills** and are in the control of the pilot and usually stored in the cockpit. The type of aircraft (passenger or cargo) will determine the type and amount of materials that can be sent via air. The most common aircraft involved with chemical spills are cargo aircraft, and in many cases exact identification is very time consuming and difficult.

FACILITY DOCUMENTS

Each facility that has chemicals above consumer quantities is supposed to have MSDS as described earlier. In SARA Title III, facilities that are covered by these requirements are also required to submit a Tier 2 form, a listing of all chemicals on site that have an MSDS, and a site plan. If they are using extremely hazardous substances (EHS) they are also required to have an emergency plan. Facilities may also be covered by other regulations and may be required to have other plans in place. Each facility should be able to provide upon request, relatively quickly and without any hassles, an MSDS for a given material. Many facilities will leave a binder of MSDS at the gate or with the security guard. Twenty-four-hour staff should have full access to information as well as emergency contact numbers. The size of the facility will determine the amount of staff available to track this process. The SARA reports and information are updated annually and must be submitted by March 1 of each year. The MSDS may not be revised annually but the facility must ensure their accuracy. Responders should meet with the facility to review these documents and to view the facility.

COMPUTER RESOURCES

Many of the chemical information texts are also available electronically, typically as a CD-ROM disk, or they are available online. Many first responders use the **Computer-Aided Management for Emergency Operations (CAMEO)** program for chemical information. CAMEO is a software package that combines chemical response information with emergency planning capability. The information within CAMEO is easily accessed and can be used by first responders. CAMEO also has the ability to provide a vapor cloud model, known as plume dispersion, as shown in **Figure 26-17.** When the local data and leak information are input, a program known as Aerial Location of Hazardous Atmospheres (ALOHA) can determine the worst-case scenario for the vapor cloud travel.

There are also CDs that have MSDS on them, and many companies, especially chemical distributors, and universities have their MSDS on the Internet. A simple Internet search of "Material Safety Data Sheets" provides several thousand results of possible MSDS locations. Other programs provide chemical information, but in most cases they are specifically designed for a hazardous materials team and may be above the first responder level. One of the greatest advantages of computer software is the ability to search for a chemical by its name and synonyms quickly.

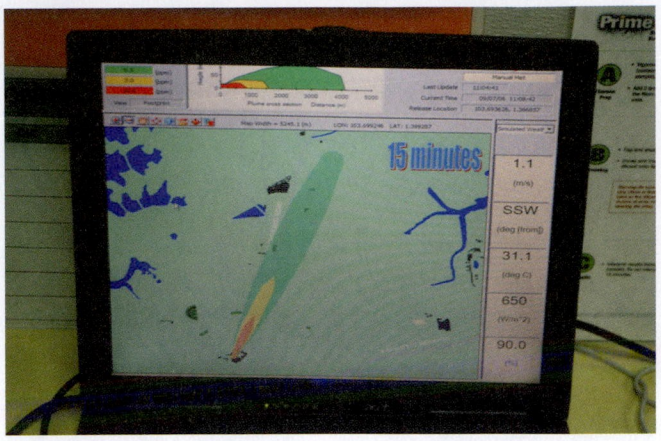

FIGURE 26-17   A toxic gas cloud projection model known as ALOHA is part of the hazardous materials software program CAMEO.

CHEMTREC

The Chemical Transportation Emergency Center, or **Chemtrec,** as it is called, is an information service provided by the American Chemistry Council (ACC). A group of chemical manufacturers established the association for several purposes, but one of the outgrowths is the Chemtrec service, which is a free service to emergency responders, paid for by the chemical manufacturers through an annual fee. Although other services also provide chemical information, Chemtrec is the largest and has been providing this service longer than any other company. When a company joins the Chemtrec system, it provides MSDS to Chemtrec as well as emergency contact information. Chemtrec can be considered a large phone book, so if a responder has a company name or chemical, Chemtrec can provide a contact name and number. Many shippers use Chemtrec as their emergency contact point.

If Chemtrec does not have an MSDS on file, they do have other chemical information databases. One of the big advantages is their ability to contact the manufacturer directly, and they will conference call with a responder so that accurate information can be obtained right from the product specialist. If Chemtrec does not have a specific manufacturer's name, they can connect the responder with other specialists who may be able to provide technical assistance. Chemtrec is well connected when dealing with chemical injuries and exposures and can provide medical information as well as a contact for further information. One thing Chemtrec does not do is make any regulatory notifications—that is the responsibility of the shipper. When calling Chemtrec, the responder should have the following information available:

- *Caller's name and phone number.* Most responders provide the dispatch phone number, because the on-scene cell phone may be tied up.

- *Name of the shipper or manufacturer.* If this information is not known, Chemtrec will contact a manufacturer for assistance, until the manufacturer is identified.

- *Shipping paper information.* This includes truck or railcar number, the carrier name, the consignee (receiver) name, type of incident, and local conditions.

The Canadian equivalent is called CANUTEC and stands for Canadian Transportation Emergency Center and provides the same services as Chemtrec. In Mexico the Emergency Transportation System for the Chemical Industry is known as SETIQ and provides the same service as Chemtrec and CANUTEC. All three services' phone numbers and other emergency contact numbers are provided in the DOT ERG.

REFERENCE AND INFORMATION TEXTS

Many texts are available from a variety of sources that provide chemical information. In the Additional Resources section at the end of the chapter there is a listing of common texts used by hazardous materials teams, like those shown in **Figure 26-18.** One thing to consider is that every piece of apparatus should carry several reference sources. If a hazardous materials team is available in a community, responders probably do not have to have any additional sources of information, because contacting the hazardous materials team by radio or phone is quick and easy. If responders are in an area where the hazardous mate-

rials team is not immediately available or it faces considerable travel time, then additional sources of information would be recommended.

The *DOT ERG* is fairly easy to use, but many of the other reference sources can be complicated and may require some knowledge of chemistry terms, many of which may be above the first-responder level. With some study of the preliminary information contained in each text, most responders can readily access the information they need. The other difference is that these texts are chemical specific and the information is for that material only. The reference texts are also slanted toward the group that develops the text. These books are generally used by hazardous materials technicians but would be recommended for first responders as well. Each book has a different perspective, and there is no one book that is the "only" book to use. The *NIOSH Pocket Guide to Chemical Hazards* shown in Figure 26-18 comes closest to being the "one" book used by hazardous materials technicians. When operating at a hazardous materials release, it is best to check more than one source of information and always confirm the information with other sources.

The National Institutes of Occupational Safety and Health (NIOSH) is the research arm to OSHA. They do not issue regulations, but recommend regulatory issues to OSHA. They conduct studies to improve worker safety and when the research demonstrates a need for new regulations or revisions to existing regulations, they provide this information to OSHA. When there are injuries or fatalities in the workplace, they conduct investigations and, again, if there is a need for new or revised regulations in this category, they also provide this information to OSHA.

They do provide a considerable number of research resources to the fire service and provide a number of sources of information. The *NIOSH Pocket Guide*

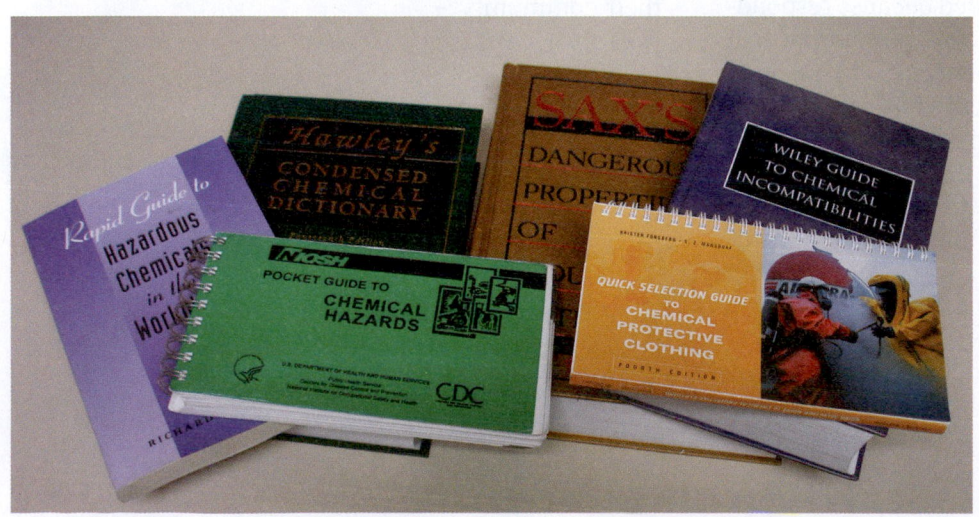

FIGURE 26-18 Several common hazardous materials reference sources, including the NIOSH pocket guide and the Sax guide to Dangerous Properties to Industrial Materials.

provides information on the commonly used industrial chemicals. The *Pocket Guide* provides information on exposure levels, chemical and physical properties, and protective equipment recommendations. It is one of the few publications that provides **ionization potentials (IP),** which are used to determine whether a **photoionization detector (PID)** will determine if a chemical is present. A PID uses an ultraviolet lamp as the mechanism to ionize gas samples. In order to be detected by a PID, the gas must have an ionization potential of less than the lamp strength in the PID. Most PIDs have a lamp strength of 10.6 electron volts (eV). When using the *NIOSH Pocket Guide,* you can determine which gases have IPs below 10.6 eV and therefore can be read by the PID.

Another commonly used book is the *Chemical Hazards Risk Information System (CHRIS),* which is published by the U.S. Coast Guard. It provides information on chemical hazards, chemical and physical properties, and water spills. The *CHRIS* book is actually a three-volume set, and it has a great synonym index. Between the *NIOSH Pocket Guide* and the *CHRIS* book, most of the common chemicals are covered.

One book that provides the most detail, and covers more than 23,000 chemicals, is *Sax's Dangerous Properties of Industrial Materials,* shown in Figure 26-18. This book is invaluable to hazardous materials teams and has considerable toxicological information, as well as chemical and physical properties. It covers trade names, synonyms, CAS numbers, and DOT numbers in the index, which is its own volume. There are a number of reference texts that are available to first responders and hazardous materials teams, many of which are listed in the suggested readings section at the end of this chapter. The books covered here are the basic texts and only represent a small number of texts that should be used to determine chemical response information.

A variety of texts is recommended because responders cannot always anticipate the type of material that they may run across. If a responder is in an area where the hazardous materials team has an extended response time, it is recommended that several persons in that department be trained to the technician level, so as to be able to provide some technical assistance to the incident commander. To function as a hazardous materials team, several persons must be trained, but one trained person could provide chemical-specific information and act as a liaison after the arrival of the hazardous materials team.

INDUSTRIAL TECHNICAL ASSISTANCE

Each community usually has a technical specialist in a given field. As an example, in a facility that ships and receives railcars, there is usually someone within that facility who is knowledgeable regarding railcars. In the event of an emergency in that community, it may be wise to use the resources of that person. Almost every town that has some type of industrial facility has a technical specialist within their community.

Many areas of the country, including Baltimore, Houston, and Baton Rouge, have industrial mutual aid groups that are designed to assist each other and the community in the event of a chemical release. In Baltimore this group is called the South Baltimore Industrial Mutual Aid Plan (SBIMAP). And it provides technical specialists who assist on the scene and also provides specialized equipment throughout a three-state area. It has been in existence for many years, and one hazardous materials team uses the services of a chemist each time it responds.

Each industrial facility usually has a person responsible for safety and health, and that person is usually a good technical resource. Within the facility there may be chemists or chemical engineers who can provide chemical information. When dealing with chemical exposures and toxicology, many facilities have industrial hygienists who work with those issues daily. These people should be contacted prior to a large incident, to find out where they live and what their availability is. When approached, many of these professionals are more than willing to assist their community.

> **NOTE**
>
> Accurate and quick information is essential to every hazardous materials incident. The more information that can be obtained, the easier the incident is to resolve.

JOB PERFORMANCE REQUIREMENT 26-1
Using the DOT Emergency Response Guide (ERG)

A What clues are available, and what portions of the DOT ERG could be used? *(Courtesy of Maryland Department of the Environment Emergency Response Division)*

B What clues are available, and what portions of the DOT ERG could be used? *(Courtesy of Maryland Department of the Environment Emergency Response Division)*

C What clues are available, and what portions of the DOT ERG could be used?

(Continued)

JOB PERFORMANCE REQUIREMENT   26-1

Using the DOT Emergency Response Guide (ERG) (*Continued*)

D   An example DOT ERG listing, which consists of two pages.

GUIDE 127   **FLAMMABLE LIQUIDS (POLAR/WATER-MISCIBLE)**   ERG2004

POTENTIAL HAZARDS

FIRE OR EXPLOSION

- **HIGHLY FLAMMABLE: Will be easily ignited by heat, sparks or flames.**
- Vapors may form explosive mixtures with air.
- Vapors may travel to source of ignition and flash back.
- Most vapors are heavier than air. They will spread along ground and collect in low or confined areas (sewers, basements, tanks).
- Vapor explosion hazard indoors, outdoors or in sewers.
- Those substances designated with a "P" may polymerize explosively when heated or involved in a fire.
- Runoff to sewer may create fire or explosion hazard.
- Containers may explode when heated.
- Many liquids are lighter than water.

HEALTH

- Inhalation or contact with material may irritate or burn skin and eyes.
- Fire may produce irritating, corrosive and/or toxic gases.
- Vapors may cause dizziness or suffocation.
- Runoff from fire control may cause pollution.

PUBLIC SAFETY

- **CALL Emergency Response Telephone Number on Shipping Paper first. If Shipping Paper not available or no answer, refer to appropriate telephone number listed on the inside back cover.**
- As an immediate precautionary measure, isolate spill or leak area for at least 50 meters (150 feet) in all directions.
- Keep unauthorized personnel away.
- Stay upwind.
- Keep out of low areas.
- Ventilate closed spaces before entering.

PROTECTIVE CLOTHING

- Wear positive pressure self-contained breathing apparatus (SCBA).
- Structural firefighters' protective clothing will only provide limited protection.

EVACUATION

Large Spill

- Consider initial downwind evacuation for at least 300 meters (1000 feet).

Fire

- If tank, rail car or tank truck is involved in a fire, ISOLATE for 800 meters (1/2 mile) in all directions; also, consider initial evacuation for 800 meters (1/2 mile) in all directions.

Page 202

D   (Continued)

ERG2004

FLAMMABLE LIQUIDS
(POLAR/WATER-MISCIBLE)

GUIDE
127

EMERGENCY RESPONSE

FIRE
CAUTION: All these products have a very low flash point: Use of water spray when fighting fire may be inefficient.

Small Fires
- Dry chemical, CO_2, water spray or alcohol-resistant foam.

Large Fires
- Water spray, fog or alcohol-resistant foam.
- Use water spray or fog; do not use straight streams.
- Move containers from fire area if you can do it without risk.

Fire involving Tanks or Car/Trailer Loads
- Fight fire from maximum distance or use unmanned hose holders or monitor nozzles.
- Cool containers with flooding quantities of water until well after fire is out.
- Withdraw immediately in case of rising sound from venting safety devices or discoloration of tank.
- ALWAYS stay away from tanks engulfed in fire.
- For massive fire, use unmanned hose holders or monitor nozzles; if this is impossible, withdraw from area and let fire burn.

SPILL OR LEAK
- ELIMINATE all ignition sources (no smoking, flares, sparks or flames in immediate area).
- All equipment used when handling the product must be grounded.
- Do not touch or walk through spilled material.
- Stop leak if you can do it without risk.
- Prevent entry into waterways, sewers, basements or confined areas.
- A vapor suppressing foam may be used to reduce vapors.
- Absorb or cover with dry earth, sand or other non-combustible material and transfer to containers.
- Use clean non-sparking tools to collect absorbed material.

Large Spills
- Dike far ahead of liquid spill for later disposal.
- Water spray may reduce vapor; but may not prevent ignition in closed spaces.

FIRST AID
- Move victim to fresh air.   • Call 911 or emergency medical service.
- Give artificial respiration if victim is not breathing.
- Administer oxygen if breathing is difficult.
- Remove and isolate contaminated clothing and shoes.
- In case of contact with substance, immediately flush skin or eyes with running water for at least 20 minutes.   • Wash skin with soap and water.
- In case of burns, immediately cool affected skin for as long as possible with cold water. Do not remove clothing if adhering to skin.
- Keep victim warm and quiet.
- Ensure that medical personnel are aware of the material(s) involved and take precautions to protect themselves.

Page 203

JOB PERFORMANCE REQUIREMENT 26-2
Determining an Action Plan Using the MSDS

A   A number of clues are present to assist with the identification of this hazardous material. *(Courtesy of Maryland Department of the Environment Emergency Response Division)*

B   This incident provides an opportunity to visualize a reactive materials release. *(Courtesy of Cambria County, Pennsylvania, Emergency Services)*

LESSONS LEARNED

First responders should be starting the information process by trying to obtain as much information as possible, so they can make safe and accurate tactical decisions. Using the recognition and identification skills, a considerable amount of information can be obtained. This information, combined with the use of reference sources, can help provide a significant amount of useful data. If waiting for a hazardous materials team, the more information that can be obtained prior to the arrival of the team will make the response task easier and less stressful. The more information that can be obtained through the use of the DOT ERG and, more important, the *NIOSH Pocket Guide,* the safer the response. Following the recommendations in the DOT ERG, the initial responders would be taking a conservative approach, which is appropriate. As more information is available, other texts, computer programs, and technical assistance can provide more specific information that can aid in a more incident-specific response. As essential as this information is, first responders should not take any risks attempting to get this information, such as entering a hazard area to get shipping papers.

KEY TERMS

ADR/RID Abbreviations used by the international shipping community and established by a European Agreement by the United Nations Economic and Social Council Committee of Experts on Transporting Dangerous Goods. The U.S. DOT is a voting member of this shipping body, which governs Regulations Concerning the International Transport of Dangerous Goods by Rail (RID) and Road (ADR)

Air Bill The term used to describe the shipping papers used in air transportation.

Chemtrec The Chemical Transportation Emergency Center, which provides technical assistance and guidance in the event of a chemical emergency. This network of chemical manufacturers provides emergency information and response teams if necessary.

Computer-Aided Management for Emergency Operations (CAMEO) Program A computer program that combines a chemical information database with emergency planning software. It is commonly used by hazardous materials teams to determine chemical information.

Consist The shipping papers that list the cargo of a train. The listing is by railcar, and the consist lists all of the cars.

Dangerous Cargo Manifest (DCM) The shipping papers for a ship, which list the hazardous materials on board.

Ionization Potential (IP) The ability of a gas or vapor to be ionized. It is most commonly used to determine whether a photoionization detector can detect a gas or vapor.

National Response Center (NRC) The location that must be called to report a spill if it is in excess of the reportable quantity.

Photoionization Detector (PID) A detection device that can detect toxic gases or vapors and is commonly used by Hazardous Materials response teams.

Polymerize A chain reaction in which the material quickly duplicates itself and, if contained, can be very explosive.

Standard Transportation Commodity Code (STCC) A number assigned to chemicals that travel by rail.

Waybill A term that may be used in conjunction with consist, but is a description of what is on a specific railcar.

REVIEW QUESTIONS

1. What are three common methods of obtaining chemical information?

2. The DOT ERG is intended to be useful for how long at an incident?

3. If given a four-digit UN identification number, which resource provides quick initial information?

4. What mode of transportation uses STCC numbers?

5. Are MSDS required in a home for household quantities of hazardous materials?

6. Can Chemtrec be used for medical information?

7. Which section of the DOT book is used if you have a shipping name?

8. Which reference book provides information related to suggested isolation distances?

9. What advantage does an industrial contact provide at a chemical spill?

10. What cargo probably will not be listed on the shipping papers?

27

Hazardous Materials: Personal Protective Equipment

Chemical protective clothing has come a long way since I was introduced to chemical hazards one evening in 1974. I had responded to a dumpster fire early one morning. We arrived to see a thick, red-brown cloud issuing from the waste container. Obviously, but unfortunately not obvious to us, we had a hazardous materials situation, not a fire. Undaunted, we went about our tasks, and, after half an hour, we eventually extinguished the fire. However, we all had burning eyes, noses, throats, and severe irritation to our wrists and necks. We called for the assistant chief on duty. He suggested our ailments were from the diesel fumes issuing from the fire truck. We thought otherwise but went back to the station, finished our tours, and went home.

The next morning I was visited by a man from the chemical company who wanted to make sure we were all right. He explained to me that the dumpster had contained discarded nitric acid bottles, and that some had still been full. They had started the fire. He was more concerned about our exposure, however, since we had not worn chemical protective clothing or even SCBA. (Hey, the fire was outside! Remember those days?) He went on to explain that nitric acid had "a delayed reaction to any acute exposure." This, he told me, would have affected the lungs. They might have developed fluid or edema, creating chemical pneumonia. In addition, my throat might have swollen and made it difficult to breathe. I remember saying to him, "You're probably glad I answered the door!" He said that, actually, he was.

Over the years I have looked back on that incident, and I know it is what got me into hazardous materials. It was the driving force behind my signing up for one of the first National Fire Academy Chemistry of Hazardous Materials courses. I know now that chemical protective clothing, and at the very least SCBA, should always be worn in any smoke, vapor, or cloud situation, but hindsight is the best sight, I guess.

A final note for anyone who might be concerned about me. Do not worry; everything works out in the end. A few months later I went to an acetylene-tank fire. I was acting officer and had decided to extinguish the fire with a sodium-bicarbonate extinguisher. Upon closer examination I noticed the fire was out. I turned to my engine crew to tell them not to charge the extinguisher. Sadly, one overanxious fellow discharged the unit right into my face—and me, once again, without SCBA! With my mouth wide open, I got a lung full of baking soda. So we can all relax, because I am balanced—chemically neutral—again!

—*Street Story by Mike Callan, President,*
Callan and Company, Middlefield, Connecticut

LEARNING OBJECTIVES

After completing this chapter, the reader should be able to:

27-1 Describe the causes of harm.

27-2 Explain the health hazards associated with chemical releases.

27-3 Discuss various chemical-related health terms.

27-4 Identify the various levels of PPE.

27-5 Demonstrate the use of SCBA and other respiratory protection at chemical releases.

27-6 Demonstrate the use of firefighting protective clothing at chemical releases.

27-7 Explain the signs and symptoms of heat stress.

*The FF I and II levels, as defined by the NFPA 1001 Standards, are identified in different colors: FF I = black, FF II = red, additional information = blue.

INTRODUCTION

Personal protective equipment, or PPE as it is commonly called, takes on many different shapes and versions. One example is provided in **Figure 27-1.** Even standard firefighter PPE has many variations and ensembles. When dealing with hazardous materials these configurations are endless.

NOTE

The use of PPE is essential to the health and safety of the first responders.

The failure to use PPE or to use it properly may cause an injury or have fatal effects.

SAFETY

Many firefighter injuries can be prevented by the proper use of PPE, most of which only takes a few seconds to put in place.

The best respiratory protection for the first responder above all else is to use the self-contained breathing apparatus (SCBA), which is described in detail in Section I, Chapter 7. SCBA offers a substantial amount of chemical protection and should be considered the minimum protection against chemical spills. Firefighter turnout gear offers limited protection against some chemicals, but it is not intended to be used for chemical spill response. Firefighter turnout gear is not tested or approved for chemical spills, and, although it may offer some protection in some chemical environments, it should only be used for immediate life-threatening rescue situations where there will be limited time in the hazard area.

As the need to be in the hazard area increases, so does the need for the proper PPE. This chapter outlines specialized types of PPE, but specific hands-on

FIGURE 27-1 PPE is used to protect the wearer from a variety of hazards, but no one type of PPE protects the wearer from all types of hazards. This Level A fully encapsulated suit protects the wearer from chemical hazards but offers little protection from heat, flame, or mechanical hazards. All PPE adds heat stress, and this garment adds significant heat stress on the wearer. Decisions on which PPE to wear are difficult to make and require additional training.

training is required prior to the use of this PPE. In general, hazardous materials technicians and specialists wear chemical-protective clothing, but persons trained to the operations level may be required to don chemical-protective clothing to perform decontamination operations or other patient-related activities after specific PPE training is provided.

HEALTH HAZARDS

Health issues are a serious concern with hazardous materials emergencies. They can affect every responder and may even be carried home. They can affect the responder immediately, or it may be years before they take a toll.

Infectious diseases are the result of the presence of microbial agents including viruses and bacteria that are able to cause disease in animals and plants. Infectious diseases are often classified as contagious. Contagious diseases, or communicable diseases, have the ability to transfer from one person to another through liquids, food, bodily fluids, inhalation, or vector borne spread.

Protecting the body from hazardous materials is easily accomplished by wearing some form of protective clothing and having a basic understanding of toxicology and how chemicals cause harm.

Toxicology

Toxicology is the study of poisons and their effect on the body, and people who study the effect of poisons on the body are known as toxicologists. Although toxicologists are typically found in the medical community, most industrial facilities have industrial hygienists on staff whose responsibility is to protect the workers' health and safety. Their primary focus is on the chemical hazards that exist within the facility. These people have extensive training in toxicology and chemical exposures and are great resources to the emergency services. Because the world of toxic exposures can be complicated and in emergency situations information is needed quickly, a quick consultation with an industrial hygienist may make the incident easier to resolve.

Types of Exposures

Being exposed to a chemical may present a risk, the level of which is typically spoken of in terms of the chemical's potential *hazard*. The hazard that a chemical presents is an indicator as to the potential harm it can cause. There are two types of exposures, acute and chronic, both of which can have serious health effects.

> **NOTE**
>
> An acute exposure is a quick, one-time exposure to a chemical.

Typically, an **acute** exposure is one where the body is subjected to a large dose over a short period of time. By definition, this exposure occurs once in a lifetime and the period of time can be from immediate to 14 days.

It should be noted that after an acute exposure the body, depending on the product and its hazards, may cause injuries ranging from minimal to major and potentially leading to death. Overall, the human body does well with short-duration exposures and recovers from them, but all chemical exposures should be avoided. A simple example of an acute exposure is that of a nonsmoker who decides to smoke one cigarette. This one-time exposure for most people is not harmful, nor would it cause any long-term health effects.

An example of a chronic exposure, however, is that of the person who smokes three packs of cigarettes a day for twenty years. This person has received doses of cigarette smoke several times a day for a long period of time and is likely to have health problems associated with this chronic exposure. (Note that abnormalities do exist; The person who smoked one cigarette may develop lung problems from that one acute exposure, and the person who chain-smokes for twenty years may never develop health problems associated with the chronic exposure.)

> **SAFETY**
>
> Hazardous materials incidents have the ability to impact humans, the environment, and property. In some cases, such as with dioxin, the risk to humans, the environment, and property can last for many years. Through immediate effects, or long-term or recurring (**chronic**) effects, hazardous materials can present a risk. Proper identification of potential hazards and the wearing of proper protective clothing are essential for responder safety.

Types of Hazards

In the realm of hazardous materials, *hazard* is defined as the category of risk that can be inflicted by exposure or contamination with a chemical. There are several methods used to identify possible hazards at a chemical release. The most common method in use today is known by the acronym **TRACEMP,** which stands for thermal, radiation, asphyxiation, chemical, **etiological,** mechanical hazards, and potential psy-

TABLE 27-1	Subcategories within TRACEMP
Thermal	Both heat and cold hazards fit into this category. If a flammable liquid ignites, it is classified as a thermal hazard. If liquefied gas contacts a person's skin, it could cause frostbite and a thermal (cold) burn. When liquefied gases are released into small spaces or rooms, the temperature can drop dramatically, presenting hypothermia and cold stress concerns.
Radiation	Types of radiation include non-ionizing and ionizing. Non-ionizing types include microwave and infrared. Ionizing types include alpha, beta, gamma, and neutron radiation and X rays. Alpha and beta are particles, while gamma, neutron, and X-ray are forms of energy.
Asphyxiation	Both simple and chemical asphyxiants fit into this category.
Chemical	This category has a number of subgroups, including poisons and corrosives. The poisons may be referred to as toxics—and some chemicals are highly toxic. There are specific levels of exposure that determine into which category a chemical fits. Also within this category are convulsants, irritants, sensitizers, and allergens. The reaction to a particular chemical varies from person to person, much in the way some people are allergic to bee stings while others are not. Chemicals affect some people and not others.
Etiological	Bloodborne pathogens and biological materials exist within this category.
Mechanical	Although not chemical in nature, mechanical hazards exist within the hazard area of a chemical spill, including the standard slip, trip, and fall hazards that one should always be concerned about. Other examples of mechanical hazards would be getting hit from blast particles, such as from a bomb or BLEVE, or a drum falling on a responder.
Psychological	When exposed to traumatic or disaster situations, some responders may suffer immediate, short-term or long-term psychological stress. Post-Traumatic Stress Syndrome (PTSS) is one example.

chological harm. Each of the individual hazards has additional hazards that fit within that classification. Much like the **risk-based response** theory, the use of TRACEMP assigns a chemical to a risk category so that tactical decisions can be based on that classification. The subcategories within TRACEMP are given in **Table 27-1.**

SAFETY

In situations where liquefied gases are released, their low temperatures will drop the temperature of the room. In some cases, the drop can be dramatic. In one situation where anhydrous ammonia had been released and there was liquid ammonia on the floor, the hazardous materials team's boots froze to the floor. Their SCBA regulators were starting to freeze and several team members had frostbitten hands. In addition to these cold hazards, the vaporization of the ammonia resulted in a vapor cloud that obstructed visibility and the crews were not able to see their team members an arm's length away, not to mention the impact of the fogged up faceshields.

Categories of Health Hazards

Within the TRACEMP categories, there are terms that responders should understand, because MSDSs or industrial contacts may describe some chemicals as fitting into one or more of these categories. One of the most commonly used terms is **carcinogen,** which refers to a material with cancer-causing potential. There are two classifications of carcinogens, known and suspected, with the majority being suspected carcinogens. According to the Chemical Abstracts Service (CAS), a division of the American Chemical Society (ACS), a group that tracks chemicals, there are 30 million chemicals in existence today. Each day the ACS adds 4,000 more chemicals to their listing. The National Toxicology Program under the Public Health Service of the U.S. Health and Human Services Administration issues an annual report on carcinogens. The eleventh annual report lists 54 chemicals that are known to cause cancer and 184 chemicals that are suspected of causing cancer. When dealing with chemical spill response, the risk of getting cancer always causes great fear. Firefighters are exposed to a large number of chemicals, many of them known cancer-causing agents, but if firefighters wear their SCBA these exposures are unlikely to cause problems. For older firefighters who worked in earlier years when SCBA was not used as extensively as it is today, cancer is still a leading cause of death, and many retired firefighters have not had the chance to enjoy retirement due to an early death.

SAFETY

Wearing SCBA and avoiding other off-duty activities that involve cancer-causing materials can prevent the development of cancer in most persons.

Another term that is commonly used is **irritant**, which is self-explanatory. An irritant is not corrosive but mimics the effect of a corrosive material in that it can cause irritation of the eyes and possibly the respiratory tract. One notable characteristic of an exposure to irritants is that the effects are easily reversed by exposure to fresh air. Mace and pepper spray are classified as irritants and may be called incapacitating agents by the military.

Sensitizer is a term used to describe a chemical that causes an effect that is in reality an allergic reaction, typically occurring through skin contact. In most cases an employee can work with a chemical for years and suffer no effects and then one day suffer a severe reaction to the material. Skin reddening, hives, itching, and difficulty breathing are possible symptoms when dealing with a sensitizing agent. Some persons, however, can become sensitive to a chemical after one exposure.

Allergens are chemicals that produce symptoms much like those of sensitizers but are the result of a reaction with an individual's immune system. In many cases, an individual may have been allergic to a substance but was unaware of it until the chemical was encountered.

A **convulsant** chemical is one that has the ability to cause convulsions upon exposure, typically by ingestion. There are a number of drugs that can cause a seizure-like response. Organophosphate pesticides make up one group of chemicals that can cause seizure-like activities. When someone is exposed to an organophosphate pesticide (OPP), the brain reacts to the chemicals and seizure-like activity results. The group of chemical warfare agents known as nerve agents is comparable to OPP compounds and causes the same seizure-like activity.

Some chemicals only affect one or more organs and are described as target organ hazards, or they may affect a body system, such as the central nervous system. The effects depend on the individual, dose, concentration, and length of exposure. Some of the target organ descriptions are provided in **Table 27-2**.

TABLE 27-2 Target Organs and Systems

Name	Target Organ or System
Central nervous system (CNS) chemicals	Affect the central nervous system and can cause short-term or long-term effects. Commonly short-term memory is lost after exposure to a CNS hazard material. Many of the hydrocarbons cause CNS effects, and people sometimes purposely expose themselves to a CNS agent to receive a "high" from the exposure. In the long term, the brain cells are damaged, never to recover. Neurotoxins essentially cause the same effects.
Peripheral nervous system (PNS) chemicals	Much like CNS chemicals, the PNS chemicals affect the body's ability to move in a coordinated fashion. Exposure to a PNS chemical causes a disruption of the brain's ability to move messages to the other body systems.
Hepatoxins	These types of materials affect the liver and if the exposure is high enough can cause severe damage to the liver.
Nephrotoxins	These materials adversely affect the kidneys.
Reproductive toxins	These toxins affect the ability to reproduce and can cause birth defects. These types of toxins can stay within the body, so they can have adverse effects on a pregnancy even if the exposure occurred a considerable time before the pregnancy.
Mutagens	An exposure to a mutagen may not cause any harm to the people who received the exposure, but the effect could be transmitted to their offspring. Mutagens cause damage to the genetic system and can cause mutations that can become hereditary.
Teratogens	These materials can affect an unborn child. The effects, however, do not happen at a cellular level and would not be passed along from generation to generation.

Radiation Hazards

Radiation has the ability to cause a number of health problems up to and including death, which can occur within weeks. Exposure to radiation can cause radiation sickness and other illnesses, some of which are shown in **Table 27-3.** Coming in direct contact with or being close to a radiation source can cause significant burns. The level of radiation and the dose determine the damage to the body. The primary issue with radiation is the fact that it can kill without any warning, since it is colorless and odorless and the danger cannot be seen. Detection devices can easily detect radiation levels and warn responders when they are entering a potentially dangerous environment. Examples of radiation action levels are provided in **Table 27-4.**

Understanding the level of radiation danger can be confusing. Equally as confusing is the process to calculate a radiation dose. There are three measurements that can be used to describe a radiation dose: absorbed dose, equivalent dose, and effective dose. The **absorbed dose** is a measure of energy transferred to a material by radiation. The absorbed dose is measured in units called gray (Gy) or rad (radiation absorbed dose). One gray is equal to 1 joule of energy absorbed by 1 kilogram of material and is the SI (System International) unit for absorbed dose. One gray is also equal to 100 rad, an older unit. To understand the impact on humans, we need to expand our thoughts on dose measurements and consider the effect of the absorbed energy on the body. The basic unit of equivalent dose in the United States is the roentgen, which is a value that is provided for the amount of ionization in air caused by X-rays or gamma radiation. Using the roentgen and applying the quality factor we can determine the dose and impact on humans, which is known as roentgen equivalent man or REM (also abbreviated as simply R). The value of

---

TABLE 27-3   Health Effects from Radiation

Exposure (REM)	Exposure (Sieverts)	Health Effect	Time to Onset
5–10	0.05–0.1	Changes in blood chemistry	
50	0.5	Nausea	Hours
50–100	0.5–1	Headache	
55	0.55	Fatigue	
70	0.7	Vomiting	
75	0.75	Hair loss	2–3 weeks
90	0.9	Diarrhea	
100	1	Hemorrhage	
100–200	1–2	Mild radiation poisoning	Within hours
200–300	2–3	Severe radiation poisoning	Within hours
400–600	4–6	Acute radiation poisoning, 60% fatal	Within an hour
600–1,000	6–10	Destruction of intestinal lining, internal bleeding, and death	15 minutes
1,000–5,000	10–50	Damage to central nervous system, loss of consciousness, and death	Less than 10 minutes
5,000–8,000	50–80	Immediate disorientation and coma, death	Under a minute
More than 8,000	More than 80	Immediate death	Immediate

TABLE 27-4	Radiation Action Levels and Assorted Doses	
Dose	**Cause or Type**	**Note**
Average Exposures to Radiation		
0.01 R annually	Chest X-ray	Per-year exposure
0.2 R annually	Radon in the home	Per-year exposure
0.081 R annually	Living at high elevations (Denver)	Per-year exposure
1.4 R	Gastrointestinal series	
Action Levels		
1 mR/hr	Isolation zone (public protection level)	Recommended exposure limit for normal activities
5 R	Emergency response	For all activities
10 R	Emergency response	Protecting valuable property
25 R	Emergency response	Lifesaving or protection of large populations
>25 R	Emergency response	Lifesaving or protection of large populations. Only on a voluntary basis for persons who are aware of the risks involved.

1 roentgen and of 1 REM are equal to each other and are used interchangeably. The rest of the world uses sieverts (Sv) to describe the equivalent dose to man. To further define how the dose impacts humans, another factor known as a weighting factor can be added. Target organs have a higher factor than target bone, for example. Radiation monitors measure in three scales: (1) REM (R) or Sieverts (S), (2) milliREM (mR) or millisieverts (mSv), and (3) microREM (μR) or microsieverts (μSv). The dose of radiation is expressed by a time factor, typically an hour. The conversions for these values are provided in **Table 27-5.**

Radiation Protection

Radiation doses should be kept "As Low As Reasonably Achievable." This is known in the nuclear industry as ALARA and is the cornerstone of radiation safety. The three factors that can influence radiation dose are time, distance, and shielding—concepts that should be applied to all types of chemical exposures as well. To minimize their dose, individuals should stay near the radiation source for as little time as possible, stand as far away from the source as possible, and place as much shielding between them and the source as possible. Naturally occurring background radiation occurs on the level of microREM (or microSieverts) per hour, such that when using a radiation detection device an accurate device will be reading this background radiation as well as any additional levels. When levels above the normal background are encountered, responders should use caution. Background radiation

readings do vary depending on the location, elevation, terrain, building type, and a number of other factors. It is when the radiation dose changes scales, such as going from microREM to milliREM, that there is a potential for health effects. Going from microREM to REM levels indicates severe danger. When encountering radiation levels that are significantly above background and have changed from microREM to milliREM levels or to REM levels, responders should move to an area of lower radiation dose.

Routes of Exposure

The four primary routes of exposure are respiratory, absorption, ingestion, and injection, **Figure 27-2.** The route that is the most commonly associated with causing health effects, both acute and chronic, is the respiratory system route, as shown in **Figure 27-3.**

SAFETY

In almost all cases the respiratory system requires some type of protection. For emergency services workers this is the easiest system to protect because they have easy access to SCBA, should be familiar with it, and are comfortable with its use.

The respiratory system can be affected by gases, vapors, and solid materials such as dust and other particles. In many cases the chemicals may not have any effect on the respiratory system itself, but may enter the body through the respiratory system and affect other organs or body systems. When dealing

TABLE 27-5 Radiation Equivalents and Conversions

Term or Amount	Equal To
rad	gray (Gy) or absorbed dose (AD)
REM	sievert (Sv) or dose equivalent (DE)
1 µR	0.01 µSv
100 µR	1 µSv
1 mR	10 µSv
100 mR	1 mSv
1 REM	10 mSv
100 REM	1 Sv
curie (Ci)	becquerel (Bq)
27 picocuries (pCi)	1 disintegration/sec (d/sec) or 1 becquerel
1 pCi	37 mBq
1 µCi	37 kBq
27 µCi	1 MBq
1 mCi	37 MBq
27 mCi	1 GBq
1 Ci	37 GBq
27 Ci	1 TBq
1 pCi	2.22 dpm
1 Bq	60 dpm
1 dpm	0.45 pCi

and carbon dioxide (CO_2) will move oxygen out of the area and cause people in the area to have difficulty breathing. If the concentrations are high enough, death could result. Chemical asphyxiants work in a different manner. They cause a chemical reaction within the body and will not allow it to use the readily available oxygen. The most common chemical asphyxiant is carbon monoxide (CO). When CO is in the air in sufficient quantities, it enters the bloodstream through the lungs. It binds with the hemoglobin in the blood, forming carboxyhemoglobin. Because hemoglobin has an attraction for CO about 225 times that of oxygen, it will not allow the oxygen molecules to bind with the blood, which causes severe health problems and often death.

The other common route of entry is via skin absorption, because the skin is the body's largest organ. Although some chemicals can cause damage to the skin and may irritate it, this does not mean that it is toxic by skin absorption. The number of chemicals that are toxic by skin absorption is relatively low, but precautions should be taken to minimize contact, and, if at all possible, have no skin contact with chemicals. Skin contact with chemicals can cause burns, rashes, or drying of the skin. The only way to provide skin protection is to wear proper protective clothing that will not allow the chemicals to get onto the skin. Firefighter turnout gear will slow the process down but will not prevent the eventual migration of the chemical to the skin. Effective decontamination is required to ensure that all of the chemical is cleaned from the skin.

The other route of entry is ingestion, which is more common than one would think. The use of SCBA generally prevents this route of exposure, at least until the rehabilitation phase of the incident.

> **CAUTION**
>
> After fighting a fire, responders are typically covered in soot and other debris. Most firefighters do not decontaminate themselves prior to eating or drinking any refreshments that may be available to them at the scene. Without proper cleaning, they will typically ingest some of these products of combustion, none of which is healthy to eat.

with respiratory hazards, there are two categories of asphyxiants, simple and chemical. Although the end result is usually the same, the manner in which a person is killed does differ significantly.

Simple asphyxiants displace oxygen in the air. It is not an adverse chemical reaction but simply a matter of something other than oxygen occupying the space in the body where the oxygen should be. Normal oxygen levels are 20.9 percent in air, and the body starts to develop difficulty breathing at less than 19 percent. A person will start to have serious problems at less than 16.5 percent oxygen. Gases such as nitrogen, halon,

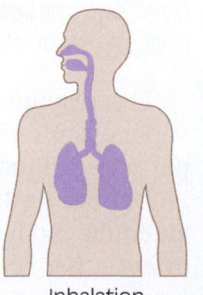  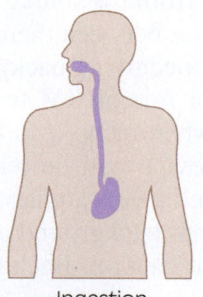 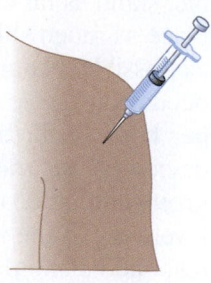

Inhalation   Absorption   Ingestion   Injection

FIGURE 27-2 Routes of exposure.

Simple Asphyxiant

Chemical Asphyxiant

FIGURE 27-3 Respiratory system route of exposure.

TABLE 27-6   Routes of Exposure/Harm by Hazard Class

Hazard Class	Primary Route of Entry	Possible Route of Entry	Harm
Explosives	Inhalation, ingestion, and absorption	Injection	Thermal, etiological, mechanical and psychological
Gases	Inhalation	Absorption	Thermal, asphyxiation, chemical, and psychological
Flammable Liquids	Inhalation, ingestion, and absorption	Injection	Thermal, asphyxiation, chemical, and psychological
Flammable Solids	Ingestion, and absorption	Inhalation and injection	Thermal, chemical, and psychological
Oxidizers and Organic Peroxides	Inhalation, ingestion, and absorption	Injection	Thermal, asphyxiation, chemical, mechanical, and psychological
Toxic or Infectious Materials	Inhalation, ingestion, and injection	Absorption	Asphyxiation, chemical, etiological, and psychological
Radioactive Materials	Inhalation, ingestion, and absorption	Injection	Thermal, radiation, and psychological
Corrosive Materials	Inhalation, ingestion, and absorption	Injection	Chemical and psychological
Miscellaneous Dangerous Goods	Inhalation, ingestion, and absorption		Thermal, radiation, asphyxiation, chemical, etiological, mechanical, and psychological

One other route of exposure is injection, although it is not considered to be one of the major routes. Many emergency services workers are exposed to hazardous materials via this route, and for this population it is probably one of the leading routes, after inhalation. The most common material that emergency services workers are exposed to is body fluids or what is referred to as bloodborne pathogens. Other methods of injection are through being near a high-pressure line when it breaks, such as a hydraulic rescue tool fluid line that would inject hydraulic fluid into a person's body. Other than standard infection control practices,

there is little protection for these types of exposures except to properly wear full PPE when working in and around situations where exposure to these fluids is possible. Some examples of the routes of exposure by hazard classes are provided in **Table 27-6.**

Factors that affect the rate of exposure, regardless of the route, are basic items such as temperature, pulse, and respiratory rate. The higher each of these items is in an individual, the more likely it is that the chemical will have some effect. The damage that chemicals have is based on the equation *Effect = Dose $\times$ Concentration $\times$ Time*. This equation relates to acute and chronic exposures. A small dose, that is, a small concentration, for a small period of time is not likely to have an adverse effect on a normal human. A large dose at a high concentration over a long period of time will in most cases have an effect. People exhibiting normal vital signs who are exposed to a chemical may not have any effects. But if they jog around the block prior to being exposed to a material, they may be affected.

> **NOTE**
>
> The increased temperature of the body will allow for faster absorption into the body, and the accompanying increased respiratory rate will cause more chemicals to enter the respiratory system.

The increased pulse rate allows for the chemicals to be spread throughout the body faster. In some confined space incidents, where chemicals may have played a factor in injuries and deaths, increased vital signs play a role. In most cases the first victim may be unconscious and therefore the body's system has slowed down, and in some cases the victim went down due to a lack of oxygen. When a person recognizes that a coworker has gone down, the vital signs increase and when trying to perform a rescue the person may actually be exposed to a higher level of the chemical than the coworker he or she is trying to rescue. This is one of the reasons why in some cases the rescuer dies, and the original victim ends up surviving.

EXPOSURE LEVELS

In industry, monitoring for exposures is commonplace and is usually a preventive action, but in the emergency services it can be an afterthought, typically after an incident has occurred. Several different types of exposure values have been issued by a variety of agencies, some of which can be very confusing. Exposure values have been established for the commonly used chemicals and for a variety of situations. The key to preventing exposures is to monitor for hazardous materials and to wear appropriate PPE. The one key agency involved with exposure values is the Occupational Safety and Health Administration (OSHA). The exposure values that they set are the ones that must be followed by all industries, including the fire service because it too is considered an industry.[1] Another organization that issues exposure values is the American Conference of Governmental Industrial Hygienists (ACGIH), a group that advocates worker safety and conducts a lot of stud-

ies regarding chemical exposures. The National Institute of Occupational Safety and Health (NIOSH), a research arm of OSHA, issues recommendations for exposure levels. OSHA is the only agency that provides legally binding exposure values; all of the others are recommendations. In some OSHA regulations the employer is required to use the lowest published exposure values, which in many cases are not OSHA's own values. When dealing with emergency situations, it is always recommended that responders follow the lowest published values.

The exposure values are based on an average male and are for an industrial application. The exposure values are typically based on an eight-hour day, forty-hour workweek with a sixteen-hour break between exposures. The values that are issued for the various substances are typically conservative; in a given population it is not atypical to find someone who is sensitive to a chemical at a lower value than the rest of the group. Each value has an extra margin for safety built in, ranging from 1 to 10,000 times the actual value. The exposure values are typically listed in parts per million (ppm) or as milligrams per meter cubed (mg/m^3). Explanations of these terms are provided in **Figure 27-4**.

The most common exposure values are expressed in ppm, but for some materials (typically solids) the values may be expressed as mg/m^3. The values are generally for a period of time, usually eight hours. OSHA refers to this eight-hour exposure value as the **permissible exposure limit (PEL).** The ACGIH refers to this eight-hour exposure as the **threshold limit value (TLV).** Both are average exposures over the eight-

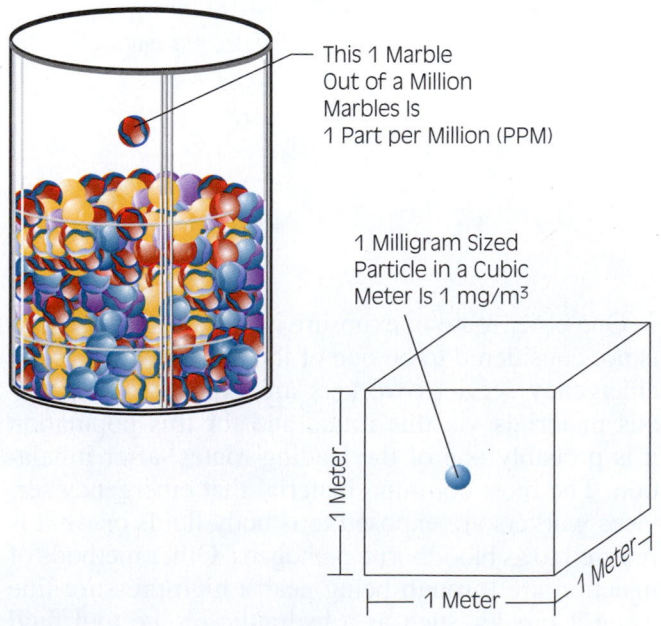

This 1 Marble Out of a Million Marbles Is 1 Part per Million (PPM)

1 Milligram Sized Particle in a Cubic Meter Is 1 mg/m^3

1 Meter

1 Meter

1 Meter

FIGURE 27-4 Explanation of units of measure.

hour period. A worker can be exposed to more than the PEL or TLV as long as at the end of the day the exposure value is less than the PEL or TLV. In some cases these exposure values are called time weighted averages (TWAs), and they may be expressed as the OSHA-TWA or the ACGIH-TWA, which is an eight-hour daily average exposure. NIOSH issues **recommended exposure limits (RELs),** which are for a ten-hour day, forty-hour workweek.

Other values that may be listed for chemicals are the **ceiling levels,** generally referred to as PEL-C or TLV-C. These provide an amount that is the highest level to which an employee can be exposed. When figuring an average, there are times when employees are going to be exposed to chemicals at a level higher than the PEL or TLV, but there are also times when they will be exposed to less than those levels. As long as their exposure average is less than the PEL or TLV, they are acceptable. Certain chemicals will cause effects at levels that may be obtained through worker exposure and would not be considered safe, but the overall exposure would fall below the PEL or TLV. To avoid unsafe levels, the safety organizations may attach a ceiling level to an exposure value, and that is the highest level that the employee can be exposed to, regardless of what the end average is.

Another exposure value is known as the **short-term exposure limit (STEL).** This value is assigned to a fifteen-minute exposure. An employee can be exposed at this level for fifteen minutes and then is required to take an hour break from the exposure. The employee can do this four times a day without any adverse effects. NIOSH is also using an excursion value that is coupled with a time limit, five to thirty minutes typically. At this level, an employee can enter an environment one time and not suffer any effects.

The last value can be confusing because it is called the immediately dangerous to life or health (IDLH) value. One would think that being exposed to a chemical at the IDLH level would mean that death may be imminent. In reality this is a value that is the maximum airborne concentration that an individual could escape and not suffer any adverse effects. The actual definition does not match the legal definition. At the IDLH level, emergency responders need to be using SCBA as an absolute minimum, and if the chemical is toxic by skin absorption then a fully encapsulated gastight suit (Level A) must be used.

Other values that a responder may see are called **lethal doses (LD$_{50}$)** or **lethal concentrations (LC$_{50}$).** The LD is for solids and liquids, and the LC is for gases. In most cases animal studies provide these values, but some are derived from human studies, suicides, and murders. The 50 attached to the LD or LC means 50 percent of the exposed population. In the studies a certain number of test subjects were exposed to a low level of chemicals. After a period of time, the test subjects were studied for any adverse effects. Another group was exposed to a higher level of the chemical. When the subjects were exposed to a cer

TOXIC VERSUS HIGHLY TOXIC

Toxic is a chemical that falls within any of the following categories:

- A chemical that has a median lethal dose (LD50) of more than 50 mg per kilogram but not more than 500 mg per kilogram of body weight when administered orally to albino rats weighing between 200 g and 300 g each
- A chemical that has a median lethal dose (LD50) of more than 200 mg per kilogram but not more than 1000 mg per kilogram of body weight when administered by continuous contact for 24 hours (or less if death occurs within 24 hours) with the bare skin of albino rabbits weighing between 2 kg and 3 kg each
- A chemical that has a median lethal concentration (LD50) in air of more than 200 parts per million but not more than 3000 parts per million by volume of gas or vapor or more than 2 mg per liter but not more than 200 mg per liter of mist, fume, or dust, when admin-

istered by continuous inhalation for 1 hour (or less if death occurs within 1 hour) to albino rats weighing between 200 g and 300 g each

Highly toxic is a chemical that falls within any of the following categories:

- A chemical that has a median lethal dose (LD50) of 50 mg or less per kilogram of body weight when administered orally to albino rats weighing between 200 g and 300 g each
- A chemical that has a median lethal dose (LD50) of 200 mg or less per kilogram of body weight when administered by continuous contact for 24 hours (or less if death occurs within 24 hours) with the bare skin of albino rabbits weighing between 2 kg and 3 kg each
- A chemical that has a median lethal concentration (LD50) in air of 200 parts per million by volume or less of gas or vapor, or 2 mg per liter or less of mist, fume, or dust, when administered by continuous inhalation for 1 hour (or less if death occurs within 1 hour) to albino rats weighing between 200 g and 300 g each

TABLE 27-7   Exposure Values

Chemical	PEL (PPM)	IDLH (PPM)	LCt_{50} (ppm) (t = 3 mins.)	ICt_{50} (ppm) (t = 3 mins.)
Sarin (a chemical warfare agent)	0.000017	0.03	12	8
Mustard (a chemical warfare agent)	0.0005	0.0005	231	21.5
Acetone	1000	2500	N/A	N/A
Acrolein	0.1	2	N/A	N/A
Ammonia	50 (ST)	300	N/A	N/A
Chlorine	1	10	N/A	N/A
Ethion (a pesticide)	0.02	N/A	N/A	N/A
Hydrogen sulfide	20 (CEILING)	100	N/A	N/A
Carbon monoxide	50 (35 NIOSH)	200 (CEILING)	N/A	N/A

Note: N/A, not applicable.

tain level and 50 percent of the test subjects died, this established the LD_{50} or LC_{50}. With terrorism, emergency responders are having to use some military data, much of which is based on LD_{50} or LC_{50} type studies. The military values are generally expressed as **LCt_{50}** or lethal concentration to 50 percent of the population, with "t" representing time, usually expressed in minutes. The military also uses **ICt_{50}**, which is the incapacitating concentration to 50 percent of the population in a certain amount of time. **Table 27-7** lists exposure values for some chemicals.

Although the exposure values can be confusing it is important to know what each of the values means, and emergency responders should be aware of the exposures they receive during incidents. It is generally recommended for emergency responders to use the TLVs or PELs as the point at which SCBA should be utilized. Emergency responders are faced with a lot of chemical exposures, and at these levels their safety can be ensured if proper PPE is worn.

SAFETY

Air monitoring of these exposure levels is a good way to ensure responder safety.

TYPES OF PERSONAL PROTECTIVE EQUIPMENT

The most common type of PPE for the firefighter is firefighter personal protective equipment (PPE). Full PPE is defined as helmet, hood, coat, pants, boots,

gloves, PASS, and SCBA, as shown in **Figure 27-5**. To be fully protected, all of this equipment must be in use. All of the snaps, zippers, and closures must be used to offer optimal protection. PPE offers protection against heat and water. The exact amount of protection depends on the type of gear used. PPE is not certified for chemical contact, nor should it be used for chemical protection. Some of the latest style gear is certified to protect against bloodborne pathogens, but to be sure, the firefighter should check for the NFPA certification.

SAFETY

Wearing protective clothing can present its own significant risk, and both OSHA and the NFPA require the use of the buddy system when entering potentially hazardous environments. When entering IDLH atmospheres, it is required that at least two firefighters enter together, and a backup team must be in place. When one partner has to leave the hazard area, the other partner should always leave at the same time. There should never be a situation where responders are in a hazard area alone.

Self-Contained Breathing Apparatus

With regard to chemical exposures, SCBA offers an assigned protection factor (APF) of 10,000. What this means is that a person wearing SCBA has an APF of 10,000 times greater than a person who is not wearing SCBA. To offer this protection, the SCBA must

FIGURE 27-5 The use of SCBA offers tremendous protection against heat and chemical hazards. Responders who enter any environment that may have a chemical present should always use SCBA.

be fitted properly to the person and must be a positive pressure device. If these two factors are not followed then the APF can be considerably less than the 10,000. A positive pressure SCBA has an airflow in the face piece all the time, and it maintains positive air pressure inside the mask to keep contaminants out. For firefighting or chemical spill response, it is imperative that the SCBA be a positive pressure unit and that it be activated automatically without any intervention on the part of the wearer.

SAFETY

A firefighter's chance for survival at a hazardous materials incident is dramatically improved when wearing SCBA, and it should be considered the minimum when dealing with chemical spills.

Although various types of SCBA equipment are available, the most common for chemical spill response is a sixty-minute type, which on an average allows twenty to thirty minutes of work time for a hazardous materials team member. When determining the use of SCBA, one has to consider the time it takes to enter the hazard area, working time, time to leave the area, and time to be decontaminated and undressed. In general, thirty- and forty-five-minute air supplies are inadequate for spill response.

Some teams and persons who deal with waste sites may use **supplied air respirators (SARs),** which have some advantages over SCBA. They do not have the weight of the SCBA, but do restrict movement somewhat due to the hoseline. Some logistical issues are associated with the use of SARs, but for long-term incidents they are of assistance.

Closed circuit SCBA's contain a cylinder of oxygen, a filter system, a regulator, and valves. They work on the principle of cleaning and filtering exhaled breath and adding pure oxygen to continue operation. The duration of the air supply is based on the filtering/cleaning and oxygen capacity of the unit.

Some response teams use **air-purifying respirators (APRs)** for minor spills, as shown in **Figure 27-6.** Although they offer some advantages and are common in industry they are not commonly used within the emergency services. With terrorism and bloodborne pathogen issues becoming more commonplace, however, the use of APRs will increase.

FIGURE 27-6 This police officer is wearing an air-purifying respirator (APR), which is filtering contaminants from the air.

Powered air purifying respirators (PAPR) are basically air purifying respirators that employ a built-in motor to blow ambient air through air purifying or filtering elements. Unlike the APR, breathing is less difficult due to the motor aiding in airflow through the filter. Since ambient air is being circulated through the mask, the PAPR must not be used in oxygen deficient atmospheres.

Protection from diseases through the air is provided by the use of a particulate respirator, which is also known as an APR. These respirators filter particles from the ambient air as the responder breathes.

APRs and PAPRs both require **fit testing** just like the SCBA to ensure that the respirator fits and will offer the wearer protection. If a responder chooses to use an APR at an incident, a decision flowchart must be followed. In general, the responder must know the chemical, and it must have good warning properties, such as a characteristic odor, and the responder has to identify the amount in the air. The amount of oxygen in the area must be verified and good levels must be maintained. In addition, a variety of cartridges can be used, and when dealing with a variety of chemicals a large stock of these cartridges must be maintained. With the exception of waste sites, minor releases, and possibly some decontamination work, SCBA is much easier to use.

Problems with SCBA include extra weight, fatigue, lack of full visibility, lack of mobility, contribution to heat stress, need to refill air supply, and other limiting factors. It is for this reason that many hazardous materials teams choose the other types of respiratory protection. The type of SCBA used will determine the amount of weight added to the responder, and in general the longer the work time the higher the weight of the apparatus. This added weight adds to the overall heat stress on the user. When using SCBA the wearer has limited vision because the face piece does not allow for a full spectrum of vision. When a chemical protective suit is added this limited field of vision gets even smaller. The use of SCBA in some applications such as confined spaces limits wearers as to how they can move about the confined space. For people who may be slightly claustrophobic with SCBA, there is an additional layer of stress placed on them when donning chemical protective clothing. The users of SCBA and chemical protective clothing need to be medically cleared to function with this type of PPE and should receive periodic medical exams, according to the applicable regulations.

Chemical Protective Clothing

There are four basic levels of chemical protective clothing, but these levels are broken down into further components. The levels are assigned letters to signify their protection levels. Both OSHA and the EPA use Level A, B, C, and D with Level A being the highest level of chemical protection.

Since the establishment of these levels, many changes have been made to PPE styles and types. When hazardous materials teams first started they had reusable protective clothing. It was used, cleaned, tested, and then reused. But the integrity of the suit became questionable after each use, so the suit manufacturers decided to switch to disposable one-time-use garments. Today, most teams use disposable suits, although some teams maintain at least one type of reusable garment, usually a Teflon fabric suit.

All chemical protective clothing must be checked prior to use for compatibility with the chemical that has been spilled. No matter the type of chemical protective clothing chosen, OSHA requires a PPE program that describes the use of protective clothing and the potential hazards employees may face. Part of the program consists of the maintenance, testing, inspection, and storage procedures. Many forms of protective clothing have specific requirements for storage and some types of clothing have finite shelf lives, some as short as five years. Level A suits require testing generally once a year or after a use. These suits are tested using a pressure test, which ensures their integrity.

CAUTION

No one suit fits all chemicals.

Some of the suits that are NFPA certified do come close, but there are a couple of chemicals that they are not to be used with. Chemical compatibility is based on the **permeation** of a chemical through the fabric of the suit. Permeation is the movement of a chemical through the fabric on the molecular level, **Figure 27-7**. In an acceptable suit, no damage to the fabric should result from a permeation test, nor should anything be seen visually. Imagine wrapping garlic in plastic wrap; initially there is no smell, but after a few hours the scent of garlic has permeated the refrigerator as if it had not been wrapped at all. Compatibility charts are provided by the manufacturer for the chemicals against which they have tested the fabric.

FIREFIGHTER FACT

Permeation is based on an 8-hour day or 480 minutes. When a department's inventory has more than one suit fabric, it is best to choose a fabric that is given a value of > 8 hours, because that means that in tests the fabric showed no breakthrough in 8 hours. After the fabric reaches 8 hours, the test is usually halted, unless specific times are indicated by the manufacturer.

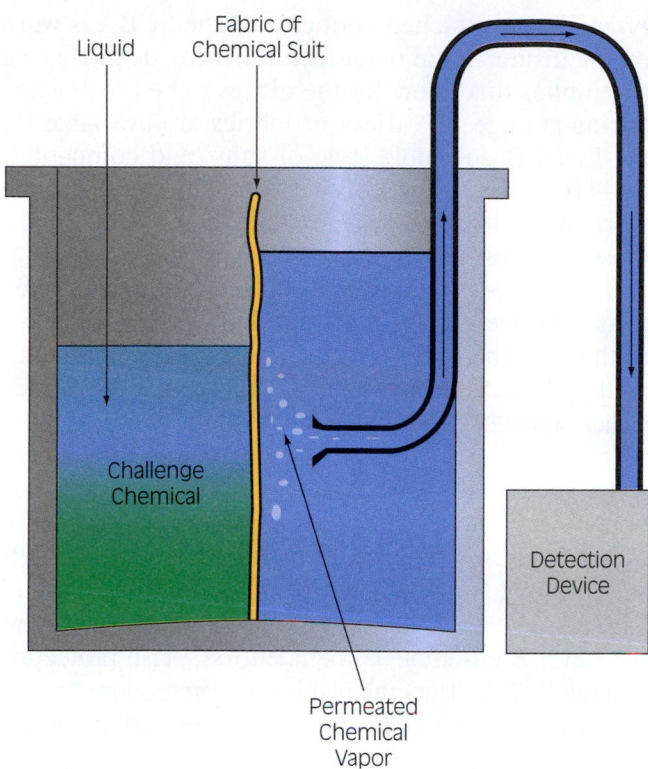

Liquid

Fabric of Chemical Suit

Challenge Chemical

Detection Device

Permeated Chemical Vapor

FIGURE 27-7 When conducting a permeation test, the fabric splits a test container, and a measurement device is used to see if the chemical goes through the fabric.

Most fabrics are tested against a battery of chemicals as specified by the American Society for Testing and Materials (ASTM) or the NFPA. This battery of chemicals represents the majority of the chemical families that responders are most likely to encounter.

Two other areas that are of concern when dealing with PPE are degradation and penetration. The degradation of a suit involves the physical destruction of the suit, leaving a hole or damage. Penetration is the movement of chemicals through natural openings such as zippers, glove/suit interface points, and other areas of the suit. It does not involve damage but is an area where chemicals may enter the suit.

Level A Ensemble Protective Clothing

The **Level A ensemble** as shown in **Figure 27-8,** is thought of as providing the highest level of protection against chemical exposure. It is a fully **encapsulated suit** and sometimes is called an encapsulated suit instead of a Level A suit.

NOTE

While it is true that a Level A ensemble offers the maximum level of chemical protection, it is also the leader as far as causing heat stress and physical and psychological stress on the responder wearing the garment.

FIGURE 27-8 This Level A ensemble is a gas/vapor-tight garment that protects against most chemicals. Although the protection offered by this ensemble is very high, it is the most stressful to wear.

To be considered a Level A ensemble the suit must have attached gloves and attached boots, and the zipper must be gastight. The suit is designed not to allow any gases to penetrate the garment, and by accomplishing this, it becomes liquid tight as well. Because materials cannot get inside the suit, including air, and gases cannot escape, the person wearing the suit needs to have an SCBA on the whole time. The suit does have relief valves that vent the exhaled air after a certain pressure buildup.

The requirement to use a Level A ensemble within the HAZWOPER regulation is when a firefighter is entering an atmosphere above the IDLH value and the chemical is toxic by skin absorption. There are several occasions, however, where the use of a Level A ensemble would be recommended for some chemicals that do not meet that definition. On occasions where a responder may be potentially covered with a toxic or corrosive material over the whole body, a Level A ensemble would be advisable.

STREETSMART TIP

When wearing a Level A ensemble prehydration is highly recommended, and dressing should take place in a cool quiet area, preferably in the shade.

During emergency operations all members must be monitored. Lack of full visibility and heat stress are major concerns when using a Level A ensemble. The Level A ensemble typically consists of the following components:

- Encapsulated suit with attached gastight gloves and boots
- Inner and outer gloves

- Hard hat
- Communication system
- Cooling system
- SCBA
- PBI/Nomex coveralls
- Overboots

To be an NFPA 1991 (Encapsulated Suit Specifications) certified Level A suit, the suit must have some flash fire resistance. To assist in compliance, newer suits use a blended fabric that offers chemical resistance as well as the flash resistance. Older style suits use a flash suit overgarment that is made of aluminized PBI/Kevlar fabric. This flash protection is not intended for firefighting, but it offers three to thirteen seconds of protection when involved in a flash fire.

Level B Ensembles

Within the **Level B ensemble** family, there is a lot of variety in suit types, as shown in **Figures 27-9 A–C**. Although the EPA and OSHA acknowledge two basic types, a large number of styles are available. The two basic types are coverall style and encapsulated, but even these have subvarieties. The encapsulated style of Level B suits is similar to the Level A style, but does not have attached gloves, nor is it vapor-tight. It

typically has attached booties, and the SCBA is worn on the inside. Some manufacturers provide glove ring assemblies that allow for the gloves to be preattached during storage. A variety of fabrics are available for the Level B ensemble style of suits, and compatibility is important. The Level B encapsulated suit is the workhorse of hazardous materials teams; it is the most common suit used.

The encapsulated Level B suit is sometimes referred to as a Bubble B or a B plus suit. These suits have some of the same heat stress issues associated with them as do the Level A ensembles, even though they are lighter. Other styles of Level B suits include a two-piece garment consisting of a jacket and pants, usually bib-overall style. The coverall style Level B suit may have attached booties or a hood, but there are a number of styles for a variety of uses. The one item that makes the Level B suit different from the lower levels of PPE is the use of SCBA. A Level A ensemble is a gastight suit, whereas the Level B ensemble is intended for splash protection and the SCBA offers the respiratory protection.

A Level B ensemble consists of these components:

- Level B suit
- Hard hat
- Inner and outer gloves
- SCBA

(A)   (B)   (C)

FIGURE 27-9   The photos here all represent Level B suits: (A) a coverall style for law-enforcement officers; (B) an encapsulated style, which is not gastight; and (C) a military-designed two-piece garment worn by tactical officers. The respiratory protection is a rebreather style, which provides a four-hour air supply.

- Communication system
- Outer boots
- Nomex/PBI coveralls

Level C Ensemble

A Level C ensemble, as shown in **Figure 27-10,** incorporates the use of an air-purifying respirator within the ensemble. For obvious reasons an APR cannot be used within an encapsulated suit, but it can be used with the other styles. A Level C ensemble can be a coverall or a two-piece garment. A Level C ensemble is used where splashes may occur, but respiratory hazards are minimal and are covered by the use of an APR.

When using an APR, an extensive listing of requirements must be met. A Level C ensemble consists of these components:

- Level C suit
- Air-purifying respirator

- Hard hat
- Inner and outer gloves
- Outer boots
- Nomex/PBI coveralls

Level D Ensemble

A Level D ensemble , shown in **Figure 27-11,** is actually regular work clothing. It is used when respiratory protection is not required and splashes are not a concern. The Level D ensemble provides no chemical protection, but does offer protection against other workplace hazards. A level D ensemble consists of these items:

- Work clothes
- Hard hat
- Chemical/work gloves
- Safety glasses
- Safety shoes/boots

High-Temperature Clothing

The two basic types of high-temperature clothing are proximity and fire entry gear. A set of proximity gear is shown in **Figure 27-12.** The most common use for

FIGURE 27-10   This Level C ensemble protects against splash hazards and low toxicity materials. It is the use of the air purifying respirator that makes this ensemble Level C.

FIGURE 27-11   The Level D ensemble is normal work clothing, this example includes Nomex coveralls, chemical gloves, safety glasses, and steel-toe shoes.

FIGURE 27-12 These firefighters are wearing proximity firefighting suits that offer a high level of thermal resistance. They are typically used by air crash rescue crews and flammable liquid firefighters. *(Courtesy of Tech Sgt. Brian E. Christiansen, USAF)*

this type of gear is in airport firefighting and flammable liquids firefighting. The entry suits may also be used in high-temperature applications within industry, such as steel making.

High-temperature gear is usually identified by its characteristic aluminized outer shell. This shell is usually attached to a PBI/Kevlar fabric that offers a higher heat resistance than normal structural firefighting clothing. This proximity gear is named for its ability to allow the wearer to get close to the burning liquid. It offers protection for temperatures up to 300 to 400°F (149 to 204°C).

Fire entry gear is designed to allow the wearer to enter a fully involved fire area for a period of thirty to sixty seconds over the life of the suit. The thought process was that it would allow for the rescue of trapped victims, which firefighters now know would not be alive if that type of garment is needed. There are certain applications for this gear involving industrial incidents where high temperatures would be found, but these are rare and most departments no longer carry this type of gear. Fire entry gear can be used in temperatures ranging up to 2,000°F (1,093°C).

Low-Temperature Clothing

When dealing with cryogenics the responder must wear protective clothing that protects the wearer against very cold temperatures. Propane, once it is released from a cylinder to the atmosphere, can reach temperatures well below 0°F (−17°C). Cryogenics by their definition are at least −150°F (−101°C) and are usually colder. Standard protective clothing is not effective against these types of materials. Responders should take precautions against cold stress, and prevent hypothermia. Although no full protective

clothing ensemble is available for dealing with cold materials, various types of gloves, **Figure 27-13,** and gauntlets and aprons are available. Layering of clothing can offer some additional protection. Unfortunately in addition to the cold, cryogenics usually present other hazards such as fire, corrosion, toxicity, and asphyxiation.

Limitations of Personal Protective Equipment

There are four basic limitations to protective clothing, and they apply across the board from EMS infection control gear, to structural firefighter clothing, to the fully encapsulated Level A ensembles. Most of these issues are only thought of when dealing with chemical protective clothing, but more thought and emphasis should be placed on these issues when using any type of protective clothing. The four major issues are heat stress, mobility, visibility, and communications problems.

> **STREETSMART TIP**
>
> Stress is a leading killer of firefighters, and heat stress plays a major factor in many of these deaths.

It is important for emergency responders to be able to recognize heat stress and its degrees of seriousness. Heat stress can lead to heat stroke, a condition that is almost always fatal. When wearing a Level A ensemble the temperature inside the suit can easily exceed 100°F (38°C), even on a cold day. When using PPE of any type, responders should take care to pre-hydrate and take their time when doffing their PPE—immediate removal of their PPE can shock the body,

FIGURE 27-13 Cryogenic gloves are used when dealing with very cold refrigerated liquids. They offer protection from a range of −150° to −450°F.

Figure 27-14, and cause serious health issues. This is important to remember when removing PPE from a hazardous materials responder, who is usually zipped up inside a very warm environment and cannot control the undressing progress.

> **CAUTION**
>
> Dehydration is a major factor when operating at a fire or a chemical spill. Frequent hydration and frequent rest breaks are important.

FIGURE 27-14 This type of PPE is very useful for USAR situations, offers chemical protection, and is well-suited for WMD response. *(Courtesy of W. L. Gore and Associates)*

The progression of heat stress is dependent on the amount of work being performed and the physical ability of the responders. To best combat any heat-related emergency, responders should hydrate and take frequent breaks, and until acclimated take extra precautions. When an incident occurs on a day in which the weather conditions would not be considered normal, personnel should pay extra attention to their activity levels. Early recognition is important for the health and safety of the responders. The levels and their warning signs are listed in **Table 27-8.**

TABLE 27-8	Health Hazards and Common Warning Signs
Heat Exhaustion	This is the real first step toward dehydration, and it is fairly common. The root cause is excess sweating, resulting in a loss of body fluid. Symptoms include dizziness, headache, nausea, diarrhea, and vomiting. Heat exhaustion affects not only hazardous materials responders but firefighters as well. Responders should have had enough fluid that they feel the need to urinate. Heat exhaustion that is not treated can lead to heat stroke, which is a very serious and often fatal condition.
Heat Stroke	This is a serious emergency and can be fatal to upward of 80 percent of the patients. In simple terms, it means the body's ability to regulate its temperature has failed and is no longer functioning. The symptoms include unconsciousness, hot and dry skin, seizures, confusion, and disorientation. Being dehydrated when approaching heat stroke levels only complicates the process and actually speeds up the patient's deterioration. Heat stroke should be prevented because once the body shuts down, the end result is usually fatal.
Cold Stress	Situations with liquefied gases or cryogenic gases present situations where cold stress could be a factor. When skin comes in contact with or in proximity to cold vapors or liquids, it can be severely damaged. The symptoms of cold stress include discoloration of the skin, numbness, and pain. The more severe the damage, the darker the skin will become. Skin severely damaged will be black from the exposure.

LESSONS LEARNED

The routes of entry and type of harm that chemicals present vary and chemicals usually present multiple hazards. Having an understanding of these health effects and routes of exposure is important for responder safety. Most chemicals present chronic hazards, but acute hazards are also to be avoided. The use of protective clothing is important for the various hazards that responders may face. The SCBA offers a high level of protection and should be the absolute minimum for respiratory protection. If a responder cannot confirm the absence of hazardous materials through the use of air monitors, then a basic level of protective clothing should be used. The decision to use chemical protective clothing is not an easy one because many factors must be examined prior to its use. With the ensuing heat stress that accompanies the wearing of chemical protective clothing, cases can arise in which the protective clothing may be more dangerous than the chemical hazard. It is for this reason that the responders must use effective risk assessment to determine the true hazard—and then dress for that hazard. Many responders have a fear of radiation, but radiation is no different than other hazardous materials; one just has to understand the hazards. With radiation and with chemicals, understanding dose is paramount to remaining safe. We are exposed to radiation every day, and just as with chemicals, the dose makes the poison.

KEY TERMS

Absorbed Dose A measure of the amount of radiation transferred to a material.

Acute A quick one-time exposure to a chemical.

Air-Purifying Respirator (APR) Respiratory protection that filters contaminants out of the air, using filter cartridges. Requires the atmosphere to have sufficient oxygen, in addition to other regulatory requirements.

Allergen A material that causes a reaction by the body's immune system.

Carcinogen A material that is capable of causing cancer in humans.

Ceiling Level The highest exposure a person can receive without suffering any ill effects. It is combined with the PEL, TLV, or REL as a maximum exposure.

Chronic A continual or repeated exposure to a hazardous material.

Convulsant A chemical that has the ability to cause seizure-like activity.

Encapsulated Suit A chemical suit that covers the responder, including the breathing apparatus. Usually associated with Level A clothing, that is gas and liquid tight, but there are some Level B styles that are fully encapsulated, but not gas or liquid tight.

Etiological   A form of a hazard that includes biological, viral, and other disease-causing materials.

Fit Testing   A test that ensures the respiratory protection fits the face and offers maximum protection.

ICt$_{50}$ The incapacitating level for time to 50 percent of the exposed group. It is a military term that is often used in conjunction with LCt$_{50}$.

Irritant   A material that is irritating to humans, but usually does not cause any long-term adverse health effects.

LCt$_{50}$ The lethal concentration for time to 50 percent of the group. Same as the LC$_{50}$, but adds the element of time. It is a military term.

Lethal Concentration (LC$_{50}$)   A value for gases that provides the amount of chemical that could kill 50 percent of the exposed group.

Lethal Dose (LD$_{50}$)   A value for solids and liquids that provides the amount of a chemical that could kill 50 percent of an exposed group.

Level A Ensemble (of Protective Clothing)   Fully encapsulated chemical protective clothing. It is gas and liquid tight and offers protection against chemical attack.

Level B Ensemble (of Protective Clothing)   A level of protective clothing that is usually associated with splash protection. Level B requires the use of SCBA. Various clothing styles are considered Level B.

Permeation   The movement of chemicals through chemical protective clothing on a molecular level; does not cause visual damage to the clothing.

Permissible Exposure Limit (PEL)   An OSHA value that regulates the amount of a chemical that a person can be exposed to during an eight-hour day.

Recommended Exposure Limit (REL)   An exposure value established by NIOSH for a ten-hour day, forty-hour workweek. Similar to the PEL and TLV.

Risk-Based Response   An approach to responding to a chemical incident by categorizing a chemical into a fire, corrosive, or toxic risk. Use of a risk-based approach can assist the responder in making tactical, evacuation, and PPE decisions.

Sensitizer   A chemical that after repeated exposures may cause an allergic-type effect on some people.

Short-Term Exposure Limit (STEL)   A fifteen-minute exposure to a chemical followed by a one-hour break between exposures. Only allowed four times a day.

Supplied Air Respirator (SAR)   Respiratory protection that provides a face mask, air hose connected to a large air supply, and an escape bottle. Typically used for waste sites or confined spaces.

Threshold Limit Value (TLV)   An exposure value that is similar to the PEL, but is issued by the ACGIH. It is based on an eight-hour day.

TRACEMP   An acronym for the types of hazards that exist at a chemical incident: thermal, radiation, asphyxiation, chemical, etiological, mechanical, and potential psychological harm.

REVIEW QUESTIONS

1. What route of entry is the easiest to protect against?
2. What material are emergency responders commonly exposed to through injection?
3. After twenty years of smoking, what type of exposure has a person received?
4. What type of exposure is a one-time event?
5. What are the three common routes of exposure?
6. What six hazards potentially exist at a chemical release?
7. Which adverse medical effect mentioned in the chapter can become hereditary?
8. Nitrogen is what type of asphyxiant?
9. What exposure value uses a fifteen-minute time limit?
10. What is the term ceiling used for with regard to exposure values?
11. Which level of chemical protective clothing offers a high level of protection against toxic chemicals that are absorbed through the skin?
12. What is a major concern as the levels of protective clothing increase?
13. Which type of respiratory protection offers the highest level of protection?

ENDNOTE

1. Not all emergency services workers are covered by the Federal OSHA regulations. Some are covered by their state OSHA or are not covered at all by OSHA regulations. This varies from state to state.

28

Hazardous Materials: Protective Actions

It was the start of what we call a normal day on the job, until later in the morning when an alarm for a chemical leak inside a beverage warehouse was sounded. The dispatch consisted of what we call a hazardous materials box: four engine companies, one truck company, a rescue squad, basic life support unit, hazardous materials company, and command officer. I was working at the hazardous materials company that day. While en route, a radio transmission by the first arriving company advised the Emergency Communications Center of a major ammonia leak inside the warehouse storage area and that they were taking protective actions. This area contained multiple storage of beverages and boxes within an enclosed and secured area that also contained valves and piping for the anhydrous ammonia refrigeration system.

Upon our arrival, the warehouse had been evacuated and a strong odor of ammonia had already consumed the entire area surrounding the warehouse. Once we had performed a hazard risk assessment and ensured that the first responders had taken appropriate protective actions, we then selected our level of protection. Three hazardous materials technicians and I entered the release area to shut off the valve to the leaking pipe. After locating the release area we located the valve and made an attempt to close the valve. While closing the valve a sudden release of gaseous and liquid ammonia covered the personnel working at and around the valve. Visibility was taken from us almost instantaneously because of the gaseous release and communications were lost between all four technicians. I was able to find my way out and noticed that my personnel were still inside the release area. Prior to making another entry to locate my personnel, I noticed a white smoke coming from my chemical boots.

After further investigation I realized that the oil-based paint from the concrete floor was causing a chemical reaction under the soles of my boots. I reentered the release area, located my personnel, and immediately withdrew from the release area to the decontamination area. Once we were refreshed, a second entry attempt was made into the release area, where we were able to locate another sectional valve and stop the leak. The hardest part of the second attempt was removing our SCBA inside our suits to squeeze past piping and valves to get to the right one.

Prior to leaving the scene we finally determined that prior to our arrival a firefighter had entered the release area and closed the valve without notifying command and/or hazardous materials personnel. When hazardous materials personnel entered the release area thinking that the valve was not closed, they actually reopened it, which caused the valve to freeze in the open position. In this event, a number of factors affected our response. First-arriving crews needed to address isolation and evacuation issues, and the type of release, but one firefighter was endangered by not taking appropriate protective actions. No personnel were injured or exposed to the ammonia, but the incident proved to be very dangerous as a result of personnel freelancing and the lack of training present at an emergency scene.

—*Street Story by Gregory L. Socks, Captain,*
Montgomery County, MD HAZMAT Team

INTRODUCTION

This chapter provides a myriad of topics for the responder and focuses on some general tactics that should be followed at a hazardous materials incident. The tactical considerations provided here are for general situations and may not apply to specific situations because each chemical spill is different, and for each spill there may be another way of handling that release.

> **NOTE**
>
> The basic concept for first responders is one of isolation—first responders should not allow other people to become part of the incident, and they should protect those involved with the incident or those who may become part of the incident in a short time.

A lot of the information in this section may not apply to specific cases. For firefighters beginning their training, it is unlikely they will be making community evacuation decisions for the next couple of years, but the material in this chapter should be kept in mind for the time when they will be making these types of decisions.

HAZARDOUS MATERIALS MANAGEMENT PROCESSES

OSHA's HAZWOPER regulation requires the use of ICS but does not state which system is required. In 2004, the Department of Homeland Security (DHS) mandated the use of the National Incident Management System (NIMS). The NIMS system was developed to merge the federal, state, and local response efforts to a disaster or terrorist attack. When developing NIMS, the various incident management systems were combined into one. For use in this text we will use Incident Command System (ICS). The reality is that for the overall incident an incident command system (ICS) will be in place, and the group of responders responsible for hazardous material will fit into that ICS using one of the processes mentioned or a combination, depending on the situation. The incident command system (ICS) is exactly what the name implies: a systematic approach for the command, control, and management of an emergency incident. It provides a command structure and designated responsibilities for the functions that must be addressed to stabilize a hazardous materials incident.

Several different management processes exist that can be used for hazardous materials incidents, many of which have been in use for many years and offer well-proven methods of organizing an incident. All of the systems have been adapted from fire service systems to fit the needs of a chemical release. The cores of all of these systems are basically the same, but they do differ in some areas. The core to all systems is the protection of life, property, and the environment.

One of the systems developed is the **8-Step Process,** which was devised by Mike Hildebrand, Greg Noll, and Jim Yvorra. Dave Lesak developed another system called the **GEDAPER process** of hazardous materials management. Another system developed early on by Ludwig Benner, Jr., is the **DECIDE process,** which is listed along with the other systems in **Table 28-1.** Regardless of which system a department chooses, it is important to choose a system that everyone understands and can use.

> **NOTE**
>
> When arriving at a suspected chemical release, it is important to isolate the area from other people who may inadvertently wander into a hazardous environment.

TABLE 28-1 Hazardous Materials Management Systems

DECIDE	8-Step Process	GEDAPER	Hazmat Strategic Goals
Detect the presence of the hazardous materials	Site management and control	Gather information	Isolation
Estimate the likely harm	Identify the problem	Estimate potential course and harm	Evacuation
Choose a response objective	Hazard and risk identification	Determine strategic goals	Notification
Identify the action	Select personal protective clothing and equipment	Assess tactical options and resources	Product identification
Do the best possible	Information management and resource coordination	Plan and implement chosen actions	Determination of appropriate personal protective equipment
Evaluate your progress	Implement response objectives	Evaluate	Decontamination
	Decontamination	Review	Spill and leak control
	Terminate the incident		Termination

Sources: The information on the 8-Step Process® is from *Managing the Incident,* Gregory Noll, Michael Hildebrand, Jim Yvorra, Fire Protection Publications, Oklahoma University, 1995. The information on the GEDAPER® is from *Hazardous Materials Strategies and Tactics,* David Lesak, Prentice Hall, 1998. Both systems reproduced with permission of the authors.

Isolation and Protection

One of the most important tasks a responder trained to the awareness level can do is to isolate the area so that others do not become part of the problem. Methods of isolation can be as simple as using barrier tape, **Figure 28-1,** to using law enforcement at traffic control points. Other methods such as traffic barriers, or even the use of emergency vehicles to block access, can be used. With a chemical hazards incident, it is important to control the incident quickly. The more people entering the suspected hazard area, the more people who may later need to be rescued or, depending on the situation, may need decontamination.

The protection of the people in a hazard area can best be accomplished by evacuation of the immediate area. This does not imply that everyone is simply told to leave, because they may need to be decontaminated or at least medically evaluated depending on the situation. A plan must be established for the holding of these people until a determination can be made as to their status. The hazardous materials team is usually the only group that can make this determination. When dealing with people in the suspected hazard area, frequent communication with the hazardous

materials team is important. The other use of protection at a chemical spill is usually associated with the adjacent community evacuation or sheltering in place. Both of these issues are further discussed later in this chapter.

If the incident is suspected of involving criminal or terrorist activity, isolation is important not only for the chemical hazards but for evidence preservation as well. First responders at the awareness level should make efforts to keep people from leaving the area and keep note of persons or vehicles that are leaving the scene. They should isolate the area and ensure that others do not enter the hazard or crime scene area. First responders should establish a hazard area and restrict entry into that area. They should also notify law enforcement and the hazardous materials team of the conditions that they have found.

Denial of Entry

If the incident is suspected of involving hazardous materials, or criminal or terrorist activity, isolation is important not only for the chemical hazards but for evidence preservation as well. First responders at the awareness level should make efforts to keep people

FIGURE 28-1 One of the first priorities should be to isolate the area so as to prevent other people from becoming involved with the incident.

from leaving the area and keep note of persons or vehicles that are leaving the scene. They should isolate the area and ensure that others do not enter the hazard or crime scene area. First responders should establish a hazard area and restrict entry into that area. They should also notify law enforcement and the hazardous materials team of the conditions that they have found. A person should be assigned to act as security for each of these zones to ensure that only authorized personnel enter these areas.

Rescue

When discussing isolation and protection, rescue is a topic that naturally follows those two important issues. The rescue of victims from a suspected hazard area can be extremely controversial. The decision to make a rescue is a personal one, because it may involve substantial risk to the rescuer. Local protocol and SOPs must be considered.

> ### SAFETY
>
> Firefighting is inherently risky. As much as firefighters would like to eliminate all risk from their occupation, it is impossible to do so. Their best hope is to safely manage that risk and use methods to identify that risk.

In reality, if responders arrive at the incident safe and sound, their chance for survival dramatically increases. One of the most dangerous parts of firefighting is responding to the incident—each year 20 to 30 percent of firefighter deaths occur while going to and from an incident.

With specific regard to Firefighters Personal Protective Equipment, scientific data published in August 2003 by the Soldiers Biological and Chemical Command (SBCCOM) provide Incident Commanders with some information as to the ability to make rescues in hazardous situations. The SBCCOM located at the Aberdeen Proving Grounds in Maryland was redesignated into several other organizations. Formerly SBCCOM, the now Research, Development, and Engineering Command (RDECOM) performed several studies involving firefighters' protective clothing, chemical protective clothing, and detection devices. The lead study, Risk Assessment of Using Firefighter Protective Ensemble (FFPE) with Self-Contained Breathing Apparatus (SCBA) for Rescue Operations During a Terrorist Chemical Agent Incident, is known as the 3/30 Rule, and researchers found that firefighters' protective clothing can offer protection where significant hazards exist to someone without protective clothing. They tested military chemical nerve and blister agent vapor, both of which present toxicity hazards; the nerve agent is extremely toxic. In situations where a nerve agent has been released and there are both live and dead victims, firefighters in full PPE and SCBA can enter this environment and make rescues with no or minimal effects. In a situation where all the victims are dead, firefighters can enter this severely toxic environment for three minutes with no or minimal effects. There are a number of factors that go into entering this type of toxic environment. The study outlines the best way for firefighters to protect themselves with PPE. There are methods discussed that use tape to add additional protection, as well as some other unique suggestions. The discussion on entering a terrorism event is controversial, and this study only offers some science behind the decision-making process. It goes without saying that in these cases emergency decontamination is a must and the gear should probably be destroyed afterward.

SAFETY

In order to understand the risks associated with the rescue of victims in a chemical warfare agent environment, responders should read the full report and understand the risks of operating in a dangerous environment. Within the report, there are several caveats to several scenarios. The report can be downloaded at http://www.ecbc.army.mil/hld/cwirp/ffpe_scba_rescue_ops_download.htm.

Once on the scene, firefighters need to evaluate the incident. Is it a rescue situation? This information should be confirmed as best it can; in some cases this information cannot be verified. What are conditions at the incident? Is there a confirmed chemical spill? If there is a release, can a person wearing no protection survive? In many cases the fire department is called to suspected chemical releases, and, after investigation, it turns out there really was not a chemical release. If the people who need to be rescued are alive, then the actual risk to the firefighters making the rescue attempt wearing full PPE is minimal. If, on the other hand, the first responders arrive at a local mall and are told that twenty people are unconscious and appear to be dead in the hardware store, the situation is different. When the responders approach the store entrance and see the twenty people lying on the floor, not moving, the responders may not want to enter that environment.

Other considerations need to be taken into account when making a rescue decision, such as response and notification time. The fact that the people in the hazardous environment have been there for a period of five to fifteen minutes, depending on the response time of the department, is critical to the decision-making process. They are not wearing any PPE and have been exposed to the material for a considerable amount of time. If they are alive when responders arrive, the risk to the rescuers is minimal. A method of emergency decontamination should be set up, and a backup crew should be standing by. The responding hazardous materials team should be consulted prior to entry so as to verify the chemical information. When rescuing the victims there should be no delay in their evacuation—a swoop-and-scoop technique is the order of the day. Stokes baskets or other methods of quick evacuation should be employed. Firefighters should not take vital signs, ask medical histories, or perform medical procedures; instead, they should quickly move the victims to a safe area. Responders should have established methods to notify the Incident Commander and other response personnel about critical conditions at the incident. Radios are the most common method, but hand signals may be needed. Some form of relay system may need to be set up to ensure communication between the response crews and the Incident Commander, so that critical information can flow both ways.

Once out of the area, decontamination should be performed if required on both victims and rescuers, and they should be isolated from the remainder of the responders. Once decontaminated, EMS personnel can begin to work on the patients using appropriate levels of PPE. Depending on the condition of the patients, decontamination may be hurried and therefore not perfect. Secondary contamination is of concern for EMS personnel. If a patient ingested a hazardous material, vomiting may result, which exposes personnel to potential contaminants in addition to the bloodborne (etiological) hazard. If the person has been contaminated with a material that is skin absorbable, then it is possible that the victim may off-gas the material or may release odors. After decontamination, the rescue team should remove their PPE and bag them. After evaluation by EMS providers and consultation with the hazardous materials team the rescuers should be sent to rehabilitation. Their turnout gear and SCBA will need to be evaluated and possibly sent for cleaning or replacement, depending on the possible contamination and the material in question.

SAFETY

The rescue of victims is made using simple risk/benefit analysis: a lot of risk is taken when a lot is at stake. No risk is taken when the benefit is little.

When handling hazardous materials incidents, encountering trapped victims is an unusual incident, but procedures should be in place to cover this contingency. The most likely scenario is a traffic accident, in which persons are trapped and chemicals are involved. This type of scenario is common if gasoline is introduced into the picture. Firefighters are used to extricating trapped victims with gasoline leaking from one or more vehicles. Protection lines are established, and a higher level of PPE is typically used by the rescue crews. The rescue crews limit the number of people in the hazard area, and a quick extrication is usually performed. If other chemicals are involved the scenario may change slightly. As hazardous materials companies arrive, personnel can be replaced by those with better chemical protection, and the hazard-

RISK-BASED RESPONSE

A risk-based response philosophy is based on the use of air monitors to guide responders safely. The monitors guide the responders as to which situations are safe to proceed into and when the use of PPE is appropriate. Unfortunately, many first responders do not have all of the required air monitors for a high level of protection. When confronted with rescue situations in potentially hazardous environments, first responders should use full protective equipment, including SCBA, as well as air monitors. They should also consult with the hazardous materials team, who can help guide them to the appropriate level of action. All emergency response involves some risk; steps need to be taken to manage (and thereby minimize) that risk.

In one situation, a truck carrying food rear-ended another tractor-trailer stopped in the middle lane of the highway as shown in **Figures 28-2, 28-3,** and **28-4.** It is estimated that the food truck was traveling at 65 mph when it hit the stopped truck. The truck it hit was carrying a mixed load consisting of mostly 55-gallon drums. The driver of the stopped truck was not hurt and was able to remove his shipping papers. The driver of the food truck was still alive but was extremely entangled in the wreckage. The driver was conscious and alert, and initial vital signs were stable. The trucks were tangled together, and the door to the truck carrying the drums was torn away from the truck. Drums had shifted onto the food truck and were laid over the front of that truck.

The first responders saw the placards on the first truck and requested a hazardous materials assignment. When they approached the food truck to evaluate the driver,

they saw the drums in a precarious position in the back of the truck. They already had turnout gear on but had also donned SCBA. They obtained the shipping papers from the driver and consulted with the hazardous materials company, which was about twenty minutes away.

FIGURE 28-3   Rescue crews are extricating the driver of a second vehicle involved in a rear-end collision. The driver remained conscious throughout the extrication and was severely entangled. Complicating the rescue was the presence of a truck full of hazardous materials containers. *(Courtesy of the Baltimore County Fire Department)*

FIGURE 28-4   Leaking drums were found during this rescue operation. Rescue crews had a foam line standing by and were air monitoring for flammable levels. When hazardous materials crews arrived, they took over air monitoring, quickly removed the leaking drums, and secured the other drums. When the dirver was extricated, hazardous materials crews took care of the remaining drums. *(Courtesy of the Baltimore County Fire Department)*

FIGURE 28-2   The truck to the right rear-ended the front truck trapping the driver in the truck on the right. In the trailer to the left are 55-gallon drums, some of which are leaking. Using PPE, air monitoring devices, and protection lines, crews continued with the rescue. *(Courtesy of the Baltimore County Fire Department)*

The contents of the drums were mostly flammable liquids and some combustible liquids. The first responders were advised to use a combustible gas indicator, continue with full PPE, establish foam lines, and begin the rescue.

Upon arrival, the hazardous materials company team members met with the incident commander (IC) to evaluate the scene. They confirmed the monitoring being done by the first responders and began to evaluate the other parts of the load. It was determined that whenever the rescue companies moved a part of the dash of the truck, one of the leaking drums would increase its flow. The hazardous materials companies secured the leak, moved the drum, and secured the remainder of the

drums. They examined the rest of the load to make sure there were no other problems and continued to monitor the atmosphere. Once the victim was removed and the rescue companies had moved away, the hazardous materials team **overpacked** and pumped the contents of the other drums into other containers.

The risk category was fire, and the PPE chosen was appropriate for the risk category. At no time were any flammable readings indicated during the rescue, although some were encountered during the transfer operation. This was an example of the various disciplines working together to rescue a victim in a hazardous situation.

ous materials company can work to control or eliminate the chemical hazard.

Making the decision to enter a hazardous environment takes training and experience. This decision should be made in direct consultation with the hazardous materials company. To make that decision, some hazardous materials teams use an approach known as risk-based response. By the heavy use of air monitors, an unknown chemical can be placed into one of four risk categories: fire, corrosive, toxic, or radiation hazards. In examining chemical and physical properties, a chemical can be placed into one or more of these categories. The major factor of concern in a potentially hazardous situation is a chemical's vapor pressure. If the chemical does not have a high vapor pressure, the risk of entering an environment containing this spilled material is very low unless a responder touches or falls into the material. One of the other items that a hazardous materials team can use to its advantage is the fact that the majority of incidents they respond to involve flammable and combustible liquids or gases. Although not designed as such, firefighter turnout gear offers ample protection for a quick rescue situation. Hazardous materials teams and now many first responders have very capable detection devices for these types of materials and can determine the true risk during an operation. Included here is a listing of the top ten chemicals spilled in this country every year. This is a hazardous materials team's bread and butter, and they should be comfortable with the handling of these materials. First responders should know the locations in which these materials are stored and used. The response to an incident involving these materials should be no different than a response to a bedroom fire. The top ten chemicals spilled are:[1]

1. Sulfuric acid
2. Hydrochloric acid
3. Chlorine
4. Ammonia

5. Sodium hydroxide
6. Gasoline
7. Propane
8. Combustible liquids
9. Flammable liquids
10. Natural gas

Site Management

The management of a hazardous materials incident can be very difficult even for a seasoned incident commander. A number of incident management strategies are available to the IC, some of which were outlined in previous chapters. The hazardous materials specific positions are outlined in **Table 28-2.** A fire department IC on a normal fire-type incident deals predominantly with resources within the IC's own agency. Occasions may arise when the IC will discuss items with the police department, and depending on the location a police officer may be in the command area. The IC may also deal with other municipal agencies such as the health department or social services. Other outside agencies such as the Red Cross may also be involved and work under the direction of the IC. Depending on the size of the community, the media may play a factor in the management of the incident. But in the overall scheme of the incident the IC deals mostly with the IC's own agency and coworkers.

When involved in a chemical release, especially one of major proportions, many more agencies get involved. In some communities, the hazardous materials team may not be associated with the local fire department; instead it may be from a county mutual aid group or from an adjoining community. In many incidents, major road closures are necessary, which increases the presence of police officers, and depending on the size of the road may bring command level officers to the scene. If the incident involves a state

TABLE 28-2   Hazardous Materials Branch Positions

- Backup: Personnel assigned to rescue the entry team if necessary. They are dressed in the same suits as the entry team and are fully prepared to make an entry, with the exception of being on air. The number of backup personnel is a minimum of two but can be expanded to more personnel depending on the incident. The backup team must be trained to the minimum of the Technician level.

- Decontamination: A person is assigned to oversee the setup and operation of the decontamination area, sometimes referred to as the decon area or the contamination reduction corridor. Other personnel will also be operating within the area performing the decon or PPE removal. People operating in this area are usually trained to the Technician level, but people trained to the Operations level may also be operating in this area. If Operations-level personnel are used, they should have received specific training in this job function.

- Entry: A minimum of a two-person team will enter the hazard area or hot zone. The type of chemical will determine the type of PPE that will be used. Prior to entry, the IC must brief the personnel who will be working in the hazard area. To be on the entry team, responders must be a Technician or a Specialist.

- Hazardous Materials Supervisor (HAZMAT Officer): This person is responsible for handling the tactical objectives, as assigned by the Incident Commander. This person is in charge of the HAZMAT team and therefore coordinates its efforts. May be referred to as HAZMAT operations.

- HAZMAT Safety Officer: This person assists the overall safety officer but is concerned mostly with HAZMAT-specific issues such as PPE selection and use. This HAZMAT safety position is usually a Technician or above. The overall safety officer is concerned with all personnel at the incident and usually focuses on other hazards not related to the chemical (e.g., slip, trip, and fall hazards).

- Information/research: This position provides information regarding the chemical and physical properties to the HAZMAT branch officer and the IC. After completion of these duties, these personnel may then assume responsibility for documenting the incident.

- Reconnaissance: These personnel use binoculars or may even enter the hazard area to determine incident severity, attempt identification of the hazardous material, and gather any additional information.

- Resources: These personnel are responsible for gathering and maintaining supplies required for the incident. May be referred to as logistics.

road or highway or has the potential to impact these roads, the state highway department may attend the incident. County or state environmental representatives or responders may also arrive at the incident to assist. In some states the Department of Natural Resources (DNR) has jurisdiction in chemical spills and, hence, may respond to an incident.

On a large spill the media will be a much larger group and may be from outside the immediate area, such as at the incident shown in **Figure 28-5.** A chemical release incident usually requires the services of a cleanup contractor, who will be arriving at the incident to assist in the cleanup effort. Depending on the incident, it is not uncommon to see insurance adjusters arriving within a few hours of the spill. On larger spills or spills that may endanger a waterway, the EPA or Coast Guard may arrive to assist with the spill.

Although this listing of agencies is nowhere near complete, it gives the reader an idea of the number of

FIGURE 28-5   The media will play a role in the handling of an incident, from evacuation instructions to minimizing any hysteria that may have occurred during the incident. A good flow of public information is essential to keeping the surrounding community at ease. *(Courtesy of Baltimore County Fire Department)*

different agencies that may assist with the mitigation of the incident. In some cases the IC may have several alarms worth of equipment of his own to manage, not to mention these other agencies. Until the incident is moved from the emergency phase to the nonemergency cleanup phase, the fire department IC is usually still in charge.[2] In some cases the IC will have a difficult time and can end up handling the "assistance" rather than actually managing the incident. In this situation the department's public information officer (PIO) and liaison officer will be of great assistance.

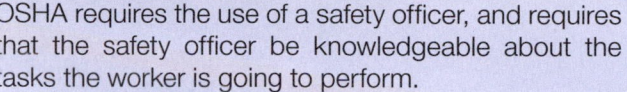

FIREFIGHTER FACT

At a hazardous materials incident, the incident commander is responsible for a number of tasks. Regardless of the local SOP, the IC has many legal obligations under the HAZWOPER regulation. The IC is responsible for the overall actions at the incident, regardless of who is performing the tasks. In most cases, the hazardous materials team performs the mitigation of the incident, and depending on the interaction of the IC and the hazardous materials team there may not be much verbal communication.

In the perfect world of ICS the IC makes all of the decisions, but the reality is that the technical group provides the suggested options. In some cases there may be only one option. If the incident goes to court it will be the IC who has to answer for the hazardous materials team's actions, so the IC must stay informed of the incident action plan and be given the opportunity to choose an appropriate response.

At a chemical release the IC is also responsible for the pre-entry briefing, which is when the entry crew is informed of the hazards that exist and what actions are expected of them. If there are no predesignated emergency signals, these must be decided on prior to entry. The IC must use a system of monitoring the progress of the incident and make changes to the system as needed. The mitigation of an incident takes the cooperation of many persons and agencies, all working toward a common goal under the guidance of the IC.

The management scenario painted did not mention victims, hospitals, or an evacuation—all of which further complicate the incident.

NOTE

The use of an incident command system is not only mandated by OSHA but is a good idea so that personnel can be tracked and the outside agencies can be managed effectively.

Regardless of the ICS system, only a small component of the overall system is typically used during routine fire operations, such as a finance branch being established during a house fire. It is likely, however, that in a major chemical release a finance branch would be established and have several persons assigned to assist in this function.

NOTE

OSHA requires the use of a safety officer, and requires that the safety officer be knowledgeable about the tasks the worker is going to perform.

In some fashion a liaison must be established with all of the responding agencies, and their specific roles also have to function within the ICS. A hazardous materials incident usually requires a minimum of two safety officers, one for overall safety and one assigned to hazardous materials specific issues.

A person trained to the operations level is not adequate to be the safety officer for the hazardous materials operation. A hazardous materials safety officer should be trained to the Technician or Specialist level to make sure that the mitigation efforts are carried out effectively and safely. Some specific hazardous materials branch functions that may be utilized during an incident are listed in Table 28-2.

Establishment of Zones

The term **zone** or **sector** is used to refer to areas that are established to identify the various isolation points, such as those shown in **Figure 28-6.** These zones are referred to as the hot, warm, and cold zones. The hot zone may also be referred to as the exclusion zone, **isolation area,** or hazard area, or may be called by a similar term. The warm zone may be known as the contamination reduction area and the cold zone as the support area. In many cases these areas are identified through colored barriers, cones, or other markings. Hot is usually identified by the color red, warm by the color yellow, and the cold zone by the color green, **Figure 28-7.**

SAFETY

The most important zone that needs to be established is the isolation area, and this needs to be established by the first arriving responder.

This area normally becomes the hot zone after the arrival of the hazardous materials team. This isolation or hot zone is the area immediately around the release

FIGURE 28-6 Proper isolation and work zones are needed to ensure the safety of responders.

and is an area that requires the use of the proper PPE. Entering the hazard area could expose personnel to the substance and possible contamination.

The minimum zone that should be established is this isolation area. The distances for this area can be determined by the use of the DOT (ERG). The minimum distance established by this book is 330 feet (101 meters), which should be the absolute minimum for an unknown material. The distance for this area is obviously determined by the local situation, so it is difficult to make blanket statements about recommended distances. If it is convenient and easy to isolate a block then it makes sense to do so, but if isolating the block will create havoc in the community, then 330 feet (101 meters) may be recommended. This distance is only valid if there are no indicators of an actual release. Any indications of a flowing spill or a vapor cloud for an unknown substance will dramatically increase this distance. If time permits, and resources are available, then an attempt should be made to set up the warm and cold zones. These zones are usually established by the hazardous materials team after arrival and setup.

First responders should position themselves in an upwind and uphill position, as shown in **Figure 28-8.** If operating at a water-based spill, they should operate upwind and upstream. The first arriving apparatus should have set up in that position when it arrived, but if not, the vehicle should be moved to that type of position as quickly as possible. Although it cannot always be accomplished, being in an upwind posi-

FIGURE 28-7 The establishment of zones is usually based on the types of hazards that may be present. For general chemical spills, the zones established are referred to as the hot, warm, and cold zones.

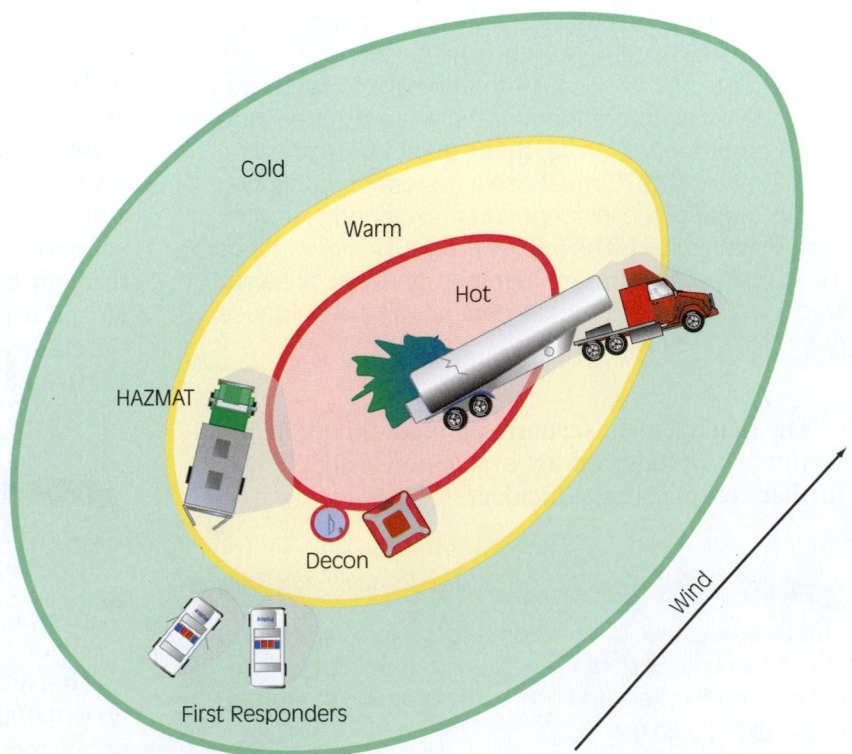

Cold

Warm

Hot

HAZMAT

Decon

First Responders

Wind

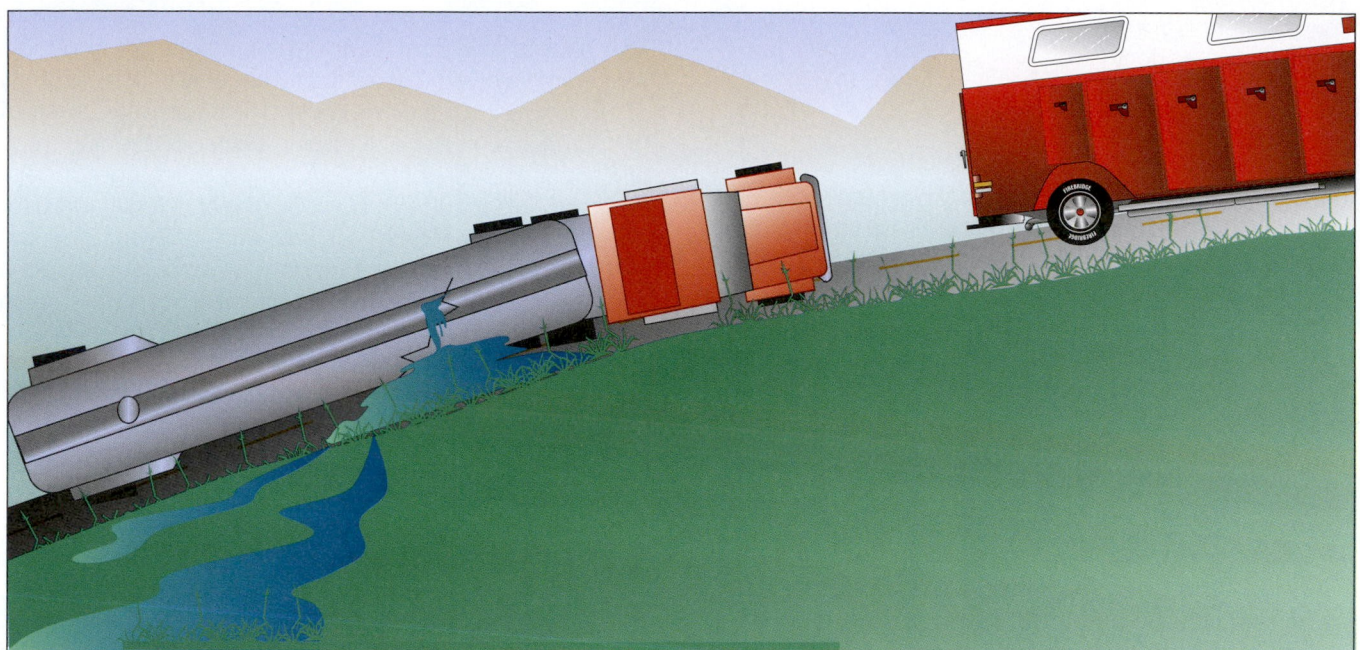

FIGURE 28-8   The best position for first responders is uphill and upwind from the release.

tion is the first priority, and uphill the second priority. If responders are upwind, but forced to set up in the downhill position, they must extend the distance away. One of the common mistakes made is that the first arriving apparatus does not communicate the on-scene wind directions and the best route of travel for other apparatus to follow as they arrive, to stage at the upwind and uphill position. All apparatus should be positioned so that they are pointed away from the incident so in the event of a catastrophic release or other emergency, the apparatus can be moved without turning around or backing.

Other items to consider for the hot zone are topography, accessibility for responding units and other resources, weather conditions, and bodies of water, including those for drinking water. Still other items to consider are setup areas for the hazardous materials team including decontamination and dress-out locations. Other exposures must be examined, and part of the isolation process is the limiting of possible ignition sources.

Public exposure potential must be considered, which can be broken down into specific time frames such as short term (minutes and hours), medium term (days, weeks, and months), and long term (years and generations). Emergency responders usually deal only with short-term and medium-term types of incidents, but the actions that first responders take can result in long-term exposures, for both the responders and the public. Sewer lines, both sanitary and storm sewers, must be taken into account as well as other utilities such as cable and phone, both aboveground and

belowground. Transportation corridors such as highways, rail lines, ports, and airports must be considered because the incident may well involve or affect these areas. For buildings, the internal areas can be broken down into zones, but items that should be considered are floor drains, ventilation ducts, and air returns.

The warm zone is set up after the arrival of the hazardous materials team and is usually where the decontamination area is established. In some cases the warm zone is also extended to allow for the hazardous materials team setup. In unusual situations the warm zone is an area where some type of PPE may be required, and depending on the circumstances may be an area that can be affected by wind shifts or a catastrophic failure. The establishment of these zones is arbitrary although it is based on the best judgment of the IC or the hazardous materials team. There is no magic line of hot and warm zones, nor could one be established. The cold zone is the area where the incident command post is established and is where the first responding companies will be positioned. All support operations such as medical and rehabilitation are set up in this area, and movement between the zones is controlled at access control points. No PPE is required for the cold zone because there should be no chance for chemical exposure in this area. A person should be assigned to act as security for each of these zones to ensure that only authorized personnel enter these areas.

Some hazardous materials teams use monitoring devices to help establish the zones. The use of air monitors to establish these isolation points is crucial, because the distances provided in the DOT ERG

THE IMPORTANCE OF ISOLATING THE SCENE

The first arriving responders can make a hazardous materials release easier to manage if they begin the isolation process. When chemicals are released, there is great potential for liability and lawsuits are probable. Persons interested in joining these lawsuits may try and gain access to the perceived hazard area. In many cases, the insurance companies will try to settle with any potential victims prior to any lawsuits and will write checks to people who were involved, even if they were only in the proximity of the incident. One of the most difficult areas to attempt to control is a mall or a shopping center where hundreds of people may be involved. A simple discharge of pepper spray will result in a number of real victims and usually generate a few more suspect victims. Everyone must be treated and possibly transported, which can create a strain on the EMS system. Early isolation and evacuation can help reduce the number of potential victims.

are very conservative and are based on the worst-case scenario. Real-time, on-scene air monitoring cannot be replaced by plume projections, estimations, or models depicted in a text.

The zone distances are based on theories established by the risk-based response system. The materials that require the greatest isolation are those that have high vapor pressure or exist as gases in their natural state. A material such as sodium hydroxide (sometimes referred to as lye or caustic) requires an isolation distance of only a few feet. Although very corrosive and contact with skin would cause some irritation and burns if not washed off immediately, it has a very low vapor pressure. The vapor pressure of sodium hydroxide is 1 mm Hg at 1,390°F (754°C), which means that the spill would have to be heated to 1,390°F (754°C), in order to produce a very small amount of vapor, considerably less than the vapor pressure of water (25 mm Hg). With this material, it is unlikely that a large isolation distance would have to be established, nor would any evacuations probably be necessary.

A chlorine release from a railcar, such as the one discussed in the Graniteville Case Study, does present an extreme risk to the community because once in the atmosphere chlorine rapidly changes to a gas. Chlorine is not only poisonous but also an oxidizer and a corrosive material with a very high vapor pressure of 5,168 mm Hg. A substantial evacuation area is required for chlorine. One advantage though is that chlorine is easily detected, so a true hazard area can be established using air monitoring.

NOTE

Making a decision to conduct an **evacuation** or to **shelter in place** can be one of the most difficult decisions for an emergency responder. Regardless of the decision, there are usually political ramifications, right, wrong, or indifferent.

Evacuations and Sheltering in Place

The IC will be bombarded by the public, media, and coworkers about the decision. To a first responder, unfortunately, not much assistance is readily available. The only resource is the DOT ERG, and that book is conservative and may not apply in a specific situation. The best way to determine whether to evacuate or shelter in place is to conduct real-time air monitoring that can determine the exact hazard area. Consultation with the hazardous materials team and the local conditions guided by the DOT ERG can establish a starting point. If the incident is at a SARA Title III facility, the jurisdiction's emergency plan should have some recommendations regarding evacuation. Some plans provide a checklist to follow to assist with the evacuation decision-making process. Close coordination with the local emergency management agency is essential to making the outcome successful.

If a decision is made to evacuate, a suitable location needs to be found, transportation may be required, and accommodations need to be established for the evacuees. This type of assistance is usually provided by the emergency management coordinator, who can be a good point of contact. The hazardous materials team can also run an ALOHA (aerial location of hazardous atmospheres) plume projection using the CAMEO (computer-aided management for emergency operations) computer program, but that is also conservative and may cause a larger evacuation than necessary. Chemical vapor plumes have characteristic shapes as determined by computer models, such as the ALOHA model. These standard plumes are to be used for worst-case scenarios and may not apply to a specific locality. Although the plumes are computed for a variety of topographies and types of weather, each local area differs, and only local weather conditions and air monitoring results can provide truly

GRANITEVILLE, SOUTH CAROLINA, CHLORINE CASE STUDY

On January 6, 2005, at 2:39 A.M., a train traveling north-bound ran into a parked train as the result of a misaligned switch. The northbound train had 42 freight cars—25 loaded and 17 empty, though chemical residue can be a factor in an otherwise empty car. Of these 42 cars, 14 held hazardous materials or residue from hazardous materials. When the trains collided, 16 cars derailed. Three of the derailed cars were tank cars carrying chlorine, one of which was leaking. The leaking tank car was carrying 13,830 gallons (52,352 liters) of chlorine, a lique-fied gas. Chlorine has an IDLH (immediate danger to life and health) value of 10 ppm and is very corrosive. Compounding the resulting chlorine release was the fact that Graniteville lies in a shallow valley, with a stream running parallel to the railroad tracks. Chlorine is heavier than air and remains low to the ground. The leaking chlorine was therefore accumulating over the streambed. Within the first few minutes, this chlorine cloud extended 2,500 feet (762 meters) to the north, 1,000 feet (305 meters) to the east, 900 feet (274 meters) to the south, and 1,000 feet (305 meters) to the west.

From the first 9-1-1 call, it was apparent that a chemical release had occurred, and the Fire Chief requested additional resources, including a hazardous materials team. When the Fire Chief arrived on scene, he developed difficulty breathing and had to abandon his immediate position. Several of the first-arriving law enforcement officers (deputy sheriffs) drove into the chlorine cloud and required medical treatment. Eleven minutes after the collision, the Fire Chief requested that the emergency notification system be activated and for citizens to shelter in place. Adjacent to the area of the derailment were a number of industrial facilities in which several persons were reported to be trapped or missing. Multiple decontamination areas were established and the local hospital was notified to expect patients. To effect rescue of persons in the hazard area, firefighters in PPE would ride on pickup trucks, which were then used to transport victims. Later that morning, the Sheriff made the decision to evacuate people in a one-mile radius around the crash site, which resulted in 5,400 persons being evacuated from their homes.

About 21 hours after the crash, hazardous materials crews were able to place a polymer patch on the leaking chlorine car. Two days later, sodium hydroxide was pumped from another of the derailed cars. During the pumping operation, the patch on the chlorine car failed, releasing vapors in proximity to the sodium hydroxide car—a potentially dangerous and reactive combination. By reducing the vapor pressure in the leaking chlorine car, the leak was reduced and plans were developed for a more permanent patch. This new patch was applied the next day, on January 12. By midnight January 18, the off-loading of the leaking rail car was complete.

At least 554 people were taken to local hospitals and 75 of them were admitted for further treatment. Including the driver, who died four months after the incident, there were nine fatalities from exposure to chlorine gas: the engineer, six workers from the adjacent milling facility to the west and north of the accident, and a resident living south of the site. Apart from the driver, only the engineer lasted more than a few moments; all of the other fatalities occurred in the first few minutes of exposure. Six fire-fighters were treated and released and one was held for treatment for several days. Two sheriff's deputies were treated and released. As a result of chlorine exposure, two engines, one ambulance, and a service truck were lost to severe damage, costing $630,000.

Think about the impact of such an incident on this small community; lives were lost, injuries occurred, and the economic impact was significant. Since the accident, a number of businesses and factories have closed. Although in this case, the chlorine exposure was the result of a violent train derailment, think about a perhaps more commonplace accidental release of a hazardous material from a faulty or loosened valve. The response to a chlorine (or comparable chemical) release must be quick, efficient, and safe because the impact can be dramatic. This incident occurred in a relatively low-population area; think about the impact in a major metropolitan community. In the event of a hazardous chemical release, using recognition and identification skills, wearing proper protective clothing, and making appropriate isolation and evacuation decisions are keys to survival for both the responders and the community.

---

accurate results. The standard shapes for plumes are provided in **Figure 28-9.**

Some studies have shown that in most cases sheltering in place is safer than evacuation. Just imagine evacuating the most populated area of a community; how would a responder notify all of the citizens? In most communities the police department may have this responsibility, but do they have the resources to accomplish this task? Is there an emergency alerting system for TV or radio in the community? In urban or metropolitan areas evacuating just a few blocks can affect thousands of people. The worst-case scenario using the DOT ERG is an evacuation distance of 4,000 feet (1,219 meters) by 7 miles (11 km), which in a city could be 25,000 to 100,000 evacuees. It would be nearly impossible to evacuate that many people,

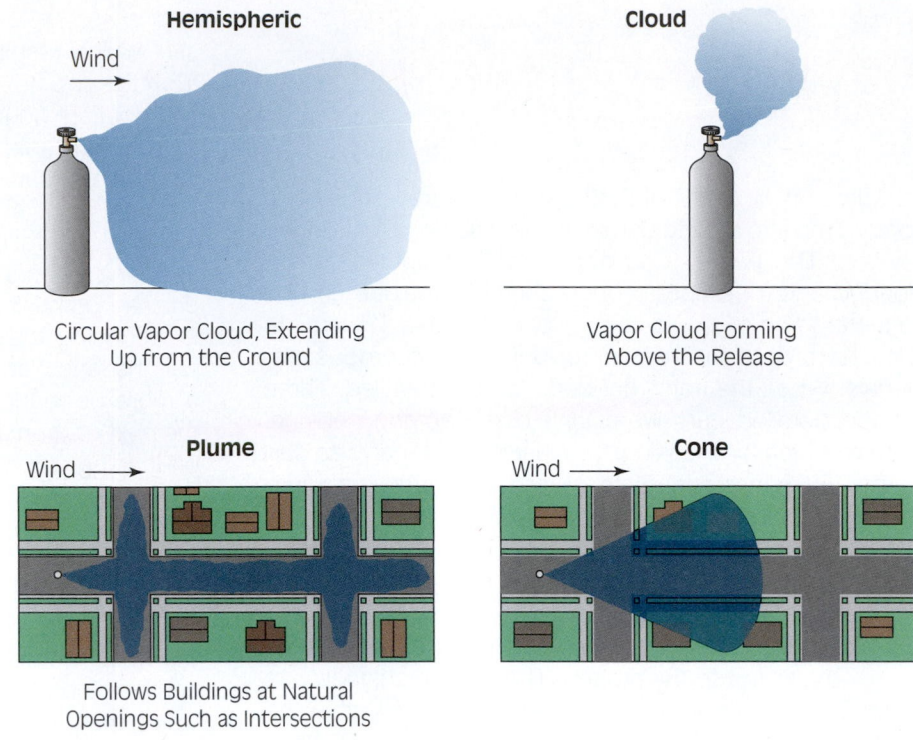

Hemispheric

Wind →

Circular Vapor Cloud, Extending
Up from the Ground

Cloud

Vapor Cloud Forming
Above the Release

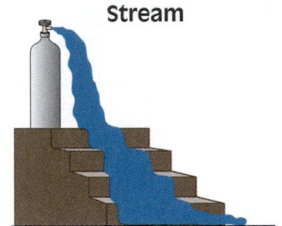

Plume

Wind →

Follows Buildings at Natural
Openings Such as Intersections

Cone

Wind →

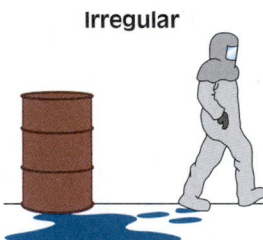

Stream

Stays Low to the Ground
Following Natural Barriers

Pool

Forms a Low-Lying Vapor
Cloud on the Ground

Irregular

Movement of the Material
by Responders or Other
Irregular Movement

FIGURE 28-9 Standard shapes for plumes or vapor clouds may form after a gas is released. The exact type varies with the topography and the buildings in the area.

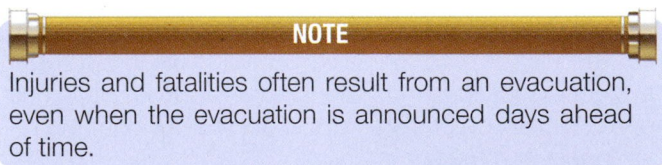

not to mention the panic and chaos that such action would bring.

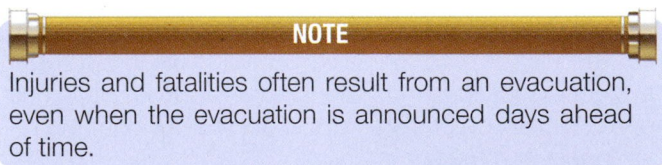

> **NOTE**
>
> Injuries and fatalities often result from an evacuation, even when the evacuation is announced days ahead of time.

There are certain times when evacuation is required such as when an explosion is probable, when explosives are involved, when a container may suffer a BLEVE or rupture with violent consequences, or when the release will continue for more than a few hours.

Sometimes an evacuation may not be recommended, such as for a hospital, nursing home, jail, or other facility in which rapid removal of the occupants is not practical. In these types of occupancies

air monitoring and control of HVAC are advisable, in addition to having a liaison remain at the facility who is in radio contact with the IC. The chemical properties of the released material will have an effect on the decision, because some materials will rise up in the air and dissipate quickly, whereas others may stay low to the ground causing evacuation problems. The type of leak must be considered: Can it be quickly and easily controlled or is there a probability that it cannot be controlled by the hazardous materials team? Will evacuating the citizens subject them to a higher level of the chemical than keeping them in place?

ACGIH has provided some planning levels for the chemicals that require planning under the Clean Air Act Amendments (CAAA). These planning levels are on three tiers and provide actual levels that can be used for emergency planning. Because the permissible exposure limits and threshold limit values were established for an eight-hour exposure during normal conditions, they do not have much applicability during an emergency release. These **emergency response planning (ERP)** levels are designed to assist with the emergency planners' preparation of the community emergency plan, and would be useful in determining the evacuation zone. When making the decision to evacuate or shelter in place, the flow of information to the public and the media is essential, because a lack of information can bring disastrous results.

When sheltering in place, the citizens should shut all windows and doors, shut off air handling systems, and stay tuned to a TV or radio station. If a continual flow of information is not available, people will become frustrated and may make attempts to go find the information for themselves. When dealing with larger facilities such as high-rises, hospitals, nursing homes, schools, or jails, it may be best to station an emergency responder at that location who acts as a direct link to the incident, so that any questions can be immediately answered and fears alleviated. The emergency management office can also establish a rumor control hotline, which can be used to answer questions about the incident from concerned citizens and family members.

COMMON INCIDENTS

This section provides an overview of common incidents and the types of releases in each of the DOT hazard classes. This listing is far from complete; it merely represents some of the most common incidents or incidents that have a great potential to impact the community. The recommendations provided here are only suggestions; local policies and procedures should be followed.

Stress, Container Breaches, and Product Releases

Hazardous materials incidents combine several factors relating to the characteristics of the container and the product and also how failures of the container can create hazards. These factors are the properties of the material, characteristics of the container, and the physics and chemical reactions. There is a direct correlation between the factors that determine the behavior of the material, the container, and the stored chemical energy of the system during an incident.

When responding to chemical incidents it is important to identify how the container was stressed, how the container was breached, and how the chemicals were released. In many cases the manner in which the breach and release occurred can help provide clues to the successful mitigation of the incident.

Stress

There are several ways of looking at the potential release of a chemical that involve stress on the container. Either the chemical itself is stressed or the container is stressed. In an incident involving a gasoline tanker that is rolled over at 55 mph (90 kph) the container is stressed, and the contents are likely to come out of the container. In an incident in which a paint waste is reacting within a drum, the material is stressed, but if the drum is sealed tight, the container will be stressed due to the chemical reaction creating pressure. There are three general types of stress.

1. *Thermal stress* is the heating of the container and its contents. The effects of this heating can result in the material reaching its boiling point inside the container, enabling existing gases to further expand inside, subsequently causing a structural weakening of the container.

2. *Mechanical stress* is the transfer of energy from one object to another. The effects of this energy transfer can be a result of abrasion, which may decrease the shell thickness of the container or be caused by a deformity of the container and/or its closures from dents, gouges, etc.

3. *Chemical stress* is the interaction between materials and/or their container. This stress can cause corrosion of the container, or generate internal temperature and pressure changes that may result in polymerization.

Breaches

Both pressurized and nonpressurized containers can breach in several ways. Breaches are any openings in a container that are unnecessary and occur when a container is stressed beyond its recovery capacity. There are five types of breaches.

1. *Disintegration* occurs when the container suffers a cataclysmic failure likely caused by an explosion.

2. *Runaway linear cracking* is a rapidly growing crack in a drum or pressure vessel that will encircle the container, violently break it into two or more pieces, and potentially cause a container failure.

3. *Closures opening up* is when attachments such as valves or pressure relief devices open up or are sheared off during the incident.

4. *Punctures* are when an object pushes through the container wall.

5. *Splits or tears* is when the container is ripped or abraded to the point of container failure.

Incidents have been quickly mitigated by the use of a drum lid. On tank trucks a common breach point is the frangible disk, which may rupture due to over-filling or a quick stop causing the liquid to slosh and rupture the disk, releasing some of the contents. The incident is quickly handled by the replacement of the disk.

Releases

Releases are the escape of material and energy from a container. There are four ways in which a container releases its contents.

1. *Detonation* is a rapid, violent ignition of a product.
2. *Violent ruptures* are when the product is released under force in less than one second.
3. *Rapid relief* is a release from a pressurized container that may last from seconds to minutes.
4. *Spill or leak.* A spill is defined as a loss of product from a naturally occurring opening such as a bung on a drum or a leaking valve. A leak is defined as a release from an unnatural opening such as a puncture in the side of a drum.

When these methods result in a chemical release, the action can be violent and can have catastrophic consequences. When a container such as a propane tank, **Figure 28-10,** detonates, it can travel up to a mile. If a container ruptures violently, it means the container was under pressure, causing the violent release. Much like a BLEVE, a violent tank rupture can send the tank or portions thereof a considerable distance.

A container that has a rapid relief valve should be able to release enough pressure to bring a margin of safety to the incident. This is only true if the cause of the increased pressure is removed and the pressure does not continue to climb within the container. If a relief valve is operating, yet still not able to relieve building pressure, a true emergency condition exists and failure to bring the pressure under control can

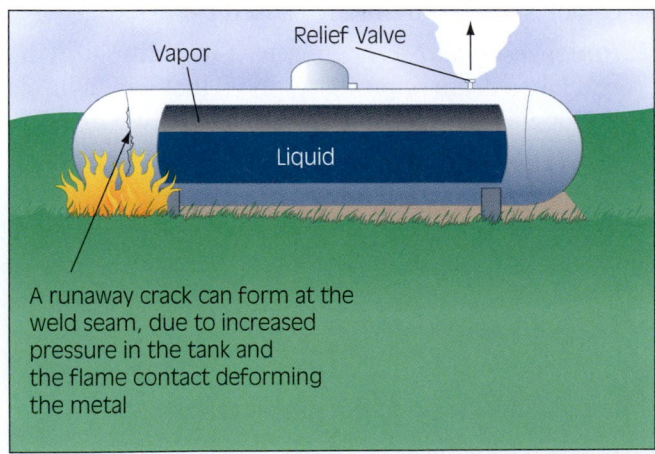

A runaway crack can form at the weld seam, due to increased pressure in the tank and the flame contact deforming the metal

FIGURE 28-10   Propane tank detonation.

have catastrophic consequences. An operational relief valve is allowing the product to escape, and if the material is flammable it may find an ignition source and ignite. If the release ignites, this may cause the internal pressure to increase even more, causing more material to ignite. It is never recommended that a relief valve be stopped, or that a fire coming from the relief valve be extinguished. The only time a fire coming from any valve, relief or otherwise, should be extinguished is when the flow of the material can be stopped with 100 percent assurance that it will be successful. Extinguishing other fire, including those that are impinging on a tank or container, should be a high priority. Hoselines should be directed so that they cool the container, which will reduce the pressure and allow the relief valve to stop operating.

Spills and leaks both result in product being lost. The end result for a spill or a leak is that product is released and can create problems for the responders. Problems may increase dramatically if an ignition source is present during a hazardous materials incident. Firefighters should take measures to eliminate these, especially if there are flammable vapors in the area. As an additional safety measure, eliminate the vapors by using a water fog to suppress or disperse them. Some typical ignition sources are:

- Open flames
- Smoking materials
- Cutting and welding operations
- Heated surfaces
- Frictional heat
- Radiant heat
- Static
- Electrical and mechanical sparks
- Lightning

Explosives

All persons must be removed from the area, and a defensive operation should be established.

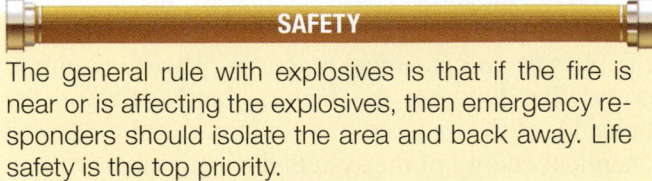

SAFETY

The general rule with explosives is that if the fire is near or is affecting the explosives, then emergency responders should isolate the area and back away. Life safety is the top priority.

Many other considerations come into play if the fire is not directly impacting the explosives. A brake fire on a truck carrying explosives is one example; as long as the fire has not reached the cargo area, firefighters may be able to extinguish it without incident.

When making the decision to fight a fire in this situation, water must be applied quickly and in large quantities to cool the cargo and then extinguish the fire. Crews should be limited, and adjoining areas evacuated. The recommendations in the DOT ERG, **Figure 28-11,** are a good starting point for isolation and evacuation.

Another incident that involves explosives arises when the fire department assists a bomb squad with a suspected explosive device. Close coordination with the bomb squad is required to make sure that the operation is conducted safely. Some examples of standoff distances are provided in **Table 28-3.** Although the distance varies with some jurisdictions, radios, cell phones, or other electronic devices should not be used within 500 to 1,000 feet (152 to 305 meters) of a suspected device. Many bomb squads detonate suspected devices in place, which may start a fire or cause a potential structural collapse. Other techniques are to "disrupt" the device or render it safe by the use

---

RESIST RUSHING IN !
APPROACH INCIDENT FROM UPWIND
STAY CLEAR OF ALL SPILLS, VAPORS, FUMES AND SMOKE

HOW TO USE THIS GUIDEBOOK DURING AN INCIDENT INVOLVING DANGEROUS GOODS

ONE IDENTIFY THE MATERIAL BY FINDING ANY **ONE** OF THE FOLLOWING:

THE 4-DIGIT ID NUMBER ON A PLACARD OR ORANGE PANEL

THE 4-DIGIT ID NUMBER (after UN/NA) ON A SHIPPING DOCUMENT OR PACKAGE

THE NAME OF THE MATERIAL ON A SHIPPING DOCUMENT, PLACARD OR PACKAGE

IF AN **ID NUMBER** OR THE **NAME OF THE MATERIAL** CANNOT BE FOUND, SKIP TO THE NOTES BELOW.

TWO LOOK UP THE MATERIAL'S 3-DIGIT GUIDE NUMBER IN EITHER:

THE ID NUMBER INDEX..(the yellow-bordered pages of the guidebook)

THE NAME OF MATERIAL INDEX..(the blue-bordered pages of the guidebook)

If the guide number is supplemented with the letter "**P**", it indicates that the material may undergo violent polymerization if subjected to heat or contamination.

If the index entry is highlighted (in either yellow or blue), it is a TIH (Toxic Inhalation Hazard) material, a chemical warfare agent or a Dangerous Water Reactive Material (produces toxic gas upon contact with water). **LOOK FOR THE ID NUMBER AND NAME OF THE MATERIAL** IN THE TABLE OF INITIAL ISOLATION AND PROTECTIVE ACTION DISTANCES (the green-bordered pages). Then, if necessary, **BEGIN PROTECTIVE ACTIONS IMMEDIATELY** (see Protective Actions on page 298). If protective action is not required, use the information jointly with the 3-digit guide.

USE GUIDE 112 FOR ALL EXPLOSIVES EXCEPT FOR EXPLOSIVES 1.4 (EXPLOSIVES C) WHERE GUIDE 114 IS TO BE CONSULTED.

THREE TURN TO THE NUMBERED GUIDE (the orange-bordered pages) **AND READ CAREFULLY.**

NOTES IF A NUMBERED GUIDE CANNOT BE OBTAINED BY FOLLOWING THE ABOVE STEPS, AND A PLACARD CAN BE SEEN, LOCATE THE PLACARD IN THE TABLE OF PLACARDS (pages 16-17), THEN GO TO THE 3-DIGIT GUIDE SHOWN NEXT TO THE SAMPLE PLACARD.

IF A REFERENCE TO A GUIDE CANNOT BE FOUND AND THIS INCIDENT IS BELIEVED TO INVOLVE DANGEROUS GOODS, TURN TO GUIDE 111 NOW, AND USE IT UNTIL ADDITIONAL INFORMATION BECOMES AVAILABLE. If the shipping document lists an emergency response telephone number, call that number. If the shipping document is not available, or no emergency response telephone number is listed, IMMEDIATELY CALL the appropriate **emergency response agency listed on the inside back cover of this guidebook.** Provide as much information as possible, such as the name of the carrier (trucking company or railroad) and vehicle number. AS A LAST RESORT, CONSULT THE TABLE OF RAIL CAR AND ROAD TRAILER IDENTIFICATION CHART (pages 18-19). IF THE CONTAINER CAN BE IDENTIFIED, REMEMBER THAT THE INFORMATION ASSOCIATED WITH THESE CONTAINERS IS FOR THE WORST CASE POSSIBLE.

Page 1

FIGURE 28-11 The DOT ERG provides some basic explosives information that provides a good margin of safety when dealing with these types of incidents.

TABLE 28-3   Bomb Threat Standoff Distances

Threat Description	Explosive Capacity	Lethal Air Blast Range	Mandatory Evacuation Distance	Desired Evacuation Distance
Pipe bomb	5 lbs. (2.2 kg)	25 ft. (8 m)	70 ft. (21 m)	850 ft. (259 m)
Briefcase or suitcase	50 lbs. (23 kg)	40 ft. (15 m)	150 ft. (46 m)	1,850 ft. (564 m)
Compact sedan	220 lbs. (100 kg)	60 ft. (18 m)	240 ft. (73 m)	915 ft. (279 m)
Sedan	500 lbs. (227 kg)	100 ft. (30 m)	320 ft. (98 m)	1,050 ft. (320 m)
Van	1,000 lbs. (454 kg)	125 ft. (38 m)	400 ft. (122 m)	1,200 ft. (366 m)
Moving van or delivery truck	4,000 lbs. (1,814 kg)	200 ft. (61 m)	640 ft. (195 m)	1,750 ft. (533 m)
Semi-trailer	40,000 lbs. (18,143 kg)	450 ft. (137 m)	1,400 ft. (427 m)	3,500 ft. (1,067 m)

Explosive capacity—Based on the maximum volume or weight of explosives (TNT equivalent) that could reasonably be hidden in the package or vehicle.

Lethal air blast range—The minimum distance personnel in the open are expected to survive from blast effects. It is based on severe lung damage or fatal impact injury from body translation.

Mandatory evacuation distance—The range to which all buildings must be evacuated. From this range to the desired evacuation distance, personnel may remain in the building (with some risk) but should move to a safe area in the interior of the building away from windows and exterior walls. Evacuated personnel must move to the desired evacuation distance.

Desired evacuation distance—The range to which personnel in the open must be evacuated and the preferred range for building evacuation. This is the maximum range of the threat from flying shrapnel/debris or flying glass from window breakage.

Source: Developed by the ATF, with technical assistance from the U.S. Corps of Engineers. Supported by the Technical Support Working Group (TSWG), a research and development arm of the National Security Council Interagency working group.

of a water cannon. These techniques could cause the device to detonate. The time to discuss these scenarios and plans of action is prior to an incident, not during it. At incidents where there is a suspected device, firefighters should be aware of the potential for a secondary device, one designed to injure the responders. When operating at the scene of an explosion, it is important to have the bomb squad search the scene for secondary devices, **Figure 28-12.** Section IV, Chapter 30 has more information on secondary devices.

Other incidents may involve a shipment of explosives that has been involved in an accident. In most cases these incidents present little risk to the responder or the community. In this case the bomb squad is a great technical resource hand should be used to evaluate the scene prior to moving any explosives. In some cases

an explosives truck, such as the one shown in **Figure 28-13,** carrying ammonium nitrate may be involved in an accident and if it has fuel oil on board, the mixture known as **ANFO** could be created. Without an initiation charge or other substantial energy source, the material is unlikely to explode, but the ammonium nitrate does present a toxicity hazard and care should be taken around spills of this material. In incidents of this type the bomb squad should also be consulted.

SAFETY

It is not uncommon for well-meaning citizens to bring explosives to the fire station. If the explosives are outside or in the citizen's vehicle, the firefighter should leave them there and call for the bomb squad.

FIGURE 28-12 A bomb technician wearing explosives protective clothing removes a potentially explosive device from a vehicle.

FIGURE 28-13 An explosives truck, used to carry explosive materials such as ammonium nitrate. These trucks are commonly used at rock quarries and in other mining situations to deliver the explosive materials. (Courtesy of Maryland Department of the Environment Emergency Response Division)

Section IV, Chapter 30 discusses some possible terrorism scenarios, and explosives are used in almost all of the incidents. Hazardous materials and bomb squad duties are becoming more and more interlaced.

Everything from distress flares and ammunition to hand grenades, pipe bombs, dynamite, and blasting caps is commonly brought to fire stations. Firefighters should not be handling these types of devices; the bomb squad should handle these types of situations. Depending on the type of device and the size, isolation and some evacuation may be necessary.

Citizens may also bring in old chemicals, some of which may be explosive and shock sensitive. With picric acid, just the simple removal of the cap or vibration of the container is enough to initiate the explosion of the container with enough force to cause fatal injuries to the person who attempts to open or set down the container. Another common item that may be brought in is containers of old ether (ethyl ether), which is outdated after a year in storage. The opening of this container may also bring fatal consequences. The job of handling these containers usually falls to the bomb squad with a hazardous materials team interface.

> **SAFETY**
>
> It is a good general practice not to bring unknown materials into the fire station.

In some cases, toxic materials may be brought in that can contaminate the person who accepts the package and cause severe health problems for the firefighters in the station. Depending on state regulations, if fire department personnel accept a package from a citizen, the department may end up owning the package and may be required to pay for the proper disposal of the item—sometimes a costly good deed.

Gases

Incidents involving gases include both flammable and nonflammable gases, with the most commonly released gases being flammable. Also in this category are poisonous gases. The poisons section provides more information on poisonous gases.

Luckily for many first responders, many departments carry gas detection devices that will warn of potentially explosive atmospheres involving flammable gases. Depending on the setup, the detector may warn of oxygen-deficient atmospheres, which may be caused by the release of a nonflammable gas. The two most commonly released flammable gases are natural gas and propane. Common propane bottles are shown in **Figure 28-14**. Although used for the same purposes, propane and natural gas do have differing characteristics that can affect a response to a gas leak. Both gases are odorless when they exist in their natural state, so when moved through a distribution system an odorant is added. In major interstate pipelines, however, the gas may be transported without an odorant added.

FIGURE 28-14 These 20-pound cylinders, found in most homes, can create large fireballs and can explode with considerable force if involved in fire.

> **NOTE**
>
> The greatest difference between natural gas and propane is the vapor density. Propane is heavier than air and will stay low to the ground trying to find an ignition source. Natural gas, on the other hand, will rise in air and should dissipate quickly.

This is weather dependent, and certain weather conditions can affect these characteristics. Under some conditions the gas may travel the length of a pipe, following the path of least resistance, usually coming up inside a building. The leak may not be detected for a few days or weeks until it surfaces. This is most common in areas of the country where the ground freezes and limits the upward movement of the gas. Many states offer training on natural gas and propane emergencies; this training is often provided by the gas companies themselves.

Incidents involving these two gases are commonplace, and for natural gas the most common incident involves a ruptured gas main. The gas system is set up on a grid, with both large distribution pipes and smaller delivery pipes. The pipes leading into a home are typically ½ to 1 inch (1.27 to 2.5 cm) diameter. Distribution pipes may be 2 to 36 inches (5 to 91 cm) depending on the region. First responder actions at these types of incidents generally involve isolation and protection, and then the team stands by until the line can be shut off. Gas detection devices should be employed to determine the true hazard area, and adjacent buildings should be checked for gas. When checking buildings, sampling should be continuous and the highest point should be checked prior to declaring a building "safe."

Another significant issue with the release of flammable gases, which also applies to vapors coming from flammable liquids, is the presence of ignition sources. Pilot lights, electrical switches, electrical contacts, and some radio and cell phone transmissions are also possible sources of ignition. Static electricity is a concern when there is movement of persons or the chemical. Attempts should be made to eliminate these sources of ignition. The electricity to a building can be cut off at the pole, which is a safer alternative than shutting off the electricity inside the hazard building. At vehicle accidents, hot engines, vehicle batteries, and static electricity are potential ignition sources.

> **CAUTION**
>
> At no point should first responders jump into a hole to attempt to shut off a leak, nor should underground valves be shut in an attempt to stop the leak.

Gas line valves may actually be keeping the amount of gas reduced, and moving the valve may increase the amount of gas escaping. If the leak is on the out-take site of the meter, it is acceptable for the valve on the meter to be shut off. The meter should be locked out and tagged out, and only the gas company should turn the flow of gas back on. Some departments carry special tags that mark the system as being out of service, and some actually lock the system in the off position. When the gas company arrives, they repair the leak and place the system back in service. When dealing with pipes and electrical systems, an OSHA regulation prescribes the procedure for shutting off a system and marking it so that the system is not accidentally turned back on prematurely. This regulation is known as the "lock out, tag out" regulation and applies to many fire service situations.

Propane releases generally involve cylinders, but leaks can also happen in a pipeline. A number of the pipelines running across many states carry propane as well as natural gas. Other than its flammability, the fact that propane sinks and stays low to the ground creates an additional hazard. If a propane cylinder is releasing liquid propane, the liquid will be very cold, sometimes as low as -90°F (-67°C). A common incident involving propane cylinders, especially those 20-pound (9 kg) cylinders designed for home barbecue grills, involves the overfilling of the cylinder. Propane cylinders are only supposed to be filled to 80 percent of the capacity of the cylinder to allow for expansion of the gas. If the cylinder is filled more than the 80 percent and the temperature increases, the gas may escape the relief valve or frangible disk.

For fires involving propane tanks, there is the potential for a BLEVE, an event that can be catastrophic to the responders. A BLEVE can occur with any flammable gas storage tank, but is usually associated with propane tanks. When pressurized tanks explode, they can travel for a considerable distance. Although no one can predict exactly when a tank will fail, some indicators that a BLEVE may be forthcoming are increased flame height, or the appearance that the flames coming from the relief valve are under high pressure. The sound of a relief valve is deafening, but when the pressure increases the pitch of the sound will get higher and may become louder.

SAFETY

If the tank becomes discolored, distended, or loses shape, or the sound from the relief valve increases, an immediate withdrawal is indicated.

When making the decision to fight a propane tank fire, a large quantity of water needs to be applied quickly and continuously. The vapor space of the tank should be concentrated on, and the fire should not be extinguished unless the responder is certain the flow of gas can be stopped. When preplanning, first responders should plan to establish a flow of water on the tank in excess of 500 gpm (1,893 lpm) within a few minutes of arrival. Each year a number of firefighters are killed by BLEVEs; in fact, in many cases whole alarm assignments have been killed and seriously injured during the BLEVE. More information in BLEVEs is found in Section IV, Chapter 24 and Chapter 25.

Some cars now are powered by natural gas or propane; the most common are cars that are part of a fleet of vehicles, including government and utility company vehicles. To determine if a vehicle is powered by one of these gases, the firefighter should look for a sticker that reads CNG (compressed natural gas), LP (liquefied petroleum), propane powered, or natural gas powered. Sample stickers are shown in **Figures 28-15** and **Figure 28-16** Other vehicles that may have compressed gas cylinders are recreational vehicles (motor homes) and work utility vehicles.

Although currently uncommon, there are cars powered by hydrogen. Hydrogen is also transported by tank truck, rail, and IMO (International Maritime Organization) style containers. This gas is extremely flammable and carries an unusual risk: it burns with an invisible flame and produces no smoke. As **Figure 28-17** and **Figure 28-18** show, this presents a major concern. Occasionally, the heat waves can be seen above a hydrogen flame, but in many cases responders cannot identify a hydrogen fire until they are too close.

FIGURE 28-15 Vehicles that use alternative fuels such as natural gas, propane, or electric are marked to indicate the fuel source. Here a CNG sticker on the rear bumper, indicates that the car is fueled by compressed natural gas.

FIGURE 28-16 This sticker is beside the front driver's quarter panel and indicates this vehicle is environmentally friendly alternatively fueled. Cars with alternative fuels typically still have a gasoline tank, and there may be gasoline present in emergencies.

FIGURE 28-17 Master streams are used to cool the hydrogen tanks on a tube trailer. *(Courtesy of Maryland Department of the Environment Emergency Response Division)*

FIGURE 28-18 Note the severe damage to this high pressure hydrogen tube trailer. It was involved in a traffic accident and caught fire. *(Courtesy of Maryland Department of the Environment Emergency Response Division)*

FIGURE 28-19 This propane tank truck had a slight rollover in a toll booth. The tank was damaged and the propane was flared off. The situation is still dangerous because the extent of the damage to the tank is unknown, and righting the tank could create stresses that could cause a catastrophic release. *(Courtesy of Maryland Department of the Environment Emergency Response Division)*

When propane tank trucks overturn, as shown in **Figure 28-19,** the truck usually will not leak, but the situation is still quite dangerous. It is impossible to determine the exact damage to the tank, and just how much increased pressure might cause a violent release of the tank contents is another unknown. If the ambient temperature is increasing, so is the pressure in the tank. It is possible that the relief valve can become damaged during a rollover and may not function as designed. It would be dangerous to right any liquefied gas tank without the product being transferred or, in the case of propane, the contents flared, as shown in **Figure 28-20.**

Other common gas releases involve carbon dioxide, chlorine, and ammonia. Although carbon dioxide (CO_2) is a common fire extinguishing agent, it is also used for the distribution of beverages and is commonly found in restaurants, bars, and convenience stores. An increase of CO_2 in a building can make people sick, and it is becoming a common **sick building chemical.** Although not commonly looked for, when dealing with emergency response to a **sick building,** CO_2 should be considered as a possible source of the problem. Both chlorine and ammonia have good warning properties. Their distinct odors are often easily identified by the people in the building. In most cases the

FIGURE 28-20 A propane tank truck was involved in a collision and rolled over. Shown are two flares that are burning off the propane, which must be done prior to the tank being righted. Many hazardous materials response teams carry flaring equipment able to flare tanks from small, grill-sized propane tanks up to rail cars. *(Courtesy of Maryland Department of the Environment Emergency Response Division)*

FIGURE 28-21 Responding to this emergency presented several challenges, due mainly to the final resting place of the overturned truck. Although hazardous materials response prefers an uphill position, the best place for the offload truck was downhill. Extra distance and additional protection lines were necessary to compensate for the disadvantaged position. *(Courtesy of Maryland Department of the Environment Emergency Response Division)*

odor is so irritating that people will self-evacuate the building prior to hazardous levels being built up.

SAFETY

Gasoline is the leading chemical when it comes to chemical accident fatalities, and transportation leads in those fatalities. Unfortunately, due to its familiarity, many emergency responders do not adequately protect themselves when responding to incidents of this type.

Flammable and Combustible Liquids

By far this is the leading category for the most common type of releases, because it includes gasoline and diesel fuel, **Figure 28-21.** Firefighters respond to these types of incidents thousands of times a day.

Gasoline has between 1 and 5 percent benzene, which is a confirmed human cancer-causing agent (carcinogen) to which firefighters are commonly exposed. A person who smells gasoline is receiving an exposure to benzene, an exposure that is repeated on a regular basis. When dealing with small spills firefighters may have a tendency to use little or no protective clothing, but their bodies are continually being exposed to this toxic material. In addition to its toxicity, gasoline is also very flammable. It can even be ignited by the static electricity generated by clothing as a responder approaches an incident.

Although of less concern than gasoline, diesel fuel, fuel oil, and kerosene do have some toxicity concerns. This lower level of concern can catch a responder unaware. Although in normal circumstances it is difficult to ignite these materials, if their temperature increases they easily ignite. On a day with a temperature below 70°F (21°C) the potential for ignition is low, but a spill onto blacktop when it is 100°F (38°C) outside adds considerable risk for a fire because a larger quantity of vapors is going to be produced on the pavement, which may have temperatures in excess of 110°F (43°C).

One common incident involves overturned or burning tanker trucks. The product found most often in these trucks is gasoline, and when one of these trucks is involved in an accident, a fire usually results.

STREETSMART TIP

To Fight or Not to Fight? When first responding companies arrive they have a tendency to try to extinguish the fire. Without large quantities of firefighting foam it is very difficult to extinguish a tanker fire. If, on arrival, the truck is well involved and is not near any exposures and is not impacting an adjacent community, it is usually best to let the fire continue to burn.

When attempting to extinguish this type of fire there is considerable runoff, which usually contains both the gasoline and firefighting foam, both of which are best kept out of waterways. Also, if the fire is extinguished, a large amount of gasoline at an elevated temperature may remain so foam must be reapplied every few minutes to ensure that the fire does not reignite, **Figure 28-22.** It is then necessary to remove the hot gasoline from the truck and pump it into another truck, a dangerous proposition. If, on the other hand, the fire is allowed to burn, nothing would remain that could cause harm to responders or the environment. The resulting black smoke, although it looks horrible, is in reality less damaging to the environment than a liquid spill. The smoke is predominantly carbon, which provides the thick black smoke. However, if the truck is impacting a community or adjacent structure, then all attempts should be made to extinguish the fire. The truck may be wrecked under a bridge, which is usually considered a critical structure in a community. The cost of a bridge can be in the millions, so responders should try to calculate the financial impact in terms of hard dollars and inconvenience if the bridge were destroyed and it took a year to rebuild. If units arrive and are confronted with a large fire, the initial water should be directed toward the bridge. When sufficient foam and water have arrived, the truck fire should be extinguished, while continually applying foam.

The application of foam at a non-fire incident usually causes friction between firefighters and environmental agencies because the foam can be damaging to the environment. Even if an "environmentally safe" foam is used, the breakdown products of the fuel are not environmentally safe, so environmental agencies prefer the use of minimal foam.

Another concern arises when a truck is overturned and it must be drilled and pumped out prior to righting the truck. The application of foam makes the situation very slippery, hence creating additional hazards. The use of air monitors to determine when the foam blanket is breaking down is recommended, and the use of as little foam as possible is recommended.

In years past, spills of flammable or combustible liquids were commonly flushed down storm drains. This practice for the most part has been discontinued because it is severely environmentally damaging, not to mention that it only moves the problem to another location. Many areas of the country collect the spilled material and dispose of it in an environmentally sound manner. The discharging of oil (includes fuels) is a violation of the federal Clean Water Act, and emergency responders could face serious fines if the material were flushed into a waterway.[3]

Flammable Solids, Water Reactives, and Spontaneously Combustible Materials

When dealing with materials in these categories, a specific identity and emergency response information are crucial. Consultation with the hazardous materials team is important, because using the wrong tactic can be devastating to the community. Most emergency responders have some experience with flammable solids, because road flares are classified as flammable solids. In most cases flammable solids are difficult to ignite, but once ignited they burn vigorously and are difficult to extinguish.

The water-reactive group can be defined in two ways. When a water-reactive material gets wet, a violent reaction, such as a fire or explosion, can result, as shown in **Figure 28-23, Figure 28-24** and **Figure 28-25.**

This is what most people think about when they hear the term water reactive. The water-reactive category also includes reactive metals that, once ignited, can create a problem, such as a violent reaction, if water is applied. Magnesium is one such reactive metal. Once ignited, magnesium burns vigorously and when water is applied it may explode, as shown in **Figure 28-26.**

Some materials such as calcium carbide are shipped as water-reactive materials. When water is applied

FIGURE 28-22 A diesel tank truck cab caught fire, impinging on the cargo tank. A quick and aggressive response by the Washington, DC, and Prince Georges County Fire Departments was able to knock down the fire before the contents were ignited. If the tank had become compromised, firefighting would have been very challenging because the tank truck was on a significant incline. Burning fuel would have traveled down the highway, possibly into storm drains. *(Courtesy of Maryland Department of the Environment Emergency Response Division)*

FIGURE 28-23 A sea box container of magnesium was involved in an accident, and the trailer was ignited by a welder trying to make some repairs to the trailer. *(Courtesy of Cambria County, Pennsylvania, Emergency Services)*

FIGURE 28-24 The magnesium caught fire and resulted in an aggressive, very hot fire. The smoke is toxic and corrosive. Adding water to this fire would have been very dangerous. *(Courtesy of Cambria County, Pennsylvania, Emergency Services)*

FIGURE 28-25 Sand, which should be very dry and stored indoors, can be used to attempt to extinguish smaller fires of flammable metals. *(Courtesy of Cambria County, Pennsylvania, Emergency Services)*

FIGURE 28-26 In this photo eight ounces of magnesium shavings were in a pool of burning diesel fuel. When the magnesium was heated, a slight water mist was sprayed over the fire. The white sparks are from the magnesium and the fireball is from the reaction as well. Relate the size of this violent reaction from a cup of magnesium to that of a truckload of magnesium.

to calcium carbide, a little bubbling occurs, which would not be considered very violent. The gas that is being released by the bubbles, however, is acetylene gas. When pure acetylene gas is produced in this manner it is unstable and reactive. If there is an ignition source nearby there will be a fire or explosion. So the actual application of water in itself does not produce a significant problem, but the action of the water may create additional concerns.

SAFETY

The use of water on flammable metals is not recommended and can cause severe injuries to the hose crew, because the metal has a tendency to explode, sending hot metal and other fragments flying.

Materials that are spontaneously combustible are usually transported in a manner designed to keep them stable. To remain stable, they may require storage at a specific temperature or have a material added to them. White phosphorus, which is usually covered with water, is an example of a spontaneously combustible material, because it ignites in the presence of air. Communities with facilities that use flammable metals should have a stockpile of metal extinguishing agents such as Metal-X or Lith-X, commonly referred to as Class D extinguishing powder. Sand that is known to be dry (e.g., it has been stored inside in a closed container) can also be used, but most sand has moisture contained within it.

The water will break down and release hydrogen gas, which will increase the amount of heat and flame being produced. Most hazardous materials teams have a limited supply of this type of extinguishing agent and usually rely on a local industrial resource for more.

Oxidizers and Organic Peroxides

Like the previous group, this is another class of materials for which help should be requested early. Oxidizers and organic peroxides can have explosive characteristics, and attempting to deal with them can lead to fatal mistakes. Organic peroxides can react explosively even if the responders have taken no action.

One of the best known oxidizers is ammonium nitrate. By itself it is relatively harmless, although it does have some toxicity associated with it. It cannot explode unless mixed with a fuel and an initiating charge added. Although all three items should not be transported together, on occasion ammonium nitrate, fuel, and an initiating charge might be on the same truck. In other situations the oxidizer does not require any oxygen to start a fire, because it will provide its own. Oxygen is placarded with a specific oxygen placard or it can be shipped with an oxidizer placard. Oxygen itself cannot be ignited, but it will greatly intensify a fire when it is involved in amounts greater than the 20.9 percent that is found in air.

Liquefied oxygen (LOX), which is a cryogenic, presents even more hazards in addition to supporting combustion. On asphalt (or other hydrocarbon material) it is shock sensitive and can detonate if compressed, such as by a firefighter walking on it. Another concern is the absorption of oxygen from a LOX release by a firefighter's turnout gear, which can present a flammability problem. Although caution should be used, and the exposed gear should be given time to air out, this would be a very rare occurrence. Most hospitals have large upright cryogenic storage tanks of liquid oxygen, as do many nursing homes, **Figure 28-27.** Smaller in-home versions are available as well, presenting a large fire risk in a residential home, not to mention a freezing and contact hazard if knocked over.

Another commonplace situation involving oxidizers is encountered when dealing with pool chemicals.

FIGURE 28-27 A leak of liquid oxygen on asphalt can present a shock-sensitivity problem in addition to the increased risk of a fire.

Most pool chemicals are oxidizers or may react with oxidizers and can be involved in a chemical reaction or fire. If the materials do not ignite, a very dangerous situation can develop, because a large vapor cloud can be created from just a handful of pool chemicals—one that can have disastrous effects on a neighborhood. Specific advice from the hazardous materials team is required prior to handling or disposing of any of these types of chemicals. There is a lessened level of concern with these types of chemicals because they are considered "household," but they do present a large hazard to the community.

Poisons

Although Class 6 is poison liquids, included here are poisonous gases from Class 2 (gases). These types of materials, which include the "Stow Away from Foodstuffs" and "Marine Pollutant" materials, are toxic to both humans and the environment. Although they are toxic in varying degrees, for the first responder they are poisonous and should be treated as such. The materials labeled as poisonous are very much so, because the DOT does not include many of the materials that are toxic in this category. When choosing a hazard class for a material, the predominant hazard determines its placement into one or another category. Some improvements have been made in the subsidiary placarding and labeling category, but there is room for more. Poisonous gases may be accompanied by a label on a bulk container that reads "Poisonous by Inhalation" and would be listed on the shipped papers as PIH. By definition, these materials present a risk to humans and the community and require extra precautions.

The most common incidents with these types of materials result from pesticides and agricultural chemicals. Although most household materials are of lesser concentrations, on occasion higher strength materials have been used. There have been several incidents in which bug-spraying companies (or individuals) have used full-strength agricultural products in residential homes, requiring evacuations lasting a few weeks and removal of some homes due to the contamination.

> **SAFETY**
>
> Technical-grade pesticides are very dangerous and should be handled with care.

Incidents involving commercial home fertilizing trucks are common, but in most cases this material is diluted and ready for application. Fertilizers do not present much risk to the responders or the environment unless found in large quantities. Pesticides and insecticides do, however, present a risk to the responders and the environment, and responders should consult with a hazardous materials team prior to taking any action. Many home pest control vehicles carry a mixed load of premixed and undiluted materials. If such a vehicle is involved in an auto accident, the use of full gear is recommended, and anyone who comes in contact with any of the liquid or solid materials should be decontaminated.

Radioactive Materials

Although considerable emphasis was placed on radiation in the early days of hazardous materials, due to a low number of incidents involving these materials, some of the emphasis has been redirected to other areas. It is true that incidents involving these types of materials are rare, but they do have a significant impact on the well-being of the community, and with the potential now for terrorism, radiation is being emphasized again.

Radioactive materials are commonly used in the community, in smoke detectors, in ground imaging equipment, and in the medical community. Although many people express concern when around radio-active materials, most radioactive materials are not harmful even for a lengthy exposure.

Radiation is divided into two categories: ionizing and nonionizing radiation. Radioactive materials that emit alpha, beta, gamma, and neutron forms of radiation are ionizing types. The ionizing forms of radiation are able to make changes in atoms, which can lead to health problems with humans. Nonionizing radiation takes the form of sunlight (visible light), microwaves, and radio waves. These forms of energy are not as strong as their ionizing counterparts, but high amounts or repeated exposure (e.g., to sunlight) can cause health problems.

Some radioactive materials, such as those coming to and from a nuclear power–generating facility, can present some risk to the community. Radioactive materials that present a risk to the community are shipped in high-strength containers designed to withstand a substantial crash and to be involved in a fire situation without any release of radioactive materials. These shipments are usually well tracked and, when involved in an incident, specialized help is usually already on the way before local responders call for it. Knowing who is the local contact for radiation emergencies is key to an effective response when dealing with radioactive materials.

> **SAFETY**
>
> Keeping the material in the container that it is intended to be in is paramount to reducing the exposure levels of the responders.

When dealing with potential radiation hazards, the adage "time, distance, and shielding" should always be followed. This basic principle of radiation exposure should be applied to all chemical exposures. Firefighters must be careful to limit their time around a radiation source, keep their distance, and wear some type of shielding; then they can be protected against most forms of radiation. Most radiation exposure guidelines are based on a one-hour exposure time. By reducing their time, firefighters reduce their exposure. By doubling their distance, they can reduce their radiation exposure to one-fourth of the original exposure. Protective clothing and SCBA can provide some shielding for some forms of radiation, while more substantial forms of shielding such as lead or concrete may be needed for more dangerous types of radiation. Persons who have been exposed to a radiation source are not radioactive. They may suffer some significant health problems, but they do not present a risk to others. Persons who have radioactive material on them, such as a radioactive powder, are contaminated, may present a risk to others, and are considered to have an external contamination. Alpha and beta radiation sources are hazards and can contaminate persons through inhalation and ingestion, which would be considered internal contamination. Neutron and gamma radiation don't create contamination issues with regards to persons. Neutron radiation sources do have the ability to irradiate some materials, but the chances of encountering a neutron source are small, except in war or a terrorist attack. Simple decontamination with soap and water is needed to reduce the damage to the person. When dealing with radioactive materials it is best to try to contain the runoff so as not to further contaminate the incident scene. The best protection against radiation is to keep the radiation source in its protective container.

Corrosives

After the flammable and combustible categories, the next most likely hazardous materials incident will probably involve a corrosive. The most common incidents occur with sulfuric acid, hydrochloric (muriatic) acid, and sodium hydroxide. Both of the acids are transported as liquids, but the sodium hydroxide may be transported wet or dry. Both sulfuric acid and sodium hydroxide have little or no vapor pressure, so they present little risk outside of the immediate spill area. Hydrochloric acid, on the other hand, does have a high vapor pressure and may require some additional isolation and evacuation. Nitric acid is less common but is easy to detect as it has a characteristic vapor cloud, as shown in **Figure 28-28.** When released, it forms a brown vapor cloud comparable to the release of bromine.

To handle any of these materials, chemical protective clothing is required, because after some time they can eat a hole in turnout gear. A responder who is splashed with any of these three materials should wash the material off as soon as possible. Note that the actual burns will take a couple of minutes to become evident, and will progress in severity if not quickly washed off. This applies to these three materials only, because some of the less common acids, such as oleum or sulfur trioxide, will cause burns immediately upon contact, but it is still a matter of time before disfiguring injuries occur.

Chemical neutralization may be the best choice for handling a corrosive spill. This type of operation needs to be performed by a hazardous materials technician who has donned appropriate levels of PPE. When neutralizing chemicals, heat and violent reactions may occur, so neutralization should take place only after consultation with a chemist.

Other Incidents

It is impossible to outline each specific action that first responders should take at a chemical release. The first rule if a responder suspects that hazardous

FIGURE 28-28   The shipping papers did not indicate the presence of nitric acid. The brown vapor cloud is a result of a chemical reaction between bromine and red fuming nitric acid. When the hazardous materials team opened the back of the truck, they were greeted with these vapors. The team members in the photograph retreated when the vapors were released from the back of the truck and changed into chemical protective clothing. *(Courtesy of Maryland Department of the Environment Emergency Response Division)*

materials are involved is to isolate the area and evacuate other persons in the immediate area until further information is available. Request the assistance of the closest hazardous materials team because their technical expertise and equipment will be needed. First responders are not trained nor equipped to provide detection or characterization of unknown materials, and to use their senses, such as smell, is to play Russian roulette.

Many toxic materials are odorless and colorless and thus will provide little warning prior to causing lethal effects. It is not the job of first responders to seek the cause of a chemical release nor to provide an identification. If the information is presented to them or can be obtained without risk, then it is of benefit, but otherwise it is the job of a hazardous materials team to perform those tasks.

Common incidents that first responders may get involved with are sick buildings, or odor complaints that can occur in buildings such as the one shown in **Figure 28-29.** The use of air monitors is the key to survival with these types of incidents, but the sense of smell is not sufficient to determine immediately dangerous levels. When involved in sick buildings or odor complaints, the best course of action for first responders is to determine the validity of the complaint. Once it is determined that the call is valid and not related to a sunny Friday afternoon, the first responders should don SCBA and request the services of a hazardous materials team. The occupants of the building should be removed and the windows and doors shut. The HVAC system should be shut down so as not to remove the unknown material. These steps are crucial to the success of the hazardous materials team's attempts to find the source of the problem.

FIGURE 28-29 It is difficult to determine the origin of unusual odors in a high rise building. There are many occupants, offices, and potential sources.

First responders may also get called to gas leaks inside a building, and their ultimate safety is determined by a suitable air monitor. It is impossible to determine the level of a gas in a building based on smell. Once in an environment that has a gas odor, the body will desensitize itself to the odor and responders will not be able to smell the gas anymore, even though the amount has been increasing, moving closer to a potential explosion.

Gas grills are prone to propane leaks, and if the leak reaches an ignition source can cause severe damage to buildings. Many hazardous materials teams carry pipe flares that burn off the gas, preventing an explosion. First responder actions should be limited to isolation and evacuation and to preparing standby hoselines. If the vapors are traveling toward other homes, then the use of hoselines with fog nozzles to move the vapors is appropriate.

DECONTAMINATION

The task of decontamination may fall to a person trained at the first responder Operations level. If a first responder is expected to perform decontamination, then specific training in that area is required, as well as training in the use of chemical PPE. The definition of **decontamination,** or decon for short, is the physical removal of contaminants from people, equipment, and the environment. It is important to note that the concepts provided here may be for one or all three of these areas of concern. Just because one type of decontaminating solution is effective for a tool does not mean that it can be used on humans. Prior to performing any decontamination procedure on humans, first responders should consult with the hazardous materials team or a chemist. Some decon solutions use very dangerous materials, especially if they come in contact with skin. For equipment, it may be acceptable to use large quantities of sodium hydroxide mixed with detergent, but this same solution would cause severe burns if placed on the skin. It is commonly recommended that no more than soap and water be used on victims.

The two ways to become contaminated are via direct contact or secondary contamination. If first responders go into a spill area and put their hands into a drum of blue paint, direct contamination has occurred. When they leave the area and shake hands with another responder, they have secondarily contaminated the other responder with the blue paint.

Types of Decontamination

With the revision to the NFPA 472 standard, there are three general types of decontamination levels: **emergency decontamination, technical decon-**

tamination, and **mass decontamination.** One other type, **fine decontamination,** is added here because it is an important type of decontamination for victims. In a large full-scale incident, all four types may be used, but in reality the majority of incidents require only minimal decon. There are two large categories of decontamination, wet and dry decontamination. The two categories should be self-explanatory: One requires the use of water or other liquid solution, whereas the other does not involve a wet process.

The process of decontamination is chemical specific so there are no absolute rules. First responders should as a minimum have a good understanding of the emergency decon procedures and develop a local procedure to handle this type of problem. When confronted with a contaminated patient who is in need of decon, there will be no time to develop the procedure. With potential terrorism in mind, responders should think about the emergency decon of not one person, but thousands of people.

Emergency Decontamination

When a victim or a responder is contaminated with a hazardous material, it is vitally important to remove the contamination as rapidly as possible. To delay the removal of the hazard could result in fatal consequences. No matter what type of emergency activity is occurring, responders and the IC should have some thought of emergency decontamination. Even responding to a medical call can result in the need for emergency decon; having a plan is essential for survival. Emergency decontamination can be as simple as taking a hoseline and washing someone off, or it can be complicated by an unconscious patient on a backboard or in a Stokes basket. For most chemicals, plain water washing is sufficient to decontaminate the person, as long as the person's clothes are removed. Removal of the patient's clothes removes 60 to 90 percent of the contamination. As toxic as nerve agents are, if the patient has been contaminated with an aerosol, then just having the person stay in fresh air for fifteen minutes can be an effective decon method. Between fresh air and clothing removal, effective emergency decon can be accomplished.

The common methods of emergency decon involve the use of a water line and some method of runoff control, such as that shown in **Figure 28-30.** Once through emergency decontamination, the victim should receive technical or fine decontamination, since emergency decontamination is not likely to remove all of the contaminant.

FIGURE 28-30   One of the simplest forms of emergency decon is the use of a hoseline.

> **NOTE**
>
> In reality when people's lives are on the line, runoff is not a large concern, but when not under those types of conditions, attempts should be made to recover any runoff. When dealing with radioactive decontamination, efforts should be taken to contain runoff.

There are a number of manufacturers who can provide decontamination containment. In the absence of formal measures, there are some practical alternatives. An inflatable children's pool, which can be easily carried on a fire truck, can be used to contain the runoff from decontamination. Another method involves the use of a tarp or plastic that is spread between two ladders. If none of these methods is possible, then using a tarp or plastic to catch as much runoff as possible is advised. If all other methods are not available, then it is best to take advantage of natural areas that would collect the runoff.

Decon should not be performed over storm drains or other environmental concerns. An industrial plant may have safety showers, and it is acceptable to use these safety showers. Prior to using the showers it is important to ask the people from the facility where the runoff goes. In most cases the runoff will go to the facility's own water treatment area and can be taken care of by that system. Caution should be used when the safety shower drains straight to the sanitary system; prior to using such a shower, the local sewer authority should be contacted.

One of the other general rules with regard to emergency decon involves the removal of corrosives from a victim. Large quantities of water should be used, because if small amounts of water are mixed with the corrosive there is the potential for a heat increase.

Massive amounts of water will eliminate this potential for further injury. Besides using large quantities of water on corrosives, it is necessary to continue the flushing for a minimum of twenty minutes. No matter how bad the temptation is for EMS providers to remove the patient, the best treatment for corrosive burns is the water flush, especially for burns caused by a base, such as sodium hydroxide.

Technical Decontamination

The first step in technical decontamination is the removal of the majority of the contaminants, and this is usually addressed by the first rinsing station within a full decon setup, as shown in **Figure 28-31.** This step is known as gross decon, and in many cases gross decon is just part of the whole process, but it can be the only step in a minor setup. Some hazardous materials teams use shower setups for gross decon, whereas others may use hoselines or pump tanks. Because this step is the most likely location for contaminants to collect, attempts should be made to recover any of the runoff. When part of a multistep process, the entry team usually performs this step on themselves without any assistance.

When dry decon is used, overgarments such as booties or extra gloves may be used, and these are placed in a designated container when leaving the hot zone. For many materials, dry decon is an acceptable use of PPE and resources.

When using any type of decon, the bottom line is to make it safe for the responder to get out of the PPE and safe for the responders assisting with the PPE removal. When using reusable PPE, decon needs to be more formal, but even extensive cleaning can be done to the PPE after the incident and removal of the responder.

Further in the technical decon process is a step in which some actual scrubbing and cleaning take place to remove any residual contaminants, **Figure 28-32.** It is usually done after the gross decon and usually has a couple of steps. Attempts to recover any runoff should be used, but this is not as critical as for the gross decon step.

FIGURE 28-31 Gross decontamination is used to remove the majority of the contamination.

FIGURE 28-32 Formal decontamination is used to remove any further contamination that may remain after gross decontamination.

It is during the scrubbing step that decon solutions are typically added to the process. These solutions are chemical specific. This process may involve showers, hoselines, or pump sprayers, and one or two people dressed in a lower level of PPE may perform this activity. When performing decontamination, it is important to pay particular attention to the areas most likely to be contaminated, the hands and feet.

Fine Decontamination

This form of decon is not performed by first responders, but by hospital-based personnel. It is a cleaning that removes all of the contaminants from the body. It would be nearly impossible to clean eyes, ears, fingernails, and other areas of the body in the street; these tasks are better accomplished in a hospital. According to OSHA the staff must have proper training to perform fine decon, and they must be trained to the Operations level, although some training to the Technician level is preferred. Some hospitals have designated decon areas with a separate entrance, separate ventilation system, and a holding tank for runoff control.

Through the Local Emergency Planning Committee (LEPC) plan, hospitals are required to identify themselves as being able to accomplish decontamination of patients. Every hospital should have a policy to cover a walk-in contaminated patient, and emergency responders should test these plans on a regular basis. Regular training and interfacing are required to keep this type of system functioning well within a community. Responders should ensure that this capability is well run and that sufficient training has taken place, because they are likely candidates to become patients someday!

Mass Decontamination

Most hazardous materials decontamination setups are designed primarily to decontaminate the members of the hazardous materials team. They are designed to handle two to four people dressed in chemical protective clothing. Although most of these systems can handle civilians, they are designed to decontaminate responders' PPE. When decontaminating civilians, as shown in **Figure 28-33,** several issues arise: clothing, privacy, valuables, weather impact, and inconvenience. Civilians may not be willing to undergo decontamination or to separate from their relatives. They may not want to remove their clothes or valuables.

The thought for mass decontamination is that plans must be developed to handle decon of thousands of people. Communication will be difficult and bullhorns

FIGURE 28-33 Mass decon should have the victims in water for as long as possible. If time permits, soap and water should be used to better the decontamination process.

or loudspeakers should be employed to address large crowds. Keeping contaminated victims informed is one method to keep them calm. Some mass decon setups are provided in **Figure 28-34** and **Figure 28-35.** Most of the mass decontamination issues arise when terrorism events are discussed. It is best to have the victims decontaminated before they get to the hospital, because contaminated victims will create additional problems at the hospital. More information on possible terrorism scenarios is discussed in Section IV, Chapter 30. If victims are contaminated with military nerve or blister agents, speed is of the essence. If decontamination is not performed on symptomatic contaminated victims, their chance for survival may be slim. The hardest part of this process is identifying those who need physical decontamination and those who need **psychological decontamination.** If victims are symptomatic and there is a credible threat presented, then decontamination should be accomplished quickly. First arriving units need to establish this decontamination process. The best decontamination solution is water, followed by soap.

Studies from the military have shown that the use of bleach can actually cause more deaths of and injuries to contaminated victims. Decontamination is chemical specific, and unless first responders can identify the specific chemical compound they should use a simple soap and water solution. A hoseline can be used to accomplish emergency victim decontamination; as more assistance arrives, the system can become more elaborate, as one example in **Figure 28-36** demonstrates. When setting up the decontamination process, it is important to avoid victim backup and if there are delays, to make sure that the victims

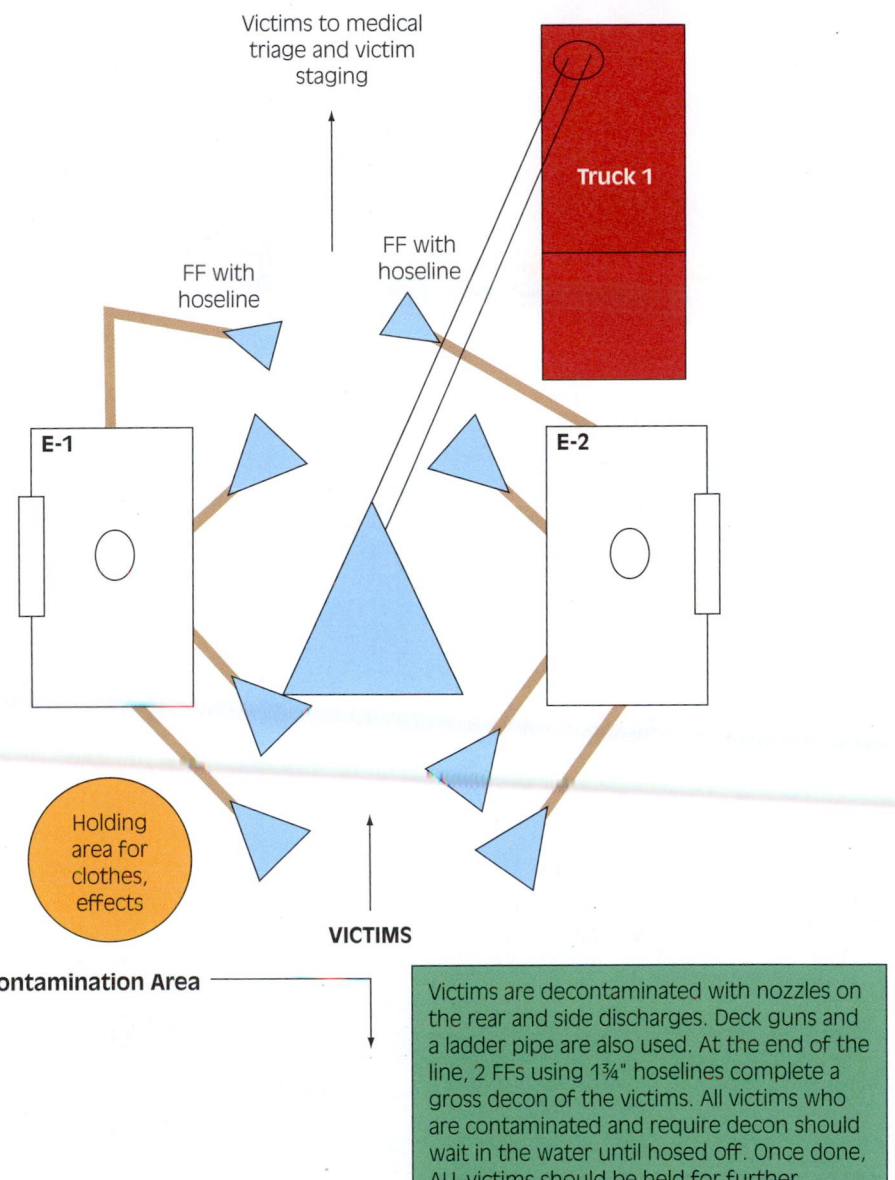

BASIC PLAN
Hazardous Materials Response Team
Mass Casualty Decontamination

Victims to medical triage and victim staging

Truck 1

FF with hoseline

FF with hoseline

E-1

E-2

Holding area for clothes, effects

VICTIMS

Contamination Area

Victims are decontaminated with nozzles on the rear and side discharges. Deck guns and a ladder pipe are also used. At the end of the line, 2 FFs using 1¾" hoselines complete a gross decon of the victims. All victims who are contaminated and require decon should wait in the water until hosed off. Once done, ALL victims should be held for further examination or questioning.

FIGURE 28-34   (A) Mass decon set-ups. *(Courtesy of Baltimore County Fire Department)*

wait in water. After the victims exit the decontamination area they should be guided to holding areas. It is important to keep symptomatic victims from nonsymptomatic victims. If these victims are mixed, the nonsymptomatic victims will tend to pick up the symptoms due to psychological reasons.

One method that is built into the engine company is shown in **Figure 28-37.** This engine company even carries robes and slippers to be distributed after decontamination. Another method of mass decontam-

ination is the use of mobile showers, shown in **Figure 28-38** and **Figure 28-39,** which can enable a number of victims to decontaminate themselves.

SAFETY

The use of a bleach and water solution in terrorism decontamination is no longer considered an acceptable practice.

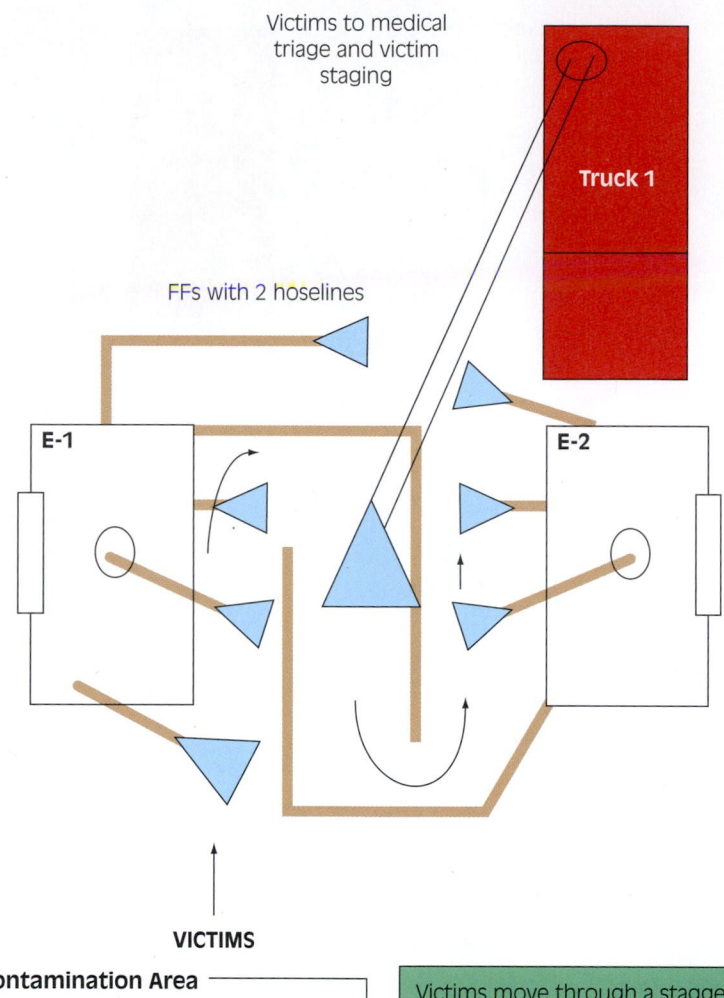

ADVANCED PLAN
Hazardous Materials Response Team
Mass Casualty Decontamination

Victims to medical
triage and victim
staging

Truck 1

FFs with 2 hoselines

E-1

E-2

VICTIMS

Contamination Area

Victims move through a staggered line,
like that at an amusement park or bank.
Gross decon is completed at the end of
the line with two 1¾" hoselines. While
waiting in line, nozzles from the rear,
side, and deck pipe are washing the
victims as well. Rope or barrier tape can
be used for the line.

FIGURE 28-34 (B) Once additional resources arrive then the advanced plan
for mass decon can be implemented.

Decontamination Process

There are several variations to a decontamination
process, but in the overall scheme of things they are
basically similar.

JPR 28-1: Basic Decontamination Steps

*(For step-by-step photos of this skill sequence, see
page 1168)*

1. Tool drop
2. Gross decon
3. Scrubbing and rinse
4. PPE removal
5. SCBA removal
6. Clothing removal, body wash, and dry off
7. Medical evaluation, including rehydration

One should consider a final step and that would
be recordkeeping. While this may sound out of the
ordinary, all exposure information (including the suit
worn and any other pertinent information) should be
logged for future use should it be needed. The variations include the use of shower systems, tents, num-

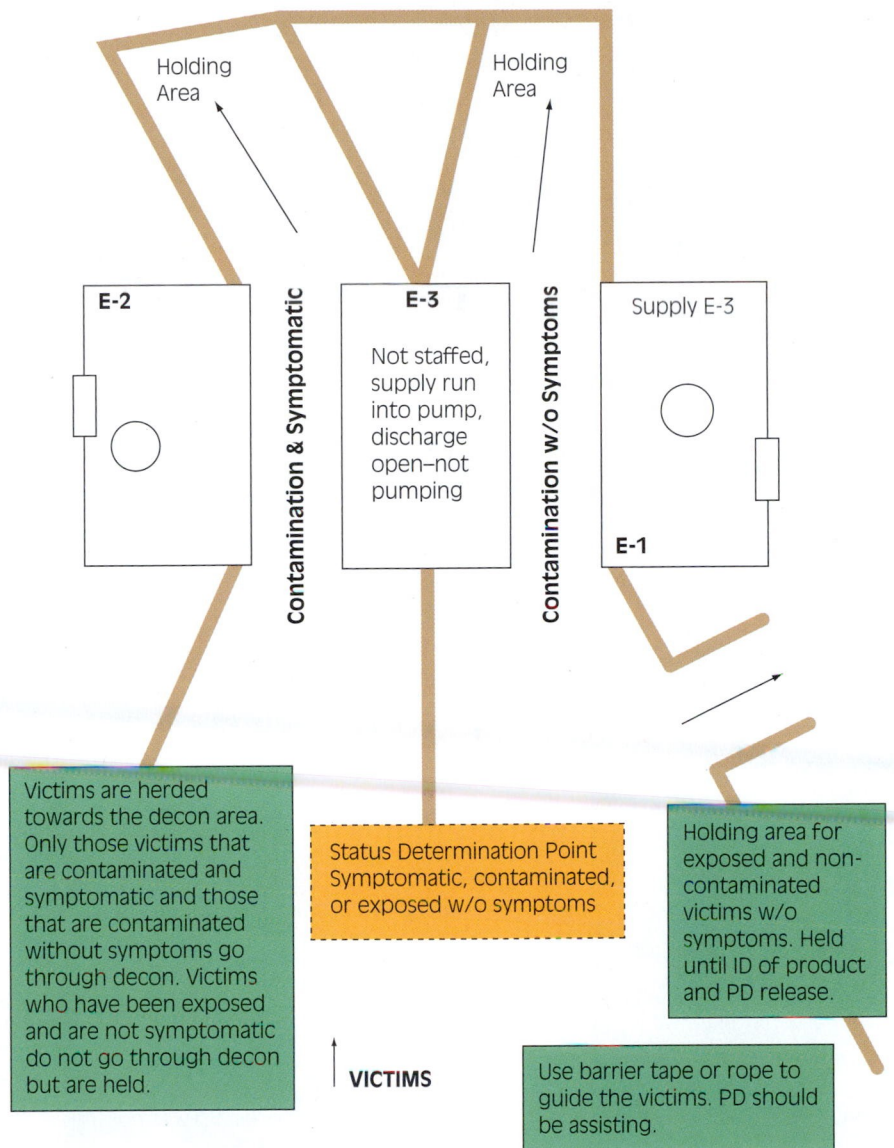

Herding Guide
Baltimore County Fire Department
Hazardous Materials Response Team

Holding Area

Holding Area

E-2

Contamination & Symptomatic

E-3

Not staffed, supply run into pump, discharge open–not pumping

Contamination w/o Symptoms

Supply E-3

E-1

Victims are herded towards the decon area. Only those victims that are contaminated and symptomatic and those that are contaminated without symptoms go through decon. Victims who have been exposed and are not symptomatic do not go through decon but are held.

Status Determination Point Symptomatic, contaminated, or exposed w/o symptoms

Holding area for exposed and non-contaminated victims w/o symptoms. Held until ID of product and PD release.

VICTIMS

Use barrier tape or rope to guide the victims. PD should be assisting.

FIGURE 28-35 Mass decon plan for herding mass numbers of potentially contaminated people.

ber and location of steps, and personnel. The method of runoff collection varies team to team as well. For most situations that involve solids and liquids a minimum type of decon should be set up.

The HAZWOPER regulation requires that the IC have a plan for decontamination, which can vary from no setup through a multistep process as depicted in the decontamination photo series. For a material such as a technical-grade pesticide, a full setup with all stations is recommended. For gases, in reality no decon is required; however, a minimum setup should be established prior to entry in the event of an emergency and for psychological purposes. Regardless of the type or size of the setup used, it should be set up and ready prior to any entry into a hazard area.

METHODS OF DECONTAMINATION

There are general methods of decontamination that apply to humans, equipment, and the environment, **Table 28-4.** Each method may not apply directly to all three categories. It is important to consult with the hazardous materials team or a chemist prior to using any of these methods on a human, because they may be more dangerous than the contaminant.

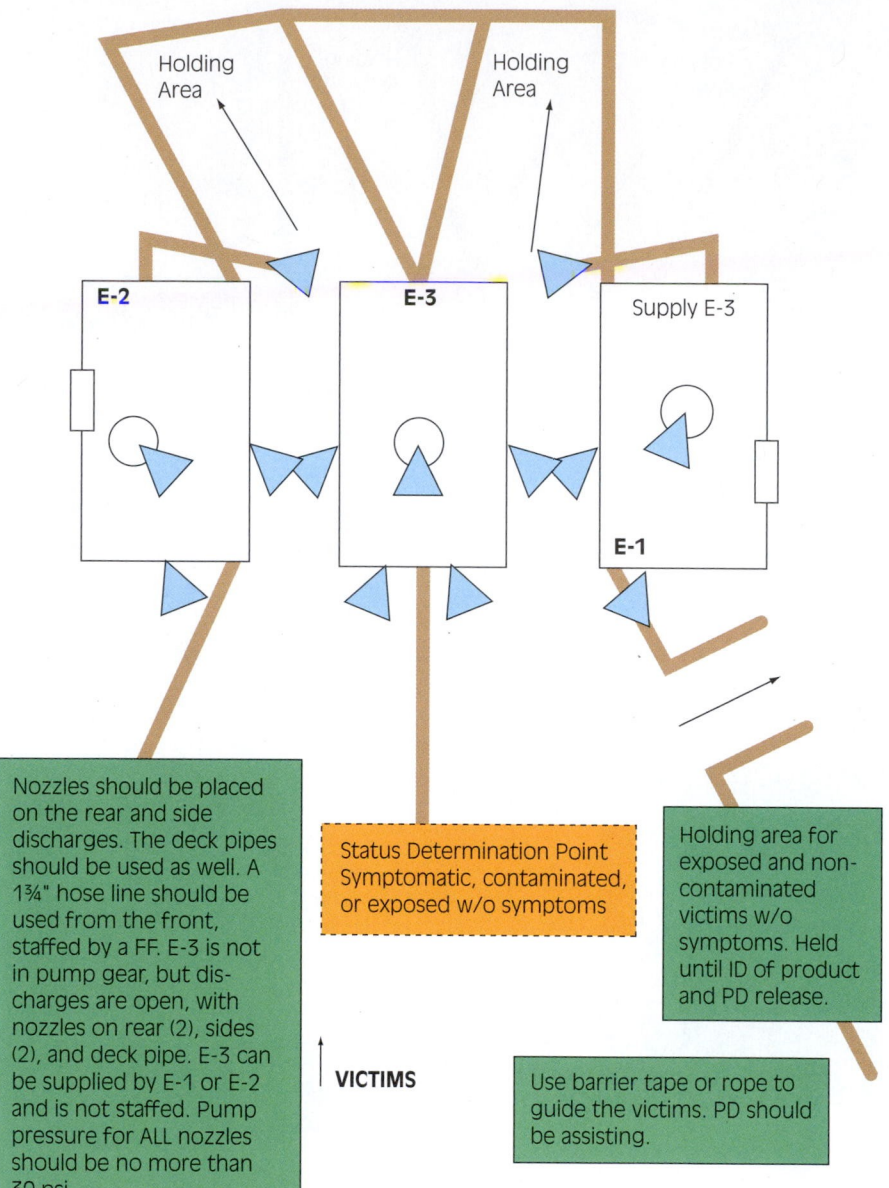

Water Flow for Herding Guide
Baltimore County Fire Department
Hazardous Materials Response Team

Holding Area

Holding Area

E-2

E-3

Supply E-3

E-1

Nozzles should be placed on the rear and side discharges. The deck pipes should be used as well. A 1¾" hose line should be used from the front, staffed by a FF. E-3 is not in pump gear, but discharges are open, with nozzles on rear (2), sides (2), and deck pipe. E-3 can be supplied by E-1 or E-2 and is not staffed. Pump pressure for ALL nozzles should be no more than 30 psi.

Status Determination Point Symptomatic, contaminated, or exposed w/o symptoms

Holding area for exposed and non-contaminated victims w/o symptoms. Held until ID of product and PD release.

VICTIMS

Use barrier tape or rope to guide the victims. PD should be assisting.

FIGURE 28-36 In addition to herding and organizing mass numbers of people, water for decontamination can be added to a plan.

Absorption

The spilled material is picked up by the absorbent material, which acts like a sponge. Common absorbent materials include ground-up newspaper, clay, kitty litter, sawdust, charcoal, and poly fiber. The absorbed material does not change, and the volume of material may increase. The whole mixture will have to be treated as waste and will not change any of the characteristics such as flammability. Compatibility needs to be researched prior to using these materials, because some materials could cause an adverse reaction.

Adsorption

The material to be picked up bonds to the outside of the adsorption medium. Activated carbon and sand are the most common adsorbents and should be available locally at a chemical facility. Many chemical facili-

FIGURE 28-37 This engine has a canopy that unfolds to provide a decontamination corridor. The side of the engine is pre-piped with shower nozzles and garden hose extensions for additional water sprays. Sides can be added for privacy.

FIGURE 28-38 An example of a decontamination vehicle that has showers and can contain any runoff. *(Courtesy of Maryland Department of the Environment Emergency Response Division)*

ties have drums, bags, or totes of activated carbon set up to receive liquids within a process, and may have some available in the event of an emergency.

Chemical Degradation

The ability to degrade a chemical varies from compound to compound and is much like neutralization. To degrade a chemical, another chemical is added or the chemical is exposed to the elements, which breaks down the chemical into less hazardous components.

Dilution

The ability to dilute a contaminant is dependent on the chemical structure of the spilled material, and it does not work on all contaminants. Imagine getting coated with used motor oil; flushing with water will not remove all of the contamination. When dealing with corrosives, large quantities of water are required, sometimes into the hundreds of thousands of gallons. In recent studies, it has been determined that plain water is as effective as some of the available bleach and water solutions. But with some types of neutralization, dilution is a possible solution.

Disinfection

When dealing with humans a 0.5 percent bleach and water solution can be used for some etiological contaminants, although contact time is needed for success. It was initially thought that bleach solutions would also be effective for biological contaminants, but it has been determined that plain water is just as effective for humans. This is due to the contact time required for effectiveness: the exposure of skin to bleach solutions causes burns if left on for more than a short time. For equipment or the environment, disinfection can be accomplished using a higher percentage of bleach or other chemical.

Evaporization

Is allowing a chemical to evaporate, changing its state of matter? If a solid or a liquid is allowed to remain in the open, depending on its vapor pressure, that solid or liquid will eventually change to a vapor. A chemical that approaches its boiling point will also evaporate, moving the material to the vapor state. The material does not disappear but merely changes its state of matter.

Isolation and Disposal

One of the easier forms of decontamination is to isolate the contaminant, collect it using appropriate protective clothing, and then dispose of the contaminant. The disposal should follow appropriate federal, state, and local regulations.

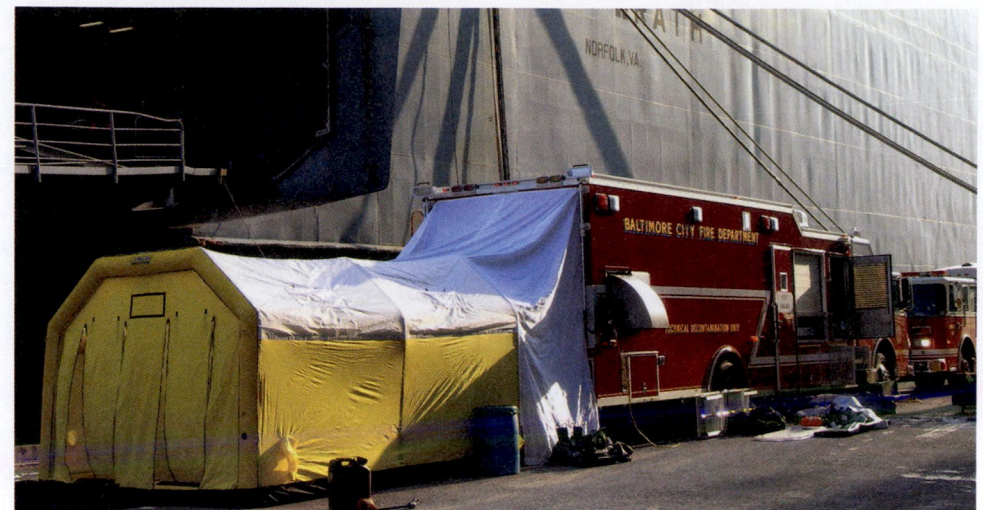

FIGURE 28-39 An example of the decontamination vehicle completely set up. The use of the tent adds an additional layer of privacy. *(Courtesy of Maryland Department of the Environment Emergency Response Division)*

TABLE 28-4   Methods of Decontamination

Type	Used on Humans	Used on Equipment	Used on Environment
Absorption	Yes	Yes	Yes
Adsorption	Yes	Yes	Yes
Chemical degradation	No	Yes	Yes
Dilution	Yes	Yes	Yes
Disinfection	Yes	Yes	Yes
Evaporization	Yes	Yes	Yes
Isolation and disposal	No	Yes	Yes
Neutralization	Under specific conditions	Yes	Yes
Sterilization	No	Yes	Yes
Solidification	No	Yes	Yes
Vacuuming	Yes	Yes	Yes
Washing	Yes	Yes	Yes

Neutralization

Neutralization is usually reserved for corrosive materials, but is also used here to describe the procedure that reduces the toxicity of a poisonous material. Responders are urged to consult with a chemist prior to performing this type of activity. The chemicals used to reduce this risk are in themselves hazardous—but in a much lesser fashion than the contaminant. An emergency responder should avoid mixing a neutralization solution using sodium hydroxide because it is very corrosive, but with some chemicals it is extremely useful in small amounts in reducing the hazard of some types of contaminants. In general neutralization it is exothermic and can produce violent reactions, although these are extremely rare.

Sterilization

There are two primary methods of sterilization, one through a combination of steam and high heat and the other chemical sterilization. The use of steam and high heat is useful for items that are contaminated with etiological materials. The use of chlorine dioxide has been used to decontaminate some buildings for anthrax contamination. Chlorine dioxide is very corrosive and would be considered a very concentrated form of bleach.

Solidification

Solidification is another method that, depending on the solidification agent used, may alter the suspect agent. In some cases, however, the solidification agent will not have any effect on the agent and will reduce the hazard and enable sampling to occur. Chemical compatibility is an item that needs to be confirmed prior to the use of a solidification material.

Vacuuming

For solid materials such as dusts or fibers, a vacuum would certainly reduce the hazard. An ordinary shop vac is not recommended for this task, however, because they are not filtered well enough to keep the exhaust from blowing the agent back into the air. Instead it is recommended to use a vacuum that is equipped with a high-efficiency particulate air (HEPA) filter. Special vacuums are made exclusively for picking up mercury without the mercury vapors being released into the air.

Washing

Typically washing is done with soap and water. This is one of the more effective decontamination solutions, but in reality the soap and water merely removes the contaminant. The contaminant may still present a risk but is removed from the person or item. Washing can be low pressure or high pressure, and can also use elevated-temperature liquids. Under the direction of a chemist, other solutions may be added to the washing solution.

To test the effectiveness of decontamination, responders can use air monitors, particularly a PID to determine if the contaminant has been removed. Other detection devices can be used as required. For the ultimate assurance, responders can take swipe samples and send the swipes off to a laboratory for analysis. As part of the termination procedures, the use of detection devices or other tests for the effectiveness of decontamination should be noted in the report. The type of decontamination and the specific type of decontamination solutions should also be recorded. The report should also include any potential exposures to hazardous materials and an activity log for the responders documenting their actions. These records should be kept for at least 30 years past the last date of employment for the responders. In addition to meeting NFPA requirements, these recordkeeping requirements are also part of the OSHA HAZWOPER regulation.

Some response teams and emergency plans outline incident levels to provide quick notifications. **Table 28-5** lists the incident levels and their associated concerns.

TABLE 28-5	Incident Levels

Level	Incident Scale
1	Small-scale incident can usually be handled by the first responders but may also require minimal additional resources. Notifications are usually local and may only be internal to the fire department. The maximum level of PPE required is firefighter turnout gear and SCBA, and even that may not be necessary. The material is not toxic by skin absorption, although it may present an inhalation hazard. The size of the spill is usually small and will have minimal environmental impact. Incidents of this type include natural gas or propane leaks and small fuel spills.
2	Incidents at this level usually require additional assistance, such as from a HAZMAT team. The incident may require additional notifications to environmental or emergency management agencies on a local or state basis. The amount of material may be larger or more hazardous than in a Level 1 incident. The type of PPE will probably be chemical protective clothing or other clothing not carried by the first responders. The first responders may have switched from an active role to a support role. A Level 2 incident may require a small evacuation and possibly a large isolation. The incident will also probably impact the ability of the jurisdiction to respond to other emergencies, depending on the resources available. An example incident would be an overturned gasoline tanker or a leaking propane tanker. A leaking drum in the back of a tractor-trailer would be classed as a Level 2 incident.
3	A Level 3 incident requires substantial local resources and the assistance of other agencies, local and state. Resources on the federal level may also be required or, at a minimum, are to be notified. The incident will require the evacuation of the affected area and a substantial isolation area. The release is large or the material is extremely toxic. Examples of a Level 3 incident include a train derailment with leaking chlorine railcars. A substantial leak from an ammonia tank truck would also be an example of a Level 3 incident.

JOB PERFORMANCE REQUIREMENT   28-1
Basic Decontamination Steps

A   The tool drop is usually the first stage in the decon setup. The tools are placed for the use of another entry team or may be collected here for later decontamination.

B   In most systems gross decon is the next step. Some response teams use showers or hoselines to accomplish this task.

C   The next step is formal decon and may involve two stations in which the responder is rinsed, scrubbed, and rinsed again. The solution used is chemical dependent and varies depending on the contaminant. This is the first step in which other responders may assist.

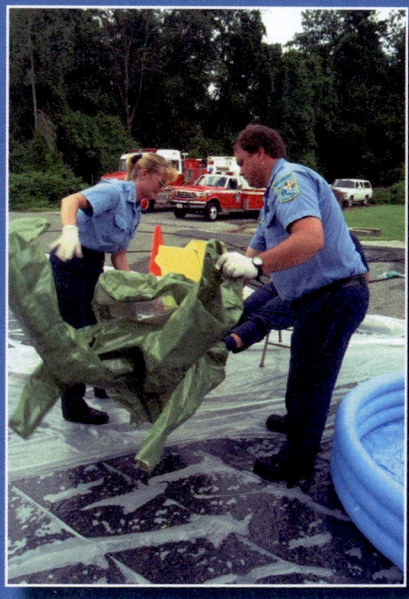

D   After the wet portion of the decon process, the entry team removes their PPE, with the assistance of other responders.

E After the removal of the PPE, the entry team will remove their SCBA. It is usually placed into some type of containment system for later cleaning.

F In some systems a fourth washing area is established for body washing. Some teams may set up a tent for this purpose.

G The last steps are hydration and medical evaluation.

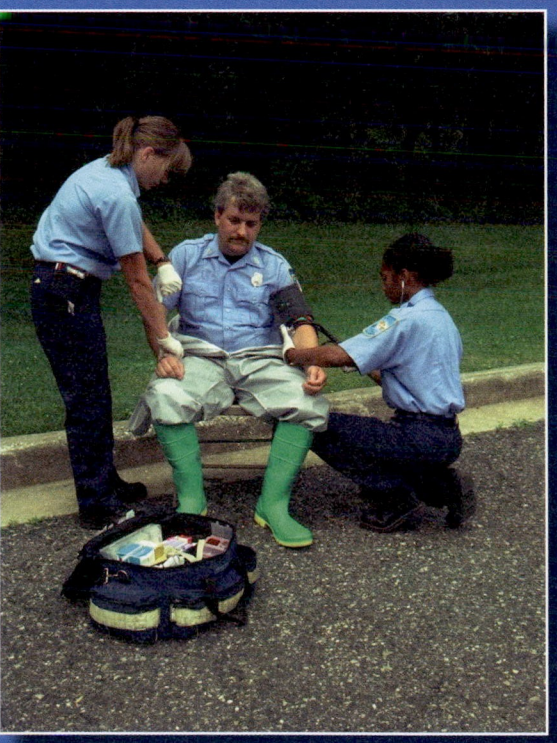

H Medical evaluation after entry is very important. Hazardous materials suit work is very stressful on the body.

LESSONS LEARNED

The chlorine railcar accident in Graniteville offers an important lesson. Many of the topics in this chapter directly relate to that incident. Although the accident had a significant impact on that community and many lives were lost, the chlorine release was not as large as it might have been. Protective actions are used for a variety of purposes, the most important of which is to protect the public and, in many cases, the responder. Management of a chemical-release incident is not an easy task, because there are multiagency jurisdictions and a variety of objectives may need to be accomplished at the same time. Unlike most fires or accidents, political pressure is usually associated with a chemical release and there is the potential that thousands of people may require evacuation and relocation.

The determination for decontamination can also be a difficult decision, because most departments do not have the ability to decontaminate large numbers of victims. Choosing the method of decontamination can be difficult as well because of the many possibilities. The best options for protection are to provide prompt isolation and prevent further escalation of the exposed population. For those people who are contaminated, a water-based decontamination method is usually recommended.

KEY TERMS

8-Step Process A management system used to organize the response to a chemical incident. The elements are site management and control, identifying the problem, hazard and risk evaluation, selecting PPE and equipment, information management and resource coordination, implementing response objectives, decon and cleanup operations, and terminating the incident.

ANFO The acronym that is used for ammonium nitrate fuel oil mixture, which is a common explosive. ANFO was used in the Oklahoma City bombing incident.

DECIDE Process A management system used to organize the response to a hazardous materials incident. The factors of DECIDE are detect, estimate, choose, identify, do the best, and evaluate.

Decontamination The physical removal of contaminants (chemicals) from people, equipment, or the environment. Most often used to describe the process of cleaning to remove chemicals from a person. Often shortened to "decon."

Emergency Decontamination The rapid removal of a material from a person when that person (or responder) has become contaminated and needs immediate cleaning. Most emergency decon setups use a single hoseline to perform a quick gross decon of a person with water.

Emergency Response Planning (ERP) Levels that are used for planning purposes and are usually associated with the preplanning for evacuation zones.

Evacuation The movement of people from an area, usually their homes, to another area that is considered to be safe. People are evacuated when they are no longer safe in their current area.

Fine Decontamination The most detailed of the types of decontamination. Usually performed at a hospital that has trained staff and is equipped to perform fine decon procedures.

GEDAPER Process A management system used to organize the response to a chemical incident. The factors are gather information, estimate potential, determine goals, assess tactical options, plan, evaluate, and review.

Isolation Area An area that is set up by responders and is intended to keep people, both citizens and responders, out. May later become the hot zone/sector as the incident evolves. Is the minimum area that should be established at any chemical spill.

Mass Decontamination A process used to decontaminate large numbers of contaminated victims.

Overpacked A response action that involves the placing of a leaking drum (or container) into another drum. There are drums made specifically to be used as overpack drums in that they are oversized to handle a normal size drum.

Psychological Decontamination The process performed when persons who have been involved in a situation think they have been contaminated and want to be decontaminated. Responders who have identified that the persons have not been contaminated should still consider what can be done to make them feel better.

Sector   An area established and identified for a specific reason, typically because a hazard exists within the sector. The sectors are usually referred to as hot, warm, and cold sectors and provide an indication of the expected hazard in each sector. Sometimes referred to as a zone.

Shelter in Place   A form of isolation that provides a level of protection while leaving people in place, usually in their homes. People are usually sheltered in place when they may be placed in further danger by an evacuation.

Sick Building   A term that is associated with indoor air quality. A building that has an air quality problem is referred to as a sick building. In a sick building, occupants become ill as a result of chemicals in and around the building.

Sick Building Chemical   When a building is referred to as a sick building, certain chemicals exist within that cause health problems for the occupants. These chemicals are referred to as sick building chemicals.

Technical Decontamination   The process used to perform decontamination of persons or equipment.

Zone   An area established and identified for a specific reason, typically because a hazard exists within the zone. The zones are usually referred to as hot, warm, and cold zones and provide an indication of the expected hazard in each zone. Sometimes referred to as a sector.

REVIEW QUESTIONS

1. What is the first priority of any incident command system?

2. What zone or sector should be established first?

3. When dealing with a flammable liquid spill, what could be used to establish the hot zone?

4. What level of incident would be used to describe a chemical release in which 2,000 people were evacuated?

5. What is the minimum amount of decontamination that should be set up for every incident?

6. What are the four types of decontamination?

7. Which decon step are hospitals involved in?

8. The decision to attempt a rescue in a hazardous materials situation depends on what information?

9. What causes the most common breach in a 55-gallon drum?

10. If fire has reached the cargo portion of a truck carrying DOT Class 1.1 materials, what tactics are used to fight the fire?

11. How can air monitors be used in the determination of hazard zones?

ENDNOTES

1. These top ten (which are in random order) are derived from the EPA, OSHA, and DOT by examining the chemical release records. Because each agency tracks releases in a different fashion, we calculated each agency's top ten and determined an average, the result of which is the list given.

2. In most states the fire department IC is in charge of the incident until command is relinquished to another agency. Check local and state regulations for more information. Although most fire departments will leave when the cleanup contractor begins to work, a problem exists if the contractor's crew is working in a hazardous environment. As long as they are, the FD should maintain a presence because the FD is solely responsible for public safety, a responsibility that cannot be delegated.

3. Any location such as a storm drain, ditch, or culvert that eventually leads to a waterway is defined as a waterway.

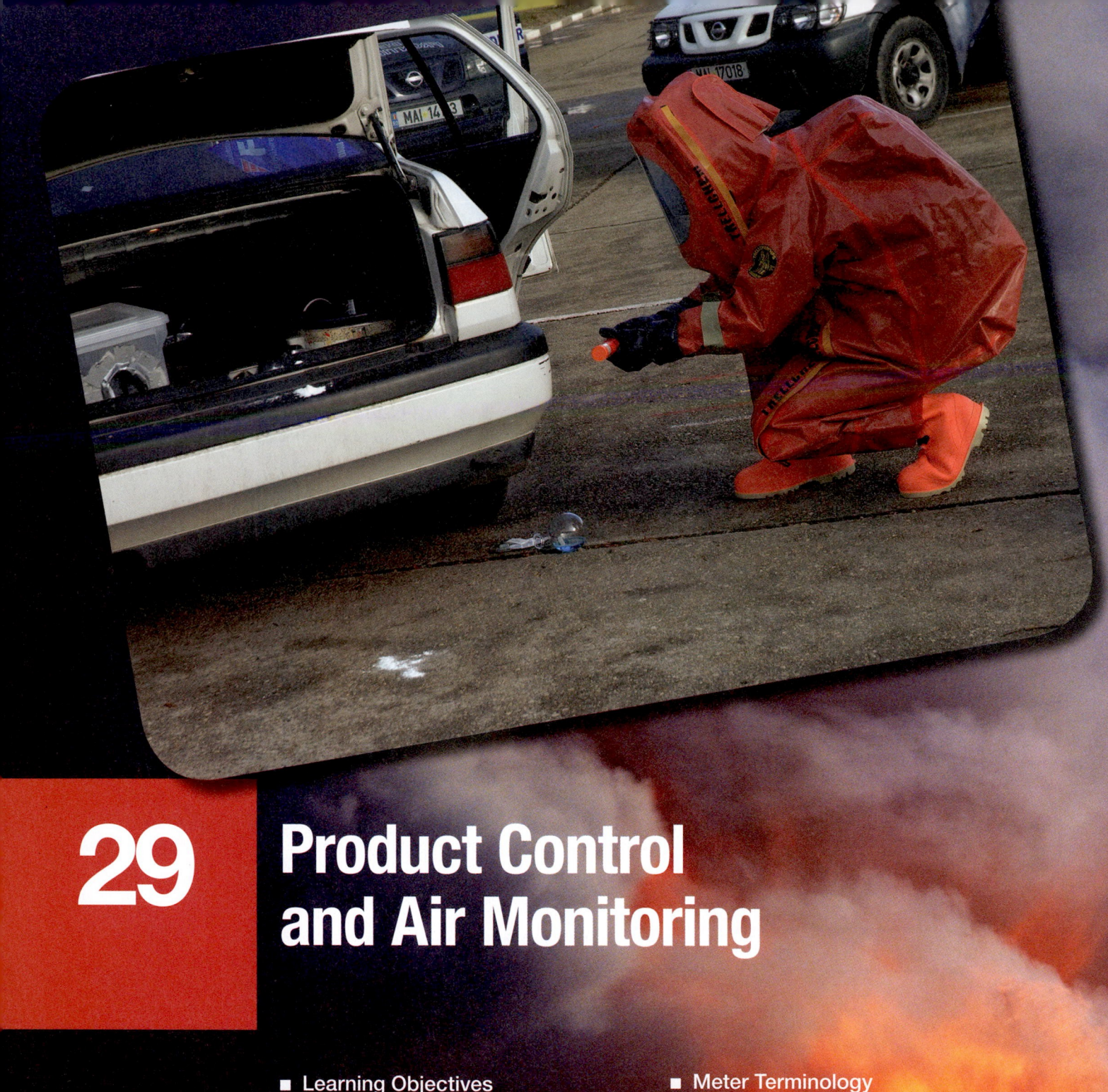

29

Product Control and Air Monitoring

On a weekday morning, our hazardous materials team was called to assist an Engine Company in the Clear Lake/NASA area of Houston, Texas. On arrival, the engine company officer advised us that a car had driven through a fence into the backyard of a one-story single-family residence and pushed a gas meter into the house. The vehicle was parked on top of the meter halfway inside the house. He did not want his crew to make entry into the house because they did not have any type of monitoring instrument to check for the level of gas in the house.

Because of the time of day and occupancy, we decided that it was necessary to do a quick primary search of the house to check for anyone stuck inside. Two of us suited up in structural firefighter protective clothing with SCBA and combustible gas indicators (CGI). The engine company was staged down the street due to the potential for explosion.

As we entered the house through a broken-down door near the car, my CGI went into alarm at 10 percent of the LEL and then into "over range" (OR). I was holding the monitor at arms length as high into the ceiling as I could reach because natural gas is lighter than air. As I looked at the monitor just past the end of my arm I saw a ceiling fan running in the high vaulted ceiling. I quickly processed the information and concluded that we were in an environment that was well above the upper explosive limit (UEL). Because we were already in the house, we did a quick primary search to ensure that no one was in the house, then quickly exited through the front door, which we left open for ventilation.

There was no one in the house. We had the gas shut off and it was safely ventilated.

It is important to carry monitoring equipment to aid in your decision making process with this very common type of call. Sometimes, firefighters have to make difficult and rapid decisions based on risk vs. benefit when dealing with natural gas emergencies. This was a very dangerous "common call" that ended with a positive outcome.

—*Street Story by Bill Hand, Houston Fire Department Hazardous Materials Response Team, Houston, Texas*

LEARNING OBJECTIVES

After completing this chapter, the reader should be able to:

29-1   Identify equipment that can determine hazardous environments and isolation areas.

29-2   Describe available defensive operations for a release.

29-3   Describe the various methods of damming, diking, diverting, and other defensive operations.

29-4   Explain what types of detection equipment are available to the first responder and equipment that a hazardous materials team might carry that could be used at a chemical release.

*The FF I and II levels, as defined by the NFPA 1001 Standards, are identified in different colors:
FF I = black, FF II = red, additional information = blue.

INTRODUCTION

The use of product control techniques can provide a quick reduction in the damage that is done to a community and the environment in the event of a spill. The reduction of the surface area over which a product has spread provides a direct reduction in the danger to responders and the community. By implementing these types of measures, first responders can provide an extreme advantage to the overall mitigation of the incident. The use of air monitoring devices is becoming more commonplace, and many first responders have some type of air monitoring device available to them. This chapter provides a brief overview of this complex subject and concentrates on the knowledge that the first responder should have. To ensure firefighter safety, basic air monitoring must be accomplished, even for basic firefighting, natural gas leaks, gasoline spills, or any other incident in which firefighters may be exposed to hazardous materials.

DEFENSIVE OPERATIONS

According to the NFPA, three large tasks fit into Operations-level tactical response: basic air monitoring, containment, and confinement. All of these can be combined into the task of product control, which is the one task that persons trained to the Operations level can perform. The tasks of containment and confinement should be performed away from the hazard area and to the smallest, safest area. To perform defensive operations, the responders do not really get close to the actual spill itself. The activities that fit into product control operations are absorption, diking, damming, diverting, retention, dilution, vapor dispersion, vapor suppression, and the use of remote shutoffs. The type and level of PPE required for each of these tasks will vary with the chemical involved, but

a minimum would be Firefighting PPE and SCBA. If making a dam a considerable distance away from the chemical release—and no contact with the material is possible—then a lesser level could be used. The tasks discussed next are common tasks assigned to responders trained to the Operations level, and they are critical tasks that can protect the community and the environment. Although it is not necessary for first responders to carry all of this type of equipment, they should know its location in the event it is needed or know how to improvise, adapt, and overcome if the equipment is not available.

Absorption

First responders are often asked to clean up a spilled material, and on a daily basis, they probably use the **absorption** technique on a large quantity of fuel and oil products using a product similar to the one shown in **Figure 29-1.** First responders should know the type of absorbent materials carried on their apparatus. Some absorbent materials will not pick up water, while others absorb any liquid with which they come in contact. When trying to pick up fuels off water, having an absorbent that does not pick up water is important. Absorbents vary from clay kitty litter, ground-up newspaper, and corn husks to synthetic fibers. Their weight and absorbent capabilities vary, and not all can be used interchangeably. They may be available in loose bulk form, as pads, **Figure 29-2,** as large rolls, or in boom style.

To hold the absorbent material, construction of a filter fence such as the one shown in **Figure 29-3** is recommended. The filter fence can be made of "rat wire" or other small-weave fencing material. Wooden stakes can be used to hold the fence in place, unless the soil is rocky, in which case metal stakes are recommended.

FIGURE 29-1 A common defensive action is the application of absorbent to a spilled material.

Also available on the market are solidification agents that encapsulate the spill, as well as microbiological oil-eating bugs. Both of these agents are expensive and although sold as absorption materials, are not generally used by emergency services due to their cost and the quantity required. It is best to carry a variety of absorbent products, as well as several different styles.

Diking and Damming

The quantity of material spilled will determine to what extent **diking** and **damming** operations are performed. The actions required at a 50-gallon spill differ from those of a 500,000-gallon release. Diking, damming, and diverting are techniques to consider using when a spill occurs on the ground or on a waterway. A spill on the ground is much easier to control than one on the water, but the principles are the same. When creating a dike or a dam, the first responder is either stopping the flow of the material or keeping it from a specific area. Except for small streams, it is nearly impossible to stop a body of water from flowing. If the spilled material is soluble with water, and the waterway has a good flow, it will be impossible to collect the spilled material, unless heavy equipment and a large area are available that could be used to create a very large dam.

First responders often use earth, sand, or rocks for dikes or dams depending on what is available. Having local contacts available around the clock that can supply these items is important, and some jurisdictions have emergency contacts for dump truck loads of sand or dirt. Local or state highway departments are a good place to start when these items are required. Heavy equipment such as front-end loaders or track hoes can also be used to create dams or dikes. When constructing a dam or a dike, it is best to construct three setups. Responders should start at the farthest point away from the spill and work back toward the spill. It may be necessary to establish the first containment miles away from the spill. If the barrier is started near the spill, responders may be overtaken by the spill and be forced to play catch-up, not to mention the need for additional PPE.

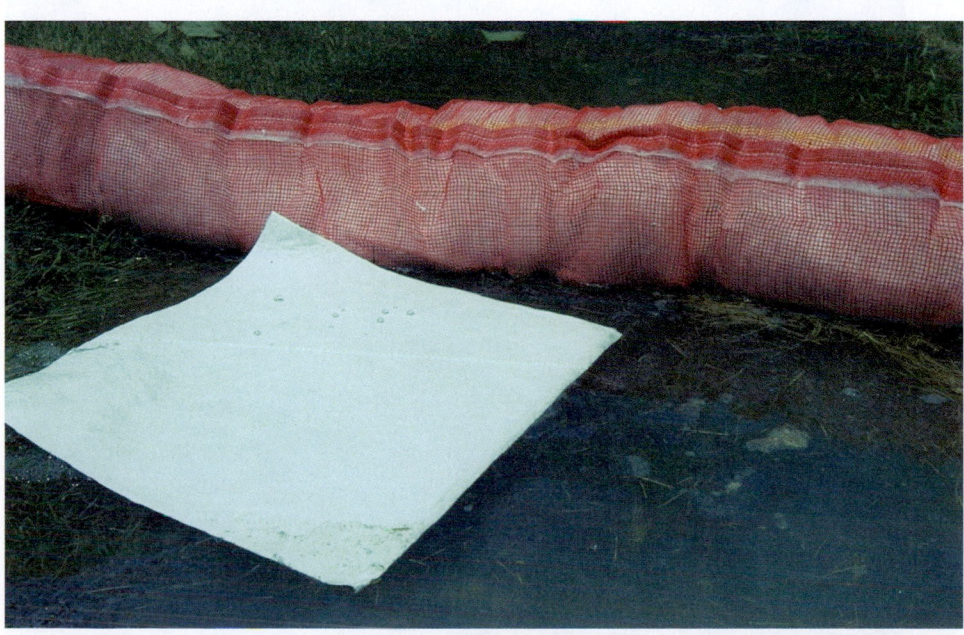

FIGURE 29-2 Absorbent pads will pick up oil from the water, but will not absorb the water, allowing them to float.

FIGURE 29-3 A screen fence can be used to collect the absorbent material floating on the water after the absorbent has collected the spilled material. The water flows through the fence below the absorbent.

Two basic types of dams can be created: overflow and underflow dams, **Figure 29-4.** The specific gravity of the spilled material dictates whether the material will float on top of the water or travel on the bottom, which determines what type of dam needs to be constructed. An overflow dam allows the water to flow over the dam and contain the spilled material at the base of the dam. An underflow dam allows the water to flow under the dam and collect the material on top of the water. Most spilled materials float on top of the water and hence require the use of an underflow dam.

The materials required to construct either type of dam are shovels, dirt or sand, and pipes. Pipes should be at least 4 inches in diameter, but larger is better. The higher the flow, the bigger the pipe and the greater the number of pipes required. A good rule of thumb to use is to have enough pipes to cover two-thirds of the width of the waterway, which would be the minimum amount required. Standard PVC pipe is adequate and, for emergency situations, the hard sleeves (suction) on the engine company are a good substitute. **Figure 29-5, Figure 29-6,** and **Figure 29-7** show both styles of dams and the use of hard sleeves, respectively. For an overflow dam, first responders tend to establish a dam, place the pipes on top, and consider the dam finished. It is best, however, to set the pipes on top of the dam, and then continue with a layer of dirt or sand on top of the pipes, because there is a tendency for the water to erode the dam, and the containment will collapse without this extra layer of dirt or sand.

Diverting

Almost anything can be used, such as dirt, sand, absorbent material, tarps, and hoselines, to divert a running spill, **Figure 29-8.** The most common use

FIGURE 29-4 The overflow dam allows the clean water to flow over the dam and collects the spilled material at the base of the dam. The underflow dam collects the spilled material on top and allows the water to flow through the bottom of the dam.

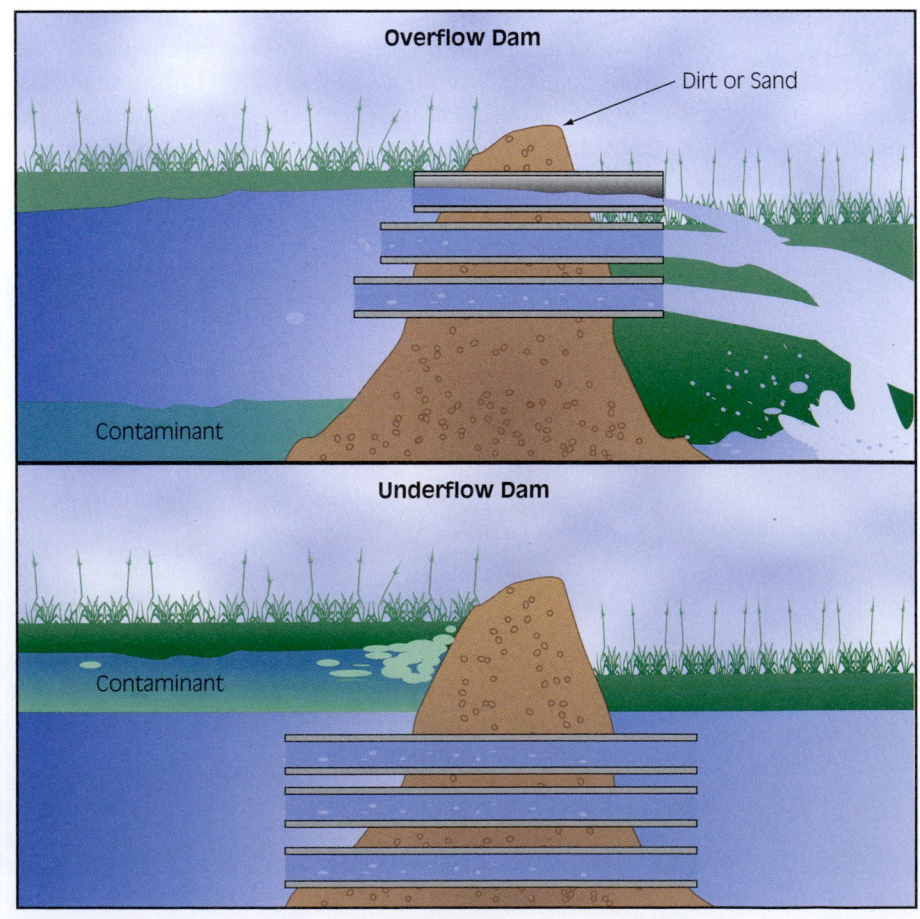

FIGURE 29-5 Overflow dam.

FIGURE 29-6 Underflow dam.

FIGURE 29-7 First responders have equipment such as hard sleeves to establish overflow and underflow dams.

FIGURE 29-8 If a spill cannot be controlled, it may be better to divert the spill around items such as sewers and allow the spill to continue down the street.

of the **diverting** technique is to keep a running spill from entering a storm drain. A ring of dirt around the storm drain will keep the spilled material out, but does not stop the flow of the product. There are times when, due to the large quantity of spilled material, stopping the flow would be impossible, but it would be possible to keep it from the storm drain.

The diversion tactic can also be used for spills on water. By using a solid boom, such as a harbor boom, **Figure 29-9,** material can be diverted into another area or even contained. Another method of diversion is to dig a trench alongside the waterway to collect the spilled material off the top of the water. This is effective for large spills if there is a collection vehicle such as a vacuum truck standing by to collect the spilled material as it runs into the diversion. These types of vehicles are available from hazardous waste cleanup contractors. The most common method is to use a floating absorbent boom that absorbs the material off the water as shown in **Figure 29-10.**

Retention

The most common method of **retention** is the digging of a hole, as shown in **Figure 29-11,** either by hand or by machine, such as with a track hoe. To catch a running spill, the best method is to dig a hole large enough to collect the spill and then divert the spill into it.

Having the ability to create a large enough containment area is paramount to success. If unable to build a large enough containment area, then first responders should concentrate on building several in a row to collect the material. As with dikes and dams, it is best to start construction of the containment area farthest from the spill and then create several other areas closer to the spill.

> **STREETSMART TIP**
>
> When digging retention areas, it is important to be aware of other underground utilities such as water, electric, sewer, cable TV, phone, or natural gas lines.

Dilution

Much like **dilution** when conducting decontamination processes, this technique is not always the solution to pollution. If a water-soluble material is in a waterway, then for all intents and purposes it is being diluted. If the waterway is small, the fire department

FIGURE 29-9 Shown is an example of using a harbor boom. The boom should have been angled so that the material being collected would be near a shoreline. In this case, the boom was stretched straight across the waterway causing the material to be collected in the middle.

FIGURE 29-10 Floating boom collects the spilled material that is floating on top of the water. Like the pads, the boom does not absorb water and should stay afloat even when saturated with the spilled material.

FIGURE 29-11 Building a retention area is a good method for holding released products. Retention areas can be dug from existing earth or created with truckloads of sand or dirt. Small holding areas can be hand dug, but for larger spills heavy equipment is required.

may need to add some water to the spilled materials. Simple flushing of a material into a waterway is not an acceptable tactic anymore and should not be considered. Items such as fuels, oils, or other hydrocarbons are not usually water soluble and cannot be diluted by water. For some specific chemicals, dilution is possible only when combined with some of the other containment tactics.

Vapor Dispersion

The topic of **vapor dispersion** can create some confusion for responders, because by their nature firefighters like to use water, usually in large quantities, and on some occasions the application of water will make spill situations safer. For instance, if a severe life threat exists because an adjacent nursing home cannot be evacuated, then using a water spray to disperse vapors would be a good idea. If, on the other hand, the release is occurring in a remote area away from any population, then a water spray to disperse vapors such as the one shown in **Figure 29-12** is not necessary. To be effective, the material that is being released must be water soluble or the vapor cloud must be able to be moved by the water streams. Although the DOT ERG states for many types of spills "use water spray to reduce vapors or divert vapor cloud," water is not recommended unless a life threat scenario exists.

Some departments use water sprays at natural gas leaks, which is not necessary and further complicates repair of the leak. Many of the leaks occur in a hole that was created by a backhoe. Water can fill the hole and make any attempted repair very dangerous and complicated. Also the use of water may actually knock down vapors that normally rise up and dissipate quickly.

FIGURE 29-12 The use of a water spray is not necessary unless the vapor cloud is impacting the public. In some cases it creates more hazards than it eliminates.

Use of a water spray can also result in the creation of another substance. When anhydrous ammonia is released into the air, and a water spray is applied, the resulting runoff is ammonium hydroxide, a very corrosive liquid that itself would require containment and cleanup. Both natural gas and propane are not water soluble and the use of water vapor will just relocate the vapors to another area.

> **NOTE**
>
> It is important to know what material is leaking prior to the application of water, because the water spray may cause more problems. It is very important to consult with the hazardous materials team prior to using this tactic.

Vapor Suppression

With the use of firefighting foams, vapor suppression is another tactic that first responders can use. The type of material spilled dictates the type of foam to be used, because not all foams will work on all products. See Section II, Chapter 11 for more information regarding foam and its application. The use of foam to suppress vapors is generally limited to flammable liquid spills, but some chemical foams are available for other types of materials.

Before the application of foam the responders should ensure that the material is contained, and that the application of foam will not cause any further problems and is compatible with the spilled material. As when using foam to extinguish a fire, responders need to make sure they have enough foam stockpiled to keep a layer of foam on the spilled material.

Remote Shutoffs

The shutting of valves is not usually a tactic used by first responders, but there are some exceptions to this rule. On most tank trucks and at some fixed facilities, there are **remote shutoffs** that could be operated by first responders. Each type of truck is different but in general an emergency shutoff is located behind the driver's side of the cab and another near the control valves. The shutoffs are both mechanically operated and in most cases are self-operating in the event of a fire. They are usually well marked as emergency shutoffs and in an easy-to-find location. The two most common locations are behind the driver's door at the tank or at the valve controls. At some facilities, typically those with loading racks, a remote shutoff may be located near the entrance so that in the event of an emergency the first responders can shut off the flow of product.

JPR 29-1: Implementing Defensive Operations

(For step-by-step photos of this skill sequence, see page 1194)

There are a number of considerations for the first responder that can help determine the overall outcome of the incident. The photos at the end of the chapter walk you through the necessary steps for developing effective defensive operations for various types of incident as a first-arriving fire department responder.

1. Scenario 1—The *visual clues* to potential hazards include:

   ■ There has been a traffic accident.

   ■ A double tank truck with the rear tank overturned is involved.

   ■ From this vantage point, there are no placards visible.

   ■ Although difficult to see, the dome on the rear tank is open, so if the tank was full, the product has been released.

Some *response priorities* include (order is dictated by on-scene conditions):

   ■ A high priority is to determine what the contents of the tank are (or were).

   ■ Update the other response units, such as the hazardous materials team, and obtain guidance from them.

   ■ Determine what type of PPE would be required.

   ■ Determine the status of the driver and determine the contents of the tank.

   ■ Determine where the police officer is and the officer's status.

   ■ Isolate the scene to remove bystanders and to keep new bystanders from becoming part of the problem.

   ■ Perform decontamination on any victims or responders who may be contaminated.

   ■ Determine where the contents of the tank went and if any control measures can be implemented.

   ■ If appropriate, use your air monitors to determine the presence of hazardous gases.

2. Scenario 2—The *visual clues* to potential hazards include:

   ■ There has been a traffic accident with a number of vehicles involved.

   ■ A tank truck is involved.

   ■ There is a Class 2 non-flammable gas placard with "2187" in the middle indicating a bulk shipment. The product name "Carbon Dioxide Refrigerated Liquid" is stenciled below the placard.

   ■ Although the piping was involved in the accident, from this vantage point there does not appear to be a release.

   ■ There is some damage to the outer shell of the insulated tank. Think about the incident priorities if the occupants of this vehicle were trapped.

Some *response priorities* include (order is dictated by on-scene conditions):

   ■ A high priority is to determine what the hazards of carbon dioxide refrigerated liquid would be, and what impact this would have on any potential rescue situations.

   ■ Update the other response units, such as the hazardous materials team, and obtain guidance from them.

   ■ Determine what type of PPE would be required.

   ■ Determine the status of the other drivers and occupants.

   ■ Determine where the highway crew is and their status.

   ■ Isolate the scene to remove bystanders and keep new bystanders from becoming part of the problem.

   ■ If appropriate, use your air monitors to determine the oxygen content. Carbon dioxide is a colorless and odorless gas that is a simple asphyxiant.

3. Scenario 3—The *visual clues* to potential hazards include:

   ■ There is a visible vapor cloud coming from the valve box on the bottom of the railcar. The box seems to be very cold.

   ■ From this vantage point there are no placards visible, but the rail car is placarded with a Class 2 "Flammable Gas" placard. The material is ethylene and liquefied flammable gas.

   ■ Although it is not known where the release is coming from, all of the transfer piping and valving is inside the box.

Some *response priorities* include (order is dictated by on-scene conditions):

   ■ A high priority is to confirm the contents of the tank.

   ■ Update the other response units, such as the hazardous materials team, and obtain guidance from them.

   ■ Determine the status of the train crew and determine the contents of the tank.

- Isolate the scene to remove bystanders and keep new bystanders from becoming part of the problem.

- Determine what downwind exposures may be at risk.

- Determine where the vapors will go and determine if any control measures can be implemented, such as water spray. Since ethylene is a flammable gas and not water soluble, this tactic would have minimal impact.

- If appropriate, use your air monitors to determine the presence and location of flammable and hazardous gases. Ethylene is one of the few gases that will rise, and it has a vapor density of 0.98, which is very close to the vapor density of air. What would change if the humidity was 85 percent or higher?

4. Scenario 4—The *visual clues* to potential hazards include:

- There is a fire involving a tank truck and some sort of traffic accident.

- From this vantage point, there are no placards visible, but there are Class 3 "Flammable Liquid" placards on the truck.

To fight the fire or not to fight the fire?

- If the truck contained gasoline, overall it would be more hazardous, but the decision to fight the fire or retreat is the same for both gasoline or diesel; the difference between the gasoline and diesel fuel are minimal. Once either is burning, comparable resources are required to extinguish the fire.

- Do you have enough resources to extinguish the fire?
  - □ If the cargo were involved, foam would be required.
  - □ If it is gasoline, do you have the appropriate foam for ethanol gasoline?
  - □ Do you have enough foam?
  - □ Can you maintain a foam blanket every five to ten minutes?

- Are there exposures or persons at risk?

- What if you did nothing?

- What is the impact on the community?

Some *response priorities* include (order is dictated by on-scene conditions):

- A high priority is to determine what the contents of the tank are (or were). Given the placarding and markings on this truck, the most likely products are gasoline or diesel fuel.

- Update the other response units, such as the hazardous materials team, and obtain guidance from them.

- Determine what type of PPE would be required.

- Determine the status of the driver and determine the contents of the tank.

- Isolate the scene to remove bystanders and keep new bystanders from becoming part of the problem.

- Perform decontamination on any victims or responders who may be contaminated.

- Determine where the contents of the tank went and if any control measures can be implemented.
  - □ The truck is sitting on a downhill incline.
  - □ There are storm drains on the downhill side.
  - □ The storm drains feed to a creek.

- If appropriate, use your air monitors to determine the presence of hazardous gases.

AIR MONITORING AT THE FIRST RESPONDER OPERATIONS LEVEL

The use of air monitoring is one of the most important tasks to accomplish when responding to chemical incidents. Air monitoring can keep responders alive. The NFPA 472 Committee recognized this and incorporated a number of air monitoring competencies at the Operations level. The new objectives are added as Mission Specific Competencies. These are additional competencies that are optional. If firefighters are going to use air monitors at the Operations level, then the mission-specific competencies should be used for training. With many departments purchasing detectors to assist with carbon monoxide alarms, first responders are experiencing associated benefits. Depending on the type of alarm purchased, it may also be used to detect flammable gases, oxygen levels, and one or two toxic gases, **Figure 29-13**. Many departments use these instruments for flammable gas leaks and for confined space entries.

When responding to situations that involve unknown materials, hazardous materials responders need pH detection for corrosives, a combustible gas detector for the fire risks, a **photoionization detector (PID)** for the toxic risks, and a radiation monitor for radiation risks. Unfortunately, many responders only rely on one or two of these detectors, and the most common detector is a three-, four-, or five-gas instru-

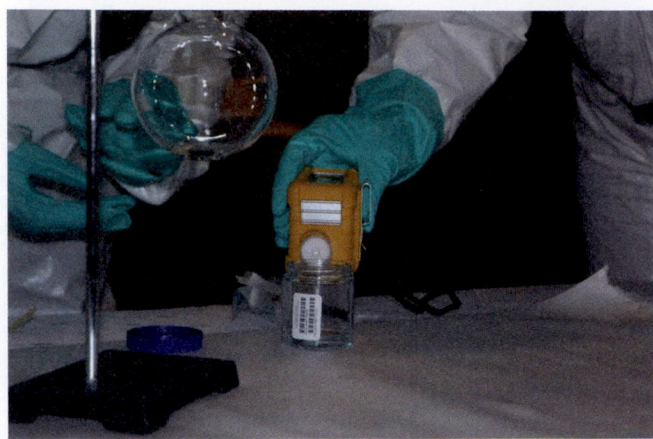

FIGURE 29-13 First responders commonly use devices that detect the presence of flammable gases, carbon monoxide, and hydrogen sulfide and determine the oxygen content. This detector also detects the presence of common toxic vapors.

ment. Other instruments are not available or not used. By not using a method of detecting toxic materials, responders (and the public) can be lulled into a false sense of security. Combustible gas detectors are not made or designed to measure toxic gases in air; they only read those types of gases when they become fire hazards. Many flammable gases are very toxic at considerably lower levels than their lower explosive limit. Many first responders have purchased air monitors, but may not have a full understanding of how they work.

NOTE

The use of air monitoring and sampling equipment can be complicated and requires practice.

Most fire service responders have a reasonable understanding of combustible gas indicators but even that knowledge can be limited. Society is becoming much more sophisticated and the citizens (the fire service's customers) expect more. This section presents some concepts on air monitoring, monitoring strategies, and information on how the monitors work and their uses. The field of air monitoring technology is ever changing and new technology emerges each year, although the basic principles of safe decision making remain the same. When purchasing air monitors it is important to understand the basic features of the instruments so that the purchase benefits the organization making the purchase.

NOTE

The general rule when purchasing an air monitor is to figure one-quarter of the purchase price into the department's annual budget for upkeep and repairs. Even though the purchaser may have been told differently, it is necessary to keep the instrument calibrated and maintained on a regular basis.

Regulations and Standards

The Occupational Safety and Health Administration (OSHA) HAZWOPER regulation (29 CFR 1910.120) does not provide a lot of specific requirements for air monitoring, even for the hazardous materials technician, but air monitoring is the principal safety element throughout the document. OSHA wants the incident commander (IC) to identify and classify the hazards that are present on a site. The use of air monitoring is the primary key to fulfilling this obligation, but in order to fulfill this obligation a hazardous materials technician must be present along with other detection devices.

The NFPA 472 document also has some requirements for air monitoring but, like the OSHA regulation, they are fairly generic. The 2008 edition of NFPA 472 includes some new Mission Specific Competencies that provide training objectives for the use of air monitors. The Technician level was also revised with some additional WMD monitoring competencies. There always have been air monitoring competencies at the Technician level, but as technology has improved, NFPA 472 needed to address these new technologies. The WMD specific monitoring competencies were added as mission-specific much like the air monitoring competencies at the Operations level.

It is important for first responders to understand how monitors work and what their deficiencies are because they may be placing their lives on the line depending on the readings received on an air monitor. First responders are more affected by the confined space regulation with regard to air monitors than they are by the hazardous materials regulation.

It is unfortunate, but it can be predicted that a lack of understanding and/or maintenance will play a factor in future firefighter fatalities and injuries based on the large number of departments using these instruments.

Air Monitor Configurations

Most departments purchase an instrument that is known as a three-, four-, or five-gas instrument, commonly referred to as a **multi-gas detector,** such as the one shown in **Figure 29-14.** In other words it samples

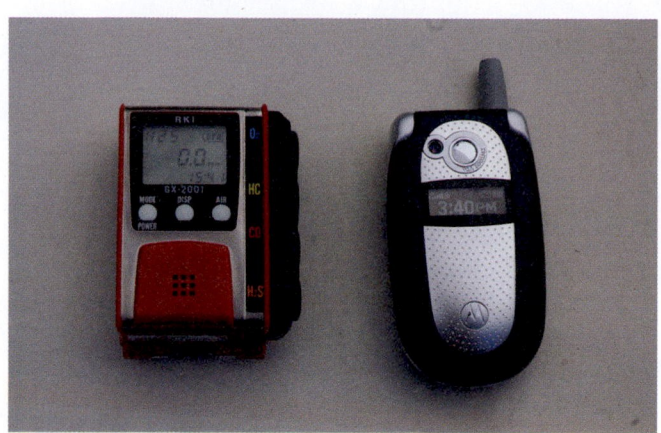

FIGURE 29-14   The miniature detector on the left detects flammable gases, carbon monoxide, hydrogen sulfide, and oxygen content. These types of instruments require calibration and regular maintenance.

for the presence of three to five different gases. Normal units sample for oxygen levels, a lower explosive limit (LEL), and two or three toxic gases. The normal "toxics" are carbon monoxide (CO) and hydrogen sulfide (H_2S). If a third gas can be sampled for, most departments choose chlorine or sulfur dioxide. For most departments, though, sampling for CO and H_2S is more than sufficient.

Although these instruments are considered expensive, in the grand scheme of things they are inexpensive compared to other detection devices used by a hazardous materials team. Many departments may purchase a device with little or no training, which can be detrimental to the successful use of the instrument. When buying a detector it is essential to purchase several maintenance items designed to keep the instrument functioning. Typically these items are included in a special kit price and include a battery-powered sample pump, calibration gas, and hardware. The calibration gas usually has an expiration date and will need to be replaced in six months to a year. The sensors in the detector need regular calibration, and the instrument should be turned on and used on a regular basis for a period specified by the manufacturer. Depending on the charging system, rechargeable batteries may need to be replaced on occasion.

The sensors also need replacement. Although most manufacturers provide two-year warranties for the sensors, in general, the LEL sensor will last four to five years, the oxygen sensor eighteen to twenty-four months, and the CO and H_2S sensors eighteen to twenty-four months. Other toxic sensors may last six to eighteen months depending on the type.

Most of the perceived problems with instruments are created by a lack of maintenance and adequate training. Regardless of what a salesperson says, no device

exists that can conclusively identify an unknown material for a first responder.

METER TERMINOLOGY

To comprehend fully the use of air monitors, the responder must understand the basic terminology that is generic to all monitors and applies across the board. More detailed information related to the specific monitors is provided in later sections.

Bump Test

Two terms that need further explanation are bump test and calibration. A **bump test,** also known as a field test, exposes a monitor to known gases, allowing the monitor to go into alarm mode, and then removes the gas. By exposing the monitor to a known quantity of gas, a person can verify the monitor's response to that gas. Most manufacturers provide bump gas cylinders, which contain the gases required to check their instrument. When bump testing, firefighters should follow the manufacturer's recommendations.

In most cases, the bump test is used to ensure that the alarms function as intended and the instrument is reading. By regularly calibrating and bump testing, the user can determine how accurate the bump test will be in the field. Some instruments react very well to bump testing and will display the levels as provided on the bump gas cylinder, while others will be off slightly. When responding to confined spaces, responders are required to bump test the instrument prior to entry into the confined space.

Calibration

Calibration is used to determine if a monitor responds accurately to exposure to a known quantity of gas, **Figure 29-15.** When new sensors are installed, they will usually read higher than intended; calibration electronically changes the sensor to read the intended value. As the sensor gets older, it becomes less sensitive and calibration electronically raises the value that the sensor displays. When a sensor fails, it cannot be electronically brought up to the correct value. Regular calibration gives a picture of the expected life of a sensor, which will deteriorate over time.

The regularity of calibration is subject to great debate. Some departments calibrate daily, others every six months. The only item found in the regulations (for anything that requires air monitoring) is in the confined space regulation, which requires calibration according to the manufacturer's recommendations. Most of the written instruction guides from the manufacturers require calibration before each

FIGURE 29-15 Calibration involves exposing the instrument on the right to a known quantity of gases to ensure that it is reading the gases correctly. This kit calibrates the monitor with a variety of gases.

use. The International Safety Equipment Association (ISEA), an industrial association, has lobbied OSHA to add regulations governing calibration. The ISEA recommends bump testing prior to use and, depending on the conditions where the device may be used, recommends calibration every thirty days. The definition of a calibration at this point is also subject to debate, because the department can verify a monitor's accuracy by exposing it to a known quantity of gas, but not perform a "full" calibration. The ISEA calls this type of test a functional test. Most response teams establish a regular schedule of calibration (weekly/monthly) and then perform "bump" or "field" checks during an emergency response. It is essential to check with manufacturers as to what calibration/bump test policy they recommend because they are all different.

Reaction Time

All monitors have a lag time or, as it is better known, reaction time. The use of a pump or not will vary the reaction time. Monitors operating without a pump are in what is known as the diffusion mode and will generally have a fifteen- to thirty-second lag time. Monitors operating with a pump have a typical reaction time of seven to thirty seconds, depending on the sensor type. Some instruments, primarily WMD detection devices, may take up to ninety seconds to react. Hand-aspirated pumps usually require ten to fifteen pumps to draw in an appropriate sample. Hand-aspirated pumps are not recommended, because the readings they provide will be too varied depending on how well the user operates the hand pump. The goal is to provide a given amount of volume across the sensors. It is important to follow the manufacturer's recommended lengths of hose to ensure that the pump operates correctly.

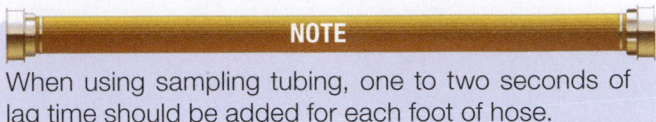

NOTE

When using sampling tubing, one to two seconds of lag time should be added for each foot of hose.

Recovery Time

Monitors also have a recovery time. Recovery time is the amount of time it takes the monitor to clear itself of the air sample. This time is affected by the chemical and physical properties of the sample, the amount of sampling hose, and the amount absorbed by the monitor. Some monitors take an extended period of time to clear if they are exposed to a large quantity of a gas. In some cases the instrument must be taken out of the environment and shut off, then started again. The reaction time will affect the overall recovery time.

Relative Response

When a gas monitor is purchased, it is set to read a specific type of gas, such as methane. If any other gas is sampled, the instrument will usually read the gas but at different values. Flammable gases have varied lower explosive limits (LELs) and therefore react differently with the flammable gas detector. The term relative response is used to describe the way the monitor reacts to a gas other than the one it was calibrated for, and is a term that is not commonly used by emergency responders.

To maintain a high level of safety, each person operating an air monitor must have a basic understanding of relative response so as not to let the monitor lead them into dangerous situations. Each detector has what is known as a relative response curve that compensates for different types of gases, **Table 29-1.** The monitor's manufacturer has tested the monitor against other gases and has provided a factor, commonly known as a correction factor, that is a relative response factor that can be used to determine the amount of gas present when sampling. The user multiplies the displayed reading by the factor to arrive at the reading for the gas that is actually present. For instance, consider a responder at a xylene spill who is using an ISC TMX-412 calibrated for pentane. The detector is reading 50 percent of the LEL. According to Table 29-1, the response curve factor for xylene is 1.3, which is multiplied by the LEL reading:

$$\text{Detector reading} \times \text{Response curve factor} =$$
$$\text{Actual LEL reading}$$

$$50 \times 1.3 = 65$$

So the actual LEL is 65 percent, a number higher than what the instrument was providing. On the other

TABLE 29-1   Correction Factors (Calibrated to Pentane)[a]

Gas Being Sampled	ISC TMX 412 Factors[b]	MSA 261 Factors[c]	RAE System Factors[d]	Molecular Weight
Methane	0.5	0.6	1	16.04
Propane	0.8	0.9	1.88	44.1
Ethanol	0.8	N/A	1.69	46.1
Pentane	1	1	1	72.2
Benzene	1	1.1	2.51	78.1
Toluene	1.1	1.2	2.47	92.1

[a]Always use the correction factors suplied by the manufacturer, keeping in mind these are laboratory estimations.

[b]Factors for Industrial Scientific Corporation LEL Sensor 1704A1856-200 calibrated with pentane.

[c]Factors for the Mine Safety Appliances MSA 360/361 calibrated with pentane.

[d]Factors are for the RAE sytem catalytic bead LEL sensor calibrated with pentane.

hand, if the responder used the same instrument for a spill of propane, the response curve factor is 0.8. Using the same scenario—that the instrument was reading 50 percent of the LEL—the actual reading is 40 percent (50 × 0.8 = 40), a safer situation than reported by the detector.

Oxygen Monitors

Oxygen is one of the most important things to sample for. Humans need it to survive, and the other instruments need it to function correctly. Normal air contains 20.9 percent oxygen; below 19.5 percent is considered oxygen deficient and is considered to be a health risk, and above 23.5 percent is considered a fire risk. If an oxygen drop is noted on the monitor, one or possibly more than one contaminants are present, causing the reduced oxygen levels, and another material (i.e., toxic, flammable, corrosive, or inert) is causing the oxygen-deficient atmosphere. When in oxygen-deficient atmospheres, any combustible gas readings will also be deficient and cannot be relied on.

If in an oxygen-enriched atmosphere, then the combustible gas readings will be increased and not accurate. Many oxygen-enriched atmospheres result from a chemical reaction involving oxidizers and typically present a dangerous situation. Monitoring for oxygen is important and the alarm values of 19.5 percent and 23.5 percent are set by OSHA regulation, but the reality is that responders can be in danger prior to these values. As an example, an oxygen drop to 20.8 percent can be a very dangerous situation depending on the contaminant. A drop to 19.9 percent would not result in an alarm, but considerable contaminants may be present in the air at this value nevertheless. Figure 29-16 provides additional information regarding this topic.

Oxygen Monitor Limitations

There are three types of oxygen sensors: two versions of lead wool technology and a solid polymer electrolyte (SPE) sensor. In all of the sensors, there is a chemical reaction and the calibrated sensors are able to determine the level of oxygen in the air. As long as the sensor is exposed to O_2, it will cause a reaction within the sensor. This is the most commonly replaced sensor and usually needs to be replaced every eighteen months, although some may last longer. Chemicals that hurt O_2 sensors are those with lots of oxygen in their molecular structure such as carbon dioxide (CO_2) and strong oxidizing materials such as chlorine and ozone. The problem with CO_2 is that it is always present in the air, and the higher the percentage the faster the sensor will deteriorate.

The optimal temperature for operation is between 32°F (0°C) and 120°F (49°C). Between 0°F (−18°C) and 32°F (0°C) the sensor slows down (electrolyte is like a slushy), and temperatures below 0°F (−18°C) can permanently damage the sensor. Operation depends on absolute atmospheric pressure, and calibration is required at the atmospheric pressure at which the user will be sampling. The sensor should also be calibrated for the temperature and weather conditions for the area being sampled.

Flammable Gas Indicators

NOTE

Flammable gas indicator (FGI) is a new term. The old terminology for these detectors described them as combustible gas indicators (CGIs), which is a misapplied term. These devices only detect the presence of flammable gases, not combustible gases. The term *combustible gases* is derived from their history and development in mining safety. Another term that could be used to describe these units is *LEL sensors,* as they are used to determine the percent of the lower explosive limit (LEL) of the material present. This text uses FGI, but LEL sensor would also apply.

All of these types of sensors work, some better than others in different situations. It is important to understand how each of these sensors works, because budgetary considerations may dictate the purchase of only one type. (Note, however, that to be able to detect a wide variety of chemicals effectively, it may be necessary to have more than one type of sensor.)

Most of the new FGIs are used to measure the LEL of the gas for which they are calibrated. The majority of FGIs are calibrated for methane (natural gas) using pentane gas. When calibrated for methane, the FGI sensor will read up to the LEL, and with some of the new units will in fact shut off the sensor when the atmosphere exceeds the LEL. This is an important consideration, because the longer the sensor is exposed to an atmosphere above the LEL, the quicker it loses its life.

The FGI reads up to 100 percent of the LEL, so that if an FGI is calibrated for methane and the user is sampling methane, the FGI will read 100 percent, but the actual concentration in that area will be 5 percent for methane (the LEL for methane is 5 percent). If the FGI displays a reading of 50 percent, then the concentration for methane is 2.5 percent. Any flammable gas sample that passes over the sensor will cause a reaction; how much of a reaction depends on the gas. To further complicate this issue, the reading on the FGI can either be higher or lower depending on the gas. If the reading is below the actual percentage the user is safe, but if the actual percentage is above what the FGI is reading then the user may no longer be in a safe atmosphere because the LEL may have been exceeded.

The basic principle of most FGIs is that a stream of sampled air passes through the sensor housing, causing a heat increase and, conversely, creating an electric charge that causes a reading on the instrument. The three combustible gas sensor types are shown in **Figure 29-16.** When purchasing or using a monitor, responders should be aware of the different sensor types. Readings can and do vary among the three, and the safety of responders is in the hands of that instrument so they must understand how it works.

Catalytic Bead

The **catalytic bead** sensor is the most common sensor in use today. A catalytic bead sensor is wire between two poles with a solid bead of metal in the

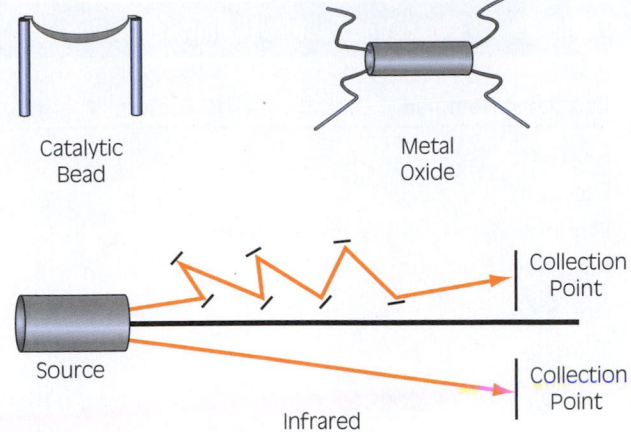

FIGURE 29-16 Examples of flammable gas indicators: catalytic bead, metal oxide, and infrared.

center. There are two wires and beads in the sensor: one burns the gas and the other is used to compare the heat increase or decrease. The beads are coated with a catalytic material that facilitates the burn off of the gas. These sensors replaced the wheatstone bridge sensor, which was a coiled piece of platinum wire between two poles.

Metal Oxide Sensor

The **metal oxide sensor** is commonly referred to as an **MOS** and is sometimes referred to as a tin dioxide sensor. Its very nature makes it a very sensitive sensor, which causes all of its "perceived" problems. If used correctly (and interpreted correctly) this sensor can provide many clues and answers at chemical incidents.

The MOS is a semiconductor in a sealed unit that has a wheatstone bridge sensor in it surrounded by a coating of a metal oxide. Heater coils provide a constant temperature. When the sample gas passes over the heated bridge, it combines with a pocket of oxygen created from the metal oxide. This reaction causes an electrical change, which causes the FGI to provide a reading. This sensor is very sensitive and will pick up almost anything that crosses it, which can be confusing to the user. It requires regular maintenance and calibration, more so than the other types of sensors. Most MOS FGIs do not provide a readout of the percentage; most provide only an audible warning or provide a number from within a range. The range extends from 0 to 50,000 units and is a relative scale. If a tablespoon of baby oil is spilled on a table and an MOS sensor passed over it, the reading will be in the range of 5 to 30. If an MOS sensor is taken into a room with 5 percent methane, the reading will be within the range for a flammable gas, probably near 45,000 to 50,000 units. Some monitors allow the

MOS to read in percentages of the LEL, in addition to a general sensing range of 0 to 50,000 units. The MOS reacts to tiny amounts, which is an outstanding feature of the monitor—most monitors are not nearly as sensitive. Note, however, that if a department can afford only one combustible gas sensor, a catalytic bead sensor is preferred because the MOS can be erratic. The MOS sensor does have great applicability for a hazardous materials team or for a department that already has another type of LEL sensor.

Infrared Sensors

The **infrared sensor** is common in industry but not all that common in the emergency response field. It uses a hot wire to produce a broad range of wavelengths, a filter to obtain the desired wavelength, and a detection device on the other side of the sensor housing. The light that is emitted from the hot wire is split, one wavelength going through the filter, the other to the detection device to be used as a reference source. When a gas is sent into the sample chamber, the gas molecules will absorb some of the infrared light and will not reach the detection device, which will read the amount of light reaching the detector. The amount of light reaching the detection device is compared and a reading is provided. The big advantage of infrared is that it does not require oxygen in which to function, that is, it can take readings in oxygen-deficient atmospheres. The device also is not affected by temperature, nor is it easily poisoned by high exposures. The disadvantages include cost and its many cross sensitivities. This is another sensor that should not be purchased as a stand-alone unit, but should be considered after the purchase of a catalytic bead or wheatstone bridge sensor.

Toxic Gas Monitors

This section describes the sensors that are commonly used in three-, four-, and five-gas units (LEL, O_2, toxic 1, toxic 2, toxic 3) and are usually used to measure carbon monoxide (CO) and hydrogen sulfide (H_2S). Toxic sensors are available for a variety of gases, however, including chlorine, sulfur dioxide, hydrogen chloride, hydrogen cyanide, and nitrogen dioxide. The most common unit sold today by far is a four-gas unit that measures LEL, O_2, CO, and H_2S. This combination is a direct result of the confined space regulation issued by OSHA. Many people understand why the first three sensors were chosen, but do not understand why H_2S was chosen because OSHA did not specify H_2S. Most confined space entries (and consequently the most deaths and injuries) take place in sewers and manholes, and H_2S is commonly found in sewers and anyplace else where things are

rotting. So when choosing one of the optional gases to sample for, it is best to choose H_2S unless the department has a specialized need for some of the other gases.

Many response teams consider choosing one of the other toxic gases mentioned in the preceding paragraph as their fourth or fifth gas, but they must weigh the cost of doing so. More cost-effective methods of detecting toxic gases such as chlorine or sulfur dioxide are available. The average cost for the sensor is $400 to $500 and is usually guaranteed for one year. The calibration gas is $300 to $400 and is only good for six months before it expires. The other problem is that if the instrument is taken into an environment with more than 20 ppm chlorine, it will be ruined.

Most toxic sensors are electrochemical sensors that have electrodes (two or more) and a chemical mixture sealed in a sensor housing. The gases pass over the sensor, causing a chemical reaction within it, and an electrical charge is created, which causes a readout to be displayed. All toxic sensors display in parts per million. Some toxic sensors use metal oxide technology and react in the same fashion as FGI metal oxide sensors.

Tactical Use of Multi-Gas Detection Devices

It is important to understand the types of sensors that are in the multi-gas detection device so that responders can make effective tactical decisions. Among the three flammable gas sensors, there are some notable differences. The metal oxide sensors (MOS) will react very quickly and will detect low levels of flammable gases, whereas the other flammable sensors have a seven to ten second delay in reading the gas levels and a detection threshold that is not as low. The MOS does not report readings that would be considered accurate, and they fluctuate considerably. In high humidity situations, the MOS will detect water vapor and give erroneous readings. MOS sensors offer an advantage in some situations and disadvantages in others. They are useful when determining if there is a contaminant in the air but are not accurate for determining LEL levels. In contrast, the catalytic bead and infrared flammable gas sensors are both accurate for determining LEL levels.

The other consideration when using flammable gas detectors lies in identifying the gas or vapor that is present. Most flammable gas sensors are calibrated for methane and will read only methane accurately. They will detect all other flammable gases, but the readings will not be accurate. Using correction factors will make the devices accurate. For instance, any reading on the flammable gas sensor indicates that there is a flammable gas or vapor present. Most detection devices alarm at 10 percent of the LEL for methane. If the device is calibrated for methane, this reading is accurate only for methane. There are correction factors that exceed a factor of 3, which means your monitor readings are off by a factor of 3. This would not be dangerous if the meter read 10 percent, since the actual reading would then be 30 percent of the LEL. If the meter reading is 40 percent, then multiplied by a factor of 3, the actual reading is 120 percent and well into the flammable range, and in some cases above the upper explosive limit (UEL) for some materials.

When an LEL sensor alarms at the 10 percent level for an unidentified material, first responders should evaluate why they are in a building or in this level of flammable vapor. They should retreat and take steps to reduce the level of flammable gas. They should immediately consult with their local hazardous materials team for additional tactical decisions. Citizens should be removed as rapidly as possible from the immediate area. Detection devices can help determine the size and boundary of the hot zone. If the gas is identified as methane and the meter is set to read methane, the reading of 10 percent is of concern and the source must be identified as rapidly as possible. If the reading at the ground floor door of a three-story apartment building is 10 percent of the LEL, no matter where the source is, the levels on the top floor will be extremely high, and probably well into the flammable range. Any spark could ignite the vapors causing a massive explosion of the building and putting responders at great risk. The electrical company should be immediately contacted to shut off the electricity to the building as rapidly as possible. The electricity should be shut off away from the building, preferably at the closest pole or transformer.

As for the toxic sensors, they are calibrated to their specific gases and are accurate for only those gases. Both CO and H_2S react to some other gases and will give readings for about thirty-five other materials. There is no way for a first responder to determine which gas may be present. Victims may be able to provide some clues; for example, CO is odorless, so if victims or bystanders reported an odor, then CO is not a likely culprit. The exception to this would be if the victims reported a vehicle exhaust odor; carbon monoxide makes up approximately 7 percent of vehicle exhaust. If H_2S is present, victims will report a rotten egg odor; if they do not report this odor, then H_2S is not the culprit. Both CO and H_2S are toxic and any exposure should be avoided. Citizens should be removed from the area if these gases are found. Carbon monoxide is only produced from incomplete combustion, so this source must be located to ensure the occupants' safety.

Other Detectors

In most cases the detectors discussed in the following subsections will be used by a hazardous materials team, but the first responder should be aware of the devices and their capabilities. In areas without a hazardous materials team or where the hazardous materials team response times are unusually long, the first responder may want to consider the purchase of these items. As is true of the monitoring devices already discussed, these instruments are complicated, and this is not by far a complete listing of the instruments that should be carried by a hazardous materials team. First responders are capable of beginning the detection process but in almost all cases further detection capability is required.

Photoionization Detectors

The photoionization detector (PID) is shown in **Figure 29-17.** Because of their ability to detect a wide variety of gases in small amounts, PIDs are becoming essential tools of hazardous materials response teams. The PID does not indicate what materials are present, much the same as the FGI will not identify the specific material that is present, but when used as a general survey instrument, the user can identify potential areas of concern and possible leaks/contamination. These are the instruments that look for toxic materials in the air, and they are essential to determining exposures to many toxic materials. The flammable gas detector will start reading toxic mate-

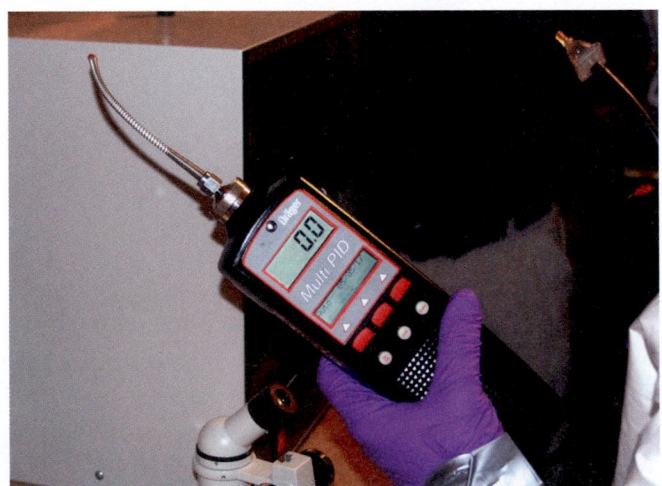

FIGURE 29-17 This instrument is a photoionizing detection device, which has the ability to detect a wide variety of potentially toxic gases or vapors. It doesn't identify the gases, it only indicates their presence. Some instruments combine this detection capability with other sensors for flammable gases, carbon monoxide, hydrogen sulfide, and oxygen.

rials in a range of 50 to 500 ppm, which for some gases is extremely high and could have immediate effects on a responder. The PID starts to read at levels near 0.1 ppm, which is very sensitive. There are PIDs that can measure in parts per billion, a very sensitive device. Because of their sensitive nature, they can detect small amounts of hydrocarbons in the soil. Sick building calls are on the increase, and the PID is a valuable tool in identifying possible hot spots within the building.

SAFETY

The dose makes the poison, and so responders should have some understanding of dose levels, as discussed in Section IV, Chapter 27. Most toxic materials report dose levels in parts per million (ppm) or parts per billion (ppb) levels. These designations typically are indicated for airborne inhalation hazards. Most first-responder toxic sensors report their readings in ppm. Some alarms are for instantaneous readings, some for a fifteen-minute exposure, and others for an eight-hour exposure. An alarm with a carbon monoxide sensor in an industrial application is different than one found in a home. Workplace exposures may be legal at certain levels, and some sensors alarm at levels lower than those required by law. Responders should consult with their local hazardous materials team when confronted with exposure levels that cause their meter to alarm.

FIGURE 29-18   This detection device monitors the radiation dose and will also identify the radioactive isotope.

Radiation Detection

Many first responders are carrying some form of radiation detection, as shown in **Figure 29-18.** There are two primary types: pagers/dosimeters and detection devices. Many agencies are using radiation pagers or comparable dosimeters. Radiation pagers such as the one shown in **Figure 29-19** provide an indication that a radiation level above background has been detected. It alerts the user that a radiation source is nearby. The dosimeter shown in **Figure 29-20** also alerts when levels above background are indicated, but it provides a continual reading of radiation dose as well. The ability to know the dose level is crucial to responder safety. Detection devices can read radiation dose rates and may also provide radiation count rates. Some devices may also indicate the type of radiation source that may be present.

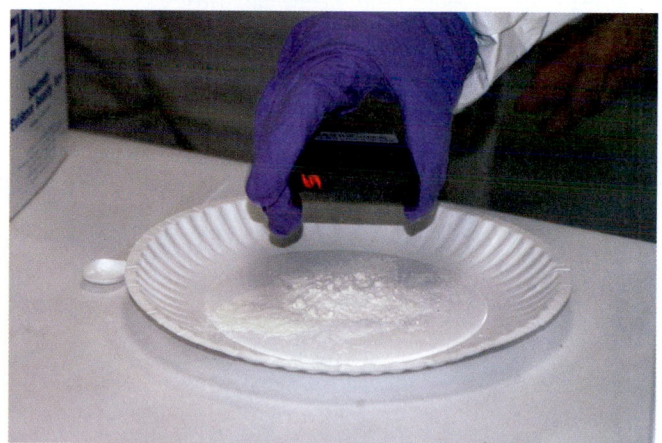

FIGURE 29-19   This radiation pager will alert the user to radiation levels above background.

Colorimetric Sampling

Colorimetric tubes are used for the detection of known and unknown vapors. With sick building calls becoming more and more frequent, the use of colorimetric tubes by hazardous materials teams has greatly increased. Colorimetric sampling consists of taking a glass tube filled with a reagent (usually a powder or crystal). The reagent is placed into a pumping mechanism that causes air to pass over the reagent. A colorimetric sampling device is shown in **Figure 29-21.** If the gas reacts with the reagent, then a color change should occur, indicating a response to the gas sample. Detection tubes are made for a wide variety of gases and generally follow the chemical family lines (i.e.,

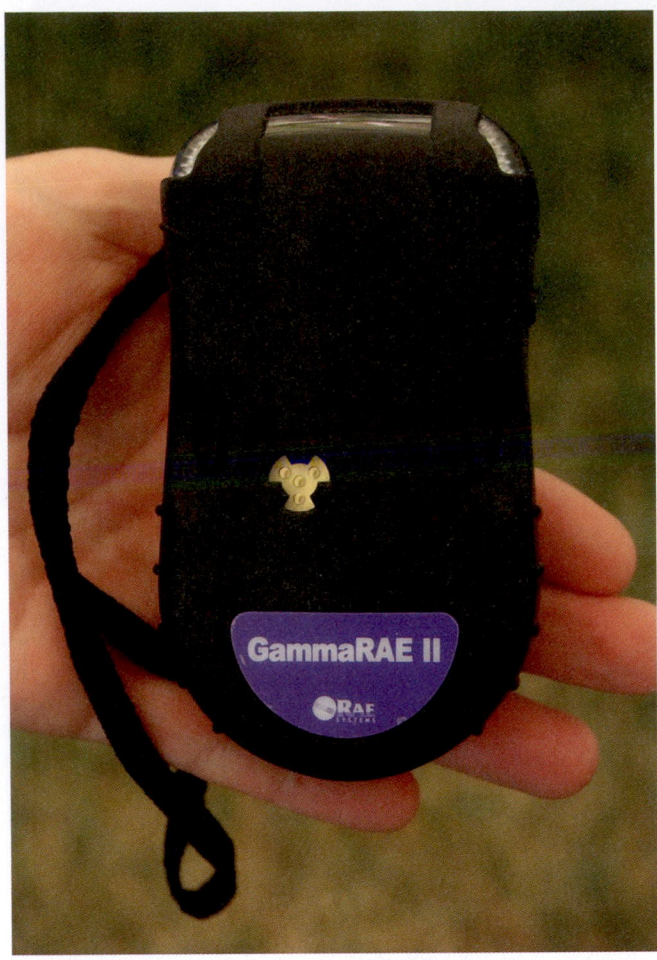

FIGURE 29-20   This radiation pager will alert when the radiation level increases, and it also provides the radiation dose rate.

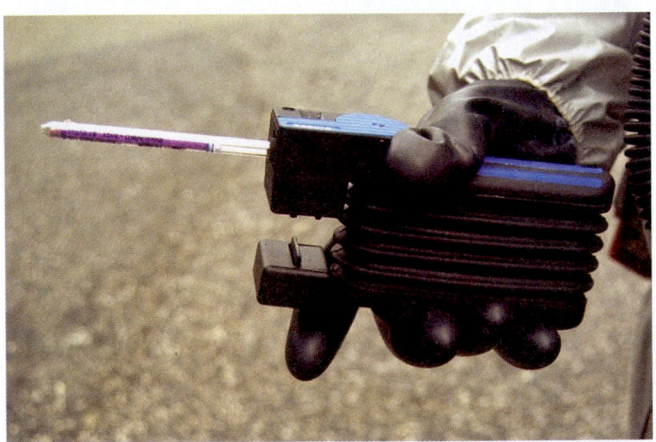

FIGURE 29-21   This is a single colorimetric tube, which samples for a specific gas or a chemical family of gases.

hydrocarbons, halogenated hydrocarbons, acid gases, amines, etc.). Although the tubes may be marked for a specific gas, they usually have cross sensitivities (react to other materials), which at times is the most valuable aspect of colorimetric sampling.

FIGURE 29-22   The chip measurement system uses a chip to analyze a gas and provides a digital readout. Standard colorimetric sampling requires an interpretation of the color change and the length of the change. The CMS takes the guesswork out of the system by providing the reading in ppm.

One system that has been developed involves the use of bar-coded sampling chips. This system, called the **chip measurement system (CMS),** involves the insertion of a sampling chip into a pump, **Figure 29-22.** The pump recognizes the chip in use and provides the correct amount of sample through the reagent. The pump, by means of optics and light transfer system reflective measurement, provides an accurate reading of the gas that may be present.

Technology related to the detection of hazardous materials is growing at a rapid rate. There are many detection devices that are available for hazardous materials teams; some are simple indicators and others are laboratory-style devices. Some of the other instruments used by a hazardous materials team include pH detection, flame-ionization detectors, a variety of warfare agent detection devices, and characterization sample kits. The detection of unknown materials is complicated even for some hazardous materials teams. Accordingly, when first responders are confronted with a possible chemical release, no matter how insignificant they think it is, they should request a hazardous materials team or at a minimum consult with the closest hazardous materials group available.

Using Air Monitoring Equipment

Air monitoring is a mission-specific skill at the NFPA Operations level. To use air monitors, responders must understand the response actions based on air monitor readings.

1. Assure battery level, and bump test the instrument.
   - If the instrument does not respond appropriately to the bump test, then it should be calibrated.

2. Understand what the flammable gas sensor detects and what it does not detect.
   - Most instruments detect flammable gases and vapors.
   - They do not identify the gas, they only indicate the presence of the gas.
   - This sensor takes seven to ten seconds to indicate.
   - The levels indicated are only accurate for the gas the sensor is calibrated for, such as methane (natural gas).
     - If other gases are detected, the readings will need to be corrected.
     - Correction factors are used to correct the readings.
     - Some instruments can be changed to read various gases.
   - Readings are usually based on the lower explosive limit for a given flammable gas based on the calibration gas.
     - Most detectors are factory set for methane, based on pentane gas calibration.
   - The lower explosive limit (LEL) for methane is 5 percent.
   - The flammable gas sensor indicates with a 0 to 100 percent scale. When the detection device reads 100 percent, there is 5 percent methane present.
     - When the device indicates 50 percent, there is 2.5 percent methane present.

3. The oxygen sensor indicates the level of oxygen.
   - Normal oxygen is 20.9 percent.
   - This sensor takes twenty to thirty seconds to indicate.

4. Alarm levels:
   - The flammable gas sensor typically alarms at 10 percent of the LEL for methane.
     - If you do not know the type of gas or vapor released, a reading of 30 percent is extremely dangerous—you do not know the correction factor.
     - Once you identify the gas or vapor, then the correction can be made.
     - An indication of 1 percent means there is a flammable gas or vapor present.
     - A reading of 10 percent or more indicates the need to isolate and evacuate the area.
   - Oxygen alarms at 19.5 percent (low) and 23.5 percent (high).
     - A reading of 20.8 percent means that there is a contaminant in the air causing an oxygen drop.
     - A reading of 19.9 percent means that there is a significant contaminant in the air causing the drop, and the meter has not yet alarmed.

CARBON MONOXIDE INCIDENTS

The CDC (Centers for Disease Control) estimates that more than 40,000 people are treated for carbon monoxide (CO) poisoning each year. In the United States, accidental CO poisoning causes the death of about 500 people each year and is responsible for 50 percent of the fatal poisonings. More than 2,000 people die each year from intentional CO poisoning. Many communities as part of the fire protection code also require the installation of carbon monoxide (CO) detectors, as shown in **Figure 29-23.** Since CO detectors first became available to homeowners in the mid-90s, the fire service is responding to an increasing number of calls involving CO detector alarms. In 1995 the city of Chicago experienced several thousand carbon monoxide detector alarms in one day, due to an inversion that kept the smog, pollution, and carbon monoxide at a low elevation within the city. Because carbon monoxide is colorless, odorless, and very toxic it is important that first responders understand the characteristics of carbon monoxide and how the detectors in the home work. As with other chemicals, carbon monoxide (CO) can be an acute or chronic toxicity hazard. It is only acutely toxic at high levels, typically at levels in excess of 100 ppm. At levels of less than 100 ppm, the hazard comes from a chronic exposure, which can be hazardous. The EPA (Environmental Protection Agency) has established an outdoor air quality level of 9 ppm for an eight-hour exposure, and 25 ppm for a one-hour exposure. The Consumer Product Safety Commission recommends that a long term (eight hours) exposure indoors be less than 15 ppm, and 25 ppm for an hour. Women who are pregnant and are exposed to CO should be encouraged to seek medical attention. Fetal hemoglobin has a higher affinity for CO, and it is possible that the mother may be asymptomatic while the unborn child suffers its ill effects. CO is also a

FIGURE 29-23 This carbon monoxide detector for home use not only monitors the CO levels but can also detect flammable gases such as propane and methane. Responders should be aware that this detection device can activate due to a number of causes, not just CO.

suspected teratogen (agent capable of inducing developmental malformations) and possible **abortifacient** (abortion-inducing agent), and chronic exposure can cause low birth weights as well. The fetus cannot clear the resulting carboxyhemoglobin as well as the mother can. First responders' CO detectors typically alarm at 35 ppm. Because of carbon monoxide's properties, it can only be detected by a CO detector. In extremely high concentrations, CO can be explosive. Exposure to CO causes flulike symptoms, headache, nausea, dizziness, confusion, and irritability. Exposure to high levels can cause vomiting, chest pain, shortness of breath, loss of consciousness, brain damage, and death.

> **NOTE**
>
> If CO is found in a home, its residents should seek medical attention because signs and symptoms of CO poisoning may be delayed for twenty-four to seventy-two hours.

Although confronted with levels of CO in a house, the residents may not be exhibiting any signs or symptoms. Levels over 100 ppm are extremely dangerous and the residents should be medically evaluated. Monitoring with a CO monitor is essential to determine the possible exposure to CO. Persons who may only show minor effects of CO poisoning and who would normally be transported to the closest hospital need to have the residence monitored. If high levels are found using a monitor, then the exposed residents will need alternative treatment such as the **hyperbaric chamber** and such treatment should not be delayed. An oxygen saturation monitor will often be used incorrectly to determine the O_2 level in a patient. Patients who have been exposed to CO will cause an oxygen saturation monitor to read 100 percent because the monitor reads the oxygen molecule in CO as being O_2. The elderly, children, or women who are pregnant are especially susceptible to CO and may have had a serious exposure without showing any effects.

If the first arriving units do not have a CO monitor and victims may still be in the residence, personnel are to have SCBA on and functioning when searching the residence. After determining no victims are present, crews are to ensure that the house is closed up and then wait outside for a CO monitor. Crews should not enter an area with a CO detector that is activated without the use of SCBA. When using a monitor, if crews find levels that exceed 35 ppm, crews should use SCBA to continue the investigation. Crews should be suspicious when responding to reports of an unconscious person or reports of "several people down" and should not enter an area without SCBA if it is possible that CO (or other toxic gases) may be present. When several persons are reported to be ill in the same area, the cause is usually chemically related, and first responders should use extreme caution. An air monitor will ensure responder safety with regard to the gases for which it samples. Failure to use an air monitor to check the atmosphere for contaminants is a dangerous practice.

Occasionally there are reports that people may be found unconscious due to a natural gas leak, but this is very unlikely. Natural gas has a distinctive odor, is a non-toxic gas, and will only asphyxiate a person by pushing oxygen out of an area. The only sign of this exposure is unconsciousness or death; any of the flulike symptoms are due to CO poisoning, not natural gas. If the level of natural gas is high enough to cause unconsciousness, then a very severe explosion hazard is present; in fact, an explosion would be imminent. Air monitoring is critical to responder safety.

When home CO detectors are activated, it is possible that a standard fire department air monitor will not pick up any CO when first responders arrive. This is because the CO detectors purchased for the homes are made to detect small amounts of CO over a long period of time, but fire service detectors provide "instant" readings and only pick up 1 ppm or more. The fact that firefighters may not pick up any readings, however, does not mean the residence's detector is defective. Many factors may cause fire service monitors not to get any readings, including if low amounts of CO are present or a momentary high level of CO activated the alarm but then dissipated prior to fire department arrival. The amount of time the residence is open will also dramatically affect fire service readings. Crews are reminded to keep the residence closed up so that the air monitor has a chance to monitor the level of CO. As a reminder, any time units respond to unknown odors or sick building calls, responders should remove any people from the building and keep it closed up. Because the amounts of toxic gases in sick building incidents are usually small, keeping them contained is very important. A patient cannot be treated for toxic gas exposure unless the source and type of exposure are known. For the patient's long-term health as well as the responder's, quick, reliable gas samples are a necessity.

The brand and type of CO detector will determine how well the device actually performs. The three basic sensing technologies are **biomimetic,** metal oxide, and electrochemical, each having advantages and disadvantages. Location, weather conditions, and the type of sensor will determine the types of readings that can be expected from a particular brand of detector.

Biomimetic

A biomimetic is a gel-like material that is designed to operate in the same fashion as the human body does when exposed to CO. This type of sensor is prone to false alarms, because it cannot reset itself unless it is placed in an environment free of CO, which in most homes is impossible. The sensor may need twenty-four to forty-eight hours to clear itself after an exposure to CO. The actual concentration of CO at the time the detector sounds may be low, but the exposure may have been enough to send the detector over the alarming threshold. If responding to an incident in which one of these detectors has activated, it will need to be placed in a CO-free environment for twenty-four to forty-eight hours to allow itself to clear. It is important for some type of detection device to be left in place for the residents until their detector clears because it is not advisable to leave them unprotected.

Metal Oxide

This is the same type of sensor that is used in the combustible gas detector, but it is designed to read carbon monoxide. This detector—although superior to the biomimetic sensor—does have some cross sensitivities and will react to other gases.

Although it is hoped that responders would be using a three- to five-gas detection device to check a home, it is possible for this type of sensor to alarm for natural gas, propane, and other flammable gases. It will even react to the flammable vapors from nail polish remover. Responders using only a CO instrument may find themselves walking into a flammable atmosphere. This type of detector can usually be identified by the use of a power cord because the sensor requires a lot of energy and, in most cases, provides a digital readout. Once activated this sensor needs some time to clear itself, but this is usually less than twenty-four hours.

Electrochemical

An electrochemical sensor is also referred to as an instant detection and response (IDR) sensor, and it is the same type of electrochemical sensor found in three- to five-gas instruments. It has a sensor housing with two charged poles in a chemical slurry. When CO goes across the sensor it causes a chemical reaction, which changes the resistance within the housing. If the amount is high enough, it will cause an alarm. It provides an instant reading of CO and does not require a buildup of CO to activate. It has an internal mechanism that checks the sensor to make sure it is functioning, which is a unique feature. Out of the three types of residential detectors, based on sensor technology, the electrochemical sensor would provide the best sensing capability.

Common sources of CO include furnaces (oil and gas), hot water heaters (oil and gas), fireplaces (wood, coal, and gas), kerosene heaters or other fueled heaters, gasoline engines running inside (basements or garages), barbecue grills burning near the residence (garage or porch), and faulty flues or exhaust pipes.

JOB PERFORMANCE REQUIREMENT 29-1
Implementing Defensive Operations

A Scenario 1: Visual clues

B Scenario 2: Visual clue *(Photo courtesy of Maryland Department of the Environment Emergency Response Division)*

C Scenario 2: Visual clue *(Photo courtesy of Maryland Department of the Environment Emergency Response Division)*

D Scenario 3: Visual clues *(Photo courtesy of Mike Austin)*

E Scenario 4: Visual clues *(Photo courtesy of Mark E. Brady, PGFD)*

LESSONS LEARNED

The use of defensive product control methods is a key component for the protection of a community and the environment. In most cases, first responders have the equipment necessary to handle these tasks. With some modification or adaptation, first responders can accomplish the control of many spills. The limiting of spills will mitigate the incident sooner and prevent its spread. If first responders cannot stockpile the necessary equipment, then contacts should be made with those facilities that may have the materials. (Under the requirements of the Oil Pollution Act [OPA] of 1990, certain facilities are required to maintain stockpiles of emergency equipment.)

First responders are also becoming more involved with air monitoring and more aware of the hazards chemicals present. Incidents involving flammable gases such as methane or propane and response to carbon monoxide incidents are the most common calls that first responders are involved in. When using air monitors, first responders are reminded that they are not all-encompassing and their use requires training and experience. Understanding the action levels are an important consideration for safety—when is safe really safe? Detection devices require testing and regular maintenance and cannot be expected to function properly without this upkeep. To determine true levels, first responders must use a range of instruments, which may be above their level. When in doubt, first responders should consult with the local hazardous materials team.

KEY TERMS

Abortifacient A chemical or material that can cause abortions.

Absorption A defensive method of controlling a spill by applying a material that absorbs the spilled chemical.

Biomimetic A form of gas sensor that is used to determine levels of carbon monoxide. It is of the type of sensors used in home CO detectors. It closely re-creates the body's reaction to CO and activates an alarm.

Bump Test Used to determine if an air monitor is working. It will alarm if a toxic gas is present. It is a quick check to make sure the instrument responds to a sample of gas.

Calibration Used to set the air monitor and to ensure that it reads correctly. When calibrating a monitor, it is exposed to a known quantity of gas to make sure it reads the values correctly.

Catalytic Bead The most common type of combustible gas sensor that uses two heated beads of metal to determine the presence of flammable gases.

Chip Measurement System (CMS) A form of colorimetric air sampling in which the gas sample passes through a tube. If the correct color change occurs, the monitor interprets the amount of change and indicates a level of the gas on an LCD screen.

Colorimetric Tubes Crystal-filled tubes that change colors in the presence of the intended gases. These tubes are made for the detection of known and unknown gases.

Damming The stopping of a body of water, which at the same time stops the spread of the spilled material.

Diking A defensive method of stopping a spill. A common dike is constructed of dirt or sand and is used to hold a spilled product. In some facilities, a dike may be preconstructed such as around a tank farm.

Dilution The addition of a material to the spilled material to make it less hazardous. In most cases water is used to dilute a spilled material, although other chemicals could be used.

Diverting Using materials to divert a spill around an item. For instance, several shovels full of dirt can be used to divert a running spill around a storm drain.

Hyperbaric Chamber A chamber that is usually used to treat scuba divers who ascended too quickly and need extra oxygen to survive. The chamber re-creates the high-pressure atmosphere of diving and forces oxygen into the body. It is also successful in the treating of carbon monoxide poisoning and smoke inhalation, because both of these problems require high amounts of oxygen to assist with the patient's recovery.

Infrared Sensor A sensor that uses infrared light to determine the presence of flammable gases. The light is emitted in the sensor housing and the gas passes through the light. If it is flammable the sensor will indicate the presence of the gas.

Metal Oxide Sensor (MOS)   A coiled piece of wire that is heated to determine the presence of flammable gases. Also called tin dioxide sensor.

Multi-gas Detector   A term used to describe an air monitor that measures oxygen levels, explosive (flammable) levels, and one or two toxic gases such as carbon monoxide or hydrogen sulfide.

Photoionization Detector (PID)   An air monitoring device used by hazardous materials teams to determine the amount of toxic materials in the air.

Remote Shutoffs   Valves that can be used to shut off the flow of a chemical. The term remote is used to denote valves that are located away from the spill.

Retention   The digging of a hole in which to collect a spill. Can be used to contain a running spill or collect a spill from the water.

Vapor Dispersion   The intentional movement of vapors to another area, usually by the use of master streams or hoselines.

REVIEW QUESTIONS

1. What type of dam would be required for a fuel spill in which the fuel has a specific gravity of less than 1?

2. What type of dam is required for a chemical spill in which the material has a specific gravity of greater than 1?

3. Describe the two key items needed to construct an underflow or overflow dam.

4. Who should be consulted prior to using vapor dispersion techniques?

5. Describe the normal configuration for a multi-gas detector.

6. If a detector is calibrated for methane and a person responds to a propane release, describe whether the instrument will detect propane and, if it will, how it does so.

7. Explain why and how often detectors should be calibrated.

8. With regard to safety and the ability to detect various toxic gases, how would you rate the use of a multi-gas detector with this combination: LEL, O_2, CO, H_2S?

9. When responding to situations involving unknown materials, which four detection devices are needed?

10. Does NFPA 472 address hazardous-materials, operations-level responders' use of detection devices?

11. When do most manufacturers recommend detection devices be calibrated?

12. How long do some WMD detection devices take to indicate?

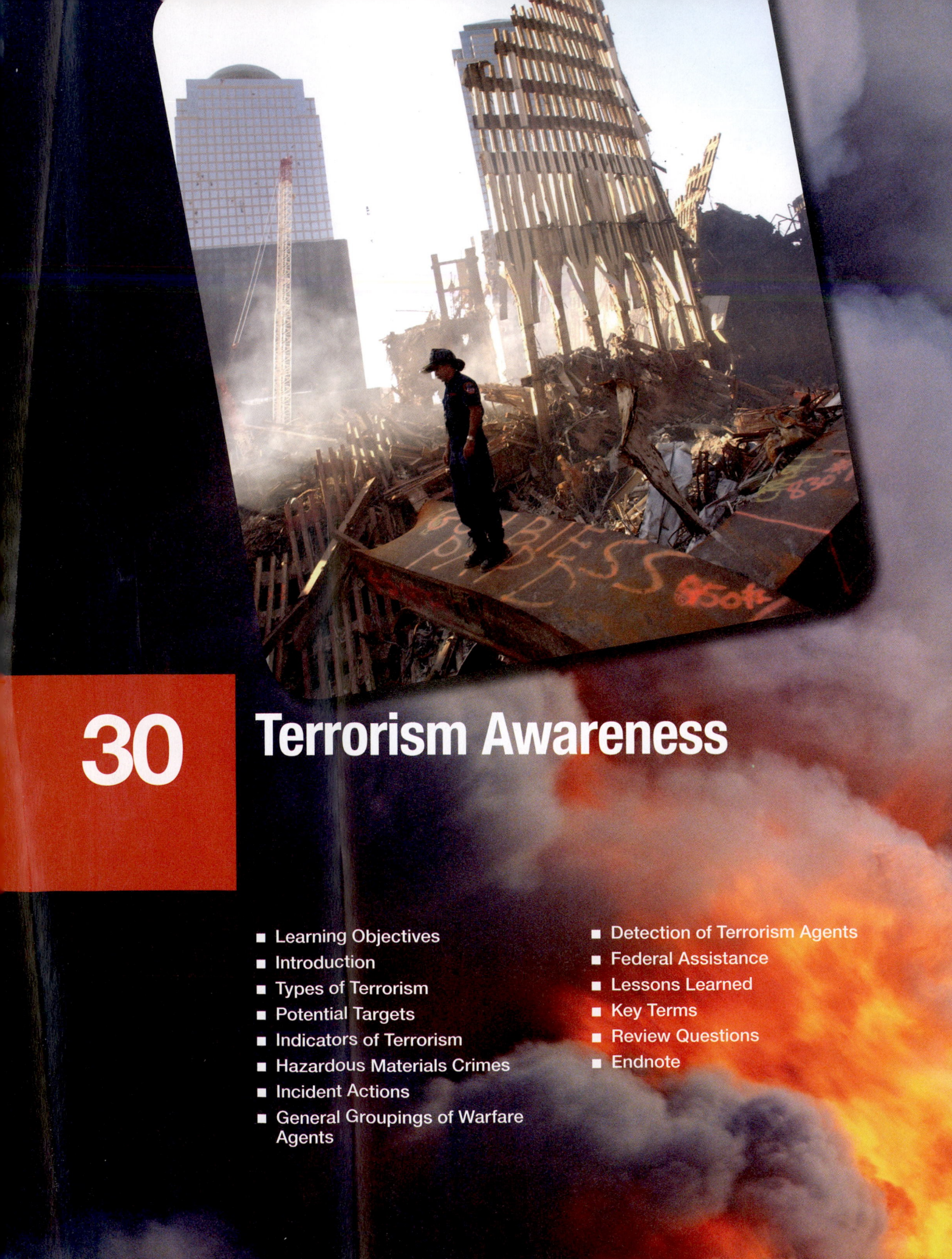

30 Terrorism Awareness

The fire service is made up of a large, tight-knit family, and the special bond we have is carried over into the other emergency services.

On September 11, 2001, our family and our nation were attacked in a manner that was unprecedented. After the attacks in New York and Arlington, Virginia, and the plane crash in Shanksville, Pennsylvania, America was at war. Terrorists had struck, and in a few precious minutes we suffered a grievous loss. America changed that day, and the emergency services took a severe hit, but we persevered. During the first few minutes of the fire in the World Trade Center, my thoughts were that it was a large fire, and that it would be a tough battle, but one the FDNY could win. When the towers collapsed, my heart sank. I knew that I had just witnessed hundreds of my family members being struck down. My thoughts were immediately for my good friends who work in FDNY and for all of the other emergency responders at that scene.

The events of that day caused the deaths of many emergency responders and an enormous amount of emotional damage to the survivors, family members, and population of the United States. But on that day, when citizens across the country dialed 9-1-1, emergency responders showed up.... We proved that we were not defeated. We are emergency responders and the keystones of our communities. We are the first line of defense against terrorism in this country. We must remain focused on our mission, which is to protect the citizens in our community. With that said, we as emergency responders must look at the real threat that terrorism presents to our communities.

There are many events that occur in this country that present substantial risks to you as a responder and to the community. The events discussed in this chapter are events that can have catastrophic consequences for your community. You must prepare for these events and become tuned in to the potential for terrorism or other criminal behavior in your community. The tactics and motivation of terrorists have changed even since September 11. Suicide/homicide bombers (SHB) are attacking commuter systems throughout the world. SHB attack locations where large numbers of people congregate. In many cases, the profiles of these bombers indicate that they are sympathetic to al Qaeda but not directly associated with them. In Iraq, terrorists are now using explosive devices combined with chlorine and other chemicals. The idea of suicide/homicide bombers is disturbing, and the use of toxic industrial chemicals adds a dangerous twist for terrorism responders. Domestic terrorism is rearing its ugly head again, and although there have been only a few arrests, these individuals present a risk as they typically act alone. Every community has the potential to be attacked or to be a base of terrorist operations. Read the listing of the case histories, and you will learn that all communities are at risk. As you respond in your community, remember those who served on September 11, 2001, and dedicate your career to being the most knowledgeable and prepared firefighter that you can be.... You never know what the future may bring.

—*Street Story by Christopher Hawley, Baltimore County Fire Department (Ret.)*

INTRODUCTION

It is unfortunate that a chapter on terrorism needs to be included in this firefighting text, and until recent times this would not have been necessary. Although everyone has seen terrorism on the evening news, it used to occur in places such as Northern Ireland, Beirut, Israel, or somewhere other than the United States. Until recently, the United States remained for the most part immune to the reign of terrorists. This changed in February 1993 when the World Trade Center was bombed in New York City, **Figure 30-1.** Even when this bombing occurred, the fire service did not pay much attention, because the bombing was looked at as the type of incident that happened only in big cities. In addition, the persons who were found to be responsible for the bombing were controlled by an influence from outside the United States, so the thought was that it was an isolated foreign attack.

However, when the Alfred P. Murrah Federal Building in Oklahoma City was devastated by a bomb on April 19, 1995, the United States fire service took notice. The attack on the federal building, **Figure 30-2,** was brought on by a person who did not have ties to another country and was a natural born citizen of the United States. It was perceived as an attack on America from one of its own, not from some unknown citizen from a foreign nation.

And perhaps the most devastating attack on the United States occurred on September 11, 2001, when the World Trade Center and the Pentagon were hit with a total of three airplanes. Then, in Shanksville, Pennsylvania, a hijacked plane crashed into a field. As a result of the impact of the planes and the resulting fire, the 110-story twin towers of the World Trade Center collapsed. The death toll from all three sites was more than 2,600 civilians and 365 emergency responders.

Shortly following three horrific crashes, in October 2001 a flood of noncredible anthrax letters hit the country. Mixed in with the thousands of noncredible letters were several letters containing real anthrax.

These letters were responsible for five deaths and up to twenty-two other illnesses. The letters were sent to news media outlets and to members of Congress.

Regardless of the origin, the potential for terrorism is here in this country and has to remain in firefighters' thoughts as they respond to any incident. Other incidents are occurring on a regular basis, and, although they do not fit the exact definition of terrorism, they involve the use of large-caliber weapons or a large number of weapons. This chapter looks at terrorism, **hazardous materials crimes,** and other potentially dangerous criminal situations that are sometimes just as deadly as terrorism. Firefighters have to think outside the box in regard to terrorism. There are many events that occur in this country that place responders at risk. Terrorism, hazardous materials crimes,

FIGURE 30-1   In 1993 a van was used to carry explosives to an underground parking garage in the World Trade Center. Six people were killed and more than 1,000 were injured. There were 50,000 people in the building, and the goal of the terrorist was to collapse the building into the adjacent tower.

FIGURE 30-2 A truck bomb caused the devastation in the Oklahoma City bombing in which 167 people were killed and 759 injured. The damage extended several blocks in each direction, and 300 buildings were damaged. Fatalities occurred in 14 separate buildings. *(Courtesy of John O'Connell)*

murders, and other criminal events all present a risk to firefighters. Someone using ricin to kill another person is committing murder, but the weapon is as deadly to the responders as it is to the intended victim. A booby-trapped drug lab is not terrorism, but it presents significant risk to responders. Many criminals are protecting themselves with body armor and fortified vehicles. Many crimes such as bank robberies involve the use of explosives and sophisticated weaponry. Street violence that can be associated with gangs is increasingly violent, and when people are killed or injured the fire service is called into action. Crimes such as assaults and murders are increasing in the school system. In recent times there have been deadly riots in Los Angeles and in St. Petersburg, Florida, usually after a sporting event. In Chicago, twenty-one people were killed and fifty were injured in a nightclub after a fight broke out. These

deaths may have been caused by a lack of exits and the use of pepper spray by security personnel, which panicked the crowd. The workplace is also becoming an increasingly dangerous place to be. A firefighter in Jackson, Mississippi, entered the headquarters fire station intent on killing the fire chief. Several persons were killed, not to mention the emotional damage that occurred.

Crimes are becoming more violent, and it is perceived that this trend will continue to increase. Emergency responders are immediately placed into dangerous situations, and can get caught—literally and figuratively—in the cross fire.

The following sections provide some information about the type of terrorism agents that currently exist. Some are unique ideas and thoughts, and some represent possible scenarios. There exists some thought within the fire service that this type of information should not be published. All of this information is readily available in the public domain and in training programs and texts throughout the United States. Most of the exact recipes and "how-to" instructions for these and any other device are easily obtained through printed texts and the Internet. This text does not provide any information that cannot be easily obtained elsewhere. If an ordinary citizen is able to obtain it in an easy, normal, and legal fashion, it is certain that the terrorist already has it. Responders need to be aware of the various chemicals and devices that someone may design to kill them and others in their community. The best defenses are education and trying to stay one step ahead of the terrorist. By being informed as to how a terrorist may operate or what some of the devices may involve, responders will be alert to a potentially fatal situation.

TYPES OF TERRORISM

The types of terrorism are divided into two distinct areas: foreign based and domestic. Until the Murrah building bombing, the fear of terrorism was aimed at a foreign source, and it was thought that any terrorist attack would be from a foreign country. To be foreign based, the motivation or supervision must come from a foreign country. Domestic terrorism originates from within the United States and is not influenced by any foreign party. From the list of terrorist acts provided earlier and from other statistics, it is clear that the largest percentage of terrorism is domestic in origin.

The FBI defines terrorism as a violent act or an act dangerous to human life in violation of the criminal laws of the United States or any segment to intimidate or coerce a government, the civilian population, or any segment thereof, in furtherance of political or social objectives. The key to this definition is the

HISTORY OF MODERN TERRORISM AND HAZARDOUS MATERIALS CRIMES

As a result of the terrorist attacks on September 11, 2001, the previous bombing of the World Trade Center in 1993, the Alfred P. Murrah Federal Building bombing in Oklahoma City, and other incidents, the fire service's response to some types of incidents will be forever changed. In the late 1990s there were two other bombings that are fairly well known to the fire service: the Atlanta Olympic Park bombing in 1996 and the Atlanta abortion clinic bombing in 1997. That particular abortion clinic bombing is significant because a secondary device was used. It is thought that it was strategically placed with the sole intention of harming the responders. The device was placed near the location where the incident command post was set up. Luckily, only minor injuries were received by the responders who were near the blast. In 1998 there was another bombing at a Birmingham, Alabama, abortion clinic in which an off-duty police officer employed as a security guard was killed by a device that was more accurately placed to target the responders. This secondary device was activated by Eric Rudolph, the person responsible as he watched the victim approach the radio-controlled device. A brief overview of some of the recent acts of terrorism, hazardous materials crimes, use of terrorism materials, and criminal acts that presented risk to responders follows.

1980s

A series of bombings targeted primarily at the Internal Revenue Service (IRS) included an attempt at a chemical bomb using ammonia and bleach. Another attempt included the use of a hot water heater as a very large pipe bomb, but the vehicle carrying the bomb caught fire while the terrorist was driving the vehicle.

1984

- Dr. Michael Swango had a long history of suspicious deaths while he moved throughout the United States and abroad. He was stripped of his license to work as a doctor and got a job as a paramedic. Although his crimes do not fit the pattern of standard terrorism, he was finally arrested for attempting to poison his paramedic coworkers with chemicals.
- The Rajneesh Foundation was responsible for poisoning 715 people with Salmonella bacteria. They poisoned the salad bars in ten restaurants. The group had used other biological agents and had targeted a water supply for attack. They had previously used raw sewage and dead rodents in an attempt to poison the system. They used nursing home patients as their test targets for some of the biological attacks. The attacks were an attempt to effect a change in a local election.

1985

Members of a militant group developed a plot to poison a water supply. They were going to use 35 gallons of cyanide, which they had in their possession. Other members of the group were arrested for church arsons and attempting to blow up gas pipelines.

1993

The World Trade Center was damaged by a van bomb. Six people were killed.

1995 (and throughout the 1990s)

The Aum Shinrikyo are the cult group that is best known for a 1995 sarin nerve agent attack in the Tokyo subway. For a long period of time the group carried off chemical and biological weapons attacks and went to great lengths to develop their weapons program. An examination of this group yields many valuable lessons on chemical and biological attacks. The group had assets of more than $300 million. The group was also home to more than a hundred scientists whose sole function was to develop chemical and biological weapons. The group used, or explored the use of, Clostridium botulinum, Bacillus anthracis (anthrax), Q fever, Ebola virus, and viral hemorrhagic fever. They carried off ten chemical attacks and nine biological attacks, not to mention numerous small and large full-scale tests. They killed seven people and injured 200 in a sarin nerve agent attack in Matsumoto in 1994. The subway attack in 1995 killed thirteen but injured more than 5,000. The attack consisted of placing sarin in three subway lines, with a total of eleven sarin bags that were pierced with umbrellas. The deaths occurred to those who came in contact with liquid or were in small confined spaces. Over 4,000 people went on their own to 278 hospitals. The Tokyo Fire Department ambulances transported 688 patients. The breakdown of the injured was seventeen critical and thirty-seven severe. There were 948 patients suffering from miosis (pinpoint pupils). All of the moderately injured patients were treated and released within six hours after they arrived at the hospital. Although 85 percent of the patients did not require any treatment, they still flooded the hospitals.

1995

- A man was arrested after manufacturing **ricin,** an extremely deadly toxin, with the intention of killing someone he was jealous of.
- The Alfred P. Murrah Federal Building in Oklahoma City was the target of a truck bomb. Several hundred people were injured and 186 people were killed in the blast.
- The members of a militia group known as the Patriots Council were arrested for the manufacture of ricin in the attempted assassination of a U.S. marshal.
- Two separate ricin incidents involved two doctors. Dr. Deborah Green killed two of her three children by burning her house, and she attempted to kill her husband three times with ricin. In Virginia, Dr. Ray Mettetal

attempted to murder his boss with a syringe filled with boric acid and saline. He also possessed ricin and a number of other materials.

- The Anaheim Fire Department in California was made aware of a potential sarin nerve agent attack at Disneyland. The chief was notified at midnight to be at the command post four hours later. Police agencies, federal law enforcement, and the military had known about the threat for five days. Up until notifying the fire chief and assembling the resources prior to the event, no other planning had occurred. The fire department was placed in charge of the incident and quickly developed a plan of action. Disneyland did not close and 30,000 to 40,000 people visited the park during the threat period. Luckily, the threat never materialized.

1996

- Members of a paramilitary group who had access to preplanning information obtained through the fire department were arrested when they attempted to blow up a Department of Justice complex in West Virginia.
- Theodore Kaczynski, known as the Unabomber, was responsible for a string of bombings that lasted eighteen years. He sent sixteen bombs, which killed three people and injured twenty-three. He wrote a manifesto that appeared in the New York Times and was recognized by his brother, who turned him in to the Federal Bureau of Investigation (FBI).
- A disgruntled firefighter in Jacksonville, Mississippi, went to his department's headquarters with the intent of killing the fire chief. The fire chief was not injured, but several other fire department personnel were killed.
- A lab worker brought in pastries that were poisoned with Shigella dysenterie to coworkers. She made twelve people ill and was found guilty on five felony assault charges. Police also learned that she had attempted on more than one occasion to poison a former boyfriend with Shigella and other biological agents.

1996 and 1997

The Atlanta area was besieged with bombings, including one at the Olympic Park that killed one person. At an abortion clinic in Birmingham, Alabama, a booby-trapped device killed an off-duty police officer and wounded a nurse.

1997

- EMS and police responded to a shooting incident, and the suspect was found to have ricin, E. coli, and a mixture of nicotine and dimethyl sulfoxide (DMSO).
- Four members of the Ku Klux Klan were arrested for plotting to blow up a hydrogen sulfide tank in order to create a diversion for an armored car robbery.

1998

- In Nairobi, Kenya, and Dar-es-Salaam, Tanzania, bombs exploded in two U.S. embassies, killing 224 people.

- Kathryn Schoonover attempted to mail 100 envelopes with cyanide disguised as diet powder all over the United States.
- Two men were arrested on the suspicion of **anthrax** possession. They were later released, since it was determined that they only possessed a possible anthrax vaccine. Larry Wayne Harris had previously been arrested in 1995 for the possession of plague, a biological toxin that he had ordered through the mail. He was a previous member of the Aryan Nation but was deemed too radical for the group.
- In Charlotte, North Carolina, a man with an explosive device held some hostages in a government building. The explosive device was thought to also contain some type of chemical agent. It was later determined that the filling agent was harmless, although the explosive was live.
- Abortion clinics in Florida, Louisiana, and Texas were affected by attacks using **butyric acid,** a material with a horrible, irritating odor.
- In Lafayette, Indiana, a pickup truck rammed into the courthouse. The bed of the pickup had flammable and combustible materials as well as several explosives.

1999

- In Colorado, two students who were armed with an array of guns and explosives attacked their own high school. The suicide attack resulted in fifteen deaths and spurred a rash of bomb hoaxes throughout the country.
- The FBI investigated hundreds of anthrax hoaxes, none of which involved the actual use of the biological agent. Abortion clinics were the targets in most of the cases.
- Ahmed Ressam was arrested in Port Angeles when he crossed over the border into the United States. He was responsible for plotting the millennium "border bomb" and had the components to a large bomb in his car. His target was the Los Angeles Airport (LAX) or the Space Needle in Seattle on New Years Eve. This plot is the responsibility of an al Qaeda sympathetic group, who have worldwide ties to other terrorist cells.

2000

- Dr. Larry Ford's house in Irvine, California, was searched after he attempted to murder his partner. Dr. Ford committed suicide, and his house was found to have held numerous chemicals and chemical agents. The search took several weeks, and chemicals, guns, and biological materials were found.
- During a genetics conference in Minneapolis, Minnesota, protestors attacked three restaurants with hydrogen cyanide.
- Members of a militia group in Sacramento, California, were arrested for plotting to blow up a 24-million-gallon propane tank.

(Continued)

2001

- Over 2,600 civilians and 365 emergency responders lost their lives at the World Trade Center in New York City; the Pentagon in Arlington, Virginia; and a field in Shanksville, Pennsylvania. Four planes were used in the attack, two hitting the World Trade Center and one hitting the Pentagon. The fourth plane crashed into a field. Members of the al Qaeda group have been held responsible for the attack.
- In October there was one death as a result of an anthrax-laden letter. Later, there was a total of seven deaths from anthrax. The emergency services across the country were deluged with calls about white powder. A very small percentage of the calls was investigated by the FBI, but they opened more than 14,000 cases regarding the white powder events. Only five cases involved real anthrax, which was sent to members of Congress, to New York City, and to Boca Raton, Florida. In December, Clayton Lee Waagner was arrested and admitted to sending more than 500 noncredible anthrax letters to women's reproductive health centers (WRHCs). He was a member of the Army of God, a group known for pursuit of the right to life cause. As of 2007, no one has been arrested for the anthrax attacks. The FBI is still actively pursuing the perpetrator. They have traveled all over the world looking for clues and have interviewed more than 9,100 persons.
- The Animal Liberation Front (ALF) and the Earth Liberation Front (ELF) committed a number of terrorist acts throughout the year. The ALF admitted to 137 illegal actions at a variety of locations. Both groups target buildings and businesses that have any connection to the use of animals or perceived violation of the environment. The acts typically create inconveniences, such as glued locks, but on occasion do include arson and other violent destruction of property. For the most part these groups take great care to avoid any potential injuries to humans, but responders could be killed or injured while responding to or handling these acts.
- Two Jewish Defense League (JDL) members were arrested for plotting to blow up a mosque and a congressman's office. Other members of the JDL have been quoted as having a desire to continue the militant work of the two men arrested.
- Richard Reid was arrested on a Paris to Miami flight for attempting to detonate a PETN explosive that was located in his tennis shoes. PETN is one of the more powerful explosives. He was overpowered by other passengers and the flight crew while trying to light the fuse. He has ties to al Qaeda and is supportive of their cause.

2002

- Two different bombers created fear in the United States, one in Philadelphia and the other in the Midwest. Preston Lit set pipe bombs off in U.S. Postal Service mailboxes in the Philadelphia area. Lucas Helder, a college student, set off pipe bombs in residential home mailboxes in five states. Helder's intention was to set off bombs so that the explosions when drawn on a map would form a smiley face. His bombings injured six people. He set a total of eighteen pipe bombs, six of which exploded.
- A fifteen-year-old stole a small plane and flew it into a high-rise building in Tampa, Florida, killing himself. His suicide note stated that he wanted to be just like Tim (McVeigh), Eric (Rudolph), and Osama (bin Laden).
- Joseph Konopka, a member of the Realm of Chaos group, had stashed potassium and hydrogen cyanide in a Chicago subway tunnel. The group is known for wanting to destroy public utility, water, sewage, and telecommunications systems.
- The leaders of two militia groups in Kentucky and Pennsylvania were arrested for possession of weapons and explosives.
- The FBI arrested Jose Padilla, also known as Abdullah al Muhajir, who is associated with al Qaeda, in Chicago. He was researching the use of, and looking for materials to detonate, a radiological dispersion device. He had traveled to Pakistan and had studied methods to pull off such an attack with al Qaeda operatives. In late 2002 the FBI was still looking for 100 al Qaeda members and investigating 150 persons and groups who may have al Qaeda ties. They made two large arrests in Detroit and Buffalo, as well as other arrests throughout the United States, apprehending a number of suspects who had al Qaeda ties.

2003

Mr. Iyman Faris was sentenced to twenty years in prison for providing assistance to al Qaeda and conspiring to commit a terrorist act. Mr. Faris was a U.S. citizen who was born in Kashmir, and last lived in Columbus, Ohio. He was accused of scouting targets, one of which included the Brooklyn Bridge. He had traveled to Afghanistan in 2000 to attend a training camp with Bin Laden. His experience as a truck driver and access to cargo planes was of interest to al Qaeda. He performed a number of research tasks for the al Qaeda leadership. In early 2003, he became a double agent for the FBI, providing information on al Qaeda. In 2004, Nuradin Abdi was charged with conspiracy to commit a terrorist act, by setting off explosives at a shopping center. When Adbi arrived in Columbus, it was Faris who picked him up.

2004

- There was a series of bombings in Madrid, Spain. The attacks, which took place on four commuter trains, killed 191 persons and injured 2,050. Although it was suspected that the attacks were carried out by al Qaeda, the actual terrorists were local Islamics and several others who were not Islamic. The terrorists used thirteen backpack explosives, of which ten detonated. The attack was very well coordinated and each of the explosions took place within several minutes of each other in different parts of the city, overwhelming the emergency response system. The attacks did create some attention among other terrorist groups, including al Qaeda, as the attack changed the outcome of the national elections. The government immediately placed blame on the Euskadi Ta Askatasuna, which is commonly called the ETA. This group has committed a number of terrorist attacks in Spain and was a likely suspect group. There were a number of misstatements made by government officials, which did not make the citizens happy. On the days following the attack, thousands of citizens protested in the streets. The demonstrations increased in size and intensity each day new facts were learned that contradicted the government's position. As a result, the party that was in power prior to the attack lost the election and government control went to the opposition party.

■ William Krar was sentenced to eleven years, along with Judith Bruey who was sentenced for just over four months in federal prison. They were charged with possession of sodium cyanide and other chemicals, which could be used in weapons of mass destruction. The chemicals were primarily acids, which could be used to produce sodium cyanide gas if mixed with the sodium cyanide. In addition to the chemicals, the FBI found hundreds of thousands of ammunition rounds, a substantial number of pipe bombs, machine guns, and a number of remote-controlled explosive devices disguised as briefcases. Mr. Krar had sent false identification documents, but the package was delivered to the wrong address. The exact motive or intended target of the weapons is not known, but Mr. Krar had a history of being anti-government and had a number of previous weapons-related arrests.

■ Demetrius Van Crocker was convicted in a domestic terrorism case. Working through intelligence, the FBI was able to work the case with an undercover FBI agent. The FBI was posing as a security guard at the U.S. Army Pine Bluff Arsenal, one of the locations that stored military chemical weapons. Van Crocker conspired with the FBI agent to purchase sarin nerve agent and C-4 explosives, which the agent purported that he could steal from the arsenal. Van Crocker was a white supremacist who disliked the government and supported Timothy McVeigh. Although Van Crocker had a low IQ, he had studied the process to manufacture military chemical weapons and other toxic industrial chemicals that could be used as effective weapons as well. Through discussions with the undercover agent, he was able to describe several scenarios involving chemical weapons, and he had more than a general interest in the topic.

2005

■ Much like the Madrid bombings, London's public subway system was the location for a series of coordinated attacks. There were two sets of bombings, a series on July 7 and July 21. Within one minute of each other, three TATP-based bombs exploded on the London Underground. An hour later a fourth backpack bomb exploded on a double-decker bus. The attack killed fifty-two, including four of the attackers, and injured more than seven hundred. The placement and timing of the explosions created hardships for the rescue crews, as the trains had left and were in between stations. The relationship with al Qaeda is not known, but it is suspected that the terrorists were sympathetic to their cause. The bombings on the 21st were similar, but the four explosive devices malfunctioned and did not detonate. Three trains and a bus were targeted, but only the detonators exploded and the main charges failed to detonate. Like the earlier bombs, these were made of TATP, which is very unstable. Two days later, another bomb was found that was similar to the other devices. The explosives in both cases had been homemade and considered very unstable. It is possible that the explosives were made at the same time, and due to the length of time intervening, the explosives deteriorated and were no longer explosive. In both bombings, a total of sixteen people were arrested and a number of others were held and released.

■ Bali, Indonesia, and two adjacent communities were the targets of a number of explosions. The bombings occurred in markets and other areas where tourists are known to congregate. In one bombing, the SHB walked into a crowded cafÈ and set off a backpack bomb. In all, twenty-three people were killed, including the three SHB. Among the 129 injured were six Americans. The attacks came before Ramadan, which is a period of significance for Muslims. A number of Australian tourists were in Bali for vacation during this period, and the victims included four Australians who died and 19 who were injured. In 2004, the Australian Embassy in Jakarta was bombed. In August 2003, an SHB detonated a car bomb outside Jakarta's Marriott hotel. This attack against a U.S. interest killed 12 and injured 150 persons. In 2002, there was another series of well-coordinated attacks in Bali, where 202 persons were killed and 209 were injured. The majority of those killed were from countries outside of Indonesia, including seven from the United States. Australia suffered the most casualties, as the area is popular with Australian tourists. The terrorists used diversionary attacks to move victims to a central location where a larger van bomb was exploded. The SHB in the van did not have to activate the bomb, as it could have been remotely controlled if the bomber had a change of mind. In all of the bombings, members of the Islamic group Jemaah Islamiyah have been arrested or suspected to be behind these years of terrorism.

(Continued)

2006

- Police in Ontario, Canada, arrested seventeen persons thought to be part of an Islamic terrorist cell. Allegedly the group had made plans to use truck bombs to target government and other high-profile buildings and to attack events where there would be large numbers of civilians. The group had ordered 6,600 pounds of ammonium nitrate, a common explosive material. As the police had intelligence information and had infiltrated the group, they were able to make the arrests prior to any terrorist actions occurring. In addition to the truck bombs, the group had made plans to storm government buildings, take hostages, and execute high-level government officials.

- The FBI was alerted that Derrick Shareef, also known as Talib abu Salam Ibn Shareef, had a desire to commit a terrorist act in Rockford, Illinois, planning to attack government buildings and commit other crimes to raise money. During discussions with an informant, Shareef changed his plan of attack to that of the Rockford Mall. Shareef discussed with an undercover FBI agent the desire to purchase four hand grenades and two pistols. He traded two stereo speakers for the grenades and one pistol, and as the U.S. Attorney stated, he fit the profile of a "wannabe" terrorist. Shareef planned his attack on the mall with the thought to maximize casualties and, it being Christmastime, he hoped to panic the citizens across the United States.

- Police in London arrested twenty-four persons for conspiring to hijack and destroy ten airplanes. The planes in question were departing London and headed to the United States, and the explosives were going to be hidden in the carry-on luggage. As a result of these arrests, liquids were thereafter banned from carry-on luggage, as the explosives were allegedly liquid-based. The explosives most likely would have been triacetone triperoxide (TATP) and hexamethylene triperoxide diamine (HMTD). The use of TATP has attracted the attention of terrorists due to its ease of manufacture using common chemicals.

2007

The American Embassy in Greece was hit with a rocket-propelled grenade (RPG) in the second such attack of the American Embassy. The Marxist group Revolutionary Struggle is suspected in this attack. There were no injuries or deaths as a result of the attack, although the Embassy suffered some minor damage. The Greek terrorist group November 17 had previously attacked the embassy in 1975.

---

intimidation of the government or the civilian population. A militant group trying to influence the local political process sprayed Salmonella bacteria on a fast-food restaurant salad bar and was successful in making more than 600 people sick. The Tokyo subway sarin attack was an attempt to destroy a good portion of the police department in an effort to prevent a raid on the terrorist compound.

> ### SAFETY
>
> The fire service will be called to many incidents that will not fit the exact definition of terrorism, but the hazards from a pipe bomb are the same regardless of the motivation of the builder.

Many responses that would have been routinely handled in the past must now be treated much differently, and responders must always be on their guard for terrorist-style devices or potential acts of terrorism.

The terrorist's motivation is to produce fear that may be aimed at the general public or the government. Fear can be provoked by large-scale actions such as the acts on September 11, 2001, the original bombing of the World Trade Center in 1993, or the bombing of the Alfred P. Murrah Federal Building in Oklahoma City. The rail attacks in Madrid affected the outcome of Spain's national election. Even acts that are not terrorism can create terror in the community. The Chicago nightclub incident, which was caused by pepper spray, and the Rhode Island nightclub fire have sparked fear and concern about safety in nightclubs. A terrorist can also incite fear just by planting the thought of potential terrorism or by devising a hoax. The latter scenario is the more likely and can be very difficult to handle from an emergency service perspective.

The thought process for determining if a threat is credible or not has five elements. If the person known or thought to be responsible is determined to have several of these capabilities, it increases the credibility factor:

1. The first of the five qualifiers involves the potential terrorist's educational ability to make a device or agent that, unlike explosives or ricin, is very difficult to attain. To truly make a biological pathogen agent, in most cases, one needs an advanced knowledge of biology. Ricin, a biological toxin, does not require any advanced knowledge compared with anthrax. Some of the threats with letters or packages have misstated the origin of the material, such as calling anthrax a virus, or have misspelled the

agent's name. If the terrorists do not know the true origin of the material or cannot spell the material correctly, they probably do not have the education necessary to make the material they state they produced. This does not take into account a person who may purchase the material.

2. The next qualifier is a person's ability to obtain the raw materials necessary to make the agents. Many of the materials necessary to make chemical warfare agents are banned for sale. Others appear on hot lists, which means they are only sold to legitimate businesses. This would not preclude someone from buying the raw materials on the black market or stealing them from a legitimate business.

3. The third qualifier is the ability to manufacture the devices or machinery required to make the agent. To manufacture chemical warfare agents requires the use of a reactor vessel, which requires about a 10-foot by 10-foot (3-meter by 3-meter) space to produce less than a gallon (3.7 liters). There are some agents that could be produced in a bathtub using backyard chemistry, but these are not the high-end agents that attract much attention. Many people who have attempted to make agents in less than ideal conditions have died during the production process. Many criminals do not take the time to follow standard industrial safety precautions.

4. One qualifier that is often overlooked is the ability to disseminate these agents. The military conducted many tests on chemical and biological warfare agents, and although they have some good methods of dissemination, even they lack a 100 percent effective method of dissemination. The Aum Shinrikyo cult in Japan is a perfect example, as they were a group with millions of dollars in assets and full chemical and biological lab and production facilities. They employed the services of 235 scientists to develop and manufacture chemical and biological agents. The Aum Shinrikyo abandoned their biological weapons program after a full-scale release of anthrax that failed. They used sarin nerve agent twice, the first time in Matsumoto, Japan, in which seven people were killed and 200 injured. The dissemination method used in the Matsumoto attack was much more effective than the one used in the Tokyo subway attack. If they had used the same dissemination method they would have greatly altered the course of events. They would have been limited by the amount of agent that could have been produced in a short period of time.

5. The last qualifier, which is the most important, is whether the person or group has the motivation to pull off the attack. The intentional killing of one person takes significant motivation, and the intentional killing of hundreds requires a whole lot of motivation. There is always the potential for an attack, but it takes considerable education, raw materials, manufacturing, and dissemination ability to pull off a chemical or biological terrorist attack. On the other hand, explosives are easy to manufacture and do not require much education, only simple tools, and the materials required are easy to assemble. It is for this reason that explosives are used in the majority of cases and are quite successful in completing an attack. In many cases the terrorist can be successful because of the hysteria associated with a potential terrorism incident. A balance must be struck between a cautious approach and one that does not allow the terrorist to win by crippling a community and causing hysteria.

Another consideration, as we have seen in England and Ireland, is the disturbing trend of viewing emergency service personnel as targets. One theory currently under examination is that the second explosion at one of the Atlanta abortion clinic bombings was aimed at the emergency responders.

SAFETY

Responders must always be alert to the potential of a secondary device. Although the use of secondary devices in the United States is not common, the loss of responders' lives would be a high priority for a terrorist. The use of secondary devices has been a common tactic and is still being used overseas. Terrorists, especially those who are patient and study methodologies and strategies worldwide, will use this tactic. It is a diversionary tactic that responders should be aware of and, although responders should not overanalyze an event, they should keep secondary devices in the back of their minds. Just as the use of toxic industrial chemicals and suicide/homicide bombers have been predicted and now have come to fruition, the use of secondary devices could endanger significant responder lives.

POTENTIAL TARGETS

Potential targets exist throughout every community in the nation and can be commercial buildings, high-rise buildings, and even residential homes. Although some incidents do not fit the definition of terrorism, the materials used are the same as those a terrorist might use. Whether the objective is murder or terrorism, the danger to the responder is the same. When looking at terrorism, potential targets can be grouped into several categories: public assembly such as the area shown in **Figure 30-3;** federal, state, and local public buildings; mass transit systems; high economic impact areas; telecommunication facilities; and historical or symbolic locations.

While obviously not an exclusive list, buildings that could be targeted include the Federal Bureau of Investigation (FBI); the Bureau of Alcohol, Tobacco, and Firearms (ATF); the Internal Revenue Service (IRS); military installations, Social Security buildings; transportation areas; city or county buildings, including fire and police stations; abortion clinics and Planned Parenthood offices; fur stores; laboratories; colleges; cosmetic production/testing facilities; banks; utility buildings; churches; and chemical storage facilities. Transportation facilities such as airports and train, bus, or subway stations are high on the potential

list of targets, given the number of people who may be potential targets and the relative ease of escape. In the southeastern United States, a large number of churches have been subjected to arson fires, and in some cases explosive devices have been used. A large number of abortion clinics have been the subject of bombings, attacks, and other threats. Any incident in or near one of these facilities should be approached with caution. Responders should know the location of these facilities in their jurisdiction. Preplans for these facilities should be thought out by the company officers, but it is not recommended that these plans be committed to paper. As the battle between pro-life and pro-choice groups continues to rage, incidents at these facilities can only be expected to rise, with emergency responders caught in the cross fire.

Many of the potential targets of terrorism have not been buildings at all but events where large numbers of people are present. The Atlanta Olympic Park bombing is an example. Other scenarios involve sports stadiums, such as the one pictured in **Figure 30-4,** public assembly locations, transportation hubs, and fairs and festivals. First responders should have some preplans for these types of locations. One possible scenario for a stadium, devised by Captain Richard Brooks (ret.) of the Baltimore County Fire Department, describes the first-in medic unit arriving at a stadium where in Section 300 there are 40 people projectile vomiting. After five minutes 200 people are projectile vomiting, and as time goes on the number increases. What happens to the responders when confronted with a situation of this nature? How many responders would be affected by this massive amount of people vomiting? How many responders would be

FIGURE 30-3 Any location is a potential target for a terrorist. Any location where large numbers of people are present, such as a mall or sports event, is a prime target.

FIGURE 30-4 Other than special events, the most common location where large numbers of people are together is at sporting events. At this stadium, if an incident were to occur, more than 50,000 people could become part of the incident.

needed to handle this incident? This act of terrorism could be accomplished by putting syrup of ipecac in the ketchup container beside one of the hot dog vendors, an easy task. Imagine the hysteria if a note was found stating that a biological agent was distributed in that section. What impact would that have on the remaining 50,000 people in the stadium if that information got out? Planners and responders involved on the national level in trying to develop response profiles to terrorism are grappling with how to plan for incidents involving 100 people, 1,000 people, 10,000 people, and 50,000 people. Terrorism incidents can very quickly overwhelm the responders and their whole emergency response system.

When dignitaries visit locations, a lengthy planning process typically takes place in which the fire department should be involved. When the Pope visited Baltimore in 1997 the planning process took more than eight months. Planning for such a large event takes the cooperation of local, state, and federal agencies. Even when dignitaries visit locations such as New York City or Chicago, advance planning occurs. Other events such as political conventions or other large political gatherings all bring the potential for an incident. When one of these events comes to a community, local responders need to be prepared for not only the people arriving at the event, but also the massive federal response that may be pre-positioned. For many special events, whole task forces of federal resources may be hidden away just in case of an incident.

Certain dates have significance to several militant groups. The date of April 19 is the anniversary of the Waco, Texas, incident in which the ATF stormed a compound that housed the Branch Davidians, a group thought to be a militia group. April 19 was also the date of the Oklahoma City bombing, and the date was chosen by the bomber as a way to retaliate for the Waco incident. Other dates provoke the potential for terrorist acts. For instance, the anniversary of *Roe v. Wade,* January 22, could incite a strike by antiabortion groups.[1] Within the United States, forty states are suspected to have members of militia, patriot, or constitutionalist groups. Membership counts vary from fifty people in some states to several thousand members in other states. Groups of concern include anarchist groups and white supremacy groups such as the Ku Klux Klan, Zionist Occupation Government, skinheads, and neo-Nazi groups, including the Aryan Nation. Other groups suspected of activity or thought to have terrorism potential include patriots, New World Order militia, constitutionalists, and tax protesters. To learn which groups are active, it is relatively simple to use a Web browser and search the Internet for many of these groups, because most have Web sites.

INDICATORS OF TERRORISM

An explosion or explosive device is the most common tool of the terrorist, and police across the country have made several arrests of persons for making or storing large quantities of explosives.

FIREFIGHTER FACT

According to an FBI source, more than 93 percent of terrorism incidents use explosives as the weapon of choice.

The most common device is a pipe bomb such as the one shown in **Figure 30-5.** Any incident where an explosion has occurred or first responders believe that an explosion has occurred should be suspected of being a terrorist incident.

If one explosion has occurred firefighters should always be aware of a possible secondary device at the incident location. Identifiers of a secondary device are as follows: containers with unknown liquids or materials, unusual devices or containers with electronic components such as wires, circuit boards, cellular phones, antennas and other items attached or exposed, devices containing quantities of fuses, fireworks, match heads, black powder, smokeless powder, incendiary materials or other unusual materials, and suspicious or out of place packages.

In this day and age any suspicious package should be suspected of containing explosives and should be dealt with by a bomb technician. Firefighters should never assume that they can handle the package or remove it from the area. Just like EMS and hazardous materials, the handling of bombs or suspicious packages is a very specialized field and should only be done by a person trained to do that type of job.

The presence of chemicals or lab equipment in an unusual location, such as a home or apartment, is an indication of possible illegal activity. When looking at chemicals or a lab there are three possibilities: drug making, bomb making, or terrorism agent production (chemical or biological). The most likely, based on statistics, is drug making, followed by bomb making. Although very common in some parts of the country, predominantly the West Coast and the Northwest, drug production labs are not commonly located throughout the whole United States. Currently the number of drug labs is increasing in the Midwest, with responders taking down more than twenty a month in some areas. This trend is slowly moving from the West Coast to the East Coast. The most likely scenario is locating someone making explosive devices, as many individuals like to make homemade devices. It is possible, but very unlikely, that firefighters would locate a facility attempting to make a terrorism agent such

FIGURE 30-5 The most common explosive device is a pipe bomb, and it is very effective. It is a very dangerous device, not only for responders but for the builder as well.

as sarin. The exception to the biological agents would be for the production of ricin, as the items required to make ricin are easily obtained and the production is just as easy. Fortunately, ricin does not have the potential to easily kill large numbers of people. It is primarily an injection hazard, although it is still very toxic through inhalation or ingestion. A responder in a small community or a rural setting is the most likely to run across any of these types of production areas. Several persons a year are arrested for the possession of ricin with the intent to use it for some type of criminal act. Most of the arrests are occurring in small towns throughout the United States. No matter what agent is located, the responder should immediately isolate the area and call for assistance. The call should go simultaneously to the police, the hazardous materials team, and the bomb squad. This request would apply to all three types of labs—drug, bomb, or warfare agent—as any of these labs should be handled as a cooperative effort.

Another indicator of potential terrorism is the intentional release of chemicals into a building or the environment. Finding a chlorine cylinder in a courthouse would be unusual and should put the responders on alert to the fact that there is a high probability that a terrorist may be at work. Finding chemical containers such as bottles, bags, cylinders, or other containers in unusual locations would also be suspect. In an industrial facility in which chemicals may be common, responders may find that there is the intentional release of a chemical.

> **NOTE**
>
> One of the best indicators of potential terrorism will be a pattern of unexplained illness or injury. If these indicators show up immediately, then the probable cause is a release of a chemical. If these illnesses start showing up a few days to a week later, then the incident is probably biological.

A response to a mall for a seizure patient is not unusual. However, a response to a mall for six people having seizures is very unlikely and could involve a chemical release from a terrorist attack. Seizures, twitching, tightness in the chest, pinpoint pupils, runny nose, nausea, and vomiting are all signs and symptoms of a warfare agent attack. EMS providers will probably be the first group to identify the use of a chemical agent. Imagine arriving at an explosion where there is a large amount of debris and twenty victims. To most the injuries would appear to be blast injuries, as they would in most cases be visible during a quick survey of a patient. Other signs and symptoms, primarily pinpoint pupils, probably would not be identified until the patient is given a more thorough exam. When confronted with victims that are unconscious or dead and now have outward signs of death, such as trauma, the responder must face the possibility that they may have been the victims of a chemical attack. This obviously has to be put in perspective. If responders in a metropolitan area of the Northeast are called to an apartment building in the winter because there are twelve unconscious people, and this happens several times each winter, it is probably carbon monoxide poisoning. However, if responders are called to a mall in the summertime because there are twelve unconscious people, it is probably not due to carbon monoxide but to a chemical release of some type. When distributing a warfare agent the explosive device will usually not have a destructive effect on the building or the surrounding area. In some cases the larger the device the more likely it is that the detonation will consume the agent. Most of the victims will not display signs or symptoms of a blast, although the one closest to the device may suffer some of those types of injuries.

Smelling unusual odors or seeing a vapor cloud may be an indicator that chemicals have been released. As may be seen with other toxic gases such as chlorine, arriving and finding dead birds or other animals should alert the responder not only to a chemical hazard but also to the potential for terrorism. It would also be unusual to respond to an office building or a house and notice security measures of the type that one would not expect in that occupancy. Items such as extra locks, bars on windows, surveillance cameras, fortified doors, guards, and other unusual protection devices may provide a clue to the responder that something out of the ordinary is at play.

> ### SAFETY
> The same hazards that exist in a chemical emergency—thermal, radioactive, asphyxiation, chemical, etiological, mechanical, and psychological—also exist at chemical or biological attacks. In some cases, the risk might be increased, since these incidents are intentional and may be intended specifically to kill or injure responders.

HAZARDOUS MATERIALS CRIMES

Although this chapter covers terrorism, emergency responders may respond to other criminal-related events. Whether the incident is terrorism related or not, the effects on responders can be the same. Criminals are using more weapons than just guns today, and the use of clandestine labs for illicit production is on the rise. These incidents are referred to as hazardous materials crimes, since chemicals are being used in an illegal fashion. Occasionally, there will be a robbery of a convenience store where the weapon is a chemical. The robbery suspect might throw a corrosive liquid on the clerk in order to rob the cash register. There have been other robberies, attacks, or attempted murders using corrosive liquids. Whether or not the law-enforcement community or the prosecutor decides the

incident is an act of terrorism, emergency responders need to recognize potentially hazardous situations and have the ability to protect themselves.

One big exposure issue for emergency responders is drug related. Many drug addicts use chemicals as part of the process to get high, and the drugs themselves are usually toxic and may present other hazards. Drug users may use a combination of chemicals to get high, and many of these items are toxic and flammable. Ether is used to assist in the heating process of several types of drugs. This material in pure form is extremely flammable, and the container may eventually become a shock-sensitive explosive. Most drugs are in solid form, which means that they present little risk unless eaten or touched with bare hands. In some cases the drugs may be stored or used with flammable and toxic liquids.

The most common situation in which emergency responders could be directly affected by drug use is when a person is huffing. While people are huffing they typically use a toxic and/or flammable material. Many common household items such as paints, glues, hair spray, and solvents provide the high that some people desire. From a hazardous materials point of view, these materials are toxic and flammable and airborne. In most cases, users will spray the material into a bag to concentrate the vapors, but when they are done the vapors remain in the room of use.

Clandestine Labs

There are several types of clandestine labs that emergency responders are likely to encounter. The most common is the drug lab, but other possible labs include explosives labs, chemical labs, and biological weapons labs.

> ### SAFETY
> All labs have inherent dangers for responders, and all are very dangerous locations to occupy. All may be booby-trapped or be set up to harm responders, and a booby trap does not know if the person coming through the door is a police officer, a firefighter, or an EMT.

For the most part a biological weapons lab, shown in **Figure 30-6,** can run unattended without any major concern and may be shut down in any number of ways without major consequence. A drug, explosives, or chemical lab, on the other hand, should only be shut down by someone trained to do so, as these labs are especially dangerous.

Drug Labs

There are many types of drug labs. Just for methamphetamines, there are more than six common methods of production. In 1973, the Drug Enforcement Admin-

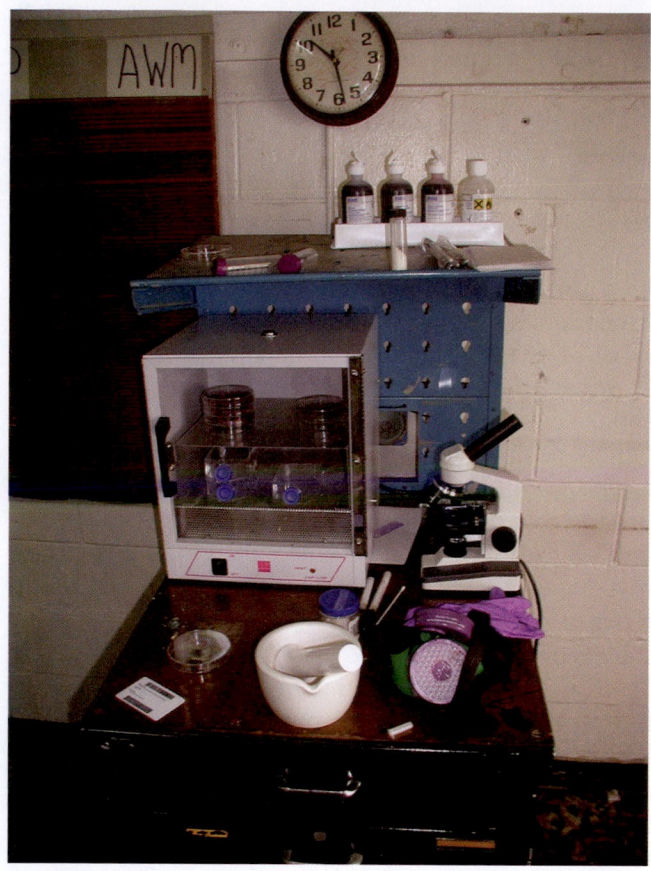

FIGURE 30-6 Example of a possible biological lab. Shown is a microscope and an incubator; both would be used in the production of biological materials.

istration (DEA) discovered 41 labs; in 1999, they discovered 2,155 labs; in 2001, they raided 12,715 labs; and in 2006, they discovered 6,435 labs. The map shown in **Figure 30-7** provides proof that the labs have moved eastward at a rapid rate; and now all of the United States has to deal with this problem. Due to its popularity, meth production is becoming common but it does involve a dangerous process. The production of drugs requires the use of many chemicals. These chemicals can be purchased outright, stolen, or manufactured using other chemicals. As many drug-producing chemicals are hot listed or cannot be purchased, the producer must resort to innovative methods to produce the chemicals. The "nazi" method of meth production involves the use of anhydrous ammonia, which is usually stolen from a chemical facility. If a 150-pound (68-kg) cylinder of anhydrous ammonia developed a leak or catastrophically failed, people for a considerable distance downwind could be affected, and those in the immediate area would be in grave danger.

Drug labs can be very complicated setups and can be found in any number of locations such as homes, barns, hotels, storage units, and even trucks. Emergency responders routinely encounter these labs inadvertently through other responses. Responders who encounter a drug lab should notify their hazardous materials team, the bomb squad, and the local office of the DEA. The shutting down of a drug lab is a complicated and very dangerous process.

FIGURE 30-7 Methamphetamine lab seizures across the United States. Note the high number of labs in the Midwest. In years past, the largest numbers of lab seizures occurred in the West. The prevalence of methamphetamine labs is moving eastward at a fast pace.

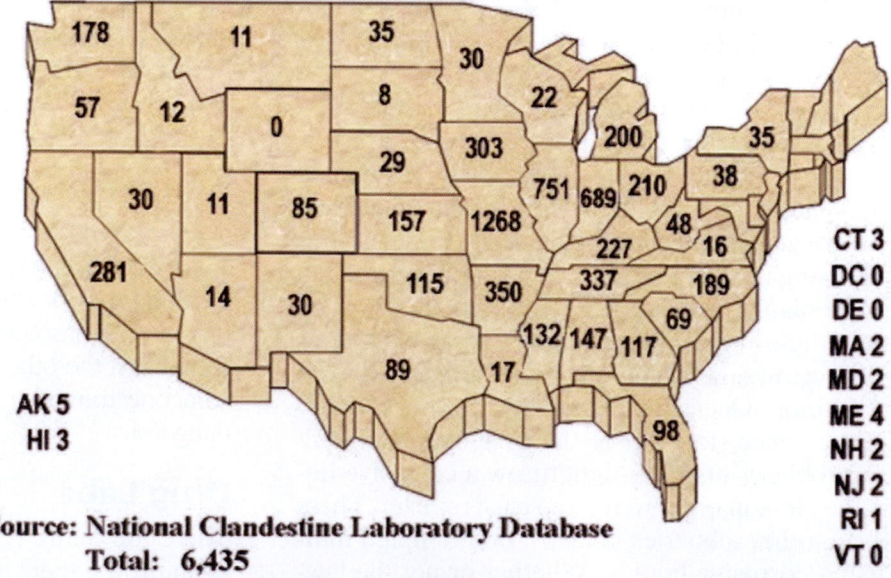

**Total of All Meth Clandestine Laboratory Incidents
Including Labs, Dumpsites, Chem/Glass/Equipment
Calendar Year 2006**

AK 5
HI 3

CT 3
DC 0
DE 0
MA 2
MD 2
ME 4
NH 2
NJ 2
RI 1
VT 0

Source: National Clandestine Laboratory Database
Total: 6,435
Dates: 01/01/2006 - 12/31/2006

Map last updated February 2007

It is this heating and cooling process that indicates the type of lab that may be present. When responders see glassware that is distilling chemicals—in other words, evaporating a certain component—and then rehydrating or condensing another portion of the original chemical, this is indicative of possible drug production. The end result of the process is a solid form, usually a powder. There will be a production line-type formation to the glassware, with some solutions being heated while others are cooled. At some point, gas cylinders may be present and the gas is allowed to mix with some part of the process. Many of the chemicals involved in the production of drugs are flammable, and although many are also toxic the predominant hazard is flammability.

The materials that are toxic cannot harm the responder as long as SCBA is worn and the materials are not touched with bare hands or eaten. It is highly recommended that personnel from the police, fire, and EMS departments receive training in drug lab awareness, with some receiving specialized training in this area.

Explosives Labs

Although not common, it is possible that emergency responders might encounter an explosives lab, which is the predominant weapon of choice for a terrorist. An explosives lab can be anything from a workbench pipe bomb builder to a full chemical production facility. A person building a pipe bomb does not need much equipment other than some simple hand tools, pipes, caps, and powder. A person making cyclotrimethylenetrinitramine, commonly referred to as RDX, or another more sophisticated explosive will need more equipment. Depending on the availability of some materials, the bomb maker may have to make some chemical components as opposed to purchasing them. It is when materials are made at home that the danger increases for a responder. The processes to make the chemical components for explosives are very dangerous and present a significant risk to the builder and responders. An explosives lab differs from a drug lab in that most of the processes do not involve heating, cooling, condensing, or distilling. However, some processes used to make the chemicals do perform some of these functions, so there are no black-and-white rules for identification. The major work at an explosives lab is usually mixing of materials, in most cases solids and liquids. If gases are used, they are usually coupled with the explosive device and may be used to increase the heat of the explosion, boosting its efficiency. People making explosives will usually have a large amount of powders in their house.

Some of the other indicators of an explosives lab are the presence of ignition devices, boosting charges, or blasting caps. Many people who manufacture explosives are doing so to make fireworks and will have cardboard tubes for the explosives to be placed into. In one case, it was originally thought that an explosives maker was just making fireworks, but some other explosive devices were found to have BBs and nails taped and glued to the outside of the explosive. Neither of these have an effect on the display ability of the explosive device and are only designed to kill or maim. When an explosives lab is discovered, the local bomb squad and the local hazardous materials team should be called in to assess the materials. As the stability of many of the materials cannot be ensured, it is usually necessary to remotely destroy many of the found materials.

Terrorism Agent Labs

These are the least likely labs to be encountered in emergency response, but responders must be aware of their existence and some of the unique features of these types of labs. The two types of terrorism agent, or **weapons of mass destruction (WMD),** labs are chemical weapons or biological weapons labs. Statistically the most likely lab is a biological toxin lab, which may be used to manufacture ricin, which is shown in **Figure 30-8.** The FBI, on average, arrests a

FIGURE 30-8   Shown is a lab that could be used to produce a biological toxin such as ricin.

few people a year for the possession of ricin, usually after the person has made a threat. Biological labs may be set up to attempt to make other biological materials. Other than ricin, botulinum toxin, and a few other biological materials, the manufacture of biological weapons is very difficult. The manufacture of ricin is a simple process that only requires a few items, such as castor beans and some readily available chemicals. The process to make some of the more advanced biological agents such as anthrax is more difficult and requires a higher level of education, sophisticated equipment, and access to raw materials. The development of biological agents involves culturing the material, usually in a petri dish. These petri or culturing dishes may be placed in an incubator or oven-like device to keep the material warm and at a constant temperature. Depending on the type of agent, there may be grinders, dryers, and sieves present to finish off the product. The major route of entry for many of these products is through touch or ingestion. The only inhalation concern would be during the grinding process, but a simple high-efficiency particulate air (HEPA) mask offers more than enough protection for a responder. Some of the materials used as part of the process may be flammable and have some toxicity, but the chemicals used to clean glassware and tools are usually highly corrosive and in most cases will be sodium hydroxide (lye) and/or bleach. A bioweapons lab will involve some chemicals, but may resemble more of a greenhouse than a chemical lab. Some biological materials are sensitive to light and will be kept in the dark, and the characteristic distilling and condensing glass apparatus will be missing.

Chemical weapons labs use two methods of production: the development of new chemical agents through standard production methods and the synthesis of existing materials. The development of a new chemical agent is the more difficult of the two processes and is nearly impossible except for someone with a chemistry background and access to some chemistry equipment. The recipes for chemical weapons are fairly sophisticated and require access to many raw materials that are hot listed. There have only been two arrests of persons in the United States for attempting to make a chemical weapon such as sarin nerve agent. Both were arrested for ordering the raw materials, and it is thought that the two individuals did not have the educational capacity to manufacture the agent. The production of these agents can be very risky to the producer and, without safety precautions, may result in death. Some of the off gases from the production of chemical warfare agents are extremely dangerous.

The most probable scenario for the development of a chemical warfare agent is the synthesis of an existing product. The criminal would take existing materials, which are usually in diluted form, and synthesize them or reduce them down to a concentrated product. Luckily most, if not all, of the existing products do not present much risk to humans, as they are strictly engineered to harm only insects. This would not stop a terrorist from attempting to use one of these products in an illegal manner, however. There are some pesticides on the market that are applicable to this type of scenario. The standard household pesticide usually has 0.05 to 0.5 percent of pure product mixed with an inert ingredient. A mixture that is used by a farmer may be on the order of 40 to 50 percent pure form and is then diluted in the farmer's tank. In order to be more harmful to humans, the pesticide must be concentrated and not diluted. The criminal must devise a method of removing one or more of the inert ingredients. As the inert ingredient is usually flammable or combustible, this is not a difficult task. A mechanism must be devised to off-gas the inert ingredient and then capture the pesticide and collect it.

One form of pesticides, known as technical grade pesticides, is already concentrated. These technical grade pesticides are in pure form and are not diluted. These materials are not in common use but can be found in and around the United States.

INCIDENT ACTIONS

A terrorist incident combines four types of emergency response into a large incident.

> ### NOTE
>
> These types of incidents have these four characteristics until proven otherwise: Mass casualty EMS incident + hazardous materials release and/or explosive devices + crime scene considerations = incident management challenges.

The first-in companies at these types of incidents can easily be overwhelmed and are going to be committed to basic actions, such as life preservation. The IC will have enormous responsibilities dealing with all of the required actions. All of the components present are difficult to handle individually, and now in this type of incident they are combined and must be handled simultaneously. The handling of a 100-person **mass casualty** incident is difficult and has the potential to overload the ICS, and that may only be one-quarter of the result of a terrorist incident. Responders should examine their response systems and determine how many patients present a concern. Some systems define a mass casualty as an incident involving five victims, while in other systems it may be ten to twelve. Obviously in a system with one EMS unit, more than one patient begins to cause system problems and may involve delays in treatment and transportation.

Imagine responding to an explosion at a mall, with 100 people injured. Such a mass casualty incident (MCI) will have EMS playing a predominant role. The police department, on the other hand, will be concerned with evidence preservation and crime scene considerations. The possibility exists that the perpetrator(s) also used a chemical or biological weapon, and the explosion was the means of distribution. This situation would be an MCI, a crime scene, and a potential chemical release situation with some, if not most, of the patients contaminated. Add to this scenario the potential for a secondary explosive device, one that is aimed at the responders! If responders can eliminate the chemical and secondary device issues, then all they have to deal with is the simple 100-person MCI and the crime scene issues. EMS will handle the patients, and the police department will handle the criminal element.

Identifying whether an incident is a crime, a terrorist act, or just an emergency situation can be difficult. In many cases, in the first few moments, there may be no way to determine a cause. Even a small explosion could be from a number of sources. Differentiating between a chemical and a biological attack can sometimes be done by observing the immediacy of the symptoms. Chemical agents or toxic industrial chemicals cause immediate reactions and symptoms. Biological agents do not cause immediate effects and may take anywhere from days to several weeks for symptoms to show. Part of the issue that responders have to deal with is the potential public hysteria regarding biological-threat agents. In some cases, the population may be extremely frightened and may exhibit some psychologically driven medical problems. For some, the symptoms may be real, and for others, they may be suspect.

The other aspect of a suspected terrorism incident will be the massive response from the federal government, even if not requested. Later in this chapter more information is provided about federal resources that are available to respond. In most cases, the minimum response to a suspected terrorism incident would come from the FBI, initially from the local agent. Field offices are located across the country, and almost every major city has a field office. Agent(s) usually arrive one to two hours into the incident. If terrorism is suspected, the FBI is the lead agency of the incident as provided by a Presidential Decision Directive known as **PDD 39.** As with hazardous materials incidents it is important to know all involved players

before an incident. Knowing who the local FBI agent is can be very important, because meeting for the first time in front of an incident is not conducive to effective scene management.

Working out the "who's in charge" concerns prior to an incident is important. The fire department should liaison with its local police, EMS, and emergency management agency prior to incidents. These and many other agencies are going to be involved and have a variety of responsibilities at an incident. In general, a unified command is recommended, and although this does not mean command by consensus, input from the various agencies should be considered. The agency with the majority of the tasks to do is generally in charge. In the initial stages the fire/EMS authority would be in charge while rescuing victims, but after the victims are removed and evidence recovery becomes the next priority, the command may switch to the police department.

The magnitude of response to a terrorism incident can be the most difficult part of the incident to manage. The IC will be overtaken by a large number of federal agencies' representatives, who on a regular basis will be replaced by later arriving supervisors. Response groups consisting of two to seventy responders may arrive uninvited, all trying to assist with the incident. Another group that can overwhelm the system is the media. In most cases the local media will react as normal, and for the most part will cooperate. Members of the national media do not know, nor do they follow, local protocol, and they will require information. A lack of accurate information can be disastrous to an incident by causing a deterioration of the incident and creating more hysteria than is already present. Media relations are very important in these types of events.

> **NOTE**
>
> If an incident occurs that involves federal responders, the national media will follow close behind.

One of the most important issues that will arise other than the life safety hazard is that of evidence preservation. As much care as possible should be taken to preserve evidence and make sure it is taken care of appropriately. The collection of evidence is primarily a law-enforcement responsibility. Unless properly trained, fire service personnel should not collect evidence but should alert the police of its presence. A whole host of issues goes along with the collection of evidence, including the chain of custody. This chain of custody, or the paper trail that follows any evidence, is crucial to the successful prosecution of the persons responsible. The failure to follow proper procedure or document the travel of evidence

SAMPLING AND EVIDENCE COLLECTION

One of the major concerns of local law enforcement and the FBI regarding the prosecution of a terrorism case is the purity of the evidence. In order for the evidence to be used in a successful prosecution, the evidence must be pure beyond all doubts. If there is any suspicion that the evidence has been tampered with, altered, or contaminated, it could affect the outcome of the prosecution. When responders enter the hazard area, there is great potential for evidence to be destroyed. The best way to proceed is to coordinate your efforts with your local FBI WMD coordinator, who can assist you with the preservation of evidence. Evidence collection is very process driven, but procedures can be implemented that will help preserve any potential evidence. The gold standard for any courtroom is laboratory analysis, so evidence must be available to be analyzed by a laboratory. This doesn't mean that every incident requires that material be sent off to the laboratory, but steps must be taken to preserve a portion of the evidence in the event that laboratory testing is required.

When there is a large amount of potential evidence, it can be overwhelming, but following a process helps eliminate error and preserve evidence. There is great potential to become involved in a court case, and following the proper procedures every time minimizes your risk in the courtroom. The one time that you don't follow evidentiary procedures will be the time that you are grilled in the courtroom for contaminating the evidence. The FBI and local law enforcement have response teams that collect evidence in these situations, and they can provide great assistance when trying to process a potential crime scene.

can result in a case being dismissed, regardless of any other evidence. There have been cases, such as the Murrah bombing, in which fire service personnel have collected evidence, but that is a very unusual occurrence. Firefighters who are assigned to collect evidence should be fire investigators or fire marshals because they are typically trained in evidence collection. Another alternative is to double up and use a firefighter and a police officer to collect evidence.

A cooperative effort is needed to combat a terrorist attack. The primary functions are rescue/life safety by fire and EMS personnel, hazard identification by the hazardous materials team, identification of possible secondary devices by a bomb technician, and incident management. It is important to communicate the hazards to all personnel, and to limit the response to essential personnel. Instead of having the whole alarm assignment report to the front of the building, it is preferable to use one or two companies to investigate while staging the other companies away from dumpsters, mailboxes, or dead-end streets. A secondary device can be hidden almost anywhere, but the key is to look for something out of the ordinary. When dealing with victims it is essential to isolate them until the cause is identified. The victims can have a large amount of information and should be questioned quickly. Questions to ask include these: What did you see, hear, or smell? Was this coming from one area or was it throughout the building? Did you see anything else suspicious? What type of signs and symptoms do you have? In addition, the police will need to conduct interviews. Documentation and preservation of the evidence are essential to the successful prosecution of the terrorist.

GENERAL GROUPINGS OF WARFARE AGENTS

Terrorists could use any of a number of possible warfare agents. They are classified into three broad areas. Weapons of mass destruction are commonly used by the military. Some of the regulations that prohibit the making, storing, or using of terrorism agents are called WMD laws or regulations. Any item that has the potential to cause significant harm or damage to a community or a large group of people is considered a WMD. The other two classifications are nuclear, biological, and chemical (NBC) and chemical, biological, radiological, nuclear, and explosive (CBRNE). Both of these are descriptions of the types of materials that could be used in a terrorism attack. Although there are slight variations, they are all used to describe the various types of agents that a terrorist could use. Most of the language differences come from funding legislation or a specific federal agency.

The military has devised a naming system for many of these agents, many of which are listed in **Table 30-1.** Responders should become familiar with these names because much of the literature and help guides refer to these agents by these names. For instance, when using a military detection device, the military name is used. When dealing with terrorism, firefighters are entering another world that has its own language. The fire service has to adopt this new language to survive in this new world. The three groupings mentioned earlier are further subdivided into the categories discussed in the following subsections.

SIGNS AND SYMPTOMS OF NERVE AGENTS

All of the nerve agents present the same types of signs and symptoms as organophosphorus pesticides, and in reality the difference is minor. Nerve agents are pesticides for humans and are a stronger, more concentrated version of commercially available pesticides. The signs and symptoms can be generally described using the acronym SLUDGEM, which stands for:

S alivation—excessive drooling

L acrimation—tearing of the eyes

U rination—loss of bladder control

D efecation—loss of bowel control (diarrhea)

G astrointestinal—nausea and vomiting

E mesis—vomiting

M iosis—pinpoint pupils

The term SLUDGEM describes all of the symptoms from the minor ones to the extreme signs and symptoms. A slight exposure to any of the nerve agents will cause pinpoint pupils, a runny nose, and difficulty breathing. A person who has come in contact with the liquid will be experiencing all the SLUDGEM signs in addition to convulsions. A person who is in convulsions needs immediate decontamination and medical treatment in order to survive. This treatment sequence has to occur in less than five to ten minutes. In addition, there must be sufficient medication available on scene to accomplish the treatment. Most paramedic units carry enough medication to treat one or two patients who have severe symptoms.

TABLE 30-1	**Military Designations for Agents**	
Name	**Military Designation**	**UN/DOT Hazard Class**
Anthrax	N/A	6.2
Biological agents	N/A	6.2
Chlorine	CL	2.3
Cyanogen chloride	CK	2.3
Distilled mustard	HD	6.1
Hydrogen cyanide	AC	6.1
Lewisite	L	6.1
Mace	CN	6.1
Mustard	H	6.1
Nitrogen mustard	HN	6.1
Pepper Spray	OC	2.2 and 6.1
Phosgene	CG	2.3
Ricin	N/A	6.2
Sarin	GB	6.1
Soman	GD	6.1
Tabun	GA	6.1
Tear gas	CS	6.1
Thickened soman	TGD	6.1
V agent	VX	6.1

Nerve Agents

Nerve agents are related to organophosphorus pesticides and include tabun, sarin, soman, and V agent. They were designed for one purpose and that is to kill people. Although very toxic, their ability to kill large numbers of people requires that the dissemination device function correctly and that a number of other critical factors be in place to be truly effective. Although several gallons of sarin agent were used in the Tokyo subway attack, the distribution method was ineffective, so out of the twelve people who died, the only people killed by the sarin itself were the two people who actually touched the liquid. The chemical and physical properties of these agents hinder their ability to be effective as a stand-alone killer. To best produce the desired effect, the agents must touch people in liquid form or be breathed in while the materials are in aerosol form. The materials will not stay in aerosol form for very long, and they have a very low vapor pressure and thus will not create vapors as standing liquid. All of the military warfare agents have a vapor pressure less than water, which means they do not evaporate quickly and unless the liquid is touched or placed on the skin it does not present a large hazard.

Incendiary Agents

For the sake of classification, **incendiary agents** are placed into the chemical classification, because chemicals are used in these devices. The most commonly

SIGNS AND SYMPTOMS OF BLISTER AGENT EXPOSURE

One of the biggest risks with the blister agents is that they may present delayed effects. If not detected this could result in victims being released only to later have problems. In general, blister agents are not designed to kill; they were designed to incapacitate the enemy, resulting in troops being assigned to assist with the wounded. It is possible to create scenarios in which fatalities could occur, but these would be unusual cases. The effects from blister agents include irritation of the eyes, burning of the skin, and difficulty in breathing. The more severe exposure results in blisters, which may be delayed. The only real street treatment for these signs and symptoms is decontamination and supportive measures. It is important to have anyone with liquid contact blot the liquid off the skin and avoid spreading the agent. Fortunately, the chemical and physical properties of these agents make them difficult to disseminate, and coming in contact with the liquid would be the primary means of injury.

SIGNS AND SYMPTOMS OF BLOOD AND CHOKING AGENTS

Many of the blood agents are commonly found in industry and may be found at normal chemical facilities. The signs and symptoms of slight exposure to blood agents include dizziness, difficulty in breathing, nausea, and general weakness. With cyanides, the breathing initially will be rapid and deep, followed by respiratory depression and usually death. The two most common choking agents are very common in industrial use. Signs and symptoms include difficulty in breathing and respiratory distress, eye irritation, and, in higher amounts, skin irritation. Phosgene may present delayed effects, while chlorine's effects are immediate.

used chemicals are flammable and combustible liquids. The standard Molotov cocktail is an example of an incendiary device that could be used by a terrorist. In some cases arsonists have used a mixture of chemicals, usually oxidizers, to create very fast high-temperature fires.

Blister (Vesicants)

The category of **vesicants,** or, as they are more commonly called, **blister agents,** includes chemical compounds called mustard, distilled mustard, nitrogen mustard, and lewisite. These materials were never designed to kill. They were designed to incapacitate the enemy so that if one person was affected by one of these agents several more would be needed to care for the affected person. Although at high concentrations these materials can be toxic, their biggest threat is from skin contact which causes severe irritation and blistering. Their chemical and physical properties make them less of a hazard than the nerve agents. One major concern with these agents is the fact that the effects from an exposure can be delayed from fifteen minutes to several hours. Quick identification of a blister agent is key to keeping the victims safe.

Blood and Choking Agents

The four chemicals discussed here in addition to being terrorism agents are also common industrial chemicals. The first category is **blood agents** and includes hydrogen cyanide and cyanogen chloride. Both of these materials are gases that disrupt the body's ability to use the oxygen within the bloodstream. They are also referred to as chemical asphyxiates. The **choking agents,** chlorine and phosgene, are very common in industry. Chlorine is present in almost every town, because it is used for water treatment processes and in swimming pools. Any community that has a water system or swimming pool has some form of chlorine. Chlorine comes in cylinders of 150 pounds (68 kg) to 90-ton (82 metric tons) railcars. It also comes in the tablet form (HTH) that is typically used in residential pools. The release of chlorine from a 90-ton (82 metric tons) railcar could result in several hundred thousand injuries and possibly an equal number of deaths,

SIGNS AND SYMPTOMS OF IRRITANTS

The signs and symptoms for a slight exposure up to a high dose are the same, with the exception of increasing severity. The signs and symptoms are eye and respiratory irritation. There is no real treatment except removal to fresh air; in fifteen to twenty minutes the symptoms will begin to disappear. Supportive care can be provided.

especially in an urban area. A small amount of chlorine can be very deadly or at a minimum create substantial panic in a community.

Irritants (Riot Control)

The most commonly used materials that are classified as potential terrorism agents are irritants and include mace, pepper spray, and tear gas. An incident that uses an irritant often impacts a large number of people, because the usual target is a school, mall, or other large place of assembly. The use of one small container can affect large numbers of people and make them immediately symptomatic. Luckily these materials are not extremely toxic—although they are extremely irritating—and the symptoms will usually disappear after fifteen to twenty minutes of exposure to fresh air. The response to one of these incidents is difficult because patients with real medical problems need treatment and the source of the irritant is often difficult to identify.

Biological Agents and Toxins

Other than explosives, the most likely agents to be used in a terrorism scenario are **biological agents** and **toxins.** Some of the materials in this grouping include anthrax, mycotoxins, smallpox, plague, tularemia, and ricin. Out of all of the agents a terrorist might make, this grouping is the easiest, especially ricin. The fatal route of entry differs with each of the agents and could occur via skin contact, inhalation, or injection. These agents are difficult to distribute effectively, and in some cases exposure to sunlight may neutralize many of these agents. Some of the potential indicators of biological terrorism include the presence of a powder, a liquid material, containers associated with biology, and in some cases a threat letter expressing the use of a biological material. The use of a dissemination device or placement so that the material is disseminated is usually more robust than with chemical agents. The visual sighting of a dust or vapor cloud being released in an unusual fashion could be indicative of a biological threat agent release. A small explosion followed by an uncharacteristic dust or vapor cloud indicates a possible dispersal device. Any unusual packages, equipment, or boxes that are out of place could be holding a biological material.

The Centers for Disease Control (CDC) have three categories of biological agents that could be used for weapons, as shown in **Table 30-2.** Category A agents are those that can be easily disseminated or transmitted person to person, have potential to cause a large-scale public health emergency, might cause public panic and social disruption, and may require special action for public health preparedness. Category B agents are those that are moderately easy to disseminate, result in moderate to low fatality rates, and require specific enhancements of CDC's diagnostic capacity and enhanced disease surveillance. Category C agents are those that might be emerging pathogens that can be used for mass dissemination in the future because of their availability and ease of production and dissemination, and have the potential for high fatalities and major health impact. Specific information about the two most popular agents is provided next.

Anthrax

Anthrax is a naturally occurring bacterial disease that is commonly found in dead sheep. It is contagious through skin contact or by inhalation of the anthrax spores. Although relatively easy to obtain, it is more difficult to culture and grow the proper grade of anthrax. To produce fatal effects, the type of anthrax required is called weapons-grade anthrax and is very difficult to produce. Even if developed it must be distributed effectively and under the right conditions.

Ricin

Although ricin is easy to make, the required distribution method leaves a lot to be desired because it must be injected to be truly effective. It is 10,000 times more toxic than the nerve agent sarin. A small amount such as one milligram, about the size of a pinhead, can be fatal. Death usually occurs several days after

TABLE 30-2 Biological Agents

Category A Agents

Name	Notes
Anthrax	Noncontagious bacteria, highly infectious through inhalation
Smallpox	Highly contagious virus, with high potential for fatalities
Botulinum toxin	Produced from *Clostridium botulinum* and one of the worst toxins known
Ebola	Also known as viral hemorrhagic fever, high fatality rates
Plague	Caused by the *Yersinia pestis* bacteria. Easy to manufacture and spread by flea bites. Can be inhaled through the spread of pneumonic plague.
Marburg	A viral hemorrhagic fever, with high fatality potential
Tularemia	Caused by the *Francisella tularensis* bacteria. Also known as rabbit fever. Incapacitates as opposed to causing fatalities.

Category B Agents

Name	Notes
Brucellosis	
Epsilon toxin	*Clostridium perfringens*
Food safety threats	*Salmonella spp., E. coli* o157:H7, Shigella
Glanders	*Burkholder mallei*
Melioidosis	*Burkholder pseudomallei*
Psittacosis	*Chlamydia psittaci*
Q fever	*Caxiella burnetii*
Ricin toxin	*Ricinus Communis*
Staphylococcal enterotoxin B	
Typhus	*Rickettsia prowazekii*
Viral encephalitis	Alphaviruses, Venezuelan equine encephalitis, eastern and western equine encephalitis
Water supply threats	*Vibrio cholerae, Cryptosporidium parvum*

Category C Agents

Name	Notes
Nipah virus	N/A
Hantavirus	N/A
Multidrug-resistant tuberculosis	N/A

injection. By other routes of entry, such as inhalation or ingestion, the most likely consequence is that many people would get sick but would eventually recover. After explosives, Ricin is one of the leading choices of domestic terrorists, and several times a year someone is arrested for possession of ricin. Some example ricin cases include:

■ 2002—Kenneth Olsen was arrested for possession of ricin, which he kept in his office cubicle. While at work at a high tech company in Spokane, Washington, he researched other poisons and explosives and left this information throughout the office.

■ 2003—A letter containing ricin was mailed to the White House. It was one of two letters from the "fallen angel" who was protesting the change of work hours being mandated by the Department of Transportation (DOT). The second letter was located at an airport postal facility in Greenville, South Carolina.

ANTHRAX SCARE 2001

The FBI is still investigating the anthrax attacks of October 2001 that started in Boca Raton, Florida, at the American Media building. It is suspected that a letter was sent and opened at this facility, which publishes the National Enquirer. One person died and several others fell ill at this building, but a letter was never recovered from this facility. The letters that were sent targeted the media and members of Congress. The NBC studios, the New York Post, and ABC all received letters. Two members of Congress, Senator Daschle and Senator Leahy, received letters. The letters killed five people and made eighteen others ill. Some of the deaths and illnesses involved people at various post offices. It is thought that the mail handling process enabled anthrax to aerosolize and get into the air. The letters contained small amounts of anthrax, which has been tested to be the Ames strain of anthrax that was produced in the United States. A lot has been learned about anthrax and its ability to be used as a weapon. The theoretical dose of anthrax in each of the recovered letters was estimated to be able to kill several hundred thousand people. The reality is that each letter typically killed one person. The other major factor in the deaths was a delay in treatment and in some cases misdiagnoses. The persons who did not seek quick treatment or who were misdiagnosed had difficult or unsuccessful recoveries. Those who sought quick medical treatment, during which the agent was recognized as anthrax, survived. The CDC reported that there were ten cases of inhalational anthrax and twelve confirmed or suspected cases of cutaneous anthrax. Of the ten inhalational cases, seven occurred to postal workers in New Jersey and Washington, DC, at mail sorting facilities. In the American Media case, one person who received the letter died, and the person sorting the mail fell ill. Six of the ten individuals with inhalational cases of anthrax survived after treatment, which is higher than originally anticipated.

The signs and symptoms of inhalational anthrax are a one- to four-day period of malaise, fatigue, fever, muscle tenderness, and a nonproductive cough followed by a rapid onset of respiratory distress, cyanosis, and sweating. The recent cases also had profound, often drenching, sweating, along with nausea and vomiting.

- 2003—FBI agents arrested Bertier Ray Riddle of Omaha, Arkansas, for sending a letter supposedly containing ricin to the local FBI office in Little Rock.

- 2004—Ricin was located in Senator Bill Frist's office in the Dirksen Senate Office building. It is not known what the source of the ricin was since no letter was found. Frist from Tennessee was the majority leader at the time. There is a possibility that the ricin is connected to several other ricin letters from the "fallen angel."

- 2004—Richard Alberg was arrested for possession of ricin after he placed a large order for castor beans, which are the main component required to make ricin. Although no specific attack was determined, Alberg fantasized about the use of biological agents to kill.

- 2005—Six persons were arrested in London for conspiring to commit terrorist acts. Inside their apartment, the police found ricin and materials that can be used to manufacture ricin. The Islamic group that was arrested had ties to Chechnya and to al Qaeda terrorists who plotted the millenium bombing attempt in the United States. The establishment of an al Qaeda and Chechnya relationship is very troublesome. The Chechnyan region of Russia is responsible for well-coordinated terrorist attacks in Russia. Their large-scale attacks have killed thousands of Russian citizens, and they are the group responsible for the Beslam school attack, the Moscow theater takeover, the downing of two planes, and a subway bombing, among many other attacks.

- 2006—Ricin along with pipe bombs and blasting caps were found in the home of William Matthews of Nashville, Tennessee. Matthews reportedly had problems at work, and his wife had sought a protection order due to his substance abuse. It is not known what he intended to do with the ricin, which was found in a baby food jar.

Radioactive Agents

Nuclear agents, unfortunately, have to be put back on the list of possibilities that a terrorist could use. There are two types of radiation events, nuclear detonation and radiological dispersion. The use of an actual nuclear detonation device is very unlikely given the security these materials have. The amount required and the specific type also make the use unlikely. A **radiological dispersion device (RDD)** disperses radiological materials, usually through an explosive device. One example would be the use of a pharmaceutical grade radioactive material attached to a pipe bomb, which could cause a large amount of radioactive material to be distributed.

RESPONSE TO ANTHRAX HOAXES

Emergency responders should establish a plan for response for noncredible anthrax events. In 2001 and 2002, these events kept hazardous materials and other emergency response teams very busy. From the outset it would be difficult for someone to manufacture and successfully pull off an anthrax threat. The use of real anthrax is most likely an isolated case. One cannot say with absolute certainty, but a new attack is not likely and is not likely to kill or injure mass numbers of people. In the hazardous materials and WMD business one should avoid the use of the words never, always, and best. The credibility factor and the technological difficulty make weaponized anthrax an unlikely candidate for use as a weapon. In many of the hoax incidents, letters proclaimed that the recipients had been exposed to anthrax. In some of these cases the only thing present in the envelope was a letter, which meant that there was no other material present. Anthrax is not invisible, and in order to improve the likelihood that the attack may be successful a quantity of the material must be present. In order to cause health effects the material must be inhaled, and in order to be inhaled the material needs to be distributed to put it into the air. The material, since it is a solid, requires a device to put it into the air. If this device is not present the risk to any persons in a building is very small. The only action required is to double- or triple-bag the envelope, package, or material in bags suitable for evidence collection. Procedures for the collection of evidence should be followed, and the local police as well as the FBI should be consulted. The persons who touched the material should be instructed to wash their hands with soap and water. They do not require full body decontamination nor are special solutions required such as a bleach and water mixture. The military advises that water is more than sufficient for humans. Anyone who was in the immediate vicinity (i.e., several feet) of the person does not require decontamination. The only reason for a full body wash would be if the material was splashed on a person from head to toe, but it is not required to be done immediately. The person can be taken to a shower or be provided privacy to take a shower. The person who did open the envelope should be entered into the health care system, advised of the signs and symptoms of exposure, and provided with emergency contact information. There is no need to start prophylactic antibiotics just because a person opened a letter with a powder in it. There is sufficient time to do lab analysis on the material before medical treatment is required. The threat from a solid material (no matter how toxic) is from inhaling the dust or touching the material. Gloves and SCBA are adequate protection for the collection of evidence. The FBI has a number of labs around the country set up to assist with the identification of WMD agents, and the local FBI office should be contacted for assistance.

The strength of the radiation source dictates how harmful an RDD would be. In many cases the RDD would present more of an investigative issue than a significant health issue. There are some radiation sources that would present a risk, but these are not in common use. The other factor is the explosive device and its limitations. Technologically an RDD is viable for a small area such as one-half acre of land. Attempting to distribute a significant amount of radioactive material takes a significant explosive device, not to mention the radiation source. The larger the RDD the more danger there is to the terrorist who has to assemble, transport, and detonate the device. There is an advantage to a noncredible RDD or a small RDD, and that is the public's reaction. The perception to the public and to many responders is that this would be a radioactive disaster. Radiation meters would indicate that there were radioactive materials spread around. The reality is that the amount of radiation would not be dangerous. As time passed, the danger would lessen as the radioactive material became less hazardous. Still, there would likely be tremendous concern in the community. Radiation causes fear because it is a big unknown, making it a prime weapon for a terrorist. Education on the hazards of radiation and the effective use of radiation monitors can ease this fear and allow responders to make an informed response.

Other Terrorism Agents

Although considerable emphasis is placed on warfare agents, in reality many other common industrial or household materials can be just as deadly, if not more so.

NOTE

Remember that a terrorist wants to create panic, not necessarily kill or injure massive numbers of people.

A likely scenario is a pipe bomb, followed by the use of ricin. Responders should not be lulled into a false sense of security if they do not find a "warfare" agent or if the device is only a small pipe bomb and not a moving truck filled full of ammonium nitrate and

FBI AND CDC LABORATORY RESPONSE NETWORK

When discussing WMD detection, it is important to note the laboratory system that the FBI and CDC have set up. The Laboratory Response Network, known as the LRN, handles the initial analysis of potential WMD materials. The LRNs have been in existence since 1999 but really came to light in 2001 after the anthrax attacks. The FBI partnered with the CDC to ensure that standardized protocols were implemented and that both public health and evidentiary concerns were addressed. The WMD coordinator can assist in getting evidence or samples to a LRN facility. The gold standard for analysis is laboratory testing, which is the desire of the FBI. Without laboratory testing, they can't make their case, and the suspect may be allowed to go free. They do not place much emphasis on street detection methods and generally discourage messing with the potential evidence. Prior to shipment they want the evidence screened for fire, corrosive, toxic, and radioactive hazards. The FBI has a number of hazardous materials response teams throughout the United States and maintains a highly technical response group known as the **Hazardous Materials Response Unit (HMRU),** who can provide assistance when potential WMD materials are found or suspected. **Figure 30-9** shows one their response vehicles.

There are 140 LRN labs set up for biological threat agents throughout the United States, and generally the local or state health departments run the labs. For chemical attacks, there are 62 laboratories that can conduct chemical threat agent analysis. These are the labs that most responders would send their samples to, through the FBI WMD coordinator.

FIGURE 30-9   The FBI's Hazardous Materials Response Unit (HMRU) has responsibility to assist in the collection of evidence at high hazard crime scenes. These hazards typically are WMD materials or other hazardous materials. They are also equipped to handle building collapses, trench scenes, and confined spaces, much like a hazardous materials or USAR team.

fuel oil. Nerve agents (or any other chemical warfare agent) have never been used in the United States, nor manufactured by anyone other than the military. The FBI Bomb Data Center reports that out of the 3,000 bombings that occur each year, there have been only two large bombings. However, these "small" bombs kill an average of 32 and injure 277 people each year.

DETECTION OF TERRORISM AGENTS

The detection of terrorism agents is difficult, and given the potential circumstances is done under severe conditions. The response to terrorism incidents has changed how many hazardous materials teams operate and has increased their capability to handle other situations. The confirmation that terrorist agents have been used is difficult because they are extremely hard to detect. The exceptions are the standard industrial materials such as chlorine, which is easy to detect and confirm its presence.

The detection of terrorism agents addresses three major categories of hazards: chemical agents, radiological materials, and biological agents. There are a number of devices that detect chemical warfare agents. They range from inexpensive paper strip tests (M-8 and M-9) to sophisticated electronic devices costing more than $125,000. The most common are direct reading devices, which detect nerve and blister agents, **Figure 30-10.** Some devices will also detect drugs, explosives, mace and pepper spray, and industrial chemicals. The radiological detection field has dosimeters that track the dose of radiation that one is receiving to handheld instruments. Radiation pagers and pager/dosimeters are very small. Pager-sized devices are available and can detect levels of radiation and alert the user to potential danger. Typical radiation monitors detect the dose a human is absorbing, while some newer models will determine and identify the exact type of radiation source present. The detection of biological agents is more difficult, and there are limited choices. One detection device that is new to the emergency response world is a polymerase

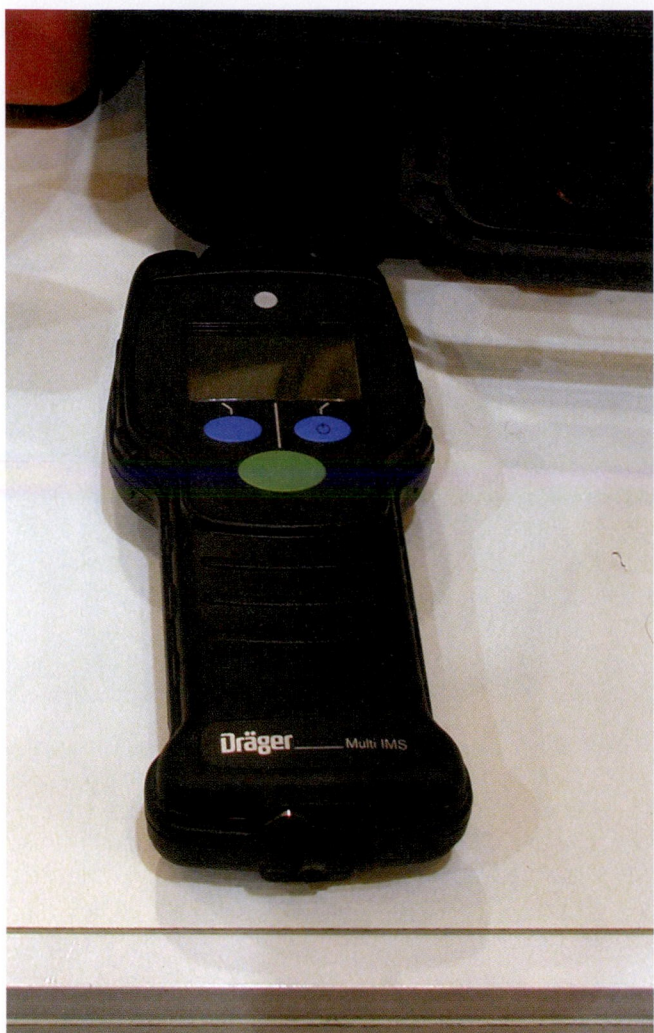

FIGURE 30-10 This detection device has the capability to detect chemical warfare agents such as sarin nerve agent and several toxic industrial chemicals.

chain reaction (PCR) unit, which detects the DNA of the sample and compares it to other DNA samples loaded in the library. It is the same method that is used in the laboratory to detect biological agents. The only other choice is handheld bioassays, which, depending on the brand, function with varying degrees of accuracy. Some work well and have accuracy rates in the 90th percentile, while others have accuracy rates in the 30th percentile. The major issue with any biological agent detection is that the sample collection method has to follow exacting standards without any deviation. The detection of any terrorism agent is difficult, and the hazardous materials team should take responsibility for doing the testing.

In some cases standard civilian detection devices such as photo-ionization detectors also play a role in the detection of terrorism agents. Other tests are used by hazardous materials teams that have applicability in the detection of terrorism agents. Due to the

many mitigating factors, the detection of these agents is going to be difficult, and to be conclusive will probably require lab tests.

FEDERAL ASSISTANCE

The federal government has established roles and responsibilities in the event an act of terrorism occurs, and these roles are provided in PDD 39. Per this document, the FBI is designated the lead agency during the emergency (crisis management) stage of an incident.

> **NOTE**
>
> At the end of November 2002, the Department of Homeland Security (DHS) was formed. This organization is designed to protect the United States against terrorist attacks and to respond to natural disasters. Its goal is to prevent, deter, and respond to terrorism and disaster situations. A number of existing federal agencies were absorbed into DHS. The one agency within DHS that has direct interaction with state and local responders is the Federal Emergency Management Agency.

The Federal Emergency Management Agency (FEMA) becomes the lead when the incident is no longer in the emergency phase (consequence management). The FBI has a Hazardous Materials Response Unit (HMRU) that provides identification, mitigation assistance, and evidence collection for potential terrorist incidents. The HMRU is a multifaceted group that not only responds to terrorism incidents, but also responds to incidents involving explosives, drug labs/incidents, and environmental crimes. The FBI HMRU has responsibility to collect evidence from potentially hazardous environments. A response from FEMA will vary with the incident, as will the number of FEMA personnel, but the response will be similar to any other disaster; FEMA assists in restoration and recovery issues.

The Urban Search and Rescue (USAR) teams fall under FEMA, and are activated by following the emergency management chain from the local level to the state level and then to FEMA for activation. At this time there are twenty-eight USAR teams across the country. They provide expertise in heavy rescue operations such as building collapses. This team of seventy people is composed of rescue specialists, dog search teams, medical specialists, hazardous materials WMD specialists, communications specialists, and an engineering and rigging component. The majority of victims are rescued by the first responders and other local specialized resources. But in some cases trapped victims may be alive for many days and

BASIC INCIDENT PRIORITIES

When dealing with an incident that involves terrorism, first responders should follow these guidelines:

- To rescue live victims, use full protective clothing including SCBA. All of the agents listed in this chapter are predominantly hazardous through inhalation. Avoid touching any unknown liquids or solids, because most are also toxic through skin contact.
- Use a quick in/quick out approach: Do not treat victims, remove them from the area. Keep in mind that the terrorist may be among the injured. Watch for secondary devices.
- Request hazardous materials and the police bomb squad. The sooner responders eliminate the potential for chemical agents or a secondary device, the better off they will be. The hazardous materials team may have the ability to detect the agents listed in this chapter, and most hazardous materials teams are working with their bomb squads on a more frequent basis.
- Limit personnel operating in the hazard area.

- Establish multiple staging areas, out of the line of sight.
- Notify the local emergency management agency so that they can mobilize the state and federal resources.
- If a building has collapsed or there is potential for a building collapse, request assistance from a tactical rescue team or a USAR team.
- Isolate all victims, separating contaminated from clean victims.
- Establish a safe triage, treatment, and transport area away from the impact or hot zone.
- Notify all area hospitals of the incident.
- Remember that the incident is a crime scene and make provisions to preserve as much evidence as possible.
- If there is reason to suspect the presence of chemical agents, use the DOT ERG or other reference sources such as the Medical Management of Chemical Casualties Handbook to suggest safety precautions and patient treatments.

may require the expertise and equipment that a USAR team has available. The unfortunate thing is that the USAR teams have a delayed response, because it takes them several hours to become airborne, not to mention travel time. If there is a possibility they may be needed, their assistance should be requested early. Like many of the teams, it is not uncommon for an advance party to arrive hours prior to the arrival of the remainder of the team.

A number of other agencies may be involved in the event of an incident. The military has a couple of units that have terrorism response capabilities and responsibilities. The National Guard has Civil Support Teams (CST) that are set up like a local hazardous material response team. They are comprised of 22 persons and carry detection devices, protective clothing, state of the art communications abilities, and decontamination equipment. There are fifty-five such teams across the United States, which are activated by the states, even though they are federally funded. The army has the Technical Escort Unit (TEU), headquartered at the Aberdeen Proving Grounds, Maryland. There are other units at Dugway Proving Grounds, Utah; Pine Bluff Arsenal, Arkansas; and Fort Belvoir, Virginia. The TEU is assigned to provide escort service for warfare agents and to be

the troubleshooting group in the event of an incident involving warfare agents or explosives. When such an incident occurs, the TEU will respond to assist with identification and the mitigation of the incident. Team members handle chemical, biological, and explosive materials as well as other hazardous materials. TEU is a self-contained unit and has the lab resources of the Research Development and Engineering Command (RDECOM) at the Aberdeen Proving Grounds available to assist them.

The United States Marines have a unit known as the Chemical and Biological Incident Response Force (CBIRF), which comes from Indian Head, Maryland. This unit responds to acts of terrorism across the world and has three main components: decontamination, medical, and security. It is a self-contained unit and requires only a water source for conducting long-term operations. CBIRF is able to respond nationwide upon request to terrorist incidents and can provide detection and mitigation as well. All of the federal resources including CBIRF, CST, and TEU can be and have been pre-positioned for certain events, such as the Olympics, political conventions, and visits by dignitaries. These resources integrate within the local system, usually hidden away from the public, and are immediately available.

Within DHS, a number of agencies in FEMA have been reorganized. As part of this reorganization, major national preparedness components and functions include The Office of Grants and Training, the United States Fire Administration, National Capital Region Coordination, Chemical Stockpile Emergency Preparedness, and the Radiological Emergency Preparedness Program transferred to FEMA, effective April 1, 2007. Other federal programs include training under the Nunn-Lugar-Domenici Legislation, which was passed in September 1996 (P.L. 104-201). This law, known as the Domestic Preparedness Training Initiative, mandated that the Department of Defense provide training to the 157 major cities and counties across the United States. It is intended to enhance the capability of the local, state, and federal response to incidents involving NBC materials. Part of this process is an assessment to determine the cities' capabilities after the training. Also set up around the country are Metropolitan Medical Response Teams (MMRTs), which are a group of 129 people on each team. They are trained and equipped to handle the medical component of a terrorism event. The plan is to have them in the 120 largest metropolitan areas in the country.

LESSONS LEARNED

Within each community there exist multiple agencies that would respond to a terrorist attack. One of the major issues would be the coordination of those resources. The response to a potential terrorism incident can be very challenging, and every responder must be alert to the possibility of such an incident. To be safe, responders should wear all PPE, not linger in the environment, and relocate to a safe area once the live victims are out. It is important to be aware of the potential for secondary devices and request hazardous materials and bomb squad assistance quickly. One to thousands of victims may be injured or killed. Scenarios involve tremendous loss of life, including a large number of responders. One of the major challenges involves the evidence collection process and collection of potentially hazardous evidence. There exists the possibility that responders may lose and the terrorist will win, a situation that can be avoided by training, planning, and preparing for such an incident.

There are a number of Federal resources that can assist with the response to a terrorist attack, or even a suspected attack. Working with and coordinating this response is an important consideration. Special response companies or teams should know who their local contacts are and should develop a relationship with the Federal liaisons. As the FBI has the responsibility and is the lead agency to investigate an act of terrorism, responders should know that the FBI has WMD coordinators who are responsible for interacting with local responders to help coordinate the overall response.

KEY TERMS

Anthrax   A biological material that is naturally occurring and is severely toxic to humans. It is commonly used in hoax incidents.

Biological Agents   Microorganisms that cause disease in humans, plants, and animals; they also cause the victims' health to deteriorate. Biological agents have been designed for warfare purposes.

Blister Agents   A group of chemical agents that cause blistering and irritation of the skin. Sometimes referred to as vesicants.

Blood Agents   Chemicals that affect the body's ability to use oxygen. If they prevent the body from using oxygen, fatalities result.

Butyric Acid   A fairly common lab acid that has been used in many attacks on abortion clinics. Although not extremely hazardous, it has a characteristic stench that permeates the entire area where it is spilled.

Choking Agent   Agent that causes a person to cough and have difficulty breathing. The terrorism agents that are considered choking agents are chlorine and phosgene, both very toxic gases.

Hazardous Materials Crime   A criminal act that uses or threatens the use of hazardous materials as a weapon.

Hazardous Materials Response Unit (HMRU) A specialized response group within the FBI Laboratory division that responds to WMD and other potentially hazardous crime scenes.

Incendiary Agents   Chemicals that are used to start fires, the most common being a Molotov cocktail.

Mass Casualty   An incident in which the number of patients exceeds the capability of the EMS to manage the incident effectively. In some jurisdictions this can be two patients, while in others it may take ten to make the incident a mass casualty.

Nerve Agents   Chemicals that are designed to kill humans, specifically in warfare. They are chemically similar to organophosphorus pesticides and cause the same medical reaction in humans.

PDD 39   Presidential Decision Directive 39, which established the FBI as the lead agency in terrorism incidents responsible for crisis management. It also established FEMA as the lead for consequence management.

Radiological Dispersion Device (RDD)   An explosive device that spreads radioactive material throughout an area.

Ricin   A biological toxin that can be used by a terrorist or other person attempting to kill or injure someone. It is the easiest terrorist agent to produce and one of the most common.

Toxins   Disease-causing materials that are extremely toxic and in some cases more toxic than other warfare agents such as nerve agents.

Vesicants   A group of chemical agents that cause blistering and irritation of the skin. Commonly referred to as blister agents.

Weapons of Mass Destruction (WMD)   A term that is used to describe explosive, chemical, biological, and radiological weapons used for terrorism and mass destruction.

REVIEW QUESTIONS

1. Describe four potential targets of terrorism.
2. Describe four indicators of potential terrorist activity.
3. What are the most readily available agents that could be used in a terrorist attack?
4. Which three main local/regional agencies or groups should be notified immediately of a suspected terrorist attack?
5. Describe the process of requesting federal assistance.
6. Explain which of the CBRNE agents is the most likely to be found at an incident.
7. Describe the second most likely agent to be found at an incident.
8. Describe which agent is designed to kill immediately.
9. What are the immediate signs and symptoms of sarin exposure?
10. Who is the lead agency to investigate acts of terrorism?
11. What is the significance of category A biological agents?
12. Which biological toxin is used most of the time in the United States?
13. Which is more toxic, ricin or sarin?
14. What are the outward indicator differences between a chemical attack and a biological attack?
15. Which federal agency should be contacted to assist with the collection of evidence in a hazardous environment?

ENDNOTE

1. The *Roe v. Wade* decision was the Supreme Court case that allowed legalized abortions, and is the case that has created a lot of the turmoil between pro-choice and pro-life forces.

Additional Resources

Building Construction

Angle, James, Michale Gala, David Harlow, William Lombardo, Craig Maciuba, *Firefighting Strategies and Tactics*, 2nd ed. Thomson Delmar Learning, Clifton Park, NY, 2008.

Brannigan, Francis and Glen Corbett. *Building Construction for the Fire Service*, 4th ed. Jones and Bartlett Publishing, Quincy, MA, 2007. (Note: While the fourth edition has several updates and improved artwork, it also pared down some of Mr. Brannigan's original text. The third edition is worth finding).

Building Construction for Fire Suppression Forces—Wood & Ordinary. National Fire Academy, National Emergency Training Center, Emmittsburg, MD, 1986.

Diamantes, David, *Fire Prevention: Inspection and Code Enforcement*, 3rd ed. Thomson Delmar Learning, Clifton Park, NY, 2007.

Dunn, Vincent, *Collpase of Burning Buildings,* Fire Engineering Books and Video, a Division of PennWell Publishing Company, Tulsa, OK, 1988.

Frechette, Leon A., *Build Smarter with Alternative Materials*, Craftsman Book Company, Carlsbad, CA, 1999.

"High Rise Office Building Fire, One Meridian Plaza, Philadelphia, PA," Technical Report Series, Report #049, U.S. Fire Administration, Washington, DC.

NIOSH Alert, *Preventing Injuries and Deaths of Firefighters due to Truss System Failures,* Cincinnati, OH, NIOSH Publications, 2005. (Publication Number 2005-132 is available at http://www.cdc.gov/niosh).

Queen, Phil, *Visual Dictionary of Firefighting Tools and Resources.* Thomson Delmar Learning, Clifton Park, NY, 2007.

"Without Warning, A Report on the Hotel Vendome Fire," Boston Sparks Association.

Communications and Alarms

Basic Telecommunicator Course. Association of Public Safety Communications Officials, South Daytona, FL, http://www.apcointl.org.

Bukowski, Richard W., and Robert J. O'Laughlin, *Fire Alarm Signaling Systems.* National Fire Protection Association, Quincy, MA, and Society of Fire Protection Engineers, Boston, MA, 1997.

Bunker, Merton W., Jr., *National Fire Alarm Code Handbook*, 1996 Edition. National Fire Protection Association, Quincy, MA.

Fire Department Communications Manual, A Basic Guide to System Concepts and Equipment. U.S. Fire Administration, Federal Emergency Management Agency, Washington, DC, 1995.

International Association of Fire Chiefs, FCC Narrowbanding Mandate—A Public Safety Guide for Compliance.

Mason, Michael R. and Jeffrey S. Pindelski, *Rapid Intervention Company Operations.* Thomson Delmar Learning, Clifton Park, NY, 2006.

National Academies of Emergency Dispatch, http://www.emergencydispatch.org.

National Emergency Number Association, http://www.nena9-l-l.org.

NFPA 1061. National Fire Protection Association, Quincy, MA.

NFPA 1221. National Fire Protection Association, Quincy, MA.

Queen, Phil, *Visual Dictionary of Firefighting Tools and Resources.* Thomson Delmar Learning, Clifton Park, NY, 2007.

Emergency Medical Services

American Association of Poison Control Centers, http://www.aapcc.org.

American Heart Association, http://www.americanheart.org.

American Red Cross, http://www.redcross.org.

American Heart Association, *American Heart Association Guidelines 2005,* http://www.americanheart.com.

Beebe, Richard, and Deborah Funk, *Fundamentals of Basic Emergency Care*, 2nd ed. Thomson Delmar Learning, Clifton Park, NY, 2005.

Burn Resource Center, http://www.burnsurvivor.com.

Centers for Disease Control, http://www.cdc.gov.

Fundamentals of BLS for Healthcare Providers (Video). American Heart Association, Dallas, Texas, 2002.

Occupational Safety and Health Administration, http://www.osha.gov.

Walter, Andrea, Marty L. Rutledge, and Christopher N. Edgar, *First Responder Handbook: Fire Service Edition*. Thomson Delmar Learning, Clifton Park, NY, 2003.

Fire Behavior

Bettelheim, Frederick, William H. Brown, Mary K. Campbell, and Shawn O. Farrell, *Introduction to General, Organic, and Biochemistry,* 8th ed. Brooks/Cole, a part of the Thomson Corporation, Belmont, CA, 2006.

Bevelacqua, Armando S., *Hazardous Materials Chemistry,* 2nd ed. Thomson Delmar Learning, Clifton Park, NY, 2006.

Cracolice, Mark S. and Edward I. Peters, *Introductory Chemistry: An Active Learning Approach,* 2nd ed. Brooks/Cole, a part of the Thomson Corporation, Belmont, CA, 2006.

Fire Protection Handbook, 20th ed. National Fire Protection Association, Quincy, MA, 2008.

Joesten, Melvin, D., Mary E. Castellion, and John L. Hogg, *World of Chemistry: Essentials,* 4th ed. Brooks/Cole, a part of the Thomson Corporation, Belmont, CA, 2006.

Klinoff, Robert, *Introduction to Fire Protection,* 3rd ed. Thomson Delmar Learning, Clifton Park, NY, 2007.

Lowe, Joseph D., *Wildland Firefighting Practices.* Delmar Thomson Learning, Albany, NY, 2001.

Masterton, William L. and Cecil N. Hurley, *Chemistry: Principles and Reactions,* 5th ed. Brooks/Cole, a part of the Thomson Corporation, Belmont, CA, 2005.

McMurry, John E., *Organic Chemistry*, 7th ed. Brooks/Cole, a part of the Thomson Corporation, Belmont, CA, 2007.

Queen, Phil, *Visual Dictionary of Firefighting Tools and Resources.* Thomson Delmar Learning, Clifton Park, NY, 2007.

Quintiere, James G., *Principles of Fire Behavior.* Delmar Publishers, Albany, NY, 1998.

Skoog, Douglas A., Donald M. West, James F. Holler, and Stanley R. Crouch, *Fundamentals of Analytical Chemistry,* 8th ed. Brooks/Cole, a part of the Thomson Corporation, Belmont, CA, 2004.

Whitten, Kenneth W., Raymond E. Davis, Larry Peck, and George G. Stanley, *General Chemistry: The Core,* 8th ed. Brooks/Cole, a part of the Thomson Corporation, Belmont, CA, 2006.

http://www.firetactics.com provides additional information regarding fire behavior and flashover events.

http://www.vincentdunn.com outlines many fire behavior events.

Fire Hose and Appliances

NFPA 1962: Standard for Care, Use and Testing of Fire House Including Couplings and Nozzles. National Fire Protection Association, Quincy, MA, 2008.

Sturtevant, Thomas, *Introduction to Fire Pump Operations,* 2nd ed. Thomson Delmar Learning, Clifton Park, NY, 2005.

Fire Department

Diamantes, David, *Fire Prevention: Inspection and Code Enforcement,* 3rd ed. Thomson Delmar Learning, Clifton Park, NY, 2007.

Incident Command System National Fire Academy (available through state training agencies).

Incident Command System National Training Curriculum I-100 and I-200 courses (available through state training agencies).

International Fire Service Accreditation Congress, http://www.if-sac.org.

Klinoff, Robert, *Introduction to Fire Protection*, 3rd ed. Thomson Delmar Learning, Clifton Park, NY, 2007.

Mason, Michael R. and Jeffrey S. Pindelski, Rapid Intervention Company Operations. Thomson Delmar Learning, Clifton Park, NY, 2006.

National Board on Fire Service Professional Qualifications, http://www.npqs.win.net.

National Fire Service Incident Management System Consortium, *Model Procedures Guide for Structural Firefighting,* Fire Protection Publications, Oklahoma State University, 1993.

Queen, Phil, *Visual Dictionary of Firefighting Tools and Resources.* Thomson Delmar Learning, Clifton Park, NY, 2007.

Smoke, Clinton H., *Company Officer,* 2nd ed. Thomson Delmar Learning, Clifton Park, NY, 2005.

United States Fire Administration, http://www.usfa.fema.gov, http://www.fema.gov/emergency/nims.

Fire Service

Bureau of Alcohol, Tobacco, and Firearms, Arson and Explosives National Repository, http://www.atf.treas.gov/aexis2/index.htm.

Damrells Fire, A Docema Film, DVD, http://www.damrellsfire.com.

Firefighters, The National Fallen Firefighters Foundation, http://www.firehero.org.

Firefighter Near Miss Program, http://www.firefighternearmiss.com.

Goodman, Edward C., *Fire! The 100 Most Devastating Fires and the Heroes Who Fought Them.* Black Dog and Leventhal Publishers, NY, 2001.

Goudsblom, Johan, *Fire and Civilization.* Penguin Press, NY, 1992.

Great Fires of America. Country Beautiful Corporation, Waukesha, WI, 1973.

Gurka, Andrew G., Hot Stuff Firefighting Collectibles. L-W Book Sales, Gas City, IN, 1994.

Into the Fire, Firemens Fund Insurance Company, DVD.

National Center for Injury Prevention and Control, http://www.cdc.gov/ncipc.

National Institute for Occupational Safety and Health, Fire Fighter Fatality Investigation and Prevention Program, http://www.cdc.gov/Niosh/fire/.

National Fire Information Council, http://www.nfic.org.

National Interagency Fire Center, http://www.nfic.gov.

Smith, Dennis, *History of Firefighting in America*, Dial Press, NY, 1978.

United States Department of Transportation, Office of Hazardous Materials Safety, http://hazmat.dot.gov.

United States Fire Administration, http://www.usfa.fema.gov.

Fire Suppression

Angle, James, Michael Gala, David Harlow, William Lombardo, Craig Maciuba, *Firefighting Strategies and Tactics*, 2nd ed. Thomson Delmar Learning, Clifton Park, NY, 2008.

California State Fire Training, Fire Command 1A and 1B. California State Fire Marshal Office.

Dodson, David W., *Fire Department Incident Safety Officer,* 2nd ed. Thomson Delmar Learning, Clifton Park, NY, 2007.

Lowe, Joseph D., *Wildland Firefighting Practices.* Delmar Thomson Learning, Albany, NY, 2001.

Mason, Michael R. and Jeffrey S. Pindelski, *Rapid Intervention Company Operations.* Thomson Delmar Learning, Clifton Park, NY, 2006.

NFPA 1001: Standard for Fire Fighter Professional Qualifications, 2007 Edition. National Fire Protection Association, Quincy, MA, 2007.

Queen, Phil, *Visual Dictionary of Firefighting Tools and Resources.* Thomson Delmar Learning, Clifton Park, NY, 2007.

Smoke, Clinton H., *Company Officer,* 2nd ed. Thomson Delmar Learning, Clifton Park, NY, 2005.

Firefighter Safety

Angle, James S., *Occupational Safety and Health in the Emergency Services,* 2nd ed. Thomson Delmar Learning, Clifton Park, NY, 2005.

Dodson, David W., *Fire Department Incident Safety Officer,* 2nd ed. Thomson Delmar Learning, Clifton Park, NY, 2007.

Firefighter Close Calls, http://www.firefighterclosecalls.com (free safety and survival downloads and links to firefighter safety sites).

National Fallen Firefighters Foundation, http://www.everyonegoeshome.org (information on the "Everyone Goes Home" program).

National Institute for Occupational Safety and Health: http://www.cdc.gov/niosh/ to obtain firefighter fatality reports and recommendations.

NFPA 1500: Standard on Fire Department Occupational Safety and Health Program, National Fire Protection Association, Quincy, MA, 2007.

Queen, Phil, *Visual Dictionary of Firefighting Tools and Resources.* Thomson Delmar Learning, Clifton Park, NY, 2007.

Risk Management Practices for the Fire Service, FA-166. United States Fire Administration, Emmitsburg, MD, 1996.

United States Fire Administration, http://www.usfa.dhs.gov/fireservice (firefighter injury and death statistics).

Firefighter Survival

Angle, James S., *Occupational Safety and Health in the Emergency Services,* 2nd ed. Thomson Delmar Learning, Clifton Park, NY, 2005.

Dodson, David W., *Fire Department Incident Safety Officer,* 2nd ed. Thomson Delmar Learning, Clifton Park, NY, 2007.

Emergency Incident Rehabilitation, FA-114. U.S. Fire Administration, Washington, DC, 1996.

EMS Safety: Techniques and Applications, FA-144. U.S. Fire Administration, Washington, DC, 1996.

Firefighter Fatality Investigations, http://www.cdc.gov/niosh.

Firefighter Near-Miss Reports, http://www.firefighternearmiss.com.

Firefighter Survival Information, http://www.firefighterclosecalls.com.

LeCuyer, John, *Designing the Fitness Program—A Guide for Public Safety Organizations,* Fire Engineering Books & Videos, a Division of PennWell Corporation, Saddlebrook, NJ, 2001.

Mason, Michael R. and Jeffery S. Pindelski, *Rapid Intervention Company Operations.* Thomson Delmar Learning, Clifton Park, NY, 2006.

NFPA 1500: Standard on Fire Department Occupational Safety and Health Program, 2007 ed.. National Fire Protection Association, Quincy, MA, 2007.

NFPA 1561: Standard on Emergency Services Incident Management System, 2007 ed. National Fire Protection Association, Quincy, MA, 2007.

NFPA 1584: Rehabilitation of Members Operating at Incident Scene Operations and Training Exercises, 2007 ed. National Fire Protection Association, Quincy, MA, 2007.

Queen, Phil, *Visual Dictionary of Firefighting Tools and Resources.* Thomson Delmar Learning, Clifton Park, NY, 2007.

Risk Management Practices for the Fire Service, FA-166. U.S. Fire Administration, Washington, DC, 1996.

Forcible Entry

Brennon, Tom, *Forcible Entry Program* (DVD in 5 parts).

Fritz, Richard, *Firefighter Hand Tools and Their Use,* #1: Cutting and Striking Tools.

Fritz, Richard, *Firefighter Hand Tools and Their Use,* #2: Prying Tools.

Queen, Phil, *Visual Dictionary of Firefighting Tools and Resources.* Thomson Delmar Learning, Clifton Park, NY, 2007.

Richman, Harold, *Ladder Company Fireground Operations,* 3rd Edition.

Hazardous Materials Response

American Association of Railroads, *American Association of Railroads Hazardous Materials Action Guides,* Washington D.C., 2006. Provides basic to advanced response information; one of the few guides that actually provides cleanup and mitigation strategies; lists common materials transported by rail.

American Association of Railroads, Emergency Handling of Hazardous Materials in Surface Transportation, *Washington, D.C., 2005. Covers materials commonly transported on rail, which applies to highway incidents as well.*

American Conference of Governmental Industrial Hygienists, ACGIH Threshold Limit Values, *2007 ed., Cincinnati, OH. Provides the threshold limit values (TLV) for all of the chemicals that have been studied by the ACGIH; targets common industrial chemicals.*

Ash, Michael; and Ash, Irene (eds.), Gardener's Chemical Synonym and Trade Names, *11th ed., Milne, Hoboken, NJ, 1999. Listing of common synonyms for industrial materials and household products; since some chemicals have more than 60 different synonyms, this text can be invaluable; no response information.*

Basics of Nuclear Radiation, Technical Note TN-176, RAE Systems, San Jose, CA, 2005.

Bevelacqua, Armando S., *Hazardous Materials Chemistry*, 2nd ed. Thomson Delmar Learning, Clifton Park, NY, 2006.

Bevelacqua, Armando S. and Richard Stilp, *Hazardous Materials Field Guide,* 2nd ed. Thomson Delmar Learning, Clifton Park, NY, 2007.

Bevelacqua, Armando S. and Richard Stilp, *Terrorism Handbook for Operational Responders,* 3rd ed. Delmar, a part of Cengage Learning, Clifton Park, NY.

Brethrick, L., *Brethricks Handbook of Reactive Substances*, 6th ed. Butterworths, Boston, MA, 2000. One of the few sources of information on what happens when chemicals combine; not slanted toward any grouping of chemicals.

Chemical Hazard Risk Information System (CHRIS). U.S. DOT/U.S. Coast Guard, Washington D.C., 2001, http://www.chris-manual.com. Provides detailed chemical information, is one of the best reference texts, and is free.

Compressed Gas Association, *Handbook of Compressed Gases,* 4th ed., Van Nostrand Reinhold, New York, 1999. Provides extensive detail on common compressed gases; includes detailed information on each type of cylinder.

Crop Protection Handbook. Meister Publishing, Willoughby, OH, 2003. Formerly Farm Chemicals Handbook; definitive source for pesticide, herbicide, and insecticide information; no other text is as up to date or in-depth.

Fosberg, K., and Mansdorf, S. Z., *Quick Selection Guide to Chemical Protective Clothing,* 4th ed. Van Nostrand Reinhold, New York, 2003. Provides chemical compatibility information on several hundred chemicals; listings are for the most common types of protective clothing.

Greingor, J. L., J. M. Tosi, S. Ruhlman, and M. Aussedat, "Acute Carbon Monoxide Intoxication During Pregnancy." *Emergency Medicine Journal,* http://emj.bmj.com, 2000.

Hawley, Chris, *Hazardous Materials Air Monitoring and Detection Devices,* 2nd ed. Thomson Delmar Learning, Clifton Park, NY, 2007.

Hawley, Chris, *Hazardous Materials Incidents*, 3rd ed. Thomson Delmar Learning, Clifton Park, NY, 2007.

Henry, Timothy V. *Decontamination for Hazardous Materials Emergencies*. Delmar Publishers, Albany, NY, 1998.

Laughlin, Jerry, and David Trebisacci, eds., *Hazardous Materials Response Handbook*, 4th ed. National Fire Protection Association, Quincy, MA, 2002.

Lesak, David, *Hazardous Materials Strategies and Tactics*. Prentice-Hall, 1998.

NIOSH Pocket Guide to Chemical Hazards. National Institute of Occupational Safety and Health, Washington D.C., 2005, http://www.cdc.gov/niosh/npg. Covers most occupational chemicals and most chemicals that responders encounter. Is an excellent source of information and has some detailed chemical and physical properties. Both hard copy and electronic versions are available. The electronic version is free, and single copies of the text are also free.

Noll, Gregory, Michael Hildebrand, and James Yvorra, *Hazardous Materials Managing the Incident*, 3rd ed. Fire Protection Publications, Oklahoma University, 2005.

Paul Gantt, *Hazardous Materials: Regulations, Response and Site Operations*, 2nd ed. Delmar, a part of Cengage Learning, Clifton Park, NY, 2009.

Radiation Basics Technical Note, Health Physics Society, McLean, VA, 2005.

Richard J. Lewis, Sr., *Dangerous Properties of Industrial Materials*, 11th ed. John Wiley & Sons Publishing, Hoboken, NJ, 2005. One of the best and most detailed reference sources. Three-volume text with extensive cross-referencing and information on 26,000 chemicals.

Stilp, Richard and Armando Bevelacqua, *Emergency Medical Response to Hazardous Materials Incidents*. Delmar Publishers, Albany, NY, 1996.

Tomaszewski, Christian, *Carbon Monoxide Poisoning*, Post Graduate Medicine Online, http://www.jpgmonline.com, 1999.

Wahlström, Björn, *Understanding Radiation*, Medical Physics Publishing, Madison, WI, 1995.

Ladders

NFPA 1931: Standard on Design and Verification Tests for Fire Department Ground Ladders, 2004 edition, National Fire Protection Association, Quincy, MA.

NFPA 1932: Standard on Use, Maintenance and Service Testing of In-Service Fire Department Ground Ladders, 2004 edition, National Fire Protection Association, Quincy, MA.

Queen, Phil, *Visual Dictionary of Firefighting Tools and Resources*. Thomson Delmar Learning, Clifton Park, NY, 2007.

Large Scale Incident Response

Buck, George, *Preparing for Biological Terrorism: An Emergency Services Guide*. Delmar Thomson Learning, Albany, NY, 2002.

Buck, George, *Preparing for Terrorism: An Emergency Services Guide*. Delmar Publishers, Albany, NY, 1998.

Buck, George, Lori Buck, and Barry Mogil, *Preparing for Terrorism: The Public Safety Communicator's Guide*. Thomson Delmar Learning, Clifton Park, NY, 2003.

Hawley, Chris, *Hazardous Material Air Monitoring and Detection Devices*, 2nd ed. Thomson Delmar Learning, Clifton Park, NY, 2007.

Hawley, Chris, Michael Hildebrand, and Gregory Noll, *Special Operations: Response to Terrorism and HazMat Crimes,* Red Hat Publishing, Chester, MD, 2002.

Federal Emergency Management Agency, http://www.fema.gov.

Department of Homeland Security, http://www.dhs.gov.

FEMA Emergency Management Institute, http://www.training.fema.gov.

FEMA Urban Search and Rescue Teams, http://www.fema.gov/emergency/usr/.

Federal Bureau of Investigation, http://www.fbi.gov.

National Response Plan, http://www.dhs.gov/xprepresp/committees/editorial_0566.shtm.

National Guard Bureau, http://www.ngb.army.mil/default.aspx.

National Emergency Management Association, http://www.nemaweb.org.

US Disaster Medical Assistance Teams (DMAT), http://www.dmat.org.

Nozzles, Fire Stream and Foam

Crapo, William F., *Hydraulics for Firefighting*, 2nd ed. Thomson Delmar Learning, Clifton Park, NY, 2008.

Gagnon, Robert M., *Design of Special Hazard and Fire Alarm Systems*, 2nd ed. Thomson Delmar Learning, Clifton Park, NY, 2008.

Klinoff, Robert, *Introduction to Fire Protection*, 3rd ed. Thomson Delmar Learning, Clifton Park, NY, 2007.

NFPA 11: Standard for Low, Medium, and High Expansion Foam Systems. National Fire Protection Association, Quincy, MA, 2005.

NFPA 1961: Standard on Fire Hose. National Fire Protection Association, Quincy, MA, 2007.

NFPA 1962: Standard for Care, Use and Testing of Fire Hose Including Couplings and Nozzles. National Fire Protection Association, Quincy, MA, 2003.

NFPA 1964: Standard for Spray Nozzles (Shutoff and Tip). National Fire Protection Association, Quincy, MA, 2003.

Queen, Phil, *Visual Dictionary of Firefighting Tools and Resources*. Thomson Delmar Learning, Clifton Park, NY, 2007.

Stern, Jeff and J. Gordon Routley, *Class A Foam for Structural Firefighting*. U.S. Fire Administration, Emmittsburg, MD, December 1996.

Sturtevant, Thomas, *Introduction to Fire Pump Operations*, 2nd ed. Thomson Delmar Learning, Clifton Park, NY, 2005.

Personal Protective Equipment

Angle, James S., *Occupational Safety and Health in the Emergency Services,* 2nd ed. Thomson Delmar Learning, Clifton Park, NY, 2005.

Dodson, David W., *Fire Department Incident Safety Officer,* 2nd ed. Thomson Delmar Learning, Clifton Park, NY, 2007.

Emergency Incident Rehabilitation, FA-114, United States Fire Administration, Emmitsburg, MD, 1992.

FEMSA Official User Information Guide, Fire and Emergency Manufacturers and Services Association, Inc., Lynnfield, MA, 1996.

Mason, Michael R. and Jeffrey S. Pindelski, *Rapid Intervention Company Operations*. Thomson Delmar Learning, Clifton Park, NY, 2006.

NFPA 1001: Standard for Fire Fighter Professional Qualifications, National Fire Protection Association, Quincy, MA 2008.

NFPA 1500: Standard on Fire Department Occupational Safety and Health Programs, National Fire Protection Association, Quincy, MA, 2007.

NFPA 1851: Standard on Selection, Care, and Maintenance of Structural Fire Fighting Protective Ensembles, National Fire Protection Association, Quincy, MA, 2008.

NFPA 1971: Standard on Protective Ensemble for Structural Firefighting and Proximity Firefighting, National Fire Protection Association, Quincy, MA, 2007.

NFPA 1977: Standard on Protective Clothing and Equipment for Wildland Firefighting, National Fire Protection Association, Quincy, MA, 2005.

Queen, Phil, *Visual Dictionary of Firefighting Tools and Resources.* Thomson Delmar Learning, Clifton Park, NY, 2007.

Risk Management Practices for the Fire Service, FA-166, United States Fire Administration, Emmitsburg, MD, 1996.

Portable Fire Extinguishers

Diamantes, David, *Fire Prevention: Inspection and Code Enforcement*, 3rd ed. Thomson Delmar Learning, Clifton Park, NY, 2007.

Gagnon, Robert, *Design of Special Hazard and Fire Alarm Systems*, 2nd ed. Thomson Delmar Learning, Clifton Park, NY, 2008.

Gagnon, Robert, *Design of Water-Based Fire Protection Systems.* Delmar Publishers, Albany, NY, 1996.

Klinoff, Robert, *Introduction to Fire Protection*, 3rd ed. Thomson Delmar Learning, Clifton Park, NY, 2007.

NFPA 10: Standard for Portable Fire Extinguishers, 1998 Edition. National Fire Protection Association, Quincy, MA.

NFPA 17: Standard for Dry Chemical Extinguishing Systems, 1998 Edition. National Fire Protection Association, Quincy, MA.

NFPA 17A: Standard for Wet Chemical Extinguishing Systems, 1998 Edition. National Fire Protection Association, Quincy, MA.

Queen, Phil, *Visual Dictionary of Firefighting Tools and Resources.* Thomson Delmar Learning, Clifton Park, NY, 2007.

Sturtevant, Thomas, *Introduction to Fire Pump Operations,* 2nd ed. Thomson Delmar Learning, Clifton Park, NY, 2005.

Prevention, Public Education, and Pre-Incident Planning

Halon Recycling Corporation, http://www.halon.org.

Klinoff, Robert, *Introduction to Fire Protection,* 3rd ed. Thomson Delmar Learning, Clifton Park, NY, 2007.

NFPA Fire Prevention Handbook, http://www.NFPAcodesonline.org.

Occupational Safety and Health Administration, http://www.osha-safety-training.net.

Queen, Phil, *Visual Dictionary of Firefighting Tools and Resources.* Thomson Delmar Learning, Clifton Park, NY, 2007.

United States Fire Administration, http://www.usfa.fema.gov.

Protective Systems

Diamantes, David, *Fire Prevention: Inspection and Code Enforcement*, 3rd ed. Thomson Delmar Learning, Clifton Park, NY, 2007.

Gagnon, Robert, *Design of Special Hazard and Fire Alarm Systems*, 2nd ed. Thomson Delmar Learning, Clifton Park, NY, 2008.

Gagnon, Robert, *Design of Water-Based Fire Protection Systems.* Delmar Publishers, Albany, NY, 1996.

Klinoff, Robert, *Introduction to Fire Protection*, 3rd ed. Thomson Delmar Learning, Clifton Park, NY, 2007.

NFPA 13: Standard for the Installation of Sprinkler Systems, 2007 Edition. National Fire Protection Association, Quincy, MA.

NFPA 13D: Standard for the Installation of Sprinkler Systems in One- and Two-Family Dwellings and Manufactured Homes, 2007 edition. National Fire Protection Association, Quincy, MA.

NFPA 13E: Guide for Fire Department Operations in Properties Protected by Sprinkler and Standpipe Systems, 2007 edition. National Fire Protection Association, Quincy, MA.

NFPA 13R: Standard for the Installation of Sprinkler Systems in Residential Occupancies up to and Including Four Stories in Height, 2007 edition. National Fire Protection Association, Quincy, MA.

NFPA 14: Standard for the Installation of Standpipe, Private Hydrants, and Hose Systems, 2007 edition. National Fire Protection Association, Quincy, MA.

NFPA 25: Standard for the Inspection, Testing, and Maintenance of Water-Based Fire Protection Systems, 2008 edition. National Fire Protection Association, Quincy, MA.

Queen, Phil, *Visual Dictionary of Firefighting Tools and Resources.* Thomson Delmar Learning, Clifton Park, NY, 2007.

Residential Fire Safety Institute, http://www.firesafehome.org.

Sturtevant, Thomas, *Introduction to Fire Pump Operations*, 2nd ed. Thomson Delmar Learning, Clifton Park, NY, 2005.

Rescue Procedures

Brown, Michael G., *Engineering Practical Rope Rescue Systems.* Delmar Thomson Learning, Albany, NY, 2000.

Browne, George J. and Gus S. Crist, *Confined Space Rescue.* Delmar Thomson Learning, Albany, NY, 1999.

Downey, Ray, *The Rescue Company.* Fire Engineering Books and Videos, 1992.

Frank, James and Jerrold Smith, *Rope Rescue Manual.* California Mountain Company Limited, Santa Barbara, CA, 1987.

Mason, Michael R. and Jeffrey S. Pindelski, *Rapid Intervention Company Operations.* Thomson Delmar Learning, Clifton Park, NY, 2006.

Queen, Phil, *Visual Dictionary of Firefighting Tools and Resources.* Thomson Delmar Learning, Clifton Park, NY, 2007.

Setnicka, Tim, *Wilderness Search and Rescue.* Appalachian Mountain Club, Boston, MA, 1980.

Technical Rescue Program Development Manual. U.S. Fire Administration, Washington, DC.

Ropes and Knots

NFPA 1983, Standard on Life Safety Rope and Equipment for Emergency Services, 2006 edition.

Queen, Phil, *Visual Dictionary of Firefighting Tools and Resources.* Thomson Delmar Learning, Clifton Park, NY, 2007.

Salvage, Overhaul and Fire Cause Determination

NFPA 921: Guide for Fire and Explosion Investigations. National Fire Protection Association, Battery March Park, MA, 2004.

Self-Contained Breathing Apparatus

Because of the special nature of SCBA, readers should also see specific manufacturers' instructions.

Bernocco, S., Phillips, C., Jose, P., and Yob, C., "Train in the Rule of Air Management," *Fire Engineering,* April 2003.

Bernzweig, D. "Expanding Time to Exit For Firefighters," *Fire Engineering*, June 2004.

Hashagen, P. "The Development of Breathing Apparatus," *Firehouse,* September 1998.

Mason, Michael R. and Jeffrey S. Pindelski, *Rapid Intervention Company Operations.* Thomson Delmar Learning, Clifton Park, NY, 2006.

Queen, Phil, *Visual Dictionary of Firefighting Tools and Resources.* Thomson Delmar Learning, Clifton Park, NY, 2007.

Thiel, A., *Prevention of Self-Contained Breathing Apparatus Failures*, Special Report, United States Fire Administration, March 1998.

Terrorism Response

Bevelacqua, Armando, *Hazardous Materials Chemistry*, 2nd ed. Thomson Delmar Learning, Clifton Park, NY, 2006.

Bevelacqua, Armando and Richard Stilp, *Hazardous Materials Field Guide*, 2nd ed. Thomson Delmar Learning, Clifton Park, NY, 2007.

Bevelacqua, Armando and Richard Stilp, *Terrorism Handbook for Operational Responders,* 3rd ed. Delmar, a part of Cengage Learning, Clifton Park, NY.

Buck, George, *Preparing for Terrorism: An Emergency Services Guide.* Delmar Publishers, Albany, NY, 1998.

Buck, George, *Preparing for Biological Terrorism: An Emergency Services Guide.* Delmar Thomson Learning, Albany, NY, 2002.

Buck, George, Lori Buck, and Barry Mogil, *Preparing for Terrorism: The Public Safety Communicator's Guide.* Thomson Delmar Learning, Clifton Park, NY, 2003.

Hawley, Chris, *Hazardous Materials Incidents*, 3rd ed. Thomson Delmar Learning, Clifton Park, NY, 2007.

Hawley, Chris, *Hazardous Material Air Monitoring and Detection Devices*, 2nd ed. Thomson Delmar Learning, Clifton Park, NY, 2007.

Hawley, Chris, Michael Hildebrand, and Gregory Noll, *Special Operations: Response to Terrorism and HazMat Crimes,* Red Hat Publishing, Chester, MD, 2002.

Henry, Timothy V., *Decontamination for Hazardous Materials Emergencies.* Delmar Publishers, Albany, NY, 1998.

Medical Management of Biological Casualties. U.S. Army Medical Research Institute of Infectious Diseases, Ft. Dietrick, Fredrick, MD, 1996.

Medical Management of Chemical Casualties Handbook. U.S. Army Chemical Casualty Care Office, Medical Research Institute of Chemical Defense, Aberdeen Proving Ground, MD, Sept. 1995.

Pickett, Mike, *Explosives Identification Guide*, 2nd ed. Thomson Delmar Learning, Clifton Park, NY, 2005.

Paul Gantt, *Hazardous Materials: Regulations, Response and Site Operations,* 2nd ed. Delmar, a part of Cengage Learning, Clifton Park, NY, 2009.

Smelby, L. Charles, Jr. (ed.), *Hazardous Materials Response Handbook*, 3rd ed. National Fire Protection Association, Quincy, MA, 1997.

Stilp, Richard and Armando Bevelacqua, *Citizen's Guide to Terrorism Preparedness.* Thomson Delmar Learning, Clifton Park, NY, 2002.

Stilp, Richard, and Armando Bevelacqua, *Emergency Medical Response to Hazardous Materials Incidents.* Delmar Publishers, Albany, NY, 1996.

Ventilation

Angle, James, Michael Gala, David Harlow, William Lombardo, Craig Maciuba, *Firefighting Strategies and Tactics*, 2nd ed. Thomson Delmar Learning, Clifton Park, NY, 2008.

Casey, James, *Fire Chief's Handbook.* Technical Publishing Company, New York, 1978.

Klinoff, Robert, *Introduction to Fire Protection,* 3rd ed. Thomson Delmar Learning, Clifton Park, NY, 2007.

Queen, Phil, *Visual Dictionary of Firefighting Tools and Resources.* Thomson Delmar Learning, Clifton Park, NY, 2007.

Water Supply

Crapo, William F., *Hydraulics for Firefighting*, 2nd ed. Thomson Delmar Learning, Clifton Park, NY, 2008.

Fire Protection Handbook, 20th ed. National Fire Protection Association, Quincy, MA, 2008.

Gagnon, Robert, *Design of Special Hazard and Fire Alarm Systems*, 2nd ed. Thomson Delmar Learning, Clifton Park, NY, 2008.

Klinoff, Robert, *Introduction to Fire Protection,* 3rd ed. Thomson Delmar Learning, Clifton Park, NY, 2007.

NFPA 22: Standard for Water Tanks for Private Fire Protection, 2003 Edition. National Fire Protection Association, Quincy, MA.

NFPA 24: Standard for the Installation of Private Fire Service Mains and Their Appurtenances, 2007 Edition. National Fire Protection Association, Quincy, MA.

NFPA 1410: Standard on Training for Initial Emergency Scene Operations, 2005 Edition. National Fire Protection Association, Quincy, MA.

NFPA 291: Recommended Practices for Fire Flow Testing and Marking of Fire Hydrants, 2007 Edition. National Fire Protection Association, Quincy, MA.

NFPA 1142: Standard on Water Supplies for Suburban and Rural Firefighting, 2007 Edition. National Fire Protection Association, Quincy, MA.

NFPA 1901: Standard on Automotive Fire Apparatus, 2003 Edition. National Fire Protection Association, Quincy, MA.

Queen, Phil, *Visual Dictionary of Firefighting Tools and Resources.* Thomson Delmar Learning, Clifton Park, NY, 2007.

Sturtevant, Thomas, *Introduction to Fire Pump Operations*, 2nd ed. Thomson Delmar Learning, Clifton Park, NY, 2005.

Acronyms

ACGIH American Conference of Governmental Industrial Hygienists

ACS American Chemical Society

AFFF aqueous film–forming foam

AHA American Heart Association

AHJ authority having jurisdiction

ALARA as low as reasonably achievable

ALOHA aerial location of hazardous atmospheres

ALS advanced life support

ANFO ammonium nitrate and fuel oil

APCO Association of Public Safety Communications Officials—International, Inc.

APR air-purifying respirators

ARFF aircraft rescue and firefighting

AST aboveground storage tanks

ASTM American Society for Testing and Materials

ATC alcohol-type concentrates

ATF (Bureau of) Alcohol, Tobacco, Firearms (and Explosives)

AVL automatic vehicle location

BLEVE boiling liquid expanding vapor explosion

BLS basic life support

BNICE biological, nuclear, incendiary, chemical, and explosive

BOCA Building Officials Conference Association

BSI body substance isolation

Btu British thermal unit

CAAA Clean Air Act Amendment

CAD computer-aided drawing; computer-aided dispatch

CAFS compressed air foam systems

CAMEO Computer-Aided Management for Emergency Operations

CANUTEC Canadian Transportation Emergency Center

CAS Chemical Abstracts Service

CBDCOM Chemical and Biological Defense Command

CBIRF Chemical and Biological Incident Response Force (Marines)

CBRNE chemical, biological, radiological, nuclear, and explosive

CDC Centers for Disease Control

CERCLA Comprehensive Environmental Response, Compensation, and Liability Act

CFR *Code of Federal Regulations*

CGI combustible gas indicators

CHEMTREC Chemical Transportation Emergency Center

CHRIS Chemical Hazards Risk Information System

CIDS Critical Incident Dispatch System

CIS critical incident stress

CISD critical incident stress debriefing

CISM critical incident stress management

CMA Chemical Manufacturers Association

CMS chip measurement system

CNS central nervous system

CO carbon monoxide

CPR cardiopulmonary resuscitation

CST (National Guard) Civil Support Team

DCM dangerous cargo manifest

DECIDE detect, estimate, choose, identify, do the best, evaluate

DHS Department of Homeland Security

DNR Department of Natural Resources

DOT Department of Transportation (U.S.)

DRD drag rescue device

EAP Employee Assistance Program

EBSS emergency escape breathing support system

EDITH Exit Drills In The Home

EHS extremely hazardous substance

EMS emergency medical services

EMT emergency medical technician

EMT-P emergency medical technician-paramedic

EPA Environmental Protection Agency

EPCRA Emergency Planning and Community Right to Know Act

ERG Emergency Response Guidebook

ERP emergency response planning

FAST firefighter assist and search team

FBI Federal Bureau of Investigation

FCC Federal Communications Commission

FD fire department

FEMA Federal Emergency Management Agency

FFFP fluoroprotein film–forming foam

FFPC firefighter protective clothing

FFPE firefighter protective ensemble

FGC Fire Ground Command

GEDAPER gather information, estimate potential, determine goals, assess tactical options, plan, evaluate, review

GFI ground fault interrupters

gpm gallons per minute

GPS global positioning system

Gy gray

HAZMAT hazardous materials

HAZWOPER hazardous waste operations and emergency response

HCS hazard communication standard

HFT hydraulic forcible entry tools

HIV human immunodeficiency virus

HMIS Hazardous Materials Information System

HMRU Hazardous Materials Response Unit (FBI)

HMTD hexamethylene triperoxide diamine

HUD heads up display

HVAC   heating, ventilation, and air conditioning

IAFF   International Association of Fire Fighters

IAP   incident action plan

IC   incident commander

ICS   incident command system

ICt$_{50}$ incapacitating concentration to 50 percent of the population, with "t" representing time, usually expressed in minutes

IDLH   immediately dangerous to life and health

IDR   instant detection and response

IFSAC   International Fire Service Accreditation Congress

IM   intermodal

IMS   (National Fire Service) Incident Management System

IR   infrared

IRIC   Initial Rapid Intervention Crew

IRS   Internal Revenue Service

ISC   Industrial Scientific Corporation

ISEA   International Safety Equipment Association

kPa   kilopascals

L/min   liters per minute

LC$_{50}$ lethal concentration to 50 percent of the population

LCES   (rooftop) lookouts, communications, escape routes, and safety zones

LCt$_{50}$ lethal concentration to 50 percent of the population, with "t" representing time, usually expressed in minutes

LD$_{50}$ lethal dose to 50 percent of the population

LEL   lower explosive (flammable) limit

LEPC   Local Emergency Planning Committee

LOX   liquid oxygen

LRN   Laboratory Response Network (of the CDC and FBI)

LSA   low specific activity (radiation)

LUST   leaking underground storage tank

MCI   mass casualty incident

MOS   metal oxide sensor

mph   miles per hour

MSA   mine safety appliances

MSDS   Material Safety Data Sheet

MSV   mobile support vehicle

MTBE   Methyl Tert-Butyl Ether

NASA   National Aeronautics and Space Administration

NBC   nuclear, biological, and chemical

NBFU   National Board of Fire Underwriters

NENA   National Emergency Number Association

NFFF   National Fallen Firefighters Foundation

NFPA   National Fire Protection Association

NHT   national hose thread

NIMS   National Incident Management System

NIOSH   National Institute for Occupational Safety and Health

NOS   not otherwise specified

NP   nozzle pressure

NPL   national priority list

NPQB   National Professional Qualifications Board

NR   nozzle reaction

NRC   National Response Center

NST   national standard thread

OPA   Oil Pollution Act

OPP   organophosphate pesticide

ORM-D   other regulated material, Class D

OS&Y   outside stem and yoke valve

OSHA   Occupational Safety and Health Administration

P   polymerization hazard

PAR   personnel accountability report

PASS   personal accountability safety system

PDD 39   Presidential Decision Directive 39

PEL   permissible exposure limit

PFD   personal flotation device

PG I, II, or III   packing group I, II, or III

PID   photo-ionization detector

PIH   poison inhalation hazard

PIO   public information officer

PNS   peripheral nervous system

PPA   positive pressure attack

ppb   parts per billion

PPC   personal protective clothing

PPE   personal protective equipment

ppm   parts per million

PPV   positive pressure ventilation

PSA   public service announcements

PSAP   public safety answering point

psi   pounds per square inch

psia   pounds per square inch absolute

psig   pounds per square inch gauge

PVC   polyvinyl chloride

QRT   Quick Response Teams

R&I   recognition and identification

rad   radiation absorbed dose

RAID   Rapid Assessment and Initial Detection (team)

RAT   Rescue Assist Team

RDECOM   Research, Development, and Engineering Command

RECEO   rescue, exposures, confinement, extinguishment, overhaul

REL   recommended exposure limit

REM   roentgen equivalent man; also abbreviated as R

REVAS   rescue, exposures, ventilation, attack, and salvage

RIC   Rapid Intervention Crew (or Company)

RIT   rapid intervention team

RPG   rocket-propelled grenade

RQ   reportable quantity

SADT   self-accelerating decomposition temperature

SAR   supplied air respirator

SARA   Superfund Amendments and Reauthorization Act

SBCCOM   Soldiers Biological and Chemical Command

SBIMAP   South Baltimore Industrial Mutual Aid Plan

SCA   surface contaminated articles (radiation)

SCBA   self-contained breathing apparatus

SERC   State Emergency Response Committee

SETIQ   Emergency Transportation System for the Chemical Industry

SHB   suicide/homicide bomber

SOG   standard operating guideline

SOP   standard operating procedure

SPE   solid polymer electrolyte (sensor)

SRS   supplemental restraint systems

SSP   standard sprinkler pendent

SSU   standard sprinkler upright

STCC   Standard Transportation Commodity Code

STEL   short-term exposure limit

TATP   triacetone triperoxide

TC   Transportation Canada

TDD   telecommunications devices for the deaf

TEU   Technical Escort Unit

TIA   Tentative Interim Amendment

TIC   thermal imaging camera

TLV   threshold limit value

TOFC   trailers on flat cars

TOG   turnout gear

TPP   thermal protective performance

TRACEMP   thermal, radiation, asphyxiation, chemical, etiological, mechanical hazards, and potential psychological harm

UEL   upper explosive (flammable) limit

UN/NA   United Nations/North America

UPS   uninterruptible power supply

USAR   Urban Search and Rescue (team)

UST   underground storage tank

UV   ultraviolet

VES   vent enter search

VOIP   voice over Internet protocol

VTR   violent tank rupture

WIV   wall indicator valve

WMD   weapons of mass destruction

Glossary

8-Step Process A management system used to organize the response to a chemical incident. The elements are site management and control, identifying the problem, hazard and risk evaluation, selecting PPE and equipment, information management and resource coordination, implementing response objectives, decon and cleanup operations, and terminating the incident.

9-1-1 Emergency telephone number that provides access to the public safety services in the community, region, and, ultimately, nation.

Abandonment Abandonment occurs when an emergency responder begins treatment of a patient, and then leaves the patient or discontinues treatment prior to the arrival of an equally or higher trained responder.

Abortifacient A chemical or material that can cause abortions.

Aboveground Storage Tank (AST) Tank that is stored above the ground in a horizontal or vertical position. Smaller quantities of fuels are often stored in this fashion.

Abrasion A scrape or brush of the skin usually making it reddish in color and resulting in minor capillary bleeding.

Absolute Pressure The measurement of pressure, including atmospheric pressure. Measured in pounds per square inch absolute.

Absorbed Dose A measure of the amount of radiation transferred to a material.

Absorption A defensive method of controlling a spill by applying a material that absorbs the spilled chemical.

Accelerator A device to speed the operation of the dry pipe valve by detecting the decrease in air pressure. It pipes air pressure below the clapper valve, speeding its opening.

Acclimation The act of becoming accustomed or used to something. Typically achieved through repeated practice within a given set of conditions.

Active Emitter Objects Objects that generate their own thermal energy, for example, animals and humans. Active emitters can vary in relation to mass and density. They can also be suppressed by barriers such as clothing and construction features inside structures.

Activity A measure of the number of decays per second that occur with the source.

Acute A quick one-time exposure to a chemical.

Adapter A device that adapts or changes one type of hose thread to another, allowing connection of two different lines. Adapters have a male end on one side and a female on the other with each side being a different thread type, for example, an iron pipe to national standard adapter.

Administrative Warrant An order issued by a magistrate that grants authority for fire personnel to enter private property for the purpose of conducting a fire safety survey.

ADR/RID Abbreviations used by the international shipping community and established by a European Agreement by the United Nations Economic and Social Council Committee of Experts on Transporting Dangerous Goods. The U.S. DOT is a voting member of this shipping body, which governs Regulations Concerning the International Transport of Dangerous Goods by Rail (RID) and Road (ADR)

Air Bill The term used to describe the shipping papers used in air transportation.

Air Monitoring Devices Used to determine oxygen, explosive, or toxic levels of gases in air.

Aircraft Rescue and Firefighting (ARFF) Of or pertaining to firefighting operations involving fixed or rotary wing aircraft.

Air-Purifying Respirator (APR) Respiratory protection that filters contaminants out of the air, using filter cartridges. Requires the atmosphere to have sufficient oxygen, in addition to other regulatory requirements.

Alcohol-Resistant Foam (AR Foam) Foam containing a polymer that will form protective layering between the burning surface and the foam. This layering prevents foam breakdown due to alcohol present in the burning fuel. Alcohol-resistant foams are used for fighting fuel fires that contain alcohol additives, such as the E85 gasoline blend.

Alcohol-Resistant Foams These foams contain a polymer that forms a protective layer between the burning surface and the foam, preventing the foam breakdown by alcohols present in the burning fuel.

Allergen A material that causes a reaction by the body's immune system.

Allergic Reaction The body's reaction to a substance to which there is an allergy.

Americans with Disabilities Act Public law that bars discrimination on the basis of disability in state and local services. Enacted in 1990.

Amputation Occurs when part of the body is severed completely as a result of an injury.

Anchor Point A safe location from which to begin line construction on a wildland fire.

ANFO The acronym that is used for ammonium nitrate fuel oil mixture, which is a common explosive. ANFO was used in the Oklahoma City bombing incident.

Anthrax A biological material that is naturally occurring and is severely toxic to humans. It is commonly used in hoax incidents.

Application Rate Amount of foam or foam solution needed to extinguish a fire. Usually expressed in gallons per minute per square foot or liters per minute per square meter.

Aqueous Film-Forming Foam (AFFF) A synthetic foam that as it breaks down forms an aqueous layer or film over a flammable liquid.

Aquifer A formation of permeable rock, gravel, or sand holding water or allowing water to flow through it.

Arterial Bleeding Bleeding from an artery.

Arteries The blood vessels, or tubes, within the body that carry blood rich with oxygen and nutrients away from the heart.

Articulating Boom Ladder An apparatus with a series of booms and a platform on the end. It is maneuvered into position by adjusting the various boom sections into place to position the platform at the desired location.

Aspect The direction a slope faces given in compass directions.

Asphyxiation Condition that causes death due to lack of oxygen and an excessive amount of carbon monoxide or other gases in the blood.

Asphyxiation Loss of consciousness or death caused by too little oxygen and too much carbon dioxide.

Association of Public-Safety Communications Officials International, Inc. (APCO) International not-for-profit organization dedicated to the advancement of public safety communications. Membership is made up of public safety professionals from around the world.

Atmospheric Pressure The pressure exerted by the atmosphere, which for Earth is 14.7 pounds per square inch at sea level.

Attack Hose Small- to large-diameter hose used to supply nozzles and other applicators or protective system for fire attack. Attack hose commonly means handheld hoselines from 1½ to 2½ inches (38 or 63 mm) in diameter.

Authority Having Jurisdiction (AHJ) The responsible governing organization or body having legal jurisdiction.

Autoextended When a fire goes out the window on one floor, up the side of the building, which is often noncombustible, and extends through the window or cockloft directly above.

Autoignition Temperature The temperature of a liquid at which it will ignite on its own without an ignition source. Can be compared to SADT (Self-Accelerating Decomposition Temperature).

Automated External Defibrillator (AED) A portable computer-driven device that analyzes a patient's heart rhythm and delivers defibrillation shocks when necessary.

Automatic or Constant Pressure Nozzle Nozzle with a spring mechanism built in that reacts to pressure changes and adjusts the flow and resultant reach of the nozzle.

Automatic or Constant Pressure Nozzle Nozzle with a spring mechanism built in that reacts to pressure changes and adjusts the flow and resultant reach of the nozzle.

Automatic Vehicle Location (AVL) System that utilizes GPS technology to pinpoint the location of an emergency incident as well as the response vehicle that may be closest according to a computerized map.

Automatic Vehicle Location (AVL) System that utilizes GPS technology to pinpoint the location of an emergency incident as well as the response vehicle that may be closest according to a computerized map.

Auxiliary Appliances Another term for protective devices, particularly sprinkler and standpipe systems.

Auxiliary Appliances Another term for protective devices, particularly sprinkler and standpipe systems.

Available Flow Amount of water that can be moved to extinguish the fire. Depends on the water supply, pump(s) and their capabilities, and the size and length of hose.

Avulsion An injury where a part of the skin is torn away, but still attached, leaving a flap or loose area hanging.

Awareness Level The basic level of training for emergency response to an incident involving hazardous materials/weapons of mass destruction (WMDs), the basis of which is the firefighters' ability to recognize a hazardous situation, protect themselves and call for trained assistance, and secure the scene.

Axial Load A load passing through the center of the mass of the supporting element, perpendicular to its cross section.

Backdraft The sudden and explosive ignition of pressurized, superheated, and oxygen-deprived gases (within a closed space) caused by the reintroduction of oxygen.

Backflow Preventers A check valve or set of valves used to prevent a backflow of water from one system into another.

Backstretch or Flying Stretch An attack line lay where the engine is at the hydrant and the line is stretched back from the engine to the fire. The flying stretch is a version of the backstretch where the engine stops in front of the fire, the attack portion is removed, and the engine proceeds to the hydrant.

Balloon Frame A style of wood frame construction in which studs are continuous the full height of a building.

Balloon Frame A style of wood frame construction in which studs are continuous the full height of a building.

Base Radio Radio station that contains all of the antennas, receivers, and transmitters necessary to transmit and receive messages. Normally located in a station or communications center.

Basic 9-1-1 Telephone system that automatically connects a person dialing the digits "9-1-1" to a predetermined answering point through normal telephone service facilities. Number and location information is not normally provided in basic systems.

Beam A structural member subjected to loads perpendicular to its length.

Beam A structural member subjected to loads perpendicular to its length.

Becquerel (Bq) A measure of radioactivity; the metric version of Curies.

Bed Ladder The nonextending part of an extension ladder.

Bight A doubled section of rope, usually made along the standing part, that forms a U-turn in the rope that does not cross itself.

Biological Agents Microorganisms that cause disease in humans, plants, and animals; they also cause the victims' health to deteriorate. Biological agents have been designed for warfare purposes.

Biomimetic A form of gas sensor that is used to determine levels of carbon monoxide. It is of the type of sensors used in home CO detectors. It closely re-creates the body's reaction to CO and activates an alarm.

Black Fire A term used to describe smoke that is high volume, turbulent velocity (hot), ultradense, and black. Black fire is a sure sign of impending autoignition and flashover.

BLEVE Boiling liquid expanding vapor explosion. Describes the rupture of a container when a confined liquid boils and creates a vapor pressure that exceeds the container's ability to hold it.

Blister Agents A group of chemical agents that cause blistering and irritation of the skin. Sometimes referred to as vesicants.

Blood Agents Chemicals that affect the body's ability to use oxygen. If they prevent the body from using oxygen, fatalities result.

Body Substance Isolation (BSI) Precautions A set of precautions for emergency responders designed to prevent exposure to any body fluid or substance.

Boiling Point The temperature to which a liquid must be heated in order to turn into a gas.

Boiling Point The temperature where a liquid will convert to a gas at a vapor pressure equal to or greater than atmospheric pressure.

Bolt Throw The distance the bolt of a lock travels into the jamb or strike plate. Usually ½ to 1½ inches.

Booster Hose Smaller diameter, flexible hard-rubber-coated hose of ¾- or 1-inch (19- to 25-mm) size usually mounted on reel that can be used for small trash and grass fires or overhaul operations after the fire is out.

Bourdon Gauge The type of gauge found on most fire apparatus that operates by pressure in a curved tube moving an indicating needle.

Box Canyon A canyon open on one end and closed on the other. They become very dangerous when wildfire enters them.

Brachial Artery A major artery in the inside of the upper arm that supplies blood to the arm. Can be used as a pressure point for controlling bleeding and for locating a pulse on an infant.

Branch The command designation established to maintain span of control over a number of divisions, sectors, or groups.

Bresnan Distributor A device used to fight fire in basements or cellars when firefighters cannot make a direct attack on the fire; has six or nine solid tips or broken stream openings designed to rotate in a circular spray pattern.

Bresnan Distributor. A device used to fight fire in basements or cellars when firefighters cannot make a direct attack on the fire; has six or nine solid tips or broken stream openings designed to rotate in a circular spray pattern.

Brush Gear Another term for a wildland personal protective ensemble.

Building Officials Conference Association (BOCA) A group that establishes minimum building and fire safety standards.

Bulk Tank A large transportable tank, comparable to a tote, but considered to be the larger of the two.

Bump Test Used to determine if an air monitor is working. It will alarm if a toxic gas is present. It is a quick check to make sure the instrument responds to a sample of gas.

Bunkers A term that is used mostly to describe the components of a structural firefighting ensemble. The original use of the term *bunkers* referred only to the pant/boot combination that firefighters wore at night and placed next to their bunks for rapid donning.

Butyric Acid A fairly common lab acid that has been used in many attacks on abortion clinics. Although not extremely hazardous, it has a characteristic stench that permeates the entire area where it is spilled.

Bypass Eductor Eductor with two waterways and a valve that allows plain water to pass by the venturi or through the venturi to create foam solution.

Calibration Used to set the air monitor and to ensure that it reads correctly. When calibrating a monitor, it is exposed to a known quantity of gas to make sure it reads the values correctly.

Cantilever Beam A beam that is supported at only one end.

Capillaries The very small blood vessels in the body that connect arteries and veins, and filter the oxygen and nutrients from the blood into the tissues of the body.

Capillary Bleeding Bleeding from a capillary.

Carbon Dioxide (CO_2) An inert colorless and odorless gas that is stored under pressure as a liquid that is capable of being self-expelled and is effective in smothering Class B and C fires.

Carbon Monoxide (CO) Colorless, odorless, poisonous gas that when inhaled combines with the red blood cells excluding oxygen.

Carcinogen A material that is capable of causing cancer in humans.

Cardiovascular System The heart, blood vessels, and blood within the body.

Carotid Pulse The pulse located on either side of the neck.

Catalytic Bead The most common type of combustible gas sensor that uses two heated beads of metal to determine the presence of flammable gases.

Ceiling Level The highest exposure a person can receive without suffering any ill effects. It is combined with the PEL, TLV, or REL as a maximum exposure.

Cellar Nozzle A device with four spray nozzles designed to rotate in a circular spray pattern for fighting fires in basements or cellars when firefighters cannot make a direct attack on the fire.

Cellar Nozzle A device with four spray nozzles designed to rotate in a circular spray pattern for fighting fires in basements or cellars when firefighters cannot make a direct attack on the fire.

Chain of Command Common fire service term that means to always work through one's direct supervisor. The fire service is viewed as a paramilitary organization and because of this all requests for information outside the assigned workplace should go through the supervisor.

Check Valve Valve installed to control water flow in one direction, typically when different systems are interconnected.

Chemical Burn A burn caused by chemical substances that come into contact with the skin or tissues of the body, creating a caustic reaction.

Chemtrec The Chemical Transportation Emergency Center, which provides technical assistance and guidance in the event of a chemical emergency; a network of chemical manufacturers that provide emergency information and response teams if necessary.

Chemtrec The Chemical Transportation Emergency Center, which provides technical assistance and guidance in the event of a chemical emergency. This network of chemical manufacturers provides emergency information and response teams if necessary.

Chimney A topographic feature on the side of a hill or mountain that naturally collects water runoff, channeling it to the bottom of the rise. Fire is attracted to this feature. Also referred to as drainage.

Chip Measurement System (CMS) A form of colorimetric air sampling in which the gas sample passes through a tube. If the correct color change occurs, the monitor interprets the amount of change and indicates a level of the gas on an LCD screen.

Choking Agent Agent that causes a person to cough and have difficulty breathing. The terrorism agents that are considered choking agents are chlorine and phosgene, both very toxic gases.

Chord The top and bottom components of a beam or truss (sometimes referred to as *flanges*). The top chord is subjected to compressive force; the bottom chord is subjected to tensile force.

Chord The top and bottom components of a beam or truss (sometimes referred to as *flanges*). The top chord is subjected to compressive force; the bottom chord is subjected to tensile force.

Chronic A continual or repeated exposure to a hazardous material.

Cistern An underground water tank made from natural rock or concrete.

Clandestine Drug Labs Illegal labs set up to manufacture street drugs.

Class A Classification of fire involving ordinary combustibles such as wood, paper, cloth, plastics, and rubber.

Class B Classification of fire involving flammable and combustible liquids, gases, and greases. Common products are gasoline, oils, alcohol, propane, and cooking oils.

Class C Classification of fire involving energized electrical equipment, which eliminates using water-based agents.

Class D Classification of fire involving combustible metals and alloys such as magnesium, sodium, lithium, and potassium.

Class K A classification of fire as of 1998 that involves fires in combustible cooking fuels such as vegetable or animal oils and fats.

Clipping Term associated with the use of two-way radios that is used to describe instances when either the first part of a message or the last part of a message is cut off as the result of either speaking before pressing the transmit key or releasing the transmit key prior to the end of a transmission.

Closed-Circuit SCBA A type of SCBA unit in which the exhaled air remains in the system to be filtered and mixed with oxygen for reuse.

Cockloft The area between the roof and the ceiling.

Cockloft The area between the roof and the ceiling.

Cold Smoke Smoke that remains after a fire is extinguished by a sprinkler system.

Collapse Zone The area around a building where debris will land when it falls. As an absolute minimum this distance must be at least 1½ times the height of the building.

Collapse Zone The area around a building where debris will land when it falls. As an absolute minimum this distance must be at least 1 ½ times the height of the building.

Colorimetric Tubes Crystal-filled tubes that change colors in the presence of the intended gases. These tubes are made for the detection of known and unknown gases.

Column A structural element that is subjected to compressive forces—typically a vertical member.

Column A structural element that is subjected to compressive forces—typically a vertical member.

Combination Attack A combined attack based on partial use of both offensive and defensive attack modes.

Combination Attack A combined attack based on partial use of both offensive and defensive attack modes.

Combination Fire Attack A blend of the direct and indirect fire attack methods, with firefighters applying water to both the fuel and the atmosphere of the room.

Combination Fire Attack A blend of the direct and indirect fire attack methods, with firefighters applying water to both the fuel and the atmosphere of the room.

Combination Nozzle A spray nozzle that is capable of providing straight stream and spray patterns, which are adjustable or variable by the operator. Most fog nozzles used today are combination nozzles.

Combination Nozzle A spray nozzle that is capable of providing straight stream and spray patterns, which are adjustable or variable by the operator. Most fog nozzles used today are combination nozzles.

Combustion A chemical reaction that includes the self-sustaining rapid oxidation of a fuel accompanied by the release of heat and light. Commonly referred to as *fire*.

Command Vehicle Typically used by operations chief officers in the fire service.

Common Terminology The designation of a term that is the same throughout an IMS.

Communicable Disease A disease that can be transmitted from one person to another.

Communications Sending, giving, or exchanging of information.

Company A team of firefighters with apparatus assigned to perform a specific function in a designated response area.

Compressed Air Foam Systems (CAFS) A foam system where compressed air is injected into the foam solution prior to entering any hoselines. The fluffy foam created needs no further aspiration of air by the nozzle.

Compression A force that tends to push materials together.

Computer-Aided Dispatch (CAD) Computer-based automated system that assists the telecommunicator in assessing dispatch information and then recommends responses.

Computer-Aided Dispatch (CAD) Computer-based automated system that assists the telecommunicator in assessing dispatch information and then recommends responses.

Computer-Aided Management for Emergency Operations (CAMEO) Program A computer program that combines a chemical information database with emergency planning software. It is commonly used by hazardous materials teams to determine chemical information.

Concentrated Load A load applied to a small area.

Confined Space A space that is large enough to be entered but is not designed for continuous occupancy.

Consent The acceptance of emergency medical treatment by a patient or victim.

Consist The shipping papers that list the cargo of a train. The listing is by railcar, and the consist lists all of the cars.

Constant or Set Volume Nozzle Nozzle with one set volume at a set pressure. For example, 60 gpm at 100 psi (227 L/min 690 kPa). The only adjustment is the pattern.

Constant or Set Volume Nozzle Nozzle with one set volume at a set pressure. For example, 60 gpm at 100 psi (227 L/min 690 kPa). The only adjustment is the pattern.

Constricted A condition of the pupils where they are much smaller than normal and may appear almost like a "pinpoint."

Continuous Beam A beam that is supported in three or more places.

Control Room A room on the ground floor of a high-rise building where all building systems controls are located.

Convulsant A chemical that has the ability to cause seizure-like activity.

Cribbing The use of various dimensions of lumber arranged in systematic stacks (pyramid, box, step, etc.) to support an unstable load.

Cryogenic Gas Any gas that exists as a liquid at a very cold temperature, always below 2150 F (2101 C)

Curies (Ci) The measure of activity level for radiation sources.

Cutting Tools The group of tools used to cut through or around materials.

Cutting Tools The group of tools used to cut through or around materials.

Damming The stopping of a body of water, which at the same time stops the spread of the spilled material.

Dangerous Cargo Manifest (DCM) The shipping papers for a ship, which list the hazardous materials on board.

Database Organized collection of similar facts.

Dead Load The weight of the building materials and any part of the building permanently attached or built in.

DECIDE Process A management system used to organize the response to a hazardous materials incident. The factors of DECIDE are detect, estimate, choose, identify, do the best, and evaluate.

Decontamination The physical removal of contaminants (chemicals) from people, equipment, or the environment. Most often used to describe the process of cleaning to remove chemicals from a person. Often shortened to "decon."

Defensive Attack A calculated attack on part of a problem or situation in an effort to hold ground until sufficient resources are available to convert to an offensive form of attack.

Defensive Attack A calculated attack on part of a problem or situation in an effort to hold ground until sufficient resources are available to convert to an offensive form of attack.

Deflagrates Rapid burning, which in reality with regard to explosions can be considered a slow explosion, but is traveling at a lesser speed than a detonation.

Deflagration Combustion rate below the speed of sound (subsonic).

Dehydration A loss of water and vital fluids in the body.

Deluge System Protective system designed to protect areas that may have a fast-spreading fire engulfing the entire area. All of its sprinkler heads are already open, and the piping contains atmospheric air. When the system operates, water flows to all heads, allowing total coverage. The system uses a deluge valve that opens when a separate fire detection system senses the fire and signals to trip the valve open.

Deluge System Protective system designed to protect areas that may have a fast-spreading fire engulfing the entire area. All of its sprinkler heads are already open, and the piping contains atmospheric air. When the system operates, water flows to all heads, allowing total coverage. The system uses a deluge valve that opens when a separate fire detection system senses the fire and signals to trip the valve open.

Density The mass per unit volume of a substance under specified conditions of pressure and temperature.

Department of Homeland Security (DHS) Federal agency created as a result of The National Strategy for Homeland Security and the Homeland Security Act of 2002 that leads the unified national effort to secure America by preventing and deterring terrorist attacks and protecting against and responding to threats and hazards to the nation. DHS ensures safe and secure borders, welcomes lawful immigrants and visitors, and promotes the free flow of commerce.

Deployment Plan Predetermined response plan of apparatus and personnel for specific types of incidents and specific locations.

Depth of Char A term commonly used by fire investigators to describe the amount of time wooden material had burned. The deeper the char, the longer the material was burning or exposed to direct flame.

Depth of Char A term commonly used by fire investigators to describe the amount of time wooden material had burned. The deeper the char, the longer the material was burning or exposed to direct flame.

Design Load A load the engineer planned for or anticipated in the structural design.

Detergent-Type Foams Use synthetic surfactants to break down the surface tension of water to create a foaming blanket.

Detonation Combustion rate above the speed of sound (supersonic).

Diffusion A naturally occurring event in which molecules travel from levels of high concentration to areas of low concentration.

Diking A defensive method of stopping a spill. A common dike is constructed of dirt or sand and is used to hold a spilled product. In some facilities, a dike may be preconstructed such as around a tank farm.

Dilated A condition of the pupils where they are much larger than normal and can take up almost the whole colored portion of the eye.

Dilution The addition of a material to the spilled material to make it less hazardous. In most cases water is used to dilute a spilled material, although other chemicals could be used.

Direct Fire Attack An attack on the fire made by aiming the flow of water directly at the material on fire.

Direct Fire Attack An attack on the fire made by aiming the flow of water directly at the material on fire.

Direct Source Emitter An object that provides the most thermal energy and can be easily visualized through thermal imaging. Examples are flame, moving fire gases, or pre-flashover thermal layering within a room or structure.

Disassembly The actual taking apart of vehicle components.

Disaster Medical Assistance Teams (DMAT) Teams that mobilize medical resources to an affected area in the event of a disaster or mass casualty incident; DMAT teams are composed of medical resources including physicians, nurses, therapists, pharmacists, EMTs, and administrative staff.

Discharge Flow Total amount of water flowing from the discharge side of the pump.

Displacement The relocating of major parts (i.e., doors, roof, dash, steering column) of a vehicle.

Distortion The bending of sheet metal or components.

Distributed Load A load applied equally over a broad area.

Distributor Pipe or Extension Pipe Devices that allow a nozzle or other device to be directed into holes to reach basements, attic, and floors that cannot be accessed by personnel. The distributor pipe has self-supporting brackets that help hold it into place when in use.

Diverting Using materials to divert a spill around an item. For instance, several shovels full of dirt can be used to divert a running spill around a storm drain.

Division Command designation responsible for operations within an assigned geographic area.

Double Female Allows the two male ends of hose to be connected.

Double Male Used to connect two female thread couplings.

Drafting The pumping of water from a static source by taking advantage of atmospheric pressure to force water from the source into the pump.

Drainage A topographic feature on the side of a hill or mountain that naturally collects water runoff, channeling it to the bottom of the rise. Fire is attracted to this feature. Also referred to as chimney or fire chimney.

Dressing The practice of making sure that all parts of a knot are lying in the proper orientation to the other parts and look exactly as the pictures herein indicate.

Dry Chemicals Dry extinguishing agents divided into two categories. Regular dry chemicals work on Class B and C fires; multipurpose dry chemicals work on Class A, B, and C fires.

Dry Hydrant A piping system for drafting from a static water source with a fire department connection at one end and a strainer at the water end.

Dry Pipe System Air under pressure replaces the water in the system to protect against freezing temperatures. The sprinkler control valve uses a dry pipe valve to keep pressurized air maintained above with the supply water under pressure below the valve.

Dry Pipe System Air under pressure replaces the water in the system to protect against freezing temperatures. The sprinkler control valve uses a dry pipe valve to keep pressurized air maintained above with the supply water under pressure below the valve.

Dry Powders Extinguishing agents for Class D fires.

Dump Site The area where tenders are unloaded or their load dumped.

Dutchman A short fold of hose or a reverse fold that is used when loading hose and a coupling comes at a point where a fold should take place or when two sets of couplings end up on top of or next to each other. The dutchman moves the coupling to another point in the load.

Dynamic A rope having a high degree of elongation (10 to 15 percent) at normal safe working loads.

Ears Elongated folds or flaps at the ends of a layer of hose to assist in pulling that layer.

Earthquakes The result of shifts in plates of the Earth's crust along geographic fault lines causing large areas of the Earth's surface to shake violently in waves of vibrations.

Eccentric Load A load perpendicular to the cross section of the supporting element that does not pass through the center of mass.

Eductor Device that siphons a liquid from a container into a moving stream.

Electronic Monitoring A method of electronic communication (data transmission) between an in-place fire extinguisher and an electronic monitoring device/system.

Emergency Call Box System of telephones connected by private line telephone, radio-frequency, or cellular technology usually located in remote areas and used to report emergency situations.

Emergency Communications Center Facility either wholly or partially dedicated to being able to receive emergency and, in some instances, nonemergency reports from citizens. Centers such as these are sometimes referred to as fire alarm, headquarters, dispatch, or a public safety answering point (PSAP).

Emergency Decontamination The rapid removal of a material from a person when that person (or responder) has become contaminated and needs immediate cleaning. Most emergency decon setups use a single hoseline to perform a quick gross decon of a person with water.

Emergency Medical Dispatch (EMD) System designed for use by telecommunicators to assist them in evaluating patient symptoms using predetermined criteria and responses.

Emergency Medical Services The delivery of prehospital medical treatment.

Emergency Medical Technician (EMT) An individual trained and certified to provide basic life support emergency medical care.

Emergency Operations Center (EOC) A communication and coordination center placed in service during major emergencies or disasters that is composed of representatives from many different agencies.

Emergency Planning and Community Right to Know Act (EP-CRA) The portion of SARA that specifically outlines how industries report their chemical inventory to the community.

Emergency Response Guidebook (ERG) Book provided by the DOT that assists the first responder in making decisions primarily at transportation-related hazardous materials incidents.

Emergency Response Planning (ERP) Levels that are used for planning purposes and are usually associated with the preplanning for evacuation zones.

Emergency Traffic Term used to signify that a priority message is to follow on the radio.

Encapsulated Suit A chemical suit that covers the responder, including the breathing apparatus. Usually associated with Level A clothing, that is gas and liquid tight, but there are some Level B styles that are fully encapsulated, but not gas or liquid tight.

Encoder Device that converts an "entered" code into paging codes, which in turn activate a variety of paging devices.

Endothermic Reaction A chemical reaction in which heat is absorbed, and the resulting mixture is cold.

Endothermic Reaction Chemical reactions that absorb heat or require heat to bond atoms or molecules.

Engine Company The unit designation of a group of firefighters assigned to a piece of apparatus designed to deliver water to the fire scene.

Enhanced 9-1-1 Similar in nature to basic 9-1-1 but with the capability to provide the caller's telephone number and address.

Etiological A form of a hazard that includes biological, viral, and other disease-causing materials.

Evacuation The movement of people from an area, usually their homes, to another area that is considered to be safe. People are evacuated when they are no longer safe in their current area.

Evaporation A process in which the molecules of a liquid are liberated into the atmosphere at a rate greater than the rate at which the molecules return to the liquid. Ultimately the liquid becomes fully airborne in a gaseous state.

Excepted Packaging Type of packaging used to transport low-risk radioactive materials; the packaging is exempted from DOT regulations.

Exhauster A device to speed the operation of the dry pipe valve by detecting the decrease in air pressure. It helps bleed off air.

Exit Drills in the Home (EDITH) A fire survival program to encourage people to practice fire drills from their home or residence.

Exothermic Reaction A chemical reaction that releases heat, such as when two chemicals are mixed and the resulting mixture is hot.

Exothermic Reaction A chemical reaction that results in heat release.

Exposure Fire Any combustible item threatened by something burning nearby that has caught on fire.

Exposure Fire Any combustible item threatened by something burning nearby that has caught on fire.

Exposure A contact with a potentially disease-producing organism; the contact does not necessarily produce the disease in the exposed individual.

Extension Ladder A ladder consisting of two or more sections that has the ability to be extended to a desired height through the use of a halyard.

External Bleeding Bleeding that is coming from an open wound on the body.

External Floating Roof Tank Tank with the roof exposed on the outside that covers the liquid within the tank. The roof floats on the top of the liquid, which does not allow for vapors to build up.

Extremely Hazardous Substances (EHS) A list of 366 substances that the EPA has determined pre sent an extreme risk to the community if released.

Extricate To set free, release, or disentangle a patient from an entrapment situation.

Federal Communications Commission (FCC) Government agency charged with administering the provisions of the Communications Act of 1934 and the revised Telecommunications Act of 1996 and is responsible for nonfederal radio-frequency users.

Federal Emergency Management Agency (FEMA) On March 1, 2003, the Federal Emergency Management Agency (FEMA) became part of the U.S. Department of Homeland Security (DHS). The primary mission of the Federal Emergency Management Agency is to reduce the loss of life and property and protect the nation from all hazards, including natural disasters, acts of terrorism, and other man-made disasters, by leading and supporting the nation in a risk-based, comprehensive emergency management system of preparedness, protection, response, recovery, and mitigation.

Femoral Artery A major artery in the lower body near the groin that supplies the leg with blood. Can be used as a pressure point for controlling bleeding in the lower extremities.

Fill Site The area where tenders are filled or attain their water.

Film-Forming Fluoroprotein Foam (FFFP) Foam that incorporates the features of AFFF and fluoroprotein foams with good resistance and the film-forming barrier.

Fine Decontamination The most detailed of the types of decontamination. Usually performed at a hospital that has trained staff and is equipped to perform fine decon procedures.

Fire Flow Capacity The amount of water available or amount that the water distribution system is capable of flowing.

Fire Flow Capacity The amount of water available or amount that the water distribution system is capable of flowing.

Fire Flow Requirement A measure comparing the amount of heat the fire is capable of generating versus the amount of water required for cooling the fuels below their ignition temperature.

Fire Flow Requirement A measure comparing the amount of heat the fire is capable of generating versus the amount of water required for cooling the fuels below their ignition temperature.

Fire Hazard Any condition, situation, or operation that could lead to the ignition of unwanted combustion or result in proper combustion becoming uncontrolled.

Fire Hose A flexible conduit used to convey water or other agent from a water source to the fire.

Fire Hose A flexible conduit used to convey water or other agent from a water source to the fire.

Fire Intensity A measurement of Btus produced by a fire. Sometimes measured in flame length in the wildland environment.

Fire Intensity A measurement of Btus produced by a fire. Sometimes measured in flame length in the wildland environment.

Fire load The amount of heat generated when the building and its contents burn.

Fire Load The amount of heat generated when the building and its contents burn.

Fire Point The lowest temperature at which a fuel off-gases an ignitable mixture that, when introduced to a spark or flame, will ignite and sustain burning.

Fire Resistive The capacity of a material to withstand the effects of fire.

Fire Resistive The capacity of a material to withstand the effects of fire.

Fire Shelter A last-resort protective device for wildland firefighters caught or trapped in an environment where a firestorm or blowup is imminent.

Fire Station Alerting System System used to transmit emergency response information to fire station personnel via voice and/or digital transmissions.

Fire Station Alerting System System used to transmit emergency response information to fire station personnel via voice and/or digital transmissions.

Fire Stopping Pieces of material, usually wood or masonry, placed in stud or joist channels to slow the extension of fire.

Fire Stopping Pieces of material, usually wood or masonry, placed in stud or joist channels to slow the extension of fire.

Fire Stream The water or other agent as it leaves the hose and nozzle toward its objective, usually the fire.

Fire Tetrahedron Four-sided pyramid-like figure used to depict the four ingredients necessary for combustion: heat, fuel, oxygen, and chemical chain reaction.

Fire Watch An organized patrol of a protected property when the sprinkler or other protection system is down for maintenance. Personnel from the property regularly check to make sure a fire has not started and assist in evacuation and prompt notification of the fire department.

Fire (see combustion)

Firefighter Assist and Support Team (FAST) A company designated to search for and rescue trapped or lost firefighters. May also be called a rapid intervention team (RIT).

Fire-Resistive Rating The time in hours that a material or assembly can withstand fire exposure. Fire-resistive ratings are usually provided for testing organizations. The ratings are expressed in a time frame, usually hours or portions thereof.

Fire-Resistive Rating The time in hours that a material or assembly can withstand fire exposure. Fire-resistive ratings are usually provided for testing organizations. The ratings are expressed in a time frame, usually hours or portions thereof.

First Responders A group designated by the community as those who may be the first to arrive at a chemical incident. This group is usually composed of police officers, EMS providers, and firefighters.

Fissile Isotopes that can be induced into fission, which creates a release of energy.

Fit Testing A test that ensures the respiratory protection fits the face and offers maximum protection.

Flammable Limits The concentration level of a substance at which it will burn. Flammable limits are expressed as a percent range mixed in air.

Flammable Range Ratio of gas to air that will sustain fire if exposed to flame or spark.

Flanks of the Fire The sides of a wildland fire running from the start point up each side to the end of the fire running into unburned areas.

Flashover A sudden event that occurs when all the contents of a container reach their ignition temperature nearly simultaneously.

Flashpoint The lowest temperature at which a fuel off-gases an ignitable mixture that, when introduced to a spark or flame, will briefly ignite, but not sustain burning.

Flow The rate or quantity of water delivered, usually measured in gallons per minute or liters per minute (1 gpm 5 3.785 L/min).

Fluoroprotein Film-Forming Foam (FFFP) Combines protein with the film-forming fluorinated surfactants of AFFF to improve on the qualities of both types of foam.

Fluoroprotein Foam Designed as an improved protein foam with a fluorinated surfactant added.

Flush or Slab Doors Doors that are flat or have a smooth surface and may be of either hollow-core or solid-core construction.

Fly Ladder That portion of a ladder that extends out from the bed ladder. Also called fly section.

Foam An aggregate of gas-filled bubbles formed from aqueous solutions of specially formulated concentrated liquid foaming agents.

Fog Nozzle Delivers either a fixed spray pattern or variable combination of straight stream and spray patterns.

Fog Nozzle Delivers either a fixed spray pattern or variable combination of straight stream and spray patterns.

Forcible Entry The fire scene task of gaining entry to a building or secured area by disabling, breaking, or going around locking and security devices.

Forcible Entry The fire scene task of gaining entry to a building or secured area by disabling, breaking, or going around locking and security devices.

Forestry Hose Specially designed hose for use in forestry and wildland firefighting. It comes in 1- and 1½-inch (25- and 38-mm) sizes and should meet U.S. Forestry Service specifications.

Fracture A medical term for a broken or cracked bone in the body.

Frangible Disk A type of pressure-relieving device that actually ruptures in order to vent the excess pressure. Once opened the disk remains open; it does not close after the pressure is released.

Freelancing The act of working alone or performing a task for which the firefighter has not been assigned.

Freelancing The act of working alone or performing a task for which the firefighter has not been assigned.

Freezing Point The temperature at which liquids become solids.

Friction Loss Measurement of friction in a system such as a hoseline.

Friction Caused by the rubbing of materials against each other while in movement and converts or robs some of the movement energy into heat energy.

Frontage The portion of a property that faces and actually touches the street.

Fuel Resistance Ability to tolerate the fuel and to avoid being saturated by or picking up the fuel.

Full Thickness Burns Burns affecting not only the skin structure but the tissues and muscles underneath. Full thickness burns may be red, white, or charred in color and will appear dry because the blood vessels in the skin are damaged extensively and are not supplying fluids to the area.

Garden Apartment A two- or three-story apartment building with common entryways and layouts on each floor, surrounded by greenery and landscaping, sometimes having porches and patios.

Gas A state of matter that describes the material in a form that moves freely about and is difficult to control. Steam is an example.

Gate Valve Indicating or nonindicating valve that is opened and closed to control water flow.

Gauge Pressure Measures pressure without atmospheric pressure. Normally fire department gauges do not measure atmospheric pressure. Gauge pressure is measured in psi or psig.

GEDAPER Process A management system used to organize the response to a chemical incident. The factors are gather information, estimate potential, determine goals, assess tactical options, plan, evaluate, and review.

Girder A large structural member used to support beams or joists—that is, a beam that supports beams.

Girder A large structural member used to support beams or joists—that is, a beam that supports beams.

Glazing The glass or other clear material portion of the window that allows light to enter.

Global Positioning System (GPS) System of twenty-four satellites used as reference points to calculate positions. The satellites were placed into orbit by the U.S. Department of Defense and were originally dedicated for military operations.

Global Positioning System (GPS) System of twenty-four satellites used as reference points to calculate positions. The satellites were placed into orbit by the U.S. Department of Defense and were originally dedicated for military operations.

Golden Hour The first hour after the onset of trauma to a person. If surgical intervention can be reached within that hour, chances for survival are greatly increased.

Gross Negligence Occurs when an individual disregards training and continues to act in a manner without regard for others.

Ground Fault Interrupter (GFI) An electrical switch that will trip when electrical current is interrupted or shorted. It may be reset when the cause of the interruption is corrected.

Ground Pads Sheets of plywood, planks, aluminum sheets, and so on, used to distribute weight over a larger area.

Guard Dogs Trained animals that will bark and attack an intruder.

Guideline/Lifeline Rope used as a crew is searching a structure to assist them in finding their way back out.

Gusset Plate A connecting plate used in truss construction. In steel trusses, these plates are flat steel stock. In wood trusses, the plates are either light-gauge metal or plywood.

Gusset Plate A connecting plate used in truss construction. In steel trusses, these plates are flat steel stock. In wood trusses, the plates are either light-gauge metal or plywood.

Hailing System A system used to help locate a victim at a collapse situation. Rescuers are placed at strategic areas of a debris pile while ordering complete silence of people and machinery. One rescuer will then call out for a response from the victim, hoping to receive an acknowledgement. If none is given, the next positioned rescuer will call out. If a reply is heard, another rescuer at a different position should call out in order to pinpoint the position of the response.

Half Life The amount of time for a given radiation source to decay, which means it is emitting radioactivity.

Half-life The time required for a drug or other substance deposited in a living organism to be metabolized or eliminated by normal biological processes.

Halligan Tool From the prying group, a 30-inch forged steel tool with three primary parts: the adz end, the pike end, and the fork end.

Halligan Tool From the prying group, a 30-inch forged steel tool with three primary parts: the adz end, the pike end, and the fork end.

Halyard A rope or cable that is used to raise the fly ladders of an extension ladder.

Hard Suction Hose A special type of hose that does not collapse when used for drafting.

Harnesses Webbing sewn together to form a belt, seat harness, or seat and chest harness combination.

Hazardous Material A substance (either matter—solid, liquid, or gas—or energy) that, when released, is capable of harming people, the environment, and property. It includes weapons of mass destruction (WMD) as defined in 18 U.S. Code, Section 2332a, as

well as any other criminal use of hazardous materials, such as illicit labs, environmental crimes, or industrial sabotage.

Hazardous Materials Crime A criminal act that uses or threatens the use of hazardous materials as a weapon.

Hazardous Materials Response Unit (HMRU) A specialized response group within the FBI Laboratory division that responds to WMD and other potentially hazardous crime scenes.

Hazardous Materials Technician An individual trained to meet the requirements of CFR OSHA 1910.120, Technician Level for Hazardous Materials Response.

Hazardous Waste Operations and Emergency Response (HAZWOPER) The OSHA regulation that covers safety and health issues at hazardous waste sites, as well as response to hazardous materials incidents.

Head of the Fire The running top or aggressive end of the fire away from the start point.

Head Pressure Measures the pressure of a column of water in feet (meters). Head pressure gain or loss results when water is being pumped above or below the level of the pump. A head of 2.31 feet (0.7 m) would equal 1 psi (6.895 kPa).

Heads Up Display (HUD) A device that displays—in the user's field of vision—the amount of air remaining in the SCBA cylinder .

Heat Resistance The ability of foam to stand up to the heat of the fire or to hot surfaces near the fire.

Helix The metal or plastic bands or rings used in hard suction hose to prevent its collapse under drafting conditions.

Higbee Cut The blunt ending of the threads of fire hose couplings that allows the threads to be properly matched, avoiding cross-threading.

Higbee Indicator A groove notched into coupling lugs to help firefighters align the Higbee cuts.

Hoistway The shaft in which an elevator or a number of elevators travel.

Hollow-Core Door Any door that is not solid, usually with some type of filler material between face panels.

Home Alerting Devices Emergency alerting devices primarily used by volunteer department personnel to receive reports of emergency incidents.

Home Alerting Devices Emergency alerting devices primarily used by volunteer department personnel to receive reports of emergency incidents.

Hook A tool with a 32-inch to 12-foot handle with a pike and hook on one end. Used for pulling ceilings or separating other materials. Also known as a pike pole.

Hook A tool with a 32-inch to 12-foot handle with a pike and hook on one end. Used for pulling ceilings or separating other materials. Also known as a pike pole.

Horizontal Ventilation Channeled pathway for fire ventilation via horizontal openings.

Horizontal Ventilation Channeled pathway for fire ventilation via horizontal openings.

Hose Bed The portion or compartment of fire apparatus that carries the hose.

Hose Bridge Device that allows vehicles to pass over a section of hose without damaging it.

Hose Cap A device used to cap the end of a hoseline or appliance to prevent water flow.

Hose Cart A handcart or flat cart modified to be able to carry hose and other equipment around large buildings. Some departments use them for high-rise situations.

Hose Clamp A device to control the flow of water by squeezing or clamping the hose shut. Some work by pushing a lever that closes the jaws of the device, and others have a screw mechanism or hydraulic pump that closes the jaws.

Hose Jacket Metal or leather device used for stopping leaks without shutting down the line that is fitted over the leaking area and either clamped or strapped together to control the leak.

Hose Roller or Hoist A metal frame, with a securing rope, shaped to fit over a windowsill or edge of a roof with two rollers to allow the hose to roll over the edge, preventing chafe.

Hose Strap A short strap with a forged handle and cinch clip attached. Used to help maneuver hose and attach hose to ladders and stair rails.

Hostile Fire Event A unique fire phenomenon (such as flashover, backdraft, and rapid fire spread) that takes place within a building.

Hurricanes Low-pressure weather systems that form in the tropics, resulting in harsh, spinning winds accompanied by thunderstorms that move across the oceans and onto land.

HVAC Acronym for heating, ventilation, and air-conditioning unit. HVACs are typically a rooftop unit on commercial buildings. Buildings may have one or dozens of these units.

HVAC Acronym for heating, ventilation, and air-conditioning unit. HVACs are typically a rooftop unit on commercial buildings. Buildings may have one or dozens of these units.

Hydrant Valve or Switch Valve Valve used on a hydrant that allows an engine to connect and charge its supply line immediately but also allows an additional engine to connect to the same hydrant without shutting down the hydrant, and increases the flow of the hydrant.

Hydrant Wrench Tool used to operate the valves on a hydrant. May also be used as a spanner wrench. Some are plain wrenches and others have a ratchet feature to speed the operation of the valve.

Hydraulic Pistons Mechanical rams that operate by pressure exerted through the use of a liquid, usually some form of oil.

Hydraulics The study of fluids at rest and in motion.

Hydrogen Cyanide (HCN) A prevalent toxin that is readily found in large quantities in smoke. Hydrogen cyanide is colorless and is produced by the combustion of natural products such as wool, silk, cotton, and paper as well as synthetics or plastics.

Hyperbaric Chamber A chamber that is usually used to treat scuba divers who ascended too quickly and need extra oxygen to survive. The chamber re-creates the high-pressure atmosphere of diving and forces oxygen into the body. It is also successful in the treating of carbon monoxide poisoning and smoke inhalation, because both of these problems require high amounts of oxygen to assist with the patient's recovery.

Hypoperfusion A serious condition caused by a problem or failure of the circulatory system that results in a decrease of oxygen and vital nutrients to the body's tissues. Also known as shock.

Hypoxia A deficiency of oxygen.

ICt$_{50}$ The incapacitating level for time to 50 percent of the exposed group. It is a military term that is often used in conjunction with LCt$_{50}$.

Ignition Temperature The lowest temperature that a fuel will off-gas an ignitable mixture that will self-ignite and continue to burn (also known as auto-ignition). May also be referred to as ignition point.

Immediately Dangerous to Life and Health (IDLH) The maximum level of a dangerous atmosphere one could be exposed to and still escape without experiencing any effects that may impair escape or cause irreversible health effects.

Impact Load A load that is in motion when it is applied.

Implied Consent The assumption of acceptance of emergency medical treatment by an unconscious patient or a child with no parents or legal guardians present.

Incendiary Agents Chemicals that are used to start fires, the most common being a Molotov cocktail.

Incident Action Plan (IAP) A strategic and tactical plan developed by the incident commander.

Incident Action Plan (IAP) A strategic and tactical plan developed by the incident commander.

Incident Command System (ICS) A command system utilized on the emergency scene that is designed to keep order and follow a sequence of set guidelines.

Incident Command System (ICS) An expandable command system used to deal with a myriad of incidents. Helps achieve the highest level of accountability and effectiveness for incident handling. Limits span of control and provides a framework to help make tasks manageable.

Incident Command System (ICS) An organized, systematic method for the command, control, and management of an emergency incident.

Incident Commander Level A training level that encompasses the operations level with the addition of incident command training. Intended to be the person who may command a chemical incident.

Incident Management Team A team of specially trained individuals that fill the command and general staff positions of the Incident Command System when an incident takes place.

Incision A cut to the skin that leaves a straight, even pattern.

Increaser Used to connect a smaller hose to a larger one.

Indirect Fire Attack An attack made on interior fires by applying a fog stream into a closed room or compartment, thus converting the water into steam to extinguish the fire.

Indirect Fire Attack An attack made on interior fires by applying a fog stream into a closed room or compartment, thus converting the water into steam to extinguish the fire.

Infection Control Procedures and practices for firefighters and emergency medical care providers to follow to prevent the transmission of diseases and germs from a patient to themselves or other patients.

Infectious Disease See *communicable disease.*

Infrared Sensor A sensor that uses infrared light to determine the presence of flammable gases. The light is emitted in the sensor housing and the gas passes through the light. If it is flammable the sensor will indicate the presence of the gas.

Initial Assessment The initial investigative action taken by care providers to determine if the patient has the basic signs of life as well as any serious, life-threatening injuries.

Initial Rapid Intervention Crew (IRIC) As outlined in NFPA 1710, two members of the initial attack crew must be assigned as a rapid deployment rescue team for the purposes of rescuing lost or trapped firefighters.

In-Line Eductor Eductor in which the waterway is always piped through a venturi.

Intake Relief Valve Required on large-diameter hose at the receiving engine that functions as a combined overpressurization relief valve, a gate valve, and an air bleed-off.

Integrated Communications The ability of all units or agencies to communicate at an incident.

Interface Firefighting Fighting wildland fire and protecting exposed structures in rural settings.

Intermodal Containers These are constructed in a fashion so that they can be transported by highway, rail, or ship. Intermodal containers exist for solids, liquids, and gases.

Internal Bleeding Bleeding within the body when no visible open wound is present.

Internal Floating Roof Tank Tank with a roof that floats on the surface of the stored liquid, but also has a cover on top of the tank, so as to protect the top of the floating roof.

Ionization Potential (IP) The ability of a gas or vapor to be ionized. It is most commonly used to determine whether a photoionization detector can detect a gas or vapor.

Irons The combination of a Halligan tool and flathead ax or maul.

Irons The combination of a Halligan tool and flathead ax or maul.

Irritant A material that is irritating to humans, but usually does not cause any long-term adverse health effects.

Isolation Area An area that is set up by responders and is intended to keep people, both citizens and responders, out. May later become the hot zone/sector as the incident evolves. Is the minimum area that should be established at any chemical spill.

Isotope A material that has a different form due to the number of neutrons that are in the nucleus.

Jacket The outer part of the hose, often a woven cloth or rubberized material, which protects the hose from mechanical and other damage.

Jamb The mounting frame for a door.

Jet Dump A device that speeds the process of dumping a load of water from a tanker/tender.

Jet Siphon A device that speeds the process of transferring water from one tank to another.

Joist A wood framing member that supports floor or roof decking.

Joist A wood framing member that supports floor or roof decking.

Kern A derivative of the term kernel, which is defined as "the central, most important part of something; core; essence."

Knockdown Speed Speed with which foam spreads across the surface of a fuel.

Laceration A cut to the skin and underlying tissues that leaves an irregular, even pattern.

Ladder Pipe An appliance that is attached to the underside of an aerial ladder for an elevated water application.

Laminated Glass Glass composed of two or more sheets of glass with a plastic sheet between them. The purpose of the plastic sheet is to hold the glass together if broken, thus reducing the hazard of flying glass.

Landing Plate The plate at the top or bottom of an escalator where the steps disappear into the floor.

Landslides Large areas of rock, earth, and/or debris that shift violently and unexpectedly down a slope.

Laws Legislation that is passed by the House and Senate, and signed by the president.

LCt_{50} The lethal concentration for time to 50 percent of the group. Same as the LC_{50}, but adds the element of time. It is a military term.

Leaking Underground Storage Tank (LUST) Describes a leaking tank that is underground.

Ledge Door Door built with solid material, usually individual boards, common in barns and warehouses.

Lethal Concentration (LC_{50}) A value for gases that provides the amount of chemical that could kill 50 percent of the exposed group.

Lethal Dose (LD_{50}) A value for solids and liquids that provides the amount of a chemical that could kill 50 percent of an exposed group.

Level A Ensemble (of Protective Clothing) Fully encapsulated chemical protective clothing. It is gas and liquid tight and offers protection against chemical attack.

Level B Ensemble (of Protective Clothing) A level of protective clothing that is usually associated with splash protection. Level B requires the use of SCBA. Various clothing styles are considered Level B.

Liability The possibility of being held responsible for individual actions.

Life Safety Line According to NFPA 1983, rope dedicated solely to the purpose of constructing lines for supporting people during rescue, firefighting, or other emergency operations, or during training evolutions.

Life Safety Term applied to the fire protection concept in which buildings are designed to allow for the escape of building occupants without injuries. Life safety usually makes the building more fire resistant, but this is not the main goal.

Lifting A term used to describe the removal of upper level smoke and heat when cool air replaces the upper level hot air that is escaping.

Light-use (one-person) orgeneral-use (two-person) rope According to NFPA 1983, a one-person rope requires a minimum tensile strength of 4,500 pounds, and a two-person rope requires a minimum tensile strength of 9,000 pounds.

Liner The inner layer of fire hose, usually made of rubber or a plastic material, that keeps the water in the tubing of the hose.

Lintel A beam that spans an opening in a load-bearing masonry wall.

Liquid A state of matter that implies fluidity, which means a material has the ability to move as water would. There are varying states of being a liquid from moving very quickly to very slowly. Water is an example.

Live Load The weight of all materials and people associated with but not part of a structure.

Load-Bearing Wall Any wall that supports other walls, floors, or roofs.

Load-Bearing Wall Any wall that supports other walls, floors, or roofs.

Loaded Stream Extinguisher Combats the water freezing problem by adding an alkali salt as an antifreezing agent.

Loading The weight of building materials or objects in a building.

Loading The weight of building materials or objects in a building.

Local Application System Designed to protect only a certain or local portion of the building, usually directly where the hazard will occur or spread.

Local Application System Designed to protect only a certain or local portion of the building, usually directly where the hazard will occur or spread.

Local Emergency Planning Committee (LEPC) A group composed of members of the community, industry, and emergency responders to plan for a chemical incident, and to ensure that local resources are adequate to handle an incident.

Locking Devices A mechanical device or mechanism used to secure a door or window.

Loop A turn in the standing part that crosses itself and results in the standing part continuing on in the original direction of travel.

Low Specific Activity (LSA) Designation that indicates that a material is emitting low levels of radiation.

Lower Explosive Limit (LEL) The lower part of the flammable range; the minimum required to have a fire or explosion.

Mantle Anything that cloaks, envelops, covers, or conceals.

Mask Confidence, or "Smoke Divers," Training Training courses designed to develop a firefighter's skills and confidence for using SCBA.

Mass Casualty An incident in which the number of patients exceeds the capability of the EMS to manage the incident effectively. In some jurisdictions this can be two patients, while in others it may take ten to make the incident a mass casualty.

Mass Casualty EMS incidents that involve more than five patients.

Mass Decontamination A process used to decontaminate large numbers of contaminated victims.

Master Stream or Heavy Appliances Non-handheld water applicator capable of flowing over 350 gallons of water per minute (1325 L/min).

Mastery The concept that an individual can achieve 90 percent of an objective 90 percent of the time.

Material Safety Data Sheet (MSDS) An information sheet that provides chemical-specific safety information.

Material Safety Data Sheet (MSDS) Information sheet for employees that provides specific information about a chemical, with attention to health effects, handling, and emergency procedures.

Matter Something that occupies space and can be perceived by one or more senses; a physical body, a physical substance, or the universe as a whole. Something that has mass and exists as a solid, liquid, or gas.

Mayday A term used *only* to signify that a person is in a life-threatening situation and needs immediate assistance. A mayday can be declared by anyone having knowledge of a person in distress.

Mayday A universal call for help. A Mayday indicates that an individual or team is in extreme danger. When declaring a mayday, it is best to repeat the word three times (Mayday, mayday, mayday).

Means of Egress A safe and continuous path of travel from any point in a structure leading to a public way. Composed of three parts: the exit access, the exit, and the exit discharge.

Medium-Diameter Hose (MDH) Either 2½- or 3-inch (63- or 75-mm) hose.

Medi-Vac An ambulance that transports patients by air. Typically, medi-vac units are helicopters with highly trained EMS personnel and nurses.

Melting Point The temperature at which solids become liquids.

Metal Oxide Sensor (MOS) A coiled piece of wire that is heated to determine the presence of flammable gases. Also called tin dioxide sensor.

Midslope An area partway up a slope. Any location not on the bottom or top of a slope, as in a midslope road crossing the slope horizontally.

Miscible Having the ability to mix with water.

Mission Statement A written declaration by a fire agency describing the things that it intends to do to protect its citizenry or customers.

Mission Vision A term used to describe a condition in which a person becomes so focused on an objective that peripheral conditions are not noticed, as if the person is wearing blinders.

Mission-specific Competencies The knowledge, skills, and abilities to perform the mission of the organization. Examples are provided in NFPA 472.

Mobile Data Computer Communications device that, unlike the mobile data terminal, does have information processing capabilities.

Mobile Data Computer Communications device that, unlike the mobile data terminal, does have information processing capabilities.

Mobile Data Terminal (MDT) Communications device that in most cases has no information processing capabilities.

Mobile Data Terminal (MDT) Communications device that in most cases has no information processing capabilities.

Mobile Radio Complete receiver/transmitter unit that is designed for use in a vehicle.

Mobile Radio Complete receiver/transmitter unit that is designed for use in a vehicle.

Mobile Support Vehicle (MSV) Vehicle designed exclusively for use as an on-scene communication center and command post.

Modular Organization The ability to start small and expand if an incident becomes more complex.

Molecule The smallest particle into which an element or a compound can be divided without changing its chemical and physical properties; a group of like or different atoms held together by chemical forces.

Mortar Mixture of sand, lime, and portland cement used as a bonding material in masonry construction.

Mounting Hardware Hinges, tracks, or other means of attaching a door to the frame or jamb.

Multi-gas Detector A term used to describe an air monitor that measures oxygen levels, explosive (flammable) levels, and one or two toxic gases such as carbon monoxide or hydrogen sulfide.

Multiple-Alarm Incident Involves the response of additional personnel.

Mutual Aid or Assistance Agreements Prearranged written agreements of the type and amount of assistance one jurisdiction will provide to another in the event of a large-scale fire or disaster. The key to understanding mutual aid is that it is a reciprocal agreement.

National Emergency Number Association Not-for-profit organization founded in 1982 and made up of more than 6,000 members. The association fosters technical advancement, availability, and implementation of a universal emergency telephone number system.

National Fire Protection Association (NFPA) A not-for-profit membership organization that uses a consensus process to develop model fire prevention codes and firefighting training standards.

National Incident Management System (NIMS) A standardized system of operating at and controlling resources during an incident where multiple agencies may interact.

National Response Center (NRC) The location that must be called to report a spill if it is in excess of the reportable quantity.

Needed or Required Flow Estimate of the amount of water required to extinguish a fire in a certain time period. Based on the type and amount of fuel burning.

Negligence Acting in an irresponsible manner, or different from the way in which someone was trained; that is, differing from the standard of care.

Nerve Agents Chemicals that are designed to kill humans, specifically in warfare. They are chemically similar to organophosphorus pesticides and cause the same medical reaction in humans.

Nested The state when all the ladders of an extension ladder are unextended.

NFPA 1001 Standard for Fire Fighter Professional Qualifications, a national consensus training standard establishing the job performance requirements of tasks to be performed by firefighters.

NFPA 1404 National Fire Protection Association standard created by the Fire Service Training Committee detailing the requirements for fire service SCBA programs, including training and maintenance procedures.

NFPA 1500 National Fire Protection Association standard created by the Technical Committee on Fire Service Occupational Safety and Health that addresses a number of issues concerning safety and protective equipment.

NFPA 1981 National Fire Protection Association standard specific to open-circuit SCBA for fire service use that contains additional requirements above the NIOSH certification.

NFPA 1982 National Fire Protection Association standard specific to Personal Alert Safety Systems (PASS).

NFPA 72 National Fire Alarm Code.

NFPA Standard 1931 Standard for manufacturer's design of fire department ground ladders.

NFPA Standard 1932 Standard on use, maintenance, and service testing of in-service fire department ground ladders.

NIOSH National Institute for Occupational Safety and Health, 42 CFR Part 84, sole responsibility for testing and certification of respiratory protection including fire service SCBA.

No-Knowledge Hardware Locking devices that require no key or special knowledge to operate.

Nozzle Flow The amount or volume of water that a nozzle will provide. Flow is measured in gallons per minute or liters per minute.

Nozzle Pressure The pressure required to effectively operate a nozzle. Pressure is measured in pounds per square inch or kilopascals.

Nozzle Reach The distance the water will travel after leaving the nozzle. Reach is a function of the pressure, which is converted to velocity or speed of the water leaving the nozzle.

Nozzle Reaction The force of that makes the nozzle move in the opposite direction of the water flow. The nozzle operator must counteract the thrust exerted by the nozzle to maintain control.

Nozzle A tapered or constricted tube used to increase the speed or change the direction of water or other fluids.

Nozzle A tapered or constricted tube used to increase the speed or change the direction of water or other fluids.

Nuclear Regulatory Agency (NRC) The government regulatory agency responsible for oversight of nuclear materials.

Occupancy Classifications The use for which a building or structure is designed.

Occupancy Classifications The use for which a building or structure is designed.

Occupant Use Hose Hose that is used in standpipe systems for building occupants to fight incipient fires. It is usually 1½-inch (38-mm) single-jacket hose similar to attack hose.

Offensive Attack An aggressive attack on a situation where resources are adequate and capable of handling the situation.

Offensive Attack An aggressive attack on a situation where resources are adequate and capable of handling the situation.

Open-Circuit SCBA A type of SCBA unit in which the exhaled air is vented to the outside atmosphere.

Operational Period The time frames for operations at an incident.

Operations Level The next level of training above awareness that provides the foundation which allows for the responder to perform defensive activities at a hazardous materials incident.

Ordinary Tank A horizontal or vertical tank that usually contains combustible or other less hazardous chemicals. Flammable materials and other hazardous chemicals may be stored in smaller quantities in these types of tanks.

OSHA 29 CFR 1910.134 Standard establishing minimum medical, training, and equipment levels for respiratory protection programs.

Outside Stem and Yoke (OS & Y) Valve Has a wheel on a stem housed in a yoke or housing. When the stem is exposed or outside, the valve is open. Also called an outside screw and yoke valve.

Overpacked A response action that involves the placing of a leaking drum (or container) into another drum. There are drums made specifically to be used as overpack drums in that they are oversized to handle a normal size drum.

Oxidation Any process in which oxygen combines with an element or substance.

Oxidizer A substance that readily releases oxygen, and by yielding oxygen, an oxidizer can easily cause or enhance the combustion of other materials. Oxidizers can dramatically increase the rate of burning when the combustible material is ignited.

Oxidizer Materials that readily release oxygen; by yielding oxygen, an oxidizer can easily cause or enhance the combustion of other materials. Oxidizers can dramatically increase the rate of burning when combustible material is ignited.

Oxygen-Deficient Atmosphere An atmosphere with an oxygen content below 19.5 percent by volume.

Packaging The bandaging and preparing of a patient to be moved from the place of injury to a stretcher.

Panel Doors Doors with a solid stile and rails with panels made of wood or glass or other materials.

Panic Hardware Hardware mounted on doors that enable them to be opened by pushing from the inside.

Paragraph q The paragraph within HAZWOPER that outlines the regulations that govern emergency response to hazardous materials incidents.

Paramedic (EMT-P) An individual trained and certified to provide advanced life support emergency medical care, including drug therapy.

Parapet The projection of a wall above the roofline of a building.

Parapet The projection of a wall above the roofline of a building.

Partial Thickness Burns Burns affecting the entire skin structure that lies over the top of the fatty tissues and muscles causing skin to turn red and blister.

Passive Emitter A inanimate object whose temperature will vary depending upon the environment and time frame that it is exposed to a heat source.

Passport A term given to a specific accountability system where crews are tracked using a card (passport) with all members listed. An accountability manager tracks the passports on an accountability board.

PDD 39 Presidential Decision Directive 39, which established the FBI as the lead agency in terrorism incidents responsible for crisis management. It also established FEMA as the lead for consequence management.

Permeation The movement of chemicals through chemical protective clothing on a molecular level; does not cause visual damage to the clothing.

Permissible Exposure Limit (PEL) An OSHA value that regulates the amount of a chemical that a person can be exposed to during an eight-hour day.

Persistence An indication of the time that a material will remain as a liquid, and is related to vapor pressure.

Personal Accountability Safety System (PASS) A small motion-sensitive device worn by firefighters. The motion device will go into an audible alarm signaling that the firefighter may be in trouble if there is no movement for a set period of time. It can also be manually activated by the firefighter if desired. Also known as a personal alert safety system.

Personal Alert Safety System (PASS) A device that emits a loud alert or warning that the wearer is motionless.

Personal Size-Up A continuous situational awareness and mental evaluation of an individual's immediate environment, facts, and probabilities.

Personnel Accountability Report (PAR) This is an organized roll call of all units assigned to an incident.

Photoionization Detector (PID) A detection device that can detect toxic gases or vapors and is commonly used by Hazardous Materials response teams.

Photoionization Detector (PID) An air monitoring device used by hazardous materials teams to determine the amount of toxic materials in the air.

Piercing Nozzle Nozzle originally designed to penetrate the skin of aircraft and now has been modified to pierce through building walls and floors.

Piercing Nozzle Nozzle originally designed to penetrate the skin of aircraft and now has been modified to pierce through building walls and floors.

Pike Pole See hook.

Pike Pole See hook.

Pipe Chase A construction term used to describe a void designed to house building water supply and waste pipes. The term electrical chase is used for wiring.

Pitot Gauge A device with an opening in its blade-shaped section that allows water to flow to a Bourdon gauge and register the flowing discharge pressure of an orifice.

Plan View A drawing or diagram of a building or area as seen from directly overhead. May include a site plan or a floor plan.

Platform Framing A style of wood frame construction in which each story is built on a platform, providing fire stopping at each level.

Platform Framing A style of wood frame construction in which each story is built on a platform, providing fire stopping at each level.

Point of No Return Refers to the firefighter's individual management of their air supply. It is determined by the amount of air consumed going into an IDLH environment versus the amount of residual air needed to safely exit the environment.

Polar Solvent Type of Foam Foam that is compatible with alcohol and/or polar solvents by creating a polymeric barrier between the water in the foam and the polar solvent.

Polar Solvent A material that will mix with water, diluting itself.

Polymeric Barrier A separation barrier made up of polymer or a chain of molecules linked in a series of long strands. This separates a polar solvent from an ATC foam blanket.

Polymerize A chain reaction in which the material quickly duplicates itself and, if contained, can be very explosive.

Portable Hydrant or Manifold Like a large water thief and may have one or more intakes and numerous outlets to allow multiple hoselines to be utilized with or without a pumper at the fire location.

Portable Water Tanks Collapsible or inflatable temporary tanks for the storage of water that is dumped from tankers or tenders. Usually carried by the tender to set up a dump site.

Positive Pressure A feature of SCBA providing a continuous supply of air, delivered by the regulator to the face piece, keeping toxic gases from entering. This pressure (1½ to 2 psi, depending on the manufacturer) is slightly above atmospheric pressure.

Post Indicator Valve (PIV) A control valve that is mounted on a post case with a small window, reading either "OPEN" or "SHUT."

Post-Incident Thought Patterns A phenomenon that describes an individual's inattentiveness following a significant incident. Post-incident thought patterns can lead to injuries or even death.

Potential Evidence Any item within and around the structure could potentially become evidence in the determination of the area of origin, fire cause, or identification of who or what may be responsible for the event.

Preaction System Similar to the dry pipe and deluge systems. The system has closed piping and heads with air under no or little pressure, but the water does not flow until signaled open from a separate fire detection system. The preaction valve then opens and allows water to flow through the system to the closed heads. When an individual head is heat activated, it opens and water attacks the fire. Usually used when water can cause a large dollar loss.

Preaction System Similar to the dry pipe and deluge systems. The system has closed piping and heads with air under no or little pressure, but the water does not flow until signaled open from a separate fire detection system. The preaction valve then opens and allows water to flow through the system to the closed heads. When an individual head is heat activated, it opens and water attacks the fire. Usually used when water can cause a large dollar loss.

Prearrival Instructions Self-help instructions intended to enhance the overall safety of the citizen until first responders arrive on the scene.

Pre-Incident Plan Advance planning of firefighting tactics and strategies or other emergency activities that can be anticipated to occur at a particular location. Also referred to as pre-incident survey or pre-plan.

Pre-Incident Survey Advance planning of firefighting tactics and strategies or other emergency activities that can be anticipated to occur at a particular location. Also referred to as pre-incident plan or pre-plan.

Pressure The force, or weight, of a substance, usually water, measured over an area.

Pressure The force, or weight, of a substance, usually water, measured over an area.

Pressure-Regulating Device Designed to control the head pressure at the outlet of a standpipe system to prevent excessive nozzle pressures in hoselines. Some are field adjustable, whereas others are preset at the factory.

Protein Foam Made from chemically broken down natural protein materials, such as animal blood, that have metallic salts added for foaming.

Prying Tools The group of tools used to separate objects by means of a mechanical advantage.

Prying Tools The group of tools used to separate objects by means of a mechanical advantage.

Psychological Decontamination The process performed when persons who have been involved in a situation think they have been contaminated and want to be decontaminated. Responders who have identified that the persons have not been contaminated should still consider what can be done to make them feel better.

Pulling Tools The group of tools used to pull away materials.

Pulling Tools The group of tools used to pull away materials.

Pulmonary Edema Fluid filling the lungs causing death by drowning.

Pump Operator A generic term to describe the person responsible for operating a fire apparatus pump. Other commonly used titles include motor pump operator, engineer, technician, chauffeur, and driver/operator.

Puncture An injury caused by an object that has stabbed the body.

Purlins A series of wood beams placed perpendicular to trusses to help support roof decking.

Pyrolysis Decomposition or transformation of a compound caused by heat.

Quantitative Decomposition The temperature at which a material begins to decompose and release quantities of gases prior to reaching its ignition temperature.

Quint A combination fire service apparatus with components of both an engine company and a truck company.

Rabbeted A door stop that is cut (rabbeted) into the door frame. On metal door frames the stop is an integral part of the frame.

Radial Pulse The pulse located in either wrist.

Radioactive Decay Process whereby, as a material emits radiation, it decays, changing its form; for example, uranium eventually becomes lead after it decays.

Radioisotope A radioactive form of an element that is unstable due to the number of neutrons.

Radiological Dispersion Device (RDD) An explosive device that spreads radioactive material throughout an area.

Radon A radioactive gas that is emitted from the earth, sometimes collecting in basement areas.

Rafter A wood joist that is attached to a ridge board to help form a peak.

Rapid Intervention Crew (RIC) A company designated to search for and rescue trapped or lost firefighters. May also be called a Firefighter Assist and Support Team (FAST).

Rapid Intervention Crew/Universal Air Connection (RIC/UAC) A special coupling that will allow an SCBA cylinder that is low on air to be transfilled from another SCBA cylinder regardless of the manufacturer's make or model.

Rapid Intervention Team (RIT) A company designated to search for and rescue trapped or lost firefighters. May also be called a Firefighter Assist and Support Team (FAST).

Rate of Spread A ground cover fire's forward movement or spread speed. Usually expressed in chains or acres per hour.

RECEO Acronym coined by Lloyd Layman standing for Rescue, Exposures, Confinement, Extinguishment, and Overhaul.

RECEO Acronym coined by Lloyd Layman standing for Rescue, Exposures, Confinement, Extinguishment, and Overhaul.

Recommended Exposure Limit (REL) An exposure value established by NIOSH for a ten-hour day, forty-hour workweek. Similar to the PEL and TLV.

Reducers Used to connect a larger hose to a smaller one.

Reel Coil Memory that wire develops from having been placed on a wooden spool as it is being manufactured.

Regulations Developed and issued by a governmental agency and have the weight of law.

Rehabilitation (Rehab) A shortened word meaning rehabilitation. Rehab typically consists of rest, medical evaluation, hydration, and nourishment.

Relief Valve A device designed to vent pressure in a tank, so that the tank itself does not rupture due to an increase in pressure. In most cases these devices are spring loaded so that when the pressure decreases the valve shuts, keeping the chemical inside the tank.

Remote Shutoffs Valves that can be used to shut off the flow of a chemical. The term remote is used to denote valves that are located away from the spill.

Reportable Quantity (RQ) Both the EPA and DOT use the term. It is a quantity of chemicals that may require some type of action, such as reporting an inventory or reporting an accident involving a certain amount of the chemical.

Rescue Company The unit designation of a group of firefighters assigned to perform specialized rescue work and/or tactics and functions such as forcible entry, search and rescue, ventilation, and so on.

Rescue Specialist A firefighter with specialized training and experience in areas such as high angle rope rescue, confined space, trench, or structural collapse rescue.

Rescue Those actions that firefighters perform at emergency scenes to remove victims from imminent danger or to extricate them if they are already entrapped.

Residential Sprinkler System Smaller and more affordable version of a wet or dry pipe sprinkler system designed to control the level of fire involvement such that residents can escape.

Residential Sprinkler System Smaller and more affordable version of a wet or dry pipe sprinkler system designed to control the level of fire involvement such that residents can escape.

Residual Pressure The pressure in a system after water has begun flowing.

Residual Pressure The pressure in a system after water has begun flowing.

Respiratory Protection Program Management program designed to ensure employee respiratory protection as required by OSHA 29 CFR 1910.134 and NFPA 1500.

Respiratory System The system of the human body that exchanges oxygen and waste gases to and from the circulatory system.

Retard Chamber Acts to prevent false alarms from a sudden pressure surge in the water supply by collecting a small volume of water before allowing a continued flow to the alarm device. The water from a surge is drained from a small hole in the bottom of the collection chamber.

Retention The digging of a hole in which to collect a spill. Can be used to contain a running spill or collect a spill from the water.

Ricin A biological toxin that can be used by a terrorist or other person attempting to kill or injure someone. It is the easiest terrorist agent to produce and one of the most common.

Ridge The land running between mountain peaks or along a wide peak. A high area separating two drainages running parallel with them.

Ringdown Circuits Telephone connection between two points. Going "off-hook" on one end of the circuit causes the telephone on the other end of the circuit to "ring" without having to dial a number.

Risk/Benefit Analysis Weighing the probable outcomes in comparison to the level of risk that rescuers will be exposed to.

Risk/Benefit An evaluation of the potential benefit that a task will accomplish in relationship to the hazards that will be faced while completing the task

Risk/Benefit An evaluation of the potential benefit that a task will accomplish in relation to the hazards that will be faced while completing the task.

Risk-Based Response An approach to responding to a chemical incident by categorizing a chemical into a fire, corrosive, or toxic risk. Use of a risk-based approach can assist the responder in making tactical, evacuation, and PPE decisions.

River Bottom Topographic feature where water runs from higher elevations to lower. Can be dry or wet depending on season or recent rains.

Rollover The intermittent ignition of lower-ignition-temperature gases within the upper thermal layer of a room. Rollover is a warning sign of impending flashover.

Rope Hose Tool About 6 feet (2 m) of 1½-inch (13-mm) rope spliced into a loop with a large metal hook at one end and a 2-inch (50-mm) ring at the other. Used to tie in hose and ladders, carry hose, and perform many other tasks requiring a short piece of rope.

Round Turn Formed by continuing the loop on around until the sections of the standing part on either side of the round turn are parallel to one another.

Run Card System System of cards or other form of documentation that provides specific information on what apparatus and personnel respond to specific areas of a jurisdiction.

Run Card System System of cards or other form of documentation that provides specific information on what apparatus and personnel respond to specific areas of a jurisdiction

Running End End of the rope that is not rigged or tied off.

Saddle A pass between two peaks that has a lower elevation than the peaks. Wind will pass through this area faster than over the peaks, so fire is drawn into this feature.

Safety Container A storage can that eliminates vapor release by using a self-closing lid. Also contains a flame arrestor in the dispenser opening.

Sea Containers Shipping boxes that were designed to be stacked on a ship, then placed onto a truck or railcar.

Search and Rescue Attempts by fire and emergency service personnel to coordinate and implement a search for a missing person and then effect a rescue.

Secondary Containment Any approved method that will prevent the runoff of spilled hazardous materials and confine it to the storage area.

Sectional View A vertical view of a structure as if it were cut in two pieces. Each piece is a cross section of the structure showing roof, wall, horizontal floor construction, and the location of stairs, balconies, and mezzanines.

Sector An area established and identified for a specific reason, typically because a hazard exists within the sector. The sectors are usually referred to as hot, warm, and cold sectors and provide an indication of the expected hazard in each sector. Sometimes referred to as a zone.

Self-Accelerating Decomposition Temperature (SADT) Temperature at which a material will ignite itself without an ignition source present. Can be compared to ignition temperature.

Self-Contained Breathing Apparatus (SCBA) A type of respiratory protection in which a self-contained air supply and related equipment are worn or attached to the user. Fire service SCBA is required to be of the positive pressure type.

Sensitizer A chemical that after repeated exposures may cause an allergic-type effect on some people.

Setting The finishing step, making sure that the knot is snug in all directions of pull.

Severance The cutting off of components (i.e., brake pedal, steering wheel) in a vehicle.

Shear A force that tends to tear a material by causing its molecules to slide past each other.

Sheetrock A trademark and another name for plasterboard.

Shelter in Place A form of isolation that provides a level of protection while leaving people in place, usually in their homes. People are usually sheltered in place when they may be placed in further danger by an evacuation.

Shock Load A load or impact being transferred to a rope suddenly and all at one time.

Shock A serious condition caused by a problem or failure of the circulatory system that results in a decrease of oxygen and vital nutrients to the body's tissues. Also known as hypoperfusion.

Shoring The use of timbers to support and/or strengthen weakened structural members (roofs, floors, walls, etc.) in order to avoid a secondary collapse during the rescue operation.

Short-Term Exposure Limit (STEL) A fifteen-minute exposure to a chemical followed by a one-hour break between exposures. Only allowed four times a day.

Shoulder Load Hose load designed to be carried on the shoulders of firefighters.

Shuttle Operation The cycle in which mobile water supply apparatus is dumped, moves to a fill site for refilling, and is returned to the dump site.

Siamese A device that connects two or more hoselines into one line with either a clapper valve or gate valve to prevent loss of water if only one line is connected.

Sick Building Chemical When a building is referred to as a sick building, certain chemicals exist within that cause health problems for the occupants. These chemicals are referred to as sick building chemicals.

Sick Building A term that is associated with indoor air quality. A building that has an air quality problem is referred to as a sick building. In a sick building, occupants become ill as a result of chemicals in and around the building.

Simple Beam A beam supported at the two points near its end.

Slab Door See flush or slab door.

Slot Loads Narrow section of a hose bed where hose is flat loaded in the slot.

Small Lines or Small-Diameter Hose Hose less than 2½ inches (63 mm) in diameter.

Smoke The products of incomplete combustion or decomposition heating (pyrolysis) that include an aggregate of solids, liquids, and gases suspended in the thermal plume.

Soft Suction Hose Large-diameter woven hose used to connect a pumper to a hydrant. Also known as a soft sleeve.

Solid Stream Nozzles Type of nozzle that delivers an unbroken or solid stream of water to the fire. Also called solid tip, straight bore, or smooth bore.

Solid Stream Nozzles Type of nozzle that delivers an unbroken or solid stream of water to the fire. Also called solid tip, straight bore, or smooth bore.

Solid A state of matter that describes materials that may exist in chunks, blocks, chips, crystals, powders, dusts, and other forms. Ice is an example.

Solid-Core Doors Doors made of solid material such as wood or having a core of solid material between face panels.

Solubility A liquid's ability to mix with another liquid.

Spalling Deterioration of concrete by the loss of surface material due to the expansion of moisture when exposed to heat.

Spalling Deterioration of concrete by the loss of surface material due to the expansion of moisture when exposed to heat.

Span of Control The ability of one individual to supervise a number of other people or units. The normal range is three to seven units or individuals, with the ideal being five.

Span of Control The ability of one individual to supervise a number of other people or units. The normal range is three to seven units or individuals, with the ideal being five.

Spanner Wrench Used to tighten or loosen couplings. It may also be useful as a pry bar, door chock, gas valve control, and so on.

Special Egress Control Device Door hardware that will release and unlock the door a maximum of 15 seconds after it has been activated by pushing on the bar.

Specialist Level A level of training that provides for a specific type of training, such as railcar specialist; someone who has a higher level of training than a technician.

Specification (Spec) Plates All trucks and tanks have a specification plate that outlines the type of tank, capacity, construction, and testing information.

Spent Nuclear Fuel Radioactive fuel that was used in a nuclear reactor.

Sprain Injury to the ligaments that hold the body's joints together and allow them to move.

Sprinkler System Protective system designed to automatically distribute water through sprinklers placed at set intervals on a system of piping, usually in the ceiling area, to extinguish or control the spread of fires.

Sprinkler System Protective system designed to automatically distribute water through sprinklers placed at set intervals on a system of piping, usually in the ceiling area, to extinguish or control the spread of fires.

Staging Part of the operations section where apparatus and personnel assigned to the incident are available for deployment within three minutes.

Stairwell An enclosed stairway attached to the side of a high-rise building or in the center core of same.

Stairwell An enclosed stairway attached to the side of a high-rise building or in the center core of same.

Standard of Care A legal term that means for every emergency medical incident, an emergency responder should treat the patient in the same manner as another emergency responder with the same training.

Standard Operating Procedure (SOP) Specific information and instruction on how a task or assignment is to be accomplished.

Standard Transportation Commodity Code (STCC) A number assigned to chemicals that travel by rail.

Standards Usually developed by consensus groups establishing a recommended practice or standard to follow.

Standing Part The part of a rope that is not used to tie off.

Standpipe System Piping system that allows for the manual application of water in large buildings.

Standpipe System Piping system that allows for the manual application of water in large buildings.

State Emergency Response Committee (SERC) A group that ensures that the state has adequate training and resources to respond to a chemical incident.

States of Matter Describes in what form matter exists, such as solids, liquids, or gases.

Static Pressure The pressure in the system with no hydrants or water flowing.

Static A rope having very little (less than 2 percent) elongation at normal safe working loads.

Staypoles The stabilizer poles attached to the sides of Bangor ladders that are used to assist in the raising of this type of ladder. Once raised, they are not used to support the extended ladder.

Steepness of Slope The degree of incline or vertical rise to a given piece of land.

Storz Couplings The most popular of the nonthreaded hose couplings.

Straight Stream A nozzle pattern that creates a hollow stream, similar in shape to the solid stream pattern, but the straight stream pattern must pass around the baffle of the nozzle. Newer fog nozzle designs, especially the automatic nozzles, only have this hollow effect from the tip on and, hence, create a solid stream with good reach and penetration abilities, some better than solid stream nozzles.

Straight Stream A nozzle pattern that creates a hollow stream, similar in shape to the solid stream pattern, but the straight stream pattern must pass around the baffle of the nozzle. Newer fog nozzle designs, especially the automatic nozzles, only have this hollow effect from the tip on and, hence, create a solid stream with good reach and penetration abilities, some better than solid stream nozzles.

Strainers Placed over the end of a suction hose to prevent debris from being sucked into the pump. Some strainers have a float attached to keep them at or near the water's surface. A different style of strainer or screen is located on each intake of a pump.

Strategic Goals The overall plan developed and used to control an incident.

Stream Shape The arrangement or configuration of the water or other agent droplets as they leave the nozzle.

Stream Straightener A metal tube, commonly with metal vanes inside it, between a master stream appliance and its solid nozzle tip. The purpose is to reduce any turbulence in the stream, allowing it to flow straighter.

Strike Plate The metal piece attached to a door jamb into which the lock bolt slides. Also called a strike or striker.

Striking Tools The group of tools designed to deliver impact forces to break locks or drive another tool.

Striking Tools The group of tools designed to deliver impact forces to break locks or drive another tool.

Sublimation The ability of a solid to go to the gas phase without being liquid.

Superficial Burns Burns affecting the outermost layer of skin, which typically cause redness of the skin, swelling, and pain.

Superfund Amendments and Reauthorization Act (SARA) A law that regulates a number of environmental issues, but is primarily for chemical inventory reporting by industry to the local community.

Supplied Air Respirator (SAR) A type of SCBA in which the self-contained air supply is remote from the user, and the air is supplied by means of air hoses.

Supplied Air Respirator (SAR) Respiratory protection that provides a face mask, air hose connected to a large air supply, and an escape bottle. Typically used for waste sites or confined spaces.

Supply Hose or Large-Diameter Hose (LDH) Larger hose (3½ inches [90 mm] or bigger) used to move water from the water source to attack units. Common sizes are 4 and 5 inches (100 to 125 mm).

Surface Contaminated Object (SCO) Materials that may be contaminated with radioactive waste and are usually low hazard.

Surface-to-Mass Ratio Exposed exterior surface area of a material divided by its weight.

Suspension System The springs, shock absorbers, tires, and so on, of a vehicle.

Tactics The specific operations performed to accomplish the strategic goals for an incident.

Tactilely Using the sense of touch to feel for any differences or abnormality.

Tag/Guide Lines Tag lines are ropes held and controlled by firefighters on the ground or lower elevations in order to keep items being hoisted from banging against or getting caught on the structure as they are being hoisted.

Tanker The term given to aircraft capable of carrying and dropping water or fire retardant. Some departments still use the term to describe land-based water apparatus.

Target Hazard An occupancy that has been determined to have a greater than average life hazard or complexity of firefighting operations. Such occupancies receive a high priority in the pre-incident survey process and often a higher level of first-alarm response assignment.

Teamwork A number of persons working together in an effort to reach a common goal.

Technical Decontamination The process used to perform decontamination of persons or equipment.

Technician Level A high level of training that allows specific offensive activities to take place, to stop or handle a hazardous materials incident.

Telecommunications Devices for the Deaf (TDD) Device that allows citizens to communicate with the telecommunicator through the use of a keyboard over telephone circuits instead of voice communications.

Telecommunicator Individual whose primary responsibility is to receive emergency requests from citizens, evaluate the need for a response, and ultimately sound the alarm that sends first responders to the scene of an emergency.

Tempered Glass Plate glass that has been heat treated to increase its strength.

Tender The abbreviated term for water tender. A water tender is defined as a land-based mobile water supply apparatus. Some departments still use the term *tender* to describe a hose-carrying support apparatus.

Tensile Strength Breaking strength of a rope when a load is applied along the direction of the length, generally measured in pounds per square inch.

Tension A force that pulls materials apart.

Terrorism Acts of violence that are arbitrarily committed against lives or property and intended to create fear and anxiety.

Tether Line A rope that is held by a team on shore during a water rescue to be used to haul the rescuer and victim back to shore.

Thermal Burns Burns caused by heat or hot objects.

Thermal Layering The stratification of gases produced by fire into layers based on their temperature.

Thermal Level A layer of air that is of the same approximate temperature.

Thermal Plume A column of heat rising from a heat source. A fully formed plume will resemble a mushroom as the upper level of the heat plume cools, stratifies, and begins to drop outside the rising column.

Thermal Protective Performance (TPP) A rating level, expressed in seconds, used to characterize the protective qualities of a PPE component before serious injury is experienced by the wearer.

Threshold Limit Value (TLV) An exposure value that is similar to the PEL, but is issued by the ACGIH. It is based on an eight-hour day.

Through-the-Lock Method A method of forcible entry in which the lock cylinder is removed by unscrewing or pulling and the internal lock mechanism is operated to open a door. Also, the family of tools used to perform this operation.

Tidal Changes The rising and falling of the surface water levels due to the gravitational effects between the Earth and the moon.

Tip Arc The path that a ladder's tip will take while being raised.

Tornados Weather events that result from thunderstorms as violent winds, rotating in a funnel-shaped cloud that can eventually make land contact at speeds as high as 300 miles per hour.

Torsion Load A load parallel to the cross section of the supporting member that does not pass through the long axis. A torsion load tries to "twist" a structural element.

Total Flooding System Used to protect an entire area, room, or building by discharging an extinguishing agent that completely fills or floods the area with the extinguishing agent to smother or cool the fire or break the chain reaction.

Total Flooding System Used to protect an entire area, room, or building by discharging an extinguishing agent that completely fills or floods the area with the extinguishing agent to smother or cool the fire or break the chain reaction.

Tote A large tank usually 250 to 500 gallons (946 1893 liters), constructed to be transported to a facility and dropped for use.

Tower Ladder An apparatus with a telescopic boom that has a platform on the end of the boom or ladder. It can be extended or retracted and rotated like an aerial ladder.

Toxicity Poisonous level of a substance.

Toxins Disease-causing materials that are extremely toxic and in some cases more toxic than other warfare agents such as nerve agents.

TRACEMP An acronym for the types of hazards that exist at a chemical incident: thermal, radiation, asphyxiation, chemical, etiological, mechanical, and potential psychological harm.

Trench An excavation that is deeper than it is wide, with 15 feet being the maximum width.

Triage A quick and systematic method of identifying which patients are in serious condition and which patients are not, so that the more seriously injured patients can be treated first.

Truck Company The unit designation of a group of firefighters assigned to perform tactics and functions such as forcible entry, search and rescue, ventilation, and so on.

Truss A rigid framework using the triangle as its basic shape to emulate a beam.

Truss A rigid framework using the triangle as its basic shape to emulate a beam.

Tsunami A series of enormous waves that are actually caused by an event that occurs underwater, such as an earthquake or volcanic eruption.

Tunnel Vision The focus of attention on a particular problem without proper regard for possible consequences or alternative approaches.

Turntable The rotating platform of a ladder that affords an elevating ladder device the ability to turn to any target from a fixed position.

Two In/Two Out The procedure of having a crew standing by completely prepared to immediately enter a structure to rescue the interior crew should a problem develop.

Type A Container A container designed to hold low-risk radioactive material and offers moderate protection against accidents.

Type A Reporting System System in which an alarm from a fire alarm box is received and retransmitted to fire stations either manually or automatically.

Type B Container A container designed to hold high-risk radioactive materials. It is a hardened transport case capable of withstanding severe accidents.

Type B Reporting System System in which an alarm from a fire alarm box is automatically transmitted to fire stations and, if used, to outside alerting devices.

Type I, Fire-Resistive Construction Type in which the structural members, including walls, columns, beams, girders, trusses, arches, floors, and roofs, are of approved noncombustible or limited combustible materials with sufficient fire-resistive rating to withstand the effects of fire and prevent its spread from story to story.

Type I, Fire-Resistive Construction Type in which the structural members, including walls, columns, beams, girders, trusses, arches, floors, and roofs, are of approved noncombustible or limited combustible materials with sufficient fire-resistive rating to withstand the effects of fire and prevent its spread from story to story.

Type II, Noncombustible Construction Type not qualifying as Type I construction, in which the structural members, including walls, columns, beams, girders, trusses, arches, floors, and roofs, are of approved noncombustible or limited combustible materials with sufficient fire-resistive rating to withstand the effects of fire and prevent its spread from story to story.

Type II, Noncombustible Construction Type not qualifying as Type I construction, in which the structural members, including walls, columns, beams, girders, trusses, arches, floors, and roofs, are of approved noncombustible or limited combustible materials with sufficient fire-resistive rating to withstand the effects of fire and prevent its spread from story to story.

Type III, Ordinary Construction Type in which the exterior walls and structural members that are portions of exterior walls are of approved noncombustible or limited combustible materials, and interior structural members, including wall, columns, beams, girders, trusses, arches, floors, and roofs, are entirely or partially of wood of smaller dimension than required for Type IV construction or of approved noncombustible or limited combustible materials.

Type III, Ordinary Construction Type in which the exterior walls and structural members that are portions of exterior walls are of approved noncombustible or limited combustible materials, and interior structural members, including wall, columns, beams, girders, trusses, arches, floors, and roofs, are entirely or partially of wood of smaller dimension than required for Type IV construction or of approved noncombustible or limited combustible materials.

Type IV, Heavy Timber Construction Type in which exterior and interior walls and structural members that are portions of such walls are of approved noncombustible or limited combustible materials. Other interior structural members, including columns, beams, girders, trusses, arches, floors, and roofs, shall be of solid or laminated wood without concealed spaces.

Type IV, Heavy Timber Construction Type in which exterior and interior walls and structural members that are portions of such walls are of approved noncombustible or limited combustible materials. Other interior structural members, including columns, beams, girders, trusses, arches, floors, and roofs, shall be of solid or laminated wood without concealed spaces.

Type V, Wood Frame Construction Type in which the exterior walls, bearing walls, columns, beams, girders, trusses, arches, floors, and roofs are entirely or partially of wood or other approved combustible material smaller than the material required for Type IV construction.

Type V, Wood Frame Construction Type in which the exterior walls, bearing walls, columns, beams, girders, trusses, arches, floors, and roofs are entirely or partially of wood or other approved combustible material smaller than the material required for Type IV construction.

Underground Storage Tank (UST) Tank that is buried under the ground. The most common are gasoline and other fuel tanks.

Undesigned Load A load not planned for or anticipated.

Unified Command The structure used to manage an incident involving multiple response agencies or when multiple jurisdictions have responsibility for control of an incident.

Unity of Command Principle that there is one designated leader or officer to command an incident.

Upper Explosive Limit (UEL) The upper part of the flammable range. Above the UEL, fire or an explosion cannot occur because there is too much fuel and not enough oxygen.

Urban Search and Rescue (USAR) Teams Groups of technical rescue specialists formed in strategic areas around the country to respond to disasters and provide technical rescue expertise and equipment; the USAR program is managed by FEMA.

Utility Rope Rope used for utility purposes only. Some of the tasks utility ropes are used for in most every fire department are hoisting tools and equipment, cordoning off areas, and stabilizing objects. Also used as ladder halyards.

Vacuum (Negative) Pressure The measurement of the pressure less than atmospheric pressure, which is usually read in inches of mercury (in. Hg or mm Hg) on a compound gauge.

Vapor Density The weight of a gas compared to air. The weight of air has been given a value of "1" (at 70 F and one atmosphere

or, equivalently, 14.7 psi [21 C and 101 kPa]). Gases that weigh more than air will have a value greater than one and those lighter will have a value less than one.

Vapor Density Weight of a gas in relation to air. Air is rated 1.

Vapor Dispersion The intentional movement of vapors to another area, usually by the use of master streams or hoselines.

Vapor Pressure The amount of force that is pushing vapors from a liquid.

Vapor Pressure The amount of force that is pushing vapors from a liquid.

Vapor Pressure The amount of force that is pushing vapors from a liquid. The higher the force the more vapors (gas) being put into the air.

Vapor Suppression Ability to contain or control the production of fuel vapors.

Vaporization The process in which liquids are converted to a gas or vapor.

Vaporization The process in which liquids are converted to a gas or vapor.

Variable, Adjustable, or Selectable Gallonage Nozzle Nozzle that allows the nozzleperson to select the flow, with usually two or three choices, and the pattern.

Variable, Adjustable, or Selectable Gallonage Nozzle Nozzle that allows the nozzleperson to select the flow, with usually two or three choices, and the pattern.

Veins The blood vessels, or tubes, within the body that carry blood lacking oxygen and nutrients back to the heart.

Velocity Pressure The forward pressure of water as it leaves an opening.

Veneer Wall A covering or facing, not a load-bearing wall, usually with brick or stone.

Venous Bleeding Bleeding from a vein.

Vent Enter Search (VES) Extremely high-risk search technique of breaking out a window and entering for a primary search versus a blind crawl through thick smoke and heat to find a stairway. Oftentimes performed prior to hoselines being put into place.

Venturi Principle A process that creates a low-pressure area in the induction chamber of the eductor and allows the foam concentrate to be drawn into and mixed with the water stream.

Vertical Ventilation Channeled pathway for fire ventilation via vertical openings.

Vertical Ventilation Channeled pathway for fire ventilation via vertical openings.

Vesicants A group of chemical agents that cause blistering and irritation of the skin. Commonly referred to as blister agents.

Visqueen A trade name for black plastic. It can be used very effectively in salvage and overhaul operations. Many examples are discussed in this chapter.

Voice Over Internet Protocol (VOIP) Technology used in telephones that converts audio signals into digital data that is transmitted over the Internet.

Voids Spaces within a collapsed area that are open and may be an area where someone could survive a building collapse.

Volcanos Mountains that open beneath the Earth's crust to the molten rock, known as lava, that lies underneath; the pressure from gases in the lava can build, resulting in an upward expulsion of lava through the center of the mountain.

Wall Indicator Valve (WIV) A control valve that is mounted on a wall in a metal case with a small window, reading either "OPEN" or "SHUT."

Watch Dogs Trained dogs that will bark and create a commotion, but will not attack.

Water Columning A condition in a dry pipe sprinkler system in which the weight of the water column in the riser prevents the operation of the dry pipe valve.

Water Curtain Nozzle Designed to spray water to protect exposures against heat by wetting the exposure's surface.

Water Curtain Nozzle Designed to spray water to protect exposures against heat by wetting the exposure's surface.

Water Hammer A sudden surge of pressure created by the quick opening or closing of valves in a water system. The surge is capable of damaging piping and valves.

Water Hammer The pressure surge that occurs when water valves are rapidly opened or closed.

Water Table The level of groundwater under the surface.

Water Tender The term given to land-based water supply apparatus.

Water Thief A variation of the wye that has one inlet and one outlet of the same size plus two smaller outlets with all of the outlets being gated. The standard water thief usually has a 2½-inch (65-mm) inlet with one 2½-inch (65-mm) and two 1½-inch (38-mm) outlets.

Waybill A term that may be used in conjunction with consist, but is a description of what is on a specific railcar.

Weapons of Mass Destruction (WMD) A term that is used to describe explosive, chemical, biological, and radiological weapons used for terrorism and mass destruction.

Web Gear The term given to a whole host of personal items carried on a belt/harness arrangement worn by wildland firefighters. Items include water bottles, a fire shelter, radio, and day sack.

Web The portion of a truss or I beam that connects the top chord with the bottom chord.

Web The portion of a truss or I beam that connects the top chord with the bottom chord.

Webbing Nylon strapping, available in tubular and flat construction methods.

Wet Chemicals Extinguishing agents that are water-based solutions of potassium carbonate-based chemicals, potassium acetate-based chemicals, potassium citrate-based chemicals, or a combination of these.

Wet Pipe Sprinkler System Has automatic sprinklers attached to pipes with water under pressure all the time.

Wet Pipe Sprinkler System Has automatic sprinklers attached to pipes with water under pressure all the time.

Wire Glass Glass with a wire mesh embedded between two or more layers to give increased fire resistance.

Working End The end of the rope that is utilized to secure/tie off the rope.

Working Length The length of the ladder that spans the distance from the ground to the point of contact with the structure. This does not include any distance the ladder might go beyond the point of contact as would be the case when the tip extends beyond the roof.

Wye A device that divides one hoseline into two or more. The wye lines may be the same size or smaller size, and the wye may or may not have gate control valves to control the water flow.

Zone An area established and identified for a specific reason, typically because a hazard exists within the zone. The zones are usually referred to as hot, warm, and cold zones and provide an indication of the expected hazard in each zone. Sometimes referred to as a sector.

Zoning A term given to the establishment of specific hazard zones, that is, hot zone, warm zone, cold zone, and also collapse zones.

Index